Wildlife-Habitat Relationships
in Oregon and Washington

Project Sponsors and Contributing Sponsors

Birds of Oregon Project
Birds of Washington Project
Confederated Tribes of the Warm Springs
Donavin A. Leckenby
Environmental Protection Agency
Federal Highways Administration
Fish and Wildlife Information Exchange
National Fish and Wildlife Foundation
Northwest Power Planning Council
Oregon Chapter of The Wildlife Society
Oregon Cooperative Wildlife Research Unit
Oregon Department of Fish and Wildlife
Oregon Department of Transportation
Oregon Forest Resources Institute
Oregon/Washington Partners In Flight
Pacific States Marine Fish Commission
Quileute Indian Tribe
Rocky Mountain Elk Foundation
Paul F. and Teresa J. Roline
Society for Northwestern Vertebrate Biology
USDA Forest Service
USDI Bureau of Land Management
USDI Fish and Wildlife Service
USGS Biological Resources Division
Washington Chapter of The Wildlife Society
Washington Community, Trade, and Economic Development
Washington Cooperative Fish and Wildlife Research Unit
Washington Department of Natural Resources
Washington Department of Transportation
Washington Forest Protection Association
Weyerhaeuser Company
Wildlife Management Institute

The Washington Department of Fish and Wildlife and the Northwest Habitat Institute
were the lead organizations on this project.

Managing Directors' Dedication

The ecological face of the Pacific Northwest has been, and continues to be, significantly altered by our human hands. In the last 200 years, there have been many changes to the wildlife and human populations that share this most wonderful place on earth. In recognition that to reverse these trends will require human intervention, we dedicate this book to you, and to those with a profound respect of nature who have come before us. It will be through your efforts and dedication to restoring and conserving the health of our natural resources that we may thrive as a people in unison with our environment.

Wildlife-Habitat Relationships in Oregon and Washington

Managing Directors

David H. Johnson

Thomas A. O'Neil

Oregon State University Press
Corvallis

The paper in this book meets the guidelines for permanence and durability of the Committee on Production Guidelines for Book Longevity of the Council on Library Resources and the minimum requirements of the American National Standard for Permanence of Paper for Printed Library Materials Z39.48-1984.

Library of Congress Cataloging-in-Publication Data
Wildlife-habitat relationships in Oregon and Washington / managing directors, David H. Johnson, Thomas A. O'Neil.— 1st ed.
    p. cm.
Includes bibliographical references and index.
  ISBN 0-87071-488-0 (alk. paper)
1. Animal ecology—Oregon. 2. Animal ecology—Washington (State). 3. Habitat (Ecology)—Oregon. 4. Habitat (Ecology)—Washington (State). I. Johnson, David H. II. O'Neil, Thomas A.
  QH105.O7 W565 2000
  591.7'09795—dc21
                        00-010729

OREGON STATE UNIVERSITY

Oregon State University Press
101 Waldo Hall
Corvallis OR 97331-6407
541-737-3166 • fax 541-737-3170
http://osu.orst.edu/dept/press

Table of Contents

CD-ROM: Matrixes for Wildlife–Habitat Relationships in Oregon and Washington
T.A. O'Neil, D.H. Johnson, C. Barrett, M. Trevithick, K.A. Bettinger, C. Kiilsgaard,
M. Vander Heyden, E.L. Greda, D. Stinson, B.G. Marcot, P.J. Doran, S. Tank, and L. Wunder

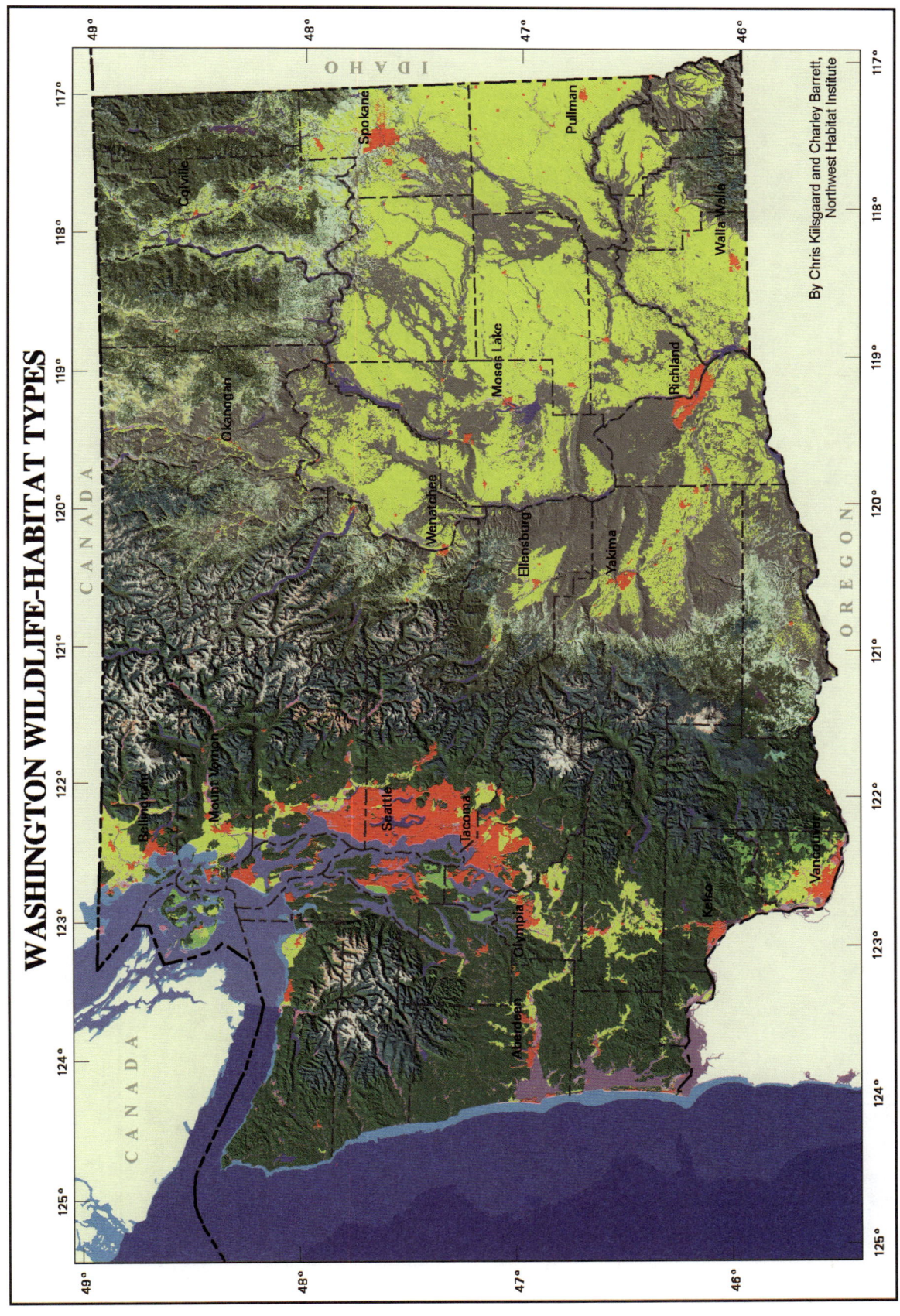

WASHINGTON WILDLIFE-HABITAT TYPES

By Chris Kiilsgaard and Charley Barrett, Northwest Habitat Institute

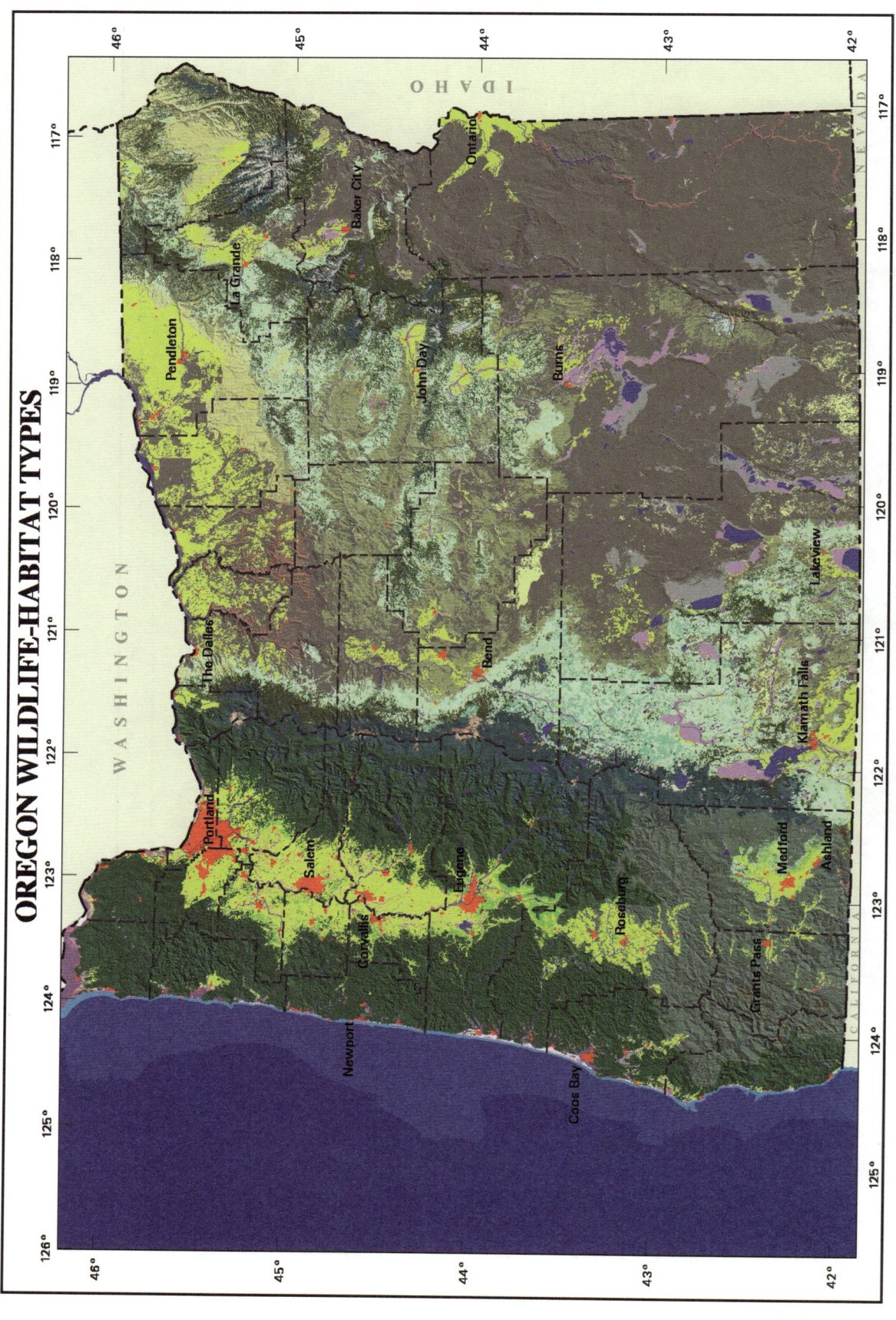

OREGON WILDLIFE–HABITAT TYPES

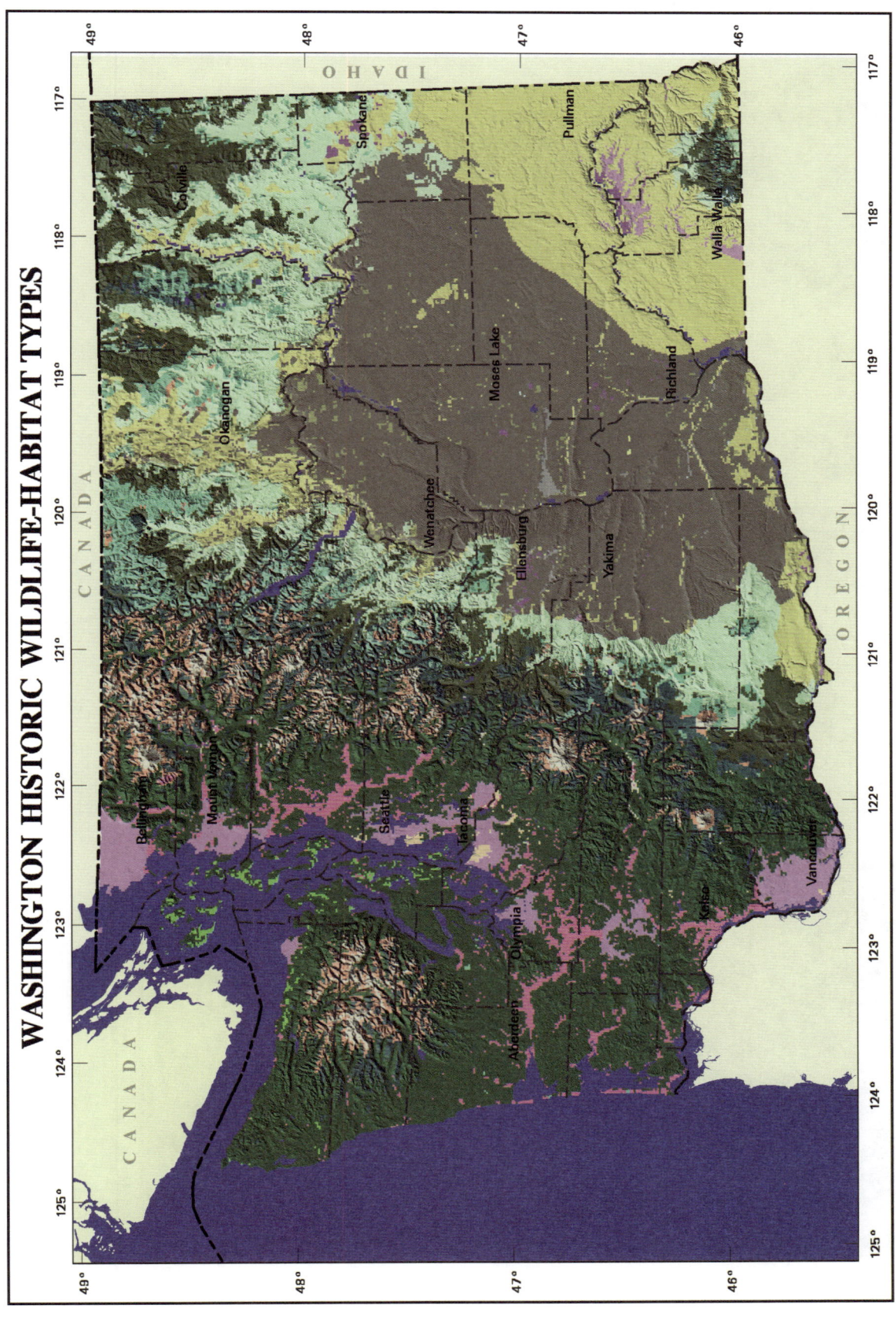

WASHINGTON HISTORIC WILDLIFE-HABITAT TYPES

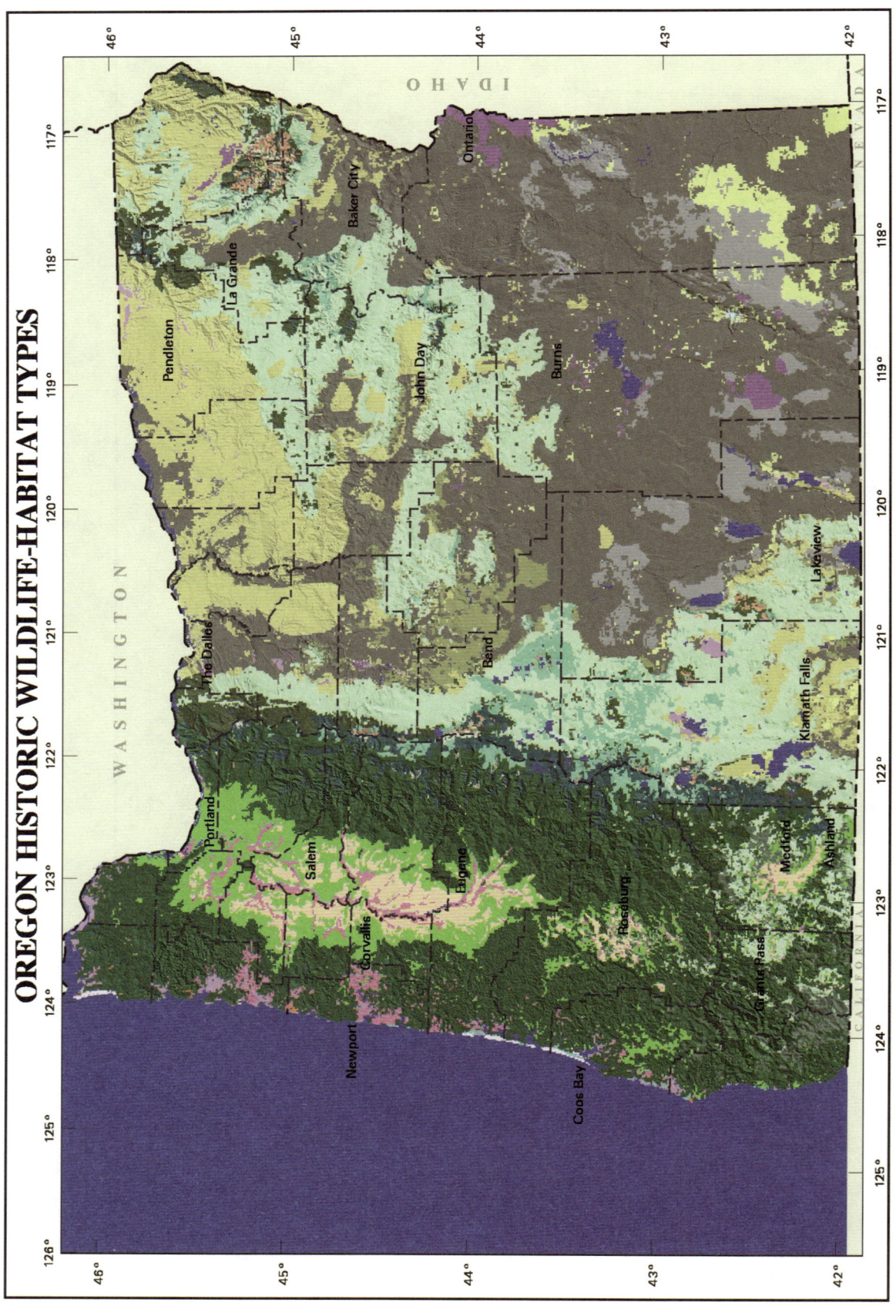

Oregon and Washington Wildlife-Habitat Types

Forest & Woodland Habitats

- Westside Lowland Conifer-Hardwood Forest
- Westside Oak & Dry Douglas-fir Forest & Woodlands
- Southwest Oregon Mixed Conifer-Hardwood Forest
- Montane Mixed Conifer Forest
- Eastside (Interior) Mixed Conifer Forest
- Western Juniper & Mountain Mahogany Woodlands
- Lodgepole Pine Forest & Woodlands
- Ponderosa Pine & Eastside White Oak Forest & Woodlands
- Upland Aspen Forest
- Subalpine Parklands

Grassland & Shrubland Habitats

- Alpine Grasslands & Shrublands
- Westside Grasslands
- Ceanothus-Manzanita Shrublands
- Eastside (Interior) Canyon Shrublands
- Eastside (Interior) Grasslands
- Shrub-steppe
- Dwarf Shrub-steppe
- Desert Playa & Salt Scrub

Developed Habitats

- Agriculture, Pasture & Mixed Environs
- Urban & Mixed Environs

Aquatic & Riparian Habitats

- Lakes, Rivers, Ponds & Reservoirs
- Herbaceous Wetlands
- Westside Riparian - Wetlands
- Montane Coniferous Wetlands
- Eastside (Interior) Riparian - Wetlands

Maritime & Coastal Habitats

- Coastal Dunes & Beaches
- Coastal Headlands & Islets
- Bays & Estuaries
- Inland Marine Deeper Waters
- Marine Nearshore
- Marine Shelf
- Oceanic

Oregon: Scale 1:3,500,000

Washington: Scale 1: 3,000,000

Current: circa 1999

Historic: circa 1850 (modeled)

Map compilation and cartography by Chris Kilsgaard and Charley Barrett, Northwest Habitat Institute

Marine contours provided by Terence P. Johnson, Washington Dept. of Fish and Wildlife

Current Eastern Washington Shrub-Steppe data provided by John E. Jacobson, Washington Dept. of Fish and Wildlife

Managing Directors' Preface

Wildlife–Habitat Relationships in Oregon and Washington and accompanying maps and digital information have been developed to synthesize and disseminate the current state of knowledge about amphibians, birds, mammals, and reptiles, and their terrestrial, freshwater, and marine habitats of Oregon and Washington. Throughout this book, we have focused our attention on scientific rigor and an honest evaluation of what we know, and do not know, about these wildlife species and their habitat relationships. Under a multiple partnership framework, we are broadening our understanding of wildlife habitats. Our wildlife habitats are made up of three components: wildlife-habitat types, structural conditions, and habitat elements. We have also added exciting new information about species' key ecological functions, management activity links (with habitat elements), and salmon-wildlife relationships. Throughout this book and the digital information, we have done our best to integrate disciplines often considered as independent: terrestrial, freshwater, and marine systems. We have also designed the information within this book to be applicable at the local, watershed, state, and regional levels. Further, we have worked hard to develop the information, and overall framework, for easy application to surrounding states and provinces, the natural extension of our ecological context. We felt this overall approach was fundamentally important, as we strove to establish a cohesive information source on which to base a common understanding for management to address ecologically and economically important issues. Based on previous works, and the combined strength of the contributions to this book, we anticipate that the information presented here will be used in natural resource planning, conservation, and education efforts for years to come.

Because so many people and organizations have a stake in the wildlife heritage of the Pacific Northwest, more than 600 people have been involved in the creation of this document and associated products. Thirty-three Project Partners and Contributing Sponsors have supported the project financially or by detailing staff, facilities, and/or equipment. The final tally shows that we received input from 246 questionnaire respondents, 225 people who attended the landscape modeling workshop in Olympia, 73 species specialists who participated in the fifteenscientific panels, 40 people who accessed and contributed to our Internet site, 88 chapter authors, and another 30 people who guided our process by participating on one of five advisory teams. Please see the full acknowledgments at the end of the book for details about the many organizations and individuals without whose contribution this project would not have been completed.

Our work was done in a spirit of cooperation and collaboration, and relationships were forged that we hope will continue into the future as we continue to learn from each other how best to meet our common goals. Our travels around Oregon and Washington offered us unique insights into our truly remarkable natural resources and people that occupy the region. Our philosophy that lead us to undertake the development of this book is best reflected within David Suzuki's *Declaration of Interdependence* (below). To the many, many people we have worked with during the last four years, we thank you. It has been an honor and true pleasure to work with so many talented and caring individuals.

October 12, 2000
David H. Johnson and Thomas A. O'Neil

Declaration of Interdependence

This We Know

We are the earth, through the plants and animals that nourish us.
We are the rains and the oceans that flow through our veins.
We are the breath of the forests of the land, and the plants of the sea.
We are human animals, related to all other life as descendants of the firstborn cell.
We share with these kin a common history, written in our genes.
We share a common present, filled with uncertainty.
And we share a common future, as yet untold.
We humans are but one of thirty million species
weaving the thin layer of life enveloping the world.
The stability of communities of living things depends upon this diversity.
Linked in that web, we are interconnected -
using, cleansing, sharing and replenishing the fundamental elements of life.
Our home, planet Earth, is finite, all life shares its resources and the energy from the sun,
and therefore has limits to growth.
For the first time, we have touched those limits.
When we compromise the air, the water, the soil and the variety of life,
we steal from the endless future to serve the fleeting present.

This We Believe

Humans have become so numerous and our tools so powerful
that we have driven fellow creatures to extinction, dammed the great rivers,
torn down ancient forests, poisoned the earth, rain and wind, and ripped holes in the sky.
Our science has brought pain as well as joy; our comfort is paid for by the suffering of millions.
We are learning from our mistakes, we are mourning our vanished kin,
and we now build a new politics of hope.
We respect and uphold the absolute need for clean air, water and soil.
We see that economic activities that benefit the few while shrinking the inheritance of many are wrong.
And since environmental degradation erodes biological capital forever,
full ecological and social cost must enter all equations of development.
We are one brief generation in the long march of time; the future is not ours to erase.
So where knowledge is limited, we will remember all those who will walk after us,
and err on the side of caution.

This We Resolve

All this that we know and believe must now become the foundation of the way we live.
At this turning point in our relationship with Earth,
we work for an evolution: from dominance to partnership;
from fragmentation to connection; from insecurity, to interdependence.

Declaration of Interdependence published with permission from the David Suzuki Foundation,
219-2211 West 4th Ave, Vancouver, British Columbia, Canada. V6K 4S2.

Foreword

The biggest challenge facing natural resource conservation efforts in the next millennium is the maintenance of viable ecosystems and the biological diversity they encompass. Addressing this imbroglio will require us to find solutions that are efficient yet effective. In meeting this challenge, we need to continue to strive and build a common foundation of knowledge that will serve to support sound and wise decisions concerning our natural resources. This book is an attempt to synthesize a substantially large and diverse body of wildlife information into an agreed foundation of knowledge. It began as a grassroots effort of the Managing Directors; with a little bit of persuasion, we were successful in initially convincing the Washington and Oregon Departments of Fish and Wildlife to offer support for the project. We realized from the beginning that simply having these agencies' endorsement and initial support was insufficient; if we were to be successful we would require participation from numerous professionals and support from a variety of groups and agencies. Eventually, 34 entities became project partners or contributing sponsors, and provided financial or in-kind support to the project. As we approached the conclusion of this effort, we found ourselves almost completely relying on our project partners' support. After four years of focused effort, numerous consensus decision-making work sessions, and involvement of more than 600 people, we present our best effort here in *Wildlife — Habitat Relationships in Oregon and Washington* in hopes of building a common understanding for management.

The information reported here is the first comprehensive effort that we are aware of to compile and synthesize the body of knowledge about Oregon and Washington wildlife. Concurrent with the advent of ecosystem-based management is the need to address multi-species requirements while addressing single-species issues, and to see the land, its habitats and wildlife communities as a system. This book is a response to that need and updates our understanding of Oregon's and Washington's forests, shrublands, grasslands, marine and aquatic habitats, and describes the wildlife relationships that are believed to exist. To achieve this we have combined the information found in the literature with the knowledge of individuals expert in wildlife, botany, fisheries, conservation biology, the fields of forest, rangeland, and marine ecology, earth-bound and satellite vegetation mapping technologies, and computer science. The presentation of this information is based on a foundation of sound theoretical and scientific perspectives, and recognizes that the final products must have practical applications. After all, the work developed here should allow for new insights, suggest new hypotheses, and offer mechanisms to address single and multi-species needs in a world that is becoming less and less wild with each passing day. Our hope for the future is that people from all walks of life can and will work together to keep those natural resources we cherish healthy throughout the next millennium. For those people who come after us and are willing to continue updating and expanding these concepts, we hope that you will experience the comradery, laughter, and passion for the resources that we experienced; may nature's forces be with you!

Introduction

Effectively managing our natural resources requires the combination of high-quality science and the availability of data-rich information. A key information need for natural resource managers, scientists, and educators is the current state of knowledge of wildlife species and their habitats. In this book, we use the term "wildlife" to refer to terrestrial vertebrates or amphibians, birds, mammals, marine mammals, and reptiles. While we support the more encompassing description of wildlife to include fishes and invertebrates, by necessity we had to narrow the focus in this book.

Wildlife habitat is a concept related to a particular wildlife species.[9] More specifically, habitat is an area with the combination of the necessary resources (e.g., food, cover, water) and environmental conditions (temperature, precipitation, presence or absence of predators and competitors) that promotes occupancy by individuals of a given species (or population), and allows those individuals to survive and reproduce.[5] The arrangement of these habitat resources and features to meet the biological needs of a species identifies the habitat niche a species occupies, and from a systems perspective, provides a framework for the ecological role or function that an individual species plays within the environment.[6] From a manager's perspective, the habitat concept needs to be tangible and scientifically supported, inasmuch as habitat features need to be plainly defined, and the wildlife species associations with them made clear. Addressing this management perspective is the primary focus of this book.

Understanding species and their habitat relationships is paramount to predicting species' responses to past, present, and future land uses within a managed landscape. In the past, one approach to habitat management could be considered similar to the motto in the movie *Field of Dreams*: "Build or protect it (habitat) and they (species) will come." But in order to build it or conserve it, resource managers need to know what the relationships are between the individual species and their habitat. Furthermore, because of disturbances, habitats are in a constant state of change, and thus wildlife communities constantly change in response. Habitats need to be evaluated throughout their entire geographic range where a species occurs to assess the influence of these habitats on the life history characteristics (such as breeding, feeding, and wintering) of that wildlife species.

Unfortunately, most wildlife relationship information has generally not been easily accessible. Information germane to this topic is often found in largely unread journals and transactions of symposia, in researchers' files, and in the minds of knowledgeable individuals.[10] Recent technological advances in data retrieval (e.g., bibliographic searches) have been beneficial, but users are still left with the daunting task of synthesis. Earlier publications have forged the basic wildlife relationship information available for Oregon and Washington,[1,2,4,7,9] and while important,

most of it is about fifteen to twenty years old and in need of updating. Also, there were geographic gaps in the information, as well as important differences in how the habitats were defined and described.

This publication builds on the work of past regional publications and subscribes to (1) taking a two-state perspective and making seamless the terrestrial, freshwater, and marine environs of Oregon and Washington; (2) providing updated information on wildlife species and their habitat relationships; (3) reviewing basic concepts and current thinking regarding these relationships; and 4) providing data and approaches that could be used in local, regional, and state planning. Importantly, we have focused a good deal of attention on the integration of terrestrial, freshwater, and marine systems. Also, this project advances the following new concepts: defining the hierarchical nature of wildlife habitats (reflected in our wildlife habitat types, structural conditions, and habitat elements) and key ecological functions; tying management activities to habitat elements; characterizing salmon—wildlife relationships; assigning associations and confidence levels to each species' relationship to wildlife habitat types and structural conditions; and developing the first wildlife habitat maps for each state. The geographic area covered by these systems is substantial: the terrestrial and freshwater areas encompass 75 counties with an area of 105,710,720 acres (42,781,128 ha); the surface area of the marine waters include an additional 65,747,200 acres (26,298,880 ha).

A fundamental paradigm shift has occurred since the initial habitat relationship books were published. The shift reflects the move away from a focus on wildlife habitats having value for few individual species, like elk, towards wildlife habitats having multiple values for the mix of species they may contain. Importantly, the shift incorporates a view of the land and its value in a systems context, giving rise to an ecosystem-based management philosophy. In 1940, Aldo Leopold[3] proposed a "land ethic" that put natural resource decisions on a non-special-interest footing. Leopold stated that a natural resource decision is "right when it tends to preserve the integrity, stability and beauty of the biotic community," and "it is wrong when it tends otherwise." Since 1960, an ethic for our natural resources has been steadily building, and it is influencing our thinking and policies by giving greater considerations to species other than our own. This ethical perspective is aimed at becoming more holistic with one's surroundings and recognizes that humans are part of a larger system. These broader views are shaping the principles that underlie the way we think about and value the environment as a whole, as well as how we approach natural resource problems.

Over the past forty years, a number of laws have played a part in shaping our landscapes and marine waters. Among the most influential federal laws are the Mining

Law of 1872, Multiple Use-Sustained Yield Act of 1960, Wilderness Act of 1964, Wild and Scenic Rivers Act of 1968, Clean Air Act of 1970, Clean Water Act of 1972, Marine Mammal Protection Act (1972), Endangered Species Act of 1973 (ESA), Forest and Rangeland Renewable Resources Planning Act of 1974, National Forest Management Act of 1976, and the Federal Land Policy and Management Act of 1976. Likewise, a number of state laws have shaped our landscape as well. Oregon's Land Use Planning Act (Oregon Revised Statute #197.005 to .860) requires counties to prepare comprehensive land-use plans that include consideration of wildlife habitat, open space needs, and the consideration of ecologically significant natural areas. The Oregon Endangered Species law has provisions that protect native vertebrates and plants on state lands only (Oregon Revised Statute #496.172 to .192; 498.026; 564.100 to .135), and the Oregon Forest Practices Act requires consideration of the impacts of forest practices on threatened, endangered, and special concern species (Oregon Revised Statute #527.610). In Washington, there is also an Endangered Species law that covers animals, but not plants (Revised Code of Washington #77.16.040, 77.16.120, 77.12.020, 77.08.010, 77.12.055 to 3, 77.21.010, 79.08.250). The Washington Growth Management Act affects land-use activities in counties with populations of 50,000 or more and requires the designation of critical areas of habitat and open space corridors (Revised Code of Washington #36.70A.010 et seq.). Washington also has a Forest Protection Act (Revised Code of Washington #76.09, enacted in 1974), which establishes minimum standards for forest practices, and provides protection for critical wildlife habitats of threatened and endangered species. The Washington Hydraulics Code (Revised Code of Washington #75.20 enacted in 1949) requires a permit for any activity that will use, divert, obstruct, or change the bed or flow of state waters. The Washington Shoreline Management Act (Revised Code of Washington #90.58, enacted in 1972) applies to all marine waters, submerged tidelands, lakes >20 acres (49.5 ha), and larger streams, marshes, bogs, and swamps and adjacent landward area; with the intent to ensure that ". . . development of these shorelines . . . will promote and enhance the public interest." The Washington State Environmental Policy Act (Revised Code of Washington #43.21C, enacted in 1971), provides for the review of environmental impacts of activities that may potentially affect Washington's air, water, soil, human health, and environment.

Whereas these laws and policies help to improve land and marine conservation, prior to 1990 most natural resource management was driven by commodity production, like the Allowable Sale Quantity (ASQ), the amount of board feet of timber from a National Forest; Animal Units per Month (AUMs) for livestock on rangelands; and Maximum Sustained Yield (MSY) for fisheries from our marine waters. But in the early 1990s these attitudes and policy directions began to change because several wildlife species became listed as threatened or endangered under the federal ESA. In April 1993, these changes, affecting the economic stability for the segment of society who were dependent on natural resources in the Northwest, became the subject of President Clinton's Forest Summit hosted in Portland, Oregon. Hence new ideas and approaches influencing our landscapes were required. Likewise, technological changes allowed us to view, and analyze, our landscapes and marine areas through an increasingly rich array of tools. The growing body of science and technology played a key role in shaping the information on which policy decisions were based.

More recently, our increasing knowledge about the environment, and about species such as the northern spotted owl and salmon, has led to a growing awareness that human society needs to adapt its activities to protect crucial ecological processes. To do so requires an integration of ecological, economic, and social values to manage biological and physical systems in a way that safeguards their long-term sustainability, natural diversity, and productivity. Thus, ecosystem-based management is a philosophy that focuses on desired states to provide the outputs sought within a framework of sustainable and viable ecological conditions, rather than management primarily for system outputs. It acknowledges the need to protect or restore critical ecosystem components, functions, and structures in order to sustain ecological systems in perpetuity. A shortcoming of this effort is that we are unable to address the ecosystem processes as a whole, and for a good reason: we do not fully understand them. This project only addresses the portion of the ecosystem that is associated with terrestrial, freshwater, and marine vertebrates; users of our information seeking to implement an ecosystem-based management approach are urged to seek additional information on the status and functions of other ecological components, such as fish and invertebrates. These other species should not be overlooked within the landscapes and aquatic environs as they contribute to fundamentally important ecological functions.

Other approaches are also emerging to address an ecosystem understanding. Organizations are moving away from traditional static approaches and towards becoming *learning institutions*. Agencies that are striving to become learning institutions focus on information management, training and education, integration and cooperation, evaluation and feedback loops, efficient and flexible processes, and empowerment of employees and the public.[8] An example of a learning approach is adopting adaptive management strategies into the management of natural resources. Adaptive management is an approach that incorporates uncertainty into managing our natural resources. It involves using a systematic process to evaluate management actions that subsequently increase our level of understanding about our natural systems so that we can make better future decisions. Key to the success of an adaptive management strategy is monitoring not only the natural communities' response to a management action, but also future management decisions, and adjusting them in accordance with results obtained. Embracing adaptive management would forge a closer tie between researchers and managers, blur boundaries between traditional funding sources, and more effectively pool resources and give a higher return on our investment.

Managing our natural resources requires the combination of science and the availability of information. Therefore, a primary emphasis of our endeavor is to develop high-quality data sets on our wildlife habitats and the species associated with them. We achieved this by first defining, describing, and depicting various component details about our wildlife habitats that are offered in this book, maps, and accompanying CD-ROM. Secondly, we brought together as much scientific talent from the Pacific Northwest as possible to review, determine, comment, and assign relationships between our wildlife species and their habitats. Our approach moves away from defining what is *primary* or *secondary* habitat for a species, and towards identifying the overall strength and context of the relationship between the wildlife species and their habitat(s). As such, the strength of the relationship is designated as *Closely Associated, Generally Associated,* or *Present* within the wildlife habitat types or structural conditions. In addition, we also assign a confidence rating to the relationship and its strength based on what we know today. This approach allows for an individual species, as well as multiple species, to be assessed across habitats. Using all matrixes will allow the reader to begin to acquire an ecosystem perspective by depicting, for example, species relationships to: one another, ecological functions, unique and common life history characteristics (like diets and reproductive sites), and potential effects from human activities. These relationships, coupled with the confidence rating for each association, should allow users to develop potential species lists with a confidence ranking for each, along with the ability to model uncertainty of wildlife use of habitats by applying the confidence rankings as a *fuzzy set logic.*[11]

As part of our updating wildlife-habitat relationships, we offer seven data sets found on the CD-ROM for the reader to use. These seven data sets are depicted as matrixes; three matrixes detail the wildlife species' associations with *Wildlife Habitat Types, Structural Conditions,* and *Habitat Elements.* Other matrixes are descriptive and illustrate *Key Ecological Functions, Life History* characteristics, *Management Activity Links* (linkages between management actions and the Habitat Elements) and *Salmon—Wildlife Relationships.* Additionally, where the literature was used to support life history or management activity information, it can also be found under a separate heading called *Citations.* The matrixes were developed in a hierarchical manner so that the user can go from a coarse level (*wildlife habitat types*) to site-specific, fine-scale features (*habitat elements*). The descriptive matrixes are developed to give the reader an idea of the role that the species play within the ecosystem and what type of activities may affect them. Each matrix is offered in a digital format and designed to give the user updated and accurate information; there are some 60,000 records of data in the seven matrixes. We recognize that not all end users

may agree with the information reported, but we have reviewed more than 100,000 pieces of literature, impaneled fifteen groups of expert specialists, cited our sources or gave comments to support a description or claim where appropriate, and performed quality assurance/quality control checks on the matrixes.

Finally, minor changes were made to the matrixes on the CD-ROM after the book chapters were finalized. Hence, some query totals found in the chapters may slightly deviate from the values reported on the CD-ROM.

Although we would have liked to include a chapter addressing the topic of "cumulative effects," one is not offered in this book. However, components of this book can be arranged into an approach to begin to address these effects, and we propose an illustration (Figure 1) of one such way of using the matrixes in this book to this end. In the past, most managers have typically thought about cumulative effects in one or two dimensions, that is, we have reviewed management activities in terms of direct impacts to species, groups of species or habitat(s). But here is where we have stopped. To assess the health of ecosystems we need to incorporate a third dimension to our thinking, which in the context of this book is framed

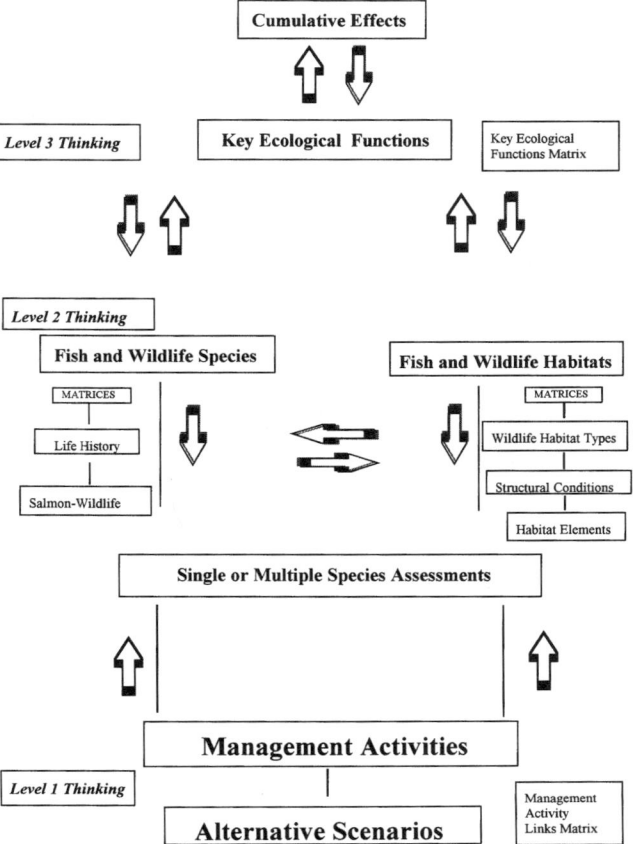

Schematic of how the Wildlife-Habitat Relationships data matrixes can be used with 3-dimensional thinking to address cumulative impacts.

by the examination of key ecological functions for each of the wildlife species. The term "Key Ecological Function" (KEF) refers to the principal set of ecological roles performed by each species in its ecosystem. KEFs directly highlight the influence of organisms on their environments, and how the presence (or absence) of a species serves to modify its own environment as well as the environments of other organisms (see Chapter 6). By using KEFs, we can begin to assess cumulative effects in a more consistent context because we can evaluate management actions in terms of their influences on ecological functions—via habitat elements and species relationships, and through the functions provided by these species—over time and space.

By design, the text within this book makes limited management recommendations. Several factors guided our rationale for this: (1) the main focus of this project was to offer credible, scientific evidence on wildlife—habitat relationships; (2) our area of interest covered a two-state region (including the marine waters out to the 200-mile exclusive economic zone) and management actions are often localized, directed by state and federal policies or regulations, and rarely does one prescription fit all situations; and (3) management recommendations can become outdated relatively quickly. Our position was that whereas management recommendations come and go, the scientific information on which management actions are based (the focus herein) will span a longer time frame. In summary, our principal intent was to make scientifically-rigorous and ecosystem-based wildlife information available to resource managers and interested others, thereby allowing them to pursue management options that fit their local, watershed, or regional situation.

Chapter Organization

The chapters that follow are designed to help the reader become aware of new concepts and approaches as well as give concrete examples using the matrixes. The maps on pages vii-xi depict historic (circa 1850) and current (circa 1999) maps of Oregon and Washington to give the reader an idea of how and where the wildlife-habitat types have changed over time. Chapters 1 and 2 describe the process of how the wildlife habitats were derived and contain a detailed description of each habitat. These chapters support and highlight the *Wildlife Habitat Types* matrix. Chapter 3 describes the structural conditions and land use/land cover classes and the importance of habitat elements in considering management of a site, and is supported by the *Structural Conditions* and *Habitat Elements* matrixes. Chapter 4 offers a discussion of how habitat elements can be designed into forest management schemes. Chapter 5 provides a review of the differing efforts useful in identifying and prioritizing areas and habitats for conservation, and offers a summary of the amounts of each of the 32 wildlife-habitat types in Oregon and Washington and the relative amounts of each under current conservation regimes. Chapter 6 is directly linked to the *Key Ecological Functions* matrix and gives a rationale and examples of how to use this data set. The next nine

chapters (7-15) discuss wildlife as part of multi-habitat communities; the habitats have been grouped because of their general ecological similarities: Westside and High Montane Forests; Eastside (Interior) Forest and Woodlands; Alpine and Subalpine; Westside Grasslands and Chaparral; Eastside (Interior) Shrubland and Grasslands; Urban; Agriculture, Pastures, and Mixed Environs; Riparian; and Coastal and Marine. Chapters 16, 17, and 18 deal with introduced species, genetic considerations for introduced and augmented populations, and an examination of the species that have been lost from the terrestrial and marine environs, respectively. Chapters 19 through 25 report on some of the most recent thinking in regard to characterizing wildlife species at risk; land-use activities (Chapter 20 directly supports the *Management Activity Links* matrix); modeling wildlife and their habitats, integrating wildlife information into the planning process; assessing snags, green trees, and down wood; and considering multi-species planning while trying to address single-species issues. Chapter 26 closely examines salmon and other aquatic resource relationships with wildlife and is linked to the *Salmon–Wildlife Relationships* matrix. The final chapter (27), a summary, gives an introduction to what can be found on the CD-ROM. A complete text of the metadata is found on the CD-ROM.

CD-ROM Organization

The digital wildlife habitat relationship information that accompanies this book is developed as seven matrixes that are found on the CD-ROM. These data sets comprised the following matrixes: wildlife habitat types, structural conditions, habitat elements, key ecological functions, life histories, management activity links, and salmon–wildlife relationships. Citations are also given to support the information in the Life History and Management Activity Links matrixes.

Layout of the information on the CD-ROM is done to report data by wildlife species or by wildlife habitat. There are canned queries that allow the user to review predetermined information, like a list of species that require a snag or are generally associated with lowland mixed conifer forests.

The information is accessible on the CD-ROM by using the Internet Explorer, version 5.0 or greater. By using the Internet browser, the user can print out any of the information that is displayed. For a description of the metadata and definition of terms, please see Chapters 1, 2, 3, 6, 20, and 26.

Conclusion

Because the primary focus of our effort has been wildlife habitat, we have tried not to lose sight of the population dynamics that are associated with habitat changes. With this in mind, we had an opportunity while convening the panels of species experts to ask them what they thought the population status was for each species by state. As a result, 558 species were assessed for their population status, with about 10% decreasing, 10% increasing, 40%

stable, and another 40% unknown. As far as we know this is the first time that there has been an attempt to survey species' experts to establish a trend estimate for each species by state.

As we face the new millennium, we recognize that ecosystems are dynamic and evolutionary, and that steady-state management solutions to environmental issues aren't always possible or appropriate. Ecosystems are organized within the hierarchical scales of time and space. They are limited in their ability to adapt to changes, which may be biophysically or socially defined, although some limits may be mitigated with the input of resources and energy. There are also limits to the predictability of ecosystem patterns and processes; some conditions and events may be predictable at some temporal and spatial scales, but not at others. Ecosystem-based management is an abstract idea, and hence is not an end in itself. Rather, it is an approach to try and achieve sustainable conditions and provide wildlife habitat, outdoor recreation, wilderness, water, wood, mineral resources, and food while retaining the aesthetic, historical, and spiritual qualities of the land. It joins the needs of people and environmental values in such a way that Oregon's and Washington's forests, grasslands, lakes, streams, and marine environments can continue to represent diverse, healthy, productive and sustainable ecosystems.

Wildlife-habitat relationship information is fundamental to our pursuit of ecosystem-based conservation efforts. We hope this book and its related products meaningfully assist you in your endeavors, and provide a springboard for the continued evolution of conservation, scientific thought, and education.

Literature Cited

1. Brown, E. R., technical editor. 1985. Management of wildlife and fish habitats in forests of western Oregon and Washington (Volume 1 & 2). U.S. Forest Service, Publication R6-F&WL-192-1985. Pacific Northwest Region, Portland, OR.

2. Capp, J., B. Carter, J. Diebert, J. Inman, and E. Styskel. 1976. Wildlife habitats relations of south central Oregon. U.S. Forest Service, Bend, OR.

3. Leopold, A. 1976. A Sand County Almanac. Oxford University Press, New York, NY.

4. Maser, C., J. W. Thomas, and R. G. Anderson. 1984. Wildlife habitats in managed rangelands—the great basins of southwestern Oregon. U.S. Forest Service, General Technical Report PNW-172. Pacific Northwest Forest and Range and Experiment Station, Portland, OR.

5. Morrison, M., B. Marcot, and R. Mannan. 1992. Wildlife—habitat Relationships—concepts and applications. University of Wisconsin Press, Madison, WI.

6. ———, ———, ———. 1998. Wildlife—habitat relationships—concepts and applications. 2nd Edition. University of Wisconsin Press, Madison, WI.

7. Proctor, C. M., J. C. Garcia, D. V. Galvin, G. C. Lewis, L. C. Loehr, and A. M. Massa. 1980. An ecological characterization of the Pacific Northwest coastal region. 5 volumes. U.S. Fish and Wildlife Service, Biological Services Program. FWS/OBS-79/11 through 79/15.

8. Senge, P. M. 1994. The Fifth Discipline. Doubleday Publications, New York, NY.

9. Thomas, J. W., technical editor. 1979. Wildlife habitats in managed forests: the Blue Mountains of Oregon and Washington. U.S. Forest Service, Agriculture Handbook 553. Washington, D.C.

10. ———. 1991. Research on wildlife in old growth forests: setting the stage. Pages 1-4 in Wildlife and vegetation of unmanaged Douglas-fir forests. U.S. Forest Service, General Technical Report PNW-GT-285. Pacific Northwest Forest and Range and Experiment Station, Portland, OR.

11. Von Altrock, C. 1995. Fuzzy logic and neurofuzzy applications explained. Prentice-Hall, NJ.

Oregon and Washington Wildlife Species and Their Habitats

Thomas A. O'Neil & David H. Johnson

Introduction

In this chapter, we address the fundamental issues of developing a common nomenclature for wildlife species, and determining their occurrence and breeding status within the two-state area. We then describe the methods used to identify the 32 wildlife-habitat types. Combined, the species list and habitats form the backbone for the rest of this book, associated maps, and CD-ROM.

Developing the Wildlife Species List

For almost two hundred years, information has been collected about Oregon's and Washington's wildlife species. These data have been obtained from field studies, surveys, and general observations. Most of this information now resides in either museum collections throughout the United States, or as records in journals, publications, and public agency files. Over time, each state has developed, mostly independently, its own wildlife species list based on these data sets. Thus, a first step for us was to bring both states' wildlife lists together into a regional data set. As we developed the joint species list, we simultaneously determined the occurrence and breeding status for each of the species. Developing the combined species list, and status categories, proved somewhat challenging because the common and scientific names for many species have changed over the decades, and the status of some species have been confusingly described as "common," "regular," "irregular," "incidental," "accidental," or "vagrant."

Once the joint species list was compiled, a variety of sources were used to establish current taxonomic nomenclature (i.e., scientific and common names) for each species. For most species, initial nomenclatures were obtained from the *Oregon Species Information System*.[36] Collins et al.,[13] Leonard et al.,[31] and Storm and Leonard[42] were used to update the amphibians and reptiles. Bird nomenclature follows the recent *Checklist of North American Birds, Seventh Edition* by the American Ornithologists' Union[2]. Verts and Carraway[45] and Wilson and Reeder[48] were used as the primary sources of scientific names for mammals; Jones et al.[30] and Hall[23] were used as secondary sources for scientific and common mammal names. Frost and Timm[19] and van Zyll de Jong[44] were consulted for bat nomenclature.

We used a number of primary sources to establish/confirm species occurrences: *Atlas of Oregon Wildlife*,[14] *Land Mammals of Oregon*,[45] *Mammals of the Pacific States*,[26] *The Mammals of North America*,[23] *The Mammals and Life Zones of Oregon*,[6] *Birds of Oregon*,[20] *Oregon Species Information System*,[36] *Oregon Wildlife Diversity Plan*,[38] *Amphibians and Reptiles of Washington State: Location Data and Predicted Distributions*,[18] *Terrestrial Mammals of Washington State: Location Data and Predicted Distributions*,[29] *Breeding Birds of Washington State: Location Data and Predicted Distributions*,[40] and *Oregon and Washington Marine Mammal and Seabird Survey*.[10] We also supplemented the above references with Alexander,[1] Aubry,[4, 5] Best,[7] and Bradley.[8]

Scientific and common names and species occurrence status by state were reviewed by Dick Johnson (Washington State University), B. J. Verts (Oregon State University), Tom O'Neil (Northwest Habitat Institute), Rolf Johnson (Washington Department of Fish and Wildlife [WDFW]), Derek Stinson (WDFW), Kelly Bettinger (WDFW), Charlie Bruce (Oregon Department of Fish and Wildlife [ODFW]), Kelly McAllister (WDFW), Bruce Mate (Oregon State University Marine Science Lab), Steven Jeffries (WDFW), and Robin Brown (ODFW). Taxonomic order follows regional publications or commonly accepted national books so that cross-referencing could be made easier. The publications that guided taxonomic order were Leonard et al.[31] for amphibians; Storm and Leonard[42] for reptiles; The American Ornithologists' Union *Checklist of North American Birds, Seventh edition*[2] for birds; and Hall[23] and Verts and Carraway[45] for mammals and marine mammals.

We settled on five occurrence status categories for the species: *Occurs, Accidental, Non-native, Reintroduced*, and *Extirpated*; the species could be listed as any one of these categories in either state. To list a species as having an *Occurs* status within each state required >15 documented observations, that is, they are considered to be regularly occurring species for the area. This figure of 15 documented observations was derived from its use by the states' ornithological groups. *Accidental* denotes those species with <15 documented occurrences, or >15 records but Oregon and Washington are not a regular part of the species' range; *Non-native* denotes species that are not native but now are found in Oregon or Washington;

Reintroduced denotes native species that were eliminated from Oregon or Washington or reduced to such low population levels that additional individuals were required to supplement or re-establish the species; and *Extirpated* refers to a native species whose populations have been completely removed from Oregon or Washington. Additionally, several categories were used to describe the breeding status of the species. *Breeds* is for those species with >5 documented breeding records by separate pairs unless professionals familiar with the species believed that breeding is probable but has not yet been documented. *Non-breeder* refers to those species that occur in the state(s) but do not breed, or have <5 documented breeding records. *Bred Historically* refers to those species that used to breed in the state(s) but currently do not.

Comparing the species lists helps us determine, for the first time, the degree of commonality and uniqueness of wildlife that exists between Oregon and Washington. The combined list of 743 wildlife species now allows us to jointly determine common and scientific names for each species (Appendix, page 687). The combined species list further allows us to characterize the occurrence and breeding status of each species by state (Table 1, at end of chapter). These steps towards standardization are fundamental for building a common understanding about our wildlife because species' names (both common and scientific), occurrence, and breeding status have fluctuated over time. Reaching agreement on these lists not only brings the information up to date, but also becomes the basis for identifying which wildlife species are associated with the various habitat types.

A total of 743 wildlife species occur in Oregon and Washington; 700 of these species occur in Oregon and 651 species in Washington. Some 593 species (80%) occur in both states, whereas 92 species are unique to Oregon and 43 are unique to Washington. In the following breakdown we list the species totals for both states, followed by the total for each state. A total of 479 species *Occur* in both states (548 OR, 501 WA), 80 species are *Accidental* in both states (120 OR, 116 WA), 22 are *Non-native* in both states (24 OR, 28 WA), no similar species have been *Reintroduced* in both states (2 OR, 2 WA), and 2 species have been *Extirpated* from both states (6 OR, 4 WA). A total of 385 species *Breed* in both states, 187 are *Non-breeders*, and 2 *Bred Historically*. A summary of the breeding status of the wildlife species in each state is shown in Figure 1.

Determining the Wildlife-Habitats in Oregon and Washington

In this book, we use the term *wildlife-habitat type* or *habitat types* to mean a group of vegetation cover types (or land use/land cover types) that were determined based on the similarity of wildlife use. Our habitat types are based on actual conditions (e.g., current vegetation), and therefore can be mapped, and we assume they contain all the essential needs for a species' maintenance and viability. Wildlife-habitat types are not species-specific because they are based on the similarity of many wildlife species using a suite of vegetation types. However, a wildlife species' "habitat" refers to an individual, species-specific use of a wildlife-habitat type. Thus, habitat is fundamentally linked to the distribution and abundance of species and underlies explanations of the factors, patterns, and processes that support the fitness of wildlife at the individual, population, and community levels, as well as their continuing evolution.

Too often the term "habitat type" is misused and is misunderstood. The term was initially used by Daubenmire[15] as a way to characterize forest vegetation associations by describing the potential of that association to reach a specific climax stage and was later refined and expanded to forest management.[16, 17, 37] This early use of the term "habitat type" described vegetative associations in relationship to seral stage, and actually had no wildlife species use clearly identified with it. Further, Daubenmire's habitat types classification dealt with the potential of a vegetative association at a site to move towards a climax stage, hence, the forest classification described future, not current, conditions. Wildlife use an area because of current conditions, not future ones. Thus, to a large extent, our past thinking revolved around vegetation classification schemes developed essentially for forest production and inventory purposes, and then overlaid with wildlife attributes.

Recently, Hall et al.[24] reviewed the literature for the appropriate use of the term "habitat type" and concluded that based on the Daubenmire definition the term should not be used when discussing wildlife-habitat relationships. While we agree in part with Hall et al.'s[24] statement, we recognize that the problem stems not from usage of the term in wildlife studies, but rather, that the term needs to be redefined by recognizing the interrelationships of wildlife with different vegetation types. Further, as Hall et al.[24] stated, there is also a need to standardize terminology. Hence, if we are willing to accept the definition of "habitat" as the resources and conditions present in an area to result in occupancy by a given wildlife

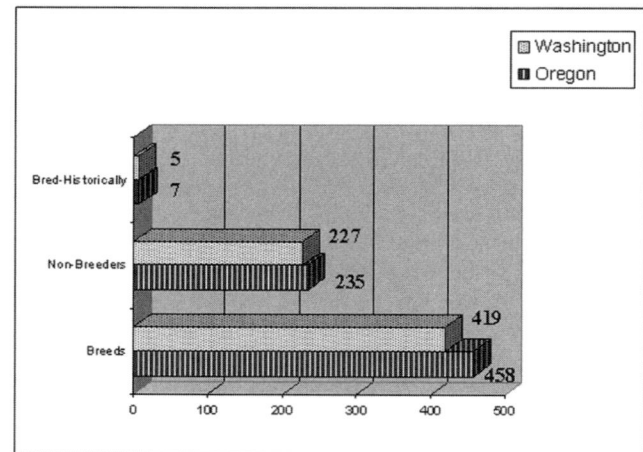

Figure 1. The total number of breeding wildlife species in Oregon and Washington.

species, then a redefinition of the term for wildlife is appropriate. Wildlife species affect, and are affected by their habitats, thus we have approached *wildlife-habitat types* as a collection of wildlife species interacting with different vegetation communities.

Several authors have advocated that wildlife-habitat relationships be placed in the proper spatial and temporal scales when they are being assessed.[34, 47] That is, we need to recognize that wildlife-habitat relationships are scale dependent. For example, Johnson[28] and Hutto[25] suggested that animals select habitat in a hierarchical manner: the first level is determined by the species' geographic range; the second is at a level where a species conducts its daily/ seasonal activities (i.e., home range); and the third level is for habitat components that are local or site-specific within their home ranges. In recognizing the scale dependency of wildlife-habitat relationships, we have also structured the wildlife-habitat data associated with this book to go from a coarse level or *wildlife-habitat type* (that includes the essential components for a species to survive and reproduce), to an intermediate spatial level or *structural condition,* to the finest level of *habitat elements* that can be found at a specific site. Our hope is that knowing the species' relationship with its habitat type, structural conditions and habitat elements will help us make better predictions for species occurrences and ecological conditions for an area. Knowing that ecological condition is based on physical parameters should also help us to identify the ecological processes that are operating (as well as missing) in an area.

Presently, our ability to identify floristic types is greater than our ability to identify associated fauna. Therefore, each identifiable vegetation type probably does not represent a unique wildlife-habitat.[33] We present an approach to classify vegetation alliances into wildlife-habitat types based on the similarity of native breeding terrestrial vertebrate species. The basis for this approach has been discussed in detail by O'Neil et al.[35] This approach has its advantage because the status of our knowledge of wildlife and their habitat relationships has increased dramatically, and the process is quantitative and repeatable. We summarize the approach as follows.

Our initial task was to identify *plant alliances,* as defined by dominant plant species that are found in Oregon and Washington. We used the national vegetation classification system of Grossman et al.[22] and Anderson et al.[3] as the basis for identifying the 287 plant alliances found in Oregon and Washington. Although there are many vegetation classification systems available, we chose this classification system because it addresses current natural vegetation cover, allowed both states a unified classification scheme, and was hierarchical in design, thus providing sufficient detail for linking it to wildlife use.

Upon review of the 287 plant alliances, we recognized that the plant alliances represented a level of detail too fine for ascribing wildlife use. To remedy this, we contracted with a member(s) from each state's Natural Heritage Program to create alliance groupings based on similar vegetative communities; this effort resulted in

pooling the 287 alliances into 85 vegetative groups[12]. To these groups, we added 5 agricultural land cover types and 1 urban classification. Twenty-eight marine classifications were also identified, based on a series of meetings with a multi-agency team of marine experts. Thus, a total of 119 vegetative/land and aquatic/marine cover types were identified, and used in the subsequent analysis steps.

We linked the associations of wildlife species with the following three steps:

1. Existing wildlife-habitat relationships matrixes[9, 11, 33, 43] were incorporated into the Oregon Species Information System that supported the Oregon Gap Analysis project.

2) The Oregon Gap data were then brought together with other regional projects that classified vegetation for wildlife purposes, namely the Washington Gap Analysis and the Interior Columbia Basin Ecosystem Management Project. Although these efforts, taken individually, only covered a portion of the two-state area, taken in composite, these efforts offered information for the entire area, and contained the desired presence/absence information on the wildlife species for the suite of the vegetation/land use groupings.

3) Crosswalks were then developed between the vegetative and land use cover types from each of the projects to the appropriate 85 vegetative groups, 5 agricultural land cover, and 1 urban types. Links to the wildlife occurrence data occurred simultaneously. We then reviewed the wildlife occurrence data (from the combined projects) for inconsistencies.

We selected 541 native breeding species to use in conjunction with the 119 classes of vegetative/land cover/ marine types for the subsequent analysis. We used native breeding species because we believe existing habitat information for breeders is more accurate and species preferences for breeding habitat are more specific than for other associations.[35] We then used the group of wildlife species associated with each of the 119 classes as a set of attributes to describe each class. This increased class specificity increased our ability to discriminate effectively among similar vegetative/land cover/marine types.

We constructed a matrix of 119 rows (vegetation/land cover/marine types) by 541 columns (wildlife species), and populated it with wildlife use data. In the matrix, a "1" indicate that the wildlife species could be found using the vegetation/land cover/marine class, and a "0" indicated absence or limited use of the class. A wildlife species could have an association with any number of the 119 classes.

Multivariate statistics were then used to undertake a cluster analysis process. Consistent patterns of wildlife association within groups suggest that wildlife species perceive the groups as a similar habitat. From this matrix we calculated wildlife species similarity between each pair of types using two different coefficients (Jaccard and Ochaia) as a measure of species' association.[21, 27, 32] The difference between the two coefficients is that the Jaccard gives equal weight to the absence or presence of a species sharing a vegetation type and the Ochaia gives weight to species sharing vegetation type(s). In general, the two coefficients gave similar results. We then created a distance matrix (species' dissimilarity) for vegetation types (119 x 119) by subtracting the Jaccard or Ochaia coefficients from 1. To groups whose vegetation types had high species' similarities, we performed a cluster analysis using the distance matrix and Ward linkage method,[46] a sequential, agglomerative, hierarchic, nonover-lapping cluster strategy,[41] which minimizes the within-cluster variance to the between-cluster variance at each step. We then reviewed each vegetation type's membership (and the wildlife species associated with it) with each set of clusters to see that it was consistent with our ecological understanding of plant and wildlife associations. If we questioned the vegetation membership within a cluster group then we also reviewed the wildlife species associated with the particular vegetation type. Experts then reviewed our initial groups and made suggestions for refining them. A total of 32 wildlife-habitats in Oregon and Washington were determined from this process. Table 2 illustrates the results that were achieved using this method. The Frontispiece reflects the mapping of these 32 wildlife habitats.

Wildlife-Habitat Data Matrix

To maximize the utility of wildlife-habitat relationship information, a digital database that links wildlife with its habitats can be found on the CD-ROM included with this book. Wildlife occurrence with a particular habitat type was determined through an expert panel process held during the summer of 1998. The expert panels while cloistered had available to them range maps for each species, wildlife-habitat types distribution maps, statewide vegetation maps, and a variety of reference materials. These sources of material helped the panelists to determine wildlife species occurrence with a particular habitat type. The codes used were: *Y—Yes the species occurs, H—Historically occurred,* and *U—Unsure.* Alongside the occurrence category, the panelist also identified the types of activity that the species performs while utilizing the habitat. The activity codes for the wildlife species within a particular habitat were: *B—Both feeds and breeds, F—Feeds only, R—Reproduces only,* and *O—Other.* The *Other* category reflects activities such as roosting/resting, hibernacula, or use of the habitat for cover (thermal and hiding) purposes.

Defining the Level of Associations Between Wildlife and the Habitats

The science of wildlife ecology continues to embrace new concepts and frameworks. The relatively new paradigm of ecosystem management challenges us to broaden traditional definitions and incorporate concepts that allow us to assess the value of landscapes to single as well as multi-species strategies. One such concept that we are building on is the *degree of association* between wildlife species and their habitats.[39] For our purpose, the panelists

used the following categories for characterizing the degree of association:

Closely Associated. A species is widely known to depend on a habitat or structural condition for part or all of its life history requirements. Identifying this association implies that the species has an essential need for this habitat or structural condition for its maintenance and viability. Some species may be closely associated with more than one habitat or structural condition, others may be closely associated with only one habitat or structural condition. Examples of species exhibiting close associations are red-winged blackbirds to wetland habitats, and spotted owls to mature and giant tree structural conditions.

Generally Associated. A species exhibits a high degree of adaptability and may be supported by a number of habitats or structural conditions. In other words, the habitats or structural conditions play a supportive role for its maintenance and viability. Examples include the black bear's association with a variety of forested habitats or black-tailed deer associated with a number of structural conditions.

Present. A species demonstrates occasional use of a habitat or structural condition. The habitat or structural condition provides marginal support to the species for its maintenance and viability. Examples are the rough-legged hawk in desert playa and salt scrub or mink in a montane mixed conifer forest. Table 3 depicts the total number of species associated with each wildlife habitat.

Finally, the expert panelists assigned an overall *confidence rating* to the occurrence and activity headings for each species within each habitat type. The confidence ratings were simply high (e.g., many peer or published accounts), moderate, and low (e.g., few or no published accounts). By ascribing a confidence rating, our objective was to offer users an evaluation of the overall strength of the scientific evidence. We understand that this is the first time a confidence rating system has been applied to wildlife-habitat relationship information of this type.

Literature Cited

1. Alexander, L. F. 1996. A morphometric analysis of geographic variation within *Sorex monticolus* (Insectivora: Soricidae). Miscellaneous Publication. University of Kansas, Natural History Museum 88: 1-54.
2. American Ornithologists' Union. 1998. Checklist of North American birds. Seventh Edition, American Ornithologists' Union, Washington D.C.
3. Anderson, M. P. Bourgeron, M. Bryer, R. Crawford, L. Engelking, D. Faber-Langendoen, M. Gallyoun, D. H. Grossman, K. Goodin, S. Landaal, K. Metzler, K. P. Patterson, M. Pyne, M. Reid, L. Sneddon, A. W. Weakley. 1998. International classification of ecological communities: terrestrial vegetation of the United States. Volume II: List of vegetation types. The Nature Conservancy, Boulder, CO.
4. Aubry, K. B. 1982. The recent history and present distribution of the red fox in Washington. Paper presented at the 63rd annual meeting, Pacific Northwest Bird and Mammal Society, May 1, 1982.
5. ———, and D. B. Houston. 1992. Distribution and status of the fisher (*Martes pennanti*) in Washington. Northwestern Naturalist 73: 69-79.
6. Bailey, V. 1936. The mammals and life zones of Oregon. U.S. Department of Agriculture, Bureau of Biological Survey, North America Fauna No. 55: 1-416.

7. Best, T. L. 1988. Morphologic variation in the spotted bat *Euderma maculatum*. American Midland Naturalist 119: 244-252.

8. Bradley, W. P. 1982. Some observations on feral horses on the Yakima Reservation. Paper presented at the 63rd annual meeting of the Pacific Northwest Bird and Mammal Society, May 1, 1982.

9. Brown, E. R., technical editor. 1985. Management of wildlife and fish habitats in forests of western Oregon and Washington (Volume 1 & 2). U.S. Forest Service, Publication R6-F&WL-192-1985. Pacific Northwest Region, Portland, OR.

10. Brueggeman, J. J. 1992. Oregon and Washington marine mammal and seabird surveys. Pacific OCS Region, Mineral Management Service, U.S. Department of Interior, Los Angeles, CA.

11. Capp, J., B. Carter, J. Diebert, J. Inman, and E. Styskel. 1976. Wildlife habitats relations of south central Oregon. U.S. Forest Service, Bend, OR.

12. Chappell, C., R. Crawford, J. Kagen, and P. Doran. 1997. A vegetation, land use, and habitat classification system for the terrestrial and aquatic systems of Oregon and Washington. Wildlife habitats and species associations in Oregon and Washington—building a common understanding. Progress Report 3. Washington Department of Fish and Wildlife, Olympia, WA.

13. Collins, J. T. 1990. Standard common and current scientific names for North American amphibians and reptiles, Third edition. Society for the Study of Amphibians and Reptiles, University of Kansas, Herpetological Circular Number 19, Lawerence, KA.

14. Csuti, B., A. J. Kimmerling, T. A. O'Neil, M. M. Shaughnessy, E. P. Gaines, M. M. P. Huso. 1997. Atlas of Oregon wildlife. Oregon State University Press, Corvallis, OR.

15. Daubenmire, R. 1952. Forest vegetation on northern Idaho and adjacent Washington, and its bearing on concepts of vegetation classification. Ecological Monographs 22: 301-330.

16. ———. 1968. Plant communities: a textbook of plant synecology. Harper and Row, New York, NY.

17. ———, and J. B. Daubenmire. 1968. Forest vegetation of eastern Washington and northern Idaho. Washington Agriculture Experimental Station, Technical Bulletin 60.

18. Dvornich, K. M., K. R. McAllister, and K. Aubry. 1997. Amphibians and reptiles of Washington State: Location data and predicted distributions in: K. M. Cassidy, C. E. Grue, M. R. Smith, and K. M. Dvornich, editors. Washington Cooperative Fish and Wildlife Research Unit, University of Washington, Seattle, WA.

19. Frost, D. R., and R. M. Timm. 1992. Phylogeny of plecotine bats (Chiroptera: Vespertilionidae): summary of the evidence and proposal of a logically consistent taxonomy. American Museum Novitate 3034: 1-16.

20. Gilligan, J., M. Smith, D. Rogers, and A. Contreras. Birds of Oregon. Cinclus Publications, McMinnville, OR.

21. Goodall, D. W. 1973. Sample similarity and species correlation. Pp. 99-149 in: R. H. Whittaker, editor. Ordination and classification of communities. W. Junk, The Hague.

22. Grossman, D. H., D. Faber-Langendoen, A. W. Weakley, M. Anderson, P. Bourgeron, R. Crawford, K. Goodin, S. Landaal, K. Metzler, K. D. Patterson, M. Pyne, M. Reid and L. Sneddon. 1998. International classification of ecological communities: terrestrial vegetation of the United States. Volume 1: The National Vegetation Classification Standard. The Nature Conservancy, Boulder, CO.

23. Hall, R. E. 1981. The mammals of North America. Second Edition. John Wiley & Sons, New York, NY.

24. Hall, L. S., P. R. Krausman, and M. L. Morrison. 1997. The habitat concept and a plea for standard terminology. Wildlife Society Bulletin 25 (1): 173-182.

25. Hutto, R. L. 1985. Habitat selection by nonbreeding, migratory land birds. Pp. 455-476 in: M. L. Cody, editor. Habitat selection in birds. Academic Press, Orlando, FL.

26. Ingles, L. G. 1965. Mammals of the Pacific States. Stanford University Press, Stanford, CA.

27. Jaccard, P. 1901. Distribution de la flore alpine dans le Bassin des Dranes et dans quelques regions voisines. Bulletin Societe Vaudoise des Sciences Naturelles 37: 241-272.

28. Johnson, D. 1980. The comparison of usage and availability measurements for evaluating resource preference. Ecology 61: 65-71.

29. Johnson, R. E. and K. M. Cassidy. 1997. Terrestrial mammals of Washington State: Location data and predicted distributions. Volume 3 in: K. M. Cassidy, C. E. Grue, M. R. Smith, and K. M. Dvornich, editors. Washington State Gap Analysis, Final Report. Washington Cooperative Fish and Wildlife Research Unit, University of Washington, Seattle, WA.

30. Jones, J. K., R. S. Hoffman, D. W. Rice, C. Jones, R. J. Baker, and M. D. Engstrom. 1992. Revised checklist of North American mammals north of Mexico, 1991. Occasional Papers, The Museum Texas Tech University, Lubbock, TX.

31. Leonard, W. P., H. A. Brown, L. L. C. Jones, K. R. McAllister, and R. M. Storm. 1993. Amphibians of Washington and Oregon. Published by the Seattle Audubon Society, Seattle, WA.

32. Ludwig, J. A., and J. F. Reynolds. 1988. Statistical ecology. John Wiley and Sons, New York, NY.

33. Maser. C. J., J. W. Thomas, and R. G. Anderson. 1984. Wildlife habitats in managed rangelands—the Great Basins of southeastern Oregon. U.S. Forest Service, Pacific Northwest Forest and Range and Experiment Station, General Technical Report PNW-172. Portland, OR.

34. Morrison, M., B. G. Marcot, and R. Mannan. 1992. Wildlife-habitat relationships: concepts and applications. Univ. Wisconsin Press, Madison, WI.

35. O'Neil, T. A., R. J. Steidl, W. D. Edge, and B. Csuti. 1995. Using wildlife communities to improve vegetation classification for conserving biodiversity. Conservation Biology, Volume 9 No. 6, 1482-1491.

36. Oregon Department of Fish and Wildlife. 1994. Oregon species information system. User manual developed by W. McKenzie, S. Olson-Edge, and T. O'Neil. Corvallis, OR.

37. Pfister, R., B. Kovalchik, S. Arno, and R. Presby. 1977. Forest habitat types of Montana. U.S. Forest Service General Technical Report INT-34. Intermountain Forest & Range Experiment Station, Ogden, UT.

38. Puchy, C., and D. Marshall. 1993. Oregon wildlife diversity plan. Oregon Department of Fish and Wildlife, Portland, OR.

39. Ruggiero, L., L. Jones, and K. Aubry. 1991. Plant and animal habitat associations in Douglas-fir forests of the Pacific Northwest: an overview. Pp. 447-450 in: Wildlife and vegetation of unmanaged Douglas-fir forests. U.S. Forest Service General Technical Report PNW-GTR-285. Portland, OR.

40. Smith, M. R., P. W. Mattocks Jr., and K. M. Cassidy. 1997. Breeding birds of Washington State: Location data and predicted distributions. Volume 4 in: K. M. Cassidy, C. E. Grue, M. R. Smith, and K. M. Dvornich, editors. Washington State Gap Analysis, Final Report. Seattle Audubon Society Publications in Zoology No. 1, Seattle, WA.

41. Sneath, P. H. A., and R. R. Sokal. 1973. Numerical taxonomy. W. H. Freeman, San Francisco, CA.

42. Storm, R. M., and W. P. Leonard. 1995. Reptiles of Oregon and Washington. Seattle Audubon Society, Seattle, WA.

43. Thomas, J. W., technical editor. 1979. Wildlife habitats in managed forests: the Blue Mountains of Oregon and Washington. U.S. Forest Service, Agriculture Handbook 553. Washington, D.C.

44. van Zyll de Jong, C. 1984. Taxonomic relationships of Nearctic small-footed bats of the *Myotis leibii* group (Chiroptera: Vespertilionidae). Canadian Journal of Zoology 62: 2519-2526.

45. Verts, B. J., and L. N. Carraway. 1998. Land mammals of Oregon. University of California Press, Berkeley and Los Angeles, CA.

46. Ward, J. H., Jr. 1963. Hierarchial grouping to optimize an objective function. Journal of the American Statistical Association 58: 236-244.

47. Wiens, J. A. 1981. Scale problems in avian censusing. Studies in Avian Biology. 6: 513-521.

48. Wilson, D. E., and D. M. Reeder. 1993. Mammal species of the world, 2nd edition. Smithsonian Institution Press, Washington, D.C.

Table 1. Wildlife species occurrence and breeding status in Oregon and Washington.

Common Name	Oregon		Washington	
	Occurrence	Breeding Status	Occurrence	Breeding Status
Amphibians				
Tiger Salamander	Occurs	Breeds	Occurs	Breeds
Northwestern Salamander	Occurs	Breeds	Occurs	Breeds
Long-toed Salamander	Occurs	Breeds	Occurs	Breeds
Cope's Giant Salamander	Occurs	Breeds	Occurs	Breeds
Pacific Giant Salamander	Occurs	Breeds	Occurs	Breeds
Olympic Torrent Salamander	Does Not Occur	Not Applicable	Occurs	Breeds
Columbia Torrent Salamander	Occurs	Breeds	Occurs	Breeds
Southern Torrent Salamander	Occurs	Breeds	Does Not Occur	Not Applicable
Cascade Torrent Salamander	Occurs	Breeds	Occurs	Breeds
Rough-skinned Newt	Occurs	Breeds	Occurs	Breeds
Dunn's Salamander	Occurs	Breeds	Occurs	Breeds
Larch Mountain Salamander	Occurs	Breeds	Occurs	Breeds
Van Dyke's Salamander	Does Not Occur	Not Applicable	Occurs	Breeds
Western Red-backed Salamander	Occurs	Breeds	Occurs	Breeds
Del Norte Salamander	Occurs	Breeds	Does Not Occur	Not Applicable
Siskiyou Mountains Salamander	Occurs	Breeds	Does Not Occur	Not Applicable
Ensatina	Occurs	Breeds	Occurs	Breeds
Clouded Salamander	Occurs	Breeds	Does Not Occur	Not Applicable
Black Salamander	Occurs	Breeds	Does Not Occur	Not Applicable
Oregon Slender Salamander	Occurs	Breeds	Does Not Occur	Not Applicable
California Slender Salamander	Occurs	Breeds	Does Not Occur	Not Applicable
Tailed Frog	Occurs	Breeds	Occurs	Breeds
Great Basin Spadefoot	Occurs	Breeds	Occurs	Breeds
Western Toad	Occurs	Breeds	Occurs	Breeds
Woodhouse's Toad	Occurs	Breeds	Occurs	Breeds
Pacific Chorus (Tree) Frog	Occurs	Breeds	Occurs	Breeds
Red-legged Frog	Occurs	Breeds	Occurs	Breeds
Cascades Frog	Occurs	Breeds	Occurs	Breeds
Oregon Spotted Frog	Occurs	Breeds	Occurs	Breeds
Columbia Spotted Frog	Occurs	Breeds	Occurs	Breeds
Foothill Yellow-legged Frog	Occurs	Breeds	Does Not Occur	Not Applicable
Northern Leopard Frog	Occurs	Breeds	Occurs	Breeds
Bullfrog	Non-native	Breeds	Non-native	Breeds
Green Frog	Does Not Occur	Not Applicable	Non-native	Breeds
Reptiles				
Snapping Turtle	Non-native	Non-Breeder	Non-native	Breeds
Painted Turtle	Occurs	Breeds	Occurs	Breeds
Western Pond Turtle	Occurs	Breeds	Occurs	Breeds
Red-eared Slider Turtle	Non-native	Breeds	Non-native	Breeds
Loggerhead Sea Turtle	Accidental	Non-Breeder	Accidental	Non-Breeder
Green Sea Turtle	Accidental	Non-Breeder	Accidental	Non-Breeder
Pacific Ridley Sea Turtle	Accidental	Non-Breeder	Accidental	Non-Breeder
Leatherback Turtle	Occurs	Non-Breeder	Occurs	Non-Breeder
Northern Alligator Lizard	Occurs	Breeds	Occurs	Breeds
Southern Alligator Lizard	Occurs	Breeds	Occurs	Breeds
Mojave Black-collared Lizard	Occurs	Breeds	Does Not Occur	Not Applicable
Long-nosed Leopard Lizard	Occurs	Breeds	Does Not Occur	Not Applicable
Short-horned Lizard	Occurs	Breeds	Occurs	Breeds
Desert Horned Lizard	Occurs	Breeds	Does Not Occur	Not Applicable
Sagebrush Lizard	Occurs	Breeds	Occurs	Breeds
Western Fence Lizard	Occurs	Breeds	Occurs	Breeds
Side-blotched Lizard	Occurs	Breeds	Occurs	Breeds
Western Skink	Occurs	Breeds	Occurs	Breeds

| Common Name | Oregon | | Washington | |
	Occurrence	Breeding Status	Occurrence	Breeding Status
Western Whiptail	Occurs	Breeds	Does Not Occur	Not Applicable
Plateau Striped Whiptail	Non-native	Breeds	Does Not Occur	Not Applicable
Rubber Boa	Occurs	Breeds	Occurs	Breeds
Racer	Occurs	Breeds	Occurs	Breeds
Night Snake	Occurs	Breeds	Occurs	Breeds
Sharptail Snake	Occurs	Breeds	Occurs	Breeds
Ringneck Snake	Occurs	Breeds	Occurs	Breeds
Night Snake	Occurs	Breeds	Occurs	Breed
Common Kingsnake	Occurs	Breeds	Does Not Occur	Not Applicable
California Mountain Kingsnake	Occurs	Breeds	Occurs	Breeds
Striped Whipsnake	Occurs	Breeds	Occurs	Breeds
Gopher Snake	Occurs	Breeds	Occurs	Breeds
Western Ground Snake	Occurs	Breeds	Does Not Occur	Not Applicable
Pacific Coast Aquatic Garter Snake	Occurs	Breeds	Does Not Occur	Not Applicable
Western Terrestrial Garter Snake	Occurs	Breeds	Occurs	Breeds
Northwestern Garter Snake	Occurs	Breeds	Occurs	Breeds
Common Garter Snake	Occurs	Breeds	Occurs	Breeds
Western Rattlesnake	Occurs	Breeds	Occurs	Breeds
Birds				
Red-throated Loon	Occurs	Non-Breeder	Occurs	Non-Breeder
Pacific Loon	Occurs	Non-Breeder	Occurs	Non-Breeder
Common Loon	Occurs	Non-Breeder	Occurs	Breeds
Yellow-billed Loon	Accidental	Non-Breeder	Occurs	Non-Breeder
Pied-billed Grebe	Occurs	Breeds	Occurs	Breeds
Horned Grebe	Occurs	Breeds	Occurs	Breeds
Red-necked Grebe	Occurs	Breeds	Occurs	Breeds
Eared Grebe	Occurs	Breeds	Occurs	Breeds
Western Grebe	Occurs	Breeds	Occurs	Breeds
Clark's Grebe	Occurs	Breeds	Occurs	Breeds
Shy Albatross	Does Not Occur	Not Applicable	Accidental	Non-Breeder
Laysan Albatross	Occurs	Non-Breeder	Occurs	Non-Breeder
Black-footed Albatross	Occurs	Non-Breeder	Occurs	Non-Breeder
Short-tailed Albatross	Occurs	Non-Breeder	Occurs	Non-Breeder
Northern Fulmar	Occurs	Non-Breeder	Occurs	Non-Breeder
Murphy's Petrel	Accidental	Non-Breeder	Accidental	Non-Breeder
Mottled Petrel	Accidental	Non-Breeder	Accidental	Non-Breeder
Cook's Petrel	Does Not Occur	Not Applicable	Accidental	Non-Breeder
Pink-footed Shearwater	Occurs	Non-Breeder	Occurs	Non-Breeder
Flesh-footed Shearwater	Occurs	Non-Breeder	Occurs	Non-Breeder
Buller's Shearwater	Occurs	Non-Breeder	Occurs	Non-Breeder
Sooty Shearwater	Occurs	Non-Breeder	Occurs	Non-Breeder
Short-tailed Shearwater	Occurs	Non-Breeder	Occurs	Non-Breeder
Manx Shearwater	Does Not Occur	Not Applicable	Accidental	Non-Breeder
Black-vented Shearwater	Accidental	Non-Breeder	Does Not Occur	Not Applicable
Wilson's Storm-petrel	Accidental	Non-Breeder	Accidental	Non-Breeder
Fork-tailed Storm-petrel	Occurs	Breeds	Occurs	Breeds
Leach's Storm-petrel	Occurs	Breeds	Occurs	Breeds
Black Storm-petrel	Accidental	Non-Breeder	Does Not Occur	Not Applicable
Red-billed Tropicbird	Does Not Occur	Not Applicable	Accidental	Non-Breeder
Blue-footed Booby	Does Not Occur	Not Applicable	Accidental	Non-Breeder
American White Pelican	Occurs	Breeds	Occurs	Breeds
Brown Pelican	Occurs	Non-Breeder	Occurs	Non-Breeder
Brandt's Cormorant	Occurs	Breeds	Occurs	Breeds
Double-crested Cormorant	Occurs	Breeds	Occurs	Breeds
Pelagic Cormorant	Occurs	Breeds	Occurs	Breeds
Magnificent Frigatebird	Accidental	Non-Breeder	Accidental	Non-Breeder
American Bittern	Occurs	Breeds	Occurs	Breeds

Common Name	Oregon		Washington	
	Occurrence	Breeding Status	Occurrence	Breeding Status
Least Bittern	Occurs	Breeds	Does Not Occur	Not Applicable
Great Blue Heron	Occurs	Breeds	Occurs	Breeds
Great Egret	Occurs	Breeds	Occurs	Breeds
Snowy Egret	Occurs	Breeds	Accidental	Non-Breeder
Little Blue Heron	Accidental	Non-Breeder	Accidental	Non-Breeder
Tricolored Heron	Accidental	Non-Breeder	Does Not Occur	Not Applicable
Cattle Egret	Occurs	Breeds	Occurs	Non-Breeder
Green Heron	Occurs	Breeds	Occurs	Breeds
Black-crowned Night-heron	Occurs	Breeds	Occurs	Breeds
Yellow-crowned Night-heron	Does Not Occur	Not Applicable	Accidental	Non-Breeder
White-faced Ibis	Occurs	Breeds	Occurs	Non-Breeder
Turkey Vulture	Occurs	Breeds	Occurs	Breeds
California Condor	Extirpated	Bred Historically*	Extirpated	Bred Historically*
Fulvous Whistling-Duck	Accidental	Non-Breeder	Accidental	Non-Breeder
Greater White-fronted Goose	Occurs	Non-Breeder	Occurs	Non-Breeder
Emperor Goose	Accidental	Non-Breeder	Accidental	Non-Breeder
Snow Goose	Occurs	Non-Breeder	Occurs	Non-Breeder
Ross's Goose	Occurs	Non-Breeder	Occurs	Non-Breeder
Canada Goose	Occurs	Breeds	Occurs	Breeds
Western Canada Goose	Occurs	Breeds	Occurs	Breeds
Giant Canada Goose	Non-native	Breeds	Non-native	Breeds
Taverner's Canada Goose	Occurs	Non-Breeder	Occurs	Non-Breeder
Aleutian Canada Goose	Occurs	Non-Breeder	Occurs	Non-Breeder
Cackling Canada Goose	Occurs	Non-Breeder	Occurs	Non-Breeder
Dusky Canada Goose	Occurs	Non-Breeder	Occurs	Non-Breeder
Vancouver Canada Goose	Occurs	Non-Breeder	Occurs	Non-Breeder
Lesser Canada Goose	Occurs	Non-Breeder	Occurs	Non-Breeder
Brant	Occurs	Non-Breeder	Occurs	Non-Breeder
Mute Swan	Non-native	Breeds	Non-native	Breeds
Trumpeter Swan	Occurs	Breeds	Occurs	Breeds
Tundra Swan	Occurs	Non-Breeder	Occurs	Non-Breeder
Whooper Swan	Accidental	Non-Breeder	Does Not Occur	Not Applicable
Wood Duck	Occurs	Breeds	Occurs	Breeds
Gadwall	Occurs	Breeds	Occurs	Breeds
Falcated Duck	Does Not Occur	Not Applicable	Accidental	Non-Breeder
Eurasian Wigeon	Occurs	Non-Breeder	Occurs	Non-Breeder
American Wigeon	Occurs	Breeds	Occurs	Breeds
American Black Duck	Accidental	Non-Breeder	Non-native	Breeds
Mallard	Occurs	Breeds	Occurs	Breeds
Blue-winged Teal	Occurs	Breeds	Occurs	Breeds
Cinnamon Teal	Occurs	Breeds	Occurs	Breeds
Northern Shoveler	Occurs	Breeds	Occurs	Breeds
Northern Pintail	Occurs	Breeds	Occurs	Breeds
Garganey	Accidental	Non-Breeder	Accidental	Non-Breeder
Baikal Teal	Accidental	Non-Breeder	Does Not Occur	Not Applicable
Green-winged Teal	Occurs	Breeds	Occurs	Breeds
Canvasback	Occurs	Breeds	Occurs	Breeds
Redhead	Occurs	Breeds	Occurs	Breeds
Ring-necked Duck	Occurs	Breeds	Occurs	Breeds
Tufted Duck	Accidental	Non-Breeder	Accidental	Non-Breeder
Greater Scaup	Occurs	Non-Breeder	Occurs	Non-Breeder
Lesser Scaup	Occurs	Breeds	Occurs	Breeds
Steller's Eider	Accidental	Non-Breeder	Accidental	Non-Breeder
King Eider	Accidental	Non-Breeder	Accidental	Non-Breeder
Harlequin Duck	Occurs	Breeds	Occurs	Breeds
Surf Scoter	Occurs	Non-Breeder	Occurs	Non-Breeder
White-winged Scoter	Occurs	Non-Breeder	Occurs	Non-Breeder

Common Name	Oregon		Washington	
	Occurrence	Breeding Status	Occurrence	Breeding Status
Black Scoter	Occurs	Non-Breeder	Occurs	Non-Breeder
Long-tailed Duck (Oldsquaw)	Occurs	Non-Breeder	Occurs	Non-Breeder
Bufflehead	Occurs	Breeds	Occurs	Breeds
Common Goldeneye	Occurs	Non-Breeder	Occurs	Breeds
Barrow's Goldeneye	Occurs	Breeds	Occurs	Breeds
Smew	Accidental	Non-Breeder	Accidental	Non-Breeder
Hooded Merganser	Occurs	Breeds	Occurs	Breeds
Common Merganser	Occurs	Breeds	Occurs	Breeds
Red-breasted Merganser	Occurs	Non-Breeder	Occurs	Non-Breeder
Ruddy Duck	Occurs	Breeds	Occurs	Breeds
Osprey	Occurs	Breeds	Occurs	Breeds
White-tailed Kite	Occurs	Breeds	Occurs	Breeds
Bald Eagle	Occurs	Breeds	Occurs	Breeds
Northern Harrier	Occurs	Breeds	Occurs	Breeds
Sharp-shinned Hawk	Occurs	Breeds	Occurs	Breeds
Cooper's Hawk	Occurs	Breeds	Occurs	Breeds
Northern Goshawk	Occurs	Breeds	Occurs	Breeds
Red-shouldered Hawk	Occurs	Breeds	Accidental	Non-Breeder
Broad-winged Hawk	Accidental	Non-Breeder	Accidental	Non-Breeder
Swainson's Hawk	Occurs	Breeds	Occurs	Breeds
Red-tailed Hawk	Occurs	Breeds	Occurs	Breeds
Ferruginous Hawk	Occurs	Breeds	Occurs	Non-Breeder
Rough-legged Hawk	Occurs	Non-Breeder	Occurs	Breeds
Golden Eagle	Occurs	Breeds	Occurs	Breeds
American Kestrel	Occurs	Breeds	Occurs	Breeds
Merlin	Occurs	Bred-Historically	Occurs	Breeds
Gyrfalcon	Occurs	Non-Breeder	Occurs	Non-Breeder
Peregrine Falcon	Occurs	Breeds	Occurs	Breeds
Prairie Falcon	Occurs	Breeds	Occurs	Breeds
Chukar	Non-native	Breeds	Non-native	Breeds
Gray Partridge	Non-native	Breeds	Non-native	Breeds
Ring-necked Pheasant	Non-native	Breeds	Occurs	Breeds
Ruffed Grouse	Occurs	Breeds	Occurs	Breeds
Greater Sage-grouse	Occurs	Breeds	Occurs	Breeds
Spruce Grouse	Occurs	Breeds	Occurs	Breeds
White-tailed Ptarmigan	Does Not Occur	Not Applicable	Occurs	Breeds
Blue Grouse	Occurs	Breeds	Occurs	Breeds
Sharp-tailed Grouse	Reintroduced	Breeds	Occurs	Breeds
Wild Turkey	Non-native	Breeds	Non-native	Breeds
Mountain Quail	Occurs	Breeds	Occurs	Breeds
Scaled Quail	Does Not Occur	Not Applicable	Non-native	Non-Breeder
California Quail	Occurs	Breeds	Non-native	Breeds
Northern Bobwhite	Non-native	Breeds	Non-native	Breeds
Yellow Rail	Occurs	Breeds	Accidental	Non-Breeder
Virginia Rail	Occurs	Breeds	Occurs	Breeds
Sora	Occurs	Breeds	Occurs	Breeds
Common Moorhen	Accidental	Non-Breeder	Does Not Occur	Not Applicable
American Coot	Occurs	Breeds	Occurs	Breeds
Sandhill Crane	Occurs	Breeds	Occurs	Breeds
Black-bellied Plover	Occurs	Non-Breeder	Occurs	Non-Breeder
American Golden-Plover	Occurs	Non-Breeder	Occurs	Non-Breeder
Pacific Golden-Plover	Occurs	Non-Breeder	Occurs	Non-Breeder
Mongolian Plover	Accidental	Non-Breeder	Does Not Occur	Not Applicable
Snowy Plover	Occurs	Breeds	Occurs	Breeds
Semipalmated Plover	Occurs	Non-Breeder	Occurs	Non-Breeder
Piping Plover	Accidental	Non-Breeder	Accidental	Non-Breeder
Killdeer	Occurs	Breeds	Occurs	Breeds
Mountain Plover	Accidental	Non-Breeder	Accidental	Non-Breeder

Common Name	Oregon		Washington	
	Occurrence	Breeding Status	Occurrence	Breeding Status
Eurasian Dotterel	Does Not Occur	Not Applicable	Accidental	Non-Breeder
Black Oystercatcher	Occurs	Breeds	Occurs	Breeds
Black-necked Stilt	Occurs	Breeds	Occurs	Breeds
American Avocet	Occurs	Breeds	Occurs	Breeds
Greater Yellowlegs	Occurs	Non-Breeder	Occurs	Non-Breeder
Lesser Yellowlegs	Occurs	Non-Breeder	Occurs	Non-Breeder
Spotted Redshank	Accidental	Non-Breeder	Does Not Occur	Not Applicable
Solitary Sandpiper	Occurs	Non-Breeder	Occurs	Non-Breeder
Willet	Occurs	Breeds	Occurs	Non-Breeder
Wandering Tattler	Occurs	Non-Breeder	Occurs	Non-Breeder
Gray-tailed Tattler	Does Not Occur	Not Applicable	Accidental	Non-Breeder
Spotted Sandpiper	Occurs	Breeds	Occurs	Breeds
Upland Sandpiper	Occurs	Breeds	Extirpated	Bred Historically
Whimbrel	Occurs	Non-Breeder	Occurs	Non-Breeder
Bristle-thighed Curlew	Accidental	Non-Breeder	Accidental	Non-Breeder
Long-billed Curlew	Occurs	Breeds	Occurs	Breeds
Hudsonian Godwit	Accidental	Non-Breeder	Accidental	Non-Breeder
Bar-tailed Godwit	Accidental	Non-Breeder	Accidental	Non-Breeder
Marbled Godwit	Occurs	Non-Breeder	Occurs	Non-Breeder
Ruddy Turnstone	Occurs	Non-Breeder	Occurs	Non-Breeder
Black Turnstone	Occurs	Non-Breeder	Occurs	Non-Breeder
Surfbird	Occurs	Non-Breeder	Occurs	Non-Breeder
Great Knot	Accidental	Non-Breeder	Accidental	Non-Breeder
Red Knot	Occurs	Non-Breeder	Occurs	Non-Breeder
Sanderling	Occurs	Non-Breeder	Occurs	Non-Breeder
Semipalmated Sandpiper	Occurs	Non-Breeder	Occurs	Non-Breeder
Western Sandpiper	Occurs	Non-Breeder	Occurs	Non-Breeder
Red-necked Stint	Accidental	Non-Breeder	Does Not Occur	Not Applicable
Little Stint	Accidental	Non-Breeder	Does Not Occur	Not Applicable
Long-toed Stint	Accidental	Non-Breeder	Does Not Occur	Not Applicable
Least Sandpiper	Occurs	Non-Breeder	Occurs	Non-Breeder
White-rumped Sandpiper	Does Not Occur	Not Applicable	Accidental	Non-Breeder
Baird's Sandpiper	Occurs	Non-Breeder	Occurs	Non-Breeder
Pectoral Sandpiper	Occurs	Non-Breeder	Occurs	Non-Breeder
Sharp-tailed Sandpiper	Occurs	Non-Breeder	Occurs	Non-Breeder
Rock Sandpiper	Occurs	Non-Breeder	Occurs	Non-Breeder
Dunlin	Occurs	Non-Breeder	Occurs	Non-Breeder
Curlew Sandpiper	Accidental	Non-Breeder	Accidental	Non-Breeder
Stilt Sandpiper	Occurs	Non-Breeder	Occurs	Non-Breeder
Buff-breasted Sandpiper	Occurs	Non-Breeder	Occurs	Non-Breeder
Ruff	Occurs	Non-Breeder	Occurs	Non-Breeder
Short-billed Dowitcher	Occurs	Non-Breeder	Occurs	Non-Breeder
Long-billed Dowitcher	Occurs	Non-Breeder	Occurs	Non-Breeder
Common Snipe	Occurs	Breeds	Occurs	Breeds
Wilson's Phalarope	Occurs	Breeds	Occurs	Breeds
Red-necked Phalarope	Occurs	Non-Breeder	Occurs	Non-Breeder
Red Phalarope	Occurs	Non-Breeder	Occurs	Non-Breeder
South Polar Skua	Occurs	Non-Breeder	Occurs	Non-Breeder
Pomarine Jaeger	Occurs	Non-Breeder	Occurs	Non-Breeder
Parasitic Jaeger	Occurs	Non-Breeder	Occurs	Non-Breeder
Long-tailed Jaeger	Occurs	Non-Breeder	Occurs	Non-Breeder
Laughing Gull	Accidental	Non-Breeder	Accidental	Non-Breeder
Franklin's Gull	Occurs	Breeds	Occurs	Non-Breeder
Little Gull	Accidental	Non-Breeder	Accidental	Non-Breeder
Black-headed Gull	Accidental	Non-Breeder	Accidental	Non-Breeder
Bonaparte's Gull	Occurs	Non-Breeder	Occurs	Non-Breeder
Heermann's Gull	Occurs	Non-Breeder	Occurs	Non-Breeder

Common Name	Oregon		Washington	
	Occurrence	Breeding Status	Occurrence	Breeding Status
Mew Gull	Occurs	Non-Breeder	Occurs	Non-Breeder
Ring-billed Gull	Occurs	Breeds	Occurs	Breeds
California Gull	Occurs	Breeds	Occurs	Breeds
Herring Gull	Occurs	Non-Breeder	Occurs	Non-Breeder
Thayer's Gull	Occurs	Non-Breeder	Occurs	Non-Breeder
Iceland Gull	Does Not Occur	Not Applicable	Accidental	Non-Breeder
Slaty-backed Gull	Accidental	Non-Breeder	Accidental	Non-Breeder
Western Gull	Occurs	Breeds	Occurs	Breeds
Glaucous-winged Gull	Occurs	Breeds	Occurs	Breeds
Glaucous Gull	Occurs	Non-Breeder	Occurs	Non-Breeder
Sabine's Gull	Occurs	Non-Breeder	Occurs	Non-Breeder
Black-legged Kittiwake	Occurs	Non-Breeder	Occurs	Non-Breeder
Red-legged Kittiwake	Accidental	Non-Breeder	Accidental	Non-Breeder
Ross's Gull	Accidental	Non-Breeder	Accidental	Non-Breeder
Ivory Gull	Does Not Occur	Not Applicable	Accidental	Non-Breeder
Caspian Tern	Occurs	Breeds	Occurs	Breeds
Elegant Tern	Occurs	Non-Breeder	Occurs	Non-Breeder
Common Tern	Occurs	Non-Breeder	Occurs	Non-Breeder
Arctic Tern	Occurs	Non-Breeder	Occurs	Breeds
Forster's Tern	Occurs	Breeds	Occurs	Breeds
Least Tern	Accidental	Non-Breeder	Accidental	Non-Breeder
Black Tern	Occurs	Breeds	Occurs	Breeds
Common Murre	Occurs	Breeds	Occurs	Breeds
Thick-billed Murre	Accidental	Non-Breeder	Accidental	Non-Breeder
Pigeon Guillemot	Occurs	Breeds	Occurs	Breeds
Long-billed Murrelet	Accidental	Non-Breeder	Accidental	Non-Breeder
Marbled Murrelet	Occurs	Breeds	Occurs	Breeds
Kittlitz's Murrelet	Does Not Occur	Not Applicable	Accidental	Non-Breeder
Xantus's Murrelet	Accidental	Non-Breeder	Accidental	Non-Breeder
Ancient Murrelet	Occurs	Non-Breeder	Occurs	Non-Breeder
Cassin's Auklet	Occurs	Breeds	Occurs	Breeds
Parakeet Auklet	Accidental	Non-Breeder	Accidental	Non-Breeder
Rhinoceros Auklet	Occurs	Breeds	Occurs	Breeds
Horned Puffin	Accidental	Non-Breeder	Accidental	Non-Breeder
Tufted Puffin	Occurs	Breeds	Occurs	Breeds
Rock Dove	Non-native	Breeds	Non-native	Breeds
Band-tailed Pigeon	Occurs	Breeds	Occurs	Breeds
White-winged Dove	Accidental	Non-Breeder	Accidental	Non-Breeder
Mourning Dove	Occurs	Breeds	Occurs	Breeds
Black-billed Cuckoo	Does Not Occur	Not Applicable	Accidental	Non-Breeder
Yellow-billed Cuckoo	Occurs	Bred Historically	Extirpated	Bred Historically
Barn Owl	Occurs	Breeds	Occurs	Breeds
Flammulated Owl	Occurs	Breeds	Occurs	Breeds
Western Screech-owl	Occurs	Breeds	Occurs	Breeds
Great Horned Owl	Occurs	Breeds	Occurs	Breeds
Snowy Owl	Occurs	Non-Breeder	Occurs	Non-Breeder
Northern Hawk Owl	Accidental	Non-Breeder	Accidental	Non-Breeder
Northern Pygmy-owl	Occurs	Breeds	Occurs	Breeds
Burrowing Owl	Occurs	Breeds	Occurs	Breeds
Spotted Owl	Occurs	Breeds	Occurs	Breeds
Barred Owl	Occurs	Breeds	Occurs	Breeds
Great Gray Owl	Occurs	Breeds	Occurs	Breeds
Long-eared Owl	Occurs	Breeds	Occurs	Breeds
Short-eared Owl	Occurs	Breeds	Occurs	Breeds
Boreal Owl	Occurs	Breeds	Occurs	Breeds
Northern Saw-whet Owl	Occurs	Breeds	Occurs	Breeds
Common Nighthawk	Occurs	Breeds	Occurs	Breeds
Common Poorwill	Occurs	Breeds	Occurs	Breeds

Common Name	Oregon		Washington	
	Occurrence	Breeding Status	Occurrence	Breeding Status
Black Swift	Occurs	Breeds	Occurs	Breeds
Vaux's Swift	Occurs	Breeds	Occurs	Breeds
White-throated Swift	Occurs	Breeds	Occurs	Breeds
Black-chinned Hummingbird	Occurs	Breeds	Occurs	Breeds
Anna's Hummingbird	Occurs	Breeds	Occurs	Breeds
Costa's Hummingbird	Accidental	Non-Breeder	Does Not Occur	Not Applicable
Calliope Hummingbird	Occurs	Breeds	Occurs	Breeds
Broad-tailed Hummingbird	Occurs	Breeds	Does Not Occur	Not Applicable
Rufous Hummingbird	Occurs	Breeds	Occurs	Breeds
Allen's Hummingbird	Occurs	Breeds	Accidental	Non-Breeder
Belted Kingfisher	Occurs	Breeds	Occurs	Breeds
Lewis's Woodpecker	Occurs	Breeds	Occurs	Breeds
Acorn Woodpecker	Occurs	Breeds	Occurs	Non-Breeder
Williamson's Sapsucker	Occurs	Breeds	Occurs	Breeds
Yellow-bellied Sapsucker	Accidental	Non-Breeder	Accidental	Non-Breeder
Red-naped Sapsucker	Occurs	Breeds	Occurs	Breeds
Red-breasted Sapsucker	Occurs	Breeds	Occurs	Breeds
Nuttall's Woodpecker	Accidental	Non-Breeder	Does Not Occur	Not Applicable
Downy Woodpecker	Occurs	Breeds	Occurs	Breeds
Hairy Woodpecker	Occurs	Breeds	Occurs	Breeds
White-headed Woodpecker	Occurs	Breeds	Occurs	Breeds
Three-toed Woodpecker	Occurs	Breeds	Occurs	Breeds
Black-backed Woodpecker	Occurs	Breeds	Occurs	Breeds
Northern Flicker	Occurs	Breeds	Occurs	Breeds
Pileated Woodpecker	Occurs	Breeds	Occurs	Breeds
Olive-sided Flycatcher	Occurs	Breeds	Occurs	Breeds
Western Wood-pewee	Occurs	Breeds	Occurs	Breeds
Eastern Wood-pewee	Accidental	Non-Breeder	Does Not Occur	Not Applicable
Willow Flycatcher	Occurs	Breeds	Occurs	Breeds
Least Flycatcher	Occurs	Non-Breeder	Occurs	Breeds
Hammond's Flycatcher	Occurs	Breeds	Occurs	Breeds
Gray Flycatcher	Occurs	Breeds	Occurs	Breeds
Dusky Flycatcher	Occurs	Breeds	Occurs	Breeds
Pacific-slope Flycatcher	Occurs	Breeds	Occurs	Breeds
Cordilleran Flycatcher	Occurs	Breeds	Occurs	Breeds
Black Phoebe	Occurs	Breeds	Accidental	Non-Breeder
Eastern Phoebe	Accidental	Non-Breeder	Accidental	Non-Breeder
Say's Phoebe	Occurs	Breeds	Occurs	Breeds
Vermilion Flycatcher	Accidental	Non-Breeder	Accidental	Non-Breeder
Ash-throated Flycatcher	Occurs	Breeds	Occurs	Breeds
Tropical Kingbird	Accidental	Non-Breeder	Accidental	Non-Breeder
Western Kingbird	Occurs	Breeds	Occurs	Breeds
Eastern Kingbird	Occurs	Breeds	Occurs	Breeds
Scissor-tailed Flycatcher	Accidental	Non-Breeder	Accidental	Non-Breeder
Fork-tailed Flycatcher	Does Not Occur	Not Applicable	Accidental	Non-Breeder
Loggerhead Shrike	Occurs	Breeds	Occurs	Breeds
Northern Shrike	Occurs	Non-Breeder	Occurs	Non-Breeder
White-eyed Vireo	Does Not Occur	Not Applicable	Accidental	Non-Breeder
Bell's Vireo	Accidental	Non-Breeder	Does Not Occur	Not Applicable
Yellow-throated Vireo	Does Not Occur	Not Applicable	Accidental	Non-Breeder
Plumbeous Vireo	Occurs	Non-Breeder	Does Not Occur	Not Applicable
Cassin's Vireo	Occurs	Breeds	Occurs	Breeds
Hutton's Vireo	Occurs	Breeds	Occurs	Breeds
Warbling Vireo	Occurs	Breeds	Occurs	Breeds
Philadelphia Vireo	Does Not Occur	Not Applicable	Accidental	Non-Breeder
Red-eyed Vireo	Occurs	Breeds	Occurs	Breeds
Gray Jay	Occurs	Breeds	Occurs	Breeds
Steller's Jay	Occurs	Breeds	Occurs	Breeds
Blue Jay	Accidental	Non-Breeder	Accidental	Non-Breeder

Common Name	Oregon		Washington	
	Occurrence	Breeding Status	Occurrence	Breeding Status
Western Scrub-Jay	Occurs	Breeds	Occurs	Breeds
Pinyon Jay	Occurs	Breeds	Accidental	Non-Breeder
Clark's Nutcracker	Occurs	Breeds	Occurs	Breeds
Black-billed Magpie	Occurs	Breeds	Occurs	Breeds
American Crow	Occurs	Breeds	Occurs	Breeds
Northwestern Crow	Occurs	Non-Breeder	Occurs	Breeds
Common Raven	Occurs	Breeds	Occurs	Breeds
Sky Lark	Does Not Occur	Not Applicable	Non-native	Breeds
Horned Lark	Occurs	Breeds	Occurs	Breeds
Purple Martin	Occurs	Breeds	Occurs	Breeds
Tree Swallow	Occurs	Breeds	Occurs	Breeds
Violet-green Swallow	Occurs	Breeds	Occurs	Breeds
Northern Rough-winged Swallow	Occurs	Breeds	Occurs	Breeds
Bank Swallow	Occurs	Breeds	Occurs	Breeds
Cliff Swallow	Occurs	Breeds	Occurs	Breeds
Barn Swallow	Occurs	Breeds	Occurs	Breeds
Black-capped Chickadee	Occurs	Breeds	Occurs	Breeds
Mountain Chickadee	Occurs	Breeds	Occurs	Breeds
Chestnut-backed Chickadee	Occurs	Breeds	Occurs	Breeds
Boreal Chickadee	Does Not Occur	Not Applicable	Occurs	Breeds
Oak Titmouse	Occurs	Breeds	Does Not Occur	Not Applicable
Juniper Titmouse	Occurs	Breeds	Does Not Occur	Not Applicable
Bushtit	Occurs	Breeds	Occurs	Breeds
Red-breasted Nuthatch	Occurs	Breeds	Occurs	Breeds
White-breasted Nuthatch	Occurs	Breeds	Occurs	Breeds
Pygmy Nuthatch	Occurs	Breeds	Occurs	Breeds
Brown Creeper	Occurs	Breeds	Occurs	Breeds
Rock Wren	Occurs	Breeds	Occurs	Breeds
Canyon Wren	Occurs	Breeds	Occurs	Breeds
Bewick's Wren	Occurs	Breeds	Occurs	Breeds
House Wren	Occurs	Breeds	Occurs	Breeds
Winter Wren	Occurs	Breeds	Occurs	Breeds
Marsh Wren	Occurs	Breeds	Occurs	Breeds
American Dipper	Occurs	Breeds	Occurs	Breeds
Golden-crowned Kinglet	Occurs	Breeds	Occurs	Breeds
Ruby-crowned Kinglet	Occurs	Breeds	Occurs	Breeds
Blue-gray Gnatcatcher	Occurs	Breeds	Accidental	Non-Breeder
Northern Wheatear	Accidental	Non-Breeder	Does Not Occur	Not Applicable
Western Bluebird	Occurs	Breeds	Occurs	Breeds
Mountain Bluebird	Occurs	Breeds	Occurs	Breeds
Townsend's Solitaire	Occurs	Breeds	Occurs	Breeds
Veery	Occurs	Breeds	Occurs	Breeds
Gray-cheeked Thrush	Accidental	Non-Breeder	Accidental	Non-Breeder
Swainson's Thrush	Occurs	Breeds	Occurs	Breeds
Hermit Thrush	Occurs	Breeds	Occurs	Breeds
Wood Thrush	Accidental	Non-Breeder	Does Not Occur	Not Applicable
American Robin	Occurs	Breeds	Occurs	Breeds
Varied Thrush	Occurs	Breeds	Occurs	Breeds
Wrentit	Occurs	Breeds	Does Not Occur	Not Applicable
Gray Catbird	Occurs	Breeds	Occurs	Breeds
Northern Mockingbird	Occurs	Non-Breeder	Occurs	Breeds
Sage Thrasher	Occurs	Breeds	Occurs	Breeds
Brown Thrasher	Accidental	Non-Breeder	Accidental	Non-Breeder
California Thrasher	Accidental	Non-Breeder	Does Not Occur	Not Applicable
European Starling	Non-native	Breeds	Non-native	Breeds
Siberian Accentor	Does Not Occur	Not Applicable	Accidental	Non-Breeder
Yellow Wagtail	Does Not Occur	Not Applicable	Accidental	Non-Breeder
White Wagtail	Does Not Occur	Not Applicable	Accidental	Non-Breeder

Common Name	Oregon		Washington	
	Occurrence	Breeding Status	Occurrence	Breeding Status
Black-backed Wagtail	Accidental	Non-Breeder	Accidental	Non-Breeder
Red-throated Pipit	Does Not Occur	Not Applicable	Accidental	Non-Breeder
American Pipit	Occurs	Breeds	Occurs	Breeds
Bohemian Waxwing	Occurs	Non-Breeder	Occurs	Non-Breeder
Cedar Waxwing	Occurs	Breeds	Occurs	Breeds
Phainopepla	Accidental	Non-Breeder	Accidental	Non-Breeder
Blue-winged Warbler	Accidental	Non-Breeder	Does Not Occur	Not Applicable
Golden-winged Warbler	Accidental	Non-Breeder	Does Not Occur	Not Applicable
Tennessee Warbler	Occurs	Non-Breeder	Accidental	Non-Breeder
Orange-crowned Warbler	Occurs	Breeds	Occurs	Breeds
Nashville Warbler	Occurs	Breeds	Occurs	Breeds
Virginia's Warbler	Accidental	Non-Breeder	Does Not Occur	Not Applicable
Lucy's Warbler	Accidental	Non-Breeder	Does Not Occur	Not Applicable
Northern Parula	Accidental	Non-Breeder	Accidental	Non-Breeder
Yellow Warbler	Occurs	Breeds	Occurs	Breeds
Chestnut-sided Warbler	Accidental	Non-Breeder	Accidental	Non-Breeder
Magnolia Warbler	Accidental	Non-Breeder	Accidental	Non-Breeder
Cape May Warbler	Accidental	Non-Breeder	Accidental	Non-Breeder
Black-throated Blue Warbler	Occurs	Non-Breeder	Accidental	Non-Breeder
Yellow-rumped Warbler	Occurs	Breeds	Occurs	Breeds
Black-throated Gray Warbler	Occurs	Breeds	Occurs	Breeds
Black-throated Green Warbler	Accidental	Non-Breeder	Accidental	Non-Breeder
Townsend's Warbler	Occurs	Breeds	Occurs	Breeds
Hermit Warbler	Occurs	Breeds	Occurs	Breeds
Blackburnian Warbler	Accidental	Non-Breeder	Accidental	Non-Breeder
Yellow-throated Warbler	Accidental	Non-Breeder	Does Not Occur	Not Applicable
Pine Warbler	Accidental	Non-Breeder	Does Not Occur	Not Applicable
Prairie Warbler	Accidental	Non-Breeder	Accidental	Non-Breeder
Palm Warbler	Occurs	Non-Breeder	Occurs	Non-Breeder
Bay-breasted Warbler	Accidental	Non-Breeder	Does Not Occur	Not Applicable
Blackpoll Warbler	Accidental	Non-Breeder	Accidental	Non-Breeder
Black-and-white Warbler	Accidental	Non-Breeder	Accidental	Non-Breeder
American Redstart	Occurs	Breeds	Occurs	Breeds
Prothonotary Warbler	Accidental	Non-Breeder	Accidental	Non-Breeder
Worm-eating Warbler	Accidental	Non-Breeder	Does Not Occur	Not Applicable
Ovenbird	Accidental	Non-Breeder	Accidental	Non-Breeder
Northern Waterthrush	Occurs	Breeds	Occurs	Breeds
Kentucky Warbler	Accidental	Non-Breeder	Accidental	Non-Breeder
Mourning Warbler	Accidental	Non-Breeder	Does Not Occur	Not Applicable
Macgillivray's Warbler	Occurs	Breeds	Occurs	Breeds
Common Yellowthroat	Occurs	Breeds	Occurs	Breeds
Hooded Warbler	Accidental	Non-Breeder	Accidental	Non-Breeder
Wilson's Warbler	Occurs	Breeds	Occurs	Breeds
Canada Warbler	Accidental	Non-Breeder	Does Not Occur	Not Applicable
Yellow-breasted Chat	Occurs	Breeds	Occurs	Breeds
Summer Tanager	Accidental	Non-Breeder	Does Not Occur	Not Applicable
Scarlet Tanager	Accidental	Non-Breeder	Does Not Occur	Not Applicable
Western Tanager	Occurs	Breeds	Occurs	Breeds
Green-tailed Towhee	Occurs	Breeds	Occurs	Breeds
Spotted Towhee	Occurs	Breeds	Occurs	Breeds
California Towhee	Occurs	Breeds	Does Not Occur	Not Applicable
American Tree Sparrow	Occurs	Non-Breeder	Occurs	Non-Breeder
Chipping Sparrow	Occurs	Breeds	Occurs	Breeds
Clay-colored Sparrow	Occurs	Non-Breeder	Occurs	Breeds
Brewer's Sparrow	Occurs	Breeds	Occurs	Breeds
Black-chinned Sparrow	Accidental	Non-Breeder	Does Not Occur	Not Applicable
Vesper Sparrow	Occurs	Breeds	Occurs	Breeds
Lark Sparrow	Occurs	Breeds	Occurs	Breeds

Common Name	Oregon		Washington	
	Occurrence	Breeding Status	Occurrence	Breeding Status
Black-throated Sparrow	Occurs	Breeds	Occurs	Breeds
Sage Sparrow	Occurs	Breeds	Occurs	Breeds
Lark Bunting	Accidental	Non-Breeder	Accidental	Non-Breeder
Savannah Sparrow	Occurs	Breeds	Occurs	Breeds
Grasshopper Sparrow	Occurs	Breeds	Occurs	Breeds
LeConte's Sparrow	Accidental	Non-Breeder	Accidental	Non-Breeder
Nelson's Sharp-tailed Sparrow	Does Not Occur	Not Applicable	Accidental	Non-Breeder
Fox Sparrow	Occurs	Breeds	Occurs	Breeds
Song Sparrow	Occurs	Breeds	Occurs	Breeds
Lincoln's Sparrow	Occurs	Breeds	Occurs	Breeds
Swamp Sparrow	Occurs	Non-Breeder	Occurs	Non-Breeder
White-throated Sparrow	Occurs	Non-Breeder	Occurs	Non-Breeder
Harris's Sparrow	Occurs	Non-Breeder	Occurs	Non-Breeder
White-crowned Sparrow	Occurs	Breeds	Occurs	Breeds
Golden-crowned Sparrow	Occurs	Non-Breeder	Occurs	Non-Breeder
Dark-eyed Junco	Occurs	Breeds	Occurs	Breeds
McCown's Longspur	Accidental	Non-Breeder	Does Not Occur	Not Applicable
Lapland Longspur	Occurs	Non-Breeder	Occurs	Non-Breeder
Chestnut-collared Longspur	Accidental	Non-Breeder	Accidental	Non-Breeder
Rustic Bunting	Accidental	Non-Breeder	Accidental	Non-Breeder
Snow Bunting	Occurs	Non-Breeder	Occurs	Non-Breeder
Mckay's Bunting	Accidental	Non-Breeder	Accidental	Non-Breeder
Rose-breasted Grosbeak	Accidental	Non-Breeder	Accidental	Non-Breeder
Black-headed Grosbeak	Occurs	Breeds	Occurs	Breeds
Blue Grosbeak	Accidental	Non-Breeder	Does Not Occur	Not Applicable
Lazuli Bunting	Occurs	Breeds	Occurs	Breeds
Indigo Bunting	Accidental	Non-Breeder	Accidental	Non-Breeder
Painted Bunting	Accidental	Non-Breeder	Does Not Occur	Not Applicable
Dickcissel	Accidental	Non-Breeder	Accidental	Non-Breeder
Bobolink	Occurs	Breeds	Occurs	Breeds
Red-winged Blackbird	Occurs	Breeds	Occurs	Breeds
Tricolored Blackbird	Occurs	Breeds	Does Not Occur	Not Applicable
Western Meadowlark	Occurs	Breeds	Occurs	Breeds
Yellow-headed Blackbird	Occurs	Breeds	Occurs	Breeds
Rusty Blackbird	Accidental	Non-Breeder	Accidental	Non-Breeder
Brewer's Blackbird	Occurs	Breeds	Occurs	Breeds
Common Grackle	Accidental	Non-Breeder	Accidental	Non-Breeder
Great-tailed Grackle	Accidental	Non-Breeder	Accidental	Non-Breeder
Brown-headed Cowbird	Occurs	Breeds	Occurs	Breeds
Orchard Oriole	Accidental	Non-Breeder	Accidental	Non-Breeder
Hooded Oriole	Accidental	Non-Breeder	Accidental	Non-Breeder
Streak-backed Oriole	Accidental	Non-Breeder	Does Not Occur	Not Applicable
Baltimore Oriole	Accidental	Non-Breeder	Accidental	Non-Breeder
Bullock's Oriole	Occurs	Breeds	Occurs	Breeds
Scott's Oriole	Accidental	Non-Breeder	Accidental	Non-Breeder
Brambling	Accidental	Non-Breeder	Accidental	Non-Breeder
Gray-crowned Rosy-Finch	Occurs	Breeds	Occurs	Breeds
Black Rosy-finch	Occurs	Breeds	Does Not Occur	Not Applicable
Pine Grosbeak	Occurs	Breeds	Occurs	Breeds
Purple Finch	Occurs	Breeds	Occurs	Breeds
Cassin's Finch	Occurs	Breeds	Occurs	Breeds
House Finch	Occurs	Breeds	Occurs	Breeds
Red Crossbill	Occurs	Breeds	Occurs	Breeds
White-winged Crossbill	Occurs	Non-Breeder	Occurs	Breeds
Common Redpoll	Occurs	Non-Breeder	Occurs	Non-Breeder
Hoary Redpoll	Accidental	Non-Breeder	Accidental	Non-Breeder
Pine Siskin	Occurs	Breeds	Occurs	Breeds
Lesser Goldfinch	Occurs	Breeds	Occurs	Breeds
Lawrence's Goldfinch	Accidental	Non-Breeder	Does Not Occur	Not Applicable

Common Name	Oregon		Washington	
	Occurrence	Breeding Status	Occurrence	Breeding Status
American Goldfinch	Occurs	Breeds	Occurs	Breeds
Evening Grosbeak	Occurs	Breeds	Occurs	Breeds
House Sparrow	Non-native	Breeds	Non-native	Breeds
Mammals				
Virginia Opossum	Non-native	Breeds	Non-native	Breeds
Masked Shrew	Does Not Occur	Not Applicable	Occurs	Breeds
Preble's Shrew	Occurs	Breeds	Occurs	Breeds
Vagrant Shrew	Occurs	Breeds	Occurs	Breeds
Montane Shrew	Occurs	Breeds	Occurs	Breeds
Baird's Shrew	Occurs	Breeds	Does Not Occur	Not Applicable
Fog Shrew	Occurs	Breeds	Does Not Occur	Not Applicable
Pacific Shrew	Occurs	Breeds	Does Not Occur	Not Applicable
Water Shrew	Occurs	Breeds	Occurs	Breeds
Pacific Water Shrew	Occurs	Breeds	Occurs	Breeds
Trowbridge's Shrew	Occurs	Breeds	Occurs	Breeds
Merriam's Shrew	Does Not Occur	Not Applicable	Occurs	Breeds
Pygmy Shrew	Does Not Occur	Not Applicable	Occurs	Breeds
Shrew-mole	Occurs	Breeds	Occurs	Breeds
Townsend's Mole	Occurs	Breeds	Occurs	Breeds
Coast Mole	Occurs	Breeds	Occurs	Breeds
Broad-footed Mole	Occurs	Breeds	Does Not Occur	Not Applicable
California Myotis	Occurs	Breeds	Occurs	Breeds
Western Small-footed Myotis	Occurs	Breeds	Occurs	Breeds
Yuma Myotis	Occurs	Breeds	Occurs	Breeds
Little Brown Myotis	Occurs	Breeds	Occurs	Breeds
Long-legged Myotis	Occurs	Breeds	Occurs	Breeds
Fringed Myotis	Occurs	Breeds	Occurs	Breeds
Keen's Myotis	Does Not Occur	Not Applicable	Occurs	Breeds
Long-eared Myotis	Occurs	Breeds	Occurs	Breeds
Silver-haired Bat	Occurs	Breeds	Occurs	Breeds
Western Pipistrelle	Occurs	Breeds	Occurs	Breeds
Big Brown Bat	Occurs	Breeds	Occurs	Breeds
Hoary Bat	Occurs	Non-Breeder	Occurs	Non-Breeder
Spotted Bat	Accidental	Non-Breeder	Occurs	Breeds
Townsend's Big-eared Bat	Occurs	Breeds	Occurs	Breeds
Pallid Bat	Occurs	Breeds	Occurs	Breeds
Brazilian Free-tailed Bat	Occurs	Breeds	Does Not Occur	Not Applicable
American Pika	Occurs	Breeds	Occurs	Breeds
Pygmy Rabbit	Occurs	Breeds	Occurs	Breeds
Brush Rabbit	Occurs	Breeds	Does Not Occur	Not Applicable
Eastern Cottontail	Non-native	Breeds	Non-native	Breeds
Nuttall's (Mountain) Cottontail	Occurs	Breeds	Occurs	Breeds
European Rabbit	Does Not Occur	Not Applicable	Non-native	Breeds
Snowshoe Hare	Occurs	Breeds	Occurs	Breeds
White-tailed Jackrabbit	Occurs	Breeds	Occurs	Breeds
Black-tailed Jackrabbit	Occurs	Breeds	Occurs	Breeds
Mountain Beaver	Occurs	Breeds	Occurs	Breeds
Least Chipmunk	Occurs	Breeds	Occurs	Breeds
Yellow-pine Chipmunk	Occurs	Breeds	Occurs	Breeds
Townsend's Chipmunk	Occurs	Breeds	Occurs	Breeds
Allen's Chipmunk	Occurs	Breeds	Does Not Occur	Not Applicable
Siskiyou Chipmunk	Occurs	Breeds	Does Not Occur	Not Applicable
Red-tailed Chipmunk	Does Not Occur	Not Applicable	Occurs	Breeds
Yellow-bellied Marmot	Occurs	Breeds	Occurs	Breeds
Hoary Marmot	Does Not Occur	Not Applicable	Occurs	Breeds
Olympic Marmot	Does Not Occur	Not Applicable	Occurs	Breeds

Common Name	Oregon		Washington	
	Occurrence	Breeding Status	Occurrence	Breeding Status
White-tailed Antelope Squirrel	Occurs	Breeds	Does Not Occur	Not Applicable
Townsend's Ground Squirrel	Occurs	Breeds	Occurs	Breeds
Merriam's Ground Squirrel	Occurs	Breeds	Does Not Occur	Not Applicable
Piute Ground Squirrel	Occurs	Breeds	Does Not Occur	Not Applicable
Washington Ground Squirrel	Occurs	Breeds	Occurs	Breeds
Wyoming Ground Squirrel	Extirpated	Bred Historically	Does Not Occur	Not Applicable
Belding's Ground Squirrel	Occurs	Breeds	Does Not Occur	Not Applicable
Columbian Ground Squirrel	Occurs	Breeds	Occurs	Breeds
California Ground Squirrel	Occurs	Breeds	Occurs	Breeds
Golden-mantled Ground Squirrel	Occurs	Breeds	Occurs	Breeds
Cascade Golden-mantled Ground Squirrel	Does Not Occur	Not Applicable	Occur	Breeds
Eastern Gray Squirrel	Non-native	Breeds	Non-native	Breeds
Eastern Fox Squirrel	Non-native	Breeds	Non-native	Breeds
Western Gray Squirrel	Occurs	Breeds	Occurs	Breeds
Red Squirrel	Occurs	Breeds	Occurs	Breeds
Douglas' Squirrel	Occurs	Breeds	Occurs	Breeds
Northern Flying Squirrel	Occurs	Breeds	Occurs	Breeds
Northern Pocket Gopher	Occurs	Breeds	Occurs	Breeds
Western Pocket Gopher	Occurs	Breeds	Occurs	Breeds
Camas Pocket Gopher	Occurs	Breeds	Does Not Occur	Not Applicable
Botta's (Pistol Riv.) Pocket Gopher	Occurs	Breeds	Does Not Occur	Not Applicable
Townsend's Pocket Gopher	Occurs	Breeds	Does Not Occur	Not Applicable
Great Basin Pocket Mouse	Occurs	Breeds	Occurs	Breeds
Little Pocket Mouse	Occurs	Breeds	Does Not Occur	Not Applicable
Dark Kangaroo Mouse	Occurs	Breeds	Does Not Occur	Not Applicable
Ord's Kangaroo Rat	Occurs	Breeds	Occurs	Breeds
Chisel-toothed Kangaroo Rat	Occurs	Breeds	Does Not Occur	Not Applicable
California Kangaroo Rat	Occurs	Breeds	Does Not Occur	Not Applicable
American Beaver	Occurs	Breeds	Occurs	Breeds
Western Harvest Mouse	Occurs	Breeds	Occurs	Breeds
Deer Mouse	Occurs	Breeds	Occurs	Breeds
Columbian Mouse	Does Not Occur	Not Applicable	Occurs	Breeds
Canyon Mouse	Occurs	Breeds	Does Not Occur	Not Applicable
Pinon Mouse	Occurs	Breeds	Does Not Occur	Not Applicable
Northern Grasshopper Mouse	Occurs	Breeds	Occurs	Breeds
Desert Woodrat	Occurs	Breeds	Does Not Occur	Not Applicable
Dusky-footed Woodrat	Occurs	Breeds	Does Not Occur	Not Applicable
Bushy-tailed Woodrat	Occurs	Breeds	Occurs	Breeds
Southern Red-backed Vole	Occurs	Breeds	Occurs	Breeds
Western Red-backed Vole	Occurs	Breeds	Does Not Occur	Not Applicable
Heather Vole	Occurs	Breeds	Occurs	Breeds
White-footed Vole	Occurs	Breeds	Does Not Occur	Not Applicable
Red Tree Vole	Occurs	Breeds	Does Not Occur	Not Applicable
Meadow Vole	Does Not Occur	Not Applicable	Occurs	Breeds
Montane Vole	Occurs	Breeds	Occurs	Breeds
Gray-tailed Vole	Occurs	Breeds	Occurs	Breeds
California Vole	Occurs	Breeds	Does Not Occur	Not Applicable
Townsend's Vole	Occurs	Breeds	Occurs	Breeds
Long-tailed Vole	Occurs	Breeds	Occurs	Breeds
Creeping Vole	Occurs	Breeds	Occurs	Breeds
Water Vole	Occurs	Breeds	Occurs	Breeds
Sagebrush Vole	Occurs	Breeds	Occurs	Breeds
Muskrat	Occurs	Breeds	Occurs	Breeds
Northern Bog Lemming	Does Not Occur	Not Applicable	Occurs	Breeds
Black Rat	Non-native	Breeds	Non-native	Breeds
Norway Rat	Non-native	Breeds	Non-native	Breeds
House Mouse	Non-native	Breeds	Non-native	Breeds
Western Jumping Mouse	Occurs	Breeds	Occurs	Breeds
Pacific Jumping Mouse	Occurs	Breeds	Occurs	Breeds
Common Porcupine	Occurs	Breeds	Occurs	Breeds

Common Name	Oregon		Washington	
	Occurrence	Breeding Status	Occurrence	Breeding Status
Nutria	Non-native	Breeds	Non-native	Breeds
Coyote	Occurs	Breeds	Occurs	Breeds
Gray Wolf	Extirpated	Bred Historically	Occurs	Breeds
Red Fox	Occurs	Breeds	Occurs	Breeds
Kit Fox	Occurs	Breeds	Does Not Occur	Not Applicable
Gray Fox	Occurs	Breeds	Does Not Occur	Not Applicable
Black Bear	Occurs	Breeds	Occurs	Breeds
Grizzly Bear	Extirpated	Bred Historically	Occurs	Breeds
Ringtail	Occurs	Breeds	Does Not Occur	Not Applicable
Raccoon	Occurs	Breeds	Occurs	Breeds
American Marten	Occurs	Breeds	Occurs	Breeds
Fisher	Occurs	Breeds	Occurs	Breeds
Ermine	Occurs	Breeds	Occurs	Breeds
Long-tailed Weasel	Occurs	Breeds	Occurs	Breeds
Mink	Occurs	Breeds	Occurs	Breeds
Wolverine	Occurs	Breeds	Occurs	Breeds
American Badger	Occurs	Breeds	Occurs	Breeds
Western Spotted Skunk	Occurs	Breeds	Occurs	Breeds
Striped Skunk	Occurs	Breeds	Occurs	Breeds
Northern River Otter	Occurs	Breeds	Occurs	Breeds
Mountain Lion	Occurs	Breeds	Occurs	Breeds
Lynx	Occurs	Breeds	Occurs	Breeds
Bobcat	Occurs	Breeds	Occurs	Breeds
Wild Burro	Non-native	Breeds	Does Not Occur	Not Applicable
Feral Horse	Non-native	Breeds	Non-native	Breeds
Feral Pig	Non-native	Breeds	Non-native	Non-Breeder
Roosevelt Elk	Occurs	Breeds	Occurs	Breeds
Rocky Mountain Elk	Occurs	Breeds	Occurs	Breeds
Black-tailed Deer	Occurs	Breeds	Occurs	Breeds
Mule Deer	Occurs	Breeds	Occurs	Breeds
Columbian White-tailed Deer	Occurs	Breeds	Occurs	Breeds
White-tailed Deer (Eastside)	Occurs	Breeds	Occurs	Breeds
Moose	Accidental	Non-Breeder	Occurs	Breeds
Mountain Caribou	Does Not Occur	Not Applicable	Occurs	Breeds
Pronghorn Antelope	Occurs	Breeds	Extirpated	Bred Historically
Bison	Extirpated	Bred Historically	Extirpated	Bred Historically*
Mountain Goat	Reintroduced	Breeds	Occurs	Breeds
Rocky Mountain Bighorn Sheep	Reintroduced	Breeds	Reintroduced	Breeds
California Bighorn Sheep	Reintroduced	Breeds	Reintroduced	Breeds
Northern Fur Seal	Occurs	Non-Breeder	Occurs	Non-Breeder
Northern (Steller) Sea Lion	Occurs	Breeds	Occurs	Non-Breeder
California Sea Lion	Occurs	Non-Breeder	Occurs	Non-Breeder
Harbor Seal	Occurs	Breeds	Occurs	Breeds
Northern Elephant Seal	Occurs	Breeds	Occurs	Non-Breeder
Sea Otter	Extirpated	Bred Historically	Reintroduced	Breeds
Gray Whale	Occurs	Non-Breeder	Occurs	Non-Breeder
Minke Whale	Occurs	Breeds	Occurs	Breeds
Sei Whale	Accidental	Non-Breeder	Accidental	Non-Breeder
Blue Whale	Occurs	Non-Breeder	Occurs	Non-Breeder
Fin Whale	Accidental	Non-Breeder	Accidental	Non-Breeder
Humpback Whale	Occurs	Non-Breeder	Occurs	Non-Breeder
Northern Right Whale	Accidental	Non-Breeder	Accidental	Non-Breeder
Striped Dolphin	Accidental	Non-Breeder	Accidental	Non-Breeder
Common Saddle-backed Dolphin	Accidental	Non-Breeder	Accidental	Non-Breeder

Common Name	Oregon		Washington	
	Occurrence	Breeding Status	Occurrence	Breeding Status
False Killer Whale	Accidental	Non-Breeder	Accidental	Non-Breeder
Short-finned Pilot Whale	Occurs	Non-Breeder	Occurs	Non-Breeder
Killer Whale	Occurs	Breeds	Occurs	Breeds
Northern Right-whale Dolphin	Occurs	Non-Breeder	Occurs	Non-Breeder
Harbor Porpoise	Occurs	Breeds	Occurs	Breeds
Dall's Porpoise	Occurs	Breeds	Occurs	Breeds
North Pacific Bottle-nosed Whale	Occurs	Non-Breeder	Occurs	Non-Breeder
Goose-beaked Whale	Accidental	Non-Breeder	Accidental	Non-Breeder
Bering Sea Beaked Whale	Accidental	Non-Breeder	Accidental	Non-Breeder
Arch-beaked Whale	Accidental	Non-Breeder	Accidental	Non-Breeder
Pygmy Sperm Whale	Accidental	Non-Breeder	Accidental	Non-Breeder
Sperm Whale	Occurs	Non-Breeder	Occurs	Non-Breeder
Beluga Whale	Does Not Occur	Not Applicable	Accidental	Non-Breeder

*Historical Breeding Unknown.

Table 2. The 32 wildlife-habitat types as determined from the cluster analysis procedure using 541 native breeding species and 119 Pacific Northwest vegetation, land use, and marine groupings.

Wildlife-Habitat Types
Vegetative/Land Use/Marine Groupings

Westside Lowland Conifer-Hardwood Forest
Alnus rubra-Acer macrophyllum Upland Forests
Picea sitchensis-Tsuga heterophylla Forests
Pseudotsuga menziesii-Alnus rubra-Acer macrophyllus Forests
Maritime Tsuga heterophylla-Thuja plicata Forests
Forested Dunes

Westside Oak and Dry Douglas-fir Forest and Woodlands
Westside Quercus garryana Forests and Woodlands
Westside Quercus garryana-Pseudotsuga menziesii Forests
Westside Dry Pseudotsuga menziesii Forests
Pseudotsuga menziesii-Arbutus menziesii Forests

Southwest Oregon Mixed Conifer-Hardwood Forest
Abies concolor Mixed Conifer Forests
Pinus jeffreii Woodlands
Pseudotsuga menziesii-Lithocarpus densiflorus Forests
Southwest Oregon Low Elevation Mixed Conifer Forests

Montane Mixed Conifer Forest
Abies amabilis-Tsuga heterophylla Forests
Abies lasiocarpa-Picea englemannii Forests
Abies magnifica var. shastensis Forests and Woodlands
Tsuga mertensiana Forests
Tsuga mertensiana-Abies amabilis Forests

Eastside (Interior) Mixed Conifer Forest
Eastside Abies grandis-Pseudotsuga menziesii Forests
Eastside Pseudotsuga menziesii-Pinus ponderosa Forests
Eastside Tsuga heterophylla-Thuja plicata Forests

Lodgepole Pine Forest and Woodlands
Pinus contorta Grass understory
Pinus contorta Shrub understory
Pinus contorta Subalpine Forests
Pinus contorta Woodlands and Forests on Pumice

Wildlife-Habitat Types
Vegetative/Land Use/Marine Groupings

Ponderosa Pine Forest and Woodlands
Pinus ponderosa Woodlands
Eastside Pinus ponderosa-Quercus garryana Forest and Woodlands
Eastside Quercus garryana Woodlands

Upland Aspen Forest
Populus tremuloides Upland Forests

Subalpine Parklands
Subapline and Alpine Wetlands
Pinus albicaulis-Abies lasiocarpa Woodlands and Parklands
Tsuga mertensiana Parklands

Alpine Grasslands and Shrublands
Subalpine and Alpine Grasslands
Alpine Dwarf Shrublands-Fellfields and Sedge Turf

Westside Grasslands
Westside Festuca idahoensis var. romeri-Danthonia californica

Ceanothus-Manzanita Shrublands
Chaparral

Western Juniper and Mountain Mahogany Woodlands
Juniperus occidentalis Scablands
Juniperus occidentalis-Artemisia tridentata Tall Shrublands
Juniperus occidentalis/Bunchgrass
Cercocarpus ledifolius

Eastside (Interior) Canyon Shrublands
Eastside Moist Deciduous Shrublands

Eastside (Interior) Grasslands
Pseudoroegneria spicata Grasslands
Eastside Low-to-Mid-elevation Festuca idahoensis Grasslands
Eastside Modified Grasslands
Sporobolus cryptandrus-Aristida puppurea var. longiseta Grasslands

Wildlife-Habitat Types
Vegetative/Land Use/Marine Groupings

Shrub-steppe
Artemisia tripartita Shrub-steppe
Artemisia cana Shrub-steppe
Artemisia tridentata ssp. tridentata and ssp.wyomingensis Shrub-steppe
Artemisia tridentata ssp. vaseyana Shrublands
Purshia tridentata Shrub-steppe
Sandy steppe and Shrub-steppe

Dwarf Shrub-steppe
Artemisia rigida/Eriogonum spp./Poa secunda Dwarf-Shrub Scabland
Artemisia arbuscula Dwarf-Shrub-steppe
Artemisia nova Dwarf-Shrublands

Desert Playa and Salt Scrub
Alkali Grasslands and Wetlands
Atriplex confertifolia Shrublands
Mixed Saltdesert Shrub-Non-Playa
Mixed Saltdesert Shrub-Playa
Sarcobatus vermiculatus Shrublands

Agriculture, Pasture, and Mixed Environs*
Cultivated Croplands
Improved Pasture
Modified Grasslands
Orchard/Vineyard/Nursery
Unimproved Pasture

Urban and Mixed Environs*
High Density
Moderate Density
Low Density

Open Water-Lakes, Rivers, Streams
Riverine
Lacustrine -Open Water

Herbaceous Wetlands
Graminoid Wet Meadow
Freshwater Aquatic Beds
Herbaceous and Sedge Wetlands

Westside Riparian - Wetlands
Alnus viridis ssp. sinuata-Acer circinatum Shrublands
Westside Riparian and Wetland Deciduous Forests
Picea sitchensis Wetland Forests and Woodlands
Tsuga heterophylla-Thuja plicata coniferous wetlands
Westside Riparian/Wetland Shrublands
Shrub/herbaceous Sphagnum Bogs
Wooded Bogs

Montane Coniferous Wetlands
Westside Montane Coniferous Wetlands
Picea engelmannii Forested Wetlands

Eastside (Interior) Riparian - Wetlands
Eastside Midmontane *Alnus incana-Salix spp.* Riparian Shrublands
Eastside Lowland Riparian Shrublands
Eastside *Populus balsamifera spp. trichocarpa*
Alnus rhombifolia Riparian

Wildlife-Habitat Types
Vegetative/Land Use/Marine Groupings

Pinus ponderosa Riparian Woodlands
Populus tremuloides Riparian/Wetland Forests and Woodlands

Coastal Dunes and Beaches
Coastal Dune Grasslands
Coastal Dune Shrublands

Coastal Headlands and Islets
Coastal Headland Shrublands and Grasslands

Bays and Estuaries*
Bays and Estuaries (includes Intertidal Marshes)

Inland Marine Deeper Waters*
Puget Sound to Strait of Juan de Fuca

Marine Nearshore*
Marine environment from shore line to 20 m depth

Marine Shelf*
Marine environment from 20 m to 200 m depth

Oceanic*
Marine environment greater than 200 m depth

*Wildlife-habitats that were determined by an expert panel process rather than through the cluster analysis effort.

Table 3. Number of wildlife species by habitat and by their association.

Wildlife-Habitat Types	Total	Closely Associated	Generally Associated	Present	Unsure
Westside Lowland Conifer-Deciduous Forest	233	48	155	29	1
Westside Oak and Dry Douglas-fir Forest and Woodlands	229	34	148	40	7
Southwest Oregon Mixed Conifer-Deciduous Forest	236	35	163	33	5
Montane Mixed Conifer Forest	221	35	138	38	10
Eastside (Interior) Mixed Conifer Forest	225	38	146	35	6
Lodgepole Pine Forest and Woodlands	166	15	95	48	8
Ponderosa Pine Forest and Woodlands	247	26	160	46	15
Upland Aspen Forest	144	4	85	42	13
Subalpine Parkland	184	19	92	69	4
Alpine Grasslands and Shrublands	132	19	35	71	7
Westside Grasslands	157	12	105	39	1
Ceanothus-Manzanita Shrublands	138	4	89	39	6
Western Juniper and Mountain Mahogany Woodlands	173	17	113	36	7
Eastside (Interior) Canyon Shrubland	147	13	84	38	12
Eastside (Interior) Grasslands	174	33	100	38	3
Shrub-steppe	184	47	100	34	3
Dwarf Shrub-steppe	149	23	86	32	8
Desert Playa and Salt Scrub	143	27	78	33	5
Agriculture and Pastures Mixed Environs	346	68	174	99	5
Urban and Mixed Environs	266	18	144	102	2
Open Water - Lakes, Rivers, Streams	162	96	53	13	0
Herbaceous Wetlands	228	105	90	31	2
Westside Riparian/Wetlands	256	74	145	35	2
Montane Coniferous Wetlands	148	17	101	28	2
Eastside (Interior) Riparian/Wetlands	271	81	149	36	5
Coastal Dunes and Beaches	155	31	79	41	4
Coastal Headlands and Islets	118	28	63	24	3
Bays and Estuaries	175	66	74	34	1
Inland Marine Deeper Waters	61	26	20	15	0
Marine Nearshore	86	38	33	14	1
Marine Shelf	72	41	20	11	0
Oceanic	44	31	8	5	0

* Historical associations are not shown.

2

Wildlife Habitats: Descriptions, Status, Trends, and System Dynamics

*Christopher B. Chappell, Rex C. Crawford, Charley Barrett,
Jimmy Kagan, David H. Johnson, Mikell O'Mealy,
Greg A. Green, Howard L. Ferguson, W. Daniel Edge,
Eva L. Greda, & Thomas A. O'Neil*

Introduction

In the previous chapter, the authors described the process by which the 32 wildlife-habitat types of Oregon and Washington (Table 1, Frontispiece) were defined. In this chapter, we offer detailed descriptions of each wildlife-habitat type to support a common understanding for their delineation, inventory, and management across the region.

Each wildlife habitat below is described as to its geographic distribution, physical setting, landscape setting, structure, and composition. Additionally, we include other information that might help managers, researchers, and others gain further insight into each habitat, such as listing other classification systems and key references, natural disturbance regimes, succession and stand dynamics, effects of management and anthropogenic impacts, and status and trends. We have also included photographs of each wildlife-habitat type to give the reader an idea of what each habitat type looks like. Multiple photographs are offered for most habitats in order to depict some of the variability that exists within each type. Each of the habitats below is numbered (Table 1); the descriptions in this chapter reflect the same numbering sequence.

The following are definitions of each category used to characterize the wildlife-habitat types:

Geographic Distribution. Describes the broad geographic range within which the habitat is located, both within Oregon and Washington and elsewhere. Major variations in dominance type are noted either here or under Composition.

Physical Setting. Describes physical features of the environment on sites where the habitat is found in Oregon and Washington. These typically include climate, elevation, soils, hydrology, geology, and topography.

Landscape Setting. Describes the landscape pattern and distribution of the habitat in relation to other habitats. Primary land use is also noted.

Structure. Describes the physical structure of the habitat, both its typical aspect and the range of variation in structure present within the habitat. Aspects of physical structure include some description of cover or density

(horizontal dimension) of vegetation; layering (vertical dimension) of vegetation; dominant growth forms, leaf phenologies (evergreen or deciduous), leaf characters (conifer or broadleaf), and vegetation persistence (annual or perennial) represented in different structural layers; and significant structural components of dead and decaying vegetation. Growth forms include trees, shrubs (>1.6 ft [0.5 m] tall), dwarf-shrubs (<1.6 ft [0.5 m] tall), graminoids (grasses, sedges, rushes), forbs, ferns, mosses, lichens, and algae. Vegetation cover categories frequently referred to include forest (>60% cover of trees), woodland (25-60% cover of trees), shrubland (>25% cover of shrubs), dwarf-shrubland (>25% cover of dwarf-shrubs), and grassland (graminoids dominant). Water-dominated habitats (e.g., marine and open water) may be described in terms of the physical aspects of the water column and the bottom substrate of the habitat.

Composition. Describes the species composition of the vegetation that creates structure. Composition is described as dominant, co-dominant (shares dominance with ≥1 species), or important indicator species by structural layer. English names for all vertebrates are used in the text and corresponding standard names are in Appendix I. The geographic distribution or physical setting is noted for those dominant species that occur only in particular physical settings or specific geographic areas of the overall habitat's range of occurrence.

Other Classifications and Key References. Notes other names that have been applied to this habitat by other classifications or major summary publications, and important references that describe the habitat or parts of the habitat in greater detail.

Natural Disturbance Regime. Describes the major natural disturbances that are important in the habitat. The regime includes the disturbance type, severity, frequency, extent, and range of variation in these characteristics.

Succession and Stand Dynamics. Describes the way in which structure and composition change over time in relation to natural disturbances.

**Table 1. The 32 wildlife habitats and their total acreage in Oregon and Washington.
The marine waters extend out to the 200-mile Exclusive Economic Zone.**

Wildlife Habitat	Oregon Total Acreage	Washington Total Acreage	Page Number
1. Westside Lowlands Conifer-Hardwood Forest	9,349,756	9,064,128	24
2. Westside Oak and Dry Douglas-fir Forest and Woodlands	433,132	425,038	26
3. Southwest Oregon Mixed Conifer-Hardwood Forest	4,020,320	Does Not Occur	28
4. Montane Mixed Conifer Forest	2,949,586	4,653,306	30
5. Eastside (Interior) Mixed Conifer Forest	4,126,957	4,662,101	32
6. Lodgepole Pine Forest and Woodlands	532,587	119,201	34
7. Ponderosa Pine Forest and Woodlands (includes Eastside Oak)	6,226,351	1,927,176	35
8. Upland Aspen Forest	19,685	100,621	37
9. Subalpine Parkland	84,240	327,442	38
10. Alpine Grassland and Shrublands	291,494	1,591,115	40
11. Westside Grasslands	133[1]	22,491	41
12. Ceanothus-Manzanita Shrublands	52,104	Does Not Occur	43
13. Western Juniper and Mountain Mahogany Woodlands	4,037,221	Does Not Occur	45
14. Eastside (Interior) Canyon Shrublands	358,250	Not Mapped[3]	46
15. Eastside (Interior) Grasslands	1,935,794	1,002,076	48
16. Shrub-steppe	17,420,753	7,144,697[3]	50
17. Dwarf Shrub-steppe	514,066	Not Mapped[3]	52
18. Desert Playa and Salt Scrub Shrublands	719,503	Not Mapped[3]	53
19. Agriculture, Pasture and Mixed Environs	6,197,887	9,251,107	55
20. Urban and Mixed Environs	575,087	1,204,680	57
21. Open Water - Lakes, Rivers, Streams	780,901	761,360	59
22. Herbaceous Wetlands	1,031,343	210,451	93
23. Westside Riparian-Wetlands	168,872	347,653	94
24. Montane Coniferous Wetlands	56,099	241,450	97
25. Eastside (Interior) Riparian-Wetlands	31,121	100,763	98
26. Coastal Dunes and Beaches	52,451	Not Mapped[2]	100
27. Coastal Headlands and Islets	9,137	7,776	102
28. Bays and Estuaries	172,748	226,336	103
29. Inland Marine Deeper Water	Does Not Occur	1,855,780	105
30. Marine Nearshore	223,371	750,329	106
31. Marine Shelf	3,905,164	4,780,625	107
32. Oceanic	33,987,189	19,845,660	108
Totals	100,263,303	70, 532,093	

[1] Because of difficulty in classifying this type using remote sensing (i.e., discerning native grasslands from pasture lands) native westside grasslands have inadvertently been classified within the agriculture habitat type. Nonetheless, there are few areas known to be native westside grasslands in Oregon.

[2] This type was not part of the vegetation classification when the Washington Gap Project mapped the state of Washington. Thus, no wildlife habitat area was determined.

[3] In Washington, Eastside Canyon Shrublands, Dwarf Shrub-steppe, and Desert Playa and Salt Scrub Shrublands were mapped as part of Shrub-steppe for the Washington GAP Project. Thus, no wildlife habitat area was determined.

Effects of Management and Anthropogenic Impacts. Describes typical changes in structure and composition observed after typical management activities (human disturbances) and widespread changes in the habitat that have occurred since Euro-American settlement. Disturbances addressed include land uses that do not necessarily convert the habitat to urban or agriculture, but have a significant influence on structure or composition, e.g., hydrologic alterations, logging, and grazing. Exotic species that have become abundant in the habitat are noted.

Status and Trends. Describes the general extent of the type in Oregon and Washington, its current ecological condition, and historical and current trends in extent and condition. Ecological condition refers primarily to how similar the current structure, composition, and disturbance regime is to natural or presettlement conditions. The total number of plant associations recognized in the habitat and the number of those that are considered globally imperiled provide some idea of the degree of loss, degradation, and threat that is associated with the habitat.

1. Westside Lowlands Conifer-Hardwood Forest

Christopher B. Chappell & Jimmy Kagan

Geographic Distribution. This forest habitat occurs throughout low-elevation western Washington, except on extremely dry or wet sites. In Oregon it occurs on the western slopes of the Cascades, around the margins of the Willamette Valley, in the Coast Range, and along the outer coast. The global distribution extends from southeastern Alaska south to southwestern Oregon.

Physical Setting. Climate is relatively mild and moist to wet. Mean annual precipitation is mostly 35-100 inches (90-254 cm), but can vary locally. Snowfall ranges from rare to regular, but is transitory. Summers are relatively dry. Summer fog is a major factor on the outer coast in the Sitka spruce zone. Elevation ranges from sea level to a maximum of about 2,000 ft (610 m) in much of northern Washington and 3,500 ft (1,067 m) in central Oregon. Soils and geology are very diverse. Topography ranges from relatively flat glacial till plains to steep mountainous terrain.

Landscape Setting. This is the most extensive habitat in the lowlands on the westside of the Cascades, except in southwestern Oregon, and forms the matrix within which other habitats occur as patches, especially Westside Riparian-Wetlands and less commonly Herbaceous Wetlands or Open Water. It also occurs adjacent to or in a mosaic with Urban and Mixed Environs (hereafter Urban) or Agriculture, Pasture and Mixed Environs (hereafter Agriculture) habitats. In the driest areas, it occurs adjacent to or in a mosaic with Westside Oak and Dry Douglas-fir Forest and Woodlands. Bordering this habitat at upper elevations is Montane Mixed Conifer Forest. Along the coastline, it often occurs adjacent to Coastal Dunes and Beaches. In southwestern Oregon, it may border Southwest Oregon Mixed Conifer-Hardwood Forest. The primary land use for this habitat is forestry.

Structure. This habitat is forest, or rarely woodland, dominated by evergreen conifers, deciduous broadleaf trees, or both. Late seral stands typically have an abundance of large (>164 ft [50 m] tall) coniferous trees, a multi-layered canopy structure, large snags, and many large logs on the ground. Early seral stands typically have smaller trees, single-storied canopies, and may be dominated by conifers, broadleaf trees, or both. Coarse woody debris is abundant in early seral stands after natural disturbances but much less so after clearcutting. Forest understories are structurally diverse: evergreen shrubs tend to dominate on nutrient-poor or drier sites; deciduous shrubs, ferns, and/or forbs tend to dominate on relatively nutrient-rich or moist sites. Shrubs may be low (1.6 ft [0.5 m] tall), medium-tall (3.3-6.6 ft [1-2 m]), or tall (6.6-13.1 ft [2-4 m]). Almost all structural stages are represented in the successional sequence within this habitat. Mosses are often a major ground cover. Lichens are abundant in the canopy of old stands.

Composition. Western hemlock (*Tsuga heterophylla*) and Douglas-fir (*Pseudotsuga menziesii*) are the most characteristic species and one or both are typically present. Most stands are dominated by one or more of the following: Douglas-fir, western hemlock, western redcedar (*Thuja plicata*), Sitka spruce (*Picea sitchensis*), red alder (*Alnus rubra*), or bigleaf maple (*Acer macrophyllum*). Trees of local importance that may be dominant include Port-Orford cedar (*Chamaecyparis lawsoniana*) in the south, shore pine (*Pinus contorta* var. *contorta*) on stabilized dunes, and grand fir (*Abies grandis*) in drier climates. Western white pine (*Pinus monticola*) is frequent but subordinate in importance through portions of this habitat. Pacific silver fir (*Abies amabilis*) is largely absent except on the wettest portion of the western Olympic Peninsula, where it is common and sometimes co-dominant. Common small subcanopy trees are cascara buckthorn (*Rhamnus purshiana*) in more moist climates and Pacific yew (*Taxus brevifolia*) in somewhat drier climates or sites.

Sitka spruce is found as a major species only in the outer coastal area at low elevations where summer fog is a significant factor. Bigleaf maple is most abundant in the Puget Lowland, around the Willamette Valley, and in the central Oregon Cascades, but occurs elsewhere also. Douglas-fir is absent to uncommon as a native species in the very wet maritime outer coastal area of Washington, including the coastal plain on the west side of the Olympic Peninsula. However, it has been extensively planted in that area. Port-Orford cedar occurs only in southern Oregon. Paper birch (*Betula papyrifera*) occurs as a co-dominant only in Whatcom County, Washington. Grand fir occurs as an occasional co-dominant only in the Puget Lowland and Willamette Valley.

Dominant or co-dominant understory shrub species of more than local importance include salal (*Gaultheria shallon*), dwarf Oregongrape (*Mahonia nervosa*), vine maple (*Acer circinatum*), Pacific rhododendron (*Rhododendron macrophyllum*), salmonberry (*Rubus spectabilis*), trailing blackberry (*R. ursinus*), red elderberry (*Sambucus racemosa*), fools huckleberry (*Menziesia ferruginea*), beargrass (*Xerophyllum tenax*), oval-leaf huckleberry (*Vaccinium ovalifolium*), evergreen huckleberry (*V. ovatum*), and red huckleberry (*V. parvifolium*). Salal, rhododendron, and beargrass are particularly associated with low nutrient or dry sites.

Swordfern (*Polystichum munitum*) is the most common herbaceous species and is often dominant on nitrogen-rich or moist sites. Other forbs and ferns that frequently dominate the understory are Oregon oxalis (*Oxalis oregana*), deerfern (*Blechnum spicant*), bracken fern (*Pteridium aquilinum*), vanillaleaf (*Achlys triphylla*), twinflower (*Linnaea borealis*), false lily-of-the-valley (*Maianthemum dilatatum*), western springbeauty (*Claytonia siberica*), foamflower (*Tiarella trifoliata*), inside-out flower (*Vancouveria hexandra*), and common whipplea (*Whipplea modesta*).

Other Classifications and Key References. This habitat includes most of the forests and their successional seres within the *Tsuga heterophylla* and *Picea sitchensis* zones.[88] This habitat is also referred to as Douglas-fir-western hemlock and Sitka spruce-western hemlock forests,[87] spruce-cedar-hemlock forest (*Picea-Thuja-Tsuga*, No. 1) and cedar-hemlock-Douglas-fir forest (*Thuja-Tsuga-Pseudotsuga*, No. 2).[136] The Oregon Gap II Project[126] and Oregon Vegetation Landscape-Level Cover Types[127] would crosswalk with Sitka spruce-western hemlock maritime forest, Douglas-fir-western hemlock-redcedar forest, red alder forest, red alder-bigleaf maple forest, mixed conifer/ mixed deciduous forest, south coast mixed-deciduous forest, and coastal lodgepole forest. The Washington Gap Vegetation map includes this vegetation as conifer forest, mixed hardwood/conifer forest, and hardwood forest in the Sitka spruce, western hemlock, Olympic Douglas-fir, Puget Sound Douglas-fir, Cowlitz River and Willamette Valley zones.[37] A number of other references describe elements of this habitat.[13, 25, 26, 40, 42, 66, 90, 104, 110, 111, 114, 115, 210]

Natural Disturbance Regime. Fire is the major natural disturbance in all but the wettest climatic area (Sitka spruce zone), where wind becomes the major source of natural disturbance. Natural fire-return intervals generally range from about 100 years or less in the driest areas to several hundred years.[1, 115, 160] Mean fire-return interval for the western hemlock zone as a whole is 250 years, but may vary greatly. Major natural fires are associated with occasional extreme weather conditions.[1] Fires are typically high-severity, with few trees surviving. However, low- and moderate-severity fires that leave partial to complete live canopies are not uncommon, especially in drier climatic areas. Occasional major windstorms hit outer coastal forests most intensely, where fires are rare. Severity of wind disturbance varies greatly, with minor events being extremely frequent and major events occurring once every few decades. Bark beetles and fungi are significant causes of mortality that typically operate on a small scale. Landslides are another natural disturbance that occur in some areas.

Succession and Stand Dynamics. After a severe fire or blowdown, a typical stand will be briefly occupied by annual and perennial ruderal forbs and grasses as well as predisturbance understory shrubs and herbs that resprout.[102] Herbaceous species generally give way to dominance by shrubs or a mixture of shrubs and young trees within a few years. If shrubs are dense and trees did not establish early, the site may remain as a shrubland for an indeterminate period. Early seral tree species can be any of the potential dominants for the habitat, depending on environment, type of disturbance, and seed source. All of these species except the short-lived red alder are capable of persisting for at least a few hundred years. Douglas-fir is the most common dominant after fire, but is uncommon in the wettest zones. It is also the most fire resistant of the trees in this habitat and survives moderate-severity fires well. After the tree canopy closes, the understory may become sparse, corresponding with the stem-exclusion stage.[168] Eventually tree density will decrease and the understory will begin to flourish again, typically at stand age 60-100 years. As trees grow larger and a new generation of shade-tolerant understory trees (usually western hemlock, less commonly western redcedar) grows up, a multi-layered canopy will gradually develop and be well expressed by stand age 200-400 years.[89] Another fire is likely to return before the loss of shade-intolerant Douglas-fir from the canopy at stand age 800-1,000 years, unless the stand is located in the wet maritime zone. Throughout this habitat, western hemlock tends to increase in importance as stand development proceeds. Coarse woody debris peaks in abundance in the first 50 years after a fire and is least abundant at about stand age 100-200 years.[193]

Effects of Management and Anthropogenic Impacts. Red alder is more successful after typical logging disturbance than after fire alone on moist, nutrient-rich sites, perhaps because of the species' ability to establish abundantly on scarified soils.[100] Alder is much more common now because of large-scale logging activities.[87] Alder grows more quickly in height early in succession than the conifers, thereby prompting many forest managers to apply herbicides for alder control. If alder is allowed to grow and dominate early successional stands, it will decline in importance after about 70 years and die out completely by age 100. Often there are suppressed conifers in the subcanopy that potentially can respond to the death of the alder canopy. However, salmonberry sometimes forms a dense shrub layer under the alder, which can exclude conifer regeneration.[88] Salmonberry responds positively to soil disturbance, such as that associated with logging.[19] Bigleaf maple sprouts readily after logging and is therefore well adapted to increase after disturbance as well. Clearcut logging and plantation forestry have resulted in less diverse tree canopies, and have focused mainly on Douglas-fir, with reductions in coarse woody debris over natural levels, a shortened stand initiation phase, and succession truncated well before late-seral characteristics are expressed. Douglas-fir has been almost universally planted, even in wet coastal areas of Washington, where it is rare in natural stands.

Status and Trends. Extremely large areas of this habitat remain. Some loss has occurred, primarily to development in the Puget Lowland. Condition of what remains has been degraded by industrial forest practices at both the stand and landscape scale. Most of the habitat is probably now in Douglas-fir plantations. Only a fraction of the original old-growth forest remains, mostly in national forests in the Cascade and Olympic mountains. Areal extent continues to be reduced gradually, especially in the Puget Lowland. An increase in alternative silviculture practices may be improving structural and species diversity in some areas. However, intensive logging of natural-origin mature and young stands and even small areas of old growth continues. Of the 62 plant associations representing this habitat listed in the National Vegetation Classification, 27 % are globally imperiled or critically imperiled.[10]

2. Westside Oak and Dry Douglas-fir Forest and Woodlands

Christopher B. Chappell & Jimmy Kagan

Geographic Distribution. This habitat is primarily found in the Willamette Valley, Puget Lowlands, and Klamath Mountains ecoregions. In the Puget Lowlands, it is common in and around the San Juan Islands and in parts of Thurston, Pierce, and Mason counties. In southwestern Oregon, it is now restricted mainly to the valleys of the Rogue and Umpqua rivers. Minor occurrences can also be found in the northeastern Olympic Mountains and western Cascades.

This habitat is comprised of several geographic variants: California black oak and ponderosa pine are important only in southwestern Oregon and the southern Willamette Valley. The latter is also found in a small area of Pierce County, Washington. Shore (lodgepole) pine is only important in the Puget Lowland, mainly in San Juan and Mason counties. Dry Douglas-fir forests (without oak or madrone) are mainly in the Puget Lowland and rarely in the Olympic Mountains, west Cascades, and Willamette Valley. Pacific madrone and Douglas-fir/Pacific madrone stands without oak are limited to the Puget Lowland and the southern Willamette Valley foothills. Mixed oak-madrone stands occur primarily in Oregon, especially southwestern Oregon.

Physical Setting. This habitat typically occupies dry sites west of the Cascades. Annual mean precipitation ranges from 17 to 60 inches (43 to 152 cm), occasionally higher. Elevation ranges from sea level to about 3,500 ft (1,069 m)

in the Olympic Mountains, but is mainly below 1,500 ft (457 m). Topography ranges from nearly level to very steep slopes, where aspect tends to be southern or western. Soils on dry sites are typically shallow over bedrock, very stony, or very deep and excessively drained. Willamette Valley soils are typically much older and have more moderate drainage and water availability. Parent materials include various types of bedrock, shallow or very coarse glacial till, alluvium, and glacial outwash.

Landscape Setting. This habitat is found in a mosaic with, or adjacent to, Westside Grasslands, Westside Lowlands Conifer-Hardwood Forest, Westside Riparian-Wetlands, Southwest Oregon Mixed Conifer-Hardwood Forest, Urban, and Agriculture. Inclusions of Open Water or Herbaceous Wetlands sometimes occur. In the Puget Lowland, this habitat is sometimes found adjacent to Puget Sound (Nearshore Marine). Land use of this habitat includes forestry (generally small scale), livestock grazing, and low-density rural residential.

Structure. This is a forest or woodland dominated by evergreen conifers, deciduous broadleaf trees, evergreen broadleaf trees, or some mixture of conifers and broadleaf trees. Canopy structure varies from single- to multi-storied. Large conifers, when present, typically emerge above broadleaf trees in mixed canopy stands. Large snags and logs are less abundant than in other westside forest habitats, but can be prominent, especially in unlogged old stands. Understories vary in structure: grasses, shrubs, ferns, or some combination will typically dominate. Deciduous broadleaf shrubs are perhaps most typical as understory dominants in the existing landscape. Early successional stand structure varies depending on understory species present and if initiated following logging or fire.

Composition. The canopy is typically dominated by one or more of the following species: Douglas-fir (*Pseudotsuga menziesii*), Oregon white oak (*Quercus garryana*), Pacific madrone (*Arbutus menziesii*), shore (lodgepole) pine (*Pinus contorta* var. *contorta*), or California black oak (*Q. kelloggii*). Ponderosa pine (*Pinus ponderosa*) is important in southwestern Oregon and the southern Willamette Valley as a subordinate or co-dominant with oak. Grand fir (*Abies grandis*) is occasionally co-dominant with Douglas-fir in the northern Puget Lowland or in the Willamette Valley. Oregon ash (*Fraxinus latifolia*) is occasionally co-dominant with white oak in riparian oak stands. Several other tree species may be present, but western hemlock (*Tsuga heterophylla*) and western redcedar (*Thuja plicata*) generally cannot regenerate successfully because of dry conditions. This lack of shade-tolerant tree regeneration, along with understory indicators like tall Oregongrape (*Mahonia aquifolium*), and blue wildrye (*Elymus glaucus*), help distinguish dry Douglas-fir forests from mid-seral Douglas-fir stands on more mesic sites, which are part of the Westside Lowlands Conifer-Hardwood Forest. Tree regeneration, when present, is typically Douglas-fir, less commonly grand fir. Sweet cherry (*Prunus avium*) and/or English hawthorn (*Crataegus monogyna*) have invaded and now dominate a subcanopy layer in many oak forests of the Willamette Valley.

Deciduous shrubs that commonly dominate or co-dominate the understory are oceanspray (*Holodiscus discolor*), baldhip rose (*Rosa gymnocarpa*), poison-oak (*Toxicodendron diversiloba*), serviceberry (*Amelanchier alnifolia*), beaked hazel (*Corylus cornuta*), trailing blackberry (*Rubus ursinus*), Indian plum (*Oemleria cerasiformis*), snowberries (*Symphoricarpos albus* and *S. mollis*), wedge-leaf ceanothus (*Ceanothus cuneatus*), and oval-leaf viburnum (*Viburnum ellipticum*). Evergreen shrubs or vines that sometimes are dominant where conifers are important in the canopy include salal (*Gaultheria shallon*), dwarf Oregongrape (*Mahonia nervosa*), Pacific rhododendron (*Rhododendron macrophyllum*), hairy honeysuckle (*Lonicera hispidula*), evergreen huckleberry (*Vaccinium ovatum*), and Piper's barberry (*Mahonia piperiana*).

Native graminoids that commonly dominate or co-dominate the understory are western fescue (*Festuca occidentalis*), Alaska oniongrass (*Melica subulata*), blue wildrye, and long-stolon sedge (*Carex inops*). Kentucky bluegrass (*Poa pratensis*) is a major non-native dominant in oak woodland understories. Swordfern (*Polystichum munitum*) or, less commonly, bracken fern (*Pteridium aquilinum*) sometimes co-dominates the understory, especially on sites that formerly supported grasslands and savannas. Forbs, many of which are characteristic of these dry sites, are often abundant and diverse, but typically do not dominate. Common camas (*Camassia quamash*), cleavers (*Galium aparine*), or other forbs are occasionally co-dominant with graminoids.

Other Classifications and Key References. This habitat has been described as oak groves and dry site Douglas-fir forest in the *Tsuga heterophylla* zone of western Washington and northwestern Oregon as well as oak woodland in the interior valleys of western Oregon.[88] It is also referred to as Oregon oakwoods No. 22 and a minor part of Cedar-hemlock-Douglas-fir forest No. 2;[136] The Oregon Gap II Project[126] and Oregon Vegetation Landscape-Level Cover Types[127] that would represent this type are Oregon white oak forest and Douglas-fir/white oak forest. The Washington Gap Project represents this habitat as part of hardwood forest, mixed hardwood/conifer forest, and conifer forest in the Woodland/Prairie Mosaic, Puget Sound Douglas-fir, and, to a minor degree, Cowlitz River, and Willamette Valley zones of Washington.[37] Other references also describe elements of this habitat.[13, 17, 40, 41, 86, 111, 115, 202, 210]

Natural Disturbance Regime. Fire is the major natural disturbance in this habitat. In presettlement times, fire frequency probably ranged from frequent (every few years) to moderately frequent (once every 50-100 years), and reflected low-severity and moderate-severity fire regimes.[1] Fire frequency has been much lower in the last 100 years. Windstorms are an occasional disturbance, most important in the San Juan Islands and vicinity. Understories are sometimes browsed heavily by deer in the San Juan Islands, thus preventing dominance by deciduous shrubs and favoring grasses and forbs.

Succession and Stand Dynamics. Many of these forests and woodlands were formerly either grasslands or savannas that probably burned frequently, thus preventing dominance by trees.[41, 54] Some portions of this habitat in the central Puget Lowlands may have formerly been dominated by shrubs (salal, beaked hazel, evergreen huckleberry, hairy manzanita [*Arctostaphylos columbiana*]) for lengthy periods, probably also because of the particular combination of fire frequency and intensity. Other areas were woodlands to semi-open forests that burned moderately frequently, as evidenced by the relict stands of old-growth Douglas-fir. The dominant trees in this habitat establish most abundantly after fire. Moderate-severity fires kill many trees but also leave many alive, creating opportunities for establishment of new cohorts of trees and increasing structural complexity.[1] Oaks and madrone resprout after fire if they are top-killed. Without periodic fire, most oak-dominated stands will eventually convert to Douglas-fir forests.[1] Animal dissemination of acorns may be important in dispersal of oaks. Shore pine, where present, is an early-seral upper canopy species that grows quickly and dies out after about 100-150 years, yielding to a mature Douglas-fir stand unless another fire intervenes before the death of the pine.

Effects of Management and Anthropogenic Impacts. Clearcut or similar logging reduces canopy structural complexity and abundance of large woody debris. Dry Douglas-fir stands are well suited to alternative silvicultural practices, such as uneven-aged management or maintaining two-storied stands. Oaks and madrone will typically resprout after logging and thus can increase in importance relative to conifers in mixed canopy stands. Selective logging of Douglas-fir in oak stands can prevent long-term loss of oak dominance. With fire exclusion, stands have probably increased in tree density and grassy understories have been replaced by deciduous shrubs.[41] Moderate to heavy grazing or other significant ground disturbance, especially in grassy understories, leads to increases in non-native invader species, many of which are now abundant in stands with grassy or formerly grassy understories. Scot's broom (*Cytisus scoparius*) is an exotic shrub particularly invasive and persistent in oak woodlands. Exotic herbaceous invaders include colonial bentgrass (*Agrostis capillaris*), common velvetgrass (*Holcus lanatus*), Kentucky bluegrass, tall oatgrass (*Arrhenatherum elatius*), rigid brome (*Bromus rigidus*), orchardgrass (*Dactylis glomerata*), hedgehog dogtail (*Cynosurus echinatus*), tall fescue (*Festuca arundinacea*), and common St. Johnswort (*Hypericum perforatum*).

Status and Trends. This habitat is relatively limited in area and is currently declining in extent and condition. With the cessation of regular burning 100-130 years ago, many grasslands and savannas were invaded by a greater density of trees and thus converted to a different habitat. Conversely, large areas of this habitat have been converted to Urban or Agriculture habitats. Most of what remains has been considerably degraded by invasion of exotic species or by logging and consequent loss of structural diversity. Ongoing threats include residential development, increase and spread of exotic species, and fire suppression effects (the latter especially in oak-dominated stands). Thirteen of 27 plant associations listed in the National Vegetation Classification are considered globally imperiled or critically imperiled.[10]

3. Southwest Oregon Mixed Conifer-Hardwood Forest

Christopher B. Chappell & Jimmy Kagan

Geographic Distribution. This upland forest and woodland habitat occurs in southwestern Oregon, northwestern California, and the Sierra Nevada. In southern Oregon, it is found at low and middle elevations in the Klamath Mountains, Cascades, Coast Range, and Eastern Cascade Slopes and Foothills ecoregions. Portions of Curry, Josephine, Jackson, Douglas, Lane, and Klamath counties are included in the range of this habitat.

Physical Setting. The climate varies from relatively dry and very warm to moderately moist and cool to slightly warm and very moist. Mean annual precipitation ranges from 20 to 140 inches (51 to 356 cm). Snow is uncommon except at the highest elevations, where a winter snow pack occurs for a few months. Summers are hot and dry. Elevation ranges from near sea level to 6,000 ft (1,829 m). Topography is mostly mountainous but also includes two fairly large valleys, and a corresponding variety of terrain. Soils are diverse as is the bedrock geology. Serpentine soils are common in portions of the Siskiyou Mountains, where they have a major effect on vegetation.

Landscape Setting. This habitat is typically bounded at its upper elevation limits by Montane Mixed Conifer Forest and at its lower limits, along the coast, by Westside Lowlands Conifer-Hardwood Forest. At lower elevations in the Rogue and Umpqua valleys it can be found in a mosaic with Westside Oak and Dry Douglas-fir Forest and Woodland, Ceanothus-Manzanita Shrublands, Urban, and Agriculture. Small inclusions of Open Water, Herbaceous Wetlands, Westside Riparian-Wetlands, and Ceanothus-Manzanita Shrublands occur scattered throughout this habitat. The predominant land use is forestry. Low-density residential is prominent in the Rogue and Umpqua valleys.

Grazing occurs on some areas, especially at lower elevations.

Structure. Conifer trees typically dominate this forest or woodland habitat. In some generally more coastal areas, a well developed subcanopy layer of smaller evergreen broadleaf trees is present. Occasionally, deciduous broadleaf trees are co-dominant. Complex multi-layered canopies are typical, though single-layered canopies also occur, especially in areas of intensive forest management. Dominant canopy trees vary from 60 to >300 ft (18 to >91 m) tall at maturity. Large woody debris (snags and logs) is typically common, although variable. Understories are mostly dominated by shrubs, but can be dominated by forbs, graminoids, or may be largely depauperate.

Composition. The tree canopy is often diverse. Douglas-fir (*Pseudotsuga menziesii*), white fir (*Abies concolor*), sugar pine (*Pinus lambertiana*), ponderosa pine (*P. ponderosa*), or incense cedar (*Calocedrus decurrens*) are typically dominant or co-dominant. Port-Orford cedar (*Chamaecyparis lawsoniana*), tanoak (*Lithocarpus densiflorus*), canyon live oak (*Quercus chrysolepis*), and Pacific madrone (*Arbutus menziesii*) are locally important. Jeffrey pine (*Pinus jeffreyi*) is dominant on serpentine parent materials in the Siskiyou Mountains, and to a lesser degree in the southwestern Cascades.

Douglas-fir is found in almost every area; ponderosa pine is also found in most stands, although it has been declining with fire suppression. White fir, incense cedar, and sugar pine are common in mixed stands in the Cascades and central and eastern Siskiyous on all but the driest sites. White fir dominates the canopy in only the moist, cool sites at higher elevations, although it is the major tree regeneration in most areas. Jeffrey pine and knobcone pine (*Pinus attenuata*) are limited primarily to serpentine soils, which they dominate. Port-Orford cedar dominates some more moist sites near the coast and riparian and wetland habitats inland. Brewer's spruce (*Picea breweri*) is an uncommon dominant at high elevations in the Siskiyous. The broadleaf subcanopy is most prominent on the western sides of the Coast Range and Siskiyous, where tanoak is most abundant, with Pacific madrone, golden chinquapin (*Castanopsis chrysophylla*), or canyon live oak also sometimes dominating the subcanopy. Coast redwood (*Sequoia sempervirens*) occurs only in a very small area near the coast in far southern Oregon.

Dominant or co-dominant evergreen shrubs include pinemat manzanita (*Arctostaphylos nevadensis*), green-leaf manzanita (*A. patula*), white-leaf manzanita (*A. viscida*), kinnikinnick (*A. uva-ursi*), Piper's barberry (*Mahonia piperiana*), dwarf Oregongrape (*M. nervosa*), tobacco brush (*Ceanothus velutinus*), squawcarpet (*C. prostratus*), salal (*Gaultheria shallon*), deer oak (*Quercus sadleriana*), huckleberry oak (*Q. vacciniifolia*), snow bramble (*Rubus nivalis*), Pacific rhododendron (*Rhododendron macrophyllum*), and evergreen huckleberry (*Vaccinium ovatum*). Major deciduous shrubs are serviceberry (*Amelanchier alnifolia*), sticky currant (*Ribes viscosissimum*), oceanspray (*Holodiscus discolor*), creeping snowberry (*Symphoricarpos mollis*), baldhip rose (*Rosa gymnocarpa*), beaked hazel (*Corylus cornuta*), Rocky Mountain maple (*Acer glabrum*), vine maple (*A. circinatum*), poison-oak (*Toxicodendron diversiloba*), big huckleberry (*Vaccinium membranaceum*), deerbrush (*Ceanothus integerrimus*), and trailing blackberry (*Rubus ursinus*). Early seral shrublands, part of this habitat, can be difficult to distinguish from Ceanothus-Manzanita Shrublands. They are best separated by their different species composition, especially the predominance in this habitat of *Ceanothus velutinus, Arctostaphylos patula,* and *A. nevadensis.*

Graminoids that are most prominent are long-stolon sedge (*Carex inops*), Idaho fescue (*Festuca idahoensis*), and California fescue (*F. californica*). Forbs that are indicative of site conditions or dominate understories include common whipplea (*Whipplea modesta*), twinflower (*Linnaea borealis*), sidebells (*Orthilia secunda*), rattlesnake plantain (*Goodyera oblongifolia*), vanillaleaf (*Achlys triphylla*), beargrass (*Xerophyllum tenax*), and starry false solomonseal (*Maianthemum stellata*).

Other Classifications and Key References. This habitat includes the conifer-dominated forests and their successional seres within the Interior Valley, Mixed-Conifer, Mixed-Evergreen, and *Abies concolor* zones of southwestern Oregon, plus Redwood forests in the *Picea sitchensis* Zone.[88] It is also referred to as Klamath Mountains mixed evergreen forests and Sierran-type mixed conifer forests,[87] *Pseudotsuga menziesii*/hardwood forests and *Abies concolor* forests,[1] Mixed conifer forest No. 5, Redwood forest No. 6, California mixed evergreen forest No. 29, and Montane chaparral No. 34.[136] The Oregon Gap II Project[126] and Oregon Vegetation Landscape-Level Cover Types[127] that would represent this type are the southwestern portion of the Douglas-fir dominant-mixed conifer forest, Jeffrey pine forest and woodland, serpentine conifer woodland, Douglas-fir-Port Orford cedar forest, Douglas-fir mixed deciduous forest, Douglas-fir-white fir/tanoak-madrone mixed forest, and Siskiyou Mountains mixed deciduous forest. Other references also describe this habitat.[13, 15, 17, 111, 117]

Natural Disturbance Regime. Fire is the predominant natural disturbance. Fire regime varies depending on environmental conditions Drier, hotter sites within this area have a low-severity fire regime. Cooler and/or moister sites typically have a moderate-severity fire regime. Presettlement mean fire return intervals vary from 10 years to about 80 years.[1, 98, 154] Lightning ignitions are more frequent here than anywhere else in the region and Native Americans probably burned some areas intentionally.[1] Wind is a somewhat important disturbance at higher elevations. Root rot fungi and insects are other important disturbances in some forests, mostly operating at small-scales.

Succession and Stand Dynamics. Most evergreen broadleaf trees, when present, are top-killed by moderate-severity fires but resprout vigorously to dominate or co-dominate after most fires.[14, 152] Mature Oregon white oak

and canyon live oak can survive fairly hot fires if the fuels do not extend into the canopy. Conifers are at a disadvantage in regeneration following stand replacement fires because of dependence on local seed-fall. Many conifers of this habitat are able to survive moderate-severity fire well, including, in decreasing order of fire resistance, Ponderosa pine, Jeffrey pine, Douglas-fir, sugar pine, coast redwood, incense cedar, and Port-Orford cedar. These species are fairly well represented throughout the successional sequence, unless a high-severity fire was closely followed by another, in which case the subcanopy broadleaf species are likely to dominate.[1] Development of complex multi-layered canopies of conifers and broadleaf evergreens are typical under a moderate-severity fire regime.

Where hardwoods are absent and white fir is prominent, succession differs from that described above. Under a low-severity fire regime with frequent fires, white fir is relatively unimportant and fire-resistant conifers, especially Douglas-fir and ponderosa pine, dominate. White fir increases in the absence of fire.[171] With a moderate-severity fire regime, i.e., less frequent fires, white fir can dominate or co-dominate, especially on cooler sites.[1] Small gaps created by moderate-severity fires, blowdown, or disease afford opportunities for regeneration of less shade-tolerant tree species, thus maintaining a diverse tree canopy for lengthy periods. Evergreen shrubs, especially tobacco brush, often dominate after high-severity fire and may persist as a cover type for decades, especially if they are reburned.[51, 88] On the driest, hottest sites in this habitat, white fir does not grow and tree regeneration is limited to Douglas-fir and ponderosa pine, with the former tending to increase in the absence of fire.

Effects of Management and Anthropogenic Impacts. Clearcut logging where hardwoods are present favors post-disturbance dominance of tanoak or madrone. Control of this competing vegetation has been a major focus of timber management in this habitat. Fire control over the last 100 years has decreased fire frequencies and altered stand structure through increases in small tree density and heavy fuels, especially where low-severity fire regimes were prevalent. As a result, most of these areas are more susceptible to stand-replacement fires. White fir has increased dramatically on drier sites where it occurs, creating dense subcanopy thickets.[1, 128] Evergreen shrubs often dominate after clearcut logging and in some cases hinder the establishment of conifers.[140] Clearcut logging tends to decrease tree species diversity, coarse woody debris loads, and structural diversity. The non-native species white pine blister rust (*Cronartium ribicola*) and *Phytophthora lateralis*, a root rot disease, have had significant negative impacts on the abundance of sugar pine and Port-Orford cedar, respectively.

Status and Trends. This habitat covers most of southwestern Oregon and has declined little in areal extent. Conditions of most communities and stands have been degraded by forestry practices and by fire suppression. The low-elevation, driest communities have been altered by grazing and invasion of exotic species. Port-Orford cedar has declined dramatically in extent from logging and *Phytophthora lateralis*.[230] Effects of fire suppression and logging-related impacts continue to be threats. Twenty-one % of 68 plant associations representing this habitat listed in the National Vegetation Classification are listed as globally imperiled.[10]

4. Montane Mixed Conifer Forest

Christopher B. Chappell

Geographic Distribution. These forests occur in mountains throughout Washington and Oregon, excepting the Basin and Range of southeastern Oregon. These include the Cascade Range, Olympic Mountains, Okanogan Highlands, Coast Range (rarely), Blue and Wallowa mountains, and Siskiyou Mountains.

Physical Setting. This habitat is typified by a moderate to deep winter snow pack that persists for three to nine months. The climate is moderately cool and wet to moderately dry and very cold. Mean annual precipitation ranges from about 40 inches (102 cm) to >200 inches (508 cm). Elevation is mid- to upper montane, as low as 2,000 ft (610 m) in northern Washington, to as high as 7,500 ft (2,287 m) in southern Oregon. On the westside, it occupies an elevational zone of about 2,500 to 3,000 vertical feet (762 to 914 m), and on the eastside it occupies a narrower zone of about 1,500 vertical feet (457 m). Topography is generally mountainous. Soils are typically not well developed, but varied in their parent material: glacial till, volcanic ash, residuum, or colluvium. Spodosols are common.

Landscape Setting. This habitat is found adjacent to Westside Lowlands Conifer-Hardwood Forest, Eastside Mixed Conifer Forest, or Southwest Oregon Mixed Conifer-Hardwood Forest at its lower elevation limits and to Subalpine Parkland at its upper elevation limits. Inclusions of Montane Forested Wetlands, Westside Riparian-Wetlands, and less commonly Open Water or Herbaceous Wetlands occur within the matrix of montane forest habitat. The typical land use is forestry or recreation. Most of this type is found on public lands managed for

timber values and much of it has been harvested in a dispersed-patch pattern.

Structure. This is a forest, or rarely woodland, dominated by evergreen conifers. Canopy structure varies from single- to multi-storied. Tree size also varies from small to very large. Large snags and logs vary from abundant to uncommon. Understories vary in structure: shrubs, forbs, ferns, graminoids or some combination of these usually dominate, but they can be depauperate as well. Deciduous broadleaf shrubs are most typical as understory dominants. Early successional structure after logging or fire varies depending on understory species present. Mosses are a major ground cover and epiphytic lichens are typically abundant in the canopy.

Composition. This forest habitat is recognized by the dominance or prominence of one of the following species: Pacific silver fir (*Abies amabilis*), mountain hemlock (*Tsuga mertensiana*), subalpine fir (*A. lasiocarpa*), Shasta red fir (*A. magnifica* var. *shastensis*), Engelmann spruce (*Picea engelmannii*), noble fir (*A. procera*), or Alaska yellow-cedar (*Chamaecyparis nootkatensis*). Several other trees may co-dominate: Douglas-fir (*Pseudotsuga menziesii*), lodgepole pine (*Pinus contorta*), western hemlock (*Tsuga heterophylla*), western redcedar (*Thuja plicata*), or white fir (*A. concolor*). Tree regeneration is typically dominated by Pacific silver fir in moist westside middle-elevation zones; by mountain hemlock, sometimes with silver fir, in cool, very snowy zones on the westside and along the Cascade Crest; by subalpine fir in cold, drier eastside zones; and by Shasta red fir in the snowy mid- to upper-elevation zone of southwestern and south-central Oregon.

Subalpine fir and Engelmann spruce are major species only east of the Cascade Crest in Washington, in the Blue and Wallowa mountains, and in the northeastern Olympic Mountains (spruce is largely absent in the Olympic Mountains). Lodgepole pine is important east of the Cascade Crest throughout and in central and southern Oregon. Douglas-fir is important east of the Cascade Crest and at lower elevations on the westside. Pacific silver fir is a major species on the westside as far south as central Oregon. Noble fir, as a native species, is found primarily in the western Cascades from central Washington to central Oregon. Mountain hemlock is a common dominant at higher elevations along the Cascade Crest and to the west. Western hemlock, and to a lesser degree western redcedar, occur as dominants primarily with silver fir at lower elevations on the westside. Alaska yellow-cedar occurs as a co-dominant west of the Cascade Crest in Washington, rarely in northern Oregon. Shasta red fir and white fir occur only from central Oregon south, the latter mainly at lower elevations.

Deciduous shrubs that commonly dominate or co-dominate the understory are oval-leaf huckleberry (*Vaccinium ovalifolium*), big huckleberry (*V. membranaceum*), grouseberry (*V. scoparium*), dwarf huckleberry (*V. cespitosum*), fools huckleberry (*Menziesia ferruginea*), Cascade azalea (*Rhododendron albiflorum*), copperbush (*Elliottia pyroliflorus*), devil's-club (*Oplopanax horridus*), and, in the far south only, baldhip rose (*Rosa gymnocarpa*), currants (*Ribes* spp.), and creeping snowberry (*Symphoricarpos mollis*). Important evergreen shrubs include salal (*Gaultheria shallon*), dwarf Oregongrape (*Mahonia nervosa*), Pacific rhododendron (*Rhododendron macrophyllum*), deer oak (*Quercus sadleriana*), pinemat manzanita (*Arctostaphylos nevadensis*), beargrass (*Xerophyllum tenax*), and Oregon boxwood (*Paxistima myrsinites*).

Graminoid dominants are found primarily just along the Cascade Crest and to the east and include pinegrass (*Calamagrostis rubescens*), Geyer's sedge (*Carex geyeri*), smooth woodrush (*Luzula glabrata* var. *hitchcockii*), and long-stolon sedge (*Carex inops*). Deerfern (*Blechnum spicant*) and western oakfern (*Gymnocarpium dryopteris*) are commonly co-dominant. The most abundant forbs include Oregon oxalis (*Oxalis oregana*), single-leaf foamflower (*Tiarella trifoliata* var. *unifoliata*), rosy twisted-stalk (*Streptopus roseus*), queen's cup (*Clintonia uniflora*), western bunchberry (*Cornus unalaschkensis*), twinflower (*Linnaea borealis*), prince's pine (*Chimaphila umbellata*), five-leaved bramble (*Rubus pedatus*), dwarf bramble (*R. lasiococcus*), sidebells (*Orthilia secunda*), avalanche lily (*Erythronium montanum*), Sitka valerian (*Valeriana sitchensis*), false lily-of-the-valley (*Maianthemum dilatatum*), and Idaho goldthread (*Coptis occidentalis*).

Other Classifications and Key References. This habitat includes most of the upland forests and their successional stages, except lodgepole pine dominated forests, in the *Tsuga mertensiana, Abies amabilis, A. magnifica* var. *shastensis, A. lasiocarpa* zones of Franklin and Dyrness.[88] Portions of this habitat have also been referred to as *A. amabilis-Tsuga heterophylla* forests, *A. magnifica* var. *shastensis* forests, and *Tsuga mertensiana* forests.[87] It is equivalent to Silver fir-Douglas-fir forest No. 3, closed portion of Fir-hemlock forest No. 4, Red fir forest No. 7, and closed portion of Western spruce-fir forest No. 15.[136] The Oregon Gap II Project[126] and Oregon Vegetation Landscape-Level Cover Types[127] that would represent this type are mountain hemlock montane forest, true fir-hemlock montane forest, montane mixed conifer forest, Shasta red fir-mountain hemlock forest, and subalpine fir-lodgepole pine montane conifer; also most of the conifer forest in the Silver Fir, Mountain Hemlock, and Subalpine Fir Zones of Washington Gap.[37] A number of other references describe this habitat.[13, 15, 17, 25, 26, 36, 38, 90, 108, 111, 114, 115, 118, 144, 148, 212, 221]

Natural Disturbance Regime. Fire is the major natural disturbance in this habitat. Fire regimes are primarily of the high-severity type,[1] but also include the moderate-severity regime (moderately frequent and highly variable) for Shasta red fir forests.[39] Mean fire-return intervals vary greatly, from ≥800 years for some mountain hemlock-silver fir forests to about 40 years for red fir forests. Windstorms are a common small-scale disturbance and occasionally result in stand replacement. Insects and fungi are often important small-scale disturbances. However, they may affect larger areas also, for example, laminated root rot (*Phellinus weirii*) is a major natural disturbance, affecting

large areas of mountain hemlock forests in the Oregon Cascades.[72]

Succession and Stand Dynamics. After fire, a typical stand will briefly be occupied by annual and perennial ruderal forbs and grasses, as well as predisturbance understory shrubs and herbs that resprout. Stand initiation can take a long time, especially at higher elevations, resulting in shrub/herb dominance (with or without a scattered tree layer) for extended periods.[3, 109] Early seral tree species can be any of the potential dominants for the habitat, or lodgepole pine, depending on the environment, type of disturbance, and seed source. Fires tend to favor early seral dominance of lodgepole pine, Douglas-fir, noble fir, or Shasta red fir, if their seeds are present.[1] In some areas, large stand-replacement fires will result in conversion of this habitat to the Lodgepole Pine Forest and Woodland habitat, distinguished by dominance of lodgepole. After the tree canopy closes, the understory typically becomes sparse for a time. Eventually tree density will decrease and the understory will begin to flourish again, but this process takes longer than in lower elevation forests, generally at least 100 years after the disturbance, sometimes much longer.[1] As stand development proceeds, relatively shade-intolerant trees (lodgepole pine, Douglas-fir, western hemlock, noble fir, Engelmann spruce) typically decrease in importance and more shade-tolerant species (Pacific silver fir, subalpine fir, Shasta red fir, mountain hemlock) increase. Complex multi-layered canopies with large trees will typically take at least 300 years to develop, often much longer, and on some sites may never develop. Tree growth rates, and therefore the potential to develop these structural features, tend to decrease with increasing elevation.

Effects of Management and Anthropogenic Impacts. Forest management practices, such as clearcutting and plantations, have in many cases resulted in less diverse tree canopies with an emphasis on Douglas-fir. They also reduce coarse woody debris compared to natural levels, and truncate succession well before late-seral characteristics are expressed. Post-harvest regeneration of trees has been a perpetual problem for forest managers in much of this habitat.[16, 97] Planting of Douglas-fir has often failed at higher elevations, even where old Douglas-fir were present in the unmanaged stand.[115] Slash burning often has negative impacts on productivity and regeneration.[186] Management has since shifted away from burning and toward planting noble fir or native species, natural regeneration, and advance regeneration.[16, 103] Noble fir plantations are now fairly common in managed landscapes, even outside the natural range of the species. Advance regeneration management tends to simulate wind disturbance but without the abundant downed wood component. Shelterwood cuts are a common management strategy in Engelmann spruce or subalpine fir stands.[221]

Status and Trends. This habitat occupies large areas of the region. There has probably been little or no decline in the extent of this type over time. Large areas of this habitat are relatively undisturbed by human impacts and include significant old-growth stands. Other areas have been extensively affected by logging, especially dispersed patch clearcuts. The habitat is stable in area, but is probably still declining in condition because of continued logging. This habitat is one of the best protected, with large areas represented in national parks and wilderness areas. The only threat is continued road building and clearcutting in unprotected areas. None of the 81 plant associations representing this habitat listed in the National Vegetation Classification are considered imperiled.[10]

5. Eastside Mixed Conifer Forest
Rex C. Crawford

Geographic Distribution. The Eastside Mixed Conifer Forest habitat appears primarily in the Blue Mountains, East Cascades, and Okanogan Highland Ecoregions of Oregon, Washington, adjacent Idaho, and western Montana. It also extends north into British Columbia.

Douglas-fir-ponderosa pine forests occur along the eastern slope of the Oregon and Washington Cascades, the Blue Mountains, and the Okanogan Highlands of Washington. Grand fir-Douglas-fir forests and western larch forests are widely distributed throughout the Blue Mountains and, less so, along the east slope of the Cascades south of Lake Chelan and in the eastern Okanogan Highlands. Western hemlock-western redcedar-Douglas-fir forests are found in the Selkirk Mountains of eastern Washington, and on the east slope of the Cascades south of Lake Chelan to the Columbia River Gorge.

Physical Setting. The Eastside Mixed Conifer Forest habitat is primarily mid-montane with an elevation range of between 1,000 and 7,000 ft (305-2,137 m), mostly between 3,000 and 5,500 ft (914-1,676 m). Parent materials for soil development vary. This habitat receives some of the greatest amounts of precipitation in the inland northwest, 30-80 inches (76-203 cm)/year. Elevation of this habitat varies geographically, with generally higher elevations to the east.

Landscape Setting. This habitat makes up most of the continuous montane forests of the inland Pacific Northwest. It is located between the subalpine portions of the Montane Mixed Conifer Forest habitat in eastern Oregon and Washington and lower tree line Ponderosa Pine and Forest and Woodlands.

Structure. Eastside Mixed Conifer habitats are montane forests and woodlands. Stand canopy structure is generally diverse, although single-layer forest canopies are currently more common than multilayered forests with snags and large woody debris. The tree layer varies from closed forests to more open-canopy forests or woodlands. This habitat may include very open stands. The undergrowth is complex and diverse. Tall shrubs, low shrubs, forbs or any combination may dominate stands. Deciduous shrubs typify shrub layers. Prolonged canopy closure may lead to development of a sparsely vegetated undergrowth.

Composition. This habitat contains a wide array of tree species (nine) and stand dominance patterns. Douglas-fir (*Pseudotsuga menziesii*) is the most common tree species in this habitat. It is almost always present and dominates or co-dominates most overstories. Lower elevations or drier sites may have ponderosa pine (*Pinus ponderosa*) as a co-dominant with Douglas-fir in the overstory and often have other shade-tolerant tree species growing in the undergrowth. On moist sites, grand fir (*Abies grandis*), western redcedar (*Thuja plicata*) and/or western hemlock (*Tsuga heterophylla*) are dominant or co-dominant with Douglas-fir. Other conifers include western larch (*Larix occidentalis*) and western white pine (*Pinus monticola)* on mesic sites, Engelmann spruce (*Picea engelmannii*), lodgepole pine (*Pinus contorta*), and subalpine fir (*Abies lasiocarpa*) on colder sites. Rarely, Pacific yew (*Taxus brevifolia*) may be an abundant undergrowth tree or tall shrub.

Undergrowth vegetation varies from open to nearly closed shrub thickets with one to many layers. Throughout the eastside conifer habitat, tall deciduous shrubs include vine maple (*Acer circinatum*) in the Cascades, Rocky Mountain maple (*A. glabrum*), serviceberry (*Amelanchier alnifolia*), oceanspray (*Holodiscus discolor*), mallowleaf ninebark (*Physocarpus malvaceus*), and Scouler's willow (*Salix scouleriana*) at mid- to lower elevations. Medium-tall deciduous shrubs at higher elevations include fools huckleberry (*Menziesia ferruginea*), Cascade azalea (*Rhododendron albiflorum*), and big huckleberry (*Vaccinium membranaceum*). Widely distributed, generally drier site mid-height to short deciduous shrubs include baldhip rose (*Rosa gymnocarpa*), shiny-leaf spirea (*Spiraea betulifolia*), and snowberry (*Symphoricarpos albus, S. mollis*, and *S. oreophilus*). Low shrubs of higher elevations include low huckleberries (*Vaccinium cespitosum*, and *V. scoparium*) and five-leaved bramble (*Rubus pedatus*). Evergreen shrubs represented in this habitat are chinquapin (*Castanopsis chrysophylla*), a tall shrub in southeastern Cascades, low to mid-height dwarf Oregongrape (*Mahonia nervosa* in the east Cascades and *M. repens* elsewhere), tobacco brush (*Ceanothus velutinus*), an increaser with fire, Oregon boxwood (*Paxistima myrsinites*) generally at mid- to lower elevations, beargrass (*Xerophyllum tenax*), pinemat manzanita (*Arctostaphylos nevadensis*) and kinnikinnick (*A. uva-ursi*).

Herbaceous broadleaf plants are important indicators of site productivity and disturbance. Species generally indicating productive sites include western oakfern (*Gymnocarpium dryopteris*), vanillaleaf (*Achlys triphylla*), wild sarsparilla (*Aralia nudicaulis*), wild ginger (*Asarum caudatum*), queen's cup (*Clintonia uniflora*), goldthread (*Coptis occidentalis*), false bugbane (*Trautvetteria caroliniensis*), windflower (*Anemone oregana, A. piperi, A. lyallii*), fairybells (*Disporum hookeri*), Sitka valerian (*Valeriana sitchensis*), and pioneer violet (*Viola glabella*). Other indicator forbs are dogbane (*Apocynum androsaemifolium*), false solomonseal (*Maianthemum stellata*), heartleaf arnica (*Arnica cordifolia*), several lupines (*Lupinus caudatus, L. latifolius, L. argenteus* ssp. *argenteus* var *laxiflorus*), western meadowrue (*Thalictrum occidentale*), rattlesnake plantain (*Goodyera oblongifolia*), skunkleaf polemonium (*Polemonium pulcherrimum*), trailplant (*Adenocaulon bicolor*), twinflower (*Linnaea borealis*), western starflower (*Trientalis latifolia*), and several wintergreens (*Pyrola asarifolia, P. picta, Orthilia secunda*).

Graminoids are common in this forest habitat. Columbia brome (*Bromus vulgaris*), oniongrass (*Melica bulbosa*), northwestern sedge (*Carex concinnoides*) and western fescue (*Festuca occidentalis*) are found mostly in mesic forests with shrubs or mixed with forb species. Bluebunch wheatgrass (*Pseudoroegneria spicata*), Idaho fescue (*Festuca idahoensis*), and junegrass (*Koeleria macrantha*) are found in drier more open forests or woodlands. Pinegrass (*Calamagrostis rubescens*) and Geyer's sedge (*C. geyeri*) can form a dense layer under Douglas-fir or grand fir trees.

Other Classifications and Key References. This habitat includes the moist portions of the *Pseudotsuga menziesii*, the *Abies grandis*, and the *Tsuga heterophylla* zones of eastern Oregon and Washington.[88] This habitat is called Douglas-fir (No. 12), Cedar-Hemlock-Pine (No. 13), and Grand fir-Douglas-fir (No. 14) forests in Kuchler.[136] The Oregon Gap II Project[126] and Oregon Vegetation Landscape-Level Cover Types[127] that would represent this type are the eastside Douglas-fir dominant-mixed conifer forest, ponderosa pine dominant mixed conifer forest, and the northeast Oregon mixed conifer forest. Quigley and Arbelbide[181] referred to this habitat as Grand fir/White fir, the Interior Douglas-fir, Western larch, Western redcedar/Western hemlock, and Western white pine cover types and the Moist Forest potential vegetation group. Other references detail forest associations for this habitat.[45, 59, 117, 118, 123, 122, 144, 148, 208, 209, 212, 221, 228]

Natural Disturbance Regime. Fires were probably of moderate frequency (30-100 years) in presettlement times. Inland Pacific Northwest Douglas-fir and western larch forests have a mean fire interval of 52 years.[22] Typically, stand-replacement fire-return intervals are 150-500 years with moderate severity-fire intervals of 50-100 years. Specific fire influences vary with site characteristics.

Generally, wetter sites burn less frequently and stands are older with more western hemlock and western redcedar than drier sites. Many sites dominated by Douglas-fir and ponderosa pine, which were formerly maintained by wildfire, may now be dominated by grand fir (a fire sensitive, shade-tolerant species).

Succession and Stand Dynamics. Successional relationships of this type reflect complex interrelationships between site potential, plant species characteristics, and disturbance regime.[228] Generally, early seral forests of shade-intolerant trees (western larch, western white pine, ponderosa pine, Douglas-fir) or tolerant trees (grand fir, western redcedar, western hemlock) develop some 50 years following disturbance. This stage is preceded by forb- or shrub- dominated communities. These early stage mosaics are maintained on ridges and drier topographic positions by frequent fires. Early seral forest develops into mid-seral habitat of large trees during the next 50-100 years. Stand replacing fires recycle this stage back to early seral stages over most of the landscape. Without high-severity fires, a late-seral condition develops either single-layer or multilayer structure during the next 100-200 years. These structures are typical of cool bottomlands that usually only experience low-intensity fires.

Effects of Management and Anthropogenic Impacts. This habitat has been most affected by timber harvesting and fire suppression. Timber harvesting has focused on large shade-intolerant species in mid- and late-seral forests, leaving shade-tolerant species. Fire suppression enforces those logging priorities by promoting less fire-resistant, shade-tolerant trees. The resultant stands at all seral stages tend to lack snags, have high tree density, and are composed of smaller and more shade-tolerant trees. Mid-seral forest structure is currently 70% more abundant than in historical, native systems.[181] Late-seral forests of shade-intolerant species are now essentially absent. Early-seral forest abundance is similar to that found historically but lacks snags and other legacy features.

Status and Trends. Quigley and Arbelbide[181] concluded that the Interior Douglas-fir, Grand fir, and Western redcedar/Western hemlock cover types are more abundant now than before 1900, whereas the Western larch and Western white pine types are significantly less abundant. Twenty percent of Pacific Northwest Douglas-fir, grand fir, western redcedar, western hemlock, and western white pine associations listed in the National Vegetation Classification are considered imperiled or critically imperiled.[10] Roads, timber harvest, periodic grazing, and altered fire regimes have compromised these forests. Even though this habitat is more extensive than pre-1900, natural processes and functions have been modified enough to alter its natural status as functional habitat for many species.

6. Lodgepole Pine Forest and Woodlands

Rex C. Crawford

Geographic Distribution. This habitat is found along the eastside of the Cascade Range, in the Blue Mountains, the Okanogan Highlands and ranges north into British Columbia and south to Colorado and California.

With grassy undergrowth, this habitat appears primarily along the eastern slope of the Cascade Range and occasionally in the Blue Mountains and Okanogan Highlands. Subalpine lodgepole pine habitat occurs on the broad plateau areas along the crest of the Cascade Range and the Blue Mountains, and in the higher elevations in the Okanogan Highlands. On pumice soils this habitat is confined to the eastern slope of the Cascade Range from near Mt. Jefferson south to the vicinity of Crater Lake.

Physical Setting. This habitat is located mostly at mid- to higher elevations (3,000-9,000 ft [914-2,743 m]). These environments can be cold and relatively dry, usually with persistent winter snowpack. A few of these forests occur in low-lying frost pockets, wet areas, or under edaphic control (usually pumice) and are relatively long-lasting features of the landscape. Lodgepole pine is maintained as a dominant by the well-drained, deep Mazama pumice in eastern Oregon.

Landscape Setting. This habitat appears within Montane Mixed Conifer Forest east of the Cascade crest and the cooler Eastside Mixed Conifer Forest habitats. Most pumice soil lodgepole pine habitat is intermixed with Ponderosa Pine Forest and Woodland habitats and is located between Eastside Mixed Conifer Forest habitat and either Western Juniper Woodland or Shrub-steppe habitat.

Structure. The lodgepole pine habitat is composed of open to closed evergreen conifer tree canopies. Vertical structure is typically a single tree layer. Reproduction of other more shade-tolerant conifers can be abundant in the undergrowth. Several distinct undergrowth types develop under the tree layer: evergreen or deciduous medium-tall shrubs, evergreen low shrub, or graminoids with few shrubs. On pumice soils, a sparsely developed shrub and graminoid undergrowth appears with open to closed tree canopies.

Composition. The tree layer of this habitat is dominated by lodgepole pine (*Pinus contorta* var. *latifolia* and *P. c.* var. *murrayana*), but it is usually associated with other montane conifers (*Abies concolor, A. grandis, A. magnifica* var. *shastensis, Larix occidentalis, Calocedrus decurrens, Pinus lambertiana, P. monticola, P. ponderosa, Pseudotsuga menziesii*). Subalpine fir (*Abies lasiocarpa*), mountain hemlock (*Tsuga mertensiana*), Engelmann spruce (*Picea engelmannii*), and whitebark pine (*Pinus albicaulis*), indicators of subalpine environments, are present in colder or higher sites. Quaking aspen (*Populus tremuloides)* may occur in small numbers.

Shrubs can dominate the undergrowth. Tall deciduous shrubs include Rocky Mountain maple (*Acer glabrum*), serviceberry (*Amelanchier alnifolia*), oceanspray (*Holodiscus discolor*), or Scouler's willow (*Salix scouleriana*). These tall shrubs often occur over a layer of mid-height deciduous shrubs such as baldhip rose (*Rosa gymnocarpa),* russet buffaloberry (*Shepherdia canadensis*), shiny-leaf spirea (*Spiraea betulifolia*), and snowberry (*Symphoricarpos albus* and/or *S. mollis*). At higher elevations, big huckleberry (*Vaccinium membranaceum*) can be locally important, particularly following fire. Mid-tall evergreen shrubs can be abundant in some stands, for example, creeping Oregongrape (*Mahonia repens*), tobacco brush (*Ceanothus velutinus*), and Oregon boxwood (*Paxistima myrsinites*). Colder and drier sites support low-growing evergreen shrubs, such as kinnikinnick (*Arctostaphylos uva-ursi*) or pinemat manzanita (*A. nevadensis*). Grouseberry (*V. scoparium*) and beargrass (*Xerophyllum tenax*) are consistent evergreen low shrub dominants in the subalpine part of this habitat. Manzanita (*Arctostaphylos patula*), kinnikinnick, tobacco brush, antelope bitterbrush (*Purshia tridentata*), and wax current (*Ribes cereum*) are part of this habitat on pumice soil.

Some undergrowth is dominated by graminoids with few shrubs. Pinegrass (*Calamagrostis rubescens*) and/or Geyer's sedge (*Carex geyeri*) can appear with grouseberry in the subalpine zone. Pumice soils support grassy undergrowth of long-stolon sedge (*C. inops*), Idaho fescue (*Festuca idahoensis*) or western needlegrass (*Stipa occidentalis*). The latter two species may occur with bitterbrush or big sagebrush and other bunchgrass steppe species. Other nondominant indicator graminoids frequently encountered in this habitat are California oatgrass (*Danthonia californica*), blue wildrye (*Elymus glaucus*), Columbia brome (*Bromus vulgaris*) and oniongrass (*Melica bulbosa*). Kentucky bluegrass (*Poa pratensis*), and bottlebrush squirreltail (*Elymus elymoides*) can be locally abundant where livestock grazing has persisted.

The forb component of this habitat is diverse and varies with environmental conditions. A partial forb list includes goldthread (*Coptis occidentalis*), false solomonseal (*Maianthemum stellata*), heartleaf arnica (*Arnica cordifolia*), several lupines (*Lupinus caudatus, L. latifolius, L. argenteus* ssp. *argenteus* var. *laxiflorus),* meadowrue (*Thalictrum* occidentale), queen's cup (*Clintonia uniflora*), rattlesnake plantain (*Goodyera oblongifolia*), skunkleaf polemonium (*Polemonium pulcherrimum*), trailplant (*Adenocaulon bicolor*), twinflower (*Linnaea borealis*), Sitka valerian (*Valeriana sitchensis*), western starflower (*Trientalis latifolia*), and several wintergreens (*Pyrola asarifolia, P. picta, Orthilia secunda*).

Other Classifications and Key References. The Lodgepole Pine Forest and Woodland habitat includes the *Pinus contorta* zone of eastern Oregon and Washington.[88] The Oregon Gap II Project[126] and Oregon Vegetation Landscape-Level Cover Type[127] that would represent this type is lodgepole pine forest and woodlands. Quigley and Arbelbide[181] referred to this habitat as Lodgepole pine cover type and as a part of the Dry Forest potential vegetation group. Other references detail plant associations with this habitat.[117, 118, 122, 123, 144, 212, 221]

Natural Disturbance Regime. This habitat typically reflects early successional forest vegetation that originated with fires. Inland Pacific Northwest lodgepole pine has a mean fire interval of 112 years.[22] Summer drought areas generally have low to medium-intensity ground fires occurring at intervals of 25-50 years, whereas areas with more moisture have a sparse undergrowth and slow fuel build-up that results in less frequent, more intense fire. With time, lodgepole pine stands increase in fuel loads. Woody fuels accumulate on the forest floor from insect (mountain pine beetle) and disease outbreaks and residual wood from past fires. Mountain pine beetle outbreaks thin stands that add fuel and create a drier environment for fire or open canopies and create gaps for other conifer regeneration. High-severity crown fires are likely in young stands, when the tree crowns are near deadwood on the ground. After the stand opens up, shade-tolerant trees increase in number.

Succession and Stand Dynamics. Most Lodgepole Pine Forest and Woodlands are early- to mid seral stages initiated by fire. Typically, lodgepole pine establishes within 10-20 years after fire. This can be a gap phase process where seed sources are scarce. Lodgepole stands break up after 100-200 years. Without fires and insects, stands become more closed-canopy forest with sparse undergrowth. Because lodgepole pine cannot reproduce under its own canopy, old unburned stands are replaced by shade-tolerant conifers. Lodgepole pine on pumice soils is not seral to other tree species; these extensive stands, if not burned, thin naturally, with lodgepole pine regenerating in patches. On poorly drained pumice soils, quaking aspen sometimes plays a mid-seral role and is displaced by lodgepole when aspen clones die. Serotinous cones (cones releasing seeds after fire) are uncommon in eastern Oregon lodgepole pine (*P. c.* var. *murrayana*). On the Colville National Forest in Washington, only 10% of lodgepole pine (*P. c.* var. *latifolia*) trees in low-elevation Douglas-fir habitats had serotinous cones, whereas 82% of cones in high-elevation subalpine fir habitats were serotinous.[4]

Effects of Management and Anthropogenic Impacts. Fire suppression has left many single-canopy lodgepole pine habitats unburned to develop into more multilayered stands. Thinning of serotinous lodgepole pine forests with fire intervals <20 years can reduce their importance over time. In pumice-soil lodgepole stands, lack of natural regeneration in harvest units has lead to creation of "pumice deserts" within otherwise forested habitats.[47]

Status and Trends. Quigley and Arbelbide[181] concluded that the extent of the lodgepole pine cover type in Oregon and Washington is the same as before 1900 and in regions may exceed its historical extent. Five percent of Pacific Northwest lodgepole pine associations listed in the National Vegetation Classification are considered imperiled.[10] At a finer scale, these forests have been fragmented by roads, timber harvest, and influenced by periodic livestock grazing and altered fire regimes.

7. Ponderosa Pine Forest and Woodlands (includes Eastside Oak)
Rex C. Crawford & Jimmy Kagan

Geographic Distribution. This habitat occurs in much of eastern Washington and eastern Oregon, including the eastern slopes of the Cascades, the Blue Mountains and foothills, and the Okanogan Highlands. Variants of it also occur in the Rocky Mountains, the eastern Sierra Nevada, and mountains within the Great Basin. It extends into south-central British Columbia as well.

In the Pacific Northwest, ponderosa pine-Douglas-fir woodland habitats occur along the eastern slope of the Cascades, the Okanogan Highlands, and in the Blue Mountains. Ponderosa pine woodland and savanna habitats occur in the foothills of the Blue Mountains, along the eastern base of the Cascade Range, the Okanogan Highlands, and in the Columbia Basin in northeastern Washington. Ponderosa pine is widespread in the pumice zone of south-central Oregon between Bend and Crater Lake east of the Cascade Crest. Ponderosa pine-Oregon white oak habitat appears east of the Cascades in the vicinity of Mt. Hood near the Columbia River Gorge north to the Yakama Nation and south to the Warm Springs Nation. Oak-dominated woodlands follow a similar distribution as Ponderosa Pine-White Oak but are more restricted and less common.

Physical Setting. This habitat generally occurs on the driest sites supporting conifers in the Pacific Northwest. It is widespread and variable, appearing on moderate to steep slopes in canyons, foothills, and on plateaus or plains near mountains. In Oregon, this habitat can be maintained by the dry pumice soils, and in Washington it can be associated with serpentine soils. Average annual precipitation ranges from about 14 to 30 inches (36 to 76 cm) on ponderosa pine sites in Oregon and Washington, and often as snow. This habitat can be found at elevations of 100 ft (30m) in the Columbia River Gorge to dry, warm areas over 6,000 ft (1,829 m). Timber harvest, livestock grazing, and pockets of urban development are major land uses.

Landscape Setting. This woodland habitat typifies the lower tree line zone forming transitions with Eastside Mixed Conifer Forest and Western Juniper and Mountain Mahogany Woodland, Shrub-steppe, Eastside Grasslands, or Agriculture habitats. Douglas-fir-ponderosa pine woodlands are found near or within the Eastside Mixed Conifer Forest habitat. Oak woodlands appear in the driest, most restricted, landscapes in transition to Eastside Grasslands or Shrub-steppe.

Structure. This habitat is typically a woodland or savanna with tree canopy coverage of 10-60%, although closed-canopy stands are possible. The tree layer is usually composed of widely spaced large conifer trees. Many stands tend towards a multilayered condition with encroaching conifer regeneration. Isolated taller conifers above broadleaf deciduous trees characterize part of this habitat. Deciduous woodlands or forests are an important part of the structural variety of this habitat. Clonal deciduous trees can create dense patches across a grassy landscape rather than scattered individual trees. The undergrowth may include dense stands of shrubs or, more often, be dominated by grasses, sedges, or forbs. Shrub-steppe shrubs may be prominent in some stands and create a distinct tree-shrub-sparse-grassland habitat.

Composition. Ponderosa pine (*Pinus ponderosa*) and Douglas-fir (*Pseudotsuga menziesii*) are the most common evergreen trees in this habitat. The deciduous conifer, western larch (*Larix occidentalis*), can be a co-dominant with the evergreen conifers in the Blue Mountains of Oregon, but seldom as a canopy dominant. Grand fir (*Abies grandis*) may be frequent in the undergrowth on more productive sites, giving stands a multilayer structure. In rare instances, grand fir can be co-dominant in the upper canopy. Tall ponderosa pine over Oregon white oak (*Quercus garryana*) trees form stands along part of the east Cascades. These stands usually have younger cohorts of pines. Oregon white oak dominates open woodlands or savannas in limited areas.

The undergrowth can include dense stands of shrubs or, more often, be dominated by grasses, sedges, and/or forbs. Some Douglas-fir and ponderosa pine stands have

a tall to medium-tall deciduous shrub layer of mallowleaf ninebark (*Physocarpus malvaceus*) or common snowberry (*Symphoricarpos albus*). Grand fir seedlings or saplings may be present in the undergrowth. Pumice soils support a shrub layer represented by green-leaf or white-leaf manzanita (*Arctostaphylos patula* or *A. viscida*). Short shrubs, pinemat manzanita (*Arctostaphylos nevadensis*) and kinnikinnick (*A. uva-ursi*) are found across the range of this habitat. Antelope bitterbrush (*Purshia tridentata*), big sagebrush (*Artemisia tridentata*), black sagebrush (*A. nova*), green rabbitbrush (*Chrysothamnus viscidiflorus*), and in southern Oregon, curl-leaf mountain mahogany (*Cercocarpus ledifolius*) often grow with Douglas-fir, ponderosa pine and/or Oregon white oak, which typically have a bunchgrass and shrub-steppe ground cover.

Undergrowth is generally dominated by herbaceous species, especially graminoids. Within a forest matrix, these woodland habitats have an open to closed sodgrass undergrowth dominated by pinegrass (*Calamagrostis rubescens*), Geyer's sedge (*Carex geyeri*), Ross' sedge (*C. rossii*), long-stolon sedge (*C. inops*), or blue wildrye (*Elymus glaucus*). Drier savanna and woodland undergrowth typically contains bunchgrass steppe species, such as Idaho fescue (*Festuca idahoensis*), rough fescue (*F. campestris*), bluebunch wheatgrass (*Pseudoroegneria spicata*), Indian ricegrass (*Oryzopsis hymenoides*), or needlegrasses (*Stipa comata, S. occidentalis*). Common exotic grasses that may appear in abundance are cheatgrass (*Bromus tectorum*), and bulbous bluegrass (*Poa bulbosa*). Forbs are common associates in this habitat and are too numerous to be listed.

Other Classifications and Key References. This habitat is referred to as Merriam's Arid Transition Zone, Western ponderosa forest (*Pinus*), and Oregon Oak wood (*Quercus*) in Kuchler,[136] and as Pacific ponderosa pine-Douglas-fir, Pacific ponderosa pine, and Oregon white oak by the Society of American Foresters. The Oregon Gap II Project[126] and Oregon Vegetation Landscape-Level Cover Types[127] that would represent this type are ponderosa pine forest and woodland, ponderosa pine-white oak forest and woodland, and ponderosa pine-lodgepole pine on pumice. Other references describe elements of this habitat.[45, 62, 88, 117, 118, 121, 122, 123, 144, 148, 209, 212, 221, 222]

Natural Disturbance Regime. Fire plays an important role in creating vegetation structure and composition in this habitat. Most of the habitat has experienced frequent low-severity fires that maintained woodland or savanna conditions. A mean fire interval of 20 years for ponderosa pine is the shortest of the vegetation types listed by Barrett et al.[22] Soil drought plays a role in maintaining an open tree canopy in part of this dry woodland habitat.

Succession and Stand Dynamics. This habitat is climax on sites near the dry limits of each of the dominant conifer species and is more seral as the environment becomes more favorable for tree growth. Open seral stands are gradually replaced by more closed shade-tolerant climax stands. Oregon white oak can reproduce under its own shade but is intolerant of overtopping by conifers. Oregon white oak woodlands are considered fire climax and are seral to conifers. In drier conditions, unfavorable to conifers, oak is climax. Oregon white oak sprouts from the trunk and root crown following cutting or burning and form clonal patches of trees.

Effects of Management and Anthropogenic Impacts. Before 1900, this habitat was mostly open and park like with relatively few undergrowth trees. Currently, much of this habitat has a younger tree cohort of more shade-tolerant species that gives the habitat a more closed, multilayered canopy. For example, this habitat includes previously natural fire-maintained stands in which grand fir can eventually become the canopy dominant. Fire suppression has lead to a buildup of fuels that in turn increase the likelihood of stand-replacing fires. Heavy grazing, in contrast to fire, removes the grass cover and tends to favor shrub and conifer species. Fire suppression combined with grazing creates conditions that support cloning of oak and invasion by conifers. Large late-seral ponderosa pine, Douglas-fir, and Oregon white oak are harvested in much of this habitat. Under most management regimes, typical tree size decreases and tree density increases in this habitat. Ponderosa pine-Oregon white oak habitat is now denser than in the past and may contain more shrubs than in presettlement habitats. In some areas, new woodlands have been created by patchy tree establishment at the forest-steppe boundary.

Status and Trends. Quigley and Arbelbide[181] concluded that the Interior Ponderosa Pine cover type is significantly less in extent than pre-1900 and that the Oregon White Oak cover type is greater in extent than pre-1900. They included much of this habitat in their Dry Forest potential vegetation group,[181] which they concluded has departed from natural succession and disturbance conditions. The greatest structural change in this habitat is the reduced extent of the late-seral, single-layer condition. This habitat is generally degraded because of increased exotic plants and decreased native bunchgrasses. One third of Pacific Northwest Oregon white oak, ponderosa pine, and dry Douglas-fir or grand fir community types listed in the National Vegetation Classification are considered imperiled or critically imperiled.[10]

8. Upland Aspen Forest
Rex C. Crawford & Jimmy Kagan

Geographic Distribution. Quaking aspen groves are the most widespread habitat in North America, but are a minor type throughout eastern Washington and Oregon. Upland Aspen habitat is found in the isolated mountain ranges of southeastern Oregon, e.g., Steens Mountain, and in the northeastern Cascades of Washington. Aspen stands are much more common in the Rocky Mountain states.

Physical Setting. This habitat generally occurs on well-drained mountain slopes or canyon walls that have some moisture. Rockfalls, talus, or stony north slopes are often typical sites. It may occur in steppe on moist microsites, but is not associated with streams, ponds, or wetlands. This habitat is found from 2,000 to 9,500 ft (610 to 2,896 m) elevation.

Landscape Setting. Aspen forms a "subalpine belt" above the Western Juniper and Mountain Mahogany Woodland habitat and below montane Shrub-steppe Habitat and Alpine Grasslands on Steens Mountain in southern Oregon. It can occur in seral stands in the lower Eastside Mixed Conifer Forest and Ponderosa Pine Forest and Woodland habitats. Primary land use is livestock grazing.

Structure. Deciduous trees usually <48 ft (15 m) tall dominate this woodland or forest habitat. The tree layer grows over a forb-, grass-, or low-shrub-dominated undergrowth. Relatively simple two-tiered stands characterize the typical vertical structure of woody plants in this habitat. This habitat is composed of one to many clones of trees with larger trees toward the center of each clone. Conifers invade and create mixed evergreen-deciduous woodland or forest habitats.

Composition. Quaking aspen (*Populus tremuloides*) is the characteristic and dominant tree in this habitat. It is the sole dominant in many stands although scattered ponderosa pine (*Pinus ponderosa*) or Douglas-fir (*Pseudotsuga menziesii*) may be present. Snowberry (*Symphoricarpos oreophilus* and less frequently *S. albus*) is the most common dominant shrub. Tall shrubs, Scouler's willow (*Salix scouleriana*) and serviceberry (*Amelanchier alnifolia*) may be abundant. On mountain or canyon slopes, antelope bitterbrush (*Purshia tridentata*), mountain big sagebrush (*Artemisia tridentata* ssp. *vaseyana*), low sagebrush (*A. arbuscula*), and curl-leaf mountain mahogany (*Cercocarpus ledifolius*) often occur in and adjacent to this woodland habitat.

In some stands, pinegrass (*Calamagrostis rubescens*) may dominate the ground cover without shrubs. Other common grasses are Idaho fescue (*Festuca idahoensis*), California brome (*Bromus carinatus*), or blue wildrye (*Elymus glaucus*). Characteristic tall forbs include horsemint (*Agastache* spp.), aster (*Aster* spp.), senecio (*Senecio* spp.), coneflower (*Rudbeckia* spp.). Low forbs include meadowrue (*Thalictrum* spp.), bedstraw (*Galium* spp.), sweetcicely (*Osmorhiza* spp.), and valerian (*Valeriana* spp.).

Other Classifications and Key References. This habitat is called "Aspen" by the Society of American Foresters and "Aspen woodland" by the Society of Range Management. The Oregon Gap II Project[126] and Oregon Vegetation Landscape-Level Cover Type[127] that would represent this type is aspen groves. Other references describe this habitat[2, 88, 119, 161, 222].

Natural Disturbance Regime. Fire plays an important role in maintenance of this habitat. Quaking aspen will colonize sites after fire or other stand disturbances through root sprouting. Research on fire scars in aspen stands in central Utah[119] indicated that most fires occurred before 1885, and concluded that the natural fire return interval was 7-10 years. Ungulate browsing plays a variable role in aspen habitat; ungulates may slow tree regeneration by consuming aspen sprouts on some sites, and may have little influence in other stands.

Succession and Stand Dynamics. There is no generalized successional pattern across the range of this habitat. Aspen sprouts after fire and spreads vegetatively into large clonal or multiclonal stands. Because aspen is shade intolerant and cannot reproduce under its own canopy, conifers can invade most aspen habitat. In central Utah, quaking aspen was invaded by conifers in 75-140 years. Apparently, some aspen habitat is not invaded by conifers, but eventually clones deteriorate and succeed to shrubs, grasses, and/or forbs. This transition to grasses and forbs occurs more likely on dry sites.

Effects of Management and Anthropogenic Impacts. Domestic sheep reportedly consume four times more aspen sprouts than do cattle. Heavy livestock browsing can adversely impact aspen growth and regeneration. With fire suppression and alteration of fine fuels, fire rejuvenation of aspen habitat has been greatly reduced since about 1900. Conifers now dominate many seral aspen stands and extensive stands of young aspen are uncommon.

Status and Trends. With fire suppression and change in fire regimes, the Aspen Forest habitat is less common than before 1900. None of the 5 Pacific Northwest upland quaking aspen community types in the National Vegetation Classification are considered imperiled.[10]

9. Subalpine Parkland
Rex C. Crawford & Christopher B. Chappell

Geographic Distribution. The Subalpine Parkland habitat occurs throughout the high mountain ranges of Washington and Oregon (e.g., Cascade crest, Olympic Mountains, Wallowa and Blue mountains, and Okanogan Highlands), extends into mountains of Canada and Alaska, and to the Sierra Nevadas and Rocky Mountains.

Physical Setting. Climate is characterized by cool summers and cold winters with deep snowpack, although much variation exists among specific vegetation types. Mountain hemlock sites receive an average precipitation of >50 inches (127 cm) in 6 months and several feet of snow often accumulates. Whitebark pine sites receive 24-70 inches (61-178 cm) per year and some sites only rarely accumulate a significant snowpack. Summer soil drought is possible in eastside parklands but rare in westside areas. Elevation varies from 4,500 to 6,000 ft (1,371 to 1,829 m) in the western Cascades and Olympic Mountains and from 5,000 to 8,000 ft (1,524 to 2,438 m) in the eastern Cascades and Wallowa mountains.

Landscape Setting. The Subalpine Parkland habitat lies above the Mixed Montane Conifer Forest or Lodgepole Pine Forest habitat and below the Alpine Grassland and Shrubland habitat. Associated wetlands in subalpine parklands extend up a short distance into the alpine zone. Primary land use is recreation, watershed protection, and grazing.

Structure. Subalpine Parkland habitat has a tree layer typically between 10 and 30% canopy cover. Openings among trees are highly variable. The habitat appears either as parkland, that is, a mosaic of treeless openings and small patches of trees often with closed canopies, or as woodlands or savanna-like stands of scattered trees. The ground layer can be composed of (1) low to matted dwarf-shrubs (<1 ft [0.3 m] tall) that are evergreen or deciduous and often small-leaved; (2) sod grasses, bunchgrasses, or sedges; (3) forbs; or (4) moss- or lichen-covered soils. Herb or shrub-dominated wetlands appear within the parkland areas and are considered part of this habitat; wetlands can occur as deciduous shrub thickets up to 6.6 ft (2 m) tall, as scattered tall shrubs, as dwarf shrub thickets, or as short herbaceous plants <1.6 ft (0.5 m) tall. In general, western Cascades and Olympic areas are mostly parklands composed of a mosaic of patches of trees interspersed with heather shrublands or wetlands, whereas, eastern Cascades and Rocky mountain areas are parklands and woodlands typically dominated by grasses or sedges, with fewer heathers.

Composition. Species composition in this habitat varies with geography or local site conditions. The tree layer can be composed of one or several tree species. Subalpine fir (*Abies lasiocarpa*), Engelmann spruce (*Picea engelmannii*) and lodgepole pine (*Pinus contorta*) are found throughout the Pacific Northwest, whereas limber pine (*P. flexilis*) is restricted to southeastern Oregon. Alaska yellowcedar (*Chamaecyparis nootkatensis*), Pacific silver fir (*A. amabilis*), and mountain hemlock (*Tsuga mertensiana*) are most common in the Olympics and western Cascades. Whitebark pine (*P. albicaulis*) is found primarily in the eastern Cascades mountains, Okanogan Highlands, and Blue Mountains. Subalpine larch (*Laryx lyallii*) occurs only in the northern Cascade Mountains, primarily east of the crest.

West Cascades and Olympic areas generally are parklands. Tree islands often have big huckleberry (*Vaccinium membranaceum*) in the undergrowth interspersed with heather shrublands between. Openings are composed of pink mountain-heather (*Phyllodoce empetriformis*), and white mountain-heather (*Cassiope mertensiana*) and Cascade blueberry (*Vaccinium deliciosum*). Drier areas are more woodland or savanna-like, often with low shrubs, such as common juniper (*Juniperus communis*), kinnikinnick (*Arctostaphylos uva-ursi*), low whortleberries or grouseberries (*Vaccinium myrtillus or V. scoparium*) or beargrass (*Xerophyllum tenax*) dominating the ground cover. Wetland shrubs in the Subalpine Parkland habitat include bog-laurel (*Kalmia microphylla*), Booth's willow (*Salix boothii*), undergreen willow (*S. commutata*), Sierran willow (*S. eastwoodiae*), and blueberries (*Vaccinium uliginosum or V. deliciosum*)

Undergrowth in drier areas may be dominated by pinegrass (*Calamagrostis rubescens*), Geyer's sedge (*Carex geyeri*), Ross' sedge (*C. rossii*), smooth woodrush (*Luzula glabrata* var. *hitchcockii*), Drummond's rush (*Juncus drummondii*), or short fescues (*Festuca viridula, F. brachyphylla, F. saximontana*). Various sedges are characteristic of wetland graminoid-dominated habitats: black (*Carex nigricans*), Holm's Rocky Mountain (*C. scopulorum*), Sitka (*C. aquatilis* var. *dives*) and Northwest Territory (*C. utriculatia*) sedges. Tufted hairgrass (*Deschampsia caespitosa*) is characteristic of subalpine wetlands.

The remaining flora of this habitat is diverse and complex. The following herbaceous broadleaf plants are important indicators of differences in the habitat: American bistort (*Polygonum bistortoides*), American false hellebore (*Veratrum viride*), fringe leaf cinquefoil (*Potentilla flabellifolia*), marsh marigolds (*Caltha leptosepala*), avalanche lily (*Erythronium montanum*), partridgefoot (*Luetkea pectinata*), Sitka valerian (*Valeriana sitchensis*), subalpine lupine (*Lupinus arcticus* ssp. *subalpinus*), and alpine aster (*Aster alpigenus*). Showy sedge (*Carex spectabilis*) is also locally abundant.

Other Classifications and Key References. This habitat is called the Hudsonian Zone,[155] Parkland subzone,[134] meadow-forest mosaic,[74] upper subalpine zone,[88] Meadows and Park, and Subalpine Parkland.[20] Quigley and Arbelbide[181] called this habitat Whitebark pine and Whitebark pine-Subalpine larch cover types. Kuchler[136] included this within the subalpine fir-mountain hemlock forest. The Oregon Gap II Project[126] and Oregon Vegetation Landscape-Level Cover Types[127] that would represent this type are whitebark-lodgepole pine montane forest and subalpine parkland. Additional references describe this habitat.[11, 49, 75, 105, 112, 114, 115, 139, 144, 221]

Natural Disturbance Regime. Although fire is rare to infrequent in this habitat, it plays an important role, particularly in drier environments. Whitebark pine woodland fire intervals varied from 50 to 300 years before 1900. Mountain hemlock parkland fire reccurrence is 400-800 years. Wind blasting by ice and snow crystals is a critical factor in these woodlands and establishes the higher limits of the habitat. Periodic shifts in climatic factors, such as drought, snowpack depth, or snow duration either allow tree invasions into meadows and shrublands or eliminate or retard tree growth. Volcanic activity plays a long-term role in establishing this habitat. Wetlands are usually seasonally or perennially flooded by snowmelt and springs, or by subirrigation.

Succession and Stand Dynamics. Succession in this habitat occurs through a complex set of relationships between vegetation response to climatic shifts and catastrophic disturbance, and plant species interactions and site modification that create microsites. A typical succession of subalpine trees into meadows or shrublands begins with the invasion of a single tree, subalpine fir and mountain hemlock in the wetter climates and whitebark pine and subalpine larch in drier climates. If the environment allows, tree density slowly increases (over decades to centuries) through seedlings or branch layering by subalpine fir. The tree patches or individual trees change the local environment and create microsites for shade-tolerant trees, Pacific silver fir in wetter areas, and subalpine fir and Engelmann spruce in drier areas. Whitebark pine, an early invading tree, is dispersed long distances by Clark's nutcrackers and shorter distances by mammals. Most other tree species are wind dispersed.

Effects of Management and Anthropogenic Impacts. Fire suppression has contributed to change in habitat structure and functions. For example, the current "average" whitebark pine stand will burn every 3,000 years or longer because of fire suppression. Blister rust, an introduced pathogen, is increasing whitebark pine mortality in these woodlands.[4] Even limited logging can have prolonged effects because of slow invasion rates of trees. This is particularly important on drier sites and in subalpine larch stands. During wet cycles, fire suppression can lead to tree islands coalescing and the conversion of parklands into a more closed forest habitat. Parkland conditions can displace alpine conditions through tree invasions. Livestock use and heavy horse or foot traffic can lead to trampling and soil compaction. Slow growth in this habitat prevents rapid recovery.

Status and Trends. This habitat is generally stable with local changes to particular tree variants. Whitebark pine is maybe declining because of the effects of blister rust or fire suppression that leads to conversion of parklands to more closed forest. Global climate warming will likely have an amplified effect throughout this habitat. Less than 10% of Pacific Northwest subalpine parkland community types listed in the National Vegetation Classification are considered imperiled.[10]

10. Alpine Grasslands and Shrublands
Christopher B. Chappell & Jimmy Kagan

Geographic Distribution. This habitat occurs in high mountains throughout the region, including the Cascades, Olympic Mountains, Okanogan Highlands, Wallowa Mountains, Blue Mountains, Steens Mountain in southeastern Oregon, and, rarely, the Siskiyous. It is most extensive in the Cascades from Mount Rainier north and in the Wallowa Mountains. Similar habitats occur throughout mountains of northwestern North America.

Physical Setting. The climate is the coldest of any habitat in the region. Winters are characterized by moderate to deep snow accumulations, very cold temperatures, and high winds. Summers are relatively cool. Growing seasons are short because of persistent snow pack or frost. Blowing snow and ice crystals on top of the snow pack at and above treeline prevent vegetation such as trees from growing above the depth of the snow pack. Snow pack protects vegetation from the effects of this winter wind-related disturbance and from excessive frost heaving. Community composition is much influenced by relative duration of snow burial and exposure to wind and frost heaving.[75] Elevation ranges from a minimum of 5,000 ft (1,524 m) in parts of the Olympics to ≥10,000 ft (3,048 m). The topography varies from gently sloping broad ridgetops, to glacial cirque basins, to steep slopes of all aspects. Soils

are generally poorly developed and shallow, though in subalpine grasslands they may be somewhat deeper or better developed. Geologic parent material varies with local geologic history.

Landscape Setting. This habitat always occurs above upper treeline in the mountains or a short distance below it (grasslands in the subalpine parkland zone). Typically, it occurs adjacent to, or in a mosaic with, Subalpine Parkland. Occasionally, it may grade quickly from this habitat down into Montane Mixed Conifer Forest without intervening Subalpine Parkland. In southeastern Oregon, this habitat occurs adjacent to and above Upland Aspen Forest and Shrub-steppe habitats. Small areas of Open Water, Herbaceous Wetlands, and Subalpine Parkland habitats sometimes occur within a matrix of this habitat. Cliffs, talus, and other barren areas are common features within or adjacent to this habitat. Land use is primarily recreation, but in some areas east of the Cascade Crest, it is grazing, especially by sheep.

Structure. This habitat is dominated by grassland, dwarf-shrubland (mostly evergreen microphyllous), or forbs. Cover of the various life forms is extremely variable, and total cover of vascular plants can range from sparse to complete. Patches of krummholz (coniferous tree species maintained in shrub form by extreme environmental conditions) are a common component of this habitat, especially just above upper treeline. In subalpine grasslands, which are considered part of this habitat, widely scattered coniferous trees sometimes occur. Five major structural types can be distinguished: (1) subalpine and alpine bunchgrass grasslands, (2) alpine sedge turf, (3) alpine heath or dwarf-shrubland, (4) fellfield and boulderfield, and (5) snowbed forb community. Fellfields have a large amount of bare ground or rocks with a diverse and variable open layer of forbs, graminoids, and less commonly dwarf-shrubs. Snowbed forb communities have relatively sparse cover of few species of mainly forbs. In the alpine zone, these types often occur in a complex fine-scale mosaic with each other.

Composition. Most subalpine or alpine bunchgrass grasslands are dominated by Idaho fescue (*Festuca idahoensis*), alpine fescue (*F. brachyphylla*), green fescue (*F. viridula*), Rocky Mountain fescue (*F. saximontana*), or timber oatgrass (*Danthonia intermedia*), and to a lesser degree, purple reedgrass (*Calamagrostis purpurascens*), downy oatgrass (*Trisetum spicatum*) or muttongrass (*Poa fendleriana*). Forbs are diverse and sometimes abundant in the grasslands. Alpine sedge turfs may be moist or dry and are dominated by showy sedge (*Carex spectabilis*), black alpine sedge (*C. nigricans*), Brewer's sedge (*C. breweri*), capitate sedge (*C. capitata*), nard sedge (*C. nardina*), dunhead sedge (*C. phaeocephala*), or western single-spike sedge (*C. pseudoscirpoidea*).

One or more of the following species dominates alpine heaths: pink mountain-heather (*Phyllodoce empetriformis*), green mountain-heather (*P. glanduliflora*), white mountain-heather (*Cassiope mertensiana*), or black crowberry (*Empetrum nigrum*). Other less extensive dwarf-shrublands may be dominated by the evergreen coniferous common juniper (*Juniperus communis*), the evergreen broadleaf kinnikinnick (*Arctostaphylos uva-ursi*), the deciduous shrubby cinquefoil (*Pentaphylloides floribunda*) or willows (*Salix cascadensis* and *S. reticulata* ssp. *nivalis*). Tree species occurring as shrubby krummholz in the alpine are subalpine fir (*Abies lasiocarpa*), whitebark pine (*Pinus albicaulis*), mountain hemlock (*Tsuga mertensiana*), Engelmann spruce (*Picea engelmannii*), and subalpine larch (*Larix lyallii*).

Fellfields and similar communities are typified by variable species assemblages and co-dominance of multiple species, including any of the previously mentioned species, especially the sedges, as well as golden fleabane (*Erigeron aureus*), Lobb's lupine (*Lupinus sellulus* var. *lobbii*), spreading phlox (*Phlox diffusa*), eight-petal mountain-avens (*Dryas octopetala*), louseworts (*Pedicularis contorta, P. ornithorhyncha*) and many others. Snowbed forb communities are dominated by Tolmie's saxifrage (*Saxifraga tolmiei*), Shasta buckwheat (*Eriogonum pyrolifolium*), or Piper's woodrush (*Luzula piperi*).

Other Classifications and Key References. This habitat is equivalent to the alpine communities and the subalpine *Festuca* communities of Franklin and Dyrness.[88] It is also referred to as Alpine meadows and barren No. 52.[136] The Oregon Gap II Project[126] and Oregon Vegetation Landscape-Level Cover Types[127] that would represent this type are subalpine grassland and alpine fell-snowfields; represented by nonforest in the alpine/parkland zone of Washington Gap.[37] Other references describe this habitat.[61, 65, 75, 80, 94, 105, 112, 123, 139, 195, 207]

Natural Disturbance Regime. Most natural disturbances seem to be small scale in their effects or very infrequent. Herbivory and associated trampling disturbance by elk, mountain goats, and occasionally bighorn sheep seems to be an important disturbance in some areas, creating patches of open ground, though the current distribution and abundance of these ungulates is in part a result of introductions. Small mammals can also have significant effects on vegetation: e.g., the heather vole occasionally overgrazes heather communities.[80] Frost heaving is a climatically related small-scale disturbance that is extremely important in structuring the vegetation[80] Extreme variation from the norm in snow pack depth and duration can act as a disturbance, exposing plants to winter dessication,[80] shortening the growing season, or facilitating summer drought. Subalpine grasslands probably burn on occasion and can be formed or expanded in area by fires in subalpine parkland.[139]

Succession and Stand Dynamics. Little is known about vegetation changes in these communities, in part because changes are relatively slow. Tree invasion rates into subalpine grasslands are relatively slow compared to other subalpine communities.[139] Seedling establishment for many plant species in the alpine zone is poor. Heath communities take about 200 years to mature after initial establishment and may occupy the same site for thousands of years.[139]

Effects of Management and Anthropogenic Impacts. The major human impacts on this habitat are trampling and associated recreational impacts, e.g., tent sites. Resistance and resilience of vegetation to impacts varies by life form.[48] Sedge turfs are perhaps most resilient to trampling and heaths are least resilient. Trampling to the point of significantly opening an alpine heath canopy will initiate a degradation and erosion phase that results in continuous bare ground, largely unsuitable for vascular plant growth.[80] Bare ground in the alpine zone left alone after recreational disturbance will typically not revegetate in a time frame that humans can appreciate. Introduction of exotic ungulates can have noticeable impacts (e.g., mountain goats in the Olympic Mountains). Domestic sheep grazing has also had dramatic impacts,[196] especially in the bunchgrass habitats east of the Cascades.

Status and Trends. This habitat is naturally very limited in extent in the region. There has been little to no change in abundance over the last 150 years. Most of this habitat is still in good condition and dominated by native species. Some areas east of the Cascade Crest have been degraded by livestock use. Recreational impacts are noticeable in some national parks and wilderness areas. Current trends seem to be largely stable, though there may be some slow loss of subalpine grassland to recent tree invasion. Threats include increasing recreational pressures, continued grazing at some sites, and, possibly, global climate change resulting in expansion of trees into this habitat. Only 1 out of 40 plant associations listed in the National Vegetation Classification is considered imperiled.[10]

11. Westside Grasslands
Christopher B. Chappell & Jimmy Kagan

Geographic Distribution. This habitat is restricted primarily to the Puget Lowland and Willamette Valley ecoregions, with most now occurring in Pierce, Thurston, and San Juan counties, Washington. It also occurs in scattered small outliers in the Coast Range of Oregon, the eastern Olympic Mountains and the Western Cascades of southern Washington and Oregon, and in adjacent southwestern British Columbia.

Physical Setting. The climate is mild and moderately dry (17-55 inches [43-140 cm] mean annual precipitation), with moist winters and dry summers. Elevation is mostly low and ranges up to a maximum of about 3,500 ft (1,067 m). Topography varies from flat, to mounded or rolling, to steep slopes. Most sites are topoedaphically dry and experience extreme soil drought in the summer. Much of what currently remains of this habitat is found on the South Puget prairies, which are underlain by very deep gravelly/sandy glacial outwash that is excessively well drained. Many other small sites, often called "balds", have shallow soils overlying bedrock and typically are on south- or west-facing slopes. The habitat also occurs rarely in Oregon on deeper soils that are not excessively drained.

Landscape Setting. This habitat occurs adjacent to or in a mosaic with Westside Riparian-Wetlands, Westside Oak and Dry Douglas-fir Forests and Woodlands, Agriculture, or Urban habitats. Westside grassland habitat occurs less commonly in a matrix of Westside Lowland Conifer-Hardwood Forest. In the San Juan Islands, the habitat sometimes occurs on bluffs or slopes adjacent to marine habitats. Currently this habitat is used for grazing, recreation, and, in the southern Puget Sound area, for military training.

Structure. This habitat is grassland or, less commonly, savanna, with <30% tree or shrub cover. Bunchgrasses predominate in native-dominated sites, with space between the vascular plants typically covered by mosses, fruticose lichens, or native forbs. Montane balds are sometimes dominated in part by short forbs (<1.6ft. [0.6 m]) or dwarf shrubs. Degraded sites are dominated by rhizomatous exotic grasses with some native herbaceous component still present. Scattered trees are either evergreen conifers or deciduous broadleaves. Shrubs may be absent, scattered, or very prominent, and include evergreen and deciduous broadleaf physiognomy.

Composition. The major native dominant bunchgrass is Roemer's fescue (*Festuca idahoensis* var. *roemeri*). Red fescue (*F. rubra*) and California oatgrass (*Danthonia californica*) are frequently dominant or co-dominant on a local basis. Long-stolon sedge (*Carex inops*) is occasionally co-dominant, especially in savannas and in the Columbia Gorge. Slender wheatgrass (*Elymus trachycaulus*), blue wildrye (*Elymus glaucus*), prairie junegrass (*Koeleria macrantha*), and Lemmon's needlegrass (*Stipa lemmonii*) can be important locally. Major exotic dominant species are colonial bentgrass (*Agrostis capillaris*), sweet vernalgrass (*Anthoxanthum odoratum*), Kentucky bluegrass (*Poa pratensis*), tall oatgrass (*Arrhenatherum elatius*), medusahead (*Taeniatherum caput-medusae*), tall fescue (*F. arundinacea*), and soft brome (*Bromus mollis*). Common camas (*Camassia quamash*) is probably the most important forb in terms of cover, but it rarely dominates. The bracken fern (*Pteridium aquilinum*) is sometimes co-dominant. A rich diversity of native forbs is typical of sites in good condition.

Roemer's fescue is distributed throughout the Puget Lowland and the Willamette Valley and in montane balds of the eastern and northeastern Olympics. Native red fescue is a major component neat saltwater in the northern

Puget Lowland and in montane balds of the Oregon Coast Range and the Columbia Gorge. Non-native varieties of red fescue can occur throughout the area, especially in degraded habitats. California oatgrass communities are found in the San Juan Islands and in the Willamette Valley. Junegrass is a co-dominant in some montane balds and the Willamette Valley prairies; it occurs less abundantly throughout the area. Lemmon's needlegrass is primarily found on shallow-soiled balds of Willamette Valley fringes and the San Juan Islands.

The most common savanna tree is Douglas-fir (*Pseudotsuga menziesii*). Oregon white oak (*Quercus garryana*) formerly was part of extensive savannas, but is now rare in that structural condition. Ponderosa pine (*Pinus ponderosa*) is very local. The most common shrub is the exotic species Scot's broom (*Cytisus scoparius*), which frequently forms open stands over the grass. Common snowberry (*Symphoricarpos albus*), Nootka rose (*Rosa nutkana*), poison-oak (*Toxicodendron diversilobum*), and serviceberry (*Amelanchier alnifolia*) are other common shrubs. The dwarf shrubs kinnikinnick (*Arctostaphylos uva-ursi*) and common juniper (*Juniperus communis*) sometimes dominate small areas in montane balds, and the former sometimes on South Puget prairies. *Racomitrium canescens* is the most common ground moss.

Other Classifications and Key References. Portions of this habitat have been referred to as prairies by many authors. Franklin and Dyrness[88] described this habitat as prairie in the Puget Sound area, grassland in the San Juan Islands and Interior Valley zone of Oregon, and grass balds in the Oregon Coast Range. The Oregon Gap II Project[126] and Oregon Vegetation Landscape-Level Cover Types[127] effort did not map this type; it was inadvertently aggregated with the agriculture classification. The Washington Gap project mapped this habitat as part of nonforested in the Woodland/Prairie Mosaic Zone. Other references describe elements of this habitat.[7, 40, 41, 54, 99, 142, 211]

Natural Disturbance Regime. Historically, fire was a major component of this habitat. In addition to occasional lightning strikes, fires were intentionally set by indigenous inhabitants to maintain food staples such as camas and bracken fern.[165] Although there is no definitive fire history information, evidence suggests that many, if not most, of these grasslands burned every few years. Annual soil drought naturally eliminated or thinned invading trees and promoted higher frequency fire regimes in the past.

Succession and Stand Dynamics. Historically, regular fires or extreme environmental conditions on the most xeric sites prevented the establishment and continued growth of most woody vegetation, thereby maintaining the grasslands and oak savannas. In some patches, scattered oaks or even Douglas-fir survived long enough to obtain some fire resistance and the frequent light fires then helped to maintain savannas.[1, 41] Oaks were also able to resprout if the above-ground stem was killed. High fire frequencies combined with digging of roots by Native Americans could have favored the abundance of forbs over that of grasses in many areas of the pre-European landscape.

Effects of Management and Anthropogenic Impacts. The exclusion of fire from most of this habitat over the last 100+ years has resulted in profound changes. Oak savanna has, for all practical purposes, disappeared from the landscape.[41] Douglas-fir encroachment, in the absence of fire, is a "natural" process that occurs eventually on the vast majority of westside grasslands, except perhaps on the very driest sites. This encroachment leads to the conversion of grasslands to forests. Fire exclusion has also resulted in increases in shrub cover and the conversion of some grasslands to shrublands. Exotic species are prominent in this habitat and generally increase after ground-disturbing activities like grazing or off-road vehicle use. Scot's broom, tall oatgrass, colonial bentgrass, sweet vernalgrass, tall fescue, common velvetgrass (*Holcus lanatus*), Kentucky bluegrass, soft brome, common St. Johnswort (*Hypericum perforatum*), and hairy catsear (*Hypochaeris radicata*) are among the most troublesome species. The dominant native grass, Roemer's fescue, can be eliminated with heavy grazing. Prescribed fire and other management tools have been used recently to control Scot's broom, Douglas-fir encroachment, and to attempt to mimic historical conditions in some areas.[78]

Status and Trends. This habitat is very rare and limited in areal extent. In the southern Puget Sound area, only about 10% of the original area of the habitat is extant, and only 3% is dominated by native species.[54] Overall decline is significantly greater than these figures suggest because the habitat is even more decimated and degraded elsewhere. Causes of the decline are fire suppression, conversion to agriculture and urban, and invasion of exotic species. Most of what remains is dominated or co-dominated by exotic species. Current trends are continued decline both in area and condition. Ongoing threats include urban conversion, increase of exotic species, ground disturbance via tracked vehicle use for military training, and effects of fire suppression. Eleven out of 12 native plant associations representing this habitat listed for the National Vegetation Classification are considered imperiled or critically imperiled.[10]

12. Ceanothus-Manzanita Shrublands
Christopher B. Chappell & Jimmy Kagan

Geographic Distribution. This habitat ranges from southwestern Oregon south through much of California. Within Oregon, it is primarily located in the Rogue and Illinois valleys in Curry, Josephine, and Jackson counties; it is also found scattered in the Siskiyou Mountains of the same counties, in Douglas County, and in the southern Cascades of Jackson and western Klamath counties.

Physical Setting. Climate is mostly very warm and relatively dry (about 17-30 inches [43-76 cm] mean annual precipitation), but extends less commonly on serpentine or extremely dry sites into somewhat cooler and moister climates (up to 50 inches [127 cm] annual precipitation). Summers are very dry; winters are only slightly cool, much warmer than shrublands on the eastside. Primary elevation range is about 1,000-2,000 ft (305-610 m), but extends up to a maximum of 5,000 ft (1,524 m). Topography is typified by mainly lower valley slopes and foothills, but extends to nearly flat valley bottoms (where before European settlement this habitat was a major type) and sporadically onto mountain slopes. This habitat tends to occur on southern aspects when it does occur outside of low valleys. Soils are typically shallow to bedrock or are derived from coarse alluvial deposits. Ultramatic bedrock is a major parent material in the western Siskiyou Mountains, whereas the eastern Siskiyou Mountains and adjacent valleys are largely volcanic.

Landscape Setting. This habitat occurs adjacent to or in a mosaic with Southwest Oregon Mixed Conifer-Hardwood Forest, Westside Oak and Dry Douglas-fir Forest and Woodlands, Agriculture, and rarely, Westside Grassland. Urban is also adjacent in a few areas. Westside Riparian-Wetlands habitat occurs as small inclusions within this habitat. This habitat covers large areas only in lower elevation valleys or on extensive areas of serpentine soils. At moderate to high elevations it is mainly small patches within a forest mosaic. Major land use of this habitat is grazing and low-density residential development.

Structure. This habitat consists mainly of shrubland dominated by sclerophyllous evergreen broadleaf shrubs, but can also include grasslands with scattered tall shrubs.

Deciduous broadleaf shrubs are less important, but are in some cases dominant. The shrubs are mostly 3.3-13 ft (1-4 m) high. Shrub canopy ranges from very open to completely closed. Herbaceous cover varies inversely with shrub canopy cover. Perennial bunchgrasses are the dominant understory at sites in good condition, whereas annual grasses dominate at sites in poorer condition. If shrubs are not too dense, forbs are abundant. Historically, many of these shrublands were probably grasslands with scattered shrubs. Occasional conifers or broadleaf trees are sometimes scattered in the habitat.

Composition. Sclerophyllous and hemi-sclerophyllous shrubs that dominate are, most commonly, wedge-leaf ceanothus (*Ceanothus cuneatus*) and white-leaf manzanita (*Arctostaphylos viscida*), and less commonly, chaparral whitethorn (*Ceanothus leucodermis*), blueblossom (*C. thyrsiflorus*), deerbrush (*C. integerrimus*), and deer oak (*Quercus sadleriana*). Wedge-leaf ceanothus is the most abundant species at low elevations in the major valleys and is the shrub most tolerant of xeric conditions. Other common, but not typically dominant shrubs include hairy manzanita (*Arctostaphylos columbiana*), pinemat manzanita (*A. nevadensis*), birchleaf mountain-mahogany (*Cercocarpus montanus* var. *glaber*), Klamath plum (*Prunus subcordata*), bitter cherry (*P. emarginata*), chokecherry (*P. virginiana*), Brewer's oak (*Quercus garryana* var. *breweriana*), huckleberry oak (*Q. vacciniifolia*), California yerba-santa (*Eriodictyon californicum*), and bearbrush (*Garrya fremontii*).

The native bunchgrasses are Idaho fescue (*Festuca idahoensis*), California fescue (*F. californica*), California oatgrass (*Danthonia californica*), Lemmon's needlegrass (*Stipa lemmonii*), western needlegrass (*S. occidentalis*), and bluegrass (*Poa secunda*). Forb diversity is often high and common genera include *Lilium*, *Calochortus*, *Fritillaria*, *Microseris*, *Monardella*, and *Erigeron*. One of several species of oak (*Quercus*) or pine (*Pinus*), Douglas-fir (*Pseudotsuga menziesii*), or incense cedar (*Calocedrus decurrens*) are sometimes present as scattered individuals, especially on less xeric sites.

Other Classifications and Key References. Franklin and Dyrness[88] referred to this habitat as sclerophyllous shrub communities in the interior valleys of Oregon. It is also called chaparral.[125, 130, 141] The Oregon Gap II Project[126] and Oregon Vegetation Landscape-Level Cover Type[127] that would represent this type is Siskiyou Mountain serpentine shrubland. Other references include describe aspects of this habitat.[69, 188]

Natural Disturbance Regime. Fire is the major natural disturbance. Fire regimes have not been studied in Oregon, but in central California the fire-return interval has been estimated at 30-60 years.[84] Fire frequency may have been greater during historic times based on chaparral fire regimes in southern California.[157] High-severity fires are typical, with most shrubs being top-killed.

Succession and Stand Dynamics. Wedge-leaf ceanothus and white-leaf manzanita are killed by fire. Some less common shrubs sprout after fire. The dominant shrub species regenerate abundantly after fire from a long-lived seedbank that is scarified by fire.[187] Many seedlings die in the first 3 years after fire.[50] Wedge-leaf ceanothus can maintain prominence for >100 years.[124] Shrub canopy cover generally increases up to 30 years or so after the last fire, and in the absence of another fire, the herbaceous understory can be reduced under dense late-successional shrub canopies. Wedge-leaf ceanothus is considered a climax, or late-successional, dominant species on low-elevation dry sites in the Rogue Valley and on some serpentine sites.[14, 69] On many other sites, this habitat seems to be maintained by occasional fires, and trees, especially oaks and ponderosa pine, will gradually increase in the absence of fire.

Effects of Management and Anthropogenic Impacts. Fire suppression has probably increased the predominance of dense, tall shrub stands versus a more open, patchy structure. It also seems to reduce the cover of bunchgrasses and forbs as stands become old. Grazing reduces native bunchgrasses in favor of exotics and/or the native rhizomatous California brome (*Bromus carinatus*). Exotic species that have successfully invaded the understory of this habitat are soft brome (*Bromus mollis*), medusahead (*Taeniatherum caput-medusae*), hedgehog dogtail (*Cynosurus echinatus*), and yellow star-thistle (*Centaurea solstitialis*).

Status and Trends. This habitat occupies a small area within this region; it has declined considerably because of conversion to agriculture, residential development, and fire suppression. Most of this habitat has been degraded by fire suppression, grazing, and exotic species invasions. This habitat is still declining in extent from development pressures. One out of 7 Oregon plant associations listed in the National Vegetation Classification is considered imperiled globally,[10] but this type of vegetation has been poorly studied in Oregon and there may be other associations.

13. Western Juniper and Mountain Mahogany Woodlands

Rex. C. Crawford & Jimmy Kagan

Geographic Distribution. This habitat is distributed from the Pacific Northwest south into southern California and east to western Montana and Utah, where it often occurs with pinyon-juniper habitat. In Oregon and Washington, this dry woodland habitat appears primarily in the Owyhee Uplands, High Lava Plains, and northern Basin and Range ecoregions. Secondarily, it develops in the foothills of the Blue Mountains and East Cascades ecoregions, and seems to be expanding into the southern Columbia Basin ecoregion, where it was naturally found in outlier stands.

Western juniper woodlands with shrub-steppe species appear throughout the range of the habitat primarily in central and southern Oregon. Many isolated mahogany communities occur throughout canyons and mountains of eastern Oregon. Juniper-mountain mahogany communities are found in the Ochoco and Blue Mountains.

Physical Setting. This habitat is widespread and variable, occurring in basins and canyons, and on slopes and valley margins in the southern Columbia Plateau, and on fire-protected sites in the northern Basin and Range province. It may be found on benches and foothills. Western juniper and/or mountain mahogany woodlands are often found on shallow soils, on flats at mid- to high elevations, usually on basalts. Other sites range from deep, loess soils and sandy slopes to very stony canyon slopes. At lower elevations, or in areas outside of shrub-steppe, this habitat occurs on slopes and in areas with shallow soils. Mountain mahogany can occur on steep rimrock slopes, usually in areas of shallow soils or protected slopes. This habitat can be found at elevations of 1,500- 8,000 ft (457-2,438 m), mostly between 4,000-6,000 ft (1,220-1,830 m). Average annual precipitation ranges from approximately 10 to 13 inches (25 to 33 cm), with most occurring as winter snow.

Landscape Setting. This habitat reflects a transition between Ponderosa Pine Forest and Woodlands and Shrub-steppe, Eastside Grasslands, and rarely Desert Playa and Salt Desert Scrub habitats. Western juniper generally occurs on higher topography, whereas the shrub communities are more common in depressions or steep slopes with bunchgrass undergrowth. In the Great Basin, mountain mahogany may form a distinct belt on mountain slopes and ridgetops above pinyon-juniper woodland. Mountain mahogany can occur in isolated, pure patches that are often very dense. The primary land use is livestock grazing.

Structure. This habitat is made up of savannas, woodlands, or open forests with 10-60% canopy cover. The tallest layer is composed of short (6.6-40 ft [2-12 m] tall) evergreen trees. Dominant plants may assume a tall-shrub growth form on some sites. The short trees appear in a mosaic pattern with areas of low or medium-tall (usually evergreen) shrubs alternating with areas of tree layers and widely spaced low or medium-tall shrubs. The herbaceous layer is usually composed of short or medium tall bunchgrass or, rarely, a rhizomatous grass-forb undergrowth. These vegetated areas can be interspersed with rimrock or scree. A well-developed cryptogam layer often covers the ground, although bare rock can make up much of the ground cover.

Composition. Western juniper and/or mountain mahogany dominate these woodlands either with bunchgrass or shrub-steppe undergrowth. Western juniper (*Juniperus occidentalis*) is the most common dominant tree in these woodlands. Part of this habitat will have curl-leaf mountain mahogany (*Cercocarpus ledifolius*) as the only dominant tall shrub or small tree. Mahogany may be co-dominant with western juniper. Ponderosa pine (*Pinus ponderosa*) can grow in this habitat and in some rare instances may be an important part of the canopy.

The most common shrubs in this habitat are basin, Wyoming, or mountain big sagebrush (*Artemisia tridentata* ssp. *tridentata*, ssp. *wyomingensis*, and ssp. *vaseyana*) and/or bitterbrush (*Purshia tridentata*). They usually provide significant cover in juniper stands. Low or stiff sagebrush (*Artemisia arbuscula* or *A. rigida*) are dominant dwarf shrubs in some juniper stands. Mountain big sagebrush appears most commonly with mountain mahogany and mountain mahogany mixed with juniper. Snowbank shrubland patches in mountain mahogany woodlands are composed of mountain big sagebrush with bitter cherry (*Prunus emarginata*), quaking aspen (*Populus tremuloides*), and serviceberry (*Amelanchier alnifolia*). Shorter shrubs such as mountain snowberry (*Symphoricarpos oreophilus*) or creeping Oregongrape (*Mahonia repens*) can be dominant in the undergrowth. Rabbitbrush (*Chrysothamnus nauseosus* and *C. viscidiflorus*) will increase with grazing.

Part of this woodland habitat lacks a shrub layer. Various native bunchgrasses dominate different aspects of this habitat. Sandberg bluegrass (*Poa sandbergii*), a short bunchgrass, is the dominant and most common grass throughout many juniper sites. Medium-tall bunchgrasses such as Idaho fescue (*Festuca idahoensis*), bluebunch wheatgrass (*Pseudoroegneria spicata*), needlegrasses (*Stipa occidentalis, S. thurberiana, S. lemmonii*), bottlebrush squirreltail (*Elymus elymoides*) can dominate undergrowth.

Threadleaf sedge (*Carex filifolia*) and basin wildrye (*Leymus cinereus*) are found in lowlands and Geyer's and Ross' sedge (*Carex geyeri, C. rossii*), pinegrass (*Calamagrostis rubescens*), and blue wildrye (*E. glaucus*) appear on mountain foothills. Sandy sites typically have needle-and-thread (*Stipa comata*) and Indian ricegrass (*Oryzopsis hymenoides*). Cheatgrass (*Bromus tectorum*) or bulbous bluegrass (*Poa bulbosa*) often dominate overgrazed or disturbed sites. In good condition this habitat may have mosses growing under the trees.

Other Classifications and Key References. This habitat is also called Juniper Steppe Woodland.[136] The Oregon Gap II Project[126] and Oregon Vegetation Landscape-Level Cover Types[127] that would represent this type are ponderosa pine-western juniper woodland, western juniper woodland, and mountain mahogany shrubland. Other references describe this habitat.[64, 79, 122, 207]

Natural Disturbance Regime. Both mountain mahogany and western juniper are fire intolerant. Under natural high-frequency fire regimes both species formed savannas or occurred as isolated patches on fire-resistant sites in shrub-steppe or steppe habitat. Western juniper is considered a topoedaphic climax tree in a number of sagebrush-grassland, shrub-steppe, and drier conifer sites. It is an increaser in many earlier seral communities in these zones and invades without fires. Most trees >13 ft (4 m) tall can survive low-intensity fires. The historic fire regime of mountain mahogany communities varies with community type and structure. The fire-return interval for mountain mahogany (along the Salmon River in Idaho) was 13-22 years until the early 1900's and has increased ever since. Mountain mahogany can live to 1,350 years in western and central Nevada. Some old-growth mountain mahogany stands avoid fire by growing on extremely rocky sites.

Succession and Stand Dynamics. Juniper invades shrub-steppe and steppe and reduces undergrowth productivity. Although slow seed dispersal delays recovery time, western juniper can regain dominance in 30-50 years following fire. A fire-return interval of 30-50 years typically arrests juniper invasion. The successional role of curl-leaf mountain mahogany varies with community type. Mountain brush communities where curl-leaf mountain mahogany is either dominant or co-dominant are generally stable and successional rates are slow.

Effects of Management and Anthropogenic Impacts. Over the past 150 years, with fire suppression, overgrazing, and changing climatic factors, western juniper has increased its range into adjacent shrub-steppe, grasslands, and savannas. Increased density of juniper and reduced fine fuels from an interaction of grazing and shading result in high severity fires that eliminate woody plants and promote herbaceous cover, primarily annual grasses. Diverse mosses and lichens occur on the ground in this type if it has not been too disturbed by grazing. Excessive grazing will decrease bunchgrasses and increase exotic annual grasses plus various native and exotic forbs.

Animals seeking shade under trees decrease or eliminate bunchgrasses and contribute to increasing cheatgrass cover.

Status and Trends. This habitat is dominated by fire-sensitive species, and therefore, the range of western juniper and mountain mahogany has expanded because of an interaction of livestock grazing and fire suppression. Quigley and Arbelbide[181] concluded that in the Inland Pacific Northwest, Juniper/Sagebrush, Juniper Woodlands, and Mountain Mahogany cover types now are significantly greater in extent than before 1900. Although it covers more area, this habitat is generally in degraded condition because of increased exotic plants and decreased native bunchgrasses. One third of Pacific Northwest juniper and mountain mahogany community types listed in the National Vegetation Classification are considered imperiled or critically imperiled.[10]

14. Eastside Canyon Shrublands
Rex. C. Crawford & Jimmy Kagan

Geographic Distribution. This habitat occurs primarily on steep canyon slopes in the Blue Mountains and the margins of the Columbia Basin in Idaho, Oregon, and Washington. This habitat also appears as isolated patches across Washington's Columbia Basin.

Physical Setting. This habitat develops in hot dry climates in the Pacific Northwest. Annual precipitation totals 12-20 inches (31-51 cm); only 10% falls in the hottest months, July through September. Snow accumulation is low (1-6 inches [3-15 cm]), persisting only a few weeks. Sites are generally steep (>60%) on all aspects but most common on northerly aspects in deep, dry canyons. Columbia River basalt is the major geologic substrate although many sites are underlain with loess deposits mixed with colluvium. Steep northerly aspects in the Palouse Hills can also support this habitat. This habitat is found from 500 to 5,000 ft (152 to 1,524 m) in elevation.

Landscape Setting. This habitat is generally found in steep canyons surrounded by the Eastside Grassland Habitat and below or in a mosaic with the Ponderosa Pine Forest and Woodland habitat. This habitat can develop near talus slopes, at the heads of dry drainages, and toe slopes in moist shrub-steppe and steppe zones. At lower elevation sites, these are more often in a mix with bluebunch wheatgrass, dry rocky grasslands, and low-elevation riparian habitats. The primary surrounding land use is livestock grazing.

Structure. The Eastside Canyon Shrubland habitat is generally a mix of tall (5 ft [1.5 m]) to medium (1.6 ft [0.5 m]) deciduous shrublands in a mosaic with bunchgrass or annual grasslands. Shrub canopies are almost always closed (>60% cover), forming a thicket of interwoven stems and branches. Shrub layers can be one or two-tiered but often are so dense they restrict the herbaceous layer to shade-tolerant rhizomatous species.

Composition. Mallowleaf ninebark (*Physocarpus malvaceus*), a major dominant, bitter cherry (*Prunus emarginata*), chokecherry (*Prunus virginiana*), oceanspray (*Holodiscus discolor*) or Rocky Mountain maple (*Acer glabrum*) are the most common tall shrubs in this habitat. In moist areas, black hawthorn (*Crataegus douglasii*) may appear and can dominate some sites as a tall shrub or small tree. Other tall shrubs such as syringa (*Philadelphus lewisii*) or serviceberry (*Amelanchier alnifolia*) often dominate sites associated with talus. Common medium-tall shrubs are common snowberry (*Symphoricarpos albus*), rose (*Rosa nutkana, R. woodsii*), smooth sumac (*Rhus glabra*), and currants (*Ribes* spp.). Basin or Wyoming big sagebrush (*Artemisia tridentata* ssp. *tridentata* or *A. t.* ssp. *wyomingensis*), along with rabbitbrush (*Chrysothamnus* spp.), may be important members of these thickets in weedy sites, dry areas, or transitions with grasslands. Scattered ponderosa pine (*Pinus ponderosa*), black cottonwood (*Populus balsamifera* ssp. *trichocarpa*) and rarely Douglas-fir (*Pseudotsuga menziesii*) trees may be found in and adjacent to this habitat.

Idaho fescue (*Festuca idahoensis*), bluebunch wheatgrass (*Pseudoroegneria spicata*), Thurber's needlegrass (*Stipa thurberiana),* and Sandberg's bluegrass (*Poa sandbergii*) found in the surrounding steppe or shrub-steppe are common but never abundant in these thickets. Basin wildrye (*Leymus cinereus*) can be locally important. Kentucky bluegrass (*Poa pratensis*) is a common introduced grass and, where grazed by livestock, is a dominant undergrowth species. Annual grasses (*Bromus tectorum, B. briziformis*) can be abundant especially on disturbed dry sites. Cleavers (*Galium aparine*) is a frequent member of the herbaceous component of this habitat. Other common forbs include red avens (*Geum triflorum*), horsemint (*Agastache urticifolia*), sticky cinquefoil (*Potentilla gracilis*), balsamroots (*Balsamorhiza* spp.), and fleabanes (*Erigeron* spp.).

Other Classifications and Key References. This habitat is called shrub garland[88] or talus thickets. The Oregon Gap II Project[126] and Oregon Vegetation Landscape-Level Cover Type[127] that would represent this type is eastside big sagebrush shrubland. Other references describe this habitat.[60, 122, 123, 207]

Natural Disturbance Regime. This habitat is within the sagebrush and bunchgrass vegetation type of Barrett et al.[22] who concluded it had a fire-return interval of 25 years. Canyon shrublands associated with talus burn less frequently but are subject to talus movement. Similar shrubfields are associated with forest landscapes and are early seral stages of the Eastside Mixed Conifer Forest Habitat.

Succession and Stand Dynamics. Many of the major shrubs sprout following fire and will be maintained with moderate fire frequency. Most thickets will increase in size without fire. This habitat has increased primarily in moist steppe and shrub-steppe habitat with fire suppression and restricted grazing. Prolonged fire suppression may lead to invasions by tree species. Apparently some representatives of this habitat could potentially support Douglas-fir or ponderosa pine woodlands after a long fire-free period.

Effects of Management and Anthropogenic Impacts. Livestock grazing in adjacent grassland or shrub-steppe habitat changes the surrounding fine-fuel matrix for fire. That, combined with fire suppression, leads to a change in habitat patch size, structure, and composition. In response to fire suppression, shrub thickets on northerly aspects near lower treeline tend to increase in patch size and height and are invaded by tree species. With heavy livestock grazing, shrubs are browsed, broken, and trampled, which eventually creates a more open shrubland with a more abundant herbaceous layer.

Status and Trends. The Eastside Canyon Shrubland habitat is restricted in range and probably has increased locally in area. Johnson and Simon[123] reported increases in common snowberry-rose communities as a response to fire suppression and heavy grazing that depleted bunchgrass cover. One of the three Eastside Canyon Shrubland community types in the National Vegetation Classification is considered imperiled.[10]

15. Eastside Grassslands
Rex. C. Crawford & Jimmy Kagan

Geographic Distribution. This habitat is found primarily in the Columbia Basin of Idaho, Oregon, and Washington, at mid- to low elevations and on plateaus in the Blue Mountains, usually within the ponderosa pine zone in Oregon.

Idaho fescue grassland habitats were formerly widespread in the Palouse region of southeastern Washington and adjacent Idaho; most of this habitat has been converted to agriculture. Idaho fescue grasslands still occur in isolated, moist sites near lower treeline in the foothills of the Blue Mountains, the Northern Rockies, and east Cascades near the Columbia River Gorge. Bluebunch wheatgrass grassland habitats are common throughout the Columbia Basin, both as modified native grasslands in deep canyons and the dry Palouse and as fire-induced representatives in the shrub-steppe. Similar grasslands appear on the High Lava Plains ecoregion, where they occur in a matrix with big sagebrush or juniper woodlands. In Oregon they are also found in burned shrub-steppe and canyons in the Basin and Range and Owyhee Uplands. Sand dropseed and three-awn needlegrass grassland habitats are restricted to river terraces in the Columbia Basin, Blue Mountains, and Owyhee Uplands of Oregon and Washington. Primary location of this habitat extends along the Snake River from Lewiston south to the Owyhee River.

Physical Setting. This habitat develops in hot, dry climates in the Pacific Northwest. Annual precipitation totals 8-20 inches (20-51 cm); only 10% falls in the hottest months, July through September. Snow accumulation is low (1-6 inches [3-15 cm]) and occurs only in January and February in eastern portions of its range and November through March in the west. More snow accumulates in grasslands within the forest matrix. Soils are variable: (1) highly productive loess soils up to 51 inches (130 cm) deep, (2) rocky flats, (3) steep slopes, and (4) sandy, gravel or cobble soils. An important variant of this habitat occurs on sandy, gravelly, or silty river terraces or seasonally exposed river gravel or Spokane flood deposits. The grassland habitat is typically upland vegetation but it may also include riparian bottomlands dominated by non-native grasses. This habitat is found from 500 to 6,000 ft (152-1,830 m) in elevation.

Landscape Setting. Eastside grassland habitats appear well below and in a matrix with lower treeline Ponderosa Pine Forest and Woodlands or Western Juniper and Mountain Mahogany Woodlands. It can also be part of the lower elevation forest matrix. Most grassland habitat occurs in two distinct large landscapes: plateau and canyon grasslands. Several rivers flow through narrow basalt canyons below plateaus supporting prairies or shrub-steppe. The canyons can be some 2,132 ft (650 m) deep below the plateau. The plateau above is composed of gentle slopes with deep silty loess soils in an expansive rolling dune-like landscape. Grasslands may occur in a patchwork with shallow soil scablands or within biscuit scablands or mounded topography. Naturally occurring grasslands are beyond the range of bitterbrush and sagebrush species. This habitat exists today in the shrub-steppe landscape where grasslands are created by brush removal, chaining or spraying, or by fire. Agricultural uses and introduced perennial plants on abandoned or planted fields are common throughout the current distribution of eastside grassland habitats.

Structure. This habitat is dominated by short to medium-tall grasses (<3.3 ft [1 m]). Total herbaceous cover can be closed to only sparsely vegetated. In general, this habitat is an open and irregular arrangement of grass clumps rather than a continuous sod cover. These medium-tall grasslands often have scattered and diverse patches of low shrubs, but few or no medium-tall shrubs (<10% cover of shrubs are taller than the grass layer). Native forbs may contribute significant cover or they may be absent. Grasslands in canyons are dominated by bunchgrasses growing in lower densities than on deep-soil prairie sites. The soil surface between perennial plants can be covered with a diverse cryptogamic or microbiotic layer of mosses, lichens, and various soil bacteria and algae. Moister environments can support a dense sod of rhizomatous perennial grasses. Annual plants are a common spring and early summer feature of this habitat.

Composition. Bluebunch wheatgrass (*Pseudoroegneria spicata*) and Idaho fescue (*Festuca idahoensis*) are the characteristic native bunchgrasses of this habitat and either or both can be dominant. Idaho fescue is common in more moist areas and bluebunch wheatgrass more abundant in drier areas. Rough fescue (*F. campestris*) is a characteristic dominant on moist sites in northeastern Washington. Sand dropseed (*Sporobolus cryptandrus*) or three-awn (*Aristida longiseta*) are native dominant grasses on hot dry sites in deep canyons. Sandberg bluegrass (*Poa sandbergii*) is usually present, and occasionally codominant in drier areas. Bottlebrush squirreltail (*Elymus elymoides*) and Thurber needlegrass (*Stipa thurberiana*) can be locally dominant. Annual grasses are usually present; cheatgrass (*Bromus tectorum*) is the most widespread. In addition, medusahead (*Taeniatherum caput-medusae*), and other annual bromes (*Bromus commutatus, B. mollis, B. japonicus*) may be present to co-dominant. Moist environments, including riparian bottomlands, are often co-dominated by Kentucky bluegrass (*Poa pratensis*).

A dense and diverse forb layer can be present or entirely absent; >40 species of native forbs can grow in this habitat including balsamroots (*Balsamorhiza* spp.), biscuitroots (*Lomatium* spp.), buckwheat (*Eriogonum* spp.), fleabane (*Erigeron* spp.), lupines (*Lupinus* spp.), and milkvetches (*Astragalus* spp.). Common exotic forbs that can grow in this habitat are knapweeds (*Centaurea solstitialis, C. diffusa, C. maculosa*), tall tumblemustard (*Sisymbrium altissimum*), and Russian thistle (*Salsola kali*).

Smooth sumac (*Rhus glabra*) is a deciduous shrub locally found in combination with these grassland species. Rabbitbrushes (*Chrysothamnus nauseosus, C. viscidiflorus*) can occur in this habitat in small amounts, especially where grazed by livestock. In moist Palouse regions, common snowberry (*Symphoricarpos albus*) or Nootka rose (*Rosa nutkana*) may be present, but is shorter than the bunchgrasses. Dry sites contain low succulent pricklypear (*Opuntia polyacantha*). Big sagebrush (*Artemisia tridentata*) is occasional and may be increasing in grasslands on former shrub-steppe sites. Black hawthorn (*Crataegus douglasii*) and other tall shrubs can form dense thickets near Idaho fescue grasslands. Rarely, ponderosa pine (*Pinus ponderosa*) or western juniper (*Juniperus occidentalis*) can occur as isolated trees.

Other Classifications and Key References. This habitat is called Palouse Prairie, Pacific Northwest grassland, steppe vegetation, or bunchgrass prairie in general ecological literature. Quigley and Arbelbide[181] called this habitat Fescue-Bunchgrass and Wheatgrass Bunchgrass and the dry Grass cover type. The Oregon Gap II Project[126] and Oregon Vegetation Landscape-Level Cover Types[127] that would represent this type are northeast Oregon canyon grassland, forest-grassland mosaic, and modified grassland; Washington Gap[37] types 13, 21, 22, 24, 29-31, 82, and 99 map this habitat. Kuchler[136] includes this within Fescue-wheatgrass and wheatgrass-bluegrass. Franklin and Dyrness[88] include this habitat in steppe zones of Washington and Oregon. Other references describe this habitat.[1, 28, 60, 159, 166, 206, 207]

Natural Disturbance Regime. The fire-return interval for sagebrush and bunchgrass is estimated at 25 years.[22] The native bunchgrass habitat apparently lacked extensive herds of large grazing and browsing animals until the late 1800s. Burrowing animals and their predators likely played important roles in creating small-scale patch patterns.

Succession and Stand Dynamics. Currently fires burn less frequently in the Palouse grasslands than historically because of fire suppression, roads, and conversions to cropland.[159] Without fire, black hawthorn shrubland patches expand on slopes along with common snowberry and rose. Fires covering large areas of Shrub-steppe habitat can eliminate shrubs and their seed sources and create Eastside Grasslands habitat. Fires that follow heavy

grazing or repeated early season fires can result in annual grasslands of cheatgrass, medusahead, knapweed, or yellow star-thistle. Annual exotic grasslands are common in dry grasslands and are included in modified grasslands as part of the Agriculture habitat.

Effects of Management and Anthropogenic Impacts. Large expanses of grasslands are currently used for livestock ranching. Deep soil Palouse sites are mostly converted to agriculture. Drier grasslands and canyon grasslands, those with shallower soils, steeper topography, or hotter, drier environments, were more intensively grazed and for longer periods than were deep-soil grasslands.[207] Evidently, these drier native bunchgrass grasslands changed irreversibly to persistent annual grass and forblands. Some annual grassland, native bunchgrass, and shrub-steppe habitats were converted to intermediate wheatgrass, or more commonly, crested wheatgrass (*Agropyron cristatum*)-dominated areas. Apparently, these form persistent grasslands and are included as modified grasslands in the Agriculture habitat. With intense livestock use, some riparian bottomlands become dominated by non-native grasses. Many native dropseed grasslands have been submerged by dam reservoirs.

Status and Trends. Most of the Palouse prairie of southeastern Washington and adjacent Idaho and Oregon has been converted to agriculture. Remnants still occur in the foothills of the Blue Mountains and in isolated, moist Columbia Basin sites. The Palouse is one of the most endangered ecosystems in the U.S.[166] with only 1% of the original habitat remaining; it is highly fragmented with most sites <10 acres. All these areas are subject to weed invasions and drift of aerial biocides. Since 1900, 94% of the Palouse grasslands have been converted to crop, hay, or pasture lands. Quigley and Arbelbide[181] concluded that Fescue-Bunchgrass and Wheatgrass bunchgrass cover types have significantly decreased in area since before1900, while exotic forbs and annual grasses have significantly increased since pre-1900. Fifty percent of the plant associations recognized as components of Eastside Grasslands habitat listed in the National Vegetation Classification are considered imperiled or critically imperiled.[10]

16. Shrub-steppe
Rex. C. Crawford & Jimmy Kagan

Geographic Distribution. Shrub-steppe habitats are common across the Columbia Plateau of Washington, Oregon, Idaho, and adjacent Wyoming, Utah, and Nevada. They extend up into the cold, dry environments of surrounding mountains.

Basin big sagebrush shrub-steppe occurs along stream channels, in valley bottoms and flats throughout eastern Oregon and Washington. Wyoming sagebrush shrub-steppe is the most widespread habitat in eastern Oregon and Washington, occurring throughout the Columbia Plateau and the northern Great Basin. Mountain big sagebrush shrub-steppe occurs throughout the mountains of the eastern Oregon and Washington. Bitterbrush shrub-steppe appears primarily along the eastern slope of the Cascades, from north-central Washington to California and occasionally in the Blue Mountains. Three-tip sagebrush shrub-steppe occurs mostly along the northern and western Columbia Basin in Washington and occasionally appears in the lower valleys of the Blue Mountains and in the Owyhee Upland ecoregions of Oregon. Interior shrub dunes and sandy steppe and shrub-steppe is concentrated at low elevations near the Columbia River and in isolated pockets in the Northern Basin and Range and Owyhee Uplands. Bolander silver sagebrush shrub-steppe is common in southeastern Oregon. Mountain silver sagebrush is more prevalent in the Oregon East Cascades and in montane meadows in the southern Ochoco and Blue Mountains.

Physical Setting. Generally, this habitat is associated with dry, hot environments in the Pacific Northwest although variants are in cool, moist areas with some snow accumulation in climatically dry mountains. Elevation range is wide (300-9,000 ft [91-2,743 m]) with most habitat occurring between 2,000 and 6,000 ft (610-1,830 m). Habitat occurs on deep alluvial, loess, silty or sandy-silty soils, stony flats, ridges, mountain slopes, and slopes of lake beds with ash or pumice soils.

Landscape Setting. Shrub-steppe habitat defines a biogeographic region and is the major vegetation on average sites in the Columbia Plateau, usually below Ponderosa Pine Forest and Woodland, and Western

Juniper and Mountain Mahogany Woodland habitats. It forms mosaic landscapes with these woodland habitats and Eastside Grasslands, Dwarf Shrub-steppe, and Desert Playa and Salt Scrub habitats. Mountain sagebrush shrub-steppe occurs at high elevations occasionally within the dry Eastside Mixed Conifer Forest and Montane Mixed Conifer Forest habitats. Shrub-steppe habitat can appear in large landscape patches. Livestock grazing is the primary land use in the shrub-steppe although much has been converted to irrigation or dry land agriculture. Large areas occur in military training areas and wildlife refuges.

Structure. This habitat is a shrub savanna or shrubland with shrub coverage of 10-60%. In an undisturbed condition, shrub cover varies between 10 and 30%. Shrubs are generally evergreen although deciduous shrubs are prominent in many habitats. Shrub height typically is medium-tall (1.6-3.3 ft [0.5-1.0 m]) although some sites support shrubs approaching 9 ft (2.7 m) tall. Vegetation structure in this habitat is characteristically an open shrub layer over a moderately open to closed bunchgrass layer. The more northern or productive sites generally have a denser grass layer and sparser shrub layer than southern or more xeric sites. In fact, the rare good-condition site is better characterized as grassland with shrubs than a shrubland. The bunchgrass layer may contain a variety of forbs. Good-condition habitat has very little exposed bare ground, and has mosses and lichens carpeting the area between taller plants. However, heavily grazed sites have dense shrubs making up >40% cover, with introduced annual grasses and little or no moss or lichen cover. Moist sites may support tall bunchgrasses (>3.3 ft [1 m]) or rhizomatous grasses. More southern shrub-steppe may have native low shrubs dominating with bunchgrasses.

Composition. Characteristic and dominant mid-tall shrubs in the shrub-steppe habitat include all three subspecies of big sagebrush, basin (*Artemisia tridentata* ssp. *tridentata*), Wyoming (*A. t.* ssp. *wyomingensis*) or mountain (*A. t.* ssp. *vaseyana*), antelope bitterbrush (*Purshia tridentata*), and two shorter sagebrushes, silver (*A. cana*) and three-tip (*A. tripartita*). Each of these species can be the only shrub or appear in complex seral conditions with other shrubs. Common shrub complexes are bitterbrush and Wyoming big sagebrush, bitterbrush and three-tip sagebrush, Wyoming big sagebrush and three-tip sagebrush, and mountain big sagebrush and silver sagebrush. Wyoming and mountain big sagebrush can codominate areas with tobacco brush (*Ceanothus velutinus*). Rabbitbrush (*Chrysothamnus viscidiflorus*) and short-spine horsebrush (*Tetradymia spinosa*) are common associates and often dominate sites after disturbance. Big sagebrush occurs with the shorter stiff sagebrush (*A. rigida*) or low sagebrush (*A. arbuscula*) on shallow soils or high elevation sites. Many sandy areas are shrub-free or are open to patchy shrublands of bitterbrush and/or rabbitbrush. Silver sagebrush is the dominant and characteristic shrub along the edges of stream courses, moist meadows, and ponds. Silver sagebrush and rabbitbrush are associates in disturbed areas.

When this habitat is in good or better ecological condition a bunchgrass steppe layer is characteristic. Diagnostic native bunchgrasses that often dominate different shrub-steppe habitats are (1) mid-grasses: bluebunch wheatgrass (*Pseudoroegneria spicata),* Idaho fescue (*Festuca idahoensis*), bottlebrush squirreltail (*Elymus elymoides*), and Thurber needlegrass (*Stipa thurberiana*); (2) short grasses: threadleaf sedge (*Carex filifolia*) and Sandberg bluegrass (*Poa sandbergii*); and (3) the tall grass, basin wildrye (*Leymus cinereus*). Idaho fescue is characteristic of the most productive shrub-steppe vegetation. Bluebunch wheatgrass is codominant at xeric locations, whereas western needlegrass (*Stipa occidentalis*), long-stolon (*Carex inops*) or Geyer's sedge (*C. geyeri*) increase in abundance in higher elevation shrub-steppe habitats. Needle-and-thread (*Stipa comata*) is the characteristic native bunchgrass on stabilized sandy soils. Indian ricegrass (*Oryzopsis hymenoides*) characterizes dunes. Grass layers on montane sites contain slender wheatgrass (*Elymus trachycaulus*), mountain fescue (*F. brachyphylla*), green fescue (*F. viridula*), Geyer's sedge, or tall bluegrasses (*Poa* spp.). Bottlebrush squirreltail can be locally important in the Columbia Basin, sand dropseed (*Sporobolus cryptandrus*) is important in the Basin and Range and basin wildrye is common in the more alkaline areas. Nevada bluegrass (*Poa secunda*), Richardson muhly (*Muhlenbergia richardsonis*), or alkali grass (*Puccinella* spp.) can dominate silver sagebrush flats. Many sites support non-native plants, primarily cheatgrass (*Bromus tectorum*) or crested wheatgrass (*Agropyron cristatum*) with or without native grasses. Shrub-steppe habitat, depending on site potential and disturbance history, can be rich in forbs or have little forb cover. Trees may be present in some shrub-steppe habitats, usually as isolated individuals from adjacent forest or woodland habitats.

Other Classifications and Key References. This habitat is called Sagebrush steppe and Great Basin sagebrush by Kuchler.[136] The Oregon Gap II Project[126] and Oregon Vegetation Landscape-Level Cover Types[127] that would represent this type are big sagebrush shrubland, sagebrush steppe, and bitterbrush-big sagebrush shrubland. Franklin and Dyrness[88] discussed this habitat in shrub-steppe zones of Washington and Oregon. Other references describe this habitat. [60, 116, 122, 123, 212, 224, 225]

Natural Disturbance Regime. Barrett et al.[22] concluded that the fire-return interval for this habitat is 25 years. The native shrub-steppe habitat apparently lacked extensive herds of large grazing and browsing animals until the late 1800s. Burrowing animals and their predators likely played important roles in creating small-scale patch patterns.

Succession and Stand Dynamics. With disturbance, mature stands of big sagebrush are reinvaded through soil-stored or windborne seeds. Invasion can be slow because sagebrush is not disseminated over long distances. Site dominance by big sagebrush usually takes a decade or more depending on fire severity and season, seed rain,

postfire moisture, and plant competition. Three-tip sagebrush is a climax species that reestablishes (from seeds or commonly from sprouts) within 5-10 years following a disturbance. Certain disturbance regimes promote three-tip sagebrush and it can out-compete herbaceous species. Bitterbrush is a climax species that plays a seral role colonizing by seed onto rocky and/or pumice soils. Bitterbrush may be declining and may be replaced by woodlands in the absence of fire. Silver sagebrush is a climax species that establishes during early seral stages and coexists with later arriving species. Big sagebrush, rabbitbrush, and short-spine horsebrush invade and can form dense stands after fire or livestock grazing. Frequent or high-intensity fire can create a patchy shrub cover or can eliminate shrub cover and create Eastside Grasslands habitat.

Effects of Management and Anthropogenic Impacts. Shrub density and annual cover increase, whereas bunchgrass density decreases, with livestock use. Repeated or intense disturbance, particularly on drier sites, leads to cheatgrass dominance and replacement of native bunchgrasses. Dry and sandy soils are sensitive to grazing, with needle-and-thread replaced by cheatgrass at most sites. These disturbed sites can be converted to modified grasslands in the Agriculture habitat.

Status and Trends. Shrub-steppe habitat still dominates most of southeastern Oregon although half of its original distribution in the Columbia Basin has been converted to agriculture. Alteration of fire regimes, fragmentation, livestock grazing, and the addition of >800 exotic plant species have changed the character of shrub-steppe habitat. Quigley and Arbelbide[181] concluded that Big Sagebrush and Mountain Sagebrush cover types are significantly smaller in area than before 1900, and that Bitterbrush/Bluebunch Wheatgrass cover type is similar to the pre-1900 extent. They concluded that Basin Big Sagebrush and Big Sagebrush-Warm potential vegetation type's successional pathways are altered, that some pathways of Antelope Bitterbrush are altered and that most pathways for Big Sagebrush-Cool are unaltered. Overall this habitat has seen an increase in exotic plant importance and a decrease in native bunchgrasses. More than half of the Pacific Northwest shrub-steppe habitat community types listed in the National Vegetation Classification are considered imperiled or critically imperiled.[10]

17. Dwarf Shrub-steppe

Rex C. Crawford & Jimmy Kagan

Geographic Distribution. Dwarf-shrub and related scabland habitats are located throughout the Columbia Plateau and in adjacent woodland and forest habitats. They are more common in southern Oregon than in Washington.

Low sagebrush steppe is common in the Basin and Range and the Owyhee Uplands in eastern Lake, Harney, and Malheur counties and is a minor type in eastern Washington and northeastern Oregon. It usually occurs on low, scabby plateaus above lake basins. Stiff sagebrush/ Sandberg bluegrass is a major type widely distributed in the Columbia Basin, particularly associated with the channeled scablands, High Lava Plains, and in isolated spots throughout the Blue Mountains and the Palouse. Black sagebrush steppe is not found in Washington and is rare in Oregon, occurring along the Nevada border in southern Lake, Harney, and Malheur counties, in the southern Basin and Range and Owyhee Uplands Physiographic Province.

Physical Setting. This habitat appears on sites with little soil development that often have extensive areas of exposed rock, gravel, or compacted soil. The habitat is characteristically associated with flats, plateaus, or gentle slopes although steep slopes with rock outcrops are common. Scabland types within the shrub-steppe area occur on barren, usually fairly young basalts or shallow loam over basalt <12 inches (30 cm) deep. In woodland or forest mosaics, scabland soils are deeper (still <26 inches [65 cm]) but too droughty or extreme soils for tree growth. Topoedaphic drought is the major process influencing these communities on ridge tops and gentle slopes around ridgetops. Spring flooding is characteristic of scablands in concave topographic positions. This habitat is found across a wide range of elevations from 500 to 7,000 ft (152 to 2,134 m).

Landscape Setting. These scabland habitats form a mosaic with Shrub-steppe habitat, Eastside Grasslands habitat, and with Western Juniper and Mountain Mahogany Woodland or Ponderosa Pine Forest and Woodland habitats. Low sagebrush stands are often extensive and occasionally occur in a mosaic with big sagebrush, stiff sagebrush, or black sagebrush steppe or within lower treeline woodlands. Stiff sagebrush stands may also be extensive, but usually occur in a mosaic with grassland, big sagebrush or occasionally in juniper (*Juniperus occidentalis*) or Ponderosa pine (*Pinus ponderosa*) woodlands. Black sagebrush stands are extensive and may occur in a mosaic with low sagebrush or mountain or Wyoming big sagebrush.

Structure. These low shrub (<1.6 ft [0.5 m] high) communities have an undergrowth of short grasses and forbs with extensive exposed rock and cryptogamic crusts. More productive sites have an open, native medium-tall bunchgrass layer with scattered low shrubs. Some scablands in the Columbia Basin have few to no dwarf shrubs and the habitat is entirely dominated by grasses and forbs. Total vegetation cover is open to sparse. Individual trees can appear among the low shrubs when this habitat appears in the forest matrix.

Composition. Several dwarf-shrub species characterize this habitat: low sagebrush (*Artemisia arbuscula*), black sagebrush (*A. nova*), stiff sagebrush (*A. rigida*), or several shrubby buckwheat species (*Eriogonum douglasii, E. sphaerocephalum, E. strictum, E. thymoides, E. niveum, E. compositum*). These dwarf-shrub species can be found as the sole shrub species or in combination with these or other low shrubs. Purple sage (*Saliva dorrii*) can dominate scablands on steep sites with rock outcrops.

Sandberg bluegrass (*Poa sandbergii*) is the characteristic and sometimes the dominant grass making up most of this habitat's sparse vegetative cover. Taller bluebunch wheatgrass (*Pseudoroegneria spicata*) or Idaho fescue (*Festuca idahoensis*) grasses may occur on the most productive sites with Sandberg bluegrass. Bottlebrush squirreltail (*Elymus elymoides*) and Thurber needlegrass (*Stipa thurberiana*) are typically found in low cover areas, although they can dominate some sites. One-spike oatgrass (*Danthonia unispicata*), prairie junegrass (*Koeleria macrantha*), and Henderson ricegrass (*Achnatherum hendersonii*) are occasionally important. Exotic annual grasses, commonly cheatgrass (*Bromus tectorum*), increase with heavy disturbance and can be locally abundant. Common forbs include serrate balsamroot (*Balsamorhiza serrata*), Oregon twinpod (*Physaria oregana*), Oregon bitterroot (*Lewisia rediviva*), big-head clover (*Trifolium macrocephalum*), and Rainier violet (*Viola trinervata*). Several other forbs (*Arenaria, Collomia, Erigeron, Lomatium,* and *Phlox* spp.) are characteristic, early blooming species. A diverse lichen and moss layer is a prominent component of these communities.

Medium-tall shrubs, such as big sagebrush (*Artemisia tridentata*), Silver sagebrush (*A. cana*), antelope bitterbrush (*Purshia tridentata*), and rabbitbrush (*Chrysothamnus* spp.) occasionally appear in these scablands.

Other Classifications and Key References. This habitat is called scabland, biscuit-swale topography, lithosolic steppe, or low shrub-steppe. Quigley and Arbelbide[181] called this habitat low sagebrush cover type and "Low

Sagebrush-Xeric" and "Low Sagebrush-Mesic" potential vegetation groups. The Oregon Gap II Project[126] and Oregon Vegetation Landscape-Level Cover Type[127] that would represent this type is low-dwarf sagebrush. Kuchler[136] did not distinguish this habitat but included it within Sagebrush Steppe. Franklin and Dyrness[88] discussed this habitat as lithosolic sites in steppe and shrub-steppe zones of Washington and as plant associations in steppe and shrub-steppe zones of central and southern Oregon. Other references describe this habitat.[60, 64, 122, 123, 207]

Natural Disturbance Regime. Scabland habitats often do not have enough vegetation cover to support wildfires. Bunchgrass sites with black or low sagebrush may burn enough to damage shrubs and decrease shrub cover with repetitive burns. Many scabland sites have poorly drained soil and because of shallow soil are prone to winter flooding. Freezing of saturated soil results in "frost-heaving" that churns the soil and is a major disturbance factor in vegetation patterns. Stiff sagebrush is a preferred browse for elk as well as livestock. Native ungulates use scablands in early spring and contribute to churning of the soil surface.

Succession and Stand Dynamics. Grazing reduces the cover of bunchgrasses and increases the abundance of common yarrow (*Achillea millefolium*), phlox species, bighead clover, serrate balsamroot, bottlebrush squirreltail and annual bromes on dwarf shrublands. Increased ground disturbing activities increases exotic plant abundance, particularly on deeper soil sites. All dwarf-shrub species are intolerant of fire and do not sprout. Consequently, redevelopment of dwarf shrub-steppe habitat is slow following fire or any disturbance that removes shrubs.

Effects of Management and Anthropogenic Impacts. Scabland habitats provide little forage and consequently are used only as a final resort by livestock. Heavy use by livestock or vehicles disrupts the moss/lichen layer and increases exposed rock and bare ground that create habitat for exotic plant invasion. Exotic annual bromes have become part of these habitats with natural soil churning disturbance.

Status and Trends. Quigley and Arbelbide[181] concluded that the low sagebrush cover type is as abundant as it was before 1900. They concluded that "Low Sagebrush-Xeric" successional pathways have experienced a high level of change from exotic invasions and that some pathways of "Low Sagebrush-Mesic" are unaltered. Twenty percent of Pacific Northwest dwarf shrub-steppe community types listed in the National Vegetation Classification are considered imperiled or critically imperiled.[10]

18. Desert Playa and Salt Scrub Shrublands

Rex C. Crawford & Jimmy Kagan

Geographic Distribution. The desert playa and salt scrub habitat centers on the Great Basin of Nevada and Utah. In the Pacific Northwest, it is most common and abundant in the larger, alkaline lake basins in southeastern Oregon, although it is represented throughout the Columbia Plateau, Basin and Range, and Owyhee Provinces.

Shadscale salt desert shrub and mixed salt desert shrub range from southeastern Oregon south into Utah and Nevada. Black greasewood salt desert scrub and alkaline/saline bottomland grasslands and wetlands appear throughout the Columbia Plateau of Washington and Oregon.

Physical Setting. This habitat typically occupies the lowest elevations in hydrologic basins in the driest regions of the Pacific Northwest. Elevation range is highly variable, from 3,000 to 7,500 ft (914 to 2.286 m) in southeastern Oregon to 500 to 5,500 ft (152-1,676 m) in central Washington. Structural and compositional variation in this habitat are related to changes in salinity and fluctuations in the water table. Areas with little or no vegetative cover have highly alkaline and saline soils and are poorly drained or irregularly flooded. Other arid soil types include desert pavement and barren ash. The wettest variants of the habitat are usually found at the mouth of stream drainages or in areas with some freshwater input into a playa. These have finer, deeper alluvial soils that occur in low alkaline dunes, around playas, on slopes above alkaline basins or in small, poorly drained basins in sagebrush steppe. Topographically, this habitat occurs on playas or desert pavement, or on low benches above playas with occasional low alkaline dune ridges.

Landscape Setting. This habitat is typically surrounded by shrub-steppe habitat. It forms a habitat mosaic of playas, salt grass meadows, salt desert shrublands and sagebrush shrublands. This habitat may be associated with Herbaceous Wetland habitat. Local land use can result in juxtaposition with Agriculture or Eastside Grasslands habitat. Most of this habitat provides rangeland for livestock, particularly as winter range. Portions of this habitat associated with water are most attractive to livestock. Other portions of this type are designated wildlife refuges.

Structure. This habitat is structurally diverse, ranging from dense shrublands to sparse grasslands to unvegetated flats. Generally, low to medium-tall alkali or saline tolerant shrubs form an open layer over a grass and annual, often succulent, forb undergrowth. Deciduous shrubs, when present, usually create <50% cover but occasionally can exceed 70% on previously disturbed ground. Ground cover between shrubs is variable, ranging from widely spaced tall, medium-tall, or short bunchgrasses to dense, short rhizomatous grasses. Other areas have no shrubs and support a fairly continuous cover of graminoids, occasionally with widely spaced bunchgrasses. Sites can have extensive bare ground, usually gravelly flats, ash, desert pavement, or low alkaline dune ridges. Typically, this habitat is a mosaic of open medium-tall to low shrubland communities, patchy grasslands or herb lands, and sparsely to unvegetated areas.

Composition. Characteristic medium-tall shrubs that dominate well-drained sites are shadscale (*Atriplex confertifolia*), bud sagebrush (*Artemisia spinescens*), and hopsage (*Grayia spinosa*). Characteristic low shrubs are greenmolly (*Kochia americana*), saltbush (*Atriplex gardneri* or *A. nuttallii*), and winter fat (*Krascheninnikovia lanata*). Other medium-tall shrubs, big sagebrush (*Artemisia tridentata*), horsebrush (*Tetradymia nuttallii* or *T. glabrata*), Mormon tea (*Ephedra viridis*), or rabbitbrush (*Chrysothamnus nauseosus* or *C. viscidiflorus*) can be co-dominant. The medium-tall shrub black greasewood (*Sarcobatus vermiculatus*) or low shrubs, iodinebush (*Allenrolfea occidentalis*) or Mojave seablite (*Suaeda moquinii*) can be dominant or co-dominant on less well drained, generally more saline parts of this habitat.

Herbaceous indicators of salt desert habitats appear in various habitats. On densely vegetated sites, native bunchgrasses, basin wildrye (*Leymus cinereus*), curly bluegrass (*Poa secunda*), and needle-and-threadgrass (*Stipa comata*) are important, usually with shrubs in this habitat. Basin wildrye is also a common and diagnostic grass in sites with less alkaline, deeper soils and some movement of water. Indian ricegrass (*Oryzopsis hymenoides*) and bottlebrush squirreltail (*Elymus elymoides*) are dominant grasses on the alkaline dunes. Introduced plants, particularly cheatgrass (*Bromus tectorum*) or halogeton (*Halogeton glomeratus*), often dominate overgrazed sites. Saltgrass (*Distichlis spicata*) is a common, diagnostic native sod-forming grass on more saline sites that often dominates large areas with and without shrubs. Pickleweed (*Salicornia virginica*) is found in wetter saline areas. Alkaline sites have mat muhly (*Muhlenbergia richardsonis*), alkali bluegrass (*Poa secunda* ssp. *juncifolia*), beardless wildrye (*Leymus triticoides*), and Lemmon's alkaligrass (*Puccinella lemmonii*). Common reedgrass (*Phragmites australis*), bulrush (*Scirpus americanus, S.*

maritimus), and creeping spikerush (*Eleocharis palustris*) are diagnostic of the wettest parts of this habitat.

Other Classifications and Key References. Popular literature refers to this habitat as shadscale, salt desert scrub, and saltflat desert. This habitat encompasses the "Desert or Salt Desert Shrub" and "*Distichlis stricta* Associations on Saline-Alkali Soils" in Franklin and Dyrness[88] and Saltbush-greasewood in Kuchler.[136] The Oregon Gap II Project[126] and Oregon Vegetation Landscape-Level Cover Types[127] that would represent this type are salt desert scrub shrubland and alkali playa. Other references describe this habitat. [29, 30, 52, 60, 123, 131, 147, 175, 184]

Natural Disturbance Regime. Fire plays a minor role over much of the distribution of the type because of sparse vegetation and lack of fuel. Many of these areas are prone to irregular flooding and prolonged droughts; both factors lead to a redistribution of component species and creation of sparsely or unvegetated areas.

Succession and Stand Dynamics. Many of the dominant shrub species sprout following fire, herbicide treatments, or heavy grazing.[4] The characteristic shrubs of this habitat increase with grazing and can invade adjacent big sagebrush communities with intense grazing.

Effects of Management and Anthropogenic Impacts. Several exotic species invade this habitat with grazing. Halogeton, a toxic exotic plant, is found most commonly in this habitat. Other noxious but nontoxic exotics that increase with grazing are Russian thistle (*Salsola kali*), tall tumblemustard (*Sisymbrium altissimum*), and cheatgrass. These can replace native grasses and change the structure of the native habitat.

Status and Trends. Agricultural development is generally not feasible; consequently, little of this habitat is converted to other uses. Most of this habitat is used for livestock grazing, which overall has increased shrub and annual cover and decreased bunchgrass cover. Quigley and Arbelbide[181] concluded that the Salt Desert Shrub cover type is less abundant now than before 1900. They further noted that the cover type has undergone a moderate level of change, so that some successional pathways have been unaltered. Approximately one third of Pacific Northwest salt desert and related community types listed in the National Vegetation Classification are considered imperiled or critically imperiled.[10]

19. Agriculture, Pastures, and Mixed Environs

W. Daniel Edge, Rex C. Crawford, & David H. Johnson

Geographic Distribution. Agricultural habitat is widely distributed at low to mid-elevations (<6,000 ft [1,830 m]) throughout both states. This habitat is most abundant in broad river valleys throughout both states and on gentle rolling terrain east of the Cascades.

Physical Setting. This habitat is maintained across a range of climatic conditions typical of both states. Climate constrains agricultural production at upper elevations where there are <90 frost-free days. Agricultural habitat in arid regions east of the Cascades with <10 inches (25 cm) of rainfall require supplemental irrigation or fallow fields for 1-2 years to accumulate sufficient soil moisture. Soils types are variable, but usually have a well developed A horizon. This habitat is found from 0 to 6,000 ft (0 to 1,830 m) elevation.

Landscape Setting. Agricultural habitat occurs within a matrix of other habitat types at low to mid-elevations, including Eastside grasslands, Shrub-steppe, Westside Lowlands Conifer-Deciduous Forest and other low- to mid-elevation forest and woodland habitats. This habitat often dominates the landscape in flat or gently rolling terrain, on well-developed soils, broad river valleys, and areas with access to abundant irrigation water. Unlike other habitat types, agricultural habitat is often characterized by regular landscape patterns (squares, rectangles, and circles) and straight borders because of ownership boundaries and multiple crops within a region. Edges can be abrupt along the habitat borders within agricultural habitat and with other adjacent habitats.

Structure. This habitat is structurally diverse because it includes several cover types ranging from low-stature annual grasses and row crops (<3.3 ft [1 m]) to mature orchards (>66 ft [20 m]). However, within any cover type, structural diversity is typically low because usually only one to a few species of similar height are cultivated. Depending on management intensity or cultivation method, agricultural habitat may vary substantially in structure annually; cultivated cropland and modified grasslands are typified by periods of bare soil and harvest whereas pastures are mowed, hayed, or grazed one or more times during the growing season. Structural diversity of agricultural habitat is increased at local scales by the presence of noncultivated or less intensively managed vegetation such as fencerows, roadsides, field borders, and shelterbelts.

Composition. Agricultural habitat varies substantially in composition among the cover types it includes. Cultivated cropland includes >50 species of annual and perennial plants in Oregon and Washington, and hundreds of varieties ranging from vegetables such as carrots, onions, and peas to annual grains such as wheat, oats, barley, and rye. Row crops of vegetables and herbs are characterized by bare soil, plants, and plant debris along bottomland areas of streams and rivers and areas having sufficient water for irrigation. Annual grains, such as barley, oats, and wheat are typically produced in almost continuous stands of vegetation on upland and rolling hill terrain without irrigation.

The orchard/vineyard/nursery cover type is composed of fruit and nut (apples, peaches, pears, and hazelnuts) trees, vineyards (grapes, Kiwi), berries (strawberries, blueberries, blackberries, and raspberries), Christmas trees, and nursery operations (ornamental container and greenhouses). This cover type is generally located on upland sites with access to abundant irrigation. Cultivation for most orchards, vineyards and Christmas tree farms includes an undergrowth of short-stature perennial grasses between the rows of trees, vines, or bushes. Christmas trees are typically produced without irrigation on upland sites with poorer soils.

Improved pastures are used to produce perennial herbaceous plants for grass seed and hay. Alfalfa and several species of fescue (*Festuca* spp.) and bluegrass (*Poa* spp.), orchardgrass (*Dactylis glomerata*), and timothy (*Phleum pratense*) are commonly seeded in improved pastures. Grass seed fields are single-species stands, whereas pastures maintained for haying are typically composed of two to several species. The improved pasture cover type is one of the most common agricultural uses in both states and produced with and without irrigation.

Unimproved pastures are predominately grassland sites, often abandoned fields that have little or no active management such as irrigation, fertilization, or herbicide applications. These sites may or may not be grazed by livestock. Unimproved pastures include rangelands planted to exotic grasses that are found on private land, state wildlife areas, federal wildlife refuges and U.S. Department of Agriculture Conservation Reserve Program (CRP) sites. Grasses commonly planted on CRP sites are crested wheatgrass (*Agropyron cristatum*), tall fescue (*F. arundinacea*), perennial bromes (*Bromus* spp.) and wheatgrasses (*Elytrigia* spp.). Intensively grazed rangelands, which have been seeded to intermediate wheatgrass (*Elytrigia intermedia*), crested wheatgrass, or are dominated by increaser exotics such as Kentucky wheatgrass (*Poa pratensis*) or tall oatgrass (*Arrhenatherum elatius*) are unimproved pastures. Other unimproved pastures have been cleared and intensively farmed in the past, but are allowed to convert to other vegetation. These sites may be composed of uncut hay, litter from previous seasons, standing dead grass and herbaceous material, invasive exotic plants (tansy ragwort [*Senecio jacobea*], thistle [*Cirsium* spp.], Himalaya blackberry [*Rubus discolor*], and Scot's broom [*Cytisus scoparius*]) with patches of native black hawthorn (*Crataegus douglasii*), snowberry (*Symphoricarpos* spp.), spirea (*Spirea* spp.), poison oak (*Toxicodendron diversilobum*), and encroachment of various tree species, depending on seed source and environment.

Modified grasslands are generally overgrazed habitats that formerly were native eastside grasslands or shrub-steppe but are now dominated by annual plants with only remnant individual plants of the native vegetation. Cheatgrass (*Bromus tectorum*), other annual bromes, medusahead (*Taeniatherum caput-medusae*), bulbous bluegrass (*Poa bulbosa*), and knapweeds (*Centaurea* spp.) are common increasers that form modified grasslands. Fire, following heavy grazing or repeated early season fires can create modified grassland monocultures of cheatgrass.

Agricultural habitat also contains scattered dwellings and outbuildings such as barns and silos, rural cemeteries, ditchbanks, windbreaks, and small inclusions of remnant native vegetation. These sites typically have a discontinuous tree layer or one to a few trees over a ground cover similar to improved or unimproved pastures.

Other Classifications and Key References. Quigley and Arbelbide[181] referred to this as agricultural and exotic forbs-annual grasses cover types. Csuti et al.[58] referred to this habitat as agricultural. The Oregon Gap II Project[126] and Oregon Vegetation Landscape-Level Cover Type[127] that would represent this type is agriculture. U.S. Department of Agriculture Conservation Reserve Program lands are included in this habitat.

Natural Disturbance Regime. Natural fires are almost totally suppressed in this habitat, except for unimproved pastures and modified grasslands, where fire-return intervals can resemble those of native grassland habitats. Fires are generally less frequent today than in the past, primarily because of fire suppression, construction of roads, and conversion of grass and forests to cropland.[159] Bottomland areas along streams and rivers are subject to periodic floods, which may remove or deposit large amounts of soil.

Succession and Stand Dynamics. Management practices disrupt natural succession and stand dynamics in most of the agricultural habitats. Abandoned eastside agricultural habitats may convert to other habitats, mostly grassland and shrub habitats from the surrounding native habitats. Some agricultural habitats that occur on highly erodible soils, especially east of the Cascades, have been enrolled in the U.S. Department of Agriculture Conservation Reserve Program. In the absence of fire or mowing, westside unimproved pastures have increasing amounts of hawthorn, snowberry, rose (*Rosa* spp.), Himalaya blackberry, spirea, Scot's broom, and poison

oak. Douglas-fir or other trees can be primary invaders in some environments.

Effects of Management and Anthropogenic Impacts. The dominant characteristic of agricultural habitat is a regular pattern of management and vegetation disturbance. With the exception of the unimproved pasture cover type, most areas classified as agricultural habitat receive regular inputs of fertilizer and pesticides and have some form of vegetation harvest and manipulation. Management practices in cultivated cropland include different tillage systems, resulting in vegetation residues during the non-growing season that range from bare soil to 100% litter. Cultivation of some crops, especially in the arid eastern portions of both states, may require the land to remain fallow for 1-2 growing seasons in order to store sufficient soil moisture to grow another crop. Harvest in cultivated cropland, Christmas tree plantations, and nurseries, and mowing or haying in improved pasture cover types substantially change the structure of vegetation. Harvest in orchards and vineyards is typically less intrusive, but these crops as well as Christmas trees and some ornamental nurseries are regularly pruned. Improved pastures are often grazed after haying or during the nongrowing season. Livestock grazing is the dominant use of unimproved pastures. All of these practices prevent agricultural areas from reverting to native vegetation. Excessive grazing in unimproved pastures may increase the prevalence of weedy or exotic species.

Status and Trends. Agricultural habitat has steadily increased in amount and size in both states since Eurasian settlement of the region. Conversion to agricultural habitat threatens several native habitat types.[166] The greatest conversion of native habitats to agricultural production occurred between 1950 and 1985, primarily as a function of U.S. agricultural policy.[96] Since the 1985 Farm Bill and the economic downturn of the early to mid 1980s, the amount of land in agricultural habitat has stabilized and begun to decline.[164] The 1985 and subsequent Farm Bills contained conservation provisions encouraging farmers to convert agricultural land to native habitats.[96, 153] Clean farming practices and single-product farms have become prevalent since the 1960s, resulting in larger farms and widespread removal of fencerows, field borders, roadsides, and shelterbelts.[96, 153, 164] In Oregon, land-use planning laws prevent or slow urban encroachment and subdivisions into areas zoned as agriculture. Washington's growth management is currently controlled by counties and agricultural land conversion to urban development is much less regulated.

20. Urban and Mixed Environs

Howard L. Ferguson

Geographic Distribution. Urban habitat occurs throughout Oregon and Washington. Most urban development is located west of the Cascades of both Oregon and Washington, with the exception of Spokane, Washington. However, urban growth is being felt in almost every small town throughout the Pacific Northwest.

Physical Setting. Urban development occurs in a variety of sites in the Pacific Northwest. It creates a physical setting unique to itself: temperatures are elevated and background lighting is increased; wind velocities are altered by the urban landscape, often reduced except around the tallest structures downtown, where high-velocity winds are funneled around the skyscrapers. Urban development often occurs in areas with little or no slope and frequently includes wetland habitats. Many of these wetlands have been filled in and eliminated. Today, ironically, many artificial "wetland" impoundments are being created for stormwater management, whose function is the same as the original wetland that was destroyed.

Landscape Setting. Urban development occurs within or adjacent to nearly every habitat type in Oregon and Washington, and often replaces habitats that are valuable for wildlife. The highest urban densities normally occur in lower elevations along natural or human-made transportation corridors, such as rivers, railroad lines, coastlines, or interstate highways. These areas often contain good soils with little or no slope and lush vegetation. Once level areas become crowded, growth continues along rivers or shores of lakes or oceans, and eventually up elevated sites with steep slopes or rocky outcrops. Because early settlers often modified the original landscape for agricultural purposes, many of our urban areas are surrounded by agricultural and grazing lands.

Structure. The original habitat is drastically altered in urban environments and is replaced by buildings, impermeable surfaces, bridges, dams, and planting of non-native species. Some human-made structures provide habitats similar to those of cavities, caves, fissures, cliffs, and ledges. With the onset of urban development, total crown cover and tree density are reduced to make way for the construction of buildings and associated infrastructure. Many structural features typical of the historical vegetation, such as snags, dead and downed wood, and brush piles, are often completely removed from the landscape. Understory vegetation may be completely absent, or if present, is diminutive and single-layered. Typically, three zones are characteristic of urban habitat.

High-density Zone

The high-density zone is the downtown area of the inner city. It also encompasses the heavy industrial and large commercial interests of the city in addition to high-density housing areas such as apartment buildings or high-rise condominiums. This zone has >60% of its total surface area covered by impervious surfaces. This zone has the smallest lot size, the tallest buildings, the least amount of total tree canopy cover, the lowest tree density, the highest percentage of exotics, the poorest understory and subcanopy, and the poorest vegetative structure.[4a, 116a, 185a] Human structures have replaced almost all vegetation.[23b, 148a] Road density is the highest of all zones. An example of road density can be seen from Washington's Growth Management Plan requiring Master Comprehensive Plans to set aside 20% of the identified urban growth area for roads and road rights-of-way. For example, Spokane's urban growth area is approximately 57,000 acres (23,077 ha); therefore >11,000 acres (4,453 ha) were set aside for road surfaces.

In the high-density zone, land-use practices have removed most of the native vegetation. Patch sizes of remaining natural areas often are so small that native interior species cannot be supported. Not only are remaining patches of native vegetation typically disconnected, but also they are frequently missing the full complement of vertical strata.[149] Stream corridors become heavily impacted and discontinuous. Most, if not all, wetlands have been filled or removed. Large buildings dominate the landscape and determine the placement of vegetation in this zone.[30a] This zone has the most street tree strips or sidewalk trees, most of which are exotics. There is virtually no natural tree replacement, and new trees are planted only when old ones die or are removed. Replacement trees are chosen for their small root systems and are generally short in stature with small diameters. Ground cover in this zone, if not synthetic or impervious, is typically exotic grasses or exotic annuals, most of which are rarely allowed to go to seed. Snags, woody debris, rock piles, and any other natural structures are essentially nonexistent. There are few tree cavities because of cosmetic pruning, cavity filling, snag removal, and tree thinning.[149]

Medium-density Zone

This zone, continuing out from the center of the continuum, is composed of light industry mixed with high-density residential areas. Housing density of 3-6 single-family homes per acre (7-15 per ha) is typical. Compared with the high-density zone, this zone has more potential wildlife habitat. With 30-59% impervious soil cover, this zone has 41-70% of the ground available for plants. Road density is less than the high-density zone.

Vegetation in this mid-zone is typically composed of non-native plant species. Native plants, when present, represent only a limited range of the natural diversity for the area. The shrub layer is typically clipped or minimal, even in heavily vegetated areas. Characteristic of this zone are manicured lawns, trimmed hedges, and topped trees. Lawns can be highly productive.[82a, 97a] Tree canopy is still discontinuous and consists of 1-2 levels, if present at all. Consequently, vertical vegetative diversity and total amount of understory are still low. Coarse and fine woody debris is minimal or absent; most snags and diseased live trees are still removed as hazards in this zone.[119a, 119b]

Isolated wetlands, stream corridors, open spaces, and greenbelts are more frequently retained in this zone than in the high-density zone. However, remnant wetland and upland areas are often widely separated by urban development.

Low-density Zone
The low-density zone is the outer zone of the urban-rural continuum. This zone contains only 10-29% impervious ground cover and normally contains only single-family homes. It has more natural ground cover than artificial surfaces. Vegetation is denser and more abundant than in the previous two zones. Typical housing densities are 0.4-1.6 single-family homes per acre (1-4 per ha). Road density is lowest of all three zones and consists of many secondary and tertiary roads. Although this zone may have large areas of native vegetation and is generally the least impacted of all three zones; it still has been significantly altered by human activities and associated disturbances.

Roads, fences, livestock paddocks, and pets are more abundant than in neighboring rural areas. With many animals and limited acreage, pasture conditions may be more overgrazed in this zone than in the rural zone; overgrazing can significantly affect shrub layers as well. Areas around home sites are often cleared for fire protection. Dogs are more likely to be loose and allowed to run free, increasing disturbance levels and wildlife harassment in this zone. Vegetable and flower gardens are widespread; fencing is prevalent.

Many wetlands remain and are less impacted. Water levels are more stable and peak flows are more typical of historical flows. Watertables are less impacted and vernal wetlands are more frequent; stream corridors are less impacted and more continuous.

This zone has the most vertical and horizontal structure and diversity of any of the three urban zones.[30a, 80a, 140a, 187a] In forested areas, tree conditions are semi-natural, although stand characteristics vary from parcel to parcel. The tree canopy is more continuous and may include multiple levels. Patch sizes are large enough to support native interior species. Large blocks of native vegetation may still be found, and some of these may be connected to large areas of native undeveloped land.[220a] In this zone, snags, diseased trees, coarse and fine woody debris, brush piles, and rock piles are widespread. Structural diversity approaches historical levels. Non-native hedges are nearly nonexistent and the native shrub layer, except for small

areas around houses, is relatively intact. Lawns are fewer and native ground covers are more common than in the previous two zones.

Composition. Remnant isolated blocks of native vegetation may be found scattered throughout a town or city mixed with a multitude of introduced exotic vegetation. As urban development increases, these remnant native stands become fragmented and isolated. The dominant species in an urban setting may be exotic or native; for example, in Seattle, the dominant species in one area may be Douglas-fir (*Pseudotsuga menziesii*), whereas a few blocks away it may be the exotic silver maple (*Acer saccharinum*). Dominant species will not only vary from city to city but also within each city and within each of the three urban zones. Nowack[167] found that in the high-density urban zone, species richness is low, and in one case, four species made up almost 50% of the cover. In the same study, exotics made up 69% of the total species.

In urban and suburban areas, species richness is often increased because of the introduction of exotics. The juxtaposition of exotics interspersed with native vegetation produces a diverse mosaic with areas of extensive edge. Also, because of irrigation and the addition of fertilizers, the biomass in the urban communities is often increased.[149] Interest in the use of native plants for landscaping is rapidly expanding,[135, 172] particularly in the more arid sites where drought-resistant natives are the only plants able to survive without water.

Across the U.S., urban tree cover ranges from 1 to 55%.[167] As expected, tree cover tends to be highest in cities developed in naturally forested areas with an average of 32% cover in forested areas, 28% in grasslands, and 10% in arid areas. Yakima, Washington, has an overall city tree cover of 18%, ranging from 10% to 12% in the industrial/commercial area to 23% in the low-density residential zone.[167] Remnant blocks of native vegetation or native trees left standing in yards and parks will compositionally be related to whatever native habitat was on site prior to development. In the Puget Sound and Willamette Valley areas, Douglas-fir is a major constituent, whereas the Spokane area has a lot of ponderosa pina (*Pinus ponderosa*).

Other Classifications and Key References. Many attempts have been made to classify or describe the complex urban environment. The Washington GAP Analysis[37] classified urban environments as "developed" land cover using the same three zones as described above: (1) high density (>60% impervious surface); (2) medium density (30-60% impervious surface); and (3) low density (10-30% impervious surface). The Oregon Gap II Project[126] and Oregon Vegetation Landscape-Level Cover Types[127] represented this type as an urban class. Several other relevant strudies characterizing the urban environment have been reported. [182, 129, 34, 70, 151]

Natural Disturbance Regime. In many instances, natural disturbances are modified or prevented from occurring by humans over the landscape and this is particularly true of urban areas. However, disturbances such as ice, wind, or firestorms still occur. The severity of these intermittent disturbances varies greatly in magnitude and their impact on the landscape varies accordingly. One of the differences between urban and nonurban landscapes is the lengthening of the disturbance cycles. Another is found in the aftermath of these disturbances. In urban areas, damaged trees are often entirely removed and if they are replaced, a shorter, smaller tree, often non-native, is selected. The natural fire disturbance interval is highly modified in the urban environment. Fire (mostly accidental or arson) still occurs, and is quickly suppressed. Another natural disturbance in many of our Pacific Northwest towns is flooding, which historically altered and rerouted many of our rivers and streams, and still scarifies fields and deposits soil on flood plains and potentially recharges local aquifers. Floods now are more frequent and more violent than in the past because of the many modifications made to our watersheds. Attempts to lessen flooding in urban areas often lead to channelization, paving, or diking of our waterways, most of which fail in their attempt to stem the flooding and usually result in increased flooding for the communities farther downstream.

Succession and Stand Dynamics. Due to anthropogenic influences found in the urban environment, succession differs in the urban area from that expected for a native stand. Rowntree[185] emphasized that urbanization is not in the same category as natural disturbance in affecting succession. He points out that urbanization is anthropogenic and acts to remove complete vegetation associations and creates new ones made of mixes of native residual vegetation and introduced vegetation. Much human effort in the city goes toward either completely removing native vegetation or sustaining or maintaining a specific vegetative type, e.g., lawns or hedges. Much of the vegetative community remains static. Understory and ground covers are constantly pruned or removed, seedlings are pulled and lawns are planted, fertilized, mowed, and meticulously maintained. Trees may be protected to maturity or even senescence, yet communities are so fragmented or modified that a genuine old-growth community never exists. However, a type of "urban succession" occurs across the urban landscape. The older neighborhoods with their mature stands are at a later seral stage than new developments; species diversity is characteristically higher in older neighborhoods as well. An oddity of the urban environment is the absence of typical structure generally found within the various seral stages. For example, the understory is often removed in a typical mid-seral stand to give it a "park-like" look. Or if the understory is allowed to remain, it is kept pruned to a consistent height. Lawns are the ever-present substitute for native ground covers. Multilayered habitat is often reduced to one or two heights. Vertical and horizontal structural diversity is drastically reduced.

Effects of Management and Anthropogenic Impacts. These additional, often irreversible, impacts include more impervious surfaces, more and larger human-made structures, large-scale storm and wastewater management, large-scale sewage treatment, water and air pollution, toxic chemicals, toxic chemical use on urban lawns and gardens, removal of species considered to be pests, predation and disturbance by pets and feral cats and dogs, and the extensive and continual removal of habitat due to expanding urbanization, and in some cases, uncontrolled development. Another significant impact is the introduction and cultivation of exotics in urban areas. Native vegetation is often completely replaced by exotics, leaving little trace of the native vegetative cover.

Status and Trends. From 1970 to 1990, >30,000 mile^2 (77,700 km^2) of rural lands in the U.S. became urban, as classified by the U.S. Census Bureau. That amount of land equals about one third of Oregon's total land area.[12] From 1940 to 1970, the population of the Portland urban region doubled and the amount of land occupied by that population quadrupled.[201] More than 300 new residents arrive in Washington each day, and each day, Washington loses 100 acres (41 ha) of forest to development.[215] Using satellite photos and GIS software, American Forests[9] discovered that nearly one third of Puget Sound's most heavily timbered land has disappeared since the early 1970s. The amount of land with few or no trees more than doubled, from 25% to 57%, an increase of >1 million acres (404,858 ha). Development and associated urban growth was blamed as the single biggest factor affecting the area's environment. This urban growth is predicted to continue to increase at an accelerated pace, at the expense of native habitat.

21. Open Water— Lakes, Rivers, and Streams

Eva L. Greda, David H. Johnson, & Thomas A. O'Neil

Lakes, Ponds, and Reservoirs

Geographic Distribution. Lakes in Oregon and Washington occur statewide and are found from near sea level to about 10,200 ft (3,110 m) above sea level. There are 3,887 lakes and reservoirs in western Washington and they total 176,920 acres (71,628 ha)[226]. In contrast, there are 4,073 lakes and reservoirs in eastern Washington that total 436,843 acres (176,860 ha).[227] There are 6,000 lakes, ponds, and reservoirs in Oregon including almost 1,800 named lakes and over 3,800 named reservoirs, all amounting to 270,641 acres (109,571 ha). Oregon has the deepest lake in the nation, Crater Lake, at 1,932 ft (589 m).[23]

Physical Setting. Continental glaciers melted and left depressions, where water accumulated and formed many lakes in the region. These kinds of lakes are predominantly found in Lower Puget Sound. Landslides that blocked natural valleys also allowed water to fill in behind them to form lakes, like Crescent Lake, Washington. The lakes in the Cascades and Olympic ranges were formed through glaciation and range in elevation from 2,500 to 5,000 ft (762 to 1,524 m). Beavers create many ponds and marshes in Oregon and Washington. Craters created by extinct volcanoes, like Battleground Lake, Washington, also formed lakes. Human-made reservoirs created by dams impound water that creates lakes behind them, like Bonneville Dam on the main stem of the Columbia River. In the lower Columbia Basin, many lakes formed in depressions and rocky coulees through the process of seepage from irrigation waters.[226]

Structure. There are four distinct zones within this aquatic system: (1) the littoral zone at the edge of lakes is the most productive with diverse aquatic beds and emergent wetlands (part of Herbaceous Wetlands habitat); (2) the limnetic zone is deep open water, dominated by phytoplankton and freshwater fish, and extends down to the limits of light penetration; (3) the profundal zone below the limnetic zone, devoid of plant life and dominated with detritivores; (4) and the benthic zone reflecting bottom soil and sediments. Nutrients from the profundal zone are recycled back to upper layers by the spring and fall turnover of the water. Water in temperate climates stratifies because of the changes in water density. The uppermost layer, the epilimnion, is where water is warmer (less dense). Next, the metalimnion or thermocline, is a narrow layer that prevents the mixing of the upper and lowermost layers. The lowest layer is the hypolimnion, with colder and most dense waters. During the fall turnover, the cooled upper layers are mixed with other layers through wind action.

Rivers and Streams

Geographic Distribution. Streams and rivers are distributed statewide in Oregon and Washington, forming a continuous network connecting high mountain areas to lowlands and the Pacific coast. There are >12,000 named rivers and streams in Oregon, totalling 112,640 miles (181,238 km) 23 in length. Oregon's longest stretch of river is the Columbia (309 miles [497 km]) that borders Oregon and Washington. The longest river in Oregon is the John Day (284 miles [457 km]) and the shortest river is the D River (440 ft [134 m]) that is the world's second shortest river. Washington has more streams than any other state except Alaska. In Washington, the coastal region has 3,783 rivers and streams totaling 8,176 miles (13,155 km)[174]. The Puget Sound Region has 10,217 rivers and streams, which add to 16,600 miles (26,709 km) in length.[223] The rivers and streams range from cold, fast-moving high-elevation streams to warmer lowland valley rivers.[223] In all, there are 13,955 rivers and streams that add up to 24,774 miles (39,861 km).[174] There are many more streams in Washington yet to be catalogued.[174] Streams reflect flowing water ≥6 feet (2 m) wied; narrower water bodies are considered within their respective habitats.

Physical Setting. Climate of the area's coastal region is very wet. The northern region in Washington is volcanic and bordered to the east by the Olympic Mountain Range, on the north by the Strait of Juan de Fuca, and on the west by the Pacific Ocean. In contrast, the southern portion in Washington is characterized by low-lying, rolling hills.[174] The Puget Sound Region has a wet climate. Most of the streams entering Puget Sound have originated in glacier fields high in the mountains. Water from melting snowpacks and glaciers provide flow during the spring and winter. Annual rainfall in the lowlands ranges from 35 to 50 inches (89-127 cm), from 75 to 100 inches (191 to 254 cm) in the foothills, and from 100 to >200 inches (254 to 508 cm) in the mountains (mostly in the form of snow).[174]

Rivers and streams in southwestern Oregon are fed by rain and are located in an area composed of sheared bedrock and thus an unstable terrain. Streams in that area have high suspended-sediment loads. Beds composed of gravel and sand are easily transported during floods. The western Cascades in Washington and Oregon are composed of volcanically derived rocks and are more stable. They have low sediment-transport rates and stable beds composed largely of cobbles and boulders, which move only during extreme events.[81] Velocities of river flow ranges from as little as 0.2 to 12 mph (0.3 to19.3 km/hr) while large streams have an average annual flow of 10 cubic feet (0.3 m3) per second or greater.[23, 169] Rivers and streams in the Willamette Valley are warm, productive, turbid, and have high ionic strength. They are characterized by deep pools, and highly embedded stream bottoms with claypan and muddy substrates, and the greatest fish species diversity. High desert streams of the interior are similar to those of the Willamette Valley but are shallower, with fewer pools, and more runs, glides, cobbles, boulders, and sand. The Cascades and Blue mountains are similar in that they have more runs and glides and fewer pools, similar fish assemblages, and similar water quality.[218]

Lakes, Rivers and Streams

Landscape setting. This habitat occurs throughout Washington and Oregon. Ponds, lakes, and reservoirs are typically adjacent to Herbaceous Wetlands, while rivers and streams typicaslly adjoin the Westside Riparian-Wetlands, Eastside Riparian-Wetlands, Herbaceous Wetlands, and Bays and Estuaries habitats.

Text continues on page 92

Westside Lowlands Conifer-Hardwood Forest

1. Goodman Creek, Oregon

2. H.J. Andrews Experimental Forest, Oregon

3. Deception Pass State Park, Island County, Washington

4. Capitol Forest, Thurston County, Washington

5. Dungeness River Valley, Clallam County, Washington

6. Lake Quinalt, Grays Harbor County, Washington

Westside Oak and Dry Douglas-fir Forest and Woodlands

1. *Orcas Island, San Juan County, Washington*

2. *James Island, San Juan County, Washington*

3. *San Juan Island, San Juan County, Washington*

4. *Fort Lewis, Pierce County, Washington*

Southwest Oregon Mixed Conifer-Hardwood Forest

1. Southwestern Oregon

2. Jackson County, Oregon

3. Ruch, Oregon

Montane Mixed Conifer Forest

1. *Mt. Pilchuck Conservation Area, Snohomish Co., WA*

2. *Pend Oreille County, Washington*

3. *Mt. Pilchuck Conservation Area, Snohomish County, Washington*

4. *Arlecho Creek, Whatcom County, Washington*

Eastside (Interior) Mixed Conifer Forest

1. *Wenatchee Mountains, Kittitas County, Washington*

2. *Alice Mae Mountain, Stevens County, Washington*

2. *Alice Mae Mountain, Stevens County, Washington*

4. *Rainbow Creek Research Natural Area, Blue Mountains, Washington*

Lodgepole Pine Forest and Woodlands

1. Loomis State Forest, Okanogan County, Washington

2. Loomis State Forest, Okanogan County, Washington

3. Loomis State Forest, Okanogan County, Washington

4. Loomis State Forest, Okanogan County, Washington

5. Loomis State Forest, Okanogan County, Washington

Ponderosa Pine Forest and Woodlands (includes Eastside Oak)

1. Barker Mountain, Okanogan County, Washington

2 Near Sisters, Oregon

3. Turnbull National Wildlife Refuge, Spokane County, Washington

4. Indian Ford, north of Sisters, Oregon

5. Briske Canyon, Washington

6. Badger Gulch Natural Area Preserve, Klickitat County, Washington

Upland Aspen Forest

1. Winthrop, Okanogan County, Washington

2. Hart Mountain National Wildlife Refuge, Oregon

3. Hart Mountain Mountain National Wildlife Refuge, Oregon

4. Steens Mountain, Oregon

Subalpine Parkland

1. Mt. Rainier National Park, Washington

2. Strawberry Mountain, Oregon

3. Mt. Pilchuck Conservation Area, Snohomish County, Washington

4. Eagle Cap Wilderness Area, Wallowa Mts., Oregon

5. Goat Rocks Wilderness Area, Lewis County, Washington

Alpine Grasslands and Shrublands

1. Mt. Rainier National Park, Washington

2. Steens Mountain, Oregon

3. Buckhorn Wilderness Area, Clallam County, Washington   *4. Mt. Rainier National Park, Washington*

Westside Grasslands

1. Scatter Creek Wildlife Area, Thurston County, Washington

2. Lane County, Oregon

3. Near Stayton, Linn County, Oregon

4. Burrows Island, Skagit County, Washington

5. Mima Mounds Natural Area Preserve, Thurston County, Washington

Ceanothus – Manzanita Shrublands

1. Southwestern Oregon

2. South of Shady Cove, Oregon

3. Southwestern Oregon

4. South of Shady Cove, Oregon

Western Juniper and Mountain Mahogany Woodlands

1. Sheldon National Wildlife Refuge, Nevada

2. Sheldon National Wildlife Refuge, Nevada

3. Jefferson County, Oregon

4. Klamath County, Oregon

Eastside (Interior) Canyon Shrublands

1. Campus Prairie Biological Study Area, Whitman County, Washington

2. Douglas Creek, Douglas County, Washington

3. Asotin County, Washington

4. Kramer Palouse Biological Study Area, Whitman County, Washington

5. Grand Ronde River, Asotin County, Washington

Eastside (Interior) Grasslands

1. Gibraltar Mountain, Ferry County, Washington

2. Palouse River, Franklin County, Washington

3. Columbia Hills Natural Area Preserve, Klickitat County, Washington

4. Arid Lands Ecology Reserve, Hanford, Washington

Shrub-Steppe

1. Horse Heaven Hills, Benton County, Washington

2. Yakima Firing Range, Washington

3. Yakima Firing Range, Washington

3. Steens Mountain, Oregon

4. Vernita, Grant County, Washington

Dwarf Shrub-Steppe

1. Castle Rock, Grand Coulee, Washington

2. Umtanum Ridge, Kittitas County, Washington

3. Saddle Mountains, Hanford, Washington

4. Sheldon National Wildlife Refuge, Nevada

Desert Playa and Salt Scrub Shrublands

1. Alvord Desert, Oregon

2. Alvord Desert, Oregon

3. Alvord Desert, Oregon

4. Harney Basin, Oregon

5. Alvord Desert, Oregon

Agriculture, Pasture, and Mixed Environs

1. Near Samish Bay, Skagit County, Washington

2. Sinlahekin Valley, Okanogan County, Washington

3. Ellensburg area, Kittitas County, Washington

4. Southeast Washington (wheat field)

5. Benton County, Oregon

6. Scotch Creek Wildlife Area, Okanogan County, Washington

Urban and Mixed Environs

1. *Example of high density urban, Seattle, Washington*

2. *Example of medium density urban, Washington*

3. *Example of low density urban, Benton County, Oregon*

4. *Example of low density urban, Wenatchee, Washington*

Open Water—Lakes, Rivers, Streams

1. Lake Ozette, Washington

2. Lower Soleduck River, Washington

3. Grant County, Washington

4. Snake River, Washington

5. Willamette River, Linn County, Oregon

Herbaceous Wetlands

1. Okanogan County, Washington

2. Methow Valley, Okanogan County, Washington

3. Pacific County, Washington

4. Lincoln County, Washington

5. Linn County, Oregon

Westside Riparian - Wetlands

1. Stequaleho Creek, Jefferson County, Washington

2. Cranberry Creek, Grays Harbor County, Washington

3. Maxfield Creek, Clallam County, Washington

4. Cranberry Creek, Grays Harbor, Washington

5. Quinalt River, Grays Harbor County, Washington

Montane Coniferous Wetlands

1. *Arlecho Creek, Whatcom County, Washington*

2. *Roger Lake, Okanogan County, Washington*

3. *Mt. Rainier National Park, Washington*

4. *Mt. Pilchuck Conservation Area, Snohomish County, Washington*

Eastside (Interior) Riparian-Wetlands

1. *Northrup Canyon, Grant County, Washington*

2. *Little Pend Oreille River, Stevens County, Washington*

3. *Douglas Creek, Douglas County, Washington*

4. *Myers Creek, Okanogan County, Washington*

5. *Eastern Klickitat County, Washington*

6. *Crimm's Creek, Washington*

Coastal Dunes and Beaches

1. Florence, Oregon

2. Whidbey Island, Washington

3. Ocean Shores, Grays Harbor County, Washington

4. Whidbey Island, Washington

5. Whidbey Island, Washington

Coastal Headlands and Islets

1. Near Cape Perpetua, Oregon

2. Heceta Head, Oregon

3. Near Yaquina Head, Oregon

4. Near Cape Perpetua, Oregon

5. Coastline near Quinalt River, Grays Harbor County, Washington

Bays and Estuaries

1. Estuary, Washington

2. Dogfish Point, Skagit County, Washington

3. Niawiakum River Natural Area Preserve, Willapa Bay, Washington

4. Nisqually Delta, Thurston County, Washington

5. Niawiakum River, Willapa Bay, Washington

6. Newport, Oregon

Marine Nearshore

1. Ben Ure Island, Island County, Washington

2. Cypress Island, Skagit County, Washington

3. Guemes Island, Skagit County, Washington

4. Near Biz Point, Skagit County, Washington

5. Near Cascade Head, Lincoln County, Oregon

Inland Marine Deeper Water

1. *Squaxin Island, South Puget Sound, Washington*

2. *San Juan Islands, Washington*

Marine Habitats

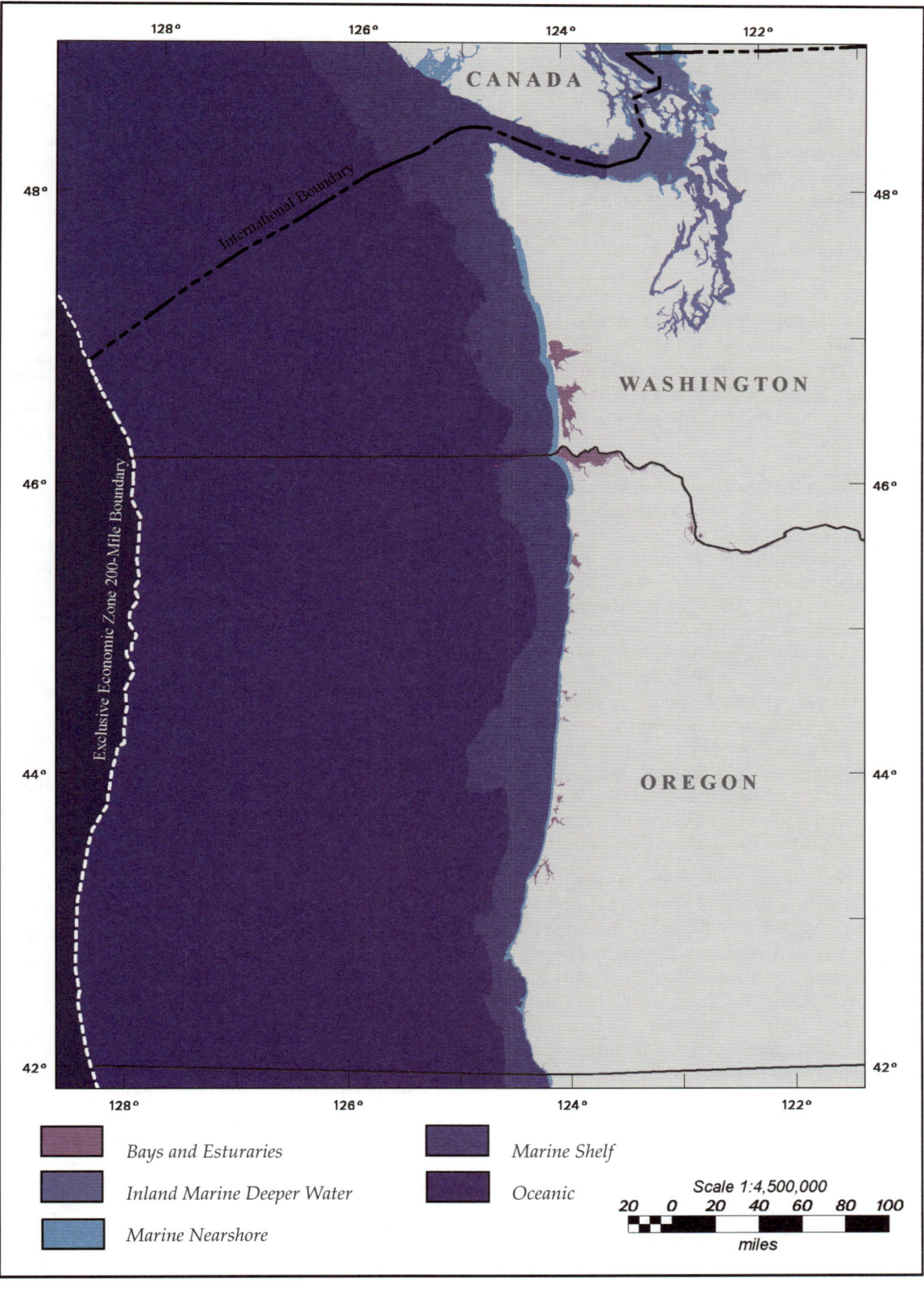

Bays and Esturaries

Inland Marine Deeper Water

Marine Nearshore

Marine Shelf

Oceanic

Scale 1:4,500,000

20   0   20   40   60   80   100

miles

Other Classifications and Key References. This habitat is called riverine and lacustrine in Anderson et al.,[10] Cowardin et al.,[53] Washington Gap Analysis Project,[37] Mayer and Laudenslayer,[150] and Wetzel.[217] However, this habitat is referred to as Open Water in the Oregon Gap II Project[126] and Oregon Vegetation Landscape-Level Cover Types.[127]

Natural Disturbance Regime. There are seasonal and decadal variations in the patterns of precipitation. In the Coast Range, there is usually 1 month of drought per year (usually July or August) and 2 months of drought once in a decade. The Willamette Valley and the Cascades experience 1 month with no rain every year and a 2-month dry period every third year. In eastern Oregon, dry periods last 2 or 3 months every year, with dry spells as long as 4-6 months occurring once every 4 years. Dry years, with <33% of normal precipitation, occur once every 30 years along the coast, every 20 years in the Willamette Valley, every 30 years in the Cascades, and every 15 years in most of eastern Oregon.[23]

Floods occur in Oregon and Washington every year. Flooding season west of the Cascades occurs from October through April, with more than half of the floods occurring during December and January. Floods are the result of precipitation and snow melts. Floods west of the Cascades are influenced by precipitation mostly and thus are short-lived, while east of the Cascades floods are caused by melting snow, and the amount of flooding depends on how fast the snow melts. High water levels frequently last up to 60 days. In 1984, heavy precipitation flooded Malheur and Harney lakes to the point where the two lakes were joined together for several years. The worst floods have resulted from cloudbursts caused by thunderstorms, like Heppner, Oregon's 1903 flood. Other "flash floods" in the region were among the largest floods in the U.S. and occurred in the John Day Basin's Meyers Canyon in 1956 and the Umatilla Basin's Lane Canyon in 1965.[23]

Effects of Management and Anthropogenic Impacts. Sewage effluents caused eutrophication of Lake Washington in Seattle, where plants increased in biomass and caused decreased light transmission. The situation was corrected, however, before it became serious as a result of a campaign of public education, and timely cleanup of the lake.[146] Irrigation projects aimed at watering drier portions of the landscape may pose flooding dangers, as was the case with Soap Lake and Lake Leonore in eastern Washington. Finally, natural salinity of lakes can decrease as a result of irrigation withdrawal and can change the biota associated with them.[92]

Removal of gravel results in reduction of spawning areas for anadromous fish. Overgrazing, and loss of vegetation caused by logging produces increased water temperatures and excessive siltation, harming the invertebrate communities such as that reported in the John Day River Basin, Oregon.[146] Incorrectly installed culverts may act as barriers to migrating fish and may contribute to erosion and siltation downstream.[174] Construction of dams is associated with changes in water quality, fish

passage, competition between species, loss of spawning areas because of flooding, and declines in native fish populations.[146] Historically, the region's rivers contained more braided multi-channels. Flood control measures such as channel straightening, diking, or removal of streambed material along with urban and agriculture development have all contributed to a loss of oxbows, river meanders, and flood plains. Unauthorized or over-appropriated withdrawals of water from the natural drainages also has caused a loss of open water habitat that has been detrimental to fish and wildlife production, particularly in the summer.[174]

Agricultural, industrial, and sewage runoff such as salts, sediments, fertilizers, pesticides, and bacteria harm aquatic species.[146] Sludge and heavy waste buildup in estuaries is harmful to fish and shellfish. Unregulated aerial spraying of pesticides over agricultural areas also poses a threat to aquatic and terrestrial life.[174] Direct loss of habitat and water quality occurs through irrigation.[130] The Oregon Department of Environmental Quality, after a study of water quality of the Willamette River, determined that up to 80% of water pollution enters the river from nonpoint sources and especially agricultural activity.[23] Very large floods (e.g., Oregon Flood of 1964) may change the channels permanently through the settling of large amounts of sediments from hillslopes, through debris flow, and through movement of large boulders, particularly in the montane areas. The width of the channel along the main middle fork of the Willamette increased over a period of 8 years. Clearcutting creates excessive intermittent runoff conditions and increases erosion and siltation of streams as well as diminishes shade, and therefore causes higher water temperatures, fewer terrestrial and aquatic food organisms, and increased predation. Landslides, which contributed to the widening of the channel, were a direct result of clearcutting. Clearcut logging can alter snow accumulation and increase the size of peak flows during times of snowmelt.[197] Clearcutting and vegetation removal affects the temperatures of streams, increasing them in the summer and decreasing in winter, especially in eastern parts of the Oregon and Washington.[24] Building of roads, especially those of poor quality, can be a major contributor to sedimentation in the streams.[82]

Status and Trends. The principal trend has been in relationship to dam building or channelization for hydroelectric power, flood control, or irrigation purposes. As an example, in 1994, there were >900 dams in Washington alone. The dams vary according to size, primary purpose, and ownership (state, federal, private, local).[214] The first dam and reservoir in Washington was the Monroe Street Dam and Reservoir, built in 1890 at Spokane Falls. Since then the engineering and equipment necessary for dam building developed substantially, culminating in such projects as the Grand Coulee Dam on the Columbia River.[214] In response to the damaging effects of dams on the indigenous biota and alteration and destruction of freshwater aquatic habitats, Oregon and Washington state governments questioned the benefits of dams, especially in light of the federal listing of several salmon species. There are now talks of possibly removing small dams, like the Savage Rapids Dam in Oregon, to removing large federal dams like those on the lower Snake River.[23]

22. Herbaceous Wetlands

Rex C. Crawford, Jimmy Kagan, and Christopher B. Chappell

Geographic Distribution. Herbaceous wetlands are found throughout the world and are represented in Oregon and Washington wherever local hydrologic conditions promote their development. This habitat includes all those except bogs and those within Subalpine Parkland and Alpine habitats.

Freshwater aquatic bed habitats are found throughout the Pacific Northwest, usually in isolated sites. They are more widespread in valley bottoms and high rainfall areas (e.g., Willamette Valley, Puget Trough, coastal terraces, coastal dunes), but are present in montane and arid climates as well. Hardstem bulrush-cattail-burreed marshes occur in wet areas throughout Oregon and Washington. Large marshes are common in the lake basins of Klamath, Lake, and Harney counties, Oregon. Sedge meadows and montane meadows are common in the Blue and Ochoco mountains of central and northeastern Oregon, and in the valleys of the Olympic and Cascade mountains and Okanogan Highlands. Extensive wet meadow habitats occur in Klamath, Deschutes, and western Lake counties in Oregon.

Physical Setting. This habitat is found on permanently flooded sites that are usually associated with oxbow lakes, dune lakes, or potholes. Seasonally to semi-permanently flooded wetlands are found where standing freshwater is present through part of the growing season and the soils stay saturated throughout the season. Some sites are temporarily to seasonally flooded meadows and generally occur on clay, pluvial, or alluvial deposits within montane meadows, or along stream channels in shrubland or woodland riparian vegetation. In general, this habitat is flat, usually with stream or river channels or open water present. Elevation varies between sea level to 10,000 ft (3,048 m), although infrequently above 6,000 ft (1,830 m).

Landscape Setting. Herbaceous wetlands are found in all terrestrial habitats except Subalpine Parkland and Alpine Grasslands and Shrublands habitats. Herbaceous wetlands commonly form a pattern with Westside and Eastside Riparian-Wetlands and Montane Coniferous Wetlands habitats along stream corridors. These marshes and wetlands also occur in closed basins in a mosaic with open water by lakeshores or ponds. Extensive deflation plain wetlands have developed between Coastal Dunes and Beaches habitat and the Pacific Ocean. Herbaceous wetlands are found in a mosaic with alkali grasslands in the Desert Playa and Salt Scrub habitat.

Structure. The herbaceous wetland habitat is generally a mix of emergent herbaceous plants with a grass-like life form (graminoids). These meadows often occur with deep or shallow water habitats with floating or rooting aquatic forbs. Various wetland communities are found in mosaics or in nearly pure stands of single species. Herbaceous cover is open to dense. The habitat can be comprised of tule marshes >6.6 ft (2 m) tall or sedge meadows and wetlands <3.3 ft (1 m) tall. It can be a dense, rhizomatous sward or a tufted graminoid wetland. Graminoid wetland vegetation generally lacks many forbs, although the open extreme of this type contains a diverse forb component between widely spaced tall tufted grasses.

Composition. Various grasses or grass-like plants dominate or co-dominate these habitats. Cattails (*Typha latifolia*) occur widely, sometimes adjacent to open water with aquatic bed plants. Several bulrush species (*Scirpus acutus, S. tabernaemontani, S. maritimus, S. americanus, S. nevadensis*) occur in nearly pure stands or in mosaics with cattails or sedges (*Carex* spp.). Burreed (*Sparganium angustifolium, S. eurycarpum*) are the most important graminoids in areas with up to 3.3 ft (1m) of deep standing water. A variety of sedges characterize this habitat. Some sedges (*Carex aquatilis, C. lasiocarpa, C. scopulorum, C. simulata, C. utriculata, C. vesicaria*) tend to occur in cold to cool environments. Other sedges (*C. aquatilis* var. *dives, C. angustata, C. interior, C. microptera, C. nebrascensis*) tend to be at lower elevations in milder or warmer environments. Slough sedge (*C. obnupta*), and several rush species (*Juncus falcatus, J. effusus, J. balticus*) are characteristic of coastal dune wetlands that are included in this habitat. Several spike rush species (*Eleocharis* spp.) and rush species can be important. Common grasses that can be local dominants and indicators of this habitat are American sloughgrass (*Beckmannia syzigachne*), bluejoint reedgrass (*Calamagrostis canadensis*), mannagrass (*Glyceria* spp.), and tufted hairgrass (*Deschampsia caespitosa*). Important introduced grasses that increase and can dominate with disturbance in this wetland habitat include reed canary grass (*Phalaris arundinacea*), tall fescue (*Festuca arundinacea*) and Kentucky bluegrass (*Poa pratensis*).

Aquatic beds are part of this habitat and support a number of rooted aquatic plants, such as yellow pond lily (*Nuphar lutea*) and unrooted, floating plants such as pondweeds (*Potamogeton* spp.), duckweed (*Lemna minor*), or water-meals (*Wolffia* spp.). Emergent herbaceous broadleaf plants, such as Pacific water parsley (*Oenanthe sarmentosa*), buckbean (*Menyanthes trifoliata*), water star-warts (*Callitriche* spp.), or bladderworts (*Utricularia* spp.) grow in permanent and semi-permanent standing water. Pacific silverweed (*Argentina egedii*) is common in coastal dune wetlands. Montane meadows occasionally are forb-dominated with plants such as arrowleaf groundsel (*Senecio triangularis*) or ladyfern (*Athyrium filix-femina*). Climbing nightshade (*Solanum dulcamara*), purple loosestrife (*Lythrum salicaria*), and poison hemlock (*Conium maculatum*) are common non-native forbs in wetland habitats.

Shrubs or trees are not a common part of this herbaceous habitat although willow (*Salix* spp.) or other woody plants occasionally occur along margins, in patches or along streams running through these meadows.

Other Classifications and Key References. This habitat is called Palustrine emergent wetlands in Cowardin et al.[53] Other references describe this habitat.[43, 44, 57, 71, 131, 132, 138, 147, 219] This habitat occurs in both lotic and lentic systems. The Oregon Gap II Project[126] and Oregon Vegetation Landscape-Level Cover Types[127] that would represent this type are wet meadow, palustrine emergent, and National Wetland Inventory (NWI) palustrine shrubland.

Natural Disturbance Regime. This habitat is maintained through a variety of hydrologic regimes that limit or exclude invasion by large woody plants. Habitats are permanently flooded, semipermanently flooded, or flooded seasonally and may remain saturated through most of the growing season. Most wetlands are resistant to fire and those that are dry enough to burn usually burn in the fall. Most plants are sprouting species and recover quickly. Beavers play an important role in creating ponds and other impoundments in this habitat. Trampling and grazing by large native mammals is a natural process that creates habitat patches and influences tree invasion and success.

Succession and Stand Dynamics. Herbaceous wetlands are often in a mosaic with shrub- or tree-dominated wetland habitat. Woody species can successfully invade emergent wetlands when this herbaceous habitat dries. Emergent wetland plants invade open-water habitat as soil substrate is exposed; e.g., aquatic sedge and Northwest Territory sedge (*Carex utriculata*) are pioneers following beaver dam breaks. As habitats flood, woody species decrease to patches on higher substrate (soil, organic matter, large woody debris) and emergent plants increase unless the flooding is permanent. Fire suppression can lead to woody species invasion in drier herbaceous wetland habitats; e.g., Willamette Valley wet prairies are invaded by Oregon ash (*Fraxinus latifolia*) with fire suppression.

Effects of Management and Anthropogenic Impacts. Direct alteration of hydrology (i.e., channeling, draining, damming) or indirect alteration (i.e., roading or removing vegetation on adjacent slopes) results in changes in amount and pattern of herbaceous wetland habitat. If the alteration is long term, wetland systems may reestablish to reflect new hydrology, e.g., cattail is an aggressive invader in roadside ditches. Severe livestock grazing and trampling decreases aquatic sedge, Northwest Territory sedge (*Carex utriculata*), bluejoint reedgrass, and tufted hairgrass. Native species, however, such as Nebraska sedge, Baltic and jointed rush (*Juncus nodosus*), marsh cinquefoil (*Comarum palustris*), and introduced species dandelion (*Taraxacum officinale*), Kentucky bluegrass, spreading bentgrass (*Agrostis stolonifera*), and fowl bluegrass (*Poa palustris*) generally increase with grazing.

Status and Trends. Nationally, herbaceous wetlands have declined and the Pacific Northwest is no exception. These wetlands receive regulatory protection at the national, state, and county level; still, herbaceous wetlands have been filled, drained, grazed, and farmed extensively in the lowlands of Oregon and Washington. Montane wetland habitats are less altered than lowland habitats even though they have undergone modification as well. A keystone species, the beaver, has been trapped to near extirpation in parts of the Pacific Northwest and its population has been regulated in others. Herbaceous wetlands have decreased along with the diminished influence of beavers on the landscape. Quigley and Arbelbide[181] concluded that herbaceous wetlands are susceptible to exotic, noxious plant invasions.

23. Westside Riparian-Wetlands
Christopher B. Chappell & Jimmy Kagan

Geographic Distribution. This habitat is patchily distributed in the lowlands throughout the area west of the Cascade Crest south into northwestern California and north into British Columbia. It also occurs less extensively at mid- to higher elevations in the Cascade and Olympic mountains, where it is limited to more specific environments.

Physical Setting. This habitat is characterized by wetland hydrology or soils, periodic riverine flooding, or perennial flowing freshwater. The climate varies from very wet to moderately dry and from mild to cold. Mean annual precipitation ranges from 20 to >150 inches (51 to >381 cm) per year. This habitat is found at elevations mostly below 3,000 ft (914 m), but it does extend up to 5,500 ft (1,676 m) in Washington and 6,500 ft (1,981 m) in Oregon in the form of Sitka alder communities. Wetlands above these elevations are generally considered part of the Subalpine Parkland habitat and are not included here. Topography is typically flat to gently sloping or undulating, but can include moderate to steep slopes in the mountains. Geology is extremely variable. Gleyed or mottled mineral soils, organic soils, or alluvial soils are typical. Flooding regimes include permanently flooded (aquatic portions of small streams), seasonally flooded, saturated, and temporarily flooded. Nutrient-poor acidic bogs, except those high in the mountains, are considered part of this habitat.

Landscape Setting. This habitat typically occupies patches or linear strips within a matrix of forest or regrowing forest. The most frequent matrix habitat is Westside Lowlands Conifer-Hardwood Forest. If not forest, the matrix can be Agriculture, Urban, or Coastal Dunes and Beaches habitats, or rarely Westside Grasslands or Ceanothus-Manzanita Shrublands. This habitat also forms mosaics with or includes small patches of Herbaceous Wetlands. Open Water habitat is often adjacent to Westside Riparian-Wetlands. The major land use of the forested portions of this habitat is timber harvest. Livestock grazing occurs in some areas. Peat mining occurs in some bogs.

Structure. Most often this habitat is either a tall (6-30 ft [2-9 m]) deciduous broadleaf shrubland, woodland or forest, or some mosaic of these. Short to medium-tall evergreen shrubs or graminoids and mosses dominate portions of bogs. Trees are evergreen conifers or deciduous broadleaf or a mixture of both. Conifer-dominated wetlands in the lowlands are included here, whereas mid-elevation conifer sites are part of Montane Coniferous Wetlands habitat. Height of the dominant vegetation can be >200 ft (62 m). Canopy height and structure vary greatly. Typical understories are composed of shrubs, forbs, and/or graminoids. Water is sometimes present on the surface for a portion of the year. Large woody debris is abundant in late seral forests and adjacent stream channels. Small stream channels and small backwater channels on larger streams are included in this habitat.

Composition. Red alder (*Alnus rubra*) is the most widespread tree species, but is absent from sphagnum bogs. Other deciduous broadleaf trees that commonly dominate or co-dominate include black cottonwood (*Populus balsamifera* ssp. *trichocarpa*), bigleaf maple (*Acer macrophyllum*), Oregon ash (*Fraxinus latifolia*) and, locally, white alder (*Alnus rhombifolia*). Pacific willow (*Salix lucida*

ssp. *lasiandra*) can form woodlands on major floodplains or co-dominate with other willows in tall shrublands. Oregon white oak (*Quercus garryana*) and California black oak (*Q. kelloggii*) can be important in the interior valleys of western Oregon. Conifers that frequently dominate or co-dominate include western redcedar (*Thuja plicata*), western hemlock (*Tsuga heterophylla*), and Sitka spruce (*Picea sitchensis*). Grand fir (*Abies grandis*) sometimes co-dominates, especially in drier climates and riverine flood plains. Douglas-fir (*Pseudotsuga menziesii*) is relatively uncommon. Shore pine (*Pinus contorta* var. *contorta*) is common in bogs and in deflation plain wetlands along the outer coast. Dominant species in tall shrublands include Sitka willow (*Salix sitchensis*), Hooker's willow (*S. hookeriana*), Douglas' spirea (*Spirea douglasii*), red-osier dogwood (*Cornus sericea*), western crabapple (*Malus fusca*), salmonberry (*Rubus spectabilis*), stink currant (*Ribes bracteosum*), devil's-club (*Oplopanax horridum*), and sweet gale (*Myrica gale*). Labrador-tea (*Ledum groenlandicum, L. glandulosum*), western swamp-laurel (*Kalmia microphylla*), sweet gale, and salal (*Gaultheria shallon*) often dominate sphagnum bogs. Vine maple (*Acer circinatum*) or Sitka alder (*Alnus viridis* ssp. *sinuata*) dominate tall shrublands in the mountains that are located on moist talus or in snow avalanche tracks.

Forests and willow, spirea, and dogwood shrublands within this habitat are limited to the area west of the Cascade Crest. Oregon ash communities occur primarily in the southern Puget Lowland (King County south), Willamette Valley, and Klamath Mountains ecoregions. White alder occurs only in the Willamette Valley and southwestern Oregon. Sitka spruce communities are mainly found in the Coast Range area and western Olympic Peninsula in areas of coastal fog influence. Western hemlock and western redcedar riparian and wetland habitats are largely absent from the southern Oregon Cascades and the Klamath Mountains. Sitka alder and vine maple communities are located in the mountains, mainly in western Washington but to a lesser degree on the east slope of the Cascades and in the Oregon Cascades. Sweet gale communities are found primarily at low elevations on the western Olympic Peninsula. Lodgepole pine-dominated communities are found as bogs in western Washington and along the outer coast of Oregon. Most sphagnum bogs are found in low elevation western Washington.

Shrubs that commonly dominate underneath a tree layer include salmonberry, salal, vine maple, red-osier dogwood, stink currant, Labrador-tea, devil's-club, thimbleberry (*Rubus parviflorus*), common snowberry (*Symphoricarpos albus*), beaked hazel (*Corylus cornuta*), and Pacific ninebark (*Physocarpus capitatus*). Understory dominant herbs include slough sedge (*Carex obnupta*), Dewey sedge (*C. deweyana*), Sitka sedge (*C. aquatilis* var. *dives*), skunk-cabbage (*Lysichiton americanus*), coltsfoot (*Petasites frigidus*), great hedge-nettle (*Stachys ciliata*), youth-on-age (*Tolmiea menziesii*), ladyfern (*Athyrium filix-femina*), oxalis (*Oxalis oregana, O. trilliifolia*), stinging nettle (*Urtica dioica*), swordfern (*Polystichum munitum*), great burnet (*Sanguisorba officinalis*), scouring-rush (*Equisetum hyemale*), blue wildrye (*Elymus glaucus*), Pacific golden-saxifrage (*Chrysosplenium glechomifolium*), and field horsetail (*Equisetum arvense*). Bogs often have areas dominated by ≥1 species of sedge (*Carex* spp.) or beakrush (*Rhynchospora alba*) and sphagnum moss (*Sphagnum* spp.) that are included within this habitat, despite their lack of woody vegetation. Sphagnum moss is a major ground cover in most bogs.

Other Classifications and Key References. This habitat includes all palustrine, forested wetlands and scrub-shrub wetlands at lower elevations on the westside as well as a small subset of persistent emergent wetlands, those within sphagnum bogs.[53] However, drier portions of this habitat in riparian flood plains may not qualify as wetlands according to Cowardin's definition.[53] They are associated with both lentic and lotic systems. Much of this habitat is probably not mapped as distinct types by the Gap projects because of its relatively small scale on the landscape and the difficulty of distinguishing forested wetlands. A portion of this habitat is mapped as the Oregon Gap II Project[126] and Oregon Vegetation Landscape-Level Cover Types[127] westside cottonwood riparian gallery, palustrine forest, palustrine shrubland, NWI (National Wetland Inventory) palustrine emergent, NWI estuarine emergent, and alder/cottonwood riparian gallery. In the Washington Gap project, this habitat occupies portions of open water/wetlands (especially riparian), hardwood forest, and mixed hardwood/conifer forest, and to a minor degree, conifer forest in the following zones: Western hemlock, Sitka spruce, Olympic Douglas-fir, Puget Sound Douglas-fir, Cowlitz River, Willamette Valley, and Woodland/prairie mosaic.[37] This habitat also occupies much of hardwood forest in the Silver fir, Mountain hemlock, portions of Subalpine fir, Interior western hemlock/redcedar, and Grand fir zones.[37] Other references describe this habitat.[41, 71, 85, 88, 90, 91, 104, 113, 114, 115, 138, 210, 220]

Natural Disturbance Regime. The primary natural disturbance is flooding. Flooding frequency and intensity vary greatly with hydro-geomorphic setting. Floods can create new surfaces for primary succession, erode existing streambank communities, deposit sediment and nutrients on existing communities, and selectively kill species not adapted to a particular duration or intensity of flood. Most plant communities are more or less adapted to a particular flooding regime,[138] or they occupy a specific time in a successional sequence after a major disturbance.[85] Debris flows/torrents are also an important, typically infrequent, and severe disturbance where topography is mountainous.[200] Fires were probably infrequent or absent because of the combination of landscape position and site moisture, although fires within the watershed would usually have effects on the habitat through impacts on flooding, sedimentation, and large woody debris inputs. Windthrow of trees can also be significant, especially near the outer coast or on saturated soils. Beavers act as important disturbances by changing the hydrology of a stream system through dams. Grazing by native ungulates, e.g., elk, can have a major effect on vegetation.

Succession and Stand Dynamics. Riparian, i.e., streamside, habitats are extremely dynamic.[162] Succession varies greatly depending on the hydro-geomorphic environment. A typical sequence on a riparian terrace on a large stream involves early dominance by Sitka willow, mid-seral dominance by red alder or cottonwood, with a gradual increase in conifers, and eventual late-seral dominance of spruce, redcedar, and/or hemlock. Such a sequence corresponds with increasing terrace height above the bankfull stream stage.[85] Some communities in bogs or depressional wetlands, as opposed to riverine, seem to be relatively stable given a particular flooding regime and environment. Successional sequences are not completely understood and can be complex. Beaver dams or other alterations of flood regime often result in vegetation changes.

Effects of Management and Anthropogenic Impacts. Intense logging disturbance in conifer or mixed riparian or wetland forests, except bogs, often results in establishment of red alder, and its ensuing long-term dominance. Salmonberry responds similarly to this disturbance and tends to dominate the understory. Logging activities reduce amounts of large woody debris in streams and remove sources of that debris.[27] Timber harvest can also alter hydrology, most often resulting in post-harvest increases in peak flows.[107] Mass wasting and related disturbances (stream sedimentation, debris torrents) in steep topography increase in frequency with road building and timber harvest.[198] Roads and other water diversion/retention structures change watershed hydrology with wide-ranging and diverse effects,[93] including major vegetation changes. The most significant of these are the major flood controlling dams, which have greatly altered the frequency and intensity of bottomland flooding. Increases in nutrients and pollutants are other common anthropogenic impacts, the former with particularly acute effects in bogs. Reed canarygrass (*Phalaris arundinacea*) is an abundant non-native species in low-elevation, disturbed settings dominated by shrubs or deciduous trees. Many other exotic species also occur.

Status and Trends. This habitat occupies relatively small areas and has declined greatly in extent with conversion to urban development and agriculture. What remains is mostly in poor condition, having experienced any of various anthropogenic impacts that have degraded the functionality of these ecosystems: channeling, diking, dams, logging, road-building, invasion of exotic species, changes in hydrology and nutrients, and livestock grazing. Current threats include all of the above as well as development. Some protection has been afforded to this habitat through government regulations that vary in their scope and enforcement with jurisdiction. Of the 77 plant associations representing this habitat in the National Vegetation Classification, almost half are considered imperiled or critically imperiled.[10]

24. Montane Coniferous Wetlands

Christopher B. Chappell & Jimmy Kagan

Geographic Distribution. This habitat occurs in mountains throughout much of Washington and Oregon, except the Basin and Range of southeastern Oregon, the Klamath Mountains of southwestern Oregon, and the Coast Range of Oregon. This includes the Cascade Range, Olympic Mountains, Okanogan Highlands, Blue and Wallowa mountains.

Physical Setting. This habitat is typified as forested wetlands or floodplains with a persistent winter snow pack, ranging from moderately to very deep. The climate varies from moderately cool and wet to moderately dry and very cold. Mean annual precipitation ranges from about 35 to >200 inches (89 to >508 cm). Elevation is mid- to upper montane, as low as 2,000 ft (610 m) in northern Washington, to as high as 9,500 ft (2,896 m) in eastern Oregon. Topography is generally mountainous and includes everything from steep mountain slopes to nearly flat valley bottoms. Gleyed or mottled mineral soils, organic soils, or alluvial soils are typical. Subsurface water flow within the rooting zone is common on slopes with impermeable soil layers. Flooding regimes include saturated, seasonally flooded, and temporarily flooded. Seeps and springs are common in this habitat.

Landscape Setting. This habitat occurs along stream courses or as patches, typically small, within a matrix of Montane Mixed Conifer Forest, or less commonly, Eastside Mixed Conifer Forest or Lodgepole Pine Forest and Woodlands. It also can occur adjacent to other wetland habitats: Eastside Riparian-Wetlands, Westside Riparian-Wetlands, or Herbaceous Wetlands. The primary land uses are forestry and watershed protection.

Structure. This is a forest or woodland (>30% tree canopy cover) dominated by evergreen conifer trees. Deciduous broadleaf trees are occasionally co-dominant. The understory is dominated by shrubs (most often deciduous and relatively tall), forbs, or graminoids. The forb layer is usually well developed even where a shrub layer is dominant. Canopy structure includes single-storied canopies and complex multi-layered ones. Typical tree sizes range from small to very large. Large woody debris

is often a prominent feature, although it can be lacking on less productive sites.

Composition. Indicator tree species for this habitat, any of which can be dominant or co-dominant, are Pacific silver fir (*Abies amabilis*), mountain hemlock (*Tsuga mertensiana*), and Alaska yellow-cedar (*Chamaecyparis nootkatensis*) on the westside, and Engelmann spruce (*Picea engelmannii*), subalpine fir (*Abies lasiocarpa*), lodgepole pine (*Pinus contorta*), western hemlock (*T. heterophylla*), or western redcedar (*Thuja plicata*) on the eastside. Lodgepole pine is prevalent only in wetlands of eastern Oregon. Western hemlock and redcedar are common associates with silver fir on the westside. They are diagnostoc of this habitat on the east slope of the central Washington Cascades, and in the Okanogan Highlands. Douglas-fir (*Pseudotsuga menziesii*) and grand fir (*Abies grandis*) are sometimes prominent on the eastside. Quaking aspen (*Populus tremuloides*) and black cottonwood (*P. balsamifera* ssp. *trichocarpa*) are in certain instances important to co-dominant, mainly on the eastside.

Dominant or co-dominant shrubs include devil's-club (*Oplopanax horridus*), stink currant (*Ribes bracteosum*), black currant (*R. hudsonianum*), swamp gooseberry (*R. lacustre*), salmonberry (*Rubus spectabilis*), red-osier dogwood (*Cornus sericea*), Douglas' spirea (*Spirea douglasii*), common snowberry (*Symphoricarpos albus*), mountain alder (*Alnus incana*), Sitka alder (*Alnus viridis* ssp. *sinuata*), Cascade azalea (*Rhododendron albiflorum*), and glandular Labrador-tea (*Ledum glandulosum*). The dwarf shrub bog blueberry (*Vaccinium uliginosum*) is an occasional understory dominant. Shrubs more typical of adjacent uplands are sometimes co-dominant, especially big huckleberry (*V. membranaceum*), oval-leaf huckleberry (*V. ovalifolium*), grouseberry (*V. scoparium*), and fools huckleberry (*Menziesia ferruginea*).

Graminoids that may dominate the understory include bluejoint reedgrass (*Calamagrostis canadensis*), Holm's Rocky Mountain sedge (*Carex scopulorum*), widefruit sedge (*C. angustata*), and fewflower spikerush (*Eleocharis quinquiflora*). Some of the most abundant forbs and ferns are ladyfern (*Athyrium filix-femina*), western oakfern (*Gymnocarpium dryopteris*), field horsetail (*Equisetum arvense*), arrowleaf groundsel (*Senecio triangularis*), two-flowered marshmarigold (*Caltha leptosepala* ssp. *howellii*), false bugbane (*Trautvetteria carolinensis*), skunk-cabbage (*Lysichiton americanus*), twinflower (*Linnaea borealis*), western bunchberry (*Cornus unalaschkensis*), clasping-leaved twisted-stalk (*Streptopus amplexifolius*), singleleaf foamflower (*Tiarella trifoliata* var. *unifoliata*), and five-leaved bramble (*Rubus pedatus*).

Other Classifications and Key References. This habitat includes nearly all of the wettest forests within the *Abies amabilis* and *Tsuga mertensiana* zones of western Washington and northwestern Oregon and most of the wet forests in the *Tsuga heterophylla* and *Abies lasiocarpa* zones of eastern Oregon and Washington.[88] On the eastside, they may extend down into the *Abies grandis* zone also. This habitat is not well represented by the Gap projects because of its relatively limited acreage and the difficulty of identification from satellite images. But in the Oregon Gap II Project[126] and Oregon Vegetation Landscape-Level Cover Types[127] the vegetation types that include this type would be higher elevation palustrine forest, palustrine shrubland, and NWI palustrine emergent. These are primarily palustrine forested wetlands with a seasonally flooded, temporarily flooded, or saturated flooding regime.[54] They occur in both lotic and lentic systems. Other references describe this habitat.[36, 57, 90, 101, 108, 111, 114, 115, 118, 123, 132, 221]

Natural Disturbance Regime. Flooding, debris flow, fire, and wind are the major natural disturbances. Many of these sites are seasonally or temporarily flooded. Floods vary greatly in frequency depending on fluvial position. Floods can deposit new sediments or create new surfaces for primary succession. Debris flows/torrents are major scouring events that reshape stream channels and riparian surfaces, and create opportunities for primary succession and redistribution of woody debris. Fire is more prevalent east of the Cascade Crest. Fires are typically high in severity and can replace entire stands, as these tree species have low fire resistance. Although fires have not been studied specifically in these wetlands, fire frequency is probably low. These wetland areas are less likely to burn than surrounding uplands, and so may sometimes escape extensive burns as old forest refugia.[1] Shallow rooting and wet soils are conducive to windthrow, which is a common small-scale disturbance that influences forest patterns. Snow avalanches probably disturb portions of this habitat in the northwestern Cascades and Olympic mountains. Fungal pathogens and insects also act as important small-scale natural disturbances.

Succession and Stand Dynamics. Succession has not been well studied in this habitat. Following disturbance, tall shrubs may dominate for some time, especially mountain alder, stink currant, salmonberry, willows (*Salix* spp.), or Sitka alder. Quaking aspen and black cottonwood in these habitats probably regenerate primarily after floods or fires, and decrease in importance as succession progresses. Lodgepole pine is often associated with post-fire conditions in eastern Oregon,[131] although in some wetlands it can be an edaphic climax species. Pacific silver fir, subalpine fir, or Engelmann spruce would be expected to increase in importance with time since the last major disturbance. Western hemlock, western redcedar, and Alaska yellow-cedar typically maintain co-dominance as stand development progresses because of the frequency of small-scale disturbances and the longevity of these species. Tree size, large woody debris, and canopy layer complexity all increase for at least a few hundred years after fire or other major disturbance.

Effects of Management and Anthropogenic Impacts. Roads and clearcut logging practices can increase the frequency of landslides and resultant debris flows/torrents, as well as sediment loads in streams.[198, 199, 229] This in turn alters hydrologic patterns and the composition and structure of montane riparian habitats. Logging typically reduces large woody debris and canopy structural complexity. Timber harvest on some sites can cause the water table to rise and subsequently prevent trees from establishing.[221] Wind disturbance can be greatly increased by timber harvest in or adjacent to this habitat.

Status and Trends. This habitat is naturally limited in its extent and has probably declined little in area over time. Portions of this habitat have been degraded by the effects of logging, either directly on site or through geohydrologic modifications. This type is probably relatively stable in extent and condition, although it may be locally declining in condition because of logging and road building. Five of 32 plant associations representing this habitat listed in the National Vegetation Classification are considered imperiled or critically imperiled.[10]

25. Eastside Riparian-Wetlands
Rex C. Crawford & Jimmy Kagan

Geographic Distribution. Riparian and wetland habitats dominated by woody plants are found throughout eastern Oregon and eastern Washington.

Mountain alder-willow riparian shrublands are major habitats in the forested zones of eastern Oregon and eastern Washington. Eastside lowland willow and other riparian shrublands are the major riparian types throughout eastern Oregon and Washington at lower elevations. Black cottonwood riparian habitats occur throughout eastern Oregon and Washington, at low to middle elevations. White alder riparian habitats are restricted to perennial streams at low elevations, in drier climatic zones in Hells Canyon at the border of Oregon, Washington, and Idaho, in the Malheur River drainage and in western Klickitat and southcentral Yakima counties, Washington. Quaking aspen wetlands and riparian habitats are widespread but rarely a major component throughout eastern Washington and Oregon. Ponderosa pine-Douglas-fir riparian habitat occurs only around the periphery of the Columbia Basin in Washington and up into lower montane forests.

Physical Setting. Riparian habitats appear along perennial and intermittent rivers and streams. This habitat also appears in impounded wetlands and along lakes and ponds. Their associated streams flow along low to high gradients. The riparian and wetland forests are usually in fairly narrow bands along the moving water that follows a corridor along montane or valley streams. The most typical stand is limited to 100-200 ft (31-61 m) from streams. Riparian forests also appear on sites subject to temporary flooding during spring runoff. Irrigation of streamsides and toeslopes provides more water than precipitation and is important in the development of this habitat, particularly in drier climatic regions. Hydrogeomorphic surfaces along streams supporting this habitat have seasonally- to temporarily-flooded hydrologic regimes. Eastside riparian and wetland habitats are found from 100- 9,500 ft (31-2,896 m) in elevation.

Landscape Setting. Eastside riparian habitats occur along streams, seeps, and lakes within the Eastside Mixed Conifer Forest, Ponderosa Pine Forest and Woodlands, Western Juniper and Mountain Mahogany Woodlands, and part of the Shrub-steppe habitat. This habitat may be described as occupying warm montane and adjacent valley and plain riparian environments.

Structure. The Eastside riparian and wetland habitat contains shrublands, woodlands, and forest communities. Stands are closed to open canopies and often multilayered. A typical riparian habitat would be a mosaic of forest, woodland, and shrubland patches along a stream course. The tree layer can be dominated by broadleaf, conifer, or mixed canopies. Tall shrub layers, with and without trees, are deciduous and often nearly completely closed thickets. These woody riparian habitats have an undergrowth of low shrubs or dense patches of grasses, sedges, or forbs. Tall shrub communities (20-98 ft [6-30 m], occasionally tall enough to be considered woodlands or forests) can be interspersed with sedge meadows or moist, forb-rich grasslands. Intermittently flooded riparian habitat has ground cover composed of steppe grasses and forbs. Rocks and boulders may be a prominent feature in this habitat.

Composition. Black cottonwood (*Populus balsamifera* ssp. *trichocarpa*), quaking aspen (*P. tremuloides*), white alder (*Alnus rhombifolia*), peachleaf willow (*Salix amygdaloides*), and, in northeast Washington, paper birch (*Betula papyrifera*) are dominant and characteristic tall deciduous trees. Water birch (*B. occidentalis*), shining willow (*Salix lucida* ssp. *caudata*), and, rarely, mountain alder (*Alnus incana*) are co-dominant to dominant mid-size deciduous trees. Each can be the sole dominant in stands. Conifers can occur in this habitat, rarely in abundance, more often as individual trees. The exceptions are ponderosa pine (*Pinus ponderosa*) and Douglas-fir (*Pseudotsuga menziesii*) that characterize a conifer-riparian habitat in portions of the shrub-steppe zones.

A wide variety of shrubs are found in association with forest/woodland versions of this habitat. Red-osier dogwood (*Cornus sericea*), mountain alder, gooseberry (*Ribes* spp.), rose (*Rosa* spp.), common snowberry (*Symphoricarpos albus*) and Drummonds willow (*Salix drummondii*) are important shrubs in this habitat. Bog birch

(*B. nana*) and Douglas spiraea (*Spiraea douglasii*) can occur in wetter stands. Red-osier dogwood and common snowberry are shade-tolerant and dominate stand interiors, while these and other shrubs occur along forest or woodland edges and openings. Mountain alder is frequently a prominent shrub, especially at middle elevations. Tall shrubs (or small trees) often growing under or with white alder include chokecherry (*Prunus virginiana*), water birch, shining willow, and netleaf hackberry (*Celtis reticulata*).

Shrub-dominated communities contain most of the species associated with tree communities. Willow species (*Salix bebbiana*, *S. boothii*, *S. exigua*, *S geyeriana*, or *S. lemmonii*) dominate many sites. Mountain alder can be dominant and is at least codominant at many sites. Chokecherry, water birch, serviceberry (*Amelanchier alnifolia*), black hawthorn (*Crataegus douglasii*), and red-osier dogwood can also be codominant to dominant. Shorter shrubs, Woods rose, spiraea, snowberry and gooseberry are usually present in the undergrowth.

The herb layer is highly variable and is composed of an assortment of graminoids and broadleaf herbs. Native grasses (*Calamagrostis canadensis*, *Elymus glaucus*, *Glyceria* spp., and *Agrostis* spp.) and sedges (*Carex aquatilis*, *C. angustata*, *C. lanuginosa*, *C. lasiocarpa*, *C. nebrascensis*, *C. microptera*, and *C. utriculata*) are significant in many habitats. Kentucky bluegrass (*Poa pratensis*) can be abundant where heavily grazed in the past. Other weedy grasses, such as orchard grass (*Dactylis glomerata*), reed canarygrass (*Phalaris arundinacea*), timothy (*Phleum pratense*), bluegrass (*Poa bulbosa*, *P. compressa*), and tall fescue (*Festuca arundinacea*) often dominate disturbed areas. A short list of the great variety of forbs that grow in this habitat includes Columbian monkshood (*Aconitum columbianum*), alpine leafybract aster (*Aster foliaceus*), ladyfern (*Athyrium filix-femina*), field horsetail (*Equisetum arvense*), cow parsnip (*Heracleum maximum*), skunkcabbage (*Lysichiton americanus*), arrowleaf groundsel (*Senecio triangularis*), stinging nettle (*Urtica dioica*), California false hellebore (*Veratrum californicum*), American speedwell (*Veronica americana*), and pioneer violet (*Viola glabella*).

Other Classifications and Key References. This habitat is called Palustrine scrub-shrub and forest in Cowardin et al.[53] Other references describe this habitat.[44, 57, 60, 131, 132, 147, 156] This habitat occurs in both lotic and lentic systems. The Oregon Gap II Project[126] and Oregon Vegetation Landscape-Level Cover Types[127] that would represent this type are eastside cottonwood riparian gallery, palustrine forest, palustrine shrubland, and National Wetland Inventory (NWI) palustrine emergent.

Natural Disturbance Regime. This habitat is tightly associated with stream dynamics and hydrology. Flood cycles occur within 20-30 years in most riparian shrublands although flood regimes vary among stream types. Fires recur typically every 25-50 years but fire can be nearly absent in colder regions or on topographically protected streams. Rafted ice and logs in freshets may cause considerable damage to tree boles in mountain habitats. Beavers crop younger cottonwood and willows and frequently dam side channels in these stands. These forests and woodlands require various flooding regimes and specific substrate conditions for reestablishment. Grazing and trampling is a major influence in altering structure, composition, and function of this habitat; some portions are very sensitive to heavy grazing.

Succession and Stand Dynamics. Riparian vegetation undergoes "typical" stand development that is strongly controlled by the site's initial conditions following flooding and shifts in hydrology. The initial condition of any hydrogeomorphic surface is a sum of the plants that survived the disturbance, plants that can get to the site, and the amount of unoccupied habitat available for invasions. Subsequent or repeated floods or other influences on the initial vegetation selects species that can survive or grow in particular life forms. A typical woody riparian habitat dynamic is the invasion of woody and herbaceous plants onto a new alluvial bar away from the main channel. If the bar is not scoured in 20 years, a tall shrub and small deciduous tree stand will develop. Approximately 30 years without disturbance or change in hydrology will allow trees to overtop shrubs and form woodland. Another 50 years without disturbance will allow conifers to invade and in another 50 years a mixed hardwood-conifer stand will develop. Many deciduous tall shrubs and trees cannot be invaded by conifers. Each stage can be reinitiated, held in place, or shunted into different vegetation by changes in stream or wetland hydrology, fire, grazing, or an interaction of those factors.

Effects of Management and Anthropogenic Impacts. Management effects on woody riparian vegetation can be obvious, e.g., removal of vegetation by dam construction, roads, logging, or they can be subtle, e.g., removing beavers from a watershed, removing large woody debris, or construction of a weir dam for fish habitat. In general, excessive livestock or native ungulate use leads to less woody cover and an increase in sod-forming grasses particularly on fine-textured soils. Undesirable forb species, such as stinging nettle and horsetail, increase with livestock use.

Status and Trends. Quigley and Arbelbide[181] concluded that the Cottonwood-Willow cover type covers significantly less in area now than before 1900 in the Inland Pacific Northwest. The authors concluded that although riparian shrubland was a minor part of the landscape, occupying 2%, they estimated it to have declined to 0.5% of the landscape. Approximately 40% of riparian shrublands occurred above 3,280 ft (1,000 m) in elevation before 1900; now nearly 80% is found above that elevation. This change reflects losses to agricultural development, roading, dams and other flood-control activities. The current riparian shrublands contain many exotic plant species and generally are less productive than historically. Quigley and Arbelbide[181] found that riparian woodland was always rare and the change in extent from the past is substantial.

26. Coastal Dunes and Beaches
Christopher B. Chappell, David H. Johnson, & Jimmy Kagan

Geographic Distribution. This habitat occurs primarily along the outer coast from southern Washington (Grays Harbor County) south to northern California. It occurs in all coastal Oregon counties, most abundantly in Tillamook County and between Florence and Reedsport. In Washington it occurs mainly in Grays Harbor and Pacific counties, and sporadically along the inland marine waters of Clallam, San Juan, Skagit, Jefferson, Whatcom, King, Pierce, Kitsap, Snohomish, and Island counties. It also occurs in British Columbia.

Physical Setting. This habitat occurs primarily in wet, mild outer coastal climates. Precipitation, almost always rain, typically averages >80 inches (203 cm) annually. Summers are relatively dry, but fog is common. Elevation is at and very near sea level, only extending as high as the highest dunes. Topography is mildly to strongly undulating in the form of mostly north-south trending dune ridges and troughs. Soils, when present, are always sandy and are underlain by deep deposits of sand, thereby creating edaphically dry sites. Soils are also very poor in nutrients and organic matter. These dunes, spits, and berms are derived from sand carried by longshore drift and wind erosion. Dunes consist of several types that differ in their physical form, including foredunes, transverse dunes, parabola dunes, and retention ridges.[220] Outlier examples away from the outer coast in the Puget Trough are small in extent, occur in a drier climate, and mainly occur in the form of sand spits and berms as opposed to dunes.

Landscape Setting. This habitat occurs in a natural mosaic with Westside Lowland Conifer-Hardwood Forest, Westside Riparian-Wetlands, and Herbaceous Wetlands. Forests adjacent to this habitat are found on stabilized dunes and are dominated by shore pine (*Pinus contorta* var. *contorta*) and Sitka spruce (*Picea sitchensis*). Wooded, shrubby, and herbaceous wetlands occur in seasonally flooded deflation plains or dune troughs. Hooker's willow (*Salix hookeriana*) and slough sedge (*Carex obnupta*) are the two most characteristic species in these wetlands. This habitat is in a mosaic with the Urban habitat, as coastal areas have been developed extensively for tourism and low-density residential uses. Recreation is a major land use and includes the use of off-road vehicles. In southern Washington and northern Oregon, the wetlands are often converted to agriculture for cranberries.

Structure. This habitat consists of a variable mosaic of structures ranging from open sand with sparse herbaceous vegetation to dense shrublands. Trees are typically absent but may be scattered. Unstabilized sand may have very little vegetation or open short grasslands or forb-dominated communities, though these are now relatively uncommon and local. Medium-tall grasslands, typically closed, are a major component in the current landscape. Tall broadleaf evergreen shrubs, typically dense, are also a significant component of the mosaic.

Composition. Where they are vegetated, unstabilized dunes or strand are typically dominated or co-dominated by American dunegrass (*Leymus mollis*), dune bluegrass (*Poa macrantha*), or Chinook lupine (*Lupinus littoralis*). Red fescue (*Festuca rubra*) was once a major dominant on more stabilized dunes but has been largely replaced by European beachgrass (*Ammophila arenaria*), an introduced species that is now the most common dune grass. Many forb species are largely confined to herb-dominated dunes or strand and may take on local importance.

Tall shrublands are dominated primarily by salal (*Gaultheria shallon*) and evergreen huckleberry (*Vaccinium ovatum*), but may also have prominent amounts of hairy manzanita (*Arctostaphylos columbiana*), kinnikinnick (*Arctostaphylos uva-ursi*), bush lupine (*Lupinus arboreus*), or California wax-myrtle (*Myrica califorica*). Coyotebrush (*Baccharis pilularis*) is abundant in southern Oregon. Both Scot's broom (*Cytisus scoparius*) and gorse (*Ulex europaeus*) are exotic shrubs that dominate disturbed areas. Scattered trees are mainly shore pine (*Pinus contorta* var. *contorta*), or, less commonly, Sitka spruce (*Picea sitchensis*).

Other Classifications and Key References. Franklin and Dryness[88] called this habitat sand dune and strand communities. The Oregon Gap II Project[126] and Oregon Vegetation Landscape-Level Cover Types[127] would crosswalk with coastal dunes habitat. This habitat is not well represented by the Washington Gap project: it takes up small percentages of several types in the Sitka spruce zone, including conifer forest, hardwood forests, and coastline, sandy beaches, and rocky islands. Other references describe this habitat.[8, 42, 137, 219, 220]

Natural Disturbance Regime. Erosion and deposition of sand are the primary natural processes controlling this habitat. Sand is deposited initially on beaches, and then moved into dunes through wind erosion.[220] Wind also maintains unstabilized dune areas. Major winter storm events may result in blowouts that create holes in existing stabilized or unstabilized dunes, creating new areas of sand deposition.

Succession and Stand Dynamics. The different structural variants of the mosaic within this habitat are primarily stages in succession from freshly deposited sand to completely stabilized shrub-dominated dunes.[220] Unstabilized sand, such as foredunes with little European beachgrass, has the most open and herbaceous vegetation. Closing of the vegetation typically results in stabilization of the sand. Recently stabilized dunes are now primarily dominated by European beachgrass. Given more time without a major disturbance, shrubs and/or trees colonize the grasslands. Shrublands are sometimes an intermediate stage in succession toward forests. Pine woodlands are another very common intermediate stage. Eventually, pine woodlands are colonized by Sitka spruce or Douglas-fir and become mixed pine-spruce or pine-Douglas-fir forests. Any one of these stages can be set back to sand by a blowout or reburial by dunes, and a cyclic successional sequence is common in many areas.[42]

Effects of Management and Anthropogenic Impacts. European beachgrass has been extensively planted for stabilization purposes and has also spread widely on its own. Unstabilized sand is now a relatively rare condition primarily because of the introduction of this species. The physical forms of dunes also have been altered by beachgrass.[55, 56] Forests are probably forming at a greater rate than they did in the past because of increased stabilization. Exotic species, especially sweet vernalgrass (*Anthoxanthum odoratum*) and common velvetgrass (*Holcus lanatus*), are now a nearly ubiquitous component of herb-dominated communities. The spread of such species may be related to past livestock grazing in many areas.[42] Scot's broom and gorse are aggressive exotic shrub invaders that were planted for stabilization and have spread widely. Since both are legumes, they result in major nitrogen increases where they establish. Off-road vehicle use has resulted in complete destruction of native herbaceous communities in some areas[220] Trampling is a potential threat in herbaceous communities.[42, 220]

Status and Trends. This habitat covers a relatively limited area and major expanses of it have been converted to other uses. The vast majority of herbaceous vegetation that remains is in poor condition, being dominated by exotic species. Current trends are probably decreasing in both extent and condition because of continued development in coastal areas and continuing expansion of exotic species into the few remaining native-dominated areas. Six of 11 plant associations currently listed in the National Vegetation Classification representing this habitat are considered imperiled or critically imperiled.[10]

27. Coastal Headlands and Islets

Christopher B. Chappell & David H. Johnson

Geographic Distribution. This shrubland, grassland, and nearshore rocky island habitat occurs along slopes and exposed headlands along the outer coast, from Cape Flattery, Clallam County, Washington, southward to (and beyond) California. On small islands it also extends into the Strait of Juan de Fuca and the inland marine waters of Puget Sound and Hood Canal, Washington. Sporadic along the Washington coast (absent between Point Grenville and Cape Disappointment), this habitat becomes most extensive on the southern Oregon coast.

Physical Setting. Wind is extreme in this environment and, in combination with abundant salt-spray, limits tree growth. Fog is common in the summer. Climate is generally mild and moist to wet, with mean annual precipitation ranging from about 70 to 120 inches (178 to 305 cm). Elevation is sea level to about 500 ft (152 m). This habitat occurs mainly on coastal headlands, bluffs, and islands with steep slopes or cliffs. Soils are typically shallow to bedrock or consist of exposed glacial deposits on steep erodable bluffs. Slopes range from gentle to very steep. In some areas, seeps create moist to wet microsites.

Landscape Setting. This habitat is always located adjacent to, or in the case of the rock islets ("sea stacks"), within the Marine Nearshore habitat. It is found mainly along the outer coastline where it typically occupies small areas between the Marine Nearshore and Westside Lowland Conifer-Hardwood Forest or on small islands. Cliffs are a common feature. In far southern Oregon (Curry County), it occupies continuous ocean-facing slopes for many miles. Land use is recreation or low-density residential.

Structure. This habitat is a shrubland, grassland, forbland, rocky island, or often a mosaic of these. The dominant shrubs may be tall or short and composed of evergreen or deciduous broadleaf shrubs. Native grasses can be short or up to 5 ft (1.5 m) in height and rhizomatous or cespitose. Forbs or ferns dominate some patches. Coniferous trees are sometimes scattered, occur in small clumps, or form dominant patches of short wind-blasted individuals.

Composition. Shrublands are dominated by salal (*Gaultheria shallon*), evergreen huckleberry (*Vaccinium ovatum*), salmonberry (*Rubus spectabilis*), black twinberry (*Lonicera involucrata*), California wax-myrtle (*Myrica californica*), thimbleberry (*Rubus parviflorus*), or the dwarf shrub, crowberry (*Empetrum nigrum*). Deer brush (*Ceanothus integerrimus*), and hairy manzanita (*Arctostaphylos columbiana*) become important on the southern Oregon coast, as does the non-native gorse (*Ulex europaeus*). Sitka spruce (*Picea sitchensis*) is the most common tree, although western hemlock (*Tsuga heterophylla*), Douglas-fir (*Pseudotsuga menziesii*), or red alder (*Alnus rubra*) also may occur. Native dominant grasses are red fescue (*Festuca rubra*) or Nootka reedgrass (*Calamagrostis nutkaensis*). Blue wildrye (*Elymus glaucus*), California danthonia (*Danthonia californica*), and Sitka brome (*Bromus sitchensis*) can also be important. A diversity of forbs occurs, with some of the most prominent being Canada goldenrod (*Solidago canadensis*), Martindale's lomatium (*Lomatium martindalei*), giant vetch (*Vicia gigantea*), giant horsetail (*Equisetum telmateia*), and coastal wormwood (*Artemisia suksdorfii*). Bracken (*Pteridium aquilinum*) is a fern that often co-dominates. Southern Oregon has a number of unique herbaceous species.

Other Classifications and Key References. Franklin and Dyrness[88] described portions of this habitat as oceanfront communities on northern Oregon headlands and the southern Oregon coast. The Oregon Gap II Project[126] and Oregon Vegetation Landscape-Level Cover Type[127] that would represent this type is coastal strand. The Washington Gap project mapped parts of this habitat as coastline, sandy beaches, and rocky islands. Other important references describe elements of the habitat.[7, 63, 177, 183]

Natural Disturbance Regime. Wind may topple trees if they do attain upright stature. Charcoal in the soil at some sites in Oregon suggests that this habitat may have had occasional fires in the past.[183]

Succession and Stand Dynamics. Little is known about the dynamics of this habitat. Trees slowly invade some areas of this habitat. As they do so, herbaceous or shrub-dominated vegetation declines. Fires would favor maintenance of grasslands or forblands.

Effects of Management and Anthropogenic Impacts. Livestock grazing of the grasslands results in decreasing importance of native grasses, especially bunchgrasses, and increasing importance of exotic species. Sweet vernalgrass (*Anthoxanthum oderatum*), common velvetgrass (*Holcus lanatus*), and orchardgrass (*Dactylus glomerata*) are major exotic grass species that dominate significant areas. Gorse has invaded large areas on the southern Oregon coast.

Status and Trends. This habitat occupies a very small area relative to other habitats in the Pacific Northwest. Condition of the grasslands is generally poor, with an abundance of non-native species. Grasslands continue to decline in condition and extent over time. Shrublands are probably more stable. Three of 5 plant associations listed in the National Vegetation Classification are considered imperiled,[10] but portions of this habitat have not been described at the association level.

28. Bays and Estuaries
Mikell O'Mealy & David H. Johnson

Geographic Distribution. This habitat reflects areas with significant mixing of salt and freshwater, including lower reaches of rivers, intertidal sand and mud flats, saltwater and brackish marshes, and open-water portions of associated bays. The habitat is distributed along the marine coast and shoreline of Washington and Oregon. There are some 21 principal bays and estuaries on the Oregon coast, and 34 in Washington.[5, 95, 178] Willapa Bay and Grays Harbor (both in Washington) are expansive and have the largest and second largest intertidal areas of the two states. The Columbia River Estuary is the largest estuary in the Pacific Northwest. This habitat does not include open water areas of Puget Sound (see Inland Marine Deeper Waters). Similar bay and estuarine habitats exist on the coasts of California and British Columbia.

The greater Puget Sound at times is considered a very large estuary;[180] for purposes of this project Puget Sound is comprised of three wildlife habitats: Bays and Estuaries, Marine Nearshore, and Inland Marine Deeper Waters.

Physical Setting. Climate is moderated by the Pacific Ocean and is usually mild. Mean temperatures at coastal stations generally range from 40 to 70°F (4-21°C) year-round with little north-south variation within Washington and Oregon. Annual rainfall along the coastal zone averages 80-90 inches (203-229 cm) and is concentrated in winter months, producing correspondingly high river runoff to bays and estuaries.[173, 176.] Elevation is at sea level to a few feet above. Coastal zone topography is characterized by long stretches of sandy beaches broken by steep rocky cliffs, rocky headlands, and the mouths of bays and estuaries. Organics, silt, and sand are the primary substrate components of this habitat and vary in specific composition and distribution with variable physical factors.[120]

Landscape Setting. This habitat is adjacent to Westside Riparian-Wetlands, Coastal Dunes and Beaches, Westside Lowland Conifer-Hardwood Forest, Coastal Headlands and Islets, Marine Nearshore, and Inland Marine Deeper Waters habitats. Major uses of bays and estuaries are recreation, tourism, the shellfish industry, and navigation. The terrestrial interface portions of this habitat have been extensively converted for agricultural crop production, livestock grazing, and residential and commercial development. Water channels of many areas have been dredged for ship navigation.

Structure. At the most seaward extent (e.g., river mouths), water depths are shallow (mostly <20 ft [6 m]) except for dredged channels. This habitat is strongly influenced by the daily tides and currents. Depending on location, mean higher high water to mean lower low water ranges from 6.1 to 10.2 ft (1.9 to 3.1 m). Tidal currents in channels of the principal estuaries typically range from 1 to 5 knots (0.5 to 2.6 m/sec) 176.

Diverse habitats result from riverine discharges and tidal fluxes, salinity, mixing, sedimentation, discharge, and insolation. Unconsolidated or consolidated tideflats are composed of rocks, gravel, sand, silt and clay as well as abundant organic material.[68, 204] Inundated by daily tidal flows, tideflats may support eelgrass, various algal species, and invertebrate communities.[5] Eelgrass meadows create protected environments and structured habitats for various wildlife species.[173] Salt marshes form at the upper tidal boundary above tideflats.[120] Salt marshes are usually open to closed graminoid or forb communities. Highly branched estuarine channels drain across salt marshes and tideflats, creating a diverse mix of structures.[192] At the most inland extent of this habitat, transitional marsh forms between salt marshes and bordering upland vegetation dominated by grass or woody vegetation.[6]

The Columbia River estuary is characterized as a partially mixed estuary and can be divided into three sections along the salinity gradient: from the mouth to about river mile 7 it is basically marine; from river mile 7 to mile 23 it is transitional (mixing); and above river mile 23 it is fluvial (fresh water). Pruter and Alverson[179] compiled available physical and biological studies at the interface between riverine and marine waters in the nearshore aspects of the Columbia River estuary and adjacent waters.

Composition. Eelgrass meadows stabilize submerged tideflats and are co-dominated by surfgrass and eelgrass species. Three diagnostic surfgrass species (*Phyllospadix scouleri*, *P. torreyi*, and *P. serrulatus*) occur on rocky substrates in exposed waters, whereas two species of eelgrasses (*Zostera marina*, *Z. japonica*) are characteristic of mud or mixed mud-sand substrates in areas sheltered from turbulent waters.[68, 173] Highly productive macroalgae that dominate estuarine channels include various blue-green algae, green algae (*Enteromorpha* spp.) and rockweed (*Fucus* spp.).[192] Tideflats bordering salt marshes often are co-dominated by pickleweed (*Salicornia virginica*), arrowgrass (*Triglochin maritima*) and three-square rush (*Scirpus americanus*).[5] The transition to higher areas of the low-marsh zone is indicated by the dominance of jaumea (*Jaumea carnosa*), saltgrass (*Distichlis spicata*), and Lyngby's sedge (*Carex lyngbyei*).[68] Major components of mid and high salt marsh areas are alkaligrass (*Puccinellia pumila*) and Canadian sand spurry (*Spergularia canadensis*).[120] Salt rush (*Juncus lesueurii*), tufted hairgrass (*Deschampsia caespitosa*), Pacific silverweed (*Argentina egedii*) and spreading bentgrass (*Agrostis stolonifera*) are salt-tolerant upland species diagnostic of high salt marshes that experience freshwater runoff or riverine discharge.[6]

Other Classifications and Key References. Cowardin et al.[53] included marine and estuarine systems of the Columbian Province. Dethier[68] described a classification for marine and estuarine habitat types in Washington State. Habitat types are defined by depth, substratum type, energy level, and a few modifiers. Species (plants and animals) are described for combinations of these physical variables. The Oregon Gap II Project[126] and Oregon Vegetation Landscape-Level Cover Types[127] that would represent this type are exposed tidal flats and estuarine emergent. Harper et al.[106] described a shore-zone sensitivity mapping system. Proctor et al.[176] described an ecological characterization of the Pacific Northwest Coastal Region, including physical and chemical environments as well as socioeconomic aspects of watersheds of the region. Schoch and Dethier[189] provided high-resolution data on the physical features and associated biota of Puget Sound's shorelines using the SCALE model (Shoreline Classification and Landscape Extrapolation). Downing[76] offered a detailed review of the geological and broad ecological development of Puget Sound.

Natural Disturbance Regime. Natural disturbance perpetuates the dynamic, transitional nature of this habitat. Tides, seasonal riverine discharges, winds, storm events, erosion, and accretion are the primary natural processes that shape this habitat. Tides are mixed, characterized by two unequal high and low tides daily, with varying intrusion into estuaries and bays at different locations along the coast. Tides and winds push salt-water wedges up through the system, causing varying degrees of mixing with incoming riverine waters and significant vertical stratification.[5] Riverine discharges and fresh-water runoff vary seasonally with precipitation and freshet regimes. Generally, a large range in annual discharge exists with high volumes of fresh water entering the system in winter and significantly reduced flows in summer.[192] Short-term storm events produce dramatic variations in physical habitat conditions. Sudden erosion or accretion may result from strong oceanic currents at the mouth of the system or from increased fresh water discharges at the head of the system. For a detailed conceptual model of disturbance regimes in estuary zones, see Proctor et al.[176]

Succession and Stand Dynamics. General successional stages reflect unconsolidated barren tideflats to stabilized high salt marshes and salt meadows. Unvegetated tideflats are colonized by pioneer plants, commonly eelgrass, that are tolerant of extended tidal inundation and vary depending on sediment type.[173] Initial colonization causes

sediment accretion and gradual rise in land elevation, changes that shift environmental conditions and permit other plants to establish. Arrowgrass, pickleweed, sand spurry, and spike rush can invade the emerging marsh, further increasing and stabilizing substrates. Saltgrass and sedge establish on higher areas of the marsh. When initial colonizers die back, tufted hairgrass and salt rush may establish.[5] Various exotic species have become naturalized in Oregon and Washington, including spreading bentgrass (*Agrostis stolonifera*) and sand spurry (*Spergularia marina*) introduced from Europe, brass buttons (*Cotula coronopifolia*) introduced from South Africa, and marsh cordgrass (*Spartina alterniflora*) introduced from the Atlantic Coast of North America.[6, 120] These successional stages can be disrupted by riverine or tidal scouring and succession can be reinitiated at any point.

Effects of Management and Anthropogenic Impacts. Management, water quality, contaminants, and land-use practices have altered significant portions of this habitat and continue to impact remaining areas.[216] The dredging and filling of marshes and tideflats to serve various human needs remove estuarine vegetation. Channel flow, tidal inundation, and fresh water discharges are disrupted by construction of seawalls, jetties, dikes, and dams. The physical and chemical conditions of these habitats are degraded by the discharge of municipal, industrial, and agricultural effluents. Functional plant and animal communities are altered by domestic and agricultural runoff of pesticides, herbicides, and fertilizers. Invasions of exotic plants (e.g., *Spartina*) and invertebrates (e.g., green crabs) pose significant, long-term ecological and economic threats to this habitat. Large tracts of habitat have been lost and converted for coastal development. Additionally, upland activities occurring throughout the watershed, including logging, mining, and hydroelectric power development, can have destructive impacts downstream in estuarine and bay environments.[192, 205]

Status and Trends. Significant quantitative and qualitative alterations of this habitat have occurred with Euro-American settlement. Although natural erosion and accretion processes continue, most habitat modification can be attributed to anthropogenic impacts.[192] Because of original diking for crop production and flood control, almost no areas of natural high marsh remain in Oregon.[120] These dikes, and other more recent barriers, prevent natural recovery and re-establishment of this habitat. Remaining examples of the bay and estuarine habitat exist in various conditions, from the more natural areas, areas undergoing active restoration, to the more prevalent polluted, degraded, or overused areas throughout Oregon and Washington. With increasing population pressures in coastal areas and the corresponding threats of habitat use and conversion, future trends will likely be continued degradation and reduction of remaining bay and estuarine areas.

29. Inland Marine Deeper Water

David H. Johnson

Geographic Distribution. This habitat is located in the northwestern portion of Washington and adjacent areas of southwest British Columbia. It includes the open waters of the Strait of Georgia, Puget Sound, Hood Canal, and the Strait of Juan de Fuca. More specifically, this habitat reflects waters >66 ft (20 m) deep, found inland from a line between the Elwha River (just west of Port Angeles) on the Washington side of the Strait of Juan de Fuca, northward to Race Rocks on the southeastern tip of Vancouver Island, British Columbia. This line was independently determined based on (1) kelp distribution, (2) marine bird distribution, and (3) fish species and abundance data. With the exception of Marine Nearshore areas, waters west of this line are considered Marine Shelf. The Inland Marine Deeper Water habitat is not found in Oregon.

Physical Setting. This habitat lies largely within the Puget Lowland and northward in Georgia Strait on the east side of Vancouver Island, British Columbia. Mean air temperatures generally range between 40 and 70°F (4-21°C) year round, with little north-south variation. Rainfall averages 20 to 80 inches (50 to 200 cm) annually and is concentrated in winter months, producing correspondingly high river runoff to bays, estuaries, and inland marine waters.

Landscape Setting. This habitat is commonly adjacent to Bays and Estuaries, Coastal Headlands and Islets, and Marine Nearshore habitats and merges with the Marine Shelf habitat in the Strait of Juan de Fuca. Inland marine waters are used extensively for navigation, commercial transport of goods, recreation, tourism, and fishery operations.

Structure. A diversity of underwater structures are created as swift tidal currents circulate waters of the Pacific Ocean through the reaches of Strait of Georgia, Puget Sound, Hood Canal and the Strait of Juan de Fuca. Aspects of geology are particularly important in understanding the structure and dynamics of this habitat. Glacial ice initially excavated several long, narrow valleys that today form

Lake Washington, Lake Sammamish, Hood Canal, and the major basins of Puget Sound. The arrangement of the present shorelines was established 13,000 years ago when glacial ice retreated from the Puget Lowland.[76] Organics, silt, and sand are the primary substrate components of this habitat and vary in specific composition and distribution with fluctuating physical factors.[120] Through deposition of sediments, major river deltas have advanced substantial distances into the deep basins of Puget Sound.[76]

Composition. Marine waters dominate fresh water influences in areas away from riverine discharges or from the shoreline. Because of the water depths involved, sunlight is diffused, and few if any plants attached to the benthic substrates are capable of growing.

Other Classifications and Key References. Cowardin et al.[53] included this region in the Columbia Province and described a hierarchical classification for wetlands and deepwater habitats in the U.S. Dethier[68] described a classification for marine and estuarine habitat types in Washington State. Habitat types were defined by depth, substratum type, energy level, and a few modifiers. Harper et al.[106] described a shore-zone mapping system for use in sensitivity mapping and shoreline countermeasures. Proctor et al.[176] described an ecological characterization of the Pacific Northwest Coastal Region, including physical and chemical environments as well as socioeconomic aspects of watershed units and of the region. Schoch and Dethier[189] provided high-resolution data on the physical features and associated biota of Puget Sound's shorelines using the SCALE model (Shoreline Classification and Landscape Extrapolation).

Natural Disturbance Regime. Seasonal and larger, periodically occurring disturbances shape this habitat. Seasonal variation in tidal regimes, precipitation and riverine discharges (winter highs), as well as periodic storm events cause changes in temperature, salinity, energy level, and gradual or sudden erosion and accretion in localized areas.

Successional and Community Dynamics. Diverse plant and invertebrate communities compete for a variety of habitats in this region. Succession occurs in each habitat area as disturbances create temporary vacancies, allowing opportunistic species to become established.

Effects of Management and Anthropogenic Impacts. Land conversion, use, and management have altered significant portions of this habitat. The physical, chemical, and biological condition of some habitats are degraded by both point and nonpoint discharges from municipal and industrial effluents. Functional plant and animal communities are altered by domestic and agricultural runoff of pesticides, herbicides, and fertilizers. Large portions of shoreline have been converted for residential, commercial, and port development, affecting inputs into the adjacent deeper waters. Benthic communities are significantly impacted by maintenance dredging done to support navigation and commerce. The transport of oil and chemical substances creates the potential for harmful spills that can affect these areas for extended periods of time. Passage of vessels from other regions increases the introduction rate of exotic species which, once established, can effectively out-compete native species.

Status and Trends. With the important exceptions of locally increased sedimentation rates and contaminant deposition/retention, the status and trends in the physical and biological aspects of this habitat are poorly known.

30. Marine Nearshore

David H. Johnson

Geographic Setting. This habitat reflects marine water areas (high tide line to depth of 66 ft [20 m]) along shorelines not significantly affected by freshwater inputs (i.e., excludes Bays and Estuaries). This includes all marine shorelines of Puget Sound, Hood Canal, San Juan Islands, Strait of Georgia, Strait of Juan de Fuca, and along the outer coastlines of Washington and Oregon. In Washington, there are 3,100 miles (4,990 km) of this nearshore habitat[85] (H. D. Berry, Department of Natural Resources, Aquatic Resources Division, Olympia, pers. comm.); in Oregon, there are 377 miles (607 km) of this nearshore habitat (C. Barrett, Northwest Habitat Institute, Corvallis, Oregon, pers. comm.). For mapping and classification purposes, this habitat does not extend into, or overlap with, shallow or intertidal areas found within Bays and Estuaries.

Physical Setting. The outer coastline of Washington and Oregon can be characterized as a series of sandy beaches interspersed with rocky headlands. This coastline is oriented in a north-south direction and is subjected to long-fetch, high-energy waves. Nearshore areas within Puget Sound, Hood Canal, and elsewhere landward from the Strait of Juan de Fuca, are more protected. With the exception of the far-reaching Columbia River plume,[194] the effects of coastal streams are generally local and seasonal.[170]

Landscape Setting. This habitat is adjacent to the Marine Shelf, Inland Marine Deeper Water, Bays and Estuaries, and a number of terrestrial-based habitats (e.g., Coastal Dunes and Beaches, Westside Lowland Conifer-

Hardwood Forest, and Urban). It occurs in a mosaic with Coastal Headlands and Islets.

Structure. Fresh waters drain from lands surrounding these inland marine waters to create estuarine environments nearshore[133] (see Bays and Estuaries habitat). Nearshore subtidal habitats are diversified by degree of wave and current action, availability of sunlight, and presence of vegetation. Submerged unvegetated habitats cover a greater area than do vegetated nearshore habitats, such as salt marshes and eelgrass beds. Various combinations of water depth, character of substrates, and exposure to tidal action create a wide range of benthic habitats. Sand, cobble, boulders, and hardpan are commonly found in areas of moderate to strong currents, whereas silt and clay settle out in protected inlets and bays.[67, 145]

Composition. This habitat supports marine organisms capable of withstanding short-term exposure to air. Bottom substrates in exposed areas are generally rock or sand, but can include cobble or gravel. The subtidal photic zone includes the region from mean low low water (MLLW or the 0 ft depth) to about -50 ft (-15 m) where water is deep enough to prevent sufficient light penetration to the marine floor for primary productivity of kelp and other marine plants. The rocky-bottom intertidal habitats support kelps (*Laminaria* spp., *Lessoniopsis* spp., *Hedophyllum sessile*), brown rockweed (*Pelvetiopsis limitata*), red algae (*Iridaea* spp.), and surfgrass (*Phyllospadix scouleri*), as well as an abundance and variety of sessile benthic invertebrates. The larger kelps, such as *Macrocystis integrifolia* and *Nereocystis leutkeana*, are found in the rocky-bottom subtidal areas. Because of constant wave action, the sandy-bottom areas of the intertidal and subtidal zones support few or no plants. The moderate to low energy intertidal and subtidal areas where sand, mud, and gravel accumulate support eelgrass (*Zostera marina* and *Z. japonica*), and the red alga (*Gracilaria pacifica*).

Other Classifications and Key References. Dethier[68] provided a detailed classification scheme for the estuary, intertidal, and shallow subtidal areas of Washington. The Cowardin et al.[53] classification scheme has several limitations with regards to adopting it for marine and estuarine systems. Levings and Thom[143] described 9 categories of nearshore habitat in Puget Sound and Georgia Basin.

Natural Disturbance Regimes. This habitat is strongly influenced by tidal rhythms, wave action, storm events, light penetration, and bottom substrate. Because of these factors, this habitat is characterized by a high degree of patchiness; this patchiness leads to differences in its faunal makeup and use. Herbivory by marine invertebrates also causes significant disturbance in plant communities, as evidenced by the direct control of kelp beds by urchin populations.

Succession and Stand Dynamics. The primary natural processes that shape the nearshore habitats include tides, erosion, accretion, and storm events. The rocky surf zone of the outer coast of the Olympic Peninsula includes some of the most complex and diverse shores in the United States.[67] Here, high wave energy provides space for habitation for species as materials are eroded away, and by increasing the capacity of algae to acquire nutrients and use sunlight. Examples of succession can be found on rocky intertidal shores where wave energy periodically disturbs established communities, or in kelp forests where herbivory or the scouring action of swift tidal currents removes vegetation.

Effects of Management and Anthropogenic Impacts. This habitat reflects the interface between land and sea, and is the site of intense commercial and navigational activities, such as seaports, marinas, ferry docks, and log booms. A significant concern identified by Broadhurst[35] is the site-by-site consideration of projects with no ability to account for and assess the cumulative environmental effects of various development activities (from small residential projects to large commercial and industrial development projects). Without the ability to measure or understand cumulative effects, managers are permitting individual activities that may result in dramatic resource losses over time. Making high-quality nearshore vegetation and shoreline characteristics inventory mapping available to land-use planners, natural resource scientists, and the public will increase opportunities to protect this habitat.

Status and Trends. Shoreline modification such as bulkheading, filling, and dredging can lead to direct habitat loss. Indirectly, it can lead to changes in the sediment and wave energy on a beach and in adjacent subtidal areas. One third of Puget Sound's shorelines, approximately 800 miles (1,287 km), has been modified.[180, 190, 191] The Central Puget Sound region, with high human population levels, shows the highest level of modification overall (52%).[180] In Washington there are 26 species of kelp, more than any other area worldwide.[77] Data on floating kelp along the Strait of Juan de Fuca suggest that while kelp areas are dynamic, the overall extent of kelp has remained stable during 1993-1997.[180]

31. Marine Shelf
David H. Johnson

Geographic Setting. This habitat consists of marine waters along the outer coast of Washington and Oregon that are 66-656 ft (20-200 m) deep. This also includes the western portion of the Strait of Juan de Fuca (excluding the Marine Nearshore areas), west of a line from the Elwha River on the Washington side of the Strait of Juan de Fuca to Race Rocks on the southeastern tip of Vancouver Island.

Physical Setting. Along the coasts of Oregon and Washington, the Marine Shelf (also called the Continental Shelf) habitat exists as a relatively shallow, flat, submerged area, which varies from about 9 to 40 miles (14 to 64 km) in width. At about the 656-ft (200-m) isobath, roughly the edge of the marine shelf, the bottom drops off more steeply to the continental slope, which is indented by several major submarine canyons. Beyond the shelf and slope are deep abyssal oceanic waters.

Landscape Setting. This habitat is located between the Nearshore Marine and Oceanic habitats; at about one third of the way into the Strait of Juan de Fuca, this habitat adjoins the Inland Marine Deeper Water habitat.

Structure. The marine shelf extends seaward from the 60-ft (20 m) to the 656-ft (200 m) isobath. It is occasionally divided into the inner (0-328 ft [100 m] deep) and outer (328-656 ft [100-200 m] deep) shelf areas, reflective of differing oceanographic influences. The bottom substrate of the shelf is mostly sand, giving way to silt on the outer edges, and is described as smooth as a result of sediment accumulation. Currents over the shelf tend to follow the seasonal pattern of the oceanic currents (i.e., northward during winter [Davidson current]). The mean surface temperatures in summer were >5°C lower, and mean salinities were 0.1-0.3% higher in upwelling areas than farther offshore. These conditions reflect an active upwelling process.[83] The shelf areas off Oregon and Washington are known for their heavy surface waves; extremes of wave heights ranging from 49 to 95 ft (15 to 29 m) have been recorded. More typical are waves of 20 to 33 ft (6 to 10 m) during storm events.

Composition. The Marine Shelf, as with other deep-water regions, does not support rooted plant life. Strickland and Chasan[194] offered a synthesis of phytoplankton; the Washington shelf has been ranked in the highest productivity category of U.S. continental shelves.

Other Classifications and Key References. Strickland and Chasan[194] offered a synthesis of information on Coastal Washington, as related to understanding impacts of offshore oil and gas exploration. The National Oceanic and Atmospheric Administration[163] reflects the Final Environmental Impact Statement/Management Plan for the Olympic Coast National Marine Sanctuary. Bottom et al.,[33] Dodimead et al.,[73] Favorite et al.,[83] Thomson,[203] and Ware and McFarlane[213] offered classification schemes and terminology for understanding the oceanic systems in the Northeastern Pacific.

Natural Disturbance Regimes. Currents over the shelf tend to follow the seasonal pattern of the oceanic currents, but also are strongly influenced by local winds, bottom and shoreline configuration, and freshwater input. On average, water flows southward in the upper 328 ft (100 m) during summer, and northward below that. Water over the shelf flows generally northward at all depths during the winter.[194] In addition to ocean currents, this habitat is heavily influenced by freshwater infusion from the Columbia River, the largest river on the Pacific west coast. The Columbia River effluent amounts to approximately 60% of the freshwater entering the Pacific Ocean between San Francisco and the Strait of Juan de Fuca in winter, and >90% the rest of the year.[21] Furthermore, the 12- to 16-mile (20 to 25 km)-wide Strait of Juan de Fuca, separating Vancouver Island from mainland Washington, is a glacially excavated channel that is the primary avenue for vigorous estuarine exchange between the shelf and the inland marine waters of Washington and British Columbia.

Effects of Management and Anthropogenic Impacts. The chief human influence on the ecology of Oregon and Washington marine waters is fishing, especially bottom and mid-water trawl fishing in shelf waters for flatfish and Pacific hake. Other significant fisheries include salmon (inner shelf), shrimp (outer shelf), and albacore (slope). The effects of overfishing have been documented. Other risks to the shelf environment include pollution, contaminants, and oil spills.

Status and Trends. Until the early 1990s, the Minerals Management Service of the U.S. Department of the Interior had planned to conduct Lease Sales for offshore oil and gas exploration in federal waters on the outer marine shelf of Oregon and Washington.[194] A moratorium on these leases is now in place. Designated in 1994, the Olympic Coast National Marine Sanctuary covers 2.11 million acres (854,251 ha) and is managed to protect its natural resources while encouraging compatible commercial and recreational uses.

32. Oceanic
David H. Johnson & Greg Green

Geographic Setting. Deep water (>656 ft [200 m] deep), open areas of the northeast Pacific Ocean extending seaward from the 656-ft (200-m) isobath along the outer coast of Washington and Oregon.

Physical Setting. This habitat includes the continental slope, which is generally found at depths between the 656- and 6,560-ft (200- and 2,000-m) isobath. In general, the 656-ft (200-m) isobath follows a north-south line.

Landscape Setting. The Oceanic habitat adjoins the shallower marine waters of the Marine Shelf habitat.

Structure. The oceanic area off Oregon and Washington includes marine waters >656 ft (200 m) deep. This region can be divided into two general habitats reflecting geomorphic features and water depth.

The continental shelf along Oregon is characterized by a series of oceanic banks including Daisy, Stonewall, Perpetua, Heceta, and Silcoos banks, and a major promontory, Cape Blanco. In contrast, the Washington shelf is furrowed by Nitinat, Juan de Fuca, Quinault, Grays, Guide, Willapa, and Astoria submarine canyons, remnants of the last glacial period.

The continental slope is usually defined as depths between 656 and 6,560 ft (200 and 2,000 m) deep. Beyond the slope is a more gradual sloping area known as the continental rise. The rise eventually terminates at the abyssal sea floor (approximately 13,120 ft [4,000 m] deep). Boundaries between the rise and abyssal plain are not clearly distinct. Both the rise and abyssal sea floor are composed largely of mud.

Marine currents define important aspects of this habitat, as wind-driven equatorward surface flow in the spring and summer results in episodic upwelling of cold nutrient-rich water; the poleward surface flow in the autumn and winter is when the downwelling prevails. The transition from poleward to equatorward flow occurs abruptly in the spring and the reverse transition occurs somewhat less abruptly in the autumn.

Other Classifications and Key References. A number of oceanographic studies have attempted to identify distinct features or parameters to define the geographical extent of the Subarctic and Subtropical Pacific waters. These studies were based mostly on temperature-salinity properties and oxygen-salinity curves.[73, 213] Favorite et al.[83] and Thomson[203] discussed new information and terminology for the oceanic systems in the Northeast Pacific. A recent review by Bottom et al.[33] synthesized existing information on the characteristics of the ocean environment and the influence of ocean variability on the capacity of the Northeast Pacific to produce salmon. Bottom et al.[31, 32] provided information on classifications of subsystems of the Coastal Upwelling Domain.

Natural Disturbance Regime. This deeper water habitat is influenced by seasonal water temperatures, winds, currents, and upwelling. The shelf slope essentially defines the pathway of the large-scale California Current, which transports water eastward from mid-ocean, then southward to California. The current is essentially driven by seasonal northwesterly winds. The relatively wide shelf off Oregon and Washington tends to isolate the California Current and its infusion of nutrients from nearshore flow. It is pushed farther offshore during the winter when southeasterly winds generate the seasonal northern flow of the more coastal Davidson Current.[141]

In addition to driving currents, seasonal winds also promote upwelling of colder, nutrient-rich waters. Coastal upwelling occurs most frequently in summer and fall, promoted by northerly and northwesterly winds. The upwelling season runs from April to October, with maximum intensity in July and August.[18] Upwelling intensity is typically greatest along the southern Oregon coast (Cape Blanco upwelling zone) and diminishes northward. However, it can occur anywhere along the Oregon-Washington coast under favorable winds. Shelf promontories such as Heceta Bank, the submarine canyons off Washington, and the shelf edge and slope create biologically productive mixing zones influenced both by large-scale currents and upwelling. Sporadic events causing large-scale changes in both nearshore and offshore habitats include extreme La Niña (cold water) and El Niño (warm water) events, and toxic red tide blooms.

Effects of Management and Anthropogenic Impacts. The main activities in this habitat are fishing and commercial transport; other than regulations to support these actions, little active management for biological resources takes place in this habitat.

Literature Cited

1. Agee, J. K. 1993. Fire ecology of Pacific Northwest forests. Island Press, Washington, DC.

2. ———. 1994. Fire and weather disturbances in terrestrial ecosystems of the eastern Cascades. U.S. Forest Service, Pacific Northwest Research Station. General Technical Report PNW-GTR-320.

3. ———, and L. Smith. 1984. Subalpine tree establishment after fire in the Olympic Mountains, Washington. Ecology 65:810-819.

4. Ahlenslager, K. E. 1987. *Pinus albicaulis*. *In* W.C. Fischer, compiler. The Fire Effects Information System (Data base). Missoula, Montana. U.S. Forest service, Intermountain Research Station, Intermountain Fire Sciences Laboratory. http://www.fs.fed.us/database/feis/plants/tree/pinalb.

4a. Airola, T. M., and K. Buchholz. 1984. Species structure and soil characteristics of five urban sites along the New Jersey Palisades. Urban Ecology 8: 149-164.

5. Akins, G. J., and C. A. Jefferson. 1973. Coastal wetlands of Oregon. Oregon Conservation and Development Commission, Portland, OR.

6. Albright, R., R. Hirschi, R. Vanbianchi, and C. Vita. 1980. Pages 449-887 *in* Coastal zone atlas of Washington, land cover/land use narratives, Volume 2. Washington State Department of Ecology, Olympia, WA.

7. Aldrich, F.T. 1972. A chorological analysis of the grass balds in the Oregon Coast Range. Ph.D. Dissertation. Oregon State University, Corvallis, OR.

8. Alpert, P. 1984. Inventory and analysis of Oregon coastal dunes. Unpublished manuscript prepared for the Oregon Natural Heritage Program, Portland, OR.

9. American Forests. 1998. Study documents dramatic tree loss in Puget Sound area. American Forest Press Release July 14, 1998.

10. Anderson, M., P. Bourgeron, M. T. Bryer, R. Crawford, L. Engelking, D. Faber-Langendoen, M. Gallyoun, K. Goodin, D. H. Grossman, S. Landaal, K. Metzler, K. D. Patterson, M. Pyne, M. Reid, L. Sneddon, and A. S. Weakley. 1998. International classification of ecological communities: terrestrial vegetation of the United States. Volume II. The National Vegetation Classification System: list of types. The Nature Conservancy, Arlington, VA.

11. Arno, S. F. 1970. Ecology of alpine larch (*Larix lyallii* Parl.) in the Pacific Northwest. Ph.D. Dissertation. University of Montana, Missoula.

12. Associated Press. 1991. Census: cities takeover U.S., *Statesman Journal*, December 18, 1991.

13. Atzet, T., and L. A. McCrimmon. 1990. Preliminary plant associations of the southern Oregon Cascade Mountain Province. U.S. Forest Service, PNW Region, Siskiyou National Forest, Grants Pass, OR.

14. ———, and D. L. Wheeler. 1982. Historical and ecological perspectives on fire activity in the Klamath Geological Province of the Rogue River and Siskiyou National Forests. U.S. Forest Service, Pacific Northwest Region, Portland, OR.

15. ———, and ———. 1984. Preliminary plant associations of the Siskiyou Mountains Province, Siskiyou National Forest. U.S. Forest Service, Pacific Northwest Region, Portland, OR.

16. ———, ———, G. Riegel, and others. 1984. The mountain hemlock and Shasta red fir series of the Siskiyou Region of southwest Oregon. FIR Report 6(1): 4-7.

17. ———, D. E. White, L. A. McCrimmon, P. A. Martinez, P. R. Fong, and V. D. Randall. 1996. Field guide to the forested plant associations of southwestern Oregon. U.S. Forest Service, Pacific Northwest Research Paper R6-NR-ECOL-TP-17-96.

18. Bakun, A. 1973. Coastal upwelling indices, west coast of North America, 1946-71. U.S. Department of Commerce, National Oceanic and Atmospheric Administration, National Marine Fisheries Service.

19. Barber, W. H., Jr. 1976. An autecological study of salmonberry (*Rubus spectabilis*, Pursh) in western Washington. M.S. Thesis. University of Washington, Seattle, WA.

20. Barbour, M. G., and W. D. Billings, editors. 1988. North American terrestrial vegetation. Cambridge University Press, New York, NY.

21. Barnes, C. A., A. C. Duxbury, and B. A. Morse. 1972. Circulation and selected properties of the Columbia River effluent at sea. Pages 41-80 *in* A. T. Pruter and D. L. Alverson, editors. The Columbia River Estuary and adjacent ocean waters, bioenvironmental studies. University of Washington Press, Seattle, WA.

22. Barrett, S. W., S. F. Arno, and J. P. Menakis. 1997. Fire episodes in the inland Northwest (1540-1940) based on fire history data. U.S. Forest Service, Intermountain Research Station. General Technical Report INT-GTR-370.

23. Bastasch, R. 1998. Waters of Oregon. A source book on Oregon's water and water management. Oregon State University Press, Corvallis, OR.

23b. Beisiinger, S. R. and D. R. Osborne. 1982. Effects of urbanization on avian community organization. Condor 84: 75-83.

24. Beschta, R. L., R. E. Bilby, G. W. Brown, L. B. Holtby, and T. J. D. Hofstra. 1987. Pages 191-232 *in* E. O. Salo and T. W. Cundy, editors. Streamside management: forestry and fishery interactions. College of Forest Resources, University of Washington, Seattle, WA.

25. Bigley, R., and S. Hull. 1992. Siouxan guide to site interpretation and forest management. Washington Department of Natural Resources, Olympia, WA.

26. ———, and ———. 1995. Draft guide to plant associations on the Olympic Experimental Forest. Washington Department of Natural Resources, Olympia WA.

27. Bilby, R. E., and J. W. Ward. 1991. Large woody debris characteristics and function in streams draining old growth, clear-cut, and second-growth forests in southwestern Washington. Canadian Journal of Fisheries and Aquatic Sciences 48:2499-2508.

28. Black, A. E., J. M. Scott, E. Strand, R. G. Wright, P. Morgan, and C. Watson. 1998. Biodiversity and land-use history of the Palouse Region: pre-European to present. Chapter 10 *in* Perspectives on the land use history of North America: a context for understanding our changing environment. USDI/USGS. Biological Resources Division, Biological Science Report USGS/BRD-1998-003.

29. Blackburn, W. H., P. T. Tueller, and R. E. Eckert Jr. 1969. Vegetation and soils of the Coils Creek Watershed. Nevada Agricultural Experiment Station Bulletin R-48. Reno, NV.

30. ———, ———, and ———. 1969. Vegetation and soils of the Cow Creek Watershed. Nevada Agricultural Experiment Station Bulletin R-49. Reno, NV.

30a. Blair, R. B. 1996. Land use and avian species diversity along an urban gradient. Ecological Applications 6: 506-519.

31. Bottom, D. K., K. K. Jones, J. D. Rodgers, and R. F. Brown. 1989. Management of living marine resources: a research plan for the Washington and Oregon continental margin. National Coastal Resources Research and Development Institute, Publication No. NCRI-T-89-004.

32. ———, ———, ———, and ———. 1993. Research and management in the Northern California Current ecosystem. Pages 259-271 *in* K. Sherman, L. M. Alexander, and B. D. Gold, editors. Large marine ecosystems: stress, mitigation, and sustainability. AAAS Press, Washington DC

33. ———, J. A. Lichatowich, and C. A. Frissell. 1998. Variability of Pacific Northwest marine ecosystems and relation to salmon production. Pages 181-252 *in* B. R. McMurray and R. J. Bailey, editors. Change in Pacific coastal ecosystems. National Oceanic and Atmospheric Administration Coastal Ocean Program Decision Analysis Series No. 11. NOAA Coastal Ocean Office, Silver Spring, MD.

34. Brady, R. F., T. Tobius, P. F. J. Eagles, R. Ohrner, J. Micak, B. Veale, and R. S. Dorney. 1979. A typology for the urban ecosystem and its relationship to large biogeographical landscape units. Urban Ecology. 4:11-28.

35. Broadhurst, G. 1998. Puget Sound nearshore habitat regulatory perspective: a review of issues and obstacles. Puget Sound Water Quality Action Team. Olympia, WA.

36. Brockway, D. G., C. Topik, M. A. Hemstrom, and W. H. Emmingham. 1983. Plant association and management guide for the Pacific silver fir zone, Gifford Pinchot National Forest. U.S. Forest Service. R6-Ecol-130a.

37. Cassidy, K. M. 1997. Land cover of Washington state: description and management. Volume 1 *in* K. M. Cassidy, C. E. Grue, M. R. Smith, and K. M. Dvornich, editors. Washington State Gap Analysis Project Final Report. Washington Cooperative Fish and Wildlife Research Unit, University of Washington, Seattle, WA.

38. Chappell, C. B. 1991. Fire ecology and seedling establishment in Shasta red fir forests of Crater Lake National Park, Oregon. M.S. Thesis. University of Washington, Seattle, WA.

39. ———, and J. K. Agee. 1996. Fire severity and tree seedling establishment in *Abies magnifica* forests, southern Cascades, Oregon. Ecological Applications 6:628-640.

40. ———, R. Bigley, R. Crawford, and D. F. Giglio. In prep. Field guide to terrestrial plant associations of the Puget Lowland, Washington. Washington Natural Heritage Program, Department of Natural Resources, Olympia, WA.

41. ———, and R. C. Crawford. 1997. Native vegetation of the South Puget Sound prairie landscape. Pages 107-122 in P. Dunn and K. Ewing, editors. Ecology and conservation of the South Puget Sound prairie landscape. The Nature Conservancy of Washington, Seattle, WA.

42. Christy, J. A., J. S. Kagan, and A. M. Wiedemann. 1998. Plant associations of the Oregon Dunes National Recreation Area, Siuslaw National Forest, Oregon. Technical paper R6-NR-ECOL-TP-09-98. U.S. Forest Service, Pacific Northwest Region, Portland, OR.

43. ———, and J. A. Putera. 1993. Lower Columbia River natural area inventory, 1992. Unpublished report to the Washington Field Office of The Nature Conservancy, Seattle, Washington. Oregon Natural Heritage Program, Portland, OR.

44. ———, and J. H. Titus. 1996. Draft, wetland plant communities of Oregon. Unpublished manuscript, Oregon Natural Heritage Program, Portland, OR.

45. Clausnitzer, R. R., and B. A. Zamora. 1987. Forest habitat types of the Colville Indian Reservation. Unpublished report prepared for the Department of Forest and Range Management, Washington State University, Pullman, WA.

46. Clemens, J., C. Bradley, and O. L. Gilbert. 1984. Early development of vegetation on urban demolition sites in Sheffield, England. Urban Ecology. 8:139-148.

47. Cochran, P. H. 1985. Soils and productivity of lodgepole pine. in D. M. Baumgartner, R. G. Krebill, J. T. Arnott, and G. F. Gordon, editors. Lodgepole pine: the species and its management: symposium proceedings, Washington State University, Cooperative Extension, Pullman, WA.

48. Cole, D. N. 1977. Man's impact on wilderness vegetation: an example from Eagle Cap Wilderness, NE Oregon. Ph.D. Dissertation. University of Oregon, Eugene, OR.

49. ———. 1982. Vegetation of two drainages in Eagle Cap Wilderness, Wallowa Mountains, Oregon. U.S. Forest Service Research Paper INT-288.

50. Conard, S. G., A. E. Jaramillo, K. Cromack, Jr., and S. Rose, compilers. 1985. The role of the genus *Ceanothus* in western forest ecosystems. General Technical Report PNW-182. U.S. Forest Service, Pacific Northwest Forest and Range Experiment Station, Portland, OR.

51. ———, and S. R. Radosevich. 1981. Photosynthesis, xylem pressure potential, and leaf conductance of three montane chaparral species in California. Forest Science 27(4):627-639.

52. Copeland, W. N. 1979. Harney Lake RNA Guidebook, Supplement No. 9. U.S. Forest Service Experiment Station, Portland, OR.

53. Cowardin, L. M., V. Carter, F. C. Golet, and E. T. LaRoe. 1979. Classification of wetlands and deepwater habitats of the United States. U.S. Fish and Wildlife Service. FWS/OBS-79.31.

54. Crawford, R. C., and H. Hall. 1997. Changes in the South Puget Sound prairie landscape. Pages 11-15 in P. Dunn and K. Ewing, editors. Ecology and conservation of the South Puget Sound prairie landscape. The Nature Conservancy of Washington, Seattle, WA.

55. Crook, C. S. 1979. An introduction to beach and dune physical and biological processes. In K. B. Fitzpatrick, editor. Articles of the Oregon Coastal Zone Management Association, Inc., Newport, OR.

56. ———. 1979. A system of classifying and identifying Oregon's coastal beaches and dunes. In K. B. Fitzpatrick, editor. Articles of the Oregon Coastal Zone Management Association, Inc., Newport, OR.

57. Crowe, E. A., and R. R. Clausnitzer. 1997. Mid-montane wetland plant associations of the Malheur, Umatilla and Wallowa-Whitman National Forests. U.S. PNW Technical Paper, R6-NR-ECOL-TP-22-97.

58. Csuti, B., A. J. Kimerling, T. A. O'Neil, M. M. Shaughnessy, E. P. Gaines, and M. M. P. Huso. 1997. Atlas of Oregon wildlife. Oregon State University Press, Corvallis, OR.

59. Daniels, J. D. 1969. Variation and integration in the grand fir-white fir complex. Ph.D. Dissertation, University of Idaho, Moscow.

60. Daubenmire, R. F. 1970. Steppe vegetation of Washington. Washington State University Agricultural Experiment Station Technical Bulletin No. 62.

61. ———. 1981. Subalpine parks associated with snow transfer in the mountains of Idaho and eastern Washington. Northwest Science 55(2):124-135.

62. ———, and J. B. Daubenmire. 1968. Forest vegetation of eastern Washington and northern Idaho. Technical Bulletin 60. Washington Agricultural Experiment Station, College of Agriculture, Washington State University, Pullman, WA.

63. Davidson, E. D. 1967. Synecological features of a natural headland prairie on the Oregon coast. M.S. Thesis. Oregon State University, Corvallis, OR.

64. Dealy, J. E. 1971. Habitat characteristics of the Silver Lake mule deer range. U.S. Forest Service Research Paper PNW-125.

65. del Moral, R. 1979. High elevation vegetation of the Enchantment Lakes Basin, Washington. Canadian Journal of Botany 57(10):1111-1130.

66. ———, and J. N. Long. 1977. Classification of montane forest community types in the Cedar River drainage of western Washington, U.S.A. Canadian Journal of Forest Research 7(2):217-225.

67. Dethier, M. N. 1988. A survey of intertidal communities of the Pacific coastal area of Olympic National Park, Washington. Prepared for the National Park Service and cooperating agencies.

68. ———. 1990. A marine and estuarine habitat classification system for Washington State. Washington Natural Heritage Program, Department of Natural Resources, Olympia, WA.

69. Detling, L. E. 1961. The chaparral formation of southwestern Oregon, with considerations of its postglacial history. Ecology 42:348-357.

70. Detwyler, T. R. 1972. Urbanization and environment. Duxbury Press, Belmont, CA.

71. Diaz, N. M., and T. K. Mellen. 1996. Riparian ecological types, Gifford Pinchot and Mt. Hood National Forests, Columbia River Gorge National Scenic Area. U.S. Forest Service, Pacific Northwest Region, R6-NR-TP-10-96.

72. Dickman, A., and S. Cook. 1989. Fire and fungus in a mountain hemlock forest. Canadian Journal of Botany 67(7):2005-2016.

73. Dodimead, A. J., F. Favorite, and T. Hirano. 1963. Salmon of the North Pacific Ocean— Part II. Review of oceanography of the subarctic Pacific region. International Commission Bulletin No. 13.

74. Douglas, G. W. 1970. A vegetation study in the subalpine zone of the western North Cascades, Washington. M.S. Thesis, University of Washington, Seattle, WA.

75. ———, and L. C. Bliss. 1977. Alpine and high subalpine plant communities of the North Cascades Range, Washington and British Columbia. Ecological Monographs 47:113-150.

76. Downing, J. P. 1983. The coast of Puget Sound: its process and development. Washington Sea Grant Publication, University of Washington. Seattle, WA.

77. Druehl, L. D. 1969. The northeast Pacific rim distribution of the Laminariales. Proceedings of the International Seaweed Symposium 6:161-170.

78. Dunn, P. V., and K. Ewing, editors. 1997. Ecology and conservation of the South Puget Sound Prairie Landscape. The Nature Conservancy, Seattle, WA.

79. Eddleman, L. E. 1984. Ecological studies on western juniper in central Oregon. In Proceedings western juniper management short course, 1984 October 15-16. Oregon State University, Extension Service and Department of Rangeland Resources, Corvallis, OR.

80. Edwards, O. M. 1980. The alpine vegetation of Mount Rainier National Park: structure, development, and constraints. Ph.D. Dissertation. University of Washington, Seattle, WA.

80a. Emlen, J. T. 1974. An urban bird community of Tucson, Arizona: derivation, structure, regulation. The Condor 76: 184-197.

81. Everest, F. H. 1987. Salmonids of western forested watersheds. Pages 3-38 in E. O. Salo and T. W. Cundy, editors. Streamside management: forestry and fishery interactions. College of Forest Resources, University of Washington, Seattle, WA.

82. ———, R. L. Beschta, J. C. Scrivener, K. V. Koski, J. R. Sedell, and C. J. Cederholm. 1987. Fine sediments and salmonid production: a paradox. Pages 98-142 in E. O. Salo and T. W. Cundy, editors. Streamside management: forestry and fishery interactions. College of Forest Resources, University of Washington, Seattle.

82a. Falk, J. H. 1976. Energetics of a suburban lawn ecosystem. Ecology 57: 141-150.

83. Favorite, F., A. J. Dodimead, and K. Nasu. 1976. Oceaonography of the subarctic Pacific region, 1960-71. International North Pacific Fisheries Commission Bulletin No. 33.

84. Florence, M. 1987. Plant succession on prescribed burn sites in chamise chaparral. Rangelands 9(3):119-122.

85. Fonda, R. W. 1974. Forest succession in relation to river terrace development in Olympic National Park, Washington. Ecology 55:927-942.

86. ———, and J. A. Bernardi. 1976. Vegetation of Sucia Island in Puget Sound, Washington. Bulletin of the Torrey Botanical Club 103(3):99-109.

87. Franklin, J. F. 1988. Pacific Northwest forests. Pages 104-130 in M. G. Barbour and W. D. Billings, editors. North American terrestrial vegetation. Cambridge University Press, New York, NY.

88. ———, and C. T. Dyrness. 1973. Natural vegetation of Oregon and Washington. U.S. Pacific Northwest Forest and Range Experiment Station, General Technical Report. PNW-8, Portland, OR.

89. ———, K. Cromack, Jr., W. Denison, A. McKee, C. Maser, J. Sedell, F. Swanson, and G. Juday. 1981. Ecological characteristics of old-growth Douglas-fir forests. U.S. Forest Service, Pacific Northwest Forest and Range Experiment Station. General Technical Report PNW-118. Portland, OR.

90. ———, W. H. Moir, M. A. Hemstrom, S. E. Greene, and B. G. Smith. 1988. The forest communities of Mount Rainier National Park. U.S. National Park Service, Scientific Monograph Series 19, Washington, D.C.

91. Frenkel, R. E., and E. F. Hieinitz. 1987. Composition and structure of Oregon ash (Fraxinus latifolia) forest in William L. Finley National Wildlife Refuge, Oregon. Northwest Science 61:203-212.

92. Frey, D. G., editor. 1966. Limnology in North America. The University of Wisconsin Press, Madison, WI.

93. Furniss, M. J., T. D. Roeloggs, and C. S. Yee. 1991. Road construction and maintenance. Pages 297-323 in W. R. Meehan, editor. Influences of forest and rangeland management on salmonid fishes and their habitats. American Fisheries Society Special Publication No. 19, Bethesda, MD.

94. Ganskopp, D. C. 1979. Plant communities and habitat types of the Meadow Creek Experimental Watershed. M.S. Thesis. Oregon State University, Corvallis, OR.

95. Gaumer, T. F., S. L. Benson, L. W. Brewer, L. Osis, D. G. Skeesick, R. M. Starr, and J. F. Watson. 1985. Estuaries. In E. R. Brown, editor. Management of wildlife and fish habitats in forests of western Oregon and Washington. U.S. Forest Service, Pacific Northwest Region, Portland, OR.

96. Gerard, P. W. 1995. Agricultural practices, farm policy, and the conservation of biological diversity. USDI, National Biological Service, Biological Science Report 4.

97. Gordon, D. T. 1970. Natural regeneration of white and red fir: influence of several factors. U.S. Forest Service, Research Paper PSW-90.

97a. Green, R. J. 1984. Native and exotic birds in a suburban habitat. Australian Wildlife Research 11:181-190.

98. Greenlee, J. M., and J. H. Langenheim. 1990. Historic fire regimes and their relation to vegetation patterns in the Monterey Bay area of California. American Midland Naturalist 124(2):239-253.

99. Habeck, J. R. 1961. Original vegetation of the mid-Willamette Valley, Oregon. Northwest Science 35:65-77.

100. Haeussler, S., and D. Coates. 1986. Autecological characteristics of selected species that compete with conifers in British Columbia: a literature review. Land Management Report No. 33. Ministry of Forests, Information Services Branch, Victoria, BC, Canada.

101. Hall, F. C. 1973. Plant communities of the Blue Mountains in eastern Oregon and southeastern Washington. U.S. Forest Service, R-6 Area Guide 3-1.

102. Halpern, C. B. 1989. Early successional patterns of forest species: interactions of life history traits and disturbance. Ecology 70:704-720.

103. Halverson, N. M., and W. H. Emmingham. 1982. Reforestation in the Cascades Pacific silver fir zone: a survey of sites and management experiences on the Gifford Pinchot, Mt. Hood and Willamette National Forests. U.S. Forest Service. R6-ECOL-091-1982.

104. ———, C. Topik, and R. van Vickle. 1986. Plant associations and management guide for the western hemlock zone, Mt. Hood National Forest. U.S. Forest Service, R6-ECOL-232A-1986.

105. Hamann, M. J. 1972. Vegetation of alpine and subalpine meadows of Mount Rainier National Park, Washington. M.S. Thesis. Washington State University, Pullman.

106. Harper, J. R., D. E. Howes, and P. D. Reimer. 1991. Shore-zone mapping system for use in sensitivity mapping and shoreline countermeasures. Proceedings of the 14th Arctic and Marine Oil spill Program (AMOP), Environment Canada.

107. Harr, R. D., and B. A. Coffin. 1992. Influence of timber harvest on rain-on-snow runoff: a mechanism for cumulative watershed effects. Pages 455-469 in M.. E. Jones and A. Laemon, editors. Interdisciplinary approaches in hydrology and hydrogeology. American Institute of Hydrology. Minneapolis, MN.

108. Hemstrom, M. A., W. H. Emmingham, N. M. Halverson, S. E. Logan, and C. Topik. 1982. Plant association and management guide for the Pacific silver fir zone, Mt. Hood and Willamette National Forests. U.S. Forest Service R6-Ecol 100-1982a.

109. ———, and J. F. Franklin. 1982. Fire and other disturbances of the forests in Mount Rainier National Park. Quaternary Research 18:32-51.

110. ———, and S.E. Logan. 1986. Plant association and management guide, Siuslaw National Forest. U.S. Forest Service Report R6-Ecol 220-1986a. Portland, OR.

111. ———, ———, and W. Pavlat. 1987. Plant association and management guide, Willamette National Forest. U.S. Forest Service. R6-ECOL 257-TP-86.

112. Henderson, J. A. 1973. Composition, distribution, and succession of subalpine meadows in Mount Rainier National Park, Washington. Ph.D. Dissertation. Oregon State University, Corvallis, OR.

113. ———. 1978. Plant succession on the Alnus rubra/Rubus spectabilis habitat type in western Oregon. Northwest Science 52(3):156-167.

114. ———, D. A. Peter, and R. Lesher. 1992. Field guide to the Forested Plant Associations of the Mt. Baker-Snoqualmie National Forest. U.S. Forest Service Technical Paper R6-ECOL 028-91.

115. ———, ———, ———, and D.C. Shaw. 1989. Forested Plant Associations of the Olympic National Forest. U.S. Forest Service Publication R6-ECOL-TP 001-88.

116. Hironaka, M., M. A. Fosberg, and A. H. Winward. 1983. Sagebrush-grass habitat types of southern Idaho. Forestry, Wildlife, and Range Experiment Station Bulletin No. 15, University of Idaho, Moscow.

116a. Hobbs, E. 1988. Using ordination to analyze the composition and structure of urban forest islands. Forest Ecology and Management 23: 139-158.

117. Hopkins, W. E. 1979. Plant associations of the Fremont National Forest. U.S. Forest Service Publication R6-ECOL-79-004.

118. ———. 1979. Plant associations of South Chiloquin and Klamath Ranger Districts— Winema National Forest. U.S. Forest Service Publication R6-ECOL-79-005.

119. Howard, J. L. 1996. Populus tremuloides. In D. G. Simmerman, compiler. The Fire Effects Information System [Data base]. U.S. Forest Service, Intermountain Research Station, Intermountain Fire Sciences Laboratory. Missoula, MT. http://www.fs.fed.us/database/feis/plants/tree/poptre.

119a. Ingold, D. J. 1996. Delayed nesting decreased reproductive success in northern flickers: implications for competition with European starlings. Journal of Field Ornithology 67: 321-326.

119b. Ingold, D. J. and R. J. Densmore. 1992. Competition between European starlings and native woodpeckers for nest cavities in Ohio. Sialia 14: 43-48.

120. Jefferson, C.A. 1975. Plant communities and succession in Oregon coastal salt marshes. Ph.D. Dissertation. Oregon State University, Corvallis, OR.

121. John, T., and D. Tart. 1986. Forested plant associations of the Yakima Drainage within the Yakima Indian Reservation. Review copy prepared for the Yakima Indian Nation—Bureau of Indian Affairs Soil Conservation Service.

122. Johnson, C. G., and R. R. Clausnitzer. 1992. Plant associations of the Blue and Ochoco mountains. U.S. Forest Service, Pacific Northwest Region, Wallowa-Whitman National Forest R6-ERW-TP-036-92.

123. ———, and S.A. Simon. 1987. Plant associations of the Wallowa-Snake Province. U.S. Forest Service R6-ECOL-TP-255A-86.

124. Keeley, J. E. 1975. Longevity of nonsprouting Ceanothus. American Midland Naturalist 93(2):504-507.

125. ———, and S. C. Keeley. 1988. Chaparral. Pages 165-208 in M. G. Barbour and W. D. Billings, editors. North American terrestrial vegetation. Cambridge University Press, New York, NY.

126. Kiilsgaard, C. 1999. Oregon vegetation: mapping and classification of landscape level cover types. Final Report. U.S. Geological Survey-Biological Resources Division: Gap Analysis Program. Moscow, ID

127. ———, and C. Barrett. 1998. Oregon vegetation landscape-level cover types 127. Northwest Habitat Institute, Corvallis, OR.

128. Kilgore, B. M. 1973. The ecological role of fire in Sierran conifer forests—its application to National Park management. Quaternary Research 3:496-513.

129. King County Park, Planning and Resource Department. 1987. Wildlife habitat profile—King County Open Space Program, Seattle, WA.

130. Knutson, K. L., and V. L. Naef. 1997. Priority habitat management recommendations: riparian. Washington Department of Fish and Wildlife, Olympia, WA.

131. Kovalchik, B. L. 1987. Riparian zone associations—Deschutes, Ochoco, Fremont, and Winema national forests. U.S. Forest Service R6 ECOL TP-279-87.

132. ———. 1993. Riparian plant associations of the National Forests of eastern Washington. A partial draft version 1. U.S. Forest Service, Colville National Forest.

133. Kozloff, E. N. 1973. Seashore life of Puget Sound, the Strait of Georgia, and the San Juan Archipelago. University of Washington Press, Seattle, WA.

134. Krajina, V. J. 1965. Bioclimatic zones and classification of British Columbia. Pages 1-17 in V. J. Krajina, editor. Ecology of western North America. Volume 1. University of British Columbia, Vancouver, British Columbia, Canada.

135. Kruckeberg, A. R. 1996. Gardening with native plants of the Pacific Northwest: an illustrated guide. University of Washington Press, Seattle. ISBN 0-295-97476-1.

136. Kuchler, A. W. 1964. Manual to accompany the map: potential natural vegetation of the conterminous United States. Special Publication. 36, American Geographic Society, New York, NY.

137. Kumler, M. L. 1969. Plant succession on the sand dunes of the Oregon coast. Ecology 50(4):695-704.

138. Kunze, L. M. 1994. Preliminary classification of native, low elevation, freshwater wetland vegetation in western Washington. Washington Natural Heritage Program, Department of Natural Resources, Olympia, Washington.

139. Kuramoto, R.T., and L. C. Bliss. 1970. Ecology of subalpine meadows in the Olympic Mountains, Washington. Ecological Monograph 40:317-347.

140. Laacke, R .J., and J. N. Fiske. 1983. Red fir and white fir. Pages 41-43 in R. M. Burns, compiler. Silvicultural systems for the major forest types of the United States. U.S. Forest Service Agriculture Handbook No. 44. Washington, D.C.

141. Landry, M. R., and B. M. Hickey, editors. 1989. Coastal oceanography of Washington and Oregon. Elsevier Science Publishing Company, New York, NY.

142. Lang, F.A. 1961. A study of vegetation change on the gravelly prairies of Pierce and Thurston counties, western Washington. M.S. Thesis. University of Washington, Seattle, WA.

143. Levings, C. D., and R. M. Thom. 1994. Habitat changes in Georgia Basin: implications for resource management and restoration. Pages 330-351 in R. C. H. Wilson, R. J. Beamish, F. Aitkins, and J. Bell, editors. Review of the marine environment and biota of Strait of Georgia, Puget Sound and Juan de Fuca Strait. Canadian Technical Report of Fisheries and Aquatic Sciences. No. 1948.

144. Lillybridge, T. R., B. L. Kovalchik, C. K. Williams, and B. G. Smith. 1995. Field guide for forested plant association of the Wenatchee National Forest. U.S. Forest Service General Technical Report PNW-GTR-359, Portland, OR.

145. Little, C., and J. A. Kitching. 1996. The biology of rocky shores. Oxford University Press, New York, NY.

146. Mac, M. J., P.A. Opler, C. E. Puckett Haecker, and P. D. Doran. 1998. Status and trends of the nation's biological resources. Volume 1. U.S. Department of the Interior, U.S. Geological Survey, Reston, VI.

147. Manning, M. E., and W. G. Padgett. 1992. Riparian community type classification for the Humboldt and Toiyabe national forests, Nevada and eastern California. Unpublished Draft Report prepared for U.S. Forest Service, Intermountain Region Ecology and Classification Program, Ogden, UT.

148. Marsh, F., R. Helliwell, and J. Rodgers. 1987. Plant association guide for the commercial forest of the Warm Springs Indian Reservation. Confederated Tribes of the Warm Springs Indians, Warm Springs, OR.

148a. Marzluff, J. M. 1997. Effects of urbanization and recreation on songbirds. Pages 89-102 in W. M. Block and D. M. Finch, editors. Songbird ecology in southwestern ponderosa pine forests: a literature review. U.S. Forest Service General Technical Report RM-292, Fort Collins, CO.

149. Marzluff, J. M., F. R. Gehlbach, and D. A. Manuwal. 1998. Urban environments: influences on avifauna and challenges for the avian conservationist. Pages 283-299 in J. M. Marzluff and R. Sallabanks, editors. Avian conservation, research, and management. Island Press, Washington D.C.

150. Mayer, K. E., and W. F. Laudenslayer, Jr., editors. 1988. A guide to wildlife habitats of California. State of California, the Resources Agency, Department of Fish and Game, Wildlife Management Division, CWHR Program, Sacramento, CA.

151. McBride, J. R., and C. Reid. 1988. Urban. Pages 142-144 in K. E. Mayer and W. F. Laudenslayer, Jr., editors. A guide to wildlife habitats of California. California Department of Forestry and Fire Protection, Sacramento, CA.

152. McDonald, P. M., and J.C. Tappeiner, II. 1987. Silviculture, ecology, and management of tanoak in northern California. Pages 64-70 in T. R. Plumb and N. H. Pillsbury, technical coordinators. Proceedings of the symposium on multiple-use management of California's hardwood resources; 12-14 November 1986; San Luis Obispo, California. U.S. Forest Service General Technical Report PSW-100.

153. McKenzie, D. F., and T. Z. Riley, editors. 1995. How much is enough? A regional wildlife habitat needs assessment for the 1995 Farm Bill. Wildlife Management Institute, Washington, D.C.

154. McNeil, R. C., and D. B. Zobel. 1980. Vegetation and fire history of a ponderosa pine-white fir forest in Crater Lake National Park. Northwest Science 54(1):30-46.

155. Merriam, C. H. 1898. Life zones and crop zones of the United States. U.S. Department of Agriculture, Division of Biological Survey, Bulletin 10.

156. Miller, T. B. 1976. Ecology of riparian communities dominated by white alder in western Idaho. M.S. Thesis. University of Idaho, Moscow.

157. Minnich, R.A. 1983. Fire mosaics in southern California and north Baja California. Science 219:1287-1294.

158. Mitchell, R., and W. Moir. 1976. Vegetation of the Abbott Creek Research Natural Area, Oregon. Northwest Science 50:42-57.

159. Morgan, P., S. C. Bunting, A. E. Black, T. Merrill, and S. Barrett. 1996. Fire regimes in the interior Columbia River Basin: past and present. Final Report RJVA-INT-94913. U.S. Forest Service, Intermountain Research Station, Intermountain Fire Sciences Lab, Missoula, MT.

160. Morrison, P., and F. J. Swanson. 1990. Fire history and pattern in a Cascade Range landscape. U.S. Forest Service General Technical Report PNW-GTR-254.

161. Mueggler, W. F. 1988. Aspen community types of the Intermountain Region. U.S. Forest Service, General Technical Report INT-250. Intermountain Research Station, Ogden, UT.

162. Naiman, R. J., H. Decamps, and M. Pollock. 1993. The role of riparian corridors in maintaining regional biodiversity. Ecological Applications 3:209-212.

163. National Oceanic and Atmospheric Administration. 1993. Olympic Coast National Marine Sanctuary, Final Environmental Impact Statement/Management Plan, November 1993 . NOAA, Sanctuaries and Reservoirs Division, Washington D.C.

164. National Research Council. 1989. Alternative agriculture. National Academy Press, Washington, D.C.

165. Norton, H. H. 1979. The association between anthropogenic prairies and important food plants in western Washington. Northwest Anthropological Research Notes 13:199-219.

166. Noss, R. F., E. T. LaRoe, and J. M. Scott. 1995. Endangered ecosystems of the United States: a preliminary assessment of loss and degradation. U.S. National Biological Service, Biological Report 28.

167. Nowak, D. J. 1994. Understanding of the structure of urban forests. Journal of Forestry October: 42-46.

168. Oliver, C. D. 1981. Forest development in North America following major disturbances. Forest Ecology and Management 3:153-168.

169. Oregon Department of Forestry. 1994. Water protection rules: purpose, goals, classification, and riparian management. OAR No.629-635-200-Water classification. Oregon Department of Forestry, Salem, OR.

170. Oregon State University. 1971. Oceanography of the nearshore coastal waters of the Pacific Northwest relating to possible pollution. Volume 1. Corvallis, OR.

171. Parsons, D. J., and S. H. DeBenedetti. 1979. Impact of fire suppression on a mixed-conifer forest. Forest Ecology and Management 2:21-33.

172. Pettinger, A. 1996. Native plants in the coastal garden: a guide for gardeners in British Columbia and the Pacific Northwest. Whitecap Books 1-55110-405-9. Vancouver, BC.

173. Phillips, R. C. 1984. The ecology of eelgrass meadows in the Pacific Northwest: a community profile. U.S. Fish and Wildlife Service, FWS/OBS-84/24.

174. Phinney, L. A., and P. Bucknell. 1975. A catalog of Washington streams and salmon utilization. Washington Department of Fisheries. Volume 2: coastal region.

175. Poulton, C. E. 1955. Ecology of the non-forested vegetation in Umatilla and Morrow counties, Oregon. Ph.D. Dissertation. State College of Washington, Pullman, WA.

176. Proctor, C. M., J. C. Garcia, D. V. Galvin, G. B. Lewis, L. C. Loehr, and A. M. Massa. 1980. An ecological characterization of the Pacific Northwest coastal region. Volume 2. U.S. Fish and Wildlife Service, Biological Services Program. FWS/OBS-79/14.

177. ———, ———, ———, ———, ———, and ———. 1980. An ecological characterization of the Pacific Northwest coastal region. Volume 3. U.S. Fish and Wildlife Service, Biological Services Program. FWS/OBS-79/14.

178. ———, ———, ———, ———, ———, and ———. 1980. An ecological characterization of the Pacific Northwest coastal region. Volume 4. U.S. Fish and Wildlife Service, Biological Services Program. FWS/OBS-79/14.

179. Pruter, A. T., and D. L. Alverson, editors. 1972. The Columbia River estuary and adjacent waters: bioenvironmental studies. University of Washington Press, Seattle.

180. Puget Sound Water Quality Authority. 1997. 1997 Puget Sound update. Seventh annual report of the Puget Sound Ambient Monitoring Program. Puget Sound Water Quality Authority, Olympia, Washington.

181. Quigley, T. M., and S. J. Arbelbide, technical editors. 1997. An assessment of ecosystem components in the interior Columbia basin and portions of the Klamath and Great Basins. Volume 2. U.S. Forest Service General Technical Report PNW-GTR-405.

182. Quinn, T. 1997. Coyote (Canis latrans) food habits in three urban habitat types of western Washington. Northwest Science 71(1):1-5.

183. Ripley, J. D. 1983. Description of the plant communities and succession of the Oregon coast grasslands. M.S. Thesis. Oregon State University, Corvallis, OR.

184. Roberts, K., L. Bischoff, K. Brodersen, G. Green, D. Gritten, S. Hamilton, J. Kierstead, M. Benham, E. Perkins, T. Pogson, S. Reed, and D.E. Kerley. 1976. A preliminary ecology survey of the Alvord Basin, Oregon. Unpublished, Final Technical Report, Eastern Oregon State College, La Grande. NSF Grant 76-08175.

185. Rowntree, R. A. 1986. Ecology of the urban forest—introduction to part II. Urban Ecology 9(3/4):229-243.

185a. Rudnicky, J. L., and M. J. McDonnell. 1989. Forty-eight years of canopy change in a hardwood-hemlock forest in New York City. Bulletin of the Torrey Botanical Club 116: 52-64.

186. Ruth, R. H. 1974. Regeneration and growth of west-side mixed conifers. In O. P. Camer, editor. Environmental effects of forest residues in the Pacific Northwest: a state-of-knowledge compendium. U.S. Forest Service General Technical Report PNW-24.

187. Sampson, A. W., and B. S. Jespersen. 1963. California range brushlands and browse plants. University of California, Division of Agricultural Sciences, California Agricultural Experiment Station, Extension Service, Berkeley, CA.

188. Sawyer, J. O., and T. Keeler-Wolf. 1995. A manual of California vegetation. Native Plant Society of California, Sacramento, CA.

189. Schoch, G. C., and M. N. Dethier. 1997. Analysis of shoreline classification and biophysical data for Carr Inlet. Washington State Department of Natural Resources. Olympia, WA.

190. Shipman, H. 1997. Shoreline armoring on Puget Sound. In T. Ransom, editor. Puget Sound Notes No. 40. Puget Sound Water Quality Action Team, Olympia, WA.

191. Shreffler, D. K., R. M. Thom, and K. B. MacDonald. 1995. Shoreline armoring effects on biological resources and coastal ecology in Puget Sound. In E. Robichaud, editor. Puget Sound Research 1995: Proceedings. Puget Sound Water Quality Action Team, Olympia, WA.

192. Simenstad, C. A. 1983. The ecology of estuarine channels of the Pacific Northwest coast: a community profile. U.S. Fish and Wildlife Services. FWS/OBS-83/05.

193. Spies, T. A., J. F. Franklin, and T. B. Thomas. 1988. Coarse woody debris in Douglas-fir forests of western Oregon and Washington. Ecology 69:1689-1702.

194. Strickland, R., and D. J. Chasan. 1989. Coastal Washington, a synthesis of information. Washington State and Offshore Oil and Gas, Washington Sea Grant, University of Washington, Seattle, WA.

195. Strickler, G. S. 1961. Vegetation and soil condition changes on a subalpine grassland in eastern Oregon. U.S. Forest Service Research Paper PNW-40, Portland, OR.

196. ———, and W. B. Hall. 1980. The Standley allotment: a history of range recovery. U.S. Forest Service, Forest and Range Experiment Station Research Paper, PNW-278.

197. Sullivan, K., T. E. Lidle, C. A. Dolloff, G. E. Grant, and L. M. Reid. 1987. Stream channels: the link between forest and fishes. Pages 39-97 in E. O. Salo and T. W. Cundy, editors. Streamside management: forestry and fishery interactions. College of Forest Resources. Contribution No. 57. University of Washington, Seattle, WA.

198. Swanson, F. J., L. E. Benda, S. H. Duncan, G. E. Grant, W. F. Megaham, L. M. Reid, and R. R. Zeimer. 1987. Mass failures and other processes of sediment production in Pacific Northwest forest landscapes. Pages 9-38 in E. O. Salo and T. W. Cundy, editors. Streamside management: forestry and fisheries interactions. College of Forest Resources Contribution No. 57, University of Washington, Seattle, WA.

199. ———, and C. T. Dyrness. 1975. Impact of clearcutting and road construction on soil erosion by landslides in the western Cascade Range, Oregon. Geology 3:393-396.

200. ———, R. L. Fredriksen, and F. M. McCorison. 1982. Material transfer in a western Oregon forested watershed. Pages 223-266 in R. L. Edmonds, editor. Analysis of coniferous forest ecosystems in the western United States. Hutchinson Ross, Stroudsburg, PA.

201. The University of Oregon's Atlas of Oregon. 1976.

202. Thilenius, J. F. 1968. The Quercus garryana forests of the Willamette Valley, Oregon. Ecology 49:1124-1133.

203. Thomson, R. E. 1981. Oceanography of the British Columbia coast.

Canadian Special Publication, Fisheries and Aquatic Sciences 56:1-292.

204. Thompson, K., and D. Snow. 1974. Fish and Wildlife Resources: Oregon coastal zone. Oregon Coastal Conservation and Development Commission, Portland, OR.

205. Tiner, R. W. 1984. Wetlands of the United States: current status and recent trends. National Wetlands Inventory. U.S. Fish and Wildlife Service.

206. Tisdale, E. W. 1983. Grasslands of western North America: the Pacific Northwest bunchgrass type. Pages 223-245 in A. C. Nicholson, A. McLean and T. E. Baker, editors. Grassland ecology and classification symposium proceedings. British Columbia Ministry of Forests, Victoria, BC.

207. ———. 1986. Canyon grasslands and associated shrublands of west-central Idaho and adjacent areas. Bulletin No. 40. Forest, Wildlife and Range Experiment Station, University of Idaho, Moscow, ID.

208. Topik, C. 1989. Plant association and management guide for the Grand Fir Zone, Gifford Pinchot National Forest. U.S. Forest Service, R6-ECOL-006-88.

209. ———, N. M. Halverson, and T. High. 1988. Plant association and management guide for the Ponderosa Pine, Douglas-fir, and Grand Fir Zones, Mount Hood National Forest. U.S. Forest Service, R6-ECOL-TP-004-88.

210. ———, ———, and D. G. Brockway. 1986. Plant association and management guide for the Western Hemlock Zone, Gifford Pinchot National Forest. U.S. Forest Service, R6-ECOL-230A-1986.

211. Turner, R. B. 1969. Vegetation changes of communities containing medusahead (Taeniatherum asperum [Sim.] Nevski) following herbicide, grazing and mowing treatments. Ph.D. Dissertation. Oregon State University, Corvallis, OR.

212. Volland, L. A. 1976. Plant communities of the central Oregon pumice zone. U.S. Forest Service R-6 Area Guide 4-2. Pacific Northwest Region, Portland, OR.

212a. Walcott, C. F. 1974. Changes in bird life in Cambridge, Massachusetts from 1960 to 1964. The Auk 91: 151-160.

213. Ware, D. M., and G. A. McFarlane. 1989. Fisheries production domains in the Northeast Pacific Ocean. Pages 359-379 in R. J. Beamish and G. A. McFarlane, editors. Effects of ocean variability on recruitment and evaluation of parameters used in stock assessment models. Canadian Special Publication, Fisheries and Aquatic Sciences 108.

214. Washington Department of Ecology. 1994. Inventory of dams. Washington Department of Ecology, Water Resources Program, Dam Safety Section. Publication No. 9

215. Washington Department of Natural Resources. 1998. Our changing nature—natural resource trends in Washington State. Washington Department of Natural Resources, Olympia, WA.

216. West, J. E. 1997. Protection and restoration of marine life in the inland waters of Washington State. Puget Sound/Georgia Basin Environmental Report Series: No. 6. Puget Sound Water Quality Action Team, Olympia, WA.

217. Wetzel, R. G. 1983. Limnology. Saunders College Publishing. New York, NY.

218. Whittier, T. R., R. M. Hughes, and D. P. Larsen. 1988. Correspondence between ecoregions and spatial patterns in stream ecosystems in Oregon. Canadian Journal of Fisheries and Aquatic Sciences 45:1264-1278.

219. Wiedemann, A. M. 1966. Contributions to the plant ecology of the Oregon Coastal Sand Dunes. Ph.D. Dissertation. Oregon State University, Corvallis, OR.

220. ———. 1984. The ecology of Pacific Northwest coastal sand dunes: a community profile. U.S. Fish and Wildlife Service, FWS/OBS-84/04.

221. Williams, C. K., B. F. Kelley, B. G. Smith, and T. R. Lillybridge. 1995. Forested plant associations of the Colville National Forest. U.S. Forest Service General Technical Report PNW-GTR-360. Portland, OR.

222. ———, and T. R. Lillybridge. 1983. Forested plant associations of the Okanogan National Forest. U.S. Forest Service, R6-Ecol-132b. Portland, OR.

223. Williams, R. W., R. M. Laramie, and J. J. Ames. 1975. A catalog of Washington streams and salmon utilization. Washington Department of Fisheries. Volume 1: Puget Sound Region.

224. Winward, A. H. 1970. Taxonomic and ecological relationships of the big sagebrush complex in Idaho. Ph.D. Dissertation. University of Idaho, Moscow.

225. ———. 1980. Taxonomy and ecology of sagebrush in Oregon. Oregon State University Agricultural Experiment Station Bulletin 642:1-15.

226. Wolcott, E. E. 1973. Lakes of Washington. Water Supply. State of Washington, Department of Conservation, Bulletin No. 14. Volume 1: Western Washington. Olympia, WA.

227. ———. 1973. Lakes of Washington. Water Supply. State of Washington, Department of Conservation, Bulletin No. 14. Volume 2: Eastern Washington. Olympia, WA.

228. Zack, A. C., and P. Morgan. 1994. Early succession on hemlock habitat types in northern Idaho. Pages 71-84 in D. M. Baumgartner, J. E. Lotan, and J. R. Tonn, editors. Interior cedar-hemlock-white pine forests: ecology and management. Cooperative Extension Program, Washington State University, Seattle, WA.

229. Ziemer, R. R. 1981. Roots and the stability of forested slopes. Pages 343-361 in Proceedings of a symposium on erosion and sediment transport in Pacific Rim steeplands. Publication 132. International Association of Hydrological Scientists. Washington, D.C.

230. Zobel, D. B., L. F. Roth, and G. L. Hawk. 1985. Ecology, pathology and management of Port-Orford cedar (Chamaecyparis lawsoniana). U.S. Forest Service General Technical Report PNW-184.

3

Structural Conditions and Habitat Elements of Oregon and Washington

Thomas A. O'Neil, Kelly A. Bettinger, Madeleine Vander Heyden, Bruce G. Marcot, Charley Barrett, T. Kim Mellen, W. Matthew Vanderhaegen, David H. Johnson, Patrick J. Doran, Laurie Wunder, & Kathryn M. Boula

Introduction

Wildlife habitat is a concept that describes attributes in the environment that serve as life requisites for wildlife allowing them to survive and reproduce. An overt assumption is that wildlife use habitats that are arranged or comprised of vital components; these arrangements result in healthy and viable wildlife populations. However, the wildlife habitat concept does not address demographic (survival and reproduction of individuals), environmental (food supply, predators, weather), natural catastrophe (flood, drought, fire), or genetic (genetic drift or inbreeding) uncertainties. These uncertainties, though important, are difficult to determine, predict, and manage. Nonetheless, we have studied and assessed wildlife life history and how species use their habitat(s) for quite some time. Thus, the wildlife habitat concept does address features in the environment that we can manipulate to enhance or discourage wildlife use of an area.

The principal purpose of this chapter is to report what we know or surmise about wildlife and its use of the landscape into features that land managers can understand, recognize, and strive to achieve. Describing and defining these features allows us to begin to establish a common understanding for management so that we can affect and influence the landscape to meet wildlife needs. What follows is an expansion of our wildlife habitats; two major features that also service wildlife species needs are structural conditions and specific habitat elements of a site. These two features should be viewed in a hierarchical manner, with structural conditions occurring on tracts of land such as at a forest stand or watershed level, whereby habitat elements are described at a site-specific or local level, like within a forest stand. Knowing something about structural conditions of an area along with the habitat elements occurring at a site will allow us to better predict what kinds of wildlife may use the site, as well as predict, if enhancements or modifications were made what wildlife might continue to use the site or area.

The collective set of environmental conditions provided by wildlife habitats, structural conditions, and habitat elements constitutes a species' overall habitat. Hence, from an ecological perspective, we are striving to determine the current and potential ecological conditions of a landscape or site. We achieve this knowledge by interpolating ecological condition from assessing wildlife habitats hierarchically, that is, from knowing something about the wildlife habitats, structural conditions, and habitat elements from area or site. Thus, current and desired ecological conditions can be assessed for wildlife and written:

Wildlife Habitats = wildlife cover type(s) + structural condition(s) + habitat element(s)

Structural Conditions

Structure is what a natural resource manager can manipulate to achieve various objectives. After all, manipulating the structural features of a forest stand is what silviculture is all about.[10] Structural conditions are often thought to follow the plant succession series. That is, if a stand, say in a forest, is left alone and given enough time it will achieve a climax or old-growth state. So the continuum starts with the earliest stages, and given a forest example, this would be at grass/forb, and then progressing through shrub/seedling; sapling/pole; small, medium, and large trees; and eventually achieving giant trees. Catastrophic events along with some management prescriptions can reset the succession stages (fire burning a tract of land mimics in many ways the effect of a clearcut). So several key structural elements for forest and shrubland/grasslands habitats are a wide range in tree and shrub sizes, a wide range in tree diameters and tree and shrub heights, and varying amounts of tree and shrub canopies. Multiple canopy layers are also considered to be significant and this includes the continuous distribution of foliar surface from the top of the crown to the ground. Such canopy distributions are thought to create greater quantities and greater diversity of animal habitats.[10, 13] That is, canopies are important not only for their physical characteristics, such as thermal qualities that influence stream temperatures and ground conditions (i.e. shade, shelter from precipitation) but also for their abiotic abilities such as fixating CO_2 or nitrogen from the atmosphere.[2, 11, 17] Finally, plant and tree understories may or may not be present depending on the structural condition of the stand;

however, their presence can influence the kind and number of species that use the stand.

Defining Structural Conditions

Knowing the wildlife-habitat type(s) of an area will allow a person to predict a list of species that may be found at that site. To refine the prediction, however, requires further information, like knowing a species range extent and condition of the habitat in the area of concern. The term condition implies, knowing the types and amounts of different structural stages and habitat elements. The condition of the habitat can predispose a species to use an area, and thus can serve as a driver for its occurrence.

Definitions of structural conditions were determined by the Species Habitat Team that was formed from multiple organizations to help clarify terms, represent field ecologist and biologist needs, and to help ensure a usable end-product. The definitions are divided into four major categories: Forest, Shrubland/Grassland, Agriculture, and Urban. The definitions for the first two categories are based on describing the characteristics of trees, shrubs, and grasses. The last two categories are based on describing land use/land cover types. This was done because of the difficulty in describing these categories from a plant community perspective and because there is very little literature available that would allow a more detailed breakdown. Table 1 was created to depict those wildlife habitats that are associated with each structural condition or land use/land cover type. A computer simulation that represents each Forest and Shrubland/Grassland structural condition can be found by reviewing the color photographs that accompany this chapter. The following are the definitions for structural conditions:

Forest Structural Conditions

The forest structural conditions are based on the following attributes: (1) tree size diameter at breast height (dbh); (2) percent canopy cover (or percent grass/forb cover); and, (3) number of canopy layers. These attributes have the following values:

Tree size (dbh)

shrub/seedling	<1 inch	<2.5 cm
sapling/pole	1-9 inches	2.5-24 cm
small tree	10-14 inches	25-37 cm
medium tree	15-19 inches	38-49 cm
large tree	20-29 inches	50-75 cm
giant tree	≥30 inches	≥76 cm

Percent canopy cover

open	10-39%
moderate	40-69%
closed	70-100%

Number of canopy layers

single story	1 stratum
multi-story	≥2 strata

The above attributes have been combined into the following structural conditions:

1. Grass/Forb—Open

Grass/Forb dominated with <70% coverage by grasses and forbs. Shrubs and small seedlings may be present, but do not dominate stand, (seedlings <10% canopy cover), and there may be remnant trees (trees remaining from the previous stand) that provide <10% canopy cover (Figure 1).

2. Grass/Forb—Closed

Grass/forb dominated with >70% coverage by grasses and forbs. Shrubs and small seedlings may be present, but do not dominate stand, (seedlings <10% canopy cover), and there may be remnant trees (trees remaining from the previous stand) that provide <10% canopy cover (Figure 2).

3. Shrub/Seedling—Open

Seedlings are large enough to add structure to the stand but are small enough that the structure is similar to shrubs and may have remnant trees (trees remaining from the previous stand) that provide <10% canopy cover. There is <70% cover of shrubs or seedlings. Tree size is <1 inch dbh, and there is 1 canopy stratum (Figure 3).

4. Shrub/Seedling—Closed

Seedlings are large enough to add structure to the stand but are small enough that the structure is similar to shrubs. Remnant trees (trees remaining from the previous stand) may provide <10% canopy cover. There is >70% cover of shrubs or seedlings. Tree size is <1 inch dbh, and there is 1 canopy stratum (Figure 4).

5. Sapling/Pole—Open

The canopy is open enough that understory vegetation may be abundant. Remnant trees (trees remaining from the previous stand) may provide <10% canopy cover. There is 10-39% cover of sapling and pole-sized trees. Tree size is 1-9 inches dbh, and there is 1 canopy stratum (Figure 5).

6. Sapling/Pole—Moderate

Understory development is hampered by available light and moisture. Remnant trees (trees remaining from the previous stand) may provide <10% canopy cover. There is 40-69% cover of sapling and pole-sized trees. Tree size is 1-9 inches dbh, and there is 1 canopy stratum (Figure 6).

7. Sapling/Pole—Closed

The understory is depauperate or absent. Remnant trees (trees remaining from the previous stand) may provide <10% canopy cover. There is >70% cover of sapling and pole-sized trees. Tree size is 1- 9 inches dbh and there is 1canopy stratum (Figure 7).

8. Small Tree—Single Story—Open

A grass/forb or shrub understory may be present. Remnant trees (trees remaining from the previous stand) may provide <10% canopy cover. There is 10-39% cover of small trees, with <10% cover of other tree sizes. Tree size is 10-14 inches dbh, and there is 1 canopy stratum (Figure 8).

Table 1. Matrix for wildlife habitats that are associated with structural conditions or land use/land cover types.

Wildlife habitats	Forest Structural Conditions[1]	Grassland/Shrubland Structural Conditions				Land Use/Land Cover Conditions	
		Grass/ Forb	Low shrub	Medium shrub	Tall shrub	Urban[2]	Agriculture[3]
Westside lowlands conifer-hardwood forest	X	—	—	—	—	—	—
Westside oak & Douglas-fir forest & woodlands	X	—	—	—	—	—	—
Southwest Oregon mixed conifer-hardwood forest	X	—	—	—	—	—	—
Montane mixed conifer forest	X	—	—	—	—	—	—
Eastside (Interior) mixed conifer forest	X	—	—	—	—	—	—
Lodgepole pine forest & woodlands	X	—	—	—	—	—	—
Ponderosa pine forests & woodlands	X	—	—	—	—	—	—
Upland aspen forest	X	—	—	—	—	—	—
Subalpine parklands	X	—	—	—	—	—	—
Alpine grasslands & shrublands	—	X	X	—	—	—	—
Westside grasslands	—	X	—	—	—	—	—
Ceanothus/manzanita shrublands	—	—	—	—	X	—	—
Western juniper & mountain mahogany woodlands	X	—	—	—	—	—	—
Eastside (Interior) canyon shrublands	—	—	—	X	X	—	—
Eastside (Interior) grasslands	—	X	—	—	—	—	—
Shrub-steppe	—	—	—	X	—	—	—
Dwarf shrub-steppe	—	—	X	—	—	—	—
Desert playa & salt scrub	—	—	—	X	—	—	—
Agriculture, pasture, and mixed environs	—	—	—	—	—	—	X
Urban and mixed environs	—	—	—	—	—	X	—
Open water—lakes, rivers, streams	—	—	—	—	—	—	—
Herbaceous wetlands	—	—	—	—	—	—	—
Westside riparian/wetlands	X	—	—	—	X	—	—
Montane coniferous wetlands	X	—	—	—	—	—	—
Eastside (Interior) riparian/wetlands	X	—	—	—	X	—	—
Coastal dunes & beaches	—	X	—	X	X	—	—
Coastal headlands & islets	—	X	—	X	X	—	—
Bays & estuaries	—	—	—	—	—	—	—
Inland marine deeper waters	—	—	—	—	—	—	—
Marine nearshore	—	—	—	—	—	—	—
Marine shelf	—	—	—	—	—	—	—
Oceanic	—	—	—	—	—	—	—

[1]Includes a variety of tree sizes and canopies

[2]Includes low, medium, and high densities of impervious areas.

[3]Includes areas like croplands, orchards, and pastures

9. Small Tree—Single Story—Moderate

Some grass/forb or shrub understory may be present. Remnant trees (green trees remaining from the previous stand) may provide <10% canopy cover. There is 40-69% cover of small trees with <10% cover of other sized trees. Tree size is 10-14 inches dbh, and there is 1 canopy stratum (Figure 9).

10. Small Tree—Single Story—Closed

Grass/Forb or shrub understory minor or absent. Remnant trees (trees remaining from the previous stand) may provide <10% canopy cover. There is >70% cover of small trees, with <10% cover of other-sized trees. Tree size is 10-14 inches dbh, and there is 1 canopy stratum (Figure 10).

11. Medium Tree—Single Story—Open

A grass/forb or shrub understory may be present. Remnant trees (trees remaining from the previous stand) may provide <10% canopy cover. There is 10-39% cover of medium trees, with <10% cover of other-sized trees. Tree size is 15-19 inches dbh, and there is 1 canopy stratum (Figure 11).

12. Medium Tree—Single Story—Moderate

Grass/Forb or shrub understory may be present. Remnant trees (trees remaining from the previous stand) may provide <10% canopy cover. There is 40-69% cover of medium trees with <10% cover of other-sized trees. Tree size is 15-19 inches dbh, and there is 1 canopy stratum (Figure 12).

13. Medium Tree—Single Story—Closed

A grass/forb or shrub understory may be present. Remnant trees (trees remaining from the previous stand) may provide <10% canopy cover. There is >70% cover of medium trees with <10% cover of other-sized trees. Tree size is 15-19 inches dbh, and there is 1 canopy stratum (Figure 13).

14. Large Tree—Single Story—Open

Grasses, shrubs, and/or seedlings may occur in the understory. There is 10-39% cover of large and/or giant size trees with <10% cover of other-sized trees. Tree size is 20-29 inches dbh, and there is 1 canopy stratum (Figure 14).

15. Large Tree—Single Story—Moderate

Some grass/forb or shrub understory may be present. There is 40-69% cover of large and/or giant trees with <10% cover of other-sized trees. Tree size is 20-29 inches dbh, and there is 1 canopy stratum (Figure 15).

16. Large Tree—Single Story—Closed

Grasses, shrubs, and/or seedlings may occur in the understory. There is >70% cover of large and/or giant trees with <10% cover of other-sized trees. Tree size is 20-29 inches dbh, and there is 1 canopy stratum (Figure 16).

17. Small Tree—Multi-story—Open

These stands have an overstory of small trees with a distinct subcanopy of saplings and/or poles. Scattered larger trees may be present but make up ≤10% canopy cover. Grass/forb or shrub understory may be present. There is 10-39% total canopy cover dominated by small trees, ≥10% canopy cover of ≥1 other smaller tree sizes. Tree size is 10-14 inches dbh, and there are ≥2 canopy strata (Figure 17).

18. Small Tree—Multi-story—Moderate

These stands have an overstory of small trees with a distinct subcanopy of saplings and/or poles. Scattered larger trees may be present but make up ≤10% canopy cover. Grass/forb or shrub understory may be present, but is probably limited. There is 40-69% total canopy cover dominated by small trees, ≥10% canopy cover of ≥1 or more other smaller tree sizes. Tree size is 10-14 inches dbh, and there are ≥2 canopy strata (Figure 18).

19. Small Tree—Multi-story—Closed

These stands have an overstory of small trees with a distinct subcanopy of saplings and/or poles. Scattered larger trees may be present but make up ≤10% canopy cover. Grass/forb or shrub understory extremely limited or absent. There is >70% total canopy cover dominated by small trees, ≥10% canopy cover of ≥1 other smaller tree sizes. Tree size is 10-14 inches dbh, and there are ≥2 canopy strata (Figure 19).

20. Medium Tree—Multi-story—Open

These stands have an overstory of medium trees with a distinct subcanopy of smaller trees. Scattered larger trees may be present but make up ≤10% canopy cover. Grass/forb or shrub understory may be present, but is probably limited. There is 10-39% total canopy cover dominated by medium trees, ≥10% or more canopy cover of ≥1 other smaller tree sizes. Tree size is 15-19 inches dbh, and there are ≥2 canopy strata (Figure 20).

21. Medium Tree—Multi-story—Moderate

These stands have an overstory of medium trees with a distinct subcanopy of smaller trees. Scattered larger trees may be present but make up ≤10% canopy cover. Grass/forb or shrub understory may be present, but is probably limited. There is 40-69% total canopy cover dominated by medium trees, ≥10% or more canopy cover of 1 or more smaller tree sizes. Tree size is 15-19 inches dbh, and there are ≥2 canopy strata (Figure 21).

22. Medium Tree—Multi-story—Closed

These stands have an overstory of medium trees with a distinct subcanopy of smaller trees. Scattered larger trees may be present but make up ≤10% canopy cover. Grass/forb understory may be present, but is probably limited. There is >70% total canopy cover dominated by medium trees, ≥10% or more canopy cover of ≥1 smaller tree sizes. Tree size is 15-19 inches dbh, and there are ≥2 canopy strata (Figure 22).

23. Large Tree—Multi-story—Open

These stands have an overstory of large or giant-sized trees with one or more distinct canopy layers of smaller trees. Stands with >40% cover of giant trees are classified in the Giant, multi-storied stage. In westside forests, stands dominated by large trees usually have giant trees scattered in the stand, with fewer in eastside forests. A grass/forb or shrub understory is often present, especially in canopy gaps. There is 10-39 % total canopy cover, with ≥10% or more canopy cover from large and/or giant trees and another ≥10% canopy cover from ≥1 or more smaller tree size classes. Tree size is 20-29 inches dbh, and there are ≥2 canopy strata (Figure 23)

24. Large Tree—Multi-story—Moderate

These stands have an overstory of large or giant-sized trees with ≥1 distinct canopy layers of smaller trees. Stands with >40% cover of giant trees are classified in the giant, multi-storied stage. In westside forests, stands dominated by large trees usually have giant trees scattered in the stand, with fewer in eastside forests. Grass/Forb or shrub understory is often present, especially in canopy gaps. There is 40-69% total canopy cover, ≥10% canopy cover from large trees with another ≥10% canopy cover from ≥1 smaller tree size classes. Tree size is 20-29 inches dbh, and there are ≥2 canopy strata (Figure 24).

25. Large Tree—Multi-story—Closed

These stands have an overstory of large or giant-sized trees with ≥1 distinct canopy layers of smaller trees. Stands with > 40% cover of giant trees are classified in the giant, multi-storied stage. In westside forests, stands dominated by large trees usually have giant trees scattered in the stand, with fewer in eastside forests. A grass/forb or shrub understory is often present, especially in canopy gaps. There is >70% total canopy cover, ≥10% or more canopy cover from large trees with another ≥10% canopy cover from ≥1 smaller tree size classes. Tree size is 20- 29 inches dbh, and there are ≥2 canopy strata (Figure 25).

26. Giant Tree—Multi-story

These stands have an overstory of giant-sized trees with ≥1 distinct canopy layers of smaller trees. Stands with <40% canopy cover are classified in the large tree multi-story open stage. There is >40% canopy cover. Tree size is >30 inches dbh, and there are ≥2 canopy strata (Figure 26a and b).

Shrubland and Grassland Structural Conditions

The shrubland and grassland structural conditions are based on the following attributes: (1) shrub height, (2) percent shrub cover (or percent grass/forb cover), and (3) shrub age class. These attributes have the following values:

Shrub height

low	<1.6 ft	<0.5 m
medium	1.6-6.5 ft	0.5-2.0 m
tall	>6.5 ft-16.5 ft	>2.0 m-5.0 m

Percent shrub cover

open	10-69% shrub cover
closed	70-100% shrub cover

Shrub age class

seedling/young	negligible crown decadence
mature	≤ 25% crown decadence
old	26-100% crown decadence

The above attributes have been combined into the following structural conditions:

1. Grass/Forb—Open

Grasslands that have <10% shrub cover and <10% tree canopy cover. Grasses and forbs cover are <70% of the ground, and bare ground is evident (Figure 27).

2. Grass/Forb—Closed

Grasslands that have <10% shrub cover and <10% tree canopy cover. Grasses and forbs cover >70% of the ground (Figure 28).

3. Low Shrub—Open Shrub Overstory—Seedling/Young

Shrublands with shrubs <1.6 ft tall, shrub canopy cover 10-69% and may have <10% tree canopy cover. Areas with <10% shrub cover are categorized as Grass/Forb. These are post-disturbance regenerating shrublands dominated by seedlings or young shrubs. Mature, legacy shrubs may persist from before the disturbance, but occur as scattered singles or widely scattered clumps. Crown decadence is negligible (Figure 29).

4. Low Shrub—Open Shrub Overstory—Mature

Shrublands with shrubs <1.6 ft tall, shrub canopy cover 10-69% and may have <10% tree canopy cover. Areas with <10% shrub cover are categorized as Grass/Forb. Crown decadence is ≤25% (Figure 30).

5. Low Shrub—Open Shrub Overstory—Old

Shrublands with shrubs <1.6 ft tall, shrub canopy cover 10-69% and may have <10% tree canopy cover. Areas with <10% shrub cover are categorized as Grass/Forb. Crown decadence is >25% (Figure 31).

6. Low Shrub—Closed Shrub Overstory—Seedling/Young

Shrublands with shrubs <1.6 ft tall, shrub canopy cover >70%, and may have <10% tree canopy cover. These are post-disturbance regenerating shrublands dominated by seedlings or young shrubs. Mature, legacy shrubs may persist from before the disturbance, but occur as scattered singles or widely scattered clumps. Crown decadence is negligible (Figure 32).

7. Low Shrub—Closed Shrub Overstory—Mature

Shrublands with shrubs <1.6 ft tall and shrub canopy cover >70%, and may have <10% tree canopy cover. Crown decadence is ≤25% (Figure 33).

8. Low Shrub—Closed Shrub Overstory—Old

Shrublands with shrubs <1.6 ft tall, shrub canopy cover >70%, and may have <10% tree canopy cover. Crown decadence is >25% (Figure 34).

9. Medium Shrub—Open Shrub Overstory—Seedling/Young

Shrublands with shrubs 1.6 - 6.5 ft tall, shrub canopy cover >10% and <70%, and may have <10% tree canopy cover. Areas with <10% shrub cover are categorized as Grass/Forb. These are post- disturbance regenerating shrublands dominated by seedlings or young shrubs. Mature, legacy shrubs may persist from before the disturbance, but occur as scattered singles or widely scattered clumps. Crown decadence is negligible (Figure 35).

10. Medium Shrub—Open Shrub Overstory—Mature

Shrublands with shrubs 1.6 - 6.5 ft tall, shrub canopy cover >10% and <70%, and may have <10% tree canopy cover. Areas with <10% shrub cover are categorized as Grass/Forb. Crown decadence is ≤25% (Figure 36).

11. Medium Shrub—Open Shrub Overstory—Old

Shrublands with shrubs 1.6 - 6.5 ft tall, shrub canopy cover >10% and <70%, and may have <10% tree canopy cover. Areas with <10% shrub cover are categorized as Grass/Forb. Crown decadence is >25% (Figure 37).

12. Medium Shrub—Closed Shrub Overstory—Seedling/Young

Shrublands with shrubs 1.6 - 6.5 ft tall, shrub canopy cover >70%, and may have <10% tree canopy cover. These are post-disturbance regenerating shrublands dominated by seedlings or young shrubs. Mature, legacy shrubs may persist from before the disturbance, but occur as scattered singles or widely scattered clumps. Crown decadence is negligible (Figure 38).

13. Medium Shrub—Closed Shrub Overstory—Mature

Shrublands with shrubs 1.6 - 6.5 ft tall, shrub canopy cover >70%, and may have <10% tree canopy cover. Crown decadence is ≤25% (Figure 39).

14. Medium Shrub—Closed Shrub Overstory—Old

Shrublands with shrubs 1.6 - 6.5 ft tall, shrub canopy cover >70%, and may have <10% tree canopy cover. Crown decadence is >25% (Figure 40).

15. Tall Shrub—Open Shrub Overstory—Seedling/Young

Shrublands with shrubs >6.5ft tall, shrub canopy cover >10% and <70%, and may have <10% tree canopy cover. Areas with <10% shrub cover are categorized as Grass/Forb. These are post- disturbance regenerating shrublands dominated by seedlings or young shrubs. Mature, legacy shrubs may persist from before the disturbance, but occur as scattered singles or widely scattered clumps. Crown decadence is negligible (Figure 41).

16. Tall Shrub—Open Shrub Overstory—Mature

Shrublands with shrubs >6.5 ft tall, shrub canopy cover >10% and <70% and may have <10% tree canopy cover. Areas with <10% shrub cover are categorized as Grass/Forb. Crown decadence is ≤25% (Figure 42).

17. Tall Shrub—Open Shrub Overstory—Old

Shrublands with shrubs >6.5 ft tall, shrub canopy cover >10% and <70%, and may have tree canopy cover <10%. Areas with <10% shrub cover are categorized as Grass/Forb. Crown decadence is >25% (Figure 43).

18. Tall Shrub—Closed Shrub Overstory—Seedling/Young

Shrublands with shrubs >6.5 ft tall, shrub canopy cover >70%, and may have tree canopy cover <10%. These are post-disturbance regenerating shrublands dominated by seedlings or young shrubs. Mature, legacy shrubs may persist from before the disturbance, but occur as scattered singles or widely scattered clumps. Crown decadence is negligible (Figure 44).

19. Tall Shrub—Closed Shrub Overstory—Mature

Shrublands with shrubs >6.5 ft tall, shrub canopy cover >70%, and may have tree canopy cover <10%. Crown decadence is ≤25% (Figure 45).

20. Tall Shrub—Closed Shrub Overstory—Old

Shrublands with shrubs >6.5 ft tall, shrub canopy cover >70%, and may have <10% tree canopy cover. Crown decadence is >25% (Figure 46).

Urban Land Use/Land Cover Conditions

The Urban land use/land cover conditions are based on the level of urban development as determined by the percent of land surface covered by impervious materials.

The Urban land use/land cover classification consists of the following conditions:

1. Low Density

Surfaces that are covered with 10-29% of impervious material. Examples would include rural residential areas, suburban housing with large lots (≥1 acre [2.4 ha]).

2. Medium Density

Surfaces that are covered by 30-59% of impervious material. Examples would include single family housing areas (lot size >1 acre [2.4 ha]), suburban development.

3. High Density

Surfaces that are covered by 60-100% of impervious material. Examples would include core downtown areas within cities (e.g., Seattle, Portland), commercial areas (e.g., shopping malls), industrial areas, high density housing (e.g., apartment buildings), and transportation corridors (e.g., highways).

Agricultural Land Use/Land Cover Conditions

1. Cultivated Cropland

Farmland used for production of annual crops such as vegetables and herbs is characterized by bare soil, and plant debris either in the field or along the periphery. The location tends to be along bottomland areas of streams and rivers and areas with a sufficient source of irrigation. Farmland used for production of annual grasses such as wheat, oats, barley, and rye is characterized by upland and rolling hill terrain, generally without irrigation. This agricultural division has similar pesticide use and/or irrigation requirements. That is, row crops are treated the same way in regard to the general application of pesticides and cultural techniques in land preparation and harvest. There is a wide range of soil conservation practices in this category.

2. Improved Pasture

Farmland used for the production of perennial grass such as grass seed and hay. Perennial grass is generally grown without irrigation. Perennial crops are treated the same way in regard to the general application of pesticides and cultural techniques.

3. Orchards/Vineyards/Nursery

Farmland used for production of tree fruits (apples, peaches, pears, hazelnuts), vineyards (grapes), berries (strawberries, raspberries, blueberries, blackberries), Christmas trees, and nursery stock (ornamental container and greenhouse operations). This cover type is generally located in upland areas with access to a high volume of irrigation. Christmas trees are characterized by upland areas, poorer soils and no irrigation. The use of chemicals in non-food crops, such as Christmas trees and nursery stock, is considerably different both in materials and time of applications.

4. Modified Grasslands

Annual or introduced perennial grasslands, including cheatgrass (*Bromus tectorum*), medusahead (*Taeniatherum caput-medusae*), and other annual bromes; moist environments, including riparian bottomlands, are often dominated by Kentucky bluegrass. Annual grasslands (and areas of introduced forbs) are usually dominated by one or two introduced annuals which comprise most of the vegetation cover. Perennial grasslands are usually dominated by a single planted bunchgrass with introduced annuals and weedy forbs between the bunches. Some environments support rhizomatous perennial grasses. These areas occur mostly on uplands but also includes riparian bottomlands that are dominated by non-native grasses. Modified grasslands can be found throughout the steppe and grasslands areas of eastern Oregon and Washington and at low elevation sites in southwestern Oregon.

5. Unimproved Pasture

Farmland that seems to have no active management such as fertilizer application, irrigation or weed control. This land might be grazed by livestock, but shows no evidence of irrigation. It can also be characterized by uncut hay, organic debris from the previous season, uncut standing dead grass, exotic plants like tansy ragwort (*Senecio jacobea*), thistle (*Cirsium spp.*), Himalaya blackberry (*Rubus discolor*) and their debris, patches of shrubs such as hawthorn (*Crataegus spp.*), snowberry (*Symphoricarpos spp.*), spirea (*Spirea spp.*), poison oak (*Rhus diversiloba*), and encroachment by various tree species. This land has usually been cleared and farmed intensively in the past. This category also includes lands that are designated within the Conservation Reserve Program (CRP) and areas planted with crested wheatgrass (*Agropyron cristatum*). Land owners use unimproved pasture for grazing livestock, otherwise it lies dormant. Thus, those lands that are not grazed either revert to brushy field or volunteer forest.

Structural Conditions Data Matrix

To maximize the utility of wildlife-habitat relationship information, a digital database that links wildlife with its structural conditions can be found on the CD-ROM included with this book. Wildlife occurrence within a particular structural condition type was determined through an expert panel process held during the fall of 1998. Table 1 was created to assist the expert panel in identifying what wildlife habitats are associated with what structural conditions. The categories that depict a wildlife species occurrence with a particular structural condition are *Y—Yes the species occurs*, *H—Historically occurred*, and *U—Unsure*. Alongside the occurrence category, we identify the types of activity that the species does while using the structural condition. The activity codes for the wildlife species within a particular condition are: *B—Both feeds and breeds*, *F/R-HE—Feeds and Reproduces when a specific habitat element is present*, *F—Feeds only*, *R—Reproduces only*, and *O—Other*. The *Other* category reflects activities such as roosting/resting, hibernating, or using the habitat for cover (thermal and hiding) purposes.

Defining the Level of Associations Between Wildlife and Structural Conditions

As mentioned in Chapter 1, we continue to embrace the new concept of *degree of association* between wildlife species and their habitats.[19] For the purposes of this project, we used the following categories for characterizing the degree of association.

Closely Associated. A species is widely known to depend on a habitat or structural condition for part or all of its life history requirements. Identifying this association implies that the species has an essential need for this habitat or structural condition for its maintenance and viability. Some species may be closely associated with more than one habitat or structural condition, others may be closely associated with only one habitat or structural condition. Examples of species exhibiting close associations are red-winged blackbirds to wetland habitats, and spotted owls to mature and giant tree structural conditions.

Generally Associated. A species exhibits a high degree of adaptability and may be supported by a number of habitats or structural conditions. In other words, the habitats or structural conditions play a supportive role for its maintenance and viability. Examples include the black bear's association with a variety of forested habitats or the black-tailed deer associated with a number of structural conditions.

Present. A species demonstrates occasional use of a habitat or structural condition. The habitat or structural condition provides marginal support to the species for its maintenance and viability. Examples are the rough-legged hawk in desert playa and salt scrub shrubland or the mink in a montane mixed conifer forest.

Finally, the expert panelists assigned an overall *confidence rating* to the occurrence and activity headings for each species within each structural condition. The confidence ratings were simply high (e.g., many peer or published accounts), moderate, and low (e.g., few or no published accounts). By assigning a confidence rating, our objective was to offer users an evaluation of the overall strength of the scientific evidence.

Habitat Elements

Site-specific habitat elements (HEs) are those components of the environment believed to most influence wildlife species' distribution, abundance, fitness, and viability (definition adapted from Marcot et al.[14] and Mayer and Laudenslayer.)[16] In this context, HEs include natural attributes, both biological and physical (e.g., large trees, woody debris, cliffs, and soil characteristics) as well as anthropogenic features and their effects such as roads, buildings, and pollution. Including these fine-scale attributes of an animal's environment when describing the habitat associations for a particular species expands the concept and definition of habitat, a term widely used only to characterize the vegetative community or structural condition occupied by a species. Failing to assess and inventory HEs within these communities and conditions may lead to errors of commission; species may be presumed to occur when in actuality they do not. Habitat elements that influence a species negatively may preclude occupancy or breeding despite adequate floristic or structural conditions.

Defining Habitat Elements

Traditionally defined, the term *habitat* is that set of environmental conditions, usually depicted as food, water, and cover, used and selected for by a given organism. Despite this broad definition, many land management agencies use the term *habitat* to denote merely the vegetation conditions and/or structural or seral stages used by a particular species. However, many other environmental attributes or features influence and affect the population viability of wildlife species. Marcot et al.[14] in their assessment of the terrestrial species of the Columbia River Basin emphasized the importance of examining all features that exert influence on wildlife by

expanding the definition of habitat to encompass all environmental correlates, naming the entirety of these attributes Key Ecological Correlates or KECs. All environmental scales, from broad floristic communities to fine-scale within-stand features, were included in their definition of a KEC. The word Key in Key Environmental Correlate refers to the high degree of influence (either positive or negative) the environmental correlates exert on the realized fitness of a given species. Nonetheless, when this information was determined, only direct relationships between an HE and a species were identified. Most of the HE-species associations refer to on the CD-ROM are mostly a positive influence between the HE and the species. Negative influence between HE and the species may be viewed as environmental stressors; however, a comprehensive list of negative influences is not presented here. The following are HE definitions.

1. Forest, Shrubland and Grassland Habitat Elements

Biotic, naturally occurring attributes of forest and shrubland communities; the information that follows is for mostly positive relationships.

1.1 *Forest/woodland vegetative elements or substrates.* Biotic components found within a forested context.

1.1.1 *Down wood.* Includes downed logs, branches, and rootwads.

1.1.1.1 *Decay class.* A system by which down wood is classified based on its deterioration.

1.1.1.1.1 *hard (class 1, 2).* Little wood decay evident; bark and branches present; log resting on branches, not fully in contact with ground; includes classes 1 and 2 as described in Thomas[22].

1.1.1.1.2 *moderate (class 3).* Moderate decay present; some branches and bark missing or loose; most of log in contact with ground; includes class 3 as described in Thomas[22].

1.1.1.1.3 *soft (class 4, 5).* Well decayed logs; bark and branches missing; fully in contact with ground; includes classes 4 and 5 as described in Thomas.[22]

1.1.1.2 *Down wood in riparian areas.* Includes down wood in the terrestrial portion of riparian zones in forest habitats. Does not refer to in-stream woody debris.

1.1.1.3 *Down wood in upland areas.* Includes downed wood in upland areas of forest habitats.

1.1.2 *Litter.* The upper layer of loose, organic (primarily vegetative) debris on the forest floor. Decomposition may have begun, but components still recognizable.

1.1.3 *Duff.* The matted layer of organic debris beneath the litter layer. Decomposition more advanced than in litter layer; intergrades with uppermost humus layer of soil.

1.1.4 *Shrub layer.* Refers to the shrub strata within forest stands.

1.1.4.1 *Shrub size.* Refers to shrub height.

1.1.4.2 *Percent shrub canopy cover.* Percent of ground covered by vertical projection of shrub crown diameter.

1.1.4.3 *Shrub canopy layers.* Within a shrub community, differences in shrub height and growth form produce multi-layered shrub canopies in the forest understory.

1.1.5 *Moss.* Large group of green plants without flowers but with small leafy stems growing in clumps.

1.1.6 *Flowers.* A modified plant branch for the production of seeds and bearing leaves specialized into floral organs.

1.1.7 *Lichens.* Any of various lower plants made up of an alga and a fungus growing as a unit on a solid surface.

1.1.8 *Forbs.* Broad-leaved herbaceous plants. Does not include grasses, sedges, or rushes.

1.1.9 *Cactus.* Any of a large group of drought-resistant plants with fleshy, usually jointed stems and leaves replaced by scales or spines.

1.1.10 *Fungi.* Mushrooms, molds, yeasts, rusts, etc.

1.1.11 *Roots, tubers, underground plant parts.* Any underground part of a plant that functions in nutrient absorbtion, aeration, storage, reproduction and/or anchorage.

1.1.12 *Ferns.* Any of a group of flowerless, seedless vascular green plants.

1.1.13 *Herbaceous layer.* Understory non-woody vegetation layer beneath shrub layer (forest context). May include forbs, grasses, ferns.

1.1.14 *Trees.* Includes both coniferous and hardwood species.

    1.1.14.1 *Snags.* Standing dead trees.

        1.1.14.1.1 *Decay class.* A system by which snags are classified based on their deterioration.

            1.1.14.1.1.1 *hard.* Little wood decay evident; bark, branches, top, present; recently dead; includes class 1 as described in Brown[3].

            1.1.14.1.1.2 *moderate.* Moderately decayed wood; some branches and bark missing and/or loose; top broken; includes classes 2 and 3 as described in Brown.[3]

            1.1.14.1.1.3 *soft.* Well decayed wood; bark and branches generally absent; top broken; includes classes 4 and 5 as described in Brown.[3]

    1.1.14.2 *Snag size.* Measured in diameter at breast height (dbh) the standard measurement for standing trees taken at 4.5 feet above the ground.

        1.1.14.2.1 *seedling* <1 inch dbh

        1.1.14.2.2 *sapling/pole* 1-9 inches dbh

        1.1.14.2.3 *small tree* 10-14 inches dbh

        1.1.14.2.4 *medium tree* 15-19 inches dbh

        1.1.14.2.5 *large tree* 20-29 inches dbh

        1.1.14.2.6 *giant tree* ≥30 inches dbh

    1.1.14.3 *Tree size.* Measured in dbh, same as 1.1.14.2 above.

        1.1.14.3.1 *seedling* <1 inch dbh

        1.1.14.3.2 *sapling/pole* 1-9 inches dbh

        1.1.14.3.3 *small tree* 10-14 inches dbh

        1.1.14.3.4 *medium tree* 15-19 inches dbh

        1.1.14.3.5 *large tree* 20-29 inches dbh

        1.1.14.3.6 *giant tree* ≥30 inches dbh

1.1.14.4 *Mistletoe brooms/witches brooms.* Dense masses of deformed branches caused by any type of broom-forming parasite (fungal or plant).

1.1.14.5 *Dead parts of live tree.* Portions of live trees with rot; can include broken tops; branches with decay; tree base with rot.

1.1.14.6 *Hollow living trees (chimney trees).* Tree bole with large hollow chambers.

1.1.14.7 *Tree cavities.* Smaller chamber in a tree; can be in bole, limbs, or forks of live or dead trees. May be excavated or result from decay or damage.

1.1.14.8 *Bark.* Includes crevices or fissures, and loose or exfoliating bark.

1.1.14.9 *Live remnant/legacy trees.* A live mature or old-growth tree remaining from the previous stand. Context is remnant trees in recently harvested or burned stands up through young forested stands. See dead parts of live trees, hollow living trees, tree cavities, and bark to see which species benefit from remnant trees with these attributes.

1.1.14.10 *Large live tree branches.* Large branches often growing horizontally out from the tree bole.

1.1.14.11 *Tree canopy layer.* Refers to the strata occupied by tree crowns.

    1.1.14.11.1 *Sub-canopy.* The space below the predominant tree crowns.

    1.1.14.11.2 *Above canopy.* The space above the predominant tree crowns.

    1.1.14.11.3 *Tree bole.* The tree trunk.

    1.1.14.11.4 *Canopy.* The more or less continuous cover of branches and foliage formed collectively by the crowns of adjacent trees and other woody growth.

1.1.15 *Fruits/seeds/nuts.* Plant reproductive bodies that are used by animals.

1.1.16 *Edges.* The place where plant communities meet or where successional stages or vegetative conditions within plant communities come together.

1.2 *Shrubland/grassland vegetative elements or substrates.*

Biotic components found within a shrubland or grassland context (these are positive influences only).

1.2.1 *Herbaceous layer.* Zone of understory nonwoody vegetation beneath shrub layer (nonforest context). May include forbs, grasses.

1.2.2 *Fruits/seeds/nuts.* Plant reproductive bodies that are used by animals.

1.2.3 *Moss.* Large group of green plants without flowers but with small leafy stems growing in clumps.

1.2.4 *Cactus.* Any of a large group of drought-resistant plants with fleshy, usually jointed stems and leaves replaced by scales or spines.

1.2.5 *Flowers.* A modified plant branch for the production of seeds and bearing leaves specialized into floral organs.

1.2.6 *Shrubs.* Plant with persistent woody stems and <16.5 feet tall; usually produces several basal shoots as opposed to a single bole.

  1.2.6.1 *Shrub size.* Refers to shrub height.
    1.2.6.1.1 small <2.0 feet
    1.2.6.1.2 medium 2.0 - 6.5 feet
    1.2.6.1.3 large 6.5 - 16.5 feet

  1.2.6.2 *Percent shrub canopy cover.* Percent of ground covered by vertical projection of shrub crown diameter.

  1.2.6.3 *Shrub canopy layer.* Within a shrub community, differences in shrub height and growth form produce multi-layered shrub canopies.
    1.2.6.3.1 *Sub-canopy.* The space below the predominant shrub crowns.
    1.2.6.3.2 *Above canopy.* The space above the predominant shrub crowns.

1.2.7 *Fungi.* Mushrooms, molds, yeasts, rusts, etc.

1.2.8 *Forbs.* Broad-leaved herbaceous plants. Does not include grasses, sedges, or rushes.

1.2.9 *Bulbs/tubers.* Any underground part of a plant that functions in nutrient absorbtion, aeration, storage, reproduction and/or anchorage.

1.2.10 *Grasses.* Members of the Graminae family.

1.2.11*Cryptogamic crusts.* Non-vascular plants that grow on the soil surface. Primarily lichens, mosses, and algae. Often found in arid or semi-arid regions. May form soil surface pinnacles.

1.2.12 *Trees* (located in a shrubland/grassland context). Small groups of trees or isolated individuals.

  1.2.12.1 *Snags.* Standing dead trees.
    1.2.12.1.1 *Decay class.* System by which snags are classified based on their deterioration.
      1.2.12.1.1.1 *hard.* Little wood decay evident; bark, branches, top, present; recently dead; includes class 1 as described in Brown.[3]
      1.2.12.1.1.2 *moderate.* Moderately decayed wood; some branches and bark missing and/or loose; top broken; includes classes 2 and 3 as described in Brown.[3]

      1.2.12.1.1.3 *soft.* Well decayed wood; bark and branches generally absent; top broken; includes classes 4 and 5 as described in Brown.[3]

  1.2.12.2 *Snag size.* Measured in dbh, as previously defined.
    1.2.12.2.1 *shrub/seedling* <1 inch dbh
    1.2.12.2.2 *sapling/pole* 1-9 inches dbh
    1.2.12.2.3 *small tree* 10-14 inches dbh
    1.2.12.2.4 *medium tree* 15-19 inches dbh
    1.2.12.2.5 *large tree* 20-29 inches dbh
    1.2.12.2.6 *giant tree* ≥30 inches dbh

  1.2.12.3 *Tree size.* Measured in dbh, as previously defined.
    1.2.12.3.1 *shrub/seedling* <1 inch dbh
    1.2.12.3.2 *sapling/pole* 1-9 inches dbh
    1.2.12.3.3 *small tree* 10-14 inches dbh
    1.2.12.3.4 *medium tree* 15-19 inches dbh
    1.2.12.3.5 *large tree* 20-29 inches dbh
    1.2.12.3.6 *giant tree* ≥30 inches dbh

1.2.13 *Edges* The place where plant communities meet or where successional stages or vegetative conditions within plant communities come together.

2. Ecological Habitat Elements

Selected interspecies relationships within the biotic community; they include both positive and negative influences.

2.1 *Exotic species.* Any non-native plant or animal, including cats, dogs, and cattle.

  2.1.1 *Plants.* This field refers to the relationship between an exotic plant species and animal species.

  2.1.2 *Animals.* This field refers to the relationship between an exotic animal species and the animal species.
    2.1.2.1 *Predation.* The species queried is preyed upon by or preys upon an exotic species.
    2.1.2.2 *Direct displacement.* The species queried is physically displaced by an exotic species, either by competition or actual disturbance.
    2.1.2.3 *Habitat structure change.* The species queried is affected by habitat structural changes caused by an exotic species, for example, cattle grazing.
    2.1.2.4 *Other.* Any other effects of an exotic species on a native species.

2.2 *Insect population irruptions.* The species directly benefits from insect population irruptions (i.e., benefits from the insects themselves, not the resulting tree mortality or loss of foliage).

  2.2.1 *Mountain pine beetle.* The species directly benefits from mountain pine beetle eruptions.

  2.2.2 *Spruce budworm.* The species directly benefits from spruce budworm irruptions.

  2.2.3 *Gypsy moth.* The species directly benefits from gypsy moth irruptions.

Text continues on page 133

Figure 1. Grass/Forb—Open

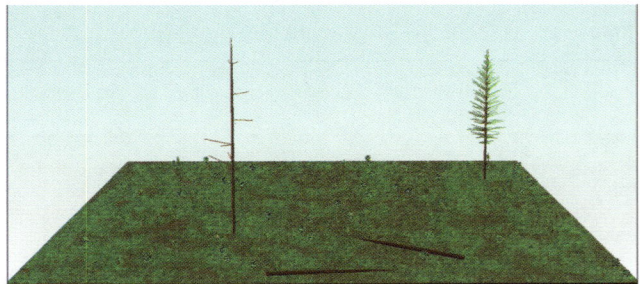

Figure 2. Grass/Forb—Closed

Figure 3. Shrub/Seedling—Open

Figure 4. Shrub/Seedling—Closed

Figure 5. Sapling/Pole—Open

Figure 6. Sapling/Pole—Moderate

Figure 7. Sapling/Pole—Closed

Figure 8. Small Tree—Single Story—Open

Figure 11. Medium Tree—Single Story—Open

Figure 9. Small Tree—Single Story—Moderate

Figure 12. Medium Tree—Single Story—Moderate

Figure 10. Small Tree—Single Story—Closed

Figure 13. Medium Tree—Single Story—Closed

Figure 14. Large Tree—Single Story—Open

Figure 15. Large Tree—Single Story—Moderate

Figure 16. Large Tree—Single Story—Closed

Figure 17. Small Tree—Multi-story—Open

Figure 18. Small Tree—Multi-story—Moderate

Figure 19. Small Tree—Multi-story—Closed

Figure 20. Medium Tree—Multi-story—Open

Figure 23. Large Tree—Multi-story—Open

Figure 21. Medium Tree—Multi-story—Moderate

Figure 24. Large Tree—Multi-story—Moderate

Figure 22. Medium Tree—Multi-story—Closed

Figure 25. Large Tree—Multi-story—Closed

Figure 26a. Giant Tree—Multi-story

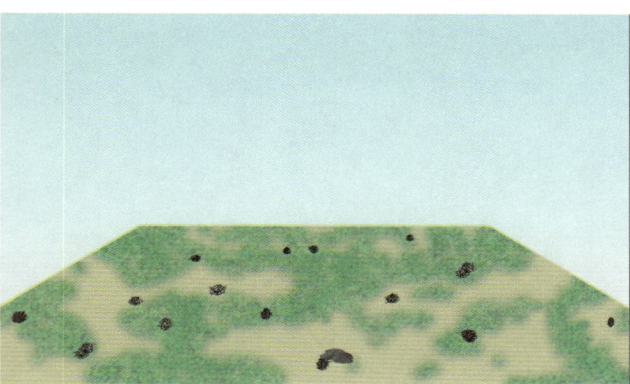

Figure 27. Grass/Forb—Open

Figure 26b. Giant Tree—Multi-story. This photograph has been included due to the limitations of the computer model.

Figure 28. Grass/Forb—Closed

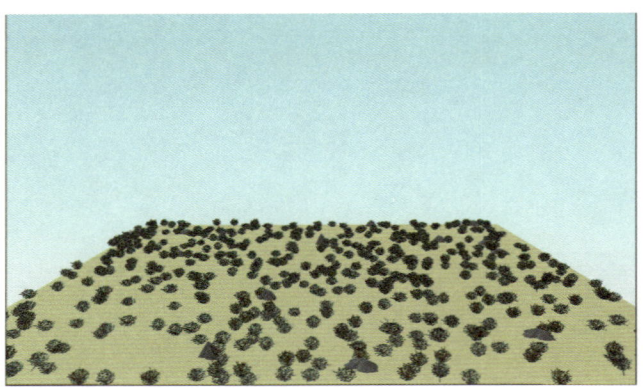

Figure 29. Low Shrub—Open Shrub Overstory—Seedling/ Young

Figure 32. Low Shrub—Closed Shrub Overstory—Seedling/ Young

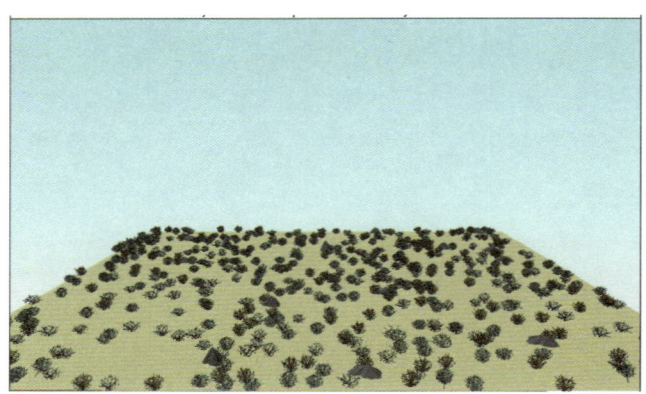

Figure 30. Low Shrub—Open Shrub Overstory—Mature

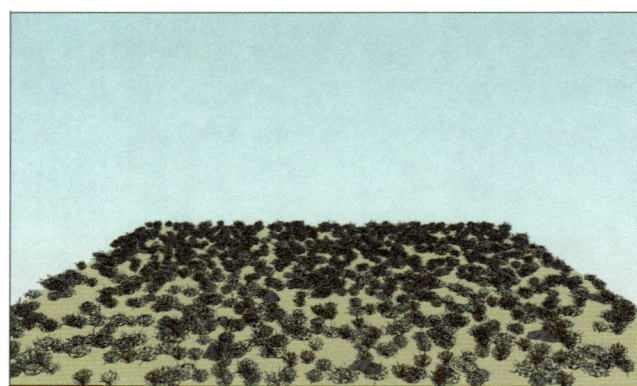

Figure 33. Low Shrub—Closed Shrub Overstory—Mature

Figure 31. Low Shrub—Open Shrub Overstory—Old

Figure 34. Low Shrub—Closed Shrub Overstory—Old

Figure 35. Medium Shrub—Open Shrub Overstory—Seedling/Young

Figure 38. Medium Shrub—Closed Shrub Overstory—Seedling/Young

Figure 36. Medium Shrub—Open Shrub Overstory—Mature

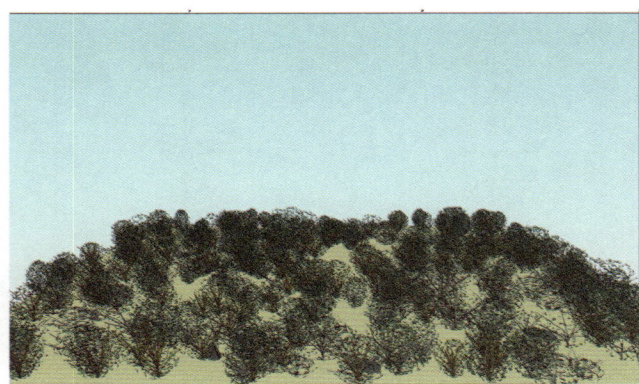

Figure 39. Medium Shrub—Closed Shrub Overstory—Mature

Figure 37. Medium Shrub—Open Shrub Overstory—Old

Figure 40. Medium Shrub—Closed Shrub Overstory—Old

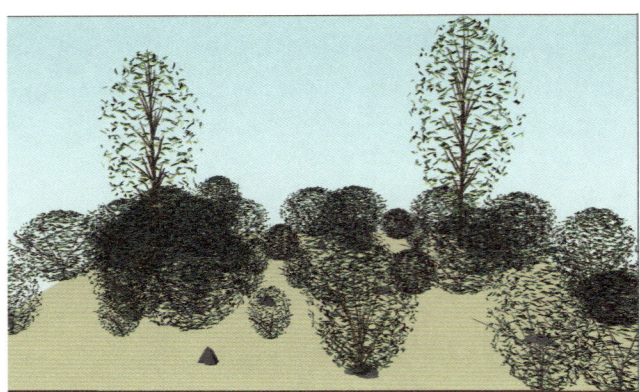

Figure 41. Tall Shrub—Open Shrub Overstory—Seedling/ Young

Figure 44. Tall Shrub—Closed Shrub Overstory—Seedling/ Young

Figure 42. Tall Shrub—Open Shrub Overstory—Mature

Figure 45. Tall Shrub—Closed Shrub Overstory—Mature

Figure 43. Tall Shrub—Open Shrub Overstory—Old

Figure 46. Tall Shrub—Closed Shrub Overstory—Old

2.3 *Beaver/muskrat activity.* The results of beaver activity including dams, lodges, and ponds, that are beneficial to other species.

2.4 *Burrows.* Aquatic or terrestrial cavities produced by burrowing animals that are beneficial to other species.

3. Non-Vegetative, Abiotic, Terrestrial Habitat Elements

Nonliving components found within any ecosystem. Primarily positive influences with a few exceptions as indicated.

3.1 *Rocks.* Solid mineral deposits.

3.1.1 *Gravel.* Particle size from 0.1-3.0 inches (0.2-7.6 cm) in diameter; gravel bars associated with streams and rivers are a separate category.

3.1.2 *Talus.* Accumulations of rocks at the base of cliffs or steep slopes; rock/boulder sizes varied and determine what species can inhabit the spaces between them.

3.1.3 *Talus-like habitats.* Refers to areas that contain many rocks and boulders but are not associated with cliffs or steep slopes.

3.2 *Soils.* Various soil characteristics.

3.2.1 *Soil depth.* The distance from the top layer of the soil to the bedrock or hardpan below.

3.2.2 *Soil temperature.* Any measure of soil temperature or range of temperatures that are key to the queried species.

3.2.3 *Soil moisture.* The amount of water contained within the soil.

3.2.4 *Soil organic matter.* The accumulation of decomposing plant and animal materials found within the soil.

3.2.5 *Soil texture.* Refers to size distribution and amount of mineral particles (sand, silt, and clay) in the soil; examples are sandy clay, sandy loam, silty clay, etc.

3.3 *Rock substrates.* Various rock formations.

3.3.1 *Avalanche chute.* An area where periodic snow or rock slides prevent the establishment of forest conditions; typically shrub and herb dominated (sitka alder [*Alnus sinuata*] and/or vine maple [*Acer circinatum*]).

3.3.2 *Cliffs.* A high, steep formation, usually of rock. Coastal cliffs are a separate category under Marine Habitat Elements.

3.3.3 *Caves.* An underground chamber open to the surface with varied opening diameters and depths; includes cliff-face caves, intact lava tubes, coastal caves, and mine shafts.

3.3.4 *Rocky outcrops and ridges.* Areas of exposed rock.

3.3.5 *Rock crevices.* Refers to the joint spaces in cliffs, and fissures and openings between slab rock; crevices among rocks and boulders in talus fields are a separate category (talus).

3.3.6 *Barren ground.* Bare exposed soil with >40% of area not vegetated; includes mineral licks and bare agricultural fields; natural bare exposed rock is under the rocky outcrop category.

3.3.7 *Playa (alkaline, saline).* Shallow desert basins that are without natural drainage-ways where water accumulates and evaporates seasonally.

3.4 *Snow.* Selected features of snow.

3.4.1 *Snow depth.* Any measure of the distance between the top layer of snow and the ground below.

3.4.2 *Glaciers, snow field.* Areas of permanent snow and ice.

4. Freshwater Riparian and Aquatic Bodies Habitat Elements

Includes selected forms and characteristics of any body of freshwater.

4.1 *Water characteristics.* Includes various freshwater attributes. Ranges of continuous attributes that are key to the queried species, if known, will be in the comments.

4.1.1 *Dissolved oxygen.* Amount of oxygen passed into solution.

4.1.2 *Water depth.* Distance from the surface of the water to the bottom substrate.

4.1.3 *Dissolved solids.* A measure of dissolved minerals in water

4.1.4 *Water pH.* A measure of water acidity or alkalinity.

4.1.5 *Water temperature.* Water temperature range that is key to the queried species; if known, it is in the comments field.

4.1.6 *Water velocity.* Speed or momentum of water flow.

4.1.7 *Water turbidity.* Amount of roiled sediment within the water.

4.1.8 *Free water.* Water derived from any source.

4.1.9 *Salinity and alkalinity.* The presence of salts.

4.2 *Rivers and streams.* Various characteristics of streams and rivers.

4.2.1 *Oxbows.* A pond or wetland created when a river bend is cut off from the main channel of the river.

4.2.2 *Order and class.* Systems of stream classification.

4.2.2.1 *Intermittent.* Streams/rivers that contain nontidal flowing water for only part of the year; water may remain in isolated pools.

4.2.2.2 *Upper perennial.* Streams/rivers with a high gradient, fast water velocity, no tidal influence; some water flowing throughout the year, substrate consists of rock, cobbles, or gravel with occasional patches of sand; little floodplain development.

4.2.2.3 *Lower perennial.* Streams/rivers with a low gradient, slow water velocity, no tidal influence; some water flowing throughout the year, substrate consists mainly of sand and mud; floodplain is well developed.

4.2.3 *Zone.* System of water body classification based on the horizontal strata of the water column.

>4.2.3.1 *Open water.* Open water areas not closely associated with the shoreline or bottom.

>4.2.3.2 *Submerged/benthic.* Relating to the bottom of a body of water, includes the substrate and the overlaying body of water within 3.2 feet (1 m) of the substrate.

>4.2.3.3 *Shoreline.* Continually exposed substrate that is subject to splash, waves, and/or periodic flooding. Includes gravel bars, islands, and immediate nearshore areas.

4.2.4 *In-stream substrate.* The bottom materials in a body of water.

>4.2.4.1 *Rocks.* Rocks >10 inches (256 mm) in diameter.

>4.2.4.2 *Cobble/gravel.* Rocks or pebbles, .1-10 inches (2.5-256 mm) in diameter, substrata may consist of cobbles, gravel, shell, and sand with no substratum type >70% cover.

>4.2.4.3 *Sand/mud.* Fine substrata <.01 inch (1mm) in diameter, little gravel present, may be mixed with organics.

4.2.5 *Vegetation.* Herbaceous plants.

>4.2.5.1 *Submergent vegetation.* Rooted aquatic plants that do not emerge above the water surface.

>4.2.5.2 *Emergent vegetation.* Rooted aquatic plants that emerge above the water surface.

>4.2.5.3 *Floating mats.* Unrooted plants that form vegetative masses on the surface of the water.

4.2.6 *Coarse woody debris in streams and rivers.* Any piece of woody material (debris piles, stumps, root wads, fallen trees) that intrudes into or lies within a river or stream.

4.2.7 *Pools.* Portions of the stream with reduced current velocity, often with water deeper than surrounding areas.

4.2.8 *Riffles.* Shallow rapids where the water flows swiftly over completely or partially submerged obstructions to produce surface agitation, but where standing waves are absent.

4.2.9 *Runs/glides.* Areas of swiftly flowing water, without surface agitation or waves, which approximates uniform flow and in which the slope of the water surface is roughly parallel to the overall gradient of the stream reach.

4.2.10 *Overhanging vegetation.* Herbaceous plants that cascade over stream and river banks and are <3.2 feet (1m) above the water surface.

4.2.11 *Waterfalls.* Steep descent of water within a stream or river.

4.2.12 *Banks.* Rising ground that borders a body of water.

4.2.13 *Seeps or springs.* A concentrated flow of ground water issuing from openings in the ground.

4.3 *Ephemeral pools.* Pools that contain water for only brief periods of time usually associated with periods of high precipitation.

4.4 *Sand bars.* Exposed areas of sand or mud substrate.

4.5 *Gravel bars.* Exposed areas of gravel substrate.

4.6 *Lakes/ponds/reservoirs.* Various characteristics of lakes, ponds, and reservoirs.

4.6.1 *Zone.* System of water body classification based on the horizontal strata of the water column.

>4.6.1.1 *Open water.* Open water areas not closely associated with the shoreline or bottom substrates.

>4.6.1.2 *Submerged/benthic.* Relating to the bottom of a body of water, includes the substrate and the overlaying body of water within one meter of the substrate.

>4.6.1.3 *Shoreline.* Continually exposed substrate that is subject to splash, waves, and/or periodic flooding. Includes gravel bars, islands, and immediate nearshore areas.

4.6.2 *In-water substrate.* The bottom materials in a body of water.

>4.6.2.1 *Rock.* Rocks >10 inches (256 mm) in diameter.

>4.6.2.2 *Cobble/gravel.* Rocks or pebbles, .1-10 inches (2.5-256 mm) in diameter, substrata may consist of cobbles, gravel, shell, and sand with no substratum type exceeding 70% cover.

>4.6.2.3 *Sand/mud.* Fine substrata <.1 inch (2.5 mm) in diameter, little gravel present, may be mixed with organics.

4.6.3 *Vegetation.* Herbaceous plants.

>4.6.3.1 *Submergent vegetation.* Rooted aquatic plants that do not emerge above the water surface.

>4.6.3.2 *Emergent vegetation.* Rooted aquatic plants that emerge above the water surface.

>4.6.3.3 *Floating mats.* Unrooted plants that form vegetative masses on the surface of the water.

4.6.4 *Size.* Refers to whether or not the species is differentially associated with water bodies based on their size.

>4.6.4.1 *Ponds.* Bodies of water <5 acre (2 ha).

>4.6.4.2 *Lakes.* Bodies of water $\geq$5 acre (2 ha).

4.7 *Wetlands/marshes/wet meadows/bogs and swamps.* Various components and characteristics related to any of these systems.

4.7.1 *Riverine wetlands.* Wetlands found in association with rivers.

4.7.2 *Context* When checked, indicates that the setting of the wetland, marsh, wet meadow, bog, or swamp is key to the queried species.

>4.7.2.1 *Forest.* Wetlands within a forest.

>4.7.2.2 *Nonforest.* Wetlands that are not surrounded by forest.

4.7.3 *Size.* When checked, indicates that the queried species is differentially associated with a wetland, marsh, wet meadow, bog, or swamp based on the size of the water body.

4.7.4 *Marshes.* Frequently or continually inundated wetlands characterized by emergent herbaceous vegetation (grasses, sedges, reeds) adapted to saturated soil conditions.

4.7.5 *Wet meadows.* Grasslands with waterlogged soil near the surface but without standing water for most of the year.

4.8 *Islands.* A piece of land made up of either rock and/or unconsolidated material that projects above and is completely surrounded by water.

4.9 *Seasonal flooding.* Flooding that occurs periodically due to precipitation patterns.

5. Marine Habitat Elements

Selected biotic and abiotic components and characteristics of marine systems.

5.1 *Zone.* System of marine classification based on water depth, and relationship to substrate.

5.1.1 *Supratidal.* The zone that extends landward from the higher high water line up to either the top of a coastal cliff or the landward limit of marine process (i.e., storm surge limit).

5.1.2 *Intertidal.* The zone between the higher high water line and the lower low water line.

5.1.3 *Nearshore subtidal.* The zone that extends from the lower low water line seaward to the 65 foot (20 m) isobath, typically within .6 miles (1 km) of shore.

5.1.4 *Shelf.* The area between the 65-650 feet (20-200 m) isobath, typically within 36 miles (60 km) of shore.

5.1.5 *Oceanic.* The zone that extends seaward from the 650 feet (200 m) isobath.

5.2 *Substrates.* The bottom materials of a body of water.

5.2.1 *Bedrock.* The solid rock underlying surface materials.

5.2.2 *Boulders.* Large, worn, rocks >10 inches (256 mm) in diameter.

5.2.3 *Hardpan.* Consolidated clays forming a substratum firm enough to support an epibenthos and too firm to support a normal infauna (clams, worms, etc.), but with an unstable surface that sloughs frequently.

5.2.4 *Cobble.* Rocks or pebbles, 2.5-10 inches (64-256 mm) in diameter, may be a mix of cobbles, gravel, shells, and sand, with no type exceeding 70% cover.

5.2.5 *Mixed-coarse.* Substrata consisting of cobbles, gravel, shell, and sand with no substratum type exceeding 70% cover.

5.2.6 *Gravel.* Small rocks or pebbles, 0.2-2.5 inches (4-64 mm) in diameter.

5.2.7 *Sand.* Fine substrata <0.2 inch (4 mm) in diameter, little gravel present, may be mixed with organics.

5.2.8 *Mixed-fine.* Mixture of sand and mud particles <0.2 inch (4 mm) in diameter, little gravel present.

5.2.9 *Mud.* Fine substrata <.002 inch (0.06 mm) in diameter, little gravel present, usually mixed with organics.

5.2.10 *Organic.* Substrata composed primarily of organic matter such as wood chips, leaf litter, or other detritus.

5.3 *Energy.* Degree of exposure to oceanic swell, currents, and wind waves.

5.3.1 *Protected.* No sea swells, little or no current, and restricted wind fetch.

5.3.2 *Semi-protected.* Shorelines protected from sea swell, but may receive waves generated by moderate wind fetch, and/or moderate-to-weak tidal currents.

5.3.3 *Partially exposed.* Oceanic swell attenuated by offshore reefs, islands, or headlands, but shoreline substantially exposed to wind waves, and/or strong-to-moderate tidal currents.

5.3.4 *Exposed.* Highly exposed to oceanic swell, wind waves, and/or very strong currents.

5.4 *Vegetation.* Includes herbaceous plants and plants lacking vascular systems.

5.4.1 *Mixed macro algae.* Includes brown, green, and red algae.

5.4.2 *Kelp.* Subaquatic rooted vegetation found in the nearshore marine environment

5.4.3 *Eelgrass.* Subaquatic rooted vegetation found in an estuarine environment

5.5 *Water depth.* Refers to the vertical layering of the water column.

5.5.1 *Surface layer.* The uppermost part of the water column.

5.5.1.1 *Tide rip.* A current of water disturbed by an opposing current, especially in tidal water or by passage over an irregular bottom.

5.5.1.2 *Surface microlayer (neuston).* The thin uppermost layer of the water's surface.

5.5.2 *Euphotic.* Upper layer of a water body that receives sufficient sunlight for the photosynthesis of plants.

5.5.3 *Disphotic.* Area below the euphotic zone where photosynthesis ceases.

5.5.4 *Demersal/benthic.* Submerged lands including vegetated and unvegetated areas.

5.6 *Water temperature.* Measure of ocean water temperature.

5.7 *Salinity.* The presence and concentration of salts; salinity range that is key to the species, if it is known, will be in the comments field.

5.8 *Forms.* Morphological elements within marine areas.

5.8.1 *Beach.* An accumulation of unconsolidated material (sand, gravel, angular fragments) formed by waves and wave-induced currents in the intertidal and subtidal zones.

5.8.2 *Off-shore islands/rocks/sea stacks/off-shore cliffs.* A piece of land made up of either rock and/or unconsolidated material that projects above and is completely surrounded by water at higher high water for large (spring) tide. Includes off-shore marine cliffs.

5.8.3 *Marine cliffs (mainland).* A sloping face steeper than 20° usually formed by erosion and composed of either bedrock and/or unconsolidated materials.

5.8.4 *Delta.* An accumulation of sand, silt, and gravel deposited at the mouth of a stream where it discharges into the sea.

5.8.5 *Dune.* In a marine context; a mound or ridge formed by the transportation and deposition of wind-blown material (sand and occasionally silt).

5.8.6 *Lagoon.* Shallow depression within the shore zone continuously occupied by salt or brackish water lying roughly parallel to the shoreline and separated from the open sea by a barrier.

5.8.7 *Salt marsh.* A coastal wetland area that is periodically inundated by tidal brackish or salt water and that supports significant (15% cover) nonwoody vascular vegetation (e.g., grasses, rushes, sedges) for at least part of the year.

5.8.8 *Reef.* A rock outcrop, detached from the shore, with maximum elevations below the high-water line.

5.8.9 *Tidal flat.* A level or gently sloping (<5°) constructional surface exposed at low tide, usually consisting primarily of sand or mud with or without detritus, and resulting from tidal processes.

5.9 *Water clarity.* As influenced by sediment load.

6. (No Data)

Formerly contained topographic information, such as elevation, that has been moved to the life history matrix.

7. Fire as a Habitat Element

Refers to species that benefit from fire. The time frame after which the habitat is suitable for the species, if known, will be found in the comments field.

8. Anthropogenic Related Habitat Elements

This section contains selected examples of human-related Habitat Elements that may be a key part of the environment for many species. These Habitat Elements may have either a negative or positive influence on the queried species.

8.1 *Campgrounds/picnic areas.* Sites developed and maintained for camping and picnicking.

8.2 *Roads.* Either paved or unpaved.

8.3 *Buildings.* Permanent structures.

8.4 *Bridges.* Permanent structures typically over water or ravines.

8.5 *Diseases transmitted by domestic animals.* Some domestic animal diseases may be a source of mortality or reduced vigor for wild species.

8.6 *Animal harvest or persecution.* Includes illegal harvest/poaching, incidental take (resulting from fishing net by-catch, or by hay mowing, for example), and targeted removal for pest control.

8.7 *Fences/corrals.* Wood, barbed wire, or electric fences.

8.8 *Supplemental food.* Food deliberately provided for wildlife (e.g., bird feeders, ungulate feeding programs, etc.) as well as spilled or waste grain along railroads and cattle feedlots.

8.9 *Refuse.* Any source of human-derived garbage (includes landfills).

8.10. *Supplemental boxes, structures and platforms.* Includes bird houses, bat boxes, raptor and waterfowl nesting platforms.

8.11 *Guzzlers and waterholes.* Water sources typically built for domestic animal use.

8.12 *Toxic chemical use.* Proper use of regulated chemicals; documented effects only.

8.12.1 *Herbicides/fungicides.* Chemicals used to kill vegetation and fungi.

8.12.2 *Insecticides.* Chemicals used to kill insects.

8.12.3 *Pesticides.* Chemicals used to kill vertebrate species.

8.12.4 *Fertilizers.* Chemicals used to enhance vegetative growth.

8.13 *Hedgerows/windbreaks.* Woody and/or shrubby vegetation either planted or that develops naturally along fencelines and field borders.

8.14 *Sewage treatment ponds.* Settling ponds associated with sewage treatment plants.

8.15 *Repellents.* Various methods used to repel or deter wildlife species that damage crops or property (excluding pesticides and insecticides).

8.15.1 *Chemical (taste, smell, or tactile).* Chemical substances that repel wildlife.

8.15.2 *Noise or visual disturbance.* Nonchemical methods to deter wildlife.

8.16 *Culverts.* Drain crossings under roads or railroads.

8.17 *Irrigation ditches/canals.* Ditches built to transport water to agricultural crops or to handle runoff.

8.18 *Powerlines/corridors.* Utility lines, poles, and rights-of-way associated with transmission, telephone, and gas lines.

8.19 *Pollution.* Human-caused environmental contamination.

8.19.1 *Chemical.* Contamination caused by chemicals.

8.19.2 *Sewage.* Contamination caused by human waste.

8.19.3 *Water.* Aquatic contamination from any source.

8.20. *Piers.* Structures built out over water.

8.21 *Mooring piles, dolphins, buoys.* Floating objects anchored out in the water for nautical purposes.

8.22 *Bulkheads, seawalls, revetment.* Retaining structures built to protect the shoreline from wave action.

8.23 *Jetties, groins, breakwaters.* Structures built to influence the current or protect harbors.

8.24 *Water diversion structures.* Structures built to funnel or direct water, including dams, dikes and levies.

8.25 *Log boom.* A raft of logs lashed together either to transport the logs or as barriers to boat traffic near marinas or dams.

8.26 *Boats/ships.* Watercraft, either motorized or nonmotorized.

8.27 *Dredge spoil islands.* Sediment deposited from dredging operations.

8.28 *Hatchery facilities and fish.* Fish that are hatched in captivity and later released into the wild. For simplicity this refers to freshwater areas, though marine birds and mammals likely feed on hatchery-released fish too. This also includes the facilities and their operation.

Habitat Elements Data Matrix

Based on Marcot's work, the Habitat Elements or HEs Digital Matrix (found on the CD-ROM) focuses on those correlates that consist of fine-scale or within-stand features. In keeping with the initial intent, HEs do not include those elements that may be used, but are nonessential to the success or viability of a population, thus they may also be thought of as the drivers of habitat selection. The HEs matrix depicts associations with all common species, including both residents and migrants, in Oregon and Washington.

The list of HEs and their definitions was derived from Marcot et al.[14] and was refined and edited based on the published literature (in Literature Cited, these are identified with an asterisk) and expert review. The final list comprises 287 HEs, including naturally occurring biological and physical elements as well as elements created or caused by human actions. Definitions are provided to characterize each element, and clarify the nature of its influence on wildlife species. The HEs are grouped into seven major categories, six of which have related subclasses (Table 2).

Category 1, entitled Habitat Elements, contains two broad subclasses: Forest/Woodland Vegetative Elements or Substrates and Shrubland/grassland Vegetative Elements or Substrates. These subclasses denote the floristic components and attributes that are found within forested and nonforested (shrubland/grassland) communities. Floristic components include plants (lichens, fungi, and ferns), and plant parts (roots, flowers, and bark); attributes include the herbaceous and shrub layers. The subclasses are nested within the broad wildlife habitats (see Chapter 1). That is, only species that occur within Forested Wildlife-habitats may be associated with Forest/Woodland Vegetative Elements or Substrates, and only species that occur within Shrubland/Grassland Wildlife-habitats may be associated with Shrubland/Grassland Vegetative Elements or Substrates. Some species that occur within both forested and nonforested habitats may be associated with HEs in both of the subclasses. Conversely, some forest or shrubland/grassland species may not be associated with any of the floristic components and attributes found in the category HEs; other HEs may play a larger role in driving the species' occurrence. Salamanders and other amphibians, for example, may occur in forested habitats but are most influenced by water characteristics, found in the category Freshwater Riparian and Aquatic Bodies discussed below. Thus, while the broader Wildlife-habitat determines which HEs (either forested or nonforested) a species may be associated with, however, it does not mandate an association. That is, when a HE from either or both of the two subclasses Forest/Woodland Vegetative Elements or Shrubland/Grassland Vegetative Elements is associated with a species, the association(s) imply that the HE is key to the species well being because it promotes the species' distribution, abundance, fitness, and viability

Category 2, Ecological Habitat Elements, contains HEs that describe interspecific relationships. Namely, these elements depict the influence of exotic species, insect population irruptions, beaver/muskrat activities, and burrows on a given species. The first two, Exotic Species and Insect Population Control may exert either positive or negative influences on a given species, whereas the latter two depict positive relationships only. For example, only species that are positively influenced by another species' burrow (e.g., secondary burrow users) will be associated with burrows. Species that are negatively influenced, such as animals that may break their legs stepping into badger burrows, are not.

Nonvegetative, Abiotic Terrestrial Habitat Elements, Category 3, contains a variety of abiotic HEs such as Rocks, Soils, Rock Substrates, and Snow; each has >2 subclasses. Habitat Elements under the classes Soils and Snow may depict either negative or positive influences. Freshwater Riparian and Aquatic Bodies Habitat Elements, Category 4 characterizes many aquatic attributes, including those associated with rivers and streams, lakes, ponds and reservoirs, and wetlands/marshes/wet meadows/bogs/swamps. These HEs depict primarily positive influences, with the exception of the class Water Characteristics, which includes Habitat Elements such as Water Temperature, Water Velocity and Water Depth, all of which could either promote or inhibit species' viability.

Marine Habitat Elements are found in Category 5. Only species that occur in the Marine Wildlife-habitats are associated with Marine Habitat Elements. Examples include the classes Zone, Substrates, Energy Vegetation, Water Depth, Water Temperature Salinity, Forms, and Water Clarity. The only HEs that may exhibit either positive or negative influences are Water Temperature and Salinity.

Category 7 consists of only one Habitat Element, Fire. This element, when associated with a given species, indicates that fire has a positive effect on the species distribution, occurrence, and viability. Some species that benefit from fire include grazers, which feed on the lush new growth initiated after burns, cavity and snag users, and species that use down wood. Panelists indicated in the comments the period of time after which the burned stand becomes beneficial to the species.

Lastly, Category 8, Anthropogenic-related Habitat Elements, contains human-made features, such as roads, buildings, and fences; activities such as toxic chemical use, animal harvest and hatchery fish releases; and outcomes, such as pollution and refuse. We included these HEs to acknowledge that most animals encounter the effects of humans in their environment, some of which may benefit certain species, many of which do not. Because of the dual nature of these anthropogenic elements, all of them may exert positive or negative influences depending on a variety of factors, including the species involved, the intensity and scale of the activity or effect, and the duration of the activity.

On completion of the HEs list, we used expert panels to indicate which HEs are associated with each species. Specifically, we posed the question, Which HEs, based on our list, most strongly influence this species' distribution, abundance, fitness, and viability? As a group, the panelists reviewed the HEs list, and indicated those elements they believed to be most relevant. When necessary, the panelists formulated comments to elucidate the relationship between the Habitat Element and the species. Clarification was particularly important for Habitat Elements that may exert both positive and negative influences. Other types of information that may be found in the comments include the following: (1) the range of values necessary for use of the HE. For example, if the element Water Depth is listed as being important to the animal, the specific water depth(s) may be described; (2) the spatial, temporal, or geographical/topographical contexts necessary for use of the HE. That is, whether the HE is used only in certain parts of the animal's range, or in specific seasons of the year, or on particular slopes, etc.; (3) whether the importance of the HE is dependent on relationships with other HEs, Structural Conditions, and/or Wildlife Habitats. If so, the relevant habitat components are listed; and (4) the activity (e.g., reproduction, foraging, etc.) associated with the HE. Some HEs may only be used for specific purposes, for example, snags during the nesting season.

Conclusion

The final Structural Conditions and Habitat Elements Matrices are the most complete characterization of intermediate and fine-scale features to date; however, because of the sheer magnitude of the exercise, some species associations may have been mistakenly omitted. We expect that as our collective knowledge about wildlife-habitat relationships evolves, additions and corrections to all three habitat matrices (Wildlife-Habitat Types, Structural Conditions, and Habitat Elements) may be identified. The matrices associated with this book are relationally tied to illustrate the fact that wildlife exist in a multi-dimensional environment. For this reason, we stress the importance of getting the big picture and using the matrices to their best advantage, that is, as inter-related databases, each providing a part of the equation that depicts in its entirety the wildlife environment. For

example, there are 130 species closely associated with the forest structural conditions, but 40 (38%) of these species require a specific Habitat Element to reproduce in the forest environment. Thus, when classifying what is or is not habitat for a particular species it is imperative to look across scales and ask, What is important at the landscape or regional scale? This information is provided in the broad Wildlife-habitat Types Matrix that depicts habitat relationships based on floristic communities (see Chapter 1). Next, one must assess the particular structures and/or seral stages within those communities to derive information on how species occurrence is influenced by vegetation of varying age, size class, and density. The Structural Conditions Matrix, defined in this chapter, elucidates wildlife-habitat relationships at what can be thought of as the "stand-level." Lastly, the fine-scale features, both physical and biological, that are essential for a species occurrence or precludes it, must be identified. The Habitat Elements Matrix, described in this chapter, delineates the with in-stand attributes that most influence each species.

Literature Cited

1. *American Fisheries Society. 1985. Aquatic habitat inventory, glossary and standard methods. Habitat Inventory Committee, American Fisheries Society, Western Division.
2. Bilby, R. 1988. Interactions between aquatic and terrestrial systems. Pages 13-29 in: K. Raedeke, editor. Streamside management-riparian wildlife and forestry interactions. University of Washington, Institute of Forest Resources, Seattle, WA. Contribution 59.
3. *Brown, E. R., editor. 1985. Management of wildlife and fish habitats in forests of western Oregon and Washington. U.S. Forest Service, Publication R6-F&WL-192-1985, Portland, OR.
4. *Buckman, H. O., and N. C. Brady. 1969. The nature and properties of soils, 7th edition. Macmillan Co., New York, NY.
5. *Bull, E. L., C. G. Parks, and T. R. Torgersen. 1997. Trees and logs important to wildlife in the interior Columbia River basin. U.S. Forest Service General Technical Report 391. Portland, OR.
6. *Cooperrider, A. Y., R. J. Boyd, and H. R. Stuart. 1986. Inventory and monitoring of wildlife habitat. U.S. Bureau of Land Management Service Center, BLM-YA-PT-87/001. Denver, CO.
7. *Cowardin, L. M., V. Carter, F. C. Golet, and E. T. LaRoe. 1979. Classification of wetland and deepwater habitats of the United States. U.S. Fish and Wildlife Service, FWS/OBS-79/31.
8. *Dethier, M. N. 1990. A marine and estuarine habitat classification system for Washington state. Washington Natural Heritage Program, Department of Natural Resources, Olympia, WA.
9. *Forest Ecosystem Management Assessment Team. 1993. Forest ecosystem management: an ecological, economic and social assessment. U.S. Forest Service, Portland, OR.
10. Franklin, J., and T. Spies. 1991. Composition, function, and structure of old-growth Douglas-fir forests. Pages 71-80 in: L. F. Ruggiero, K. B. Aubry, A. B. Carey, and M. H. Huff, editors. Wildlife and vegetation of unmanaged Douglas-fir forests. U.S. Forest Service, Pacific Northwest Research Station, Portland, OR. General Technical Report PNW-GTR-285.
11. Hattenschwiler, S., F. Miglietta, A. Raschi, and C. Korner. 1997. Thirty years of in situ tree growth under elevated CO_2, a model for future forest responses? Global Change Biology 2: 377-387.
12. *Howes, D. E., J. R. Harper, and E. H. Owens. 1994. Physical shore-zone mapping system for British Columbia. British Columbia Ministry of Environment, Lands and Parks, Victoria, BC, Canada.

13. Kauffman, B. 1988. The Status of Riparian Habitats in Pacific Northwest Forests. Pages 45-55 *in:* K. Raedeke, editor. Streamside management!riparian wildlife and forestry interactions. University of Washington, Institute of Forest Resources, Seattle, WA. Contribution 59.

14. Marcot, B. G., M. A. Castellano, J. A. Christy, L. K. Croft, J. F. Lehmkuhl, R. H. Naney, K. Nelson, C. G. Niwa, R. E. Rosentreter, R. E. Sandquist, B. C. Wales, and E. Zieroth. 1997. Pages 1497-1713 *in:* T. M. Quigley and S. J. Arbelbide, editors. Terrestrial ecology assessment: an assessment of ecosystem components in the interior Columbia Basin and portions of the Klamath and Great Basins. Volume 3. US Forest Service General Technical Report PNW-GTR-405. U.S. Forest Service Pacific Northwest Research Station, Portland, OR.

15. *Maser, C., J. M. Geist, D. M. Concannon, R. Anderson, and B. Lovell. 1979. Wildlife-habitats in managed rangelands. The Great Basin of southeastern Oregon: geomorphic and edaphic habitats. U.S. Bureau of Land Management, General Technical Report PNW-99.

16. *Mayer, K. E., and W. F. Laudenslayer, Jr., editors. 1988. A guide to wildlife-habitats of California. California Department of Forestry and Fire Protection, Sacramento, CA.

17. Norby, R.J., C. A. Gunderson, S. D. Wullschleger, E. G. O'Neill, and M. K. McCracken. 1992. Productivity and compensatory responses of yellow-poplar trees in elevated CO_2, Nature 357: 322-324.

18. *Parks, C. G., E. L. Bull and T. R. Torgerson. 1997. Field guide for the identification of snags and logs in the interior Columbia River basin. U.S. Forest Service General Technical Report 390. Portland, OR.

19. Ruggiero, L. F., L. C. Jones, K. B. Aubry. 1991. Plant and animal habitat associations in Douglas-fir forests of the Pacific Northwest: an overview. Pages 447-462 *in:* L. F. Ruggerio, K. B. Aubry, A. B. Carey, and M. H. Huff, editors. Wildlife and vegetation of unmanaged Douglas-fir forests. U.S. Forest Service Pacific Northwest Research Station, Portland, OR. General Technical Report PNW-GTR-285.

20. *Smith, R. L. 1966. Ecology and field biology. Harper and Row, New York, NY.

21. *Stevens, V. 1995. Wildlife diversity in British Columbia: distribution and habitat use of amphibians, reptiles, birds, and mammals in biogeoclimatic zones. Research Branch, British Columbia Ministry of Forestry, Wildlife Branch, British Columbia Ministry of the Environment, Lands and Parks, Victoria, BC, Canada. Work Paper 04/1995.

22. *Thomas, J. W. 1979. Wildlife-habitats in managed forests—the Blue Mountains of Oregon and Washington. U.S. Forest Service Agricultural Handbook No. 553, Portland, OR.

23. *Washington Department of Fish and Wildlife. 1993. Guidelines for using lake and stream survey forms. Washington Department of Fish and Wildlife, Fisheries Management Division. Olympia, WA.

24. *West, N. R. 1990. Structure and function of microphytic soil crusts in wildland ecosystems of arid to semi-arid regions. *In* M. Begon, A. H. Fitter and A. Macfadyen, editor. Advances in Ecological Research, Academic Press, Volume 20. San Diego, CA.

4

Management of Within-stand Forest Habitat Features

William C. McComb

Introduction

Managing forests to produce a desirable mix of forest resources, including timber and wildlife, requires an understanding of how animals respond to habitat within and among stands over a landscape. Management strategies aimed at long-term population change are most likely to succeed if they alter habitat quantity, quality, and/or distribution. In this chapter, I focus on characteristics of habitat that could be manipulated through silviculture or vegetation management within stands. Habitat also includes abiotic factors such as soils, geology and climate, but vegetative characteristics represent the greatest opportunity for manipulation.

I provide a conceptual framework for wildlife habitat management at the stand scale that silviculturists can consider when taking actions to manage wildlife habitat and, ultimately, wildlife populations and communities. That framework builds on a common understanding of several terms: *habitat* is the place where a species lives. It includes the physical and biological resources necessary to support a viable population over space and through time. Each species and each population has its own habitat requirements. *Populations* are self-sustaining assemblages of individuals of one species over space and through time. *Communities*, by contrast, are assemblages of populations over space and through time.

There are two common approaches to management of forest habitat for wildlife: management of individual species and management of communities. Occasionally managers choose to focus management on indicator species or guilds of species in order to meet the needs for many other species. This approach should be avoided, because each species has its own habitat requirements and each will respond differently to management activities.[31] Forest disturbances change the abundance of individuals in most populations, and those changes also affect the composition of wildlife communities. Management of individual species, therefore, has consequences for other species in the community occupied by the species that is being managed. Forest disturbances that benefit black-tailed deer, for example, probably would benefit creeping voles and orange-crowned warblers but not Douglas squirrels or pileated woodpeckers.

Vegetation management by forest-land managers is probably the greatest factor influencing the abundance and distribution of animals in our forests. I will introduce concepts of habitat function, population change, and habitat patterning. These concepts are used to illustrate how silvicultural systems and habitat patterning can influence demographic patterns of animals over large forest areas. The result, finally, is a framework for meeting forest-wildlife goals through habitat manipulation.

Habitat Function

A population gains energy from food resources and conserves energy by exploiting cover resources (Figure 1). The rate of net primary production is fixed over large areas and time, because there is a solar constant and because climate changes are relatively slow. Herbivores in forests exist in a sea of energy, but food quantity may not be as important as food quality.[36] Animals require the digestible energy in food, but indigestible portions of food (e.g., cellulose, lignin, chiton, or bones) or compounds in the plants that inhibit digestion (tannins and other phenols) reduce food quality. Net energy available after metabolism increases the animal's fitness depending on the quality of the cover available to the animal.

Many browse species contain high levels of phenols that reduce their digestibility for many herbivores,[19] but mule deer saliva contains a substance (prolene) that binds with the phenols and reduces their effectiveness.[49] Some plants, therefore, may have high levels of compounds that

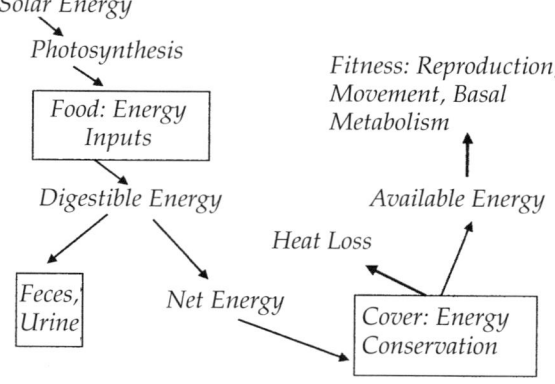

Figure 1. Conceptual diagram of acquisition and conservation of energy by an animal and the effect on individual fitness.

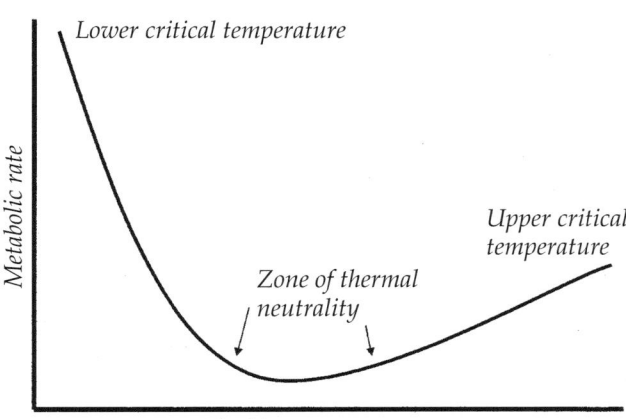

Figure 2. The zone of thermal neutrality is the ambient temperature where a homeothermic organism expends the least amount of energy maintaining its body temperature.

reduce digestibility for some wildlife species, but the co-evolution of plants and herbivores has resulted in plant defense mechanisms that are less effective for herbivores such as deer.

It is advantageous for an animal to conserve any energy that it acquires. Mammals and birds that maintain a constant body temperature expend a large amount of energy to maintain that temperature. Energy expenditures to maintain body temperature are minimized in an animal's *thermal neutral zone* (Figure 2). Any departure from the thermal neutral zone results in increased expenditure of energy, so animals often select habitat that reduces climatic extremes. There are upper and lower critical temperatures beyond which is lethal. Cover from overheating is especially important to large animals such as elk that may find it particularly difficult to release excess heat unless water is available to aid in evaporative cooling. Cover from severe cold is especially important to a species with a high surface area to mass ratio (e.g., small bird). Cover that allows an animal to stay within an acceptable range of temperatures (particularly those that approach the thermal neutral zone) is important to maintaining a positive balance of net energy and hence influences animal fitness.

Cover also can refer to the portion of habitat that an animal uses for nesting and escape from predators. The effectiveness of nest box programs for wood ducks, bluebirds and other secondary cavity-using species, demonstrates that manipulation of the quantity, quality, and availability of cover resources can be an effective management technique.[39] Land managers can manage habitat for a species by altering food quality, quantity, and/or availability while also altering the quality, quantity, and/or availability of cover. This strategy can lead to drastic changes in habitat quality for the species.

Water is differentially important to animal species. Some require free water or high humidity (mountain beaver, for example, have a primitive uretic system). Other species obtain most of their water from their food (e.g.,

pocket gophers). Some species use water as a form of cover to enhance evaporative cooling (e.g., elk) or to escape predators (e.g., white-tailed deer). Still others such as amphibians require free water or moist environments for reproduction.

The size of habitat is an important determinant of its suitability for a species. A patch of habitat must be sufficiently large to provide energy inputs and energy conservation features to sustain a population. Habitat may occur in one large unit, but more commonly it is distributed in patches through other less suitable habitat. If these habitat patches are too widely distributed, then the animal expends more energy moving among patches than it receives from those patches. The amount of habitat, therefore, and its quality and distribution are interrelated. Increasing any or all of these attributes of habitat increases the net energy available to animals that use this energy to maintain body temperature, move to food and cover, and reproduce.

Population Growth

Populations can be manipulated by modifying habitat and thereby influencing animal fitness. The linkages between animal demography and habitat are complex, but some understanding of these relations is necessary for successful habitat management. Each species has its own potential for population increase, and this potential is described as the *intrinsic rate of natural increase*. There is a solar constant, so energy available to animals is limited (Figure 1). Food, therefore, becomes scarcer or of poorer quality as the population grows. As populations grow, cover is occupied by other individuals and the risk of disease and parasitism increases. Interspecific competition for resources causes some subordinate individuals to use suboptimal cover.[18] As food, cover, or other resources become limiting, the population growth rate decreases, because mortality increases or because reproduction decreases. This process is termed *logistic growth* (Figure 3). If we assume that resources are constant, then the population reaches a point where births equal deaths and growth becomes 0. This point is termed the *carrying capacity* of the habitat for the population.

But resources are not constant; they change daily, seasonally, and annually. Birth rates, death rates, and movement rates are variable over both space and time as habitat changes through disturbances and succession. Carrying capacity, consequently, is always changing. The concept of a *dynamic carrying capacity* is useful to land managers, because it provides the link between habitat quality and population growth. Manipulating habitat to change carrying capacity is a particularly effective approach to long-term manipulation of wildlife populations.

Populations do not always reach carrying capacity in relation to habitat quality. Some species, such as voles, snowshoe hares, and ruffed grouse, follow a "boom and bust" population pattern. Populations grow for about 3-6 years and then rapidly decline for another 3-6 years. High-quality habitat usually increases the highs and decreases

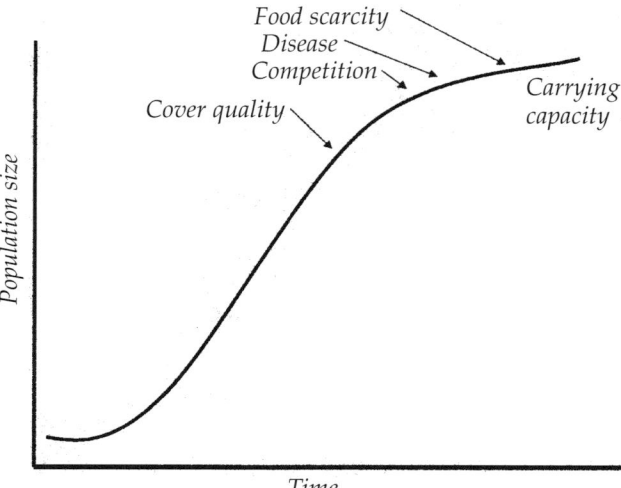

Figure 3. Logistic population growth occurs as resources become more limited with increasing populations.

the lows of a population cycle, but habitat probably does not directly mediate these cycles, because they occur throughout much of the geographic range of the species.[26]

Movement and Dispersal

Animals move for various reasons, but movement to and from food and cover is the most obvious daily requirement of any species. This distance is very short for species such as clouded salamanders, while others such as American marten may move among stands over a landscape. The arrangement of food and cover in an area can affect energy costs associated with movements from one resource to another. Another type of movement, dispersal, may result from overcrowding. As a population approaches carrying capacity, it can be energetically advantageous for some individuals search for more acceptable habitat.[18] Dispersal can be influenced by habitat quality. The rate of movement and the time- and habitat-specific probability of survival in various habitat types during dispersal influence the probability that an organism will survive its dispersal to a good patch of habitat.[62]

If we assume that species have evolved and persisted within a range of natural variability from natural disturbances in forest systems, then we might better be able to understand how human-induced disturbances may influence habitat availability and population dynamics.

The following section describes natural disturbances common in the Pacific Northwest and the habitat features that seem to change dramatically depending on the type of disturbance.

Natural Disturbance in Western Coniferous Forests

Species inhabiting forests often are adapted to the natural forest disturbance regimes in the region. Some species colonize after large, recent disturbances; others recolonize after the forest develops; and still others occupy late successional stages. Dead wood can be an important component of each of these seral stages (Table 1). Some colonizers of early seral stages are more abundant now than ever before. They include many herbivores that can influence the trajectory of ecological succession within a stand,[45] particularly where the size and distribution of various seral stages optimize rates of population growth and colonization for each species.

Knowledge of natural disturbances in western coniferous forests can help when developing silvicultural systems that might meet the needs of forest-associated wildlife.[40] Coarse-scale disturbances typically occur over tens or hundreds of acres (hectares) both within stands as well as among them. Fine-scale disturbances occur within stands at a scale of <1 tree height in width. Coarse- and fine-scale disturbances have affected the establishment, development, and destruction of unmanaged western coniferous forests.

The term *patch* is used to describe a homogeneous unit of resources that is different from the matrix in which it occurs.[42] Stands and patches can be synonymous or patches can occur within stands, depending on the resources under consideration. Patches may occur at any scale; a canopy gap may be a patch within a stand and a stand may be a patch within a landscape.

Coarse-scale disturbances

Except in coastal forests, fire was the most frequent and widespread coarse-scale disturbance occurring at scales up to 250,000 acres (100,000 ha). The fire regime of the Douglas-fir (*Pseudotsuga menziesii*) region in western Oregon and Washington is complex and includes fires of varying frequencies, intensities, sizes, and patterns. For instance, return intervals for stand-replacement fires at Mount Rainier National Park in the Washington Cascades

Table 1. Examples of animal species that would use four types of dead or dying trees in four stages of stand development in Pacific Northwest forests.

	Tree decay stage			
Seral stage	Live cavity-tree	Hard snag	Soft snag	Log
Stand reinitiation	House wren	American kestrel	Western bluebird	Alligator lizard
Stem exclusion	Winter wren	Hairy woodpecker	Chestnut-backed chickadee	Ensatina salamander
Transition	Red-breasted nuthatch	Northern flicker	Red-breasted sapsucker	Clouded salamander
Shifting-gap phase	Spotted owl	Pileated woodpecker	Northern flying squirrel	American marten

are 400-500 years.[24] Stand-replacement fires in the central Oregon Cascades have return intervals of >200 years, while smaller ground fires that kill some green trees have return intervals of about 100-150 years.[44, 58]

Currently several different scenarios of forest development and succession are recognized for Oregon and Washington.[1, 44, 57] If a fire kills most of the trees in a stand and initiates establishment of Douglas-fir and western hemlock (*Tsuga heterophylla*) seedlings, then the stand will probably develop old-growth structure at a stand age of 150-250 years in the absence of management (depending on site quality). Snags and logs in such stands would probably follow a pattern of increase, decrease, and increase over a 400- to 800-year cycle[55] (Figure 4).

Lower intensity fires kill fewer trees in a patchy pattern (2-100 acres [1-40 ha]). A stand of various tree sizes and large dead trees produced a multi-layered canopy sooner than in the first scenario. Inputs of dead wood are more constant compared to the first scenario,[55] although fires may recur and delay forest development (Figure 5).

Fire return intervals in ponderosa pine (*Pinus ponderosa*) forests are much shorter (every 5-15 years), but return intervals increase with increasing elevation in eastside forests.[1] Fire intensity also often increases with elevation in these eastside areas.

Wind may cause either coarse- or fine-scale disturbances. Areas of high wind, such as the Pacific coast and the Columbia River Gorge may produce large windfall areas following the opening of a small part of a stand as the stand progressively blows down, in a domino effect. Elsewhere, wind disturbance usually consists of single trees or small patches (<10 acres [<4 ha]) of trees that are broken off or blown over. Rarely are all trees in a patch blown over, and large amounts of dead wood remain after the event (Figure 6).

Fine-scale disturbances

Fine-scale (<2 acres [<1 ha]) disturbances influence stand dynamics in natural stands. Fine-scale disturbances such as suppression mortality, root rots (e.g., *Phellinus* spp. and *Armillaria* spp.), localized windthrow, and light ground fires can lead to the death of individual or small groups of trees (Figure 7). Formation rates of these gaps are relatively low in old-growth Douglas-fir stands[56] (0.1-0.8% of the stand/year) compared to other old-growth forest types.[51] However, these canopy gaps are the sites of tree regeneration in mature stands,[56, 58] particularly for shade tolerant species. Considerable mortality and replacement of dominant trees occurred during 36 years in 500-year-old Douglas-fir/western hemlock forest.[16] Rates of dominant tree mortality were 0.75%/year, but the mortality was balanced by recruitment of shade-tolerant tree species. Gaps may be slow to fill because >70% of old-growth canopy trees die without disrupting the forest floor.[16, 56] In these cases, existing shade-tolerant shrubs dominate gaps.

In eastern Oregon and Washington, fire control has led to homogenization of large landscapes.[1] Especially in ponderosa pine forests, fire exclusion has caused many

Figure 4. A stand replacement fire kills most trees in a stand, but leaves considerable amounts of dead wood, which is used by many species of cavity-nesting wildlife. (Photograph by W.C. McComb)

Figure 5. A partially burned stand leaves many live green trees to grow after the disturbance as well as many dead trees that provide habitat for cavity nesters. A two-story stand is likely to develop after this disturbance. (Photograph by J. Tappeiner)

Figure 6. Windthrow leads to a dramatic increase in coarse woody debris in a stand and these structures are used by many species of mammals and amphibians in northwest forests. (Photograph by J. Tappeiner)

Figure 7. A canopy gap provides a site for tree and shrub regeneration and can aid in development of vertical structure in the stand. (Photograph by J. Tappeiner)

stands to develop outside of the range of natural variability in terms of species composition and structure. Clearly fire or some other disturbance will need to be reintroduced to recover these stands.

All of these scenarios have occurred in the region. The landscape patterns prior to timber management were a combination of stands created by disturbances of various frequencies, intensities, and sizes. Since timber management began in Pacific Northwest forests, the managed stands represent a much reduced range of variability in frequency, intensity, and size of disturbance from the natural disturbance regime.

Plant-Animal Interactions

Biotic interactions also can affect forest development and consequently the structure and composition of habitat for a variety of species. Certainly herbivory can play a significant role in directing successional pathways. Some plants have chemical and physical defenses against herbivory.[14] Plants also respond to animal herbivory by altering growth rates. Among many grasses, forbs, and some shrubs, moderate levels of herbivory can actually stimulate growth above the levels of either undisturbed or heavily grazed or browsed plants.[2, 13] It is widely assumed that this compensatory growth occurs at the expense of reproduction and that herbivory, therefore, results in decreased seed production or smaller seed sizes.

Herbivores affect forest systems in ways other than consumption. They aid in the dissemination of seeds, and they may help maintain site quality. Some plants are well adapted to dispersal on animals (for example, bedstraws [*Gallium* spp.]). Other plant species (e.g., dogwoods [*Cornus* spp.] and cherries [*Prunus* spp.]) are well adapted to scarification that results from passing through animal digestive systems and "direct-seeding" in a packet of fertilizer. Many fencerows, consequently, are dominated by cherries, hawthorns (*Crataegus* spp.), and dogwoods because birds often perch on fences after eating the fruits of these plant species.

Mycorrhizal fungi aid plants in the uptake of water and nutrients, and they can be particularly important to early plant growth and survival on harsh sites.[47] These fungi produce fruits underground, unlike most other fungi, and they do not, therefore, rely on aerial spore dispersal, as do other fungi. They seem, instead, to be well adapted to animal dispersal. Some fungi known as truffles are important components of the diets of some small mammals, particularly red-backed voles. These animals eat fruits and ingest spores, which then pass through the digestive system in a few days and are deposited at a new site. A new fungal mat may then grow from this site and ensure the presence and widespread distribution of mycorrhizae in the soil.[11] Mixing organic matter in the soil by these burrowing animals also is likely to influence decomposition rates and soil function.

The activities of some herbivores can have tremendous impact on habitat for other species.[45] The activities of American beaver, for example, create early seral-stage riparian forest and pool habitat that can be important to other species. For example, Suzuki[59] found very different amphibian and mammal communities associated with areas in the Oregon Coast Range that were impacted by beavers compared to similar areas where beavers did not build dams. Other examples include black bears that kill patches of trees in plantations, gophers that eliminate regeneration in patches, or elk herds that browse heavily next to riparian zones. All these activities create patchiness or heterogeneity in affected sites, and such patches can be important resource areas for other species.

Spatial and Temporal Scales of Habitat

Habitat typically is characterized by the physical setting (soils, aspect, elevation, and climate) and by the structure and composition of vegetation occupying the site. One attribute of habitat is its size. A 2-acre (1-ha) patch of habitat may be more than sufficient for a California red-backed vole but clearly is inadequate for its predator, the northern spotted owl. Habitat is also dynamic. It is affected by the geomorphic setting that slowly changes, by the weather that changes daily, seasonally, yearly, and over centuries, and by plant-community succession, which is rapid in the early stages and slower in the later stages. Vegetation management can alter the composition and structure of habitat and the direction of plant succession. Forest managers, therefore, must think of habitat over a range of spatial and temporal scales. One convenient way of dealing with these scales is to visualize a hierarchy of small patches of habitat for a species that serve as building blocks within the species' home range, which in turn serve as building blocks for populations (Figure 8). I discuss how habitat for a species might function at the stand level (a unit of forest land that will receive a silvicultural treatment), and then how habitat functions in aggregates of stands at the forest or landscape scales.

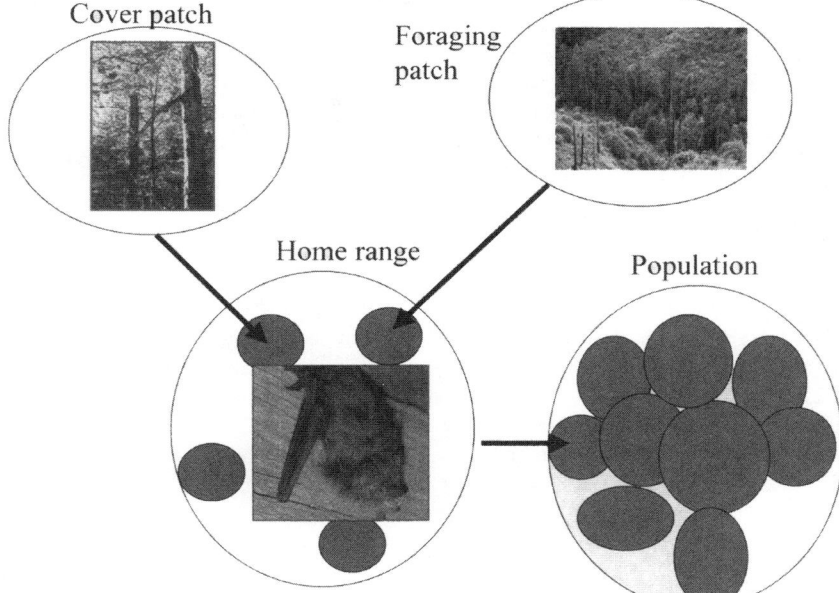

Figure 8. Adequate foraging and cover patches must be present within a species home range for it to be occupied, and sufficient home ranges must be available to support a population.

Spatial Scales of Habitat

Stands do not exist in isolation. Each stand is surrounded by other stands, corridors of vegetation, water courses, and other features (the *context* for the stand). All these features comprise all or a portion of a *landscape*. A landscape typically is considered the mosaic of vegetation and geomorphic features that occur across an area.[15] The elements of the landscape pattern that are important to some species of wildlife include the types of patches that make up the landscape, the sizes of those patches, the length of surrounding edge, and the nature of corridors or barriers between patches. Many of these elements interact to determine the suitability of habitat over the landscape for various plant and animal species. Large carnivores use very large areas for foraging, necessitating a landscape perspective in habitat management.[29] The total amount of habitat patches across a landscape, accordingly, strongly influences the occurrence or abundance of a species on that landscape.[37] Managers can manipulate stand structure and distribution across the landscape to influence the abundance and distribution of species within a planning area.

Species may be influenced by stand arrangement on the landscape.[28] Species that require forest-interior (core) patches may be adversely impacted by the proliferation of edges within the landscape that may result from disturbance.[34, 41] As more of the landscape is affected by disturbance, animals that move among remaining patches of habitat expend increasing amounts of energy, and availability of forest-interior habitat decreases.[17, 60] This process of habitat loss and fragmentation can be mediated by managing stands within the landscape to meet the needs of managed species.

Natural Levels of Variability in Managed Forests

Estimating parameters characteristic of natural disturbance regimes can facilitate prediction of forest recovery through the direction of succession and the subsequent development of vegetation structure.

Disturbance size and shape can influence the animal species that either remain in or recolonize a disturbed area.[50, 60] Organisms would likely be displaced by severe disturbances of a size larger than a home range, but they may not be displaced if the disturbance is sufficiently small relative to the home range size. I suspect that species inhabiting mature forests have been able to persist (1) by including fine-scale disturbances within their home ranges, or (2) by recolonizing stands of sufficient size that regrow to maturity following coarse-scale disturbances. *Disturbance shape* influences the amount of edge and hence microclimatic characteristics of the stand. The characteristics of stand edges can have profound effects on a number of bird and mammal species.[24, 42]

Disturbance intensity influences the amount of organic material destroyed and redistributed by the disturbance and hence the amount and form (living or dead) of material that remains after the disturbance. These residual structures may directly or indirectly be habitat for mature forest species. Gaps in mature and old-growth forests produce the snags and logs in the stand, and enhance its vertical complexity.[40] The residual structures following coarse-scale disturbances may persist into the next stand and provide the large trees and snags used by some species.[38, 46, 53] The residual organic material that remains after disturbance can influence the direction of succession and the rate of subsequent development following the disturbance.[57]

Disturbance frequency will influence the tree species composition and the amount of living and dead organic material present on the site over time. Frequent coarse-scale disturbances can delay the onset of forest development, or may preclude it. Infrequent fine-scale disturbance may delay the development of vertical structure in a stand.[56]

Disturbance density is the percentage of the stand or landscape occupied by a patch type resulting from a disturbance. The density of fine-scale disturbances also influences the abundance of structural features produced by the disturbance. A stand in which 1% of the stand is in canopy gaps <30 years old will have a less well developed vertical structure and fewer hard snags than one in which 15% is in gaps. Finally, the *disturbance pattern* may also influence habitat quality for some species. Clumped distributions of fine-scale disturbances may result in a cumulative decrease in habitat availability within an individual's home range. A more random or uniform distribution of disturbances may allow that individual to tolerate the same disturbance density because only a small portion of any individual's home range in the stand would be affected.

No single stand management system will precisely match the variability inherent in natural forests that resulted from a variety of disturbance regimes. But some of the variation can be incorporated into the managed forest landscapes of the region by using a variety of silvicultural systems. The choice of these systems will depend on the biological, social, and economic objectives for the stand and the landscape, and they will imitate natural disturbances to varying degrees (Figure 9). Indeed, the basis for development of existing silvicultural systems for timber objectives was that these systems reflect the regeneration and growth strategies of the commercially important tree species in a region. Intensive timber management as currently practiced leaves less dead wood

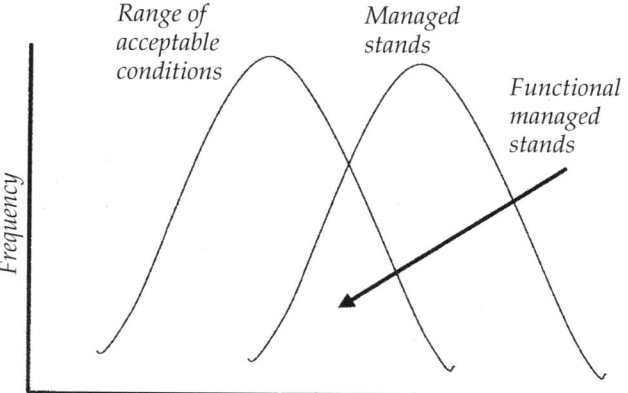

Figure 9. Species require certain habitat features over a range of sizes or amounts. The degree to which managed stands can provide these conditions will influence the proportion of the landscape that can be used by each species of wildlife.

and noncommercial plant species than natural disturbances,[23] so it may not imitate natural disturbances for other forest resources as well as it does for timber.

Habitat Characteristics Associated with Forest Wildlife

I assume that animals in forests use certain key structural and compositional features over home range-sized areas or larger. However, each species differs in its requirements for the magnitude and condition of these structures. To meet the needs of forest wildlife species, these habitat features should be considered as part of silvicultural systems employed in stands over areas at least as large as the largest home range size among the species under consideration multiplied by the desired population (Figure 8).

Large Trees

Some trees within a stand should be designated as legacy trees and left to grow to maturity and die through natural processes. Douglas-firs >50 inches dbh (>125 cm) are used by marbled murrelets,[53] red tree voles, and northern spotted owls. Large trees add to the vertical structure within old forests. Vertical structure may be associated with bird community structure in some forest types,[30] but it is unclear how important vertical structure is in western coniferous forests. Large trees also add large surfaces of deeply-fissured or scaly bark that is used by bark-foraging birds such as brown creepers[32] and they support lichens, an important food source for northern flying squirrels.[33] Designating Douglas-fir, western redcedar (*Thuja plicata*), bigleaf maple (*Acer macrophyllum*), Pacific madrone (*Arbutus menziesii*), red alder (*Alnus rubra*) and western hemlock as legacy trees in coastal and Cascades forests would provide a range of growth rates and bark surfaces and contribute to complexity in the stand. See chapter 24 for information on guidelines for numbers of living trees per acre (ha).

Some hardwoods, shrubs, and noncommercial conifers should be allowed to grow to maturity and die within these stands. Hardwoods often are considered competing vegetation in conifer stands. However, species such as Oregon oak (*Quercus garryanna*), Pacific madrone, and bigleaf maple (especially trees >20 inches [> 50 cm] dbh) produce many natural cavities and dead limbs (effectively, elevated snags), that are used by cavity nesting animals.[21,48] Hardwoods and shrubs also provide forage, fruits, and foliar insects that are food resources for many forest vertebrates[40] and hardwood composition in unmanaged stands in western Oregon influenced the abundance and occurrence of several bird species.[25]

Dead Wood

Species vary in their use of dead wood size and decay classes.[38] Although some quantitative information is available on snag use by some species, such as red-breasted sapsuckers, pileated woodpeckers, and hairy woodpeckers,[3,46,52] we have little quantitative information for other species, especially log-users.[38] Hence, setting stand goals for dead wood for species such as bats, shrew-moles, tailed frogs, and northwestern salamanders is difficult, but must be done to develop management prescriptions. McComb and Lindenmayer[38] offer one approach to addressing this problem.

Forest Floor Litter

Forest floor and below-ground conditions influence habitat quality for ground-foraging and burrowing species. Some terrestrial mammals and amphibians remain active below-ground during the summer. For instance, rough-skinned newts use logs and burrow systems of voles and shrews as summer daytime refugia (W.C. McComb and C.L. Chambers, Northern Arizona University, unpubl. data). Burrow systems of mountain beavers and pocket gophers are used by other species.[35] Except for anecdotal descriptions, we have no quantitative information that could be used to set goals for species associated with forest floor litter and burrows (e.g., northwestern salamanders, Pacific shrews, and ensatina salamanders).

Identifying Desired Future Conditions

Because habitat is dynamic, it is important to plan for a series of desired future conditions over space and through time for a stand. The first step is to develop diameter distributions (and associated levels of variance) for living and dead wood from stands that are functioning for the species of wildlife that are desired. The goal of the silvicultural system would be to provide the structure, composition, and as nearly as possible, the dead wood distribution in the managed stand to fall within the range of variability acceptable to the species. Models that can help the manager understand the dynamics of trees and dead wood should be used when setting goals (see chapter 24).

Effects of Silvicultural Treatments on Habitat

Silvicultural activities have partially replaced natural disturbances in managed forests. Forest managers can select the types and rates of disturbances that will meet specific resource objectives. Some habitat management issues that foresters and wildlife biologists face in the Pacific Northwest may result from insufficient consideration of the size, frequency, intensity and patterning of silvicultural disturbances on the landscape. A range of management decisions can be made on any given site that will result in stand conditions and plant communities that support only certain species (Figure 10). Consider, for example, an Oregon Coast Range site managed with the following combination of decisions: clearcut, no legacy (retention of logs, snags or green trees), no site preparation, rely only on natural regeneration, no vegetation management, and no precommercial thinning. The result would probably be a hardwood-shrub plant-community of small stature. Now consider the same site managed with the following decisions: clearcut, snag

legacy, plant Douglas-fir and western hemlock, seedling release, and thin to 300 trees/acre (750 trees/ha). The result would probably be a conifer-dominated stand condition with a grass-forb understory (sparse shrubs) during the early stages of development. Each condition is habitat for a different suite of species on the same site managed in one of two ways. Many possible decisions could be made early in stand development that could produce a wide range of stand conditions (Figure 10).

Traditional Management Approaches

Even-age Management

Clearcut and plant regeneration has been the system most frequently used in the westside forests of the Pacific Northwest, especially on private lands, but there are other options. Seed-tree and shelterwood systems are useful on certain sites and with certain tree species. The deferred rotation system described by Smith et al.[54] is an alternative that benefits certain species of animals by leaving large trees to grow through ≥2 rotations. Whichever system is chosen, the stand usually proceeds through site preparation, stand re-establishment, vegetation management, and stand-density management before it is ready for harvest at the end of the rotation. Decisions at each stage influence stand character and affect habitat quality for the wildlife species present at various stages of stand development (Figure 10).

Regeneration System. The selection of a regeneration system affects stand structure during the early stages of stand development. Seed-tree and shelterwood systems leave vertical structures (important to some birds and arboreal mammals) that remain in the stand until the seed trees and overwood are removed. Duration of the grass-forb stand condition is usually shorter and shrub conditions in the stand are more predominant with these systems than with traditional clearcutting, because of the difficulties associated with vegetation management when residual trees and snags are present. Deferred rotation systems offer the advantage of some vertical structure remaining in the stand throughout the rotation. Even-age management, however, often results in the creation of sharp, induced edges (depending on the stature of adjacent stands). These high contrast edges are beneficial to some species,[41] but not to others, especially some species of amphibians.[34] Plant species associated with these regeneration systems are often shade intolerant, although this can be adjusted with shelterwood systems and artificial regeneration. Partial-cut and clearcut stands function similarly when <10 trees/acre (<25/ha) are retained in partial cut stands.[7] Nonetheless, partial cut stands seem to provide habitat for more species of birds than clearcut stands, at least in the breeding season.[7, 61] More species were detected in uncut or patch-cut stands than in clearcut or partial cut stands during the winter, however.[6]

Legacy. At harvest and before site preparation, the land manager may decide to leave certain structural components of the previous stand on the site and into the next rotation. Snags are the most visible type of legacy left or created on many sites and they can significantly influence occurrence of a number of species.[4, 5, 38, 52] Logs and living conifers and hardwoods (and the lichens, bryophytes, and other less vagile species associated with them) left on the site can provide structural and compositional features that create conditions in the new stand more typical of those found after natural disturbances. Animal communities associated with stands that include these features would be more complex than communities in stands that lack similar components.

Site preparation. Site preparation ranges from very intense (mechanical scarification) to none. Mechanical scarification may significantly affect the below ground structure of the stand by temporarily removing burrows and compacting soils. It also may affect plant communities that develop after the disturbance. Intense scarification, burning, or some herbicides may reduce shrub development in subsequent stages of stand development. Light fires may proliferate the sprouting of shrubs and reduce the presence of grasses and forbs in an early stand. The choice of site preparation influences not only the trajectory of the plant community that develops on the site, but also the level of residual "legacy" that remains after the treatment. An intense burn or mechanical scarification, for example, reduces levels of dead wood on the site. This may be a desirable method for manipulating the habitat of some species that may influence regeneration success (such as mountain beaver), but it also may have adverse consequences for other species. Nonetheless, species such as Oregon voles and vagrant shrews increase after intense site preparation, but others such as Pacific and Trowbridge's shrews, ensatina salamanders, and Pacific giant salamanders decrease after site preparation.[9, 10]

Stand re-establishment. Land managers can determine the composition of developing stands by deciding which plant species will be re-established after site preparation. With natural advance regeneration, that decision may be unnecessary. Planting seedlings, however, can greatly affect the structure and composition of stands. A stand that includes several tree and shrub species, with different growth rates, will be more diverse than a stand planted with a single tree species. Density and spacing of the regeneration also could be affected. The size distribution of trees in the stand can be narrow (uniform spacing) or broad (irregular spacing) in an even-aged stand. Mixed-species planting or variable-density planting can provide heterogeneity in an otherwise homogenous system.

Vegetation Management. Management of competing vegetation can significantly affect the availability of certain plant species as food and cover. Herbicide applications that release conifers can temporarily decrease the availability of shrubs for shrub-nesting birds.[43] For small mammals and amphibians, site preparation seems to have a greater effect on animal abundance than spraying of glyphosate.[9, 10] Either chemical or manual vegetation control can influence the heterogeneity of developing stands. Spot control of competing vegetation could lead to a more heterogeneous stand than possible with broadcast application of a treatment. Manual control of many shrub species can lead to a proliferation of sprouts that increase amounts of available browse.

Perhaps the most profound effect of vegetation management is the influence that such activities have on the composition and structure of developing stands. Lack of any vegetation management in many Coast Range sites may lead to stands with a large component of shrubs or hardwoods that will benefit some species of wildlife.[25] Intensive vegetation management may lead to a conifer-dominated stand with little shrub development unless stand density is manipulated as the stand develops.

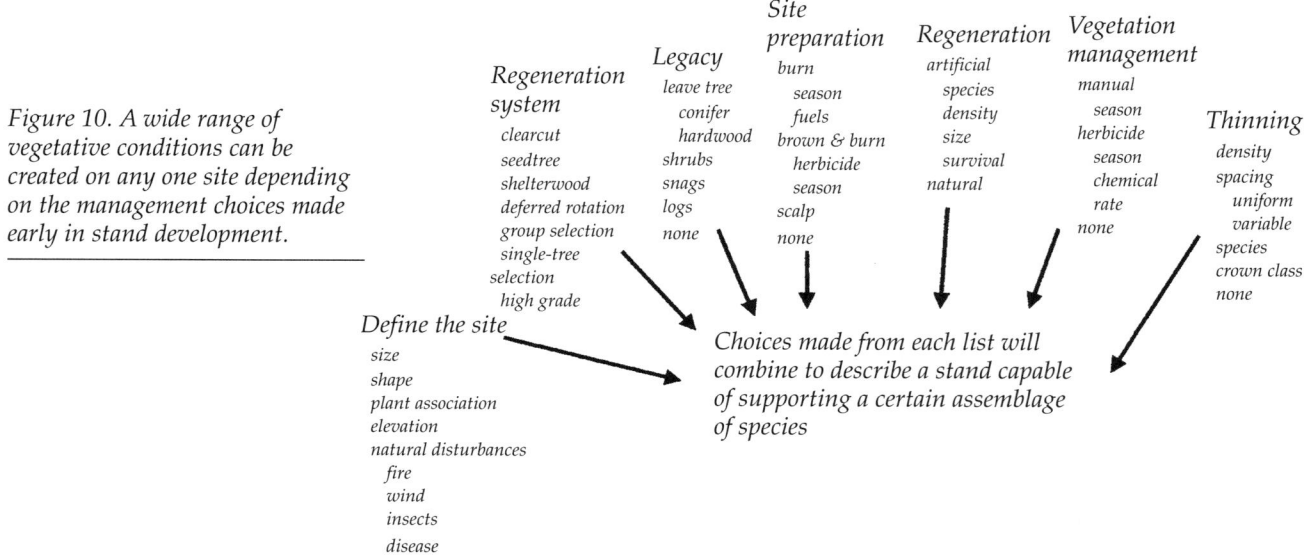

Figure 10. A wide range of vegetative conditions can be created on any one site depending on the management choices made early in stand development.

Intermediate Treatments. Precommercial and commercial thinning influences the structure and composition of the understory and may influence, consequently, the vertical complexity and quantity of browse in a stand. The plant composition of the stand is unlikely to be influenced to any great degree at this point in stand development. Opening the crowns by thinning also influences habitat quality for those species that find cover and food (cones) in tree crowns. Also, those species that feed among tree crowns, such as Hammond's flycatchers may benefit from thinning.[22] Variable-intensity thinning can produce a wide range of tree diameters and greatly influence the production of small snags early in stand development. It also can influence the distribution of food and cover in young, pole-stage stands. Small openings <0.2 acre (<(0.1 ha) in second-growth stands of spruce-hemlock (*Picea-Tsuga*) in southeast Alaska allowed access to food close to cover from snow for Sitka black-tailed deer and allowed certain bird species to occur in otherwise unsuitable habitat.[12]

Uneven-age Management

Uneven-age management usually involves group selection or individual tree selection. These regeneration systems cause a fine-scale disturbance, so stand-level vertical structure is usually high, edge and fragmentation effects are usually low, and stand heterogeneity is usually high compared to even-aged systems. The decision to leave a legacy of dead wood depends on the objectives of the land manager; retention of green trees normally is part of the silvicultural system, but the allocation of large reserve trees in the stand is also feasible.

Site preparation is usually minimal in these systems, because they usually rely on natural advance regeneration. Artificial regeneration is feasible, however, and especially within the group selection system. Chemical or mechanical site preparation, therefore, may help ensure establishment and survival of tree seedlings in these harvest groups. Mechanical scarification of the harvest groups and compaction of the soils along designated skid trails can significantly affect below-ground habitat by reducing the availability of burrow systems and restricting the ability of animals to burrow in the surface soil. Careful harvest planning and use of designated skid trails is essential.[27]

Artificial regeneration in harvest patches with a group selection system provides a significant opportunity to influence the composition and structure of the stand. More shade-intolerant species can be regenerated with group selection than with single-tree selection. Small patch, group-selection systems or single-tree selection systems that rely on existing advance-regeneration or large planting stock, however, may eliminate grass-forb stand conditions and reduce the length of time that a stand remains in a shrub stage. Indeed, Chambers et al.[7] reported that many bird species found in clearcut and green-tree retention stands did not occur in stands managed using small group selection. Many of the species using the group selection system were also found in uncut mature forest stands.[6, 8] Cutting cycle length, target tree size, and thinning

intensity all affect the structure and composition of uneven-aged stands.

The choice of which silvicultural system to use should be determined by the plant community, site conditions, logging constraints, and species. Activities in even-aged stands that could enhance conditions for species typical of late seral stages include:

- allocation of dead and living trees as structural legacy after harvest,
- variable-density planting and thinning, and
- harvesting techniques that minimize soil disturbance.

Uneven-age management strategies that could enhance habitat quality for many species that inhabit late seral-stage conditions include:

- establishing a large target tree size,
- lengthening cutting cycles,
- minimizing disturbance to the stand during logging with designated skid trails,
- harvesting with small-group or single-tree selection systems where they are appropriate,
- managing for shade-tolerant tree species,
- maintaining high-density groups of regeneration, and
- allocate dead or large, living trees.

Non-traditional Management Approaches

The following systems represent management approaches that differ from those traditionally used for timber management. They are only a few of the possible strategies to produce structural characteristics typical of those following natural disturbances within western Oregon and Washington.[40] The first system, modified single-storied stands, represents a slight modification of current stand management practices that are designed to produce relatively uniform stands for timber production using the clearcut regeneration system. Rather than complete removal of all trees from the unit, 3-10 live trees/ acre (8-25/ha) and large dead wood are left on the unit to provide some of the organic carry-over important to organisms that use these features in early seral stages (Figure 11). The second system, few-storied stands, is the creation of stands with 2 or 3 layers of canopy trees as well as retention of large dead wood (Figure 12). The third option, many-storied stands, uses small group selection cutting to create a stand that is composed of >3 layers of canopy trees in a mosaic of gaps (Figure 13). Again, large legacy trees and snags are maintained in the stand. Each of these systems represents elements found in natural disturbance regimes, and each is applicable to many forest types throughout the Pacific Northwest. The first two systems imitate coarse-scale disturbances. Stands that develop from these systems may be recolonized by species associated with mature forests as the stand develops, but would not provide adequate habitat for mature forest species until the regeneration develops to a sufficient size. Disturbance associated with the single-story system is more intense

Figure 11. A managed stand with a legacy of living and dead trees retained to provide habitat for species that are adapted to post-fire conditions. (Photograph by W.C. McComb)

Figure 12. This two-story stand can provide habitat for species adapted to stands that develop after partial burns. (Photograph by W.C. McComb)

Figure 13. An uneven-aged stand managed by group selection cutting. Species typical of mature forest conditions continue to use this stand despite repeated entries. (Photograph by W.C. McComb)

and leaves fewer large living and dead trees than most natural disturbances, except when fires have occurred within a stand at high frequency. This system represents the low end of within-stand natural variation in patch size, canopy cover, and large live tree survival. The few-storied system also has elements of coarse-scale disturbances, but produces greater variability in vertical structure and tree diameter classes than the single-story system. Both systems may be smaller in size and more frequent in occurrence than some naturally occurring coarse-scale disturbances on the landscape. Long rotations and large harvest units would more nearly imitate natural disturbances in western Oregon and Washington.

The many-storied system is patterned after fine-scale natural disturbances. Cut gaps may have to be larger than most natural canopy gaps to allow natural regeneration of shade intolerant species and to make harvesting more efficient. On gentle terrain, harvesting costs are not significantly higher than for clearcutting.[27] This system would have high within-stand variability in tree size and vertical complexity. This system might provide acceptable habitat for mature forest species while allowing some small but regular timber removal.

Natural disturbances occur in a range of sizes, shapes, frequencies, intensities, and patterns across landscapes. These parameters can be varied within both coarse- and fine-scale silvicultural systems by varying managed stand sizes and shapes, the frequency of entry, the levels of residual living and dead wood, and the arrangement of stands on the landscape, respectively. Diversity can be retained within landscapes by adopting a broader range of silvicultural systems and applying them to reflect the variety in size, intensity, and frequency of the natural disturbance regimes.

Density management during stand development might enhance structural complexity in these stands as well as in existing plantations. Typically a stand is precommercially thinned and then may receive 1 or 2 commercial thins prior to harvest. Usually thinning leaves the remaining trees uniformly spaced with adequate room among crowns for continued volume growth.[22] Thinnings that vary in intensity within the stand provide the opportunity for rapid growth by some trees in the stand and reduced growth or death (snags) by others within the stand. These patterns imposed over several thinnings can result in a range of tree sizes within the stand. If some of these trees were designated at final harvest for retention into the next rotation, and variable density planting and thinning were used, then an even more diverse forest structure could result and the stand would become two-storied. These stands might provide habitat for most species typical of mature forest within 80-120 years.

Landscapes by Design

Currently we are inheriting landscapes created by past disturbances and timber-driven management objectives on mixed ownerships that do not consider large-scale habitat patterning. Combining different silvicultural systems on the landscape in a manner that considers size and connectedness of mature forest habitat over large areas through time would be one step toward designing forest patterns. The silvicultural systems proposed above should be considered only examples; they will not apply in all situations nor will they meet all wildlife objectives. Silviculturists will have to work with wildlife biologists, forest ecologists, harvesting specialists, and other resource managers to:

1. identify clear objectives for habitat (for 1 or many species) and timber,

2. identify existing examples of conditions that meet the objectives,

3. design prescriptions specific to local conditions that represent desired future conditions to meet the objectives, and

4. plan the location of the stands on the landscape to create landscape patterns that meet the objectives.[7]

As an example, consider how several of the approaches described above might be distributed on a landscape to meet a management objective of maintaining mature forest species while allowing some timber removal. Areas known to support mature-forest associated species could be managed using the many-storied management to provide habitat for some mature-forest species and provide higher connectivity among mature forest patches than young plantations. Modified single-storied and few-storied stands could be used on steep side slopes where harvesting costs for uneven-aged management may be prohibitive.[27] The few-storied stands may be most appropriate on steep slopes adjacent to riparian buffer strips because they would provide at least a low contiguous canopy cover and a source of large logs for the riparian system. The resulting landscape would be a mosaic of many-storied stands interspersed with single- and few-storied stands in various seral stages. To be effective, managers would have to be committed to seeing the landscape designed to meet the needs of various wildlife species over time. This landscape design may not meet management objectives for other resources. Considerations for aesthetics, water and fisheries resources should be included. Although early seral stage habitats will be created during initial stages of development following single-storied and few-storied system harvests, species associated with young stands may not be as abundant as with traditional even-aged management systems.

Regardless of the landscape design, there would be continual (though low) volume removal until the areas came into a fully managed condition, and then there would be an increase in timber production after this time.[20] Volume removals would be less than from stands managed solely for timber production. The level of volume reduction will depend on the commitment made by the forest managers to providing habitat for selected species.

Summary

Habitat is a species-specific concept. Habitat represents the resources necessary to support a population over space and through time. Habitat manipulation represents the best opportunity for long-term management of populations. Within forests, key structures such as large trees, species composition of trees and shrubs, dead wood, and forest litter layers should be given particular attention when designing stand prescriptions. Each species has its own habitat requirements and these should be estimated as well as possible when developing goals for stand prescriptions.

Structure and composition of a developing stand can be dramatically influenced by a wide range of management decisions. If managers wish to provide habitat for selected species, decisions should be made based on the potential of the site to develop the magnitude and composition of required habitat features over time. It is important to recognize that even currently common species should be considered during management to ensure that both common and rare species continue to be represented across managed landscapes over time. Further, these habitat features should be distributed over space so that the species' ability to use resources throughout its home range is considered. It is this planning over space and time at biologically meaningful scales that will be required to meet the needs for wildlife in managed forests.

Literature Cited

1. Agee, J. 1999. Disturbance effects on landscape fragmentation in Interior West Forests. Chapter 3 in Forest Fragmentation: Wildlife and Management Implications. J. P. Rochelle, L. A. Lehman, and J. Wisniewski, editors. Forest fragmentation: Wildlife and Management Implications. Brill Press, Netherlands.

2. Belsky, J. 1986. Does herbivory benefit plants? A review of the evidence. American Naturalist 127: 870-892.

3. Brown, E. R., technical coordinator. 1985. Management of wildlife and fish habitats in western Oregon and Washington. Publication R6-F&WL-192-1985. U.S. Forest Service. 2 volumes.

4. Bull, E. L., and A. D. Partridge. 1986. Methods of killing trees for use by cavity-nesters. Wildlife Society Bulletin 14:142-146.

5. Chambers, C. L., T. Carrigan, T. Sabin, J. Tappeiner, and W. C. McComb. 1997. Use of artificially created Douglas-fir snags by cavity-nesting birds. Western Journal of Applied Forestry 12(3):93-97.

6. Chambers, C. L., and W. C. McComb. 1997. Effects of silvicultural treatments on wintering bird communities in the Oregon Coast Range. Northwest Science 71(4):298-304.

7. Chambers, C. L., W. C. McComb, J. C. Tappeiner II, L. D. Kellogg, R. L. Johnson, and G. Spycher. 1999. CFIRP: What we learned in the first ten years. The Forestry Chronicle 74:431-434.

8. Chambers, C. L., W. C. McComb, and J. C. Tappeiner. 1999. Breeding bird responses to three silvicultural treatments in the Oregon Coast Range. Ecological Applications 9(1):171-185.

9. Cole, E. C., W.C. McComb, M. Newton, C. L. Chambers, and J. P. Leeming. 1997. Response of amphibians to clearcutting, burning, and glyphosate application in the Oregon Coast Range. Journal of Wildlife Management 61:656-664.

10. Cole, E. C., W. C. McComb, M. Newton, J. P. Leeming, and C. L. Chambers. 1998. Response of small mammals to clearcutting, burning, and glyphosate application in the Oregon Coast Range. Journal of Wildlife Management 62:1207-1216.

11. Cork, S. J. and G. J. Kenagy. 1989. Rates of gut passage and retention of hypogeous fungal spores in two forest-dwelling rodents. Journal of Mammalogy 70:512-519.

12. DellaSala, D.A., J. C. Hagar, K.A. Engel, W. C. McComb, R. L. Fairbanks, and E. G. Campbell. 1996. Effects of silvicultural modifications of temperate rainforest on breeding and wintering bird communities, Prince of Wales Island, Southeast Alaska. Condor 98:706-721.

13. du Toit, J. T., J. P. Bryant, and K. Frisby. 1990. Regrowth and palatability of acacia shoots following pruning by African savanna browsers. Ecology 71:149-154.

14. Farentinos, R. C., P. J. Capretta, R. E. Kapner, and V. M. Littlefield. 1981. Selective herbivory in tassel-eared squirrels: role of monoterpenes in ponderosa pines chosen as feeding trees. Science 213:1273-1275.

15. Forman, R. T. T. and M. Godron. 1986. Landscape ecology. New York: John Wiley & Sons, New York, NY.

16. Franklin, J. F. and D. S. DeBell. 1988. Thirty-six years of tree population change in an old-growth *Pseudotsuga-Tsuga* forest. Canadian Journal of Forest Research 18:633-639.

17. Franklin, J. F. and R. T. T. Forman. 1987. Creating landscape patterns by forest cutting: ecological consequences and principles. Landscape Ecology 1:3-18.

18. Fretwell, S. D., and H. L. Lucas, Jr. 1969. On territorial behavior and other factors influencing habitat distribution in birds. I. Theoretical development. Acta Biotheorologica 19:16-36.

19. Friesen, C. A. 1991. The effect of broadcast burning on the quality of winter forage for elk, western Oregon. Thesis, Oregon State University, Corvallis, OR.

20. Gagliuso, R. and W. C. McComb. 1992. Designing landscapes for wildlife and timber, pages 379-388 in G. Wood and B. Turner, editors. Integrating forest information over space and time. ANUTECH Pty Ltd, Canberra, Australia.

21. Gumtow-Farrior, D. L. 1991. Cavity resources in Oregon white oak and Douglas-fir stands in the mid-Willamette valley, Oregon. Thesis, Oregon State University, Corvallis, OR.

22. Hagar, J., W. C. McComb and W. H. Emmingham. 1996. Bird communities in commercially thinned and unthinned Douglas-fir stands of western Oregon. Wildlife Society Bulletin 24:353-366.

23. Hansen, A. J., T.A. Spies, F. J. Swanson and J. L. Ohmann. 1991. Lessons from natural forests. BioScience 41:382-392.

24. Hemstrom, M. A., and J. F. Franklin. 1982. Fire and other disturbances of the forests in Mount Rainier National Park. Quaternary Research 18:32-51.

25. Huff, M. H., and C. M. Raley. 1991. Regional patterns of diurnal breeding bird communities in Oregon and Washington, pages 177-206 in L.F. Ruggiero, K. B. Aubrey, A. B. Carey, and M. H. Huff, editors. Wildlife and vegetation of unmanaged Douglas-fir forests. U.S. Forest Service General Technical Report 285.

26. Keith, L., and L.A. Windberg. 1978. A demographic analysis of the snowshoe hare cycle. Wildlife Monographs 58.

27. Kellogg, L. D., P. Bettinger, and R. M Edwards. 1996. A comparison of logging planning, felling, and skyline yarding between clearcutting and five group-selection harvesting methods. Western Journal of Applied Forestry 11:90-96.

28. Leopold, A.S. 1933. Game management. New York: Charles Scribner's Sons, New York, NY.

29. Lindstedt, S. L., B. J. Miller, and S. W. Buskirk. 1986. Home range, time, and body size in mammals. Ecology 67:413-418.

30. MacArthur, R. H., and J. W. MacArthur. 1961. On bird species diversity. Ecology 42:594-598.

31. Mannan, R. W., M. L. Morrison, and E. C. Meslow. 1984. Comment: the use of guilds in forest bird management. Wildlife Society Bulletin 12:426-430.

32. Mariani, J. M. 1987. Brown creeper (*Certhia americana*) abundance patterns and habitat use in the southern Washington Cascades. Thesis, University of Washington, Seattle, WA.

33. Martin, K. J. 1994. Movements and habitat associations of northern flying squirrels in the central Oregon Cascades. Thesis, Oregon State University, Corvallis, OR.

34. Martin, K. J. 1999. Habitat associations of small mammals and amphibians in the central Oregon Coast Range. Dissertation, Oregon State University, Corvallis, OR. .

35. Maser, C., B. R. Mate, J. F. Franklin, and C. T. Dyrness. 1981. Natural history of Oregon Coast mammals. U.S. Forest Service General Technical Report PNW-133.

36. Mautz, W.W. 1978. Nutrition and carrying capacity. Pages 321-348 in J. L. Schmidt and D. L. Gilbert, D.L., editors. Big game of North America. Stackpole Books, Harrisburg, PE.

37. McComb, W. C. 1999. Forest fragmentation: wildlife and management implications – synthesis of the conference. Chapter 17 in J. P. Rochelle, L. A. Lehman, and J. Wisniewski, editors. Forest fragmentation: Wildlife and Management Implications. Brill Press, Netherlands.

38. McComb, W. C., and D. Lindenmayer. 1999. Dying, Dead, and Down Trees, Chapter 10, in Hunter, M. L., Jr., Maintaining Biodiversity in Forest Ecosystems. Cambridge University Press, Cambridge, England.

39. McComb, W. C., and R. E. Noble. 1981. Nest box and natural cavity use in three mid-South forest habitats. Journal of Wildlife Management 45:92-101.

40. McComb, W. C., T.A. Spies, and W. H. Emmingham. 1993. Stand management for timber and mature-forest wildlife in Douglas-fir forests. Journal of Forestry 91(12):31-42.

41. McGarigal, K., and W. C. McComb. 1995. Relationships between landscape structure and breeding birds in the Oregon Coast Range. Ecological Monographs 65:235-260.

42. McGarigal, K., and W. C. McComb. 1999. Forest fragmentation and breeding bird communities in the Oregon Coast Range. Chapter 12 in J. P. Rochelle, L. A. Lehman, and J. Wisniewski, editors. Forest fragmentation: Wildlife and Management Implications. Brill Press, Netherlands.

43. Morrison, M. L., and E. C. Meslow. 1984. Effects of the herbicide glyphosate on bird community structure, western Oregon. Forest Science 30:95-106.

44. Morrison, P. H. and F. J. Swanson. 1990. Fire history and pattern in a Cascade Range landscape. U.S. Forest Service General Technical Report PNW-GTR-254.

45. Naiman, R. J. 1988. Animal influences on ecosystem dynamics. BioScience. 38(11):750-752.

46. Nelson, S. K. 1989. Habitat use and densities of cavity-nesting birds in the Oregon Coast Range. Thesis, Oregon State University, Corvallis, OR.

47. Perry, D. A., M. P. Amaranthus, J. G. Borchers, [and others]. 1989. Bootstrapping in ecosystems. BioScience. 39:230-237.

48. Raphael, M. G. 1987. Use of Pacific madrone by cavity-nesting birds. Pages 198-202 in Symposium on multiple-use management of California's hardwood resources. U.S. Forest Service General Technical Report PSW-100.

49. Robbins, C. T., S. Mole, A. E. Hagerman, and T. A. Hanley. 1987. Role of tannins in defending plants against ruminants: Reduction in dry matter digestion? Ecology 68:1606-1615.

50. Rosenberg, K. V., and M. G. Raphael. 1986. Effects of forest fragmentation on vertebrates in Douglas-fir forests. Pages 263-272 in J. Verner, M. L. Morrison, and C. J. Ralph, editors. Wildlife 2000: Modeling habitat relationships of terrestrial vertebrates. The University of Wisconsin Press, Madison, WI.

51. Runkle, J. R. 1985. Disturbance regimes in temperate forests. Pages 17-34 in S. T. A. Pickett and P. S. White, editors. The ecology of natural disturbance and patch dynamics. Academic Press, New York, NY.

52. Schreiber, B. 1988. Diurnal bird use of snags on clearcuts in central coastal Oregon. Thesis. Oregon State University, Corvallis, OR.

53. Singer, S. W., N. L. Naslund, S. A. Singer, and C. J. Ralph. 1991. Discovery and observations of two tree nests of the marbled murrelet. Condor 93:330-339.

54. Smith, H. C., N. I. Lamson, and G. W. Miller. 1986. An esthetic alternative to clearcutting. Journal of Forestry 87:14-18.

55. Spies, T. A., and J. F. Franklin. 1988. Old-growth and forest dynamics in the Douglas-fir region of western Oregon and Washington. Natural Areas Journal 8:190-201.

56. Spies T. A., J. F. Franklin, and M. Klopsch. 1990. Canopy gaps in Douglas-fir forests of the Cascade Mountains. Canadian Journal of Forest Research 20:649-658.

57. Spies, T. A., J. F. Franklin, and T. B. Thomas. 1988. Coarse woody debris in Douglas-fir forests of western Oregon and Washington. Ecology 69:1689-1702.

58. Stewart, G. H. 1986. Forest development in canopy openings in old-growth *Pseudotsuga* forests of the western Oregon Cascade Range, Oregon. Canadian Journal of Forest Research 16:558-568.

59. Suzuki, N. 1992. Habitat classification and characteristics of small mammal and amphibian communities in beaver-pond habitats of the Oregon Coast Range. Thesis, Oregon State University, Corvallis, OR.

60. Temple, S. A. 1986. Predicting impacts of habitat fragmentation on forest birds: a comparison of two models. Pages 301-303 in J. Verner, M. L. Morrison, and C. J. Ralph, editors. Wildlife 2000: modeling habitat relationships of terrestrial vertebrates. University of Wisconsin Press, Madison, WI.

61. Vega, R. M. S. 1993. Bird communities in managed conifer stands in the Oregon Cascades: habitat associations and nest predation. Thesis, Oregon State University, Corvallis, OR.

62. With, K. 1999. Is landscape Connectivity necessary for wildlife management? Chapter 7 in J. P. Rochelle, L. A. Lehman, and J. Wisniewski, editors. Forest fragmentation: Wildlife and Management Implications. Brill Press, Netherlands.

5

Conservation of Biodiversity: Considerations and Methods for Identifying and Prioritizing Areas and Habitats

Margaret M. Shaughnessy & Thomas A. O'Neil

Introduction

Conservationists have advocated two approaches to counter the accelerating loss of biodiversity. One operates in a crisis mode, rescuing species in danger of extinction. The other focuses on protecting communities of plants and animals not yet in serious jeopardy, but which are likely to be driven towards extinction with increased habitat loss. These two complementary approaches have been described as "fine filter" and "coarse filter" strategies for maintaining biodiversity.[26] Implementation of the "fine filter" approach is exemplified in actions taken in support of the Endangered Species Act. In implementing the "coarse filter" approach, conservationists and natural resource organizations often include wildlife-habitat associations in their assessment of biological diversity.[37, 43] In doing so, these organizations often look to conserve vegetation types in hopes that they will also protect the plant, invertebrate, and vertebrate species associated with them. This premise implies that vegetation serves as a satisfactory indicator of the environmental variables that interact on a particular site[41, 43] and as such, assumes that plant communities can serve as substitutes for ecosystems and the elements of biodiversity.[27] Given the complex ways in which species interact with their ecosystem, the combination of proactive fine and coarse filter strategies present a greater hope for conserving biodiversity.[37]

Until recently, many conservation strategies have focused on identifying areas that will function as reserves, or areas that could be set aside to conserve some aspect of biodiversity. However, reserves are vulnerable to evolutionary processes, catastrophic events, and global warming. In addition, no area or group of areas likely will be sufficient to support the full range of native species and ecological processes that contribute to biodiversity at ecoregional, state, and continental scales.[14, 35] Therefore, the identification and establishment of reserves represents a relatively short-term solution, and the focus of conservation efforts has expanded to include managing multiple-use public and private lands for biodiversity.[27] Several recent conservation efforts have demonstrated this broader focus by (1) identifying and mapping locations of priority habitats, species, and areas; (2) identifying lands currently managed for biodiversity values and then developing methods to identify and conserve areas that will complement that existing conservation network of private and public lands; and (3) developing easily accessible information sources that can be used as tools by natural resource managers and the public to identify priority habitats, conserve native species, and increase awareness and understanding of (a) habitats at risk; (b) function, distribution, and abundance of habitats; and (c) effects of land management activities.[4, 7, 48]

This chapter will review the methods currently used to prioritize terrestrial habitats and areas, focusing on the data sets used, important considerations and inclusions, and qualitative and quantitative aspects of each. In addition, several statewide strategies for identifying and prioritizing conservation areas are summarized to provide examples of strategies for addressing this complex issue. Finally, total and protected acreages for each of the 32 habitats found in Oregon and Washington are provided.

Methods for Identifying and Prioritizing Sites

Methods for identifying and prioritizing areas important to conservation values range from subjective techniques based on the knowledge of experts to more quantitative methodologies. Generally, to achieve the desired conservation goal, most processes rely on both subjective and quantitative techniques, with the latter often used as a tool to aid in the decision-making process.[4, 7] An advantage of using quantitative techniques is that the methods and data sets can be made available to individuals and organizations who, in turn, can learn how decisions were made and gain understanding about how their values and land-management decisions affect large-scale conservation decisions.[4, 52] Important components of quantitative techniques include well-defined goals, an appropriate spatial scale for the analyses, analyses conducted at several scales (e.g., local, ecoregion, and state), an understanding of the limitations of the maps that are used (e.g., resolution, accuracy, and habitats and habitat elements that may be excluded), appropriate units for the analyses, and indices of viability and threat. The goals of the conservation program will determine the type and quality of the data required (e.g., resolution of vegetation map) and the spatial scale for the analysis.[42, 52]

Maps and Units of Analysis

Critical to most conservation efforts is our ability to (1) map habitats and species distributions, (2) identify the habitat associations of our wildlife species, and (3) understand the changes that have occurred in the landscape over time. Therefore, several types of maps and digital information are useful to conservation efforts, including maps of vegetation, species distribution, and historic vegetation.

Vegetation Maps. Vegetation maps developed though interpretation of satellite imagery provide a good source of information about vegetation and land forms currently existing on the landscape. Generally, these maps do not adequately represent small features or linear features such as wetlands, riparian areas, and small areas of specific vegetation types such as Romer's fescue grasslands also are not represented well.[4, 6, 7, 37, 39] In addition, although they are important biological characteristics for predicting distributions of vertebrate and invertebrate species, many fine-scale features such as structure, habitat elements, and presence of individual plant species generally are not included in maps developed from remotely sensed data.[37, 39] Nevertheless, maps developed from remotely sensed data provide some of our best and most cohesive sources of information with which to conduct analyses and identify conservation priorities.

Species Distribution Maps. Maps of species distributions for plants, vertebrates, and invertebrates also are important components of most fine and coarse filter strategies for conservation. For rare, threatened, and endangered species, species distribution maps generally can be developed from databases that track occurrences of those species. For more common species, distributions can be modeled using information such as museum records, verified sighting records, maps of current vegetation, and information on habitat associations and elevation.[6, 38] Accuracy of distribution maps is difficult to determine; however, the accuracy will be related to the resolution, vegetation classification, and attributes of the vegetation map as well as the predictive capabilities of the habitat associations.

Because our knowledge of species and habitat associations is limited, O'Neil et al.[28] developed a method for grouping vegetation types into more generalized habitat types based on use by similar groups of wildlife species. The generalized habitat classifications allowed investigators to predict with greater certainty the associations between species and habitats.[28] However, databases that provide only general associations between species and generalized habitat features provide weak inferences for predicting the presence and distribution of species,[39] and may not be the best choice for the units of biodiversity analyses since information about vegetation type, structure, successional stage, and habitat elements will be lacking.

Historic Vegetation Maps. Maps depicting the historic distribution of vegetation are useful tools that allow the user to gain understanding of the changes that have occurred in the relative amounts and general distribution of vegetation types, and thereby help identify changes relevant to conservation efforts.[7] For example, comparisons between maps of current and historic vegetation provide a context in which changes in vegetation from native species to exotic vegetation types or those associated with human uses (e.g., urban and agricultural land cover types) can be evaluated.[7]

Units of Analysis. Conservation areas have been identified and prioritized using various units of analysis including (1) sampling grids or hexagons of uniform size that have been superimposed over the landscape;[5, 16, 49, 52] (2) mapped vegetation or habitat polygons; and (3) existing conservation reserves or areas known to be important, rare, or limiting. All of these methods have benefits as well as limitations, and the goals of the project will help determine which method is most appropriate. For example, sampling grids represent a convenient way to divide the landscape and are useful tools for large-scale regional or national assessments because attributes such as species presence or absence can be assigned easily to each cell of the grid. However, because the cells or hexagons defined by the sampling grids represent arbitrarily defined areas that bear no relationship to features on the ground and often encompass a wide range of elevations and a variety of vegetation and habitat types, the cell may provide only a small piece of the habitat needed by many of the species attributed to it.[7] Therefore, the cell identified through an analysis may not contain the characteristics or considerations that are important or necessary to achieve the desired conservation goal.[4] Field surveys may be conducted to confirm that the cell contains the desired biological characteristics;[6] however, the biological characteristics of the cells not chosen through the analysis would remain unverified and therefore unknown even though they may contain characteristics that would be considered more desirable to the conservation goal.

Vegetation patches or polygons identified through various mapping methodologies with remotely sensed data also have been used as units of analysis for identification of conservation priorities.[4, 6, 7] An assumption of these analyses is that the vegetation polygons represent a single vegetation type, although the degree to which that is true is a function of the vegetation classifications, map accuracy, mapability, minimum mapping unit (mmu, the smallest polygon or unit of classification), and scale. Until recently, most statewide vegetation maps produced from satellite imagery were created with a relatively large mmu (e.g., 350 acres or 140 ha). However, advances in technology have enabled investigators to more easily create maps with smaller mmu, making these vegetation maps more useful for natural resource managers and local-scale conservation assessments. Other limitations of vegetation maps were discussed previously.

Quantitative methods

Generally, most quantitative methodologies for identifying and prioritizing conservation sites can be categorized into three types, (1) species richness, (2) scoring techniques and indices, and (3) complementarity or iterative analyses.

Species Richness. Species richness analyses are simple approaches used by many investigators to evaluate the relative value of sites as wildlife habitat. In its most basic form, species richness methodologies rank the value of a site based on the total number of species that are predicted to occur in that area.[5, 15, 16, 36, 45, 52] Several investigators have expressed concern over the use of species richness methodologies to identify or prioritize sites for conservation values because areas of species richness for a particular group (e.g., mammals, threatened and endangered species, plants) rarely overlap with areas of species richness for other groups.[6, 20, 30, 36] However, other investigators have found that richness analyses for species with restricted ranges may identify areas that are more biologically diverse and include areas important to a variety of taxa.[12, 44]

Additional concern about species richness methodologies has arisen because often the analyses are based on predicted distributions of species or do not include an estimate or index of abundance for each species.[6, 16, 45, 52] Therefore, a site that contains only 1 individual of each of 30 species would have a ranking similar to a site with 100 individuals of each of 30 species. Similarly, a site ranked as poor in terms of species richness may contain many individuals of each species. In addition, habitats of intermediate quality frequently are ranked as good in terms of species richness, but may contain many urban or non-native species.[33] To address these concerns and ensure that viable populations are identified and conserved, several investigators have included estimates of population density or indices of abundance into the analyses.[45, 52] In addition, the use of species richness analyses now usually is limited to identification of areas important for specific species or groups of species rather than being used for more general coarse-filter approaches to identification and prioritization of conservation areas.

Scoring Techniques and Indices. A widely used approach to evaluate areas in terms of their contribution to conservation goals involves scoring techniques and indices that enable investigators to rank or prioritize areas based on various selection criteria.[1, 3, 4, 24, 51, 53] Areas may be ranked for criteria such as species richness, species diversity, species rarity, species priority, endemism, total number of individuals of each species or an index of abundance, naturalness, size, or economic and scientific value.[4, 9, 23, 31, 45, 46, 53] Selection criteria often are weighted in complex multi-criteria indices that are used to evaluate and prioritize the potential conservation areas.[3, 4, 24, 51] However, the relationships between criteria often are too complex to allow for the development of a meaningful index,[9, 21, 40, 46] and the factors contributing to a site's importance become hidden in a complex formula and are not easily identifiable to the investigators themselves or

to the general public. Therefore, some authors recommend the use of single criterion rankings that then can be used to subjectively evaluate and rank sites for conservation value.[45]

Complementarity or Iterative Analyses. A more recently developed method for identifying and prioritizing areas involves an iterative process in which an algorithm is used to identify sites with the largest number of species from a target group (e.g., threatened species, amphibians, species with restricted ranges) and then iteratively identifies additional sites that contribute the most species not already represented in the site selected previously.[6, 16, 19, 22, 25, 31, 32] The iterative process continues until a subset of sites is identified in which all species of the target group are represented. An advantage of this approach is that, unlike many scoring and ranking methodologies that simply select the highest ranking sites, replicating many species and excluding other species,[19] the iterative selection process takes into consideration the attributes of other sites, and then identifies and selects complementary areas.[32] As with other prioritization methodologies, it is important that these analyses include information on the total number of individuals of a species or an index of abundance to ensure that viable populations are identified and conserved.[45]

Factors that should be considered for inclusion in any of the subjective or quantitative analyses include the following:
1. management needs of a habitat or area (e.g., fire in Oregon White Oak communities);
2. abundance or total numbers of each species;
3. important breeding habitat, seasonal range, movement corridors;
4. limited abundance or distribution of habitats;
5. vulnerability to disease, habitat alteration, proximity to threat, development, and land-use activities;
6. unique or dependent species;
7. uniqueness of plant or wildlife community;
8. function of plant, invertebrate, or vertebrate species;
9. function of habitat or some component of habitat for plant, invertebrate, or vertebrate species;
10. status of plant, vertebrate, and invertebrate species;
11. endemism or species with restricted ranges;
12. land ownership;
13. protection status or management goals for site; and
14. species richness, species rarity, and species priority.

Current Approaches for Evaluating and Identifying Priority Areas and Habitats

At the local, state, regional, and national levels, the task of identifying and conserving priority areas and habitats is daunting. Until recently, few states have had the data sets needed to develop quantitative approaches to this conservation challenge. However, the National Gap Analysis Program has provided vision and start-up funding for most states to begin development of these basic data sets. With the completion of some of these data sets, several states have begun to develop approaches for identifying and conserving priority habitats and areas. In addition to the nation-wide approach developed by the National Gap Analysis Program, three statewide approaches are reviewed in this section. All three approaches are quite different, reflecting differing goals, confidence or availability of data sets, and available technology. Of the three statewide approaches, Washington's program was developed the earliest and consequently has the simplest design. The approaches for Oregon and Colorado were developed in 1997 and 1998, respectively. Both of theses approaches are complex, using recently developed data sets and technology and incorporating quantitative analyses as well as subjective evaluations. All three approaches take important steps to begin addressing conservation needs at local, ecoregion, and statewide scales.

National Gap Analysis Program

The National Gap Analysis Program is a nationwide program managed by the United States Geological Survey, Biological Resources Division. The program was developed in the mid-1980s and uses a coarse-filter approach to assess conservation needs at state, regional, and national scales. The program focuses on working with each state in the United States and developing digital data layers that can be used with GIS to identify "gaps" or natural land cover types and native vertebrate species not adequately represented in the existing network of conservation lands. Specific objectives of the National Gap Program include conducting statewide, regional, and national level analyses to:

1. Determine the area occupied by vertebrate species and land cover/vegetation types that are within lands managed for conservation values.
2. Identify public lands having vertebrate species and land cover/vegetation types that are not well represented in the existing conservation network.
3. Identify gaps in the natural reserve system of the United States.

The National Gap Analysis Program uses several data layers to meet their program goals including maps depicting (1) current vegetation and land cover in each state, (2) distribution of terrestrial vertebrates, and (3) land ownership and management. For each state, vegetation is mapped from satellite imagery to the alliance level (dominant/co-dominant cover types) using the National Vegetation Classification System.[8] The mmu for each of the statewide maps is ≤250 acres (100 ha). Using a variety of techniques and data sets, maps of vertebrate distribution also are developed for each state. Because few states have been systematically surveyed for all species, vertebrate distribution maps generally identify the predicted distribution of each species rather than the known

distribution of the species. Finally, the National Gap Analysis Program also relies on maps depicting land ownership and land management. These maps identify the major patterns of land ownership and management, aiding in the identification of lands managed for conservation values and in identification of "gaps" in the current network of conservation lands.

Washington: Priority Habitats and Species Program

The Washington Department of Fish and Wildlife Habitat Program (WDFW) has developed a Priority Habitats and Species list that identifies habitats and species considered priorities for conservation and management.[48] The WDFW program has a broad focus and includes information for terrestrial, aquatic, and marine species and habitats. The program identifies (1) priority habitats, (2) priority species, and (3) priority areas. Habitats, species, or areas may be considered priorities throughout the state or that designation may be restricted to specific geographic areas. Biologists use the criteria described below to identify priority habitats, species, and areas; these occurrences are then mapped, and locational and descriptive data are recorded in a GIS. WDFW then develops management recommendations for Washington's priority habitats, species, and areas. These documents can be obtained from the WDFW Habitat Program Internet site at www.wa.gov/wdfw/hab/phspage.htm. The Priority Habitats and Species program currently has identified approximately 160 priority species and 20 priority habitats.[48]

Priority Habitats. Priority habitats are defined as habitat types or elements with unique or significant value to a diverse assemblage of species. Priority habitats may consist of a unique vegetation type or dominant plant species, a specific successional stage, or specific habitat elements (e.g., talus, caves). For a habitat type or element to be considered a priority habitat, it must have at least oneof the following characteristics:

1. relatively high fish and wildlife density or species diversity;
2. important fish and wildlife breeding habitat, seasonal range, or movement corridor;
3. rare or of limited availability;
4. high vulnerability to habitat alteration; or unique or dependent species.

Priority Species. Priority Species are defined as species that require protective measures for their perpetuation because of their (1) population status; (2) sensitivity to habitat alteration; or (3) recreational, commercial, or tribal importance. Priority species include:

1. all state listed (threatened, endangered and sensitive) and candidate species;
2. vulnerable aggregations: species or groups of species susceptible to population decline because of their tendency to aggregate (e.g., heron rookeries, sea bird concentrations, marine mammal haul outs, shellfish beds, fish spawning and rearing areas); and

3. native and non-native species with recreational, commercial, or tribal importance that are at risk due to habitat loss or degradation.

Priority Areas. Priority areas are defined as specific areas or locations that are a priority because they support relatively high numbers of individuals (e.g., heron rookeries, locations of rare species) or are important to the life history and ecology of the species. Examples of priority areas include the following:

1. breeding, rearing, and hibernation sites;
2. leks;
3. areas commonly or traditionally used by individuals of a species or a group of animals;
4. migration corridors; or
5. foraging areas.

Oregon:
The Oregon Biodiversity Project

Currently in Oregon, there is not a state-sponsored program to identify or prioritize habitats or areas important to conservation values. However, a nonprofit organization, The Defenders of Wildlife, has begun to try to address these concerns for Oregon.[7] The Oregon Biodiversity Project (OBP), a project overseen by Defenders of Wildlife, recently completed an analysis of Oregon's biodiversity and identified processes and opportunities for conserving the State's native species and habitats.[7] Both statewide and ecoregional analyses were conducted, resulting in the identification of the network of conservation lands existing currently in Oregon and the development of a process for identifying conservation needs and strategies for addressing those needs.

Data Layers

The OBP relied on both quantitative and qualitative processes to meet its goals. Data layers and digital data sets used for both the statewide and ecoregional analyses included the following:

Statewide Vegetation Map. This map was developed in cooperation with the Oregon Gap Analysis Program[13] and depicts the current vegetation and land cover of Oregon. The map was created through visual interpretation of satellite photos (1:250,000 scale) taken in the late 1980s and has a mmu of 350 acres (140 ha). One hundred thirty-three vegetation and land cover types were identified based on dominant species (e.g., Oregon White Oak, Ponderosa Pine) and basic structure (e.g., woodland, grassland, forest). This large mmu resulted in several vegetation and land cover types not being represented (e.g., small stands of native forest types) or being under-represented on the map (e.g., wetlands, riparian areas). The map was used for "coarse filter" analyses to assess the distribution, abundance, and status of Oregon's biodiversity. Two statewide vegetation maps of Oregon have been completed since the ORB completed its assessment. One map was developed by the Oregon Gap Analysis Program and has a mmu of 250 acres (100 ha). The second map was developed in 1999 by the Northwest

Habitat Institute and has a mmu of 10 acres (4 ha). Both maps have 62 vegetation and land cover types and were developed by Killsgaard and Barrett.[18]

Historic Vegetation Map. This map was created by the OBP, and represents a blending of several regional data sets including those developed by the Interior Columbia River Basin Ecosystem Management Project, U.S. Forest Service vegetation maps, and General Land Office maps. The historic vegetation map was not intended to identify the vegetation types that existed at a specific place and time; rather it was developed as an indicator of the relative amount and general distribution of vegetation, enabling investigators to identify historic vegetation patterns and land-cover changes that have occurred over the past century. As with the map of current vegetation, the scale of the historic vegetation map limits the detail that could be shown, under-representing or omitting several vegetation and land-cover types.

At-risk Species. At-risk species are defined as plant and animal (aquatic and terrestrial vertebrates and invertebrates) species considered highly vulnerable to extinction. The OBP used the Natural Heritage Program's database and ranking system to identify approximately 200 species considered to be at risk in Oregon. The Heritage Program's database incorporates information on factors such as known occurrences, threats, sensitivity, and area occupied, as well as species status at state, national, and global levels.

Land Ownership, Administration, and Management. The OBP refined the existing data layer that identified current land ownership and administration for Oregon. In addition, the OBP developed a ranking system, the Biodiversity Management Rating System, to identify the degree to which lands were managed for biodiversity values. Lands managed by state and federal agencies and those managed by The Nature Conservancy were rated subjectively on a 10-point scale to indicate their contribution to long-term conservation of native biodiversity. Land managers assessed and rated their lands for the following criteria:

1. Management Objectives: is biodiversity conservation or protection of natural values the primary objective for the area?
2. Security: is the area formally designated for long-term protection of natural values?
3. Biodiversity Values: does the area include land that has high-quality habitat or does it make an important contribution to ecosystem function?
4. Size: is the land a large block, and is it managed as a unit?

Current Conservation Network. The land ownership data layer and the ratings developed though the Biodiversity Management Rating System were used to identify lands in Oregon that are devoted to the long-term protection of biodiversity, or a "Network of Conservation Lands." Only lands that were rated at the high end of the scale (i.e., 8-10) by the Biodiversity Management Rating System were included in the conservation network. Using those guidelines, the OBP determined that almost 10% of Oregon's total area, not including tribal or most private lands, was included in the current conservation network.

Salmon Core Areas. Salmon Core Areas were identified by the Oregon Coastal Salmon Restoration Initiative as areas of critical importance to sustain salmon populations in individual basins. Maps identifying the historic distribution and current status of Oregon's anadromous fish species[50] also were used to identify areas important to salmon populations.

Aquatic Diversity Areas. These areas were identified by the Oregon Chapter of the American Fisheries Society as being important to the conservation of aquatic diversity[2] because they represent Oregon's best remaining aquatic habitats and include connecting corridors and at-risk fish species.

Statewide Analysis

The OBP's statewide analyses consisted of coarse-filter analyses that provided an overview of current conditions in Oregon. To conduct the analyses, OBP used information about current and historic vegetation, aquatic ecosystems, at-risk species, human population and land development (e.g., population growth, road networks), and land ownership and administration. General assessments included:

1. Identification and description of Oregon's network of conservation lands.

- Greater than 1/3 of Oregon's native vegetation types have <5% of their distribution within the existing network of conservation lands.

- Largely due to wilderness designations and the President's Northwest Forest Plan, current management for biodiversity is most extensive in alpine habitats and Westside forests.

- Westside ecoregions (Coast Range, Klamath, and West Cascades) have large federal ownerships and approximately 25% of the land is included in the network of conservation lands.

- Willamette Valley and Columbia Basin Ecoregions have a large proportion of land in private ownership and have less than 2% of land in the network of conservation lands.

- Eastside ecoregions (Owyhee Uplands, East Cascades, Blue Mountains, Basin and Range, and High Lava Plains) have large federal ownership, but only 2-7% of the land is included in the network of conservation lands.

2. Statewide and ecoregional analyses identifying vegetation types represented in the existing conservation network and how well each type was represented. Examples of results of these analyses include the following:

- Over 90% of subalpine and alpine meadows are included in the conservation network.

- About 1.8% of the Big Sagebrush-Bunchgrass type is included in the conservation network.

- About 3.1% of Oregon white oak woodlands are included in the conservation network.

- Only about 0.1% of bitterbrush steppe is included in the conservation network.

3. Identification of statewide priorities for conservation based on widespread decline of habitat types and significance of the habitat in multiple ecoregions. Results of the analyses indicated the following habitat types were priorities for conservation in Oregon:

- Oak savanna and woodlands,
- Wetlands,
- Riparian,
- Bottomland hardwood forests,
- Old-growth conifer, and
- Native grasslands and prairies.

4. Summary of the abundance and distribution of non-native habitats. Non-native land cover types account for greater than 16% of Oregon's landscape.

- 11% of Oregon's land is in farmland and developed pastures;

- 4.5% of Oregon's land is dominated by exotics such as cheatgrass;

- 0.7% of Oregon's land is in urban, industrial, and residential classes or cover types;

- Native habitats most affected by conversion to non-native types include grasslands, prairies, wetlands, and bottomland hardwood forests.

Ecoregional Analyses

As a complement to the statewide analyses, OBP conducted analyses at the ecoregional level. The goal of these analyses were to (1) identify the elements of biodiversity that required additional protection and those that were protected by the current conservation network, and (2) identify areas with the potential to address statewide and ecoregional priorities. The results of the ecoregional analyses included a general description of each ecoregion, its current and historic vegetation, and information about climate, land forms, habitats, industry, and human population levels. Analyses included the following:

1. Vegetation Analysis: identification of vegetation types that are conservation priorities based on current management status and an assessment of changes in abundance and distribution over the past century;

2. At-risk Species: analysis of abundance, distribution, risks/threats, and representation in the current conservation network;

3. Ecosystem changes: assessment of changes in ecosystem processes and vegetation structure not addressed by analyses of the coarse-scale vegetation map, and the results and impacts of management practices such as fire suppression, timber harvest, and grazing;

4. Analyses of aquatic species;

5. Summary of conservation issues for the ecoregion, such as:

• threats, human population growth, economic development, pollution, conversion to non-native habitats;

• changes in management practices (e.g., fire suppression, grazing, timber harvest, conversion of wetlands);

• changes in natural disturbance regimes (e.g., fire suppression, flooding);

• invasions of non-native plant and animal species;

• habitat fragmentation; and small private ownerships.

6. Identification of "Conservation Opportunity Areas," or areas with the potential to address statewide and ecoregional conservation priorities. Generally, characteristics of these areas included the following:

• large blocks of native habitats,

• vegetation or habitats that have declined,

• vegetation types not well represented in the conservation network,

• at-risk species, and

• potential to complement or connect the existing conservation network.

The OBP used a two-step approach to identify Conservation Opportunity Areas. The first step used GIS data layers as a quantitative approach that enabled the project to (1) identify gaps in the existing conservation network, (2) assess changes in the conservation network, (3) assess changes from historic vegetation patterns, and (4) display areas identified as having significant biodiversity values. The second step consisted of a subjective assessment of the potential of different areas to enhance the existing conservation network. This assessment included evaluations of land ownership, current management, existing and potential programs for conservation, pending public policy discussions, and potential threats to the elements of biodiversity.

Defenders of Wildlife published a book, *Oregon's Living Landscape: Strategies and Opportunities to Conserve Biodiversity*,[7] outlining the process followed and the results achieved by the OBP. The book is written in a style that is appealing to the general public, and describes the components of biodiversity as well as a process that can be used to identify conservation needs and opportunities in Oregon.

Colorado: The High Priority Habitat System

Recently, the Colorado Division of Wildlife (CDOW) developed a system, the High Priority Habitat system, to rank habitats in terms of their priority for conservation.[4] This approach to identification and conservation of priority habitats and areas is the most recently developed and complex approach for the western states. The system

is unique in that it was designed to be used by many user-groups, accommodating geographic and species-specific queries and thus the differing priorities of those groups. For example, the system can be used by the CDOW to help identify the habitats that should be protected in a specific resource area to maintain native fauna or to protect areas important to a single species. Additionally, the system can be used by county planners interested in identifying and protecting habitats for economically important species in their county. The High Priority Habitat system is available through an interactive Internet site, the Colorado Natural Diversity Information Source (http://www.ndis.nrel.colostate.edu/), a site maintained through a cooperative effort of state and private agencies. The goal of the Natural Diversity Information Source is to provide easy access to biological, geopolitical, and demographic data needed to understand potential impacts for changes in land use on wildlife and natural communities.

Data Sets

To develop the High Priority Habitat System, the CDOW used existing data sets, including a statewide vegetation map, species distribution maps, and the Colorado Vertebrate Ranking System, an information system that incorporates information from several existing databases to identify species at risk.

Statewide Vegetation Map, Fifty-two vegetation types were identified and mapped by the Colorado Gap Analysis Project. The mmu for the vegetation map was 250 acres (100 ha), and was determined by the National Gap Analysis Program. Basin-wide classifications and maps provided more detailed information.

Species Distributions Maps. CDOW used two sources of information to delineate vertebrate distribution. For rare species or economically important species, CDOW used data sets that tracked the occurrence of each species to develop maps identifying the known distribution of each species. For most other species, the project used maps that predicted species distributions based on the model created by the Colorado Gap Analysis Project. The Gap model used the vegetation map, vegetation associations, and elevation constraints to predict the distribution of each wildlife species.

Colorado Vertebrate Ranking System. The CDOW developed the Colorado Vertebrate Ranking System[10] (COVERS) to provide information necessary to evaluate the conservation needs of Colorado's native vertebrates. The system was modeled after a process developed by the Florida Non-game Wildlife Program,[24] and is linked to and incorporates information from several statewide databases including the Colorado Natural Heritage Program's ranking databases, the Colorado Wildlife Species Database, and the Colorado Bird Observatory database.

The COVERS ranking process is designed to identify species at risk, and uses a two-step process to accomplish that task. First, all species are ranked for variables that indicate (1) the degree of biological imperilment of the species and (2) the state of knowledge about the species' distribution, abundance, and ecology. Variables ranked in this step include the following:

1. Biological imperilment
- global population size
- population concentration
- largest global population size
- global population trend
- state population trend
- global distribution
- global distribution trend
- ecological specialization
  dietary specialization
  reproductive specialization
  other specialization
- sensitivity to exotic/invasive organisms
- sensitivity to human-induced factors

2. Current state of knowledge and management
- knowledge of state distribution
- knowledge of state populations trend
- knowledge of state population limitations
- ongoing management in Colorado
- number of protected sites or populations in Colorado

Species identified as "imperiled" though the first step of the process underwent an additional rating process to determine the degree of imperilment and to evaluate factors that could influence conservation of the species. Variables ranked at this stage of the process included:

3. Additional biological variables
- condition of largest population
- current distribution in Colorado
- reproductive potential for recovery
  number of offspring per year
  years to first reproduction
- magnitude of habitat threat
- immediacy of habitat threat
- direct threat to species
- portion of available habitat currently occupied

4. Importance of Colorado populations
- systematic significance at the state level
- percent of total range in Colorado
- percent of global population in/visiting Colorado
- relation of Colorado distribution to global range
- importance of species to other organisms

5. Social considerations
- economic consequences of population decline
- economic benefits of population increase
- current listing status
  other considerations

The scores for each variable were recorded in the COVERS database, and can be accessed to address or evaluate specific conservation needs of each species. The COVERS database also calculates summary scores for each species that can be used to indicate (1) the biological imperilment of the species, (2) the state of knowledge, protection, and management for the species, and (3) the importance of the species to the health of the global population and to biodiversity in Colorado.

Ranking Habitat Value

The High Priority Habitat system is designed so that the user can specify the geographic area and group(s) of species that are of interest to them. Possible geographic areas include the entire state of Colorado, any county or hydrounit, Bureau of Land Management District or Resource Area, National Forest or Grassland, and CDOW Regions or Game Management Units. The system continually is being modified and improved, and eventually will allow for queries of user-defined groups of species and for evaluations of single species. Presently, the system is capable of accommodating queries for the following groups of species:

- Rare species
  Federally Listed Threatened and Endangered
  State Listed Threatened and Endangered
  State Species of Special Concern
  Colorado Natural Heritage Program Imperiled Species
- Taxonomic groups
  amphibians
  reptiles
  birds
  mammals
  native fish
  mollusks
- Other Groupings
  all species
  economically important species

The High Priority Habitat system ranks the vegetation polygons identified in the Colorado Gap Analysis statewide vegetation map. These polygons were identified from remote sensing techniques and each polygon was assumed to represent a single vegetation type. Therefore, the High Priority Habitat system serves as a tool to help identify the value and importance of each vegetation polygon identified in the project's vegetation maps.

Three methods were developed for ranking vegetation polygons. The first two methods were developed to rank vegetation polygons according to their importance to (1) economically important species and (2) rare species. These methods used the distribution maps developed by CDOW for rare species and economically important species. Unlike the maps developed through the Colorado Gap Analysis Program, which predict the distribution of species, the maps developed by CDOW for rare species and economically important species identify the known distribution of each species. The third method for ranking vegetation polygons can be used for all species, including rare species and economically important species, and uses the maps of potential distribution of each species predicted by the Colorado Gap Analysis model.

Ranking Habitat Values for Economically Important Species. For each economically important species, "activity areas," or areas used by the species for a specific purpose (e.g., elk calving areas) or during a specific season (e.g., elk winter range), were mapped and then rated subjectively in terms of their importance to the species. For example, winter range was rated as relatively less important to a species' survival than severe winter range. A composite map for all economically important species was developed by incorporating ratings for the importance of activity areas from the individual species maps. For the composite map, values assigned to vegetation polygons were not additive, but reflected the highest score assigned to the polygon for any of the economically important species. The composite map can be used to identify the vegetation polygons that are relatively more important to economically important species.

Ranking Habitat Values for Rare Species. Similar to the methods followed for economically important species, for each rare species, activity areas were mapped and then rated according to their importance to the species. In addition, activity areas were weighted according to the rarity of the species (e.g., Federally Endangered, Federally Threatened or State Endangered, State Threatened, or Federal Category 1). A composite map for all rare species was developed by incorporating ratings for species rarity and the importance of activity areas. For the composite map, values assigned to vegetation polygons were not additive, but reflected the highest score assigned to the polygon for any of the rare species. The composite map can be used to identify the vegetation polygons that are relatively more important to rare species.

Ranking Habitat Values for All Species. For each species, vegetation polygons within the species' distribution were assigned the value from the COVERS database indicating the degree of biological imperilment. This score provides an indication of the biological imperilment of the species in Colorado and ranks the species in terms of its needs for conservation efforts and habitat protection. Composite maps were developed for rare species, different taxonomic groups, and economically important species by assigning each vegetation polygon a score calculated as the sum of the scores for biological imperilment for all species predicted to occur in that polygon. Therefore, the overall value of a vegetation polygon incorporates information on species richness and species imperilment, and provides an indication of the importance of each polygon to the conservation of wildlife species.

Conservation and Stewardship Incentives

Many programs exist nationally as well as regionally and locally to encourage landowners and managers to adopt alternative land management practices and conserve biodiversity. Examples of incentives include relief from natural resource regulations, land acquisitions and exchanges, easements, management agreements, zoning and planning designations, tax incentives, technical assistance, and cost sharing. The OBP published a booklet that discusses these incentives, their strengths and weaknesses, and provides suggestions for modifications or additional incentives that will contribute to biodiversity conservation on lands managed primarily for commodity production and other human uses.[47] Another source of information about conservation incentives has been compiled by the Oregon Coordinated Resource Management (OCRM) Task Group,[29] a group of federal and state agencies supporting coordination of renewable natural resource planning and management on state, federal, and private lands. The OCRM Task Group has identified tax incentives and public funding sources for individual landowners as well as public and private organizations. Projects funded by incentive programs include but are not limited to the following:

1. Watershed improvement projects
2. Fish and wildlife habitat improvement projects
3. Wetland/Riparian Area improvement projects
4. Timber stand development and improvement projects
5. Soil projection projects
6. Landowner stewardship plan development projects
7. Education projects

Environmental Report Cards

The fundamental goal of conservation programs is to maintain ecosystems and the biodiversity contained within them. Incorporating the concept of "ecosystems" into conservation efforts serves to broaden our view of the environment by recognizing that management of our natural resources must integrate ecological relationships with social and political values so that our natural systems can be protected and maintained over time. To be successful in our attempt to maintain functional and viable ecosystems, we will need to measure and monitor our progress. One way of monitoring the success of our achievements is to report our progress in terms of our goals and objectives. An ideal way to do this is to adopt a "report card" that routinely apprizes the general public, scientific community, and decision-makers about our progress.

Criteria for Developing a Report Card

The report card should document our progress towards achieving our desired conditions and should be written so that it is understandable by a wide audience. This is because clear communication of the desired conditions will allow the public to follow and aspire to the same vision. In addition, the report card should define terminology as well as identify our desired conditions, measures used to evaluate performance, the scientific basis for assigning grades, and the current conditions/processes of the abiotic and biotic resources. Other information that is important to incorporate is identified by Harwell et al.[11] and includes but is not limited to: (1) habitat quality—landscape mosaic; spatial extent; landscape and community diversity; connectivity, fragmentation, structural diversity; (2) integrity of biotic community—biodiversity; trophic structure; exotic, invasive or noxious species; threatened and endangered species; economic or aesthetically important species; (3) ecological processes—primary and secondary productivity; succession; key ecological functions nutrient cycling; species dispersal and migration; (4) water quality—biological characteristics, physical characteristics, chemical characteristics; (5) hydrological system—hydroperiod; surface and groundwater flow; water storage; channel complexity or sinuosity; sediment transport; (6) disturbance regime—frequency and intensity of fire, flooding, storms, and drought; disease or pest outbreaks; anthropogenic disturbances; (7) soil quality—biological, physical, and chemical characteristics; erosion and accumulation of soil and sediment; and (8) air quality—biological, physical, and chemical characteristics.

In addition to reporting on the progress being made towards the project's environmental goals, it also would be helpful to identify 5-10 questions that the project is striving to answer. This step would be a tremendous help with initiating research coordination and inter- and intra-agency cooperation. Finally, any grade assigned should be based on a scientific rationale that gives the reader a clear understanding of the current status of the project and what steps are needed to achieve its goals.

For most projects, involving a variety of interest groups in the development of the desired conditions will facilitate and promote shared expectations and ownership of the ideas and goals. For example, since Oregon and Washington share several diverse ecoregions, a regional assessment of both state's ecoregions would make a large contribution to conservation efforts by allowing the region and multiple agencies to address multi-jurisdictional issues in a cohesive and unified manner. Agencies like the Northwest Power Planning Council, Bonneville Power Administration, U.S. Forest Service, and U.S. Fish and Wildlife Service would be able to set their program goals in a manner that would meet a regional vision and would transcend political boundaries.

Two recently completed efforts that have used ecoregion-level analysis to address conservation concerns were the OBP and World Wildlife Fund's review of the Terrestrial Ecoregions of North America.[34] Both approaches took a landscape-level view of the ecoregion and discussed its biological uniqueness, current conservation status, and priority actions. However, conservation efforts and assessments should not be limited to regional and ecoregional analyses. Assessments need to be conducted at several scales including local scales and at the level of hydrologic units comprising basins and watersheds. A multi-scale approach will depict results of a shared vision along with noting the implemental achievements at the region to local levels. Because specific goals and objectives may vary at these different levels, it is important to address the time requirement to meet strategies across all levels.

Wildlife Habitat Maps for Oregon and Washington

As a starting point for the development of quantitative processes for identifying and prioritizing areas important to conservation values, we have summarized existing digital data for Oregon and Washington. The frontispiece of this book depicts the wildlife habitats for each state and Tables 1, 2, and 3 (which follow the Literature Cited) list the acreage of each wildlife habitat and the amounts that are protected within each ecoregion of each state. We hope these will provide a simple overview of the abundance and distribution of our wildlife habitat and land cover types, generating ideas and contributing to the development of quantitative processes for identifying priority areas and habitats and conserving the biodiversity of the region.

Literature Cited

1. Anselin, A., P. M. Meire, and L. Anselin. 1989. Multi-criteria techniques in ecological evaluation: an example using the analytical hierarchy process. Biological Conservation 49: 215-229.
2. Bottom, D., S. Beckwitt, J. Christy, S. Clarke, J. Dambacher, C. Rissell, R. Hughes, H. Li, D. McCullough, A. McGee, K. Moore, R. Nawa, and S. Thiele. 1993. Oregon critical watersheds database. Watershed Classification Subcommittee of the Natural Production Committee, Oregon Chapter American Fisheries Society, Corvallis, OR.
3. Bolton, M. P. and R. L. Specht. 1983. A method for selecting conservation reserves. Australian National Parks and Wildlife Service Occasional Paper 8.
4. Colorado Division of Wildlife. 1998. High priority habitats. Natural Resource Ecology Laboratory, Colorado State University, Fort Collins, CO.
5. Csuti, B., P. J. Kennelly, S. M. Meyers, M. M. P Huso, and J. Sifneos. 1998. Multiscale biodiversity conservation: a prototype process for Oregon. Final report: Task 2A; Develop and refine methodology for analyzing biodiversity at the state scale. U.S. Environmental Protection Agency, Corvallis, OR.
6. ———, A. J. Kimerling, T. A. O'Neil, M. M. Shaughnessy, E. P. Gaines, and M. M. P. Huso. 1997. Atlas of Oregon wildlife: distribution, habitat, and natural history. Oregon State University Press, Corvallis, OR.
7. Defenders of Wildlife. 1998. Oregon's living landscape: Strategies and opportunities to conserve biodiversity. Oregon State University Press, Corvallis, OR.
8. Federal Geographic Data Committee. 1996. FGDC vegetation classification and information standards. U.S. Geological Survey, Reston, VA.
9. Gotmark, F., M. Ahlund, and M. O. G. Eriksson. 1986. Are indices reliable for assessing conservation value of natural areas? An avian case study. Biological Conservation 38: 55-73.
10. Gross, J. E., and Melcher. 1997. Colorado vertebrate ranking system. Natural Resource Ecology Laboratory, Colorado State University, Fort Collins, CO.
11. Harwell, M., V. Myers, T. Young, A. Barruska, N. Gassman, J. Gentile, C. Harwell, S. Appelbaum, J. Barko, B. Causey, C. Johnson, A. McLean, R. Smola, P. Templet, and S. Tosini. 1999. A framework for an ecosystem integrity report card. Bioscience 49:7: 543-556.
12. International Council for Bird Preservation. 1992. Putting biodiversity on the map: priority areas for global conservation. Cambridge, England.
13. Kagan, J. S., and S. Caicco. 1992. Manual of Oregon actual vegetation. Idaho Fish and Wildlife Research Cooperative Unit, University of Idaho, Moscow, ID.
14. Karr, J. R. 1990. Biological integrity and the goal of environmental legislation: lessons for conservation biology. Conservation Biology 4: 244-250.

15. Kershaw, M., G. M. Mace, and P. H. Williams. 1995. Threatened status, rarity, and diversity as alternative selection measures for protected areas: a test using Afrotropical antelopes. Conservation Biology 9: 324-334.

16. Kiester, A. R., J. M. Scott, B. Csuti, R. F. Noss, B. Butterfield, K. Sahr, and D. White. 1996. Conservation prioritization using GAP data. Conservation Biology 10: 1332-1342.

17. Kiilsgaard, C. 1999. Oregon Vegetation: Mapping and Classification of Landscape Level Cover Types. Final Report. U.S. Geological Survey-Biological Resources Division: Gap Analysis Program. Moscow, ID.

18. Kiilsgaard C., and C. Barrett. 1999. Map of: Oregon Vegetation Landscape-Level Cover Types. Published by Northwest Habitat Institute, Corvallis, OR.

19. Kirkpatrick, J. B. 1983. An iterative method for establishing priorities for the selection of nature reserves: an example from Tasmania. Biological Conservation 25: 127-134.

20. Lawton, J. H., J. R. Prendergast, and B. C. Eversham. 1994. The numbers and spatial distributions of species: analyses of British data. Pages 177-195 in: P. L. Forey, J. Humphries, and R. I. Vane Wright, editors. Systematics and conservation evaluation. Clarendon Press, Oxford, England.

21. Margules, C. R. 1989. Introduction to some Australian developments in conservation evaluation. Biological Conservation 50: 1-11.

22. ———, C. R., A. O. Nicholls, and R. L. Pressey. 1988. Selecting networks of reserves to maximize biological diversity. Biological Conservation 43: 63-76.

23. ———, and M. B. Usher. 1981. Criteria used in assessing wildlife conservation potential: a review. Biological Conservation 21: 79-109.

24. Millsap, B. A., J. A. Gore, D. E. Runde, and S. I. Cerulean. 1990. Setting priorities for the conservation of fish and wildlife species in Florida. Wildlife Monographs 111.

25. Nicholls, A. O., and C. R. Margules. 1993. An upgraded reserve section algorithm. Biological Conservation 64: 165-169.

26. Noss, R. F. 1987. From plant communities to landscapes in conservation inventories: a look at The Nature Conservancy (U.S.A.). Biological Conservation 41: 11-37.

27. ———, and A. Y. Cooperrider. 1994. Saving nature's legacy: protecting and restoring biodiversity. Island Press, Washington, D.C.

28. O'Neil, T. A., R. J. Steidl, W. D. Edge, and B. Csuti. 1995. Wildlife habitat clusters: using wildlife communities to improve vegetation classification for conserving biodiversity. Conservation Biology 9: 1482-1491.

29. Oregon Coordinated Resource Management Task Group. 1996. Public funding sources for landowner assistance. Oregon CRM Task Group, Portland, OR.

30. Prendergast, J. R., R. M. Quinn, J. H. Lawton, B. C. Eversham, and D. W. Gibbon. 1993. Rare species, the coincidence of diversity hotspots and conservation strategies. Nature 365: 335-337.

31. Pressey, R. L., C. J. Humphries, C. R. Margules, and R. I. Vane-Wright. 1993. Beyond opportunism: key principles for systematic reserve selection. Trends in Ecology and Evolution 8: 124-128.

32. ———, and A. O. Nicholls. 1989. Application of a numerical algorithm to the selection of reserve in semi-arid New South Wales. Biological Conservation 50: 263-278.

33. Rapoport, E. H., G. Borioli, J. A. Monjeau, J. E. Puntieri, and R. D. Oviedo. 1986. The design of nature reserves: a simulation trial for assessing specific conservation value. Biological Conservation 37: 269-290.

34. Ricketts, T., E. Dinerstein, D. Olson, C. Loucks, W. Eichbaum, D. DellaSala, K. Kavanagh, P. Hedao, P. Hurley, K. Carney, R. Abell, and S. Walters. 1999. Terrestrial ecoregions of North America, a conservation assessment. Supported by World Wildlife Fund. Island Press, Washington, D.C.

35. Rojas, M. 1992. The species problem and conservation: what are we protecting? Conservation Biology 6: 170-178.

36. Saetersdal, M., J. M. Line, and H. J. B. Birks. 1993. How to maximize biological diversity in nature reserve selection: vascular plants and breeding birds in deciduous woodlands, western Norway. Biological Conservation 66: 131-138.

37. Scott, J. M., F. Davis, B. Csuti, R. Noss, B. Butterfield, C. Groves, H. Anderson, S. Caicco, F. D'Erchia, T. C. Edwards, Jr., J. Ulliman, and G. Wright. 1993. Gap analysis: a geographical approach to protection of biological diversity. Wildlife Monographs 123.

38. ———, and M. D. Jennings. 1998. Large-area mapping of biodiversity. Annals of the Missouri Botanical Garden 85:34-47.

39. Short, H. L. and J. B. Hestbeck. 1995. National biotic resource inventories and GAP analysis. BioScience 45: 535-539.

40. Smith, P. G. R., and J. B. Theberge. 1987. Evaluating natural areas using multiple criteria: theory and practice. Environmental Management 11: 447-460.

41. Specht, R. L. 1975. The report and its recommendation. Pages 11-21 in: F. Fenner, editor. A national system of ecological reserves in Australia. Report 19. Australian Academy of Science, Canberra, Australia.

42. Stoms, D. M. 1994. Scale dependence of species richness maps. Professional Geographer 46: 346-358.

43. Thomas, J. W., technical editor. 1979. Wildlife habitats in managed forests: the Blue Mountains of Oregon and Washington. Agriculture Handbook 553. U.S. Forest Service, Washington D.C.

44. Thomas, C. D., and H. C. Mallorie. 1985. Rarity, species richness and conservation: butterflies of the Atlas Mountains in Morocco. Biological Conservation 33: 95-117.

45. Turpie, J. K. 1995. Prioritizing South African estuaries for conservation: a practical example using waterbirds. Biological Conservation 74: 175-185.

46. Usher, M. B. 1986. Wildlife conservation evaluation: attributes, criteria and values. In: M. B. Usher, editor. Wildlife conservation evaluation. Chapman and Hall, London, England.

47. Vickerman, S. 1998. Stewardship incentives: conservation strategies for Oregon's working landscape. Defenders of Wildlife, Lake Oswego, OR.

48. Washington Department of Fish and Wildlife. 1996. Priority habitats and species list. Habitat Program, Washington Department of Fish and Wildlife, Olympia, WA.

49. White, D., A. J. Kimerling, and W. S. Overton. 1992. Cartographic and geometric components of a global sampling design for environmental monitoring. Cartography and Geographic Information Systems 19: 5-22.

50. Wilderness Society, The. 1993. Pacific salmon and federal lands, a regional analysis. The Wilderness Society, Bolle Center for Forest Ecosystem Management, Washington D.C.

51. Williams, G. 1980. An index for ranking of wildfowl habitats, as applied to eleven sites in west Surrey, England. Biological Conservation 18: 3-99.

52. Williams, P., D. Gibbons, C. Margules, A. Rebelo, C. Humphries, and R. Pressey. 1996. A comparison of richness hotspots, rarity hotspots and complementary areas for conserving diversity of British birds. Conservation Biology 10: 155-174.

53. Wright, D. F. 1977. A site evaluation scheme for use in the assessment of potential nature reserves. Biological Conservation 11: 293-305.

Table 1. The total acreage of wildlife habitat types that occur and are under conservation-oriented protection strategies in Oregon and Washington.

Wildlife Habitat	Oregon		Washington	
	Total	Protected[1]	Total	Protected[1]
Westside Lowlands Conifer-Hardwood Forest	9,349,756	1,544,745	9,064,128	469,747
Westside Oak and Dry Douglas-fir Forest and Woodlands	433,132	2,596	425,038	13,832
Southwest Oregon Mixed Conifer Forest	4,020,321	900,642	Does Not Occur	Does Not Occur
Montane Mixed Conifer Forest	2,949,586	933,681	4,653,306	2,061,798
Eastside (Interior) Mixed Conifer Forest	4,126,957	279,223	4,662,101	400,429
Lodgepole Pine Forest and Woodlands	532,587	13,834	119,201	17,720
Ponderosa Pine Forest and Woodlands	6,226,351	98,349	1,927,176	98,907
Upland Aspen Forest	19,685	7,783	100,621	3,823
Subalpine Parkland	84,240	48,410	327,442	190,593
Alpine Grassland and Shrublands	291,494	163,170	1,591,115	1,300,415
Westside Grasslands	133[2]	Not Mapped	22,491	1,320
Ceanothus-Manzanita Shrublands	52,104	22,016	Does Not Occur	Does Not Occur
Western Juniper and Mountain Mahogany Woodlands	4,037,221	105,708	Does Not Occur	Does Not Occur
Eastside (Interior) Canyon Shrublands	358,250	68,709	Not Mapped[3]	Not Mapped
Eastside (Interior) Grasslands	1,935,794	132,070	1,002,076	65,535
Shrub-steppe	17,420,753	490,879	7,144,697	473,622
Dwarf Shrub-steppe	514,066	26,777	Not Mapped[4]	Not Mapped
Desert Playa and Salt Scrub Shrublands	719,503	48,105	Not Mapped[4]	Not Mapped
Agriculture, Pasture and Mixed Environs	6,197,887	28,134	9,251,107	43,991
Urban and Mixed Environs	575,087	487	1,204,680	4,967
Open Water - Lakes, Rivers, Streams	780,901	96,437	761,360	46,896
Herbaceous Wetlands	1,031,343	135,121	210,451	12,966
Westside Riparian-Wetlands	168,872	10,339	347,653	38,287
Montane Coniferous Wetlands	56,099	2,373	241,450	25,551
Eastside (Interior) Riparian-Wetlands	31,121	2,604	100,763	9,285
Coastal Dunes and Beaches	52,451	1,821	Not Mapped[4]	Not Mapped
Coastal Headlands and Islets	9,137	987	7,776	1,831
Bays and Estuaries	172,748	31,332	226,336	3,6896
Inland Marine Deeper Water	Does Not Occur	Does Not Occur	1,855,780	2,9076
Marine Nearshore	223,371	1,350	750,329	1,0446
Marine Shelf	3,905,164	Not Determined	4,780,625	Not Determined[5]
Oceanic	33,987,189	Not Determined	19,845,660	Not Determined[5]
Totals	100,263,303		70, 532,093	

[1] Oregon's values were determined by using the Defenders of Wildlife's biodiversity management rating (Defenders of Wildlife 1998). Specifically, only those lands rated with a high degree of protection were used, and these were areas that had a rating of 8, 9, or 10. Washington's values were determined using the Washington Gap Analysis protected areas data layer; specifically, only lands with a high degree of protection were used, and these were areas that had a status of 1 or 2 (Cassidy 1997:13). Marine protected areas were determined from the following publications: Murray, M.R. 1998. The status of marine protected areas in Puget Sound. Volumes I and II Puget Sound/Georgia Basin Environmental Report Series 8. Puget Sound Water Quality Action Team, Olympia, Washington, USA; and Robinson, M.K. 1999. The status of Washington's coastal marine protected areas. Washington Department of Fish and Wildlife, Olympia, USA.

[2] Because of difficulty in classifying this type using remote sensing (i.e., discerning native grasslands from pasture lands) native westside grasslands have inadvertently been classified within the agriculture habitat type. Nonetheless, there are few areas known to be native westside grasslands.

[3] This type was only recognized along the Oregon and Washington border, otherwise it was not part of the vegetation classification when the Washington Gap Project mapped the state of Washington. Thus, no wildlife habitat area was determined.

[4] This type was not part of the vegetation classification when the Washington Gap Project mapped the state of Washington. Thus, no wildlife habitat area was determined.

[5] While not included in this analysis, there are 102 Marine Protected Areas (MPAs) in Puget Sound (Murray 1998); an additional 33 MPAs occur along the outer Washington coastline (Robinson 1999). While most MPAs are quite small in size, the Olympic Coast National Marine Sanctuary along the outer coast of northwest Washington offers protection to some 2,112,000 acres of marine nearshore, marine shelf, and oceanic habitat.

Table 2. Amount of protected[1] wildlife habitat types in Oregon by ecoregion.

Wildlife-Habitat[2]	Ecoregion[3]												
	1	2	3	4	5	6	7	8	9	10	11	12	13
1	790,718	8,005	98,483	628,149	0	19,389	0	0	0	0	0	0	0
2	85	1,757	0	662	0	92	0	0	0	0	0	0	0
3	3,714	8,143	608,870	248,819	31,095	0	0	0	0	0	0	0	0
4	0	0	7,941	247,269	391,740	77,026	42,611	167,093	0	0	0	0	0
5	0	0	0	0	19,802	81,002	150,958	26,769	74	0	0	185	0
6	0	0	0	316	5,317	8,201	0	0	0	0	0	0	0
7	0	850	0	8,519	15,733	26,319	39,466	641	0	0	0	130	0
8	0	0	0	0	0	0	0	59	1,102	0	0	0	0
9	0	0	185	712	40,540	2,712	0	0	0	0	0	0	0
10	0	0	133	0	16,424	0	3,988	141,516	0	0	0	0	0
11	(this habitat not mapped, thus no acreage figures currently available)												
12	0	20,549	1,459	0	0	0	0	0	0	0	0	0	
13	0	0	0	0	0	808	15,624	0	9,750	0	50	9,715	0
14	0	0	0	0	0	0	0	0	0	0	2,351	0	0
15	0	0	0	0	0	2,206	125,246	4,798	0	0	0	0	0
16	0	0	0	0	0	679	3,045	44	226,119	0	3,045	7,569	0
17	0	0	0	0	0	231	0	0	25,027	0	0	0	0
18	0	0	0	0	0	0	0	0	10,323	0	0	0	0
19	575	6,655	259	2,6230	2,769	1,142	0	129	0	0	0	0	
20	0	294	0	0	0	120	0	0	0	0	0	0	0
21	128	468	63	1,413	1,215	2,268	110	802	40,618	0	102	0	0
22	885	834	688	203	343	46,320	1,310	0	7,008	0	28	0	0
23	7,324	2,330	69	315	0	0	0	0	0	0	0	0	0
24	0	0	0	988	585	692	106	0	0	0	0	0	0
25	0	0	0	0	0	0	0	0	238	0	0	0	0
26	371	0	1,451	0	0	0	0	0	0	0	0	0	0
27	369	0	621	0	0	0	0	0	0	0	0	0	0
28	31,030	13	26	268	0	0	0	0	0	0	0	0	0
30	731	0	619	0	0	0	0	0	0	0	0	0	0
31	0	0	0	0	0	0	0	0	0	0	0	0	0
32	0	0	0	0	0	0	0	0	0	0	0	0	0

[1] Oregon's values were determined by using the Defenders of Wildlife's biodiversity management rating. Specifically, only those rated with a high degree of protection were used, and these were areas that had a rating of 8, 9, or 10.

[2] Wildlife Habitats: (1) Westside Lowlands Conifer-Hardwood Forest, (2) Westside Oak and Dry Douglas-fir Forest and Woodlands, (3) Southwest Oregon Mixed Conifer Forest, (4) Montane Mixed Conifer Forest, (5) Eastside (Interior) Mixed Conifer Forest, (6) Lodgepole Pine Forest and Woodlands, (7) Ponderosa Pine Forest and Woodlands, (8) Upland Aspen Forest, (9) Subalpine Parkland, (10) Alpine Grassland and Shrublands, (12) Ceanothus-Manzanita Shrublands, (13) Western Juniper and Mountain Mahogany Woodlands, (14) Eastside (Interior) Canyon Shrublands, (15) Eastside (Interior) Grasslands, (16) Shrub-steppe, (17) Dwarf Shrub-steppe, (18) Desert Playa and Salt Scrub Shrublands, (19) Agriculture, Pasture and Mixed Environs, (20) Urban and Mixed Environs, (21) Open Water-Lakes, Rivers, Streams, (22) Herbaceous Wetlands, (23) Westside Riparian-Wetlands, (24) Montane Coniferous Wetlands, (25) Eastside (Interior) Riparian-Wetlands, (26) Coastal Dunes and Beaches, (27) Coastal Headlands and Islets, (28) Bays and Estuaries, (30) Marine Nearshore, (31) Marine Shelf, (32) Oceanic.

[3] Ecoregion: (1) Oregon Coast Range, (2) Western Oregon Interior Valley, (3) Siskiyou Mountains, (4) Westside Oregon Cascades, (5) High Oregon Cascades, (6) Eastside Oregon Cascades, (7) Oregon Blue and Wallowa Mountains, (8) High Oregon Blue and Wallowa Mountains, (9) Basin and Range, (10) Columbia Basin, (11) Owyhee Uplands, (12) High Lava Plains, (13) Snake River Plains.

Table 3. Amount of protected[1] wildlife habitat types in Washington by ecoregion.

Wildlife Habitat[2]	Ecoregion[3]												
	1	2	3	4	5	6	7	8	9	10	11	12	13
1	101,073	21,038	0	0	0	746	2,979	0	0	93,120	58,577	857	191,353
2	0	3,532	0	0	17	0	8,921	0	0	0	1,222	139	0
4	273,608	0	11,984	0	17	195,521	91,259	32,658	277,140	677,505	336,052	0	166,054
5	0	0	71,349	23	1,248	178,324	17,704	44,307	69,145	11,775	6,554	0	0
6	0	0	384	0	11	4,776	2,296	0	5,049	3,375	1,829	0	0
7	0	0	18,053	474	8,940	25,270	11,362	18,677	15,433	693	5	0	0
8	0	0	190	63	29	297	0	0	2,680	564	0	0	0
9	0	0	135	0	0	44,703	7,793	0	27,015	50,477	23,838	0	36,632
10	75,390	0	4,298	0	0	153,578	12,187	5,872	203,560	582,801	145,585	0	117,143
11	0	1,276	44	0	0	0	0	0	0	0	0	0	0
14	(this habitat not mapped, thus no acreage figures currently available)												
15	0	0	2,826	2,240	3,784	19,492	6,140	13,723	14,737	2,593	0	0	0
16	0	0	9,120	2,042	420,826	11,292	4,329	17,077	8,927	10	0	0	0
19	287	1,658	3,770	557	34,256	441	196	137	1,378	37	25	1,123	126
20	273	3,074	405	897	0	0	0	0	0	0	0	112	206
21	5,743	2,072	3,011	665	8,640	4,252	441	57	2,175	11,162	6,204	0	2,473
22	1,890	1,456	111	0	6,212	0	0	0	0	595	0	1,148	1,554
23	21,230	2,723	0	0	0	0	0	0	0	5,971	2,501	1,129	4,733
24	225	0	3,790	60	0	4,140	4,655	20	6,038	2,751	3,870	0	0
25	0	0	0	0	3,726	2,418	60	860	2,220	0	0	0	0
26	(this habitat not mapped, thus no acreage figures currently available)												
27	1,831	0	0	0	0	0	0	0	0	0	0	0	0
28	2,166	882	0	0	0	0	0	0	0	0	0	478	163
29	0	2,907	0	0	0	0	0	0	0	0	0	0	0
30	718	326	0	0	0	0	0	0	0	0	0	0	0
31	0	0	0	0	0	0	0	0	0	0	0	0	0
32	0	0	0	0	0	0	0	0	0	0	0	0	0

[1] Washington's values were determined using the Washington Gap Analysis protected areas data layer. Specifically, only those lands with a higher degree of protection were used, and these were areas that had a status of 1 or 2.

[2] Wildlife Habitats: (1) Westside Lowlands Conifer-Hardwood Forest, (2) Westside Oak and Dry Douglas-fir Forest and Woodlands, (4) Montane Mixed Conifer Forest, (5) Eastside (Interior) Mixed Conifer Forest, (6) Lodgepole Pine Forest and Woodlands, (7) Ponderosa Pine Forest and Woodlands, (8) Upland Aspen Forest, (9) Subalpine Parkland, (10) Alpine Grassland and Shrublands, (11) Westside Native Grasslands, (14) Eastside (Interior) Canyon Shrublands, (15) Eastside (Interior) Grasslands, (16) Shrub-steppe, (19) Agriculture, Pasture and Mixed Environs, (20) Urban and Mixed Environs, (21) Open Water-Lakes, Rivers, Streams, (22) Herbaceous Wetlands, (23) Westside Riparian-Wetlands, (24) Montane Coniferous Wetlands, (25) Eastside (Interior) Riparian-Wetlands, (26) Coastal Dunes and Beaches, (27) Coastal Headlands and Islets, (28) Bays and Estuaries, (29) Inland Marine Deeper Water, (30) Marine Nearshore, (31) Marine Shelf, (32) Oceanic.

[3] Ecoregion: (1) Outer Olympic Peninsula, (2) Puget Sound, (3) Northeastern Corner, (4) Okanogan Highlands, (5) Columbia Basin, (6) East Central Cascades, (7) Southeast Cascades, (8) Blue Mountains, (9) Northeast Cascades, (10) Northwest Cascades, (11) Southwest Cascades, (12) Willamette Valley, (13) Inner Olympic Peninsula.

6

Key Ecological Functions of Wildlife Species

Bruce G. Marcot & Madeleine Vander Heyden

Introduction

An ecosystem is the set of all its component organisms and populations and their ecological interactions with each other and with the abiotic world. Traditionally, wildlife-habitat relationships (WHR) programs, models, and databases have focused on how the presence of terrestrial vertebrates is influenced by environmental conditions, and have mostly ignored ecological interactions. WHR approaches have assumed that wildlife (W) basically is a function of habitat (H), or W = f(H). Further, most evaluations of biodiversity patterns and of imperiled or scarce species have focused only on mapping species counts or evaluating the role of habitat on species' presence and abundance.

In most of these worthy pursuits, community interactions and ecological roles of species—how organisms change their environments by what they do—have been largely absent. That is, as much as organisms are a function of their habitat, environmental conditions, including habitat and resources available for other species, are influenced by the ecological roles of organisms. This chapter provides a foundation for explicitly considering the ecological roles and interactions of vertebrate wildlife and for integrating those roles with the more traditional habitat-based WHR focus.

The Functional Foundation of Sustainable Ecosystems

Much has been written on sustainable management of natural resources and conservation of wildlife populations and natural ecosystems.[6] Sustainability may be defined as resource use habits that do not outstrip the capacity of an ecosystem to produce desirable conditions and commodities, and often has been seen as a *de facto* conservation goal. But what influences sustainability? Sustainable use of resources is made possible by ensuring that rates of production are not exceeded by rates of loss plus extraction.[43] Further, natural ecosystems are sustainable only when their native biodiversity (the variety of life and *its processes*) and the functional basis of productivity are maintained. Ultimately, it is the set of ecological roles played by its component organisms, including humans, that influences ecosystem biodiversity, productivity, and sustainability.

To ensure sustainable wildlife populations, conserving threatened or endangered species should not stop at addressing only their individual habitat needs, that is, the W = f(H) relation. In fact, a primary purpose of the Endangered Species Act is ". . . to provide a means whereby the ecosystems upon which endangered species and threatened species depend may be conserved . . . ." (Sec. 2[b]). As well, the U.S. Forest Service has moved to an "ecosystem management" basis for planning and management on national forests and grasslands.[34] To conserve and manage ecosystems means understanding their dynamics and processes, including the ecological functions of species.

Key Ecological Functions of Wildlife Species

The term *key ecological functions* (KEFs) refers to the principal set of ecological roles performed by each species in its ecosystem. KEFs refer to the main ways organisms use, influence, and alter their biotic and abiotic environments. "Key" refers to the main roles played by each species. As we show in this chapter, categories of KEFs can be depicted for each species and used in multiple-species planning, biodiversity conservation, and other facets of ecosystem analysis and management.

Background and Theory

Expanding the WHR Paradigm

The explicit use of functional categories for evaluating effects of land use planning on ecological communities appeared in the terrestrial ecology science assessments of the Interior Columbia Basin Ecosystem Management Project.[23, 24, 28] In that work, the traditional WHR paradigm was expanded by depicting the environmental and functional influences on a species in a "species influence diagram" (Figure 1). The species influence diagram illustrates how the distribution and abundance of a species (or species group) are influenced by key environmental correlates; that a species performs specific KEFs; that KEFs in turn influence the biodiversity, productivity, and sustainability (BPS) of the ecosystem; that management

Key environmental correlates   *Key ecological functions*

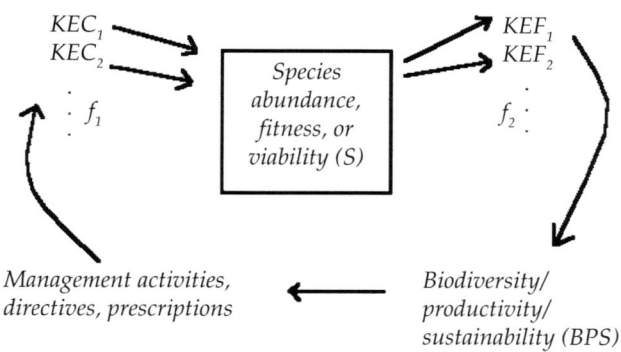

Figure 1. A generalized species influence diagram, showing (1) the relations of species (S) to their key environmental correlates (KECs or habitat elements; f1 functions) and key ecological functions (KEFs); (2) that species' KEFs influence ecosystem biodiversity, productivity, and sustainability (BPS; f2 functions); and (3) that management activities can be based on BPS and can influence KECs and habitats, in turn influencing the viability of species. The matrices on the CD-ROM provide data on each of these diagram elements. Arrows represent rates of influence among the elements.

goals for BPS can help establish management guidelines; and that management activities influence species' key environmental correlates. The term "key environmental correlate" refers to wildlife habitats, habitat elements, and other nonhabitat influences on the distribution and abundance of organisms.

We distinguish *KEFs of species*, which are the biotic and ecological roles of organisms, from abiotic *ecosystem processes*, which include weather disturbance events and geochemical cycles. In this chapter we focus on the link between species and their environments through KEFs and demonstrate how other components in this overall species influence diagram can be used in evaluations.

Ecological Functions of Vertebrates

It has long been recognized that the ecological roles of vertebrate species influence ecosystems. Only recently, however, has this been integrated into theory and practice.[7, 42, 44] Examples of some ecological functions of vertebrate species include how:

- browsing or grazing by ungulates can change plant communities;[1, 3, 14, 41]
- animals can act as "environmental engineers" and influence geomorphology[5] and ecosystem processes;[17]
- frugivores can support viable fruit-bearing plants;[25]
- pollinators can support plant diversity;[36]
- seed dispersers[8] and frugivores[13] can influence forest succession and regeneration;
- carrion feeding can support the trophic structure of a community;[12]
- carnivore predation can influence populations of ungulate prey species;[48]

- rodents can serve to disseminate beneficial mycorrhizal fungi in western U.S. forests;[18] and
- reptilian primary burrow excavators can provide for avian secondary burrow users.[46]

Many other examples can be found in the literature, especially on ecological functions of plants and invertebrates.

What is the manager to do with all this information? The types and ramifications of ecological functions and interactions in an ecosystem can be overwhelming. This is why we have developed a structured classification of KEFs and provide here a way of considering functions when formulating and evaluating effects of land use management actions. Through the species influence diagram and use of the matrixes on the CD-ROM, management activities can be linked to habitat elements and structures, and thence to species, their KEFs, and their influence on ecosystem BPS.

Ecological Functions Central to the Trophic Structure of Wildlife Communities

Ecological functions of organisms support the trophic structure of ecosystems, that is, energy flows, food webs, and nutrient cycling. More biodiverse systems support wider arrays of ecological functions,[16, 37] and thus might support a broader or deeper trophic structure. Ecosystems differ in their rates and patterns of primary production (photosynthesis by plants), which largely supports vertebrate primary consumers and much of the rest of the food web and trophic pyramid. Trophic structures differ among ecosystems in that different sets of vertebrate species participate in various ways in providing major standing crop biomass, types and rates of consumption and decay or recycling functions, energy transfer and nutrient cycling, and interspecific interactions including competition and predation. We provide some examples below.

Building a Database of Species' Key Ecological Functions

Purpose

The purpose of building a matrix of KEFs of vertebrates is to provide a consistent framework from which to consider the ecological roles of wildlife in the management of populations, habitats, and ecosystems. A related purpose is to provide a means of posing working hypotheses of the ecological roles of wildlife and effects of management actions on those roles.

Methods

We developed the KEF matrix by first refining the existing hierarchical classification system of KEFs developed by Marcot et al.[23] and then populating the KEF matrix by denoting the pertinent KEF categories for each species. Marcot et al.'s KEF classification was developed for use with soil microorganism groups, lichens, bryophytes, vascular plants, and invertebrates, as well as vertebrates. To maintain some similarity, we retained their major KEF categories and numbering system, trimmed out subcategories not pertinent to vertebrates, and expanded some subcategories. As with the original system, our revised KEF classification is strictly hierarchical so that one can query the matrix at various levels of specificity.

In populating the KEF species matrix, we started with that of the Interior Columbia Basin Ecosystem Management Project.[21] We added pertinent wildlife species not included in that database, including coastal and marine birds and mammals, and deleted species not found in Washington and Oregon. We also reviewed published literature, used our own experience and professional judgment, and sought peer review of the species or KEF categories we added. We lacked data by which to describe KEFs for a species for each wildlife habitat, habitat element, or habitat structure individually.

Database Results

The KEF classification system includes some 85 categories of ecological functions, including major headings and subheadings (Table 1). The 8 major headings are trophic relations, nutrient cycling, organismal relations, disease vectors, soil relations, wood structure relations, water relations, and vegetation relations.

The KEF matrix on the CD-ROM includes 598 wildlife species out of the 755 species identified for the project area. The 157 species not included are considered "accidental" or are subspecies of KEF-coded species. The KEF matrix has 7,319 records, each record being a specific KEF category for a species. Thus, species average 12 KEF categories each (including KEF categories and subcategories, so there is some duplication in this count) and range from 3 (California condor and turkey vulture, both carrion-feeder specialists, the former of only historical occurrence), to 30 (American black bear) KEF categories.

As might be expected, the frequency of species by number of KEF categories roughly follows a normal distribution that is slightly skewed to the right (Figure 2). That is, few species are coded with very few, or very many, KEF categories, and there are slightly more species with many KEF categories than predicted by a normal distribution. The species with few KEF categories are *functional specialists*; like the condor and vulture, they perform only a very few functions within their ecosystems. The species with many KEF categories tend to be *functional generalists*; they perform many functions. We discuss the ecological implications of such functional patterns below.

Patterns of Key Ecological Functions of Vertebrate Species in Washington and Oregon

Following are some examples of various functional patterns of wildlife communities in Washington and Oregon.

Trophic Structure of Wildlife Communities

The KEF matrix can be used to depict general trophic structures of communities by summing number of species according to their trophic and diet relations. In Washington and Oregon as a whole, 50% of all wildlife species are primary consumers (herbivores), 87% are secondary consumers (primary predators), and only 1% are tertiary consumers (secondary predators), with bird species playing a proportionally greater role across this spectrum (Figure 3). Other minor trophic categories include carrion-feeders (6%, mostly birds and mammals), cannibalistic feeders (1%, amphibians, birds, and mammals), and coprophagous feeders (2%, all mammals).

The main (>50 species) primary-consumer categories are spermivores (seed-eaters), frugivores (fruit-eaters), grazers (grass/forb-eaters), foliovores (leaf-eaters), and aquatic herbivores (Figure 4). Amphibians play the main role of feeding on plant material in water on decomposing

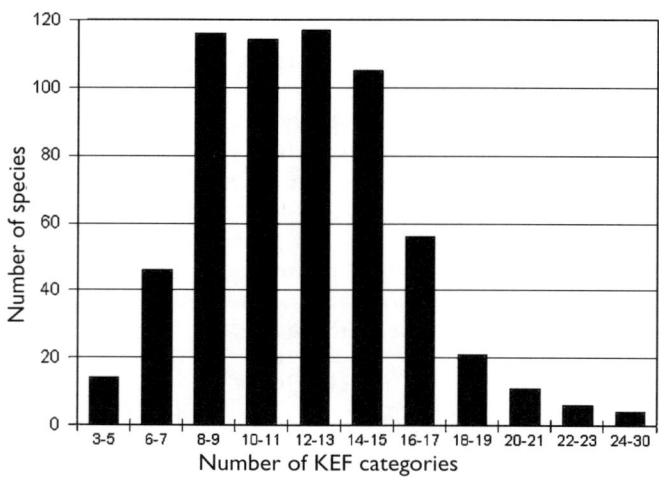

Figure 2. Frequency histogram of number of vertebrate wildlife species in Washington and Oregon by number of categories of key ecological functions (KEFs) they perform, as denoted in the KEF matrix.

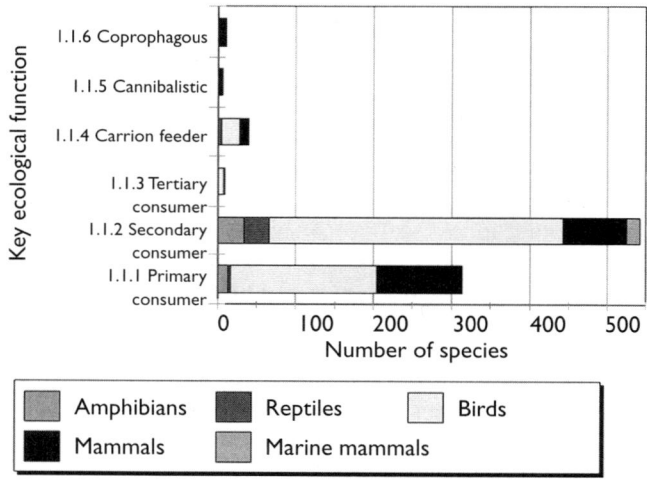

Figure 3. Trophic level functions of wildlife in Oregon and Washington.

Table 1. The classification of key ecological functions of wildlife in Washington and Oregon.

1 Trophic relationships *
    1.1 heterotrophic consumer (an organism that is unable to manufacture its own food and must feed on other organisms)*
        1.1.1 primary consumer (herbivore; an organism that feeds primarily on plant material) (also see below under Herbivory) *
            1.1.1.1 foliovore (leaf eater) *
            1.1.1.2 spermivore (seed eater) *
            1.1.1.3 browser (leaf, stem eater)
            1.1.1.4 grazer (grass, forb eater)
            1.1.1.5 frugivore (fruit eater) *
            1.1.1.6 sap feeder
            1.1.1.7 root feeders *
            1.1.1.8 nectivore (nectar feeder)
            1.1.1.9 fungivore (fungus feeder) *
            1.1.1.10 flower/bud/catkin feeder
            1.1.1.11 aquatic herbivore
            1.1.1.12 feeds in water on decomposing benthic substrate (benthic is the lowermost zone of a water body)
            1.1.1.13 bark/cambium/bole feeder
        1.1.2 secondary consumer (primary predator or primary carnivore; a carnivore that preys on other vertebrate or invertebrate animals, primarily herbivores) *
            1.1.2.1 invertebrate eater
                1.1.2.1.1 terrestrial invertebrates
                1.1.2.1.2 aquatic macroinvertebrates (e.g., not plankton)
                1.1.2.1.3 freshwater or marine zooplankton
             1.1.2.2 vertebrate eater (consumer or predator of herbivorous or carnivorous vertebrates) *
                1.1.2.2.1 piscivorous (fish eater) *
             1.1.2.3 ovivorous (egg eater)
        1.1.3 tertiary consumer (secondary predator or secondary carnivore; a carnivore that preys on other carnivores)
        1.1.4 carrion feeder (feeds on dead animals)
        1.1.5 cannibalistic (eats members of its own species)
        1.1.6 coprophagous (feeds on fecal material)
        1.1.7 feeds on human garbage/refuse
            1.1.7.1 aquatic (e.g., offal and bycatch of fishing boats)
            1.1.7.2 terrestrial (e.g., garbage cans, landfills)
    1.2 prey relationships
        1.2.1 prey for secondary or tertiary consumer (primary or secondary predator)
2 Aids in physical transfer of substances for nutrient cycling (C,N,P, etc.) *
3 Organismal relationships *
    3.1 controls or depresses insect population peaks *
    3.2 controls terrestrial vertebrate populations (through predation or displacement) *
    3.3 pollination vector
    3.4 transportation of viable seeds, spores, plants, or animals (through ingestion, caching, caught in hair or mud on feet, etc.) *
        3.4.1 disperses fungi
        3.4.2 disperses lichens
        3.4.3 disperses bryophytes, including mosses
        3.4.4 disperses insects and other invertebrates (phoresis)
        3.4.5 disperses seeds/fruits (through ingestion or caching)
        3.4.6 disperses vascular plants *
    3.5 creates feeding, roosting, denning, or nesting opportunities for other organisms *
        3.5.1 creates feeding opportunities (other than direct prey relations) *
            3.5.1.1 creates sapwells in trees
        3.5.2 creates roosting, denning, or nesting opportunities *
    3.6 primary creation of structures (possibly used by other organisms) *
        3.6.1 aerial structures (typically large raptor or squirrel stick or leaf nests in trees or on platforms, or barn swallow/cliff swallow nests)*
        3.6.2 ground structures (above-ground, non-aquatic nests and ends and other substrates, such as woodrat middens, nesting mounds of swans, for example)*
        3.6.3 aquatic structures (muskrat lodges, beaver dams)*
    3.7 user of structures created by other species
        3.7.1 aerial structures (typically large raptor or squirrel stick or leaf nests in trees or on platforms, or barn swallow/cliff swallow nests)

3.7.2 ground structures (above-ground, non-aquatic nests and ends and other substrates, such as woodrat middens, nesting mounds of swans, for example)

3.7.3 aquatic structures (muskrat lodges, beaver dams)

3.8 nest parasite

3.8.1 interspecies parasite (commonly lays eggs in nests of other species)

3.8.2 common interspecific host (parasitized by other species)

3.9 primary cavity excavator in snags or live trees (organisms able to excavate their own cavities)

3.10 secondary cavity user (organisms that do not excavate their own cavities and depend on primary cavity excavators or natural cavities)

3.11 primary burrow excavator (fossorial or underground burrows)

3.11.1 creates large burrows (rabbit-sized or larger)

3.11.2 creates small burrows (less than rabbit-sized)

3.12 uses burrows dug by other species (secondary burrow user)

3.13 creates runways (possibly used by other species; runways typically are worn paths in dense vegetation)

3.14 uses runways created by other species

3.15 pirates food from other species

3.16 interspecific hybridization (species known to regularly interbreed)

4 Carrier, transmitter, or reservoir of vertebrate diseases

4.1 diseases that affect humans *

4.2 diseases that affect domestic animals

4.3 diseases that affect other wildlife species

5 Soil relationships *

5.1 physically affects (improves) soil structure, aeration (typically by digging) *

5.2 physically affects (degrades) soil structure, aeration (typically by trampling) *

6 Wood structure relationships (either living or dead wood) *

6.1 physically fragments down wood *

6.2 physically fragments standing wood *

7 Water relationships *

7.1 impounds water by creating diversions or dams *

7.2 creates ponds or wetlands through wallowing

8 Vegetation structure and composition relationships *

8.1 creates standing dead trees (snags) *

8.2 herbivory on trees or shrubs that may alter vegetation structure and composition (browsers)

8.3 herbivory on grasses or forbs that may alter vegetation structure and composition (grazers)

---

* Key ecological functions of *Homo sapiens*.

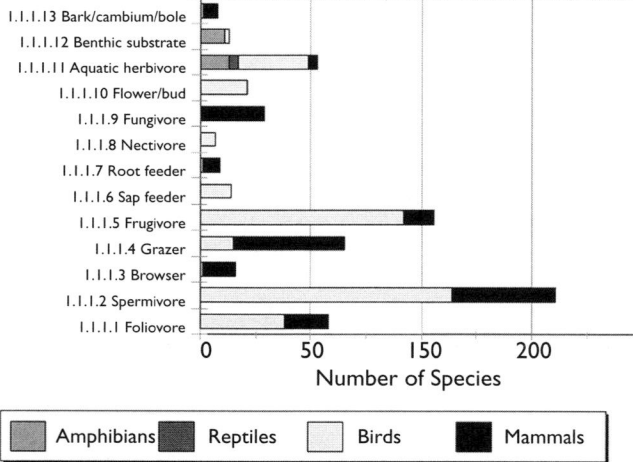

Figure 4. Primary consumption (herbivory) functions of wildlife in Washington and Oregon.

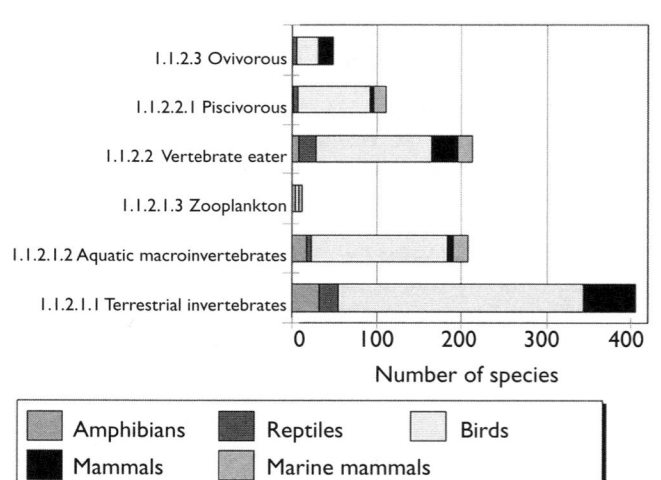

Figure 5. Secondary consumption (carnivory) functions of wildlife in Washington and Oregon.

benthic substrates. Birds play main roles as frugivores, sap feeders, nectivores (nectar-feeders), and flower, bud, and catkin feeders. Mammals play main roles as browsers, grazers, root feeders, and fungivores (fungus-feeders). Both birds and mammals participate more equally as foliovores and spermivores.

The main (>100 species) secondary consumer categories are feeders on terrestrial invertebrates, vertebrates, aquatic macroinvertebrates, and fish (piscivores) (Figure 5). Amphibians participate in all secondary consumer categories, but proportionally are the greatest consumers of freshwater invertebrates. Birds proportionately dominate all categories except that of consuming freshwater invertebrates. Mammals participate in most categories with their greatest proportionate representation as terrestrial invertebrate-feeders, vertebrate feeders, and ovivores (vertebrate egg-eaters). Tertiary consumers include a few birds and mammals only.

These patterns differ among native wildlife habitats. For example, avian primary consumers tend to be most numerous (total number of species and proportion of all species) in both eastside (east of the Cascade Mountains crest) and westside wetlands and in dry forest types, whereas avian secondary consumers are most numerous in westside (wetland, forest, woodland, and riparian) types. Mammalian primary consumers are most numerous in higher elevation or eastside grassland, shrubland, or forest types, whereas mammalian secondary consumers are most numerous in westside types (riparian, wetland, grassland, woodland, and forest types alike). Much of this can be explained by the kinds of climates and vegetation physiognomies present in various geographic locations, such as the dry eastside rainshadow effects where grasslands, shrublands, and grazing and browsing mammals (mammalian primary consumers) are more common.

Understanding such trophic patterns and their biogeographic relations may be important to managers at a broad scale, as land ownership patterns and management activities might affect habitats differently, favoring one trophic category over another in different areas. For example, the two anthropogenic (human-created) habitats—Agriculture, Pastures, and Mixed Environments, and Urban and Mixed Environments— provide poorly for mammalian secondary consumers. Comparing historical to current habitat patterns may reveal broad-scale and longer-term shifts in trophic structures of wildlife communities, which in turn can disclose differential repercussions on other ecosystem functions and biodiversity patterns. It may be of interest to compare current and potential future conditions to extrapolated historical conditions, to see if and how wildlife communities have changed in terms of their trophic patterns and to help set broad-scale planning objectives. Such objectives might include restoring or maintaining all historical trophic categories by geographic area or habitat, or mitigating the adverse reduction of trophic categories in anthropogenically disturbed environments.

Organismal Relations Within Wildlife Communities

Some 26 categories of organismal relations within wildlife communities can be evaluated with the KEF matrix. Patterns presented here are for all communities in Washington and Oregon (Figures 6-8); patterns among specific wildlife habitats may differ.

A number of examples of symbiosis can be demonstrated. Seven species of birds (hummingbirds and orioles) serve as pollination vectors for plants. Among terrestrial vertebrates, mammals are the sole dispersers of fungi and lichens; among terrestrial vertebrates, primarily birds disperse invertebrates and vascular plant parts; and both birds and mammals disperse seeds and fruits (Figure 6). Birds create sapwells for feeding by other species. Both birds and mammals create roosting, denning, or nesting structures in aerial, ground, and aquatic situations that other amphibian, reptile, bird, and mammal species also use (Figure 7).

Three bird species act as nest (brood) parasites (although only the brown-headed cowbird regularly parasitizes nests of other species within the Oregon-Washington area) and 58 bird species serve as their hosts. Eighteen birds and 1 mammal species act as primary cavity excavators, serving 32 bird and 24 mammal secondary cavity-using species. Twenty-one mammal species dig large (rabbit size) burrows and 9 bird and 53 mammal species dig small burrows, which also provide for 7 amphibian, 14 reptile, 1 bird, and 59 mammal secondary burrow-using species. Forty-six mammal species create runways in terrestrial vegetation that also can be used by 2 amphibian and 41 other mammal species (Figure 8).

Managers might wish to determine which species of such symbiotic, organismal relations occur in the wildlife habitats and communities under their charge to ensure that the habitat requirements of all members are duly addressed, and that full ramifications of modifying habitat for one set of species on their symbiotic partners are acknowledged.

Soil, Wood, Water, and Vegetation Relations

Some examples of soil, wood, water, and vegetation relations are shown in Figure 9. Seventy-five species of amphibians, reptiles, birds, and mammals may help improve or maintain soil structure and aeration through tunneling, digging, and turning over soil. This also has the benefit of increasing aerobic bacterial decay in soils and incorporating coarse decaying wood as organic matter into the soil. One bird and 20 mammal species physically help fragment down wood, and 1 bird and 1 mammal species fragment standing wood; these functions, too, help hasten the wood decay process.

Seven mammal species, including 2 elk subspecies, impound water by creating diversions or dams, or create ponds or wetlands through wallowing. Such ponds and wetlands can then provide for a wide array of other wildlife species; the habitat elements database lists 242 species of amphibians, reptiles, birds, and mammals as being associated with ponds and wetlands. Three

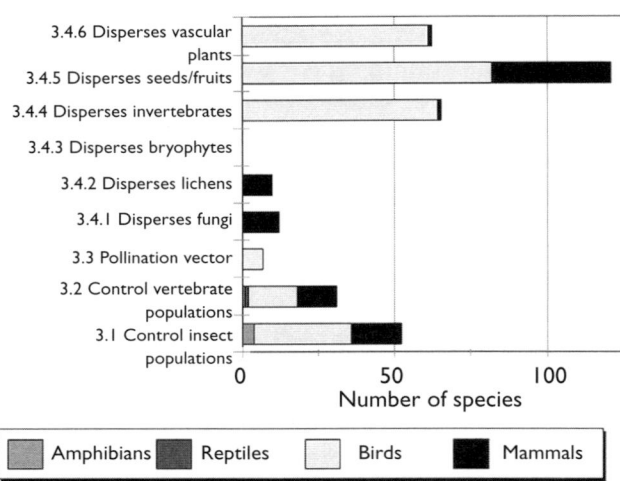

Figure 6. Organismal functional relations of wildlife in Washington and Oregon: population control, and pollination and dispersal vectors.

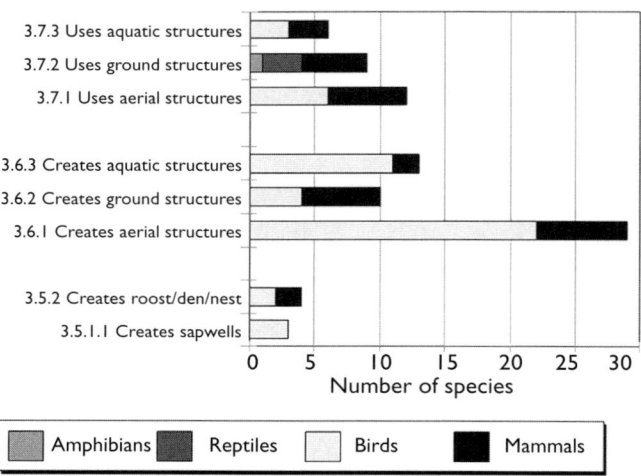

Figure 7. Organismal functional relations of wildlife in Washington and Oregon: creation and use of feeding, breeding, and resting structures.

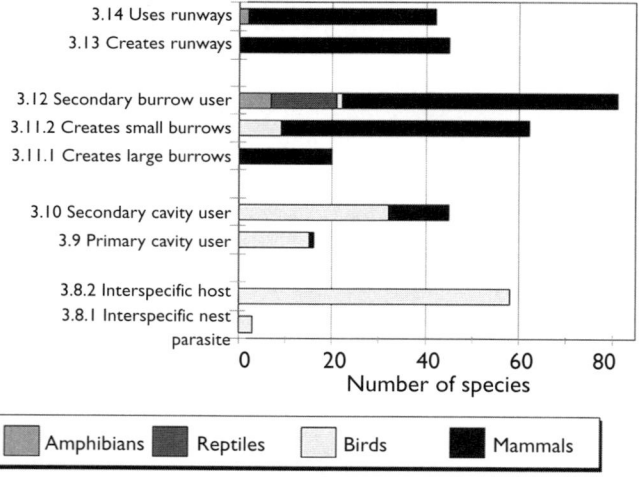

Figure 8. Organismal functional relations of wildlife in Washington and Oregon: nest parasitism and hosts, and cavity, burrow, and runway excavation and use.

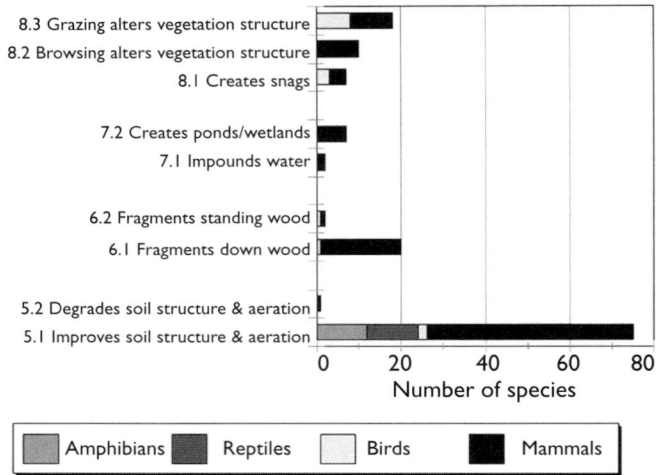

Figure 9. Soil, wood, water, and vegetation functional relations of wildlife in Washington and Oregon.

mammal species create snags through shredding, digging into, or girdling of trees. The habitat elements matrix lists 131 species of amphibians, birds, and mammals as being associated with snags. Browsing by 10 mammal species and grazing by 1 bird and 7 mammal species can alter vegetation structure and composition. This in turn can change habitat conditions for a wide array of other wildlife associated with grass, forb, and shrub species and cover.

Taxonomy and Definitions of Functional Patterns

Beyond the simple species tallies discussed above, what are the functional characteristics of ecosystems that can be evaluated by using the KEF matrix? We suggest a taxonomy of 15 types of patterns of ecological functions (Table 2, follows Literature Cited at end of chapter) and provide some examples here of using the matrix for evaluating them. Some of these patterns are discussed in the literature but many we pose here for the first time as working hypotheses of how wildlife communities operate and can be evaluated. All of the types of patterns need empirical study to determine actual rates of influence of the functional roles of animals.

Patterns to Watch

Community patterns. Community patterns of KEFs in wildlife communities include functional redundancy, functional richness, total functional diversity, functional webs, functional profiles, and functional homologies.

Functional redundancy is the number of wildlife species performing a specific ecological function, and is determined by tallying species in the KEF matrix for specific KEF categories for specific habitats conditions. *Functional richness* is the number of KEF categories in a community. *Total functional diversity* is functional richness weighted by functional redundancy,[4] analogous to species diversity.

A number of wildlife habitats in Washington and Oregon are more or less equally functionally rich except

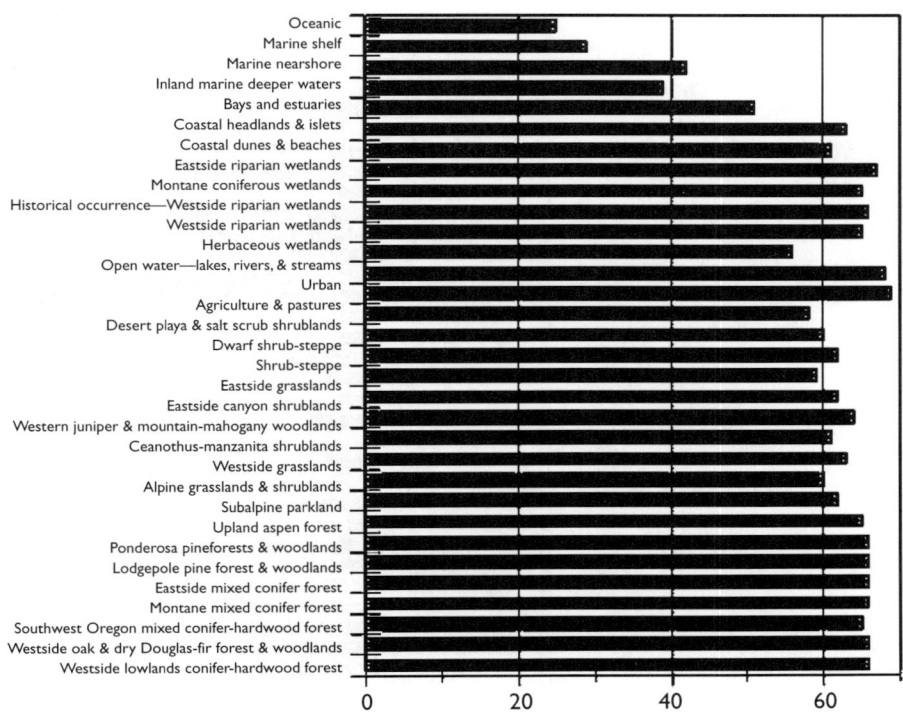

Figure 10. Total functional richness (number of key ecological function categories) by wildlife habitat in Washington and Oregon.

for the less rich marine and coastal habitats (Figure 10). Agriculture, Pastures, and Mixed Environments, and Urban and Mixed Environments are the most functionally rich, but the matrixes might overstate the viability of wildlife species within these anthropogenic environments. Among native wildlife habitats, a number of forest, woodland, and wetland environments rank as the most functionally rich.

Total functional diversity differs more among wildlife habitats in Washington and Oregon than does functional richness, although it follows much of the same ordinal pattern (Figure 11). An exception is with various alpine and shrubland wildlife habitats that have low functional diversity. This means that even though a relatively high number of categories of KEFs are represented in alpine and shrubland wildlife communities, there is lower functional redundancy (fewer wildlife species) performing many of those functions than found in forest, woodland, and wetland communities; there are far fewer still in marine and coastal communities. This may imply a greater sensitivity of alpine, shrubland, coastal, and marine communities to change than with forest, woodland, and wetland communities.

This illustrates the premise that functionally redundant, rich, and diverse communities may be more resistant or resilient to perturbations[20, 29, 33] and can support greater levels of biodiversity[45] than less functionally redundant, rich, or diverse communities. Managers may wish to develop functional profiles (see below) not just of local wildlife habitats but also of structural stages and combinations of habitat elements. Then they can identify those communities that are naturally rich, diverse, and redundant in their ecological functions, and those that are not, to help prioritize further evaluations and to focus

conservation attention on the latter and ensure the continued integrity of the former.

A *functional web* is the full array of all KEFs associated with a set of species that may be specified by some habitat element or structure. An example given in Morrison et al.[28] from the interior Columbia Basin illustrates an array of 22 KEF categories just of mammals associated with down wood in eastside ponderosa pine forest communities. Many of these functions, such as dispersal of fungi or primary burrow excavation, extend well beyond the confines of down wood substrates per se. This demonstrates the concept of how functions supported in part by specific habitat elements can influence parts of the ecosystem well beyond those habitat elements.

Marcot et al.[24] provided *functional profiles*, or graphs showing functional redundancy for each KEF category, and of total functional diversity of all KEF categories, across all vegetation communities of the interior Columbia Basin. The concept of *functional homologies*, or how communities compare in their functional categories and redundancies, is new but can be easily evaluated using the KEF database (Table 2) and similarity analyses. Managers may wish to evaluate functional webs, profiles, and homologies to help describe the unique functional roles of communities in their charge.

Geographic Patterns. Species' range maps or distributions by wildlife habitats can be used to depict geographically many of the functional patterns discussed here. This opens the door to a new way to display functional patterns spatially, and may be termed "geographic functional ecology". For example, stacking the range maps of species with a particular KEF category produces a map of functional redundancy. Areas of lowest redundancy that serve to link areas of high redundancy can be delineated

Figure 11. Total functional diversity (functional richness times mean number of species per function) by wildlife habitat in Washington and Oregon.

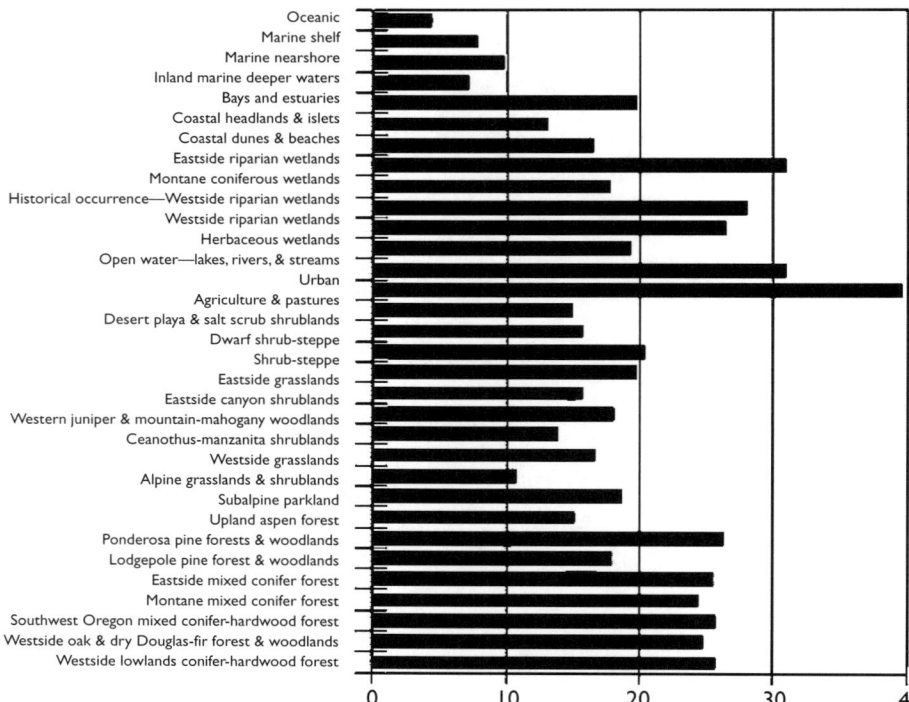

as *functional bottlenecks or cold spots*, just as areas of highest redundancy can be identified as *functional linkages or hot spots*. In this way, many other aspects of KEFs can be mapped. It may be of interest and importance to managers to map such geographic functional patterns, to identify areas and communities worthy of greater conservation attention.

Species' Functional Roles. Two major patterns of the functional roles of species are functional keystones and functional specialization. *Functional keystones* are species whose removal would most alter and degrade the structure or function of the community. If the species is the only one that performs a particular ecological function within a community, it is a *critical functional link species.* The function it performs is then a *critical function. Functional breadth* is the number of ecological functions performed by a species. When a species performs very few functions, it is a *functional specialist.*

Functional keystones, functional link species, and functional specialists can be identified by querying the KEF database. For example, the turkey vulture was identified above as being an extreme functional specialist (it performs the fewest total functions). Its function is KEF category 1.1.4 (see Table 1), which means that it is a heterotrophic consumer and, more specifically, a carrion feeder. It has no other trophic categories, which means that it is a specialist in this function. Loss of carrion means loss of this species.

But is the reverse also true, that is, is the turkey vulture critical to the carrion-feeding function? This is determined by identifying the functional redundancy of that function. Querying the KEF database shows that there are 39 other wildlife species also coded as carrion feeders in Washington and Oregon. Granted, many of these species

feed on different kinds of carrion in different habitats and on different substrates than does the turkey vulture. Also, there may be fewer carrion-feeding wildlife species in any given wildlife habitat; this is determined by linking the KEF database to the wildlife habitats database. For example, there are only 17 carrion-feeding species found in Westside Lowland Conifer-Hardwood Forest. In fact, the fewest number (8 species) occurs in Inland Marine Deeper Waters and Marine Nearshore habitats, but the turkey vulture does not occur in either of these habitats. Where it does occur, the carrion-feeding function is shared by other species. Therefore, although the turkey vulture is a functional specialist, its function is usually performed by at least some other species, so, although specialized, its carrion-feeding role, in general, is not entirely a keystone role. It may be a keystone role, however, for specific kinds of carrion on specific substrates or in specific seasons; there are many dimensions to a species' niche.

Some KEFs, however, have few participants, even if the associated species are not functional specialists. An example is the American beaver, which is the only species coded in the database with KEF code 7.1. "impounds water by creating diversions or dams". So for this function, this species is a critical functional link and a keystone. American beaver is also coded with 24 other KEF categories (these include subcategories as well), so it is not much of a functional specialist. However, it *is* the only species that performs KEF 7.1, which is a function that influences many other species. In fact, the habitat elements database lists 49 other wildlife species directly and positively associated with habitat element 2.3, "Beaver/muskrat activity, including dams, lodges, and ponds" in Washington and Oregon.

The lesson for the manager here is to pay attention to both the array of functions by species, as well as the array of species by function. The concept of a keystone species or keystone function has many dimensions and potential management implications. We caution against careless use of the general term "keystone" without clarifying its context and significance.[26] For example, species that are functional specialists are probably highly vulnerable to changes in their environment if they depend on that function for survival. A species that is a functional generalist with broad functional breadth, that is, that performs the widest array of functions-might greatly influence an ecosystem upon its removal, but this depends on what and how other species also perform such functions. A function with one or few species may itself be vulnerable and of management interest for ensuring its continuance in an ecosystem.

Functional Responses of Communities. The categories discussed above can be used to characterize the functional responses of communities. Community functional responses refer to how communities respond functionally to disturbances. Responses can be categorized as functional resilience, resistance, attenuation, and shifting, and imperiled functions.

Functional resilience is the capacity of a community to rebound to a starting functional state after a perturbation, such as a natural or human-caused disturbance event. *Functional resistance* is the degree of inertia of a community to retain its initial functional state and not change under stressors. *Functional attenuation* is the degree to which a community loses functions after a disturbance, and *functional shifting* is the difference in functions before and after disturbance when the community has finally stabilized in species composition. *Imperiled functions* are those functions that are represented by few species that are themselves at risk.

Some Hypotheses on Ecological Functions

Ecological implications of functional patterns of species and communities, as discussed above, can be taken as testable hypotheses about the roles of wildlife and how ecosystems work (Table 1). Some additional hypotheses include the following:

1. Functional switching imparts functional resistance: a greater degree of functional switching by species imparts a greater functional resistance to the community.
2. Functional redundancy imparts resilience: for a particular function, the higher the functional redundancy, the greater is the functional resilience (or resistance) for that function.
2. Functional diversity imparts low attenuation: the greater the functional richness and diversity of a community, the lower is the degree of functional attenuation (loss of functions).
3. The more functionally diverse a community, the greater is its productivity and its biodiversity.
4. The greater the functional redundancy, the more sustainable is the set of resources that the function provides.

5. Functional breadth is not necessarily related to critical functional links. That is, species, which are the sole (or few) performer(s) of a given function in a community might also perform very few, or many other, functions.
6. Critical functional links cannot be predicted from functional breadth.

Of course, working hypotheses also can be stated as null hypotheses for the purpose of critical testing of their veracity. For example, the first working hypothesis can be stated as the following null hypothesis: there is no influence on functional resistance of a community from degree of functional switching of its component species. Then, an appropriate research study can be devised that compares the functional constitution of communities before and after a disturbance, for communities that differ in their degree of functional switching of their component species.

Lost Functions of Extirpated Species

Thirteen species are recognized in this effort as extirpated from either Oregon and/or Washington (see Chapter 18). Extirpated species performed 49 ecological functions (including subcategories) that were potentially reduced or eliminated as a result of extirpation. Because ecosystem sustainability is dependent on the maintenance of biodiversity and native ecological functions, the loss of these species may have affected the functional integrity of the communities in which they formerly occurred. The degree of such influence, however, is largely unstudied.

Key Ecological Functions of Extirpated Species in Oregon and Washington

Trophic Relations. The 13 extirpated species varied widely in their degree of functional specialization, with most species performing >1 trophic-related KEF. The grizzly bear, for example, had the widest functional breadth of trophic KEFs (7) among the extirpated species, whereas California condor was a functional specialist (1), feeding exclusively on carrion. The upland sandpiper and yellow-billed cuckoo also were narrow in their trophic relations, feeding solely on terrestrial invertebrates and serving as prey for secondary or tertiary consumers. Overall, trophic relations of extirpated species included spermivores (grizzly bear, Wyoming ground squirrel, trumpeter swan, and sharp-tailed grouse), grazers (pronghorn, bison, bighorn sheep, and trumpeter swan) frugivores (gray wolf, grizzly bear, and sharp-tailed grouse), root-feeders (grizzly bear), and foliovores (trumpeter swan and sharp-tailed grouse). Eight species were secondary consumers, including 4 species that consumed vertebrates; 1 species was a tertiary consumer (gray wolf); 2 were carrion-feeders; 1 was cannibalistic, and 9 were prey for other species.

Organismal Relations. The 13 extirpated species performed 6 categories of KEFs related to organismal relations. Three mammal species were primary burrow excavators; 3 mammals were creators or users of runways;

2 mammals potentially controlled vertebrate populations; 2 birds and 3 mammals dispersed insects and/or seeds; and 2 mammals created feeding opportunities for other species. Other functions of extirpated species included carriers of disease (4 mammals), and functions related to soil (1 mammal), wood (3 mammals), and water relations (1 mammal).

To assess the potential impact of losing these functions, we calculated functional redundancy by tallying the number of species with selected KEFs by wildlife habitat. These species function profiles depict the degree of functional redundancy among habitats, thus serving to identify imperiled or lost functions. From among the set of KEFs performed by each extirpated species, we selected for analysis those functions that might be expected to vary among communities, as we discuss next.

Carrion-feeding. Feeding exclusively on carrion, the California condor was an extreme functional specialist. Historically, it inhabited the Westside Grasslands, Westside Riparian-Wetlands, and Coastal Dunes and Beaches of Oregon and Washington. Querying the matrices shows that 24 individual species perform this function in these 3 habitats. Twelve species perform this function in Westside Grasslands, 20 in Westside Riparian-Wetlands, and 16 in Coastal Dunes and Beaches. Similar to the turkey vulture, the condor was not a keystone species for the maintenance of the carrion-feeding function. Nonetheless, functional specialists may be highly sensitive to changes in their environment, thus increasing their vulnerability to extinction or extirpation.[31] The extreme degree of specialization (for carrion as well as perhaps other needs) of the California condor may have contributed to its demise.

Control of Vertebrate Populations. This KEF was examined for redundancy within the range of habitats historically inhabited by gray wolves in Oregon. Thirty other species potentially perform this function, including 1 amphibian, 1 reptile, 16 birds, and 11 mammals (excluding the grizzly bear, also extirpated from Oregon). Obviously, not all of these species perform this KEF in comparable ways. Among the assemblage of mammals associated with this function, coyotes, mountain lions, and black bears might perform this KEF similarly to gray wolves, although they may selectively prey on other species.

Importantly, the loss of top predators from a community may have reverberating effects on the rest of the community. Predators lost from a complex food web "release" their herbivore prey, thus potentially leading to significant changes in vegetation structure, cover, and species from increases of the prey population.[31] In the Greater Yellowstone Ecosystem, for example, the extirpation of wolves led to an unprecedented growth in the moose population, which had profound effects on willow communities, and consequently, on neotropical migrants associated with those communities (Berger et al., unpubl. data). Furthermore, wolves are dominant over coyotes and aggressively exclude or displace them. With the extirpation of wolves, coyotes increased dramatically.[27] Studies currently are assessing the impacts and dynamics of wolf recolonization on coyotes in the Greater Yellowstone Ecosystem.

Provision of Feeding Opportunities. The grizzly bear, now rare in Washington and extirpated from Oregon, performed the function of providing carrion in 11 habitats. The species function profile for the KEF "provides feeding opportunities" revealed that 6 other species (minus the gray wolf, which was also extirpated from Oregon) also perform this function. However, only 1, the mountain lion performs this KEF in essentially the same manner, by providing carrion. It does so in all 12 of the habitats that the grizzly bear once occupied. This is an example of a potentially imperiled function. Although this function was not completely lost, the elimination of grizzly bears and gray wolves from their historical range may have greatly constricted the occurrence and rate of this carrion-providing function. Because providing carrion is a symbiotic, organismal relation, other species that use carrion might be affected by the scarcity of this function.

Creation of Ponds or Wetlands Through Wallowing. Wallowing bison likely once created small ponds and wetlands in 8 habitats in Oregon and Washington. Four ungulate species currently perform this function in subsets of these habitats. Of the 4 species, the Rocky Mountain elk is the only ungulate that can create small ponds and wetlands in all 8 of the habitats formerly occupied by bison, and it is the only ungulate that performs this function in Lodgepole Pine Forests and Woodlands. Although not imminent, extirpation of the Rocky Mountain elk from this habitat would result in the loss of this potentially imperiled function.

The Rocky Mountain elk, therefore, is a critical functional link species for the critical function of creating small ponds and wetlands in Lodgepole Pine Forests and Woodlands. Twenty wildlife species that occur in this habitat are associated with ephemeral pools, including 8 species of amphibians. Interestingly, 4 species of frogs that occur in Brazil breed only in peccary wallows and other small permanent ponds; conserving these amphibians necessitates preserving the wallowing function of peccaries.[40] This is an example of how the loss of one species (bison or peccary) can influence the distribution of another.

We found no evidence that the extirpation of the pronghorn, Wyoming ground squirrel, sea otter, or upland sandpiper resulted in the loss or imperilment of any specific ecological function in their former habitats. The trophic KEF categories of these species (e.g., grazer, spermivore, invertebrate-eater) were and are shared by other species; thus the habitats they occupied were highly redundant for these functions. However, we recognize that the level of specificity in our database is insufficient to characterize the complete set of ecological functions for every species; simply performing the same general KEF in the same habitat does not necessarily confer substitutability. At some level, one species is never completely interchangeable with another, regardless of functional similarity.

Evaluating the KEFs of extirpated species may help land managers understand the influences that historically contributed to ecosystem function. From this understanding, restoration goals can then be formulated to restore lost functions and to protect rare ones. Species that perform functions with low functional redundancy in a particular habitat could receive priority to ensure perpetuation of that function in the community. Because most species are involved in several to many trophic, organismal, and symbiotic relations, the conservation of overall biodiversity can surely lend to community stability. By minimizing species loss, the integrity of ecosystem function may be maintained.[45] Understanding how species contribute to community function therefore may be imperative to crafting "ecosystem management" guidelines and defining ecosystem restoration objectives. Also, KEF analyses can be used to formulate and test hypotheses related to the effects of function loss on community stability and ecosystem integrity. The manager can apply this approach at a variety of scales. That is, a species need not go extinct in a two-state area before its more local reduction or extirpation of its populations and associated functions are addressed.

Competing Functions of Exotic Species

Thirty-one non-native (exotic) species occur in Oregon and Washington, including 2 amphibians, 3 reptiles, 14 birds, and 12 mammals (see Chapter 16). Collectively, they perform 54 KEFs in 30 wildlife habitats, and thus exotic species are able to exploit niches occupied by native animals. The most successful non-native species have KEFs that confer a competitive advantage, often to the detriment of native species. Exotic species threaten biodiversity by displacing native species either directly through predation or displacement, or indirectly through the alteration of ecosystems, and have been described as one of the greatest threats to native species and human-disturbed ecosystems in the world.[35]

The exotic species found in Oregon and Washington compete with native species through trophic and organismal relations; soil, wood, and water relations; and by altering vegetation structure, and composition. Trophic relations involve competition for food or predation. Most exotic primary consumers are foliovores (9 bird species and 2 mammals), spermivores (12 bird species and 6 mammals), and frugivores (8 bird species and 5 mammals). Seven exotic species are grazers, including the wild burro and feral horse, which compete with native ungulates for forage. Seven exotic species are aquatic herbivores, and a few are root feeders, browsers, flower feeders, and decomposing benthic substrate feeders (2 species each). None of the exotic species consumes sap, nectar, or bark. Of the 25 exotic species that are secondary consumers, most consume invertebrates (24 species) and 12 prey on vertebrates, including 3 exotic species that feed on fish (bullfrog and 2 reptiles) and 2 species that are ovivorous (Eastern fox squirrel and Virginia opossum). Three species

feed on carrion (2 reptiles and Virginia opossum), 2 mammals are coprophagous, and 23 are prey for secondary or tertiary consumers.

Twenty-three exotic species participate in organismal relationships including the potential control of terrestrial vertebrate populations (bullfrog), and transportation of seeds (9 birds and 4 mammals). Two exotic squirrel species create structures that are used by other species, and 3 exotic species use structures created by other species. Exotic secondary cavity users, the eastern fox squirrel, eastern gray squirrel, English or house sparrow, and European starling, compete for cavities in their respective habitats with up to 40 native species. Declines of Lewis' woodpecker and purple martin have been attributed, in part, to competition with starlings for cavities. Three exotic mammals are primary burrow excavators and 7 are secondary burrow users, with 2 exotic rats (Norway rat and black rat) alone competing with up to 51 native secondary burrow users. Five exotic mammal species create runways and 5 other exotic mammal species use runways created by others.

Exotic species can affect gene pools (e.g., eastern cottontail hybridizes with native species) and spread disease (at least 5 exotic mammal species carry diseases that may affect humans). Soil-related KEFs include potentially physically improving soil structure by digging or burrowing (4 exotic mammals and 1 exotic reptile) and physically degrading soil structure by compaction (wild burro, feral horse). Three exotic species fragment down wood (wild burro, feral horse, and feral pig) and the same 3 species can create small ponds or wetlands through wallowing. Lastly, the eastern cottontail, feral pig, and wild burro can alter vegetation structure and composition through grazing.

Understanding how exotic species compete with native wildlife may help land managers predict which native species are most likely to be affected, and in which habitats. Although the mechanisms by which systems resist introductions are not well understood, they seem to be related, in part, to trophic complexity.[31] Introductions of more trophically generalized mammals to species-poor, trophically simple systems (e.g., islands) have had extreme effects on the native species. Conversely, trophically complex, species-rich systems may be more resilient. Maintaining the functional web, or full set of KEFs, within a community may, therefore, increase resistance or resilience to the effects of exotic species whether introduced intentionally or not.

The Functional Roles of Marine Species

Marine Birds

There are 83 species of coastal and pelagic birds associated with the 4 oceanic wildlife habitats: Inland Marine Deeper Waters, Marine Nearshore, Marine Shelf, and Oceanic habitats. Collectively, they comprise 53 KEF categories. They are all secondary consumers; 13 are also primary consumers, mostly including aquatic herbivores, but a few (mostly waterfowl) are also foliovores, spermivores, browsers, and grazers. Many consume invertebrates (76 bird species) and/or vertebrates (63 bird species), including fish (60 bird species) and eggs (12 bird species). Seventeen bird species (skua, eagle, gulls, and jaegers) pirate food from other species, an adaptation found much less in terrestrial habitats; 13 species (waterfowl, fulmar, eagle, gulls, storm-petrels, and shearwaters) are carrion feeders; and 12 species (skua, fulmar, gulls, albatrosses, shearwaters, and jaegers) feed on human garbage or refuse in the ocean, principally offal and bycatch of fishing boats. Some help disperse invertebrates (28 bird species) or plants (25 bird species).

Because most coastal and pelagic bird species nest on stacks, sea cliffs, and other coastal substrates, and some are fossorial and nocturnal, there are 7 species (guillemot, auklets, kingfisher, puffin, and storm-petrels) that are primary excavators of small burrows and 11 species (grebes, 1 eagle, 1 gull, 1 tern, 1 pelican, and osprey) that create structures used by other species. There are even 4 secondary cavity users (waterfowl) and 3 species (cormorants) that can kill trees and create snags. Again, many of these are examples of how species' ecological roles extend beyond just the realm of their primary habitat, in this case, coastal and open ocean.

Marine Mammals

The matrix includes 18 species of marine mammals. Collectively, they comprise some 20 KEF categories. All 18 species consume aquatic invertebrates and 3 (whales) consume marine zooplankton. Seventeen species consume vertebrates (15 of these eat fish) and 7 serve as prey for other predators. Four whale species create feeding opportunities for other species, and a number of species aid in nutrient cycling and carry or transmit disease, although these functions are poorly studied in marine environments.

The Functional Roles of Humans

Although not explicitly included in the KEF matrix, *Homo sapiens* nonetheless is one of the most effective ecological change agents that can adversely affect BPS.[9, 19] Anthropogenic change of habitats is a major disturbance event in native ecosystems of Washington and Oregon (see chapter 20 on Management Activities).

Humans perform some 35 categories of KEFs (Table 1) (this tally includes category headings and subheadings, so there is some repetition in this value). This exceeds the influence of the two most functionally broad native wildlife species, the American black bear (30 KEF categories) and the American beaver (24), and helps explain how humans have so greatly altered ecosystem functions and processes. Humans certainly rate as a functional keystone species, in that their removal from a system would likely result in major redistributions of other species and their functions. However, none of the ecological functions performed by humans are necessarily critical link functions, which are not performed in at least some general way by other wildlife species.

So, what is it about the ecological functions of humans that so greatly influence native ecosystems and wildlife communities? It is not that humans play any critical functional roles, or that humans are functional specialists. Rather, it seems to be our amazing functional breadth—that is, the number of functions we perform and our resounding influence on the rest of the functional web of natural communities. Such functional influences should be considered when evaluating effects of human activities. Managers, in fact, may wish explicitly to include the KEF categories of humans in their evaluations of functional patterns of wildlife species and communities.

Management Implications

Throughout the text above, we have offered some ideas on how managers might use this new approach to describe the functional patterns of wildlife species and communities and influences of human activities. More specifically, the manager can link the KEF matrix with the other matrixes, especially those depicting wildlife habitats, habitat elements, habitat structures, and management activities.

For example, one could begin by specifying a particular management activity, such as using herbicides in shrubland or grassland management (Management Activities 5c; see Chapter 20) within a particular wildlife habitat such as Eastside Grasslands (Wildlife Habitat 15; see Chapter 20). Linking the management activities, wildlife habitats, and habitat elements matrixes, one can determine which wildlife species are potentially influenced by this activity in this habitat, that is, which species are coded for using habitat elements that are affected by this activity in this wildlife habitat. In this case, it is a set of 2 amphibians, 7 reptiles, 75 birds, and 44 mammals.

Next, one can link this species set to their KEFs and determine which KEF categories might be most influenced by this management activity by comparing their functional redundancies to those of all wildlife species in this habitat. For example, the ecological function of fungi dispersal is performed overall by two wildlife species in Eastside Grasslands: the deer mouse and Rocky Mountain elk. These two species both are also potentially influenced by use of herbicides in this habitat, because herbicide use is linked to at least one habitat element used by each of these species, namely the habitat elements of herbaceous layer (Habitat Element code 1.2.1), fruits/seeds (1.2.2), shrubs (1.2.6), fungi (1.2.7), forbs (1.2.8), grasses (1.2.10), and plants (2.1.1). Herbicide use may have a less salient influence on some other KEFs depending on how many wildlife species perform the functions that are *not* influenced by the activity. In this and other ways, the manager can use the matrixes to evaluate the potential influence of specific management activities on ecological functions and determine the functions that may be most sensitive to the activities, and perhaps those worthy of special mitigation or conservation attention. We suggest using the categories and analysis procedures listed in Table 2 to address this and other patterns of ecological function.

Ultimately, a goal of wildlife conservation may be to help ensure fully functional ecosystems. To this end, the manager could use the approaches we offer in this chapter to determine, for individual wildlife habitats:

1. the full set of native ecological functions,
2. the full functional redundancy of each KEF category and total functional diversity,
3. habitat elements contributing to maintaining functional diversity of wildlife communities, and
4. which management activities simplify, extirpate, or restore functions.

The manager can use functional profiles to compare among communities, or a community over time under different management scenarios, to determine: (1) which scenario would provide for the fullest array of native KEF categories; (2) the degree to which full functionality of an ecosystem can be maintained; (3) which functions might be at higher risk; and (4) which species, habitat elements and structures, and management activities account for reduction or loss of specific functions. This provides a basis for identifying threats to KEFs as influenced by human (or other) disturbance events. Linking to maps of species and functions as discussed above, the manager can then determine spatially where such threats might occur.

Caveats

We emphasize that the information in the KEF database largely represents a collective expert judgment and is mostly not based on scientific field studies. As well, the KEF information is coded largely as categorical data and not quantified as rates of influence. Thus, patterns generated by querying the KEF matrix should be considered as tentative working hypotheses requiring empirical validation and refinement. We view the major value of the KEF matrix as providing (1) a way of explicitly and repeatably evaluating the functional patterns of species and communities and the potential influences from human activities, and (2) generating working hypotheses on such influences for management consideration and testing.

Following Marcot,[22] we offer the following caveats when using the KEF matrix:

Completeness of the KEF Database

Although all vertebrate species are included, there may be many ways to represent their ecological functions, and some functions likely are incompletely depicted. An example is KEF category 2 on nutrient cycling relations. This category was originally developed more for plants and invertebrates[23] but could be more fully developed for vertebrates. Some KEF categories require additional attention. An example is KEF category 4, disease vector. Species listed under this category need review and refinement by wildlife disease experts. There are likely to be other errors of omission and commission in the matrix and other needs to revise or expand the KEF classification system (Table 1).

Need for Validation and Primary Field Studies

Ecological implications of the KEF pattern categories discussed in this chapter should be viewed as working hypotheses. Further, the relations depicted among the elements in the species-influence diagram may be quantified as specific rates (e.g., rates of soil digging and turnover, rates of pollination, etc.). To do this for the entire set of species, KECs, KEFs, and their relations to BPS and management activities would entail a major research program. Thus, the matrixes can be used to prioritize critical relations and identify functions most in need of empirical, quantitative study.

Scientific knowledge of some functions and species is more complete than for other functions or species. We did not denote this variation in the database per se, but wildlife specialists or systems ecologists should be able to determine areas needing further, or primary, empirical study.

Resolution of the KEF Categories

Some KEF categories are broadly generalized whereas others are very specific. This is inevitable, given the variation in scientific knowledge and understanding across species and functional categories. The user of the KEF matrix should not expect it to accurately predict ecological functions within small or local geographic areas such as single forest stands; if linked to the wildlife habitats and habitat elements matrixes, the KEF matrix will likely err on the side of commission of functions rather than omission. The KEF matrix might serve the manager better at broader geographic scales.

Ecological Functions of Other Taxonomic Groups

The user needs to integrate the information provided in the KEF matrix with some understanding of the functions of other taxonomic groups, namely algae, plankton, plants, plant allies (cryptogams), and invertebrates. The user should not expect to be able fully to describe (and prescribe!) ecosystem processes only in terms of vertebrate wildlife species. This is especially true with ecological functions that are poorly represented by vertebrates, such as those pertaining to soil productivity, wood decay, carbon sequestration, filtering of xenobiotics, and many other functions.[23]

When managing wildlife in an ecosystem context, it is important not to ignore the functional influence of plants and invertebrates, which really perform most ecological functions in ecosystems. For example, plants and their allies—including algae, fungi, lichens, bryophytes (mosses and liverworts), and cryptogamic soil crusts, as well as vascular plants—provide the foundation of net primary production, and trophic and physical structures of ecosystems (except for rare autotrophic, chemosynthetic and other extremophile-based systems found in a few locations). Invertebrates, including soil microorganisms such as mites, springtails, nematodes, rotifers, protozoa, and bacteria, also play central functional roles in ecosystems, especially in soil and decaying wood; E.O. Wilson called invertebrates the "little things that run the world."[47]

Ecological functions of plants and invertebrates may best be addressed in functional groups of species,[10, 30, 32] for example, as suggested for wetland plants,[2] exotic plant species,[11, 15] plants central to ecosystem process and global change,[39] and invertebrates that play ecological roles in managed forests.[38]

Vertebrates also take a back seat to other organisms in terms of species richness. In the interior Columbia River Basin in the U.S., vertebrates constituted only 4% of all known species and 1% of all expected (known and unknown) species of macroorganisms, and even far less if microorganisms are included.[23] Nonetheless, the relation of plants, plant allies, and invertebrates to vertebrate species is pertinent and may be of interest to managers for maintaining overall ecological functions and ecosystem processes. Plants support vertebrate primary consumers including browsers and grazers; invertebrates support insectivores and other secondary consumers. Plants and plant allies provide many kinds of physical habitat structures, help filter water and detoxify soils, support soil structures for burrowing, and many other functions of use to vertebrates. Invertebrates or their byproducts (e.g., honey) serve as a food source; they also foster wood decay, create snags and down wood, serve as disease transmission vectors, and play many other ecological roles affecting vertebrates.

Conclusion

Ecosystem managers can address the biotic component of ecosystems by explicitly including key ecological functions of species (including humans) in assessments and in formulating management direction. Many ecosystem management approaches to wildlife and land use planning are already considering ecosystem processes of disturbance events (e.g., fire, wind, drought, and flooding). The picture needs to be made complete by also considering the functional role of organisms, and the influence of those functions on overall ecosystem biodiversity, productivity, and sustainability. The framework we offer here provides a rigorous, repeatable means of evaluating at least the qualitative influence of wildlife on their ecosystems, and should be viewed as starting hypotheses from which to focus conservation or research attention. We fully expect that many KEF categories and even their qualitative relations with wildlife species may change given empirical study.

Acknowledgments

We thank Tom O'Neil, Kelly Bettinger, and Sally Olsen-Edge for helpful comment on the manuscript. We gratefully acknowledge the contributions of Marla Trevithick in developing the databases. We benefited from ecological discussions with Mikell O'Mealy and Susan Tank.

Literature Cited

1. Augustine, D. J., and S. J. McNaughton. 1998. Ungulate effects on the functional species composition of plant communities: herbivore selectivity and plant tolerance. Journal of Wildlife Management 62: 1165-1183.

2. Boutin, C., and P. A. Keddy. 1993. A functional classification of wetland plants. Journal of Vegetation Science 4: 591-600.

3. Brelsford, M. J., J. M. Peek, and G. A. Murray. 1998. Effects of grazing by wapiti on winter wheat in northern Idaho. Wildlife Society Bulletin 26: 203-208.

4. Brown, J. H. 1995. Macroecology. The University of Chicago Press, Chicago, IL.

5. Butler, D. R. 1995. Zoogeomorphology: animals as geomorphic agents. Cambridge University Press, New York, NY.

6. Callicott, J. B., L. B. Crowder, and K. Mumford. 1999. Current normative concepts in conservation. Conservation Biology 13: 22-35.

7. Chapin, F. S., III, and others. 1996. The functional role of species in terrestrial ecosystems. In B. Walker and W. Steffen, editors. Global change and terrestrial ecosystems. Cambridge University Press, New York, NY.

8. Duncan, R. S., and C. A. Chapman. 1999. Seed dispersal and potential forest succession in abandoned agriculture in tropical Africa. Ecological Applications 9: 998-1008.

9. Forester, D. J., and G. E. Machlis. 1996. Modeling human factors that affect the loss of biodiversity. Conservation Biology 10: 1253-1263.

10. Golluscio, R. A., and O. E. Sala. 1993. Plant functional types and ecological strategies in Patagonian forbs. Journal of Vegetation Science 4: 839-846.

11. Gordon, D. R. 1998. Effects of invasive, non-indigenous plant species on ecosystem processes: lessons from Florida. Ecological Applications 8: 975-989.

12. Green, G. I., D. J. Mattson, and J. M. Peek. 1997. Spring feeding on ungulate carcasses by grizzly bears in Yellowstone National Park. Journal of Wildlife Management 61: 1040-1055.

13. Green, R. J. 1995. Using frugivores for regeneration: a survey of knowledge and problems in Australia. Pages 651-657 in J. A. Bissonette and P. R. Krausman, editors. Integrating people and wildlife for a sustainable future. The Wildlife Society, Bethesda, MD.

14. Hobbs, N. T. 1996. Modification of ecosystems by ungulates. Journal of Wildlife Management 60: 695-713.

15. Horvitz, C. C., J. B. Pascarella, S. McMann, A. Freedman, and R. H. Hofstetter. 1998. Functional roles of invasive non-indigenous plants in hurricane-affected subtropical hardwood forests. Ecological Applications 8: 947-974.

16. Huston, M. A. 1997. Hidden treatments in ecological experiments: re-evaluating the ecosystem function of biodiversity. Oecologia 110: 449-460.

17. Jones, C. G., J. H. Lawton, and M. Shachak. 1996. Organisms as ecosystem engineers. Pages 130-147 in F. B. Samson and F. L. Knopf, editors. Ecosystem management: selected readings. Springer, New York, NY.

18. Li, C. Y., C. Maser, Z. Maser, and B. A. Caldwell. 1986. Role of three rodents in forest nitrogen fixation in western Oregon: another aspect of mammal-mycorrhizal fungus-tree mutualism. Great Basin Naturalist 46: 411-414.

19. Mack, M. C., and C. M. D'Antonio. 1998. Impacts of biological invasions on disturbance regimes. Trends in Ecology and Evolution 13: 195-198.

20. MacNally, R. C. 1995. Ecological versatility and community ecology. Cambridge University Press, New York, NY.

21. Marcot, B. G. 1997. Species-environment relations (SER) database. U.S. Forest Service, Pacific Northwest Research Station, Portland, OR. Interior Columbia Basin Ecosystem Management Project database available on-line at URL: http://www.icbemp.gov/spatial/metadata/databases/dbase.html

22. Marcot, B. G. 1997. The species-environment relations (SER) database: an overview and some cautions in its use. U.S. Forest Service, Pacific Northwest Research Station, Portland, OR. Interior Columbia Basin Ecosystem Management Project database documentation available on-line at URL: http://www.icbemp.gov/spatial/metadata/databases/792aux2.html

23. Marcot, B. G., M. A. Castellano, J. A. Christy, L. K. Croft, J. F. Lehmkuhl, R. H. Naney, K. Nelson, C. G. Niwa, R. E. Rosentreter, R. E. Sandquist, B. C. Wales, and E. Zieroth. 1997. Terrestrial ecology assessment. Pages 1497-1713 in: T. M. Quigley and S. J. Arbelbide, editors. An assessment of ecosystem components in the interior Columbia Basin and portions of the Klamath and Great Basins. Volume III. U.S. Forest Service General Technical Report PNW-GTR-405.

24. Marcot, B. G., L. K. Croft, J. F. Lehmkuhl, R. H. Naney, C. G. Niwa, W. R. Owen, and R. E. Sandquist. 1998. Macroecology, paleoecology, and ecological integrity of terrestrial species and communities of the interior Columbia River Basin and portions of the Klamath and Great Basins. U.S. Forest Service, General Technical Report PNW-GTR-410.

25. Mawdsley, N. A., S. G. Compton, and R. J. Whittaker. 1998. Population persistence, pollination mutualisms, and figs in fragmented tropical landscapes. Conservation Biology 12: 1416-1420.

26. Mills, L. S., M. E. Soule, and D. F. Doak. 1993. The keystone-species concept in ecology and conservation. BioScience 43: 219-224.

27. Moore, G. C., and G. R. Parker. 1992. Colonization by the eastern coyote (Canis latrans). Pages 23-37 in A. H. Boer, editor. Ecology and management of the eastern coyote. Wildlife Research Unit, University of New Brunswick, Fredericton, New Brunswick, Canada.

28. Morrison, M. L., B. G. Marcot, and R. W. Mannan. 1998. Wildlife-habitat relationships: concepts and applications. Second edition. University of Wisconsin Press, Madison, WI.

29. Naeem, S. 1998. Species redundancy and ecosystem reliability. Conservation Biology 12: 39-45.

30. Parueo, J. M., and W. K. Lauenroth. 1996. Relative abundance of plant functional types in grasslands and shrublands of North America. Ecological Applications 6: 1212-1224.

31. Pimm, S. L. 1994. Biodiversity and the balance of nature. Pages 347-359 in E. D. Shulze and H. A. Mooney, editors. Biodiversity and ecosystem function. Springer-Verlag, Berlin, Heidelberg, Germany.

32. Pugnaire, F. I., and F. Valladares, editors. 1999. Handbook of functional plant ecology. Marcel Dekker, New York, NY.

33. Rastetter, E. B., L. Gough, A. E. Hartley, D. A. Herbert, K. J. Nadelhoffer, and M. Williams. 1999. A revised assessment of species redundancy and ecosystem reliability. Conservation Biology 13: 440-443.

34. Rauscher, H. M. 1999. Ecosystem management decision support for federal forests in the United States: a review. Forest Ecology and Management 114: 173-198.

35. Reid, W. V., and K. R. Miller. 1989. Keeping options alive-the scientific basis for conserving biodiversity. World Resources Institute, Washington, D.C.

36. Robertson, A. W., D. Kelly, J. J. Ladley, and A. D. Sparrow. 1999. Effects of pollinator loss on endemic New Zealand mistletoes (Loranthaceae). Conservation Biology 13: 499-508.

37. Schlapfer, F., and B. Schmid. 1999. Ecosystem effects of biodiversity: a classification of hypotheses and exploration of empirical results. Ecological Applications 9: 893-912.

38. Schowalter, T., E. Hansen, R. Molina, and Y. Zhang. 1997. Integrating the ecological roles of phytophagous insects, plant pathogens, and mycorrhizae in managed forests. Pages 171-189 in K. A. Kohm and J. F. Franklin, editors. Creating a forestry for the 21st century: the science of ecosystem management. Island Press, Washington, D.C.

39. Smith, T. M., H. H. Shugart, and F. I. Woodward, editors. 1997. Plant functional types: their relevance to ecosystem properties and global change. Cambridge University Press, New York, NY.

40. Soule, M. and K. A. Kohm. 1989. Research Priorities for Conservation Biology. Island Press, Washington, D.C.

41. Stohlgren, T. J., L. D. Schell, and B. Vanden Heuvel. 1999. How grazing and soil quality affect native and exotic plant diversity in Rocky Mountain grasslands. Ecological Applications 9: 45-64.

42. Stone, R. 1995. Taking a new look at life through a functional lens. Science 269: 316.

43. Urbanska, K., N. Webb, and P. Edwards, editors. 1998. Restoration ecology and sustainable development. Cambridge University Press, New York, NY.

44. van Gardingen, P. R., G. M. Foody, and P. J. Curran, eds. 1997. Scaling-up: from cell to landscape. Cambridge University Press, New York, NY.

45. Walker, B. H. 1992. Biodiversity and ecological redundancy. Conservation Biology 6: 18-23.

46. Williams, J. B., B. E. Tieleman, and M. Shobrak. 1999. Lizard burrows provide thermal refugia for larks in the Arabian desert. Condor 101: 714-717.

47. Wilson, E. O. 1987. The little things that run the world (the importance and conservation of invertebrates). Conservation Biology 1: 344-346.

48. Young, D. D., Jr, and T. R. McCabe. 1997. Grizzly bear predation rates on caribou calves in northeastern Alaska. Journal of Wildlife Management 61: 1056-1066.

Table 2. A taxonomy of patterns of key ecological functions (KEFs) of wildlife species and communities, and how to evaluate them using the KEF database.*

Functional pattern	Definition	Ecological implications	How to evaluate
Community patterns			
Functional redundancy	The number of species performing the same ecological function in a community.	High redundancy imparts greater resistance of the community to changes in its overall functional integrity. Low redundancy suggests critical functions to watch.	Tally number of species by KEF category for specific wildlife habitats, comparing changes over time or among habitats.
Functional richness	The total number of KEF categories in a community.	Denotes degree of functional complexity; greater functional richness means more functionally diverse systems. Also denotes the degree to which the full "functional web" of a community would be provided or conserved.	Tally number of KEF categories among all species present in a wildlife habitat and, optionally, habitat structure. Compare such tallies resulting from changes in habitats and structures.
Total functional diversity	The total array of KEF categories weighted by their redundancy, i.e., the number of functions times the mean functional redundancy across all functions.	Denotes total functional capacity of a community. High functional diversity means many functions and even redundancy among functions; low functional diversity means few functions and skewed redundancy (some functions with few species).	Tally number of species by KEF category for a given wildlife habitat; calculate mean number of species per KEF category. Can assign weights to some KEF categories if they pertain to specific management objectives.
Functional web	The set of all KEFs within a community and their connections among species and thence to habitat elements.	Depicts how habitat elements provide for species, and the array of ecological functions performed by those species. Functions typically extend well beyond the specific habitat elements.	Identify habitat elements of management interest; list all species within a wildlife habitat that are associated with those habitat elements; list all KEF categories associated with those species. Compare changes in habitat elements.
Functional profile	The degree of functional redundancy compared across communities.	Identifies communities with low (or high) redundancy of particular KEF categories. This can help prioritize habitat management, e.g., to ensure continuance of low-redundancy functions.	Tally number of species by KEF category and by wildlife habitat. Identify wildlife habitats with lowest tallies for each KEF category.
Functional homologies	The functional similarity of communities even if species composition differs.	Two communities are functionally homologous if they have similar functional profiles and patterns of functional redundancy, even if the species performing the functions differ. Functionally homologous communities can be expected to operate in similar ecological ways.	Produce functional profiles across all KEF categories for several communities or for a community over time based on its expected changes in habitat elements, habitat structures, etc. Compare the profiles (e.g., via contingency analysis) and identify statistically similar (functionally homologous) communities.

* Many of these categories are unstudied in wildlife communities of Washington and Oregon; thus, ecological implications should be viewed as working hypotheses. (See Table 1 for KEF categories.)

Functional pattern	Definition	Ecological implications	How to evaluate
Geographic patterns			
Functional bottlenecks or cold spots	Geographic locations with very low functional redundancy of an otherwise widely-distributed functional category.	Functional bottlenecks denote areas of higher risk of severing functionally connected communities across the landscape. Severing functions might set the stage for degradation of functional ecosystems.	Map wildlife habitats and/or distributional ranges of wildlife species. For a given KEF category, map the number of wildlife species in each habitat or overlay their range maps. Identify locations with lowest species richness bordering higher richness on each side; these are geographic functional bottlenecks ("cold spots").
Functional linkages or hot spots	Geographic locations with very high functional redundancy.	Functional hot spots denote areas where many species provide a specific ecological function; such communities may be more resilient to changes in environment or habitat for that function.	Map species richness for a particular KEF category as above. Identify locations with highest species richness; these are functional linkages ("hot spots"). Determine which species occur in a given hot spot and their wildlife habitats, habitat elements, and habitat structures, and how changes might influence the persistence of the species and thus the redundancy of the function.
Species' functional roles			
Functional keystone species, critical functional link species, and critical functions	Functional keystone species are species whose removal would most alter the structure or function of the community. One form of this may be critical functional links, which are species that are the only ones that perform a specific ecological function in a community. A critical function therefore is the associated functional category represented by only 1 (or very few) species within a community.	Reduction or extirpation of populations of functional keystone species and critical functional links may have a ripple effect in their ecosystem, causing unexpected or undue changes in biodiversity, biotic processes, and the functional web of a community.	For a given wildlife habitat, tally the number of species for each KEF category (functional redundancy). For KEF categories with only 1 species, determine which species performs this function. This is a critical functional link species for this particular function in this habitat.
Functional breadth and functional specialization of species	The number of ecological functions performed by a species.	Species with the narrowest functional breadth (i.e., fewest functions) are functional specialists and may be more vulnerable to extirpation from changes in conditions supporting that function.	For a given wildlife habitat, tally the number of KEF categories for each species. Identify the species with the fewest number of categories. These are functional specialists. Determine their habitat elements and structures and thus their potential vulnerability to changes thereof. Functional specialists that are also functional keystones may be of high priority for conservation attention.

Functional pattern	Definition	Ecological implications	How to evaluate
Functional responses of communities			
Functional resilience	The capacity of a community to return to a starting pattern of total functional diversity, richness, and redundancy, following a disturbance event.	Functionally resilient communities are better able to maintain their biotic processes in the face of disturbances. Conversely, it is important to know how far a community can be changed by some anthropogenic disturbance event and still be able to return to its starting functional pattern.	Determine the total functional diversity, functional richness, and functional redundancy of a pre-disturbance community. Then determine the types and rates of recovery of its wildlife habitats, habitat elements, and habitat structures following some disturbance; the wildlife species associated with such recovery stages; and the species' KEF categories and functional diversity, richness, and redundancy for each recovery stage. Compare stages for functional similarity and thus resilience.
Functional resistance	The ability of a community to resist changing its functional diversity, richness, and redundancy, following a disturbance event.	Functionally resistant communities can be counted on to continue to provide specific ecological functions in spite of and during disturbances. They may provide a bastion for a specific desired function in a disturbed or managed landscape.	Analyze as above and determine the degree to which functional diversity, richness, and redundancy do not change for each post-disturbance stage. This is a measure of functional resistance.
Functional attenuation	The degree to which the set of ecological functions within a community simplify following a disturbance event.	Functionally attenuated communities provide fewer or lower redundancies of ecological functions. It may be particularly important to know the degree of functional attentuation to be expected following anthropogenic disturbances.	Analyze as above and determine which KEF categories likely drop out and which remain over post-disturbance recovery stages as compared with initial conditions. Calculate functional diversity, richness, and redundancy for each stage to determine the rate of functional attenuation. Compare final stage to initial conditions to determine overall functional attenuation.
Functional shifting	The degree to which a community changes to a new, stable, functional constitution following a disturbance event.	Communities with low functional resilience or resistance may end up with a new array of functions and a new pattern of functional diversity, richness, and redundancy. It may be particularly important to know how a community might functionally shift following anthropogenic disturbances and thus which functions might be weakened, strengthened, lost, or gained.	Analyze as above and compare KEF categories and functional diversity, richness, and redundancy between predisturbance and final, stable stages. Identify which functions are lost or gained, and which change in redundancy.
Imperiled functions	A function that is represented by very few species (critical functional link species) or by species that are themselves scarce, declining, or moribund, where extirpation of the species would mean loss of the function. Imperiled functions also can be identified geographically.	Loss of imperiled functions serve to degrade overall ecosystem integrity. Even seldom-performed ecological functions might be critical to maintaining ecosystems, such as occasional dispersal of plant seeds in the face of shifting climates.	For a given wildlife habitat, determine KEF categories with the lowest functional redundancy, and the risk level of the associated species. Imperiled functions are those with 1 or few species which are themselves at risk.

7

Wildlife of Westside and High Montane Forests

*Deanna H. Olson, Joan C. Hagar, Andrew B. Carey, John H. Cissel, &
Frederick J. Swanson*

Introduction

More than 300 vertebrate species are associated with western forests of Oregon and Washington (Table 1). Western and montane conifer-hardwood forests and oak woodlands are some of the more species-rich areas within the two states. Both the productivity and the mosaic of conditions within western forests contribute to the higher vertebrate diversity. These forests are exceeded in richness only by habitats encompassed by riparian-wetlands, urban, and agriculture and pasture designations. Interestingly, these are habitats that either border or are found nested within western forests. A high overlap of species occurs between these habitat types and western forests, especially along their interfaces. Forest species include those taxa that are obligates to forested habitats for all or part of their life history, more generalist species that occur in the forest matrix but also in other nonforest types, and transient species that are found incidentally in forests because of their proximity to other habitats.

This chapter provides an overview of the broad- and fine-scale patterns of western forest wildlife assemblages, emphasizing the main faunal habitat associations with forest conditions. Drivers of the geographic distributions of many taxa include climate conditions, the legacy of past natural disturbances, and vegetation types. This mix of physical and biological conditions has been elegantly consolidated into ecoregion designations for western forests. Ecoregions provide a context for broad-scale species richness pattern assessment. At finer spatial scales,

site, microhabitat, and microclimate conditions, and recent disturbance events contribute to explanations of species distribution patterns. Habitat assessments of forest-associated wildlife conducted for this volume are compiled to summarize species-habitat relations. Across spatial scales, species life history, behavior, and intra- and interspecific interactions are significant elements for our understanding of wildlife habitat associations. These abiotic and biotic components are outlined to more fully conceptualize their roles in the organization of forest faunal assemblages. Hotspots of wildlife diversity are addressed at both broad landscape and finer forest stand spatial scales.

Although all vertebrate classes have some coverage in this chapter, amphibians are used for many lead examples. Whereas amphibians do not comprise a large percentage of the overall wildlife fauna in these habitats (about 13%), western forests are key habitats for many amphibians. Of 31 native amphibian species occurring in Oregon and Washington, 29 (93%) occur in western forests (Table 1) and 22 (71%) are restricted to this region.[39] These restricted species are either endemic to western Oregon and Washington or have ranges within the larger ecological extent of the western forest landscape.

There are many unique amphibian taxa reliant on these forested habitats. Among these are (1) the tailed frog (Figure 1); (2) the largest terrestrially occurring salamander, the Pacific giant salamander (Figure 2); (3)

Table 1. Species occurrences (no. species) in western and montane forest habitat types of Washington and Oregon.*

Taxon	n	Montane mixed conifer	SW Oregon mixed conifer-hardwood	Westside oak and dry Douglas-fir & woodlands	Westside lowlands conifer-hardwood
Amphibians	29	27	19	18	26
Reptiles	20	5	19	18	14
Birds	164	107	131	119	120
Mammals	104	82	67	74	73
Total	317	221	236	229	233

* Data are compiled from expert panel assessments for this book, and no historic occurrences are included (e.g., bison, gray wolf, merlin). *n* = total no. species evaluated.

Figure 1. Tailed frogs (Ascaphus truei) are primitive anurans strongly associated with conditions of western Oregon and Washington old-growth forests. J. S. Applegarth, photographer.

Figure 2. Pacific giant salamanders are stream breeders, but after metamorphosis they are terrestrial, occurring in upslope forests. J. S. Applegarth, photographer.

the torrent salamanders, Rhyacotriton spp.; and (4) several endemic plethodontid salamanders[34] (Figure 3, page 197). Western forests are recognized as a region of phylogenetic radiation for amphibians, yet our understanding of the amphibian biodiversity is incomplete. As a result of ecological, morphometric, and genetic studies, formal designations of distinct populations (e.g., evolutionarily significant units [ESUs][30]) or species across the western forest landscape are expected to change. Within this vertebrate class, the tendency for philopatry and limited dispersal capability has resulted in isolation of subpopulations, that now seem to have diverged ecologically, morphologically, and potentially taxonomically. Support for newly recognized species, or proposals to recognize distinct populations or species, have developed for Rhyacotriton, Dicamptodon, several Plethodon salamanders, 2 Rana frogs, and 1 Aneides salamander. The potentially hidden biodiversity within amphibians is expected to be a topic of research for some years to come. As species are examined in this light, concern for the status of unique subpopulations or species is heightened. This has ramifications for species conservation, and consequently western forest management policies.

Final sections of this chapter address forest management and its role in maintaining the persistence of wildlife populations in the western forest region. Anthropogenic disturbances over the last 100 years have reshaped forest habitats in Oregon and Washington. A current pivot in trajectories for public land management is introducing new standards for landowner stewardship of biological resources. Consequences of both the past management policies that emphasized economic resources and the current new management directions balancing socioeconomics with ecosystem integrity are significant

for species. These have and will mold habitats, and incidentally or by design, wildlife distribution and abundance patterns. We are learning how to be stewards for the long-term economic productivity and ecological sustainability of these forested landscapes. As our knowledge gaps are filled and standards for biological conservation are adjusted, we can expect an adaptive framework for forest management policies for the next millennium.

Broad Scale Patterns

Western forests in Oregon and Washington harbor a diverse and unique fauna. From a broad regional perspective, the habitats provided by western coniferous forests across the landscape can be easily distinguished. Their high productivity and structural complexity are captured by remote sensing images (Figure 4, page 200). This is not a homogeneous landscape, however, as might seem apparent at this broad spatial scale. It is the habitat heterogeneity, both across the region and within landscapes, that is pivotal to understanding the wildlife diversity of the western forests.

The large-scale mosaic of habitats across westside and montane forests is exemplified by the multiple physiographic provinces or ecoregions for this area. Physiographic province designations take primarily physical factors into consideration, including soils, geomorphology, natural disturbance history, and climate. When discussed in light of wildlife distributions, Nussbaum et al.[37] recognized 3 physiographic provinces over the western forest landscape (Cascade Mountains, Coast Ranges, and Klamath Mountains). These are subdivided into 8 physiographic provinces in the Recovery Plan for the northern spotted owl[56] (Cascade Mountains into 4 provinces: eastern and western Cascade Ranges in both Oregon and Washington; Coast Ranges into 3 provinces: Olympic Peninsula, Washington lowlands, Oregon Coast Range; and the Klamath Mountains). Ecoregions are geographic descriptors of environmental conditions; they are based on spatial patterns of physiography, climate, disturbance, vegetation, and wildlife. Omernik[40] recognized 5 Level III ecoregions for the area of interest in this chapter (Coast Range, Cascades, Eastern Cascades and foothills, North Cascades, Klamath Mountains); these are analogous to some of the physiographic provinces above. Pater et al.[42] provided a further subdivision of the Level II ecoregions into 42 Level IV regions (Figure 5, page 198).

Whereas general wildlife patterns were criteria for ecoregion determination, correspondence of wildlife species distributions with ecoregions is not fully investigated. Evaluation of species' associations with the newly developed Level IV ecoregions for western Oregon and Washington have not been conducted. However, some initial pattern assessments have developed using Oregon Level III ecoregions. The Oregon Biodiversity Project[41] presented species richness categories on a hexagonal grid for the State, with 150,000-acre (60,750-ha) hexagons, for butterflies, amphibians, fish, reptiles, and mammals (Figure 6, page 201).

Across Oregon, fish species richness was greatest in the western portion of the state coincident with western forests (many hexagons with >20 species and >29 species), compared to east-side areas (<8 species in much of the southeastern portion of the state, and patches from <8 to 20-29 species in the northeastern portion). In the Cascades, fish species richness decreased with increasing elevation. Mammal and butterfly richness patterns were similar to each other, although mammals were less variable in richness numbers across the western forested landscape. These 2 taxa had greatest diversity in the southern Cascade Range and up the ridge of the Cascade crest (including portions of the Klamath, and eastern and western Cascade ecoregions; >102 butterfly and >64 mammal species in much of this area). In contrast, among the western forested regions, butterfly and mammal richness were least in the Coast Range. Amphibian patterns in this analysis were extremely patchy. Like fish, diversity was greater in the western forests, with the least richness in the eastern Cascades. Among the other 3 western forested ecoregions, each had some hexagons with higher amphibian diversity (>15 species), most with intermediate (7-15 species) richness categories, and a number of patches with <5 amphibians. Reptile richness showed the most remarkable concordance with ecoregion boundaries. Among western forested regions, reptile richness was highest (>15 species) in the Klamath region. Much of the Coast Range had the lowest reptile richness category (<6 species), and the western Cascades showed a decreasing richness with elevation, with 12-15 species in the foothills and 6-8 species in the higher landscape.

Although Oregon bird species richness patterns were not compiled for the Oregon Biodiversity Project,[41] bird species ranges correspond to several patterns related to western forest ecoregions. Nine general species range patterns were derived from avian range maps in Csuti et al.[16] (Figure 7, >240 species evaluated, shorebirds restricted to coastal shores were excluded). Proportion of species per pattern and example species are (1) 2%, northern spotted owl and harlequin duck; (2) 8%, MacGillivray's warbler and varied thrush; (3) 3%, purple finch and chestnut backed chickadee; (4) 22%, American dipper and dark-eyed junco; (5) 2%, plain titmouse and black phoebe; (6) 7%, boreal owl and three-toed woodpecker; (7) 6%, calliope hummingbird and flammulated owl; (8) 9%, northern flicker and common raven; and (9) 39%, western meadowlark and yellow-headed blackbird. Half of the bird species in Oregon and Washington have the majority of their ranges in westside forests (patterns 1-7, Figure 7.) Most western forest birds have some occurrences east of the Cascade Range (e.g., patterns 2, 4, 5, and 7), and relatively few have ranges restricted to westside forests (e.g., patterns 1, 3, and 6).

Such a species range assessment provides a slightly shifted perspective of the spatial arrangements of species in comparison to species richness maps (e.g., Figure 6). For comparison among wildlife vertebrate taxa and between Oregon and Washington, additional range patterns were assessed (Figure 8). To facilitate comparisons, patterns 1 through 4 were combined to consolidate species with general west-side forest relationships, and patterns 6 and 7 were combined to more easily represent species restricted to the Cascade Range. In Oregon, forests in the Cascade Range and westward (Figures 7 and 8, general range patterns 1 through 7) are focal portions of the ranges of >50% of the species in each of these vertebrate classes. However, for all vertebrate classes assessed, east-side landscapes provide significant habitats for some species (Figures 7 and 8, general range pattern 9). This fact might be overlooked if only species richness maps were examined. For example, amphibian species richness (Figure 6) is clearly tied to western forests, yet east-side landscapes are of critical importance for the ranges of about 20% of Oregon and Washington amphibian species (Figure 8, general range pattern 9). Interestingly, reptiles in Washington are much more reliant on east-side habitats than those in Oregon (Figure 8). In Oregon, the west-side Klamath ecoregion contributes a significant reptile species "hotspot."

Herpetological distribution patterns in the Klamath ecoregion are further addressed by Bury and Pearl.[4] They reported that this region has the most species-rich herpetofauna of any similar-sized mountain range in the Pacific Northwest (38 native species of amphibians and reptiles; this number reflects total richness for the region, as opposed to richness per hexagon in the previous discussion). As already noted, this is largely attributable to the higher reptile richness in the Klamath region. Bury and Pearl[4] provided explanation for this patterning based on organism ecology, life history and behavior, and the common data elements that go into defining ecoregions (e.g., elevation, latitude, climate, legacy of past natural disturbance events, vegetative structure and composition).

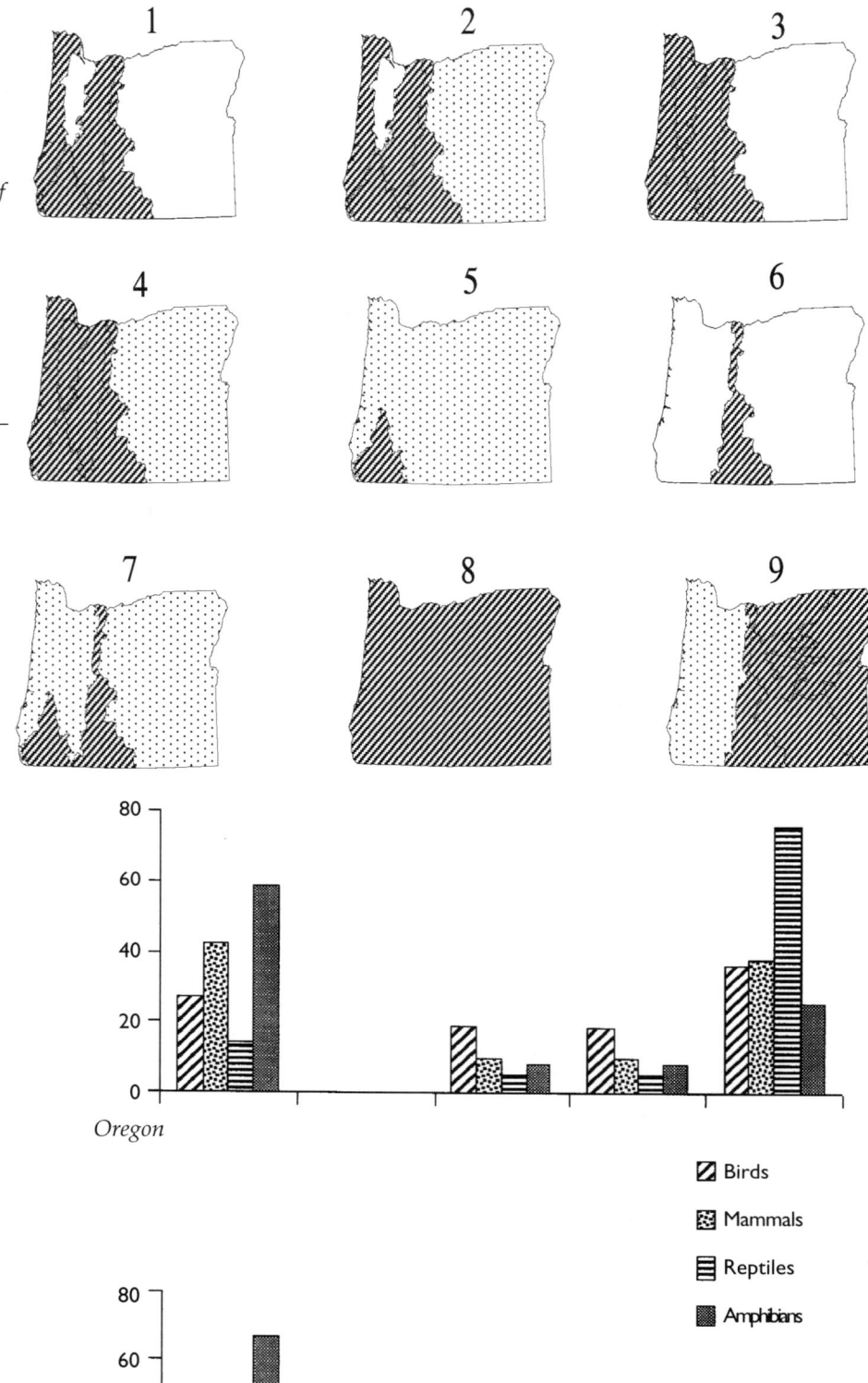

Figure 7. General range patterns of Oregon birds, derived from species range maps in Csuti et al.[16] Dark shading indicates the main portion of species ranges (areas in which most localities or habitats are known), light stippling indicates some occurrences, white indicates no or very few localities are known. Level III ecoregion boundaries (Figure 5) are used for general range boundaries.

Figure 8. General range patterns (1-9 from Figure 7) of Oregon wildlife and extension of those geographic patterns in Washington. Patterns 1-4 from Fig. 7 are combined to represent taxa with predominantly west-side ranges, and patterns 6 and 7 are combined to represent species occurring primarily in the Cascade Range. Data are compiled from range maps in Csuti et al.,[16] Johnson and Cassidy,[22] Leonard et al.,[25] Smith et al.,[46] and Storm and Leonard.[48]

These are the puzzle pieces that must be evaluated to understand the broad-scale species richness trends for any wildlife taxa. This rationale for herpetofauna of the Klamath ecoregion can be extended to demonstrate probable drivers of regional patterns in the western forests of Oregon and Washington.

Across western Oregon and Washington forests, many wildlife species diversity patterns are established with latitude, elevation, and climate, some of the basic data elements of ecoregions. These factors interact to produce different biotic and abiotic environments for wildlife, and species composition changes markedly with these factors. Taxa reliant on temperature and moisture regimes, such as reptiles and amphibians, show strong physiographic associations.[37] For example, species richness decreases with increases in latitude and elevation for reptiles and amphibians in western forests as temperature and moisture tolerance limits are encountered. Among amphibians occurring in the Cascade Range, approximately 20 species generally occur at elevations <4,000 ft (1,219 m), 13 species at 4,000-5,000 ft (1,219-1,524 m), 9 species at 5,000-6,000 ft (1,524-1,829 m), and 7 species at >6,000 ft (1829 m, range limits compiled from Leonard et al.[25]). Latitudinally, reptile species distributions become much more restricted to the north: only 3 of about 18 species retain broad distributions, 8 species are found primarily within the inland valleys (e.g., Willamette Valley), and 7 species' ranges end with the Klamath Mountains ecoregion. In southern Oregon, amphibian species gradients are found with distance from the coast. Presumably the cooler, moister climates of the coastal zone may explain this; conditions become xeric rapidly inland. Of 9 salamanders occurring at the coast in the extreme southwestern corner of the state, only 5 have likely ranges extending through Jackson County, about 100 miles (161 km) inland (Pacific giant salamanders, roughskin newts, Ensatina, clouded salamanders, and Del Norte salamanders are "replaced" by Siskiyou Mountain salamanders; Figure 3).

Current broad distributions of many taxa also reflect a legacy effect of past natural disturbances or environmental conditions. In particular, less vagile taxa with specific habitat requirements are not as resilient to disturbances, and their distributions may retain a signature of past events for extended time periods. The development of the Pacific Northwest herpetofauna and current herpetofaunal distributions were summarized by Nussbaum et al.[37] in light of the regional-to-continental extent of the northern temperate rain forests comprising the Arcto-Cretaceo-Tertiary Geoflora, and subsequent orogenic, glaciation, and drought events. They surmised that some currently depauperate areas can be accounted for by both insufficient time for reinvasion and the current relatively cold, harsh environments restricting colonization. They proposed that reptiles with broader distributions elsewhere and peripherally distributed in the Pacific Northwest may be relatively recently introduced to this region subsequent to the last glaciation (25,000-10,000 years ago). Many of the region's reptiles are derived from

species in the southwestern U.S., which presumably moved north with the retreating glaciers and establishment of warmer, more xeric conditions.

Likewise, amphibians responded to the same combination of events, the broad-ranging temperate forests subject to subsequent mountain-building, glaciation, and drought. Yet rather than these events resulting in colonization from adjoining regions, as in reptiles, many amphibian populations became isolated as the temperate rain forests contracted from their historical extent, becoming fragmented, with inhospitable conditions developing between remnants. Both glaciation and drought exacerbated isolation of such species dependent on milder conditions. Welsh[51] elaborated on this scenario in his presentation of relictual amphibian species adapted to the ancient primeval coniferous forest ecosystem. Several species now associated with old-growth forest conditions may be relics of this historical forest landscape, now strongly tied to stands retaining a semblance of the microhabitat and microclimate conditions of yore. Welsh's[59] analyses from southwestern Oregon and northwestern California identified the Del Norte salamander, the tailed frog, and the southern torrent salamander as such relics that generally do not tolerate conditions in highly managed forest stands. As presented by Welsh,[59] relatively small sedentary organisms with restricted distributions and narrow habitat requirements and climatic tolerances would be sensitive to disturbance or environmental change. Many of the 22 native amphibian species restricted to the western forest landscape occur in association with old-growth forest conditions, and are in essence phylogenetically symbolic of a past legacy of events and conditions in this region. Isolation scenarios based on disturbance and changing environments can be conceived for many of them (e.g., the plethodontids in Figure 3). The current distribution of the Larch Mountain salamander (Figure 3) likely reflects volcanic disturbance events from Mount Saint Helens, Mount Rainier, and Mount Hood, in addition to spatio-temporal fluctuations in microhabitat and microclimate conditions. Such dot map distributions also reflect biases of survey locations (i.e., remote areas are less sampled).[34]

Synthesizing the data contributing to ecoregion designations can lead to a better understanding of "hotspots" of wildlife diversity at the broad spatial scale of western and montane forests. As discussed, reptile diversity is highest in the more xeric Klamath Mountain region. The margins of several species ranges occur here, at their northern or western extent, whereas those that are found to the north often show more restricted distributions to lowland areas. Birds mirror this pattern (see Case History 1, below). DellaSala et al.[17] referred to the Klamath-Siskiyou ecoregion as an area of "extraordinary biodiversity", ranking its biodiversity "among the world's most outstanding temperate coniferous forests". Of 2,377 terrestrial animals they analyzed (snails, butterflies, birds, mammals, reptiles and amphibians), 168 (7%) occurred nowhere else. They found high endemism among aquatic animals: 42% of fish (n=33), and 60% of mollusks (*n*=235). At the subspecies level, 281 (8%) of >3,500 plants were cited as endemic. A dominant contributor to the Klamath-Siskiyou wildlife diversity hotspot is the convergence of several mountain ranges, in an "H" pattern. As a result, there is juxtaposition of several highly dissected stream drainage networks, extensive elevational and microclimatic gradients, and relative proximity of different life zones or ecoregions: the high desert of the Great Basin (east); California mountains (i.e., Sierra Nevada), plains and valleys (south); Coast (west); and the inland wet-temperate river valley systems (in the Klamath region and north to the Willamette Valley). In addition to the diverse array of physical conditions and vegetation types from the intersection of these zones, the survival of relictual flora and fauna due to reduced glacial impacts and, consequently, remnant habitats in this area adds a legacy effect to its biological diversity. For amphibians, at a provincial or ecoregion level, a broader ranging diversity hotspot occurs in the more productive wet temperate forests of the Coast Range.[37] Broad-scale diversity hotspots for mammals are less easily identified. Diversity hotspots may be distinguished at finer spatial scales, and are discussed below relative to patterns resulting from more fine-grained habitat elements.

Case History I
Patterns of Bird Communities in Coniferous Forests of Western Oregon and Washington
Joan C. Hagar

The majority of bird species occurring in conifer-dominated forests west of the Cascade crest in Oregon and Washington are widely distributed within this zone. Chestnut-backed chickadees and varied thrushes are notable residents of moist, low elevation (<5,000 ft [1,524 m]) coniferous forests west of the Cascade crest because they are broadly endemic to the Pacific Northwest and coastal southern California. Other year-round resident species that are characteristic of these forests include northern spotted owl, pileated woodpecker, Steller's jay, winter wren, and golden-crowned kinglet. During the breeding season, migrant species swell the ranks of the avifauna in terms of both numbers of species and density of individuals. Among the most broadly distributed and abundant migrant species in these conifer forests are the Swainson's thrush, Pacific-slope flycatcher, and hermit warbler. The breeding distribution of hermit warblers is largely restricted to Washington, Oregon, and California, defining this species as an endemic.

Avian community composition is influenced by elevation, seral stage, vegetation structure and composition, presence of water and other special features, and the interaction of all these factors. Total bird abundance in western Oregon and Washington is generally negatively correlated with elevation.[21] A negative relationship between elevation and bird species richness in old-growth forests also has been reported.[24]

Changes in riparian bird communities along an elevational gradient contribute to the decline in species richness with increasing elevation. Several riparian obligate species, such as great blue heron, wood duck, osprey, and kingfisher, are most abundant at low elevations, where forests or trees adjacent to large streams, rivers, and estuaries provide habitat for nesting, roosting, and foraging. At higher elevations, fewer species are strongly associated with steep, highly constrained streams and riparian vegetation that typically is only subtly distinguished from surrounding upslope forest. Dippers, however, specialize in foraging and nesting along such clear, swift mountain streams throughout western Oregon and Washington.

Bird community composition varies with seral stage, with differences being most pronounced between very early open canopy (i.e., grass-forb-shrub) and closed canopy stages. Species richness of birds tends to be similar in early and late stages of forest development, and lowest in the structurally simple mid-seral stages of managed forests.[2] Although relations between abundance and seral stage vary for many species on a geographic scale,[44] species that are typically associated with early seral conditions west of the Cascade Range include willow flycatchers, white-crowned sparrows, song sparrows, and spotted towhees.[2, 11, 57] Species that typically are more abundant in old forests include Pacific-slope flycatcher, varied thrush, and many members of the bark-foraging guild (e.g., brown creeper, chestnut-backed chickadee, red-breasted nuthatch, hairy woodpecker[44]). The avian species that are most closely associated with old-growth forests are marbled murrelets, northern spotted owls, and Vaux's swifts.[44] However, most species that reach their greatest abundance in older forests nonetheless also will use early seral habitats as long as key structural features are present. For example, species that forage on bark and nest in cavities, such as chestnut-backed chickadees and red-breasted nuthatches, occur in recent harvest units where green trees and snags have been retained.[11, 57]

Hardwood trees and shrubs may be one of the most important factors influencing bird community composition in the conifer-dominated landscape of the Pacific Northwest. The abundance and diversity of birds has been correlated positively with the abundance and distribution of hardwoods.[7, 19, 20, 21, 26, 32, 43] Deciduous hardwoods provide different resources for foraging and nesting than conifers, and thus provide unique habitat with which several bird species are strongly associated. Warbling vireos are predictably found in alder groves, and several species of neotropical migrant warblers (e.g., MacGillivray's, orange-crowned, Wilson's) typically forage and nest in thickets of deciduous shrubs.[16, 31] Unique associations between individual avian species and either hardwoods or conifers result in high bird diversity where hardwoods are mixed with or adjacent to conifers. Examples include the margins of large valleys where Oregon white oak occurs and the Klamath region.

Provincial Bird Patterns

Few bird species have a close affinity to only one physiographic province or ecoregion, but species richness and abundances of many species vary among provinces. Although direct comparisons of avian species richness across all provinces are not available, some general patterns are evident. Total bird abundance in western Oregon and Washington is generally correlated negatively with latitude and positively with longitude.[21] Thus, mean abundance and species richness of diurnal breeding birds is higher in the Oregon Coast Range than the Cascades of Oregon and southern Washington,[21] and species richness is highest in the Klamath region. Characteristics of the avifauna that distinguish each province are described below.

The avifauna of the tall, dense forests of the Coast Range and Puget lowlands shows some influence of the proximity to ocean, bays, and estuaries. The marbled murrelet is a unique example of this coastal influence, being the only seabird that nests in forest habitats. Limited by the necessity to feed at sea, this species nests only within 60 miles of the ocean, and predominantly in old-growth forests where branches of sufficient diameter provide nest platforms. No bird species is particularly associated with the coastal band of Sitka spruce, but wrentits are notable for the area because they do not regularly occur west of the coastal scrub along the southern and central Oregon coast.

As in the Coast Range, dense coniferous forests dominate habitats of the lower west slopes of the Cascade Range, so it is not surprising that the avifaunas of these two ecoregions are very similar. In Oregon, a slightly lower average avian species richness on the lower west slopes of the Cascades[21] may be partially attributable to a lower abundance of broad-leafed deciduous trees, such as big-leaf maple and red alder, than in the Coast Ranges.[35] On the other hand, some habitats that are unique to the Cascades contribute some distinctive members to the bird community. For example, two duck species, bufflehead and Barrow's goldeneye, breed on lakes in the high Cascades, and are considered forest associates because they use cavities, usually in snags, for nesting. Harlequin ducks are associated with fast-moving mountain streams, and find ideal breeding habitat along drainages of the west slope of the Cascades. Breeding populations of these three duck species are patchy and local throughout the Cascade Range in Oregon and Washington. Other species that are more likely to occur in the Cascades than in other provinces include golden eagle and goshawk.

Changes in bird communities that can be attributed to elevation are more obvious in the Cascades and Olympics than in the Coast Ranges. For example, Clark's nutcrackers do not occur below the high elevation spruce-fir forests in the Cascades, and mountain chickadees replace chestnut-backed chickadees above the Douglas-fir zone.[16] The presence of species such as great gray owls, boreal owls, black-backed woodpeckers, and boreal chickadees in the high Cascades (>5,000 ft [1,524 m]), North Cascades, and high Olympics indicates the boreal influence on these communities. In addition, merlins, northwestern crows, black swifts, and pine grosbeaks distinguish the avifauna of the north Cascades in Washington from those south of the Columbia River. Although both hermit warblers and Townsend's warblers breed in both states, hermit warblers reach the northern extent of their contiguous range in Washington, and Towsend's warblers do not breed south of northern California. Thus, Townsend's warblers are more abundant in the Washington Cascades, while hermit warblers are prevalent in Oregon, and the entire region comprises a zone of hybridization for these two species.

The Klamath region in southern Oregon stands out from the rest of the Pacific Northwest for two reasons: high plant species diversity, and the convergence of several ecological zones and their associated fauna. The highest avian species richness west of the Cascade crest in Oregon and Washington occurs in the Klamath Mountains. The high diversity of birds in this region has been attributed to the diversity of vegetation, and in particular the abundance of hardwoods.[43] Several avifaunas converge in the Siskiyou-Klamath mountains. Species such as Allen's hummingbird, black phoebe, oak titmouse, and blue-gray gnatcatcher reach the northwestern extent of their geographic ranges in this region. Species that are typical of pine habitats to the south and east (e.g., Lewis' woodpecker, white-headed woodpecker), and of Great Basin habitats (e.g., calliope hummingbird, green-tailed towhee, ash-throated flycatcher) occur in forested areas within this region. Other representatives of southern or arid regions are more abundant here than further north (e.g., acorn woodpecker), but the full compliment of species that are characteristic of moist coastal forests also occur (e.g., varied thrush, marbled murrelet, hermit warbler).

Patterns within Landscapes

Narrowing our focus from the broad regional scale perspective of species distribution patterns to a finer within-landscape approach (i.e., within the five Level III western forest ecoregions) allows a more concise discussion of wildlife-habitat relations. Whereas coarse-grained elements such as legacy disturbance effects and regional climate gradients may help our understanding of regional taxonomic diversity trends, individuals of a species survive and reproduce at finer scales. In the western and montane forests of Oregon and Washington, micro- to macrohabitat conditions at the forest stand level are of critical importance to the individual. Aggregating up in biological organization, populations similarly function within these narrower bounds. In this section, the dominant finer scale habitat associations of western forest wildlife within landscapes are presented, including forest plant species and stand structural conditions. Key habitat elements within western forests to which wildlife have strong ties (e.g., logs, rock substrates, litter, snags, and large trees) are distinguished. Old growth, young seral stages, riparian forests, and forest edges are highlighted because of their roles as wildlife habitat hotspots for various taxa within our current managed landscapes. At this finer spatial scale, heterogeneity among these habitat types and microhabitat features remains a dominant driver of the western wildlife species diversity.

Current knowledge of wildlife species' use of habitats and general ecology in Oregon and Washington was compiled by panels of species-experts and is presented in several matrixes on the CD-ROM accompanying this book. This expert knowledge includes data supported by research, personal observations, or expert opinion. In the tables below, these data are summarized to show wildlife relationships with forest habitat type, structural conditions, habitat elements, and trophic and organismal relations.

Forest Habitat Types and Structural Conditions

Western forests are a complex mix of vegetative conditions. Herb, shrub, and canopy tree structure and composition are key predictors of the occurrence of various wildlife species. Wildlife species were assessed relative to 4 western forest habitat types: (1) Westside Lowlands Conifer-deciduous Forest; (2) Westside Oak and Dry Douglas-fir Forest and Woodlands; (3) Southwest Mixed Conifer-Hardwood Forest; and (4) Montane Mixed Conifer Forest. Fairly similar numbers of total species occurred in these habitats (Table 1), with differences reflecting some of the larger scale patterns already discussed (e.g., reptiles).

Table 2. Western forest wildlife habitat associations with vegetation height and successional stage.[1]

Taxon	Grass/forb	Shrub/seedling	Sapling/pole	Small trees	Medium trees	Large & giant trees
Amphibians[2]	19	19	20	21	28	28
Reptiles[3]	19	16	16	18	18	17
Birds[3]	61	92	98	130	140	140
Mammals[3]	66	65	64	72	79	78
Total	165	192	198	241	265	263

[1]Data are compiled from expert panel assessments for this book, see accompanying CD-ROM. Numbers of species are indicated.

[2]These species require specific habitat elements to occur within these structural conditions.

[3]These species may require specific habitat elements to occur within these structural conditions.

Table 3. Numbers of western forest wildlife species associated with tree size (large includes giant trees) and canopy complexity (single = single story canopies, multi = multiple story canopies).*

Taxon	Small trees		Medium trees		Large trees	
Canopy Cover	single	multi	single	multi	single	multi
Amphibians						
Open	21	21	22	22	24	22
Moderate	21	21	22	28	24	28
Closed	20	21	22	28	28	28
Reptiles						
Open	17	18	16	15	17	16
Moderate	10	9	10	9	10	9
Closed	6	6	6	6	6	8
Birds						
Open	106	96	115	104	121	114
Moderate	89	85	97	101	102	104
Closed	68	67	74	76	76	96
Mammals						
Open	61	62	68	58	71	73
Moderate	54	54	65	60	69	68
Closed	33	41	43	47	53	65

* Data are compiled from expert panel assessments for this book.

Table 4. Numbers of species using (use) and closely associated (close assoc.) with old-growth forests in western Oregon and Washington, and their relative dispersal ability (stand is within approximately 60 acres [24 ha]; range = across provinces).[1]

Taxon	Use	Close assoc.	Close assoc. in Oregon	Close assoc. in Washington	Dispersal of close assoc.
Amphibians	31[2]	16	16	10	Stand
Reptiles	10	0	—	—	—
Birds	119	38	38	37	Stand to Range
Mammals	67	26	21	21	Stand to Range
Total	227	80	75	68	

[1]Data compiled from Thomas et al.,[51] and includes species from northwestern California forests.

[2]The former Olympic salamander is represented here as the 4 torrent salamander species.

Within these 4 habitats, forest/woodland structural conditions were identified. Three components were used to classify structure: vegetation height and successional stage (i.e., grass/forb, shrub/seedling, sapling/pole, and small-to-giant tree categories); number of canopy layers (single or multiple); and canopy cover (open, moderate, closed). Several patterns of wildlife associations with these components are apparent (Tables 2 and 3). Total species richness increases with vegetation height, and this holds true for each taxon except reptiles (Table 2). Bird species show a particularly dramatic response to vegetation height, successional stage, and canopy layers. Mammals and amphibians show a similar but dampened pattern. All 4 wildlife groups show a pattern with canopy closure (Table 3). Greater numbers of birds, mammals and reptiles are associated with open rather than closed canopies. The reverse is apparent for amphibians.

Associations of western forest fauna with forest age-size or successional stage categories have been analyzed in numerous studies and assessments across the region. Results from these studies and existing knowledge of species' forest associations were compiled for the Report of the Scientific Analysis Team[51] for areas within the range of the northern spotted owl. The particular assignment for this assessment was to identify those species likely to be closely associated with late-successional and old-growth forest conditions, from the longer list of those that used forest habitats (Table 4). Close associates met one of several criteria, such as having greater abundance in old-growth forest than in mature or pole stands and requiring habitat components that are contributed by old-growth forest (Table 5-1 in Thomas et al.[51]). From their data compilation, about a third of the vertebrates using western forests were identified as likely close associates with the older forest conditions (Table 4). No reptiles and about half the amphibians were close associates. These patterns may be reflected in the current habitat assessments. About a third more forest wildlife species occurred in association with large trees than the earliest successional stages. However, there was only about a 10% increase between small and large trees (Table 2). Amphibians were more associated with closed canopies. Structural conditions of old-growth forests include components in Table 3 that are related to vertical stratification: large tree size, multiple story, and closed canopies. These create habitats or conditions to which some old-growth associates are reliant (see Habitat Elements below).

Some old-growth associated species do not spend their entire life in these forests. For example, neotropical migrant birds spend their winters in Mexico or Central America (7 species), and some waterfowl migrate in winter to lowland bays, lakes, and surf zones (6 species). Similarly, many aquatic-breeding amphibians reproduce in lentic and lotic waters within the forested landscape, and move to the upslope forest matrix after metamorphosis from their aquatic stages.

Early successional stands are used by many western forest-dependent wildlife. From Table 2, herb, shrub and sapling conditions are used by a high percentage of species in every taxonomic group. Open conditions across tree size categories are especially important for birds and mammals (Table 3). Similarly, 13 (25%) of 53 mammal species were associated with early seral stages in the Augusta Creek watershed, in the western Oregon Cascade Range.[18] McGarigal and McComb[27] reported about a third of bird species (n=99) used early seral forests in western Oregon.

Riparian forests are critical to the life history of numerous vertebrates in western Oregon and Washington. The western forest landscape in the Pacific Northwest is highly dissected by stream channels, and the stream-upslope interface is not far from any locale in the forest matrix. Riparian areas are distinguished by their cool, moist environments, and by their multitude of conditions across a watershed. Small headwater streams are ecologically distinct from their downstream mainstem counterparts. Along this entire aquatic network, riparian areas are prone to small-scale disturbances, such as flooding, bank and slope slippage, landslides, and treefalls. These add heterogeneity to riparian forest conditions. Heterogeneity also is added as slope gradients vary, as streams flow through constrained and unconstrained reaches, and with temporal variation in foliage cover (i.e., deciduous trees) and rainfall patterns. Unconstrained reaches and tributary junctions, in particular, are proposed as higher species diversity areas

Table 5. Primary habitat element associations of species with close ties to old-growth forest conditions.[1]

Taxon	n	Talus	Logs	Duff/litter	Large snags	Large trees	Riparian
Amphibians	16	12	11	9	0	0	13[2]
Birds	38	0	7	0	25	10	8
Mammals	26	4	14	7	14	10	11
Total	80	16	32	16	39	20	32

[1] n = total number of species per taxon. Species may be represented in multiple habitat element columns. Compiled from Thomas et al.[51] (includes northwestern California species).

[2] The former Olympic salamander is represented here as the 4 torrent salamander species.

Table 6. Habitat elements used by forest wildlife in western Washington and Oregon, as determined during the panel assessments for this book.

Taxon	Surface rock	Logs	Duff/litter	Snags	Live trees	Moss	Cavities	Shrubs
Amphibians	3	12	10	4	3	1	0	1
Reptiles	10	5	4	0	0	0	0	0
Birds	27	18	5	57	72	5	38	21
Mammals	35	50	19	22	37	4	22	11
Total	75	85	38	83	112	10	60	33

for several aquatic-dependent taxa. As habitat complexity may beget diversity, riparian forests are predictable species hotspots in western forests because they may encompass numerous habitat and microhabitat conditions.

As discussed for old-growth forests above, whereas some taxa use riparian areas for their entire lives, others are users for critical life history functions. For example, several myotis bats use riparian corridors for foraging habitat and travel corridors, but roost upslope. The terrestrial stages of aquatic-breeding amphibians may rely on riparian habitats for foraging or refuge. Assessments in Thomas et al.[51] determined that 29 (36%) of 80 old-growth forest associated vertebrates were strongly associated with riparian areas (Table 5). These include 53% of the amphibians (10 species, primarily stream- and pond-breeders, and bank associates), 42% of the mammals (7 bats, Pacific shrew, shrew-mole, deer mouse, marten), and 21% of the birds (marbled murrelet, bald eagle, and 6 waterfowl species) that were assessed. As can be seen in Table 5, riparian-associated species include users of several forest habitat elements (talus, logs, snags, large trees).

Extending the diversity hotspot concept to within-landscape, forest stand-scale areas, both old-growth and riparian forests are areas of higher richness and diversity for many taxa. However, overall reduced breeding bird diversity was found along streams in comparison to upslope forest habitats by McGarigal and McComb[26] in the Oregon Coast Range. They suggested differences in vegetation structure and composition could explain their results: fewer large conifers and snags were noted along their study streams. More generally, riparian areas within late-successional or old-growth conditions are particularly high in wildlife numbers. An understanding of the habitat elements and microclimate components of these areas is important to fully grasp the habitat complexities of these areas. Riparian-dependent wildlife are discussed further in Chapter 13.

Habitat Elements

Several forest habitat elements are used for foraging, refuge, or reproduction by western forest wildlife. These include features such as downed logs, snags, duff or litter, and rocky substrates. They also include biotic or live vegetative components, including shrubs, live tree branches, or live remnant trees. Many of the old-growth forest associated species may occur in younger stands if these critical habitat elements are found there. Table 6 summarizes habitat elements associations for forest wildlife. Table 5 shows the primary habitat elements of old-growth associated vertebrates, compiled from Thomas et al.[51] Each element in Tables 5 and 6 can be considered a dominant contributor to western forest wildlife habitats across taxa. Microhabitats provided by logs and snags, rock, duff and litter, cavities, shrubs, and large trees are key elements for forest-dwelling species. It is interesting to note that there are several differences between the results of these two assessments (Tables 5 and 6). Table 6 shows that many more species *use* these habitat elements in western forests than the more limited pool of species that are both closely tied to old-growth forest conditions and *associated* with these elements (Table 5). Criteria for use versus association are different. Among elements considered in both assessments, logs and snags were the dominant habitat elements represented in Table 5, whereas surface rock and live trees also are dominant in Table 6.

Dead and downed wood occurs in numerous forms (Figure 9). In western forests, downed wood includes logs, rootwads and stumps, wood piles and slash, roots, branches, loose bark, and bark piles. Different species tend to be associated with the different downed wood types. Among amphibians, for example, both ensatina and 57. Vega, R. M. S. 1993. Bird communities in managed conifer

Text continues on page 200

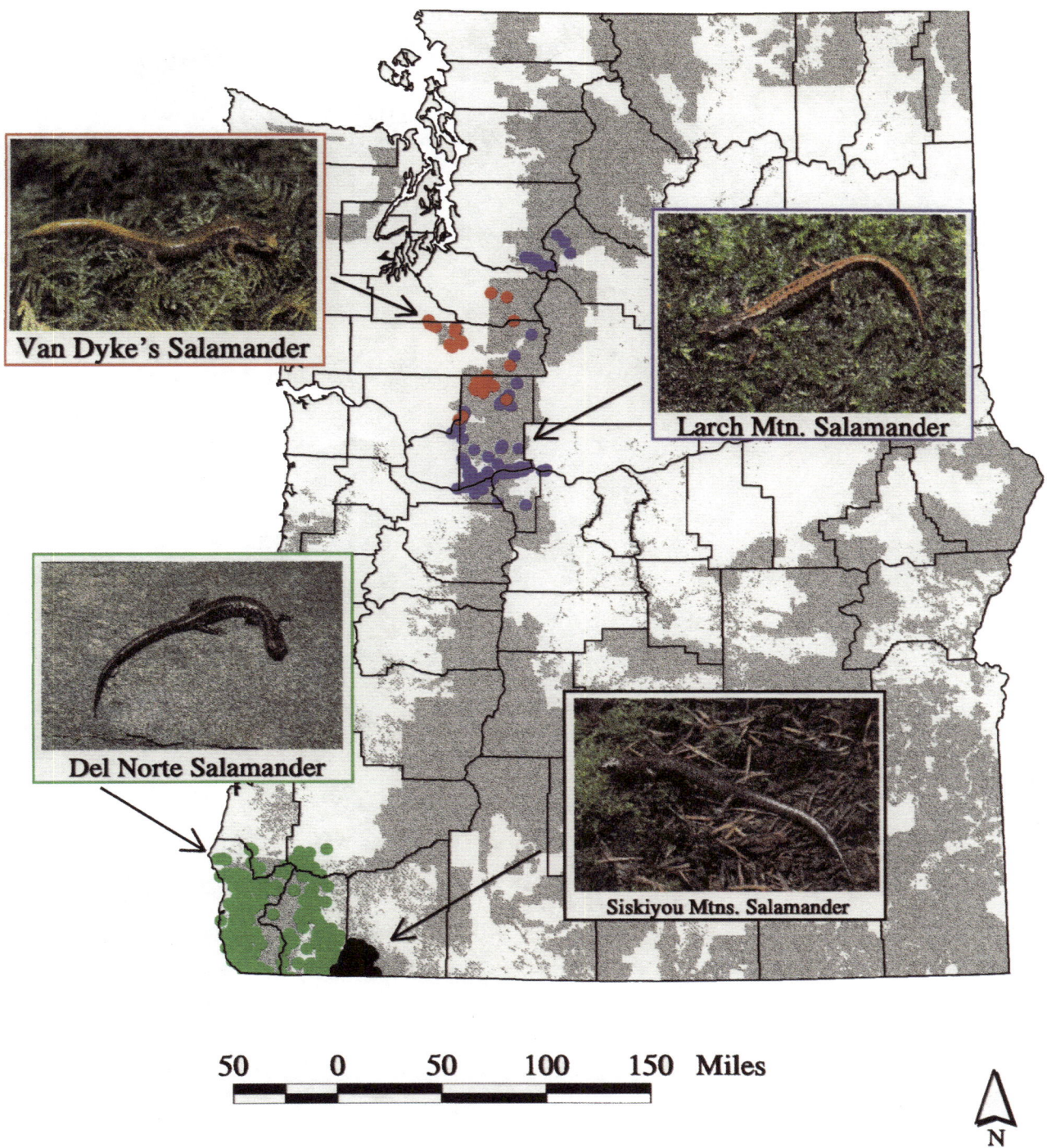

Figure 3. Four terrestrial salamanders restricted to forested habitats of the Pacific Northwest. These species are covered under the federal Survey and Manage provision of the Northwest Forest Plan, for which surveys are conducted prior to ground disturbing activities and known sites are currently managed for salamander persistence.[34, 54, 55] W. P. Leonard, photographer. R. S. Nauman, GIS technician.

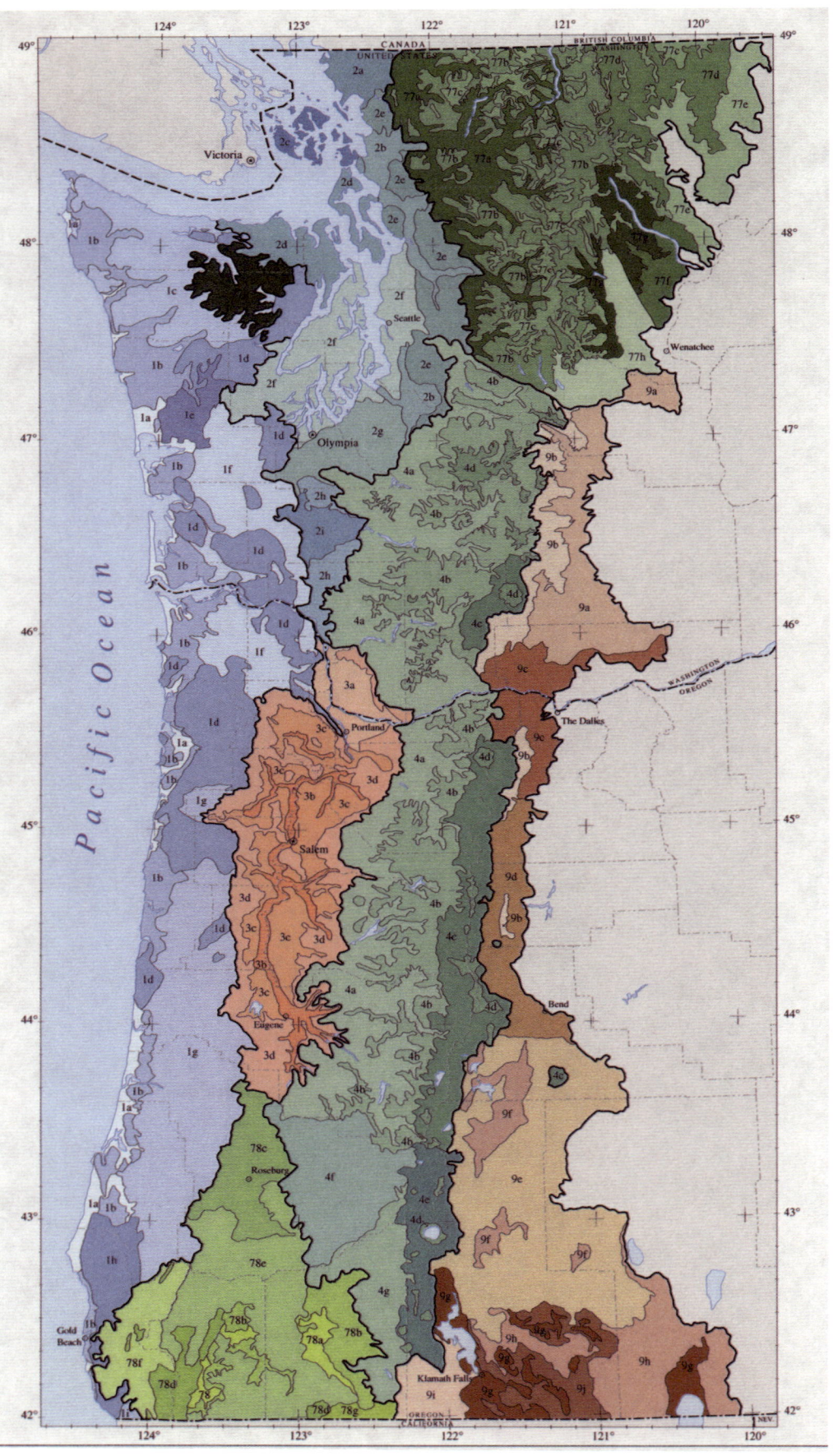

Figure 5. Level III and IV ecoregions for western Oregon and Washington, compiled at a scale of 1:250,000 by Pater et al.[42]

1 Coast Range
1a Coastal Lowlands
1b Coastal Uplands
1c Low Olympics
1d Volcanics
1e Outwash
1f Willapa Hills
1g Mid-Coastal Sedimentary
1h Southern Oregon Coastal Mountains
1i Redwood Zone

2 Puget Lowland
2a Fraser Lowland
2b Eastern Puget Riverine Lowlands
2c San Juan Islands
2d Olympic Rainshadow
2e Eastern Puget Uplands
2f Central Puget Lowland
2g Southern Puget Prairies
2h Cowlitz/Chehalis Foothills
2I Cowlitz/Newaukum Prairie Floodplains

3 Willamette Valley
3a Portland/Vancouver Basin
3b Willamette River and Tributaries Gallery Forest
3c Prairie Terraces
3d Valley Foothills

————— Level III ecoregion
————— Level IV ecoregion
· · · · · County boundary
— · — · — State boundary
— — — — International boundary

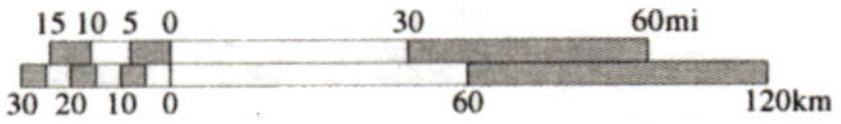

15 10 5 0   30   60mi
30 20 10 0   60   120km
Universal Transverse Mercator Projection, Zone 10

4 Cascades
4a Western Cascades Lowlands and Valleys
4b Western Cascades Montane Highlands
4c Cascade Crest Montane Forest
4d Cascades Subalpine/Alpine
4e High Southern Cascades Montane Forest
4f Umpqua Cascades
4g Southern Cascades

9 Eastern Cascades Slopes and Foothills
9a Yakima Plateau and Slopes
9b Grand Fir Mixed Forest
9c Oak/Conifer Eastern Cascades Columbia Foothills
9d Ponderosa Pine/Bitterbrush Woodland
9e Pumice Plateau Forest
9f Cold Wet Pumice Plateau Basins
9g Klamath/Goose Lake Warm Wet Basins
9h Fremont Pine/Fir Forest
9i Southern Cascades Slope
9j Klamath Juniper/Ponderosa Pine Woodland

77 North Cascades
77a North Cascades Lowland Forests
77b North Cascades Highland Forests
77c North Cascades Subalpine/Alpine
77d Pasayten/Sawtooth Highlands
77e Okanogan Pine/Fir Hills
77f Chelan Tephra Hills
77g Wenatchee/Chelan Highlands
77h Chiwaukum Hills and Lowlands
77i High Olympics

78 Klamath Mountains
78a Rogue/Illinois Valleys
78b Siskiyou Foothills
78c Umpqua Interior Foothills
78d Serpentine Siskiyous
78e Inland Siskiyous
78f Coastal Siskiyous
78g Klamath River Ridges

Principal authors: David E. Pater (Dynamac Corporation), Sandra A. Bryce, (Dynamac Corporation), Thor D. Thorson (NRCS), Jimmy Kagan (Oregon Natural Heritage Program), Chris Chappell (Washington DNR), James M. Omernik (USEPA), Sandra H. Azevedo (OAO Corporation), and Alan J. Woods (Dynamac Corporation).

Collaborators and contributors: Terry L. Aho (NRCS), Duane Lammers (USFS), Thomas Atzet (USFS), Robert Meurisse (USFS), Kenneth Radek (USFS), Carl Davis (USFS), Thomas Loveland (USGS), M. Frances Faure (OAO Corporation), and Jeffrey A. Comstock (OAO Corporation).

This project was partially supported by funds from the USEPA – Office of Research and Development – Regional Applied Research Effort (RARE) program.

but the clouded salamander is frequently found under the bark of logs and the ensatina is often found in bark piles associated with decaying snags. Coarse woody debris, particularly large down logs, are the habitat elements most frequently used by amphibians and mammals. Large wood decays gradually, its rate dependent on the tree size, ambient conditions, and tree species. Wildlife species may have affinities for specific wood decay classes.

For vertebrates, wood provides foraging, cover, and sites for reproduction. For some species within each vertebrate class, large downed wood provides thermal refugia, buffering temperature and moisture extremes. Blessing et al.[1] demonstrated the temperature buffering capacity of a log, 20 in x 13 ft (50 cm x 4 m), containing a Van Dyke's salamander nest site. While ambient air temperatures ranged 43-76°F (6.3 to 24.7°C) in the shade 6.6 ft (2 m) from the log, temperatures inside the log cavity at the salamander nest ranged 46-63°F (7.7-17°C). For several days in the summer, the maximum nest temperature was cooler than the minimum outside air temperature. For amphibians, logs provide cool, moist, and stable microhabitats suitable for their physiological temperature and moisture requirements. Some

plethodontid salamander species have limited home ranges, remaining at log sites for indefinite periods. They find sufficient foraging opportunities at logs, in addition to using them for cover and reproduction. For mammals, downed wood habitats similarly provide resting, nesting and denning, and foraging habitat for numerous species (Table 6). Of the 14 log-associated mammals listed in Table 5, 2 species are bats that potentially use logs for roosting, and the remaining 12 species, 10 rodents and insectivores and 2 forest carnivores, use the downed wood for multiple life history functions. Although fewer birds rely on logs, their use of downed wood includes perching and lookout in addition to foraging sites, cover, and nesting. The sharptail snake is one of the few western forest reptiles associated with downed wood, and often is found in logs.

Snags are used extensively by birds and mammals (Tables 5 and 6). Cavities, cracks, crevices, and loose bark on or in snags are used by numerous species as cover and resting, roosting and nesting sites. Snag decay class is important for cavity excavators, and for species using snags for foraging. Some snag-users have preferences for snag size. Protection from predation is considered a selective force of snag use, and again, thermal buffering is thought to be a critical component of these habitat elements. Standing dead trees within intact stands will provide suitable thermal refugia for species sensitive to temperature extremes. However, some species tend to use snags in open conditions, or in a variety of closed and open forest types. Many bats have roosting and hibernacula in larger snags, some preferring snags with loose bark. Fishers use cavities for denning and resting, preferring large snags. Cavity-nesting birds include many waterfowl, owl, and woodpecker species.

Duff, litter, live trees, and surface rock are additional western forest components required by various wildlife species (Table 5 and 6). Many amphibians in western forests use duff, litter and surface rock, but only 1 species (clouded salamander, in California) has been found in trees. Rock associates include some plethodontid salamanders, such as the Del Norte salamander. Surface rock can be covered by litter and not readily identified as a likely site.[38] Mammals use all of these habitat elements, yet relatively little is known about the ecology of some arboreal mammals. The northern red tree vole is thought to live almost exclusively within the forest canopy. Numerous birds use live trees. Cavities and crevices are a main refuge and nesting place whereas the various canopy layers are used for foraging, cover, and nesting. Large conifers may be of particular importance to several species.

Forest Wildlife Assemblages

Sorting wildlife species by their main habitat associations helps us understand their roles in western forest ecosystems. Such assemblages or communities can be derived from the classifications of taxa relative to the habitat parameters presented here, including forest type, structural components, and habitat elements. The scale at which assemblages are identified might vary with the context that is being examined. The reptile assemblage of

Figure 4. Satellite image of the Pacific Northwest. Western and montane forests are shown as darker areas on the left, and are coincident with the range of the northern spotted owl and the Northwest Forest Plan. M. Fiorella, GIS technician.

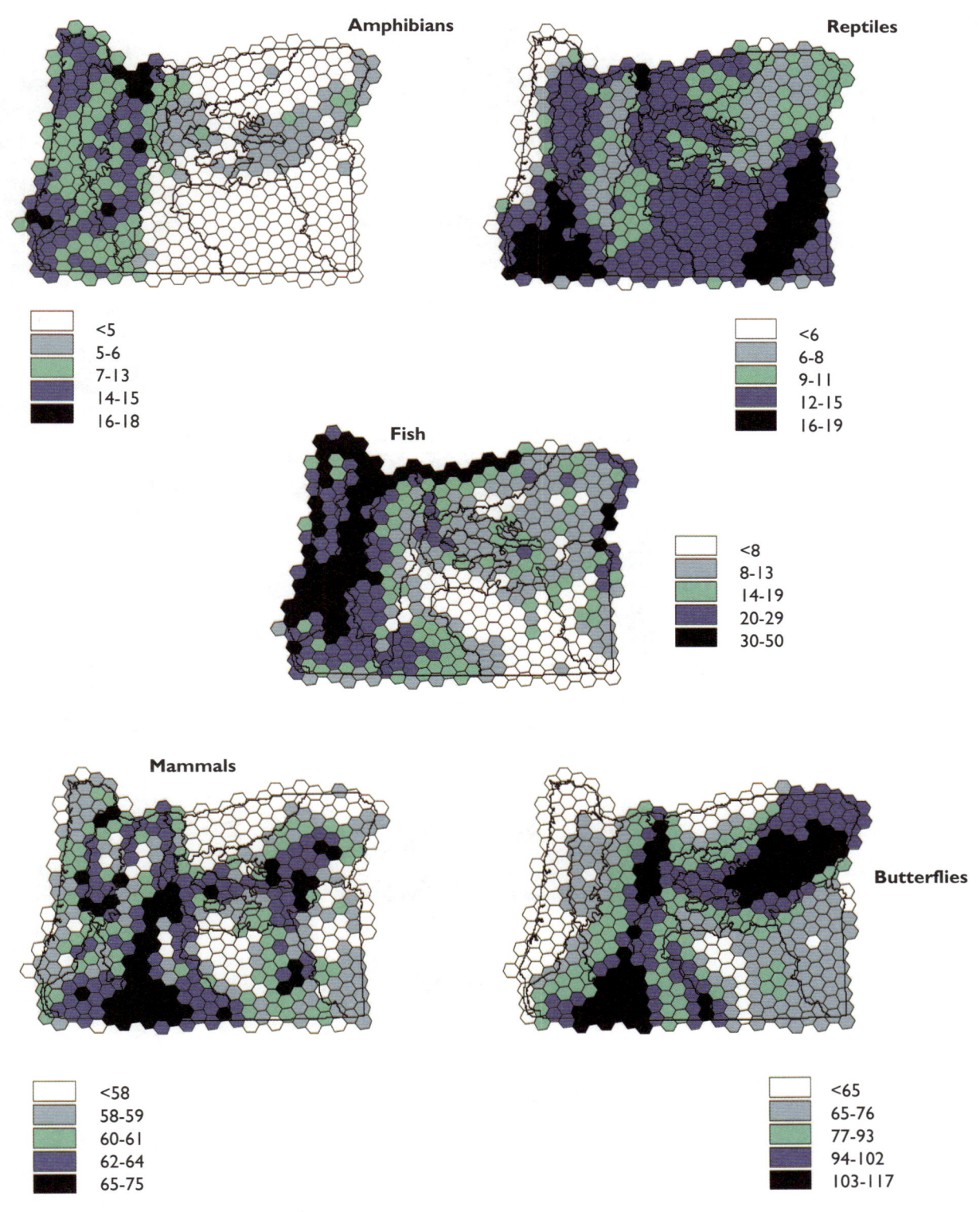

Figure 6. Species richness maps for Oregon, derived from data compiled for the Oregon Biodiversity Project[41] CD-ROM. Species richness (numbers of species) categories per taxonomic group per hexagon (approximately 150,000 acres [60,750 ha]) are shown. Bird data were not available. (Courtesy of R.S. Nauman, GIS technician)

Figure 12. Satellite images of the eastern Oregon Cascade Range, showing the Three Sisters Wilderness from an eastern perspective. Upper image shows forests of the eastern Cascade ecoregion (red) and areas of downslope forest management. Lower image shows the likely impact areas of recreation (within 328 ft [100 m])of roads and trails, focussed along riparian areas). M. Richmond, GIS technician.

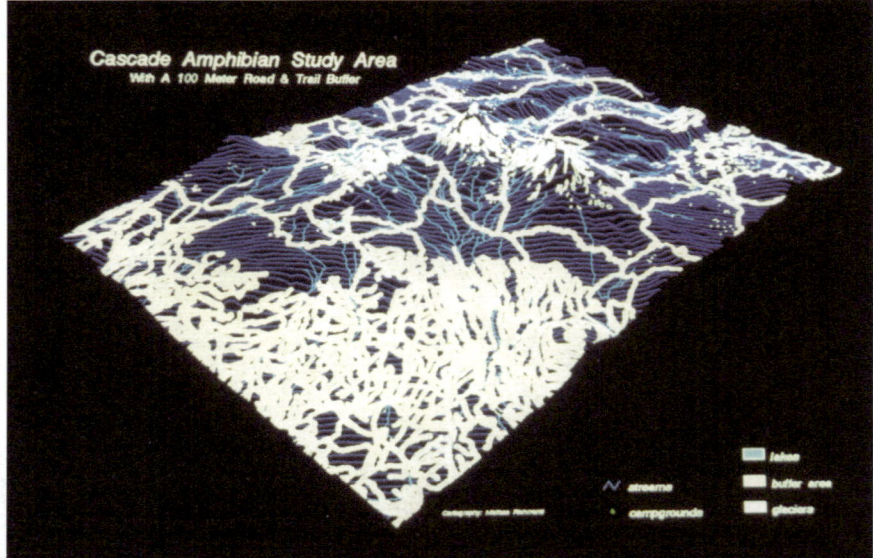

Figure 9. Downed wood provides habitat for multiple vertebrate species. H. J. Andrews Experimental Forest, western Oregon Cascades. J. Means, photographer.

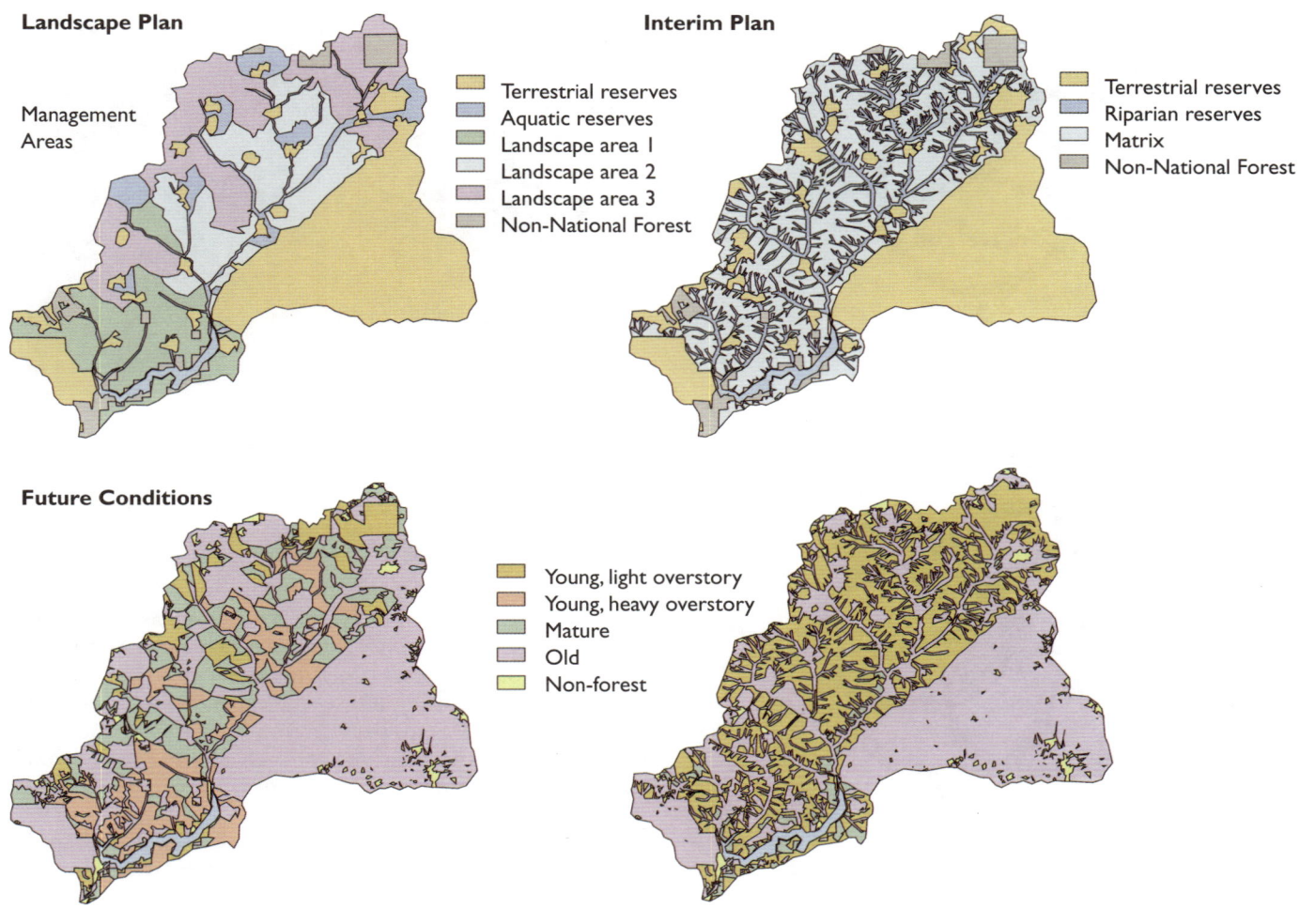

Landscape Plan

Management
Areas

- Terrestrial reserves
- Aquatic reserves
- Landscape area 1
- Landscape area 2
- Landscape area 3
- Non-National Forest

Interim Plan

- Terrestrial reserves
- Riparian reserves
- Matrix
- Non-National Forest

Future Conditions

- Young, light overstory
- Young, heavy overstory
- Mature
- Old
- Non-forest

Figure 13. Management areas and projected future landscape structures for the Blue River watershed managed under Matrix and Riparian Reserve designations of the Northwest Forest Plan[54] ("Interim Plan"), and for the Blue River landscape management strategy ("Landscape Plan").[14] T. Turner, GIS technician.

the Klamath ecoregion might be identified for 1 purpose, whereas in another vein Klamath woodland snakes might be distinguished. Ground-dwelling small mammals and forest canopy mammals are examples of 2 assemblages around which hypotheses of ecological function may be developed (Case History 2, below). Given the large numbers of species using snag or tree cavities, cavity-nesting species of birds and mammals is another useful assemblage in this context. Assemblages have increasing validity as distinct ecological entities when they are populated by alternative members across landscapes that span multiple species' ranges.

For amphibians, 3 main assemblages are generally partitioned for the Pacific Northwest (Figure 10). These are separated by breeding habitat: terrestrial; "pond" (inclusive of all lentic habitats; e.g., lakes and wetlands) (e.g., Figure 11); and "stream" (inclusive of all lotic waters; e.g., streams and seeps). From a finer-grained habitat

assessment of species in western and montane forests (lower portion of Figure 10), we have added complexity to this model. In western forests, distinct assemblages of stream-breeders are found in association with stream size. In headwater streams and seeps, *Rhyacotriton* torrent salamanders dominate assemblages. Cope's giant salamanders and tailed frogs may occur in some headwater channels and seeps as well. Larger downstream channels are dominated by Pacific giant salamanders, co-occurring with cottid and salmonid fishes in many systems. The stream bank community is distinct, often comprised of terrestrial-breeding Dunn's salamanders, western red-backed salamanders, and Van Dyke's salamanders. We have split the terrestrial assemblage into two groups: the rock and downed wood associates. Both groups are highly fossorial, spending much of their time subsurface. When temperature and moisture regimes at the forest surface are suitable, they can be found in

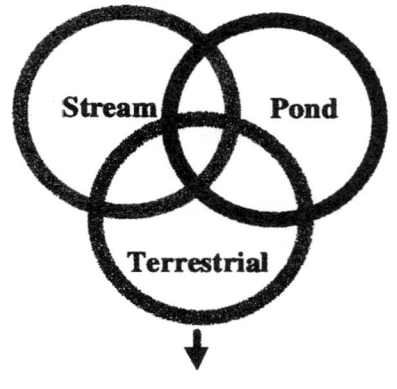

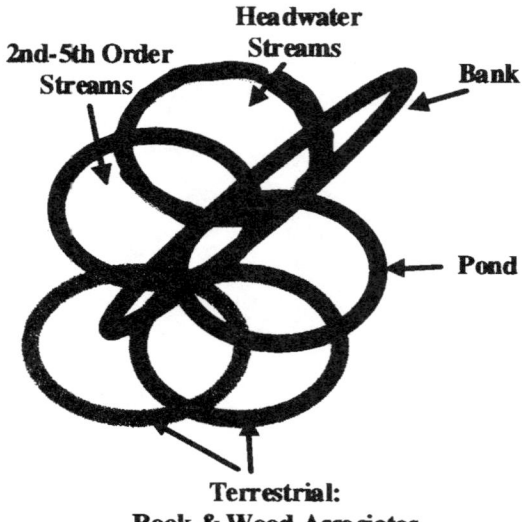

Figure 10. Fundamental (top) and finer-grained (bottom) habitat assemblages of western and montane forest amphibians.

Figure 11. Red legged frog (Rana aurora) is a pond breeder in western forests. W. P. Leonard, photographer.

inhospitable warm, dry surface conditions away from aquatic habitats. Pond-breeding amphibians such as Cascades frogs often are found along montane creeks in the summer. Foraging opportunities may be enhanced in such locations, and they may function as dispersal corridors. The relatively high species richness found along banks also might result from this being an edge between habitat types. As a boundary between habitats, you might be more likely to find members of neighboring assemblages along its interface. Or alternatively, as an edge, this region may represent a barrier to further movement.

Among terrestrial forms, rock and wood associates have the potential for relatively high habitat overlap as individuals opportunistically use cover as it is available and as suitable microclimate conditions warrant. This occurs in two ways. First, within populations, there might be use of both wood or rock, depending on its availability and suitability as habitat cover. Individuals in such populations might use the different cover types for different functions, such as dispersal cover, foraging areas, or reproduction. Dispersal cover, in particular, might be used more opportunistically. Second, for some species, there seems to be segregation of habitat use geographically among populations. For example, the Larch Mountain salamander is found in association with surface rock along the Columbia River Gorge, but is found associated with downed wood and loamy soils elsewhere.[15] Similarly, the Del Norte salamander appears to be a rock associate at inland locations, and can be found with downed wood at coastal sites.[38] The critical factor in this might still be habitat availability and suitability. Suitable microclimate and microsite conditions for these terrestrial salamanders might occur through combinations of either of these cover types.[15, 38]

We discuss western forest wildlife assemblages again below. First, as trophic relations are identified, assemblages may link to ecological functions and processes (see also Case History 2). Second, assemblages often are used during the development of protective measures for wildlife when forest management activities are proposed.

association with wood or rock cover. At some locations, these animals may be found year-round if suitable surface refugia are available. Wood associates include the slender salamanders, the black salamander, and those species associated with logs that were mentioned previously. Rock associates include the plethodontids in Figure 3. As stream or pond breeders move into upslope forests, they often opportunistically use both wood and rock microhabitats, as well as subsurface retreats.

Overlap among assemblages is considerable for amphibians because of their complex life history. Adults are generally not restricted to breeding habitats. Thus, stream and pond breeders venture from the aquatic and riparian forest landscapes into upslope forests. Likewise, terrestrial forms occur in riparian zones and may be found streamside. The bank seems to be the primary area of assemblage overlap, as members of all assemblages are found in this near-water riparian zone. This may occur because bank conditions in western and montane forests may be almost ideally suited for amphibians, having cool moist microclimates. During summer, in particular, banks might offer surface refugia for amphibians from

Species' Life History, Behavior, and Biotic Interactions

Distribution patterns of western forest wildlife are highly dependent on several aspects of their behavior and ecology. In particular, species' life history, behavior, and species interactions may need to be understood to fully explain species-habitat relations.

Dispersal limitations seem to contribute to the isolation of many amphibian populations. Amphibians generally are reliant on relatively narrow windows of temperature and moisture conditions for surface activity. Spring and fall rainy seasons are primarily when dispersal occurs for most taxa. Compounded by their relatively low mobility, amphibians' survival during migrations across heterogeneous forested landscapes can be affected. However, few studies on western forest amphibian dispersal have been conducted. Some taxa may move hundreds of feet to several miles (*Ambystoma, Taricha, Dicamptodon,* and *Rana* spp.). The terrestrial plethodontid salamanders are considered less vagile. Studies of some log-associates found movement of only a few feet over seasons to years.

A tendency for site fidelity may interact with amphibian dispersal capabilities. There may be philopatric tendencies among both the more and less mobile amphibians. Many toads known to trek miles in montane forests have high breeding site fidelity. Traditionally used breeding sites are common in pond breeding salamanders and frogs, and may occur in the other forest amphibian assemblages.

Mammals and birds seem to be less restricted in their movements (Table 3), yet many have relatively small home ranges (i.e., <10 acres [4 ha][27]). Even those with large home ranges may be restricted in their movements by perceived barriers on the landscape. Clearcuts and roads are suspected barriers for some species.

Interactions among species are the threads that weave the fabric of the integrated living forest ecosystem. Intraspecific interactions are often dominated by competition for food, space, and mates, and interspecific interactions involve both competition and predation. Both intra- and interspecific interactions may displace individuals or populations, affect survivorship, and result in altered distribution and abundance patterns across the landscape.

Trophic relations are established for most western and montane forest wildlife species (Table 7). Typical of food webs, most numerous among forest wildlife are the primary consumers (herbivores) and secondary consumers (primary predators). Primary consumers eat a host of plant material (e.g., leaves, seeds, sap, roots, bark, fruit). Primary consumers are birds (58% of this taxon) and small-to-large bodied mammals (e.g., rodents, deer, elk; 67% of mammals), and larval pond-breeding amphibians. Secondary consumers prey on invertebrates, vertebrates, and eggs. In western forests, some mammals, most birds and all reptile and amphibian species are secondary consumers (Table 7). Many of these are gape-limited predators that often change diet opportunistically with their size or age. Sharptail snakes are noted as feeding

Table 7. Distribution of species by trophic relationships in western Washington and Oregon.*

Taxon	Primary consumer	Secondary consumer	Tertiary consumer	Carrion feeder
Amphibians	10	29	0	0
Reptiles	1	20	0	2
Birds	96	161	6	8
Mammals	70	62	1	12
Total	177	272	7	22

* Data are compiled from expert panel assessments for this book.

on only small slugs, and most small salamanders prey on various invertebrates.

Amphibians and mammals are likely key vertebrate connectors within the western forest system. Both groups are the chief prey for secondary predators (= tertiary consumers), and as primary consumers and primary predators are conduits of energy from the lower trophic levels, particularly the diverse arthropod and fungal communities. In eastern U.S. forests, amphibians may comprise a major component of the vertebrate biomass;[3] however, such a biomass estimation has not been done for western Oregon and Washington forests. Rodents comprise about half the mammal species of the region, functioning as prey for numerous species. Bats are believed to consume enormous quantities of insect prey, mainly over streams, ponds, and riparian areas. Shrews and moles are carnivorous, eating predominantly arthropods. The fungi-feeding mammals link the vertebrate trophic network with the diverse fungal community, of pivotal importance in the forest ecosystem.[52] Fungi also are food for microorganisms and invertebrates, and have critical forest functions as decomposers, nutrient cyclers, and through their mutualistic or symbiotic relationships with other forest species.[28]

Competition and trophic relationships are only a part of the interspecific interactions among wildlife in forests (Table 8). Wildlife mediate the reproduction of numerous forest species as dispersers or pollinators. Birds and mammals are dominant players in organismal relationships (Table 8). Their actions can alter habitats significantly, making them either suitable or unsuitable for other species. For example, cavity, burrow, and runway creators provide critical habitats for many other wildlife species.

The stability of vertebrate communities can be dependent on the balance of intraspecific, trophic, and other organismal relations with species' habitat conditions. Measures of biotic integrity attempt to determine the status of communities by assessing indicator conditions or species. These can be examined relative to disturbance impacts, such as forest management activities, to monitor forest system functions. In western forests, arboreal rodent and forest-floor small mammal assemblages are suggested to be ideal indicators of forest biotic integrity (Case History 2).

Table 8. Number of species by organismal relationships in western Washington and Oregon.*

Organismal relationship	Amphibians	Reptiles	Birds	Mammals
Controls insect populations	8	0	29	13
Controls terrestrial vertebrate populations	1	1	14	12
Pollination vector	0	0	6	0
Transportation of seeds, spores, plants or animals	0	0	65	36
Nest parasite	0	0	40	0
Creates feeding, roosting, denning, or nesting sites for other species	0	0	4	4
Creates structures	0	0	13	11
Uses structures created by other species	1	3	5	11
Cavity excavator	0	0	14	1
Cavity user	0	0	29	12
Burrow excavator	0	0	0	44
Burrow user	0	9	0	38
Runway creator	0	0	0	32
Runway user	0	0	0	33
Pirates food from others	0	0	4	0
Interspecific hybridization	0	3	12	6

* Data are compiled from expert panel assessments for this book.

Case History 2
Biotic Integrity

Andrew B. Carey

How can the effectiveness of ecosystem and landscape management be evaluated? One attractive method is to measure the integrity of select vertebrate communities.[10] Ideally, these communities would consist of a limited number of year-round resident species, common enough to be found in most, if not all, patches of suitable habitat, yet sensitive enough to management that ≥1 species would be absent or severely reduced in abundance in unsuitable or low-quality environments. A limited number of species is desirable because operational practicality requires use of one technique of high reliability. Year-round resident species are desirable because population levels of migrants incorporate variability due to wintering and migration factors that are independent of the area being managed. Relatively high abundance is required to ensure that all species can be measured effectively when present. Two communities fit these requirements: forest-floor small mammals and arboreal rodents.

Forest-floor small mammals are interesting because complete communities with moderately high abundances of each species depends on Hutchinsonian preinteractive niche diversification—in other words, biocomplexity in the structure and processes of the forest floor. Thus, the community represents diverse forest-floor functions, abundance of coarse woody debris, and understory development. On the Olympic Peninsula in Washington, for example, the ranked relative abundances of the species in the community are Trowbridge's shrew (5), southern red-backed vole (4), montane shrew (4), deer mouse (4), forest deer mouse (3), shrew-mole (3), creeping vole (3), and vagrant shrew (2). Assemblages in other provinces

are listed in Carey and Johnson.[8] This community can be described with various techniques: live trapping, pitfall trapping, and snap trapping.

Another interesting community is the arboreal rodent community. In Washington, the community consists of northern flying squirrel, Douglas' squirrel, and Townsend's chipmunk. In Oregon, one has the option of adding bushy-tailed woodrat, dusky-footed woodrat, and red tree vole. Individually and collectively, the biomass of these species is indicative of carrying capacities for a variety of vertebrate predators including mustelids, hawks, and owls. In addition, in managed forests, community arrangement will diverge from the high abundances of all members found in old growth, to communities dominated by 1 or 2 species when management has failed to adequately address key ecosystem components.[6] In particular, the arboreal rodent community measures ecological productivity—the energy that the system of fungi, understory plants, and overstory trees diverts to reproduction (truffles, mushrooms, fruits, seeds, and nuts). This community integrates production of fruit with the decadence process that produces cavities and affects spatial arrangements of habitat elements. This results in niche diversification. Thus, the sum of the population sizes of species in the arboreal rodent community measures, in large part, the energy the system is putting into reproduction.[9] This community (with the exception of red tree vole) can be effectively described with live trapping.[5]

Arboreal rodents and forest-floor small mammals can be used both to monitor effectiveness[6] but also to predict results of forest management strategies through modeling.[10] The use of biotic integrity, when supported by basic research, offers an alternative approach for populations of rare, cryptic, or other species that are difficult to study, survey, and manage.

Management Issues

Western and montane forest management policies in Oregon and Washington are undergoing tremendous changes, largely to address the long-term persistence of species and to ensure the maintenance of ecosystem integrity. The main human uses of western and montane forests include timber production, water resources, special forest products, recreation, mining, and the associated support systems of roads and trails (Figure 12, page 202). These forests also maintain the treasuries of biophysical legacies, functions, and processes having aesthetic, ethical, and cultural values within our society. However, across this landscape, timber management has been the dominant focus on both public and private lands for >100 yrs. By the 1970s and 1980s, concern for high profile species such as the northern spotted owl grew as late-successional and old-growth forests were diminished and fragmented because of logging, and as the U.S. Endangered Species Act (ESA) of 1973 prohibited harm to species listed under the Act, and to their habitats. Studies were initiated to understand main habitat associations of the northern spotted owl, as well as the requirements of numerous other suspected obligates to the ancient forests of this region. By 1994, >1000 taxa were identified as likely associates of old-growth forest habitats.

Federal Forest Management

The range of the northern spotted owl has close resemblance to the western and montane forests considered in this chapter (Figure 4), but also extends latitudinally beyond Oregon and Washington. In the U.S., 42% of this landscape is federally administered (24.3 million acres [9.8 million ha][53]). This landscape across 3 states was used to develop the first ecosystem management plan for the nation which spanned land ownerships: the federal Northwest Forest Plan.[54] The Plan's goals included providing for the highest contribution to the socioeconomic needs of the region while ensuring the long-term viability of the old-growth forest ecosystem and associated species. Although the Plan has been in place for 6 years, the balance of socioeconomic productivity and protection of biological integrity is a challenge that is still being developed as we enter the 21st century. The Plan is based on adaptive management, and thus the adaptive phase is expected to continue as new knowledge and management tools develop for western forests.

For species protective measures, the foundation of this Plan relies on several land allocations, habitat provisions, and species-specific mitigations.[54] A backbone of reserved lands was created across the region for the maintenance of well-distributed populations of most of the broad ranging taxa considered. For habitats of fish and riparian-dependent species of concern in the region, an Aquatic Conservation Strategy[45, 54] was formulated, including protections of key watersheds, development of watershed analysis, watershed restoration, and identification of Riparian Reserves, primarily along streams and around unstable areas. Forest habitat provisions included coarse woody debris management, specified snag and green tree retention levels, and forest management for red tree vole and northern spotted owl dispersal corridors. For species that were not adequately protected by these series of measures, a "Survey and Manage" provision was created.[53, 54, 55] For those of the most concern, sites proposed for ground disturbance would be surveyed for the species of interest, and if found, managed to maintain the persistence of that species[54, 55] (Figure 3). These protection measures are additive, together addressing the long-term persistence of old-growth associated species on the federal forest landscape in this region.

Since implementation of this Plan, alternative landscape management plans have developed. The measures developed for the Northwest Forest Plan might be considered experimental, since nothing of that scope has ever before been attempted in this region. The Plan may not be the sole means to maintain ecosystem integrity and biological legacies in western forests, while also providing timber products and other socioeconomic values. For the federally administered Augusta Creek watershed, an alternative scenario of forest management was developed using the natural disturbance processes (i.e., fire) of the landscape as criteria for forest management.[13] Forest rotations and green tree retention levels were matched to fire frequency intervals (100, 200, and 300 years) and intensities (15-50% retention). The Plan's Riparian Reserve system was reduced to provide buffers along only the mainstem fish-bearing streams; however, tree retention would be weighted along other stream channels in harvested units. Aquatic reserves also were placed in small basins for species or areas of concern. The resulting landscape was modeled and evaluated after several hundred years and found to have advantages over the Northwest Forest Plan landscape at the watershed scale. Importantly, the fragmented and edgy "spaghetti" landscape of the Plan (i.e., spatial pattern resulting from the highly dendritic stream network and its accompanying Riparian Reserves, between which harvested units are located) is consolidated into larger contiguous forest blocks. Benefits for wildlife included improved habitat connectivity and maintenance of interior old-growth conditions in much of the landscape because of reduced edge effects on microclimate.[12] This scenario, with some adjustments, is being tested in a neighboring watershed of the Willamette National Forest in the western Oregon Cascade Range (Case History 3).

Case History 3
The Blue River Landscape Study
John H. Cissel and Frederick J. Swanson

A team of scientists and managers working on the H. J. Andrews Experimental Forest and the Blue River Ranger District of the Willamette National Forest have been cooperating for most of this decade to develop and test a landscape management approach based on natural disturbance regimes.[13, 14] The team has been motivated to a significant degree by concern over the loss and fragmentation of older forests, and the lack of a coherent long-term strategy for conservation of older forest systems in managed landscapes. The underlying assumption of this approach is that by approximating key aspects of important disturbance regimes in management regimes, risks posed to native species and ecological processes are reduced as compared to other historical and contemporary landscape management approaches.[23, 29, 49]

The Blue River Landscape Study is intended to evaluate the potential effects of implementing a landscape plan based on historical landscape dynamics. The landscape management approach used in the study is intended to meet the same general objectives underlying the Northwest Forest Plan:[54] provide habitat to sustain species associated with late-successional forests, maintain and restore aquatic ecosystems, and provide a sustainable supply of timber. A combination of effectiveness monitoring, long-term plots, retrospective studies, and modeling assessments are being used to evaluate and adjust this landscape management approach. The Blue River watershed provides an ideal setting for the study due to its size (approximately 56,790 acres [23,000 hectares]) and the presence of the H. J. Andrews Experimental Forest and numerous long-term studies within the watershed. In addition, the Blue River watershed is a part of the Central Cascades Adaptive Management Area, a federal land allocation in the Northwest Forest Plan that encourages development and evaluation of new approaches.

The dynamics of historical landscapes in this area were heavily influenced by fire of varying frequency, severity and spatial extent. General patterns of past fire behavior were interpreted into three fire regimes based on a 500-year dendrochronological record.[33, 50, 58] For example, wet, cool sites burned infrequently while warmer, drier sites burned more frequently. Characteristics of these three regimes were used to establish timber and fire management regimes in actively-managed portions of the landscape. Timber harvest frequency and rotation age (100-260 years) were based upon historical fire frequency, timber harvest intensity (15-50% overstory canopy retention) was based upon historical fire severity, and the spatial patterns of timber harvest were based upon the spatial patterns of historical fires. Implementation guidelines are intended to reflect natural disturbance patterns to the extent feasible while protecting ecological values.

An aquatic reserve system also was established to help meet the aquatic ecosystem objectives in the Northwest Forest Plan.[45, 54] These reserves are of two types: small-watershed reserves and corridor reserves. Small-watershed reserves are strategically located throughout the watershed to encompass areas of particular importance to aquatic ecosystems and spotted owls. In addition, corridor reserves are established on all fish-bearing streams. Figure 15 depicts the landscape management plan ("Landscape Plan"), and, for comparison, a literal implementation of the Northwest Forest Plan as if it were applied to the Blue River watershed ("Interim Plan"; e.g., Riparian Reserves occur on all streams, and 80-year rotation regeneration harvests with 15% retention occur in the "Matrix"[54]).

A watershed restoration strategy is an integral component of the Blue River landscape management approach. Restoration activities are intended to reestablish a resilient, interconnected aquatic network capable of maintaining aquatic habitats and processes while management activities are occurring in the watershed. Road restoration activities are planned to occur first in areas where risks to aquatic ecosystems are high.

Future timber harvest and forest successional patterns were projected across the watershed for the next 200 years for both plans (Figure 13, page 203) and analyzed. Results show that the landscape plan will produce more late-successional habitat (71% of the watershed versus 59%) in a less fragmented landscape as compared to the interim plan.[14] Larger patches in the landscape plan create more interior habitat, thought to benefit some wildlife species such as the northern spotted owl. Less edge between old and young forests in the landscape plan reduces edge effects such as altered microclimates and increased plant mortality, and may reduce habitat for certain species that favor edges, such as elk. More complex stand structures are present in the landscape plan due to generally higher overstory canopy retention levels. Retention of live and dead trees in young stands has been found to favor cyanolichens, certain fungi and invertebrates associated with older forests, amphibians with life histories requiring both stream/riparian and upland habitats, provide more options for protection of rare species, and to moderate understory environments. The landscape plan also maintains a substantial component of mature forest (80-200 years old). In contrast, the interim plan nearly eliminates the mature forest component because almost all lands are either in a reserve, where all stands grow old and large-scale disturbance is eliminated, or in matrix lands where a relatively short rotation (approximately 80 years) prevents regrowth of mature forest. We feel that the absence of mature forest in the interim plan poses substantial risk when mortality due to disturbance, climate change or senescence eliminates older Douglas-firs in the reserves.

Landscape structures resulting from both the landscape management plan in this study[14] and from the interim plan are historically unprecedented. For that reason we feel it is critical that an adaptive management approach be followed for both plans. We are pressing ahead with implementation, monitoring, modeling, and research to better define and evaluate a historically-rooted approach in the Blue River watershed based on the landscape dynamics inherent to the area. We hope these concepts can be tested in other provinces in the region, and that the Matrix and Riparian Reserve approach of the interim plan[54] can be similarly tested.

57. Vega, R. M. S. 1993. Bird communities in managed conifer stands in the

State and Private Forest Management

In the last decade, several forest management plans at watershed to landscape scales have been designed by state agencies and industrial land owners. These alternative approaches to forest management reflect the diverse alignment of roles as wildlife stewards of these land owners. On federal lands, the more conservative standard has been set for species maintenance or restoration as a priority equal to or greater than providing economic returns. On state lands, timber revenue is an identified priority, and consequently a more intensive timber management program is implemented. Although species persistence is addressed by states and many rare species protective measures are implemented, a relatively greater risk to native habitats and species is perceived with their more intensive timber harvest practices, reduced reserved lands, and narrower riparian buffers. Whether or how states might alter their role as ecosystem and biodiversity stewards is currently a debated issue regionally and nationally. Private and industrial forest land owners seek to maximize timber returns, and while they actively design provisions to maintain biodiversity as legally required, their provisions may minimally protect species habitats, rarely identify all taxa associated with the forest landscape, and do not necessarily restore habitats to allow rare species to [re]colonize their lands. Industrial land owners with more extensive land holdings have been more proactive for species protections, yet on the broader spectrum, they seem to be held less accountable for species persistence than state and federally administered lands. Although there is acknowledgment of the different roles of land owners for ecosystem, habitat or species stewardship, a good model for a managed alignment of these diverse roles has not been developed for a landscape with multiple ownerships. The "Coastal Landscape Analysis and Management System" project is hoped to advance such a model for the Oregon Coast Range province.[47]

Standards for species and particularly wildlife conservation are changing and we are still mid-pivot. "Precautionary principles" are more often cited as rationale for conservative forest management decisions, and there has been a shift in the burden of proof for species and habitat protections: we've gone from needing to prove a value needs protection before providing it, to proving it has adequate protection before lifting it.[36] As mentioned above, there also has been a switch from addressing a few threatened and endangered wildlife species to a broader spectrum of species (e.g, fungi, lichens, bryophytes, mollusks) and assemblages (e.g, arthropod functional groups).[53, 54, 55] Increased public concern and review add complexities to processes that now seem to require full consensus, whereas they were more authoritatively controlled previously. Litigation or the threat of litigation has been an effective driver of these changing land management ethics. As adaptive management approaches are being advocated, long-term contracts for state and private Habitat Conservation Plans are becoming more and more difficult to achieve. And finally, while the policy arena is embroiled in controversy over how much wildlife protection is needed in different portions of the western forest region, by ownership and location, the science of forest management is rapidly changing.

Sustainable forestry techniques currently being tested across western Oregon and Washington forests are numerous and innovative. Forest density management and alternative silviculture is being examined for both restoration and regeneration harvests by interagency collaborative partners. Selective harvests are being more broadly implemented. Mosaics of thinning levels, clearcut islands, and green tree leave islands may achieve multiple forest objectives, retaining localized patches of rare species or species hotspots while opening other patches for regeneration of shade intolerant tree species and production of greater wood volumes. Such practices are more costly, involving greater site reconnaissance and site preparation, and more complex logging directions, but they also may attain goals for compatible wood production and biological resource protection at the site level. The role of leave islands for vertebrate species persistence within a managed forest landscape needs further study.

Several forest management approaches and provisions are being tested, or are in need of being tested. Riparian management approaches are being examined at the site level. Alternative stream buffer widths are being examined (e.g., Figure 14). Concurrently, forest biological resources are being investigated. Are there habitats or taxa that require special consideration in various portions of the stream network? If so, what are their responses to alternative forest management designs? Stream buffers may not be the only mechanism for aquatic and riparian resource maintenance. Those in current use do not provide interior old-growth microclimate conditions along streams, but rather mitigate for slope slippage, water temperatures, wood inputs to streams, or perhaps development of late-successional structural components (e.g., large tree size). Patch reserves along streams have been proposed but not tested. Patches and buffers might be used together, like beads on a string, to provide intermittent riparian habitats with interior conditions and narrower intervening sections designed to retain stream temperatures, limited downed wood recruitment, and near-stream habitat.

Management of downed wood recruitment is another topic in need of additional attention. As presented in this chapter, many wildlife habitat associations rely on dead and downed wood. Yet are we maintaining and managing for the recruitment of sufficient large logs and snags? Examination of log decay classes in managed forests reveals a paucity of hard logs, and in some locations, mostly just legacy large wood from high-grading harvests of half a century or more ago. Loss of coarse woody debris has implications for mammals and amphibians, key taxa in food webs, linking producers to consumers.

Spatial scale of protection is an issue that needs to be addressed for multiple wildlife species and habitats across Oregon and Washington. What resources should be maintained at the site, at the watershed or land-ownership block, or at the landscape and region? Is the intent to

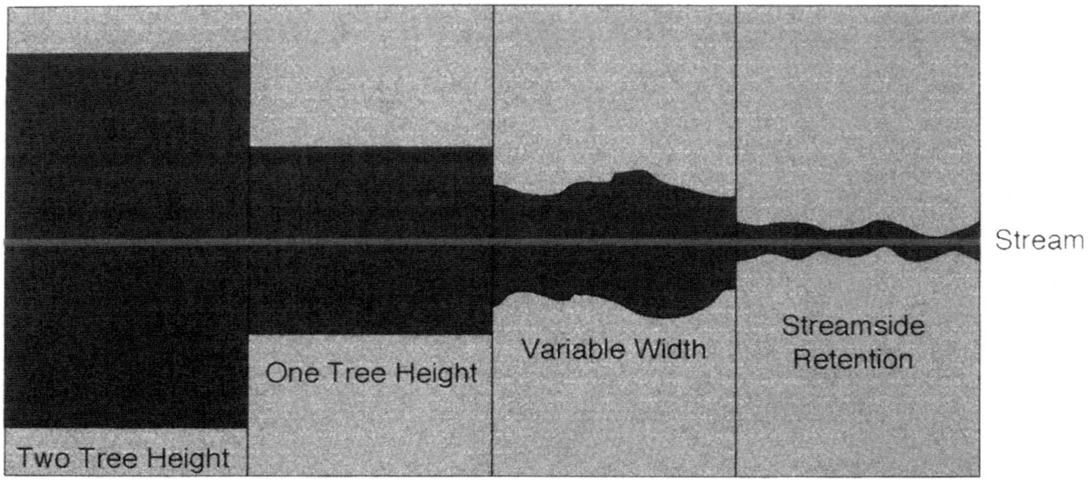

Figure 14. Alternative riparian buffer widths being examined in headwaters, with upslope density management. Tree heights correspond to federal interim Riparian Reserves for the Northwest Forest Plan.[54] The variable buffer has a 50-feet (15 m) minimum and varies with topographic or vegetation conditions, and the streamside retention buffer retains only streamside trees for bank stability, those within about 20 feet (6 m) of the stream (D. Olson, Riparian Buffer Study, interagency research conducted by the Bureau of Land Management and the U.S. Forest Service in western Oregon forests). K. Ronnenberg, graphics artist.

maintain all rare species and key habitats at all localities? At what level of rarity can sites be prioritized for maintenance at larger scales, such as a watershed or land ownership, so that losses at individual sites are acceptable if larger scale persistence is assured? What levels of risk are acceptable for these different species-rarities, habitats, and spatial scales? Can protective mitigations be nested among sites, watersheds, provinces, landscapes, and ownerships? Mechanisms and processes for multiscale and inter-landowner management approaches need advancement. These need to be tied to effective monitoring strategies and adaptively managed.

Conclusion

Several common themes are presented for new directions in western and montane forest management. These are sure to develop further in the next few years and decades.

1. The burden of proof is shifting to demonstrable stewardship of species and their habitats prior to implementation of west-side forest management activities, with increased use of the "precautionary principle" and conservative approaches to hedge uncertainties and risk to species persistence. Increased public involvement and oversight forest land management activities is partly responsible for this trend, resulting in greater land-owner accountability for the maintenance of natural resources.

2. Well-defined goals are needed for wildlife management in western forests among federal and state agencies, private and industrial land ownerships.

3. Integrated habitat-based and species-specific management approaches are being designed by federal, state, and private landowners.

4. Collaborative efforts are being initiated to investigate alternative management approaches to achieve multiple resource production and protection across landscapes.

Literature Cited

1. Blessing, B. J., E. P. Phoenix, L. L. C. Jones, and M. G. Raphael. 1999. Nests of Van Dyke's salamander (*Plethodon vandykei*) from the Olympic Peninsula, Washington. Northwestern Naturalist 80: 77-81.

2. Brown, E. R., technical editor. 1985. Management of wildlife and fish habitats in forests of western Oregon and Washington. Part 2, Appendices. U.S. Forest Service R6-F&WL-192-1985, Portland, OR.

3. Burton, T. A., and G. E. Likens. 1975. Salamander populations and biomass in the Hubbard Brook Experimental Forest, New Hampshire. Copeia 1975: 541-546.

4. Bury, R. B., and C. A. Pearl. 1999. Klamath-Siskiyou herpetofauna: biogeographic patterns and conservation strategies. Natural Areas Journal 19: 341-350.

5. Carey, A. B. 1995. Sciurids in Pacific Northwest managed and old-growth forests. Ecological Applications 5: 648-661.

6. Carey, A. B. 2000. Effects of new forest management strategies on squirrel populations. Ecological Applications 10:248-257.

7. Carey, A. B., M. M. Hardt, S. P. Horton, and B. L. Biswell. 1991. Spring bird communities in the Oregon Coast Range. Pages 123-142 in: L. F. Ruggiero, K. B. Aubry, A. B. Carey, M. H. Huff, technical coordinators. Wildlife and vegetation of unmanaged Douglas-fir forests. U.S. Forest Service General Technical Report PNW-GTR-285, Portland, OR

8. Carey, A. B., and M. L. Johnson. 1995. Small mammals in managed, naturally young, and old-growth forests. Ecological Applications 5: 336-352.

9. Carey, A. B., J. Kershner, B. Biswell, and L. D. de Toledo. 1999. Ecological scale and forest development: squirrels, dietary fungi, and vascular plants in managed and unmanaged forests. Wildlife Monographs 142: 1-71.

10. Carey, A. B., B. R. Lippke, and J. Sessions. 1999. Intentional Ecosystem Management: managing forests for biodiversity. Journal of Sustainable Forestry 9: 83-125.

11. Chambers, C. L., W. C. McComb, and J. C. Tappeiner II. 1999. Breeding bird responses to three silvicultural treatments in the Oregon Coast Range. Ecological Applications 9(1): 171-185.

12. Chen, J., J. F. Franklin, and T. A. Spies. 1995. Growing-season microclimatic gradients from clearcut edges into old-growth Douglas-fir forests. Ecological Applications 5:74-86.

13. Cissel, J. H., F. J. Swanson, G. E. Grant, D. H. Olson, S. V. Gregory, S. L. Garman, L. R. Ashkenas, M. G. Hunter, J. N. Kertis, J. H. Mayo, M. D. McSwain, S. G. Swetland, K. A. Swindle, and D. O. Wallin. 1998. A landscape plan based on historical fire regimes for a managed forest ecosystem: the Augusta Creek study. U.S. Forest Service General Technical Report PNW-GTR-422, Portland, OR.

14. Cissel, J. H., F. J. Swanson, and P. J. Weisberg. 1999. Landscape management using historical fire regimes: Blue River, OR. Ecological Applications 9(4): 1217-1231.

15. Crisafulli, C. M. 2000. Survey protocol for the Larch Mountain salamander (Plethodon larselli). Chapter VII pages 253-310 in: D. H. Olson, editor. Survey protocols for amphibians under the Survey and Manage provision of the Northwest Forest Plan. Interagency publication of the Regional Ecosystem Office, Portland, OR. U.S. GPO [2000-589-124/04022 Region No. 10] available at http://www.or.blm.gov/surveyandmanage/

16. Csuti, B., A. J. Kimerling, T. A. O'Neil, M. M. Shaughnessy, E. P. Gaines, and M. M. P. Huso. 1997. Atlas of Oregon wildlife: distribution, habitat, and natural history. Oregon State University Press, Corvallis, OR..

17. DellaSalla, D. A., S. B. Reid, T. J. Frest, J. R. Strittholt, and D. M. Olson. 1999. A global perspective on the biodiversity of the Klamath-Siskiyou ecoregion. Natural Areas Journal 19: 300-319.

18. Garman, L. R., and M. G. Hunter. 1998. Phase 4-Evaluation: Birds and mammals. Pp. 59-61 in: J. H. Cissel, F. J. Swanson, G. E. Grant, D. H. Olson, S. V. Gregory, S. L. Garman, L. R. Ashkenas, M. G. Hunter, J. N. Kertis, J. H. Mayo, M. D. McSwain, S. G. Swetland, K. A. Swindle, and D. O. Wallin. A landscape plan based on historical fire regimes for a managed forest ecosystem: the Augusta Creek study. U.S. Forest Service General Technical Report PNW-GTR-422, Portland, OR.

19. Gilbert, F. F., and R. Allwine. 1991. Spring bird communities in the Oregon Cascade Range. Pages 145-158 in: L. F. Ruggiero, K. B. Aubry, A. B. Carey, and M. H. Huff, technical coordinators. Wildlife and vegetation of unmanaged Douglas-fir forests. U.S..Forest Service General Technical Report PNW-GTR-285, Portland, OR.

20. Hagar, J. C., W. C. McComb, and W. H. Emmingham. 1996. Bird communities in commercially thinned and unthinned Douglas-fir stands of western Oregon. Wildlife Society Bulletin 24(2): 353-366.

21. Huff, M. H., and C. M. Raley. 1991. Regional patterns of diurnal breeding bird communities in Oregon and Washington. Pages 177-205 in: L. F. Ruggiero, K. B. Aubry, A. B. Carey, and M. H. Huff, technical coordinators. Wildlife and vegetation of unmanaged Douglas-fir forests. U.S. Forest Service General Technical Report PNW-GTR-285, Portland, OR.

22. Johnson, R. E., and K. M. Cassidy. 1997. Terrestrial mammals of Washington state: location data and predicted distributions. Volume 3 in: K. M. Cassidy, C. E. Grue, M. R. Smith, and K. M. Dvornich, editors. Washington State Gap Analysis Project-final report, Washington cooperative fish and wildlife research unit, University of Washington, Seattle, WA.

23. Landres, P. B., P. Morgan, and F. J. Swanson. 1999. Overview of the use of natural variability concepts in managing ecological systems. Ecological Applications 9(4): 1179-1188.

24. Lehmkuhl, J. F., L. F. Ruggiero, and P. A. Hall. 1991. Landscape-scale patterns of forest fragmentation and wildlife richness and abundance in the southern Washington Cascade Range. Pages 425-442 in: L. F. Ruggiero, K. B. Aubry, A. B. Carey, and M. H. Huff, technical coordinators. Wildlife and vegetation of unmanaged Douglas-fir forests. U.S. Forest Service General Technical Report PNW-GTR-285, Portland, OR.

25. Leonard, W. P., H. A. Brown, L. L. C. Jones, K. R. McAllister, and R. M. Storm. 1996. Amphibians of Washington and Oregon. The Trailside Series, Seattle Audubon Society, Seattle, WA.

26. McGarigal, K., and W. C. McComb. 1992. Streamside versus upslope breeding bird communities in the central Oregon Coast Range. Journal of Wildlife Management 56: 10-23.

27. McGarigal, K., and W. C. McComb. 1993. Research problem analysis on biodiversity conservation in western Oregon forests. Special Research Report, Bureau of Land Management, The Pacific Forest and Basin Rangeland Systems Cooperative Research and Technology Unit, Corvallis, OR.

28. Molina, R., T. O'Dell, S. Dunham, and D. Pilz. 1999. Biological diversity and ecosystem functions of forest soil fungi: management implications. Pages 45-58 in: R. T. Meurisse, W. G. Ypsilantis, and C. Seybold, technical editors. Proceedings: Pacific Northwest forest & rangeland soil organism symposium. U.S. Forest Service General Technical Report PNW-GTR-461, Portland, OR.

29. Morgan, P., G. H. Aplet, J. B. Haufler, H. C. Humphries, M. M. Moore, and W. D. Wilson. 1994. Historical range of variability: a useful tool for evaluating ecosystem change. Journal of Sustainable Forestry 2: 87-111.

30. Moritz, C. 1994. Defining "evolutionarily significant units" for conservation. Trends in Ecology and Evolution 9(10): 373-375.

31. Morrison, M. L. 1981. The structure of western warbler assemblages: analysis of foraging behavior and habitat selection in Oregon. Auk 98: 578-588.

32. Morrison, M. L., and E. C. Meslow. 1983. Avifauna associated with early growth vegetation on clearcuts in the Oregon Coast Ranges. U.S. Forest Service General Technical Report PNW-GTR-305, Portland, OR.

33. Morrison, P., and F. J. Swanson. 1990. Fire history and pattern in a Cascade Range landscape. U.S. Forest Service General Technical Report PNW-GTR-254, Portland, OR.

34. Nauman, R. S., and D. H. Olson. 2000. Survey and Manage salamander known sites. Chapter II in: D. H. Olson, editor. Survey protocols for amphibians under the Survey and Manage provision of the Northwest Forest Plan. Interagency publication of the Regional Ecosystem Office, Portland, OR. U.S. GPO [2000-589-124/04022 Region No. 10] available at http://www.or.blm.gov/surveyandmanage/

35. Niemiec, S. S., G. R. Ahrens, S. Willits, and D. E. Hibbs. 1995. Hardwoods of the Pacific Northwest. Forest Research Laboratory, Oregon State University, Research Contribution 8. Corvallis, OR.

36. Noss, R. F., M. A. O'Connell, and D. D. Murphy. 1997. The science of conservation planning: habitat conservation under the Endangered Species Act. Island Press, Washington, D.C.

37. Nussbaum, R. A., E. D. Brodie, Jr., and R. M. Storm. 1983. Amphibians and reptiles of the Pacific Northwest. University of Idaho Press, Moscow, ID.

38. Ollivier, L. M., and H. H. Welsh, Jr. 2000. Survey protocol for the Del Norte salamander (Plethodon elongatus). Chapter V pages 163-200 in: D. H. Olson, editor. Survey protocols for amphibians under the Survey and Manage provision of the Northwest Forest Plan. Interagency publication of the Regional Ecosystem Office, Portland, OR. U.S. GPO [2000-589-124/04022 Region No. 10] available at http://www.or.blm.gov/surveyandmanage/

39. Olson, D. H. and W. P. Leonard. 1997. Amphibian inventory and monitoring: a standardized approach for the Pacific Northwest. Chapter 1 in: D. H. Olson, W. P. Leonard, and R. B. Bury, editors. Sampling amphibians in lentic habitats: methods and approaches for the Pacific Northwest. Northwest Fauna 4:1-22.

40. Omernik, J. M. 1987. Ecoregions of the conterminous United States. Map Supplement (Scale 1:7,500,000). Annals of the Association of American Geographers 77: 118-125.

41. Oregon Biodiversity Project. 1998. Oregon's living landscape: strategies and opportunities to conserve biodiversity. A Defenders of Wildlife Publication, Oregon State University Press, Corvallis, OR.

42. Pater, D. E., S. A. Bryce, T. D. Thorson, J. Kagan, C. Chappell, J. M. Omernik, S. H. Azevedo, and A. J. Woods. 1998. Ecoregions of western Oregon and Washington. Interagency poster produced by U.S. Geological Survey, Reston, VA.

43. Ralph, C. J., P. W. C. Paton, and C. A. Taylor. 1991. Habitat association patterns of breeding birds and small mammals in Douglas-fir/hardwood stands in northwestern California and southwestern Oregon. Pages 379-393 *in:* L. F. Ruggiero, K. B. Aubry, A. B. Carey, and M. H. Huff, technical coordinators. Wildlife and vegetation of unmanaged Douglas-fir forests. U.S. Forest Service General Technical Report PNW-GTR-285, Portland, OR.

44. Ruggiero L. F., L. L. C. Jones, K. B. Aubry. 1991. Plant and animal habitat associations in Douglas-fir forests of the Pacific Northwest: an overview. Pages 447-462 *in:* L. F. Ruggiero, K. B. Aubry, A. B. Carey, and M. H. Huff, technical coordinators. Wildlife and vegetation of unmanaged Douglas-fir forests. U.S. Forest Service General Technical Report PNW-GTR-285, Portland, OR.

45. Sedell, J. R., G. H. Reeves, and K. M. Burnett. 1994. Development and evaluation of Aquatic Conservation Strategies. Journal of Forestry 92(4): 28-31.

46. Smith, M. R., P. W. Mattocks, Jr., and K. M. Cassidy. 1997. Breeding birds of Washington state. Volume 4 *in:* K. M. Cassidy, C. E. Grue, M. R. Smith, and K. M. Dvornich, editors. Washington State Gap Analysis Project-final report, Seattle Audubon Society Publications in Zoology No. 1, Seattle, WA.

47. Spies, T. A., G. H. Reeves, K. M. Burnett, W. C. McComb, K. N. Johnson, G. Grant, J. L. Ohmann, S. L. Garman, and P. Bettinger. In press. Assessing the ecological consequences of forest policies in a multi-ownership province in Oregon. *In:* J. Liu and W. W. Taylor, editors. Integrating landscape ecology into natural resources management. Cambridge University Press, NY.

48. Storm, R. M., and W. P. Leonard, coordinating editors. 1995. Reptiles of Washington and Oregon. The Trailside Series, Seattle Audubon Society, Seattle, WA.

49. Swanson, F. J., J. A. Jones, D. O. Wallin, and J. H. Cissel. 1994. Natural variability - implications for ecosystem management. Pp. 89-106 *in:* M. E. Jensen, and P. S. Bourgeron, technical editors. Eastside forest ecosystem health assessment - Volume II: ecosystem management: principles and applications. U. S. Forest Service General Technical Report PNW-GTR-318, Portland, OR.

50. Teensma, P. D. A. 1987. Fire history and fire regimes of the central western Cascades of Oregon. Ph.D. Dissertation, University of Oregon, Eugene, OR.

51. Thomas, J. W., M. G. Raphael, R. G. Anthony, E. D. Forsman, A. G. Gunderson, R. S. Holthausen, B. G. Marcot, G. H. Reeves, J. R. Sedell, and D. M. Solis. 1993. Viability assessments and management considerations for species associated with late-successional and old-growth forests of the Pacific Northwest. Report of the Scientific Analysis Team. U. S. Forest Service, Portland, OR.

52. Trappe, J. M., and D. L. Luoma. 1992. The ties that bind: fungi in ecosystems. Chapter 2 *in:* G. C. Carroll, and D. T. Wicklow, editors. The fungal community: its organization and role in the ecosystem, Second edition. Marcel Dekker, Inc., NY.

53. U.S. Department of Agriculture, Forest Service, U. S. Department of Commerce (NOAA), U.S. Department of Interior (Bureau of Land Management, National Park Service and U.S. Fish and Wildlife Service) and the Environmental Protection Agency. 1993. Forest ecosystem management: an ecological, economic, and social assessment. Report of the Forest Ecosystem Management Assessment Team [FEMAT]. U. S. GPO 1993-793-071. Available at: Regional Ecosystem Office, Portland, OR.

54. U.S. Department of Agriculture, Forest Service, and U.S. Department of Interior, Bureau of Land Management. 1994. Record of decision for amendments for Forest Service and Bureau of Land Management planning documents within the range of the northern spotted owl (Northwest Forest Plan). U.S. D.A. Forest Service and U.S. D.I. Bureau of Land Management, Portland, OR.

55. U.S. Department of Agriculture, Forest Service, and U.S. Department of Interior, Bureau of Land Management. 1999. Draft-Supplemental environmental impact statement: for amendment to the survey and manage, protection buffer, and other mitigating measures standards and guidelines. [Draft SEIS]. U.S. Forest Service and U.S. Bureau of Land Management, Portland, OR.

56. U.S. Department of Interior. 1992. Recovery plan for the northern spotted owl—final draft. U.S. Department of the Interior, 2 volumes. Portland, OR.

57. Vega, R. M. S. 1993. Bird communities in managed conifer stands in the Oregon Cascades: habitat associations and nest predation. M.S. thesis. Oregon State University, Corvallis, OR.

58. Weisberg, P. J. 1996. Blue River fire regime analysis and description. Unpublished report, available at the Blue River Ranger District Office, U.S. Forest Service, Blue River, OR.

59. Welsh, H. H., Jr. 1990. Relictual amphibians and old-growth forests. Conservation Biology 4: 309-319.

8

Wildlife of Eastside (Interior) Forests and Woodlands

Rex Sallabanks, Bruce G. Marcot, Robert A. Riggs,
Carolyn A. Mehl, & Edward B. Arnett

Introduction

Eastside (interior) forests and woodlands are those forest and woodland environments that occur in Oregon and Washington east of the crest of the Cascade Mountain Range. Eastside (interior) forests and woodlands are broadly classified into 4 forest habitats based on conifer composition: (1) Eastside (Interior) Mixed Conifer Forest (EMCF), (2) Lodgepole Pine (*Pinus contorta*) Forest and Woodlands (LPFW), (3) Ponderosa Pine (*Pinus ponderosa*) Forest and Woodlands (PPFW), and (4) Upland Aspen (*Populus tremuloides*) Forest (UAF). We specifically do not address forested riparian habitats (see Kauffman et al.[87]). In this chapter, we first highlight the principal characteristics of Eastside (interior) forests and woodlands, describe key habitat features that make them suitable for wildlife, provide an overview of the wildlife assemblages that inhabit them, and discuss aspects of biological diversity and wildlife residency. With an emphasis on ponderosa pine, we then focus on management issues relevant to Eastside (interior) forests and woodlands, consider historical disturbance regimes, threats, and trends, and then relate these issues to wildlife. Finally, we provide several case histories that illustrate specific management issues and/or wildlife-habitat relationships and provide some suggestions for future research and monitoring.

A broad discussion of all Eastside (interior) forests and woodlands and their associated wildlife throughout Oregon and Washington is beyond the scope of both this chapter and our collective working knowledge. Throughout this chapter, therefore, our comments are often based on our own research experiences and reflect certain biases. Most notably, these biases are nongame landbirds and the Blue Mountains of northeastern Oregon. Where appropriate, however, we focus on other ecologically significant habitats (e.g., ponderosa pine) or wildlife taxa (e.g., large herbivores). For more specific information on UAF, see DeByle and Winokur[48a] and DeByle et al.[48b] In summary, our intent is to describe key aspects of Eastside (interior) forests and woodlands, highlight significant management issues, and by way of specific examples, illustrate wildlife-habitat relationships and potential interactions with land-use practices.

Habitat Characterizations

Eastside (interior) forests and woodlands are characterized by highly variable climate, landscape topography, elevational gradients, and varying degrees of natural and anthropogenic disturbances. Mountain ranges, rolling hills, and deep canyons characterize much of eastern Oregon and Washington where Eastside (interior) forests and woodlands dominate the landscape. Elevation ranges from 100 feet (30 m) above sea level (asl) (e.g., PPFW in the Columbia River Gorge) to 9,500 feet (2,896 m) asl (e.g., UAF).[38] As a result of this broad elevational gradient, and the extensive geographic distribution of Eastside (interior) forests and woodlands, physical site conditions are highly variable. In general, however, mature vegetation conditions of Eastside (interior) forests and woodlands are controlled over broad regions by climate, with soils and topography playing secondary roles; that is, most mature vegetation is zonal in nature. Variation in disturbance regimes also has influenced species composition and density in forests of the inland West.[42]

Historically, the presence and structure of forests dominated by Douglas-fir (*Pseudotsuga menziesii*) and grand fir (*Abies grandis*) throughout much of the inland West were influenced by frequent, low-intensity ground fires that reduced densities of trees and surface vegetation. For low elevation cover types (primarily PPFW), this disturbance regime produced open, park-like stands of all ages but predominantly large ponderosa pine, with a grass-dominated understory. These forests were regularly interrupted by nonforest openings (e.g., sclerophyllous shrubs, shrubsteppe, grassy parks, or balds, and herbaceous- or broadleafed-dominated riparian areas)[57, 153] (Figure 1). As a result, low elevation forests of the inland West, including Eastside (interior) forests and woodlands, were historically fragmented, at least with respect to forest cover.[138]

Eastside (Interior) Mixed Conifer Forest currently is dominated by grand fir, Douglas-fir, western larch (*Larix occidentalis*), and ponderosa pine. These forests occur on the periphery of the maritime climatic region that supports cedar-hemlock forests.[41, 146] Mixed conifer forest is more diverse along eastern slopes of the Cascade Mountains than elsewhere in the interior Columbia Basin. For example, in the very southern Cascades and northern Sierras and Klamath mountain convergence areas within Oregon, white fir (*A. concolor*), incense cedar (*Calocedrus decurrens*), Jeffrey pine (*Pinus jeffreyi*), and sugar pine (*P. lambertiana*) also can be found in association with Douglas-fir and ponderosa pine.[42, 118] However, these species do not typically occur in most EMCF. Extensive dry forests of PPFW occur at elevations below the mixed conifer forest zone. Lodgepole Pine Forest and Woodlands occur throughout the mixed conifer forest zone, primarily in cold, relatively dry environments and usually where winter snowpack is persistent. Pumice soil lodgepole pine habitat (see case history 4) is typically intermixed with PPFW and lies in between EMCF and either western juniper (*Juniperus occidentalis*) woodland or shrub-steppe habitat.[38] Finally, UAF, found from 2,000 to 9,500 feet (610 to 2,896 m) can occur as seral stands in both EMCF and PPFW or as climax forest on well-drained mountain slopes or canyon walls that have some moisture.

Some wildlife species are more or less closely associated with each of these forest types in Washington and Oregon. Examples of species largely confined to each of these types (i.e., that also occur within 3 or fewer other habitat types, according to the Wildlife-habitat Relationship [WHR] matrixes [see CD-ROM]) include (in increasing order of number of other habitat types also used): Cordilleran flycatcher, pygmy shrew, red-tailed chipmunk, boreal owl, and gray catbird in EMCF; pygmy nuthatch, pygmy shrew, pinyon jay, and gray flycatcher in PPFW; boreal owl and gray flycatcher in LPFW; and yellow warbler, American redstart, and boreal owl in UAF. None of these species is confined to only one habitat type.

Other wildlife species found within the Eastside (interior) forest and woodland array have evolved associations with specific substrates or environmental conditions. Examples include the northern three-toed woodpecker, which invades recently burned-over, high-

elevation lodgepole pine forests for feeding and nesting. This woodpecker selects for smaller-diameter snags over larger-diameter ones,[130] but most other Eastside cavity nesters select for the largest available snag (e.g., pygmy nuthatch and white-headed woodpecker). The pinyon jay is well known to cache viable seeds of pinyon and other pine species, thereby acting as a pine dispersal agent.

Distributional Aspects of Eastside Environments

Within Oregon and Washington, EMCF occurs primarily throughout the east Cascades, Blue Mountains (northeast Oregon), and Okanogan Highlands (northeast Washington). Lodgepole Pine Forest and Woodlands have a similar geographic distribution to EMCF;[38] in particular, this forest type is common in the pumice zone of southcentral Oregon from near Mt. Jefferson south to the Crater Lake area (also see case history 4).

Ponderosa Pine Forest and Woodlands are also widespread throughout foothills of the Blue Mountains, east Cascades, Okanogan Highlands, and central Oregon pumice zone, as well as in the Columbia Basin in northeastern Washington. Upland Aspen Forest is primarily found in the Steens Mountains (southeast Oregon) and in the northeastern Cascades of Washington. Mixed conifer, lodgepole pine, and ponderosa pine habitats are widespread and relatively common, often occurring in large contiguous tracts throughout the Blue Mountains, eastern slopes of the Cascades, and Okanogan Highlands. In contrast, UAF is relatively uncommon and highly patchy in its distribution across the Eastside (interior) forest landscape. At least currently, where it occurs as a seral stage within other forest types, aspen stands are small and isolated from one another (Figure 2). Whereas many wildlife species that inhabit Eastside (interior) forests and woodlands might be considered "associates" of specific forest communities (see above),

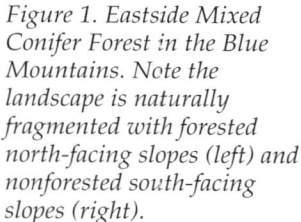

Figure 1. Eastside Mixed Conifer Forest in the Blue Mountains. Note the landscape is naturally fragmented with forested north-facing slopes (left) and nonforested south-facing slopes (right).

Figure 2. Patches of Upland Aspen Forest, often intermixed with other Eastside forest and woodland habitats, are currently small and isolated.

they are not often considered "obligates." Thus, we do not generally consider the distributional aspects of any of the Eastside (interior) forest and woodland habitats to be limiting for wildlife. Possible exceptions to this might exist for species that inhabit upland aspen and old-growth ponderosa pine, because of the patchy distribution of these habitats across the landscape. Species associated with aspen habitat include the red-naped sapsucker[46] and several species of bats, including the big brown bat.[44, 86] Species known to be associated with old-growth ponderosa pine include the white-headed woodpecker,[60] white-breasted nuthatch,[80a] and flammulated owl.[74]

Associations Between Habitat Elements and Wildlife

Here we emphasize wildlife associations with 4 habitat features (elements) that seem to be especially significant in Eastside (interior) forests and woodlands. The Habitat Element matrixes (see CD-ROM) can be used to list habitat elements, including substrates and microhabitat and conditions, that are used by species associated with each of the 4 Eastside (interior) forests and woodlands.

Snags. Dead standing trees or snags are widely recognized to be an essential forest component for numerous wildlife species.[130] In the 4 Eastside (interior) forest and woodland types, 77 species of vertebrates associate with snag substrates during some part of the annual cycle, including 2 amphibians, 51 birds, and 24 mammals, which use specific snag characteristics for diverse purposes. Primary cavity excavators include 15 species (2 chickadees, 3 nuthatches, 9 woodpeckers, and black bear). A few species use bark piles at the base of snags, including the northwestern salamander, which also uses down logs, and the Larch Mountain salamander, which uses bark piles particularly during dry seasons and on sites with moist talus and low soil content.[105] Partially decomposed standing wood provides nest, feeding, and roost sites for many other species; decomposing down wood is also used by various other species.

Most species of bats occurring in Eastside (interior) forests and woodlands use snags and large trees with structural defects (e.g., broken tops) for roosting.[17, 35, 104] Bats generally select tall, large-diameter snags for roosting.[18] However, the influence of forest structure and landscape pattern surrounding roost sites is not well understood for most species in this region. Snags and trees selected for roosting by bats occur in more open habitats (e.g., less canopy closure and understory vegetation) or near edges, and are close to water.[17, 18, 35] Anecdotal information on local declines of snag-associated bats may be related to reduction in the number of large snags.[105] Species experiencing habitat losses over the historical period were closely associated with ponderosa pine habitats (especially single-story old-forest stands).

Down Wood. Sixty-eight species of wildlife found in Eastside (interior) forests and woodlands have correlations with down wood, including 6 amphibians, 5 reptiles, 13 birds, and 44 mammals. A few amphibian species, such as the Larch Mountain salamander, may lay eggs in moist habitats but are associated with down wood or talus as adults, rather than aquatic environments. The distribution and abundance of many amphibian species are more closely associated with specific substrates (such as down wood) and microhabitat conditions (such as deep, moist talus), than with general vegetation cover types and structural stages.[105] Collectively, these amphibians are most sensitive to changes in down wood, litter, and duff depths and characteristics, still and flowing water quality and quantity, and precipitation quality and weather patterns.

Bull et al.[32] provide details of use of down wood by numerous other wildlife species in the interior Columbia River Basin. Some species, such as the pileated woodpecker in the Blue Mountains of northeastern Oregon, frequently forage on down wood for carpenter ants[30] (*Camponotus* spp.) (Figure 3).

Closed-canopy Forest. Some 233 wildlife species found in Eastside (interior) forests and woodlands are generally associated with closed-canopy forest structures, including single- and multi-story structures of shrub/seedling, sapling/pole, small-tree, medium-tree, and large-tree

Figure 3. Down wood is an important habitat element for many wildlife species in Eastside forests and woodlands. The cavities seen in this photograph were likely excavated by the pileated woodpecker in search of carpenter ants.

forest stages. No wildlife species within this set has an obligate association with closed-canopy conditions of shrub/seedling, sapling/pole, small-tree, medium-tree or large-tree forest structures. Such associates use 10 other structural conditions (see CD-ROM matrixes). Examples include the least flycatcher, willow flycatcher, dusky-footed woodrat, fox sparrow, green-tailed towhee, and yellow-pine chipmunk. Most of these species occur more frequently in sapling/pole and small-tree stages. Species frequenting medium- and large-tree stages include only the least flycatcher and dusky-footed woodrat. The western red-backed vole and western wood-pewee use closed-canopy conditions of medium- or large-tree structures and also a dozen other structural conditions. In EMCF of northeastern Oregon, Townsend's warbler is highly associated with closed-canopy conditions, but not exclusively so (see case history 1). All of the species listed here constitute the closest set of closed-canopy associates in medium- and large-tree stages of Eastside (interior) forests and woodlands.

Grass/forb and Shrub/seedling. In contrast to closed-canopy forest, many wildlife species are closely associated with the closed-canopy conditions of grass/forb and shrub/seedling stages of Eastside (interior) forests and woodlands. Species nearly tied (that is, that use only 1 other structure) to such closed grass/forb structures include the killdeer, meadow vole, American pika, California ground squirrel, northern bog lemming, and Brazilian free-tailed bat. Species closely tied (that is, that use only 3 other structures; none uses less) to closed shrub/sapling structures include the rough-legged hawk, barn swallow, and Brewer's sparrow. A lack of shrub cover in EMCF of Oregon may actually influence the abundance of some shrub-nesting songbirds, many (83%) of which are Neotropical migrants[34] (R. Sallabanks, 1995 Annual Report, Sustainable Ecosystems Institute, Meridian, ID).

One can speculate on why more wildlife species seem to be more closely associated with closed structures of grass/forb and shrub/seedling structures than with older forest structures; evolutionary conditions and adaptive advantages of these associations are largely unstudied. Most Eastside (interior) forests did not commonly occur historically as closed-canopy structures (with high volumes of dead wood), which developed only in recent decades with the advent of fire suppression management (see below), whereas early successional grass/forb and shrub/seedling stages did occur naturally following intermittent natural fires.

Wildlife Assemblages

In this section we provide an overview of all vertebrate wildlife taxa that inhabit Eastside (interior) forests and woodlands and discuss species residency (i.e., migratory status). This section is intended to be general, although in some cases little information is available on distribution, abundance, and habitat relationships for wildlife species, like reptiles. Some species, such as birds, are relatively well studied.

Overview of Wildlife

According to the WHR data matrices, there are 287 vertebrate wildlife species that currently inhabit Eastside (interior) forests and woodlands (8 salamanders, 10 frogs/toads, 2 turtles, 6 lizards, 13 snakes, 98 mammals, and 150 birds). This species richness is roughly equivalent to Eastside shrubland and grassland habitats (285 species), coastal and marine environments (278 species), and urban environments (265 species), subtantially greater than alpine and subalpine habitats (173 species), and less than freshwater, riparian, and wetland habitats (429 species), agriculture and pastures (344), and westside and high montane forest habitats (320). Three species occurring historically in Eastside (interior) forests and woodlands are now locally extirpated: northern leopard frog, bison, and sharp-tailed grouse (recently reintroduced). Much of the following information has been summarized from previous publications.[105, 159]

Amphibians. Amphibians of Eastside (interior) forests and woodlands are largely unstudied. For many species, even those listed as sensitive, such as the tailed frog in Oregon,[28] data on populations, distribution, and habitat requirements are scarce to nonexistent. In general, the 4 Eastside (interior) forest and woodland habitats were considered to play a supportive (but not essential) role in species' maintenance and viability (i.e., most amphibian species were classified as being generally "associated" with each habitat) (Figure 4a).

Within Eastside (interior) forests and woodlands, standing or flowing water is required for egg laying and larval development for most amphibian species. Frogs are predominantly associated with forested habitats; riparian areas provide essential habitat for most adult forms. At lower elevations, in dryer forest types characterized by ponderosa pine and Douglas-fir, several amphibian species occur, most of which are locally endemic and more common west of the Cascade Range. Ponderosa Pine Forest and Woodlands and EMCF provide habitat for 13 and 12 amphibian species, respectively (Table 1). Within EMCF, the moister cedar-hemlock and grand fir habitat types support the richest amphibian communities because of the damp climate and greater abundance of aquatic habitats.

At higher elevations, in colder forest habitat types such as Engelmann spruce (*Picea engelmannii*), lodgepole pine, and subalpine fir (*A. lasiocarpa*), temperatures are generally too cold and the breeding season too short to support diverse amphibian communities. Lodgepole Pine Forest and Woodlands and UAF support only 9 and 4 amphibian species, respectively (Table 1).

In some cases, exotic fish species introduced into high elevation lakes have impacted indigenous amphibian species, such as the Cascades frog. Elsewhere, exotic amphibian species (e.g., the bullfrog) may be directly responsible for the decline or elimination of native amphibian species (e.g., the northern leopard frog[99]). Other species, such as the southern race of the Columbia spotted frog and the western toad have declined or become locally extirpated for unknown reasons.[105] Little is documented about the effects of livestock grazing on amphibian

populations. Whereas some species seem to be negatively impacted by excessive grazing (e.g., Cascade frog and Columbia spotted frog), others seem tolerant (e.g., Pacific chorus frog). Other land-use practices also may affect amphibians. Irrigation canals or field flooding can provide adequate habitat for egg laying and larval development, but if water is shut off prior to hatching or metamorphosis, species will fail to reproduce.

Reptiles. Distribution and abundance of reptile species have been poorly studied. Habitat selection for snakes and lizards is driven more by the need for warm climates, rocks, talus, and soils than by the presence of general vegetation types. Most reptile species occur in the drier PPFW (21 species) at lower elevations (Table 1) than in the other 3 Eastside (interior) forest and woodland habitats. Moister forest habitats at intermediate elevations also have many reptile species, but to a lesser extent compared to PPFW. At higher elevations, in cold forest habitats, reptiles are relatively rare.

Like amphibians, most reptile species are considered generally associated with Eastside (interior) forest and woodland habitats; none are considered "closely associated" (dependent on a habitat or structural condition for part or all of its life history requirements) and few are considered "present" (known to occasionally use a habitat or structural condition) (Figure 4b).

Birds. Eastside (interior) forests and woodlands provide habitat for many bird species, most of which are known to occur in PPFW and EMCF (Table 1). Riparian vegetation within Eastside (interior) forests and woodlands is of particular importance to many species, especially Neotropical migratory songbirds.[136] The distribution and abundance of many birds seems to be more heavily influenced by the presence of key structural attributes (habitat elements) within forested stands rather than habitat types or seral stages per se (see case histories 1 and 4). In general, the 4 Eastside (interior) forest and woodland habitats were considered to play a supportive

Table 1. Numbers of vertebrate wildlife species known to occur in Eastside forest and woodland habitats.*

Taxonomic class	Mixed conifer	Lodgepole pine	Ponderosa pine	Upland aspen
Amphibians	12	9	13	4
Reptiles	11	12	21	5
Birds	116	83	131	77
Mammals				
Small mammals	43	26	31	24
Bats	11	9	15	5
Carnivores	18	13	14	10
Ungulates	9	8	7	5
All mammals	81	56	67	44
All species	220	160	232	130

* Information derived from a query of Wildlife-Habitat Relationships Project data matrixes.
Species with historical or unknown occurrences are excluded.

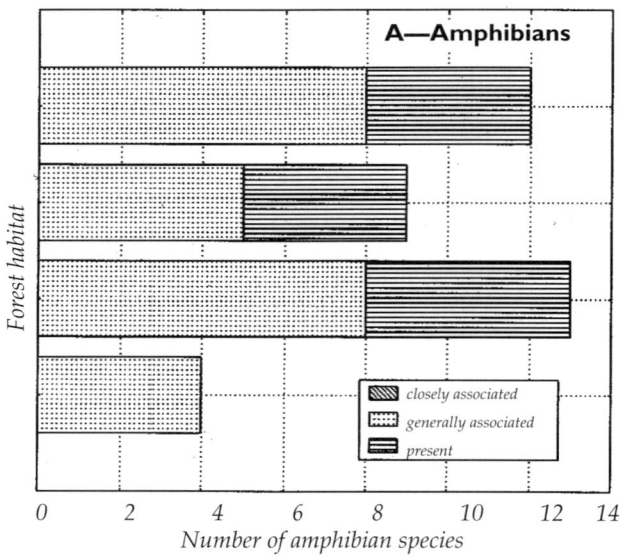

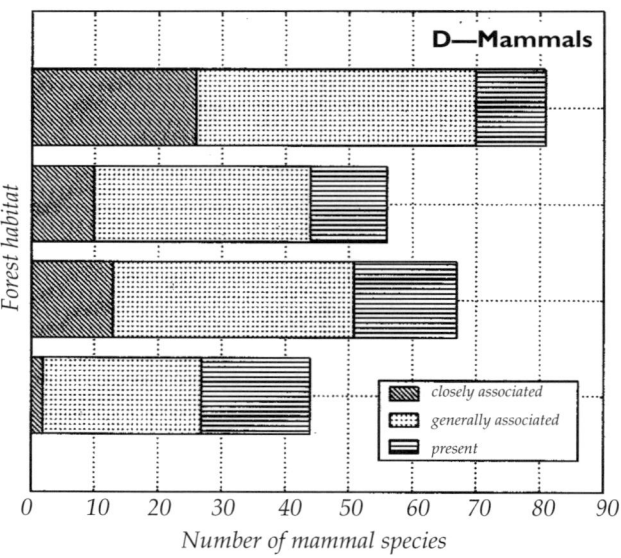

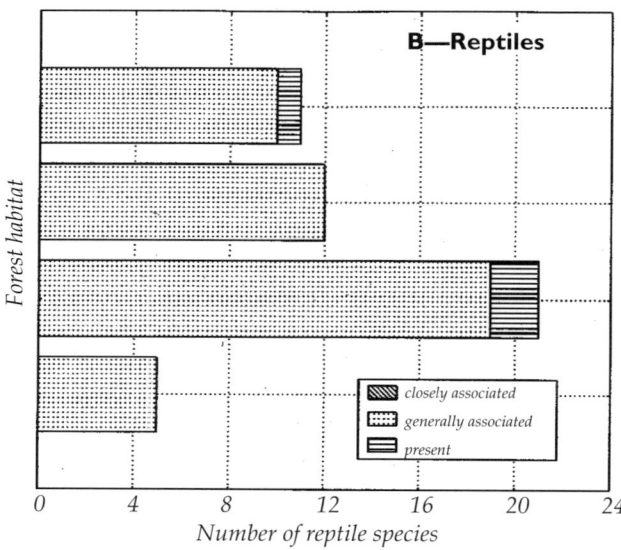

Figure 4. Degree of association of amphibian (A), reptile (B), bird (C), and mammal (D) species among the 4 Eastside forest and woodland habitats (top: Mixed Conifer; next: Lodgepole Pine; next: Ponderosa Pine; bottom: Upland Aspen). Closely associated—a species is widely known to depend on a habitat or structural condition for part or all of its life history requirements; generally associated—a species exhibits a high degree of adaptability and may be supported by a number of habitats or structural conditions; present—a species demonstrates occasional use of a habitat or structural condition.

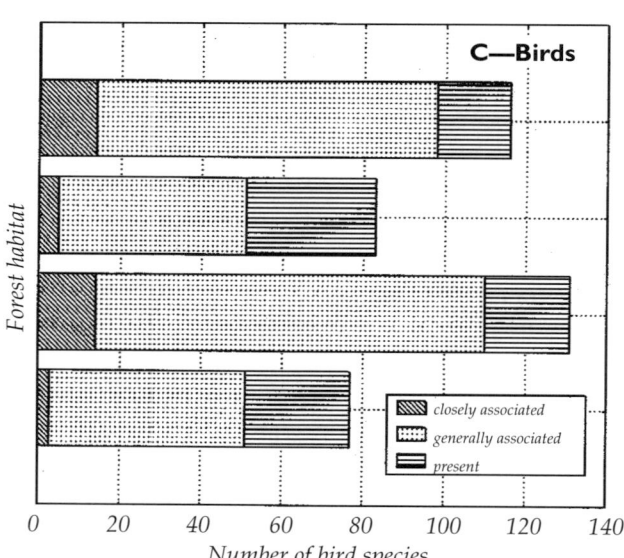

(but not essential) role in species' maintenance and viability (Figure 4c).

Most bird species are wide-ranging and use several, if not all, of the 4 Eastside (interior) forest and woodland habitats. Relatively few species were classified as being closely associated with any of the forest habitats, suggesting that few species depend on any 1 specific forest habitat for maintenance and viability (Figure 4c). Examples of such "closely associated" species include cordilleran flycatcher and EMCF, boreal owl and LPFW, northern saw-whet owl and PPFW, and red-naped sapsucker and UAF.

Most birds using Eastside (interior) forest and woodlands are "foliage-gleaners" (Figure 5a). Such species (e.g., warblers and chickadees) forage primarily by gleaning insects or fruit from vegetation (foliage and occasionally branches), not from the ground. Many foliage-gleaning bird species occur in PPFW and EMCF. Other foraging guilds (swooping divers, ground-gleaners, aerial insectivores, and bark-gleaners) are relatively evenly distributed among Eastside (interior) forest and woodland habitats in terms of species numbers (Figure 5a). Nesting guilds are dominated primarily by open-cup (e.g., thrushes and warblers) and cavity (e.g., woodpeckers and owls) nesters; platform (e.g., accipiters) and ground (e.g., grouse and quail) nesters are less common (Figure 5b). Many open-cup nesting bird species occur in PPFW and EMCF. Cavity nesters seem to be distributed similarly among the 4 Eastside (interior) forest and woodland habitats (Figure 5b).

Small Mammals. Few data exist relative to distribution and abundance of small mammal species in Eastside (interior) forests and woodlands. Several are exotic to the region, including Virginia opossum, which was recorded in Umatilla County, northeast Oregon, as recently as 1994.[161] There are more small mammal species in EMCF than the other 3 habitats (Table 1). Most use a wide variety of forest types, however, and like many species in this assessment, use key habitat elements such as burrows, litter, down wood, rock outcrops, and forest openings in all seral stages. Many squirrels, mice, woodrats, and other species depend on seeds from trees, especially large ponderosa pine seeds. Some species, especially ground squirrels, have benefited from agricultural intensification of dryer, low-elevation forest communities. Others, such as the northern pocket gopher and common porcupine have benefited from forestry practices.[105]

More mammal species exhibit close associations with Eastside (interior) forest and woodland habitats (especially EMCF) than any other taxonomic group considered in this assessment (Figure 4d). Examples include bushy-tailed woodrat and EMCF, snowshoe hare and LPFW, western pocket gopher and PPFW, and least chipmunk and UAF. Many species are considered generally associated with Eastside (interior) forests and woodlands; relatively few are considered present (Figure 4d).

Bats. Of all vertebrate wildlife groups, we perhaps know least about bats. Bats occupy a unique and important niche in Eastside (interior) forests and woodlands, as they are the only mammal with powered flight capability and the only nocturnal predator of many forest insects. Bats spend more than half their lives subjected to the selective pressure of roosting, and availability and quality of roost sites are thought to be critical factors influencing population size and distribution of some species.[95] Proximity to drinking and foraging sites also influences bat use of habitat and may limit populations.[35, 39, 45, 95, 165] In general, forest-dwelling bats use large-diameter snags and trees as roosts, and frequently switch roosts, which may be related to the type and permanency of roost occupied, and inversely related to the availability of roosts.[100] High levels of bat activity in mature and old-growth forests[123, 151, 152] are thought to be partially a consequence of availability of older trees for roosting in these stands.[35, 39, 95, 152] However, western long-eared myotis are flexible in selection of roost sites, and their patterns of use of roost structures may differ with landscape characteristics[165] (e.g., habitat pattern, snag availability). Bats also use various human-made structures such as bridges, buildings, and mines,[95] thus human development may favor some species of bats by providing roost sites, hibernacula, and nurseries.[52, 104]

Information on basic ecology, abundance, distribution, and habitat selection of bats in Eastside (interior) forests and woodlands is generally rare.[17, 35, 122] Bats reach their richest communities in PPFW where 15 species are known

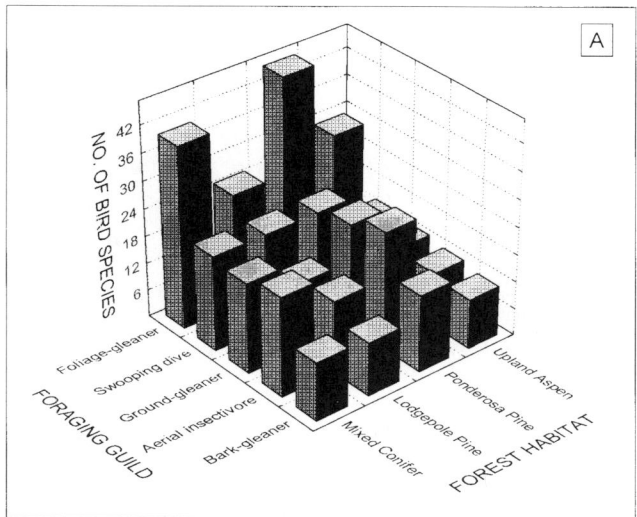

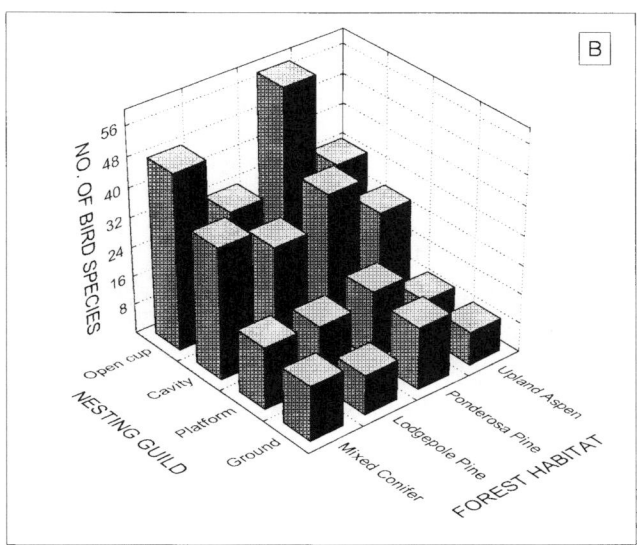

Figure 5. Distribution of bird species, summarized by foraging guild (A) and nesting guild (B), among the 4 Eastside forest and woodland habitats.

to occur (Table 1). Important management issues relative to bats include timber harvest, particularly removal of snags, and disturbance at mine and cave sites. Campbell et al.[35] suggested that retention and recruitment of snags in managed forests are important for the conservation of bats in this region. Selective harvest techniques that retain trees and snags distributed across the landscape could provide habitat to maintain populations of bats.[35, 163] Restricting human entry to caves and mines where bats are known to occur may reduce potentially adverse effects from recreation.

Carnivores. Eastside (Interior) Mixed Conifer Forest provides habitat for 18 species of carnivores; among the 3 other Eastside (interior) forest and woodland habitats, carnivores are relatively evenly distributed (Table 1). Many of the carnivores that inhabit Eastside (interior) forests and woodlands are habitat generalists, dependent more on availability of prey and lack of disturbance from humans (e.g., black bear). Others, such as the American marten and Pacific fisher, are closely associated with late-successional conifer forests and riparian habitats or with interior forest environments;[61] these 2 mustelids also use down logs for den sites.[105] However, marten also use intensively managed lodgepole pine forests in the central Oregon Pumice Zone[126] (see case history 4).

The distribution and abundance of some carnivores have changed dramatically from historical conditions.[117] Species such as the grizzly bear, gray wolf, and mountain lion were (and still are) considered a threat to human safety and livestock production. As a result, they were pursued by trappers from the mid-1800s through the mid-1900s. Grizzly bear and gray wolf populations were extirpated from Eastside (interior) forests and woodlands within Oregon and significantly reduced in Washington. Although likely reduced during the trapping era, mountain lion populations have increased toward the end of the century subsequent to changes in harvest regulations. Other carnivorous species such as coyotes, foxes, and skunks have apparently increased because of reduced competition and removal of their major predators[105] as well as the encroachment of agricultural lands at lower elevations.

Ungulates. Ungulates are significant components of Eastside (interior) forests because of their profound influence on vegetation (see case history 3). Ungulate species are similarly distributed among EMCF, LPFW, and PPFW. Upland Aspen Forest has the fewest species of ungulates (Table 1). Vegetation communities at mid-elevations, especially open plant communities in ponderosa pine and Douglas-fir zones, and mixed conifer communities at higher elevations, support the largest numbers of herbivores.[82] These forest zones dominate the Blue Mountains landscape, particularly at mid-elevations where animals must gravitate to meet nutritional demands. Hence it is the strategic position of these types and their broad distribution that dictate their importance to ungulates. Highly productive habitats (e.g., UAF) can be extraordinarily important to large herbivores, despite their limited distribution because of their high productivity.

Deer and elk have expanded their ranges in recent times, and where densities have exceeded the capacity of remaining wildlands, they have become depredators of crops in some agricultural areas, particularly on private lands near the interface of valley and montane zones. In some forest settings, elk and deer use dense stands of shade-tolerant understory trees for cover, which would not have been as available under natural fire regimes. Within UAF, aspen and its associated understory plants are heavily used forage sources, sometimes throughout the year. Many aspen communities have become degraded as a result of persistent heavy foraging by seasonal concentrations of high ungulate numbers (both wild and domestic).

Like elk and deer, bighorn sheep are popular for hunting and viewing, and at higher elevations within EMCF, frequently use cliff and rock walls. Whereas some populations of bighorns seem to be maintaining current numbers, others are generally declining, perhaps due to widespread habitat changes[164] such as replacement of grass, forbs, and low shrubs with tall shrubs and trees, which bighorns generally avoid.[128] Fire suppression has contributed significantly to these changes. Bighorn sheep benefit from management activities that maintain early-seral vegetative conditions, such as prescribed fire[9, 78, 121a, 127a] and timber harvest.[10]

Aspects of Wildlife Residency

Among wildlife taxa considered in this assessment, only bird species migrate entirely out of the region seasonally. In fact, there are almost as many Neotropical migratory birds in Eastside (interior) forests and woodlands as there are year-round resident birds (Figure 6). Most Neotropical migrants spend the winter in Central and South America, returning to Oregon and Washington in April and May to breed. Fall migration to wintering grounds typically occurs in August and September. Numbers of migratory bird species vary among habitat communities in relation to overall numbers of bird species; no particular habitat seems to have disproportionately more or less (Figure 6). The status, trends, and general habitat relationships of Neotropical migratory birds in the interior Columbia River Basin, which includes Eastside (interior) forests and woodlands, have been reviewed elsewhere.[136]

In general, however, wildlife species that inhabit Eastside (interior) forests and woodlands are primarily year-round residents. Of the 287 species that currently inhabit these environments, 227 (79%) remain in Oregon

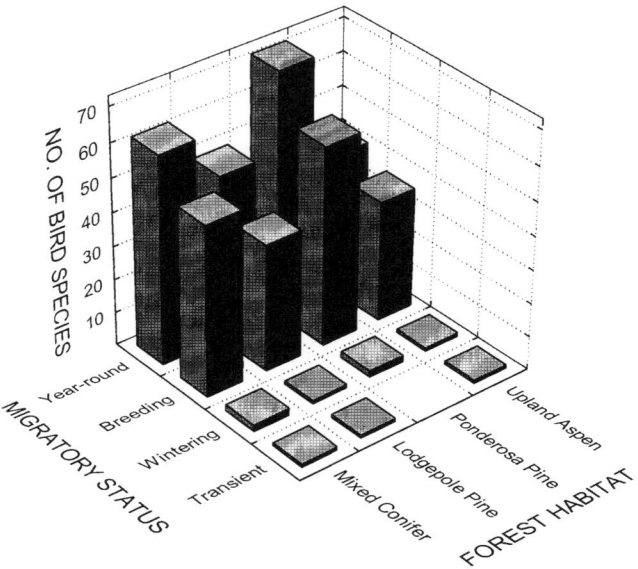

Figure 6. Distribution of bird species, summarized by migratory status, among the 4 Eastside forest and woodland habitats.

and Washington year-round (although not necessarily in Eastside [interior] forests and woodlands). There is some seasonal movement within eastern Oregon and Washington by several ungulate species, but there is no migration out of the region. Some bat species are thought to migrate southward during the winter months (e.g., hoary bat and spotted bat), but this has yet to be confirmed.

Many bird species that are considered permanent (i.e., nonmigratory) residents exhibit "migrations" between breeding and nonbreeding locations within Eastside (interior) forests and woodlands. Such species can be summarized using the classification system derived by Marcot.[103] Elevational migrants include (1) species that breed locally, but move to lower elevations in winter (e.g., Cooper's hawk); (2) species that breed at higher elevations, but move into local areas in winter (e.g., ruby-crowned kinglet and Lincoln's sparrow); and (3) species that breed at low elevations/bottomlands, but move to local areas post-breeding (e.g., American goldfinch, Brewer's blackbird, and brown-headed cowbird). Latitudinal migrants include "displacement" species that move latitudinally within the region (e.g., sharp-shinned hawk, band-tailed pigeon, and winter wren). Finally, there are several nomadic permanent residents that exhibit less predictable movements within Eastside (interior) forests and woodlands throughout the year (e.g., cedar waxwing and red crossbill).

Ecological Functions of Wildlife Species in Eastside (Interior) Forests and Woodlands

The CD-ROM matrixes were queried to determine the key ecological functions (KEFs) performed by wildlife associated with Eastside (interior) forests and woodlands. All 4 Eastside (interior) forest and woodland habitats contained the same categories of KEFs.[107] A high degree of functional similarity was found in patterns of species redundancy (richness) by KEF category, with a few interesting exceptions. There were fewer foliovores and browsers in PPFW, and EMCF (more in UAF and LPFW) than expected by overall patterns of species richness among the 4 forest types. There were fewer grazers, fungivores, and bark feeders in PPFW than expected.

Ponderosa Pine Forest and Woodlands were also depauperate in species that disperse fungi and lichens. This was also true for species that fragment down wood and impound water. This might be related to the greater aridity of PPFW than the other 3 forest habitats. However, overall trophic structures (percent of all species in primary, secondary, and tertiary consumer categories) were nearly identical among all 4 Eastside (interior) forest and woodland habitats. This seems to be an ecosystem constant.

Management Issues in Eastside (Interior) Forests and Woodlands

Wildfire is a driving force in shaping the structure and function of all Eastside (interior) forests and woodlands.[2, 3] In the last century, alteration of the natural fire regime that operates in drier, low- to mid-elevation Eastside (interior) forests has been a primary impact of modern land-management activities in the region. This low to mid-elevation range corresponds primarily with the historical occurrence of ponderosa pine forest[2] and woodlands in eastern Washington.[97] Fire exclusion policies, coupled with and accelerated by selective harvest of the seral tree component and livestock grazing, have allowed many PPFWs to progress farther along successional pathways than would have occurred under historical fire regimes.[3, 43, 97] At the same time, a return to a more natural disturbance regime would result in the loss of some old-forest characteristics (e.g., closed-canopy, dead wood) that have developed from fire exclusion.

Whereas management issues exist in many or all Eastside (interior) forests and woodlands (i.e., clearcutting, roads), the primary anthropogenic impacts of the last 100-150 years have resulted in extensive changes in distribution, structure, and species composition of ponderosa pine forests,[2] and to a lesser extent (because of limited historical range in eastern Washington) woodlands.[97] The corresponding impacts to associated wildlife are poorly understood and opportunities for future research are limited because of the extensive loss of these habitats in eastern Washington.

Efforts to restore and maintain biodiversity within targeted landscapes often concentrate on the most significantly altered environment relative to historical distribution and conditions.[114] For this reason, we confine most of our discussion of management impacts within Eastside (interior) forests and woodlands to PPFW.

Natural Disturbance Regimes

Historically, landscape vegetation patterns in the Pacific Northwest and in many Eastside (interior) forest and woodland habitat types were primarily influenced by fire at return intervals ranging from 7-20 years (ponderosa pine forests), 50-100 years (Douglas-fir and grand fir forests), and >100 years (subalpine fir forests).[2, 3] In dry, low- to mid-elevation PPFW, the frequency of fire kept fuel loading of the forest floor and understory to a minimum which, in turn, prevented fire from reaching the overstory and killing trees.[2, 97, 172] The result was an overall reduction in fire intensity, dependent on the fire return interval. One response of vegetation was to favor more fire-tolerant tree species (e.g., ponderosa pine and/or white oak [*Quercus garryana*] in the overstory and bunchgrasses in the understory) that were well adapted to a frequent low-severity fire regime.[5, 13, 172]

Whereas many of these fires were ignited naturally, Native Americans used Eastside (interior) forests and woodlands, especially PPFW, for food and building materials for thousands of years prior to Euro-American settlement[14] and their most significant influence on these forests was their propensity to intentionally start fires. Possible reasons for setting fires were to improve desired wildlife habitat, increase desired plant species, drive game animals into traps, and open transportation routes.[67] Historically then, these fires generally burned extensively

Figure 7. Fire exclusion policies throughout Eastside forests and woodlands during the past 100 years have created densely stocked stands of trees. Note the high fuel load on the forest floor, which in combination with the crowded overstory, would increase the likelihood of a stand-replacing wildfire in this forest.

throughout low- to mid-elevation Eastside (interior) forests and woodlands, perhaps only being extinguished by fall rains or lack of fuel due to previous fires. Under such a natural disturbance regime, the potential for high-severity wildfires,[2] insect outbreaks, and disease events were low.[168] Volume of dead wood was probably low as well.[127b, pages 308-9]

Fire Suppression. Along with logging and grazing, fire suppression has been responsible for significant changes in forest structure and species composition of Eastside (interior) forests and woodlands; this is especially true for PPFW.[2, 13] Such management action has lengthened the fire return interval and has significantly altered the species composition and structure, compared to forests existing under a low-severity fire regime.[3] Today, many PPFWs no longer could support a low-severity fire regime; many of these forests have not burned since the 1800s and are characterized by considerable fuel loading in the understory.[3, 172]

Fire exclusion policies during the past 100 years have interrupted the developmental pattern of clumped groups of trees by allowing regeneration to survive across an entire site and not just in openings.[5] As a result, extensive areas of dry, low- to mid-elevation Eastside (interior) forest landscapes, including EMCF and LPFW, have become increasingly homogeneous, with "dog-hair thickets" of trees in many areas (Figure 7).[3] Most, if not all, low- to mid-elevation potential vegetation types are believed to be experiencing tree densities outside the historical range of conditions for these sites.[3, 144, 172] Dense overstory and understory conditions provide vertical continuity of fuels that effectively creates a "ladder" for fire to spread from the understory to tree crowns. As a result, fires now kill more trees through loss of overstory in addition to understory, as well as being more severe in terms of temperature and site impacts. Consequently, Eastside (interior) forest landscapes are now characterized by infrequent fire return intervals and large expanses of high-severity, stand-replacing fires.[3] Research on historical fire return intervals throughout the Intermountain West has uncovered little to no evidence of infrequent fire return

intervals (say, >50 years) and high-severity fire regimes occurring within the dry, low-elevation ponderosa pine zone.[12, 54, 145, 170, 172] On similar ponderosa pine sites in central Idaho, Steele[144] found these high-severity fires to have dysfunctional effects on the ecosystem through reduced loss of site production, seed sources, nutrients, and even soil.

Timber Harvest. Effects of timber harvest on Eastside (interior) forests and woodlands have changed over the years.[3, 96] Early timber harvests usually targeted the largest trees, which primarily included ponderosa pine and western larch. More recently, species such as Douglas-fir, grand fir, lodgepole pine, and subalpine fir have also been used.[3] Historical harvest regimes, coupled with fire suppression, allowed smaller, more shade tolerant, late successional species to capture the growing space of ponderosa pine forests.[140] The result has been a rapid shift on many sites from forests dominated by seral species to those dominated by late successional species, and from relatively open stands of old-growth trees to relatively closed stands of young trees which have high fuel and large woody debris loads.[3, 172]

More recently, timber management programs have used more intensive harvest practices. Clearcut areas in low-elevation forests generally recover slowly from logging disturbance, and efforts to reforest clearcuts have been generally unsuccessful.[146] Today, selective harvest that favors retention of early seral species, while also thinning the stand and reducing fuels, is considered the most ecologically responsible harvest method in ponderosa pine systems, especially for those sites where under-burning is considered a viable management practice over the long term.[144]

Eastside white oak woodlands have relatively little economic value as a timber species and are more commonly harvested for firewood purposes.[97]

Grazing. Greater livestock densities at lower elevations have led to greater grazing impacts in PPFW relative to the other Eastside (interior) forest and woodland habitats.[3] Grazing began in the late 1800s when enormous herds of

Figure 8. Livestock grazing, in addition to fire suppression and timber harvest, is a management activity that has played a major role in changing historical fire regimes of Eastside forests and woodlands.

domestic sheep and cattle were allowed to graze freely throughout PPFW. By 1860, in addition to feral horses and sheep, there were 200,000 cattle in Oregon.[59] Steele et al.[146] observed and documented substantial damage to soils and vegetation on ponderosa pine sites in Idaho, especially where herds were concentrated. Perennial bunchgrasses, in particular, are still recovering from severe overgrazing of the early days.[3, 96, 146]

Grazing pressure at this level can also contribute to fire suppression by reducing continuity of understory vegetation and preventing low intensity fires from spreading in their normal pattern across the landscape.[43] Today, livestock grazing continues throughout all Eastside (interior) forests and woodlands, but at much lower densities (Figure 8). Localized damage to vegetation and soils may still occur where animals concentrate, particularly in riparian areas and forest openings.[64] On the other hand, grazing by wild herbivores increased dramatically during the twentieth century.[82]

Grazing continues to affect forest ecology in terms of succession, species composition, and species diversity.[127c] Specific influences on forest structure have been increased tree numbers, decreased native grasses, increased accumulation of downed woody material, increased spread of exotic and noxious weeds, and increased forest floor duff.[69, 174] However, changes in the densities of wild and domestic herbivores, relative to one another, suggest that the focus of future research on grazing should extend beyond merely the effects of livestock grazing.[127c]

Status, Threats, and Trends of Habitats and Communities

Collectively, fire suppression, grazing, and logging have contributed to the loss of most fire-maintained old-growth Eastside (interior) forests and woodlands. Ponderosa Pine Forest and Woodlands have been more heavily impacted by these factors relative to the other three Eastside (interior) forest and woodland habitats. Today, the two greatest immediate threats to future viability of ponderosa pine forests are high-severity fire occurrences and increased site-specific competition for nutrients and

moisture that results in reduced ponderosa pine regeneration and increased mortality over the long-term.[144] Little forest remains today that represents historical old-growth (>200 years[68]) ponderosa pine forest.[74, 114] Where stands still contain old-growth ponderosa pine, tree densities and fuel accumulations present a significant risk to long-term survival and future restoration of this forest type.[144] Lightening and accidental fire starts are expected to cause future burns of unprecedented and uncontrollable intensity and magnitude.[4]

Allowing these forests to burn under a high-severity fire regime to "reset the balance" is considered by some to be a nonviable alternative for restoration of these forests.[144] Such fires could potentially eliminate what little old-growth remains, as well as have detrimental effects on wildlife.[96] Replacing these old forests would take >200 years, thereby foregoing any immediate restoration options and flexibility for spatial distribution. To complicate matters, the occurrence of high-severity fires on these sites often damages the soil or promotes species better adapted to a "hot" fire regime, making these landscapes more prone to future high-severity fires.[3] The result can be delayed recolonization by all species where soil is damaged, or recolonization by shrubs that are able to shade out seral tree species and delay their establishment for many years to come (R. Steele, forestry consultant, pers. comm.). The end result is the same, however: the extended loss of the old-growth ponderosa pine ecosystem from the landscape.

Kay[88] concludes that, historically, Native Americans commonly determined the structure of entire plant and animal communities by hunting and by setting fires (also see papers cited in Knowles and Knowles [93]), and that a current "natural regulation" approach to management does not recognize, and thus probably would not replicate, such historical conditions.[106] Perhaps with careful thinning, along with prudent reintroduction of fire into these ecosystems, old-growth ponderosa pine forests can be extensively restored. Given the dynamic nature of ecological communities in Eastside (interior) forests and woodlands, particularly regarding potential effects of fire, perhaps the very concept of defining "desired future conditions" for planning could be replaced with a concept of describing "desired future dynamics." That is, long-term evolutionary potentials can be met only by accounting for potential future changes in conditions. No condition can be static for long. Impending changes in regional climates, too, have the capacity for causing great shifts in composition of ecological communities.

Relationships Among Management Issues, Wildlife, and Disturbance Regimes

Fire suppression has resulted in conifer encroachment on certain shrub habitats and has contributed to loss of habitat for many reptiles.[105] Little information is available on the effects of fire and fire suppression, season of burn, and changes in vegetation pattern resulting from burns and fire suppression on birds in Eastside (interior) forests and woodlands.[156a] Although general trends have been reported, species-specific effects on bird population density and trends from livestock grazing, changes in distribution and density of nest parasites (principally the brown-headed cowbird), changes in insect populations, and availability of free water are poorly understood. Livestock grazing is well known to indirectly influence some nongame bird populations via its effects on vegetation.[148, 149]

Resource management, including fire suppression and timber management, and resulting changes in the structure and distribution of vegetation communities, has also influenced the distribution and abundance of many avian species. For example, the amount of old-growth ponderosa pine forest that has been maintained by frequent, low-severity fires has declined by approximately 85% from historical conditions to present across the entire region; in some areas, such as the Klamath Plateau and eastern slopes of the Cascade Mountains, <5% remains today.[74] Species associated with this community, such as the white-headed woodpecker and flammulated owl, likely have declined in abundance and are considered sensitive by federal and state agencies. Lewis', black-backed, and three-toed woodpeckers, along with

mountain and western bluebirds, are associated with sites where fires have killed the forest overstory (also see case history 2). Black-backed and three-toed woodpeckers also associate with mid- to high-elevation forests following irruptions of insect populations such as bark beetles.[22]

Hutto[80b] documented differential responses by bird species to stand-replacement fires in the northern Rocky Mountains, and noted that one species, the black-backed woodpecker, was nearly restricted to sites with standing fire-killed trees. He also listed other bird species that may benefit over the long term from stand-replacement fires. (For more details on relationships between severity of wildfire and bird communities, see case history 2).

Other western studies, although not conducted in Oregon or Washington, provide data on the response of bird communities to snag harvest following fire (i.e., salvage logging). Raphael[125] explored the response of birds to reduced densities of snags, simulating various snag harvest levels immediately following fire using simulation models. He predicted that optimum snag densities under the constraints tested should be between 17-37 snags/acre (7-15 snags/ha). Also using models, Raphael and White[127] predicted drastic declines in cavity-nesting bird density following removal of snags from burned areas in the Sierra Nevada.

In ponderosa pine forests of southwestern Idaho, Saab and Dudley[135] found Lewis' woodpecker to be the most abundant and successful cavity nester in 2- to 4-year-old burns. At least during the first 4 years post-fire, nesting densities of cavity-nesting birds continued to increase. Lewis' woodpeckers seemed to be positively affected by salvage logging, experiencing highest nesting success in salvage-logged units. All bird species selected nest sites with higher tree densities than those measured at random sites, and cavity nesters as a group selected clumps of snags as opposed to snags retained in uniform, evenly spaced distributions. Retaining snags in clumps during salvage logging operations is therefore a management recommendation. Another is to maintain snags (especially broken-topped snags) in forests that are susceptible to

Figure 9. Salvage logging of fire-killed trees is a common management practice in Eastside forests and woodlands today. More studies are needed on the effects of this type of forestry on wildlife species.

stand-replacement wildfires to provide nest trees during post-fire years when most trees cannot yet be excavated.

Hutto[80b] stated that post-fire salvage cutting (Figure 9) might be conducted too frequently to be justified on the basis of sound ecosystem management because some bird species require burned forests for maintenance of viable populations. He pointed out that some bird species differ in the microhabitats that they occupy within a burn; salvage prescriptions that tend to "homogenize" forest structure (e.g., selective removal of all trees of a certain size) are therefore unlikely to maintain the necessary variety of microhabitats within a burned forest. Hutto[80b] suggested that rather than implementing a selective tree removal salvage operation, it may be better to take trees from one part of a burn and leave another part completely untouched, dependent on management objectives. Unfortunately, the paucity of empirical data on effects of snag harvest on wildlife populations following fire makes it difficult to suggest viable management alternatives.

Research Activities in Eastside (Interior) Forests and Woodlands

To provide practical examples of relationships between wildlife species and management issues in Eastside (interior) forests and woodlands, we present the following four case histories. These case histories offer details on specific wildlife species, habitat relationships, and effects of management activities. With one exception (case history 3), we emphasize birds and/or the Blue Mountains to illustrate wildlife use of Eastside (interior) forests and woodlands. Based on our review of literature and current state of knowledge of wildlife communities in Eastside (interior) forests and woodlands, we have identified several information needs. We conclude with suggestions for future research.

Case History I
Bird-habitat Relationships in Grand Fir Forests of the Blue Mountains, Oregon

Habitat selection by birds has been well studied, both empirically[40, 84, 169] and theoretically.[58, 131, 160] Since Thomas,[153] however, few additional data have been added to our knowledge of forestry-avifauna relationships in the interior Pacific Northwest. Prior to the study reported here, Mannan and Meslow[102] and Bull et al.[33] provided the only data on songbirds in the Blue Mountains. This case history summarizes a three-year study of bird populations in grand fir forests of the Blue Mountains. The objectives of the study were to (1) assess how forest stand attributes, such as canopy cover, snag density, and understory structure influence habitat selection by birds; (2) investigate bird use of forest structural classes in Eastside mixed conifer (grand fir) forests; and (3) relate habitat use by birds to forest management options and the potential for timber harvest practices to provide suitable breeding habitat for birds. We present data on avian community composition, relationships with specific structural attributes of forest stands, and on how birds were distributed among forest structural classes (i.e., developmental stages). Other potential drivers of habitat use by birds, such as landscape features and abiotic factors, were controlled for wherever possible. To enhance the relevance of this study to current management, we describe avian abundance with respect to a forest classification scheme formulated for the Interior Columbia Basin Ecosystem Management Project (ICBEMP).[124]

This study was conducted in the Marine Physiographic Zone[36] of the northern Blue Mountains Ecological Province[57] in northeastern Oregon. Study sites were located in forests managed by Boise Cascade Corporation ($n = 59$), and the U.S. Forest Service, Umatilla and Wallowa-Whitman National Forests (NF) ($n = 24$). All sample stands were in the *Abies grandis* series of forest habitat types.[85] Sample stands ranged from 40-124 acres (16-50 ha). By limiting the study to grand fir forests,

Figure 10. Stem exclusion, open canopy forest is a common structural class in Eastside Mixed Conifer Forests of the Blue Mountains. Our research has found this structural class, which is often management-induced (note stumps in the foreground), to have the richest avifauna of all grand fir forest types in the region.

variation among study plots with respect to abiotic factors such as climate, soil conditions, and potential evapotranspiration was minimized. Grand fir forests are part of EMCF.

Data were collected in 83 forest stands between 1994 and 1996. Data were collected annually in each stand and vegetation data were collected in the first year that a particular stand was sampled (either 1994 or 1995). The same 45 stands were sampled in all three years and the same 70 stands were sampled in both 1995 and 1996. All stands were classified into 1 of 7 structural classes according to structure[115]: (1) stand initiation (SI); (2) stem exclusion, open canopy (SEOC) (Figure 10); (3) stem exclusion, closed canopy (SECC); (4) understory reinitiation (UR); (5) young forest, multistory (YFMS); (6) old forest, single-story (OFSS); and (7) old forest, multistory (OFMS). This scheme has been adopted by ICBEMP, and it essentially represents an expansion of the 4 "stand development stages" first described by Oliver.[116] Bird communities were censused using standard point count methodology.[81] Vegetation was sampled using protocols modified from James and Shugart.[83] For more details see R. Sallabanks (1996 Annual Report, Sustainable Ecosystems Institute, Meridian, ID).

In total, 36,602 detections of 90 bird species were recorded. Rank order of abundance changed only slightly from year to year. Overall, the red-breasted nuthatch, Townsend's warbler, and dark-eyed junco dominated the pooled sample. Among the 30 species for which data were analyzed, 29 exhibited significant correlations between their abundance and one or more structural attributes, in at least 1 year of study. Only abundance of the pine siskin was apparently unrelated to the structural attributes measured. Adjusted $R2$ ranged from 0.07 (hermit thrush in 1995) to 0.63 (chipping sparrow in 1994) with an overall mean of 0.24. Four species (American robin, chipping sparrow, orange-crowned warbler, and Townsend's warbler) showed significant correlation with the same structural attribute(s) in all 3 years; habitat functions were subsequently developed for these species by pooling census data among years.

Abundance of 23 bird species (64%) varied significantly among forest structural classes. All species except ruffed grouse, brown creeper, Swainson's thrush, Townsend's warbler, house wren, and mountain bluebird were present in all structural classes. The house wren and mountain bluebird were found almost exclusively in SI, and Hammond's flycatcher was far more abundant in SECC than in other structures. The following species were most abundant in, but not necessarily restricted to, particular single structural classes: dusky flycatcher, orange-crowned warbler, chipping sparrow, northern flicker, house wren, dark-eyed junco, and mountain bluebird (SI); western tanager (SEOC); Hammond's flycatcher, yellow-rumped warbler, pine siskin, and golden-crowned kinglet (SECC); hermit thrush (UR); ruffed grouse, Swainson's thrush, and brown-headed cowbird (YFMS); and brown creeper (OFMS).

Several of the relationships observed are consistent with previous studies. The relative abundance of Townsend's warbler, for example, was consistently correlated with canopy cover across years, and canopy cover explained almost 50% of the variation in Townsend's warbler numbers among structural classes when data were pooled among years. Mannan and Meslow[102] also reported that Townsend's warblers nested in sites that had high canopy volumes of grand fir and Douglas-fir, and that canopy volume of grand fir was the primary attribute separating nest sites from random sites.

In general, however, research of this nature has not occurred in Eastside (interior) forest and woodlands and most of the habitat relationships we report have not been previously described. Only 4 species showed sufficiently consistent relationships with structural attributes to warrant development of habitat functions to predict relative abundance. Such functions may be used to predict the abundance of a species if we know the values of the different parameters in that species' habitat model. Thus, the potential for a particular forest stand to provide suitable habitat for a species can be evaluated for different management scenarios. Overstory removal, for example, can reduce numbers of Townsend's warblers at the stand level (note, however, that any stand-level effects should not be considered without respect to landscape-level modification and resulting dynamics). Similarly, the orange-crowned warbler is a ground-nesting songbird that was more abundant in stands with well-developed understories (>1,236 shrubs, seedlings, and saplings/acre [>500/ha]) and substantial ground cover (>60% ground cover). Nest predators presumably have more difficulty locating nests in such stands, where nests built on the ground are well hidden beneath shrubs, grasses, forbs, and conifer seedlings. Thus, where bird densities are strongly linked to shrub density and/or ground cover, it is possible that prescribed burning and/or grazing by large ungulates (i.e., wild or domestic) may be expected to influence nesting or foraging habitat, and thereby influence bird density or reproductive success. Overstory manipulation is relatively unimportant in this context. On the other hand, understory suppression may improve habitat for the American robin, a species found to be more common in stands with little understory (<988 shrub stems, seedlings, and saplings/acre [<400/ha]). This ground-foraging thrush requires flat, open forest floors in which to search for invertebrates.[139] Hence, what is potentially deleterious for one species may be beneficial for another (see also Sallabanks et al.[137]). The chipping sparrow would apparently benefit from silviculture that produces open forest canopy (10-30%) and removal of larger trees (>89 feet [>27 m]) in height and 12.6 inches [>32 cm] diameter at breast height [dbh]). This species was consistently found to be more common in stands with short trees (89 feet [27 m]) and open canopies (<40%). In fact, for the chipping sparrow, we found that canopy height explained the greatest amount of variation (65%) in bird abundance.

On average, SEOC had the richest avifauna; YFMS was also relatively species rich. Those forest types at the extremes of the structural class spectrum (SI and OFMS) tended to have the most depauperate bird assemblages. These results are the first to document relative bird abundance among these structural classes; they can also be used to assess the degree to which different bird species are "associates" of any one structural class (i.e., found more or less exclusively in only one class). The implication for management of such structural class associates is that landscape planners may use decision rules for projecting bird abundance and thereby evaluate the potential effects of proposed forestry practices on particular bird species. Despite the fact that fewer bird species were found in SI than elsewhere, this class had more structural class associates than any other class studied. Two ground nesters (dark-eyed junco and orange-crowned warbler), 2 low-mid canopy nesters (chipping sparrow and dusky flycatcher), and 3 cavity nesters (house wren, northern flicker, and mountain bluebird) were all significantly more common in this structural class than elsewhere. Two species (house wren, mountain bluebird) were almost exclusively observed in SI. Clearly, the loss of this particular structural class from the Blue Mountains' landscape, at least in the *Abies grandis* zone, would leave species like the house wren and mountain bluebird with few alternatives. The only other clear structural class specialist was the Hammond's flycatcher, in SECC.

Only one species (brown creeper) was found to be significantly more common in OFMS, but not to the extent to be considered an old forest associate. The brown creeper has been found to be associated with structures characteristic of old forest by others.[1, 108] Other species in the Blue Mountains have also been referred to as "old forest specialists," primarily because of their apparent dependency on large snags (e.g., pileated woodpecker[30, 31, 102] and Vaux's swift[27, 29, 102]). This was not so in our study, where the presence of pileated woodpecker did not differ significantly among structural classes.

The fact that so few species exhibited significant "selection" for structural class suggests that the structural attributes and habitat elements required by most species are present in several structural classes rather than being restricted to just one. Stand Initiation possibly contained the most structural class associates because it differed the most from the other structural classes with respect to key habitat elements and structural attributes that influence the distribution and abundance of birds, such as canopy cover. Unless a species requires an entirely open canopy (e.g., house wren) or large decadent trees with flaking bark for nesting cover (e.g., brown creeper), any number of the intermediate structural classes from SEOC to YFMS provide suitable habitat. Those structural classes that represent the extremes of the successional spectrum (SI and OFMS) are likely to be the most different structurally, have the most specialized avifauna, and the fewest generalists. When attempting to maintain avian biodiversity, therefore, structural classes per se may not be relevant and are not likely to be limiting for most

species. Instead, to provide suitable habitat for most species, a rather broad range of the structural continuum may be required, as long as the key habitat elements (e.g., snags) and structural attributes (canopy cover, understory) are provided at different levels throughout that range.

Case History 2
Response of Breeding Bird Communities to Severity of Wildfire in Subalpine Fir Forests of the Blue Mountains, Oregon

Fire is the dominant disturbance agent in the interior Pacific Northwest[2] and especially in the Blue Mountains of northeastern Oregon.[3] Hence, a better understanding of how fire influences ecosystem structure and function is necessary if we are to improve management and conservation of natural systems in this region. The Blue Mountains province is a fire-dominated ecosystem[3] where forest health issues such as effects of stand-replacement wildfire on wildlife populations are much needed areas of study. In particular, management of forested ecosystems throughout the interior Pacific Northwest would greatly benefit from knowledge of how changes in forest habitat following wildfire influence the wildlife species inhabiting them, especially for those species perceived to be at risk or in decline. For example, the decline of some Neotropical migratory bird species, such as the olive-sided flycatcher, is perhaps linked to a century of fire suppression in western forests.[6]

Early studies of the effects of fire on bird populations focused primarily on game species such as quail and grouse.[16, 50, 132] More recently, research has focused on snag use by cavity-nesting birds in forested systems[48, 135] and general effects of fire on songbird habitat.[79, 101] In their examination of post-fire succession of avifauna in the Olympic Mountains of western Washington, Huff et al.[79] found bird communities to be similar in forest 1 year post-fire compared with nearby old-growth (preburn) forest. Avian breeding density and diversity did decline, however, in burned forest in years 2 and 3 post-fire. Studying avian community structure during early post-fire succession in the Sierra Nevada, Bock et al.[25] found species richness, diversity, and evenness all to be highest on a burned plot 8 years post-fire, but to be lowest on the same plot 15 years post-fire. Blake[21] reported reduced species richness of nonbreeding bird communities in burned compared with unburned ponderosa pine habitat in Arizona. Other studies, in other habitat types, also found bird populations declining following fire.[98, 154]

In general, however, most studies report few, if any, differences between the avifaunal composition of recently burned and unburned forests.[16, 51, 101, 147, 150] A few studies report increases in avifaunal richness following fire.[23, 24] In a survey of breeding bird communities from 34 burned forest sites in western Montana and northern Wyoming, Hutto[80b] discovered that within 2 years after stand-replacement fire, the relative abundance of woodpeckers, flycatchers, and seed-eaters (a total of 15 species) was significantly higher, presumably due to increased foraging and nesting opportunities. Given the influence of fire and current forest health issues, the Blue Mountains provide us with an ideal setting in which to study this topic.

In 1994, the Twin Lakes Fire burned approximately 22,000 acres (8,900 ha) of high elevation EMCF in the Blue Mountains. This case history reports briefly on the influence of the Twin Lakes Fire on structure and composition of breeding bird communities during the first 3 years (1995-97) following the fire (for more details on study design and sampling protocols, see R. Sallabanks, 1995 Annual Report, Sustainable Ecosystems Institute, Meridian, ID).

To address the effects of wildfire on bird communities, 4 treatments were selected: (1) unburned forest (i.e., control); (2) low severity burn forest (defined as 0-33% tree mortality with few differences in ground and/or understory vegetation compared with unburned forest); (3) medium severity burn forest (defined as 34-66% tree mortality with approximately 50% of live ground and/or understory vegetation absent); and (4) high severity burn forest (defined as 67-100% tree mortality with little or no live ground and/or understory vegetation present) (Figure 11). Five replicate plots of each of the no burn, low severity burn, and medium severity burn forest, and 10 replicate plots of the high severity burn forest were located in Spring 1995 and 25 permanent plots were established. "Plots" can best be described as forest patches, which may incorporate multiple stands that were homogeneous with respect to severity of burn. All plots were at least 40 acres (16 ha). Most plots ($n = 20$) were located in the Pine Ranger District, southeast of the Eagle Cap Wilderness Area, of the Wallowa-Whitman NF. The remaining plots ($n = 5$) were in the Hells Canyon National Recreation Area, and 2 of these 5 were within the boundary of the Wilderness Area.

Each year from 1995 through 1997, breeding bird communities were sampled using standard fixed-radius point count censusing techniques. In 1995 and 1997, habitat characteristics were measured using a relatively intensive vegetation sampling protocol. These data helped facilitate an evaluation of the effects of wildfire on forest habitat structure and a better understanding of the causal mechanisms underlying changes in the bird community. Due to space limitations, however, this case history focuses on the response of avifauna only; changes in habitat features and their relationships with the distribution and abundance of bird species will not be discussed.

During the 3-year study, 8,399 bird detections representing 74 bird species were recorded. The pine siskin, dark-eyed junco, and chipping sparrow dominated the avian community in the recently burned Blue Mountains forest landscape. Overall, wildfire had a somewhat negative effect, reducing bird species richness wherever burn intensity was the highest. Hutto[80b] argues that many previous studies of wildfire effects may not have found a response of bird communities to fire because analyses were based on composite statistics such as total bird density, species richness, and within-guild abundances; these measures may "hide more than they reveal" in terms of the biological effects of fire on specific

Figure 11. High intensity wildfire can cause 100% mortality in overstory and understory vegetation, such as seen in this photograph taken 1 year post-fire. Many Eastside forests and woodlands are now characterized by large, infrequent, stand-replacing fire regimes.

bird species. We concur with Hutto and emphasize the need for species-specific relationships to be assessed in all studies of this nature.

Species-specific analyses were conducted on all birds that were detected 10 times (within 164 feet [50 m] of count stations) in 2 of the 3 years of this study ($n = 23$ species). For analysis we used repeated measures ANOVA, with "treatment" (degree of burn) as the independent variable, "year" (1995, 1996, 1997) and "visit" (1, 2, 3) number as the repeated measures, and "relative abundance" (mean no. birds detected per point count per visit per species) as the dependent variable. Fourteen of the 23 species analyzed were found to vary significantly with severity of burn. Most species, including those with the most significant effects, such as the golden-crowned kinglet and mountain chickadee, were negatively affected. "Tree foliage-searching" species such as these also were found to be more common in unburned compared with burned plots by Bock and Lynch[24] in conifer forests in the Sierra Nevada. Lyon and Marzluff[101] found similar results for this guild, reporting that foliage-insect feeders, in particular, declined to only one-third of the numbers prefire. In mixed conifer forests of the Grand Teton and

Yellowstone national parks, foliage-feeding bird species also were more abundant in unburned and moderately burned compared with heavily burned sites.[147]

Although significant year effects were found for 12 species, significant interaction effects between treatment and year were only found for 3 species, suggesting that relationships between severity of burn and species' abundance were generally consistent among years. Cavity nesting birds in general, but especially the hairy woodpecker, mountain bluebird, and black-backed woodpecker, responded positively to the fire. Woodpecker species such as the black-backed and three-toed in particular rapidly colonize recently burned forests.[20, 71, 80b, 101, 147, 173] The increased availability of snags and open foraging habitat are identified as potential causal mechanisms for these species.

Two species of management concern,[53] the black-backed woodpecker and olive-sided flycatcher, warrant closer attention. Both species are associated with recently burned forest.[80b] The relative abundance of the black-backed woodpecker increased significantly in the most severely burned forest of the Blue Mountains in 1996 compared with 1995, but numbers were back down again in 1997 (R. Sallabanks, unpubl. data). The olive-sided flycatcher, on the other hand, continued to increase in abundance throughout the 3-year study, reaching its highest numbers 3 years post-fire.

The Twin Lakes Fire had mixed effects on the bird community in the Oregon Blue Mountains. In general, fewer species used the most severely burned forest compared with unburned and less severely burned forest. However, several species seemed to be opportunistically fire-dependent and clearly benefited from the 1994 burn. Changes in nesting and foraging opportunities simultaneously triggered positive responses from some species (e.g., cavity nesters and aerial feeders) and negative responses from others (e.g., overstory nesters and foliage gleaners). One relatively rare species (black-backed woodpecker) and 1 rapidly declining species (olive-sided flycatcher) were both found to be positively affected by the fire. These species, it seems, are precariously dependent on stand-replacement wildfires occurring on the landscape, even though their use of a recently burned forest may be short term (1-3 years). Additional research to better understand the role of wildfire as a natural disturbance event on bird communities in forested landscapes is needed.

Case History 3
Modification of Forest Vegetation by Wild and Domestic Ruminant Herbivores in Mixed Conifer Forests of the Blue Mountains, Oregon

Ruminant herbivores usually manipulate the composition of vegetation to some degree, especially when their density is high relative to food supply.[8, 37, 94, 129, 155] Ruminants also may regulate community processes by modifying cycling and availability of nutrients.[55, 110, 120] Their dung and urine contribute directly to the pool of nitrogen that is immediately available for plant growth and increase nitrogen to carbon ratio.[56, 111, 120, 134] This accelerates decomposition of litter by microbes and mineralization of nitrogen,[120, 133, 142] thereby accelerating nutrient cycling and increasing production of some plants. Ruminants also regulate community processes by foraging selectively. Selective feeding directly alters the composition of plant communities, influencing the amount and quality of plant litter that becomes available to decomposers over time.[120]

Through nutrient processing and selective suppression of plants, ruminants influence forest attributes such as wood production,[62, 63, 120, 166] herbivore carrying capacity,[82, 109] and biodiversity.[26, 49, 112] Their propensity to stimulate plant growth has been documented in grasslands[110] and in subarctic tree-line communities.[113] Their ability to alter the succession and stability of aspen ecosystems also is well-documented.[15, 89, 90, 91] Where herbivores suppress deciduous plants in forests, they advance succession, increase nutrient flow to the conifer overstory, and thereby may reduce forest productivity over time.[119, 120, 121]

Influences of herbivores can be most perplexing when they interact with episodic disturbance. In *Abies* and *Pseudotsuga* forests, for example, successional potential is predisposed by site characteristics, past disturbance, and by the reproductive strategies and abundance of individual plants (i.e., seed banks, rootstocks). Secondary succession is initiated by episodic agents (e.g., silviculture and fire), and its starting point and trajectory are influenced by the attributes that define each disturbance agent's regime (e.g., intensity and size of disturbance).[72, 73, 92, 141, 156] Subsequent episodic disturbances and herbivory may then further modify succession. Under these circumstances, the influence of any disturbance agent can be contingent, or dependent, on the influences of other disturbance agents. Such contingent influences have been described for grasslands[77] and in aspen forests,[15] and obviously have implications for any land management strategy that is based on disturbance management. Nonetheless, multi-agent contin-gencies have not been deeply explored in forests.

Herbivore exclosures have been used to study the influences of herbivory on vegetation development in various settings, and where several exclosures are used to document a range of episodic disturbance, they can provide insight on the role of disturbance-agent interactions, and on the relative importance of individual agents to development of communities over time. The study on which this case history is based[127c] explored influences of ruminants on community composition, biomass accumulation, and nutrient pools under varied episodic-disturbance regimes that were defined by variation in logging history, fire regime, and grass seeding.

Seven exclosures, built in the mid-1960s, were studied in the following 3 localities of the Blue Mountains Ecological Province:[57] Mottet (n = 1), Hoodoo (n = 3), Hall Ranch (n = 3). All but 1 of the sites were dominated by vegetation types within the *Abies grandis* forest series and had long histories of use by wild and domestic herbivores. Although undocumented, the sites may have been grazed by the horses of Native Americans in the eighteenth and early nineteenth century. Elk and mule deer were indigenous, but were nearly extirpated between 1870 and 1900. Domestic ruminants increased dramatically during the same period.[82] All 7 sites probably were grazed to some extent by domestic sheep in the first half of the twentieth century. Sheep grazing extended to present day in the vicinity of Mottet, but no evidence of sheep was observed at the exclosure itself during the 27 years of the experiment. During experiments, cattle were the dominant livestock at all sites except Mottet. All 7 study sites were also subject to use by elk and both white-tailed and mule deer. Whereas livestock were more prevalent than wild herbivores early in the century, wild ruminants were variously more common during the exclosure experiments summarized here.

Plant communities were sampled periodically between 1965 and 1995 inside and outside of all 7 exclosures. Similarly, total biomass and nutrients of nonconiferous understory and forest floor were estimated in 1995. Composition of herbivore diets was assessed using hand-reared elk in 1992.[127c]

Throughout the experiments, several key influences of herbivory were documented, including (1) suppressed development of shrub canopies; (2) suppressed accumulation of standing biomass and litter; and (3) reduced nutrient accumulations at forest floors (total N, total and exchangeable calcium [Ca] and magnesium [Mg]). Suppression of deciduous shrubs is a common result of ruminant herbivory in forest ecosystems.[7, 75, 89, 120, 155, 171] Regression and cluster analyses suggested a continuum of vegetation response to elk dietary preferences on the study sites. Deciduous shrubs predominated at one extreme, whereas unpalatable herbs dominated at the other. Forage preferences of elk correlated with development of plant taxa over time (P<0.002), and several taxa were predisposed to suppression by their high palatability and low abundance. These included several indicators of climax community (e.g., *Acer glabrum*, *Taxus brevifolia*, *Linnaea borealis*), shrubs that contribute vertical structure and leaf litter during forest succession (*Acer glabrum*, *Amelanchier alnifolia*, *Salix* spp., *Sorbus sitchensis*), and plants that actively replenish N pools after episodic loss (*Astragalus canadensis*, *Ceanothus sanguineus*, *C. velutinus*). The consequences of "overgrazing" by herbivores at or near maximum carrying capacity are universally recognized. However, the fact that

herbivores feed selectively when given choices implies that some taxa are always predisposed to suppression, even when grazing pressure is low on average across the plant community.[143]

Results of this study also implicated herbivores in controlling understory biomass accumulation and nutrient dynamics. At the end of the experiment, total understory and forest-floor biomasses respectively averaged 2.1 and 1.5 times greater inside exclosures than outside. Amounts of total and exchangeable Ca at the forest floor were greater inside exclosures than outside (18% and 11%, respectively). Amounts of total and exchangeable Mg at the forest floor were also greater inside than outside (20% and 16%, respectively). Concentration of total N at the forest floor was 26% higher inside exclosures than outside, but concentration of mineral N did not differ significantly. No differences were detected for nutrients in soil.

Shrubs were major contributors to the litterfall inside the exclosures, particularly in clear-cuts (where potential for shrub expression was greatest). Also, shrub foliage (and that of forbs) contained relatively high concentrations of N, phosphorus (P), sulfur (S), Ca, and Mg compared to graminoids. Greater amounts of these nutrients inside exclosures probably reflected the contribution of shrub foliage to forest-floor biomass as well. Greater accumulation of forest-floor Ca inside exclosures probably reflected translocation of Ca from the lower soil profile by some shrubs.[19]

In conclusion, herbivores influence forests by selectively suppressing plant taxa and by accelerating the cycling of nutrients. Based on results in some shrub-dominated riparian communities,[66, 148] one can logically hypothesize that both of these influences have implications for ecosystem management. First, the herbivory regime can influence the trajectory of succession and change its terminus, thereby confounding the seral and climax vegetations that managers use in forest planning.[66, 127c] Second, herbivores can suppress the forest's carrying capacity for fauna linked to shrubs in forest understories.[49, 148] This may have its greatest relevance in predicting avifaunal composition because many birds typically nest on the ground and in shrub canopies, rather than in forest overstories. Finally, by suppressing some plants via selective feeding, herbivores can reduce the rate at which certain nutrients are replenished on sites following episodic disturbance. This observation probably has its greatest relevance in the context of frequent burning. Burning volatizes nutrients, especially N and S, thereby contributing to their loss.[60a, 154a] The stability of the burned site's nutrient pool is thus influenced by the rate of loss via volitization (as functions of the frequency and intensity of burning) and by the rate of nutrient replenishment (as a function of passive input and N fixing, the latter being potentially controlled by herbivores through their selective herbivory). In this study, the most profound influence of wild and domestic ruminant herbivores may have been their modification of site production potential, via selective suppression of N-fixing plants.

Case History 4
A Review of Wildlife-habitat Relationships in Managed Lodgepole Pine Forests of the Central Oregon Pumice Zone

Perhaps one of the least understood habitats in Eastside (interior) forests and woodlands is the lodgepole pine forest of the central Oregon Pumice Zone, particularly with respect to wildlife-habitat relationships. This area lies east of the Cascade crest and encompasses >2.7 million acres (>1.1 million ha) within the Deschutes, Fremont, and Winema National Forests.[162] The topography of this region strongly reflects its volcanic origin, with numerous cinder cones and volcanic domes arising from broad basins formed by basalt flows.[162] The entire region is blanketed by recent pumice eruptions from Mount Mazama (Crater Lake), Newberry Caldera, and other sources; pumice soil sustains a climax forest of lodgepole pine with an antelope bitterbrush (*Purshia tridentata*) shrub component.[57, 162] Wildfires and mountain pine beetle (*Dendroctonus ponderosae*) epidemics were historical disturbances associated with lodgepole pine forests in central Oregon.[76] A history of fire suppression and recent mountain pine beetle epidemics, coupled with drought conditions, have resulted in extensive mortality in lodgepole and ponderosa pine forests in this region.[76] Subsequently, salvage logging of these forests has been implemented in recent years on several national forests east of the Cascade Mountains.

Salvage logging of forests subjected to fire, insects, and forest pathogens has increased over the past two decades. This forestry practice typically influences vegetative structure by removing snags and down wood, but generally retains varying densities of green trees and snags. Salvage harvest prescriptions vary and can mimic clearcut or selective harvest techniques. Government agencies often use a "pay-as-cut" timber sale contract, where purchasers pay only for wood volume removed. This differs from purchasing a fixed volume estimate, with economic incentives to remove as much of the available wood volume as possible, unless logistics and/or costs are prohibitive. With a pay-as-cut sale, contractors typically seek areas with concentrations of wood volume (dead trees in this case) to reduce operational expenses, and do not generally spend time and money to search for individuals and small groups of trees. Consequently, snags and down wood can be retained throughout harvest units in varying densities and distributions. There are few published studies on wildlife-habitat relationships from the central Oregon Pumice Zone,[126] and available data generally are limited to work in progress (e.g., E. B. Arnett, B. Altman, W. P. Erickson, and K. A. Bettinger, 2000 Final Report, Weyerhaeuser Company, Springfield, OR), agency reports,[65] and other unpublished "gray literature." In this case history, we summarize available information on wildlife-habitat relationships in lodgepole pine forests of the central Oregon Pumice Zone, focusing on studies of American marten[126] and forest songbirds (E. B. Arnett, B. Altman, W. P. Erickson, and K. A. Bettinger, 2000 Final Report, Weyerhaeuser Company, Springfield, OR).

American Marten. In summer 1993, the U.S. Forest Service Pacific Northwest Research Station began a five-year study on American marten in the central Oregon Pumice Zone (U.S. Forest Service, 1999 Annual Report, U.S. Forest Service, Olympia, WA). Objectives of this study were to (1) obtain basic natural history and habitat selection information on martens in lodgepole pine-bitterbrush communities; (2) evaluate movement and behavior of martens during winter; (3) evaluate marten response to different density, size, and configuration of slash piles at the stand- and landscape-scale; and (4) determine the influence of salvage logging on marten ecology. The study area is on the Chemult Ranger District of the Winema NF, approximately 12.4 miles (20 km) northeast of Chemult, OR. Martens were livetrapped and fitted with radiocollars to evaluate home range size, use of resting and denning sites, and habitat selection patterns. One hundred and fifteen martens (71 M, 44 F) were captured during the study and 6,251 radio relocations were obtained from marked individuals. Martens were found resting or denning on 1,851 occasions (30% of all telemetry locations) in 821 resting sites, 70 denning sites, and 27 resting and denning sites. Slash piles were most commonly used for resting and denning (44 and 40%, respectively). Down logs, live trees, and snags also were used extensively by marten for resting and denning. Raphael and Jones[126] reported seasonal differences in structures used for rest sites by martens; 60% of resting sites were in slash piles during snow-free periods, whereas only 17% were in slash during times of snow. Additionally, martens in this area generally used down logs more for denning (39%) than resting (29%). Raphael and Jones[126] also reported that live trees, snags, and logs with diameters >19.7 inches (>50 cm) dbh were selected significantly more than predicted from availability.

These data support the importance of managing structural woody features in forests to provide habitat for marten and their prey. Retaining slash piles following timber harvest seems to provide important habitat for martens[126] (U.S. Forest Service, 1999 Annual Report, U.S. Forest Service, Olympia, WA), and other species of wildlife (U.S. Forest Service, Chemult Ranger District, unpubl. data). Retaining large (>19.7 inches [>50 cm] dbh) snags and down logs also is important for providing resting and denning sites for marten. Raphael and Jones[126] estimated that 0.5 live trees >19.7 inches dbh, 0.7 snags >19.7 inches dbh, 1.5 logs, and 3.2 slash piles per acre (0.2 live trees >50 cm dbh, 0.3 snags >50 cm dbh, 0.6 logs, and 1.3 slash piles per ha) retained across the landscape would provide habitat for resting and denning, and suggested that these estimates of structure distribution could be relatively easy to accommodate in managed forests.

Forest Songbirds. Habitat relationships of forest songbirds occupying lodgepole pine forests of the central Oregon Pumice Zone are poorly understood. Herein, we summarize results of bird-habitat relationships as studied in this region from 1995 to 1999 (E. B. Arnett, B. Altman, W. P. Erickson, and K. A. Bettinger, 2000 Final Report, Weyerhaeuser Company, Springfield, OR). Objectives of this study were to (1) estimate bird species composition and relative abundance in stands of lodgepole pine; (2) model relationships between structural habitat features and bird species relative abundance and probabilities of use in lodgepole pine stands; (3) monitor nests of key species to evaluate reproductive success and nesting habitat relationships; and (4) evaluate relationships between bird species abundance and nesting success and salvage logging of lodgepole pine forests. The study area lies in the northeast corner of the Chemult Ranger District of the Winema NF,[157] and in the northwest portion of the Silver Lake Ranger District of the Fremont NF.[158]

Forty-two bird species were recorded within 164 feet (50 m) during point count surveys on both the Fremont and Winema national forests. Among species detected,17, 15, and 10 species were permanent residents, Neotropical migrants, and short-distance migrants, respectively. The mountain chickadee, yellow-rumped warbler, dusky flycatcher, dark-eyed junco, and chipping sparrow were the most common species detected on the Fremont NF, whereas the mountain chickadee, yellow-rumped warbler, chipping sparrow, dark-eyed junco, and American robin were most common on the Winema NF. Results indicated that salvage logged stands consistently had a higher total number of species and mean number of species per stand compared to reference stands. In general, relative abundance for the most common species, all species combined, and mean number of species per stand did not differ ($P > 0.10$) between reference and treatment stands on both forests. For species having >20 detections, the hermit thrush and red-breasted nuthatch were significantly more abundant in reference stands compared to treatment stands on the Fremont NF, whereas dark-eyed juncos, American robins, and Cassin's finches were significantly more abundant in treatment stands on this forest. The dusky flycatcher, hairy woodpecker, and red-breasted nuthatch all were significantly more abundant in treatment stands on the Winema NF, while the gray flycatcher was more abundant in reference stands on this forest.

Between 1997 and 1999, 420 nests of 21 different species were monitored within 12 100-acre (40-ha) nest plots (6 in salvage-logged stands and 6 in reference stands). Eighty-one per cent of these nests ($n = 336$) were from 6 species: mountain chickadee ($n = 83$), yellow-rumped warbler ($n = 63$), dark-eyed junco ($n = 58$), chipping sparrow ($n = 57$), American robin ($n = 44$), and dusky flycatcher ($n = 31$). Average Mayfield estimates of nest success for the 6 most common species ranged from 30% (dark-eyed junco) to 70% (yellow-rumped warbler). Among nesting guilds, mean nest success was lowest for ground nesters (28.8%, $n = 2$ species, $n = 72$ nests), highest for cavity nesters (66.1%, $n = 8$ species, $n = 131$ nests), and moderate for open-cup foliage nesters (55.3%, $n = 11$ species, $n = 217$ nests). Nest success was similar between reference and treatment stands for 5 of the 6 common species: American robin nest success was greater in treatment stands compared to reference stands. Though not specifically quantified, causes of nest failure included predation by small mammals and gray jays, abandonment, and possibly weather-related mortality. Cowbird parasitism was not observed during nest monitoring, though cowbirds were detected during point counts.

To illustrate wildlife-habitat relationships of birds in lodgepole pine forests in this region, we present summary results of nest and relative abundance-habitat selection models for 3 species: chipping sparrow, dark-eyed junco, and hermit thrush (E. B. Arnett, B. Altman, W. P. Erickson, and K. A. Bettinger, 2000 Final Report, Weyerhaeuser Company, Springfield, OR). The probability of presence of male chipping sparrows and dark-eyed juncos was greater in areas containing fewer total live trees and saplings, less canopy closure, and greater shrub cover. Conversely, presence of male hermit thrushes was positively related to the number of live trees and saplings, percent canopy closure, and percent seedling cover. When comparing nest-site habitat characteristics with those at random points (37.1 feet [11.3 m] fixed-radius plot), all three species selected nest sites with greater percent canopy closure. Chipping sparrow nest sites also were positively associated with greater percent seedling cover and number of saplings. Dark-eyed juncos selected nest sites in areas having greater percent cover of grasses and forbs, greater percent seedling cover, and less bare ground cover.

Results of this study generally support the characterization of avian community guilds and species composition of lodgepole pine forests. Because these forests are often single-story, dense, mono-typical forests with limited structural complexity and a sparsely vegetated understory, species richness in lodgepole pine is typically low and species dominance is often high. Despite geographic and climatic differences, avian species composition in lodgepole pine forest seems to be relatively consistent throughout the range of this forest type. Although no species is known to be dependent exclusively on lodgepole pine forests, the abundance and consistency in occurrence suggest the importance of this forest type for several bird species, such as black-backed woodpeckers.

Preliminary results of habitat selection models generally corroborate nesting and foraging strategies of avian species highlighted in this case history. Chipping sparrows and dark-eyed juncos usually are ground foragers that are often associated with more open forest conditions, while hermit thrushes appear more associated with increasing density of trees and saplings, percent canopy closure, and shrub cover. Birds in these lodgepole pine forests selected nest sites with specific structural habitat features related to cover for protection from predators and weather, often features that were not selected for by singing males at larger spatial scales. Thus, identifying structural habitat components important to birds at multiple spatial scales will be valuable to managers as they incorporate the needs of birds in management planning.

It may be possible to maintain forest bird relative abundance and reproductive success in lodgepole pine stands that have been salvage logged under certain prescriptions (E. B. Arnett, B. Altman, W. P. Erickson, and K. A. Bettinger, 2000 Final Report, Weyerhaeuser Company, Springfield, OR). The same authors hypothesized that more diverse and structurally complex harvest stands have a much greater probability of maintaining bird species relative abundance, diversity, and reproductive success. Thus,we believe that structure retained after this particular harvest treatment (e.g., live tree and snag density, down logs, understory vegetation) could explain findings presented in this case history.

Future Research Needs

Inevitably, our synthesis of wildlife communities of Eastside (interior) forests and woodlands generated more questions than it answered. Research in Eastside (interior) forests and woodlands, as elsewhere, has been heavily biased toward charismatic megafauna, birds, and ungulates (especially game species). Nongame species, especially smaller vertebrates such as bats, reptiles, and amphibians, have been relatively unstudied. In many cases we have yet to describe accurately the distribution and abundance of a species, let alone study its relationship with habitat structure or management activities. In general, more information is needed on relationships between wildlife species and the 3 predominant management issues in Eastside (interior) forests and woodlands: fire, timber harvest, and grazing. In addition, forest fragmentation, a process found to severely impact wildlife populations elsewhere in the U.S., has not yet been studied in Eastside (interior) forests and woodlands.[138] Relative to EMCF and PPFW, UAF and LPFW have been poorly studied. Many research needs identified by forest managers in the Pacific Northwest (Oregon, Washington, and Idaho) are relevant to Eastside (interior) forests and woodlands and have been discussed elsewhere.[11]

Below we offer a few specific examples of research needs:

1. The implications of prescribed burning on plant and subsequent invertebrate diversity are unknown. Factors such as fire interval, intensity, duration, season, patchiness, and spatial extent need to be examined to determine their effects on plant community composition and diversity and abundance of invertebrate herbivores and predators;
2. Assessments of the viability of species, both in space and time, that inhabit landscapes managed for the natural range of variability;
3. Studies on effects of fire occurrence, fire suppression, season of burn (see A. L. Turner and R. Sallabanks, 1999 Annual Report, Sustainable Ecosystems Institute, Meridian, ID), and changes of vegetation pattern resulting from burns and fire suppression on ground-nesting birds and their habitats (see Figure 12);
4. Studies to determine the effects of herbivory on forest understory composition and associated bird communities;
5. Effects of salvage logging of fire- and insect-killed trees on predator-prey dynamics, behavior, habitat suitability, and productivity and survivorship of wildlife species;
6. Studies that address the effects of unevenaged management (e.g., thinning from below) in PPFW and EMCF on habitat selection and population dynamics of wildlife species;
7. Restoration efforts, such as management action, to restore aspen or old-growth ponderosa pine to Eastside (interior) forests and woodlands. Such efforts should incorporate a research component to examine the response of wildlife habitat restoration over the long-term; and
8. Studies that address how manipulation of natural processes, such as fire and ecological succession, can influence genetics and viability of populations clearly dependent on those processes.

Figure 12. Ponderosa Pine Forest and Woodlands have been heavily impacted by fire suppression, grazing, and logging. Restoration efforts today include controlled burning to reduce fuels. In this stand on the Umatilla National Forest in northeast Oregon, smoke still lingers from a recent fire prescription.

Acknowledgments

The research discussed here has been funded by the U.S. Forest Service, Boise Cascade Corporation, National Council for Air and Stream Improvement, Blue Mountains Natural Resources Institute, Weyerhaeuser Company, Oregon Department of Fish and Wildlife, U.S. Fish and Wildlife Service, Oregon Agricultural Experiment Station, and Sustainable Ecosystems Institute. We are especially grateful to D. H. Johnson and K. A. Bettinger for providing technical and logistical support throughout this project. A. L. Turner provided Figure 1, and S. Edge made numerous editorial comments on an earlier version of this manuscript. All other photographs courtesy of R. Sallabanks.

Literature Cited

1. Adams, E. M., and M. L. Morrison. 1993. Effects of forest stand structure and composition on red-breasted nuthatches and brown creepers. Journal of Wildlife Management 57:616-629.

2. Agee, J. K. 1993. Fire ecology of pacific northwest forests. Island Press, Covelo, CA.

3. ———. 1996. Fire in the Blue Mountains: a history, ecology, and research agenda. Pages 119-145 in R. G. Jaindl and T. M. Quigley, editors. Search for a solution: sustaining the land, people, and economy of the Blue Mountains. American Forests, in cooperation with the Blue Mountains Natural Resources Institute, Washington, D.C.

4. ———. 1997. The severe weather wildfire: too hot to handle? Northwest Science 71:153-56.

5. ———. 1998. The landscape ecology of western forest fire regimes. Northwest Science 72:24-34.

6. Altman, B., and R. Sallabanks. Olive-sided flycatcher (Contopus cooperi). In A. Poole, and F. Gill, editors. The birds of North America, No. 502. 2000. The Birds of North America, Inc., Philadelphia, PA.

7. Ammer, C. 1996. Impact of ungulates on structure and dynamics of natural regeneration of mixed mountain forests in the Bavarian Alps. Forest Ecology and Management 88:43-53.

8. Anderson, R. C., and A. J. Katz. 1993. Recovery of browse-sensitive tree species following release from white-tailed deer (Odocoileus virginianus Zimmerman) browsing pressure. Biological Conservation 63:203-208.

9. Arnett, E. B. 1990. Bighorn sheep habitat selection patterns and response to fire and timber harvest in south-central Wyoming. Thesis, University of Wyoming, Laramie, WY.

10. ———, L. L. Irwin, F. G. Lindzey, and T. J. Hershey. 1990. Use of clearcuts by Rocky Mountain bighorn sheep in south-central Wyoming. Proceedings of the Northern Wild Sheep and Goat Council 7:194-205.

11. ———, and R. Sallabanks. 1998. Land manager perceptions of avian research and information needs: a case study. Pages 399-413 in J. M. Marzluff and R. Sallabanks, editors. Avian conservation: research and management. Island Press, Covelo, CA.

12. Arno, S. F., H. Y. Smith, and M. A. Krebs. 1997. Old-growth ponderosa pine and western larch stand structures: influences of pre-1900 fires and fire exclusion. U.S. Forest Service, Research Paper INT-RP-495.

13. Barnhardt, S. J., J. R. McBride, C. Cicero, P. da Silva, and P. Warner. 1987. Vegetation dynamics of the northern oak woodland. Pages 53-58 in T. R. Plumb and N. H. Pillsbury, technical coordinators. Proceedings of the Symposium on multiple-use of California's hardwood resources. U.S. Forest Service, General Technical Report PSW-100.

14. Barret, S., and S. Arno. 1982. Indian fires as an ecological influence in the Northern Rockies. Journal of Forestry 80:647-651.

15. Bartos, D. L., J. K. Brown, and G. D. Booth. 1994. Twelve years of biomass response in aspen communities following fire. Journal of Range Management 47:79-83.

16. Bendell, J. F. 1974. Effects of fire on birds and mammals. Pages 73-138 in T. T. Kozlowski and C. E. Ahlgren, editors. Fire and ecosystems. Academic Press, New York, NY.

17. Betts, B. J. 1996. Roosting behaviour of silver-haired bats (Lasionycteris noctivagans) and big brown bats (Eptesicus fuscus) in northeast Oregon. Pages 55-61 in R. M. R. Barclay and R. M. Brigham, editors. Bats and forests symposium, Victoria, British Columbia, Canada. BC Ministry of Forests, Victoria, BC.

18. ———. 1998. Roosts used by maternity colonies of silver-haired bats in northeastern Oregon. Journal of Mammalogy 79:643-650.

19. Binkley, D., and D. L. Husted. 1983. Nitrogen accretion, soil fertility, and Douglas-fir nutrition in association with redstem ceanothus. Canadian Journal of Forest Research 13:122-125.

20. Blackford, J. L. 1955. Woodpecker concentration in burned forest. Condor 57:28-30.

21. Blake, J. G. 1983. Influence of fire and logging on nonbreeding bird communities of ponderosa pine forests. Journal of Wildlife Management 46:404-415.

22. Bock, C. E., and J. H. Bock. 1974. On the geographical ecology and evolution of the three-toed woodpeckers, Picoides tridactylis and P. arcticus. American Midland Naturalist 92:397-405.

23. ———, and ———. 1983. Responses of birds and deer mice to prescribed burning in ponderosa pine. Journal of Wildlife Management 47:836-840.

24. ———, and J. F. Lynch. 1970. Breeding bird populations of burned and unburned conifer forest in the Sierra Nevada. Condor 72:182-189.

25. ———, M. Raphael, and J. H. Bock. 1978. Changing avian community structure during early post-fire succession in the Sierra Nevada. Wilson Bulletin 90:119-123.

26. Bowers, M. A. 1997. Influence of deer and other factors on an old-field plant community. Pages 310-326 in W. L. McShea, H. B. Underwood, and J. H. Rappole, editors. The science of overabundance: deer ecology and population management. Smithsonian Institution Press, Washington, D.C.

27. Bull, E. L. 1991. Summer roosts and roosting behavior of Vaux's swift in old-growth forests. Northwestern Naturalist 72:78-82.

28. ———, and B. E. Carter. 1996. Winter observations of tailed frogs in northeastern Oregon. Northwestern Naturalist 77:45-47.

29. ———, and J. E. Hohmann. 1992. The association between Vaux's swift and old-growth forests in northeastern Oregon. Western Birds 24:38-42.

30. ———, and R. S. Holthausen. 1993. Habitat use and management of pileated woodpeckers in northeastern Oregon. Journal of Wildlife Management 57:335-345.

31. ———, and E. C. Meslow. 1977. Habitat requirements of the pileated woodpecker in northeastern Oregon. Journal of Forestry 75:335-337.

32. ———, C. G. Parks, and T. R. Torgersen. 1997. Trees and logs important to wildlife in the Interior Columbia River Basin. U.S. Forest Service, General Technical Report PNW-391.

33. ———, T. R. Torgersen, A. K. Blumton, C. M. McKenzie, and D. S. Wyland. 1995. Treatment of an old-structure stand and the effect on birds, ants, and large woody debris: a case study. U.S. Forest Service, General Technical Report PNW-353.

34. Bunnell, F. L., L. L. Kremaster, and R. W. Wells. 1997. Likely consequences of forest management on terrestrial, forest-dwelling vertebrates in Oregon. Report M-7 of the Centre for Applied Conservation Biology, University of British Columbia, Vancouver, BC.

35. Campbell, L. A., J. G. Hallett, and M. A. O'Connell. 1996. Conservation of bats in managed forests: use of roosts by Lasionycteris noctivagans. Journal of Mammalogy 77:976-984.

36. Caraher, D. L., J. Henshaw, F. Hall (and others), panel members. 1992. Restoring ecosystems in the Blue Mountains: a report to the regional forester and the forest supervisors of the Blue Mountain forests. U.S. Forest Service, Pacific Northwest Region.

37. Chadde, S. W., and C. E. Kay. 1991. Tall-willow communities on Yellowstone's northern range: a test of the "natural-regulation" paradigm. Pages 231-262 in R. B. Keiter and M. S. Boyce, editors. The Greater Yellowstone ecosystem: redefining America's wilderness heritage. Yale University Press, New Haven, CT.

38. Chappell, C. B., R. C. Crawford, C. Barrett, J. Kagan, D. H. Johnson, M. O'Mealy, G. A. Green, H. L. Ferguson, W. D. Edge, E. L. Greda, and T. A.

O'Neil. 2001. Wildlife habitats: descriptions, status, trends, and system dynamics. Pages 22-114 in D. H. Johnson and T. A. O'Neil, managing directors. Wildlife-habitat relationships in Oregon and Washington. Oregon State University Press, Corvallis, OR.

39. Christy, R. E., and S. D. West. 1993. Biology of bats in Douglas-fir forests. U.S. Forest Service, General Technical Report PNW-GTR-308.

40. Cody, M. L., editor. 1985. Habitat selection in birds. Academic Press, New York, NY.

41. Cooper, S. V., K. E. Neiman, and D. W. Roberts. 1991. Forest habitat types of northern Idaho: a second approximation. U.S. Forest Service, General Technical Report INT-236.

42. Covington, W. W., R. L. Everett, R. Steele, L. L. Irwin, T. A. Daer, and A. N. D. Auclair. 1994. Historical and anticipated changes in forest ecosystems of the inland west of the United States. Pages 13-63 in R. N. Sampson and D. L. Adams, editors. Assessing forest ecosystem health in the Inland West. The Haworth Press, New York, NY.

43. ———, and M. M. Moore. 1994. Postsettlement changes in natural fire regimes and forest structure: ecological restoration of old growth ponderosa pine forests. Pages 153-181 in R. N. Sampson and D. L. Adams, editors. Assessing forest ecosystem health in the Inland West. The Haworth Press, New York, NY.

44. Crampton, L. H., and R. M. R. Barclay. 1998. Selection of roosting and foraging habitat by bats in different-aged aspen mixed-wood stands. Conservation Biology 12:1347-1458.

45. Cross, S. P. 1986. Bats. Pages 497-517 in A. Y. Cooperrider, R. J. Boyd, and H. R. Stuart, editors. Inventory and monitoring of wildlife habitat. U.S. Bureau of Land Management, BLM Service Center, Denver, CO.

46. Csuti, B., A. J. Kimerling, T. A. O'Neil, M. M. Shaughnessy, E. P. Gaines, and M. M. P. Huso. 1997. Atlas of Oregon wildlife: distribution, habitat, and natural history. Oregon State University Press, Corvallis, OR.

48. Davis, J. W., G. A. Goodwin, and R. A. Ockenfels, technical coordinators. 1983. Snag habitat management: proceedings of the Symposium. U.S. Forest Service, General Technical Report RM-99.

48a. DeByle, N. V., and R. P. Winokur, editors. 1985. Aspen: Ecology and management in the western United States. U.S. Forest Service, General Technical Report RM-119.

48b. ———, C. D. Bevins, and W. C. Fischer. 1987. Wildlife occurrence in aspen in the interior western United States. Western Journal of Applied Forestry 2:73-76.

49. deCalesta, D. S. 1994. Effects of white-tailed deer on songbirds within managed forests in Pennsylvania. Journal of Wildlife Management 58:711-718.

50. Doerr, P. D., L. B. Keith, and D. H. Rusch. 1971. Effects of fire on a ruffed grouse population. Tall Timbers Fire Ecology Conference Proceedings 10:25-46.

51. Emlen, J. T. 1970. Habitat selection by birds following a forest fire. Ecology 51:343-345.

52. Fenton, M. B. 1997. Science and the conservation of bats. Journal of Mammalogy 78:1-14.

53. Finch, D. M. 1992. Threatened, endangered, and vulnerable species of terrestrial vertebrates in the Rocky Mountain Region. U.S. Forest Service, General Technical Report RM-215.

54. Finch, R. B. 1984. Fire history of selected sites on the Okanogan National Forest. U.S. Forest Service, Pacific Northwest Region, Okanogan National Forest, Okanogan, WA.

55. Frank, D. A., and R. D. Evans. 1997. Effects of native grazers on grassland N cycling in Yellowstone National Park. Ecology 78:2238-2248.

56. ———, R. S. Inouye, N. Huntly, G. W. Minshall, and J. E. Anderson. 1994. The biogeochemistry of a north-temperate grassland with native ungulates: nitrogen dynamics in Yellowstone National Park. Biogeochemistry 10:163-180.

57. Franklin, J. F., and C. T. Dyrness. 1988. Natural vegetation of Oregon and Washington. Oregon State University Press, Corvallis, OR.

58. Fretwell, S. D., and H. L. Lucas. 1970. On territorial behavior and other factors influencing habitat distribution in birds. Acta Biotheoretica 19:16-36.

59. Galbraith, W. A., and E. W. Anderson. 1991. Grazing history of the Northwest. Rangelands 13:213-218.

60. Garrett, K. L., M. G. Raphael, and R. D. Dixon. 1996. White-headed woodpecker (Picoides albolarvatus). In A. Poole and F. Gill, editors. The birds of North America, No. 252. The Birds of North America, Inc., Philadelphia, PA.

60a. Geist, J. M. 1974. Chemical characteristics of some forest and grassland soils of northeastern Oregon. U.S. Forest Service, Research Note PNW-217.

61. Gibilisco, C. J. 1994. Distributional dynamics of modern Martes in North America. Pages 59-71 in S. W. Buskirk, A. S. Harestad, M. G. Raphael, and R. A. Powell, editors. Martens, sables, and fishers biology and conservation. Cornell University Press, Ithaca, NY.

62. Gill, R. M. A. 1992a. A review of damage by mammals in north temperate forests: 1. Deer. Forestry 65:145-169.

63. ———. 1992b. A review of damage by mammals in north temperate forests: 3. Impact on trees and forests. Forestry 65:363-388.

64. Gillen, R. L., W. C. Krueger, and R. F. Miller. 1985. Cattle use of riparian meadows in the Blue Mountains of northeastern Oregon. Journal of Range Management 38:205-209.

65. Goggans, R., R. D. Dixon, and L. C. Seminara. 1987. Habitat use by three-toed and black-backed woodpeckers. Oregon Department of Fish and Wildlife, Nongame Report 87-3-02, Bend, OR.

66. Green, D. M., and J. B. Kauffman. 1985. Succession and livestock grazing in a northeastern Oregon riparian system. Journal of Range Management 48:307-314.

67. Gruell, G. E. 1985. Indian fires in the Interior West: a widespread influence. Pages 68-74 in Proceedings: Symposium and workshop on wilderness fire. U.S. Forest Service, General Technical Report INT-182.

68. Hamilton, R. C. 1993. Characteristics of old growth forest in the Intermountain Region. U.S. Forest Service, Intermountain Resource Station, Ogden, UT.

69. Harrod, R. J., R. J. Taylor, W. L. Gaines, T. Lillybridge, and R. Everett. 1996. Pages 107-117 in R. G. Jaindl and T. M. Quigley, editors. Search for a solution: sustaining the land, people, and economy of the Blue Mountains. American Forests, in cooperation with the Blue Mountains Natural Resources Institute, Washington, D.C.

70. Hein, D. 1980. Management of lodgepole pine for birds. Pages 238-246 in R. M. DeGraaf, technical coordinator. Management of western forests and grasslands for nongame birds. U.S. Forest Service, General Technical Report INT-86.

71. Heinselman, M. L. 1973. Fire in the virgin forests of the Boundary Waters Canoe Area, Minnesota. Quarterly Research (NY) 3:329-382.

72. ———. 1978. Fire in wilderness ecosystems. Pages 248-278 in J. C. Hendee, G. H. Stankey, and R. C. Lucas, editors. Wilderness management. U.S. Forest Service, Miscellaneous Publication No. 1365.

73. ———. 1981. Fire intensity and frequency as factors in the distribution and structure of northern ecosystems. Pages 7-57 in H. A. Mooney, T. M. Bonnicksen, N. L. Christensen, J. E. Lotan, and W. A. Reiners, technical coordinators. Fire regimes and ecosystem properties. U.S. Forest Service, General Technical Report WO-26.

74. Henjum, M. G., J. R. Karr, D. L. Bottom, D. A. Perry, J. C. Bednarz, S. G. Wright, S. A. Beckwitt, and E. Beckwitt. 1994. Interim protection for late-successional forests, fisheries, and watersheds: national forests east of the Cascade crest, Oregon and Washington. The Wildlife Society, Bethesda, MD.

75. Hernandez, M. P. G., and F. J. Silva-Pando. 1996. Grazing effects of ungulates in a Galician oak forest (northwest Spain). Forest Ecology and Management 88:65-70.

76. Hessburg, P. F., R. G. Mitchell, and G. M. Filip. 1994. Historical and current roles of insects and pathogens in eastern Oregon and Washington forested landscapes. U.S. Forest Service, General Technical Report PNW-327.

77. Hobbs, N. T., D. S. Schimel, C. E. Owensby, and D. S. Ojima. 1991. Fire and grazing in the tallgrass prairie: contingent effects on nitrogen budgets. Ecology 72:1374-1382.

78. ———, and R. A. Spowart. 1984. Effects of prescribed fire on nutrition of mountain sheep and mule deer during winter and spring. Journal of Wildlife Management 48:551-560.

79. Huff, M. H., J. K. Agee, and D. A. Manuwal. 1985. Postfire succession of avifauna in the Olympic Mountains, Washington. Pages 8-15 in Fire's effects on wildlife habitat-Symposium proceedings. U.S. Forest Service, General Technical Report INT-186.

80a. Hutto, R. L. 1995. 1995 Annual Report. University of Montana, Missoula, MT.

80b. ———. 1995. Composition of bird communities following stand-replacement fires in northern Rocky Mountain (U.S.A.) conifer forests. Conservation Biology 9:1041-1058.

81. ———, S. M. Pletschet, and P. Hendricks. 1986. A fixed-radius point count method for nonbreeding and breeding season use. The Auk 103:593-602.

82. Irwin, L. L., J. G. Cook, R. A. Riggs, and J. M. Skovlin. 1994. Effects of long-term grazing by big game and livestock in the Blue Mountains forest ecosystems. U.S. Forest Service, General Technical Report PNW-325.

83. James, F. C., and H. H. Shugart, Jr. 1970. A quantitative method of habitat description. Audubon Field Notes 24:727-736.

84. ———, and N. O. Warner. 1982. Relationships between temperate forest bird communities and vegetation structure. Ecology 63:159-171.

85. Johnson, C. G., Jr., and R. R. Clausnitzer. 1992. Plant associations of the Blue and Ochoco mountains. U.S. Forest Service, Pacific Northwest Region, R6-ERW-TP-036-92.

86. Kalcounis, M. C., and R. M. Brigham. 1998. Secondary use of aspen cavities by tree-roosting big brown bats. Journal of Wildlife Management 62:603-611.

87. Kauffman, J. B., M. Mahrt, L. Mahrt, and W.D. Edge. 2001. Wildlife of riparian habitats. Pages 361-388 in D.H. Johnson and T.A. O'Neil, managing directors. Wildlife-habitat relationships in Oregon and Washington. Oregon State University Press, Corvallis, OR.

88. Kay, C. E. 1995a. Aboriginal overkill and native burning: implications for modern ecosystem management. Proceedings of the 8th George Wright Society Conference on research and resource management on public lands.

89. ———. 1995b. Browsing by native ungulates: effects on shrub and seed production in the Greater Yellowstone Ecosystem. Pages 310-320 in B.A. Roundy, E. D. McArthur, J. S. Haley, and D. K. Mann, compilers. Proceedings: wildland shrub and arid land restoration symposium. U.S. Forest Service, General Technical Report INT-315.

90. ———. 1997. The condition and trend of aspen, Populus tremuloides, in Kootenay and Yoho national parks: implications for ecological integrity. Canadian Field-Naturalist 111:607-616.

91. ———, and S. Chadde. 1992. Reduction of willow seed production by ungulate browsing in Yellowstone National Park. Pages 92-99 in W. P. Clary, E. D. McArthur, D. Bedunah, and C. L. Wambolt, compilers. Proceedings of Symposium on ecology and management of riparian shrub communities. U.S. Forest Service, General Technical Report INT-289.

92. Kilgore, B. M. 1981. Fire in ecosystem distribution and structure: western forests and scrublands. Pages 58-59 in H.A. Mooney, T. M. Bonnicksen, N. L. Christensen, J. E. Lotan, and W.A. Reiners, technical coordinators. Fire regimes and ecosystem properties. U.S. Forest Service, General Technical Report WO-26.

93. Knowles, C. J., and P. R. Knowles. 1993. A bibliography of literature and papers pertaining to presettlement wildlife and habitat of Montana and adjacent areas. FaunaWest Wildlife Consultants, Boulder, MT.

94. Kroll, J. C., W. D. Goodrum, and P. J. Behrman. 1986. Twenty-seven years of over-browsing: implications to white-tailed deer management on wilderness areas. Pages 294-303 in D. L. Kulhavy and R. N. Conner, editors. Wilderness and natural areas in the eastern United States: a management challenge. Center for Applied Studies, School of Forestry, Stephen F. Austin University, Nacagdoches, TX.

95. Kunz, T. H. 1982. Roosting ecology of bats. Pages 30-55 in T. H. Kunz, editor. Ecology of bats. Plenum Press, New York, NY.

96. Langston, N. 1995. Forest dreams, forest nightmares. University of Washington Press, Seattle, WA.

97. Larsen, E. M., and J. T. Morgan. 1998. Management recommendations for Washington's priority habitats: Oregon white oak woodlands. Washington Department of Fish and Wildlife, Olympia, WA.

98. Lawrence, G. E. 1966. Ecology of vertebrate animals in relation to chaparral fire in the Sierra Nevada foothills. Ecology 47:278-291.

99. Leonard, W. P., H. A. Brown, L. L. C. Jones, K. R. McAllister, and R. M. Storm. 1993. Amphibians of Washington and Oregon. Seattle Audubon Society, Seattle, WA.

100. Lewis, S. E. 1995. Roost fidelity of bats: a review. Journal of Mammalogy 6:481-496.

101. Lyon, L. J., and J. M. Marzluff. 1985. Fire's effects on a small bird population. Pages 16-22 in Fire's effects on wildlife habitat—Symposium proceedings. U.S. Forest Service, General Technical Report INT- 186

102. Mannan, R. W., and E. C. Meslow. 1984. Bird populations and vegetation characteristics in managed and old-growth forests, northeastern Oregon. Journal of Wildlife Management 48:1219-1238.

103. Marcot, B. G. 1985. Habitat relationships of birds and young-growth Douglas-fir in northwestern California. Dissertation, Oregon State University, Corvallis, OR.

104. ———. 1996. An ecosystem context for bat management: a case study of the interior Columbia River Basin, U.S.A. Pages 19-36 in R. M. R. Barclay and R. M. Brigham, editors. Bats and forests symposium, Victoria, British Columbia, Canada. BC Ministry of Forests, Victoria, BC, Canada.

105. ———, M.A. Castellano, J. A. Christy, L. K. Croft, J. F. Lehmkuhl, R. H. Naney, K. Nelson, C. G. Niwa, R. E. Rosentreter, R. E. Sandquist, B. C. Wales, and E. Zieroth. 1997. Terrestrial ecology assessment. Pages 1497-1713 in T. M. Quigley and S. J. Arbelbide, editors. An assessment of ecosystem components in the interior Columbia Basin and portions of the Klamath and Great Basins. Volume III. U.S. Forest Service, General Technical Report PNW-405.

106. ———, L. K. Croft, J. F. Lehmkuhl, R. H. Naney, C. G. Niwa, W. R. Owen, and R. E. Sandquist. 1998. Macroecology, paleoecology, and ecological integrity of terrestrial species and communities of the interior Columbia River Basin and portions of the Klamath and Great Basins. U.S. Forest Service, General Technical Report PNW-410.

107. ——— and M. Vander Heyden. 2001. Key ecological functions of wildlife species. Pages 168-186 in D.H. Johnson and T.A. O'Neil, managing directors. Wildlife-habitat relationships in Oregon and Washington. Oregon State University Press, Corvallis, OR.

108. Mariani, P., and D. A. Manuwal. 1990. Factors influencing brown creeper (Certhia americana) abundance patterns in the southern Washington Cascade Range. Studies in Avian Biology 13:53-57.

109. McCullough, D. R. 1979. The George Reserve deer herd. The University of Michigan Press, Ann Arbor, MI.

110. McNaughton, S. J. 1976. Serengeti migratory wildebeest: facilitation of energy flow by grazing. Science 191:92-94.

111. ———. 1992. The propagation of disturbance in savannas through food webs. Journal of Vegetation Science 3:301-314.

112. McShea, W. J., and J. H. Rappole. 1997. Herbivores and the ecology of forest understory birds. Pages 298-309 in W. J. McShea, H. B. Underwood, and J. H. Rappole, editors. The science of overabundance: deer ecology and population management. Smithsonian Institution Press, Washington, D.C.

113. Molvar, E. M., R. T. Bowyer, and V. Van Ballenberghe. 1993. Moose herbivory, browse quality, and nutrient cycling in an Alaskan treeline community. Oecologia 94:472-470.

114. Noss, R. F., E. T. LaRoe, III, and J. M. Scott. 1995. Endangered ecosystems of the United States: a preliminary assessment of loss and degradation. Biological Report 28, U.S. Department of the Interior, National Biological Service, Washington, D.C.

115. O'Hara, K. L., P.A. Latham, P. F. Hessburg, and B. G. Smith. 1996. A structural classification for inland northwest forest vegetation. Journal of Applied Forestry 11:97-102.

116. Oliver, C. D. 1981. Forest development in North America following major disturbances. Forest Ecology and Management 3:153-168.

117. Paquet, P., and A. Hackman. 1995. Large carnivore conservation in the Rocky Mountains: a long-term strategy for maintaining free-ranging and self-sustaining populations of carnivores. World Wildlife Fund, Washington, D.C.

118. Parsons, D. J., and S. H. DeBenedetti. 1979. Impact of fire suppression on a mixed-conifer forest. Forest Ecology and Management 2:21-33.

119. Pastor, J., and Y. Cohen. 1997. Herbivores, the functional diversity of plant species, and the cycling of nutrients in ecosystems. Theoretical Population Biology 51:165-179.

120. ———, and R. J. Naiman. 1992. Selective foraging and ecosystem processes in boreal forests. American Naturalist 139:691-705.

121. ———, ———, B. Dewey, P. McInnes, and Y. Cohen. 1993. Moose browsing and soil fertility in the boreal forests of Isle Royale National Park. Ecology 74:467-480.

121a. Peek, J.M., D.A. Demarchi, and D.E. Stucker. 1985. Bighorn sheep and fire: seven case histories. Pages 36-43 in Fire's effects on wildlife habitat—Symposium proceedings. U.S. Forest Service, General Technical Report INT-186.

122. Perkins, J. M., J. M. Barss, and J. Peterson. 1990. Winter records of bats in Oregon and Washington. Northwestern Naturalist 71:59-62.

123. ———, and S. P. Cross. 1988. Differential use of some coniferous forest habitats by hoary and silver-haired bats in Oregon. The Murrelet 69:21-24.

124. Quigley, T. M., R. W. Haynes, and R. T. Russell, technical editors. 1996. Integrated scientific assessment for ecosystem management in the interior Columbia Basin and portions of the Klamath and Great Basins. U.S. Forest Service, General Technical Report PNW-382.

125. Raphael, M. G. 1983. Cavity-nesting bird response to declining snags on a burned forest: a simulation model. Pages 211-215 in J.W. Davis, G. A. Goodwin, and R.A. Ockenfels, technical coordinators. Snag habitat management: Proceedings of the Symposium. U.S. Forest Service, General Technical Report RM-99.

126. ———, and L. L. C. Jones. 1997. Characteristics of resting and denning sites of American martens in central Oregon and western Washington. Pages 146-165 in G. Proulx, H. N. Bryant, and P. M. Woodard, editors. Martes: taxonomy, ecology, techniques and management. The Provincial Museum of Alberta, Edmonton, Alberta, Canada.

127. ———, and M. White. 1984. Use of snags by cavity-nesting birds in the Sierra Nevada. Wildlife Monographs 86:1-66.

127a. Riggs, R.A., and J.M. Peek. 1980. Mountain sheep habitat use patterns related to post-fire succession. Journal of Wildlife Management 44:933-938.

127b. ———, S. Bunting, and S.E. Daniels. 1996. Prescribed fire. Pages 295-319 in P.R. Krausemann, editor. Rangeland wildlife. Society for Rangeland Management, Denver, CO.

127c. ———, A.R. Tiedemann, J.G. Cook, T.M. Ballard, P.J. Edgerton, M. Vavra, W.C. Krueger, F.C. Hall, L.D. Bryant, L.L. Irwin, and T. DelCurto. In press. Modification of mixed-conifer forests by ruminant herbivores in the Blue Mountains ecological province. U.S. Forest Service, Research Paper PNW-RP-527.

128. Risenhoover, K. L., and J.A. Bailey. 1985. Foraging ecology of mountain sheep: implications for habitat management. Journal of Wildlife Management 49:797-804.

129. Rose, A. B., and K. H. Platt. 1987. Recovery of northern fiordland alpine grasslands after reduction in the deer population. New Zealand Journal of Ecology 10:23-33.

130. Rose, C., J. Ohmann, K. Waddell, D. Schreiber, D. Lindley, K. Mellen, and B. G. Marcot. 2001. Decaying wood in Pacific Northwest forests: Concepts and tools for habitat management. Pages 580-623 in D.H. Johnson and T.A. O'Neil, managing directors. Wildlife-habitat relationships in Oregon and Washington. Oregon State University Press, Corvallis, OR.

131. Rosenzweig, M. L. 1985. Some theoretical aspects of habitat selection. Pages 517-540 in M. L. Cody, editor. Habitat selection in birds. Academic Press, New York, NY.

132. Rowe, J. S., and G. W. Scotter. 973. Fire in the boreal forest. Quarterly Research (NY) 3:444-464.

133. Ruess, R. W. 1987. The role of large herbivores in nutrient cycling of tropical savannas. Pages 93-141 in B. H. Walker, editor. Determinants of tropical savannas. International Union of Biological Scientists, Paris, France.

134. ———, and S. J. McNaughton. 1987. Grazing and the dynamics of nutrient and energy regulated microbial processes in the Serengeti grasslands. Oikos 49:101-110.

135. Saab, V.A., and J. G. Dudley. 1998. Responses of cavity-nesting birds to stand-replacement fire and salvage logging in ponderosa pine/Douglas-fir forests of southwestern Idaho. U.S. Forest Service, Research Paper RM-11.

136. ———, and T. D. Rich. 1997. Large-scale conservation assessment for Neotropical migratory land birds in the Interior Columbia River Basin. U.S. Forest Service, General Technical Report PNW-399.

137. Sallabanks, R., E. B. Arnett, T. B. Wigley, and L. L. Irwin. In press. Accommodating birds in managed forests of North America: A review of bird-forestry relationships. National Council for Air and Stream Improvement Technical Bulletin: In press.

138. ———, P. J. Heglund, J. B. Haufler, B.A. Gilbert, and W. Wall. 1999. Forest fragmentation of the Inland West: issues, definitions, and potential study approaches for forest birds. Pages 187-199 in J.A. Rochelle, L.A. Lehmann, and J. Wisniewski, editors. Forest fragmentation: wildlife and management implications. Brill Academic Publishers, Leiden, The Netherlands.

139. ———, and F. C. James. 1999. American robin (Turdus migratorius). In A. Poole and F. Gill, editors. The birds of North America, No. 462. The Birds of North America, Inc., Philadelphia, PA.

140. Sampson, R. N., D. L. Adams, S. S. Hamilton, S. P. Mealey, R. Steele, and D. Van De Graaff. 1994. Assessing forest ecosystem health in the Inland West. Pages 3–10 in R. N. Sampson and D. L. Adams, editors. Assessing forest ecosystem health in the Inland West. The Haworth Press, New York, NY.

141. Schimmel, J., and A. Granstrom. 1996. Fire severity and vegetation response in the boreal Swedish forest. Ecology 77:1436-1450.

142. Seagle, S. W., S. J. McNaughton, and R. W. Ruess. 1992. Simulated effects of grazing on soil nitrogen and mineralization in contrasting Serengeti grasslands. Ecology 73:1105-1123.

143. Smith, A. D. 1965. Determining common use grazing capacities by application of the key species concept. Journal of Range Management 18:196-201.

144. Steele, R. 1994. The role of succession in forest health. Pages 183–190 in R. N. Sampson and D. L. Adams, editors. Assessing forest ecosystem health in the Inland West. The Haworth Press, New York, NY.

145. ———, S. F. Arno, and K. Geier-Hayes. 1986. Wildfire patterns change in central Idaho's ponderosa pine-Douglas-fir forest. Western Journal of Applied Forestry 1:16-18.

146. ———, R. D. Pfister, R. A. Ryker, and J. A. Kittams. 1981. Forest habitat types of central Idaho. U.S. Forest Service, General Technical Report INT-114.

147. Taylor, D. L., and W. J. Barmore, Jr. 1980. Post-fire succession of avifauna in coniferous forests of Yellowstone and Grand Teton national parks, Wyoming. Pages 130-145 in Workshop Proceedings: management of western forests and grasslands for nongame birds. U.S. Forest Service, General Technical Report INT-86.

148. Taylor, D. M. 1986. Effects of cattle grazing on passerine birds nesting in riparian habitat. Journal of Range Management 39:254-258.

149. ———, and C. D. Littlefield. 1986. Willow flycatcher and yellow warbler response to cattle grazing. American Birds 40:1169-1173.

150. Theberge, J. B. 1976. Bird populations in the Kluane Mountains, southwest Yukon, with special reference to vegetation and fire. Canadian Journal of Zoology 54:1346-1356.

151. Thomas, D. W. 1988. The distribution of bats in different ages of Douglas-fir forests. Journal of Wildlife management 52:619-626.

152. ———, and S. D. West. 1991. Forest age associations of bats in the southern Washington Cascade and Oregon Coast Ranges. Pages 295-303 in L. F. Ruggiero, K. B. Aubry, A. B. Carey, and M. H. Huff, technical coordinators. Wildlife and vegetation of unmanaged Douglas-fir forests. U.S. Forest Service, General Technical Report PNW-285.

153. Thomas, J. W., technical editor. 1979. Wildlife habitats in managed forests: The Blue Mountains of Oregon and Washington. U.S. Forest Service, PNW Region, Agricultural Handbook No. 553.

154. Tiagwad, T. E., C. M. Olson, and R. E. Martin. 1982. Single-year response of breeding bird populations to fire in a curlleaf mountain mahogany-big sagebrush community. Pages 101-110 in E. E. Starkey, J. F. Franklin, and J. W. Matthews, editors. Ecological research in national parks of the Pacific Northwest. Forest Research Laboratory, Oregon State University, Corvallis, OR.

154a. Tiedemann, A. R., R. R. Mason, and B. E. Wickman. 1998. Forest floor and soil nutrients five years after urea fertilization in a grand fir forest. Northwest Science 72:88-95.

155. Tilghman, N. G. 1989. Impacts of white-tailed deer on forest regeneration in northwestern Pennsylvania. Journal of Wildlife Management 53:524-533.

156. Turner, M. G., W. H. Romme, R. H. Gardner, and W. W. Hargrove. 1997. Effects of fire size and pattern on early succession in Yellowstone National Park. Ecological Monographs 67:411-433.

156a. Turner, A. L., and R. Sallabanks. 1999 Annual Report, Sustainable Ecosystems Institute, Meridian, ID.

157. U.S. Forest Service. 1992a. Environmental assessment for the First Timber Sale. U.S. Forest Service, Winema National Forest, Klamath Falls, OR.

158. ———. 1992b. Antelope salvage planning area: environmental assessment. U.S. Forest Service, Fremont National Forest, Lakeview, OR.

159. ———, and U.S. Bureau of Land Management. 1997. Eastside Draft Environmental Impact Statement, Volume 1. Interior Columbia Basin Ecosystem Management Project (ICBEMP), Walla Walla, WA.

160. Verner, J., M. L. Morrison, and C. J. Ralph, editors. 1986. Wildlife 2000: Modeling habitat relationships of terrestrial vertebrates. University of Wisconsin Press, Madison, WI.

161. Verts, B. J., and L. N. Carraway. 1998. Land mammals of Oregon. University of California Press, Berkeley, CA.

162. Volland, L. 1985. Plant associations of the central Oregon Pumice Zone. U.S. Forest Service, Pacific Northwest Region, R6-ECOL-104-1985, Portland, OR.

163. Vonhof, M. J. 1996. Roost-site preferences of big brown bats (Eptesicus fuscus) and silver-haired bats (Lasionycteris noctivagans) in the Pend d'Oreille Valley in southern British Columbia. Pages 62-80 in R. M. R. Barclay and R. M. Brigham, editors. Bats and forests symposium, Victoria, British Columbia, Canada. BC Ministry of Forests, Victoria, BC, Canada.

164. Wakelyn, L. A. 1987. Changing habitat conditions on bighorn sheep ranges in Colorado. Journal of Wildlife Management 51:904-912.

165. Waldien, D. L., J. P. Hayes, and E. B. Arnett. 2000. Day-roosts of female long-eared myotis in western Oregon. Journal of Wildlife Management 64:785-796.

166. Weigand, J. F., R. Haynes, A. R. Tiedemann, R. A. Riggs, and T. M. Quigley. 1993. Ecological and economic impacts of ungulate herbivory on commercial stand development in forests of eastern Oregon and Washington. Forest Ecology and Management 61:137-155.

168. Wickman, B. E. 1992. Forest health in the Blue Mountains: the influence of insects and diseases. U.S. Forest Service, General Technical Report PNW-295.

169. Wiens, J. A., and J. T. Rotenberry. 1981. Habitat associations and community structure of birds in shrubsteppe environments. Ecological Monographs 51:21-41.

170. Wischnofske, M. G., and D. W. Anderson. 1983. The natural role of fire in the Wenatchee Valley. U.S. Forest Service, Pacific Northwest Region, Wenatchee National Forest, Wenatchee, WA.

171. Woodward, A., E. G. Schreiner, D. B. Houston, and B. B. Moorhead. 1994. Ungulate-forest relationships in Olympic National Park: retrospective exclosure studies. Northwest Science 68:97-110.

172. Wright, C. S. 1996. Fire history of the Teanaway River drainage, Washington. Thesis, University of Washington, Seattle, WA.

173. Yunick, R. P. 1985. A review of recent irruptions of the black-backed woodpecker and three-toed woodpecker in eastern North America. Journal of Field Ornithology 56:138-152.

174. Zimmerman, G. T., and L. F. Neuenschwander. 1984. Livestock grazing influences on community structure, fire intensity, and fire frequency within the Douglas-fir/Ninebark habitat type. Journal of Range Management 37:104-110.

9

Wildlife in Alpine and Subalpine Habitats

Kathy M. Martin

Introduction

The landscape of western North America is defined by vast areas of striking mountainous terrain. Since settlement, humans have been fascinated and frustrated with these imposing habitats that offer more vertical than horizontal relief. In the old world, humans have used alpine areas for hunting and agriculture since the beginning of recorded history. Europeans and Asians have practiced alpine agriculture for over 10 centuries, cultivating crops and moving livestock up to alpine pastures in summer and down to lower elevations in fall.[9] In North America, our relationship with alpine areas is much more recent. Mountains posed serious barriers to exploration and the development of agriculture in the previous century. Today, however, alpine areas are valued for their intrinsic beauty and wildness, for their recreational potential, and as a refuge from dense urban areas.

The alpine zone consists of rugged, partially vegetated terrain with snowfields and rocky ridges, above the natural treeline. Alpine ecosystems are structurally simple with few plant species compared to most lower elevation habitats. High elevation habitats are characterized by high winds, prolonged snow cover, steep terrain, extremes of heat and cold, and intense ultra-violet radiation. With increasing elevation, time for breeding decreases and environmental stochasticity increases; at the highest elevations, hypoxic conditions add additional energetic living costs. These factors result in short, intense breeding seasons for wildlife, and the need for seasonal movements to and from patchy breeding habitats and wintering areas.

Despite being highly valued for their intrinsic beauty and wildness, alpine vertebrates and their high elevation habitats are a neglected area for research and management. Alpine ecosystems recently have experienced large increases in amount and kinds of human use, and some areas now show significant deterioration. In this chapter, I describe the wildlife communities inhabiting the alpine zone above natural treeline. I describe what is special about high elevation communities, and what is changing. About one third of the vertebrate fauna in the Pacific Northwest is connected to alpine and sub-alpine habitats across one or more seasons. Alpine habitats are essentially vertical islands. Some animals, such as white-tailed ptarmigan, hoary marmots and mountain goats, remain in their high

elevation 'islands' year-round, and leave only to travel to other alpine patches. However, the majority of species move to lower elevation habitats at some life history stage. Thus, connectivity is a key ecological process to maintain for alpine wildlife. Anthropogenically-induced changes at both high and low elevations constitute potential threats for alpine animals which are well adapted for extreme conditions, but not so well for increased warming, competition or predation. Alpine sites have the potential to serve as natural experiments that allow 'space for time' perspectives in predicting animal responses to global climate change.

1. The Nature, Distribution, and Diversity of Alpine Habitats—Global to Local

A mountain is a landmass arising above the general landscape that induces a change in climate that affects vegetation and animal life[111] The word *alpine* comes from the Alps, and refers to the zone above the natural treeline, with persistent or permanent snowfields, rocky ridges, occasional wind-shaped trees and continuous to scattered tundra vegetation.[92] The treeline, the lower boundary of the alpine life zone, is often fragmented over several hundred meters of altitude.[81] Several factors define the alpine zone, including elevation, aspect and high relief, but climate is probably the best determinant of where alpine zones begin.[81, 111] Alpine climates are characterized by high winds, low temperatures, low effective moisture, and short growing seasons.[11, 21] Alpine zones increase in elevation from north to south and from coastal areas to interior. Northward across the mid-latitudes of North America, treeline decreases about 152 m (500 ft) in elevation per 160 km (100 miles).[43] In the Cascades, treeline increases in elevation from about 2000 m (6560 ft) on the west side to 2500 m (8200 ft) on the eastern continental side.[5] In this chapter, the alpine zone includes two habitat types, alpine treeless and partially vegetated areas at the top, and the sub-alpine, which is the zone between closed upper montane forest and the upper limits of small trees in open parklands.

The global alpine landmass comprises about 4 million km², about 3% of the global landmass,[81] an area about half the size of the continental United States (~7 million km², 2.98 million sq. mi.). About 30% of this alpine landmass is vegetated.[81] In Washington, the total area of alpine and sub-alpine habitat is about 4.4% of the landbase and in Oregon, about 0.6%.[76] Further north in the Pacific Northwest, 17% of British Columbia is classified as alpine (BC Ministry of Forests Database). By global and continental standards, the mountains in Washington and Oregon are modest in elevation. Mt. Everest 29,029 ft (8,848 m), on the Tibet-Nepal border is the highest elevation on earth. Mt. Denali in Alaska 20,320 ft (6194 m) is the highest peak in North America. In the Pacific Northwest, Mt. Rainier is the highest mountain in Washington at 4,409 ft (4392 m) (Figure 1), while Mt. Hood is the highest in Oregon at 11,239 ft (3426 m).

The North Cascades Range provides a rugged set of young mountains, geologically complex, and heavily glaciated. Mt. Baker, Glacier Peak and Mt. Rainier, high ice-covered volcanoes, are the most conspicuous mountains scattered along the crest of the Range. Below the ice, these "island" mountains support well-developed alpine plant communities in moist and cool environments.[11, 46] The Cascades Range is distinguished by being one of the snowiest places on the planet. Mt. Rainier and Mt. Baker have the right combination of winter precipitation and oceanic air currents, as well as steep temperature and elevational gradients to generate impressive snowfall. During the winter of 1998-99, Mt. Baker recorded a world record snowfall of 95.0 ft (28.9 m)(U.S. Dept. of Commerce and the National Oceanic and Atmospheric Administration, Press release, August 3, 1999).

Alpine and sub-alpine habitats are among the most undisturbed habitats remaining in the Pacific Northwest. The amount or distribution of high elevation habitat in Washington and Oregon has changed little since presettlement times. Alpine ecosystems, however, are sensitive to sustained and heavy use. Olympic and Mt. Rainier National Parks show significant deterioration of alpine fellfields and sub-alpine meadows caused by recreational activities, grazing, and air-borne contaminants.[15, 69, 78, 114] This region has also experienced several impressive natural disturbance events. The Mt. St. Helens volcano in Washington erupted in 1980, scorching or leveling 500 km[2] (200 mi[2]) of surrounding forest up to a distance of 25 km (15 mi).[59] Recovery of the mountain flora and fauna after this natural disturbance is of special ecological interest to biologists. Shortly after Mt. St. Helens cooled, wildlife species began to recolonize and, recovery continues. Species with the most complete recovery tend to be associated with standing dead trees, stream ecosystems, or living beneath the soil surface.[41, 94]

The Cascades Mountains, the major mountain range in the region, extend from British Columbia south through Washington and Oregon to California. The Selkirk Mountains run from southern British Columbia to northeastern Washington. The Olympic Mountains lie west, and the Okanogan Highlands lie east of the Cascades Range. The major mountain ranges in Oregon are the Wallowa Mountains, Blue Mountains, Steens Mountain, and the Siskiyous.

2. High Elevation Wildlife Habitats in Washington and Oregon

In Washington and Oregon, the two main high elevation wildlife habitats classified were Sub-alpine Parklands (#9) and Alpine Grasslands and Shrublands (#10).[36, 76] These habitats comprise about 2.2% of the Washington and Oregon land base (Table 1), and include a broad diversity of alpine habitats and conditions. Coastal or maritime alpine located on the western side of the mountains differs dramatically from interior mountain ranges on the eastern side.[11, 55] On the western side, there is more coastal alpine, connected to Upper Aspen habitat, montane conifers such as Engelmann spruce (*Picea engelmannii*), and sub-alpine fir (*Abies lasiocarpa*), and coniferous wetland habitat types. On the eastern side, it is drier, alpine habitats are at higher elevations, and these are connected to lodgepole pine (*Pinus contorta*) and high shrub Steppe habitat types.[35] Locally, high elevation habitats also differ in climate, vegetation, and biological processes such as treeline limit and resilience to impacts depending upon location, slope, elevation, and aspect.

Figure 1. Sub-alpine flowers in the foreground, open montane forest with hints of tree islands, and a spectacular view of snow-capped Mt Rainier in the background. Slide by Steve Ogle (#17), mid to late August 1999.

Sub-alpine Parkland habitats occur below alpine or krummholz (the zone where growth of trees and shrubs is stunted and deformed) and above continuous montane forest. The parklands are mosaics of patches of herbaceous or dwarf-shrub vegetation and tree islands or scattered trees with 10 to 30% canopy cover. Sub-alpine Parklands occur throughout the high mountain ranges of Washington and Oregon, extending north into Canada, and south to the Sierra Mountains. Mountain hemlock (*Tsuga mertensiana*) sub-alpine parklands occur along the Cascades Crest and in the Olympic Mountains. In the west Cascades and Olympic Mountains, Sub-alpine Parklands are mosaics of tree patches and heath shrublands or wetlands. Sub-alpine Parklands in the east Cascades and Wallowa Mountains, occurring at slightly higher elevations (up to 2438 m, 8000 ft; Table 1), contain Whitebark pine (*Pinus albicaulis*)-subalpine fir with ground cover typically dominated by sedges and grasses, and with less heath than Alpine Grasslands and Shrublands.

Alpine Grasslands and Shrublands include all vegetated areas above the upper treeline in the highest mountains, as well as significant expanses of grassland just below the upper treeline within the sub-alpine zone. Upper treeline is defined as the elevation above which trees are unable to grow in an upright form. Alpine vegetation is dominated by sedge species, grasses, hardy forbs and/or dwarf shrubs such as heathers. This habitat type includes the krummholz. Alpine Grasslands and Shrublands occur in the high mountains throughout the Cascades, Olympic Mountains, Okanogan Highlands, Wallowa Mountains, Blue Mountains, Steens Mountain in southeast Oregon, and occasionally in the Siskiyous. Alpine heath communities are found primarily along the Cascade crest and west, especially from Mt. Rainier north. Sub-alpine and alpine wetland habitat occurs throughout the range, and is more common in the high mountain ranges of Washington.[35]

Most natural disturbances in Alpine Grasslands and Sub-alpine Parklands involve weather events or animal activities, and usually have small-scale impacts. Frost heaving can have small-scale but important effects on alpine vegetative communities.[49] Wind blasting and extreme variation in snow pack between years in high elevation habitats can kill and desiccate plants or shorten growing seasons. Because seedling establishment is generally poor, most alpine plants also propagate by clones or stolons.[81] Trees often invade from wind-dispersed seeds, but seed dispersal by mountain birds such as Clark's nutcrackers may determine upper treeline, especially for whitebark pine.[87] Tree invasion rates into Sub-alpine Grasslands are slow compared to sub-alpine tree or shrub communities.[56, 83] Herbivory and associated trampling disturbance by elk and mountain goats creates patches of open ground.[69] Avalanches and snow-slumping convert coniferous forest to open meadows or deciduous forests providing nutritious forage for breeding or migrating birds, rodents, bears, and ungulates.[82] Sub-alpine grasslands burn on occasion, but since 80-90% of sub-alpine plant biomass is underground, fire does not affect the structure of sub-alpine grasslands greatly. In Sub-alpine Parklands, fire suppression has contributed to changes in habitat structure and function. During wet climatic cycles, reduced fire frequency can lead to tree islands coalescing and, parklands becoming a more closed forest.[1] Area of alpine grassland, however, may be increased by fires in Sub-alpine Parklands.[83] Periodic shifts in climatic factors such as drought, and depth or duration of snow pack may either lower treeline or allow tree invasions into meadows and shrublands creating more parkland habitat.

Status and trends. There has been little change in abundance of alpine habitat over the past 150 years. Most areas are dominated by native species and are still in good condition.[76] Current trends for most alpine grasslands are considered stable, but there are increasing threats from recreational pressures and livestock grazing, and possibly some slow loss of sub-alpine grassland to recent tree invasion. Conditions are changing, however, as over the past half century, mountain recreational activities have increased markedly. Some alpine areas on the eastside are degraded physically. Recreational impacts are noticeable in some national parks and wilderness areas. Mt. Rainier National Park gets two million visitors per year, most of whom converge on Paradise and Sunrise sub-alpine

Table 1. Areas of high elevation habitats, and elevational ranges for Washington and Oregon[a]

	Washington				Oregon			
	Acres	Sq. mi	Km²	% Protected[b]	Acres	Sq. mi	Km²	% Protected[b]
#9. Sub-alpine Parklands[c]	327,442	512	1325	58.2	84,240	132	341	57.5
#10. Alpine Grasslands and Shrublands[d]	1,599,115	2,499	6,471	81.3	291,494	455	1,180	56.0
Total high elevation area	1,926,557	3,010	7,796	77.4	395,734	618	1,601	56.3
Total state land base	43,164,632	67,443	174,678		61,974,831	96,834	250,800	

[a] Kiilsgaard (1999).

[b] Percent of habitat type protected in Category 1 and 2 (Shaughnessy and O'Neil 2000).

[c] Elevation west of Cascade Crest (1372-1829 m; 4500-6000 ft), East of Cascade Crest (1524-2438 m; 5000-8000 ft).

[d] Elevation: 1524 m (5000 ft) to > 3050 m (10,000 ft).

Table 2. Living and breeding in high elevation environments: constraints, consequences and wildlife adaptations

A. Environmental Constraints
Cold and extreme temperatures
High winds
Open habitats
Strong temporal resource gradient (melting snow fields)
Strong spatial resource gradient (food phenology)
Aridity
UVB light
Hypoxia
Airborne toxins
Fragmented habitats

B. Ecological consequences
High energetic costs for living and breeding
Cooling and warming adaptations required
Patchy distribution and strong seasonality of resources
Low parasitism levels, good health
Delayed breeding schedules
Increased reproductive synchrony
Increased reproductive stochasticity
Fewer broods/litters per season
Longer development times
High predation risk on young and adults
Small, low density populations
Need to disperse across unsuitable habitats

C. Biological adaptations
Physiological
Increased visual acuity
Large thermal neutral zone for heat and cold
Biochemical adaptations to hypoxia
Molting patterns for cryptic plumage
Night-time torpor
Flexible reproductive timing
High tolerance for environmental toxins
Cold-tolerant embryos, antifreeze solutions
Structural
Larger body size
Thicker, warmer pelage/plumage
Appendages modified for wind stabilization
Behavioral
Cryptic behaviour
Non-directional vocalizations
Social and or sub-nivean roosting
Energy-minimizing behaviors such as gliding and walking uphill

D. Life history adaptations
Cryptic plumage/pelage
Extensive molting patterns
Increased parental care
Monogamous mating system
Strong seasonality in habitat use
Hibernation or dispersal in winter
Increased longevity with elevation
Strong age dependence and senescence
Excellent dispersal/migration abilities
External recruitment for rescue

meadows. Sedge turfs are the most resilient to trampling and heaths the least resilient. When alpine heath is opened up to bare ground it typically does not revegetate for decades, unless active restoration programs are initiated. Sunrise Meadow in Mt. Rainier National Park has had success in restoration of the sub-alpine meadows.[78, 114] The major human impacts to high alpine grasslands are trampling and associated impacts caused by tent sites on Mt. Rainier National Park.[78] Exotic ungulates, such as the mountain goats introduced to the Olympic Mountains, can have profound impacts on high elevation grasslands.[69] Grazing by domestic animals also has negative impacts,[19] but these effects are not well studied in the Pacific Northwest. Sub-alpine Parkland habitat trends are believed to be generally stable. Whitebark pine may be declining due to effects of blister rust or fire suppression.[83] For Alpine Grasslands and Shrublands, only one of 40 plant associations listed in the National Vegetation Classification is considered threatened, and less than 10% of the Pacific Northwest Sub-alpine Parkland community types are identified as threatened.[65]

3. Alpine Environments and Wildlife Adaptations for High Altitudes

Alpine environments are characterized by low mean temperatures, high winds, prolonged snow cover, and intense ultra-violet radiation[12] (Table 2A). Topography varies from steep to gentle, and winter snow pack ranges from several meters in gullies to exposed windswept ridges. Although the vegetation structure is simple, variation in slope and aspect results in rapid changes in habitat types with only modest changes in elevation, thus generating significant structural complexity to alpine habitats. Steep local gradients in snow cover and soil moisture often govern the productivity and distribution of alpine plants.[11] Precipitation is more important than temperature for alpine plant growth, as most plant species are cold tolerant and can carry on photosynthesis down to -6°C.[12]

Wildlife living at high elevations must be able to cope with high winds, cold temperatures, and desiccation, since often little precipitation originates from rainfall and it drains quickly. Although alpine soils are normally cold, daily temperatures on the ground vary over a range of 47°C[144] (Box 1). Thus, during mid day, overheating can be a problem for alpine wildlife species. On elevational gradients, resources can be patchily distributed in narrow

1. Getting by on high: Avian adaptations to high altitude and life history adjustments

Alpine and arctic birds face similar problems of living in habitats characterized by low temperatures, high winds, short growing seasons and delayed breeding schedules. At high altitudes, hypoxia and aridity further elevate energetic costs for survival and reproduction. Physiological adaptations and life history traits of white-tailed ptarmigan in the Colorado Rocky Mountains (3600-4400 meters) are compared with arctic willow ptarmigan (0.5-800 m). Ptarmigan embryos in the alpine must develop at cellular oxygen tension levels that are normally lethal for tissue maintenance.[32, 33] To deal with hypoxic conditions, white-tailed ptarmigan embryos have elevated levels of blood hematocrit and citrate synthase enzyme in muscle tissue at a much earlier stage of development than willow ptarmigan that show similar levels to other avian embryos at low elevations. Hematocrit and red blood cell organic phosphate levels decline when white-tailed ptarmigan clutches are incubated at lower altitudes, indicating that high elevation levels are a physiological acclimatization to hypoxia, rather than a genetic adaptation.[33, 47] Daily temperatures in the alpine range from -2°C to >45°C and, females adjust incubation schedules based on nest site type and ambient temperature.[145] Cooling as well as generating heat to maintain homeothermy is likely a problem for most alpine vertebrates. Embryo viability does not differ between the two species, but living at high elevation likely imposes significant life history costs. Alpine ptarmigan have smaller clutches, slower egg laying rates, longer incubation periods and higher predation than arctic ptarmigan.[97] Alpine birds show stronger age-dependent effects on survival and reproduction compared to arctic ptarmigan.[145]

bands of diverse habitat types that vary sharply in time within a season (e.g., plant phenology or insect emergence; Table 2B). Spatial and temporal variation in resources can extend the availability of food and cover to wildlife, but also requires good mobility and longer migration distances from patchy breeding habitats to winter areas. Alpine environments also show significant stochasticity in environmental conditions annually. Some years have low snow cover leading to an absence of cover for sub-nivean species and, in other years, like 1999 in the Pacific Northwest, much breeding habitat remained snow-covered for the entire season (K. Martin unpubl. data). At the highest elevations, hypoxic conditions add an additional energy cost.[26, 97] These factors result in short, intense breeding seasons for wildlife.

Animals respond to living in extreme environments in several ways. They can migrate to warmer environments, hibernate, or stay active all year. Alpine animals that stay active all year have developed biological and life history adaptations to survive in these extreme habitats (Table 2). Vertebrate species living at high elevations have developed impressive physiological, behavioral and morphological adaptations to conserve energy (Table 2C). The type of adaptation may increase with elevation, or with increased time spent at high elevation. For example, species that use low elevation alpine, or high alpine for only short periods, may require only behavioral adjustments like moving to more benign or sheltered sites to conserve energy.[38, 86] Hummingbirds exploiting the rich resources in sub-alpine meadows during migration go into nighttime torpor if their energy reserves at nightfall are below a critical set point.[34] With less vegetative cover for concealment, many alpine mammals and birds have developed cryptic appearances and behaviors to enhance blending into the landscape. Although ptarmigan are renowned for their cryptic plumage, larger mammals such as bighorn sheep and coyotes blend exceptionally well in these open landscapes. Often species living in the alpine such as marmots, pika and ptarmigan have territorial and social contact calls that are hard to localize, which may reduce risk of detection by predators when vocalizing.[22] Like arctic wildlife, animals living at high elevations develop fat deposits, extra feathers or thicker fur to increase insulation. Red foxes in the alpine develop a thick pelage and a faded cryptic gray coat color.

True alpine species may develop structural adaptations. Species living at the highest elevations for extended periods make biochemical adjustments such as increasing their blood haematocrit concentrations[26, 32, 33] (Box 1). In the Himalayas, alpine passerines tend to have long pointed wings for efficient flight to scattered resources and improved flight performance in strong and variable winds. Finches and chats at the highest elevations have square-ended or only shallow forked tails for flight stability in wind, and strong hind limbs with small feet for ground foraging.[84, 85, 86] Winds may pose problems for smaller animals, but larger animals such as raptors use wind to remain aloft and move efficiently.

The reduced time for breeding within a season may alter life history traits. Generally, vertebrate species living at high elevations exhibit lifestyles where it takes longer to achieve independence, there is lower expected annual fecundity and, at least for some species, accompanied by longer life spans and more developed social behaviors. White-tailed ptarmigan, a high alpine grouse, showed slower laying rates, smaller clutches, longer incubation periods with consequent higher reproductive failure compared to arctic breeding willow ptarmigan[97] (*L. lagopus*; Box 1). A study of golden-mantled ground squirrels across five elevations in the Sierra Nevada in California showed reduced time above ground for adults and juveniles, later age of maturity, lower litter size and greater survivorship of females with increasing elevation.[24] Parental care patterns vary with elevation in marmots.

2. Sleeping through winter with family members: Marmot hibernation and sociality

Marmots are large diurnal, ground dwelling rodents that occupy middle and upper elevation alpine and sub-alpine mountains across North America and Eurasia. At the high elevations in Washington and Oregon, the three marmot species living in alpine and sub-alpine habitats have non-overlapping ranges. Yellow-bellied marmots occur in habitats #9 and #10 only in Oregon, and in Washington are replaced in the high elevation habitats by hoary marmots, which live only in the Cascades. The Olympic marmot is restricted to the Olympic Peninsula. Marmots are the largest true mammalian hibernators, with species spending from 4.5 to 8.5 months per year underground.[4] Marmots thus have a short active season above ground, on average 4.8 months, when they must grow, reproduce, and prepare for hibernation. Marmots have complex social systems, delayed maturity of young, and flexible patterns of dispersal and recruitment as a consequence of their large body size and living in extreme and patchy environments.[4, 6, 7] The degree of sociality across the 14 marmot species correlates with environmental harshness and increases with altitude and latitude.[8] The low elevation woodchuck (*Marmota monax*) is solitary, young of the year disperse from natal burrows, and individuals usually reproduce after their first hibernation.[51] The yellow-bellied marmot, the smallest marmot, lives at medium to high altitudes[57] and is moderately social. The basic social units are polygynous groups usually developed through recruitment of daughters. Dispersal of young typically occurs after the first hibernation.[4] European alpine marmots (*M. marmota*), living at the highest elevations, are the most social species with up to 20 individuals in a group defended by a dominant male and female.[8] Juveniles typically do not disperse for two or more years, and their over-winter survival is enhanced by increased group size in joint hibernacula.[6, 7, 8] Like birds, the principal life history costs incurred by marmots living in alpine and sub-alpine environments appear to be time delays. Young require more (longer) parental care at higher elevations and thus remain with their parents or extended family for one or more years after they reach maturity.[4, 8] Delayed dispersal results in groups of high relatedness, which in some species leads to reproductive suppression of subordinate members in the group.[4] Reproductive skipping between years occurs in hoary and Olympic marmots due to extreme conditions, but not in yellow-bellied marmots.[4] Higher longevity is observed in high elevation marmots, and this may be due to delayed dispersal, reproductive skipping, or higher sociality that may reduce predation risk.[4]

Alpine marmots have prolonged parental care, live and hibernate in larger groups and take longer to reach maturity compared to marmot species living at lower elevations[4, 6, 7] (Box 2).

Among mountain ungulates, there is considerable variation in patterns of parental investment and time to offspring independence in high and low elevation populations, but it is not clear what influence alpine habitats have in explaining these patterns. In order to maximize maternal survival, mountain goats may reduce their investment in reproduction with increasing altitude, but species spending more time at high elevations may prolong the period of maternal care to young.[52, 60] Delayed primiparity may be the life history trait most likely to vary with elevation for ungulates (M. Festa-Bianchet, pers. comm.). One expects other life history tactics such as increased probability of bet hedging against the regular but stochastic reproductive failure,[24] but such demographic parameters are usually not available for species living at high elevation.

Despite the tendency for small population sizes, birds adapted to live in naturally fragmented alpine habitats appear to have well developed dispersal abilities (Box 3). Pikas also are excellent dispersers,[118] but generally dispersal from and to alpine habitats is more challenging for alpine mammals than birds. The frequent speciation that occurred in the high elevation marmots (Olympic, Hoary, Yellow-bellied, and Vancouver Island) and ground squirrels (Beldings, Columbian, golden-mantled and Cascade golden-mantled) may have been in response to isolation imposed by dispersal barriers.

3. Advantages to Living and Breeding in Alpine Habitats

Why would any animal live at high elevation given the extreme environmental conditions and the high energetic costs to living there? Is it worth it? The answer is that there are definite advantages to living in high elevation environments. In winter, despite extensive snow pack, most alpine areas have wind-swept ridges with exposed herbaceous stems and seeds for foraging, and as winter progresses, increasing snow levels allow herbivores access to new layers of vegetation. In spring, vast numbers of cold-numbed insects are swept up from lower elevations to land on high elevation snowfields. These insects provide a nutritious and abundant food supply for alpine-breeding birds and mammals.[104,122] Snowfields melting through the season create a gradient in plant phenology that provides

3. Patterns and mechanisms of rescue and dispersal in alpine ptarmigan

The periodic "rescue" of populations that are declining with recruitment from productive populations may be an important feature of population biology for many species. Dispersal and recruitment are crucial life history parameters for persistence of populations or metapopulations, but data on patterns and mechanisms are lacking for most vertebrates. We studied four populations of white-tailed ptarmigan breeding in highly fragmented alpine habitats, with corresponding small populations, in the Rocky Mountains of Colorado, from 1987 to 1996. Multiple populations allowed us to distinguish between local and regional events. Populations showed dramatic variation in offspring production and local survival (return) of adults across years and sites, yet populations remained relatively stable. Variation in reproductive success and survival were unlinked across sites and appeared to be driven mainly by internal ecological processes such as depredation of eggs and young.[22] Sites varied temporally in predation risk,[22] perhaps because the generalist predators that prey on ptarmigan live for several years and their hunting skills likely improve with age. Computer simulation models of individual populations predicted that all populations should go extinct in 2 to 10 years, assuming no linkage between them. Treating the four populations as different fragments of a single large closed population resulted in predicted persistence times of about nine years.[98] Yet all populations have persisted for at least 30 years and likely substantially longer. In one site, the population remained stable despite zero production the previous year and no return of breeding females.

Ptarmigan have adapted to breeding in highly fragmented and stochastic alpine habitats with a system of extensive external recruitment that functions at a landscape scale. Local populations were able to avoid extinction due to external recruitment, since about 95% of females and 75% of males recruiting to sites did not originate from any of the studied populations. Excellent dispersal abilities allowed recruitment from populations during a productive episode to "rescue" populations at risk of collapse were they dependent solely on internal recruitment. Studies of radio-marked juveniles in Colorado revealed a spatial scale of about 35 km (22 mi), and movement patterns of juveniles were not related to locations or movements of their mothers.[95] The enabling factors for this extensive external recruitment appear to be low costs to dispersal and low benefits to philopatry. Observations and experiments on mate and territory replacement showed no apparent reproductive costs to females that switch mates or territories[66] (K. Martin unpubl. data), suggesting that dispersal to unfamiliar sites is not costly. Populations will be maintained as long as there is a balance of populations producing recruits in reasonable proximity to populations requiring rescue. Rescue by external recruitment appears to apply to other grouse such as red grouse (*Lagopus lagopus scoticus*) , sage grouse and capercaillie (*Tetrao urogallus*),[109, 125] and perhaps generally to other taxonomic groups. For small populations in stochastic environmentals, maintaining connectivity between populations is crucial because populations producing recruits must be within reasonable proximity to those requiring rescue.[95]

an extended supply of high quality food for herbivores that can migrate along the green-up line. In late summer, leaf budding, flowering, and fruiting of huckleberry (*Vaccinium deliciosum*) or bearberry (*Arctostaphylos uva-ursi*) may co-occur in close proximity grading away from the edge of a snowfield. Avalanche chutes at high elevation provide lush vegetation adjacent to forest cover for bears, marmots and songbirds. Availability of spring forage may be crucial for breeding in a number of species such as bighorn sheep and mountain goats.[52, 110] Generally, compared to other habitats, there are few intraspecific or interspecific competitors relative to food supplies.

Predation risk in the alpine may or may not differ from lower elevation sites, but it appears easier for prey to detect aerial and mammalian predators in open alpine habitats. It is a challenge for medium and large predators such as foxes, mountain lion and wolves to approach prey

undetected, although they are clearly successful in doing so. In northern British Columbia, caribou give birth to their calves at alpine sites because they are considered to be refugia from predators, and maybe also from biting insects.[117]

Alpine habitats appear to be healthy environments with low levels of parasitism or disease, at least for alpine grouse, which have few blood infections or intestinal parasites.[22, 123] Blood parasite infections in white-tailed ptarmigan varied with degree of fragmentation in the low elevation coastal alpine on Vancouver Island as 84% of 25 birds sampled on isolated mountains in the north and south had blood parasites compared to only 57% of 53 individuals in the more continuous alpine in Strathcona Provincial Park (K. Martin and N. J. K. Braun, unpubl. data). The higher rates on smaller, more isolated mountains were likely related to ptarmigan mixing more with blue grouse at these sites. In contrast to birds, parasites and infectious viruses can be a problem for alpine marmots[8] and other mammals. Alpine ungulates, particularly bighorn sheep, are susceptible to a range of infectious diseases including pneumonia and blood and organ parasites, especially when animals are in contact with domestic livestock.[54, 107, 113]

Relative to the arctic, where long distance migrations to winter sites are required for most seasonal breeding species, winter habitat for alpine species often occurs in close proximity to breeding sites. In alpine habitats in the Pacific Northwest, animals may simply descend one or a few kilometers to habitats that offer greater shelter or more benign weather. In the low elevation coastal alpine in central Vancouver Island, white-tailed ptarmigan moved around the mountain to a southern or western aspect to winter in openings in the montane forest or parkland only 100 to 400 m (300-1000 ft) lower in elevation than their breeding habitat. In the Cascades Range, ptarmigan likely make similar movements in winter.

Despite their reputation for unfavorable thermal conditions, alpine habitats offer thermal advantages for wildlife species in both winter and summer. Winter is reliably cold, usually remaining below freezing for periods of seven to eight months. Thus arousal for hibernating animals is likely to occur at appropriate times. There is generally a reliable supply of snow to provide well-insulated and safe sub-nivean habitats. Summers are reliably cool and breezy and, a number of birds and mammals escape the late summer heat at lower elevations by moving up to alpine meadows to forage and to escape insects. Temperature inversions are common in mountain landscapes and provide an exception to the general rule of decreased temperature with increased elevation. A thermal belt results in warmer temperatures at mid-slope than either the valley bottom or the upper slopes.[111] On the north slope of Mt. Hood in Oregon, a sharply delimited thermal belt between the Hood River valley bottom and the upper slopes allows suitable growing conditions for such crops as cherries.

In winter, thermal inversions occur in high elevation areas, and temperatures may remain high for several weeks. In the Northeastern Asian highlands, temperatures at timberline from December to February can be 25-30°C warmer than the low forest conditions of -50°C. Moose migrate to high elevation habitats during thermal inversions, possibly to capitalize on the favorable temperatures, although they may also be attracted to sub-alpine shrublands (A. Andreev, pers. comm.).

In summary, there may be high ecological costs to living in open habitats at high elevations as the need to move for food or cover may result in increased risk of detection by predators. However, alpine animals likely experience lower levels of inter-specific competition and less habitat degradation than wildlife occupying lower elevation habitat types.

4. Resident Wildlife and Migrant Visitors to Alpine and Sub-alpine Habitats

Year-round and summer residents. Only a few North American bird species, white-tailed ptarmigan, American pipit, black-crowned rosy finches, gray-crowned rosy finches, the alpine sub-species of horned lark (*Eremophila alpestris alpinus*), and golden-crowned sparrow breed exclusively in alpine and sub-alpine habitats. Of these, only ptarmigan are year round residents (Figures 2, 3). Mammals, such as hoary marmots, Olympic marmots, Cascade golden-mantled ground squirrels, mountain goats, pikas and water voles, associate primarily with high elevation habitats year-round. Pikas can live at low elevations if appropriate habitat features occur, such as talus, but high elevation populations are non-migratory. Overall, more species of mammals than birds are closely associated with high alpine habitats (Table 3). There are no true alpine amphibians or reptiles in the Pacific Northwest, and only the Cascade frog is closely associated with sub-alpine parkland habitats (Table 3).

Alpine songbirds face special challenges to survive and breed in alpine environments. With their small body size they must cope with cold temperatures and high winds as well as incubate small eggs that cool rapidly in near freezing ambient temperatures. All alpine and sub-alpine passerines provide bi-parental care to their altricial young. True alpine passerines (birds breeding only at high

Figure 2. White-tailed ptarmigan in winter plumage and in winter habitat at Guanella Pass, Colorado, with Mount Evans (breeding habitat) in the background. Photo by K. Martin, January 1998.

Figure 3. Male (left) and female (upper right) white-tailed ptarmigan in cryptic spring breeding plumage feeding on alpine territory, Loveland Pass, Colorado. Photo by Robert E. Bennetts.

elevations) are hardy and appear to survive well despite storms and inclement weather. They take shelter from wind beside or under rocks, and sit on dark rocks to warm up in cold sunny weather (K. Martin, unpubl. obs.). A study of alpine finches showed ecological and morphological differentiation across an elevational gradient in the Himalayas with the heaviest species occupying the highest elevations.[84, 85, 86] Both sexes of the alpine sub-species of horned lark, which lives only in the Washington Cascades and Olympic Mountains, have longer wings and tarsi than low elevation sub-species in the same region.[10]

Most alpine passerines are both granivorous and insectivorous.[136] Within alpine sites, species tend to use different microhabitats and foraging modes. Rosy finches, associated with steep rocky slopes, are slow searchers for insects and seeds, while horned larks and pipits are active searchers in more grassy areas.[38] Wheatears (*Oenanthe oenanthe*) frequent open slopes and plains and pounce from boulder perches to capture ground and low-flying insects. Food supplies in alpine sites for songbirds are plentiful generally, but of short duration.[38] Receding snowfields in summer continually uncover new food for rosy finches and pipits.[133] Many alpine passerines forage on chilled and dead wind-blown arthropods deposited on snowfields.[122] Verbeek[135] emphasized the importance of having snowfields to provide an insect food supply within territories of pipits in Wyoming. However, horned larks and pipits foraged more frequently in snow-free areas than on snow on Mt. Baker in Snoqualmie National Forest, although both species did use snowfields in late afternoon after updrafts from low elevations deposited insects on snow.[104] The proportion of energy budget provided by chilled insects on snowfields was not determined in any of these studies.

With such limited vertical structure in the alpine, passerines need to choose nest sites that are protected as much as possible from predators and extremes of wind, precipitation, or temperature. Rosy finches and mountain bluebirds nest in inaccessible cliffs or rock crevices, while pipits and horned larks nest on open grassy tundra slopes that are potentially more exposed to predators and climatic extremes. Johnson[74] reported 81% of 139 rosy finch nests were in cliff crevices, and suggested that rodents might limit rosy finch nesting to cliffs. Nest sites in cliff crevices and under rocks are colder thermal environments than open cup nest sites in the alpine and arctic tundra.[93, 144] Water pipits (*Anthus spinoletta*) in Austria spent three to four weeks choosing potential nest sites within their territories.[14] Nest sites were on steep slopes and oriented NW-NE, providing shelter from drifting snow or rain. All nests were placed below ground level with a solid nest roof resembling a nest cavity more than a nest cup and providing considerable insulation against the cold alpine temperatures.[14, 136] American pipit nests are sunk into the ground on steep grassy slopes.[99, 135]

The breeding schedules of alpine bird species are three to six weeks later than those for related species at low elevations. Initiation of egg-laying in white-tailed ptarmigan populations above 3200 m (10,500 ft) in Colorado may begin on 25 May and extend to 11 July (mean of 8 June; 5 years),[22] about six weeks later than ruffed grouse, a low elevation ground nesting tetraonid, that commences egg laying by 10 April and extends to 19 July in British Columbia.[28] Some alpine species extend breeding later into the season to compensate for their late start in spring. There was no overlap in laying dates for

Table 3. Vertebrate species use of sub-alpine and alpine habitats in Washington and Oregon[1]

	Mammals				Birds				Herptiles			
	A[2]	B	C	D	A	B	C	D	A	B	C	D
Sub-alpine parklands (#9)	11	39	15	65	7	48	47	102	1	7	3	11
Alpine grasslands (#10)	13	22	14	49	5	16	52	73	0	4	1	5
Number of species				65				112				11

1. Associations derived from Expert Panels of the Species Habitat Project and supplemented by atlases from Washington (Smith et al 1997), Oregon (Csuti et al 1997), British Columbia (Campbell et al 1990, 1996), and Colorado (C. E. Braun 1969), as well as unpublished field data from K. Martin (University of British Columbia—Website: http://www.forestry.ubc.ca/alpine/index.htm), and field counts from L. Steiner, Washington.

2. A = Closely Associated; B = Generally Associated; C = Present; D = Total.

rosy finches in the Montana Rocky Mountains (egg laying 16 June to 6 July), over seven weeks later than on Amchitka Island, Alaska (28 April to 26 May).[75] The alpine sub-species of horned lark at 3200 m in Wyoming laid eggs between 25 June and 10 July, up to three months later than low elevation populations.[10, 134]

Some mountain goat and bighorn sheep herds remain at high elevation year-round, subsisting in winter on the windswept alpine grasslands and meadows. Herds consist of small groups of bachelors or female with young. Mountain goat populations may be limited by quality and quantity of winter food as suggested by Houston and Stevens,[69] but recent work indicates that spring weather or timing of access to new plant growth in spring is more important than winter conditions[110] (M. Festa-Bianchet, pers. comm.).

Species breeding across wide elevational gradients. Some birds and mammals range naturally across wide elevational gradients with their breeding habitats sometimes extending from coastline into the alpine zone (Table 3). Allee and Schmidt[2] referred to this group as 'alpine-tolerant'. Examples include Canada geese, blue grouse, rufous hummingbird, killdeer, dipper, mountain bluebird, rock wren, and Townsend's solitaire, and a variety of mammals such as long-legged *Myotis* bat, white-tailed jackrabbit, yellow-bellied marmot, Beldings and Columbian ground squirrels, northern pocket gopher, western jumping mouse, heather vole, long-tailed vole, fox, coyote, black bear, grizzly bear, wolverine, mountain lion, and big horn sheep. At high elevations, Columbian ground squirrels have lower litter size, lower female body weight, lower proportion of young females breeding, and higher adult survival relative to low elevation populations.[45] Some within-species differences in ground squirrel life history with elevation, such as litter size and proportion of young females with litters, disappeared when high elevation populations were food-supplemented.[44] They suggested that ground squirrel populations showed a phenotypically plastic life history response to variation in food availability with elevation. Pika range from high alpine to sea level where there are grasses and forbs close to talus. Pika show similar life history patterns to Columbian ground squirrels as high alpine populations have smaller litters and greater longevity of adults than those in low elevation talus.[119, 120]

There are few directed studies for 'alpine tolerant' birds on whether and how life history or behavior shifts with increasing elevation. However, we can make some inferences. Breeding schedules such as arrival dates and initiation of breeding are significantly later with increases in elevation. Garden warblers (*Sylvia borin*) breeding in Switzerland at 4,921 ft (1500 m) arrived on territories about three weeks later than birds breeding at 656 ft (200 m).[142] The high elevation population of garden warblers did not extend their egg laying period beyond lower elevation populations and thus, had one or two fewer broods in a season than low elevation pairs.[141] If restricted gene flow results in local adaptation, clutch size or offspring survival is predicted to be higher in high elevation environments

to compensate for reduced number of breeding attempts per season. However, clutch size did not increase with elevation in garden warblers.[141] Despite less favorable thermal conditions, both incubation and nestling periods were one day shorter in high elevation garden warblers compared to lowland populations; this was attributed to higher levels of male parental care and thicker nests.[141] Nest structures in the high elevation population weighed almost twice as much as those at 200 m and presumably conferred greater insulative warmth. Success of individual nesting attempts was good, but the frequency of total failure for a season was higher at increased elevation.[141]

For most 'alpine-tolerant' species, fewer broods per season are produced at higher elevations. For example, American robins breeding at low elevation in British Columbia can produce up to five clutches and three broods[28] (J. N. M. Smith, pers. comm.), but in the sub-alpine they may only have time to produce one brood, or possibly two clutches. Dippers and horned larks also produce multiple broods per season at low elevations and single broods at high elevations.[77, 134]

Population densities of species that breed across a range of elevations are expected to decrease with increasing elevation, but this was not found for garden warblers.[142] Densities of territorial male black redstarts (*Phoenicuros ochruros*) in Austria increased with elevation.[79] Densities of alpine chats also remained high with increasing elevation in the Himalayas.[84] However, these studies had small study plots and reported densities might be inflated as a result of plots being located in optimal habitats. Across elevational gradients, individuals inhabiting higher elevations were not younger or less competitive birds unable to obtain good territories at lower elevations. High elevation populations of garden warblers in Switzerland and redstarts in Austria did not contain a higher than average proportion of first time breeders, and natal philopatry and breeding site fidelity of birds in the upper elevations was high.[79, 142]

Poikilothermic (cold-blooded) animals such as frogs, salamanders, and snakes have special problems moving and developing eggs in cold alpine climates as low temperatures result in prolonged development times for amphibians and reptiles. In alpine habitats, frogs commonly pass the first winter as tadpoles and require an extra year to achieve full development,[2] in contrast to lower elevation individuals that usually achieve full maturity in their first year. In the Alps, the three species of reptiles, green lizard (*Lacerta vivipara*:), common viper (*Vipera berus*) and the blindworm (*Angius fragilis*) that reach the alpine zone are all viviparous (bear live young).[2] Reptile eggs cannot develop and hatch in such cold climates, but when viviparous female snakes or lizards at high elevations retain eggs in their body, they can speed embryonic development by basking in the sun during the day, and moving to sheltered locations at night.[111] Interestingly, the green lizard is oviparous (egg laying) throughout most of its lower elevational range in Europe. Snakes in the European Alps show delayed sexual maturity and reproduce only every two to four years in the alpine, rather than annually as in low elevation populations.[31, 116] Cold-blooded animals living in mountain environments, including amphibians, reptiles and invertebrates are almost universally dark-colored. Melanism apparently contributes to heat absorption and helps protect against ultraviolet radiation.[2, 111] For hibernation, cold-blooded animals move under rocks or into animal burrows. They tend to aggregate in multi-species groups (insects, snakes, and lizards), and one can find predators and prey congregated under one rock for winter hibernation.[111]

In Washington and Oregon, five species of amphibians and reptiles occur in alpine grasslands and 11 species of salamanders, frogs, toads, and garter snakes live in sub-alpine parklands if appropriate habitat features such as standing water are available. Western toads and Pacific Treefrogs breed in high elevation ponds.[39] Western toads occur at elevations up to 7,400 ft (2255 m)[39] in Washington, and in Colorado to 12,000 ft (3658 m, C. E. Braun, pers. comm.). Garter snakes occur in the alpine and sub-alpine on southern and central Vancouver Island at elevations ranging from 5,056 to 5,904 ft (1,540 to 1,800 m, K. Martin, unpubl. data). Little is known about the reproductive success and specific adaptations to life in alpine and sub-alpine habitats of amphibians and reptiles in the Pacific Northwest.

In sum, the ecology, behavior, and life history of the upper elevational ranges for 'alpine-tolerant' vertebrates is poorly understood. Quite possibly there is more restricted gene flow and greater local adaptation than we realize. Other species, in addition to horned lark, marmots and ground squirrels, may have well differentiated high elevation sub-species or eco-types yet to be discovered. We cannot calculate relative reproductive success and survival for wildlife species living at high and low elevations, mainly because longitudinal studies of individuals at high elevation sites have not been done.

Seasonal migration of wildlife to high elevation habitats. The seasonal migration of vertebrates to alpine and sub-alpine habitats has been noted,[108, 111] and naturalists acknowledge the use of alpine habitats by migrating birds.[30] Mammalian wildlife also move up to high elevations seasonally. However, the importance of the phenomenon is not widely acknowledged and, the ecological processes involved have not been determined. At Chinook Pass and Lake Valhalla, north of Stevens Pass in the Cascades Mountains, Washington, 32 bird species were observed during 8 surveys in July, September and October (Les Steiner, unpubl. data). During field studies in central and southern British Columbia, over 113 bird species were recorded foraging or resting in high elevation habitats over a three-month period (Box 4). About 80% of bird species observed in late summer do not breed at high elevation. Some species were clearly migrants from higher latitudes using alpine sites as migration corridors, including arctic shorebirds like Baird's sandpipers and greater yellowlegs, and northern raptors such as sharp-shinned hawks and northern goshawks. We suspect a high proportion of birds arriving in the alpine were elevational

Table 4. Characteristics of alpine and sub-alpine habitats in late summer for migrants and residents

1. Rich food resource gradients—plants, insects and small prey
2. Open habitats to detect predators
3. Energetically more favorable for ascent (updrafts)
4. Habitat features similar to high latitude environments

migrants of local origin. Some forest birds, such as nuthatches and pine siskins, certainly moved up the mountain to track emerging food resources (insects/flower/fruit) from snow melt and green-up. For North America, over 200 species of birds were recorded using alpine and sub-alpine habitats in late summer (literature and field surveys, web site: http://www.forestry.ubc.ca/alpine/index.htm).

There are a number of reasons why alpine habitats are suitable for fall migrants (Table 4). In late summer, food resources have declined at lower elevations whereas in the alpine, late July to early September is the peak period of flowering and invertebrate emergence.[81] Migrants from the north use alpine and sub-alpine habitats as refueling points. The proximity of alpine habitats to forest, grasslands, shrub steppe or coastal habitats at lower elevations means that animals can move between these habitats in a short time. In late summer, few robins are observed on low elevation coastal breeding areas, a time when large numbers are observed in nearby alpine and sub-alpine habitats, as well as in lowland berry-rich wetlands (J. N. M. Smith, pers. comm.). Rufous hummingbirds move up to sub-alpine meadows and parklands in late summer and defend territories around flower patches.[128] Territorial boundaries of migrant hummingbirds shift daily in relation to temporal patterns of flowering plants such as paintbrush (*Castilleja* spp.) or columbine (*Aquilegia* spp.).[61, 62] From mid August to early September, alpine grasslands support large numbers of grasshoppers that are eaten by kestrels, mountain bluebirds and Townsend's solitaire (S. Ogle, pers. obs.). Yellow-rumped warblers, common passerine migrants in the alpine, fly-catch from cliff edges in late September. The increasing abundance of prey attracts predators. Migrating raptors use the open mountain terrain to hunt for land birds and small mammals, often using updrafts along cliff faces and ridges. The duration that individual altitudinal migrants spend in the alpine is unknown, but temporal patterns are most likely related to resource availability and may vary with species. Some altitudinal migrants could move to alpine and sub-alpine areas on a daily basis and spend their nights at lower elevations. Band-tailed pigeon could be one such species where availability of late summer fruits in the sub-alpine, such as huckleberry, crowberry (*Empetrum nigrum*), and bearberry, may help compensate for the loss of traditional fruit and nutbearing shrubs due to habitat loss at lower elevations.

In summary, about one-third of the vertebrate fauna in Washington and Oregon use high elevation habitats at some period in their life history. The chief period of occupation for many species is between late July and late

4. Elevational migration: use of high elevation habitats in late summer

Studies in British Columbia suggest that alpine areas support a high diversity and abundance of birds and mammals during late summer and fall. Since 1996, information was collected on temporal and spatial patterns of use of high elevation habitats by wildlife, especially migrating birds, on Vancouver Island and mainland British Columbia (K. Martin, University of British Columbia and Environment Canada). In 1996 and 1997, we collected data opportunistically on species use while censusing ptarmigan on 25 mountains across Vancouver Island, including relatively contiguous alpine in the central Island and more fragmented alpine habitats to the north and south. During this time, we observed 53 avian species with peak migration in these coastal alpine and sub-alpine habitats starting in late August and remaining high through September. Only 11 (21%) of these species regularly breed in high elevation habitats.[28, 130]

In 1998 and 1999, we surveyed alpine, sub-alpine and upper montane parkland habitats in mainland British Columbia ranging from 58° N latitude to the Washington border. We recorded a total of 107 species and 8,969 detections, using line transect sampling, over a total of 82 km (51 mi) in

alpine habitat, 69 km (43 mi) in sub-alpine, and 52 km (32.3 mi) in montane parkland (approximately 270 survey hours on 62 days, years pooled). Species richness was similar in all three habitats in 1998, a warm and relatively dry summer (Figure 4). Species richness was much reduced in 1999, a cold, wet summer with greatly delayed snowmelt and early onset of winter conditions. Except during peak migration (5-19 September), proportionately fewer species used alpine habitats than either the sub-alpine or montane in 1999 (Figure 4). In 1999, one interior site that was snow-free by late June had the greatest number of species (9 more species than in 1998). One coastal site, where the snow-line extended down to 1100 m (3610 ft.) in late August, had only 26 detections/km surveyed in 1999, compared to 93 detections/km in 1998. Overall, during a 10-week period in late summer, 113 species from a diverse array of avian families used coastal and interior high elevation habitats on Vancouver Island and the central and southern mainland British Columbia. Data summaries are available on Centre for Alpine Studies web site: http://www.forestry.ubc.ca/alpine/index.htm.

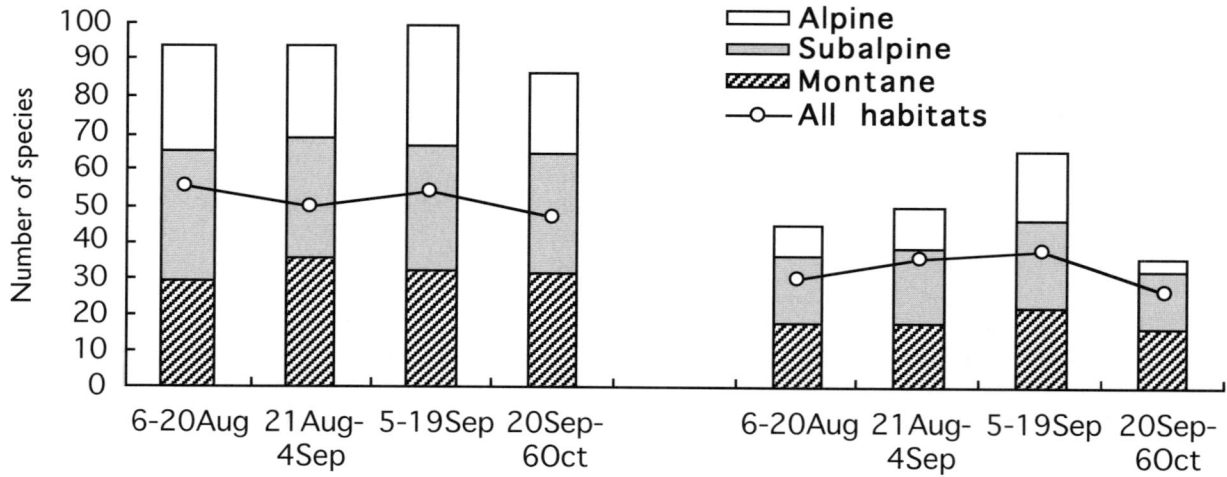

Figure 4. Avian use of high elevation habitats during migration in south and central British Columbia. Ten mountains were surveyed in 1998 (left), and five sites were re-censused in 1999 (right). Bars represent total numbers of species observed by habitat type for 10-day intervals. Since many species were observed in more than one of alpine, sub-alpine, or montane parkland habitats, lines represent the total species counted during each period.

October (Figure 4). Thus, the biodiversity of high elevation ecosystems is often greatly underestimated, and, clearly, we need to consider life history periods outside the breeding season.

Winter residents. Few wildlife species winter in alpine areas, and fewer still remain active in winter. Some birds, including white-tailed ptarmigan, raven, and rosy finch, remain at the highest elevations during winter, and a few arctic or northern species like snowy owls and snow buntings migrate south to use alpine areas in the Pacific Northwest. Rosy finches forage for exposed seeds on windswept alpine ridges.[58] Gray jays, raven, Clark's nutcracker, and hardy songbirds like golden-crowned kinglets and dark-eyed juncos use tree islands in sub-alpine parklands in winter. A number of mammals make use of alpine and sub-alpine habitats in winter, including pika, white-tailed jackrabbits, mountain caribou, and mountain goats, and their predators, foxes, coyotes, pine marten, wolverine, lynx, bobcat, and mountain lion, hunt along or above treeline. Mammals develop thick insulated fur coats and are adept at finding shelter from alpine winds and winter temperatures. Three *Myotis* bats hibernate in sub-alpine winter caves or mines in Washington and Oregon. Most species use snow roosts or snow burrows for sleeping or during storms. White-tailed jackrabbits winter as individuals at the interface of wind swept ridges and the krummholz.[18] For marmots and ground squirrels that hibernate overwinter at high elevations, choice of burrow and amount of snow cover can be crucial to overwinter survival.[4, 7, 8] Pika, weasels, and voles remain active in the sub-nivean layer where they have temporary reprieve from life-threatening predators and winter storms.

5. Threats/Status for Wildlife Species

Most local and regional scale threats to wildlife species in alpine and sub-alpine habitats relate to human activities (Table 5). These include on and off-site impacts due to ski resorts, hiking, mountain biking, climbing, fishing, group camping, off-road use of all-terrain vehicles, and helicopter-supported skiing, hiking, fishing and sight-seeing activities. In Mt. Rainier National Park, 12% of plants (107 of 894 species) and 14% of birds (21 of 147) are exotic species.[53] Other potential threats to ecological conditions in alpine areas include livestock grazing, mining, montane forestry, ground source contaminants, and barriers imposed by roads and trails.[19, 88] Except in a few parks, anthropogenically induced impacts on alpine

Table 5. Land uses and possible threats in alpine and sub-alpine habitats

1. Ski area developments, on-site and off-site impacts
2. Hiking, fishing, and hunting, especially with helicopter access
3. Mountain biking, rock climbing, para-gliding
4. Livestock grazing, especially sheep grazing
5. Mining, montane forestry
6. Airborne toxins, ground source contaminants

wildlife are not well monitored in Washington and Oregon.

Local Impacts
Wildlife conflicts in winter—downhill skiing and other alpine recreation activities. Winter may be a period of great sensitivity, with human disturbances possibly causing the most adverse impacts on wildlife species. At this time, grouse and ungulates occur in social groups or flocks and live at about the same elevational range as downhill ski facilities. On Vancouver Island, British Columbia, for example, white-tailed ptarmigan move from an average summer elevation of 5,500 ft (1,676 m) down to treeline at 4,500 ft (1,372 m) in winter (K. Martin, unpubl. data), while blue grouse move up to treeline.[121] Winter ptarmigan flocks can include individuals from a distance of 14 mi (23 km).[67] Thus activities that impact flocks adversely may be observed over considerable distances.

Ski resort developments impose both on- and off-site impacts. Tourist activities in winter cause serious disturbance to black grouse (*Tetrao tetrix*), now found mainly in mountain forests due to extensive habitat loss at low elevations in central Europe.[126] The construction of ski stations in the Alps between 1950 and 1980 removed or reduced the quality of many black grouse winter habitats in France and Germany. Ski stations also cause major disturbances in winter feeding behavior, winter range and disturbance on leks, resulting in dramatic population declines.[100, 148] Downhill and back country skiers searching for deep powder snow cause additional disturbance as these sites are also sought by grouse for snow burrows to escape the cold and predators (A. Zeitler, pers. comm.). The use of explosives to precipitate controlled avalanches causes ptarmigan to flush from snow burrows on ski areas in the Alps and Pyrenees.[100] Mountain goats are particularly sensitive to disturbance, as they do not habituate to alpine recreational activities such as helicopter hiking or skiing.[40]

Adverse effects from ski hill facilities can occur year-round. After a road and chairlifts were constructed in 1960-1961 at Cairngorm Mountain in the Scottish highlands, allowing easier human access, numbers of crows and gulls on the mountain increased, largely sustained by food scraps,[137, 139] but they also preyed on ptarmigan (*L. mutus*) eggs and dotterel (*Charadrius morinellus*).[102, 131] Density and breeding success of red grouse (*L. l. scoticus*) and ptarmigan on Cairngorm Mountain did not change in relation to skier activities, but ski lift cables caused significant mortality for both species when birds flew into them, and harassment by free-running dogs was reported.[138] In France, 91 dangerous ski cables were identified, and measures proposed to improve visibility of these cables.[106] Mortality of grouse due to ski area activities is almost certainly additive since it occurs in late winter and early spring, after the period when highest overwinter mortality in grouse populations was recorded in Norway (G.E. Frilund and H. Chr. Pedersen, pers. comm.) and in Iceland (O.K. Nielsen, pers. comm.).

Wildlife conflicts in summer. Alpine species may be heavily impacted directly by recreational pressures during the breeding season. Recreational activities like rock climbing or hang gliding can disturb cliff-nesting raptors or ungulates with young. No differences were observed in reproductive success of black grouse inside and outside areas visited by tourists in the French Alps.[101] However, black grouse use of habitat close to a heavily frequented path in a French National Park was reduced by 40 to 50%, perhaps due to the presence of wandering dogs.[68] Alpine marmots in the Swiss and French Alps adjust well to human presence once they become habituated to large numbers of hikers.[91, 103] However, chamois (*Rupicapra rupicapra*) and mouflon (*Ovis orientalis*) that are hunted in the Alps and Pyrenees retreat to more inaccessible alpine areas when they encounter humans, even when there are high numbers of hikers. Mouflon came down to forage on alpine grasslands when there were few hikers or when fog reduced visibility, but they quickly retreated upslope when disturbed (J.-L. Martin and K. Martin, pers. obs.). Thus, alpine ungulates are less likely to habituate to disturbance from tourists where they are hunted than where they are protected, and the impact of disturbance will be greatest when high alpine refuges offer only low quality forage.

Visitors feeding wildlife in parks and recreation areas can cause local negative impacts. Wildlife feeding can change habitat use patterns and inappropriate food can be injurious to health of the animals. Wildlife feeding may result in naivete, particularly of young animals. Mountain goats and bighorn sheep have been killed by cars on mountain roads because they were unwary and associated cars with food (K. Martin, pers. obs.), or because they were attracted to road salt (M. Festa-Bianchet, pers. comm.). Densities of such generalist predators as chipmunks, corvids or canids, or herbivore species such as marmots or golden-mantled ground squirrels may increase due to accidental or intentional feeding of wildlife. This might change predation risk regimes for alpine species due to a direct increase in predator abundance or the amount of time predators spend in the alpine.[138]

Grazing by domestic species in alpine sites, particularly by sheep, reduces habitat cover for predator avoidance and may reduce forage quality for native herbivores.[21] If areas are heavily grazed and trampled, species composition can change dramatically, with many herbaceous plants being extirpated locally. Large areas of the Okanogan Highlands in the Paysaten Wilderness have been heavily grazed by thousands of domestic sheep since the late 1800s. Recently, the U.S. Forest Service reduced the number of sheep on this area to a few hundred. Studies on long-term effects of sheep grazing in alpine and sub-alpine grasslands are needed for this region. However, other studies have suggested that domestic livestock have a greater impact on native ungulates through transmission of exotic diseases rather than by grazing competition.[54, 107, 113]

Regional Impacts

Air and water quality. Alpine areas are considered pristine habitats where the air is pure and water is clean and both are continually renewed by wind and snow melt. However, emerging information suggests that alpine sites are not as pristine as we think. Mt. Rainier National Park in western Washington consistently had the highest average weekly concentrations of ozone.[15] Although a natural component of the earth's trophosphere, in high concentrations ozone can injure vascular plants and be potentially hazardous to human health.[90] Ozone concentrations are higher at high elevations, which is partly a natural atmospheric phenomenon, but also partly due to upwind transport of pollutants from lower elevation urban areas. Thus, alpine plants and animals in Washington are exposed to high ozone concentrations on warm and sunny days.[15] Blais et al.[13] found high levels of organochlorines in mountain parks in western Canada. Airborne toxins may influence snow melt patterns in alpine sites and thus change environmental conditions for wildlife species as well as quality of the water supply for nearby urban sites.

Alpine habitats are similar to arctic ecosystems because both have shallow soils that are impervious much of the year and have low levels of biological activity (e.g., soil bacteria);[37, 81] thus neither are effective at filtering toxic materials. They differ in that the alpine atmosphere is less resistant to airborne transport of particulate matter. Air and water quality in many mountain wilderness sites in western North America are also relatively more compromised than arctic systems, given their proximity to large urban centers such as Seattle, Vancouver, and Denver.[15]

Ground source contaminants are also a potential risk to wildlife at high elevations (Table 5). In the metal-rish ore belt of the Colorado Rocky Mountains, high concentrations of cadmium, an extremely toxic but naturally occurring heavy metal, were found in white-tailed ptarmigan tissue and also in their foods.[88] High cadmium concentrations were associated with kidney damage, particularly for overwintering female ptarmigan. This study also reported a possible reduction in ptarmigan survival and lower population densities in the ore belt regions. Since willow was shown to biomagnify cadmium, other herbivores such as elk and deer living in these alpine regions could also be at risk of cadmium toxicity.[88] Effects of changes in the quality of air, water, and soils on alpine wildlife merit further study.

The increasing value of high elevation habitats. Alpine and sub-alpine areas have increased in value as wildlife habitat, given changes in habitats or processes at low elevations. Wildlife species that formerly used habitats over a broad range of elevations are becoming restricted to the upper elevations of their former range. The problem of upward shifting is well advanced in central Europe. Here most forest grouse species that originally occupied a range of habitats from low elevation bogs, heaths, and moorlands up to the sub-alpine treeline have been extirpated from the low elevations in the past century, due to intensive agricultural and forestry activities. Black grouse in central Europe are now 'de facto' sub-alpine parkland species due to the current unsuitability of traditional low elevation habitats.[126] The relative productivity for wildlife species in high and low elevation habitats was not established previous to extirpation at low elevation. However, high elevation habitats regularly experience total reproductive failure and many species are endangered.[126] Formerly, high elevation populations may have been rescued in years of low production by low elevation populations, but this option is less and less possible. Predation risk might be increasing at high elevations if predators have increased food availability as a result of human presence. In Europe, wolves and brown bears are largely restricted to high latitudes and mountain habitats.[129] For large carnivores, the retreat to high elevation habitat patches results in species distributions being more fragmented than formerly.[23] Eurasian lynx (*Lynx lynx*) in Switzerland may travel over 100 km to an adjacent mountain group to circumvent barriers imposed by highways and densely settled areas.[23]

On first consideration, it might appear that we need not worry about mean elevational increases for species in the Pacific Northwest. However, in western Washington and Oregon, low elevation coniferous forests have been converted to intensive agricultural, industrial, or high density urban areas, and many low elevation wetlands have disappeared.[48, 50] Hence, wildlife species that traditionally occupied a broad elevational range of habitats here have also experienced significant range reduction, accompanied with an 'upward shift', including such species as white-tailed jackrabbit, western toad, other amphibians, blue grouse, ruffed grouse, wolf, and other large predators. For example, blue grouse numbers have declined in many low elevation areas of Washington and Oregon, while high elevation populations remain unchanged (M.A. Schroeder, Washington Department of Fish and Wildlife, pers. comm.). In the case of grizzly bear, the 'upward shift' in population distribution is also facilitated by greater conflicts with humans at low elevations, in addition to low elevation habitat loss.

For wildlife species living at higher elevations, we need to determine which species have experienced the largest elevational shifts and whether any appear in difficulty. One problem relates to potential increased energetic costs of living at higher elevations as discussed in Section 4, as well as the impact on population viability of more frequent and greater stochasticity in annual production.[141] We should establish whether wildlife populations at low elevation rescue high elevation populations following years of reproductive failure at high elevations. Risks to wildlife population viability in spatially separated high elevation 'islands' that result from removal of low elevation populations will increase due to habitat fragmentation and reduced effective population size.

Ultimate factors—climate change and global warming

Climate change has strong impacts on wildlife species at high elevations.[27, 70, 73] Increased concentrations of CO_2 available for photosynthesis and changes in temperature and moisture availability increase growth rates, abundance and elevational range of many alpine plants at higher elevations.[81] In Austria, surveys of alpine plants growing 15-20 m from the peak in 1992 and 1993 found 26 (87%) of 30 mountains had higher species richness than surveys done in identical sites 40 to 100 years earlier.[64, 112] Upslope movement rates for alpine plants for these mountains were calculated as 3.2 to 13 ft (1 to 4 m) over a 10-year period.[64] In the context of global warming, the small-scale patterns of environmental conditions in the alpine zone, allow 'space for time' studies, because vegetation has had no time to adjust to these temperature gradients, and thus is likely to produce a more realistic picture of longterm responses to changing climates.[81]

Research on wildlife responses to climate change is beginning to emerge.[70] With climate change, upper limits of plant growth will increase, and food availability per unit area for alpine herbivores may increase. However, with the increasing altitude of treeline, another consequence of climate change, alpine habitats will become more fragmented, with smaller and more isolated patches. Animals living in these patches will have smaller populations and be required to disperse longer distances to other alpine patches, or pay the consequences of not dispersing.[95, 115]

Alpine animals are well adapted for living in cold climates, but most have not developed sophisticated mechanisms for coping with warm temperatures.[119] Daytime ground temperatures in the Colorado alpine regularly exceeded 45°C, and thus cooling can be a concern for alpine birds in summer.[144] Alpine animals cool themselves by going underground or by moving to snowfields. In the low elevation alpine of Washington and Oregon, an increase of 1 to 2°C in mean temperature could have a large impact on the extent of snow pack and

Figure 5. Cattle Creek in Stein Park shows range of habitat types from riparian valley to mountaintop with treeline going up much higher on north side (left) than on the drier south side. Photo by Steve Ogle, August 1998.

snowmelt patterns. Faster melting snowfields might not persist through summer (Figure 5). Fewer or smaller snowfields leave animals with restricted habitat availability and longer dispersal distances to other habitats. With less area to search, predators may be more successful in capturing prey.

In response to increasing temperatures in spring, pied flycatchers (*Ficedula hypoleuca*) increased their egg size over a 19-year period in the mountains of northern Finland; larger eggs had higher hatching success, but there was no improvement in numbers of young raised.[71, 72] Increasing spring temperatures over a 28-year period were also correlated with earlier initiation of egg laying and larger clutches.[71] However, duration of abundant food resources in late summer will decrease if plant phenology gradients are reduced by faster snowmelt. Global warming is thus expected to have large consequences for high elevation wildlife species, including reduced habitat availability and suitability, reduced duration of food abundance, but perhaps greater per unit area of food availability and possibly increased predation risk. Unfortunately, empirical data for alpine plants and climate change have been ambiguous with responses by plant species being quite specific.[81] Overall responses will be difficult to predict until the key ecological or physical factors are determined. For example, temperature may be less important than moisture in determining the distribution of mountain flora and fauna.[147] Long-term monitoring programs for alpine and sub-alpine wildlife will have high efficacy, as even small increments in warming have significant impacts on habitat quantity and quality for breeding and migration.

6. Information Needs and Conservation Recommendations

Need to develop concerns for alpine conservation and management. Probably the largest concern about alpine habitats is *the lack of concern* about high elevation conservation. The idea prevails that much alpine area is already protected, maybe too much, and that most alpine areas are inaccessible and thus not at great risk. Alpine areas are given lower priority than low elevation habitats because they have modest diversity of breeding species. Such attitudes account for a lack of concern and knowledge regarding status and trends of alpine wildlife populations, in particular for non-game species. Most research for alpine fauna has been done in summer and, thus our knowledge of seasonal use of alpine areas is limited. We are not in a good position to detect early population declines, especially if ecological processes such as predation are shifting at large cross-habitat scales.

Wildlife research is needed to determine the important ecological processes and habitat functions required by both 'alpine obligate' and 'alpine tolerant' vertebrate species. Research is also needed to improve predictions on how wildlife species will cope with changing climatic and environmental conditions. Life history studies are needed for high elevation populations of 'alpine tolerant' species that are experiencing extensive habitat loss at low elevations. Better understanding is required about the importance of maintaining connectivity across time and space for species with seasonal altitudinal shifts. For ungulates and large carnivores, we need to ensure that alpine species have access to valley lowlands and other important wintering habitats. In national parks and wilderness sites, more research and application is required on restoration of impacted alpine grasslands and sub-alpine meadows, particularly in the North Cascades Range.[114] Finally, to detect and predict the ecological effects of global climate change, it will be crucial to establish long-term alpine habitat and wildlife monitoring programs similar to programs in the Austrian Alps,[64] or in northern Finland,[71, 72] or in the Rocky Mountains.[70]

Migration corridors in the alpine: the need to consider habitat connectivity. Alpine habitats function as seasonal migration and dispersal corridors for wildlife. High latitude birds use alpine and sub-alpine habitats as migration highways to move from northern breeding territories to southern winter areas (Box 4). Nocturnal and diurnal migrants may use mountain ridges for navigation.[25] The continued loss of low elevation migration habitat such as riparian and coastal areas strengthens the need to understand and manage for such ecological functions as alpine migration corridors. Increasing use of alpine areas is expected by large area-sensitive animals, particularly predators, where landscapes have been fragmented.

The issue of connectivity between high and low elevation habitats is an important biological and management question. Parks in the Rocky Mountains have a disproportionate amount of their habitat at high elevation, and most parks do not contain year-round habitat requirements for a number of wildlife species.[140] For example, some elk are year-round residents in Banff National Park, but others are migratory; some migratory elk use up to six different management jurisdictions in a single year, including both hunted and non-hunting zones in two provinces.[146] For alpine residents, we need to ensure that connectivity is maintained from alpine habitats to adjacent lower elevation forests and valley bottoms (Figure 5). In mountain habitats, landscape context can be critical as access to large areas may contain bottlenecks.[105] Urban areas and roads in valley bottoms can sub-divide populations or restrict access to seasonally used habitats. Mountain passes are also likely important for seasonal access or population connectivity, and these tend to be where roads, trails and recreational huts are situated.

The most severe wildlife-human conflicts occur in mountain landscapes with narrow valleys that are densely settled or developed. Unfortunately, mammalian wildlife in western North America often require access to valley bottoms to move between mountain chains, since alpine areas tend to be closed off by cliffs. Thus the presence of urban or highway development is almost guaranteed to interrupt wildlife movements, act as a dispersal barrier and fragment populations of large ungulates and predators. All U.S. Interstate highways in mountains impede movements of large mammals (C.E. Braun, pers. comm.). Also, this is a particularly important problem in large mountain parks such as Banff National Park where urban developments and major highways in the narrow Bow Valley have greatly impeded movements of large mammals and fragmented their populations.[105] Species vary in their willingness to cross such dispersal barriers and their wariness when doing so.[140] Determining the ecological value of the alpine as migration habitat and minimizing interruptions of wildlife movements in mountains are two research priorities.

Changes in ecological parameters at high elevations. Increases in predator density or the amount of time predators spend in the alpine could significantly increase the predation risk for alpine and sub-alpine species, as well as for 'upward shifted' alpine tolerant species. Medium and large-sized predators such as grizzly, wolf or lynx that do not persist easily in high-density urban areas have probably moved higher. For them, ungulates provide a reliable food resource through the winter period. Forestry activities at high elevation may also result in changes to wildlife. Research is needed on the influence of montane forest cutting in the Pacific Northwest on distribution and abundance of generalist predators, and consequent impacts on survival and reproduction of wildlife in adjacent alpine and sub-alpine habitats.

Increases in recreational use of high elevation areas increases food availability for generalist predators. High elevation habitats are generally unsuitable for crows or Steller's jays, but populations can exist if there are dwellings or dumps nearby.[121] Along the Cascade Crest, crows occur in ski areas and towns. Raven densities may increase along roads as a result of road kills. A study during the breeding season in northern Finland found more magpies and crows around three ski hill centers than in pristine forest of the same type and age in the region. Also, predation of artificial nests was much higher for the ski hill centers than the undisturbed forest (J. Jokimaki, unpubl. data, jukka.jokimaki@urova.fi).

Human activities at lower elevations might also influence ecological processes in the alpine. Mid-sized predators like ravens and coyotes have substantial home ranges and, could easily travel 5 to 10 kilometers from lower elevations to the alpine, but it is unknown whether they do so. Over-winter survival of generalist predators is improved by road-kills, dumps in adjacent urban areas, or from feeding in backyards at holiday homes or ranchettes in montane forest. Generalist predators no longer depend year-round on traditional prey to survive when anthropogenic food supplies are available, but if they kill the natural prey they encounter even in just one season, they could profoundly impact alpine prey population densities. Given that many high elevation wildlife species are long-lived, a reduction in adult survival has a relatively higher demographic impact than for r-selected species. We need to conduct field research on possible disturbance factors that increase generalist predator densities in or near high elevation areas, whether these predators spend more time at high elevation, and whether these increases significantly impact alpine wildlife populations. At the same time, we need to manage

garbage and educate people to avoid feeding wildlife in and adjacent to alpine areas to ensure natural population densities of generalist predators.

Coexistence of alpine wildlife and winter recreation activities. Ski areas and adjacent resorts have negative ecological impacts on alpine habitats such as soil erosion from sub-alpine forest cutting, which can in turn impact water quality. Wildlife may be displaced or their movements restricted due to cleared areas for ski runs or building facilities. Snow making for fall skiing occurs in many North American ski operations, and can draw down water tables in the drier interior mountain conditions.[89] Little is known about impacts on alpine plants and animals of additives (ice-nucleating bacteria, fertilizer) to water to improve snowmaking efficiency.[89] Before approval of skiing and other recreational developments in alpine and montane habitats, careful surveys should be done in winter to determine which areas to avoid to minimize wildlife conflicts. Patterns of winter habitat use by high elevation wildlife species (e.g., blue grouse, ptarmigan, elk, bighorn sheep) should be determined before facilities are approved. For example, after it was determined that a proposed expansion of ski runs and lifts disrupted an elk migration route in Colorado, the Aspen Skiing Company revised their plans in consultation with the Pitkin County Commission, Aspen town council, Colorado Division of Wildlife, and The Sierra Club local chapter[63] (web site: http://www.skiaspen.com/environment). Construction of ski runs and buildings was done in winter. Trees were cut, but no bulldozer grading was done, to avoid disturbing sub-alpine vegetation. Little is known, or being investigated, regarding potential impacts to alpine or sub-alpine species from other disturbances associated with ski facilities, such as large volumes of sewage and garbage, compacted sub-alpine vegetation, light pollution from night skiing, and the consequences of later spring green-up caused by delayed snow melt from compacted snow and snow making.[89]

In the past two decades, there has been a dramatic increase in alpine cross-country ski touring, and in the use of helicopters to access remote areas for skiing, fishing, and hiking. Such activities do not result in many visits per mountain; however, the overall size of area impacted has increased dramatically. The impacts of such disturbances on wildlife are difficult to monitor, control or to predict. Ski touring groups in Bavaria cause considerable disturbance to wintering populations of grouse and ungulates.[148] Proactive multi-partner consultations have had good success in designing ski touring routes with leave areas for wintering wildlife in Bavaria.[127, 148]

The increased use of alpine facilities in fall, particularly with late summer hiking and mountain biking, may result in conflicts with seasonal movements of wildlife from late summer habitats to winter sites. Careful evaluation of habitat use should be made, and seasonal wildlife travel corridors determined in advance of designing and approving new ski developments or expanding existing operations. The effects of ski developments cover a much larger area than proposed development sites, particularly since they are often positioned at the headwaters of watersheds.[89] A secondary effect of ski area developments is the urbanization of mountain valleys. Increasingly, ski resort areas are becoming amenities to support luxury second home communities. Reviews of proposed new developments or expansions of ski resort areas should consider off-site impacts from this secondary urbanization, including all concerns relevant to urban areas near alpine habitats. Potential impacts should be evaluated at a sufficiently large scale to include connectivity issues discussed earlier for maintaining viable mountain wildlife populations. Proposed alpine recreational developments should demonstrate that they have allowed for maintenance of normal ecological processes.

7. Summary

The alpine zone consists of rugged, partially-vegetated terrain with snowfields and rocky ridges, above the natural treeline. Alpine ecosystems are structurally simple with few plant species compared to most lower elevation habitats. In Washington and Oregon, the main wildlife habitats are (1) alpine grasslands and shrublands, and (2) sub-alpine parklands. These habitats are characterized by high winds, prolonged snow cover, steep terrain, extremes of heat and cold, and intense ultra-violet radiation. With increasing elevation, time for breeding decreases and environmental stochasticity increases; at the highest elevations, hypoxic conditions add additional energetic living costs. Other ecological costs may also be high, such as high predation risk in the open alpine habitats, and the need for seasonal movements from patchy breeding habitats to wintering areas. These factors result in short, intense breeding seasons for wildlife.

About one third of the vertebrate fauna (188 species) in Washington and Oregon use alpine or sub-alpine habitats at some period in their life history. These include 65 species of mammals, 112 birds and 11 reptiles and amphibians. Only a few species such as white-tailed ptarmigan, hoary marmots and mountain goats live exclusively in the alpine, while the majority breed in both alpine and lower elevation habitats or use high elevation habitats during migration. For true alpine breeders, some life history traits may vary with altitude as reproductive rates are often lower, development times longer, with accompanying increases in longevity. For wildlife species breeding across elevational gradients, reproductive success at high elevation appears similar to those for populations at lower elevations. However, with compressed reproductive seasons, wildlife at high elevation produce fewer offspring in a season.

An unexamined aspect of biodiversity for high elevation ecosystems is their use by migrating wildlife during late summer and fall. From late July through October., alpine areas support a rich diversity and abundance of birds and mammals that move up from lower elevations, as well as high latitude avian migrants. This is a time when alpine habitats offer rich food resources, while productivity in many low elevation habitats has declined. Thus, we need to include life history periods outside the breeding season when evaluating high elevation habitats for wildlife. Connectivity is a key ecological process to maintain high elevation wildlife populations. Connectivity needs to be maintained (1) among patchy alpine habitats, (2) along mountain corridors for north-south migrants, and perhaps most importantly (3) between alpine and adjacent lower elevation habitats and valley bottoms.

Despite being highly valued for their intrinsic beauty and wildness, alpine vertebrates and high elevation habitats are a neglected area for management and research. Even though currently most alpine species appear secure in mostly stable habitat conditions, change within and adjacent to alpine and sub-alpine habitats is happening rapidly and over extensive areas. Alpine ecosystems have experienced large increases in amount and kinds of human use, with some areas showing significant deterioration caused by recreational activities, livestock grazing, mining, and air-borne contaminants. High elevation habitats are vulnerable to erosion, especially near melting snowfields or when soil crusts are broken above volcanic ash, and vegetation recovers slowly after such disturbances. To prevent elevated predation risk for alpine wildlife, we should avoid inadvertent increases in food supplies for generalist predators in and near high elevation habitats. On a landscape scale, developed valley bottoms constitute dispersal barriers for many wildlife species with seasonal vertical movements. Since we lack comprehensive understanding of what is required to maintain ecological processes in alpine habitats and have limited experience in restoring these sensitive ecosystems, the precautionary principle is strongly advised for managing alpine and sub-alpine communities. Finally, we need to establish long-term monitoring programs to detect and predict the ecological effects of climate change on alpine habitats and wildlife populations.

Acknowledgments

I thank Environment Canada for allowing me time to develop the ideas for this chapter. I thank I. Storch, The Munich Wildlife Society, Germany, and J. L. Martin, cefe-CNRS, Montpellier, France for providing office logistics to produce this paper. Numerous colleagues from Europe and North America provided insights and discussed ideas concerning alpine ecology and conservation: W. Arnold, A. Bernard-Laurent, C. Böhn, C. E. Braun, U. Breitenmoser, Ch. Breitenmoser-Würsten, R. W. Campbell, E. R. Dunn, L. E. Ellison, M. Festa-Bianchet, C. L. Gass, H. Gossow, M. Guentert, A. Landmann, J.-L. Martin, M. A. Schroeder, I. Storch, D. B. A. Thompson, H. Zeiler, and A. Zeitler. D. H. Johnson and K. Bettinger provided vital local material and guidance during the chapter production. K.E.H. Aitken, C. L. Hitchcock, and M.D. Mossop provided logistical assistance. I thank C. E. Braun, M. Festa-Bianchet, C. M. I. Meslow, B. K. Sandercock, M. A. Schroeder, J. N. M. Smith, and N.A.M. Verbeek for reviewing various drafts.

Literature Cited

1. Agee, J.K. and L. Smith. 1984. Subalpine tree establishment after fire in the Olympic Mountains, Washington. Ecology 65: 810-819.

2. Allee, W. C. and K. P. Schmidt. 1951. Ecological Animal Geography. John Wiley and Sons, Inc. New York, NY.

3. Armitage, K. B. 1986. Individuality, social behavior, and reproductive success in yellow-bellied marmots. Ecology 67: 1186-1193.

4. ———. 1999. Evolution of sociality in marmots. Journal of Mammalogy 80: 1-10.

5. Arno, S. F. and R.P. Hammerly. 1984. Timberline: Mountains and arctic forest frontiers. The Mountaineers, Seattle, WA.

6. Arnold, W. 1990a. The evolution of marmot sociality: I. Why disperse late? Behavioural Ecology and Sociobiology 27: 229-237.

7. ———. 1990b. The evolution of marmot sociality: II. Costs and benefits of joint hibernation. Behavioural Ecology and Sociobiology 27: 239-246.

8. ———. 1993. Social evolution in marmots and the adaptive value of joint hibernation. Verh. Deutsch Zool. Ges. 86: 79-93.

9. Bätzing, W. 1991. Die alpen: entstehung und Gefährdung einer europäischen kulturlandschaft. Beck, Munich, Germany.

10. Beason, R.C. 1995. Horned lark (Eremophila alpestris). Pages 1-24 in A. Poole and F. Gill, editors. The Birds of North America, No. 195. The Academy of Natural Sciences; Philadelphia.

11. Billings, W. D. 1989. Alpine vegetation. Pages 392-420 in W. D. Billings and R. K. Peet, editors. Terrestrial vegetation of North America. Paragon Books.

12. ———, and H.A. Mooney. 1968. The ecology of arctic and alpine plants. Biological Review 43: 481-529.

13. Blais, J. M., D. W. Schindler, D. C. G. Muir, L. E. Kimpes, D. B. Donald, B. Rosenburg. 1998. Accumulation of persistent organochlorine compounds in mountains of western Canada. Nature 395: 585-588.

14. Böhn, C., and A. Landmann. 1995. Nest-site selection and nest construction in the water pipit (Anthus spinoletta). Journal für Ornithologie 136: 1-16.

15. Brace, S. and D. L. Peterson. 1998. Spatial patterns of tropospheric ozone in the Mount Rainier region of the Cascade Mountains, U.S.A. Atmospheric Environment, 32: 3629-3637.

16. Braun, C. 1969. Population dynamics, habitat, and movements of White-tailed Ptarmigan in Colorado. Ph.D. thesis, Colorado State University, Fort Collins, CO.

17. Braun, C. E. 1971. Habitat requirements of Colorado white-tailed ptarmigan. Proceedings of Western Association of State Game and Fish Commissioners. 51: 284-292.

18. Braun, C. E., and R. G. Streeter. 1968. Observations on the occurrence of white-tailed jackrabbits in the alpine zone. Journal of Mammalogy 49: 160-161.

19. Braun, C. E., R. W. Hoffman, and G. E. Rogers. 1976. Wintering areas and winter ecology of white-tailed ptarmigan in Colorado. Special Report No. 38, Colorado Division of Wildlife, Denver, CO.

20. Braun, C. E., R. K. Schmidt, and G. E. Rogers. 1973. Census of Colorado white-tailed ptarmigan with tape-recorded calls. Journal of Wildlife Management 37:90-93

21. ———. 1980. Alpine bird communities of western North America: implications for management and research. Pages 280-291 in R. M. DeGraff and N. G. Tilghman, compilers. Workshop Proceedings: Management of western forests and grasslands for nongame birds. U.S. Department Agriculture Forest Service General technical report INT-86.

22. ———, K. Martin, and L.A. Robb. 1993. White-tailed ptarmigan (Lagopus leucurus). Pages 1-24 in A. Poole, and F. Gill, editors. The birds of North America, No. 68. Philadelphia: The Academy of Natural Sciences; Washington, D.C.

23. Breitenmoser, U. 1998. Large predators in the Alps: the fall and rise of man's competitors. Biological Conservation 83: 279-289.

24. Bronson, M. T. 1979. Altitudinal variation in the life history of the golden-mantled ground squirrel (Spermophilus lateralis). Ecology 60: 272-279.

25. Bruderer, B. 1982. Do migrating birds fly along straight lines? Pages 3-14 in F. Papi and H. G. Wallraff, editors. Avian Migration. Springer, Heidelberg, Germany.

26. Bullard, R. W. 1972. Vertebrates at altitudes. Pages 209-226 in M. K. Yousef, S. M. Horvath, and R. W. Bullard editors. Physiological adaptations, desert and mountain. Academic Press, New York, NY.

27. Burton, J. F. 1995. Birds and climate change. Helm, London, UK.

28. Campbell, R. W., N. K. Dawe, I. McTaggart-Cowan, J. M. Cooper, G. W. Kaiser, M. C. E. McNall, G. E. J. Smith. 1990 and 1996. The Birds of British Columbia. Volumes 1-3. University of British Columbia Press, Vancouver, BC, Canada.

29. Cannings, R. A., R. J. Cannings, and S. G. Cannings. 1987. The birds of the Okanagan Valley. Royal BC Museum, Victoria, BC, Canada.

30. Cannings, R. J., and S. G. Cannings. 1996. British Columbia, A Natural History. Douglas and McIntyre, Vancouver, BC, Canada.

31. Capula, M, L. Luiselli, and C. Anibaldi. 1992. Complementary study on the reproductive biology in female adder, Vipera berus (L.) from eastern Italian Alps. Vie Milieu 42: 327-336.

32. Carey, C. 1980. Adaptation of the avian egg to high altitude. American Zoologist, 20: 449-459.

33. ———, and K. Martin. 1997. Physiological ecology of incubation of ptarmigan eggs at high and low altitudes. Wildlife Biology 3: 211-218.

34. Carpenter, F. L., and M. A. Hixon. 1988. A new function for torpor: fat conservation in a wild migrant hummingbird. The Condor 90: 373-378.

35. Cassidy, K. M., C. E. Grue, M. R. Smith, and K. M. Dvornich, editors. 1997. Washington State Gap Analysis - Final Report. Volumes 1-5. Washington Cooperative Fish and Wildlife Research Unit, University of Washington, Seattle, WA.

36. Chappell, C.B., R.C. Crawford, J. Kagan, D.H. Johnson, M. O'Mealy, G.A. Green, H.L. Ferguson, and W.D. Edge. 2001. Wildlife habitats: descriptions, status, trends, and system dynamics. Pages 22-114 in D.H. Johnson and T.A. O'Neil, managing directors. Wildlife-habitat relationships in Oregon and Washington. Oregon State University Press, Corvallis, OR.

37. Chapin, F. S. III, and C. Körner, editors. 1994. Arctic and alpine biodiversity: patterns, causes and ecosystem consequences. Springer-Verlag, New York, NY.

38. Cody, M. L. 1985. Habitat selection in grassland and open country birds. Pages 191-226 in Habitat selection in birds. Academic Press. New York, NY.

39. Corkran, C.C. and C. R. Thoms. 1996. Amphibians of Oregon, Washington, and British Columbia. Lone Pine, Vancouver, BC, Canada.

40. Côte, S. D. 1996. Mountain goat responses to helicopter disturbance. Wildlife Society Bulletin 24: 681-685.

41. Crisafulli, C. M., and C. P. Hawkins. 1998. Ecosystem recovery following a catastrophic disturbance: Lessons learned from Mount St. Helens. in M. J. Mac, P.A. Opler, C. E. Puckett Haecker, and P. D. Doran, editors. Status and trends of the nations biological resources. Volumes 1-2. U.S. Department of Interior, U.S. Geological Survey, Reston, VA.

42. Csuti, B., A.J. Kimerling, T.A. O'Neil, M.M Shaughnessy, E.P. Gaines, and M. M. P. Huso. 1997. Atlas of Oregon wildlife: Distribution, habitat and natural history. Oregon State University Pres, Corvallis, OR.

43. Daubenmire, R. 1954. Alpine timberlines in the Americas and their interpretation. Butler University Botanical Studies 11: 119-136.

44. Dobson, F. S. and J.O. Murie. 1987. Interpretation of intraspecific life history patterns: evidence from Columbian ground squirrels. American Naturalist 129: 382-397.

45. ———, R. M. Zammuto, and J.O. Murie. 1986. A comparison of methods for studying life history in. Columbian ground squirrels. Journal of Mammalogy 667: 154-158.

46. Douglas, G.W. and L.C. Bliss. 1977. Alpine and high subalpine plant communities of the north Cascades Range, Washington and British Columbia. Ecological Monographs 47: 113-150.

47. Dragon, S., and in vivo C. Carey, K. Martin, and R. Bauman. 1999. Effect of high altitude and in vivo adenosine/β-adrenergic receptor blockade on ATP and 2,3BPG concentrations in red blood cells of avian embryos. Journal of Experimental Biology 202:2787-2795

48. Edge, W.D. 2001. Wildlife of agriculture, pastures, and mixed environs. Pages 342-360 in D.H. Johnson and T.A. O'Neil, managing directors. Wildlife-habitat relationships in Oregon and Washington. Oregon State University Press. Corvallis, OR.

49. Edwards, O. M. 1980. The alpine vegetation of Mount Rainier National Park: structure, development and constraints. Ph.D. dissertation, Univ. of Wash., Seattle, WA.

50. Ferguson, H.L., K. Robinette, K. Stenberg. 2001. Wildlife of urban habitats. Pages 317-341 in D.H. Johnson and T.A. O'Neil, managing directors. Wildlife-habitat relationships in Oregon and Washington. Oregon State University Press. Corvallis, OR.

51. Ferron, J., and J. P. Ouellet. 1989. Temporal and intersexual variations in the use of space with regard to social organization in the woodchuck *Marmota monax*. Canadian Journal of Zoology 67: 1642-1649.

52. Festa-Bianchet, M., M. Urquhart, and K.G. Smith. 1994. Mountain goat recruitment: kid production and survival to breeding age. Canadian Journal of Zoology 72: 22-27.

53. Filley, B. 1996. The big fact book about Mount Rainier. Dunamis House Publishing, Issaquah, WA.

54. Foreyt, W. J., K. P. Snipes, and R. W. Kasten. 1994. Fatal pneumonia following innoculation of healthy bighorn sheep with *Pasturella haemolytica* from healthy domestic sheep. Journal of Wildlife Disease 30: 137-145.

55. Franklin, J. F., and C. T. Dyrness. 1973. Natural Vegetation of Oregon and Washington. Oregon State University Press, Corvallis, OR.

56. Franklin, J. F., W.H. Moir, M.A. Hemstrom, S. E. Greene, and B. G. Smith. 1988. The forest communities of Mount Rainier National Park. U.S.D.I. National Park Service Scientific Monograph Series 19, Washington, D.C.

57. Frase, B.A. and R. S. Hoffman. 1980. *Marmota flaviventris*. Mamalian Species 135: 1-8.

58. French, N. R. 1959. Life history of the black rosy finch. Auk 76: 158-180.

59. Frenzen, P. M. and C. M. Crisafulli. 1990. Mount St. Helens ten years later: past lessons and future promise. Northwest Science 64: 263-267.

60. Gaillard, J. -M., M. Festa-Bianchet, and K. G. Smith. 1998. Population dynamics of large herbivores: variable recruitment with constant adult survival. Trends in Research in Evolution and Ecology 13: 58-63.

61. Gass, C. L., and K. P. Lertzman. 1980. Capricious mountain weather: a driving variable in hummingbird territorial dynamics. Canadian Journal of Zoology 58: 1964-1968.

62. ———, and G. Sutherland. 1985. Specialization by territorial hummingbirds on experimentally enriched patches of flowers: energetic profitability and learning. Canadian Journal of Zoology 63: 2125-2133.

63. Giezentanner, K. 1994. Wildlife and biodiversity analysis for the Snowmass Ski Development expansion. U.S. Forest Service report, CO.

64. Grabherr, G. and M.G. H. Pauli. 1994. Climate effects on mountain plants. Nature 369: 448.

65. Grossman, D.H., D. Faber-Langendoen, A.W. Weakley, M. Anderson, P. Bourgeron, R. Crawford, K. Goodin, S. Landaal, K. Metzler, K.D. Patterson, M. Pyne, M. Reid and L. Sneddon. 1998. International classification of ecological communities: terrestrial vegetation of the United States. Vol. 1: The National Vegetation Classification Standard. The Nature Conservancy.

66. Hannon, S.J. and K. Martin. 1996. Mate fidelity and divorce in ptarmigan: polygyny on the tundra. Pages 192-210 in J.M. Black, editor. Partnerships in birds, the study of monogamy. Oxford University Press, Oxford, England.

67. Hoffman, R.W. and C. E. Braun. 1975. Migration of a wintering population of white-tailed ptarmigan in Colorado. Journal of Wildlife Management 39: 485-490.

68. Houard, T., and M. Mure. 1997. Les tetras-lyre des vallons de Salese et Mollieres. Parc National du Mercantour. Domaine vital et influence du tourisme. Revue d'Ecologie (La Terre et la Vie), Supplemente 4: 165-197.

69. Houston, D. B., and V. Stevens. 1988. Resource limitation in mountain goats: a test by experimental cropping. Canadian Journal of Zoology 66: 228-238.

70. Inouye, D.W., B. Barr, K.B. Armitage and B.D. Inouye. 2000. Climate change is affecting altitudinal migrants and hibernating species. Proceedings of the National Academy of Science 97: 1630-1633.

71. Järvinen, A. 1994. Global warming and egg size of birds. Ecography 17: 108-110.

72. ———. 1995. Effects of climate change on mountain bird populations. Pages 73-74 in A. Guisan et al., editors. Potential Ecological impacts of climate change in the Alps and Fennoscandian Mountains. Conservatoire et Jardin botaniques de la Ville de Geneve: Geneva, Switzerland.

73. Jenik, J. 1997. The diversity of mountain life. Pages 199-231 in B. Messerli, and J. D. Ives, editors. Mountains of the world: a global priority. Parthenon Publishing Group, London, UK.

74. Johnson, R. E. 1965. Reproductive activities of rosy finches, with special reference to Montana. Auk, 82: 190-205.

75. ———. 1983. Nesting biology of the rosy finches on the Aleutian Islands, Alaska. Condor 85: 447-452.

76. Kiilsgaard, C.W. 1999. Oregon vegetation: mapping and classification of landscale level cover types. Final report submitted to the U.S. Geological Survey, Biological Resources Division: GAP Analysis Program, Moscow, ID.

77. Kingery, H. E. 1996. American Dipper (*Cinclus mexicanus*). Pages 1-28 in A. Poole, and F. Gill, editors. The birds of North America, No. 229. The Academy of Natural Sciences; Philadelphia. Washington, D.C.

78. Kirk, R. 1999. Sunrise to paradise: the story of Mount Rainier National Park. University of Washington Press, Seattle, WA.

79. Kollinsky, C., and A. Landmann. 1996. Altitudinal distribution of male black redstarts: are there age-dependent patterns? Bird Study 43: 103-107.

80. Körner, C. 1994. Alpine plant diversity, a global survey and functional interpretation. Pages 45-64 in F. S. Chapin III, and C. Körner, editors. Arctic and alpine biodiversity: patterns, causes and ecosystem consequences. Springer-Verlag, New York, NY.

81. ———. 1999. Alpine Plant Life: functional plant ecology of high mountain ecosystems. Springer-Verlag, Heidelburg, Germany.

82. Krajick, K. 1998. Animals thrive in an avalanche's wake. Science 279: 1853.

83. Kuramoto, R.T. and L. C. Bliss. 1970. Ecology of subalpine meadows in the Olympic Mountains, Washington. Ecological Monographs 40:317-347.

84. Landmann, A. and N. Winding. 1993. Niche segregation in high-altitude Himalayan chats (Aves, Turdidae): does morphology match ecology? Oecologia 95: 506-519.

85. ———, and ———. 1995a. Adaptive radiation and resource partitioning in Himalayan high-altitude finches. Zoology Analysis of Complex Systems 99: 8-20.

86. ———, and ———. 1995b. Guild organization and morphology of high-altitude granivorous and insectivorous birds: convergent evolution in an extreme environment. Oikos 73: 237-250.

87. Lanner, R. M. 1988. Dependence of Great Basin Bristlecone pine on Clark's nutcracker for regeneration at high elevations. Arctic and Alpine Research, 20: 358-362.

88. Larison, J.R., G.E. Likens, J.W. Fitzpatrick, and J.G. Crock. 2000. Cadmium toxicity among wildlife in the Colorado Rocky Mountains. Nature 406: 181-183.

89. Legault, S. 1997. Down a slippery slope: ski resorts and their effect on ecological, social and economic health. Unpublished Discussion paper, UTSB Research Inc., Banff, Alberta, Canada.

90. Lippman, M. 1989. Health effects of ozone: a critical review. Journal of the air and waste management association 39: 672-695.

91. Louis, S., and M. Le Berre. 1997. Visitor impact on alpine marmot foraging behavior. Journal of Wildlife Research 2: 133-136.

92. Love, D. 1970. Subarctic and subalpine: where and what? Arctic and alpine research 2: 63-73.

93. Lyon, B. E. and R. D. Montgomerie. 1987. Ecological correlates of incubation feeding: a comparative study of high arctic finches. Ecology 68: 713-722.

94. Manuwal, D.A., M. H. Huff, M. R. Bauer, C. B. Chappell, and K. Hegstad. 1987. Summer birds of the upper subalpine zone of Mount Adams, Mount Rainier, and Mount St. Helens, Washington. Northwest Science, 61: 82-92.

95. Martin, K., C. E. Braun, and P. B. Stacey. 2000. Recruitment, dispersal, and demographic rescue in spatially structured white-tailed ptarmigan populations. Condor 102:503-506.

96. ———, C. Doyle, F. Mueller and S. J. Hannon. 2001. Forest grouse and ptarmigan. Chapter 11 *In* C.J. Krebs, S. Boutin, and R. Boonstra, editors. Ecosystem dynamics of the Boreal Forest: The Kluane Project. Oxford University Press, in press.

97. ———, R. F. Holt and D. W. Thomas. 1993. Getting by on high: ecological energetics of arctic and alpine grouse. Pages 33-41 *in* C. Carey, G. L. Florant, B. A. Wunder, and B. Horwitz, editors. Life in the Cold: Ecological, physiological, and molecular mechanisms. Westview Press, Boulder, CO.

98. ———, P. B. Stacey, and C. E. Braun. 1997. Demographic rescue and maintenance of population stability in grouse - beyond metapopulations. Wildlife Biology 3: 295-296.

99. Medin, D. E. 1987. Breeding birds of an alpine habitat in the southern Snake Range, Nevada. Western Birds 18: 163-168.

100. Menoni, E., and Y. Magnani. 1998. Human disturbance of grouse in France. Grouse News, No. 15: 4-8.

101. Miquet, A. 1986. Contribution a la etude des relations entre Tetras-lyre *Tetrao tetrix* L. Tetraonidae et tourisme hivernal en Haute-Tarentaise. Acta Oecologia, Oecologia Applications 7: 325-355.

102. Nethersole-Thompson, D., and A. Watson. 1981. The Cairngorms. Melven Press, Perth, U.K.

103. Neuhaus, P. and B. Mainini. 1998. Reactions and adjustment of adult and young alpine marmots *Marmota marmota* to intense hiking activities. Wildlife Biology, 4: 119-123.

104. Norvell, J. R., and P. D. Creighton. 1990. Foraging of horned larks and water pipits in alpine communities. Journal of Field Ornithology 61: 434-440.

105. Noss, R. F., H. B. Quigley, M. G. Hornocker, T. Merrill, and P. C. Paquet. 1996. Conservation biology and carnivore conservation in the Rocky Mountains. Conservation Biology, 10: 949-963.

106. Observatoire des Galliformes de Montagne. 1997. Rapport annuel de l'Observatoire des Galliformes de Montagne.

107. Onderka, D. K., S. A. Rawluk, and W. D. Wishart. 1988. Susceptibility of Rocky Mountain bighorn sheep and domestic sheep to pneumonia induced by bighorn and domestic livestock strains of *Pasturella hemolytica*. Canadian Journal of Veterinary Research 52: 439-444.

108. Pattie, D. L. and N. A. M. Verbeek. 1966. Alpine birds of the Beartooth Mountains. Condor 68: 167-176.

109. Piertney, S. B., A. D. C. MacColl, P. J. Bacon, and J. F. Dallas. 1998. Local genetic structure in red grouse *(Lagopus lagopus scoticus)*: evidence from microsatellite DNA markers. Molecular Ecology, 7: 1645-1654.

110. Portier, C., M. Festa-Bianchet, J. M. Gaillard, J. T. Jorgenson and N. G. Yoccoz. 1998. Effects of density and weather on survival of bighorn sheep lambs *(Ovis canadensis)*. Journal of Zoology, London 245: 271-278.

111. Price, L. W. 1981. Mountains and man: a study of process and environment. University of California, Berkeley, CA.

112. Price, M. F. and R. G. Barry. 1997. Climate change. Pages 409-445 *in* B. Messerli, and J. D. Ives, editors. Mountains of the world: a global priority. Parthenon Publishing Group, London.

113. Pybus, M. J., S. Groom, and W. M. Samuel. 1996. Meningeal worm in experimentally infected bighorn and domestic sheep. Journal Wildlife Disease 32: 614-618.

114. Rochefort, R. M. and S. T. Gibbons. 1992. Mending the meadow: high altitude meadow restoration in Mount Rainier National Park. Restoration and Management Notes 10: 120-126.

115. Roland, J., N. Keyghobadi, and S. Fownes. 2000. Alpine *Parnassius* butterfly dispersal: effects of landscape and population size. Ecology 81: 1642-1653.

116 Saint Girons, H. 1992. Strategies reproductrices des viperidae dans les zones temperees fraiches et froides. Bulletin de la Société zoologique de France 117: 267-278.

117. Shackleton, D. 1999. Hoofed mammals of British Columbia. University of British Columbia Press, Vancouver, BC, Canada.

118. Shaughnessy, M. M. and T. A. O'Neil. 2001. Conservation of biodiversity: Considerations and methods for identifying and prioritizing areas and habitats. Pages 154-167 in D.H. Johnson and T.A. O'Neil, managing directors. Wildlife-habitat relationships in Oregon and Washington. Oregon State University Press. Corvallis, OR.

119. Smith, A. T. 1974. The distribution and dispersal of pikas: influences of behavior and climate. Ecology 55: 1368-1376.

120 Smith, A. T. 1978. Comparative demography of pikas (Ochotona): effect of spatial and temporal age-specific mortality. Ecology 59: 133-139.

121. Smith, M. R., P. W. Mattocks, Jr., and K. M. Cassidy. 1997. Breeding birds of Washington State. Volume 4 in K. M. Cassidy, C. E. Grue, M. R. Smith, and K. M. Dvornich, editors. Washington State Gap Analysis—Final report. Seattle Audubon Society Publications in Zoology No.1, Seattle, WA.

122. Spalding, J. B. 1979. The aeolian ecology of White Mountain Peak, California: wind-blown insect fauna. Arctic and Alpine Research 11: 83-94.

123. Stabler, R. M., N. J. Kitzmiller, and C. E. Braun. 1974. Hematozoa from Colorado birds, IV. Galliformes. Journal of Parasitology 60: 536-537.

124. Storch, I. 1993. Patterns and strategies of winter habitat selection in alpine capercaillie. Ecography 16: 351-359.

125. ———. 1997. The role of the metapopulation concept in conservation of European woodland grouse. Wildlife Biology, 3: 272.

126. ———, compiler. 2000. Grouse Status Survey and Conservation Action Plan 2000-2004. WPA/Birdlife/SSC Grouse Specialist Group. IUCN, Gland. Switzerland and Cambridge, UK and the World Pheasant Association, Reading, UK

127. Suchant, R., and R. Roth. 1998. Tourism in the Black Forest - danger for the capercaillie. Grouse News, No. 15: 13-17.

128. Sutherland, G. D., C. L. Gass, P. A. Thompson, and K. P. Lertzman. 1982. Feeding territoriality in migrant rufous hummingbirds: defense of yellow-bellied sapsucker *(Sphyrapicus varius)* feeding sites. Canadian Journal of Zoology 60: 2046-2050.

129. Sutherland, W. J. and J. D. Reynolds. 1998. Sustainable and unsustainable exploitation. Pages 90-115 *in* W. J. Sutherland, editor. Conservation science and management. Blackwell, Oxford, UK.

130. Taylor, K. A. 1994. Pages 165-185 *in* A Birder's Guide to Vancouver Island. Keith Taylor Birdfinding Guides, Victoria, BC, Canada.

131. Thompson, D. B. A. and D. P. Whitfield. 1993. Research on mountain birds and their habitats. Scottish Birds 17: 1-8.

132. ———, A. Watson, S. Rae, and G. Boobyer. 1996. Recent changes in breeding bird populations in the Cairngorms. Botanical Journal of Scotland. 48: 99-110.

133. Twining, H. 1940. Foraging behavior and survival in the Sierra Nevada rosy finch. Condor 42: 64-72.

134. Verbeek, N. A. M. 1967. Breeding biology and ecology of the horned lark in alpine tundra. Wilson Bulletin 79: 208-217.

135. ———. 1970. Breeding ecology of the water pipit. Auk 87: 425-451.

136. ———. 1981. Nesting success and orientation of water pipit *Anthus spinoletta* nests. Ornis Scandinavica 12: 37-39.

137. Watson, A. 1979. Bird and mammal numbers in relation to human impact at ski lifts on Scottish hills. Journal of Applied Ecology, 16: 753-764.

138. ———. 1982. Effect of human impact on ptarmigan and red grouse near ski lifts in Scotland. Page 51 *in* Annual Report of ITE for 1981, Cambridge, UK.

139. ———. 1991. Increase of people on Cairngorm plateau following easier access. Scottish Geographical Magazine 107: 99-105.

140. Weaver, J. L., P. C. Paquet and L. F. Ruggiero. 1996. Resilience and conservation of large carnivores in the Rocky Mountains. Conservation Biology 10: 964-976.

141. Widmer, M. 1993. Breeding biology of the garden warbler *Sylvia borin* in a subalpine habitat in the central Swiss Alps. Der Ornithologische Beobachter 90: 83-115. German with English abstract.

142. ———. 1996. Phänologie, Siedlungsdichte und Populationsökologie der Gartengrasmucke *Sylvia borin* in einem subalpinen Habitat der Zentralalpen. Journal für Ornithologie, 137: 479-501.

143. ———. 1999. Altitudinal variation of migratory traits in the garden warbler *Sylvia borin*. Ph. D. Thesis, University of Zurich, Zurich, Switzerland.

144. Wiebe, K. L., and K. Martin. 1997. Effects of predation, body condition and temperature on incubation rhythms of ptarmigan. Wildlife Biology 3: 219-227.

145. ———, and ———. 1998. Age-specific patterns of reproduction in white-tailed and willow ptarmigan *(Lagopus leucurus* and *L. lagopus)*. Ibis 140:14-24.

146. Woods, J.G. 1991. Ecology of a partially migratory elk population. Ph.D. Thesis. University of British Columbia, Vancouver, BC, Canada.

147. Woodward, F. I. (editor). 1992. Advances in Ecological Research. Volume 22. The ecological consequence of global climate change. Academic Press, London.

148. Zeitler, A. and U. Glanzer. 1998. Skiing and grouse in the Bavarian Alps. Grouse News, 15: 8-12.

Web sites:

Centre for Alpine Studies (UBC): http://www.www.forestry.ubc.ca/alpine/index.htm

Centre of Alpine ecology (Italy): http://www.itc.it/cea/index_e.html

10

Wildlife of Westside Grassland and Chaparral Habitats

*Bob Altman, Marc Hayes, Stewart Janes,
& Richard Forbes*

Introduction

Westside Grasslands and Chaparral (Ceanothus-Manzanita Shrublands) are among the most distinctive and endangered wildlife habitats in Oregon and Washington. They are characterized by a limited distribution either naturally (i.e., chaparral) or from extensive losses associated primarily with post-European settlement (i.e., grassland). Although not as species rich or biologically diverse as many habitats, they contain an array of vertebrates that occur rarely elsewhere in Oregon and Washington. These habitats also occur where most of the human population lives, thus they have been extensively altered since European settlement.

Loss of westside native grassland habitat has been especially dramatic; >90% in the Puget Lowlands[27] and >99% in the Willamette Valley.[23] Remaining parcels are invariably small, moderately to heavily disturbed, and geographically disjunct. Consequences for grassland-associated wildlife have been local and regional extirpation of some species, and severe declines among other species with a concomitant increase in exotic species.

Habitat Types

Grasslands

Throughout this chapter we use the term grassland to include both prairie and savanna habitats (see Chapter 2 for descriptions). We emphasize to the extent possible native grasslands, which occur as scattered historical remnants. We also acknowledge agricultural grasslands (e.g., hayfields, pasture, annual and perennial grass fields, fallow agricultural fields) because these dominate the current landscape, and some wildlife associated with them were integral components of native grasslands.

Westside grasslands are open landscapes characterized by herbaceous plants, particularly graminoids. Periodic fire, the degree of seasonal flooding, and local variation in soils likely created a mosaic of vegetation and successional conditions within native grasslands. Perennial bunchgrasses dominated native sites,[1] but exotic annual grasses are prevalent in many remnant grasslands.[43, 123] Some sites, particularly wet prairies, are rich in forbs. Vegetation height, density, cover, and species composition are dependent upon several factors including soil type, aspect, climate, and land use and management.

Westside valley grasslands include glacial outwash prairies of the Puget Lowlands of Washington, particularly the South Puget Sound area; and early seral fire-maintained grasslands of the Willamette, Rogue, and Umpqua valleys in Oregon.[43] Most discussion in this chapter focuses on these grasslands, which comprise the largest area of westside grasslands. Limited reference is made to coastal headland and dune grasslands and shrublands, montane grass balds, and balds in the San Juan Islands, which are minor components of westside grasslands and for which few data exist on wildlife relationships. Coastal headland grasslands and shrublands include flat coastal plain grasslands (e.g., Quillayute Prairie, Washington) and the rocky, herbaceous and low shrub-dominated exposed headlands along the ocean (e.g., Heceta Head, Oregon; Cascade Head, Oregon). These grasslands are generally minor components within the Sitka spruce (*Picea sitchensis*) vegetation zone. Montane balds are grass-dominated, south-facing hillsides of Coast Range and Olympic Peninsula summits. They occur as scattered but conspicuous openings in an otherwise forest-dominated landscape.[43] Balds in the San Juan Islands are low elevation (<2,000 feet [610 m]) dry sites on south to west aspects, moderate to steep slopes, with shallow to bedrock soils[22] (C. Chappell, Washington Natural Heritage Program, pers. comm.).

We categorized grasslands into three types: wet prairie, dry prairie, and savanna. In the Willamette Valley, wet prairies occur primarily on poorly drained (high clay fraction) soils[134] (see Chapter 2 this volume—herbaceous wetlands). Dry prairies occur on better drained lowland and valley soils, especially along valley margins.[134] Savanna refers to dry prairie with singular widely scattered trees or small open-canopy tree groves with <30% canopy cover.[1] The tree species is usually savanna-form Oregon white oak (*Quercus garryanna breweri*), but in some areas it may be ponderosa pine (*Pinus ponderosa*) or Douglas-fir (*Pseudotsuga menziesii*).[43, 103, 111]

Chaparral

Chaparral occurs in southwestern Oregon in the interior valley and foothills below 6,800 feet (2,100 m) in the Rogue River basin and patchily northward into the Umpqua River basin (see Chapter 2—ceanothus-manzanita shrublands). Valley chaparral, a diluted extension of the California chaparral types,[33] occurs mostly below 3,600 feet (1,100 m). A higher elevation montane chaparral with a different species composition occurs between 3,600-6,800 feet (1,100-2,100 m) in the Siskiyou Mountains and southern Cascades. Some chaparral communities may be climax (e.g., wedgeleaf ceanothus [*Ceanothus cuneatus*] on the Rogue Valley floor), and others are believed to be fire-maintained. Chaparral consists of a shrubland with a highly variable canopy and herbaceous understory, both of which may depend on fire history. Dominant plants are multi-stemmed shrubs usually 1.6-13 feet (0.5-4 m) tall. Chaparral can either be a climax or a successional community. Regardless of the seral condition, a tree component may be present. Although gradients exist between shrub- and tree-dominated communities, chaparral comprises the shrub-dominated end of this spectrum.

Habitat Distribution and Impacts

Grasslands

Before European settlement, the landscape of the Willamette Valley, particularly south of Salem, was largely an open expanse of prairie and savanna.[43] An estimated 1,030,604 acres (417,249 ha) of prairie and 539,938 acres (218,598 ha) of savanna existed within the Willamette Valley at the time of European settlement (E. Alverson, The Nature Conservancy, pers. comm.). Among prairie habitats, approximately one-third, 301,472 acres (122,053 ha), was wet prairie. Native prairie and savanna habitats in the Willamette Valley have been reduced from being the most abundant vegetative communities, covering approximately 45% of the landscape (30% prairie and 15% savanna), to a few small, scattered parcels of semi-natural remnants amid farmland and urban and rural residential development totaling approximately 2,000-3,000 acres (800-1,200 ha) or <1% of the historical extent (E. Alverson, The Nature Conservancy, pers. comm.).

In the South Puget Lowlands, prairies were a less significant part of the pre-European landscape than in the Willamette Valley. Assessment of pre-European vegetative conditions based on soil type indicates 149,360 acres (60,470 ha) of prairie habitat, or about 10% of the landscape.[27] Current estimates of prairie habitat are 12,582 acres (5,094 ha), and most of this (91%, 11,500 acres [4,656 ha]) occurs on Fort Lewis Military Installation.[27] Only 3% (2,993 acres [1,212 ha]) of the pre-European total is considered "intact prairie" and again most of this (71%, 2,130 acres [862 ha]) occurs on Fort Lewis. Thus, pre-settlement prairie vegetation in the South Puget Lowlands has sustained losses >90%. Losses also have been extreme for grasslands and balds in the northern Puget Lowlands (C. Chappell, Washington Natural Heritage Program, pers. comm.).

In the Rogue and Umpqua valleys, no estimates of the extent of pre-European grassland habitat are available. Most of these grasslands, and essentially all of them at the lowest elevations, are thought to have been modified by agriculture. Most grasslands with some resemblance to the historical condition are small pockets on tablelands north of Medford.

Key factors that changed historical conditions in westside grasslands were:
1. alteration of natural disturbance processes (fire and flooding), which shaped and maintained the pre-European structure of plant communities;
2. agricultural development (livestock grazing and cropland), which accelerated the conversion of native grasslands; and
3. urbanization, which resulted in the irretrievable loss of native grasslands.

Fire suppression became significant in the mid-1800s[1] when the expanding population became dominated more by Europeans than Native Americans. Before that, the extent and persistence of native grasslands were closely tied to the regular disturbance of fire, either wildfire or fires set by Native Americans.[1, 14, 124, 130] Europeans deemed that fires were a threat to property. Fire suppression allowed succession in fire-maintained grasslands, resulting in widespread shrub (including exotics) and tree (particularly Douglas-fir) encroachment into native grasslands and savannas.[1, 73] For example, in the South Puget Lowlands, 32% of prairie loss has been attributed to forest invasion and conversion.[55]

Agricultural development resulted in the loss and degradation of native grasslands from conversion to cropland and livestock grazing, respectively. The effect of conversion to cropland was the immediate and permanent loss of native prairie on a local scale. Until the early 1900s, largely animal-dependent farming limited the size of areas that could be cultivated. Thus, the landscape remained a mosaic of cultivated areas and native prairie habitats. Mechanical agriculture and fossil-fueled machinery resulted in progressive conversion of native prairie and savanna to agricultural land. By the mid-1970s, only tracts too small, tracts otherwise poorly suited for cultivation, or tracts under some protection, were untouched. These are among the few scattered areas that remain today as refugia for native prairie plants. Many savanna-form oaks also were lost in the conversion to cropland, and those that remain are relicts in the midst of a sea of cultivation.[66]

Livestock grazing has altered grassland habitat primarily through degradation. However, moderate to high levels of grazing can also result in the permanent loss of the more palatable native grasses and forbs (E. Alverson, The Nature Conservancy, pers. comm.). Nearly all remnant grassland parcels not converted to cropland have been grazed. Most grazing was open range until the late 1800s, when a shift to fenced pastureland occurred.[66] This practice, in conjunction with repeated disturbance of soil and vegetative communities from cultivation, importation of selected forage grasses and forbs, and the seasonal movement of livestock between parcels, set the stage for the invasion of exotic plants. Many remnant prairie parcels are exotic species-dominated. Heavily grazed sites tend to have the highest proportion of exotic species.[123] Colonial bentgrass (*Agrostus capillaris*), sweet vernalgrass (*Anthoxanthum odoratum*), Scot's broom (*Cytisus scoparius*), medusa-head wildrye (*Elymus caput-medusae*), and reed canarygrass (*Phalaris arundinacea*) are prominent among exotic species present. Consequences of exotic plant invasion were the displacement or local extirpation of native plant species, which also may have disrupted food webs for wildlife.

Attempts at quantifying prairie loss to agriculture are rare, but these losses are significant. In the South Puget Lowlands, 30% of prairie losses have been attributed to agricultural uses.[55] In addition, factors that reduced native ungulate populations (e.g., habitat fragmentation [e.g., fencing], hunting) altered successional relationships between native browsers and grazers and grassland species. In Washington, declining recruitment of savanna oaks after 1870 is thought to be linked to heavy grazing.[1, 95]

Urbanization resulted in local and permanent losses of grassland,[1] particularly in the South Puget Lowlands and northern Willamette Valley. Beyond the near-complete loss of native grassland habitat, urbanization continues to reduce the availability of agricultural grassland habitats. This pattern is most common in the central and southern Willamette Valley and central Rogue Valley, where urban growth boundaries have extended into active and passive agricultural lands, resulting in loss of habitat for several grassland-associated species that have been able to persist in those areas.[3] Attempts to quantify losses resulting from urban development are also rare, but the levels of loss identified are significant. In the South Puget Lowlands, one-third of prairie losses were attributed to urban development.[55]

Chaparral

Little is known about the pre-European distribution of chaparral habitat in the Rogue and Umpqua valleys.[43] The only available current estimate has identified 38,412 acres (15,551 ha) of chaparral occurring in the Rogue and Umpqua river basins[96] (T. O'Neil, Northwest Habitat Institute, pers. comm.). However, this figure is an underestimate because it excludes the montane subtype and successional stages of other communities that have a chaparral structure.

Chaparral habitat has declined in extent primarily because of conversion to agriculture and urban/residential development, and the quality of chaparral habitat has been degraded because of fire suppression. Subsequent to European settlement, much chaparral was converted to pasture and cropland, particularly on the lower slopes and valley bottoms. In the Rogue Valley, little chaparral remains on the valley floor. Steeper slopes have been less affected. Where chaparral remains, fire suppression probably has increased the predominance of dense, tall, shrub stands instead of the more open, patchy structure. Where grazing has been excluded, savannas or open woodlands have developed a chaparral-like structure.[1] The latter would account for minor local increases in chaparral habitat.

More recently, population growth in the Rogue Valley has further encroached on chaparral habitat. New subdivisions and clearing associated with low density rural development present a different threat to chaparral habitat. As development of private lands abuts public lands, land managers are increasingly required to manage lands to reduce fire hazard to adjacent residences. Typically this involves manual removal of chaparral plants to minimize the likelihood of fire.

The chaparral community is adapted to recurring fire. Fire suppression inevitably will lead to profound changes in the distribution and abundance of plant species.[10] For example, the recruitment of new generations of shrubs likely will be affected. The germination of many chaparral plants is triggered by the heat of fire or the chemicals found in smoke. Without wildfire, new recruitment is limited, especially among chaparral herbaceous plants, and as chaparral shrubs become decadent, their productivity is reduced.[11, 57] Given our limited knowledge of the functioning of this ecosystem along the northern margin of its range, many changes that might arise as a consequence of fire suppression cannot be anticipated.

Wildlife Communities

Wildlife communities in grassland, and to a lesser extent chaparral habitats, tend to be ecologically simple, with relatively low species diversity and richness. Within each habitat type, the more uniform the vegetative structure, the lower the species diversity. As different structural components are added (e.g., shrubs in grassland, trees in chaparral), the wildlife community becomes more diverse. However, extensive encroachment of these structural elements may make the habitat less suitable to its highly associated species.

Numerous site-specific factors result in differences in vegetative composition within grassland and chaparral habitats. However, from a wildlife standpoint, coarse-level vegetative structural similarities exist that are linked to wildlife use. The principal feature in grasslands is the dominance of herbaceous vegetation, with low, but varying numbers of shrubs and/or scattered trees. The principal feature in chaparral is the dominance of shrubs, interspersed with varying amounts of herbaceous vegetation and hardwood or, less frequently, coniferous trees. Differences in wildlife use within grassland and chaparral habitats are often related to differences in vegetation structure within physiognomic layers other than the dominant layer. Additionally, microhabitat features, such as rocks or water, often account for the presence/absence of many amphibians, reptiles, and small mammals.

Extensive loss of grasslands has resulted in a fragmented habitat pattern that presents significant obstacles to the persistence of wildlife populations. Two important issues are (1) genetic isolation and dispersal, particularly for some amphibians, reptiles, and small mammals with limited mobility; and (2) the inability of small patches of habitat to meet the areal requirements of some birds, reptiles, and medium-sized to large mammals. Consequences of genetic isolation may be local losses of genetic diversity and greatly increased susceptibility to local extirpation. Pronounced fragmentation may result in habitat patches with varying amounts of a depauperated fauna, whose populations may have lost their genetic flexibility.

Historical Knowledge and Trends

Information on wildlife communities in native habitats at the time of European settlement are sparse, and evidence of population status and trends is mostly anecdotal. Additionally, historical information on the vertebrate fauna is disproportionate; greatest for large mammals and birds, and almost nonexistent for small mammals and herpetofauna. Additionally, population trend information is available only for birds, and only for the last thirty years from the Breeding Bird Survey (BBS). Only recently have systematic studies been conducted on wildlife use of westside grassland and chaparral habitats (for example, see Case Histories).

Where data are unavailable on historical wildlife populations, our knowledge of wildlife habitat relationships is based on current conditions within these habitats. For example, in grasslands, an understanding of wildlife species relationships with native habitats is partly based on current knowledge of species relationships with agricultural grassland habitats. This is largely a consequence of native grasslands having nearly disappeared, and the fact that some wildlife species have been able to persist in agricultural grasslands.

Mammals. Most information on the mammal fauna is for medium-to-large mammals, and a significant portion of it is based on hunting and trapping data. Suspected population changes for most mammal species are based on an evaluation of potential conflicts between species' habitat requirements and types of habitat change, and our relatively limited knowledge about a species' ability to tolerate change.[68, 90, 128]

Mammals associated with native grasslands were probably reduced with post-settlement habitat losses and alteration, but suggestions of this pattern are based entirely on large mammals. For example, at the time of settlement, elk foraged in valley grasslands, but hunting and cultivation drove them into more secluded forested areas.[58, 119, 128] Declines in elk and other large mammals in lowland grassland habitats undoubtedly reflected loss and alteration of their requisite habitat. However, the pattern is difficult to disentangle from concomitant spatial and temporal changes in burning, climate variation, forestry

Table 1. Bird species regularly associated with westside grassland and chaparral habitat with significantly declining population trend estimates based on Breeding Bird Survey data (Sauer et al. 1997).

Species	Habitat association[a]		BBS population trend[b]		Comments
	Grassland	Chaparral	1966-98	1980-98	
American kestrel	x		**-7.3****	**-9.6****	
Ring-necked pheasant	x	x	**-5.8*****	-0.7	introduced species
Mourning dove	x	x	**-3.3*****	**-4.3*****	
Common nighthawk	x	x	**-6.8****	8.6	
Vaux's swift		x	**-3.9***	-1.8	
Northern flicker	x	x	**-1.2***	-0.6	
Acorn woodpecker	x	x	-1.8	**-2.2*****	Oregon only (i.e., westside)
Western wood-pewee	x	x	**-2.9*****	-0.8	
Ash-throated flycatcher	x	x	-0.8	**-3.5*****	Rogue and Umpqua valleys only
Bushtit		x	**-12.1*****	**-9.9*****	
Oak titmouse		x	**-3.8***	-2.2	Rogue valley only
Western bluebird	x	x	**-3.6****	-7.7	
House wren		x	-3.4	**-6.8***	
Wrentit		x	-0.9	**-1.7*****	Oregon only
Lazuli bunting	x	x	**-4.0****	-0.6	
Chipping sparrow	x	x	**-8.3*****	**-4.5****	
Lark sparrow		x	**-5.2****	-4.2	Rogue and Umpqua valleys only
Song sparrow	x		**-1.1*****	**-1.0****	
Dark-eyed junco		x	**-2.7*****	**-2.2****	
Western meadowlark	x	x	**-7.0*****	**-4.9***	
Brown-headed cowbird	x	x	**-2.8*****	**-1.8****	
Brewer's blackbird	x	x	-3.3	**-2.7****	
Bullock's oriole		x	**-3.0****	-2.8	
American goldfinch	x	x	**-2.3***	-1.6	

[a] Grassland includes dry and wet prairie and savanna; chaparral encompasses valley and montane subtypes; x = species association with this habitat type.

[b] Trend expressed as percent annual change. Only includes species that have occurred on ≥14 routes. Bold indicates significantly declining trends. Significance values are * P <0.10, ** P <0.05, and *** P <0.01.

practices, and hunting. Although Columbian black-tailed deer have been influenced by the same factors, they seem significantly more tolerant to land conversion and human-caused disturbances than elk.[128]

Birds. Populations of bird species associated with native grasslands were undoubtably reduced to some extent with post-settlement losses of native grassland habitats. Some species such as the streaked horned lark, western meadowlark, and Oregon vesper sparrow were apparently able to persist in newly created agricultural grasslands such as hayfields and pasture. These species were referred to as "abundant or common" in the Willamette Valley and Puget Lowlands by all authors from the time of European settlement[122] through the late 1800s and early 1900s[5, 30, 65, 100] up until the mid-1900s.[44, 54, 72] Since the late 1960s, however, many westside grassland bird species have experienced substantial population declines (Table 1). Of the 44 species with either long-term (1966-98) or short-term (1980-98) significantly declining trends in the BBS Region encompassing westside grasslands and chaparral (Southern Pacific Rainforest), 24 are regularly (closely and generally) associated with grasslands and/or chaparral. A few of these species also use forest habitats (e.g., western bluebird, Bewick's wren), thus declines may be affected by populations in forest habitats. Several grassland species that lack sufficient BBS data for analysis (e.g., short-eared owl, northern harrier, streaked horned lark, Oregon vesper sparrow) are considered to be declining in western Oregon and Washington based on anecdotal information.

Statistically descriptive indications of population declines can also be made for some grassland species based on BBS relative abundance data. In the Willamette Valley, western meadowlark abundance declined from a mean of approximately 13 birds/route ($n = 11$) in the early 1970s to <1 bird/route in the mid 1990s.[4] The chipping sparrow, a savanna-associate, declined in abundance on Willamette

Western meadowlark. (Photograph by Richard B. Forbes)

Valley BBS routes from a mean of approximately 11 birds/route in the early 1970s to approximately 2 birds/route in the mid 1990s.

In contrast to declining species, only 16 species have significantly increasing trends.[106] Only 2 of those are grassland associates (common yellowthroat and red-winged blackbird in wet prairie), and 2 are chaparral associates (California towhee and American robin). The remaining 12 species are forest, riparian, or urban associates.

Concurrent trends in land use during the BBS time frame (1966-96) include extensive suburban and residential development, and conversion to row crop agriculture and large, intensively cultivated grass fields, particularly in the Willamette Valley. Row crop agriculture provides almost no habitat for grassland birds, and although cultivated grass fields (e.g., hayfields, annual and perennial grass fields) can provide habitat for a few native species (e.g., savannah sparrow, western meadowlark), they can also function as ecological traps when harvesting aborts nesting attempts and reduces reproductive success.

Two broad patterns of change that have occurred in contemporary grassland avifauna worldwide are declines in native and endemic species and increases in exotic species.[75] These patterns also are evident in the Willamette Valley based on censusing of agricultural grassland habitats in 1996.[3] Among the 12 most abundant species, only the savannah sparrow, the most abundant species, is considered highly associated with native grassland habitat (Table 2). The second most abundant species, the European starling, is an exotic, and the third, fourth, and twelfth most abundant species (Brewer's blackbird, red-winged blackbird, and killdeer, respectively) are characteristic of agricultural lands. Of the remaining species, 3 swallows (barn, cliff, and violet-green) are open-country aerial foragers, and 4 other species (American robin, American goldfinch, song sparrow, and common yellowthroat) are relatively common edge or generalist species that use grasslands opportunistically. Species highly associated with native grasslands such as the western meadowlark,

Table 2. The 12 most abundant species during point-count censusing in Willamette Valley agricultural grasslands, May-June, 1996.[a]

Species	Birds/point count
Savannah sparrow	1.62
European starling	1.11
Brewer's blackbird	0.80
Red-winged blackbird	0.54
Barn swallow	0.53
American goldfinch	0.51
Song sparrow	0.47
American robin	0.46
Common yellowthroat	0.39
Cliff swallow	0.37
Violet-green swallow	0.20
Killdeer	0.18

[a]Based on detections within 328 feet (100 m) of 544 point-count stations visited once in May and once in June.[3]

Table 3. Historical records of herpetofauna of the Southern Puget Trough outwash prairies (modified from Leonard and Hallock[83]).

Species	Selected sources	Reproductive mode[a]	Occurrence[b] Historical	Occurrence[b] Current
Northwestern salamander	79	Sb	C	C
Long-toed salamander	79	Sb	C	C
Rough-skinned newt	79	Sb	C	C
Western toad	107, 109	Sb	C	U
Pacific treefrog	79, 109	Sb	C	C
Northern red-legged frog	79, 109	Sb	C	C
Oregon spotted frog	109	Sb	U	R
Bullfrog	79, 109	Sb	C	C
Western pond turtle	109	El	C	X
Northern alligator lizard	109	Lb	C	C
Western fence lizard	109	El	C	X
Rubber boa	109	Lb	U	U
Racer	2	El	R	X
Gopher snake	122	El	R	X
Western terrestrial garter snake	79, 109	Lb	C	C
Northwestern garter snake	79, 109	Lb	C	C
Common garter snake	79, 109	Lb	C	C

[a] Sb = stillwater breeding, El = egg laying, Lb = live bearing

[b] C = common, U = uncommon, R = rare, X = functionally extirpated

Oregon vesper sparrow, and streaked horned lark were uncommon to rare.

Herpetofauna. Leonard and Hallock[83] provided an overview of the historical information on herpetofauna in the South Puget Lowlands prairie landscape. Most records are from museum collections and personal field notes of amateurs and professional biologists. These records date back approximately 150 years and provide at least qualitative documentation of the relative abundance of several species. It is noteworthy that several species documented as occurring historically are now extremely rare or extirpated (Table 3). Additionally, a number of species seem to be much less abundant than described in these historical accounts.

Historical data addressing grassland-associated Willamette Valley herpetofauna were erratically collected, both geographically and temporally. Early surveys provided distributional data from selected areas, e.g., Portland,[70] Benton and Linn Counties,[50] or anecdotal information as part of larger regional compendiums,[49, 52, 108, 125] but data on habitat relationships were generally sparse (although see Graf[50]). Because few historical grasslands remain, reduction or disappearance of at least some of the herpetofauna characteristic of those systems is strongly suspected. Historical data on grassland-associated herpetofauna for the Umpqua Valley are lacking. Some data on herpetofauna exist for the Rogue Valley,[41] but only a few of these data apply to grassland-associated herpetofauna.

Current Populations and Habitat Relationships

A few mammal studies have been recently conducted in westside grassland and chaparral habitats. Research on the Western pocket gopher in the Puget Lowlands has emphasized biology and habitat use,[135] and species distribution relative to the degree of soil rockiness in prairies.[115] Several studies have been conducted on the ecology of gray-tailed voles in the Willamette Valley,[136, 137, 138] including effects of habitat fragmentation,[139] mowing,[34] and pesticides.[35, 94] An examination of the relationship of small mammals and fire in wet prairie was conducted at Finley National Wildlife Refuge in the Willamette Valley (see Case Histories).

Several recent bird community studies, which include work done in the Willamette Valley,[3] Umpqua Valley,[28] and Puget Lowlands[101, 104, 105] have examined habitat relationships of grassland and savanna species. Additionally, species-specific studies include the blue-gray gnatcatcher in the Rogue Valley[112] and the western bluebird in the Willamette Valley.[36] Among the grassland studies, Altman[3] focused on agricultural grasslands because native grasslands are nearly extirpated in the Willamette Valley, and Rogers[104] focused on a few of the remaining parcels of native prairie in the South Puget Lowlands. The three Puget Lowlands studies focused on remnant grasslands on military installations at Fort Lewis and McChord.

Several recent surveys, some of which occurred on grasslands, have attempted to assess the current status and habitat relationships of herpetofauna in the South

Puget Lowlands.[18, 56, 82] Most have focused on state-listed or federal candidate species such as the western pond turtle[42] and the peripherally grassland-associated Oregon spotted frog.[47, 92] Several of these studies were conducted on remnant prairies found on military installations at Fort Lewis and McChord.

Studies focused on grassland-associated herpetofauna in Oregon are limited (R. Goggans, Oregon Department of Fish and Wildlife, pers. comm.). St. John[116, 117, 118] provided some habitat data on grassland herpetofauna in the Rogue, Umpqua, and Willamette valleys, respectively. Additionally, there is an ongoing effort to characterize the snake fauna of several habitats in the Willamette hydrographic basin, although only some of this effort addresses grasslands or grassland-containing assemblage[91] (R. Mason, Oregon State University, pers. comm.).

Grasslands: Mammals. The mammal fauna of westside grasslands includes a diverse array of approximately 55 species that occupy a variety of ecological niches (Table 4). This fauna includes a large suite of grazing, browsing, or seed-eating species, represented by exposed surface-active (deer, elk), vegetation tunneling (selected rabbits, voles), or burrowing forms (pocket gophers, selected squirrels). The mammal fauna also includes a smaller group of aerial (bats) and terrestrial (shrews, moles) insectivores, and a species rich and diverse group of carnivores. This fauna includes forms that range from being closely tied to grassland habitats (e.g., gray-tailed vole, Camas pocket gopher), to generalized species that occur across a range of habitat types (e.g., bobcat, coyote, weasels, skunks).

Mammals most characteristic of westside grasslands are all small species dependent on an herbaceous structure or the invertebrate fauna supported by that structure. These include Camas pocket gopher, gray-tailed vole, Townsend's mole, and local forms of the Western pocket gopher in the Puget Lowlands. Regionally, some of these species are replaced by ecologically equivalent but more generalized species. For example, Botta's pocket gopher and California vole occupy grasslands in the Rogue and Umpqua valleys, whereas the Camas pocket gopher and gray-tailed vole are restricted to Willamette Valley grasslands. In a similar pattern, the broad-footed mole rather than Townsend's mole is the more common mole in grasslands in the Rogue Valley.

Several additional insectivores and rodents are common in westside grasslands, but are widespread in distribution and occur in several habitat types. These include the creeping vole, Townsend's vole, and Western pocket gopher in western Oregon. The only large mammal that could be viewed as a grassland specialist, the pronghorn, was extirpated historically across the western limit of its distribution in the Rogue Valley;[7] evidence is lacking that it was once more widespread in westside grasslands.

Several groups of mammals, although frequent and widespread in westside grasslands, lack representative species specialized for westside grassland existence *per se*. These include bats, carnivores, rabbits, shrews, and selected ungulates (elk, deer). All species in these groups either use grassland habitat only for foraging or require some structural element not present in exclusively herbaceous grassland (e.g., shrub or tree structure, an aquatic habitat, or both). Shrews and most bats require proximity to a mesic association or an aquatic habitat, whereas rabbits, many carnivores, and ungulates need some woody structure for concealment. For example, black-tailed jackrabbits require the relatively open matrix found in many grasslands to effectively use running as an anti-predator strategy. However, they also require scattered shrub clumps to conceal themselves following a predator-escape run. Additional species are common in habitats outside grasslands (oaks, woodland, or forest margins), but also may be abundant on grassland margins (e.g., California ground squirrel).

Some of the aforementioned patterns contribute to a generally increasing species richness gradient among mammals with decreasing latitude (Table 4), a frequent pattern observed in nearly all faunal groups on a broader (global) scale. For example, of the 3 grassland-specialized mammals previously discussed, only 1, Townsend's mole, is widespread in Washington; another, Camas pocket gopher, does not occur in the state; and the third, gray-tailed vole, is only known from Clark County.

Grasslands: Birds. The breeding avifauna in westside grasslands is characterized by relatively low species diversity, and a unique assemblage of cryptically colored ground nesters. Species diversity is greater in savanna habitat, where several cavity nesters are associated with the tree component.

Grassland birds have evolved several adaptations to accommodate to life in an herbaceous, open landscape. Most nest on the ground, and most nests are well concealed in vegetation for protection from predators. Some species, such as the western meadowlark, enhance nest concealment by building a domed nest and a short runway to the nest. Others, such as the burrowing owl, nest underground in mammal burrows. Many passerines developed behavioral adaptations to avoid nest predation,

Gray-tailed vole. (Photograph from W. Daniel Edge)

Table 4. Mammal distribution in westside grassland and chaparral habitats.

	Grassland[a,b]				Chaparral[a,b]	
	P	W	U	R	U	R
Insectivores						
Fog shrew[c]					px[d]	px
Vagrant shrew	x	x	x	x		
Broad-footed mole	x	px				
Coast mole	x	x	x	x		
Townsend's mole	x	x	x	x		
Bats						
California myotis	x	x	x	x	px	px
Long-eared myotis	x	x	px	x	x	x
Little brown myotis[c]	x	x	px	x		
Long-legged myotis	x	x	px	px	px	x
Yuma myotis[c]	x	x	x	x		
Hoary bat	m	m	m	x	px	px
Silver-haired bat	x	px	px	x	x?	x?
Big brown bat	x	x	px	x	x	x
Townsend's big-eared bat	x	x	x	x	px	x
Pallid bat		m	x	x	px	x
Brazilian free-tailed bat		x	x	px?	x?	
Rabbits						
Brush rabbit		x	x	x	x	x
Black-tailed jackrabbit		x	x	x		
Rodents						
Siskiyou chipmunk						x
California ground squirrel		x	x	x	x	x
Golden-mantled ground squirrel					m	x
Botta's pocket gopher		m	x	x	px	x
Camas pocket gopher		x				
Western pocket gopher[f]	m	m	x	x	x	x
California kangaroo rat						x
Deer mouse	x	x	x	x	x	x
Pinon mouse						x
Dusky-footed woodrat		x	x	x	x	x
California vole		m	x	x		x
Gray-tailed vole	m	x				
Long-tailed vole	x	m	px	x		
Creeping vole	x	x	x	x	x	x
Townsend's vole	x	x	x	x		
Pacific jumping mouse[c]	x	x	x	m		
Common porcupine	m	m	px	x	m	x
Carnivores						
Coyote	x	x	x	x	x	x
Gray wolf	h	h	h	h	h	h
Gray fox		x	x	x		
Red fox[e]	x	x	x	x		
Black bear	x	h	x	x	x	px
Grizzly bear	h	h	h	h	h	h
Ringtail					x	x
Raccoon	x	x	x	x	x	x
Ermine	x	x	px	px	px	x
Long-tailed weasel	x	x	px	x	x	x
American badger				h		h
Western spotted skunk	x	x	x	x	x	x
Striped skunk	x	x	x	x	x	x
Mountain lion	x	x	x	x	x	x
Bobcat	x	x	x	x	x	x
Ungulates						
Elk	x	x	x	x	x	x
Black-tailed deer	x	x	x	x	x	x
Columbian white-tailed deer	x	x	x			
Pronghorn	h					
Totals	38	46	43	42	31	33

[a] Grasslands include dry and wet prairie and savanna; chaparral encompasses valley and montane subtypes (see text for details).

[b] P = Puget Lowlands, W = Willamette Valley, U = Umpqua Valley, R = Rogue Valley.

[c] Species typically needs some kind of aquatic habitat beyond that found in seasonally wet prairie (e.g., pond, stream, marsh) and thus, is only weakly linked to grasslands as defined here.

[d] x = known occurrence in the geographic areas indicated for each habitat type (p preceding the designation indicates probable occurrence); m = present, but of marginal occurrence with a greater distribution outside the region; and h = historically present, but functionally extirpated. Data are based on a combination of Ingles[68], Maser et al.[90], Verts and Carraway[128], other sources, and personal observations.

[e] Data on natural range expansions and introduced populations in Oregon are confusing (Verts and Carraway[128]).

[f] Portions of what is known as Western pocket gopher is treated as Mazama pocket gopher by some investigators.[115]

including landing away from the nest and running concealed along the ground to get to the nest, injury-feigning distraction displays (killdeer), and acoustical mimicry of rattlesnakes (juvenile burrowing owls). Grassland birds often have shorter incubation and nestling periods than other passerines, and young leave the nest prior to being able to fly, but can run effectively along the ground until capable of flight. Other ecological adaptations unique to grassland birds with limited or no perches available include flight songs (horned lark), singing from the ground (many species), and hovering in the air as an aerial perch for hunting (American kestrel).

Approximately 42 bird species are regularly associated breeding species in westside grassland habitats (Table 5). Species richness increases from wet prairie (16 species), to dry prairie (23), to savanna (27). Some species occur only within 1 type of habitat along the grassland continuum (e.g., common snipe in wet prairie, acorn woodpecker in savanna). Others are equally associated with multiple grassland types (e.g., western meadowlark, streaked horned lark). Wet prairie is characterized by species associated with mesic sites such as common snipe,

Text continues after Table 5, on page 271

Table 5. Breeding bird species regularly associated with grassland and chaparral habitats in western Oregon and Washington.[a]

Species[b]	Status[c]	Guilds		Grassland[d]			Chaparral[d]			Grasslands[e]				Chaparral[e]	
		Forage[f]	Nest[g]	Wet Prairie	Dry Prairie	Savanna	Chaparral Grassland	Chaparral Shrub	Oak Chaparral	P	W	U	R	U	R
White-tailed kite	T	G,C	F	x		x				m	m		m		
Northern harrier	T	G,C	G	X	x	x				x	x	m	m		
American kestrel	T	G,C	C			x				x	x	x	x	x	x
Red-tailed hawk	T	G,C	F		x	x	x	x		x	x	x	x	x	x
Wild turkey (I)	R	G,G	G			x		x	x	m	m	x	x	x	x
Ring-necked pheasant (I)	R	G,O	G		x		x			x	x	x	x		
Northern bobwhite (I)	R	G,G	G		x					m	m				
Mountain quail	R	G,H	G					x	x					x	x
California quail	R	G,G	G			x	x	x		x	x	x	x	x	x
Mourning dove	T	G,G	F			x	x	x		x	x	x	x	x	x
Sandhill crane	T	G,O	G	x						h	h				
Killdeer	R	G,I	G	x	x	x	x			x	x	x	x	x	x
Western screech owl	R	G,C	C											x	x
Burrowing owl	T	G,C	G		X				x		h	h	h		
Short-eared owl	T	G,C	G	X	x					m	m	m	m	m	m
Common snipe	T	G,O	G	X						x	x	x	x	x	x
Common nighthawk	N	A,I	G		x		x		x	m	m	m	m	m	x
Common poorwill	N	A,I	G					x	x	m	m	m	m	m	x
Vaux's swift	N	A,I	C						x					x	x
Anna's hummingbird	T	F,O	F			X		x	x				x	x	x
Lewis' woodpecker	T	A,I	C			x			x	h	h	m	m	m	m
Acorn woodpecker	R	B,G	C			x			x		m			x	x
Northern flicker	R	G,I	C			x			x	x	x	x	x	x	x
Western wood-pewee	N	A,I	F			x			x	x	x	x	x	x	x
Ash-throated flycatcher	N	A,I	C			x		x	x				x	x	x
Dusky flycatcher	N	A,I	F					X	x					m	m
Western kingbird	N	A,I	F			x	x		x	m	m	x	x	m	x
Say's phoebe	N	A,I	C							h	h	h	h		
Streaked horned lark	R	G,O	G	x	X	x	x			x	x	x	x		
Tree swallow	N	A,I	C			x			x	x	x	x	x	x	x
Violet-green swallow	N	A,I	C			x			x	x	x	x	x	x	x
Western scrub jay	R	G,O	F			x	x	x	x	m	m	x	x	x	x
Oak titmouse	R	F,O	C						X					x	x
Bushtit	R	F,I	F					x	x					x	x
White-breasted nuthatch	R	B,I	C						x	h	h		x	x	x
Bewick's wren	R	G,I	C			x		x	x		h	h	h	x	x
House wren	N	F,I	C						x					x	x
Blue-gray gnatcatcher	N	F,I	F						X						x

Species[b]	Status[c]	Forage[f]	Nest[g]	Grassland[d] Wet Prairie	Dry Prairie	Savanna	Chaparral[d] Chaparral Grassland	Chaparral Shrub	Oak Chaparral	Grasslands[e] P	W	U	R	Chaparral[e] U	R
Western bluebird	T	G,I	C			X			×			×	×	×	×
American robin	T	G,V	F					×	×					×	×
Wrentit	R	F,I	F				×	X	×					×	×
European starling (I)	R	G,O	C	×	×	×	×			×	×	×	×	×	×
Common yellowthroat	N	F,I	F	×			×			×	×	×	×		×
Black-headed grosbeak	N	F,I	F			×		×	×	×		×		×	m
Lazuli bunting	N	F,O	F		×		×	×	×		×		m	×	
Spotted towhee	R	G,O	G				×	×	×					×	×
California towhee	R	G,O	G				×	×	×					×	×
Green-tailed towhee	N	G,O	G				×	×	×					×	m
Chipping sparrow	N	G,O	G		×	×	×		×	×	×	×	×	×	×
Oregon vesper sparrow	N	G,O	G		×	×				×	×	m	m		
Lark sparrow	N	G,O	G		×	×	×			×	h	m	m		
Savannah sparrow	T	G,O	G	×						×	×	×	×		
Grasshopper sparrow	N	G,O	G		×					×	×	m	m		
Song sparrow	R	G,O	G	×			×			×	×	×	×		
Fox sparrow	T	G,O	F				×	×						×	×
Dark-eyed junco	T	G,O	G				×	×	×					×	×
Western meadowlark	R	G,I	G	×	X	×	×	X	×	×	×	×	×	×	×
Red-winged blackbird	R	G,O	F	×		×	×		×	×	×	×	×	×	×
Brewer's blackbird	T	G,O	F	×		×	×	×	×	×	×	×	×	×	×
Brown-headed cowbird	T	G,O	—	×	×	×	×		×	×	×	×	×	×	×
Bullock's oriole	N	F,I	F				×		×					×	×
House finch	R	G,F	F	×	×	×	×	×	×		×	×	×	×	×
American goldfinch	T	G,G	F		×	×	×		×	×	×	×	×	×	×
Lesser goldfinch	R	G,G	F			×	X		×					×	×
Total Species				16	23	27	27	22	38	35	41	39	41	43	47

[a] The list includes only species that *regularly* nest in grassland and chaparral habitats, currently or historically. Species that only forage in these habitats during the breeding season, such as rufous hummingbird (wet prairie) and rock dove (dry prairie, savanna), some raptors and other species that require specific structural elements for nesting, such as common raven and peregrine falcon (cliffs) or barn swallows and cliff swallows (outbuildings) are not included.

[b] I = Introduced species.

[c] Status refers to the predominant pattern for the majority of individuals of that species that breed in western Oregon and Washington. It is based on information presented in Puchy and Marshall[99] with professional opinion from others. Populations of some species considered temperate migrants may be resident in the Rogue Valley (e.g., mourning dove, American robin, western bluebird, Anna's hummingbird, and American goldfinch), but the majority of the population in westside grassland and chaparral habitats is migratory. R = resident; N = nearctic (long-distance) migrant; T = temperate (short-distance) migrant

[d] X = species has a high degree of association with this habitat during the breeding season (obligate or semi-obligate) such that species abundance is significantly higher in this habitat than any other in western Oregon and Washington; x = species has a relatively strong association with this habitat during the breeding season; i.e., it is as abundant in this habitat as others, but it is not obligate to this habitat for breeding.

[e] x = regular breeding species in the geographic areas indicated for each habitat type; m = marginal presence or in low abundance as a breeding species in the geographic areas indicated for each habitat type (i.e., much greater distribution outside the geographic area); h = historically present as a breeding species, but believed to be extirpated in the geographic areas indicated for each habitat type.

fFirst letter is predominant foraging zone during the breeding season: G = ground, A = aerial; F = foliage; B = bark. Second letter is predominant food type during the breeding season: C = carnivore (vertebrates); F = frugivore (fruits); G = granivore (nuts, seeds); H = herbivore (plants); I = insectivore (insects); O = omnivore (plant and animal matter); V = vermivore (worms).

gNesting zones: G = ground (includes low shrub <1 m off ground); F = foliage (open-cup in shrub or tree >1 m off ground); C = cavity or mostly enclosed structure (e.g., ledge).

northern harrier, and short-eared owl. Dry prairie is characterized by species such as Oregon vesper sparrow, burrowing owl, and grasshopper sparrow. Where a scattered shrub component is able to persist in dry prairie, species diversity may increase to include the lazuli bunting and American goldfinch, and species abundance may increase for some obligate birds such as the Oregon vesper sparrow and western meadowlark. The presence of singular or small groves of scattered oaks, Douglas-fir, or ponderosa pine (savanna) results in an entirely new group of highly associated species such as American kestrel, western bluebird, chipping sparrow, western kingbird, and acorn woodpecker. Many of these species forage on the ground or in the air above the grassland, but are responding to availability of cavities for nesting (e.g., American kestrel, western bluebird), presence of acorns as a food source (e.g., acorn woodpecker), elevated perches for singing (e.g., chipping sparrow), and/or the availability of branch structure for nesting substrate (e.g., western kingbird).

As expected in an herbaceous vegetative community, grassland bird assemblages are dominated by ground foragers (71%, *n* = 30) (Table 5). Nesting guilds are less dominated by ground nesters (41%, *n* = 17), and more equally represented by cavity nesters (29%, *n* = 12) and foliage nesters (27%, *n* = 11). However, if those species exclusively associated with the tree component in savanna (*n* = 13) are not considered, ground foraging and ground nesting guilds become more dominant (89% and 57%, respectively). It is in the savanna community that bird species of other foraging and nesting guilds become more representative. This includes foraging guilds such as aerial insectivores (western kingbird) and granivores (acorn woodpecker), and nesting guilds such as cavity nesters (American kestrel, western bluebird) and open-cup foliage-nesters (chipping sparrow, western kingbird).

The composition of breeding bird species in westside grasslands is predominantly migrants (64%, *n* = 27) (Table 5). When distinguishing type of migrant (temperate vs. nearctic) there is relatively equal representation among resident (36%, *n* = 15), temperate migrant (26%, *n* = 15), and nearctic migrant (29%, *n* = 13) species. Further delineation excluding savanna-only species results in similar increases in species richness from nearctic migrants (27%, *n* = 8) to residents (20%, *n* = 9) to temperate migrants (43%, *n* = 13).

The most distinctive geographic pattern is that the Puget Lowlands grassland breeding avifauna is less species rich than the 3 Oregon valleys (Table 5). Several westside grassland or savanna species (i. e., white-tailed kite, acorn woodpecker, ash-throated flycatcher, lark sparrow, grasshopper sparrow, lesser goldfinch) reach the northern limit of their geographic ranges in Oregon. There are no northern grassland species that reach the southern limit of their breeding range in the Puget Lowlands, although historically sandhill cranes maintained a small breeding population in Puget Lowland prairies.[72]

The nonbreeding-season bird community in grasslands differs considerably from that of the breeding season. Species diversity can be highly variable, dependent on several factors such as weather, field type and management, fluctuating invertebrate populations, degree of seasonal surface water, and timing of spring green-up. Use is characterized by flocking species and aerial hunting behavior. Breeding species present in winter flocks include western meadowlark, streaked horned lark, and killdeer, but the dominant avian presence is nonbreeding flocking species such as Canada geese, sandhill cranes, American pipits, and several duck, gull, and shorebird species (e.g., dunlin and dowitchers). There also is an influx of raptors that hunt for small mammals and birds in native and agricultural grasslands.

A few species of note occur in the coastal and dune grassland and low shrubland types. Streaked horned lark breeds in coastal dune grasslands,[110] which are also important habitat for several wintering raptors and for small populations of migrant and wintering longspurs and snow buntings. The fox sparrow subspecies *fuliginosa* breeds in coastal headland shrublands north of Grays Harbor in Washington.[110] Yellow-rumped warblers concentrate in coastal dune shrublands in winter, especially in California wax myrtle (*Myrica californica*) (C. Chappell, Washington Natural Heritage Program, pers. comm.).

Grasslands: Herpetofauna. No amphibians and few reptiles can be considered grassland dependent; most use these habitats opportunistically. All are facultatively associated with grasslands through the presence of wet prairies and aquatic habitats within the grassland matrix. Facultative use of grasslands may occur for various aspects of a species life history, such as foraging, breeding, thermoregulation, and refuge/overwintering sites (Table 6). In general, grassland habitat use by herpetofauna requires the presence of at least one key feature on which these species are dependent, either within or next to grasslands. For many amphibians, the key feature is usually some type of surface water, but other species require rocks, sizable woody debris, or some kind of tree canopy. Westside grasslands possess temporary pools similar to the more expansive grasslands in California, but the amphibian fauna characteristic of the latter (e.g., spadefoot toads and tiger salamanders) is absent from the grassland pools of the Pacific Northwest.

Text continues after Table 6 on page 274

Table 6. Relationships of herpetofauna with grassland and chaparral habitats in western Oregon and Washington.

Species	Reproduction[a]	Food[b]	Refuge/Overwintering[c]	Grasslands[d] WP	DP	S	Chaparral[e] V	M	Comments
Amphibians									
Northwestern salamander	E-P Stlw B: BR, ST	L:Aq Mc, Inv, Am PM:Inv, Sl, Ea	L:AV, SS, WDs PM: MB, RC	x[f]	x	x			Ephemeral sites must have a 4.5 month or more hydroperiod
Long-toed salamander	E-P Stlw B:VA	L:Aq Mc, Inv PM:Inv, Sl, Ea	L:SS, AV, WDs PM: MB, RC	x	x	x		x	Ephemeral sites can have a 2.5-3 month hydroperiod
Rough-skinned newt	P Stlw B:AV	L:Aq Mc, Inv PM:Inv, Sl, Ea	L:AV, SS, WDs PM: MB, RC	x	x	x	x	x	Oviposition typically on dense submerged aquatic vegetation
Ensatina	Terr MB, WD	Co, Inv	UB, WDt, MB, RC			x	x	x	Not in WA grasslands; prefers woody debris
Black salamander	Terr UN	Co, Inv	RK, MB?, RC?				x		Extreme SW OR only; open shrubland near Ashland
Western toad	P Stlw NB	L:Al, Po, Oa, Car PM:Inv	L:N, AV PM:MB, WDt	x	x	x	x	x	Large woody debris may be preferred for overwintering
Pacific treefrog	E-P Stlw B:VA	L:Al, Oa, Car PM:Inv	L:AV, SS PM:WD, UB, LI	x	x	x	x	x	Ephemeral sites can have a 2.5-3 month hydroperiod
Northern red-legged frog	E-P Stlw B: EV, BR	L:Al, Oa, Car PM:Inv, Sl, Ea, Sm Am	L:AV, SS PM:TV, MB, PD	x	x	x			Ephemeral sites must have a 4.5 month or more hydroperiod
Foothill yellow-legged frog	P Flw B: RK, WDs	L:Al, Oa, Car PM:Inv, Sm Am	L:RS, WDs PM: CM, O	x	x	x	x	x	Most frequently in medium-sized (3rd to 6th order) stream
Oregon spotted frog	E-P Stlw NB	L:Al, Oa, Car PM:Inv, Sm Am	L:AV, SS PM: SP, O PM:SP, O	x	x	x			Adult habitat P Stlw; reproductive site may be ephemeral
Reptiles									
Western pond turtle	Egg SO	Inv, Car, AqVg	R:AV, SS, WD, RK O: LI, AV	xR		xR	x		Active-season habitat in permanent water, but highly variable in type
Long-nosed leopard lizard	Egg SO	Inv, Sm Lz	R:SH O: MB?		x		x		SW OR only; western fringe of range; few occurrences known
Sagebrush lizard	Egg SO	Inv	R:TV, SH, RK, MB O: MB?				x		SW OR only; western fringe of range; fragmented occurrence
Western fence lizard	Egg MB, SO	Inv	R:TV, SH, RK, MB O: MB, UB, O	x		x	x	x	Coastal Puget Sound non-grassland in WA; in chaparral abundant; needs exposed vertical structure for basking
Western skink	Egg UR, MB	Inv	R:MB, RK, LI, WD O: MB?	x	x		x		Prefers rocky habitat; Columbia Gorge non-grassland in WA

Species	Reproduction[a]	Food[b]	Refuge/Overwintering[c]	Grasslands[d] WP	DP	S	Chaparral[e] V	M	Comments
Northern alligator lizard	Live	Inv	R:TV,SH,WD, RK O: UN		×	×	×		Present in low-elevation grassland only in WA
Southern alligator lizard	Egg MB,WD,SO	Inv, NMa	R:TV, SH, WD O: MB, WD		X	X	X		Not in western WA grasslands; dense vegetation matrix key
Rubber boa	Live	NMa, Nb	R: UB, WD O:WD, O			×	×	×	Needs sizable woody debris
Ringneck snake	Egg UR, WD	Sm Am, Sl, Ea	R: RK, WD, MB, LI O: UN			×	×		Preferred refuges beneath well-insolated rocks
Sharptail snake	Egg WD	Sl, Sm Sa	R:WD, LI O: UN			×	×		Associated with woody debris; historic only in WA
Racer	Egg MB	Sm Lz, Be, Nb, Sm Ma, Inv	R:TV, MB O: MB?		×	×	X		Prefers a shrub-level matrix; historic only in WA
Striped whipsnake	Egg UN	Sm Lz, Be, Nb, Sm Ma	R:TV, MB O: MB?			×	×		SW OR only; western fringe of range; few occurrences known
Gopher snake	Egg MB	Sm Ma, Sm Lz, Be, Nb	R: MB, TV, RK O: MB, O	×	X	X	×	×	Presumed extirpated in western WA
Common kingsnake	Egg MB	Sm Lz, Sm Ma, Be, Nb, Sm, Sn	R: MB, WD O: MB, O		×	X	×		Often near riparian corridors; SW OR only
California mountain kingsnake	Egg UN	Sm Lz, Be, Nb, Sm Ma, Sm, Sn	R: RK, WD, MB O: UN				×		In mixed decid-conif forest or riparian; SW OR only
Pacific coast aquatic garter snake	Live	F, Sm Am, AmL	R: RK, TV O: UN		×	×	×		SW OR only; fishing snake; needs coarse substrate stream
Western terrestrial garter snake	Live	Sm Ma, Sm Am, F, Sl, Nb	R:TV, MB, WD O:WD, O	×	×	×	×	×	Most common along low, but well-vegetated aquatic margins
Northwestern garter snake	Live	Sl, Ea, Sm Sa	R:TV, LI, WD O: MB, O		×	×	×		Common where habitat favors native slugs—primary prey
Common garter snake	Live	Sm Am, Am L, F, Sl	R: AV,TV, MB, WD O: RK, O	×	×	×	×	×	Most common along stillwater aquatic margins
Western rattlesnake	Live	Sm Ma, Sm Lz	R: RK,TV, MB, WD O: RK, MB		×	×	×		Most common in rocky talus or fractured rock areas

Where multiple codes are given, they are listed in order of decreasing importance.

[a] Reproduction codes: Mode: E = Ephemeral, Flw = flowing water, P = Permanent, Stlw = stillwater, Terr = terrestrial, Egg = egg laying, Live = Live bearing; Location (where known): B = oviposition brace, NB = no oviposition brace, BR = branches (live woody vegetation), AV = aquatic vegetation, EV = emergent vegetation, MB = mammal burrow, RK = rock, SO = soil excavation (not in burrow or under rock), ST = sticks (dead woody vegetation), UN = unknown, UR = under rock, VA = highly variable, WD = woody debris, WDs = woody debris (submerged), WDt = woody debris (terrestrial).

[b] Food codes: For amphibians with a free-living larval stage, larva (L) and post-metamorphic (PM) diet is partitioned; non-egg stages of terrestrial amphibians are considered collectively. Post-hatching or post-partum stages of reptiles are considered collectively: Food types: Al = algae, Am L = amphibian larvae, Aq Mc = aquatic microcrustaceans, Aq Vg = aquatic vegetation (non-algal forms), Be = bird eggs, Car = carrion, Co = springtails, Ea = earthworms, F = small fish, Inv = invertebrates, Nb = nestling birds, NMa = nestling mammals, Oa = Other aquatic microorganisms (not microcrustaceans), Po = pollen, Sl = slugs, Sm Am = small amphibians, Sm Ma = small mammals, Sm Lz = small lizards, Sm Sa = small salamanders, Sm Sn = small snakes. Only major food categories are listed.

Notes continue on next page

c Refuge/Overwintering site codes: For amphibians with a free-living larval stage, larval and postmetamorphic sites are partitioned; too little is known to partition refuge (active-season) and overwintering sites categories; Notations for refuge (R) versus overwintering (O) sites are partitioned; AV = aquatic vegetation, CM = channel margin (instream), LI = litter (leaves), MB = mammal burrow, N = none, O = other (this is noted wherever other site type(s) besides the one(s) listed is(are) suspected), PD = pond, RC = root channel (terrestrial), RK = under rocks (terrestrial), RS = rocky substrate (aquatic only), SH = shrub matrix, SP = spring, SS = soft substrate (aquatic only), TV = terrestrial vegetation, UB = under bark, UN = unknown, WDT = woody debris (terrestrial; not under bark of debris), WD = woody debris, WDs = woody debris (submerged), ? = indicated site type is suspected.

d WP = wet prairie, DP = dry prairie, S = savanna,

e V = valley, m = Montane

f Occurrence codes: x = present, X = strong association, xR = used as reproductive habitat because active season habitat does not provide requisite conditions.

Southern alligator lizard. (Photograph by Richard B. Forbes)

An evaluation of historical and current data for the Puget Lowlands listed 1 amphibian (Oregon spotted frog) and 3 reptiles (northwestern pond turtle, yellow-bellied racer, and gopher snake) as obligate in prairie habitat.[83] An obligatory association for the 3 reptiles is easily justified—all three lay eggs in terrestrial nests and need thermally adequate oviposition sites.[15, 66] Moreover, the 2 snakes forage preferentially in upland habitat with a low vegetative structure, and the turtle uses terrestrial overwintering locations with Oregon white oaks.[66]

In contrast, justifying the Oregon spotted frog as a grassland obligate is difficult. This frog is a warmwater marsh specialist,[59, 60, 62] and its occurrence within grassland habitat seems dependent on the presence of warmwater marsh and appropriate overwintering sites (e.g., springs or selected flowing water channels). Moreover, available data indicate that Oregon spotted frog use of upland habitats is not only limited, but its focal habitat, warmwater marsh, may occur within matrices of different upland habitat types without grassland habitat being present. At best, the Oregon spotted frog should be viewed as only generally associated with grassland habitats.

The Willamette Valley encompasses the northern limit (Pacific slope portion) of the main body of the geographic range for 10 species of reptiles (western pond turtle, western fence lizard, western skink, southern alligator lizard, ringneck snake, sharptail snake, racer, gopher snake, western rattlesnake, and western terrestrial garter snake).[97, 114] Most (8 of 10) are egg-layers and all are known or presumed to deposit eggs in a terrestrial nest. The only 4 of these egg-laying reptiles recorded from the greater Portland area (northern Willamette Valley) are known from fewer than a dozen records,[84] and all were from well insolated, south-facing slopes that at least historically consisted of a grassland or Oregon white oak savanna association. These reptiles seem to have a patchy or

fragmented distribution along their northern range margins, which may make them vulnerable to local extirpation. Puchy and Marshall[99] viewed populations of western rattlesnakes in the southern Willamette Valley as relict because of their confinement to a few south-facing foothills and buttes. However, whether the western rattlesnake was ever common on the valley floor is unknown because of the few historical records.

Because of their terrestrial life histories, reptiles exhibit stronger links to grassland habitats than amphibians. However, only 2 reptiles, the gopher snake and southern alligator lizard, can be described as having an obligatory grassland link; the others are best regarded as generally associated with grasslands. The gopher snake has high insolation requirements, which make it among the most surface active of Pacific Northwest snakes;[98] insolation needs likely restrict the gopher snake to grasslands at higher latitudes. However, lack of burrowing mammals can also exclude this species from grasslands. Burrowing mammals are not only prey, but also create most refuge and oviposition sites for gopher snakes in a grassland landscape. The other potential obligate, the southern alligator lizard, favors a high-density herbaceous matrix, through which it literally "swims," pushing off closely spaced stems with its small limbs.[40] Two additional snakes, the racer and western terrestrial garter snake, seem to be most common in grassland associations in the northern parts of their ranges in the Pacific Northwest, also likely because of their insolation requirements. However, despite the fact that both have relatively diverse diets,[113] their highest densities are attained in grasslands with some shrub structure and an aquatic margin, respectively. These patterns may be related to favored prey achieving their highest densities in these grassland subtypes. The racer favors a scattered shrub matrix for foraging on eggs, nestling birds, and lizards; whereas the western terrestrial garter snake robs the nests of voles.

Latitudinal patterns linked to insolation requirements result in southwestern Oregon (Rogue and Umpqua valleys) having a greater species richness of reptiles than the Puget Lowlands (Table 7). Some egg-laying snakes (e.g., kingsnakes) have their northern limits of distribution in southwestern Oregon, whereas others (e.g., ringneck snake, sharptail snake) are geographically highly restricted outside this region in the Pacific Northwest.[41, 97, 113, 114] Nonetheless, except for the gopher snake and southern alligator lizard, all other reptiles occurring in Rogue and

Table 7. Herpetofaunal species distribution in westside grassland and chaparral habitats.

	Grassland[a]				Chaparral[a]	
	P[b]	W	U	R	U	R
Frogs and toads[c]						
Western toad	m[d]		x	x		
Pacific treefrog	x	x	x	x	x	x
Northern red-legged frog	x	x	x	m	m	h
Foothill yellow-legged frog		m	x	x	x	x
Bullfrog (E)	x	x	x	x	x	x
Oregon spotted frog	x	h				
Salamanders						
Northwestern salamander[c]		x	x		x	
Long-toed salmander[c]	x	x	x	x		
Rough-skinned newt[c]	x	x	x	x	x	x
Ensatina	x	x	x	x	x	x
Black salamander						x
Lizards						
Long-nosed leopard lizard						m
Sagebrush lizard						m
Western fence lizard	h	x	x	x	x	x
Western skink	m	x	x	x	x	x
Northern alligator lizard	x	m			m	
Southern alligator lizard		m	x	x	x	x
Snakes						
Rubber boa	x	x	x	m	x	x
Ringneck snake		x	x	x	x	x
Sharptail snake	h	x	x	x	x	x
Racer	h	x	x	x	x	x
Striped whipsnake				m		m
Gopher snake	h	x	x	x	x	x
Common kingsnake			x	x	m	m
California mountain kingsnake				m	x	x
Pacific coast aquatic garter snake			x	x	x	x
Western terrestrial garter snake	x	x	x	x	x	x
Northwestern garter snake	x	x	x	m	x	x
Common garter snake	x	x	x	x	x	x
Western rattlesnake		m	x	x	x	x
Turtles						
Western pond turtle	h	x	x	x	x	x
Totals	20	23	24	25	24	26

[a] Grasslands include dry and wet prairie and savanna; chaparral encompasses valley and montane subtypes (see text for details).

[b] P = Puget Lowlands, W = Willamette Valley, U = Umpqua Valley, R = Rogue Valley.

[c] Species typically needs some kind of aquatic habitat beyond that found in seasonally wet prairie (e.g., pond, stream, marsh) and thus, is only weakly linked to grasslands as defined here.

[d] x = known occurrence in the geographic areas indicated for each habitat type; m = present, but of marginal occurrence with a greater distribution outside the region; and h = historically present, but functionally extirpated without or with few recent records. Data are based on a combination of Corkran and Thoms[26], Leonard and Hallock[83], numerous other sources, and personal observations.

Umpqua grasslands require that some structural or habitat element be present for these species to occur within a grassland landscape. For example, the western skink and ring-necked snake are typically not found in grasslands lacking rocks (J. Applegarth, Bureau of Land Management, pers. comm.), because these seem to lack appropriate thermal requirements for nest sites. Sharptail snakes, specialized slug consumers, are typically not found in grasslands lacking a large woody matrix generator, typically oaks,[25] probably because the slugs they consume, mostly members of the tail-dropping slug genus *Prophysaon*, are tied to this habitat element.

Chaparral: Mammals. Approximately 39 species of mammals occur in chaparral habitats of southwestern Oregon (Table 4). Species richness is slightly less than in grasslands. Few mammals are tied exclusively to the westside chaparral of the Rogue and Umpqua basins. The best indicators of chaparral in Oregon are mammals that have a fundamental tie to the shrubby structure that characterizes this habitat. The California kangaroo rat, fairly common on brushy hillsides in the upper Rogue River Valley, is likely the best candidate, although its distribution covers only part of what may be characterized as chaparral in Oregon. Other candidates include the dusky-footed woodrat, which often builds stick nests in a shrub matrix, and brush rabbit, usually no more than a few feet away from a shrubby refuge.[128] However, dusky-footed woodrats attain their highest densities where tree cover is >90% and has not been fire-disturbed,[19] and brush rabbits attain highest densities in more mesic coastal sites,[141] so both species seem to respond to habitat features that chaparral does not provide. Many other species occur in varying abundance in chaparral, but are common only in the presence of selected trees (conifers: long-eared myotis, golden-mantled ground squirrel); riparian areas (several bats and shrews); or rocks or rocky outcrops (California ground squirrel). One of the striking absences from Oregon chaparral is the brush or chaparral mouse, an indicator species for California chaparral,[68] recorded within 2 miles (3.2 km) of the Oregon border.[53, 128]

Chaparral: Birds. The chaparral habitat of southwestern Oregon harbors a unique assemblage of bird species, some of which are found in few places elsewhere in the Pacific Northwest. Shrublands, including the chaparral of southwestern Oregon, are intermediate in structural complexity between grasslands and forest, and elsewhere have been demonstrated to support intermediate numbers of bird species.[9, 37, 88] However, Pacific region chaparral contains more bird species than many other shrublands, including the shrub-steppe community of eastern Oregon and Washington.[133]

The bird assemblage of Oregon chaparral is similar to that observed in chaparral to the south.[6] The most noticeable difference is the absence of the California thrasher in Oregon. Also, Bewick's wren seems to be numerically more important in California chaparral, where it is one of the most frequently detected species in both chamise and mixed-chaparral. Two species, the wrentit and lesser goldfinch, have a restricted distribution in the Pacific Northwest, but attain their highest densities in chaparral.

Approximately 48 bird species regularly breed within chaparral habitats of southwestern Oregon (Table 5). Species composition changes markedly as structure of the chaparral changes. Some species are limited to chaparral of a specific configuration, whereas others occur across the full range of habitat types. In general, bird species richness is lowest in chaparral shrub (Table 5), which lacks trees and extensive herbaceous openings. Only 22 species regularly breed among wedgeleaf ceanothus and whiteleaf manzanita (*Arctostaphylos vicida*) in chaparral shrub. Some of the most common species include the western scrub-jay, spotted towhee, California towhee, and lesser goldfinch. Diversity is slightly higher if chaparral contains substantial amounts of grassland. This feature adds species such as western meadowlark and lark sparrow. Species richness climbs sharply on chaparral sites that include hardwood trees. On the most mesic sites, which include madrone (*Arbutus menziesii*) and oak, common species include mountain quail, oak titmice, white-breasted nuthatch, black-headed grosbeak, chipping sparrow, and dark-eyed junco. A number of factors contribute to this increase in diversity. Trees provide a broader range of foraging opportunities because of increased structural complexity and nesting opportunities for cavity users. A different group of birds uses high-elevation chaparral sites. Some of the more closely associated species are the green-tailed towhee, fox sparrow, and dusky flycatcher.

Migrant species represent a greater portion of the breeding bird species in chaparral (62%, $n = 29$) than residents (38%, $n = 18$) (Table 5). Among migrants, there are more nearctic migrant species (55%, $n = 16$) than temperate migrant species (45%, $n = 13$). Nearctic migrants particularly contribute to the increased species richness in chaparral with hardwood trees. Nearly absent from treeless shrublands, nearctic migrants represent a significant proportion of the bird fauna in chaparral with hardwood trees, comprising 42% ($n = 14$) of the breeding species (Table 5).

Ground foragers dominate (58%, $n = 27$) chaparral bird assemblages (Table 5). The percentage of ground foragers declines with increasing site moisture due, in part, to additional bird species exploiting alternative foraging opportunities. Bark foragers, foliage and twig gleaners, and flycatchers all increase in species richness along the moisture gradient. Among nesting guilds, foliage nesters (41%, $n = 19$), cavity nesters (30%, $n = 14$), and ground nesters (28%, $n = 13$) are roughly similar in number (Table 5).

The winter bird community displays many similarities with the breeding bird community. Chaparral shrub contains the fewest species in both seasons. Likewise, more species occur along the interface with grasslands, and the assemblage is richest in the presence of hardwoods. In winter, several species leave the drier sites, but other migrants largely compensate for the loss, including the golden-crowned and white-crowned sparrows from the north as well as altitudinal migrants such as dark-eyed juncos. Nearctic migrants leaving the mesic habitats are replaced by fewer migrants in the winter (kinglets and chickadees), resulting in a net loss of about 3-8 species. In winter, bird densities are greatest near the valley floor and decrease with increasing elevation.

Somewhat surprisingly, foraging guild composition changes relatively little seasonally. Ground foragers dominate in both seasons and show the same relative decline with increasing moisture. In winter, the decline is both absolute as well as relative. The contribution of foliage and twig gleaners also changes little on a seasonal basis. Changes that do occur include the disappearance of flycatchers in winter, and increases in bark foragers and drillers.

Besides their importance to breeding and wintering birds, chaparral habitats are important to migrants, especially in spring. Species such as the warbling vireo, and yellow, Wilson's, MacGillivray's, and yellow-rumped warblers stop to feed before continuing north to breed in other habitats. Northbound migrants typically travel through the lower elevations, taking advantage of the transient abundance of canopy-dwelling arthropods in chaparral shrubs and hardwoods. In late April, wedgeleaf ceanothus experiences a brief flush of arthropods that drops quickly to very low levels for the rest of the year. Oregon white oak experiences a similar pulse at about the same time, but arthropod volume does not fall as precipitously and persists at a moderate level into the fall. The ephemeral abundance of arthropods also helps explain the relatively low numbers of foliage gleaners breeding in chaparral, particularly in chaparral lacking hardwoods.

Chaparral: Herpetofauna. The herpetofauna of chaparral habitats includes a diverse array of 24 species, including 4 frogs, 4 salamanders, 3 lizards, and 13 snakes (Table 6). No reptile or amphibian species can be considered exclusively tied to chaparral. The western whiptail, perhaps the reptile species known to be most strongly linked to chaparral in California,[114] does not occur in chaparral in southwestern Oregon, although it occurs in a large area of high desert shrub-steppe in eastern Oregon.[97] However, among the habitat associations found in southwestern Oregon, the western fence lizard and southern alligator lizard reach their highest densities in chaparral, especially wedgeleaf ceanothus- dominated areas. Chaparral shrubs, especially in combination with rocks, provide the complex, vertical structure for foraging, and elevated vantage sites for predator-detection characteristic of western fence lizard habitat.[87] Wedgeleaf ceanothus also provides the dense stem matrix within which the southern alligator lizard can forage and maneuver to escape predators.[40] The sharptail snake and western rattlesnake may also attain their highest densities in wedgeleaf ceanothus chaparral, but data on variation in snake assemblages in southwestern Oregon are lacking.

Several reptiles occur or are more common in chaparral if selected structural elements are present: rocks (ringneck snake and western skink); and Oregon white oak and madrone (rubber boa, western skink, and common kingsnake). Three garter snakes (common, Pacific coast aquatic, and western terrestrial garter snake) occur within chaparral largely because the requisite aquatic habitat (stillwater or stream) occurs within the chaparral habitat matrix. How these species use upland habitat such as chaparral is not well known, but the most significant use is probably for overwintering or short-term retreat sites.[66] Both kingsnakes seem to be more common in chaparral where a riparian zone exists, which likely reflects more suitable levels of prey; this probably also applies to the western skink and northern alligator lizard.

In the Pacific Northwest, no amphibian species exhibits strong ties to chaparral. The ensatina salamander is the amphibian most likely to be found in chaparral, especially where oaks are present. However, this species is significantly more abundant in more mesic westside associations. Black salamanders have been recorded patchily from chaparral-like habitats on the north slopes of the Siskiyou Mountains, but this species also has been most frequently recorded along permanent riparian corridors in the region (D. Clayton, Rogue River National Forest, pers. comm.). Remaining amphibians that occur within the chaparral landscape are there because the aquatic habitat they require is present; their use of the upland with a chaparral structure is likely restricted to seasonally important refuge sites. This applies mostly to the Pacific chorus (tree) frog, foothill yellow-legged frog, and rough-skinned newt.

Endemics, Extirpations, and Peripheral Species

Wildlife species population patterns in westside grassland and chaparral habitats reveal several features:
1. losses are prominent, but gains have been rare,
2. losses have been greater in grasslands than in chaparral, which reflects the relative magnitude of change in each habitat type,
3. no species inhabiting grasslands or chaparral has become extinct, but a number of species have been extirpated over large areas, a pattern especially pronounced in grasslands,
4. regional endemics and species that are distributionally peripheral or naturally rare seem to have sustained the greatest losses, and
5. a few species have expanded their ranges in westside grasslands.

Grasslands: Mammals. The Camas pocket gopher and gray-tailed vole are endemic to the Willamette Valley *sensu lato*.[128] The gray-tailed vole formerly was associated with prairie habitats, and now occurs in agricultural grasslands, particularly pastures, hayfields, grass-seed fields, and grain fields. The Camas pocket gopher occurs primarily in cultivated cropland such as grain fields. A well-drained field and the presence of plants with reserves stored in bulbs or tap roots characterizes suitable habitat.[128]

Local forms of the Western pocket gopher are declining in the Puget Lowlands,[115] and this species is patchily distributed in westside grasslands. In the Puget Lowlands, Western pocket gopher occurrence is negatively associated with increasing levels of rocky soils.[115]

At the time of European settlement, gray wolves occurred in low densities throughout westside grasslands,[17, 122, 128] grizzly bears were present in grasslands in western Oregon,[128] and the pronghorn, a species with a strong grassland tie, was present in the upper Rogue Valley.[7] Disappearance of grizzly bears, gray wolves, and pronghorn from westside grasslands occurred shortly after settlement, probably a function of hunting and/or incompatability with humans, rather than habitat alteration, because most habitat alteration was subsequent to their disappearance.[90, 128]

Black-tailed jackrabbits have declined substantially in the Willamette Valley.[89] Bailey[7] reported them as "commonly reaching as far north as the country about Salem, and more rarely to the Columbia River." On one day in November in 1930, he observed 9 road-killed black-tailed jackrabbits on a trip between Salem and Eugene. In addition to loss of grassland habitat, sport hunting and disease may have adversely impacted this species.[119] A few were seen in the 1970s,[128] and jackrabbits were seen at approximately 15 locations from 1996 to 1999 (BA; MH). Most sightings were in Christmas tree farms, and occurred throughout the valley south of Portland.

Columbian white-tailed deer occurred throughout grasslands of the Willamette and Umpqua valleys in Oregon at the time of European settlement.[128] They now are restricted to remnant populations in Umpqua Valley near Roseburg, and a few sites in bottomland forests and agricultural grasslands along the Lower Columbia River.[46] This form is federally endangered, although the Douglas County populations are proposed for delisting (D. Peterson, U.S. Fish and Wildlife Service, pers. comm.). Loss of habitat, increased human density, and hunting have reduced use of grassland habitats by Columbian white-tailed deer.

Grasslands: Birds. Two subspecies of birds, the Oregon vesper sparrow and streaked horned lark, are endemic to westside grasslands. At the time of European settlement, the Oregon vesper sparrow was "rather abundant on the Nisqually plains, Puget Sound".[122] In the late 1800s, it was "abundant summer resident, found everywhere in open country" in Washington County, Oregon;[5] in the early 1900s, "fairly common on the prairies and grassy fields of western Oregon and Washington;"[65] in the 1930s, "abundant in Willamette Valley native and agricultural grasslands;"[44] in the 1940s, "common summer resident" in the southern Willamette Valley;[54] and in the early 1950s, "numerous about pastures and prairies of the Puget Sound Region."[72] Currently, it is rare to locally uncommon in widely scattered areas of the Willamette Valley.[3] In the Puget Lowlands, it regularly occurs only in the San Juan Islands[85] and on prairies in the South Puget Lowlands, especially at Fort Lewis Military Installation.[104] In the Willamette Valley, it occurs almost exclusively in light- to moderately-grazed pastures and weedy Christmas tree farms in the valley foothills.[3]

At the time of European settlement, the streaked horned lark was a "very abundant summer resident on gravelly prairies of Puget Sound".[122] In the late 1800s, it was "a rather common summer resident" in Washington County, OR;[5] in the early 1900s, "common resident in western Oregon and southwestern Washington;"[65] in the 1930s, "a common breeding bird of the open fields in western Oregon;"[44] and in the 1940s, "common permanent resident" in the southern Willamette Valley.[54] Currently, it is rare to locally uncommon, with scattered small populations in the Willamette Valley, the largest breeding population occurring on and north of Baskett Slough National Wildlife Refuge.[3] Breeding habitat is prairie and most types of agricultural grasslands where the vegetation is short, <1 ft (<0.3 m), with patches of bare or sparsely vegetated ground.[3] In Washington, it has been extirpated from the San Juan Islands and much of the Puget Lowlands.[85] A few scattered breeding populations exist in the South Puget Lowlands, the largest on prairies at Fort Lewis Military Installation.[104] A few pairs also occur in coastal dune habitats on the southern Washington coast[110] (C. Chappell, Washington Natural Heritage Program, pers. comm.).

Several bird species have been extirpated as breeding species from all or some westside grasslands since European settlement. The burrowing owl was considered "a familiar sight"[44] and a "common breeding species"[16] in the 1920s and 1930s in the Rogue Valley, and may have nested in the Willamette Valley[50] and Puget Lowlands.[110] The last suspected nesting in westside grasslands was in the early 1980s (O. Swisher, retired, pers. comm.). An experimental reintroduction program in the Rogue Valley in the mid 1980s was unsuccessful.[89] Burrowing owls occasionally winter in grassland habitats in the Willamette, Rogue, and Umpqua valleys.

Lewis' woodpeckers formerly nested in oak savannas of the Puget Lowlands[72] and Willamette Valley.[44] Prior to 1965, they were fairly common residents in Columbia River bottomlands,[45] and nested in a few oak savanna locations in the Willamette Valley (e.g., Finley National Wildlife Refuge) into the early 1970s (A. Contreras, University of Oregon, pers. comm.). A drastic reduction in the population occurred in the late 1960s and early 1970s, and the last documented breeding record in the Willamette Valley was in 1977 near Scapoose along the Columbia River.[48] They have been extirpated primarily because of (1) competition with the European starling for nest cavities since the arrival of that species in Oregon in the 1950s and its proliferation in the 1960s, and (2) loss of large savanna-form oaks to development and agriculture. Lewis' woodpeckers may nest infrequently in Rogue and Umpqua valley savanna habitats, but have declined in those areas also.[89] It occurs as a regular wintering species in the Rogue Valley, and as an occasional migrant and wintering bird in the other valleys.

The common nighthawk was "very abundant on the prairies near Puget Sound,"[122] and nested in grasslands in the Willamette Valley,[44] but may be extirpated there.[3] The loss of nesting nighthawks may have resulted from reductions in the insect prey base, and increases in populations of urban and rural predators, such as the American crow. Nests have been recently reported in Christmas tree farms and clearcuts in the foothills of the Willamette Valley.[3]

Two species, Say's phoebe and the sandhill crane, occurred in small local breeding populations in westside grasslands, but disappeared as breeding species many years ago. Say's phoebe was considered a "regular resident of Rogue, Umpqua, and Willamette valleys."[44] It maintained a small breeding population near Corvallis at least through the 1940s,[39] but has not been reported as a nesting species since. Sandhill cranes formerly were a "summer resident in the prairies of western Washington."[72] At the time of European settlement they were considered a "common summer resident."[122]

The lark sparrow has been extirpated as a breeding species in the Willamette Valley, but still maintains breeding populations in the Rogue and Umpqua valleys. Shortly after European settlement, lark sparrows were reported as "common breeders in the Willamette Valley,"[74] and into the early 1900s were considered an "uncommon summer resident at Corvallis and Dayton."[140] It has not been reported as a breeding species in the Willamette Valley for approximately 50 years.

One obligate grassland species, the grasshopper sparrow, is considered naturally rare in westside grasslands. It maintains a small, tenuous breeding presence in the Willamette, Umpqua, and Rogue valleys. This inconspicuous bird was first reported in the Rogue Valley in 1963.[102] It has been reported periodically since then, but only a few birds at most. It was not recorded in the Willamette Valley until the early 1970s.[93] Generally, 1-3 birds have been reported most years since, until an extensive study in 1996-97 reported 20-25 singing males, mostly on private pasture lands.[3] Habitat in the Willamette Valley includes relatively short grass (<18 in [<0.5 m]) pasture lands with little to no shrub cover.[3]

The white-tailed kite and acorn woodpecker are species that have expanded their breeding range in westside savanna habitats. The acorn woodpecker has always been a common to uncommon species in oak savanna of the Rogue and Umpqua valleys, but has expanded its range in the Willamette Valley in the last 50 years.[48] Prior to 1940, there were only a few records of this species near Eugene,[39, 44] but now it occurs north to Portland. Its distribution in the Willamette Valley is spotty, and it is less abundant further north. It does not occur in the Puget Lowlands. Only a few incidental nonbreeding season records of white-tailed kite existed in westside grasslands prior to the 1960s and 1970s.[48] Sightings have increased since then, and kites are now a regularly occurring wintering species throughout westside valley grassland habitats. They are also a rare breeding species in grasslands of all of the interior valleys of Oregon.[48]

Grasslands: Herpetofauna. No amphibian or reptile species that occurs in westside grasslands in Oregon and Washington is endemic to this habitat, and no species is known to be completely extirpated from this habitat type.

However, 2 frogs with a peripheral link to grasslands have experienced significant regional extirpation. The Oregon spotted frog is thought to be extirpated on the Willamette Valley floor and has been nearly extirpated in the Puget Lowlands, where only 2 populations about 10 miles (16 km) apart are known to remain in the Black River watershed in Thurston County[83, 92] (K. McAllister, Washington Department of Fish and Wildlife, pers. comm.). Surveys for the Oregon spotted frog conducted in 1993-94 failed to detect the species at any of the 10 verifiable historical sites on the Willamette Valley floor,[60, 62] and surveys of historical sites in Washington have similarly failed to detect the species.[92] Similarly, the foothill yellow-legged frog is thought to be nearly extirpated in the Willamette Valley, as the species was found at only 1 of 10 verifiable historical sites.[13] Both species are presumed to have declined because of changes in their aquatic habitats, rather than changes in the associated grassland matrix.

Three of the 4 egg-laying reptiles (western pond turtle, racer, and gopher snake) that occurred historically in the South Puget Lowland prairies are presumed extirpated (Table 3). Confusion exists regarding the fourth species, the western fence lizard. Leonard and Hallock[83] listed it as historically common, but indicated that the association of this species with prairie appears to be in error. Moreover, current reports may involve misidentifications. Uncertainty about the historical distribution arising from comments by Suckley and Cooper[122] make it best to conservatively interpret western fence lizard as functionally extirpated (K. McAllister, Washington Department of Fish and Wildlife, pers. comm.). Regardless of their historical abundance, all egg-laying grassland reptiles in the South Puget Lowlands may be extirpated.

Based on knowledge of the magnitude of habitat change on the Willamette Valley floor, selected reptiles may have sustained major reductions in geographic range in this region. The western terrestrial garter snake that occurs in the Willamette Valley is a morphologically distinct variant that currently lacks taxonomic recognition (A. St. John, Independent consultant, pers. comm.). This snake occurred across the southern half of the Willamette Valley with a narrow extension north along the toe of the Coast Range to the vicinity of Forest Grove.[97, 114] It is now uncommon in the southern Willamette and has not been encountered at the northern end of the western extension despite extensive searches (J. Applegarth, Bureau of Land Management, pers. comm.; A. St. John, Independent consultant, pers. comm.).

Chaparral: Mammals. Only 1 mammal, the California kangaroo rat, may be viewed as endemic to chaparral in Oregon, and no mammals known to occur in Oregon chaparral are extinct. However, small mammal surveys focused in Oregon chaparral to detect endemics like the brush mouse have not been conducted.

Chaparral: Birds. Three bird species are endemic or near-endemic to chaparral habitats of southwestern Oregon. All have a broader geographic distribution in California.

The blue-gray gnatcatcher occurs in the zone where Oregon white oak and chaparral shrubs coexist. It rarely occurs in chaparral without oaks, and is not encountered in oak savanna lacking wedgeleaf ceanothus. Since the first reported breeding in 1963,[102] blue-gray gnatcatchers have been found at scattered chaparral sites throughout Jackson and Josephine counties. Densities are relatively low as might be expected for a peripheral population. Between 1.4 and 2.7 individuals/25 ac (10 ha) have been recorded on Lower Table Rock (SJ). Of 8 nests monitored by Speer and Felker,[112] one fledged a brown-headed cowbird and 3 successfully fledged young gnatcatchers. Nests are constructed on a horizontal limb usually in an oak, and birds forage in both oak and ceanothus.

Oak titmice have similar habitat requirements and a similar range to the blue-gray gnatcatcher. Being cavity nesters, they are further restricted to habitat containing mature Oregon white oaks with dead limbs and enough rot to permit excavation of a nest site. Densities of between 4.1-7.3 individuals/25 ac (10 ha) were recorded on Lower Table Rock (SJ). Titmice display a broad range of foraging activities. Not only do they forage among both oaks and ceanothus, but they also regularly feed on the ground.

Another species endemic to chaparral is the California towhee. Unlike the blue-gray gnatcatcher and oak titmouse, it occurs over a broad range of chaparral habitats and as far north as the Umpqua Valley. The California towhee is a terrestrial forager that forages in openings between shrubs, usually ceanothus or manzanita. Despite feeding in more open areas than the spotted towhee, it is seldom found far from dense cover.

Chaparral: Herpetofauna. No amphibians or reptiles are endemic to chaparral habitats of southwest Oregon, and no amphibians or reptiles recorded from Oregon chaparral habitats are extinct. Several infrequently observed species of snakes (California mountain kingsnake, common kingsnake, ringneck snake, and sharptail snake) that occur in chaparral are thought to have a restricted distribution in Oregon. Although these snakes are generally infrequently encountered and may be considered rare (e.g., sharptail snake),[89] the extent of their distribution is unknown because few systematic surveys have been conducted when these species are near-surface active. Moreover, recent surveys in the Umpqua Basin suggest that the sharptail snake has a much broader altitudinal range than previously thought[61, 63] (R. Hoyer, retired, pers. comm.). Since similar biases probably exist with the remainder of these rarely observed snakes, their status will remain ambiguous until enough surveys are completed. Regardless of how abundant they may be, these snakes are not restricted to chaparral habitats.

Three species, long-nosed leopard lizard, sagebrush lizard, and striped whipsnake, have been recorded from few localities in southwestern Oregon.[77, 97] These are members of the Great Basin herpetofauna that reach the western fringe of their geographic ranges in the upper Rogue Valley. Distributional information on these species is lacking, but it is unlikely that any of these species are widespread in the upper Rogue; thus, they can be considered distributionally peripheral and naturally rare.

The black salamander is distributionally the most peripheral chaparral-associated species, occurring only across a small area of the Siskiyou Mountains of southwestern Oregon. However, recent surveys show it to be somewhat more widespread than previously thought (D. Clayton, Rogue National Forest, pers. comm.).

Exotic Species

Several introduced species of wildlife maintain populations in westside grassland and chaparral habitats. At least 8 introduced vertebrates occur in these habitats (Virginia opossum, eastern cottontail, house mouse, Norway rat, wild turkey, ring-necked pheasant, northern bobwhite, and European starling), and at least 4 other introduced vertebrates occur in aquatic habitats to which grasslands or chaparral are peripherally linked (nutria, bullfrog, common snapping turtle, and red-eared slider). Additionally, ambiguous data suggest that the Willamette Valley—Puget Lowland populations of red fox were introduced around the turn of the century.[128] Some of these exotic species are well established in grassland and chaparral habitats, but others, (i.e., house mouse and Norway rat) are only rarely dissociated from human dwellings and highly modified human habitats into adjacent grasslands or chaparral. Data regarding whether exotic species impact native species vary greatly. Evidence of negative impacts exist, and positive effects have not been found. Most data suggesting some kind of exotic species effects are confounded with alternative factors that have potential to influence native species.[20, 64]

The Virginia opossum, first introduced to Oregon between 1910 and 1921,[71] is now established throughout western Oregon and Washington.[128] Opossums are characterized as highly opportunistic carnivores,[67] but seem to be too slow to be efficient predators on even moderately mobile prey. However, immobile or poorly mobile prey, including bird's eggs and nestlings and immature and nursing small mammals are vulnerable. Because opossum diet in western Oregon and Washington has not been evaluated, the impact of this exotic on populations of ground nesting birds and mammals is unknown.

The eastern cottontail was introduced into Benton and Linn counties in 1937 and 1941 from Ohio and Illinois, respectively.[51] From these introductions, it has spread at least through the mid-Willamette Valley.[127, 129] The source of eastern cottontails in the northern Willamette Valley and Puget Lowlands is unknown, but they may be descendants from Missouri stock originally introduced in 1933 near Battle Ground, Washington.[29] Chapman and Verts[21] provided evidence that eastern cottontails were behaviorally dominant over native brush rabbits. Verts and Carraway[126] offered evidence that eastern cottontails and brush rabbits hybridize under natural conditions at least occasionally. However, Chapman and Trethewey[20] indicated that the ecological and economic impacts of the introduction of eastern cottontails into Oregon were unknown. Despite available evidence, the expansion of eastern cottontails in the Willamette Valley, and contraction of brush rabbit distribution in the same region has not been demonstrated to have a direct connection.

The date of introduction of the house mouse to Oregon and Washington is not known, but it was probably early considering its near ubiquitous association with humans. The infrequent capture of house mice in the field, and their almost uniform occurrence near human habitation suggests that populations in open fields are typically temporary, near human structures, and limited to warmer and drier seasons.[128] The apparent failure of house mice to become established in the field in the Pacific Northwest may be because native voles interfere with their reproduction,[12] recruitment,[32] or a combination of population processes.[86]

The Norway rat, like the house mouse a native of China and Siberia, is almost ubiquitous in its association with humans.[128] The first record of the species in the Pacific Northwest was an individual collected at Astoria by Lieutenant Trowbridge in 1855.[8] The Norway rat nearly always resides close to human dwellings, but in California, noncommensal populations of Norway rats occurred in riparian woodlands and adjacent fallow agricultural fields several miles from human habitation.[121] As Norway rats have been trapped in agricultural and fallow lands in the Willamette Valley, noncommensal populations may also occur there. Nonetheless, data are lacking regarding interactions between Norway rats and native small mammals in westside grasslands.

At least 3 grassland and/or chaparral bird species were introduced as game species: the wild turkey, ring-necked pheasant, and northern bobwhite. Wild turkey, introduced to the Pacific Northwest just after 1900, are now most visible in the rolling oak savannas and, to a lesser extent, adjacent chaparral of the interior lowland Umpqua and Rogue valleys. Several different forms of the wild turkey met with poor to variable success until Rio Grande turkeys from Texas were introduced during the early 1980s (T. Farrell, Oregon Department of Fish and Wildlife, pers. comm.). This form has proven extraordinarily successful, largely because adults attempt to renest after nesting failures even very late in the season.

The oldest of the 3 game bird introductions is the ring-necked pheasant. Ring-necked pheasants were first brought from China in 1881 by Judge Owen Denny. This introduction failed, but subsequent Denny-encouraged introductions resulted in its establishment by 1884.[81] By 1891, it was well enough established in Willamette Valley agricultural grasslands that Oregon declared the first open hunting season on pheasants, during which an estimated 50,000 individuals were taken. Even before that, pheasants were being live-trapped for transport to other areas, and western Washington was among the first to receive translocated pheasants. Soon after 1900, the ring-necked pheasant was established over most of the Pacific Northwest agricultural grasslands. Increasingly manicured agricultural practices (e.g., reduction of hedgerows) that began during the mid-1970s have resulted

in local declines in ring-necked pheasants. However, the ring-necked pheasant remains a visible part of the fauna, particularly in the agricultural grasslands of western Oregon. No data exist regarding interactions between ring-necked pheasant and the native fauna in westside grasslands.

The northern bobwhite was introduced to agricultural areas of Oregon during 1900-1960, but has been largely extirpated from most areas where it once seemed to have been established, such as the Willamette Valley.[48] Nonetheless, a few local sightings continue, some of which are attributable to escaped birds used in dog training, private introductions, and in some cases, populations that have persisted for decades without reintroduction.

Attempts to introduce the European starling to the Pacific Northwest were first made at Portland, Oregon in 1889 and 1892.[81] Both of these early efforts resulted in brief establishment, but eventually failed in 1901 or 1902.[48] The eventual permanent introduction can be traced to Eugene Scheifflin, a New York drug manufacturer whose bird and Shakespeare hobbies engrossed him. Scheifflin introduced starlings in New York in 1906, where they began nesting the same year. The species spread like a wave westward and by 1928, was usually spoken of as anywhere east of the Mississippi.[81] The first starlings recorded in the westside Pacific Northwest were at Eugene and Meadow View (both Lane County, Oregon) on 26 December 1947.[48] Today, the starling is common to abundant year-round in urban areas, towns, and agricultural lands through the westside Pacific Northwest. Starlings are aggressive cavity nesters and compete directly with many native birds for a limited cavity supply.[48] Three westside grassland species are thought to have declined due at least in part to the European starling: the western bluebird, American kestrel, and Lewis' woodpecker.

Nutria (also known as coypu), native to the Patagonian subregion of South America, were imported to North America for a fur-farm at Elizabeth Lake (California) in 1899.[38] Unknown numbers of nutria were liberated from a fur farm in Tillamook County during a flood (S. Jewett *in* Larrison[80]). From this release, a colony was established in the vicinity of Garrison Lake. Subsequently, feral nutrias were observed along the Nestucca River and appeared to be doing well (F. Wire *in* Larrison[80]). By 1946, nutrias were established across the westside interior valleys, including the lowland Willamette.[76] Nutria use of grassland habitat adjacent to wetlands is limited to movements between aquatic habitats and marginal foraging. Based on this pattern of habitat use, effects of nutria are likely to be greater in wetland habitats than in the adjacent upland (e.g., grasslands).

The eastern bullfrog was originally introduced to westside habitats in the Pacific Northwest by Matt Ryckman, who brought bullfrogs to the McKenzie Fish Hatchery in 1921.[78] Several subsequent introductions led to the bullfrog-farming craze of the 1930s, which saw a large number of entrepreneurial bullfrog ventures appear through the West.[69] All of these ventures failed and the purported frog farmers either left their stock in situ or actually seeded them in aquatic sites nearby. By 1950, the bullfrog was well established in westside habitats across the Pacific Northwest (R. Storm, Oregon State University, pers. comm.). Like nutria, bullfrogs do not use upland habitats frequently; such use is restricted to some foraging adjacent to and movements between aquatic sites, typically under wet or high humidity conditions. As a consequence, much of the potential impact of bullfrogs has focused on native aquatic species such as the Oregon spotted frog and foothill yellow-legged frog,[13, 59, 60, 62] and the effects bullfrogs have on these species is arguable.[64] What impact bullfrogs have, if any, on those species associated with grassland or chaparral habitat proper, which they only occasionally encounter, is unknown.

The common snapping turtle is a large aquatic turtle native to most of the eastern U.S.[24] The original introduction is not known and likely represents a series of releases by pet owners. Similar to the red-eared slider, its use of upland habitat is restricted to nesting-related activities by females and to a lesser extent, movement between aquatic sites and to overwintering sites. Like the bullfrog, a diverse diet and imposing feeding mode have been the basis of suggestions that it might negatively influence native species.[66] However, the potential negative impacts of this species have not been investigated, and the impacts it may have on species in grassland or chaparral adjacent to aquatic sites, if any, are unknown.

The date of introduction of the red-eared slider, an aquatic turtle native to and widespread in the eastern U.S,[24] is not known. The species appeared in the pet trade for many years, and similar to the common snapping turtle, it is likely that scattered point introductions from pet turtle owners are responsible for its establishment in the westside Pacific Northwest. Red-eared sliders now exist at scattered points in lowland drainages in western Oregon and Washington. Evidence of nesting has been found at several sites, and indirect evidence of reproduction (e.g., appearance of tiny hatchlings) also exists. As with the previous aquatic species, use of upland habitats is limited, being largely restricted to nesting and nest site assessment activities by females as well as movements to alternative aquatic sites or to overwintering sites. Further, any evidence of influence on native species is restricted to aquatic forms,[66] so if red-eared sliders affect species that use grasslands or chaparral proper, those interactions have not been addressed.

Management and Conservation

Conservation of wildlife in westside grassland and chaparral habitats will require integration of a suite of activities on both private and public lands. These activities, which may or may not be interdependent, include:

1. inventory of habitat types and their degree of modification, particularly where data are lacking or limited (i.e., Rogue and Umpqua valleys);
2. protection of high-quality sites and restoration of degraded habitats; and
3. management to maintain habitat quality and wildlife populations.

Protection and restoration may be approached through several modes. These include acquisition, conservation agreements, landowner outreach and incentive programs, legislative enactments, and management agency policies. Where opportunities occur, acquisition by land management agencies or organizations may facilitate long-term protection and management. Mechanisms to accomplish this include purchase, land-swaps, and deed-trusts. Frequently, extensive acquisitions are economically unfeasible, so alternative approaches must be considered.

Most land ownership of westside grassland and chaparral habitat is private, so actions to protect or enhance habitat for wildlife must be in partnership with private landowners, many of whom make their living on these same lands. This approach is especially important where protected (public) and private land interdigitate, and the management of private land has opportunity to significantly influence system dynamics on public land. For these reasons, land management agencies and organizations (e.g., U.S. Fish and Wildlife Service, The Nature Conservancy), and private landowner government outreach agencies (e.g., Natural Resources Conservation Service, Division of State Lands) must play a major role in wildlife conservation. A significant portion of this role must be cooperative development of guidelines for conservation with private landowners. For grassland wildlife in particular, managing existing and future agricultural lands to provide suitable habitat is crucial because of the near disappearance of native grasslands, and the low likelihood that extensive areas adequate for supporting viable populations can be restored to native conditions.

On public lands, efforts must be directed towards including all wildlife in habitat management plans, and minimizing human-wildlife conflicts when allowing multiple-use activities. This must include institutionalizing policy that directs management plans to encompass conservation of the entire spectrum of wildlife species. Some public lands also may offer the best potential for large areas of quality habitat where population viability for area-sensitive wildlife species (e.g., western meadowlark, several large mammals) is more likely. Public lands such as State Parks and Federal Wildlife Refuges should be a conservation priority, since these lands potentially represent stable, long-term commitments.

Land use on westside agricultural lands often changes yearly, and this instability likely adversely impacts grassland wildlife populations. Thus, the establishment of designated conservation areas on public and private lands may be important to maintain stable populations at multiple locations. Examples of these efforts include the Oregon Biodiversity Project[31] and a Partners in Flight Landbird Conservation Plan for Westside Lowlands and Valleys.[4]

Beyond protection and restoration through cooperative efforts, site-specific management requires attention to system dynamics. Fire, the primary natural disturbance process that shaped westside grassland and chaparral habitats, occurs rarely relative to historical conditions. Thus appropriately, current restoration and management of grasslands has focused on restoring fire back into the system as well as removal and control of exotic plants, and seeding of native species.[135] Significant attempts also have been made to identify and acquire remaining parcels of prairie habitat (E. Alverson, The Nature Conservancy, pers. comm.).

The expanding human population of western Oregon and Washington has resulted in continual loss and degradation of grassland and chaparral wildlife habitat. Current conservation emphasis has focused on rare and declining species on public lands. A significant need exists for more community-based conservation and wildlife partnerships with private landowners. Without these components, the future of wildlife in westside grassland and chaparral habitats will probably be relegated to protection and management of highly disjunct populations and sites.

Case Histories

1. Small Mammals and Fire in a Wet Prairie of Western Oregon
James Faulkner

Periodic fire and flooding, and variations in microtopography, soil types, and hydrology most likely created a mosaic of vegetation composition and successional stages in the historical prairies of the Willamette Valley. Current prairie restoration and management efforts use prescribed fire to set back succession and attempt to recreate historical prairie conditions. This case history investigates the influence of prescribed fire on the small mammal community of a native wet prairie in western Oregon.

Study Site. The study site was a 339 acre (137 ha) section of native wet prairie in the Willamette Floodplain Research Natural Area (RNA). The RNA is located at W. L. Finley National Wildlife Refuge, 10 miles (16 km) south of Corvallis, Benton County, Oregon. Elevation ranges from 269-289 feet (82-88 m). Soils are silty clay loams of alluvial origin. The impermeable clay holds winter precipitation, causing shallow flooding of low areas until late spring. The prairie is surrounded by fields cultivated for commercial grass seed and the production of goose forage. A program of prescribed burning was initiated in 1990 on the RNA prairie to control invasive alien plant species and reduce shrub and tree cover.[120] The small mammal study site consisted of a 69-acre (28 ha) section of prairie burned annually since 1990 (Section A), a 91-acre (37 ha) section burned triennially since 1991 (Section B), and a 178-acre (72 ha) section unburned since 1979 (Section C). Burning during this study took place in September 1996 and 1998 in Section A, and September 1998 in Section B.

The RNA wet prairie has a high diversity of plant species. Of the most common species, graminoids include *Deschampsia cespitosa, Agrostis tenuis, Holcus lanatus,* and various *Carex* and *Juncus* species; forbs include *Hypericum perforatum, Myosotis laxa, Veronica scutellata, Plagiobothrys figuratus,* and *Camassia quamash*; shrubs include *Rosa* spp. and *Spirea douglasii*; and trees include *Fraxinus latifolia* and *Crataegus douglasii.*[120] Vegetation sampling during trap periods showed that the height of woody shrubs was substantially reduced in burned sections, but percent cover was not greatly affected. Diversity and percent cover of grasses were greater in the burned sections than the unburned. Litter cover and depth increased with time since fire and was consistently greater in the unburned section.

Small Mammals. Trapping surveys for small mammals were conducted for 5 consecutive days in late summer prior to burning in 3 years (1996-98), with an additional trap period 2 weeks after burning in 1998. Small mammals were live-trapped along 9 permanent transects, 3 in each prairie section.

In decreasing order of abundance, small mammal species captured on the prairie were the deer mouse, vagrant shrew, Townsend's vole, gray-tailed vole, short-tailed weasel, dusky-footed woodrat, and Pacific jumping mouse. Numbers of individuals captured over 3 trap periods differed between prairie sections for most species (Table 8). The deer mouse, a habitat generalist, was fairly abundant in all prairie sections, but was not evenly distributed among sections ($X^2 = 35.3, df = 2, P < 0.01$), having greatest abundance in Section C. The vagrant shrew was not evenly distributed among sections ($X^2 = 17.1, df = 2, P < 0.01$), having greatest abundance in Section B and least in Section A. The distribution of Townsend's voles did not differ among prairie sections ($X^2 = 4.7, df = 2, P > 0.08$), nor did the distribution of gray-tailed voles ($X^2 = 0.2, df = 2, P > 0.25$). Dusky-footed woodrats were captured only in Section C, and no nests were observed outside of that section.

Small mammal species had different direct responses to fire, as indicated by the pre- and post-fire trap periods in 1998. The chance of recapturing an individual deer mouse from the pre-fire trap period during the post-fire trap period did not differ between the combined burned sections (A and B) and the unburned section ($P = 0.49$). Likewise, the chance of capturing a vole (species combined) in a burned area before fire did not differ from after fire ($P = 0.45$). This suggests little direct fire mortality

Table 8. Numbers of individuals of common small mammal species captured during 3 trap periods (1996-98) within the 3 (A, B, and C) differentially burned sections of prairie.

Species	A	B	C	Total
Deer mouse	73	62	136	271
Vagrant shrew	11	37	17	65
Townsend's vole	9	16	21	46
Gray-tailed vole	10	11	9	30
Dusky-footed woodrat	0	0	6	6

or fire-induced emigration from burned areas for deer mice or voles. Conversely, 7 of the 11 vagrant shrews captured in the pre-fire trap period were in the sections to be burned, whereas 0 of 11 were trapped in those burned sections following fire. This suggests that fire-induced mortality and/or emigration may be significant for vagrant shrews.

Discussion. The wet prairies of the Willamette Valley historically were shaped by fire and water. Grassland is not the climax state in this system, as fire suppression results in shrubland and ultimately ash woodland in some areas. The combination of prairie sections with different burn frequencies on the RNA prairie creates different stages of succession, which may allow for a greater diversity of small mammal species. In this study, each species had different direct responses to fire, as well as different associations with the prairie sections (i.e., stages of succession). Despite this variability in species associations, none of the species were ever most abundant in the annually burned section. A period of 2-3 years may be needed to overcome the effects of fire. Therefore, a prairie management scheme that takes the diversity and population characteristics of small mammal species into account may benefit by providing a variety of successional stages, while allowing adequate time between fire events for populations of fire-sensitive species to recover.

2. Managing Agricultural Grasslands for Landbirds: The Role of Scale and Context
Bob Altman & Rebecca Goggans

In 1996, the Oregon Department of Fish and Wildlife initiated a 2-year project to develop management and conservation strategies for several state sensitive grassland-associated bird species in the Willamette Valley, particularly the western meadowlark, Oregon vesper sparrow, and streaked horned lark. All 3 species were formerly common breeding species in native grasslands, but have suffered extensive population declines.

Methods. Data collection integrated several methods at extensive and intensive scales. Broad-scale (extensive) censusing was conducted at 544 roadside point-count stations throughout the Willamette Valley in 1996 to characterize the grassland breeding bird community, and quantify species abundance and habitat associations. Territory-mapping (intensive scale) was conducted primarily in 1997 at selected sites to quantify species area-requirements and habitat relationships.

Habitat characterization was conducted within a 328 feet (100 m) radius of each point-count station. Among habitat variables described were estimates of percent cover of habitat type and percent cover of vegetation height classes in 6-inch (15.2 cm) increments (e.g., 0-6 in [0-15.2 cm], 6-12 in [15.2-30.5 cm], etc.). Habitat data were collected within each territory along three 164-foot (50 m) transects randomly located from the center of the territory. Habitat variables described included estimates of percent cover by growth form, and percent cover of vegetation

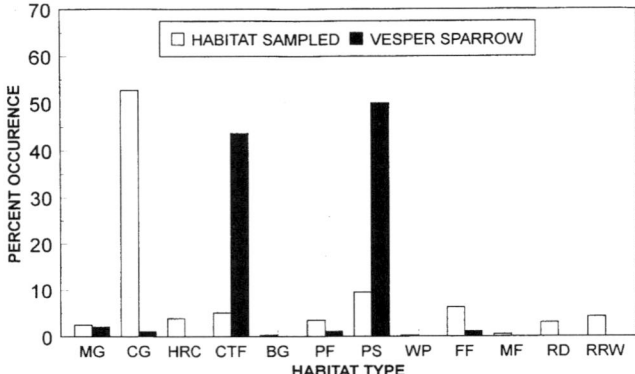

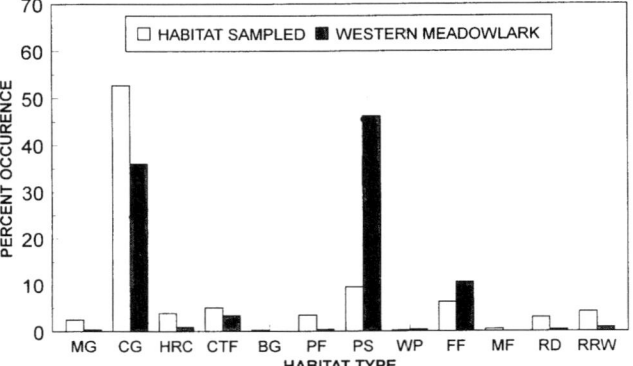

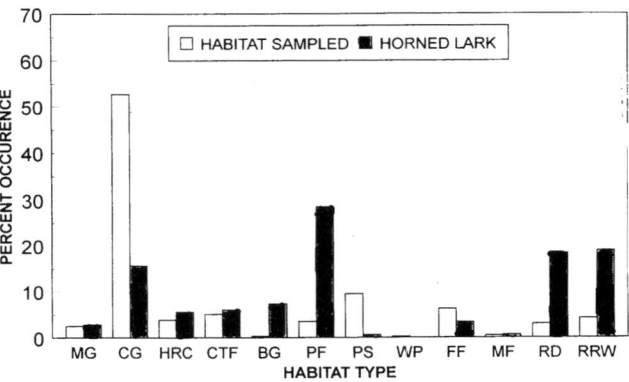

Figure 1. Occurrence of three grassland-associated bird species based on point-count censusing in the Willamette Valley, May - June 1996.

MG – mowed grass
CG – cultivated grass seed
HRC – herbaceous row crop
CTF – Christmas tree farm
BG – bare ground
PF – plowed field
PS – pasture
WP – wet prairie
FF – fallow field
MF – maintained field
RD – gravel/dirt road
RRW – roadside right-of-way

height classes as described above. Territories were delineated using the "repeat-flush" technique.[131] Territory size was generated from boundary point data collected with a Global Positioning System (GPS) and downloaded into Arc View software for calculations.

Results: Censusing. Patterns of species occurrence by habitat type indicated a relatively high degree of specialization for vesper sparrows and western meadowlarks (each predominantly in 2 habitat types), and a relatively wide range of habitat types used by horned larks (Figure 1). Pasture was a predominant habitat for the western meadowlark and vesper sparrow, accounting for nearly 50% of the detections of both species, despite <10% of the sampling effort in pasture. Christmas tree farms were the other high-use habitat for vesper sparrow (approximately 45% of the detections). Grass-seed fields, which comprised 53% of the sampling effort, were not used by vesper sparrows, but were the second primary habitat for western meadowlarks (approximately 35% of the detections). Percent detections (10.6) of western meadowlarks in fallow fields were also greater than the percent of this habitat type sampled (6.3). The 4 principal habitats used by horned larks were gravel and dirt roads, roadside rights-of-way, plowed fields, and grass-seed fields. Horned lark detections in bare ground, Christmas tree farms, and herbaceous row crops were also greater than the percent sampling effort in those habitat types.

Western meadowlark and horned lark occurrence in grass-seed fields was further defined by the type of field (i.e., annual or perennial). Western meadowlarks ($n = 80$) were detected nearly equally in annual and perennial grass fields, but horned larks ($n = 29$) were substantially more abundant in annual grass fields (Figure 2). This difference is likely because of the presence of more bare ground in annual grass fields in the spring due to planting, whereas perennial grass fields are vegetated to a greater extent during this time period.

Western meadowlark and vesper sparrow occurrence in pastures was further defined by the amount of vertical structure provided by shrubs and trees. Both species occurred primarily in entirely herbaceous pastures or pastures with <10% shrub-tree cover, although western meadowlarks were most often associated with predominantly herbaceous pastures, and vesper sparrows with a higher percentage of shrub/tree cover (Figure 3). Pastures with >10% shrub-tree cover were moderately used by vesper sparrows, but rarely used by western meadowlarks. Shrubs and trees are important to both species for singing perches.[132]

Patterns of species occurrence relative to grass height were different for all 3 species. Most horned lark detections were in short grass (0-6 inches [0-15.2 cm]), vesper sparrows occurred in relatively short grass (1-12 inches [2.5-30 cm]), and western meadowlarks were most tolerant of different grass heights (up to 36 inches [91.4 cm]) depending on the type of field.

Results: Territory Mapping. Mean territory size for western meadowlark ($n = 21$) was 14.3 acres (5.8 ha). Sixty-

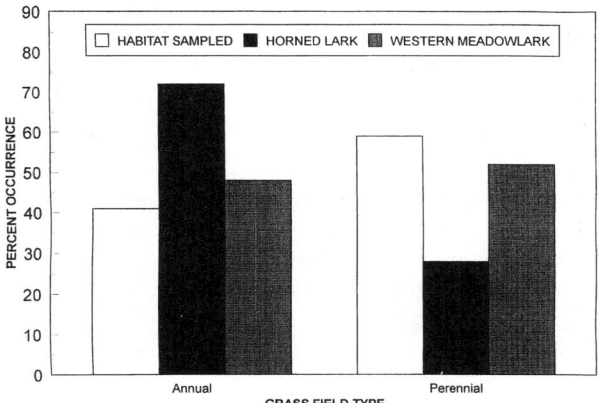

Figure 2. Horned lark and western meadowlark occurrence in annual and perennial grass fields based on point-count censusing in Willamette Valley, May - June 1996.

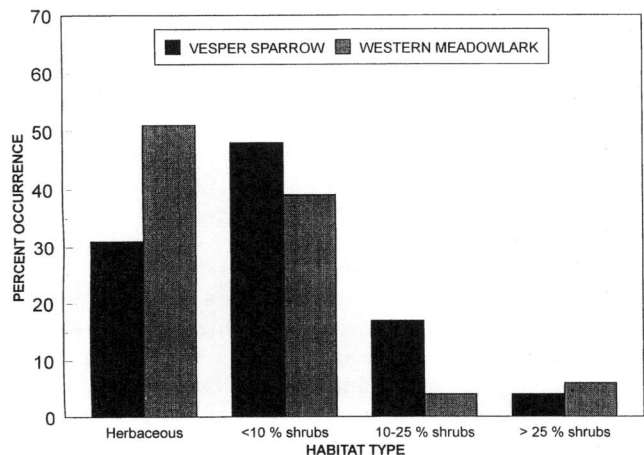

Figure 3. Vesper sparrow and western meadowlark occurrence in pasture types based on point-count censusing in Willamette Valley, May - June 1996.

seven percent ($n = 14$) of the territories were 10-19 acres (4-7.7 ha). Only 19% ($n = 4$) were <10 acres (4.7 ha), and 14% ($n = 3$) >20 acres (8.1 ha). Only 24% ($n = 5$) of western meadowlark territories included grass-seed fields, and none were >50% grass-seed fields, despite efforts to locate territories in areas that were exclusively or near-exclusively grass-seed fields. Territory sizes for vesper sparrows and horned larks were considerably smaller than those of western meadowlarks. Mean territory size for vesper sparrow ($n = 38$) was 3.1 acres (1.3 ha). However, all but 1 of the territories were >6 acres (2.4 ha). Mean territory size for the horned lark was 1.9 acres (0.8 ha), although the sample size was only 3 territories.

Mean percent occurrence of each grass height class in western meadowlark territories was relatively similar. However, individual territories were highly variable, such that each height class was dominant or codominant in at least 1 territory. This suggests habitat selection by western meadowlarks may not have been dependent on grass height. Vesper sparrow territories were distinctive for particular grass height classes, which suggests some degree of habitat selection by grass height. Over 80% of mean percent occurrence of each height class was in the 3 classes <18 inches (45.7 cm). Vesper sparrow territories were dominated (46%) by grass heights of 6-12 inches (15.2-30.5 cm).

Conservation and Management. Habitat issues of greatest importance to the western meadowlark were size and context. Western meadowlark was relatively tolerant of various local habitat conditions (e.g., grass height) and field types (e.g., annual or perennial grass fields, pasture, fallow fields), but more area-sensitive than the other species, and more sensitive to the context and juxtaposition of habitat types. Thus, the management focus needs to be on coarse-level conditions at landscape scales. With a relatively large territory size for a passerine, western meadowlarks require relatively large areas of suitable contiguous habitat to maintain populations. However, since cultivated grass fields apparently do not meet their habitat requirements exclusively, large patches of habitat must include at least half their preferred habitat types (pasture and fallow fields).

Habitat issues of greatest importance to vesper sparrows were type and condition. They regularly occurred in small patches (e.g., <10 acres [4 ha]) of grassland habitat, but only within 1 grassland type (pastures) and with relatively specific local conditions for grass height and shrub-tree cover. Thus, management does not need to be conducted at landscape scales, but must be within the context of a specific agricultural grassland type, and be designed to create and maintain relatively specific structural conditions within that habitat type.

Horned larks used a wide variety of habitat types, but were relatively specific at the microhabitat scale for grass height and the presence of bare ground. Thus, habitat management can be accomplished at small scales by creating or maintaining bare ground or sparsely vegetated areas within or adjacent to a wide variety of suitable habitat types. It also seems that suitable habitat can support relatively dense populations because of their small territory size and their propensity to forage communally during the breeding season at good foraging sites.

The near extirpation of native grasslands in the Willamette Valley leaves few opportunities to maintain populations of grassland bird species on these lands. Thus, conservation of declining and state sensitive grassland birds must occur primarily within the context of these species' use of non-native agricultural grasslands. Habitat requirements of the western meadowlark, Oregon vesper sparrow, and streaked horned lark share some similarities, but also have unique differences that can be managed for in agricultural grasslands. However, conservation of these species in Willamette Valley agricultural grasslands will require integrated management actions at several different scales. Decision making will be dependent on opportunities available, and the context of type of land use occurring.

3. Squaw Flat:
A Fire Link between Grasslands and Reptiles
Marc Hayes

In southwestern Oregon, grasslands intermix with a complex of conifer and deciduous tree-dominated assemblages. At low elevations, relatively continuous grasslands dominate in an areal sense. Forested patches cover progressively larger areas as elevation increases. At mid-elevations, grasslands of a lowland character are reduced to scattered patches. At higher elevations yet, this type of grassland vanishes from the forest matrix. Regions of grassland-forest contact are not static. These areas change in a manner that can markedly influence the fauna associated with their adjoining vegetation assemblages. The example provided here uses the reptile fauna to illustrate the grassland maintenance element of such a dynamic for a mid-elevation site in the South Umpqua Basin.

Study site. The study site is the proposed Squaw Flat Research Natural Area (hereafter simply Squaw Flat). Squaw Flat comprises 558 acres (226 ha) of mixed conifer-deciduous forest at the intersection of Jackson and Squaw creeks in the South Umpqua Basin 40 miles (64 km) southeast of Roseburg, Oregon. The site ranges from 1,740 feet (533 m) to 2,450 feet (747 m) in elevation, slopes mostly west and northwest, and has a complex volcanic substrate altered through substantial landflows. Except for 2 tiny ponds (<1.7 acre [0.7 ha]), scattered seeps, a few ephemeral drainages, and the riparian margins of Jackson and Squaw creeks, Squaw Flat is uplands. Uplands can be fundamentally divided into forest and prairies. Forest, the dominant upland category, has a continuous but highly variable conifer-dominated matrix that covers about 95% of the upland area. The forest ranges from mesic (Douglas-fir and western hemlock as dominants) to xeric (ponderosa pine dominated) in a diffuse west-to-east gradient. The other 5% (ca. 20 acres [8 ha]) of upland is 7 small prairies (0.6-10.3 acres [0.2-4.2 ha]) imbedded in the forest matrix (Figure 4; A-G). Native grasses, graminoids, and forbs are dominant on these prairies. Prairies are fundamentally open with low-stature vegetation, but Oregon white oaks ring most of their peripheries.

Squaw Flat is in as close to an undisturbed condition as a site in this elevation range can be expected to be. Timber harvest, which has occurred at points along its periphery, has not occurred on the site, recently or historically (Figure 4). As a consequence, much of the forest is old growth that exceeds a few hundred years in age. The only significant disturbance to Squaw Flat over the last 30 years has been the irregular trespass of livestock from adjacent Umpqua National Forest land; appropriate exclusion barriers are lacking.

Reptile Fauna. Squaw Flat has a rich reptile fauna with 14 species recorded (northern alligator lizard, southern alligator lizard, western skink, western fence lizard, rubber boa, racer, western rattlesnake, ringneck snake, sharptail snake, Pacific coast aquatic garter snake, western terrestrial

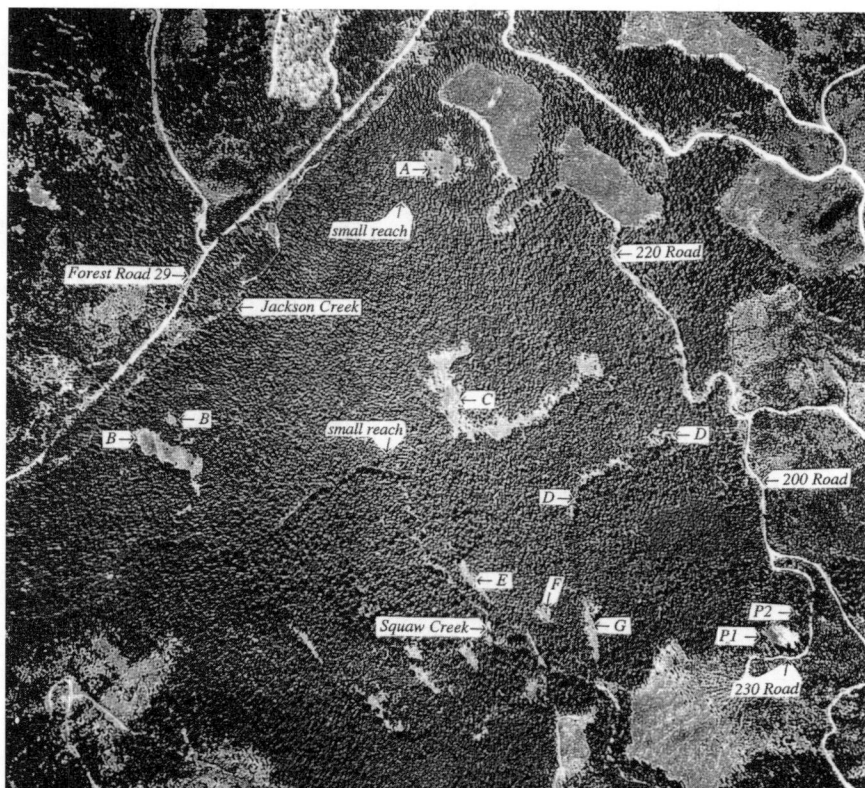

Figure 4. Squaw Flat aerial photograph. Based on a 19 July 1992 overflight. Letters indicate prairies; ponds are P1 and P2 (see text). True north is 12 degrees left of the vertical axis of the page.

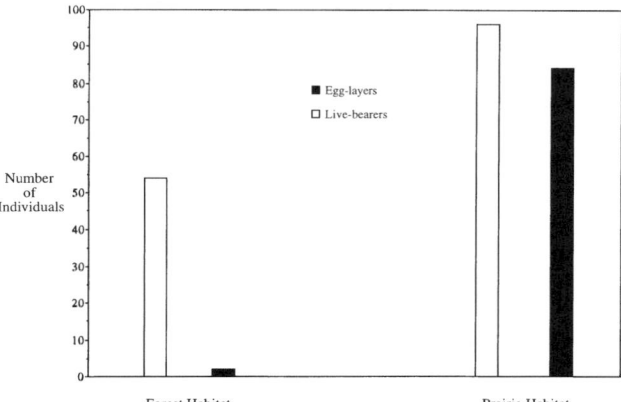

Figure 5. Number of reptiles encountered at Squaw Flats during surveys with effort-adjusted for the area of major habitat categories.

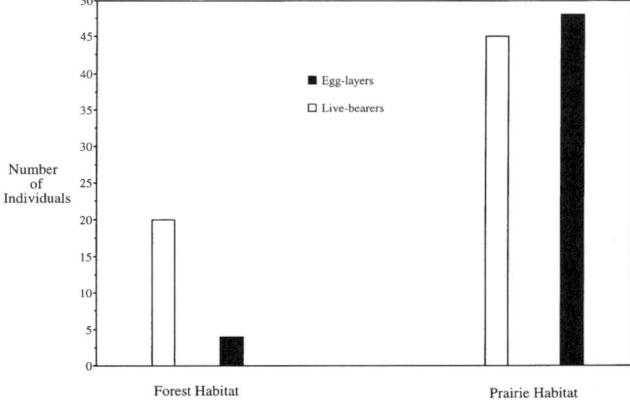

Figure 6. Number of reptiles encountered at Squaw Flats during surveys with equal effort in forest and prairie habitats.

garter snake, northwestern garter snake, common garter snake, western pond turtle) but as many as 3 other species potentially present. Preliminary surveys of this fauna were completed during 1 May-15 June 1996. Surveys were habitat proportional; effort was adjusted to the area of each habitat type. Despite the enormous asymmetry in effort that favored forested habitat because of its much greater area, reptiles were found on prairies in much greater numbers than those observed in forested habitats (Chi-square test: $X2 = 16.7$, $P <0.005$; Figure 5).

Subsequent equal effort surveys between forest and prairie habitats completed during 15 April—15 September in 1997 and 1998 showed that this distributional asymmetry was even more skewed (Chi-square test: $X2 = 32.3$, $P <0.0001$; Figure 6). Egg-laying and live-bearing reptiles both contributed to the pattern, but the pattern was more pronounced among egg layers. Twenty-two reptile nests were found in prairie or prairie-edge locations; 5 different species were represented (western pond turtle, $n = 8$; western fence lizard, $n = 2$; western skink, $n = 10$; gopher snake, $n = 1$; racer, $n = 1$). A female western pond turtle was observed in the process of nesting on the northernmost prairie (A in Fig. 4). No nests of egg-laying reptiles were found in the forested habitats despite over 4 times the search effort allocated to these areas.

Vegetation Changes. Aspects of Squaw Flat vegetation reveal that changes have occurred during the last 60 years.

First, young true firs, primarily of the grand/white fir complex (*Abies grandis*/*A. concolor*), and young western hemlocks dominate the forest understory. Over most of the area, saplings and seedlings of true firs and western hemlock outnumber the younger trees of other canopy dominants by 10 to 1. This contrasts sharply with an overstory dominated by Douglas-fir, ponderosa pine, and in a few areas western hemlock. Second, the forested side of the white oak fringe of most prairies consisted of a melange of healthy conifers and deciduous trees, including frequently senescent white oaks. The latter have partially or completely dead crowns. This pattern is consistent with change inferable from historical photographs. Sixty-year-old aerial photographs indicate that, except for 1 site, the area of prairies has been reduced 10-20%. The sole exception was prairie F (Figure 4), which was landflow-created since 1960. Observed vegetation changes are consistent with the fire suppression management policy that the Umpqua National Forest has engaged in during this period. Fire suppression across this elevation range in this region of the Pacific Northwest results in the characteristically successful recruitment of fire-intolerant, shade-tolerant conifers that progressively encroach on other vegetation.

Future Outlook. Egg-laying reptiles depend on prairies for nesting sites and development. This is probably because prairies may be the only significant areas available on Squaw Flat with the appropriate temperature characteristics for incubating eggs, basking, or warming surface objects needed to gain heat by convection. Estimation of the rate of encroachment on the Squaw Flat prairies indicates that >50% of prairie area will be lost over the next 120 years. The level of closure that may alter relative abundance or species composition of the egg-laying reptiles present is not known. However, if current trends continue, the abundance and species richness of egg-laying reptiles is likely to be reduced. Maintaining the high diversity of amphibians and reptiles at Squaw Flat is a basic RNA objective. The meeting of that objective cannot be accomplished with passive management, it will require adopting a management policy that can reverse the existing prairie-reduction trend. The most direct way to meet that objective is to engage in a trend-reversing fire management policy: a prescribed burn schedule that will limit encroachment of fire-intolerant conifers.

Grassland encroachment at forest interfaces is not unique to Squaw Flat. Mid-elevations in southwestern Oregon and prominent latitudinal vegetation changes in central western Oregon abound with grassland-forest transitions. Soil and hydrology can be responsible for grassland occurrence in such interface areas, but Squaw Flat illustrates that fire may be fundamental in grassland patch maintenance. That a suite of reptiles, including most egg-layers have their northern and altitudinal limits of distribution in this grassland-forest interface-rich region is no coincidence. As an important factor in the maintenance of grassland patches, fire is probably equally fundamental to the maintenance of reptile diversity where these patches exist.

Literature Cited

1. Agee, J. K. 1993. Fire ecology of Pacific Northwest forests. Island Press, Washington, D.C.
2. Alcorn, G. D. 1935. The western yellow-bellied racer in western Washington. Copeia 1935(2):135.
3. Altman, B. 1999a. Status and conservation of grassland birds in the Willamette Valley. Final report submitted to Oregon Dept. of Fish and Wildlife, Corvallis, OR.
4. ———. 1999b. Conservation strategy for landbirds in the lowlands and valleys of western Oregon and Washington. Version 1.0. American Bird Conservancy and Oregon-Washington Partners in Flight.
5. Anthony, A. W. 1886. Field notes on the birds of Washington County, Oregon. Auk 3:161-172.
6. Avery, M. L., and C. van Riper. 1990. Evaluation of wildlife-habitat relationships data base for predicting bird community composition in central California chaparral and blue oak woodlands. California Fish and Game 76(2):103-117.
7. Bailey, V. 1936. The mammals and life zones of Oregon. U.S. Department of Agriculture, Bureau of Biological Survey, North American Fauna (55):1-416.
8. Baird, S. F. 1857[1858]. Mammals. Pages 1-757 in Reports of explorations and surveys, to ascertain the most practicable and economical route for a railroad from the Mississippi River to the Pacific Ocean. Beverly Tucker, Printer, Washington, D.C., 8 (Part 1).
9. Balda, R. P. 1975. Vegetation structure and breeding bird density. Pages 59-80 in Proceedings of the symposium on management of forest and range habitats for nongame birds. U.S. Forest Service General Technical Report WO-1.
10. Barbour, M. G., and J. Major. 1977. Terrestrial vegetation of California. John Wiley and Sons. New York, NY.
11. Biswell, H. H. 1974. The effects of fire on chaparral. Pages 321-364 in T. T. Kozlowski and C. E. Ahlgren, editors. Fire and ecosystems. Academic Press, New York, NY.
12. Blaustein, A. R. 1980. Behavioral aspects of competition in a three-species rodent guild of coastal southern California. Behavioral Ecology and Sociobiology 6:247-255.
13. Borisenko, A. N., and M. P. Hayes. 1999. Status of the foothill yellow-legged frog (*Rana boylii*) in Oregon. Final report prepared for The Nature Conservancy under contract with the U.S. Fish and Wildlife Service with assistance from the Oregon Department of Fish and Wildlife, the U.S. Army Corps of Engineers, and the U.S. Geological Survey.
14. Boyd, R. 1986. Strategies of Indian burning in the Willamette Valley. Canadian Journal of Anthropology 5:65-86.
15. Brodie, E. D., Jr., R. A. Nussbaum, and R. M. Storm. 1969. An egg-laying aggregation of five species of Oregon reptiles. Herpetologica 25(3):223-227.
16. Browning, M. R. 1975. The distribution and occurrence of the birds of Jackson County, Oregon, and surrounding areas. North American Fauna 70.
17. Burroughs, R. D., editor. 1995. The natural history of the Lewis and Clark Expedition. Michigan State University Press, East Lansing, MI.
18. Bury, R. B. 1992. Fort Lewis biodiversity study: western pond turtle and herpetofauna study. National Ecology Resource, U.S. Fish and Wildlife Service, U.S. Forest Service, PNW Research Laboratory, Olympia, WA.
19. Carraway, L. R., and B. J. Verts. 1991. *Neotoma fuscipes*. Mammalian Species (386):1-10.
20. Chapman, J. A., and D. E. C. Trethewey. 1972. Movements within a population of introduced eastern cottontail rabbits. Journal of Wildlife Management 36:1221-1226.
21. ———, and B. J. Verts. 1969. Interspecific aggressive behavior in rabbits. The Murrelet 50:17-18.
22. Chappell, C. B., and R. C. Crawford. 1997. Native vegetation of the south Puget Sound prairie landscape. Pages 107-122 in P. Dunn and K. Ewing, editors. Ecology and conservation of the South Puget Sound prairie landscape. The Nature Conservancy, Seattle, WA.

23. Christy, J., E. Alverson, M. Dougherty, S. Kolar, L. Ashkenas, and P. Minnear. 1999. Pre-settlement vegetation map for the Willamette Valley, Oregon, compiled from records of the General Land Office Surveyors (c. 1850). Oregon Natural Heritage Program, Portland, OR. .

24. Conant, R., and J. T. Collins. 1991. A field guide to reptiles and amphibians: Eastern and Central North America. Houghton Mifflin, Boston.

25. Cook, S. F., Jr. 1960. On the occurrence and life history of *Contia tenuis*. Herpetologica 16(2):163-173.

26. Corkran, C. C. and C. Thoms. 1996. Amphibians of Oregon, Washington, and British Columbia: a field identification guide. Lone Pine Press, Edmonton, Alberta, Canada.

27. Crawford, R. C., and H. Hall. 1997. Changes in the South Puget prairie landscape. Pages 11-16 in P. V. Dunn and K. Ewing, editors, Ecology and conservation of the South Puget Sound prairie landscape. The Nature Conservancy. Seattle, WA.

28. Cross, S. P., and J. K. Simmons. 1983. Bird populations of the mixed-hardwood forests near Roseburg, Oregon. Oregon Department of Fish and Wildlife Technical Report 82-2-05.

29. Dalquest, W. W. 1941. Distribution of cottontail rabbits in Washington state. Journal of Wildlife Management 5:408-411.

30. Dawson, L. W., and J. H. Bowles. 1909. The birds of Washington. Occidental Press, Seattle, WA.

31. Defenders of Wildlife. 1998. Oregon's living landscape: strategies and opportunities to conserve biodiversity. Defenders of Wildlife, Lake Oswego, OR and The Nature Conservancy, Portland, OR.

32. DeLong, K. T. 1966. Population ecology of feral house mice: interference by *Microtus*. Ecology 47:481-487.

33. Detling, L. E. 1961. The chaparral formation of southwestern Oregon, with considerations of its postglacial history. Ecology 42:348-357.

34. Edge, W. D., J. O. Wolff, and R. L. Carey. 1995. Density-dependent responses of gray-tailed voles to mowing. Journal Wildlife Management 59:245-251.

35. ———, R. L. Carey, J. O. Wolff, L. M. Ganio, and T,. Manning. 1996. Effects of Guthion 2S on *Microtus canicaudus*: a risk assessment validation. Journal Applied Ecology 32:269-278.

36. Elzroth, E. K. 1987. Population study and breeding biology of the western bluebird (*Sialia mexicana*). Oregon Department of Fish and Wildlife, Nongame Wildlife Program, Technical Report 86-1-01.

37. Emlen, J. T. 1972. Size and structure of a wintering avian community in southern Texas. Ecology 53:317-329.

38. Evans, J. 1970. About nutria and their control. U.S. Department of the Interior, Bureau of Sport Fisheries and Wildlife, Resource Publication 86:1-65.

39. Evenden, F. G., Jr. 1949. Habitat relations of typical austral and boreal avifauna in the Willamette Valley, Oregon. Dissertation, Oregon State College, Corvallis, OR.

40. Fitch, H. S. 1935. Natural history of the alligator lizards. Transactions of the Academy of Science of Saint Louis 39(1):3-38.

41. Fitch, H. S. 1936. Amphibians and reptiles of the Rogue River Basin, Oregon. American Midland Naturalist 17:634-652.

42. Forrester, B. and R. Storre. 1992. Western pond turtle survey: Ft. Lewis, Washington. Unpublished report, Department of Army. David Evans Associates., Bellevue, WA.

43. Franklin, J. F., and C. T. Dyrness. 1973. Natural vegetation of Oregon and Washington. U.S. Forest Service General Technical Report PNW-8.

44. Gabrielson, I. N., and S. G. Jewett. 1940. Birds of Oregon. Oregon State Monograph No. 2. Oregon State College, Corvallis, OR.

45. Galen, C. 1989. A preliminary assessment of the status of the Lewis' woodpecker in Wasco County, Oregon. Oregon Department Fish and Wildlife, Nongame Technical Report 88-3-01.

46. Gavin, T. A. 1984. Whitetail populations and habitats: Pacific Northwest. Pages 487-496 in L. K. Halls, editor. White-tailed deer: ecology and management. Stackpole Books, Harrisburg, Pennsylvania.

47. Gilbert, B., T. Williams, and J. Bottorff. 1991. 1991 spotted frog survey Ft. Lewis Military Reservation, Pierce County, Washington. U.S. Fish and Wildlife Service, Olympia, WA.

48. Gilligan, J., M. Smith, D. Rogers, and A. Contreras. 1994. Birds of Oregon: status and distribution. Cinclus Publishing, McMinnville, OR.

49. Gordon, K. L. 1939. The Amphibia and Reptilia of Oregon. Oregon State College Monographs 1:1-81.

50. Graf, W. 1939. The distribution and habitats of amphibians and reptiles in Lincoln, Benton, and Linn counties. Thesis, Oregon State College, Corvallis, OR.

51. Graf, W. 1955. Cottontail rabbit introductions and distribution in western Oregon. Journal of Wildlife Management 19:184-188.

52. Graf, W., S. G. Jewett, Jr., and K. L. Gordon. 1939. Records of amphibians and reptiles from Oregon. Copeia 1939(2):101-102.

53. Grinnell, J. 1933. Review of the recent mammal fauna of California. University of California Publications in Zoology 40:71-234.

54. Gullion, G. W. 1951. Birds of the southern Willamette Valley, Oregon. The Condor 53:129-149.

55. Hall, H. L., R. C. Crawford, and B. Stephens. 1995. Regional inventory of prairies in the southern Puget Trough: Phase I. File report. Washington Department Natural Resources, Natural Heritage Program, Olympia, WA.

56. Hallock, L. A., and W. P. Leonard. 1997. Herpetofauna of the Ft. Lewis Military Reservation, Pierce and Thurston Counties, Washington. Washington Natural Heritage Program.

57. Hanes, T. L. 1971. Succession after fire in the chaparral of southern California. Ecological Monographs 41:27-52.

58. Harper, J. A. and numerous authors. 1987. Ecology and management of Roosevelt elk in Oregon. Revised edition. Oregon Department of Fish and Wildlife, Portland, OR.

59. Hayes, M. P. 1994a. The spotted frog (*Rana pretiosa*) in western Oregon. Oregon Department of Fish and Wildlife, Wildlife Diversity Program, Technical Report 94-0-01:1-19 + appendices.

60. ———. 1994b. Current status of the spotted frog in western Oregon. Oregon Department of Fish and Wildlife, Wildlife Diversity Program, Technical Report 94-0-01:1-11.

61. ———. 1996. Assessment of the sensitive amphibian and reptile fauna on selected culvert, pond, road, slide, and stream sites within the Tiller Ranger District (Umpqua National Forest). Final report prepared for the Umpqua National Forest under subcontract to Resources Northwest, Inc.

62. ———. 1997. Status of the Oregon spotted frog (*Rana pretiosa*) in the Deschutes Basin and selected other systems in Oregon and northeastern California with a rangewide synopsis of the species status. Final report prepared for the U.S. Fish and Wildlife Service with assistance from the Oregon Department of Fish and Wildlife.

63. ———. 1998. Assessment of the sensitive amphibian and reptile fauna on Apple Jack and Johnnie Springs salvage units, Wildcat Timber Sale Units, and ERFO sites within the Tiller Ranger District (Umpqua National Forest). Final report prepared for the Umpqua National Forest under subcontract to EA Engineering, Science, and Technology, Inc.

64. ——— and M. R. Jennings. 1986. Decline of ranid frog species in western North America: are bullfrogs (*Rana catesbeiana*) responsible? Journal of Herpetology 20(4):490-509.

65. Hoffman, R. 1927. Birds of the Pacific States: containing brief biographies and descriptions of about four hundred species with special reference to their appearance in the field. Boston, MA.

66. Holland, D. C. 1994. The western pond turtle: habitat and history. Final report to the U.S. Department of Energy, Bonneville Power Administration, Division of Fish and Wildlife, and Wildlife Diversity Program, Oregon Department of Fish and Wildlife.

67. Hopkins, D. D., and R. B. Forbes. 1980. Size and reproductive patterns of the Virginia opossum in northwestern Oregon. The Murrelet 60:95-98.

68. Ingles, L. G. 1965. Mammals of the Pacific States: California, Oregon, and Washington. Stanford University Press, Stanford, CA.

69. Jennings, M. R., and M. P. Hayes. 1985. Pre-1900 overharvest of California red-legged frogs (*Rana aurora draytonii*): the inducement for bullfrog (*Rana catesbeiana*) introduction. Herpetologica 41(1):94-103.

70. Jewett, S. G., Jr. 1936. Notes on the amphibians of the Portland, Oregon, area. Copeia 1936(1):71-72.

71. ———, and H. W. Dobyns. 1929. The Virginia opossum in Oregon. Journal of Mammal. 10:351.

72. ———, W. P. Taylor, W. T. Shaw, and J. W. Aldrich. 1953. Birds of Washington State. University of Washington Press, Seattle, WA.

73. Johannessen, C. L., W. A. Davenport, A. Millet, and S. McWilliams. 1971. The vegetation of the Willamette Valley. Annals of the Association of American Geography 61:286-302.

74. Johnson, O. B. 1880. List of the birds of the Willamette Valley, Oregon. American Naturalist 14:485-491, 635-641.

75. Knopf, F. L. 1994. Avian assemblages on altered grasslands. Studies in Avian Biology 15:247-257.

76. Kuhn, L. W., and E. P. Peloquin. 1974. Oregon's nutria problem. Proceedings Vertebrate Pest Conference 6:101-105.

77. Lais, M. C. 1969. Distribution of amphibians and reptiles in Jackson County, Oregon. Thesis, Southern Oregon State College, Ashland, OR.

78. Lampman, B. H. 1946. The coming of pond fishes. Binfords and Mort, Portland, OR.

79. Lardie, R. L. 1963. Checklist of the amphibians and reptiles of Pierce County, Washington, with special reference to those found on McChord Air Force Base. Tricor 3(1):10-17.

80. Larrison, E. J. 1943. Feral coypus in the Pacific Northwest. The Murrelet 24:3-9.

81. Laycock, G. 1966. The alien animals: the story of imported wildlife. The Natural History Press, Garden City, New York, NY.

82. Leonard, W. P. 1995. Survey of amphibians and reptiles at McChord Air Force Base, Pierce County, Washington. Washington Natural Heritage Program.

83. Leonard, W. P., and L. A. Hallock. 1997. Herpetofauna of the South Puget Sound prairie landscape. Pages 65-74 in P. V. Dunn and K. Ewing, editors, Ecology and conservation of the South Puget Sound prairie landscape. The Nature Conservancy, Seattle, WA.

84. Lev, E., J. Fugate, M. P. Hayes, D. Smith, L. Wilson, and R. Wisseman. 1994. The biota of Smith and Bybee Lakes Management Area. Final report prepared for Metro Regional Parks and Greenspaces, Portland, OR.

85. Lewis, M. G., and F. A. Sharpe. 1987. Birding in the San Juan Islands. The Mountaineers, Seattle, WA.

86. Lidicker, W. Z., Jr. 1966. Ecological observations on a feral house mouse population declining to extinction. Ecological Monographs 36:27-50.

87. Lillywhite, H. B., G. Friedman, and N. Ford. 1977. Color matching and perch selection by lizards in recently burned chaparral. Copeia 1977(1):115-121.

88. MacArthur, R. H., and J. W. MacArthur. 1961. On bird species diversity. Ecology 42:594-598.

89. Marshall, D. B., M. Chilcote, and H. Weeks. 1996. Species at risk: sensitive, threatened, and endangered vertebrates of Oregon. Second edition. Oregon Department of Fish and Wildlife, Portland, OR.

90. Maser, C, B. R. Mate, J. F. Franklin, and C. T. Dyrness. 1981. Natural history of Oregon Coast mammals. U.S. Forest Service; and U.S. Department of the Interior, Bureau of Land Management, General Technical Report PNW-133:1-496.

91. Mason, R. T., D. T. Lerner, and I. T. Moore. 1997. Snake communities in the central Willamette Valley: effects of changing habitat. Presented at the Annual Meeting of the Society of Northwestern Vertebrate Zoologist, Olympia, WA.

92. McAllister, K. M., W. P. Leonard, and R. M. Storm. 1993. Spotted frog surveys in the Puget Trough of Washington, 1989-1991. Northwestern Naturalist 74:10-15.

93. McQueen, L. 1979. The grasshopper sparrows of the Willamette Valley. Oregon Birds 5:32-35.

94. Meyers, S. M., and J. O. Wolff. 1994. Comparative toxicity of azinphos-methyl to house mice, laboratory mice, deer mice, and gray-tailed voles. Archives of Contamination and Toxicology 26:478-482.

95. Murray, K. 1968. The pig war. Washington Historical Society, Tacoma, WA.

96. Northwest Habitat Institute. 1999. Oregon: Landscape-scale vegetation map. Northwest Habitat Institute, Corvallis, OR.

97. Nussbaum, R. A., E. D. Brodie, Jr., and R. M. Storm. 1983. Amphibians and reptiles of the Pacific Northwest. University of Idaho Press, Moscow, ID.

98. Parker, W. S., and W. S. Brown. 1980. Comparative ecology of two colubrid snakes in northern Utah. Milwaukee Public Museum, Publications in Biology and Geology (7):1-104.

99. Puchy, C. A., and D. B. Marshall. 1993. Oregon Wildlife Diversity Plan. Oregon Department of Fish and Wildlife, Portland, OR.

100. Rathburn, S. F. 1902. A list of the land birds of Seattle, Washington, and vicinity. Auk 19:131-141.

101. Resources Northwest, Inc. and Pentec Environmental Inc. 1995. Neotropical migratory bird survey, Fort Lewis Military Reservation. Unpublished report submitted to U.S. Army Corps of Engineers and HQ, I Corps and Fort Lewis Directorate of Engineering and Housing.

102. Richardson, C., and F. W. Sturges. 1964. Bird records from southern Oregon. Condor 66:514-515.

103. Riegel, G. M., B. G. Smith, and J. F. Franklin. 1992. Foothill oak woodlands of the interior valleys of southwestern Oregon. Northwest Science 66:66-76.

104. Rogers, R. 1999. Microhabitat selection and status of streaked horned lark, western bluebird, Oregon vesper sparrow, and western meadowlark in western Washington. Thesis, Evergreen State College, Olympia, WA.

105. Rolph, D. N. 1998. Assessment of Neotropical migrant landbirds on McChord Air Force Base, Washington. Unpublished report, The Nature Conservancy of Washington, Seattle, WA.

106. Sauer, J. R., J. E. Hines, I. Thomas, J. Fallow, and G. Gough,. 1999. The North American Breeding Bird Survey: results and analysis 1966-1998. Version 98.1. U.S.G.S. Patuxent Wildlife Research Center, Laurel, MD.

107. Slater, J. R. 1955. Distribution of Washington amphibians. Occasional Papers of the College of Puget Sound 24:212-242.

108. Slevin, J. R. 1928. The amphibians of western North America: An account of the species known to inhabit California, Alaska, British Columbia, Washington, Oregon, Idaho, Utah, Nevada, Arizona, Sonora, and Lower California. Occasional Papers of the California Academy of Sciences 16:1-151.

109. Slipp, J. W. 1940. The mammals, reptiles, and freshwater fishes of the Tacoma area. Thesis, College of Puget Sound, Tacoma, WA.

110. Smith, M. R., P. W. Mattocks, Jr., and K. M. Cassidy. 1997. Breeding birds of Washington State. Volume 4 in Washington State Gap Analysis —Final Report in K. M. Cassidy, C. E. Grue, M. R. Smith, and K. M. Dvornich, editors. Seattle, Audubon Society Publication In Zoology No I, Seattle, WA.

111. Smith, W. P. 1985. Plant associations within the interior valleys of the Umpqua River basin, Oregon. Journal of Range Management 38:526-530.

112. Speer, F. and J. Felker. 1991. Blue-gray gnatcatcher (Polioptila caerulea) study: Lower Table Rock Preserve. Unpublished report submitted to The Nature Conservancy, Medford, OR.

113. Stebbins, R. C. 1954. Amphibians and reptiles of western North America. University of California Press, Berkeley, CA.

114. Stebbins, R. C. 1985. A field guide to the amphibians and reptiles of western North America. Houghton Mifflin, Boston, MA.

115. Steinberg, E. and D. Heller. 1997. Using DNA and rocks to interpret the taxonomy and patchy distribution of pocket gophers in western Washington prairies. Pages 43-51 in P. V. Dunn and K. Ewing, editors, Ecology and conservation of the South Puget Sound prairie landscape. The Nature Conservancy, Seattle, WA.

116. St. John, A. D. 1984. The herpetology of Jackson and Josephine Counties, Oregon. Oregon Department of Fish and Wildlife, Nongame Wildlife Program, Technical Report 84-2-05:1-78.

117. St. John, A. D. 1985. The herpetology of the Interior Umpqua River Drainage, Douglas County, Oregon. Oregon Department of Fish and Wildlife, Nongame Wildlife Program, Technical Report 85-2-02:1-69.

118. St. John, A. D. 1987. The herpetology of the Willamette Valley, Oregon. Oregon Department of Fish and Wildlife, Nongame Wildlife Program, Technical Report 86-1-02:1-79.

119. Storm, R. M. 1941. Effect of white man's settlement on wild animals in the Mary's River valley. Thesis. Oregon State University, Corvallis, OR.

120. Streatfeild, R., and R. E. Frenkel. 1997. Ecological survey and interpretation of the Willamette Floodplain Research Natural Area, W. L. Finley National Wildlife Refuge, Oregon, U.S.A. Natural Areas Journal 17:346-354.

121. Stroud, D. C. 1982. Population dynamics of *Rattus rattus* and *norvegicus* in a riparian habitat. Journal of Mammalogy 63:151-154.

122. Suckley, G., and J. G. Cooper. 1860. The natural history of Washington Territory and Oregon. Final Reports. On the survey of the Northern Pacific Railroad route. Bailliere Brothers, New York, NY.

123. Sugihara, N. G., L. J. Reed, and J. M. Lenihan. 1987. Vegetation of the Bald Hills oak woodlands, Redwood National Park, California. Madrono 34:193-208.

124. Taylor, R. J., and T. R. Boss. 1975. Biosystematics of *Quercus garryana* in relation to its distribution in the state of Washington. Northwest Science 49:49-57.

125. Van Denburgh, J. 1922. Reptiles of North America: an account of the species known to inhabit California, Alaska, British Columbia, Washington, Oregon, Idaho, Utah, Nevada, Arizona, Sonora, and Lower California. Volumes 1 and 2. California Academy of Sciences, San Francisco, CA.

126. Verts, B. J., and L. R. Carraway. 1980. Natural hybridization of *Sylvilagus bachmani* and introduced *S. floridanus* in Oregon. The Murrelet 61:95-98.

127. ———, and ——— 1981. Dispersal and dispersion of an introduced population of *Sylvilagus floridanus*. The Great Basin Naturalist 41:167-175.

128. ———. and ———. 1998. Land mammals of Oregon. University of California Press, Berkeley, CA.

129. ———., S. A. Ludwig, and W. Graf. 1972. Distribution of eastern cottontail rabbits in Linn County, Oregon, 1953-1970. The Murrelet 53:25-26.

130. White, R. 1980. Land use, environment, and social change: The shaping of Island County, Washington. University of Washington Press, Seattle, WA.

131. Wiens, J. A. 1969. An approach to the study of ecological relationships among grassland birds. Ornithology Monograph 8:1-93.

132. ———. 1973. Interterritorial habitat variation in grasshopper and savannah sparrows. Ecology 54:877-884.

133. ———, and M. I. Dyer. 1975. Rangeland avifaunas: Their composition, energetics and role in the ecosystem. Pages 146-181 *in* Proceedings of the symposium on management of forest and range habitats for nongame birds. U.S. Forest Service General Technical Report WO-1.

134. Wilson, M. V., E. R. Alverson, D. L. Clark, R. H. Hayes, C. A. Ingersoll, and M. B. Naughton. 1995. The Willamette Valley Natural Areas Network. Restoration and Management Notes 13(1):26-28.

135. Witmer, G. W., R. D. Sayler, and M. J. Pipas. 1996. Biology and habitat use of the Mazama pocket gopher (*Thomomys mazama*) in the Puget Sound Area Washington. Northwest Science 70(2): 93-98.

136. Wolff, J. O., and E. M. Schauber. 1996. Space use and juvenile recruitment in gray-tailed voles in response to intruder pressure and food abundance. Acta Theriologica 41:35-43.

137. Wolff, J. O., W. D. Edge, and R. Bentley. 1994. Behavioral and reproductive biology of the gray-tailed vole. Journal Mammal. 75:873-879.

138. Wolff, J. O., T. Manning, S. M. Meyers, and R. Bentley. 1996. Population ecology of the gray-tailed vole (*Microtus canicaudus*). Northwest Science 70:334-340.

139. Wolff, J. O., E. M. Schauber, and W. D. Edge. 1997. Effects of habitat loss and fragmentation on the demography of gray-tailed voles. Conservation Biology 11:945-956.

140. Woodcock, A. R. 1902. An annotated list of the birds of Oregon. Oregon Agricultural Experiment Station Bulletin 68.

141. Zoloth, S. R. 1969. Observations of the population of brush rabbits on Ano Nuevo Island, California. The Wasmann Journal of Biology 27:149-161.

11

Wildlife of Eastside Shrubland and Grassland Habitats

W. Matthew Vander Haegen, Scott M. McCorquodale, Charles R. Peterson, Gregory A. Green, & Eric Yensen

Introduction

The rain shadow of the Cascade Mountains gives rise to a suite of arid and semi-arid habitats that differ substantially from those of the surrounding forest. O'Neil and Johnson[100] have classified these habitats into 6 different types: shrub-steppe, dwarf shrub-steppe, desert playa and salt scrub, western juniper and mountain mahogany woodlands, eastside canyon shrublands, and eastside grasslands. Most of these communities are dominated by shrubs and herbaceous vegetation (grasses and forbs) and typically have a microbiotic crust of lichens and mosses binding the upper surface of the soil. With the exception of the western juniper and mountain mahogany woodlands habitat type, trees in these communities are limited mainly to riparian zones and ecotones with forested habitats and are entirely absent from extensive areas. Much of the historical vegetation in these habitat types, particularly in Washington and north central Oregon, has been converted to agricultural crops.[27, 113] In some areas, the only remaining native communities are on rocky soils or steep slopes unfit for agriculture. The dominant land use in these shrubland and grassland habitats is livestock grazing, and few examples of undisturbed stands exist, limited primarily to sites where topography or remoteness from water has made access for livestock grazing impractical.[22]

Although pristine climax communities do exist for eastside grassland and shrubland habitats, the majority of sites have been shaped by a legacy of past land uses that includes continuous grazing by livestock and range improvements to increase livestock forage and that in turn has facilitated invasion by exotic vegetation. This legacy has modified the vegetation community in many areas, with some changes occurring so long ago that they are not apparent to the present day observer.[22] Changes in the herb community brought about by excessive grazing and exotic invaders are particularly damaging in these arid habitats where the herb layer often contains the most vegetation biomass. Moreover, the successional trajectory of vegetation communities in arid habitats can be modified by influences such as grazing and fire, resulting in present day "zootic" climax communities that differ greatly from those which occurred historically.[22, 128] Sites in southcentral Washington that were dominated by exotic annuals in the 1950s still have not been colonized by native plants some four decades later.[115]

The low vertical structural diversity inherent in these habitats provides fewer habitat layers for wildlife, resulting in lower diversity in some taxa. There are, for example, no arboreal mammals or canopy nesting birds.

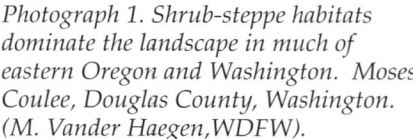

Photograph 1. Shrub-steppe habitats dominate the landscape in much of eastern Oregon and Washington. Moses Coulee, Douglas County, Washington. (M. Vander Haegen, WDFW).

Habitats with a shrub component tend to have more diverse wildlife communities than grass dominated habitats, a function of increased nesting and foraging strata. For example, the shrub-steppe habitat has 49 closely associated species, whereas eastside grassland has only 34. Sites dominated by native plants have more closely associated species than sites dominated by exotics (34 species closely associated with eastside grasslands vs. 2 with modified grasslands).

Available water is a defining factor in these arid and semi-arid habitats, and this strongly shapes the composition of plant communities[22] and influences the ecology and behavior of associated wildlife. In Washington, precipitation occurs primarily during late autumn and winter with annual totals ranging from 5.9 inches (150 mm) in the lowest parts of the Columbia Basin to 21.7 inches (550 mm) in the higher elevations near the forest ecotone. Annual snowfall can be substantial at higher elevations and snow can remain in colder areas into spring. Growth of vegetation in spring is affected by available soil moisture, a result of "bioyear" precipitation—water that falls as rain or snow from October-April and is stored in the soil.[22] Bioyear precipitation varies widely from year to year, affecting plant growth[69] and influencing both forage for herbivorous wildlife and populations of herbivorous insects that form the food base for many reptiles and breeding birds. The marked seasonality in precipitation creates a flush of available food in spring and early summer that is capitalized on by breeding birds. In the lower rainfall zones most vegetative growth is completed by early summer and many plants senesce in preparation for the dry, hot months ahead.

Adaptations to Arid, Seasonally Hot and Cold Environments

Hot and arid conditions that prevail in many of these habitat types in summer play a part in determining the animal life that can persist there. Daytime temperatures can exceed 113°F (45°C) and free water can be scarce. Some species are physiologically adapted to survive in such an environment, whereas others modify their behavior— some survive by a combination of both strategies (Table 1). Species that require daily access to free water (e.g., bats, elk) must restrict their daily use areas to include open water. Mourning doves must drink daily and frequently fly great distances to reach free water.[78] Other species drink infrequently and conserve body water by minimizing water lost through evaporation, respiration, and excretion. These species meet their daily water requirement through moisture contained in food and from metabolic water produced from oxidation as part of digestion.[17] The grasshopper mouse and the sage sparrow are examples of species that spend considerable time far from free water. Black-throated sparrows are particularly well adapted to life in arid environments; experiments have shown that they can survive on a diet of dry seeds without access to free water and without restricting their activity.[78]

Birds and reptiles have some physiological characteristics that allow them to tolerate arid conditions. Birds excrete nitrogenous waste as uric acid, a characteristic that they share with reptiles and that uses <10% of the water required by mammals excreting urea.[78] Birds also operate at a higher body temperature than do most other animals, allowing them to cool themselves by convection at most ambient air temperatures. Some arid-land birds (e.g., black-throated sparrow) can drink and process water with a high saline content.[78] Reptiles are particularly well adapted to arid environments. Their dry, scaley integument and lack of exocrine glands reduces water loss to the environment[17] and, unlike amphibians, they can reproduce independent of standing water.

Many arid-land species modulate their activities to avoid temperature extremes, seeking shade during the heat of the day or, in the case of the rock wren, seeking the cooler environment offered by rock crevices.[78] Fossorial species find cooler temperatures and reduced water loss underground. Seeking milder conditions in underground

Table 1. Living and breeding in arid and semi-arid environments: constraints, consequences, and animal adaptations.

A. Environmental Constraints
High and low temperature extremes
High winds
Open habitats
Aridity
Strong temporal resource gradients
Fragmented habitats
High annual variability in rainfall

B. Ecological Consequences
Cooling and warming adaptations required
Increased variability in annual reproductive output
Need to disperse across unsuitable habitats
Water conservation required

C. Adaptations
- Physiological
    Facultative breeding
    Flexible reproductive timing
    Ability to process hypersaline water
    Ability to survive without free water
    Large ears for cooling (jackrabbits)
    Speed for eluding prey (pronghorn)
- Behavioral
    Cryptic behavior
    Shade-seeking
    Burrow into soil to avoid heat extremes
    Daily trips to available water

D. Life History Adaptations
Cryptic plumage/pelage
Strong seasonality in habitat use
Colonial living
Accelerated larval development
Hibernation or migration during winter
Estivation during summer

burrows can be a temporary tactic used during the heat of the day, or it can be a way to escape harsh conditions for much longer periods. Some fossorial mammals, including ground squirrels and pocket mice, may estivate during the driest months and emerge from their burrows when the winter rains renew vegetation growth. Great Basin spadefoot toads are one of few amphibians to live in arid habitats, surviving the arid summer months by estivating underground. Members of this genus minimize water loss by burrowing underground and by allowing urea levels in the blood to rise as the soil around them dries. Sufficiently high internal osmotic pressures create water potentials that should allow spadefoot toads to absorb water from even very dry soil.[111] Spadefoot toads are also capable of tolerating considerable loss of body water (up to nearly 50%).[138]

Spadefoot toads may spend many months burrowed deep in the soil, emerging to feed or to breed on rainy nights. Their facultative breeding system is well adapted to arid habitats, allowing them to track local conditions and breed after rain events or, in more recent times, after irrigation has wetted the soil.[98] They are adapted to temporary breeding ponds by needing only 2-3 days for egg hatching and only a few weeks for development to metamorphosis. Some tadpoles may feed on carrion as well as vegetation, which also may result in more rapid development. Tadpoles of some other species of spadefoot toads are known to be cannibalistic.[98]

Cold conditions may limit the activity times of ectotherms daily and seasonally; however, their low energetic requirements (typically <10% of those of similar sized endotherms) allow amphibians and reptiles to survive for extended periods of time without eating. For example, an adult rubber boa, maintained in captivity under simulated field thermal gradients, voluntarily fasted for 23 months before eating, with no apparent long-term ill effects (Michael E. Dorcas, Davidson College, pers. comm.). Such fasting abilities enable these animals to survive conditions such as droughts when food availability may be low. The low metabolic rates of amphibians and reptiles also result in low rates of respiratory water loss.[111]

Habitat Elements— Special Habitat Features

Some wildlife species of eastside grassland and shrubland habitats are strongly associated with special features (habitat elements) on the landscape and are unlikely to be found in their absence. Many species are associated with geological formations such as rock outcrops, cliffs, and talus slopes. Peregrine falcons, cliff swallows, and golden eagles nest on cliffs and rock faces and are among the 35 species associated with these structures. All of the snake species and about half of the lizard species in shrub-steppe habitats are associated with rocky features (individual rocks, talus slopes, outcrops, cliffs, ridges, caves, crevices, etc.) that serve a variety of functions including providing foraging locations (e.g., side-blotched lizards), retreat sites from predators (e.g., western fence lizards), vantage points

Photograph 2. Cliffs provide valuable roost sites for bats as well as nest sites for numerous species of shrub-steppe wildlife. Grand Coulee, Douglas County, Washington. (M. Vander Haegen, WDFW).

within territories (e.g., Mojave black-collared lizards), nesting sites (e.g., ringneck snakes), and gestation sites (e.g., western rattlesnakes). Rocky features are often needed to provide the temperature gradients required for effective behavioral thermoregulation.[56] Rocky features such as talus slopes and crevices within lava flows are especially important as overwintering sites for snakes, and their availability may limit the distribution of some species.[43] Often, overwintering sites are used by multiple species (e.g., striped whipsnakes, gopher snakes, western terrestrial garter snakes, and western rattlesnakes). Talus slopes and talus-like structures (rock piles, lava stringers) are associated with 22 species and provide refuge for small mammals like the least chipmunk, and hibernacula for a variety of snakes including the western rattlesnake. Some amphibians (e.g., long-toed salamanders and Pacific treefrogs) may be associated with talus. Rocky outcrops provide nest sites for ferruginous hawks and habitat for rock wrens, yellow-bellied marmots, and 42 other species. Caves are used by 18 species, including bobcats and common ravens, and are critical habitats for bats for roosting and hibernation.

Burrowing owls and fossorial mammals like the Columbian ground squirrel and badger require deep soils for constructing nests and dens. Washington ground squirrels used sites where soils were deeper, weaker, and contained less clay than adjacent, unoccupied sites.[4] Soil type also can affect the persistence of underground burrows. In an Oregon study, nest burrows of burrowing owls were more likely to be reused in subsequent years when constructed in silty-loam soil; burrows in loamy-sand soils were often silted in and thus unusable by the next breeding season.[38] Loose soils are important to burrowing species like Great Basin spadefoot toads, sagebrush lizards, and horned lizards. Several species of reptiles (e.g., long-nosed leopard lizards) bury their eggs in loose soils.

The burrows of other animals are often used by burrowing owls, many species of lizards and snakes,[134] and some species of amphibians (e.g., tiger salamanders and western toads). Burrows may serve a variety of functions, including providing retreats from predators (e.g., long-nosed leopard lizards), foraging sites (e.g., rubber boas), egg deposition sites (e.g., collared lizards), and thermal gradients for regulating body temperature. The use of burrows by reptiles and amphibians in dry shrub-steppe habitats is particularly important for minimizing evaporative water loss.

All amphibians in eastside shrubland and grassland communities are associated with standing or slow-moving water sometime during their life cycle, because it is required for breeding and larval development. Several species of reptiles also are associated with water (i.e., marshes, pools, ponds, lakes, streams, or rivers). Western skinks and rubber boas are often found near moist areas. Garter snakes often forage in the water or along shorelines. Painted turtles overwinter, breed, and forage in water. Human-made stock ponds have probably increased the amount of suitable habitat for some of these water associated species.

Anthropogenic structures are frequently used for nesting sites or as shelter by a variety of wildlife. Common ravens, American crows, and several buteonid hawks nest on power transmission towers. Construction of such towers was responsible for an increase in the number of breeding ferruginous hawks on the Hanford Site in southcentral Washington in the 1980s.[32] Cement road culverts and bridges are used as shelter or as nest sites by cliff swallows and 13 other species; they provide appropriate nesting structure previously missing in many areas. Garter snakes often use the foundations of bridges and buildings for nocturnal retreat sites and overwintering sites. Western fence lizards and racers use stone walls for retreat sites and foraging areas.

Abandoned homesteads and farm buildings provide unique vertical structure in many areas of shrubland and grassland habitat. Old buildings are used as nest sites by kestrels, barn owls, deer mice, and numerous other species. Abandoned buildings and trash piles may attract small mammals and thus reptiles that prey on them (e.g., western rattlesnakes). Gopher snakes may use wells for estivation sites (C.R. Peterson, Idaho State University, pers. obs.). Trees planted as windbreaks around homesteads and orchards provide nesting structure for red-tailed hawks, black-billed magpies, orioles, and numerous other birds. Abandoned farm equipment serves as nesting sites for European starlings, domestic pigeons, western kingbirds, and other birds as well as various small mammals.

Bird Communities in Shrub-steppe

The following wildlife community profiles focus on shrub-steppe habitat, with references to other eastside shrubland and grassland habitats where appropriate. We selected shrub-steppe as a focal habitat based on its dominance in eastside landscapes and because of the severe management issues that this habitat currently faces.

Shrub-steppe bird communities are characterized by a relatively small number of breeding species. The

Photograph 3. Talus slopes and talus-like structures provide hibernacula for western rattlesnakes. Grant County, Washington. (M. Vander Haegen, WDFW).

assemblage of passerines, for example, typically totals 4-8 birds and often is dominated by a single species.[156] Extensive surveys including many sites may generate a much larger species list;[27] however, many species on these longer lists occur infrequently or breed in other habitats and are tallied on surveys as they forage or migrate through the shrub-steppe. For example, three years of surveys on one site in the shrub-steppe of eastern Washington produced a list totaling 28 species; however, only 5 species were documented to nest on the site (Washington Department of Fish and Wildlife [WDFW] unpubl. data). Fragmentation of formally extensive shrub-steppe and new habitats associated with agriculture and irrigation contribute species that are not typical of the shrub-steppe community type.

What the shrub-steppe bird community may lack in variety, it makes up for in specificity. Several species associated with shrub-steppe are so dependent on sage cover that they are termed sagebrush obligates. Sage and Brewer's sparrows, sage thrashers, and sage grouse are considered obligates, whereas vesper sparrows and green-tailed towhees are classified as near-obligates.[8] These species often characterize the big sagebrush (Artemisia tridentata) shrub-steppe community, although other shrub-steppe or grassland-associated species may dominate at specific sites. In big sagebrush communities in northcentral Oregon and southcentral Washington, the breeding bird community was dominated by sage sparrows and western meadowlarks[54, 146] (WDFW unpubl. data). In northcentral Washington, the breeding bird community was dominated by Brewer's sparrows and vesper sparrows (WDFW unpubl. data). Disturbed sites with few shrubs frequently are dominated by western meadowlarks, horned larks, grasshopper sparrows, and long-billed curlews.[54, 146]

Most passerines that breed in shrub-steppe eat insects and other arthropods at some period during the year. Some birds are primarily insectivorous, consuming a variety of invertebrates that they capture on the wing, or glean from shrubs or herbaceous vegetation. Even granivorous birds that feed mostly on seeds and plant material as adults feed invertebrates to their young to supply critical nutrients. Total precipitation during the bioyear affects primary and secondary production and varies widely among years in this ecosystem.[22] This yearly variation in rainfall can affect reproductive success of shrub-steppe nesting birds, probably through its influence on availability of arthropod prey.[121]

The community of breeding birds in shrub-steppe is largely comprised of migrants, many (n = 16) wintering south of the United States and therefore termed Neotropical migrants. Spring migrants that pass through the shrub-steppe on their way to more northern breeding grounds and that spend time in the sagebrush communities include the white-crowned sparrow and ruby-crowned kinglet. Resident species are largely gallinaceous birds (e.g., greater sage-grouse and sharp-tail grouse) and corvids (e.g., common raven and black-billed magpie). The winter bird community is supplemented by species that breed in more

northern sites but spend all or part of the winter in the shrub-steppe, including rough-legged hawks and northern shrikes.

Two native galliforms, greater sage-grouse and sharp-tailed grouse, live in shrub-steppe communities of Oregon and Washington. Both are ground nesters and require large areas for their annual home ranges, including open areas for leks in the spring. Both grouse are listed as threatened in Washington, and the sharp-tailed grouse was only recently reintroduced to Oregon after having been extirpated in the 1960s.[99] Loss of habitat and fragmentation by agriculture are believed to be primary causes for the decline of these species.[16, 124] Exotic galliforms that frequently can be found nesting in shrub-steppe include the chukar, gray partridge, and ring-necked pheasant.

Numerous raptors use shrub-steppe for nesting and foraging, preying on a variety of small mammals, reptiles, birds, and insects. Burrowing owls and short-eared owls nest on the ground in shrub-steppe and can be found nesting in stands of big sage, antelope bitterbrush (Purshia tridentata), or open, low grasslands. Northern harriers often forage in shrub-steppe, especially along edges with agriculture, but they require wetlands or similar areas of thick reeds or grasses for their ground nests. Most raptors require elevated nesting sites and historically nested on buttes, cliffs, and in riparian-associated trees. In northcentral Oregon, juniper trees were important nesting platforms for ferruginous and Swainson's hawks, and long-eared owls.[39, 54] Juniper trees also were important as nest sites for ferruginous hawks in southcentral Washington.[31] Artificial planting of trees and the proliferation of power transmission lines during the last century have increased the availability of suitable nest sites and likely have boosted raptor populations in some areas.[119]

A recent analysis of population trends using the Breeding Bird Survey (BBS)[122] identified 8 shrub-steppe-associated species that are declining in the interior Columbia River Basin. Four of these species, the Brewer's sparrow, lark sparrow, loggerhead shrike, and western meadowlark, are closely associated with shrub-steppe. Numerous other birds, including the sage sparrow, grasshopper sparrow, and burrowing owl, likely are not monitored adequately by the BBS and will require specialized monitoring to detect changes in their populations.[122]

Organizing Principles: Birds

Shrub-steppe communities extend from the northern border of Washington to the southern border of Oregon. Although some shrub-steppe-associated birds are common to much of this area, there are differences in avian species assemblages across this latitudinal gradient. For example, black-throated sparrows and green-tailed towhees reach the northern extent of their range near Oregon's northern border and occur only sporadically in Washington. Sharp-tailed grouse occur in several locations in northcentral and eastern Washington, but were extirpated from Oregon and have only recently been

Photograph 4. Sage sparrows are sagebrush obligates and nest in extensive tracts of shrub-steppe. Moses Coulee, Douglas County, Washington. (M. Vander Haegen, WDFW).

reintroduced.[99] Bird communities also may change along an elevational gradient in shrub-steppe, as rainfall increases with elevation and changes the vegetation community. Vesper and Brewer's sparrows occur rarely in sage communities on the low elevation Columbia River Plain, whereas they are regularly found on ridges only a few kilometers to the west in sage communities above 5,900 feet (1,800 m).[146] As one proceeds north in the Columbia Basin, elevation increases and these species become common at all elevations. In fact, these two sparrows, along with the sage thrasher, typify sage communities in northcentral Washington (WDFW unpubl. data).

Presence and abundance of individual bird species vary with a range of local and landscape variables. On a gross scale, vegetation structure determines what species can breed in a community through presence of suitable nesting and foraging strata. A suite of grassland-associated birds that breed in shrub-steppe includes the grasshopper sparrow, horned lark, western meadowlark, and long-billed curlew. These birds nest on the ground and depend on grasses and forbs to conceal their nests. Native perennial grasses generally dominate the ground layer in undisturbed shrub-steppe communities, and presence of several ground-nesting species, including the grasshopper sparrow and horned lark, increased with percent cover of these grasses.[147] In contrast, long-billed curlews in northcentral Oregon[104] and southeast Washington[2] seemed to prefer short-statured annual grasslands for nesting and foraging, an apparent adaptation to this recent, disturbance-related vegetation community.

Shrub cover is a requirement for the suite of shrub-nesting birds that nest in Washington and Oregon's shrub-steppe. Sage, Brewer's, and lark sparrows, along with sage thrashers, and loggerhead shrikes, nest in or immediately beneath shrubs. They also use shrubs for singing and foraging perches, and Brewer's sparrows forage extensively within the foliage of shrubs.[162] Presence of trees, either along riparian areas or planted as windbreaks or near homesteads, provides nesting platforms for black-billed magpies and a host of raptors, including red-tailed and Swainson's hawks, and long-eared owls. Without the vegetation structure to support nests and foraging activities, these species are unlikely to establish territories and breed successfully.

Floristics also play a part in determining the composition of the avian community in shrub-steppe. In their extensive study of shrub-steppe birds that included sites in Washington, Oregon, and Nevada, Wiens and Rotenberry[158] found that whereas habitat structure played a part in determining species occurrence at a biogeographic scale, presence of particular plant species was important to some birds at a regional scale. Sage sparrows, true to their name, prefer a shrub community dominated by sagebrush over other species of shrub.[158] Even within the sagebrush family, sage sparrows show a preference for stands of big sagebrush. Recent work in Washington has shown that both sage sparrows and Brewer's sparrows occur at greater abundance in communities of big sagebrush than in stiff sagebrush (*A. rigida*) communities typical of rocky soils.[147] Wiens and Rotenberry[158] found abundance of Brewer's sparrows to be negatively correlated with cover of spinescent shrubs such as hopsage and budsage. These authors suggested that some shrub-steppe birds key on particular shrubs to take advantage of arthropod foods particular to those shrub species.

Although we have only recently begun to examine spatial components of community structure, there is evidence that landscape characteristics influence the occurrence of some shrub-steppe species. In Idaho, the probability of finding sage sparrows and Brewer's sparrows in patches of sagebrush increased with size of the patch.[63] Recent work in Washington has shown patch size to be very important to sage sparrows, with males establishing territories and nesting only on sites many times larger than an average territory (WDFW unpubl. data). The landscape context in which these spatial effects are examined also may have a bearing on how species react. In extensive shrub-steppe in Idaho, where sagebrush

Photograph 5. Young burrowing owls near their nest burrow. Burrowing owls in the Columbia Basin depend on fossorial mammals both for nest sites and for much of their diet. Grant County, Washington. (D. Hoyt, WDFW).

communities were fragmented by fire and subsequent cheatgrass (*Bromus tectorum*) invasion, sage thrashers occurred more frequently in spatially similar sites with low fragmentation of sage.[63] However, in Washington, where sagebrush stands are fragmented by agriculture, thrashers were regularly found in small fragments of sage and were more likely to occur in fragmented than extensive sites.[147] For some species, these landscape features likely interact with local vegetation characteristics to determine suitability of a site. Landscape characteristics such as patch size and spatial similarity seem to have little effect on occurrence of some of the more generalist birds such as western meadowlarks and horned larks, at least at the scales studied.[63, 147]

Another physical variable that influences the bird community in shrub-steppe is soil type. Texture and depth of the soil can affect its suitability for foraging and nesting, and composition of the vegetation community is influenced by soil characteristics. Burrowing owls nest in underground burrows and require deep, friable soils. Sage and Brewer's sparrows in Washington were more abundant in deep soil communities characterized by big sagebrush than in shallow soil communities dominated by low-growing stiff sagebrush.[147] Size of the sage shrubs affects their suitability as nest sites and perhaps also determines availability of insect prey. In this same analysis, loggerhead shrikes occurred in greatest numbers, and western meadowlarks in lowest numbers, in sandy soil communities typical of the Columbia River Plain. Mourning doves were the only species that occurred at greater abundance in shallow soil communities.

Occurrence of some shrub-steppe birds may be related to the presence of other animal species. Burrowing owls in the West are closely tied to populations of fossorial mammals and the vegetation communities and soil types that support them.[38, 49] Burrowing owls may depend on badgers for nest sites in the Columbia Basin.[38] Although burrowing owls eat a variety of prey, small mammals are a key component and western populations often are associated with colonies of ground squirrels and other small, burrowing mammals. In the Columbia Basin of Washington and Oregon, small mammals, particularly

Great Basin pocket mice, comprised the majority of the biomass in pellets collected at nest sites.[40]

Brown-headed cowbirds may use different habitats for feeding and for breeding and selection of both may be driven by the presence of other species. Cowbirds likely evolved with large, grazing animals (i.e., bison) of the short-grass prairies of the Midwest, following the animals as they grazed and eating seeds and insects exposed by trampling hooves.[74] In the shrub-steppe of Oregon and Washington, cowbirds are typically associated with livestock and can be found foraging among cattle and horses in pastures and feedlots. Cowbirds are "nest parasites" who lay their eggs in the nests of other birds, and therefore require suitable populations of "host" species to reproduce successfully. In big sagebrush communities in eastern Washington, cowbirds typically parasitized sage, Brewer's, and vesper sparrows, although rates of parasitism were low compared to other communities.[148] Cowbirds also parasitize birds that nest in riparian communities within shrub-steppe,[33, 102] and their distribution likely is tied to that of suitable hosts in these communities as well.

Mammal Communities in Shrub-steppe

Species richness is typically related to the structural complexity of dominant vegetation.[18, 116] Not surprisingly, the diversity of mammals in the shrub-steppe of western North America is lower than that typical of structurally more complex habitats of the region. For example, 40 small mammal species are closely associated with forested habitats of Oregon and Washington, whereas only 20 are closely associated with shrub-steppe habitat. Ten carnivore species are closely associated with forested habitats compared to 2 in shrub-steppe. Because of low precipitation, high incident solar radiation and wide fluctuations in seasonal temperatures, the shrub-steppe is a challenging environment for homeotherms. Not surprisingly, many shrub-steppe mammals are relatively specialized for life in this arid region.

In addition to the relatively few mammals that are

Photograph 6. Pygmy rabbits are sagebrush obligates, building their burrows in deep soils beneath sagebrush plants and feeding on sagebrush foliage. Douglas County, Washington. (WDFW).

clearly associated with true shrub-steppe vegetation, numerous species may be associated with very specialized habitats that occur as minor components of shrub-steppe systems. For example, black bears, beavers, and muskrats are not considered shrub-steppe species, but all may occur in larger riparian corridors that extend from forested communities into shrub-steppe.

Small and Meso-sized Mammals

Small mammals of the shrub-steppe include ubiquitous species such as the white-footed deer mouse and several species that are relatively restricted to shrub-steppe and other arid biotic associations. Hedlund and Rogers[50] trapped Great Basin pocket mice, northern grasshopper mice, sagebrush voles, western harvest mice, and deer mice in shrub-steppe habitat of the Hanford Site of southcentral Washington. Marr et al.,[79] trapping another area of the Hanford Site, caught deer mice, western harvest mice, grasshopper mice, montane voles, Great Basin pocket mice, and northern pocket gophers. Rogers and Rickard[120] also listed vagrant shrews as occurring in shrub-steppe habitat of southcentral Washington. Although small mammal communities of the shrub-steppe commonly contain several of these species, a few species, notably Great Basin pocket mice and deer mice, numerically dominate most assemblages in eastern Washington.[36, 50, 79] Although uncommon in Washington, kangaroo rats, such as Ord's kangaroo rat and the less common chisel-toothed kangaroo rat, inhabit shrub-steppe in eastern Oregon.[19] The least chipmunk and dark kangaroo mouse also occur in some areas of shrub-steppe.[19]

Meso-sized mammals of the shrub-steppe include several lagomorphs, such as white-tailed and black-tailed jackrabbits, mountain cottontails, and pygmy rabbits. The 2 jackrabbit species are sympatric; however, white-tailed jackrabbits are generally less common and tend to be distributed in smaller, isolated populations, at least in recent years.[19, 80] Both species are prone to cyclic population growth,[44, 64, 97] and variation between high and low population levels in black-tailed jackrabbits can be dramatic.[3] When the 2 jackrabbit species occur together, white-tailed jackrabbits tend to occupy more open, grass-dominated areas and black-tailed jackrabbits predominate in areas with a strong shrub component.[19, 143] Areas occupied by jackrabbits are obvious due to their practice of clipping vegetation to make runways, which they travel repeatedly.[97] Pygmy rabbits are uncommon and typically occur only in areas dominated by tall, dense stands of Great Basin or big sagebrush, which provide preferred forage.[19, 41, 153] Pygmy rabbits also excavate burrows in which they den, and therefore also require areas with friable soil. Mountain cottontails often are associated with distinct microhabitats such as riparian areas or rocky ravines and also occur in areas developed by humans.[19] Yellow-bellied marmots are not associated with shrub-steppe associations except where rock piles occur as special, localized habitat features. Bushy-tailed wood rats, Columbian ground squirrels, Belding's ground squirrels, and a complex of smaller ground squirrels (discussed below) also occur in shrub-steppe habitats of eastern Oregon and Washington.

Carnivores of the Shrub-Steppe

Because diversity and biomass tend to decline with increasingly higher order consumers,[116] the diversity and abundance of shrub-steppe carnivores is a fraction of that of small and meso-sized mammals. Among the common carnivores are the ubiquitous coyote, the more habitat restricted badger, and the widespread long-tailed weasel. The kit fox reaches its northern range limit in arid plant associations of southeastern Oregon.[19]

Coyotes, badgers, and weasels consume a diverse array of small mammals, insects, birds, and reptiles. Carnivore densities are related not only to densities of their prey but also to complex social factors.[92, 126, 130] Coyote, weasel, and badger occupancy of shrub-steppe habitats is probably linked to vegetation indirectly via effects of vegetation on prey diversity and abundance. The bobcat, another widespread species, occurs in shrub-steppe where rock outcrops, ravines, or lava formations are available for suitable den sites.[62] Mountain lions are the largest carnivores that use shrub-steppe habitat and typically occur only where mountainous terrain and shrub-steppe vegetation are contiguous.[19, 20] Bobcats and mountain lions prefer prey at least as large as a hare, and are found only where suitable prey biomass is relatively high. Not surprisingly, bobcat and mountain lion reproductive success and spatial use of the landscape varies with cyclic changes in densities of primary prey, whether lagomorphs or ungulates.[62] Mountain lions tend to be found only where large mammal prey also are available.

Large Herbivores

Native species of large herbivores in the shrub-steppe include generalists such as the mule deer and elk, as well as more specialized ungulates such as the pronghorn antelope. Bighorn sheep are not a true shrub-steppe species and have been historically rare in arid mountain ranges of eastern Oregon and Washington, but bighorns did occur prehistorically in some mountainous areas adjacent to shrub-steppe habitats and have been reintroduced to several locations in Oregon[101] and Washington.[151] All of these large herbivores require free water and therefore are not found in extensive tracts of shrub-steppe that lack surface water such as springs or ponds. Large mammal ecology in the shrub-steppe is discussed in greater detail in a later section.

Organizing Principles: Mammals

The determinants of habitat suitability for mammals in the shrub-steppe undoubtedly varies across species. For example, floristics may be important for some species because of diet specialization. For some species, however, the vegetative structure of the association is probably more important than the specific assemblage of plant species that comprise the association.

Secondary and tertiary consumers are usually linked to vegetation associations indirectly via habitat affinities of their prey species. But because most predators in the shrub-steppe will consume a variety of prey organisms,[62] their generalized affinities for plant communities reflect the diversity of habitats used by their smaller mammalian prey. However, as noted, several shrub-steppe carnivores directly select structural and geological habitat features because of their importance to other life history needs (e.g., den availability). Accordingly, these important structural habitat features seem to be strong determinants of habitat suitability for some shrub-steppe carnivores.

Occurrence and density of primary consumers would normally be expected to be directly linked to community floristics. However, generalists such as the larger herbivores seem to have broad tolerances for plant associations as long as acceptable forage biomass is above critical limits that relate to foraging efficiency thresholds. Primary consumers with specialized diets, such as pygmy rabbits, may be very selective of plant associations.[41] Black-tailed jackrabbits are seasonally selective of forage type, but will consume a variety of plant species[143] and therefore display more generalized affinities for plant associations than pygmy rabbits. Black-tailed jackrabbits will generally not consume cheatgrass, and therefore largely avoids pure cheatgrass swards.[143] Where cheatgrass invasion is occurring, jackrabbits foraging on native bunchgrasses can convey a competitive advantage to cheatgrass and facilitate further conversion towards an annual grassland.[132]

Mammal species vary considerably in their affinity to associations with a strong sagebrush component. Pygmy rabbits are strongly associated with sagebrush, rarely occurring where sagebrush is a minor component or lacking from the plant community. Sagebrush voles have a strong affinity for sagebrush but occur in stands lacking a sagebrush overstory if grass understories are thick enough to provide cover.[19] Pronghorns are the only large herbivore in the shrub-steppe that frequently forage on sagebrush. Pronghorns often show an affinity for sagebrush[81, 135] and are most successful where sagebrush species are available for winter forage, although they can occupy areas without sagebrush if other acceptable forage is available.[125]

The Role of Ground Squirrels in Shrub-Steppe Communities

In pre-European times, shrub-steppe habitats in eastern Oregon and Washington and adjacent states were

Photograph 7. Badgers feed extensively on ground squirrels and are ecologically linked to another ground squirrel predator, the burrowing owl, that uses badger dens for nest sites. Morrow County, Oregon. (G. Green, Foster Wheeler Environmental Corporation).

inhabited by the smaller ground squirrels of the genus *Spermophilus*. Like other ground squirrels they were sedentary, and rivers and other barriers were significant obstacles to gene flow.[24, 37] Consequently, they diverged into a complex of 7 species, each with a separate range but all with similar ecological roles. Larger species of this genus, such as Columbian and Belding's ground squirrels, also inhabited shrub-steppe habitats, but were generally in deeper soils, and frequently in more mesic habitats, and are not considered here. Each species occupied a separate geographic area. In Washington, Townsend's ground squirrel lived between the Cascade Mountains and the Yakima River,[47, 52] whereas a similar, but chromosomally distinct form occurred between the Yakima and Columbia Rivers.[96] The Washington ground squirrel formerly occupied the Columbia Basin between the Columbia River and the range of the Columbian ground squirrel to the east, and extended south into the Columbia Basin of Oregon.[4] Merriam's ground squirrel (*S. canus canus*) occupied the shrub-steppe communities in central Oregon, except in Malheur County where it was replaced by the similar *S. c. vigilis*.[47, 55, 149] These small ground squirrels remain poorly studied, but available information indicates that they prefer deep, well-drained sandy to loamy soils[4, 55] but otherwise occur in a variety of native vegetation types.[129, 149, 167]

There is little information available on the abundance of ground squirrels in pre-European times. It is likely that their distribution was patchy, with subpopulations acting as sources and sinks. Their distribution may have been driven by foraging needs, as small herbivores have a low digestive capacity and therefore depend on increasingly higher quality forage as body size decreases.[25] Ground squirrels depend upon high quality forage and an abundant supply of seeds in order to store the fat necessary to survive 8 months of hibernation. High quality forage in turn depends on wildfire and disturbance providing young, succulent, productive growth. In presettlement times, wildfires were patchy and post-burn succession generally maintained a mosaic of patches in various successional stages[165] that would have allowed ground squirrels to move about the landscape monopolizing new growth as it appeared.

Ground squirrels are important as a prey base for many of the predators in their ecosystems. In Idaho, Paiute ground squirrels are a keystone species, providing a critical food source for prairie falcons and an important prey source for red-tailed and ferruginous hawks, badgers, western rattlesnakes, and gopher snakes. They also are prey for long-tailed weasels, ravens, and others.[167]

As burrowers, ground squirrels are important in mixing soils.[1, 53, 137] For example, arctic ground squirrels were documented to move 19 tons of soil/acre/yr.[112] They are important in soil aeration,[57] and fertilize the soil with their feces and urine,[42, 112, 136, 137, 141] resulting in significantly greener vegetation in the vicinity of their burrows.[112] The burrows significantly increase water infiltration into the soil, which increases plant productivity. In an Idaho study, ground squirrel burrows increased productivity of bunchgrasses by about 20%.[67, 68] In shrub-steppe environments, ground squirrels are important to the very plants they eat.

Badgers feed extensively on ground squirrels, and areas with high ground squirrel densities usually have high densities of badger digs.[167] These badger digs also increase water infiltration and aeration of the soil[53] and provide nest sites and shelter for a variety of wildlife. Thus, the presence of ground squirrels has positive effects on other animal species, as well as on soil and vegetation.

Invasion of exotic annuals has changed both the fire regimes and successional patterns in shrub-steppe habitats. Range fires burn much hotter in exotic annual infested rangelands, killing shrubs and allowing exotic annuals to out-compete native species. In many areas, native communities of shrubs, bunch grasses, and forbs have been replaced by annual grasses (especially cheatgrass) and dicots (particularly mustards). Ground squirrels will eat exotic plants,[166] but the productivity of these invasive plants varies annually with precipitation, providing an unstable food base for ground squirrels. In southwestern Idaho, ground squirrel populations are unstable in areas dominated by exotic annuals and are prone to extinction.[167] This instability is compounded by the large amounts of indigestible silica in cheatgrass and especially in medusahead (*Taeniatherium asperum*) seeds, making them a poor food source regardless of the quantity.[123]

Fire also affects ground squirrels by reducing shrub cover. Although ground squirrels can survive in areas without shrubs as long as the herbaceous layer remains unchanged,[106] in the long term their density decreases when shrub cover drops below 11%, probably due to loss of protective cover, moisture changes, or other factors. Ground squirrel density also drops when shrub cover exceeds 20%, apparently due to reduced productivity in the herbaceous layer.[59]

Agricultural conversion has had an even more drastic effect than exotic annuals on ground squirrel populations. In the Columbia Basin, agricultural conversion has reduced the Washington ground squirrel to a handful of isolated populations with low probability of long-term survival[5] (E. Yensen, Albertson College, pers. obs.). Agricultural conversion likely has had similar effects on the Townsend's ground squirrel.[45] No systematic survey has been done on Merriam's ground squirrel, but it seems to be extirpated over much of its range (E. Yensen, Albertson College, unpubl. data). In this case, vegetation changes favoring the larger Belding's ground squirrel may be responsible.

Despite their importance in shrub-steppe ecosystems, ground squirrels generally have not been appreciated by humans. Ground squirrels do invade agricultural fields, eat alfalfa, grain and other produce, and dig holes in irrigation ditches. In return, ground squirrels have been poisoned, trapped, used for target practice, and reviled for over a century with little understanding of, or interest in, their ecological roles. Ironically, their demise has unfortunate ecological consequences for shrub-steppe ecosystems that may portend economic losses as well. The decline of ground squirrels in some areas implies reduced productivity of the rangeland and, thus, lower value to grazing cattle.

Ecology of Large Herbivores in Shrub-Steppe

Living in warm, arid climates presents unique challenges to large mammals in terms of water conservation and maintenance of homeothermy.[94] As previously noted, the availability of drinking water is an important component of suitable habitat for large herbivores in the shrub-steppe.[30, 86] Large herbivores also have relatively high forage intake demands; this intake rate is strongly correlated with the biomass of acceptable forage.[15, 155] Accordingly, minimum forage biomass would also be expected to limit the types of arid environments that large herbivores can effectively exploit.[84] Among ruminant herbivores, smaller animals tend to have lower intake demands, but because of morphological and digestive constraints, require higher quality forage relative to larger ruminants.[48, 58] Smaller ruminants can afford the costs of selective herbivory typically required to obtain high quality diets. Larger ruminants have high intake demands and must maintain high intake rates in order to meet these demands within the amount of time they invest in foraging. Therefore, it has generally been theorized that smaller ruminants are typically forage-quality limited, in contrast with larger ruminants that are expected to be constrained more by forage quantity.[58]

Because of low primary productivity in the shrub-steppe relative to more mesic habitats of the region, it would seem that the shrub-steppe would be a more suitable environment for small ruminant herbivores. However, many shrub-steppe associations are dominated by bunchgrasses and nonpalatable shrubs such as big sagebrush. Because of high silica content and relatively thick cell walls, grasses are not easily digested by small herbivores.

Contemporary shrub-steppe habitats are exploited by 3 large ruminant herbivores: pronghorn antelope, mule deer, and elk. Bison occurred historically in the shrub-steppe of southeast Oregon,[145] and there is archaeological evidence that they once occurred in the shrub-steppe of the Columbia Plateau.[103] Of the 3 contemporary shrub-steppe ruminants, only the pronghorn is a steppe specialist. Pronghorns are widespread in steppe-like habitats of western and midwestern North America,[135] but the shrub-steppe of eastern Washington and Oregon represents the fringe of their distribution. The pronghorn is well suited for occupation of open steppe-like country.[12] As previously noted, pronghorns have a pronounced affinity for plant associations with a strong sagebrush component,[135] and their distribution conforms relatively well to the distribution of areas occupied by *Artemisia* species. Pronghorns are the only ruminant herbivore in North America that forages heavily on sagebrush, a relatively unpalatable and poorly digested shrub. Pronghorns also seasonally consume a diverse array of forbs, although they generally consume little grass.[135, 168, 169] Pronghorns are relatively common in eastern Oregon but have been extirpated in eastern Washington.[135] The factors relating to their disappearance in Washington are not well understood, but competition with livestock for palatable forb species and the generally large-scale conversion of sagebrush-dominated rangelands to other uses are implicated.

Mule deer are generalists and occur in a relatively broad array of habitats of western North America from densely forested coastal areas to arid deserts of the southwest.[150] Mule deer are relatively common in the shrub-steppe of eastern Oregon and Washington, although their distribution is far from uniform.[70] On the Hanford Site of southcentral Washington, mule deer densities are highest near riparian areas such as the Columbia River shoreline, waste ponds, and perennial springs.[29, 144] Individual Hanford Site mule deer also make disproportionate use of areas near water.[30] Carson and Peek[13] found that mule

Photograph 8. The Rattlesnake Hills elk herd resides entirely in shrub-steppe habitat in eastern Washington. Elsewhere, shrub-steppe serves as winter habitat for some migratory populations of elk. Arid lands ecology reserve, Benton County, Washington. (S. McCorquodale, Yakama Nation).

1: The Rattlesnake Hills Elk Herd

Until recently, shrub-steppe communities were not thought to be year-round habitat for elk.[10, 95] Although elk were known to use shrub-steppe habitat contiguous with forested habitats during severe winters, elk were generally thought to be precluded from year-round occupation of shrub-steppe due to low forage productivity and the inability to tolerate high thermal loads. In 1972, however, a small group of elk colonized the shrub-steppe habitat of the Hanford Site in southcentral Washington.[114] This group, commonly known as the Rattlesnake Hills elk herd, has been studied extensively. The original colonizing group of 4-7 elk grew rapidly because of high reproductive output and extremely high first-year survival.[88] Despite the fact that shrub-steppe had been considered unsuitable year-round habitat for elk, individual fitness, indexed by reproductive success in females and age-specific antler growth in males, has been exceptional in the Rattlesnake Hills elk population.[86, 87, 88] Currently this population numbers >700 individuals (B. Tiller, Battelle Memorial Institute unpubl. data).

Although primary production is low in shrub-steppe relative to forested environments, energetic modeling has indicated that shrub-steppe communities may rival or exceed more productive environments in terms of forage energy availability because all production occurs in the herbaceous and shrub layers.[84] Grass-dominated associations typical of eastern Oregon and Washington shrub-steppe seem to be well suited to exploitation by elk; forage biomass typical of many shrub-steppe communities is above that considered minimum for elk.[84, 155] In grass-dominated communities, elk seem to be competitively superior to mule deer.

Elk in the shrub-steppe use a broad array of plant associations. Telemetry data indicate that stands with a sagebrush overstory are used heavily by elk for bedding, particularly during the heat of summer.[89] Foraging elk, however, have shown an affinity for grass swards where historical fires have eliminated sagebrush.[83] Elk diets are diverse in the shrub-steppe, with grasses dominating spring and winter diets, and forbs dominating summer and fall diets.[82, 85]

The Rattlesnake Hills elk population has shown little dependence on shrub forage, even during winter, but the predominant shrubs in the herd range, sagebrush, greasewood (*Sarcobatus vermiculatus*), and spiny hopsage (*Atriplex spinosa*) are not considered good elk forage.

Other factors have contributed to the success of the Rattlesnake Hills elk population. The portion of the Hanford Site occupied by elk is topographically diverse,[82] and access restrictions associated with Hanford Site security are believed to limit harassment to levels tolerable to elk.[89] The presence of perennial springs in the southern portion of the Hanford Site are probably essential to the thermoregulatory strategy of elk in high thermal loading environments.[89, 105]

---

deer in northcentral Washington preferred riparian cover types and areas with some topographic diversity as opposed to flat expanses of shrub-steppe vegetation. Mule deer in the shrub-steppe consume diverse diets typical of a generalist herbivore, but forbs and the early growth of shrubs are particularly important,[144] consistent with energetic predictions. Large areas dominated by bunchgrasses and lacking palatable shrubs do not seem to be vegetation types exploited effectively by mule deer.[70]

Although mule deer and elk may be residents of the shrub-steppe region (see Box 1), some migratory populations also use shrub-steppe exclusively as winter habitat. Where shrub-steppe occurs adjacent to low-elevation xeric forest, suitable arrays of forage species and biomass and favorable climatological features make shrub-steppe rangeland highly valuable winter range, especially during severe winters. In eastern Oregon and Washington, large numbers of migratory deer and elk display this seasonally intense use of shrub-steppe winter range, although during the warmer months they use higher elevation forests.

Reptile and Amphibian Communities in Shrub-Steppe

Relative to other classes of terrestrial vertebrates, diversity of amphibians in shrub-steppe habitat of Oregon and Washington is low (10 of the 32 species occurring in these two states). Only 3 of 21 species of salamanders occur in shrub-steppe habitat: long-toed salamander, tiger salamander, and roughskin newt. Seven of 11 native species of anurans occur in shrub-steppe. Great Basin spadefoot toad, western toad, and Woodhouse's toad are the anuran species most likely to be found in shrub-steppe away from standing water. In southeastern Idaho, Great Basin spadefoot toads may be found in shrub-steppe as far as 5 km from standing water (S. Cooper Doering and C. Peterson, Idaho State University unpubl. data). Ranid frogs are most closely associated with wetlands. Columbia spotted frogs in southwestern Idaho, for example, spend most of their time within several meters of water (T. Carrigan, Bureau of Land Management unpubl. data).

The occurrence and distribution of some anuran species has changed within the last 30 years. Northern leopard frogs have disappeared from many sites in Oregon and Washington where they occurred previously.[131] Bullfrogs have been introduced into Oregon and Washington and pose a threat to native species such as northern leopard frogs, spotted frogs, and western pond turtles.[72, 131]

In contrast to amphibians, reptile diversity in shrub-steppe habitat is relatively high (21 of 28 native species in Oregon and Washington). Lizards are the group of reptiles most associated with shrub-steppe (9 of 11 species in Oregon and Washington). No lizard species occurs exclusively in shrub-steppe, but 3 species (Mojave black-collared lizard, long-nose leopard lizard, and desert horned lizard) occur only in shrub-steppe, dwarf shrub-steppe, and desert playa/salt scrub shrublands.

Ten of the 15 (67%) snake species in Oregon and Washington occur in shrub-steppe habitat. The ground snake occurs only in shrub-steppe, dwarf shrub-steppe, and desert playa/salt scrub shrublands. Striped whipsnakes occupy these three habitats plus juniper and mountain mahogany woodlands, whereas night snakes are found in the same habitats as striped whipsnakes plus eastside canyon shrublands. Three species (racer, gopher snake, and western rattlesnake) occur in a wide variety of habitats, including shrub-steppe. Three more species (rubber boa, western terrestrial garter snake, and common garter snake) also occur in a wide variety of habitats, including shrub-steppe, especially near water.

Although both species of freshwater turtles (painted turtle and western pond turtle) occur more frequently in other habitats, they use shrub-steppe if near permanent water (marshes, slow rivers and streams, ponds, or lakes with abundant aquatic vegetation). These turtles venture onto land to bask or to disperse, and they may lay their eggs in nests up to 150 and 800 m, respectively, away from water.[98, 134]

Although species richness of amphibians and reptiles is lower than that of birds and mammals in Oregon and Washington shrub-steppe, amphibians and reptiles can be very important ecologically. Because their long-term conversion efficiencies are many times higher than those of birds and mammals, they can contribute disproportionately to biomass production and make large amounts of energy available to other trophic levels.[110] For example, Turner at al.[140] found that the annual biomass produced by side-blotched lizards in the Nevada desert was equal to or greater than that of birds and mammals in desert and grassland habitats.

General Organizing Principles:
Reptiles and Amphibians

The presence of water is an essential habitat feature for the amphibians occurring in Oregon and Washington shrub-steppe. All of these amphibians breed in lentic environments (e.g., marshes, pools, ponds, side-channels, or oxbows). Consequently, the presence of standing or slow-moving water for at least the time required for eggs to hatch and larvae to complete metamorphosis is required for successful reproduction. However, because many of these species can live for considerable lengths of time (>10 years in some species), suitable breeding habitat does not have to be present every year. Temperature variation among shrub-steppe communities does not seem to explain differences in the occurrence of amphibian species within Oregon and Washington.

Because reptiles are ectothermic, thermal conditions play a key role in determining the occurrence, distribution, and numbers of reptiles that will be found in shrub-steppe habitats. In general, reptile species richness decreases with increasing latitude and altitude.[60] The number of reptile species decreases by over 40% between southern Oregon and northern Washington. For example, Mojave black-collared lizards and ground snakes do not occur in Washington, and side-blotched lizards and striped whipsnakes do not occur north of central Washington. Within a landscape, the distribution and abundance of reptiles will vary with topography. For example, snake dens are usually located on south facing slopes[43, 61] (S. Cooper Doering and C. R. Peterson, Idaho State University unpubl. data).

For many species of reptiles, it is likely that the time available for embryo development at appropriate temperatures is the condition that limits distribution. For example, biophysical analyses indicate that adult and juvenile desert iguanas could survive in shrub-steppe habitat in Washington, but that soil temperatures are too low to allow successful incubation of eggs.[109] Gravid rubber boas and western rattlesnakes have been observed at den sites in the late fall of cool years in southeastern Idaho, indicating that summer temperatures were not sufficiently high to allow the embryos to develop (M. E. Dorcas, Davidson College, pers. comm., J. Lee, Idaho State University pers. comm.). These field observations, combined with laboratory studies of the thermal dependence of embryo development[14] and observations of the inability of embryos to survive simulated hibernation (M. Dorcas, Davidson College pers. comm.), suggest that the amount of time at suitable temperatures available for embryo development may play an important role in setting the distributional limits of snakes. Indeed, the evolution of viviparity in reptiles is generally viewed as an adaptation to cool conditions.[111]

Vegetation structure and floristics seem to explain less of the variation in occurrence and distribution of reptiles than in birds and mammals. In developing GIS models for predicting reptile distributions for the Idaho Gap Analysis project, Butterfield et al.[11] found that factors other than vegetation appear to limit the distribution of many reptiles. These factors may include temperature, moisture, and the special habitat features previously described. Nevertheless, vegetation does influence reptiles in several ways. Vegetation structure may influence reptiles directly; for example, sagebrush lizards in northcentral Oregon use habitats with tall shrubs (thermal cover and protection from avian predators) and sparse ground cover (ease in detecting and pursuing prey) (G. Green, Foster Wheeler Environmental Corporation, unpubl. data). Because most reptiles inhabiting shrub-steppe are carnivorous, differences in vegetation probably influence reptile occurrence indirectly via the habitat affinities of their prey.

Management Issues in Shrub-Steppe

In this section we consider some of the management issues that have had (and likely will continue to have) considerable impact on shrub-steppe communities at a regional scale. There are certainly other threats to shrub-steppe communities, particularly at the local scale (e.g., real estate development and inundation by water projects), but we limit our discussion here to more widespread issues.

Conversion to Agriculture

There is little doubt that the conversion of native plant communities to agricultural uses has had, and continues to have, profound effects on shrub-steppe habitats in the Columbia Basin. Beginning with the westward migration of Euro-Americans in the mid-1800s that brought farmers into the deserts of Oregon and Washington, and accelerated by the damming of the Columbia River that made large-scale irrigation possible, about 14.8 million acres (6 million ha) of shrub-steppe have been converted to wheat fields, row crops, and orchards.[113] Agricultural development has been most pronounced in Washington where >50% of the land originally in shrub-steppe has been converted[27] (Figure 1). This large-scale displacement of one habitat type for another has substantially reduced the area available to native shrub-steppe wildlife. Moreover, the addition of new, human-related habitats (agricultural and rural development) has elevated the food base for some predators (e.g., magpies and gulls) and likely their populations as well, with unknown impacts on shrub-steppe wildlife. The addition of cattle feedlots, pastures, and lawns to the landscape has enhanced the suitability of the area for brown-headed cowbirds, a nest parasite that lays its eggs in the nests of other birds and thereby depresses the host bird's reproductive success.

Agricultural conversion has not occurred randomly across the landscape, but instead has focused on the most arable, deep soil communities. This has resulted in a disproportionate loss of these communities and an increase in the proportion of shallow soil shrub-steppe habitats on the landscape (Figure 2). Some species of shrub-steppe wildlife, such as badgers, ground squirrels, and burrowing owls, depend on deep soil communities. The pygmy rabbit, listed as endangered in Washington, is found only in deep, loamy soil sagebrush stands. Furthermore, some shrub-steppe passerines occur in greater abundance in loamy soil communities than in other soil types.[147] Conversion of deep soil shrub-steppe communities to irrigated agriculture will likely continue in the foreseeable future, making this one of our most endangered arid land communities.

Although loss of native plant communities should be avoided, some habitats associated with agricultural development can have values for wildlife. Wetlands associated with agricultural development provide breeding and feeding areas for species not typically associated with shrub-steppe. Wetlands created as part of numerous irrigation projects provide habitat for various nesting waterfowl and marshland birds, as well as amphibians and aquatic mammals. These areas also serve as migration stop-over sites for waterfowl. Wetlands with a woodland component provide stop-over sites for passerines that historically used naturally occurring wooded riparian habitats, a resource that has been greatly depleted.

Farm fields enrolled in the Conservation Reserve Program (CRP) can have considerable value to shrub-steppe birds. These fields are taken out of production for ≥10 years and planted to tame grasses such as crested wheatgrass, providing nesting habitat for greater sage-grouse (M. Schroeder, WDFW pers. comm.) and grasshopper sparrows[107] on lands that offered few such values when under cultivation. As sagebrush colonizes CRP fields via seeds from adjacent shrub-steppe, habitat value increases because of added structure and food for sage-dependent species. The value of CRP fields to shrub-steppe wildlife could be increased by planting them with native vegetation and extending the period of enrollment.

Habitat Fragmentation

The pattern of agricultural conversion within the shrub-steppe of eastern Washington and northcentral Oregon has resulted in a highly fragmented landscape (Figure 1). Where once native grasslands and shrublands stretched unbroken for thousands of square miles, there exists now only fragments of native habitats in a matrix of agricultural fields. This breakup of formerly contiguous habitats can have detrimental effects on species occurrence and population dynamics. Much of the research documenting fragmentation effects has examined avian communities in forested ecosystems, although some recent work has focused on grasslands and shrublands.[6, 63, 147]

Some forest bird species are area sensitive and will not inhabit habitat patches below a minimum size.[34, 117] Extensive surveys in Washington suggest that sage sparrows are most likely to occur in blocks of shrub-steppe >2,470 acres (1,000 ha) and that male sage sparrows found singing in small fragments are unlikely to maintain a territory or attract a mate (WDFW unpubl. data). Numerous studies have documented greater rates of nest predation[28, 152, 163] and nest parasitism[9, 118] in fragmented landscapes. Elevated rates of nest predation and parasitism may result from an increase in the number of predators and brown-headed cowbirds in fragmented landscapes and an increase in habitat edge. In Washington, 3 shrub-steppe birds (Brewer's sparrow, lark sparrow, and sage thrasher) showed evidence of lower nesting success in fragmented shrub-steppe compared to continuous, unbroken tracts (WDFW unpubl. data). Cameras monitoring artificial nests baited with quail eggs revealed that black-billed magpies and common ravens likely were responsible, at least partly, for increased predation on nests in fragmented landscapes (WDFW unpubl. data). As remnant habitat becomes smaller and more fragmented, it is under greater influence of the surrounding landscape[157] and more susceptible to external influences, be they predators, nest parasites, potential competitors, or the wind-blown seeds of exotic weeds.

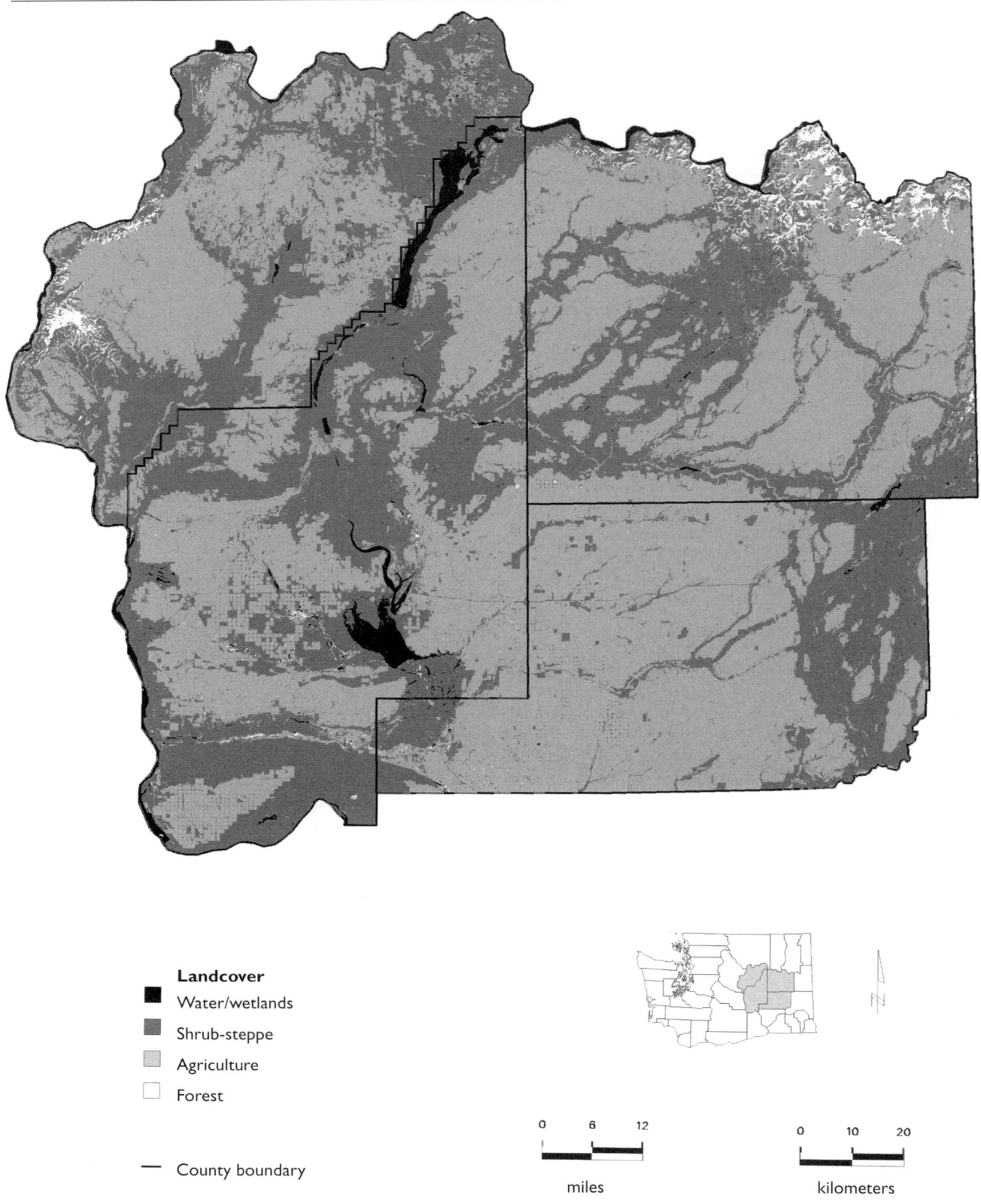

Landcover

- ■ Water/wetlands
- ■ Shrub-steppe
- ■ Agriculture
- □ Forest

— County boundary

miles

kilometers

Figure 1. Current landcover of four counties in eastern Washington, illustrating the degree of fragmentation in this formerly shrub-steppe-dominated landscape. Landcover classes were derived from Landsat Thematic Mapper data using multi-temporal analysis (scene dates: May 1993 and August 1994). Counties illustrated are (clockwise from the bottom right) Adams, Grant, Douglas, and Lincoln.

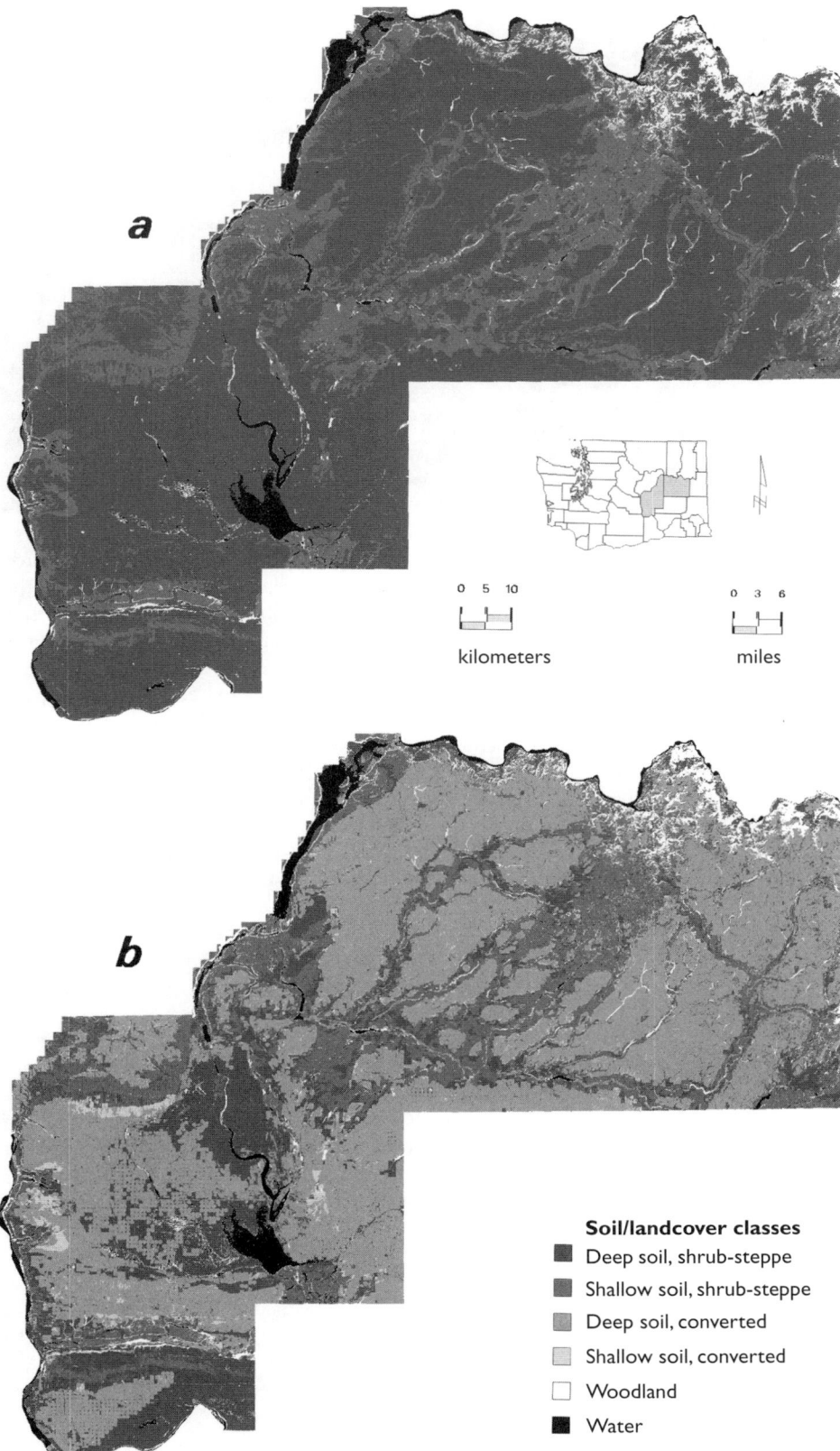

Figure 2. (a) Historical distribution of deep soil and shallow soil shrub-steppe communities in Grant and Lincoln Counties, Washington. (b) Current distribution of deep soil and shallow soil shrub-steppe communities in Grant and Lincoln Counties, illustrating the extensive conversion of deep-soil shrub-steppe (primarily to agricultural fields) and minimal conversion of shallow-soil shrubsteppe.

Soil/landcover classes
- Deep soil, shrub-steppe
- Shallow soil, shrub-steppe
- Deep soil, converted
- Shallow soil, converted
- Woodland
- Water

Photograph 9. Shrub-steppe communities of eastern Washington and northcentral Oregon have been fragmented by agricultural conversion. Douglas County, Washington. (M. Vander Haegen, WDFW).

Fragmentation of the shrub-steppe landscape very likely has disrupted the dynamics of dispersal and immigration that allows populations to persist over large areas. Stochastic events may cause the extirpation of a species from one habitat patch, necessitating recruitment from nearby patches to reestablish a population. Highly fragmented landscapes have lower connectivity, meaning that dispersing individuals must cross unfavorable lands (in this case agricultural fields or developed lands) to move from one habitat patch to another. In theory, the smaller the patch and the more distant other population sources, the lower the probability that recolonization will occur.[75, 164] Species with small home ranges and limited dispersal capabilities, such as many small mammals and reptiles, are most likely to be affected (see Box 2). For species that normally occur at relatively low densities, such as the northern grasshopper mouse and pygmy rabbit, small breeding assemblages could become genetically isolated and vulnerable to extirpation.[65, 66] Likewise, for populations that are prone to dramatic cyclic fluctuations in population size, such as lagomorphs, fragmentation may increase the probabilities of local extinctions associated with decline phases in isolated populations. Although research has addressed the consequences of genetic isolation and the probability of persistence for small populations in fragmented forest communities,[71, 93] little work has been done in shrub-steppe.

Livestock Grazing

With the rise of the Cascade Range in the early Pliocene, maritime influences east of the range diminished and the Intermountain West became drier, with rainfall patterns for the most part centered in the fall and winter. Grasses capable of estivating during the dry summer months, such as *Agropyron, Poa,* and *Festuca,* dominated the landscape.[23, 77] In turn, the lack of moisture in the plants during the summer months, coupled with poor distribution of drinking water, may have imposed severe constraints on the establishment of significant numbers of large

herbivores, especially bison.[77] For whatever reason, herds of native ungulates never reached the numbers in the shrub-steppe regions of Oregon and Washington that they did east of the Rocky Mountains. Consequently, the *Agropyron, Poa,* and *Festuca* dominated grasslands of Oregon and Washington may have been poorly adapted to withstand the grazing pressures of European livestock introduced over the last 200-300 years.

The legacy of livestock grazing in the shrub-steppe regions of Oregon and Washington began about 1700 when the Shoshone brought horses into southeastern Oregon from the Spanish missionaries at Santa Fe.[46] By 1730, horses had reached the Columbia Basin, where the Nez Perce and Cayuse built herds into the thousands by 1800.[35] The impact of these horses on the local grassland ecology is unrecorded. Cattle grazing as an industry did not begin east of the Cascades until the 1860s, but quickly expanded, reaching its zenith in the late 1870s. At about the same time cattle began competing with hundreds of thousands of sheep and the thousands of horses needed for cattle raising, plus thousands more unattended Indian ponies called "cayuses." By 1885 the range was showing signs of deterioration. Farming in the latter part of the century not only fueled greater competition on less and less range, but it brought with it exotic seed contaminants such as cheatgrass, Jim Hill mustard (*Sysimbrium altissimum*), and Russian thistle (*Salsola kali*), that facilitated further deterioration of the range. As a result of public outcry about poor range conditions, the Federal government finally gained control of all unclaimed rangelands with the Taylor Grazing Act in 1934.

Today, grazing management is dictated by the science of range management defined by Stoddart et al.[133] as optimizing the returns from rangelands in those combinations most desired by and suitable to society through the manipulation of range ecosystems. Manipulation in the Intermountain West mostly has involved reseeding of deteriorated rangelands with non-native grasses (largely because seed sources for many

2: Spatial Scale

The scale at which different species use a landscape varies widely, as does our knowledge of this component of wildlife ecology. Sage and sharp-tailed grouse, both resident species, may cover a considerable area over the course of a year as they fill their various needs for breeding, food, and cover. A female sharp-tailed grouse must visit a male on his lek in the spring, then locate suitable nesting cover to lay and incubate a clutch, rear her young in suitable brood cover, perhaps move to another area as she tracks changing food resources in the fall, and then complete the year in riparian cover where hardwood trees provide buds as a source of winter food. Her annual home range may cover 1,000 acres (405 ha) and include a multitude of vegetation communities.[16] A female greater sage-grouse may include >10 miles^2 (26 km^2) in her breeding season home range alone.[124] In contrast, a Brewer's sparrow arrives in the shrub-steppe after having spent the winter in South America, establishes a <2.5 acre (1 ha) territory in a big sagebrush stand[160] and may spend the entire breeding season nesting and rearing young on this small territory. Fragmentation and other changes in the landscape likely will affect these two species quite differently. However, our knowledge of the habitat needs of most shrub-steppe passerines is meager and is focused primarily on the breeding territory. There may be other needs, such as critical premigratory habitats or post-fledging habitats, that we are not aware of and that would expand the landscape use of species like Brewer's sparrows.

Large mammals have relatively large home ranges, and in shrub-steppe they may require access to large tracts of habitat to be successful.[29, 84, 86, 169] Large mammals such as elk have remarkable dispersal abilities, however, and thus are less prone than smaller mammals to becoming genetically isolated by fragmentation unless the distance between patches becomes very large.

Meso-sized and small mammals have smaller home ranges, and therefore patch scale has different implications relative to large mammals. Small mammals can, in theory, continue to exploit small patches of shrub-steppe.[64] However, because of small home range sizes and limited dispersal abilities, fragmentation of shrub-steppe habitat may effectively isolate many populations of meso-sized and small mammals, with undetermined consequences.

Although reptiles generally respond to their environments on a finer spatial scale than birds or most mammals, a wide range of variation exists in the size of areas used by different species. Territorial, sit-and-wait predators, such as side-blotched lizards, move on a scale of feet and have home ranges of <5,000 feet^2 (500 m^2).[139] In contrast, active, widely foraging species, such as the western whiptail lizard, have home ranges several times larger.[90] Some species of snakes that use communal overwintering sites may seasonally migrate several miles in one direction and have home ranges of hundreds of acres (e.g., western rattlesnakes).[14] Species with smaller spatial requirements may be better able to persist for short periods of time in fragmented habitats; however, it is unclear whether their long-term persistence will be lesser or greater than those species that require large home ranges but are better able to recolonize areas where populations have gone locally extinct.

---

native grasses were simply not available), and establishing seasonal grazing regimes designed to prevent further deterioration of rangeland.[51, 133] However, these present-day systems are still designed to maximize livestock production, often to the detriment of shrub-steppe wildlife.

Do present day grazing practices benefit or adversely affect shrub-steppe wildlife? The answer is mixed. Some wildlife species, such as long-billed curlews[104] and burrowing owls[38] may actually benefit from reduced vegetational structure (grazed perennials or low-statured annuals), and Great Basin pocket mice will attain high population numbers in pure cheatgrass stands.[127] However, ground-nesting birds and small mammals that require protective cover from vegetation may not benefit, especially if they become more susceptible to predation.

The greatest impact from grazing, however, is probably the perpetuation of the weed legacy from livestock trampling of the soil. The friable soils of the shrub-steppe zone, especially in the drier areas, are held together by layers of cryptogamic mosses and lichens. This cryptogamic crust can prevent establishment of annual weeds and provide a moisture cap that reduces soil evaporation. However, the trampling action of livestock, especially horses and cattle, can degrade these layers and provide seed beds for cheatgrass and weedy forbs. Dense stands of cheatgrass not only outcompete native bunchgrasses (especially for moisture in the early stages of growth), but are also susceptible to hot wildfires that can virtually eliminate sagebrush.[22] Consequently, whereas the proximate effect of livestock grazing on wildlife may be the removal of grass and forb biomass important as forage and cover to many wildlife species, the ultimate effect may be perpetuation of weedy annuals that out-compete native plants that local wildlife have adapted to use.

Photograph 10. Grazing by livestock has profoundly influenced the vegetation in many shrub-steppe communities. Grand Coulee, Douglas County, Washington. (M. Vander Haegen, WDFW).

The Fire/Cheatgrass Cycle

Another human-caused agent of change that threatens to degrade shrub-steppe habitats, related in part to agricultural practices and to livestock grazing in particular, is the conversion of extensive areas to simplified annual grasslands. Cheatgrass is an exotic annual grass that was introduced to the Intermountain West sometime in the late 1800s, probably as an agricultural pest.[76] Overgrazing of native bunchgrass communities by livestock led to deterioration of the range and opened the door for the widespread invasion of this exotic annual grass.[76] Each of the arid and semi-arid grassland and shrubland habitat types in eastern Oregon and Washington is susceptible to cheatgrass invasion.[113] Characteristics of cheatgrass's life history and physical structure, coupled with a native flora ill-equipped to compete with this new invader, have allowed it to change the composition of shrub-steppe communities and even alter ecosystem processes.[21]

Cheatgrass is a winter annual that goes to seed early in the year and generally desiccates in spring or early summer. A native of the Mediterranean region, it is adapted to the climate of eastern Oregon and Washington and, unlike the native bunch grasses, can thrive under heavy livestock grazing. Cheatgrass rapidly colonizes bare soil and moves readily into disturbed sites. Intact communities with native bunchgrasses and a healthy cryptogamic crust can keep the invader in check; however, heavy grazing of native grasses and mechanical breakdown of the cryptogamic crust through trampling can provide a window for cheatgrass to spread and exert dominance. Wildfire also can provide a window of opportunity by killing shrubs and making valuable soil moisture available to cheatgrass for germination and growth. Cheatgrass recovers quickly following wildfires and can out-compete native grasses.[91]

The most extreme changes that cheatgrass has caused in the West have resulted from repeated wildfires and the response of cheatgrass to post-fire conditions.[21] Native bunchgrasses generally grow sparsely, with forbs and bare soil or cryptograms between plants. This discontinuous fuel layer does not carry a fire well, and wildfires in these communities typically burn patches, creating a mosaic of burned and unburned areas.[154] Cheatgrass, in contrast, forms continuous stands that desiccate early in the season and carry fires well[154] (Photograph 11). As a result, burns in cheatgrass are larger and more frequent. As cheatgrass biomass increases, the ability of the community to carry fire also increases, resulting in a positive-feedback loop where fire promotes cheatgrass dominance, leading to more frequent fires.[21] On Idaho's Snake River plain, sagebrush communities evolved with a fire-return interval of 35-100 years; following invasion by cheatgrass, fire-return intervals in some shrub-steppe communities are now as low as 3-5 years.[154]

Increased fire frequency in steppe and shrub-steppe vegetation leads to a vegetation community with lower species richness.[154] On frequently burned areas of Idaho's Snake River plain, almost all of the vegetation was introduced annuals—primarily cheatgrass.[154] Vegetation life forms react differently and predictably to the fire/cheatgrass cycle; essentially, annuals increase dramatically and all other plants decrease.[154] Wildfire kills big sagebrush, and because most species of sagebrush do not resprout from root crowns after fires, regeneration depends on the existing seed bank. Big sagebrush seeds are short-lived, and if fire returns before the new seedlings reach reproductive age (4-6 years) the species can be eliminated from the community.[154] Even rabbitbrush, a common successional shrub in sage communities that readily sprouts after fire, can be lost from the community with fire-return intervals of 2-4 years.[154] As cheatgrass-fueled fires lead to more extensive burns, seed sources from adjacent, unburned areas become more distant and the patch dynamics that likely promoted revegetation in these communities historically no longer functions. This change in the fire frequency leads to a change in the trajectory of plant succession and represents an alteration of ecosystem processes.[21]

Photograph 11. Cheatgrass can form continuous stands that burn frequently, preventing reestablishment of the former vegetation community. Hanford Site, Benton County, Washington. Photo: M Vander Haegen, WDFW.

Research Needs

A host of wildlife species have received little attention in ecological studies, yet have great potential to be affected by changes in shrub-steppe landscapes. This list would include many of the reptiles, amphibians, and small mammals. We need studies examining demography and habitat affinities of the Great Basin spadefoot toad, long-nosed leopard lizard, night snake, sagebrush vole, and the full complex of ground squirrels, among others. We need a better understanding of the relationship between fossorial mammals and the species that depend on them for burrows and as prey. Fourteen species of bats are known to use shrub-steppe, yet we know little about their population trends, population dynamics, and habitat needs, or how disturbance by humans affects breeding or roosting activities. We need studies examining the effects of habitat fragmentation on the distribution and demography of shrub-steppe wildlife. We need to develop and implement surveys for sage sparrows, grasshopper sparrows, burrowing owls, and other birds that are not covered sufficiently by the Breeding Bird Survey, as well as for reptiles and mammals that are not easily observed.

We know little about how the condition of extant shrub-steppe varies across the landscape. Mapping shrub-steppe communities across a gradient of conditions, from pristine to highly degraded, would provide a more realistic assessment of the current status of the resource and would be invaluable for future modeling of wildlife distribution. Livestock grazing has altered the vegetaion in many shrub-steppe communities, yet we lack information on how these changes affect most wildlife species. We need research on the role of microbiotic rusts in maintaining ecosystem function and as an indicator of ecosystem integrity. Are microbiotic crusts an important component for some species?

Expanding habitat linkages in rfagmented landscapes and increasing the amount of deep soil shrub-steppe for species such as pygmy rabbits and ground squirrels will require restoration of converted agricultural lands. How do we restore native vegetation on former agricultural sites and maximize the chance of regaining the full range of ecological function? How do we restore native vegetation and ecological function to highly degraded rangelands? Agricultural lands enrolled in CRP are a significant part of the landscape in eastern Washington. Research on the value of CRP to wildlife will help us assess its place in the shrub-steppe ecosystem and guide future enrollments to benefit native wildlife.

Applying the Data Matrixes to a Management Example:

Changes in the Wildlife Community Following Wildfire and Conversion to Annual Grasslands

Loss of Shrubs

Little published work has examined the effects of repeated wildfires and conversion to annual grasslands on shrub-steppe wildlife communities. Loss of the shrub layer through repeated fires eliminates habitat for shrub nesting birds, including some key shrub-steppe obligates (sage sparrow, Brewer's sparrow, and sage thrasher). Although fidelity to their breeding site will sometimes bring adults back in the year following a fire[108] or other catastrophic loss of the shrub layer,[159, 161] without suitable nesting shrubs such returns are likely to be a short-term phenomenon. Loss of shrubs would be detrimental to some lizard species, reducing the availability of shaded sites needed for thermoregulation and cover used to avoid predators. When fire removes shrubs from a community, ungulates that browse on sagebrush, bitterbrush, and other shrubs lose valuable winter habitat. Updike et al.[142] documented a decline in big game winter range following fire, cheatgrass invasion, and suppression of shrub regeneration by cheatgrass. Species that depend on sagebrush for forage during all or part of the year, such as

Table 2. Numbers of species associated with shrub-steppe habitat type, modified grassland structural condition, and shrubs as a key habitat element, in Oregon and Washington.

Species group	Shrub-steppe		Modified Grasslands		Shrubs as a key element
	Generally associated	Closely associated	Generally associated	Closely associated	
Birds	44	22	31	2	22
Mammals	26	27	34	0	12
Reptiles	20	0	7	0	6
Amphibians	9	0	10	0	1
Totals	103	49	82	2	41

greater sage-grouse and pygmy rabbits, likely will be excluded as sagebrush communities degrade to annual grasslands.

The Habitat Element Matrix (contained on the CD-ROM) can be used to determine which species depend on shrubs as an important component of shrub-steppe habitat and to predict which species would be lost from the community as fire removes the shrub layer. A query of the matrix revealed 22 birds, 12 mammals, 6 reptiles, and 1 amphibian that are associated with shrub-steppe habitat and require shrubs for some life function (Table 2).

Conversion of the Herb Layer to Cheatgrass

Changes in structure and composition of the herb layer that follow cheatgrass invasion also affect the wildlife community. Conversion of native forbs and bunchgrasses to exotic annuals results in a less stable food base for small herbivores like Townsend's ground squirrel, increasing the amplitude of their population fluctuations and the potential for localized extinctions.[167] Long-billed curlews seem to prefer cheatgrass dominated sites for nesting in southcentral Washington[2] and northcentral Oregon,[104] whereas some other ground-nesting birds have been found to occur at abnormally low densities in cheatgrass.[7] Birds that prefer open ground for foraging (e.g., sage sparrow and loggerhead shrike) avoid sites with dense cheatgrass.[27, 73] Conversion to cheatgrass probably would decrease availability of prey for lizards and make it more difficult for lizards to move about. Some snakes also would be negatively affected by such a conversion; night snakes and striped whipsnakes prey largely on lizards and thus would likely be negatively affected by the loss of lizards due to habitat conversion. Preliminary results from a comparative trapping study conducted in the Snake River Birds of Prey Area in 1978-1979[26] and 1997-1998 (J. Cossel, Jr. and C. Peterson, Idaho State University unpubl. data) are generally consistent with these predictions. Located in southwestern Idaho, much of this area has been converted from natural shrubland to non-native grassland. Although there is considerable year-to-year variation in the occurrence and trapping rates of reptiles, side-blotched lizards and night snakes seem to have declined whereas more generalist species like racers have increased.

Changes in the wildlife community that might be expected following conversion of shrub-steppe to annual grasslands can be derived from the matrixes. Sites dominated by cheatgrass are classified as *modified grasslands* in the Structural Condition Matrix. A query of this matrix revealed 82 species "generally associated" with annual grasslands and only 2 species classified as "closely associated." A query of the Wildlife-Habitat Matrix for species in shrub-steppe found 103 species "generally associated" and 49 species "closely associated" with this habitat type (Table 2). From the results of these queries we can conclude that the conversion of shrub-steppe communities to annual grasslands through the fire/cheatgrass cycle can cause dramatic changes in the wildlife community. The change may be particularly severe in the breeding bird community, where 20 species closely associated with shrub-steppe (several of them obligates) are predicted to be excluded. There are limitations to such an analysis, and one must give careful thought to other factors that may influence species occurrence at a particular site.

Acknowledgments

We thank the following individuals for reviewing previous drafts of the manuscript: S. Burton, J. Cossell, Jr., J. Lee, W. Leonard, M. Schroeder, and P. Sherman. We also thank B. J. Verts for providing information on Merriam's ground squirrel and J. Jacobson for preparing Figures 1 and 2.

Literature Cited

1. Abaturov, B. D. 1972. The role of burrowing animals in the transport of mineral substances in the soil. Pedobiologia 12:261-266.

2. Allen, J. N. 1980. The ecology and behavior of the long-billed curlew in southeastern Washington. Wildlife Monographs 73.

3. Bailey, V. 1936. The mammals and life zones of Oregon. U. S. Department of Agriculture, Bureau of Biological survey, North American Fauna 55:1-416.

4. Betts, B. J. 1990. Geographic distribution and habitat preferences of Washington ground Squirrels (Spermophilus washingtoni). Northwestern Naturalist 71:27-37.

5. ———. 1999. Current status of Washington ground squirrels in Oregon and Washington. Northwestern Naturalist 80:35-38.

6. Bolger, D. T., T. A. Scott, and J. T. Rotenberry. 1997. Breeding bird abundance in an urbanizing landscape in coastal southern California. Conservation Biology 11:406-421.

7. Brandt, C. A., and W. H. Rickard. 1994. Alien taxa in the North American shrub-steppe four decades after cessation of livestock grazing and cultivation agriculture. Biological Conservation 68:95-105.

8. Braun, C. E., et al. 1976. Conservation committee report on effects of alteration of sagebrush communities on the associated avifauna. Wilson Bulletin 88:165-171.

9. Brittingham, M. C., and S. A. Temple. 1983. Have cowbirds caused forest songbirds to decline? Bioscience 33:31-35.

10. Bryant, L. D., and C. Maser. 1982. Classification and distribution. Pages 1-60 in J. W. Thomas and D. E. Toweill, editors. Elk of North America: ecology and management. Stackpole Books, Harrisburg, PA.

11. Butterfield, B. R., B. Csuti, and J. M. Scott. 1994. Modeling vertebrate distributions for Gap Analysis. Pages 53-68 in R. I Miller, editor. Mapping the diversity of nature. Chapman and Hall, London, England.

12. Byers, J. A. 1997. American pronghorn: social adaptations and the ghosts of predators past. University of Chicago Press, Chicago, IL.

13. Carson, R. G., and J. M. Peek. 1987. Mule deer habitat selection patterns in northcentral Washington. Journal of Wildlife Management 51:46-51.

14. Cobb, V. A. 1994. The thermal ecology of pregnancy in free-ranging Great Basin rattlesnakes (Crotalus viridis lutosus). Dissertation, Idaho State University, Pocatello, ID.

15. Collins, W. B., and P. J. Urness. 1983. Feeding behavior and habitat selection of mule deer and elk on northern Utah summer range. Journal of Wildlife Management 47:646-663.

16. Connelly, J. W., M. W. Gratson, and K. P. Reese. 1998. Sharp-tailed grouse (Tympanuchus phasianellus). In A. Poole and F. Gill, editors, The Birds of North America, No. 354. The Birds of North America, Inc., Philadelphia, PA.

17. Costa, G. 1995. Behavioural adaptations of desert animals. Springer-Verlag, New York, NY.

18. Cox, C. B., and P. D. Moore. 1993. Biogeography: an ecological and evolutionary approach. Blackwell Science, Oxford, England.

19. Csuti, B., A. J. Kimerling, T. A. O'Neil, M. M. O'Shaughnessy, E. P. Gaines, and M. M. P. Huso. 1997. Atlas of Oregon wildlife. Oregon State University Press, Corvallis, OR.

20. Currier, M. J. P. 1983. Felis concolor. Mammalian Species 200:1-7, American Society of Mammalogists.

21. D'Antonio, C. M., and P. M. Vitousek. 1992. Biological invasions by exotic grasses, the grass/fire cycle, and global change. Annual Review of Ecology and Systematics 23:63-87.

22. Daubenmire, R. 1970. Steppe vegetation of Washington. Washington Agricultural Experiment Station Technical Bulletin 62. Washington State University , Pullman, WA.

23.———. 1975. Floristic plant geography of eastern Washington and northern Idaho. Journal of Biogeography 2:1-18.

24. Davis, W. B. 1939. The recent mammals of Idaho. Caxton Printers, Ltd., Caldwell, ID.

25. Demment, M. W., and P. J. Van Soest. 1985. A nutritional explanation for body-size patterns of ruminant and nonruminant herbivores. American Naturalist 125:641-672.

26. Diller, L. V., and D. R. Johnson. 1982. Ecology of reptiles in the Snake River Birds of Prey Area. U. S. Bureau of Land Management Snake River Birds of Prey Research Project, Boise, ID.

27. Dobler, F. C., J. Eby, C. Perry, S. Richardson, and M. Vander Haegen. 1996. Status of Washington's shrub-steppe ecosystem: extent, ownership, and wildlife/vegetation relationships. Research Report. Washington Department of Fish and Wildlife, Olympia, WA.

28. Donovan, T. M., F. R. Thompson III, J. Faaborg, and J. R. Probst. 1995. Reproductive success of migratory birds in habitat sources and sinks. Conservation Biology 9:1380-1395.

29. Eberhardt, L. E., E. E. Hanson, and L. L. Cadwell. 1982. Analysis of radionuclide concentrations and movement patterns of Hanford Site mule deer. Report BNWL-4420, Battelle, Pacific Northwest Laboratory, Richland, WA.

30.———, S. M. McCorquodale, and G. A. Sargeant. 1989. Elk and deer studies related to the Basalt Waste Isolation Project. Report PNL-6798, Battelle, Pacific Northwest Laboratory, Richland, WA.

31. Fitzner, R. E., D. Berry, L. L. Boyd, and C. A. Rieck. 1977. Nesting of ferruginous hawks (Buteo regalis) in Washington, 1974-75. Condor 79:245-249.

32.———, and R. L. Newell. 1989. Ferruginous hawk nesting on the U.S. DOE Hanford Site: case history of a recent invasion caused by transmission lines. Pages 125-132 in Proceedings IV: issues and technology in the management of impacted wildlife. Thorne Ecological Institute, Boulder, CO.

33. Freeman, S., D. F. Gori, and S. Rohwer. 1990. Red-winged blackbirds and brown-headed cowbirds—some aspects of a host-parasite relationship. Condor 92:336-340.

34. Freemark, K. E., and H. G. Merriam. 1986. Importance of area and habitat heterogeneity to bird assemblages in temperate forest fragments. Biological Conservation 36:115-141.

35. Galbraith, W. A., and E. W. Anderson. 1971. Grazing history of the Northwest. Journal of Range Management 24:6-12.

36. Gano, K. A. 1979. Analysis of small mammal populations inhabiting the environs of a low-level radioactive waste pond. Report PNL-2479, Battelle, Pacific Northwest Laboratory, Richland, WA.

37. Gavin, T. A., P. W. Sherman, E. Yensen, and B. May. 1999. Population genetic structure of the northern Idaho ground squirrel. Journal of Mammalogy 80:156-168.

38. Green, G. A., and R. G. Anthony. 1989. Nesting success and habitat relationships of burrowing owls in the Columbia Basin, Oregon. Condor 91:347-354.

39.———, and M. L. Morrison. 1983. Nest-site characteristics of sympatric ferruginous and Swainson's hawks. Murrelet 64:20-22.

40.———, R. E. Fitzner, R. G. Anthony, and L. E. Rogers. 1993. Comparative diets of burrowing owls in Oregon and Washington. Northwest Science 67:88-93.

41. Green, J. S., and J. T. Flinders. 1980. Brachylagus idahoensis. Mammalian Species 125:1-4, American Society of Mammalogists.

42. Greene, R. A., and C. Reynard. 1932. The influence of two burrowing rodents, Dipodomys spectabilis spectabilis (kangaroo rat) and Neotoma albigula albigula (pack rat) on desert soils in Arizona. Ecology 13:73-80.

43. Gregory, P. T. 1982. Reptilian hibernation. Pages 53-154 in C. Gans and F. H. Pough, editors. Biology of the Reptilia, Volume 13, Physiology D, Physiological Ecology. Academic Press Inc., New York, NY.

44. Gross, J. E., L. C. Stoddart, and F. H. Wagner. 1974. Demographic analysis of a northern Utah jackrabbit population. Wildlife Monographs 40.

45. Hafner, D. J., E. Yensen, and G. L. Kirkland, Jr. 1998. North American rodents. Status survey and conservation action plans. IUCN, Gland, Switzerland.

46. Haines, F. 1938. The northward spread of horses among the Plains Indians. American Anthropology 40:429-436.

47. Hall, E. R. 1981. Mammals of North America, 2nd edition. John Wiley & Sons, New York, NY.

48. Hanley, T. A. 1980. Nutritional constraints of food and habitat selection by sympatric ungulates. Dissertation, University of Washington, Seattle, WA.

49. Haug, E. A., B. A. Millsap, and M. S. Martell. 1993. Burrowing owl (Speotyto cunicularia). In A. Poole and F. Gill, editors, The Birds of North America, No. 61. The Birds of North America, Inc., Philadelphia, PA.

50. Hedlund, J. D., and L. E. Rogers. 1976. Characterization of small mammal populations inhabiting the B-C Crib environs. Report BNWL-2181, Battelle Pacific Northwest Laboratory, Richland, WA.

51. Heitschmidt, R. K., and J. W. Stuth. 1991. Grazing management, an ecological approach. Timber Press, Portland, OR.

52. Hoffmann, R. S., C. G. Anderson, R. W. Thorington, Jr., and L. R. Heany. 1993. Family Sciuridae. Pages 419-465 in D. E. Wilson and D. M. Reeder, editors. Mammal species of the world: a taxonomic and geographic reference. Smithsonian Institution Press, Washington, D.C.

53. Hole, F. D. 1981. Effects of animals on soil. Geoderma 25:75-112.

54. Holmes, A. L., and G. R. Geupel. 1998. Avian population studies at naval weapons systems training facility Boardman, Oregon. Final report. Point Reyes Bird Observatory, Stinson Beach, CA..

55. Howell, A. H. 1938. Revision of North American ground squirrels, with a classification of Sciuridae. North American Fauna 56:1-256.

56. Huey, R. B., C. R. Peterson, S. J. Arnold, and W. P. Porter. 1989. Hot rocks and not-so-hot rocks: retreat-site selection by garter snakes and its thermal consequences. Ecology 70:931-944.

57. Inouye, R. S., N. J. Huntly, D. Tilman, and J. R. Tester. 1987. Pocket gophers (Geomys bursarius), vegetation, and soil nitrogen along a successional sere in east central Minnesota. Oecologia 72:178-184.

58. Jarman, P. J. 1974. The social organization of antelope in relation to their ecology. Behaviour 48:215-267.

59. Johnson, K. H., R. A. Olson, and T. D. Whitson. 1996. Composition and diversity of plant and small mammal communities in tebuthiuron-treated big sagebrush (Artemisia tridentata). Weed Technology 10:404-416.

60. Kiester, A. R. 1971. Species density of North American amphibians and reptiles. Systematic Zoology 20:127-137.

61. Klauber, L. M. 1956. Rattlesnakes. Volume 1. University of California Press, Berkeley, CA.

62. Knick, S. T. 1990. Ecology of bobcats relative to exploitation and a prey decline in southeastern Idaho. Wildlife Monographs 108.

63. ———, and J. T. Rotenberry. 1995. Landscape characteristics of fragmented shrub-steppe habitats and breeding passerine birds. Conservation Biology 9:1059-1071.

64. ———, and D. L. Dyer. 1997. Distribution of black-tailed jackrabbit habitat determined by GIS in southwestern Idaho. Journal of Wildlife Management 61:75-85.

65. Lacy, R. C. 1997. Importance of genetic variation to the viability of mammalian populations. Journal of Mammalogy 78:323-335.

66. Lande, R. 1988. Genetics and demography in biological conservation. Science 241:1455-1460.

67. Laundré, J. W. 1993. Effects of small mammal burrows on water infiltration in a cool desert environment. Oecologia 94:43-48.

68. ———. 1998. Effect of ground squirrel burrows on plant productivity in a cool desert environment. Journal of Range Management 51:638-643.

69. Le Houerou, H. N., R. L. Bingham, and W. Skerbek. 1988. Relationship between the variability of primary production and the variability of annual precipitation in world arid lands. Journal of Arid Environments 15:1-18.

70. Leckenby, D. A., D. P. Sheehy, C. H. Nellis, R. J. Scherzinger, I. D. Luman, W. Elmore, J. C. Lemos, L. Doughty, and C. E. Trainer. 1982. Wildlife habitats in managed rangelands of the Great Basin of southeastern Oregon: mule deer. General Technical Report PNW-139, Pacific Northwest Forest and Range Experiment Station, Portland, OR.

71. Lehmkuhl, J. F., S. D. West, C. L. Chambers, W. C. McComb, D. A. Manuwal, K. B. Aubry, J. L. Erickson, R. A. Gitznen, and M. Leu. 1999. An experiment for assessing vertebrate response to varying levels and patterns of green-tree retention. Northwest Science 73:45- 63.

72. Leonard W. P., H. A. Brown, L. L. C. Jones, K. R. McAllister, and R. M. Storm. 1993. Amphibians of Washington and Oregon. Seattle Audubon Society, Seattle, WA.

73. Leu, M. 1995. The feeding ecology and the selection of nest shrubs and fledgling roost sites by loggerhead shrikes (Lanius ludovicianus) in the shrub-steppe habitat. Thesis, University of Washington, Seattle, WA.

74. Lowther, P. E. 1993. Brown-headed cowbird (Molothrus ater). In A. Poole and F. Gill, editors, The Birds of North America, No. 47. The Birds of North America, Inc., Philadelphia, PA.

75. MacArthur, R. H., and E. O. Wilson. 1967. The theory of island biogeography. Princeton University Press. Princeton, NJ.

76. Mack, R. N. 1981. Invasion of Bromus tectorum L. into western North America: an ecological chronicle. Agro-Ecosystems 7:145-165.

77. ———, and J. N. Thompson. 1982. Evolution in steppe with few large, hoofed mammals. American Naturalist 119:757-773.

78. Maclean, G. L. 1996. Ecophysiology of desert birds. Springer-Verlag. New York, NY.

79. Marr, N. V., C. A. Brandt, R. E. Fitzner, and L. D. Poole. 1988. Habitat associations of vertebrate prey within the Controlled Area study zone. Report PNL-6495, Battelle Pacific Northwest Laboratory, Richland, WA.

80. Marshall, D. B. 1992. Sensitive vertebrates of Oregon. Oregon Department of Fish and Wildlife, Portland, OR.

81. Mason, E. 1952. Food habits and measurements of Hart Mountain antelope. Journal of Wildlife Management 16:387-389.

82. McCorquodale, S. M. 1985. The ecology of a shrub-steppe elk (Cervus elaphus) population. Thesis, University of Washington, Seattle, WA.

83. ———. 1987. Fall-winter habitat selection by elk in the shrub-steppe of Washington. Northwest Science 61:171-173.

84. ———. 1991. Energetic considerations and habitat quality for elk in arid grasslands and coniferous forests. Journal of Wildlife Management 55:237-242.

85. ———. 1993. Winter foraging behavior of elk in the shrub-steppe of Washington. Journal of Wildlife Management 57:881-890.

86. ———, and L. E. Eberhardt. 1993. The ecology of elk in an arid environment: an overview of the Hanford elk project. Pages 56-63 in R. L. Callas, D. B. Koch, and E. R. Loft, editors. Proceedings of the 1990 Western States and Provinces Elk Workshop. California Department of Fish and Game, Sacramento, CA.

87. ———, ———, and G. A. Sargeant. 1989. Antler characteristics in a colonizing elk population. Journal of Wildlife Management 53:618-621.

88. ———, L. L. Eberhardt, and L. E. Eberhardt. 1988. Dynamics of a colonizing elk population. Journal of Wildlife Management 52:309-312.

89. ———, K. J. Raedeke, and R. D. Taber. 1986. Elk habitat use patterns in the shrub-steppe of Washington. Journal of Wildlife Management 50:664-669.

90. McCoy, C. J. 1965. Life history and ecology of the Cnemidophorus tigris septentrionalis. Dissertation, University of Colorado, Boulder, CO.

91. Melgoza, G., R. S. Nowak, and R. J. Tausch. 1990. Soil water exploitation after fire: competition between Bromus tectorum (cheatgrass) and two native species. Oecologia 83:7-13.

92. Messick, J. P., and M. G. Hornocker. 1981. Ecology of the badger in southwestern Idaho. Wildlife Monographs 76.

93. Mills, L. S. 1996. Fragmentation of a natural area: dynamics of isolation for small mammals on forest remnants. Pages 199-219 in R. G. Wright, editor. National parks and protected areas: their role in environmental protection. Blackwell Science, Cambridge, MA.

94. Moen, A. N. 1973. Wildlife ecology: an analytical approach. W. H. Freeman, San Francisco, CA.

95. Murie, O. J. 1951. The elk of North America. Stackpole Books, Harrisburg, PA.

96. Nadler, C. H. 1968. The chromosomes of Spermophilus townsendi (Rodentia: Sciuridae) and report of a new subspecies. Cytogenetics 7:144-157.

97. Nowak, R. M. 1991. Walker's mammals of the world. Fifth edition. Johns Hopkins University Press, Baltimore, MD.

98. Nussbaum, R. A., E. D. Brodie, and R. M. Storm. 1983. Amphibians and reptiles of the Pacific Northwest. University of Idaho Press, Moscow, ID.

99. Olsen, B. 1976. Status report: Columbian sharp-tailed grouse. Oregon Wildlife 34:10.

100. O'Neil, T. A. and D. H. Johnson. 2001. Oregon and Washington wildlife species and their habitats. Pages 1-21 In: D.H. Johnson and T.A. O'Neil, managing directors. Wildlife-habitat relationships in Oregon and Washington. Oregon State University Press, Corvallis, OR.

101. Oregon Department of Fish and Wildlife. 1994. Oregon Species Information System. Wildlife Research, Corvallis, OR.

102. Orians, G. H., E. Roskaft, and L. D. Beletsky. 1989. Do brown-headed cowbirds lay their eggs at random in the nests of red-winged blackbirds? Wilson Bulletin 101:599-605.

103. Osborne, D. 1953. Archaeological occurrence of pronghorn antelope, bison, and horse in the Columbia Plateau. Scientific Monthly 77:914-917.

104. Pampush, G. J., and R. G. Anthony. 1993. Nest success, habitat utilization and nest-site selection of long-billed curlews in the Columbia Basin, Oregon. Condor 95:957-967.

105. Parker, K. L., and C. T. Robbins. 1984. Thermoregulation in mule deer and elk. Canadian Journal of Zoology 62:1409-1422.

106. Parmenter, R. R., and J. A. MacMahon. 1983. Factors determining the abundance and distribution of rodents of a shrub-steppe ecosystem: the role of shrubs. Oecologia 59:145-156.

107. Patterson, M. P., and L. B. Best. 1996. Bird abundance and nest success in Iowa CRP fields: the importance of vegetation structure and composition. American Midland Naturalist 135:153-167.

108. Petersen, K. L., and L. B. Best. 1987. Effects of prescribed burning on nongame birds in a sagebrush community. Wildlife Society Bulletin 15:317-329.

109. Porter, W. P,. and C. R. Tracy. 1983. Biophysical analyses of energetics, time-space utilization, and distributional limits. Pages 55-83 in R. B. Huey, E. R. Pianaka, and T. W. Schoener, editors. Lizard ecology. Harvard University Press, Cambridge, MA.

110. Pough, F.A. 1980. The advantages of ectothermy for tetrapods. American Naturalist 115: 92-112.

111. Pough, F. H., J. B. Heiser, and W. N. McFarland. 1996. Vertebrate life. Fourth edition. Prentice Hall, Upper Saddle River, NJ.

112. Price, L. W. 1971. Geomorphic effect of the arctic ground squirrel in an alpine environment. Geografiska Annaler 53A:100-106.

113. Quigley, T. M., and S. J. Arbelbide. 1997. An assessment of ecosystem components in the interior Columbia Basin and portions of the Klamath and Great Basins. General Technical Report PNW-GTR-405. U.S. Forest Service, Pacific Northwest Research Station, Portland, OR.

114. Rickard, W. H., J. D. Hedlund, and R. E. Fitzner. 1977. Elk in the shrub-steppe of Washington: an authentic record. Science 196:1009-1010.

115. ———, and B. E. Vaughan. 1988. Characteristics of contrasting shrub-steppe plant communities. Pages 109-179 in W. H. Rickard, L. E. Rogers, B. E. Vaughan, and S. F. Liebetrau, editors. Shrub-steppe: balance and change in a semi-arid terrestrial ecosystem. Elsevier, Amsterdam, The Netherlands.

116. Ricklefs, R. E. 1983. Ecology (second edition). Chiron Press, New York, NY.

117. Robbins, C. S., D. K. Dawson, and B.A. Dowell. 1989. Habitat area requirements of breeding forest birds of the Middle Atlantic states. Wildlife Monographs 103.

118. Robinson, S. K., F. R. Thompson III, T. M. Donovan, D. R. Whitehead, and J. Faaborg. 1995. Regional forest fragmentation and the nesting success of migratory birds. Science 267:1987-1990.

119. Rogers, L. E., R. E. Fitzner, L. L. Cadwell, and B. E. Vaughan. 1988. Terrestrial animal habitats and population responses. Pages 182-250 in W. H. Rickard, L. E. Rogers, B. E. Vaughan, and S. F. Liebetrau, editors. Shrub-steppe: balance and change in a semi-arid terrestrial ecosystem. Elsevier, Amsterdam, The Netherlands.

120. ———, and W. H. Rickard. 1977. Ecology of the 200 Area plateau waste management environs: a status report. Report PNL-2253, Battelle, Pacific Northwest Laboratory, Richland, WA.

121. Rotenberry, J. T., and J. A. Wiens. 1991. Weather and reproductive variation in shrub-steppe sparrows: a hierarchical analysis. Ecology 72:1325-1335.

122. Saab, V. A., and T. D. Rich. 1997. Large-scale conservation assessment for Neotropical migratory land birds in the interior Columbia River Basin. General technical report PNW-GTR-399. U.S. Forest Service, Pacific Northwest Research Station, Portland, OR.

123. Savage, D. E., J. A. Young, and R. A. Evans. 1969. Utilization of medusahead and downy brome caryopses by chukar partridges. Journal of Wildlife Management, 33:975-978.

124. Schroeder, M.A., J. R. Young, and C. E. Braun. 1999. Sage Grouse (Centrocercus urophasianus). In A. Poole and F. Gill, editors, The Birds of North America, No. 425. The Birds of North America, Inc., Philadelphia, PA.

125. Schwartz, C. C., and J. G. Nagy. 1976. Pronghorn diets relative to forage availability in northeastern Colorado. Journal of Wildlife Management 40:469-478.

126. Seidensticker, J. C., IV, M. G. Hornocker, W. V. Wiles, and J. P. Messick. 1973. Mountain lion social organization in the Idaho Primitive Area. Wildlife Monographs 35.

127. Small, R.J., and B.J. Verts. 1983. Responses of a population of Perognathus parvus to removal trapping. Journal of Mammalogy 64:139-143.

128. Smith, E. L., et al. 1995. New concepts for assessment of rangeland condition. Journal of Range Management 48:271-282.

129. Smith, G. W., and D. R. Johnson. 1985. Demography of a Townsend ground squirrel population in southwestern Idaho. Ecology 66:171-178.

130. Springer, J. T. 1982. Movement patterns of coyotes in south central Washington. Journal of Wildlife Management 46:191-200.

131. Stebbins, R. C., and N. W. Cohen. 1995. A natural history of amphibians. Princeton University Press, Princeton, NJ.

132. Stewart, G., and A. C. Hull. 1969. Cheatgrass, Bromus tectorum L.—an ecological intruder in southern Idaho. Ecology 30:58-74.

133. Stoddart, L.A., A.D. Smith, and T.W. Box. 1975. Range Management. McGraw-Hill Book Company, New York, NY.

134. Storm, R. M., W. P. Leonard, H. A. Brown, R. B. Bury, D. M. Darda, L. V. Diller, and C.R. Peterson. 1995. Reptiles of Washington and Oregon. Seattle Audubon Society, Seattle, WA.

135. Sundstrom, C., W. G. Hepworth, and K. L. Diem. 1973. Abundance, distribution, and food habits of the pronghorn. Bulletin 10, Wyoming Game and Fish Commission, Cheyenne, WY.

136. Taylor, W.P. 1935. Some animal relations to soil. Ecology 16:127-136.

137. Thorp, J. 1949. Effects of certain animals that live in soils. Scientific Monthly 68:180-191.

138. Thorson, T., and A. Svihla. 1943. Correlation of the habits of amphibians with their ability to survive the loss of body water. Ecology 24:374-381.

139. Turner, F. B., R. I. Jennrich, and J. D. Weintraub. 1969. Home ranges and body size of lizards. Ecology 50:1076-1081.

140. ———, P. A. Medica, and B. W. Kowalewsky. 1976. Energy utilization by a lizard (Uta stansburiana). U.S. International Biological Programme Monograph No. 1. Utah State University Press, Logan, UT.

141. Turner, G. T., R. M. Hansen, V. H. Reid, H. P. Tietjen, and A. L. Ward. 1973. Pocket gophers and Colorado mountain rangeland. Colorado State University Experiment Station Bulletin, 554S. Fort Collins, CO.

142. Updike, D. R., E. R. Loft, and F.A. Hall. 1990. Wildfires on big sagebrush/antelope bitterbrush range in northeastern California: implications for deer populations. Pages 41-46 in E. S. McArthur, R. M. Romney, S. D. Smith, and P.T. Tueller, editors. Proceedings—symposium on cheatgrass invasion, shrub die-off, and other aspects of shrub biology and management. General Technical Report INT-GTR-276. U. S. Forest Service, Ogden, UT.

143. Uresk, D. W., J. F. Cline, and W. H. Rickard. 1975. Diets of black-tailed hares on the Hanford Reservation. Report BNWL-1931, Battelle, Pacific Northwest Laboratory, Richland, WA.

144. ———, and V.A. Uresk. 1980. Diets and habitat analyses of mule deer on the 200 Areas of the Hanford Site in southcentral Washington. Report PNL-2461, Battelle, Pacific Northwest Laboratory, Richland, WA.

145. Van Vuren, D., and M. P. Bray. 1985. The recent geographic distribution of Bison bison in Oregon. Murrelet 66:56-58.

146. Vander Haegen, W. M. 1996. Survey of breeding bird communities on BRMaP sites, Hanford Site, 1996. Project completion report. Washington Department of Fish and Wildlife, Olympia, WA.

147. ———, F. C. Dobler, and D. J. Pierce. 2000. Shrubsteppe bird response to habitat and landscape variables in eastern Washington, USA. Conservation Biology 14:1145-1160.

148. ———, and B. Walker. 1999. Parasitism by brown-headed cowbirds in the shrub-steppe of eastern Washington. Studies in Avian Biology 18:34-40.

149. Verts, B. J., and L. N. Carraway. 1998. Land mammals of Oregon. University of California Press, Berkeley, CA.

150. Wallmo, O. C. 1981. Mule and black-tailed deer distribution and habitats. Pages 1-26 in O. C. Wallmo, editor. Mule and black-tailed deer of North America. University of Nebraska Press, Lincoln, Nebraska.

151. Washington Department of Fish and Wildlife. 1995. Washington State management plan for the bighorn sheep. Game Division, Washington Department of Fish and Wildlife, Olympia, WA.

152. Weinberg, H. J., and R. R. Roth. 1998. Forest area and habitat quality for nesting wood thrushes. Auk 115:879-889.

153. Weiss, N. T., and B. J. Verts. 1984. Habitat and distribution of pygmy rabbits (Sylvilagus idahoensis) in Oregon. Great Basin Naturalist 44:563-569.

154. Whisenant, S. G. 1990. Changing fire frequencies on Idaho's Snake River Plains: ecological and management implications. Pages 4-10 in E. S. McArthur, R. M. Romney, S. D. Smith, and P.T. Tueller, editors. Proceedings—symposium on cheatgrass invasion, shrub die-off, and other aspects of shrub biology and management. U. S. Forest Service, Ogden, UT.

155. Wickstrom, M. L., C. T. Robbins, T. A. Hanley, D. E. Spalinger, and S. M. Parish. 1984. Food intake and foraging energetics of elk and mule deer. Journal of Wildlife Management 48:1285-1301.

156. Wiens, J. A. 1985. Habitat selection in variable environments: shrub-steppe birds. Pages 227-251 in M. Cody, editor. Habitat Selection in birds. Academic Press, New York, NY.

157. ———, C. S. Crawford, and J. R. Gosz. 1985a. Boundary dynamics: a conceptual framework for studying landscape ecosystems. Oikos 45:421-427.

158. ———, and J. T. Rotenberry. 1981. Habitat associations and community structure of birds in shrub-steppe environments. Ecological Monographs 51:21-41.

159. ———, and ———. 1985. Response of breeding passerine birds to rangeland alteration in a North American shrubsteppe locality. Journal of Applied Ecology 22:655-668.

160. ———, ———, and B. Van Horne. 1985b. Territory size variations in shrub-steppe birds. Auk 102:500-505.

161. ———, ———, and ———. 1986. A lesson in the limitations of field experiments: shrub-steppe birds and habitat alteration. Ecology 67:365-376.

162. ———, B. Van Horne, and J. T. Rotenberry. 1987. Temporal and spatial variations in the behavior of shrub-steppe birds. Oecologia 73:60-70.

163. Wilcove, D. S. 1985. Nest predation in forest tracts and the decline of migratory songbirds. Ecology 66:1211-1214.

164. ———, C. H. McLellan, and A. P. Dobson. 1986. Habitat fragmentation in the temperate zone. Pages 237-256 in M. E. Soule, editor. Conservation biology: the science of scarcity and diversity. Sinauer Associates, Sunderland, MA.

165. Yensen, D. L. 1980. A grazing history of southwestern Idaho with emphasis on the Snake River Birds of Prey Area. U. S. Bureau of Land Management, Boise District, Boise, ID.

166. Yensen, E. and D. L. Quinney. 1992. Can Townsend's ground squirrels survive on a diet of exotic annuals? Great Basin Naturalist 52:269-277.

167. ———, ———, K. Johnson, K. Timmerman, and K. Steenhof. 1992. Fire, vegetation changes, and population fluctuations of Townsend's ground squirrels. American Midland Naturalist 128:299-312.

168. Yoakum, J. 1958. Seasonal food habits of the Oregon pronghorn. Interstate Antelope Conference Transactions 10:58-72.

169. ———. 1980. Habitat management guides for the American pronghorn antelope. Technical Note 347, Bureau of Land Management, Denver, CO.

12

Wildlife of Urban Habitats

Howard L. Ferguson, Kevin Robinette,
& Kate Stenberg

Description, History, and Status

People often believe that when land is urbanized, it loses its value for wildlife. The development of cities does not eliminate all wildlife habitats—some are lost, some are altered, and some are maintained. In the process of urbanization, some species of wildlife are favored while others are stressed or eliminated. Over the past two decades researchers have studied wildlife in urban areas; most of this urban research has been conducted in eastern North America and Europe, with little having been conducted in the Pacific Northwest. More needs to be learned in this region because the issue of urbanization and the protection of our natural resources, including wildlife and their habitats, is being raised daily in Oregon and Washington.

Attempting to characterize urban habitats and urban wildlife is challenging. Urban ecosystems are in perpetual change.[43] They are a mix of natural and human-made activities and disturbances managed and unmanaged lands, native and non-native plants, and natural and human-made structures all merged together.[198] Instead of one or two management goals, as is common in a forest stand, each urban landowner has a different goal and a different strategy for achieving that goal.

An urban area may be viewed as a continuum ranging from the highly developed inner city out to the relatively undeveloped rural areas.[18, 30, 43, 50, 67, 69, 137, 208] The position on the urban-rural gradient has a significant impact on land-cover regimes and can actually induce substantially different landscape structures.[208] This concept implies that environmental variability is spatially structured and that this pattern influences ecological processes and annual population dynamics.[18, 137] We have chosen to follow the concept of an urban gradient in this chapter. This continuum is based on the level of urban development as determined by the percent of the land surface covered by impervious materials. At the center of this continuum is the highly developed area of the inner city, what we call the high-density zone, having >60% impervious surface (area covered by roads, buildings, parking lots, sidewalks, and driveways). As you move out from the inner city, you encounter the middle zone or the medium-density zone, the area having 30-59% impervious surface. Farthest away from the inner city is the low-density zone, with 10-29% impervious surface.

The low-density zone generally has the most native wildlife habitat available, the largest blocks of unfragmented land, and the most diverse assemblage of native plants and animals. This same trend is true for habitat structure. The high-density zone typically has little understory, and few, if any, snags, downed woody material, and brush or rock piles.[2, 102, 170] These key habitat structures increase as you move out from the inner city.[198] And, as expected, both the abundance and the diversity of wildlife present for a particular area depend on its location along this continuum. The terms "urban" and "suburban" refer to concentrated human development in and around towns and cities. Our use of the word "urban" encompasses all 3 urban zones as described above and in Chapter 2. Rather than being circular, the 3 urban zones typically appear as zones of urbanization along highways, rivers, or valleys (often looking like isotherms). Further complicating the continuum is the ontogeny of urban "sprawl," the irregular or unplanned spread of urbanization. It is caused by people, and in some cases, businesses, in the inner suburbs moving farther and farther out as the city continues to grow. State growth management laws attempt to address and control urban sprawl (see section on "Growth Management").

The difference between urban conversions and many other types of land conversions is the permanence of the changes made to the landscape—changes are maintained by humans and are likely to persist for hundreds of years. Typically, urbanization removes, alters, or replaces natural or existing vegetation with impervious surfaces.[18, 179] Plant species composition is changed by the culturing of some native species at the expense of others and by the introduction of exotic species. Selected, non-native plants flourish in urban settings because of the increased availability of water, fertilizers, and cultivation. Exotic shrubs and trees outnumber native shrubs and trees 6:1 in the understory layers.[17]

With the influx of exotics and the overall intensive maintenance of urban areas, succession is altered and the community equilibrium changes.[18, 134] The protection of human structures and property has led to intense fire suppression in and around urban areas, disrupting the natural fire disturbance regime.[133, 134, 157] Another conspicuous consequence of urbanization is the fragmentation of the natural environment into a mosaic of patches of different size, shape, composition, and age. Patches become surrounded by areas influenced by humans—cleared areas, roads, buildings, and other human-made structures—and, in general, become smaller and more isolated toward the urban center[46] (see section on "Habitat Loss and Fragmentation"). Succession is further altered by human landscaping and maintenance activities.

Patches of native presettlement vegetation or special or unique geological habitats, such as riparian areas, canyons, rock outcrops, cliffs, and lakes, are often left undeveloped within urban zones. Wildlife found in and around these remnant habitats or inclusions are usually a subset of the wildlife normally expected for each habitat. The species assemblage will be determined by the size of the remnant patch as well as the degree and amount of urbanization surrounding the remnant. Often, these features are the reason for initial human settlement, and that is why many of our Northwest cities have such habitats: Portland and Spokane are close to rivers; Tacoma and Seattle are near Puget Sound waterways (Figure 1). These features can greatly enhance the wildlife abundance and diversity normally expected to be found in an urban area. These inclusions are the reason coyotes, bald eagles, ospreys, kingfishers, and even moose and cougars may sometimes be found in the urban sections of our Pacific Northwest cities. Wildlife habitat, and the urban environment, can be enhanced through the preservation of remnant forests, riparian areas, wetlands, cliffs, meadows, fields, and orchards as cities expand.[48]

One factor that may influence the use of urban habitats by wildlife in Oregon and Washington is the relatively young age of our urban areas. Urbanization has been extremely rapid in North America in comparison to Europe, Asia, and Africa. With few exceptions, events that took place slowly over many centuries in the Old World have been accomplished in just a few decades in the New World; animals have had less time to adjust to urbanization in North America. In the western U.S., most of the

Figure 1. Spokane River in Spokane, Washington. (Photograph by Howard L. Ferguson)

urbanization that has caused significant habitat alteration has occurred during the last 50-100 years. Blondel[19] stated: "Urban settlements (in the Old World) have existed for at least 3,500 years, especially in the eastern Mediterranean", and he described the progressive adjustments and habituation of rural birds to urban life in Old World cities. This can be contrasted with birds found in North American cities, most of which have been occupied for <200 years.[48, 104] One example of a bird that seems to have adjusted to urban life in Europe but not in North America is the northern goshawk. It is recognized as an urban species in Germany[63, 64] and is increasing in human-influenced habitats in Poland,[22] whereas in the U.S. it is considered an interior forest species that requires mature/old-growth trees.

Urbanization is a widespread and ever-increasing process. In 1989, 74% of the U.S. population (203 million people) lived in urban areas, and that number is expected to increase to >80% by 2025.[81, 99] This increase throughout the country has resulted in the conversion of cropland, pastures, and forests into urban and suburban environments.[72] Between 1960 and 1970, urban land in the U.S. increased by 9 million acres (3,643,725 ha), and between 1970 and 1980 it increased by 13 million acres (5,263,158 ha).[83]

In the Puget Sound area of Washington, between 1979 and 1989, suburban zone area increased by 7.3%, and urban zone area increased by 15.7%. For example, nearly 47% of the private land in Snohomish and Kitsap counties is now suburban.[130] Between 1945 and 1970, 257,000 acres (104,049 ha) of Washington timberland were converted to urban use,[24] and these westside forests are among the most productive in the U.S.[24] From 1970 to 1992, nearly 10% of all forests in the Puget Sound Region were converted to other uses: roads, suburbs, cities, and farms.[130] An estimated 30,000-80,000 acres (12,146-32,389 ha) of wildlife habitat is lost each year in Washington state; >90% of its wetlands in urban areas and 90-98% of its urban coastal wetlands are already gone.[206] The greatest losses of Washington freshwater wetlands have probably resulted from urban development.[37] From 1992 to 1998, >16,000 acres (6,478 ha) of natural areas were lost to development and other changes in the Portland metropolitan area (J. Budhabhatti, Senior Planner, Portland Metro, pers. comm.). The need for immediate conservation and regional and statewide planning is essential, considering that by 2045 our population is expected to nearly double.[206]

Common Ecological Features of Urban Habitats

Buildings and Human Structures

Characteristic of urban areas are buildings, streets, roads, parking lots, and other artifacts of human construction.[18, 134] These structures occupy most of the ground surface and form a largely impermeable and sterile covering of the soil, which once supported native vegetation. The water runoff from these impervious surfaces is high and fast, with little infiltration into underlying strata, and results in a reduced rate of recharge to natural groundwater reservoirs, aquifers, and a lowering of the water table.[126]

Emlen[73] noted that the special features and artifacts of human origin contributed importantly to the physiognomy and diversity of the urban habitat. Rooftops and their superstructures, such as air-conditioner units and television antennae, provide song and resting perches for birds. Supplemental nesting sites provided by human structures include vent holes between bricks, vent holes for dryers, openings around eaves and eave troughs, edges of tile roofs, ledges under roofs, window shutters, and decorative boxes.[23, 54] Overhead wires and other high song perches may enhance the value of the urban area for territorial species, whereas traffic hazards, cat predation, and disturbance or harassment by dogs and humans are selectively detrimental to ground nesting and low-shrub foraging birds.[23, 54]

The major direct contribution of the human-built environment to bird communities may be an increased supply of holes, crevices, and ledges for nesting.[73] The degree to which these human structures positively impacted birds was questioned by Lancaster and Rees[123] when they surmised that the major contributing elements of buildings and other human-made structures in urban environments are, essentially, closed volumes and barren surfaces with limited potential for diversification of niches. They noted that even the song and resting perches associated with buildings, poles, and wires are insignificant when compared with those made available by comparable cover by trees and bushes.[123] However, when natural features and structures are not present, such as in downtown urban areas, these human structures do provide some potential nesting and roosting sites for animals able to adjust, or for those few commensal animals already adapted to urban areas.

Climate

The climate in large urban centers can differ substantially from the surrounding exurban areas. Cities are warmer, with higher average daily temperatures and less variable daily and annual temperatures, than the surrounding areas.[68, 74, 191, 201] In San Francisco, differences of up to 20°F (11°C) have been recorded between the densely built-up business districts of the city and the undeveloped urban parkland of Golden Gate Park.[69] In central California, this temperature differential increased with increasing city size.[69] Warmer temperatures may be a factor that has allowed, at least indirectly, some species to overwinter farther north than they did historically (e.g., blue jay[182] and merlin[110]). Studies elsewhere have revealed some evidence that breeding may begin earlier and end later in the year partially due to this temperature differential in cities.[74, 88, 143]

More precipitation occurs over highly urbanized areas compared to similar rural areas of the same region.[107] Additional heat generated by cities causes air above cities to rise more rapidly than from the surrounding countryside, resulting in more rapid cooling and rainfall. Consequences of this are largely unknown for the Pacific Northwest, but it may benefit wildlife in the more arid areas of Oregon and Washington.

Light

In general, there is greater illumination in urban areas because of increased reflective surfaces and widespread artificial lighting. Artificial lighting prolongs the day length in cities, which may affect diurnal cycles and trigger earlier reproductive activity in some species.[74, 200] It may also influence migratory behavior and interfere with reproductive behavior of nocturnal species. Artificial lighting may be a major factor contributing to the problem of bird collisions with buildings.[116, 117]

Habitat Loss and Fragmentation

The most obvious consequence of urbanization is the loss of rural or agricultural habitat. Much of the land converted for urbanization may be considered loss of habitat because of the permanence of the impervious surfaces covering these lands. With urban development and loss of habitat comes fragmentation (the division of habitat into smaller, more isolated patches of natural vegetation). As the landscape matrix shifts from natural vegetative cover types to urban land uses, the natural habitats become remnants or patches—smaller areas that are isolated and disconnected from each other. The ability of these patches to support specific wildlife species is a function of their size and isolation, as well as the neighboring vegetative type.[4, 167]

The impact of fragmentation depends on the surrounding habitat;[167] this may explain the different conclusions being reported for eastern U.S. and Pacific Northwest forests. Forest fragmentation in the eastern U.S., at both the landscape and site level, results in serious impacts on wildlife.[60, 211, 213] It can result in genetic isolation of less-mobile species (e.g., reptiles and amphibians).[175] Because of species' area needs, population isolation, and spatial heterogeneity, fragmentation further reduces the amount of habitat available for certain species.[34, 49, 202] In southern California, Bolger[21] found that in areas surrounded by urban development, fragmentation seemed to cause more isolation than that reported for the forested landscapes of eastern and mid-western U.S. He reported that local extinctions in these isolated habitat fragments were common, and that recolonization of these areas was rare. Studies reporting significant negative impacts from fragmentation had either agriculture or urban habitats as the surrounding or neighboring land use, whereas the patches of the Northwest are typically surrounded by different-aged forests—not agriculture or urban. These differences may explain why the few studies from the Northwest have reported fewer impacts from fragmentation on wildlife than the eastern U.S. Just what impact does fragmentation have on urbanized areas of the Pacific Northwest? More research is needed to determine the impact that fragmentation has on wildlife in urban areas of the Northwest.

Another effect of fragmentation is the creation of more edge, the interface where different plant communities meet. Moderate levels of urban development fragment formerly continuous habitat, increasing the amount of edge. If development occurs in non-continuous small blocks, the amount of edge increases even more. As time goes on, and urbanization and development increase, more and more native habitat is eliminated, and edge habitat will eventually decrease, to the point where all native habitats are eliminated and no edge habitat exists.[18] Whereas edges are beneficial to some wildlife species, they also can negatively impact others.[4, 165, 213, 216] Higher nest predation and parasitism rates have been reported for small, fragmented eastern U.S., where edge was bounded by either urban or agriculture land use.[131, 156, 211, 212]

Roads

Roads are a major component of urban areas and have major impacts on wildlife, wildlife habitat quality, and connectivity. In urban areas, roads or road corridors function ecologically as (1) habitat for animals, their predators, and competitors; (2) sources and collection points of contaminants; (3) sources of human disturbance; (4) partial or complete barriers to animal movement; (5) conduits for animal movement or range expansion; (6) barriers to surface water percolation (lowering groundwater tables); (7) sources of fragmentation of wildlife habitats; (8) facilitators of the spread of non-native species; and (9) mortality agents.[80, 206] Roads divide once large populations into smaller groups, which are at greater risk of extinction and genetic stagnation.[80, 124, 125] Roads often cut directly across long-established animal trails to water and feeding grounds,[169] and with increased road density, human access increases, which can result in more disturbances.[80] Additional information on roads and the impacts to wildlife can be found at http://www.fhwa.dot.gov/environment/wildlife crossings/.

Human Disturbance, Noise, and Predation

Urban wildlife species are constantly exposed to human-related disturbances, including noise and disturbance related to human activity around homes and buildings, road traffic, construction activities, and aircraft activity.[207] Some wildlife species become accustomed to regular and predictable disturbances. Watson and Pierce[207] found that irregular noise and unpatterned disturbances—especially human pedestrian disturbances—had more negative impact on bald eagles than disturbances caused by autos, boats, and aircraft. In Puget Sound, human activity altered the normal feeding guild relationships of crows, gulls, and bald eagles.[180] Eagles were least tolerant of human activity and gulls were most tolerant, even feeding more in the presence of humans. This behavior may negatively impact the less tolerant eagles in urban areas during years with limited food resources.[180]

For birds, noise may interfere with intraspecific communications required for successful reproduction, and noise may have contributed to the elimination of some key interior species in the eastern U.S.[115, 118, 161] In his book on noise and wildlife, Busnel[33] concluded that noise, at least indirectly, might cause bodily injury, energy loss, decreased feeding, habitat avoidance and abandonment, and reproductive losses. Long-term exposure to noise causes excessive stimulation to the nervous system, resulting in chronic stress that may harm the overall health of wildlife and their reproductive fitness.[79]

Predation rates are higher in urban areas[16, 68, 94, 194, 196] in comparison to similar exurban areas; with an increase in edge comes an increase in nest predation and brood parasitism.[181, 212] Avian nest predators and brood parasites are associated with strip corridors and edges between fields and woodlands, resulting in decreased nesting success for some bird species. One explanation for this is that predators can penetrate deeper[211, 212] or are able to search smaller areas more efficiently.[131] However, Martin[132] believes nest predation cannot fully explain the losses of bird species from fragments, and believes other ecological requirements, such as nest site preference, in combination with increased predation may explain loss of species. Results from a study in the eastern U.S.[31] attributed the lack of nesting success of ovenbirds in fragmented forests to reduced food supplies.

Several authors have attributed increased predation in urban areas to human pets—cats and dogs.[44, 114] Domestic cats can be efficient small predators and may occur in high numbers in urban areas.[40] Prey of free-ranging domestic cats in one study was approximately 70% small mammals, 20% birds, and 10% various other animals including reptiles and invertebrates.[78] Cats often outnumber other mid-sized native predators and may compete with them for many of the same food sources.[44] Buildings and barren ground reduce and simplify vegetation within patches, and provide hunting areas for domestic cats and dogs that may effectively reduce the local abundance of vertebrate prey.[60] For more information on feral cats, see http://www.dfg.ca.gov/hcpb/feralcat.html.

Food

The type and quantity of foods available in the urban environment differ considerably from those found in undisturbed or more natural areas. Many authors have commented on the abundance of food available in the urban environment.[32, 55, 56, 74, 94] This abundance is limited to a few food types; quantities of seeds, fruits, and refuse are unintentionally provided through spillage, waste, disposal, and collection and accumulation of food at refuse sites and landfills. Even the ubiquitous urban lawn can be highly productive for certain types of food, e.g., insects and worms,[76, 82, 94] if chemical use is limited.

Considerable quantities of food are also provided throughout the year by people. This food is primarily seeds and nuts, the exception being nectar (hummingbird) feeders. In 1994, Americans spent >$2 billion on birdseed alone; this figure does not include the cost of feeders and other equipment.[188] The urban environment provides these artificial seed sources in much greater quantities and concentrations than would ever occur in natural or undisturbed habitats. Another center of human-supplied food is parks, where quantities of bread and other human foods are provided to wildlife (Figure 2).

Figure 2. Riverfront Park, Spokane, Washington. (Photograph by Howard L. Ferguson)

Replanting with non-native plants similarly alters the availability of foods.[17] Native plant food sources, including native berries, fruits, and seeds, are typically unavailable or present in limited amounts in urban areas. Exotic plants support fewer insects than do native plants,[133] and the urban practice of controlling insects with broad-spectrum pesticides further limits the availability of insects in the urban environment. Removal of all down and dead wood, diseased trees, and snags from the urban landscape curtails the food base for those species that depend on insects or fungi from these sources.[187] Consequently, the urban food available to wildlife contributes to reduced species diversity while it sustains high densities of favored, often exotic, species.[20, 123] The visible increase of the abundance of a few species of ground feeding, cavity-nesting omnivores (e.g., house sparrows and starlings) is consistent with the nature and availability of urban food, which also corresponds to the gradient of increasing urbanization.[123]

Wildlife Use of Urban Habitats

Most of the ecological research on wildlife in urban areas has been conducted in the eastern and southwestern U.S. and Europe, and most of the data and references presented in this chapter are results from those studies. Little research has been done in the urban areas of the Northwest, but the basic impacts of urbanization are consistent.[43, 199] Clergeau et al.,[43] in a study comparing Quebec, Canada with Rennes, France, found similar patterns for species abundance and compositions in both cities. They concluded that the physical setting of a city does not greatly influence the structure of the urban bird community.[43] Most of the results from these urban studies are likely to be true for the Northwest, and if not, can at least serve as starting points for studies in our area.

Overall, as urbanization increases, species diversity declines and species densities increase, primarily because of high numbers of a few exotic species.[53, 123] These exotic species may make total abundance and biomass actually higher in urban areas than in adjacent, more natural areas.[73, 123] The decline in species diversity has been observed in ants,[201] click beetles,[151] birds,[16, 73, 93, 94, 136] reptiles and amphibians,[140, 155] and mammals.[198]

A significant impact of urbanization is the removal, alteration, and replacement of the natural vegetation.[17] The resultant urban vegetation differs from presettlement vegetation in several ways: species composition is altered, sterile varieties are common, and most importantly, woody plants are sparsely distributed. Habitats are lost for those species adapted to the naturally occurring vegetation types and structures of an area. In urban areas, bird and small mammal diversity is correlated with both the abundance and diversity of vegetation, and also with habitat heterogeneity;[123] however, highest species richness is not always in the most natural habitats, but often occurs in moderately disturbed ones.[157] Urban landscapes support a variety of wildlife species that are adapted to, or have adjusted to, urban areas.[73, 199] Refer to Table 1 for a summary of the environmental conditions and constraints confronting urban wildlife and the ecological consequences of these conditions. See Table 2 for a summary of adjustments made by wildlife in response to the urban environment.

Birds

A common result of increased urbanization is a decline in number of species and a simultaneous increase in total bird density; i.e., an avifauna comprised of a few exotic species that become very abundant.[14, 56, 73, 123] The proportion of native bird species, especially shrub dwelling, ground nesting, or insectivorous canopy feeders, progressively declines as urban development intensifies.[53] Flocks of gregarious, weakly territorial, omnivorous, and granivorous species replace territorial[73] and insectivorous migrant species.[51, 53] The tempering of the urban climate may reduce prey fluctuation, increase prey biomass, and reduce bird energy requirements, enabling urban birds to attain higher densities, breed for longer periods of the year, produce more young, and have higher annual survival.[134]

The granivorous and omnivorous guilds make up nearly 93% of the total avian biomass of urban areas.[17, 73, 215] Insectivorous species are common in the low-density zone, but are nearly absent from the high- and medium-density zones, especially in winter.[43, 51] Bark drillers and gleaners are reduced in the high-density zone as well.[17] Differences in nesting guilds are also evident. Species nesting on tree branches or buildings are more abundant in the high-density zone, whereas ground, cavity, shrub, and tree-twig nesters either are absent or occur at low densities in this same zone.[56] Scavenging omnivores (e.g., gulls, corvids, blackbirds, and European starlings) benefit from spilled waste.[134] Emlen[73] characterized native species in urban areas as often having wide geographic ranges and broad ecological tolerances, and by definition are opportunistic species (e.g., mourning dove and house finch).

In Canada and France, Clergeau et al.[43] found that the pattern of breeding bird abundance, unlike species diversity, is not directly associated with the urbanization gradient. Savard[172] found a similar pattern in Toronto, with

Table 1. Environmental conditions and constraints of urban areas and the ecological consequences.

Environmental conditions and constraints	Ecological consequences
Altered disturbance regime—	
Landscaping with non-natives	Alteration of type and available timing of food resource. Change in cover type. Constant or varied maintenance.
Suppression of fire	Altered succession, with unpredictable seral stages. Alteration of equilibrium.
Altered phenology	Different pollination schedules. Provides extended duration of season.
Concentrated human populations	Human and associated pet predation. Reduction in large predators. Large amount of human waste. High potential for human disturbance.
Elimination of snags	Elimination of habitat for primary and secondary cavity nesters, and food for many insectivores.
Extended photoperiod—artificial lighting	Increased day length—earlier breeding cycles, multiple broods. Lights act as effective barriers to movements and dispersal.
Extensive lawn or "savannah" habitats	Improved foraging for some species like robins and starlings. Elimination of foraging and nest sites for ground- and shrub- dependent wildlife.
Fragmented landscapes	Small isolated patches of habitat. Small, isolated, and high-density populations, more vulnerable to extinction. Decreased mobility and dispersal potential.
Higher temperatures	Lower energetic requirements. Allows multiple broods. May allow some birds to "winter" over.
Human byproducts—garbage	Increased food sources. Concentration of feeding animals. This dependence may interrupt seed dispersal and pollination.
Human byproducts—sewage ponds	Increased habitat and extended ranges for wetland-associated wildlife.
Human structures—barns, bridges, buildings, chimneys, eaves, culverts	Increased nesting and roosting structures. Increased danger for flying animals.
Human structures—bird boxes	Temporary supplement or substitute for snags and associated cavities. Also supports aggressive exotic birds.
Human structures—bird feeders	Sustains or supplements birds and food. Seed eaters concentrated in these areas. Provide prey concentrations (e.g., for accipiters).
Increased impervious surfaces	Concentration of pollutants, decrease in stormwater percolation, increased sedimentation, etc.
Increased fertilizers	High NPK, increased algal blooms impact other beneficial vegetative growth. Low oxygen, can cause fish kill.
Mobility barriers—lights, roads, vehicles, structures	Wildlife forced to disperse across unsuitable and dangerous habitats. High predator risk.
Power poles, towers, and power lines	Increased nesting and roosting. Provides travel routes (e.g., squirrels). Increased mortality.
Presence of human pets	Increased predation, particularly ground nesters and feeders
Roads	Decreased mobility, increased mortality, isolation. Increase in roadside habitats.
Simplification of habitat—little or no understory or mid layer (conversion to lawn)	Decrease of ground- and shrub-nesting species. Increase in shrub nesting height.

Table 2. Ecological adjustments made by urban wildlife.

Physiological	Behavioral	Life history
Greater longevity	Changes in nesting habits—higher nests.	Reduced migrations, winter residency in areas formerly vacated.
Prolonged breeding season	Changes in feeding and foraging behavior.	Changes in food/diet—human waste products.
More and/or multiple broods	Tolerance of humans. Increased intraspecific tolerance. Change in aggressiveness—increased and decreased.	Changes in nesting habits.
Prolonged circadian cycles	Higher population densities. Increased nocturnal activity, and extended mating and breeding.	Higher population densities.

a higher bird abundance in residential areas adjacent to downtown than in the downtown area itself, a decrease in abundance in old residential neighborhoods with tall trees, and an increase in more open and young residential neighborhoods, followed by a decrease in the least urbanized sectors.

It is the type and volume of vegetative cover, particularly the shrub and canopy layer, and habitat patchiness, that influences avian diversity and numbers rather than species composition.[16, 73] In urban areas, the amount of woody vegetation has been singled out as the most important factor for promoting a diverse native bird assemblage.[199] Numerous studies, mostly from the eastern U.S., have shown a positive linear relationship between bird species diversity and foliage height diversity across the urban gradient.[53, 123] The diversity of birds in urban areas was also influenced by the age of the neighborhood,[127] type of housing,[89] and degree of urbanization.[13] However, in California, Blair[18] found that bird diversity, bird density, and bird biomass peaked at intermediate levels of urbanization compared to the most natural sites. McDonnell et al.[137] predicted that species richness should peak at an intermediate level of development because biotic limitations are high at the rural end, whereas physical limitations are high at the urban end.

Unlike in forested regions, urban development in deserts,[73, 96] shrub, or steppe typically results in increased insectivorous species because of the growth and proliferation of planted and artificially watered woody vegetation. This increase seems to be true only for the short-term; if urban development becomes denser over time, diversity declines. In the arid southwest, Soulé et al.[186] found that the xeric chaparral-obligate species began disappearing a few decades after being isolated by urban development.

The heights of birds' nests in urban areas in the eastern U.S. are higher than in rural areas in response to increased predation, human disturbance, or to decreased amounts of lower-level vegetation.[32, 52, 173] DeGraaf et al.[54] found that various species in the eastern U.S. differed in nesting height in response to housing density, which may be interpreted as a measure of disturbance. They also found that shrub-nesting species respond to available vegetation rather than to the level of disturbance in an area.

In Colorado, Cringan and Horak[48] found that urbanization had a complex impact on raptors. Solitary species with large home ranges were negatively impacted, whereas small-mammal eaters were positively impacted at first, but ultimately their populations decreased as urbanization increased. Bird-eating raptors benefited from the large prey populations in cities, if minimum habitat resources and sufficient protection were available.[48] Cringan and Horak[48] further speculated that as biomass of the urban bird population increased, a behavioral response would probably first occur, with migrant and vagrant raptors feeding on the abundant city birds, particularly during winter. If nesting habitat was suitable, this might be followed by a nesting response. Falcons,

Table 3. Total number of species by association in each urban zone. The number in parenthsis is the percent of non-native species in each category.

	High-density	Medium-density	Low-density
Closely associated	12 (67)	13 (62)	15 (53)
Associated	10 (0)	68 (3)	145 (6)
Present	32 (3)	77 (7)	104 (2)

especially merlins, and accipiters seem to be taking advantage of the increased prey base.[48] The urbanization of the merlin in Saskatoon, Saskatchewan, has been well-documented.[110]

Prey of free-ranging domestic cats was approximately 20% birds in several studies.[40, 78] Coleman et al.[44] estimated between 8 and 217 million birds were killed each year by cats in Wisconsin. Other avian predators that seem to increase in urban areas are sharp-shinned hawks, Cooper's hawks, and merlins.[133] Human predators can also seriously deplete local songbird populations.[133] Yet Martin[131] believes that nest predation is probably the most limiting factor for songbirds, and that nest predators may be more abundant in urban areas than in native habitats, having increased dramatically in the western U.S. during the last century.[133] In Arizona, Emlen[73] concluded that 2 ground-nesting species had been eliminated by disturbances in an urban area.

Mammals

Urban mammals have been studied much less than urban birds. Mammalian species richness seemed to decline rapidly with increasing human disturbance and increasing barren ground,[148] and decreasing distance to human habitation. Species richness increased with increasing density of vegetation per patch, especially in the shrub mid-layer.[60] Small mammal populations in urban areas, like birds, seem to be positively correlated with density of vegetation—residual patches of natural and semi-natural vegetation support the most dense populations.[61, 62, 197] In Oxford, Dickman and Doncaster[61] found that remnant patches of natural or semi-natural vegetation contained the highest density of small mammals. Intensively managed and maintained patches, like parks and golf courses, contained fewer small mammals than expected.[62]

Whereas many specialized native mammalian species are excluded from urban areas, a few native omnivorous and scavenging species benefit from the food and shelter provided by humans. Several non-native species thrive in the urban setting and are more abundant in urban areas than in more natural habitats.[29] As Obara et al.[152] noted, the urban environment is often characterized by the replacement of native by non-native species (Table 3). Small native mammals that do survive seem to show a low degree of territoriality and high intraspecific tolerance. These behavioral characteristics facilitate aggregations of individuals in favorable microhabitats and movements between habitat patches (buildings) that help the small mammals exploit the urban environment. Yet any disruption of these small remnant urban patches will likely affect these concentrated species more than the larger, less concentrated species that move frequently between patches.[61, 197] Soulé et al.,[185] in their study of urban San Diego, found that rodent populations seemed to be particularly susceptible to local extinctions in a fragmented urban environment.

A limiting factor of small mammals in urban areas may be the presence of predators, such as coyotes,[158, 159] cats,[35] dogs,[15] and foxes.[65, 98] Small mammals may comprise up to 70% of the total prey mass of free-ranging domestic cats.[78] These urban predators may depress the distribution and abundance of many small mammal populations and further reduce the availability of prey for small native carnivores.[3]

In many urban areas of the eastern U.S., deer are so common they often become pests.[204] High deer densities have been attributed to the availability of abundant forage, lack of large predators, and hunting restrictions within urban areas. In addition, these large ungulates have become habituated to urban disturbances and tolerate human presence and disturbance.[204] Deer may occasionally attract large predators, like cougars or bears, into the urban area, which can cause human-wildlife conflicts.

There are basically 2 types of bats found in urban environments in Oregon and Washington: the colonial bats including the little brown (probably the most abundant), big brown, pallid, Brazilian free-tailed, and western pipistrelle, and the solitary types including the hoary and silver-haired bats. In urban areas, most bat species are found in fewer numbers and at fewer locations when compared with rural sites.[86] In Michigan, species diversity was twice as high in rural areas as in the city.[122] In this same study, big brown bats were the most common bats of the urban areas, but their numbers dropped significantly in the high-density commercial zone. This may be the result of the lower insect numbers and diversity found in urban areas as compared to rural areas.[82, 122, 187] Although increased foraging activity around artificial lighting in urban areas has been reported by some,[11, 77] in Michigan no difference in foraging activity was detected.[122] Bat foraging behavior may change in urban areas as well.[84] Urban animals may spend more time foraging than rural ones, perhaps reflecting differences in prey density.[86] The biggest problem bats may face in urban areas is disturbance of day and night roost sites, maternity sites, and hibernacula by humans and cats.

Reptiles and Amphibians
Modification of habitat is probably the chief threat of urbanization to reptiles and amphibians.[36, 189] Urbanization may destroy or severely alter upland habitats that are essential for land-dwelling stages or species, and consequently, feeding and refuge (e.g., estivation and hibernation) patches may be destroyed or fragmented.[162] Such impacts increase the risk of extinction by hindering

movements between these upland patches and breeding sites (e.g., wetlands), and the resultant isolation of individuals from these environments and each other.[162] In watersheds with urban development, amphibian abundance decreases and the quality of the remaining breeding habitat is severely impacted by runoff.[162, 163] Clearly, reptiles or amphibians restricted to water for breeding, egg laying, and development are harmed by wetland loss.[97, 140] Even the smallest wetlands are important habitats for some amphibians.[91, 163] Another complicating factor in the conservation of amphibians is their strong philopatry to breeding site wetlands—adults may return to breeding sites several years after the sites have been destroyed.[162, 189] Females of some species of snakes are more prone to be road-killed than are males, because females travel more.[36] Burrowing snakes are better able to survive in urban areas than are surface foragers, whereas larger or venomous snakes suffer great declines from urbanization.[36]

There seems to be a strong correlation of terrestrial-breeding amphibian distribution with large woody debris, dead and decaying wood and organic matter, and other habitat conditions favorable to thermoregulation, foraging, and resting.[162] However, reptiles and amphibians are not as closely associated with vegetative complexity as are birds and mammals. Instead, thin-stemmed emergents, including rushes, sedges, herbs, and grasses seem more important for lentic breeding species.[164] Richter et al.[164] also found a strong inverse correlation between species richness and water-level fluctuation and degree of urbanization. Increases in the duration and frequency of flooding, and the changes in discharge rates resulting from the magnification and increasing frequency of existing storm peaks in urban watersheds decreased the success of lentic breeding amphibians.[27]

Urban development often cuts off and isolates wetlands or riparian areas from formerly connected terrestrial environments. Amphibians are particularly vulnerable to this type of urban habitat fragmentation.[6, 162] An example in Washington is the western toad, which was recently listed as a candidate species by Washington Department of Fish and Wildlife. Culverts may block fish and wildlife passage, whereas road crossings and other developments may pose barriers to wildlife following riparian corridors.[145, 162] Streams in urban areas sometimes are

Table 4. Total number of species occurring in high-, medium-, and low-density urban areas in Oregon and Washington taken from Wildlife Habitats Data Matrix.[154]

	High-density	Medium-density	Low-density
Amphibians	0	5	18
Reptiles	1	12	21 (+1 unsure)
Birds	39	100	149 (+1 unsure)
Mammals	14	41	76 (+1 unsure)
Total species	54	158	264 (+3 unsure)

completely buried in pipes or bisected by roads, making them unavailable or unsuitable for both aquatic and terrestrial species.

Wildlife Profiles of the Urban Zones

Wildlife in the High-density Zone

Fifty-four wildlife species could potentially occur within this high-density zone (Table 4), with 12 species classified as closely associated, 10 as associated, and another 32 species as present.[154] These numbers are derived from a list of all species that may occur in any high-density zone anywhere in either Oregon or Washington. Furthermore, each species will only be present if and when the appropriate habitat is available.

Large, wide-ranging or ecologically specialized species and those that cannot tolerate high levels of human activity and disturbance are excluded from this high-density urban zone. Species that do well are gregarious, weakly territorial, human disturbance tolerant, and dietary generalists.[39, 73, 98] Exotic wildlife species often are most abundant and conspicuous of all wildlife in this zone.[199] See Table 3 for a summary of non-native species in all density zones.

Some ecological features faced by wildlife in this densest zone are the presence of more concrete than soil, an extensive stormwater system, a multitude of roads and associated vehicles, ever-present background noise, artificial lighting, and a highly maintained and manicured landscaping of non-native vegetation (Figure 3). Wildlife dispersal or movement is limited and highly restricted, and often dangerous.[169] As a result, most of the species found in this zone are either birds, or small mammals with small home ranges.

Most native vegetation has been eliminated, and much of the wildlife habitats consist of human-made structures. Rooftops, cornices, eaves, and arches—analogs of cavities, rocky flats, rocky outcrops, cliffs and caves[23]—may provide roosting and nesting habitats for a few species of birds and bats, species adapted to such habitat features. In addition, most of the species found in this zone have adjusted their behavior and foraging habits or are already adapted to take advantage of the waste products of humans.

Birds. Seedeaters and ground-foraging omnivorous species—starlings, house sparrows, and rock doves—are the most common birds found in the high-density zone.[73, 111] These exotics are tolerant of human activity and other environmental disturbances found in urban areas, which allows them to exploit artificial urban structures and out-compete less aggressive native species. All 3 of these exotic species are known to feed on human handouts. The rock dove is a cliff dweller that uses buildings in the downtown area for roosting and nesting.[111] Rock doves are more common in this high-density zone than any other, favoring the industrial and commercial areas where there is an abundance of roosting and nesting sites, especially in older

buildings.[123] The English house sparrow and the European starling nest in almost any cavity they can find, aggressively chasing off other cavity nesters. Savard and Falls[173] found that house sparrows preferred human structures to natural sites for nests. They seemed to prefer crevices, especially holes in eaves, to open nests.[173]

Gulls feed at landfills and trash bins, and use parking lots, schoolyards, and tops of buildings for roosting. Mallards are often seen in parks with ponds or a river, and gulls often ring the shoreline. Nectar (hummingbird) feeders in this zone may attract rufous hummingbirds. Bridges in the high-density zone often are used as nesting structures for several bird species including white-throated swifts, cliff swallows, crows, ravens, and peregrine falcons. The American robin, house finch, and black-capped chickadee tend to avoid commercial and industrial areas found in this zone.[123] Both killdeer and common nighthawks have historically nested on rooftops in urban areas, but there are no recent records of rooftop nesting for the Northwest.[154]

Unusual as it may seem, an observer in this high-density zone may observe a merlin in winter[110] or a peregrine falcon in summer. These two raptors have adjusted to the conditions of this high-density zone, and take advantage of the modified food sources, and in the case of the peregrine falcon, nesting structures of downtown urban areas.[48] Merlins feed on house sparrows[153, 183] whereas many urban peregrine falcons may be targeting rock doves.[9, 113]

The high-rise buildings in this high-density zone have killed millions of birds; Mesure[139] estimated that approximately 100 million birds are killed each year in North America. Klem[116] reported the annual mortality to be between 98 and 976 million birds. Why birds are unable to avoid urban skyscrapers is unknown, but it seems that mortality can be significantly reduced if the lights of these urban skyscrapers are extinguished at night.[116, 117]

Mammals. Extreme urbanization has heavily impacted mammals; there are no native species of mammals listed as either "closely associated" or " generally associated" within the high-density zone.[154] Instead, the black rat, Norway rat, and house mouse are the most common

Figure 3. High-density urban zone. Neighborhoods in Spokane, Washington. (Photograph by Howard L. Ferguson)

mammals of the high-density urban areas; they are more abundant here than in more natural habitats.[29] They are all Old World exotics that have coexisted with humans for centuries. Only in coastal areas may the black rat outnumber the Norway rat.[135, 210] These small mammals use small cavities associated with the plumbing and electrical pathways of buildings or stormwater pipes as burrows, and power and telephone lines as travel corridors,[135] and they have been observed feeding at dumpsters.[95] These exotic urban dwelling species are well adapted for urban areas by having small or reduced home ranges.[39, 98] The house mouse may have a total home range of 10.8 yard^2 (9 m^2).[135] Another common behavior of urban-dwelling species is their ability to tolerate extremely high population densities, densities that may be caused by inhibited dispersal,[120] or because of widespread availability of building debris and rubbish that provides abundant shelter and protection from predators and weather.[61, 62] Newton[147] remarked that adequate forage and protection from predation might permit urban animals to use habitats that would be inferior under natural conditions. Small mammals in this zone escape extinction or genetic attrition because habitat patches are probably close enough to allow limited colonization and gene flow.[60] However, disruption of these small habitat patches will likely affect these small species more than larger species because their small home ranges keep them isolated in one patch, unlike larger species that move more frequently between patches.[61]

Squirrels are common inhabitants of parks or older neighborhoods where there is an open canopy composed of large deciduous trees. In urban parks, squirrels take advantage of human handouts.[95] Raccoons and opossums may be seen in this zone if there is some greenspace or corridor (e.g., underground sewer or stormwater channel) that allows them safe travel to and from the city.

Big brown bats often form maternity colonies inside structures such as houses or churches in downtown urban areas.[10] Brazilian free-tailed bats, which occur primarily in southwestern Oregon, nest under tile roofs and hibernate inside buildings.[135] Bats often use bridges in this high-density zone, if the structure contains appropriate crevices. Hoary bats and little brown bats also have been observed foraging around artificial lights and roosting under bridges in this zone. Bat habitat often can be enhanced in urban areas by ensuring that bridges are designed to accommodate bats. For more information, visit Bat Conservation International's bridge web page at http://www.batcon.org/bridge/ambatsbridges/index.html.

Amphibians and Reptiles. Amphibians generally are absent from high-density urban areas because of the lack of wetlands. According to Stebbins and Cohen[189] and Campbell,[36] modification of habitat (i.e., wetland loss) is probably the chief threat of urbanization to reptiles or amphibians restricted to water for breeding, egg laying, and development.[97, 140] No native amphibians or reptiles are listed as either "closely associated" or "generally associated" within the high-density zone.[154] Only 1 exotic reptile—the red-eared slider—is listed in this zone. It is often inappropriately released in urban ponds or lakes.

Wildlife in the Medium-density Zone

Larger remnant patches of native vegetation and less fragmentation result in a 3-fold increase in the number of wildlife species potentially occurring in this zone compared to the high-density zone. A total of 158 wildlife species potentially occur within this medium-density zone (Table 4); 13 species are classified as closely associated, 68 as generally associated, and another 77 species as potentially present. These numbers are derived from a list of all species that may occur in any medium-density zone anywhere in Oregon or Washington.[154] Each species will only be present if and when the appropriate habitat is available.

Wildlife diversity is greater in this zone than in the high-density zone. Most non-native generalists that were present in the inner city are still present here. However, with increased vegetation and patch size, many native species appear in this zone. Instead of only ecological generalists of the inner city, now seed- (grass-) eaters, and a few insectivorous species are present, especially in older neighborhoods.[203]

Birds. House sparrows and starlings often reach maximum densities in this zone.[123] Starlings occur wherever cover ranges between 35 and 85%, especially near lawns. Rock dove densities are considerably less in this zone than in the high-density zone, but they may occur in such high densities that action is required to reduce their numbers. One solution is to design buildings without holes, cavities, or ledges, and to reduce them on existing structures. Improved methods for food waste disposal and regulation of feeders will also help in their control.[123]

With more snags,[7] a small amount of woody debris, and increased ground cover, many new species can be found in this zone compared to the high-density zone.[134] Primary cavity nesters begin to appear in this zone, as well as an increased variety of secondary cavity nesters. Northern flickers might be seen feeding on ants, and swallows on insects shortly after dawn or before dusk. With high densities of house sparrows and starlings (both cavity nesters) and a limited number of snags, there is often intense competition between these aggressive exotics and native cavity nesters.[108, 109]

Lawns are commonly found in this zone. Flock-feeding species like American robins, Brewer's blackbirds, and starlings may be seen foraging on lawns anytime during the day. Robins are common in this zone—lawns with shade trees seem to provide ideal habitat.[143] In addition, watering of lawns mimics precipitation, causing earthworms to rise to the soil surface.[143] Older neighborhoods in this zone, with expansive lawns and large older-age shade trees, typically support more insectivores and more shrub and tree branch nesters than do younger neighborhoods.[56, 127] The occasional presence of towhees, juncos, and sparrows are evidence of a more dense and diverse understory compared to the high-density zone. Mourning doves may appear if there is enough woody vegetation for them to nest in and some unmowed lawns or grassy edges to forage in.

With the many individual human residences or lots in this zone, most remnant patches of native habitat are isolated, and the amount of edge is often at a maximum. This increased edge now attracts brown-headed cowbirds (nest parasites) and more nest predators, such as jays, crows, skunks, and raccoons. These nest parasites and predators, operating near the edge of an urban area, may have enough impact on birds of small habitat patches to create population sinks (specific areas of population declines) where local reproduction is insufficient to offset adult mortality.[166] Additionally, cat predation is high in this zone and may negatively impact bird populations to the point where these urban residential zones might be considered ecological traps.[41, 90, 106]

Crows are human tolerant and omnivorous, and do well in all zones: >656 different food items have been identified from stomach samples of the American crow.[160]

Bird feeders in this zone are common. Many seed-eating birds are attracted to these feeders in summer and in winter, when snow may be on the ground and food is scarce. Jays may be present, along with white-crowned sparrows, dark-eyed juncos, and evening grosbeaks. The house finch, a new closely associated species of this zone, often monopolizes sunflower seed feeders. Sharp-shinned and Cooper's hawks may be frequent visitors, preying on the concentrated passerines, especially in the winter.[71] Anna's hummingbirds seem to take advantage of the increased urban temperatures and the abundance of artificial feeders in this zone;[218] they can now be found in Portland and Seattle in midwinter.

Because of the increased prey base found in this zone, several raptor species may be observed. Red-tailed hawks, kestrels, and great-horned owls all tolerate human disturbance, allowing them to exploit the available small-mammal prey base.[28, 48] Red-tailed hawks and great-horned owls are generalist predators that seem (if enough habitat is available)to benefit from forest fragmentation and urbanization.[28, 47] Screech owls, also tolerant of people, show up in this zone in forested parks or older residential neighborhoods that have large deciduous trees.[87, 89] Kestrels are able to survive in this zone as well, using the extensive array of telephone wires as hunting perches,[48] and will breed, if cavities are available for nesting.

In this zone, collisions with human-made objects were the greatest indirect cause of avian deaths, accounting for approximately 32% of deaths.[8] The Cornell Laboratory of Ornithology's Feeder Watch Program suggested that approximately 100 million birds are killed annually by striking windows in urban residential areas in the U.S.[70]

Mammals. With less impervious surface in the medium-density zone, more earth is available for vegetative growth and for the digging and burrowing of small mammals, such as moles (Townsend's coast, and broad-footed), rabbits, and gophers. In western Oregon and Washington, a predator of these small mammals is the red fox, a dietary generalist that seems to have adjusted well to human development.

More available habitat, larger patch size, or a combination of the two is evidenced by the wider array of larger mammalian omnivores recorded in this zone. Eastern gray squirrels, raccoons, skunks, and opossums (west of the Cascades) can be quite abundant. Higher densities of these mammals seem to occur because their home ranges are smaller in urban areas than in more rural areas. This social behavioral change has been noted in raccoons,[6, 39, 98, 174] whose densities in urban areas may be twice as high as in comparable rural areas. Raccoons are versatile and take advantage of urban areas; they use sewers and human-made structures for travel while searching for human-supplied foods. Buildings and culverts provide additional den sites when natural dens are unavailable in urban areas. A successful omnivore, the raccoon is known to eat >46 different species of plants, insects, crustaceans, birds, bird eggs, and several small mammal species (including rabbits and Norway rats),[103] and can easily open trash containers.

Another dietary generalist taking advantage of the medium-density environments is the opossum. A native of the southeastern U.S., it was introduced in Oregon in the early twentieth century, and now has spread to all areas west of the Cascades in both Oregon and Washington, with a few being recorded east of the Cascades.[135] One was reported as far east as Spokane, Washington, in 1993.[112] The coyote is present in all medium- and low-density urban zones throughout the U.S..[5, 129] Although coyotes seem to prefer relatively undisturbed habitats in urban environments,[159] they are able to adapt and survive in the highly modified medium-density zone. Coyotes, and other native carnivores, may even help reduce the abundance of house cats and other small mammalian carnivores that prey on songbirds, thereby indirectly contributing to the maintenance of a native avifauna.[158, 186] Ecological specialists such as shrews, moles, and mice may persist in small urban patches, if human disturbance is limited and appropriate habitat components are retained.[61] However, shrews and voles often are absent or rare in urban areas, perhaps because their diurnal behavior may inhibit movements across open spaces.[217] Perhaps this is why only one species, the montane vole, is listed as potentially being present in this zone.[154] The Townsend's mole may be found in lawns, moist meadows, and fields.

With the presence of larger and older trees, more insects, and more water, there is a corresponding increase in the number of bat species. Several additional *Myotis* species (Yuma, long-eared, little brown, and the solitary silver-haired bat) may occupy this medium-density zone, as well as the big brown and Brazilian free-tailed bats that are present in the high-density zone. Bat boxes in this zone may help to supplement natural roosts and cavities. Refer to Bat Conservation International's bat box project for more information: http://www.batcon.org.

Amphibians and Reptiles. Some remnant wetlands persist in this zone. In some cases, reptiles and amphibians are facilitated by the human infrastructure; snakes, frogs,[145] and salamanders[100] may travel in sewers, mains, and pipes. Reservoirs and artificial ponds can be beneficial for turtles and toads.[145] Salamanders and turtles—both native and non-native—may be found in this zone. Reflecting the increased small-mammal prey base, snakes are more plentiful, with garter snakes being the most likely to be seen. The many roads and widespread use of culverts in this zone has had significant detrimental impacts on amphibians and reptile.[162] The ubiquitous introduced bullfrog has taken over many small ponds and lakes in this zone. It requires warm-water ponds, marshes, and backwaters for breeding. The Pacific tree frog may be heard in this zone if small ponds, seasonal pools, temporarily filled depressions, or slow streams are present for it to breed in.

The many raccoons in this zone may negatively impact reptiles and amphibians;[12] great blue herons also help regulate their numbers in this zone. The collecting of herpetofauna by humans can severely impact small isolated populations. The impact of cat and dog predation on reptile and amphibian populations in this zone is largely unknown and needs to be investigated.

Wildlife in the Low-density Zone

This zone most closely approaches predevelopment conditions of the native flora (Figure 4). As expected, species diversity (but not density) also approaches predevelopment levels. A total of 264 wildlife species potentially occur within this low-density zone (Table 4), with 15 species classified as closely associated, 145 as generally associated, and another 104 species as potentially present. These numbers are derived from a list of all species that may occur in any low-density zone anywhere in either Oregon or Washington; each species will only be present if and when the appropriate habitat is available.[154] The actual number of species will vary according to the native background matrix, be it forest, shrub-steppe, or grassland, more than in either of the other two urban zones. Most of the wildlife species found in this zone are native species (Table 3).

Birds. Numbers of avian specialists, including insectivores, may approximate those in undisturbed areas. Swallows, woodpeckers, warblers, and chickadees are numerous. House sparrows may be uncommon, and instead, chickadees, nuthatches, flickers, robins, juncos, and native sparrows are the most common birds observed. Few rock doves are present except around farms or granaries. A more natural ground or forb layer is apparent by the presence of quail and sparrows feeding. Thrushes, indicators of nearly natural ground cover and shrub layers, may be widespread. Gray catbirds may be found in this zone on the eastside and indicate the presence of a well-developed dense understory as well. Western tanagers are present where there is forest cover of large trees with a full canopy.[168] Large trees in this zone also account for the presence of brown creepers and golden-crowned kinglets. With the retention of snags, most of the woodpecker species occur in this zone, including white-headed and Lewis's woodpeckers. Because of the presence of these primary cavity nesters in this zone, new cavities are created each year, promoting use by secondary cavity nesters, e.g., chickadees, nuthatches, bluebirds, and wrens.

Figure 4. Low-density urban zone. Benton County, Oregon. (Photograph by Kelly A. Bettinger)

English house sparrow populations decrease sharply when total vegetation cover reaches 60%.[123] This is the zone where the ground-nesting guilds of birds begin to approach predevelopment numbers, not only for foraging, but for breeding as well. Thrushes and towhees are abundant in many areas. In dormant agricultural fields or active hay fields, savannah, vesper, and song sparrows may be seen and heard, with juncos, golden-crowned, and white-crowned sparrows living on the edge of these fields or in the bordering hedgerows.

Bird feeders are quite common, and bird variety seen at these feeders can be much greater than in the medium-density zone: purple finches, pine siskins, black-capped or chestnut-sided chickadees, red- and white-breasted nuthatches, and even red-winged blackbirds may appear, especially in the winter. At nectar feeders all of the local hummingbirds are likely to occur, with some, like Anna's, even over-wintering.[218] Nest boxes are quite common in this zone, and help supplement natural cavities, attracting western and mountain bluebirds,[101] swallows, chickadees, nuthatches, wrens, flickers, and even pygmy and screech owls.

At dawn and dusk, swallows (violet-green, tree, cliff, bank, and barn) take to the air. Near the coast, purple martins may be present. In wetland areas with dense shrubs, diversity will be high, and birds present may include common yellowthroats, Wilson's, orange-crowned, and sometimes yellow warblers, great blue herons, killdeers, osprey, and rails. Song sparrows occur in most brushy spots, whereas northern orioles favor large open-canopied deciduous trees, along with warbling and red-eyed vireos, black-throated gray warblers, and Hutton's vireo.

Brown-headed cowbird populations may be quite high in this zone because of abundant livestock, and high nest parasitism may occur. However, nest predation may be less common in this zone than in other zones, probably because of better habitat for the host species, more dense cover, and less edge.[181, 212] In this zone, ravens begin to outnumber crows, and in the winter, large groups of ravens may be seen together at road-kills, along with the occasional bald eagle.

With the native prey base at near natural levels, so too will be the predator populations. Hawks and owls are present and occur in this zone for foraging as well as for nesting. Northern harriers may also occur here, especially in and around agricultural or hay fields.

Mammals. Most mammal species present before urban growth occurred will be present in this low-density zone, except large carnivores. All of the large ungulates occur in this zone, including deer, elk, and even moose (e.g., Spokane, Washington). Deer are often most abundant in this zone because of the mix of agricultural fields and reduced hunting pressures. Because of high prey densities, cougars may occur in this zone.

Mustelids recorded in this zone include the long-tailed weasel and mink. The deer mouse is now the most common mouse, replacing the house mouse. Raccoons, skunks and opossums are still present but in lower, more natural densities than in the inner zones. Now, with the presence of downed wood, chipmunks and native squirrels occur. Rarely will the non-native Eastern gray squirrels be found in this zone, outside of parks and golf courses. Instead, Douglas and red squirrels become more common. Ground squirrels may be visible along the road banks. Perhaps, reflecting the amount of soil available to diggers in this zone compared to the two other zones, 2 shrews, 2 voles, and 3 moles may occur. Pocket gophers are regionally common.

With the increase in prey, coyotes also become increasingly common in this zone. More large areas of native vegetation are now present that can be used as cover by coyotes, who travel in and around the scattered homes and search for food.[158, 159] Black bears may be found foraging. Populations of bobcats will be present unless pets are overabundant or overgrazing is prevalent.

With more wetlands, muskrats, beavers, and mink may occur. Bats depend on these small wetlands for food and water, especially the mosquito loving little brown and Yuma myotis bats. Preservation of maternity and hibernacula sites is a high management priority for these animals.

Amphibians and Reptiles. Instead of seeing only the ubiquitous tree frog or the aggressive non-native bullfrog, both the spotted frog or western toad may be found in wet areas. Most of the expected native salamanders and lizards may also be found in this zone in appropriate habitats. Toads may be seen in and around the many lawns of residential areas where landscaping and irrigation practices provide adequate food and shelter for survival. They rely heavily on Hymenoptera found in the lawns.[12]

Within the urban environs, this is the zone most suited for amphibians and reptiles; Richter[162] pointed out that areas with <15% impervious surfaces are best. Concern here should be to limit runoff, maintain large buffers around wetlands, stabilize water level fluctuations of wetlands, and protect shorelines where spawning occurs, particularly northwestern shorelines that are preferred.

Faunal diversity declines sharply in streams in basins with >10-15% impervious surface. Macroinvertebrate diversity is consistently rated "poor" when watershed imperviousness exceeds 10-15%.[176] Species that employ feeding strategies such as filtering organic matter or preying on insects are especially sensitive to changes triggered by increases in impervious surfaces. Fish diversity also drops sharply at this level of impervious surface.[27, 176] Abundance and spawning success of sensitive species are also affected at >10% impervious surface. Trout and other salmonids are particularly susceptible to changes triggered by increases in impervious surfaces.[176] Even with complete retention of streamside buffers, basins with >10% impervious cover show measurable levels of stream degradation.[190]

Wildlife in the Urban Fringe Areas

The urban fringe area may be defined as the region of transition between the rural area and the low-density urban zone.[30, 171] Nilon and VanDruff[148] characterize the urban fringe as being predominately vacant with agricultural lands dominated by large amounts of growing space. Only higher use of rural roads may distinguish this area from the neighboring rural zone. Unlike the rural zone, neither agriculture nor the timber industry is the primary employer, and as a result, the use and management of these lands is dissimilar to the intensely farmed fields or harvested woodlots of the rural areas. In this urban fringe, compared to rural areas, hedgerows are often more lush, fields more often idle, pastures unoccupied, timber rotations longer, and tree harvests more selective. In the urban fringe, farmland left idle before its urban development may temporarily improve the forage base available to small mammals, which can subsequently support raptors and other small mammal predators.[48, 148] Most important, it is this area that is in imminent danger of being developed and subdivided.

Land in this fringe area is less fragmented and often less expensive than in any of the three urban zones. Protection and conservation of wildlife habitat can often be achieved at a lower cost, and often with less impact to the citizens of the region. Typically many large tracts of private and public land remain that can be identified and preserved as important wildlife corridors or core areas. Although planners and biologists often overlook the urban fringe area, it may offer the greatest opportunities for conservation and landscape-level management.

Key Urban Planning Issues

Growth Management

Throughout the U.S., urban growth and development is one of the largest threats to wildlife diversity. Oregon and Washington are two of only about a dozen states with comprehensive statewide "growth management" legislation. Oregon's Comprehensive Land Use Planning Program has its roots in laws passed in 1973. Oregon measures its land use planning successes by attempting to meet 19 statewide planning goals. Washington began taking a proactive approach to managing growth with the passage of the Growth Management Act in 1990. Growth management has changed local planning in many ways, and provides opportunities for managing fish and wildlife resources.

Controlling Urban Sprawl. Growth management requires local jurisdictions to control urban sprawl. Urban sprawl uses land inefficiently, is typically damaging to natural resources, and is expensive to provide for public services. Urban sprawl has been defined as:

> *Unplanned, uncontrolled, and uncoordinated single-use development that does not provide for an attractive and functional mix of uses and/or is not functionally related to the surrounding land use and which variously appears as low density, ribbon or strip, scattered, leapfrog, or isolated development.*[146]

Oregon and Washington both address controlling urban sprawl in their statewide goals under their growth management legislation. The intent of Oregon's Statewide Planning Goal 14 is "To provide for an orderly and efficient transition from rural to urban land use". In Washington, two of the planning goals are to "encourage development in urban areas where adequate public facilities and services exist or can be provided in an efficient manner "and to "reduce the inappropriate conversion of undeveloped land into sprawling, low-density development". Both states attempt to do this by requiring local jurisdictions to identify Urban Growth Boundaries (Oregon) and Urban Growth Areas (Washington). These are essentially the same thing, an administrative boundary that marks where urban development ends and rural areas begin. Urban Growth Boundaries/Areas (UGB/A) are included as part of a local jurisdiction's comprehensive plan and are designated using population growth projections (Figure 5). For the purposes of the Urban Wildlife Manager or Urban Planner, the UGB/A marks where the high- and medium-density zones end and the low-density zone begins.

Protecting Fish and Wildlife Habitat. Before the passage of growth management legislation, fish and wildlife habitat was given a backseat when local jurisdictions planned for various developments. Notable exceptions to this were habitats for species protected under the Federal Endangered Species Act, and wetland resources protected under the Federal Clean Water Act. Oregon and Washington now require local jurisdictions to consider and protect habitat areas for non-listed species as well as special habitat areas including riparian areas.

Oregon's Statewide Planning Goal 5 seeks "to protect natural resources and conserve scenic and historic areas and open spaces." Local governments are required to go through a process in which they inventory and evaluate Goal 5 resources, which include riparian corridors, wetlands, and other wildlife habitat. After the inventory and evaluation process, the local jurisdiction, in consultation with the appropriate state agencies, must decide whether to protect the resource, fully allow conflicting uses, or limit conflicting uses in some fashion. An example from Oregon is the Zoning Code in Deschutes

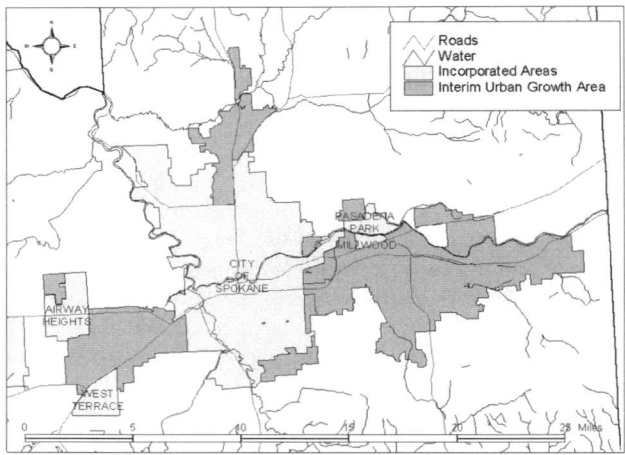

Figure 5. Interim urban growth boundaries for Spokane County, Washington, 1998.

County that includes a Wildlife Area Combining Zone. It applies to areas identified in Deschutes County's comprehensive plan as deer winter range, significant elk habitat, antelope range, or deer migration corridor. The wildlife area zone functions as an overlay zone that changes the permitted uses in the underlying zone. For example, golf courses and dude ranches, while permitted conditionally in the rural residential zone, are not permitted in the wildlife area zone. The zoning code also includes an overlay zone that protects sensitive birds, mammals, and wildlife areas like great blue heron rookeries and Townsend's big-eared bat sites.

Washington's Growth Management Act (GMA) requires all local jurisdictions to identify and protect critical areas, and identify natural resource lands. Critical areas include wetlands, aquifer recharge areas, fish and wildlife habitat conservation areas, frequently flooded areas, and geologically hazardous areas. The GMA directs the State Department of Community Trade and Economic Development (DCTED) to develop minimum guidelines for identifying critical areas. Local jurisdictions must then identify their critical areas at least to the minimum standards identified by DCTED. Then they must draft and implement regulations that protect critical areas. These protective regulations are often referred to as Critical Areas Ordinances (CAOs).

Spokane County's CAO provides additional protection for wetlands, geologically hazardous areas, and fish and wildlife habitat conservation areas. The county has adopted the Washington Department of Fish and Wildlife's (WDFW) Priority Habitats and Species (PHS) program to designate fish and wildlife habitat conservation areas.[205] The PHS program includes distribution maps of species and habitats considered to be priorities for conservation and management. When a proposed development occurs within a priority habitat or within a certain distance of a priority species nest site, the developer is required to submit a Habitat Management Plan that documents how the development will avoid and mitigate the habitat or species in question. Mitigation measures can include establishing buffer areas, clustering development and preserving open space, seasonal restriction of construction activities, and preservation of critically important trees and plants. Habitat Management Plans are reviewed by WDFW. Spokane County grants final approval.

Planning for Open Space. Growth management legislation in Washington and Oregon requires local jurisdictions to consider and plan for open space. Historically, policies discussing the need for open space for recreation and wildlife habitat were included in local comprehensive plans. However, these policy statements were typically not implemented. Planning for open space is required in both Oregon and Washington. In Oregon, "open space" is included as one of the Goal 5 resources that must be inventoried and evaluated. Washington's GMA requires that local jurisdictions planning under the act include a Land Use Element in their comprehensive plan. Among others, "open space" is identified as one of the land uses to be designated and described. There may be some disadvantages to open space systems. They can be expensive to develop and local tax revenues (or potential tax revenues) may be reduced. However, the benefits derived from open space by both wildlife and humans normally greatly outweigh any disadvantages.

Land-use Planning Tools. Usually, growth management makes use of existing land-use planning tools like comprehensive plans and development regulations. However, under growth management legislation, policies identified in comprehensive plans are required to be consistent with the ordinances that implement them. Wildlife habitat goals can be achieved using many of the planning tools currently in use by local jurisdictions—both regulatory and voluntary. Traditional land-use planning involves the use of different land-use zones. Zoning regulations dictate permissible uses in the zones and building densities.

Transfer of Development Rights programs allow a developer or landowner to transfer their "right" to develop property from an area where development is not wanted to a more appropriate area. These areas are often referred to as "sending" and "receiving" zones.

Current Use Taxation programs give landowners the ability to reduce the amount of property tax they pay on a parcel if they agree to keep it in open space or wildlife habitat. Local jurisdictions can develop rating systems where a landowner would receive a greater tax break for more valuable wildlife habitat, e.g., riparian areas or wetlands.

Educational Programs can protect or enhance urban wildlife habitat through voluntary efforts on the part of the landowner. The National Wildlife Federation's Backyard Wildlife Habitat Program provides information for urban homeowners on bird feeding, landscaping for wildlife, and other small projects that enhance habitat. The WDFW's Backyard Wildlife Sanctuary Program provides similar information specific to the Pacific Northwest.

Purchase of the land or easement rights, either outright, partial, or fee simple is often the most realistic way to protect valuable wildlife habitat. This can be achieved through outright acquisition, or with conservation easements, which allow the landowner to retain ownership of the land while prohibiting certain activities that would be harmful to wildlife. Easements can be written in many different ways. Activities that are typically prohibited in conservation easements include building, grazing, and clearing of vegetation. Public access can often be written into an easement but is not a requirement. Local governments often seek the help of land trusts or organizations with land trust skills.

Clustering is often used in planned developments to integrate open space and/or wildlife habitat. Buildings or homes are grouped, or clustered, in a small area of a development while the remaining land is retained in common ownership. Local jurisdictions occasionally allow for density bonuses when developers use this technique.

Open Space, Greenways, and Corridors

One of the most politically acceptable ways to preserve and protect wildlife habitat in and around urban areas is through an open space program. Greenspaces, open spaces, wildlife corridors, natural parks, and trails—all of these terms refer to parcels of open land scattered throughout a metropolitan region. Greenspace not only benefits wildlife but is also recognized by the public as providing important aesthetic and recreational benefits to the community. As a result, communities are often very supportive of greenspaces and open spaces. As Ludwig[128] noted, properly planned open spaces help reduce the isolation and fragmentation of habitat patches, communities, and ecosystems. They also help to ensure and enhance the movement and dispersal of animals, which promotes genetic exchange and population stability.[128]

Selection and Design Basics. When defining the vision and priorities for a green or open space program, it is important to look at the structure and use of the surrounding landscape and how each natural area fits within the region as a whole. Several authors have contributed guidelines for the design of open spaces.[58, 59, 214] Noss[149] admitted that in the absence of detailed autecological, species-specific information, some empirical generalizations for reserve design stand out. The following is a compendium of ideas from Thomas et al.,[195] Noss,[149] and Noss and Cooperrider:[150]

1. Maintain large, intact patches of native vegetation by preventing fragmentation of those patches by development. Remember that blocks of habitat close together are better than blocks far apart.
2. Maintain connections among wildlife habitats by identifying and protecting corridors for movement, remembering that habitat in contiguous blocks is better than fragmented blocks and that interconnected blocks are better than isolated blocks. Focus on the idea of connectivity.

3. Minimize edge. Round patches are better than linear patches.
4. If possible, establish priorities for species protection and protect habitats that constrain the distribution and abundance of those species.
5. Protect rare landscape features or ecological hotspots. Guide development toward areas of landscape containing "common" widespread features.
6. Maintain significant ecological processes in protected areas. Provide enough land for processes to continue.
7. Contribute to the regional persistence of rare species by protecting some of their habitat locally.
8. Balance the opportunity for recreation by the public with the habitat needs of wildlife. Remember open spaces that are roadless or otherwise inaccessible to humans are better than roaded and accessible habitat blocks for wildlife.

Open space selection should also be done on a landscape-level scale. An open space system should be planned on a regional, multi-county scale, ensuring connection of high-quality habitats with all adjoining counties and protection of these habitats throughout the region. Take advantage of all lands, state or federal, already protected, incorporating them into the open space system. A tool to help design and carry out an open space plan is the Geographical Information System (GIS). Most states have already completed a GAP analysis—a GIS-based mapping inventory of land ownership and protection assessment.[38, 177] This analysis and the associated data can be used for a more detailed analysis required for a localized open space plan.

Riparian, Wetland, and Stormwater Systems

Urbanization affects the natural drainage systems of an area, its streams, wetlands, and associated riparian areas and buffers. This section is broken into three main parts: direct impacts, indirect impacts, and management. Stormwater control structures are included in the management section, as these systems are the primary tools available to mitigate the impacts of urbanization on aquatic and riparian systems. In some cases, zoning and local ordinances, for example, those specifying buffers, can also be instrumental in protecting local water resources.

Urbanization results in two basic types of impacts on drainage systems—direct and indirect. Direct impacts include dredging, straightening, bank stabilization, piping, filling, buffer removal and reduction, and fragmentation or isolation from other components of the landscape. Indirect impacts result when the pervious to impervious surface ratios of the watershed are altered. As little as 10% impervious surface in a watershed can change the hydroperiod of the natural drainage systems.[26, 27, 176] Altered hydroperiods result in habitat changes that affect the fish and wildlife in these systems. Impervious cover has especially important impacts on storm surges and concentrations of pollutants in streams.[192, 209] The direct and indirect impacts are interrelated and rarely occur separately. This distinction forms a useful way to frame the discussion.

Unfragmented aquatic systems with large (>656 feet [>200 m]) buffers will contain many wildlife species typically found in these habitats in non-urbanized watersheds, such as kingfishers, eagles, waterfowl, beavers, salmon, otters, and amphibians.

Direct Impacts
Direct impacts of urbanization on water systems include (1) loss or reduction of riparian habitat or buffers, (2) loss of woody debris and other instream structures, (3) degradation or loss of stream channels and wetlands, (4) reduction in water quality, (5) habitat fragmentation, and (6) introduction of non-native species and pets. Many of these direct impacts will result in secondary changes in the aquatic environment.

Loss or Reduction of Riparian Habitat and Buffers. In Washington, an estimated 50-90% of the riparian habitat has been lost or modified.[119] Remnant riparian habitat may provide the only remaining connectivity, refuges, and reserves of habitat within an urban area. The function and value of urban riparian zones can vary widely depending on their width and connectivity to habitat areas outside the urban zone.

Urbanization often results in a direct loss of riparian zones and aquatic buffers as the natural vegetation is either trimmed or removed and converted to impervious surfaces and human structures. Bank stabilization activities also contribute to the loss of aquatic buffers. The loss of riparian vegetation results in a direct reduction in habitat for terrestrial wildlife species. Removal of riparian vegetation, as well as alteration of organic inputs and hydrologic regimes by expanding urbanization, generally results in increased erosion, increased algal production, changes to temperature regimes, and reduced concentrations of dissolved oxygen.[209] Sedimentation may be increased as well as nutrients tied to sedimentation. Smaller buffers result in less filtering of pollutants, such as artificial fertilizers and pesticides from runoff, resulting in increased nutrients and other pollutants in the aquatic system.[119]

Loss of Woody Debris and Other Instream Structures. Typically, large woody debris on the ground and standing snags and other large trees are removed. Hazard tree removal and reduction of the riparian zone directly removes potential large woody debris from both the buffer zone and from the aquatic system; this results in the loss of pools, riffles, backwaters, and off-channel habitats. In addition, large woody debris is removed directly from streams and rivers for navigation enhancement and flood control.

Degradation or Loss of Stream Channels and Wetlands. Direct impacts include channelization, dredging, filling, bank stabilization, culverts, and piping, which may all result in stream degradation. Wetlands are frequently filled in urban areas, resulting in habitat losses and further alteration of the water regime in a basin. Rivers and streams are dredged and straightened, which removes and alters the stream bottom conditions and habitats.

Straightening results in greater stream velocities and increased stream channel instability. Streams may also be buried completely in pipes or bisected by roads, affecting the connectivity of the whole system. Bank stabilization, loss of riparian buffers, dredging, channelization, and filling can all increase the amount of fine sediments present in a stream system. Fine sediments can smother bottom invertebrates and salmon spawning gravels.

Many of our regulatory systems assume that small wetlands are expendable. However, most natural wetlands are small and are vital to maintaining ecological diversity.[178] Richter and Azous[163] found no correlation between wetland size and amphibian diversity. The authors concluded that the loss of most small wetlands could severely impede source-sink processes and place remaining wetlands at increased probabilities of local amphibian population extinction.

Reduction in Water Quality. Impervious surfaces collect pollutants that wash off during storm events and may be deposited in large pulses directly into aquatic systems unless there is some sort of filtration or other water quality control structure present. Improperly applied herbicides and pesticides may directly affect riparian and aquatic flora and fauna. Reduction of riparian buffers reduces the capacity of the natural system to filter out these pollutants and nutrients and exacerbates the problem. The size of the pollutant load is directly related to the level of impervious surfaces in the watershed.[119] Fertilizers applied to lawns, gardens, golf courses, and other urban landscaping are carried to surface waters in stormwater runoff. Land clearing, which exposes soil, can result in high levels of phosphorus in stormwater runoff. These excess nutrients can enhance the growth of invasive aquatic plant species. Decomposition of excess plant growth can rob the water of dissolved oxygen, further reducing its suitability for aquatic fauna.

Runoff from parking lots and other impervious surfaces tends to be warmer, which will increase the overall temperature of the receiving waters regardless of buffer size. Combined with the loss of trees that would have provided shade, temperatures in impervious surface areas may be 10-12°F higher than in the forests and fields that they replace.[176] Increased summer temperatures in urban streams are directly related to the level of impervious surfaces found in the basin. Lack of riparian cover and stormwater detention ponds that warm water before releasing it into streams and wetlands can amplify the observed increases in water temperatures.

Indirect Impacts
With urbanization, runoff rates increase, resulting in significant increases in the frequency and the duration and frequency of flood events. Changes in runoff rates and volumes may be more severe in western Washington and Oregon, where stormwater is normally stored and carried off with subsurface flows, than in eastern towns where stormwater in more typically carried across the land surface. Peak discharges increase and urbanized basins may experience entirely new peak runoff events.[25]

Urbanization alters the volume, velocity, duration, and frequency of runoff events that can alter the structural integrity of aquatic systems that may not be otherwise impacted by urban developments.

Channel Instability and Runoff. Increased runoff in urban areas results in additional soil erosion, stream channel cutting, and water quality deterioration. Channel instability triggers streambank erosion and habitat degradation. Channel instability also expresses itself in the loss of instream habitat structures such as pools and riffles, large woody debris, and overhead cover.[27, 119] Channel cutting and increased erosion can result in channels becoming straightened, with meanders, back channels, and overflow channels being eliminated or cut off from the main channel. As channel scouring increases there is loss of substrates needed for salmonid spawning and aquatic invertebrates. Stream channel erosion downstream of culverts (plunge pools) can create barriers to fish passage where they did not exist when the structures were installed. Sedimentation and deposition upstream of culverts similarly impacts aquatic fauna by burying spawning gravel beds and destroying pools and riffles.

Water Table Alterations. More rapid runoff of stormwater results in decreased infiltration and groundwater recharge. Wetlands and streams may transform to drier, more intermittent, or seasonally flooded conditions, impacting native plant and animal species adapted to a longer wet season. Increased runoff may also result in wetlands holding more water for a shorter period than under undisturbed conditions. These changes in the water regime with urbanization alter the vegetative community of wetland areas.[6, 45, 75, 105]

Water Level Fluctuations. Increased runoff changes the natural pattern of water level fluctuations. In the greater Seattle area, water level fluctuations >7.9 inches (20 cm) are commonly seen in basins with >50% total urbanization. These alterations in the natural pattern of water level fluctuations are correlated with decreased amphibian species diversity. Amphibian species that attach their egg masses to vegetation are extremely vulnerable to water level changes. Some researchers believe that such species will not be able to survive in basins with moderate to high levels of urbanization.[6, 162, 163]

Management
Management of urban riparian and aquatic systems should take a three-pronged approach. First, manage the amount of impervious surfaces and vegetation clearing throughout the basin. Second, work to keep the riparian systems as natural as possible through maintenance of connectivity and adequate buffer widths. Third, manage stormwater runoff to control the impacts of altered hydroperiods.

The amount of impervious surface in a basin is closely linked to the health of the aquatic systems in the basin. Whereas maintaining <10% impervious surface is not a realistic goal for most urban areas, urban managers should try to minimize impervious surfaces as much as possible through such techniques as landscaping requirements, clearing restrictions, use of pervious surfacing materials, and under-structure parking. Basin-wide clearing restrictions that control both the amount and the timing of clearing can significantly mitigate the impacts to water quality and hydroperiod.

Maintaining connectivity and adequate buffer widths will help riparian systems retain natural functions and values. Connectivity should include continuity between instream structures, the connectivity of the riparian corridor throughout the urban area, and the connectivity of wetlands to protected upland habitats as well as other aquatic systems. Landscape-level planning should account for these habitats before development. Strategies might include using bridges rather than culverts, daylighting impacted stream corridors, targeted acquisitions and incentive programs for riparian protection, and restrictions on development in riparian areas. Although McGuckin and Brown[138] suggested integrating municipal stormwater management facilities into urban greenways, this is not a good solution. These stormwater detention ponds experience extreme water level fluctuations and consequently become amphibian mortality sinks.[6, 163] To minimize this effect: (1) separate stormwater detention ponds from natural areas, (2) do not allow vegetation with stem diameters of 0.12-0.20 inches (3-5 mm) to grow in detention ponds, (3) construct sides of ponds to be steep to minimize areas of appropriate water depth for breeding, and (4) construct curbing or fencing around the ponds to discourage use by amphibians.[6, 162, 163, 164]

Typical regulatory buffer widths of 50-100 feet (15.2-30.5 m) are designed for maintenance of water quality.[119] Vegetated buffers of 75 feet (23 m) do a good job at filtering pollutants and nutrients from stormwater runoff and moderately good job of temperature regulation.[119] Buffers 100 feet (30.5 m), however, do a poor job of providing wildlife habitat, large woody debris recruitment, sediment filtration, or maintenance of natural hydroperiods.[119] Buffers should also be wide enough to allow for natural channel migration.

Most urban buffers will benefit from active vegetation management. Noxious and invasive weed species should be removed or controlled. Management may include long-term replacement of non-native vegetation with natives to lessen erosion or impact on wildlife. A strong component of conifer trees should be maintained to provide the best source for long-lasting large woody debris in both the aquatic and terrestrial portions of the system. "Hazard" tree removal should be carefully monitored. Where trees are determined to be hazards, every attempt should be made to shorten them to a safe height while still maintaining a standing wood component. Portions of "hazard" trees that are removed should be placed either in the aquatic or the terrestrial portion of the system. Large trees of any type should be maintained in the buffers for bank stabilization and temperature regulation.

The primary emphasis should be on protection of existing natural systems. Replacement and restoration, although commonly attempted, have had limited success. Of 40 wetland mitigation projects monitored in King County, 11 were not installed and 79% of the rest were unsuccessful by current performance standards. Only 1 site successfully replaced the functions of the impacted watershed.[141]

Management of stormwater runoff in urban basins is critical to maintaining the functions of natural wetland and riparian systems. Stormwater detention systems must be big enough to accommodate the increased runoff that is generated from urban basins. They also need to be designed to moderate the changes in velocity, volume, duration, and frequency of discharges to natural systems. Stormwater systems in urban areas should also be designed to filter pollutants and oils and to remove excess sediments from stormwater. The ideal stormwater control system would also provide some method to moderate discharge temperatures and dissolved oxygen levels. Stormwater detention facilities should not be confused with natural wetland systems. Stormwater detention ponds that are functioning properly will catch and experience the severe water level fluctuations generated in urbanized basins. If the stormwater system is catching sediments properly, it will need to be dredged on a regular basis, which will prevent more than herbaceous vegetation from becoming established. Stormwater basin maintenance schedules should be timed to have the least impact on wildlife species that may be using the limited habitats provided by the facility. Stormwater control structures are not intended to provide or replace wildlife habitat; they are intended to protect existing wildlife habitat from the impacts generated by urban development.

Urban Parks, Golf Courses, Cemeteries, and Open Space

Urban parks and open spaces are rapidly assuming a central role in the protection of native wildlife from urban-related disturbances.[42] In addition, the popularity of golf has increased greatly in the past 20 years resulting in many new and expanded golf courses.[144] Many of these are at the interface between expanding population centers and diminishing rural lands[142] (Argyle et al., Greater Vancouver Regional District, Development Services Department, Burnaby, British Columbia, Canada, unpublished data). Cemeteries increase as the urban population increases. In many urban settings, parks, golf courses, and cemeteries often contain the last remaining patches of native vegetation. Unfortunately, not all wildlife species can exist within these small islands. The diversity of species present in these areas is generally lower than in the original habitat.[60, 92, 184, 186] Little information is available concerning the effects of urbanization on small mammals and the ability of parks to maintain community structure in urban areas.[1, 148] Dickman and Doncaster[61] found that intensively managed and maintained patches of habitat, like parks and golf courses, contained fewer small mammals than expected. However, with design changes and different management approaches, golf courses, parks, and cemeteries can be significantly improved in order to provide habitat for many native wildlife species.

The following factors are important in determining the richness, diversity, and abundance of bird species in an urban park setting:[42] (1) size of area, (2) amount of edge, (3) presence of natural vegetation, (4) amount of shrub cover, (5) average canopy height and canopy cover, (6) density of adjacent buildings, (7) presence of water, and (8) availability of snags.

When designing these areas and during construction, it is important to minimize the impact on the natural vegetation of the area. Large spatial scale or landscape-level planning is essential. Try to protect the largest habitat patches possible, especially those containing landscape-level habitat elements like rivers, lakes, and woodlots. Riparian corridors should always remain undisturbed and protected with appropriate buffers. Attempt to make golf course fairways as narrow as possible leaving undisturbed native vegetation on the sides for "roughs". Route the course through existing openings.[92] In urban parks, leave as much native vegetation as possible around the perimeter. Better yet, design in connecting corridors to larger native patches offsite, even if they are small. If only small native patches are available, an array of small patches can potentially serve the function similar to a single, large patch, by providing the same habitat components. The key is that they be close together and connected with similar habitat. Consider clustering of facilities, buildings, roads, and parking lots; a similar technique is being used in residential development. In addition, position these areas of disturbance in low-quality habitat patches, leaving high-quality patches undisturbed.

Pesticide and insecticide use in these areas can be greatly reduced with the adoption of Integrated Pest Management (IPM) practices. IPM is a management strategy for dealing with insects, weeds, and plant diseases that establishes acceptable levels of infestation or damage. Treatment only occurs if these levels are exceeded. IPM treatment uses cultural practices and biological controls that start with the least harmful to the environment and progresses to harmful chemicals only as a last resort.[92] Where pesticide levels are low, birds become a natural ally for managers, helping to maintain insect levels below outbreak levels.

Another factor that can contribute to increased wildlife use, while at the same time fit into IPM practices, is the use and propagation of native plants rather than exotics. Exotic plants typically require more water, are less resistant to insect infestations, and are often sterile, providing no pollen or seeds for wildlife. By using natives, the escape and spread of exotics are also minimized. The removal of native shrubs and ground covers in parklands alters the nesting habits of birds.[32] Altering the native forest in Seattle through planting of formal gardens, or clearing natural brush and reducing park size, were associated with decreased overall bird diversity, fewer regularly occurring species, and a greater proportion of species typically associated with urban environments.[85]

Specialized habitats like butterfly or hummingbird gardens can be established even in the smallest patches of habitat in our parks, golf courses, and cemeteries. Many of the same flowering plants that attract hummingbirds will also attract butterflies.[57, 121] These gardens will generate much public interest and can serve as education points.

Wildlife use, especially by wading birds, waterfowl, and shorebirds, can be greatly enhanced by the presence of a pond or lake. Especially important for shorebirds is the practice of timing drawdowns with the spring and fall migration flights of these birds. During migration, these birds search for food in areas of shallow water (<7 inches deep [17.8 cm]), especially where there are exposed mudflats. Drawdowns should occur in spring roughly from late March through early June, while fall migration peaks during August and September.[92] Other management considerations include: (1) winter flooding protects the invertebrate eggs and larvae (food for birds), (2) slow drawdowns maintain soil moisture and encourage establishment of vegetation, and (3) working vegetation into the soil releases nutrients for invertebrates. Drawing down the water level gradually will continually expose new areas of saturated soil and keep a steady supply of invertebrates available to birds.[92] To protect wildlife nesting and roosting in these aquatic environments, good management practices dictate that trails and roads should not be built completely around these natural features. Continual disturbance by humans and their pets will chase away all but the most tolerant wildlife from the area.

Snags are used by >85 North American bird species, including many owls, woodpeckers, chickadees, swallows, and other songbirds as places to find food, watch for prey, build homes, and attract mates.[193] Snags are so valuable to wildlife that their protection needs to be an integral part of any forest habitat. Snags to preserve should be >6inches (15.25 cm) in diameter—the larger the snag the more opportunities for nesting and foraging. Snags can be made more attractive by trimming back the branches; however, it is best to leave 1-3 feet (0.3-0.9 m) on each limb for perching. If a snag is close enough to be a threat to a building or structure, it can be topped to a height to eliminate the threat. Snags are more beneficial when left in clumps of >3. Recommended snag densities are 3-16/acre (7.4-39.5/ha). Eventually a snag will rot to the point that it will fall. Let fallen snags lie, they are still valuable to wildlife and the environment.

Gavareski[85] emphasized the importance of size and vegetative condition of parks for attracting and sustaining wildlife; the larger the park and the more natural the vegetation (fewer exotics, natural understory), the more wildlife species will be found. Cicero[42] also pointed out the importance of buffers in isolating species from the negative influences of urbanization, and in providing avenues of faunal interchange among ponds and peripheral habitats. The number of trails should be limited in these areas and judiciously placed away from sensitive breeding or roosting areas.

Managing Urban Habitat for Wildlife

Backyard Wildlife in Residential Areas

The key to attracting wildlife is providing proper habitat. Generally, this entails increasing the abundance and variety of native vegetation on your property. The basic requirement for supporting wildlife in your backyard is to provide food, water, shelter, and places to protect young.

To determine management activities that can either enhance, or in some cases, discourage or repel a particular species, consult the management activity links matrix.[154] Using this matrix, you can determine the habitat elements that you can affect. Or when planning, this database can be used to predict potential losses and gains of species when different land use or management actions are proposed.

Food. Imitate nature. Provide a variety of grasses, forbs, shrubs, and trees that provide fruits, nuts, and seeds through the year. Plant a variety of annuals and perennials, and evergreen and deciduous trees of varying height. Use plants that naturally occur in your area and are not introduced by humans. Native plants are adapted to your local environment and the local wildlife species have evolved along with these plants. Benefits include less maintenance, less water, and less disease, with better survival over time and food that wildlife can use. Keep lawns to a minimum.

Water. A year-round supply of water will help to initially attract and retain wildlife in your local area. The source of water can vary from a pie plate to a stream, a birdbath, or a pond.

Shelter. Wildlife need a place to rest, escape from enemies, and seek shelter from inclement weather. Evergreen trees and shrubs or thorny bushes add year-round protective cover from predators. Deciduous shrubs offer effective summer cover. Rock, log, slash, and mulch piles offer other cover. Clumping your plantings will help provide cover. Vines on a fence or wall or snag can provide other forms of shelter. Don't cut down dead trees; leave them standing, and when they do fall, let them lie. Not only do they provide shelter, but food and places to raise young as well. Work with your neighbors to provide as large as possible parcels of native habitat.

Places to Raise Young. Snags provide places to raise young for birds, snakes, small mammals, and bats. Rock, log, and slash piles provide nursery space for rabbits, snakes, birds, mice, and shrews. A variety of heights and density of vegetation provides the diversity required for diverse wildlife. Wetlands and ponds provide proper nesting habitat for frogs, salamanders, dragonflies, and other insects.

For more information, join WDFW's Backyard Wildlife Program: http://www.wa.gov/wdfw/wlm/byw_prog.htm, Oregon's Naturescaping Program: http://www.dfw.state.or.us/ODFWhtml/Education/Naturescaping.html, or the National Wildlife Federation's Backyard Wildlife Habitat Program: http://www.nwf.org/.

Research Needs

There are many questions that need to be answered for urban areas of the Northwest. How has growth management affected wildlife? Are urban metapopulations sources or sinks, and for what types of animals and why? Do urban areas serve as sources of exotic plants and animals, spreading these species to rural areas? How effective are wildlife corridors in sustaining target populations? How can they be designed to be more effective? How does clustered housing impact wildlife? How do wildlife communities (e.g., birds, small mammals, bats) change with increasing urbanization? What impact does dispersed settlement (urban fringe) have on wildlife? Can we accurately predict the impact that future urbanization will have on wildlife; can we adequately assess different alternatives?

Research is needed to determine the impact that fragmentation has on wildlife in urban areas of the Pacific Northwest. Is fragmentation different in urban areas because of exotics, size of patches, composition of urban zones? How does edge impact wildlife in the urban area? Does it increase brood parasitism and nest predation? Are cowbird populations increasing in urban areas of the Northwest?

What is the best measure of urbanization? Is the "urban gradient" concept the best and most appropriate paradigm to use when conducting research?[137] How far from the urban center do the impacts of urbanization extend into the exurban area? Quantify the rural-urban exchange (through corridors). Determine the ease through which wildlife can travel in urban areas. How far from the urban centers do the positive and negative effects of development extend into surrounding wildlands? How does the amount of juxtaposition of varying types of settlement affect birds? Determine the foraging difference along the urban rural gradient for different bird species and guilds.

See Marzluff et al.[134] for a thorough discussion on avian research needs in an urban environment. They recommend that research should share some of the following features: (1) long-term investigation, (2) use of rigorous experimental design allowing simultaneous comparison of population viability along several points of the urban-wildland continuum, (3) identification of mechanisms responsible for the effects of urbanization on birds (all wildlife), and (4) quantification of avian (all wildlife) demography.

Literature Cited

1. Adams, L. W., and L. E. Dove. 1989. Wildlife reserves and corridors in the urban environment. National Institute of Urban Wildlife, Columbia, MD.
2. Airola, T. M., and K. Buchholz. 1984. Species structure and soil characteristics of five urban sites along the New Jersey Palisades. Urban Ecology 8: 149-164.
3. Alcock, I., and P. Warsop. 1982. Diet, distribution and habitat preferences of stoats and weasels in Sheffield. Sorby Record 20: 5-10.
4. Andre, H. 1994. Effects of habitat fragmentation on birds and mammals in landscapes with different proportions of suitable habitat: a review. Oikos 71: 355-366.
5. Atkinson, K. T., and D. M. Shackleton. 1991. Coyote, Canis latrans, ecology in a rural-urban environment. Canadian Field-Naturalist 105: 49-54.

6. Azous, A. L., and K. O. Richter. 1995. Amphibian and plant community responses to changing hydrology in urban wetlands. Pages 156-162 in E. Robichaud, editor. Puget Sound Research ?95 Proceedings. Puget Sound Water Quality Authority, Olympia, WA.

7. Balda, R. P. 1975. The relationships of secondary cavity-nesters to snag densities in western coniferous forests. U.S. Forest Service Wildlife Habitat Technical Bulletin 1, Albuquerque, NM.

8. Banks, R. C. 1979. Human-related mortality of birds in the United States. U.S. Fish and Wildlife Service Special Scientific Report 215.

9. Barber, J. C., and M. M. Barber. 1983. Prey of an urban peregrine falcon. Maryland Birdlife 39: 108-110.

10. Barbour, R. W., and W. H. Davis. 1969. Bats of America. University Press of Kentucky, Lexington, KY.

11. Barclay, R. M. R. 1984. Observations on the migration, ecology, and behavior of bats at Delta Marsh, Manitoba. Canadian Field-Naturalist 98: 331-336.

12. Barrentine, C. D. 1991. Food habits of western toads (*Bufo boreas halophilus*) foraging from a residential lawn. Herpetological Review 22: 84-87.

13. Batten, L. 1972. Breeding birds species diversity in relation to increasing urbanization. Bird Study 19: 157-166.

14. Batten, L. 1974. Blackbird boom in suburbia. Wildlife 16: 274-277.

15. Beck, A. M. 1974. The ecology of urban dogs. Pages 57-59 in J. Noyes and D. Progulske, editors. Wildlife in an urbanizing environment. The Wildlife Society, Bethesda, MD.

16. Beissinger, S. R. 1978. Avian community organization and habitat structure in a residential area and a virgin beech-maple forest. M.S. Thesis. Miami University, Oxford, OH.

17. Beissinger, S. R. and D. R. Osborne. 1982. Effects of urbanization on avian community organization. Condor 84: 75-83.

18. Blair, R. B. 1996. Land use and avian species diversity along an urban gradient. Ecological Applications 6: 506-519.

19. Blondel, J. 1985. Mediterranean bird faunas in the light of anthropic pressures since the Neolithic. Pages 594-607 in V. D. Ilyichev and V.M. Gavrilov, editors. Acta 18th International Ornithological Congress, Moscow, Russia.

20. Boersma, P. D., and K. N. Almasi. 1991. Can birds tell the difference between native and introduced conifers in the arboretum? Washington Park Arboretum Bulletin 54: 7-9.

21. Bolger, D. T. 1999. Fragmentation effects on birds in southern California: contrast to the paradigm. Abstract from 69th Annual Meeting of the Cooper Ornithological Society, March 29-April 3, Portland, OR.

22. Bogumile, O. 1986. Breeding ecology of goshawk in forest-agricultural habitats of central Poland. Pages 7-9 in H. Ouellet, editor. Proceedings of XIX Congressus Internationalis Ornithologici, Ottawa, Canada.

23. Bolen, E. G. 1991. Analogs: a concept for the research and management of urban wildlife. Landscape and Urban Planning 20: 285-289.

24. Bolsinger, C. L. 1973. Changes in commercial forest area in Oregon and Washington, 1945-70. U.S. Forest Service Resource Bulletin PNW-46, Portland, OR.

25. Booth, D. B. 1991. Urbanization and the natural drainage system - impacts, solutions, and prognoses. Northwest Environmental Journal 7: 93-118.

26. Booth, D. B. and C. R. Jackson. 1994. Urbanization of aquatic systems-degradation thresholds and limits of mitigation. Pages 425-434 in R.A. Marston and V. R. Hasfurther, editors. Proceedings of symposium on effects of human induced changes on hydrologic systems. American Water Resources Association, Jackson Hole, WY.

27. Booth, D. B. and L. E. Reinelt. 1993. Consequences of urbanization on aquatic systems – measured effects, degradation thresholds, and corrective strategies. Pages 545-550 in Watershed 93, Proceedings of a national conference on watershed management, Alexandria, VA.

28. Bosakowski, T., and D. G. Smith. 1997. Distribution and species richness of a forest raptor community in relation to urbanization. Journal of Raptor Research 31: 26-33.

29. Brooks, J. E. 1973. A review of commensal rodents and their control. Critical Reviews in Environmental Control 3: 405-453.

30. Bryant, C. 1982. The rural real estate market. Department of Geography Publication Series No. 18, University of Waterloo, Ontario, Canada.

31. Burke, D. M., and E. Nol. 1998. Influence of food abundance, nest-site habitat, and forest fragmentation on breeding ovenbirds. The Auk 115: 96-104.

32. Burr, R. M., and R. E. Jones. 1968. Influence of parkland habitat management on birds in Delaware. Transactions North American Wildlife and Natural Resources Conference 33: 299-306.

33. Busnel, R. G., and John Fletcher, editors. 1978. Effects of noise on wildlife. Academic Press, New York, NY.

34. Butcher, G. S., W. A. Niering, W. J. Barry, and R. H. Goodwin. 1981. Equilibrium biogeography and the size of nature preserves; an avian case study. Oecologia (Berlin) 49: 29-37.

35. Calhoon, R. E., and C. Haspel. 1989. Urban cat populations compared by season, subhabitat and supplemental feeding. Journal of Animal Ecology 58: 321-328.

36. Campbell, C. A. 1973. Survival of reptiles and amphibians in urban environments. Pages 61-66 in J. Noyes and D. Progulske, editors. Wildlife in an urbanizing environment. The Wildlife Society, Springfield, MA.

37. Canning, D. J., and M. Stevens. 1989. Wetlands of Washington: a resource characterization. Environmental 2010 Project, Washington State Department of Ecology, Olympia, WA.

38. Cassidy, K. M., M. R. Smith, C. E. Grue, K. M. Dvornich, J. E. Cassady, K. R. McAllister, and R.E. Johnson. 1997. Gap analysis of Washington State: an evaluation of the protection of biodiversity. Volume 5 in K. M. Cassidy, C. E. Grue, M. R. Smith, and K. M. Dvornich, editors. Washington State Gap Analysis—Final Report. Washington Cooperative Fish and Wildlife Research Unit, University of Washington, Seattle, WA.

39. Cauley, D. L. 1974. Urban habitat requirements of four wildlife species. Pages 143-147 in J. Noyes and D. Progulske, editors. Wildlife in an urbanizing environment. The Wildlife Society, Springfield, MA.

40. Christian, D. P. 1975. Vulnerability of meadow voles, *Microtus pennsylvanicus*, to predation by domestic cats. American Midland Naturalist 93: 498-502.

41. Churcher, P. B., and J. H. Lawton. 1987. Predation by domestic cats in an English village. Journal of Zoology 212: 439-455.

42. Cicero, C. 1989. Avian community structure in a large urban park: controls of local richness and diversity. Landscape Urban Planning 17: 221-240.

43. Clergeau, P., J. L. Savard, G. Mennechez, and G. Falardeau. 1998. Bird abundance and diversity along an urban-rural gradient: a comparative study between two cities on different continents. The Condor 100: 413-425.

44. Coleman, J. S., S. A. Temple, and S. R. Craven. 1997. Cats and wildlife: a conservation dilemma. Extension Publications, Madison, WI.

45. Cooke, S. S., and A. L. Azous. 1993. Effects of urban stormwater runoff on palustrine wetland vegetation. A Report to the U.S. Environmental Protection Agency, Region 10. Seattle, WA.

46. Cousins, S. H. 1982. Species size distributions of birds and snails in an urban area. Pages 99-109 in R. Bornkamm, J.A. Lee, and M. R. D. Seaward, editors. Urban Ecology. Blackwell Scientific Publications, Oxford, England.

47. Craighead, J. J., and F. C. Craighead, Jr. 1956. Hawks, owls and wildlife. Stackpole Publishing, Harrisburg, PA.

48. Cringan, A. T., and G. C. Horak. 1989. Effects of urbanization on raptors in the western United States. Pages 219-228 in Proceedings of western raptor management symposium and workshop. National Wildlife Federation, Washington, D.C.

49. Cronon, W. 1983. Changes in the land. Hill and Wang, New York, NY.

50. Dansereau, P., and G. Par. 1977. Ecological grading and classification of land-occupation and land-use mosaics. Geography Paper No. 58. Fisheries and Environment Canada, Ottawa, Ontario, Canada.

51. DeGraaf, R. M. 1978. Avian communities and habitat associations in cities and suburbs. Pages 7-24 in C. M. Kirkpatrick, editor. Proceedings of the John S. Wright Forestry Conference, Wildlife and People. Purdue University, West Lafayette, IN.

52. DeGraaf, R. M. 1985. Residential forest structure in urban and suburban environments: some wildlife implications in New England, USA. Journal of Arboriculture 11: 236-241.

53. DeGraaf, R. M. 1986. Urban bird habitat relationships: application to landscape design. Transactions North American Wildlife and Natural Resources Conference 51: 232-248.

54. DeGraaf, R. M., H. R. Pywell, and J. W. Thomas. 1975. Relationships between nest heights, vegetation, and housing density in New England suburbs. Transactions of Northeast Fish and Wildlife Conference 32: 130-150.

55. DeGraaf, R. M. and J. W. Thomas. 1976. Wildlife habitat in or near human settlements. Pages 54-62 in J. W. Anderson, editor. Trees and forests for human settlements. Proceedings of Pl.05-00 Symposia in Vancouver, BC, Canada.

56. DeGraaf, R. M. and J. M. Wentworth. 1981. Urban bird communities and habitats in New England. Transactions North American Wildlife and Natural Resources Conference 46: 396-413.

57. Dennis, J. V. and M. Tekulsky. 1991. How to attract hummingbirds and butterflies. Ortho Books, San Ramon, CA.

58. Diamond, J. M. 1976. Island biogeography and conservation: strategy and limitations. Science 193: 1027-1029.

59. Diamond, J. M. and R. M. May. 1976. Island biogeography and the design of natural reserves. Pages 163-186 in R. M. May, editor. Theoretical ecology: principles and applications. W. B. Saunders, Philadelphia, PA.

60. Dickman, C. R. 1987. Habitat fragmentation and vertebrate species richness in an urban environment. Journal of Applied Ecology 24: 337-351.

61. Dickman, C. R. and C. P. Doncaster. 1987. The ecology of small mammals in urban habitats. Volume I. Populations in a patchy environment. Journal of Animal Ecology 56: 629-640.

62. Dickman, C. R. and C. P. Doncaster. 1989. The ecology of small mammals in urban habitats. Volume II. Demography and dispersal. Journal of Animal Ecology 58: 119-127.

63. Dietrich, J. 1980. Der habicht (Accipiter gentilis) im stadtverband Saarbrucken. Verhandlungen der Gesellschaft für Ökologie. 10: 301-305.

64. Dietrich, J. and H. Ellenberg. 1981. Aspects of goshawk urban ecology. Pages 163-175 in R. E. Kenward and I. M. Lindsay, editors. Understanding the goshawk. International Association of Falconry and Conservation of Birds of Prey, Oxford, England.

65. Doncaster, C. P. 1985. The spatial organization of urban foxes (Vulpes vulpes) in Oxford. Unpublished Thesis, University of Oxford, England.

66. Dorney, R. S. 1985. Urban wildlife populations: a look at uptown, downtown, and suburban residents. In Wildlife survivors in the human niche. National Zoological Park, Smithsonian Institute, Washington, D.C.

67. Dorney, R. S. 1986. Bringing wildlife back to the cities. Technology Review October: 48-54.

68. Dorney, R. S. and P. W. McClellan. 1984. The urban ecosystem: its spatial structure, its scale relationships, and its subsystems attributes. Environments 16: 9-20.

69. Duckworth, F. S., and J. S. Sandberg. 1954. The effect of cities upon horizontal and vertical temperature gradients. Bulletin American Meteorological Society 35: 198-207.

70. Dunn, E. H. 1993. Bird mortality from striking residential windows in winter. Journal of Field Ornithology 64: 02-309.

71. Dunn, E. H. and D. L. Tessaglia. 1994. Predation of birds at feeders in winter. Journal of Field Ornithology 65: 8-16.

72. Ehrenfeld, D. W. 1970. Biological conservation. Holt, Rienhart and Winston, New York, NY.

73. Emlen, J. T. 1974. An urban bird community of Tucson, Arizona: derivation, structure, and regulation. The Condor 76: 184-197.

74. Erz, W. 1966. Ecological principles in the urbanization of birds. Ostrich Supplement 6: 357-363.

75. Ewing, K. 1996. Tolerance of four wetland plant species to flooding and sediment deposition. Environmental and Experimental Biology 36: 131-146.

76. Falk, J. H. 1976. Energetics of a suburban lawn ecosystem. Ecology 57: 141-150.

77. Fenton, M. B., H. G. Merriam, and G. L. Holryod. 1983. Bats of Kootenay, Glacier, and Mount Revelstoke national parks in Canada: identification by echolocation calls, distribution, and biology. Canadian Journal of Zoology 61: 2503-2508.

78. Fitzgerald, B. M. 1988. Diet of domestic cats and their impact on prey populations. Pages 123-147 in D. C. Turner, and P. Bateson, editors. The domestic cat: the biology of its behaviour. Cambridge University Press, Cambridge, England.

79. Fletcher, J. L. 1990. Review of noise and terrestrial species: 1983-1988. Pages 181-188 in B. Berglund and T. Lindvall, editors. Noise as a public health problem. Volume 5: new advances in noise research part II. Swedish Council For Building Research, Stockholm, Sweden.

80. Forman, R. T. 1995. Land mosaics: the ecology of landscapes and regions. Cambridge University Press, Cambridge, England.

81. Fox, R. 1987. Population images. United Nations Fund for Population Activities.

82. Frankie, G. W., and L. E. Ehler. 1978. Ecology of insects in urban environments. Annual Review Entomology 23: 367-387.

83. Frey, H. T. 1984. Expansion of urban area in the United States: 1960-1980. U.S. Department of Agriculture Economic Research Service Staff Report AGES830615, Washington, D.C.

84. Furlonger, C. L., H. J. Dewar, and M. B. Fenton. 1987. Habitat use by foraging insectivorous bats. Canadian Journal Zoology 65: 284-288.

85. Gavareski, C. A. 1976. Relation of park size and vegetation to urban bird populations in Seattle, WA. The Condor 78: 375-382.

86. Geggie, J. F., and M. B. Fenton. 1985. A comparison of foraging by Eptesicus fuscus (Chiropter: Vespertilionidae) in urban and rural environments. Canadian Journal Zoology 63: 263-266.

87. Gehlbach, F. 1986. Odd couples of suburbia. Natural History 95: 56-66.

88. Gehlbach, F. 1988. Population and environmental features that promote adaptation to urban ecosystems: the case of eastern screech-owls (Otus asio) in Texas. Pages 1809-1813 in H. Ouellet, editor. Proceedings of XIX Congressus Internationalis Ornithologici, Ottawa, Canada.

89. Geis, A. D. 1974. The new town bird quadrille. Natural History 83: 53-61.

90. George, W. G. 1974. Domestic cats as predators and factors in winter shortages of raptor prey. Wilson Bulletin 86: 384-396.

91. Gibbs, J. P. 1993. Importance of small wetlands for the persistence of local populations of wetland-associated animals. Wetlands 13: 25-31.

92. Gillihan, S. W. 2000. Bird conservation on golf courses: a design and management manual. Sleeping Bear Press, Chelsea, MI.

93. Goldstein, E. L., M. Gross, and R. M. DeGraaf. 1986. Breeding birds and vegetation: a quantitative assessment. Urban Ecology 9: 377-385.

94. Green, R. J. 1984. Native and exotic birds in a suburban habitat. Australian Wildlife Research 11: 181-190.

95. Guth, R. W. 1979. The junk food guild: birds and mammals on picnic grounds and in residential areas. Illinois Audubon Bulletin 189: 3-7.

96. Guthrie, D. A. 1974. Suburban bird populations in southern California. American Midland Naturalist 92: 461-466.

97. Hall, R. J. 1980. Effects of environmental contaminants on reptiles: a review. U.S. Fish and Wildlife Service, Washington, D.C., Special Scientific Report—Wildlife 228: 1-12.

98. Harris, S. 1977. Distribution, habitat utilization and age structure of a suburban fox (Vulpes vulpes) population. Mammal Review 7: 25-39.

99. Haub, C., and M. M. Kent. 1989. 1989 world population data sheet. Population Reference Bureau, Washington, D.C.

100. Heck, O. A. 1971. Eastern tiger salamanders found in a window well. Engelhardtia 4: 35.

101. Herlugson, C. J. 1978. Comments on the status and distribution of western and mountain bluebirds in Washington. Western Birds 9: 21-32.

102. Hobbs, E. 1988. Using ordination to analyze the composition and structure of urban forest islands. Forest Ecology and Management 23: 139-158.

103. Hoffmann, C. O., and J. L. Gottschang. 1977. Numbers, distribution and movements of a raccoon population in a suburban residential community. Journal of Mammalogy 58: 623-636.

104. Hooper, R. G., E. F. Smith, H. S. Crawford, B. S. McGinnes, and V. J. Walker. 1975. Nesting bird populations in a new town. Wildlife Society Bulletin 3: 111-118.

105. Houck, C. A. 1996. The distribution and abundance of invasive plant species in freshwater wetlands of the Puget Sound lowlands, King County, Washington. M.S. Thesis. University of Washington, Seattle, WA.

106. Hubbs, E. L. 1951. Food habits of feral house cats in the Sacramento Valley. California Fish and Game 37: 177-189.

107. Huff, F. A. 1977. Effects of the urban environment on heavy rainfall distribution. Water Resources Bulletin 13: 807-816.

108. Ingold, D. J. 1996. Delayed nesting decreased reproductive success in northern flickers: implications for competition with European starlings. Journal of Field Ornithology 67: 321-326.

109. Ingold, D. J. and R. J. Densmore. 1992. Competition between European starlings and native woodpeckers for nest cavities in Ohio. Sialia 14: 43-48.

110. James, P. C., A. R. Smith, L. W. Oliphant, and L. G. Warkentin. 1987. Northward expansion of the wintering range of Richardson's merlin. Journal of Field Ornithology 58: 112-117.

111. Johnsen, A. M., and L. W. VanDruff. 1987. Summer and winter distribution of introduced bird species and native bird species richness within a complex urban environment. Pages 123-127 in L. W. Adams and D. L. Leedy, editors. Integrating man and nature in the metropolitan environment. Proceedings of a National Symposium on Urban Wildlife, Chevy Chase, MD.

112. Johnson, R. E., and K. M. Cassidy. 1997. Terrestrial mammals of Washington State: location data and predicted distributions. Volume 3 in: K. M. Cassidy, C. E. Grue, M. R. Smith, and K. M. Dvornich, editors. Washington State Gap Analysis—Final Report. Washington Cooperative Fish and Wildlife Research Unit, University of Washington, Seattle, WA.

113. Jordheim, S. 1966. Recollection of a peregrine falcon-pigeon pursuit. Blue Jay 24: 20.

114. Jureck, R. M. 1994. A bibliography of feral, stray, and free-ranging domestic cats in relation to wildlife conservation. California Department of Fish and Game, Wildlife Management Division, Sacramento, CA.

115. Klein, M. L. 1993. Waterbird behavioral responses to human disturbances. Wildlife Society Bulletin 21: 31-39.

116. Klem, D., Jr. 1989. Bird-window collisions. Wilson Bulletin 101: 606-620.

117. Klem, D., Jr. 1990. Collisions between birds and windows: mortality and prevention. Journal Field Ornithology 61: 120-128.

118. Knight, R. L., and K. J. Gutzwiller. 1995. Wildlife and recreationists: coexistence through management and research. Island Press, Washington, D.C.

119. Knutson, K. L., and V. L. Naef. 1997. Management recommendations for Washington's priority habitats: riparian. Washington Department of Fish and Wildlife, Olympia, WA.

120. Krebs, C. J., M. S. Gaines, B. L. Keller, J. H. Myers, and R. H. Tamarin. 1973. Population cycles in small rodents. Science 179: 35-41.

121. Kruckeberg, A. 1982. Gardening with native plants of the Pacific Northwest; an illustrated guide. University of Washington Press, Seattle, WA.

122. Kurta, A., and J. A. Teramino. 1992. Bat community structure in an urban park. Ecography 15: 257-261.

123. Lancaster, R. K., and W. E. Rees. 1979. Bird communities and the structure of urban habitats. Canadian Journal of Zoology 57: 2358-2368.

124. Leedy, D. L. 1975. Highway-wildlife relationships, Volume 1. A state-of-the-art report. U.S. Department of Transportation, Federal Highway Administration; FHWA-RD-76-4. Washington, D.C.

125. Leedy, D. L. 1975. Highway-wildlife relationships, Volume 2. An annotated bibliography. U.S. Department of Transportation, Federal Highway Administration; FHWA-RD-76-5. Washington, D.C.

126. Leedy, D. L. and L. W. Adams. 1986. Wildlife in urban and developing areas: an overview and historical perspective. Pages 8-20 in: K. Stenberg, and W. W. Shaw, editors. Wildlife conservation and new residential developments, Tucson, AZ.

127. Lucid, V. J. 1974. Bird utilization of habitat in residential areas. M.S. Thesis. Virginia Polytechnic Institute, Blacksburg, VA.

128. Ludwig, D. R. 1995. Assessment and management of wildlife diversity in an urban setting. Natural Areas Journal 15: 353-361.

129. MacCracken, J. G. 1982. Coyote food in a southern California suburb. Wildlife Society Bulletin 10: 80-281.

130. MacLean, C. D., and C. L. Bolsinger. 1997. Urban expansion in the forests of the Puget Sound region. U.S. Forest Service, Resource Bulletin PNW-RB-225. Portland, OR.

131. Martin, T. E. 1988. Habitat and area effects on forest bird assemblages: is nest predation an influence? Ecology 69: 74-84.

132. Martin, T. E. 1993. Nest predation and nest sites. Bioscience 43: 523-532.

133. Marzluff, J. M. 1997. Effects of urbanization and recreation on songbirds. Pages 89-102 in W. M. Block, and D. M. Finch, editors. Songbird ecology in southwestern ponderosa pine forests: a literature review. U.S. Forest Service General Technical Report RM-292, Fort Collins, CO.

134. Marzluff, J. M., F. R. Gehlbach, and D. A. Manuwal. 1998. Urban environments: influences on avifauna and challenges for the avian conservationist. Pages 283-299 in: J. M. Marzluff, and R. Salabanks, editors. Avian conservation, research, and management. Island Press, Washington, D.C.

135. Maser, C. 1998. Mammals of the Pacific Northwest: from the coast to the high Cascades. Oregon State University Press, Corvallis, OR.

136. Mason, P. 1985. The impact of urban development on bird communities of three Victorian towns, Lilydale, Coldstream, and Mr. Evelyn, Victoria, Australia. Corella 9: 14-21.

137. McDonnell, M. J., and S. T. A. Pickett. 1990. Ecosystem structure and function along urban-rural gradients: an unexploited opportunity for ecology. Ecology 7: 1232-1237.

138. McGuckin, C., and R. Brown. 1995. Stormwater management and urban greenways. Landscape and Urban Planning 33: 227-246.

139. Mesure, M. 1996. Bird collisions. World Wildlife Fund, Canada, Ontario, Canada.

140. Minton, S. A., Jr. 1968. The fate of amphibians and reptiles in a suburban area. Journal of Herpetology 2: 113-116.

141. Mockler, A. 1998. Results of monitoring King County wetland mitigations. King County Department of development and environmental sciences, Renton, WA.

142. Moore, K. E. 1990. Urbanization in the lower Fraser Valley, 1980-1987. Technical Report 120. Canadian Wildlife Service, Delta, BC, Canada.

143. Morneau, F., C. Lépine, R. Décarie, M. Villard, and F. DesGranges. 1995. Reproduction of American robin (Turdus migratorius) in a suburban environment. Landscape and Urban Planning 32: 55-62.

144. Moul, I. E., and J. E. Elliott. 1994. The bird community found on golf courses in British Columbia. Northwestern Naturalist 75: 88-96.

145. Neill, W. T. 1950. Reptiles and amphibians in urban areas of Georgia. Herpetologica 6: 113-116.

146. Nelson, A. C., and J. B. Duncan. 1995. Growth management principles and practices. American Planning Association, Chicago, IL.

147. Newton, I. 1986. The sparrowhawk. T. and A. D. Poyser, Ltd., Calton, Staffordshire, England.

148. Nilon, C. H., and L. W. VanDruff. 1987. Analysis of small mammal community data and applications to management of urban greenspaces. Pages 53-59 in L. W. Adams and D. L. Leedy, editors. Integrating man and nature in the metropolitan environment. National Institute Urban Wildlife, Columbia, MD.

149. Noss, R. F. 1992. The Wildlands Project: land conservation strategy. Wild Earth (Special Issue): 10-25.

150. Noss, R. F. and A. Cooperrider. 1994. Saving nature's legacy: protecting and restoring biodiversity. Defenders of Wildlife and Island Press, Washington, D.C.

151. Nowakowski, E. 1982. Influence of urbanization on the structure of wireworm communities (Coleoptera, Elateridae). Pages 79-87 in M. Luniak and B. Pisarski, editors. Animals in urban environment. Polish Academy of Science, Institute of Zoology, Warsaw, Poland.

152. Obara, H., H. Hirata, and M. Okuzaki. 1977. An aspect of the existence of mammals in urban ecosystems. Pages 173-187 in M. Numata, editor. Tokyo project: interdisciplinary studies of urban ecosystems in the metropolis of Tokyo, Tokyo, Japan.

153. Oliphant, L. W., and S. McTaggart. 1977. Prey utilized by urban Merlins. Canadian Field-Naturalist 91: 190-192.

154. O'Neil, T. A., D. H. Johnson, C. Barrett, M. Trevithick, K. A. Bettinger, C. Kiilsgaard, M. Vander Heyden, E. L. Greda, B. G. Marcot, P. J. Doran, L. Wunder, and S. Tank. 2001. Matrixes for wildlife-habitat relationships in Oregon and Washington. In: D. H. Johnson and T. A. O'Neil, managing directors. Wildlife-habitat relationships in Oregon and Washington. Oregon State University Press, Corvallis, OR.

155. Orser, P. N., and D. J. Shure. 1972. Effects of urbanization on the salamander Desmognathus fuscus fuscus. Ecology 53: 1148-1154.

156. Paton, P. W. C. 1994. The effect of edge on avian nest success: how strong is the evidence? Conservation Biology 8: 17-26.

157. Petraitis, P. S., R. E. Latham, and R. A. Niesenbaum. 1989. The maintenance of species diversity by disturbance. Quarterly Review of Biology 64: 393-418.

158. Quinn, T. 1997. Coyote (Canis latrans) food habits in three urban habitat types of western Washington. Northwest Science 71: 1-5.

159. Quinn, T. 1997. Coyote (Canis latrans) habitat selection in urban areas of western Washington via analysis of routine movements. Northwest Science 71: 289-297.

160. Reaume, T. 1987. Selective feeding by the American crow. Ontario Birds 5: 71-72.

161. Reijnen, R. 1995. Disturbance by car traffic as a threat to breeding birds in the Netherlands. M.S. Thesis. DLO-Institute for Forestry and Nature Research, Wageningen, The Netherlands.

162. Richter, K. O. 1997. Criteria for the restoration and creation of wetland habitats of lentic-breeding amphibians of the Pacific Northwest. Pages 72-94 in K. B. MacDonald and F. Weinmann, editors. Wetland and riparian restoration: taking a broader view. U.S. Environmental Protection Agency, Publication 910-R-97-007, Region 10, Seattle, WA.

163. Richter, K. O., and A. L. Azous. 1995. Amphibian occurrence and wetland characteristics in the Puget Sound Basin. Wetlands 15:305-312.

164. Richter, K. O., and A. L. Azous, S. S. Cooke, R. W. Wisseman, and R. R. Horner. 1991. Effects of stormwater runoff on wetland zoology and wetlands soils characterization and analysis. Washington State Department of Ecology Report G0091019, Olympia, WA.

165. Robbins, C. S. 1979. Effects of forest fragmentation on bird populations. Pages 198-212 in Management of northcentral and northeastern forests for nongame birds. U.S. Forest Service General Technical Report NC-51, Minneapolis, MN.

166. Robinson, S. K., F. R. Thompson III, T. M. Donovan, D. R. Whitehead, and J. Faaborg. 1995. Regional forest fragmentation and the nesting success of migratory birds. Science 267:1987-1990.

167. Rochelle, J. A. 1998. Forest fragmentation: wildlife and management implications. Conference Summary Statement, 18-19 November 1998. Portland, OR.

168. Rosenberg, K. V., A. A. Dhondt, and J. D. Lowe. 1996. Lessons from the landscape. Birdscope 10: 3-5.

169. Rost, G. R., and J. A. Bailey. 1979. Distribution of mule deer and elk in relation to roads. Journal of Wildlife Management 43: 634-641.

170. Rudnicky, J. L., and M. J. McDonnell. 1989. Forty-eight years of canopy change in a hardwood-hemlock forest in New York City. Bulletin of the Torrey Botanical Club 116: 52-64.

171. Russwurm, L. H. 1977. The surroundings of our cities. Community Planning Associated Press, Ottawa, Ontario, Canada.

172. Savard, J. P. L. 1978. Birds in metropolitan Toronto: distribution, relationships with habitat features, and nesting sites. Ph.D. Dissertation, University of Toronto, Ontario, Canada.

173. Savard, J. P. and J. B. Falls. 1981. Influence of habitat structure on the nesting heights of birds in urban areas. Canadian Journal of Zoology 59: 924-932.

174. Schinner, J. R. 1969. Ecology and life history of the raccoon (Procyon lotor) within the Clifton suburb of Cincinnati. M.S. Thesis. University of Cincinnati, Cincinnati, OH.

175. Schlauch, F. L. 1975. Agonistic behavior in a suburban Long Island population of the smooth green snake (Opheodrys vernalis). Engelhardtia 6: 25-26.

176. Schueler, T. 1994. The importance of imperviousness. Watershed Protection Techniques 1: 100-111.

177. Scott, J. M., F. Davis, B. Csuti, R. Noss, B. Butterfield, C. Groves, H. Anderson, S. Caicco, F. D'Erchia, T. C. Edwards Jr., J. Ulliman, and R. Wright. 1993. Gap Analysis: a geographic approach to protection of biological diversity. Wildlife Monographs 123: 1-41.

178. Semlitsch, R. D., and J. R. Bodie. 1998. Are small, isolated wetlands expendable? Conservation Biology 12: 5.

179. Sharpe, D. M., F. Stearns, L. A. Leitner, and J. R. Dorney. 1986. Fate of natural vegetation during urban development of rural landscapes in northeastern Wisconsin. Urban Ecology 9: 267-287.

180. Skagen, S. K., R. L. Knight, and G. H. Orions. 1991. Human disturbance of an avian scavenging guild. Ecological Applications 1: 215-225.

181. Small, M. F., and M. L. Hunter. 1988. Forest fragmentation and avian nest predation in forested landscapes. Oecologia 76: 62-64.

182. Smith, K. G. 1978. Range extension of the blue jay into Western North America. Bird Banding 49: 208-214.

183. Sodhi, N. S., and L. W. Oliphant. 1993. Prey selection by urban-breeding merlins. The Auk 110: 727-735.

184. Soulé, M. 1987. Viable populations for conservation. Cambridge University Press, Cambridge, England.

185. Soulé, M. A. C. Alberts, and D. T. Bolger. 1991. The effects of habitat fragmentation on chaparral plants and vertebrates. Oikos 63: 39-47.

186. Soulé, M. D. T. Bolger, A. C. Alberts, J. Wright, M. Sorice, and S. Hill. 1988. Reconstructed dynamics of rapid extinctions of chaparral-requiring birds in urban habitat islands. Conservation Biology 2: 75-92.

187. Southwood, T. R. E. 1961. The number of species of insects associated with various trees. Journal of Animal Ecology 30: 1-8.

188. Spencer, L. 1994. Fastest growing hobby in U.S. is for the birds. Seattle Times, January 10, 1994.

189. Stebbins, R. C., and N. W. Cohen. 1995. A natural history of amphibians. Princeton University Press, Princeton, NJ.

190. Steedman, R. J. 1988. Modification and assessment of an index of biotic integrity to quantify stream quality in Southern Ontario. Canadian Journal of Fisheries and Aquatic Sciences 45: 492-501.

191. Sukopp, H., H. Elvers, and H. Mattes. 1982. Studies in urban ecology of Berlin (West). Pages 115-130 in: M. Luniak and B. Pisarski, editors. Animals in Urban Environment. Polish Academy of Science, Institute of Zoology, Warsaw, Poland.

192. Swank, W. T., and P. V. Bolstad. 1994. Cumulative effects of land use practices on water quality. Pages 409-421 in N. E. Peters, R. J. Allan, and V. V. Tsirkunov, editors. Proceedings of the Rostov-on-don Symposium on Hydrochemistry 1993: hydrological, chemical, and biological processes affecting the transformations and the transport of contaminants in aquatic environments. IAHS Publication 219, Oxfordshire, England

193. Thomas, J. W., R. B. Andersen, C. Maser, and E. L. Bull. 1979. Snags. Pages 60-70 in J. W. Thomas, editor. Wildlife habitats in managed forests, the Blue Mountains of Oregon and Washington. U.S. Forest Service Agriculture Handbook 553, Washington, D.C.

194. Thomas, J. W., R. M. DeGraaf, and J. C. Mawson. 1977. Determination of habitat requirements for birds in suburban areas. U.S. Forest Service Research Paper NE-357.

195. Thomas, J. W., C. Maser, E. D. Forsman, J. B. Lint, E. C. Meslow, B. R. Noon, and J. Verner. 1990. A conservation strategy for the northern spotted owl. Report of the interagency committee to address the conservation strategy of the northern spotted owl. U.S. Forest Service, Portland, OR.

196. Tomialoj‰, L. 1985. Urbanization as test of adaptive potentials in birds. Pages 608-614 *in:* XVIII Congresses Internationalist Ornithologici, Moscow, Russia.

197. Valden, D. W. 1980. Urban small mammals. Journal of Zoology 191: 403-406.

198. Van Druff, L. W., and R. N. Rowse. 1986. Habitat association of mammals in Syracuse, New York. Urban Ecology 9: 413-434.

199. Van Druff, E. G. Bolen, and G. J. San Julian. 1994. Management of urban wildlife. Pages 507-530 *in:* T. A. Bookout, editor. Research and management techniques for wildlife and habitats. The Wildlife Society, Bethesda, MD.

200. Vasic, V., and V. Stevanovic. 1970. Characteristics of the avifauna of the inner city area of Belgrade. Larus 24: 107-118.

201. Vepsalainen, K., and B. Pisarski. 1982. The structure of urban ant communities along a geographical gradient from north Finland to Poland. Pages 155-168 *in:* M. Luniak and B. Pisarski, editors. Animals in urban environment. Polish Academy of Science, Institute of Zoology, Warsaw, Poland.

202. Vizyova, A. 1986. Urban woodlots as islands for land vertebrates: a preliminary attempt on estimating the barrier effects of urban structural units. Ekologia (CSSR) 5: 407-419.

203. Walcott, C. F. 1974. Changes in bird life in Cambridge, Massachusetts from 1960 to 1964. The Auk 91: 151-160.

204. Warren, J. W. 1997. Deer overabundance. Wildlife Society Bulletin Special Issue 25: 2.

205. Washington Department of Fish and Wildlife. 1999. Priority habitats and species list. Washington Department of Fish and Wildlife, Olympia, WA.

206. Washington State Department of Natural Resources. 1998. Our changing nature. Washington Department of Natural Resources, Olympia, WA.

207. Watson, J. W., and D. J. Pierce. 1998. Bald eagle ecology in western Washington with an emphasis on the effects of human activity. Final Report. Washington Department of Fish and Wildlife, Olympia, WA.

208. Wear, D. N., M. G. Turner, and R. J. Naiman. 1998. Land cover along an urban-rural gradient: implications for water quality. Ecological Applications 8: 619-630.

209. Welch, E. B., J. M. Jacoby, and C. W. May. 1998. Stream quality. Pages 69-94 *in:* R. J. Naiman and R. E. Bilby, editors. River ecology and management: lesson from the Pacific Coastal Ecoregion. Springer-Verlag, New York, NY.

210. Whitaker, J. O., Jr. 1996. National Audubon Society field guide to North American mammals. Alfred A. Knopf, New York, NY.

211. Whitcomb, R. F., D. S. Robins, J. F. Lynch, B. L. Bystrak, M. K. Limkiewitz, and D. Bystrak. 1981. Effects of forest fragmentation on avifaunas of the eastern deciduous forest. Pages 125-205 *in:* R. L. Burgess and D. M. Sharpe, editors. Forest island dynamics in man-dominated landscapes. Springer-Verlag, New York, NY.

212. Wilcove, D. S. 1985. Nest predation in forest tracts and the decline of migratory songbirds. Ecology 66: 1211-1214.

213. Wilcove, D. S., C. P. McLellan, and A. P. Dobson. 1986. Habitat fragmentation in the temperate zone. Pages 237-256 *in:* M. E. Soulé, editor. Conservation biology: the science of scarcity and diversity. Sinauer Associates, Sunderland, MA.

214. Wilson, E. O., and E. O. Willis. 1975. Applied biogeography. Pages 522-534 *in:* M. L. Cody, and J. M. Diamond, editors. Ecology and evolution of communities. Harvard University Press, Cambridge, MA.

215. Woolfenden, G. E., and S. A. Rohwer. 1969. Breeding birds in a Florida suburb. Florida State Museum Bulletin 13.

216. Yahner, R. H. 1988. Changes in wildlife communities near edges. Conservation Biology 2: 333-339.

217. Yalden, D. W. 1980. Urban small mammals. Journal of Zoology 191: 403-406.

218. Zimmerman, D. A. 1973. Range expansion of Anna's hummingbird. American Birds 27: 827-835.

13

Wildlife of Agriculture, Pastures, and Mixed Environs

W. Daniel Edge

Composition, Structure, and Distribution

The Agriculture, Pasture, and Mixed Environs habitat (hereafter agricultural habitat) in Oregon and Washington is highly diverse in part because of the large number of agricultural commodities produced in the two states, but also because this habitat classification includes other linear landscape components that could not be classified individually (e.g., road sides, shelterbelts, etc.). This habitat includes managed and unmanaged pasture, cultivated cropland, orchards, vineyards, nurseries, and associated scattered dwellings and intervening areas of vegetation. There are >410 agricultural crops produced in the two states. This habitat is maintained across a wide range of ecoregions and climatic conditions typical of both states. Climate constrains crop production at upper elevations where there are <90 frost-free days. Agricultural habitat in arid regions east of the Cascades with <10 inches (25 cm) of rainfall requires supplemental irrigation or fields lying fallow for 1-2 years to accumulate sufficient soil moisture. This habitat is found from 0 to 6,000 feet (0-1,829 m) in elevation. Soil types are variable, but soils usually have a well-developed A horizon.

Agricultural habitat is widely distributed at low- to mid-elevations (<6,000 feet [<1,829 m]) throughout both states. This habitat is most abundant in broad river valleys in both states and on gentle rolling terrain east of the Cascades. Agricultural habitat occurs within a matrix of other habitat types at low- to mid-elevations, including Eastside (Interior) Grasslands, Shrub-steppe, Westside Lowlands Conifer-Hardwood Forest, and other low- to mid-elevation forest and woodland habitats. Agricultural habitat often dominates the landscape where terrain is flat or gently rolling, on well-developed soils, broad river valleys, and areas with access to abundant irrigation water. Unlike other habitat types, agricultural habitat is often characterized by regular landscape patterns (squares, rectangles, and circles) and straight borders because of ownership boundaries and multiple crops within a region. Edges can be abrupt along the habitat borders within agricultural habitat and with adjacent habitats.

Agricultural habitats have two dominant characteristics that separate them from the other wildlife habitats described in this book: regular disturbance and ownership pattern, both of which are important in managing these habitats for wildlife. The most common characteristic is a regular pattern of disturbance. Depending on the crop produced and the cultural practices employed in these areas, agricultural habitats may be substantially modified (e.g., cultivated, grazed, harvested, hayed, mowed, pruned, tilled, etc.) at least once per year. Because of these disturbances, many agricultural habitats are important for wildlife only on a seasonal basis, whereas others may be ecological traps during the breeding season.[7] Unlike most of the other wildlife habitats covered in this book, almost all agricultural habitats are privately owned. Only the urban habitat classification contains as much privately owned property as agricultural habitat. As a consequence, management benefitting wildlife in agricultural habitats occurs predominately because of personal interest of the landowner, stewardship, other conservation benefits, the recreational value associated with wildlife in these areas, or because of financial disincentives of U.S. Department of Agriculture (USDA) programs under the Farm Bill.[30, 53]

Because of the wide distribution and regular disturbance of agricultural habitats, wildlife that occur in these areas display a number of adaptations. Many of the species that use agricultural habitats are habitat generalists, adapted for using several cover types for both feeding and breeding. In many areas of both states, pastures and cultivated cropland have replaced native prairie habitats. However, many of these areas (e.g., annual and perennial grass seed fields, wheat, barley, Conservation Reserve Program fields, and hay pastures) are structurally similar to the native prairie habitats they replaced, and often are used by habitat specialists of native prairie habitats (e.g., western meadowlark, grasshopper sparrow, bobolink).[8, 70]

Important Habitat Elements

Agricultural habitats in Oregon and Washington include several habitat elements, all of which are anthropogenic in origin. Because this landscape is human-dominated, there are no habitat features that are considered unique—those habitat features that cannot be created. However, some agricultural habitat is adjacent to unique features such as rimrocks with cliffs and caves. Important habitat elements in these areas include ephemeral wetlands, wells

Table 1. Wildlife benefits provided by habitat elements in agricultural habitats in Oregon and Washington.

Habitat feature	Wildlife benefits						
	Water	Food	Shelter	Breeding sites	Structural diversity	Edge habitat	Corridors
Farmed or ephemeral wetland	X	X		X		X	
Water developments	X	X		X	?[a]		
Buildings and farm structures			X	X	X		
Shelterbelts and windbreaks		X	X	X	X	X	X
Hedge and fence rows		X	X	X	X	X	X
Roadsides		X	X	X	X	X	X
Field borders		X	X	X	X	X	X
Odd areas		X	X	X	X	X	?[a]

[a]Unknown or uncertain benefit.

and water developments, deserted dwellings and other buildings, shelterbelts, hedge and fence rows, roadsides, and field borders (Table 1).

Several habitat elements, including shelterbelts, windbreaks, hedge and fence rows, field borders, and roadsides are important components of the agricultural landscapes because they provide permanent cover and structural diversity and may function as corridors.[5] Furthermore, these habitat features, because of their narrow, linear nature, are composed entirely of edge habitat and typically have high species diversity.[10] However, depending on their abundance and distribution in the agricultural landscape, these elements may function as ecological traps because of high rates of predation or brood parasitism.[6, 70, 100]

Ephemeral or Farmed Wetlands

Agricultural habitats include numerous ephemeral or farmed wetlands. In many areas, especially along broad floodplains, native wetlands were drained for agriculture, but most of these sites still retain two of the three characteristics that define wetlands (hydrology and hydric soils).[21, 86] Consequently, these areas have standing water on them ≥1 month of the year. Depending on the residual vegetation or the crop from the previous growing season, these sites can be extremely important wildlife habitats.[55] Ephemeral and farmed wetlands that were used to produce small grains (i.e., sorghum, rice, wheat, barley, oats, grass seed) during the previous growing season are critically important as stop-over and wintering habitat for millions of migratory waterfowl in the Pacific Flyway. These sites provide both wetland habitat and abundant food sources because of the waste grain typical of small grain production practices. During the northward migration in the spring, the abundant food in these wetlands is especially important for energy conservation and nutritional conditioning for egg production.[45]

Ephemeral or farmed wetlands may also be important breeding sites for amphibians. For example, many native hay pastures east of the Cascade Mountains in both states are flood-irrigated during the spring months, resulting in large areas of grass vegetation with 2-8 in (5-20 cm) of standing water—conditions that provide ideal breeding sites for some amphibians.

Wells and Water Developments

Wells and water developments are habitat elements that are common in agricultural areas, especially east of the Cascade Mountains in Oregon and Washington. Water is scarce in many of these areas, and water developments become an important habitat feature for wildlife. Intensive livestock management, characteristic of many agricultural areas in the arid regions of both states, requires a reliable, uniform distribution of water in grazing pastures.[44, 89] Spring developments, reservoirs, and precipitation catchments or guzzlers are water developments used for livestock production that also benefit wildlife. These structures can be designed or modified to increase wildlife use and to prevent accidental drowning in steep-sided structures.[17, 22, 72, 101] Especially when protected from livestock grazing, these sites usually develop vegetation characteristic of riparian areas and become oases for wildlife.[15, 43, 44, 88]

Buildings and Farm Structures

Deserted dwellings and other buildings are also habitat elements in agriculture habitats. Buildings provide structure in agricultural areas often characterized by homogeneous landscapes of low vegetation stature; these structures provide habitat features that may otherwise be unavailable.[20, 50] These structures are used as nest sites for swallows and barn owls, roosts and hibernacula for several species of bats, and shelter for many small mammals, reptiles, and amphibians.[3, 46, 50]

Shelterbelts and Windbreaks

Shelterbelts are vital habitat features of agricultural habitats. Shelterbelts and windbreaks are terms that have been used interchangeably for human-made habitats created by planting rows of trees and shrubs along the border of agricultural fields (Figure 1). Shelterbelts have been promoted for a number of conservation benefits including (1) reducing wind erosion; (2) shelter for growing plants, farmsteads, and livestock; (3) snow fences and living screens; (4) wood and fiber production; and (5) aesthetics. They are also excellent wildlife habitat and have been promoted as such by both state and federal conservation agencies. In large areas of eastern Oregon

Figure 1. Shelterbelts and windbreaks, created by planting rows of trees and shrubs, are important wildlife habitat in agricultural areas. (Photograph by W. Daniel Edge)

Figure 2. Hedge and fence rows are rows of shrubs and trees bordering agricultural fields that provide food, cover, and dispersal habitat for wildlife. (Photograph by W. Daniel Edge)

and Washington, shelterbelts and riparian areas account for most of the tree habitat over hundreds of square miles. Shelterbelts are disproportionally important wildlife habitats, especially for neotropical migratory landbirds,[9, 70, 75, 99] reptiles, and small mammals.[98] Shelterbelts are used for year-round cover and feeding sites, for nesting,[9, 70] and as dispersal corridors[32] by birds and mammals. Shelterbelts that incorporate several plant species and provide ground, mid-story, and canopy-level vegetation structure have higher wildlife species diversity than single-species, single-structure shelterbelts.[75, 99]

Hedge and Fence Rows

Hedge and fence rows are habitat elements of agricultural habitats that enhance these areas for wildlife; they are continuous or near-continuous shrub and tree vegetation communities that border agricultural fields (Figure 2). Fence lines are common in agricultural landscapes, although not as common as they used to be. Historically, when most farms ran some livestock, fences were much more common than they are now; they have declined in agricultural landscapes since the advent of "clean farming."[30, 70] Nevertheless, many fields, especially pastures used for grazing, are separated or surrounded by fences. Fence and hedge rows provide shelter, food, and structural complexity to agricultural landscapes (Table 1) and typically have high species richness and wildlife abundance relative to the agricultural fields they border.[9, 59, 70]

Field Borders

Field borders are narrow strips of permanent vegetation planted between fields and around the perimeters of fields. Field borders benefit wildlife by providing permanent cover, forage, increased edge habitat, and increased horizontal and vertical diversity (Table 1). Field borders also control erosion and can protect field edges that are used as "turn-arounds" or travel lanes for farm equipment. Another farmer benefit includes reduced planting and

Figure 3. Roadsides can provide important year-round cover for wildlife in agricultural areas. (Photograph by W. Daniel Edge)

farming costs. If used adjacent to riparian areas, wetlands, etc., field borders also function as filter strips and improve water quality. Few studies have looked at wildlife community responses to different field borders, but wildlife use is likely to be dependent on the width, length, and the structural diversity of the borders.[9, 59, 70]

Roadsides

Roadsides are another important habitat feature of agricultural habitats, and are a common attribute of all terrestrial ecosystems except in wilderness areas. Roads abound in agricultural landscapes because the culture and harvest of crops typically require producers to visit fields multiple times, with different pieces of equipment (Figure 3). Roads in agricultural ecosystems vary from simple dirt tracks between fields managed by the landowner to four-lane Interstate Highways managed by the state Department of Transportation (DOT). High-standard roads such as county roads and state and interstate highways are typically constructed by borrowing dirt from the roadside to raise the roadbed between the borrow pits. Thus, a road has the roadbed, the borrow pit, and the fore and back slopes of the borrow pits. The fore and back slope and the borrow pits are potentially good wildlife habitat in agricultural areas, especially compared to the adjacent crop fields. The borrow pits act as drainage systems and often function as ephemeral or permanent wetlands. The slopes are almost always covered with permanent grass or shrub vegetation, but may be gravel or bare soil.

The amount of habitat provided by roadsides depends on the road standard and ownership. Rights-of-way, the areas set aside in the road easement, vary in width from jurisdiction to jurisdiction. Roads on private land are not required to set aside areas and many roads on farms are immediately bordered by the field (i.e., the field is plowed right up to the roadside). Roadsides adjacent to county roads and state and federal highways have rights-of-way that are ≥16 feet (5 m) wide on either side of the road, and for some Interstate highways, may be >328 feet (100 m) wide in places. Thus, roadsides managed by the county road departments or the state DOT represent more potential habitat than is managed by any other state agency (i.e., state or county parks, state wildlife areas). For example, DOT manages 7,500 miles (12,060 km) in Oregon and 7,035 miles (11,320 km) in Washington of state and federal highways. If we apply the minimum of 16 feet (5 m) on either side of the road as right-of-way, roadsides managed by the DOT represent a minimum of 45 and 43 miles^2 (121 and 113 km^2) of habitat in Oregon and Washington, respectively. County road departments manage almost three times the amount of habitat as the DOT in the two states.

Most data on wildlife use of roadsides or road rights-of-way come from higher-standard roads (county road up to interstates). Roadsides provide shelter, food, and nesting sites in agricultural landscapes (Table 1). Adams and Geis[1] reported that small mammal communities along roadsides differ depending on the standard of road—interstate highways had greater species richness than did county roads. Camp and Best[13] reported higher avian species richness along roadsides than in adjacent row crop fields, and noted that the number of species was greater than the number of species reported for most grasslands. They also reported that bird abundance was related to the structural development (height and density) of roadside vegetation. Camp and Best[14] found more nests in backslope areas than in foreslope areas. Roadsides in intensively farmed areas may be particularly good habitat for some species. Warner et al.[92] found a greater density of pheasant nests in roadsides than in any other cover type, and reported that nest success was 29%, which was comparable to hayfields.

Wildlife Diversity in Agricultural Habitat

Agricultural habitats in Oregon and Washington support diverse wildlife communities. Agricultural habitat in the two states is used by 342 species—more than any other habitat. This high species richness is a function of the broad distribution of the habitat type across the two states, but more importantly, it is a function of the wide variety of habitat conditions, land uses, and crops that are included within this habitat classification. For example, agricultural habitat in eastern Oregon or Washington might include unimproved and improved pastures, orchards, row crops, annual grains, and shelterbelts all relatively close to one another. However, any one of these crops or cover types alone is likely to support a much more depauperate wildlife community than a combination of types. For example, unimproved pastures on the eastside of the two states might be expected to support a wildlife community similar in diversity to the eastside grasslands, which are used by only 172 species.

Because of their high levels of disturbance, agricultural habitats are prone to invasion by exotic organisms. Regular soil disturbance typical of many agricultural practices provides opportunities for colonization by invasive weeds. Intensive grazing may reduce the competitiveness of some plant species in pastures, allowing invasion of exotic plant species. Wildlife communities can be affected by conversion of native habitats to non-native plant communities.[96] However, in most areas intensively managed for agricultural production, invasive weeds are aggressively controlled through the use of herbicides or integrated pest management. Invasive plant species more commonly become a problem in special habitats within agricultural areas such as road sides, field borders, fence rows, and shelterbelts. Scotch broom (*Cytisus scoparius*) and Himalaya blackberry (*Rubus procerus*) are two examples of invasive weedy species that regularly colonize these special habitats in both states.

Agricultural habitats are commonly inhabited by exotic wildlife species. Nutria, Virginia opossum, and bullfrogs commonly occur in agricultural habitats along major river valleys on the westside of both states. The rock dove, house sparrow, European starling, black rat, Norway rat, and house mouse occur in many agricultural areas throughout both states. Furthermore, several species, including the chukar, gray partridge, ring-necked pheasant, wild turkey, and eastern cottontail, have been introduced for sport in both states. The ring-necked pheasant and gray partridge are closely associated with agricultural habitats and have been widely introduced for the specific purpose of establishing gamebird populations in these non-native habitats because agricultural areas do not support native gamebird populations in huntable numbers.

Because of the non-natural condition of agricultural habitats, they are typically not critical habitats for threatened, endangered, or sensitive species. However, conversion to agriculture has directly caused the decline of some native ecosystems, including wetlands, riparian areas, native prairie, and oak woodlands, and the species that depend on these systems. Several federally listed plant species occur in these declining ecosystems, and restoration of these ecosystems will likely require the conversion of agricultural habitats to native vegetation types. Only one federally listed wildlife species (Columbia white-tailed deer) is associated with agricultural habitats in the two states. Ten mammal species of concern, mostly bats, occur in agricultural areas.

Wildlife Use of Agricultural Habitat

Most wildlife species using agricultural habitat are either seasonal migrants or use these areas in conjunction with other habitat types. Most amphibians, reptiles, birds, and mammals are only partially associated with or present in agricultural habitats (Figure 4). No amphibian or reptile species is closely associated with agricultural habitats, but 53 bird and 15 mammal species are closely associated with these areas. The most common activity for all wildlife in agricultural habitats is breeding and feeding (Figure 5). However, most of these species breed and feed in other habitats as well, and are not exclusively found in agricultural habitats. Most breeding and feeding activity in agricultural habitat occurs in habitat elements such as shelterbelts, windbreaks, fence rows, and field borders rather than in crop fields. Although many species do breed or nest in crop fields, feeding is the most common activity of wildlife in cropland.[9, 10, 70] Some closely associated breeders, such as the sandhill crane, short-eared owl, western meadowlark, bobolink, grasshopper sparrow, and savannah sparrow breed in agricultural habitats because they are structurally similar to the native grassland ecosystems that they replaced.[9]

Many wildlife species make only seasonal use of agricultural habitats. Neotropical migratory landbirds make extensive use of agricultural habitats, primarily as stop-over habitat during spring and fall migrations and

to a lesser extent as breeding habitat.[70] Typically, neotropical migrants use habitat elements within agricultural areas such as shelterbelts, windbreaks, fence rows, and field borders for stop-over and breeding habitat. Waterfowl make extensive use of cropland during spring and fall migrations and as winter habitat because of waste grain in many crop fields. Canada geese extensively use grass seed fields and pastures during winter stop-overs in both states. Although many large mammals such as deer, elk, and pronghorn antelope may use agricultural habitats year round, use is most pronounced during fall and winter.

Wildlife use of agricultural habitat is dependent on the crop being produced and the tillage method used to cultivate the crop. Crops vary substantially in respect to the amount of food, shelter, structure, and frequency of disturbance. In general, unimproved pastures that are grazed rather than hayed offer some of the best habitat among agricultural crops, especially if fall regrowth is sufficient to offer winter and early spring cover. Fruit and nut orchards provide the most long-term habitat structure among agricultural crops, and are used for nesting and foraging by many songbird species. A few wildlife species will nest in most row crops, but these habitats are ephemeral and may be harvested during the nesting season (see section on ecological traps below). Most vegetable and root crops offer little habitat for wildlife, especially after the crop has been harvested. Depending on how crop residues are handled, many small grains may offer both food (waste grain) and shelter (stubble) for several months after harvest. The fallow system used for producing wheat on large areas in the eastern portion of both states may provide stubble and waste grain for up to a year every other year.

The tillage practice used also determines the residue available for winter and early spring cover and the availability of food. Conventional tillage is used to produce

Figure 4. Percent of species by vertebrate group closely associated, associated, or present in agricultural habitats. Number in parenthesis above bars is the number of species.

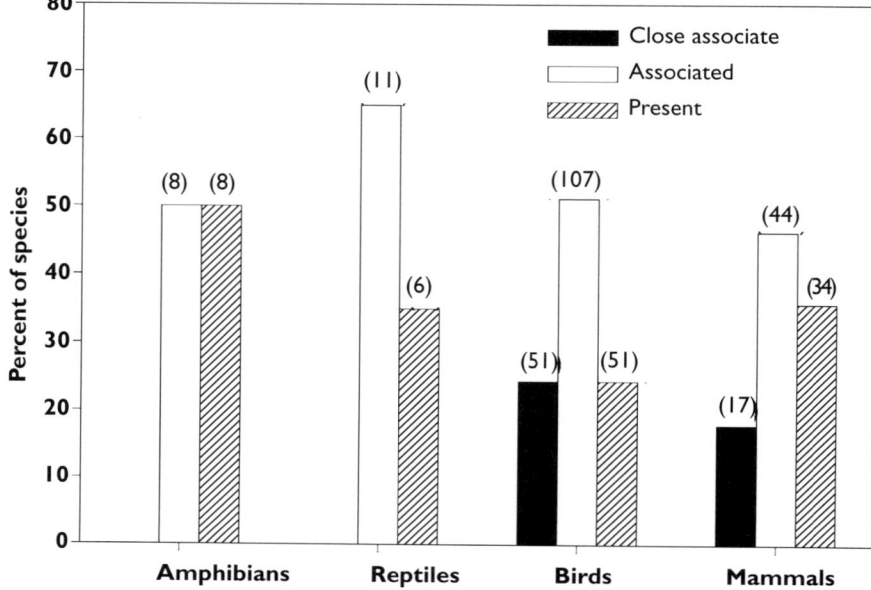

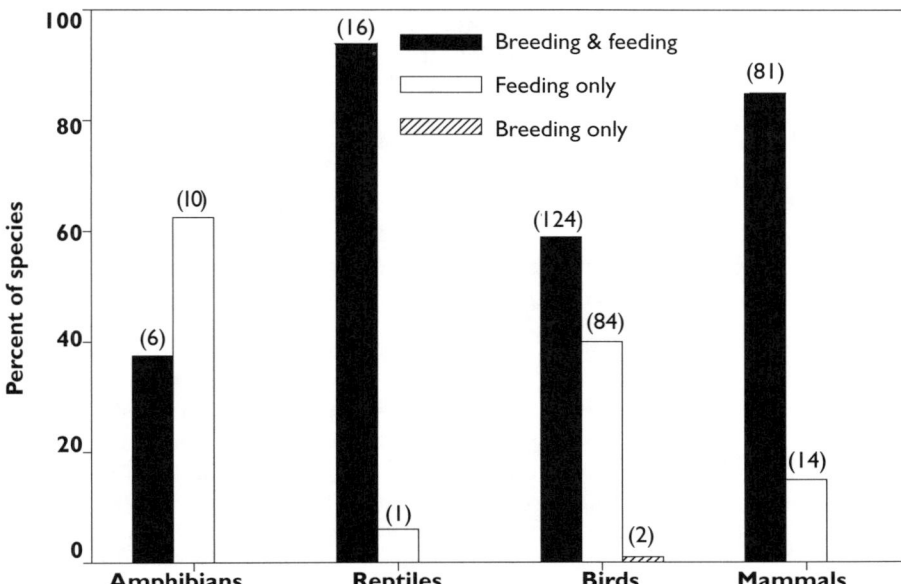

Figure 5. Percent of species by vertebrate group using agricultural habitats for breeding, feeding, or other activities. Number in parenthesis above bars is the number of species.

many crops in both states. Under conventional tillage practices, crop residues from the previous crop are plowed under in the late summer or fall, followed by disking and harrowing in the spring to prepare the seed bed for planting (Figure 6). Usually at least one pass with cultivating equipment is used to control weeds while the new crop is growing. Conventional tillage provides few wildlife benefits from the time the fields are plowed until the new crop is established. Waste grain is typically plowed under and there are no crop residues to provide cover for ≤6 months of the year. However, arthropods and small mammals may be temporarily available for wildlife following plowing and disking. Many crows, gulls, and even coyotes are often seen following plows or using fields shortly after plowing.

Conservation tillage includes a number of practices in which ≥30% of the crop residues protect the soil surface between harvest and planting. Reduced till and no-till are two of the more common conservation tillage practices. Under reduced tillage, a disk is used for the primary tillage in the spring; residues from the previous crop provide overwinter cover and waste grain is not plowed under until disking. Reduced tillage relies on cultivation to control weeds once the crop is established. Arthropods associated with the previous crop's residue may be available for wildlife under reduced tillage practices. No-

till uses special planters to plant the crop in the residue from the previous season, and crop residues are therefore present year round. No-till provides overwinter cover, waste grain is not incorporated into the soil, and arthropod communities associated with plant residues are not destroyed. However, because no-till substitutes herbicides for cultivation to reduce weeds, seeds from annual plants, which are important food sources for some wildlife, become unavailable. Crop production under all tillage practices may be an ecological trap during the nesting season.[7]

Status, Threat, and Trends in Agricultural Habitats

Agricultural habitats have changed substantially over the past two centuries. The most pronounced change has been the steady increase in the conversion to agricultural habitat until the 1950s (Figure 7). Before European settlement, native American tribes cultivated some areas along major streams and rivers, but the total area under cultivation was minute compared to the extent of modern agriculture. Although trade in agricultural products was important for some tribes, agriculture was practiced on a subsistence scale by native Americans. European and Asian immigrants introduced the horse-drawn plow beginning

Figure 6. Conventional tillage practices remove over-winter cover and bury waste grain, providing few benefits for wildlife.

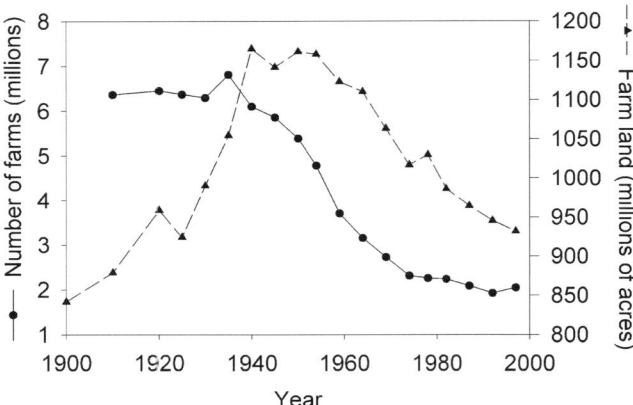

Figure 7. The number of farms and amount of farmland in the United States, 1900-1997.[84]

in the early 1800s, and throughout the rest of the century much of the land on the flood plains of the broad river valleys in both states was converted to agriculture. This expansion of agriculture resulted directly in substantial declines in natural habitats, especially wetlands, native prairies, and riparian woodlands. The advent of steam and gasoline-powered machinery accelerated the rate of conversion of native habitats to agriculture, and by the 1930s the land best suited for agriculture had already been converted. Between the 1950s and 1980s improvements in irrigation technology and USDA programs encouraged conversion of less suitable areas to farmland, primarily arid areas in the eastern portions of both states. However, recessions in the 1950s, 1960s, and early 1980s resulted in a decline in the number of farms and total harvested acres.

Since 1985, U.S. agricultural policy has changed the amount and character of agricultural habitat in both Oregon and Washington. The Food Security Act of 1985 (the Farm Bill) contained the first important conservation provisions from a wildlife standpoint.[30] The 1985 Farm Bill contained disincentives for further conversion of wetlands and highly erodible soils to agriculture under the Swampbuster and Sodbuster programs. More importantly, the 1985 Farm Bill introduced the Conservation Reserve Program (CRP), which paid farmers to take highly erodible soils out of production under 10-year contracts. The CRP has resulted in significant improvements of wildlife habitat in farmlands, especially for grassland-dependent species.[8, 30, 37, 52, 94] Building on the success of the 1985 Farm Bill, the 1995 Farm Bill enhanced the CRP with a continuous program and introduced several other significant conservation programs including the Wetland Reserve, Environmental Quality Incentives, and Waterbank programs (see McKenzie[53] for a description of these programs). The net result of these programs has been a decline in the total land area under crop production and enhancement of wildlife habitat on farms and ranches, especially wetlands and riparian areas.

Landscape and Farm Diversity

Since the beginning of the twentieth century agricultural habitats have changed dramatically in respect to diversity and spatial configuration, because of changes in agricultural markets and technology, food choice, urbanization and economic development, tax policy, farm policy, and agricultural research, etc.[30, 56] Farm diversity has declined dramatically since the 1960s. Before this, most farms produced multiple products and many were diversified crop-livestock operations. Many farmers produced forage and feed grains for their livestock in addition to other crops. Forage and hay pastures, which required longer crop rotations and in general were better habitat for wildlife than row crops, were more abundant than they are now. Since the 1960s the trend in agricultural production has been toward specialization in crop production on individual farms as well as regionally.[56] The result is extensive agricultural habitat that is spatially and structurally homogeneous. This decline in habitat diversity has resulted in declining wildlife diversity in agricultural habitats.[9, 30, 70]

Changes in farm and field size also have resulted in less diverse agricultural habitats. Although the number of farms has declined by >50% since the 1950s, the number of harvested acres has remained relatively constant,[84] and the average farm size has almost tripled. Coupled with increased product specialization and operation scale, average field size has also increased substantially. This increase in field size decreases the amount of edge habitat around fields, which are important sources of wildlife diversity in agricultural habitats.[9, 10, 66, 100] Furthermore, increased field size also decreases the abundance of habitat elements such as shelterbelts, fence rows, and field borders. Wildlife response to field size and adjacent habitats is species-specific,[10, 35, 66, 100] and dependent on land-use patterns.[27, 74, 81] Heckert[35] characterized grassland bird species as edge species, area-sensitive, and vegetation-restricted. Edge species such as the song sparrow, red-winged blackbird, and American goldfinch might be expected to decline as field size increased. However, larger field sizes will be beneficial for area-sensitive species such as the western meadowlark, grasshopper sparrow, savannah sparrow, and bobolink[35] (see Chapter 20, Westside Grasslands and Chaparral).

Changes in farming practices also have resulted in declines in habitat diversity. The advent of "clean farming" promoted by the Land Grant universities in both states caused further declines in the availability of habitat elements in agricultural habitats. Field borders, fence rows, and road sides were removed in an effort to eliminate refugia for weeds and pest species. Thus "clean farming" has resulted in further simplification of agricultural landscapes.

Ecological Traps

Agricultural habitats have a high potential to become ecological traps because of farm operations and the abundance and distribution of habitat features in agricultural landscapes. Ecological traps are human-made areas that, based on physical or vegetation characteristics, appear suitable for nesting but which, by virtue of some confounding factor(s), result in population sinks rather than sources for species that use those sites.[7, 29]

As previously described, most agricultural habitats are characterized by a regular pattern of disturbance, which varies with the crop being produced and the production methods used to grow, cultivate, and harvest the crop. Production of most crops requires multiple field operations that may include plowing, disking, harrowing, planting, cultivating, applying herbicides and pesticides, as well as harvesting (Figure 8). The production of some crops may require ≥10 field operations during the growing season, and most agricultural crops produced in Oregon and Washington require at least four field operations. The number of field operations depend on the tillage method being used to produce the crop. Reduced or no-till production methods substitute at least one herbicide application for ground-disturbing cultivation (Figure 8).

The natural history characteristics of wildlife nesting in agricultural fields determine the level of disturbance from field operations. According to Best,[7] severity of nest disturbance by field operations depends on the (1) nest position in relation to the crop rows, (2) length of the nesting cycle, (3) duration of the breeding season and the species' propensity to renest after nest failure, and (4) timing of the breeding season. Nest position within a crop field is important because field operations can destroy nests; destruction is more likely for between-row nesters such as ring-necked pheasants, killdeer, grasshopper sparrows, and western meadowlarks than for within row nesters such as mourning doves and vesper sparrows. Duration of nesting season and the propensity to renest is important for some species; mourning doves have a long nesting season and will renest multiple times. Ring-necked pheasants have a long nest cycle and typically only raise one brood per season (but some will renest). Other species will renest ≥1 time. Timing of the breeding season is also important. Horned larks begin nesting almost 1 month before field operations begin whereas mourning dove and vesper sparrow nesting extends beyond when fields are cultivated. Best[7] also points out that another form of ecological trap occurs when hayfields and pastures are converted into no-till row crops. Species that are strongly philopatric and nest in grasslands (western meadowlarks and grasshopper sparrows) may not perceive the extent of habitat alteration until later in the nesting season and are unlikely to establish new territories elsewhere. Mammal populations are also susceptible to ecological entrapment. Edge et al.[26] reported that mowing of alfalfa caused an approximately 50% decline in gray-tailed vole populations, regardless of the initial density.

Ecological traps in agricultural habitats also may occur because of the amount and distribution of edge habitat and habitat elements. Edges as ecological traps have received a substantial amount of research.[66, 100] However, most of this work is related to forest-edge habitats, and may only be relevant to agricultural ecosystems where woodlots of varying size are intermixed in farmlands. As

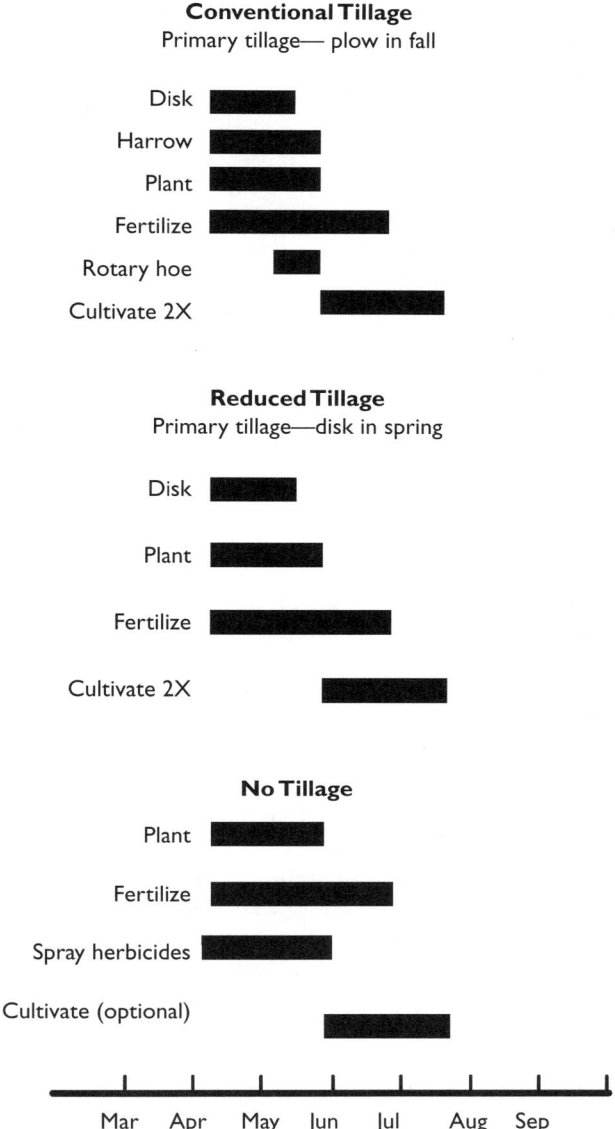

Figure 8. Sequence of field operations used in the production of a crop is dependent on the tillage method used for crop production. This diagram illustrates the production of corn and soybeans (after Best,[7] used with permission).

previously mentioned, agricultural landscapes have changed over the past quarter-century, resulting in larger field sizes and fewer edges and habitat elements such as shelterbelts, fence rows, and field borders. As the abundance of edge and special habitats declines, those that remain in the landscape become fragmented and successively isolated. Survival and reproductive success of wildlife using these remaining habitats may be low because of depredation from omnivorous predators.[74, 79, 100] Populations of many omnivorous predators including corvids, gulls, canids, raccoons, opossums, skunks, and rodents are enhanced by the interspersion of cropland food sources (i.e., crops, waste grains, arthropods, and fruits in fence rows, hedgerows, or shelterbelts). Many of these predators are highly mobile and preferentially search edge habitats, and as a result, nests in or near predator travel lanes may have a low likelihood of success.

Brood parasitism by brown-headed cowbirds may be pronounced in agricultural landscapes. Cowbirds are well adapted to agricultural landscapes[51] and exploit a variety of host species.[69] Agricultural landscapes with abundant edge habitats composed of shrub and tree communities are well suited to cowbirds because of the species' practice of spatially partitioning their breeding and feeding activities.[23, 24, 71] Cowbird feeding areas typically are open fields, livestock facilities and corrals, and other human-altered areas,[2, 23] whereas breeding sites are host-rich areas such as riparian areas,[2] sparse forests,[71] and edge habitats between forests and meadows.[29] Thus, the distribution and pattern of shrub and woody edge habitats in agricultural landscapes may increase brood parasitism and create ecological traps for species nesting in those areas.[11, 38, 69]

Agricultural Chemicals

Modern agriculture typically requires extensive chemical inputs in the form of pesticides and fertilizers. These chemicals used in agricultural areas have both direct and indirect effects on wildlife living in farm landscapes, and have been a concern of wildlife biologists.[16, 41, 63] The impact of chemical pollutants on nontarget species, primarily birds and mammals, is of special concern for threatened or endangered species[85] or species that exist in small isolated populations.[54, 93]

Pesticides. Pesticides, including herbicides, insecticides, fungicides, and rodenticides, are widely used to control agricultural pests. For example, the 1997 fall potato crop alone required 333 tons (302 metric tons) of pesticides (herbicides, insecticides, and fungicides) in Oregon and 634 tons (575 metric tons) in Washington.[84] Depending on the crop being produced, pesticides often are applied several times during the year; some crops may require the use of pesticides every 10 days to 2 weeks throughout the growing season. Furthermore, pesticides are often applied in mixtures or cocktails of ≥2 chemicals. Because they are widespread and toxic, pesticides have the potential to adversely affect wildlife throughout both states. The environmental impact of pesticides is an extensive subject that would require an entire book for complete coverage. I provide only a cursory review of some of the main points in this section.

The U.S. Environmental Protection Agency (EPA) attempts to minimize the effects of pesticides on wildlife through implementation of the Federal Insecticide, Fungicide, and Rodenticide Act of 1988 (FIFRA), which sets regulations on the acceptance and use of chemical pesticides. Even with these regulations, some chemicals are used that result in detrimental effects to nontarget birds and mammals.[41] One criterion originally required by FIFRA was to field test certain pesticides prior to registration. However, in 1992 the regulations were revised and field tests were no longer required[58, 87] (L. Fisher, U.S. Environmental Protection Agency, administrative memorandum). One reason for this revision was the cost of field testing a pesticide, which often exceeds $1 million. However, without field tests, greater uncertainty exists regarding the potential impact of pesticides on nontarget species.[42, 65, 82] Furthermore, the models the EPA uses to predict effects of a chemical on wildlife often perform poorly under field conditions[25, 73, 90] because of assumptions inherent in the use of these models.[82]

A pesticide's capacity to harm wildlife is dependent on several factors, including the chemical's characteristics and toxicity, and the timing, duration, and dose of exposure.[62; 133, 80] The characteristics of a pesticide include its propensity to bioaccumulate and its degradation rate. Toxicity of a pesticide refers to how poisonous the chemical is to wildlife. Pesticides vary substantially in their toxicity to wildlife; some are relatively nontoxic even at high application rates whereas others are extremely toxic at normal application rates. Typically, insecticides are much more toxic to wildlife than are herbicides or fungicides. Furthermore, toxicity of a chemical varies substantially among wildlife species and even within a species depending on the age, sex, reproductive condition, and health of the exposed animals. The timing of a chemical application may determine its impact on wildlife—some wildlife are more susceptible during nesting, migration, or periods of low food availability. The duration of exposure can be important because some species are affected by a pesticide only if the exposure is extended. Finally, the effect of a pesticide is dependent on the dose or concentration of the chemical in the environment. Some pesticides can be applied within a range of applications rates for a pest in a particular crop. A pesticide applied at a higher application rate will almost always have a greater potential to impact wildlife than the same chemical applied at a low application rate.

Wildlife exposure to pesticides occurs in several ways. Primary routes of exposure are through consumption, absorption through the skin, and inhalation of pesticide vapors when the chemicals are applied. Wildlife that nest or feed in recently sprayed crops may be subject to primary routes of exposure. Animals may consume pesticides directly, e.g., through consumption of contaminated vegetation, rodenticide baits, or pesticides granules as grit. Secondary exposure to pesticides occurs when wildlife consume other animals that have been exposed. Birds and other wildlife are attracted to insects dying on the soil surface after insecticide treatments, and may be exposed when they eat these insects or feed them to their young. Secondary poisoning can be a concern for some predators that consume rodents poisoned by rodenticides.

Effects of pesticides on wildlife can be both direct and indirect. Direct effects include both lethal and sublethal exposures. Not all wildlife exposed to pesticides receive a lethal dose and immediately die; smaller sublethal doses may be more common, but typically are difficult to verify. Sublethal doses can lead to changes in behavior, weight loss, impaired reproduction or predator avoidance, and inability to thermoregulate.[31] Pesticides can indirectly affect wildlife by altering habitat and reducing food resources. Although less toxic to wildlife, herbicides can have profound impacts on wildlife by altering habitat. Many seed-producing "weed" species are important components of wildlife diets. Furthermore, weed control can destroy insect habitats and result in less diverse insect communities, which in turn may reduce the availability of some important foods for some wildlife.[80] Herbicides may also modify wildlife cover and reduce vegetation complexity. Widespread use of herbicides in fence rows, field borders, roadsides, and shelterbelts to control weedy vegetation has reduced the value of these habitat areas for wildlife.

Fertilizers and Livestock Waste. Modern agricultural practices depend heavily on the application of fertilizers to maintain soil productivity and to grow many of the crop varieties that have high nutrient demands. Hundreds of million pounds of fertilizer are used in Oregon and Washington annually. For example, in 1997, Oregon wheat farmers used 37,650 tons (34,150 metric tons) of nitrogen and 2,350 tons (2,130 metric tons) of phosphate fertilizers, while Washington wheat farmers used 72,700 tons (65,940 metric tons) of nitrogen and 8,250 tons (7,480 metric tons) of phosphate.[84] Fertilizers cause few direct impacts on wildlife, but indirectly impact aquatic organisms because of reduced water quality. Nonpoint source (NPS) pollution from fertilizers may adversely impact amphibian reproduction, but that has not been documented. Phosphorus and nitrogen are the primary elements in NPS pollution in agricultural areas. Phosphorus pollution causes eutrophication of lakes and streams. The rapid growth of algae in response to phosphorus can limit oxygen availability in aquatic systems and cause fish die-offs, and presumably kills developing eggs and larvae of amphibians. Nitrogen is primarily a human health concern.

Livestock waste in agricultural habitats also may result in water quality problems. Livestock confinement operations such as dairies and feedlots produce massive amounts of waste that can result in reduced water quality where wastes are improperly stored or recycled. Livestock waste disposal during the past 20 years has moved from being primarily a problem with lack of or improper storage to one of inadequate land base for efficient reuse of manure nutrients.[49]

Animal Damage

Wildlife in agricultural habitats cause significant damage to agricultural crops and production systems. This damage may strain relationships between landowners and wildlife and habitat management agencies, and may cause some landowners to be reluctant to improve wildlife habitat on their properties. Loss to wildlife damage represents a significant portion of the agricultural production in Oregon and Washington. According to a 1997 survey conducted by the Oregon Agricultural Statistics Service,[83] wildlife caused Oregon farmers and ranchers $158 million in damage to crops, livestock, injury, and prevention expenses. Approximately 47% of Oregon's farms reported some type of wildlife damage; this loss represents 3.3% of the total value of agricultural production in Oregon.[83]

Some wildlife damage occurs in almost all agricultural crops and in facilities, production equipment, and systems. A thorough review of the types of damage caused by wildlife is beyond the scope of this chapter, but is presented in Hygnstrom et al.[36] Big game cause substantial damage in hay, pasture, alfalfa, and grain crops, especially in the eastern half of both states. Waterfowl, particularly Canada geese, cause widespread losses to pastures and grass seed fields during winter. Passerines damage a wide variety of fruit, nut, and grain crops in both states. Rodents, primarily voles, house mice, ground squirrels, and pocket gophers, damage alfalfa, hay, grain, fruit, and nut crops. Mice and rats cause significant damage to stored agricultural products, especially grains, in both states. Rodents also damage irrigation systems, and mounds of dirt from burrowing rodents can damage harvesting equipment. Predators, primarily coyotes and mountain lions, cause losses to livestock operators in both states.

Managing animal damage problems is a difficult and controversial subject. Biologists from the wildlife agencies in both states and the USDA Wildlife Services spend substantial amounts of time and agency resources in resolving these problems. Animal damage specialists have a wide array of options in their tool kits that generally fall in the categories of habitat modification, alternative crop production systems, hazing and scaring, repellents, and population control. Specific control techniques for each wildlife species are presented in Hygnstrom et al.[36] Although often discussed, birth control is generally not an option because of expense and ineffective delivery systems.[28] Lethal forms of control receive increased scrutiny in the highly urbanized areas in both states, and management agencies are continually searching for and testing alternative control methods. However, regardless of the public's perception of animal damage management, it will always be a necessary component of wildlife management in Oregon and Washington, where we have both intensive agricultural production and abundant and diverse wildlife communities.

Managing Agricultural Habitat for Wildlife

Although agricultural habitats have high wildlife diversity on a statewide basis, diversity is typically low at the farm level, and opportunities abound for enhancing wildlife habitats and increasing wildlife diversity on farms and ranches. State and federal agencies in both states have numerous assistance programs (Table 2) to help landowners manage their properties for improved environmental quality. Interested landowners should contact the agency in charge of the program. All of these programs are voluntary and contain varying levels of incentives or disincentives to encourage participation. However, biologists working with private landowners must recognize that successful programs for enhancing wildlife habitats on private lands will depend on the landowners' personal interest and sense of stewardship.[48] The level of comfort and participation in habitat

Table 2. Landowner assistance programs administered in 1999 by federal agencies in Oregon and Washington that aid private landowners in managing properties for improved environmental quality.

USDA Natural Resources Conservation Service and Farm Services Agency

The Conservation Farm Option—a pilot program intended to foster innovative conservation in whole-farm management.

Conservation of Private Grazing Land—provides technical assistance for conservation needs on pasture, hayland, and range.

Conservation Reserve Program—a land retirement program that establishes permanent grass or tree cover on environmentally valuable cropland or pasture.

Environmental Quality Incentives Program—provides incentives for addressing conservation needs on cropland or areas used for livestock operations.

Flood Risk Reduction Program—encourages less intensive uses of frequently flooded cropland.

Highly Erodible Land Conservation Compliance Program—requires participants who receive USDA assistance to implement a soil conservation system on highly erodible farmed land.

Resource Conservation and Development Program—helps local groups plan and conduct rural development and natural resource conservation activities.

Water Bank Program—compensates landowners for protecting regionally important migratory bird wetlands.

Watershed Protection and Flood Prevention Program—a cost-share program to implement watershed work plans that protect watershed functions.

Wetland Conservation Compliance Program—a disincentive program that links wetland conservation to receipt of USDA farm program benefits.

Wetland Reserve Program—a land-retirement program for restoring and protecting wetlands.

Wildlife Habitat Incentives Program—assists landowners to plan and pay for habitat improvements associated with farming operations.

U.S. Fish and Wildlife Service

Partners for Wildlife—a cost-share program for enhancing wildlife habitats, especially wetlands.

enhancement programs will vary from landowner to landowner. Significant changes in wildlife habitats on private lands may take many years and will depend on local biologists building trust with private landowners.

Odd Areas

Odd areas are areas on farms that are not cultivated because of location, topography, erosion, geomorphology, or soil conditions. Odd areas typically include sites such as rock piles, rock outcrops, borrow and gravel pits, eroded areas in fields, bare knobs, sinkholes, gullies, and good cropland cut off from the rest of the field by a stream, ditch, gully or other obstruction.[61] Center-pivot irrigation systems typically leave odd areas between the irrigated fields. Most farms have some odd areas, which represent

Table 3. Suggestions for managing odd areas on farms or ranches.[61, 78]

Protect from fire and grazing, or if grazed, control timing and intensity to prevent damage or loss of shrubs and trees.

Develop food plots adjacent to odd areas.

Plant fruit or mast-producing shrubs and trees—cherry, hawthorne, wild plum, wild pear, serviceberry, crabapple, and oaks are good candidates depending on the site or region.

Enlarge odd areas during crop rotation by adjusting field boundaries.

Plan for connectivity among odd areas and other habitat patches.

Manipulate vegetation with controlled fire, mowing, or a brush-hog to stimulate new growth of understory and enhance structural diversity.

In areas where snow drift or high winds may be a problem, develop the site as a shelterbelt or windbreak to provide cover needed.

Add brush piles from pruning of shrubs and trees.

potential sites for habitat development. Many odd areas may already be good habitat and need little in the way of enhancement. Management of odd areas and plant species used to enhance wildlife habitat depends on the region (i.e., what plants are adapted to soil and climate, and what is available) and the type of site. Snyder[78] and Payne and Bryant[61] suggested a number of practices that landowners might consider implementing on odd areas (Table 3). Typically, the landowner's desire for particular wildlife species and other habitats available on a farm would guide what practices are used.

Food Plots

Food plots are sites cultivated specifically to provide food for wildlife species, and have been widely recommended by wildlife agencies around the country. Food plots can be used to supplement natural food supplies where they are likely to be limited, or designed to attract wildlife to specific areas for viewing. They are also commonly used in conjunction with recreational hunting or fee hunting. Food plots may lead to better over-winter survival and spring conditioning for egg laying. For example, Robel et al.[68] showed that bobwhite quail with access to food plots during winter in Kansas maintained higher body weights and fat reserves, both of which are assumed to lead to better over-winter survival and early conditioning for egg production.

Food plots are not preferred to other forms of habitat management, but are typically used by landowners who wish to rapidly enhance wildlife habitats on their properties. Payne and Bryant[61:436] listed several disadvantages to food plots: (1) they concentrate wildlife, thereby increasing vulnerability to disease and predation; (2) food plots attract undesirable wildlife and feral animals (i.e., hogs, starlings, or other damage-causing species); (3) they represent a direct expense to the farmer; (4) food plots may result in high concentrations of wildlife or populations above carrying capacity; and (5) in arid regions plantings are likely to fail during drought years, and are probably unneeded during wet years.

Table 4. Metabolizable energy in seeds of plants commonly used for food plots for seed-eating birds.

Seed	Metabolizable energy content (kcal/g)
Excellent (>4.14 kcal/g)	
Giant ragweed	4.32
Good (3.49-4.14 kcal/g)	
Western ragweed	3.88
Corn	3.87
Soybean	3.78
Sunflower	3.65
Sorghum	3.59
Low (2.61-3.48 kcal/g)	
German millet	3.47
Prostrate lespedeza	3.42
Korean lespedeza	3.14
Wheat	3.06
Thistle	2.70
Shrub lespedeza	2.69
Poor (<2.61 kcal/g)	
Partridgepea	2.42
Smartweed	2.30
Multiflora rose hips	2.02
Switchgrass	1.86
Smooth sumac	1.48

After Robel et al.,[67] used with permission.

Food plants recommended for wildlife vary from region to region like the wildlife species the plots are intended to benefit, but grain sorghum, corn, legumes, and cereal grains are the most common. Payne and Bryant[61] listed plants commonly used for food plots for several wildlife species. Sorghum is typically ranked over corn because it is easier to establish in areas without irrigation, the smaller seed is acceptable to more wildlife species, and because it has energy content similar to corn. Energy content of the more common wildlife food plants varies substantially (Table 4).

The size, shape, location, and distribution of food plots are as important as the particular plant foods that are cultivated. Size depends on the target wildlife species. For most birds and small game species, plots that are 0.5-5 acres (0.2-2.0 ha) are usually sufficient. Ungulates may require 10- to 20-acre (4- to 8-ha) plots or larger, depending on the size of the herd being fed. In general, food plots are linear, which maximizes edge, access, and interspersion with other habitat types, and distributes animals to reduce vulnerability to disease and predation. Width of these linear strips also depends on the target species. For most small animals 10-20 feet (3-6 m) is wide enough. Often, this is limited more by the equipment used to cultivate a food plot. Leaving 4-5 rows of a crop unharvested is a commonly recommended practice that leaves a strip approximately 10-20 feet (3-6 m) wide depending on the crop and harvesting equipment. Payne and Bryant[61] suggested that plots should not exceed 328 feet (100 m) for ungulates and 164 feet (50 m) for some medium to small animals.

The spacing and distribution of food plots is another consideration. Usually only 1-3% of an area might need to be intensively managed for food plots.[61: 434] The distribution of food plots depends on the mobility of the target species and constraints imposed by normal farm operations. Typically, recommendations have been developed only for game species. For pheasants, one 3.7-acre (1.5-ha) plot for every 123 acres (50 ha) would insure

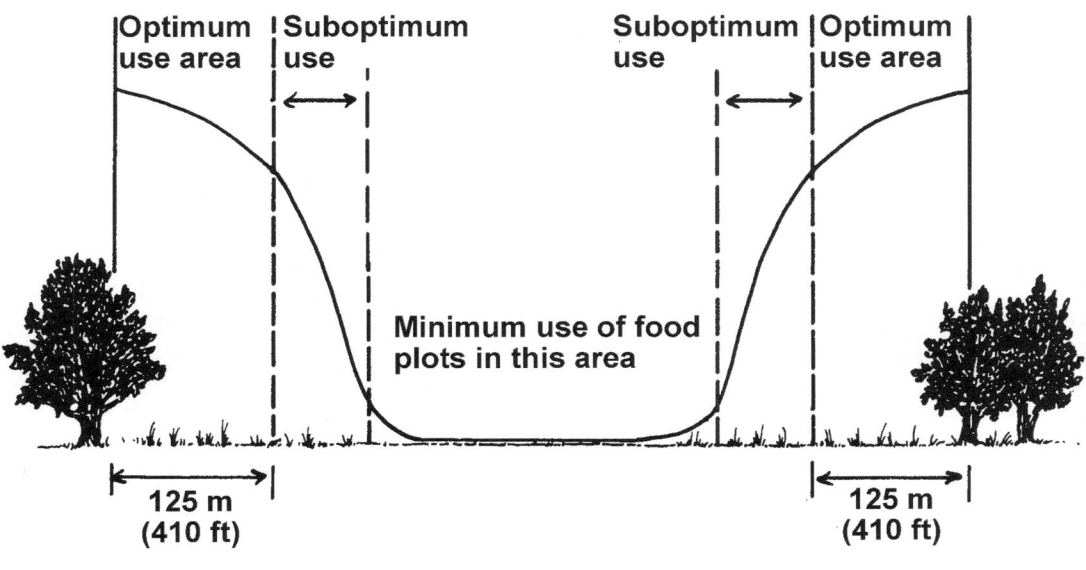

Figure 9. Deer use of food plots will be greatest in the optimal use area within 410 feet (125 m) of cover; food plots >655 feet (200 m) from cover will receive little use by deer (after Leckenby et al.[47]).

that a food plot was always within 1,640 feet (500 m). For deer, two 15- to 20-acre plots per 1,235 acres (6-8 ha per 500 ha) has been recommended.[61; 434]

Location of food plots relative to other habitat attributes is most important; proximity to cover is a prime consideration. For example, deer will be unlikely to use food plots >738 feet (225 m) from cover (Figure 9). Geese and waterfowl, conversely, will be less likely to use food plots that are within 82-164 feet (25-50 m) of cover that might conceal predators. In areas with high winds, food plots are more likely to be used if they are on the leeward side of cover.

Shelterbelts and Windbreaks

The management of shelterbelts and windbreaks will vary, depending on the location and the purpose of the shelterbelt. Typically, shelterbelts are designed and planted for specific conservation purposes other than wildlife habitat enhancement (i.e., reduced soil erosion, visual or noise screening, snow fences, livestock protection). However, because shelterbelts provide important structural and cover components in otherwise structurally simple landscapes, these habitat features can greatly enhance wildlife diversity and abundance on a farm. Wildlife managers and landowners should expect about 20 years before a shelterbelt begins to function in the manner for which it was designed. However, these sites will be used by wildlife immediately.

Numerous plant species have been used for shelterbelt plantings; they should be suitable for the purpose and adapted to soil and climate. The USDA maintains a website (http://www2.nrcs.usda.gov/Netdynamics/Vegspec/pages/HomeVegspec.htm) with software that helps users identify plant materials for use in restoration efforts. The species that should be planted will vary from region to region depending on availability of seed or stock. Species used should be tolerant of herbicides that may be used in areas being protected. Planting a variety of species is also important because it reduces the likelihood of losing the shelterbelt to a species-specific pest (e.g., Dutch elm disease), and the greater structural diversity will increase wildlife diversity.[75, 99] Payne and Bryant[61] developed a list of plants commonly used in shelterbelts. Because many sites where shelterbelts are desired have severe climatic conditions (low precipitation, high winds), only high-quality planting stock should be used. It may be necessary to reduce competition from other plants during the establishment phase by using herbicides or mechanical means. In arid regions, supplemental water may also be required during the establishment phase. Shelterbelts also need to be protected from fire and grazing.

Shelterbelt width depends on the number of rows of plantings, and typically the number of rows used in conservation practices depends on the thermal protection required and the need to control snow drift. Podol[64] suggested that any windbreak <108 feet (33 m) wide in northern areas would not create snow-free areas in winter and would have little value for winter cover. For wildlife purposes, the wider the better. When planting rows,

Yahner[99] suggested that rows should be 16 feet (5 m) apart to allow for herbaceous and shrub development in the understory.

Planting configuration may also be important for wildlife. Plantings that are irregularly spaced or that are not straight would increase the amount of edge habitat and would reduce line-of-sight distances from predators and other sources of disturbance.[57]

Hedge and Fence Rows

Plant species used in fence and hedge rows vary from region to region and should be adapted to local climate and soil conditions. Multiple species within fence and hedge rows will increase wildlife diversity.[59] Species that produce fruit or berries and that have dense growth forms are good candidates for fence rows. In Oregon and Washington, wildrose (Rosa spp.), wild pear, hawthorn (Crataegus spp.), blackberry, and cherry (Prunus spp.) are commonly used. However, in some areas these species may be considered invasive weeds.

Hedge and fence rows can be planted or allowed to grow from natural regeneration. If planted, success is best where high-quality planting stock is used. Where possible, hedge and fence rows should be at least 10 feet (3 m) wide. Creating mosaics through selective mowing and woodcutting, and retaining 1.7-2.4 snags (>18 inches diameter at breast height [dbh], 33-49 feet tall)/mile (1-2 snags [>46 cm dbh, 10-15 m tall]/km) of fence row have also been suggested for hedge and fence row management.[76] These sites, like shelterbelts, will need to be protected from fire and grazing. O'Connor and Shrubb[59] suggested that the least amount of hedge row management as possible will make the most favorable conditions for wildlife. Hedge and fence rows can be created relatively easily by plowing or disking the site where they are to be created, placing fence posts in a staggered configuration, and running a wire or string for perching by birds. Seeds in the bird droppings rapidly establish preferred plants in the plowed or disked areas.

Field Borders

Field borders are typically seeded with fast growing grasses or legumes that are adapted to the site. Field borders should be at least 16 feet (5 m) wide. In areas where they function as filter strips around wetlands and riparian areas, they should be 20-98 feet (6-30 m) wide. Field borders may be hayed for forage, but as in all haying operations where wildlife are a concern, haying should be delayed as late as possible to avoid the breeding season (usually after 15 July). Field borders, like most grass or grass/legume mixtures, occasionally need to be mowed to maintain plant vigor or to prevent invasion by woody plants. Maintenance mowing of this type can be done once every two or three years. Another consideration for both haying and mowing is to allow some regrowth before the fall/winter dormant period. Weeds are a common concern among farmers that have herbaceous field borders. Many of the "weed" species are actually good sources of food for wildlife. However, most farmers will want to control invasive weeds that are likely to interfere with crop production. Weeds in field borders can usually be controlled with spot application of herbicides.

Roadsides

Because of the prevalence of roadsides in both states, management of these habitats can profoundly affect wildlife. Roadside management varies a great deal depending on jurisdiction. The DOTs in Oregon and Washington are charged with management of interstate, U.S., and state highways. County governments typically manage county roads, while farmers manage the roads on their own property. Most roadsides managed under the first two jurisdictions are managed with safety as a primary concern. Roadside vegetation is mowed to reduce visual obstructions, and in some cases to discourage wildlife, especially deer, use adjacent to the road. Many roads on private lands do not have roadside vegetation. Rather, the field is plowed or cultivated right up to the road. Encouraging private landowners to leave a 16-feet (5-m) uncultivated strip of vegetation on each side of a road will provide wildlife habitat as well as other conservation benefits including reducing sedimentation in streams and off-road movement of petrochemicals. Warner et al.[92] noted the most attractive roadsides for pheasant nesting were well established areas seeded with a grass-legume mix. Roadsides on private lands are typically managed to prevent encroachment of woody species and to control invasion by noxious weeds.

Because of the linear nature of roadsides and the regular mowing, these areas may become ecological traps.[6, 14] Birds nesting in roadsides, as in other linear habitats, may suffer high rates of predation. Bergin et al.[6] reported 23% nest loss overall on artificial nests, but predation varied from area to area. Furthermore, they reported significantly greater predation on the backslope than the foreslope (30% vs. 16%). Backslopes may either have more predators moving through them, or predators preferentially search backslopes. Foreslopes in their study were more likely to be mowed (12% vs. 0%), and the shorter vegetation or mowing activity might make predators wary of searching these habitats. Mowing also directly destroys nests. Because of these concerns mowing should be delayed until mid-July where feasible, to allow successful nesting by ground-nesting birds and small mammals. In most cases, one-pass management is sufficient to address safety and woody vegetation encroachment concerns. One-pass management involves a single-mower-width pass on the foreslope adjacent to the road each year, beginning after mid-July. Woody vegetation on the remainder of the foreslope and the backslope can be controlled by mowing once every two to three years. Invasive weeds along the roadsides in all jurisdictions can usually be controlled with spot application of herbicides.

Vehicle collisions with deer cause over $1 billion in damage and >200 human fatalities annually in the U.S.,[19] and can be an important source of mortality for some wildlife populations. Although such collisions occur throughout both states, problems are most pronounced where roads intersect traditional migration routes or big game winter ranges. For some less mobile species such as reptiles and amphibians, roads may be a substantial barrier to movement. Construction of wildlife-proof fences or underpasses have proven to be effective for reducing the number of wildlife-vehicle collisions. In cases where the road is a barrier to movement or along a migration route, underpasses allow these movements to occur relatively unhindered if they are correctly located.[77, 91]

Reducing Pesticide Impacts

Reducing the effects of pesticides on wildlife can be accomplished in a number of ways. Reducing the total amount of pesticides is a general goal of alternative or sustainable agriculture[56] that would benefit wildlife. Adopting integrated pest management (IPM) strategies has been widely promoted as a means of reducing the reliance on pesticides.[56] IPM is a crop management system that includes all aspects of crop production in developing strategies for reducing crop damage from pests and disease. These strategies might include but are not limited to genetic engineering of the crop, residue management, tillage practices, fertilizer management, pest monitoring, pesticide applications, and introduction of pest predators. In most cases, IPM will result in a reduction in the amount of pesticides applied.[56] In some cases, wildlife predators might be part of the IPM strategy. For example, encouraging raptor use of crop fields may help reduce rodent populations.[34, 40, 96]

Chemical selection is another way to reduce pesticide impacts on wildlife. Several pesticides are usually available for controlling a particular pest in a particular crop. Most farmers select the cheapest chemical that gets the job done, but many are willing to pay more for a chemical that would cause fewer environmental impacts. Pesticides vary considerably in their toxicity (Table 5) and persistence. Selecting the chemical with the lowest toxicity to the group of animals that might be exposed to it and chemicals with shorter half-lives would help reduce pesticide impacts on wildlife. The pesticide information profile system (http://ace.orst.edu/info/extoxnet/pips/ghindex.html) is an easy way to determine the relative toxicity and the persistence of pesticides that a farmer may consider using.

Controlling the timing and location of a pesticide application is another means by which impacts on wildlife can be reduced. Where flexibility is possible, applying a pesticide when wildlife are not in the area would reduce wildlife exposure to the toxicant. For example, waterfowl commonly use croplands during the short spring migration period. If a pesticide application can be delayed until after the birds have moved through the area, there would be fewer potential impacts. This type of flexibility is often not possible. An alternative is to avoid the use of pesticides where wildlife are known to occur. Habitat features such as shelterbelts, fence rows, field borders, and odd areas should be protected from pesticides with a no-spray buffer zone of at least 16 feet (5 m) in the crop field adjacent to the habitat (Figure 10). Wetlands and riparian

Table 5. Toxicities to birds, mammals, and fish of insecticides and miticides used on fruit crops.

Pesticide (brand name)	Birds[a]	Mammals[a]	Fish[b]
azinphos-methyl (Guthion)	H	H	EH
carbaryl (Sevin)	L	L	H
chlorpyrifos (Lorsban)	H	L	EH
diazinon (Diazinon)	H[c]	M	EH
dicofol (Kelthane)	L	L	EH
dimethoate (Defend)	H[c]	M	M
dinocap (Karathane)	M	L	EH
endosulfan (Thiodan)	M	H	EH
esfenvalerate (Asana)	L	L	EH
formetanate (Carzol)	M	H	M
fenbutatin-oxide (Vendex)	L	L	EH
malathion	L	L	EH
methidathion (Supracide)	H	M	EH
methomyl (Lannate)	H	H	H
methyl parathion (Penncap-M)	H[c]	H	H
oxamyi (Vydate)	H	H	M
oxythioquinox (Morestan)	—	L	EH
permethrin (Ambush, Pounce)	L	L	EH
phosmet (Limidan)	L	M	EH
propargite (Omite)	L	L	EH

After Palmer and Bromley,[60] used with permission.

[a]Wildife hazard is based on the following toxicities: H (highly toxic) = LD_{50} <30 mg/kg and LC_{50} <500 ppm; M (moderately toxic) = LD_{50} >30 and <100 mg/kg and/or LC_{50} >500 and <1,000 ppm; L (low toxicity) = LD_{50} >100 mg/kg and LC_{50} >1,000 ppm; NT (not toxic).

[b]Fish 96-hour LC_{50} toxicities are as follows: EH (extremely toxic) <0.1 ppm; H (highly toxic) 0.1-1.0 ppm; M (moderately toxic) 1.0-10.0 ppm; L (low toxicity) >10.0 ppm.

[c]Active ingredient (not necessarily a specific product) has caused wildlife deaths.

Figure 10. Protect sensitive habitat such as wetlands, shelterbelts, fencerows, etc., from pesticides with a no-spray buffer (dashed line). A better alternative would also include a permanent grass field border that would prevent movement of chemicals during runoff. (Photograph by W. Daniel Edge)

areas should be protected with buffer zones composed of grass and herbaceous vegetation to reduce the amount of chemicals (pesticides and fertilizers) and sediment that reaches the water during periods of runoff. Depending on the steepness of the slope adjacent to wetlands and riparian areas, buffer zones of up to 98 feet (30 m) may be required. Grassed waterways in crop fields are recommended for preventing chemical runoff from reaching water sources.

Controlling chemical drift is another means of reducing the impacts of pesticides on wildlife. Chemical drift is the movement of a chemical off the intended site of application, and is a principal cause of wildlife exposure to pesticides. Both physical (i.e., chemical droplets) and vapor drift may occur when a pesticide is applied, depending primarily on wind velocity. Pesticides should not be applied when wind speeds exceed 5 mph (8 kph) or when there are gusts of up to 10 mph (16 kph), and never when sensitive habitats or wildlife are down wind. Droplet size is another major factor in chemical drift. Modern spray nozzles are designed to produce specific droplet sizes with a particular pressure. Using the most modern spray equipment that is properly calibrated will help reduce the amount of drift.

Case Study

The data matrixes in this book provide a powerful tool for assessing the impacts of management practices and landuse policies on wildlife populations. However, predicting changes in wildlife communities that result from changes in agricultural habitats is problematic because of the widespread distribution of agricultural habitats throughout the two states and because many wildlife species are only associated with agricultural habitats for feeding or occur in those habitats because of habitat features (e.g., shelterbelts, field borders, ephemeral wetlands, etc.). Such predictions should be considered only for relatively small areas within the two states (i.e., one to a few counties) and should consider the ecological roles of the species in those habitats.

As an example, I queried the database for the species that occur in Agricultural and Westside Grassland habitats in Linn and Benton counties, Oregon. Such a query might be used to assess the impact on wildlife communities of agricultural policies encouraging farmers to convert farmland to native grassland habitat, or vice versa. This query generated a list of 241 species that are associated with the two habitats in the two-county area. Approximately 42% (101) of the species occur in agricultural habitat and not in Westside Grasslands, but only 4 species are exclusive to Westside Grasslands (Table 6). Without consideration of the abundance of these two habitats and the ecological role of individual species associated with the habitats, one might easily conclude that conversion of Westside Grasslands to agricultural habitat would generally be beneficial to more species of wildlife. It is difficult to separate Westside Grasslands from unimproved pastures with remote sensing data, and therefore, an accurate estimate of the land area covered by these two habitats in Linn and Benton counties is not possible. However, agricultural habitat makes up >640 miles^2 (1,657 km^2) of the land area in these two counties, whereas unimproved pastures account for about 80 miles^2 (207 km^2); Westside Grasslands are only a small proportion of that (<5%) (T. O'Neil, Northwest Habitat Institute, personal communication). Furthermore, the species exclusively associated with agricultural habitats are exotic species (e.g., house sparrow, house mouse, Norway rat), or more often are associated with some special habitat features, such as ephemeral wetlands (e.g., shorebirds and waterfowl) or shelterbelts and fence or hedge rows (e.g., vireos and warblers).

Research Needs in Agricultural Habitats

A review of the literature cited section of this chapter reveals that little research related to wildlife in agricultural ecosystems has been conducted in Oregon or Washington; most research on this topic has been conducted in the Midwest. Although the relationships I report in this chapter are probably valid in our area, there are many research questions that will need to be addressed in order to effectively manage wildlife on farms and ranches. These research questions include responses of wildlife to crop management practices, integrated pest management, landscape design, human dimensions, and the impacts of agricultural policy on wildlife resources.

Most crops can be produced with several alternative crop management systems, and each system will provide a different array of wildlife benefits and costs. The growing

Table 6. Wildlife habitat association matrix for wildlife species that are found in western grassland and agricultural habitats in Linn and Benton counties, Oregon.

	Westside Grasslands			
	Closely associated	Associated	Present	Does not occur
Agricultural				
Closely associated	8	21	7	15
Associated	2	58	13	48
Present	0	13	14	38
Does not occur	0	2	2	

interest in sustainable or alternative agriculture among farmers will likely produce additional management systems providing additional opportunities for wildlife to exist on farms. Residue management and tillage practices are two areas of crop management that directly affect the year-round value of croplands to wildlife, and additional research is needed to identify those practices that are most beneficial to wildlife.[9, 70] Because of the many agricultural commodities produced in the two states, we will be unable to document wildlife responses to alternative practices for most crops. Therefore, research should focus on the most broadly produced crops, such as wheat, grass seed, hay pastures, and orchard fruits and vineyards.

Integrated pest management is another component of crop management systems where little research has been done on the direct effects on wildlife. Methods for managing animal damage to crops, ways of reducing impacts of pest management on wildlife, and a better understanding of how wildlife can be used to control crop pests all need to be examined. Wildlife damage to crops is a perennial problem that will become more complicated with the increasing concern for animal welfare. Alternative management systems that discourage wildlife use of commonly damaged crops need to be researched. For example, living mulches in orchards, vineyards, and other crops help to control runoff and invasive weeds, but depending on which grass species is used and how the mulch is managed, damage by voles may vary substantially. IPM also includes the use of pesticides, and although a great deal is known about wildlife response to individual pesticides based on laboratory studies, additional studies are needed to determine population-level effects under field conditions.[25, 82] The effects of using mixtures of pesticides on wildlife are virtually unknown. Finally, wildlife eat many arthropod and vertebrate pest species and can be part of the pest management strategy. Encouraging predators to use crop fields, and the effectiveness of predators in controlling crop pests need additional research. For example, Wolff et al.[96] documented an 11-fold increase in American kestrel use of grass fields with supplemental perches, but increased use by kestrels did not result in a decline in vole populations at high densities.

Understanding how agricultural production affects wildlife at the landscape scale is another important research need. The abundance and distribution of crops and habitat elements such as shelterbelts, riparian buffers, fence rows, and field borders at watershed and regional scales have important implications for the long-term persistence of wildlife in agricultural landscapes. How and to what extent some of these habitat features are used for dispersal is poorly understood.[5] Habitat elements may become ecological traps and function as population sinks depending on their abundance and distribution within a landscape. Determining how predation and brood parasitism are influenced by the spatial and temporal distribution of habitat components is a vital research need in agricultural habitats.

Human dimensions research is needed to identify a broad range of social and economic factors that affect how farmers manage their properties. Research needs in this area cover the gamut of the people side of wildlife management. Farmers' perceptions regarding wildlife damage[97] and habitat enhancement practices have been documented,[18] but research is also needed to identify alternative management systems that might be adopted on broad scales. As farmers and ranchers move to diversify operations, there is likely to be an increasing interest in wildlife-based alternative income opportunities such as fee hunting and ecotourism.[4, 12] Although some aspects of fee hunting have been documented in Oregon and Washington,[12, 39] the demand and willingness to pay for ecotourism in the two states has not been determined. Finally, human dimensions research is needed to identify processes and mechanisms by which successful conservation practices are implemented at the watershed scale.[33]

The final area of research is the impact of agricultural policy on wildlife resources. USDA farm programs may directly affect how farmers manage their properties from local to regional scales.[30] Land retirement programs such as the Conservation Reserve Program can affect wildlife diversity and abundance on broad spatial scales.[8, 52] Continued support of these programs in the next Farm Bill and when the 10-year contracts expire will depend on documenting their wildlife and ecological benefits.

Literature Cited

1. Adams, W. L., and A. D. Geis. 1979. Roads and roadside habitat in relation to small mammal distribution and abundance. Pages 54-1 to 54-17 in: D. Arner and R. E. Tillman, editors. Environmental concerns in rights-of-way management. EPRI WS-78-141, Electric Power Research Institute, Palo Alto, CA.

2. Airola, D. A. 1986. Brown-headed cowbird parasitism and habitat disturbance in the Sierra Nevada. Journal of Wildlife Management 50: 571-575.

3. Barbour, R. W., and W. H. Davis. 1969. Bats of America. University Press of Kentucky, Lexington, KY.

4. Benson, D. E. 1998. Enfranchise landowners for land and wildlife stewardship: examples from the western United States. Human Dimensions of Wildlife 3: 59-68.

5. Beier, P., and R. F. Noss. 1998. Do habitat corridors provide connectivity? Conservation Biology 12: 1241-1252.

6. Bergin, T. M., L. B. Best, and K. E. Freemark. 1997. An experimental study of predation on artificial nests in roadsides adjacent to agricultural habitats in Iowa. Wilson Bulletin 109: 437-448.

7. Best, L. B. 1986. Conservation tillage: ecological traps for nesting birds? Wildlife Society Bulletin 14: 308-317.

8. ———, H. Campa III, K. E. Kemp, R. J. Robel, M. R. Ryan, J. A. Savidge, H. P. Weeks, Jr., and S. R. Winterstein. 1997. Bird abundance and nesting in CRP fields and cropland in the Midwest: a regional approach. Wildlife Society Bulletin 25: 864-877.

9. ———, K. E. Freemark, J. J. Dinsmore, and M. Camp. 1995. A review and synthesis of habitat use by breeding birds in agricultural landscapes of Iowa. American Midland Naturalist 134: 1-29.

10. ———, R. C. Whitmore, and G. M. Booth. 1990. Use of cornfields by birds during the breeding season: the importance of edge habitat. American Midland Naturalist 123: 84-99.

11. Brittingham, M. C., and S. A. Temple. 1983. Have cowbirds caused forest songbirds to decline? BioScience 33: 31-35.

12. Butler, L. D., and J. P. Workman. 1990. A comparison of fee hunting on private rangelands in central Oregon and the Texas Trans Pecos. Pages 269-276 in: L. A. Renecker and R. J. Hudson, editors. Wildlife production: conservation and sustainable development. University of Alaska Extension Service Miscellaneous Publication 91-6, Fairbanks, AK.

13. Camp, M., and L. B. Best. 1993. Bird abundance and species richness in roadsides adjacent to Iowa rowcrop fields. Wildlife Society Bulletin 21: 315-325.

14. ———, and ———. 1994. Nest density and nesting success of birds in roadsides adjacent to rowcrop fields. American Midland Naturalist 131: 347-358.

15. Candelaria, L. M., and M. K. Wood. 1981. Wildlife use of stockwatering facilities. Rangelands 3: 194-196.

16. Carson, R. 1962. Silent spring. Houghton-Mifflin, Boston, MA.

17. Chilgren, J. D. 1979. Drowning of grassland birds in stock tanks. Wilson Bulletin 91: 345-346.

18. Conover, M. R. 1998. Perceptions of American agricultural producers about wildlife on their farms and ranches. Wildlife Society Bulletin 26: 597-604.

19. ———, W. C. Pitt, K. K. Kessler, T. J. DuBow, and W. A. Sanborn. 1995. Review of human injuries, illnesses, and economic losses caused by wildlife in the United States. Wildlife Society Bulletin 23: 407-414.

20. Cooperrider, A. 1986. Terrestrial physical features. Pages 587-601 in: A. Y. Cooperrider, R. J. Boyd, and H. R. Stuart, editors. Inventory and monitoring of wildlife habitat. U.S. Department of Interior, Bureau of Land Management, U.S. Government Printing Office, Washington, D.C.

21. Cowardin, L. M., V. Carter, F. C. Golet, and E. T. LaRoe. 1979. Classification of wetlands and deepwater habitats of the United States. U.S. Fish and Wildlife Service. FWS/OBS-79/31. Washington, D.C.

22. Craig, T. H., and L. R. Powers. 1976. Raptor mortality due to drowning in a livestock watering tank. Condor 78: 412.

23. Darley, J. A. 1982. Territoriality and mating behavior of the male brown-headed cowbird. Condor 84: 15-21.

24. Dufty, A. M., Jr. 1982. Movements and activities of radio-tracked brown-headed cowbirds. Auk 99: 316-327.

25. Edge, W. D., R. L. Carey, J. O. Wolff, L. M. Ganio, and T. Manning. 1996. Effects of Guthion 2S on *Microtus canicaudus*: a risk assessment validation. Journal of Applied Ecology 33: 269-278.

26. ———, J. O. Wolff, and R. L. Carey. 1995. Density-dependent responses of gray-tailed voles to mowing. Journal of Wildlife Management 59: 245-251.

27. Fitzgibbon, C. D. 1997. Small mammals in farm woodlands: the effects of habitat, isolation and surrounding land-use patterns. Journal of Applied Ecology 34: 530-539.

28. Garrott, R. A. 1995. Effective management of free-ranging ungulate populations using contraception. Wildlife Society Bulletin 23: 445-452.

29. Gates, J. E., and L. W. Gysel. 1978. Avian nest dispersion and fledging success in field-forest ecotones. Ecology 59: 871-883.

30. Gerard, P. W. 1995. Agricultural practices, farm policy, and the conservation of biological diversity. U.S. Department of the Interior, National Biological Service, Biological Science Report 4.

31. Grue, C. E., W. J. Fleming, D. J. Busby, and E. F. Hill. 1983. Assessing the hazards of organophosphate pesticides to wildlife. North American Wildlife and Natural Resources Conference 48: 200-219.

32. Haas, C. A. 1995. Dispersal and use of corridors by birds in wooded patches on an agricultural landscape. Conservation Biology 9: 845-854.

33. Habron, G. 1999. An assessment of community-based adaptive watershed management in three Umpqua Basin watersheds. Dissertation, Oregon State University, Corvallis, OR.

34. Hall, T. R., W. E. Howard, and R. E. Marsh. 1981. Raptor use of artificial perches. Wildlife Society Bulletin 9: 296-298.

35. Heckert, J. R. 1994. The effects of habitat fragmentation on mid-western grassland bird communities. Ecological Applications 4: 461-471.

36. Hygnstrom, S. E., R. M. Timm, and G. E. Larson, editors. 1994. Prevention and control of wildlife damage. University of Nebraska Cooperative Extension, University of Nebraska, Lincoln, NE.

37. Johnson, D. H., and L. D. Igl. 1995. Contributions of the Conservation Reserve Program to populations of breeding birds in North Dakota. Wilson Bulletin 107: 709-718.

38. Johnson, R. G., and S. A. Temple. 1990. Nest predation and brood parasitism of tallgrass prairie birds. Journal of Wildlife Management 54: 106-111.

39. Johnson, R. L. 1989. Market fee hunting opportunities in the presence of abundant public land. Western Journal of Applied Forestry 4: 24-26.

40. Kay, B. J., L. E. Twigg, T. J. Korn, and H. I. Nicol. 1994. The use of artificial perches to increase predation on house mice (*Mus domesticus*) by raptors. Wildlife Research 21: 95-106.

41. Kendall, R. J., and J. Ackerman. 1992. Terrestrial wildlife exposed to agrochemicals: an ecological risk assessment perspective. Environmental Toxicology and Chemistry 11: 1727-1749.

42. ———, and T. E. Lacher, Jr. 1994. Wildlife toxicology and population modeling: integrated studies of agroecosystems. Lewis Publishers, Boca Raton, FL.

43. Kindschy, R. R. 1986. Rangeland vegetative succession: implications to wildlife. Rangelands 8: 157-159.

44. ———. 1996. Fences, waterholes, and other range improvements. Pages 369-381 in: P. R. Krausman, editor. Rangeland Wildlife. Society for Range Management, Denver, CO.

45. Krapu, G. L., and K. J. Reineke. 1992. Foraging ecology and nutrition. Pages 1-29 in: B. D. J. Batt, A. D. Afton, M. G. Anderson, C. D. Ankney, D. H. Johnson, J. A. Kadlec, and G. L. Krapu, editors. Ecology and management of breeding waterfowl and wetlands. University of Minnesota Press, Minneapolis, MN.

46. Kunz, T. H. 1982. Roosting ecology. Pages 1-46 in: T. H. Kunz, editor. Ecology of bats. Plenum Publication Corporation, New York, NY.

47. Leckenby, D. A., D. P. Sheehy, C. H. Nellis, R. J. Scherzinger, I. D. Luman, W. Elmore, J. C. Lemos, L. Doughty, and C. E. Trainer. 1982. Wildlife habitats in managed rangelands—the Great Basin of southeastern Oregon—mule deer. U.S. Forest Service and Bureau of Land Management, General Technical Report PNW-139.

48. Leopold, A. 1945. The outlook for farm wildlife. Transactions of the North American Wildlife Conference 10: 165-168.

49. Logan, T. J. 1993. Agricultural best management practices for water pollution control: current issues. Agriculture, Ecosystems and Environment 46: 223-231.

50. Maser, C., J. W. Thomas, I. D. Luman, and R. Anderson. 1979. Wildlife habitats in managed rangelands—the Great Basin of southeastern Oregon—man-made habitats. U.S. Forest Service and Bureau of Land Management, General Technical Report PNW-86.

51. Mayfield, H. F. 1965. The brown-headed cowbird with old and new hosts. Living Bird 4: 128.

52. McCoy, T. D., M. R. Ryan, E. W. Kurzejeski, and L. W. Burger, Jr. 1999. Conservation Reserve Program: source or sink habitat for grassland birds in Missouri? Journal of Wildlife Management 63: 530-538.

53. McKenzie, D. F. 1997. A wildlife manager's field guide to the Farm Bill. Wildlife Management Institute, Washington, D.C.

54. National Academy of Sciences. 1981. Testing for effects of chemicals on ecosystems. National Academy Press, Washington, D.C.

55. National Audubon Society. 1996. Oases for wildlife: small and farmed wetlands. National Audubon Society, Washington, D.C.

56. National Research Council. 1989. Alternative agriculture. National Academy Press, Washington, D.C.

57. Norrgard, R. 1989. Woody cover plantings for wildlife. Minnesota Department of Natural Resources, St. Paul, MN.

58. Norton, S. B., D. J. Rodier, J. H. Gentile, W. H. van der Schalie, W. P. Wood, and M. W. Slimak. 1992. A framework for ecological risk assessment at the EPA. Environmental Toxicology and Chemistry 11: 1663-1672.

59. O'Connor, R. J., and M. Shrubb. 1986. Farming and birds. Cambridge University Press, Cambridge, England.

60. Palmer, W. E., and P. T. Bromley. 1992. Wildlife and pesticides: fruit trees. North Carolina Cooperative Extension Service, North Carolina State University AG-463-7, Raleigh, NC.

61. Payne, N. F., and F. C. Bryant. 1994. Techniques for wildlife habitat management of uplands. McGraw-Hill, Inc., New York, NY.

62. Peterle, T. J. 1991. Wildlife toxicology. Van Nostrand Reinhold, New York, NY.

63. Pimentel, D. S., H. Acquay, M. Biltonen, P. Rice, M. Silva, J. Nelson, V. Lipner, S. Giordano, A. Horowitz, and M. D'Amore. 1992. Environmental and economic costs of pesticide use. Bioscience 42: 750-760.

64. Podol, E. B. 1979. Utilization of windbreaks by wildlife. Pages 121-127 in: Windbreak management workshop proceedings. Great Plains Agriculture Council Publication 92. Lincoln, NE.

65. Rattner, B., and A. Fairbrother. 1991. Biological variability and the influence of stress on cholinesterase activity. Pages 89-108 in: P. Mineau, editor. Chemicals in agriculture, Volume 2. Cholinesterase inhibiting insecticides, their impact on wildlife and the environment. Elsevier, New York, NY.

66. Reese, K. P., and J. T. Ratti. 1988. Edge effect: a concept under scrutiny. Transactions of the North American Wildlife and Natural Resources Conference 53: 127-136.

67. Robel, R. J., A. R. Bisset, T. M. Clement, Jr., A. D. Dayton, and K. L. Morgan. 1979. Metabolizable energy of important foods of bobwhites in Kansas. Journal of Wildlife Management 43: 982-987.

68. ———, R. M. Case, and A. R. Bisset. 1974. Energetics of food plots in bobwhite management. Journal of Wildlife Management 38: 563-564.

69. Robinson, S. K., F. R. Thompson III, T. M. Donovan, D. R. Whitehead, and J. Faaborg. 1995. Regional forest fragmentation and the nesting success of migratory birds. Science 267: 1987-1990.

70. Rodenhouse, N. L., L. B. Best, R. J. O'Conner, and E. K. Bollinger. 1993. Effects of temperate agriculture on neotropical migrant landbirds. Pages 280-295 in: D. M. Finch and P. W. Stangel, editors. Status and management of neotropical migratory birds. U.S. Forest Service, General Technical Report RM-229.

71. Rothstein, S. I., J. Verner, and E. Stevens. 1984. Radio-tracking confirms a unique diurnal pattern of spatial occurrence in the parasitic brown-headed cowbird. Ecology 65: 77-88.

72. Sanderson, H. R., T. M. Quigley, E. E. Swan, and L. R. Spink. 1990. Specifications for structural range improvements. U.S. Forest Service, General Technical Report PNW-250.

73. Schauber, E. M., W. D. Edge, and J. O. Wolff. 1996. Insecticide effects on small mammals: influence of vegetation structure and diet. Ecological Applications 7: 136-150.

74. Schmitz, R. A., and W. R. Clark. 1999. Survival of ring-necked pheasant hens during spring in relation to landscape features. Journal of Wildlife Management 63: 147-154.

75. Schroeder, R. L., T. T. Cable, and S. L. Haire. 1992. Wildlife species richness in shelterbelts: test of a habitat model. Wildlife Society Bulletin 20: 264-273.

76. Shalaway, S. D. 1985. Fencerow management for nesting birds in Michigan. Wildlife Society Bulletin 13: 302-306.

77. Singer, F. J., and J. L. Doherty. 1985. Managing mountain goats at a highway crossing. Wildlife Society Bulletin 13: 469-477.

78. Snyder, W. D. 1984. Ring-necked pheasant nesting ecology and wheat farming on the high plains. Journal of Wildlife Management 48: 878-888.

79. Stallman, H., and L. B. Best. 1996. Bird use of an experimental strip intercropping system in northeast Iowa. Journal of Wildlife Management 60: 354-362.

80. Stinson, E. R., and P. T. Bromley. 1991. Pesticides and wildlife: a guide to reducing impacts on animals and their habitat. Virginia Cooperative Extension Communications, Blackburg, VA.

81. Tiebout, H. M., III, and R. A. Anderson. 1996. A comparison of corridors and intrinsic connectivity to promote dispersal in transient successional landscapes. Conservation Biology 11: 620-627.

82. ———, and K. E. Brugger. 1995. Ecological risk assessment of pesticides for terrrestrial vertebrates: evaluation and application of the U.S. Environmental Protection Agency's quotient model. Conservation Biology 9: 1605-1618.

83. U.S. Department of Agriculture. 1998. 1997 Oregon Wildlife damage survey. U.S. Department of Agriculture, Oregon Agricultural Statistics Service, Portland, OR.

84. ———. 1999. Agricultural statistics 1999. U.S. Department of Agriculture, National Agricultural Statistics Service, U.S. Government Printing Office, Washington, D.C.

85. U.S. Environmental Protection Agency. 1991. Endangered species protection program as it relates to pesticide regulatory activities. Report to Congress. EPA 540-09-91-120. EPA Office of Pesticides and Toxic Substances, Washington, D.C.

86. ———. 1991. Federal manual for identifying jurisdictional wetlands, proposed revisions. U.S. Environmental Protection Agency, U.S. Army Corps of Engineers, U.S. Fish and Wildlife Service, and U.S. Soil Conservation Service. Washington, D.C.

87. ———. 1992. Framework for ecological risk assessment. EPA/630/R-92/001. Washington, D.C.

88. Uresk, D. W., and K. Severson. 1988. Waterfowl and shorebird use of surface-mined and unfenced livestock water impoundments on the northern Great Plains. Great Basin Naturalist 48: 353-357.

89. Vallentine, J. F. 1971. Range development and improvements. Brigham Young University Press, Provo, UT.

90. Wang, G., W. D. Edge, and J. O. Wolff. 1999. A field test of the quotient method for predicting risk to Microtus canicaudus in grasslands. Archives of Environmental Contamination and Toxicology 36: 207-212.

91. Ward, A. L. 1982. Mule deer behavior in relation to fencing and underpasses on Interstate 80 in Wyoming. Transportation Research Record 859: 8-10.

92. Warner, R. E., G. B. Joselyn, and S. L. Etter. 1987. Factors affecting roadside nesting by pheasants in Illinois. Wildlife Society Bulletin 15: 221-228.

93. Weinstein, D. A., and E. M. Birk. 1989. The effects of chemicals on the structure of terrestrial ecosystems: mechanisms and patterns of change. Pages 181-209 in: S. A. Levin, M. A. Harwell, J. R. Kelly, and K. D. Kimball, editors. Ecotoxicology: Problems and approaches. Springer-Verlag, New York, NY.

94. Wildlife Management Institute. 1994. The Conservation Reserve Program: a wildlife conservation legacy. Wildlife Management Institute, Washington, D.C.

95. Wilson, S. D., and J. W. Belcher. 1989. Plant and bird communities of native prairie and introduced Eurasian vegetation in Manitoba, Canada. Conservation Biology 3: 39-44.

96. Wolff, J. O., T. Fox, R. R. Skillen, and G. Wang. 1999. The effects of supplemental perch sites on avian predation and demography of vole populations. Canadian Journal of Zoology 77: 535-541.

97. Wywialowski, A. P. 1994. Agricultural producers' perceptions of wildlife-caused losses. Wildlife Society Bulletin 22: 370-382.

98. Yahner, R. H. 1982. Microhabitat use by small mammals in farmstead shelterbelts. Journal of Mammalogy 63: 440-445.

99. ———. 1983. Small mammals in farmstead shelterbelts: habitat correlates of seasonal abundance and community structure. Journal of Wildlife Management 47: 74-84.

100. ———. 1988. Changes in wildlife communities near edges. Conservation Biology 2: 333-339.

101. Yoakum, J., W. P. Dasmann, H. R. Sanderson, C. M. Nixon, and H. S. Crawford. 1980. Habitat improvement techniques. Pages 329-410 in: S. D. Schemnitz, editor. Wildlife management techniques manual, fourth edition. The Wildlife Society, Inc., Washington, D.C.

14

Wildlife of Riparian Habitats

J. Boone Kauffman, Matthew Mahrt, Laura A. Mahrt, & W. Daniel Edge

Introduction

What is a Riparian Zone?

In many respects it is easier to describe the multitude of functional roles and values of riparian zones than to provide a precise definition. Natural riparian zones are some of the most diverse, dynamic, and complex biophysical habitats on the terrestrial portion of the planet.[158] Riparian zones, like many interfaces, edges, or ecotones, possess relatively high degrees of resources, control energy and material flux, are sites of biological and physical interaction at the terrestrial/aquatic interface, maintain critical habitats for rare and threatened species, and are refuges and source areas for pests and predators.[158]

In the Pacific Northwest, riparian zones are remarkably varied in their composition, size, and structure. As such, definitions vary depending on the discipline and experience of those who derive them. The word "riparian" is derived from the Latin word "ripa," which means the bank of a stream. Therefore a simple, albeit an incomplete definition of riparian zone would be vegetation rooted at the water's edge.[28]

One unifying feature of riparian zones is their association with streams or rivers (i.e., lotic systems). Like the riverine network in which they are associated, riparian zones form a continuum from headwaters to estuaries. Whereas communities commonly found within riparian zones are often wetlands (i.e., wet meadows, seasonally inundated floodplain forests, etc.), not all riparian zones are wetlands or even contain wetland-obligate plants.[184] For example, the streamside vegetation in constrained riparian reaches within montane forests may be comprised wholly of upland vegetation. *Wetlands* are defined as an ecosystem that depends on constant and recurrent shallow inundation or saturation at, or near, the surface of the substrate. The minimum essential characteristics of a wetland are recurrent, sustained inundation or saturation at or near the surface, and the presence of physical, chemical, and biological features reflective of recurrent, sustained inundation or saturation. Common diagnostic features of wetlands are hydric soils and hydrophytic vegetation. These features will be present except where specific physiochemical, biotic, or anthropogenic factors have removed them or prevented their development.[162]

Wetlands include bogs, swamps, marshes, mangroves, estuaries, and some riverine (riparian) ecosystems.[151]

Riparian zones have been investigated from many perspectives, which has created a diverse and often confusing array of definitions based on hydrologic, topographic, edaphic, and vegetative criteria. For example, Kauffman and Krueger[107] defined riparian zones as those assemblages of plant and animal communities whose presence can either be directly or indirectly attributed to factors that are stream-induced or stream-related. Naiman and Décamps[158] stated that riparian zones encompass the stream channel between the high and low water marks, and that portion of the terrestrial landscape from the high water mark towards uplands where vegetation may be influenced by elevated water tables, flooding, and the ability of the soils to hold water. These definitions describe the influences of hydrologic processes and the increased availability of moisture on the streamside or floodplain biota, but do not include the multiple functional roles that encompass how the terrestrial biota influences the geomorphology, hydrology, or stream processes (e.g., beavers, or vegetation, and how they influence the channel morphology, biogeochemistry, or biotic composition of the aquatic ecosystem). Interactions between terrestrial and aquatic ecosystems include modifications of microclimate (e.g., light, temperature, and humidity), alteration of nutrient inputs from hill slopes, contribution of organic matter to streams and floodplains, and retention of inputs.[68] From an ecosystem perspective, riparian zones are defined in terms of their multiple functional roles as the interface between aquatic and terrestrial environments. Therefore, *riparian zones* are defined as the three-dimensional zones of direct physical and biotic interactions between terrestrial and aquatic ecosystems; boundaries of the riparian zone extend outward to the limits of flooding and upward into the canopy of streamside vegetation.[68, 214]

Why Riparian Zones Have a High Degree of Biological Diversity

Globally, riparian zones are "hotspots" of biological diversity. Even within areas such as tropical rainforests, indigenous peoples use floodplain communities for much of their food gathering activities (i.e., hunting, fishing, fruits, cultivation, etc.). The riparian zones of the Pacific Northwest are no exception. Whereas they may occupy as little as 0.5-2.0% of the landscape, they contain more plant, mammal, bird, and amphibian species than do the surrounding uplands. A variety of physical and biotic features unique to the terrestrial-aquatic interface contribute to this high degree of biological diversity (Table 1).

Typically, the composition of biotic communities within riparian zones changes along the continuum from headwaters to mouth. This compositional change is in response to the elevational gradient as well as changes in hydrology, geomorphology, and disturbance regimes.[122] For example, along the elevational continuum, composition will vary due to factors associated with valley bottom width. In constrained reaches (steep canyons), riparian plant communities are narrow and may closely resemble those of upland forests, shrublands, or grasslands.[68] In contrast, riparian zones of unconstrained reaches are often composed of a heterogeneous mosaic of plant communities of differing successional stages and dominated by a diverse mix of herbs, grasses, deciduous shrubs and trees, or conifers (Figure 1).

Among the features of the riparian zone that result in high biodiversity is its unique position on the landscape. Because riparian zones are at the lowest point in the landscape, gravitational forces contribute to the input of sediments and organic matter. This includes the mineral elements that form soils (both alluvium and colluvium, ranging in size from clay particles to boulders) as well as organic matter (from bits of lichen to old-growth logs). These may be deposited at any time (i.e., objects rolling downhill), during storms (wind-throw, overland flow, etc.), or during extreme events such as landslides or debris flows, or storms occurring after fires.

Also related to topographic position is the microclimate of riparian zones, which results in the presence of unique habitats. This microclimate is related to the proximity of open water that influences temperature and relative humidity. In steep canyons, topographic shade can be important. Relative to ridgelines, disturbance from high winds and the frequency of lightning may be lower in riparian zones. Fire-return intervals can be longer and fire severity can be lower. Also, these areas may be cold air drainages during the winter.

The complexity and high frequency of natural disturbances leads to a greater species diversity in riparian zones than upslope habitats.[68] Flooding and ice floes are unique to riparian zones and their variable frequency, magnitude, and extent result in plant communities with variable composition, age, and structure. Whereas floods may destroy established riparian communities, they may also deposit the substrates necessary for many keystone

Table 1. Features of riparian zones that contribute to the biological diversity that characterizes intact or functioning ecosystems.

Geographic location-gravitational forces
Lateral inputs of organic matter of all sizes, soil colluvium, and seeds and propagules.

Disturbance regimes
Flooding, ice floes, landslides, debris flows, and fires that result in a mosaic of plant communities of variable composition, age, and structure.

Diverse geomorphology
Valley bottom width, channel/floodplain gradient, and diverse soils/substrates that provide habitat for a multitude of different plant communities.

High productivity
Deep soils, availability of water and nutrients resulting in high production of plants, invertebrates, and vertebrates.

Microclimate
Position on the landscape (topographic shade, wind breaks, cold air drainages) and interaction with water create diverse and unique microclimates relative to the uplands.

Unique biogeochemical cycles
Dynamic changes in aerobic/anaerobic conditions of soils, groundwater movements of nutrients, and overland deposition of nutrients.

Proximity to water
Source of surface water for terrestrial biota.

Animals
Herbivory, seed dispersal, nutrient inputs, trampling and dam building that modifies vegetation composition, structure, nutrient and hydrological cycles.

riparian species to establish. This includes the deposition of nurse logs for conifer tree species or mineral substrates for the establishment of stands of willows (Salix spp.) and cottonwoods (Populus spp.). In addition, willows and cottonwoods often disperse seeds in phase with seasonally high flows, facilitating germination and establishment on moist alluvial substrates.[204]

The interface of water and land strongly and positively influence biodiversity. Within the floodplain, this influence is apparent both at the streambank interface as well as at the water table/aerated soil interface. The location of these interfaces constantly changes with streamflow, greatly influencing the biogeochemistry of the riparian ecosystem. Soils may be saturated and anaerobic (highly reduced) for part of the year, and aerobic (oxidized) during other times. This creates unique habitats for organisms that must have adaptations to survive both the stress associated with waterlogged soils for part of the year and drought-like conditions at other times of the year.[61, 65]

In addition to the geomorphic and climatic features, biotic influences such as competition, herbivory, and disease are significant forces positively influencing biological diversity in riparian zones.[158] Herbivory by both

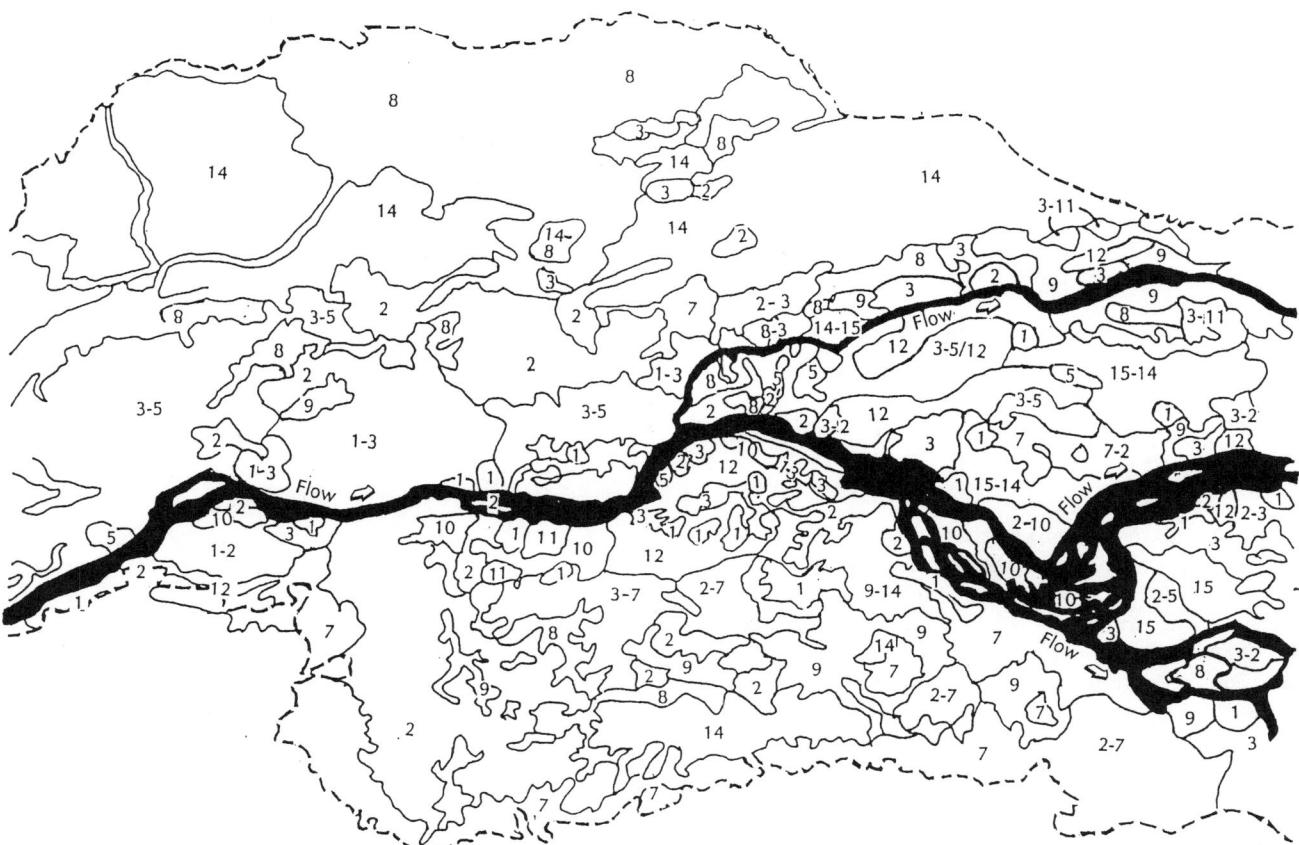

Figure 1. *The riparian zone along rivers such as Rush Creek, California (1996) is typified by a diverse structure and composition of plant species and communities. The high biological diversity is related to a diverse suite of soil substrates, geomorphic surfaces, water availability, biotic impacts (e.g., beaver, elk, humans, etc.), and disturbance history (e.g., fire, floods, etc.). This map is derived from low level aerial photography (1"=200') and extensive ground truthing. The minimum stand area in this map is about 100* square feet (3mX3m). *The numbers within plant communities denote stand dominance: 1= black cottonwood (*Populus trichocarpa*), 2=coyote willow (*Salix exigua*), 3=yellow willow (*Salix lutea*), 4= arroyo willow (*Salix lasiolepis*), 5= whiplash willow (*Salix lasiandra*), 7= Wood's rose (*Rosa woodsii*), 8=wet meadows, 9=dry meadows, 10=gravel bars, 12= unconsolidated alluvium, 14= big sagebrush/antelope bitterbrush (*Artemisia tridentata/Purshia tridentata*), and 15= Rabbitbrush (*Chrysothamnus visiciflorus and nauseosus*).*

insects and large animals is an important influence on riparian composition and structure.[31, 122] In addition, activities such as wallowing, burrowing, dam building, and trampling influence riparian zones. The net result is that the heterogeneity of riparian habitats is increased and the cycling rates of elements (such as nitrogen and phosphorus) are modified.[158] Because of the presence of water, many animals are attracted to the riparian zone simply to drink.

Viewing Riparian Zones from an Ecosystem Perspective

The conservation, management, or restoration of riparian zones requires recognition of how ecosystems function, and what attributes are responsible for their composition, structure, and productivity. A riparian zone and its associated values arise as the product of an infinite number of complex interactions between four fundamental ecosystem features: (1) soils/ geomorphology; (2) hydrology; (3) biota; and (4) climate-microclimate (Figure 2). Humans are also important influences on the vast majority of riparian zones. Hydrologic, geomorphic, and climatic factors are strong determinants of the composition, and therefore the structure and function of the riparian biota. Yet the biota, particularly riparian vegetation, also have strong influences on hydrologic, geomorphic, and microclimatic processes within the riparian zone. The soils/ geomorphology features include the floodplain, active channel, channel gradient, geologic substrates influencing soil and channel composition, and subsoil features of the floodplain. Hydrological features include the frequency, magnitude, and temporal distribution of streamflow (including peakflows and lowflows), sediment availability and transport, subsurface hydrology, and water quality. Climatic features include temperature and precipitation regimes, and microclimatic characteristics of the ecosystem. Biotic features include vegetation, vertebrates,

Figure 2. The structure and function of the riparian zone is the result of complex interactions between the climatic, biotic, hydrologic, and geomorphic components of the ecosystem. Human or natural activities that alter any of the infinite number of interactions between these 4 components will have feedback influences on all other components, thereby altering ecosystem composition, structure, function or productivity (adapted from Kauffman et al.[103]).

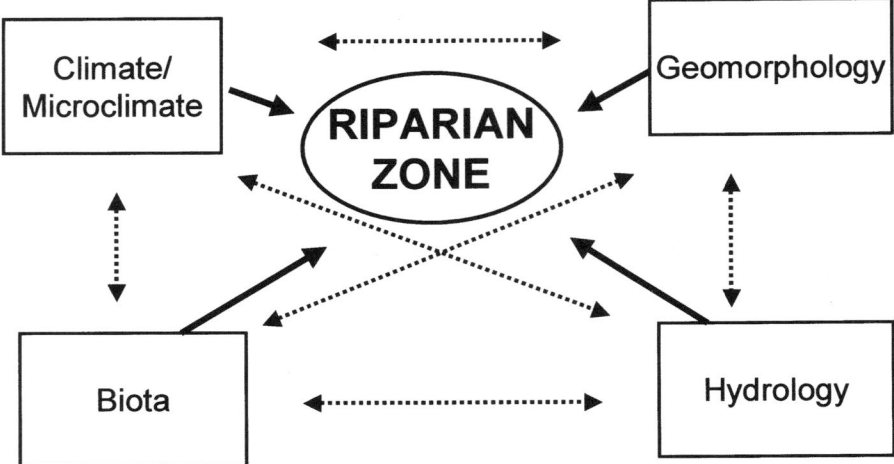

invertebrates, and microorganisms. In addition to the live biota, this component also includes dead materials (necromass) such as snags, fallen logs, fine organic debris (litter), and animals, especially salmon carcasses. Ecosystem functioning is disrupted when human land use activities alter these components or sever the interactions that link them.[46, 103] Such ecosystem alterations may ultimately be expressed as a loss in productivity, a decline in biodiversity, and/or the extinction of species.

The distribution and composition of riparian plant and animal communities reflect histories of both fluvial disturbance from floods and non-fluvial disturbance regimes of adjacent upland areas such as fire, wind, plant disease, and insect outbreaks. As for any disturbance event(s), the frequency, timing, and magnitude of the disturbance will influence the structure and composition of the biotic community. Floods result in both the erosion of established floodplains and their biota as well as the deposition of varied substrates where succession or stand establishment begins anew. These events have created complex patterns of soil morphology and groundwater dynamics that influence riparian plant and animal communities.[53, 68, 176]

Hydrologic and geomorphic conditions necessary for the establishment of many riparian-obligate trees and shrubs have been described throughout North America. Successful establishment of cottonwood trees and willows commonly occurs on point bars of coarse-textured, well-aerated soils and within the 2- to 10-year high water levels on floodplains.[57, 138] Seed dispersal and germination coincide with late spring flows when water tables are high and fresh alluvium has been deposited.[17] Successful establishment may also be limited to areas where the rate and extent of water table decline does not exceed the biological capacity of root growth.[125] At the lower limits of the floodplain, establishment is often not possible because high water destroys the young plants or their habitat.[17, 26] Geomorphic features and hydrologic events that result in long periods with waterlogged soils (e.g., mountain meadows) create conditions where the biota must be adapted to anaerobic conditions. Under these conditions, the natural plant communities are dominated by adapted species such as sedges (*Carex* spp.), rushes (*Juncus* spp.), bulrushes (*Scirpus*, and *Eleocharis* spp.) or hydrophytic grasses.[176]

Mosaics of landforms influence spatial patterns of riparian plant communities, but riparian vegetation also influences the evolution of geomorphic surfaces. Root networks of riparian stands increase resistance to erosion. Aboveground stems of streamside vegetation increase channel roughness during overbank flow, thereby decreasing the erosive action of floods and physically retaining materials in transport.[68] Riparian plant communities also contribute large woody debris to channels, a major geomorphic feature in streams and rivers.[72] Complex, naturally occurring pools and other important features of fish habitat can form as a result of interactions of hydrologic disturbance regimes, substrates, and streamside vegetation. If hydrologic patterns, sediment availability, or streamside vegetation are altered by land use activities, then channel morphology will subsequently adjust to these new conditions.

Riparian vegetation, including roots, downed wood, stems and graminoid leaves, provides physical structure that slows water, decreases stream power, and holds materials in place.[158] Land uses that reduce the structure or productivity of vegetation can result in significant adjustments of hydrological or geomorphic features with feedbacks that further negatively affect ecosystem composition and function (e.g., channel incision that severs many of the interactions between the floodplain and associated aquatic environment).

Functions and Values of Riparian Zones

The value of riparian zones to fish and wildlife is well known. In the western U.S., riparian environments provide habitats for a disproportionate number of birds, reptiles, and mammals. The aquatic biota—insects, amphibians, and fish—are strongly influenced by the composition and structure of riparian zones. In terms of management, riparian zones are considered the most critical of wildlife habitats in rangelands of southeastern

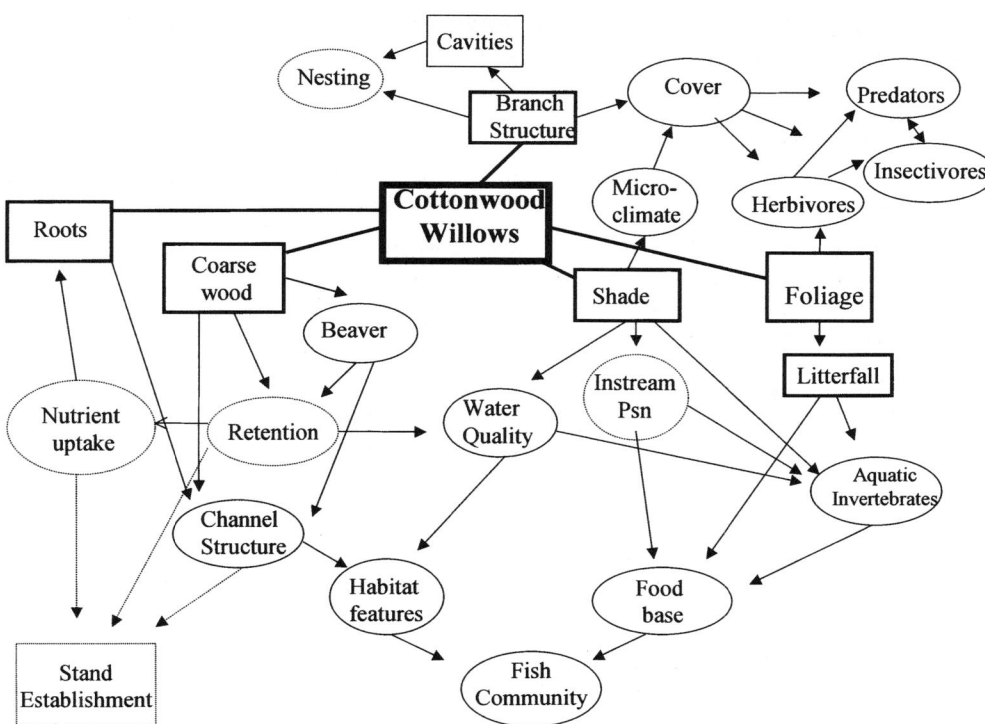

Figure 3. The multiple ecological roles and influences of riparian obligates—cottonwoods and willows—that are keystone features of both aquatic and terrestrial ecosystems. These species perform multiple functional roles that strongly influence biotic composition and diversity. Strctural features of plants are in boxes, whereas ovals contain other species or ecosystem processes.

Oregon,[219] the Blue Mountains,[218] and in western Oregon and Washington.[168] In the Blue Mountains of Oregon there are 378 terrestrial vertebrate species, of which 285 are directly dependent on riparian habitats or use them more than other habitats.[219] Of the 363 terrestrial species of wildlife known to occur in the Great Basin of southeastern Oregon, 79% are either directly dependent on riparian zones or use them more than any other habitat.[219] In northeastern Oregon >80 bird species used riparian habitats along a ≈1.5-mile (2.5-km) stream reach (Catherine Creek).[108] Of 120 native California herpetofauna, 83% of amphibians and 40% of reptiles were found in riparian habitats.[18] In this book, it is estimated that 253 species use westside and 265 species use eastside riparian habitats. Out of 593 wildlife species that occur in Oregon and Washington, 319 (53%) use riparian zones. Considering that riparian zones and wetlands cover only 1-2% of the landscape, their value associated with biological diversity is evident.

Riparian zones are important for wildlife and fisheries for many reasons:[107, 168, 220]

1. Riparian zones have a high diversity of plant species and vegetation structure, thereby providing niches for numerous fish and wildlife species.
2. Riparian zones often contain unique vegetation assemblages both in composition and structure. For example, riparian zones may be the only herbaceous-dominated communities (e.g., mountain meadows) in forested ecosystems, or the only tree-dominated communities in desert or grassland environments.
3. The linear shape of riparian zones creates high edge: area ratios with uplands on the outer margin and aquatic ecosystems (the stream) on the inner margin.

4. Riparian zones modify the environment (microclimate) for both terrestrial and aquatic organisms. In addition, riparian vegetation influences water chemistry and temperature through shade, sediment retention, and nutrient transformation.
5. Riparian zones serve as natural corridors or migration routes. They may serve as connecting corridors between areas of suitable habitats in fragmented environments (e.g., between islands of old-growth forests).
6. Riparian zones provide many habitat features necessary for many fish and wildlife species to survive. This includes forage, nesting/breeding habitat, and cover for terrestrial wildlife. Riparian zones affect the aquatic biota through inputs of litterfall and coarse wood, provision of shade, influences on channel diversity by roots and large wood, and nutrient sequestration of ground and surface water (Figure 3).

Keystone species or groups of species are those that play some unique role in ecosystem structure and function; they are species that have an inordinate effect on ecosystem structure and process.[179] Keystone species have a disproportionately large impact on ecological systems relative to their abundance on the landscape. Beaver, salmon, and cottonwoods all are considered keystone species (and all have been greatly reduced throughout their range). By damming streams and creating pools, beaver create habitats for aquatic species and raise water tables, resulting in wetlands. They also influence the structure of adjacent plant communities through harvest of woody plants.[157] Salmon are considered keystone

Table 2. Some functional attributes of the riparian biota that characterize the multiple linkages between biological and physical processes of intact riparian ecosystems.*

Biological Component	Ecological Function	Ecosystem Response or Influence
Fine organic debris (leaves, needles, cones, lichens, fine wood, etc.,)	• Allochthonous inputs of energy and nutrients	• Instream productivity • Water chemistry
Coarse wood	• Channel structure • Hydraulic roughness • Sediment/nutrient retention • Nutrients/energy source	• Aquatic habitat/diversity • Erosion control • Water quality
Roots	• Nutrient uptake • Channel morphology	• Aquatic habitat • Diminution of erosion • Sediment retention
Vegetation canopy	• Energy absorption • Thermal buffer • Influences microclimate (relative humidity, winds, vapor pressure deficits and temperature)	• Modification of temperature extremes • Influences instream productivity and biotic composition • Modification of disturbances (ice floes, fire, etc.)
Vegetation structure	• Channel/floodplain roughness • Nutrient uptake, transformation, and sequestration • Wildlife habitat • Influences on soil processes	• Dissipation of stream energy • Deposition of sediments and nutrients • Water chemistry—conversion of mineral to organic forms • Sources of foraging, reproduction and resting habitat • Soil organic matter, water holding capacity, fertility, and nutrient turnover
Macro invertebrates/vertebrates	• Detritivores • Herbivores • Predators	• Change in vegetation structure • Seed/stem dispersal • Nutrient redistribution and cycling • Channel processes and morphology • Competition • Hydrological influences—trampling, dam building
Microbial populations	• Reduction-oxidation processes • Denitrification • Mineralization • N fixation • Decomposition	• Nutrient transformations that influence water quality and habitat conditions for adapted vegetation • Biogenic emissions of greenhouse gases • Site productivity
The riparian ecosystem/landscape	• Migration/dispersal corridors • Human benefits	• Longitudinal migrations • Elevational migrations • Propagule dispersal (plant and animal) • Economic, spiritual, and ecological values

* Modified from Kauffman et al. (1996) and Swanson et al. (1982).

species because of their role as a food source to a wide variety of predators and scavengers as well as a nutrient source that drives ecosystem productivity in headwater streams.

A widespread keystone feature of many western riparian zones is that of the willows and cottonwoods (*Salicaceae*). They play multiple functional roles in riparian ecosystems, strongly influencing both the aquatic and terrestrial biota (Figure 3). Roots and wood influence channel morphology (pools, undercuts, channel width), nutrient transformations, and organic matter and sediment retention. Shade and litterfall influence the aquatic invertebrates, instream photosynthesis, and water quality (i.e., nutrients, temperature, oxygen concentration, etc.). These physical and physiological processes ultimately have strong influences on native fishes such as salmonids (Figure 3). Cottonwoods are also recognized for their importance as terrestrial wildlife habitat. Johnson et al.[93] suggested that the highest breeding densities of non-colonial birds in North America were in southwestern cottonwood habitats. Cottonwoods and willows are also very palatable to a variety of insect and wildlife species including beaver and large herbivores.[31] Management of keystone riparian vegetation communities such as willow and cottonwoods is of paramount importance in terms of managing a wide diversity of fish and wildlife species. This was recognized over a century ago by Van Cleaf[225] who observed: "the destruction of streamside trees has resulted in destruction of the natural hiding places of trout and until these natural harbors are restored it will be useless to hope for any practical benefit from restocking them".

The Linkages of Riparian Zones with the Instream Biota

Riparian vegetation exerts strong controls on habitats and food resources of aquatic vertebrates (Figure 3). Streams are directly influenced biologically, physically, and chemically by aboveground and belowground components of streamside vegetation[68, 214] (Table 2). Riparian vegetation controls the physical structure of low-order streams in coniferous forests as well as in unconstrained herbaceous-dominated reaches in mountain meadows. The composition of riparian communities determines the quality, timing, and quantity of organic matter contributed to the aquatic environment.[19] Riparian vegetation functions as a nutrient filter by retaining sediments from overland flow, and by the uptake of nutrients in groundwater. Vegetation alters the chemistry of the soil water and likely the stream water.[53] Shade influences the degree of solar radiation that reaches the water column, and hence instream photosynthesis and water temperature. Through these influences, the riparian zone also regulates the composition of the aquatic community in terms of relative importance of functional groups.[44] Abundance and composition of the detritivore assemblage in streams are determined in large measure by the composition of riparian vegetation.[226] Riparian zones also directly control the food resources of herbivorous and detritivorous fishes.[68]

Inputs of Organic Matter

Riparian plant communities offer an abundant and diverse array of food resources for both terrestrial and aquatic consumers. Riparian vegetation provides much of the food base for the aquatic biota. In headwater forest ecosystems 90% of the organic matter and 99% of the stream energy input may arise from the riparian zone, with only 1% derived from instream photosynthesis by attached algae or periphyton.[45] Naiman and Décamps[158] reported that organic matter inputs from riparian vegetation (allochthonous inputs) varied from 0.13 to 0.60 lbs/ft² (200 to 900 g/m²) in small and medium streams to about 0.010 to 0.011 lbs/ft² (15-17 g/m²) in larger rivers. Most studies of terrestrial litter inputs have focused on leaves and needles of trees, but herbaceous and shrub litter represent a significant input of litter for the aquatic ecosystem. In the Blue Mountains of northeast Oregon, total aboveground biomass (TAGB) of headwater riparian forests are 13.64 to 17.53 lbs/ft² (20,300 to 26,100 g/m²).[30] In contrast, TAGB on floodplain meadows of the same reaches were 300-1500 g/m².[53, 108, 176] However, during peak flows when floodplains are inundated, virtually all aboveground biomass in floodplain meadows is incorporated into the aquatic food webs. In contrast, only a small fraction of the forest biomass is input into streams as litterfall or lateral inputs in a given year.[19] The quantity of allochthonous inputs arising from meadows flowing through unconstrained reaches in an intact condition (not incised or grazed by cattle) may be similar to or exceed that of constrained forested reaches.

Once input into a stream, leaves of herbaceous plants and many shrubs are processed quickly (30-50 days).[68] This is because of the high nutrient concentration of many riparian plant species and a relatively low C:N ratio. For example, Case[30] found that the nitrogen concentration of black cottonwood (*Populus trichocarpa*) leaves was ≈2.0% and that of willows and thin-leaf alder (*Alnus incana*) was 2.0-2.6%. Tree leaves that are high in nutrients and not structurally resistant to breakdown require <4-6 months for complete processing while conifer needles may persist in streams for 1-2 years.[68] In addition to particulate organic matter, riparian zones may significantly contribute to the nutrient balance of a stream via dissolved organic nutrient inputs through groundwater/hyporheic movements.[158] Dwire[53] found elevated concentrations of dissolved organic carbon (DOC) in the groundwater of streamside riparian communities relative to the surface water in the channel. Riparian vegetation could potentially be a source of DOC via movements through macropore spaces.

Large Wood

Large wood greatly influences the physical and biological characteristics of small, forested streams. Wood dissipates energy, retains nutrients and sediments, and forms habitat (pools, hiding cover, backwaters, secondary channels, etc.) for both aquatic vertebrates and invertebrates.[158, 199] Wood can also redirect water currents, resulting in major channel changes and alterations in riparian community composition. Sedell et al.[199] reported that large wood debris densities in old-growth forest streams of the Pacific Northwest ranged from 18 to 45 pieces per 100m of stream reach. Biomass of large wood was 39 to 139 tons/ac (88 to 312 Mt/ha) in headwater streams. Large wood in streams may last for >200 years. In small headwater streams, large wood debris may create pools that are suitable rearing habitat for salmonids. These are sites where food is plentiful and may be obtained by juvenile fish with minimal energy expenditure. Deep pools created by large wood or accumulations of smaller pieces may afford cover from predators.[199]

Organic matter and inorganic sediments must be retained within a stream to serve as either nutritional resources or habitat for aquatic organisms.[68] Large logs and accumulations of finer wood (as well as standing stems and herbaceous vegetation) can trap materials in transport. These large and fine wood debris accumulations will slow the transport of water and dissolved solutes, increasing the potential for biological uptake, and hence ecosystem productivity. Activities that decrease the input and quantity of wood within a stream also decrease the retentive capacity of the stream, resulting in declines in productivity, increases in nutrient and organic export, and increases in erosion rates.

Downed wood plays multiple functional roles in riparian zones as well in streams and rivers. Downed wood provides habitat for a variety of plant and animal species. Acting as a nurse log, downed wood enhances germination and establishment for many plant species of the Pacific Northwest. Downed wood also affords protection and habitat for birds, amphibians, reptiles, and small mammals. It is important to recognize that downed wood is not necessarily an important ecological factor in unconstrained meadow reaches where trees are not part of the composition. Stream enhancement activities that place wood debris into these reaches may result in more ecological harm than good.[103]

Shade

Riparian vegetation blocks or filters solar radiation, having profound effects on microclimate, water quality, instream productivity, and hence both the aquatic and understory terrestrial biota (Figure 3). Shading by riparian vegetation reduces the amount of primary production in streams, because less light is available for photosynthesis. Of great importance is the role of shade in influencing stream temperatures; this in turn strongly influences on the composition of the aquatic biota. Land use activities (e.g., logging, livestock grazing, etc.) that reduce stream shading result in increased stream temperature.[118]

The degree and influence of shading on ecosystem processes is a function of the structure and composition of the riparian vegetation as well as stream size. In headwater streams, the trees, shrubs, and even herbaceous vegetation can effectively shade the stream channel. For example, sedges and grasses provide shade for >65% of the wetted channel in intact mountain meadow-dominated reaches of northeastern Oregon. The role of shading declines with large rivers. However, it is important to note that in many western stream networks, >85% of the stream length is in first or second order tributaries. Direction of the channel as well as topography (i.e., topographic shade from canyons) is also important.

Influences on Channel Morphology

Channel structure is strongly determined by streamside vegetation particularly by roots and wood debris accumulations. Root biomass can be remarkable in riparian ecosystems. For example, Otting[176] found that intact sedge meadows have a root biomass of 14 tons/acre (32 Mt/ha) where aboveground biomass was only 2.2 to 3.6 tons/acre (5-8 Mt/ha). This root biomass is similar to that of tropical rainforests, where aboveground biomass may be as high as 178 tons/acre (400 Mt/ha). Well-vegetated streambanks capture and retain sediments, resulting in decreased width-to-depth ratios, increased sinuosity, and increased nutrient retention, within the riparian zone.[10] Vegetation enhances channel diversity (i.e., formation of pools, undercuts, etc.) and hence is a critical

component of fisheries habitat. In addition to fish, numerous other vertebrates (e.g., belted kingfishers, rough-winged swallows, dippers, beaver, river otter, etc.) use streambanks as rearing or feeding cover.

Nutrient Dynamics

Numerous features of riparian zones influence the biogeochemistry of instream ecosystems. Shade, nutrient retention, sediment trapping, and organic (litterfall) inputs all are important biogeochemical functions in stream ecosystems. In addition, riparian areas can be effective in trapping nutrients and sediment-bound pollutants carried in surface runoff. This function may be of particular importance in agricultural landscapes, where riparian zones can buffer aquatic systems from excessive nutrient or toxic inputs. Riparian areas removed 80-90% of the sediments leaving agricultural fields in North Carolina.[37, 120] Grassy riparian areas trapped >50% of the sediments from uplands when overland water flows were <5 cm deep,[158] underscoring the functional values of unconstrained floodplain reaches as sites of sediment/nutrient retention.

Riparian zones also modify the chemistry of the groundwater. Oxygen diffusion is very slow in saturated soils compared to aerated soils. As such, oxygen is rapidly consumed by aerobic microbes once riparian soils flood. Prolonged saturation in the riparian zones can result in reduced (anaerobic or anoxic) soils where many important chemical transformations (e.g., denitrification) occur. Nutrients that have escaped the rooting zones of upslope vegetation or agricultural fields may be transformed via these microbial processes or incorporated into this last terrestrial site by plant uptake.

Owing to the reduced nature of many riparian soils and relatively higher organic matter contents, rates of denitrification are usually greater in riparian zones than uplands. For denitrification to occur, there must be nitrate, a C source, and anaerobic conditions. Denitrification rates of 8.9 to 49 lbs/ac/yr (10-55 kg/ha/yr) have been reported in riparian zones.[38, 86] These plant and microbial processes can greatly influence instream productivity, water quality, and hence, the aquatic composition. Rhodes et al.[187] found that in an undisturbed Sierra Nevada watershed, >99% of the incoming nitrate was denitrified in the riparian zone.

Nutrient uptake by plants is another form of nutrient removal in riparian zones. Plant uptake can represent a short-term accumulation of nutrients in non-woody biomass, or a long-term accumulation in woody biomass.[158] Given that riparian woody species are relatively fast-growing, can be found in high densities, and are relatively high in nutrient concentration, the level of nutrient sequestration can be significant. For example, following the cessation of livestock grazing, Case[30] found that mean aboveground biomass of willows increased from 2- to 5-fold in only two years. Litterfall inputs can also be viewed as part of a plant-mediated nutrient transformation. Plants uptake dissolved mineral forms of nutrients in groundwater and deposit nutrients in organic forms via litterfall and lateral inputs.

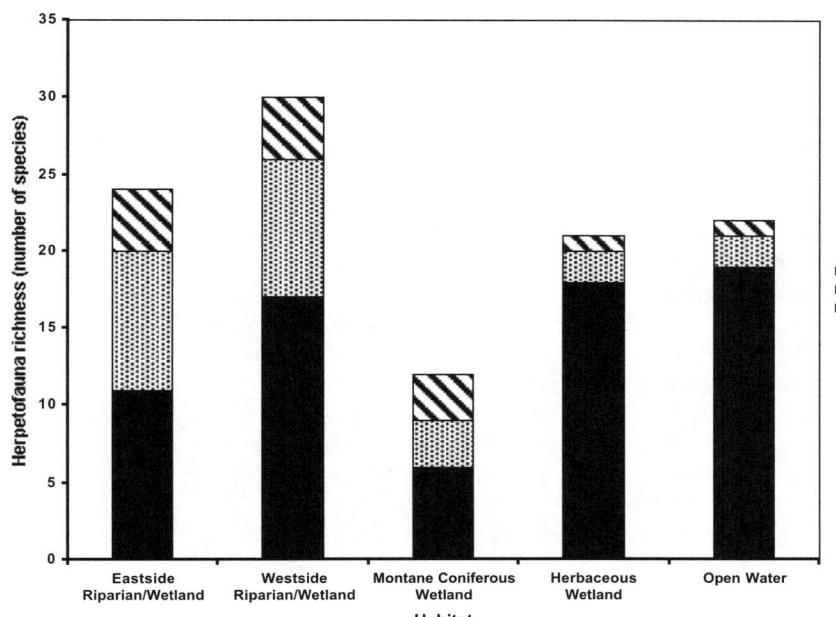

Figure 4. The richness of the herpetofauna in 5 riparian habitats types of Oregon and Washington.

Reptiles and Amphibians

Amphibians and reptiles (herptiles) represent an important biotic component of riparian ecosystems. They comprise important components of riparian and aquatic food webs as detritivores, herbivores, insectivores, and carnivores. In many landscapes herptiles comprise the largest proportion of total vertebrate biomass.[25] Compared to birds and mammals, the herpetofauna has received less attention regarding their distribution, density, diversity, and ecological function in riparian habitats. Because herptiles lack the "charisma" of mammals and birds, the public is largely unaware of the ecological importance and functions of amphibians and reptiles in riparian ecosystems.

Herptiles are good indicators of the health of aquatic systems. These animals are especially sensitive to pollution and loss of habitat through land use or land cover change.[8, 70] Declines in amphibians may cause shifts in patterns or abundance of their prey species. For example, an increase in mosquitoes in Asia has been linked to heavy cropping of frogs for food.[185] Because of their abundance, herptiles are prey species for many carnivorous mammals and raptors[43] and therefore, declines in their density could affect these predators.

Amphibians are long-lived compared to mammals and birds of similar size. For example, many species of urodeles and anurans can survive up to 20 years, whereas shrews and songbirds rarely live 5 years. These differences are related to the amphibian's poikilotherm physiology.[7] Long-lived adults can tie up considerable amounts of nutrients, making them unavailable for transfer through the ecosystem, and therefore have a large effect on ecosystem nutrient processes. Herbivorous tadpoles can have influences on ecosystem functions in freshwater habitats. In sufficient densities, tadpoles regulate the abundance and cover of algae and other aquatic plants. In their absence, algal growth could lead to dramatic shifts in ecosystem structure and environment through changes in processes such as eutrophication.[8, 209]

Herpetofauna Diversity in Riparian Areas

The herpetofauna of Oregon and Washington includes 59 native species; reptiles and amphibians are widespread throughout all riparian zones and wetland types. A total of 22 herptiles use both the eastside and westside deciduous riparian habitats (Figure 4). Herbaceous wetlands provide habitats for 32% of all herptiles in the two states. Montane coniferous riparian areas and open water provide habitat for 20% and 15% of the herptiles respectively, in Oregon and Washington.

All of the herptiles (100%) found in eastside deciduous and montane coniferous habitats reproduce in these riparian zones; 91% of the species found in westside deciduous, and 89% of the species found in herbaceous wetlands, use these riparian zones for breeding. About 60% of the species present in open water habitats breed there. Other uses of riparian zones include foraging, overwintering, and migration. Reptiles also use riparian areas extensively for foraging because of the high density of prey species, including aquatic insects and other invertebrates, fish, adult and larval amphibians, small mammals, and young birds. Because amphibians have limited mobility and dispersal capabilities, continuous riparian zones are important pathways to colonize suitable, yet unoccupied habitats.

Most amphibians require an aquatic habitat for part of their life cycle. The exceptions to this rule are the fully terrestrial salamanders of the Plethodontidae family. Plethodonts require moist rotting logs and litter for both egg development and cutaneous respiration.[39] Land uses that decrease downed log abundance would decrease numbers of amphibians that use downed wood for all or part of their life cycles.

Microhabitats within riparian zones have important effects on thermoregulation. Because the environment of riparian zones can be more moderate than surrounding uplands at both cold and warm extremes, reptiles may more efficiently thermoregulate in these habitats. For example, snakes will bask on, and hide under, the rocks along creeks, streams, and rivers.[193] Western pond and painted turtles, which are the only two native turtles in Oregon and Washington, are riparian/wetland obligates, and frequently use wood in streams and lakes for basking.

Influences of Land Use on the Herpetofauna

Disturbances that affect the riparian herpetofauna include timber harvesting, water diversion, agricultural practices, spraying of pesticides and herbicides, and livestock grazing. The tailed frog is highly specialized for cold, fast moving streams that are often heavily shaded.[39] Larvae prefer water temperatures at or below 59°F(15°C) and can require up to 5 years to metamorphose.[49] Removal of the forest overstory has resulted in declines or the disappearance of tailed frogs.[24, 40, 166] Bull and Carter[21] found significant differences in numbers of larvae or adults in streams adjacent to areas of low, moderate, and heavy amounts of timber harvest. They observed an inverse correlation between the degree of timber of harvest and tailed frog abundance. Similarly, the density and biomass of torrent salamanders, Siskiyou Mountains salamanders, and giant salamanders were lower in logged compared to unlogged streams.[18, 40, 74, 181] Because amphibians have a limited capacity to re-invade riparian zones and headwater streams following their extirpation from a logged reach, the probability of population fragmentation and local extinction is high.

Water diversions such as dam building or channelization have resulted in declines in amphibian and reptile populations. Studies of the herpetofauna in southwestern riparian ecosystems have shown that dam building resulted in decreases in population numbers or even local extinctions.[88, 99, 100, 119, 215, 228] The spotted frog was abundant throughout Oregon until the mid-1970s. A recent survey of all historical localities in western Oregon indicates that it has disappeared from 90% of its former range.[76]

The leopard frog in both Oregon and Washington has experienced significant declines and local extinctions throughout much of its historical range.[209, 211, 212] These losses are thought to be related to the conversion of wetlands to agriculture as well as pesticide control for mosquitoes.

Livestock grazing has been found to negatively impact amphibian and reptile populations, although few studies have been conducted in the Pacific Northwest. In the southwestern U.S., Szaro et al.[216] found that numbers of wandering garter snakes were significantly higher where cattle grazing was excluded. They found that the snakes were responding to losses of streamside vegetation and organic debris that provided microhabitat for preferred food items. Desert tortoise populations in the southwest U.S. have also declined due to cattle grazing. Grover and

DeFalco[69] found that livestock trample eggs and burrows and compacted the soil making it difficult for adults to dig burrows and females to dig nest cavities. Cattle also compete with desert tortoises for forage. Agriculture practices (i.e., haying, and plowing) have resulted in declines in western pond turtle populations in Oregon.[174] Lizard populations have also declined in southwestern U.S. riparian zones due to cattle grazing.[15, 98] Similar responses have been observed in heavily grazed Great Basin rangelands (L. Mahrt, Eastern Oregon University, pers. comm.). Other declines in native herptiles are the result of predation and competition by introduced species such as bass, bullfrogs, nutria, red-eared sliders, and domestic dogs and cats. Habitat modification through the establishment of exotic plant species (i.e., reed canary grass *Phalaris arundinacea*) has also affected native herptiles.[79, 174]

Birds

Whereas riparian zones only comprise 1-2% of the western landscapes, they provide breeding habitat for more species of birds than any other vegetation type in North America.[114, 115] There is a disproportionate diversity and abundance of birds in riparian and wetland ecosystems in every major biome and in every western state.[56, 60, 168, 169, 171, 196, 219, 220, 222] For example, 82% of bird species breeding in northern Colorado use riparian ecosystems.[112] In the southwest U.S., >75% of all bird species nest primarily in streamside vegetation. If all riparian habitats in the southwest were lost, 47% of the regional avifauna would face extinction.[93]

Avifauna of freshwater, riparian, and wetland environments are taxonomically diverse. However, these habitats are particularly significant to certain ecological groups. For example, all 23 species of waterfowl that breed regularly in the western U.S. south of Alaska do so in riparian and wetland environments.[9, 48] Similarly, all 14 western species of waders, a group consisting of cranes, rails, herons, and ibises, depend on riparian and wetland habitats for most of their life cycles.[48]

Shorebirds, which include stilts and avocets, sandpipers, and plovers, are typically dependent on freshwater, riparian, and wetland habitats. These environments provide nesting habitat for 10 of the 14 species that breed in the contiguous western states.[48, 55] Interior wetlands also provide crucial stopover habitat for 37 species during migration.[175, 177, 202, 206]

Riparian habitats are critical to small nongame landbirds, particularly those that breed in North America and winter south of the U.S. border (e.g., Neotropical Migratory Birds [NTMB]). In the western U.S., 60–85 % of NTMB species breed primarily in woody, deciduous riparian vegetation.[50, 58, 93, 112, 116, 169, 196] Declines of several endangered NTMB are directly linked to modification or loss of deciduous riparian woodland.[63, 223, 224] In addition to their importance as breeding sites, riparian habitats are important to NTMB during migration. Abundance of migrating NTMB may be 10 times greater in riparian zones than in surrounding uplands.[210] Similarly, richness of

Table 3. Avian species richness (extant species) in freshwater, riparian, and wetland habitats of Oregon and Washington relative to richness in all habitats.

Habitat(s)	Number of species	
	Occur	Breed
All	367	273
Inland species†	324	264
Freshwater, riparian, and wetland	266	204

† Species associated primarily with coastal and marine habitats are excluded.

Table 4. Avian species richness in freshwater, riparian, and wetland habitats of Oregon and Washington relative to type of use and degree of association.

Degree of association	Type of use			
	Breed & forage	Forage only	Other	Total
Closely Associated	103	27	2	132
Generally Associated	89	14	0	103
Present	12	16	3	31
Total	204	57	5	266

migrating species can be as much as 14 times higher in riparian than in non-riparian habitats.[77] Migrating Wilson's warblers that use native, riparian willow habitat have higher rates of fat deposition and shorter stopover periods than those using agricultural or field edge habitat.[229] Finally, riparian zones may also be critical wintering habitats for resident land birds as well as migratory species that winter north of the U.S. border.[114, 115, 186]

The status of many bird species that rely on riparian and wetland habitats is sensitive and of priority management concern. Finch[59] found that riparian woodlands and low-elevation wetlands contained more species of listed and vulnerable birds than any other habitat in the Rocky Mountain region. Similarly, 29 (62%) of 47 nongame, migratory bird species considered species of concern in the Pacific Northwest, breed in riparian and wetland habitats.[55, 223] Among the western states, 26-79% of the priority bird species listed by Partners in Flight (PIF) were associated with riparian and wetland vegetation.[55] Of these, 22 were among the 90 bird species of highest conservation priority in North America.[29]

Avian Diversity in Riparian and Wetland Habitats

Avian diversity in riparian and wetland ecosystems of Oregon and Washington is quite high relative to upland ecosystems. Overall, 266 (72%) of the 367 species of birds in Oregon and Washington use freshwater, riparian, and wetland habitats (Table 3). If one only includes inland species (i.e., excludes coastal and marine birds) the proportion using riparian areas increases to 82% (Table 3). More striking, 204 (77%) of the 266 species of inland

birds that breed in Oregon and Washington do so in riparian and wetland environments (Table 4). Of these species, 103 are closely associated with riparian and wetland habitats, whereas 89 are generally associated with riparian and wetland systems.

We define migratory land birds (MLB) as small, terrestrial birds that spend only part of the year in Oregon or Washington for nesting, as winter residents, or as transients during migration. Resident land birds (RLB) are species of small terrestrial birds that are present throughout the year. Galliforms (GB) consist of pheasants, grouse, ptarmigans, turkeys, and Old World quail, along with New World quails. Kites, hawks, eagles, falcons, owls, osprey, and New World vultures comprise birds of prey (BP). Waterbirds (WB) consist of grebes, pelicans, gulls and terns, and loons. Shorebirds (SB), waders (WA), and waterfowl (WF) were defined earlier. For 7 of these 8 ecological groups, at least 75% of the inland species use riparian and wetland habitats (Figure 5a). Similarly, for 6 of the ecological groups, >75% of the species that occur in riparian and wetlands habitats use them for breeding (Figure 5b). SB and WB groups have the smallest percentages of riparian/wetland breeders, a function of the large number of species that occur in Oregon and Washington only as winter residents or transients. The proportions of MLB and RLB, BP, and GB that are closely associated with riparian and wetland ecosystems in Oregon and Washington are much smaller than for aquatic specialists (Figure 5c). Nevertheless, >50% of the GB that occur in riparian and wetland environments are close associates, and MLB and RLB comprise a large percentage of all species that are closely associated with riparian and wetland habitats.

Mirroring trends throughout the West, riparian and wetland ecosystems in Oregon and Washington harbor more species of sensitive birds than do other habitats. Approximately 80% of bird species listed as sensitive in Oregon or Washington occur in riparian and wetland habitats.[126] This trend is consistent across ecological groups; only for MLB do <70% of sensitive species occur in riparian and wetland environments. Riparian zones function primarily as breeding habitat for most sensitive bird species. Similarly, many sensitive species are closely associated with riparian and wetland habitats. Fewer RLB and BP rely on riparian zones; most sensitive species of these groups found in riparian and wetland environments are woodpeckers, falcons, and owls. Whereas these species can reach high breeding densities in riparian and wetland ecosystems,[168, 220] most are more closely tied to conditions in the surrounding uplands.[126]

Comparison of Avian Diversity in Riparian and Wetland Habitats

Of the wildlife habitats in Oregon and Washington, Eastside Riparian Wetland has the greatest bird species richness (155 species) (Figure 6). Species richness is also high in Herbaceous Wetland (143 species) and Westside Riparian Wetland (140 species). The structurally diverse deciduous habitats dominated by cottonwoods, alder,

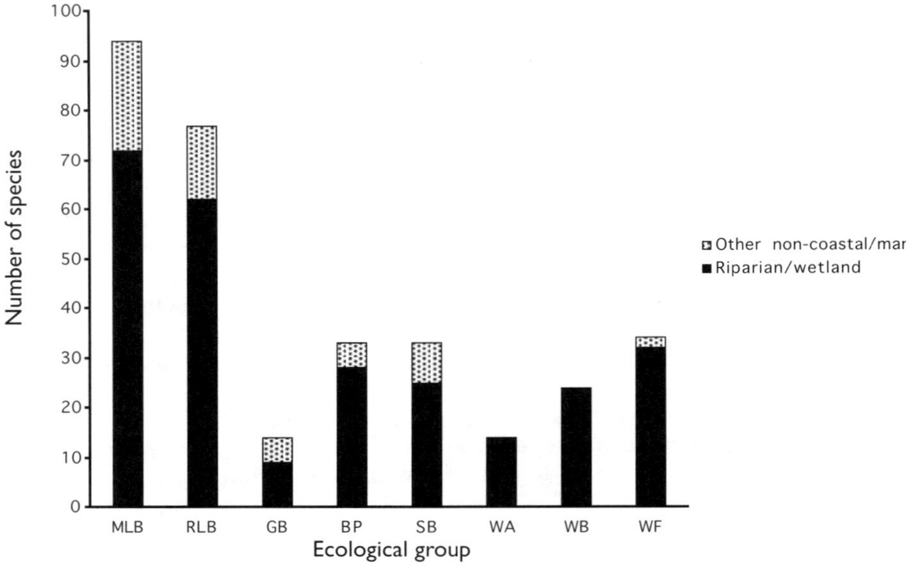

Figure 5a. Numbers of coastal/ marine and inland birds (i.e., non-coastal/marine species) that occur in riparian habitats of Oregon and Washington. Species richness is reported for the 8 major ecological groups of birds: MLB—migratory land birds, RLB—resident landbirds, GB—galliforms, BP – birds of prey, SB—shorebirds, WA—water birds, WB—wading birds, and WF— waterfowl.

Figure 5b. Numbers of inland birds (i.e., non-coastal/marine species) that use riparian zones by habitat use categories in riparian habitats of Oregon and Washington. Species richness is reported for the 8 major ecological groups of birds: MLB— migratory land birds, RLB—resident landbirds, GB—galliforms, BP – birds of prey, SB—shorebirds, WA— water birds, WB—wading birds, and WF—waterfowl.

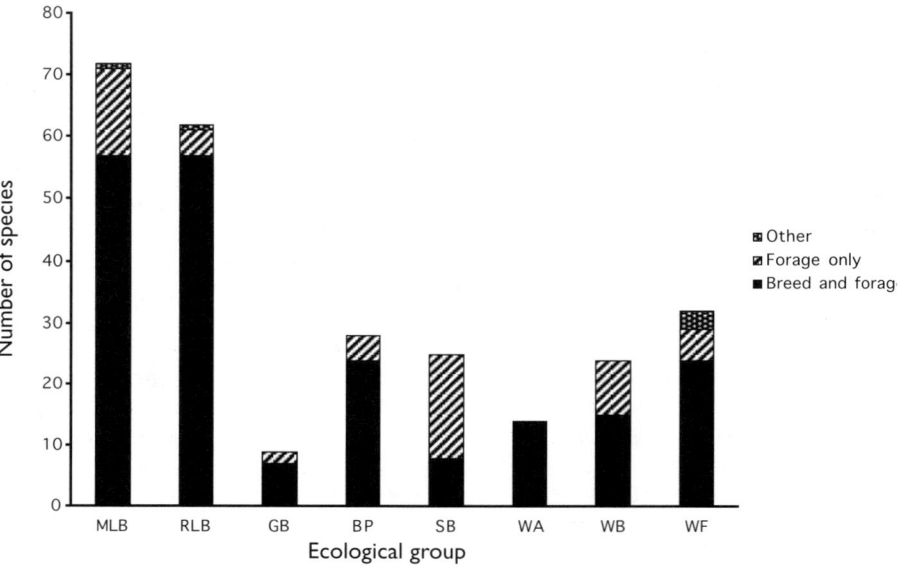

Figure 5c. Degree of association of inland birds (i.e., non-coastal/marine species) in riparian habitats of Oregon and Washington. Species richness is reported for the 8 major ecological groups of birds: MLB— migratory land birds, RLB—resident landbirds, GB—galliforms, BP – birds of prey, SB—shorebirds, WA— water birds, WB—wading birds, and WF—waterfowl.

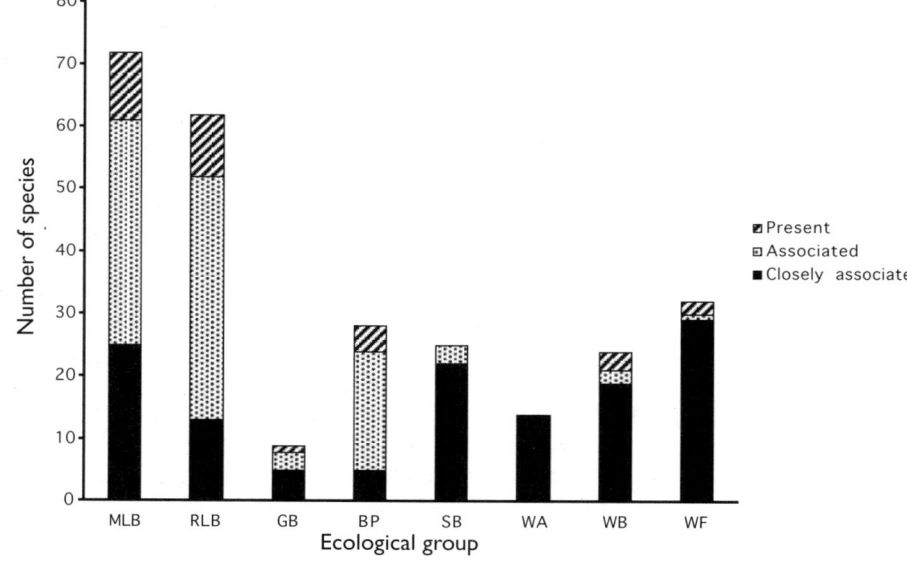

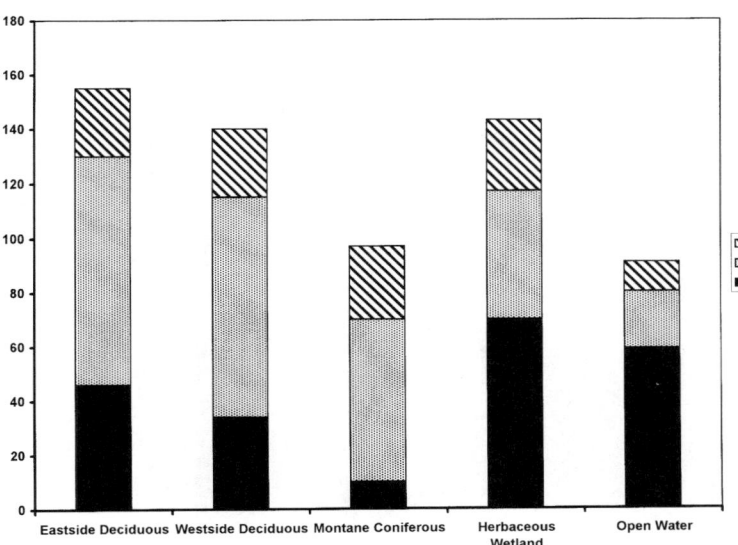

Figure 6. Numbers of inland birds (i.e., non-coastal/marine species) that use 5 riparian habitats types of Oregon and Washington by habitat use categories.

willows, and maples (*Acer* spp.) are heavily used by birds, particularly for breeding. At least 80% of all birds that occur in East- and Westside Riparian Wetland habitats breed there. In contrast, only 55% of the species that occur in Herbaceous Wetlands and only 7% of the birds that use Open Water habitats use them for breeding. However, there is often a remarkable plant and invertebrate productivity in herbaceous wetlands and open waters (i.e., marshes, estuaries, mountain meadows, emergent wetlands, and wet prairies). These habitats are largely used for foraging, particularly by WF, SB, WB, WA, BP, and many land birds.

Interestingly, more species are "closely associated" with Herbaceous Wetland and aquatic environs habitats than any other. Almost 50% of the bird species that occur in Herbaceous Wetlands, and 65% that occur in Open Water habitats are close associates. Montane Coniferous Wetlands have the lowest species richness (97 species), as well as the lowest proportion of close associates. This is likely because of their similarity to upland forests. The relatively few species that are closely associated with the deciduous riparian habitats, coupled with their high richness of bird use, implies that numerous species most associated with other habitats can also find suitable niches in these deciduous riparian communities.

East- and Westside Riparian Wetland habitats are of great importance to terrestrial MLB and RLB. For example, Eastside Riparian Wetland provides habitat for 52 (55%) and 55 (71%) species of MLB and RLB, respectively in Oregon and Washington. Approximately 75% of the MLB, and 90% of RLB that occur in Eastside Riparian habitat use them as breeding habitat. In addition, many landbirds considered associated with Eastside Riparian Wetland habitat reach their greatest breeding densities in this habitat (i.e., they are riparian-dependent). Riparian-dependents place 60 to 90% of their nests in riparian vegetation, or 60 to 90% of their abundance occurs in riparian vegetation during the breeding season. Species such as calliope hummingbird, western wood-pewee, and Macgillivray's warbler are considered riparian dependents in most of their western range. Whereas many landbirds

present in Eastside Riparian do not meet the criteria for close association, their population viability would decline significantly without this habitat.

Eastside Riparian Wetland habitat provide habitat for 8 of the 14 species of GB, and 21 of the 33 species of BP that occur in Oregon and Washington. Most galliforms that occur in Eastside Riparian Wetland use this habitat for breeding, and most are also close associates. In contrast, whereas most BP present in Eastside Riparian breed in this habitat, few are close associates.

With minor differences, the same patterns noted for Eastside Riparian Wetland occur in Westside Riparian Wetland. The same is not true for Montane Coniferous Wetlands. Although a significant number of landbirds and BP occur and breed in Montane Coniferous Wetlands, few species are closely associated with this habitat. This is likely because of the structural and compositional similarity of Montane Coniferous Wetlands to adjacent upland habitats. Indeed, Montane Coniferous Wetlands in some westside landscapes may have lower avian diversity and abundance than adjacent uplands.[142]

Herbaceous Wetlands and Open Water ecosystems provide habitats for many species of birds adapted to aquatic/wetland habitats (WA, SB, WF, etc.). Herbaceous Wetlands are of particular importance as breeding habitats for these birds. All 14 species of WA that occur in Oregon and Washington breed in Herbaceous Wetlands and are closely associated with these habitats. Twenty-seven (79%) of the 34 non-coastal/marine species of waterfowl present in Oregon and Washington use Herbaceous Wetlands. Eighteen (67%) of these species use Herbaceous Wetlands for breeding. Similarly, 18 (75%) of the 24 non-coastal/marine species of WA present in Oregon and Washington occur in Herbaceous Wetlands. Fifteen (83%) of these species breed in Herbaceous Wetlands, whereas 12 (67%) are close associates. Only 8 of the 33 non-coastal/marine species of SB that occur in Oregon and Washington breed in herbaceous wetlands. However, this number represents 80% of all such species known to breed in the two states. Fifteen species of SB are closely associated with Herbaceous Wetlands. Six of these species breed there; the

Figure 7. Degree of association of inland birds (i.e., non-coastal/marine species) in the 5 riparian habitat types of Oregon and Washington.

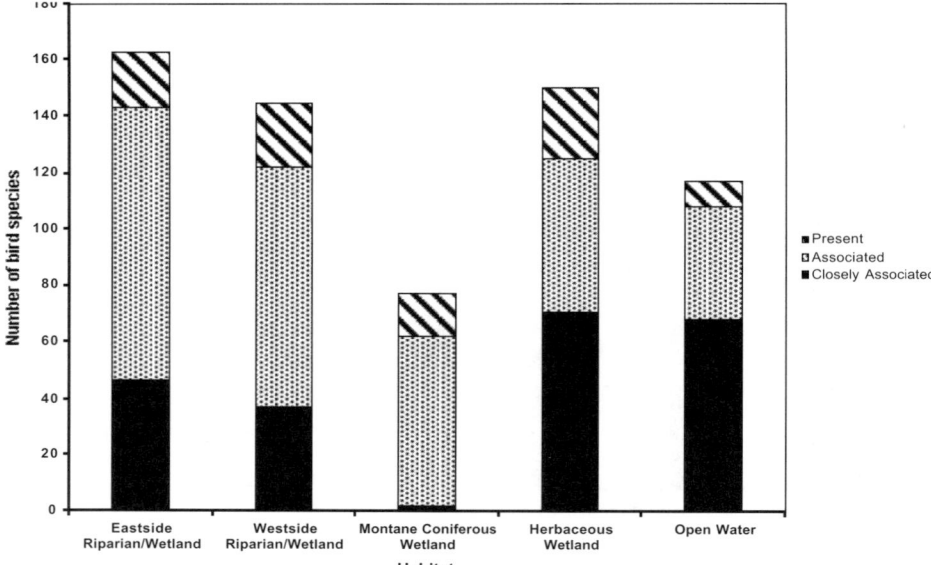

remainder depend on Herbaceous Wetlands as foraging/refuge habitat during migration and in winter.

The Open Water habitat is most significant to birds as foraging and/or refuge habitat. Across ecological groups, >80% of all species use lakes, ponds, reservoirs, and rivers only for foraging or roosting (Figure 6). However, lakes, ponds, reservoirs, and rivers are still critical to a number of aquatic birds; >60% of all non-coastal/marine WB and WF are closely associated with these habitats.

Factors Influencing Avian Diversity

The diversity and abundance of birds in riparian and wetland habitats is but another component of the extraordinary biological and physical diversity that characterizes riparian and wetland ecosystems. The high levels of plant and insect productivity in riparian zones are among the factors contributing to the exceptional diversity of the avifauna. These levels of productivity diminish competition for food;[15] high productivity can also influence avian diversity via its effects on behavior and use of space. Abundant, high-quality food decreases time and energy costs incurred in food acquisition.[127] For territorial species, such decreases permit contraction of territories.[164] Contraction of territories in competitively superior species may promote existence of ecologically similar species by mitigating exclusion via direct interference.[36] Similarly, contraction of territories reduces overlap among territories of ecologically similar species, thus alleviating depression of food resources through exploitation.[127] In addition, diverse feeding sites promote diversity within feeding guilds, allowing ecologically similar species to partition food resources and avoid interspecific competition.[36, 123, 124] The high productivity of most riparian and wetland ecosystems also provides important resources to non-breeding avifaunas. Migrating species require stopover sites with abundant, high-quality foods to replenish energy and nutrient reserves.[1, 152] Thus, interior wetland complexes provide critical habitats for an abundance and diversity of migrating SB.[47, 154] Similarly,

abundant, energy rich (insect and plant) foods in riparian zones are important to MLB during migration.[82, 83]

Vegetation diversity and complexity also provides a diversity of potential nest sites. A cottonwood forest with senescent trees and snags furnishes substrates for primary and secondary cavity nesters. Streambanks provide sites for species that nest in burrows or under overhanging ledges, whereas open nesting species find sites on the ground, in shrubs, and in the canopy. Martin[128] demonstrated that the relationship between complexity of vegetation and avian diversity was better explained by nest site diversity than by forage site diversity. Close associations between bird and plant species may derive from specialization in use of nesting substrates, which is usually narrower than specialization in use of foraging sites.[132, 133] Such diversity in nest sites may minimize nest site competition and decrease nest predation.[129, 130, 132]

Although their general importance is unclear, other resources such as escape cover or singing perches/display sites might also influence diversity of birds in riparian and wetland ecosystems. This would be particularly true in grasslands, deserts and shrublands, where riparian zones are the only tree-dominated communities on the landscape. High observation points from which to spot prey are also important structural features of tree-dominated riparian zones in non-forested landscapes. Availability of suitable singing perches or other kinds of display sites might influence occurrence of certain species (and hence overall diversity) by influencing selection of habitat by males[201] and/or selection of males by females.

Management Considerations

The maintenance of avian diversity in riparian and wetland environments ultimately depends on maintenance of natural hydrologic and disturbance regimes. In addition, because physical and biotic processes in the floodplain do not operate independently of inputs from the surrounding landscape, management must adopt a landscape-level approach. These considerations are general; they apply to all riparian and wetland habitat types and to all types of birds. In general, disruption of natural hydrologic and disturbance processes, whether by direct perturbations such as damming or channelization, or indirect perturbations, such as livestock grazing, fragment and simplify woody, riparian vegetation.[33, 52, 80, 107, 116, 138, 192] Such effects can exclude individual species or nesting guilds by eliminating nesting substrates (e.g., cottonwoods or other large trees required for cavity nesters). Similarly, simplification of woody vegetation can negatively influence individual species, nesting guilds, or the entire breeding assemblage by increasing nest predation/brood parasitism. Such an increase can occur via a number of mechanisms acting at multiple spatial scales.

At the smallest scale, land uses that decrease vegetation cover may increase susceptibility to predators by decreasing visual, auditory, and/or olfactory concealment of nests.[23, 111, 117, 131, 153] Land management activities that reduce vegetation density in the patch surrounding the nest may increase the predator's ability to locate even well concealed nests.[16] Similarly, decreased vegetation cover (e.g., willow density) can increase susceptibility to predation by reducing the number of potential prey sites a predator must search.[132, 134, 135]

The effects of brood parasitism by the brown-headed cowbird on riparian nesting species pose significant concerns in avian management. This is an example of the cascading effects associated with land cover changes in concert with the introduction of a non-native species.[190, 191] The highest rates of parasitism by brown-headed cowbirds occur in landscapes marked by human habitation.[217] Before European settlement, brown-headed cowbirds did not occur or were of low abundance in the western U.S.[136] Introduction of domestic livestock and conversion of western landscapes to agriculture apparently facilitated range expansion by providing abundant, high-quality foraging habitats for brown-headed cowbirds.[137, 194, 227] In search of nests in which to place their eggs, cowbirds seek out habitats with a high density (and potentially diversity) of MLB and RLB.[194] Because they have the highest abundance and diversity of landbirds, and because livestock concentrations, farmhouses, and other feeding sites are often close by, riparian ecosystems represent optimal breeding habitats for brown-headed cowbirds. For uncommon riparian-nesting species such as the Bell's vireo and willow flycatcher, the effects of parasitism contribute significantly to their threatened or endangered status in parts of their western range.[63, 224] Riparian-nesting species such as the yellow warbler and warbling vireo also experience significant loss of breeding productivity.[217]

Mammals

Riparian zones in Oregon and Washington are disproportionately important for mammals for many of the same reasons that they are critical to many herptiles and birds. As previously noted, riparian zones typically have higher structural diversity compared to adjacent upland habitats. Riparian zones are subject to relatively high levels of natural disturbance because of the dynamic nature of their adjacent aquatic habitats,[68] and therefore typically have high spatial heterogeneity.[165, 173] Because of their linear form, riparian zones have high edge-to-interior ratios and are functionally edge habitats, offering mammals that use these areas access to two or more habitats in close proximity. Riparian zones offer a predictable source of water, favorable microclimates, and high plant diversity that increase structural diversity and offer a varied and abundant forage supply.[107, 142, 168] Because of their diverse fauna, riparian habitats are important areas from the standpoint of mammal conservation and management. Riparian zones are used by 27 threatened, endangered or mammal species of special interest or concern in Oregon and Washington. All of the furbearers in both states use riparian areas, and all but one of the big game animals (bighorn sheep) in both states rely on riparian areas for part of their habitat requirements.

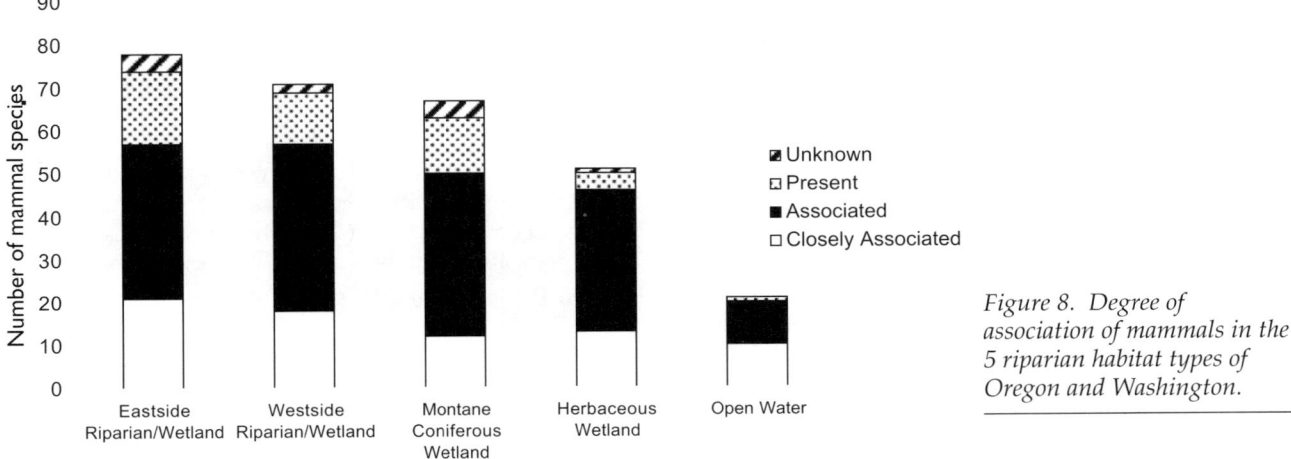

Figure 8. Degree of association of mammals in the 5 riparian habitat types of Oregon and Washington.

Figure 9. Numbers of mammal species that use 5 riparian habitats types of Oregon and Washington by habitat use categories.

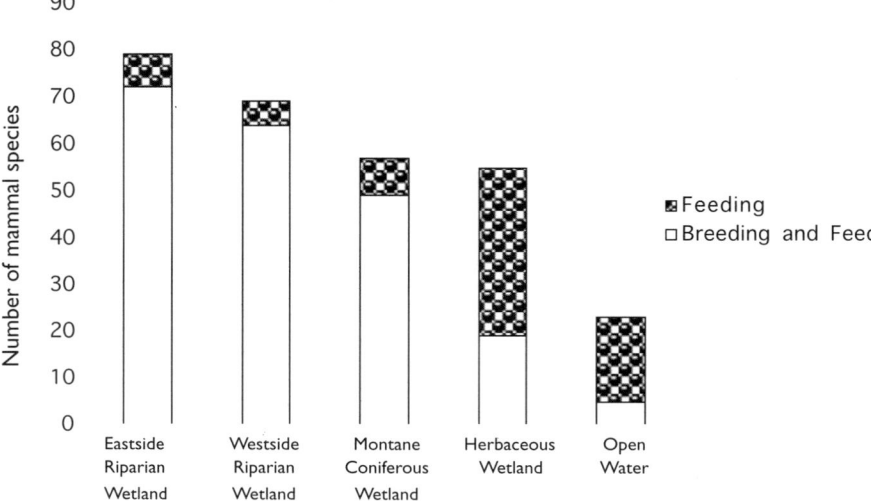

Mammal Diversity in Riparian Areas

Mammalian diversity is high in riparian zones of Oregon and Washington. Of the 147 mammal species in the two states, 95 species (65%) use riparian areas. Species richness varies among the 5 riparian habitats delineated within this book (Figure 8). Eastside and Westside Riparian Wetland habitats have the highest mammalian species diversity (79 and 69 species, respectively), whereas Open Water areas have the lowest (21 species). However, as previously noted, these numbers are somewhat misleading because of the high level of spatial heterogeneity in the distribution of riparian habitats. Eastside and Westside Riparian Wetland habitats are often closely juxtaposed with Montane Coniferous Wetland, Herbaceous Wetland, and Open Water habitats (Figure 1), and thus mammal species richness within a riparian reach containing all types would be higher than in single habitat types.

Mammal fauna in riparian areas are particularly diverse when compared to adjacent upland habitats.[168, 219] Cross[41] reported that mixed deciduous and conifer riparian areas in southwestern Oregon had greater small mammal species richness and biomass than did upland sites. The trans-riparian gradient is often less dramatic in forests in western Oregon and Washington than in other areas within the two states because upslope environments are relatively moist.[75] Nevertheless, riparian areas in western Oregon and Washington characteristically have greater species richness than upland sites, and provide habitat for some species that occur nowhere else.[64, 139, 140, 142, 183] Riparian zones in the Cascades Mountains[5, 51] and along the Columbia River[78] typically have greater mammalian richness and abundance than do upland sites. Most studies of riparian zones in arid landscapes typical of the eastern portions of the two states compared mammal abundance and diversity in grazed and ungrazed deciduous and wet meadow habitat types, rather than riparian versus upland sites. Nevertheless, these studies[108, 146, 147, 203] suggest that vegetation in well developed riparian areas supports greater mammalian abundance and richness than does upland vegetation.

Mammalian Use of Riparian Zones

Riparian areas in Oregon and Washington are used by mammals for food, shelter, a source of water, and as corridors for dispersal and movement. Most mammal species using riparian habitats in the two states use them for breeding and feeding (Figure 9). Except for open water habitats, about 50% of the species using the 5 riparian habitats breed and feed in those areas.

Several groups of mammals rely heavily on riparian areas for meeting life requirements.[143] Bats of several species preferentially use riparian areas for foraging because of the abundant insect prey emerging from the aquatic environments.[42, 121] Furthermore, in heavily forested areas streams and ponds provide open flight corridors for bats. Many species of small mammals, including the white-footed vole, Pacific jumping mouse, western jumping mouse, western harvest mouse, Richard's water vole, Pacific water shrew, montane shrew, Pacific shrew, water shrew, broad-footed mole, and shrew-mole are closely associated with riparian zones in Oregon and Washington.[4, 42, 43, 139, 143] All of these small mammals are associated with dense vegetation and/or downed wood.[143]

Several mammalian predators specialize on aquatic prey or forage almost exclusively in riparian zones. Mink are generalist predators and prey on locally available food sources.[54] Mink occupying coastal areas forage primarily in the intertidal zone,[73, 92] whereas mink in freshwater habitats forage primarily along shorelines and in emergent vegetation.[54] Like the mink, river otters are adapted to a diverse range of aquatic habitats from marine to freshwater and from coastal to high mountainous areas.[149] River otters are much more of a prey specialist than mink, feeding primarily on fish, although a wide variety of invertebrates and vertebrates have been documented in river otter diets.[149] Although not as closely tied to riparian areas as mink and river otters, raccoon populations are usually most abundant near riparian areas.[197]

Little is known about the life history characteristics of most riparian obligate mammals. Dispersal is a fundamental population process thought to be important in population regulation, maintenance of genetic diversity, meta-population dynamics, and colonization of habitats.[84, 183] Dispersal has important management implications, and is a fundamental process that needs to be understood in order to assess the effects of management activities on wildlife populations. Habitat fragmentation is thought to be a major factor contributing to the decline of wildlife populations. Riparian obligates are particularly susceptible because of the small area of the landscape occupied by riparian zones, and their high probability of fragmentation due to land and water uses. Patch area (size of suitable habitats) and degree of isolation (distance to the nearest suitable habitat and probability of migration to that site) are considered major factors determining the species dynamics of a patch. Riparian habitats and buffers are assumed to function as dispersal corridors for wildlife species,[183] but little empirical evidence is available to support this assumption.[64, 143] Small mammals adapted to Eastside Riparian Wetland and Herbaceous Wetland habitats may be more vulnerable to land management practices that fragment and isolate their habitats because matrix vegetation would be less suitable for promoting successful movements in and out of the fragmented patches.[198]

Influence of Mammals on Riparian Ecosystems

Of the vertebrate species using riparian ecosystems, mammals are the group most capable of modifying ecosystem structure and function. These influences can extend to both the terrestrial and aquatic components of riparian areas.[101, 155, 178] Mammals can alter riparian areas through predation, herbivory, trampling, and by directly modifying characteristics of stream channels.[109, 156] These activities can increase habitat complexity by altering the composition and structure of riparian vegetation, and the spatial heterogeneity of riparian and aquatic habitats. Conversely, when densities of mammals grow to unnatural levels due to management, introduction, or protection, over-utilization (by large ungulates and beavers) and trampling (by large ungulates) has resulted in ecosystem simplification or degradation.

Beaver. American beavers are the keystone vertebrate species in riparian areas where they naturally occur (Figure 3). The importance of beavers in these ecosystems is only partly understood, because their heavy exploitation during the 1800s practically eliminated them from many stream systems in Oregon and Washington. Although beavers have re-colonized most stream systems in both states, many of these systems have not been influenced by the presence of this keystone species for long time periods. Even in areas where populations have recovered, exploitation often reduces beaver numbers. Unexploited beaver populations can alter riparian ecosystem structure and function dramatically.[157, 160] Beaver alter riparian areas by (1) creating and maintaining wetlands; (2) influencing the rate, timing, and volume of downstream water and sediment movement; (3) modifying channel hydrology and geomorphology; (4) altering the dynamics of nutrient cycling and decomposition; (5) retaining sediments and organic matter; and (6) modifying plant community composition and structure, and plant physiognomy.[157, 160, 170] The signature of these habitat alterations may persist for centuries.[85, 163, 195]

Water impoundments from beaver dams are a significant impact on riparian areas, aquatic habitats, and wetlands. Beaver dams along streams can increase ecosystem complexity across the landscape.[95, 96] Flooding of alluvial terraces and valley floors causes long-term changes in plant community composition and dynamics.[94] Beaver ponds are storage and processing sites for nutrients, carbon, and energy that enters the stream as litterfall.[159] In addition, beaver dams aid in nutrient retention and transformations that affect the composition and productivity of the aquatic biota.[46] The trapping of sediments is another important function of beaver ponds. Smith[208] reported up to 90% reduction in sediment loads in streamflows below beaver dams.

Beavers profoundly affect plant communities through the impounding of water and selective foraging.[97] Beavers are generally restricted to areas within 328 ft (100 m) of water,[97] and in some regions may cut more than 2200 lbs (1 Mt) of wood annually.[144] Beavers are central place foragers, coming on land to cut vegetation that they haul back to the water to consume or store for later use.[90, 91] This leads to selective exploitation of woody vegetation in riparian areas, thereby altering tree density, height, and other components of riparian systems.[34] Beavers ingest only a small portion of the biomass they cut, contributing large amounts of woody debris to both stream channels and riparian zones.[156] Their placement of stems into streams facilitates downstream transport and deposition of living materials onto gravel bars or streambanks, and leads to the establishment of new willow, alder, cottonwood, or aspen (*Populus tremuloides*) stands.

Beavers are important in the creation of habitat for fish and other wildlife species and play an important role in riparian conservation.[2] Ponds behind beaver dams are suitable habitat for lentic species as well as stream-dwelling (lotic) species that prefer low velocity currents.[156] Beaver ponds are important overwinter areas for some coastal fish species.[14] Biomass of invertebrates in beaver ponds may be 2-5 times higher than in un-dammed streams.[141, 159] The structural complexity of vegetation around beaver ponds attracts many bird and mammal species.[167] Good river otter habitat has been associated with beaver activity.[150] Medin and Clary[148] reported that relative densities of small mammals were 3 times higher in willow-dominated habitats around a beaver pond than in an adjacent non-willow riparian habitat.

Native Ungulates. Large herbivores can have major impacts on the composition and structure of riparian vegetation, thereby altering riparian functions. Although a great deal of research has focused on domestic herbivores, comparatively little is known about the impacts of native herbivores on riparian structure and function.[172, 207] Native herbivores, especially elk, moose, and white-tailed deer, preferentially use riparian areas because of the abundant and diverse forage, and the favorable microclimate in these areas. Use of riparian areas by native herbivores can occur year-round, depending on the location of the riparian zone. Because of their status as big-game animals, native herbivore populations are often maintained at relatively high densities. This is particularly true in national parks and areas where hunting and predation pressures are low. When ungulates overpopulate their range, selective herbivory may eliminate palatable plant components such as willows or cottonwoods.[3, 26, 32, 71, 110] Deer and elk populations at high densities can degrade community composition and ecosystem function,[3, 32] or slow natural recovery rates.[30] Similar to overgrazing by domestic livestock, excessive use by wild ungulates can affect physical features of the environment such as soil and stream channel structure (R.L. Beschta, Oregon State University, pers. comm.). These effects will ultimately negatively affect fish (aquatic) habitats as well as bird and small mammal habitats.

Elk are widely distributed throughout both Oregon and Washington. Elk densities in many areas in both states are at or near all-time highs. In addition to high population levels, elk group size is largest of the native ungulates in both states. Elk preferentially use riparian areas especially during the summer months.[180, 207] This combination of preferential use of riparian areas, high densities, and large group size may result in significant impacts on both the physical components (stream channel structure) as well as biotic components (vegetation structure and diversity) of riparian habitats. In the Blue Mountains of eastern Oregon, crown volume following cessation of cattle grazing increased 550% in two years in wild ungulate (mostly elk) exclosures while recovery was only 195% outside of exclosures.[31] Furthermore, fewer willows produced catkins outside of exclosures (2%) than inside exclosures (34%). Likewise, Singer et al.[205] documented significant decreases in biomass of several willow species in Yellowstone and Rocky Mountain National Parks subject to browsing, primarily by elk. In northeastern Oregon, Skovlin[207] reported that elk and deer accounted for approximately one-third of the total amount of browsing that occurred on riparian trees and shrubs (the rest was by domestic livestock).

Human Influences on Riparian Zones

The first widespread Euro-American perturbation in western riparian/stream ecosystems (if not the introduction of the domestic horse to native American Indian tribes) was the exploitation of the beaver. The harvesting of riparian forests for fuel, shelter, and land clearing closely followed. Livestock grazing, water diversions, wetland drainage, mining, over-fishing, logging, and associated impacts (roads, splash dams, railroads), began soon after the Euro-American occupation in the Pacific Northwest. Finally, urbanization, highways, channelization, hydroelectric dams, and industrial development have contributed to the decline of riparian habitats and their associated wildlife in the Pacific Northwest. Exotic invasions, pollution, global climate change, and extinctions are increasing and are irreversible threats to the biodiversity of riparian and wetland ecosystems. All of these human influences have dramatically affected the structure and diversity of riparian zones, wetlands, and the wildlife and fish that use these ecosystems.

Many of the same attributes that contribute to the high productivity and biodiversity of wildlife species are of great economic value to human society. Hence there is competition for these ecosystems between human society's needs for resource exploitation and its needs for preservation of wildlife and fish. For example, broad floodplains formed through the millennia are productive not only for their complex wildlife habitats and linkages to the aquatic biota, but also for their nutrient-rich soils. The most productive lands in terms of tilled agriculture, forage for livestock, and forest growth for wood are riparian zones and wetlands. In addition, mining concentrated along riparian corridors in the Pacific Northwest has permanently destroyed many riparian wildlife habitats (as well as their potential use for agriculture). A balance between the needs of human society and the responsibility for maintaining biological diversity is needed.

In addition to the proximity of water and the high biotic productivity, the geographical position of riparian zones within the landscape results in their disproportionate use by both humans and wildlife. Riparian zones are often the most level areas within watersheds. As riparian zones are the low ground, plant propagules, soils, nutrients, and water all move in their direction. Because of this topographic position, even land use activities limited to uplands can have significant influences within the riparian zone. Landslides caused by logging, altered flow regimes by roads, degraded water quality due to upslope erosion by agriculture, and nutrient and fecal coliform inputs by animal-based agriculture are but a few examples.

Livestock grazing is likely the most extensive land use in the Pacific Northwest and livestock have significantly influenced the structure of riparian zones throughout this region.[52] In western Oregon and Washington, the broad floodplain riparian zones, formerly complex mosaics of coniferous and deciduous forest, marsh, and wet prairies, have been converted to simple agroecosystems of croplands, pastures, dairies, or grass fields. Similarly, almost all productive floodplain riparian zones in central and eastern Washington and Oregon have been converted to agriculture uses. In headwater reaches, range livestock production (often in concert with logging) becomes the dominant economic use of riparian zones. The degree to which livestock impact ecosystems depends on the degree of utilization, season of use, frequency of use, and inherent site characteristics such as species composition, soil textures, slope, and climate.[109]

Domestic cattle prefer to be near water and on gentle topography. Where uplands are steep or arid, cattle congregate principally in riparian zones. For example, on a Blue Mountain, Oregon, grazing allotment, the riparian zone comprised only 2% of the area, yet accounted for 81% of all forage consumed by cattle.[189] Without controls on animal numbers, timing, and duration of use, cattle can rapidly and severely degrade riparian areas through the effects of forage removal, soil compaction, streambank trampling, and the introduction of exotics.[35, 113, 182, 221] These factors have been defined as the direct effects of domestic livestock grazing on ecosystems.[109]

The direct effects of livestock combine to create indirect secondary biotic and hydrologic influences on the riparian zone, ultimately resulting in losses of ecosystem structure and composition, particularly in deciduous woodland riparian stands of cottonwood, alder, or willow. In the short term, canopy loss through the heavy use of shrubs and saplings influences the structure of these stands, thereby decreasing their value as wildlife habitat, particularly for avian nesting. Heavy use by herbivores can also decrease reproductive success of riparian willows.[31] Herbivory causes a corresponding depression in root production.[89] At stream edges, the combination of root loss and trampling weakens and collapses banks.[221] Bank loss and the resulting sediment loads contribute to downcutting, channel widening, and degradation of fisheries habitats. As the channel downcuts, over-bank flows cease, and hyporheic connectivity (subsurface stream/floodplain water exchange) is lost. Floodplain forests often develop following flood events and channel downcutting eliminates the potential for this gallery forest establishment. Loss of the riparian forests negatively affects not only the terrestrial wildlife but the aquatic biota as well. Loss of the shade provided by riparian vegetation results in increased stream temperature, altered water quality, and a change in composition and abundance of the aquatic biota.[118]

Logging riparian forests is likely deleterious to more habitats and wildlife species than logging in any other forest community or landscape in the Pacific Northwest. The harvesting of streamside forests leads directly to loss of habitats used by a disproportionate number of wildlife species. In addition, logging has reduced water quality through siltation and increases in stream water temperatures due to loss of shade.[11] Removal of streamside trees also eliminates litterfall and nutrient inputs into streams. Recruitment of stream wood is eliminated, which results in channel simplification, and possibly channel incision. Soil erosion and incidences of debris flow events are often exacerbated after logging. All of these combine to degrade habitat quality for both the aquatic and terrestrial biota.

Efficient transportation systems for commerce and human movements have often come at the expense of riparian/wetland habitats. Before 1900, splash dams were built on many mountain streams to provide sufficient flow for transporting logs to mills; such log drives are believed to have considerably simplified stream channels.[200] Today the Snake and Columbia rivers are used to transport agricultural products. This form of transportation created

via the construction of mainstem dams has come at the expense of free-flowing reaches, their associated riparian zones, and among the largest runs of salmon to have ever existed. When railroad transport of logs replaced instream log drives early in this century, many railroad grades were built adjacent to channels or within the floodplain.[145] At mid-century, road building increased as logging extended into more remote areas; many of these logging roads were also located in floodplains and riparian areas. Modern highways are often built adjacent to streams where riparian forests once existed. Roads located mid-slope or on ridgetops can alter drainage patterns and contribute to the occurrence of mass soil movements in steep mountain terrain.[20] In some instances, the drainage systems (including the road surfaces, ditches, and culverts) of a road network may alter storm hydrographs. This can occur where overland and subsurface flows are intercepted by high-density road networks such that runoff reaches the stream more quickly.[145]

Dams, water diversions, and stream channelization are commonly used in irrigation, mining, and milling operations, flood control, municipal water supplies, and conversion of wetlands for agriculture and urbanization. These engineering activities constitute an extreme perturbation of fluvial processes and riparian conditions across the landscape. Dams block the migration of fish and other aquatic organisms. Young fish returning to the ocean are often killed in turbines or from other dangers created through reservoir construction. Transport of sediment, organic matter, and nutrients is disrupted. The flow regime (peak flows and base flows) can be drastically altered, which eliminates most floods and channel forming processes. Diversion of water from streams by ditching or damming has the immediate effect of reducing or eliminating water-dependent vegetation as the water table is lowered and floodplain/stream linkages are lost. Recruitment of new individuals is diminished because many riparian plant species depend on fluvial processes to transport seeds to favorable sites created by naturally occurring highflows. The widespread loss of cottonwoods in the American West is attributed to cumulative effects of the dewatering of streams, in combination with heavy grazing.[26, 80, 213]

Flow regulation and the elimination of floods are commonly regarded as beneficial for the advancement of human civilization. However, the elimination of these natural disturbance processes has profound ecological effects on biotic diversity and production.[6] The response of vegetation, fish, and wildlife to water diversion varies with regional climate and geology and with local variations in reach type. Where flows are eliminated or dramatically reduced in channels downstream of dams, a change from riparian-obligate vegetation to exotic or even xeric vegetation can result. Impacts of these changes are reflected in the loss of native wildlife and fish species resulting from the changes in habitat structure, stream temperature, water chemistry, and alterations in food supply.

Figure 10. A conceptual approach to wetlands and riparian restoration. Figure is adapted from Kauffman et al.[103]

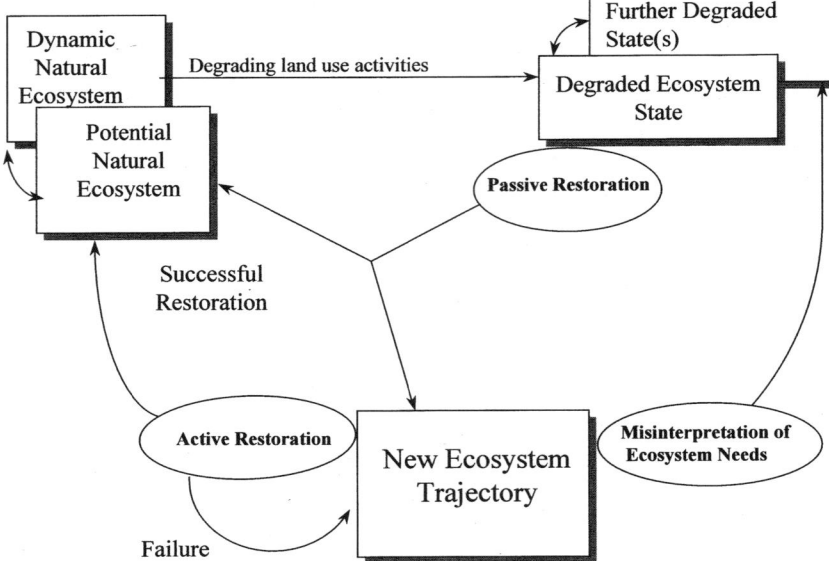

Restoration of Riparian Wildlife Habitats

Finally, few watersheds are affected by only one land use activity. Multiple land use activities often have cumulative or synergistic effects on riparian/stream habitats. Programs that take the affirmative steps toward ecological restoration must recognize the multiple effects of current and historical uses that shape the composition and structure of riparian zones today.

Ecological restoration of riparian habitats (riparian restoration) is defined as the reestablishment of predisturbance riparian functions and related chemical, biological, and hydrological characteristics.[161] Ecological restoration is the process of repairing damage caused by humans to the diversity and dynamics of indigenous ecosystems.[87] Ecological restoration attempts to return riparian zones, as closely as possible, to fully functional conditions (i.e., the potential natural range of variability in composition, structure, and ecosystem processes). However, it is important to recognize that ecosystems and landscapes are in a constant state of flux due to ever changing environmental conditions. These changes, coupled with human-caused global environmental changes (e.g., soil degradation, exotic invasions, desertification, and altered atmospheric chemistry and climate change), preclude our capability to recreate precisely the ecosystem structure and functions that previously existed. Thus the goal of a restoration project is to insure that the natural dynamic ecosystem processes are again operating sufficiently so that ecosystem structure and functions can be recovered.[161]

Ecological restoration results in the reconnection of the linkages between organisms and their environment (Figure 10). Because ecosystems are comprised of an infinite number of organisms and physical features and processes, a species-only or single process approach to restoration often results in failure.[13, 87] For example, the

Table 5. An example of steps involved in an adaptive management approach to the implementation of riparian and wetlands habitat restoration. Whereas restoration activities generally occur at the site or stream reach scale, projects should begin by examining conditions at the landscape/watershed scale. This facilitates management actions geared to preserving intact areas (more successful and less costly than restoration) as well as identifying recovering areas where passive restoration may have already been implemented. Failure to follow a step-wise and adaptive ecological approach to restoration has often resulted in failure or even further degradation of the resources, as well as the loss of public confidence.

Step	Action or approaches
1.	Recognition of a need for restoration.
2.	Delineation of ecological condition of sites within the landscape and determination of appropriate management actions — preservation, restoration, maintenance.
3.	Determination of the type, degree, and causes of degradation, resources lost, or those endangered.
4.	Determination of site potential, ecosystem attributes, and functions of areas to be restored.
5.	Determination of barriers to restoration (biological and social).
6.	Implementation of passive restoration.
7.	Evaluation and monitoring for adaptive management and determination if additional measures are necessary (appropriate time frames).
8.	Implementation of active restoration approaches.
9.	Evaluation and monitoring for adaptive management, and determination if additional measures are necessary (appropriate time frames).

reintroduction of an extirpated fish species into a degraded stream reach, or the installation of log weirs or large boulders into a mismanaged steam reach, does not constitute restoration. Whereas attempts to revive a single species are likely to target only a few more obvious habitat requirements, less obvious habitat needs or other important processes and functions are often ignored. By shifting the focus of restoration to the ecosystem or watershed, we are more likely to achieve success for both the habitat and the species of interest.

Ecological restoration also needs to be distinguished from "landscaping" or "architectural" approaches to ecosystem repair. Such "restoration" methodologies often call for the systematic reconstruction of a stream according to specific human perspectives of what the stream should look like. Merely recreating a form without the function or the functions in an artificial configuration does not constitute restoration.[161] Neither will be sustainable, and if improperly designed may create conditions that exacerbate ecological degradation. For example, the placement of artificial structures (boulders, rock gabions) does not replace the multiple functions of large woody debris, nor does it insure or promote the future recruitment of woody debris into the stream system. Such structures are commonly engineered and constructed with the false perception that they and the stream channel are static. In reality, highflows can be increasingly destructive when rigid structures are placed in degraded stream channels.[12, 67] Similarly, the creation of permanent surface water wetlands in landscapes where natural wetlands were seasonal or ephemeral may create optimal habitats for exotics (e.g., bullfrogs) to the detriment of native species.

Natural resource management, including the maintenance of biological diversity, will be most successful when decisions and actions are made at a landscape or watershed perspective. Ecological restoration is but one of a number of land management activities that should be implemented at appropriate communities or stream reaches within the landscape. Recognition that there are areas in the watershed (and perhaps its entirety) in a degraded condition and that species and/or important ecosystem processes are at risk is the first step in successful restoration (Table 5). Restoration activities should occur only in sites where appropriate, whereas intact sites should be managed under preservation/conservation strategies.

Preservation is simply the maintenance or management of intact ecosystems. The structure and function of ecosystems warranting protection are those that exist in a desired natural state.[161] The protection and preservation of intact ecosystems is of great importance both environmentally and economically. Intact sites are valuable references to judge the efficacy of restoration activities. They are also sources of natural genetic material for the reintroduction or reinvasion of nearby areas in need of restoration. Protection is a management strategy that often entails more than the prevention of human-induced alterations. It also includes management actions necessary to maintain natural functions and characteristics (e.g., prescribed fire, weed control). Measures to protect intact

ecosystems (preservation) are important because they are often easier to implement than restoration. Preserving intact ecosystems may also be less expensive than restoring them, just as preventative medicine is nearly always less expensive than corrective medicine.[27, 161]

Once decisions have been made on the causes of degradation and where to implement restoration activities, the first and most critical step is to halt those activities that are causing degradation or preventing their recovery. This has been termed *passive* or *natural restoration*.[102, 104] Many riparian species are capable of rapid recovery following cessation or removal of human perturbations because the riparian biotas have evolved adaptations to survive and even reproduce following the frequent natural disturbance events typical of streamside environments.

The two most common examples of successful passive ecological restoration in western riparian zones include the re-watering of streams following years of withdrawal for agricultural or municipal purposes and the cessation of abusive livestock grazing. Throughout the West, long periods of streamflow diversion, combined with heavy livestock grazing, have resulted in severe degradation of riparian/stream ecosystems. With the cessation of livestock grazing and the return to perennial instream flows, the recovery of riparian vegetation has been dramatic.[103]

Livestock grazing has been the most prevalent cause of ecological degradation of riparian/stream ecosystems in the Intermountain west.[62, 101, 107] Beschta et al.[12] and Kauffman et al.[102] suggested that the cessation of livestock grazing in riparian zones of eastern Oregon was the single most ecologically and economically effective approach for restoring salmonid habitats. In central Oregon, Busse[26] found a severe lack of willow and cottonwood reproduction in grazed riparian zones of the Crooked River National Grassland. Following the construction of corridor fencing, she quantified a widespread and rapid rate of willow and cottonwood establishment. In northeastern Oregon, similar rates of recovery following cessation of cattle grazing have been quantified.[30, 66] Three years after the cessation of cattle grazing, Case[30] reported that the average crown volume of willows increased nearly 300%. Average crown volume of black cottonwood and alder increased 800% and 200%, respectively. Comparing 10 years of no grazing with light-to-moderate late season grazing use, Green and Kauffman[66] reported significant increases in both the structure and density of willows and cottonwoods in the ungrazed exclosures in Northeast Oregon. Although positive trends in willow density and height in the light-to-moderately grazed areas (3 weeks annually late in the season) also occurred, recovery rates were significantly less than in the ungrazed areas.

Responses to passive restoration via corridor fencing have also been quantified in riparian wetland meadow communities. Water infiltration rates and water holding capacities were much higher in ungrazed (9 years) compared to grazed meadows (season-long) in an eastern Oregon study.[106] In addition, soil organic matter and rates of nitrogen mineralization were higher in the exclosures compared to grazed areas.

Reviews of many instream restoration and enhancement projects throughout the western U.S. clearly reveal that passive restoration was the critical first step in successful riparian restoration programs.[12, 13, 102] In many cases, this was all that was needed for the restoration of riparian ecosystems. Because of the high costs and potential for failure with active restoration/manipulation, project managers should monitor and observe the natural recovery process for an appropriate period of time (e.g., 5-10 years) following implementation of passive restoration.[103] After this time period, if recovery is not occurring, implementation of active restoration projects might begin.

A severely degraded riparian ecosystem may not recover even after implementation of passive restoration. To achieve ecological restoration in such situations, active manipulations will likely be needed. *Active restoration* is defined as the reintroduction of extirpated species, ecosystem processes, and other features of ecosystem structure; chemical clean-up of the environment; and the elimination of exotic species or conditions that allow for their persistence. Many factors may prevent a return to a natural dynamic system state when only passive restoration is implemented. This may include species extinctions (particularly keystone species such as beaver or cottonwoods), the presence of exotic predators (e.g., bullfrogs or small-mouth bass), the presence of exotic plants (e.g., reed canary grass *Phalaris arundinacea* or knapweeds [*Centaurea spp.*]), loss of hydrologic function and alterations of hydrologic disturbance regimes (e.g., diversions, regulated flows by dams, and upland disruptions of groundwater flow patterns), and alteration of geomorphic features (e.g., channel incision, soil erosion, compaction). Whereas some of these barriers to recovery can be easily ameliorated, others are much more severe in their impact and persistence, and create situations where restoration at any cost is not technologically feasible. Social, political, and economic limitations to recovery are also a common hurdle in many cases.

Active restoration, where undertaken, must be implemented in such a way as to facilitate the recovery of natural ecosystem processes.[104] Restoration activities should result in a self-sustaining, functional ecosystem. Stream and streambank manipulations under the guise of "enhancement programs" that neglect to consider or ignore the natural ecosystem dynamics and site potential usually result in failure, or even exacerbate degradation of the ecosystem. This implementation of riparian or in-channel activities that further degrades ecosystem features (structure, function, and/or composition) is defined as the *misinterpretation of ecosystem needs*.[103] The results of these misinterpretations are activities that either limit or further degrade the hydrological, geomorphic, or biotic function of riparian/stream ecosystems. These often occur when program managers fail to address the following questions:

1. What is the biotic and physical potential of this site?
2. What are the functional attributes of the riparian zone or stream now and following restoration?

3. Can I achieve restoration via passive restoration alone and/or has it already been implemented?
4. Could there be negative consequences associated with the proposed approaches to restoration, and what would be the consequences of failure?

The most common examples of misinterpretation of ecosystem needs include unnatural species introductions or habitat manipulations out of context with the natural environment. The former includes out-planting hatchery fish of non-indigenous genetic strains, introductions of non-native plant species, or planting species in sites where they would not normally occur. The creation of wetlands with yearlong surface water to mitigate the loss of natural ephemeral wetlands would not be considered ecological restoration. Some examples of misinterpretations of the natural environment include blasting pools within bedrock channels, the addition of logs and boulders in channels associated with degraded meadow systems (i.e., where large instream features would not occur), in-channel engineering approaches that are heavily anchored by cable, metal rods, geotextile fabrics, or boulders, armoring of streambanks with large boulders (riprap), and the placement of excavated sediments on streambanks and floodplains.[103] Many of these approaches not only severely limit the capacity for streams to undergo natural adjustments in channel morphology and stream sinuosity over time, but they may also suppress the recovery of riparian vegetation and hence the habitat for the associated terrestrial biota. It should be clear that riparian/stream ecosystems degraded from off-channel sources cannot be restored by focusing only on activities within the channel.

Finally it is important to recognize that success of riparian restoration cannot be based on a time frame relevant to that of a construction project. Restoration is a continuing adaptive management activity that occurs over a long time period. Some components of riparian zones rapidly recover following implementation of passive restoration (i.e., herbaceous regrowth) whereas other important riparian features may require decades to centuries for recovery (e.g., gallery forests of cottonwood

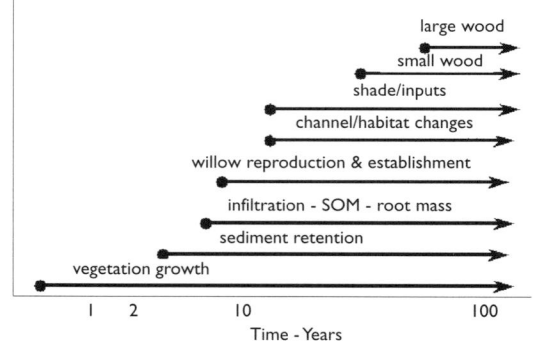

Figure 11. Recovery rates of selected leading edge and keystone features of riparian habitats following the implementation of passive restoration. Passive restoration, defined as the termination of those activities that cause degradation or prevent recovery, should be the first step in riparian/stream restoration projects.

and large wood, incised channels) (Figure 11). Recognition of the temporal requirements and logical steps of ecological restoration is an extremely important component in restoration planning and implementation.

Riparian zones are ecosystems rich in productivity, biological diversity, and those unique biogeochemical processes that influence water quality and the aquatic biota. For humans, riparian/stream ecosystems are foci of commodity, recreational, and spiritual values. The maintenance or restoration of intact riparian ecosystems is of importance to societies at local, regional, and global scales. Complex ecosystems and associated habitat features cannot be achieved via simple and artificial constructions of selected components. Ecological restoration is a holistic approach not achieved through isolated manipulations of individual elements but through approaches that ensure natural ecological processes occur.[161] If society is to use, enjoy, or benefit from the natural resource potential of western riparian/stream ecosystems, concerted efforts of ecological restoration should begin before their potential is forever lost through the extinction of their various components.

Whereas it is recognized that ecological restoration sometimes comes at a high cost, it is also important to recognize that ecological restoration is an investment into the natural capital of riparian and stream ecosystems, and hence the environmental wealth of the region. Conversely, degradation is equivalent to squandering the natural wealth and the productive capacity of natural ecosystems. It is clearly in the best environmental as well as long-term economic interests of the nation to restore once-productive ecosystems in the Pacific Northwest. Restored riparian/aquatic systems will be essentially self-maintaining, and therefore, useful in perpetuity.[27] Because these ecosystems are a fundamental component of our life-support system, restoration should represent an important priority for both public and private landowners.

Literature Cited

1. Alerstam, T., and Å. Lindström. 1990. Optimal bird migration: the relative importance of time, energy, and safety. Pages 331-351 in: E. Gwinner, editor. Bird migration: physiology and ecophysiology. Springer-Verlag, Berlin, Germany.

2. Allred, M. 1980. A re-emphasis on the value of the beaver in natural resource conservation. Journal of Idaho Academy of Science 16:3-10.

3. Alverson, W. S., D. M. Waller, and S. L. Solheim. 1988. Forests too deer: edge effects in northern Wisconsin. Conservation Biology 12:348-358.

4. Anthony, R. G., E. D. Forsman, G. A. Green, G. Witmer, and S. K. Nelson. 1988. Small-mammal populations in riparian zones of coniferous forest in western Oregon. Pages 163-165 in: K. Raedeke, editor. Streamside management: riparian wildlife and forestry interactions. Contribution No. 59. University of Washington, Institute of Forest Resources, Seattle, WA.

5. ———, E. C. Meslow, and D. S. deCalesta. 1987. The role of riparian zones for wildlife in westside Oregon forests—what we know and don't know. Pages 5-12 in: National Council on Air and Stream Improvement (NCASI) Technical Bulletin No. 514, Corvallis, OR.

6. Bayley, P. B. 1995. Understanding large river-floodplain ecosystems. BioScience 45:153-158.

7. Beebee, T. J. C. 1995. Amphibian breeding and climate. Nature 374:219-220.

8. ———. 1996. Ecology and conservation of amphibians. Chapman & Hall, London, United Kingdom.

9. Bellrose, F. C. 1980. Ducks, geese, and swans of North America. Stackpole Books, Harrisburg, PA.

10. Beschta, R. L. 1995. Restoration of riparian and aquatic systems for improved aquatic habitats in the upper Columbia River Basin. Pages 475-491 in: Stouder, P.A. Bisson, and R.J. Naiman, editors. Pacific Salmon and their ecosystems: Status and future options. Chapman & Hall, New York, NY.

11. Beschta, R.L., R.E. Bilby, G.W. Brown, L.B. Holtby, and T.D. Hofstra. 1987. Pages 191-232 in: E. O. Salo, and T.W. Cundy, editors. Streamside management: Forestry and fishery interactions. Contribution 57, Institute of Forest Resources, University of Washington, Seattle, WA.

12. ———, W. S. Platts, and J. B. Kauffman. 1991. Field review of fish habitat improvement projects in the Grande Ronde and John Day River Basins of eastern Oregon. USDOE-Bonneville Power Administration Report. Division of Fish and Wildlife. Portland, OR.

13. ———, ———, ———, and M. T. Hill. 1994. Artificial stream restoration—money well spent or money down the river. Proceedings. University Council of Water Resources, Big Sky, MT.

14. Bisson, P.A., R. E. Bilby, M. D. Bryant, C.A. Dolloff, G. B. Grette, R.A. House, M. L. Murphy, K.V. Koski, and J. R. Sedell. 1987. Large woody debris in forested streams in the Pacific Northwest: past, present, and future. Pages 143-190 in: E. O. Salo and T.W. Cundy, editors. Streamside management: forestry and fisheries interactions. Contribution 57, Institute of Forest Resources, University of Washington, Seattle, WA.

15. Bock, C. E., A. Cruz, J. R., M. C. Grant, C. S. Aid, and T. R. Strong. 1992. Field experimental evidence for diffuse competition among southwestern riparian birds. American Naturalist 140: 815-828.

16. Bowman, G. B., and L. D. Harris. 1980. Effect of spatial heterogeneity on ground nest depredation. Journal of Wildlife Management 44: 806-813.

17. Bradley, C. E., and D. G. Smith. 1986. Plains cottonwood recruitment and survival on a prairie meandering river floodplain, Milk River, southern Alberta and northern Montana. Canadian Journal of Botany 64:1433-1442.

18. Brode, J. M. and R. B. Bury. 1984. The importance of riparian systems to amphibians and reptiles. Pages 30-36 in: R. E. Warner and K. E. Hendrix, editors. California riparian systems: Ecology, conservation, and productive management. University of California Press, Berkeley, CA.

19. Brookshire, E. N. 2000. Allochthonous organic matter dynamics in headwater stream riparian areas. M.S. Thesis Oregon State University, Corvallis, OR.

20. Brown, G.W. 1985. Forestry and water quality. Second edition. Forest Research Laboratory, Oregon State University, Corvallis, OR.

21. Bull, E. L. and B. E. Carter. 1996. Tailed frogs: distribution, ecology, and association with timber harvest in Northeastern Oregon. U.S. Forest Service Report. PNW-RP-497.

22. Bureau of Land Management. 1998. Birds as indicators of riparian vegetation condition in the western U. S. Bureau of Land Management. Boise, ID.

23. Burhans, D. E. 1997. Habitat and microhabitat features associated with cowbird parasitism in two forest edge cowbird hosts. Condor 99: 866-872.

24. Bury, R. B. 1968. The distribution of Ascaphus truei in California. Herpetologica 24:39-46.

25. ———, and M.G. Raphael. 1983. Inventory methods for amphibians and reptiles. Proceedings International Conference on Renewable Resource Inventories for Monitoring Changes and Trends. Oregon State University, Corvallis, OR.

26. Busse, C.G. 1989. Ecology of the Salix and Populus species of the Crooked River National Grassland. Thesis, Oregon State University, Corvallis, OR.

27. Cairns, J. Jr., 1993. Is restoration ecology practical? Restoration Ecology. 1:3-6.

28. Campbell, A. G. and J. F. Franklin. 1979. Riparian vegetation in Oregon's western Cascade Mountains: composition, biomass, and autumn phenology. Coniferous forest biome, ecosystems analysis studies. U.S. International Biosphere Program. Bulletin Number 14.

29. Carter, M., G. Fenwick, C. Hunter, D. Pashley, D. Petit, J. Price, and J. Trappe. 1996. For the future-watchlist 1996. Field Notes 50:3.

30. Case, R.L. 1995. Structure, biomass, and successional dynamics of forested riparian ecosystems of the Upper Grande Ronde Basin. Thesis, Oregon State University, Corvallis, OR.

31. ———, and J.B. Kauffman. 1997. Wild ungulate influences on the recovery of willows, black cottonwood and thin-leaf alder following cessation of cattle grazing on Northeastern Oregon. Northwest Science. 71:115-126.

32. Chadde, S.W. and C. E. Kay. 1991. Tall-willow communities on Yellowstone's northern range: a test of the "natural regulation" paradigm. In: R. B. Keiter. and M. S. Boyce, editors. The greater Yellowstone ecosystem: redefining America's wilderness heritage. Yale University Press, New Haven, CT.

33. Chaney, E., W. Elmore, and W. S. Platts. 1993. Livestock grazing on western riparian areas. U.S. EPA NW Research. Information Center. Inc., Eagle, ID.

34. Clements, C. 1991. Beavers and riparian ecosystems. Rangelands 13:277-279.

35. Clifton, C. 1989. Effects of vegetation and land use on channel morphology. Pages 121-129 in: R. E. Gresswell, B.A. Barton, and J. L. Kershner, editors. Practical approaches to riparian resource management: an educational workshop. USDI Bureau of Land Management, Billings, MT.

36. Cody, M. L. 1974. Competition and the structure of bird communities. Princeton University Press, Princeton, NJ.

37. Cooper, J. R., J. W. Gilliam, R. B. Daniels, and W. P. Robarge. 1987. Riparian areas as filters for agricultural sediment. Soil Science Society of America Journal 51:416-420.

38. Cooper, D. J. 1990. Ecology of wetlands in Big Meadows, Rocky Mountain National Park, Colorado. Biological Report 90(15). U.S. Department of Interior, Washington, D.C.

39. Corkran, C. C. and C. Thoms. 1996. Amphibians of Oregon, Washington and British Columbia. Lone Tree Publishing, Renton, WA.

40. Corn, P. S. and R. B. Bury. 1989. Logging in western Oregon: responses of headwater habitats and stream amphibians. Forest Ecology and Management 29:39-57.

41. Cross, S. P. 1985. Responses of small mammals to forest riparian perturbations Pages 269-275 in: R. R. Johnson, C. D. Ziebell, D. R. Patton, R. F. Ffolliott, and R. H. Hamre, technical coordinators. Riparian ecosystems and their management: reconciling conflicting uses. Proceedings of the First North American Riparian Conference. U.S Forest Service General Technical Report RM-120.

42. ———. 1988. Riparian systems and small mammals and bats. Pages 93-112 in: K. Raedeke, editor. Streamside management: riparian wildlife and forestry interactions. Contribution No. 59. University of Washington, Institute of Forest Resources, Seattle, WA.

43. Csuti, B., A. J. Kimerling, T.A. O'Neil, M. M. Shaughnessy, E. P. Gaines, and M. M. Huso. 1997. Atlas of Oregon Wildlife: Distribution, habitat, and natural history. Oregon State University Press, Corvallis, OR.

44. Cummins, K.W. 1974. Structure and function of stream ecosystems. Bioscience. 24:631-641.

45. Cummins, K.W. and G. L. Spengler. 1978. Stream ecosystems. Bioscience: 10:1-9.

46. Dahm, C. N., E. H. Trotter, and J. R. Sedell. 1987. The role of anaerobic zones and processes in stream ecosystem productivity. Pages 157-178 in: R. C. Averett and D. M. McNight, editors. Chemical quality of water and the hydrologic cycle. Lewis Publishers, Chelsea, MI.

47. Davis, C.A., and L. M. Smith. 1998. Ecology and management of migrant shorebirds in the playa lakes region of Texas. Wildlife Monographs 140: 1-45.

48. DeGraaf, R. M., V. E. Scott, R. H. Hamre, L. Ernst, and S. H. Anderson. 1991. Forest and rangeland birds of the United States: natural history and habitat use. U.S. Forest Service Agriculture Handbook 688.

49. deVlaming, V. L. and R. B. Bury. 1970. Thermal selection in tadpoles of the tailed frog, Ascaphus truei. Journal of Herpetology 4:179-189.

50. Dobkin, D. S., and B.A. Wilcox. 1986. Analysis of natural forest fragments: riparian birds in the Toyabee Mountains, Nevada. Pages 293-299 in: J. Verner, M. L. Morrison, and C. J. Ralph, editors. Wildlife 2000:

modeling habitat relationships of terrestrial vertebrates. University of Wisconsin Press, Madison, WI.

51. Doyle, A. T. 1990. Use of riparian and upland habitats by small mammals. Journal of Mammalogy 71:14-23.

52. Dwire, K.A., B.A. McIntosh, and J. Boone Kauffman. 1999. Ecological influences of the introduction of livestock on Pacific Northwest Ecosystems. In: D. Gobel, editor. Environmental History of the Pacific Northwest. University of Washington Press, Seattle, WA.

53. Dwire, K. 2000. Connectivity of riparian meadows and low-order streams in the Blue Mountains, Oregon: interactions among hydrology, soil, and vegetation. Dissertation. Oregon State University, Corvallis, OR.

54. Eagle, T. C., and J. S. Whitman. 1987. Mink. Pages 615-642 in: M. Novak, J. A. Baker, M. E. Obbard, and B. Malloch, editors. Wild furbearer management and conservation in North America. Ontario Ministry of Natural Resources, Toronto, Ontario, Canada.

55. Ehrlich, P. R., D. S. Dobkin, and D. Wheye. 1988. The Birders Handbook. Simon and Schuster Inc., New York, NY.

56. England, A. S., L. D. Foreman, and W. F. Laudenslayer. 1984. Composition and abundance of bird populations in riparian systems of the California deserts. Pages 694-705 in: R. E. Warner and K. M. Hendrix, editors. California riparian systems: ecology, conservation, and productive management. University of California Press, Berkeley, CA.

57. Everitt, B. L. 1968. Use of the cottonwood in an investigation of the recent history of a floodplain. American Journal of Science 266:417-439.

58. Farley, G. H., L. M. Ellis, J. N. Stuart, and N. J. Scott, Jr. 1994. Avian species richness in different-aged stands of riparian forest along the middle Rio Grande, New Mexico. Conservation Biology 8:1098-1108.

59. Finch, D. M. 1992. Threatened, endangered, and vulnerable species of terrestrial vertebrates in the Rocky Mountain Region. U.S. Forest Service General Technical Report RM-215.

60. ———, and L. F. Ruggiero. 1993. Wildlife habitats and biological diversity in the Rocky Mountains and Northern Great Plains. Natural Areas Journal 13:191-203.

61. Finley, K. K. 1994. Hydrology and related soil features of three Willamette Valley wetland prairies. M.S. Thesis Oregon State University, Corvallis, OR.

62. Fleischner, T. L. 1994. Ecological costs of livestock grazing in western North America. Conservation Biology 8:629-644.

63. Franzreb, K. E. 1987. Endangered status and strategies for conservation of the Least Bell's Vireo (Vireo bellii pusillis) in California. Western Birds 18:43-49.

64. Gomez, D. M., and R. G. Anthony. 1996. Amphibian and reptile abundance in riparian and upslope areas of five forest types in western Oregon. Northwest Science 70:109-119.

65. Green, D. M., and J. B. Kauffman. 1989. Nutrient cycling at the land-water interface: the importance of the riparian zone. Pages 61-68 in: R. E. Gresswell, B.A. Barton, and J. L. Kershner editors. Practical approaches to riparian resource management: an educational workshop. U.S. Bureau of Land Management, Billings, MT.

66. Green, D. M. and J. B. Kauffman. 1995. Succession and livestock grazing in a Northeast Oregon riparian ecosystem. Journal of Range Management. 48:307-313.

67. Gregory, S.V. 1993. The basis for integrated watershed and stream restoration. National Habitat Restoration Workshop. Annual Meeting of the American Fisheries Society. August, 1993. Portland, OR.

68. ———, F. J. Swanson, W.A. McKee, and K.W. Cummins. 1991. An ecosystem perspective of riparian zones. Bioscience 41: 540–550.

69. Grover, M. C. and L .A. DeFalco. 1995. Desert tortoise (Gopherus agassizii): status of knowledge outline with references. U.S. Forest. Service. IRS. INT-GTR-316.

70. Hall, R. J. 1980. Effects of environmental contaminants of reptiles: a review. U.S. Fish and Wildlife Service Special. Science. Report 228. Washington, D.C.

71. Hanley, T.A., and R. D. Taber. 1980. Selective plant species inhibition by elk and deer in three conifer communities in western Washington. Forest Science 26:97-107.

72. Harmon, M. E., J. F. Franklin, F. J. Swanson, P. Sollins, S. V. Gregory, G. D. Lattin, N. H. Anderson, S. P. Cline, N. G. Aumen, J. R. Sedell, G. W. Lienkaemper, K. Cromack, Jr., and K. W. Cummins. 1986. Ecology of coarse woody debris in temperate ecosystems. In: A. MacFadyen, and E. D. Ford, editors. Advances in Ecological Research 15:133-302.

73. Hatler, D. F. 1976. The coastal mink on Vancouver Island, British Columbia. Ph.D. Thesis, University of British Columbia, Vancouver, BC, Canada.

74. Hawkins, C. P., M. L. Murphy, N. H. Anderson, and M .A. Wilzbach. 1983. Density of fish and salamanders in relation to riparian canopy and physical habitat in stream of the northwestern United States. Canadian Journal of Fish and Aquatic Sciences 40:1173-1185.

75. Hayes, J. P., M. D. Adam, D. Bateman, E. Dent, W. H Emmingham, K. G. Maas, and A. E. Skaugset. 1996. Integrating research and forest management in riparian areas of the Oregon Coast Range. Western Journal of Applied Forestry. 11:85-89.

76. Hayes, M. P. 1994. Current status of the spotted frog (*Rana prestiosa*) in western Oregon. Report to Oregon Department of Fish and Wildlife. Portland, OR.

77. Henke, M., and C. P. Stone. 1979. Value of riparian vegetation to avian populations along the Sacramento River System. Pages 228-235 *in*: R. R. Johnson and J. F. McCormick, editors. Strategies for protection and management of floodplain wetlands and other riparian ecosystems. U.S Forest Service. General Technical Report WO-12. Washington, D.C.

78. Hinschberger, M. A. 1978. Occurrence and relative abundance of small mammals associated with riparian and upland habitats along the Columbia River. Thesis, Oregon State University, Corvallis, OR.

79. Holland, D.C. 1994. The western Pond turtle: habitat and history. U.S. Department of Energy, Bonneville Power Administration, Portland, OR.

80. Howe, W. H. and F. L. Knopf. 1991. On the imminent decline of Rio Grande cottonwoods in central New Mexico. Southwest Naturalist 36:218-224.

81. Hunter, M. L., Jr. 1990. Wildlife, forests, and forestry: principles of managing forests for biological diversity. Prentice-Hall, Inc. Englewood Cliffs, NJ.

82. Hutto, R. L. 1985a. Habitat selection by nonbreeding migratory landbirds. Pages 455-476 *in*: M. L. Cody, editor, Habitat selection in birds. Academic Press, Orlando, FL.

83. ———. 1985b. Seasonal changes in the habitat distribution of transient insectivorous birds in southeastern Arizona: competition mediated? Auk 102: 120-132.

84. Ims, R. A., and N. G. Yoccoz. 1997. Studying transfer processes in metapopulations: emigration, migration, and colonization. Pages 247-266 *in*: I. A. Hanski and M. E. Gilpin, editors. Metapopulation biology: ecology, genetics, and evolution. Academic Press, San Diego, CA.

85. Ives, R. L. 1942. The beaver-meadow complex. Journal of Geomorphology 5:191-203.

86. Jacobs, T.C. and J. W. Gilliam. 1985. Nitrate from agricultural drainage waters. Journal of Environmental Quality. 14: 472-478.

87. Jackson, L.L., N. Lopoukhine, and D. Hillyard. 1995. Ecological Restoration: A definition and comments. Restoration Ecology 3:71-75.

88. Jakle, M. D. and T. A. Gatz. 1985. Herpetofaunal use of four habitats of the middle Gila River drainage, Arizona. *In*: R.R. Johnston, technical coordinator. Riparian ecosystems and their management: Reconciling conflicting uses. U.S. Forest Service. General Technical Report. RM-120.

89. Jameson, D.A. 1963. Responses of individual plants to harvesting. Botanical Review 29:532-594.

90. Jenkins, S. H. 1980. A size-distance relation in food selection by beavers. Ecology 61:740-746.

91. ———. 1975. Food selection by beavers: a multidimensional contingency table analysis. Oecologia 21:157-173.

92. Johnson, C. B. 1985. Use of coastal habitat by mink on Prince of Wales Island, Alaska. M.S. Thesis, University of Alaska, Fairbanks, AK.

93. Johnson, R. R., L. T. Haight, and J. M. Simpson. 1977. Endangered species vs. endangered habitats: a concept. Pages 68-79 *in*: R. R. Johnson and D. A. Jones, editors. Importance, preservation, and management of riparian habitat: a symposium. U.S Forest Service. General Technical Report RM-43.

94. Johnston, C. A., and R. J. Naiman. 1987. Boundary dynamics at the aquatic-terrestrial interface: the influence of beaver and geomorphology. Landscape Ecology 1: 47-57.

95. ———, and ———. 1990. Aquatic patch creation in relation to beaver population trends. Ecology 71:1617-1621.

96. ———, and ———. 1990. Browse selection by beaver: effects on riparian forest composition. Canadian Journal of Forest Research 20:1036-1043.

97. ———, and ———. 1990. The use of a geographic information system to analyze long-term landscape alteration by beaver. Landscape Ecology 4:5-19.

98. Jones, K. B. 1981. Effects of grazing on lizard abundance and diversity in western Arizona. Southwest Naturalist. 26:107-115.

99. ———. 1988. Comparison of herpetofaunas of a natural and altered riparian ecosystem. U.S Forest Service. General Technical Report RM-166.

100. ———, L.P. Kepner, and T. E. Martin. 1985. Species of reptiles occupying habitat islands in western Arizona: a deterministic assemblage. Oecologia 66:595-601.

101. Kauffman, J. B. 1988. The status of riparian habitats in Pacific Northwest forests. Pages 45-55 *in*: K. Raedeke, editor. Streamside management: riparian wildlife and forestry interactions. Contribution 59. University of Washington, Institute of Forest Resources, Seattle, WA.

102. ———, R. L. Beschta and W. S. Platts. 1993. Fish habitat improvement projects in the Fifteenmile Creek of Trout Creek Basins of central Oregon: field review and management Recommendations. U.S. Department of Energy-Bonneville Power Administration Report, Division of Fish and Wildlife. Portland, OR.

103. ———, ———, N. Otting, and D. Lytjen. 1997. Ecological approaches to riparian restoration in the United States. Fisheries 22:12-24.

104. ———, R. L. Case, D. Lytjen, N. Otting and D. L. Cummings. 1995. Ecological approaches to riparian restoration in northeast Oregon. Restoration Ecology Notes 13:12-15.

105. ———, N. Otting, D. Lytjen, and R. L. Beschta. 1996. Ecological principles and approaches to riparian restoration in the western United States. *In*: Healing the Watershed: A Guide to Watershed and Natural Fisheries Restoration. Workbook Number 2, Healing the Watershed Series, Pacific Rivers Council, Eugene, OR.

106. ———, L. Boeder, A. Thorpe, and J. Brookshire. 1998. Vegetation and soil properties of grazed and ungrazed riparian montane meadows. Abstracts. Ecological Society of America. 83rd Annual meeting. Baltimore, MD.

107. ———, and W. C. Krueger. 1984. Livestock impacts on riparian ecosystems and streamside management implications: a review. Journal of Range Management 37:430-437.

108. ———, W. C. Krueger, and M. Vavra. 1982. Impacts of a late season grazing scheme on nongame wildlife habitat in a Wallowa Mountain riparian ecosystem. Pages 208-220 *in*: L. Nelson and J. M. Peek, editors. Wildlife livestock relationships symposium, Proceedings 10, University of Idaho, Forest and Range Experiment Station, Moscow, ID.

109. ———, and D. A. Pyke. 2000. Global Influences of livestock on the biological diversity of rangeland and forest ecosystems. *In*: S. Levin, editor. Encyclopedia of Biological Diversity. Academic Press.

110. Kay, C. E. and S. Chadde. 1992. Reduction of willow seed production by ungulate browsing in Yellowstone National Park. *In*: Symposium on ecology and management of riparian shrub communities. Sun Valley, ID.

111. Kelly, J. P. 1993. The effects of nest predation on habitat selection by dusky flycatchers in limber pine-juniper woodland. Condor 95: 83-93.

112. Knopf, F. L. 1985. Significance of riparian vegetation to breeding birds across an altitudinal cline. Pages 105-111 *in*: R. R. Johnson, C. D. Ziebell, D. R. Patten, P. F. Folliot, and R. H. Hamre, technical coordinators. Riparian ecosystems and their management: reconciling conflicting uses. U.S. Forest Service. General Technical Report. RM-120.

113. ———, and R.L. Cannon. 1982. Structural resilience of a willow riparian community to changes in grazing practices. *In*: Proceedings of the wildlife-livestock relationships symposium. Forest, Wildlife, and Range Experiment Station, University of Idaho. Moscow, ID.

114. ———, R. R. Johnson, T. Rich, F. B. Sampson, and R. C. Szaro. 1988. Conservation of riparian ecosystems in the United States. Wilson Bulletin. 100:272-284.

115. ———, and F. B. Sampson. 1994. Scale perspectives on avian diversity in western riparian ecosystems. Conservation Biology 8:699-676.

116. Kreuper, D. J. 1993. Effects of land use practices on western riparian ecosystems. Pages 321-330 in: D. M. Finch and P. W. Stengel, editors. Status and management of neotropical migratory birds. U.S. Forest Service. General Technical Report RM-229.

117. Larison, B., S. A. Layman, P. L. Williams, and T. B. Smith. 1998. Song sparrows vs. cowbird brood parasites: impacts of forest structure and nest-site selection. Condor 100:93-101.

118. Li, H. W., G. A. Lamberti, T. N. Pearsons, C. K. Tait, J. L. Li, and J. C. Buckhouse. 1994. Cumulative effects of riparian disturbances along high desert trout streams of the John Day Basin, Oregon. Transactions of the American Fisheries Society 123:627-640.

119. Lowe, C. H. 1986. Amphibians and reptiles in southwest riparian ecosystems. In: R .R. Johnston, technical coordinator. Riparian ecosystems and their management: Reconciling conflicting uses. U.S. Forest Service. General Technical Report RM-120.

120. Lowrance , R. R., R. L. Todd, and L. E. Asmussen. 1984. Nutrient cycling in an agricultural watershed: I. phreatic movement. Journal of Environmental Quality 13:22-27.

121. Lunde, R. E., and A. S. Harestad. 1986. Activity of little brown bats in coastal forests. Northwest Science 60:206-209.

122. Lytjen, D. 1998. Ecology of woody riparian vegetation in tributaries of the Upper Grande Ronde River Basin, Oregon. Thesis Oregon State University, Corvallis, OR.

123. MacArthur, R. H. 1972. Geographical ecology. Harper & Row, New York, NY.

124. ———, J. W. MacArthur, and J. Peer. 1962. On bird species diversity. II. Prediction of bird census from habitat measurements. Ecology 39: 599-612.

125. Mahoney, J. M., and S. B. Rood. 1992. Response of a hybrid poplar to water table decline in different substrates. Forest Ecology and Management 54:141-156.

126. Marshall, D. B., M. W. Chilcote, and H. Weeks. 1996. Species at risk: sensitive, threatened, and endangered vertebrates of Oregon. Second edition. Oregon. Department of Fish and Wildlife. Portland, OR.

127. Martin, T. E. 1986. Competition in breeding birds: on the importance of considering processes at the level of the individual. Current Ornithology 4:181-210.

128. ———. 1988. Habitat and area effects on forest bird assemblages: is nest predation an influence? Ecology 69:74-84.

129. ———. 1988. On the advantage of being different: nest predation and the coexistence of bird species. Proceedings of the National Academy of Science 85: 169-2199.

130. ———. 1988. Processes organizing open-nesting bird assemblages: competition or nest predation? Evolutionary Ecology 2:37-50.

131. ———. 1992. Breeding productivity considerations: what are the appropriate habitat features for management? Pages 455-473 in: J. M. Haggan III and D. W. Jonston, editors. Ecology and conservation of neotropical migrant landbirds. Smithsonian Institution Press, Washington, D.C.

132. ———. 1993. Nest predation and nest sites: new perspectives on old patterns. Bioscience 43: 523-532.

133. ———. 1995. Avian life history evolution in relation to nest sites, nest predation, and food. Ecological Monographs 65:101-127.

134. ———. 1998. Are microhabitat preferences of coexisting species under selection and adaptive? Ecology 79: 656-670.

135. ———, and J. J. Roper. 1988. Nest predation and nest-site selection of a western population of hermit thrush. Condor 90:51-57.

136. Mayfield H. F. 1965. The Brown-headed cowbird with old and new hosts. Living Bird 4: 13-28.

137. ———. 1977. Brown-headed cowbird: agent of extermination? American Birds 31:107-113.

138. McBride, J. R., and J. Strahan. 1984. Establishment and survival of woody riparian species on gravel bars of an intermittent stream. American Midland Naturalist 112:235-245.

139. McComb, W. C., and J. C. Hagar. 1992. Riparian wildlife habitat literature review. Oregon Department of Forestry, Salem, OR.

140. ———, K. McGarigal, and R. G. Anthony. 1993. Small mammal and amphibian abundance in streamside and upslope habitats of mature Douglas-fir stands, western Oregon. Northwest Science 67:7-15.

141. McDowell, D. M., and R. J. Naiman. 1986. Structure and function of a benthic invertebrate stream community as influenced by beaver (Castor canadensis). Oecologia 68:481-489.

142. McGarigal, K., and W. C. McComb. 1992. Streamside versus upslope breeding bird communities in the central Oregon Coast Range. Journal of Wildlife Management 56:10-23.

143. ——— and ———. 1993. Research problem analysis on biodiversity conservation in western Oregon forests. U.S. Bureau of Land Management, Pacific Forest and Basin Rangeland Systems Cooperative Research and Technology Unit, Corvallis, OR.

144. McGinley, M. A., and T. G. Whitham. 1985. Central place foraging by beavers (Castor canadensis): a test of foraging predictions and the impact of selective feeding on the growth form of cottonwoods (Populus fremontii). Oecologia 66:558-562.

145. McIntosh, B. A., J. R. Sedell, J. E. Smith, R. C. Wissmar, S. E. Clarke, G. R. Reeves, and L. A. Brown. 1994. Management history of eastside ecosystems: changes in fish habitat over 50 years, 1935 to 1992. U.S Forest Service, General Technical Report PNW-GTR-321. Portland, OR.

146. Medin, D. E., and W. P. Clary. 1989. Small mammal populations in a grazed and ungrazed riparian habitat in Nevada, U.S. Forest Service, Intermountain Research Station Research Paper, INT-413.

147. ——— and ———. 1990. Bird and small mammal populations in a grazed and ungrazed riparian habitat in Idaho. U.S. Forest Service, Intermountain Research Station Research Paper. INT-425. Ogden, UT.

148. ——— and ———. 1991. Small mammals of a beaver pond ecosystem and adjacent riparian habitat in Idaho. U.S. Forest Service, Intermountain Research Station Research Paper. INT-445. Ogden, UT.

149. Melquist, W. E., and A. E. Dronkert. 1987. River otter. Pages 627-641 in: M. Novak, J. A. Baker, M. E. Obbard, and B. Malloch, editors. Wild furbearer management and conservation in North America. Ontario Ministry of Natural Resources, Toronto, Ontario, Canada

150. ——— and M. G. Hornocker. 1983. Ecology of river otters in west central Idaho. Wildlife Monograph 83.

151. Mitsch, J. M. and J. G. Gosselink. 1993. Wetlands, 2nd edition. Van Nostrand Reinhold, New York, NY.

152. Moore, F. R., S. A. Gauthreaux, P. Kerlinger, and T. R. Simmons. 1995. Habitat requirements during migration: important link in conservation. Pages 121-144 in: T. E. Martin and D. M. Finch, editors. Ecology and management of neotropical migratory birds: a synthesis and review of critical issues. Oxford University Press, New York, NY.

153. Murphy, M. T. 1983. Nest success and nesting habits of eastern kingbirds and other flycatchers. Condor 85:208-219.

154. Myers, J. P. 1983. Conservation of migrating shorebirds: staging areas, geographic bottlenecks, and regional movements. American Birds 37:23-25.

155. Naiman, R. J. 1988. How animals shape their ecosystems. BioScience 38:750-800.

156. ———, T. J. Beechie, L. E. Benda, D. R. Berg, P. A. Bisson, L. H. MacDonald, M. D. O'Conner, P. L. Olson, and E. A. Steel. 1992. Fundamental elements of ecologically healthy watersheds in the Pacific Northwest Coastal Ecoregion. Pages 127-188 in: R. J. Naiman, editor. Watershed management: balancing sustainability and environmental change. Springer, New York, NY.

157. ———, C. A. Johnston, and J. C. Kelley. 1988. Alteration of North American streams by beaver. BioScience 38:753-762.

158. ——— and H. Décamps. 1997. The ecology of interfaces: riparian zones. Annual Review of Ecology and Systematics 28:621-658.

159. ——— and J. M. Melillo. 1984. Nitrogen budget of a subarctic stream altered by beaver (Castor canadensis). Oecologia 62:150-155.

160. ———, ———, and J. E. Hobbie. 1986. Ecosystem alteration of boreal forest streams by beaver (*Castor canadensis*). Ecology 67:1254-1269.

161. National Research Council (US) Committee on restoration of Aquatic ecosystems-Science Technology and public policy. 1992. Restoration of Aquatic ecosystems. National Academy Press. Washington, D.C.

162. National Research Council—Committee on Characterization of Wetlands. 1995. Wetlands—characteristics and boundaries. National Academy Press, Washington, D.C.

163. Neff, D. J. 1957. Ecological effects of beaver habitat abandonment in the Colorado Rockies. Journal of Wildlife Management 21:80-84.

164. Newton, I. 1998. Population limitation in birds. Academic Press Inc., San Diego, CA.

165. Nilsson, C., G. Grelsson, M. Johansson and U. Sperens. 1989. Patterns of species richness along riverbanks. Ecology 70: 77-84.

166. Nobel, G. K. and P. G. Putnam. 1931. Observations on the life history of *Ascaphus truei*. Copeia 1931:97-101.

167. Novak, M. 1987. Beaver. Pages 282-313 in M. Novak, J. A. Baker, M. E. Obbard, and B. Malloch, editors. Wild furbearer management and conservation in North America. Ontario Ministry of Natural Resources, Toronto, Ontario, Canada.

168. Oakley, A. L., J. A. Collins, L. B. Everson, D. A. Heller, J. C. Howerton, and R. E. Vincent. 1985. Riparian zones and freshwater wetlands. Pages 57-80 in E. R. Brown, technical editor. Management of wildlife and fish habitats in forests of western Oregon and Washington. U.S. Forest Service Publication No. R6-F&WL-192.

169. Ohmart, R. D. 1994. The effects of human induced changes on the avifauna of western riparian habitats. Pages 273-285 in: J. R. Jehl. Jr., and N. K. Johnson, editors. A century of avifaunal change in western North America. Studies in Avian Biology 15.

170. ———. 1996. Historical and present impacts of livestock grazing on fish and wildlife resources in western riparian habitats. Pages 246-279 in: P. R. Krausman, editor. Rangeland wildlife. Society for Range Management, Denver, CO.

171. ——— and B. W. Anderson. 1982. North American desert riparian ecosystems. Pages 433-479 in: G. L. Bender, editor. Reference handbook on the deserts of North America. Greenwood Press, Westport, CT.

172. ——— and ———. 1986. Riparian habitat. Pages 169-199 in A.Y. Cooperrider, R. J. Boyd, and H. R. Stuart, editors. Inventory and monitoring of wildlife habitat. U.S Bureau of Land Management, U.S. Government Printing Office, Washington, D.C.

173. Oliver, C. D., and T. M. Hinckley. 1987. Species, stand structures, and silvicultural manipulation patterns for the streamside zone. Pages 259-276 in: E. O. Salo and T. W. Cundy, editors. Streamside management: forestry and fisheries interactions. Contribution 57, Institute of Forest Resources, University of Washington, Seattle, WA.

174. Oregon Department of Fish and Wildlife. 1996. Sensitive, threatened and endangered vertebrates in Oregon. DBM 3/18/96.

175. Oring, L. W., and J. M. Reed. 1998. Shorebirds of the Western Great Basin of North America: overview and importance to continental populations. International Wader Studies 9: 6-12.

176. Otting, N. J. 1999. Ecological characteristics of montane floodplain plant communities in the upper Grande Ronde Basin, Oregon. Thesis, Oregon State University, Corvallis, OR.

177. Page, G. W., and R. E. Gill, Jr. 1994. Shorebirds in western North America: late 1800s to late 1900s. Studies in Avian Biology 5:147-160.

178. Pastor, J., R. J. Naiman, B. Dewey, and P. McInnes. 1988. Moose, microbes, and the boreal forest. BioScience 38:770-777.

179. Perry, D. A. 1994. Forest ecosystems. The Johns Hopkins University Press, Baltimore, MD.

180. Pederson, R. J., A. W. Adams, and J. Skovlin. 1979. Elk management in Blue Mountain habitats. Oregon Department of Fish and Wildlife, Portland, OR.

181. Petranka, J. W. 1994. Response to impact of timber harvesting on salamanders. Conservation Biology 8: 302-304.

182. Platts, W. S. and R. L. Nelson. 1989. Characteristics of riparian plant communities and streambanks with respect to grazing in northeastern

Utah. Pages 73-81 in: R. E. Gresswell, B. A. Barton, and J. L. Kershner, editors. Riparian resource management: an educational workshop. Billings, MT.

183. Raedeke, K., editor. 1988. Streamside management: riparian wildlife and forestry interactions. Contribution 59. University of Washington, Institute of Forest Resources, Seattle, WA.

184. Reed, P. B. 1997. Revision of the 1988. National list of plant species that occur in wetlands: Northwest (Region 9). US Fish and Wildlife Service, Office of Biological Services, Washington, D.C.

185. Regier, H. A. and G. L. Baskerville. 1986. Sustainable redevelopment of regional ecosystems degraded by exploitive development. In: W. C. Clark and R.E. Munn, editors. Sustainable development of the biosphere. Cambridge University Press, Cambridge, England.

186. Rice, J. B., W. Anderson, and R. D. Ohmhart. 1980. Seasonal habitat selection by birds in the lower Colorado River valley. Ecology 61:1402-1411.

187. Rhodes, J., C. M. Skau, D. Greenlee, and D. Bown. 1985. Quantification of nitrate uptake by riparian forests and wetlands in an undisturbed headwaters watershed. In: R.. Johnson, C. D. Ziebell, D. R. Patton, P. F. Ffolliott and R. H. Hamre, coordinators. Riparian ecosystems and their management: reconciling conflicting uses. First North American Riparian Conference. U.S. Forest Service General Technical Report RM-120. Fort Collins, CO.

188. Ricklefs, R. E. 1969. An analysis of nesting mortality in birds. Smithsonian Contributions in Zoology 9.

189. Roath, L. R. and W. C. Krueger. 1982. Cattle grazing influences on a mountain riparian zone. Journal of Range Management 35:100-104.

190. Robinson, S. K., S. I. Rothstein, M. C. Brittingham, L. J. Petit, and J.A. Grzybowski. 1995a. Ecology and behavior of Brown-headed Cowbirds and their impact on host populations. Pages 428-460 in: T. E. Martin and D. M. Finch, editors. Ecology and management of neotropical migratory birds. Oxford University Press, New York, NY.

191. ———, F. R. Thompson III, T. M. Donovan, D. R. Whitehead, and J. Faaborg. 1995b. Regional forest fragmentation and the nesting success of migratory birds. Science 267: 1987-1990.

192. Rood, S. B., and J. M. Mahoney. 1990. Collapse of riparian poplar forests downstream from dams in western prairies: probable causes and prospects for mitigation. Environmental Management 14:451-464.

193. Rossman, D. A, N. B. Ford, and R. A. Siegel. 1996. The garter snakes: evolution and ecology. University Oklahoma Press, Norman, OK.

194. Rothstein, S. I., J. Verner, and E. Stevens. 1984. Radio-tracking confirms a unique diurnal pattern of spatial occurrence in the parasitic brown-headed cowbird. Ecology 65:77-88.

195. Rudemann, R., and W. J. Schoonmaker. 1938. Beaver dams as geologic agents. Science 88:523-525.

196. Saab, V. A., and C. Groves. 1992. Idaho's migratory landbirds: description, habitats, and conservation. Idaho Wildlife 12: 1-26.

197. Sanderson, G. C. 1987. Raccoon. Pages 487-499 in: M. Novak, J. A. Baker, M. E. Obbard, and B. Malloch, editors. Wild furbearer management and conservation in North America. Ontario Ministry of Natural Resources, Toronto, Ontario, Canada.

198. Schroeder, R. L., and A. W. Allen. 1992. Assessment of habitat of wildlife communities on the Snake River, Jackson, Wyoming. U.S. Fish and Wildlife Service, Resource Publication 190, Washington, D.C.

199. Sedell, J. R., P. A. Bisson, F. J. Swanson, and S. V. Gregory. 1988. What we know about large trees that fall into streams and rivers. In: From the forest to the sea, a story of fallen trees, C. Maser, R. F. Tarrant, J. M. Trappe, and J. F. Franklin, editors. U. S. Forest Service General Technical Report PNW-229. Pacific Northwest Research Station, Portland, OR.

200. Sedell, J. R., Leone, F. N., and W. S. Duval, 1991. Water transportation and storage of logs. American Fisheries Society Special Publication 19:325-367.

201. Sedgewick, J. A., and F. L. Knopf. 1992. Describing willow flycatcher habitats: scale perspectives and gender differences. Condor 94:720-733.

202. Shuford, W. D., G. W. Page, and J. E. Kjelmyr. 1998. Patterns and dynamics of shorebird use of California's Central Valley. Condor 100:227-224.

203. Shultz, T. T., and W. C. Leininger. 1991. Nongame wildlife communities in grazed and ungrazed montane riparian sites. Great Basin Naturalist 51:286-292.

204. Sigafoss, R. S. 1964. Botanical evidence of floods and floodplain deposition. U.S. Geological Survey. Professional Paper 485-A.

205. Singer, F. J., L. C. Zeigenfuss, R. G. Cates, and D. T. Barnett. 1998. Elk, multiple factors, and persistence of willows in national parks. Wildlife Society Bulletin 26:419-428.

206. Skagen, S. K. 1997. Stopover ecology of transitory populations: the case of migrant shorebirds. Ecological Studies 125:244-269.

207. Skovlin, J. M. 1984. Impacts of grazing on wetlands and riparian habitat: a review of our knowledge. Pages 1001-1104 in: Committee on developing strategies for rangeland management. National Research Council and National Academy of Science. Westview Press, Boulder, CO.

208. Smith, B. H. 1980. Not all beaver are bad: or an ecosystem approach to stream habitat management, with possible software applications. Pages 32-37 in: Proceedings 15th Annual. Meeting American Fisheries Society, Colorado-Wyoming Chapter, Fort Collins, CO.

209. Stebbins, R.C. and N.W. Cohen. 1995. A natural history of amphibians. Princeton University Press, Princeton, NJ.

210. Stevens, L., B. T. Brown, J. M. Simpson, and R. R. Johnson. 1977. The importance of riparian habitat to migrating birds. Pages 156-164 in: R. R. Johnson and D. A. Jones, editors. Importance, preservation, and management of riparian habitat: a symposium. U.S. Forest Service. General Technical Report Rep. RM-166.

211. St. John, A. D. 1985. The herpetology of the Owyhee River drainage, Malheur County, Oregon. Oregon Department of Fish and Wildlife, Technical Report 85-5-03.

212. Storm, R.M. 1986. Current status of Oregon amphibians and reptiles: a brief review. Appendix 8 in Oregon nongame wildlife management plan. Oregon Department of Fish and Wildlife, Portland, OR.

213. Stromberg, J. C. and D. T. Patten. 1992. Mortality and age of black cottonwood stands along diverted and undiverted streams in the eastern Sierra Nevada, California. Madrono 39:205-223.

214. Swanson, F. J., S. V. Gregory, J. R. Sedell, and A.G. Campbell. 1982. Land-water interactions: the riparian zone. Pages 267-291 in: Edmonds, R.L. editor. Analysis of coniferous forest ecosystems in the western United States. US/IBP Synthesis Series 14. Hutchinson Ross Publishing, Stroudsburg, PA.

215. Szaro, R. C. and S. C. Belfit. 1986. Herpetofaunal use of a desert riparian island and its adjacent scrub habitat. Journal of Wildlife Management 50:752-761.

216. ———, ———, K. Aitkin, and J. N. Rinne. 1985. Impact of grazing on a riparian garter snake. R .R. Johnston, technical coordinator. Riparian ecosystems and their management: reconciling conflicting uses. U.S. Forest Service. General Technical Report RM-120.

217. Tewksbury, J. A., S. J. Heil, and T. E. Martin. 1998. Breeding productivity does not decline with increasing fragmentation in a western landscape. Ecology 79: 2890-2903.

218. Thomas, J. W. 1979. Wildlife habitats in managed forests of the Blue Mountains of Oregon and Washington. U.S. Forest Service Agricultural Handbook Number 5532. Washington, D.C.

219. Thomas, J. W., C. Maser, and J. E. Rodiek. 1979. Riparian zones. Pages 40-47 in: J.W. Thomas editor. Wildlife habitats in managed forests: the Blue Mountains of Washington and Oregon. U.S. Forest Service, Agriculture Handbook Number 553. Washington, DC.

220. ———, ———, and ———. 1979. Wildlife habitats in managed rangelands: the great basin of southeastern Oregon. U.S. Forest Service. General Technical Report PNW-80. Portland, OR.

221. Trimble, S. W. 1994. Erosional effects of cattle on streambanks in Tennessee, USA. Earth Surface Processes and Landforms 19:451-464.

222. Tubbs, A. A. 1980. Riparian bird communities of the Great Plains. Pages 413-433 in: R. M. DeGraff and N. B. Tilghman, technical coordinators. Management of western forests and grasslands for nongame birds. U.S. Forest Service. General Technical Report INT-86. Ogden, UT.

223. U.S. Fish and Wildlife Service. 1995. Migratory nongame birds of management concern in the United States: the 1995 list. Office of Migratory Bird Management, Washington, D.C.

224. Unitt, P. 1987. Empidonax trailli extimus: an endangered subspecies. Western Birds 18:137-162.

225. Van Cleaf, J. S. 1885. How to restore our trout streams. American Fisheries Society. 14:51-55.

226. Vannote, R. L. G. W. Minshall, K. W. Cummins, et al. 1980. The river continuum concept. Canadian Journal of Fisheris and Aquatic Sciences. 37:130-137.

227. Verner, J., and L. V. Ritter. 1983. Current status of the brown-headed cowbird in the Sierra National Forest. Auk 100:355-368.

228. Warren, P.L. and C. R. Schwalbe. 1985. Herpetofauna in riparian habitats along the Colorado River in Grand Canyon. R. R. Johnston, technical coordinator, Riparian ecosystems, and their management: Reconciling conflicting uses. U.S. Forest Service General Technical Report RM-120. Ft. Collins, CO.

229. Yong, W., D. M. Finch, F, R, Moore, and J. F. Kelly. 1998. Stopover ecology and habitat use of migratory Wilson's Warblers. Auk 115:829-842.

15

Wildlife of Coastal and Marine Habitats

Joseph B. Buchanan, David H. Johnson, Eva L. Greda,
Gregory A. Green, Terence R. Wahl, & Steven J. Jeffries

Introduction

The marine environment contains perhaps the most dynamic collection of habitats in the Pacific Northwest. The habitats included in this chapter—coastal dunes and beaches, bays and estuaries, coastal headlands and islets, nearshore, inland marine deeper waters, marine shelf, and oceanic waters—are uniquely moderated by movement of marine water. Coastal upwellings, tides, currents, and waves all influence habitat use by wildlife.

The total area of individual habitats covered in this chapter varies substantially. Although estuaries, beaches, dunes, and headlands occupy comparatively smaller areas than other habitats, the area of marine water (out to 200 miles offshore) is obviously immense (about 65.5 million acres [26.5 million ha]). Many of the habitats are remote areas—coastal headlands, marine shelf, and oceanic waters—with low levels of human activity, whereas others (beaches, estuaries, inland marine deeper waters) are important commercial areas or popular recreation sites. At least 284 regularly occurring species (219 birds, 58 mammals, 7 reptiles, and 3 amphibians) are associated with these habitats.

In this chapter we present information useful to those with an interest in the ecology, conservation, and management of the marine environment. We begin by generally describing the habitats and features of the habitats important to wildlife. After a summary of information on species richness, adaptations of marine wildlife, and threats to the wildlife community, we describe the distribution and abundance of representative species in the different habitats. Acknowledging the importance of oceanic warming, we describe the phenomenon of El Niño/Southern Oscillation events (and oceanic warming in general), and conclude with a section on conservation issues in the marine environment.

Habitat Descriptions

Estuaries represent the dynamic interface of riverine and marine systems. The complex ecological processes that occur in estuaries are beyond the scope of this chapter and are described elsewhere.[54, 118, 158, 161] There are four commonly recognized types of estuaries in the coastal Pacific Northwest: bar-built estuaries develop where accumulated sand restricts the flow of water to the ocean and thereby influences sediment transport (e.g. Willapa Bay, Netarts Bay); blind estuaries occur where sediments close the outlet to the ocean, primarily during summer months (e.g., Elk River, Sixes River); drowned river estuaries occur where a rise in sea-level floods a former river valley (e.g. Coos Bay, Yaquina Bay); many Puget Sound estuaries are considered fjord estuaries due to the glacial origin of Puget Sound.[158] The major estuaries in Washington and Oregon are indicated in Figure 1.

Salt marsh is associated with many of the estuaries, although tremendous amounts of salt marsh habitat have been either destroyed or degraded through diking and conversion to agriculture or urbanization, primarily during the late nineteenth and twentieth centuries.[19, 50] For example, approximately 90% of the salt marsh in Coos Bay was destroyed between about 1880 and 1980.[158] Similarly, about 60% of the salt marsh in Puget Sound was lost following European settlement of that area.[158] To a large extent, high quality (native) salt marsh is currently found only in very limited areas.

Vast intertidal mudflats are present in the protected bays and inlets where sediment carried by the riverine system is deposited. Some bays, particularly Coos Bay, Grays Harbor and sites in Puget Sound, are important shipping ports, and shipping channels in some of these areas are maintained by regular dredging of accumulated sediments; small islands within the Columbia River mouth, Grays Harbor, and Coos Bay, for example, have been created or altered by deposited dredge spoils.

The conditions in bays and estuaries, and along their shorelines, are quite variable, typically reflecting each site's unique geological and disturbance history. Akins and Jefferson[1] recognized eight types of salt marsh in Oregon estuaries. These types vary as a function of marsh elevation, slope, substrate, drainage, and vegetation cover. These conditions influence the species composition associated with a particular salt marsh. Highly disturbed sites often lack or have much reduced salt marsh areas, an important habitat component of estuarine ecosystems. Some salt marshes are important habitat areas for roosting shorebirds and gulls, and are also used as haul-out areas (e.g. in Washington's Hood Canal) by harbor seals. Estuarine tide flats and salt marshes are important hunting grounds for a variety of raptor species including falcons, buteos, harriers, and owls. These birds benefit from the presence of drift logs, used as hunting or resting perches.

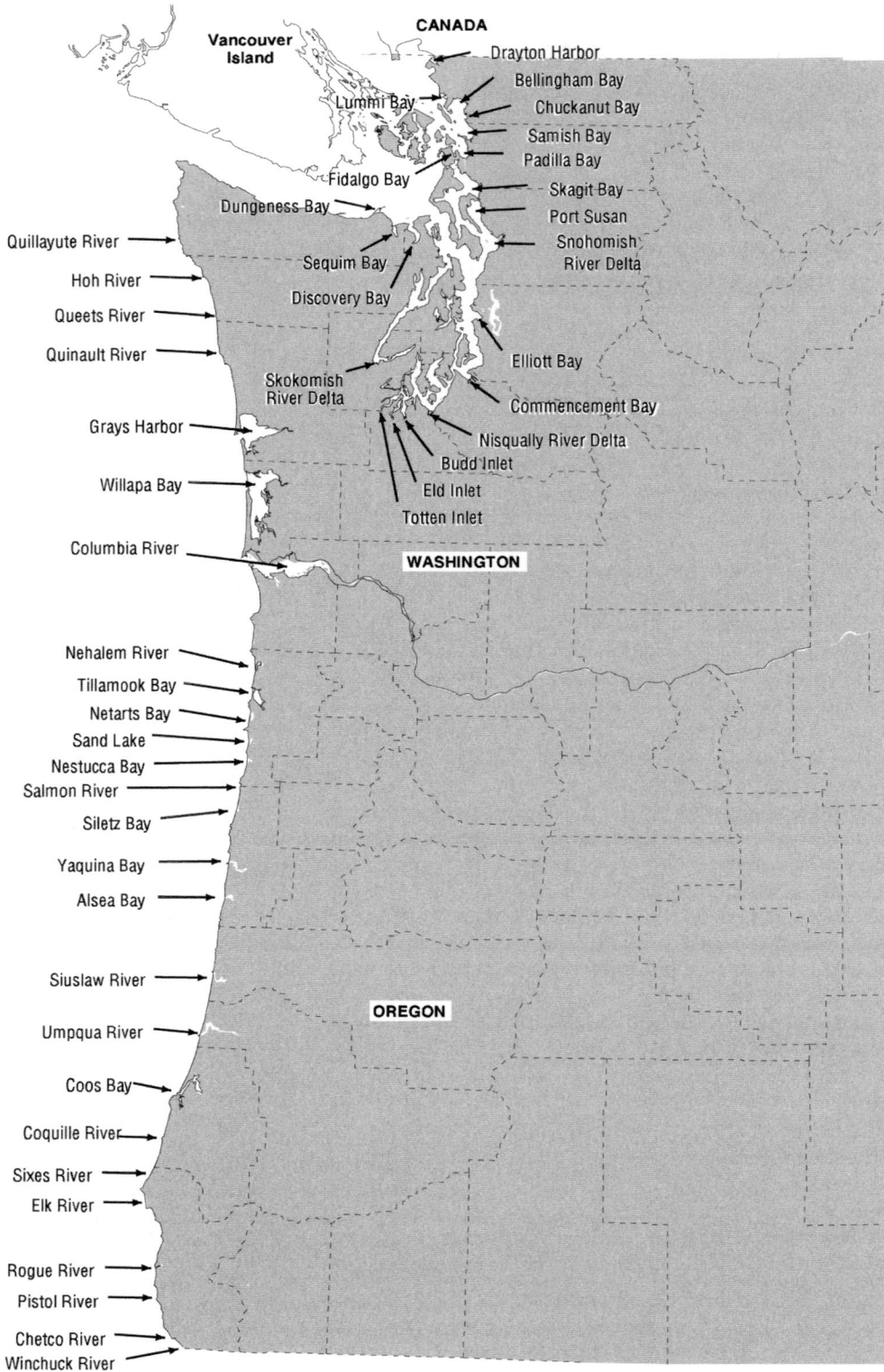

Figure 1. Location of estuaries in the marine regions of Washington and Oregon. Note that some estuaries (Elliott Bay, Commencement Bay, and Budd Inlet in Washington) have been largely converted to industrial or other human-development purposes and have therefore lost most estuarine function.

Beaches can vary from broad, relatively flat expanses of sand to narrow, fairly steep slopes dominated by pebbles or larger rocks. Pacific beaches are strongly influenced by long-shore drift, the seasonal currents that transport sand and sediment that physically alters the width or other characteristics of a beach.[103] The dominant feature of the beach is the sand or pebble substrate. Grain size can be quite variable within and among beaches; the most important areas for nesting, roosting, and foraging shorebirds are typified by fine-grained sand. In many areas, the top several cm of sand is obviously rich in invertebrate life and sustains large concentrations of overwintering and migrant sanderlings.[29, 120] Other than the actual substrate, the most prominent features of coastal beaches are the streams (or rivers) and drift logs. The streams and rivers contribute to the dynamic movement of sand and silt in this environment. Larger rivers carry huge amounts of silt from uplands and deposit the material in embayments or offshore. Much of this material is eventually transported ashore by currents and wave action.[103] Small watercourses carry less sediment but often create new channels through low dunes and influence local erosion. These small streams appear to influence the distribution of shorebirds along the beach[29] by providing a source of fresh water important for bathing.

Large numbers of drift logs wash up on beaches, particularly during storms. These logs often persist on beaches for years until they are carried away or moved to higher dunes by strong storms; buried in sand; or used as firewood, either on site or after being cut and removed. Current amounts of drift logs on beaches tend to be far lower than were present decades ago.[79] Drift logs on beaches are used as foraging and resting perches by falcons. Very small pieces of wood are occasionally used as wind breaks by roosting shorebirds, particularly the snowy plover, in windy weather.

Situated immediately landward of the beaches, sand dunes are found along 307 of the 764 km (40%) of outer coastline from Cape Flattery, Washington, to the California border.[192] There are 3 dune locations in Washington and

26 in Oregon (see Figure 2). Some dunes are associated with rivers or bays whereas others are on broad coastal plains or terraces.[192] Many dunes are vegetated with European beach grass, an exotic species used to stabilize dune movement.[194] Vegetation associated with dunes can be quite diverse, with 21 plant communities identified.[192] Most dunes in Washington rise no more than 4-6 m, whereas some dune systems in Oregon attain heights over 25 m above the interdune base.[192] In some areas, lower foredunes develop in the areas between the main body of dunes and beaches (seaward) or salt marsh (bayward). With the exception of Clatsop Spit, Oregon, where rapid progradation was responsible for the creation of foredunes (A. Wiedemann, pers. comm.), the foredunes along the coast are associated with stabilizing vegetation. In fact, Cooper[47] found no foredunes along the coast prior to the establishment of European beachgrass. Very small dune formations (embryo dunes) occur seaward of larger dunes (e.g. Leadbetter Point, Washington) although it should be noted that these dunes are not persistent (i.e., they are regularly destroyed by wave action) and are not considered foredunes.[193]

The dominant features of the coastal dunes, other than the shape and function (i.e., amount and direction of movement[192]), vary depending on the vegetation community present. Typically, the vegetation has a strong herbaceous, grass or shrub component.[76, 192] Shrubs are important habitats for overwintering passerines such as fox sparrow and yellow-rumped warbler. Raptors such as the northern harrier use trees as perch sites. Small mammals and amphibians and reptiles associated with coastal dunes find protective cover for foraging and denning in and under large drift logs deposited by winter storms. Such logs also are used as resting and feeding sites by falcons.

Another common feature of coastal dune systems, particularly those influenced by the introduction of European beachgrass, is the presence of very small, often secluded pools formed by storm- or rainwaters. The use of these pools by wildlife has not been well documented, although they are known to be used by waterfowl and certain shorebirds (e.g. greater yellowlegs; J. Buchanan, unpubl. data).

Each of the habitats above is influenced by the movement of water, either through periodic storms or the regular pattern of the tidal cycle. Beaches and dunes can be scoured or displaced by regular currents or wave action during particularly severe storms. Similarly, mudflats can also be scoured by storm runoff or regular strong tidal currents. Sand beaches can recede or accrete depending on the placement of new sediments in the system and the strength of currents.[103] Salt marsh progradation is occurring in some coastal estuaries, but in the region as a whole has occurred at a lesser rate than salt marsh loss.[158]

Beaches and mudflats are alternately covered or exposed for hours at a time due to daily fluctuations in the region's tidal regime, which has both diurnal and semi-diurnal characteristics (there are generally two high and two low tides, of varying height, each day). In larger

California Gulls at Fort Canby State Park, Washington. (Photo by Terence R. Wahl)

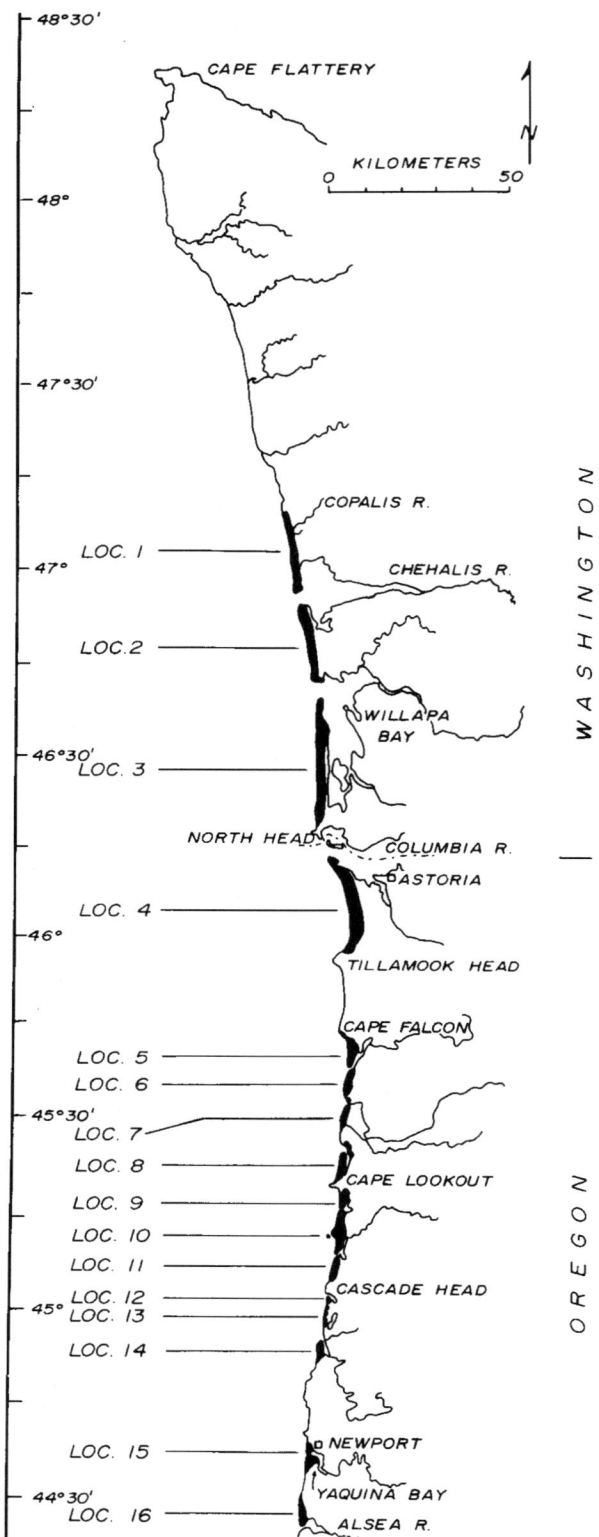

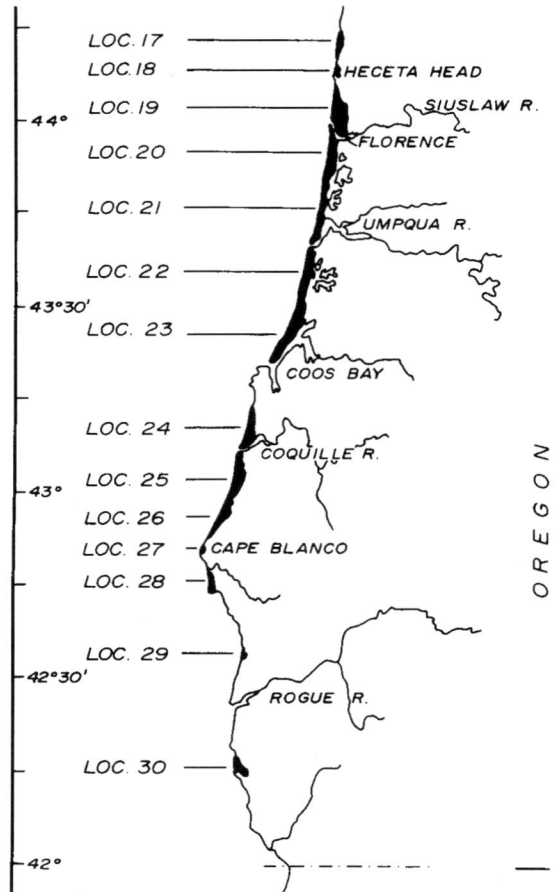

Figure 2. Distribution of dune locations in Washington and Oregon (map from Wiedemann;[192] see Cooper[47]).

estuaries such as Grays Harbor and Willapa Bay, >50 km² of tide flats may be exposed during low tides.

A number of human-made structures are used by wildlife associated with coastal beaches, dunes, bays, and estuaries. Perhaps the most permanent of these structures are the jetties used to maintain shorelines and harbor entrances. These sites are often used by roosting gulls and cormorants, and are important sites for rock sandpiper, surfbird, black turnstone, and ruddy turnstone. Shellfish beds in some areas are also used by wildlife such as turnstones, although the value of the areas is not clear. In fact, a recent study indicated that dunlins and western sandpipers were less common in bivalve management areas than in unmodified areas.[99]

Other structures used by a wide variety of species include buoys, channel markers, pilings, and docks. Buoys, channel markers, and docks are often used as haul-out sites by harbor seals or sea lions. These structures are also used as resting or foraging perches by a variety of birds including gulls, peregrine falcons, shorebirds, and pelagic cormorants (occasionally used for nesting). Pilings are used in some areas as high tide roost sites by black-bellied plovers and are used for nest sites by cavity-dependent species such as the purple martin.

The Coastal Headlands and Islets habitat includes the offshore rocks and islands, and coastal headlands and bluffs dominated by salal or evergreen huckleberry. Coastal headlands are defined as rocky cliffs of non-glacial origin on mainlands facing marine waters. Examples of coastal headlands include some of the San Juan Islands, the area between Cape Flattery and Point Grenville, and much of the coastline of Oregon. Some headlands are fronted by sand beaches, whereas others rise directly from open water. Most bluffs in Washington are fronted by cobble beach (areas of Puget Sound), while bluffs in Oregon rise from open water or beach. Headlands tend to be much taller (up to hundreds of meters) whereas bluffs tend to be smaller than about 20-50 m.

There are several key habitat features associated with coastal headlands. Most coastal headlands and the larger offshore islands have at least a thin layer of soil that supports grasses and evergreen shrubs, and in many areas, evergreen forest. It is these areas, primarily on offshore islands, but occasionally on mainland headlands, that support great breeding colonies of seabirds. Consequently, soil and vegetation properties that allow seabirds to dig long but shallow burrows are essential for many of these birds.[78] Most of the headlands and bluffs have very steep or vertical cliff faces. The areas are used by ledge-nesting seabirds and species such as the peregrine falcon. Other species, such as belted kingfisher and pigeon guillemot, nest in coastal bluffs where burrows can be excavated in the face of the escarpment.[66] Snags and large trees present on these headland faces are used as perch locations and nest sites by species such as the bald eagle. The shrub vegetation provides habitat conditions for ground- and shrub-nesting species such as winter wren and song sparrow.

Along much of the coastline there are rocky benches or accumulations of large boulders at the base of the headlands. Low-lying bench areas are used as foraging sites by rock-shoreline shorebirds and as haul-out sites by harbor seals. Higher elevation sites (those above the wave-splash zone) are used as nesting areas by the black oystercatcher.

Harbor Seals hauling out at low tide along Oregon coast. (Photo by Steven J. Jeffries)

The Nearshore Marine habitats include those marine water areas (high tide line to a depth of 66 ft [20 m]) along shorelines that are not significantly affected by freshwater inputs (i.e., excludes Bays and Estuaries). Adjacent to Marine Shelf, Inland Marine Deeper Waters, Bays and Estuaries, and a number of terrestrial-based habitats, Nearshore Marine habitat occurs in a mosaic with Coastal Headlands and Islets. A wide range of benthic habitats support marine organisms capable of withstanding short-term exposure to air. Larger kelps, such as *Macrocystis integrifolia* and *Nereocystis leutkeana* can be found in rocky-bottom subtidal areas, whereas rocky-bottom intertidal areas support smaller marine vegetation such as brown rockweed (*Pelvetiopsis limitata*) and surfgrass (*Phyllospadix scouleri*). Strongly influenced by tidal rhythms, wave action, storm events, and light penetration, the Nearshore Marine habitat is characterized by a high degree of patchiness resulting in differences in its faunal makeup and use. Examples of succession may be found in kelp forests where herbivory or the scouring action of tidal currents removes vegetation, and on rocky intertidal shores where wave action periodically disturbs established communities.

The Inland Marine Deeper Water habitat occurs in northwestern portions of Washington State and British Columbia, and does not occur in Oregon. This habitat includes waters (>66 ft [>20 m] deep) of the Strait of Georgia, Puget Sound, Hood Canal, and the Strait of Juan de Fuca. Bays and Estuaries, Coastal Headlands and Islets, and Marine Nearshore habitats adjoin Inland Marine Waters Habitat. This habitat further merges with the Marine Shelf habitat in the Strait of Juan de Fuca and is used extensively for navigation, commercial transport of goods, recreation, tourism, and fishery operations. Varying in specific composition and distribution with fluctuating physical factors, organics, silt, and sand are the primary substrate components of this habitat. Inland marine deeper waters do not support a significant number of plants due to light diffusion resulting from the depth of water involved. Seasonal variation in tidal regimes, precipitation and riverine discharges (winter highs), as well as periodic storm events cause changes in temperature, salinity, energy level, and erosion and accretion that effectively shape this habitat.

The Marine Shelf and Oceanic habitats are generally described topographically as the shelf (waters 20-200 m in depth), slope (200-2,000 m), and offshore or rise (greater than 2,000 m in depth) (Figure 3). The shelf can be further subdivided into inner (20-100 m) and outer 100-200 m) regions, especially since the oceanography and biology can change considerably closer to the shelf edge. Off Oregon and Washington the continental shelf varies in width from 15 to 40 km, and is relatively uniform compared to other coastal regions of North America. Nevertheless, there are numerous geomorphological features that influence the physical and biological processes. The continental shelf along Oregon is characterized by a series of oceanic banks, including Daisy, Stonewall, Perpetua, Heceta, and Siltcoos Banks, and a

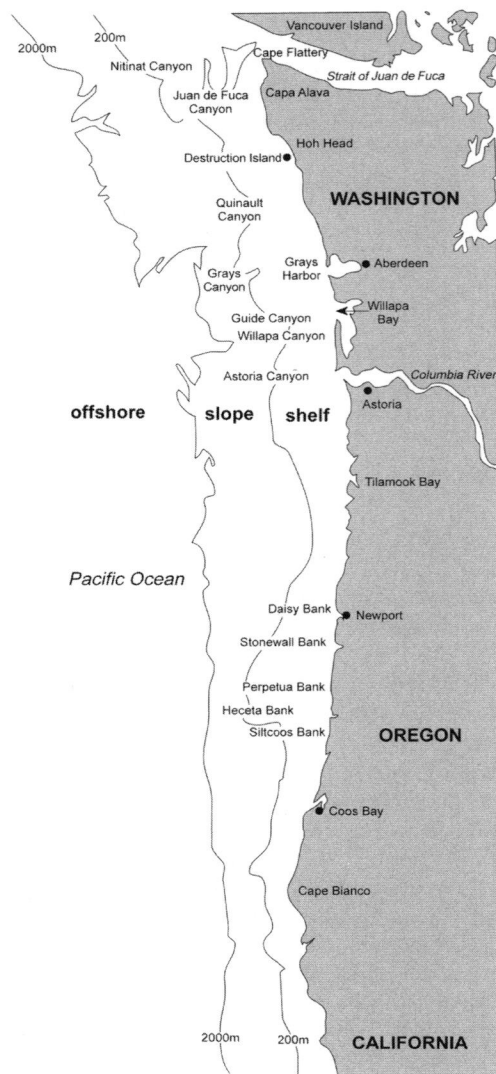

Figure 3. Major bathymetric features of the offshore region of Washington and Oregon.

major promontory, Cape Blanco. In contrast, the Washington shelf is furrowed by the Nitinat, Juan de Fuca, Quinalt, Grays, Guide, Willapa, and Astoria submarine canyons. The 20- to 25-km-wide Strait of Juan de Fuca, separating Vancouver Island from mainland Washington, is a glacially excavated channel that is the primary avenue for vigorous estuarine exchange between the coastal ocean and the inland marine waters of Washington and British Columbia. The Columbia River discharges through a relatively narrow and deep river mouth that defines the coastal boundary between Oregon and Washington.

Unlike California, and except for the southern part of Oregon, most of the region is not directly affected by the California Current system away from the more oceanic areas.[68, 108] Consequently, the nearshore biological productivity is primarily due to wind-driven coastal upwelling. In contrast, offshore nutrification results largely from water mixing due to the interaction of the California

Current waters with shelf edge geomorphological features. Coastal upwelling occurs most frequently in summer and fall, promoted by northerly and northwesterly winds. Upwelling intensity is typically greatest along the southern Oregon coast (the Cape Blanco upwelling zone). The Columbia River effluent amounts to 60% to 90% of the freshwater entering the Pacific Ocean between the Strait of Juan de Fuca and San Francisco Bay.

Marine wildlife occupying the outer shelf, slope, and offshore waters concentrate where deep and shallow waters mix. Some of the highest concentrations occur along the shelf slope influenced by the California Current, and the very edge of the outer shelf where coastal and current waters mix. These mixing zones are especially prominent at Heceta Bank where its unique bending shape and proximity to the Cape Blanco upwelling zone infuse the Bank with deepwater nutrients. Other important mixing zones occur along the headwalls of Washington's submarine canyons.

Additional oceanographic features include the Columbia River plume on the upper layers of the shelf and offshore waters (contributing to the "dilute domain"[68]) but while the effects of this on bird distribution are unknown,[23] the high concentration of freshwater within this plume may limit its attraction to marine life. This outflow extends northward along the Washington coast in winter and shifts to offshore and south in summer. More dramatic when encountered but not always present over the outer shelf or slope, the "blue water" boundary suggests oceanic water mass boundaries like the Subarctic Boundary across the North Pacific,[68] where water color, salinity, sea surface temperatures, and seabird species composition and abundance change dramatically;[180] these conditions should not be confused with warm water incursions. One noticeable change in seabird abundance at this boundary area, where albacore are often found, is the replacement of foraging fork-tailed storm-petrels by Leach's storm-petrels on the warmer, offshore side.[177] In addition, patterns of marine bird occurrence off Washington and Oregon probably reflect events or conditions in the Arctic, the southern hemisphere or elsewhere; reproductive success or failure and feeding opportunities at staging areas or during migration may determine populations of visiting species here.

Species Richness and Adapatations of Marine Wildlife

The marine habitats differ markedly in their attributes and their associated species. Evaluation of the information presented in Table 1 reveals a number of patterns of interest. Species richness was greatest in the bays and estuaries, followed by dunes and beaches, the marine nearshore zone, the marine shelf area, and the oceanic zone. Consistent with this pattern, there was a strong inverse relationship between species richness and the proportion of closely associated species found in the habitat. Not surprisingly, the oceanic habitat, with only 42 species, had the highest proportion (69%) of closely

Table 1. Species richness in the different marine habitats based on the degree of association. Because some species are associated with more than one habitat, totals within rows are not additive. Species of uncertain association (n = 9) and of historical occurrence (n = 1) were not included in the table.

Species Groups	Dunes/ Beaches	Headlands	Bays/ Estuaries	Inland Marine	Marine Nearshore	Marine Shelf	Oceanic
Closely Associated[a]							
Marine Mammals	0	3	2	5	6	11	10
Terrestrial Mammals	1	0	4	0	0	0	0
Birds	29	25	60	21	32	28	18
Reptiles	1	0	0	0	0	0	1
Amphibians	0	0	0	0	0	0	0
Total for Group	31	28	66	26	38	39	29
Generally Associated[b]							
Marine Mammals	2	0	1	2	1	5	1
Terrestrial Mammals	11	20	5	0	1	0	0
Birds	61	38	67	18	31	15	7
Reptiles	4	4	1	0	0	1	0
Amphibians	1	1	0	0	0	0	0
Total for Group	79	63	74	20	33	21	8
Present[c]							
Marine Mammals	0	0	2	2	2	1	2
Terrestrial Mammals	14	7	2	1	0	0	0
Birds	27	17	30	12	12	10	3
Reptiles	0	0	0	0	0	0	0
Amphibians	0	0	0	0	0	0	0
Total for Group	41	24	34	15	14	11	5
Total for Habitat	151	115	174	61	85	67	42

[a]Closely associated species: A species is widely known to depend on a habitat for part or all of its life history requirements. Identifying this association implies that the species has an essential need for this habitat for its maintenance and viability.

[b]Generally associated species: A species exhibits a high degree of adaptability and may be supported by a number of habitats. In other words, the habitats play a supportive role for its maintenance and viability.

[c]Present species: A species demonstrates occasional use of a habitat. The habitat provides marginal support to the species for its maintenance and viability.

associated species. Conversely, the bays/estuaries and dunes/beaches had the lowest proportions of closely associated species (38% and 20%, respectively). The two latter habitats also had the highest proportions of generally associated species (42% and 51%, respectively). These numbers make sense intuitively; generalists are less common in the oceanic and predominantly marine zones and are more prevalent in the more diverse habitats associated with the marine/upland interface.

The wildlife species associated with the marine environment have adapted to prevalent conditions along the coasts of Washington and Oregon. Certainly, the characteristics of migrant species are also uniquely influenced by conditions throughout the range of each species. All species are constrained in some way by biotic or abiotic conditions. Some of these constraints, and the associated ecological consequences and adaptations, are presented in Table 2.

Threats to Marine Habitats and Wildlife Communities

There are many threats to the species and habitats associated with the marine environment. Because an in-depth discussion of this topic would be too lengthy for this chapter, we have summarized the major issues and provided references so readers can locate additional pertinent information (Table 3).

In addition to the threats we summarized, a noteworthy threat is now emerging as a result of decades of human inability to co-exist with wildlife populations on a sustained basis. This threat is best described as the presence of conflicting management objectives. As efforts to manage wildlife populations become more difficult due to reduced or eliminated options, coupled with ever-increasing pressures from an expanding human population, the potential for management conflict becomes more likely and more significant. A current example is the issue of Caspian tern predation on salmon

Table 2. Examples of constraints, ecological consequences and animal adaptations in the marine environment.

Environmental Constraints	Ecological Consequences	Adaptations	
		Physiological/Structural	Behavioral
Temporal changes in prey availability (daily or seasonal); depleted prey populations	Higher energetic costs associated with avoiding predators, moving between foraging and roosting locations, searching for prey; density-dependent competition; mortality	Long-legged birds can forage in varying water depths to extend the foraging period; utilize lipid deposits to support emigration; fat storage as blubber; replenish lipid reserves and store as fat; use high-energy intake to fuel migration; body shape (and wing structure) adapted to avoid predators (e.g. shorebirds) and sustain long migratory movements	Nocturnal foraging allows greater access to tidal areas; periods of inactivity (roosting) to minimize energy expenditure; prey switching, habitat switching; minimize energetic costs by reducing time spent flying (e.g. western grebes)
Adverse weather conditions (flooding of foraging, roosting, or nesting areas; high winds)	Increased energetic demands; less time spent roosting, foraging, and other maintenance requirements; reduced foraging efficiency; failed reproduction	Utilize lipid deposits to support emigration; insulating plumage, pelage, and blubber	Temporal flexibility in reproductive timing
Spatially limited habitats	Greater physiological costs associated with migration; density-dependent competition		Strong seasonality in habitat use; excellent dispersal/ migration abilities; colonial breeding; breeding rookeries
Predation or predator density	Disturbance; increased energetic costs; increased vigilance and predator avoidance behavior suited to the habitat; mortality	Body shape and wing structure adapted to avoid predators; cryptic plumage	Balancing food intake and vigilance with predation risk; predator avoidance strategies effective in open habitat; flocking behavior enhances foraging behavior and predator detection; nest placement (on islands, burrow into soil); colonial breeding; breeding rookeries
Cold marine water temperatures	Warming adaptations required	Insulating plumage, pelage and blubber; small ears and robust body parts	
Marine waters as habitat	Required ability to subsist in marine environment	Tolerance of saline conditions; webbed feet and wing shape facilitate swimming under water by waterbirds; flippers	Ability of waterbirds to swim and obtain food under water surface

smolts near the tern nesting colony in the lower Columbia River estuary. The terns' appetite for salmon smolts concerns fisheries biologists and others engaged in management and recovery of salmon stocks. Potential solutions to the problem proposed by various parties have included tern population control and colony relocation. Certainly there is hope for successful colony relocation, given the species' known ability to respond to dynamic habitat changes. However, the take-home message is that a naturally established Caspian tern colony (although located on a dredge spoil island), one of the largest in the world, and therefore of great conservation significance, is in conflict with salmon conservation efforts. The solution, unfortunately, is active engagement and intervention.

Wildlife Communities

The species groups associated with marine habitats exhibit a great range of seasonal occurrence patterns in the region. With the exception of the leatherback sea turtle, all reptiles, amphibians, and terrestrial mammals are considered year-round residents. Similarly, most marine mammals are year-round residents; however, seven species (e.g. humpback whale, northern fur seal) occur only or primarily during spring/summer or fall/winter (Table 4). It is among the birds that the greatest variability in seasonal occurrence exists (Table 4). Although year-round residents are well represented (31% of the birds), they are outnumbered by winter residents (37%). Other species breed only or are exclusively migrants. A small number

Table 3. Summary of actual or potential threats to habitats or species in the marine environment.

Threat	Oceanic	Marine shelf	Marine nearshore	Inland marine	Bay/ Estuary	Coastal headland	Beach/ Dune	References
			Major Habitat Associations					*References*
Habitat Loss[a]								
Habitat Loss			X[b]	X[b]	X			19, 20
Habitat Degradation								
Habitat alteration					X		X	27, 194
Climate change	X	X	X	X	X	X	X	187
Exotic vegetation				X	X		X	27, 99, 125
Exotic invertebrates	?	?	?	X	X	?	?	27
Exotic vertebrates					X	X		78, 94
Pollution								
Agricultural					X			34, 49, 157,
Industrial/Municipal				X	X			34, 38, 100, 157,
Oil	X	X	X	X	X	X	X	27, 32, 73, 165, 100
Heavy metal					X		?	27, 34, 49, 3100
Plastic particle	X	X	X	X	X		X	27, 55, 149
Human Disturbance								
Pedestrian					X	X	X	27
Motorized vehicles				X	X	X	X	27
Non-motorized watercraft				X	X			27
Pets					X	X	X	27, 62
Hunting					X			27
Military activities						X		166
Mortality Factors								
Collisions with utility lines					X			27
Market hunting					X		X	46, 71, 83
Collisions with vehicles or watercraft				X	X		X	27
Gillnet and longline fisheries activities entanglement	X	X	X	X				57, 117, 126
Changes in predator or predation levels						X		144

[a]Habitat loss is here defined as permanent conversion of suitable habitat. Habitat degradation refers to a range of conditions, from slightly to completely degraded; by definition, it is possible to restore degraded habitats.

[b]Loss of reef structure on ocean bottom due to trawling (J. Hodder, personal communication).

of far ranging, non-breeding seabirds and marine mammals occur at various seasons, but do not occur year-round.

Habitat-associated patterns of seasonal occurrence are evident as well. Higher proportions of species are found in coastal dunes and coastal headlands during summer (year-round residents or summer only in Table 5) than in other habitats. With the exception of oceanic waters, in which 48% of the species occur in winter (year-round and winter columns in Table 5), all habitats support high proportions of species in winter. Bays/estuaries, inland marine deeper waters, and nearshore waters support the highest proportions of species which overwinter but do not breed locally (Table 5).

Many of the species found in these habitats engage in lengthy migrations or dispersal movements. With the exception of the leatherback sea turtle, all reptiles, amphibians, and terrestrial mammals generally move less than 1,000 km (Table 4). On the other hand, 69% of the marine mammals and 85% of the birds regularly travel distances greater than 1,000 km. In fact, the majority of strictly migrant species travel over 10,000 km (16% of the marine mammals and 23% of the birds travel over 10,000 km) (Table 4).

Coastal Dunes, Beaches, Bays, and Estuaries
The beach habitat supports very large numbers of vertebrates, and several areas in the region are recognized

Table 4. Summary of seasonal occurrence and migration/dispersal capability of birds and marine mammals in coastal and marine habitats of Washington and Oregon.

Seasonal Status			Migration/Dispersal Capability			Number of Species/
	<10 km & sedentary	<100 km	<1,000 km	<10,000 km	>10,000 km	Seasonal Category
Birds						
Year-round	7 (10%)	3 (4%)	16 (24%)	41 (61%)	0	67 (31%)
Summer only	0	0	2 (8%)	16 (64%)	7 (28%)	25 (12%)
Winter only	0	0	5 (6%)	65 (82%)	9 (11%)	79 (37%)
Transient	0	0	0	12 (33%)	24 (67%)	36 (17%)
Other	0	0	0	0	9 (100%)	9 (4%)
Total	7 (3%)	3 (1%)	23 (11%)	134 (62%)	49 (23%)	216
Marine Mammals						
Year-round	0	1 (8%)	5	6	0	12 (63%)
Summer only	0	0	0	1	3	4 (21%)
Winter only	0	0	0	3	0	3 (16%)
Transient	0	0	0	0	0	0
Other	0	0	0	0	0	0
Total	0	1 (5%)	5 (26%)	10 (53%)	3 (16%)	19

Table 5. Summary of seasonal residency of species associated with marine habitats in the Pacific Northwest.

Habitat	Number & Percentage of Species in each Residency Category					Notes
	Year-round	Summer only	Winter only	Spring/ Autumn only	Other	
Coastal Dunes and Beaches	80 (52%)	16 (10%)	36 (23%)	22 (14%)	0	Does not include 4 uncertain, one extirpated species
Coastal Headlands and Islets	68 (58%)	22 (19%)	19 (16%)	8 (7%)	0	Does not include 3 uncertain status species
Bays and Estuaries	64 (36%)	16 (9%)	65 (37%)	32 (18%)	1	Does not include 1 uncertain status species
Inland marine deeper waters	23 (37%)	5 (8%)	25 (40%)	9 (14%)	1	
Marine nearshore	27 (30%)	8 (9%)	37 (42%)	15 (17)	2	Does not include 1 uncertain status species
Marine shelf	25 (35%)	5 (7%)	20 (28%)	13 (18%)	9 (13%)	
Oceanic	17 (39%)	4 (9%)	4 (9%)	10 (23%)	9 (20%)	

as internationally important sites. For example, the northern coast of Oregon and the beaches adjacent to Grays Harbor and Willapa Bay support some of the highest densities of spring (472/km[120]) and autumn (214/km[31]) migrant sanderlings in North America. The three Washington beaches (Ocean Shores/Copalis Beach, Grayland Beach, North Beach [also known as Long Beach]) supported an average of 50.2, 43.3, and 46.6 sanderlings/km during winter between 1982-83 and 1989-90.[29] These beaches are also important areas for roosting shorebirds during daily high tides that inundate the nearby embayments. Winter counts of dunlins averaged 408.7, 464.4, and 618.7/km on the three beaches. North Beach is also an important area for roosting black-bellied plovers[29] (high count of 3848 in winter 1994-95).[31] Although many Oregon beaches support sanderlings, only a few (e.g., those adjacent to Bandon Marsh and Tillamook Bay) support other species of roosting shorebirds during higher

tides (J. Gilligan, pers. comm.). Additionally, the area between Coquille Point and Cape Blanco also supports flocks of roosting shorebirds, primarily during brief migratory stopovers (D. Lauten, pers. comm.). Ocean Shores/Copalis Beach, North Beach, and Clatsop/Sunset Beach (Oregon) qualify as sites of regional importance[62] in the Western Hemisphere Shorebird Reserve Network[84] (WHSRN). Eight beaches in the region are considered critical habitat for the threatened snowy plover.[172] Sand islands in large estuaries (e.g. Grays Harbor, Willapa Bay, Columbia River mouth) also support large breeding colonies of gulls and terns,[142, 167, 173] and are used as haulout sites by harbor seals (see below).

Estuarine sites in the region support substantial winter populations of Arctic- and sub-arctic-nesting shorebirds.[31, 141] Important wintering sites for shorebirds include Willapa Bay (>70,000 birds),[31] Grays Harbor (mean=about 20,000 between 1979 and 198),[141] Columbia River estuary

(20,000),[141] Port Susan Bay (>30,000),[65] Skagit Bay (>29,000),[65] and Coos Bay (11,000).[62, 141] The most abundant winter resident is the dunlin, which typically comprises about 90-95% of the winter shorebird community in western Washington.[31, 65] Other common winter resident shorebirds of estuaries (although often local in distribution) include black-bellied plover, greater yellowlegs, and western sandpiper.[30, 31, 65, 124, 141]

The region's estuaries are crucial links in the Pacific Flyway for migrating shorebirds, waterfowl, and other waterbirds. Grays Harbor is one of the three most important shorebird stopover sites on the Pacific Flyway south of Alaska; estimates of about 1 million birds were made in spring 1981.[88] High counts of >100,000 migrant shorebirds have been made at Willapa Bay[31] and the Columbia River estuary.[132, 133, 134, 135, 136, 137] Other WHSRN sites for shorebirds include Port Susan Bay (>50,000 in both spring and autumn migration), Skagit Bay (>20,000 in spring), Padilla Bay (>30,000 in spring[65]), and Tillamook Bay (nearly 100,000 in autumn 1979, although the site now apparently supports far fewer birds) and Coos Bay (perhaps 20,000 birds).[62, 84] The western sandpiper is by far the most abundant migrant in the region; other abundant species include dunlin and short-billed dowitcher. Black-bellied plover, semipalmated plover, red knot (primarily in Washington), sanderling, and greater yellowlegs are common during migration.[31, 65, 88, 139, 141, 191] At least 26 estuarine sites in Washington and Oregon have supported at least 4,000 shorebirds during the non-breeding period (see Figure 4) and 22 additional sites have supported at least 1000 shorebirds.[62]

The most abundant waterfowl restricted primarily to marine water habitats are diving ducks. Based on systematic aerial surveys in Puget Sound, these include scoters (primarily surf scoter and white-winged scoter), bufflehead, and goldeneyes (Barrow's goldeneye and common goldeneye).[127] Densities (number of birds/km^2) of these Arctic- and sub-arctic-nesting species in the nearshore zone can be quite high and vary substantially among years (Table 6). In fact, annual variability in abundance (density) is evident in all marine birds in the

Spring migrant shorebirds, primarily Western Sandpipers, at high tide roost, Bowerman Basin, Grays Harbor, Washington. (Photo by Joseph B. Buchanan)

Harbor Seals hauled out at low tide, Gertrude Island, Washington. (Photo by Steven J. Jeffries)

zone nearest shore (Table 6), and density can vary substantially over broad areas such as Puget Sound (Figures 5, 6).

Because of their shallow waters and extensive intertidal sand bars and mud flats, estuaries provide excellent habitat for harbor seals. This habitat provides numerous low tide haulout areas along channels and exposed tide flats which are used year round as resting areas. Harbor seals also select many of the larger estuaries (e.g. Tillamook Bay, Willapa Bay, Grays Harbor and Padilla Bay) as preferred haulout locations during pupping season. Estuaries provide a variety of important prey for harbor seals ranging from flatfishes to salmon. In addition to harbor seals, some estuaries such as Grays Harbor provide excellent shallow water feeding areas for gray whales migrating through Oregon and Washington coastal waters.

Inland Marine Deeper Water

Marine Birds. A relatively low number of species utilize the deeper inland marine waters. These are mostly mid-water or surface feeders, often found in large flocks which concentrate at convergences or "tide rips" in channels and passages, in offshore, deeper waters of large embayments, and near baitfish schools in the Strait of Juan de Fuca. Opportunistic gulls make little apparent distinction between deep water (>20 m depth) and nearshore (<20 m) waters (see CD-ROM in this book [Wildlife Habitats matrix]).

Some birds using deep waters often commute considerable distances between night roosts on rocks and islands or water "roost" areas. Cormorants and common murres do this in large numbers. Cormorants must roost ashore nocturnally and for periods during daylight hours. Daily movements between roosts along the Strait of Juan de Fuca and Rosario Strait and the interior channels and bays in the San Juan Islands and Whatcom, Skagit and Island counties, for example, are consistently observed. Cormorants are limited by nesting colony site requirements during summer, and are not found much south of Admiralty Inlet, unlike the winter pattern when numerous cormorants move into all portions of Puget Sound.

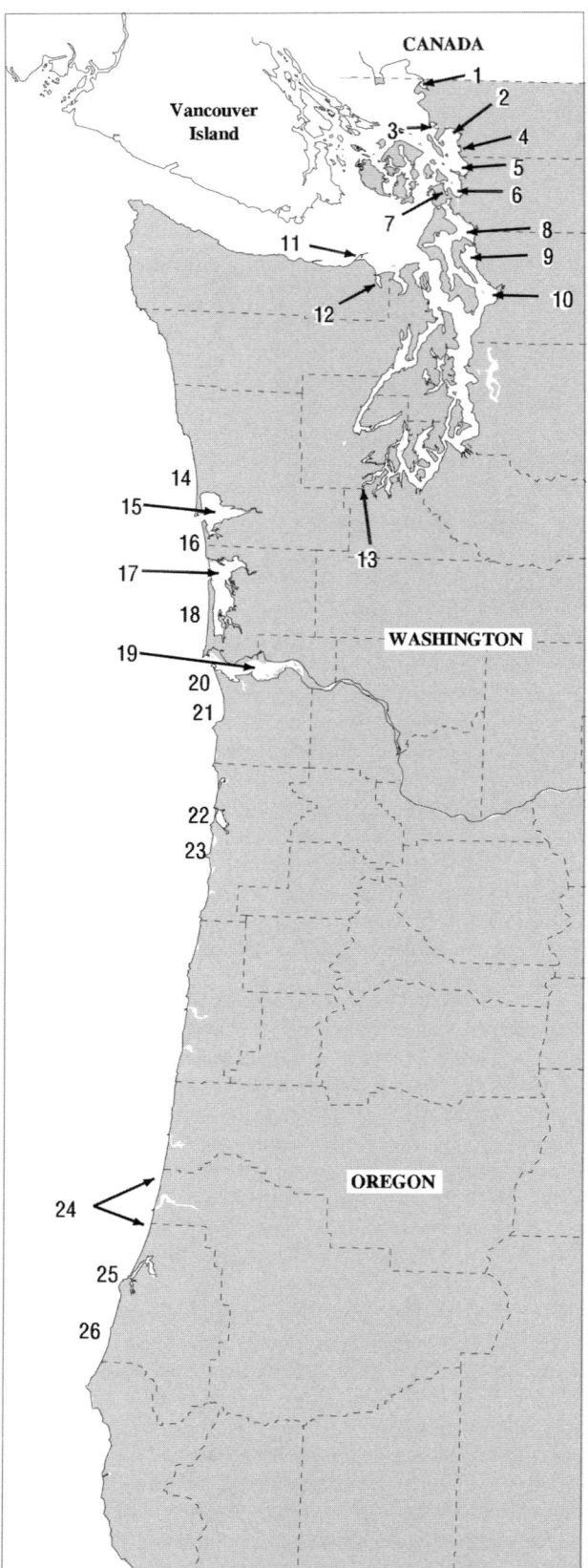

Figure 4. Location of estuarine and beach sites that have supported >4,000 shorebirds in at least one season.[62] *Sites referenced by numbers on map are: 1) Drayton Harbor, 2) Bellingham Bay, 3) Lummi Bay, 4) Chuckanut Bay, 5) Samish Bay, 6) Padilla Bay, 7) Fidalgo Bay, 8) Skagit Bay, 9) Port Susan Bay, 10) Snohomish River delta, 11) Dungeness Bay, 12) Sequim Bay, 13) Totten Inlet, 14) Ocean Shores/ Copalis Beach, 15) Grays Harbor, 16) Grayland Beach, 17) Willapa Bay, 18) North Beach, 19) Columbia River estuary, 20) Clatsop Beach, 21) Sunset Beach, 22) Tillamook Bay, 23) Netarts Bay, 24) Oregon Dunes Beach, 25) Coos Bay, and 26) Coquille River estuary and vicinity, and beach south to Cape Blanco.*

Table 6. Densities (birds/km²) of various waterfowl species, species groups, and all marine waterbird species combined, in the Bays and Estuaries of Puget Sound during winter (December to February, 1992-93 to 1998-99).*

Species or Species Group	Median Density	Range
Goldeneye (2 species)	21.97	17.27-38.34
Scaup (2 species)	20.88	13.62-29.05
Scoter (3 species)	56.34	53.56-70.43
Bufflehead	44.38	34.12-64.32
Diving Ducks	165.11	144.45-239.86
Dabbling Ducks	233.78	69.15-722.48
All Marine Birds	621.82	335.84-1284.93

* Data from Nysewander and Evenson[128])

Species diversity and populations are larger in winter than in summer. Resident species are joined in winter by large numbers of birds from coastal areas to the north and south and from inland North America. Briggs et al.[23] point out that relatively high densities of wintering common murres in inland marine waters of Washington may reflect response to seasonally rougher sea conditions (and poorer foraging possibilities) over the shelf. Seasonal numbers vary between years, with populations of species breeding on the ocean coast—Brandt's cormorant, common murre and ancient murrelet—particularly subject to oceanographic conditions. Likewise, winter feeding conditions here can affect populations returning to breeding areas.

Population status and change in inland waters is incompletely known and apparent changes lack explanation. Aerial surveys comparing present-day populations with 1978-79 characterizations of waters north of Puget Sound[183] and surveys of areas south of that in the early 1980s (Washington Department of Fish and Wildlife, unpubl. data) indicate that scoters and western grebes have declined as much as 50% or more in some areas.[146] Data from Christmas Bird Counts and unquantified observations suggest similar changes and also apparent increased seasonal abundances of phalaropes and above-average occurrences of offshore species like storm-petrels and auklets. These apparent patterns could reflect ocean productivity failure in recent years.

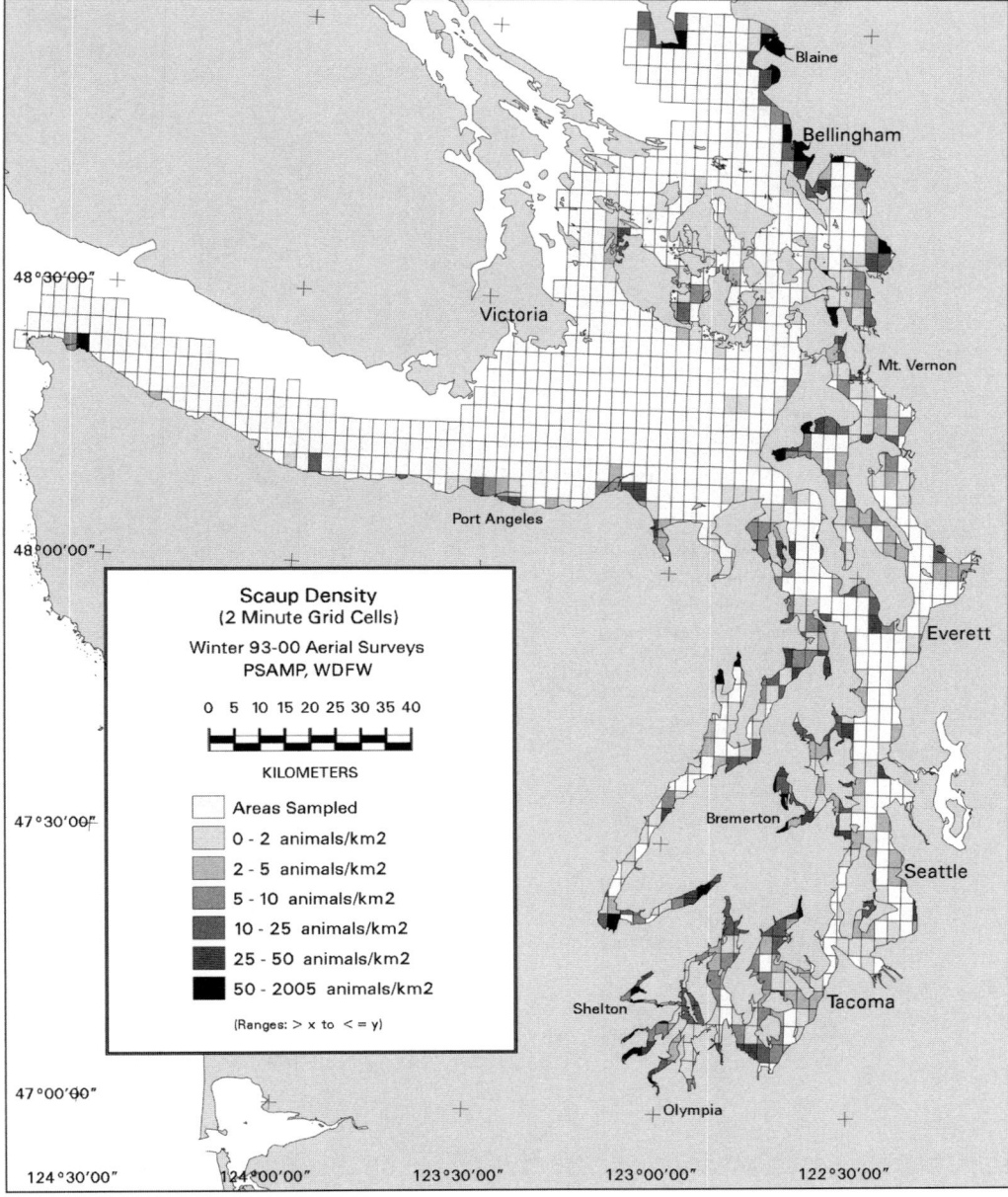

Figure 5. Density of scaups in Puget Sound, Washington, derived from aerial transect counts. Data and figure from Nysewander and Evenson.[128]

In summer, overall populations are relatively low in the inland marine deeper waters. Cormorants, gulls, and alcids concentrate in active tidal convergence areas like northern Admiralty Inlet, off Point-No-Point, in the Narrows near Tacoma, Case and Carr inlets, in Rosario, Haro and Georgia straits, Deception Pass, and in the San Juan Islands passages like San Juan and Spieden channels.[183] Numbers of birds in Puget Sound, Hood Canal, and waters east of Whidbey Island in summer 1982 were very low.[185] Deep waters are especially important in the eastern Strait of Juan de Fuca and Admiralty Inlet where a significant proportion of the Washington-Oregon breeding population of rhinoceros auklets forages.[186] The auklet concentration in summer results in Admiralty Inlet numbers peaking then,[183] unlike other inland locations where summer populations are the lowest of the year. Rhinoceros auklets and variable numbers of non-breeding

Brandt's cormorants and common murres comprise nearly all of the seabird community in the area during summer.

There are no sizable "natural" marine bird colony sites south of the eastern end of the Strait of Juan de Fuca. Human populations and development preclude the establishment of gull and tern colonies except at a few human-built locations, including abandoned piers and building roofs. There are relatively few scattered cormorant and pigeon guillemot nests.[66, 167] Nesting "deep water" species are confined almost exclusively to Protection and Smith Islands in the eastern Strait of Juan de Fuca. Population estimates prior to 1989 included 35,000 rhinoceros auklets and a small number of tufted puffins. Pre-1990s numbers of other nesting species which forage in both nearshore and deeper waters were 10,600 cormorants, 20,000 gulls and 2,400 pigeon guillemots.[167] Except for rhinoceros auklets, inland marine nesting

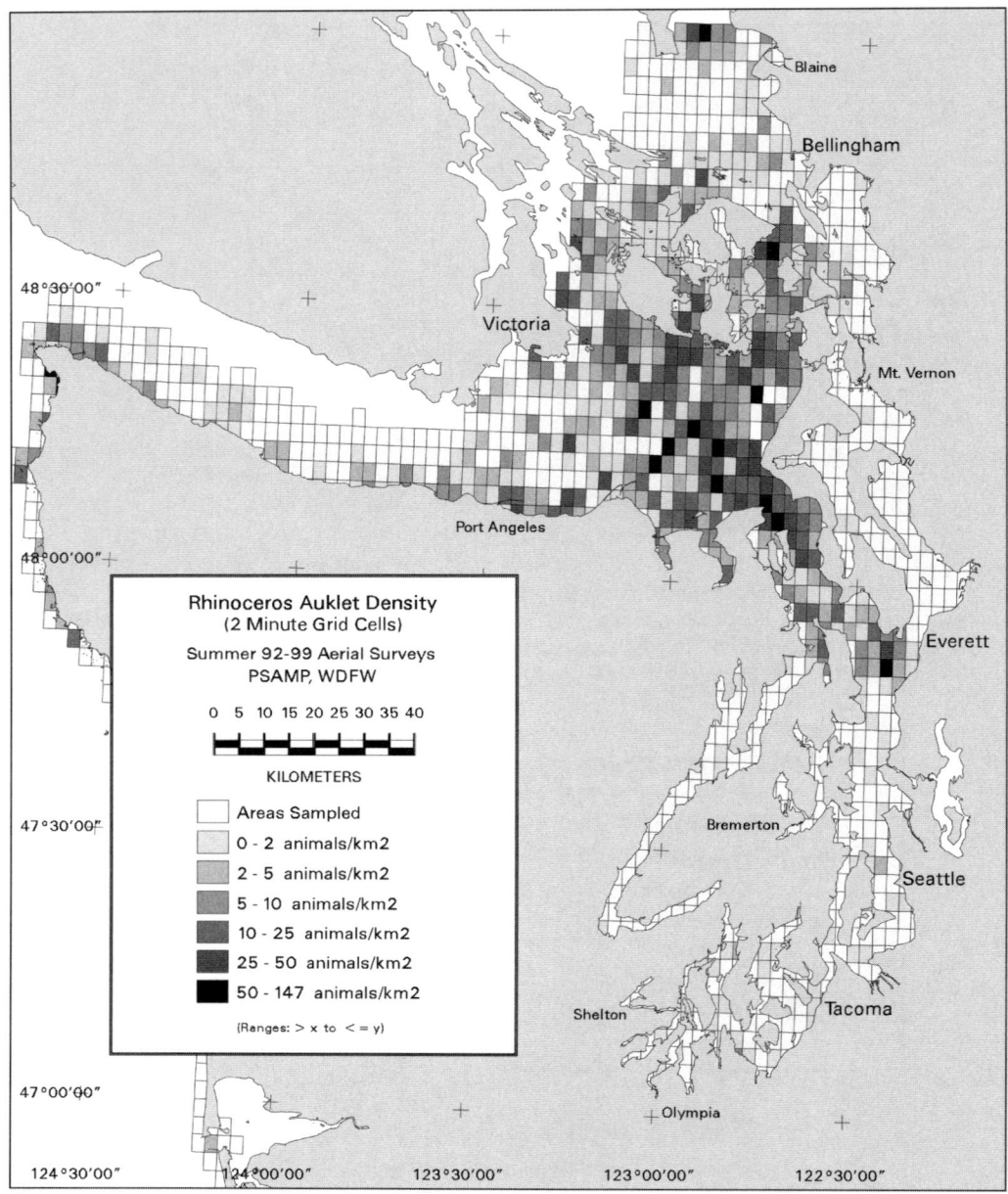

Figure 6. Density of rhinoceros auklets in Puget Sound, Washington, derived from aerial transect counts. Data and figure from Nysewander and Evenson.[128]

populations of "deep-water" foragers are essentially non-existent. The regional auklet population may benefit in their use of the area because it appears that inland marine colonies are less susceptible to oceanic food shortages and breeding failure.

Maximum seasonal populations in northern inland marine waters were not estimated for 1978-79.[183] Projected numbers for Hood Canal-Puget Sound of wintering deepwater alcids (almost all common murres) in December 1982 and February 1983 were 14,000-19,000; the estimate for grebes, mostly western grebes, was 27,000-40,000.[184] One indication of the importance of the inland area is a rough calculation of about 200,000 common murres in the eastern Strait of Juan de Fuca in early fall migrating into inland waters to the south and north in 1978-79 (T. Wahl, unpubl. data). Similarly, wintering flocks of 1,600 Pacific loons and 2,000 Brandt's cormorants were

observed several times in the San Juan Islands and Georgia Strait areas[179] (T. Wahl unpubl. data). Though inside the 20 m depth contour, up to 25,000 scoters at herring spawn events in the eastern Georgia Strait region (in 1979) indicate the importance of this prey source and why large flocks of Pacific loons were present in deep water just offshore (T. Wahl, unpubl. data).

Marine Mammals. Washington's inland water with depths greater than 20 m include most of Puget Sound, Hood Canal, Strait of Georgia, and eastern Strait of Juan de Fuca. These deeper marine waters are used primarily as foraging areas for 6 species of marine mammals including the harbor seal, California sea lion, harbor porpoise, Dall's porpoise, minke whale and killer whale.

The harbor seal population in Washington's inland waters currently exceeds 17,000 animals, and forages on a variety of seasonally abundant marine fish found in the

Strait of Juan de Fuca, Strait of Georgia, San Juan Islands, Hood Canal and Puget Sound. Primary prey of harbor seals includes Pacific whiting, herring, salmon, Pacific sandlance, tomcod, plainfin midshipman, various flatfish, and sculpins. Harbor seals are considered year-round residents in these waters and appear to select prey such as Pacific whiting, herring, or salmon, which concentrate in local areas to spawn. In the other cases they feed on the variety of less abundant but widely distributed prey such as Pacific midshipman, flatfish, or sculpins.[97, 123, 130, 155]

Dispersing into the inland waters of Washington and British Columbia in late summer from their breeding grounds in California, California sea lions forage primarily on returning adult salmon at river mouths in the fall, spawning stocks of Pacific whiting in Port Susan and the Strait of Georgia in fall and winter, and herring staging and spawning areas off Cherry Point and in the Strait of Georgia. Other important prey of California sea lions in Puget Sound includes market squid and dogfish. In many cases relatively large groups of California sea lions (500-1,000 animals) use haulout areas in close proximity to deeper waters used as spawning grounds for preferred prey (i.e., use of log-booms at Everett for feeding on spawning Pacific whiting stocks in Port Susan).[121, 122, 123]

As with the pinnipeds, the common cetaceans (harbor porpoise, Dall's porpoise, minke whale and killer whale) of Washington's deeper marine waters use these areas for foraging as well as for reproduction and social interactions. Harbor porpoise, while once considered common throughout Washington's inland waters, are rarely seen today outside of the San Juan Islands, Strait of Georgia, and Strait of Juan de Fuca.[105, 138, 156] Causes suggested for the disappearance of harbor porpoise from Puget Sound include mortality in fishing gear, effects of environmental contaminants, and increased vessel activity.[138] Within Washington inland waters (Strait of Juan de Fuca, Strait of Georgia, and San Juan Islands), the harbor porpoise stock was recently estimated at just over 3,500 animals.[105] Important prey of harbor porpoise includes squid, shrimp, and small schooling fish, such as herring and Pacific sandlance.

Dall's porpoise are the small black and white cetacean distributed throughout the inland waters of Washington. This is the species frequently encountered riding the bow wave of vessels in northern Puget Sound and the San Juan Islands. In open water this species leaves a characteristic "rooster tail" as it cavorts about. No comprehensive population assessment has been conducted for the species in Washington's inland waters, but the species is considered common.[138] Dall's porpoise feed on small schooling fish such as herring and Pacific sandlance.

Minke whales are the smallest and most common baleen whale found in northern Puget Sound, San Juan Islands, Strait of Juan de Fuca, and Strait of Georgia.[138] No population estimate is available for this species in Washington's inland waters. As with the other cetaceans found in inland waters, the minke whale's diet consists primarily of small schooling fish.

The killer whale, or orca, is the best known and most spectacular cetacean found in Washington's inland marine deeper waters. Studied in San Juan Islands, Strait of Juan de Fuca, and Strait of Georgia since the early 1970s using photo identification, the population status, natural history and behavior patterns of the southern resident killer whales (pods J, K and L) are well known. Currently numbering a total of 90 individuals, these pods move throughout the inland waters, feeding primarily on the various runs of salmon returning to rivers which flow into Puget Sound and the Strait of Georgia.[13, 72, 131, 138] In addition to the southern resident pods, which are fish eaters, additional "transient" pods of killer whales occasionally enter Washington's inland waters from the Pacific Ocean. Although very little is known about these "transient" pods, they appear to forage primarily on other marine mammals including harbor seals and gray whales.[7, 8]

Coastal Headlands, Islets, and Marine Nearshore

Marine Birds. Although most species associated with headlands and islets are found in low densities along much of the coast, high concentrations of seabirds are found in localized nesting colonies (Figures 7, 8, 9). Estimates of total breeding populations are available for some species, but for many there is considerable uncertainty. Moreover, estimates from the period prior to documented population declines in the 1990s may no longer be accurate (e.g. common murre populations in the region have varied substantially over the past two decades).[11, 43, 195]

In Oregon, the most abundant breeding species are Leach's storm-petrel and common murre. Leach's storm-petrel is found at several major colonies (1966-1967 surveys): Goat Island (high estimate of 535,800 birds), Hunter's Island (208,000), Whalehead Island (118,300), Crook Point Island (101,500), and Saddle Rock (47,900).[173] The common murre is also abundant in Oregon but has a wider distribution: Since 1966, 23 sites have supported at least 5,000 birds (131,481 at Three Arch Rocks, 28,531 at Cascade Head complex, 27,658 at Bird Rocks complex, 24,316 at Whalehead Island, 24,057 at Gull Rock [Curry County], 23,667 at Orford Reef, 23,220 at Cat and Kittens, 19,146 at Yaquina Head, 15,117 at Hubbard Mound Reef, 14,377 at Gull Rock [Lincoln County], 14,018 at Redfish Rock, 13,839 at Mack Arch, 12,865 at Island Rock, 10,512 at Twin Rocks, 8,031 at Cape Meares [south face], 7,679 at Castle Rock [Clatsop County], 7,488 at Tillamook Rock, 7,079 at Pillar Rocks, 6,647 at North Coquille Point Rock, 6,000 at both Tower Rock and Mack Reef, 5,940 at Pyramid Rock, 5,342 at Gull Rock [Clatsop County], and 7 sites have supported at least 1,000 birds (3,198 at Face Rock, 2,968 at Goat Island, 2,855 at Rogue River Reef, 2,694 at Sea Lion Rock, 2,506 at Brown Rock, 2,091 at an unnamed rock near Island Rock, 1,011 at Cape Blanco area).[43, 112, 173] It should be noted that the common murre experienced a dramatic population decline along the Washington coast associated with oceanic warming in the northeastern Pacific Ocean.[195] Although murre numbers have increased in the last few years, their abundance remains below that prior to the

Figure 7. Location of seabird nesting colonies in Washington (see text for details). Sites referenced by numbers on map are: 1) Smith Island, 2) Protection Island, 3) Seal and Sail Rocks, 4) Tatoosh Island, 5) Bodelteh Island, 6) Carroll Island, 7) Jagged Island, 8) Petrel Rock, 9) Cakesosta, 10) Rounded Island, 11) Alexander Island, 12) Destruction Island, 13) Willoughby Rock, 14) Split Rock, 15) Point Grenville, and 16) Big Stack and Grenville Arch.

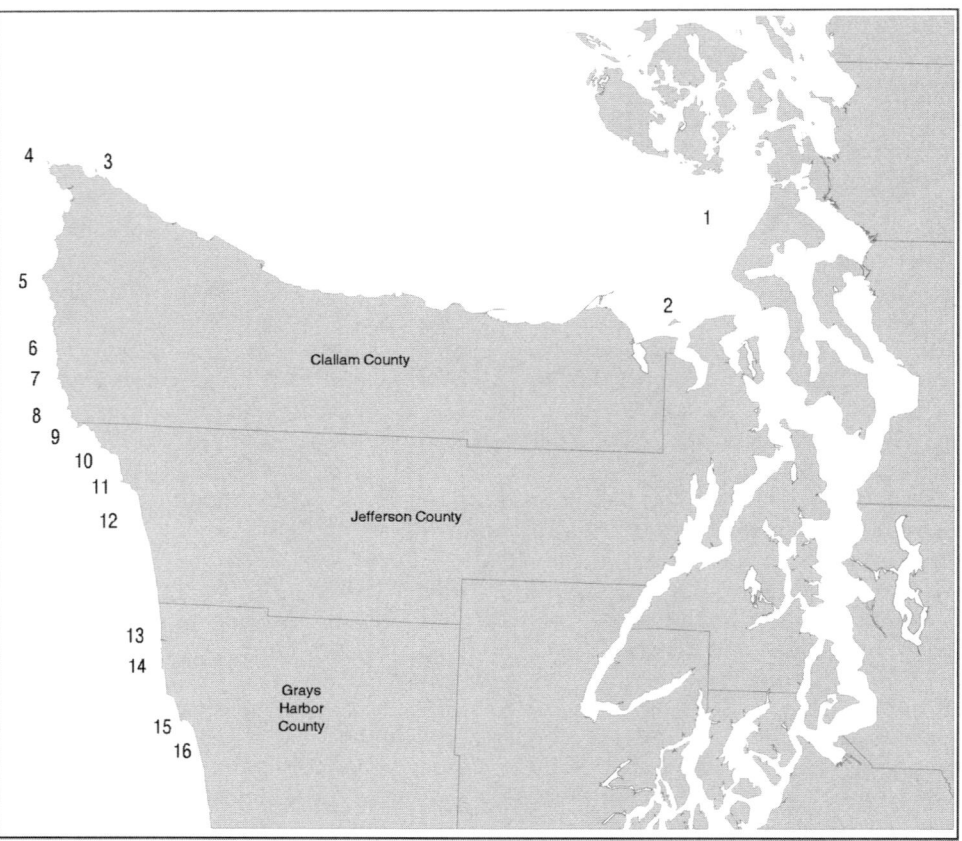

1982-83 El Niño event. In Oregon, the population of common murres has been stable in recent years, although colony attendance levels have fluctuated and productivity has been generally poor[43] (R. Lowe, pers. comm.). All other seabird species associated with coastal headlands and nearshore habitats in Oregon are found in far lower numbers than murres.[173]

The available data indicate that seabird colonies in Washington support comparatively fewer birds than those in Oregon. The most abundant breeding seabird species in Washington (from baseline data 1978-1982) is the Cassin's auklet (estimated population is 87,600 birds, of which an estimated 55,000 occur at Alexander Island; the species' breeding range is limited to 8 currently known locations between Point Grenville and Seal and Sail Rocks).[167] Other abundant species include fork-tailed storm-petrel (1,600 at Carroll Island, 1,900 on two islands in the Bodelteh Islands group), Leach's storm-petrel (20,000 at Jagged Island, 10,000 at Carroll Island, 2,600 at Petrel Rock, 2,000 at Alexander Island), double-crested cormorant (about 1,100 at Protection Island), pelagic cormorant (7 of 61 colonies support a total of about 2,000 birds), western gull (perhaps 6,000-8,000 occur south of Destruction Island), glaucous-winged gull (over 10,000 at Protection Island), rhinoceros auklet (34,216 at Protection Island, 23,600 at Destruction Island, 2,588 at Smith Island), tufted puffin (7,800 at Jagged Island, 4,000 at Alexander Island), and common murre, 1996-1998 (10,400 at Split Rock, 5,300 at Willoughby Rock, 5,000 at Grenville Arch, 4,214 at Tatoosh Island, 3,610 at Carroll Island, 2,355 at

Huntington Island, 2,200 at Rounded Island, 2035 at Big Stack [Grenville Pillar], 1135 at Cakesota).[140, 167, 196, 197]

Other species, primarily associated with cliff faces, are generally less common. Surveys conducted in greater Puget Sound in 1999 found at least 10,633 breeding pigeon guillemots at 367 locations, primarily in southern Puget Sound, the Admiralty Inlet/eastern Strait of Juan de Fuca region, and the San Juan Islands and vicinity.[66] Belted kingfishers also nest in burrows in coastal bluffs but are found in individual territories. A notable resident of this habitat is the peregrine falcon, which nests along the northern outer coast of Washington (26 known sites of which 18 were reproductive at least one year between 1997 and 1999[198] (Washington Department of Fish and Wildlife database), the northern Puget Sound and San Juan Islands area (18 sites of which 14 were reproductive between 1997 and 1999; WDFW database), and along the Oregon coast (20 sites, of which 7 were reproductive between 1997 and 1999; Oregon Department of Fish and Wildlife database; J. Pagel, pers. comm.).

Densities of birds foraging in nearshore and shelf habitats usually are higher than those in oceanic waters and include at least 30 species which are abundant at one or more seasons during the year (see Life History matrix). Depth associations of birds may be more pronounced or more obvious in Washington, where the shelf is broad, than in Oregon, where the shelf is relatively narrow. Many species concentrate to forage at coastal outflows of major rivers, embayments and the mouth of the Strait of Juan de Fuca. Depending on tides, weather and season, distribution in these habitats varies greatly.

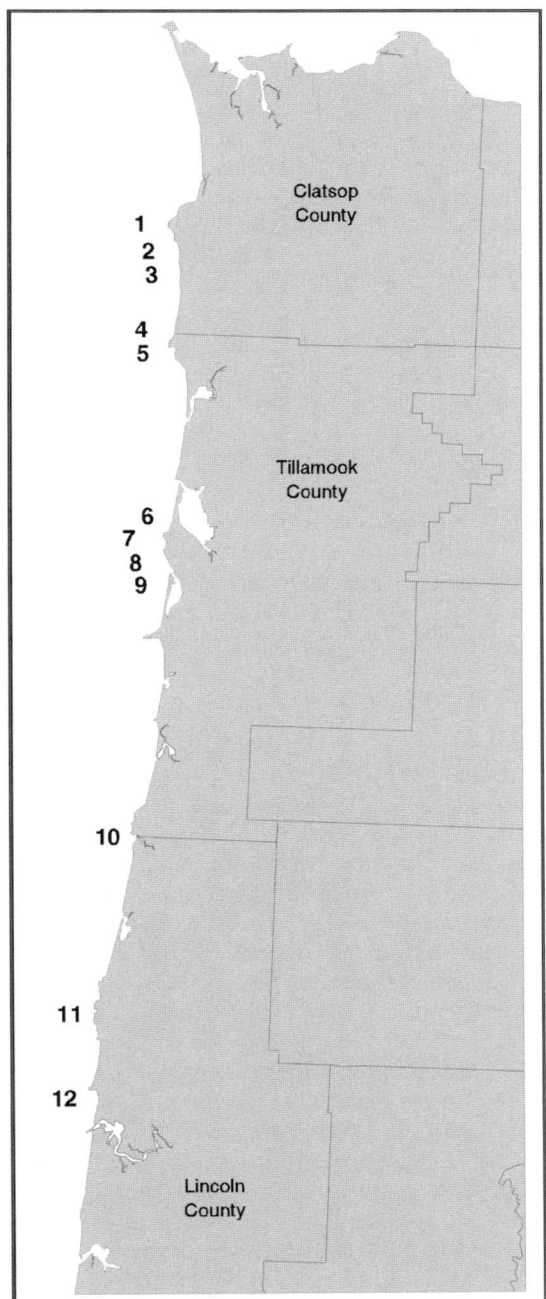

Figure 8. Location of seabird nesting colonies on the northern Oregon coast (see text for details). Sites referenced by numbers on map are: 1) Tillamook Rock, 2) Sea Lion Rock, 3) Bird Rocks, 4) Castle Rock, 5) Gull Rock, 6) Pyramid Rock, 7) Cape Meares, 8) Pillar Rocks, 9) Three Arch Rocks, 10) Cascade Head, 11) Gull Rock, and 12) Yaquina Head.

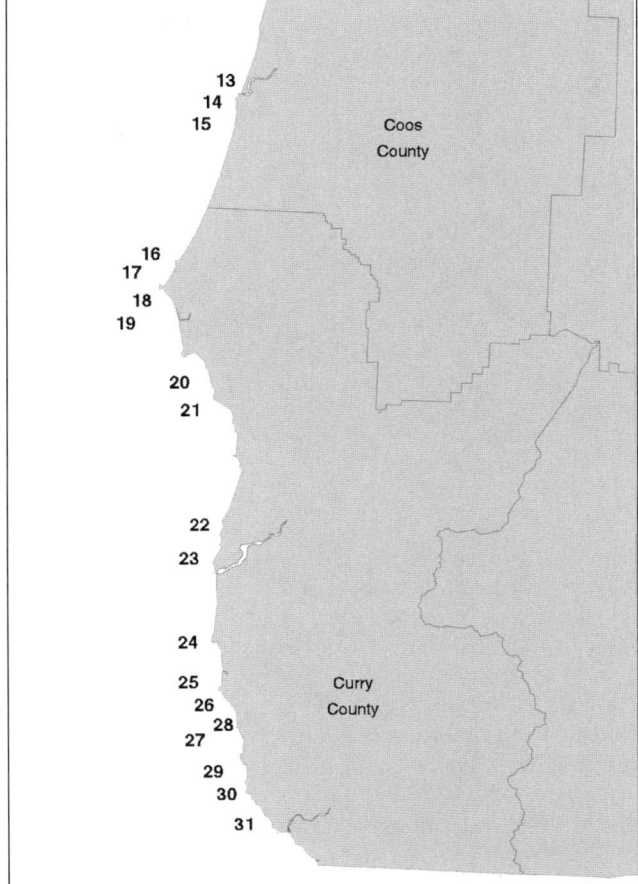

Figure 9. Location of seabird nesting colonies on the southern Oregon coast (see text for details). Sites referenced by numbers on map are: 13) North Coquille Point Rock, 14) Face Rock, 15) Cat and Kittens, 16) Tower Rock, 17) Gull Rock, 18) Cape Blanco, 19) Orford Reef, 20) Redfish Rock, 21) Island Rock, 22) Hubbard Mound Reef, 23) Rogue River Reef, 24) Hunter's Island, 25) Crook Point Island, 26) Saddle Rock, 27) Mack Reef, 28) Mack Arch, 29) Whalehead Island, 30) Twin Rocks, and 31) Goat Island.

Patterns of marine bird occurrence off the Washington and Oregon coasts likely reflect conditions in the Arctic, the southern hemisphere or elsewhere; reproductive success or failure and feeding opportunities at staging areas during migration or winter residence may determine populations of species in the region. There is great interannual variation in seasonal populations, some of which may be due to variations in oceanic productivity within the region, or to large-scale factors such as El Niño or general oceanic warming.[187] Much of the variation noted is not fully explained, as in the case of fluctuating numbers of regular transoceanic migrants like phalaropes and jaegers in nearshore waters.

Brandt's cormorants, brown pelicans, Heermann's gulls and Caspian terns concentrate in embayments or at estuarine outflows and harbor entrances but also forage in nearshore waters. In addition to these, large numbers of gulls and alcids utilize nearshore areas. In particular

Brandt's Cormorants at roost on Whale Rocks off Lopez Island, Washington. (Photo by Terence R. Wahl)

common murres and rhinoceros auklets may concentrate in very large numbers seasonally, especially in summer when anchovies are abundant. This season also draws very large numbers of sooty shearwaters to the nearshore, with historical estimates of 100,000 to 1 million present in Grays Harbor and Willapa Bay and at the mouth of the Columbia River.[9] As in other marine habitats, piscivores forage in multi-species flocks, with gulls typically locating prey schools, and then, depending on the season, other species like shearwaters, cormorants, pelicans, alcids and jaegers arrive to forage until prey schools are dispersed or decimated.

The nearshore area is an important migration corridor for a number of common seabirds (e.g., phalaropes, other shorebirds, and waterfowl), which nest in the Arctic, coastally, and inland in western North America. Additionally, brown pelicans and Heermann's gulls move north in early summer and return south in the fall. In late summer-early fall Brandt's cormorants and common murres move north for the winter, many enroute to the Strait of Juan de Fuca and inland marine waters of Washington and British Columbia; these species return south in the spring to colonies in Oregon and California (T. Wahl, unpubl. data). In winter, the nearshore waters are utilized by loons, some grebes, cormorants, and gulls and especially by large numbers of diving ducks, with scoters representing 50% of total numbers in November.[23]

Large winter and migrant populations of 14 species concentrate at deep water tidal fronts,[183, 184] and lower numbers of other nearshore species forage opportunistically in the deeper waters offshore. Almost all regions have highest numbers in winter, with loons, western grebes, Brandt's cormorants, several species of gulls, 5 alcids (especially common murre) dominating the seabird community. One diving duck, the oldsquaw, forages locally over deep-water banks, often in flocks of hundreds.[179] Three planktivore specialists—red-necked phalarope, Bonaparte's gull, and common tern—are widespread during migrations.

Nearshore waters are occasionally visited by large concentrations of species normally associated with deeper waters farther offshore. Variations due to events like El Niño may mask "natural" variation, and subsequent analyses of cause and effect are extremely complicated. Common murres historically have been widely dispersed over the shelf, but flocks in Washington concentrate over the inner shelf in less than 100 m depth[177, 178] (to 150 m depth).[23] Similarly, summer-fall flocks of sooty shearwaters often are found in the same area. These species are attracted in to nearshore waters seasonally.

At present, breeding populations of a number of species are incompletely known. The following estimates (prior to reproductive failures in the 1990s) may no longer accurately reflect the number of birds present.[11, 43, 195] Washington and Oregon breeding numbers were estimated to be 400,000 for Leach's storm-petrel and 5,000 for fork-tailed storm-petrel;[15] 5,700 double-crested cormorants, 16,600 Brandt's cormorants and 12,000 pelagic cormorants;[160] 25,000 glaucous-winged and western gulls;[176] up to 20,000 Caspian terns (in southwest Washington and the lower Columbia River);[111] 7,660 pigeon guillemots[67, 162] (although a recent estimate for Puget Sound was 10,633);[66] 457,000 common murres;[33] 8,000-12,000 marbled murrelets;[67] 87,600+ Cassin's auklets;[167, 168] 25,000 rhinoceros auklets;[33] and 28,000 tufted puffins.[33]

Marine Mammals. A total of 31 marine mammal species have been reported from the marine waters of Washington and Oregon.[12, 138, 156] A number of these species are considered pelagic offshore species, which are rare visitors to nearshore waters during El Niño events, or come ashore as strandings along coastal beaches. The species composition of marine mammals found at coastal headlands, islets and nearshore waters (<20 m) of Puget Sound, Hood Canal, San Juan Islands, Strait of Georgia, Strait of Juan de Fuca, and along the outer coasts of Washington and Oregon is dominated by 3 species of pinnipeds: the harbor seal, California sea lion and Steller sea lion. Additionally, elephant seals migrate into Washington and Oregon coastal waters with a small breeding rookery recently reported on the southern Oregon coast at Cape Arago.[90] Two species of cetaceans (gray whale and harbor porpoise) are commonly found distributed in Washington and Oregon waters from just outside breaker line to the inner continental shelf. In addition, the sea otter is an important component of the nearshore marine environment of the northern Washington (Olympic Peninsula) coast as well.

As of 1996, the harbor seal population in Washington was estimated at over 34,000 animals. These animals use 319 haulout areas, primarily on intertidal islets, rock, reefs, sandbars and shoals.[123] Areas of high harbor seal densities include Willapa Bay, Grays Harbor, Destruction Island, Giants Graveyard, Cape Alava, San Juan Islands, Skagit Bay, Gertrude Island and Hood Canal. The largest haulout sites include Cape Alava/Bodelteh Island area for harbor seals, California sea lions and Steller sea lions; Carroll Island for Steller and California sea lions; and Giant's Graveyard/Toleak Point area, Grays Harbor, Willapa Bay, and Columbia River for harbor seals.[98] An example of the

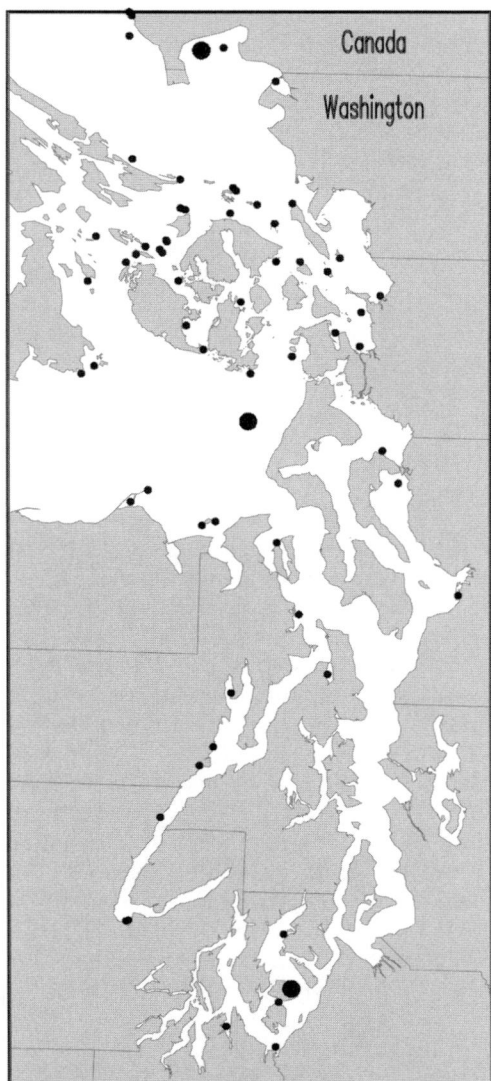

Figure 10. Distribution of harbor seal haulout sites in Puget Sound, Washington (small circles: 100-500 seals, large circles: >500 seals; data from Washington Department of Fish and Wildlife).

distribution of harbor seal haulout sites is shown in Figure 10.

Bays and estuaries along the outer coast and inland waters of Washington provide important pupping and breeding areas for harbor seals. Harbor seal pups are present along the outer Washington coast from late April through July; pups are present in the inland waters of Washington from late June through September.[12, 107, 155, 170] Important prey of harbor seals in Washington waters include Pacific whiting, herring, salmonids, smelts, plainfin midshipmen and various flatfishes.[12, 123, 130]

Harbor seals are the most abundant pinnipeds in nearshore marine waters of Oregon, with a population of over 9,000 animals using 101 haulout areas along the Oregon coast.[123] Areas of high harbor seal densities in Oregon include coastal headlands, reefs and islets at Rogue Reef, Orford Reef, Cape Arago, and Tillamook Head.

Oregon coastal estuaries (Coos Bay, Alsea Bay, Tillamook Bay and Columbia River) also provide important habitat for harbor seals. These estuaries provide protected areas on intertidal sandbars and shoals used for pupping and breeding, with pups present from early April through June.[12, 24, 107, 170] The largest haulout sites include Tillamook Bay for harbor seals; Cascade Head for California sea lions and Steller sea lion; Sea Lion Caves (Florence, OR) for Steller and California sea lions; Cape Arago for harbor seals, elephant seals, Steller sea lions and California sea lions; Orford Reef for Steller sea lions (breeding rookery) and California sea lions; and Rogue Reef for Steller sea lions (breeding rookery), California sea lions, and harbor seals.[24]

The nearshore waters of Oregon provide a variety of food for harbor seals with important prey items in their diet consisting of schooling fishes (Pacific whiting, smelts, herring and Pacific mackerel), various flatfish, lamprey, and salmonids.[123, 148] During the winter, eulachon smelt provide a seasonally abundant prey for harbor seals as well.

California sea lions are seasonal migrants into nearshore marine waters of Washington and Oregon that results from a northward dispersal of subadult and adult males from their rookeries in California and Baja Mexico following their annual breeding season in June.[12, 24, 114, 123] A wave of 5,000-6,000 animals move northward into Oregon, Washington and British Columbia nearshore waters in late summer and early fall, remaining in Northwest waters until late spring, when the majority of these animals return south to their breeding rookeries.[123] California sea lions use haulout sites at or near the mouths of most of the major rivers along the Oregon and Washington coasts. In Oregon, haulout areas are located at Rogue Reef, Orford Reef, Cape Arago, Sea Lion Caves, Cascade Head, and Columbia River. In Washington, haulout areas are located at Split Rock, Carroll Island, Bodelteh Island, Tatoosh Island, Waadah Island and near Everett in Puget Sound. The diet of California sea lion in Washington and Oregon consists primarily of seasonally abundant schooling species such as Pacific whiting, herring, Pacific mackerel, eulachon, salmon, and squid.[123, 148] Movements and distribution of California sea lions through nearshore waters of Oregon, Washington and British Columbia has been correlated with spawning aggregations of their prey, and indicate the ability of California sea lions to cue into locally abundant concentrations of these species.[123] This species has been reported to move up the Columbia and Willamette rivers following spawning runs of salmon and eulachon as well.[12, 123]

Steller sea lions occur year round in Washington and Oregon nearshore waters, and include both breeding and non-breeding animals. The main breeding rookeries for Steller sea lions along the Oregon coast are located at Rogue and Orford reefs; relatively small breeding rookeries are reported at Sea Lion Caves and Three Arch Rocks.[24, 121] Additional haulout locations in Oregon are located at Cape Arago, Cascade Head, and Columbia

River. Seasonal abundances in Oregon nearshore waters range from 1,000-4,000 animals, with peak counts occurring during the breeding season in early summer. Although both adult males and females are present in Washington, no breeding rookeries occur. Haulout locations are found along the outer Washington coast at Split Rock, Carroll Island, Bodelteh Island, Cape Alava and Tatoosh Island. Seasonal abundances range from 500-1,500 animals along the outer Washington coast. Relatively small numbers of Steller sea lions occur at haulout locations in the inland waters of Washington, although 500-1,000 animals move through the Strait of Juan de Fuca and into British Columbia waters annually to feed on herring spawning in the Strait of Geogia north of Nanaimo. Similar to California sea lions, the primary diet of Steller sea lions in the nearshore waters of Washington and Oregon consists of Pacific whiting, herring, salmon, lamprey, rockfish, flatfish, and squid.[148]

The gray whale is one of the best known cetaceans because of its annual migration along the west coast of North America. Every year, the Eastern North Pacific stock of gray whales migrates through the nearshore and coastal waters of Washington and Oregon traveling from breeding grounds in Baja Mexico to feeding areas to the north along the west coast of North America and Alaska. This stock of gray whales has recovered from whaling and over-exploitation in the late nineteenth century, and is estimated in excess of 26,000 animals. Because of this recovery, the Eastern North Pacific stock of gray whales was removed from the List of Endangered Species in 1994.[153] The southward migration takes the gray whale population from their feeding grounds to breeding lagoons in Baja Mexico, with whales passing along the Oregon and Washington coasts from late November to early January.[89, 153] The northward dispersal and migration of gray whales from their breeding ground brings them into the nearshore waters of Oregon and Washington waters from late February to late May.[89, 153] Although the majority of the population migrates to feeding grounds in the Bering and Chukchi Seas, a small proportion of the gray whale population remains in nearshore areas along the Oregon, Washington, and British Columbia coasts to feed during summer months.[51, 153] Foraging aggregations usually are found along the Oregon coast from near Yaquina Head to Depoe Bay. Along the Washington coast, feeding areas are located in shallow water areas of Grays Harbor, Cape Alava, Cape Flattery, western Strait of Juan de Fuca, and along Whidbey Island (Saratoga Passage). Gray whales are benthic feeders, and usually forage in areas where they find dense concentrations of amphipod crustaceans, polychaete worms, sand shrimp, and herring spawn (roe).[37, 53, 138, 153, 190] Individual gray whales in nearshore waters of Washington and British Columbia have been reported returning to the same areas to feed year after year.[37, 53, 153]

The smallish, olive gray harbor porpoise is the most common small cetacean found in nearshore marine waters of Oregon and Washington. Shy by nature, this species is found along the Pacific Ocean coast from just outside the breaker line to out over the continental shelf.[36, 81, 106] The Oregon and Washington coastal stock was estimated at over 44,000 animals.[106]

In the nearshore waters off the coast of Washington from Destruction Island to Neah Bay is found a growing population of sea otters. Extirpated from the coast of Washington in the early 1900s, the sea otter was translocated to the Olympic Peninsula coast in 1969 and 1970 from Amchitka Island, Alaska.[96] The species has reestablished itself off the coast with a growing population, which (as of July 1999) is estimated at 605 animals. Sea otters can be expected to play a significant role in shaping the nearshore urchin/kelp community along the Olympic Peninsula coast, with a diet consisting of sea urchins, clams, Dungeness crab, octopus, and other benthic invertebrates.[104]

Marine Shelf and Oceanic

Marine Birds. Seabird populations are characterized by great seasonality as well as distribution by general depth "habitats" over shelf, slope, and offshore/oceanic waters. Off Washington, there is a general change in species composition over the shelf at about 100 m depth, with albatrosses, fulmars, shearwaters (other than sooty shearwater), and storm-petrels becoming more numerous offshore over the deeper "outer" shelf. Off the Oregon coast these changes may be less apparent where the continental shelf is relatively narrower. The spring population includes a large number of coastal breeders, such as fork-tailed storm-petrel, Leach's storm-petrel, common murre, and Cassin's auklet. In late spring, large numbers of shearwaters arrive from the southern hemisphere for the summer, and birds from Alaska and Canada arrive as well. Diversity increases during fall migration when migrant phalaropes, jaegers, terns and alcids, in addition to summer residents, not only substantially increases the diversity of abundant species to 19 but also accounts for the largest overall seabird populations of the year.[23, 182, 184] By winter, many of the northern birds have moved south, with large numbers of northern-breeding gulls and fulmars remaining. In general, the bird community associated with shallower water is dominated by diving or pursuit plunging seabirds that prey on pelagic fish such as herring, anchovy, eulachon, and smelt (e.g., murres, auklets, puffins, and sooty shearwaters), while the deeper oceanic waters support surface-feeding birds (e.g., gulls, albatrosses, fulmars, other shearwaters, and storm-petrels) that feed largely on squid and plankton (including jellyfish).[6, 145, 180, 189] Jaegers and skuas are generally found in deeper waters, where they pirate prey from offshore gulls.

For offshore populations especially, variability is one of the most consistent features at temporal scales ranging from the short-term[187] to long-term cycles, which may persist for decades. Even more so than in nearshore waters, there is great inter-annual variation in seasonal populations present in offshore waters. Some of this is due to variations in oceanic productivity within the region and some may be due to large-scale variations (El Niño, Pacific

Decadal Oscillation). Some variations are due presumably to smaller-scale variations outside the region, such as at Arctic breeding grounds. Much of the variation is not well understood.

The shelf area, like the nearshore coastal waters, is a regular migration route for a number of species, which, though they do not forage in numbers in the habitats, are noted resting or foraging on occasion. Loons, particularly Pacific loons, many geese, dabbling and diving ducks, shorebirds and lost landbirds transit the shelf area. Pigeon guillemots, apparently migrating north for the winter, are noted in August (T. Wahl unpubl. data), and small numbers of Xantus' murrelets dispersing from southern breeding areas to Oregon-Washington offshore waters[177] are examples of less-obvious occurrences.

Birds concentrate at current boundaries and topographically induced upwellings.[23] These oceanic fronts or 'rips' at the edges of coastal upwellings concentrate small zooplankton and their larger predators. These attract planktivorous storm-petrels, phalaropes and terns and, subsequently, jaegers and gulls. Nocturnal foraging, which is known for a number of species, is apparently more important to birds over the shelf and oceanic waters than to birds inshore. Oceanic birds are characterized by patchy distribution, which reflects prey distribution and prey detection, and a response to activities like commercial fishing, which attracts birds.

The water mixing zones of the shelf slope and outer shelf edge support a high diversity of seabirds capitalizing on the biological richness of this area. Over 20 species of birds characterize these regions. Five breed along the Oregon and Washington coasts (fork-tailed storm-petrel, Leach's storm-petrel, Cassin's auklet, rhinoceros auklet, and tufted puffin); while 3 other breeders (western gull, glaucous-winged gulls and common murre) are found here, they occur in much smaller numbers than closer to shore.[23, 177] Conspicuous southern hemisphere breeders (from Australia, New Zealand, and the Antarctica region) found here include the pink-footed shearwater, Buller's shearwater, sooty shearwater, and south polar skua. Hawaiian breeders include both the black-footed albatross and Laysan albatross. The California gull breeds inland. All the remaining dominant species breed in Alaska and Canada, and include the northern fulmar, red phalarope, red-necked phalarope, Pomerine jaeger, long-tailed jaeger, Arctic tern, and 5 gulls (Bonaparte's, herring, Thayer's, Sabine's, and black-legged kittiwake).

The highest seabird densities are found in shelf waters, due to the high numbers of local breeding birds and nearshore migrants. Briggs et al.[23] found shelf water densities of between 40 and 50 birds/km² during the summer months, due mainly to the nearly 500,000 common murres that nest in Oregon[23] and the approximately 88,000 Cassin's auklets nesting along the outer coast of Washington.[167] Also, tremendous flocks (up to 100,000) of sooty shearwaters invade the shelf waters to feed on northern anchovies during the summer months, with local densities[23] exceeding 1,000 birds/km². Wahl[177] found the inner shelf (20-100 m deep) seabird community

Sabine's Gulls migrating off Grays Harbor, Washington. (Photo by Terence R. Wahl)

also to be dominated by rhinoceros auklets, tufted puffins, and 4 species of gulls (western, herring, glaucous-winged, and California).

Compared to oceanic waters, seabird densities are considerably higher[23] along the shelf edge and slope, with seasonal densities ranging from about 2 to 7 birds/km². Seasonal densities were highest during the summer (when again tubenoses dominated), but relatively high densities were also found winter (when gulls, fulmars, and Cassin's auklets dominated) and spring (when Leach's storm-petrels were prevalent). Seabirds here tended to concentrate into locally high densities, especially along the seaward edges of Heceta and Swiftsure banks, the Cape Blanco upwelling center, and behind shrimp and factory trawlers.[23, 181] In these waters, Briggs et al.[23] found local densities of sooty shearwaters exceeding 2,000 birds/km².

Farther offshore over oceanic waters (>2000 m deep), Sanger[154] found the seabird composition to change further. Black-footed albatrosses predominated, except during the winter when most of the population of this species had returned to its Hawaiian breeding grounds. Fork-tailed storm-petrel and Leach's storm-petrels are common during the summer months. By winter, northern fulmars, herring gulls, glaucous-winged gulls, and black-legged kittiwakes comprise the bulk of the population.

Few estimates have been made of seasonal offshore-shelf populations. Briggs et al.[23] estimated 1.8 million birds present at one time in late summer for an oceanic, shelf and nearshore waters study area off Washington and Oregon. Wahl[178] roughly estimated a seasonal total of 1.2 million off Washington. Maximum densities were in August-September, with highest densities over the continental shelf and lower numbers offshore in oceanic waters.[23, 178] Foraging breeding birds aggregated to within about a 50 km radius from colonies.[23] Estimates of breeding population size of a number of offshore species may not be very accurate, particularly following the widespread reproductive failures in the 1990s (see "Coastal Headlands, Islets, and Nearshore" for other population information).

Marine Mammals. At least 24 species of cetaceans and 5 pinnipeds are known to occur in the marine waters off Oregon and Washington.[18, 81, 115, 156] However, some species are rarely seen, or at least rarely identified. These include offshore species such as Sei whales, pygmy sperm whales, and Stejneger's, Hubb's, and Cuvier's beaked whales. Most beaked whale records are from strandings. Other species of beaked whales (especially mesoplodonts) may also occur in the deeper offshore waters of Oregon and Washington. Sei whale records derive mostly from old whaling logs. Blue and northern right whales are also very rare, but both have been observed off Oregon and Washington in recent years.[82, 152] Others, such as false killer whales, short-finned pilot whales, striped dolphins, and common dolphins, may intrude into Oregon and Washington waters during years of very warm water, although pilot whales may have been much more common off the coasts than they are now. A record of a single belukha whale[156] is obviously extralimital.

Extralimital strandings of ringed and ribbon seals have been reported in California.[115] Although neither has been recorded for Oregon or Washington, it is likely that these Arctic seals passed through Oregon and Washington waters before reaching California.

Gray whales, harbor porpoise, harbor seals, California sea lions, and Steller sea lions are largely denizens of the coastal and inner shelf waters. All but the gray whale are occasionally observed in shelf edge waters, though most of the at-sea sightings of pinnipeds occur near coastal rookeries or haulouts.[18]

Eleven species of marine mammals are characteristically found in the shelf edge and slope waters. The three most common species are the Pacific white-sided, Risso's, and northern right whale dolphins. Green et al.[81] estimated the spring Pacific white-sided dolphin population to be nearly 40,000 animals (including offshore animals), a number far exceeding any other population of marine mammal occurring off Oregon and Washington. Green et al.'s[81] estimate for spring and summer Risso's dolphins was nearly 8,000. Both species were most common during the spring and summer, less so during the fall, and rare during the winter months. Northern right whale dolphins, on the other hand, were found to be most common during the fall and absent during the winter.[81] Risso's dolphins were far more common over the shelf slope than the shelf edge (and entirely absent in offshore waters). These dolphins feed exclusively on the squid found in abundance in the slope currents. Interestingly, all three species of southern-originating dolphins were rarely seen off southern Oregon in the vicinity of Cape Blanco. It is likely these temperate-water dolphins avoid the colder, upwelled water found off Cape Blanco. The other smaller cetacean using shelf edge and slope waters is the Dall's porpoise. Overall, these porpoise seem to be equally common in outer shelf and slope waters, although they shift onto the shelf during the summer and fall, following schooling fish and squid,[70] and move offshore during the winter.[81] Although they are easily observed, Green et al.[81] estimated that a little over 2,000 were found year-round.

The larger whales found near the shelf edge or over slope waters include the humpback, fin, Baird's beaked, and killer whales. Green et al.[81] found all 4 species most common during the summer months, although fins were absent during the spring; humpbacks were absent during the winter, and Baird's beaked whales were missing during the fall and winter. Killer whales were found year-round. Most of the humpback whale sightings by Green et al.[81] occurred immediately south of Hecata Bank, on Swiftsure Bank near the Canadian border, or off the mouth of the Columbia River. Nearly all the fin whales were observed in a lower to mid-slope region called Newport Valley approximately 85 to 90 km west of Newport, Oregon.[81] Similarly, most of the killer whale sightings (3 pods) occurred near the shelf edge about 35 to 40 km off Newport. Other sightings of pods (2) occurred near the shelf edge off Washington. Four of the 5 groups of Baird's beaked whales observed by Green et al.[81] occurred near the lower edge of the slope. A single group of 7 mesoplodont whales was observed by Green et al.[81] in slope waters of Washington.

Photographic identification studies on humpback whales have revealed that the animals seen off Oregon and southern Washington are part of a distinct feeding aggregation that spends the spring, summer, and fall along the coast of California, Oregon, and Washington, and migrates primarily to Mexico and Central America in winter.[39, 40] This population numbers just over 900 and has been increasing at about 7-8% per year.[35] Humpback whales identified off northern Washington near the British Columbia border show a low rate of interchange with locations to the north and south and consist of a relatively small number of whales that show strong seasonal site fidelity and migrate in winter to breeding grounds in Mexico, Hawaii, and off Japan.[41, 52, 169]

Most (68%) of the 19 elephant seals observed at-sea by Bonnell et al.[18] were found in shelf edge or slope waters. These deep diving animals spend little time at the surface (thus, are rarely seen) when occurring in deeper waters, and may be more common than they appear off Oregon and Washington. Elephant seals equipped with radio transmitters have been tracked by satellite from California, through Oregon and Washington, to the Gulf of Alaska.[60] In contrast, migrating fur seals are easily seen as they rest at the surface. When standardized for differences in survey effort, Bonnell et al.[18] estimated that about 27% of the northern fur seals occurred in outer shelf or slope waters. In total, Bonnell et al.[18] recorded 172 sightings of fur seals, most (87%) between January and May, with an estimated April peak abundance of about 7,000 animals.

Only 5 species of marine mammals dominate the deeper offshore waters: Pacific white-sided dolphins, northern right whale dolphins, Dall's porpoise, sperm whales, and northern fur seals.[18, 81] However, the rarely-seen mesoplodont whales are thought to prefer these waters, and Risso's dolphins have been seen far offshore during NOAA ship-board surveys (M. Dahlheim, NOAA/National Marine Mammal Laboratory, pers. comm.).

As with seabirds, the distribution of marine mammals is linked to their diet and foraging habitats. Marine mammals that feed on schooling pelagic fish or shallow-water squid (e.g., seals, sea lions, harbor porpoise, and Dall's porpoise) are found closer to shore, while squid-eating marine mammals (e.g., dolphins, Dall's porpoise, sperm whales, beaked whales, and fur seals) are found in the deeper waters. Humpback whales are generally found where plankton or herring concentrate, regardless of water depth, and elephant seals feed on deep-water fish, especially hagfish and squid.[45]

Since 1985, 4 species of sea turtles have been found stranded along the Oregon and Washington coasts: leatherback, loggerhead, green, and olive ridley (J. Scordino, NOAA, pers. comm.). It is likely that records of the latter three warm-water species represent extralimital sightings, possibly of disoriented animals, or those sickened by the cooler temperate waters. Leatherback turtles, however, appear to be a regular summer visitor to shelf and slope waters. Bowlby et al.[21] reported on 16 leatherback turtles sighted during the marine mammal and seabird surveys conducted under Brueggeman et al.[25] Turtles were sighted during the months of June, July, and September, with the 9 July sightings occurring farther inshore (shelf waters).

El Niño and La Niña Events

The effects of El Niño have been well documented and publicized, perhaps due to the severity and frequency of these events in the past two decades. The phenomenon influences weather conditions in marine environments, which in turn produce powerful storms and adverse weather conditions that impact humans and wildlife populations over vast areas of the planet. Although first documented in 1726, El Niño has occurred for tens of thousands of years. The ocean warming events vary in their intensity and timing, although in recent years the events have occurred, on average, every 3.8 years, with very strong episodes occurring every 9.9 years.[147] The two greatest El Niño events in recorded history occurred in 1982-83 and 1997-98.[102]

The cold water counterpart of El Niño is La Niña. The conditions in the Pacific region during La Niña episodes are exactly opposite those encountered during an El Niño event. Sea surface temperatures are anomalously high in the eastern equatorial Pacific and the east-to-west trade winds are very strong. In the Pacific Northwest, winters are colder and wetter in a La Niña year than in other years. The fluctuations between El Niño and La Niña states are often called the El Niño Southern Oscillation (ENSO).

A recent long-term warming of oceanic waters in the northeast Pacific may have intensified the severity of El Niño episodes. Some scientists believe this change in temperature is related to an increase in global surface temperatures. In the last century, average temperatures have increased by about 0.5°C and are expected to rise another 1.0 to 3.5°C by 2100.[116] Some argue that the increase in temperature is the result of anthropogenic factors, especially greenhouse gases, which trap rising heat. Other scientists believe that the oceanic warming is due to decadal oscillations and that the current conditions will not persist indefinitely.[92] Significant temperature changes of as much as 10°C in a relatively short period (several years) have been documented in some parts of the world.[101] Significantly, the Pacific Decadal Oscillation (PDO) has closely interacted with the major El Niño episodes of the past two decades. Every few decades, the North Pacific basin shifts from being unusually cold water temperatures in western and central parts of the basin and warm in the east to an essentially opposite pattern. The PDO phase in effect since the 1970s, which corresponds to the recent severe El Niño events, is characterized by warmer waters in the eastern North Pacific basin. The relationship between El Niño episodes and the PDO is not well-understood.[101, 113, 143]

The conditions that influence the occurrence of El Niño are best described in relation to typical weather and oceanic conditions. Under normal conditions, cooler sea surface temperatures, a shallower thermocline depth, and a lower sea level prevail in the eastern tropical Pacific Ocean relative to areas further east. This unequal state is influenced and maintained by the Pacific trade winds. The western Pacific Ocean absorbs more heat and this warmer air rises and moves east, where it sinks. The resulting easterly trade winds occur at lower altitudes than westerly winds, and they push sea water (which is warmed by the sun) westward via the Equatorial Current. Cold marine waters in the eastern tropical Pacific Ocean are maintained by the Peru Current and equatorial as well as coastal upwellings, which transport cold subsurface water to the surface. The resulting temperature and pressure gradients between the east and the west are maintained and further strengthened by a positive feedback.[10]

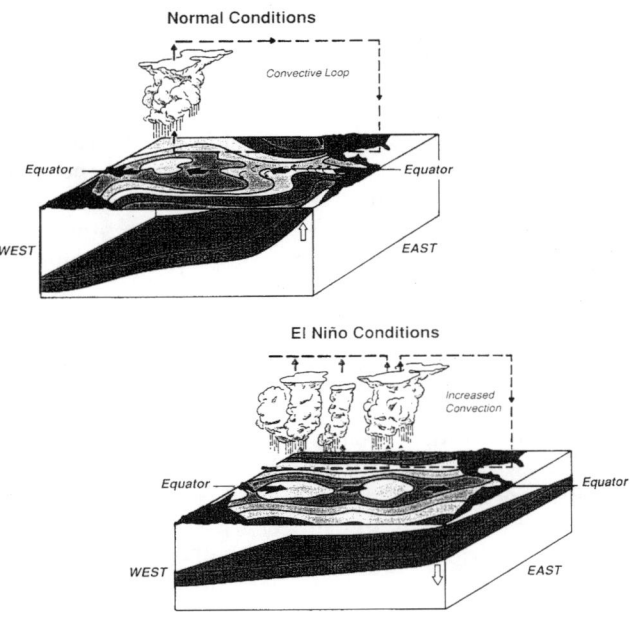

Figure 11. Schematic of the normal and El Niño conditions prevailing in the equatorial Pacific; from Frankignoul.[75]

During El Niño, the pool of warm water in the western Pacific expands eastward, changing wind patterns. Normally strong, easterly winds may weaken due to change in air pressure, and the normally weak westerly winds strengthen. With the decrease in easterly wind speed, the accumulated warm water on the west is released in the form of a Kelvin wave, readjusting sea level. As a result, warm water is transported eastward, the equatorial upwelling process decreases, sea surface temperatures and thermocline differences diminish, and sea level in the eastern Pacific rises by 0.2-0.4 m. As higher temperatures shift to the east, so do patterns of precipitation. Usually arid middle Pacific islands and coastal regions of Ecuador and Peru are drenched with rains and tormented by winds[10, 75, 150] (Figure 11). The magnitude of each El Niño episode depends on many factors, including timing, location, magnitude of oceanic warming, and phase of interdecadal climatic oscillations. Consequently, it is not yet possible to predict the onset of such an episode. Once El Niño starts, however, it is possible to determine its spatial and temporal distribution with current climate models.[159]

How a given animal is affected by El Niño depends on a variety of factors. The most important factors are the severity of a particular warming episode, and its geographical extent. Not all El Niños influence conditions as far north and east as the Oregon and Washington coasts. During less intense episodes, for example, coastal waters off California are the northern limit of El Niño conditions. Only a few, severe and long-lasting El Niños, particularly the 1940-41, 1957-58, 1982-83, and 1997-98 events, widely affected many marine species across the Pacific Ocean and along the coasts of the Americas.[10]

Of all the changes related to El Niño, the deepening of the thermocline has by far the greatest impact on the marine fauna of the eastern and central Pacific Ocean. The eastern Pacific Ocean contains some of the richest waters on the planet in terms of biological productivity. Nutrients from the breakdown of organic materials sink and accumulate in the cold waters below the thermocline. Phytoplankton has ready access to nutrients and takes advantage of sunlight for photosynthesis when the thermocline is shallow. The Kelvin wave causes the mixed layer of water above the thermocline to increase and the productivity to decrease, thereby affecting the entire food web of the eastern Pacific. Upwelling, if it occurs at all, can no longer replenish cold, saline, and nutrient-rich water from subsurface layers, because the thermocline is lower, extending the warm layers of water too deeply. During upwelling, water is transported from depths of between 40 and 80 m, but during severe El Niño episodes, the thermocline deepens by another 100-150 m, effectively negating the upwelling process.[10] At the peak of the 1982-83 El Niño in March there was a 5-fold reduction in absolute productivity at a cross-equatorial transect, and a 20-fold decrease near the Galapagos Islands. A normally 400 km wide nutrient-rich section of the Pacific ocean along the coast of Peru and Ecuador was diminished to a 30-50 km wide band in November 1982, and became even narrower in March 1983.[10] Sea surface temperatures, which normally range from 7°C to 11°C, reached a maximum of 18°C in July 1983, extending out to at least 200 km from the Oregon coast, and reaching depths of over 100 m.[80]

Marine Birds. The most common responses by marine birds to El Niño or other oceanic warming conditions have been altered species distribution, and changes in species abundance, reproductive output, and behavior. For marine birds off the Washington and Oregon coasts, these changes likely reflect reductions in the availability of suitable prey. Ocean temperatures off the California coast during the 1982-83 El Niño were 1-4°C higher than normal. Concomitant with this temperature increase, zooplankton densities declined by 90% below normal during spring 1983; phytoplankton declined as well and was distributed much deeper in the water column. In addition, rockfish and northern anchovy populations, which support large and diverse populations of seabirds, experienced lowered reproductive success.[22]

Perhaps the most significant and well-documented avian response to El Niño is population change. Populations of tufted puffins and Cassin's auklets in central coastal California were reduced by 50% in response to El Niño conditions.[4, 63] Over the period 1972-1998, and particularly during the 1990s, a decade of prolonged decline in ocean productivity, Wahl and Tweit[187] found significant decreases in abundance of a number of species off southwest Washington, apparently reflecting a decline in biological productivity. Most dramatic was the 90% decline in sooty shearwaters shown for Washington and California (see also Veit et al.).[174] Also decreasing were flesh-footed and Buller's shearwaters and regionally breeding Cassin's auklet, tufted puffin, and common murres. Especially noticeable were changes in bird densities before and after the "step," a widespread, subtle change in environmental variables that occurred in the 1970s,[64] and before and after about 1990 when ocean productivity decreased and stayed low for years. Habitat switching was evident during the 1990s off Grays Harbor, with rhinoceros auklets moving more inshore where prey abundance was maintained, while offshore productivity declined. Common murres also moved inshore, though to a lesser extent, while sooty shearwaters declined in numbers there as well as offshore.[187]

Populations of rock sandpipers demonstrated an unusual trend following the 1982-83 El Niño episode. Their abundance in coastal Christmas Bird Count circles was higher prior to the 1982-83 El Niño than it was after that event.[28] It was speculated that the species underwent a range contraction during a prolonged period of oceanic warming.[28]

Brandt's and pelagic cormorants, western gulls, common murres, pigeon guillemots, and Cassin's auklets experienced low nesting success during El Niño years.[63] In 1983, the hatching success of common murre chicks on Farallon Islands, California, was 31.7% lower than in other years (i.e., 1972-1983) and the fledging success rate was only 15%, compared to 94% in other years.[14] Brandt's cormorants abandoned their nests in large numbers during the 1983 breeding season along the Oregon coast from Coquille Point Rocks to Sea Lion Caves. Some cormorants managed successfully to raise chicks by breeding earlier in the season.[91] Chick production by murres on the Oregon coast was lower than during pre-El Niño years.[80] Pelagic cormorants suffered low reproductive success all along the Oregon coast from Cape Meares to Coos Bay. Brandt's cormorants had lower reproductive success during 1982-83 and during the El Niño of 1987-88, but recovered in subsequent years.

The diet of cormorants and gulls at the Farallones, California during the 1976 and 1978 warm water episodes and the 1982-83 El Niño was more diverse and differed in composition from the diet in non-El Niño years. Seabirds were exploring different locations in search of food.[2] The very few cormorants, common murres, and gulls that were seen at the Farallones during 1983 spent more time in search for food than during normal years. As a result, the feeding frequencies decreased and adults brought fewer fish to chicks. With a decrease in the abundance of juvenile rockfish (*Sebastes* spp.) during the warm-water episodes in 1973, 1976, 1978, and 1983, common murres and pigeon guillemots switched to other prey species, such as anchovies, squid and smelt.[5] The change in diet and an increase in foraging time likely decreased the caloric intake of nestlings, contributing to the lowered reproductive success of the seabirds.

Egg laying among common murres during the 1983 breeding season was not as synchronized as it was in normal years, exposing chicks to predators such as western gulls.[14, 163] Food shortage during the 1982-83 episode contributed to conspecific predation on western gulls chicks.[164] Some of the birds arrived at their breeding grounds later in the season than usual (e.g. pigeon Guillemots arrived in mid-July instead of June and early July at Farallon Islands, California) and did not breed at all,[2] or in much lower numbers.[163] During the 1982-83 event, double-crested cormorants produced less than half the normal number of nestlings.

Habitat switching was evident during the 1990s off Grays Harbor, presumably in response to declining prey abundance offshore. Common murres, as one conspicuous example, have historically been widely dispersed over the shelf with flocks concentrating over the inner shelf (100 m to 150 m depth).[23] Numbers of murres decreased during years of low productivity in the 1990s, and birds were relatively closer inshore, usually in nearshore waters.[187] Even more noticeable was the shift inshore of rhinoceros auklets; their numbers increased throughout the time period of 1972-1998. Sooty shearwaters declined in overall numbers nearshore as well as offshore.[174, 187] Observations and anecdotal reports in the 1990s suggested that a number of species may have concentrated in inland marine waters (Strait of Juan de Fuca) during migrations or, in the case of regional breeders, during failed breeding seasons. These included storm-petrels, brown pelicans, phalaropes, jaegers and gulls (T. Wahl, unpubl. data).

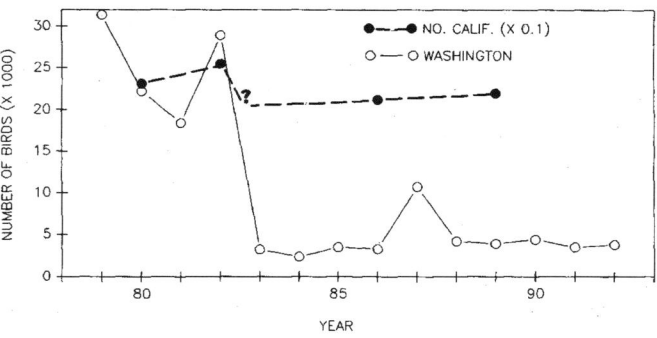

Figure 12. The number of breeding common murres in 5 California colonies and 28 colonies on the outer coast of Washington, 1979-1990. California counts are reduced by a factor of 10 to complement the y-axis scale, fitted to the Washington counts (from Ainley et al.[3]).

During the 1982-83 El Niño, a number of warm-water species, such as black-vented shearwater (*Puffinus opisthomelas*), black storm petrel (*Oceanodroma melania*), Craveri's murrelet (*Endomychura craveri*), and brown booby (*Sula leucogaster*) were present near the Farallon Islands in California. Unusually high numbers of Laysan albatrosses and black-footed albatrosses were seen off the coast of British Columbia during the 1992-93 El Niño episode.[77] A breeding range expansion occurred in California when the Heerman's Gull expanded north into southern California, apparently in response to large numbers of red crabs (*Pleuroncodes planipes*) that migrated to that region from the south during El Niño.[56]

Populations of some seabirds have been slow to recover from the impacts of El Niño events. Prior to the 1982-83 event, Brandt's cormorant numbers at the Farallon Islands, California, ranged between 15,000 and 25,000 birds. The population crashed, and had increased to only 3,000 - 10,000 by 1994[3] (Figure 13). The common murre population in California has increased by only 1-2% per year since 1982-83 in comparison to the 7.8% annual increase prior to that El Niño event[3] (Figure 12). Common murres were not as fortunate in Washington. It took until 1987 to recover from a significant drop in population (from 18,335-31,520 birds breeding during 1979-1982 to only 3,190 in 1983), and then the population was decimated to levels similar to those in the 1982-83 episode by the 1987-88 El Niño[3, 110, 195] (Figures 12, 13). On the other hand, rhinoceros auklets recovered quickly after the 1982-83 and 1986 El Niños during which they experienced only a minor decrease in numbers.[3] Double-crested cormorants along the Washington coast appear to have recovered from El Niño episodes[195] (Figure 13).

Some resident marine species do not appear to be impacted by oceanic warming. Boersma and Wheelwright[16] suggested that storm petrels nesting along the Washington and Oregon coasts were better able to cope with the fluctuating environmental conditions through their ability to leave their eggs unattended. Resistant to chilling and largely protected in underground burrows, petrel eggs may be left unattended for 31 days in

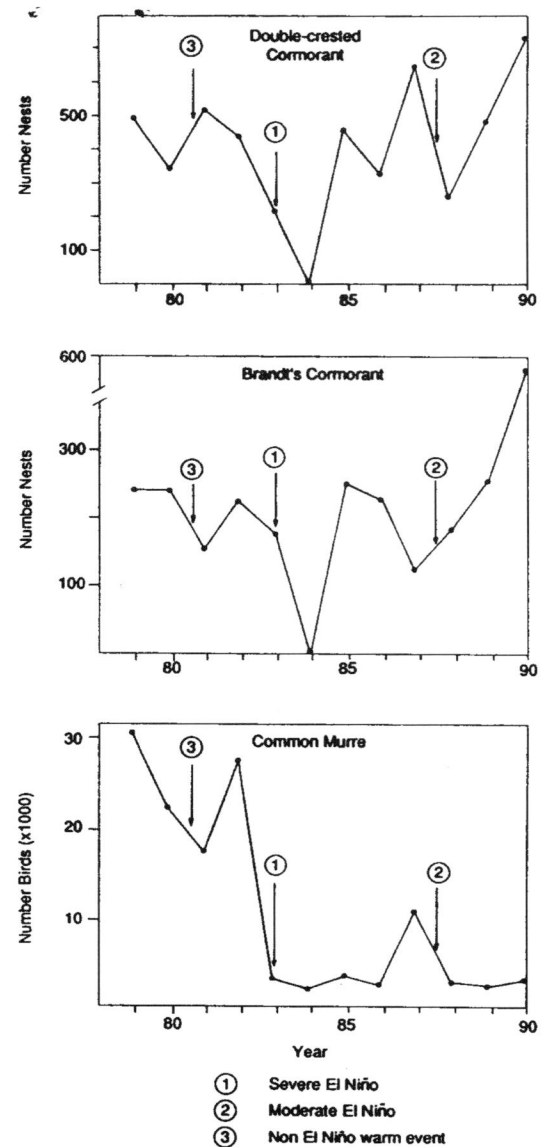

Figure 13. Seabird fluctuations and the occurrence of warm water events on the Washington outer coast, 1979-1990 (from Wilson[195]).

southeastern Alaska[16] and up to 17 days on the Farallon Islands, California.[2] This adaptation enables the adult birds to spend more time in search of food. Similarly, Leach's storm petrel feed further offshore, in areas not influenced by upwelling.[80] In fact, numbers of breeding petrels increased off southern California during periods of warmer ocean temperatures.[175] Of 5 species which increased in abundance (birds/km) off the Washington coast, only albatrosses, fulmar, and large gulls, species readily attracted to fishing vessels, increased offshore. In nearshore waters, however, two species that feed on anchovies and other fish increased significantly: brown pelican in Washington in the 1990s, and rhinoceros auklet since at least 1983.[187] Western gulls were able to switch to alternate food sources during the 1982-83 El Niño, and

they may have benefited from salmon hatchery smolts released in the Yaquina estuary the summer of 1983.[11] Although adult mortality of pigeon guillemots increased substantially and resulted in a nesting rate 19% below normal, there were no changes in the number of chicks that hatched and fledged per nest.[91] In contrast to common murres and cormorants, pigeon guillemots feed more on epibenthic fish so they were less affected by a decrease in primary productivity.

Marine Mammals. Despite a number of studies conducted on marine mammal responses to El Niño conditions in California[58, 59, 109] and Alaska,[199] very little information is available pertaining to marine mammals in the coastal areas of Washington and Oregon. The summary below is drawn from information relating to California sea lions at their California breeding areas. The California sea lion is distributed along the Pacific coast of North America from British Columbia to the southern tip of Baja California and Gulf of California, Mexico. The primary breeding areas are in the Channel Islands off California. There are no breeding areas for this species north of California. Although it is not clear whether the magnitude of impacts to sea lion populations noted in California occur in marine mammal populations in Washington and Oregon, this seems possible during severe El Niño events when oceanic productivity is altered off the coasts of Washington and Oregon.

Birth rates at the breeding rookeries in the Channel Islands declined substantially following the 1982-83 El Niño. The decrease in number of births at all rookeries there ranged from 30-62%. By 1984, birth rates had returned to pre-El Niño levels at San Nicolas Island, but lower birth rates prevailed at other islands through 1986.[59] The reduced birth rates were at least partially due to a lower number of females present at breeding sites.[74]

There was no apparent increase in female mortality during the 1982-83 El Niño. Females have high fidelity to breeding grounds, but the very strong El Niño of 1982-83, in combination with human disturbances, apparently forced them to disperse.[74] Indeed, counts of sea lions were much higher at their non-breeding rookeries, the Farallon Islands, during the summers of 1983 and 1984. Adult males, which fast during the breeding season and migrate north afterward, were not affected.[93]

Prenatal growth of California sea lion pups did not appear to be affected. Females were already pregnant at the onset of El Niño, and the mass of pups born that year did not differ significantly from non-El Niño years.[17] Growth during the first month of life did not differ significantly either, but pups began to loose mass when they reached two months of age.[17, 95] The number of foraging trips made by females from San Miguel Island increased on average by 37% in 1983, and pups received less milk than normal (female pups 3%; male pups 12%). Lower milk intake resulted in a 23% reduction in growth rates of pups and a decrease in body mass for pups 2 months of age and older.[95] "Sneak suckling", not commonly seen in California sea lions, increased, as did the proportion of yearlings suckling. Females that lost pups or did not breed that season shifted their attention to raising older offspring. Body mass of female and male pups declined by 25-35%, resulting in increased mortality at San Nicolas and San Clemente (but not on San Miguel).[59] The greater investment in foraging effort, which was already considerable,[48] therefore appeared to be related to the increase in pup mortality. The individual trips were not longer, but dives were deeper than normal as sea lions tried to compensate for more dispersed or deeper localized prey.[69] It appears that food habits changed dramatically during and shortly after the El Niño.[59, 85]

Population Status

The habitats of the marine-terrestrial interface support a substantial number of threatened, at-risk, vulnerable, or rare species. Threatened species include the snowy plover and marbled murrelet, the former occurring on sand beaches and low foredunes and the latter in coastal marine waters where it forages for fish. The snowy plover was listed as a threatened species because of habitat destruction and human disturbance in nesting areas.[172] At-risk species include the common murre, western grebe, scaups, and scoters, species that have experienced population declines in all or part of the region over the past two decades.[127, 195] Some of the species (e.g. western grebe, surf scoter) are known to carry high concentrations of environmental pollutants.[86, 87] Vulnerable species are those that meet certain life requisites only in certain habitats or where key habitat attributes are present; the destruction or degradation of those habitats could pose a substantial threat to such species. An example is the sanderling, which forages and roosts primarily on sand beaches and is vulnerable to the harmful effects of oil spills in that habitat.

Another way to define vulnerability is in terms of the proportion of the species' population that uses an area; areas that represent a substantial part of a species' range or support a significant proportion of the population might be considered vulnerable.[44, 62] Sanderlings, western sandpipers, dunlins, and Leach's storm-petrel, despite being abundant in coastal Washington and/or Oregon, easily qualify in this category because such large proportions of the species' overall or regional populations depend on habitats along our coasts.[120, 139, 167, 173] Finally, rare species include those that are numerically uncommon or rare in the region, although their populations may be stable. Examples of such species include the gyrfalcon and snowy owl.

Many other species associated with these habitats have maintained high levels of abundance or are increasing over previous population levels. For example, snow geese, brant, many "puddle" ducks, and loons appear to be common and stable in Washington estuaries (D. Kraege, pers. comm., D. Nysewander, pers. comm.). Although trend data are lacking, it appears that dunlin and sanderling populations may be stable over the past few decades.[141] Sea otter populations have increased substantially on the Washington coast following reestablishment efforts there in 1969-1970.[96] Bald eagle and peregrine falcon populations in the region have increased

in abundance (H. Allen, pers. comm.) following successful efforts to limit the use of harmful pollutants that formerly limited populations. Other species with stable or increasing populations include double-crested cormorant and California gull.[128, 187]

Conservation Mechanisms

Continued human population growth and the concomitant demands on natural resources for public commodities and recreation are challenges that require dedicated management attention. For example, recent estimates indicate that the population of Washington could double by 2045.[188] Despite the successes in recovering populations of endangered species, many other species are at risk or will become so in the next several decades. The reasons are many (Table 2) and the "solutions" are becoming more complicated and expensive.

The primary means for addressing these threats to species and habitats are 1) more proactive and innovative management (including growth management) that is designed to provide long-term solutions to protect wildlife and habitats and allow greater flexibility and options, 2) restoration and recovery of degraded habitats and wildlife populations, and 3) protection of the remaining important habitats. Although a detailed discussion of the first two topics is beyond the scope of this chapter, a few relevant points are in order.

The development of conservation initiatives in marine habitats has increased dramatically in recent years. Recognizing that many species must be managed and protected at very large spatial scales, management and conservation plans have been developed for waterfowl[42, 171] and shorebirds.[62] A conservation plan is being developed for colonially nesting birds. Important aspects of the shorebird plan include public education and identification of voluntary conservation measures (e.g. conservation easements) that could be implemented on private lands. Such actions may be crucial given that many important estuarine sites in Washington and Oregon contain areas of privately owned land.[62]

Human activities in the region have greatly influenced the extent and quality of important wildlife habitats.[19, 50] With the tremendous growth in the field of wildlife ecology in recent decades, particularly in our understanding of species-habitat relationships, there now appear to be opportunities to restore destroyed or degraded habitats. Although restoration ecology should not be viewed as a panacea for unwise or unplanned growth and development, it does offer hope that degraded habitats can be recovered to the extent that they can function for wildlife. A variety of restoration activities that would benefit shorebirds have been described.[27, 62]

Marine Protected Areas

Protection of marine habitats is an important component of marine conservation efforts. Certain marine areas in Washington and Oregon are now recognized as Marine Protected Areas (MPAs). The concept of the MPA is relatively recent, being defined, as "any area of intertidal or subtidal terrain, together with its overlying water and associated flora, fauna, historical and cultural features, which has been reserved by law or other effective means to protect part or all of the enclosed environment."[119] The number of MPAs has grown from 118 worldwide in 1970 to over 1,300 in the 1990s.[119] Important underlying principles of the MPA program are the restoration of depleted stocks of fish through the establishment of no-harvest areas, and protection of both endangered species and critical habitats.

MPAs in Washington and Oregon are managed in a variety of administrative contexts. Parks, marine sanctuaries, wildlife refuges, wilderness areas, and conservation areas have been established and administered by state or federal government, local and tribal governments, and private organizations. The MPAs vary in size, management intent, and the degree to which they are protected. Some MPAs are non-contiguous, composed of many elements such as rocks, reefs, or islands. Although there are no standard regulations used to manage the body of MPAs, a number of MPAs were established and are managed as a result of cooperation among multiple agencies and organizations.[119, 129, 151]

The need to establish and manage areas to protect marine habitats and species was recognized in Washington only recently. Increasing awareness of degradation of marine habitats led to research and monitoring in the shared inland marine waters of Washington and British Columbia and eventually to the signing of an Environmental Cooperation Agreement between British Columbia and Washington State in 1992. The goal of the agreement was to address habitat degradation and declines in fish and shellfish populations in Puget Sound, Hood Canal, Strait of Juan de Fuca and the Strait of Georgia. The recognition of the importance of MPAs was one of the outcomes of the agreement.[119] Since the founding of the Washington State Seashore Conservation Areas in 1969 and the Washington Marine Protected Areas Work Group in 1995, the entire intertidal coast of Washington has become a single MPA.[119, 151] There are 33 additional MPAs on the outer coast and 102 in Puget Sound (Figure 14a).[61, 119, 151]

MPAs in Washington and Oregon are owned and managed by a diverse group of organizations. These include Washington Department of Natural Resources (Natural Area Preserves and Natural Resources Conservation Areas), Washington Department of Fish and Wildlife (Wildlife Areas), Washington State Parks and Recreation Commission (State Parks), Washington Department of Ecology (Padilla Bay National Estuarine Research Reserve [this site is cooperatively managed with other agencies]), University of Washington (Marine Biological Preserve in waters of San Juan County and waters surrounding Cypress Island in Skagit County), Oregon Department of Fish and Wildlife (Marine Gardens, Research Reserves), Oregon Parks and Recreation Department (State Parks, Habitat Refuges), U.S. Fish and Wildlife Service (National Wildlife Refuges), United Nations Educational, Scientific and Cultural Organization

(UNESCO; Man and the Biosphere Reserve), National Oceanic and Atmospheric Administration (Olympic Coast National Marine Sanctuary), U.S. Bureau of Land Management (various protected natural areas and reserves), Environmental Protection Agency (National Estuary Plan sites), and local governments, Native American tribes, and private organizations (e.g., The Nature Conservancy, San Juan Preservation Trust).

Various state laws (e.g. ORS 196 and ORS 197), statewide Planning Goals, the Territorial Sea Plan, other agency rules and statutes, as well as federal laws, apply to Oregon ocean resources. Oregon Policy Advisory Council,[129] state agencies, local governments, and federal agencies implement the above policies and laws. One of the elements of the ocean-resource management is the Rocky Shores Strategy program, which addresses shoreline and offshore rocks and reefs. Oregon rocky shores belong to the public, with the Oregon Division of State Lands (ODSL) being the trustee on the public's behalf. ODSL shares management of the rocky shores with Oregon Parks and Recreation Department. Marine life is under the jurisdiction of the ODFW, and the "dry" part of the offshore rocks and islands are managed by the U. S. Fish and Wildlife Service as National Wildlife Refuges. Designated sites are managed by numerous state and federal agencies or private non-profit organizations. MPAs are found along much of the Oregon coast (Figure 14b).

The primary goals of the MPAs are varied, but they typically emphasize protection of rare or sensitive marine species and/or habitats. Some of the MPAs also place great importance on the scientific, educational, and recreational values unique to the site. Species protection ranges from bottom fish, shellfish, and marine invertebrates, to wildlife such as harbor seal, California and Steller's sea lions, waterfowl, seabirds, shorebirds, bald eagle, and peregrine falcon. Harvest of certain food species and full access are allowed in most MPAs, although access may be difficult in some areas due to terrain features or the remote nature of the site; access is prohibited (e.g., some Natural Area Preserves) or restricted to particular areas (e.g., Padilla Bay National Estuarine Research Reserve, breeding sites on refuges) at very few MPAs. Further details regarding MPAs in Washington and Oregon are available elsewhere.[61, 119, 129, 151]

Research Needs

There is a great need for additional research regarding the species and habitats of the marine environment. Some of the research topics are species or habitat-specific, whereas many other topics apply to multiple species and habitats. General research topics that apply to multiple species and habitats include: 1) response of marine bird and mammal populations to the effects of oil and other pollutants (essentially all habitats), various fisheries activities (estuaries, inland marine, nearshore, and shelf), and El Niño (essentially all habitats), 2) population demography and monitoring of status or trends (relates to species or communities in all habitats), 3) development of methods to reduce impacts of fisheries activities on non-target

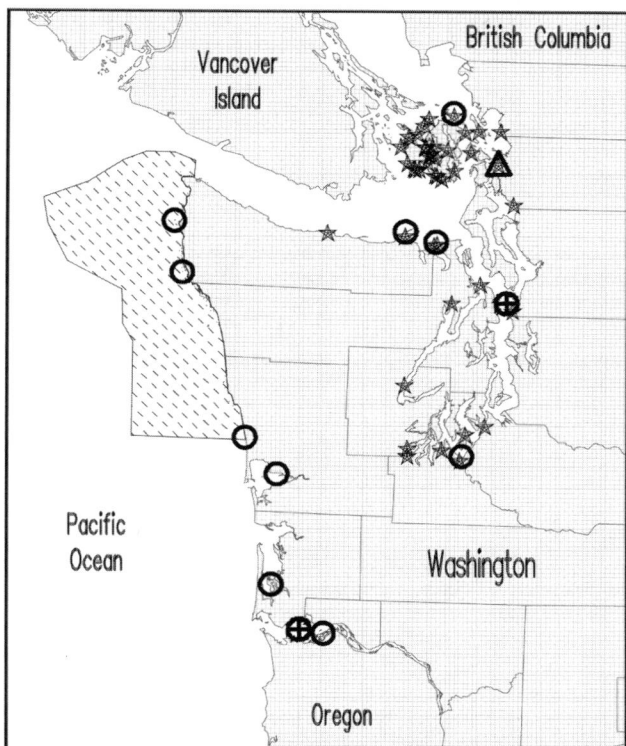

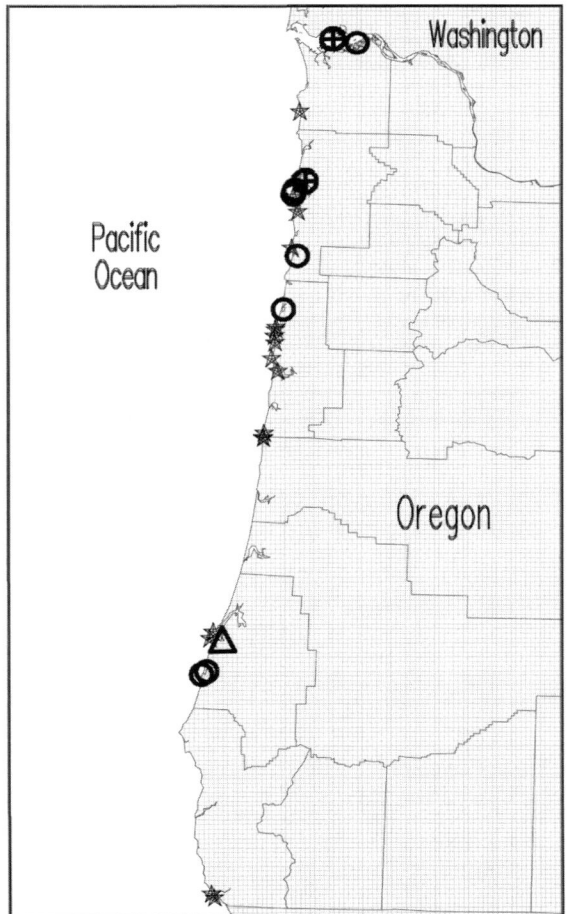

Figure 14. Distribution of selected Marine Protected Areas in Washington and Oregon.

⊕ National Estuary Program

△ National Estuarine Research Preserve

○ National Wildlife Refuge

✲ State Marine Protected Areas

⬚ National Marine Sanctuary

species (estuaries, inland marine), and 4) responses to human disturbances (beaches, estuaries, headlands). Examples of species- or habitat-specific research needs include: behavioral, physiological, or other responses to restored habitats (e.g. snowy plover on beaches after beachgrass removal, shorebirds at restored roost sites in an estuary such as Tillamook Bay, shorebird or waterfowl foraging behavior in cordgrass removal areas); more general studies of habitat use (foraging habitat preferences of waterbirds in estuaries as a function of tidal phase); studies that enable inference to be made about habitat quality (e.g. studies of food habits and roost site selection by waterfowl, waterbirds, shorebirds, colonially-nesting gulls in estuaries); and the effects of sedimentation on estuarine productivity.

Acknowledgments

We thank Janet Hodder and Dave Nysewander for making comments that improved the manuscript; John Calambokidis, Al Didier, Jr., Roy Lowe, and Ken Warheit for providing important information and references; and Jeff Gilligan and Dave Lauten for sharing their expertise.

Literature Cited

1. Akins, G. J., and C.A. Jefferson. 1973. Coastal wetlands of Oregon. Oregon Coastal Conservation and Development Commission. Salem, OR.

2. Ainley, D. G., R. J. Boekelheide, S. H. Morrell, and C. S. Strong. 1990. Pigeon Guillemot. Pages 277-305 *in:* D. G. Ainley, and R. J. Boekelheide, editors. Seabirds of the Farallon Islands: Ecology, dynamics, and structure of an upwelling-system community. Stanford University Press, CA.

3. ———, W. J. Sydeman, S. A. Hatch, and U. W. Wilson. 1994. Seabird population trends along the west coast of North America: cause and the extent of regional concordance. Studies in Avian Biology 15: 119-133.

4. ———, L. B., Spear, and S. G. Allen. 1996. Variation in the diet of the Cassin's auklet reveals spatial, temporal, seasonal, and decadal occurrence patterns of euphasids off California, USA. Marine Ecology Progress Series 137: 1-10.

5. ———, L. B. Spear, S. G. Allen and C.A. Ribic. 1996. Temporal and spatial patterns in the diet of the common murre in California waters. Condor 98: 691-705.

6. Ashmole, N. P. 1971. Seabird ecology and the marine environment. Pages 112-286 in: D. S. Farner and J. R. King, editors. Avian biology, Vol. 1. Academic Press, New York, NY.

7. Baird, R. W., and L. M. Dill. 1995. Occurrence and behavior of transient killer whales: seasonal and pod-specific variability, foraging behavior, and prey handling. Canadian Journal of Zoology 73: 1300-1311.

8. ———, and ———. 1996. Ecological and social determinants of group size in transient killer whales. Behavioral Ecology 7: 408-416.

9. Baldridge, A., and J. B. Crowell, Jr. 1966. The fall migration: northern Pacific coast region. Audubon Field Notes 20: 82.

10. Barber, R. T., and F. P. Chavez. 1983. Biological consequences of El Niño. Science 222: 1204-1210.

11. Bayer, R. D. 1986. Breeding success of seabirds along the mid-Oregon coast concurrent with the 1983 El Niño. Murrelet 67: 23-26.

12. Beach, R. J., A. C. Geiger, S. J. Jeffries, S. D. Treacy, and B. L. Troutman. 1985. Marine mammals and their interactions with fisheries of the Columbia River and adjacent waters, 1980-1982. NMFS-AFSC Processed Rept. 85-04, 316 pp. Alaska Fisheries Science Center, NMFS, NOAA, 7600 Sand Point Way NE, Seattle, WA.

13. Bigg, M. A., P. F. Olesiuk, G. M. Ellis, J. K. B. Ford, and K. C. Balcomb. 1990. Social organization and genealogy of resident killer whales (Orcinus orca) in the coastal waters of British Columbia and Washington State. Reports to the International Whaling Commission, Special Issue 12: 383-405.

14. Boekelheide, R. J., D. G. Ainley, S. H. Morrell, H. R. Huber, and T. J. Lewis. 1990. Common murre. Pages 245-275 in: D. G. Ainley, and R. J. Boekelheide, editors. Seabirds of the Farallon Islands: Ecology, dynamics, and structure of an upwelling-system community. Stanford University Press, CA.

15. Boersma, P. D., and M. J. Green. 1993. Conservation of storm-petrels in the North Pacific. Pages 112-121 in: K. Vermeer, K. T. Briggs, K. H. Morgan, and D. Siegel-Causey, editors. The status, ecology, and conservation of marine birds of the North Pacific. Canadian Wildlife Service Special Publication, Ottawa.

16. ———, and N. T. Wheelwright. 1979. Egg neglect in the procellariiformes: reproductive adaptations in the fork-tailed storm-petrel. Condor 81: 157-165.

17. Boness, D. J., O. T. Oftedal, and K. A. Ono. 1991. The effect of El Niño on pup development in the California sea lion (Zalophus californianus). Early postnatal growth. Pages 173-179 in: F. Trillmich, and K. A. Ono, editors. Pinnipeds and El Niño: Responses to environmental stress. Springer-Verlag, Berlin, Germany.

18. Bonnell, M. L., C. E. Bowlby, and G. A. Green. 1992. Pinniped distribution and abundance off Oregon and Washington, 1989-1990. Chapter 2 in: J. J. Brueggeman, editor. Oregon and Washington marine mammal and seabird surveys. OCS Study MMS 91-0093. Ebasco Environmental, Bellevue, WA and Ecological Consulting, Inc., Portland, OR.

19. Boule, M., N. Olmsted, and T. Miller. 1983. Inventory of wetland resources and an evaluation of wetland management in western Washington. Washington Department of Ecology, Olympia, WA.

20. ———, and K. F. Bierly. 1987. History of estuarine wetland development and alteration: What have we wrought? Northwest Environmental Journal 3: 43-61.

21. Bowlby, C. E., G. A. Green, and M. L. Bonnell. 1994. Observations of leatherback turtles offshore of Washington and Oregon. Northwestern Naturalist 75: 33-35.

22. Briggs, K. T., and E. W. Chu. 1987. Trophic relationships and food requirements of California seabirds: updating models of trophic impact. Pages 279-304 in: J. P. Croxall, editor. Seabird feeding ecology and role in marine ecosystems. Cambridge University Press, Cambridge, England.

23. ———, D. H. Varoujean, W. W. Williams, R. G. Ford, M. L. Bonnell, and J. L. Casey. 1992. Seabirds of the Oregon and Washington OCS, 1989-1990. Chapter 3 in: J. J. Brueggeman, editor. Oregon and Washington marine mammal and seabird surveys. OCS Study MMS 91-0093. Ebasco Environmental, Bellevue, WA and Ecological Consulting, Inc., Portland, OR.

24. Brown, R. F. 1988. Assessment of pinniped populations in Oregon, April 1984 to April 1985. NMFS-NWAFC Processed Rept. 88-05, 44 p. Alaska Fisheries Science Center, NMFS, NOAA, 7600 Sand Point Way NE, Seattle, WA.

25. Brueggeman, J. J. ed. 1992. Oregon and Washington Marine Mammal and Seabird Surveys—Final Report. OCS Study MMS 91-0093. Ebasco Environmental, Bellevue, WA and Ecological Consulting, Inc., Portland, OR.

26. Reference deleted in proof.

27. Buchanan, J. B. In press. Shorebirds: plovers, oystercatchers, avocets and stilts, sandpipers, snipes, and phalaropes. In E. M. Larsen, and N. Nordstrom, editors. Management recommendations for priority bird species. Washington Department of Fish and Wildlife, Olympia, WA.

28. ———. 1999. Recent changes in the winter distribution and abundance of rock sandpipers in North America. Western Birds 30: 193-199.

29. ———. 1992. Winter abundance of shorebirds at coastal beaches of Washington. Washington Birds 2: 12-19.

30. ———. 1988. Migration and winter populations of greater yellowlegs, Tringa melanoleuca, in western Washington. Canadian Field-Naturalist 102: 611-616.

31. ———, and J. R. Evenson. 1997. Abundance of shorebirds at Willapa Bay, Washington. Western Birds 28: 158-168.

32. Burger, A. E., and D. M. Fry. 1993. Effects of oil pollution on seabirds in the northeast Pacific. Pages 254-263 in: K. Vermeer, K. T. Briggs, K. H. Morgan, and D. Siegel-Causey, editors. The status, ecology and conservation of marine birds of the North Pacific. Canadian Wildlife Service Special Publication, Ottawa, Canada.

33. Byrd, G. V., E. C. Murphy, G. W. Kaiser, A. Y. Kondratyev, and Y. V. Shibaev. 1993. Status and ecology of offshore fish-feeding alcids (murres and puffins) in the North Pacific. Pages 176-198 in: K. Vermeer, K. T. Briggs, K. H. Morgan, and D. Siegel-Causey, editors. The status, ecology and conservation of marine birds of the North Pacific. Canadian Wildlife Service Special Publication, Ottawa, Canada.

34. Calambokidis, J., J. B. Buchanan, G. H. Steiger, and J. R. Evenson. 1991. Toxic contaminants in Puget Sound wildlife: Literature review and recommendations for research and monitoring. Contract report 68-D8-0085 to U.S. Environmental Protection Agency, Seattle, WA.

35. ———, T. Chandler, K. Rasmussen, G. H. Steiger, and L. Schlender. 1999. Humpback and blue whale photographic identification research off California, Oregon, and Washington in 1998. Final report to Southwest Fisheries Science Center, Olympic Coast National Marine Sanctuaries, and University of California at Santa Cruz. Cascadia Research Collective, Olympia, WA.

36. ——— J. C. Cubbage, J. R. Evenson, S. D. Osmek, J. L. Laake, P. J. Gearin, B. J. Turnock, S. J. Jeffries, and R. F. Brown. 1993. Abundance estimates of harbor porpoise in Washington and Oregon waters. Final Report to National Marine Mammal Laboratory, AFSC, NMFS, Seattle, WA. Cascadia Research Collective, Olympia, WA.

37. ———, J. R. Evenson, G. H. Steiger, and S. J. Jeffries. 1994. Gray whales of Washington State: Natural history and photographic catalog. Cascadia Research Collective, Olympia, WA.

38. ———, J. Peard, G. H. Steiger, J. C. Cubbage, and R. L. DeLong. 1984. Chemical contaminants in marine mammals from Washington State. NOAA Technical Memorandum NOS OMS 6, Rockville, MD.

39. ———, G. H. Steiger, J. R. Evenson, K. R. Flynn, K. C. Balcomb, D. E. Claridge, P. Bloedel, J. M. Straley, C. S. Baker, O. von Ziegesar, M. E. Dahlheim, J. M. Waite, J. D. Darling, G. Ellis, and G. A. Green. 1996. Interchange and isolation of humpback whales off California and other North Pacific feeding grounds. Marine Mammal Science 12: 215-226.

40. ———, ——— K. Rasmussen, J. Urbán R., K. C. Balcomb, P. Ladrón de Guevara P., M. Salinas Z., J. K. Jacobsen, C. S. Baker, L. M. Herman, S. Cerchio, and J. D. Darling. 2000. Migratory destinations of humpback whales that feed off California, Oregon and Washington. Marine Ecology Progress Series 192: 295-304

41. ———, ———, J. M. Straley, L. M. Herman, S. Cerchio, D. R. Salden, R. J. Urbán, J. K. Jacobsen, O. von Ziegesar, K. C. Balcomb, C. M. Gabriele, M. E. Dahlheim, S. Uchida, G. Ellis, Y. Miyamura, P. Ladron de Guevara,

M. Yamaguchi, F. Sato, S. A. Mizroch, L. Schlender, K. Rasmussen, and J. Barlow. In press. Movements and population structure if humpback whales in the North Pacific Basin. Marine Mammal Science.

42. Canadian Wildlife Service and U.S. Fish and Wildlife Service. 1998. North American waterfowl management plan (1998 update).

43. Carter, H. R., U. W. Wilson, R. W. Lowe, M. S. Rodway, D. A. Manuwal, J. E. Takekawa, and J. L. Yee. In press. Population trends of the common murre (*Uria aalge californica*). *In*: D. A. Manuwal, H. R. Carter, and T. Zimmerman, editors. Biology and conservation of the common murre in California, Oregon, Washington, and British Columbia. U.S. Geological Survey, Information and Technology Report, Washington, D.C.

44. Carter, M. F., W. C. Hunter, D. N. Pashley, and K. V. Rosenberg. 2000. Setting conservation priorities for landbirds in the United States: The Partners in Flight approach. Auk 117:541-548.

45. Condit, R., and B. J. Le Boeuf. 1984. Feeding habits and feeding grounds of the northern elephant seal. Journal of Mammalogy 65: 281-290.

46. Cooke, W. W. 1910. Distribution and migration of North American shorebirds. United States Biological Survey Bulletin No. 35.

47. Cooper, W. S. 1958. Coastal sand dunes of Oregon and Washington. Geological Society of America. Memoir 72.

48. Costa, D. P., G. A. Antonelis, and R. L. DeLong. 1991. Effects of El Niño on the foraging energetics of the California sea lion. Pages 156-165 *in*: F. Trillmich, and K. A. Ono, editors. Pinnipeds and El Niño. Responses to environmental stress. Springer-Verlag, Berlin, Germany.

49. Custer, T. W., and J. P. Myers. 1990. Organochlorines, mercury, and selenium in wintering shorebirds from Washington and California. California Fish and Game 76: 118-125.

50. Dahl, T. E. 1990. Wetlands of the United States 1780s to 1980s. U.S. Fish and Wildlife Service, Washington, D.C.

51. Darling, J. D. 1984. Gray whales off Vancouver Island, British Columbia. Pages 267-287 *in*: M. L. Jones, S. L. Swartz, and S. Leatherwood, editors. The Gray Whale, *Eschrictius robustus*. Academic Press, Inc., Orlando, FL.

52. ———, J. Calambokidis, K. C. Balcomb, P. Bloedel, K. R. Flynn, A. Mochizuki, K. Mori, F. Sato, H. Suganuma, and M. Yamaguchi. 1996. Movement of a humpback whale (*Megaptera novawangliae*) from Japan to British Columbia and return. Marine Mammal Science 12: 281-287.

53. ———, K.E. Keogh, and T. E. Steeves. 1998. Gray whale (*Eschrichtius robustus*) habitat utilization and prey species off Vancouver Island, British Columbia. Marine Mammal Science 14: 692-720.

54. Day, J. W., Jr., C. A. S. Hall, W. M. Kemp, and A. Yáñez-Arancibia. 1989. Estuarine ecology. John Wiley and Sons, New York, NY.

55. Day, R. D., H. S. Wehle, and F. C. Coleman. 1985. Ingestion of plastic pollutants by marine birds. Pages 344-386 *in*: R. S. Shomura, and H. O. Yoshida, editors. Proceedings of the workshop on the fate and impact of marine debris. U.S. Department Committee, NOAA Technical Memo, NMFS, NOAA-TM-NMFS-SWFC-54.

56. Dayton, P. K., and M. J. Tegner. 1990. Bottoms beneath troubled waters: benthic impacts of the 1982-1984 El Niño in the temperate zone. Pages 433-472 *in*: P. W. Glynn, editor. Global ecological consequences of the 1982-1983 El Niño-Southern Oscillation. Elsevier Oceanography Series, 52. University of Miami, Miami, FL.

57. DeGange, A. R., R. H. Day, J. E. Takekawa, and V. M. Mendenhall. 1993. Losses of seabirds to gill nets in the North Pacific. Pages 204-211 *in*: K. Vermeer, K. T. Briggs, K. H. Morgan, and D. Siegel-Causey, editors. The status, ecology, and conservation of marine birds of the North Pacific. Canadian Wildlife Service Special Publication, Ottawa, Canada.

58. DeLong, R. L., and G. A. Antonelis. 1991. Impact of the 1982-1983 El Niño on the northern fur seal population at San Miguel Island, California. Pages 75-83 *in*: F. Trillmich, and K. A. Ono, editors. Pinnipeds and El Niño. Responses to environmental stress. Springer-Verlag, Berlin, Germany.

59. ———, C. W. Oliver, B. S. Stewart, M. C. Lowry, and P. K. Yochem. 1991. Effects of the 1982-83 El Niño on several population parameters and diet of California sea lions on the California Channel Islands. Pages 166-172 *in*: F. Trillmich, and K. A. Ono, editors. Pinnipeds and El Niño. Responses to environmental stress. Springer-Verlag, Berlin, Germany.

60. ———, B. S. Stewart, and R. D. Hill. 1992. Documenting migrations of northern elephant seals using day length. Marine Mammal Science 8:155-159.

61. Didier, A. J., Jr. 1998. Marine Protected Areas of Washington and Oregon. Pacific Fishery Management Council. Contract No. 98-08.

62. Drut, M. S. and J. B. Buchanan. 2000. U.S. National shorebird conservation plan: Northern Pacific coast working group regional management plan.

63. Duffy, D. C. 1990. Seabirds and the 1982-1983 El Niño-Southern Oscillation. Pages 395-415 *in*: P. W. Glynn, editor. Global ecological consequences of the 1982-83 El Niño Southern Oscillation. Elsevier Oceanography Series, 52. University of Miami, FL.

64. Ebbesmeyer, C. C., D. R. Cayan, D. R. McLain, F. H. Nichols, D. H. Peterson, and K. T. Redmond. 1991. 1976 step in the Pacific climate: forty environmental changes between 1968-1975 and 1977-1984. Pages 129-141 *in*: J. L. Betancourt, and V. L. Tharp, editors. Proceedings of the Seventieth Annual Pacific Climate (PACLIM) Workshop. California Department of Water Resources. Interagency Ecological Studies Program Technical Report 26.

65. Evenson, J. R., and J. B. Buchanan. 1997. Seasonal abundance of shorebirds at Puget Sound estuaries. Washington Birds 6:34-62.

66. ———, D. R. Nysewander, and M. Mahaffy. 1999. Progress report of the 1999 pigeon guillemot breeding colony census of Washington inland marine waters. Washington Department of Fish and Wildlife.

67. Ewins, P. J., H. R. Carter, and Y. V. Shibaev. 1993. The status, distribution, and ecology of inshore fish-feeding alcids (*Cepphus* guillemots and *Brachyramphus* murrelets in the North Pacific. Pages 164-175 *in*: K. Vermeer, K. T. Briggs, K. H. Morgan, and D. Siegel-Causey, editors. The status, ecology and conservation of marine birds of the North Pacific. Canadian Wildlife Service Special Publication, Ottawa, Canada.

68. Favorite, F., A. J. Dodimead, and K. Nasu. 1976. Oceanography of the subarctic Pacific region 1960-71. International Northern Pacific Fisheries Commission Bulletin No. 33.

69. Feldkamp, S. D., R. L. DeLong, and G. A. Antonelis. 1991. Effects of El Niño 1983 on the foraging patterns of California sea lions (*Zalophus californianus*) near San Miguel Island, California. Pages 146-155 *in*: F. Trillmich, and K. A. Ono, editors. Pinnipeds and El Niño. Responses to environmental stress. Springer-Verlag, Berlin, Germany.

70. Fiscus, C., and K. Niggol. 1965. Observations of cetaceans off California, Oregon, and Washington. U.S. Fish and Wildlife Service, Special Science Report-Fisheries No. 498.

71. Forbush, E. H. 1912. A history of the game birds, wildfowl, and shore birds of Massachusetts and adjacent states. Massachusetts Board of Agriculture.

72. Ford, J. K. B., G. M. Ellis, L. G. Barrett-Lennard, A. B. Morton, R. S. Palm, and K. C. Balcomb. 1999. Dietary specialization of two sympatric populations of killer whales (*Orcinus orca*) in coastal British Columbia and adjacent waters. Canadian Journal of Zoology 76: 1456-1471.

73. Ford, R. G., D. H. Varoujean, D. R. Warrick, W. A. Williams, D. B. Lewis, C. L. Hewitt, and J. L. Casey. 1991. Final report: seabird mortality resulting from the Nestucca oil spill incident, winter 1988-89. Report for Washington Dept. of Wildlife.

74. Francis, J. M., and C. B. Heath. 1991. The effects of El Niño on the frequency and sex ratio of suckling yearlings in the California sea lion. Pages 193-210 *in*: F. Trillmich, and K. A. Ono, editors. Pinnipeds and El Niño. Responses to environmental stress. Springer-Verlag, Berlin, Germany.

75. Frankignoul, C. 1994. Equatorial ocean dynamics: observations and modeling. Pages 157-236 *in*: P. Malanotte Rizzoli, editor. The central circulation of the Oceans. Instituto Veneto Di Scienze, Lettere Ed Arti, Venice, Italy.

76. Franklin, J. F., and C. T. Dyrness. 1973. Natural vegetation of Oregon and Washington. U.S. Forest Service Gen. Tech. Rep. PNW-8. Reprinted 1988, Oregon State University Press, Corvallis, OR.

77. Gaston, A. J. 1994. Unusual number of Laysan albatrosses, *Diomedea immutabilis*, off the west coast of Haida Gwaii, Queen Charlotte Islands, British Columbia. Canadian Field-Naturalist. 108: 373.

78. ———, and S. B. C. Dechesne. 1996. Rhinoceros auklet (*Cerorhinca monocerata*). Pages 1-20 *in:* A. Poole, and F. Gill, editors. The birds of North America. The Academy of Natural Sciences, Philadelphia, PA, and The American Ornithologists' Union, Washington, D.C.

79. Gonor, J. J., J. R. Sedell, and P. A. Benner. 1988. What we know about large trees in estuaries, in the sea, and on coastal beaches. Pages 83-112 *in:* C. Maser, R. F. Tarrant, J. M. Trappe, and J. F. Franklin, editors. From the forest to the sea: A story of fallen trees. USDA General Technical Report PNW-GTR-229. Portland, OR.

80. Graybill, R., and J. Hodder. 1985. Effects of the 1982-83 El Niño on reproduction of six species of seabirds in Oregon. Pages 205-210 *in:* W. S. Wooster, and D. L. Fluharty, editors. El Niño North: El Niño effects in the eastern subarctic Pacific Ocean. Washington Sea Grant Program, University of Washington, Seattle, WA.

81. Green, G. A., J. J. Brueggeman, R. A. Grotefendt, C. E. Bowlby, M. L. Bonnell, and K. T. Balcomb, III. 1992. Cetacean distribution and abundance off Oregon and Washington, 1989-1990. Chapter 1 *in:* J. J. Brueggeman, editor. Oregon and Washington Marine Mammal and Seabird Surveys. OCS Study MMS 91-0093. Ebasco Environmental, Bellevue, WA and Ecological Consulting, Inc., Portland, OR.

82. ———, R. A. Grotefendt, M. A. Smultea, C. E. Bowlby, and R. A. Rowlett. 1993. Delphinid aerial surveys in Oregon and Washington offshore waters. Final Report to National Marine Mammal Laboratory, National Marine Fisheries Service, Seattle, WA. Ebasco Environmental, Bellevue, WA.

83. Grinnell, J., H. C. Bryant, and T. I. Storer. 1918. The game birds of California. University of California Press, Berkeley, CA.

84. Harrington, B., and E. Perry. 1995. Important shorebird staging sites meeting Western Hemisphere Shorebird Reserve Network criteria in the United States. USDI Fish and Wildlife Service.

85. Heath, C. B., K. A. Ono, D. J. Bonese, and J. M. Francis. 1991. The influence of El Niño on female attendance patterns in the California sea lion. Pages 138-145 *in:* F. Trillmich, and K. A. Ono, editors. Pinnipeds and El Niño. Responses to environmental stress. Springer-Verlag, Berlin, Germany.

86. Henny, C. J., L. J. Blus, and R. A. Grove. 1990. Western grebe, *Aechmophorus occidentalis*, wintering biology and contaminant accumulation in Commencement Bay, Puget Sound, Washington. Canadian Field-Naturalist 104: 460-472.

87. ———, ———, ———, and S. P. Thompson. 1991. Accumulation of trace elements and organochlorines by surf scoters wintering in the Pacific Northwest. Northwestern Naturalist 72: 43-60.

88. Herman, S. G., and J. B. Bulger. 1981. Distribution and abundance of shorebirds during the 1981 spring migration at Grays Harbor, Washington. U.S. Army Corps of Engineers, Contract Report DACW67-81-0936. Seattle, WA.

89. Herzing, D. L., and B. R. Mate. 1984. Gray whale migrations along the Oregon coast, 1978-1981. Pages 289-307 *in* M. L. Jones, S. L. Swartz, and S. Leatherwood, editors. The Gray Whale, *Eschrictius robustus*. Academic Press, Inc., Orlando, FL.

90. Hodder, J., R. F. Brown, and C. Cziesla. 1998. The northern elephant seal in Oregon: A pupping range extension and onshore occurrence. Marine Mammal Science 14: 873-880.

91. ———, and M. R. Graybill. 1985. Reproduction and survival of seabirds in Oregon during the 1982-83 El Niño. Condor 535-541.

92. Holden, C. 1999. Pumping up the greenhouse. Science 285: 2057.

93. Huber, H. R., C. Beckham, and J. Nisbet. 1991. Effects of the 1982-83 El Niño on northern elephant seals on the South Farallon Islands, California. Pages 219-233 *in:* F. Trillmich, and K. A. Ono, editors. Pinnipeds and El Niño. Responses to environmental stress. Springer-Verlag, Berlin, Germany.

94. Huntington, C. E., R. G. Butler, and R. A. Mauck. 1996. Leach's storm-petrel (*Oceanodroma leucorhoa*). Pages 1-32 *in:* A. Poole, and F. Gill, editors. The birds of North America. The Academy of Natural Sciences, Philadelphia, PA, and The American Ornithologists' Union, Washington, D.C.

95. Iverson, S. J., O. T. Oftedal, and D. J. Boness. 1991. The effect of El Niño on pup development in the California sea lion (*Zalophus californianus*).

Milk intake. Pages 180-184 *in:* F. Trillmich, and K. A. Ono, editors. Pinnipeds and El Niño. Responses to environmental stress. Springer-Verlag, Berlin, Germany.

96. Jameson, R. J., K. W. Kenyon, S. Jeffries, and G. A. Vanblaricom. 1986. Status of a translocated sea otter population and its habitat in Washington. Murrelet 67: 84-87.

97. Jeffries, S. J., R. F. Brown, H. R. Huber, and R. L. DeLong. 1997. Assessment of harbor seals in Washington and Oregon, 1996. Pages 83-94 *in:* P. S. Hill, and D. P. DeMaster, editors. MMPA and ESA Implementation Program, 1996. AFSC Processed Rept. 97-10. Alaska Fisheries Science Center, NMFS, NOAA, 7600 Sand Point Way NE, Seattle, WA.

98. ———, P. J. Gearin, H. R. Huber, D. L. Saul, and D. A. Pruett. 2000. Atlas of seal and sea lion haulout sites in Washington. Washington Department of Fish and Wildlife, Olympia, WA.

99. Kelly, J. P., J. G. Evens, R. W. Stallcup, and D. Wimpfheimer. 1996. Effects of aquaculture on habitat use by wintering shorebirds in Tomales Bay, California. California Fish and Game 82: 160-174.

100. Kennish, M. J. 1992. Ecology of estuaries: Anthropogenic effects. CRC Press, Boca Raton, FL.

101. Kerr, R. A. 1998. Warming's unpleasant surprise: Shivering in the greenhouse? Science. 281: 156-158.

102. ———. 1999. Big El Niños ride the back of slower climate change. Science 283: 1108-1109.

103. Komar, P. D. 1997. The Pacific Northwest Coast: Living with the shores of Oregon and Washington. Duke University Press, Durham, NC.

104. Kvitek, R. G., D. Shull, D. Canestro, E. C. Bowlby, and B. L. Troutman. 1989. Sea otters and benthic prey communities in Washington State. Marine Mammal Science 5: 266-280.

105. Laake, J. L., R. L. DeLong, J. Calambokidis, and S. Osmek. 1997. Abundance and distribution of marine mammals in Washington and British Columbia inside waters, 1996. Pages 67-73 *in:* P. S. Hill, and D. P. DeMaster, editors. MMPA and ESA Implementation Program, 1996. AFSC Processed Report 97-10. Alaska Fisheries Science Center, NMFS, NOAA, 7600 Sand Point Way NE, Seattle, WA.

106. ———, J. Calambokidis, and S. Osmek. 1998. Survey report for the 1997 aerial surveys for harbor porpoise and other marine mammals of Oregon, Washington and British Columbia outside waters. Pages 77-97 *in:* P. S. Hill, and D. P. DeMaster, editors. MMPA and ESA Implementation Program, 1997. AFSC Processed Report 98-10. Alaska Fisheries Science Center, NMFS, NOAA, 7600 Sand Point Way NE, Seattle, WA.

107. Lamont, M. M., J. T. Vida, J. T. Harvey, S. Jeffries, R. Brown, H. R. Huber, R. DeLong, and W. K. Thomas. 1996. Genetic substructure of the Pacific harbor seal (*Phoca vitulina richardsi*) off Washington, Oregon, and California. Marine Mammal Science 12: 402-413.

108. Landry, M. L. and B. M. Hickey, editors. 1989. Coastal oceanography of Washington and Oregon. Elsevier Press, Amsterdam, Netherlands.

109. Le Boeuf, B. J., and J. Reiter. 1991. Biological effects associated with El Niño-Southern Oscillation, 1982-83, on northern elephant seals breeding at Año Nuevo, California. Pages 206-218 *in:* F. Trillmich, and K. Ono, editors. Pinnipeds and El Niño. Responses to environmental stress. Springer-Verlag, Berlin, Germany.

110. Lowe, R. 1993. El Niño hard on Oregon seabirds. Pacific Seabird Group Bulletin 20: 62.

111. ———. 1998. Regional reports, Washington and Oregon. Pacific Seabirds 25: 84.

112. ———, and D. S. Pitkin. 1996. Replicate aerial photographic censuses of Oregon Common Murre colonies 1995. USFWS, Newport, OR.

113. McPhaden, M. J. 1999. Genesis and evolution of the 1997-98 El Niño. Science 283: 950-954.

114. Mate, B. R. 1975. Annual migrations of the sea lions *Eumetopias jubatus* and *Zalophus californianus* along the Oregon coast. Rapports et Proces-Verbaux des Reunions. Conseil International pour l'Exploration Scientifique de la Mer Mediterranee (Monaco) 169: 455-461.

115. ———. 1981. Marine mammals. Pages 372-492 *in:* C. Maser, B. R. Mate, J. F. Franklin, and C. T. Dyrness, editors. Natural history of Oregon coast mammals. USDA Forest Service, General Technical Report PNW-133.

116. Melillo, J. M. 1999. Warm, warm on the range. Science 283: 183-4.

117. Melvin, E. F., J. K. Parrish, and L. L. Conquest. 1999. Novel tools to reduce seabird bycatch in coastal gillnet fisheries. Conservation Biology 13: 1386-1397.

118. Mitsch, W. J., and J. G. Gosselink. 1993. Wetlands (second edition). Van Nostrand Reinhold, New York, NY.

119. Murray, M. R. 1998. The status of Marine Protected Areas in Puget Sound. Puget Sound/Georgia Basin Environmental Report Series No. 8. Olympia, WA.

120. Myers, J. P., C. T. Schick, and C. J. Hohenberger. 1984. Notes on the 1983 distribution of sanderlings along the United States' Pacific coast. Wader Study Group Bulletin 40: 22-26.

121. National Marine Fisheries Service (NMFS). 1992. Final recovery plan for Steller sea lions (*Eumetopias jubatus*). Prepared by the Steller Sea Lion Recovery Team for the National Marine Fisheries Service, Office of Protected Resources, Silver Springs, MD.

122. ———. 1995. Environmental assessment on protecting winter-run wild steelhead from predation by California sea lions in the Lake Washington ship canal. NMFS Environmental Assessment Report. Alaska Fisheries Science Center, NMFS, NOAA, 7600 Sand Point Way NE, Seattle, WA.

123. ———. 1997. Investigation of scientific information on the impacts of California sea lions and Pacific harbor seals on salmonids and on the coastal ecosystems of Washington, Oregon, and California. NOAA Tech. Memo. NMFS-NWAFSC-28. Northwest Region, NMFS, NOAA, 7600 Sand Point Way NE, Seattle, WA.

124. Nehls, H. B. 1994. Oregon shorebirds: Their status and movements. Oregon Department of Fish and Wildlife, Technical Report 94-1-02.

125. Nichols, F. H., J. K. Thompson, and L. E. Schemel. 1990. Remarkable invasion of San Francisco Bay (California, USA) by the Asian clam *Potamo-corbula amurensis*. II. Displacement of a former community. Marine Ecology Progress Series 66: 95-101.

126. Northridge, S. P. 1991. Driftnet fisheries and their impacts on nontarget species: A worldwide review. Fisheries Technical Paper 320. Food and Agriculture Organization of the United Nations, Rome, Italy.

127. Nysewander, D. R., and J. R. Evenson. 1998. Status and trends for selected diving duck species examined by the marine bird component, Puget Sound Ambient Monitoring Program (PSAMP), Washington Department of Fish and Wildlife. Pages 847-867 *in*: R. Strickland, editor. Puget Sound Research '98. Puget Sound Water Quality Action Team, Olympia, WA.

128. ———, and ———. 2000. Puget Sound Ambient Monitoring Program: Report of the marine bird, waterfowl, and marine mammal monitoring project for July 1992 to December 1999. Washington Department of Fish and Wildlife, Olympia, WA.

129. Ocean Policy Advisory Council. 1994. State of Oregon Territorial Sea Plan. Ocean Policy Advisory Council, Portland, OR.

130. Olesiuk, P. F. 1993. Annual prey consumption by harbor seals (*Phoca vitulina*) in the Strait of Georgia, British Columbia. Fishery Bull., U.S. 91: 491-515.

131. ———, M. A. Bigg, and G. M. Ellis. 1990. Life history and population dynamics of resident killer whales (*Orcinus orca*) in the coastal waters of British Columbia and Washington State. Reports to the International Whaling Commission, Special Issue 12: 209-243.

132. Oregon Department of Fish and Wildlife. 1992. Shorebird survey of the Lower Columbia Estuary: August 1992. Unpublished table of survey results on file with U.S. Fish and Wildlife Service, Portland, OR.

133. ———. 1993. Shorebird survey of the Lower Columbia Estuary: August 22, 1993. Unpublished report to U.S. Fish and Wildlife Service, Portland, OR.

134. ———. 1994a. Shorebird survey of the Lower Columbia Estuary: April 24, 1994. Unpublished report to U.S. Fish and Wildlife Service, Portland, OR.

135. ——— 1994b. Shorebird survey of the Lower Columbia Estuary: August 21, 1994. Unpublished report to U.S. Fish and Wildlife Service, Portland, OR.

136. ———. 1994c. Shorebird survey of the Lower Columbia Estuary: April 1994. Unpublished report to U.S. Fish and Wildlife Service, Portland, OR.

137. ———. 1994d. Shorebird survey of the Lower Columbia Estuary: August 1994. Unpublished report to U.S. fish and Wildlife Service, Portland, OR.

138. Osbourne, R., J. Calambokidis, and E. M. Dorsey. 1988. A guide to marine mammals of Greater Puget Sound. Island Publishers, Anacortes, WA.

139. Page, G. W., L. E. Stenzel, and J. E. Kjelmyr. 1999. Overview of shorebird abundance and distribution in wetlands of the Pacific coast of the contiguous United States. Condor 101: 461-471.

140. Parrish, J. K. 1998. Attendance and reproductive success of Tatoosh Island common murres: Draft final report 1997. University of Washington, Seattle, WA.

141. Paulson, D. 1993. Shorebirds of the Pacific Northwest. University of Washington Press, Seattle, WA.

142. Penland, S. 1981. Natural history of the Caspian tern in Grays Harbor, Washington. Murrelet 62: 66-72.

143. Philander, S. G. H. 1990. El Niño, La Niña and the Southern Oscillation. Academic Press, San Diego, CA.

144. Pitkin, D. S., and R. W. Lowe. 2000. Replicate aerial photographic censuses of Oregon common murre colonies, 1996-1997. Unpublished report to the Tenyo Maru Trustee Council. U.S. Fish and Wildlife Service, Newport, OR.

145. Prince, P. A., and R. A. Morgan. 1987. Diet and feeding ecology of Procellariiformes. Pages 135-171 *in*: J. P. Croxall, editor. Seabirds: feeding ecology and role in marine ecosystems. Cambridge University Press, Cambridge, England.

146. Puget Sound Water Quality Action Team. 1999. 2000 Puget Sound update: seventh report of the Puget Sound Ambient Monitoring Program. Second draft.

147. Quinn, W. H., and V. T. Neal. 1987. El Niño occurrences over the past four and a half centuries. Journal of Geophysical Research 92 (C13): 14449-14461.

148. Reimer, S. D., and R. F. Brown. 1997. Prey of pinnipeds at selected sites in Oregon identified by scat (fecal) analysis, 1983-1996. Oregon Dept. of Fish and Wildlife, Wildlife Diversity Program, Tech. Rept. 97-6-02.

149. Ribic, C. A., S. W. Johnson, and C. A. Cole. 1997. Distribution, type, accumulation, and source of marine debris in the United States, 1989-1993. Pages 35-47 *in*: J. M. Coe, and D. B. Rogers, editors. Marine debris: sources, impacts, and solutions. Springer-Verlag, New York, NY.

150. Richmond, R. H. 1989. The effects of the El Niño/Southern Oscillation on the dispersal of corals and other marine organisms. Pages 127-140 *In*: P. W. Glynn, editor. Global Ecological Consequences of the 1982-83 El Niño Southern Oscillation. Elsevier Oceanography Series, 52. University of Miami, FL.

151. Robinson, M. K. 1999. The Status of Washington's coastal Marine Protected Areas. Washington Department of Fish and Wildlife. Interjurisdictional Resource Management. Olympia, WA.

152. Rowlett, R. A., G. A. Green, C. E. Bowlby, and M. A. Smultea. 1994. The first photographic documentation of a northern right whale off Washington State. Northwestern Naturalist 75: 102-104.

153. Rugh, D. J., M. M. Muto, S. E. Moore, and D. P. DeMaster. 1999. Status review of the Eastern North Pacific stock of gray whales. NOAA Tech. Memo. NMFS-AFSC-103. Alaska Fisheries Science Center, NMFS, NOAA, 7600 Sand Point Way NE, Seattle, WA.

154. Sanger, G. A. 1970. The seasonal distributions of some seabirds off Washington and Oregon, with notes on their ecology and behavior. Condor 72: 339-357.

155. Scheffer, V. B., and J. W. Slipp. 1944. The harbor seal in Washington State. American Midland Naturalist 32: 373-416.

156. ———, and ———. 1948. The whales and dolphins of Washington State with a key to the cetaceans of the west coast of North America. American Midland Naturalist 39: 257-337.

157. Schick, C. T., L. A. Brennan, J. B. Buchanan, M. A. Finger, T. M. Johnson, and S. G. Herman. 1987. Organochlorine contamination in shorebirds from Washington State and the significance for their falcon predators. Environmental Monitoring and Assessment 8: 1-17.

158. Seliskar, D. M., and J. L. Gallagher. 1983. The ecology of tidal marshes of the Pacific Northwest coast: A community profile. USDI Fish and Wildlife Service, FWS/OBS-82/32.

159. Shukla, J. 1998. Predictability in the midst of chaos: A scientific basis for climate forecasting. Science 282:728-731.

160. Siegel-Causey, D., and N. M. Litvinenko. 1993. Status, ecology, and conservation of shags and cormorants of the temperate North Pacific. Pages 122-130 in: K. Vermeer,, K. T. Briggs, K. H. Morgan, and D. Siegel-Causey, editors. The status, ecology and conservation of marine birds of the North Pacific. Canadian Wildlife Service Special Publication, Ottawa, Canada.

161. Simenstad, C. A. 1983. The ecology of estuarine channels of the Pacific Northwest coast: A community profile. U.S. Fish and Wildlife Service, FWS/OBS-83-05.

162. Sowls, A. L., A. R. DeGange, J. W. Nelson, and G. S. Lester. 1980. Catalog of California seabird colonies. U.S. Fish and Wildlife Service. FWS/OBS 80/37.

163. Spear, L. B. 1993. Dynamics and effects of western gull feeding in a colony of guillemots and Brandt's cormorants. Journal of Animal Ecology 62: 399-414.

164. ———— T. M. Penniman, J. F. Penniman, H. R. Carter, and D. G. Ainley. 1987. Survivorship and mortality factors in a population of western gulls. Studies in Avian Biology 10: 44-56.

165. Speich, S. M., and S. P. Thompson. 1987. Impacts on waterbirds from the 1984 Columbia River and Whidbey Island, Washington, oil spills. Western Birds 18: 109-116.

166. ———— B. L. Troutman, A. C. Geiger, P. J. Meehan-Martin, and S. L. Jeffries. 1987. Evaluation of military flight operations on wildlife of the Copalis National Wildlife Refuge, 1984-1985. Department of the Navy, West. Div., Naval Facilities Engineering Command, San Bruno, CA.

167. ————, and T. R. Wahl. 1989. Catalog of Washington seabird colonies. U.S. Dept. Interior, Fish and Wildl. Serv., Biol. Rept. 88(6). MMS 89-0054.

168. Springer, A. M., A. Y. Kondratayev, H. Ogi, Y. V. Shibaev, and G. B. van Vliet. 1993. Status, ecology, and conservation of Synthliboramphus murrelets and auklets. Pages 187-203 in: K. Vermeer, K. T. Briggs, K. H. Morgan, and D. Siegel-Causey, editors. The status, ecology and conservation of marine birds of the North Pacific. Canadian Wildlife Service Special Publication, Ottawa, Canada.

169. Steiger, G. H., J. Calambokidis, D. K. Ellifrit, K. C. Balcomb, J. D. Darling, and G. A. Green. 1999. Distribution and population structure of humpback whales off Oregon and Washington. in: Abstracts Thirteenth Biennial Conference on the Biology of Marine Mammals, Maui, HI 28 November—3 December 1999. Society for Marine Mammalogy, Lawrence, KS.

170. Temte, J. L. 1993. Population differentiation of the Pacific harbor seal: cranial morphology parallels birth timing. Ph.D. dissertation, University of Wisconsin, Madison, WI.

171. U.S. Fish and Wildlife Service. 1997. Pacific Coast Joint Venture strategic plan. Portland, OR.

172. ————. 1999. Endangered and threatened wildlife and plants; Designation of critical habitat for the Pacific coast population of the Western Snowy Plover; Final rule. Federal Register 64: 68507-68544.

173. Varoujean, D. H. 1979. Seabird colony catalog: Washington, Oregon and California. U.S. Fish and Wildlife Service, Portland, OR.

174. Veit, R. R., J. A. McGowan, D. G. Ainley, T. R. Wahl, and P. Pyle. 1997. Apex marine predator declines ninety percent in association with changing oceanic climate. Global Change Biology 3: 23-28.

175. ————, P. Pyle, and J. A. McGovan. 1996. Ocean warming and long-term change in pelagic bird abundance within the California current system. Marine Ecology Progress Series 139: 11-18.

176. Vermeer, K., D. B. Irons, E. Velarde, and Y. Watanuki. 1993. Status, conservation and management of nesting Larus gulls in the North Pacific. Pages 131-139 in: K. Vermeer, K. T. Briggs, K. H. Morgan, and D. Siegel-Causey, editors. The status, ecology and conservation of marine birds of the North Pacific. Canadian Wildlife Service Special Publication, Ottawa, Canada.

177. Wahl, T. R. 1975. Seabirds in Washington's offshore zone. Western Birds 6: 117-134.

178. ————. 1984. Distribution and abundance of seabirds over the continental shelf off Washington (including information on selected marine mammals). Washington State Department of Ecology.

179. ————. 1996. Waterbirds in Washington's inland marine waters: some high counts from systematic censusing. Washington Birds 5: 29-50.

180. ————, D. G. Ainley, A. H. Benedict, and A.R. DeGange. 1989. Associations between seabirds and water masses in the northern Pacific Ocean in summer. Marine Biology 103: 1-17.

181. ————, and D. Heinemann. 1979. Seabirds and fishing vessels: co-occurrence and attraction. Condor 8: 390-396.

182. ————, K. H. Morgan, and K. Vermeer. 1993. Seabird distribution off British Columbia and Washington. Pages 39-47 in: K. Vermeer, K. T. Briggs, K. H. Morgan, D. Siegel-Causey, editors. The status, ecology and conservation of marine birds of the North Pacific. Canadian Wildlife Service Special Publication, Ottawa, Canada.

183. ————, S. M. Speich, D. A. Mauwal, K. V. Hirsch, and C. Miller. 1981. Marine bird populations of the Strait of Juan de Fuca, Strait of Georgia and adjacent waters in 1978 and 1979. U.S. Environmental Protection Agency, DOC/EPA interagency Energy/Environment R&D Program Report EPA/600/f-81/156.

184. ————, and ————. 1983. First winter survey of marine birds in Puget Sound and Hood Canal, December 1982 and February 1983. Report for the Washington Department of Game. Nongame Wildlife Program.

185. ————, and ————. 1984. Survey of marine birds in Puget Sound, Hood Canal and waters east of Whidbey Island, Washington, in summer 1982. Western Birds 15: 1-14.

186. ————, and ————. 1994. Distribution of foraging Rhinoceros Auklets in the Strait of Juan de Fuca, Washington. Northwestern Naturalist 75: 63-69.

187. ————, and B. Tweit. In press. Seabird abundances off Washington, 1972-1998. Western Birds.

188. Washington Department of Natural Resources. 1998. Our changing nature: Natural resource trends in Washington state. Olympia, WA.

189. Weins, J. A., and J. M. Scott. 1975. Model estimation of energy flow in Oregon coastal seabird populations. Condor 77: 439-452.

190. Weitkamp, L. A., R. C. Wissman, C. A. Simenstad, K. L. Fresh, and J. G. Odell. 1992. Gray whale foraging on ghost shrimp (Callianasa californiensis) in littoral sand flats of Puget Sound, USA. Canadian Journal of Zoology 70: 2275-2280.

191. Widrig, R. S. 1979. The shorebirds of Leadbetter Point. Independently published.

192. Wiedemann, A. M. 1984. The ecology of Pacific Northwest coastal sand dunes: A community profile. USDI Fish and Wildlife Service, FWS/OBS-84/04.

193. ————. 1998. Coastal foredune development, Oregon, USA. Journal of Coastal Research, Special Issue 26: 45-51.

194. ————, and A. Pickart. 1996. The Ammophila problem on the northwest coast of North America. Landscape and Urban Planning 34: 287-299.

195. Wilson, U. W. 1991. Responses of three seabird species to El Niño events and other warm episodes on the Washington coast, 1979-1990. Condor 93: 853-858.

196. ————. 1996. Washington common murre colony surveys 1996. USFWS, Sequim, WA.

197. ————. 1999. Common murre colony surveys, Washington Islands National Wildlife Refuge, 1997-1998. USFWS, Port Angeles, WA.

198. ————, A. McMillan, and F. C. Dobler. 2000. Nesting, population trend and breeding success of peregrine falcons on the Washington outer coast, 1980-98. Journal of Raptor Research 34:67-74.

199. York, A. E. 1995. The relationship of several environmental indices to the survival of juvenile male northern fur seals (Callorhinus ursinus) from the Pribilof Islands. Pages 317-327 in: R. J. Beamish, editor. Climate change and northern fish populations. Canadian Special Publication for the Fisheries and Aquatic Science No. 121.

16

Introduced Wildlife of Oregon and Washington

Gary W. Witmer & Jeffrey C. Lewis

Introduction

Each species of wildlife occurs as part of an ecosystem, interacting in many ways with other plant and animal species in that system as well as with the abiotic components such as soil, air, water, and other substrates. The array of wildlife species around the globe has been shaped by geological and climatological events as well as by eons of evolution and natural selection. Species have come and gone and those remaining have, in most cases, co-evolved or co-adapted with many other species so that relatively stable, and often complex, relationships exist. Usually, a great many niches have been carved out and occupied, creating a distinct flora and fauna in each region of the globe that is maintained under conditions of relative stability over time. Natural disturbances (wind, fire) and large-scale events (volcanic eruptions, drought) may occasionally alter that stability and the relationships between species, but an overall homeostatis "a return to the climatic community steady state" usually prevails. These and other concepts of biogeography have been discussed in detail.[31, 162]

Species we refer to as "native" or "indigenous" naturally occur in a particular area and have been there for a very long time. However, events can occur that bring individuals of a new species into a region where they come into contact with many species with which they are neither co-adapted nor co-evolved. In most cases, these newly arrived individuals soon succumb, but some may survive long enough to interact with, or disturb, normal relationships in the community. In a few cases, the newcomers may survive, reproduce, and become established in the ecosystem, permanently altering relationships among or between species. These newcomers are usually referred to as "introduced," "exotic," "non-native," or "non-indigenous." Species that are very successful at this are sometimes called "invasive" species. These species are often capable of spreading unchecked, increasing to high population levels, and comprising a significant portion of the total biota. In cases where the species has thrived in the new location for a relatively long period of time (in terms of human generations), they are considered "naturalized" and are essentially considered a regular part of the local flora and fauna (species complex).

In this chapter, we present information on wildlife introduced to Oregon and Washington. While other terms could be used, we will refer to these relatively new members of the fauna of Oregon and Washington as "introduced" species. Occasionally, the term "exotic" will be used, especially in the context of legal terminology, such as state or federal laws and regulations. We will not include species that are expanding their range on their own without the direct intervention of humans; examples of these species include the cattle egret and the barred owl. Also, we will not include the reintroduction or population augmentation of native species, although much of this activity is occurring in the Pacific Northwest for conservation and biodiversity purposes.

Additionally, we will only include introduced species of birds, mammals, amphibians, and reptiles. In additional to at least 125 species of vertebrates, it has been estimated that over 2,000 species of plants and over 1,100 species of invertebrates have been introduced into the U.S., along with 111 species of fish and over 50 plant pathogens.[116] There are many introduced species of plants, invertebrates, and fish that occur in the Pacific Northwest, and many are of major concern with regard to ecosystem integrity, natural resource management, crop protection, or human health and safety. Detailed discussions on the rapid and destructive spread of various noxious plant species that have been introduced to the Pacific Northwest have been presented by Peck,[120] Stein and Flack,[139] and Toney et al.[146] Most plant species introduced to the Pacific Northwest have been perennial forbs originating from Eurasia, although there has been a trend towards woody species introductions in more recent years.[146] It is ironic that many plant species were purposefully introduced for wildlife habitat enhancement.[130] Introduced invertebrate species and their impacts have been discussed,[68, 139] as well as introduced fish species and their impacts to aquatic systems.[30, 139]

In this chapter we will discuss why wildlife introductions occur; the benefits and problems associated with introductions; regulation of introductions; the introductions that have occurred in Oregon and Washington; the known or potential interactions between introduced species and native species; and the management of introduced species. We will also include several case histories that characterize favorable and adverse aspects of wildlife introductions.

How Introductions Occur

Wildlife species can be introduced to new areas through a variety of mechanisms, both accidental and purposeful (Table 1). Accidental introductions can result from animals escaping captivity, as has occurred with fox, mink, monk parakeet (*Myiopsitta monachus*), various livestock species, and an array of wild ungulates such as fallow (*Dama dama*) and axis (*Axis axis*) deer. As stowaways on ships, trains or other vehicles, some rodent species (Norway rat, black rat, house mouse) and bird species (house sparrow) have achieved worldwide distribution. Finally, human alteration of habitats or native species ranges, after an initial introduction elsewhere, has resulted in the altered and often expanded range of number of species such as the opossum into regions of the country in which they did not historically occur.

Purposeful introductions have occurred for many reasons (Table 1). The desire to have bird species from the countries of their European heritage, hence aesthetics, led the Portland Song Bird Club to attempt many songbird introductions, including the starling.[46, 76] Similarly, eastern gray and fox squirrels have been released in many urban/suburban areas of the west.[154]

Many "game" species, as well as some domestic livestock species, have been introduced to provide some combination of recreational hunting or economic benefit (fur, food, or clothing). These species include many species of upland game birds, turkeys (*Meleagris gallopavo*), foxes, eastern cottontail rabbits, nutria, and various species of deer. Domestic species have been released to provide a future source of food or transportation (pig, *Sus scrofa*; goat, *Capra hircus*; horse, *Equus caballus*; burro, *Equus asinus*).

In some cases, species were introduced to fill a perceived vacant niche, as with upland game birds, carnivores on islands, and herbivores on islands. In actuality, in some of those cases, the populations of native species occupying those niches had been greatly reduced by over-harvest or by human-induced changes in habitats or predator-prey relationships (e.g., many native grouse species). Carnivores (such as fox; mongoose, *Herpestes* spp.; European ferret, *Mustela putorius*; and domestic cat, *Felis catus*) have been introduced, especially on islands, to help control rodent or rabbit populations, many of which were also introduced accidentally or purposefully at an earlier date. This form of biological control has rarely, if ever, been successful in its intended purpose.

In some cases, animals kept in captivity were released because the owners no longer cared or could afford to maintain the animals, or because the economic incentive to raise the animals had declined, as with bullfrogs, nutria, mink, fox, exotic deer, "road-side zoo" animals, and some species of livestock. In a few cases, the animals released from captivity may have been rehabilitated animals or problem animals. Some persons may release animals from

Table 1. Some reasons why wildlife introductions occur.

A. Accidental introductions
1. Escaped captivity
2. Stowaways
3. Expanded range of species after introduction elsewhere

B. Purposeful introductions
4. Aesthetics
5. Economics
6. Recreation
7. Source of food
8. Filling "vacant" niches
9. Biological control
10. Released from captive population
11. Release of rehabilitated or problem animals
12. Whimsy: "what the heck, let's see what happens"
13. Gifts

captivity on a whim: "what the heck, let's see what happens." This may have occurred with some parakeet/parrot species as well as with some reptile and amphibian species. Finally, persons have given animals (wild or domestic) away as gifts, which later escaped or were released and established free-ranging populations.

What Makes Introductions Succeed or Fail?

Most wildlife introductions, whether accidental or purposeful, fail to establish free-ranging and sustained populations.[59, 163] There are many reasons why this is the most likely outcome of an introduction: inadequate numbers of animals, poor health or genetic quality of animals, predation, disease or parasites, inadequate habitat, competition with native species, poor planning, and others.[59, 163]

On the other hand, certain characteristics of a species or population make it more likely to be successful at "invading" a new area and becoming established.[44, 112] These include a large native range, high mobility, broad diet, short generation time, high genetic variability, gregariousness, larger size than most closely related species, few predators, association with humans, association with freshwater habitats, and ability to function under a wide range of physical conditions. Often these species are "habitat generalists" and have a "broad ecological amplitude."[44]

It is important to recognize that many factors are involved in the success or failure of a wildlife introduction. Even chance and timing play a role.[33] Disturbance of a site or community, often human-induced, may make the area more susceptible to invasion.[51, 117, 125] As such, it is difficult to predict whether or not a given introduction will succeed or fail.[129, 136, 137] There have been some efforts to construct predictive models of the likelihood of successful establishment of an introduced species.[136, 137] Unfortunately, there is much to be learned in this area of ecology. In terms of regulation of wildlife introductions,

Red fox. (Photograph by Jeffrey C. Lewis)

Table 2. Potential adverse ecological consequences of introduced wildlife species.

A. Effects on physical environment
1. Water quality, quantity
2. Soil compaction
3. Soil erosion
4. Nutrient balance

B. Effects on flora
5. Species composition
6. Species abundance
7. Vegetative structure
8. Plant succession
9. Species endangerment

C. Effects on fauna
10. Competition
   a. food
   b. habitats
   c. interference
11. Predation
12. Disease/parasite transmission
13. Hybridization
14. Species endangerment

D. Direct effects to humans
15. Disease/parasites to humans, livestock, pets
16. Crop damage
17. Structural damage
18. Livestock predation
19. Livestock forage competition
20. Human food consumption and contamination
21. Human safety
22. Aesthetics

E. Major ecosystem disruption or alteration
23. Combinations of the above effects

this situation has historically resulted in an "innocent until proven guilty" attitude, and species introductions are not prohibited until it is known that they are likely to cause adverse effects; and once these occur, they may be impossible to reverse.

Potential Benefits and Adverse Effects of Introductions

A large number of introductions of plants and animals has already occurred, and continues to occur, in the United States.[116] Some past introductions have benefited the public for the reasons listed in Table 1. Consider, for example, domestic livestock and upland game birds. In Oregon, the introduction of ring-necked pheasant in 1881 resulted in large economic and recreational benefits—so much so, that captive-reared birds were soon being exported to many other states.[35] Upland game species (both bird and small mammal) continue to provide large revenues and extensive recreation to many states.[73]

There are many potential or realized ecological consequences of wildlife introductions (Table 2). To date, the most visible effects of introductions to Pacific Northwest ecosystems appear to be from plant and invertebrate species introductions, although overgrazing by domestic livestock has affected some dryland areas. We note that the effects of an introduced wildlife species, however, may take hundreds of years to become evident: the "blink of an eye" in ecological time. The effects can be to the physical environment, the flora, the fauna, and humans directly, or more often, to a combination of these ecosystem elements. Perhaps the most common effects are from herbivory, competition, or predation. However, many other types of effects can occur, such as hybridization with native species[133] and disease transmission.[52] Numerous examples of ecological effects were presented by MacDonald et al.[107] and Simberloff.[134] In some cases, a major disruption of the ecosystem can occur, but this has not yet been well documented from wildlife introductions in terrestrial ecosystems in Oregon and Washington, with the possible exceptions of San Juan Island (European rabbit),[28] Destruction Island (European rabbit),[6] and the Olympic Peninsula (mountain goat),[67] all

in Washington. Major disruptions are most common on islands where rats,[165] carnivores,[164] or feral livestock[148] have been introduced. Erosion and community changes (species composition, abundances, biodiversity, and species loss) have occurred in these situations. On the North American mainland, similar effects have occurred in the Great Smoky Mountains National Park from the introduction of feral pigs.[10] Feral horses and burros have had substantial impacts on some southwestern ecosystems.[41] Species-specific examples of realized or (more often) potential ecological effects of wildlife introductions in Oregon and Washington are presented later in this chapter.

Regulation of Wildlife Introductions

The regulation, policies, and guidelines for wildlife introductions in the United States have had a long and convoluted history. A large number of governmental agencies—at federal, state, and local levels—have played roles; the net effect often being inconsistent, inadequate, or contradictory efforts among agencies, or policies that changed dramatically over time within an agency.[116, 131] There is a strong need for not only more regulation of species introductions, but also better coordination of regulation across jurisdictional boundaries and governmental levels.[88, 116]

Regulations and practices have evolved from encouraging the importation and release of animals to improve agricultural resources, hunting opportunities, or local economies to restricting importations because of disease hazards, threats to agricultural resources or human health and safety, or potential disruption of natural ecosystems. As early as 1923, Taylor[142] discussed benefits, adverse effects, methods, and regulations for the introduction of upland game birds in the Pacific Northwest.

The U. S. Department of Interior's (USDI) Federal—State Cooperative Foreign Game Program of 1948 added an element of central authority at a time when the importation of game species into the United States was being strongly pursued. This program was guided by three objectives: to provide an ecological and life history data base to individuals or agencies, to discourage introductions when the data base suggested an introduction was unwise, and to fill vacant or understocked habitats with foreign species as an alternate course of action following appropriate testing and trial introductions.[128]

In 1966, the USDI published guidelines and recommendations for the importation of wildlife.[128] These guidelines incorporated eight conditions:

1. Critically determine that a need exists, with desirable ecological, recreational, and economic impacts.
2. A definite niche is available and unsuited for a native species.
3. Introductions should not be considered if they threaten the reduction or displacement of native populations; nor should existing or proposed land uses be in conflict with an exotic species transplant.
4. Introductions should be preceded by ecological studies of both the animal and the habitat proposed at the release site.
5. Disease relationships require special study as well as the steps assuring appropriate quarantine leading to disease-free stock.
6. Exotic species with close relatives in the United States should be avoided, to preclude hybridization with native wildlife.
7. Small-scale experiments and a thorough evaluation of these should precede larger introductions.
8. Before an exotic species is released, methods for controlling its abundance and expansion must be available.

These guidelines resulted in eight recommend-ations from the USDI. These were meant to apply to federal lands or federal actions and included: 1) no decisions until a full assessment is at hand, 2) no exotic species placed in national parks or lands devoted to the preservation of native biota, 3) no exotic species placed in the vicinity of rare or uncommon native species, 4) no exotic grazers placed on federal lands devoted to domestic grazers, 5) no exotic big game placed in areas devoted to intensive land uses, 6) no introductions on federal lands except under a permit and a commitment from the state wildlife agency, 7) treatment of exotics leaving federal land as trespassing livestock with the responsible party held liable, and 8) periodic review of public policy regarding exotic species.

The Wildlife Society published a policy statement on species introductions in 1975. This policy included the following three considerations:[91]

1. Support the introduction of exotic species only after competent scientists have thoroughly studied the situation and potential effects and quarantine requirements have been met.
2. Urge that no state, provincial, or national agency introduce an exotic species or permit such an introduction unless that species can be contained within its jurisdiction, or unless surrounding jurisdictions have sanctioned the introduction.
3. Exclude from the provisions of this policy the importation of exotic species by officially recognized scientific and educational organizations, and the institutional exchange of such species provided that the exotics are maintained in captivity at all times.

President Carter signed Executive Order 11987 in 1977. This document, in part, restricted federal agencies from introducing species to lands they administer, encouraged the prevention of introductions by other levels of government and by private citizens, and restricted federal support of introductions outside the United States.[128] These limitations applied unless either the Secretary of Interior or the Secretary of Agriculture determined that the introduction would not have an adverse effect on natural ecosystems.

An international position statement, containing policies and guidelines similar to those above, was developed and approved by the International Union for Conservation of Nature and Natural Resources (IUCN) in 1987.[116] The IUCN is an organization comprised of scientific experts and government officials involved in conservation around the world.

The concern about "invasive alien" species continues to generate activity by the federal government. On June 17, 1997, Vice President Gore directed the preparation of a strategy to combat the introduction and spread of non-native plants and animals in the United States that are causing great economic and ecological harm to the nation. A draft document has been prepared that reviews the situation, makes recommendations, and provides an action plan.[1] Based on the results of the Task Force, President Clinton signed an Executive Order on Invasive Species on February 3, 1999. Its goals are to prevent the introduction of invasive species, to provide for their control, and to minimize the economic, ecological, and human health impacts that they cause. It was estimated that invasive species cost the U.S. economy about $123 billion per year. The Order establishes Invasive Species Council assigned the task of setting up an Advisory Committee and preparing and implementing an Invasive Species Management Plan. The Plan will 1) detail and recommend performance-oriented goals and objectives,

2) review existing and prospective approaches and authorities, 3) identify pathways of introductions and ways to minimize risks of introduction, 4) identify research needs, 5) be science-based, 6) recommend and implement measures to reduce introductions and control those that have occurred, 7) identify requirements to achieve goals and objectives, and 8) evaluate and report on the success in achieving goals and objectives.

We have reviewed some of the long history of introduced wildlife concerns, policies, and recommendations. The groundwork has been set for a vigorous effort to reduce introductions and to manage existing introductions. It remains to be seen what level of success will be achieved towards this goal.

Current Federal and State Regulations

The two main federal agencies regulating wildlife species introductions in the United States are the U.S. Department of Agriculture's (USDA) Animal and Plant Health Inspection Service (APHIS) and the USDI Fish and Wildlife Service (FWS).[116, 131] A major function of APHIS is to protect United States agriculture (both plants and animals) from diseases or plant and animal "pests" that might gain access to the country or be transported between states. Border inspections, quarantines, disease testing, and eradication programs are some of its routine functions, and most of APHIS's pest exclusion occurs at ports of entry. APHIS does not expressly prohibit species-specific imports, but requires adequate quarantine and veterinarian inspection before such imports or transportations are allowed (9 CFR Ch. 1). The agency is particularly strict regarding hedgehogs (*Erinaceus* spp.) and the brush-tailed possum (*Trichosurus vulpecula*). A major concern of APHIS is to prevent the entry of Newcastle's disease, chlamydiosis, foot-and-mouth disease, rinderpest, bovine tuberculosis, and other communicable diseases of livestock and wildlife. APHIS is also active in management and research to prevent entry of the brown tree snake (*Boiga irregularis*) into Hawaii and the mainland United States.[17]

The FWS protects threatened and endangered species by, among other activities, restricting the importation and exportation of federally listed species under the Endangered Species Act and the Convention on International Trade in Endangered Species (CITES). Except under permit and various restrictions, the FWS expressly prohibits the importation and release of individuals, progeny, or eggs of many species of vertebrates into the United States, to protect national resources (50 CFR Ch. 1). These species include flying fox (*Pteropus* spp.), mongoose, European rabbit, wild dog (genus *Cuon*), multi-mammate rat (*Mastomys* spp.), raccoon dog (*Nyctereutes procyonoides*), starling, quelea (*Quelea qualea*), Java sparrow (*Padda oryzivora*), red-whiskered bul-bul (*Pycnonotus jocosus*), all species of amphibians, and all species of reptiles. Additionally, the importation and transportation of birds of the family *Psittacidae* (parrots, parakeets, macaws, etc.) are regulated by the U.S. Public Health Service because of disease hazards (42 CFR Parts 71 and 72).

Bullfrog. (Photograph by Jeffrey C. Lewis)

For many decades, many states, including Oregon and Washington, had few regulations regarding the importation or keeping of "exotic" wildlife or the protection of native biodiversity from exotics.[21, 116] Many states in the late 1800s and early 1900s, including Oregon and Washington, encouraged—or were directly involved with—the propagation or release of many game species, including exotic species. These practices have largely been curtailed in recent decades, with notable exceptions such as with wild turkey and Sichuan pheasant, *P. c. suehschanensis*.[116] The Southeastern Cooperative Wildlife Disease Study provided a "model law" in 1988 to help guide states in regulating animal imports that addressed veterinary, humane, public safety, ecological, and other concerns.[116] It recommended a permit requirement for introduced species, that certain common domestic and naturalized species be exempt from the regulations, that criteria and a list be developed for "environmentally injurious animals," and that a technical advisory committee be formed to provide advice. Both Oregon and Washington legislatures and wildlife agencies have enacted detailed and specific regulations on the importation and keeping of introduced wildlife.

Oregon Administrative Rules (OAR 635-056-0000 to -0150) prohibit the importation or keeping of numerous vertebrate species, including hedgehog, tri-colored squirrel (*Callosciurus* spp.), brush-tailed possum, bats of any species, mongoose, wild pig, chamois (*Rupicapra* spp.), non-domestic goat (*Capra* spp., except *C. hircus*), wildebeest (*Connochaetes* spp.), gazelles (*Procapra* spp.), capybara (*Hydrochaeris hydrochaeris*), prairie dogs (*Cynomys* spp.), any species of wild canid (except fox), Egyptian goose (*Alopochen aegyptiacus*), African clawed frog (*Xenopus* spp.), bullfrogs (*Pyxicephalus* spp.), alpine newt (*Triturus* spp.), brown tree snakes, snapping turtle (Chelydridae, all spp.), pond sliders (*Pseudemys* spp.), Chinese pond turtle (*Chinemys* spp.), pond turtle (*Chrysemys* spp.), painted turtle (*Chrysemys* spp.), map turtle (*Graptemys* spp.), North American (*Apalone* spp.) and African (*Trionyx* spp.) soft shell turtles, European pond turtle (*Emys orbicularis*), Blanding's turtle (*Emydoidea blandingii*), common mud turtle (*Kinosteron subrunum*), common musk turtle (*K. odoratum*), and Asian pond turtle (*Mauremys* spp.). Many fish species are also prohibited. Exceptions, by permit, are made for zoos and research facilities if they are escape-proof and are staffed and equipped to provide adequate care. There is also a long list of domestic or otherwise exempt species, including dogs (*Canis familiaris*), cats, burros, horses, swine, European rabbits, ferrets, and parrots and parakeets (*Psittacidae*, all spp.). The State has specific requirements involving the sale, transportation, and holding of exotic animals, to help prevent escape of, or disease transmission by, introduced animals. There are strict reporting rules that apply when an introduced species escapes captivity.

Washington has similar laws (Chapter 77.08 RCW) and prohibits the importation or keeping of mute swan, mongoose, wild pig, collared peccary (*Tayassu tajacu*), and various species of exotic bovids and cervids. Many species of fish are also prohibited.

Much of the importation of introduced wildlife into the United States is because of the enormous pet industry.[116] About 23 percent of the vertebrate species of foreign origin that currently live in the wild in the United States were originally imported as cage birds or other wildlife pets.[116] There is a growing concern about the depletion of animals from exportation countries (hence the species listings in CITES, including its appendices), but also about the significant hazard that introduced species pose to native species, ecosystems, agriculture, and forestry.[1, 116, 124] It has even been speculated that the liberalization of international free trade may increase species introductions around the globe.[74] The probability of accidental release of non-native wildlife as well as disease transfer and other hazards can be reduced by the improvement of existing programs and the implementation of specific actions as presented by the Ad Hoc Federal Invasive Alien Species Task Force[1] and the Office of Technology Assessment.[116] Their recommendations include: acknowledge that the prevention of introductions is paramount, encourage governments to take an active role by establishing national and regional councils, develop new scientific expertise for dealing with introductions, develop a white-black-gray list to assist in regulating exotic species, develop a comprehensive program to prevent unintentional introductions by identifying major pathways and methods to interdict and reduce impacts, develop and implement an international regime for control and support cooperation through development assistance, develop a Web-based network of information, convene educational workshops, and consult with the United States Congress regarding new regulations and funding authority. The white-black-gray list would delineate species that are automatically allowed, never allowed, or allowed only after thorough investigation, respectively.

The implementation of adequate programs to prevent the accidental, or purposeful but prohibited, release of introduced diseases, plants, invertebrates, and vertebrates is especially important because of the difficulty and expense of eradication once an introduced species becomes dispersed and established.[36, 135] Furthermore, it is important to increase public awareness of the risks posed by wildlife introductions. A good public education program on this subject should lead not only to more public support for the prevention of future introductions and the management or eradication of past introductions (in terms of supporting appropriate legislation, management practices, and requisite budgets), but also for better public compliance with federal and state regulations.

Introduced Wildlife Species in Oregon and Washington

There have been a large number of wildlife introductions to Oregon and Washington, dating back to the 1700s (e.g., horses and burros). Most attempted introductions, whether accidental or purposeful, have failed. For example, Portland Bird Club attempted but failed between 1889 and 1907[46, 76] to introduce many species of songbirds, including Eurasian skylark, *Alauda arvensis*; wood lark, *Lullula arborea*; blackcap, *Sylvia atricapilla*; European robin, *Erithacus rubecula*; nightingale, *Luscinia megarhynchos*; Eurasian blackbird, *Turdus merula*; song thrush, *T. philomelos*; parrot crossbill, *Loxia pytyopsittacus*; Eurasian siskin, *Carduelis spinus*; Eurasian goldfinch, *C. carduelis*; linnet, *Acanthis cannabina*; Eurasian bullfinch, *Pyrrhula pyrrhula*; chaffinch, *Fringilla coelebs*; house sparrow, and European starling.

We have compiled a list of 42 wildlife species introductions to Oregon and Washington that have established free-ranging populations at least on a localized scale (Table 3).[7, 20, 27, 34, 37, 46, 54, 70, 75, 76, 94, 97, 98, 99, 104, 152] The information we provide on the 42 species includes common and scientific names, the date, and location of the introduction(s), the reason for the introduction, the current status and distribution (in general terms), and the country of origin (Table 3). We note that many of the species were introduced over a period of time and at several locations. The list includes 18 birds, 19 mammals, 3 reptiles, and 2 amphibians. About half (19 of 42) of the species listed have achieved widespread distribution in Oregon or Washington.

Most bird species were introduced for hunting or aesthetic purposes, although several arrived by range expansion after being introduced elsewhere. Many of the mammal introductions were escapees or animals released when no longer needed or economically valuable. Several were introduced for hunting or fur farming. The Old World rodents arrived as stowaways. Most amphibian and reptile introductions were for aesthetic, pet, or food purposes.

Additionally, there are many other non-native wildlife species that have been observed or reported in Oregon or Washington, but information on their occurrences is very limited and we cannot be sure whether or not those species are established (Table 4). We have included this species list because of the potential ecological consequences if they do become established and more widespread in distribution.

Table 3. Wildlife species introduced to Oregon or Washington. [7, 20, 27, 34, 37, 46, 54, 70, 75, 76, 94, 97, 98, 99, 104, 152]

Species	Place/Date	Reason	Status	Origin
Trumpeter swan, *Cygnus buccinator*	Harney Co. OR 1939-58; Spokane Co. WA 1963	aesthetics; hunting	very limited, small numbers in OR and WA	SE Alaska, NW Canada and somewhat south
Mute swan, *Cygnus olor*	Lincoln Co. OR 1950s; Deschutes Co. OR 1960s	aesthetics, escapees?	very limited, small numbers in OR and WA	Eurasia
American black duck, *Anas rubripes*	Snohomish Co. WA, date unknown	hunting	small localized popn in Puget Sound	E United States
Chukar, *Alectoris chukar*	Lake Co. OR 1951; Deschutes Co. OR 1952; Klamath Co. OR 1960s; E. WA 1930s	hunting, brood stock sale	scattered popns in E OR and E WA	Eurasia
Gray partridge, *Perdix perdix*	Linn Co. OR 1900; 23 counties OR 1913-14; Spokane Co. WA 1906; Columbia Co. WA 1908	hunting, brood stock sale	scattered popns NE OR and E WA	Eurasia
Ring-necked pheasant, *Phasianus colchicus*	Linn Co. OR 1881-82; Protection Is. WA 1883	hunting, brood stock sale	widespread, common	Eurasia
White-tailed ptarmigan, *Lagopus leucurus*	Wallowa Co. OR 1967-69	aesthetics?, hunting?	localized, small numbers in OR; native to WA	SE Alaska, W Canada into WA
Wild turkey, *Meleagris gallopavo*	OR 1899 (failed); many OR counties 1961-83; E. WA 1970s	hunting	widespread, moderate numbers in E OR and E WA; some on San Juan Is.	E United States, Southcentral United States
California quail, *Callipepla californica*	Thurston Co. WA 1857; many OR counties 1914	hunting, brood stock sale	widespread, common; native to S OR	SW United States just into OR
Scaled quail, *Callipepla squamata*	Yakima Co. WA 1913	hunting, brood stock sale, escapees	localized, small numbers; extirpated?	Southcentral United States
Northern bobwhite quail, *Colinus virginianus*	Walla Walla WA 1865 & 1920; Whidbey Is. WA 1871; Linn Co. OR 1882	hunting, brood stock sale	localized, small numbers	E United States
Mountain quail, *Oreortyx pictus*	W WA 1880	hunting, brood stock sale	localized in WA; native in OR	SW United States into SE WA
Rock dove, *Columba livia*	many OR counties <1900; W WA <1940	aesthetics, racing, messengers, then range expansion	widespread, common	Eurasia
Monk parakeet, *Myiopsitta monachus*	Multnomah Co. OR 1969	escapees	small numbers, Portland area; some in WA?	South America
Skylark, *Alauda arvensis*	Portland OR 1889 (failed?); Vanc. Is. BC 1903	aesthetics, then range expansion	small numbers on San Juan Is.	Eurasia
European starling, *Sturnus vulgaris*	Portland OR 1889 (failed); arr. on own 1940s OR & WA	aesthetics, then range expansion after introd. to E United States	widespread, common OR and WA	Eurasia
Crested mynah, *Acridotheres cristatellus*	Vancouver BC 1894	aesthetics, then range expansion	localized, small numbers in Seattle, Bellingham areas	SE Asia
House sparrow, *Passer domesticus*	Spokane Co. WA 1895; King Co. WA 1897; Portland OR 1889	aesthetics and insect control to E United States, range expansion	widespread, common OR and WA	Eurasia
Virginia opossum, *Didelphis virginiana*	Umatilla Co. OR 1910; WA <1941	aesthetics, pets escapees, fur trapping	locally common, esp. W WA, W OR & NE OR	E United States
Red fox, *Vulpes vulpes*	many places W WA by 1909, W OR by 1915	fox hunting, fur farming, escapees	widespread W OR & WA; less so in E OR & WA	Holarctic
European ferret, *Mustela putorius*	San Juan Is. WA	rabbit control	small population remains?	Europe

Species	Place/Date	Reason	Status	Origin
House cat, *Felis catus*	<1800	escapees, pest control	widespread, OR & WA?	Eurasia, Africa
Domestic dog, *Canis familiaris*	<1800	escapees	occasional occurrences	Eurasia
Burro, *Equus asinus*	E OR late 1700s	escapees or released when no longer needed	small population in SE OR	Africa
Horse, *Equus caballus*	E OR late 1700s	escapees or released when no longer needed	moderate population in SE OR	Asia
Feral pig, *Sus scrofa*	SW OR late 1800s; Skagit Co. WA 1981	hunting, escapees?	Very small, localized populations or extirpated	Eurasia
Axis deer, *Axis axis*	Pierce Co. WA >1980	aesthetics or escapees	small, localized population	India
Fallow deer, *Dama dama*	King Co. WA >1980	aesthetics or escapees	small, localized population	Europe
Mountain goat, *Oreamos americanus*	NE OR 1950 & Columbia Gorge OR 1969 (failed); Olympic Mtns. WA early 1900s	aesthetics, hunting	moderately abundant in Olympic mtns., native to N Cascade & Rocky Mtns.	Alaska to WA, Cascade and Rocky Mtns.
Eastern gray squirrel, *Sciurus carolinensis*	King Co. WA 1925; W OR 1919	aesthetics	localized, urban/suburban areas of W & NE OR and W WA	E United States
Fox squirrel, *Sciuris niger*	W OR & WA <1940; Baker Co. OR 1950s	aesthetics	localized, urban/suburban areas of E WA, W and NE OR	E United States
House mouse, *Mus musculus*	OR and WA late 1700s	stowaway, then range expansion	widespread, urban/suburban areas OR & WA	Europe
Norway rat, *Rattus norvegicus*	OR and WA <1850	stowaway, then range expansion	localized, urban/suburban areas OR & WA	Asia
Black rat, *Rattus rattus*	OR and WA <1800	stowaway, then range expansion	localized, urban/suburban areas OR & WA	Asia
Nutria, *Myocastor coypus*	King Co. WA 1930s; Lincoln and Tillamook Cos. OR 1937	fur farming, escapees, vegetation control?	localized, mostly W OR & WA	South America
European rabbit, *Oryctolagus cuniculus*	San Juan Co.. WA 1929; Destruction Is. WA 1970	aesthetics, hunting	Island populations persist, other localized populations in WA?	Europe
Eastern cottontail, *Sylvilagus floridanus*	Whitman Co. WA 1926; Linn and Benton Cos. OR 1940s	hunting	widespread, locally abundant	E United States
Bullfrog, *Rana catesbeiana*	many places OR & WA 1914 on	insect control, aesthetics, hunting, food, then range expansion	widespread, locally abundant	E and Central United States
Green frog, *Rana clamitans*	King, Stevens and Whatcom Cos. WA	aesthetics, hunting?	very localized, small populations	E. United States
Snapping turtle, *Chelydra serpentina*	many places W OR & W WA 1950s on	aesthetics, hunting?, pets, food	localized, small populations	E and Central United States
Red-eared slider turtle, *Trachemys scripta elegans*	many places OR and WA	aesthetics?, pets, escapees	Locally common in W and Central OR	SE United States
Plateau striped whiptail, *Cnemidophorus velox*	Jefferson Co. OR	aesthetics?	Localized, small population	SW United States

Habitat Use by Introduced Wildlife Species

All general habitat categories that occur in Oregon or Washington are used by at least one of the 42 introduced species, although few introduced species use alpine or marine habitats (Table 5). Only 12 of 42 (29%) introduced species are affiliated with only one general habitat category; most of those 12 species are restricted to freshwater/riparian systems. Most species (71%) can be considered habitat generalists, using several general habitat categories.

Human-disturbed areas (agriculture lands, urban/suburban areas) are used by a large number of the introduced wildlife species, 19 and 18 species, respectively (Table 5). This group of species includes most introduced upland game birds, songbirds, and mammals. Forests

Table 4. Other introduced wildlife species that have been occasionally observed or reported in Oregon or Washington.*

Birds	Mammals	Amphibians/reptiles
Domestic goose, *Anser cygnoides*	Wolf-dog hybrid, *Canis lupus x familiaris*	Eastern mud turtle, *Kinosternon subrubrum*
Egyptian goose, *Alopochen aegyptiacus*	Domestic cow, *Bos taurus*	Stinkpot, *Sternotherus odoratus*
Graylag goose, *Anser anser*	Domestic goat, *Capra hircus*	Painted turtle (non-natives), *Chrysemys picta*
Domestic mallards, *Anas platyrhynchos*	Domestic sheep, *Ovis aries*	Eastern box turtle, *Terrapene carolina*
Muscovy duck, *Cairina moschata*	Barbary sheep, *Ammotragus lervia*	Ornate box turtle, *Terrapene ornata*
Red-legged partridge, *Alectoris rufa*	Mouflon sheep, *Ovis musimon*	Malayan box turtle, *Cuora amboinensis*
Sichuan pheasant, *Phasianus colchicus suehschanensis*		Desert tortoise, *Gopherus agassizii*
Golden pheasant, *Chrysolophus pictus*		Texas tortoise, *Gopherus berlandieri*
Peafowl, *Pavo cristatus*		Gopher tortoise, *Gopherus polyphemus*
Guineafowl, *Numida meleagris*		Hermann's tortoise, *Testudo hermanni*
Psittacines (misc. Parrots, cockatoos, macaws)		Reeve's turtle, *Chinemys reevesi*
		Spiny softshell turtle, *Apalone spinifera*
		Florida softshell turtle, *Trionyx ferox*
		Big-headed turtle, *Platysternon megacephalum*
		Caiman, *Caiman crocodilus*

* Little is known about the status of most of these species; most probably do not comprise free-ranging, self-sustaining populations and have not expanded their range beyond the release site(s) in either state, however, these events could occur in the future.

Table 5. Use of general habitat categories by 42 wildlife species introduced to Oregon or Washington.

Wildlife Group	No. (% of Group) of Species by General Habitat Category[a]						
	Forest	Shrub/grass	Agriculture	Urb/suburb	Freshw/rip	Marine	Alpine
Birds (18 spp.)	8 (44%)	8 (44%)	11 (61%)	7 (39%)	4 (22%)	1 (6%)	1 (6%)
Mammals (19 spp.)	11 (58%)	4 (21%)	8 (42%)	11 (58%)	5 (26%)	0 (0%)	1 (5%)
Amphibians/reptiles (5 spp.)	1 (20%)	1 (20%)	0 (0%)	0 (0%)	5 (100%)	0 (0%)	0 (0%)
Total: (42 spp.)	20 (48%)	13 (31%)	19 (45%)	18 (43%)	14 (33%)	1 (2%)	2 (5%)

[a]Most species use more than one general habitat category.

(especially open, deciduous or mixed forests) are used by about half (48%) of the introduced species. Freshwater/riparian habitats are used by approximately equal numbers of introduced bird, mammal, and amphibian/reptile species groups, with all introduced reptiles and amphibians using those habitats. Shrub/grass habitats are used by introduced bird species and eastern cottontail rabbits, but especially upland game birds.

Specific habitat associations for many of the introduced wildlife species have not been well defined. Some insight for some species can be gained from Brown,[12] Guenther and Kucera[60] and Thomas.[144] That information, along with species-specific literature and expert opinion, has been used to complete the wildlife habitat matrixes of this book. These matrixes can be used, to some extent, to project the potential competition between native species and introduced species. For example, the introduced eastern gray and fox squirrels use oak woodlands as do native western gray squirrels; all three species use variously-aged forest stands and all use snags. White-tailed ptarmigans, introduced to Oregon, may compete with native blue and spruce grouse in the use of alpine meadows, subalpine fir

habitats, and grass-forb dominated areas. On the other hand, there is little potential for competition between white-tailed ptarmigans and ruffed grouse. Competition may also occur between introduced upland game birds and native sage and sharp-tailed grouse; all of these species use shrub-steppe and sagebrush-steppe habitats, as well as grass-dominated areas and riparian areas. Most of these species use agricultural lands, too. Likewise, in some situations, the introduced eastern cottontail rabbit may compete with native rabbits (pygmy, and brush rabbits; Nuttall's cottontail) in grass/sedge meadows and alder bottomlands as well as riparian areas, agricultural lands, brushpiles (including downed woody materials), and burrows. For many introduced species, we do not know enough about what specific habitats they could use, given the opportunity and time for populations to occupy those habitats. For example, axis and fallow deer could potentially occupy many of the same habitats as the native black-tailed deer and the endangered Columbia white-tailed deer. Interested persons are referred to the matrixes for further investigation of potential habitat competition between native and introduced species.

Ecological Consequences of Wildlife Introductions

It is very important to recognize the potential ecological consequences of wildlife introductions. While considerable effort and expense have been invested in dealing with introduced plants and insects in the Pacific Northwest, introduced wildlife also has caused, or has the potential to cause, substantial harm to Pacific Northwest ecosystems and agricultural resources (Table 2). There are many potential or realized ecological consequences for each of the 42 wildlife species introductions that have occurred in Oregon and Washington (Table 6).

Several points need to be emphasized. The code "NK" (none known) appears frequently in Table 6. With wildlife introductions, we are often presented with ecological "situations" with which we have little or no experience; hence, our predictive powers are very limited. Additionally, serious effects may occur long after the introductions. The rather cavalier attitude of the past ("let's do it and see what happens") is no longer acceptable, given the many legal mandates and policies for species and biodiversity protection, healthy ecosystem maintenance, and the protection of human health and agricultural resources. We must not only deal with existing introduction problems, but must strive to prevent future introductions that have significant potential for adverse consequences. Both require a greater ecological and managerial knowledge base than we now possess. On the other hand, great strides in agriculture, recreation, local economies, biological control of pests, and even in the medical profession, have been made as a result of species introductions. Obviously, a careful and deliberate analysis, on a case-by-case basis, must be made before proceeding with any wildlife introduction.

Introduced wildlife species have the potential for several, if not numerous, adverse ecological consequences (Table 6). Most introduced birds have the potential to adversely affect native birds, especially through forage and nest site competition. A classic example is the ability of starlings to usurp nest sites from wood ducks, bluebirds, woodpeckers of many species, and many other songbirds.[57, 155] Additionally, some hybridization problems exist; for example, black ducks hybridize with mallards,[133] some upland game species hybridize with native species[8] and eastern cottontail rabbits hybridize with the brush rabbit.[151] Avian diseases, such as avian tuberculosis, Newcastle's disease, salmonellosis, and chlamydiosis, can be transmitted to native species.[52] Members of the parrot family make popular pets, but they have the potential to spread avian tuberculosis, a disease transmissible to not only other wildlife, but to pets, livestock, and humans.[52] Congregations of introduced bird species such as house sparrows, rock doves, and starlings at roosts or feeding stations have produced significant disease hazards (e.g., histoplasmosis, ornithosis, salmonellosis).[52] Several of the introduced bird species (starling, house sparrow, rock dove, skylark, crested mynah, *Acridotheres cristatellus*) have a great potential to damage crops, contaminate foodstuffs, and cause aesthetic problems.[52, 55] Additionally, the

Wild burro. (Photograph by Jeffrey C. Lewis)

introduced rodent species cause many types of structural damage to human dwellings, livestock facilities, and constructed features such as dikes, dams, levees, transmission lines, and irrigation systems.[69] Considerable effort is expended each year to reduce the negative effects of these species. On the other hand, the introduction of upland game birds may have increased the prey base for native predators, both avian and mammalian, as well as having provided recreation as intended.

Most introduced mammal species have the potential to adversely affect various species of native plants and animals, as well as humans and their resources, through herbivory or predation. Species such as red foxes, ferrets, and feral dogs and cats can inflict high levels of mortality on ground nesting birds and have been implicated in the endangerment of numerous species, such as snowy plovers, least terns (*Sterna antillarum*), and clapper rails (*Rallus longirostris*).[164] Similar results have been reported for introduced carnivores in Europe.[82] Feral or free-ranging cats kill large numbers of songbirds every year in urban/suburban settings.[25] Essentially, all introduced herbivorous mammals can cause plant damage and may even impede regeneration of some plant species. Usually the amount of damage is related to the density of introduced mammals; hence, existing small, introduced populations of axis and fallow deer are probably not causing significant impact to the native flora. On the other hand, high densities or concentrated use can cause substantial impacts on native flora, as occurs with European rabbits, mountain goats, and feral pigs, horses, and burros. In some cases, endangerment of native plant species may occur, as with herbivory by introduced mountain goats on the Olympic Peninsula.[67] Many of the introduced mammal species, especially carnivores and Old World rodents, have been implicated in the transmission of disease to native wildlife, livestock, pets, or humans; these diseases include rabies, plague, distemper, relapsing fever, and leptospirosis.[38]

Introduced amphibians and reptiles have been implicated in the decline of many native aquatic fauna through predation or competition.[29, 111] The introduced bullfrog is a classic example and will be considered in more detail in the case histories at the end of the chapter.

Table 6. Potential or realized ecological consequences of wildlife species introduced to Oregon or Washington.

Species	Physical Environment	Flora	Fauna	Human
Trumpeter swan, *Cygnus buccinator*	NK	NK	FC, NC, AB	AB
Mute swan, *Cygnus olor*	NK	NK	FC, NC, AB	AB
American black duck, *Anas rubripes*	NK	NK	H	NK
Chukar, *Alectoris chukar*	NK	NK	D, FC, NC, IP	CD
Gray partridge, *Perdix perdix*	NK	NK	D, FC, NC, IP	CD
Ring-necked pheasant, *Phasianus colchicus*	NK	NK	D, FC, NC, IP, H	CD
White-tailed ptarmigan, *Lagopus leucurus*	NK	NK	FC, NC, AB	NK
Wild turkey, *Meleagris gallopavo*	NK	PR	FC, IP	NK
California quail, *Callipepla californica*	NK	NK	D, FC, NC, IP, H	CD
Scaled quail, *Callipepla squamata*	NK	NK	D, FC, NC, IP, H	CD
Northern bobwhite quail, *Colinus virginianus*	NK	NK	D, FC, NC, IP, H	CD
Mountain quail, *Oreortyx pictus*	NK	NK	D, FC, NC, IP, H	CD
Rock dove, *Columba livia*	NK	NK	FC, NC, D	A, HD, CF
Monk parakeet, *Myiopsitta monachus*	NK	PD?	FC, NC, D	CD
Skylark, *Alauda arvensis*	NK	PD?	FC, NC, D?	CD?
European starling, *Sturnus vulgaris*	NK	PD?	FC, NC, D	CF, A, HD
Crested mynah, *Acridotheres cristatellus*	NK	PD?	FC, NC, D?	CF, A, HD?
House sparrow, *Passer domesticus*	NK	PD?	PC, FC, NC, D	CF, A, HD
Virginia opossum, *Didelphis virginiana*	NK	NK	P, D	A, HD
Red fox, *Vulpes vulpes*	NK	PD?	P, PC, E, H, D	CD, LP, HD
European ferret, *Mustela putorius*	NK	NK	P, E, H?, D?	AB, HD?
House cat, *Felis catus*	NK	NK	P, PC, E, H?, D	HD
Domestic dog, *Canis familiaris*	NK	NK	P, PC, E, H, D	AB, LP, HD
Burro, *Equus asinus*	S, W	PD, E	FC, D	LC, HD
Horse, *Equus caballus*	S, W	PD, E	FC, D	LC, HD
Feral pig, *Sus scrofa*	S, W	PD, PR	FC, D, IP	AB, LC
Axis deer, *Axis axis*	NK	PD, PR	H?, D?	NK
Fallow deer, *Dama dama*	NK	PD, PR	H?, D?	NK
Mountain goat, *Oreamos americanus*	S	PD, PR, E	FC, IP	NK
Eastern gray squirrel, *Sciurus carolinensis*	NK	PD, PR	FC, NC, AB, D	SD
Fox squirrel, *Sciuris niger*	NK	PD, PR	FC, NC, AB, D	SD
House mouse, *Mus musculus*	NK	NK	FC?, D	HD, A, CF, SD
Norway rat, *Rattus norvegicus*	NK	NK	FC?, D	HD, A, CF, SD
Black rat, *Rattus rattus*	NK	PD?	FC?, D	HD, A, CF, SD
Nutria, *Myocastor coypus*	S, W	PD, PR	FC?, D	HD, SD
European rabbit, *Oryctolagus cuniculus*	S	PD, PR, E	FC, NC?, D, IP?	CD, LC, HD
Eastern cottontail, *Sylvilagus floridanus*	NK	PD, PR?, E?	FC, NC?, D, IP	CD, LC, HD
Bullfrog, *Rana catesbeiana*	NK	NK	PC, P, E	NK
Green frog, *Rana clamitans*	NK	NK	PC, P?, E?	NK
Snapping turtle, *Chelydra serpentina*	NK	NK	PC, P, E?, AB, D	AB
Red-eared slider turtle, *Trachemys scripta*	NK	PD?	P?, PC?, FC?, D, AB	NK
Plateau striped whiptail, *Cnemidophorus velox*	NK	NK	PC?	NK

A=aesthetics, AB=aggressive behavior, CD=crop damage, CF=contamination of foods, D=disease/parasites, E=species endangerment, FC=forage competition, H=hybridization, HD=human/livestock/pet disease/parasites, IP=increase prey base, LC=livestock forage competition, LP=livestock predation, NC=nest competition, NK=none known, P=predation, PC=prey competition, PD=plant damage, PR=plant regeneration, S=soil erosion, SD=structural damage, W=water quality/quantity.

English (house) sparrow. (Photograph by Jeffrey C. Lewis)

Management of Introduced Wildlife Species

It is better to prevent the introduction of an unwanted species rather than deal with the management or attempted eradication of the species once it becomes established.[36, 135] On the other hand, many introduced wildlife species (upland birds, cottontail rabbits, nutria, red fox) are managed as "game" species by state wildlife agencies, using traditional methods of harvestable wildlife management. Usually, a harvest license is required; seasons, bag limits, methods of take, and other regulation are set each year; and, in some cases, populations and harvests are monitored.

There are many wildlife management methods available to assist us in the management of "undesirable" introduced wildlife species (Table 7). These physical, mechanical, chemical, biological, and cultural methods are used to reduce the carrying capacity of the area for the species, to reduce population density, or to keep animals out of certain areas. Reduction of populations by lethal means may only provide a temporary "fix" unless habitats can be modified to reduce their carrying capacity for the introduced species.[149] On the other hand, commercial exploitation or bounties on introduced wildlife has been used in some situations as a way to keep population levels down while generating local income.[22, 118] Pathogenic agents are rarely used to control vertebrate populations because of the need for specificity, start-up costs and potential hazards, although efforts continue in Australia.[114] Often a variety of methods are employed, as in an integrated pest management (IPM) approach.[13, 17, 118, 165] If a new species is released in the area, it is important to restrict its spread as soon as possible.[9] Research is underway on chemosensory and reproductive inhibition devices that may provide valuable tools in the future management of introduced species.[48]

Eradication of an introduced species is often the management goal, but is difficult to achieve.[9, 24, 127] Nonetheless, eradication has been achieved in some places, especially on islands.[127, 148, 165] An entirely different philosophy is to "let nature take its course" and assume that eventually introduced species will drop out on their own, will fit in satisfactorily, or will result in a worldwide homogenization of the planet's flora and fauna. While this may be the ultimate fate of the global flora and fauna, we, as resource managers and concerned citizens, should not take such a defeatist attitude.[88]

Certain introduced species have the potential for substantial ecosystem disruption, the final and highly significant category in Table 2. Of the 42 wildlife species already introduced into Oregon or Washington, we would include feral livestock (pigs, horses, burros), mountain goats (in areas where they are not native), nutria, and European rabbits in this category. Other species (such as the Old World rodents) cause substantial ecological disruption in tropical ecosystems, but are not as damaging to temperate ecosystems. Major ecosystem disruption can occur when these species seriously impact the physical environment (soil parameters, erosion, water quantity and quality), achieve relatively high densities, exert heavy grazing pressure, or successfully compete with native fauna. Mountain goat impacts on the Olympic Peninsula and the difficulties of resolution have been described,[18, 67] and so have impacts and management of feral horse and burro.[19, 41, 98] Feral pigs have been studied extensively around the world because of their very significant impacts to ecosystems.[10, 98] The pros and cons of nutria introductions in the Pacific Northwest and elsewhere have also been discussed.[89, 93, 98] Similar discussions for European rabbits were presented as well.[6, 28, 63, 98] Because of the significant potential for ecological disruption by these species, there have been extensive efforts to eradicate them after introduction, especially from islands, and most are banned from import at the federal or state level.

Table 7. Examples of methods for the management of introduced wildlife species by category.

Cultural/Habitat	Physical	Chemical	Biological	Other
Crop selection	Barriers	Repellents	Predators	Bounties
Cover reduction	Traps	Toxicants	Disease/Parasites	Insurance
Water removal	Electrocution	Reproductive inhibition	Resistant plants	Harvest
Sanitation	Flooding	Aversive conditioning	Lethal genes	Acclimation
Buffer crops	Shooting/Frightening devices	Tranquilizers, other drugs	Biosonics	Acceptance

In most cases, an integrated management approach will be required to control most introduced species, using a problem assessment, action plan, several methods, and monitoring. Adequate surveillance and control at the point of origin are important. Additionally, adequate budgets, public support, and access to private lands will be essential to the successful management or eradication of most introduced wildlife species.[9, 36, 165] Using introduced rodents as an example; Witmer et al.[164, 165] discussed the many considerations of introduced species management and eradication.

There are many socio-political, economic, and ecological issues associated with introduced species.[88, 143] Realizing this, and involving the appropriate and interested parties in the decision-making process, will be essential to the successful resolution of current and future wildlife introductions.

Case Histories

We conclude this chapter with several case histories of wildlife introductions in Oregon and Washington. As we have mentioned, some introductions can be considered "positive" while many are considered "negative" for a number of reasons. In reality, most introductions have the potential for both positive and negative effects. Only time and our concerted efforts will determine the future status of native and introduced wildlife species and of ecosystems in Oregon and Washington. In many cases, we will continue to live with these naturalized species.

1. Ring-necked Pheasant

History, Distribution, and Status. Ring-necked pheasants are native from the Caucasus Mountains of Eurasia through Southeast Asia to Northern Japan where they are closely associated with river valleys, bamboo stands, and agricultural lands. They have been widely introduced throughout the world, primarily for upland game hunting, but also for viewing.[99, 104] Substantial revenues have been generated for state wildlife agencies and for local economies from upland game seasons.[73] The first successful introduction to the United States occurred in 1882 when Owen Denney, an Oregon attorney and judge, had 28 birds from China delivered to Portland.[138] He began breeding the birds, and they did so well that he was soon shipping them to other parts of Oregon, into Washington in about 1883, and, eventually, to other states. The first pheasant hunting in the United States occurred in Oregon in 1891. In 1911, the State of Oregon opened the first large-scale, state-operated game bird farm.[95] The facility in the Willamette Valley achieved peak production in 1950 when over 70,000 pheasants were reared and released. The State of Washington followed with the development of extensive game bird farms.[142] It has been estimated that about 100,000 pheasants were harvested in Washington in 1922,[142] and that number tripled by 1950.[95] More recently, however, many of the rearing facilities have been shut down because of increased costs, low survival rate of pen-reared birds, and other problems.[32] Interestingly, this situation has resulted in attempts to introduce another subspecies, the Sichuan pheasant, that is better adapted to wooded or shrubby habitats.[138] Pheasants are currently widespread in Oregon and Washington. The natural history of the pheasant in the United States and the Pacific Northwest has been recounted.[2, 99, 104, 138]

Ecological Implications. Although the introduction of pheasants and other upland game birds (see Table 3) has largely been considered positive, they are not without some adverse ecological effects. Concerns were expressed as early as 1923 in Washington that pheasants may damage crops (sprouting corn, potatoes) and gardens.[142] They compete for food and nest sites with native grouse species, especially because they are more adaptable and tolerant of disturbed landscapes and because they may be released in large numbers on a regular basis.[138, 150, 159] Their aggressive behavior can displace other birds. Nest parasitism with blue grouse, ruffed grouse, and other upland game bird species has been reported.[83, 132, 142, 150, 158, 159]

There is the potential for hybridization between pheasants, native grouse, and other upland game bird species,[8, 78] but the extent and seriousness of this effect is not known. The potential transfer of diseases, such as avian tuberculosis, to other bird species and even humans, pets, and livestock has been noted.[52]

Management and Research Needs. In Oregon and Washington, pheasants are classified as upland game birds and are managed with season lengths and bag limits to regulate the number of harvested birds. During most years, female pheasants have been protected from harvest to increase recruitment. Wild bird populations have been supplemented with pen-reared birds to increase harvest, although the use of this strategy has greatly declined. There have often been state and federal efforts to encourage agricultural crop producers to manage their lands for the benefit of pheasants and other upland game species. The Conservation Reserve Program is an example of one such program. Activities involve establishing or maintaining areas of woody or herbaceous vegetation; in drier areas, water sources ("guzzlers") may be provided.

It is likely that pheasants will be a less significant element of the Oregon and Washington introduced avifauna in the future. There are many reasons for this, including long-term declines in wild populations, the reduced emphasis on pen-rearing and release of birds, human encroachment on pheasant habitats, clean farming practices, and an increased interest in improving conditions for native upland game bird species. That being said, there is still substantial interest in this naturalized member of the Pacific Northwest fauna, and a wish to assure that its regional presence will continue.

2. European Starling

History, Distribution, and Status. The European starling is a palearctic species that originally ranged throughout Europe and east to Lake Baikal, Siberia, and the Middle East.[99] It has since become naturalized, via numerous introductions, to most of North America, South Africa, Australia, and New Zealand.[99] Starlings were purposefully and successfully introduced to Central Park, New York City, in 1890-91, although the Portland Songbird Club attempted an introduction in Portland, Oregon, in 1889 that failed.[77] The species range expansion in North America is nothing less than amazing, reaching Mexico in the 1940s and Alaska in the 1970s.[16, 85] While numbers of starlings appear to have stabilized over much of North America, they are now one of the most numerous bird species in North America.[16] The species was introduced for aesthetic purposes, to bring a little of the Old World to the New World. Starlings first appeared in Oregon and Washington in 1943.[77] They are now abundant throughout most of Oregon and Washington, especially in urban/suburban and agricultural settings. The natural history of the species has been described.[16, 49, 104]

Ecological Implications. Few benefits have been attributed to the introduction of starlings in North America. Cabe[16] noted, however, that much basic research on avian biology has been done using starlings. Starlings have also been attributed with high levels of insect consumption, may be hunted, and may provide food for humans in some situations.[49] Finally, starlings occur in highly disturbed settings that might otherwise have few birds present.

Feare[49] reviewed the many adverse effects of starlings, including plant damage; food and nest competition with native bird species; disease and parasite transfer to wildlife, livestock, pets, and humans; fruit consumption and damage; livestock food consumption and contamination; aircraft strikes; and aesthetic problems (droppings, odors, noise). Much of the concern about ecological effects of starlings seems to involve their highly competitive ability to usurp nest cavities (both natural and man-made) and thus contribute to the declines in populations of native cavity nesters. Adverse effects have been noted for bluebirds,[39, 122] purple martins,[11] tree swallows,[122] northern flickers,[72] various species of woodpeckers,[71, 84, 147, 156] and various cavity nesting duck species.[50, 57, 105] Not only is it difficult for these species to find and hold nest cavities in the presence of starlings, but starlings may also parasitize the nests of other species by destroying eggs or hatchlings.[50, 57, 122] Brush[15] noted, however, that significant cavity competition probably only occurs where natural cavities are very limited.

Economic losses and damage to planted crops (corn, winter wheat), fruits (grapes, peaches, blueberries, strawberries, figs, apples, and cherries) and livestock feedlots have been described,[49, 80] as well as the disease problems caused by starlings.[49, 52, 80] Constantin and Floyd[26] discussed the hazards of starlings and other birds at airports. Typically, starling problems are quite localized.

Management and Research Needs. The manage-ment of starlings is problematic at best because of their exceptional ability to exploit human-altered landscapes.[49] Many methods are used to reduce starling numbers, the damage they cause, or to disperse aggregations. These include attempts at exclusion from buildings, ledges, and trees; cultural and habitat modifications that reduce food, water,

and roost availability; the use of frightening devices based on chemicals, sounds, or objects; the use of repellents and sticky substances; the use of toxicants; shooting; and trapping, with or without live bird decoys.[80] Although some of these methods have been moderately successful for a while, most are of limited effectiveness and must be repeated on a regular and long-term basis.[16, 49] The difficulty of dealing with starlings at high density roosts has been documented by Glahn et al.[56] In all likelihood, a combination of methods and the alteration of crop and livestock production practices would be most likely to provide damage reduction or population reduction.[49]

Starlings are here to stay and can be expected to continue to impact some native bird species. We need to better understand the interactions of starlings with food sources, habitats, and other species and with the control measures that we employ. We also need to develop more effective damage management methods; research is underway on avian repellents and on immunocontraception. There has been some effort to develop specific methods to reduce the ability of starlings to usurp nest sites from other species; Grabill[57] attempted to increase wood duck nesting success by placing starling nest boxes near wood duck nest boxes. He relied on the agonistic behavior of starlings during nesting to keep other starlings from using the wood duck nest boxes. Fielder et al.[50] reduced starling use of wood duck nest boxes by covering the opening from the end of the wood duck nesting season until just before the initiation of the next wood duck nesting season. Lumsden[105] and McGilvrey and Uhler[109] also presented designs to reduce nest box use by starlings. Knowledge of starling flocking behavior was instrumental in the development of Avitrol, a chemical frightening agent. A few birds are allowed to feed on treated bait. They become sick, fly erratically, and give warning cries that frighten other starlings from the area.[80] Most starlings ingesting Avitrol will eventually die; therefore, the chemical must be used carefully to minimize secondary poisoning hazards.[66] Geis[53] noted that starling and house sparrow numbers could be kept at lower densities by the careful design of urban structures; latticework on apartment buildings, for example, was very attractive to starlings and house sparrows. These examples illustrate the value of a thorough knowledge of the biology and ecology, including behavior, of a species to assist in resolving conflicts.

3. Nutria

History, Distribution, and Status. The nutria is a large semi-aquatic rodent native to South America, which has been introduced into a number of areas in North America since the 1930s. Oregon and Washington are among 15 or more states with feral nutria populations.[161] Nutrias were first brought to the Northwest in the 1930s in the expectation that nutria farming would become a lucrative endeavor.[61, 89, 94] Inflated breeding stock prices, poor reproduction, large farming expenses, and little economic return for nutria pelts (~$1.00 per pelt) resulted in the collapse of an industry whose boom was short-lived.[47, 87,]

[89, 161] More than 600 nutria farms existed in Oregon from the 1930s to the 1950s,[89] and a number of farms existed in Washington at this time.[61,93] Flooding and storms damaged holding structures and allowed some nutrias to escape from fur farms, however, farmers often released their stock when farming became uneconomical. By the 1940s, nutrias had been captured by trappers or collected on both sides of the Cascades in Oregon and Washington, but most nutrias were found in the Puget Sound area, the Willamette Valley, along coastal Oregon rivers, and along the Columbia River.[70,79,89,93,108] Only the Yakima River drainage in southcentral Washington supports substantial numbers east of the Cascade Mountains.

The nutria is an unclassified wildlife species in Oregon and Washington, and it can be harvested in unlimited numbers at any time of the year. The records indicate fluctuating harvest levels of nutrias, which may reflect fluctuating pelt prices[152] rather than fluctuating population densities. Nutria harvest data also indicate a relatively stable population, in that nutrias are consistently captured in the same counties (i.e., nutrias do not appear to be spreading to previously unoccupied counties in appreciable numbers). Short-term stability, however, does not necessarily mean that all habitats suitable for nutrias have been colonized or that a range expansion will not occur in the future. The natural history of the nutria has been described in detail.[47,87,161]

Ecological Implications. Through foraging, nutrias can denude expanses of vegetation, eliminating vegetative structure.[87,157] While nutrias are generally opportunistic vegetarians, Wentz[157] found that broadleaf arrowhead (*Saggittaria latifolia*) and smartweed (*Polygonum* spp.) were selected by nutrias in Oregon, and these plants may be locally reduced or extirpated by foraging nutrias. Nutrias construct resting and feeding platforms of compacted vegetation in wet areas, form trails between these platforms through vegetation, and also create grooming areas, dens, and runs or slides at the water's edge.[47,87] These activities can significantly impact vegetative communities.[87,157] The clearing of vegetation by nutrias may alter plant succession, and convert marsh ecosystems to more open-water environments.

In Louisiana, increasing nutria harvests in the mid-1900s coincided with decreasing muskrat harvests.[87] The apparent decline in the muskrat population could have been the result of many factors, but the nutria irruption was considered among the most significant. Apparent declines in muskrat numbers have also been observed in areas where nutrias are abundant on the Finley National Wildlife Refuge in western Oregon (H. Brunkal, U. S. Fish and Wildlife Service, pers. comm.). Alteration of the vegetative community would be expected to have a significant influence on native fauna, especially sensitive amphibians and species that have niches similar to the nutria (e.g., muskrat, some waterfowl). Unfortunately, little information is available on the direct or indirect impacts of nutrias on other fauna.

Nutrias cause direct and indirect impacts to humans by their foraging and burrowing activities, which result in damage to agriculture, drainage systems, earthen structures (dikes, levees, embankments), and vegetative communities.[47,89,96] Burrowing can disintegrate and weaken these structures, and may cause them to fail.[96,161] Ironically, nutrias were introduced in some areas to help control marsh vegetation.[47] Kuhn and Peloquin[89] reported nutria damage to agricultural crops in the Willamette Valley and estimated losses of thousands of dollars per year. Humans, livestock and pets are vulnerable to a number of diseases and parasites carried by nutrias, including equine encephalomyelitis, leptospirosis, hemorragic septicemia, paratyphoid, salmonellosis, giardiasis, tapeworms, and liver flukes.[96,161] The aggressive behavior of nutria also poses a hazard to pets that approach them too closely (J. Tabor, Washington Department of Fish and Wildlife, pers. comm.).

Personnel with the USDA Wildlife Services and state wildlife officers respond to nutria damage complaints. Although a number of damage prevention and control methods exist for nutrias,[96] commercial trapping appears to be the most common method used in Oregon and Washington. Some trappers have certainly benefited from the introduction of nutrias, although the monetary benefits appear limited as nutria pelts are not highly valued for fur.[152] Low pelt prices offer little incentive to most trappers and consequently, commercial trapping may be limited as a management tool for nutria populations. Conversely, control of pest nutria can be a source of income for some trappers and pest control professionals.

Management and Research Needs. Trapping and localized control efforts have been used to manage nutria populations since they were first introduced, and these techniques will likely continue to provide for nutria management in the future. Trapping records indicate a relative stable nutria population in the Pacific Northwest. Until new information indicates that nutria impacts are particularly severe to certain species, ecological communities, or geographic areas, it is unlikely that current management methods will be altered or replaced. Lobbying efforts to ban trapping or outcries for nutria eradication could alter the status quo, but these do not appear to be immediate issues in Oregon or Washington. With the exception of research by Peloquin[121] on growth and reproduction and Wentz[157] on nutria density and impacts to marsh vegetation, little study of the nutria has been conducted in the Pacific Northwest, and none has been published from Washington. Future research should focus on how the nutria's alteration of aquatic environments and its physical presence (i.e., potential competition and disease transmission) could impact sensitive fauna and vegetative communities. This research may also prompt study into alternative management techniques for nutrias.

4. Red Fox

History, Distribution, and Status. Native to North America, Europe, Asia, and northern Africa,[103] the red fox has the largest geographic range of any terrestrial carnivore with the possible exception of the domestic cat.

European red foxes were introduced to the East Coast in the 1600s and 1700s[23, 126] and Australia in the mid 1800s[106] for fox hunting; actions that confused the taxonomy of red foxes in eastern North America and greatly expanded the range of the red fox. Non-native red foxes were brought to Oregon and Washington in the early 1900s for fur farming[40] and fox hunting or trapping.[4, 108, 160] In the 1910s, the fur industry was rapidly spreading west across the continent, when choice breeding stock and pelts from red foxes (predominantly the silver phase foxes) were sold for thousands of dollars.[40] By 1915, the first fox farms were established in Oregon and Washington.[40] Many introductions occurred when foxes were released or escaped from farms[3, 4, 40] or evaded hounds and hunters, forming free-roaming populations of non-native red foxes in both states. More recent introductions of red foxes in North America have included the release of pet foxes, the illegal release of farm foxes by animal rights activists, and the translocation of non-native foxes into previously unoccupied areas by pet owners, wildlife rehabilitators, and animal control officials.[102]

Non-native red foxes occur throughout many of the lowland areas in western Oregon and Washington[4, 5] and in several disjunct populations in eastern Oregon and Washington. The distribution of non-native red foxes has largely been determined as those areas where red foxes occur outside the historical ranges of native Cascade (*V. v. cascadensis*) and Rocky Mountain red foxes (*V. v. macroura*).[4, 5, 7, 62] The Cascade red fox historically occurred in the high-elevation meadows and parklands of the Cascade Range, whereas the Rocky Mountain red fox occupied similar habitats in northeastern Washington, and in the Blue and Wallowa Mountains.[7, 37, 62] Because relatively little information is available on the locations and operation of fox farms, especially in Oregon, and because there are no known means of visually distinguishing native from non-native red foxes, it is not known if introductions have occurred within the ranges of the native red foxes.

Red foxes are considered a furbearing species in Oregon and a furbearing game animal in Washington with no administrative distinctions made between native and non-native red foxes. Red foxes can be trapped in most areas of Oregon during a regulated season. In Washington, red foxes can be trapped during a regulated season except in Whatcom, Skagit, and Island Counties, and a portion of Cowlitz County. Because market prices strongly influence the harvest of most furbearer species, harvest data are not a good indication of fox population trends. While little information is available on population trends, there are no indications that fox populations in either state are increasing or decreasing dramatically. The natural history of the red fox has been described in detail.[103, 106, 141, 152, 153]

Ecological Implications. With the exception of excavating holes for dens and prey items, and leaving some uneaten prey remains scattered about, red foxes probably have little effect on their physical environment. Non-native red foxes feed on a variety of fruit-producing plants.[100] They also eat leafy vegetation, some of which may be ingested incidentally with other foods. Foxes may be important seed dispersers of both native and non-native plants. Den-site excavation and other digging could minimally disrupt flora, but would also expose a medium for seed germination.

Red foxes commonly prey on insects, earthworms, small- to medium-sized birds and mammals, and herpetofauna; predation on crustaceans and fish and the use of carrion has also been documented.[64, 100, 103] Red foxes are noted predators of species valued by humans as pets,[64] livestock,[58, 115] game birds and mammals,[45, 160] and endangered species.[164] Conservation of the snowy plover in Oregon and Washington could be hindered by red foxes should they become established near nesting colonies along the coast, as has happened in California. Interference with the reproductive behavior of native fauna, especially ground-nesting birds, can be significant.[166] Aubry[4, 5] suggested that non-native red foxes might not be physiologically or behaviorally capable of surviving in high-elevation habitats. However, an introduction of non-native red foxes within the historical ranges of native red foxes could result in resource competition, interbreeding and disease transmission.[5, 101] It is unknown if interbreeding with non-native foxes would reduce the fitness of native red fox populations. The transmission of diseases, including sarcoptic mange, rabies, canine distemper, parvovirus, and leptospirosis, is a threat that red foxes pose to other mammals.[103] Additionally, resource competition,[81] disease transmission, and interbreeding[145] would be expected to negatively affect native kit foxes should red foxes become established in southeastern Oregon.

Foxes may negatively impact humans in several ways, including livestock depredation, crop damage, disease transmission to humans and their pets, and predation or injury of pets. More indirectly, non-native red foxes may negatively impact humans by affecting species valued by the public (e.g., game or protected species). Positive impacts to the public include recreational and economical opportunities of trapping, hunting, and fur farming, and the recreational opportunities of feeding and watching wild foxes, along with the enjoyment from having foxes as pets; the latter is strongly discouraged by wildlife professionals.

Management and Research Needs. Relatively little is known about the populations of non-native red foxes in Oregon and Washington. Non-native red foxes provide additional harvest opportunities for trappers, and management in this regard comprises season restrictions and harvest regulations. Management has also involved communication with the public about occasional livestock kills and concerns about fox predation on domestic cats. Damage prevention and control methods for red foxes were reviewed by Phillips and Schmidt.[123] The wide distribution of red foxes in western Oregon and Washington reflects the potential for disease transmission to pets, livestock, other wildlife, and humans, with rabies being of particular concern. Research on non-native red foxes to determine distribution and densities, identify disease prevalence, and characterize food habits is needed to better understand populations and potential impacts in Oregon and Washington.

5. Bullfrog

History, Distribution, and Status. Bullfrogs are native to North America east of the Rocky Mountains; however, their range has greatly expanded due to introductions by humans in western North America, South America, Europe, and Asia.[14] Bullfrogs were first introduced into the Northwest in the early 1900s to provide food (i.e., frog legs), opportunities for frog hunting, and stock for frog farms.[92, 140] Being the largest frog in North America, bullfrog legs were a highly prized food, and frog farming to supply the demand for bullfrog legs was undertaken but rarely succeeded.[14] Lampman[92] states that bullfrogs from Idaho were first brought to eastern Oregon in 1914, and subsequent introductions in western Oregon in 1921 involved releasing additional Idaho bullfrogs and bullfrogs from the previously introduced populations in eastern Oregon. Nussbaum et al.[113] reported the release of 18 bullfrogs in the Grant's Pass area in 1931 and that bullfrogs were soon well established in the upper Rogue River Valley. In Washington, Dvornich et al.[43] reported the first specimens collected in the 1930s, which suggests that the first successful introductions occurred in the 1920s and early 1930s.

Bullfrogs are largely aquatic and occur in lower-elevation freshwater habitats on both sides of the Cascades in Oregon and Washington and along much of the Columbia River.[97] Within this range, the bullfrog has become widely established and locally abundant because it is a capable colonizer of a wide variety of habitats and a prolific breeder.[14] They are classified as a game fish in Oregon and as a game species in Washington; however, these classifications may soon change in both states. Fishing or hunting license (in Oregon and Washington, respectively) is required to harvest bullfrogs and there are season restrictions, but no bag limits. Bullfrogs may still be expanding their range as suitable habitats are colonized by invading or introduced individuals. Humans continue to introduce bullfrogs into new, previously unoccupied areas. Water-garden and pond stores in Vancouver and Portland recently sold bullfrog tadpoles from California and North Carolina (at $3.00 per tadpole) to individuals interested in stocking their ponds with bullfrogs (selling or possessing live bullfrogs is illegal in Washington without a permit). Also, some summer festivals include a frog-jumping contest where captured frogs (often bullfrogs) may be released after the contest into previously unoccupied habitats. The natural history of bullfrogs has been described in detail.[14, 113]

Ecological Implications. Both tadpoles and adults feed on vegetation,[14, 113] although plant consumption by adults is likely the result of incidental ingestion while capturing a prey item. Consumption of vegetation by bullfrog tadpoles could constitute an impact on local flora and a significant indirect impact on native species that use this flora for food or cover,[90] but this consumption could make a substantial contribution to nutrient cycling in aquatic ecosystems.

Much of the literature on the bullfrog in western North America has been concerned with the effect of bullfrog predation on native fauna, especially other ranids.[42, 65, 86, 111] Kupferberg[90] demonstrated that bullfrogs negatively affected the growth of developing yellow-legged frogs by outcompeting them for food resources. Although many reports have implicated the bullfrog as a major cause of declines in some native species, Hayes and Jennings[65] argued that this has not been clearly proven and a number of other factors may be at work, such as predation by introduced fish, habitat alteration, commercial exploitation, and the effects of toxicants. However, until proven otherwise, it may be wise to consider bullfrog predation and competition as detrimental to a number of vulnerable, sensitive, or listed species in Oregon and Washington, including the Oregon spotted frog, leopard frog, red-legged frog, foothill yellow-legged frog, and western pond turtle. Bullfrog predation on hatchling western pond turtles has prompted management efforts to protect remaining populations in Oregon and Washington. Management efforts involve collecting hatchling turtles from western pond turtle nests and placing them in captivity until they are too large to be eaten by bullfrogs (R. Goggans, Oregon Department of Fish and Wildlife, and K. Slavens, Washington Department of Fish and Wildlife, pers. comm.). These efforts, which started in the early 1990s in Oregon and Washington, have been successful at recruiting young turtles into resident populations, and some female recruits from the first "head-started" cohorts are expected to be large enough to breed in 1999.

Bullfrogs are beneficial to some people as a source of food, sport, and economic gain. Universities and schools have created a significant demand for bullfrogs for use in classroom and laboratory study. Others simply enjoy the sound of bullfrogs or stock them for the pleasure of having bullfrogs on their property; the latter is strongly discouraged by wildlife professionals.

Bullfrog predation on and competition with native species are impacts that cause concern among many people. This concern may prompt a modification of the legal status of the bullfrog to allow for more effective bullfrog management and protection of native species vulnerable to bullfrog predation and competition.

Management and Research Needs. The predation and competition threats posed by bullfrogs to native species have prompted consideration of bullfrog eradication in some localized areas. Removing egg masses, killing adults, or promoting harvests of bullfrogs may act to reduce their impacts on native species. However, their local abundance, widespread distribution, ability to disperse and recolonize habitats, and the tendency for people to transplant bullfrogs, makes eradication difficult at best.[110] Perpetual bullfrog control may be required where management is important to protect or restore native species. Research efforts that determine the degree to which bullfrogs threaten native species, relative to other causes of species decline, will help us focus our management actions on the most critical problems. Research into ways of controlling or eradicating bullfrogs without harming native species would also be valuable.[110] Studies that focus on single species or communities that may be impacted by bullfrog presence should also focus on obtaining data to credibly address the impacts of bullfrogs.

Conclusions

At least 42 introduced species of wildlife (birds, mammals, amphibians, reptiles) occur in Oregon and Washington. Introductions have occurred for many reasons, both accidental and purposeful. Some have greatly contributed to outdoor recreation, local economies, and state wildlife agency revenues, while others have had adverse ecological consequences through direct or indirect mechanisms (e.g., resource competition, displacement, predation, hybridization, and disease transmission). Economic losses to human-valued resources and public health hazards have been documented. There has been more stringent regulation of introduced species at the federal and state levels in recent years, in part due to the increased concern about potential harm to native flora and fauna. Options for resource managers include prevention of entry, eradication, and management after dispersal and establishment; the first is perhaps the most practical, while the latter is the most commonly employed option. Eradication is difficult and expensive in most situations. More surveillance and control at the point of origin is needed. A sustained effort, using a variety of methods and the principles of integrated pest management, is needed to limit adverse effects. Each situation must be assessed on a species- and site-specific basis. New methods are needed to improve the monitoring and management of introduced wildlife.

Acknowledgments

We would like to acknowledge the many people that provided information or review comments on a portion or all of this chapter: Keith Aubry, Charles Bruce, Richard Bruggers, Alan Clark, Bruce Coblentz, Larry Cooper, John Crawford, Richard Dolbeer, Joe Engles, James Glahn, Rebecca Goggans, Marc Hayes, David Johnson, Kelly McAllister, Charles Meslow, Thomas O'Neil, Michael Pipas, and Kate Slavens.

Literature Cited

1. Ad Hoc Federal Invasive Alien Species Task Force. 1997. Campaign against invasive alien species: an action plan for the nation. Washington, D.C.

2. Allen, D. L. 1956. Pheasants of North America. Stackpole Books, Harrisburg, PA.

3. Ashbrook, F. G. 1923. Silver fox farming. U. S. Dept. Agric. Bull. No. 1151. Washington, D.C.

4. Aubry, K. B. 1983. The Cascade red fox: distribution, morphology, zoogeography and ecology. Ph.D. Dissertation. University of Washington, Seattle, WA.

5. ———. 1984. The recent history and present distribution of the red fox in Washington. Northwest Science 58:69-79.

6. ———, and S. D. West. 1984. The status of native and introduced mammals on Destruction Island, Washington. The Murrelet 65:80-83.

7. Bailey, V. 1936. The mammals and life zones of Oregon. Bureau of Biological Survey. North American Fauna No. 55. U.S. Department of Agriculture, Washington, D.C.

8. Blackburn, D. 1977. An apparent ring-necked pheasant X blue grouse hybrid. Murrelet 58:78.

9. Bomford, M. and P. O'Brien. 1995. Eradication or control for vertebrate pests. Wildlife Society Bulletin 23:249-255.

10. Bratton, S. P. 1975. The effect of the European wild boar, Sus scrofa, on gray beech forest in the Great Smoky Mountains. Thesis, Cornell University, Ithaca, New York, NY.

11. Brown, C. R. 1981. The impact of Starlings on Purple Martin populations in undamaged colonies. American Birds 35: 266-268.

12. Brown, E. R., editor. 1985. Management of wildlife and fish habitats in forests of western Oregon and Washington. U.S. Department of Agriculture, Forest Service, Pacific Northwest Region. Portland, OR.

13. Bruggers, R. L., E. Rodriguez, and M. Zaccagnini. 1998. Planning for bird pest problem resolution: a case study. International Biodeterioration and Biodegradation 42:173-184.

14. Bury, R. B., and J. A. Whelan. 1984. Ecology and management of the bullfrog. U.S. Fish and Wildlife Service. Resource Publ. No. 155. Washington, D.C.

15. Brush, T. 1983. Cavity use by secondary cavity-nesting birds and responses to manipulations. Condor 85:461-466.

16. Cabe, P. R. 1993. European Starling. The Birds of North America 48: 1-24.

17. Campbell, E. W., G. Rodda, T. Fritts, and R. Bruggers. 1999. An integrated management plan for the brown tree snake on Pacific islands. Pages 423-433 in G. Rodda, Y. Sawai, D. Chiszar, and H. Tanaka, editors. Problem snake management. Comstock, Ithaca, New York, NY.

18. Carlquist, B. 1990. An effective management plan for the exotic mountain goats in Olympic National Park. Natural Areas Journal 10:12-18.

19. Carothers, S. W., M. Stitt, and R. Johnson. 1976. Feral asses on public lands: an analysis of biotic impact, legal considerations and management alternatives. Proceedings of North American Wildlife Conference 41:396-406.

20. Cassidy, K. M., C. Grue, M. Smith, and K. Dvornich, editors. 1997. Washington State Gap Analysis—Final Report. Volumes 1-5. Washington Cooperative Fish and Wildlife Research Unit, University of Washington, Seattle, WA.

21. Center for Wildlife Law. 1996. Saving biodiversity: a status report on state laws, policies and programs. Defenders of Wildlife, Albuquerque, New Mexico.

22. Choquenot, D., P. O'Brien, and J. Hone. 1995. Commercial use of pests: can it contribute to conservation objectives? Pages 251-258 in G. C. Grigg, P. Hale, and D. Lunney, editors. Conservation through sustainable use of wildlife. Centre for Conservation Biology, The University of Queensland, Brisbane, Australia.

23. Churcher, C. S. 1959. The specific status of the New World red fox. Journal of Mammalogy 40:513-520.

24. Coblentz, B. E. 1990. Exotic organisms: a dilemma for conservation biology. Conservation Biology 4:261-265.

25. Coleman, J. S., S. Temple, and S. Craven. 1997. Cats and wildlife: a conservation dilemma. University of Wisconsin Cooperative Extension Service, Madison, WI.

26. Constantin, B., and J. Floyd. 1992. An avian/airport study for Standiford Airport, Louisville, Kentucky: results and management implications. Proceedings Eastern Wildlife Damage Control Conference 5:166-170.

27. Corkran, C. C., and C. Thoms. 1996. Amphibians of Oregon, Washington, and British Columbia. Lone Pine, Renton, WA.

28. Couch, L. K. 1929. Introduced European rabbits in the San Juan Islands, Washington. Journal of Mammalogy 10:334-336.

29. Coulter, M. W. 1957. Predation by snapping turtles upon aquatic birds in Maine marshes. Journal of Wildlife Management 21:17-21.

30. Courtenay, Jr., W. R., and C. Kohler. 1986. Exotic fishes in North American fisheries management. Pages 401-413 in Richard H. Stroud, editor. Proceedings of The Role of Fish Culture in Fisheries Management at Lake Ozark, Missouri. American Fisheries Society, Bethesda, MD.

31. Cox, C. B., I. N. Healy, and P. D. Moore. 1976. Biogeography: an ecological and evolutionary approach. Second edition. Blackwell Scientific, Oxford, England.

32. Crawford, J. A., and S. Meyers. 1980. Survival of ring-necked pheasant hens in the Willamette Valley, Oregon. Unpublished report. Oregon Department of Fish and Wildlife, Portland, OR.

33. Crawley, M. J. 1989. Chance and timing in biological invasions. Pages 407-423 in J. A. Drake, editor. Biological invasions: a global perspective. John Wiley & Sons, New York, NY.

34. Csuti, B., A. J. Kimerling, T. A. O'Neil, M. M. Shaughnessy, E. P. Gaines, and M. M. P. Huso. 1997. Atlas of Oregon wildlife: distribution, habitat, and natural history. Oregon State University Press, Corvallis, OR.

35. Dahlgren, R. B. 1987. The Ring-Necked Pheasant. Pages 305-311 in Harmon Kallman, editor. Restoring America's wildlife- 1937-1987: the first 50 years of the Federal Aid in Wildlife Restoration (Pittman-Robertson) Act. U. S. Department of the Interior, Fish and Wildlife Service, Washington D.C.

36. Dahlsten, D. L. 1986. Control of invaders. Pages 275-302 in Harold A. Mooney, editor. Ecology of biological invasions of North America and Hawaii. Springer-Verlag, New York, NY.

37. Dalquest, W. W. 1948. Mammals of Washington. University of Kansas Publ. Natural History No. 2. University of Kansas Press, Lawrence, KS.p

38. Davis, J. W., L. Karstad, and D. Trainer. 1981. Infectious diseases of wild mammals. Iowa State University Press, Ames, IA.

39. Davis, W. H., and W. C. McComb, 1989. Bluebirds and Starlings: competition for nest sites. Sialia 11:124-126,138.

40. Dearborn, N. 1915. Silver fox farming in eastern North America. Bulletin No. 301. U. S. Department of Agriculture. Washington, D.C.

41. Douglas, C. L., and D. Leslie, Jr. 1996. Feral animals on rangelands. Pages 281-292 in P. Krausman, editor. Rangeland wildlife. Society for Range Management. Denver, CO.

42. Dumas, P. C. 1966. Studies of the Rana species complex in the Pacific Northwest. Copeia 1966:60-74.

43. Dvornich, K. M., K. R. McAllister, and K. B. Aubry. 1997. Amphibians and reptiles of Washington State: location data and predicted distributions. Volume 2, Washington State Gap Analysis, Final Report. Washington Cooperative Fish and Wildlife Research Unit, University of Washington, Seattle, WA.

44. Ehrlich, P. R., 1989. Attributes of invaders and the invading process: vertebrates. Pages 315-328 in J. A. Drake, editor. Biological invasions: a global perspective. John Wiley & Sons, New York, NY.

45. Epstein, M. B., G. A. Feldhamer, and R. L. Joyner. 1983. Predation on white-tailed deer fawns by bobcats, foxes, and alligators: predator assessment. Proceedings of the Annual Conference of the Southeastern Association of Fish and Wildlife Agencies 37:161-172

46. Evanich, J. 1986. Introduced birds of Oregon. Oregon Birds 12:156-186.

47. Evans, J. 1970. About nutria and their control. Resource Publ. No. 86. U. S. Department of Interior, Fish and Wildlife Service, Washington, D.C.

48. Fall, M. W., and W. B. Jackson. 1998. A new era of vertebrate pest control? International Biodeterioration and Biodegradation 42:85-91.

49. Feare, C. J. 1984. The starling. Oxford University Press, New York, NY.

50. Fielder, P. C., B. G. Keesee, and P. A. Popushinsky. 1990. Wood duck use of nesting structures in central Washington. Pages 265-267 in L. H. Fredrickson, G. V. Burger, S. P. Havera, D. A. Graber, R. E. Kirby, and T. S. Taylor, editors. Proceedings of North American Wood Duck Symposium, St. Louis, MI.

51. Fox, M. D., and B. J. Fox. 1986. The susceptibility of natural communities to invasion. Pages 57-66 in R. H. Groves and J. J. Burdon, editors. Ecology of biological invasions. Cambridge University Press, Cambridge, England.

52. Friend, M., editor. 1987. Field guide to wildlife diseases. Resource Publication No. 167. U. S. Department of the Interior, Fish and Wildlife Service, Washington, D.C.

53. Geis, A. D. 1973. Effects of urbanization and type of urban development on bird populations. Pages 97-105 in J. H. Noyes and D. R. Progulske, editors. Proceedings of a symposium on wildlife in an urbanizing environment. University of Massachusetts, Amherst, MA.

54. Gilligan, J., editor. 1994. Birds of Oregon. Cinclus, McMinnville, OR.

55. Glahn, J. F., and D. L. Otis. 1986. Factors influencing blackbird and European starling damage at livestock feeding operations. Journal of Wildlife Management 50:15-19.

56. ———, A. R. Stickley, Jr., J. F. Heisterberg, and D. F. Mott. 1991. Impact of roost control on local urban and agricultural blackbird problems. Wildlife Society Bulletin 19:511-522.

57. Grabill, B. A., 1977. Reducing starling use of wood-duck boxes. Wildlife Society Bulletin 5:69-70.

58. Graf, W. 1947. Mouse populations in relation to predation by foxes and hawks. The Murrelet 28:18-21.

59. Griffith, B., J. M. Scott, J. W. Carpenter, and C. Reed. 1989. Translocation as a species conservation tool: status and strategy. Science 245:477-480.

60. Guenther, K., and T. Kucera. 1978. Wildlife of the Pacific Northwest: occurrence and distribution by habitat, BLM district, and National Forest. U.S. Department of Agriculture, Forest Service, Portland, OR.

61. Guenther, S. E. 1950. Nutria. Washington State Game Department, Olympia, Washington, USA. Game Bulletin 2:5.

62. Hall, E. R. 1981. Mammals of North America. J. Wiley and Sons, New York, NY.

63. Hall, L. 1977. Feral rabbits on San Juan Island, Washington. Northwest Science 51:293-297.

64. Harris, S. 1981. On the food of suburban foxes (Vulpes vulpes), with special reference to London. Mammal Review 11:151-168.

65. Hayes, M. P., and M. R. Jennings. 1986. Decline of ranid frog species in western North America: are bullfrogs responsible? Journal of Herpetology 20:490-509.

66. Holler, N. R., and E. W. Schafer, Jr. 1982. Potential secondary hazards of avitrol baits to sharp-shinned hawks and American kestrels. Journal of Wildlife Management 46:457-462.

67. Houston, D. B., E. G. Schreiner, and B. B. Moorhead. 1994. Mountain goats in Olympic National Park: biology and management of an introduced species. Scientific Monograph NPS/NROLYM/NRSM-94/25. U.S. Department of Interior, National Park Service, Washington, D.C..

68. Human, K. G., and D. M. Gordon, 1997. Effects of Argentine ants on vertebrate biodiversity in Northern California. Conservation Biology 11:1242-1248.

69. Hygnstrom, S. E., R. M. Timm, and G. E. Larson. 1994. Prevention and control of wildlife damage. University of Nebraska Cooperative Extension, Lincoln, NE.

70. Ingles, L. G. 1965. Mammals of the Pacific States: California, Oregon, Washington. Stanford University Press, Stanford, CA.

71. Ingold, D. J. 1994. Influence of nest-site competition between European starlings and woodpeckers. Wilson Bulletin 106: 227-241.

72. ———. 1996. Delayed nesting decreases reproductive success in northern flickers: implications for competition with European starlings. Journal of Field Ornithology 67:321-326.

73. International Association of Fish and Wildlife Agencies. 1997. The economic importance of hunting: economic data on hunting throughout the entire United States. International Association of Fish and Wildlife Agencies, Washington, D.C.

74. Jenkins, P. T. 1996. Free trade and exotic species introductions. Conservation Biology 10:300-302.

75. Jewett, S. G., W. Taylor, W. Shaw, and J. Aldrich. 1953. Birds of Washington State. University of Washington Press, Seattle, WA.

76. Jobanek, G. A. 1987. Bringing the old world to the new: the introduction of foreign songbirds into Oregon. Oregon Birds 13:59-74.

77. ——— 1993. The European starling in Oregon. Oregon Birds 19:93-96.

78. Johnsgard, P. A. 1973. Grouse and quail of North America. University of Nebraska Press, Lincoln, NE.

79. Johnson, R. E., and K. M. Cassidy. 1997. Terrestrial mammals of Washington State: location data and predicted distributions. Volume 3, Washington State Gap Analysis- Final Report. Washington Cooperative Fish and Wildlife Research Unit, University of Washington, Seattle, WA.

80. Johnson, R. J., and J. F. Glahn. 1994. European starlings. Pages E109-E120 in S. Hygnstrom, R. Timm, and G. Larsen, editors. Prevention and control of wildlife damage. University of Nebraska Cooperative Extension Service, Lincoln, NE.

81. Johnson, W. E., T. K. Fuller, and W. L. Franklin. 1996. Sympatry in canids: a review and assessment. Pages 189-218 in J. L. Gittleman, ed. Carnivore behavior, ecology, and evolution. Volume 2. Cornell University Press, Ithaca, NY.

82. Kauhala, K. 1996. Introduced carnivores in Europe with special reference to central and northern Europe. Wildlife Biology 2:197-204.

83. Kebbe, C. E. 1958. Deposition of pheasant eggs in ruffed grouse nest. The Murrelet 39:10.

84. Kerpez, T. A., and N. S. Smith. 1990. Competition between European starlings and native woodpeckers for nest cavities in Saguaros. Auk 107:367-375.

85. Kessel, B. 1953. Distribution and migration of the European starling in North America. Condor 55:49-67.

86. Kiesecker, J. M., and A. R. Blaustein. 1998. Effects of introduced bullfrogs and smallmouth bass on microhabitat use, growth, and survival of native red-legged frogs. Conservation Biology 12:776-787.

87. Kinler, N. W., G. Linscombe, and P. R. Ramsey. 1987. Nutria. Pages 326-343 in M. Nowak, J. A. Baker, M. E. Obbard, and B. Malloch, edsitors. Wild furbearer management and conservation in North America. Ontario Ministry of Natural Resources, Toronto, Ontario, Canada.

88. Knopf, F. L. 1992. Faunal mixing, faunal integrity, and the biopolitical template for diversity conservation. Transactions of North American Wildlife and Natural Resources Conference 57:331-343.

89. Kuhn, L. W., and E. P. Peloquin. 1974. Oregon's nutria problem. Proceedings of the Vertebrate Pest Conference 6:101-105.

90. Kupferberg, S. J. 1997. Bullfrog (Rana catesbeiana) invasion of a California river: the role of larval competition. Ecology 78:1736-1751.

91. Labisky, R. F. 1975. Ecopolicies of The Wildlife Society. Wildlife Society Bulletin 3:36-43.

92. Lampman, B. H. 1946. The coming of the pond fishes. Binfords and Mort, Portland, OR.

93. Larrison, E. J. 1943. Feral coypus in the Pacific Northwest. The Murrelet 24:3-9.

94. ———. 1976. Mammals of the northwest: Washington, Oregon, Idaho, and British Columbia. Seattle Audubon Society. Seattle, WA.

95. Lauckhart, J. B., and J. W. McKean. 1956. Chinese pheasants in the Northwest. Pages 43-89 in Durward L. Allen, editor. Pheasants of North America. Stackpole Company, Harrisburg, PA.

96. LeBlanc, D. J. 1994. Nutria. Pages B-71 - B-80 in S. E. Hygnstrom, R. M. Timm, and G. E. Larsen, editors. Prevention and control of wildlife damage. University of Nebraska Cooperative Extension Service, Lincoln, NE.

97. Leonard, W. P., H. A. Brown, L. L. C. Jones, K. R. McAllister, and R. M. Storm. 1993. Amphibians of Washington and Oregon. Seattle Audubon Society. Seattle, WA.

98. Lever, C. 1985. Naturalized mammals of the world. Longman Scientific and Technical, NY.

99. ———. 1987. Naturalized birds of the world. Longman Scientific and Technical, New York, NY.

100. Lewis, J. C., K. L. Sallee, and R. T. Golightly, Jr. 1993. Introduced red fox in California. Non-game Bird and Mammal Section Report 93-10. California Department of Fish and Game, Sacramento, CA.

101. ———, R. T. Golightly, Jr., and R. M. Jurek. 1995. Introduction of non-native red foxes in California: implications for the Sierra Nevada red fox. Transactions of the Western Section of The Wildlife Society 31:29-32.

102. ———, K. L. Sallee, R. T. Golightly, Jr., and R. M. Jurek. 1998. Social and biological aspects of non-native red fox management in California. Proceedings of the Vertebrate Pest Conference 18:150-155.

103. Lloyd, H. G. 1980. The red fox. B. T. Batsford, London, England.

104. Long, J. L. 1981. Introduced birds of the world: the worldwide history, distribution, and influence of birds introduced to new environments. Universe Books, New York, NY.

105. Lumsden, H. G. 1986. Choice of nest boxes by tree swallows, house wrens, eastern bluebirds, and European starlings. Canadian Field-Naturalist 100:343-349.

106. Macdonald, D. W. 1987. Running with the fox. Facts On File, New York, NY.

107. MacDonald, I. A. W., L. Loope, M. Usher, and O. Hamann. 1989. Wildlife conservation and the invasion of nature reserves by introduced species: a global perspective. Pages 215-255 in J. A. Drake, editor. Scope 37: Biological invasions, a global perspective. John Wiley & Sons, New York, NY.

108. Mace, R. U. 1970. Oregon's furbearing animals. Oregon State Game Commission Wildlife Bulletin Number 6. Portland, OR.

109. McGilvrey, F. B., and F. M. Uhler. 1971. A starling-deterrent wood duck nest box. Journal of Wildlife Management 35:793-797.

110. Moler, P. E. 1994. Frogs and toads. Pages F-9 - F-11 in S. E. Hygnstrom, R. M. Timm, and G. E. Larsen, editors. Prevention and control of wildlife damage. University of Nebraska Cooperative Extension Service, Lincoln, NE.

111. Moyle, P. B. 1973. Effects of introduced Bullfrogs, Rana catesbeiana, on the native frogs of the San Joaquin Valley, CA. Copeia 1973:18-22.

112. Newsome, A. E., and I. R. Noble. 1986. Ecological and physiological characters of invading species. In: R. H. Groves and J. J. Burdon, editors. Ecology of biological invasions. Cambridge University Press, Cambridge, England.

113. Nussbaum, R. A., E. D. Brodie, Jr., and R. M. Storm. 1983. Amphibians and reptiles of the Pacific Northwest. University of Idaho Press, Moscow, ID.

114. O'Brien, P., and S. Thomas. 1998. Rabbit calicivirus: update on a new biological control for pest rabbits in Australia. Proceedings of the Vertebrate Pest Conference 18:397-405.

115. O'Gara, B. W., K. C. Brawley, J. R. Munoz, and D. R. Henne. 1983. Predation on domestic sheep on a western Montana ranch. Wildlife Society Bulletin 11:253-264.

116. Office of Technology Assessment. 1993. Harmful non-indigenous species in the United States. OTA-F-565. U.S. Government Printing Office, Washington, D.C.

117. Orians, G. H., 1986. Site characteristics favoring invasions. Pages 133-148 in Harold A. Mooney and James A. Drake, editors. Ecology of biological invasions of North America and Hawaii. Springer-Verlag, New York, NY.

118. Parkes, J. P. 1996. Integrating the management of introduced mammal pests of conservation values in New Zealand. Wildlife Biology 2:179-184.

119. ———, G. Nugent, and B. Warburton. 1996. Commercial exploitation as a pest control tool for introduced mammals in New Zealand. Wildlife Biology 2:171-177.

120. Peck, M. E., 1948. Invasion of exotic plants and their economic significance in Oregon. Northwest Science 22:126-130.

121. Peloquin, E. P. 1969. Growth and reproduction of the feral nutria Myocator coypus (Molina) near Corvallis, Oregon. M. S. Thesis, Oregon State University, Corvallis, OR.

122. Peterson, B., and G. Gauthier. 1985. Nest site use by cavity-nesting birds of the Cariboo Parkland, British Columbia. Wilson Bulletin 97:319-331.

123. Phillips, R. L., and R. H. Schmidt. 1994. Foxes. Pages C-83 - C-88 in S. E. Hygnstrom, R. M. Timm, and G. E. Larsen, editors. Prevention and control of wildlife damage. University of Nebraska Cooperative Extension Service, Lincoln, NE.

124. Pimentel, D. 1986. Biological invasions of plant and animals in agriculture and forestry. Pages 149-162 in Harold A. Mooney, editor. Ecology of biological invasions of North America and Hawaii. Springer-Verlag, New York, NY.

125. Pimm, S. L. 1989. Theories of predicting success and impact of introduced species. Pages 351-367 in J. A. Drake, editor. Biological invasions: a global perspective. John Wiley & Sons, New York, NY.

126. Presnall, C. C. 1958. The present status of exotic mammals in the United States. Journal of Wildlife Management 22:45-50.

127. Rainbolt, R. E. and B. E. Coblentz. 1997. A different perspective on eradication of vertebrate pests. Wildlife Society Bulletin 25:189-191.

128. Robinson, W. L., and E. G. Bolen. 1989. Wildlife ecology and management. Second edition. Macmillan, New York, NY.

129. Roughgarden, J., 1986. Predicting invasions and rates of spread. Pages 179-188 in Harold A. Mooney and James A. Drake, editors. Ecology of biological invasions of North America and Hawaii. Springer-Verlag, New York, NY.

130. Scheffer, T. H. 1935. Introduced food plants for waterfowl. The Murrelet 16:66-69.

131. Schmitz, D. C., and R. G. Westbrooks. 1997. The federal government's role. Pages 329-337 in Daniel Simberloff, editor. Strangers in paradise: impact and management of nonindigenous species in Florida. Island Press, Washington, D.C.

132. Schmutz, J. A. 1988. Ring-necked pheasant parasitism of wild turkey nests. Wilson Bulletin 100:508-509.

133. Simberloff, D. 1996. Hybridization between native and introduced wildlife species: importance for conservation. Wildlife Biology 2:143-150.

134. ———. 1997a. The biology of invasions. Pages 3-17 in Strangers in paradise: impact and management of nonindigenous species in Florida. D. Simberloff, editor. Island Press, Washington, D.C.

135. ———. 1997b. Eradication. Pages 221-228 in Strangers in paradise: impact and management of nonindigenous species in Florida. D. Simberloff, editor. Island Press, Washington, D.C.

136. Smallwood, K. S., and T. P. Salmon. 1992. A rating system for potential exotic bird and mammal pests. Biological Conservation 62:149-159.

137. ———. 1994. Site invasibility by exotic birds and mammals. Biological Conservation 69:251-259.

138. Spreyer, M. 1995. Pheasant obsession. Birder's World 9:44-48.

139. Stein, B. A., and S. R. Flack, editors. 1996. Americas least wanted: alien species invasion of U.S. ecosystems. The Nature Conservancy, Arlington, VA.

140. Storer, T. I. 1933. Frogs and their commercial use. California Fish and Game 19:203-213.

141. Storm, G. L., R. D. Andrews, R. L. Phillips, R. A. Bishop, D. B. Siniff, and J. R. Tester. 1976. Morphology, reproduction, and mortality of mid-western fox populations. Wildlife Monographs 49.

142. Taylor, W. P. 1923. Upland game birds in the state of Washington. The Murrelet 4:3-15.

143. Temple, S. A. 1990. The nasty necessity: eradicating exotics. Conservation Biology 4:113-115.

144. Thomas, J. W., editor. 1979. Wildlife habitats in managed forests; the Blue Mountains of Oregon and Washington. Agricultural Handbook No. 553. U.S. Department of Agriculture, Forest Service, Portland, OR.

145. Thornton, W. A., G. C. Creel, and R. E. Trimble. 1971. Hybridization in the fox genus Vulpes in west Texas. Southwest Naturalist 25:423-434.

146. Toney, J. C., P. M. Rice, and F. Forcella. 1998. Exotic plant records in the Northwest United States 1950-1996: an ecological assessment. Northwest Science 72:198-209.

147. Troetschler, R. G., 1976. Acorn woodpecker breeding strategy as affected by starling nest-hole competition. Condor 78:151-165.

148. Van Vuren, D. 1992. Eradication of feral goats and sheep from island ecosystems. Proceedings of the Vertebrate Pest Conference 15:377-381.

149. Van Vuren, D. 1998. Manipulating habitat quality to manage vertebrate pests. Proceedings of the Vertebrate Pest Conference 18:383-390.

150. Vance, D. R., and R. L. Westemeier. 1979. Interactions of pheasants and prairie chickens in Illinois. Wildlife Society Bulletin 7:221-225.

151. Verts, B. J., and L. Carraway. 1980. Natural hybridization of Sylvilagus bachmani and introduced S. floridanus in Oregon. The Murrelet 61:95-98.

152. ———, and ———. 1998. Land mammals of Oregon. University of California Press, Berkeley, CA.

153. Voigt, D. R. 1987. Red Fox. Pages 379-392 in M. Nowak, J. A. Baker, M. E. Obbard, and B. Malloch, editors. Wild furbearer management and conservation in North America. Ontario Ministry of Natural Resources. Toronto, Ontario, Canada.

154. Washington Department of Wildlife. 1993. Status of the western gray squirrel, Sciurus griseus, in Washington. Unpublished report. Washington Department of Wildlife, Olympia, WA.

155. Weitzel, N. H. 1988. Nest-site competition between the European starling and native breeding birds in Northwestern Nevada. Condor 90:515-517.

156. Welsh, J. E., and R. A. Howard, Jr. 1983. Characteristics of snags influencing their selection by cavity-nesting birds. Transactions of the Northeast Fish and Wildlife Conference 40:177.

157. Wentz, W. A. 1971. The impact of nutria (Myocastor coypus) on marsh vegetation in the Willamette Valley, Oregon. M. S. Thesis. Oregon State University, Corvallis, OR.

158. Westemeier, R. L., and T. L. Esker. 1989. An unsuccessful clutch of northern bobwhites with hatched pheasant eggs. Wilson Bulletin 101:640-642.

159. Westerskov, K. 1990. Partridges and pheasants: competitors or sharers? Pages 183-201 in Kevin E. Church, Richard E. Warner, and Stephen J. Brady, editors. Gray partridge and ring-necked pheasant workshop. Kansas Department of Wildlife and Parks, Topeka, KS.

160. Wilcomb, M. J. 1948. Fox populations and food habits in relation to game birds in the Willamette Valley, Oregon. M. S. Thesis. Oregon State University, Corvallis, OR.

161. Willner, G. R. 1982. Nutria. Pages 1059-1076 in J. A. Chapman and G. A. Feldhamer, editors. Wild mammals of North America: biology, management, and economics. Johns Hopkins University Press, Baltimore, MD.

162. Wilson, E. O., and F. M. Peters, editors. 1988. Biodiversity. National Academy Press, Washington, D.C.

163. Witmer, G. W. 1990. Re-introduction of elk in the United States. Journal of the Pennsylvania Academy of Science 54:131-135.

164. ———, J. L. Bucknall, T. H. Fritts, and D. G. Moreno. 1996. Predator management to protect endangered avian species. Transactions of the North American Wildlife and Natural Resources Conference 61:102-108.

165. ———, E. W. Campbell, and F. Boyd. 1998. Rat management for endangered species protection in the United States Virgin Islands. Proceedings of the Vertebrate Pest Conference 18:281-286.

166. Zembal, R. 1992. Status and management of light-footed clapper rails in coastal southern California. Transactions of the Western Section of The Wildlife Society 28:1-5.

17

Genetic Considerations for Introduced and Augmented Populations

Susan M. Haig & R. Steven Wagner

Introduction

Habitat fragmentation has occurred at a record pace throughout many areas of the Pacific Northwest. As a consequence, wildlife and land managers in this region are forced to deal with recovery of small populations. Frequently, these efforts have focused on politically charged issues related to spotted owls, marbled murrelets, or salmon (*Oncorhynchus* spp.). Increasingly, however, there are more species facing extirpation that receive little attention and fall outside the protection provided by more recognized species.

Direct conservation of the land is usually the best approach to dealing with habitat loss for most species. However, habitat conservation efforts often come too late for natural recovery of specific populations or species and more drastic actions are needed to assure their survival and viability. To achieve this goal, individuals may need to be translocated to/from other populations or, if a species or population has gone extinct in one locale, a reintroduction from captivity or another wild population may be necessary.

This seemingly simple concept, of moving individuals among populations or establishing new ones, has had variable success and is quite controversial. For example, the success rate was only 11% (16 cases) among 145 reintroduction programs that released captive animals back to the wild.[6] The success rate among 80 translocation programs was better, but less than half (46%) resulted in establishing self-sustaining populations of mammals and birds.[19, 60] Similar efforts for reptiles and amphibians were even less successful—only 5 of 25 programs were successful.[9] Thus, reintroduction and translocation programs should only be carried out as a last resort.[56]

There are many explanations for the lack of success of these programs (reviewed in Chapter 18), however, in this chapter we will outline genetic factors that need to be carefully considered if a decision is reached to establish a new population or augment an existing population. As in any population recovery effort, there are numerous overlapping factors to consider[51] which include: *Demographic factors, environmental factors, catastrophic events,* and *genetic factors* (see Box 1). The least understood and/ or addressed concern in new population management is the consideration of *genetic factors.* There are numerous ways in which "genetic factors" are defined[7] and used in the design of new populations. Molecular techniques, pedigree analyses, or viability models are tools used to evaluate "genetic factors."

In the broadest sense, small populations should be managed to avoid short-term demographic disasters and to insure long-term adaptability (i.e., maximize genetic diversity). Loss of genetic diversity diminishes the flexibility of populations to adapt to changing conditions, resulting in a decreased ability to cope with parasites, disease, competitors, changing climatic conditions, etc. There has been much debate over prioritization of genetic and demographic concerns. Some suggest that over the short term, concern for demographic factors should take precedence over genetic issues.[37] However, the interactions between genetic and demographic factors are not mutually exclusive. The demographic structure of a population affects the genetic structure and vice versa. Consequently, simple management decisions can be made when establishing a new population (either via reintroduction or augmentation) that will improve both demographic and genetic structure of the population.[20] Thus, both factors should be considered from the start of a program.

---

1. Factors Influencing Population Viability

Demographic factors—factors that affect individual birth and death rates. Demography is most critical to consider if a population is very small (approximately less than 40 individuals), because stochastic events can be severe in small populations making them susceptible to Allee effects.[2]

Genetic factors—factors that influence small populations through random genetic drift (random loss of alleles) and inbreeding which can reduce genetic diversity.

Environmental factors—factors that can impact a population of any size. They include variations in weather, food availability, and other factors that affect survival, reproduction and abundance.

Catastrophic factors—significant random events that can affect a population of any size. They include flood, fire, hurricanes, etc.

---

2. Factors Related to Random Genetic Drift

Random Genetic Drift[61, 62]—mechanism by which populations lose genetic diversity. It is the result of random changes in gene frequency from generation to generation.

Effective Population Size (N_e[61])—is inversely proportional to the rate of loss of genetic diversity. It is equal to the size of an ideal population.

Ideal Population—defined as a population meeting the following assumptions: random mating (e.g. no inbreeding), 1:1 breeding sex ratio, equal family sizes, large population size, and non-overlapping generations.

Population Bottleneck—a drastic decline in N_e resulting in a small number of survivors passing on a fraction of genes from the original population.

The most critical genetic threats facing small or new populations are the loss of diversity and the potential for inbreeding depression.[3] Genetic diversity is lost primarily through *random genetic drift* and, as a consequence, this increases *inbreeding* and decreases *effective population size* (see Box 2). In general, populations lose genetic diversity over time, with the rate of loss largely dependent on the number of individuals in the population. This loss can be balanced by the number of founders or migrants into the population—factors that are most affected early on in a new population. *Random genetic drift* is the mechanism by which small populations lose genetic diversity, while *effective population size* is a way to measure this rate of loss. Thus, the appropriate strategy for new/small population management is to minimize random genetic drift and maximize the effective size of the population.

Random genetic drift reduces two components of genetic diversity in small populations: *allelic diversity* and *heterozygosity*[1] (see Box 3). Maintaining both factors becomes increasingly difficult as populations become smaller, because the effect of drift becomes greater due to the random sampling of fewer genes, resulting in greater fluctuations in gene frequency. Therefore, while a number of factors can affect genetic diversity, random genetic drift is an overriding factor in controlling loss of genetic variation.[36]

Substantial genetic drift can result in increased *inbreeding*. Inbreeding can be simply defined as the mating of close relatives, but is often confused with *inbreeding depression*, which is the loss of fitness resulting from inbreeding. Generally, demographic factors, such as fecundity, fertility, development rate, age of sexual maturity, and litter size can be affected by increased inbreeding. Thus, from a conservation perspective, inbreeding is typically considered detrimental and to be avoided.[49]

A contrary view is that some amount of inbreeding allows for the maintenance of co-adapted genes[52, 53, 59] (i.e., adaptation to the local environment). Especially in new populations, this potential benefit is generally outweighed by the importance of establishing populations with unrelated individuals so as to increase genetic diversity as quickly as possible. Consequently, there is some concern that managers, in an effort to mitigate the effects of drift, may introduce genetically-distant individuals into the population. This type of introduction could lead to the disruption of co-adapted gene complexes or the ability to adapt to the local environment. The result would be *outbreeding depression* which could have consequences as severe as inbreeding depression[59] (Box 3). Populations should be monitored beyond the first generation (F1) for the effects of outbreeding. Usually the first generation shows increased fitness but this can quickly erode in subsequent generations as co-adapted complexes are broken up.[41] Overall, breeding strategies should mitigate between the effects of inbreeding and outbreeding.

An important measure for determining the rate of loss of genetic diversity in a population is effective population size[61] (N_e, Box 2). Effective population size is inversely proportional to the loss of genetic diversity. The concept of N_e can be confusing, but it is an important concept for managing genetic and demographic factors in a population—especially as new populations are established. Most simply, effective population size is equal to the size of an *ideal population,* which is defined based on the following assumptions: random breeding (e.g., no inbreeding), 1:1 breeding sex ratio, equal family sizes, large population size, non-overlapping generations, etc. The effective size is usually smaller than the *censused population size* (N) because most natural populations do not meet all of the assumptions of an ideal population. The ratio N_e/N, the comparison of effective to censused population size, gives an indication of how different the censused size is from the ideal size. Thus, it can be used by population managers to assess the status of a population and to judge progress toward achieving a more genetically diverse and demographically stable (i.e., birth rates at least equal death rates and emigration) population. Moreover, it can be used

3. Population Genetic Terminology

Allelic diversity—number of unique alleles that occur in a population.

Heterozygosity—frequency of alleles per locus.

Inbreeding—mating of close relatives.

Inbreeding depression—loss of fitness resulting from inbreeding.

Outbreeding—mating of genetically distant individuals.

Outbreeding depression—loss of fitness resulting from the mating of genetically distant individuals, usually associated with the disruption of co-adapted gene complexes.

as a means of assessing the relative effects of factors that might contribute to the viability of a new population.

A sudden or drastic decline in N_e or a *population bottleneck* results when a small number of survivors pass on a fraction of genes from the original population. By definition, bottlenecks occur when populations suffer drastic declines, such as when animals are brought into captivity, or a new population is established. Taken to an extreme, this could result in population extinction. Thus, it is critical to consider problems associated with bottlenecks in planning new populations. That is, selection of individuals to establish or enhance a population must be undertaken with extreme caution. Also, managers should try to minimize the effect of a bottleneck by increasing the number of individuals as quickly as possible. This can be achieved by introducing more breeding age individuals—with consideration to mating system and current sex ratio; providing extra-protective measures for eggs, nests, or juveniles; creating artificial nest sites or breeding areas; considering some type of predator management measures, etc.

Establishing Viable Populations

If a decision is made to establish a new population or enhance an existing population, some practical decisions should be made in view of the above discussion. Often genetic considerations are quite simple and can profoundly affect a program's success, yet they are frequently ignored. For example, in a world-wide review of 94 translocation or reintroduction programs, Kalmer[35] found genetic factors were only considered in 47 cases. Listed below are a few questions that should be considered for both the new and donor population:

1. What is the population viability goal we are trying to reach in the new population and maintained in the donor population?
2. How many individuals are needed/could be given up?
3. What is the relatedness of individuals to be introduced/to be lost?

Viability goals. As with any recovery planning effort, an assessment of needs and risks must be carried out prior to moving individuals. Managers need a realistic idea of what they have (genetically and demographically) and what they can hope for, given careful planning. Unrealistic goals will insure the failure of any program. Molecular methods can be used to assess genetic diversity, the genetic distance among various populations, and identify relatedness of individuals in a population.[20] Pedigree analyses can also be used to assess the genetic status of donor populations or evaluate breeding options.[21, 22, 24] Finally, population viability models can be used to evaluate various strategies for long and short-term survival.[23]

Several short-term and long-term population goals have been proposed in the conservation literature. Primarily, the maintenance of genetic diversity will increase with more individuals in a population and an increased N_e.[11, 12, 13] More specifically, Franklin[14] suggested

a short-term management goal of N_e=50 and a long-term management goal of N_e=500 in order to maintain population viability. Long-term goals proposing an optimum effective population size are controversial: recently Franklin and Frankham[15] revised the target to include a range of N_e between 500-1000, while others contend the N_e should be higher, between 1000-5000.[38, 42] However, as Lande and Barrowclough[38] point out, these goals are unrealistic for many populations because they are too general (e.g., an N_e of 50 for many rodents would be much easier to attain than a N_e of 50 for grizzly bears) and do not take into account population-specific life history characters and demography. Thus, population viability goals should be viewed with respect to effective population size but specific goals should be developed for each situation in order to maintain genetic diversity in a realistic fashion.

How many/which individuals to translocate. While more is usually better, often there are limited choices regarding the number of individuals available to start a population. Also a balance must be struck between offsetting the effects of drift, preserving local adaptation and providing enough individuals to overcome genetic and demographic threats. This is illustrated by a small population of red-cockaded woodpeckers (*Picoides borealis*) in South Carolina, where managers wanted to increase the population of 5-6 birds to a viable population.[23] Pre-translocation modelling indicated that annual translocation of three females and two males for a ten-year period would achieve the same genetic and demographic goals as moving many more individuals—which were not available from donor populations anyway. Thus, the modelling was able to give managers some realistic goals. Some general guidelines for increasing N_e and assuring a genetically fit and demographically stable population include the following:[39, 48]

1. Begin with at least 20-30 unrelated founders.
2. Increase carrying capacity quickly.
3. Equalize founder contribution/mean kinship.
4. Minimize inbreeding.
5. Equalize family size.
6. Establish a demographically stable population.

Meeting any or all of these management objectives may be difficult or impossible for many species or situations; however, effectively addressing any of them will increase N_e and result in a more stable population.

Monitor progress towards recovery. Genetic diversity should be measured during various phases of population growth as a means of diagnosing progress toward viability.[17, 58] For example, during the founder phase, the number of founders and their relatedness has a critical effect on genetic diversity. As the population begins to increase, a slow population growth rate, an extended duration of growth phase, or a small carrying capacity can reduce genetic diversity. Once the carrying capacity has been reached, a small carrying capacity or short duration of the capacity phase can also result in loss of diversity.

During any phase of population establishment, inbreeding, outbreeding, or the loss of founder lines can become a problem. Thus, managers need to carefully monitor productivity of adults and survivorship of young. Inbreeding can be examined in a variety of ways, but fundamental to any analysis will be establishment of a pedigree from the beginning of the population. If reduced survivorship is observed and appears to have a genetic basis, setting up test pairings may confirm the problem. Even if this is not possible, separating pairs of suspected relatives may be advised. If new populations are too small or it is too intrusive to alter pairings, introduction of new individuals may help. In general, inbreeding can best be avoided by increasing the population size quickly and selecting pairs to minimize mean kinship and equalize founder contribution.

Donor Populations and Metapopulation Management

Establishment of new populations cannot be carried out in a vacuum; that is, without regard to the effect on the structure of donor populations and the overall metapopulation. One of the first factors to consider is the source of donors to establish a new population. In general, choosing from populations that are genetically and ecologically similar will result in individuals that have adapted to the area's particular characteristics. For example, in re-establishing a viable population of red-cockaded woodpeckers in South Carolina, molecular analyses were performed on samples from throughout the species range to determine the most genetically similar donor population.[24, 26] Donor birds were then chosen from populations that were genetically and geographically close, would not suffer as a result of loss of individuals, and had similar habitat conditions.

The recommendation to move individuals from the most similar population is not always an alternative. In the case of the Florida panther (Felis concolor coryi), the genetically closest individuals were from a Texas (F. c. stanleyana) subspecies. In this case, outbreeding depression might have been an issue, however, there were no other options. Modelling the potential impact on the Florida population led biologists to conclude that the benefits would outweigh the risks.[29] In another unique situation,

Hedrick and Parker[31] evaluated stocking options for the endangered Gila topminnow (Poeciliopsis o. occidentalis) by examining genes from the major histocompatibility complex (MHC). MHC genes are thought to have adaptive significance, particularly for resistance to parasites and pathogens, hence their diversity within populations may be informative relative to their adaptive potential. They found significant differences among individuals from the four watersheds they sampled, and recommended that individuals not be mixed among watersheds.

As new populations are established, consideration should be given to the overall metapopulation structure. While many strategies have been debated, it usually comes down to consideration of the SLOSS model[55] (i.e., managing for a *single large or several small* populations). That is, how do we balance the benefits of a single large population with those of several small? Genetically and demographically, a large population is generally more viable over a longer period of time, whereas small populations face greater risk of immediate extinction. However, small populations can serve as reservoirs for unique alleles, provide a buffer if catastrophe hits the large population, and provide links to other populations.[40] Thus, a mix of both large and inter-linked small populations may be the most effective scenario. In translocation/reintroduction programs, maintenance of the metapopulation may be carried out via natural or human-induced movements of individuals among populations. Since the 1930s, the rule of thumb has been that movement into a population of one individual per generation would offset the negative effects of drift.[61] Reconsideration of this rule led Mills and Allendorf[44] to recommend that one individual was a minimum and movement of up to ten individuals per local population per generation would better assure maintenance of a viable metapopulation.

Genetic Tools to Assess and Monitor New Populations

Molecular techniques are increasingly employed to aid in designing population enhancement strategies.[20] These analyses are most helpful in defining founder relationships, monitoring genetic diversity, and determining relationships among populations in which translocations and augmentations are being considered.

Table I. Application of molecular techniques in resolving issues of small population conservation.

Application	Allozymes	DNA fingerprints (RFLPs)	Microsatellites	RAPDs	mtDNA	MHC
Parentage	(X)[1]	X	X	(X)		X
Mating system	X	X	X	(X)		X
Pedigree definition		X	X			X
Population structure	X	(X)	X	X	X	X
Population differentiation	X		X	X	X	X
Hybridization	X		X	X	X	X
Phylogeny	(X)				X	

1. () indicates that while appropriate, another technique would most likely yield more helpful information.

Each technique is best suited for a limited range of questions, hence several may be necessary in order to get a complete picture of a population structure[46] (Table 1). For example, DNA fingerprints or restriction fragment length polymorphisms (RFLPs) have a very high mutation rate and can show substantial variability among closely-related individuals. However, this variability reaches saturation among genetically distant individuals. Therefore, they are better at resolving questions related to parentage, pedigree, and population structure than for resolving higher level analyses such as addressing questions regarding population differentiation or phylogeny.

Protein electrophoresis or allozyme analyses are one of the most widely used techniques used to assay genetic diversity[34] as they can be used to address a number of questions ranging from parentage to phylogeny. Even with the advent of new techniques, they remain one of the most powerful, quick, and inexpensive techniques for certain types of questions. Typically, any type of tissue can be used (i.e., blood, muscle, heart, liver, feather pulp), and 5-10 variable loci are usually screened. While historically allozymes have been used for parentage, population structure, and population differentiation studies, depending upon the taxa and scale of the question, there might not be adequate variability to resolve some of these questions. For example, for many avian species, allozyme variability is too low to address many population-level questions.[5]

Often in new populations, relatedness of individuals in the wild donor population is not known, making it difficult to choose individuals to translocate. DNA fingerprinting (minisatellites) and microsatellite loci are two techniques increasingly employed to aid in this question. For example, DNA fingerprinting was used to determine the relatedness of founders to the captive population of Guam rails (*Rallus owstoni*) prior to their reintroduction to the wild.[21] These fingerprinting profiles tested five hypothesis proposed to explain relatedness among the nine founders. Band-sharing among the profiles was correlated with relatedness calculated from hypothesized pedigrees. This relationship was used to determine the most plausible founder hypothesis, and to construct a breeding/management scheme that maximized genetic diversity in the new population.

Microsatellites have a broader range of application than DNA fingerprints as they can address questions related to population differentiation. For example, microsatellites were used to determine the appropriateness of moving Marianas crows (*Corvus kubaryi*) from Rota to Guam.[10] Microsatellites also have an advantage over DNA fingerprinting and random amplified polymorphic DNA (RAPDs) because heterozygotes can be scored directly. Since the microsatellite technique is based upon the polymerase chain reaction[45] (PCR), any type of tissue may be used with small amounts of tissue (e.g., blood, hair, feathers, scat, bone, tissue from museum specimens, etc.). These studies usually involve screening 5-15 variable loci per individual.

This technique can be very quick and cost effective if primers have been previously designed. However, these primers are often species-specific and can be extremely difficult to elucidate, especially in birds.[47] Thus, without pre-designed primers this technique can be one of the most costly and time consuming techniques to develop.

Use of randomly amplified polymorphic DNA (RAPDs) is increasingly applied to problems of population structure, differentiation, and hybridization.[24, 25, 26, 27] There have also been a limited number of studies that have employed RAPDs to infer parentage and pedigrees. RAPDs are one of the easiest, fastest and most cost effective of all the molecular techniques. The development phase of screening for variable primers is minimal and straight forward. It is also possible to screen a large number of individuals (20-30) per population for a number of loci (15-20) relatively quickly. RAPDs are also a PCR based technique so almost any type of tissue may be used. One drawback of RAPDs is that they are a dominant marker, therefore heterozygotes can not be directly scored. RAPDs have been used to evaluate translocation strategies for the Australian woody perennial *Corrigan grevillea* (*Grevillea scapigera*), a plant that has been reduced to four populations (total of 27 plants) by habitat destruction.[50] The RAPD data indicated little difference among populations, hence they recommended that translocations would not have negative genetic effects.

Mitochondrial DNA (mtDNA) markers are some of the most widely-used in conservation biology.[4] Their best use is in addressing questions of population structure, hybridization, and phylogeny. For small populations, this marker can be useful for identifying populations from which to translocate individuals. MtDNA is a maternally-inherited, extrachromosomal small molecule of DNA (~16kb for vertebrates) found in the mitochondria. The molecule is useful because it evolves at a rapid rate and different regions have variable mutation rates. Furthermore, since it is uniparentally inherited and non-recombining there is little sequence heterogeneity within an individual. MtDNA studies involve two kinds of analyses: DNA sequencing or restriction site analysis (RFLPs). DNA sequencing usually involves sequencing of 400-1000 bases of various genes for intraspecific phylogeny studies. For example, the cytochrome b, COI, and NADH genes have been sequenced for intraspecific phylogeny studies and the highly variable d-loop region has been used to explore both intraspecific phylogeny and population stucture. Given the availability of universal primers for many mtDNA genes, these kinds of studies are moderately priced. Phylogenetic analysis among populations of the wet forest-restricted frog *Litoria pearsoniana* using the COI gene (460bp), indicated that conservation efforts should be made at the scale of major rainforest isolates. Furthermore, investigators suggested that the conservation status of the species should be considered independently for northern and southern populations due to significant differentiation.[43]

The major histocompatibility complex (MHC) is increasingly used as a marker to assess genetic variability in captive populations.[28, 30] MHC has a number of advantages over other markers. First, MHC is suggested to have alleles of selective significance, unlike the other the markers which we assume to be neutral. Second, MHC has important immune functions and is studied widely to understand autoimmune diseases, tissue transplantation, and immune response to diseases. A study of two populations of pocket gophers found low variability at the DQα locus and each fixed for unique alleles.[54] These results were also consistent with the rejection of reciprocal skin grafts, therefore, the authors suggested that these populations should not be mixed.

In addition to choosing a technique, the feasibility of obtaining adequate samples for each type of study must be considered in the planning stages. For example, parentage/pedigree studies require comparisons of highly variable genetic loci between individuals. For these types of questions sampling complete families (i.e., mother, father, all offspring, other potential fathers, and helpers; n = ~20 families) and multiple generations will be necessary. Population structure/differentiation studies involve relative comparisons among populations, so there must be adequate sampling of both individuals per population (~20-40 per population). Hybridization and phylogeny studies generally involve comparisons at higher taxonomic levels and markers that have a slower rate of mutation, therefore, depending on the marker, fewer individuals per group (~1-5) need to be sampled.

Genetic Considerations for Northwest Populations

The information presented here attains the greatest significance when we consider issues close to home. Game animals have been translocated for years within and among sites in the Northwest. More recently, however, population restoration strategies have been warranted for small populations in the region. The following examples illustrate species where restoration efforts will be enhanced if genetic factors are considered along with the other issues.

Western Pond Turtles

Western pond turtles have been rapidly declining throughout their Pacific Coast range for nearly a century and are extirpated from most of Washington and Oregon.[8, 33, 57] There are numerous hypotheses for their decline which include disease, human disturbance, and predation by introduced largemouth bass (*Micropterus salmoides*) and bullfrogs. In the Northwest, there are less than 250 individuals among six sites on the Washington side of the Columbia River Gorge and one site on the Oregon side of the Gorge. Another site, in Puget Sound, is located at an artificial pond and originated from 3 founders.[18] It is not known if these individuals are native to the Puget Sound basin because each founder was found in a semi-rural environment where there were no known populations.

From a genetic perspective, phylogenetic studies using mitochondrial DNA would help define the relationships among these populations and help determine if translocating among them is advisable. For example, because origin of the Puget Sound turtles is not clear, augmentation of the Gorge sites with Puget Sound turtles could break up co-adapted gene complexes and lead to outbreeding depression. Currently, the Washington Department of Fish and Wildlife has been "headstarting" juveniles in captivity. They have mixed juveniles across the Gorge sites but not with Puget Sound or other sites. This approach of increasing the population size but not mixing populations appears prudent until further molecular analysis can be completed.

Upland Sandpipers

Upland sandpipers are a shorebird species suffering from drastic population decline and fragmentation in the Northwest due to habitat deterioration, often as a result of overgrazing and agriculture.[8, 16, 32] They have been extirpated from Washington and less than 100 remain in eastern Oregon and Idaho. The nearest viable populations are in the prairies of the Dakotas and Manitoba, but even these are declining. Thus some of the genetic issues include whether translocating birds from the Midwest would result in an outbreeding problem, whether the Midwest populations would be further jeopardized if they donated individuals, and what population structure would be most viable for the new populations. As with pond turtles, mtDNA or microsatellite analyses of Northwest and Midwest birds will help assess differences between populations. Some population viability modelling will help determine the effect of loss of individuals on the donor populations as well as what will be needed for the new or enhanced populations.

Pygmy Rabbits

Pygmy rabbits are a Columbia Basin species whose native sagebrush habitat is now so reduced that only five populations remain in southeastern Washington (Washington Department of Fish and Wildlife, unpub. data). With less than 100, if not less than 50, individuals per population, biologists suspect that there is little genetic or demographic interchange among populations. Furthermore, the dynamic nature of the Columbia Basin often renders these small populations susceptible to catastrophic events such as floods, fires, and disease. Thus, since all populations are small, a decision to manipulate any population may result in its extinction, yet no action may have the same result. Therefore, a starting point for species recovery would be to assess the genetic distance among populations with mtDNA or microsatellites. If genetic distances were not significant, consideration of moving several individuals among populations may be warranted given concern about loss of genetic diversity due to drift. A more complex decision could be then be considered regarding population consolidation. If some populations are too small to survive and there are no donor populations outside Washington and Oregon, it may be best to move the remaining individuals to a nearby population with enough suitable habitat to sustain them.

Clearly, this would be a difficult decision as moving an entire population may result in loss of protective efforts for the habitat they occupied. However, if it stabilized another population on the brink of extinction, it may be a worthwhile endeavor. Revisiting the pros and cons suggested by the SLOSS model would be in order prior to undertaking such an action.

Oregon Silverspot Butterfly

Oregon silverspot butterfly (*Speyeria zerene hippolyta*) is a species extirpated from Washington and has only three remnant populations in Oregon. To restore the butterflies to their native Washington habitat, translocated individuals from Oregon would have to be chosen from areas that are geographically closest and occur in habitats most similar to what the original Washington populations occupied. Prior to any movements, Oregon populations need to be assessed to determine if they could sustain loss of a few individuals without significant loss of viability. Next, genetic distance should be assessed. If there is no significant genetic difference among populations, it might be best to choose individuals from habitats as similar to the new habitats as possible. This will speed their adaptation to the new site and more quickly establish the new populations.

Conclusions

In this chapter we have reviewed the significance of considering genetic factors in augmenting or establishing new populations. Because single individuals can have a significant impact on population structure and genetic diversity can be lost quickly in small populations, new populations can be helped most effectively if factors such as the number of founders, relatedness of founders, and genetic distances between donor and recipient populations are considered at the beginning of recovery efforts, not as an afterthought. These factors affect both the genetic and demographic stability of new populations, and can often be measured most efficiently with genetic techniques such as molecular tools. Overall, genetic management is often overlooked due to managers unfamiliarity with terms or techniques, yet it can have a significant impact on overall population viability.

Literature Cited

1. Allendorf, F.W., and R.F. Leary. 1986. Heterozygosity and fitness in natural popualtions of birds and mammals. Pages 57-76 in M.E. Soule, ed. Conservation Biology: the Science of Scarcity and Diversity. Sinauer Assoc., Sunderland, MA.

2. Andrewartha, H.G., and L.C. Birch. 1954. The Distribution and Abundance of Animals. University of Chicago Press, Chicago, IL.

3. Avise, J.C. 1994. Molecular Markers, Natural History, and Evolution. Chapman and Hall, New York, NY.

4. ———. 1995. Mitochondrial DNA polymorphism and a connection between genetics and demography of relevance to conservation. Conservation Biology 9: 686-690.

5. Barrowclough, G.F. 1983. Biochemical studies of microevolutionary processes. Pages 223-261 in A.H. Brush and G.A. Clark, editors. Perspectives in Ornithology. Cambridge University Press, New York, NY.

6. Beck, B.B., L.G. Rapaport, M.S. Price, and A. Wilson. 1994. Reintroduction of captive born animals. Pages 265-284 in P.J.S. Olney, G.M. Mace, and A.T.C. Feistner, editors. Creative Conservation: Interactive Management of Wild and Captive Animals. Chapman and Hall, London.

7. Beissinger, S.R., J.R. Walters, D.G. Catanzaro, K.G. Smith, J.B. Dunning, Jr., S.M. Haig, B.R. Noon, and B.M. Smith. The use of models in avian conservation. Current Ornithology (in press).

8. Csuti, B., A.J. Kimmerling, T.A. O'Neil, M.M. Shaughnessy, E.P. Gaines, and M.M.P. Huso. 1997. Atlas of Oregon Wildlife. Oregon State University Press, Corvallis, OR.

9. Dodd, C.K., and R.A. Seigel. 1991. Relocation, repatriation, and translocation of amphibians and reptiles: are they conservation strategies that work? Herpetologia 47: 336-350.

10. Duckworth, W.D., S.R. Beissinger, S.R. Derrickson, T.H. Fritts, S.M. Haig, F.C. James, J.M. Marzluff, and B.A. Rideout. 1997. Scientific Bases for Preservation of the Marianas Crow. National Academy of Science. Washington D.C.

11. Frankel, O.H., and M.E. Soule. 1981. Conservation and Evolution. Cambridge University Press, New York, NY.

12. Frankham. R. 1996. Relationship of genetic variation to population size in wildlife. Conservation Biology 10:1500-1508.

13. ———, and I.R. Franklin. 1998. Response to Lynch and Lande. Animal Conservation 1:73.

14. Franklin, I.R. 1980. Evolutionary change in small populations. Pages 135-149 in M. E. Soule and B.A. Wilcox, editors. Conservation Biology—An Evolutionary Perspective. Sinauer Assoc. Sunderland MA.

15. ———, and R. Frankham. 1998. How large must populations be to retain evolutionary potential? Animal Conservation 1: 69-73.

16. Gilligan, J., M. Smith, D. Rogers, and A. Contreras. 1994. Birds of Oregon: Status and Distribution. Cinclus Publications, McMinnville, OR.

17. Gompper, M.E., P.B. Stacey, and J. Berger. 1997. Conservation implications of the natural loss of lineages in wild mammals and birds. Conservation Biology 11: 857-867.

18. Gray, E.M. 1995. DNA fingerprinting reveals a lack of genetic variation in northern populations of the Western Pond Turtle (*Clemmys marmorata*). Conservation Biology 9:1244-1255.

19. Griffith, B., J.M. Scott, J.W. Carpenter, and C. Reed. 1989. Translocations as a species conservation tool: status and strategy. Science 245: 477-480.

20. Haig, S.M. 1998. Molecular contributions to conservation. Ecology 79: 413-425.

21. ———, J.D. Ballou, and N.J. Casna. 1994. Identification of kin structure among Guam Rail founders: a comparison of pedigrees and DNA profiles. Molecular Ecology 5:109-119.

22. ———, ——— and S.R. Derrickson. 1990. Management options for preserving genetic diversity: reintroduction of the Guam Rail to the wild. Conservation Biology 4: 290-300; 464.

23. ———, J.R. Belthoff, and D.H. Allen. 1993. Population viability analysis for a small population of Red-cockaded Woodpeckers and an evaluation of population enhancement strategies. Conservation Biology 7:289-301.

24. ———., R. Bowman, and T.D. Mullins. 1996. Population structure of Red-cockaded Woodpeckers in south Florida: RAPDs revisited. Molecular Ecology 5: 725-734.

25. ———, C.L. Gratto-Trevor, T.D. Mullins, and M.A. Colwell. 1997. Population identification of western hemisphere shorebirds throughout the annual cycle. Molecular Ecology 6: 413-427.

26. ———, J.M. Rhymer, and D.G. Heckel. 1994. Population differentiation in randomly amplified polymorphic DNA of Red-cockaded Woodpeckers. Molecular Ecology 3: 581-595.

27. ———, R.S. Wagner, E.D. Forsman, and T.D. Mullins. Geographic variation and genetic structure in Spotted Owls. Conservation Genetics (in review).

28. Hedrick, P.W. 1994. Evolutionary genetics of the major histocompatibility complex. American Naturalist 143: 945-964.

29. ———. 1995. Gene flow and genetic restoration: the Florida Panther as a case study. Conservation Biology 9:996-1007.

30. ———, and P.S. Miller. 1996. Rare alleles, MHC and captive breeding. Pages 187-204 *in* V. Loeschcke, J. Tomiuk, and S.K. Jain, editors. Conservation Genetics. Birkhauser, Basel, Switzerland.

31. ———, and K.M. Parker. 1998. MHC variation in the endangered Gila Topminnow. Evolution 52: 194- 199.

32. Herman, S.G., J.W. Scoville, and S.G. Waltcher. 1984. The Upland Sandpiper in Bear Valley and Logan Valley, Grant County, Oregon. Report to Oregon Department of Fish and Wildlife.

33. Holland, D.C. 1993. A synopsis of the distribution and current status of the Western Pond Turtle (*Clemmys marmorata*). Report to Oregon Department of Fish and Wildlife.

34. Hubby, J.L. and R.C. Lewontin. 1966. A molecular approach to the study of genetic heterozygosity in natural populations, I: The number of alleles at different loci in *Drosophila psuedoobscura*. Genetics 54:577-94.

35. Kalmer, A. 1995. Genetic assessments in animal translocation programmes: re-introductions, re-stocking, and conservation introductions. M.S. Thesis, University of Kent, Canterbury, England.

36. Lacy, R.C. 1987. Loss of genetic diversity from managed populations: interacting effects of drift, mutation, immigration, selection, and population subdivision. Conservation Biology 1:143-158.

37. Lande, R. 1995. Mutation and conservation. Conservation Biology 9: 782-791.

38. ———, and G.F. Barrowclough. 1995. Effective population size, genetic variation, and their use in population management. Pages 87-123 in Viable Populations for Conservation. M.E. Soule, editor. Cambridge University Press, Cambridge, England.

39. Leberg, P. L. 1990. Genetic considerations in the design of introduction programs. Trans. 55th N.A. Wildl. & National Research Conference: 609-619.

40. Lesica, P. and F.W. Allendorf. 1995. When are peripheral populations valuable for conservation? Conservation Biology 9: 753-760.

41. Lynch, M. 1996. A quantitative-genetic perspective on conservation issues. Pages 471-501 in J. C. Avise and J. L. Hamrick, editors. Conservation Genetics: Case Histories from Nature. Chapman Hall, New York, NY.

42. ———, and R. Lande. 1998. The critically effective size for a genetically secure population. Animal Conservation 1: 70-72.

43. McGuigan, K., K. McDonald, K. Parris, and C. Moritz. 1998. Mitochondrial DNA diversity and historical biogeography of a wet forest-restricted frog (*Litoria pearsoniana*) from mid-east Australia. Molecular Ecology 7: 175-186.

44. Mills, L.S., and F.W. Allendorf. 1996. The one-migrant-per-generation rule in conservation and management. Conservation Biology 10:1509-1518.

45. Mullis, K., F. Faloona, S. Scharf, R. Saiki, G. Horn, and H. Erlich. 1986. Specific enzymatic amplification of DNA in vitro: the polymerase chain reaction. Cold Springs Harbor Symposium on Quantitative Biology 51: 263-273.

46. Parker, P.G., A.A. Snow, M.D. Schug, G.C. Booton, and P.A. Fuerst. 1998. What molecules can tell us about populations: choosing and using a molecular marker. Ecology 79: 361-382.

47. Primmer, C.R., T. Raudsepp, B.P. Chowdhary, A.P. Moller, and H. Ellegren. 1997. Low frequency of microsatellites in the avian genome. Genome Research 7: 471-482.

48. Ralls, K. and J.D. Ballou. 1992. Managing genetic diversity in captive breeding and reintroduction programs. Trans. 57th N.A. Wildl. & Nat. Res. Conf.: 263-282

49. ———, P.H. Harvey, and A.M. Lyles. 1986. Inbreeding in natural populations of birds and mammals. Pages 35-36 in M. E. Soule, editor. Conservation Biology: The Science of Scarcity and Diversity. Sinauer Assoc. Sunderland, MA.

50. Rossetto, M., P.K. Weaver, and K.W. Dixon. 1995. Use of RAPD analysis in devising conservation strategies for the rare and endangered *Grevillea scapigera* (Proteaceae). Molecular Ecology 4:321-329.

51. Shaffer, M.L. 1981. Minimum viable population sizes for conservation. BioScience 31: 131-134.

52. Shields, W.M. 1982. Philopatry, inbreeding, and the evolution of sex. State University of New York Press, Albany, NY.

53. ———. 1993. The natural and unnatural history of inbreeding and outbreeding. Pages 143-172 *in* N.W Thornhill, editor. The Natural History of Inbreeding and Outbreeding. University of Chicago Press, Chicago, IL.

54. Sanjayan, M.A., K. Crooks, G. Zegers, and D. Foran. 1996. Genetic variation and the immune response in natural populations of pocket gophers. Conservation Biology 10:1519-1527.

55. Simberloff, D.S. and L.G. Abele 1976. Island biogeography theory and conservation practice. Science 191: 285-286.

56. Snyder, N.R., S.R. Derrickson, S.R. Beissinger, J.W. Wiley, T.B. Smith, W.D. Toone, and B. Miller. 1996. Limitations of captive breeding in endangered species recovery. Conservation Biology 10: 338-348.

57. Stebbins, R.C. 1985. A Field Guide to Western Reptiles and Amphibians. Second Edition, Houghton Mifflin Co. Boston, MA.

58. Stockwell, C.A., M. Mulvey, and G.L. Vinyard. 1996. Translocations and the preservation of allelic diversity. Conservation Biology 10: 1133-1141.

59. Templeton, A.R. 1986. Coadaptation and outbreeding depression. Pages 105-116 in M.E. Soule, editor. Conservation Biology: the science of scarcity and diversity. Sinauer Assoc., Sunderland, MA.

60. Wolf, C.M., B. Griffith, C. Reed, and S.A. Temple. 1996. Avian and mammalian translocations: update and reanalysis of 1987 survey data. Conservation Biology 10: 1142-1154.

61. Wright, S. 1931. Evolution in Mendelian populations. Genetics 16: 97-159.

62. ———. 1951. The genetical structure of populations. Annals of Eugenics 15: 323-354.

18

Extirpated Species of Oregon and Washington

Constance Iten, Thomas A. O'Neil, Kelly A. Bettinger,
& David H. Johnson

Introduction

Extinction is forever, but extirpation is not! Extinction is an incremental process of loss from local to regional and, ultimately, to global disappearance of a species. Since all extinctions are incremental, the best place to try and conserve a species is at the local level. Causes of local extinctions have been proposed to be the result of small to catastrophic random events that occur in populations, genetics, or our environment. Events that create a loss of integrity within our ecosystems have also been surmised to enhance local extinction.[120, 147] One to many stochastic events may actually affect a species. We felt it would be of value to give short historical reviews of those species that have undergone local extirpation from Oregon and/or Washington. We have provided as best as possible estimates of historical ranges and abundances. But the reader should realize that levels of interest in and knowledge of different species varied greatly in the past, and therefore the historical information on some species may be incomplete.

What follows are short descriptions of 13 wildlife species that are (or formerly were) extirpated from Oregon and/or Washington. Table 1 was developed after reviewing 593 wildlife species that occur within the terrestrial, freshwater, and marine environments of the two states. Of the 14 species listed, only one, the California condor, was near the brink of global extinction before substantial efforts were initiated to rescue the species. Four species, the bighorn sheep, sea otter, sharp-tailed grouse, and trumpeter swan have undergone reintroduction effort(s) into the region with varying levels of success. Efforts to reintroduce the trumpeter swan have been limited to breeding populations, as wintering populations have rebounded due to protection measures. The pronghorn has been reduced in Oregon and extirpated from Washington. Efforts to reintroduce them in Washington have been unsuccessful. The bison has been brought back in limited numbers by private ranching ventures. The gray wolf, grizzly bear, and Wyoming ground squirrel have undergone regional or statewide extirpations without efforts to reintroduce them. The breeding population of the merlin in Oregon is considered nearly extirpated but wintering populations continue to be present. The final two species, upland sandpiper and yellow-billed cuckoo, are also considered extirpated,

although a few individuals may occasionally still be found during the breeding season. No efforts have been made to reintroduce or augment populations of the latter three species. Thus, most of the species listed here can be considered a "short list" for either Oregon and/or Washington, each having different potential for being reintroduced to their former range.

One additional species that was nearly extinct in the 1940s, the short-tailed albatross, will not be covered in this chapter, as it breeds only on one or two islands off the coast of Japan. Formerly, the albatross was abundant enough in shallow coastal waters of the North Pacific (probably as post-breeding foragers) that it regularly occurred in coastal Indian middens.[49] It is unknown if it was ever completely absent from Oregon and Washington waters, and in recent years as the breeding population has increased in Japan, there have been sightings off our coasts.

Table 1. A list of 14 species extirpated from Oregon and/or Washington.

Wildlife Species	Oregon	Washington
Bighorn sheep	X[a]	X[a]
Bison	X	X
California condor	X	X
Sea otter	X	X[a]
Trumpeter swan	X[a]	X[b]
Yellow-billed cuckoo	X	X
Gray wolf	X	
Grizzly bear	X	
Merlin	X[c]	
Sharp-tailed grouse	X[d]	
Wyoming ground squirrel	X	
Pronghorn antelope		X
Upland sandpiper		X
Short-tailed albatross	X	X

a = successfully reintroduced

b = reintroduced breeding population but not re-established; winter population is more secure

c = breeding population likely extirpated; winter population more secure

d = reintroduced but not re-established

Extirpation of these species has occurred for a variety of reasons. Some were pursued and killed for bounty because they endangered humans and their livestock, some were shot for their meat and hides, and others were poisoned and trapped because they were considered agricultural pests. Extirpation of several of the species that were initially abundant was influenced by multiple factors, while other species had small populations that quietly slipped from view. Loss of habitat was a key determinant in the declines of most of the species. In most cases, the loss of these 13 species was in some manner or another human caused. Conversely, re-establishing viable populations of any of these species will require active human management.

Species Extirpated from Oregon and Washington

Bighorn Sheep *(Ovis canadensis)*
Other Common Names: Mountain sheep, California bighorn, rimrock bighorn, Rocky Mountain bighorn.

Former Distribution
Bighorn sheep were formerly found from southeastern British Columbia and southwestern Alberta south along the Cascade and Sierra Nevada ranges into Baja California.

In Oregon, they occurred throughout much of the non-forested region east of the Cascade Range, with Rocky Mountain bighorn occupying the northeastern corner of the state and California bighorn the remainder of the range. Rocky Mountain bighorn sheep were numerous in the Wallowa Mountains, Snake River, and Grande Ronde River Canyons.[8] California bighorn ranged over southeast and south-central Oregon and through much of the John Day and Deschutes River drainages in the north-central part of the state.[24] Bighorns were extirpated from Oregon by the mid-1940s.[22]

In Washington, California bighorns were probably found scattered on the eastern slopes of the Cascade Mountains. The last surviving populations of the California bighorn were seen in the north-central part of the state and the Blue Mountains of southeast Washington in the early 1900s. Rocky Mountain bighorns were found in the northeastern and southeastern corners of Washington at the edges of the major bighorn sheep populations of Oregon and Idaho. Bighorns were extirpated from Washington by 1935.[65]

Current Status and Distribution
Bighorn sheep have been successfully re-introduced into both Oregon and Washington. The first attempt to restore bighorn sheep in Oregon was the release of 23 Rocky Mountain bighorns at Hart Mountain in 1939. The transplant failed, however, with the last survivor seen in 1947. Although the cause of the failure has not been established, the subspecies used may have been a factor.[24] Successful restoration of Rocky Mountain bighorns began in 1971. As of 1996, there were approximately 500 Rocky Mountain Bighorn sheep in 10 established herds in

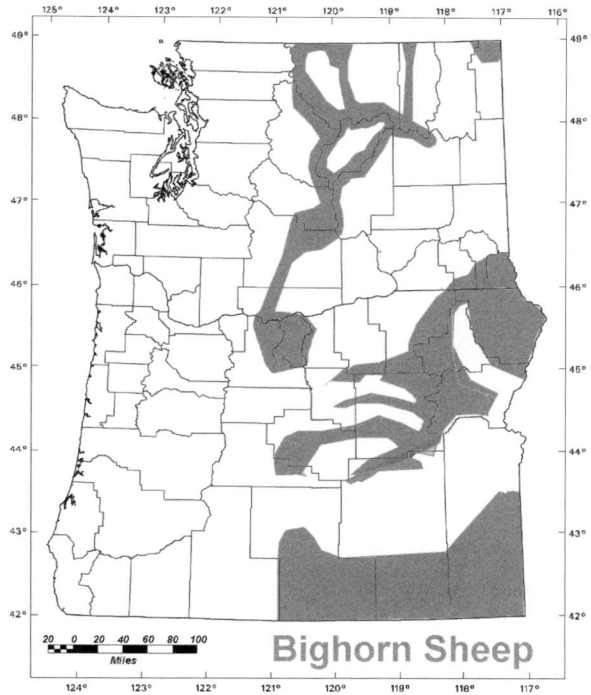

Bighorn Sheep

northeastern Oregon.[23] The first transplant of California bighorn sheep at Hart Mountain occurred in 1954, between 1954 and 1996 a total of 976 California bighorn sheep have been moved within, into, or out of Oregon. Currently, there are approximately 2500 California bighorn sheep in Oregon in 26 herd ranges throughout the state (Van Dyke, Oregon Department of Fish and Wildlife, pers. comm.).

In 1957, California bighorn sheep were re-introduced into Washington State. Over the next 16 years, California bighorns were released into 10 areas of eastern Washington that had been within the ranges of the original native populations. In 1999, the California bighorn population was estimated at 786 in 7 herds located throughout central, north central, northeastern, and southeastern Washington (Rice, Washington Department of Fish and Wildlife, pers. comm.).

Rocky Mountain bighorn sheep were reintroduced in Washington beginning in 1972.[65] In 1999 there were a total of 237 Rocky Mountain bighorns in 3 herds in southeastern Washington and 1 herd in northeastern Washington (Cliff Rice, Washington Department of Fish and Wildlife, pers. comm.).

Discussion
Although many bighorn sheep were probably extirpated due to indiscriminate hunting practices prior to established regulations, disease is considered the principal cause of bighorn declines. Additionally, unregulated grazing by domestic livestock is believed to have contributed to the overall decline of bighorn sheep.[141]

Bighorn sheep are very susceptible to a variety of diseases. The introduction of and proximity to domestic livestock as European settlement expanded provided the vector for many of these diseases. Bighorns have no natural resistance to bacteria and parasites carried by domestic stock.[65]

Pasturella spp. is the most common type of bacteria found in the respiratory tracts of bighorns. Certain types of *Pasturella* produce an acute bacterial pneumonia, which in many cases is stress-related and brought on by factors such as severe winter weather.[36, 117, 127] Domestic sheep have been implicated in the transmission of *Pasturella* spp. to bighorns.[144]

Lungworm-pneumonia complex is another major source of mortality in bighorn sheep. Areas of high sheep densities and poor forage can predispose animals with heavy lungworm infestations to a bacterial infection which causes pneumonia.[144]

Scabies is a contagious skin disease of bighorn sheep. Infected sheep stop eating, which leads to weight loss and death from environmental influences or other disease. Outbreaks of scabies have resulted in significant population declines of both California and Rocky Mountain bighorns in California, Colorado, Idaho, Montana, Oregon, and Wyoming.[19, 121] Scabies was responsible for a die-off of more than half of the Cottonwood Creek herd in Washington state in 1988.[144]

Other infectious diseases important to bighorn populations include contagious ecthyma, paratuberculosis (Johne's disease), actinomycosis (lumpy jaw), foot rot, necrobacillosis, keratoconjunctivitis, leptospirosis, caseous lymphadenitis, and listeriosis.[144]

The bighorn sheep management goal for Washington State is to "increase bighorn sheep populations of both subspecies statewide to a level where all available habitat within their historic distribution is filled and each herd is self-sustaining."[144] Management in Washington centers on several issues. Noxious weed control is important for maintaining quality forage habitat; restoration and reintroduction are priorities, as several herds may require augmentation; herds in the Blue Mountains require monitoring after a *Pasturella* die-off in 1995; and coordination and cooperation with tribes will be necessary to manage potential tribal hunting impacts.[145]

Oregon is attempting to increase bighorn sheep numbers as well as broaden the genetic diversity of the California bighorn sheep in the state and throughout its range, through cooperative efforts with other states and provinces. Trapping and transplanting are used to control population size in established herds thought to be large enough to allow removal without impacting population health. A ram to ewe ratio of 1:1 is maintained in those herds through controlled hunting.[24]

The Hells Canyon Initiative is a long-term project to restore Rocky Mountain bighorn sheep to the over 5 million acres of the Snake River Drainage of Washington, Oregon, and Idaho.[51] The Initiative represents a cooperative effort among agencies, organizations, and many private individuals with formal agreement between the state fish and wildlife agencies, the U.S. Forest Service, the Bureau of Land Management, and the Foundation for North American Wild Sheep. The overall goal of the Initiative is to "restore self-sustaining bighorn sheep herds to suitable habitat in the Hells Canyon area."

Under the Initiative, state and federal agencies will increase efforts to reintroduce bighorn sheep and manage habitat and populations to establish new herds and increase the size of existing herds. Data will be collected on bighorn sheep ecology and factors limiting population size to be analyzed and incorporated into management strategies. The area will serve as a landscape level model for bighorn sheep restoration.[51]

Bison (*Bos bison*)
Other Common Names: buffalo, American bison, plains bison, prairie bison, tatanka.

Former Distribution
Prehistorically, the bison was found in the central grasslands and northern parklands of North America, in habitats that ranged from semidesert to boreal forest with suitable grazing. Bison have been known to occur throughout North America and as far north as Alaska and south to Mexico and from the Cascades and Sierra Nevada ranges to the Atlantic Coast.[46, 84, 141] Native bison that were free roaming have never been observed in Oregon or Washington and were initially thought to be present in low numbers and a rare visitor to each state. However, according to Verts and Carraway,[141] "records in Idaho, Oregon and Nevada indicate that the bison were widespread, although likely not abundant." The Snake River valley probably served as a corridor to lead them westward. By crossing the Snake River or by following its southern bank, they could enter Oregon and establish populations in the eastern and central region of Oregon. Archeological evidence of bison occurring in the Columbia Plateau of Oregon and Washington has been provided by Osborne[93] and Van Vuren and Bray[140] who recap which museums contain bison remains (i.e., several bison skulls are housed at Oregon State University's Horner Museum).

Other written and verbal accounts of bison occurring in Oregon and Washington have been found in the Washington Historical Quarterly as reported by Kingston[68] and Pacific Northwest Quarterly as reported by Haines,[45] and are as follows:

• In February 1875, Professor O.C. Marsh noted that "the most western point at which I have myself observed remains of buffalo was in 1873 on Willow Creek, Eastern Oregon among the foothills on the eastern side of the mountains."

• The drying of Malheur Lake in 1930 disclosed the skeletal remains of more than 40 buffalo scattered over an area of 2,000 acres. It was in this region that Peter Skeene Ogden saw evidence of buffalo in 1826.

• Regarding the Columbia River region, David Thompson in 1811 noted he was told by Indians that he met at the confluence of the Snake and Columbia Rivers that they could go 3 days to where the buffalo were. This would mean that the buffalo came within 75 to 100 miles of the Snake River.

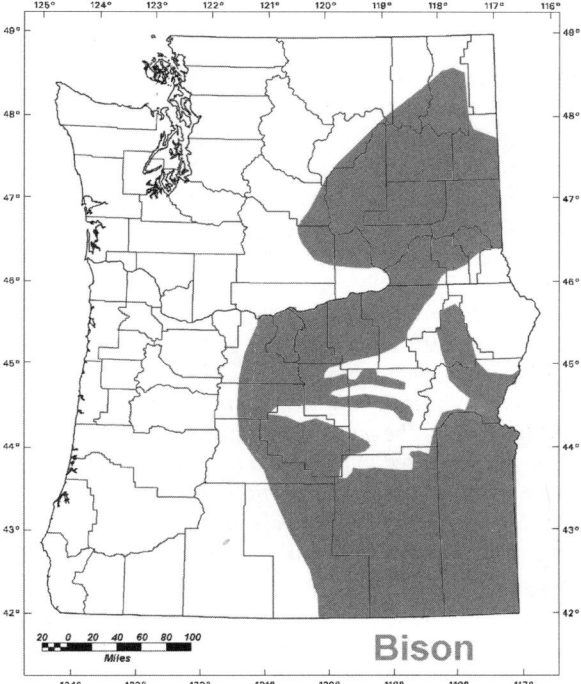

Bison

• In Washington, George Gibbs, who accompanied the Stevens surveying party of 1853 as a geologist and ethnologist, said that an Indian hunter told him that a lost buffalo bull had been killed 25 years before in the Grand Coulee. Further, Dr. Charles Pickering who accompanied the party of Lieutenant Johnson of the Wilkes expedition in 1841, which crossed the Cascades and visited Colville, Lapwai, and other places mentioned that "a single instance was on record of a stray animal having been seen in the vicinity of Colville." Mr. W. W. Lewis stated that years ago older Spokane Indians told him that their fathers surrounded and killed the last buffalo in the Spokane Valley somewhere near the Idaho border. He figured that the time of reference was about 1810 to 1820. Also, the Spokane Indians told Duncan McDonald that older members of their tribe killed a number of buffalo north of Moses Lake or the Grand Coulee.

Current Status and Distribution
The bison had been extirpated from Oregon and Washington prior to European settlement of the region. Today, bison occur in isolated populations usually within a park, preserve, or on private ranches. Most free ranging herds occur in Canada in either Mackenzie Sanctuary or Wood Buffalo National Park, and in the United States in Yellowstone National Park or the National Bison Range.[84]

By 1903, there were thought to be less than 1,700 bison in all of North America and most of these were in zoos or private herds.[42] By 1983, they had flourished from relocation and other conservation efforts to more than 75,000.[61] As of 1995, there were about 2,000 bison in 15 private ranches in Oregon.[78] As of 2000, there were 750-800 bison on 10 ranches in Washington (Jones, Northwest Bison Assoc., pers. comm.).

Discussion
The decline of the bison in Oregon and Washington appears to have been caused by a combination of events, but primarily drought, hunting, and overexploitation. Other sources of bison mortality include severe weather and on occasion drowning. Today, there are only a few instances of bison hunting. In Montana, along the northern boundary of Yellowstone Park, bison are viewed as problematic and can be shot when they leave the park. In Utah, bison are considered a game species and can be legally hunted.

Currently, bison are making a comeback primarily with the support of private interests. Game ranching of bison is profitable, and thus a viable commodity that helps expand their numbers and locations. Reintroducing bison to the wild, however, may be somewhat problematic because they show little respect for traditional livestock fencing and are known to carry brucellosis. Brucellosis may cause abortions in bison and cattle, which is against economic interests of the cattle industry.[84] However, Native American people have a strong connection with bison, and reintroduction for cultural reasons is given careful consideration. Thus, the subject of reintroduction is likely to continue. Finally, bison can coexist with other ungulates (like deer and elk) on the same range because there appears to be very little interspecific competition between them due to different habitat use and food habits.[83] Therefore, we may see more bison in the future in Washington and Oregon.

California Condor (*Gymnogyps californianus*)
Other Common Names: condor, California condor, spirit bird.

Former Distribution
California condors, although never abundant in recent centuries, were historically reported as year-round residents from British Columbia south to Baja California,[96] and were thought to have nested as far north as southern British Columbia.[146] The condor is considered a relic of the ice age, and formerly was very widespread—bones were found from New York and Florida—but well prior to European discovery of the continent its range had begun to shrink westward [31].

Bones of at least 63 California condors were found at the "Five Mile Rapids" archeological site along the Columbia River, five miles east of the Dalles, Oregon. Carbon-14 dating made it possible to date the remains between 10,000 and 7,500 years before present. According to Miller,[84a] the birds were attracted to the site by the presence of abundant living and carcasses of salmon and human refuse resulting from fishing. Another bone fragment positively identified as that of California condor was found in an Indian shell mound, 6 miles north of Brookings, Coos County, Oregon.[84b]

Some of the earliest records of condors found in the Pacific Northwest were those in the published journals of Lewis and Clark and of Patrick Gass, made while on their expedition to the lower Columbia River. A photostat of the original journal of Meriwether Lewis for February 17,

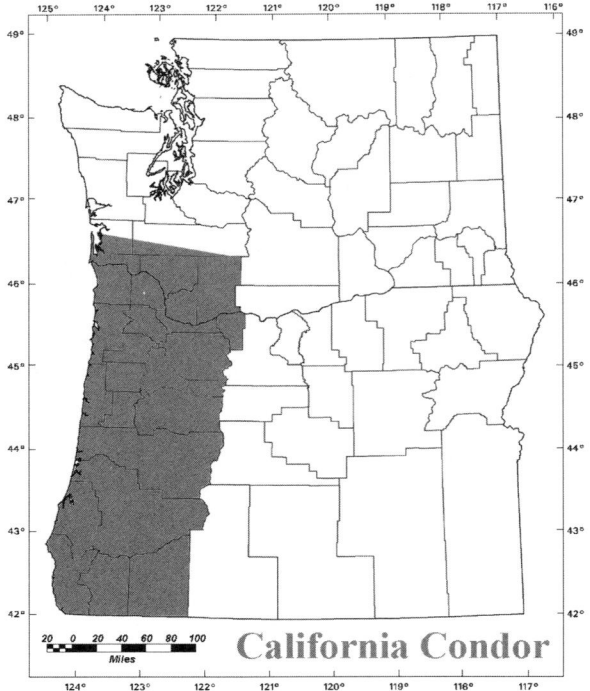

California Condor

The spring of 1827 was severe, and much snow had fallen. The consequence was that many horses died at Fort Vancouver, and we were visited by the various species of beasts and birds of prey that abound in that country. Most conspicuous among these were the California vulture. This magnate of the air was ever hovering around, wheeling in successive circles for a time, then changing the wing as if wishing to describe the figure 8; the ends of the pinions, when near enough to be seen, having a bend waving upwards, all his movements, whether soaring or floating, ascending or descending, were lines of beauty. In flight he is the most majestic bird I have seen. One morning a large specimen was brought into our square, and we had all a hearty laugh at the eagerness with which the Botanist [Douglas] pounced upon it. In a very short time he had it almost in his embraces fathoming its stretch of wings, which not being able to compass, a measure was brought, and he found it full nine feet from tip to tip.[35]

Roselle Putnam,[102] a young woman who immigrated to the Oregon territories in 1843, wrote to her mother and sister in 1852 describing her surroundings in the Umpqua Valley. In her letter she includes a description of the condor:

. . . the largest wild bird in the country is the vulture, which is only an overgrown buzzard—it only preys on the dead carcasses—I saw one measured which I think was between ten & eleven feet from the point of one wing to the point of the other.

Suckley[129] noted that:

The Californian vulture [=California condor] visits the Columbia River in fall, when its shores are lined with great numbers of dead salmon, on which this and other vultures, besides crows, ravens, and many quadrupeds, feast for a couple of months.

Swainson[133] made specific reference to the black vulture [=California condor] as common on "the plains of the Multnomah" [=Willamette River]. By 1850 they were rarely found north of California.[69]

Finley[33] mentions sightings of condors in 1903 and again in 1904 by George and Henry Peck at Drain in Douglas County, Oregon as well as in the report by Henry Peck of a condor killed on the southern Oregon coast a "number of years" earlier. Finley concluded:

These records seem to show that if the California condor was formerly found in the region of the Columbia River the numbers have decreased and the last of these northern birds seem to have taken refuge in the rough mountains regions of southern Oregon.

This represents the last published sighting of California condor in Oregon and Washington.

For many years precise population numbers were unknown, although they were considered to be a declining species since the 1890s. No longer found outside of California, the estimated total population dropped to 25 to 30 birds by the late 1970s.[139]

1806, published by Harris,[48] includes an excellent description of a living condor and a sketch of the head, leaving no doubt as to the identity of the bird. In addition to sightings of the birds, several condors were captured or killed by the Lewis and Clark expedition, as shown in their journal entries:

1805 October 30: mouth Wind River, some seen [76:vol 3 p.174]

1805 November 18: mouth Columbia River, one killed [76:vol 3 p. 232]

1805 November 30: mouth Columbia River, seen [76:vol 3 p. 259]

1806 January 3: Fort Clatsop, near mouth Columbia River, seen [76:vol 3 p. 309]

1806 February 17: near Fort Clatsop, one wounded and captured [76:vol 4 p. 81]

1806 March 15: near Fort Clatsop, two killed [43:p. 203]

1806 March 28: Deer Island, Columbia River, hunters said eagles and vultures devoured four deer [76:vol 4 p. 211]

Northwest Company fur trader Alexander Henry made record in his journal of seeing a condor in the Willamette Valley on January 25, 1814.[63:Elliott Coues, p. 112] David Douglas,[29] a Scottish botanist, referred to the range of the California condor north of California as follows:

I have met with them as far to the north as 49°N latitude, in the summer and autumn months, but nowhere so abundant as in the Columbia Valley between the Grand Rapids [=The Dalles, Wasco county] and the sea.

Douglas[30] noted in 1826 that "the Large Buzzard, so common on the shores of the Columbia, is also plentiful here [=eastern Linn County, OR[69]]: saw nine in one flock."

George Barnston, a close friend of Douglas, recounted an event which took place at Fort Vancouver:

Current Distribution

California condors are not currently found in Oregon and Washington. As of 1999, there are only 160 California condors in the world, including 47 in the wild and 113 in captive breeding facilities.[97] The birds in the wild were liberated at release sites in California and Arizona.

Discussion

Factors in the decline of the California condor include climatic as well as human-related activities.[96] Although scientists cannot pinpoint the exact reason for the reduction in the condor population, random, wanton shooting has generally been considered the single most serious cause of the condor's decline.[139] Other factors such as lead contamination from the ingestion of lead bullet fragments in carcasses,[13] egg and specimen collection, poisoning (both intentional and inadvertent), eggshell thinning from DDT,[31] and habitat modifications[96] have also contributed to their decline. The birds in the Pacific Northwest also may have been members of an isolated population that suffered from human exploitation. Such exploitation occurred between 1805 and 1850, during which time at least 13 condors were killed.[21, 29, 35, 69, 102]

Condors do not reach breeding capability until at least 6 years of age. That fact, coupled with an extremely low reproductive rate, laying only one egg per clutch, and not always laying in consecutive years, makes it difficult for the condor to recover from a significant reduction in population.[146] If, as proposed by Koford,[69] the Pacific Northwest population was a discrete one of only two or three-dozen individuals, the known mortality in the early 1800s would have been enough to jeopardize it seriously.[146]

The U.S. Fish and Wildlife Service listed the California condor as endangered in 1967. In an effort to study and preserve the few remaining condors, birds were captured and fitted with tags or radio transmitters to learn about their feeding, mating, and rearing habits. After confirming condors would lay a second or third time after losing an egg, in 1983 biologists began removing eggs laid in the wild to hatch in captivity to increase production. Young birds were also caught and bred in captivity. Despite these efforts, the numbers of condors in the wild continued to decline precipitously to a point in 1987, when the last wild condor was captured and brought into the captive breeding program to preserve the species.[139] In 1992, the first condors were reintroduced into the Los Padres National Forest's Sespe Condor Sanctuary.

From 1992 to 1999, a total of 55 birds have been released in California and Arizona as a "non-essential/experimental population" under section 10j of the Endangered Species Act. The recovery plan for condors includes continuing releases and the investigation of additional release sites in areas within the historical range of the California condor.[97]

Sea Otter (*Enhydea lutris*)

Other Common Names: southern sea otter, otter.

Former Distribution

Sea otters once occupied a nearly continuous range along the coastlines bordering the North Pacific Ocean, from Baja California to the northern islands of Japan.[14] Except for a small remnant population in Central California, sea otters were extirpated from the majority of their former eastern Pacific Ocean range between Prince William Sound, Alaska and Baja California by the early 1900s.[67, 104]

In Washington, they were once found in abundance along the central portion of the outer coast between Grays Harbor to Point Grenville.[14] Major concentrations were found between Point Grenville and Grays Harbor, and at North Head.[16]

Along the Oregon coast sea otters were originally found in areas of exploitable abundance. In 1857 Baird wrote: "They [sea otters] are found so abundantly near Cape Mendocino [California] to Port Orford [Oregon] that several companies have been organized and equipped in San Francisco expressly for their capture."[129]

Current Distribution

Despite re-introduction attempts, the sea otter population in Oregon became extinct in the mid-1970s, and otters are no longer found in the state.[104]

In Washington, 1999 surveys found a total population of 605 ranging from Tatoosh Island off Cape Flattery south to Destruction Island. The survey results indicate that the population in Washington has been growing at a finite rate of about 11% since 1989, and is expanding southward.[60]

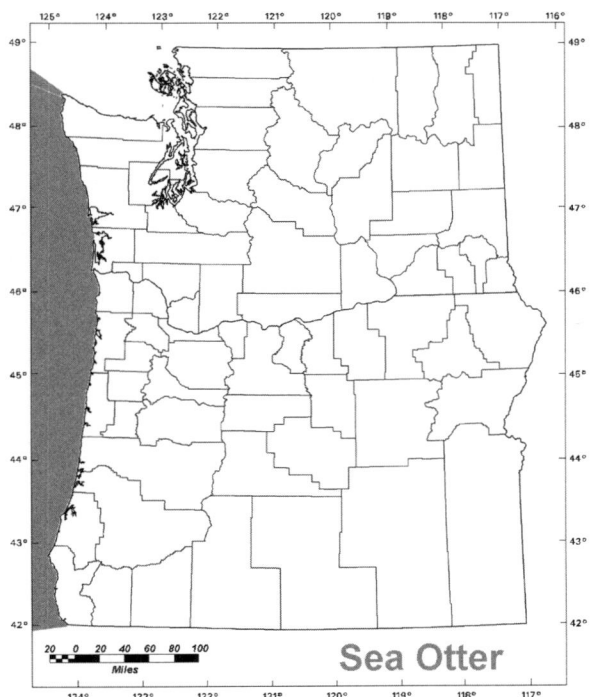

Sea Otter

Discussion

Sea otters were once plentiful throughout their range. In the early 1600s the Chinese were already receiving sea otter pelts from the Kuril Islands, harvested by Ainu hunters.[32] Discovered by the Russians in the 1740s at the Pribilof Islands of Alaska, they were hunted southward until they were all but exterminated in most of their former range. By 1870, hundreds of thousands had been harvested for their luxuriant fur, considered the finest and most valuable of any fur-bearing animal.[15] The purchase of Alaska by the United States in 1867 was based largely on the anticipated revenue from further fur harvests of both sea otters and fur seals. Russia, realizing that these resources were spent, sold the Alaska territory for a mere 7.2 million dollars.[104] After the purchase, sea otter harvest intensified and by the early 1900s the species was nearly extinct. In 1911, when unregulated killing was finally stopped, there were only a few remaining animals in 13 known locations.[67]

Between 1965 and 1972, 708 sea otters were translocated to unoccupied habitats in Alaska, British Columbia, Washington, and Oregon. In Washington, despite the 55% mortality of the initial transplant, sea otters became established and have begun to expand their range along the outer coast.[59, 60]

In Oregon, the translocated population declined drastically throughout the 1970s and is now considered extinct.[59, 104] Failure of the Oregon population has been attributed largely to emigration and mortality, resulting in a population too small to assure continued population growth.[59]

Sea otters require habitat along exposed saltwater coastlines with rocky islands and points of rock to provide shelter. Shallow (less than 180 feet of water) shorelines with underwater reef formations and extensive kelp beds are preferred, particularly those with abundant invertebrate bottom fauna. Areas such as these are available along the coastline of both Oregon and Washington.

Successful reintroduction of sea otters in Oregon may require consideration of the source of the translocated animals. Otters captured in Alaska are within the Aleutian zoogeographical province, while Oregon lies within a transitional zoogeographic province that separates the Californian and Aleutian provinces. It has been suggested that the more southerly habitat found in Oregon may have been less suitable for sea otters from northern populations.[59] Additionally, the number of otters transplanted must be capable of reproducing at a rate greater than that of the combined rate of emigration and mortality to establish a successful population.[59]

In Washington, with an expanding sea otter population,[60] management considerations may soon involve the need for the resolution of conflicts with commercial, tribal, and recreational fisheries. Populations of sea otter prey items form the basis of many commercial and recreational fisheries, including sea urchins, razor clams, mussels, and crabs.[75] Another commercial fishery interaction involves the by-catch of sea otters in net fisheries. Potential management solutions include: establishing zonal management units and the creation of take guidelines or closures for net fisheries.[75]

Trumpeter Swan (*Cygnus buccinator*)

Other Common Names: white swan, wild swan, bugler.

Former Distribution

Historically, the trumpeter swan's breeding range was extensive. The swan occurred from central Alaska east to Nova Scotia, New Brunswick, and Newfoundland south to the Carolinas, western Tennessee, northwestern Mississippi, eastern Arkansas, and Missouri, and west to Nebraska, Wyoming, southern Idaho, northeastern Oregon, and possibly California.[85] The breeding range shown by Banko[9] is more conservative and does not include Oregon, eastern Canada, or the southern United States. In winter, the trumpeter swan was found from southern Alaska south to southern California, possibly east through Arizona and New Mexico[98] to the lower Rio Grande Valley and the Gulf Coast of Texas,[10] east along the Gulf Coast to central Florida.[105]

The first and perhaps only evidence of breeding in Oregon appears to be a report in the journal *Oologist* of trumpeter swans at Malheur Lake in Harney County between 25 May and 15 June 1921.[63] Apparently no swans were known to breed in Oregon again until attempts to establish a breeding population at Malheur National Wildlife Refuge began in 1939, and the first nesting occurred in 1958.[58]

In Oregon, this species was noted by multiple authors as common in migration and/or winter along the lower Columbia River, in the Willamette Valley, and east of the Cascades in the 1800s.[44, 63] A writer by the name of John M. Murphy, who visited Oregon in the late 1800s,

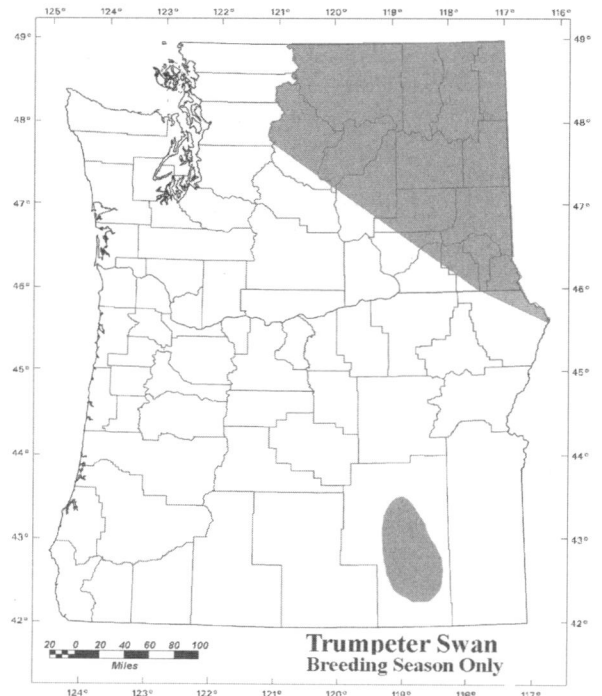

Trumpeter Swan
Breeding Season Only

described both whistling (=tundra swan) and trumpeter swans as so abundant at the mouth of the Columbia River that they appeared as "cumulus clouds" in late fall.[118] He also noted that both swans were "exceedingly common" in the Klamath Lake region, and along the Willamette, Columbia, and Snake Rivers. In 1891, Thomas G. Farrell wrote of an interview with an old resident of Seavies Island (=Sauvie Island) who described "immense flocks" of tundra and trumpeter swans that formerly visited Columbia and Multnomah Counties.[63] He goes on to say that "Of late years very few of these birds [trumpeter swans] visit the valley of the Columbia, and….it has never been my good fortune to meet with a member of the species."[63:T. G. Farrell p. 132] This decline in abundance is echoed again more than a decade later by John Minto, a pioneer who arrived in the Willamette Valley in the 1840s and noted that waterfowl populations in general had plummeted in numbers by the early 1900s, such that newcomers to the area hardly believed accounts of the earlier abundance.[118] By the time Gabrielson and Jewett published *Birds of Oregon* in 1940, they considered the species gone from Oregon and stated that there had been "no authentic records for many years."

The only evidence that the trumpeter swan may have formerly nested in the state of Washington is a report by C. F. Yokum of an account by an "old settler" of breeding at Cherry Lake in Whitman County until about 1918.[9] No nesting was confirmed until birds were introduced at Turnbull National Wildlife Refuge in Spokane County in 1963 and successful nesting occurred from 1967 to 1969.[122]

In migration and winter, trumpeter swans were reported by Suckley in 1860 to be more abundant on the Columbia River than in Puget Sound.[62] He describes immense flocks of swans along the shores of the river and spread out along the margin of the water for a distance varying from an eighth to a quarter of a mile. By the 1920s, the only known trumpeter swans in the state were a small flock of about 18 birds that for several years moved from wintering areas in British Columbia into Okanogan County to winter somewhere in eastern Washington or Oregon.[62]

Current Status and Distribution
Currently, the natural breeding range of the trumpeter swan is much reduced. The Pacific Coast population, which breeds in Alaska and winters from southern Alaska south to Washington and Oregon, numbered about 13,000 individuals in 1990.[85] The Rocky Mountain population which breeds in the Rocky Mountains from Yukon and Northwest Territories south through eastern British Columbia, Alberta, Montana, Wyoming, and Idaho numbered about 2,200 in 1992. It includes restored flocks in Oregon and Nevada. The Interior population was estimated at 629 birds in 1993 and includes restored flocks in South Dakota, Minnesota, Wisconsin, Michigan, and Ontario.

Attempts to re-establish a breeding population at Malheur National Wildlife Refuge in Oregon began in 1939 with birds from Red Rocks Lake National Wildlife Refuge in Montana, but the first nesting did not occur until 1958.[58] The breeding population at Malheur NWR peaked in the early 1980s at 19 breeding pairs and a total of 77 individuals. Between 1990 and 1999, the number of young fledged each year ranged from 3 to 18, and counts of fall adults ranged from 19 to 75 birds (the count of 75 was inflated by the release of 52 birds in 1992). Numbers have been steadily dropping since 1992. In 1991, an effort was begun to establish a separate breeding population at Summer Lake Wildlife Area in Lake County with birds from Montana, Idaho, and Malheur NWR.[58] In 1991, birds occurred only in Harney County in the summer, whereas in 1999, trumpeter swans were summering in Harney, Lake, Klamath, Crook, Grant, and Baker Counties. In addition, 4 juvenile swans were recently released on the Deschutes River in Bend as part of a plan to replace breeding exotic mute swans with the native trumpeter (Ivey, U.S. Fish and Wildlife Service, pers. comm.). The birds in eastern Oregon winter locally, sometimes wandering elsewhere in Oregon, northeastern California, or western Nevada.[57] The most regular winter location for trumpeter swans in western Oregon is in western Polk County, where a small flock has occurred since the mid-1980s.[44] In recent winters, small numbers have also been reported between late October and March along the Columbia River in Multnomah, Columbia, and Clatsop Counties.

In Washington, 6 trumpeter swan cygnets were introduced to Turnbull National Wildlife Refuge in Spokane County in 1963.[122] Nesting was attempted in 1966 and successful from 1967 to 1969. No mated pairs or nesting occurred again until 1994 when a pair was again seen, but as of 1997 that pair had not successfully bred. The only other recent report of a trumpeter swan in Washington in summer was of an adult seen 17 June 1991 at Calispell Lake in Pend Oreille County.[122] Trumpeter swans winter in Skagit and Snohomish counties (north Puget Sound) where high counts include 100 at Skagit Flats on 31 October 1996, 160 at Stanwood in Snohomish County on 16 November 1996, and 480 at Swan Reserve in Skagit County on 14 February 1998.[106, 111] Outside of the typical wintering areas, individual birds or groups of up to 10 are on occasion reported from Island, Grant, Clallam, Clark, Benton, Jefferson, Walla Walla, Klickitat, Kittitas, Pacific, Cowlitz, Grays Harbor, and Thurston Counties.[106, 107, 108, 111, 113, 114, 115, 116]

The trumpeter swan is not listed as an endangered, threatened, or sensitive species in Oregon or Washington. In 1989, the U.S. Fish and Wildlife Service was petitioned to list the Rocky Mountain population (which includes restored flocks in Oregon) because of low numbers and winter vulnerability, but the petition was denied.[85]

Discussion
The initial cause of the trumpeter swan decline was over-hunting for food, feathers, and skins.[9, 85] Factors negatively affecting trumpeter swans include lead poisoning from consuming lead shot or fishing weights, collisions with power lines and barbed wire fences, accidental shooting

by hunters mistaking the birds for other species of swans during hunting season, illegal shooting, and competition with the exotic and more aggressive mute swan.[9, 85] In Oregon, the greatest cause of mortality between 1958 and 1999 was due to power line collisions.[58] The second most important mortality factor during that time in Oregon was predation, but losses may be exaggerated, as many swans were wing-clipped and flightless. Once flying age is reached, adult trumpeter swans have few natural predators, though they are occasionally taken by golden eagles and very rarely by coyotes.[9]

Most breeding habitat is considered secure, though there are still local problems with pollution, recreation, mining, and development.[85] Swans nest in freshwater areas characterized by shallow, stable levels of unpolluted, fresh water; accessible forage; room for take off (approximately 100 m); emergent vegetation, muskrat or other platforms for nest sites; and low human disturbance.[85] Similar areas of ice-free water are used in fall and winter by migrants and wintering birds. The inadequate quality and quantity of winter habitat is considered one of the most important limiting factors in trumpeter swan restoration. Most winter habitat has been lost to urban development.

The trumpeter swan often responds well to reintroduction efforts, and active restoration projects to former breeding areas in Minnesota, Wisconsin, Michigan, and Ontario, as well as Oregon, are ongoing.[85] An additional challenge related to reintroduction that needs to be addressed is the loss of migratory traditions in some flocks.

Trumpeter swans are long-lived—up to 24 years in the wild and 32 years in captivity—and they do not typically begin to breed until 4-7 years old.[85]

Yellow-billed Cuckoo (*Coccyzus americanus*)
Other Common Names: rain bird, rain cuckoo, kow-kow.

Former Distribution
Some authorities recognize two subspecies; others consider the species to be monotypic.[56] *Coccyzus americanus occidentalis*, the California cuckoo, formerly bred from southwest British Columbia, Washington, Oregon, northern Utah, central Colorado, and west Texas south and west to southern Baja California, Sinaloa, and Chihuahua in Mexico. *Coccyzus americanus americanus* is found breeding in the remainder of the range throughout eastern North America south to eastern Mexico, the Greater Antilles, and possibly Central America and northwestern South America. The northern range limit is roughly equivalent to the northern tier of U.S. states, with populations extending into some areas of extreme southern Canada. Yellow-billed cuckoos winter in South America as far south as northern Argentina.[2]

Though common and widespread in the east, western populations of yellow-billed cuckoos probably always had a limited distribution, as they were mostly restricted to deciduous forested lowland riparian habitats. They may, however, have been locally abundant in favorable habitat,

as they were described by John K. Townsend as abundant in the area of present day Vancouver, Washington in the 1830s.[64] In 1916 R. Bruce Horsfall described yellow-billed cuckoos as "rare birds throughout most parts of Oregon."[54] In June 1924 Ira N. Gabrielson described encountering "Twelve or fifteen birds" in an area along the Columbia River near Portland, Multnomah County, where only one bird had been noted in the previous 5 years.[39] He further stated that they "continued to be found there in numbers all summer." By the 1940s, Gabrielson and Jewett[40] considered the yellow-billed cuckoo "not a common bird anywhere in Oregon" but did consider it fairly common in some past years (1920s) along the Columbia and Willamette Rivers. Jewett et al.[62] described it as a "rare summer resident in the humid Transition Zone of western Washington" and noted earlier reports from the 1920s and 1930s that described the cuckoo as "fairly plentiful" near Lake Washington and "a regular summer resident, although not numerous" in Whatcom County.

In Washington, the yellow-billed cuckoo was formerly an uncommon breeder throughout the Puget Trough and along major river systems in eastern Washington[122] and considered extirpated as a breeder by 1934.[56] In Oregon, it was formerly an abundant or common breeder along the Columbia River west of the Cascades, and in the Willamette and Rogue River Valleys.[44, 56, 81] It also occurred along major river systems such as the Grande Ronde, Umatilla, Columbia, Owyhee, and Snake in eastern Oregon, and was considered extirpated as a breeder by 1945.[56]

Current Status and Distribution
Since the 1930s, individual birds have been reported in Washington from Grays Harbor, King, Snohomish, Whatcom, Okanogan, Douglas, Grant, Franklin, Benton,

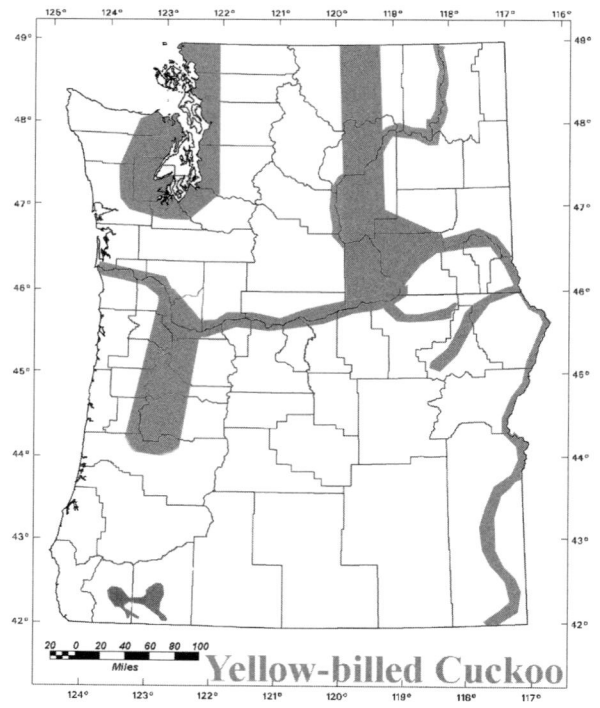

Yellow-billed Cuckoo

and Walla Walla counties. The most recent confirmed sighting is from Elma, Grays Harbor County on 3 August 1996 (Rogers, Washington Ornithological Society, pers. comm.). No breeding has been confirmed in Washington for more than 60 years.

In Oregon, there were only two records between 1940 and 1970, both from La Grande, Union County.[44] Since 1970 there have been more than 40 records of individual birds or pairs from Union, Harney, Malheur, Klamath, Deschutes, Jefferson, Umatilla, Wallowa, Baker, Lake, Linn, Clackamas, and Multnomah counties. All records have occurred between 20 May and 12 September, with most in June and July. The most recent records are from Malheur National Wildlife Refuge in May and June 1999.[126, 131] Since the 1940s, no breeding was confirmed in Oregon until 1992, when a nesting pair was found in La Grande, Union County. A female with a brood patch hit a window in Bend, Deschutes County on 18 June 1990.

The western population was formerly listed as a Category 2 (candidate for listing), but as of 1990 the western population was listed as a Category 3B (former federal candidate for listing) because it was not considered a valid subspecies.[37, 56] The U.S. Fish and Wildlife Service (USFWS) was petitioned again in 1999 to consider listing the cuckoo as endangered, and made a determination in February 2000 that listing the western subspecies may be warranted (Harvey, U.S. Fish and Wildlife Service, pers. comm.). A status review was begun and the USFWS is currently funding DNA research to determine the validity of the subspecies. As of 1999, the yellow-billed cuckoo is listed as a Species of Special Concern in Washington, and as a Species of Special Concern—Critical in Oregon.

Discussion

Several factors have been implicated in the decline of the yellow-billed cuckoo in the western U.S., but the most important one is habitat loss. Cuckoos in the western U.S. breed in extensive deciduous riparian thickets or in cottonwood and willow forests. Areas near backwaters such as sloughs and oxbows are particularly important. Size of the habitat patch appears to be crucial as well, as very few cuckoos were found in areas where suitable habitat was less than 100 m wide and less than 10 ha in size in California.[41] Laymon and Halterman[72] speculate that cuckoos may require intact riparian woodlands of more than 40 ha in size for breeding in California. The strict ties to river bottoms, sloughs, damp thickets and swampy places in the west appears to be tied to a requirement of high humidity needed to keep cuckoo eggs from desiccating; deciduous forests in the eastern U.S. where the species probably evolved are consistently humid in the summer.[41] Activities that have contributed to the loss of floodplain forests include channelization of river systems, water withdrawal for irrigation, dam building, flood control projects, livestock grazing, and clearing of areas for firewood, agriculture, and development. It is thought that once populations were stressed because of lack of suitable habitat, the widespread spraying of DDT and other pesticides in the 1940s and

1950s was the proverbial final nail in the coffin.[41] The yellow-billed cuckoo is a 100% insectivorous bird that feeds mainly on large insects such as caterpillars (including forest tent caterpillars), katydids, cicadas, gypsy moths, grasshoppers, and crickets.[56] These larger insects typically have cycles of abundance from year to year and as a result, cuckoos are thought to be nomadic just prior to breeding as they appraise local food resources. Areas which one year had numerous nesting pairs may have few or none the next, depending on food resources. This ties back to the available amount of habitat, in that protecting a limited area of riparian forest may not be adequate to sustain cuckoo populations in a region, since they need to be able to wander and find not only areas of suitable habitat, but habitat with an abundance of cyclical insects that season.

Changes in habitat on the wintering grounds are not considered an issue in the yellow-billed cuckoo's decline because their preferred winter habitat is the shrubby second-growth that results from cutting mature and old-growth forests.

No specific measures are being taken to improve habitat conditions for cuckoos in Oregon and Washington at this time, although they may benefit from efforts to restore riparian areas along major lowland river systems for other resource goals if such projects result in the formation of extensive stands of cottonwood and willow, and the preservation or restoration of seasonal flooding patterns that promote sloughs and oxbows.

Species Extirpated from Oregon

Gray Wolf (*Canis lupus*)

Other Common Names: grey wolf, timber wolf.

Former Distribution

Wolves were once found in all habitats of the northern hemisphere, with the exception of tropical forests and arid deserts.[90] In western North America, wolves ranged from Alaska south through Canada to northern California.[141] The northern Rocky Mountain gray wolf subspecies (*C. l. irremotus*) was listed as endangered by the USFWS in 1973; the entire species, *C. lupus*, was listed as endangered in the conterminous United States, except Minnesota in 1978.[1]

In 1857 Newberry reported:

Though less common than the "coyote," the large gray wolf is found in all the uninhabited parts of California and Oregon. Very few were seen by members of our party, none were killed; and we had everywhere evidence that this species was much less numerously represented on the Pacific Coast than on the Upper Missouri. In the Cascade Mountains we saw tracks of some of these wolves of most portentous size. All the large wolves seen by any of our party were gray, and all the skins, which I saw in possession of Indians or whites were also gray, and it is probable that the white and black varieties are never found in California. On the upper Columbia, in Oregon and Washington Territories, where the wolves are more numerous and

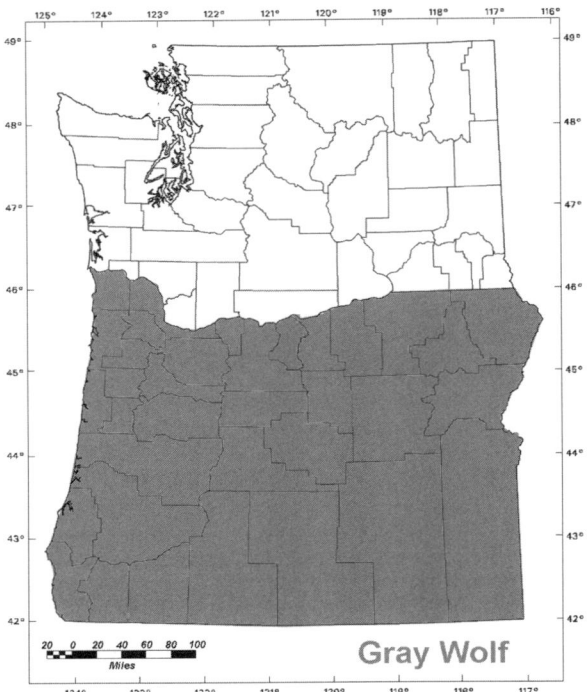

Gray Wolf

the winters are colder, the same variations occur which are common on the upper Missouri.[88]

The "Cascades wolf" *Canis lupus fucus* was historically found throughout the Cascade Range in both Washington and Oregon. It was described as a medium-sized wolf, with "a dark cinnamon or cinnamon-buffy" coloration with the dorsal area "profusely overlaid with black" and was referred to by early white settlers as "red legs." Lewis and Clark were the first white men to record the Cascades wolf, in 1805-06. They referred to it as the "large brown wolf" said to inhabit "California and the banks of the Columbia" river.[150]

The Cascades wolf is considered to have become extinct throughout its former range in the northwestern United States by 1977.[38, 99] In Oregon, wolves were considered to have been extirpated by 1972.[92]

Current Status and Distribution

Today, wolves are found in less than one percent of their former range in the contiguous United States. In Washington and Oregon, there have been a number of sightings of wolf-like canids in the last 25 years. In Washington, no recent live sightings have been confirmed as a wild wolf, and to date neither an identified pack nor mating pair has been found. Two wolves found dead in the Selkirk Mountains of northeastern Washington were confirmed to have strayed from the re-introduced 9-mile and Magic packs of Montana (John Almack, Washington Department of Fish and Wildlife, pers. comm.). However, there have been several documented family units of wolves in Washington in the last 12 years, indicating some level of reproduction.[1] In 1999, a wolf was captured in Oregon after traveling hundreds of miles from the Nez Perce Indian reservation, where efforts are being made to establish packs of Canadian wolves in central Idaho.

Currently, the only known wolf packs in the greater Pacific Northwest region are in British Columbia in the Elaho River drainage, near Squamish, B.C., in central Idaho, and in Yellowstone National Park in Montana.[71]

Discussion

Prior to the introduction of the horse in the 1700s, the human population was considered to have been "too low" to exert much pressure on wildlife populations.[101] Some native peoples of the Pacific Northwest region revered the wolf and had a taboo against killing them. Other tribes hunted wolves for their pelts for use in robes or blankets. In total, pressure on wildlife populations in the area was minimal. With the arrival of the horse, the impact on wildlife increased significantly, as native hunters were able to travel longer distances and transport more game.[71] In addition to intensive settlement by Europeans, the establishment of the Hudson's Bay Company in the area to ". . . make a profit for the stockholders by means of exploitation of the fur trade. . . ." initiated the decline of the wolf population. Between 1827 and 1859, 7,761 wolf pelts were traded at four posts in the Cascades area, from Thompson River, B.C. south to Fort Vancouver on the Columbia River, and east to Fort Colville. An additional 8,234 wolf pelts were traded at Fort Nez Perce at the confluence of the Walla Walla and the Columbia Rivers.[71]

As settlers moved into the area, livestock depredation by wolves became a major concern. In 1841, Admiral Charles Wilkes wrote about a wolf problem at Fort Vancouver, which necessitated bringing in from the field "large numbers of cattle. . . for in consequence of the numerous wolves that are prowling about; in some places it becomes necessary for the keeper to protect his beasts even in the daytime."[150] In 1843, the first "wolf meeting" was called and a $5.00 bounty on wolves was established.[71]

By 1871, the Washington Territorial Predator Control Act provided bounties on most predators in the Washington territory. In 1915, the Department of Agriculture designated federal funds for "predator control" on national forests and other public lands. In 1916, the U.S. Biological Service expanded the program to include private lands. Records from Washington State in 1924 described the program:

during September, 23 federal, State and Cooperative hunters worked a total of 653 days. We expect every man to put out all the poison baits possible. . . to cover the greatest possible area. . . to use it in all sections where the work can be carried on without endangering dogs and domestic animals. . . our first duty is to get rid of all the predatory animals possible.[7]

The last bountied wolf in Oregon was recorded in 1946. Since that time only two specimens have been collected, one in 1974 in southern Baker County and one in 1978 in eastern Douglas County. In Washington, the U.S. Biological Service records indicate that only 2 wolves were taken by hunters between 1915 and 1929.[137] The most recent confirmed wolf sighting in Washington was recorded in 1975. Identified as a British Columbia wolf *Canis lupus columbianus*, it was killed in Douglas County, after it had killed three calves. The Washington State Department of Game concluded in their 1975 Status Report that:

> *wolves may still be present in Washington, but that presence is represented by very small numbers, if at all. The wolf killed in Douglas county may have been a wild animal, but Canadian authorities cannot verify the possibility of its coming from adjacent ranges in British Columbia.*[142]

Currently, there are no efforts being made to re-introduce wolves to Washington or Oregon. Both Washington and Oregon are considered within the Northern Rocky Mountain wolf recovery plan area. The plan is currently focusing on re-introduction efforts in Yellowstone National Park. Budget constraints allow for little more than investigation of sightings outside the recovery area (Almack, Washington Department of Fish and Wildlife, pers. comm.).

Grizzly Bear (*Ursus arctos*)

Other Common Names: brown bear, big brown bear, griz, grisly, silvertip, white bear, Moccasin Joe.

Former Distribution

Grizzlies once roamed most of western and central North America and could be found from the Arctic Ocean to Mexico and from the prairies of the Great Plains to the Pacific coast.[95] There is also some evidence that the earlier range of grizzlies may have extended through parts of eastern North America.[141] The historical distribution of the grizzly bear is best summarized by Rausch.[103] Since the settlement of Europeans, the range of the grizzly bear has been steadily retreating, especially from the south and eastern portions of the United States.[26] The greatest number of bears could be found in the western portions of the United States and Canada. Along the Pacific coast, the grizzly bear used to range from Alaska to Mexico with viable populations still existing only in Alaska, Canada, and a small population in the interior of Washington. In the southern portion of the Pacific coast, the grizzly once roamed through Mexico, California, and Oregon. Most of the grizzly population in Mexico was thought to have been removed early, suggesting that the population dynamics in that area were weak. However, grizzly bear populations in California appeared viable, with a population estimated at 10,000; nonetheless, all had vanished by 1924.[26]

The grizzly bear has been extirpated from Oregon for almost 70 years, with the last verified animal shot on Chesnimnus Creek in Wallowa County on 14 September, 1931. Earlier accounts of the grizzly bear in Oregon suggest that the species occurred throughout the state prior to European settlement in the Cascade Range, Siskiyou, Blue, Steens, and Wallowa mountains; the Klamath, Rogue, Umpqua, and Willamette valleys; and some of the high desert east and south of Bend.[8, 141] The subspecies that used to occur in Oregon was *Ursus arctos horribilis*.[103] Grizzly bears also ranged over most of Washington. Today, a small population remains in the Northern Cascades, but its viability is unknown.

Current Status and Distribution

Grizzly populations are not threatened in Canada or Alaska, but in September 1975 the grizzly bear was designated as "threatened" south of the Canada border. As of 1999, grizzly bear populations could be found in 6 disparate ecosystems in Idaho, Montana, Washington, and Wyoming.[95] More specifically, they are found in the North Cascades of Washington, Selkirk Mountains of Idaho, Cabinet-Yaak of Montana, North Continental Divide of Montana and Canada, Bitterroot Mountains of Idaho and Montana, and the Greater Yellowstone of Idaho, Montana, and Wyoming. Critical habitat for the bear in the lower 48 states was designated in November 1976 and includes all of the current bear population centers except the Bitterroot Mountains.

Grizzly bears have persisted in the lower 48 states only in areas where extensive remote habitat has prevented or reduced human-caused mortality. Typically, these areas are protected by the Wilderness Act of 1964.[26] Population estimates have been very difficult to obtain, but based on a long-term study conducted by Craighead et al.[27] the Yellowstone population varied from 172 to 309 (1 bear per 80.3 km[2]).

Discussion

Grizzly bears occur in a variety of habitats from alpine to lowland valleys. They use early seral stages, riparian habitats, grasslands, and shrublands, but mostly live in rugged, largely inaccessible areas. Travel corridors are important to grizzly bear conservation because the current population centers are disjunct. The primary factors affecting population declines are loss of habitat, loss of food sources, and direct mortality. Historically, grizzlies were killed by humans defending their property and personal safety. They were also trapped and hunted.

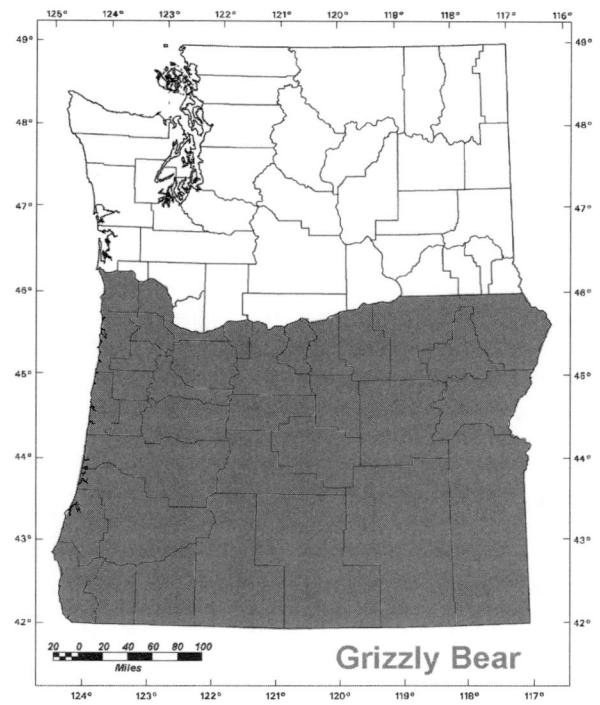

Grizzly Bear

Sullivan[130] discussed early trapping activities and noted that 382 grizzly bear hides passed through Fort Colville in 1849. However, Verts and Carraway's[141] review of the species in Oregon found no population estimates prior to their extirpation from the state. Only anecdotal information exists that suggests that populations were scarce by the early 1900s.[8]

The principal challenge in managing the grizzly bear is that large areas are required that involve multi-agency jurisdictions. Therefore, different mandates, laws, and concerns arising from various federal, state, and private organizations transcend the local politics of an area. Often the biggest concern revolves around personal safety, specifically when there is a potential for bears to interact with humans. The bears are long lived, have a low reproductive rate, and the cubs can spend several years with their mothers. Thus, direct mortality of a small number of bears can greatly influence the population dynamics of an area. Reintroducing grizzlies to Oregon is not without its own unique challenges, but is possible, as the state does have some large areas that are still remote. Likely places to reintroduce the grizzly bear in Oregon include the John Day Wilderness, or Hells Canyon Wilderness and Recreational Area—Eagle Cap Wilderness ecosystems.

Merlin (*Falco columbarius*)
Other Common Names: pigeon hawk, pigeon falcon.

Former distribution
The merlin is a circumboreal species breeding in Alaska, most of Canada, parts of the northwestern and northeastern U.S., Britain, Scandinavia, Finland, Iceland, Siberia, and north central Russia.[123] North American populations winter in southwestern Alaska, coastal western Canada, western and southern U.S., south into Panama, West Indies, Central America, northern South America, Venezuela, Colombia, Ecuador, and northern Peru.

Of the three subspecies that occur in North America, two breed in Washington and one formerly bred in Oregon.[44, 123] *Falco columbarius columbarius* (taiga merlin) is a rare breeder in high elevation forests of the north Cascades and northeastern Washington, and formerly bred in eastern Oregon; *F. c. suckleyi* (black merlin) is a rare breeder along the outer coast, Hood Canal, and Puget Sound areas of Washington.[44, 122] The black merlin is considered mostly sedentary while the taiga merlin is mostly migratory.

In Oregon, Gabrielson and Jewett[40] considered the black merlin a "Rare winter resident of [the] coast" and "only of casual occurrence inland." In Washington, Jewett et al.[62] considered the black merlin a "permanent resident coastwise and in tide flats and open prairie country of the Sound region of western Washington; in fall and winter some individuals wandering to points east of the Cascade Mountains." Only 2 confirmed nests were known at the time in Washington, both in cavities high in dead trees (although a subsequent review of the original reports on

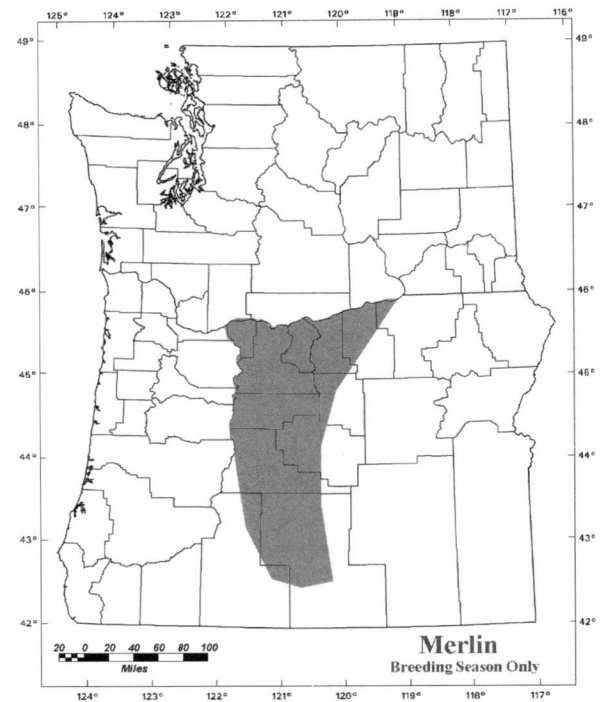

Merlin
Breeding Season Only

these nests suggested that at least one was not in a cavity; Gleason, pers. comm.). One nest was located in the Puyallup Valley of Pierce County; no location was given for the other. Black merlins were reported on several occasions hunting house sparrows and harassing pigeons in the cities of Portland and Seattle.[40, 62]

The taiga merlin was first reported as nesting in Oregon in 1857 near Klamath Lake.[40] Gabrielson and Jewett[40] describe the taiga merlin as "Very rare breeding bird and uncommon migrant and winter resident in eastern Oregon. Very rare straggler west of Cascades." They hypothesized that the species was more common than reports indicate because of the potential difficulty in distinguishing it from young sharp-shinned hawks. In Washington, Jewett et al.[62] describe the taiga merlin as a "Permanent resident in eastern Washington, rare migrant west of the Cascades."

Current Status and Distribution
The merlin is still found throughout its former range (see above) and in past decades actually expanded its range in the northern Great Plains and in New England.[123]

In Oregon, there are no confirmed breeding records from recent decades, but there are regular reports of merlins in the summer (June-July) from the 1970s, 1980s, and 1990s, including at least 33 sighting reports during the 1995-1999 Breeding Bird Atlas effort[44] (Adamus, Oregon State University, pers. comm.). The breeding season reports come from throughout most of western and eastern Oregon, lending uncertainty to the theory that only the taiga merlin breeds or bred in the state. One of the strongest pieces of evidence of possible breeding in recent years comes from Potamus Point in Morrow County where a bird appeared to be on territory 10-11 July 1998.[125] In addition, there are 6 unconfirmed reports by different observers of possible nests: a cavity nest near Klamath Falls, date unknown; a stick nest and a cavity nest on the upper Metolius River, date unknown; a nest under loose bark near Reedsport in 1966; a cavity nest near Cottage Grove in 1972; and a cavity nest near Roseburg in 1988 or 1989 (Fenske, pers. comm.). Merlins are more common in Oregon in winter, as migrants and wintering birds move into the state. They can be found throughout the state, but are most regular near major estuaries, lakes, reservoirs, and occasionally in cities where food supplies are reliable.[25] Unusually high counts were reported from two Audubon Christmas Birds Counts (CBC) in December 1997 when 6 birds were tallied on the Sauvie Island CBC in Multnomah County and 4 birds were tallied on the Dallas CBC in Polk County.[136]

In Washington, up to 13 nests and an additional 5 territorial pairs of the black merlin were found in the 1980s and 1990s in Skagit, Snohomish, Clallam, Grays Harbor, San Juan, and Jefferson Counties[122] (Gleason, pers. comm.). Of 9 nests monitored in the late 1990s, all were in old corvid nests and 8 were in Douglas fir or western hemlock trees with diameters ranging from 24 inches to 13 feet (Gleason, pers. comm.). There are two records of possible breeding by taiga merlins during the Washington Breeding Bird Atlas period (1987-1996): an adult seen north of Mt. Adams in Yakima County flying over old-growth forest and a pair seen in Stevens County[122] (Tom Gleason, pers. comm.). A third record was reported of an adult at Sanpoil, Ferry County on 19 June 1997.[109] As in Oregon, merlins are more common and widespread in Washington during winter when birds from the north move into the region.

As of 1999, the merlin has no federal or Oregon State status, but in Washington the breeding population is considered a Species of Concern—Candidate.

Discussion

Merlins hunt a variety of small to medium-sized birds, and also will feed to a lesser extent on larger flying insects such as dragonflies, and small mammals and reptiles.[123] They do not build nests of their own; instead they use old crow, raven, hawk, or magpie nests, tree cavities, or cliff ledges. At least in western Washington, merlins seem to prefer "waterfront property" and nest near lakes, rivers, and along the Puget Sound, perhaps because these areas provide breaks in otherwise continuous forest and/or because prey species are more abundant.

Washington and Oregon are at the edge of the species' breeding range and the density of breeding pairs has probably always been less than in core areas of their range. But there are probably more breeding pairs of merlins in western and eastern Oregon and Washington than the numbers of confirmed nests indicate, particularly given the number of sightings of birds during the breeding season. One explanation for the paucity of confirmed nest sites is the effort it requires to locate a pair and find the nest—it took Tom Gleason five years and many hours of patient searching and watching to find just 8 nests (Tom Gleason, pers. comm.). Wintering populations are more abundant as birds move into the region from Alaska and Canada, and tend to be easier to see as they occupy more habitats than would be used for breeding.

As with the peregrine falcon, DDT use caused eggshell thinning in the merlin, and studies in the 1960s in the Great Plains and Canada documented a 30% decline in merlin reproductive success when compared to pre-1950 data.[123] Though most merlin populations are currently thought to be reproducing well, a study of eggs collected between 1980-1988 in the Canadian Prairie Provinces showed that 35.7% of the eggs still contained levels of DDT likely to negatively affect reproductive success.[89] Though no merlin eggs have been tested in Oregon or Washington, eggshells from some peregrine falcon nests in the region still contain levels of DDT that warrant concern, and some pairs continue to lose their clutches by sitting on and breaking the eggs (Pagel, U.S. Forest Service, pers. comm.). A sample of 350 eggshell fragments collected in recent years in Oregon, Washington, and northern California were a mean 17.4% thinner than eggshells from pre-DDT use (Pagel, U.S. Forest Service, unpubl. data). The greatest thinning occurred in southwestern Oregon and northwestern California where eggshells were a mean 24% thinner. Given that merlins and peregrines have somewhat similar ranges, migratory habitats, and diets (both are bird eaters) in Oregon and Washington, it is possible that merlins also could still be experiencing eggshell thinning.

In most areas of their range, merlin populations currently appear to be increasing, with the possible exception of the northeastern U.S., where acid rain is thought to be a factor.[123] In the prairie regions, merlins are declining or have disappeared from areas where nest trees are being lost from woodlots and shelterbelts and where vegetation is cleared from around prairie potholes, but the decline may be offset by increases in nesting populations in some cities in the region.[123] It is unknown what the effects might be from habitat loss and change on wintering grounds in Central and South America.

Documented threats to breeding populations in Oregon and Washington—at least for those birds not nesting in protected areas such as the Olympic National Park— include direct persecution and nest tree loss. Of the 9 nests monitored by Tom Gleason in western Washington in the 1990s, adults at two nests were shot and an additional two nest trees were cut down by landowners. All 9 nests were on private land and many were near homes in low-density waterfront neighborhoods with mature conifer trees.

Sharp-tailed Grouse
(*Tympanuchus phasianellus*)

Other Common Names: Kennicott's grouse, western prairie chicken, prairie sharp-tailed grouse, willow grouse.

Former Distribution

At the time of European settlement, Columbian sharp-tailed grouse were found throughout much of the inland Pacific Northwest, as well as locations in western Colorado, northern Utah, northern New Mexico, western Montana, and western Wyoming.[77]

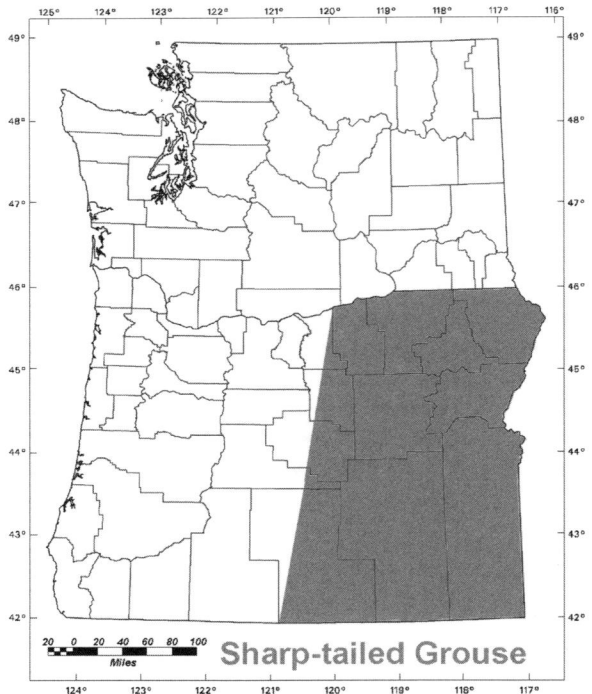

Sharp-tailed Grouse

Sharp-tailed grouse were once abundant in eastern Washington and Oregon. Early European settlement, farming practices, and hunting pressure contributed to their decline. By the late 1920s they had been extirpated from at least six counties in Washington. In 1940 Columbian sharp-tailed grouse were described as: "now being scarce and apparently in danger of early extinction in Oregon."[28] The last sighting of Columbian sharp-tails in Oregon was in the late 1960s in Baker County. They were considered extirpated from Oregon by the early 1970s (Sands, The Nature Conservancy, pers. comm.).

Current Status and Distribution
Sharp-tailed grouse persist in 8 scattered remnant populations in Douglas, Lincoln, and Okanogan counties in Washington State. They are confined to small, severely fragmented populations, some of which are located on degraded habitat. Much of the remaining potential sharp-tail habitat in Washington is threatened by further alteration and fragmentation.[50] In 1997 the total population was estimated to be 1,000 birds (Mike Schroeder, Washington Department of Fish and Wildlife, pers. comm.).

In Oregon, a small, stable sharp-tail population has been re-established in Wallowa County. Three leks currently support approximately 100 birds (Vic Coggins, Oregon Department of Fish and Wildlife, pers. comm.).

Discussion
Once plentiful throughout their range, sharp-tailed grouse were harvested in abundance by early explorers and settlers. Populations were further impacted by the conversion of grassland and sagebrush habitats to agriculture, livestock grazing in shrub/meadow steppe and increased settlement in their former range.

In the late 1800s, sharp-tailed grouse were still abundant in both Washington and Oregon. Early settlers in eastern Washington found the Columbian sharp-tailed grouse to be the most abundant and best-known game bird in the area.[149] They were also abundant in the Grande Ronde and Powder River valleys in Baker and Union Counties of Oregon. In 1881, an article on hunting in that area recounted:

This is the only part of Oregon where prairie chicken [=sharp-tailed grouse] may be found, and this fall I saw my first specimen of this beautiful bird, and as I have always wished, this, the first chicken I ever beheld, in its peculiar flight fell clean to my gun, and as I picked up the noble grouse it seemed a shame to kill so handsome a bird. . . . Great numbers of birds could be killed here daily, but with me sport loses its charm when it approaches slaughter, so what might have been scores I was content to make dozens.[119]

Settlers in Washington harvested wagonloads of sharp-tails in a single day in the 1880s and 1890s, presumably in high-concentration areas such as the shrub/meadow steppe bordering river tributaries of eastern Washington.[70]

By 1895, sharp-tailed grouse were declining from just such hunting practices. In 1895, an article in the *Oregon Naturalist* stated:

this bird [sharp-tailed grouse] once so abundant on the plains of eastern Oregon, Washington and 'Big bend of the Columbia,' is slowly and surely repeating the history of the pinnated grouse [=lesser prairie chicken] of the Western states. Where 15 years ago they could be seen in winter in flocks of fifty or more; flocks of a dozen are now uncommon. Then their great enemy was the coyote, who robbed their nest and caught their young. Now it is man with his traps and guns.[6]

In 1899, L. B. Quimby in an unpublished report for the Oregon Department of Fish and Wildlife recommended hunting restrictions in light of the rapidly declining numbers of sharp-tailed grouse attributed to over-harvest, particularly in the winter months. After the establishment of hunting seasons and bag limits for upland game birds, sharp-tailed grouse experienced a short-lived recovery. But by the turn of the century populations were once again decreasing (Crawford, Oregon State University, pers. comm.).

Additionally, farming practices were acknowledged as adding to the decline. The 1914 *Oregon Sportsman* stated:

The method of farming in this section [Umatilla County], which is almost entirely summer fallowing and burning the stubble in the spring, has nearly destroyed the prairie chicken [=sharp-tailed grouse]. . . .[100]

There were early warnings that efforts would need to be undertaken to protect the grouse, such as the account of the Columbian sharp-tailed grouse in 1915, which stated:

This grouse is one of our best game birds, but it has decreased very rapidly with the change of conditions because it does not prosper in the vicinity of man. It has always held its own in the sagebrush with the coyotes and other natural enemies, but like the sage grouse, it has suffered a great deal on account of the extensive pasturing of sheep. . . . The only way this grouse can be saved from extinction in many of the regions where it lives is by the setting aside of certain areas not fit for agricultural purposes as wild bird refuges where hunting is forbidden and the birds are well protected. Otherwise, the history of this western grouse will be the same as that of its cousin, the pinnate grouse or heath hen [=greater prairie chicken].[34]

By the mid-1920s sightings of the sharp-tailed grouse in Oregon were reduced to "a very few" in Wallowa county, considered one of the leading counties in the state as to the number of game birds.[4]

At that same time in Washington, hunting was restricted in all counties to a 4-month sharp-tail season (August-November) with a daily bag limit of 10, replacing the previous 6-month season and daily bag limit of 20 of the late 1800s. In 1909, Whitman County reduced the season to 3 months and a daily bag limit of only 5, and closed the season entirely in 1919. Between 1933 and 1952, a moratorium was placed on sharp-tailed grouse hunting in Washington. In 1953, hunting for sharp-tailed grouse was re-opened until 1988, when population declines resulted in a statewide closure.[50] Sharp-tailed grouse were placed on the state list as a threatened species in Washington in 1998 (Schroeder, Washington Department of Fish and Wildlife, pers. comm.).

In Oregon, Columbian sharp-tailed grouse were listed as an endangered species in 1969.[79] The last known sighting in Oregon occurred in 1968 or 1969 in Baker County.[91] It is now considered extirpated throughout the state of Oregon (Crawford, Oregon State University, pers. comm.). The U.S. Fish and Wildlife Service considered the sharp-tailed grouse in the Northwest a sensitive species in 1982.[138]

The Washington Department of Fish and Wildlife has been coordinating the acquisition of upland habitat for sharp-tailed grouse through the Bonneville Power Administration and the Washington Wildlife and Recreation Coalition. Approximately 22,250 acres have been purchased for sharp-tailed grouse. Additionally, WDFW has been working with private landowners to increase the benefits of CRP lands for sharp-tailed grouse. Ongoing research, coordination, and partnerships are focused on improving sharp-tailed grouse habitat, and evaluating potential sites for possible reintroduction or augmentation of populations.[50]

In Oregon, re-introduction efforts began in 1991 in Wallowa County near Enterprise. As a result of successive annual transplants with birds from southeastern Idaho, two small populations have been successfully established in areas of former distribution. In 1999, it was estimated that the existing population consisted of 100 birds. Future plans include re-introductions into other areas of the state formerly occupied by sharp-tailed grouse (Sands, The Nature Conservancy; Coggins, Oregon Department of Fish and Wildlife, pers. comm.).

Wyoming Ground Squirrel
(*Spermophilus elegans*)

Other Common Name: Richardson's ground squirrel.

Former Distribution

Fossil remains of the *Spermophilus richardsonnii* complex—of which the Wyoming ground squirrel is a member—are known from Alberta to Kansas, Texas, and Oklahoma, and as far east as Minnesota.[46, 141] The oldest fossils of the *Spermophilus richardsonni* complex were found in Southern Alberta, Canada.[47] Of Washington and Oregon, the Wyoming ground squirrel only occurred in southeastern Oregon.[141] Trying to be more exact as to its previous distribution is problematic because of taxonomic classification issues. In 1863, this species was first identified as a subspecies (*Spermophilus richardsonni elegans*) by Kennicott;[66] in 1928 Howell[55] classified it as a species (*Citellus elegans nevadensis*); in 1938 it was reclassified by Howell (*Citellus richardsonnii nevadensis*); and in 1971 Nadler et al.[87] finally classified the species as *Spermophilus elegans nevandensis*.

Current Status and Distribution

As of 1999, the Wyoming ground squirrel occurs in 3 distinct areas with separate subspecies: *nevadensis* in southeastern Oregon, southwestern Idaho and northern Nevada; *aureus* in southwestern Montana and eastern Idaho; and *elegans* in the southern half of Wyoming and northwestern third of Colorado with small extensions into Utah and Nebraska.[151] In Oregon, *Spermophilus elegans nevadensis* is only known from 7 specimens collected at 2 sites: (1) in Malheur County, Oregon and (2) near McDermitt, Nevada. In the 1970s, the Wyoming ground squirrel in Oregon was viewed as endangered if present

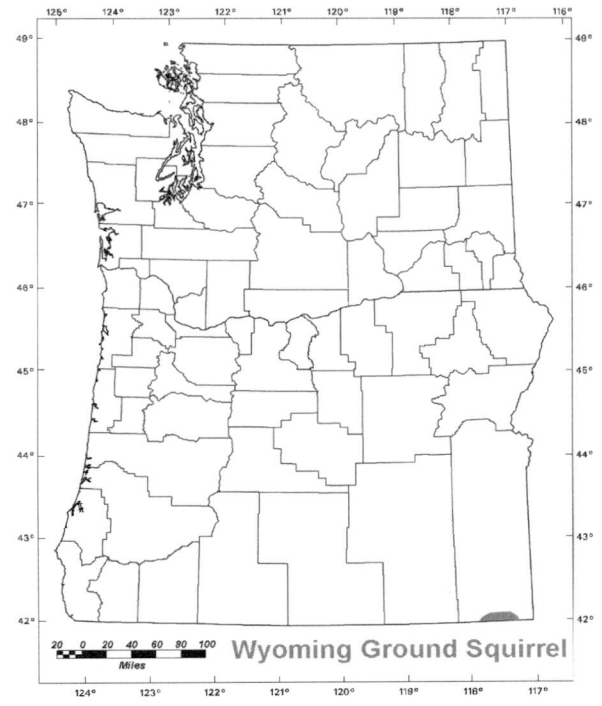

Wyoming Ground Squirrel

in the state,[141] because there had been no records for the species since 1927. Efforts in the 1980s to confirm presence of the Wyoming ground squirrel within southwestern Idaho proved unsuccessful. In 1986, the first Oregon Non-game Wildlife Management Plan was developed, and this species was listed as extirpated, but identified as the Richardson's ground squirrel.[80]

Discussion

The Wyoming ground squirrel is a member of the unspotted ground squirrels and lives underground in burrows for most of the year. In areas where it is more common, like northern Nevada, the Wyoming ground squirrel can be found in meadows or habitats that emulate meadow characteristics, as well as on occasion in sagebrush or bottom lands.[46]

The Wyoming ground squirrel can occupy mountain meadows from about 5000 feet (~1500 m) elevation to above timberline.[5, 20, 73] This ground squirrel is preyed upon by snakes, coyotes, foxes, weasels, badgers, and hawks.

No one really knows the reason(s) for the decline of the Wyoming ground squirrel in Oregon. The Oregon population is at the northwest extent of its range; hence, the Wyoming ground squirrel population may not ever have been very great. Competition with other ground squirrels may have played an initial role in limiting their distribution, but the disjunct population of *Spermophilus elegans nevadensis* is considered a relict one that has been out-competed by the Belding's ground squirrel in more mesic habitats, and by the Merriam's and Piute ground squirrels in arid habitats. The viability of the Wyoming ground squirrel has been negatively affected by shooting, poisoning, trapping, and conversion or loss of habitat.[3]

Species Extirpated from Washington

Pronghorn Antelope (*Antilocarpa americana*)
Other Common Names: pronghorn, antelope, prongbuck, berrendos, little pale deer, small caribou.

Former Distribution

Pronghorns formerly occurred from southwestern Manitoba, western Minnesota, eastern Texas, and Hidalgo west to southwestern Alberta, eastern Washington, southwestern Oregon, California (except coastal regions), and Baja California.[141]

In Oregon, early reports indicated pronghorns were abundant and widely distributed.[8] Pronghorns originally occurred throughout the area east of the Cascade Range except for the Blue and Wallowa mountains. They also occurred west of the Cascade Range in the Rogue River valley and at the headwaters of the North Umpqua River. From the beginning of European settlement through the mid-nineteenth century, pronghorns rapidly declined and became less widely distributed, although they were never completely extirpated from Oregon.[141]

Lewis and Clark mention antelope occurring on the great plains of the Columbia, though by no means as abundantly as east of the Rocky Mountains. Baird

suspected them to be "nearly, if not quite extinct there. . ." in 1857, having never "met them." He observed that pronghorn were found sparingly in Oregon; but felt, unless stragglers occurred on the Spokane plains, it was doubtful whether they entered the Washington Territory. They were said by the Indians to have been formerly quite plentiful at the Dalles of the Columbia, but were nearly exterminated in that locality by 1857.[129]

Pronghorn antelopes occurred on the Columbia Plateau from north central Washington through southeastern Washington from about 8000 B.C. They were extirpated in Washington prior to most European settlement, although archeological and ethnographic records substantiate the presence of a few pronghorn into the early 1800s.[17, 93] The spatial distribution of the archeological and ethnographic records corresponds with the semi-arid sagebrush habitat preferred by pronghorns.[17]

Current Distribution

After suffering declines through the early 1900s, pronghorn numbers have steadily increased in Oregon, in part due to the establishment of the Sheldon-Hart Mountain National Wildlife refuges in Oregon and Nevada.[141] The current population of pronghorn in Oregon is approximately 20,000 animals (Van Dyke, Oregon Department of Fish and Wildlife, pers. comm.).

In Washington, re-introductions were attempted beginning in 1938. Initially, 38 of 138 fawns brought into the state survived to be released in various areas including the Squaw Creek Refuge (the present day Yakima Training Center), Adams County, the Colockum Game Range in Kittitas County, and Grant County.[86] By 1950 the population had reached 200. Additional transplants of pronghorns were made in 1968. Seven adults and 15 fawns were released in Grant and Kittitas Counties. Despite these

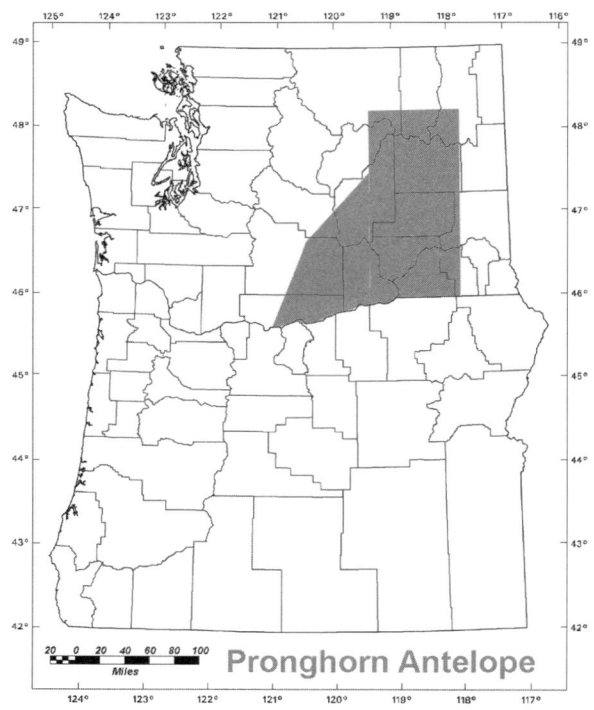

Pronghorn Antelope

introductions, the population continued to decline, to 125 in 1971,[94] and to 60 in 1978.[143] None of the introduced animals survived to establish self-sustaining herds, and at present, pronghorn are not found in Washington (R. Johnson, Washington Department of Fish and Wildlife, pers. comm.).

Discussion
Colonization of western rangelands by European settlers caused a more than a 99% decline in pronghorn numbers.[148] Fencing, habitat loss, competition with domestic livestock, and un-regulated hunting all contributed to the decline.[74] Osborne[93] suggested that climatic changes throughout the Columbia Plateau were primarily responsible for displacing pronghorn from their former pre- and proto-historic range in Washington. He suggested: "colder winters, deeper, longer lasting snows, and an influx of northern predators" as aspects of a climatic pulsation beginning as much as 4,000 years ago might well have upset the balance that made the eastern part of Washington habitable for the antelope.

Archeological sites where concentrations of pronghorn remains were found indicate a large number of individuals were acquired near those sites.[17] Ethnographic records confirm group hunting methods; pronghorn driven into coulees toward waiting hunters[134] as well as the use of group surrounds and natural enclosures.[132] C. L. Brown[18] felt the introduction of the horse and gun to the Plateau Indians accelerated the extirpation of the pronghorn antelope. Large number of furs and hides were traded to obtain horses, while both guns and horses made it easier for hunters to acquire antelopes from greater distances. Economic reasons rather than subsistence needs began to dictate hunting choices, making certain hunted species greater targets.[18]

Although it is unknown how important a role disease may have played in early pronghorn declines, losses from disease have had serious impacts on present day populations. Bluetongue, carried by cattle, caused the death of at least 3,200 pronghorn in eastern Wyoming in 1976, and another 300 deaths in northeastern Wyoming in 1984.[135] Other serious disease vectors include parasitic worms, which can be the cause of high fawn mortality.[11] Bever[12] found that areas over-grazed by domestic sheep resulted in high parasitic loads in pronghorn.

Management of pronghorn in Oregon includes managing the Sheldon-Hart Mountain National Wildlife refuges for pronghorn, in efforts to improve range conditions and hunting regulations that promote herd health and vigor.[141] In Washington, after unsuccessful attempts to re-establish herds in their former range on the Columbia Plateau, pronghorn management efforts have not been pursued (R. Johnson, Washington Department of Fish and Wildlife, pers. comm.).

The Pronghorn Antelope Workshop meets biennially; attendees represent western state and provincial wildlife agencies, federal land and wildlife agencies, universities and colleges, consultants, and private interests from Canada, Mexico, and United States. The Workshop's goals are: "to exchange information and encourage the perpetuation of sustainable wild herds of pronghorn as an ecological, aesthetic, and recreational natural resource on public and private western rangelands at their most productive levels consistent with other land uses."[74] Wildlife managers from both Oregon and Washington participate in the Workshop, and have contributed to the development of the Pronghorn Management Guides, a "compendium of suggested practices and techniques for managing pronghorn and their habitat."[74]

Upland Sandpiper (*Bartramia longicauda*)
Other Common Names: upland plover, Bartram's sandpiper, highland plover, field plover.

Former Distribution
The original status of upland sandpipers in Oregon—and likely Washington as well—before the introduction of livestock is unknown.[81] It is thought that the potential breeding range of the upland sandpiper included much of eastern Oregon and Washington, as scattered sightings have occurred from throughout the area. Experience from Oregon, however, has shown that it takes intensive surveys by experienced observers in order to confirm nesting.[82]

Though a number of references from the early 1900s included Oregon in descriptions of the range of the upland sandpiper, they all appear to be based on J. C. Merrill's 1888 "Notes on the birds of Fort Klamath, Oregon" and a few scattered other specimens and sightings.[63] Citing uncertainty in Merrill's report, S. G. Jewett placed the species on Oregon's hypothetical list in 1929. He appeared to reverse himself by 1931 after breeding was confirmed in Umatilla County, and more birds were seen in Grant County. By 1940, Gabrielson and Jewett list the upland sandpiper as a rare summer resident in the vicinity of Ukiah, Umatilla County, and Bear and Logan Valleys, Grant County.[40]

Upland sandpipers were first reported in Washington in 1905, but no other evidence of their occurrence was found until 1928 when it was confirmed as breeding in Spokane County.[62] In 1953, Jewett et al. describe the species as an "uncommon and local summer resident in eastern Washington" with at least one small nesting population in Spokane County.

Current Status and Distribution
The breeding range of the upland sandpiper stretches from north-central Alaska, northern Yukon, central and northeastern British Columbia, southwestern Mackenzie through the Canadian prairies to southern Quebec, central Maine, and New Brunswick, southeast to West Virginia, central Virginia, and Maryland, and west throughout the American Midwest to northeastern Oregon and eastern Washington.[2,53] Though the range covers a broad area, upland sandpipers are restricted to scattered local areas of suitable habitat, and are particularly scarce west of the Rocky Mountains. Populations in Oregon, Washington, and Idaho are considered disjunct and peripheral to the main range. Upland sandpipers winter in eastern South

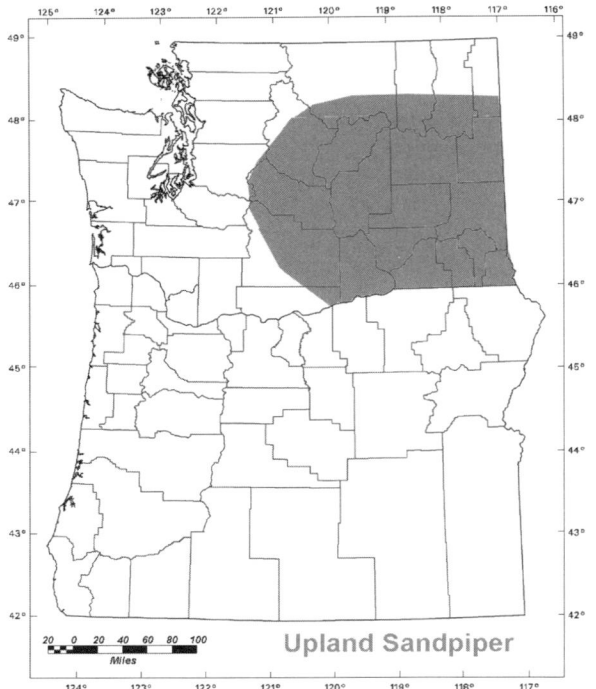

Upland Sandpiper

America from Surinam and northern Brazil south to central Argentina and Uruguay.[2]

Although it is very small, Oregon has the largest known nesting population of upland sandpipers west of the Rockies, and the status of the species in Oregon today is likely similar to what it was in the 1930s and 1940s. Surveys conducted in 1984 estimated a population of about 100 birds in Oregon, with over half occurring in Bear and Logan valleys in Grant County.[52] Subsequent surveys up through 1993 suggest a significant decline in numbers, however, from the mid-1980s.[82] Breeding has been confirmed in Bear and Logan Valleys, Grant County; Bridge Creek Wildlife Area and Ukiah Meadows, Umatilla County; Campbell Flat on the Starkey Experimental Forest, Union County; and Sycan Marsh, Lake County.[81, 128] Regular sightings of upland sandpipers, including territorial behavior and display flights and song, were reported up through summer 1999 from Grant County.[124, 125, 126]

In Washington, the last known record from the breeding area near Spokane was of two adults seen on 22 September 1993, after intensive surveys turned up no birds during the breeding season[82] (R. Rogers, Washington Ornithological Society, pers. comm.). Since then, individual sightings have been reported from Pacific, Douglas, Grays Harbor, King, Okanogan, and Clallam Counties, all in August or September[110, 112] (R. Rogers, Washington Ornithological Society, pers. comm.). No sightings were reported for the state of Washington in 1994, 1996, or 1999, even though intensive surveys were conducted in the east Spokane Valley in 1994[82] (R. Rogers, Washington Ornithological Society, pers. comm.).

The upland sandpiper was listed as Endangered by the state of Washington in 1981[82] and is a Species of Special Concern—Critical in Oregon. Though listed as Threatened or Endangered in 10 other states, and as a candidate for designation as Threatened or Endangered in British Columbia, as of 1999 the species currently has no special classification under the federal Endangered Species Act.[53, 82] A recovery plan for the upland sandpiper in Washington State was published in February 1995.

Discussion

The first threat to upland sandpiper populations occurred in the late 1800s and early 1900s when market hunters turned their attention from the disappearing passenger pigeon to the upland sandpiper.[53] With no bag limits or closed seasons, and birds being shipped by boxcar loads to market, declines were noticed by the 1870s. Birds were also hunted on their South American wintering grounds. Legal hunting of the upland sandpiper in North America ended with the Migratory Birds Convention Act of 1916, and the birds are now also protected by law in Argentina, though illegal harvest likely still occurs in South America.[53]

The biggest threats to populations now are habitat loss and alteration. Upland sandpipers are a grassland species, requiring a juxtaposition of tall, dense vegetation for nesting and shorter, more open grassy areas for resting, feeding, and brood rearing.[53] Platforms for courtship and territorial displays such as rocks or fence posts and a lack of rugged topography appear to be important too. In Oregon, nests have been found on the borders of seasonally wet mountain meadows, and near forest-grassland or sagebrush-grassland edges.[82] No descriptions are available for the two known Washington nest sites.

Development of residential housing, commercial business areas, and gravel pits has contributed to permanent habitat loss. Conversion of native grasslands to agricultural crop fields has likely resulted in "sink populations," because although upland sandpipers nest in such areas, there is probably no reproductive gain, due to repeated disturbances associated with crop management practices. Hayfield and pasture mowing have been implicated in chick mortality. Exotic plant invasion by species such as the spotted knapweed has also contributed to habitat loss and alteration. Heavy grazing results in the reduction of grass cover to the point that upland sandpipers no longer use such areas for nesting, and chicks and eggs are trampled and destroyed by cattle. It may be possible, however, to carefully control stocking, timing, and duration of grazing in order to provide the required juxtaposition of all cover conditions from short to tall grasses.[82]

While predators are an expected source of mortality, agricultural areas likely attract more or different predators than would be found in large intact areas of native grasslands (foxes, skunks, American crows, coyotes, raccoons). Studies have also shown that the upland sandpiper is a loosely colonial nester where populations are healthy, and mobbing behavior by nesting adults helps reduce losses to predators.[82]

Upland sandpipers are largely insectivorous, and over half of their diet is comprised of crickets, locusts, grasshoppers, and weevils.[53] Heavy and widespread insecticide use both on the wintering grounds and breeding grounds, particularly in the 1940s, is also thought to be partly responsible for population declines.

Literature Cited

1. Almack, J. A., and S. H. Fitkin. 1998. Grizzly bear and gray wolf investigations in Washington State 1994-1995, final progress report. Washington Department of Fish and Wildlife, Olympia, WA.

2. American Ornithologists' Union. 1998. Check-list of North American Birds. 7th edition. American Ornithologists' Union, Washington, D.C.

3. Andelt, W. F. and S. N. Hopper. 1998. Managing Wyoming ground squirrels. Colorado State University Cooperative Extension, No. 6.505. Colorado State University, Fort Collins, CO.

4. Anonymous. 1926. Wallowa County rich in birds. Oregon Sportsman. 2(3):22.

5. Armstrong, D. M. 1972. Distribution of mammals in Colorado. Monographs of the Museum of Natural History, University of Kansas 3:1-415.

6. Averill, A. B. 1895. The Sharp-tailed grouse. Oregon Naturalist 2:81-82.

7. Bach, G. R., and H. W. Dobyns. 1924. Hunter's newsletter—Washington District. Washington State Department of Agriculture, Bureau of Biological Survey. Olympia, WA.

8. Bailey, V. 1936. The mammals and lifezones of Oregon. North American Fauna, No. 55. U.S. Department of Agriculture, Bureau of Biological Survey, Washington, D.C.

9. Banko, W. E. 1960. The trumpeter swan. University of Nebraska Press, Lincoln, NE.

10. Banko, W. E., and A. W. Schorger. 1976. Trumpeter swan. Pages 5-71 in R. S. Palmer, editor. Handbook of North American birds. Volume 2, Waterfowl. Yale University Press, New Haven, CT.

11. Bever, W. 1950. Parasites and diseases of South Dakota antelope. Pages 12-16 in Pittman-Robertson Job Completion Report Project 12-R-7. South Dakota Fish and Game Department, Pierre, SD.

12. ———. 1957. The incidence and degree of the parasitic load among antelope and the development of field techniques to measure such parasitism. Pittman-Robertson Project 12-R-14. No. 1-5, 2. South Dakota Fish and Game Department, Pierre, SD.

13. Bloom, P. H., J. M. Scott, O. H. Pattee, and M. R. Smith. 1989. Lead contamination of golden eagle *Aquila chrysaetos* within the range of California condor *Gymnogyps californianus*. In B. U. Meyburg and R. D. Chancellor, editors. Raptors of the modern world. WWGBP: Berlin, London, and Paris.

14. Bowlby, C. E., B. L. Troutman, and S. J. Jefferies. 1988. Sea otters in Washington: distribution, abundance, and activity patterns. Washington Department of Wildlife. Marine Mammal Investigations, Final Report for National Coastal Resources, Research and Development Institute, Newport, OR.

15. Bonnot, P. 1951. The sea lions, seals, and sea otter of the California coast. California Fish and Game 37:371-389.

16. Brittell, J. D., Brown, J. M., and R. L. Easton. 1976. Marine shoreline fauna of Washington. Volume 2, Habitat requirements of selected species. Washington Departments of Game and Ecology. Olympia, WA.

17. Brown, C. L. 1977. An examination and evaluation of the archeological occurrence of the American pronghorn antelope in the Columbia Plateau. M.A. Thesis. Washington State University, Pullman, WA.

18. ———, 1981. Note to Washington Department of Game 7 April 1981.

19. Buechner, H. K. 1960. The bighorn sheep in the United States, its past, present and future. Wildlife Monographs Number 4.

20. Burnett, W. L. 1931. Life history studies of the Wyoming ground squirrel (*Citellus elegans*) in Colorado. Bulletin of Colorado Agriculture Experiment Station 373:1-23.

21. Burroughs, R. D. 1961. The natural history of the Lewis and Clark Expedition. Michigan State University Press, East Lansing, MI.

22. Coggins, V. L. 1980. Present status of Rocky Mountain bighorn sheep in Northeast Oregon. Biennial Symposium North. Wild Sheep and Goat Council 2:90-105.

23. ———, and P. E. Matthews. 1996. Rocky Mountain bighorn sheep in Oregon, history, and present status. Biennial Symposium North. Wild Sheep and Goat Council 10:87-92.

24. ———, Matthews, P. E., and W. Van Dyke. 1996. History of transplanting mountain goats and mountain sheep - Oregon. Biennial Symposium North. Wild Sheep and Goat Council 10:190-195.

25. Contreras, A. 1997. Northwest birds in winter. Oregon State University Press, Corvallis, OR.

26. Craighead, J. J., and J. Mitchell. 1982. Grizzly bear. In J. Chapman and G. Feldhamer, editors. Wild mammals of North America - biology, management, and economics. John Hopkins University Press, Baltimore, MD.

27. ———, J. R. Varney, and F. C. Craighead. 1974. A population analysis of the Yellowstone grizzly bears. Bulletin 40, Forest and Conservation Experimental Station, School of Forestry, University of Montana, Missoula, MT.

28. Cushing, J. E. Jr. 1941. The Columbian sharp-tailed grouse in Lake County, Oregon. Condor 43:75.

29. Douglas, D. 1829. Observations on the *Vultur californianus* of Shaw. Zoological Journal 4:328-330.

30. Douglas, D. 1914. Journal kept by David Douglas during his travels in North America 1823-1827. Antiquarian Press, New York, NY.

31. Ehrlich, P. R., D. S. Dobkin, and D. Wheye. 1988. The birder's handbook. Simon and Schuster, New York, NY.

32. Farr, A. C. M., and F. L. Bunnell. 1980. The sea otter in British Columbia - a problem or opportunity. Pages 110-128 in R. Stace-Smith, L. Johns, and P. Joslin, editors. Threatened and endangered species and habitats in British Columbia and the Yukon. British Columbia Ministry of Environment, Fish and Wildlife Branch, Victoria, BC.

33. Finley, W. L. 1908. Life history of the California condor: Part II- historical data and range of the condor. Condor 10:5-10.

34. Finley, W. L. 1915. Columbian sharp-tailed grouse. Oregon Sportsman. 3:138-140.

35. Fleming, J. H. 1924. The California condor in Washington: another version of an old record. Condor 26:111-112.

36. Foreyt, W. J. and D. A. Jessup. 1982. Fatal pneumonia of bighorn sheep following association with domestic sheep. Journal of Wildlife Diseases 18:163-168.

37. Franzreb, K. E., and S. A. Laymon. 1993. A reassessment of the taxonomic status of the yellow-billed cuckoo. Western Birds 24:17-28.

38. Friss, L. K. 1985. An investigation of subspecific relationships of the gray wolf, *Canis lupus*, in British Columbia. M.S. Thesis. University of Victoria, British Columbia, Canada.

39. Gabrielson, I. N. 1924. The season: Portland (Oregon) region. Bird-Lore 26:343-344.

40. Gabrielson, I. N., and S. G. Jewett. 1940. Birds of Oregon. Oregon State College, Corvallis, OR.

41. Gaines, D. and S. A. Laymon. 1984. Decline, status and preservation of the yellow-billed cuckoo in California. Western Birds 15:49-80.

42. Garrietson, M. S. 1938. The American bison. New York Zoological Society, New York, NY.

43. Gass, P. 1904. Gass's journal of the Lewis and Clark expedition. Reprint of edition of 1811. A. C. McClurg, Chicago, IL.

44. Gilligan, J., M. Smith, D. Rogers, and A. Contreras. 1993. Birds of Oregon. Cinclus Publications, McMinnville, OR.

45. Haines, F. D. 1940. The western limits of the buffalo range. Pacific Northwest Quarterly 31: 389-398.

46. Hall, R. E. 1981. The mammals of North America. Second edition. John Wiley & Sons, New York, NY.

47. Harrington, C. R. 1978. Quaternary vertebrate faunas of Canada and Alaska and their suggested chronological sequence. Syllogeus 15:1-105.

48. Harris, H. 1941. The annals of Gymnops to 1900. Condor 43: 3-55.

49. Hasegawa, H. and A. R. DeGange. 1982. The short-tailed albatross, *Diomedea albatrus*, its status, distribution and natural history. American Birds 36: 806-814.

50. Hays, D., M. Tirhi, and D. Stinson. 1988. Washington state status report for the sharp-tailed grouse. Washington Department of Fish and Wildlife, Olympia, WA.

51. Hells Canyon Bighorn Sheep Restoration Committee. 1997. Restoration of Bighorn Sheep. The Hells Canyon Initiative. Idaho Department of Fish and Game, Lewiston, ID.

52. Herman, S. G., J. W. Scoville, and S. G. Waltcher. 1985. The upland sandpiper in Bear Valley and Logan Valley, Grant County, Oregon. Unpublished report to Oregon Department of Fish and Wildlife, Portland, OR.

53. Hooper, T. D. 1997. Status of the upland sandpiper in the Chilcotin-Caribou Region, in British Columbia. Wildlife Working Report WR-86, Ministry of Environment, Lands and Parks, Williams Lake, BC, Canada.

54. Horsfall, R. B. 1916. California cuckoo. Oregon Sportsman 4:195-196.

55. Howell, A. H. 1938. Revision of the North American ground squirrels, with a classification of the North American Sciuridae. North American Fauna 56:1-256.

56. Hughes, J. M. 1999. Yellow-billed cuckoo (Coccyzus americanus). In A. Poole and F. Gill, editors. The Birds of North America, Number 418. The Birds of North America, Philadelphia, PA.

57. Ivey, G. L. 1990. Population status of trumpeter swans in southeastern Oregon. Pages 118-122 in D. Compton, editor. Proceedings and Papers of the Eleventh Trumpeter Swan Society Conference. The Trumpeter Swan Society, Maple Plain, MN.

58. Ivey, G. L., M. J. St. Louis, and B. D. Bales. 2000 (In press). The status of the Oregon trumpeter swan program. In R. Shea, editor. Proceedings and Papers of the Seventeenth Trumpeter Swan Society Conference. The Trumpeter Swan Society, Maple Plain, MN.

59. Jameson, R. J., K. W. Kenyon, A. M. Johnson, and H. M. Wright. 1982. History and status of translocated sea otter populations in North America. Wildlife Society Bulletin 10:100-107.

60. Jameson, R. J., and S. Jeffries. 1999. Results of the 1999 survey of the reintroduced sea otter population in Washington State. Unpublished report.

61. Jennings, D. C., and J. Hebbring. 1983. Buffalo management and marketing. National Buffalo Association, Custer, SD.

62. Jewett, S. A., W. P. Taylor, W. T. Shaw, and J. W. Aldrich. 1953. Birds of Washington State. University of Washington Press, Seattle, WA.

63. Jobanek, G. A. 1997. An annotated bibliography of Oregon bird literature published before 1935. Oregon State University Press, Corvallis, OR.

64. Jobanek, G. A., and D. B. Marshall. 1992. John K. Townsend's 1836 report of the birds of the lower Columbia River region, Oregon and Washington. Northwest Naturalist 73:1-14.

65. Johnson, R. L. 1983. Mountain goats and mountain sheep of Washington. Biological Bulletin Number 18. Washington State Game Department, Olympia, WA.

66. Kennicott, R. 1863. Descriptions of four new species of Spermophilus, in the collection of the Smithsonian Institution. Proceedings of the Academy of Natural Sciences of Philadelphia 15:157-158.

67. Kenyon, K. W. 1969. The sea otter in the eastern Pacific Ocean. North American Fauna 68:1-352.

68. Kingston, C. S. 1932. Buffalo in the Pacific Northwest. The Washington Historical Quarterly 23:163-173.

69. Koford, C. B. 1953. The California condor. National Audubon Society Research Report Number 4.

70. Larrison, E. J., and K. G. Sonnenberg. 1968. Washington birds: their location and identification. The Seattle Audubon Society, Seattle, WA.

71. Laufer, J. R., and P. T. Jenkins. 1989. Historical and present status of the gray wolf in the Cascade Mountains of Washington. Northwest Environmental Journal 5:313-327.

72. Laymon, S. A., and M. D. Halterman. 1987. Can the western subspecies of yellow-billed cuckoo be saved from extinction? Western Birds 18:19-25.

73. Lechleitner, R. R. 1969. Wild mammals of Colorado: their appearance, habits, distribution, and abundance. Pruett Publication Company, Boulder, CO.

74. Lee, R. M., J. D. Yoakum, B. W. O'Gara, T. M. Pojar, and R. A. Ockenfels, editors. 1998. Pronghorn Management Guides. Biological and management principles and practices to sustain pronghorn populations

and habitat from Canada to Mexico. 18th Biennial Pronghorn Antelope Workshop, Prescott, AZ.

75. Lentfer, J. W. 1988. Selected marine mammals of Alaska. Species accounts with research and management recommendations. Marine Mammal Commission, Washington, D.C.

76. Lewis, M., and W. Clark. 1905. Original journals of the Lewis and Clark expedition 1804-1806. R. G. Thwaites, ed. Volume 3, part 2; Volume 4, parts 1 and 2. Dodd, Mead, New York, NY.

77. Linn, R. M., editor. 1979. Proceedings of the 1st Conference on Scientific Research in the National Parks. National Park Service Transactions and Proceedings Series No. 5.

78. Manning, R. 1995. The buffalo is coming back. Defenders Magazine 71:6-15.

79. Marshall, D. B. 1969. Endangered plants and animals of Oregon. III. Birds. Oregon Agricultural Experiment Station Special Report 278.

80. Marshall, D. B. 1986. Oregon Nongame Wildlife Management Plan. Oregon Department of Fish and Wildlife, Portland, OR.

81. Marshall, D. B., M. W. Chilcote, and H. Weeks. 1996. Species at risk: sensitive, threatened and endangered vertebrates of Oregon. 2nd edition. Oregon Department of Fish and Wildlife, Portland, OR. .

82. McAllister, K. R. 1995. Washington state recovery plan for the upland sandpiper. Washington Department of Fish and Wildlife, Olympia, WA.

83. McCullough, Y. B. 1980. Niche separation of seven North American ungulates on the National Bison Range, Montana. Ph.D. Dissertation, University of Michigan, Ann Arbor, MI.

84. Meagher, M. 1986. Bison bison. Mammalian Species Number 266. The American Society of Mammalogists.

84a. Miller, L. 1957. Bird remains from an Oregon Indian Midden. Condor 59:59-63.

84b. Miller, A. H. 1942. California condor bone form the coast of southern Oregon. Murrelet 23:77.

85. Mitchell, C. D. 1994. Trumpeter swan (Cygnus buccinator). In A. Poole and F. Gill, editors. The Birds of North America, Number 105. The Birds of North America, Philadelphia, PA.

86. Moreland, R. 1969. Special Report. Antelope introductions. Washington Department of Game Status Report 1969. Olympia, WA.

87. Nadler, C. F., R. S. Hoffman, and K. R. Greer. 1971. Chromosomal divergence during evolution of ground squirrel populations (Rodentia: Spermophilus). Systematic Zoology 20:298-305.

88. Newberry, J. S. 1857. Report on the zoology of the route. Chapter 2. Report upon the birds. Pages 73-110 in Reports of explorations and surveys, to ascertain the most practicable and economical route for a railroad from the Mississippi River to the Pacific Ocean...1854-55. Volume 6. Part 4. Zoological report. Number 2. Beverly Tucker, printer, Washington D.C.

89. Noble, D. G., and J. E. Elliot. 1990. Levels of contaminants in Canadian raptors, 1966 to 1988, effects, and temporal trends. Canadian Field-Naturalist 104:222-243.

90. Nowak, R. M., and J. L. Pardiso. 1983. Walker's mammals of the world. John Hopkins University Press, Baltimore, MD.

91. Olson, B. 1976. Status report of the Columbian sharp-tailed grouse. Oregon Wildlife 31(4):10.

92. Olterman, J. H., and B. J. Verts. 1972. Endangered plants and animals of Oregon IV. Mammals. Oregon State University, Agricultural Experiment Station, Special Report, 364:1-47.

93. Osborne, D. 1953. Archeological occurrences of the pronghorn, bison, and horse in the Columbia Plateau. Scientific Monthly 77:5 260-269.

94. Parsons, Z. 1971. Unpublished report on pronghorn antelope transplants in Washington. Washington Department of Game, Olympia, WA.

95. Pasitschniak-Arts, M. 1993. Ursus arctos. Mammalian Species, Number 439. The American society of Mammalogists.

96. Pattee, O. H. and S. R. Wilbur. 1989. Turkey vulture and California condor. In B. G. Pendelton, editor. Proceedings of the Western Raptor Management Symposium and Workshop.

97. The Peregrine Fund. 1999. California Condors to be released on December 7, 1999. The Peregrine Fund. http://www.peregrinefund.org/press/con1299.html.

98. Phillips, A., J. Marshall, and G. Monson. 1964. The birds of Arizona. University of Arizona Press, Tucson, AZ.

99. Pisano, R. 1977. Status of the wolf in North America. International Union for Wildlife Conservation.

100. Plamondon, J. B. 1914. Chinese pheasants in Umatilla County. Oregon Sportsman 2(9):20.

101. Presnall, C. C. 1946. Wildlife conservation as affected by American Indian and Caucasian concepts. Journal of Mammalogy 27:458.

102. Putnam, R. 1928. The letters of Roselle Putnam. Oregon Historical Quarterly 29:242-264.

103. Rausch, R. L. 1963. Geographic variation in size of North American brown bears, Ursus arctos Linnaeus, as indicated by condylobasal length. Canadian Journal of Zoology 41:33-45.

104. Riedman, M. L., and J. A. Estes. 1987. A review of the history, distribution, and foraging ecology of sea otters. In G. R. VanBlaricom and J. A. Estes, editors. The community ecology of sea otters. Springer-Verlag, Berlin.

105. Rogers, P. M., and D. A. Hammer. 1978. Ancestral breeding and wintering ranges of the Trumpeter Swan (Cygnus buccinator) in eastern United States. Tennessee Valley Authority, Knoxville, TN.

106. Rogers, R. 1997a. Washington field notes. WOSNews 48:8-10.

107. Rogers, R. 1997b. Washington field notes. WOSNews 50:7-10.

108. Rogers, R. 1997c. Washington field notes. WOSNews 51:8-10.

109. Rogers, R. 1998a. Washington field notes. WOSNews 52:8-11.

110. Rogers, R. 1998b. Washington field notes. WOSNews 53:8-10.

111. Rogers, R. 1998c. Washington field notes. WOSNews 56:8-10.

112. Rogers, R. 1999a. Washington field notes. WOSNews 59:9-15.

113. Rogers, R. 1999b. Washington field notes. WOSNews 60:11-15.

114. Rogers, R. 1999c. Washington field notes. WOSNews 61:8-11.

115. Rogers, R. 1999d. Washington field notes. WOSNews 62:8-11.

116. Rogers, R. 1999e. Washington field notes. WOSNews 63:8-14.

117. Ryder, T. J., E. S. Williams, K. W. Mills, K. H. Bowles, and E. T. Thorne. 1992. Effect of pneumonia on population size and lamb recruitment in Whiskey Mountain on bighorn sheep. Proceedings of the Biennial Symposium of the North American Wild Sheep and Goat Council 8:136-146.

118. Rymon, L. M. 1969. A critical analysis of wildlife conservation in Oregon. Ph.D. Dissertation, Oregon State University, Corvallis, OR. .

119. S., J. 1881. Shooting in Oregon. Forest and Stream 16:9.

120. Shaffer, M. L. 1981. Minimum population sizes for species conservation. BioScience 31:131-134.

121. Smith, D. 1954. The bighorn sheep in Idaho. Idaho Department of Fish and Game, Wildlife. Bulletin No. 1.

122. Smith, M. R., P. W. Mattocks, Jr., and K. M. Cassidy. 1997. Breeding birds of Washington State. Volume 4 in K. M. Cassidy, C. E. Grue, M. R. Smith, and K. M. Dvornich, editors. Washington State Gap Analysis—Final Report. Seattle Audubon Society Publications in Zoology Number 1, Seattle, WA.

123. Sodhi, N. S., L. W. Oliphant, P. C. James, and I. G. Warkentin. 1993. Merlin (Falco columbarius). In A. Poole and F. Gill, editors. The Birds of North America, Number 44. The Birds of North America, Philadelphia, PA.

124. Spencer, K. T. 1998. Field notes: Eastern Oregon summer 1997. Oregon Birds 24:29-31.

125. Spencer, K. T. 1999. Field notes: Eastern Oregon June-July 1998. Oregon Birds 25:13-20.

126. Spencer, K. T. 2000. Field notes: Eastern Oregon summer 1999. Oregon Birds 26:138-143.

127. Spraker, T. R., and C. P. Hibler. 1982. An overview of the clinical signs, gross and histological lesions of the pneumonia complex of bighorn sheep. Proceedings of the Biennial Symposium of the North American Wild Sheep and Goat Council 3:163-172.

128. Stern, M. A., and G. A. Rosenberg. 1985. Occurrence of a breeding upland sandpiper at Sycan Marsh, Oregon. Murrelet 66:34-35.

129. Suckley, G., and J. G. Cooper. 1860. The natural history of Washington Territory and Oregon with much relating to Minnesota, Nebraska, Kansas, Utah and California, between the thirty-sixth and forty-ninth parallels of latitude, being those parts of the final reports on the survey of the Northern Pacific Railroad route, relating to the natural history of the regions explored, with full catalogues and descriptions of the plants and animals collected from 1853 to 1860. Baillière Brothers, New York, NY.

130. Sullivan, P. T. 1983. A preliminary study of historic and recent reports of grizzly bears, Ursus arctos, in the north Cascades area of Washington. Washington Department of Fish and Wildlife, Olympia, WA.

131. Sullivan, P. T. 1999. Field notes: eastern Oregon spring 1999. Oregon Birds 25:101-106.

132. Suphan, R. J. 1974. Ethnological report on the Wasco and Tenino relative to socio-political organization and land use. Garland, New York, NY.

133. Swainson, W., and J. Richardson. 1831. Fauna boreali-Americana; or the zoology of the northern parts of British America. Part II. The birds. J. Murray, London.

134. Teit, J. A. 1928. The Salishan tribes of the western Plateau. Bureau of American Ethnology Annual Report 45.

135. Thorne, E. T., E. S. Williams, T. R. Spraker, W. Helms, and T. Segerstrom. 1988. Bluetongue in free-ranging pronghorn antelope (Antilocapra americana) in Wyoming 1976 and 1984. Journal of Wildlife Diseases 24:113-119.

136. Tice, B. 1998. Field notes: Western Oregon winter 1997-1998. Oregon Birds 24:96-99.

137. United States Congress. 1929. Control of predatory animals. 70th Congress, 2nd session. House Document No. 497. Washington D.C., U.S. Government Printing Office.

138. U.S. Fish and Wildlife Service. 1982. Sensitive bird species, Region One. U.S. Fish and Wildlife Service, Portland, OR.

139. U.S. Fish and Wildlife Service. 1998. California Condor: Gymnogyps californianus. Wildlife fact sheet series 1. California condor. USFWS, Arlington, VA. http://www.fws.gov.

140. Van Vuren, D., and M. Bray. 1985. The recent geographic distribution of Bison bison in Oregon. Murrelet 66:56-58.

141. Verts, B. J., and L. Carraway. 1998. Land mammals of Oregon. University of California Press, Berkeley, CA.

142. Washington Department of Game. 1975. 1974-1975 big game status report, wolf section. Washington Department of Game. Olympia, WA.

143. Washington Department of Game. 1978. 1978 big game status report. Washington Department of Game, Olympia, WA.

144. Washington Department of Fish and Wildlife. 1995. Washington State Management Plan for bighorn sheep. Washington Department of Fish and Wildlife, Olympia, WA.

145. ———, 1998. 1998 game status and trend report. Washington Department of Fish and Wildlife, Olympia, WA.

146. Wilbur, S. R. 1973. The California condor in the Pacific Northwest. Auk 90:196-198.

147. Wilcox, B. A., and D. D. Murphy. 1983. Conservation strategy: the effects of fragmentation on extinction. American Naturalist 125:879-887.

148. Yoakum, J. D. 1968. A review of the distribution and abundance of American pronghorn antelope. Antelope States Workshop Proceedings 3:4-14.

149. Yocom, C. F. 1952. Columbian sharp-tailed grouse (Pediocetes phasianellus columbianus) in the state of Washington. American Midland Naturalist 48:185-192.

150. Young, S. P. 1944. The wolves of North America, Part I. American Wildlife Institute, Washington D.C. Condor 43(1):75.

19

Characterizing Species at Risk

John F. Lehmkuhl, Bruce G. Marcot, & Timothy Quinn

Introduction

For at least the past three decades, researchers and managers have used various forms of risk and decision analysis for assessing the status and effects of management practices on populations or habitat. Early methods that continue to be practiced today use habitat relationships information to assess changes in macro- or microhabitat area from land management practices or other human activities.[36, 147, 171] Other forms use risk analysis to identify and rank species of conservation concern.[58, 77, 107, 120] Quantitative population viability models have received extensive attention for analyzing probabilities of extinction for single species under current or future conditions,[1, 80] but they are just one of many tools in the risk analysis toolbox. Decision models have used risk analysis to compare extinction likelihood of wildlife populations to the economic cost of their conservation.[49] Many other examples can be found in the literature.

In this chapter, we attempt to describe some of those risk analysis tools, but also strive to provide readers with a basic understanding of the underlying theory and criteria for identifying species at risk. We focus on analytical tools that might be used with available datasets, like the CD-ROM in this book, to meet the risk analysis challenge at different levels of geographic and ecological scale. We define the components of risk analysis, describe the criteria for identifying species at risk, review methods of species risk analysis, examine the attributes of species currently considered at risk, and work several examples of risk analysis using the life history and habitat relationships information in this book.

Defining Risk

In much of the biological literature, the term "risk" really means "likelihood" of some event or condition occurring.[1] However, formally, ecological risk analysis typically differentiates the estimation of outcome likelihood (possibility of change or loss) from management values (or utilities) placed on those outcomes. Outcome likelihood can be estimated quantitatively (e.g., 90% persistence of a nonzero population over 100 years) from simulation modeling (rarely from empirical study), qualitatively from simpler habitat or life history models, or from expert judgment as an ordinal scale (rank order) of outcomes. "Risk" is the likelihood that some

management objective will not be met. Risk is the likelihood of an outcome occurring multiplied by the value given to an outcome. This means that to fully evaluate the "risk" of any course of action, management objectives first must be articulated in terms that can be clearly analyzed.

Distinguishing Risk Analysis from Risk Management

In risk analysis, researchers list possible outcomes, estimate their likelihood under one or more management alternatives, and then weight each outcome with a value. Outcome values are assigned by managers based on social, economic, political, or other goals. Defining what is desired (outcome values) and the likelihood of achieving it (outcome likelihood) helps determine the risk of a decision (Figure 1). In risk management, the manager first describes the values they place on possible outcomes, ideally in conjunction with the risk analysis. Then the manager describes their risk attitude towards seeking or avoiding particular kinds of outcomes. Then, they apply this attitude to the risk analysis results and choose an acceptable course of action that best fits their risk attitude and the outcome values.

Data that fuel risk analyses can be uncertain in several ways. There can be errors of measurement, sampling errors, and natural variation. Data errors of these kinds can result in greater variances in the estimation of the biological factor of interest, which can add some degree of confusion to the risk management phase. How these errors add up and interact is typically ignored in risk analyses, but the problem of error propagation is not to be neglected. Methods to negate the effects include stochastic simulation analysis and sensitivity analysis,[144] use of meta-analysis techniques,[34, 38, 56] partitioning or bootstrapping samples,[111] and, when the problem is simple enough, outright analytic solutions.[115]

A common erroneous view of uncertainty (variance) in risk analysis outcomes is that it appears as a lack of evidence of adverse effects, or even as evidence of no effect. This can be a biologically, and politically, hazardous approach leading, in the worst cases, to species extirpation or extinction. What can the risk analyzer do in the face of such unknowns? True data unknowns mean that no statistical estimation or population projection is even possible, even by using expert judgment. The risk analyzer can skirt the issue by borrowing data from a related species, time period, or geographic location, or can conduct a sensitivity analysis, letting an unknown variable vary over some biologically reasonable range of values and determining the degree to which the outcome (e.g., population persistence likelihood) is influenced by that variable. A more formal analysis of the value of additional information also can be conducted to determine the worth of expending time and money in seeking values of unknown variables, or refining estimates of ones known.

Criteria for Identifying Species at Risk

If current trends in human behavior continue, the twenty-first century may come to be known as a mass extinction event rivaling the extinction event of 65 million years ago. Some estimates suggest that up to one quarter of all species now living may be lost in the next fifty years,[102] mainly through habitat alteration, but also through introduction of invasive species, pollution, and other human activities. Clearly, to avoid such a catastrophe, natural resource professionals must better understand the process of extinction, how humans can affect species' extinction risks, and the attributes that make species particularly extinction-prone. In particular, decision makers and risk analyzers need theory and predictive tools to make wise land-use decisions in short time frames and otherwise data-poor environments, particularly in the next fifty years, when the world's human population is expected to peak. Below, we outline some of what we know and do not know about the process of extinction.

The viability of populations is driven by two basic types of processes: systematic and stochastic. Systematic processes generally reduce the size of populations,[155] and result from the removal of something essential from the environment, such as space, shelter, or food, or from the insertion of something lethal, such as a new predator or poison. Systematic processes tend to reduce populations sizes to levels that make them susceptible to stochastic processes that lead to extinction. Stochastic processes, on the other hand, are random events or a series of events that can dramatically affect birth and death rates within populations.

Systematic Extinction Processes

Primary processes. Of the 485 animals and 584 plant species that have been certified as extinct since 1600,[102] approximately 61% of the animal extinctions and 27% of the plants are island endemics. Most of these island extinctions resulted from Diamond's[29] "evil quartet" of systematic processes: habitat destruction and fragmentation, introduction of exotic species, over-exploitation by humans, and chains of extinction. Although this same evil quartet affects mainland species, island endemics, having evolved in the absence of predators or competitors including humans, are particularly vulnerable to extirpation when exotic predators or competitors are introduced. The spread of

exotic species will only increase with the growth of the global economy.

Landscape change (habitat loss, degradation, and fragmentation) is the primary cause of endangerment for two-thirds of the world's vertebrates currently considered at risk of extinction.[128] Some major habitat types such as the North American tallgrass prairie have been virtually eliminated (converted to other uses) while others such as the heathlands in the United Kingdom and thorn scrub in Sri Lanka have been reduced to less than 30% of their original extent.[141] Currently some of the highest rates of forest conversion to agricultural and urban land are found in the humid tropics,[186] our great storehouses of biodiversity. Other habitat types are changed in less obvious ways. For example, while the amount of forest cover is increasing in Europe and the former USSR, the clearing of virgin forest worldwide has increased exponentially since the 1500s.[99] Of the 4 billion hectares estimated to be in forest cover today, 1.7 billion are secondary forests or plantations.[185]

As an area is converted from natural vegetation to some other use, the remaining habitat is commonly divided into smaller parcels. The effects of these actions on biodiversity are diverse,[32, 131, 146, 181] but often discussed in terms of three related effects: fragmentation, edge, and isolation. It is important to note that although human alterations to the environment are often considered deleterious to most species, the effects of landscape change including fragmentation, isolation, and the creation of edge habitat are species-specific. What constitutes a patchy

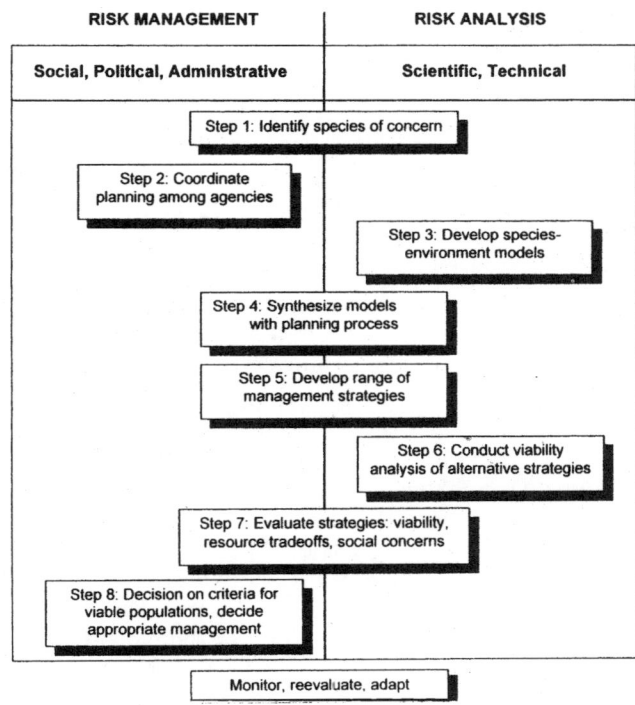

Figure 1. Example of risk management and risk analysis as applied to population viability planning (from Marcot et al. 1986).

environment for one species may constitute a homogeneous environment for another species. Likewise, degraded habitat for one species may constitute ideal habitat for another. Moreover, some species will benefit from human activities either directly by the creation of habitat (edge for example) or through indirect effects such as reducing competition or predation pressure.

Fragmentation. Habitat loss results in the reduction of habitat and in discontinuities in the distribution of remaining habitat, that is, fragmentation. Fragmentation can be viewed as the creation of remnant vegetation patches surrounded by a matrix composed of different habitat or land/uses.[146] For illustrative purposes, consider a single species on a landscape where each parcel of land is either classified as habitat or non-habitat. At one end of the fragmentation spectrum and in the simplest case, patches of habitat may be so small that they support no individuals (i.e. smaller than a single home range, or missing some important habitat feature such as nesting substrate). In areas that are slightly less fragmented, small populations may persist in some of the larger habitat patches. However, these small populations may be at increased risk of extirpation due to stochastic processes described below.

A collection of small populations (or subpopulations) each inhabiting its own patch may constitute a metapopulation. Although there are many variations on the metapopulation theme, the distinguishing characteristic of metapopulations is that population dynamics within a patch are relatively independent of dynamics in other patches. That is, dispersal among patches is a relatively rare event that has little effect on the extant population's growth rate. In theory, persistence of a metapopulation is related to the rate at which each patch (subpopulation) goes extinct and is recolonized by individuals from another patch. The metapopulation persists if recolonization is greater than extinction rate.

Edge effects. As patches of habitat become smaller (and the matrix expands), the ratio of edge habitat to interior habitat increases. Edges can be thought of as areas of abiotic and biotic gradients between patch habitat and the matrix. For example, in a clear-cut matrix with forest patches, the outside edge of the forest patch has conditions (e.g. light humidity, temperature, plant species composition) more like those in the clear-cut than in any other part of the forest patch. Edges essentially reduce the amount of habitat for species requiring interior conditions of the patch. Shade tolerant species may become restricted to the interior parts of the habitat patches, with different species requiring different distance from the edge.[73, 136] Edge habitat may have different types of species interactions than either the matrix or the patch interior. For example, nest predators and avian brood parasites are more common along forest edges than in patch interiors.[169, 180] The importance of this phenomenon depends on the scale of edge effects: some outside influences penetrate only a few meters from the boundary (e.g. sunlight),[82] whereas others infiltrate hundreds of meters in to the interior (e.g., wind,[25] nest predators).[180]

Isolation. Isolation refers to the accessibility of habitat patches to organisms, and mostly is related to two measures: distance between patches and the composition of matrix lands surrounding those patches. All else being equal, as distance between patches increases, or suitability of the matrix land as habitat decreases, the more isolated a patch becomes to a particular species. Demography within and dispersal among patches, which is a function of isolation, will often determine the response of individual species to fragmentation.[63]

Upon isolation, a patch is likely to have more species than it can sustain in the long-term. Species may be lost from the patch in a process called species relaxation.[112] The most rapid extinctions are likely to occur in species that depend entirely on native vegetation, require large (relative to the patch size) territories, and those that exist at low densities and are thus vulnerable to stochastic processes.[146]

Stochastic Extinction Processes

Demographic stochasticity. Stochastic, or random, processes are usually the proximate causes[155] of extinction for populations already small from systematic processes. There are four types of stochasticity: demographic, environmental, natural catastrophes, and genetic.[37, 160, 161, 164] Demographic stochasticity is the result of chance variation in vital rates. This form of stochasticity affects individuals randomly and increases the variance in mortality and/or natality rates and thus increases the probability of extinction.[43, 74, 101] For example, a species consisting of very few breeding pairs by chance alone could produce offspring all of the same sex, effectively reducing the rate of successful reproduction in the next generation. Alternatively, consider a species where the annual probability of mortality based on the population as a whole is 50%. If we represent the fate of individuals as the result of a coin toss, it is easy to demonstrate that 5 tails in row (where tails equal death) are easier to obtain than 100 tails in a row. The smaller the number of coin tosses (smaller the population) the more likely they will all come up tails. Demographic uncertainty is an important consideration usually only for relatively small populations[152]—on the order of 100s.

Environmental stochasticity. Environmental stochasticity is the variation in mean fecundity and survivorship rates across space and time, although it is most commonly thought of as temporal variation at a single site. It results from changes of environmental or habitat quality and interspecific interactions with competitors, predators, parasites and diseases.[101] In contrast to demographic stochasticity, environmental stochasticity can affect individuals in the population more or less equally to increase the variation in birth and death rates, and thus variation in abundance.[43, 74] Natural catastrophes, such as floods, fires, and droughts, often occur at random intervals and can have dramatic and usually adverse effects on small populations. Also, because some catastrophes such as hurricanes are very large, they can adversely affect even very large populations. However, as the geographic range of a species increases (i.e. larger population) there is less likely to be spatial correlation of environmental variation across that range. Thus, large populations may be buffered against environmental stochasticity occurring at relatively small spatial scales.

Genetic stochasticity. Genetic stochasticity is the variation in frequency of alleles within a gene pool. The type of genetic stochasticity that conservation biologists are most concerned with is loss of allelic diversity. This arises from a reduction in the "effective population size" (that is, the effective number of breeding adults) with consequential deleterious short-term effects of random allele fixation, founder effects, inbreeding or outbreeding on reproductive success, and long-term losses of genetic variation (heterozygosity) and adaptability to environmental change from genetic drift.[37, 148] Inbreeding has been shown to reduce reproductive success in small or isolated populations, including rainbow trout,[70] ungulates,[6, 133, 134] red-cockaded woodpeckers,[48] and pit vipers.[84] In the past few years, the emphasis on genetics in conservation biology has decreased for a variety of reasons, the most important being the conclusion that populations large enough to be demographically self sustaining are likely to remain genetically viable indefinitely.[55, 74, 110] However, the relationship between long-term viability and genetic diversity remains unclear and is an active area of research.

Interactive effects on populations. The population size at which stochastic processes become important is likely to vary over time and space according to factors affecting successful reproduction and dispersal.[44] For example, given two populations of equal size, high rates of reproduction or immigration in one population would allow faster recovery from disturbance than where reproduction or immigration are slow. Such population size thresholds also are species-specific and largely determined by the interaction among the particular characteristics of a species life history and population structure, and environmental conditions.[42] Birth and death models of population demography indicate that the risk of extinction declines rapidly with increasing population size.[44] As an example, Morrison et al.[115] analyzed the exponential rate of decline of native Hawaiian bird populations, and demonstrated that species with very small populations decline at a far faster rate than do larger populations. Extinction risk as a function of environmental stochasticity and large-scale catastrophes also will decrease with increasing population size, but at a slower rate than with demographic effects. In reality, extinction of small populations is probably the result of interactions among these four processes,[42] although their relative contribution to extinction is largely unknown and possibly quite complex.

For example, consider the extinction of the heath hen (*Tympanuchus cupido cupido*), a species that was fairly common in the New England States at the time of European settlement.[157] By 1900, there were fewer than 100 birds, all of which were found on Martha's Vineyard

(systematic processes). In 1907, a portion of Martha's Vineyard was set aside as a refuge for the bird and a program of predator control was implemented. The population responded, and by 1916 had reached the size of more than 800 birds. In 1916 a fire (natural catastrophe) destroyed most of the heath hens' nests and during the following winter the birds suffered unusually heavy predation (environmental stochasticity) from a relatively high concentration of a common heath hen predator, the goshawk. The combined effect of these events reduced the population to 100-150 individuals. In 1920, after the population had increased to about 200, disease (environmental stochasticity) reduced the population to below 100. Though the species persisted a while longer, by 1932 the last survivor was gone. In the final stages of the population decline, the birds appeared to have become increasingly sterile (inbreeding depression) and the proportion of males increased (demographic stochasticity). Which of these events served as the *coup de grace* is unknown. Theory suggests that in general, the probability of extinction for small populations is highest from environmental effects, followed by demographic stochasticity, longer-term genetic effects, and lastly catastrophes,[42, 72] although this hierarchy itself varies by species and condition.

In contrast to systematic pressures, the magnitude of stochastic threats depends more on population size than on life history traits of a species. In part, this is because life history traits are generally selected traits imparting persistence of the species within particular bounds of environmental conditions. Thus, a general theory of stochastic processes has been developed based on population size alone. Systematic processes, on the other hand, are likely to be as different from each other as the life history traits of species affected by those processes. It may not be possible to construct a universal theory of population decline comparable to the stochastic theory of small populations.[24] However, understanding the mechanisms of extinction (i.e. why some species and not others decline in the face of systematic extinction processes) is fundamentally important to conservation biologists and wildlife managers.

Criteria of Species at Risk

Population vulnerability analysis. The basic question is: why do some populations decline and others do not? We can use the conceptual "population vulnerability analysis" framework to determine criteria and processes for assessing extinction risk.[42] The risk of extinction due to systematic processes is likely a function of the interaction of: (1) the physical and biotic environment of the individual and population; (2) the population phenotype, including the species life history, behavior, morphology, physiology, and habitat requirements; and (3) how the first two factors interact to determine population distribution and individual fitness.[42, 174] The relative importance of these three factors is a matter of debate and probably depends on the particular blend of species characteristics and environmental characteristics. Despite the potential complexity of understanding the mechanisms of extinction, a number of studies have attempted to reduce the number of factors into an easily understood and measurable suite of characteristics. In our case, we are interested in determining *a priori* whether a particular species is more extinction prone than is another species. The question becomes: how do certain population distribution patterns and life history characteristics enable species to persist in the face of the most common threats to their viability and the ones which humans can most control—habitat loss and fragmentation?

The probability that a population will persist with habitat loss is largely a function of population density or abundance: the greater abundance or density the lower the risk of local extinction.[27, 28, 119, 125, 163, 170, 182] That is, larger populations, and populations distributed in a metapopulation pattern (with low correlation of stochastic factors and environmental conditions among population centers) are less vulnerable to stochastic processes than are small populations or populations lacking metapopulation structures, all else being equal (see discussion above).

The interaction of population distribution and geographic scale of management concern also is an important issue in determining and managing risk. For example, species that are globally common, but regionally or locally rare, may not be of concern at the large scale, but may be important for local management to preserve subspecies and the evolutionary potential of the species at the edge of its range.[95] On the other hand, for species that are locally common, but rare at broader scales, risk management might be important at both local and higher geographic scales to maintain "source" populations or strongholds of viability.

In addition to population structure, life history characteristics are important attributes of the "population phenotype" that may affect persistence.[96, 106, 153, 156] Vulnerability to extinction in isolated habitat patches may increase with large body size;[27, 28, 125] low fecundity;[125] specialization on patchy resources;[64, 168] behavioral patterns requiring the formation of large groups and thus concentrated resources, dependence on keystone or link species, or occurrence at the edge of a species range;[170] temporal variation in population size;[28, 43, 44, 66] and metabolic rate.[182] Many of these life history traits are related to body size.

Based on his study of colonization and extinction rates of birds on islands, Pimm et al.[125] summarized the theoretical underpinnings of extinction risk in four predictions. All other things being equal, extinction rate will be greater:
- In small populations than in large populations (see above),
- At low population densities, in small-bodied, fast growing, short-lived species than in large-bodied, slow-growing, long-lived species,
- At high population densities in large bodied, slow-growing, long-lived species than in small-bodied, fast growing, short-lived species, and

- In populations with high rather than low variability in numbers (but see Tracy and George).[174]

Interactive effects. While these predictions can help us formulate testable hypotheses, they say nothing about the processes which drive extinction, or how these life history and other factors work to increase risk. Moreover, the relationship between factors that limit population size and distribution and chances of extinction are complex, at least for species that have existed in small numbers within small geographic ranges for millions of years. Burgman et al.[18] suggest that factors that lead to relative rarity or restricted geographic ranges may be independent of those that cause extinction. Even those species that are tolerant of a wide range of environmental conditions may lack available habitat.

In addition to these general characteristics, Akçakaya et al.[1] identified several properties that may make species especially susceptible to human impacts:
- Habitat overlap: species that occur in the same types of habitats that people or introduced superior competitors prefer are more at risk than other species occurring in less preferred areas. Humans tend to populate areas with fertile soils and benign climates such as major rivers valleys and coastlines. These areas are subject to waste disposal and urban and agricultural development.
- Harvesting: species that taste good or otherwise valuable to humans are often at risk from hunting. Human exploitation may exacerbate species' properties, which put them at risk. For example, sea turtles require 10-40 years to reach sexual maturity, so removal of breeding adults rapidly depletes a population and slows the rate of recovery when exploitation ends (e.g., populations stay small long periods of time).[71]
- Large home range requirements: animals with extensive home ranges often occur at low densities and are susceptible to the reduction and fragmentation of habitat.
- Limited adaptability and resilience: species with limited dispersal capabilities, limited reproductive capacity (e.g. sea turtles), or narrow and inflexible habitat requirements may be incapable of rapid recovery from disturbance making them prone to extinction in human dominated landscapes.

To this list we can add dispersal ability and factors influencing successful dispersal and recolonization: species with poor dispersal ability are subject more to local stochastic influences than are species with broad dispersal ability; that is, dispersal ability influences the distribution patterns of a species.[53]

Lessons for management. There are two important lessons from small-population theory. 1) managing small populations requires enormous effort, as does conducting research on them to determine causes for decline. The best time to ensure viability is when populations are still large. 2) Managers should attempt to reverse or mollify adverse effects of systematic threats to persistence and increase the effective size of threatened populations as rapidly as possible.[55] Additionally, in some cases, it is advantageous to establish more than one population center to avoid losing the entire population to single environmental stochastic influences or catastrophes. In extreme cases, captive propagation followed by reintroduction into restored habitats, although terribly expensive, may be one of the best recourses.[149]

Methods for Determining Species at Risk

Methods for wildlife risk assessment vary in taxonomic scope, procedural complexity, and objective. The scope of assessments ranges from single-species assessments to multi-species ecosystem or biodiversity assessments in which many species are considered in the same effort. Procedural complexity varies from simple deterministic estimates of risk (e.g., consistent loss of habitat area, and use of ordinal risk rankings) to complex probabilistic estimates for explicit time periods that incorporate uncertainty related to demographic and environmental stochasticity, or management processes (Figure 2). Risk assessments also vary in their objectives, from estimates only of extinction risk to estimations of management risk that integrate extinction risk with social, economic, or other biological considerations. Several of the methods to assess risk of habitat or population loss described below, such as habitat relationships assessments and models, are covered in detail elsewhere in this book; here, we give an overview of these and other methods for analyzing risk of loss.

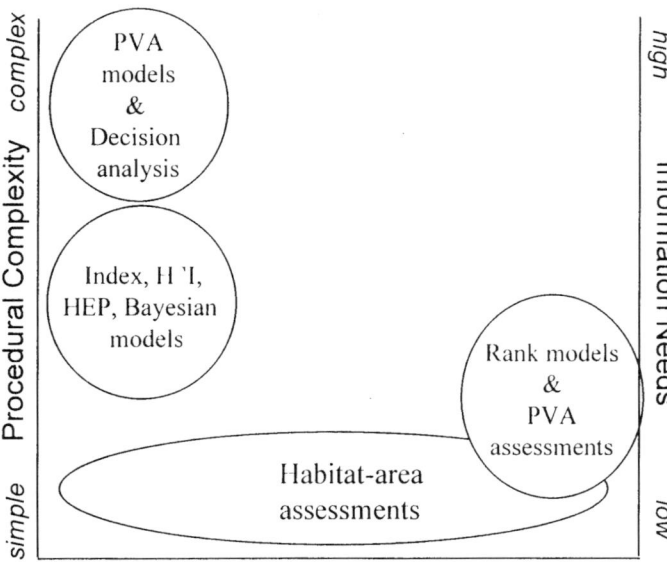

Figure 2. Methods of risk assessment for animal and plant species vary along gradients of taxonomic scope and procedural complexity. Additionally, methods vary in objective from estimating ecological risk of extinction to management risk.

Habitat Relationships Assessments

Wildlife habitat relationships (WHR) assessments or models use a variety of procedures to estimate the effects of changes in habitat quantity or quality under management alternatives. The objective of WHR models is to assess how changes in vegetation composition or structure from management might result in habitat loss for on or more species. Multiple species typically are analyzed in guilds or groups of species with similar habitat relationships, foraging or nesting behavior, movement, or other life-history attributes. Habitat relationship assessments can be applied at stand, landscape, and regional scales, although Raphael and Marcot[90] found such models most accurate and appropriate for use with some species at landscape or larger scales.

The methodological roots of wildlife habitat relationships programs for vertebrates are in the work of Patton,[122] Thomas et al.,[171] Lehmkuhl and Patton,[75] and Brown,[14] among others. Indexed WHR models include the habitat suitability models (HSI),[147] the habitat evaluation procedures (HEP),[36, 178] and habitat effectiveness models (e.g., elk models by Thomas et al.,[171] Wisdom et al.,[184] Thomas et al.)[173] Morrison et al.[115] review these models.

Habitat is typically defined as the sum of area in plant communities and their structural variants (macrohabitats). Adding to the suitability of an environment are habitat elements (microhabitats) used for feeding, nesting, and resting. Analysis usually determines the change in area of available habitats under management alternatives and describes the implications for associated species. Versatility indices of habitat use are the number, or breadth, of macrohabitats or habitat elements used by individual species. Versatility indices are used to assess implications of habitat change for species that use few habitats (low versatility) or several to many habitats (higher versatility). Index models (HEP, HSI) independently describe or index the value (typically scaled 1 to 10 or 0 to 1) of several macrohabitat and microhabitat attributes (e.g., westside conifer forests and snag density therein) then calculate a summary index of habitat value. Risk is determined from HEP and HSI approaches by comparing model index values against an independently determined threshold of habitat change considered to put the species at risk.

It is problematic, however, in HEP and HSI models to identify such thresholds. Thus, HEP and HSI type models are better used to qualitatively assess alternative conditions in a rank order comparison, rather than to estimate absolute population response. In some cases, historical and projected trends in area of forest habitats have been related to population density in habitat types to depict potential trends in the abundance of vertebrate taxa and functional groups over large areas.[137, 139] Similar trends and analyses pertain to other, non-forest environments.

Rank Models

Rank models work within a population vulnerability framework to evaluate risk.[9, 42, 150] These models provide a ranking of extirpation or management risk, using attributes of population structure and distribution, life history, habitat-use versatility, or habitat area and spatial pattern. For ordinal-rank models, each factor is usually assigned an integer score, or rank, relative to its contribution to extirpation or management risk, for example, on a scale of 1 to 10. Quantitative data on habitat area or population size, as well as qualitative information on life history attributes, can be used to develop the scores. In this approach, multiple species typically are evaluated at the same time using the same variables. The scale of application is most often regional or larger, but could be applied down to watershed or smaller scales. Ordinal-rank models have been developed to assess the impacts of large-scale habitat loss and fragmentation on potential population persistence.[51, 61, 77]

The International Union for the Conservation of Nature and Natural Resources (IUCN) developed a categorical rank system[57] for their Red Lists[58, 59] to assess worldwide species extinction risk based on population size and distribution (see Mace and Collar[83] for an application) (Figure 3). The IUCN system has been further developed for local application to assess extinction and associated management risk within individual countries or other artificial management units.[41, 124]

The Nature Conservancy (TNC) developed a global categorical-rank extinction and management rating system for the Natural Heritage Program that is based primarily on qualitative estimates of plant and animal population abundance and distribution.[98] The system also ranks "site biodiversity significance" and "site protection urgency" on a rank scale of 1-5 for use in developing management priorities. Other ordinal-rank models have

been developed that integrate extinction and management risk attributes to set state-wide or global management priorities for individual species.[4, 20, 62, 81, 114, 120, 135]

Population Viability Analyses and Assessments

Population viability analysis (PVA) is a formal modeling process for quantifying the likelihood that a species will persist for a given time into the future.[1, 9, 42, 150, 152] In PVA, attributes of population structure and demography, life history, genetic structure, and habitat quality and quantity typically are used in models combining population demography and geographically referenced metapopulation dynamics to assess viability likelihood and extinction risk. A very significant innovation that distinguishes this form of risk assessment from that based on WHR models and rank methods is a consideration of uncertainty arising from natural variation or unpredictability in population and environmental processes. Sometimes, uncertainty is addressed by depicting viability as likelihoods of extinction over a specified period of time, rather than as a single deterministic estimate for an unspecified time period as with rank methods, sensitivity analysis of population parameters, or estimating the variance of modeling analysis outcomes. However, the formal PVA approach seldom accounts for propagation of error terms and general lack of scientific knowledge on the ecology of a species.

For some authors, the process of PVA has been associated with determining the "minimum viable population" (MVP), which is presumed to be the minimum number of individuals that ensure a population's persistence for a specified length of time.[150] The value of the MVP concept has proven very limited, however, because the "minimum" population size is not a simple number, but varies with population parameters, the acceptable probability of persistence, the length of planning period,[42, 79] and how populations and their environments change unpredictably. Also, acceptable persistence likelihoods and planning period durations are functions of the amount of risk managers are willing to accept in trading species viability for other values; thus, the "minimum" viable population size largely is a function of conservation policy instead of pure science.

Typically, quantitative PVA models are developed for single species because of model complexity and the need for detailed data on population dynamics and habitat relationships. Several less-quantitative approaches have been developed to assess viability for multiple species— these may be termed Population Viability "Assessments" to distinguish them from the more stochastic and demographic modeling-based PVA process. The scale of analysis in population viability assessments can vary from regions for wide-ranging species with sparse populations, such as the northern spotted owl[94, 177] to local site conditions for less-mobile species with locally disjunct populations.

Population viability assessment was developed to accommodate the varying availability of species data and

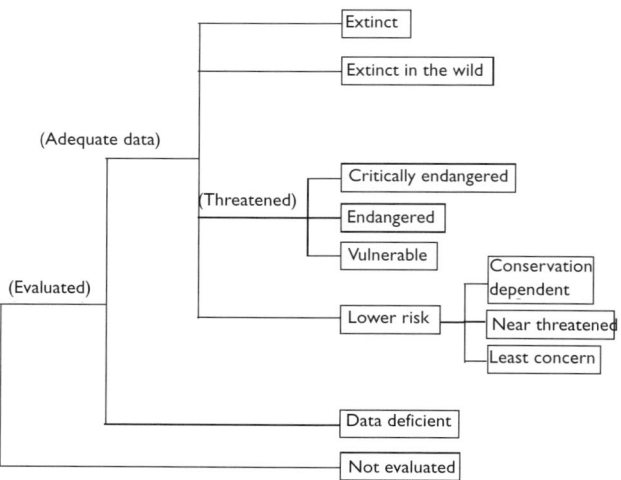

Figure 3. Rank categories of ecological risk developed by the International Union for the Conservation of Nature and Natural Resources (IUCN 1994) for the Red List of threatened and endangered species.

the needs and constraints of scientists and managers while meeting the basic criterion for estimating the likelihood of persistence over a specified period of time. For example, expert-opinion approaches were used to estimate the likelihood of persistence of the northern spotted owl and other species associated with late-successional forests in the coastal Pacific Northwest.[33, 61, 94, 172, 177] Subsequent regional assessments of multiple species for the interior Columbia River basin,[76] and southeast Alaska[154] adapted and modified methods developed for the Pacific Northwest.[33]

These approaches typically have had expert panelists score the relative abundance and distribution of habitats or populations at particular points in time under alternative management scenarios. The scores are described by several outcome classes to depict the likelihood of persistence under management alternatives. Uncertainty of viability likelihood in one approach[76] was estimated as the variance among and within viability scores of expert panelists. In a current effort, however, this "black box" of expert opinion is being opened by using Bayesian belief network models of species' interactions with their key environmental correlates as a way to measure habitat or population potential across landscapes over time (B. Marcot, pers. comm., USDA Forest Service, Portland, Oregon).

Model forms of PVA quantify the probability of persistence for single species by using mathematical and simulation models. Those forms explicitly model the probability of persistence as a function of population structure and distribution, genetic structure or processes, habitat quality and quality, or life history attributes. Uncertainty can be evaluated with sensitivity analysis that varies parameter values within a range of known or hypothesized values.[1, 79, 104, 167] Random, or stochastic, effects are modeled from hypothesized (or, less commonly, empirically fit) probability density functions. Several programs have been developed to model PVA, each with particular strengths and weaknesses.[78, 79, 113] Examples of PVA can be found in the literature for birds,[3, 7, 11, 12, 13, 23, 45, 48, 89, 103, 142] mammals,[5, 16, 21, 22, 39, 52, 67, 69, 72, 80, 97, 121, 159, 187] reptiles,[162] fish,[8, 26, 31, 100, 140] invertebrates,[116] and plants.[10, 17, 19, 109, 117]

It is our considered opinion that the quantitative PVA approach is useful primarily in uncommon cases where species have been previously screened to be at high risk, and where data on their population demography and genetics, and variations in their environment and habitats are known and can be adequately modeled. For most cases, however, the more qualitative population viability assessment approach is likely to be adequate for policy decision-making at broad scales.

Decision Analysis and Knowledge-Based Systems

Decision analysis integrates extinction risk assessment with management risk assessment. Biological and management uncertainty, in terms of probabilities and their variation, are used to calculate expected values of species abundance under several management alternatives.[79, 87, 132] Decision and knowledge-based systems differ from quantitative PVA models by using rigorous steps to capture and quantify expertise, and by explicitly incorporating management (as opposed to ecological) risk. As with PVA, decision analysis models are usually too complex to use with more than 1 species at a time, and are useful at watershed to regional scales.

Decision-tree risk analysis is the most commonly used procedure, and has been done for grizzly bear,[88] black-footed ferrets,[85] tigers,[86] and Sumatran rhinos.[87] Expected value of perfect and sample information, and Bayesian statistics[46, 168] are other methods that may be useful for wildlife risk analysis.[93] Expert systems and other expert advisory approaches are gaining acceptance as important tools when used to analyze risk related to specific problems of wildlife resource management.[93, 108]

Toxicological risk assessments

The literature on toxicological risk assessments for wildlife is relatively well established.[124, 143, 175, 176] Procedural standards have been set by the U.S. Environmental Protection Agency for laboratory and field risk assessment of environmental contaminants on wildlife.[35, 176] Risk is quantified[175] as (1) dose-response evaluated by, (a) the ratio of environmental concentration over lethal concentration resulting in 50% mortality of test individuals (LC_{50}), and (b) the ratio of toxicant consumption over the lethal dose resulting in 50% mortality of individuals (LD_{50}); and (2) by the environmental exposure quantified as the amount of toxicant in the environment over the amount required for LD_{50}. Environmental exposure is a function of population and life history attributes, as used for other risk assessments. Risk increases as these ratios approach 1. Those metrics of risk do not quantify, however, the interaction of toxicants with other factors or the important population consequences of sub-lethal effects.

Single species are the focus of toxicological assessments. Application is usually at small scales of individual contaminant sites or treatment areas, as in the case of pesticides. Watershed or regional applications would sum the lower-order effects at individual sites. Population models are increasingly being used to assess toxicological risk.[30, 40, 50, 68, 127, 158, 179]

Advantages and Disadvantages of Methods

WHR and rank models. Simple WHR habitat models or rank models that rate species risk to develop priorities for conservation or assess generalized risk from specific threats (e.g. forest fragmentation) have the advantages of being broadly applicable, rapid, and incorporating ecological data, theory, and local expert knowledge when research data are few. Simple ranking techniques usually are easily understood by managers in other disciplines, and can be applied to estimate risk for many species over large areas.

However, the generality of the procedure brings with it disadvantages. Information based on expert opinion may be biased by personal experience or tacit motivational bias of the contributors. When a common set of life history, population, or habitat attributes are used to examine risk for a group of species, there is the potential that an attribute that is critical for the persistence of one species might be missing because it was generally considered not important. Likewise, a model developed for one place might fit poorly in other areas. Simple WHR and rank models do not incorporate variation in risk associated with variability of species populations, environmental conditions, or uncertainty in species responses that is necessary for application to small scales of management, such as stands and sub-watersheds.

PVA models. Quantitative risk analyses, such as single-species PVA models, require detailed life history and population information, consider variation in assessing risk, and often examine effects of management alternatives. Variation in management and species responses to risk factors is specifically estimated as likelihood, or through sensitivity analysis. PVA models can help clarify assumptions, create testable hypotheses relative to model parameters and processes, synthesize and integrate knowledge, and demand explicit and rigorous reasoning.[9]

These potential advantages of PVA models are often offset by poor documentation, high cost, or limitations of the methodology, which have been amply discussed in the literature.[2, 11, 24, 47, 66, 74, 104, 105, 113, 167] At the heart, too, persistence likelihood is not strictly an empirically verifiable parameter—you cannot conduct a field study or experiment to verify the risk levels stated in a PVA without putting at risk the very populations you are trying to conserve. Thus, PVA is used under the presumption that its mosaic of models and concepts, which individually can be derived from sound theoretical or empirical evidence on other species, together constitute a useful approach for describing changes in some scarce species of interest.

Decision analysis. Alternatively, highly structured procedures of risk assessment using decision analysis have the advantage of explicitly stating key assumptions and possible management alternatives in a standard format.[85, 87, 90, 92] Management actions and species responses are expressed as probabilities with a defined range of expected values and variation. Different types, sources, and units of management and ecological information can be incorporated to facilitate management decisions.

However, there are several disadvantages of decision analysis when used in population viability assessments.[92, 118] Probabilities are often difficult to accurately assess by use of expert opinion, and assumptions to quantitatively estimate probabilities may inject too much additional uncertainty. Also, small changes in probabilities may greatly affect the results of the analysis. Not all management alternatives and future environmental conditions may be foreseen, thus biasing the management decision. Also, it may be difficult to define specific management objectives for individual species and to quantify outcomes with respect to habitat conditions from biological, and especially social and political, viewpoints.

Improvements Needed in Assessing Wildlife Risk

Improvements in wildlife risk assessment can be made in several areas. There is an enduring need for better information on species' population structure and the range of variability and response to environments over space and time. Also needed for model development are data on habitat selection and the impacts of management on both population structure and habitat selection. Data also are needed for verification of models—a critical and often overlooked aspect of model building because of the difficulty of obtaining independent data at appropriate spatial or temporal scales. Long-term datasets, such as Breeding Bird Surveys, waterfowl surveys, and other data from formal monitoring programs may provide useful data for model verification. Data for model building and verification can be gained through conventional research, but also through better monitoring of management outcomes within the framework of adaptive management.[54] Both approaches aim to provide reliable knowledge using basic scientific methods; hence, the perception that research is for scientists and adaptive management for managers is a false dichotomy that hinders progress in the field.

Just as important is the need for methods to better estimate economic and social values of wildlife, and explicitly incorporate the costs incurred by alternative management practices.[118, 165] Central to the goal of ecosystem management is defining essential ecosystem components, linking processes, and endpoints or desired future conditions of those properties.[118]

Characterizing Risk Species and Their Attributes

Below are several examples of methods for characterizing species at risk, determining the specific attributes that distinguish high-risk species from other species, and comparing risk from several management alternatives. The examples are not the only ways risk analysis might be done (see the methods section of this chapter), but are intended to illustrate approaches to the problems. The first two examples use different but related procedures to differentiate and classify imperiled, secure, and intermediate-risk species, and develop a rule set for classifying other species. A third example uses principles discussed in the section of this chapter on criteria for identifying species at risk to score and rank potential risk based on life history characteristics, when prior knowledge of risk is unavailable. That same approach could be used to rank management priorities. A fourth example describes a population viability assessment used to assess land management alternatives.

For the first three examples, we developed a database for a sample of 60 species that were selected based on criteria shown in Appendix 1. A sample of 20 threatened, endangered, or sensitive (TES) species in Oregon and Washington (e.g., Photo 1) were combined with a sample of 20 species we considered secure (e.g., Photo 2) and 20

Photo 1. Bobolink (Dolichonyx oryzivorous) in Shrub-steppe habitat. An example of a "Group I" species, i.e., species known to be imperiled. (Photo: Bruce G. Marcot)

Photo 2. Ensatina (Ensatina eschscholtzii) in Westside Lowlands Conifer-Hardwood Forest habitat. An example of a "Group II" species, i.e., species very likely secure. (Photo: Bruce G. Marcot)

Photo 3. Sharptail snake (Contia tenuis) in Southwest Oregon Mixed Conifer-Hardwood Forest habitat. An example of a "Group III" species, i.e., species with intermediate security. (Photo: Bruce G. Marcot)

species that had potentially intermediate viability risk (e.g., Photo 3).

For the species in the example database, we extracted a subset of the life history and habitat relationships fields from the matrixes (CD-ROM with this book) that correspond to attributes associated with variation in viability, as discussed in the section of this chapter on criteria for identifying species at risk. We combined some matrix values for a few fields to somewhat simplify the example analyses. Where field values were given as "unknown" we used the most likely value, based on our knowledge of species, to completely fill the data matrix for the example analysis. We designated primary fields in the example database as those most likely to account for risk levels based on our experience, and some secondary life history fields that could be important attributes of species at risk. Other fields were purely for information to interpret the analysis. We calculated two "versatility" indices of habitat cover and structural-type use from the data given in the habitat matrixes. The indices were calculated as the percentage of all habitat cover or structural types that the animal is known to use.

Using Classification Trees to Determine Influences on Species Risk

This first example shows how a classification tree procedure[183] can be used to determine the relative contribution of life history, population distribution, and versatility attributes in the matrixes to known risk-group membership. The results of that analysis, then, can be used to develop a rule-set to predict risk-group membership for species with unknown risk potential. Our example concentrates on comparing the attributes of TES species with species in secure or potential intermediate-risk categories. It is important to remember that this and the following examples are merely illustrations of what can be done with the larger list of species and attributes that might be used in a real situation.

The classification tree procedure worked in a way similar to a dichotomous key used for plant identification: it started with all the species in a group, then sequentially split each subsample based on the most discriminating variable until no significant splits could be performed. The "proportional reduction in error" (PRE) statistic described the goodness-of-fit, or strength, of the model, similar to the conventional R^2 statistic, varying between 0 and 1, with 1 indicating a perfect fit of the model. The classification tree figure showed the splits and relative number of species within each subsample. Crosstabulation and t-tests also were used to examine the relationship between each life history and versatility field and risk group

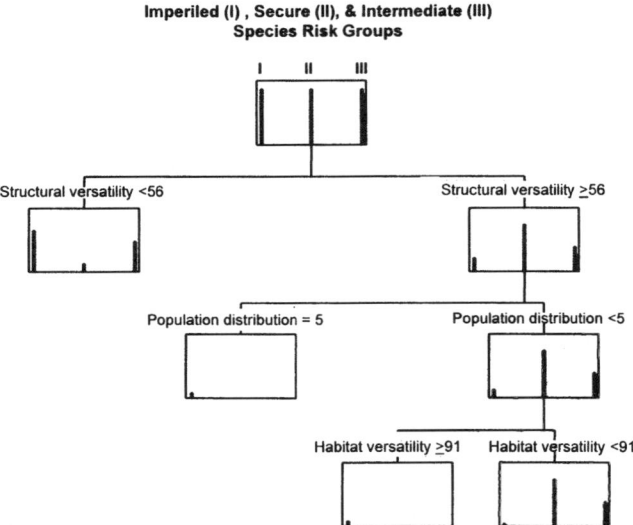

Figure 4. *Classification tree of 60 vertebrate species in Oregon and Washington known to be imperiled, secure, or of potential intermediate viability risk. The classification used life history and habitat versatility attributes from the matrices. See text for an explanation of the species and variable selection processes. Risk groups are represented by bars in the order of imperiled, secure, and intermediate; the height of the bars indicates the number of species in each group and box, starting with 20 species in each group.*

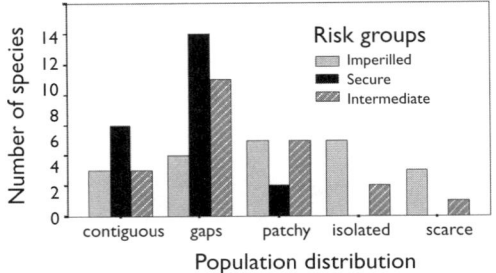

Figure 5. *Numbers of vertebrate species in the risk analysis sample dataset by population distribution category and viability risk group.*

Other than habitat and structural versatility, population distribution was the only other variable that differed among risk groups (Figure 5). Imperiled species largely had patchy, isolated, or scarce distributions; whereas, secure species mostly were contiguously distributed or had gaps in their distribution. Distributions of intermediate species varied between those 2 extremes.

A Rule Induction Approach: SARA—Species At Risk Advisor

This example discusses the use of a rule induction approach developed by B. Marcot to identify species at risk based on their life history attributes. Rule induction is a knowledge-base modeling approach for devising an optimal rule to distinguish among categories based on a set of example cases. For example, Stockwell et al.[166] used rule induction to predict different density levels of gliders (*Petauroides volans*) in Australia. In our example, the rule distinguishes among the three risk levels (imperiled, secure, and intermediate) and the examples are the 60 sample species used in the data set.

There are many kinds of rule induction algorithms, particularly in the area of knowledge-base and expert systems programming and statistical analysis.[60] Used here is the Quinlan ID-3 rule-induction algorithm.[123, 129] This approach is similar to the classification tree example presented above in that it identifies an optimal division of the known cases based on successively identifying variables and their values that best distinguish among remaining cases. What results is an optimal rule, much like a botanical key, that leads to final categories of the three species risk levels. The ID-3 rule-induction algorithm and other related knowledge-base approaches, however, differ from the classification tree analysis in that they can handle both categorical and ordinal data, as well as continuous (ratio-scale) data.

The knowledge-base expert system programming package 1st Class (1st Class Expert Systems, Inc., Wyland, MA; Release 3.65) was used to do the rule-induction analysis. In general, rule induction and use of knowledge base modeling differ from classification tree analyses in several ways. First, a knowledge base programming approach can produce a user-friendly interactive query system based on the optimal rule. In this example, the query system produced is named SARA—the Species At

membership. Differences among groups were reported if significant at P≤0.05.

Imperiled species in our sample dataset were best distinguished from the other species by their use of few structural habitat types, as reflected by low structural habitat versatility (PRE=0.125; Figure 4). Structural habitat versatility of imperiled species (mean=41%) was almost half that of secure species (mean=75%), and somewhat less than secure and intermediate species combined (mean=55%). Intermediate-risk species when considered alone could not be neatly separated from imperiled or secure species. That pattern was evident in the classification tree (Figure 4) where intermediate-risk species were split nearly equally between imperiled and secure groups on the basis of structural versatility. Although not appearing critical for separating risk groups, habitat cover-type versatility had the same pattern of differences among groups, which might be expected with a high correlation between structural and habitat versatility (r=0.76).

In some cases, population distribution and habitat cover versatility could be used to distinguish imperiled from other species. Imperiled and intermediate-risk species with low structural versatility on the left side of tree appeared to have similar life history and distribution traits and could not be easily separated by additional attributes from the sample dataset (Figure 4). However, among the species with high versatility on the right side of the tree, scarce population distribution and relatively low habitat cover versatility distinguished most of the few imperiled species from secure or intermediate species.

Risk Advisor. Second, the knowledge base programming approach can account for uncertainty and unknowns that cannot be included in a classification tree analysis. Uncertainty is handled by depicting likelihood of assigning each example to each outcome category. It then uses these likelihoods to calculate a probability of an unknown case (e.g., a species not included in the example data set) belonging to each outcome category (each risk level). Third, during the interactive query, a knowledge base model such as produced by 1st Class can allow the user to enter "do not know" or "no information" to any question, and from there the program determines the next best sequence of as-yet-unasked questions to optimally determine the outcome category. Fourth, because it is not limited to statistical assumptions inherent in the parametric classification tree approach, the ID-3 rule-induction algorithm is not limited to a specific number of variables given a fixed number of example cases (e.g., 5-10 example cases are needed for each variable considered in the classification tree approach). Fifth, "classification trees," being a strictly parametric statistical analysis procedure, provides confidence values describing the degree to which each factor accounts for classification; knowledge base approaches typically do not. Thus, rule induction and classification trees type classification analyses are complementary approaches (see Conclusions section).

SARA uses 29 life history attributes, 1 taxonomic attribute, and 60 example species. Its optimal rule is displayed in Figure 6. It should be remembered that this analysis is only an example of the kind of assessment that can be done to determine how life history attributes might contribute to viability risk levels. Similar analyses can be done on other species' attributes such as population size, trend, density, and so on, if data are available. Results of this analysis, and the SARA model, should not be interpreted as definitive analyses of the influence of life history attributes on risk levels; at best, this analysis poses testable hypotheses, and urges a fuller assessment of a fuller set of species.

Analysis results suggest that 9 life history attributes and 1 taxonomic attribute can account for, and differentiate among, the 3 viability risk levels for the 60 example species. These attributes are population distribution, foraging substrate, elevational range, age at first reproduction, structural versatility, habitat versatility, landscape use, migration distance, and taxonomic order. It should be noted that rule induction algorithms generate results whether or not they are statistically valid, so the relative power with which each of these life history attributes contributes to predicting viability risk levels is not known. This is where the parametric statistical analysis of classification trees can complement this analysis.

An example of how SARA can be used is for identifying a species of unknown risk level. For example, suppose a species has a population distribution consisting of gaps with habitat broadly distributed but with interruptions causing some population isolation (PopnDistrb = pop2, as in Figure 6 and Appendix 3), that it is an owl (Order =

Strigiformes), and that it uses only 20% of all available habitat types (HabVers < 34.5%). The SARA rule set, and the knowledge base expert system, would key this species out to belonging to Group I, that is, a potentially imperiled species. This is because at least 1 known case from the set of 60 example species had these conditions and was known to be imperiled (state or federally listed), the northern spotted owl. Whether another owl with these same attributes would be similarly imperiled is postulated by the SARA analysis but would bear further evaluation. So the purpose of the SARA model, particularly if developed among a broader set of species, is to provide a tool for initially screening species for further evaluation.

An analysis was also done that combined the secure and intermediate categories and compared this combination to the imperiled category. This more simply determined which life history attributes might be used to identify imperiled species although it could not differentiate between secure and intermediate categories. In this analysis, a somewhat different set of attributes were identified as key differentiating factors: habitat versatility, structural versatility, clutch or litter size, summer site fidelity, breeding status, population distribution, and taxonomic order. That a different set of factors resulted in this analysis as compared with the above analysis is not unexpected when the same example species cases were combined in a different way. In this example, one possible combination of attributes that would lead to identifying the imperiled category is if a species had a structural versatility < 54% (StrucVers<54), a mean litter or clutch size of 3 or less (NLitt_Clut=nlit3), and documented or suspected breeding in both Washington and Oregon (BreedStatus=bs3). Again, this should be taken only as an example of the kind of analysis possible; a fuller evaluation with more species would likely produce more reliable results that could be evaluated against a larger set of species with known population status.

A Rank-Model Example

This example shows how population and life history attributes can be used to develop a simple rank model (see methods section of this chapter) to evaluate potential extirpation risk when there is no *a priori* knowledge of risk, as was known in the previous examples. We scored risk for each value of the primary life history and versatility fields in the example dataset on a scale of 1 to 10 (Appendix 2), with 10 being the highest risk, using the principles discussed in the section of this chapter on criteria for identifying species at risk. For example, species with high annual production of offspring or numbers of litters were scored lower risk for that attribute than species with low productivity. The rationale was that a population of species with high fecundity likely would more easily recover from disturbance in their environment, i.e., be more resilient, and less susceptible to local extirpation than species with low productivity. We then summed the scores of each field to calculate a total score for each species. Species with high scores would potentially be at greater risk of extirpation than species with low scores. Differences

A. Population distribution is contiguous.
  B. The organism forages underwater or aerially, or foraging substrate is unknown.
    C. The upper elevation range of typical or regular occurrence is up to 1000 ft. (no identification)
    CC. The upper elevation range of typical or regular occurrence is up to 3000 ft. (no identification)
    CCC. The upper elevation range of typical or regular occurrence is up to 5000 ft. Group III
    CCCC. The upper elevation range of typical or regular occurrence is >5000 ft. Group I
  BB. The organism does not forage underwater or aerially, and foraging substrate is not unknown.
    C. The average age at first breeding (females) is <6 months. (no identification)
    CC. The average age at first breeding (females) is 1 year. Group II
    CCC. The average age at first breeding (females) is 2 years.
      D. The structural versatility of the species is <99. Group II
      DD. The structural versatility of the species is >=99. Group III
    CCCC. The average age at first breeding (females) is 3 years. Group III
    CCCCC. The average age at first breeding (females) is 4+ years. Group II

AA. Population distribution consists of gaps.
  B. The taxonomic order is Caudata.
    C. The structural versatility of the species is <90.50. Group II
    CC. The structural versatility of the species is >=90.50. Group III
  BB. The taxonomic order is Anura. (no identification)
  BBB. The taxonomic order is Squamata. Group III
  BBBB. The taxonomic order is Falconiformes.
    C. The average age at first breeding (females) is <6 months. (no identification)
    CC. The average age at first breeding (females) is 1 year. Group II
    CCC. The average age at first breeding (females) is 2 years. Group III
    CCCC. The average age at first breeding (females) is 3 years. Group I
    CCCCC. The average age at first breeding (females) is 4+ years. (no identification)
  BBBBB. The taxonomic order is Charadriiformes. Group II
  BBBBBB. The taxonomic order is Strigiformes.
    C. The habitat versatility of the species is <34.50. Group I
    CC. The habitat versatility of the species is >=34.50 and < 50.00. Group III
    CC. The habitat versatility of the species is >=50.00. Group II
  BBBBBBB. The taxonomic order is Apodiformes.
    C. The habitat versatility of the species is <53.50. Group II
    CC. The habitat versatility of the species is >=53.50. Group III

Figure 6. The example, optimized rule from the SARA (Species At Risk Advisor) example knowledge base model that determines species' viability risk levels based on their life history attributes. This rule was induced from analysis of 60 example wildlife species in Washington and Oregon, and is only an example of the type of analysis possible by using knowledge base, rule-induction analysis. The decision points in the rule correspond to the questions and states shown in Appendix 3. Group I = imperiled species; Group II = secure species; Group III = intermediate species. See text for explanation of methods.

among species or species groups in risk score or life history attributes were considered significant at P≤0.05.

The distribution of risk scores was very normally distributed, ranging from 33 to 94 points with a mean of 65 (Figure 7). A preponderance of low scores would have indicated few species likely to be negatively affected by management; whereas a preponderance of species with high risk scores would indicate a potentially large risk of action. Ranking species by their risk score from high to low would indicate priorities for further analysis or field work to accurately determine the impacts of management.

How well does this risk scoring and ranking method coincide with the species risk group, as defined in the sample dataset and analyzed in earlier examples? Risk scores for species showed a similar relationship among risk groups to those found with the classification and SARA procedures. Imperiled species on average had higher risk scores (mean=72) than secure species (mean=56) (Figure 8). Intermediate-risk species (mean=68) did not differ from imperiled species in average risk score, but did have higher scores than secure species. That pattern was largely a result of similarly high values for

BBBBBBBB. The taxonomic order is Piciformes.
    C. The habitat versatility of the species is <50.00. Group III
    CC. The habitat versatility of the species is >=50.00. Group II
BBBBBBBBB. The taxonomic order is Passeriformes.
    C. It is a "patch" species, likely using only 1 homogenous habitat patch during the life cycle. Group III
    CC. It is a "mosaic" species, likely using an aggregate of habitat patches but 1 structural stage.
        D. The migration or seasonal movement is <100 km. (no identification)
        DD. The migration or seasonal movement is 100 - 1000 km. Group I
        DDD. The migration or seasonal movement is >1000 km. Group II
        DDDD. The species is non-migratory. Group II
    CCC. It is a "generalist" species, likely using all or many patch types, & >1 structural stage. Group III
    CCCC. It is a "contrast" species, likely requiring contrast between 2 structural stages in close proximity. (no identification)
BBBBBBBBBB. The taxonomic order is Rodentia.
    C. The structural versatility of the species is <28.50. Group I
    CC. The structural versatility of the species is >=28.50. Group III
BBBBBBBBBBB. The taxonomic order is Carnivora. Group II

AAA. Population distribution consists of patchily distributed populations.
    B. The average age at first breeding (females) is <6 months. Group II
    BB. The average age at first breeding (females) is 1 year. Group I
    BBB. The average age at first breeding (females) is 2 years. Group III
    BBBB. The average age at first breeding (females) is 3 years.
        C. The habitat versatility of the species is <34.50. Group I
        CC. The habitat versatility of the species is >=34.50. Group II
    BBBBB. The average age at first breeding (females) is 4+ years. Group III

AAAA. Population distribution consists of isolated population(s).
    B. The migration or seasonal movement is <100 km. Group III
    BB. The migration or seasonal movement is 100 - 1000 km. Group III
    BBB. The migration or seasonal movement is >1000 km. Group I
    BBBB. The species is non-migratory. Group I

AAAAA. Population distribution is scarce.
    B. The habitat versatility of the species is <16.00. Group III
    BB. The habitat versatility of the species is >=16.00. Group I

both imperiled and intermediate species in the life history field describing population distribution, as noted in the classification tree example.

Could we use risk scores to create our own, perhaps better, risk groups instead of using existing legal or administrative definitions of risk, which might reflect political as much as ecological conditions? We used cluster analysis to create three new "risk" clusters, or groups, of species based only on our risk scores, and then we compared the clusters with the previously defined risk groups to see how well they matched the old groups.

Cluster analysis identified low- (mean=51), intermediate- (mean=68), and high-risk (mean=84) species clusters (Table 1); but, cluster membership did not always match the *a priori*-assigned risk groups. That result might be expected considering the inability to perfectly separate species in imperiled, intermediate-risk, and secure risk groups in the earlier examples. The low risk cluster included about 60% of the secure species, 20% of the intermediate-risk species, and 10% of the imperiled species. Most species clustered as moderate-risk, with nearly equal numbers of imperiled and intermediate-risk

species, and some secure species. Only 9 of the 60 species clustered as high-risk, with equal numbers of imperiled and intermediate risk species. The poor match between cluster and *a priori* risk group, in some cases, indicates that legal definitions of risk may not always be adequate to define risk, and unlisted species might also need to be analyzed.

Population Viability Assessments: The Interior Columbia Basin

A final example is taken from the population viability assessment of draft management alternatives for the Interior Columbia Basin Ecosystem Management Project.[76] The purpose of the assessment was to determine the degree to which habitat conditions on lands administered by the Forest Service and Bureau of Land Management within the interior Columbia River basin contribute to long-term persistence (at least 100 years) of select plant and animal species of conservation concern. Secondarily, they examined the extent to which other lands and other influences might affect populations.

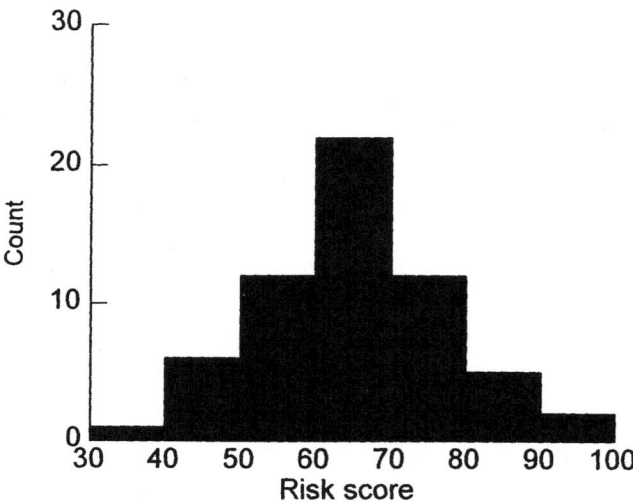

Figure 7. The distribution of risk scores for 60 species in the example dataset using the methodology described for rank model procedures. Risk scores were estimated by scoring each life history, population, or habitat attribute in the example dataset for its potential contribution to extirpation risk, then summing the attribute scores for a single risk estimate.

Table 1. Crosstabulation of risk group memberships based on clustering risk scores of life history attributes and habitat versatility, compared with risk groups in the sample dataset based on legal or administrative designations of threatened, endangered, or sensitive status and expert opinion.

| Risk group (sample dataset) | Risk Score Cluster | | | Total |
	1 (low)	2 (high)	3 (moderate)	
I (Imperiled)				
Count	2	4	14	20
% within Group I	10%	20%	70%	100%
II (Secure)				
Count	12	1	7	20
% within Group II	60%	5%	35%	100%
III (Intermediate)				
Count	4	4	12	20
% within Group III	20%	20%	60%	100%
Total				
Count	18	9	33	60
% within Group	30%	15%	55%	100%

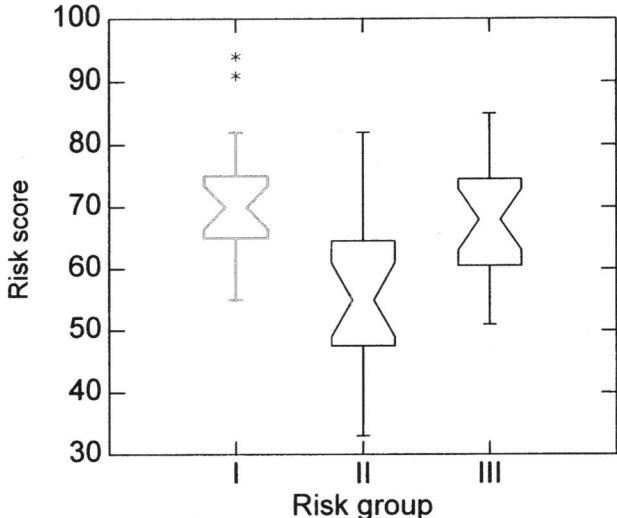

Figure 8. The distribution of risk scores, estimated by the rank model procedure, by viability group (imperiled [I], secure [II], intermediate [III]) determined a priori for the example dataset of 60 species. Boxes indicate the mid-range (50% quartile) of data values around the median; notches in the box are the confidence interval around the median. Whiskers on the boxes show the outside quartiles, and asterisks show outliers.

The assessment was not a quantitative population viability analysis, as it did not employ an explicit model of genetic or demographic risk to species persistence. Rather, the qualitative assessment was a structured and reasoned series of judgments about projected amounts and distributions of habitat and the likelihood that such habitat would allow populations to persist over the long term. Thus, it met the essential criterion of a population viability assessment to provide an estimate of the likelihood that a population will persist to some arbitrarily chosen future time.

Species assessments were based on expert opinion about the likely outcome for species and their habitats under a variety of possible management alternatives. Expert judgments were solicited from eight expert panels. The panels were provided with information on the species distributions, habitat relationships, and known population trends, and with information on the effects of management alternatives on species macro- and micro-habitat elements. Based on that information, the experts were asked to make two judgments. The first judgment rated the species' likely distribution based only on habitat conditions on the federal lands and the natural history characteristics of the species. The second was a cumulative effects analysis of the likely condition of species populations across all ownerships. Factors considered included demographic characteristics, responses to varying qualities of habitat for specific life functions, types and ranges of seasonal and permanent movements, genetic characteristics, and biotic interactions (e.g., competition, predation, herbivory).

Expert judgments were registered through a process of likelihood voting using a structured outcome scale. The

Table 2. Viability outcomes used for the population viability assessment of draft management alternatives for the Interior Columbia Basin Ecosystem Management Project.[76]

Outcome 1.
Habitat is broadly distributed across the planning area with opportunity for continuous or nearly continuous occupation by the species and little or no limitation on population interactions.

Outcome 2.
Habitat is broadly distributed across the planning area but gaps exist within this distribution. Disjunct patches of habitat are typically large enough and close enough to other patches to permit dispersal among patches and to allow species to interact as a metapopulation.

Outcome 3.
Habitat exists primarily as patches, some of which are small or isolated to the degree that species interactions are limited. Local subpopulations in most of the species' range interact as a metapopulation but some patches are so disjunct that subpopulations in those patches are essentially isolated from other populations.

Outcome 4.
Habitat is typically distributed as isolated patches, with strong limitation in interactions of populations among patches and limited opportunity for dispersal among patches. Some local populations may be extirpated and rate of recolonization will likely be slow.

Outcome 5.
Habitat is very scarce throughout the area with little or no possibility of interactions among local populations, strong potential for extirpations, and little likelihood of recolonization.

outcome scale depicted 5 distinct possible outcomes for the habitat or population, each representing points along a gradient ranging from a broadly distributed condition with high likelihood of persistence to a poorly distributed condition with high likelihood of extirpation (Table 2). For each judgment, each expert spread 100 likelihood votes across these 5 outcomes (Table 3). Placing 100 votes on a single outcome indicated certainty in that outcome; whereas, spreading the votes among several outcomes indicated less certainty in any one outcome. Consensus among panelists was not an objective of the process; moreover, the independence of experts' judgments was necessary to assess the uncertainty (standard deviation) of viability likelihood. Uncertainty included two components, variation of likelihood distributions among panelists and the spread of likelihood points among outcomes by each panelist.

There were 2 primary analyses performed on the data derived from the panels. First, the mean likelihood scores for all experts for each of the outcomes was calculated. For example, if there were 4 experts on a panel, and their likelihood votes for outcome 2 for a particular species were 30, 30, 60, and 40, then the mean likelihood score was 40. Next, to summarize mean likelihood for each species, a

weighted mean outcome was also calculated. This was calculated by assigning a value to each of the outcome categories (Outcome 1 value=1, Outcome 2 value=2, etc.), multiplying the mean likelihood of that outcome by its assigned value, adding these products for all outcomes, and then dividing by 100. Outcomes were considered improving or declining from the historical to current, or from current to future, periods if they changed at least 0.5, a value corresponding to one standard deviation of the mean outcome. The assessment did not provide a simple determination of what does and does not constitute a "viable" population. This was considered a strength of the process rather than a weakness, as there are no simple thresholds for viability, particularly when assessments are done on a broad array of taxa. Rather than providing a simple determination, this assessment described likely future conditions for species and habitats and provided for comparison of those conditions to current and historical conditions. Lack of a simple determination, however, added complexity to the job of interpreting the results and using them in a decision-making framework. The authors recommended that interpretation of the results emphasize comparison of the projected future conditions under the alternatives to historical and current conditions.

A variety of cautions must be applied to the interpretation of this form of assessment. These cautions fall into four areas: (1) the broad geographic and temporal scale of the analysis limits local inference; (2) the resolution of the data and planning guidance define the level of confidence in results; (3) limitations on the ability to infer population results from habitat analysis; and (4) gaps in knowledge limit confidence or geographic scope of inference. Lehmkuhl et al.[76] give a detailed discussion of these cautions and assumptions.

Conclusions Drawn from the Example Analyses

The examples were meant to illustrate a practical range of risk analysis methods that would be relatively easy to implement using the data matrixes in this book and conservation theory. The examples focused on frequently encountered situations when the taxonomic scope is fairly broad (i.e., many species need to be analyzed), and when procedural, or analytical, complexity may be limited by available data. The geographic scale (extent) of application would usually be large—at watershed or larger scales.

The first two examples used data in related procedures to differentiate and classify imperiled, secure, and intermediate-risk species, and to develop a rule set for classifying other species of unknown risk. The third example used species risk theory and data to score and rank potential risk based on attributes of life history and population distribution, when prior knowledge of risk was unavailable. That same type of model could be used to rank management priorities based on criteria other than extirpation risk, e.g. social or economic value. The fourth example described a qualitative population viability assessment that linked in the data matrixes with expert

Table 3. Example of the likelihood voting system used to assess viability outcomes for selected species of conservation concern for the Interior Columbia Basin Ecosystem Management Project.*[76]

Species: Flammulated Owl			Management Alternative		
Outcome	Historic	Current	Traditional	Restore	Reserve
1 contiguous	1	0	0	0	0
2 gaps	75	2	1	26	15
3 patchy	20	40	20	60	50
4 isolated	4	38	39	13	27
5 scarce	0	20	40	1	8
Total score	100	100	100	100	100

* One hundred likelihood points are distributed across the five outcomes for each time period or alternative. The distribution of points reflects the certainty of a particular outcome.

opinion to assess land management alternatives.

Other more quantitative methods, such as demographic PVA models, are valuable tools to address some of those risk issues and should not be ruled out. They were not illustrated because their complexity demands information far more sophisticated than is available in the data matrixes, and that is nonexistent for the vast majority of species that the manager has to consider. Quantitative PVA models will be most appropriate when the focus of management is a single species suspected or known to be at risk, and when reliable data are available to model population and habitat dynamics under management alternatives.

It is first and foremost necessary to remember that the examples were meant simply to illustrate some of the ways information in the data matrixes can be used to determine factors that might contribute to risk levels of species, or to rank species in terms of risk. Further and more complete analyses, using these methods or others described in this chapter, are really needed to verify our initial findings on how life history or taxonomic categories could predispose some species to various risk levels. Those conclusions should be taken as tentative working hypotheses.

The following is an example of how one might draw conclusions from the example risk analyses. Conclusions from these types of analyses might differ if the focus of such analyses shifted to specific taxonomic or functional groups. That given, it could be significant that the best distinguishing features of the three species groups analyzed in the first three examples mostly refer to habitat selection, habitat breadth, and population distribution, and not as much to inherent life history attributes per se. At the start of this analysis, we expected at least some life history attributes, such as age at first reproduction, reproduction rates, site fidelity, and movement distances, to predict risk level. Life history attributes still may be important to distinguish risk levels among species within genera or families, which remains to be formally tested with an expanded analysis of the entire species dataset.

However, the importance of life history attributes for characterizing risk within taxonomic groups does not seem to hold equally across the 4 taxonomic classes. That also was the general finding of Russell et al.,[145] who found that distributions of extinctions and threat classifications were clustered unevenly within certain genera and families, particularly in taxa that contain few species. Whether taxonomic affiliation and diversity alone predisposes species to certain risk levels, or patterns of habitat use and population distribution are themselves directed by life history characteristics, is unclear and needs further analysis.

The first three analyses were developed mostly as examples of the kinds of evaluations that could be done with data to test how life history attributes might contribute to risk levels. The rank model, the third example, also ranked species by potential risk. More specific and more thorough analyses can follow these examples by: (1) focusing on species within specific individual habitats, structural conditions, physiographic provinces, taxonomic groups, known risk categories, or geographic areas; (2) rerunning such analyses for all such species within the data matrixes; (3) better determining the mechanisms by which life history traits influence habitat selection and population distribution; and (4) better determining the influence of life history traits on risk levels among species within genera and families.

Following more complete analysis, the results can be integrated into a broader risk management framework. The risk analysis portion of a decision evaluation can include those life history traits and habitat selection behavior of species found to portend risk levels, and thereby determine how alternative management actions might influence the species' habitat use and population response. In the risk management phase, the decision-maker then would have explicit information on which species might be at greater risk.

The qualitative PVA example for the Columbia River Basin showed how information from the first three types of analyses could be melded with matrix data and expert opinion to assess and score the effects of different land management alternatives at a broad regional scale. That information, then, would inform decision makers on how best to manage risk. The qualitative expert-opinion process used in the example could be developed further by formalizing the relationships between life history, habitat, and population status in a decision analysis or knowledge-based system. Then, the risk analysis process could be easily refined or repeated as management alternatives are modified from public input or changing agency priorities. Formalizing such expert models also has the advantage of explicitly stating the relationships between species attributes, habitat, and risk, thus, opening the scientific "black box" for review and agreement among interested parties that the best available science is being used for management.

Conclusions

Wildlife managers do risk analyses that vary in spatial scale, taxonomic scope, and complexity of information needs and methods—it is a daunting task. Moreover, the public is scrutinizing resource management actions that can affect wildlife viability, and is asking for the best available science to analyze the effects of those actions. Armed with a basic understanding of the theory and criteria for identifying species at risk, managers and scientists can use available datasets and analytical tools to meet that challenge at different levels of geographic and ecological scale.

Species are at risk from a combination of systematic and stochastic processes that affect population size and distribution. Systematic processes are usually the ultimate cause of extirpation, and occur when something essential is removed from the environment, such as habitat, or when something lethal is inserted into the environment, such as a new predator or poison. Stochastic, or random, processes are usually the proximate cause of extirpation in populations already made small by systematic pressures. The vulnerability, or risk, for species is a function of how systematic and stochastic processes interact with habitat selection behavior, and the demographic and life history attributes of the species. Habitat selection, habitat breadth, and population distribution likely are among the most important influences on risk level.

Understanding the processes underlying species risk allows managers to select the right analysis tools. While much of the conservation literature on wildlife risk analysis focuses on data-intensive quantitative PVA models, there are many situations for which that tool is unsuitable in terms of number of species to be analyzed, the data available for the analysis, or the time and money available to do it. Fortunately, there are datasets and analytical tools that can meet the risk analysis needs of managers for different levels of taxonomic and procedural complexity. The data matrixes included in this book are an important basic resource for risk analysis that can be linked with analytical methods in this chapter. Results of such risk assessments will provide the essential information for decision-makers about the alternatives for managing risk.

Literature Cited

1. Akçakaya, H. R., and J. L. Atwood. 1997. A habitat-based metapopulation model of the California gnatcatcher. Conservation Biology 11:422-434.
2. ———, M. A. Burgman, and L. R. Ginzburg. 1997. Applied population ecology: principles and computer exercises using RAMAS EcoLab 1.0 Applied Biomathematics, New York, NY.
3. ———, M. A. McCarthy, and J. L. Pearce. 1995. Linking landscape data with population viability analysis: Management options for the helmeted honeyeater Lichenostomus melanops cassidix. Biological Conservation 73:169-176.
4. Allendorf, F. W., D. Bayles, D. L. Bottom, K. P. Currens, C. A. Frissell, D. Hankin, J. A. Lichatowich, W. Nehlsen, P. C. Trotter, and T. H. Williams. 1997. Prioritizing Pacific salmon stocks for conservation. Conservation Biology 11:140-152.
5. Armbruster, P., and R. Lande. 1993. A population viability analysis for African elephant (Loxodonta africana): how big should reserves be? Conservation Biology 7:602-610.
6. Ballou, J., and K. Ralls. 1982. Inbreeding genetics and juvenile mortality in small populations of ungulates: a detailed analysis. Biological Conservation 24:239-72.
7. Beissinger, S. R. 1995. Modeling extinction in periodic environments: Everglades water levels and snail kite population viability. Ecological Applications 5:618-631.
8. Botsford, L. W., and J. G. Brittnacher. 1998. Viability of Sacramento River winter-run chinook salmon. Conservation Biology 12:65-79.
9. Boyce, M. S. 1992. Population viability analysis. Annual Review of Ecology and Systematics 23:481-506.
10. Bradstock, R. A., M. Bedward, J. Scott, and D. A. Keith. 1996. Simulation of the effect of spatial and temporal variation in fire regimes on the population viability of a Banksia species. Conservation Biology 10:776-784.
11. Brook, B. W., L. Lim, R. Harden, and R. Frankham. 1997a. Does population viability analysis software predict the behaviour of real populations? A retrospective study on the Lord Howe Island Woodhen (Tricholimnas sylvestris, Sclater). Biological Conservation 82:119-128.
12. ———, ———, ———, ———. 1997b. How secure is the Lord Howe Island Woodhen? A population viability analysis using VORTEX. Pacific Conservation Biology 3:125-133.
13. Brooker, L. C., and M. G. Brooker. 1994. A model for the effects of fire and fragmentation on the population viability of the Splendid Fairy-wren. Pacific Conservation Biology 1:344-358.
14. Brown, E. R., editor. 1985. Management of wildlife and fish habitats in forests of western Oregon and Washington. Pub No. R6-F&WL-192-1985, 2 vols. USDA Forest Service, Pacific Northwest Region, Portland, OR.
15. Burgman, M., D. Cantoni, and P. Vogel. 1992. Shrews in suburbia an application of Goodman's extinction model. Biological Conservation 61:117-123.
16. ———, S. Ferson, and D. Lindenmayer. 1995. The effect of the initial age-class distribution on extinction risks: implications for the reintroduction of Leadbeater's possum. Pages 15-19 in M. Serena, editor. Reintroduction biology of Australian and New Zealand fauna. Surrey Beatty, Chipping Norton, England
17. Burgman, M. A. 1990. A stage-structured stochastic population model for the giant kelp Macrocystis-pyrifera. International Colloquim on Dynamical Models in Biology, Lausanne, Switzerland, September 13-16, 1988. Memoires De La Societe Vaudoise Des Sciences Naturelles 18:355.
18. ———, S. Ferson, and H. R. Akçakaya. 1993. Risk assessment in conservation biology. Chapman and Hall, London, England.
19. ———, and B. B. Lamont. 1992. A stochastic model for the viability of Banksia-cuneata populations environmental demographic and genetic effects. Journal of Applied Ecology 29:719-727.
20. Burke, R. L., and S. R. Humphrey. 1987. Rarity as a criterion for endangerment in Florida's fauna. Oryx 21:97-102.
21. Burke, R. L., J. Tasse, C. Badgley, S. R. Jones, N. Fishkein, S. Phillips, and M. E. Soulé. 1991. Conservation of Stephen's kangaroo rat (Dipodomys stephensi): planning for persistence. Bulletin of the Southern California Academy of Science 90:10-40.

22. Burton, M. P. 1994. Alternative projections of decline of the African elephant. Biological Conservation 70:183-188.

23. Bustamante, J. 1996. Population viability analysis of captive and released bearded vulture populations. Conservation Biology 10:822-831.

24. Caughley, G. 1994. Directions in conservation biology. Journal of Animal Ecology 63:215-244.

25. Chen, J. 1991. Edge effects: microclimatic pattern and biological responses in old growth Douglas-fir forests. Ph.D. dissertation, University of Washington, Seattle, WA.

26. Cisneros-Mata, M. A., L. W. Botsford, and J. F. Quinn. 1997. Projecting viability of *Totoaba macdonaldi*, a population with unknown age-dependent variability. Ecological Applications 7:968-980.

27. Diamond, J., K. D. Bishop, and S. Van Balen. 1987. Bird survival in an isolated Javan woodland: island of mirror? Conservation Biology 1:132-133.

28. Diamond, J. M. 1984. "Normal" extinctions of isolated populations. Pages 191-246 in M. H. Nitecki, editor. Extinctions. University of Chicago Press, Chicago, IL.

29. ———. 1989. The present, past and future of human-caused extinctions. Philosophical Transactions of the Royal Society of London B325:469-477.

30. Emlen, J. M. 1989. Terrestrial population models for ecological risk assessment: a state-of-the-art review. Environmental Toxicology and Chemistry 8:831-842.

31. ———. 1995. Population viability of the Snake River chinook salmon (*Oncorhynchus tshawytscha*). Canadian of Journal of Fisheries and Aquatic Sciences 52:1442-1448.

32. Fahrig, L., and J. Paloheimo. 1988. Effects of spatial arrangement of habitat patches on local population size. Ecology 69:468-475.

33. FEMAT. 1993. Forest ecosystem management: an ecological, economic, and social assessment. Report of the Forest Ecosystem Management Assessment Team. U.S. Government Printing Office, Washington, D.C.

34. Fernandez-Duque, E., and C. Valeggia. 1994. Meta-analysis: a valuable tool in conservation research. Conservation Biology 8:555-561.

35. Fite, E., L. W. Turner, N. J. Cook, C. Stunkard, and R. M. Lee. 1988. Guidance document for conducting terrestrial field studies. EPA 540/09, 88-109. U.S. Environmental Protection Agency, Office of Pesticide Programs, Washington, D.C.

36. Flood, B. S., M. E. Sangster, R. D. Sparrowe, and T. S. Baskett. 1977. A handbook for habitat evaluation procedure. USDI Fish and Wildlife Service, Washington, D.C.

37. Frankel, O. H., and M. E. Soulé. 1981. Conservation and evolution. Cambridge University Press.

38. Franklin, A. B., and T. M. Shenk. 1995. Meta-analysis as a tool for monitoring wildlife populations. Pages 484-487 in J. A. Bissonette, and P. R. Krausman, editors. Integrating people and wildlife for a sustainable future. The Wildlife Society, Bethesda, MD.

39. Fritts, S. H., and L. N. Carbyn. 1995. Population viability, nature reserves, and the outlook for gray wolf conservation in North America. Restoration Ecology 3:26-38.

40. Gagne, J. A. 1994. The future application of ecological models in environmental risk assessment. Pages 497-499 in Ronald J. Kendall, and Thomas E. Lacher, editors. Wildlife toxicology and population modeling: integrated studies of agroecosystems. Lewis Publishers, Boca Raton, FL.

41. Gardenfors, U. 1996. Application of Red List categories on a regional scale. *In* IUCN Red List of Threatened Animals. Gland, Switzerland.

42. Gilpin, M. E., and M. E. Soulé. 1986. Minimum viable populations: processes of species extinction. Pages 19-34 in M. E. Soulé, editor. Conservation biology: the science of scarcity and diversity. Sinauer Associates, Sunderland, MA.

43. Goodman, D. 1987a. How do species persist? lessons for conservation biology. Conservation Biology 1:59-62.

44. ———. 1987b. The demography of chance extinction. Pages 11-34 in M. E. Soulé, editor. Viable populations for conservation. Cambridge University Press,

45. Green, R. E., M. W. Pienkowski, and J. A. Love. 1996. Long-term viability of the re-introduced population of the white-tailed eagle *Haliaeetus albicilla* in Scotland. Journal of Applied Ecology 33:357-368.

46. Grubb, T. G. 1988. Pattern recognition—a simple model for evaluating wildlife habitat. RM-GTR-487. USDA Forest Service, Rocky Mountain Forest and Range Experiment Station, Fort Collins, CO.

47. Grumbine, R. E. 1990. Viable populations, reserve size, and federal land management: a critique. Conservation Biology 4:127-134.

48. Haig, S. M., J. R. Belthoff, and D. H. Allen. 1993. Population viability analysis for a small population of red-cockaded woodpeckers and an evaluation of enhancement strategies. Conservation Biology 7:289-301.

49. Haight, R. G. 1995. Comparing extinction risk and economic cost in wildlife conservation planning. Ecological Applications 5:767-775.

50. Hallam, T. G., and R. R. Lassiter. 1994. Individual-based mathematical modeling approaches in ecotoxicology: a promising direction for aquatic population and community ecological risk assessment. Pages 531-542 in Ronald J. Kendall, and Thomas E. Lacher, editors. Wildlife toxicology and population modeling: integrated studies of agroecosystems. Lewis Publishers, Boca Raton, FL.

51. Hansen, A. J., and D. L. Urban. 1992. Avian response to landscape patterns—the role of species' life histories. Landscape Ecology 7:163-180.

52. Harcourt, A. H. 1995. Population viability estimates: theory and practice for a wild gorilla population. Conservation Biology 9:134-142.

53. Hokit, D. G., B. M. Stith, and L. C. Branch. 1999. Effects of landscape structure in Florida scrub: a population perspective. Ecological Applications 9:124-134.

54. Holling, C. S. 1984. Adaptive environmental assessment and management. John Wiley & Sons, New York, NY.

55. Holtsinger, K. E., and P. Vitt. 1997. The future of conservation biology: what's a geneticist to do? Pages 202-216 in S. T. A. Pickett, R. S. Ostfeld, M. Shachak, and G. E. Likens, editors. The ecological basis of conservation: heterogeneity, ecosystem, and biodiversity. Chapman and Hall, New York, NY.

56. Hunter, J. E., and F. L. Schmidt. 1990. Methods of meta-analysis. Sage Publications, Thousand Oaks, CA.

57. IUCN. 1994. Red list categories. Gland, Switzerland.

58. ———. 1996a. Red list of threatened animals. Gland, Switzerland.

59. ———. 1996b. Red list of threatened plants. Gland, Switzerland.

60. Jeffers, J. N. R. 1991. Rule induction methods in forestry research. AI Applications 5:37-44.

61. Johnson, K. N., J. Franklin F., J. W. Thomas, and J. Gordon. 1991. Alternatives for management of late-successional forests of the Pacific Northwest: a report to the Agriculture Committee and The Merchant Marine and Fisheries Committee of the U.S. House of Representatives. Scientific Panel on Late-Successional Forest Ecosystems, USDA Forest Service, Washington, D.C.

62. Julin, K. R., compiler. 1997. Assessments of wildlife viability, old-growth timber volume estimates, forested wetlands, and slope stability. PNW Gen. Tech. Rep. PNW-392. USDA Forest Service, Pacific Northwest Research Station, Portland, OR.

63. Kareiva, P. 1987. Habitat fragmentation and the stability of predatory-prey interactions. Nature 326:288-290.

64. ———, D. Skelly, and M. Ruckelshaus. 1997. Reevaluating the use of models to predict the consequences of habitat loss and fragmentation. Pages 156-166 in S. T. A. Pickett, R. S. Ostfeld, M. Shachak, and G. E. Likens, editors. The ecological basis of conservation: heterogeneity, ecosystems, and biodiversity. Chapman & Hall, New York, NY.

65. Karr, J. R. 1982a. Avian extinction on Barro Colorado Island, Panama: a reassessment. American Naturalist 119:220-239.

66. ———. 1982b. Population variability and extinctions in the avifauna of a tropical land-bridge island. Ecology 63:1975-1978.

67. Kelly, P. A., K. D. Allred, H. P. Possingham, and D. F. Williams. 1996. Extinction risk assessment for the San Joaquin kit fox. Bulletin of the Ecological Society of America 77:228.

68. Kendall, R. J., and J. Akerman. 1992. Terrestrial wildlife exposed to agrochemicals an ecological risk assessment perspective. Environmental Toxicology and Chemistry 11:1727-1749.

69. Kenney, J. S., J. L. D. Smith, A. M. Starfield, and C. W. McDougal. 1995. The long-term effects of tiger poaching on population viability. Conservation Biology 9:1127-1133.

70. Kincaid, H. L. 1976. Effects of inbreeding on rainbow trout populations. Transactions of the American Fisheries Society 2:273-80.

71. King, F. W. 1987. Thirteen milestones on the road to extinction. Pages 7-18 in R. Fitter, and M. Fitter, editors. The road to extinction. IUCN, Gland, Switzerland.

72. Kinnaird, M. F., and T. G. O'Brien. 1991. Viable populations for an endangered forest primate, the Tana River Mangabey (*Cercocebus galeritus galerius*). Conservation Biology 5:203-213.

73. Klein, B. C. 1989. Effects of forest fragmentation on dung and carrion beetle communities in central Amazonia. Ecology 70:1715-1725.

74. Lande, R. 1988. Genetics and demography in biological conservation. Science 241:16-241.

75. Lehmkuhl, J. F., and D. R. Patton. 1984. User's manual for the RUN WILD III data storage and retrieval system. USDA Forest Service, Southwestern Region, Wildlife Unit, Albuquerque, NM.

76. ———, M. G. Raphael, R. S. Holthausen, J. R. Hickenbottom, R. H. Naney, and J. S. Shelly. 1997. Chapter 4. Effects of planning alternatives on terrestrial species in the interior Columbia River basin. Pages 539-730 in T. M. Quigley, K. M. Lee, and S. J. Arbelbide, technical editors. Evaluation of EIS alternatives by the science integration team. PNW-GTR-406. USDA, Forest Service, Pacific Northwest Research Station, Portland, OR.

77. ———, and L. F. Ruggiero. 1991. Forest fragmentation in the Pacific Northwest and its potential effects on wildlife. Pages 35-46 in L. F. Ruggiero, K. B. Aubry, A. B. Carey, and M. H. Huff, editors. Wildlife and vegetation of unmanaged Douglas-fir forests. PNW-GTR-285. USDA Forest Service, Pacific Northwest Research Station, Portland, OR.

78. Lindenmayer, D. B., M. A. Burgman, H. R. Akçakaya, R. C. Lacy, and H. P. Possingham. 1995. A review of the generic computer programs ALEX, RAMAS-space and VORTEX for modeling the viability of wildlife metapopulations. Ecological Modeling 82:161-174.

79. ———, T. W. Clark, R. C. Lacy, and V. C. Thomas. 1993. Population viability analysis as a tool in wildlife conservation policy: with reference to Australia. Environmental Management 17:745-758.

80. ———, and H. P. Possingham. 1995. Modeling the impacts of wildfire on the viability of metapopulations of the endangered Australian species of arboreal marsupial, Leadbeater's possum. Forest Ecology and Management 74:197-222.

81. ———, and ———. 1996. Ranking conservation and timber management options for Leadbeater's possum in southeastern Australia using population viability analysis. Conservation Biology 10:235-251.

82. Lovejoy, T. E., R. O. Bierregaard, Jr., A. B. Rylands, J. R. Malcom, C. E. Quintels, L. H. Harper, K. S. Brown, Jr., A. H. Powell, G. V. N. Powell, H. O. R. Schubart, and M. Hays. 1986. Edge and other effects of isolation on Amazon forest fragments. Pages 257-285 in M. E. Soulé, editor. Conservation biology: the science of scarcity and diversity. Sinauer Associates, Sunderland, MA.

83. Mace, G. M., and N. J. Collar. 1995. Extinction risk assessment for birds through quantitative criteria. IBIS 137:240-246.

84. Madsen, T., B. Stille, and R. Shine. 1996. Inbreeding depression in an isolated population of adders *Vipera berus*. Biological Conservation 75:113-8.

85. Maquire, L. A., T. W. Clark, R. Crete, J. Cada, C. Groves, M. L. Shaffer, and U. S. Seal. 1988. Black-footed ferret recovery in Montana: a decision analysis. Wildlife Society Bulletin 16:111-120.

86. ———, and R. C. Lacy. 1990. Allocating scarce resources for conservation of endangered subspecies: partitioning zoo space for tigers. Conservation Biology 4:157-166.

87. ———, U. S. Seal, and P. F. Brussard. 1987. Managing critically endangered species: the Sumatran rhino as a case study. Pages 141-158 in M. E. Soulé, editor. Viable populations for conservation. Cambridge University Press, Cambridge, U.K.

88. ———, and C. Servheen. 1992. Integrating biological and sociological concerns in endangered species management: augmentation of grizzly bear populations. Conservation Biology 6:426-434.

89. ———, G. F. Wilhere, and Q. Dong. 1995. Population viability analysis for red-cockaded woodpeckers in the Georgia Piedmont. Journal of Wildlife Management 59:533-542.

90. Marcot, B. G. 1986. Concepts of risk analysis as applied to viable population assessment and planning. Pages 89-102 in B. A. Wilcox, P. F. Brussard, and B. G. Marcot, editors. The management of viable populations: theory, applications, and case studies. Center for Conservation Biology, Stanford, CA.

91. ———, R. S. Holthausen, and H. Salwasser. 1986. Viable population planning. Pages 49-62 in B. A. Wilcox, P. F. Brussard, and B. G. Marcot, editors. The management of viable populations: theory, applications, and case studies. Center for Conservation Biology, Stanford, CA.

92. ———. 1987. Use of decision tree analysis for assessing wildlife-silviculture relationships. USFS Unpublished Report

93. ———. 1992. Putting data, experience and professional judgment to work in making land management decisions. Pages 140-161 in J. B. Nyberg and W. B. Kessler, editors. Integrating timber and wildlife in forest landscapes: a matter of scale. B.C. Ministry of Forests, Victoria, B.C., Canada.

94. ———, and R. Holthausen. 1987. Analyzing population viability of the spotted owl in the Pacific Northwest. Transactions of the North American Wildland and Natural Resources Conference 52:333-347.

95. ———, L. K. Croft, J. F. Lehmkuhl, R. H. Naney, C. G. Niwa, W. R. Owen, and R. E. Sandquist. 1998. Macroecology, paleoecology, and ecological integrity of terrestrial species and communities of the interior Columbia River Basin and portions of the Klamath and Great Basins. PNW Gen. Tech. Rep. PNW-410. USDA Forest Service, Pacific Northwest Research Station, Portland, OR.

96. Margules, C., A. J. Higgs, and R. W. Rafe. 1982. Modern biogeographic theory: are there any lessons for nature reserve design? Biological Conservation 24:115-128.

97. Marmontel, M., S. R. Humphrey, and T. J. O'Shea. 1997. Population viability analysis of the Florida manatee (*Trichechus manatus latirostris*), 1976-1991. Conservation Biology 11:467-481.

98. Master, L. L. 1991. Assessing threats and setting priorities in conservation. Conservation Biology 5:559-563.

99. Mather, A. S. 1990. Global forest resources. Belhaven Press, London, England.

100. Matsuda, H., T. Yahara, and Y. Uozumi. 1997. Is tuna critically endangered? Extinction risk of a large and overexploited population. Ecological Research 12:345-356.

101. May, R. M. 1973. Stability in randomly fluctuating versus deterministic environments. American Naturalist 107:621-650.

102. ———, J. H. Lawton, and N. E. Stork. 1995. Assessing extinction rates. Pages 1-24 in J. H. Lawton and R. M. May, editors. Extinction rates. Oxford University Press, New York, NY.

103. McCarthy, M. A. 1995. Population viability analysis of the helmeted honeyeater: risk assessment of captive management and reintroduction. Pages 21-25 in M. Serena, editor. Reintroduction biology of Australian and New Zealand fauna. Surrey Beatty, Chipping Norton, England.

104. ———, and M. A. Burgman. 1995. Coping with uncertainty in forest wildlife planning. Forest Ecology and Management 74:23-36.

105. ———, ———, and S. Ferson. 1995. Sensitivity analysis for models of population viability. Biological Conservation 73:93-100.

106. McCoy, E. D. 1982. The application of island-biogeographic theory to forest tracts: problems in the determination of turnover rates. Biological Conservation 22:217-227.

107. McElroy, D. M., J. A. Shoemaker, and M. E. Douglas. 1997. Discriminating *Gila robusta* and *Gila cypha*: risk assessment and the Endangered Species Act. Ecological Applications 7:958-967.

108. McNay, R. S., R. E. Page, and A. Campbell. 1987. Application of expert-based decision models to promote integrated management of forests. Transactions of the North American Wildland and Natural Resources Conference 52:82-91.

109. Menges, E. S. 1990. Population viability analysis for an endangered plant. Conservation Biology 4:52-62.

110. ———. 1991. The application of minimum viable population theory to plants. Pages 45-61 in D.A. Falk and K. E. Holtsinger, editors. Genetics and conservation of rare plants. Oxford University Press, New York, NY.

111. Meyer, J. S., C. G. Ingersoll, L. L. McDonald, and M. S. Boyce. 1986. Estimating uncertainty in population growth rates: jackknife vs. bootstrap techniques. Ecology 67:1156-1166.

112. Miller, R. I., and L. D. Harris. 1977. Isolation and extirpation in wildlife reserves. Biological Conservation 12:311-315.

113. Mills, L. S., S. G. Hayes, C. Baldwin, M. J. Wisdom, J. Citta, D. J. Mattson, and K. Murphy. 1996. Factors leading to different viability predictions for a grizzly bear data set. Conservation Biology 10:863-873.

114. Millsap, B. A., J. A. Gore, D. E. Runde, and S. I. Cerulean. 1990. Setting priorities for the conservation of fish and wildlife species in Florida. Wildlife Monographs 111:1-57.

115. Morrison, M. L., B. G. Marcot, and R. W. Mannan. 1998. Wildlife-habitat relationships: concepts and applications. Second edition. University of Wisconsin Press, Madison, WI.

116. Murphy, D.D., K.E. Freas, and S.B. Weiss. 1990. An environment-metapopulation approach to viability analysis for a threatened invertebrate. Conservation Biology 4:41-51.

117. Nantel, P., D. Gagnon, and A. Nault. 1996. Population viability analysis of American ginseng and wild leek harvested in stochastic environments. Conservation Biology 10:608-621.

118. Nash, S. 1991. What price nature? Future ecological risk assessments may chart the values, and the odds. BioScience 41:677-680.

119. Newmark, W. D. 1987. A land-bridge island perspective on mammalian extinctions in western North American parks. Nature 325:430-432.

120. Niemi, G. J. 1982. Determining priorities in nongame management. Loon 54:28-54.

121. Nolet, B. A., and J. M. Baveco. 1996. Development and viability of a translocated beaver Castor fiber population in the Netherlands. Biological Conservation 75:125-137.

122. Patton, D. R. 1978. RUN WILD: a storage and retrieval system for wildlife habitat information. RM-GTR-51. USDA Forest Service, Rocky Mountain Forest and Range Experiment Station, Fort Collins, CO.

123. Pazzani, M., and D. Kibler. 1992. The role of prior knowledge in inductive learning. Machine Learning 9:54-97.

124. Peterle, T. J. 1991. Wildlife toxicology. Van Nostrand Reinhold, New York, NY.

125. Pimm, S. L., H. L. Jones, and J. Diamond. 1988. On the risk of extinction. American Naturalist 132:757-785.

126. Pinchera, F., L. Boitani, and F. Corsi. 1997. Application to the terrestrial vertebrates of Italy of a system proposed by IUCN for a new classification of national Red List categories. Biodiversity and Conservation. 6:959-978.

127. Power, M., D. G. Dixon, and G. Power. 1994. Modeling population exposure-response functions for use in environmental risk assessment. Journal of Aquatic Ecosystem Health 3:45-58.

128. Prescott-Allen, R., and C. Prescott-Allen. 1978. Sourcebook for a world conservation strategy: threatened vertebrates. IUCN, Gland, Switzerland.

129. Quinlan, J. R. 1986a. Induction of decision trees. Machine Learning 1:81-106.

130. ———. 1986b. Simplifying decision trees. International Journal of Man-Machine Studies 27:221-234.

131. Quinn, J. F., and A. Hastings. 1987. Extinction in subdivided habitats. Conservation Biology 1:198-209.

132. Raiffa, H. 1968. Decision analysis: introductory lectures on choices under certainty. Addison-Wesley, Reading, MA.

133. Ralls, K, J. D. Ballou, and A. Templton. 1988. Estimates of lethal equivalents and the cost of inbreeding in mammals. Conservation Biology 2:185-93.

134. ———, K. Brugger, and J. Ballou. 1979. Inbreeding and juvenile mortality in small populations of ungulates. Science 206:1101-3.

135. Ranjit Daniels, R. J., M. Hegde, N.V. Joshi, and M. Gadgil. 1991. Assigning conservation value: a case study in India. Conservation Biology 5:464-475.

136. Ranney, J. W., J. M. C. Bruner, and J. B. Levenson. 1981. The importance of edge in the structure and dynamics of forest islands. Page 67-95 in R. L. Burgess and D. M. Sharpe, editors. Forest island dynamics in man-dominated landscapes. Springer, New York, NY.

137. Raphael, M. G. 1988. Long-term trends in abundance of amphibians, reptiles, and mammals in Douglas-fir forests of northwestern California. Pages 23-31 in Management of amphibians, reptiles, and small mammals in North America. RM-GTR-166. USDA Forest Service, Flagstaff, AZ.

138. ———, and Marcot, B. G. 1986. Validation of a wildlife-habitat-relationships model: vertebrates in a Douglas-fir sere. Pages 129-138 in J. Verner, M. L. Morrison, and C. J. Ralph. editors. Wildlife 2000: modeling habitat relationships of terrestrial vertebrates. University of Wisconsin Press, Madison, WI.

139. ———, K. V. Rosenberg, and B. G. Marcot. 1988. Large-scale changes in bird populations of Douglas-fir forests, northwestern California. Pages 63-83 in Bird conservation. University of Wisconsin Press, Madison, WI.

140. Ratner, S., R. Lande, and B. B. Roper. 1997. Population viability analysis of spring chinook salmon in the South Umpqua River, Oregon. Conservation Biology 11:879-889.

141. Redford, K. H., A. Taber, and J. A. Simonetti. 1990. There is more to biodiversity than the tropical rainforests. Conservation Biology 4:328-330.

142. Reed, J. M., C. S. Elphick, and L. W. Oring. 1998. Life-history and viability analysis of the endangered Hawaiian stilt. Biological Conservation 84:35-45.

143. Rodier, D. J., and M. G. Zeeman. 1994. Ecological risk assessment. Pages 581-604 in Lorris G. Cockerham and Barbara S. Shane, editors. Basic environmental toxicology. CRC Press, Boca Raton, FL.

144. Rossi, R. E., P. W. Borth, and J. J. Tollefson. 1993. Stochastic simulation for characterizing ecological spatial patterns and appraising risk. Ecological Applications 3:719-735.

145. Russell, G. J., T. M. Brooks, M. M. McKinney, and C. G. Anderson. 1998. Present and future taxonomic selectivity in bird and mammal extinctions. Conservation Biology 12:1365-1376.

146. Saunders, D. A., R. J. Hobbs, and C. R. Margules. 1991. Biological consequence of ecosystem fragmentation: a review. Conservation Biology 5:18-32.

147. Schamberger, M., A. H. Farmer, and J. W. Terrell. 1982. Habitat suitability index models: introduction. FWS/OBS-82/10. USDI Fish and Wildlife Service, Washington, D.C.

148. Schonewald-Cox, C. M., S. M. Chambers, B. MacBryde, and L. Thomas, editors. 1983. Genetics and conservation: a reference for managing wild animal and plant populations. Benjamin/Cummings, Menlo Park, CA.

149. Seal, U. S. 1986. Goals of captive propagation programmes for the conservation of endangered species. International Zoo Yearbook 24/25:174-179.

150. Shaffer, M. L., 1981. Minimum population sizes for species conservation. BioScience 31:131-134.

151. ———. 1983. Determining minimum viable population sizes for the grizzly bear. International Conference on Bear Research and Management 5:133-139.

152. ———. 1987. Minimum viable populations: coping with uncertainty. Pages 69-86 in M. E. Soulé, editor. Viable populations for conservation. Cambridge University Press, Cambridge, U.K.

153. ———, and F. B. Samson. 1985. Population size and extinction: a note on determining critical population sizes. American Naturalist 125:144-152.

154. Shaw, C. G., III. 1999. Use of risk assessment panels during revision of the Tongass Land and Resource Management Plan. PNW Gen. Tech. Rep. PNW-460. USDA Forest Service, Pacific Northwest Research Station, Portland, OR.

155. Simberloff, D. 1988. The contribution of population and community biology to conservation science. Annual Review of Ecology and Systematics 19:473-511.

156. ———, and L. G. Abele. 1982. Refuge design and island biogeographic theory: effects of fragmentation. American Naturalist 120:41-50.

157 Simon, N., and P. Geroudet. 1970. Last survivors. World Publishing Co., New York, NY.

158. Skalski, J. R., and S. G. Smith. 1994. Risk assessment in avian toxicology using experimental and epidemiology approaches. Pages 467-488 in Ronald J. Kendall and Thomas E. Lacher, editors. Wildlife toxicology and population modeling: integrated studies of agroecosystems. Lewis Publishers, Boca Raton, FL.

159. Song, Y. L. 1996. Population viability analysis for two isolated populations of Haianan Eld's deer. Conservation Biology 10:1467-1472.

160. Soulé, M. E. 1986. The effects of fragmentation. Pages 233-236 in M. E. Soulé, editor. Conservation biology: the science of scarcity and diversity. Sinauer Associates, Sunderland, MA.

161. ———, editor. 1987. Viable populations for conservation. Cambridge University Press, Cambridge, U.K.

162. ———. 1989. Risk analysis for the concho water snake. Endangered Species Update 6:19, 22-25.

163. ———, D. T. Bolger, A. C. Alberts, R. Sauvajot, J. Wright, M. Sorice, and S. Hill. 1988. Reconstructed dynamics of rapid extinctions of chaparral-requiring birds in urban habitat islands. Conservation Biology 2:75-92.

164. ———, and Wilcox, B. A., editors. 1980. Conservation biology: an evolutionary-ecological approach. Sinauer Associates, Sunderland, MA.

165. Starfield, A. M., and A. M. Herr. 1991. A response to Maguire. (Letters). Conservation Biology 5:435.

166. Stockwell, D. R. B., S. M. Davey, J. R. Davis, and I. R. Noble. 1990. Using induction of decision trees to predict greater glider density. AI Applications in Natural Resource Management 4:33-43.

167. Taylor, B. L. 1995. The reliability of using population viability analysis for risk classification of species. Conservation Biology 9:551-558.

168. ———, P. R. Wade, R. A. Stehn, and J. F. Cochrane. 1996. A Bayesian approach to classification criteria for spectacled eiders. Ecological Applications 6:1077-1089.

169. Temple, S. A., and J. R. Cary. 1988. Modeling dynamics of habitat: interior bird populations in fragmented landscapes. Conservation Biology 2:340-347.

170. Terborgh, J. W., and B. Winter. 1980. Some causes of extinction. Pages 119-133 in M. E. Soulé, editor. Conservation biology: the science of scarcity and diversity. Sinauer Associates, Sunderland, MA.

171. Thomas, J. W., H. Black, Jr., R. J. Scherzinger, and R. J. Pedersen. 1979. Deer and elk. Pages 104-127 in J. W. Thomas, editor. Wildlife habitats in managed forests—the Blue Mountains of Oregon and Washington. U.S. Gov. Printing Office, Washington, D.C.

172. ———, E. D. Forsman, J. B. Lint, E. C. Meslow, B. R. Noon, and J. Verner. 1990. A conservation strategy for the spotted owl. U.S. Government Printing Office, Portland, OR.

173. ———, D. A. Leckenby, M. Henjum, R. J. Pedersen, and L. D. Bryant. 1988. Habitat-effectiveness index for elk on Blue Mountain winter ranges. PNW-GTR-218. USDA Forest Service, Pacific Northwest Research Station, Portland, OR.

174. Tracy, C. R., and T. L. George. 1992. On the determinants of extinction. The American Naturalist 139:102-121.

175. Urban, D. J. 1990. The use of terrestrial field data in the practical application of ecological risk assessment principles. Pages 319-334 in L. A. Somerville and C. H. Walker, editors. Pesticide effects on terrestrial wildlife. Taylor & Francis, London, England.

176. ———, and Cook, N. J. 1986. Ecological risk assessment - hazard evaluation division standard evaluation procedure. EPA 540/9-86-167. U.S. Environmental Protection Agency, Office of Pesticide Programs, Washington, D.C.

177. USDA. 1988. Final supplement to the Environmental Impact Statement for and Amendment to the Pacific Northwest Regional Guide. USDA Forest Service, Pacific Northwest Region, Portland, OR.

178. U.S. Fish and Wildlife Service. 1980. Habitat evaluation procedures (HEP). Ecological Services Manual No. 102. U.S. Government Printing Office, Washington, D.C.

179. Veerkamp, W., and C. Wolff. 1996. Fate and exposure models in relation to risk assessment: developments and validation criteria. Environmental Science and Pollution Research International 3:91-95.

180. Wilcove, D. S. 1985. Nest predation in forest tracts and the decline of migratory songbirds. Ecology 66:1211-1214.

181. ———, C. H. McLellan, and A. P. Dobson. 1986. Habitat fragmentation in the temperate zone. Pages 237-256 in M. E. Soulé, editor. Conservation biology: the science of scarcity and diversity. Sinauer Associates, Sunderland, MA.

182. Wilcox, B. A. 1980. Insular ecology and conservation. Pages 95-117 in M. E. Soulé and B. A. Wilcox, editors. Conservation biology: an evolutionary-ecological perspective. Sinauer Associates, Sunderland, MA.

183. Wilkinson, L. 1997. Classification and regression trees. Pages 13-30 in SYSTAT 7.0: new statistics. SPSS Inc., Chicago, IL.

184. Wisdom, M. J., L. R. Bright, C. G. Carey, W. W. Hines, R. J. Pedersen, D. A. Smithey, J. W. Thomas, and G. W. Witmer. 1986. A model to evaluate elk habitat in western Oregon. USDA Forest Service PNW Region, and USDI Bureau of Land Management and Oregon Department of Fish and Wildlife, Portland, OR.

185. World Resources Institute (WRI). 1991. World Resources Report 1991-1992. A guide to the global environment. Oxford University Press, New York, NY.

186. ———. 1992. World Resources Report 1992-1993: Toward sustainable development. Oxford University Press, New York, NY.

187. Xu, H., and H. Lu. 1996. A preliminary analysis of population viability for Chinese water deer (*Hydropotes inermis*) lived in Yancheng. Acta Theriologica Sinica. 16:81-88.

Appendix I.

Criteria and procedure for selecting species to demonstrate viability risk analyses for this chapter.
To demonstrate the several examples of species viability risk analysis presented in this chapter, we selected a representative set of species from the Oregon-Washington species-environment database. At the time of the analysis, the databases were not completed, so we selected 20 species in each of 3 risk groups. The 3 risk groups were: Group I, species known to be imperiled; Group II, species very likely secure; and Group III, species with intermediate security.

To keep the comparisons at some level of parity across groups, in each group we included the same number of species from each selected taxonomic class and order. The number of species by class and order represented in each of the 3 groups was as follows:

Amphibians—3 total
Salamanders—2; Frogs—1

Reptiles—2 total
Lizards—1; Snakes—1

Mammals—5 total
Voles—1; Canids—1; Weasels or raccoons—2; Cats—1

Birds—10 total
Non-owl raptors—2; Shorebirds—1; Owls—1; Swifts—1; Woodpeckers—1; Flycatchers—1; Passerines—3

To reduce the variation in these small samples, we excluded from these lists:
(a) marine or pelagic species,
(b) obligate shorebirds, and species wholly dependent on inland freshwater aquatic environments, such as most waterfowl, although some amphibians were included,
(c) introduced or exotic species, such as wild turkey,
(d) species at the edge of their distributional ranges (e.g., Tennessee warbler),
(e) species that occur in the WA-OR area only sporadically or irruptively (e.g., snowy owl),
(f) species that occur in only a small portion of the geographic area, particularly in only one state (e.g., Olympic marmot); an exception is lynx, which historically ranged into Oregon and for which there are intermittent (and largely unconfirmed) sightings in Oregon.

We used this nonrandom selection process because our aim was to demonstrate selected types of risk analysis models across as wide array of species types as possible, and not to develop a thorough risk analysis for all species. The species chosen for the analyses were as follows.

Group I. Species Known To Be Imperiled
This group includes 20 Federally listed threatened or endangered species, as well as species listed by Washington or Oregon State as candidates, sensitive, critical, or in decline. Where multiple species matched these criteria, where possible we chose the species with the distribution that included both states.

Dunn's Salamander
Van Dyke's Salamander
Northern Leopard Frog
Plateau Striped Whiptail
Striped Whipsnake
Bald Eagle
Peregrine Falcon
Snowy Plover
Northern Spotted Owl
Vaux's Swift
White-Headed Woodpecker
Gray Flycatcher
Western Bluebird
Bobolink
Grasshopper Sparrow
Gray-Tailed Vole
Gray Wolf
Wolverine
Fisher
Lynx

Group II. Species Very Likely Secure
This group includes 20 species known to be widespread, common, or abundant, and thus highly likely to be secure.

Roughskin Newt
Ensatina
Pacific Treefrog
Western Fence Lizard
Gopher Snake
Turkey Vulture
American Kestrel
Killdeer
Barn Owl
White-Throated Swift
Hairy Woodpecker
Cordilleran Flycatcher
Cedar Waxwing
White-Crowned Sparrow
Purple Finch
Meadow Vole
Coyote
Raccoon
Long-Tailed Weasel
Bobcat

Group III. Species With Intermediate Security
This group includes 20 species with viability security
judged to be intermediate between that of Groups I and
II.

> Tiger Salamander
> Larch Mountain Salamander
> Columbian Spotted Frog
> Western Whiptail
> Sharptail Snake
> Cooper's Hawk
> Merlin
> Black-Bellied Plover
> Northern Saw-Whet Owl
> Black Swift
> Pileated Woodpecker
> Willow Flycatcher
> Bohemian Waxwing
> Sage Sparrow
> Gray-Crowned Rosy Finch
> White-Footed Vole
> Kit Fox
> Ringtail
> American Marten
> Mountain Lion

Appendix 2.

**Life history and habitat versatility fields used in examples
of risk analysis procedures. Values for some fields were simplified from the original values in the
matrixes. Risk score for values of some fields was estimated from the literature for the examples.**

SHP Field Name	Field Type[a]	Field Values[b]	Risk Score[c]
Common name	3	common name	na
Taxonomic class	3	A=amphibian; B=bird; M=mammal; R=reptile	na
Taxonomic order	3	name of taxonomic order	na
Taxonomic family	3	name of taxonomic family	na
Breeding status	3	I=breeds in OR; 2=breeds in WA; 3=breeds in both states; 4=non-breeder	na
Occurrence status by state	3	0=does not occur in the state; I=native; 2=non-native; 3=reintroduced	na
Risk category[d]	I	I=Group I—species known to be imperiled; 2=Group II—species very likely secure; 3=Group III—species with intermediate security	na
Type of seasonal activity	2	I=hibernation; 2=estivation; 3=both (hibernation & estivation) 4=none	na
Migration/seasonal movements	I	I=latitudinal	7
		2=altitudinal	4
		4=year round resident	I
		6=both I and 2	10
Migration/seasonal movement distance class	I	I=< 10 km & <100 km	4
		2=< 1,000 km	7
		3=< 10,000 & >10,000 km	10
		4=none	I
Forms aggregations	I	y=yes	10
		n=no	0
Juvenile dispersal distance class	I	I=<100 m & < I km	10
		2=<10 km	5
		3=< 100 km & >100 km	0

SHP Field Name	Field Type[a]	Field Values[b]	Risk Score[c]
Average # of offspring per litter, or eggs per clutch	1	1=1-2 & 3-6	10
		2=7-10 & 11-15	5
		3=16-20 & 21-50 & >50	0
# Litters or clutches / year	1	1 ≤ 1	10
		2 ≤ 2	7
		3 ≤ 3	4
		4 > 3	1
Average age at first breeding (females)	1	1=<6 mo.	0
		2=1 yr.	1
		3=2 yr.	4
		4=3 yr.	7
		5=4+ yr.	10
Mating system	2	1=polygamy; 2=lifetime monogamy; 3=lek ; 4=promiscuous	na
Home range size class	1	1=<1 ha & 1-10 ha	1
		2=11-50 ha & 51-100 ha	4
		3=101-500 ha & 501-1000 ha	7
		4=1001-10,000 ha & > 10,000 ha	10
Fidelity to summer range	2	1=high; 2=medium; 3=low	na
Fidelity to winter range	2	1=high; 2=medium; 3=low	na
Geographic range	1	1=locally endemic	10
		2=regionally endemic	7
		3=moderately widespread, or widespread	1
		4=WA peripheral or OR peripheral	10
Population distribution	1	1=contiguous	0
		2=gaps	1
		3=patchy	4
		4=isolated	7
		5=scarce	10
Landscape use	2	1=patch; 2=mosaic; 3=generalist; 4=contrast	na
Elevational range	2	1=0-1000 feet; 2=<3000 feet; 3=<5000 feet; 4=>5000 feet	na
Diet	1	1=primary consumer (herbivore)	2
		2=secondary or tertiary consumer (carnivore)	8
		3=primary invertebrates	4
		4=both 1 & 3 (largely, insectivorous and eats seeds and fruits)	0
Foraging location	2	1 terrestrial (underground, on ground, in down wood); 2=lower vegetation (shrub or tree bole/bark); 3=upper vegetation (tree canopy); 4=other (unknown, in or underwater, or aerial)	na
Mass (body size)	1	1=<500 gm	0
		2=<1 kg	2
		3=<10 kg	4
		4=<100 kg	6
		5=<1000 kg	8
		6=>1000 kg	10
Habitat versatility[d]	1	Continuous variable from 1-100 representing the percentage of the 32 habitats used by the species ("Y" code in occurrence field of habitat matrix)	
		0-25%	10
		26-50%	7
		51-75%	4
		76-100%	0
Structural versatility[d]	1	Continuous variable from 1-100 representing the percentage of the 54 structural types used by the species ("Y" code in occurrence field of habitat matrix)	
		0-25%	10
		26-50%	7
		51-75%	4
		76-100%	0

Notes to Appendix 2

[a] 1=primary field; 2=secondary field; 3=information field. The example classification tree analysis used primary fields in the calculations, and the secondary and information fields to aid *post hoc* interpretations. The example rule-induction analysis used all 3 types of fields in the analysis. The differences in which types of fields were used pertained to meeting the statistical assumptions of the two analyses.

[b] Values found in the matrix for some fields were combined to simplify the example analyses.

[c] Created for this analysis and not found in the matrixes. Relative risk score (1-10) for each field value. Values were estimated for primary fields only for use in the rank-model example.

[d] Field created for this analysis and not found in the matrixes.

Appendix 3.

Running the Species At Risk Advisor (SARA): factors and their values used in the example rule-induction knowledge base model that provides most likely viability risk levels of wildlife species.

Welcome to SARA: Species at Risk Advisor

Version 0.30, 4 January 1999, author: Bruce G. Marcot

This is an expert advisory model that demonstrates how life history characteristics can be used to infer potential risk levels of terrestrial vertebrate wildlife species. This model is based on a selected sample of 60 species—20 species in each of 3 known risk categories (imperiled/listed, secure, and intermediate). The model is based on an optimized rule set using the Quinlan Q3 rule-induction algorithm, that produces the most efficient use of the example information for inferring risk levels. This model is ONLY A DEMONSTRATION of the kind of risk level analysis possible. It is NOT intended to provide definitive predictions of risk levels of all species. NOTE: You may respond to each question by choosing the most appropriate answer, or typing "?" (without the quotes) to denote that you don't know. Let's begin!

TaxClas [What is the taxonomic class?]
amphibian
reptile
bird
mammal

TaxOrder [What is the taxonomic order?]
Caudata
Anura
Squamata
Falconiform [Falconiformes]
Charadriifo [Charadriiformes]
Strigiforme [Strigiformes]
Apodiformes
Piciformes
Passeriform [Passeriformes]
Rodentia
Carnivora

BreedStatus [What is the breeding status?]
bs1 [the species is documented or suspected as breeding in Oregon]
bs2 [the species is documented or suspected as breeding in Washington]
bs3 [the species is documented or suspected as breeding in both states]
bs4 [the species is documented or suspected of being a non-breeder in both states]

OccStatusOR [What is the occurrence status in Oregon?]
oso1 [native]
oso2 [non-native (introduced accidentally or purposefully, or self-invader)]
oso3 [reintroduced (native species extirpated then reestablished w/ introductions)]
oso0 [does not occur in the state]

OccStatusWA [What is the occurrence status in Washington?]
osw1 [native]
osw2 [non-native (introduced accidentally or purposefully, or self-invader)]
osw3 [reintroduced (native species extirpated then reestablished w/ introductions)]
owo0 [does not occur in the state]

OccStatusBo [What is the occurrence status in both Oregon and Washington?]
osb1 [native]
osb2 [non-native (introduced accidentally or purposefully, or self-invader)]

TypSeaInact [What is the type of seasonal inactivity?]
tsi1 [hibernation (dormancy associated with cold period of the year)]
tsi2 [estivation (dormancy associated with warm/dry period of the year)]
tsi3 [both hibernation & estivation]
tsi4 [none]

Mig_SeasMov [What is the type of migrational or seasonal movement?]
migs1 [latitudinal (change in latitude, north/south)]
migs2 [altitudinal (change in elevation)]
migs4 [year-round resident (stays in same vicinity throughout the year)]
migs6 [both latitudinal and altitudinal]

MigDist [What is the migration or seasonal movement distance class?]
migd1 [< 100 km]
migd2 [< 1000 km]
migd3 [> 1000 km]
migd4 [none (non-migratory)]

Aggreg [Does the species form aggregations or groups?]
yes [yes (forms concentrations of individuals during some periods or activities)]
no

JuvDispDist [What is the juvenile dispersal distance class?]
juv1 [< 1 km]
juv2 [< 10 km]
juv3 [> 10 km]

NOffsprg [What is the average number of offspring per litter, or eggs per clutch?]
noff1 [1-6]
noff2 [7-15]

noff3 [>15]

NLitt_Clut [What is the number of litters or clutches per year?]
nlit1 [< or = 1]
nlit2 [< or = 2]
nlit3 [< or = 3]
nlit4 [> 3]

Age1st [What is the average age at first breeding (females)?]
age1 [< 6 mo.]
age2 [1 yr]
age3 [2 yrs]
age4 [3 yrs]
age5 [4+ yrs]

MatingSys [What is the mating system?]
mat1 [polygamy (seasonal monogamy, polyandry, or polygyny)]
mat2 [lifetime monogamy]
mat3 [lek]
mat4 [promiscuity]

HRSizeClass [What is the home range size class?]
hr1 [< or = 10 ha]
hr2 [< or = 100 ha]
hr3 [< or = 1000 ha]
hr4 [> 1000 ha]

SiteFidSumm [What is the site fidelity to summer range (i.e., home range overlap among years)?]
sw1 [high (>50% of popn returns to same home range or territory the next year)]
sw2 [medium (>10% of popn returns)]
sw3 [low (<10% of popn returns)]

SiteFidWint [What is the site fidelity to winter range (i.e., home range overlap among years)?]
sw1 [high (>50% of popn returns to same home range or territory the next year)]
sw2 [medium (>10% of popn returns)]
sw3 [low (<10% of popn returns)]

GeogRng [What is the geographic range?]
geog1 [locally endemic (occurs only in OR &/or WA, or a small area thereof)]
geog2 [regionally endemic (occurs only in Pacific Northwest or Great Basin region)]
geog3 [moderately widespread, or widespread (occurs thru western states/No. Amer.)]
geog4 [WA peripheral or OR peripheral (popn in WA or OR is on edge of its range)]

PopnDistrb [What is the population distribution?]
pop1 [contiguous (habitat is broadly & contiguously distributed over sp. range)]
pop2 [gaps (habitat broadly distributed but w/ gaps causing some popn isolation)]
pop3 [patchy (habitat exists mostly as disjunct patches, moderate recolonization)]
pop4 [isolated (habitat occurs only as locally isolated patches, low recolonization)]
pop5 [scarce (habitat is very scarce thru OR/WA, little/no recolonization)]

LndscpUse [What is the landscape use category?]
lnd1 [patch (sp. likely uses only 1 homogeneous habitat patch during life cycle)]

lnd2 [mosaic (sp. uses aggregates of patches of habitat, but 1 structural stage)]
lnd3 [generalist (sp. uses all or many patch types, & >1 structural stage)]
lnd4 [contrast (sp. requires contrast betw. 2 struc. stages in close proximity)]

ElevRng [What is the upper elevational range of typical or regular occurrence?]
ele1 [up to 1000 ft]
ele2 [up to 3000 ft]
ele3 [up to 5000 ft]
ele4 [> 5000 ft]

Diet [What is the diet category?]
diet1 [primary consumer (herbivore)]
diet2 [secondary or tertiary consumer (carnivore)]
diet3 [primarily invertebrates]
diet4 [omnivore (both herbivore and carnivore, mostly insects & seeds/fruits)]

Forag1 [Does the organism forage on terrestrial substrates (underground, on ground, or on/in down wood)?]
yes
no

Forag2 [Does the organism forage in lower vegetation (shrub or understory, or tree bole/bark)?]
yes
no

Forag3 [Does the organism forage in upper vegetation (tree canopy)?]
yes
no

Forag4 [Does the organism forage underwater or aerially; or is foraging substrate unknown?]
yes
no

BodyMass [What is the body mass class (weight of larger sex)?]
bm1 [< 500 gm]
bm2 [< 1 kg]
bm3 [< 10 kg]
bm4 [< 100 kg]
bm5 [< 1000 kg]
bm6 [> 1000 kg]

HabVers<?>
StrucVers<?>

RESULT [The risk level of the species is:]
Grp1 [Group I—Species known to be imperiled (there is at least one example of a species with these attributes that is Federally or State listed)]
GrpII [Group II—Species very likely secure (there is at least one example of a species with these characteristics that is widespread or common and not in any danger)]
GrpIII [Group III—Species with intermediate security (there is at least one species with these characteristics that is intermediate in security status)]

20

Terrestrial and Marine Management Activities: Links to Habitat Elements and Ecological Processes

Madeleine Vander Heyden & Bruce G. Marcot

Introduction

Much has been published on the effects of human activities on the environment,[5, 9, 16] and recent efforts have attempted to synthesize the literature into comprehensive, easily accessible, digital formats. For example, several state wildlife agencies have produced Web pages that include information on the effects of human activities on individual wildlife species. We believe that a weakness in this approach is its inability to capture the complexity of impacts beyond the individual species level. Ecosystem management suggests that scientists and managers should also determine the effects of management activities on ecological processes and functions, and on the long-term sustainability, diversity, and productivity of resources and environments. This chapter discusses a Washington-Oregon "Management Activities Matrix" (located on the CD-ROM) and illustrates how it can be used with the other data matrices described in this volume to ask more advanced questions than suggested by existing information bases. For example, the Management Activities Matrix allows the manager not only to determine which species may be affected by a particular land use, but also to assess what ecological functions are involved, and which Habitat Elements (HEs).[14] In this chapter we present and illustrate this unique perspective and methodology. Our approach is similar to the concept of the food web: effects of management actions are not only direct and linear but also can have surprisingly indirect and nonlinear implications. To gauge the full extent of an activity's impact, it becomes necessary to examine the entire web of influences.

Development of the Management Activities Matrix

Producing the Management Activities Matrix presented many challenges. Environmental impacts are complex and defy simple categorization. Therefore we have limited the Management Activities Matrix only to those relationships that were identified either in the literature or by an expert panel, and for the most part, we have reported only direct impacts. We did not attempt to describe in detail the exact nature of every possible influence, because they depend heavily on many factors including local site characteristics and the specific nature of the proposed activity. Any *potentially* affected HE, regardless of the scale or intensity

of the activity that would be necessary to produce an effect, is linked in the Matrix. For example, one could ask: overall, do road management activities have the potential to affect soil structure and soil organic matter? The Management Activities Matrix can be used as a guide to depict or predict influences on HEs and ecological processes, and to pose concepts, frameworks, and hypotheses regarding the effects of management activities. It is necessarily not a definitive model that precisely predicts site-specific impacts resulting from the described activities.

The Land Use and Management Activities

We wanted to include activities that we thought were most relevant to Oregon and Washington, and were most useful to characterize across all land ownerships. We limited the activities addressed to those that affect inland and nearshore-marine HEs only. Our activity list was refined and edited several times with the assistance of many resource scientists and managers expert in particular areas (see Acknowledgements). Thirteen broad categories of management activities were identified, ranging from nearshore marine resource management to urban development, and encompassing activities occurring on all land ownerships, locations, and jurisdictions in both Oregon and Washington. These broad categories contain 152 specific activities (see Table 1).

Information on CD-ROM

Under the Data Query section on the CD-ROM, there are two submenus under Management Activities; one to query for related habitat elements, the other to query the citations that support the Management Activity Matrix. The user can query these to either obtain very specific information on individual HEs or to obtain general information from the literature about the potential effects of an activity. These two queries are described below.

Querying for Related Habitat Elements. Also based on the published literature, this table identifies which specific HEs are potentially affected by each management activity (either positively or negatively). In addition, under the heading "Query for Related Habitat Elements," users can query each management activity to obtain a comprehensive list of all HE-management activity links. This list was developed using the literature cited in the table and expert panels (discussed below under Data Sources).

Querying for Citations. Based on the published literature, this table describes the effects of a specific activity, e.g. "thinning" or "impounding water," on various ecological processes such as biological diversity or ecosystem function.

Linking HEs to management activities allows the manager to query the various data matrixes in this volume, asking such questions as: What species are associated with the HEs linked to this specific management activity, and, furthermore, what are their Key Ecological Functions (KEFs)?[11] We illustrate this approach in detail in the section "How to Use the Matrix: Example Queries."

Data Sources

We used two sources of data to populate the Management Activities Matrix: the published literature and an expert panel. During the literature review, we used several reference databases to search titles, keywords, and full abstracts for information, using the management activities as key words. We searched literature on wildlife, ornithology, fisheries, mammalogy, and zoology. The wildlife database search included all papers published during 1937-97 in the *Journal of Wildlife Management, Wildlife Monographs,* and the *Wildlife Society Bulletin* (>8,400 records). The ornithology database search included all papers published during 1955-97 in the *Auk, Ornithological Monographs, Condor, Studies in Avian Biology, Wilson Bulletin,* and the *Journal of Field Ornithology* (>18,000 records). The mammalogy database search included all papers published during 1950-97 in the *Journal of Mammalogy, Mammalian Species,* and *American Society of Mammalogists Special Publications* (>10,900 records). The ecological database search included papers published during 1945-97 in *Ecology, Ecological Applications,* and *Ecological Monographs,* plus some additional papers published in the *Journal of Vegetation Science* (1900-97), for a total of >12,000 records. We also obtained references from our personal reference databases, project library, and the Cambridge Scientific Review.

Whenever a link was found in the published literature that associated a particular management activity with an ecological process or a HE, we coded the information in our database. If the information was from research, we briefly described the results, where the study took place, at what time of year, and in what habitat. When the referenced links were not the results of research, the above information was filled out to the extent possible. We listed habitats as they were described, and did not attempt to "crosswalk" the habitat classes to those used in this volume (see Wildlife Habitats).[3]

To ensure that all potentially affected HEs were properly identified for each activity, we convened an expert panel consisting of scientists and managers with backgrounds in the various activity areas (see Acknowledgements). The panelists reviewed each activity and indicated which HEs were potentially linked to it. Links obtained from the panels were incorporated into the database along with links identified from the literature to form a comprehensive list of all HEs potentially associated with each activity. Individual activities may be queried to obtain this information (see Query for Related Habitat Elements on the CD-ROM).

Major Influences of Activities on Wildlife

Throughout Washington and Oregon, management activities affecting the most wildlife species (>500) pertain to conversion of habitat for development of human habitation, recreation, mineral extraction, forestry, water supply development, controlling and prescribing fire, and livestock grazing (Table 1). The management activities affecting the least wildlife species (<100) include mushroom harvesting, snow-related recreation, creation of artificial nest sites, and providing dead and down wood (Table 1). If the dead and down wood category is a surprise,[15] note that this activity still influences some 96 wildlife species and 9 HEs.

The ranked order of management activities having the most and least influence on number of wildlife species and HEs varies somewhat among specific wildlife habitats. However, the general pattern of land-development activities having the greatest effect remains mostly the same across most terrestrial habitats, including coastal environments. In addition, not surprisingly, other management activities related to aquatic resource management (Management Activity 2; see Table 1) and marine activities (Management Activity 10) show up as potentially influencing a large number of species and HEs in coastal environments. For example, in Bays and Estuaries (Wildlife Habitat 28), the most influential management activity is recreational development (potentially influencing 173 wildlife species in that habitat); other activities of major influence (potentially influencing >150 wildlife species) pertain to conversion of native habitats; road and building construction; controlling water pollution; dredging; harbor, marina, and ferry terminal development; wastewater treatment; and water level management.

How to Use the Matrix: Example Queries

Using a relational database such as Access or Paradox, the following is an example of a query on fire management that illustrates specifically how the Matrix may be used. Let's say we're interested in the influence of the management activity "low- to moderate-intensity burns" (Management Activity 1B), as may be used during prescribed burning to help restore some grasslands and forests east of the crest of the Cascade Mountains in Washington and Oregon, and how such influences may differ between grassland and forest environments.

We first linked the Management Activities Matrix to the Habitat Elements (HEs) Matrix and counted 83 HEs potentially influenced by this activity (the HE tally here refers to HE category headings and subheadings, so there is some redundancy in these figures). Next, we narrowed this by linking the number of HEs affected by this activity to those only associated with wildlife species that occur

Table 1. Management activities potentially affecting the greatest, and least, number of wildlife species, and the associated number of potentially affected Habitat Elements (HEs) across all wildlife habitats in Washington and Oregon.*

Management activity	No. wildlife species affected	No. habitat elements affected
Greatest influence (>500 species)		
Conversion of native habitats	577	74
Recreational developments	570	128
Road construction and obliteration	555	70
Mineral exploration	553	123
Building houses and businesses	552	149
Surface/strip mining and processing	544	141
Forest management (in general)	532	64
Conversion of shrubland to native or non-native grassland (for livestock management)	528	51
Conversion of shrubland to native or non-native grassland (for shrubland or grassland management)	528	51
Clearcutting	525	88
Establishing/maintaining greenways and greenbelts	517	105
Suppressing wildfire	514	83
Prescribed/controlled high intensity burns	503	91
Increasing water supply	503	66
Decreasing water supply	503	66
Livestock grazing	500	70
Least influence (<100 species)		
Retaining/providing dead/down wood	96	9
Creating/maintaining islands or rafts within impoundments	61	3
Providing artificial nest sites (for agricultural activities)	59	2
Providing artificial nest sites (for forest habitat management activities)	59	2
Snowshoeing/snow skiing/sledding	42	2
Snowmobiling	42	2
Harvesting wild mushrooms	26	1

* See Appendix for management activity codes and definitions.

in Habitat 15, Eastside (Interior) Grasslands. This asks the question, what array of HEs for wildlife are potentially influenced by low- to moderate-intensity burns, specifically in Eastside Grasslands? The result was a list of 73 HE categories. We next compared this with use of low- to moderate-intensity burns in Wildlife Habitat 7, Ponderosa Pine Forests and Woodlands, and this produced a list of 78 HE categories.

However, even though the number of HEs is similar between these two habitats, the actual HE categories and the associated array of wildlife species may differ. We tested this by listing wildlife species rather than HEs in each query, saving those results, and comparing the results between the two habitats. It turns out that the HEs potentially affected by low- to moderate-intensity burns are virtually identical in these two habitats (72 of the potentially affected HE categories are shared by these two habitats). However, of the 163 wildlife species in Eastside Grasslands and 218 species in Ponderosa Pine Forests and Woodlands that have HEs potentially influenced by this activity, only 99 of these species occur in common between these two habitats. That is, 39% of the potentially fire-affected wildlife species in the Grasslands habitat and 55% of those in the Forest habitat occur uniquely in these

habitats. Thus, as should be expected, the potential influence of this kind of burn on wildlife species is different in eastside grasslands and forests, and a greater proportion of affected species in the forest habitat would be uniquely influenced.

Discussion

Stressors and Indicators. Why should the manager be concerned about evaluating influences of management activities on Habitat Elements and wildlife? Beyond the obvious reasons—that is, environmental and biological assessments to meet legal and regulatory mandates—the Management Activities Matrix provides a basis for explicitly and repeatedly describing the potential influences of stressors on wildlife communities. The model of identifying such stressors (e.g., see the species influence diagram [11]) can be central to identifying the most influential management activities for mitigation, and prioritizing potential wildlife responses for monitoring.[10, 16, 18] What needs empirical work is validating, refining, and quantifying the linkages between management activities and habitat elements and associated wildlife populations.

Although we have purposely avoided the problematic concept of management indicator species, wildlife species most susceptible to particular stressors and management activities might be good candidates for use as bioindicators and "early warning" signals of impending changes to other aspects of their ecosystem.[13] One example is stream amphibians that indicate levels of aquatic ecosystem stress.[20] However, cryptogams, plants, and invertebrates often serve as more sensitive indicators,[1, 12, 19] or, in many cases, it may be simpler to more directly monitor specific biochemical responses to management activities.

Considerations and Caveats for Using the Matrix

The management activities listed in the Management Activities Matrix are necessarily described in general ways, that is, at broad geographical scales. However, because the effects of management activities act as "stressors" to native systems, they vary by degree of perturbation based on duration, frequency, intensity (amplitude), and combination of activities. Furthermore, the specific influence of a management activity on ecological processes and HEs will vary according to the scale, intensity, and the action of the effect (direct or indirect). Referring to the management activity, *scale issues* include the level of spatial resolution (the pervasiveness of the activity), overall geographic extent and context (over how large an area does the activity occur, and in what type of environment), and duration of the activity.

Specifically, *spatial resolution* refers to the geographic extent over which the activities and associated effects occur. For example, some activities affect entire watersheds, whereas others affect only the stream environment. Identifying the level of spatial resolution of the activity determines how finely the Management Activities Matrix can be applied. The Matrix should not be used to predict effects at the scale of individual vegetation stands or point locations, but it can be used to predict overall, general effects averaged at the scale provinces, habitats, and habitat structures.

Geographic extent and context influence an effect's intensity. The geographic extent of agricultural conversions in eastern Washington and Oregon, for example, is quite widespread; agricultural activities in that region may have different effects on wildlife community composition, structure, and function than will small inclusions of the same activities in a matrix of native grassland or forest. Likewise, agricultural activities adjacent to a wetland will impact wildlife communities differently than agricultural development in an upland context.[2]

Time duration or temporal scale refers to the influence or persistence of effects over time, in terms of how long a management activity has persisted in an area and how long wildlife communities and populations have had to respond. It also refers to the immediacy of an influence; some effects might be time-delayed. Two contrasting examples are faunal relaxation (time-delayed loss of species from isolated native habitats), and recolonization of restored native environments.

Obviously, activities vary in their effects, depending on which HEs are involved. Some HEs will be directly affected and others indirectly. Both direct and indirect effects can result from a single management activity, and can influence wildlife populations through the HEs.

Although all of the above-mentioned factors will determine the ultimate impact of an activity, they are not explicitly depicted in the Matrix because of the difficulties of addressing these complex factors in a simple, qualitative matrix format. Managers need to consider the issues described above and apply the findings of the Matrix to their own geographic area and project scope. Because the influence of a management activity depends largely on local conditions and circumstances, variations in specific effects should be expected.

When evaluating effects of activities, managers might wish to consider the following:

1. The influence of management activities on HEs can vary from those depicted in this simple Management Activities Matrix in several ways: some depicted effects may not occur, some may occur more saliently than others, and there may be some effects not depicted in the Matrix;
2. Effects can vary over space and time, including time lags and off-site influences; and
3. Not all effects are negative; identifying the positive ones could be most useful for some conservation objectives such as habitat restoration.

Conclusions and Suggestions for Further Developments

The Management Activities Matrix provides a rigorous, repeatable basis for considering the influence of land and resource management activities on wildlife habitats and species. By querying the Matrix in conjunction with the other habitat- and species-based databases offered in this volume, the manager can be prompted to consider effects across a full range of environmental conditions and wildlife species groups. Certainly, some activities and wildlife species warrant far more detailed attention than can be provided here.

Spatially, the Matrix might prove most useful when applied to broad geographic areas such as National Forest Ranger Districts, or to general land use allocations. At finer spatial scales, such as individual vegetation stands or point locations, the Matrix might overestimate the number of HEs and wildlife species affected by given management activities.

Temporally, the Matrix does not specify how quickly or for how long an effect can be expected. Stressors to ecosystems operate in different ways depending on their initial intensity as well as duration. Also, effects of some management activities may be indirect or incur time lags. The manager could query the Management Activities Matrix and determine likely HEs and species affected (positive and negative), and then sort those according to expected levels of intensity, duration, and time lag.

Local conditions, site history, and the range of specific activities greatly influence how any given management activity will affect habitats and wildlife. Thus, use of the Management Activities Matrix may best be seen as helping the manager to develop working hypotheses of effects warranting expert review and, where needed, empirical validation through local testing. At best, the Matrix can be used as a basis for devising and prioritizing adaptive management monitoring studies to better discern and refine the initial, crude estimates of effects.

The next level of development of the Management Activities Matrix could then entail refining the activity categories for greater detail, and quantifying the specific influence of activities on HEs and wildlife species. Several tools may be useful for quantifying the influence of management activities into the "causal web" of wildlife communities, including use of Bayesian belief networks[8] and sequential Bayes statistics,[7] decision-aiding programming,[6] and dynamic, stochastic simulation models.[4] The influence diagram approach presented in Chapter 6 can provide an overall framework for such further modeling exploits.[17] In the end, the challenge to the manager is to consider the influence of management activities in the full web of ecological interactions within the ecosystem.

Acknowledgments

Many people contributed their expertise to the development of the Management Links Matrix. We are indebted to the following biologists and land managers:

Database development: Susan Tank, Marla Trevithick, and Thomas O'Neil.

Classification, definition, and review of management activities: Susan Tank, Thomas O'Neil, David H. Johnson, Pat Chapman, Curt Leigh, Randy Carman, Cheryl Friesen, Donavin Leckenby, and Doug Runde.

Literature Review: Susan Tank, Derek Stinson, and Kelly Bettinger.

Expert panelists: Susan Tank, Cheryl Broyles, Paul Wagner, David H. Johnson, Kelly Bettinger, E. Charles Meslow, and Thomas O'Neil.

Literature Cited

1. Anderson, E. W. 1986. Plant indicators of effective environment. Rangelands 8:70-73.

2. Boutin, C., and B. Jobin. 1998. Intensity of agricultural practices and effects on adjacent habitats. Ecological Applications 8:544-557.

3. Chappell, C. B., R. C. Crawford, C. Barrett, J. Kagan, D. H. Johnson, M. O'Mealy, G. A. Green, H. L. Ferguson, W. D. Edge, E. L. Greda, and T. A. O'Neil. 2001. Wildlife habitats: descriptions, status, trends, and system dynamics. Pages 22-114 *in*: D. H. Johnson and T. A. O'Neil, managing directors. Wildlife-habitat relationships in Oregon and Washington. Oregon State University Press, Corvallis, OR.

4. Costanza, R., F. H. Sklar, and M. L. White. 1990. Modeling coastal landscape dynamics. BioScience 40:91-107.

5. Dale, V. H. 1997. The relationship between land-use change and climate change. Ecological Applications 7:753-769.

6. Engle, D. M., D. J. Bernardo, T. D. Hunter, J. F. Stritzke, and T. G. Bidwell. 1996. A decision support system for designing juniper control treatments. AI Applications 10:1-11.

7. Gazey, W. J., and M. J. Staley. 1986. Population estimation from mark-recapture experiments using a sequential Bayes algorithm. Ecology 67:941-951.

8. Haas, T. C. 1991. A Bayesian belief network advisory system for aspen regeneration. Forest Science 37:627-654.

9. Kryuchkov, V. V. 1993. Extreme anthropogenic loads and the northern ecosystem condition. Ecological Applications 3:622-630.

10. Maltby, L. 1999. Studying stress: the importance of organism-level responses. Ecological Applications 9:431-440.

11. Marcot, B. G., and M. Vander Heyden. 2001. Key ecological functions of wildlife species. Pages 168-186 *in*: D. H. Johnson and T. A. O'Neil, managing directors. Wildlife-habitat relationships in Oregon and Washington. Oregon State University Press, Corvallis, OR.

12. McGoech, M. A. 1998. The selection, testing, and application of terrestrial insects as bioindicators. Biological Review 73:181-201.

13. McLaren, M. A., I. D. Thompson, and J. A. Baker. 1998. Selection of vertebrate wildlife indicators for monitoring sustainable forest management in Ontario. Forestry Chronicle 74:241-248.

14. O'Neil, T. A., K. A. Bettinger, M. Vander Heyden, B. G. Marcot, C. Barrett, T. K. Mellen, W. M. Vander Haegen, D. H. Johnson, P. J. Doran, L. Wunder, and K. L. Boula. 2001. Structural conditions and habitat elements of Oregon and Washington. Pages 115-139 *in*: D. H. Johnson and T. A. O'Neil, managing directors. Wildlife-habitat relationships in Oregon and Washington. Oregon State University Press, Corvallis, OR.

15. Rose, C. L., B. G. Marcot, J. L. Ohmann, K. L. Waddell, B. Schreiber, T. K. Mellen, and D. L. Lindley. 2001. Decaying wood in Pacific Northwest forests: concepts and tools for habitat management. Pages 580-623 *in*: D. H. Johnson and T. A. O'Neil, managing directors. Wildlife-habitat relationships in Oregon and Washington. Oregon State University, Corvallis, OR.

16. Schindler, D. W. 1987. Detecting ecosystem responses to anthropogenic stress. Canadian Journal of Fisheries and Aquatic Science 44:6-25.

17. Schlapfer, F., and B. Schmid. 1999. Ecosystem effects of biodiversity: a classification of hypotheses and exploration of empirical results. Ecological Applications 9:893-912.

18. Sibly, R. M. 1999. Efficient experimental designs for studying stress and population density in animal populations. Ecological Applications 9:496-503.

19. Stolte, K., D. Mangis, R. Doty, and K. Tonnessen. 1993. Lichens as bioindicators of air quality. U.S. Forest Service General Technical Report RM-224. Rocky Mountain Forest and Range Experiment Station, Fort Collins, CO.

20. Welsh, H. H., Jr, and L. M. Ollivier. 1998. Stream amphibians as indicators of ecosystem stress: a case study from California's redwoods. Ecological Applications 8:1118-1132.

Appendix

Land use and management activities depicted in the Management Activities Matrix.

Code	Activity	Definition
I	**Fire Management**	
IA	Suppressing wildfire	Actively extinguishing or preventing wildfires.
IB	Low- to moderate -intensity burns	Fires that are usually intentionally lit (or natural fires that are allowed to burn) for a specific management objective. The extent, intensity, and timing are either planned or controlled. Low-intensity burns usually are repeated at regular time intervals.
IC	High-intensity burns	Usually natural fires that are allowed to burn for a specific management objective. The extent, intensity, and timing are either planned or controlled. High-intensity burns are usually a one-time event or occur very infrequently.
2	**Freshwater Wetland, Riparian, and Aquatic Resource Management**	
2A	Creating and maintaining impoundments	Activities include the construction of main and retention dams that create impoundments >4 ha.
2B	Controlling water levels	The effects of raising and lowering water levels within an impoundment, assuming the high and low water marks are already well established, and considering only those effects that occur within the variable zone.
2C	Creating/maintaining islands or rafts within impoundments	Naturally occurring islands that result from high water levels cutting off peninsulas, and human-made rafts created from a variety of materials. Both rafts and islands are <0.8 ha. Also includes dredge spoil islands.
2D	Draining wetlands, marshes, ponds, lakes	Effects associated with the draining of fully functional aquatic systems.
2E	Increasing water supply	Assumes, within the context of a stream, wetland, or small lake (<4 ha), that flooding results in an increase in water supply that is sustained for 2-3 months (or more) over several years.
2F	Decreasing water supply	Flow withdrawal occurs within the context of a stream, wetland, or small lake (<4 ha) and these water bodies normally contain open water for most of the year.

Code	Activity	Definition
2G	Burning wetlands to maintain successional stages	Periodic, low intensity burning of wetlands that occurs in association with rivers, lakes, and streams. The extent, intensity, and timing are either planned or controlled.
2H	Restoration of wetlands	Revegetation with native wetland species, and the maintenance of water levels for the majority of the year.
2I	Wetland management techniques	Describes a variety of methods by which wetland ecosystem function is maintained.
2J	Flooding fields and wetlands	Flooding that would, without human intervention, occur normally in agricultural habitats and other nonforest environments due to site conditions and the water table.
2K	Removing riparian vegetation	Removal of trees and shrubs within 30 m of a waterway.
2L	Livestock grazing of riparian areas	The effects of primary grazers (cattle, horses, and sheep) within 366 m of a waterway.
2M	Adding coarse woody debris and boulders to streams and rivers	Intentional addition of materials to enhance aquatic habitat conditions within streams and rivers.
2N	Removing coarse woody debris from streams and rivers	Intentional removal of materials from streams and rivers resulting in degradation of aquatic habitat conditions.
2O	Restoring/maintaining beaver populations	Restoring or maintaining beaver populations will retain the primary function of beavers: to deliver down wood to aquatic systems and produce small impoundments (<0.8 ha).
2P	Retaining riparian buffer strips	Activities associated with maintaining trees and shrubs within 30 m of a waterway.
2Q	Armoring banks for erosion control	Enhancing bank stability within streams and rivers (e.g., riprap).
2R	Controlling sediment-ation by revegetation of banks with grass-sedge-forb mixtures	Bank stabilization technique in streams and rivers.
2S	Controlling water pollution	Controlling point-source pollution discharge into lakes, streams, rivers, or nearshore marine waters.
2T	Disposing/assimilating wastewater	Controlling waste water effluent discharge into lakes, streams, rivers, or nearshore marine waters.
2U	Dredging	Periodic dredging and deposition of spoils by large ships or barges within large rivers.
2V	Locating/constructing stream crossings	The construction of roads and bridges across creeks and small rivers; includes heavy equipment, blasting, and landscape alteration.

Code	Activity	Definition
2W	Controlling aquatic plants	Activities, including herbicide application and water drawdowns, to reduce or remove emergent or submergent plants usually associated with reservoirs or impoundments.
2X	Channelization	Creating passageways to direct the flow of water.
3	**Road Management**	**Roads that are engineered and maintained; the surfaces of which can be pavement, gravel/rock/cinder, or dirt. Does not include skid roads.**
3A	Road and bridge construction/ obliteration	The actual construction (or obliteration) of roads and bridges, which includes heavy equipment, blasting, and landscape alteration.
3B	Operational aspects of road maintenance and use	Activities associated with the maintenance and use of roads and bridges, which includes roadside vegetation management (mowing, herbicides, ditch cleaning, revegetating roadsides, introducing exotic vegetation, removing hazard snags), removing beaver dams that cause road flooding, spreading oil for dust abatement, and runoff management (e.g., culverts) to reduce erosion, turbidity, and contamination of waterways by heavy metals.
3C	Road closures	Limiting road use (seasonal and yearly closures with gates or some other system, but road is still maintained).
3D	Bridges (in general)	Literature search resulted in article(s) that described the general effects of bridges.
3E	Roads (in general)	Literature search resulted in article(s) that described the general effects of roads.
4	**Agricultural Activities**	
4A	Applying fertilizers	Periodic application of fertilizers to agricultural habitats such as row crops, orchards, nurseries, etc.
4B	Applying pesticides	Periodic application of pesticides to agricultural habitats such as row crops, orchards, nurseries, etc.
4C	Applying herbicides	Periodic application of herbicides to agricultural habitats such as row crops, orchards, nurseries, etc.
4D	Applying fungicides	Periodic application of fungicides to agricultural habitats such as row crops, orchards, nurseries, etc.
4E	Haying/mowing	Vegetation removal on row crops and pasturelands.

Code	Activity	Definition
4F	Maintaining grasses and forbs within orchards, Christmas tree farms, etc.	Leaving vegetation around desired crops to provide habitat for wildlife.
4G	Providing/maintaining vegetation along field and ditch margins	Includes providing cover in the form of hedge rows, shelterbelts, or other vegetated corridors.
4H	Retaining crop residue	Practicing harvest methods that leave crop remains on the ground over the winter.
4I	Implementing farm-land conservation programs	Activities focused on the restoration and maintenance of predominantly native vegetation and erosion control measures on lands formerly managed for agricultural commodity production.
4J	Irrigating	Routine application of water to row crops or pastureland.
4K	Altering drainage	Includes ditching and tiling on a recurrent basis on lands used in the production of agricultural commodities.
4L	Decreasing water supply: flow withdrawal	Pumping water out of streams, rivers, and wetlands for irrigation, resulting in decreased water supply in aquatic habitats.
4M	No-till or minimum-till farming	Crop production techniques that minimize soil disturbance.
4N	Clean farming	Intensive agriculture that uses all available land surface, leaving no crop residues (often due to tilling and burning post-harvest).
4O	Strip intercropping	Harvest technique to conserve soil by removing alternate rows of vegetation.
4P	Conversion of native habitats	Replacing native forest or shrubland/grassland habitats with agriculture.
4Q	Control of verte-brates considered to be agricultural pests	Use of repellents, including chemical, visual, and noise, also includes trapping to remove animals causing damage to crops.
4R	Providing artificial nesting sites	Nest boxes placed along roads and near ponds to enhance passerine and waterfowl productivity.
4S	Agriculture (in general)	Literature search resulted in article(s) that described the general effects of agriculture.
5	**Shrubland and Grassland Management**	
5A	Mechanical vegetation management	Physical removal of vegetation, including chaining.
5B	Burning	Fires that are intentionally lit (or natural fires that are allowed to burn) for a specific management objective. The extent, intensity, and timing are either planned or controlled.
5C	Use of herbicides	Periodic application of herbicides to shrubland/grassland habitats.

Code	Activity	Definition
5D	Restoration	Activities used to recreate or enhance native grassland or shrubland habitats.
5E	Conversion of shrubland to native or non-native grassland	Conversion to grassland to provide forage for livestock.
5F	Livestock grazing	Allowing cattle, horses, or sheep to forage on open rangeland.
5G	Shrubland management (in general)	Literature search resulted in article(s) that described the general effects of shrubland management.
5H	Grassland management (in general)	Literature search resulted in article(s) that described the general effects of grassland management.

6   Livestock Management (cattle, sheep, and horses)

Code	Activity	Definition
6A	Livestock grazing	Allowing cattle, horses, or sheep to forage over wide areas.
6B	Conversion of shrubland to native or non-native grassland	Conversion to grassland to provide forage for livestock.
6C	Creating or providing stockponds	The presence of stockponds and their influence on livestock impacts to the environment (e.g., may reduce impacts to naturally occurring streams).
6D	Excluding livestock from riparian areas	Removing or preventing livestock from access to streams and the resultant restoration of riparian habitats.

7   Fencing

Code	Activity	Definition
7A	Fencing to control or direct wildlife access	Includes fencing to (1) exclude predators from other wildlife or livestock, (2) exclude ungulates from orchards, hay stacks, and seedlings, and (3) control wildlife access and movements along roadways.
7B	Fencing to protect or restore habitat	Nonriparian, includes aspen and special botanical areas.
7C	Fencing to exclude livestock from riparian areas	Preventing livestock access to streams and the resultant restoration of riparian habitats.

8   Mining Activities

Code	Activity	Definition
8A	Site reclamation	The purpose of reclamation is to return the disturbed areas to a stabilized and productive condition following mining and milling activities to protect long-term land, water, and air resources in the area. This most often involves modifying the final grade of gravel and substrate materials, providing for soil stability, planting vegetative cover, and addressing water flow and quality aspects.
8B	Surface/strip mining and processing	Surface and strip mining are techniques that allow the extraction of shallow ores and coal. Processing associated with these techniques involves the extraction of valuable materials from mixed ore and can include heap leach, vat leach, flotation, and other techniques. These techniques involve the removal of overburden; ore processing; waste rock disposal; tailings disposal and embankment construction; water supply development, storage, and runoff management; and power supply development.
8C	Underground mining and processing	Underground mining is the extraction of mineralized zones by underground methods. Processing associated with these techniques involves the extraction of valuable materials from mixed ore and can include heap leach, vat leach, flotation, and other techniques. These techniques involve site preparation; ore processing; waste rock disposal; tailings disposal and embankment construction; and water supply development, storage, and runoff management.
8D	Maintaining access to abandoned subsurface mines and tunnels	Leaving openings allows for wildlife use of mines and tunnels.
8E	Placer prospecting and mining	Searching for, and recovery of, minerals from streamborne deposits. Usually involves the use of water to aid recovery from these deposits. These activities include excavation of materials from these deposits and processing with various equipment including pans, sluice boxes, suction dredges, and highbankers. The primary impacts of these activities are the redistribution of existing in-stream sediment, introduction of new sediment from adjacent uplands, negative impacts to in-stream invertebrates, fish, and fish spawning areas, and alteration of the stream channel dynamics.
8F	Mineral exploration	This activity reflects the exploration of potential mineral deposits and primarily involves significant amounts of road construction (often in a grid pattern), the drilling of test holes, and associated land-clearing activities by heavy equipment.

Code	Activity	Definition
8G	Sand, gravel (aggregate), and peat mining	Surface mining to extract building and construction materials including sand, gravel, and rock. Activities are typified by rock and gravel quarries. Aggregate mining does not involve the chemical processing of materials. Peat mining is included here as it reflects similar methods for the extraction of subsurface materials.
8H	Mining (in general)	Literature search resulted in article(s) that described the general effects of mining.
8I	Mining activities involving blasting	Literature search resulted in article(s) that described the general effects of blasting.
8J	Oil and gas extraction	Literature search resulted in article(s) that described oil and/or gas extraction.

9 **Forest Management**[a]
9A *Harvest Operation Activities*

Code	Activity	Definition
9A1	Clearcutting	The harvesting of all standing trees in a given area at the same time.
9A2	Shelterwood cuts	Harvesting in which trees on a site are removed in a series of cuts over time to create an even-aged stand.
9A3	Seed tree cuts	Harvesting in which a cut removes almost all the trees in an area, but leaves a few scattered mature trees of good genetic stock to produce seed to regenerate the stand.
9A4	Group selection	The selective removal of small groups of trees in a system of uneven-age management.
9A5	Selective harvest across all tree sizes	The selective removal of single trees in a system of uneven-age management.
9A6	Selective harvest of specific tree sizes, conditions, or species	Includes varied silvicultural prescriptions, including salvage harvests.

9B *Silvicultural/Stand Improvement Activities*

Code	Activity	Definition
9B1	Precommercial thinning	Removal of young trees to increase the growth of remaining trees in a stand. The cut trees are often left as slash on the forest floor.
9B2	Commercial thinning	Release cuttings to manipulate stocking densities for enhancement of dominant or codominant trees; cut trees have commercial value and are removed from the stand.
9B3	Pruning	The removal of lower limbs to create higher quality wood and enhance tree growth rates.
9B4	Simplifying species composition and/or structure	Genetically selecting trees, favoring one or few commercially desired species.
9B5	Type conversion	Changing grasslands to tree farms, or hardwood stands to conifer stands.

Code	Activity	Definition
9B6	Prescribed burning	Fires that are intentionally lit for a specific management objective such as forest health or site preparation. The extent, intensity, and timing are either planned or controlled.
9B7	Applying insecticides	Periodic application of insecticides to prevent loss of tree rigor and mortality.
9B8	Forest management (in general)	Literature search resulted in article(s) that described the general effects of forest management.

9C *Site Preparation/Tree Establishment Activities*

Code	Activity	Definition
9C1	Applying herbicides	Application of herbicides to reduce competition to seedlings from encroaching vegetation.
9C2	Fertilizing plantations	Application of fertilizers to increase tree growth and site productivity.
9C3	Removing slash	Includes use of heavy equipment and piling, burning, or hauling off of slash for chipping; all slash is removed from site.
9C4	Planting/seeding	Includes the use of vexar tubing, shade cards, and plastic sheeting.
9C5	Tilling prior to planting	Mechanical preparation of the ground to facilitate tree planting.

9D *Habitat Management Activities*

Code	Activity	Definition
9D1	Maintaining mature/ old growth	Includes maintaining forest corridors.
9D2	Grazing livestock	Allowing horses, sheep, or cattle to freely forage within forested stands.
9D3	Retaining medium green trees	Leaving 28-48 cm dbh trees from prior stand.
9D4	Retaining large green trees	Leaving >51 cm dbh trees from prior stand.
9D5	Retaining defective trees	Deformities include cavities, broken tops, heart rot, conks, multiple tops, etc.
9D6	Creating/maintaining edges	Many harvesting activities result in contrasting boundaries between forest stands of various successional stage or species composition.
9D7	Retaining mast trees	Leaving hardwood species that produce soft or hard mast for wildlife use.
9D8	Retaining forest openings	Pertains to the creation of small forest openings (gaps), or the prevention of tree encroachment into natural meadows.
9D9	Retaining brush/ slash piles	Leaving the unwanted vegetation left from a harvest operation in piles on the site for wildlife use.
9D10	Retaining/providing dead/down wood	Leaving or providing dead wood from harvest operations on the forest floor for wildlife use (instead of collecting or burning it).

Code	Activity	Definition
9D11	Retaining/creating snags	Allowing standing dead trees to remain in the stand after harvesting, or topping/blasting live green trees to create new snags.
9D12	Retaining riparian buffers	Activities associated with maintaining trees and shrubs within 30 m of a waterway.
9D13	Providing artificial nest sites	Employing nest boxes to enhance productivity of selected forested species that use cavities.
9D14	Creating/maintaining corridors	Deliberately providing or retaining connective habitat to facilitate wildlife travel and use.
9E	*Incidental Activities*	
9E1	Introducing exotic vegetation	An example is elk forage mix.
9E2	Creating water sources	Digging pumper ponds or wildlife ponds.
9E3	Removing hazard trees	Removal of trees deemed to be hazardous to human safety from roadsides and campgrounds to comply with federal regulations.
9E4	Building skid roads and landings	The creation of skid roads and landings opens up forest canopies, increasing the amount of edge and early successional habitats.
9E5	Forest vertebrate pest control	Activities to prevent animal damage or to remove offending individuals.
9F	*Special Forest Products*	
9F1	Firewood cutting	Removal of live trees or snags for the purpose of obtaining fuel.
9F2	Harvesting wild mushrooms	Commercial harvest of wild mushroom species.
9F3	Bough collection	The pruning of lower limbs for decorative purposes.
9F4	Special forest products (in general)	Literature search resulted in article(s) that described the general effects of collecting special forest products.

10　Marine Activities

Code	Activity	Definition
10A	Marine dredging and filling	Mechanically or hydrologically removing bed materials (sand, gravel, mud) and moving them to a new location to provide increased depth for boat and ship navigation. Filling is the placement of dredged material or upland materials in marine aquatic areas. In Puget Sound, fill materials are typically placed to create uplands for commercial purposes (e.g., marina, port developments). Fill material has been used to create dredge spoil islands along the lower Columbia River.

Code	Activity	Definition
10B	Harbor, marina, and ferry terminal development	Includes both the development and subsequent use of harbors, marinas, and ferry terminals. This category reflects both fresh and saltwater environments. Includes commercial shipping, associated cargo handling, and ferry transport. Recreational boat marinas and associated infrastructure (e.g., parking lots, floats, breakwaters, fueling stations). Commercial harbors and ferry terminals are typified by Elliott Bay, Port Angeles, and Bellingham Bay, Washington; and Newport, Coos Bay, and Portland, Oregon. Recreational marinas are typified by Olympia and Des Moines, Washington; and Astoria, Oregon. Impacts extend to include bilgewater and wakes from large ships.
10C	Residential docks in marine and freshwaters	Floating and fixed docks, piers, and associated pilings in marine and freshwater environments. Physical dimensions of docks tend to be about 2 m wide and 15-30 m long. Typical dock structures have associated pilings and deck surfaces.
10D	Toxic spills in fresh and saltwater	Spills or depositions of chemicals into freshwater and marine habitats. This is typified by, but not limited to, petroleum spills, railroad car incidents, semi-truck turnovers, and marine Superfund sites (e.g., Commencement Bay, Elliott Bay, Washington). Chemicals are primarily represented by hydrocarbons, dioxins, petrochemicals, fertilizers, pesticides, and heavy metals. This category does not include spill and deposition sites that are entirely terrestrial-based.
10E	Marine shoreline armoring	Placement of rock, wood, or concrete at the water's edge to prevent shoreline erosion or bank failure. Bulkheads are sometimes placed in no-eroding areas.
10F	Developing underwater marine structures	The active creation of underwater structures, normally involving placement of large concrete and rock substrates. Objective is to provide vertical relief to create habitat structures for various marine fish and shellfish. These underwater reef structures could be 15 m wide, 61 m long, and 3 m tall. The structures are located primarily in Puget Sound,

Code	Activity	Definition
		Washington. Note: oil exploration and associated drilling platforms are currently prohibited off the Oregon and Washington coastlines, and thus are not considered in this assessment.
10G	Marine fisheries	This activity reflects marine-based harvest and processing of fish. Primary effects are derived from trawlnet, purse-seine, and gillnet fishing techniques. Primary wildlife issues in Oregon and Washington reflect the bycatch of marine seabirds (and marine mammals to a much lesser degree) in active or lost fishing gear (e.g., "ghost nets").
10H	Aquaculture	Commercial production and harvest of fish (i.e., grown in net pens) and shellfish (e.g., oysters, geoducks, clams, and mussels). This also includes impacts associated with recreational harvest of shellfish.
11	**Urban Development**	
11A	Paving	Creation of impervious surfaces, (e.g., concrete or asphalt), and the subsequent impacts associated with the loss of natural vegetation and substrates.
11B	Building houses and businesses	Converting natural habitats for human occupation, including, for example, single homes, apartments, businesses, subdivisions, shopping malls or industrial parks, and the subsequent impacts associated with the loss of natural vegetation.
11C	Presence of domestic animals	Refers to the disturbance and impacts caused by dogs and cats.
11D	Urban aquatic habitat management	Activities associated with the restoration and modification of this habitat, including paving, ditching, and channelization of urban watercourses. Also includes the development and maintenance of ponds and lakes in an urban context, for example, in golf courses or parks.
11E	Landscaping and vegetation management	The conversion of native vegetation to exotic and ornamental plant species, including lawns.
11F	Water quality and stormwater management	Activities associated with the prevention of water contamination from runoff.
11G	Establishing and maintaining greenways/greenbelts	Providing undeveloped areas of primarily natural vegetation within the urban matrix.

Code	Activity	Definition
12	**Recreational Activities**	
12A	Trail use and camping	All activities associated with trails and camping. Includes the use of pack animals and mountain bikes.
12B	Snowshoeing/snow skiing/sledding	Nonmotorized snow travel.
12C	Mountain/rock climbing	Disturbance caused to wildlife from climbing of all kinds.
12D	Motorized boating	Disturbance by and effects of motor boats on wildlife and aquatic habitats.
12E	Nonmotorized boating	Nonmotorized boat traffic including rafts, canoes, sailboats, and rowboats.
12F	Swimming	The effects of humans in waterbodies.
12G	Off-road driving	Includes the effects of all motorized off-road vehicles, including ATVs, four-wheel drive trucks, and dune buggies.
12H	Snowmobiling	Motorized snow travel.
12I	Aircraft use	Includes planes, helicopters, and other motorized aircraft.
12J	Recreational developments	Includes ski areas, and other resorts.
12K	Fish stocking	Providing fish for recreational use.
13	**Right-of-way Management**	
13A	Utility corridors	Linear rights-of-way including power lines, telephone lines, oil pipelines, etc.

[a]Some of the Forest Management Activities were defined by D. Patton, 1992, in Wildlife Habitat Relationships in Forested Ecosystems, published by Timber Press, Portland, Oregon.

21

An Overview of Models and Their Role in Wildlife Management

Gary J. Roloff, George F. Wilhere, Timothy Quinn, & Steven Kohlmann

Introduction

Much of ecology is concerned with the search for repeated patterns. When expressed in a structured, meaningful representation of natural systems, these patterns form the foundation of modeling in wildlife management. Models are built to define problems, clarify ideas, organize concepts, communicate information, develop and test hypotheses, and to make predictions. Thus, models are crucial to the resource decision-making process.[180, 183] Models should be viewed not as scientific laws that represent absolute truth, but rather, as hypotheses that offer a "purposeful representation."[182] The models we discuss here are used to characterize how animals respond to their abiotic (e.g., landform, soils) and biotic (vegetation, species interactions) environment. Animal response can take the form of occurrence, distribution, abundance, movements, productivity, or survival. The intent of this chapter is to:

1. Introduce wildlife managers to common types of resource management models.
2. Illustrate the use of models in resource management by offering simple examples.
3. Offer guidance in selecting an appropriate model for a specific use.
4. Provide references where managers can get more information.

This chapter is divided into four primary sections. The first section presents modeling concepts, terminology, an overview of how a simple model works, and a discussion of the role of models in resource management. The second section discusses three general approaches to wildlife modeling: habitat modeling, population modeling, and spatial population modeling. We provide examples for each modeling approach and offer decision trees to assist wildlife managers in selecting the type of model(s) best suited for their needs (Figures 1-3). Also in this section we compare and contrast different model types in terms of their practical application to resource management issues. The third section of this chapter presents a case study that uses a modeling approach for wildlife management. We discuss factors that should be considered prior to initiating a modeling effort and demonstrate the use of the decision trees for selecting a particular model type. Finally, we surmise on the future of modeling in wildlife management.

Wildlife Modeling Concepts, Terms, and Mechanics

When formulating a model we: (1) identify the system of interest, (2) state specific objectives, (3) hypothesize relationships between system inputs and outputs, and (4) construct a useful mechanism (e.g., an equation, a simulation, etc.) that represents these relationships. To illustrate the model building process, assume that we wanted to construct a model for canopy closure. Canopy closure is defined as the percentage of ground area shaded by overhead foliage.[47] It is a forest attribute associated with habitat quality for a number of wildlife species and it has been incorporated into some forest practices regulations. Unfortunately, canopy closure is notoriously difficult to measure precisely.[27, 42, 198] Hence, a model that accurately relates canopy closure to some other forest attribute might serve as a cost-efficient tool for assessing canopy closure. Thus far in our modeling process, we have identified a system that we wish to understand better and we have stated an objective—develop a tool for efficiently assessing canopy cover.

Models describe systems. A system is "any phenomenon, either structural or functional, having at least two separable components and some interaction between these components . . . ."[74] For the purposes of modeling, systems are decomposed into input variables, output variables, and the relationships among them. Inputs and outputs are also known as independent and dependent variables, respectively. Typically, "relationships" describe output variables as functions of the input variables, but if feedback exists in the system, then outputs are a function of both inputs and outputs. The complexity of relationships varies greatly from a single mathematical equation to a multitude of equations, inequalities, and logical expressions. Equations, inequalities, and logical expressions contain variables and parameters. As their name implies, variables do not have a constant value. The simplest model has only two variables—input and output. More complex models may have multiple inputs or output variables, and may also have intermediate variables that represent steps in the translation of input to output. Parameters are quantities that mediate the relationship(s) among variables. Parameter values are estimated from data and are often treated as constants. In simulation models, however, certain parameter values may vary according to a statistical distribution, but the distribution is described by other parameters that are constants.

For this example, assume that a forest inventory has been completed that measured canopy closure. After summarizing our data, we infer that stand basal area is the independent (or input) variable most highly correlated with canopy closure, the dependent (or output) variable. Based on *a priori* knowledge of canopy closure, we make two modeling assumptions—a stand with nonzero basal area must have nonzero canopy closure, and a stand with zero basal area must have zero canopy closure. This forces all mathematical equations through the zero-zero point of our canopy closure-basal area relationship. Upon further analysis of the data, we infer that the relationship between input and output variables is best described by the equation for a line:

Canopy Closure = K * Basal Area

The relationship has one parameter, K (although the intercept could be considered a parameter with value zero), and the value for K is derived from the data. Using other statistics, which are beyond the scope of this review, we can evaluate the reliability of our model. The validity of the model should also be tested by collecting new data on canopy closure and basal area and then comparing the data with the model's predictions. Validity of a model also depends on the "scope of inference". For example, if data for our canopy closure-basal area relationship were collected only from western hemlock (*Tsuga heterophylla*) types in western Washington the scope of inference is limited to similar types. Therefore, the model would be invalid for Douglas-fir (*Pseudotsuga menziesii*) types in eastern Washington. Finally, keep in mind that the mathematical relationship we have developed is only a hypothesis. Our model is the "best" relationship we could infer given our data. More data or more precise data might show that the relationship is not a straight line but actually a curvilinear function.

There is a great deal of literature on using model outputs and testing model validity.[34, 71, 165, 180] Model validity refers to a broad spectrum of performance standards and criteria including credibility, realism, generality, precision, breadth, and depth.[137] The common theme among validation studies is that the utility of model output depends on input data quality and realism of hypothesized relationship(s). However, the utility of modeling is not limited to generating outputs. One of the greatest benefits of a model is derived from the thought process required to build it. The modeling process requires us to explicitly state assumptions, helps us envision the system's relationships, and increases our understanding of the critical components of systems.[75] While nearly all wildlife professionals know they are dealing with complex systems, often the holistic thought processes are lost at the decision-making level where attention is focused on a "problem." Building a model highlights aspects of systems that must be considered in making decisions.[181]

One challenge in developing meaningful models is selecting the proper inputs. In the absence of empirical data, inputs are identified from expert opinion, personal experiences, or existing literature (which may be based on empirical data). If empirical data exist, there are many statistical techniques that can assist modelers in choosing the "best" inputs (where "best" can be defined as the set of inputs that explains the most variability in the output). Consider our simple canopy closure model. The choice to use a single input variable (i.e., basal area) and a linear relationship implies that we have identified the most important variable affecting canopy closure. However, we know that other attributes can affect canopy closure such as tree species and topography. So, did we create a good model? The equivocal and correct answer is "it depends." Simple models are more general, easier to understand, and easier to communicate than more complex models, but may lack realism when applied to specific cases. Including more detail in our model (e.g., adding input variables to account for tree species or topography) may make it more realistic but would also require more data and possibly more assumptions. The challenge for wildlife modelers is to evaluate the tradeoffs between realism, cost, and applicability. This evaluation depends on the complexity of the system under study, availability of data, desired outputs, and most importantly, the risk associated with making a wrong prediction. For example, a model that predicts canopy cover ±20% of the true canopy cover 80% of the time is probably useful to wildlife managers that need to predict the distribution of a common bird species in a strategic forest plan. Conversely, this type of model may not be acceptable for a manager who may be required by law to provide >70% canopy cover around the nest of an endangered species.

The Roles Models Play in Wildlife Management

We believe models are an important part of wildlife management and thus, the question is not whether to model but how to do it effectively.[183] Starfield[183] lists several misconceptions regarding the use of models and notes that these often act as impediments to model use. Nonetheless, decisions that affect resources are continually made despite a seemingly perpetual lack of data and understanding. Models built for wildlife management will never perform to the standards of physicists or chemists because it is unlikely that ecological models can ever be tested conclusively.[166, 180] We use models in wildlife management to help define or reduce risk associated with complex decisions when data are scarce (see chapter 19).

The need for models in wildlife management has increased in recent years for several reasons. First, models can help quantify tradeoffs, including the risks, of alternative management scenarios even as natural resource issues grow increasingly complex. Second, models can produce easily communicable results that transcend disciplinary boundaries that are essential if all stakeholders are to be involved in decision-making. Third, many models are specifically designed to help us predict the long-term consequences of our actions, a rare feature of past natural resource management. Finally, models help provide a framework for formulating hypotheses and research designs, which are essential parts of adaptive management.

Types of Wildlife Models

To help build understanding, we separated common wildlife models (hereafter models) into three classes: habitat models, population models, and spatial population models. This classification is based primarily on the characteristics of model inputs and outputs. Habitat models typically have habitat quality and quantity as model outputs rather than population performance measures. Output from habitat models is an index of population response to its environment. Habitat models can be designed to represent species' occurrence, density, productivity, or survival. Population models, alternatively, often use historic estimates of population performance to predict future population performance. Both inputs and outputs of population models are measures of the demographic rates (e.g., fecundity, mortality, density, and persistence). Spatial population modeling integrates information about the geometry of habitats in space and time, how organisms interact to position themselves in their habitat and how they move. Like population modeling, the output of spatial population modeling is often demographic. Spatial population modeling can also generate a variety of spatial outputs including immigration and emigration rates and success, probabilities of population persistence, movement patterns, and "source-sink" dynamics.

Habitat Modeling

The underlying ecological concept for most habitat models is niche theory. Elton[61] is credited with defining the niche as "the functional role and position of the organism in its community." Many wildlife modelers attempt to quantify important niche components and to predict a population's response to changes in those components. This can be a daunting task, considering that niches for most species are extremely complex.[93] A challenge for wildlife modelers is to mathematically represent this complexity in a context suitable for practical use and application.

Texts on habitat modeling date back to the 1960s[8, 209] although most work on modeling wildlife-habitat relationships started in the 1970s[75, 130, 180] For general reviews of wildlife-habitat relationships modeling see Starfield and Bleloch,[180] Verner et al.,[204] and Morrison et al.[136, 137] Starfield and Bleloch[180] and Verner et al.[204] wrote and edited some of the first comprehensive works relating habitat modeling to wildlife management issues. Starfield and Bleloch[180] focused on the mechanics of applying models to wildlife management. They described several model types and supplemented their descriptions with examples. The work in Verner et al.[204] consists of 60 chapters (mostly case studies) that cover theory, techniques, and the validation of habitat models. This also contains one of the most thorough literature reviews

available on habitat modeling through 1986. Morrison et al.[136, 137] combined habitat theory and the practical applications of research on animals and their habitats in a modeling context.

Wildlife modelers use habitat to model the occurrence, abundance, distribution, or population performance of animals. Habitat models can be categorized as theoretical and empirical.[137] Theoretical habitat models are based on inputs and relationships *hypothesized* as important in describing the system. The modeler says "This is how I think my system is."[74] This model building process may be based on logic, experience, or on a model constructed for some other system.[75] In contrast, empirical models use data collected specifically for constructing that model.

Following Morrison et al.,[137] we developed a decision tree to assist managers in identifying model(s) appropriate for their management issue (Figure 1). The decision branch points in Figure 1 are related to data availability and the questions to be answered. There are undoubtedly exceptions to the logic presented in Figure 1. For example, habitat suitability index models have been applied to multiple species situations[171] and empirical data sets may be used to develop the more theoretical guild and life form models. Thus, the decision tree should not be viewed as a rigid rule set but rather as a guide for reducing the number of choices and for selecting the appropriate model.

Theoretical Habitat Models

Theoretical habitat models include modeling approaches that use expert opinion, personal experience, qualitative descriptions, and data that were collected extraneous to the modeling effort to portray wildlife-habitat relationships. They range from simple, qualitative depictions of wildlife-habitat systems[137: 312] to complex, multi-scale simulation models.[12] These models are typically inexpensive to develop, easy to understand, and easy to communicate (Table 1). Theoretical habitat models include the habitat evaluation procedures and habitat effectiveness models; Bayesian and pattern recognition models; habitat suitability index and habitat capability models; species-habitat matrixes; and coarse-filter/fine-filter and guild/life form models (Figure 1).

Habitat Evaluation Procedures and Habitat Effectiveness Models. One group of theoretical habitat models includes the habitat evaluation procedures [196] and habitat effectiveness models (Figure 1). The procedures are based on two primary components: 1) an index to habitat condition, i.e., quality, and 2) a measure of habitat quantity. Habitat quality is determined from input data on vegetation structure, composition, and spatial arrangement of vegetation types (e.g., distance between suitable forage and cover). Habitat quantity is determined by multiplying the quality index by the corresponding amounts of that habitat type. The result is termed a "habitat unit."[196] The assumption is that greater numbers of habitat units represent more favorable conditions (i.e., higher carrying capacity) for the wildlife species under consideration.[157] Although not intended to estimate animal

Wildlife Habitat Modeling

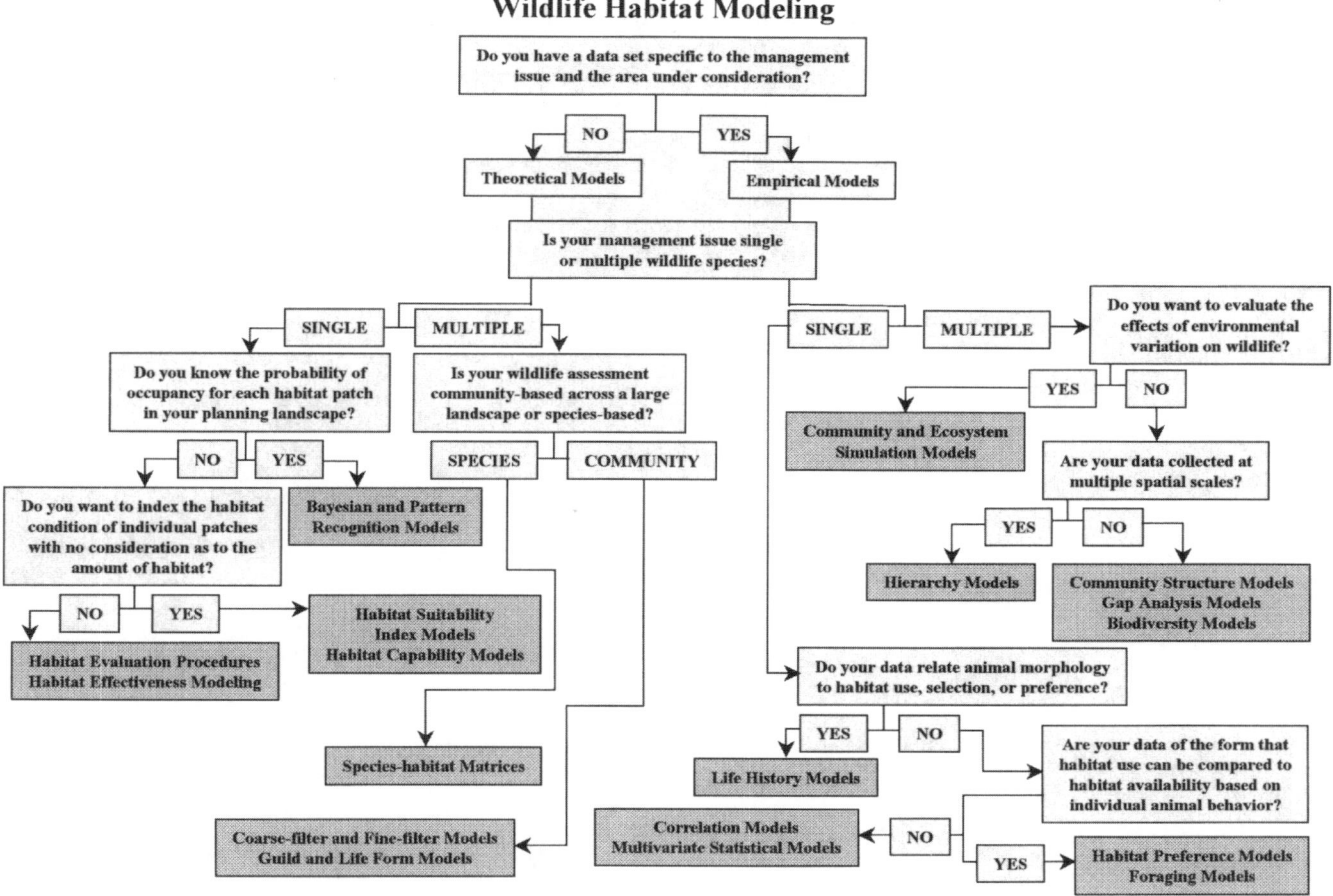

Figure 1. Decision tree for selecting a wildlife habitat modeling approach.

numbers,[6] both modeling procedures must be tested using demographic data.[159, 200] Once tested, these models can be a useful tool for population viability planning.[157]

Habitat evaluation procedures are appealing and have been used extensively by regulatory agencies in the United States to assess the adequacy of mitigation projects.[196] Habitat effectiveness models can be developed for low costs, apply to a range of similar systems, and readily associated to real-world systems (Table 1). Both of these model types have been criticized for their high levels of subjectivity and lack of perceived validity (Table 1).

There are slight differences in how the habitat evaluation procedures and habitat effectiveness models are used.[137] This difference depends on the use and interpretation of model outputs. In habitat evaluation procedures, outputs are habitat units (the product of habitat quality index and the amount of that habitat type available). In habitat effectiveness models outputs are indices that represent habitat conditions relative to some biologically meaningful optimum[213] (e.g., maximum use, carrying capacity). Habitat evaluation procedures use habitat units as a representation of carrying capacity whereas habitat effectiveness models score entire evaluation areas with a single index.[213] Also, habitat effectiveness indices typically represent habitat potential

that is discounted by human activity[185, 213] (e.g., roads, disturbance). For example, Suring et al.[185] portrayed habitat potential for brown bears (*Ursus arctos*) on the Kenai Peninsula using a habitat suitability index. Subsequently, they developed a "human activities" sub-model based on locations of urban areas, roads, hunting camps, and livestock grazing. The "human activities sub-model" was used to discount habitat potential. The modeling effort resulted in a single index of habitat effectiveness for brown bear on the Kenai Peninsula.[185]

Habitat Suitability Index Models and Habitat Capability Models. Another common theoretical habitat modeling approach includes habitat suitability index models and habitat capability models. This approach produces single patch indices of habitat quality (Figure 1). Developed by the United States Fish and Wildlife Service, habitat suitability modeling is part of the habitat evaluation procedures.[196] These models are typically single-species focused and represent "within-patch" habitat conditions (Figure 1). Habitat suitability index models have also received limited use for community assessments.[172] Habitat suitability models generate a numerical value that indexes the capacity of a habitat patch to support a species. Inputs to habitat suitability models include measures of vegetation structure, composition, and spatial arrangement. These inputs are mathematically combined

**Table 1. A coarse comparison of habitat-based wildlife modeling approaches.
Each model was ranked as unfavorable (U), moderately favorable (MF), or favorable (F),
according to the practical application of the modeling approach.
Rankings are subjective evaluations as to the historic use, performance, and perception of the models.**

	Model Evaluation Criteria									
	Applicability					*Model Structure*		*Model Output*		
	Appeal	Common data	Cost of data acquisition	Cost of model development	Training required	Level of subjectivity	Complexity	Generality	Face validity	Communicability
Theoretical models										
Bayesian & pattern recognition models	F	MF	MF	F	MF	U	MF	MF	MF	U
Habitat evaluation procedures	F	MF	MF	F	MF	U	MF	F	U	MF
Habitat capability modeling	U	MF	MF	F	MF	U	MF	F	U	F
Habitat effectiveness modeling	MF	MF	MF	F	MF	U	MF	F	U	F
Habitat suitability index models	F	MF	MF	F	F	U	F	F	U	F
Species-habitat matrixes	F	F	F	F	F	F	F	F	MF	F
Coarse- & fine-filter models	U	U	U	U	U	MF	U	MF	MF	MF
Guild and life form models	MF	MF	MF	F	MF	U	MF	F	U	MF
Empirical models										
Community & ecosystem simulation models	F	F	U	U	U	F	U	F	MF	U
Hierarchy models	MF	MF	U	MF	MF	MF	MF	MF	MF	MF
Community structure models	MF	F	U	MF	U	F	U	U	MF	U
Gap analysis models	MF	U	U	U	MF	MF	MF	F	MF	F
Biodiversity models	U	U	U	MF	U	F	MF	U	F	MF
Life history models	U	U	MF	MF	F	F	MF	F	F	MF
Correlation models	NF	MF	MF	MF	F	F	MF	U	F	F
Multivariate statistical models	MF	U	U	U	U	F	U	U	F	U
Habitat preference models	MF	U	U	U	U	F	U	U	F	MF
Optimal foraging models	U	U	U	U	U	F	U	U	F	U

aSome criteria adapted from Marcot et al.[124] where *Appeal* = How much the model framework has been used in the past and practically applied; *Common data* = Data that are collected as part of routine inventories (e.g., rimber cruises, game check stations, remote sensing); *Cost of data acquisition* = Cost to acquire the data for model development and use; *Cost of model development* = Cost (monetary and time) of developing the model; *Training required* = Amount of training required to use and understand the model; *Level of subjectivity* = Degree that subjective evaluations are used to develop the model relationships; *Complexity* = Complexity of the modeling framework, synonymous with Marcot et al.'s "wholeness"; *Generality* = How well the model represents a broad range of similar systems; *Face validity* = Perceived credibility of a modeling framework; *Communicability* = Ability to associate model output to real world systems.

into an index that ranges from 0.0 to 1.0 representing totally unsuitable areas to optimum habitat, respectively.[197] In generating this index, a linear relationship between carrying capacity and the index value is assumed.[197] The linear relationship implies that a change in index score will always correspond to the same proportionate change in carrying capacity.

Habitat suitability models have been applied to numerous wildlife management situations,[204] and their performance has varied.[189] Because of the relationship between habitat suitability modeling and the habitat evaluation procedures, terminology for the two modeling approaches is sometimes used interchangeably in the literature.[111] The important difference between these approaches is reflected in model output. Habitat suitability models generate an index of habitat condition for a patch. The habitat evaluation procedures, in contrast, generate habitat units that are based on habitat quality and quantity (index score times the amount of the habitat available). Both types of models can be tested by comparing outputs

to abundance,[108, 147] capture success,[63] abundance of nesting or denning structures,[156] animal locations,[107] or demographic performance.[22]

Habitat suitability index models are appealing as indicated by their historic use and common application, have low development costs, require minimal training for implementation, are structured in an understandable manner, can typically be applied to a broad range of similar systems (but see Roloff and Kernohan),[159] and are readily communicated to resource managers (Table 1). Drawbacks of using habitat suitability index models include a high level of subjectivity and lack of perceived validity (Table 1).

Habitat capability models are closely related to habitat suitability index models.[137] In contrast to habitat suitability index models that generate an index to habitat potential, output from habitat capability modeling has been used to represent population levels.[135] Habitat capability models typically have low development costs, can be applied to a broad range of similar systems, and are readily communicable to managers (Table 1). This model type has not been used extensively and has a high level of subjectivity and lack of perceived validity (Table 1).

Bayesian and Pattern Recognition Models. Bayesian and pattern recognition models are another form of theoretical habitat model that focuses on single species (Figure 1). The underlying framework of these models is Bayes's Thereom, which is a statistical approach that integrates both empirical data and subjective judgement.[64] Simply, Bayes's Thereom states that the probability of an event occurring is related to the degree of belief that the event will occur. Knowledge and data influence this degree of belief. As our knowledge regarding a system grows, our knowledge of system behavior improves. For example, assume that we believe, a priori, that a grassland habitat patch becomes suitable for sparrows when vegetation reaches a height of 2 m. The a priori probability for these conditions, i.e., the event of sparrows occupying habitat when it is grassland and vegetation is 2 m is 1.0 (or 100% probability of occupancy). We later observe that this sparrow only occurs in habitats meeting this description 80% of the time, thus, we may use this knowledge to adjust our a priori probability down (perhaps to 0.8). This process is iterated repeatedly to refine a priori probabilities and to create descriptions of new events. Iversen[95] and Carlin[32] provide thorough descriptions of Bayes's Theorem and Williams et al.[212] describe the adaptation of Bayesian theory to wildlife habitat assessments.

In terms of wildlife habitat modeling, Bayesian theory most often occurs in the form of pattern recognition models. Pattern recognition models were initially developed for the medical profession to assist in the diagnosis of disease.[120] Pattern recognition models integrate Bayesian probabilities across habitat space. Three factors must be understood to use the pattern recognition approach.[101] First, categories of habitat suitability into which a patch can be classed must be identified. Second, the habitat attributes that dictate habitat suitability must be identified. Lastly, the set of probabilities that reflect the association between individual habitat attributes and each suitability class must be assigned. In the sparrow example above, we classified "grasslands" as the most suitable habitat type. We identified vegetation height as the attribute that determined patch quality. Our a priori probability was 1.0 if habitat was grassland and had vegetation 2 m high (i.e., we were certain that sparrows would occur if these conditions were provided). In pattern recognition models, this process is repeated for multiple habitat patches across space, thus, these models permit the incorporation of landscape configuration into the probability estimates. For a conceptual review of pattern recognition theory, see Flather and Hoekstra.[64] Examples of using pattern recognition for modeling habitats can be found for bighorn sheep,[89] bald eagles,[72] and pileated woodpeckers.[154]

Bayesian and pattern recognition models can be developed for low costs, and they are easily updated or modified (Table 1). This model type is used less than other comparable model types but has been linked to adaptive

resource management (see below). Drawbacks of using Bayesian and pattern recognition models include the subjectivity often associated with assigning initial probabilities, and difficulties in communicating the outputs (Table 1).

Species-habitat Matrixes. Two classes of theoretical habitat models address multiple wildlife species (Figure 1). Species habitat matrices (such as those outlined in this book) are tables that list vegetation types, seral stages of ecological communities, or environmental conditions with which wildlife species associate. This group of models is designed in part to provide a rapid assessment methodology for predicting the presence or relative abundance of multiple species within particular ecological communities.[16, 60, 137] Species-habitat matrix information may also be useful in crafting habitat evaluation procedures.

The habitat rankings in species habitat matrices are often based on expert opinion or existing information. Thus, species-habitat matrices can be built at relatively low costs (Table 1). Based on historical use and application, these models are one of the most favorable types (Table 1), however, their lack of objectivity has caused some concern.[16, 48, 124, 153] Also, since habitat rankings are by community type, species-habitat matrices typically offer only within-habitat patch assessments and their validity decreases for species relying on spatial arrangements of habitat patches.

Some of the more intensive species-habitat matrix work has been conducted as part of the California Wildlife Habitat Relationships program.[1] Other models include Thomas'[191] and Brown's[23] database for the Blue Mountains of Oregon, DeGraaf and Rudis's[49] work on New England wildlife habitat, Marcot's[123] Wildlife Habitat Information Matrix Program for northwestern California, Verner and Boss's[202] model for the western Sierra Nevada, and this book.

Coarse-filter/Fine-filter Models and Guild and Life Form Models. The other group of theoretical habitat models that address multiple wildlife species function at the community-level, i.e., they either use ecological communities as surrogates for population response or they group organisms into guilds or life forms (Figure 1). Coarse-filter and fine-filter models are one approach for simultaneously assessing multiple species.[137] Coarse-filter models use the occurrence, abundance, and locations of ecological communities to predict animal responses.[84, 85] When applying a coarse-filter model, considerable attention must be directed at defining ecological communities. Haufler et al.[84, 85] outline a process for stratifying landscapes into "ecological land units" that are based on similar bio-geo-climatic conditions, existing vegetation structure, site potential, and historical disturbance regimes. These units are used to describe floral and faunal diversity for planning landscapes.[84, 85] It is often unclear, however, how to determine the amount and spatial distribution of ecological land units that best function as a coarse-filter. Thus, fine-filter models (or species assessments) are used to check the adequacy of the coarse-filter.[84, 85] A check is necessary because it is reasonable to assume that some species will "slip through the cracks" of the coarse-filter.[137] Any of the previously discussed single species habitat models that index species' response (e.g., survival, fecundity) could be used as a fine-filter.

Because of data requirements and complexity, use of the coarse-filter and fine-filter approaches has been limited (Table 1). Recent advances in quantifying existing vegetation from remotely sensed data have facilitated the use of this approach. Development of large-scale models that predict ecological site potentials and simulate historical disturbance regimes have also made the coarse-filter more user-friendly.

Guild or life form models are another multi-species, community-based type of theoretical habitat model (Figure 1). These models group wildlife species with similar habitat requirements, morphology, or behaviors into guilds or life forms. For example, Inger and Colwell[94] grouped amphibians and reptiles based on measured associations to microhabitat characteristics. Because members of the groups share common characteristics, they are assumed to respond to changes in environmental conditions in a similar way. Guild and life form models can be viewed as coarse-filter approaches and thus the concerns of individual species "slipping through the cracks of the coarse-filter" apply. The "coarseness" of a guild or life form model depends on the specificity of the animal grouping and the complexity of the environment being described. For example, Mannan et al.[122] found that guilds with >4 bird species in older forests in northeastern Oregon had less predictive power than individual species models. The advantages of these types of models are that they can be constructed from existing literature, they condense a large amount of information into tabular form, and they are appealing (Table 1). Drawbacks of using guild or life form models include the level of subjectivity and low perceived validity (Table 1).

Considerable research on the theory of guilds and life forms has been conducted.[15, 122, 174, 175] Verner[203] provides a clear understanding of a guild approach for the pine-oak woodlands of California. Most often, guild models are used to evaluate the effects of changes in the environment on songbirds.[25, 102, 160] Inger and Colwell[94] and Hariston[80] applied the approach to amphibians and reptiles. The most familiar examples of the life form approach include Thomas[191] and Brown[23] for depicting habitat relationships in the western United States.

Empirical Habitat Models. Empirical habitat models include approaches that are developed from data specifically collected for model building (Figure 1). Empirical models can be purely descriptive, as derived from case studies, or statistical, as based on data sampling.[137] These models, in comparison to theoretical habitat models, are often more difficult and expensive to apply and communicate but they lend greater scientific credibility to the modeling process and outputs (Table 1).

Empirical habitat models include life history models; correlation and multivariate statistical models; community and ecosystem simulation models; hierarchical models; community structure, gap analysis, and biodiversity models; and habitat preference and foraging models (Figure 1).

Life History Models. Empirical habitat models can be loosely categorized into single and multi-species approaches (Figure 1). A relatively simple, empirically derived single species approach is life history modeling. Life history models have been proposed for understanding how phenotypic traits evolved and how these traits can be used to predict habitat requirements.[137] The underlying premise of life history models is that the combination of ecology and morphology provide consistent expressions of ecological and evolutionary adjustments between phenotype and the environment.[155] Here, "ecology" refers to the manner in which an organism relates to its environment and "morphology" refers to the physical structure of an organism. Life history models are based on the premise that an organism's morphology is an expression of its environment, i.e., the physical structure of an organism has evolved to optimize fitness in particular environments. For example, research has demonstrated relationships between organism body size and the sizes of prey.[154] The development of reliable life history models requires fairly detailed data on animal behavior or morphology and associated habitat use information.

Life history models often require relatively little training to use, have low subjectivity, and are perceived as valid (Table 1). These models can be applied to a broad range of similar systems but only for species for which data are collected (Table 1). Life history models require specific data often not collected as part of routine inventories and their practical application may be limited (Table 1).

Correlation Models and Multivariate Statistical Models. Another group of empirical habitat models involves the use of correlation and multivariate statistics (Figure 1). These models typically portray relationships between environmental characteristics and measures of animal abundance or distribution.[137] Correlation models are most useful for demonstrating trends in wildlife habitat relationships. For example, Steeger and Hitchcock[184] used a correlation analysis to demonstrate that red-breasted nuthatch nests were positively associated with the density of trees infected with *Armillaria* root disease. These types of models are common in the habitat literature and can be adapted to other models that require habitat quality information (e.g., habitat suitability index models, pattern recognition models). Their ease of creation and communication of results make this model type one of the more favorable empirically based approaches for practical application although their generality may be limited (Table 1).

Multivariate statistical models are also common in the literature. These models simultaneously relate multiple environmental attributes to animal abundance and distribution.[137] They often take the form of "exploratory data techniques" for examining which input variables (habitat characteristics) had the greatest effect on output variables (population response). For example, if songbird abundance and several habitat variables were collected at multiple sample points, multivariate models could be used to identify which habitat characteristics appear to explain the most variation in bird abundance. Volumes of literature exist on different multivariate techniques and how they can be used as predictive tools for resource managers. Sources for more generalized information on multivariate methods and their application to wildlife management include Capen,[30] Dillon and Goldstein,[54] and Digby and Kempton.[53]

The complexity and breadth of multivariate wildlife-habitat models are virtually limitless. Multivariate models all use more than two input (or independent) variables in various mathematical formulations to predict output (or dependent) variables. One of the most common forms of multivariate habitat analysis involves the use of multiple regression. For example, Maurer[125] used multiple regression to develop predictive models of bird density using multiple environmental input variables. Similarly, Puttock et al.[151] used an expanded form of regression modeling, termed log-linear modeling, to estimate moose density based on vegetation characteristics and spatial arrangements of habitats. Logistic regression and discriminate function models are also two forms of multivariate models that are commonly used in wildlife analyses. The output of these model types is categorical, most commonly probability of presence or absence. For example, Nadeau et al.[139] used logistic modeling to determine the presence or absence of muskrat burrows along stream reaches based on vegetation characteristics and stream morphology. Capen et al.[31] used discriminate function analysis to categorize songbird habitats as "used" or "unused" based on plot samples of vegetation characteristics. Verner et al.[204] provide some excellent papers on the design and improvement of regression models.

Principal components analysis is yet another common form of multivariate modeling approach. Principal components is used to identify important components of a multivariate data set. For example, Harestad and Keisker[79] used principal components analysis to identify which tree attributes were important to primary cavity-nesting birds. Although principal components analysis does not yield a predictive model *per se*, Maurer[125] used the output of principal components analysis (namely the principal components) to generate a regression model. The primary difference between standard regression approaches (i.e., multiple and log-linear) and discriminate function and principal components analyses is that the former typically use a subset of input variables to develop the predictive equation whereas the latter use all of the input variables to generate the model.

Recently, classification and regression tree analysis and fuzzy logic modeling have received attention by wildlife modelers as useful multivariate approaches. Classification and regression tree analysis (also called CART) is used to recursively partition data sets on the basis of a set of independent variables.[142] The technique is commonly used to develop classification trees for modeling and mapping species distributions,[37, 90] biodiversity,[141] and vegetation types (Mike McGrath, National Council for Air and Stream Improvement, Corvallis, Oregon, pers. comm.). Classification and regression tree analysis tests several independent variables and identifies which of those variables most effectively partitions variation in the dependent variable. The modeling approach uses this subset of independent variables to identify threshold values that constitute splits in the decision tree.[142] Each branch of the decision tree is subjected to this procedure until certain stopping rules are encountered.[142] Once the decision tree is completely structured, the branches are pruned to an optimal fit using cross-validation procedures.[142] The result of classification and regression tree analysis is a hierarchically organized decision tree.[142]

Fuzzy logic is relatively new to wildlife modeling but it has been used for decision-making in a multitude of applications, including chemistry, engineering, financial forecasting, and mineral exploration. Fuzzy logic allows one to express a degree of uncertainty around model parameters and variables.[39] The appeal of fuzzy logic to wildlife modelers is that answers to wildlife problems are typically qualified. For example, we can seldom predict a dependent wildlife variable with absolute certainty. More often, we qualify our answers with fuzzy terms like "sometimes", "probably", or "most often". Wildlife modelers use fuzzy logic to portray these qualitative terms. Advantages of using fuzzy logic to model wildlife systems include: 1) fewer values, rules, and decisions, 2) the use of linguistic instead of numerical variables, 3) a direct relationship between inputs and outputs without having to understand all the relationships, 4) easy design and prototyping; and 5) often a few rules encompass great complexity.[132] Disadvantages of using fuzzy logic for wildlife modeling are: 1) more simulation and fine-tuning is required before models are operational, and 2) the cultural bias that our profession has for mathematical precision may make these models less appealing.[132] Because the very nature of fuzzy logic is imprecise, the use of fuzzy logic models for modeling wildlife species with regulatory impacts may be limited.

Other, less common multivariate methods exist and a complete discussion of these is beyond the scope of this chapter. Multivariate statistical models are appealing because they have a low level of subjectivity relative to other model types and they are perceived as valid (Table 1). Applicability of multivariate models is often limited by data availability, cost of model development, and training that is required to interpret the model (Table 1). These models are also complex, difficult to communicate, and often apply to a narrow range of environmental conditions (Table 1). For more information on the use of multivariate models for wildlife habitat assessments see Capen,[30] Verner et al.,[204] (Chapter 2) and Morrison et al.[136] (Chapter 7).

Habitat Preference Models and Foraging Models. Another category of empirical habitat models that emphasizes single species includes two model types that use data on habitat use and availability (Figure 1). These models are termed habitat preference models and foraging models (Figure 1). Habitat preference models rely on statistical inference on habitat selection, i.e., the differential use of habitats as compared with their availability.[137] These models are most frequently based on individual animal selection as monitored via radio telemetry or mark-recapture. The idea behind these models is that habitat preference can be inferred from the amount of time (or frequency of visits) an animal spends in a particular habitat patch relative to the amount of the habitat available to that animal. For example, Quinn[152] used radio-telemetry to show how coyotes in urban areas of Washington selected certain habitat types during routine movements.

Foraging models are closely related to habitat preference models. These models predict animal welfare or fitness based on the bioenergetics of forage intake. Optimal foraging theory suggests that when resources are not limiting, species should concentrate foraging on the best types of food.[38] In other words, organisms chose habitat to optimize their energy budgets. Inputs to foraging models include nutritional content of forages and the organism's energetic balance. Hobbs's[87] foraging model for mule deer simulates energy flow from forage resources to the animal and the allocation of that energy to individual expenditures. The output generated by Hobbs's[87] model portrays the energy balance of mule deer and he uses this information to predict deer condition and mortality. Foraging models have been developed for other species including caribou,[17] black-tailed jackrabbits,[40] insectivorous birds,[126] and waterfowl.[67]

Both habitat preference and foraging models are analytical approaches that require extensive field data. These characteristics and their overall complexity often make them less appealing to practitioners (Table 1). However, these model types can be some of the most statistically robust and scientifically credible of the habitat-based models (Table 1). Also, data from these types of studies can be used to construct other, more simplistic models (e.g., habitat effectiveness, habitat evaluation procedures).

Community and Ecosystem Simulation Models. We divided empirical habitat models for multiple species into three groups (Figure 1). One group of models can be used to project environmental conditions or evaluate the effects of environmental stochasticity on wildlife habitat. These models are categorized as community and ecosystem simulation models.[143, 145, 146, 190] These models are based on systems analysis and consist of inputs, outputs, and feedback loops that describe processes affecting that system. Community and ecosystem simulation models were designed to evaluate 1) population and system responses to changing environmental conditions, 2) the effects of catastrophic changes on populations, and 3) overall ecosystem function.[137] The simulation component of these models permits the integration of randomness into the analysis. Thus, community and ecosystem simulation models have been used to optimize systems based on the re-occurrence of desired patterns from iterative model runs.[21] The startup and maintenance costs, complexity, and difficulty in communicating the results reduce the utility of community and ecosystem simulation models for practitioners. But once a process is established, these models offer powerful planning capabilities, particularly in combination with coarse-filter models.

Community and ecosystem simulation models are appealing, often use readily available data, reduce the level of subjectivity by incorporating randomness, and may apply to a broad range of conditions (Table 1). The primary drawbacks of using these model types are the costs associated with data collection, model development, and training (Table 1). Community and ecosystem simulation models are also complex and can be difficult to communicate (Table 1).

Hierarchy Models. The second group of multi-species empirical habitat models is called hierarchy models. These models portray information at multiple spatial scales (Figure 1). Although several of the previously discussed modeling approaches consider relationships across multiple scales (e.g., coarse-filter and fine-filter models, community and ecosystem simulation models), hierarchy models retain scale relationships and depict these relationships as a series ranking one above the other. The purest hierarchy models include the models developed for ecological land classification,[59, 128] however the hierarchy concept is implicit in many of the previously discussed modeling approaches. For example, With and Crist[214] used hierarchical concepts to model grasshopper response to different habitat types and Roloff and Haufler[157] used hierarchy theory in combination with the habitat evaluation procedures to model population viability. Morrison et al.[137] noted the relationship of hierarchy models to many of the community and ecosystem simulation models. Hierarchical models are gaining favor throughout the natural resource sciences because they offer a means of logically structuring complex ecosystems. The strength of hierarchical models is in combination with other modeling approaches. By building and presenting some of the more complex models in a hierarchical fashion they often become more appealing to practitioners.

Community Structure, Gap Analysis, and Biodiversity Models. The final category of empirical habitat models portrays multiple species habitat relationships at a particular spatial scale at one point in time (Figure 1). This category includes community structure models, gap analysis models, and biodiversity models (Figure 1). Community structure models describe species distribution, abundance, or diversity as functions of environmental complexity.[137] Like single species multivariate statistical models, community structure models portray responses of wildlife communities or guilds based on empirical data. James[96] used a community structure model to display a three-dimensional ordination of a bird community in Arkansas. Each dimension represented a habitat gradient. For example, the first dimension, which accounted for 65% of the total variation in bird distribution, ranged from areas with high ground cover and few trees and shrubs to areas with mature forests and low ground cover.[96] For another example of a community structure model see James and Wamer.[97] Community structure models often use commonly collected data and have low subjectivity (Table 1). These models can be expensive, complex, site-specific, and difficult to communicate (Table 1).

Gap analysis models are another form of multi-species habitat model. These models are based on geographic distribution of many species, and can thus depict gaps in species distributions or identify areas of relatively high species richness.[173] The input data on vegetation are derived from satellite imagery and wildlife species habitat associations are assigned in a manner similar to the species-habitat matrices discussed earlier. The United States Fish and Wildlife Service initially established the gap analysis program at the state-level, however, use of the concept has spread[18] (e.g., into Mexico). First generation gap models identified patches of habitat in which wildlife species were known to occur and the locations of similar patches were used to infer species distributions.[173] Second generation gap models use a refined vegetation classification scheme and integrate minimum area requirements and landscape configuration into habitat assessments (Richard Minnis, Cooperative Fish and Wildlife Research Unit, pers. comm.).

An abundance of published literature exists on the implementation of gap analysis models; citations are provided at http://www.gap.uidaho.edi/gap. This site also contains an overview of the gap analysis process, a handbook for applying gap analysis, and a guide to the status of gap analysis for each State in the United States. Other than startup costs of data acquisition and compilation, gap analysis models are one of the more appealing multi-species, coarse-filter modeling approaches for portraying habitat relationships across large areas (Table 1) but see Morrison et al.[137: 332] for a discussion on the limitations of the approach.

Biodiversity models are similar to gap analysis models; they map species richness based on vegetation attributes. Biodiversity models have most commonly appeared in the form of logarithmic series, Shannon-Wiener functions, Simpson's diversity indices, Brillouin's indices, and evenness measures.[104] These approaches assess community structure (e.g., richness, evenness, diversity) based on the number of species in a sample and the number of individuals of each species in a sample. Although the measurement of species richness has dominated community ecology in the past, species richness is only a partial expression of biodiversity and thus, there is a need to further develop these models to expand their breadth.[137] For further reading on the topic of biodiversity models, refer to Maurer[127] and Rosenzweig.[162] Biodiversity models have low subjectivity and thus are perceived as valid (Table 1). Drawbacks of these models include historically minimal use in land management decisions, costs, and their site-specificity (Table 1).

Habitat models have a number of appealing attributes from which they derive their utility. First, theory, practice, and common sense tell us that habitat quality, quantity, and the spatial distribution of habitats are important determinants of population performance. Second, wildlife-habitat relationship data (this book) and habitat data (e.g., vegetation inventories, remotely sensed data) are frequently the only type of information available to wildlife professionals. Third, inputs and outputs of habitat models are typically easy to understand and act upon. Finally, several habitat models assess management impacts on a large number of species for which there often exists little more than life history or coarse-grained distribution information. These multi-species models are appealing because they provide information on numerous species across larger landscapes and are consistent with contemporary perspectives on ecosystem management and landscape-level planning.

Population Modeling

Population modeling forms the basis of numerous harvest management plans world wide.[119, 144, 170] A population is a group of individuals that is spatially, temporally or genetically separated from other groups.[210] Population models use demographic information such as fecundity, mortality, immigration, emigration, density, and age structure as inputs to portray population dynamics.[69] Population models can also be used to estimate which parts of a population (e.g., juveniles, adult males, females) have the greatest effect on population growth.

The extensive history and theory of population modeling is beyond the scope of this introductory chapter (see Andrewartha[4] and Slobodkin[178] for general reviews). General texts on general wildlife biology and management often address population ecology and various modeling approaches.[5, 19, 38] Dempster[50] and Moss et al.[138] discuss basic demographic theory and environmental influences on population growth, with Dempster[50] providing examples from long-term data sets. Sutherland[186] discusses how certain population ecology aspects of vertebrates can be related to behavior. Fryxell and Lundberg[69] address predator-prey modeling.

Population models can be single- or multi-species and can integrate multiple demographic and environmental factors (Figure 2). Single-species models project population dynamics for individual species typically as an extension of historic population trends[170] or in relation to environmental variables.[144] Multiple species models represent interactions between two or more populations of different species, most commonly interactions between predator and prey. Population models and spatial population models (below) can also be categorized as

Wildlife Population Modeling

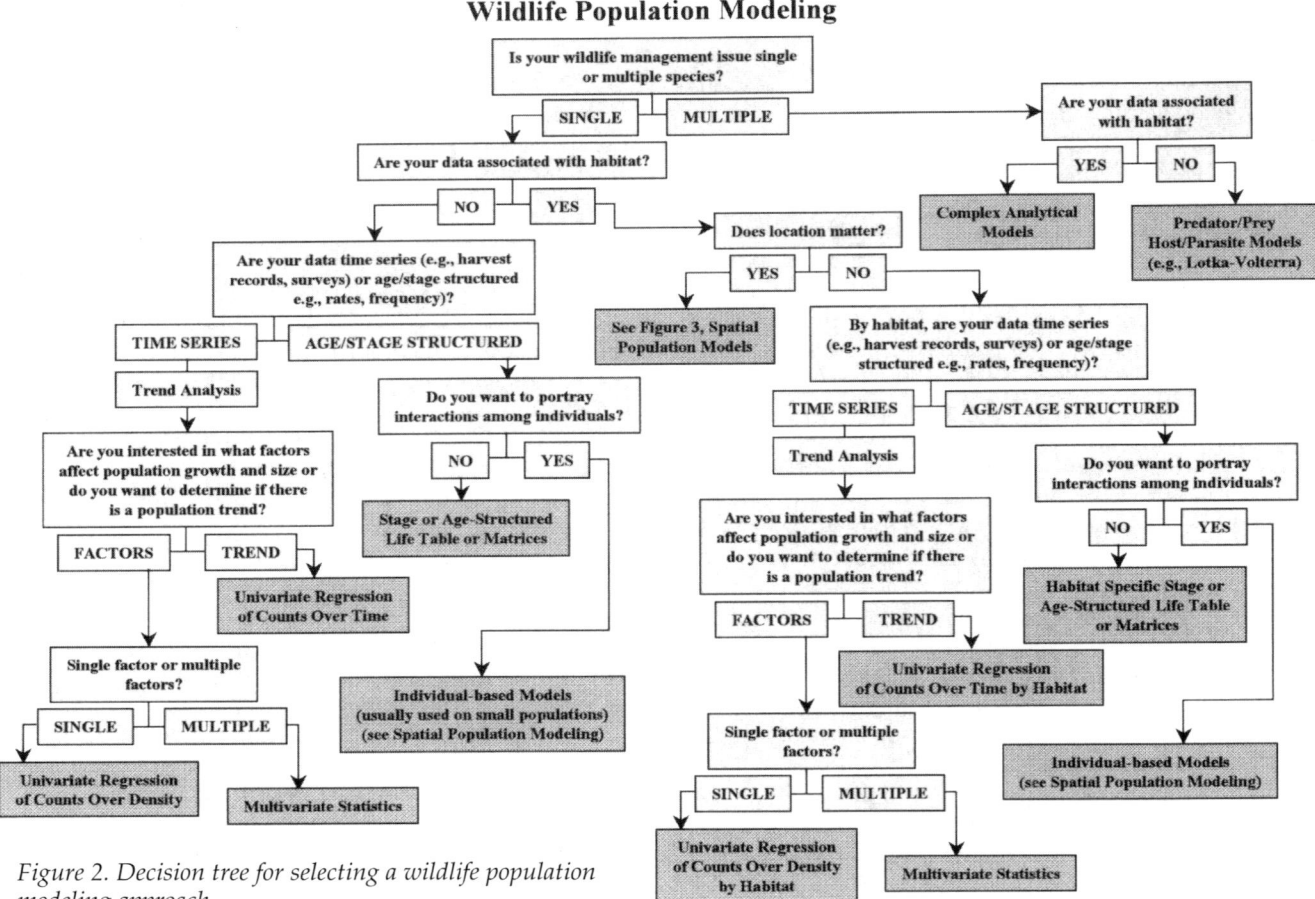

Figure 2. Decision tree for selecting a wildlife population modeling approach.

deterministic and stochastic. Deterministic models assume that parameter values are constant throughout the projections. For example, in a deterministic Leslie matrix (see below), demographic rates are treated as constants for each age- or stage-class.[11, 58] Stochastic models incorporate variability into the parameter estimates.[86] Data re-sampling techniques (e.g., Monte Carlo methods, bootstrapping) are often used to incorporate this variability. For example, in a stochastic Leslie matrix demographic rates by age- or stage-class are generated multiple times by sampling from underlying data distributions. Deterministic and stochastic methods exist for the three common approaches to modeling populations: trend analysis models, life-table or matrix models, and multiple species modeling.

Trend Analysis Models

Trend analysis models are the simplest and often most useful form of population model. Trend analysis uses a dependent variable (e.g., number of animals harvested, number of waterfowl surveyed) to index population size. Measures of dependent variables over time are used as inputs to trend models and population changes (trends) over time are estimated. Trend models portray average rates of population change, require few assumptions, and can incorporate data that are readily available from standard wildlife surveys. Pascual et al.[144] evaluated three

different forms of trend models (logistic, Ricker, and Beverton-Holt) on data from a Serengeti wildebeest (*Connochaetes taurinus*) population. Each of the three models require different assumptions on the feedback between population size and population growth. Thus, each portrayed different population dynamics and demonstrated the importance of understanding a model's assumptions and structure to interpreting model output.[144]

Univariate Regression. To determine statistical significance of a population trend, a simple linear regression applied to population time series (i.e., a series of estimates of population size taken at regular intervals) is often appropriate. Transformation of population count data (e.g., taking the natural logarithm of the population measure) and various smoothing techniques (to help discern a trend from highly variable data) are available to estimate population change over time.[168] These types of models provide an estimate or prediction of population size through time and can be used to infer if the population is stable (i.e., the trend line has a slope = 0), declining (negative slope), or increasing (positive slope). For example, McCullough[131] used univariate regression to determine the relationship between adult population size and recruitment of young for grizzly bears. Using the magnitude of the regression coefficients and the regression slope, McCullough[131] argued that young grizzly bear

**Table 2. A coarse comparison of population modeling approaches.
Each model was ranked as unfavorable (U), moderately favorable (MF), or favorable (F),
according to the practical application of the modeling approach.
Rankings are subjective evaluations as to the historic use, performance, and perception of the models.**

| | Model Evaluation Criteria | | | | | | | | | |
| | Applicability | | | | | Model Structure | | Model Output | | |
	Appeal	Common data	Cost of data acquisition	Cost of model develop-ment	Training required	Level of subjectivity	Complexity	Generality	Face validity	Communicability
Single species										
Trend analysis										
Univariate regression	F	F	F	F	F	MF	F	F	F	F
Multivariate techniques	MF	F	F	F	F	U	U	F	MF	U
Age/Stage based										
Life tables or matrixes	F	MF	MF	F	F	F	F	F	F	F
Multiple species	U	MF	U	MF	U	U	U	U	MF	F

[a]Some criteria adapted from Marcot et al.[124] where *Appeal* = How much the model framework has been used in the past and practically applied; *Common data* = Data that are collected as part of routine inventories (e.g., rimber cruises, game check stations, remote sensing); *Cost of data acquisition* = Cost to acquire the data for model development and use; *Cost of model development* = Cost (monetary and time) of developing the model; *Training required* = Amount of training required to use and understand the model; *Level of subjectivity* = Degree that subjective evaluations are used to develop the model relationships; *Complexity* = Complexity of the modeling framework, synonymous with Marcot et al.'s "wholeness"; *Generality* = How well the model represents a broad range of similar systems; *Face validity* = Perceived credibility of a modeling framework; *Communicability* = Ability to associate model output to real world systems.

recruitment into populations was highly dependent on adult population size. Univariate regression is one of the most practical approaches to population modeling (Table 2).

Multivariate Techniques. To further account for variance in population rates of change from trend data, it is possible to examine the contribution of additional environmental variables that may regulate the population. Additional independent variables (e.g. weather, harvest, predation) can be input to construct a multiple regression model that expresses population growth as a function of several independent variables including multiple time lags. Time lags are a means of capturing the effects of independent variables that appear after some period (lag) of time.[164] For example, a summer drought might not immediately influence adult mortality, but rather, the effects of the drought may be expressed 10 months later in the number of offspring produced by females.

Caution should be exercised when projecting future population size based on historic trends. This is akin to predicting the dependent variable outside the range of the independent variable (time). Moreover, relationships between population growth rate and time are not cause and effect. That is, simple trend analysis does not indicate if populations are regulated in a predictable fashion, perhaps by intrinsic (density-dependent) factors.[14] One of the most commonly used hypotheses to detect density-dependent population regulation is based on the logistic model.[51, 163, 167] The logistic model often takes the form of the familiar S-shaped curve. The logistic model can incorporate a wide variety of density-dependent effects that can imitate different forms (e.g., scramble competition,

contest competition) and intensities of intra-specific competition.

Multivariate techniques to population modeling often use commonly collected data, can be developed for low costs, and require minimal training to apply (Table 2). These models also can be applied to a broad range of similar systems (Table 2). Drawbacks of these models include their level of subjectivity and complexity (Table 2). Outputs can also be difficult to communicate (Table 2).

Age/Stage-Based Models

If structure or composition of a population is of interest, then age- or stage-based models are usually employed. These models contain elements that correspond to data collected on individuals that are extrapolated to the population using matrix algebra.[86] Matrix elements can include survival of individuals, probability of transition from one stage to the next, or fecundity.[201] Most life tables or matrix models focus on life cycle components with high potential impacts on population growth and how these components interact.[201]

Leslie[113] introduced matrix methods to wildlife biologists in the mid-1940s. Textbooks by Caswell[35] and Burgman et al.[28] advanced matrix theory in biology and the tools are widely used for population projections on a multitude of species. Life tables and matrix models were developed from matrix theory and they represent age- or stage-specific summaries of mortality or birth rates that operate on populations. These tables have been presented in several formats (e.g., Leslie matrix) and often use a specialized notation (reviewed in Krebs).[104: 413] The original work of Leslie[113] laid the foundation for age-based analysis, whereby individuals in a population are projected through different age classes. Lefkovitch[112] recognized that Leslie's work was a specialized version of a more general matrix. He noted that Leslie's[113] matrix ideas could be used to project the abundance of individuals in a particular life stage at a certain time and he established the foundation for stage-based analysis. In stage-based analysis, the assumption is made that all individuals in a particular stage (e.g., body size or development classes) exhibit the same mortality, fecundity, and growth schedules, regardless of age. Stage-based models are useful when the age of individual organisms cannot be exactly determined, but when individuals can be classified according to their developmental stage (e.g., subadults-adults, larvae-pupae-adults). For example, van Tienderen[201] identified four life cycle stages (yearlings, juveniles, reproductive females, and post-reproductive females) regulated by three probabilities (the probability of going to the next life stage, the probability of remaining in the current life stage, and fecundity) in his matrix population projection of the demography of killer whales. Van Tienderen[201] demonstrated the importance of correlation among inputs on the population projection. These considerations become especially important when matrix models are used to project populations subjected to alternative management activities, particularly if the management activity cascades through the life table because of interrelationships in the demographic parameters.

Stage-based modeling can also be performed on habitat-structured populations.[20] Since animals that occupy habitats of different quality will survive and reproduce at different rates (especially in the absence of density dependent effects), it is logical to consider that a more preferable approach to life table or matrix modeling is to structure populations according to habitat quality. Bowers[20] compared age-structured versus habitat-structured population projections using computer simulation. He demonstrated that the demography of a habitat-structured population could change fitness by changing habitats. In age-structured populations fitness could only be changed through aging. Realistically, a combination of age- or stage- and habitat-based approaches probably provides the best model. Some modelers have taken this advice and integrated habitat and demographics into population viability analyses.[2, 161]

Age- and stage- models can employ key-factor analysis, which is often used to compare the effects of different survival or fecundity rates among the different classes on population dynamics.[199] In other words, by using key-factor analysis, age- and stage-models can be used to determine the age or stage at which conservation actions have the greatest effect on population size.[44] This is often useful for evaluating the effects of density-dependence and environmental variation on population growth rates.[24, 176, 199] The key-factor analysis is data intensive in that it requires age- or stage-specific mortality and fecundity data.[14]

Life table and matrix models have been used to project populations for a wide array of wildlife species including killer whales,[201] semipalmated sandpipers,[86] spotted owls,[29, 110] marbled murrelets,[10] loggerhead sea turtles,[43] and fur seals,[58, 68] to name a few. These models are some of the most widely used and practical of the population modeling tools (Table 2).

Multiple Species Models. Often, a key question in wildlife population management relates to the role that predation plays in creating prolonged prey suppression.[134] To determine if predation is causing low prey density, researchers must quantify the response of predators to changing prey density.[134] A diversity of models exist that describe population dynamics in communities with interacting species.[129] Two-species interaction models are created when the dynamics of two interacting populations (i.e., predator-prey, herbivore-plant, competition for a resource) are of interest.

Generally, the mathematics of multiple species population models can be grouped into those that project continuous population growth rates (i.e., use differential equations) and those that project discrete population growth rates (i.e., use difference equations). Differential models are used with wildlife species that exhibit continuous reproduction, i.e., there is no "break" in the potential for population growth. In contrast, difference-equation models (originally developed by entomologists) have application to wildlife species that exhibit notable generation breaks in the birth process (e.g., salmon, insects). Generally, difference equations tend to be less stable than the corresponding differential equations because the time lapse between generations of growth in the difference models has the destabilizing effects associated with any time lag in an interactive system.[129] These models are not simply restricted to the growth equations for predator and prey, other environmental influences can also be integrated (e.g., habitat quality).[134]

Multi-species trend models are often based on the logistic growth model, where the population growth rate of a species is not only a function of its own density, but also is influenced by the density of a second species. The most well-known multi-species trend models are the predator-prey models (more commonly called Lotka-Volterra models) which estimate the demographics of two species in a predator and prey relationship. These models predict prey recruitment, prey mortality, predator recruitment, and predator mortality as functions of each other.[69] By using ratio-dependent interspecific effects (i.e., the ratio of the predator per prey) instead of a pure predator density effect the realism of the model is increased.[13] This approach is analogous to incorporating extrinsic covariates into the logistic model (see above). This generates an interesting conclusion, that carrying capacity is not a fixed entity, but depends on the covariates. In this case, the covariate is the ratio of predator to prey. Such covariates mediate and alter the strength of the density-dependent affects on carrying capacity. Examination of covariate affects can greatly increase our understanding of complex systems and enhance our ability to manage these systems successfully.[103]

Age- or stage-structured multi-species models are rarely constructed in wildlife management due to their extensive data requirements (i.e., age- or stage-specific demographic rates for predators and prey; Table 2), but they occasionally find application in the field of entomology and pest management. Age-structured multi-species models can be implemented as linked Leslie-models.[73] In these models, a realistic representation of the order of biological process (i.e., the sequence of events within the population) is of critical importance.

Spatial population modeling

Widespread recognition of space as an important consideration for both conservation theory and management practice can be traced to the equilibrium theory of island biogeography.[56, 177] MacArthur and Wilson[121] proposed that a dynamic equilibrium between extinction and immigration could explain species richness on oceanic islands, and that species richness was a function of an island's area and distance from the mainland. An analogy between oceanic islands and patches of habitat surrounded by inhospitable areas has been used to advance concepts in wildlife habitat management.[52, 70, 188] Levins[115, 116] is credited with establishing the contemporary foundation for the study of spatially structured wildlife populations. His metapopulation concept pertains to a collection of interacting populations that occur in a distinct habitat patch separated from other patches by non-habitat. The number of populations that occur within the metapopulation (or the number of occupied habitat patches) is determined by the rates of population extinction and patch colonization. There are obvious similarities between island biogeography and metapopulations, but for a number of reasons, the latter has supplanted the former as a dominant paradigm in conservation biology.[78] Hanski and Gilpin,[77] Hastings and Harrison,[83] and Hanski and Simberloff[78] provide explanations of metapopulation theory, its history, and models.

While the theoretical import of the metapopulation concept is undeniable, the practical application of the concept is limited, particularly with respect to large or mobile animals for which demographic isolation is less likely. Rarely have managers faced the problem of managing a "population of populations", which was Levins's[116] original definition for a metapopulation. Managers usually deal with a single population or a sub-population. Members of that population will distribute themselves across space according to the spatial distribution of habitats. This pattern of habitat across landscapes may have profound affects on population viability (i.e., the population's survival). "Habitat fragmentation" has become a well-known *cause célèbre* for conservation biologists, and many of their concerns are well-justified.[169] This is an important motivation behind many of the models used for investigating the relationship between the spatial pattern of habitats and population dynamics across landscapes.

Space complicates population models in three fundamental ways: (1) non-uniform patterns of habitat quality, (2) non-uniform patterns of animal density, and (3) animal movements from one location to another.[114] These complications can be modeled with various degrees of realism, and models can be classified accordingly—spatially implicit, spatially explicit, or spatially realistic (Figure 3). The taxonomy of spatial models can be further refined by classifying models according to their population structure, i.e., either as single population or metapopulation models, and according to their formulation—analytical, matrix, or individual-based (Figure 3). Analytical models occur in the form of an equation or equations. Classic examples of analytical models are the logistic and Lotka-Volterra population models. The appeal of analytical models stems from their mathematical elegance, basis in ecological principles, and tractability, either through exact solutions or numerical methods. Most analytical models describe the dynamics of idealized populations (e.g., closed-population, constant carrying capacity), hence their importance lies in contributing to conservation theory which in turn forms the foundation of management practice. While the generality of analytical models makes them less useful for quantitative answers to specific management questions, these models can provide qualitative insights.

The formulation of matrix models was previously discussed in the "Population Modeling" section of this chapter. In that section, we noted that the dimensions of non-spatial matrix models were determined by the number of population classes. The dimensions of spatial matrix models are determined by both the number of population classes and types of habitat patches.

Individual-based models track all the individuals in a population; hence they can incorporate local interactions between organisms and habitat.[92, 98] Every individual in the simulated population is represented numerically, and data on every individual is "observed" and tabulated for birth, death, reproduction, location, and movement. The behavior and fate of individuals are stochastic. That is, actions or events occur randomly according to probability distributions that are either empirically estimated from the real population or based on plausible assumptions. Rules direct the reproductive, foraging, social, and movement behaviors of individuals. The collective success or failure of all individuals determines the fate of the population. In general, individual-based models are not mathematically sophisticated, but they are computationally complex. Hence, well-honed computer programming skills are indispensable for tackling the task of model development. There are no standards for model structure, but modular, top-down, object-oriented programming is highly recommended.

Figure 4 contains examples of different analytical, matrix, and individual-based spatial population models. Figure 4(a) is an analytical spatially implicit model.[109, 110] The equation represents the territory dynamics of a population containing only females at demographic equilibrium. Figure 4(b) is a spatially implicit model

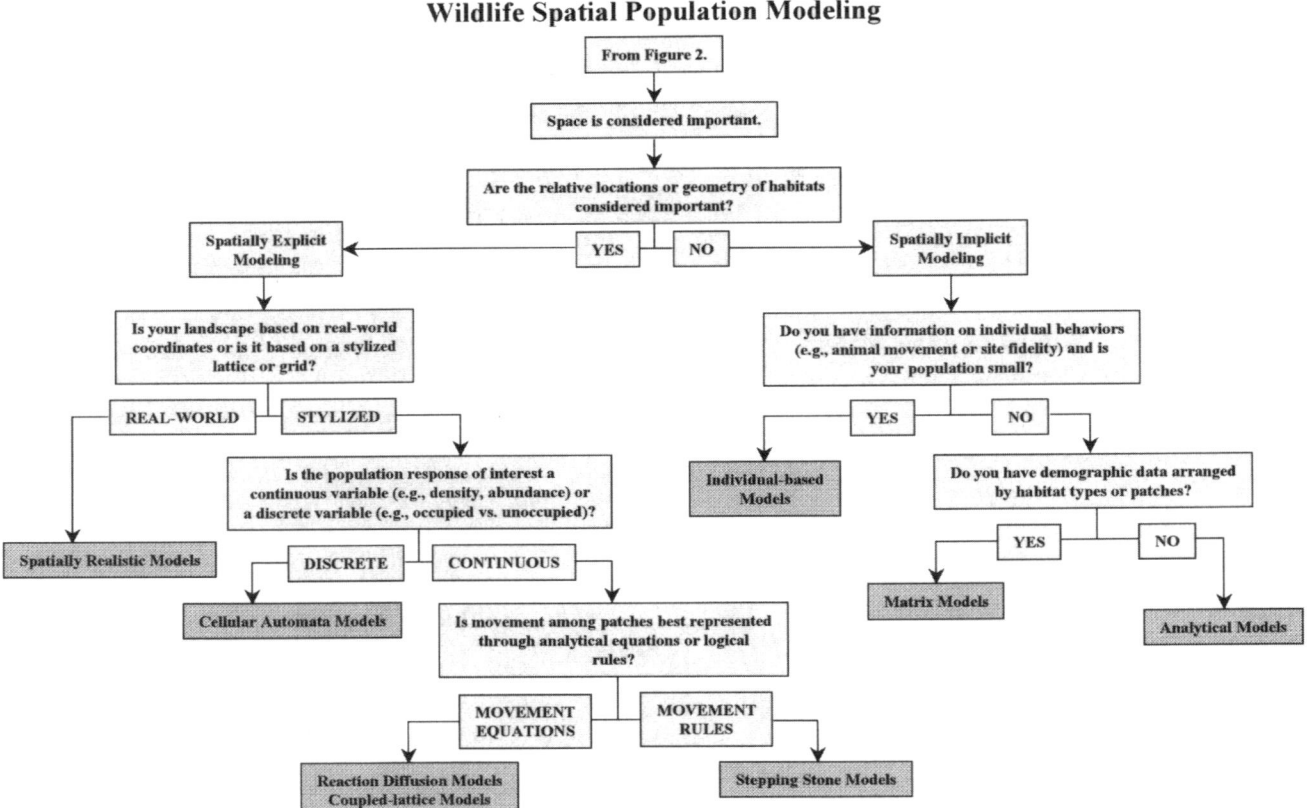

Figure 3. Decision tree for selecting a wildlife spatial population modeling approach.

formulated as a matrix. The matrix represents movement among four habitat patches where F_1^i is the fecundity of individuals in age class 1 and patch i; P_1^i is the probability of individuals in age class 1 surviving and staying in patch i; and d_1^{ij} is the probability of individuals in age class 1 surviving and dispersing to patch j from patch i (Figure 4b). Figure 4(c) is the stylized landscape model developed by Lamberson et al.[106] Each circle contains a territory cluster and arrows depict the dispersal paths between clusters. It is assumed that each cluster is identical and contains the same number of territories. Movement within a cluster is governed by an equation similar to that presented in Figure 4(a). Figure 4(d) is an example of a spatially realistic model. The input to the model is Landsat Thematic Mapper imagery classified into four forest cover categories (left); a non-habitat class and three levels of habitat quality. A hexagonal grid was superimposed on the image and each hexagon was assigned a territory quality based on its habitat composition. This type of model was a two-sex, individual-based model.[208]

The classification system for spatial population modeling we propose here should impart some order on a rapidly evolving branch of population ecology, but it undoubtedly fails to convey the variety of spatial models that exist today. The problem we faced was well expressed by Kareiva[99] who wrote, "The different ways of representing space, keeping track of populations, dealing with environmental variability, describing dispersal, and portraying dynamics within patches, together combine to yield an assortment of models that is overwhelming in its variety."

Spatially Explicit Models. Spatially explicit models are those that "keep track of the exact positions of plants and animals"[100] or "have a structure that specifies the location of each object of interest".[57] Hanski and Simberloff[78] regard the key feature distinguishing spatially explicit models from spatially implicit models to be localized interactions, i.e., the population dynamics within a patch are influenced only by the state of nearby patches in explicit models. A wide variety of modeling approaches are considered spatially explicit including reaction-diffusion models, coupled lattice models, cellular automata models, stepping stone models, and spatially realistic models (Figure 3).[78] These approaches can be applied to single populations or metapopulations.

Reaction-Diffusion Models and Coupled-Lattice Models. Reaction-diffusion models are based on the mathematics used to describe the kinetics of reacting and diffusing chemicals.[9, 114] In population modeling, "reaction" refers to within-patch population dynamics and "diffusion" refers to the dispersal of individuals between patches. A reaction-diffusion model consists of a set of partial differential equations (Figure 3) hence they are continuous in both time and space. Each equation represents a patch's population density and consists of

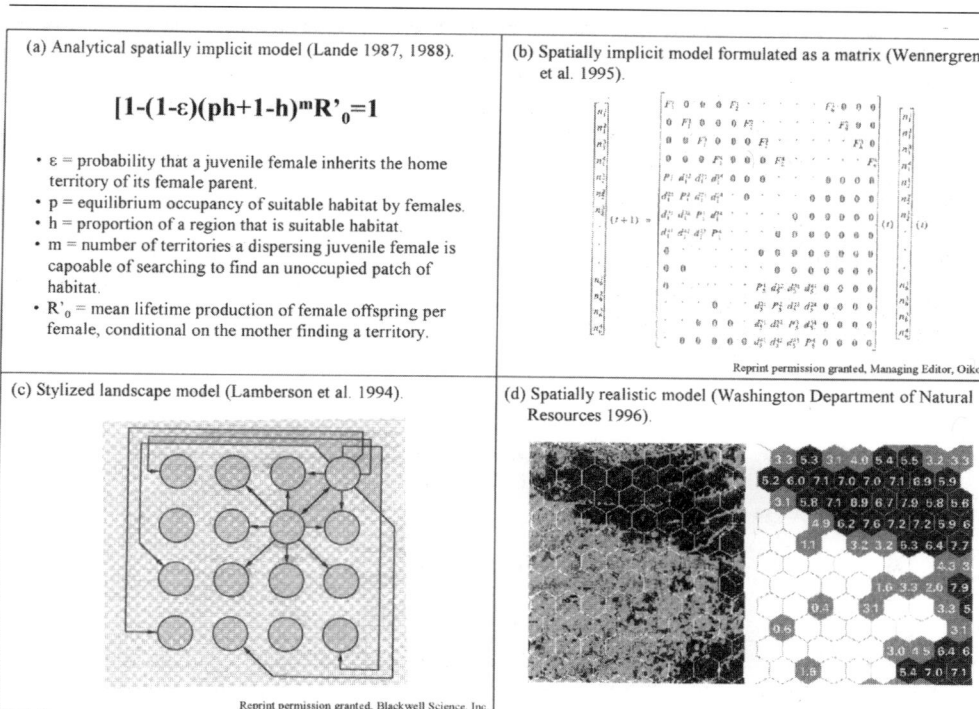

Figure 4. Examples of spatial population modeling approaches.

In the figure:

(a) Analytical spatially implicit model (Lande 1987, 1988).

$$[1-(1-\varepsilon)(ph+1-h)^m R'_0 = 1$$

- ε = probability that a juvenile female inherits the home territory of its female parent.
- p = equilibrium occupancy of suitable habitat by females.
- h = proportion of a region that is suitable habitat.
- m = number of territories a dispersing juvenile female is capaoble of searching to find an unoccupied patch of habitat.
- R'_0 = mean lifetime production of female offspring per female, conditional on the mother finding a territory.

(b) Spatially implicit model formulated as a matrix (Wennergren et al. 1995).

Reprint permission granted, Managing Editor, Oikos.

(c) Stylized landscape model (Lamberson et al. 1994).

Reprint permission granted, Blackwell Science, Inc.

(d) Spatially realistic model (Washington Department of Natural Resources 1996).

three terms: (1) a function describing within-patch population dynamics, (2) a function describing net exchange with other patches, and (3) a function describing net exchange with the region surrounding a patch.[114] To date, the use of reaction-diffusion models has produced mostly theoretical results (Table 3); practical applications have dealt mainly with invasions by pests or the spread of disease.[99]

Coupled lattice, cellular automata, and stepping stone models all arrange habitat patches, or local populations, into a regular two-dimensional array of cells. Coupled lattice models are basically a discrete time, discrete space simplification of reaction-diffusion models;[9] hence they are analytical models that take the form of difference equations. Each difference equation represents population density within a particular cell. Cells within the lattice are "coupled" through diffusion processes that represent animal dispersal. Like reaction-diffusion models, the results from coupled lattice models have been theoretical (Table 3). They are becoming well known for portraying complex dynamics and chaotic behaviors that evoke intriguing speculations about real ecosystems.[9]

Reaction-diffusion models tend to have low appeal because these models have, in general, been confined to the realm of theory (Table 3). Costs and complexity of these models can also limit their practical application (Table 3). The theoretical foundation and mathematical rigor associated with reaction-diffusion models reduce subjectivity, increase generality, and increase perceived validity

Cellular Automata Models. A cellular automaton consists of a regular array of interacting cells. Unlike a coupled lattice model that describes the state of each cell by a continuous variable (e.g., population density), the state of each cell in an automaton model is a discrete variable (Figure 3). For example, Caswell and Etter[36] describe a two-species cellular automaton in which each cell could possess one of four states: species S_1 present, species S_2 present, both species present, or neither species present. Discrete states are ideal for many metapopulation analyses where patches may be either occupied or unoccupied. The state of each cell is determined by a set of rules that account for the current state of the cell and the states of surrounding cells.[9, 45] The rules may be mathematical equations or logical operations or some combination of both. Cellular automata are difficult to solve analytically, so they are often implemented as computer simulations.[36] We are unaware of any applications of cellular automata models to address real wildlife management issues and their practical application is limited (Table 3).

Stepping Stone Models. Because of their conceptual simplicity, stepping stone models have proven to be the most practical approach for incorporating spatial structure into population models. Stepping stones, or patches of habitat, are simply arranged in a regular grid. Patch dependent demographics or local population dynamics are defined by a set of rules (Figure 3). Similar to cellular automata models, these rules can consist of mathematical equations or logical operations or some combination of both. Most stepping stone models simplify the space between patches by assuming it consists of entirely uniform non-habitat. Stepping stone models played an important role in analyses of federal management plans for spotted owl habitat.[192] Like Doak,[55] Thomas et al[192] (and see Lamberson et al.[106]) modeled owl habitat as territory clusters, but they added explicit spatial structure by arranging clusters in a rectangular array. Territories within clusters were all equally accessible, but movement from

Table 3. A coarse comparison of spatial population modeling approaches. Spatial population models can be applied to both individual populations or metapopulations. Each model was ranked as unfavorable (U), moderately favorable (MF), or favorable (F), according to the practical application of the modeling approach. Rankings are subjective evaluations as to the historic use, performance, and perception of the models.

| | Model Evaluation Criteria | | | | | | | | | |
| | Applicability | | | | | Model Structure | | Model Output | | |
	Appeal	Common data	Cost of data acquisition	Cost of model develop-ment	Training required	Level of subjectivity	Complexity	Generality	Face validity	Communicability
Spatially explicit models										
Reaction-diffusion models	U	U	MF	MF	U	F	U	F	F	MF
Coupled lattice models	U	U	MF	MF	U	F	U	F	F	MF
Cellular automata	U	U	MF	MF	U	MF	MF	F	F	MF
Stepping stone models	F	U	MF	MF	MF	U	MF	U	MF	F
Spatially realistic models	MF	U	U	U	U	U	U	U	MF	U
Spatially implicit models										
Analytical models	F	MF	MF	F	F	F	MF	F	F	F
Matrix models	F	U	MF	F	F	F	F	MF	F	F
Individual-based models	MF	U	MF	MF	MF	U	U	U	MF	MF

ªSome criteria adapted from Marcot et al.[124] where *Appeal* = How much the model framework has been used in the past and practically applied; *Common data* = Data that are collected as part of routine inventories (e.g., rimber cruises, game check stations, remote sensing); *Cost of data acquisition* = Cost to acquire the data for model development and use; *Cost of model development* = Cost (monetary and time) of developing the model; *Training required* = Amount of training required to use and understand the model; *Level of subjectivity* = Degree that subjective evaluations are used to develop the model relationships; *Complexity* = Complexity of the modeling framework, synonymous with Marcot et al's "wholeness"; *Generality* = How well the model represents a broad range of similar systems; *Face validity* = Perceived credibility of a modeling framework; *Communicability* = Ability to associate model output to real world systems.

cluster to cluster was limited to adjacent clusters. Dispersal between clusters was more hazardous than dispersal within clusters because regions between clusters contained non-habitat. This model was an individual-based model. A stepping stone model may also be implemented as a transition matrix.[35] Spatial structure can be incorporated into a matrix model through elements that represent the probability of successful dispersal between patches.[211] Demographic parameters would be patch-dependent, i.e., different vital rates can be assigned to patches according to habitat quality. Stepping stone models have high appeal as a compromise between the complexity of spatially realistic models (see below) and the mathematical rigor of reaction-diffusion and cellular automata models (Table 3).

Spatially Realistic Models. The stylized landscapes of the aforementioned models have been eclipsed by the empirically derived landscapes of spatially realistic models. Spatially realistic models use habitat data from actual landscapes that are tied to real-world spatial coordinates (Figure 3), and these habitat data can be manipulated to represent habitat conditions likely to result from management. Spatially realistic population models arose from a fortuitous convergence of theory and technology. A critical mass of research in landscape ecology, albeit predominately theoretical, has been attained during the past 15 years.[66, 81, 193, 195] This theory has led wildlife managers to recognize that the spatial arrangement and quality of habitat can have a significant impact on population viability.[157] Concurrently, an array

of powerful new technologies for the collection and analysis of spatial data, e.g., satellite imaging, global positioning systems, improved microelectronics for radio-telemetry, and geographic information systems, have provided the wherewithal to conduct large-scale empirical studies.

The first published results obtained from applying a spatially realistic model to a wildlife population are attributed to Fahrig and Paloheimo.[62] They analyzed the effects of habitat patch arrangement on the population density of cabbage butterflies (*Pieris rapae*) in Ontario. Since then, interest in such models has continually increased. Spatially realistic models have been used to analyze the effects of habitat change on forest-interior birds in Wisconsin;[187] Bachman's sparrow (*Aimophila aestivalis*) in the southeastern United States;[118, 150] an endangered possum in Australia;[117] the endangered helmeted honeyeater (*Lichenostomus melanops cassidix*) in Australia;[2] an endangered butterfly in Finland;[205] the California gnatcatcher[3] (*Polioptila c. californica*); the California spotted owl[105, 140] (*S. o. occidentalis*); and northern spotted owl.[88, 91, 133, 208]

Hanski[76] developed an analytical spatially realistic metapopulation model. This model, which she called an incidence function, could accurately model the patch turnover rate and minimum patch sizes of several butterfly species, and was described as a practical tool for management problems.

Hanski's[76] spatially realistic models are typically stochastic and individual-based. For many applications, spatially realistic individual-based models will require large computer memory and huge on-line data storage because: (1) information on every living individual in the population is maintained during a simulation, and (2) geographic information system data, particularly rasterized data, can occupy large amounts of hard-drive space. High computer processing speeds are also desirable.

Data required by spatially explicit models fall into three separate categories: (1) habitat data, (2) demographic data, (3) and movement data. Habitat maps may be generated through remote sensing methods (e.g,. aerial photography or satellite imaging), on-the-ground habitat surveys, habitat-based modeling, or most likely some combination of both. The most important demographic data are estimates of survivorship and fecundity. Ideally, demographic data would also provide information on the non-breeding portion of the population, such as the fraction of the population that is not breeding. Both demographic data and movement data should be collected such that parameter estimates can be related to habitat quality.

Generally, agreement exists that spatially explicit models have utility in land management decisions.[57, 100, 194] However, concerns have been expressed about model reliability.[7, 41, 56, 165, 211] These concerns stem from two fundamental problems with spatially explicit models. First, such models require data that are often difficult to collect, in particular data relating demographics to habitat quality and data on movements (Table 3). These data requirements are expensive (Table 3). Therefore, such models will be feasible only for species that can garner substantial commitments to data collection. Second, because of the large spatial and long temporal scales to which the models are typically applied, there is no possibility of validating the model. In other words, validation is impossible because true validation requires measuring population dynamics and landscape patterns that are replicated and randomized. These concerns can be assuaged by intelligent model design, thorough model evaluation, appropriate model application, and prudent interpretation of model outputs.[11, 26, 41, 183]

Spatially Implicit Models. Population models can incorporate aspects of space but not possess spatial structure. For example, the classic logistic model of population dynamics implicitly incorporates the amount of habitat available to a population through density dependence. The logistic model, however, does not address the spatial distribution of habitats, and hence, it is a spatially implicit model. Spatially implicit models assume that habitat is spatially structured, but simplify the structure by ignoring the geometry or relative locations of habitat (Figure 3). Typically, spatially implicit models assume that habitats are arranged as discrete patches and that each patch is equally connected to all other patches.[78] The original metapopulation model of Levins[115, 116] was an analytical spatially implicit model.

Hastings[82] developed a more detailed analytical metapopulation model that incorporated the local population dynamics of each patch. Lande[109] used metapopulation concepts to model the effects of habitat loss on a single population of a territorial species and he used this analytical model to examine management of the northern spotted owl.[110] An analytical spatially implicit population model was also used to explore the concept of "source" and "sink" habitats by Pulliam,[148] who claimed that the model could be useful for natural resource management problems.[149]

Lamberson et al.[106] (and see Thomas et al.,[192] Appendix M) developed a spatially implicit model to analyze the effects of habitat fragmentation on the northern spotted owl. All suitable territories were equally accessible to dispersing owls, but the probability of successfully finding a territory was a function of: (1) the proportion of territories unoccupied, and (2) a search efficiency parameter, which was assigned different values for juveniles and adults. Matrix models can also incorporate implicit spatial structure.[20] Unlike typical matrix models that divide populations into age classes, Bowers[20] divided a population into habitat-type classes, according to the habitat types in which individuals bred.

Spatially implicit models can possess a hierarchical structure. Doak[55] represented habitat as clusters of spotted owl territories that imparted a two-level hierarchy to its spatial structure. At the lower level, all territories within a cluster were equally accessible to owls in that cluster, but other clusters were much less accessible. At the higher level, all clusters were equally accessible to all other clusters. In other words, the relative locations of clusters were ignored, and within clusters the relative locations of territories were ignored. His model was an individual-based model—each year the location and age of each owl was known. Carroll and Lamberson[33] developed a theoretical analytical model that also had a hierarchical spatial structure.

Among spatially implicit models, individual-based formulations scored the most "unfavorable" (Table 3). This can be explained, in part, by their relative newness. Practitioners have some exposure to analytical and matrix formulations of population models, but individual-based models are still foreign. Therefore, individual-based models score "unfavorable" for appeal, training required, perceived validity, and communicability (Table 3).

Wildlife Modeling: A Case Study

Implementation of a modeling approach to address a wildlife management issue involves an explicit statement of the issue and a review of relevant information. The issue can be characterized in terms of how output will be used, risks associated with producing imprecise or biased model output, and data availability or resources available to collect data. In addition, scale is important and includes such things as the plan's time frame, planning landscape extent, and species involved. More often that not, the "best model approach" is not selected because limited time and resources (i.e., data and knowledge of the system) constrain the scope and detail of the modeling effort.

In this section, we demonstrate the process of selecting and building a model to address a real resource management issue. We frame the management issue, review the information pertinent to addressing the issue, select a modeling approach using the decision trees presented in figures 1-3, and discuss how the model output was used.

Frame the Management Issue. The State of Washington designated lynx as a threatened species in 1993. As part of the listing, state agencies are responsible for establishing forest practice rules that protect lynx or its critical habitat. State agencies have jurisdiction over state-owned and large, privately owned lands. In lieu of forest practice rules for lynx, landowners within critical habitat areas and the state agencies agreed to develop and implement conservation plans to protect lynx and their habitat. The process was appealing to landowners because it provided an opportunity to develop conservation strategies consistent with their landscape's ecology and compatible with their land management objectives and information systems. The conservation plans were subject to scientific review and approval by Washington State's Forest Practices Board.

The first step in developing the plan was for each landowner to state their goal and objectives. The goal was to craft a scientifically credible lynx conservation plan that could be integrated into the landowner's long-term resource management plan. For one landowner, objectives to achieve this goal included:
- Create a lynx conservation plan that was compatible with the landowner's resource information and forest growth models.
- Maintain or enhance lynx habitat condition over the life of the plan.
- Monitor the lynx population in compliance with the State's recovery plan.
- Maintain a reasonable flow of timber from the land.
- Use existing inventory and mapping systems whenever possible.

Review the Information Pertinent to the Management Issue. Based on this, the landowners decided that modeling was an appropriate tool to assist them in addressing the management issue. The next step in the modeling process was to determine which model type (i.e., habitat-based, population-based, or spatial population modeling) was most suited to the situation and to decide on a specific modeling approach (i.e., use the decision trees in figures 1-3 to select a model). To make these decisions, information and data pertinent to the management issue were reviewed. Planning capabilities influenced the landowner's ability to generate a conservation plan. The landowner had a timber-based land classification scheme coarsely grouped by major plant association type,[46] overstory tree species, forest structure, and vegetation density. This scheme was catalogued in an extensive stand inventory database and mapped to a resolution of 5 acres (2 ha) in a geographic information system. The geographic information system also contained transportation and hydrology map layers. Since the conservation plan needed to be completed in one year, and no empirical data on the

relationships between lynx and habitat existed for this area, the time frames were too short to initiate a large data collection effort. Thus, the project was single species focused, supported by vegetation inventories and a geographic information system, and needed to be completed in a short time frame. A theoretical habitat modeling approach was selected (Figure 1).

Select the Modeling Approach. Several single species habitat models could have been used for this project (Figure 1). The rarity of lynx in the planning area and lack of occurrence data precluded assigning probabilities of occupancy by habitat patch. Thus, the landowners did not select Bayesian or pattern recognition models. Since the amount of habitat was deemed important, they chose to index the contribution of habitat patches towards lynx conservation using the habitat evaluation procedures (Figure 1). Part of the allure of the habitat evaluation procedures was the notion that smaller patches of high quality habitat could function similar to larger patches of marginal quality habitat.[157] This tradeoff was attractive to the landowner because it allowed for operational flexibility. Another appealing characteristic of the habitat evaluation procedures was that they could be used to assess habitat at scales ranging from a single forest stand to a landscape composed of hundreds of stands.[157] Spatially explicit output at the stand-level was appealing in that it permitted the evaluation of individual silvicultural prescriptions that linked to the operational harvest plans. Similarly, a spatially explicit, multi-stand view of habitat was also important to address lynx movement requirements and the effects of harvest patterns on within-home range habitat conditions. Finally, output from the habitat evaluation procedures offered a currency that foresters, agency biologists, and the public could readily understand.

How the Model Output Was Used. The final conservation plan consisted of a 5-year harvest schedule that outlined the stands to be managed, silvicultural prescriptions, post-harvest site preparation, and lynx mitigation measures (Figure 5). The short-term effects of these activities on lynx habitat quality were assessed (Figure 5). In addition to the spatially explicit harvest plan and lynx habitat assessment, a 25-year trend in lynx habitat condition was estimated using a wildlife decision support system that consisted of a tree growth and yield model, forest planning model, and a spatial disaggregation tool.[158] In this decision support system, the lynx habitat model served several purposes. It provided the landowner a means to quantitatively assess habitat in a repeatable, timely manner. The landowner used this information to evaluate different silvicultural prescriptions (e.g., tree spacing in thinning operations); to provide home-range-level assessments that permitted an analysis of habitat configuration; to position human-made den sites, travel corridors, and road closures; and to organize management activities in time and space.

The model and its output were easily communicable to state biologists and regulators and the landowner's

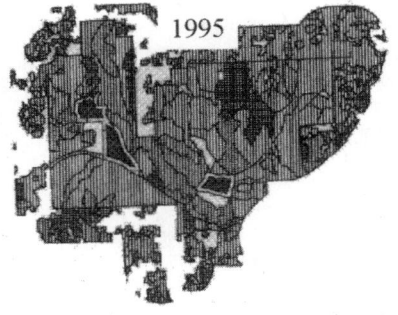

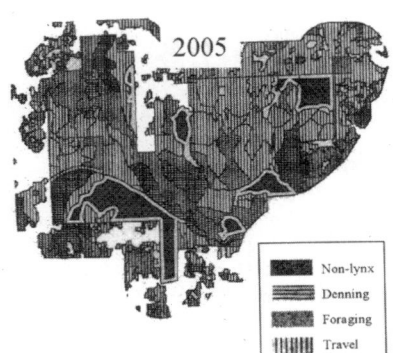

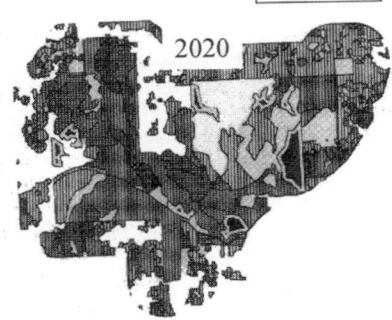

1997 Harvests

1. Commercially thinned trees >12 cm dbh.
   Retain 500 trees/ha.

2. 12 cm diameter cut leaving 10-13 hardwoods per ha. Slashed and planted.

3. Regeneration cut that left 25-37 conifers and 7-13 hardwoods per ha. Slashed and planted.

Lynx Home Range Habitat

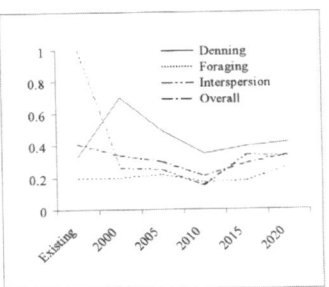

Figure 5. Components of a lynx conservation plan that was developed using a wildlife decision support system that included a tree growth and yield model, forest planning model, spatial disaggregation tool, and a habitat evaluation procedures model for lynx.

foresters and management. Considering the multiple stakeholders associated with this project (stage agencies, Forest Practices Board, landowner management) this characteristic of the model was critical. Also, because the model was habitat-based, the model and its output were easily translated to on-the-ground management practices. The model was run on map layers that corresponded to operational stand boundaries thereby permitting a straightforward crosswalk to the forest plan. The habitat evaluation procedures used in this project are also conducive to adaptive improvements that may result from a more refined lynx habitat model, modifications in how the landowner surveys and maps their forests, or from an expanded view of habitat across multiple land ownerships.

Here, wildlife modeling provided a tool for addressing regulatory issues on private lands management. A defensible, meaningful, and practical plan was developed and accepted by the Washington Forest Practices Board in 1996. The landowners have integrated lynx habitat conservation into their resource planning process and they continue to collect data and refine the relationships contained in the wildlife model.

Future Directions of Wildlife Modeling

More and more, wildlife managers are being asked to apply their expertise toward holistic, often intangible, management goals such as "sustainability," "biodiversity," and "ecological integrity." To be effective, wildlife managers cannot work within the confines of conventional wildlife biology. Rather, managers must also be broadly trained and versed in statistics, ecology, law, sociology, and economics. Managers are also expected to adapt to changing circumstances brought about by dynamic budgets, new information, or changes to regulations or public opinion. These new expectations will shape the future directions of wildlife modeling as a management tool.

Modeling will play a key role in adaptive management. Adaptive management is a process that uses feedback for improving the management process. Models can fit neatly into this process in that they help inform management decisions. The adaptive management process was originally invented for fisheries based on control system theory.[179, 206, 207] In practice, a model of the fisheries system was developed and initial management decisions would be based on information derived from the model.

Management actions were viewed as perturbations to the fishery and by measuring the response of the system to these perturbations data were acquired to improve the model. The adaptive management process would use the new data to revise the model and parameter estimates with the intent of improving the information for management decisions. This cycle continues indefinitely.

Adaptive management is complex. The information feedback mechanisms for wildlife systems may be direct (as in the case of animal harvest), but more often indirect (as is the case with habitat manipulation). For models designed to predict indirect responses of wildlife, the link between management action and system response is subtle and hence, inferences from the model may be weak. Also, wildlife managers often deal with management issues that range from decades to centuries in length. Thus, feedback for model improvement may be delayed beyond the life of the management issue. To alleviate this time lag in receiving feedback, modelers have turned to iterative modeling approaches that attempt to account for the different combinations of environmental and population events. Iterative or simulation models tend to mimic stochastic events and provide insight to the sensitivity and breadth of model output to the range of inputs. For example, iterative modeling is commonly used in extinction models[65] to understand minimum population numbers required for population persistence under a range of environmental stochasticities. Ruckelshaus et al.[165] used iterative modeling to test the sensitivity of spatially explicit population models to parameter error. Bowers[20] used simulation modeling to evaluate the dynamics of age- and habitat-structured populations. These are just a few examples of the use of simulation modeling in an adaptive improvement mode. These simulations provided feedback that facilitated the understanding of system dynamics and model behaviors.

Clearly modeling will remain an important part of wildlife management and undoubtedly a component of a good decision-making process. As wildlife practitioners, we often become preoccupied with rejecting specific models or questioning the assumptions of a particular modeling approach when we should be evaluating whether or not a model allows us to make a more informed decision.[165] This is not to say that models should be taken at face value. It is critical for model users to understand and communicate the limitations and meaning of a model's assumptions, relationships, and outputs relative to the "real world." The complexity of wildlife management dictates that no one model will suffice. Rather, we must develop, apply, evaluate, and improve all models in our toolbox as we see fit, tailoring each to meet the demands of the particular situation.

Literature Cited

1. Airola, D. A. 1988. A guide to the California wildlife habitat relationship system. California Department of Fish and Game, Sacramento, CA.

2. Akçakaya, R. H., M. A. McCarthy, and J. L. Pearce. 1995. Linking landscape data with population viability analysis: management options for the helmeted honeyeater. Biological Conservation 73:169-176.

3. ———, and J. L. Atwood. 1997. A habitat-based metapopulation model of the California gnatcatcher. Conservation Biology 11:422-434.

4. Andrewartha, H. G. 1971. Introduction to the study of animal populations. University of Chicago, Chicago, IL.

5. Bailey, J. A. 1984. Principles of wildlife management. John Wiley & Sons, New York, NY.

6. Bart, J., D. R. Petit, and G. Linscombe. 1984. Field evaluation of two models developed following the habitat evaluation procedures. Transactions of the North American Wildlife and Natural Resources Conference 49:489-499.

7. ———. 1995. Acceptance criteria for using individual-based models to make management decisions. Ecological Applications 5:411-420.

8. Bartlett, M. S. 1960. Stochastic population models in ecology and epidemiology. Wiley & Sons, New York, NY.

9. Bascompte, J., and R. V. Solé. 1995. Rethinking complexity: modelling spatiotemporal dynamics in ecology. Trends in Ecology and Evolution 10:361-366.

10. Beissinger, S. R. 1995. Population trends of the marbled murrelet projected from demographic analyses. Pages 385-393 in C. J. Ralph, G. L. Hunt, Jr., M. G. Raphael, and J. F. Piatt, editors. Ecology and conservation of the marbled murrelet. United States Forest Service General Technical Report PSW-GTR-152.

11. ———, and M. I. Westphal. 1998. On the use of demographic models of population viability in endangered species management. Journal of Wildlife Management 62:821-841.

12. Benson, G. L., and W. F. Laudenslayer, Jr. 1986. DYNAST: Simulating wildlife responses to forest management strategies. Pages 351-356 in J. Verner, M. L. Morrison, and C. J. Ralph, editors. Wildlife 2000: modeling habitat relationships of terrestrial vertebrates. University of Wisconsin, Madison, WI.

13. Berryman, A. A. 1990. POPSYS Series 2 - two-species analysis. User manual, Ecological Systems Analysis, Pullman, WA.

14. ———. 1994. Population dynamics: forecasting and diagnosis from time-series. Pages 119-128 in S. R. Leather, A. D. Watt, N. Mills, and K. F. A. Walters, editors. Individuals, populations and patterns in ecology. Intercept, Andover, England.

15. Block, W. M., L. A. Brennan, and R. J. Gutiérrez. 1986. The use of guilds and guild-indicator species for assessing habitat suitability. Pages 109-113 in J. Verner, M. L. Morrison, and C. J. Ralph, editors. Wildlife 2000: modeling habitat relationships of terrestrial vertebrates. University of Wisconsin, Madison, WI.

16. ———, M. L. Morrison, J. Verner, and P. N. Manley. 1994. Assessing wildlife-habitat-relationships models: a case study with California oak woodlands. Wildlife Society Bulletin 22:549-561.

17. Boertje, R. D. 1985. An energy model for adult female caribou of the Denali herd, Alaska. Journal of Range Management 38:468-472.

18. Bojorquez-Tapia, L. A., I. Azuara, E. Ezcurra, and O. Flores-Villela. 1995. Identifying conservation priorities in Mexico through geographic information systems and modeling. Ecological Applications 5:215-231.

19. Bookout, T. A. 1994. Research and management techniques for wildlife and habitats. The Wildlife Society, Bethesda, MD.

20. Bowers, M. A. 1994. Dynamics of age- and habitat-structured populations. Oikos 69:327-333.

21. Boyce, S. G. 1985. Forestry decisions. United States Forest Service General Technical Report SE-35.

22. Breininger, D. R., V. L. Larson, B. W. Duncan, and R. B. Smith. 1998. Linking habitat suitability to demographic success in Florida scrub-jays. Wildlife Society Bulletin 26:118-128.

23. Brown, E. R. 1985. Management of wildlife and fish habitats in forests of western Oregon and Washington. Part I. United States Forest Service, Portland, OR.

24. Brown, D., N. D. E. Alexander, R. W. Marrs, and S. Albon. 1993. Structured accounting of the variance of demographic change. Journal of Animal Ecology 62:490-502.

25. Brush, T., and E. W. Stiles. 1986. Using food abundance to predict habitat use by birds. Pages 57-63 in J. Verner, M. L. Morrison, and C. J. Ralph, editors. Wildlife 2000: modeling habitat relationships of terrestrial vertebrates. University of Wisconsin, Madison, WI.

26. Bunnell, F. L. 1989. Alchemy and uncertainty: what good are models? United States Forest Service General Technical Report PNW-232.

27. ———, and D. J. Vales. 1990. Comparison of methods for estimating forest overstory cover: differences among techniques. Canadian Journal of Forest Research 20:101-107.

28. Burgman, M. A., S. Ferson, and R. H. Akçakaya. 1993. Risk assessment in conservation biology. Chapman and Hall, London, England.

29. Burnham, K. P., D. R. Anderson, and G. C. White. 1996. Meta-analysis of vital rates of the northern spotted owl. Studies in Avian Biology 17:92-101.

30. Capen, D. E. 1981. The use of multivariate statistics in studies of wildlife habitat. United States Forest Service General Technical Report RM-87.

31. ———, J. W. Fenwick, D. B. Inkley, and A. C. Boynton. 1986. Multivariate models of songbird habitat in New England forests. Pages 171-175 in J. Verner, M. L. Morrison, and C. J. Ralph, editors. Wildlife 2000: modeling habitat relationships of terrestrial vertebrates. University of Wisconsin, Madison, WI.

32. Carlin, B. P. 1996. Bayes and empirical Bayes methods for data analysis. Chapman & Hall, New York, NY.

33. Carroll, J. E., and R. H. Lamberson. 1993. The owl's odyssey: a continuous model for the dispersal of territorial species. SIAM Journal of Applied Mathematics 53:205-218.

34. Caswell, H. 1976. The validation problem. Pages 296-308 in H. H. Shugart and R. V. O'Neill, editors. Systems ecology. Dowden, Hutchinson and Ross, Stroudsburg, PA.

35. ———. 1989. Matrix population models. Sinauer, Sunderland, MA.

36. ———, and R. J. Etter. 1993. Ecological interactions in patchy environments: from patch occupancy models to cellular automata. Pages 93-109 in S. A. Levin, T. M. Powell, and J. H. Steele, editors. Patch dynamics. Springer-Verlag, New York, NY.

37. Caughley, G., J. Short, G. C. Grigg, and H. Nix. 1987. Kangaroos and climate: an analysis of distribution. Journal of Animal Ecology 56:751-761.

38. ———, and A. R. E. Sinclair. 1994. Wildlife ecology and management. Blackwell Scientific Publications, Boston, MA.

39. Chang, C.-L. 1997. Fuzzy-logic-based programming. World Scientific, Singapore, Singapore.

40. Clark, W. R., and G. S. Innis. 1982. Forage interactions and black-tailed jack rabbit population dynamics: a simulation model. Journal of Wildlife Management 46:1018-1035.

41. Conroy, M. J., Y. Cohen, F. C. James, Y. G. Matsinos, and B. A. Maurer. 1995. Parameter estimation, reliability, and model improvement for spatially explicit models of animal populations. Ecological Applications 5:17-19.

42. Cook, J. G., T. W. Stutzman, C. W. Bowers, K. A. Brenner, and L. L. Irwin. 1995. Spherical densiometers produce biased estimates of forest canopy cover. Wildlife Society Bulletin 23:711-717.

43. Crouse, D. T., L. B. Crowder, and H. Caswell. 1987. A stage-based population model for loggerhead sea turtles and implications for conservation. Ecology 68:1412-1423.

44. Crowder, L. B., D. T. Krouse, S. S. Heppell, and T. H. Martin. 1994. Predicting the impact of turtle excluder devices on loggerhead sea turtle populations. Ecological Applications 4:437-445.

45. Czárán, T., and S. Bartha. 1992. Spatiotemporal dynamics models of plant populations and communities. Trends in Ecology and Evolution 7:38-42.

46. Daubenmire, R. 1952. Forest vegetation of northern Idaho and adjacent Washington, and its bearing on concepts of vegetation classification. Ecological Monographs 22:301-330.

47. ———. 1959. A canopy-coverage method of vegetation analysis. Northwest Science 33:43-64.

48. Dedon, M. F., S. A. Laymon, and R. H. Barrett. 1986. Evaluating models of wildlife-habitat relationships of birds in black oak and mixed-conifer habitats. Pages 115-119 in J. Verner, M. L. Morrison, and C. J. Ralph, editors. Wildlife 2000: modeling habitat relationships of terrestrial vertebrates. University of Wisconsin, Madison, WI.

49. DeGraaf, R. M., and D. D. Rudis. 1986. New England wildlife: habitat, natural history, and distribution. United States Forest Service General Technical Report NE-108.

50. Dempster, J. P. 1975. Animal population ecology. Academic Press, New York, NY.

51. Dennis, B., and M. L. Taper. 1994. Density dependence in time series observations of natural populations: estimation and testing. Ecological Monographs 64:205-224.

52. Diamond, J. M. 1975. The island dilemma: lessons of modern biogeographical studies for the design of natural reserves. Biological Conservation 7:129-146.

53. Digby, P. G. N., and R. A. Kempton. 1987. Multivariate analysis of ecological communities. Chapman and Hall, London, England.

54. Dillon, W. R., and M. Goldstein. 1984. Multivariate analysis: methods and applications. John Wiley & Sons, New York, NY.

55. Doak, D. 1989. Spotted owls and old growth logging in the Pacific Northwest. Conservation Biology 3:389-396.

56. Doak, D. F., and L. S. Mills. 1994. A useful role for theory in conservation. Ecology 75:615-626.

57. Dunning, J. B., D. J. Stewart, B. J. Danielson, B. R. Noon, T. L. Root, R. H. Lamberson, and E. E. Stevens. 1995. Spatially explicit population models: current forms and future uses. Ecological Applications 5:3-11.

58. Eberhardt, L. L. 1990. A fur seal population model based on age structure data. Canadian Journal of Fisheries and Aquatic Sciences 47:122-127.

59. Ecomap. 1993. National hierarchical framework of ecological units. United States Forest Service, Washington, D.C.

60. Edwards, T. C., Jr., E. T. Deshler, D. Foster, and G. G. Moisen. 1996. Adequacy of wildlife habitat relation models for estimating spatial distributions of terrestrial vertebrates. Conservation Biology 10:263-270.

61. Elton, C. 1927. Animal ecology. Macmillan Company, New York, NY.

62. Fahrig, L., and J. Paloheimo. 1988. Effect of spatial arrangement of habitat patches on local population size. Ecology 69:468-475.

63. Fitzgerald, E. C., and C. H. Nilon. 1993. Testing the accuracy of an HSI model in an urban county. Transactions of the North American Wildlife and Natural Resources Conference 58:117-123.

64. Flather, C. H., and T. W. Hoekstra. 1985. Evaluating population-habitat models using ecological theory. Wildlife Society Bulletin 13:121-130.

65. Foley, P. 1997. Extinction models for local populations. Pages 215-246 in I. Hanski and M. E. Gilpin, editors. Metapopulation biology: ecology, genetics, and evolution. Academic Press, San Diego, CA.

66. Forman, R. T. T., and M. Godron. 1986. Landscape ecology. John Wiley and Sons, New York, NY.

67. Frederick, R. B., W. R. Clark, and E. E. Klass. 1987. Behavior, energetics, and management of refuging waterfowl: a simulation model. Wildlife Monographs 96:1-35.

68. Frisman, E. Y., E. I. Skaletskaya, and A. E. Kuzin. 1982. A mathematical model of the population dynamics of a local northern fur-seal herd. Ecological Modelling 16:151-172.

69. Fryxell, J. M., and P. Lundberg. 1998. Individual behavior and community dynamics. Chapman & Hall, New York, NY.

70. Goeden, G. B. 1979. Biogeographic theory as a management tool. Environmental Conservation 6:27-32.

71. Grant, W. E. 1986. Systems analysis and simulation in wildlife and fisheries sciences. John Wiley & Sons, New York, NY.

72. Grubb, T. G. 1988. Pattern recognition-a simple model for evaluating wildlife habitat. United States Forest Service Research Note RM-487.

73. Gutierrez, A. P. 1996. Applied population ecology - a supply-demand approach. John Wiley and Sons, New York, NY.

74. Hall, C. A. S., and J. W. Day, Jr. 1977a. Systems and models: Terms and basic principles. Pages 6-36 in C. A. S. Hall and J. W. Day, Jr., editors. Ecosystem modeling in theory and practice. John Wiley & Sons, New York, NY.

75. ———, and J. W. Day, Jr. 1977b. Ecosystem modeling in theory and practice. John Wiley & Sons, New York, NY.

76. Hanski, I. 1994. A practical model of metapopulation dynamics. Journal of Animal Ecology 63:151-162.

77. ———, and M. Gilpin. 1991. Metapopulation dynamics: brief history and conceptual domain. Biological Journal of the Linnean Society 42:3-16.

78. ———, and D. Simberloff. 1997. The metapopulation approach, its history, conceptual domain, and application to conservation. Pages 5-26 in I. Hanski and M. E. Gilpin, editors. Metapopulation biology: ecology, genetics, and evolution. Academic Press, San Diego, CA.

79. Harestad, A. S., and D. G. Keisker. 1989. Nest tree use by primary cavity-nesting birds in south central British Columbia. Canadian Journal of Zoology 67:1067-1073.

80. Hariston, N. G. 1981. An experimental test of a guild: salamander competition. Ecology 62:65-72.

81. Harris, L. D. 1984. The fragmented forest: island biogeography theory and the preservation of biological diversity. University of Chicago, Chicago, IL.

82. Hastings, A. 1991. Structured models of metapopulation dynamics. Biological Journal of the Linnean Society 42:57-71.

83. ———, and S. Harrison. 1994. Metapopulation dynamics and genetics. Annual Review of Ecology and Systematics 25:167-188.

84. Haufler, J. B., C. A. Mehl, and G. J. Roloff. 1996. Using a coarse-filter approach with species assessment for ecosystem management. Wildlife Society Bulletin 24:200-208.

85. ———, C. A. Mehl, and G. J. Roloff. 1998. Conserving biological diversity using a coarse-filter approach with a species assessment. Pages 107-125 in R. K. Baydack, H. Campa III, and J. B. Haufler, editors. Practical approaches to the conservation of biological diversity. Island Press, Washington, D.C.

86. Hitchcock, C. L., and C. Gratt-Trevor. 1997. Diagnosing a shorebird local population decline with a stage-structured population model. Ecology 78:522-534.

87. Hobbs, N. T. 1989. Linking energy balance to survival in mule deer: development and test of a simulation model. Wildlife Monographs 101:1-39.

88. Hof, J., and M. G. Raphael. 1997. Optimization of habitat placement: a case study of the northern spotted owl in the Olympic Peninsula. Ecological Applications 7:1160-1169.

89. Holl, S. A. 1982. Evaluation of bighorn sheep habitat. Desert Bighorn Council Transactions, 47-49.

90. Hollander, A. D., F. W. Davis, and D. M. Stoms. 1994. Hierarchical representations of species distributions using maps, images, and sighting data. Pages 71-88 in R. I. Miller, editor. Mapping the diversity of nature. Chapman and Hall, London, England.

91. Holthausen, R. S., M. G. Raphael, K. S. McKelvey, E. D. Forsman, E. E. Starkey, and D. E. Seaman. 1995. The contribution of federal and nonfederal habitat to persistence of the northern spotted owl on the Olympic Peninsula, Washington: report of the reanalysis team. United States Forest Service General Technical Report PNW-352.

92. Huston, M., D. DeAngelis, and W. Post. 1988. New computer models unify ecological theory. Bioscience 38:682-691.

93. Hutchinson, G. E. 1957. Concluding remarks. Cold Spring Harbor Symposium on Quantitative Biology 22:415-427.

94. Inger, R. F., and R. K. Colwell. 1977. Organization of contiguous communities of amphibians and reptiles in Thailand. Ecological Monographs 47:229-253.

95. Iversen, G. R. 1984. Bayesian statistical inference. Sage Publications, Beverly Hills, CA.

96. James, F. C. 1971. Ordinations of habitat relationships among breeding birds. Wilson Bulletin 83:215-236.

97. ———, and N. O. Wamer. 1982. Relationships between temperate forest bird communities and vegetation structure. Ecology 63:159-171.

98. Judson, O. P. 1994. The rise of the individual-based model in ecology. Trends in Evolution and Ecology 9:9-14.

99. Kareiva, P. 1990. Population dynamics in spatially complex environments: theory and data. Pages 53-68 in M. P. Hassell and R. M. May, editors. Population regulation and dynamics. Royal Society, London, England.

100. Kareiva, P., and U. Wennergren. 1995. Connecting landscape patterns to ecosystem and population processes. Nature 373:299-302.

101. Kling, C. L. 1980. Pattern recognition for habitat evaluation. Thesis, Colorado State University, Fort Collins, CO.

102. Knopf, F. L., J. A. Sedgwick, and R. W. Cannon. 1988. Guild structure of a riparian avifauna relative to seasonal cattle grazing. Journal of Wildlife Management 52:280-190.

103. Kohlmann, S. G., and M. J. Hedrick. In Press. Density dependence in two Great Basin desert ungulates: implications for adaptive management. Journal of Wildlife Management.

104. Krebs, C. J. 1989. Ecological methodology. HarperCollins Publishers, New York, NY.

105. Lahaye, W. S., R. J. Gutiérrez, and H. R. Akçakaya. 1994. Spotted owl metapopulation dynamics in southern California. Journal of Animal Ecology 63:775-785.

106. Lamberson, R. H., B. R. Noon, C. Voss, and K. S. McKelvey. 1994. Reserve design for territorial species: the effects of patch size and spacing on the viability of the northern spotted owl. Conservation Biology 8:185-195.

107. Lancia, R. A., S. D. Miller, D. A. Adams, and D. W. Hazel. 1982. Validating habitat quality assessment: an example. Transactions of the North American Wildlife and Natural Resources Conference 47:96-110.

108. ———, and D. A. Adams. 1985. A test of habitat suitability index models for five bird species. Proceedings of the Annual Conference of the Southeastern Association of Fish and Wildlife Agencies 39:412-419.

109. Lande, R. 1987. Extinction thresholds in demographic models of territorial populations. American Naturalist 130:624-635.

110. ———. 1988. Demographic models of the northern spotted owl (Strix occidentalis caurina). Oecologia 75:601-607.

111. Laymon, S. A., and R. H. Barrett. 1986. Developing and testing habitat-capability models: pitfalls and recommendations. Pages 87-91 in J. Verner, M. L. Morrison, and C. J. Ralph, editors. Wildlife 2000: modeling habitat relationships of terrestrial vertebrates. University of Wisconsin, Madison, WI.

112. Lefkovitch, L. P. 1965. The study of population growth in organisms grouped by stages. Biometrics 21:1-18.

113. Leslie, P. H. 1945. On the use of matrices in certain population mathematics. Biometrika 33:183-212.

114. Levin, S. A. 1976. Population dynamic models in heterogeneous environments. Annual Review of Ecology and Systematics. 7:287-310.

115. Levins, R. 1969. Some demographic and genetic consequences of environmental heterogeneity for biological control. Bulletin of the Entomological Society of America 15:237-240.

116. ———. 1970. Extinction. Pages 77-107 in M. Gerstenhaber, editor. Some mathematical problems in biology. American Mathematical Society, Providence, RI.

117. Lindenmayer, D. B., and H. P. Possingham. 1996. Ranking conservation and timber management options for Leadbeater's possum in southeastern Australia using population viability analysis. Conservation Biology 10:235-251.

118. Liu, J., J. B. Dunning, and H. R. Pulliam. 1995. Potential effects of a forest management plan on Bachman's sparrow (Aimophila aestivalis): linking a spatially explicit model with GIS. Conservation Biology 9:62-75.

119. Lubow, B. C., G. C. White, and D. R. Anderson. 1996. Evaluation of a linked sex harvest strategy for cervid populations. Journal of Wildlife Management 60:787-796.

120. Lusted, L. B. 1968. Introduction to medical decision making. Charles C. Thomas, Springfield, IL.

121. MacArthur, R. H., and E. O. Wilson. 1967. The theory of island biogeography. Princeton University, Princeton, NJ.

122. Mannan, R. W., M. L. Morrison, and E. C. Meslow. 1984. Comment: the use of guilds in forest bird management. Wildlife Society Bulletin 12:426-430.

123. Marcot, B. G. 1979. California wildlife/habitat relationships program: north coast/Cascades zone. Volumes 1-4. United States Forest Service, Eureka, CA.

124. ———, M. G. Raphael, and K. H. Berry. 1983. Monitoring wildlife habitat and validation of wildlife-habitat relationships models. Transactions of the North American Wildlife Conference 48:315-329.

125. Maurer, B. A. 1986. Predicting habitat quality for grassland birds using density-habitat correlations. Journal of Wildlife Management 50:556-556.

126. ———. 1990. Extensions of optimal foraging theory for insectivorous birds: implications for community structure. Studies in Avian Biology 13:455-461.

127. ———. 1994. Geographical population analysis: tools for the analysis of biodiversity. Blackwell Scientific Publications, Oxford, England.

128. Maxwell, J. R., C. J. Edwards, M. E. Jensen, S. J. Paustian, H. Parrott, and D. M. Hill. 1995. A hierarchical framework of aquatic ecological units in North America (Nearctic zone). United States Forest Service General Technical Report NC-176.

129. May, R. M. 1973. On relationships among various types of population models. American Midland Naturalist 107:46-57.

130. Maynard-Smith, J. 1974. Models in ecology. Cambridge University Press, Cambridge, England.

131. McCullough, D. R. 1981. Population dynamics of the Yellowstone grizzly bear. Pages 173-196 in C. W. Fowler and T. D. Smith, editors. Dynamics of large mammal populations. John Wiley & Sons, New York, NY.

132. McNeill, F. M., and E. Thro. 1994. Fuzzy logic: a practical approach. Academic Press, Cambridge, MA.

133. McKelvey, K. 1992. A spatially explicit life-history simulator for the northern spotted owl. Appendix 4-P in Eugene District Resource Management Plan & EIS. United States Bureau of Land Management, Eugene, OR. 2 vols.

134. Messier, F. 1994. Ungulate population models with predation: a case study with the North American moose. Ecology 75:478-488.

135. Mills, T. R., M. A. Rumble, and L. D. Flake. 1996. Evaluation of a habitat capability model for nongame birds in the Black Hills, South Dakota. United States Forest Service Research Paper RM-RO-323.

136. Morrison, M. L., B. G. Marcot, and R. W. Mannan. 1992. Wildlife-habitat relationships: concepts and applications. University of Wisconsin, Madison, WI.

137. ———, B. G. Marcot, and R. W. Mannan. 1998. Wildlife habitat relationships: concepts and applications. Second edition. University of Wisconsin Press, Madison, WI.

138. Moss, R., A. Watson, and J. Ollason. 1982. Animal population dynamics. J. W. Arrowsmith Limited, Bristol, England.

139. Nadeau, S., R. Décarie, D. Lambert, and M. St-Georges. 1995. Nonlinear modeling of muskrat use of habitat. Journal of Wildlife Management 59:110-117.

140. Noon, B. R., and K. McKelvey. 1992. Stability properties of the spotted owl metapopulation in southern California. United States Forest Service General Technical Report PSW-133.

141. O'Connor, R. J., M. T. Jones, D. White, C. T. Hunsaker, T. Loveland, B. Jones, and E. Preston. 1996. Spatial partitioning of environmental correlates of avian biodiversity in the conterminous United States. Biodiversity Letters 3:97-110.

142. ———, and M. T. Jones. 1997. Using hierarchical models to index the ecological health of the nation. Transactions of the North American and Natural Resources Conference 62:501-508.

143. Pacala, S. W., C. D. Canham, and J. A. Silander, Jr. 1993. Forest models defined by field measurements. I. The design of a northeastern forest simulator. Canadian Journal of Forest Research 23:1980-1988.

144. Pascual, M. A., P. Kareiva, and R. Hilborn. 1997. The influence of model structure on conclusions about the viability and harvesting of Serengeti wildebeest. Conservation Biology 11:966-976.

145. Pastor, J., and W. M. Post. 1985. Development of a linked forest productivity-soil process model. Publication of Oak Ridge National Laboratory, Environmental Sciences Division, Number 2455, Oak Ridge, TN.

146. Poiani, K. A., and W. C. Johnson. 1993. A spatial simulation model of hydrology and vegetation dynamics in semi-permanent prairie wetlands. Ecological Applications 3:279-293.

147. Prosser, D. J., and R. P. Brooks. 1998. A verified habitat suitability index for the Louisiana waterthrush. Journal of Field Ornithology 69:288-298.

148. Pulliam, R. H. 1988. Sources, sinks, and population regulation. American Naturalist 132:652-661.

149. ———, and B. J. Danielson. 1991. Sources, sinks, and habitat selection: a landscape perspective on population dynamics. American Naturalist 137:550-566.

150. Pulliam, R. H., J. B. Dunning, and J. Liu. 1992. Population dynamics in complex landscapes: a case study. Ecological Applications 2:165-177.

151. Puttock, G. D., P. Shakotko, and J. G. Rasaputra. 1996. An empirical habitat model for moose, *Alces alces*, in Algonquin Park, Ontario. Forest Ecology and Management 81:169-178.

152. Quinn, T. 1997. Coyote (*Canis latrans*) habitat selection in urban areas of western Washington via analysis of routine movements. Northwest Science 71:289-297.

153. Raphael, M. G., and B. G. Marcot. 1986. Validation of a wildlife-habitat-relationships model: vertebrates in a Douglas-fir sere. Pages 129-138 in J. Verner, M. L. Morrison, and C. J. Ralph, editors. Wildlife 2000: modeling habitat relationships of terrestrial vertebrates. University of Wisconsin, Madison, WI.

154. Renken, R. B. 1988. Habitat characteristics related to pileated woodpecker densities and territory size in Missouri. Dissertation, University of Missouri, Columbia, MO.

155. Ricklefs, R. E., and D. B. Miles. 1994. Ecological and evolutionary inferences from morphology: an ecological perspective. Pages 13-41 in P. C. Wainwright and S. M Reilly, editors. Ecological morphology. University of Chicago, Chicago, IL.

156. Robel, R. J., L. B. Fox, and K. E. Kemp. 1993. Relationship between habitat suitability index values and ground counts of beaver colonies in Kansas. Wildlife Society Bulletin 21:415-421.

157. Roloff, G. J., and J. B. Haufler. 1997. Establishing population viability planning objectives based on habitat potentials. Wildlife Society Bulletin 25:895-904.

158. ———, B. Carroll, and S. Scharosch. 1999. A decision support system for incorporating wildlife habitat quality into forest planning. Western Journal of Applied Forestry 14:91-99.

159. ———, and B. J. Kernohan. 1999. Evaluating the reliability of habitat suitability index models. Wildlife Society Bulletin 27:973-985.

160. Root, R. B. 1967. The niche exploitation pattern of the blue-gray gnatcatcher. Ecological Monographs 37:317-350.

161. Root, K. V. 1998. Evaluating the effects of habitat quality, connectivity, and catastrophes on a threatened species. Ecological Applications 8:854-865.

162. Rosenzweig, M. L. 1995. Species diversity in space and time. Cambridge University Press, New York, NY.

163. Rotella, J. J., J. T. Ratti, K. P. Reese, M. L. Taper, and B. Dennis. 1996. Long-term population analysis of gray partridge in eastern Washington. Journal of Wildlife Management 60:817-825.

164. Rothery, P., I. Newton, L. Dale, and T. Wesolowski. 1997. Testing for density dependence allowing for weather effects. Oecologia 112:518-523.

165. Ruckelshaus, M., C. Hartway, and P. Kareiva. 1997. Assessing the data requirements of spatially explicit dispersal models. Conservation Biology 11:1298-1306.

166. Rykiel, E. J., Jr. 1996. Testing ecological models: the meaning of validation. Ecological Modelling 90:229-244.

167. Saitoh, T., N. C. Stenseth, and O. N. Bjornstadt. 1997. Density dependence in fluctuating gray-sided vole populations. Journal of Animal Ecology 66:14-24.

168. Sauer, J. R., and S. Droege. 1990. Survey designs and statistical methods for the simulation of avian population trends. United States Fish and Wildlife Service Biological Report 90(1).

169. Saunders, D.A., R. J. Hobbs, and C. R. Margules. 1991. Biological consequences of ecosystem fragmentation: a review. Conservation Biology 5:18-32.

170. Schmutz, J.A., R. F. Rockwell, and M. R. Petersen. 1997. Relative effects of survival and reproduction on the population dynamics of emperor geese. Journal of Wildlife Management 61:191-201.

171. Schroeder, R. L. 1986. Habitat suitability index models: wildlife species richness in shelterbelts. United States Fish and Wildlife Service Biological Report 82(10.128).

172. ———, and S. L. Haire. 1993. Guidelines for the development of community-level habitat evaluation models. United States Fish and Wildlife Service Biological Report 8, Washington, D.C.

173. Scott, J. M., F. Davis, B. Csuti, R. Noss, B. Butterfield, C. Groves, H Anderson, S. Caicco, F. D'erchia, T. C. Edwards, Jr., J. Ulliman, and R. G. Wright. 1993. Gap analysis: a geographic approach to protection of biological diversity. Wildlife Monographs 123:1-41.

174. Severinghaus, W. D. 1981. Guild theory development as a mechanism for assessing environmental impacts. Environmental Management 5:187-190.

175. Short H. L., and K. P. Burnham. 1982. Technique for structuring wildlife guilds to evaluate impacts on wildlife communities. United States Fish and Wildlife Service Special Science Report on Wildlife, Number 244, Washington, D.C.

176. Sibly, R. M., and R. H. Smith. 1998. Identifying key factors using lambda contribution analysis. Journal of Animal Ecology 67:17-24.

177. Simberloff, D. 1988. The contribution of population and community biology to conservation science. Annual Review of Ecology and Systematics 19:473-511.

178. Slobodkin, L. B. 1980. Growth and regulation of animal populations. Dover Publishing, New York, NY.

179. Smith, A. D. M., and C. J. Walters. 1981. Adaptive management of stock-recruitment systems. Canadian Journal of Fisheries and Aquatic Sciences 38:690-703.

180. Starfield, A. M., and A. L. Bleloch. 1986. Building models for conservation and wildlife management. Macmillan Publishing Company, New York, NY.

181. ———, and A. L. Bleloch. 1991. Building models for conservation and wildlife management. Burgess International Group, Edina, MN.

182. ———, K.A. Smith, and A. L. Bleloch. 1994. How to model it: Problem-solving for the computer age. Burgess International Group, Edina, MN.

183. ———. 1997. A pragmatic approach to modeling for wildlife management. Journal of Wildlife Management 61:261-270.

184. Steeger, C., and C. L. Hitchcock. 1998. Influence of forest structure and diseases on nest-site selection by red-breasted nuthatches. Journal of Wildlife Management 62:1349-1358.

185. Suring, L. H., K. R. Barber, C. C. Schwartz, T. N. Bailey, W. C. Shuster, and M. D. Tetreau. 1998. Analysis of cumulative effects on brown bears on the Kenai Peninsula, southcentral Alaska. Ursus 10:107-117.

186. Sutherland, W. J. 1996. From individual behaviour to population ecology. Oxford University Press, New York, NY.

187. Temple, S.A., and J. R. Cary. 1988. Modeling dynamics of habitat-interior bird populations in fragmented landscapes. Conservation Biology 2:340-347.

188. Terborgh, J. 1974. Preservation of natural diversity: the problem of extinction prone species. Bioscience 24:715-722.

189. Terrell, J.W., and J. Carpenter. 1997. Selected habitat suitability index model evaluations. United States Geological Survey Information and Technical Report USGS/BRD/ITR-1997-0005.

190. Tester, J. R., A. M. Starfield, and L. E. Frelich. 1997. Modeling for ecosystem management in Minnesota pine forests. Biological Conservation 80:313-324.

191. Thomas, J.W. 1979. Wildlife habitats in managed forests: the Blue Mountains of Oregon and Washington. United States Forest Service Agriculture Handbook Number 553.

192. ———, E. D. Forsman, J. B. Lint, E. C. Meslow, B. R. Noon, and J. Verner. 1990. A conservation strategy for the northern spotted owl. Report of the Interagency Scientific Committee to address the conservation of the northern spotted owl. Appendix M. United States Forest Service, Portland, OR.

193. Turner, M. G. 1989. Landscape ecology: the effect of pattern on process. Annual Review of Ecology and Systematics 20:171-197.

194. ———, G. J. Arthaud, R. T. Engstrom, S. J. Hejl, J. Liu, S. Loeb, and K. McKelvey. 1995. Usefulness of spatially explicit population models in land management. Ecological Applications 5:12-16.

195. Urban, D. L., R. V. O'Neill, and H. H. Shugart. 1987. Landscape ecology. Bioscience 37:119-27.

196. United States Fish and Wildlife Service. 1980. Habitat evalaution procedures (HEP). United States Fish and Wildlife Service, Division of Ecological Services, ESM 102, Washington, D.C.

197. ——— 1981. Standards for the development of habitat suitability index models. United States Fish and Wildlife Service Division of Ecological Services, ESM 103, Washington, D.C.

198. Vales, D. J., and F. L. Bunnell. 1988. Comparison of methods for estimating forest overstory cover. I. Observer effects. Canadian Journal of Forest Research 18:606-609.

199. Varley, G. C., G. R. Gradwell, and M. P. Hassell. 1973. Insect population ecology—an analytical approach. Blackwell Press, Oxford, England.

200. Van Horne, B., and J. A. Wiens. 1991. Forest bird habitat suitability models and the development of general habitat models. United States Fish and Wildlife Service Research Number 8, Washington, D.C.

201. Van Tienderen, P. H. 1995. Life cycle trade-offs in matrix population models. Ecology 76:2482-2489.

202. Verner, J., and A. S. Boss. 1980. California wildlife and their habitats: western Sierra Nevada. United States Forest Service General Technical Report PSW-37.

203. ———. 1984. The guild concept applied to management of bird populations. Environmental Management 8:1-14.

204. ———, M. L. Morrison, and C. J. Ralph. 1986. Wildlife 2000: modeling habitat relationships of terrestrial vertebrates. University of Wisconsin, Madison, WI.

205. Wahlberg, N., A. Moilanen, and I. Hanski. 1996. Predicting the occurrence of endangered species in fragmented landscapes. Science 273:1536-1538.

206. Walters, C. J., and R. Hilborn. 1976. Adaptive control of fishing systems. Journal of the Fisheries Research Board, Canada 33:145-159.

207. ———, and R. Hilborn. 1978. Ecological optimization and adaptive management. Annual Review of Ecology and Systematics 9:157-188.

208. Washington Department of Natural Resources. 1996. Draft environmental impact statement: habitat conservation plan. Washington State Department of Natural Resources, Olympia, WA.

209. Watt, K. E. F. 1966. Systems analysis in ecology. Academic Press, New York, NY.

210. Wells, J.V., and M. E. Richmond. 1995. Populations, metapopulations, and species populations: what are they and who should care? Wildlife Society Bulletin 23:458-462.

211. Wennergren, U., M. Ruckelshaus, and P. Kareiva. 1995. The promise and limitations of spatial models in conservation biology. Oikos 74:349-356.

212. Williams, G. L., K. R. Russell, and K. R. Seitz. 1977. Pattern recognition as a tool in the ecological analysis of habitat. Pages 521-531 in Classification, inventory, and analysis of fish and wildlife habitat. United States Fish and Wildlife Service FWS/OBS-78/76.

213. Wisdom, M. J., L. R. Bright, C. G. Carey, W.W. Hines, R. J. Pedersen, D. A. Smithey, J.W.Thomas, and G.W.Witmer. 1986. A model to evaluate elk habitat in western Oregon. United States Forest Service, Pacific Northwest Region, Publication Number R6-F&WL-216-1986.

214. With, K.A., and T. O. Crist. 1996. Translating across scales: simulating species distributions as the aggregate response of individuals to heterogeneity. Ecological Modeling 93:125-137.

22

Five Case Studies in Wildlife Modeling

Timothy Quinn & David H. Johnson, editors

Case Study 1
A spatially realistic population model for informing forest management decisions
G.F. Wilhere, N.H. Schumaker, & S.P. Horton

Case Study 2
A model to assess potential vertebrate habitat at landscape scales: HABSCAPES
M. Huff, T.K. Mellen, & R. Hagestedt

Case Study 3
Cross-scale classification trees for assessing risks of forest practices to headwater stream amphibians
G.D. Sutherland & F.L. Bunnell

Case Study 4
Applying Gap analysis to county and regional land use planning
M.R. Stevenson

Case Study 5
A model to determine potential Northern spotted owl nesting areas
N.W. Darby & T. Young

Introduction to the Case Studies

The five case studies that make up this chapter were presented at the *Landscape Management of Pacific Northwest Forest Workshop,** held in Olympia, Washington in February 1998. Designed for land mangers and other natural resource professionals, the purpose of the workshop was to help bridge the gap between the theory and practice of conservation biology. In particular, as workshop organizers, we (T. Quinn and D.H. Johnson) were interested in studies describing practical techniques and tools that provide insights or solutions to complex, large-scale or multi-species conservation issues.

The five studies presented below represent an accurate range, and sampling, of the many projects presented at the workshop. They are unique in the sense that each case study presents a methodology used in a real conservation setting, and one that also addresses common problems with data quality and quantity. In other words, they used data they had "at hand." We present these case studies as examples of creative approaches rather than solutions to specific conservation issues. There are no perfect answers here, but there is abundant creative energy about important, and ever-improving, work by practitioners.

Case Study #1 by Wilhere et al. is an example of a potentially powerful management tool to address a fundamental management issue: predicting the effect of landscape-level habitat changes to wildlife species or populations. The single-species, spatially realistic population model described here requires movement and demographic data (see previous chapter), which typically are available only for the best-studied species. To address data limitations, Wilhere et al. introduce a process called *parameter tuning*. Parameter tuning is a way of estimating specific parameter values for which no specific data exist by iteratively solving for those values during model simulations. By treating known parameters as constants, Wilhere et al. could search for appropriate values of the unknown parameter by comparing the characteristics of the simulated and real populations. Model output tended to become more concordant with empirical data, as increasingly realistic values are used to estimate the unknown parameter. While this process does not lessen the need for good data, it may provide managers with better ways of using existing data.

---

*The sponsors of the workshop were: Washington State Timber, Fish, and Wildlife Program; Wildlife-Habitat Relationships Project; Washington Dept. Of Fish and Wildlife; U.S. Fish and Wildlife Service; Washington and Oregon Chapters of The Wildlife Society; Organization of Fish and Wildlife Information Managers; and Environmental Systems Research Institute (ESRI).

Case Study #2 by Huff et al. describes a process called HABSCAPES for determining distributions of suitable habitat for a species or collections of species (guilds), which are grouped by resource use. The process uses vegetation cover type data (including vegetation structural characteristics) in conjunction with data on species-habitat associations, home range or territory size, and species life history, to identify all patches of potential habitat. Users can specify criteria, such as the size and composition of patches that comprise habitat to quickly delineate potential habitat at relatively large spatial scales. One unique feature of HABSCAPES, and an obvious place to issue caution, is that it can streamline multi-species assessments by guilding species. It is best thought of as a coarse filter approach to habitat assessments.

Case Study #3 by Sutherland and Bunnel explored the use of classification and regression tree analysis (CART) for determining habitat relationships of tailed frogs at different spatial scales. They argue that many of the spotted frog studies are contradictory because they were done at different places, or they explained frog response using independent variables representing different spatial scales. Sutherland and Bunnel suggest that CART can be more appropriate than multiple or logistic regression at teasing out relationships occurring at different spatial scale when only inventory data (relative abundance or occurrence) are available. While noting that it is heuristic in nature, they concluded that CART has four important strengths as demonstrated in their study: 1) it reveals pathways by which environmental variation may determine species distribution, 2) it can use commonly available information comprised of numerical and nominal data, 3) it can facilitate the use of separate data without making assumptions about how environmental variables interact at all locations, and 4) it produces outputs that are easy to interpret and to convert into hypotheses.

Case Study #4: Stevenson describes a process of applying data from Washington Gap and other sources to help Spokane County plan for human population growth and the protection of biodiversity. He outlines five steps to the planning process: 1) identify the minimum amount of land for each species, 2) establish species richness thresholds to located candidate reserves, 3) delineate reserves as an iterative process to account for local knowledge, economics and politics, 4) delineate corridors connecting reserves, and 5) evaluate success. While the steps reflect common approaches to conservation, this case study demonstrates how local knowledge and multiple data sets can be combined in useful ways during planning processes. Perhaps the most important thing about this case study is that someone despite the potential problems with data and bureaucracy took the time to work with local jurisdictions that were struggling with managing human population growth.

Case Study #5: Darby and Young discuss how to build a geographical information system (GIS) model to help managers make more informed decisions about conservation. Similar to HABSCAPES, the authors outline a process for assessing the distribution of potential northern spotted owl (NSO) nesting sites using vegetation cover type data (including vegetation structural data) and autecological information about the NSO. The model is built using routines commonly found in GIS software. This case study is not included here because we believe it has just the right recipe for spotted owls, but rather because it reminds us of the power of GIS as an integrator of information from the fields such as forestry, biology, and economics.

Case Study 1

A Spatially Realistic Population Model for Informing Forest Management Decisions

George F. Wilhere, Nathan H. Schumaker, & Scott P. Horton

Introduction

Spatially realistic population models (SRPMs) address a fundamental problem commonly confronted by wildlife managers—predicting the effects of landscape-scale habitat management on an animal population. SRPMs typically consist of three submodels: (1) a habitat submodel, (2) a movement submodel, and (3) a demographic submodel. We describe the submodels and data requirements for the typical SRPM. The most frustrating problem with SRPMs is the lack of data needed to relate movement and demographic parameters to habitat quality. We developed a SRPM to evaluate the relative effects of different habitat management strategies on the spotted owl subpopulation of the Olympic Peninsula. This case study documents some plausible assumptions we made to circumvent the data problem, and explains an approach called "parameter tuning" that we used to generate parameter values.

Focus of Case Study

The Washington State Department of Natural Resources (DNR) has developed a habitat conservation plan[7] for 107,000 ha of state lands on the western Olympic Peninsula that emphasizes the habitat needs of the Northern spotted owl. The plan will alter landscape conditions across more than 4000 km^2 of inter-mixed state, federal, and private land, and thus influence the fate of the Peninsula's spotted owl subpopulation. The critical question was how much of an influence? Specifically, would the conservation plan appreciably reduce the likelihood of survival and recovery of spotted owls on the Peninsula? In addition, since Washington's State Environmental Policy Act dictates that state agencies shall consider more than one alternative when planning large projects, DNR needed to answer questions about the relative effects of different habitat management strategies. During development of the plan, spatially realistic population models (SRPMs) were emerging as a practical approach for answering these types of questions,[24, 29] so the DNR funded the development of a new SRPM that offered some advantages over existing models.[6, 7, 31] We will refer to the model of Schumaker[31] as SOS for "Spotted Owl Simulator". SOS is a pre-cursor to the model known as PATCH.[32]

An Introduction to Spatially Realistic Population Models

Spatially realistic population models address a fundamental problem commonly confronted by wildlife managers—predicting the effects of landscape-scale habitat management on an animal population. SRPMs are a type of spatially explicit population model.[16] Spatially explicit models "keep track of the exact positions of plants and animals,"[39] but most types of spatially explicit models are based on simplistic, stylized landscapes (Roloff et al., this volume). In contrast, a SRPM uses empirically derived habitat data that are referenced to real geographic coordinates. These "realistic" landscapes can be manipulated to represent conditions likely to result from future habitat management. SRPMs are usually individual-based models. That is, location, birth, death, reproduction, and movement are "observed" and recorded for every individual in the simulated population. The behavior and fate of individuals are stochastic. This means that most actions or events occur randomly according to probability distributions that are either estimated from empirical data or based on plausible assumptions. SRPMs are typically comprised of three submodels: (1) a habitat submodel, (2) a movement submodel, and (3) a demographic submodel.

The habitat submodel creates an abstract representation of a real landscape. SRPMs are usually limited to two spatial dimensions (X and Y coordinates), a third dimension related to habitat quality, and a fourth dimension is time, which implies dynamic habitat quality. The habitat submodel must interface with the demographic and movement submodels. That is, the submodel should delineate landscape units that have demographic significance, such as territories, or that may be relevant to movement behavior, such as habitat corridors. Also, a habitat submodel must provide metrics of habitat quality that can be used by habitat dependent functions in the other submodels.

A common approach to representing landscapes uses a hexagonal grid to define landscape units.[25, 28, 32] A hexagonal structure is often adopted for its geometric simplicity, and because, unlike a rectangular or triangular grid, the distances between a polygon center and centers of all adjacent polygons are equal. This simplifying assumption—that the spatial distribution and connectivity of habitat can be adequately represented by a grid—obviously imposes some limitations on realism of the habitat submodel. The most significant limitations are that (1) the number of hexagons sets the carrying capacity of the entire landscape, (2) the hexagon size determines the maximum local territory density, and (3) and habitat quality is usually characterized by one only variable: an index calculated for each hexagon.

The data requirements of a habitat submodel correspond with its multiple dimensions. An intermediate data product must be a habitat map showing the location of various vegetation classes or habitat categories. The simplest map could consist of two categories: habitat and nonhabitat.[20, 29] Such maps may be generated through remote sensing methods (e.g,. aerial photography or satellite thematic mapper imaging), through on-the-ground habitat surveys, or most likely some combination of both. The habitat map is digitized using a geographic information system and the habitat submodel translates the digitized map into a simulated landscape. Forest growth models or information about future land use patterns can be used to create maps of likely changes in habitat conditions over time.

Ideally, a movement submodel would simulate realistic movement behavior. That is, metrics of simulated movement, such as total distance moved, net distance moved, net direction moved, linear auto-correlation (i.e., the tendency to move in a straight line), and habitat preference would be statistically representative of real movements. Certainly, habitat quality must influence movement behavior, as must interactions with conspecifics, but unfortunately, our understanding of such behavior is crude and the cost of improving our understanding has been prohibitive. Consequently, a common approach has been to simply intuit the cognitive mechanisms, such as "memory," habitat "selectivity," and intraspecific conflict "aversion" that reputedly manifest movement behaviors.[25, 28]

If a habitat submodel uses a hexagonal grid to define landscape units, then movement paths consist of a sequence of steps from hexagon to hexagon. At each hexagon, an individual can move in one of six directions or not move at all. Submodel parameters define how "cognitive mechanisms" influence individual movement decisions. Movement submodels in SRPMs usually allow a range of behaviors from totally random to completely directed by habitat quality and the presence of conspecifics. Because long-distance dispersal is considered a risky behavior, some SRPMs incorporate dispersal related mortality into the movement submodel.

A frustrating problem with SRPMs has been a lack of movement data.[30] Movement data should be collected such that parameter estimates can be related to habitat quality. Mark-recapture studies can yield some information on animal movements, but the most efficient means of studying movement is radio-telemetry. Understanding how conspecific interactions influence movement behavior would require detailed knowledge of a local population.

The demographic submodel defines the population structure and deals with changes in population size over time. SRPMs can be two-sex or one-sex. The population is often broken into stage classes, such as juvenile, subadult, and adult.[25] The birth and death of individuals are usually modelled as stochastic processes with the probabilities of nesting success or producing a certain number of offspring or death being functions of local habitat quality.[20, 28] The crux of the demographic submodel is developing these quantitative relationships between demographic parameters and habitat quality.

The most important data for the demographic submodel are those needed for estimates of stage-specific mortality and maternity parameters. These data are obtained through mark-recapture studies and observations on the number of offspring produced per breeding female (or breeding pair), respectively. Ideally, demographic data would also provide information on the nonbreeding portion of the population, e.g., floaters. Demographic data should be collected such that parameter estimates can be related to habitat quality.

The basic output of SRPMs is the same as other population models: the number of individuals over time. SRPMs generate an output that other populations' models do not: the spatial distribution of individuals over time. One can compare the likely outcomes of different habitat management strategies, which could be useful for optimizing land management.[18] An optimal strategy might be one that maintains a desired population density for the minimum cost.

In general, SRPMs are not mathematically sophisticated, but they are computationally complex. Well-honed computer programming skills are indispensable for tackling the task of model development. There are no standards for model structure, but modular, top-down, object-oriented programming is highly recommended. For most applications, a spatially realistic, individual-based model will require large computer memory and huge on-line data storage because: (1) information on every living individual in the population is maintained during a simulation, and (2) GIS data, particularly rasterized data, can occupy gigabytes of nonvolatile memory. High CPU speeds are also very desirable.

An Example: The Spotted Owl Simulator

The Spotted Owl Simulator[6, 31] was developed to evaluate the effects of different habitat management strategies on the spotted owl subpopulation of the Olympic Peninsula. With respect to the abundance and richness of data, the Olympic Peninsula was an excellent place to attempt analyses using a SRPM. The Peninsula's owl subpopulation has been the subject of intensive study and the results of numerous investigations were available.[4, 10, 11, 12, 23, 26, 33, 34] Demographic parameter estimates had been published.[3, 14] Habitat use versus habitat availability studies (E. D. Forsman, U.S. Forest Service, unpubl. data), and juvenile dispersal studies (E.D. Forsman, U. S. Forest Service, pers. comm.) had been conducted. A gross estimate of population size had been issued in a draft report.[19] Data for habitat maps were also available. Landsat Thematic Mapper (TM) data for the western Peninsula, the actual planning area, had been classified into land cover categories,[38] and classified Landsat TM data also existed for the remainder of the Peninsula.[27]

Even with all this information, other essential information was missing. There were no data that could be used to relate mortality and maternity parameters to habitat quality and no data to relate the movement of owls to habitat quality. The movement data available to us only described the minimum, maximum, and mean distances (with standard error of the mean) for dispersing juveniles.

The Habitat Submodel

SOS uses a hexagonal grid to represent the spatial distribution of owl habitat. Six parameters govern the translation of rasterized habitat maps to a hexagonal landscape: hexagon size, "territory expansion," a "territory threshold," and the relative habitat values of three land cover categories. The hexagon size corresponds to the minimum home range area of an owl pair. We assumed that the smallest home ranges on the Peninsula would exist in the high-quality habitat of Olympic National Park. The density of owl pairs in the low-elevation old-growth forests of the Park was estimated to be 0.08/100 ha.[34] This density is equivalent to an exclusive home range of 1,250 ha/pair, and we equated the size of the exclusive home range to the minimum home range. The minimum home range is probably not equivalent to the exclusive home range, but theoretically, in a sparsely populated region composed of high-quality habitat it should be nearly so.

The classified Landsat TM data[38] had eight discrete forest categories, but only three categories (old-growth, large saw, and small saw) were assumed to have value as spotted owl habitat—an assumption based on Forsman et al.[13] and Carey et al.[4] A habitat utilization index (HUI) was calculated for each forest category. It is defined by a ratio: (percent of owl relocations within a category)/ (percent of home range area in that category).

HUI describes habitat preferences. If HUI is greater than 1.0, then owls expressed a preference for that forest category. The ratio of two HUI quantifies the relative habitat value of two forest categories. HUIs were calculated using radio-telemetry data collected from 20 owl home ranges on the western Olympic Peninsula (E. D. Forsman, U. S. Forest Service, unpublished data). The shortcomings of indices based on habitat use versus habitat availability were known to us,[21, 37] but the data available to us precluded any other approach.

The hexagon habitat scores were calculated with the following equation:

$$HS = HV_{OG}A_{OG} + HV_{LS}A_{LS} + HV_{SS}A_{SS}$$

where HV_x is the relative habitat value of either old-growth (OG), large-saw (LS), or small-saw (SS) forest, and A_x is the proportion of a hexagon covered by that forest category. Hexagons having scores above a threshold value are classified as potential owl territories. The threshold value was determined by applying the same equation to the 20 owl home ranges studied by Forsman, where A_x refers to the proportion of a home range. We found that setting the threshold value to the mean habitat score of these 20 home ranges (which was less than the median score) yielded an acceptable spatial distribution of potential territories when compared to the known locations of territorial owls on the Olympic Peninsula (E. D. Forsman, U.S. Forest Service, unpubl. data; D.E.

Seaman, National Biological Service, unpubl. data; S. P. Horton, DNR, unpubl. data). The minimum and first quartile scores each yielded a worse correspondence between predicted potential territories and known locations.

The maximum home range of an owl pair is determined through an expansion parameter. This parameter represents two aspects of owl behavior. First, in areas of fragmented habitat, owls expand the size of their home range, and second, the home ranges of neighboring owls may overlap.[4] The expansion parameter represents the maximum amount of neighboring hexagons that may be included in a home range. For 20 owl home ranges studied on the western Olympic Peninsula (E. D. Forsman, U. S. Forest Service, unpubl. data), the maximum home range size was 11,248 ha, but the third quartile size was 6301 ha. With a hexagon size of 1250, the maximum home range that can be modeled is 8750 ha (a hexagon plus its six neighboring hexagons). The expansion parameter value was assigned its maximum value. The difference between the observed maximum home range and the maximum size that can be modeled will cause the number of potential territories to be underestimated.

Hexagon "expansion" is a novel aspect of SOS. Expansion does not actually increase the score or size of hexagons. It is simply a mechanism that allows hexagons with low scores to function as territories by laying claim to unallocated habitat present in adjacent hexagons. The expansion of a sub-threshold hexagon continues until it reaches the potential territory threshold or until it reaches the expansion limit. Only hexagons that are below the threshold expand, and hexagons with low scores require more expansion to become potential territories. This process does not affect the score of any hexagons. The habitat submodel produces three types of hexagons: supra-threshold potential territory, sub-threshold potential territory, and not a potential territory.

The Movement Submodel

The simulated landscape in SOS is a hexagonal grid. From each location, owls can move in only six directions and every move must be the same length, from hexagon center to hexagon center. For hexagons of 1,250 ha, the "step" size is about 3.80 km. Owls stop moving when a territory or mate is found, or a maximum number of steps have been taken.

The submodel requires specification of the minimum number of steps that must be moved by dispersing juveniles. Moving adults ignore the minimum movement distance. This assumption was made because no data on minimum movement distances exist for adults. The ratio of total distance to net distance for owl dispersal equals about 2,[35] and the minimum net dispersal distance observed for juveniles on the Peninsula is 8.67 km (E.D. Forsman, U.S. Forest Service, unpubl. data). So, the minimum number of steps equaled two times the net dispersal distance divided by the mean step size, or about four hexagons. Likewise, the maximum net dispersal distance observed for juveniles on the Peninsula is 58.25 km, and so the maximum number of steps was set to 31.

The degree to which owl movements are guided by habitat quality is specified by the "Bias to Quality" parameter. This parameter determines the frequency at which owls move to the adjacent hexagon with the best habitat quality. If Bias to Quality equals 0.60, then owls move to the hexagon with the highest habitat score about 60 percent of their steps. If an owl does not consider habitat quality when moving, then the "Linearity" parameter specifies the probable direction of the next step. It determines the tendency to move in a forward direction. When Linearity equals zero, there is an equal probability of moving to any of the neighboring six hexagons. As Linearity increases from zero, the simulated owls have a greater tendency to move in a straight line. The values of Bias to Quality and Linearity were determined through parameter tuning (described below).

Little is known about the behavioral mechanisms that direct the movement of individual animals across landscapes. Hence, the strategy for developing the movement submodel was to avoid parameters that are behavioral. This is contrary to the modeling strategy recommended by others[5] who believe that parameters should be biologically meaningful. At least one parameter, namely Bias to Quality, was needed to link habitat quality to movement. The other movement parameters—minimum and maximum number of steps and Linearity—are basically statistical descriptors of the movement path. Our current understanding is that males, females, juveniles, and adults probably exhibit different movement behaviors.[15, 22] However, besides the minimum movement distance, all movement behavior was modeled the same way, regardless of sex or stage class. A lack of data motivated this simplification.

No mortality occurs during movement. When juvenile owls are dispersing substantial mortality probably occurs, and it presumably is a function of habitat quality and dispersal distance. But, having no data with which to develop such a relationship, we elected to lump all death into a single annual mortality.

The Demographic Submodel

The Spotted Owl Simulator has a two-sex, three stage-class population structure. As with other spotted owl population models,[3, 24] the stage classes are juvenile, subadult, and adult. Estimates for demographic parameters were available,[14] but the crux was developing quantitative relationships between these demographic parameters and habitat quality—a challenging task with no habitat data. The first step was to select the form of the relationships. A logistic function is intuitively appealing, but it requires five parameters to define completely. A logistic function may be approximated with a three-segment, piece-wise linear function that requires only four parameters. We assumed that the least important domain of this function was the flat segment over lower quality habitat. Hence, we eliminated this segment and adopted a two-segment, piece-wise linear function with only three parameters. The shape of this function was based on the plausible assumption that above some value of habitat quality, survivorship (the complement of mortality) and maternity do not increase appreciably and can be considered constants. The same assumption was adopted by Holthausen et al.[20] The value at which survivorship and maternity become constants is known as the "saturation value" of habitat quality.

Adult and subadult stage classes had separate habitat dependent functions for the probability of death and the probability of nesting success. The values for the minimum and maximum parameters were determined through parameter tuning (described below). Juvenile survivorship was not a function of habitat quality because there is considerable uncertainty about the value of this parameter.[14] Instead, separate simulations were run using different constant values of juvenile survivorship. These runs represented a range of plausible scenarios from worst case to best case, and were a form of sensitivity analysis for juvenile survivorship.

Parameter Tuning

The values of some model parameters, such as fledgling sex ratio, came directly from published studies of spotted owls.[13] The value of other model parameters, such as minimum number of steps for movement were calculated using available data. Unfortunately, SRPMs require types of information that are often difficult to acquire, such as the relationships between habitat quality and demographic parameters. SRPMs also entail modeling processes that are poorly understood, such as dispersal behavioral. For these situations there were no data with which to derive parameter values directly, but there were data that allowed us to iteratively search for values such that the average characteristics of the simulated population approximated known characteristics of the real population. This is parameter tuning. For example, to determine values for the movement parameters, simulations were run on SOS with all other parameters assigned fixed values. Upon completion of a simulation, statistical descriptors of the simulated dispersal paths were compared to empirical data. Based on this comparison, Bias to Quality and Linearity were adjusted, and the process was repeated until the mean net dispersal distance was approximately equal to the mean observed on the Peninsula (24 km; E.D. Forsman, U.S. Forest Service, unpubl. data) and the ratio of mean net distance to mean total distance was approximately 2. To enable tuning of the dispersal parameters, mean net distance and mean total distance had to be model outputs.

To establish the relationship between demographic parameters and habitat quality, a maximum value was assumed using available information[20] and the minimum value was determined through parameter tuning. The minimum parameter value was iteratively adjusted until the realized mean parameter value of the simulated population was approximately those reported by Forsman et al.[14] To enable tuning of the demographic parameters, adult and subadult mortality and mean maternity had to be outputs of the model.

Conclusions

Spatially realistic population models are an exciting new technology in wildlife management, but the excitement is somewhat tempered by their extraordinary data requirements. Part of the excitement surrounding SRPMs undoubtedly arises from the array of powerful new technologies for the collection and analysis of spatial data: satellite thematic mapper imaging, global positioning systems, improved microelectronics for radio-telemetry, and geographic information systems. Yet, even with this technology, the extraordinary data requirements of most SRPMs are difficult to satisfy, and this has elicited some concerns about model reliability.[1, 5, 8, 30, 39] Many of these concerns can be assuaged by thorough explication of model assumptions, rigorous model evaluation, thoughtful interpretation of results, and cautious use of the results for decision making. But, for those who choose to develop a SRPM for analyzing their wildlife management problems, grappling with insufficient data remains a vexing challenge.

Developing a SRPM will invariably require assumptions about the population or species autecology. Assumptions will be necessary to circumvent insufficient information or to make computer programming a practical task, but each assumption or simplification made during model development can introduce poor judgement or bias. Nearly all other categories of models familiar to wildlife biologists—habitat suitability indices, matrix models, equations derived from statistical regression, etc.—have well-established principles or methods for their development, and much of their credibility as decision-making tools derives from the process of model formulation that reduces subjective judgements. Likewise, the credibility of a SRPM depends upon the handling of subjective judgements, i.e., assumptions, by the modeller. The most satisfactory approach to dealing with model assumptions is collecting the data needed to eliminate them. When this is impractical, each assumption should be explicitly stated and justified, and a thoughtful consideration of its potential effects on model output should be documented. The effects of each assumption can be rigorously evaluated through sensitivity analyses.

Models which link habitat quality to population demographics are a relatively new tool. This may account for the scarcity of critical data needed by SRPMs. In fact, considering the needs of nongame species at large spatial scales is a relatively recent intellectual development in wildlife management. A critical mass of research in landscape ecology, albeit predominately conceptual or theoretical, has been attained during the past fifteen years.[2, 9, 17, 36] As is often the case, empirical research has not matched the pace of theoretical research, nor has it matched the pace of the technological developments that have enabled SRPMs. Nevertheless, the outlook is promising. Awareness of important processes that take place at the landscape scale has begun to direct field research in wildlife biology and data needed by SRPMs will accumulate.

Acknowledgments

We gratefully acknowledge the reviewers of this paper, Eric Forsman and John Dunning Jr., whose comments motivated major revisions. Timothy Quinn added considerable polish through his conscientious editing. We thank David Johnson for the opportunity to present our work.

Literature Cited

1. Bart, J. 1995. Acceptance criteria for using individual-based models to make management decisions. Ecological Applications 5:411-420.

2. Bissonette, J.A., editor. 1997. Wildlife and Landscape Ecology: effects of patterns and scale. Springer-Verlag, New York, NY.

3. Burnham, K.P., D.P. Anderson, and G.C. White. 1994. Estimation of vital rates of the northern spotted owl. Appendix J, pp. 1-26 in Final supplemental environmental impact statement on management of habitat for late-successional and old-growth forest related species within the range of the northern spotted owl. USDA Forest Service, Portland, OR.

4. Carey, A.B., S.P. Horton, B.L. Biswell. 1992. Northern spotted owl: influence of prey base and landscape character. Ecological Monographs 62:223-250.

5. Conroy, M.J., Y. Cohen, F.C. James, Y.G. Matsinos, and B.A. Maurer. 1995. Parameter estimation, reliability, and model improvement for spatially explicit models of animal populations. Ecological Applications 5:17-19.

6. Department of Natural Resources. 1996. Draft Environmental Impact Statement: Habitat Conservation Plan. Washington State Department of Natural Resources, Olympia, WA.

7. ———. 1997. Final Habitat Conservation Plan. Washington State Department of Natural Resources, Olympia, WA.

8. Doak, D.F., and L.S. Mills. 1994. A useful role for theory in conservation. Ecology 75:615-626.

9. Forman, R.T.T., and M. Godron. 1986. Landscape Ecology. John Wiley and Sons, New York, NY.

10. Forsman, E.D. 1990. Habitat use and home range characteristics of spotted owls on the Olympic Peninsula, Washington. pages numbers unknown in Wildlife habitat relationships in western Washington and Oregon. PNW Annual Report—Fiscal Year 1990. USDA Forest Service, Pacific Northwest Research Station, Wildlife Ecology Team, Olympia, WA.

11. ———. 1991. Habitat use and home range characteristics of spotted owls on the Olympic Peninsula, Washington. pages 12-16 in Wildlife habitat relationships in western Washington and Oregon. PNW Annual Report—Fiscal Year 1991. USDA Forest Service, Pacific Northwest Research Station, Wildlife Ecology Team, Olympia, WA.

12. ———. 1992. Demographic characteristics of spotted owls on the Olympic Peninsula, Washington, 1987-1992. pages 18-27 in Wildlife habitat relationships in western Washington and Oregon. PNW Annual Report—Fiscal Year 1992. USDA Forest Service, Pacific Northwest Research Station, Wildlife Ecology Team, Olympia, WA.

13. ———, E. C. Meslow, and H. M. Wight. 1984. Distribution and biology of the spotted owl in Oregon. Wildlife Monographs 87:1-64.

14. ———, S. G. Sovern, E. D. Seaman, K.J. Maurice, M. Taylor, and J.J. Zisa. 1996. Demography of the northern spotted owl on the Olympic Peninsula and east slope of the Cascade Range, Washington. In E.D. Forsman, S. DeStefano, M.G. Raphael, and R.J. Gutierrez, editors. Demography of the Northern Spotted Owl. Studies in Avian Biology, vol. 17. Cooper Ornithological Society, Camarillo, CA.

15. Greenwood, P.J., and P.H. Harvey. 1982. The natal and breeding dispersal of birds. Annual Review of Ecology and Systematics. 13:1-21.

16. Hanski, I., and D. Simberloff. 1997. The metapopulation approach, its history, conceptual domain, and application to conservation. Pages 5-26 in I. Hanski and M.E. Gilpin. Metapopulation Biology: Ecology, Genetics, and Evolution. Academic Press, San Diego, CA.

17. Harris, L.D. 1984. The Fragmented Forest: Island Biogeography Theory and the Preservation of Biological Diversity. University of Chicago Press, Chicago, IL.

18. Hof, J., and M.G. Raphael. 1997. Optimization of habitat placement: a case study of the northern spotted owl in the Olympic Peninsula. Ecological Applications 7:1160-1169.

19. Holthausen, R.S., M.G. Raphael, K.S. McKelvey, E.D. Forsman, E.E. Starkey, and D.E. Seaman. 1994. The contribution of federal and nonfederal habitat to persistence of the northern spotted owl on the Olympic Peninsula, Washington. Draft Report of the Reanalysis Team, U.S. Forest Service, Portland, OR.

20. ———, M.G. Raphael, K.S. McKelvey, E.D. Forsman, E.E. Starkey, and D.E. Seaman. 1995. The contribution of federal and nonfederal habitat to persistence of the northern spotted owl on the Olympic Peninsula, Washington: report of the Reanalysis Team. U.S. Forest Service, General Technical Report PNW-352.

21. Hobbs., N.T., and T.A. Hanley. 1990. Habitat evaluation: do use/availability data reflect carrying capacity? Journal of Wildlife Management 54:515-522.

22. Johnson, M.L., and M.S. Gaines. 1990. Evolution of dispersal: theoretical models and empirical tests using birds and mammals. Annual Review of Ecology and Systematics 21:449-480.

23. Lehmkuhl, J.F., and M.G. Raphael. 1993. Habitat pattern around northern spotted owl locations on the Olympic Peninsula, Washington. Journal of Wildlife Management 57:302-315.

24. McKelvey, K. 1992. A spatially explicit life-history simulator for the northern spotted owl. Appendix 4-P in Eugene District Resource Management Plan & EIS. U.S. Department of Interior, Bureau of Land Management, Eugene, OR.

25. ———, B.R. Noon, and R.H. Lamberson. 1993. Conservation planning for species occupying fragmented landscapes: the case of the northern spotted owl. Pages 424-450 in P.M. Kareiva, J.G. Kingsolver, and R.B. Huey, editors. Biotic Interactions and Global Change. Sinauer Associates Inc., Sunderland, MA.

26. Mills, L.S., R.J. Fredrickson, and B.B. Moorhead. 1993. Characteristics of old-growth forests associated with northern spotted owls in the Olympic National Park. Journal of Wildlife Management 57:315-321.

27. PMR. 1993. Washington State forest cover classification and cumulative effects screening for wildlife and hydrology. Pacific Meridian Resources, Portland, OR.

28. Pulliam, R.H., J.B. Dunning, and J. Liu. 1992. Population dynamics in complex landscapes: a case study. Ecological Applications 2:165-177.

29. Raphael, M.G., J.A. Young, K. McKelvey, B.M. Galleher, and K.C. Peeler. 1994. A simulation analysis of population dynamics of the northern spotted owl in relation to forest management alternatives. Final environmental impact statement on management of habitat for late-successional and old-growth forest related species within the range of the northern spotted owl. Vol. 2, Appendix J-3. U.S.D.A. Forest Service, Portland, OR.

30. Ruckelshaus, M., C. Hartway, and P. Kareiva. 1997. Assessing the data requirements of spatially explicit dispersal models. Conservation Biology 11:1298-1306.

31. Schumaker, N.H. 1995. Habitat Connectivity and Spotted Owl Population Dynamics. Dissertation. University of Washington, Seattle, WA.

32. ———. 1998. A user's guide to the PATCH model. U.S. EPA/600/R-98/135, Environmental Protection Agency. Environmental Research Laboratory, Corvallis, OR.

33. Seaman, D.E., R.J. Fredrickson, D.B. Houston, B.B. Moorhead, and R.A. Hoffman. 1992. Northern spotted owl inventory, Olympic National Park. Unpublished progress report.

34. ———, S.A. Gremel, S. L. Roberts, and D.W. Smith. 1996. Spotted Owl Inventory-Monitoring in Olympic National Park, Final Report. National Biological Service and Olympic National Park, Port Angeles, WA.

35. Thomas, J.W., E.D. Forsman, J.B. Lint, E.C. Meslow, B.R. Noon, and J. Verner. 1990. A conservation strategy for the northern spotted owl. Report of the Interagency Scientific Committee to Address the Conservation of the Northern Spotted Owl. Appendix M. U.S. Forest Service, Portland, OR.

36. Turner, M.G. 1989. Landscape ecology: the effect of pattern on process. Annual Review of Ecology and Systematics 20:171-197.

37. Van Horne, B. 1983. Density as a misleading indicator of habitat quality. Journal of Wildlife Management 47:893-901.

38. WDW. 1993. Experimental forest classification project. Unpublished report to the Washington Department of Natural Resources, Washington Department of Wildlife, Olympia, WA.

39. Wennergren, U., M. Ruckelshaus, and P. Kareiva. 1995. The promise and limitations of spatial models in conservation biology. Oikos 74:349-356.

Case Study 2

A Model to Assess Potential Vertebrate Habitat at Landscape Scales: HABSCAPES

Mark Huff, T. Kim Mellen & Rich Hagestedt

Introduction

Our objective was to develop a model and analysis procedure targeted for land managers that (1) assesses the amount and distribution of "potential" habitat of all terrestrial vertebrates and aquatic amphibians relative to large landscape patterns (e.g., 1 million acres) and (2) provides options for developing maps and databases to display and analyze "potential" habitat distribution patterns. Our modeling approach, termed HABSCAPES, takes the user's spatially explicit vegetation data (i.e., vegetation maps linked to electronic coordinates), which have been classified into mapped areas for wildlife habitat purposes, and links it to stand- and landscape-level species-habitat relations and life history information. We used the life history information as a surrogate for detailed demographic data that are largely unavailable for most vertebrate species in any given planning area. We developed a series of computer programs and queries to access this information, and to determine distributions of suitable habitat for individual species and groups of species (hereafter, guilds) with similar habitat needs relative to their spatial requirements (e.g., home range area). The results are spatially referenced in a database and stored for visual display and further analyses.

Methods

Development and use of HABSCAPES follows the five general steps described below. Two separate databases are needed to implement HABSCAPES: a vertebrate habitat relations database (Step 1) and a spatially-referenced vegetation database with sufficient information to characterize vegetation features into habitat conditions for animal species (Step 3). HABSCAPES version 1.0 operates in a DOS environment for 486 MHz or higher, IBM compatible PCs. In May 2000, HABSCAPES version 2.0 was completed to run in an ARC/INFO platform.

Process

Step 1. Define vertebrate stand-level habitat relationships and distribution patterns. A series of species-habitat and life history matrixes[1,7] have to be developed for each species. These matrixes organize information, for example, on structural conditions and specific habitat features used by species, and spatial relations of species including home range size class and use at a landscape scale. These matrixes are provided for species found in Washington and Oregon in the CD-ROM included in this book. The distribution of each species also needs to be identified for the analysis area.

Step 2. Group species into guilds based on known habitat use at the stand and landscape scales. HABSCAPES was designed to use a guild approach, where all vertebrate species are grouped by similar uses of resources.[3] To assign a species to a single guild, we first divided all of the vertebrate species in our analysis area (Mt. Hood National Forest) into three groups: special and unique habitat obligates (e.g., scree slopes or springs and seeps), riparian habitat obligates, and terrestrial habitat users. Only species belonging to the riparian habitat obligate and terrestrial habitat groups were placed in guilds; only the terrestrial guild is presented here. The special and unique habitat obligate species were difficult to place into groups and therefore did not fit well into our guild approach. In addition to guilds, HABSCAPES can be used to assess potential habitat of individual species using specific habitat and spatial attributes.

Guilds for the terrestrial habitat group were based on combinations of three attributes: home range size, landscape-use categories, and structural habitat conditions. Home range sizes were divided systematically into three categories: small (<60 acres [24 ha]), medium (60-1,000 acres [24-400 ha]), and large (>1,000 acres [400 ha]). These categories were selected based on observed gaps in home range sizes for species reported in the literature and on the size of management units within the planning area. Landscape-use categories were assigned to each species based on interpreting the literature on how species use habitat within their home range and professional judgment. Four categories were developed: patch (tend to use one homogeneous habitat patch), mosaic (aggregate patches of homogeneous habitat), contrast (use of patches of two different structural stages in close proximity), and generalist (use a variety of structural stages). The third attribute, structural habitat conditions, was derived from the habitat relations information in the matrices, but simplified considerably into three categories: open, small tree (8-21 inches [20-53 cm] dbh), or large tree (≥ 21 inches [53 cm] dbh) conditions. In developing HABSCAPES using Mt. Hood National Forest data, we had 36 possible terrestrial guilds (combinations of 3 home range, 4 landscape use, and 3 structural habitat categories); species were represented in 18 of the possible guilds.

Step 3. Develop a habitat relations-based classification of vegetation using composition and structure. The vegetation database stores the spatially referenced information on which habitat relationship assessments are made. At a minimum, these spatially explicit data provided by the user should be organized into composition-classified communities (e.g., *Pinus ponderosa* series) and into successional or structural development (e.g., late seral, open canopy) types. Vegetation must be classified accurately. At large spatial scales these data are usually obtained from interpreted aerial photographs and/or satellite imagery and stored as raster-based (pixel) maps. The attributes of the vegetation (e.g., canopy cover) are stored in a database with specific variables that are spatially linked to vertebrate species distribution (e.g., range maps). As we developed HABSCAPES, county boundaries were used to delineate distribution (i.e., potential occurrence) for vertebrates. To complete the linkage to vegetation; county boundaries were integrated into the classified vegetation database.

Step 4. Delineate groups of like vegetation pixels into patches. Interpretation of vertebrate habitat relations is done usually at the patch or "patchiness" scale. To make predictions of suitable habitat for different species and guilds, patches need to be created. Typically, patches are areas of relatively homogeneous conditions (e.g., tree sizes), with fixed boundaries and delineated for a specific purpose other than animal habitat relations. To provide users of HABSCAPES with flexibility to delineate their own patches for determining suitable habitat, we developed a program called PATCH that aggregates habitat-related data stored at the pixel scale into patches. PATCH "grows" patches through a pixel by pixel

neighborhood analysis that aggregates like pixels together based on the user's rules (e.g., minimum patch size for given set of objectives and analysis requirements and the type and amount of adjoining habitat surrounding each pixel).[5] PATCH provides the user with a consistent and repeatable method for delineating portions of a map as potential habitat. The output from the PATCH program is a spatially-referenced database used to create individual "patch" maps for specific sets of conditions. In developing HABSCAPES, for example, we created separate patch maps for three structural states (open, small tree, and large tree structural habitats) that were used as the basis to determine suitable habitat for different guilds.

Step 5. Evaluate suitability of patches for different guilds using landscape-level characteristics and habitat relationships. Habitat suitability can be assessed for each guild once patch maps are made. Suitability classes are assigned using programs that evaluate patch size, neighborhood relationships of patches, and the amount of area within fixed home ranges for each guild (or species) that have similar patch-vegetation characteristics. Examples of the parameters we used for our guild-level analysis are shown in Table 1.

The generalist guild is the easiest one to interpret for potential habitat: all vegetation patches are considered suitable. For the guilds with species confined to a single

Table 1. Parameters used to run HABSCAPES programs for each guild. Blanks indicate parameter was not needed to assess habitat for the guild.

Guild Code[1]	Home Range	Patch Configuration Size	Structure Stage Type	Min. Patch	% Home Range in Size	Total Adjacent Habitat	Buffer Width Habitat[2]	% total habitat of each type[3]
TSPO	52 (Small)	Patch	Open	20	50			
TSPST	52 (Small)	Patch	Small tree	20	50			
TSPLT	52 (Small)	Patch	Large tree	20	50			
TSMO	52 (Small)	Mosaic	Open	4.8	50	30		
TSMST	52 (Small)	Mosaic	Small tree	4.8	50	30		
TSGOS	52 (Small)	Generalist	Open/small tree	4.8	50	30		
TSGSL	52 (Small)	Generalist	Small/large tree	4.8	50	30		
TSGG	52 (Small)	Generalist	Generalist	4.8				
TMPO	1000 (Med)	Patch	Open	500	50			
TMMO	1000 (Med)	Mosaic	Open	20	50	700		
TMMLT	500 (Med)	Mosaic	Large tree	20	50	350		
TMGG	500 (Med)	Generalist	Generalist	4.8				
TLMO	3000 (Large)	Mosaic	Open	40	50	2100		
TLMLT	3000 (Large)	Mosaic	Large tree	40	50	2100		
TLGG	3000 (Large)	Generalist	Generalist	4.8				
TSC	52 (Small)	Contrast	Open/large tree	4.8	20		100	>=25
TMC	500 (Med)	Contrast	Open/large tree	10	20		200	>=25
TLC	3000 (Large)	Contrast	Open/large tree	20	20		400	>=25

[1]Guild codes all begin with "T" to indicate terrestrial guilds. The remainder "first letter" code sequentially indicates home range size (S, M, and L), patch configuration(P, M, G, and C) and structure stage (O, ST, and LT); contrast guild codes do not include a structure stage.

[2]Used for mosaic guild only to determine Adjacent-Neighborhood habitat. Acres are 70% of home range size.

[3]Used for contrast guilds only; a type 1 and type 2 contrast habitat must each be at least 25% of total contrast to ensure adequate contrasting environments on the landscape.

patch (patch species), suitable habitat is assessed by determining if a patch created by the PATCH program exceeds the designated minimum patch size for the given structure conditions. We developed specific programs to assess potentially suitable habitat for groups of species that tend to aggregate multiple patches together to make suitable habitat, and groups of species that tend to aggregate patches based on contrasting conditions; they are termed Mosaic and Contrast species, respectively. The landscape-level habitat relationships for the different guilds of Mosaic and Contrast species are conceptual and based on the limited information available for the few species studied at landscape scales. Consequently, much professional judgment was needed to build these types of models.[5]

The SUIT program assesses habitat suitability for the Mosaic species (guilds) by centering a guild-defined, home range-sized circle on each pixel in each patch. For each pixel, a series of menu-based questions are asked (i.e., set of analysis rules) to determine the amount and distribution of patches within the circle.[5] This process has seven possible outcomes: (1) unsuitable patch, not the correct structural stage; (2) suitable large patch, a single patch that is larger than some specified portion of the guild home range size (e.g., 50%); (3) suitable aggregated patch, a habitat patch which meets minimum patch size and which falls within a home range size area of which some specific portion (e.g., 50%) consists of suitable habitat; (4) suitable adjacent-neighborhood patch, a habitat patch which meets minimum patch size and is made suitable when patches on the edge of the home range extend beyond the home range size are combined with patches within the home range size meet some specific portion of suitable habitat (e.g., 70%); (5) unsuitable isolated-contributing patch, a patch which meets the minimum patch size criteria but is too isolated from other patches to be aggregated into suitable habitat and does contribute to making another set of patches suitable; (6) unsuitable isolated patch, meets

minimum patch size criteria but does not contribute to the suitability of another patch; and (7) unsuitable small patch, does not meet the minimum patch size.

The CONTRAST program assesses the distribution of contrasting habitats within a home range by merging two different (structure stage) patch maps (e.g., mature forest adjacent to open/early successional conditions classified together as a patch). Rules are defined by the user for the width, amount, and proportions of contrast conditions between the different patch types. The program produces four potential outcomes of patches that are suitable contrasts or contribute to making a patch a suitable contrast.

Examples
Below we illustrate uses of HABSCAPES using the programs SUIT, PATCH, and CONTRAST, and demonstrate guild and single species approaches.

Bull Run Watershed Analysis
HABSCAPES was used in the Bull Run Watershed Analysis on the Mt. Hood National Forest to assess potential change in amount of habitat for different terrestrial guilds between historic and current conditions. Only guilds with species occurring in the watershed were analyzed. The historic vegetation condition was based on 1940 county forest surveys. The survey provided stand structure information prior to most timber management activities.

All three HABSCAPES programs were used for the terrestrial guild assessment. See Table 1 for guild codes and the parameters used for each guild. The assessment showed declines in habitat for TSPLT, TMMLT, and TLMLT guilds (Table 2). All these guilds use large tree habitat. Consequently, the guild analysis was followed up with an analysis of two individual species of concern within guilds experiencing declining habitat, northern spotted owl and red tree vole. The other guild showing a

Table 2. Historic and Current Habitat Available for Terrestrial Guild

Guild Code	Home Range	Patch Association	Structure Stage	Historic % of Watershed	Current % of Watershed	% Change from Historic Habitat
TSPO	Small	Patch	Open	6	13	+117
TSPLT	Small	Patch	Large tree	58	31	-46
TSMO	Small	Mosaic	Open	6	13	+117
TSGOS	Small	Generalist	Open/small tree	35	65	+86
TSGSL	Small	Generalist	Small/large tree	59	83	+41
TSGG	Small	Generalist	Generalist	95	98	+3
TMMO	Medium	Mosaic	Open	5	9	+125
TMMLT	Medium	Mosaic	Large tree	20	13	-35
TMGG	Medium	Generalist	Generalist	95	98	+3
TLMO	Large	Mosaic	Open	5	2	-60
TLMLT	Large	Mosaic	Large tree	58	28	-52
TLGG	Large	Generalist	Generalist	95	98	+3
TSC	Small	Contrast	Open/large tree	2	7	+250
TMC	Medium	Contrast	Open/large tree	4	9	+125
TLC	Large	Contrast	Open/large tree	6	14	+133

decline in habitat was associated with habitat of large open areas (TMLO), however, this habitat was rare even historically. The guild analysis also revealed that habitat increased substantially for the following guilds: TSPO, TSMO, TSGOS, TSGSL, TMMO, TSC, TMC, and TLC (Table 2).

North Willamette Late Successional Reserve (LSR) Assessment

The Record of Decision for Amendments to Forest Service and Bureau of Land Management Planning Documents within the Range of the Northern Spotted Owl (ROD) (USDA and USDI, 1994) required the two agencies to develop LSR Assessments. The following two examples come from the North Willamette LSR Assessment.[10]

Northern Spotted Owl Habitat. HABSCAPES was used to evaluate the current distribution, amount and quality of late-successional habitat within and outside LSRs. The mosaic guild representing species with large home ranges and associated with late-successional habitat (TLMLT) was used as a "first screening" of potential late-successional

habitat.[9] Specific analyses was done on the northern spotted owl, a threaten species belonging to the TLMLT guild. Analysis attributes for locating suitable nesting and foraging habitat on Mt. Hood National Forest included patches below 5,000 feet (1,610 m) elevation with >60% canopy closure and trees >21 inch dbh or a mix of trees >21 and 8-21 inches (13-53 cm) dbh. Patches with greater than 40% canopy closure and a mix of trees 5-21 inches dbh were classified as suitable dispersal habitat. Other analysis parameters specific to the 1990 guidelines to avoid a "taking"[11] were home range size of 2,995 acres (1198 ha), minimum patch size of 40 acres (16 ha), and minimum acres of suitable habitat within a home range of 40% (1,182 acres [474 ha]). The "Large Patch" polygons shown in Map 1 from Program SUIT represent the largest, least fragmented blocks of suitable habitat. "Aggregated Patches" are more fragmented but still represent areas where at least 40% of the home range area is in suitable habitat. Other patch types do not meet this 40% assumption. The majority of the "Large Patch" polygons are within LSRs or wilderness areas. Additional "Large

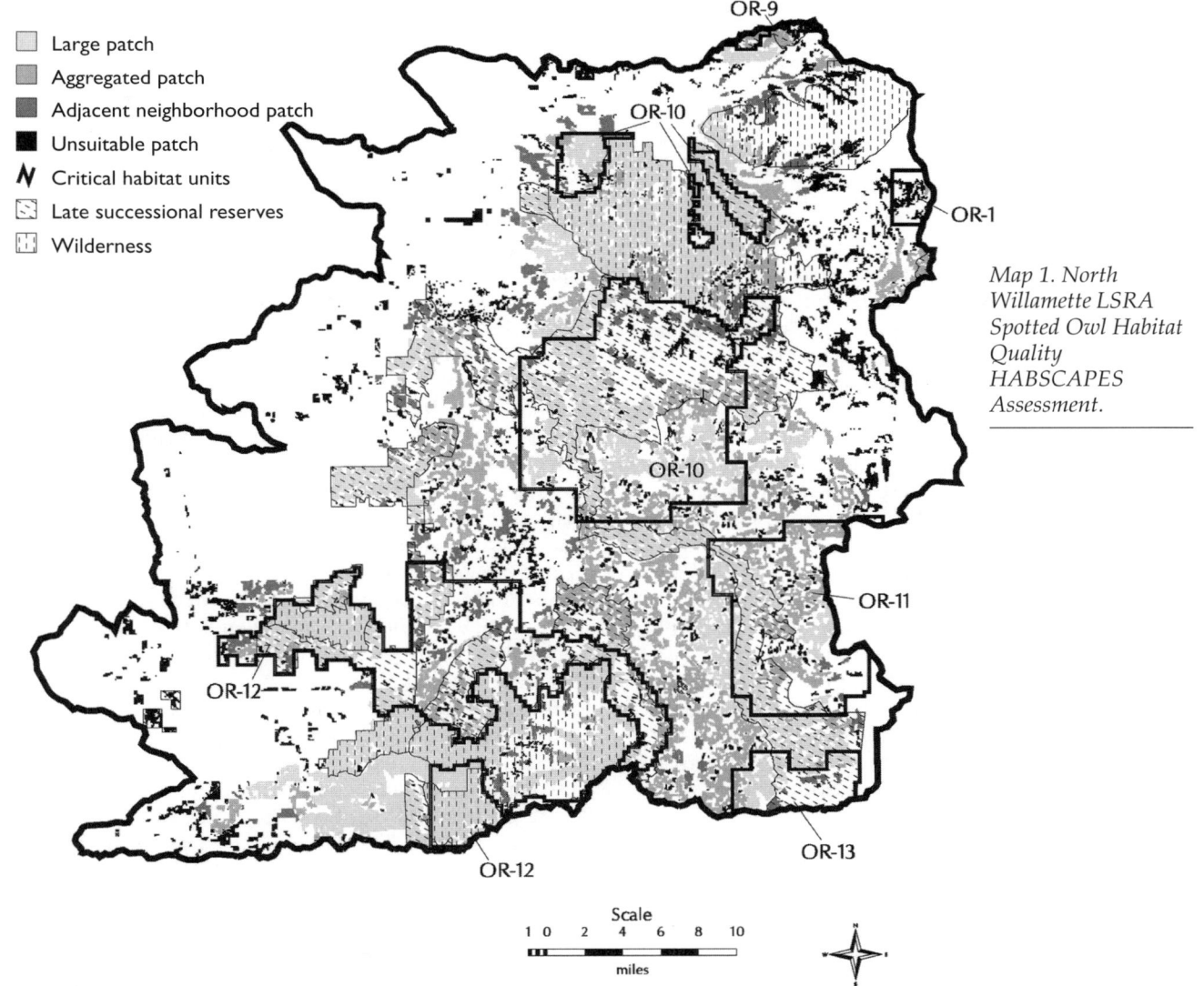

Map 1. North Willamette LSRA Spotted Owl Habitat Quality HABSCAPES Assessment.

Legend:
- Large patch
- Aggregated patch
- Adjacent neighborhood patch
- Unsuitable patch
- *N* Critical habitat units
- Late successional reserves
- Wilderness

Scale

1 0 2 4 6 8 10

miles

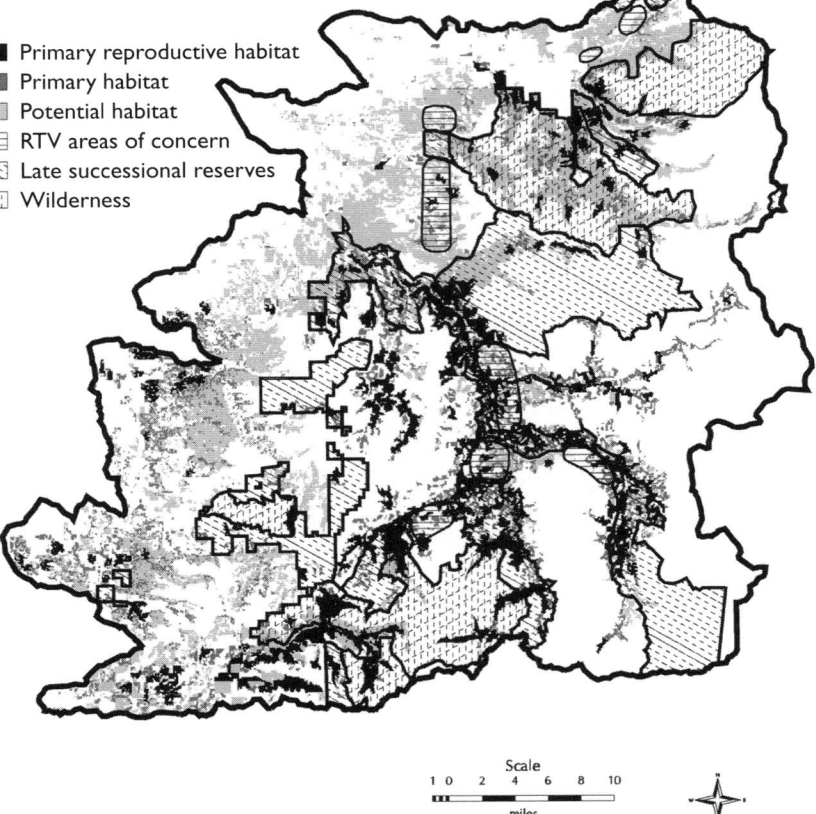

- ■ Primary reproductive habitat
- ▨ Primary habitat
- ▨ Potential habitat
- ▤ RTV areas of concern
- ▨ Late successional reserves
- ⊡ Wilderness

Map 2. North Willamette LSRA Red Tree Vole Habitat and Connectivity Areas of Concern.

Scale
1 0 2 4 6 8 10
miles

Patch" polygons occur in the northern part of the assessment area due to differences in the definition of suitable habitat. There are large amounts of "Aggregated and Adjacent-Neighborhood Patches" habitat outside these protected areas. Dispersal habitat was added to the map after suitable habitat was run through the model.

The condition of spotted owl habitat in Critical Habitat Units (CHUs) was also a management concern. In the western Oregon Cascade Range, overlap is low between LSRs and CHUs,[8] and more site-specific analysis was needed for Mt. Hood National Forest. Output from HABSCAPES was compared to assess the quality of habitat between LSRs/wilderness areas and CHUs based on distribution of patch types (Map 1). There were more acres of "Large and Aggregated Patches" in CHUs than in LSRs/wilderness areas primarily because of more total acres in the CHUs, however, the proportion of habitat in the different patch categories was similar between CHUs and LSRs/wilderness areas. The proportion of suitable habitat in large, contiguous patches was highest in areas where CHUs and LSRs/wildernesses areas overlap, indicating that areas with the least amount of fragmented habitat are covered by both designations.

Red Tree Vole. The red tree vole is a species of management concern because of its close association with late-successional forests, it lives almost exclusively in the canopy of conifers, and its need for habitat connectivity in landscapes because of limited dispersal capability.[8, 9] In the ROD for the Northwest Forest Plan, the red tree vole was identified as a Survey and Manage Species requiring surveys before initiating "ground disturbing" management activities.[8] Because of these circumstances, a suitable habitat assessment was done for red tree voles as part of the LSR assessment on Mt Hood National Forest.

The PATCH program in HABSCAPES was used to assess amount and distribution of red tree vole habitat. Habitat descriptions and model parameters were based on Huff et al.[4] Red tree voles are most abundant in late-successional forests; however, they do occur in younger conifer forest with a closed canopy,[2] which provides dispersal habitat. Habitat was divided into primary reproductive, late-successional patches >100 acres (40 ha); primary other than reproduction, late-successional patches of any size; and potential habitat, patches with closed-canopy conifers and tree diameters 8-21 inches. Forests above about 3,000 feet (970 m) elevation were considered unsuitable habitat.

Suitable red tree vole habitat identified using HABSCAPES was limited primarily by elevation (Map 2). High elevations create a barrier to the vole movement between the Salmon-Huckleberry Wilderness and the Roaring River LSR in the north-central portion of the assessment area. The HABSCAPES maps helped to identify areas to sustain connectivity for the voles at lower elevations to the west and for four areas to the south between LSRs.

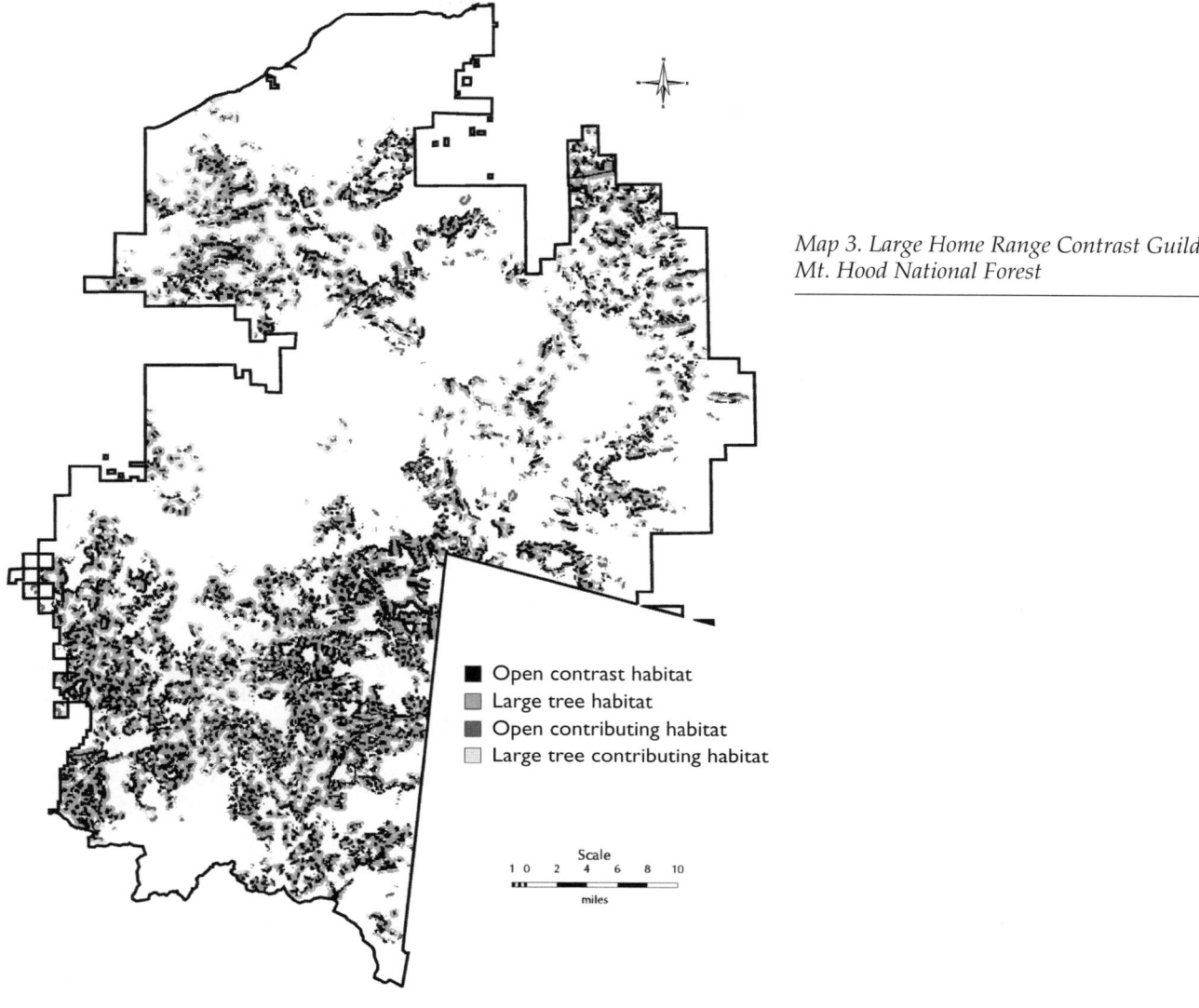

Map 3. Large Home Range Contrast Guild, Mt. Hood National Forest

■ Open contrast habitat
▨ Large tree habitat
■ Open contributing habitat
☐ Large tree contributing habitat

Scale
1 0   2   4   6   8   10
miles

Large Home Range, Contrast Guild (TLC)

The CONTRAST program in HABSCAPES was used to create a map of habitat for the TLC guild, large tree and open habitat within 400 m of each other.[12] Species in this guild include elk, great horned owl, and red-tailed hawk. Minimum patch size was set at 20 acres (8 ha), percent habitat in the home range needed to be at least 20 percent, and each type of contrast habitat needed to contribute at least 25 percent to total contrast habitat (based on professional judgement) to ensure adequate contrast between seral stages. The results (Map 3) indicated where the best potential elk habitat could be found on the Mt. Hood National Forest. Although not shown, separate queries on winter and summer range attributes can done and overlain together to assess the relations of seasonal habitat patterns. Mapping the TLC guild also predicts potential great horned owl habitat. This map could be compared with the map of suitable spotted owl habitat to identify potential places where predation of great horned owls on spotted owls[6, 11] is probable.

Conclusions

HABSCAPES is a "coarse filter" approach to assess potential habitat and to develop conservation strategies for many species simultaneously. HABSCAPES assists biologists and resource managers with complex multi-scale assessments, such as watershed analyses required in the Northwest Forest Plan. Our approach of grouping species into similar habitat guilds, which considers species' spatial requirements, and of providing suitability indices of potential habitat gives biologists a tool to make relatively quick habitat assessments for large numbers of vertebrate species in a planning area.

HABSCAPES is versatile, as shown from our case studies and assessments of individual species and guilds that addressed many natural resource planning issues. HABSCAPES provides users with options to create patches based on specific objectives, and to change patch shape, size, and composition for different objectives using program PATCH. As PATCH aggregates like pixels into patches, it stores information about patches at the pixel level, thus giving the user flexibility to assign a pixel to as many patches as warranted. It is the user who defines the rule sets for the analysis programs PATCH, SUIT, AND CONTRAST and for the queries to determine potential habitat.

Tradeoffs have to be made to assess multi-species issues at large landscape scales, as done with HABSCAPES. Generalizations are made about species habitat relations and spatial requirements to fit groups of species into guilds. The amount of habitat for certain species unknowingly can be over- or under-estimated, especially for those species that are closely associated with specific habitat features that are not evaluated. Therefore, the guild approach adopted in HABSCAPES should not be used to evaluate population and species viability issues associated with habitat. Further, HABSCAPES projections of suitable habitat are dependent on estimates of minimum home range sizes, however many species have not been studied sufficiently for different habitats and life history functions. Currently few tests have been done to validate the HABSCAPES suitable habitat predictions; though, we did examine how well spotted owl nest occurrence corresponded with suitable habitat projects. We placed the known spotted owl nest sites for Mount Hood National Forest over habitat suitability types in Map 1. Sixty-five percent of the nests were concentrated in the "most" suitable types for spotted owls, while 22 and 18% were in other suitable habitat patch types and dispersal habitat, respectively. We expected a higher percentage of nests in the most suitable habitat; we suspect it will be higher as the vegetation database is refined (e.g., narrower tree diameter classes).

Literature Cited

1. Brown, E.R., technical editor. 1985. Management of wildlife and fish habitats in forests of western Oregon and Washington. Part 2—Appendices. Publ. R6-F&WL-192-1985. U.S. Department of Agriculture, Forest Service, Portland, OR.

2. Carey, A.B. 1991. The biology of arboreal rodents in Douglas-fir forests. In Biology and management of old-growth forests. M.H. Huff, R.S. Holthausen, and K.B. Aubry (technical coordinators). PNW-GTR-276. USDA Forest Service, Pacific Northwest Research Station, Portland, OR.

3. Hale, W.G., and J.P. Margham. 1991. The HarperCollins dictionary of biology. HarperCollins Publishers, New York, NY.

4. Huff, M.H., R.S. Holthausen, and K.B. Aubry. 1992. Habitat management for red tree voles in Douglas-fir forests. In: Biology and management of old-growth forests. M.H. Huff, R.S. Holthausen, and K.B. Aubry (technical coordinators). PNW-GTR-302, USDA Forest Service, Pacific Northwest Research Station, Portland, OR.

5. Mellen, K., M. Huff, and R. Hagestedt. 1995. "HABSCAPES" Interpreting landscape patterns: a vertebrate habitat relationships approach. Pages 135-144 in Thompson, J.E., compiler. Analysis in support of ecosystem management. Analysis workshop III, April 10-13, 1995, Fort Collins, Colorado. USDA Forest Service, Ecosystem Manage. Cntr., Washington, D.C.

6. Miller, G.S. 1989. Dispersal of juvenile northern spotted owls in western Oregon. M.S. thesis, Oregon State University, Corvallis, OR

7. Thomas, J.W., editor. 1979. Wildlife habitats in managed forests: the Blue Mountains of Oregon and Washington. Washington DC: U.S. Department of Agriculture, Forest Service.

8. USDA Forest Service and USDI Bureau of Land Management. 1994. Final supplemental environmental impact statement on management of habitat for late successional and old growth forest related species within the range of the northern spotted owl. Forest Service, Bureau of Land Management. Portland, OR. 3 vols.

9. ———, and ———. 1994. Record of Decision for Amendments to Forest Service and Bureau of Land Management planning Documents Within the Range of the Northern Spotted Owl, and Standards and Guidelines for Management of Habitats for Late Successional and Old Growth Forest Related Species Within the Range of the Northern Spotted Owl. Forest Service, Bureau of Land Management. Portland, OR. 3 vols.

10. ———, and ———. 1998. North Willamette LSR Assessment. Mt. Hood National Forest, Sandy, OR. Unpublished report.

11. USDI Fish and Wildlife Service. 1992. Recovery plan for the northern spotted owl—draft. Portland, OR.

12. Wisdom, M.J., L.R. Bright, C.G. Carey, W.W. Hines, R.J. Pedersen, D.A. Smithey, J.W. Thomas, and G.W. Witmer. 1986. A model to evaluate elk habitat in western Oregon. Pub. No. R6-F&WL-216-1986, USDA Forest Service, Pacific NW Region, Portland, OR.

Case Study 3

Cross-scale Classification Trees for Assessing Risks of Forest Practices to Headwater Stream Amphibians

Glenn D. Sutherland & Fred L. Bunnell

Introduction

Because organisms respond to environmental conditions and processes operating at different scales, extrapolation of specific research findings to guide general forest practices can be misleading.[3] We illustrate the use of classification and regression tree (CART) methods to reveal relationships occurring at different scales when only inventory data are available. Our primary objectives are to: 1) identify the range of habitat attributes permitting the species presence; 2) reveal potential differences in habitat relationships within the species range; and 3) identify sites where conservation measures would be most effective. We chose the tailed frog for five reasons: 1) it is designated "at risk" or "of special concern" in most of its range; 2) available data are contradictory, suggesting that broader scale phenomena are acting;[2, 5, 11, 14] 3) generalized conservation measures could have enormous impact on timber harvest; 4) similar approaches can be developed for other species in headwater streams; and 5) given the number of studies on the species, it is worthwhile developing an approach that permits analysis of geographically scattered data.[3]

Methods

Bunnell et al.[4] interpreted the disparate results of studies evaluating impacts of forest practices on tailed frogs as a function of different physical settings. Common approaches to describing environmental influences on species response variables—multivariate habitat models (e.g. logistic or multiple regression)—can be misleading if different factors limit a species in different portions of its range, or if the interactions between variables are non-linear. Moreover, spatial autocorrelation within and between environmental variables can alter or disguise the apparent response.[8]

First, we developed empirical multivariate species-habitat models from extensive inventory data, unstratified by management activities or ecological gradients, using CART analysis. CART models create hierarchical trees by recursive partitioning of sets of numeric or categorical habitat predictor variables into mutually exclusive subsets, which are most homogeneous with respect to a response variable (species occurrence or abundance).[1] Each step finds the variable most important in reducing remaining variation, dependent on all previous steps. The output tree diagram represents a nested set of ecological dependencies among habitat factors, exposing how key environmental variables can act to constrain the ranges of other variables, given the observed species response. That permits inference about the consequences of management or conservation actions in different environmental settings.

Because broad-scale variability due to climatic, geologic, and cumulative effects of disturbance were potential covariates of tailed frog responses, we then assessed potential spatial autocorrelation between variables and sites to help interpret model structure. We used a multivariate approach based on Mantel tests, which quantifies the degree of concordance between two matrixes.[9] We compared a matrix of abiotic and biotic site descriptors using the CART node structure as a classifier of ecological similarity between sites, with a matrix of geographic distances obtained from UTM coordinates.

For practitioners seeking empirical species-habitat models, CART methods have advantages over logistic and generalized linear methods.[15] They are easier to interpret when both continuous and categorical predictor variables are used. They assume no specific multiplicative relationship between predictor variables, so resulting models are robust to both the shape of frequency distributions of predictor variables, and the presence of outliers. They can be tested on other datasets, even when some variable values are missing. In short, they are appropriate to the kinds of data commonly available.

Data Requirements

We used both occurrence and relative abundance data routinely collected in broad-scale inventory surveys, and density data summarized from site-specific and longer-term research studies. For predictor variables, we chose *site-specific habitat variables* that could be associated with local abundance of tadpoles, including: substrate composition, channel disturbance, stream width, water temperature, and recovery of streambeds from disturbance.[2, 14] We added summary variables of *climate, topography and forest structure and/or disturbance effects* from existing GIS databases, standard climate records, or maps.

Analytical outputs

Primary outputs of the CART approach are heuristic prediction rules that can be used to classify site suitability across a wide range of environmental conditions for a given species. The rules provide guidance on key environmental factors (physical, climatic, habitat structure) determining population status, and expose ecological and environmental thresholds that can be used in GIS software to delineate protection zones or possible

risks to the species from management activities. Combined with spatial analyses they can help to identify the appropriate spatial scale for applying management guidelines. The approach also provides probability rankings that help to classify new sites where only some predictor variables have been measured, and to identify potential classifications for sites in areas not yet surveyed.

Once such empirically-based habitat models are developed, they can be used to make predictions in two ways. Using the original data set, the likelihood of any rule in the tree applying in similar environmental conditions elsewhere can be estimated from the proportion of points falling into each node. This statistic should be used cautiously, because it will be imprecise when samples are not representative of the new environmental conditions.[13] One also can use the original tree to examine new data, and use the misclassification rate as an estimate of the goodness-of-fit of the tree to the new data.[6] O'Connor et al.[12] suggested that while tree-based models tend to be robust to moderate sampling biases, prediction accuracy degrades significantly if key environmental variables are missing or are undersampled.

Example

We sought to determine where tailed frogs occurred and which attributes of headwater stream habitats were necessary to classify, map, and evaluate potential protection measures. We georeferenced 846 records of tailed frog occurrence, abundance, and habitat from three sources: an extensive survey across the range of the tailed frog in British Columbia[7] (L. Dupuis, unpubl. data), and two intensive studies of habitat association in replicated forest management treatments.[14] Most records were of tadpoles, but adults were recorded when found. Although these sources had different objectives and methods of data collection, they could be combined in CART analysis.

Details of data collection are found in the sources noted. Access to sites was usually by road, restricting most samples to watersheds with some harvesting history and encouraging opportunistic sampling. At each site, occurrence of tailed frogs (tadpoles or adults) was determined. For 59.9% of the sites, estimates were also made of tadpole density (area constrained searches; $n = 134$) or relative abundance (time constrained searches; $n = 374$). Up to 20 stream characteristics were recorded at a site, but many sites ($n = 263$) recorded fewer than 10 of these.

To this basic dataset, we added broad-scale climate, geological and topographic attributes of sub-basins potentially relevant to disturbance regimes and stream productivity. We used GIS databases to estimate harvesting attributes of each sub-basin, including: mean road density within 100m of streams inventoried in the provincial stream database, percent area of recent harvest (since 1978) at different elevations (300m, 300-800m and above 800m), and area harvested on slopes >50% and within 100m of streams. From the BC Ministry of Forests climate database, we extracted mean precipitation and air temperature variables representing long-term climate averages for each biogeoclimatic subzone in our samples. We measured stream gradients, and determined parent lithology from Geological Survey of Canada maps. We eliminated redundant variables (correlations >0.75 with other variables), and samples with many missing predictor variables.

To assess whether different environmental relationships were acting in different portions of the tailed frog range, we divided sites into regional subsets reflecting influences of coastal and inland climates (Coast Mountain Range and Kootenay Mountains of southeastern BC). We derived models for occurrence (Coast: n=180; Kootenays: n=113) and, on sites where frogs occurred, relative abundance (Coast n=31; Kootenays: n=27), and density (Coast: n=92). Here we excluded sites outside these two regions.

We first developed initial classification trees using CART[10] for each response variable and sample subset, by recursively splitting each node until either the proportion of the deviance remaining for sample sites in the node was less than 5% of the original deviance, or there were fewer than 5 sample sites left in the node. For these trees we used only sites with no missing data for all predictor variables. We then "pruned back" each initial tree using a ten-fold cross-validation analysis.[6] That process iteratively removes random subsets of the original observations, and uses the remaining observations to refit the classification tree, allowing the analyst to select the number of nodes that provides a parsimonious fit to the data.

Much can be inferred from CART analysis. We note five major points. First, environmental variables operating at broad spatial scales (macro- and meso scales in Table 1) accounted for >54% of the explained deviance in all models. Micro-scale characteristics of streams contributed less to explaining occurrence and abundance patterns, although they were associated with patterns of tadpole density in coastal sites (Table 1). Parent bedrock was the dominant predictor of distribution and abundance on the coast, and relative abundance in the Kootenays (Table 1; Figures 1 and 2). Variables reflecting extent and location of harvests and roads near streams at the sub-basin scale also predicted distribution and abundance in both regions (Table 1). Second, positive spatial autocorrelations between geographic distances separating sites and their classification by CART was apparent in patterns of occurrence (Mantel tests; coast: $P<0.001$; Kootenays: $P=.036$), and in the relative abundance data for the coast ($P < 0.036$). Coastal densities and Kootenay relative densities showed no spatial autocorrelation ($P>0.5$). Underlying geology governed occurrence, but variation in stream characteristics related to tailed frog abundance were independent of geographic location. Third, while there is overlap in habitat variables predicting occurrence in the two regions (Table 1), the relative importance of each variable differed between the two regions. In particular, variables linked to forest harvesting (areal extent and location of recent harvest, and stream substrate) were more important in the Kootenays than in coastal sites (Table 1; Figures 1 and 2). The % cover of fines and stream temperature helped distinguish sites of high and low tadpole density on the coast. Differences in the importance

Table 1. Percentages of original deviance explained by the CART models for tailed frog occurrence (Figures 1 and 2) and abundance (see text) among the various classes of predictor variable. Scales defined by range of detectable spatial autocorrelation in the original variables (macro = 80-120 km; meso = 25-60 km; micro = <25 km). Sample sizes indicate the number of sites used in each model.

Predictor class	Spatial scale	Coast			Kootenays	
		Occurrence (n = 180)	Relative abundance (#/ 20 min.) (n = 31)	Density (#/ m²) (n = 92)	Occurrence (n = 113)	Relative abundance (#/ 20 min.) (n = 27)
Climate[1]	macro	-	-	5.4	-	-
Topography[2]	meso	9.5	-	11.1	3.5	9.9
Bedrock Lithology	meso	21.1	-	22.5	2.3	24.5
Sub-basin Harvesting and Road Density[3]	meso	9.6	7.5	2.1	18.8	9.9
Stand Type[4]	micro	-	-	-	-	-
Stream Substrate[5]	micro	6.9	-	37.0	13.8	10.6
Stream Temperature	micro	-	-	9.1	6.0	-
Stream Width	micro	-	35.2	-	12.1	-
Fine Organic Debris	micro	2.1	-	-	-	-
Reach Slope (%)	micro	-	-	-	-	-

[1] means for: annual precipitation, summer precipitation (in mm), annual temperature, temperature of warmest month (°C),

[2] elevation (m), aspect (°), stream gradient (%)

[3] % sub-basin area logged since 1978; % logged area within 100m of streams, road density within 100m of streams (km/ha)

[4] age class of adjacent stand (0-10 yrs; 11-20 yrs; 21-60 yrs; 61-100 yrs, > 100 yrs), presence of riparian buffers (yes/no)

[5] % cover of boulders, cobbles, and fines

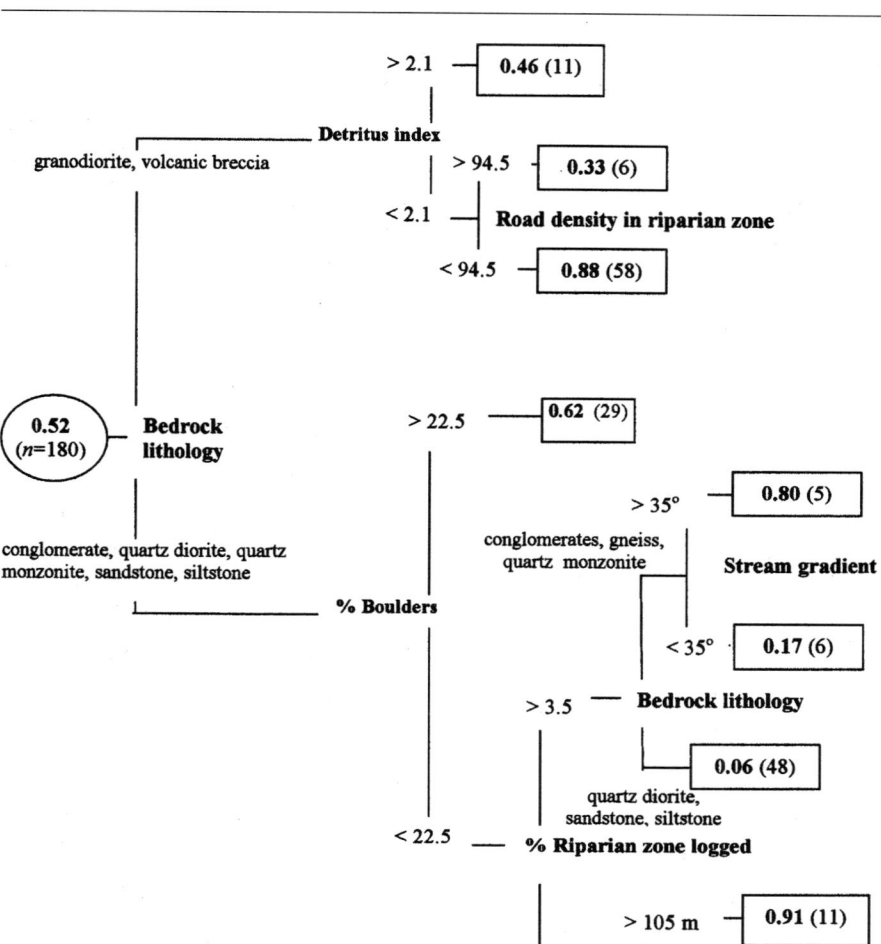

Figure 1. Classification and Regression Tree (CART) model for tailed frog occurrence in the western coastal mountain zone of British Columbia. Numbers inside the rectangles (end nodes) are the proportion of sampled sites (sample size in brackets) with tailed frogs present. The splitting predictor variable and its threshold value is shown for each branch of each node. The length of each branch (growing to the right) is proportional to the percent of deviance explained by the splitting variable at each node.

of macro-scale environmental variables relative to micro-scale variables may reflect differences in regional hydrology regimes or histories of forest management. They could also reflect the restricted spatial distribution of occurrence records in the Kootenays,[7] which limits the ability to adequately separate components of variation. Fourth, important dependencies between habitat variables were exposed by the occurrence trees (Figures. 1 and 2). On the coast, sites with the highest probability of tailed frog occurrence were associated with relatively undeveloped watersheds in regions of highly competent bedrock (granodiorites or volcanic breccias) or, in regions of less competent bedrock types, with boulder-dominated streams, higher elevations, or steep stream gradients if a moderate amount of logging had occurred in the riparian zone. In the Kootenays, sites with the highest probability of tailed frog occurrence were associated with watersheds having extensive harvesting history, in smaller boulder-dominated streams, but, in less developed watersheds, with sites having cobble-producing bedrock (argillite; quartzite). Finally, variables directly reflecting forest management practices adjacent to sample sites (e.g. presence or absence of buffers) had no apparent predictive capacity.

The process can derive qualitative and quantitative criteria for locating conservation measures and refining site-specific forest practices for species at risk. Physical setting had a greater influence on occurrence and abundance of larval frogs than did adjacent forest practices. We offer two cautions when interpreting these models: 1) our source data for extent of recent logging may be at too coarse a scale to reliably detect an influence, and 2) our models are directly relevant only to the in-stream life stage. Nonetheless, managers can use classification trees to identify habitat protection options in several ways. For example, if the goal is maintenance of current habitat, then all sites meeting the environmental criteria with a >0 probability of occurrence become candidates for protection measures. Our analyses suggest that practices designed to conserve existing tailed frog tadpole populations (e.g., buffers around streams) will be most successful if: 1) they are situated in intrusive or metamorphic bedrock formations; 2) they target streams on intermediate gradients with significant boulder cover;

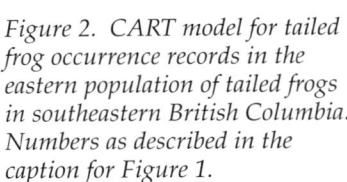

Figure 2. CART model for tailed frog occurrence records in the eastern population of tailed frogs in southeastern British Columbia. Numbers as described in the caption for Figure 1.

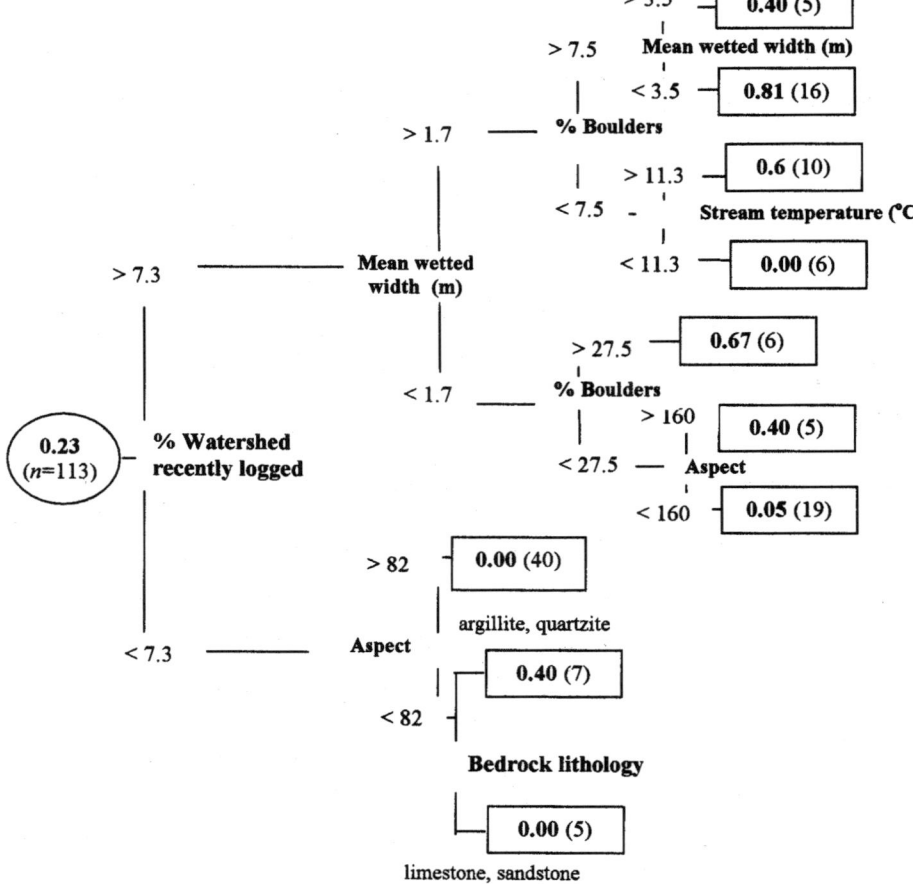

and 3) they are in watersheds with low or moderate levels of historical harvest. Conservation efforts will be less effective where weathering products from surrounding geology produces fine-grained or rubbly stream sediments, at low elevation sites, and in sites with increased chances of dry periods during the summer. In portions of the tailed frog range, entirely natural factors have much greater influence than human-induced changes. Such areas are not good candidates for expensive protective measures. More complex and quantitative rules can be derived from the trees (see Methods) to develop stream classifications for species sensitivity, or compare predictions with other datasets.

Conclusions

Hierarchical classification trees can expose complex species-environmental relationships operating at different scales, using field inventory samples overlaid with readily available data describing climate, topography, and geology. Protection measures for tailed frogs must account for spatial variation in the sensitivity of sites to be effective. More generally, for species in which the effects of forest practices act most strongly through altered gradients affecting growth, fecundity and movements, a classification tree approach yields practical design criteria for management more quickly and with less effort than do standard linear models.

There are four strengths to this approach in evaluating species-environment relationships. First, CART techniques complement landscape-scale methods for predicting presence and diversity of species (e.g., Gap analysis) by revealing pathways by which environmental variation may be determining species' distribution. Second, CART techniques make efficient use of minimal, low cost data sets involving combinations of nominal and numerical data. For relatively unstudied species, robust estimation techniques like CART can quickly explore existing habitat databases to yield provisional rules to guide management options. Third, CART models permit data of several independent, small-scale studies to be "scaled up" into a more spatially extensive analysis without making restrictive assumptions about how the same environmental variables interact in all locations. Finally, CART models are easy to interpret, leading to hypotheses that may predict observed distribution and abundance, and ranking the likelihood of responses of species to combination of environmental variables. Our models can easily be tested and used in other areas where standard inventory data is routinely collected (e.g., stream surveys).

There are weaknesses in the approach. Most importantly, classification tree approaches are heuristic in nature—processes inferred from them must be treated as hypotheses until their predictions are confirmed by more detailed experimental studies. Secondly, interpretations of model outputs are constrained to the precision of the datasets used as input. Finally, classification tree approaches by themselves cannot infer effects of specific habitat protection measures on future extirpation probabilities, because of basic uncertainties in species' life history.

Acknowledgments

We thank Linda Dupuis and Pierre Friele for the tailed frog survey database for British Columbia. Malcolm Gray (BC Ministry of Environment, Lands and Parks) provided his Watershed Ranking database for coastal BC, while Matthew Craig provided detailed GIS data for the Kootenay region. Arnold Moy assisted with analyses. Comments by Linda Dupuis, David Huggard and two anonymous reviews improved the manuscript. Research support provided by Forest Renewal BC, the BC.Ministry of Environment, Lands and Parks, the BC Ministry of Forests, and Wildlife Habitat Canada to FLB. This is contribution M-10 of the Centre for Applied Conservation Biology, UBC.

Literature Cited

1. Brieman, L., J.H. Friedman, R.A. Olshen, and C.J. Stone. 1984. Classification and regression trees. Wadsworth International Group, Belmont, CA.

2. Bull, E.L., and B.E. Carter. 1996. Tailed frogs: distribution, ecology and association with timber harvest in northeastern Oregon. USDA Forest Service Research Paper PNW-RP-497.

3. Bunnell, F.L., and D.J. Huggard. 1999. Biodiversity across spatial and temporal scales: problems and opportunities. Forest Ecology and Management 115: 113-126.

4. ———, L.L. Kremsater, and R.W. Wells. 1997. Likely consequences of forest management on terrestrial, forest-dwelling vertebrates in Oregon. Oregon Forest Resources Institute, Portland, OR.

5. Bury, R.B. 1968. The distribution of *Ascaphus truei* in California. Herpetologica 24: 39-46.

6. Clark, L.A., and D. Pregibon. 1992. Tree-based models. Pages 377-419 in J.M. Chambers and T.J. Hastie, editors. Statistical Models in S. Wadsworth & Brooks/Cole Computer Science Series, Pacific Grove, CA.

7. Dupuis, L.D., and F.L. Bunnell. 1997. Status and distribution of the tailed frog in British Columbia. Unpublished report for Forest Renewal British Columbia.

8. Legendre, P. 1993. Spatial autocorrelation: trouble or new paradigm? Ecology 74: 1659-1673.

9. ———, and M.-J. Fortin, 1989. Spatial pattern and ecological analysis. Vegetation 80: 107-138.

10. Mathsoft, Inc. 1998. SPLUS 4 Guide to Statistics. Mathsoft, Seattle, WA.

11. Metter, D.E. 1964. A morphological and ecological comparison of two populations of the tailed frog, *Ascaphus truei* Stejneger. Copeia 1964: 181-195.

12. O'Connor, R.J., M.T. Jones, D. White, C. Hunsaker, T. Loveland, B. Jones, and E. Preston. 1996. Spatial partitioning of environmental correlates of avian biodiversity in the conterminous United States. Biodiversity Letters 3: 97-110.

13. Thompson, J.D., G. Weiblen, B.A. Thomson, S. Alfaro, and P. Legendre. 1996. Untangling multiple factors in spatial distributions: lilies, gophers, and rocks. Ecology 77: 1698-1715.

14. Wahbe, T.R. 1996. Tailed frogs (*Ascaphus truei*) in natural and managed coastal temperate rainforests of Southwestern British Columbia, Canada. Thesis, University of British Columbia, Vancouver, BC, Canada.

15. Verbyla, D.M. 1987. Classification trees: a new discrimination tool. Canadian Journal of Forest Research 17: 1150-1152.

Case Study 4
Applying Gap Analysis to County and Regional Land Use Planning

Matthew R. Stevenson

Introduction

Biodiversity is the range of living organisms and the processes which created and sustain them. Human activities on the landscape in the form of resource extraction, agricultural conversion, and urban development have had a substantial impact on biodiversity. In fact, biological diversity is being lost to such a degree that the present rate of extinction rivals the catastrophic loss of species at the end of the Paleozoic and Mesozoic eras. The rate at which species are becoming extinct is 1,000 to 10,000 times higher than before human intervention.[20] The Nature Conservancy estimates that 200 plant species and 71 species and subspecies of vertebrates have gone extinct in North America (excluding Mexico) since European settlement.[9] The condition of ecosystems is similar: according to the Office of Technology Assessment, "twenty-three ecosystem types that once covered about half the coterminous United States now cover about seven percent."[21]

The reasons for this ecological catastrophe are complex, but to a large degree the responsibility can be attributed to the way in which cities, counties and regions have chosen to locate, organize, and regulate land uses. Perhaps in partial recognition of this, the State of Washington adopted the Growth Management Act (RCW 36.70A) in 1990, which includes in its requirement that jurisdictions planning under the act "conserve fish and wildlife habitat" and "protect the environment." An additional legal imperative facing counties and regions is the Federal Endangered Species Act (ESA), which can trigger severe limitations upon local activities once a species (or a population of a species) has been listed.[17]

In addition to meeting these regulatory requirements, cities, counties, and regions should be concerned with preserving biodiversity because only properly functioning ecosystems can provide the free ecosystem services civilization requires. However, one problem confronting jurisdictions interested in protecting biodiversity is the lack of information describing the distribution of biologically important areas. In an attempt to provide this information and improve upon previous fragmented approaches to single species preservation, the National Gap Analysis Program (NGAP) was charged with conducting the largest effort ever to map the distribution of biological resources in the United States.

Gap analysis, which uses land cover, wildlife habitat relationship models, and other data to predict the distribution of terrestrial vertebrates, provides information which can be applied to landscapes and regions in order to ensure that land management and land use plans more effectively incorporate the protection of biodiversity. This case study details one approach for applying these data to land use planning.

Methods

Description of Process and Data Requirements

The process developed for applying Washington GAP data to county and regional land use planning involves five steps.

1. Finding the minimum amount of land required to represent each species predicted to occur in the study area.

This first step is intended to ensure that every species predicted to occur in the County is represented at least once by any system of protected lands. This concept, commonly referred to as "representation," does not necessarily mean that all life requisites of any given species can be accommodated within the representative land area, but does guarantee that at least some habitat for every species predicted to occur in the County is included.[3, 11] Unfortunately, WAGAP did not determine home range size for each species, which is necessary to determine the amount of habitat required to maintain minimum viable populations for all vertebrates. Consequently, representation means simply that at least one vegetation polygon per species is included within the proposed network of reserves and corridors.

Finding representation begins with clipping raw WAGAP data to the study area. The WAGAP data are organized into three discrete groups: land cover, which consists of 100 ha minimum mapping unit (MMU) land cover polygons delineated using 1991 Landsat TM data; vertebrate distributions, which were derived using wildlife/habitat relationship models in conjunction with the land cover map; and stewardship, which rates the management of large land holdings on a scale of 1 to 4 based upon the level of protection of biodiversity.

First, using the clipped species distribution data, the polygon or set of polygons with the highest species richness (greatest number of species) for each given taxonomic group is located. At the same time, species which are not predicted to occur within this initial polygon or set of polygons are also identified. Next, these non-represented species within each group are ranked in descending order according to the number of occurrences within the study area. Then, starting with the polygons for the non-represented species with the highest predicted number of occurrences, polygons are eliminated by adding species to the representative set.

In other words, out of 180 total bird species in a given study area, 100 species may be present in the richest polygon, meaning 80 species are not represented. Of these 80 non-represented species, the number of polygons in which each occurs will be different. As an example, if the Brewer's blackbird were not represented in the richest polygon but occurred most frequently among all non-represented species, its distribution would form the initial set of polygons for the second round of selection. If the Brewer's blackbird were to occur in 600 polygons, then 600 polygons would make up the largest set of polygons that would appear in the second round of selection. The next most representative species from that set—for instance, the violet-green swallow—might occur in 480 polygons where the Brewer's blackbird has also been predicted to occur. Selecting the polygons where both the blackbird and the violet-green swallow occur reduces the number of polygons to 480 needed to represent both species. Each additional species added to the set reduces the number of polygons in a similar fashion until finally the addition of one more species results in the selection of no polygons because no occurrences overlap. This process is repeated until all species in each taxonomic group are represented. The approach is similar (though less sophisticated) to the representative set solutions discussed by Pressey and others.[2, 3, 11, 15, 19]

2. Establishing species richness thresholds to locate candidate areas for habitat reserves.

Establishing species richness thresholds by taxonomic group is necessary before areas of high species richness can be identified as potential locations for reserves. This process requires the creation of a decision rule for which polygons should be selected or "turned on," based upon the number of species predicted to occur in each. Determining a universally applicable richness threshold is not as straightforward as is representation, since richness is relative to the total number of vertebrates (by group) within the study area. For this analysis, threshold levels were arbitrarily set at 75 percent of possible richness for mammals and birds and 50 percent of possible richness for reptiles and amphibians. A lower threshold for reptiles and amphibians was used because their populations tend to be concentrated in discrete habitat patches such as wetlands, rivers, and lakes, particularly for amphibians.

3. Delineating potential habitat reserves.

I considered all polygons with relatively high species richness for designation as a habitat reserve area. Factors used in the selection of habitat reserves included polygon size, land cover, level of human development and intensity of human activity within the area, adjacent development, relationship to Priority Habitats and Species (PHS) mapped designations, and proximity to other polygons satisfying the richness requirement. The Washington Department of Fish and Wildlife (WDFW) established and maintains the PHS database, which was developed to help guide management decisions regarding biologically important areas. PHS differs from WAGAP in two important ways: first, it does not use wildlife-habitat relationship models to predict species distributions; and second, it does not encompass all terrestrial vertebrates within Washington.[18] Since PHS information is limited to priority species (priority species include threatened, endangered, or sensitive species; vulnerable aggregations; and species of recreational, commercial, and/or tribal importance) it should be used in conjunction with WAGAP data to corroborate areas indicated as important for establishing representation, corridors, and habitat reserves.

The polygons selected for potential habitat reserves were large polygons with high species richness, natural land cover, low internal and adjacent human development, and high PHS polygon and point coverage overlap.[6] Additionally, heron rookeries, waterfowl concentrations, and other vulnerable wildlife aggregations received special consideration when potential reserves were delineated.[18]

4. Refining polygon sets to determine corridor alignments.

A considerable amount of land was selected during the process of identifying key habitat areas based on representation and richness of species, and the next step was to refine this area into a comprehensive system of habitat reserves and wildlife corridors. Three kinds of data were used to reduce and refine the area resulting from steps 1-3: predicted species distributions from WAGAP for all counties surrounding the study area county; PHS data from WDFW, and a variety of thematic data unique to the County, such as large-scale land cover, topography, wetlands and hydrology, land use, zoning, growth area boundaries, critical areas, digital orthophotography, and air photo stereo pairs.

The larger WAGAP data set for the counties surrounding the study area county were used during the process of refining the selected set to ensure that reserves and corridors connected to potential reserves in adjacent counties. The PHS database was used to ensure that important habitats were retained as the larger WAGAP polygons were refined into corridors. The corridors themselves were delineated using the county-specific data in conjunction with the WAGAP and PHS data.

The most important county-specific data set was a large-scale land cover map with a minimum mapping unit (MMU) that is substantially smaller than the WAGAP MMU of 100 ha. The smaller MMU is necessary because it lends greater accuracy to the process of delineating potential corridor alignments, particularly through highly developed or disturbed areas.[9, 13] I created this map in the summer of 1997 by interpreting and digitizing land use and land cover from digital orthophotos supplied by WDFW. Unfortunately, this type of mapping was expensive, time-consuming, and did not cover the entire county. Outside the mapped area, a parcels coverage and primary roads coverage were used as proxies for the location and intensity of development and human activity.

A USGS digital elevation model (DEM) was used to ensure that corridors follow realistic gradients (i.e. corridors should not run over cliffs) and that the complete system "maintain(s) natural ecosystems and biodiversity across the full extent of environmental gradients".[9] Encompassing all environmental gradients will become increasingly important as global temperatures continue to rise and ecological communities slowly shift northward and upward to compensate.[10]

Additionally, wetlands, rivers, and streams data were extremely important for delineating corridors in general but especially for amphibians and birds. As demonstrated by Klaus Richter in a study of 19 wetlands in King County, Washington, wetlands are used disproportionately by birds and amphibians and are therefore critical habitat areas for these taxa.[Richter, cited in 1]

Data on land use, zoning, growth area boundaries, and critical areas are maintained by all counties and can be used as an indication of how the landscape will likely change in the future and can also be used to detect areas of potential conflict between planned or zoned development and areas that may be important as habitat reserves or corridors. Digital orthophotos and stereo pairs were used to examine actual land cover conditions during the process of refining boundaries for reserves and corridors.

5. Establishing an ecological and political context for system evaluation.

The reality of land management is political, so it is useful to establish a context for decision making by analyzing how policy decisions may affect stewardship status within both ecologically and politically determined boundaries. I chose the statewide extent of the vegetation zones passing through the study area (a county) as the ecological unit of analysis for the relative levels of preservation of vegetation zones and vertebrates and I used the county as the politically determined boundary. The stewardship status of vegetation zones can serve as a rough indicator of how well a county is protecting biodiversity relative to the entire extent of the vegetation zones that pass through the county. I used four measurements to establish the political and ecological context. First, I determined the percentage of land within the statewide extent of vegetation zones which is managed primarily for the protection of biodiversity. Second, I determined the percentage of land within all vegetation zones found within the study area. Third, I determined the percentage of study area vegetation zones managed primarily for the protection of biodiversity. Lastly, I determined the new percentage of study area vegetation zones that would be managed primarily for the protection of biodiversity within any proposed system of corridors and reserves.

Process Output

Using the methods described above, this process will produce four primary products: an inventory of the predicted distributions of terrestrial vertebrates and their level of protection within the study area and the state; an inventory of vegetation zones and their level of protection within the study area and the state; a set of potential habitat reserves and wildlife corridor alignments within the study area; and lastly, a "before and after" description of the effect of any new system of corridors and reserves upon the protection of vertebrates and vegetation zones.

It is important to remember that WAGAP data was designed for landscape-scale analyses; thus, it may be too coarse to use directly for acquiring conservation lands. Additional field work, ground truthing, mapping, and surveying are critical before any specific land management decisions are made.[12] However, it is also important to note that the WAGAP data base is the only one of its kind in existence, it is readily available (for a nominal fee) to any county or city that is interested in using it, and it can be analyzed and manipulated on a PC within a realistic time frame and budget.

Example:
Spokane County Comprehensive Plan

The process described above was developed in Spokane County, Washington through an interdisciplinary studio class in the Department of Urban Design and Planning at the University of Washington in the fall of 1997. The initial motivation for conducting the analysis came from staff in the WDFW Region 1 office, who were working with the Spokane County Division of Long Range Planning to locate "wildlife corridors" and "landscape linkages," two categories within the County's recently adopted Critical Areas Ordinance (CAO) which had been defined but not yet identified on the ground. In order to map these specific critical areas, the County was willing to serve as the pilot project for the application of WAGAP data. I went on to further develop this application in my Master's thesis at the University of Washington.[14]

The first step in the analysis was to examine patterns of vertebrate richness by taxonomic group (in this case mammals, birds, reptiles, and amphibians). The second step involved the establishment of representation for all vertebrates (i.e., the minimum configuration of polygons that captures all vertebrate taxa in at least one polygon). Output from the richness analysis played a key role in the selection of WAGAP polygons considered as starting points for reserves. These three data sets were combined with a variety of additional data sets to produce the potential habitat reserve locations and wildlife corridor alignments.

The initial system of habitat reserves was determined as described above in step three. In Spokane County, areas of habitat with high species richness, natural land cover, low internal and adjacent human development, and high PHS polygon and point coverage overlap were located in the northeast, west, and southwest portions of the county. These potential habitat reserve areas encompass such well-known places as Mt. Spokane State Park, Turnbull National Wildlife Refuge, and portions of the Spokane River.

The most crucial piece of data used in refining the initially-selected reserves and corridors was the fine-grained land-use land-cover map I created of the Interim Urban Growth Area (IUGA) surrounding greater metropolitan Spokane. This map was created with a five acre (2.5 ha) MMU, substantially smaller than the 100 ha MMU used by WAGAP. The IUGA does not cover the entire county, so we used a parcels coverage and primary roads coverage as indicators of human development and activity in areas lacking higher resolution land cover information. A parcels coverage can serve as an effective proxy because smaller parcels typically indicate higher densities and intensities of development, impervious surfaces, non-native vegetation, fences, and automobile traffic. Conversely, larger parcels tend to have lower densities, less impervious surface, and more native vegetation (with the exception of agricultural areas). Based upon field visits, we determined that parcels smaller than approximately 20 acres were too disturbed to be included in potential corridor alignments.

Roads (particularly interstates and highways) were an equally important consideration when determining potential reserve locations and corridor alignments, because they may impact all of the taxonomic groups modeled by WAGAP (see for example *transportation and wildlife: reducing wildlife mortality and improving wildlife passageways across transportation corridors*[16]).

The refined corridor system is based primarily upon the existing riparian network and utilizes upland connections to complete and complement this foundation. Riparian corridors were selected as the backbone for the system in Spokane County because of their disproportionately significant contribution to biodiversity.[8] Wherever current land use permitted, corridors were drawn at a minimum of $^1/_4$-mile to $^1/_2$-mile wide on center in order to protect riparian vegetation and provide realistic adjacent upland movement corridors (particularly in forested areas) that would be wide enough to avoid the "edge" effects, which can be deleterious to some species.

Results

If the refined reserve and corridor system were to be fully implemented within Spokane County, the amount of additional land managed primarily for the protection of biodiversity would increase from 4% to 30%. As shown in the last column of Table 1, all vegetation zones within the County would gain at least 10% in protected area in addition to currently protected land. These increases are important within the County but also address statewide conservation needs. As described by WAGAP in their final report summary, "The vegetation zones with the highest

Table 1. Conservation status (acres, percent) of vegetation zones in Spokane County under proposed system of corridors and reserves.

Vegetation zone	Acres currently protected	Percent of vegetation zone	Acres in proposed system	Additional percent of vegetation zone within system
Ponderosa pine	21,917	1.9	165,821	26.5
Palouse	907	<0.01	36,404	10.6
Interior Doug fir	17,441	1.5	121,125	55.7
Three-tip sage	0	0	13,188	23.9
Grand fir	47	<0.01	2,025	97.3
Interior redcedar	0	0	49	89.2

Table 2. Protected habitat as a percentage of total habitat for terrestrial vertebrates in Spokane County before and after proposed system of reserves and corridors.

Taxonomic group	Count	Mean (%)		Median (%)	
		Before	After	Before	After
Mammals	66	5	41	4	39
Birds	161	6	37	6	38
Reptiles/Amphibians	17	5	44	6	43
All vertebrates	244	6	39	6	39

Table 3. Test results comparing the original and modified reserve and corridor system to field data specific to Spokane County (results given in % of original database points or area [if polygons]).*

System	PHS	Heritage	Herp	Sensitive Plants	Heron Rookeries	Elk	Public Input
Original	<1% omitted	31% omitted	33% omitted	30% omitted	None omitted	3% omitted	3.5% omitted
Modified	<1% omitted	7.8% omitted	7.7% omitted	7.5% omitted	None omitted	<1% omitted	<1% omitted

* (Adapted from Ferguson, Robinette, and Stevenson 1999)

Conservation Priority Index (CPI) are steppe zones (which include Palouse and Three-tip Sage)." Additionally, WAGAP describes the Ponderosa Pine and Oak zones as "zones of high reptile and avian diversity" which have "moderately high CPIs."[7] The Ponderosa Pine zone and both of the steppe zones found in Spokane County would be substantially more protected under the system designed with this process. Furthermore, the majority of existing protected areas in Washington State are located at high elevations in the Cascade and Olympic mountains. The system proposed for Spokane County would improve statewide conservation not just by increasing the area of protected land in relatively under-represented low elevation vegetation zones, but also by connecting important habitats in an area of the state with very little biodiversity protection.

Table 2 shows current habitat protection in Spokane County for all terrestrial vertebrates at a mean and median of 6% of predicted distributions. If the entire system of reserves and corridors were implemented with no substantial modifications, the median area of protected habitat for all species would increase to 39%, more than six times the existing level of protection. This additional protection occurs by increasing the amount of protected land to just 30% of the County's total land area.

Based upon interviews I conducted with the Spokane County Division of Long Range Planning, WDFW, and the Inland Northwest Land Trust (INLT), the information produced through this application of WAGAP is unique and valuable to all of them. These results were finalized in January 1998, and so far Spokane County is using the information to update its Critical Areas Ordinance and evaluate a proposed system of extensive, high capacity roads encircling the greater Spokane area. The INLT has used and will continue to use the results as the foundation for their Threads of Hope campaign, which serves as the backbone for their entire conservation strategy. WDFW has used the data to evaluate parcels nominated for a recently re-approved and taxpayer funded Conservation Futures Program, and has collaborated with Spokane County to evaluate and further refine the proposed system into a Final Open Space Plan.

As shown in Table 3, the WDFW evaluation of the original system of reserves and corridors was made with field data, WDFW Heritage points, WDFW reptile and amphibian (herp) points, DNR sensitive plant species points, great blue heron rookery locations, elk habitat data, and public input. WDFW biologists detected relatively high omission rates for Heritage points, herps, and sensitive plant species, but fairly good coverage of all other evaluation criteria. By using these data to increase the land area included in the system of reserves and corridors by just 2%, omission rates were substantially reduced, as indicated in the second row of Table 3. According to Ferguson et al,[4] "this final version appears to have satisfied our goals—a biologically defensible open space plan with a reasonable amount of land being identified in just over two months."

Conclusions

Strengths and Weaknesses

The methodology described above has the capability of informing the process by which land use plans are created and land management policies are formulated. However, this process should be used with an understanding that it is only a tool for developing the solution required to conserve biodiversity in the face of expanding human population and human activity. Additionally, it is important to realize that although there are many strengths to this approach, there are also shortcomings which must be acknowledged and addressed accordingly.

Strengths. Gap data predicts distributions for all vertebrates in the state, not just rare, threatened, or endangered species. Combining richness, representation, and connectivity ensures that every species is contained within any proposed system of protected or specially managed areas. Working at a variety of resolutions allows land planners and managers to see patterns that would otherwise not be visible, and integrating the results of this process into land use planning can be an effective way of preventing additional populations of vertebrates and other organisms from becoming threatened or endangered. Identifying the most important areas of habitat can provide some degree of certainty to the development and resource extraction industries. The successful adoption and implementation of the results of this process can lead to a significant increase in the amount of land managed primarily for the protection of biodiversity. A number of management and policy-making bodies are necessarily involved in the process and are therefore more likely to find the outcome valuable. Finally, creating a connected system of habitat areas maintains ecosystem services, enhances the aesthetic qualities of the landscape, raises property values, and can provide recreational opportunities to the burgeoning Washington population.

Weaknesses. The bulk of the process is based upon Gap analysis, which being predictive in nature contains errors of commission and omission, and overlaying these distribution maps to create composite richness maps can lead to additional error accumulation.[5] The satellite imagery used to conduct WAGAP was collected in 1991, and substantial changes may have occurred in the landscape in the intervening eight years. Vertebrate distributions are not always the most accurate proxy for the distributions of other taxa. The various data sets used in this analysis were created at different scales for different purposes and do not always work well in conjunction with one another. Capturing the presence of any given taxon does not guarantee long-term persistence (i.e., geographic units or polygons do not necessarily contain sustainable populations). The coarse scale of the analysis necessitates a substantial amount of follow up fieldwork and additional fine-scale analysis, as detailed above. Lastly, classification of vegetation may be inadequate for certain taxa, and habitat relationship models may be weak for certain taxa.

Relevance to the Landowner

This process is relevant to landowners in that it can provide a considerably higher degree of certainty concerning the use of their land. The current ad-hoc, reactionary, species-by-species approach to protecting biodiversity is divisive and counter-productive. Land owners will probably be more interested in actively participating in the stewardship of their land when they realize it is biologically significant on a regional scale. However, it is important that landowners be engaged and involved through a process relying on incentives for cooperation, not government command regulations.

Additional Research

The Spokane County pilot project provided a tremendous opportunity for the development of a methodology for applying WAGAP to real problems. The process described above can provide important information to decision makers. However, there are several key areas of additional research which would enhance the quality, accuracy, and usability of the finished product. Regional coordination is indispensable because the effectiveness of any system of corridors and reserves will be severely constrained if similar biodiversity planning activities in adjacent jurisdictions are not coordinated. Ground truthing should be conducted in parallel with the expanded mapping of land cover to ensure that what is being mapped actually exists on the ground. Developing (or applying existing) set coverage and optimization algorithms would allow for the evaluation of a variety of approaches to representation, such as comparing the differences between representing each taxonomic group and adding the results versus representing all groups simultaneously. Land use modeling would be beneficial for determining which of those areas within the proposed system are also the most likely to come under intense development pressure and should consequently be top priorities for protection. Lastly, public involvement is extremely important, because the level of support for this approach will be related to the degree to which the people affected by it understand, appreciate, and acknowledge its usefulness.

Literature Cited

1. Azous, A.L. and R.R. Horner. 1997. Wetlands and urbanization: implications for the future. Final report of the Puget Sound Wetlands and Stormwater Management Research Program, Washington State Department of Ecology, Olympia, WA.

2. Church, R.L., D.M. Stoms, and F.W. Davis. 1996. Reserve selection as a maximal covering location problem. Biological Conservation 76:105-112.

3. Csuti, B., J.D. Camm, B. Downs, M. Huso, R. Hamilton, M. Kershaw, A.R. Kiester, S. Polasky, R.L. Pressey, K. Sahr, and P.H. Williams. 1996. A comparison of reserve selection algorithms using data on terrestrial vertebrates in Oregon. Biological Conservation. 80:83-97.

4. Ferguson, H.L., K. Robinette, and M.R. Stevenson. 1999. Searching for the best science available: a method for identifying a corridor-open space system for land use planning. Proceedings, Fourth International Symposium on Urban Wildlife Conservation, Tucson, AZ.

5. Flather, C.H., K.R. Wilson, D.J. Dean, and W.C. McComb. 1997. Identifying gaps in conservation networks: of indicators and uncertainty in geographic-based analyses. Ecological Applications 7:531-542.

6. Forman, R.T.T. 1995. Land mosaics: the ecology of landscapes and regions. Cambridge University Press, Cambridge, England.

7. Grue, C.E., K.M. Cassidy, and K.M. Dvornich. 1998. GAP Bulletin Number 6: Final report summary, Washington GAP analysis project. U.S. Department of the Interior, U.S. Biological Service, Washington D.C.

8. Naiman, R.G., H. Decamps, and M. Pollock. 1993. The role of riparian corridors in maintaining regional biodiversity. Ecological Applications 3:209-212.

9. Noss, R.F., and A.Y. Cooperrider. 1994. Saving nature's legacy: protecting and restoring biodiversity. Island Press, Washington, D.C.

10. Peters, R.L., and T.E. Lovejoy, eds. 1992. Global warming and biological diversity. Yale University Press, New Haven, CT.

11. Pressey, R.L., H.P. Possingham, and J.R. Day. 1997. Effectiveness of alternative heuristic algorithms for identifying indicative minimum requirements for conservation reserves. Biological Conservation 80:207-219.

12. Scott, J.M., F. Davis, B. Csuti, R.F. Noss, B. Butterfield, C. Groves, H. Anderson, S. Caicco, F. D'Erchia, T.C. Edwards Jr., J. Ullman, and R.G. Wright. 1993. Gap analysis: a geographic approach to protection of biological diversity. Wildlife Monographs 123.

13. Stine, P.A. 1995. A multi-scale conservation assessment of plant communities in southern California. Dissertation, University of California at Santa Barbara, Santa Barbara, CA.

14. Stevenson, M.R. 1998. Protecting biodiversity: applying gap analysis in Spokane County, Washington. Thesis. Department of Urban Design and Planning, University of Washington, Seattle, WA.

15. Stokland, J.N. 1997. Representativeness and efficiency of bird and insect conservation in Norwegian boreal forest reserves. Conservation Biology 11:101-111.

16. United States Department of Transportation, Federal Highway Administration, State of Florida Department of Transportation. 1996. Transportation and wildlife: reducing wildlife mortality and improving wildlife passageways across transportation corridors. Proceedings of the Florida Department of Transportation/Federal Highway Administration Transportation-Related Wildlife Mortality Seminar. Federal Highway Administration, Washington, D.C.

17. United States Fish and Wildlife Service. 1988. Endangered Species Act of 1973. As amended through the 100th Congress. United States Department of the Interior Washington, D.C.

18. Washington Department of Fish and Wildlife. 1996. Priority habitats and species list. Olympia, WA.

19. Williams, P., D. Gibbons, C. Margules, A. Rebelo, C. Humphries, and R. Pressey. 1996. A comparison of richness hotspots, rarity hotspots, and complementary areas for conserving diversity of British birds. Conservation Biology 10:155-174.

20. Wilson, E.O., and F.M. Peter, editors. 1988. Biodiversity. National Academy Press, Washington, D.C.

21. Winckler, S. 1992. Stopgap Measures. The Atlantic Monthly January: 74-81.

Case Study 5
A Model to Determine Potential Northern Spotted Owl Nesting Areas

Neal Darby and Tim Young

Introduction

Private, State and Federal wildlife managers need planning tools that assist in determining the potential occurrence of sensitive, threatened and endangered species on their respective lands. These planning tools are particularly needed when wildlife managers lack the physical or financial resources to conduct field surveys. To meet this need a Geographical Information System (GIS) model was developed to synthesize existing GIS vegetation maps and create a separate landscape level map depicting areas with a high potential to contain nesting northern spotted owls. This model was designed for landscape level planning in areas where surveys for northern spotted owls may no longer be practicable. The model uses desktop GIS technology and existing forest inventory GIS data representing various northern spotted owl habitat quality parameters (e.g., tree size, percent canopy closure, tree species composition, etc.). Threshold levels used to delineate high potential nesting areas were based on spatial analyses of habitat located around known northern spotted owl locations and nest sites as identified in the literature. The ability to identify areas that have a high potential to support northern spotted owl nesting

over large landscapes could provide natural resource managers with greater flexibility in the design of forest management plans that promote the conservation of northern spotted owls and their habitat. Finally, the model can be adapted for other species of concern by identifying other vegetation attributes and spatial parameters that describe habitat conditions conducive for that species' needs.

Methods

ArcView GIS 3.0a with the Spatial Analyst Extension (ESRI 1997) was used to process the model. The model consists of several iterative steps. In each step, specific features in a GIS map are selected for and modified to create a new GIS map, until the final GIS map depicting high potential nesting areas is created. ArcView GIS commands used in each step are all found within the default menus, requiring no special programming. This model process was developed using the Gifford Pinchot National Forest's (GPNF) GIS vegetation map and interim vegetation table (I-Veg). Both the GPNF GIS vegetation map and I-Veg table are publicly available on the World Wide Web (http://www.fs.fed.us/gpnf/gis/index.html).

To begin the model process, a GIS vegetation map with forest stand attribute information detailed enough to allow the delineation of suitable northern spotted owl habitat is required (Figure 1, Step A). The organization and detail of forest stand attribute information can vary widely within and between federal, state and private agencies. Thus, a biologist needs to review the GIS vegetation map database and, using local knowledge and professional judgment, determine if it is adequate to confidently identify the extent of stands containing suitable northern spotted owl habitat throughout the area of interest. It is not important to have precise attribute information, only that all GIS vegetation map features representing suitable habitat be distinguishable from unsuitable habitat.

The GPNF GIS vegetation map and I-Veg table were used to establish a base GIS vegetation map (Figure 1, Step A). The I-Veg table is a database of forest stand attribute information (i.e., average tree diameter at breast height (DBH), trees per acre, stand age, etc.). The I-Veg table is directly joined to the GIS vegetation map feature table by matching fields (table columns) that contain identical information, such as a stand or polygon identifier number. Other GIS vegetation maps may not require a separate table be joined, having the forest stand attribute information in its feature table. Some GIS vegetation map feature tables may have a field designating spotted owl habitat. In fact, the I-Veg table contains a field titled "Spotted Owl Habitat." Under the "Spotted Owl Habitat" field, each polygon representing a forest stand in the GIS vegetation map is given one of four habitat descriptions: Unsuitable, Dispersal, Foraging or Nesting. Each habitat description is based on forest stand attribute information found under other fields in the I-Veg table for each respective polygon. Suitable northern spotted owl habitat, as defined here, consisted of stands designated as Foraging and Nesting in the GPNF I-Veg table and included forest

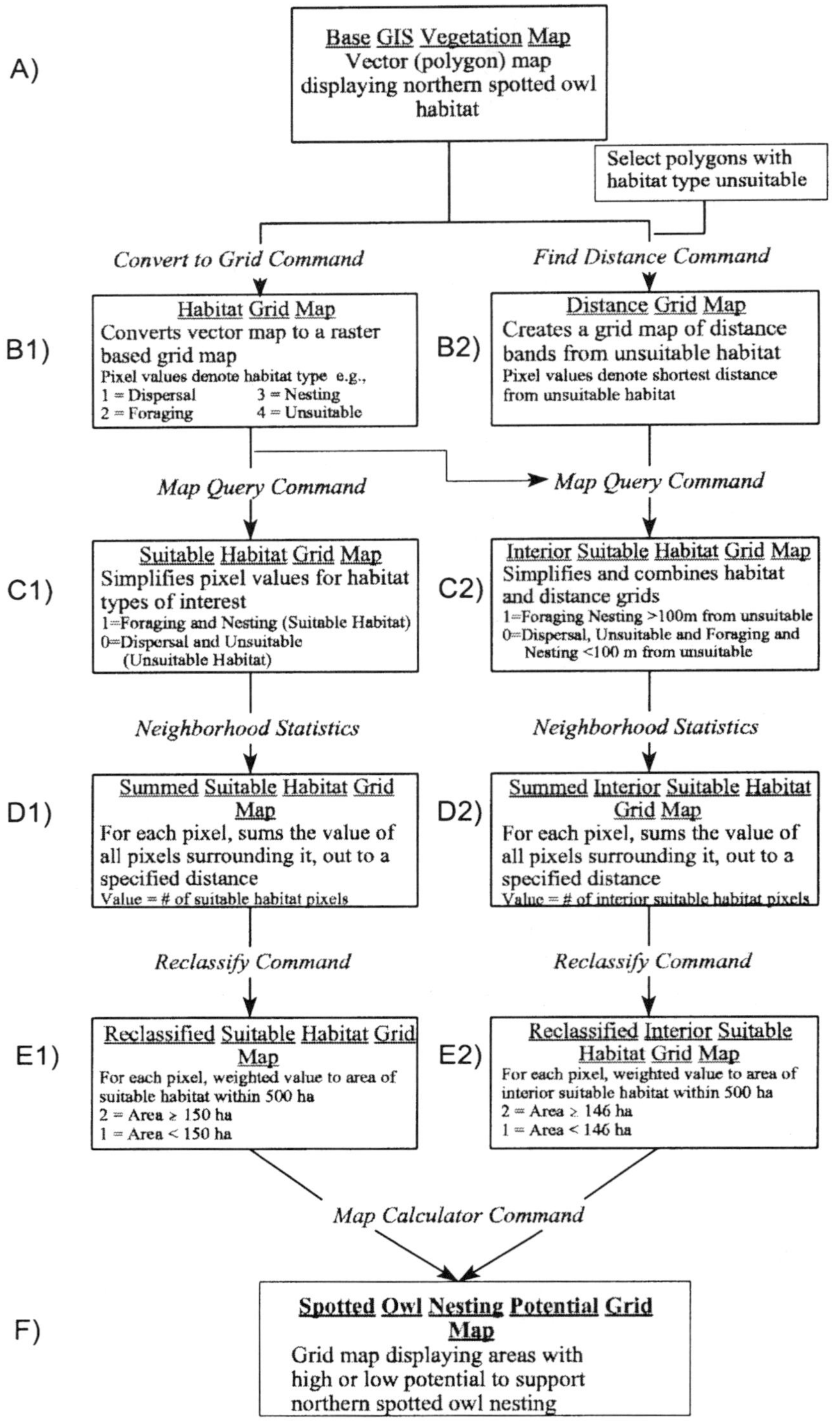

A)

Base GIS Vegetation Map
Vector (polygon) map
displaying northern spotted owl
habitat

Select polygons with
habitat type unsuitable

Convert to Grid Command

Find Distance Command

B1)

Habitat Grid Map
Converts vector map to a raster
based grid map
Pixel values denote habitat type e.g.,
1 = Dispersal   3 = Nesting
2 = Foraging   4 = Unsuitable

B2)

Distance Grid Map
Creates a grid map of distance
bands from unsuitable habitat
Pixel values denote shortest distance
from unsuitable habitat

Map Query Command

Map Query Command

C1)

Suitable Habitat Grid Map
Simplifies pixel values for habitat
types of interest
1=Foraging and Nesting (Suitable Habitat)
0=Dispersal and Unsuitable
(Unsuitable Habitat)

C2)

Interior Suitable Habitat Grid Map
Simplifies and combines habitat
and distance grids
1=Foraging Nesting >100m from unsuitable
0=Dispersal, Unsuitable and Foraging and
Nesting <100 m from unsuitable

Neighborhood Statistics

Neighborhood Statistics

D1)

**Summed Suitable Habitat Grid
Map**
For each pixel, sums the value of
all pixels surrounding it, out to a
specified distance
Value = # of suitable habitat pixels

D2)

**Summed Interior Suitable Habitat
Grid Map**
For each pixel, sums the value of
all pixels surrounding it, out to a
specified distance
Value = # of interior suitable habitat pixels

Reclassify Command

Reclassify Command

E1)

**Reclassified Suitable Habitat Grid
Map**
For each pixel, weighted value to area of
suitable habitat within 500 ha
2 = Area ≥ 150 ha
1 = Area < 150 ha

E2)

**Reclassified Interior Suitable
Habitat Grid Map**
For each pixel, weighted value to area of
interior suitable habitat within 500 ha
2 = Area ≥ 146 ha
1 = Area < 146 ha

Map Calculator Command

F)

**Spotted Owl Nesting Potential Grid
Map**
Grid map displaying areas with
high or low potential to support
northern spotted owl nesting

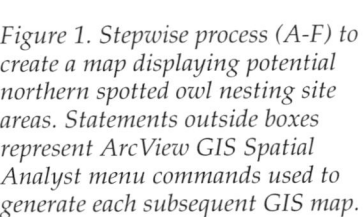

Figure 1. Stepwise process (A-F) to create a map displaying potential northern spotted owl nesting site areas. Statements outside boxes represent ArcView GIS Spatial Analyst menu commands used to generate each subsequent GIS map.

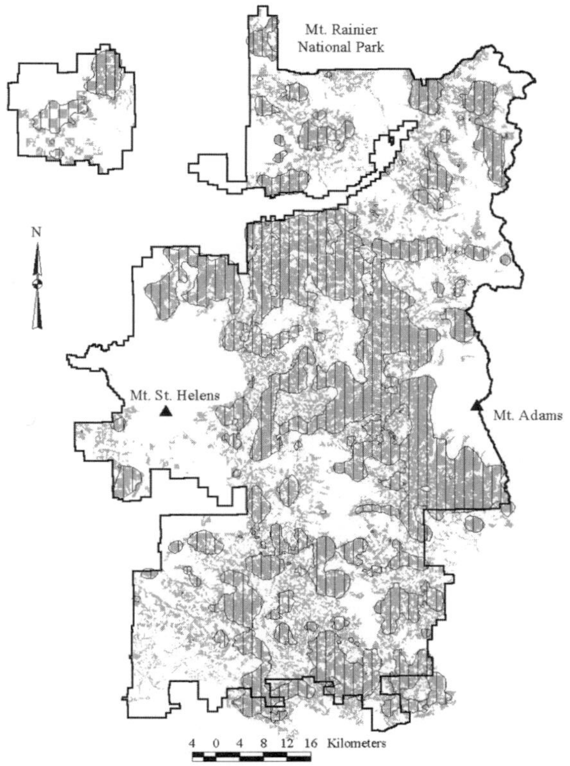

Figure 2. Extent of spotted owl suitable habitat on the Gifford Pinchot National Forest (solid gray pattern) and areas modeled as a high potential to support spotted owl nesting sites (crosshatch).

stands approximately 80-years-old and older, an average DBH >41 cm, one or more canopy layers, and an average canopy closure >40 percent.

Once the base GIS vegetation map displaying all spotted owl habitat attributes was established, it was then converted into a grid map using Spatial Analyst's *Convert to Grid* command (Figure 1, Step B). The *Convert to Grid* command involved overlaying the vector (polygon) base GIS vegetation map with a grid pattern. Each pixel in the grid is assigned a value (a whole number beginning with 1) representing the attribute it mostly overlays on the vector based map. Thus, the command created a Habitat Grid Map with pixel values of 1 for Dispersal; 2 for Foraging; 3 for Nesting; and, 4 for Unsuitable spotted owl habitat.

In addition to the Habitat Grid Map described above, a grid map depicting interior forest habitat is needed (Figure 1, StepB2). Interior forest habitat is defined here as an area of suitable northern spotted owl habitat beyond a specified edge distance from Unsuitable habitat. Creating an interior forest habitat grid map involved selecting all Unsuitable habitat that consisted of openings or early seral forest stands on the base GIS Vegetation Map and then executing Spatial Analyst's *Find Distance* command (Figure 1, Step B2). As with the *Convert to Grid* command, the *Find Distance* command overlays the vector based GIS Vegetation Map with a grid pattern. But, instead of

assigning pixel values based on overlaying polygon attribute information, pixel values in the Distance Grid Map depict the actual distance of the pixel from selected Unsuitable habitat. As a result, distance bands are created from the edges of unsuitable habitat, consisting of openings and early seral forests, into mid seral forests and suitable habitat. This allows for the elimination of edge-affected areas, defined by a specified edge width, to highlight interior forest habitat on the landscape. Also, by having distance bands radiating through unsuitable habitat that consist of mid seral forest, a GIS edge is not created between mid and late seral forests.

The display of forest interior patches is a better indicator of forest fragmentation because it quantifies landscape composition and configuration.[9] The higher the amount of fragmentation the less interior forest that is available. Research on forest landscape patterns around known spotted owl nest sites has indicated that the quantity of forest interior habitat patches is important for continued nest site use.[3, 6, 7, 8, 10]

The next step simplified the values of the Habitat Grid Map and combined the Habitat and Distance Grid Maps using the *Map Query* command (Figure 1, Step C1-2). The *Map Query* command allows for selecting specific values from one or more grid maps and combining them to create a new grid map with pixel values of 0 and 1. The value of 1 represents selected values from the Habitat and Distance Grid maps and the value 0 represents the rest of the Habitat and Distance Grid maps' extent. Using the *Map Query* command on the Habitat Grid Map, pixels valued as 2 (Foraging) and 3 (Nesting) were selected and queried creating a Suitable Habitat Grid map (Figure 1, Step C1). This created a grid map whose pixel value of 1 outlined northern spotted owl suitable habitat across the GPNF landscape (Figure 2).

To create a grid map displaying interior northern spotted owl suitable habitat, the *Map Query* command was used to combine selected values from the Habitat Grid map and Distance Grid map. Pixel values 2 (Foraging) and 3 (Nesting) from the Habitat Grid Map were selected and, from the Distance Grid Map, pixel values representing distances >100 meters from an edge were selected. After completing the query a new Interior Suitable Habitat Grid Map was created depicting only suitable habitat >100 meters from an edge. The 100 m forest edge buffer was based on work by Johnson.[6] He noted that spotted owl nest sites were predominately located >100 m from forest edges.

Step D involves summarizing or, in this case, totaling the number of pixels valued as 1 within a specified distance from each individual pixel for both the Suitable Habitat Grid Map and Interior Suitable Habitat Grid Map (Figure 1, Step D). This involves using the *Neighborhood Statistics* command in Spatial Analyst. The specified distance to evaluate on the GPNF was a 1.26 km radius (500 ha). Research has noted the importance of the quantity of northern spotted owl suitable habitat within an approximate 500 ha area surrounding a nest site. The quantity of suitable habitat within a 500 ha area appears to influence nest site location at landscape scales.[5, 7, 8, 10] In addition, this area receives concentrated use as a foraging site for the female and as a staging area for the owlets after they leave the nest but prior to dispersal.[4]

Two grid maps were created following steps D1 and D2: a Summed Suitable Habitat Grid Map and a Summed Interior Suitable Habitat Grid Map. The value for a pixel in each new grid map being the sum of the values for all pixels surrounding it within a 1.26 km radius.

Each pixel value in the Summed Suitable Habitat and Interior Suitable Habitat Grid Maps was then converted to display the total area in hectares of suitable habitat and interior suitable habitat that surrounds the respective pixel within the 1.26 km radius. To determine the area represented by the pixel value, the pixel value was multiplied by the area of the pixel. Back in steps B1 and B2 when the *Convert to Grid* and *Find Distance* commands were executed the user is prompted for pixel size. Thus, if a pixel size of 50 m was chosen: each pixel would represent an area of 2500 m^2 or about 0.25 ha. If a summed pixel value from the Summed Suitable Habitat Grid Map is 100 (100 pixels representing suitable habitat within 1.26 km), multiply 100 by 0.25 ha to obtain the area (25 ha) of suitable habitat within 1.26 km of that pixel.

With this understanding of the pixel values in the Summed Suitable Habitat and Summed Interior Suitable Habitat Grid Maps, they can be reclassified using the *Reclassify* command (Figure 1, Step E). The *Reclassify* command allows grouping of pixel values to represent ranges of specified values. The specified values are provided by the user and should, for purposes of delineating high potential nesting areas, represent values greater than the minimum area of the habitat parameters (e.g., suitable habitat and interior suitable habitat) shown to support northern spotted owl nesting. Reclassified pixel values need to be weighted when reclassified so that the grid maps can be joined in the final step. For the analysis on the GPNF, summed pixel values <150 ha in the Summed Suitable Habitat Grid Map were reclassified to a weighted pixel value of 1, while summed pixel values >150 ha were reclassified to a weighted pixel value of 2, creating a Reclassified Suitable Habitat Grid Map. Likewise, summed pixel values in the Summed Interior Suitable Habitat Grid Map representing <146 ha and >146 ha were reclassified to weighted pixel values of 1 and 2, respectively, creating a Reclassified Interior Suitable Habitat Grid Map.

In the example for the GPNF, 150 ha and 146 ha were the threshold values that delineated low potential nesting areas from high potential nesting areas. The selective use of areas containing a higher proportion of mature and old-growth forest (suitable habitat) by nesting spotted owls is supported in scientific literature.[2, 3, 6, 7, 8, 10] Bart and Forsman,[2] Bart[1] and Ripple et al.[10] have further shown that reproductive success may be depressed when the proportion of suitable habitat falls below 30 percent of a 1.26 km radius circle, as well as larger radius circles. The threshold value of 150 ha used to create the Reclassified Suitable Habitat Grid Map for the GPNF approximates

30 percent of the 500 ha area summarized in step D1. The threshold value of 146 ha used to create the Reclassified Interior Suitable Habitat Grid Map was derived from Johnson[6] and Ripple et al.[10] Johnson[6] stated that spotted owl nest sites in central Oregon were predominately greater than 100 m from the forest edge, indicating the importance of interior forest for nest site location. Furthermore, Johnson[6] found the majority of spotted owl responses to survey calls occurred in areas that contained >146 ha of interior suitable habitat. Johnson[6] defined interior forests as mature and old-growth stands buffered 100 m from an opening. In addition, Johnson[6] noted that areas with <146 ha of interior suitable habitat contained the majority of great horned owls. Carey et al.[3] also noted an increased occurrence of great-horned owls in fragmented landscapes as compared to larger intact, forested landscapes.

For the final step, step F, the Reclassified Suitable and Reclassified Interior Suitable Habitat Grid Maps were joined together using the *Map Calculator* command (Figure 1, Step F). This step involved adding the weighted pixel values of each overlaying pixel from each reclassified grid map and dividing by the number of grids being added together. This was a simple step since only two grids were combined. The grid with the most restricted area (a pixel value of 2 for each map) would be the result (Figure 2). Because only two grids were joined with weighted values of 1 and 2, the final Spotted Owl Nesting Potential Grid Map exhibited pixel values of 1 and 2. Pixels with a value of 1 on the Nesting Potential Grid Map represented an area with a low potential to support a spotted owl nest site due to inadequate amounts of suitable habitat and interior suitable habitat in a surrounding 500 ha area. Pixels with a value of 2 would represent an area with a high potential to support spotted owl nest sites due to sufficient quantities of suitable habitat and interior suitable habitat within the surrounding 500 ha (Figure 2). The Spotted Owl Nesting Potential Grid Map was converted to a shapefile using the *Convert to Shapefile* command, so it could be overlain with other GIS vector map layers.

Examples

An example of how the model's Spotted Owl Nesting Potential Grid Map could be used is as follows. A timber sale is proposed and the GIS vegetation map depicts the entire planning area with extensive coverage of suitable habitat for the spotted owl, though it has been subjected to past harvest activities. Spotted owl surveys have not been conducted and are not planned for the area. Therefore, it is not known where spotted owls are nesting and it must be assumed that all suitable habitats are occupied. Without further information, measures designed to minimize adverse impacts to spotted owls, such as timing restrictions on when logging could occur, would be implemented across the entire planning area. The impact of these restrictions could make the timber sale more difficult and costly to implement.

However, after running the model for the planning area, a wildlife manager could overlay the Spotted Owl Nesting

Potential Grid Map with the proposed project map. After assessing the degree of overlap of High Potential Nesting Areas to proposed project sites and, accounting for personal knowledge of habitat conditions and spotted owl use, the wildlife manager could suggest protective measures for the spotted owl be applied specifically to those areas within High Potential Nesting Areas. No protective measures to protect spotted owl nesting would be provided in Low Potential Nesting Areas. This process would allow more flexibility in planning and conducting the timber sale while still considering the needs of spotted owls. This is clearly shown by comparing the extent of the Spotted Owl Nesting Potential Grid Map to the GPNF GIS vegetation map of spotted owl habitat types (Figure 2). An analysis of the GPNF shows 107,030 ha, or approximately 49 percent of the total area of suitable spotted owl habitat (218,070 ha) shown in the GPNF GIS vegetation map was identified as having a high potential to support spotted owl nesting (Figure 2).

For a second example, another timber sale is planned in an area known to have supported spotted owls in the past. After selecting the general area in which to conduct the sale, the Spotted Owl Nesting Potential Grid Map could be used as a planning and project design tool. One alternative would be to place units that fall outside of areas depicted as having a high potential to support spotted owl nesting. A second alternative could be to place units in forest patches that constitute primarily edge, to maintain the highest quantity of interior suitable habitat that defines high potential spotted owl nesting areas. A third alternative would be to propose silvicultural prescriptions that would maintain the high quantity of suitable habitat. Such silvicultural prescriptions could include higher canopy retention levels and protection of large trees and snags. Thus, harvest units could be placed and designed to minimize impacts to both suitable habitat and the spotted owl.

As a final example, the Interior Suitable Habitat Grid Map derived in Step C2 of the model (Figure 1, Step C2) could be used to analyze effects of proposed projects on interior forests and the extent of forest fragmentation that may occur. Forest fragmentation is a common issue brought out during scientific and public review. By overlaying proposed project areas on the interior suitable habitat grid map, the amount of interior suitable habitat that would be lost or affected and the amount of edge created as a result of the proposed project could be easily assessed.

Conclusions

The High Potential Spotted Owl Nesting Area Model was designed for use on ArcView GIS software with the Spatial Analyst extension. ArcView is commonly used by numerous agencies to meet their GIS needs so there is no need to download a specific software or program and train for its use. All commands to run the iterative steps in the model are located in ArcView's default menus.

The only data requirement is a GIS vegetation map layer that contains sufficient forest stand attribute information to differentiate unsuitable and suitable spotted owl habitat. However, the model is flexible in that additional GIS map layers can be incorporated to further define suitable habitat. For example, a GIS layer showing barred owl, a competitively advantaged species over the spotted owl, distribution could be processed through each iterative step to develop a Reclassified Barred Owl Grid Map with weighted pixel values. This layer could then be joined with the Reclassified Suitable Habitat and Reclassified Interior Suitable Habitat Grid Maps under step F. Some iterative steps could be skipped to accommodate map layer modifications. The final GIS map can then be saved and used as an overlay for project planning.

It is recommended that any GIS map layer used in the model should be reviewed based on local knowledge, personal judgment, and field reconnaissance. It is also recommended that the model be run every time the GIS vegetation maps are updated.

Threshold levels used in the High Potential Spotted Owl Nesting Area Model to differentiate low and high potential nesting areas were derived from published scientific literature.[2, 3, 6, 7, 8, 10] These studies exhibited the common occurrence of a high proportion of suitable habitat and large suitable habitat patch sizes (interior habitat) around spotted owl nest sites.

To determine if modeled results were realistic, known spotted owl activity center locations on the GPNF were compared with the Spotted Owl Nesting Potential Grid Map to see how well they correlated. A Chi-square test of independence showed highly significant distributions with modeled areas containing both >30 percent suitable habitat ($p < 0.0001$) and >146 ha of interior suitable habitat ($p < 0.01$) within a 1.26 km radius circle. Still, more activity centers were located in areas depicting low potential for spotted owl nesting. This was because since the establishment of most activity centers, timber harvest had reduced suitable habitat below these thresholds.

Furthermore, scattered surveys for spotted owls on the GPNF have not established any nest sites or activity centers within areas modeled as low potential nesting areas. Three new activity centers, including a nest site, were all located within modeled high potential nesting areas. Another weakness involves the exclusion of contiguous suitable habitat stands along the periphery of high potential spotted owl nesting areas. This is because of the nature of the *Neighborhood Statistic* command. As neighborhood statistics are derived for each pixel, fewer pixels valued as habitat are summed as it goes further from a habitat patch. Therefore, the edges of a large suitable habitat patch would be delineated as low potential spotted owl nesting areas due to insufficient quantities of suitable and interior suitable habitat on one side, even though they are the edges of a large contiguous, interior patch.

Another similar situation involved the delineation of a circular high potential area that, when overlaid with the suitable habitat map, mostly consist of unsuitable habitat with suitable habitat located around the periphery. This occurs in highly fragmented areas. Since the *Neighborhood*

Statistics command summarizes each pixel regardless of whether it is suitable or unsuitable, the high potential area is situated over the landscape that maximizes the greatest amount of suitable habitat in a 1.26 km radius.

These two situations can be overcome by selecting suitable habitat polygons that intersect high potential spotted owl nesting areas, but only when the polygons do not cover more than approximately 100 ha. If polygons cover extensive areas, then buffering the high potential spotted owl nesting area designation by 0.25 mile appears to incorporate the majority of stands that extend beyond the boundary. It would be advised that biologists determine the status of suitable habitat stands along the periphery of modeled high potential spotted owl nesting areas before accepting any decision of it being a low potential spotted owl area.

Finally, this process need not be restricted to an analysis for spotted owls. Other habitat parameters for other species could be developed and modeled much like were done here. An array of similarly modeled spatial relationships of wildlife and their habitats could greatly assist in ecosystem level planning and project design.

Literature Cited

1. Bart, J. 1995. Amount of suitable habitat and viability of northern spotted owls. Conservation Biology 9:943-946.

2. Bart, J. and E. D. Forsman. 1992. Dependence of northern spotted owls on old-growth forests in the western USA. Conservation Biology 62:95-100.

3. Carey, A. B., S. P. Horton, and B. L. Biswell. 1992. Northern Spotted Owls: Influence of prey base and landscape character. Ecological Monographs 62:223-250.

4. Forsman, E. D., E. C. Meslow, and H. M. Wight. 1984. Distribution and biology of the spotted owl in Oregon. Wildlife Monographs 87:1-64

5. Hanson, E., D. Hays, L. Hicks, L. Young, and J. Buchanan. 1993. Spotted owl habitat in Washington: A report to the Washington Forest Practices Board. Final Report.

6. Johnson, D. H. 1993. Spotted owls, great horned owls, and forest fragmentation in the central Oregon Cascades. M.S. Thesis. Oregon State University, Corvallis, OR.

7. Lehmkuhl, J. F. and M. G. Raphael. 1993. Habitat pattern around northern spotted owl locations on the Olympic Peninsula, Washington. Journal of Wildlife Management 57:302-315.

8. Meyer, J. S., L. L. Irwin and M. S. Boyce. 1992. Influence of habitat fragmentation on spotted owl site location, site occupancy, and reproductive status in western Oregon. Progress Report.

9. McGarigal, K., and B. J. Marks. 1995. FRAGSTATS: Spatial pattern analysis program for quantifying landscape structure. General Technical Report PNW-GTR-351. USDA, Forest Service, Pacific Northwest Research Station, Portland, OR.

10. Ripple, W. J., P. D. Lattin, K. T. Hershey, F. F. Wagner and E. C. Meslow. 1997. Landscape composition and pattern around northern spotted owl nest sites in southwest Oregon. Journal of Wildlife Management 61:151-158.

23

Integrating Wildlife Species Habitat Goals and Quantitative Land Management Planning Processes

Pete Bettinger, Kevin Boston, John Sessions, & William C. McComb

Introduction

Land management planning is an iterative process. Goals are developed and adjusted during planning processes, and desired outcomes are displayed in a management plan. Wildlife habitat goals are becoming more common in management plans, and resource allocation models are used to attempt to resolve conflicts among management goals. For example, one goal may be to provide an adequate supply of habitat for one or more selected species, while another may be to provide a minimum amount of timber harvest volume. Quantitative land management planning techniques can be used to represent multiple resource goals, and to plan activities that are compatible with resource sustainability.

Quantitative planning approaches commonly have helped guide silviculture and the biological and economic management of forests. Using quantitative goals or objectives implies that desired conditions have been defined for a planning area. These goals and objectives may change over time, as societal, regulatory, or organizational goals change. For example, in the Pacific Northwest we have seen a shift during the last two decades from commodity production objectives to forest structure objectives on public lands. Quantitative planning methods can be applied to a wide range of landowner goals (e.g., social, biological, economic) to provide managers with management alternatives.

Planning should be viewed as a continuous and incremental process that assists managers in making decisions. The decisions faced by each organization could require short- or long-term plans, or strategic (forest-wide) or tactical (site-specific) plans. The potential limitations to planning generally fall into four categories: organizational, personnel, database, and technological. Despite the limitations, planning helps managers (1) make decisions, (2) establish credibility, (3) accomplish tasks efficiently, (4) achieve continuity in management, and (5) provide a standard for comparison against other plans and monitoring results.

The spatial arrangement of wildlife habitat and forest management activities is important for a number of reasons, such as adhering to regulatory rules, complying with organizational goals and policies, and maintaining aesthetic conditions. Forest regulations, for instance, are placing increasingly restrictive limits on the size and spatial relationships of harvest units.[6] As a result of managing forest land within regulatory frameworks, forest management planning now often requires multiple resource goals and the use of spatial rules for the selection of timber harvest units.[17]

In this chapter we illustrate how to develop and communicate wildlife habitat goals and provide an example of how they can be included in land management planning processes. To accomplish this, we first briefly describe a generic, yet typical, planning process. Next we describe several types of wildlife species habitat goals that could be used in natural resource planning. We then define several common terms used in quantitative management planning and illustrate their use with the species habitat goals. Several management plans are developed with these goals in mind. Finally, we discuss four important areas of planning and their influences on management planning.

Quantitative Management Planning Processes

We undertake planning processes every day. We decide what we want to do, express some impression of what the desired goals should be, examine a few ways to achieve the goals, pick an alternative, implement the alternative, and perhaps assess the results. Several examples come to mind: driving to work (how do I get there? when should I leave? which route do I take?); shopping for groceries (what do I need to purchase? what do I want to purchase? what is my budget? what groceries do I already have at home?); and writing a report (when is it due? how long should it be? what other tasks must I accomplish during this same period?).

Organizations may undertake planning processes less frequently, yet the steps in planning are similar, and certainly more formal. Formal planning processes can take many forms, but generally consist of the following tasks: (1) define the scope of the planning problem, (2) articulate future goals, (3) isolate alternative solutions to the problem, (4) set criteria for evaluating the alternatives, (5) evaluate the alternatives, (6) choose an alternative for the plan, (7) monitor the plan, and (8) reevaluate alternatives after implementation.

The nature of planning problems ranges from those with relatively simplistic goals (e.g., habitat management for ruffed grouse in aspen forests), to problems that may include many resources and goals over large areas (e.g., national forest planning). During planning processes, the goals, objectives, and constraints should become apparent, and may be based on an organization's mission, goals, or objectives. Goals include outcomes such as providing habitat for certain wildlife species. An objective may be to maximize the amount of habitat produced, timber volume produced, or profit, or to minimize costs to achieve habitat goals.

Constraints may include minimum harvest ages for timber stands or budget limitations. These parts of planning processes are very important and will be discussed in more detail later in this chapter. The isolation of alternatives and the determination of criteria for evaluating alternatives often involve the use of decision models. The objectives should be quantitative, providing criteria by which plans are judged and their implementation monitored.

Much of the remainder of this chapter focuses on three types of wildlife goals that could be used in planning problems. These goals are categorized as either spatial or nonspatial. After discussing the goals in general terms, we define common terminology that could be used throughout planning process discussions. This terminology is directed toward the professional involved with natural resource management.

Wildlife Species Habitat Goals

Species habitat goals can be either qualitatively or quantitatively defined. Whereas qualitative goals may be important for certain organizations, this chapter focuses on quantitative goals. Quantitative goals for wildlife species can be categorized in many ways: do the goals require a measurement of snags per unit area; do the goals involve land classification (uplands vs. riparian areas); are the goals expressed as linear or nonlinear equations? One categorization on which we concentrate is whether the goals require spatial information to measure their attainment. We describe three wildlife species habitat goals below; two of them require spatial information in their computations. Nonspatial goals may consist of assembling a number of acres in certain age classes, or strata. Spatial goals may include configurations such as requiring minimum patch sizes, or adjacent habitat types.

Nonspatial Goals

These types of planning goals generally are based on achieving the *amount* of some resource in a planning area. For example, goals could be developed with the criteria that some amount of habitat, such as mature closed canopy forest, will be achieved. Examples of nonspatial goals related to the type of nesting habitat required for four wildlife species in the Pacific Northwest include:
1. Sharp-shinned hawk prefers 25- to 50-year-old even-aged conifer stands with a mean tree density of 477 trees/acre (1,180 trees/ha).

2. Cooper's hawk prefers 30- to 70-year-old even-aged conifer stands with a mean stand density of 367 trees/acre (907 trees/ha).
3. Northern goshawk prefers ≥150-year-old conifer stands with a mean tree density of 195 trees/acre (482 trees/ha).
4. Red tree vole prefers old-growth forests that are ≥195 years old.

A planning team would define how much area with these characteristics is desired (e.g., 100 acres, 1000 acres, as much as possible) in order to use these goals as objectives or constraints in a quantitative planning process, although they still could simply be measured. A relaxation of the tree density requirements would permit increased flexibility in the achievement of forest goals. For example, three of the goals could be restated as:
1. Sharp-shinned hawk prefers 25- to 50-year-old even-aged conifer stands with a mean tree density of 405-525 trees/acre (1000-1300 trees/ha).
2. Cooper's hawk prefers 30- to 70-year-old even-aged conifer stands with a mean stand density of 324-405 trees/acre (800-1000 trees/ha).
3. Northern goshawk prefers ≥150-year-old conifer stands with a mean tree density of 174-214 trees/acre (430-530 trees/ha).

Spatial Goals: Minimum Patch Size

Some forest planning goals use spatial characteristics of the landscape to determine their value to particular wildlife species. One type of spatial goal might be to require patches to be a minimum size before they contribute positively toward the achievement of habitat. For example, the varied thrush, winter wren, and Hammond's flycatcher may need intact stands of single story open canopy forests ≥49.4 acres (≥20 ha) in size.

From a planning perspective, the definitions of single story open canopy forest are important. While structural attributes of forests (i.e., variation in tree diameter classes, level of coarse woody debris, vertical complexity, etc.) could be used to define these conditions, we will assume for discussion purposes that these are forested stands ≥90 years old. The minimum size requirements could then be met with individual stands that are ≥49.4 acres (≥20 ha) and ≥90 years old. The goal could also be met with groups of contiguous stands, where the sum of the area is ≥49.4 acres (≥20 ha), and all stands have forests ≥90 years old. Finally, the planning team would need to determine, for the analysis area, how many of these patches would be required (e.g., provide at least 3 distinct patches/mile² [7.7/km²]), or how much area with these characteristics is required (e.g., 100 acres, 1000 acres, as much as possible).

Spatial Goals: Adjacency Requirements

Goals for wildlife species may also be developed that indicate, for optimal benefit to a particular species, that one type of habitat should be placed next to another. For example, the great gray owl prefers early seral stage forests (clearcuts) for foraging, yet these areas should be adjacent to single story open canopy forests containing snags or large trees with broken tops. Alternatively, great gray owls may show a preference for older forest stands adjacent to permanent meadows.

Further clarification of this goal is important. For example, do all early seral stage forests need to be adjacent to single story open canopy forests, or just some portion of them? Conversely, the amount of single story open canopy forests adjacent to early seral stands might be of interest. At some point a quantitative measure or threshold must be defined before the goal can be used in a planning process. We will proceed with a rather simple assumption: acceptable habitat for great gray owls requires ≥49.4 acres (≥20 ha) of single story open canopy forest adjacent to ≥24.7 acres (≥10 ha) of early seral habitat.

Placing Wildlife Species Habitat Goals in a Management Planning Context

The process we suggest in the development of a management plan consists of describing the planning problem in words, then converting the descriptions to quantitative relationships. In the early stages of planning it may be beneficial for those involved in developing management plans to identify the decision-maker. The decision-maker is a person or group who has designated stewardship responsibilities, who chooses the policies or actions that will be undertaken, and thereby accepts responsibility for the choice of management plan.

The applied economics and management science fields have defined several terms that are commonly associated with quantitative management planning processes. We place the goals we discussed above into context using some of these terms. First we will restate the goals, and then develop goal criteria. Then we will determine the activities and decision variables that could contribute to goal achievement. Finally, we describe the objective function and constraints of the resulting problems.

Goals. Goals are the outcomes of a planning process; they are relevant to the planning problem and are expected to be achieved (e.g., increase habitat for marbled murrelets). These outcomes are useful when they can be stated in quantitative terms, for example: (1) to provide the greatest amount of varied thrush habitat, or (2) to provide the greatest amount of habitat for all species, each species given equal value.

Goal Criteria. Goal criteria measure goal achievement, and help to determine whether or not the activities employed are, or will, move a landscape toward or away from goals. For example: (1) goal: to provide the greatest amount of varied thrush habitat, and (2) goal criterion: the amount of land considered to be varied thrush habitat in each planning period. Multiple goals could tend to compete against one another (e.g., sharp-shinned hawk and red tree vole). With multiple goals, it may be hard to determine whether a particular management plan achieves, in sum, more of the multiple goals than another plan does, and if so, by how much. How the multiple goals and goal criteria are designed is a major factor in evaluating which goal may be more influential in the development of a management plan.

Activities and Decision Variables. Activities include the suite of actions that can be used to achieve the goals (e.g., silvicultural activities), and are essentially the types of projects that can be implemented to achieve the goals. Management activities could include a broad variety of silvicultural prescriptions. For example, we could thin a timber stand at age 30 years, and remove 25 ft^2/acre (5.7 m^2/ha) of tree basal area, or, we could remove 30 ft^2/acre (6.9 m^2/ha) of tree basal area, or, we could create 10 snags/acre (24.7 snags/ha), etc. Fertilization treatments, the creation of snags, leaving trees on a harvest site at the time of regeneration harvest, and other silvicultural treatments are all activities that can be combined to produce a single forest management prescription.

Decision variables represent the level or levels to which each activity is assigned (e.g., land units, roads). For example, an activity (thinning) may be applied to all, or part of, a decision variable (a timber stand), to open up the canopy and promote vertical structure complexity.

Objective Function. An objective function is a mathematical statement of a goal, or goals, combining the criterion for measurement of goal achievement and the decision variables specified for the problem. Although many problems in business and engineering adopt a single goal (often a statement to maximize profit or minimize cost), a broader definition is often useful in wildlife management planning. The solution with the best objective function value, all else being equal, will be chosen as the best solution. The objective function value serves as a decision guide for sorting through the possible feasible decision variable combinations to determine what combination is best according to the criteria used to evaluate the objective function.

Goals for wildlife management planning can be expressed using one of several forms: (1) maximize or minimize 1 goal with the other goals expressed as constraints, (2) minimize deviations from ≥1 goal with other goals expressed as constraints, or (3) a hybrid or combination of the previous 2 forms. In form 1, the goals represented as constraints *must* be met and the goal in the objective function might be viewed as the residual goal that is achieved to the extent possible given the goals represented as constraints. In form 2, the objective function is expressed as the sum of the deviations from targets for ≥1 goal and the remaining goals are expressed as constraints. Using form 2 the decision-maker must develop targets for the goals and the optimization attempts to minimize the deviations between the targets and the goal achievement while satisfying any remaining goals represented as constraints. Form 2 is sometimes referred to as goal programming. Goal programming can involve either linear or nonlinear programming models. In Form 3 the objective function maximizes or minimizes 1 goal minus the sum of the deviations from targets of ≥1 goal. The remaining goals are represented as constraints.

Constraints. Anything that limits the achievement of a goal (i.e., a minimum amount of marbled murrelet habitat must be maintained) is a constraint. Usually constraints arise from resource limitations (e.g., cannot schedule activities to more than what you manage), decision-maker goals (e.g., need to produce X amount of timber volume/year), or externally imposed policies or regulations (e.g., do not harvest within Y feet of an owl location). A management plan is assumed to be feasible when all constraints are met. Of course, the level of the objective function after meeting the constraints may not be acceptable, and new activities and a revision of the constraints may be needed.

Next we will define the goals, goal criteria, activities, decision variables, objective function, and constraints for a planning problem that uses the species goals we have described above. Then we will make some assumptions about a planning exercise for a hypothetical forest. Finally, we illustrate several management plans that emphasize the achievement of the maximum amount of each of the three types of goals.

Nonspatial Goals

The nonspatial goals we have defined for the four wildlife species in the Pacific Northwest are to provide the greatest amount of habitat for each over the length of the planning horizon (Table 1). The criteria to evaluate the goals are to measure the amount of land considered to be habitat for each species, in each planning period, and divide by the total number of acres on the forest, to arrive at a percentage of land in habitat for each species. The tree age and species composition on these land units can be used to define whether habitat goals have actually been achieved for each species. To simplify our example, we assume that the age of the stand is the most important factor, and assume that the composition of trees within the stand meet the requirements for each species. A more detailed analysis could incorporate growth and yield estimates to determine the estimated number of trees/acre (ha), or the structural composition of stands (to determine whether they meet a criterion for old growth).

The activities we will illustrate include just two options: regeneration or no harvest. Activities that affect the status of decision variables, and either contribute positively or negatively to the condition of the decision variables (with respect to the goals) could include management actions such as thinning, regeneration harvests, fertilization treatments, and snag creation, although here we model regeneration harvests only. The decision variables for sharp-shinned hawk, Cooper's hawk, northern goshawk, and red tree vole are units of land allocated to various age classes. If the decision variables and the activities are combined, we can track the amount of land, within a land unit, that is scheduled to receive a silvicultural treatment. This implies that not all of a timber stand, for example, could be harvested. We assume, however, that if a land unit is scheduled for an activity, the activity will occur on the entire land unit (i.e., it is a binary decision). This assumption makes choosing the size of the average timber stand important.

The objective function for this problem is to maximize the amount of land, for all four species, that can be

Table 1. Problem formulation for the nonspatial species habitat goals.

Goals:
To provide the greatest amount of sharp-shinned hawk habitat over time.
To provide the greatest amount of Cooper's hawk habitat over time.
To provide the greatest amount of northern goshawk habitat over time.
To provide the greatest amount of red tree vole habitat over time.

Goal criteria
Amount of land considered to be sharp-shinned hawk habitat in each planning period.
Amount of land considered to be Cooper's hawk habitat in each planning period.
Amount of land considered to be northern goshawk habitat in each planning period.
Amount of land considered to be red tree vole habitat in each planning period.

Activities and decision variables
Silvicultural activities include no harvest or regeneration harvest.
Decision variables are individual land units allocated to silvicultural activities.

Objective function
Maximize the percent of land considered to be habitat for each species in each planning period.

$$\text{maximize} \left(\sum_{j=1}^{10} \sum_{i=1}^{n} \sum_{k=1}^{m} A_i H_{ijk} \right) / \left(\sum_{k=1}^{m} A_i \right)$$

Where: i = land unit
  j = time period
  k = wildlife species
  A_i = acres in land unit i
  H_{ijk} = habitat for species k, on land unit i, during planning period j (binary, 0 for no, 1 for yes). The determination of habitat is a function of average tree species age.

Constraints:
Only 1 regeneration harvest can occur during the planning horizon.

$$\sum_{j=1}^{10} X_{ij} = 1 \quad \forall i$$

Where: X_{ij} = binary (0 for no, 1 for yes) variable indicating a regeneration harvest on land unit i during planning period j

Timber volume produced must be above a minimum volume.

$$\sum_{j=1}^{10} A_i X_{ij} V_{ij} \geq \text{minimum harvest value} \quad \forall j$$

Where: V_{ij} = timber volume on land unit i during planning period j

considered habitat for each species, while satisfying the constraints (Table 1).

Spatial Goals:
Minimum Patch Size

The minimum patch size spatial goals for varied thrush, winter wren, and Hammond's flycatcher are to provide the greatest amount of habitat over time (Table 2). The criteria we use to measure these goals consist of measuring the percentage of land, in each planning period that meets the habitat requirements. The activities and decision variables are similar to those described for nonspatial goals. The objective function is also similar to that of nonspatial goals, yet only maximizes the percentage of land in patches ≥49.4 acres (≥20 ha), and ≥90 years old. The definition of habitat could, in this case, be considered a constraint, because the aggregation of land units ≥90 years old may be required before they are (in aggregate) ≥49.4 acres (≥20 ha) in size. The other constraints follow those of the nonspatial goals.

Spatial Goals:
Adjacency Requirements

The adjacent patch spatial goal for the great gray owl is to provide the greatest amount of habitat over time (Table 3). The criteria we use to measure these goals consist of measuring the percentage of land, in each planning period that meets the habitat requirements. The activities and decision variables are similar to those described for nonspatial goals. The objective function is also similar to that of nonspatial goals, yet only maximizes the percentage of land in patches ≥49.4 acres (≥20 ha), and ≥90 years old, and adjacent to patches ≥24.7 acres (≥10 ha) and ≥10 years old. The definition of habitat could, in this case, be considered a constraint, because to be eligible to be considered, the aggregation of land units ≥90 years of age may be required before they are (in aggregate) ≥49.4 acres (≥20 ha) in size, and similarly an aggregation of patches may be required for early seral forested stands before the goal can be achieved. In addition, we add one more level of complexity to this problem by limiting the size of clearcuts to ≥120 acres (≥48.6 ha). We add this last constraint to illustrate an increasing complexity in planning goals. The other constraints follow those of the nonspatial goals.

Developing a Management Plan

We will now develop a management plan with the wildlife goals we have developed in the preceding sections. Described below are the assumptions we use regarding a hypothetical forest landscape, including a description of the initial wildlife habitat classes, how habitat types transition after regeneration harvests, and how we make decisions concerning the activities that are included in the management plan.

Case Study Landscape

We use a hypothetical forest (Figure 1) to illustrate the implementation of the wildlife species habitat goals. This forest is 2,500 acres (1,012 ha), and has an initial age class distribution illustrated in Figure 2. There are 74 analysis

Table 2. Problem formulation for the minimum patch size species habitat goal.

Goals:
To provide the greatest amount of varied thrush habitat over time.

Goal criteria:
Amount of land considered to be varied thrush habitat in each planning period.

Activities and decision variables:
Silvicultural activities include no harvest or regeneration harvest. Decision variables are individual land units allocated to silvicultural activities.

Objective function:
Maximize the percent of land considered to be habitat for each species in each planning period.

$$\text{maximize} \left(\sum_{j=1}^{10} \sum_{i=1}^{n} \sum_{k=1}^{m} A_i H_{ijk} \right) \Big/ \left(\sum_{k=1}^{m} A_i \right)$$

Where: i = land unit
  j = time period
  k = wildlife species
  A_i = acres in land unit i
  H_{ijk} = habitat (binary, 0 for no, 1 for yes) for species k, on land unit i, during planning period j. The determination of habitat is a function of average tree species age and size of contiguous habitat.

Constraints:
Only 1 regeneration harvest can occur during the planning horizon.

$$\sum_{j=1}^{10} X_{ij} = 1 \quad \forall i$$

Where: X_{ij} = binary (0 for no, 1 for yes) variable indicating a regeneration harvest on land unit i during planning period j

Timber volume produced must be above a minimum volume.

$$\sum_{j=1}^{10} A_i X_{ij} V_{ij} \geq \text{ minimum harvest value} \quad \forall j$$

Where: V_{ij} = timber volume on land unit i during planning period j

Habitat patches must be larger than a minimum size.
H_{ijk} = 1 if size of contiguous habitat of ≥90 year-old forest is ≥49.4 acres (≥20 ha), otherwise H_{ijk} = 0. The determination of habitat is made using an area restriction model, similar to that described in Murray (1999), which is a recursive function that evaluates all adjacent units y to unit i, and all adjacent units z to unit y, and so on.

units (polygons) defining the forest, with an average size of 33.8 acres (13.7 ha). The forest is considered to be primarily a conifer-dominated forest, with hardwoods dominating in the riparian areas.
Management Activities. As mentioned earlier, we simplify the management choices for this discussion. The

Table 3. Problem formulation for the adjacent patch species habitat goal.

Goals:

To provide the greatest amount of great gray owl habitat over time.

Goal criteria:

Amount of land considered to be great gray owl habitat in each planning period.

Activities and decision variables:

Silvicultural activities include no harvest or regeneration harvest. Decision variables are individual land units allocated to silvicultural activities.

Objective function:

Maximize the percent of land considered to be habitat for each species in each planning period.

$$\text{maximize} \left(\sum_{j=1}^{10} \sum_{i=1}^{n} \sum_{k=1}^{m} A_i H_{ijk} \right) / \left(\sum_{k=1}^{m} A_i \right)$$

Where: i = land unit

j = time period

k = wildlife species

A_i = acres in land unit i

H_{ijk} = habitat (binary, 0 for no, 1 for yes) for species k, on land unit i, during planning period j. The determination of habitat is a function of average tree species age and size of contiguous habitat.

Constraints:

Only one regeneration harvest can occur during the planning horizon.

$$\sum_{j=1}^{10} X_{ij} = 1 \quad \forall i$$

Where: X_{ij} = binary (0 for no, 1 for yes) variable indicating a regeneration harvest on land unit i during planning period j

Timber volume produced must be above a minimum volume.

$$\sum_{j=1}^{10} A_i X_{ij} V_{ij} \geq \text{minimum harvest volume } \forall j$$

Where: Vij = timber volume on land unit i during planning period j

Habitat patches must be of certain sizes, and adjacent to each other.

H_{ijk} = 1 if size of contiguous habitat of ≥90-year-old forest ≥49.4 acres (≥20 ha), is adjacent to a contiguous area of ≥10-year-old forest that is ≥24.7 acres (≥10 ha) (or vice versa), otherwise H_{ijk} = 0. The determination of habitat is made using two area restriction models, similar to that described in Murray (1999), which are recursive functions that evaluate all adjacent units y to unit i, and all adjacent units z to unit y, and so on.

Regeneration harvests must be equal to or smaller than a maximum size. An area restriction model is used to control the size of regeneration harvests (Murray 1999).

$$X_{ij} \sum_{y \in N_i \cup S_i} A_i X_{yj} \leq \text{maximum clearcut size}$$

Where: X_{yj} = binary (0 for no, 1 for yes) variable indicating a regeneration harvest on land unit y during planning period j

N_i = set of land units adjacent to unit i

S_i = subset of regenerated land units containing all units adjacent to the neighbors of land unit i, and all land units adjacent to neighbors of neighbors, and so on.

timing of the regeneration harvest will be the only silvicultural choice. The quantitative management planning technique will determine when to apply the regeneration harvest. Of course, one choice is that no harvest will occur during the planning horizon. In summary, we will allow the following types of management activities: (1) regeneration harvest in any land unit with an age ≥40 years; and (2) grow only, no harvest activity.

Vegetation Transition after Regeneration Harvest. In order to emulate a dynamic forested landscape, we could develop and use vegetation type transition probabilities, which we could then use to determine a new wildlife habitat type after harvest. The decision on vegetation type transitions would be based on such factors as previous forest type, and distance from the stream system. We have chosen, however, to use very simple transition rules for this chapter's examples. If regeneration harvests occur, we will assume that the resulting forest will return to the vegetation type that existed prior to harvesting. This is, of course, a simplification of forest succession, especially if intensive stand establishment activities occur.

Length of Planning Horizon. Planning horizons vary from a few years to several decades, depending on the objectives of the organization(s) guiding the planning process. Here, we assume that the planning horizon will be 50 years, and that the analysis will be divided into ten 5-year planning periods.

Activity Scheduling Considerations. Most forest resource organizations have developed or adapted mathematical programming techniques to enable them to produce management plans. On public lands in the western United States, land management planning has been increasingly emphasizing both economic efficiency (or commodity production) and wildlife goals,[3] such as the goals we have described that require the achievement of certain levels of wildlife habitat. Forest planning algorithms that optimize the spatial arrangement of forest resources to meet a set of management goals vary from the more traditional optimizations techniques, such as linear or mixed integer programming,[10] to heuristic techniques such as binary search and simulated annealing.[15] These types of problems are generally considered to be difficult to solve with traditional mathematical programming techniques.

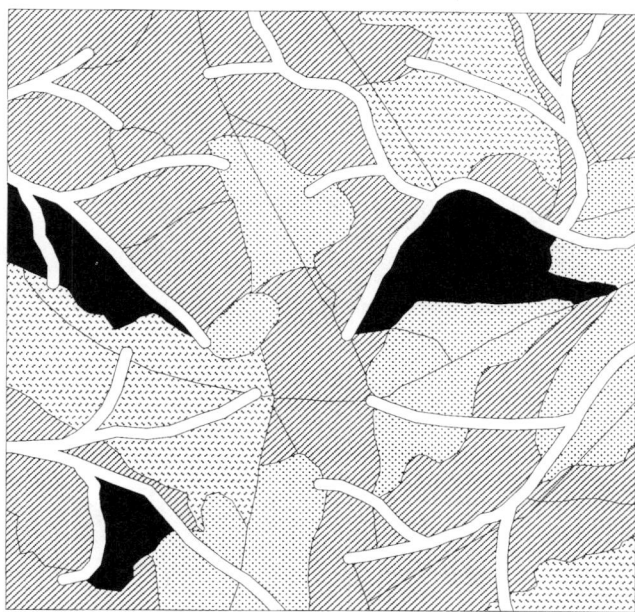

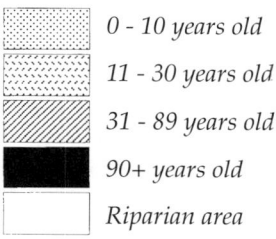

0 - 10 years old

11 - 30 years old

31 - 89 years old

90+ years old

Riparian area

Figure 1. The hypothetical forest that is used in the planning exercise.

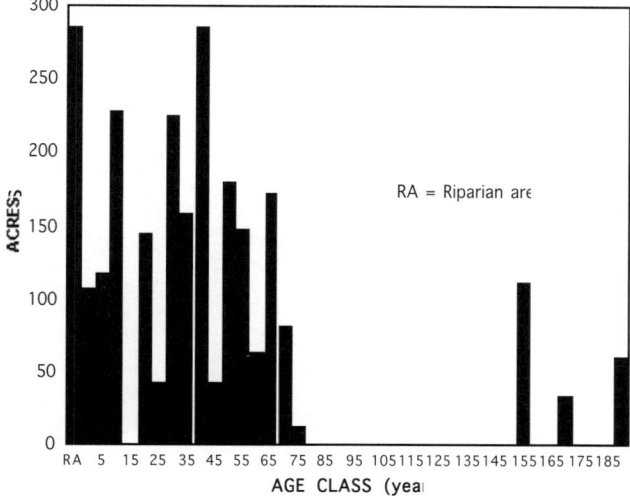

Figure 2. Age class distribution of the hypothetical forest

Table 4. A general heuristic procedure.

Step	Process
1	Number all land units from 1 to n.
2	Number all activities j eligible for each land unit i from 1 to k_i.
3	Set the initial value of the objective function to a large negative number.
4	Choose a land unit randomly by choosing a random number between 1 and n.
5	Choose an activity to assign to that land unit by choosing a random number between 1 and k_i.
6	Calculate the value of the objective function.
7	If this value is better than the last, continue to Step 10, otherwise return to Step 4.
8	Update the solution by incorporating the activity from Step 5.
9	Have we reached a stopping criterion? In other words, has the objective function value stabilized, or have we run out of time?
10	If we want to continue, return to Step 4, or else stop.

For example, mixed integer and integer programming techniques have been used to produce management plans with adjacency concerns, but these types of techniques have severe limitations (directly related to problem size) when applied to large combinatorial problems.[13]

The potential types of mathematical programming techniques that are available to develop management plans with spatial goals are therefore limited, because of the nonlinear, and integer nature of the resulting planning problem. This should not, however, discourage the use of complex habitat goals. Heuristic programming techniques have been applied to traditional forest harvest scheduling problems[11] as well as to forest transportation problems,[15, 16, 18, 20, 21] wildlife conservation and management,[1, 4, 9] aquatic system management,[5] and biological diversity.[12] Although a heuristic technique does not guarantee that a global optimum solution can be identified for a particular problem, it can produce a feasible, efficient solution to a complex problem in a reasonable amount of time.

The nature of the planning goals will be a major factor in determining how difficult it may be to perform planning exercises with each technique. This chapter does not delve into the differences between these techniques, but on the implementation of the goals. We have chosen a heuristic programming technique (tabu search) to use as the demonstration tool to help develop a management plan. The general structure of tabu search has been previously described.[2, 5, 7, 8] A traditional optimization technique (e.g., integer programming) can be used to develop management plans for the nonspatial goals, however the use of traditional optimization techniques is limited for more spatially complex planning problems. For comparison, we developed a management plan for the nonspatial goals using an integer programming technique as well as tabu search, and found the tabu search produced a result which was within 0.02% of the integer programming result.

Generally, heuristic techniques operate on a set of rules (see Table 4). Rule-based procedures allow the development of a solution by moving from 1 location in a "solution space" to another; in our example, this happens by a single change in the status of a land unit. This procedure will lead to "good" solutions, but may become stalled at local optima and may not be able to reach other areas of the solution space, where even better solutions might be found. These procedures can be modified to allow the acceptance of changes (in solutions) that lead temporarily to decreases in the objective function value. Procedures like this allow movement out of local optima of the solution space, and thus allow the search process to explore other arrangements of activities across a landscape, which may prove to be better overall. Tabu search, simulated annealing, and Monte Carlo simulation[19] all have more sophisticated rules to allow search processes to do just that.

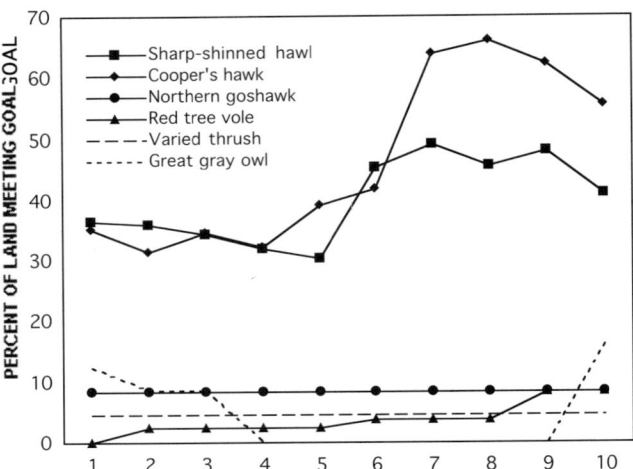

Figure 3. Results from maximizing nonspatial wildlife species habitat goals.

Management Plan Alternatives

The management plans that we will illustrate include the following three scenarios:

1. Maximize the nonspatial goals for the sharp-shinned hawk, Cooper's hawk, northern goshawk, and red tree vole, subject to the following: minimum harvest age constraints (40 yr), timber volume goals (3 million board feet per 5-yr period [approximately 19,000 m³ per 5-yr period]), and only one regeneration harvest per land unit during the planning horizon. Both of the spatial goals will be measured in a posterior (after-the-fact) manner on the resulting "best" solution found for the nonspatial goals.

2. Maximize the spatial goal requiring a minimum patch size for the varied thrush, subject to the following: minimum harvest age constraints (40 yr), timber volume goals (3 million board feet per 5-yr period [approximately 19,000 m³ per 5-yr period]), and only one regeneration harvest per land unit during the planning horizon. Both the nonspatial goals and the spatial goal requiring adjacent patch types will be measured in a posterior (after-the-fact) manner on the resulting "best" solution found for this spatial goal.

3. Maximize the spatial goal requiring adjacent patch types for the great gray owl, subject to the following: minimum harvest age constraints (40 yr), timber volume goals (3 million board feet per 5-yr period [approximately 19,000 m³ per 5-yr period]), and only one regeneration harvest per land unit during the planning horizon. In addition, clearcut sizes must be ≥120 acres (48.6 ha). Both the nonspatial goals and the spatial goals requiring minimum patch sizes will be measured in a posterior (after-the-fact) manner on the resulting "best" solution found for this spatial goal.

In addition to illustrating increasingly complex wildlife goals, these three plans will serve to illustrate the trade-offs planners encounter when the emphasis of a plan is

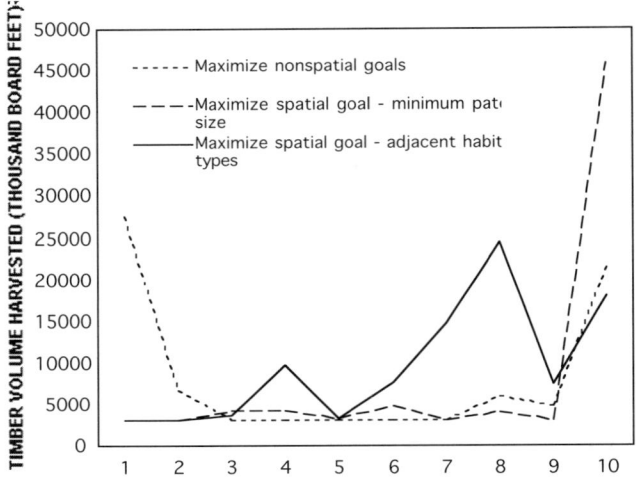

Figure 4. Timber volume produced for three planning problems.

Figure 5 (right). Cooper's hawk habitat (shaded area) in the initial landscape condition (top), after period 5 (middle), and after period 10 (bottom), when nonspatial goals are maximized.

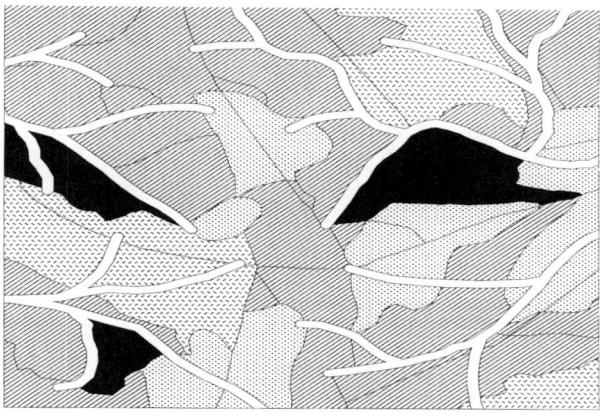

0 - 10 yrs old
11 - 30 yrs old
31 - 89 yrs old
90 + yrs old
Riparian area

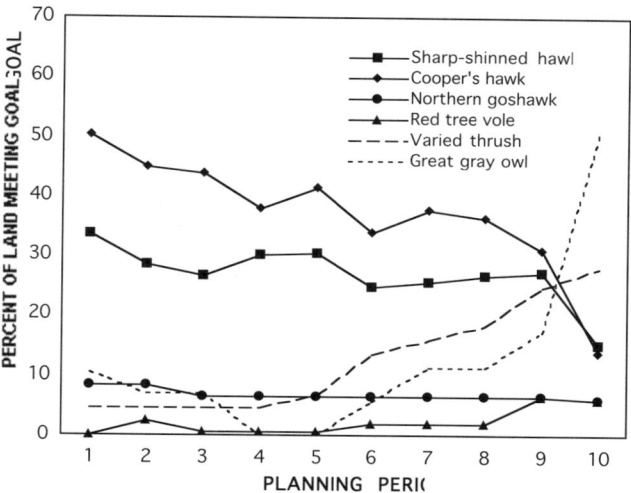

Figure 6. Results from maximizing the spatial goal requiring minimum patch sizes for varied thrush habitat.

placed on goals that may not necessarily be complementary, but perhaps competitive.

Results from Case Study

We developed feasible solutions to the three planning problems described earlier in this chapter. The solutions can be considered good feasible solutions to the problem, but not necessarily optimal solutions, since heuristic techniques do not guarantee optimal solutions will be located during the search process.

Maximizing the Nonspatial Goals

When we used the heuristic technique to maximize the nonspatial goals, we found that the percentage of land in sharp-shinned hawk and Cooper's hawk habitat was quite high, averaging 40% and 46%, respectively and tended to increase over the last half of the planning horizon (Figure 3). Northern goshawk and red tree vole habitat types averaged 8% and 4% per period, respectively, with the goshawk goal remaining constant and the red tree vole goal increasing gradually over time. The percentage of land in varied thrush habitat (spatial goal requiring minimum patch sizes) remained constant over time, and averaged 4.5% of the landscape, whereas the percentage of land suitable for great gray owl habitat (spatial goal requiring adjacent habitat types) was relatively high in the first and last periods, yet was negligible in the intermediate periods, as older forest patch size was limited and little harvesting occurred (creating few young seral forests. The average great gray owl goal was 4.5%. Timber volume per 5-year planning period averaged 8.2 million board feet (approximately 51,500 m^3) (Figure 4), although it ranged from about 3 million (approximately 18,900 m^3) to about 27.6 million board feet (approximately 173,900 m^3) per period. The large harvest area in the first period was beneficial to both the sharp-shinned and Cooper's hawk goals in the later periods. The increase in harvest volume in period 10 had little effect on any of these goals,

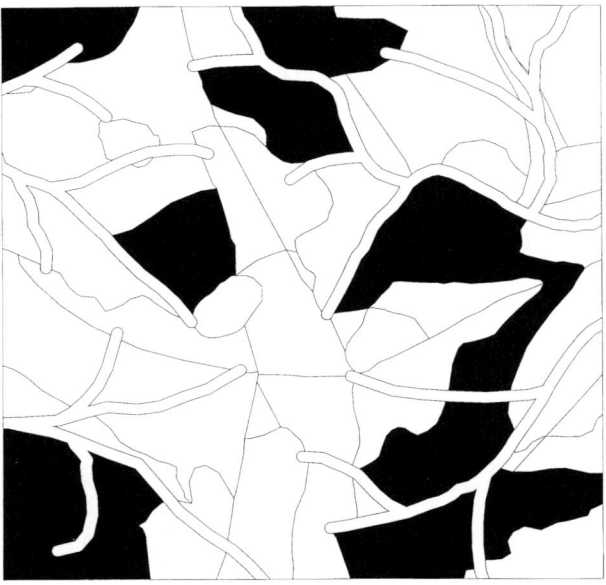

since the stands harvested were greater than 70 and less than 150 years old. The spatial arrangement of Cooper's hawk habitat over time is illustrated in Figure 5.

Maximizing the Spatial Goal Requiring Minimum Patch Sizes

When we used the heuristic technique to maximize the spatial goal requiring minimum patch sizes, we found that the average percentage of land in sharp-shinned hawk and Cooper's hawk habitat decreased about 10%, and dropped dramatically in period 10. Fewer stands were harvested in period 1 of this scenario than in the previous scenario, thus as the forest gets older (>70 yrs) these habitat goals decline. Their level of goal achievement is illustrated in Figure 6. The percentage of land in varied thrush habitat (spatial goal requiring minimum patch sizes) gradually increased over time, and was about 28% in period 10 of this scenario. Therefore about 23% more land is in this habitat type during period 10 than in the previous scenario, because we attempted to maximize the amount of varied thrush habitat here. The percentage of land in great gray owl habitat (spatial goal requiring adjacent habitat types) increased dramatically in period 10, as more stands were clearcut, yet averaged about 10% per period. Timber volume per 5-year planning period averaged 7.9 million board feet (approximately 49,500 m^3) (Figure 4), and ranged from 3 million (approximately 18,900 m^3) to 45.9 million board feet (approximately 288,600 m^3) per period. Harvest volume is quite high in the last period because some cutting here did not constrain the achievement of varied thrush habitat, and because we did not control harvest levels with a maximum harvest level constraint. The spatial arrangement of varied thrush habitat is illustrated in Figure 7.

Maximizing the Spatial Goal Requiring Adjacent Patch Types

When we used the heuristic technique to maximize the spatial goal requiring adjacent patch types, we found that the percentage of land in sharp-shinned hawk and Cooper's hawk habitat were lower than when the nonspatial goals were maximized (23 and 32%, respectively), with both generally declining over time (Figure 8). Northern goshawk, and red tree vole habitat types (the nonspatial goals) were approximately 50-70% lower here than when these nonspatial goals were maximized. The percentage of land in varied thrush habitat (spatial goal requiring minimum patch sizes) gradually increased over time, yet was about 1% lower here than when we attempted to maximize this spatial goal. The percentage of land in great gray owl habitat averaged 27% over the planning horizon, with dramatic increases in the last 3 planning periods. One reason for

Figure 7 (left). Varied thrush habitat (shaded area) in the initial landscape condition (top), after period 5 (middle), and after period 10 (bottom), when the spatial goal requiring minimum patch sizes is maximized.

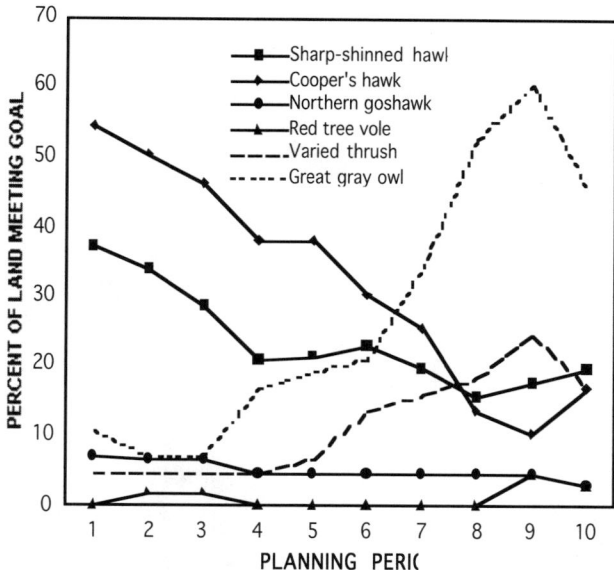

Figure 8. Results from maximizing the spatial goal requiring adjacent patch types for great gray owl habitat.

these patterns is that the planning model tries to develop both old and young forests simultaneously, so it allows some of the older stands to continue to grow, while the younger ones are harvested between the ages of 40 and 50, in order to create as much early seral forest as possible. Therefore, the amount of land in Cooper's hawk habitat types (stands between the ages of 30 and 70 yr) gradually decreases over time. The spatial arrangement of great gray owl habitat is illustrated in Figure 9. Timber volume per 5-year planning period averaged 9.5 million board feet (approximately 60,000 m³) (Figure 4), and ranged from 3 million (approximately 18,900 m³) to 24.6 million board feet (approximately 154,500 m³) per period. Harvest volume is high in the last few periods because some cutting is required to create early seral patches to complement the older forest patches, and because we did not control harvest levels with a maximum harvest level constraint. The landscape that results at the end of the planning horizon is much different than those that result from maximizing the other goals (Figure 10).

Discussion

Formulating management plans with a variety of goals (e.g., economic and wildlife species habitat) can be challenging. We have illustrated a process in which wildlife species habitat goals can be identified, and criteria for their measurement developed. This type of process can be used by planning teams to facilitate the incorporation of these goals into planning. The complexity of the goal criteria should not be a barrier to this process, as we have shown that there are techniques to incorporate nonlinear

Figure 9 (right). Great gray owl habitat in the initial landscape condition (a), after period 5 (b), and after period 10 (c), when the spatial goal requiring adjacent patch types is maximized.

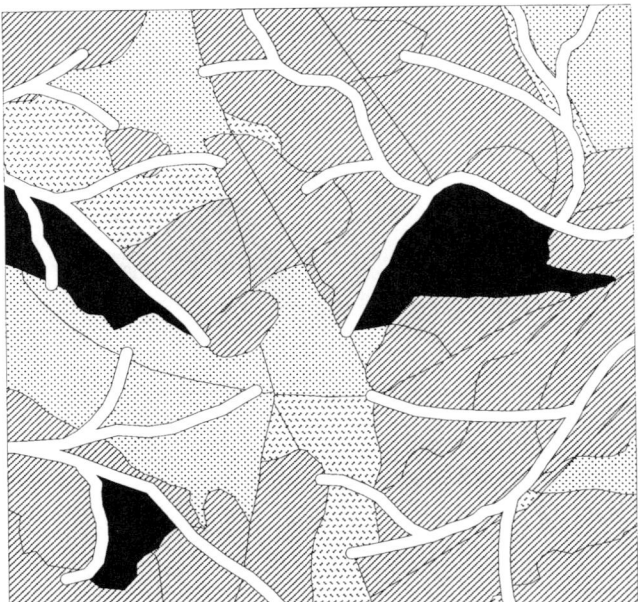

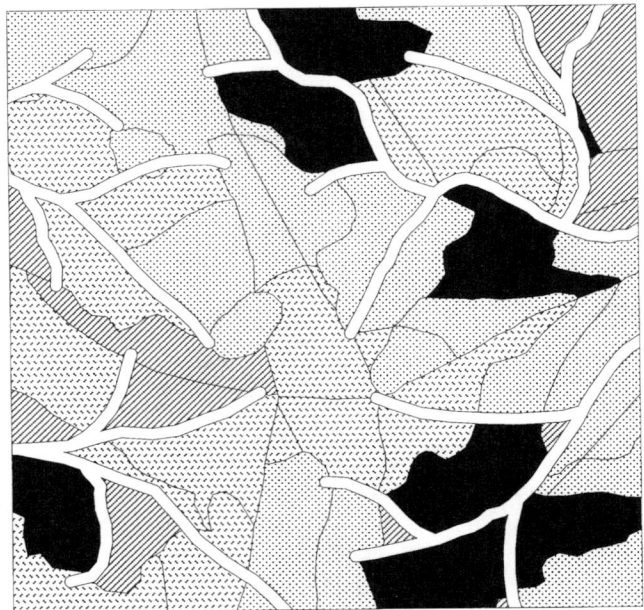

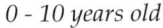	0 - 10 years old
	11 - 30 years old
	31 - 89 years old
	90+ years old
	Riparian area

	0 - 10 years old
	11 - 30 years old
	31 - 89 years old
	90+ years old
	Riparian area

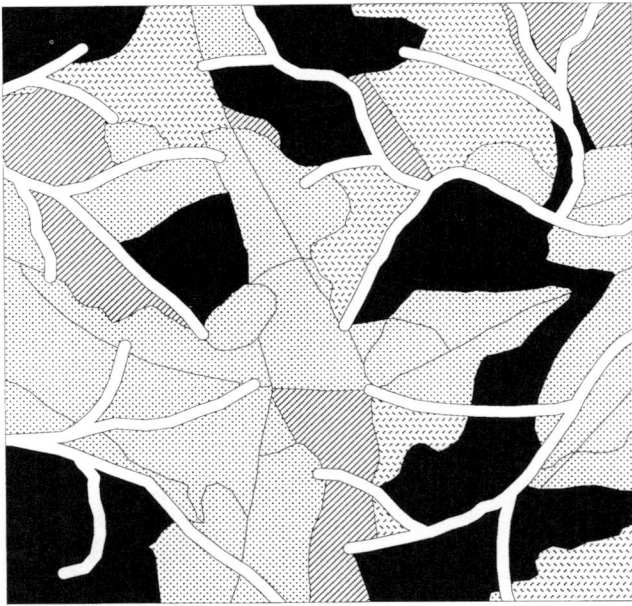

	0 - 10 years old
	11 - 30 years old
	31 - 89 years old
	90+ years old
	Riparian area

Fig. 10. Landscape condition in planning period 10 when maximizing the nonspatial goals (a, top left), maximizing the spatial goal requiring minimum patch sizes (b, lower left), maximizing the spatial goal requiring adjacent patch types (c, above).

integer type variables in a planning process. Spatial goals require information about the locations of resources, and can be represented by linear or nonlinear relationships. For example, a nonlinear relationship could require one type of habitat patch to be adjacent to another type of habitat. A linear relationship could be a variable such as distance from the ocean, which essentially can be assumed to be a constant.

Whereas the development of goals and goal criteria are important for planning processes, the specialists involved in developing goal criteria, and the planners whose task it is to implement the goals in a planning process, encounter potential challenges in four focus areas; those dealing with people, databases, technology, or organizational commitment. People are the instrument by which databases and technology are joined to help develop management alternatives. The types of tools planners possess will generally guide a process toward the use of a particular set of tools and data structures. Databases are constantly evolving, as are the computer and software technology we use to manipulate data. Organizations should match the planning objectives (e.g., the types of goals to be measured, how they will be measured, the spatial extent of the planning area, etc.) with the databases and the technology at their disposal (or which could be acquired). In doing this, estimates of budget levels and cash flow needs, and other institutional limitations (e.g., personnel), will become apparent. The reaction of the leaders of the organization to the planning needs will be one indication of the commitment of the organization to the planning process. If ≥ 1 of these four essential focus areas (people, databases, technology, organizational commitment) is limiting, an organization may consider re-examining the expected results of the planning process.

Conclusions

The central role of forest managers is good land stewardship, which is achieved through decision making, or choosing among alternative courses of action. Quantitative planning techniques allow managers to examine variations in goals, goal criteria, and activities, and to develop alternative management plans. We are fortunate that advances in applied economics and management science have allowed the natural resources fields to use quantitative planning tools, and to say with some confidence that recommended land management plans are efficient and best meet the objectives of forest managers. Managers are then relied on to balance the divergent goals of an organization and appraise the reliability and meaning of information developed in plans that are developed and designed to meet those goals.

Four major aspects of planning projects are people, databases, technology, and commitment by the organization to the development of the plan. Assuming databases, technology, and organizational commitment are available at appropriate levels to facilitate a planning process, the limiting factor (if there is one) is people. The communication process among individuals can be enhanced with a common understanding of the pieces of a planning process, including the basic steps planners and biologists must take to understand the goals:

1. develop quantitative description of species goal;
2. establish criteria to measure the goals;
3. select the activities that can be used to help achieve the goals;
4. work with planners to develop mathematical statements of the goals, in the form of an objective function and constraints, to help guide the selection of a combination of activities that will comprise a management plan.

Literature Cited

1. Arthaud, G.J., and D. Rose. 1996. A methodology for estimating production possibility frontiers for wildlife habitat and timber value at the landscape level. Canadian Journal of Forest Research 26:2191-2200.
2. Bettinger, P. and K. Boston. 1999. Intensifying a heuristic forest harvest scheduling search procedure with 2-opt decision choices. Canadian Journal of Forest Research 29:1784-1792.
3. Bettinger, P., K.N. Johnson, and J. Sessions. 1996. Forest planning in an Oregon case study: defining the problem and attempting to meet goals with a spatial-analysis technique. Environmental Management 20:565-577.
4. Bettinger, P., J. Sessions, and K. Boston. 1997. Using tabu search to schedule timber harvests subject to spatial wildlife goals for big game. Ecological Modeling 94:111-123.
5. Bettinger, P., J. Sessions, and K.N. Johnson. 1998. Ensuring the compatibility of aquatic habitat and commodity production goals in eastern Oregon with a tabu search procedure. Forest Science 44:96-112.
6. Daust, D.K., and J.D. Nelson. 1993. Spatial reduction factors for strata-based harvest schedules. Forest Science 39:152-165.
7. Glover, F. 1989. Tabu search—Part I. ORSA Journal of Computing 1:190-206.
8. Glover, F. 1990. Tabu search—Part II. ORSA Journal of Computing 2:4-32.
9. Haight, R.G., and L.E. Travis. 1997. Wildlife conservation planning using stochastic optimization and importance sampling. Forest Science 43:129-139.
10. Hof, J., M. Bevers, L. Joyce, and B. Kent. 1994. An integer programming approach for spatially and temporally optimizing wildlife populations. Forest Science 40:177-191.
11. Hoganson, H.M., and D. Rose. 1984. A simulation approach for optimal timber management scheduling. Forest Science 30:220-238.
12. Kangas, J., and T. Pukkala. 1996. Operationalization of biological diversity as a decision objective in tactical forest planning. Canadian Journal of Forest Research 26:103-111.
13. Lockwood, C., and T. Moore. 1993. Harvest scheduling with spatial constraints: a simulated annealing approach. Canadian Journal of Forest Research 23:468-478.
14. Murray, A.T. 1999. Spatial restrictions in harvest scheduling. Forest Science 45:45-52.
15. Murray, A.T., and R.L. Church. 1995. Heuristic solution approaches to operational forest planning problems. OR Spektrum [Operations Research] 17:193-203.
16. Nelson, J., and J.D. Brodie. 1990. Comparison of a random search algorithm and mixed integer programming for solving area-based forest plans. Canadian Journal of Forest Research 20:934-942.
17. O'Hara, A.J., B.A. Faaland, and B.B. Bare. 1989. Spatially constrained timber harvest scheduling. Canadian Journal of Forest Research 19:715-724.
18. Pulkki, R. 1984. A spatial database—heuristic programming system for aiding decision-making in long-distance transport of wood. Acta Forestalia Fennica 188.
19. Reeves, C.R. and J.E. Beasley. 1993. Introduction. Pages 1-19 in C.R. Reeves, editor. Modern heuristic techniques for combinatorial problems. John Wiley & Sons, Inc., New York, NY.
20. Weintraub, A., G. Jones, A. Magendzo, M. Meacham, and M. Kirby. 1994. A heuristic system to solve mixed integer forest planning models. Operations Research 42:1010-1024.
21. Weintraub, A., G. Jones, M. Meacham, A. Magendzo, A. Magendzo, and D. Malchuk. 1995. Heuristic procedures for solving mixed-integer harvest scheduling—transportation planning models. Canadian Journal of Forest Research 25:1618-1626.

24

Decaying Wood in Pacific Northwest Forests: Concepts and Tools for Habitat Management

Cathy L. Rose, Bruce G. Marcot, T. Kim Mellen, Janet L. Ohmann,
Karen L. Waddell, Deborah L. Lindley, & Barry Schreiber

Introduction

Decaying wood has become a major conservation issue in managed forest ecosystems.[16, 64, 69a, 149, 201] Of particular interest to wildlife scientists, foresters, and managers are the roles of wood decay in the diversity and distribution of native fauna, and ecosystem processes. Numerous wildlife functions are attributed to decaying wood as a source of food, nutrients, and cover for organisms at numerous trophic levels.[231, 232, 234, 346, 369] Principles of long-term productivity and sustainable forestry include decaying wood as a key feature of productive and resilient ecosystems.[10, 229, 291, 293, 386] In addition to a growing appreciation of the aesthetic, spiritual, and recreational values of forests, society increasingly recognizes ecosystem services of forests as resource "capital" with tangible economic value to humans, such as air and water quality, flood control, and climate modification.[15, 262, 290]

The ecological importance of decaying wood is especially evident in coniferous forests of the Pacific Northwest. In this region, the abundance of large decaying wood is a defining feature of forest ecosystems, and a key factor in ecosystem diversity and productivity.[127] Native forests west of the Cascade Crest are highly productive and accumulate large amounts of live and dead wood—a result of the temperate climate that favors tree growth.[125, 126, 395] Large accumulations of decaying wood provide wildlife habitat and influence basic ecosystem processes such as soil development and productivity, nutrient immobilization and mineralization, and nitrogen fixation.[85, 115, 218, 233] Forests east of the Cascade Crest are also strongly influenced by accumulations of decaying wood that set the stage for ecosystem disturbances from fire, insects, and disease.[56, 137, 390]

Decaying wood has a pivotal role in both estuarine and coastal marine ecosystems, supporting complex trophic webs from benthos to higher vertebrates. More than half of the total organic carbon content of Washington's off-shore, midshelf sediment originates from coniferous forests.[139] Inputs of decaying wood are crucial to most aspects of stream processes, such as channel morphology, hydrology, and nutrient cycling.[233, 260, 262, 354] Although knowledge of these processes is incomplete, decaying wood in freshwater and marine food webs illustrates the potentially far-reaching implications of terrestrial wood management.

Historical interest in decaying wood has centered mainly on impediments to reforestation and on economic consequences of decay to timber production. In the first half of this century, wood decay or "decadence" was viewed simply as an undesirable attribute of overmature forests at increased risk of damage by fire, insects, or disease. Recognition of the diverse ecological benefits derived from decaying wood in aquatic and terrestrial ecosystems has grown appreciably over the past two decades.[56, 198, 201, 234] Wood decay in forests of the Pacific Northwest has recently become a topic of renewed interest at national and global scales, regarding the role of terrestrial carbon storage in the reduction of atmospheric CO_2 (a greenhouse gas). In light of recent projections for accelerated climatic warming during the next several decades,[255] regulatory targets for terrestrial carbon storage are being considered.[16]

New research over the past three decades has emphasized the significance of decaying wood to many fish and wildlife species,[48, 56, 230, 233, 234, 369] and to overall ecosystem function.[389] The importance of decaying wood to ecosystem biodiversity, productivity, and sustainability is a keynote topic in two recent regional ecosystem assessments in Oregon and Washington.[114, 225] These, and other publications address both the specific roles of wood decay in ecosystem processes and functions, as well as ecological functions of wildlife species associated with wood decay.[13, 68, 216, 250]

Interactions among wildlife, other organisms, and decaying wood substrates are essential to ecosystem processes and functions. In the process of meeting their needs, animals accomplish ecosystem "work" with respect to transformation of energy and cycling of nutrients in wood. For example, chipmunks and squirrels disperse mycorrhizal fungi which play key roles in nutrient cycling for tree growth; birds, bats, and shrews consume insects that decompose wood or feed on invertebrates and microbes; beavers and woodpeckers create habitats by modifying physical structures; arthropods build and aerate soil by decomposing wood material. Relations between wood decay and wildlife have been examined in several recent analyses.[56, 225, 282, 283, 284, 286] Species-specific associations have been identified for some tree species, types of decay fungi, and different forms of wood, such as bark piles at the base of snags, hollow living trees, and broomed trees. Simplified classification schemes and inventory procedures have been developed for decaying wood, particularly wood habitat structures relevant to wildlife.[56]

Wildlife species associated with wood decay, and their ecosystems are affected by management activities. Intensively managed forest plantations have replaced old-growth throughout most of the commercial forest land base in Oregon and Washington.[41, 42] Intensive forest management regimes have substantially altered the abundance and composition (species, size, decay class) of decaying wood in forest ecosystems in the Pacific Northwest. Managed forests, on average, have lower amounts of large down wood and snags than do natural forests.[59, 81, 82, 114, 154, 225, 276, 344, 368] Furthermore, in forests east of the Cascade crest, fire suppression has altered stand dynamics and produced accumulations of fine fuels conducive to stand-replacement fires.[5, 211, 302] Forest health problems and declining populations of some vertebrate and invertebrate wildlife species have coincided with changes in forest structure. These changes have raised concerns about the future biodiversity, productivity, and sustainability of the region's forests, particularly in coastal and eastside forests.[21, 170, 187, 222, 367]

Since the publication of Thomas et al.[369] and Brown,[48] new research has indicated that more snags and large down wood are needed to provide for the needs of fish, wildlife, and other ecosystem functions than was previously recommended by forest management guidelines in Washington and Oregon. For example, the density of cavity trees selected and used by cavity-nesters is higher than provided for in current management guidelines.[53, 102] Reductions in the wood content of estuarine and marine environments, and consequences to assorted aquatic organisms have been documented.[139] Reductions in forest productivity linked to management practices such as stem removal, slash burning, and soil disturbance have also been recognized.[10, 13, 44, 198, 199]

Critical ecosystem functions of wood, coupled with incomplete knowledge for management, make the topic of decaying wood a priority for future research and adaptive management.[78, 149, 164] Effective approaches to managing decaying wood require that dead wood components of wildlife habitats be viewed within the context of the larger interacting ecosystem. To help managers achieve the goal of effective management of decaying wood, this chapter seeks to provide a focus on the ecological context for wood decay and associated wildlife in forests of the Pacific Northwest. Emphasis on

Wood Legacies in Managed Forests

John Hayes

Legacies are structures or components of ecosystems that exist prior to a disturbance and are "inherited" by the post-disturbance community. Legacies can provide important temporal connectivity within a stand, allowing organisms present in a pre-disturbance community to persist in an area following disturbance. In addition, legacy wood can provide structural elements and complexity in a stand that would otherwise require very long periods of time to develop. In managed forests, wood legacies, including large diameter trees, snags, and down wood, are ecologically important structures that play central roles in diverse ecosystem processes and functions, such as geomorphic processes, hydrology, nutrient cycling, and habitat for fish and wildlife. The ecological value of wood legacies has begun to gain widespread recognition only within the past two decades.[122, 164]

As a result of a variety of operational, safety, and economic considerations, application of intensive forest management practices often results in removal of legacy structures from stands and minimal retention of future legacy structures. Growing replacement structures with similar characteristics (e.g., large diameter trees with large diameter branches, thick and deeply-furrowed bark, and complex crown structure) requires decades or longer. Moreover, unless special provisions are made, large diameter trees, snags, and logs with these

characteristics may never be produced in forests managed intensively on short- to moderate-length rotations. Habitat quality for species that depend upon or are closely associated with these structures can be seriously diminished with their loss from forest stands. The ecological importance of wood legacies combined with the difficulties of creating replacement structures provide convincing reasons to conserve legacy structures during management activities.

Managing wood legacies through time in managed forests is a multi-staged process. Existing structures that will serve as legacy structures in the post-disturbance environment should be identified prior to a disturbance event, such as logging. In some cases, it may be adequate to rely on the timber sale administrator or loggers to identify appropriate structures and implement the management strategy in the field. Since one intent of legacy structures is to provide various functions through time, it will often be valuable to either individually mark important legacy structures, or to document their location and purpose so that future managers can take the structures into account. Of equal importance, plans for recruitment of future legacy structures should be prepared to ensure that legacy structures will be available in future stands. Innovative silvicultural practices can be employed to create conditions favorable to development of future legacy structures.

concepts of long-term productivity in this chapter reflects an underlying principle that habitat functions of decaying wood are inextricably linked to ecosystem processes. Careful attention to the whole ecosystem is a prerequisite to successful management of decaying wood for wildlife.

This chapter provides a synthesis of knowledge on processes and functions of wood decay in forest productivity and wildlife habitat, and summarizes available information on the current regional status of decaying wood. It then offers managers a stepwise assessment process to set goals and objectives, and select silvicultural tools to manage wood decay for desired results. Although the primary emphasis is on upland forests, the chapter also highlights wood decay structures and functions in forested wetlands, riparian forests, and aquatic systems. The analysis addresses a wide range of woody plant structures, with emphasis on large down wood and snags that form 'legacy' structures linking successive generations of stands.

Ecological Significance of Decaying Wood

Wood Decay Processes

Decay processes in wood form the basis for understanding ecological roles and relationships among wildlife and decaying wood. Familiarity with these relationships facilitates the translation of these concepts into successful management strategies. Down wood, snags, and other persistent wood structures (such as stumps, root wads, and coarse roots) occur in most forest ecosystems. In moist

---

Terminology of Decaying Wood

Decaying wood includes all portions of a tree, standing or down, that are dead and in the process of decay. This is a broad definition that seeks to address wildlife uses[56] and other ecological functions of decaying wood. It includes the following wood structures:

• living trees with decay (e.g., heart-rot, root-rot, broken or decaying tops, large decaying branches, broken branch stubs, bole wounds, loose bark, stem cavities, and old trees with deep, fissured bark and thick, bulky branches)
• living hollow trees (advanced heartwood decay)
• broomed trees (diseased branches)
• snags (dead trees or portions of trees that remain upright)
• down wood (decaying trees or portions of trees that have fallen to the forest floor; includes other commonly used terms, such as logs, coarse woody debris, and large woody debris)
• stumps and other tree parts (rootwads, bark piles, coarse roots, and fine litter)

Ecosystem Definition

Ecosystems can be described at a variety of spatial scales and the spatial hierarchy of ecosystems is helpful for understanding the ecological functions of wood decay. Therefore, we use the "macrosystem" definition of ecosystems,[382] which recognizes the functional continuum of terrestrial, stream, estuarine and coastal ocean environments. Functional linkages include both the physical and biological processes and the interactions between organisms and decaying wood that contribute to nutrient, energy, and material cycling this continuum.[233] Smaller sub-units of ecological organization[273] are recognized within this macrosystem, such as the forest floor and riparian communities.

---

forests, biological activity is the primary determinant of wood decay rates.[164, 346, 404] In drier forest regions, fire is often the primary agent of wood breakdown. In forest ecosystems with infrequent fire, the principal mechanisms of wood breakdown include leaching, fragmentation, transport, collapse, settling, seasoning, respiration, and biological transformation.[164] Decay processes in wood can be divided into two groups depending on whether the mode of action is mechanical fragmentation, or chemical breakdown. In the following section, we describe biologically mediated chemical breakdown of wood, and various physical and biological processes that fragment wood structure.

Knowledge of wood structure and composition is useful for understanding processes of wood decay. Wood tissues of tree stems include the outer bark, cork cambium, inner bark (phloem), vascular cambium, outer xylem (living sapwood), and the inner xylem (non-living heartwood). Outer bark is a non-living barrier between the inner tree and harmful factors in the environment, such as fire, insects, and diseases. The cork cambium (phellogen) produces bark cells. The vascular cambium produces both the phloem cells (principal food-conducting tissue) and xylem cells of the sapwood (the main water storage and conducting tissue) and heartwood. Wood properties are determined mainly by plant cell walls, consisting of cellulose fibrils impregnated within a cement-like matrix of lignin and other organic polymers in the approximate proportions: cellulose (40-50%), hemicellulose (20-35%), and lignin (15-35%).

Wood Fragmentation

Mechanical disruption of wood structure may be initiated abiotically by natural events, such as heavy snows, rime ice, avalanches, mass soil movement, floods, and windstorms that uproot or break trees.[86, 230, 270, 301, 316] Thermal expansion and contraction of wood, and water in woody tissues also promotes the physical disruption of wood structure. These disturbance agents affect wood decay variously by: 1) directly or indirectly inducing mortality of wood tissues, 2) altering the microclimate surrounding the affected wood structure, 3) increasing the surface area of wood exposed to the environment, and 4) increasing wood accessibility to microbes.

Fragmentation of decaying wood during the early stages of decomposition is accomplished by wood-boring insects such as beetles, carpenter ants, and termites.[23, 89, 230] As fissures develop in wood, invertebrates from minute mites to centipedes, millipedes, slugs and snails find suitable habitat, and contribute to further fragmentation. In the soil, an abundance of soil animals (including nematodes, earthworms, and microarthropods) contribute to the fragmentation of decaying wood and increase the surface area for microbial attack.[186] Functions and interactions between these soil fauna are varied and complex.[230]

A variety of vertebrate species contribute to the fragmentation of decaying wood in both live and dead trees. Most notable are the primary cavity excavators, such as woodpeckers and nuthatches that break wood fibers in the process of cavity excavation[216] (Figure 1). Other animal activities that fragment wood include limb pruning or breakage, foraging, denning, and burrowing. As decaying wood softens, tunneling and burrowing by other vertebrates such as salamanders, shrew, shrew-moles, and voles continue the fragmentation process in rotten wood and bark, and promote favorable conditions for further decomposition by invertebrates, fungi, and bacteria.

Principal Biological Agents of Wood Decay

Biologically mediated chemical breakdown of wood can be grouped into two basic categories: primary decay that establishes on living trees, and secondary decay that commences after tree mortality caused by other factors. Of the biological agents of wood decay, insects and fungi are the principal players in coniferous forest ecosystems. Wood-boring insects contribute to the early stages of wood decomposition by consuming and chemically digesting

Figure 1. The pileated woodpecker, a primary excavator. Photo: David H. Johnson.

Figure 2. Conk of the fungus Fomes pini. Photo: Washington Department of Natural Resources.

wood and by the enzymatic action of microbes. Insects also serve as vectors for other decay organisms and increase wood surface area accessible to both microbial and fungal attack.[23, 47, 89, 118, 230] Fungi accomplish wood decay by enzymatically decomposing the cell wall. Wood decay fungi can be separated into two main groups—the pathogenic heart-rot and root-rot fungi that colonize primarily the heartwood or roots of living trees (primary decay), and the saprobic fungi that colonize decaying wood (secondary decay) (Figure 2).

On live trees, vulnerable locations for primary decay include heartwood, branch nodes, roots, dead and broken tops or branches, and wounds from fire, logging activities, windthrow, and lightning. Wounding is instrumental to primary decay because germinating fungal spores cannot penetrate intact bark or living sapwood, but must land directly on exposed dead wood connected to the heartwood. Generally, decay fungi located in the upper stems of living trees are heart-rot fungi.[220] Heart-rot fungi are not restricted to the heartwood, but usually infect through wounds in the sapwood of living trees. Every tree species is susceptible to at least one heart-rot fungus, but few fungi can cause heart-rot in living trees.[56, 175, 220]

Root rot is caused by a group of primary decay fungi that infect roots and spreads to other trees through root contact. These root pathogens act by causing nutrient deficiency after disrupting root tissue, altering root morphology, parasitizing the root cambium, or functioning as heart-rotters of the roots and lower tree bole.[220, 322] By weakening the root structure, root rot may directly kill trees, or indirectly contribute to mortality from windthrow and insects.

Processes of secondary decay are similar in both dead trees and in dead portions of live trees. Heart-rot fungi already present in a tree at the time of death usually contribute little to further decay of the dead tree. Instead they are replaced by other fungi termed saprobes, that are better adapted to degrade dead wood. Sapwood rotting fungi are quick to colonize a dead tree, particularly if tree mortality occurred without depletion of nutrient and carbohydrate stores. In general, once the sapwood is fully decayed, the sapwood fungi are replaced by saprobes that continue to degrade the sapwood and proceed into the heartwood. Saprobes may also invade larger patches of dead tissue on live trees, however.

Live plant roots contribute to wood decay by producing enzymes and other exudates that accelerate cellulose decomposition and catalyze nutrient release from organic matter, and by forming symbioses with mycorrhizal fungi, N-fixing bacteria and actinomycete fungi to aid lignin decomposition.[1]

Stages and Structures of Decaying Wood

Various woody tissues in a tree have inherent differences in decay susceptibility. Decay normally begins at the exterior of a tree and progresses inward, though heart rot may degrade the interior of a tree with intact sapwood. The outer bark disappears mainly by fragmentation and sloughing from the top and sides of down trees. Despite a high concentration of nitrogen, decay of outer bark is inhibited by a high content of lignin and other anti-microbial compounds.[176, 203, 230] Inner bark rapidly decays due to its small volume and high nutritional quality (high digestibility and nutrient content). The outer ring of living sapwood is usually highly resistant to decay despite its high nutrient content, and may remain so even when adjacent tissues are extensively decayed. Dead sapwood decays rapidly, however. Compared to sapwood, heartwood decays slowly because of its low nutrient quality and limited digestibility to microbes (high lignin and low nitrogen).[342]

All dead wood progresses through discreet stages of decay characterized by progressive changes in wood attributes, decomposition rates, and decomposer species. This chronology has been described in more detail for down wood than for snags. At each stage, decay is influenced by conditions in the physical environment, and by the structural and chemical composition of wood. As decomposition advances, changes in wood properties include decreased density and digestibility (higher ratio of lignin to cellulose), and increased water content and nutrient concentrations (including nitrogen, phosphorus, calcium, and magnesium in microbial tissue). Microbial decomposers vary in their nutrient requirements and in their inherent ability to degrade the structural matrix of wood (cellulose and lignin). Altered nutrient and structural attributes of wood with advancing decay thus creates sequential niches for different decomposer species.[220] Decay rates for dead wood are thus logarithmic rather than linear, because the more readily-decomposed fraction of wood is digested first, leaving the less digestible fraction,[184] and because nitrogen availability to microbes decreases over time.

Classification Systems For Decaying Wood

To facilitate the description and quantification of habitat attributes provided by decaying wood, several systems to classify decay conditions in conifers have been developed. New studies are needed to adapt such systems to hardwoods, which typically do not follow the same decay phases as conifers. A five-class system has been commonly used to describe the extent of decay in down wood.[28, 234, 339, 376] Decay classes are useful to ecological studies of wood decomposition because they describe

functional and structural properties of wood, and eliminate the need to determine the date of tree mortality. The classes are based on a variety of tree characteristics, including the presence or absence of bark and fine twigs, bole shape, wood texture, and extent of log contact with soil. The decay status of a tree when it falls to the forest floor depends on its condition as a live tree or snag. For example, a live or recently dead tree that falls to the forest floor becomes a Class 1 log, however, a fallen snag may create a Class 1 to a Class 4 log.

A classification system for Douglas-fir snags describes a continuum of nine decay stages.[234, 370] The nine stages of snag decay correspond to the five decay classes for logs described above, with four additional stages: two earlier stages for live, but declining trees, and two later stages for progressive breakage and decay of the lower bole.[230] This nine-stage system was modified and refined into a five-stage system by eliminating Stages 1 and 2, and combining stages 7-9.[82, 263]

A simplified system of three structural classes each for conifer logs and snags has been developed on the premise that fewer classes are sufficient to describe wildlife use and easier to apply in the field (Figure 3). Visual, chemical, and physical properties of conifer wood were evaluated in relation to degree of decay. Wood density, cellulose, and lignin content were found to be the best indicators of decay, whereas visual characteristics were relatively poor indicators.[182] Consequently, although the three class system of visually classifying wood decay may be satisfactory for assessing wood conditions important to many vertebrate wildlife species, the five-class system may better describe subtle changes in wood properties (density, moisture) important to other organisms, and to ecosystem functions.

Ecological Functions of Decaying Wood

Wildlife Habitat Roles

Down wood, snags, and live trees with decay serve vital roles in meeting the life history needs of wildlife species in Oregon and Washington. Literature describing habitat relationships for wildlife and decaying wood in the Pacific Northwest are most comprehensive for vertebrate foraging ecology and cavity-nesting relations. Knowledge is more extensive for snags than for down wood, and for westside than for eastside ecosystems. Major regional ecosystem assessments recently completed in Oregon and Washington have examined the roles of decaying wood in habitat and in ecosystem biodiversity, productivity and sustainability.[114, 225, 302]

Recent significant advancements have defined wildlife species-specific relationships with particular characteristics and components of decaying trees, both standing and fallen,[56, 95, 185, 284, 351, 373, 386, 402] and implications for management.[13, 68, 223, 226, 250, 327] Much of this recent research has improved our understanding of poorly documented wildlife uses of microhabitats. Some of these are represented in the Habitat Element matrixes on the CD-ROM accompanying this book, and in the DecAID

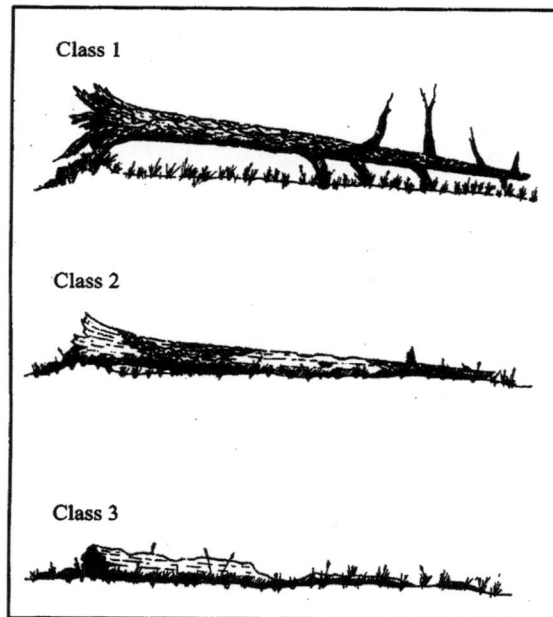
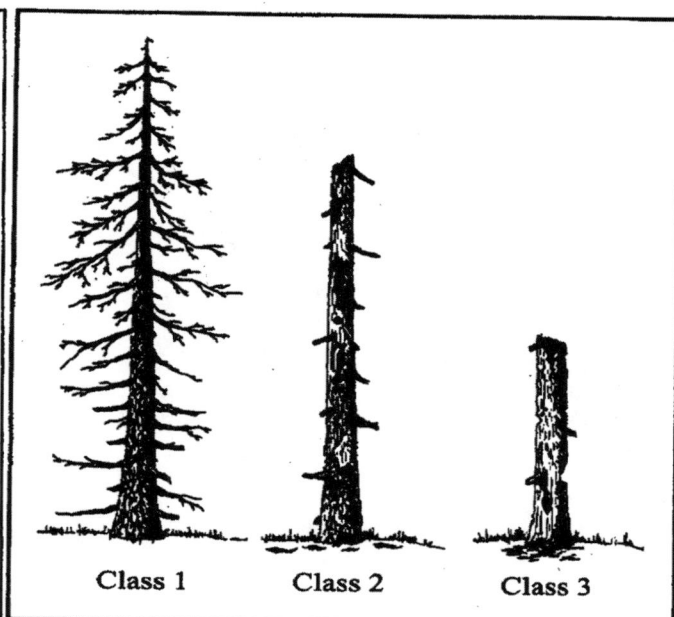

Figure 3. Structural classes of snags and down wood, from Bull et al.[56]

model.[226] For example, research on amphibians previously known to be associated with down wood, elucidated ways in which these species use distinctly different features of down wood.[22, 59a] For example, ensatinas occur primarily in bark piles at the base of snags, redback salamanders are found most often under moderately decayed logs, clouded salamanders prefer habitat under bark on logs, and Oregon slender salamanders inhabit the interior of logs.[22, 59a]

Up-to-date and specific information on species' habitat associations and key ecological functions is available in the matrixes on the CD-ROM accompanying this book and in the DecAID model.[226] The remainder of this section presents a sampling of new findings about resource partitioning among various wildlife communities in Pacific Northwest forests, with examples of current information available in the matrices and in the DecAID model.

Tree Species. Few studies have demonstrated the causal relationship between tree species per se, and the habitat functions of decaying wood. However, some habitat functions of wood are provided primarily by certain tree species. Woodpeckers, sapsuckers, and nuthatches are highly specific in their selection of tree species for nesting and roosting, and this selectivity is attributed to the presence of decay fungi.[29, 56, 95, 351] Tree species also influences other life requisites, such as prey production. For example, carpenter ants, a primary prey species of pileated woodpeckers in northeastern Oregon, prefer down larch logs (*Larix laricina*).[373] Decay characteristics, size, and wood density may account for more variation in wildlife use of wood than tree species. These parameters are likely to be highly correlated; that is, some tree species

are more readily decomposed by organisms creating suitable conditions for cavity excavators.[284, 323] Tree species may also be correlated with environmental conditions that influence habitat suitability, such as microclimate, canopy position, geographic location, etc.

Decaying Wood Habitat Structures. In a recent review, decaying wood structures were grouped into five categories representing key habitat elements for wildlife, including living trees with decay, hollow trees, trees with brooms, dead trees (snags), and down woody material (logs).[56] An overview of these habitat structures is provided below, with reference to the three-class systems for classifying wood decay in snags and logs.[56]

• *Standing Trees.* Living trees often have pockets of decay, dieback, wounds, or broken tops and limbs that serve as wildlife habitat. To be useful to most cavity excavators, live trees usually must contain wood in a Class 2 stage of decomposition.[56, 238] For example, strong excavators, such as Williamson's sapsuckers, pileated woodpeckers, and black-backed woodpeckers, select trees with a sound exterior sapwood shell and decaying heartwood to excavate their nest cavities.[238, 284, 307] *Armillaria* root rot and other saprophytic fungi produce softer exterior wood suitable for cavity construction by weaker excavators such as the red-breasted nuthatch and chickadee.[216] Trees with decaying tops or large dead or deformed branches provide roosting and drumming habitat for birds, and nesting platforms for murrelets, owls, and raptors (Figure 4).[102, 114, 153, 269] In addition, localized wood decay resulting from tree wounding and scarring (particularly at the tree base) provides substrate for ants and other insects consumed by foraging birds.[53]

Figure 4. Spotted owl roost. Photo: Deborah Lindley.

Figure 5. Hollow snag of ponderosa pine. Photo: Deborah Lindley.

A hollow tree (live or dead) is formed when advanced heartwood decay (Class 3) causes the heartwood cylinder to detach from the sapwood, leaving a hollow core surrounded by a reinforcing band of sapwood. The softened heartwood of trees colonized by heart-rot fungi provides suitable conditions for excavating a nest chamber, and the living sapwood functions to maintain the tree's structural integrity (Figure 5).[56] The structural elements produced by advancing heartwood decay include hollow cavities from small patches to larger cylinders, to entire hollow trunks. These decay legacies and resulting habitat elements are described in detail by Bull et al.[56]

Hollow trees larger than 20 inches (51 cm) in diameter at breast height (dbh) are the most valuable for denning, shelter, roosting, and hunting by a wide range of animals.[7, 52, 55, 56, 265, 282] Hollow chambers are used as dens by black bears, as night roosts by woodpeckers, and as dens, shelter, roosts, and hunting sites by a variety of animals, including flying squirrels, wood rats, bats, American marten, northern flickers and Vaux's swift.[7, 51, 52, 55, 56, 265, 282] Hollow trees and down wood are formed from only a few tree species that can maintain bole structural integrity as the heartwood decays.[56] Western redcedar is especially valuable in providing hollow trees because the decay-

resistant sapwood remains structurally sound for centuries. In the Interior Columbia Basin, grand fir and western larch form the best hollow trees for wildlife uses.

Broomed trees caused by mistletoe, rust, or needlecast fungi may remain alive for decades, and have attributes distinct from decay patches in live trees. Abundant forage is produced from mistletoe shoots and fruits. Regardless of the extent of decay, broom infections provide various habitat functions to wildlife depending on how and where they form along the bole.[371, 372] For example, mistletoe brooms form platforms used for nesting, roosting, and resting sites by owls, hawks, and song birds; roosting by grouse; and resting cover by squirrels, porcupines, and marten.[56, 114] Little information is published on the use of rust or needlecast brooms by wildlife, but it is likely they serve similar functions as dwarf mistletoe. Documented uses include rest sites for American marten and nest sites for squirrels and great horned owls.[50, 56, 256] Because the dieback of infected tree parts may occur over a protracted time frame, a single tree infected with these types of pathogens may contain discrete patches of dead wood in various stages of decay.[51, 256]

Recent studies have provided valuable insight on wildlife uses of snags (dead trees).[21, 56, 314, 402] Snags provide essential habitat features for many wildlife species (Figure 6). The abundance of cavity-using species is directly related to the presence or absence of suitable cavity trees. Habitat suitability for cavity-users is influenced by the size (diameter and height), abundance, density, distribution, species, and decay characteristics of snags.[307] In addition, the structural condition of surrounding vegetation determines foraging opportunities.[402]

The Habitat Elements matrix on the CD-ROM with this book lists a total of 96 wildlife species associated with snags in forest (93 species) or grassland /shrubland (47 species) environments. Most of these species use snags in both environments. In forests, this includes 4 amphibian, 63 bird, and 26 mammal species. Additionally, 51 wildlife species are associated with tree cavities, 45 with dead parts of live trees, 33 with remnant or legacy trees (which may have dead parts), 28 with hollow living trees, 21 with bark crevices, and 18 with trees having mistletoe or witch's brooms. Habitat uses include nesting, roosting, preening, foraging, perching, courtship, drumming, and hibernating (Figure 7).

Of the 93 wildlife species associated with snags in forest environments, 21 are associated with hard snags (Stages 1 and 2), 20 with moderately decayed snags (Stage 3), and 6 with soft snags (Stages 4-5) in the five-stage classification system. According to the matrixes,[188] most snag-using wildlife species are associated with snags >14.2 inches (36 cm) diameter at breast height (dbh), and about a third of these species use snags >29.1 inches (74 cm) dbh.

This query of the Habitat Elements matrix illustrates the breadth of updated information about wildlife and snag habitat relations. Research results have expanded the number and variety of decaying wood categories over what was previously presented in Thomas[366] and Brown.[48] For example, the portion of the query (above) about cavity-

Figure 6. Snags provide essential habitat for wildlife. Photo: Deborah Lindley.

Figure 7. Brown creeper nest in sloughing bark. Photo: Harry Hartwell.

nesters includes subsets of species associated with different types of cavities, such as bark crevices, hollow trees, cavities in dead trees (snags), and cavities in living trees. Further differences in matrix summaries of species' relations may result from changes in the taxonomic classification of species, updated species' range and geographic distribution data, as well as data limitations for some species' associations with specific habitat elements. It should be noted that the Habitat Elements matrix denotes wildlife species' associations with the various habitat elements, but does not distinguish species' uses of habitat elements for life history needs (eg., nesting, feeding, hibernating, etc.), nor does it indicate the degree of association or species' versatility. However, species' habitat use information is available in the DecAID advisory model.[226]

Stumps provide a variety of wildlife habitats. Stumps with sloughing bark (Class 2) provide sites for bat roosts,[387] and foraging sites for flickers, and downy, hairy and pileated woodpeckers. In openings, tall stumps with advanced decay (Class 3) provide nest sites for flickers, and subsequently for blue birds and other secondary cavity-nesters associated with openings. Squirrels and chipmunks also use stumps as lookouts and platforms for cone-shredding. The matrixes on the CD-ROM with this book do not include stumps as a habitat element.

• *Down Woody Material (logs).* Down wood affords a diversity of habitat functions for wildlife, including foraging sites, hiding and thermal cover, denning, nesting, travel corridors, and vantage points for predator avoidance.[56, 64, 230] Larger down wood (diameter and length) generally has more potential uses as wildlife habitat. Large diameter logs, especially hollow ones are used by vertebrates for hiding and denning structures.[214, 230] Bears forage for invertebrates in logs during summer and fall. Fishers use large logs to a limited degree as den sites.[297]

Lynx select dense patches of downed trees for denning.[200] Jackstrawed piles of logs form a habitat matrix offering thermal cover, hiding cover, and hunting areas for species such as marten, mink, cougar, lynx, fishers, and small mammals[305, 313] (Figure 8). Smaller logs benefit amphibians, reptiles, and mammals that use wood as escape cover and shelter. Small mammals use logs extensively as runways[234] (Figure 9). California red-backed voles use Class 2-3 down logs for cover, and feed on fungi (especially truffles) and lichens growing in close association with down wood.[378] The orientation of down wood also influences wildlife habitat use.[56] Logs oriented along slope contours may be useful travel lanes for wildlife, whereas logs oriented across contours impede travel.

The moist environment beneath loose bark, bark piles and in termite channels of logs with advanced decay provides a protected area for foraging by salamanders.[22] The cool, moist environment of rotten wood may be required for some species of salamanders to survive heat stress during summer. Decaying wood also provides habitat for invertebrates on which salamanders and other foraging vertebrates feed (e.g., collembolans, isopods, millipedes, mites, earthworms, ants, beetles, flies, spiders and snails).[230] The folding-door spider constructs a silk tube within the cracks and crevices of wood with advanced decay.

Generalizations regarding decay classes of habitat structures must be made cautiously. Habitat functions may encompass several decay classes or be specific to particular patterns and classes of decay.[56, 233]

Habitat structures in upper layers of the forest floor (soil, litter, duff) result from processes involving organic material (litter, decaying roots, vertebrate and invertebrate carrion, and fecal matter) and a diverse community of organisms, including bacteria, fungi, algae, protozoa, nematodes, arthropods, earthworms, amphibians, reptiles,

Figure 8. Jackstrawed logs offer hunting areas and den sites for lynx. Photo: Gary Koehler.

Figure 9. Townsend's chipmunk, Photo: D. Anderson, National Museum.

and small mammals[167, 185, 207, 230, 249, 376] The complex trophic web supported by nutrient and moisture conditions within the litter and duff layers transforms plant material into a variety of degradation products, thereby storing and releasing nutrients within the ecosystem.

According to the Habitat Elements matrix on the CD-ROM with this book, in forest environments, 86 vertebrate wildlife species are associated with down wood, 38 with litter (undecomposed fine wood), and 14 with duff (decomposed litter and other vegetation matter underlying the litter layer). Of these species, 58 are associated exclusively with down wood, 10 exclusively with litter, and none exclusively with duff. Wildlife uses include reproduction, hibernation, feeding, resting, sunning, drumming, preening, dusting, lookout and travel. All decay stages of down wood are used by wildlife.

Decaying wood forms many habitat structures in riparian forests. Accumulations of large wood on stream banks provide habitat for small mammals and birds that feed on stream biota,[103, 104] and provide structural diversity in streamside forests.[233, 261, 262, 376] The matrices on the CD-ROM with this book list 27 wildlife species associated specifically with down wood in riparian areas.

The role of down wood in salmon habitat has received much attention over the past two decades. Large wood is a key component of salmonid habitat both as a structural element and as cover and refugia from high flows. Large wood serves key functions in channel morphology, as well as sediment and water routing.[33, 38, 262] The importance of wood to salmon habitat varies from headwater to stream mouth.[235] As stream order increases and gradient decreases in third- to fifth-order streams, down wood is a dominant channel-forming feature. Larger wood deflects water and increases hydraulic diversity, producing a range of pool conditions that serve as habitats for juvenile salmonids in summer.[37] Diverse channel margins are a primary aspect of rearing habitat. Flow obstructions created by large wood provide foraging areas for young salmonid fry that are not yet able to swim in fast currents,[40] and provide refugia to juvenile salmonids at high flow.[257] In higher order streams, flow deflections created by large wood trap sediments and nutrients, and enhance the quality of gravels for spawning. Down wood is less of a channel-forming feature along large rivers, but defines meander cutoffs and provides cover and increased invertebrate productivity for juvenile salmonids.[392]

Long-Term Productivity

Long-term productivity is technically founded on the concept of net primary productivity—the conversion of solar energy to plant biomass, expressed as grams of organic matter per unit area per year. At any point in time, ecosystem productivity may vary depending on the successional stage of vegetation, disturbances, and variation in climate and other environmental factors. In terms of present-day forest management objectives, long-term ecosystem productivity typically refers to the production, in perpetuity, of diverse resources, including timber, fish, wildlife, and other ecosystem services, such as air and water quality. Processes that sustain the long-term productivity of ecosystems have become the centerpiece of new directives in ecosystem management and sustainable forestry.[78, 229, 291, 320] Given the key role of decaying wood in long-term productivity of forest ecosystems in the Pacific Northwest,[122, 169, 261, 302] the topic should remain of keen interest to scientists and managers during the coming decade.[149] Below, we highlight functions of decaying wood directly linked to long-term productivity, including influences on the frequency and severity of disturbances such as fire, disease, and insect outbreaks.[5, 6, 133, 137]

Nutrient Cycling and Soil Fertility. Decaying wood has been likened to a savings account for nutrients and organic matter,[376] and has also been described as a short-term sink, but a long-term source of nutrients in forest ecosystems.[164] The total amount and distribution of nutrients in different woody tissues varies from region to region and among forest types. Coarse wood contains 0.3-4.4% of the total nitrogen and 4.2-10.6% of the total phosphorus in westside forests. Nutrient cycling via foliage and fine litter has been well-described.[1, 161, 388, 408] Substantial amounts of nitrogen are returned to the soil from coarse wood inputs, yet even where annual rates of wood input are high, 4 to 15 times more nitrogen is returned to the forest floor from foliage than from large wood.[164] This is a consequence of higher nutrient concentrations and shorter turnover times of leaf litter compared to wood.[164, 300] The relative contribution of large wood to the total nutrient pool in an ecosystem depends to a large extent, on the size of other organic pools in the system.

The proportion of tree nutrition derived from large wood during normal stand development is unknown. The low nutrient content in wood, small mass of tree boles relative to foliar litterfall, and slow rates of wood decay suggest that large wood plays a minor role in forest nutrition.[18, 159, 162] After large scale disturbance such as fire and blowdown, however, the large nutrient pool stored in woody structures of trees (bole, branches, twigs, roots) becomes available to the regrowing forest. Large down wood may thus be an ample source of nutrients throughout secondary succession.[281]

The slow rate of nutrient release from decomposing wood may serve to synchronize nutrient release with nutritional demands in forests, and also to minimize nutrient losses via leaching to the ground water.[8, 34, 83, 174, 341] In addition to nitrogen bound chemically within wood, down wood reduces nutrient losses from ecosystems by intercepting nutrients in litterfall and throughfall. Favorable temperature and moisture conditions also makes large decaying wood sites of significant nitrogen inputs via N-fixation.

Chronosequence studies indicate that large decaying wood accumulates nitrogen, calcium, and magnesium over time, but loses phosphorus and potassium.[76, 117, 146, 164, 205, 340] The time frame for nutrient immobilization and release (mineralization), as well as the specific processes responsible for changes in nutrient content of down wood have not been fully elucidated. Mineralization of nitrogen has been associated with a critical carbon-to-nitrogen ratio of approximately 100. That is, nitrogen and other elements stored in microbial tissues are gradually released to plants as ratios decline below 100.[105, 340] Despite growing recognition of the roles of decaying wood in soil productivity, details of many processes are poorly understood.[125, 131, 164, 376] Harmon et al.[164] describe patterns in the accumulation and release of major nutrients in decaying wood through time.

Recent studies indicate that wood may release nutrients more rapidly than previously thought through a variety of decay mechanisms mediated by means other than microbial decomposers, i.e. fungal sporocarps, mycorrhizae and roots, leaching, fragmentation, and insects.[107, 158, 159, 162, 339, 405] Harmon et al.[162] found that during early stages of decomposition, fungal sporocarps concentrated nitrogen, potassium and phosphorus from down wood. Annual dieback of the sporocarps returned nutrients to the available soil nutrient pool.

Soil is the foundation of the forest ecosystem.[68, 348] Large wood is a major source of humus and soil organic matter that improves soil development.[164, 242] An estimated 9-68% of the forest floor in Douglas-fir/western hemlock (*Tsuga heterophylla*) clearcuts in the Pacific Northwest is derived from decaying wood.[213] On the H. J. Andrews Experimental Forest of western Oregon, 20-30% of the soil volume consists of decaying wood dispersed throughout a matrix of litter and duff.[294] Because wood is a relatively inert substance, it may help to stabilize pools of organic matter in forests by slowing soil processes and buffering against rapid changes in soil chemistry. Humus and soil organic matter are critical to site productivity on dry sites, and may limit site productivity on excessively wet and cold sites.[169, 377, 381]

Litter decomposition processes in the upper soil are the major locus of nutrient cycling within forests of the Pacific Northwest.[1] Soil organic matter and wood can abiotically (chemically) retain large amounts of nitrogen and other elements in forest ecosystems. A significant fraction of the nitrogen is believed to be in soil organic matter pools.[58, 189, 219, 266, 355, 385] Down wood and soil organic matter thus, may regulate productivity in some forest ecosytems by limiting nutrient availability.[106] Numerous studies have demonstrated that losses in soil productivity often are closely linked to losses in soil organic matter.[298]

Moisture Retention. Water stored in large decomposing wood accelerates microbial decay rates by stabilizing temperature and preventing dessication during the summer.[11, 160, 376] Moist conditions within the wood favor decay by attracting burrowing and tunneling mammals and invertebrates that improve aeration of wood, and by providing colonization substrate and moisture for mycorrhizae and other fungi.[88, 168, 206] Moist "nurse logs" also provide excellent sites for seedling establishment and production of sporocarps. These processes increase retention and cycling of nutrients within ecosystems and contribute to higher biodiversity and biomass production.

Mycorrhizae. Mycorrhiza, meaning "fungus-root", is a symbiotic association of fungi with plant roots. The fungus improves nutrient and water availability to the host in exchange for energy derived from plant sugars. Mycorrhizae are necessary for the survival of numerous tree families, including pine, hemlock, spruce, true fir, Douglas-fir, larch, oak, and alder. Mycorrhizal associations are a source of nutrients to promote wood decay. By the time a log reaches more advanced stages of decomposition (Class 3) fungal colonization leads to the accumulation of nutrients in hyphae, rhizomorphs and sporocarps,[23, 87] especially for ectomycorrhizal fungi, where >90% of the fungal activity is associated with organic material.[119, 168, 310] Ectomycorrhizal fungi decrease the ratio of carbon to nitrogen in decomposing wood, and mediate nutrient availability to plants while improving nutrient retention by forest ecosystems.[13, 206, 251]

Mass Wasting and Surface Erosion. In the Pacific Northwest, combined effects of steep slopes, high rainfall, a history of tectonic uplift, and rapid weathering of weak rocks make unstable slopes a dominant erosion process. Large wood helps to anchor snowpacks, limit the extent of snow avalanches, and may even stabilize debris flows, depending on the depth of the unstable area.[125, 356, 358] The energy derived from falling or flowing water is the driving force behind erosion processes in Pacific Northwest forests. By covering soil surfaces and dissipating energy in flowing and splashing water, logs and other forms of coarse wood significantly reduce erosion.[357] Large trees lying along contours reduce erosion by forming a barrier to creeping and raveling soils, especially on steep terrain. Material deposited on the upslope side of fallen logs absorbs moisture and creates favorable substrates for plants that stabilize soil and reduce runoff.[230]

Stand Regeneration and Ecosystem Succession. Decomposing wood serves as a superior seed bed for some plants because of accumulated nutrients and water, accelerated soil development, reduced erosion, and lower competition from mosses and herbs.[160, 376] In the Pacific Northwest, decaying wood influences forest succession by serving as nursery sites for shade-tolerant species such as western hemlock, the climax species in moist Douglas-fir habitat.[80, 123, 160, 163, 244] Wood that covers the forest floor also modifies plant establishment by inhibiting plant growth, and by altering physical, microclimatic, and biological properties of the underlying soil. For example, elevated levels of nitrogen fixation in *Ceanothus velutinus* and red alder[35, 88] have been reported under old logs.

Streams and Riparian Forests. Long-term productivity in streams and riparian areas is closely linked to nutrient inputs, to attributes of channel morphology, and to flow dynamics created by decaying wood.[144, 233, 360] Small wood contributes to nutrient dynamics within streams and provides substrates to support biological activity by microorganisms, as well as invertebrates and other aquatic organisms.[145, 262] Much of the organic matter processed by the aquatic community originates in riparian forests and is stored as logs.[90, 259] Down wood also helps to retain nutrients in streams, by trapping carcasses of dead salmon and increasing carcass availability to terrestrial scavengers.[72]

Large wood is the principal factor determining the productivity of aquatic habitats in low- and mid-order forested streams.[262] Large wood stabilizes small streams by dissipating energy, protecting streambanks, regulating the distribution and temporal stability of fast-water erosional areas and slow-water depositional sites, shaping channel morphology by routing sediment and water, and by providing substrate for biological activity.[361] The influence of large wood on energy dissipation in streams influences virtually all aspects of ecological processes in aquatic environments, and is responsible for much of the habitat diversity in stream and riparian ecosystems.[262, 376] The stair-step gradients produced by wood in small stream basins supports higher productivity and greater habitat diversity than that found in even-gradient streams lacking wood structure.[32, 40, 142, 173, 254, 357]

Key Ecological Functions of Wildlife Species Associated With Decaying Wood

As previously described, decaying wood provides habitat elements for many wildlife species. Beyond this, the associated wildlife species play diverse ecological roles in their ecosystems, which in turn can influence other species. These roles are termed "key ecological functions" or KEFs, which are categorized and denoted for each species in the Key Ecological Functions matrix on the CD-ROM with this book.[224]

Detailed descriptions of the patterns of KEFs of wildlife species associated with wood decay have been summed from the Key Ecological Functions matrix.[223] An associated "functional web" summarizing the ecological roles of wildlife associated with down wood included 86 wildlife species in Washington and Oregon and traced various functional categories.

Similarly, a functional web can be described here for the 96 vertebrate species associated with snags in forest or grassland/shrubland environments, as denoted by the Habitat Elements matrix on the CD-ROM with this book. For example, a query of the KEF matrix for these species shows that 40% of snag-associated species are primary consumers, 95% are secondary consumers, and 8% are carrion feeders. The remaining >5% are tertiary consumers, cannibalistic feeders, coprophages (feed on fecal matter), or feed on human refuse. The primary consumers include spermivores or seed-eaters (63% of all primary consumers associated with snags), frugivores or fruit-eaters (50%), sap feeders (18%). The others include (each <15%) fungivores or fungus-eaters, foliovores or leaf-eaters, grazers, and feeders on flowers or buds, aquatic plants, bark or cambium, and roots. Secondary consumers associated with snags include insectivores (90% of all secondary consumers associated with down wood), vertebrate predators (36%), ovivores or egg-eaters (13%), and piscivores or fish-eaters (8%). Percentages may sum to >100% because some species have multiple roles.

Various symbiotic relations can be described for the 96 snag-associated species. Sixteen species are primary cavity excavators and 35 are secondary cavity users; 8 are primary burrow excavators and 11 are secondary burrow users; 5 are primary terrestrial runway excavators and 6 are secondary runway users. Nine snag-associated species create nesting or denning structures and 8 use created structures. Sixteen species might influence vertebrate population dynamics and 22 might influence invertebrate population dynamics. Snag-associated species also contribute to dispersal of other organisms including seeds and fruits (21 snag-associated wildlife species perform this function), invertebrates (8 species), plants (8 species), fungi (2 species), and lichens (1 species). Six snag-associated species can improve soil structure and aeration through digging, 2 species fragment standing wood, and 2 species fragment down wood. One snag-associated species creates snags, and at least 1 can alter vegetation structure and succession through herbivory.

The ecological roles of wildlife associated with various decaying wood structures have both many similarities and salient differences. For example, both snag- and down wood-associated wildlife more or less equally participate in dispersal of seeds and fruits (although the particular species they disperse may differ); however, snag-associated wildlife play a greater role in dispersal of invertebrates and plants, and down wood-associated wildlife play a greater role in dispersal of fungi and lichens. Down wood-associated species might contribute more to improving soil structure and aeration through digging, and to fragmenting wood.

This is one example of the far greater differentiating power afforded by a well-constructed set of matrixes than was previously available in Thomas[366] and Brown.[48] That is, a combined query of the Habitat Elements matrix and the Key Ecological Functions matrix can produce a list of species with unique combinations of habitat element associations and particular ecological functions. For example, other decaying wood elements, such as hollow live trees, cavities in dead trees, dead parts of live trees, bark crevices and bark piles, support different arrays of wildlife species; their ecological roles (KEFs) may also differ from those discussed here.

It should be recognized, too, that invertebrates play critical major roles in wood decay ecology in both standing and down wood.[327] The full array of ecological roles of invertebrates associated with wood decay and related soil formation processes has been studied little, but is likely to be substantial. Included in 11 functional invertebrate groups described by FEMAT,[114] based on ecological roles, were coarse wood chewers, litter and soil dwellers, understory and forest gap herbivores, canopy herbivores, epizootic forest species, pollinators, riparian herbivores, and riparian predators. A square meter of undisturbed forest soil is estimated to contain from 200 to 250 species of arthropods, including up to 75 species of fungivorous mites reaching densities of 200,000 individuals.[249]

Current Regional Patterns of Decaying Wood

Factors Influencing Regional Abundance of Decaying Wood

Quantities and characteristics of decaying wood in an ecosystem represent a balance between additions through tree death, breakage, transport, and losses through processes of decomposition and fire consumption, all of which can be influenced by forest management activities.[164] Management effects on the abundance of decaying wood can be accurately assessed only within the context of underlying natural patterns of wood dynamics. The following section describes predominant influences on accumulations of decaying wood in forests of the Pacific Northwest, including inputs (primary productivity and mortality), and outputs (decay and combustion), and interactions with various disturbances of natural and human origin.

Wood Input Rates

Primary Productivity. Regionally, ecosystem accumulations of decaying wood are closely related to gradients in net primary productivity as a result of moisture and temperature limitations. Consequently, dead wood abundance in the Pacific Northwest declines along an easterly gradient from the highly productive Douglas-fir zone[120] to the high desert. A summary of existing studies in Washington and Oregon[164] showed greater input rates of decaying wood biomass in mature and old-growth Douglas-fir and Sitka spruce/western hemlock forests (0.20 - 12.14 Mg/ac/yr) than in higher elevation Pacific silver fir (0.12 Mg/ac/yr). No data were available for eastside forests; however, lower accumulations of dead wood in juniper forests are closely tied to low biomass productivity, as well as fire. Detailed discussions of regional patterns in forest productivity have been provided elsewhere.[138, 148, 208, 274, 315, 396, 398, 399]

Successional Development. Pulses in tree mortality occur normally in stand regeneration and development. Self thinning from competition-induced suppression, for example, reduces stand density during the course of even-aged stand development in western Oregon and Washington.[277, 343] Suppressed trees are also more susceptible to insects and diseases. Suppressed trees are typically of small diameter and often remain standing until blown down by wind. Compared to young regenerating stands, natural mortality in old-growth stands usually affects a smaller number of trees, usually older and larger trees with slower inherent rates of decay and longer residence times compared to small trees. Natural aging processes may directly cause tree mortality by impairing tree physiological and hydraulic functioning, or indirectly by predisposing trees to other agents of mortality, such as insects, diseases, and windthrow.[194, 317, 318 393, 394, 400] Consequently, older trees with a higher ratio of heartwood to sapwood are more susceptible to heart-rot fungi, root-rotting fungi, sapwood fungi, and insects such as bark beetles.[79, 108, 237, 400]

The input rates and average piece size of dead wood generally increase with stand age,[164] although the amount of decaying wood can follow a U-shaped pattern if young forests inherit large amounts of decaying wood and live trees from preceding stands.[346] Age-related inputs and accumulation of wood throughout stands succession in the Pacific Northwest have been summarized.[71, 164, 277, 346] Information is lacking to describe natural patterns of decaying wood inputs and accumulation during stand development in forests east of the Cascade Range.

Fire. Prior to European settlement, natural disturbance regimes dominated by fire had a major influence on input rates and accumulations of dead wood in forests of the Pacific Northwest,[2, 5, 390] but other factors such as wind, pathogens, and insects also influenced forest development to varying degrees. Fire effects on ecosystem inputs of wood are varied and complex, depending on specific ecosystem attributes of soil, vegetation characteristics, and historic disturbance patterns.[1] Fire plays a major role in the dynamics of decaying wood by altering the rates and composition (size, species) of dead wood inputs, as well as combustions losses. The historic frequency and severity of fire events varies widely across the region. East of the Cascade Range, pre-settlement forests were subject to frequent low-intensity fires that killed only a small percentage of living trees, but consumed much of the decaying wood (85%), due to low moisture contents of fuel wood during late summer.[5, 24, 95] In moist forests west of the Cascade crest, by comparison, fire disturbances were infrequent, but of high intensity and magnitude, producing high inputs of wood from tree mortality.[2, 345, 346] For example, Agee and Huff[6] reported a 10-fold increase in snags and a 150% increase in down wood in old-growth hemlock-Douglas-fir forests after wildfire.

Insects and Pathogens. Insects and pathogens deserve special mention because they have had a major influence on historic inputs of dead wood in the Pacific Northwest. Although these disturbance agents are natural components of the ecosystems, their prevalence and degree of influence on ecosystem structure and function may be greatly modified by human activities. Insects and pathogens play a key role in maintaining diverse and productive forests by creating habitat and stimulating nutrient cycling. Catastrophic insect and pathogen infestations, though, may be symptomatic of imbalances causing reduced ecosystem productivity.

Insects and pathogens of Pacific Northwest forests are too numerous to describe in entirety here, but detailed information is available elsewhere.[133, 220, 322] Insects can be separated into general groups of defoliators, sap-suckers, bark beetles, wood borers, cone and seed insects, and excavators of seasoned wood. Insects may be primary agents of tree mortality or incidental agents that take advantage of trees weakened by other stresses. Bark beetle species represent both groups and may inhabit trees that are healthy, dying, or recently dead. Of the numerous varieties of tree pathogens, those that cause brooming in conifers induce substantial inputs of decaying wood in Pacific Northwest forests. These include the dwarf mistletoes (*Arceuthobium* spp.), rust fungi (*Chrysomyxa* spp. or *Melampsorella* spp.), or the needle cast fungus (*Elytroderma deformans*). Dwarf mistletoes are perennial parasitic plants that derive water and nutrition from their hosts, progressively killing branches, trees, and sometimes entire stands of vulnerable species.[171] Broom rusts are obligately parasitic pathogenic fungi that form witches' brooms on their coniferous hosts.[407] Needle cast fungus causes early needle drop in pines, and can form witches' brooms similar in appearance to dwarf mistletoe.[77]

Other Natural Disturbances. Flooding has diverse effects on ecosystems processes, thus specific effects on net accumulation of decaying wood may be difficult to discern. Streamside forests are often perturbed and even destroyed by flooding. Periodic flooding produces pulsed inputs of wood to forest ecosystems, but chronic flooding may eventually cause conversion to non-timber vegetation by impairing site productivity for trees. Reduced aeration in flooded soils can inhibit decay rates of wood in close contact with the soil, such as soil humus, small diameter wood, and large wood in advanced stages of decay. Poor aeration inhibits nitrogen mineralization and induces denitrification, reducing nitrogen supply to wood decomposers.[179] Prolonged flooding that eliminates beneficial populations of soil decomposer organisms thus, can also inhibit decay rates of wood.[353]

Wind is one of the most common disturbances to forests, with damage ranging from single trees to many hectares. In historic times, wind was the major disturbance factor controlling inputs of decaying wood to forests west of the Cascade crest in Washington and Oregon, and a primary influence in maintaining soil productivity.[46, 399] Wind produces pulsed inputs of decaying wood, often selecting for larger trees, though smaller trees in dense stands may also be highly vulnerable to wind, especially along natural topographic breaks, and on the windward side of harvest edges. Windthrow frequently also triggers outbreaks of bark beetles. Windthrow risk is difficult to predict, as it is is influenced by a complex interactions at scales ranging from individual trees to stands, to the surrounding landscape. Excellent discussions of windthrow processes in ecosystems have been presented elsewhere.[270, 335, 349] A recent study by Sinton et al.[337] is the first to evaluate the landscape context of forests as well as environmental factors contributing to historic windthrow patterns.

The Pacific Northwest is a region of active volcanic influence, most recently impacted by the explosion of Mt. St. Helens.[128, 129] Volcanic events variably contribute to large input, redistribution, and consumption of dead wood to ecosystems depending on proximity to the blast zone, pyroclastic flows, and debris flows. Thus, effects of volcanic activity on the subsequent abundance and decay rates of wood may be highly variable and difficult to predict. Mudflows from volcanic events can completely cover down wood and create a sanitized soil surface with reduced populations of decomposers.[101] However, plant succession after volcanic events may include many N-fixing species that serve to increase N availability and accelerate wood decomposition.

Landslides are a natural but infrequent occurrence (estimated interval of 500-1500 years) on steeper slopes within the region.[31, 190, 361] Altered slope hydrology and soil stability due to activities such as road building and logging, however, can increase the intensity, magnitude, and frequency of such events.[262] Landslides in the upper basins of steep mountainous terrain form debris flows that transport most of the sediment and wood in first- and second-order (i.e. low-order) channels.[359] Landslides and debris flows also contribute most of the sediment and coarse material into larger third to fifth-order channels (i.e. mid-order). These geomorphic events increase and redistribute large amounts of wood in Northwest ecosystems.

Wetlands Perpetuate Wood Legacies
Richard Bigley

Forested wetlands in the Pacific Northwest often contain a disproportionally high concentration of snags and down wood compared to most upland forests. Inundation promotes the accumulation of down wood by limiting decay and creating snags. Decaying wood contributes to many wetland functions and exerts considerable influence on water quality and flow routing in the forest.

Environmental conditions for wood decay in wetlands. Flooding often produces anaerobic soil conditions that curtail the decomposition of wood and other organic matter for many years. Buried wood thousands of years old has been recovered from deep peat and other anaerobic environments. Anaerobic conditions in wetlands inhibit wood decomposition rates by favoring facultative and obligate anaerobes that are less efficient dcomposers than areobic bacteria. Many fungal decomposers become dormant when water-saturated, and continue to grow only when oxygen is restored.

Clues as to the frequency, duration, and seasonality of wetland conditions can be inferred from soil features. Short periods of flooding may not be sufficient to inhibit wood decomposition in upland forests. Soils in areas without prolonged flooding usually have evenly-distributed red and yellow colors characteristic of oxidized iron. In contrast, forests with prolonged flooding often have deep organic surface horizons and underlying mineral soils lacking oxidized iron. Fluctuating water tables can usually be identified by the degree of "mottling" (red and yellow flecks) resulting from variable oxygen levels (and hence iron oxidation) in soils.

Wood structures contribute to wetland ecological function. Water storage in wetlands serves to reduce flooding at lower elevations, and to moderate seasonal variability in streamflow. Hydrological functions and surface runoff thus are major concerns in managing forests in and around wetlands. In forested wetlands where ground vegetation can be sparse, down wood is often a dominant influence in water routing and channel morphology.

Down wood significantly reduces erosion by dissipating flow energy and trapping sediments. Stable wood structures in wetlands create effective barriers to downslope transport of unconsolidated sediments, and facilitate vegetation establishment. Major fluvial erosion and deposition processes are influenced by overland flow resulting from heavy rainfall events and the limited infiltration capacity of some forest soils in the Pacific Northwest.

As with upland sites, down wood in wetlands provides sites for plant germination and establishment, protects vegetation from grazing, and serves as travel corridors for wildlife.

Wetland management and wood legacies. Forest management in and around wetlands is primarily concerned with soil disturbance and windthrow. Selective harvest and salvage of valuable western red cedar (*Thuja plicata*) has been the main impetus for management activities in forested wetlands of the Pacific Northwest. Management objectives to retain snags and down wood in forested wetlands require more detailed consideration of the different ecological functions of fast-growing hardwoods like popular (*Populus tremulodes*) and more decay-resistant but slower-growing species like western red cedar.

Wood Output Rates

Aside from wood removal in commercial harvesting, decay and combustion processes are the primary factors influencing the ouputs of decaying wood over time. Wood decay rates have been estimated for a number of forest types, but are highly sensitive to wood structure and chemistry as described earlier, and to climatic conditions, especially temperature, moisture, and aeration.[18, 105, 113, 146, 164, 300, 340] Decay rates of wood are most rapid under conditions of warm temperature and good aeration that stimulate the metabolic rates of decomposer organisms. Adequate moisture also is required to soften wood structure.[376]

Seasonal patterns of temperature and moisture in decaying wood have been reviewed in detail.[164, 230] In the western Cascades, decay is limited more by temperature during the winter and by moisture in the summer. Aeration may limit decay where poor drainage results in saturated soil and a perched water table. The brief warm season limits decay processes more in the eastern

Cascades. Mesoscale atmospheric phenomena (e.g. orographic effects) that introduce finer-scale climate patterns also produce more variability in wood decay rates within ecosystems. Microclimatic variability at the scale of stands and individual trees is also affected by forest type, successional stage, aspect, slope, elevation, soil drainage, shading, and other factors.[131, 141, 146, 234] Wood size, shape, and placement affect decomposition rates by changing microclimates and wood exposure to decomposers.[56, 1:5, 205, 218, 341]

In addition to stimulating inputs of dead wood, fire eliminates dead wood via processes of decay and combustion. Decay rates are influenced by climatic and nutritional conditions for decay organisms in the post-fire environment.[193, 245, 379] Higher rates of wood decay in the warmer temperatures of forest openings created by fire, may be counterbalanced by moisture limitation. Similarly, changes in nutrient availablility after fire, may either stimulate or inhibit decay. Cool-burning fires consume less wood and produce a pulse of nutrients to decomposers.

Conversely, high intensity fires consume more wood and may cause nutrient losses via volatilization and leaching, or by depressing populations of beneficial microbial and invertebrate decomposer organisms in soil.[2, 5, 43] In addition, charring and case-hardening of wood surfaces by fire can retard decay processes.[218]

Effects of Human Disturbances on Wood Dynamics

Timber Management. Intensive forest management for timber objectives has been shown to simplify stand and landscape structure.[154] Harvest and burn cycles of 60-100 years in the western Cascades have been estimated to be equivalent to a 4- to 5 -fold increase in fire frequency compared to natural conditions, in terms stand structure.[368] Two rotations of managed Douglas-fir have been estimated to reduce the abundance of dead wood by 90%, compared to levels in natural old-growth systems. In addition to removing wood, management activities also influence wood decay rates. For example, warmer microclimates after harvesting stimulate decay. Fertilization conducted to improve wood production in commercial forest lands[312] can accelerate decomposition rates of down wood by supplying decomposer organisms with limiting nutrients, particularly nitrogen. Conversely, intensive practices that deplete soil nitrogen, such as short-rotation, whole tree harvesting[198] may inhibit rates of wood decomposition over the longer term. Models estimate that between 50 to 100 years are required to replace nitrogen losses after timber harvest.[1]

Carbon Storage and Climate Change. Douglas-fir forests in low elevation temperate rain forests have the capacity to store more carbon in the form of biomass and dead wood than any forest ecosystem on earth.[197, 258] Consequently, these forests provide another benefit of global significance—the ability to sequester carbon from the atmosphere, and potentially to counteract global warming.[16] Intensive forestry practices that released large quantities of carbon stored as biomass and dead wood in Northwest forests during the 1980s have been implicated as a contributing factor in the rising atmospheric concentrations of carbon dioxide.[19, 161] The relative contribution of carbon storage in terrestrial forests to climatic warming remains equivocal, however.[16] The potential for warmer climates to accelerate decomposition of decaying wood in terrestrial systems remains an issue of concern.[294]

Furthermore, ecosystem accumulation of total wood (live and dead) has been postulated to increase in response to temporary stimulation of biomass productivity by higher temperatures, carbon dioxide, and nitrogen availability.[19, 308, 309] The increased rates of wood accumulation may level off, however, as nitrogen availability to forests stabilizes.[100] As a result of global warming, complex linkages between temperature, carbon dioxide, plant growth, decomposition, and nutrient cycling could alter the size and relative distributions of carbon and nutrient pools in forest soils and vegetation.

Hence, wood accumulation and decay processes may exhibit considerable variation from one ecosystem to another.

Anthropogenic Pollution. Anthropogenic pollution can affect stand microclimates, soil chemistry, populations of decay organisms, and a variety of basic ecosystem processes, thus it is not surprising that pollutants alter processes of wood decay.[1, 280, 329] Ecological effects on forest ecosystems are highly specific to the source, type, and magnitude of the pollutants. Of the diverse pollutants affecting forest ecosystems, nitrogen pollution is the most widespread and has been thoroughly studied.[166, 219, 264, 288, 329] Nitrogen pollution is of concern because biomass production and storage in many forest ecosystems is tightly regulated by nitrogen availability. One of the most comprehensive studies of pollution impacts to date documented inhibited decomposition rates of organic material at a site receiving chronic inputs of nitrogen and sulfur.[166] Elevated nitrogen loading has been documented in coastal watersheds and estuarine environments;[380] consequently, wood dynamics may be altered in these environments, as well.

Fire Suppression. In the eastern Cascades and through much of the intermountain area, extensive forest insect and disease problems have resulted from decades of fire suppression in combination with selective harvesting of pines.[177, 194, 236, 401, 403] An analysis of landscape dynamics in the Interior Columbia River Basin[302, 379] revealed that fire suppression resulted in a decreased abundance of large-diameter trees, and caused fuel accumulations that predisposed forests to stand-replacement fires. As mentioned previously, more intense fires not only consume more wood, but can inhibit wood decay by reducing nitrogen availability (and other elements) through volatilization and leaching, especially for wood in close association with the soil.[245] Wood decay in post-fire regenerating forests also may be exacerbated by a decline in symbiotic nitrogen-fixing plant species in stands subject to prolonged fire suppression.[169]

Summary of Regional Patterns of Abundance for Snags and Down Wood

Regional variation in dead wood abundance reflects strong underlying gradients in physical environment, disturbance, and biological processes that affect community composition and structure, forest dynamics, and rates of dead wood input and output, as reviewed earlier in this chapter. These factors interact in complex ways in influencing abundance of dead wood on a given site. Basic information about regional ecological patterns of dead wood can provide context for management decisions at a variety of scales, and for analyzing forest policy at regional and national levels. A wealth of information on snags and down wood has recently become available from grids of field plots established by regional forest inventories across Washington and Oregon. This section summarizes findings of a recent analysis of these data by Ohmann and Waddell.[276]

Ohmann and Waddell[276] analyzed data on dead wood collected on 16,867 field plots in 9 forested wildlife habitat types.[188] The plots represent about 49 million ac (20 million ha) of forest within Washington and Oregon, about 54 percent of which is publicly owned, 24 percent is owned by timber industry, and 22 percent is owned by nonindustrial private landowners.[296] The field plots were measured from 1984-1997 as part of inventories conducted by the BLM (Natural Resource Inventory, NRI); the USDA Forest Service, National Forest System (Current Vegetation Survey, CVS); and the USDA Forest Service, Pacific Northwest Research Station (Forest Inventory and Analysis, FIA). Dead wood data were unavailable for national and state parks, and down wood data were unavailable for nonfederal lands in Oregon and western Washington. Snags on plots were sampled on fixed- and variable-radius circular plots, and down wood was sampled on line transects. See Ohmann and Waddell[276] for information about inventory design, field methods, and data summary.

Plots in upland forest were classified into 1 of 9 wildlife habitat types based on potential natural vegetation and ecoregion.[75] Each plot was also classified into 1 of the following 3 successional stages, which are groupings of the structural stages based on current vegetation structure[188]: early (tree stocking [sensu[217]] <10 percent, or tree stocking >=10 percent and quadratic mean diameter [QMD] 1.0-9.8 in [2.5-24.9 cm]), middle (tree stocking >=10 percent and QMD 9.8-19.6 in [25.0-49.9 cm]), late (tree stocking >=10 percent and QMD >=19.7 in [50.0 cm]). The QMD is the diameter of the tree of average cross-sectional area at breast height (4.5 ft; 1.37 m) on the plot.

Differences in Dead Wood Among Wildlife Habitat Types

The abundance of snags and down wood varied substantially across the region. The greatest differences in dead wood abundance were among the wildlife habitat types, although differences among successional stages within wildlife habitat types also were significant in many cases. Total snag densities were greatest at higher elevations: 15.1/ac (37.2/ha) in montane mixed-conifer forest and 14.6/ac (36.0/ha) in subalpine parks (Table 1). Snags were least dense in the drier wildlife habitat types on the eastside: 0.3/ac (0.8/ha) in western juniper woodland and 2.0/ac (5.0/ha) in eastside ponderosa pine (Table 1). Large snags were most abundant in montane mixed-conifer forest (3.8/ac; 9.6/ha) and in westside conifer-hardwood forest (2.2/ac; 5.5/ha), and least abundant in western juniper woodland (0.1/ac; 0.2/ha) and ponderosa pine (0.4/ac; 1.0/ha) (Table 1). The volumes of both total and large down wood were greatest in westside conifer-hardwood forest and lowest in western juniper woodland (Table 2, 3). Total down wood volume among wildlife habitat types ranged from 105.6 to 2,619.7 ft³/ ac (7.4 to 183.3 m³/ha), and large wood from 64.3 to 1,883.7 ft³/ac (4.5 to 131.8 m³/ha). Differences in total dead wood generally were more pronounced among wildlife habitat types on the west side of the Cascades than among the eastside types, and the amounts of total snags and

down wood in montane mixed-conifer forests were significantly different from almost all other wildlife habitat types.

Much of the regional variation in dead wood abundance, expressed as differences among the wildlife habitat types, probably can be attributed to strong gradients in net primary productivity. For example, the large amount of dead wood in westside conifer-hardwood forests probably can be explained by high rates of input within these forests, which are the most productive of the wildlife habitat types.[120] In contrast, the large amount of dead wood in montane mixed-conifer forest also may be explained by slow rates of decomposition in the cold temperatures at high elevations. The high density of snags in the subalpine parkland and montane mixed-conifer types may be a function of high mortality rates and low fall rates in these habitats.

Successional Patterns of Dead Wood

Snag density generally increased with stand age. Within wildlife habitat types, total snag density always was lowest in the early successional stage and usually was highest in the late stage, although no differences were detected among stages of subalpine parkland, ponderosa pine, and western juniper (Table 1). Large snag abundance increased with successional development in all of the wildlife habitat types except western juniper woodland, where no trends were evident (Table 1).

The volume of total and large down wood also generally increased with forest development, but successional patterns differed somewhat among the wildlife habitat types (Table 2, 3). Late successional stages contained the largest concentrations of both total and large down wood in 7 of the 9 wildlife habitat types (Table 2, 3). In the westside wildlife habitats and in montane mixed-conifer forest, the volume of total and large down wood in the late stage usually was significantly different from the early and mid stages, but early and mid stages usually were not significantly different from one another (Table 2, 3). Large down wood volumes differed significantly between the early and middle successional stages in all of the eastside wildlife habitat types except western juniper woodland (Table 3).

No wildlife habitats exhibited a U-shaped pattern in snag abundance, which can occur if large amounts of wood is inherited from a preceding stand.[346] Down wood also most often increased with succession, but this pattern was less consistent than for snags, and some wildlife habitat types did exhibit a U-shaped pattern. The lack of a U-shaped successional pattern for snags is not surprising. Snags have much shorter lag times in the forest than down wood. Natural processes of fragmentation and decomposition begin much sooner, and they disappear as recognizable structures much faster.[164] In addition, much of the dead wood in westside forests is input directly as down wood rather than snags.[164] Snags also are much more likely than down wood to be damaged or intentionally removed by humans through the course of forest management and harvest activities. In a previous analysis of regional plot data, Hansen et al.[154] found that large snags

Table 1. Weighted mean (standard error) density of snags by wildlife habitat type, successional stage, and snag size, Oregon and Washington.

	Early Total	Early Large	Middle Total	Middle Large	Late Total	Late Large	All stages Total	All stages Large
Westside conifer-hardwood	2.1[a] (0.1)	0.8[a] (0.0)	2.4[b] (0.1)	7.2[b] (0.2)	5.8[c] (0.2)	12.7[c] (0.4)	5.8 (0.1)	2.2 (0.1)
Westside white oak-Douglas-fir	2.5[a] (0.5)	0.6[a] (0.2)	1.1[a] (0.1)	4.6[a] (0.4)	2.4[b] (0.4)	6.9[b] (1.0)	4.1 (0.3)	1.0 (0.1)
SW OR mixed conifer-hardwood	3.8[a] (0.4)	1.0[a] (0.1)	2.0[b] (0.1)	6.9[b] (0.4)	3.8[c] (0.2)	8.5[b] (0.4)	6.2 (0.2)	2.1 (0.1)
Montane mixed-conifer	7.2[a] (0.6)	1.2[a] (0.1)	4.2[b] (0.2)	20.0[b] (0.5)	8.8[c] (0.3)	16.3[c] (0.6)	15.1 (0.4)	3.9 (0.1)
Subalpine parkland	14.1 (3.0)	0.7[a] (0.2)	2.3[b] (0.4)	15.1 (1.7)	NA	NA	14.6 (1.7)	1.5 (0.2)
Eastside mixed-conifer	6.0[a] (0.4)	0.8[a] (0.1)	8.7[b] (0.2)	1.7[b] (0.0)	8.4 (0.6)	3.2[c] (0.2)	7.9 (0.2)	1.5 (0.0)
Lodgepole pine	6.7[a] (0.6)	0.3[a] (0.0)	11.2[b] (1.1)	0.9[b] (0.2)	NA	NA	8.0 (0.5)	0.5 (0.0)
Ponderosa pine (eastside)	2.0 (0.2)	0.4 (0.0)	2.0 (0.2)	1.1 (0.0)	2.1 (0.3)	0.6 (0.1)	2.0 (0.1)	0.4 (0.0)
Western juniper	0.3 (0.1)	0.1 (0.0)	0.6 (0.2)	0.2 (0.1)	0.3 (0.2)	0.1 (0.1)	0.3 (0.1)	0.1 (0.0)
All wildlife habitat types	3.7[a] (0.1)	0.8[a] (0.0)	8.6[b] (0.1)	2.1[b] (0.0)	11.5[c] (0.2)	5.4[c] (0.1)	7.0 (0.1)	2.0 (0.0)

NA: Not applicable—sample size <10 plots.

Total = snags >=10.0 in DBH, decay classes 1-5, and >=6.6 ft tall

Large" = >=19.7 in DBH, decay classes 1-5, and >=6.6 ft tall).

Significantly different means (alpha=0.05) within rows (among successional stages for a given snag size-class) are indicated by different letter footnotes.

Modified from Ohmann and Waddell (in press).

were 3-5 times more abundant in stands that had never been clearcut than in stands that had been clearcut at least once. These factors taken together suggest that snag levels would more closely track recent disturbance and forest succession, while down wood amounts would be more strongly influenced by the long-term history and productivity of the site.

Dead Wood in Wilderness Areas

Over all wildlife habitat types, large snags were more than twice as dense in wilderness areas than outside wilderness (Figure 10 a). The strongest differences were for westside conifer-hardwood forest (2.06/ac [5.1/ha] outside wilderness vs. 6.15/ac [15.2/ha] within wilderness), eastside mixed-conifer forest (1.30/ac [3.2/ha] vs. 3.84/ac [9.5/ha]), and lodgepole pine (0.3/ac [0.8/ha] vs. 1.1/ac [2.7/ha]). In contrast, large down wood was more abundant outside wilderness than within wilderness in all of the wildlife habitat types except eastside ponderosa pine (Figure10 b), although the differences usually were not significant. The most pronounced differences in down wood volume were in southwest Oregon mixed conifer-hardwood (918.9 ft³/ac [64.3 m³/ha] outside wilderness

vs. 323.1 ft³/ac [22.6 m³/ha] inside wilderness) and montane mixed-conifer (1,061.9 ft³/ac [74.3 m³/ha] vs. 502.9 ft³/ac [35.2 m³/ha]).

The wilderness stratification of the plots was intended to separate plots with different likelihoods of having been disturbed by timber harvest and management. However, comparisons of dead wood within and outside of wilderness areas must be interpreted with caution. Plots in wilderness are strongly biased towards higher elevations and lower productivities, which may account for much of the higher amounts of down wood outside wilderness. In addition, even though wilderness areas are off-limits to future timber harvesting, they have been affected by other human activities to some degree (e.g., roads, recreation, exotic species introduction, fire suppression). Furthermore, many plots outside wilderness areas sample old growth and younger natural forest.

Nevertheless, if snags are more strongly influenced by timber management activities than down wood, then wilderness areas would be more likely to contain greater amounts of snags than areas outside wilderness, as the data show. In fact, OSHA standards historically have

Table 2. Weighted mean (standard error) volume, percent cover, and density of total[1] down wood by wildlife habitat type and successional stage, Oregon and Washington.

Wildlife habitat type	Successional Stage											
	Mean (SE) cubic feet per acre				Mean (SE) percent cover				Mean (SE) pieces per acre			
	Early	Middle	Late	All stages	Early	Middle	Late	All stages	Early	Middle	Late	All stages
Westside conifer-hardwood	2,169 (109)	2,396ᵇ (89)	3,234ᵇ (120)	2,619 (61)	4.4ᵃ (0.2)	4.5ᵃ (0.1)	5.5ᵇ (0.2)	4.8 (0.1)	132.1ᵃ (5.6)	41.6ᵃ (1.1)	101.5ᵇ (2.8)	110.0 (2.1)
Westside white oak-Douglas-fir	772 (167)	632 (87)	1,017 (176)	733 (73)	1.8 (0.3)	1.5 (0.1)	2.2 (0.3)	1.7 (0.1)	70.5 (11.1)	16.6 (1.7)	51.9 (7.5)	48.9 (3.8)
SW OR mixed conifer-hardwood	1,213ᵃ (123)	983ᵃ (66)	1,695ᵇ (114)	1,227 (54)	2.5ᵃ (0.2)	2.2ᵃ (0.1)	3.2ᵇ (0.2)	2.5 (0.1)	64.3ᵃ (5.9)	24.2ᵃ (1.3)	69.7ᵇ (3.8)	63.5 (2.3)
Montane mixed-conifer	1,459ᵃ (91)	1,608ᵃ (47)	2,837ᵇ (151)	1,769 (46)	3.8ᵃ (0.2)	4.3ᵇ (0.1)	5.1ᶜ (0.2)	4.3 (0.1)	102.8ᵃ (4.1)	41.0ᵇ (1.1)	90.0ᶜ (3.8)	99.8 (1.9)
Subalpine parkland	502 (104)	789 (149)	NA	629 (90)	1.6 (0.3)	2.1 (0.4)	NA	1.8 (0.2)	45.8 (7.0)	21.1 (3.2)	NA	47.8 (5.1)
Eastside mixed-conifer	672ᵃ (31)	780ᵇ (19)	840ᵇ (70)	753 (16)	2.0 (0.1)	2.2ᵃ (0.0)	1.9ᵇ (0.1)	2.1 (0.0)	63.9 (1.2)	23.4ᵃ (0.4)	41.3ᵇ (2.7)	58.5 (0.9)
Lodgepole pine	714 (36)	792 (60)	NA	734 (31)	2.9 (0.1)	2.8 (0.2)	NA	2.9 (0.1)	81.6 (3.4)	26.4 (2.0)	NA	77.3 (2.8)
Ponderosa pine (eastside)	304 (19)	406ᵇ (20)	282 (43)	362 (14)	0.8ᵃ (0.0)	1.0ᵇ (0.0)	0.7ᵃ (0.1)	0.9 (0.0)	28.4ᵃ (1.6)	12.9ᵃ (0.5)	18.1ᵃ (2.8)	29.7 (0.9)
Western juniper	107 (44)	104 (27)	107 (67)	106 (27)	0.2 (0.1)	0.3 (0.1)	0.3 (0.2)	0.2 (0.0)	6.6 (1.7)	3.8 (1.0)	10.9 (6.6)	7.8 (1.4)
All wildlife habitat types	1000ᵇ (27)	1,109ᵇ (19)	2,738ᶜ (19)	1,255 (17)	2.6ᵃ (0.1)	2.7ᵇ (0.0)	4.1ᶜ (0.1)	2.9 (0.0)	74.0ᵃ (1.4)	27.4ᵇ (0.3)	77.5ᶜ (1.6)	71.0 (0.7)

NA: Not applicable—sample size <10 plots.

1. Total = >=4.9 in large end diameter, decay classes 1-4, and >=6.6 ft long.

Significantly different means (alpha=0.05) among successional stages are indicated by different letter footnotes.

Modified from Ohmann and Waddell (in press).

Table 3. Weighted mean (standard error) volume, percent cover, and density of "large" down wood* by wildlife habitat type and successional stage, Oregon and Washington.

Wildlife habitat type	Successional Stage											
	Mean (SE) cubic feet per acre				Mean (SE) percent cover				Mean (SE) pieces per acre			
	Early	Middle	Late	All stages	Early	Middle	Late	All stages	Early	Middle	Late	All stages
Westside conifer-hardwood	1,408[a] (97)	1,709[a] (81)	2,469[b] (114)	1,883 (57)	1.9[a] (0.1)	2.2[a] (0.1)	3.1[b] (0.1)	2.4 (0.1)	23.9[a] (1.6)	19.7[a] (0.9)	24.1[b] (1.1)	22.3 (0.7)
Westside white oak-Douglas-fir	442 (147)	373 (79)	609 (154)	432 (64)	0.6 (0.2)	0.6 (0.1)	0.9 (0.2)	0.6 (0.1)	9.0 (2.4)	5.9 (1.1)	6.5 (1.8)	6.6 (0.9)
SW OR mixed conifer-hardwood	766[a] (107)	609[a] (57)	1,197[b] (106)	803 (49)	0.9[a] (0.1)	0.9[a] (0.1)	1.6[b] (0.1)	1.1 (0.1)	7.9[a] (1.2)	7.4[a] (0.8)	12.0[b] (1.2)	8.9 (0.6)
Montane mixed-conifer	699[a] (80)	660[a] (34)	2,071[b] (137)	906 (40)	1.0[a] (0.1)	1.0[a] (0.0)	2.9[b] (0.2)	1.3 (0.1)	10.3[a] (0.9)	8.7[a] (0.4)	23.4[b] (1.5)	11.6 (0.4)
Subalpine parkland	104 (47)	319 (91)	NA	207 (54)	0.2 (0.1)	0.5 (0.1)	NA	0.4 (0.1)	2.6 (1.2)	4.4 (1.4)	NA	3.4 (0.9)
Eastside mixed-conifer	247[a] (26)	329[b] (14)	513[c] (60)	317 (13)	0.4[a] (0.0)	0.5[b] (0.0)	0.7[c] (0.1)	0.5 (0.0)	3.6[a] (0.2)	4.2[b] (0.2)	6.4[c] (0.7)	4.2 (0.1)
Lodgepole pine	80[a] (13)	209[b] (40)	NA	116 (14)	0.2[a] (0.0)	0.3[b] (0.1)	NA	0.2 (0.0)	1.7[a] (0.3)	3.2[b] (0.7)	NA	2.1 (0.3)
Ponderosa pine (eastside)	147[a] (16)	221[b] (17)	160 (34)	191 (11)	0.2[a] (0.0)	0.3[b] (0.0)	0.3 (0.1)	0.3 (0.0)	2.3[a] (0.2)	2.9[b] (0.2)	2.5 (0.6)	2.7 (0.2)
Western juniper	71 (40)	50 (23)	89 (66)	64 (24)	0.1 (0.0)	0.1 (0.0)	0.2 (0.2)	0.1 (0.0)	2.5 (0.2)	0.8 (0.4)	4.5 (3.3)	0.7 (0.2)
All wildlife habitat types	496[a] (23)	579[b] (16)	1,695[c] (59)	720 (14)	0.7[a] (0.0)	0.8[b] (0.0)	2.2 (0.1)	1.0[c] (0.0)	7.1[a] (0.3)	6.9[b] (0.2)	17.5[c] (0.6)	8.5 (0.2)

NA: Not applicable—sample size <10 plots.

* Large = >=19.7 in large end diameter, decay classes 1-4, and >=6.6 ft long.

Significantly different means (alpha=0.05) among successional stages are indicated by different letter footnotes.

Modified from Ohmann and Waddell (in press).

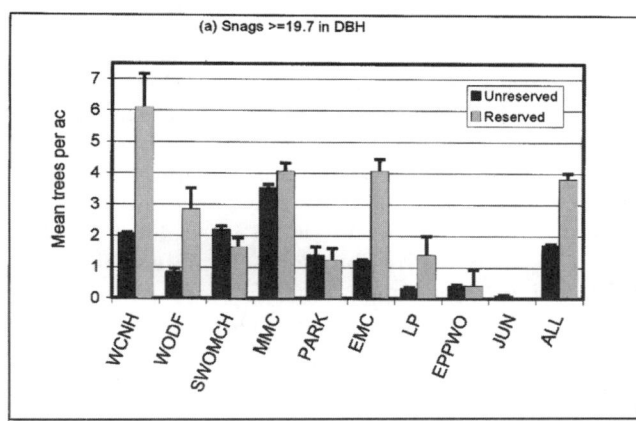

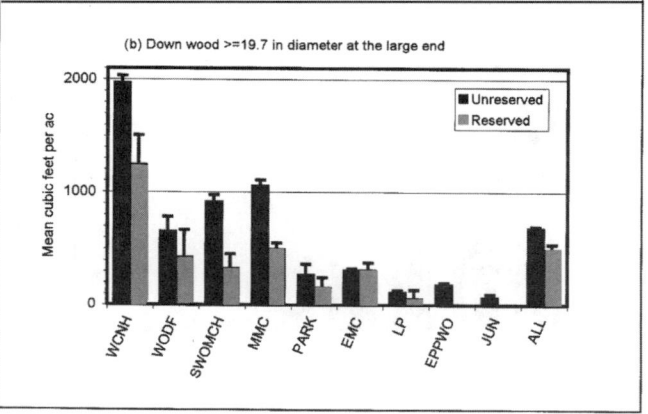

Figure 10. Abundance of dead wood by habitat type and reserved status, Oregon and Washington. (a) Weighted mean density of snags >=19.7 in DBH, decay classes 1-5, and >=6.6 ft tall. (b) Weighted mean volume of down wood >=19.7 in diameter at the large end, decay classes 1-4, and >=6.6 ft long. Error bars indicate one standard error of the mean. WCNH=westside conifer—hardwood, WODF=white oak—Douglas-fir, SWOMCH=southwest Oregon mixed conifer-hardwood, MMC=montane mixed-conifer, PARK=subalpine parkland, EMC=eastside mixed-conifer, LP=lodgepole pine, EPPWO=eastside ponderosa pine—white oak, JUN=western juniper, ALL=all habitats. There were <10 plots in reserved western juniper.

required the removal of most snags from harvest units for worker safety. Therefore, fewer snags were expected in managed stands outside wilderness. If snags are cut and left on site, this would contribute to the larger amount of down wood we observed outside wilderness areas. High snag densities in the high elevation habitats (subalpine parkland and montane mixed conifer) also result from inaccessibility for timber and firewood cutting.

Distribution of Dead Wood Abundance

Estimates of wood abundance in Ohmann and Wadell[276] are regional averages within wildlife habitats. The standard errors of these estimates were fairly low because of the very large sample sizes for most of the wildlife habitat types and successional stages. In reality, the plot-level amounts of dead wood within the wildlife habitat types were extremely variable. This variability reflects the high spatial and temporal variability in the many interacting environmental and disturbance factors that influence dead wood on a site. All of the wildlife habitat types examined had similar patterns: distributions were non-normally distributed and strongly skewed to the right. A large proportion of the plots contained no snags or down wood, and a very small proportion of the plots contained extremely large accumulations of dead wood. Mean values for these skewed distributions must be interpreted with caution. The distribution of snags for the conifer alliance of westside conifer-hardwood forest (Figure 11) illustrate this pattern. In this wildlife habitat type, 39 percent of the area sampled had no snags.

Comparisons with Other Studies

Very few estimates of dead wood abundance at broad geographic scales are available for comparison with results from Ohmann and Waddell.[276] Direct comparisons with published studies are extremely difficult to make because of differences in geographic location; the vegetation types, stand ages, and disturbance histories sampled; sampling design; definitions (e.g., dead wood sizes and decay classes); and units of measure (numbers of trees, volume, density, cover, or linear meters). Other regional studies of dead wood in Washington and Oregon have been restricted either to federal or to nonfederal lands, which usually represent very different ecological conditions.[274] A study by Ohmann et al.[275] was limited to snags on nonfederal lands, because data were unavailable for dead wood on federal lands and down wood on nonfederal lands at that time. The study by Spies et al.[346] was confined to natural Douglas-fir forests >40 years old on federal lands on the westside. Published information for eastside forests is not available,[111] or consists of summaries of a few local studies.[56] Scientists for the Interior Columbia River Basin Ecosystem Management Project relied on expert opinion and local studies to estimate current and historical amounts of decaying wood.[202] Harmon et al.[164] did not include any studies from eastern Washington or eastern Oregon.

Estimates of down wood volume presented by Ohmann and Waddell[276] are somewhat lower than other published numbers, but this is expected for several reasons: the minimum diameters are larger than in many other studies; the data describe both managed and natural forests of all ages, not just older natural forests originating after fire; down wood of decay class 5 is excluded; the values represent means across many stands, including stands where no dead wood was observed, and maximum values are not presented. Estimates of percent cover of down wood also may be lower than in other studies that used plot sampling or total tallies, as percent cover calculated from line intersect sampling has been shown to underestimate true values.[27]

Although estimates of mean total down wood volume in successional stages of westside conifer-hardwood forest ranged from 2,169.2 to 3,233.8 ft³/ac (151.8 to 226.3 m³/ha) (Table 2), the maximum value on a plot was 30,625.9 ft³/ac (2,142.9 m³/ha). This compares to a range of 4,416

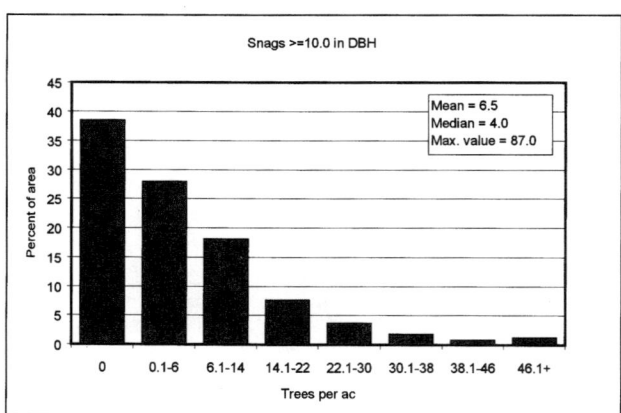

Figure 11. Density of snags >=10.0 in DBH, decay classes 1-5, and >=6.6 ft tall across plots in the conifer alliance of westside conifer-hardwood forest, Oregon and Washington, displayed as a percent of the sampled area.

to 20,306 ft³/ac (309 to 1,421 m³/ha) in various studies in westside Douglas-fir-western hemlock summarized by Harmon et al.,[164] and to 2,115 to 4,473 ft³/ac (148 to 313 m³/ha) reported by Spies et al.[346] The large snag densities in westside conifer-hardwood forest (Table 1) were substantially less than those reported by Spies et al.;[346] the estimate of 0.9 large snags/ac (2.1 large snags/ha) in early stages probably represent stands younger than the 40 yr minimum sampled by Spies et al.;[346] the estimate for middle-successional stages of 2.4/ac (6.0/ha) compares to 10.9/ac (27/ha) in their young stands; and the estimate for late stages of 5.70/ac (14.3/ha) compares to their mature 6.48/ac (16/ha) and old growth 10.9/ac (27/ha) classes.

Estimates of decaying wood presented by Ohmann and Waddell[276] are not directly comparable to those reported in most wildlife studies.[226] Wildlife studies typically describe habitat conditions around nest sites, where decaying wood abundance may be substantially higher than in surrounding stands because many wildlife species select nest sites within clumps of snags.[226] Limited evidence suggests that decaying wood is most often distributed randomly within stands, but sometimes is clumped,[82, 215] (Marcot et al. in prep., unpublished data). Twenty-five percent of stands sampled in the Oregon Coast Range were found to contain patches of 5-10 trees that died simultaneously.[82]

Management Considerations

Management Ramifications of Snag and Down Wood Abundance

Wildlife Species

By querying the CD-ROM matrixes, some generalizations can be drawn about the implications to wildlife of the current status of decaying wood as detailed above. Unfortunately, such inventory data are not available for historic conditions, so little can be quantified about potential changes in wildlife communities.

It is clear that wilderness and parks generally provide substantial standing dead wood habitat, but this provides for wildlife associated more with montane through alpine environments than with lower elevation environments. This would favor such species as black-backed and northern three-toed woodpeckers. The apparent dearth of large snags in Ponderosa pine may mean lower suitability for the 54 wildlife species associated with large snags (20+ in or 51+ cm dbh) in that wildlife habitat.

Intensive forest management activities that have decreased the density of large snags in early forest successional stages (sapling/pole and small tree stages) may have had adverse impacts on the 61 associated wildlife species (Figure 12). Similarly, the lesser amount of large down wood in early forest successional stages may not provide as well for the 24 associated wildlife species. Such results suggest the continuing need for specific management guidelines to provide large standing and down dead wood in all successional stages.

Ecological Processes

Natural patterns of wood distribution and legacy retention are characterized by high variability through space and time. The lack of data on historical conditions of snag and down wood abundance, combined with highly altered disturbance regimes, limits conclusions that may be drawn regarding temporal changes in decaying wood across the region.[164, 178] The interpretation of inventory data will be improved in future years only as more detailed background information on stand conditions and disturbance histories becomes available. Until then, it may be more productive for managers to assess future trends in dead wood abundance and composition using new models that evaluate effects of specific management practices on wood habitat functions, such as DecAID.[223, 226] To assist with the assessment process, the following section examines evidence regarding impacts of past silvicultural practices on wood dynamics in the Pacific Northwest.

During the past century, management practices in the Douglas-fir region have favored even-aged stands, although there was some selective cutting prior to the 1950s (see reviews[94, 365]) Clear-cut harvesting on short rotations was the dominant silvicultural system, depending on site productivity, market preferences, and accessibility. To facilitate site preparation, reserved trees were usually absent and standing snags were felled and slash was burned. Management priorities emphasized a shift in stand composition from species mixtures to monocultures of the most valuable and fast-growing species, such as Douglas-fir. The economic rotation age varied from 40 years to over 100 years.[59] In addition to fire suppression, forest management east of the Cascade crest emphasized even-aged silviculture via clear-cutting, and uneven-aged silviculture using selective removal of overstory dominants.[59, 137] Piling and burning of slash was common in even-aged management regimes. These silvicultural practices clearly altered the abundance and recruitment of large down wood and snags in managed forests of the Pacific Northwest,[1, 2, 5, 59, 94, 112, 122, 130, 154, 198, 228, 230, 261, 262, 275, 276, 279, 302, 330, 344, 346, 357, 376] including:

1. Lower abundance of large diameter snags and down wood "legacies" in managed forests (and streams); e.g. lack of the U-shaped pattern; higher accumulation of smaller-diameter fuels in eastside forests.
2. Reduced recruitment and retention of large trees to provide future "legacies"
3. Shorter mean residence time for down wood (i.e. faster decomposition as a function of reduced log diameter).
4. Altered species composition of forests (westside: more Douglas-fir, less western red cedar; eastside: less pine, more true fir species).

Depletion of Large Wood. The loss of large wood structures has numerous potential impacts on ecological functions of forests, although available information is inadequate for a definitive assessment. The lack of large logs on steep slopes can decrease water percolation into soil, impair slope stability, accelerate soil erosion and

Figure 12. Management practices of the past have simplified forest structure. Note the lack of snags in this regenerating forest on the right side of this hillslope near Satsop, Washington. Photo: Deborah Lindley.

sediment input to streams, and increase nutrient losses in litter.[164, 358, 359, 360, 361] Some data support a linkage between intensive management (especially depletion of decaying wood) and reduced forest biomass productivity, particularly on less productive sites. Lower productivity is attributed to nutrient losses from managed forests, reduced nutrient availability in older stands, and decreased nutrient storage, particularly in the soil.[272, 383, 384] Depletion of soil organic matter has been cited as a primary factor contributing to declining forest productivity and biodiversity in the Pacific Northwest and elsewhere.[17, 137, 198, 199, 228, 292, 293, 298, 299]

One study of a western hemlock stand in Oregon determined that after 150 years of intensive utilization and short rotations (Forcyte-10 model), reduced timber yields were linked to depleted soil organic matter and lower nutrient availability.[319] Such imbalances frequently reduce tree growth in early stages, and impair ecosystem resilience to disturbance in later stages.[9, 291, 292, 293] In contrast, Curtis et al.[94] review evidence that forest management does not degrade site productivity in moist temperate forests. Management effects on site productivity are likely to be highly variable, depending on site-specific attributes of climate, soil, and vegetation.

Small Wood Accumulation. In eastside forests subject to fire suppression, large accumulations of small wood fuels have been shown to improve site productivity due to increased soil organic matter and nutrient retention. These nutritional benefits to tree growth may be short-lived, however, as the forests are predisposed to high intensity fires that consume wood and accelerate nutrient losses.[137, 170]

Riparian Forests. Human alteration of riparian and aquatic ecosystems in the Pacific Northwest over the past century has proceeded with limited understanding of the ecosystems and the effects of human disturbances. The few studies that have examined streams as ecological systems (see review in Naiman et al.[262]) highlight the spatial and temporal complexity of riparian ecosystems, and the importance of connectivity to diverse ecosystem functioning. A balanced ecosystem is one in which biodiversity, productivity, biogeochemical cycles are in balance with the geological and climatic conditions of the region.[191, 192] The delivery of woody material to stream channels has been identified as a key process determining the ecological functioning of watersheds in the Pacific Northwest coastal ecoregion.[33, 40, 260, 262] This statement may be extended to most areas west of the Cascade crest in Oregon and Washington. Far-reaching effects of the absence of large wood structures in streams include: 1) simplification of channel morphology, 2) increased bank erosion, 3) increased sediment export and decreased nutrient retention, 4) loss of habitats associated with diversity in cover, hydrologic patterns, and sediment retention.[33, 144, 262] In coastal environments and estuaries, the loss of large wood may disrupt trophic webs and alter coastal sediment dynamics.[233]

Other Management Effects. In forests throughout the region, multiple stand entries and stump residues have spread diseases, particularly root pathogens. Root diseases initiate changes in stand composition, structure, and growth that alter nutrient and biomass dynamics of ecosystems.[137, 220, 237] Root diseases also predispose trees to windthrow. Injury to residual trees during partial cutting contributes to the spread of stem decays that influence input rates of dead wood and provide wood structures for wildlife habitat. Silvicultural practices that produce multi-layered stands of susceptible species may also affect ecosystem processes by favoring the spread of mistletoe. In general, it is important to recognize that stresses imposed by management have the potential to alter the abundance and dynamics of decaying wood by altering ecosystem energy and nutrient cycles, stand microclimates, and forest susceptibility to disturbance.

Lessons Learned During the Last Fifteen Years

What new developments have ensued since publication of Thomas[366] and Brown[48] in our applied knowledge of wildlife-wood decay relations? Much basic research has been conducted on terrestrial vertebrates associated with snags, although much remains to be learned of wildlife associated with the other forms of decaying wood. Major research initiatives such as the Demonstration of Ecosystem Management Options (DEMO) study by USDA Forest Service are helping to quantify relations of wildlife presence to down wood. Nothing can replace field zoology and ecology in providing such basic understanding of wildlife relations with decaying wood.

Several models have been introduced that help track the demography of snags,[60, 183, 221, 286, 306] and several other unpublished models provide further evidence. More recently, limited capabilities for handling snag dynamics have been incorporated into the Forest Service's Forest Vegetation Simulator (FVS) model. Some of these models track numbers of snags over time by dbh and decay class, and variously incorporate creation, decay, and fall rates of snags. The only available down wood dynamics model to date is the Coarse Wood Dynamics Model for the western Cascade Mountains of Oregon and Washington.[247]

No model, however, had been advanced to replace the "biological potential" models[263, 369] until the DecAID model was developed for this book[188] and USDA Forest Service (see next section). Several major lessons have been learned in the period 1979-1999 that have tested critical assumptions of these earlier management advisory models:

- Calculations of numbers of snags required by woodpeckers based on assessing their "biological potential" (that is, summing numbers of snags used per pair, accounting for unused snags, and extrapolating snag numbers based on population density) is a flawed technique. Empirical studies are suggesting that snag numbers in areas used and selected by some wildlife species are far higher than those calculated by this technique.[226]

- Setting a goal of 40% of habitat capability for primary excavators, mainly woodpeckers,[369] is likely to be insufficient for maintaining viable populations.
- Numbers and sizes (dbh) of snags used and selected by secondary cavity-nesters often exceed those of primary cavity excavators.
- Clumping of snags and down wood may be a natural pattern, and clumps may be selected by some species, so that providing only even distributions may be insufficient to meet all species needs.
- Other forms of decaying wood, including hollow trees, natural tree cavities, peeling bark, and dead parts of live trees, as well as fungi and mistletoe associated with wood decay, all provide resources for wildlife, and should be considered along with snags and down wood in management guidelines.
- The ecological roles played by wildlife associated with decaying wood extend well beyond those structures per se, and can be significant factors influencing community diversity and ecosystem processes.

We have also learned that managing forests with decay processes should be done as part of a broader management approach to stand development, with attention paid to retaining legacies of large trees and decaying wood from original or prior stands. Further lessons have been learned in the area of technical and operational developments; some of these are discussed below.

Habitat Assessment

Regional summaries of the current abundance of snags and down wood presented by Ohmann and Waddell[276] have several potential applications to forest management, planning, and policy. One use is in broad-scale assessments of wildlife habitat. Managers and planners can compare the inventory information on current dead wood resources to characteristics selected and used by wildlife and other organisms in developing management guidelines for federal lands, or in evaluating forest practice regulations or incentive programs for state and private lands. Such guidelines currently are based on very limited scientific data.

Comparisons of inventory estimates to those reported in most wildlife studies are complicated, however, by the fact that inventory estimates represent average conditions within a wildlife habitat type at the regional level rather than around nest sites (see earlier discussion), and few wildlife studies report both values (nest site and stand average) for comparison.[226] Furthermore, although the analysis of inventory data presents data on decaying wood abundance, management actions at the local level may best be focused on the ecological processes that lead to development of these forest structures rather than on the abundance of structures themselves. Management decisions also may require information on the spatial distribution (landscape pattern) of decaying wood, which cannot be estimated from sample-based inventories.

Information on regional patterns of decaying wood is currently being incorporated into the DecAID model[248] depicting the range of natural conditions. This information is intended to help guide managers in considering dead wood and processes of decomposition in forest management. The regional inventory database contains information on occurrence of pathogens such as stem decays and root diseases that contribute dead wood. In addition, the data contain new information about the range of variability in decaying wood—both historically and in the current landscape. The range of variability in decaying wood abundance among plots in the region can help guide management decisions regarding the desired distribution of decaying wood within a large landscape or watershed, but not spatial variation within stands.

Caution must be exercised in using the regional plot data to describe the historical range of conditions in dead wood. The regional plot data sample only current conditions and lack data on site history, as discussed earlier. Even if plots in "natural" forest could be identified, current levels of decaying wood have been altered to an unknown degree by fire suppression and other human influences. On the eastside in particular, current levels of decaying wood may be elevated above historical conditions due to fire suppression and increased mortality, and may be depleted below historical levels in local areas burned by intense fire or subjected to repeated salvage and firewood cutting. Plot data from natural forests on the westside, where fire return intervals are longer,[5] may provide a reasonable approximation of historical conditions.

Template for a Stepwise Assessment Process

The following template presents a stepwise process for habitat assessment and management planning to achieve objectives for down wood and snags. These recommendations are based on existing knowledge of wood dynamics and decay processes in ecosystems, and insights developed by the inventory analysis of Ohmann and Waddell.[276] This process also can be applied in managing for other forms of decaying wood or wood legacies.

Set Quantitative Objectives. To develop effective management plans for down wood and snags, it is imperative to evaluate: 1) land ownership and desired biodiversity, in terms of wildlife species to be supported, 2) specific land allocation, 3) site capability and history, and 4) management scale. Consideration of these factors assists in identifying a range of target values as guideposts for management activities.

Evaluate Current and Historical Status of Decaying Wood. Information on the current and historical status of decaying wood is helpful in determining how the existing patterns in abundance, composition, and distribution of down wood and snags relate to objectives identified for the area of interest. If detailed data on the current and historical range of natural conditions is lacking (which is likely), it may be preferable to substitute functional target values for specific wildlife species. For example, to provide maximum habitat elements for specific cavity-nesting species, a designated quantity and distribution of snags

of a given size and decay class may be the identified objective. The functional objectives should be determined to be within the productive capability of the ecosystem. Productive capability may be evaluated by examining information on primary productivity for the ecoregion and forest type (described earlier) and if available, data summaries on the historical range of natural conditions of decaying wood.

Wood status may include various descriptors of the amount (volume or weight), size class, decay class, and perhaps species of wood structures. Current inventory methods for quantifying the status of decaying wood, including plot design and sampling techniques, have been reviewed.[56, 71, 164, 165, 276, 285] Field inventories should be conducted at the scale at which management guidelines are intended to be developed and implemented, such as among watersheds within basins.

Incorporate Wood Dynamics Into Planning. The extended residence time for decaying wood legacies necessitates that snag and log dynamics models be incorporated into the process of long-term management planning. New and existing wood dynamics models are reviewed below. It is important to remember that such models typically operate at the stand level, whereas management guidelines for snags and down wood are directed at broader scales. Thus, model output should be averaged across stands within watersheds, or management guidelines should address stand- (meaning successional stage) specific conditions.

Monitor Management Effects on Wood Dynamics. Due to the general lack of information guiding management of decaying wood, wood dynamics and associated wildlife response is a priority topic for monitoring and adaptive ecosystem management.[149] Approaches for monitoring natural resources and for modifying management activities must be carefully planned and evaluated, with consideration of appropriate landscape scale.

How To Set Quantitative Objectives Using DecAID

How should the manager determine appropriate objectives for managing decaying wood? This chapter is not intended to prescribe standards for wood decay management. Instead, we offer the following set of questions that the land manager could use to develop overall objectives and specific guidelines for wood decay management.

What is the land ownership and overall expectation for biodiversity conservation? It may be socially acceptable to have different levels of expectation for conservation and maintenance of biological diversity, including decaying wood and associated wildlife, on different land ownerships and land use allocations. Land ownerships are subject to different sets of laws and regulations regarding biodiversity and habitat conservation. For example, commercial forestry landowners may be regulated by the appropriate state forest practices act, as well as any habitat conservation plans, but are not subject to the more stringent wildlife habitat regulations found in the National Forest Management Act that regulate national forests. Consequently, commercial tree farms and commercial forestry lands may call for conservation of decaying wood and associated wildlife at moderate statistical levels presented in the cumulative species curves of the DecAID advisory model, whereas some Federal lands such as national forests and national parks may strive for the high statistical levels.

What is the specific land use allocation? For a given land ownership, specific land use allocations may guide objectives for the abundance of decaying wood. For example, on commercial forestry lands, intensive timber-use zones on upland slopes may provide for lower levels of decaying wood than in riparian buffer zones, and on some Federal lands, old-forest conservation areas (such as Managed Late-Successional Reserves) may warrant even higher levels.

What is the site capable of producing overall and given its particular site history? For a given site, such as a particular forest stand, the manager may wish to determine what the site is capable of producing in terms of sizes and densities of decaying wood structures such as snags, hollow live trees, and down wood. Then, considering current stand conditions and recent site history, the manager can craft reasonable management direction to provide for appropriate levels of decaying wood as integrated into other site management activities such as tree pruning, precommercial thinning, final tree harvests, etc.

What is the appropriate spatial scale for managing decaying wood? The tools discussed in this chapter pertain to helping prescribe or predict effects of the abundance of decaying wood at the scale watersheds or larger, rather than for particular sites such as individual forest stands. However, this does not mean that attending to individual stands is unimportant. Studies suggest that wood habitat structures function best for wildlife when they are broadly distributed as well as occurring in locally-dense clumps, such as with scattered snag or down wood patches. The manager might use some of the concepts, suggestions, and tools in this chapter at stand levels but would need to validate them at broader geographic scales.

Even broader geographic areas such as subbasins might need to be evaluated when considering or projecting natural and human-caused disturbance regimes. Disturbances can radically change local conditions. Borrowing from landscape ecology, one rule of thumb that the manager might follow is to evaluate, as analysis units, a land area large enough to fully encompass the occurrence and full extent of the infrequent (e.g., 50-year) disturbance event such as a major wildfire or harvest schedule rotation period. Planning should be targeted for average levels of decaying wood across the analysis unit realizing that individual sites within the unit might be subject to greater uncertainty of specific conditions at various times.

The manager can use tools such as the DecAID model and matrixes on the CD-ROM to perform a risk analysis of the likelihood of providing for wildlife species and ecosystem functions associated with particular amounts and sizes of snags and down wood and other forms of decaying wood, in specific wildlife habitats, over time, and at several statistical levels. Using such tools, the manager may consider alternative scenarios for retention, recruitment, and restoration of decaying wood for wildlife and ecosystem function. Uncertainty in wildlife response, expected levels of decaying wood on site, likelihoods of fire, insect, and disease conditions, and future disturbances, may factor into specific management guidelines, as well as into monitoring programs to help evaluate efficacy of the guidelines and objectives.

How To Evaluate The Current And Historical Status of Decaying Wood

To determine a quantitative objective for dead wood, it is necessary to evaluate the functional target for wood relative to the historic range of natural consditions and current status of dead wood in the area of interest. A new modeling tool named DecAID is available to assist with this task. DecAID (as in "decayed" or "decay aid") is a new Decayed Wood Advisory Model being developed to address some of the recent lessons learned.[226, 247] DecAID is based on a thorough review of literature, available research and inventory data, and expert judgement. It broadens the paradigm for wildlife species and habitat assessment by considering the key ecological functions of wildlife (see below) as well as the ecosystem context of wood decay in terms of secondary effects on forest productivity, fire, pest insects, and diseases.

DecAID has four components:

(1) Species-habitat associations, specifically, the use and selection by terrestrial wildlife species of snag dbh, snag density (number per unit area), down wood diameter, and down wood percent cover (percent of forest floor covered by down wood or wood litter). This information is based on a synthesis of empirical data, and interpretations made by the authors and experts.[226] DecAID presents cumulative species curves for combinations of wildlife habitats and habitat structures, showing the size and density of snags (or diameter and percent cover of down wood) corresponding with each species as reported in field studies. Where data permit, these curves are shown for low (30% tolerance interval), moderate (mean), and high (80% tolerance interval) statistical levels. Some studies pertain to snag (or down wood) densities averaged across study stands, whereas others pertain to locally high densities in snag (or down wood) clumps centered on nest/breeding, den/roost, or foraging sites.

(2) Key ecological functions of wildlife species associated with decaying wood, and full lists of *wildlife species associated with decaying wood*. This information is based on the species functions database of the matrices on the CD-ROM with additional information from the literature and other databases.[223, 224]

(3) Range of natural conditions. This information is an analysis of current and recent past inventory data on snags and fallen trees.[276] The inventory data are taken from several sources, including the Forest Inventory and Analysis of USDA Forest Service, Pacific Northwest Research Station; the Current Vegetation Survey of USDA Forest Service, National Forest System; and the Natural Resource Inventory of USDI Bureau of Land Management. Collectively, these inventory data cover all national forests and BLM lands, and all other non-reserved lands. Use of inventory data is limited for eastside, fire-dominated systems.

(4) Ecosystem processes and productivity. This information is based on a synthesis of publications and research identifying how wood decay provides for ecosystem productivity.

The manager will be able to use DecAID for advice on the following topics by first specifying wildlife habitat, structural stage, and statistical (confidence) level: 1) wildlife species associated with particular sizes and densities of snags and down wood, or, conversely, the sizes and densities required to meet specified wildlife management objectives, at three levels of confidence; 2) the array of key ecological functions of wildlife associated with decaying wood; 3) the recent-historic and current range of natural conditions of snags and fallen trees; 4) advice on fire risk assessment and mitigation; 5) advice on the roles of insects and diseases associated with various amounts of decaying wood; 6) and the influence of the abundance of decaying wood on ecosystem processes and productivity.

As an example of using Component 1 and 2 of DecAID, a manager might be interested in the role of snags in the Upland Aspen Forest wildlife habitat type. A query of the CD-ROM portion of DecAID reveals that 30 bird and 10 mammal species are coded as being associated with snags in this forest type. DecAID has compiled all available empirical data on aspen-related species, which totals 9 birds (8 primary and 1 secondary cavity-nester) for snag density data, and 23 bird (16 primary and 6 secondary cavity-nesters, the primaries including 4 sapsucker hybrids) and 1 mammal species for snag dbh data. For Upland Aspen Forest, the cumulative species curves suggest that providing for all nesting or breeding species as found in field studies entails locally dense clumps of snags averaging 32.4 snags per acre larger than 33 in (80 snags per ha) with at least some snags 20.3 in (52 cm) or greater in dbh.

If the manager had a known or expected condition different from this, the expected associated species could also be determined. An example is an Upland Aspen Forest stand with locally dense snag clumps averaging 20.3 snags per ac (60 snags per ha) of 15.7 in (40 cm) dbh. In this case, the snag density corresponds to providing for 6 of the 9 studied bird species, and the snag dbh corresponds to providing for all of the 21 species (100%) at low statistical levels, 14 species (67%) at average, and 6 species (35%) at high statistical levels; DecAID also lists the specific species in each case. The manager can use the information on statistical levels as a risk analysis and also as a means of setting objectives, where the high statistical level might correspond with highly conservatory objectives such as for Late-Successional Reserves.

DecAID also ties into the CD-ROM matrix on wildlife species and their key ecological functions. The 30 bird and 10 mammal species associated with snags in Upland Aspen Forest collectively perform some 60 categories of key ecological functions. These include primary consumers or herbivores (16 species), secondary consumers or primary carnivores (38 species), and tertiary consumers or secondary carnivores (3 species). Other ecological functions performed by this set of species include potential control of insect populations (10 species) or vertebrate populations (7 species); dispersal of fungi, lichens, insects and other invertebrates, seeds, fruits, or vascular plants (11 species); creating sapwells (2 species), creating nesting structures (4 species), excavating cavities (11 species), digging burrows (3 species), and creating terrestrial runways (2 species) potentially used by other species, some of which are also associated with snags in Upland Aspen Forest. Other ecological functions of snag-associated wildlife in Upland Aspen Forest are potential improvement of soil structure and aeration through digging and tunneling (2 species), physically fragmenting standing or down wood (2 species), creating snags (1 species), and potentially altering vegetation structure or succession through herbivory (1 species). Using the CD-ROM, the manager also can determine the specific ecological roles of just those species expected to be provided at given sizes and amounts of snags and down wood, and trade-offs with other functions. Overall, through such lists, DecAID provides the manager with insights into the array of ecological roles performed by wildlife species associated with decaying wood structures in specific wildlife habitats. The model also indicates how those functions might respond to changing attributes of decaying wood. Thus, the model offers an indication of how different management practices affect the relative "functionality" of ecosystems.

How To Incorporate Wood Dynamics Models Into Planning

Once an appropriate objective is determined with quantitative targets, the next step is to integrate wood dynamics into the planning process. This is a critical step because of the extended time frames required to replace large wood structures. Dynamics models predict the general rate of fall of snags and decay of snag and logs on a site to determine the amount of decaying wood occurring through time (e.g. a rotation). These models can also be used to estimate the number of green replacement trees to leave on a site and when to convert them to snags or logs. Mortality estimates from growth and yield models (Forest Vegetation Simulator, Organon, DFSIM, etc.) can be input into all the models reviewed here to track snag and log recruitment in forest stands. The following are three dynamics models available for Washington and Oregon.

Snag Recruitment Simulator (SRS)—Marcot. [221] SRS is based on the snag life-table approach.[263] The model was built to generalize snag decay and falling rates, and can be parameterized with any such rates. Data limitations, however, necessitated that the westside version of the model (i.e., Oregon and Washington west of the Cascades crest) predict fall and decay of Douglas-fir snags based on data from the Oregon Coast Range,[82] and the eastside version predict fall rates of ponderosa pine snags from the Blue Mountains of Oregon and decay rates from the Coast Range Douglas-fir data. SRS consists of a series of compiled Lotus spreadsheets and runs in DOS on a PC. It is the easiest of the three models to use. Several people have parameterized SRS for specific geographic areas. A similar model based on a Leslie Matrix was later published.[253]

Snag Dynamics Projection Model (SDPM)—McComb and Ohmann.[240] SDPM uses logistic regression analysis to predict the probability of a snag falling over a 10-year period. A straight rate is used for the probability of a snag changing from hard to soft over the same period. Forest Inventory and Analysis (FIA) remeasurement data from western Washington was used to develop the model. SDPM models Douglas-fir, western hemlock and western redcedar snags. SDPM incorporates site factors such as climate, slope, and aspect which affect snag fall and decay rates. The model is an executable DOS-based program written in C language.

Coarse Wood Dynamics Model (CWDM)—Mellen and Ager,[246] Mellen and Ager,[247] Mellen et al.[248] CWDM analyzes the dynamics of both snags and down logs in forests on the westside of the Cascade Mountains. The model assesses snag fall, height loss and decay, and log decay for Douglas-fir and western hemlock. Snag fall and height loss rates are from FIA remeasurement data from western Oregon and Washington. Decay rates are based on work in the western Oregon Cascades.[140] A single-exponential equation[164] was used to model the dynamics of snags and logs. CWDM is available in both DOS and Windows versions. An east-side version of CWDM will be developed as data on decay rates of snags and logs in eastern Washington and Oregon become available.

Management Tools and Opportunities

Silvicultural Methods

Traditional forestry practices in the Pacific Northwest[94] have been based on the simplification and homogenization of forests to achieve economical wood fiber production.[122, 228, 289] As a result, the amount of old-growth forests has declined over 50% in the last 60 years,[41] and remaining stands are highly fragmented.[347] Such practices have been at the center of intense controversy over the past three decades.[271, 272] Concerns center on the loss of biological "legacies" resulting from natural disturbances—surviving organisms and wood structures that contribute to resilient ecosystems with structural, compositional, and functional diversity.[291]

New management paradigms, originally coined "New Forestry,"[121] have shifted the emphasis from maximized fiber production over the short-term to diverse forest values and maintenance of long-term forest productivity (see also [201]). Alternative forest management systems are required to meet these new objectives, particularly the goal to produce and retain legacies of decaying wood.[122, 228, 292] Management goals for dead wood, thus, are inextricably embedded within the broader goals for diverse forest resources. To accomodate multiple objectives, Franklin[122] has suggested that new management should do the following:

1. Emphasize adding new silvicultural tools to the old.
2. Address stand and landscape-level objectives.
3. Maintain stands that are structurally and compositionally diverse (practices will differ from stand to stand, depending on forest type, condition, environment and specific objectives).

New silvicultural practices are a high priority in regenerating stands where one or more harvest cycles have simplified stand structure and composition and removed wood legacies. In young stands, Franklin[122] recommends that management should:

1. Aggressively create stands of mixed composition to maintain habitat for a broad array of species (and to achieve diversity in quality and timing of nutrient inputs to streams).
2. Delay the process of early canopy closure (wide spacings, pre-commercial thinning etc.).
3. Provide for adequate amounts and a continuous supply of large wood, including snags and down logs, for maintaining structural divesity in forests and streams and maintaining all other ecosystem processes associated with wood.

The basic theme of these revisions of intensive forestry practices is to retain the higher levels of complexity found in natural forests, and in so doing, to protect processes and structures that retain future options for ecosystem management. Effective management of decaying wood must do more than simply provide for inputs of dead trees. Rather, management should strive to provide for diversity of tree species and size classes, in various stages of decay and in different locations and orientations within the stand and landscape.[122, 233] Examples of new practices include retention of large woody structures at the time of harvest, long rotations, and creation of snags and logs from green trees. An overview of different silvicultural practices to regain (i.e. retain and create) wood legacies in managed stands is provided below. Detailed discussions of alternative silvicultural practices are available.[94, 132] Through the integrated application of silvicultural practices described below, it becomes possible to conserve biodiversity in managed forests by retaining legacy structures and accelerating the development of structural complexity.[61, 69a]

Structural Retention. Structural retention refers to any practice that retains significant structural elements from a harvested stand for incorporation into a new stand. The focus of this practice has been on harvesting in old and mature stands. Objectives in structural retention include maintenance of organisms and processes on harvested areas, structural enrichment of the regenerating forest, and enhanced connectivity across managed landscapes. To a limited degree, structural retention mimics the effects of variable disturbance patterns to create legacy structures. Retention of wood structures from a harvested stand for carryover into a new stand offers an infinite array of alternatives to clearcutting and other regeneration harvest systems.[132, 364] Variables in structural retention include the type, size, number of structures, and spatial distribution of the retention. Large, old, and decadent trees, standing dead trees (snags), and down wood are examples of structures selected for retention (Figure 13).

Retention of snags and large logs is a particularly effective practice in maintaining large wood when harvesting stands that already contain significant wood legacies—such as young and mature stands of natural origin and old-growth forests. Practices including the removal or burning of unmerchantable trees and down wood in forests and near stream channels should be minimized or eliminated. Retention of snags provides numerous habitat benefits.[154, 239, 402] However, safety and liability issues associated with snag retention have posed an operational barrier to management objectives for structural retention. Two approaches useful in reducing hazards associated with snags are: 1) to cluster snags in patches rather than wide dispersal, and 2) to create snags from green trees after cutting.[122]

Many questions remain regarding the most desirable density and spatial distribution of snags and down wood for wildlife.[94] New information suggests that snags and down wood may follow a naturally "clumped" distribution. Therefore, managers can take opportunistic advantage of site-specific occurrences of snags and down wood without having to match a particular spatial distribution pattern of clumps. This offers managers broad flexibility to provide varying local densities of snags and down wood across the ground, within and among stands. Managers must also consider the temporal dimension to decaying wood, to ensure that sufficient sufficient snag and down wood densities are provided through time. Franklin et al.[132] review information on the type, amount, and distribution of structural retention for various objectives. Additional examples of wildlife benefits from structural retention have been described.[20, 21, 63, 64, 69, 152, 154, 210, 314, 326]

Live (Green) Tree Retention. Retention of living trees on cutover areas is one form of structural retention that can provide for future recruitment of snags and down wood (created artificially or through natural processes).[122] Other terms commonly used for this method include partial cutting or partial retention to distinguish it from selection cutting. Green tree retention involves reserving a significant percentage (10-40%) of living trees, including dominants through the next rotation. The density, composition, condition, size classes, and spatial distribution of the retained trees varies according to management objectives, stand and site conditions, and other constraints. The objective is to maintain a more structurally diverse stand than could be achieved through even-aged management.

Green trees function as a refugium of biodiversity in forests. For example, many species of invertebrate fauna in soil, stem, and canopy habitats of old-growth forests do not disperse well, and thus, do not readily recolonize clear-cut areas.[207, 326] The same concept holds for many mycorrhizae-forming fungal species.[293] Added benefits of green tree retention include moderated microclimates of the cutover area, which may increase seedling survival, reduce additional losses of biodiversity on stressed sites,[293] and facilitate movement of organisms through cutover patches of the landscape. Green trees retained across harvest cycles can also be used to grow very large trees for either ecologic or economic goals. This may be an especially valuable practice in providing large wood to riparian forests subject to harvest: green tree retention can be implemented to favor species such as Douglas-fir and redcedar that produce larger and more persistent wood. Other benefits of green tree retention include reduced hazards of landslides via maintenance of root strength and reduced potential for rain-on-snow flood events.

Green tree retention offers many benefits to wildlife. For example, the higher structural diversity in young stands that contain legacy trees from previous stands provides much improved habitat values to late successional species such as the northern spotted owl, as well as other vertebrates that use late-successional stands for some elements of their life history.[69, 122, 314] Such stands may provide wildlife habitat as early as age 70-80 years rather than 200-300 years, the approximate time interval required for old-growth conditions to develop after secondary succession. Green tree retention on a harvest cycle of 120 years has been proposed as a method to provide habitat for late successinal species in only 40-50 years.[278] Different scenarios for green tree retention have been offered.[122, 278] To meet needs of all species likely to occur in area, it is critical to identify tree species for retention. Updates on wildlife and vegetation response to

Figure 13. Retention of old-growth "legacy" snags in a commercial thinning unit in western Washington. Photo: Deborah Lindley.

green tree retention are recently available for the DEMO program in Oregon and Washington, and other silvicultural trials.[20, 30, 152, 155, 156, 210, 268, 311, 406]

Variable Retention Harvest Systems. Franklin et al.[132] recommended the term "variable retention harvest systems" to refer to harvesting practices that allow a continuous spectrum of removal and retention in mature and older stands, depending on objectives. The development and maintenance of structurally complex managed forests is the primary rationale for retaining structural elements of the harvested stand. Retention of various decaying wood structures through variable retention harvest provide many benefits to many wildlife species and functions,[20, 26, 64, 70, 73, 156, 326] as well as other forest resource commodities.[12]

More recently, there has been increasing interest in aggregated or "patch" retention—the maintenance of small forest patches, instead of dispersion of retained structures.[64] These patches can provide refugia, while also providing microclimatic gradients for more sensitive species and functions.[151] By providing for the maintenance of refugia, tree and patch retention may benefit species limited by slow dispersal rates, rather than by particular habitat structures.[13, 239, 343] Franklin et al.[132] review major issues in developing harvest prescriptions based on the variable retention harvest concept, including the type, amount, and distribution of structural retention for various objectives. Franklin[122] described several alternative management approaches for maintaining wood production and complex forest structure suggested by fire history research. In a multi-aged management strategy, selective cutting practices can be used to sustain complex stand structure and composition for long periods. It may be beneficial that the management system mimic the natural disturbance regime to the extent possible, so that the site can accomodate future natural disturbances as they interact with managed stands.

Long Rotations. Long rotations involve the use of rotation ages that are significantly longer than that defined by the culmination of mean annual increment.[91, 92, 93, 94] Long rotations provide for the recruitment of larger and more complex wood structures by allowing tree to grow to a larger size and by eliminating logging disturbances that damage or remove wood legacies. When coupled with a series of silvicultural treatments, long rotations can produce complex managed forests, increase commodity yields, and address cumulative issues where too much of the landscape is in a recently-harvested condition. Long rotations also may be applied to patches of trees at a scale smaller than individal stands. As an added benefit, extended rotations can reduce the need for permanent transportation systems.[196]

Long rotations may not provide adequate protection for all structural elements, processes, and organisms, particularly those most sensitive to disturbance, or with more rapid turnover rates. For example, large-diameter, moderately-decayed snags removed at the time of harvesting (via clearcutting) may not be replaced within the rotation cycle. Rather, retention of snags at harvest, combined with long rotations would be necessary to provide for present and future recruitment of such snags. Long rotations are unlikely to provide forests with structural, functional, and compositional features comparable to late-successional reserves.[12, 132, 156]

Thinning. Thinning in plantations can be used to decrease the time required to develop larger trees and multiple-canopy forests for species associated with late-successional forests, while producing ecomonic benefits from thinnings and shorter rotations.[94, 364] It may not produce desired amounts of down wood or heavy-limbed tree crowns. However, benefits of accelerating the rate of stand development must be weighed against the detrimental effects of logging disturbances on species, particularly those with limited dispersal capabilities.[343] Variable density thinning is a variant of the traditional unifrom spacing that shows promise for accelerating structural and compositional diversity.[66] In general, thinning is beneficial to the development of more structural diversity in young stands, and to a variety of wildlife species, particularly in more complex and patchy stand structures, with legacy structures.[26, 70, 73, 150, 155]

Restoration Techniques

Restoration with respect to decaying wood involves silvicultural manipulations to develop and create wood legacies in stands lacking suitable existing structures. In addition to retention of existing snags and down wood at the time of harvest, and young stand management to favor structural complexity (retain green trees), intentional methods have been developed and tested to create snags, down wood, cavities and other habitat niches.[96, 132] Techniques include girdling, injection with herbicide, topping, explosives, fungal innoculation, and use of pheromones to attract beetles.[25, 54, 57, 84, 212, 284] Various techniques for creating artifical cavities have been tested, including creation of cavities by den routing,[63, 65] or by cutting a hole with a chainsaw and covering with a faceplate.[62]

Many of these techniques have produced favorable habitat for wildlife. However, characteristics of artificially created snags may not always be comparable to natural snags. Additional research and monitoring is needed to better evaluate the attributes and habitat uses of snags and down wood created by artificial methods. Costs for creating artificial snags and down wood are important management considerations, thus it is desirable to retain existing legacy structures, as well as trees for future snag recruitment. In addition, it may be both biologically and operationally beneficial to create and preserve snags within patches of uncut trees, rather than to distribute them uniformly across stands.

Few long-term data are available to assess the effectiveness of different methods of snag creation. Results of a recent study[328] monitoring natural and artificially created snags on the Siuslaw National Forest in the Oregon Coast Range provide useful guidance to managers. The study tracked longevity and wildife use in 150 green leave trees, 91 natural (Class 1) snags, 27 intentionally-topped Douglas-fir trees, and 23 hardwood leave trees (big-leaf maple and alder) and snags in variable retention harvest units from 1987 to 1998 (Figure 14).

The study identified several interesting facts about topped leave trees and residual snags:

1. Topped leave trees (blasted or cut) are far more windfirm than natural snags or green leave trees (Figures 15, 16). Greater windfirmness of topped trees (0.7% rate of windthrow) compared to natural snags (11%) or green leave trees (17%) is attributed to both lower rates of root failure and stem breakage. Windthrow of green leave trees can be substantial, particularly in wind-prone areas. Numerous cavities and other defects in the lower bole indicate that bole breakage could increase in the future. Thus, long-term cavity habitat requires periodic topping, rather than a single entry, especially for small-diameter trees that decay more rapidly.

2. Live leave trees experience high rates of windthrow and breakage due to increased exposure in clearcuts. Thus, live green trees may not provide for snag recruitment for an extended period post-harvest.

3. Degradation and loss of natural snags is high due to root throw, but more significantly to bole breakage. Breakage is most frequent near cavities, areas of advanced decay, and structural deformities. The longevity of Class 1 snags is limited. Continued recruitment of new snags requires snags to be created periodically from green trees during the course of stand development.

4. Trees topped above two branch whorls survive and develop new tops. Continued diameter growth in these trees provide higher values as wildlife snags. Large crooks formed in these trees also provide platform nest sites and create future breaking points to form a tall snag. The greater longevity of these live-topped trees should reduce the need to cause intentional mortality in leave trees in the future.

5. Methods for topping trees by either blasting or chainsawing produce similar results for both snags and live-topped trees. Blasting provides a more natural look, but the chainsaw method allows for directional felling if the salvage of tops is planned.

6. Natural and created snags show high levels of use by cavity-nesting birds. Topped trees rapidly develop cavities throughout the bole. Live-topped trees develop cavities ten years after topping, with cavities forming first near the upper bole. The creation of live trees with cavity habitat is highly desirable, as it allows cavity habitat to be maintained over longer periods.

7. Big-leaf maple has relatively high survival and provides a high density of cavity sites. The rapid diameter growth of big-leaf maple allows a tree size suitable for cavity production to be developed within thirty years.

Figure 14. Stand treated to restore wildlife snags. Treatments included topping with and without retention of live branch whorls. Photo: Barry Schreiber.

Topped trees present fewer conflicts with harvest and silvicultural activities than snags, and if trees are topped prior to harvest, some of the tops may be salvaged to offset the cost of topping. In planning for the desired density of leave trees, managers need to consider the vulnerability of green leave trees to wind and fire, and degree of competition to living topped trees by the regenerating forest. Thinning may be desirable to extend the lifespan of green trees or live-topped trees. Tree defects will affect the quality and longevity of leave trees as snag habitat. For example, bole crooks are weak points that can determine where stem breakage will occur, hence this feature can be selected to provide future snag habitat of desired height. Trees with butt rot, hollow stems, large bole sweeps, forks at the lower bole, or leaning trees are less suitable for retention, as they are less stable, and succumb more rapidly to wind and gravity. Live-topping of green trees has been found to have great potential to provide long-term cavity habitat in managed stands. This possibility should be more thoroughly investigated.[328] Additional considerations for active management of decaying wood are available in a recent review.[69a]

Management of Wood in Streams. In riparian areas, the development and maintenance of large trees is required to provide inputs of large wood to streams and rivers[33, 143] The forest adjacent to channels is particularly critical in small streams, since wood inputs from other sources may be negligible.[241] Natural stable wood in streams should be undisturbed[376] and supplementary wood recruited through management of riparian forests to the maximum extent possible. The delivery and routing of wood to streams is important to the vitality of watersheds and their component drainages in the Pacific Northwest.[262] Integrity of the lateral, longitudinal and vertical components, as well as the temporal and spatial characteristics of wood cycling within basins must be maintained. Thus, managers must pay attention to geomorphic considerations and connectivity between parts of the larger system when planning for inputs of large wood in riparian areas.

Figure 15. Snag created by blasting. Photo: Barry Schreiber.

Figure 16. Snag created by topping with a saw. Photo: Barry Schreiber.

Aquatic functions dependent upon connectivity include: 1) requirements for habitat access during discrete life history stages of fish and wildlife,[304, 321] 2) synchronization of emergence and migration of aquatic organisms to stream temperature,[303, 362] 3) regulation of nutrient and material exchanges between forests and streams,[295, 361] 4) and maintenance of hydraulic regimes within boundaries of evolutionary adaptation for specific organisms.[350] For example, in coastal Oregon, debris flows from tributaries provide most of the large wood inputs.[39] However, in southeastern Alaska, floodplain forests adjacent to the stream are the primary source of wood to the stream system.[233]

New silvicultural methods being developed and tested in riparian forests[143] include re-establishing conifer species, retaining snags, down logs, and green trees, and avoiding salvage. Underburning may be used to re-establish conifers, if necessary, to reduce competition with understory shrubs. Thinning and underplanting may also be used to create snags and provide large wood. In heavily degraded riparian areas, forest complexity can be improved by retaining large, and broken trees in the adjacent harvest unit. Where harvesting is allowed, group selection or single tree selection is preferred to clear-cutting.[3] Thinning should strive to leave trees dispersed in irregular patches.

Summary of Management Recommendations

The information presented in this chapter emphasizes several properties of decaying wood in forest ecosystems: (1) each structure formed by decaying wood helps support a different functional web in the ecosystem; (2) no one decaying wood structure supports all functions equally; and (3) all decaying wood habitats together support the widest array of ecological functions and associated wildlife species. The CD-ROM with this book in combination with the DecAid model provides managers with a powerful tool that makes it possible to assess the degree of "full functionality" of ecosystems as supported by the various decaying wood structures, and which functions are strengthened, diminished, or lost through alternative silvicultural management practices.

Lessons for managers are:

1. Examine forestry practices for how they influence the distribution and abundance of down wood and snags in relation to forest landscape patterns. In situations where forest management objectives extend beyond wood production to broader biological and human values, intensive forestry practices by themselves may inadequately maintain or restore biodiversity, especially in early and late successional forest development phases. Species, processes, and values associated with older stages of stand development (transition and and shifting gap stages)[277] are likely impaired or absent from intensively managed stands.[343] Species and processes associated with the early establishment phase also have shorter duration than may occur naturally. This does not mean that intensive forest management practices are incompatible with multiple forest objectives at a landscape scale, but rather that species and processes associated with early and late stages of forest development should be assessed over large areas such as landscapes, subregions, and regions.[291, 343] Management for certain species must consider habitat requirements at different spatial and temporal scales.[181] It may then be possible to modify silvicultural practices at the stand scale to meet multiple objectives at landscape and larger scales. The landscape perspective also is pertinent to managing riparian systems,[143] where the role of wood decay in riparian environments varies according to the type and geography of the associated water body.

2. Emphasize retention of wood legacies, and secondarily promote restoration where legacies are deficient to meet stated objectives. The decline of species associated with late-successional forest structures, as well as the prolonged time needed to produce wood legacies, suggests that it is both ecologically and economically advantageous to retain legacy structures across harvest cycles wherever possible, rather than attempt to restore structures that have been depleted. This is especially obvious for slow-growing tree species and very large wood structures. Retention of old-growth structural legacies has been identified as critical to conservation of biodiversity between large reserves and conservation areas.[222, 267]

3. Use an adaptive management approach to assess management options where possible. Our ability to sustain forest ecosystem values while producing commodities is uncertain. Given the imperfect state of our knowledge regarding management effects on biodiversity and long-term productivity of forests, prudence calls for an adaptive management approach that spreads environmental risk across a range of management strategies. Management must seek a blend of practices to meet biological and social objectives.[343] Guidance on approaches to adaptive management is available in several publications.[45, 110, 157, 338, 391]

4. Address management objectives and data needs to protect functions of decaying wood in basic processes of forest ecosystems, not solely to fish and wildlife habitat. At the forest policy level, broad-scale assessments of down wood are needed to address Criteria and Indicators for the Conservation and Sustainable Management of Temperate and Boreal Forests, developed through the Montreal Process. Although dead wood was not considered in the first national-level assessment, dead wood abundance will be addressed in the first assessment of forest sustainability to be conducted by any state in the U.S., by the Oregon Department of Forestry.[36]

Operational Considerations

Management for decaying wood will require new and innovative operational approaches. Structural retention and restoration pose many new challenges to forest operations, both in current and future harvest activities. To be successful, any plan for provision of wood decay elements should address operational aspects of implementation. This includes scoping operational approaches for plan implementation, and identifying strategies for resolving potential conflicts ahead of time. Two common sources of operational difficulties include safety concerns and associated administrative costs.

Worker safety

From 1980-1989, the average annual fatality rate for workers in the logging industry was more than 23 times that for all U.S. workers. More than half of these fatalities occurred when workers were struck by falling or flying objects or were caught in or between objects; most of these fatalities involved trees, logs, snags or limbs.[264a] An analysis of claims data from 1990-1997, ranked workers in the logging industry as having the highest risk of traumatic head and brain injuries in Washington state.[401b]

In recognition of the need for tree retention in harvests, OSHA revised the federal Logging Standard (29 CFR 1910.266) in 1995, to clarify its intent that danger trees may be avoided, rather than being removed or felled.[72a] A danger tree is any standing tree (live or dead) that poses a hazard to workers, from unstable conditions such as deterioration, damage, or lean. The revised rule allows some discretion in determining the hazard area around a danger tree, by "...allowing work to commence within two tree lengths of a marked danger tree, provided that the employer demonstrates that a shorter distance will not create a hazard for an employee."(OSHA Logging Preamble, Section V). Determining a safe working distance requires a case-by-case "...evaluation of various factors such as, but not limited to, the size of the danger tree, how secure it is, its condition, the slope of the work area, and the presence of other employees in the area." The employer is responsible for marking hazard trees and making this determination. Washington State guidelines for reserve tree selection provide definitions of hazard areas and examples of operational techniques that are compatible with safe work practices.[401a]

Oregon (Chapter 437, Oregon Administrative Rules) and Washington (Chapter 296-54, Washington Administrative Code) state safety regulations require employers to have a site-specific safety plan before logging activities begin. The operator can include strategies for safe retention of reserve trees in the safety plan. A safety consultation program, completely separate from OSHA's inspection program, is available through state safety agencies. The operator may request free, confidential, on-site assistance with safety planning, without risk of citations or penalties. State safety plan requirements and OSHA consultation services are two key mechanisms to resolve operational safety issues associated with structural retention and restoration. Reserve tree requirements are best implemented through descriptive criteria in contract or application conditions, rather than marking individual trees and snags. Because the employer is responsible for identifying hazards and preventing worker exposure, pre-marked trees could reduce the operator's flexibility to deal with unforeseen hazards which arise in the course of logging activities. If a tree marked by a timber sale administrator or forest practices officer turned out to be a danger tree, confusion over safety compliance authority also might result.

Potential safety conflicts can be addressed in advance by requiring a pre-work conference with the operator and a state safety consultant to review the operator's site-specific strategies for meeting reserve tree objectives in compliance with safety regulations. At this time, the types, locations, and distributions of leave trees can be specified, considering the logging system to be used, and site-specific topography, unit layout and decay conditions. Once landings and tail holds are identified, reserve trees may be mapped; as an example, clumps of reserve trees may be left between cable roads. For the variety of situations that may be encountered during operations, it is important to devise alternative strategies. This is a good way for managers to improve operational knowledge and expedite administrative compliance time.

Concerns frequently arise where high public use creates a risk of third party liability. Considerations include the proximity of reserve trees to roads, trails, campgrounds, ski areas, and other recreation areas and public access points. Methods for addressing these concerns include signage and clear delineation of potential hazard areas, fencing and other barriers to discourage public access, snag height reduction and use of setbacks to minimize exposure.

Costs and Forest Management Complexity

Legacy retention generally results in higher logging costs than clearcutting and may require alternative logging technology. Compared to clearcutting, logging costs have been estimated to be significantly greater for dispersed retention, but only slightly higher for aggregated retention.[201,406a] Aggregated retention has also been recommended to minimize interference with aerial application of pesticides, fertilizers, and herbicides.[201] Other concerns regarding tree retention are effects on stocking density and genetic composition of the regenerating forest. Compared to live trees, the retention of snags, decadent trees, and down wood within a harvest unit have little influence on future stand genetics or stocking density. Retention of live trees and decaying wood may reduce wood yields due to volume in wood structures permanently retained on a site, or reduced growth in the regenerating stand due to shading.[1b, 311, 311a,401c, 406] Conversely, decaying wood may increase future wood yields by improving seedling survival and growth, and site productivity.[164]

Long Range Planning Considerations

Long-range plans should provide criteria for tree selection and distribution that are flexible enough to account for ongoing tree decay processes, and changes in harvest plans. For instance, a time lag between unit layout and harvest may result in changes in reserve tree decay classes. It is also important to consider how reserve trees will fit into plans for other forest practices after harvest, such as thinning operations, fertilization, vegetation management, and fire control. This is particularly important in uneven-age management schemes where multiple entries are planned. When developing conceptual long range plans and site-specific plans, it is thus important to consider whether a hazard would exist by the time future activities take place.

New Approaches

Few studies have analyzed operational challenges to managing for decaying wood. However, a variety of forest managers are currently testing various methods to retain and restore decaying wood in various forms as snags, decadent trees, and down wood. Recent publications offer useful discussions of concepts and silvicultural approaches to managing for structurally complex forest stands, including operational considerations; they also discuss recent lessons from experimental forestry operations.[31a, 69a, 201, 288a, 328, 406a]

Literature Cited

1. Aber, J. D., and J. M. Melillo. 1991. Terrestrial ecosystems. Saunders College Publishing, of Holt, Rinehart and Winston, Orlando, FL.

1b. Acker, S. A., E. K. Zenner, and W.H. Emmingham. 1998. Structure and yield of two-aged stands on the Willamette National Forest, Oregon: implications for green tree retention. Canadian Journal of Forest Research 28:749-758.

2. Agee, J. K. 1981. Fire effects on Pacific Northwest forests: flora, fuels, and fauna. Pages 54-66 in Northwest Fire Council Proceedings, Northwest Fire Council, Portland, OR.

3. ———. 1988. Successional dynamics of forest riparian zones. Pages 31-43 in K. J. Raedecke, editor. Streamside management: riparian wildlife and forestry interactions. Institute of Forest Resources, Contribution Number 59. University of Washington, Seattle, WA.

4. ———. 1993. Fire ecology of the Pacific Northwest forests. Washington, D.C. Island Press.

5. ———. 1994. Fire and weather disturbances in terrestrial ecosystems of the eastern Cascades. U.S. Forest Service, Pacific Northwest Research Station, Portland, Oregon. General Technical Report PNW-GTR-320.

6. ———, and M. H. Huff. 1987. Fuel succession in a western hemlock / Douglas-fir forest. Canadian Journal of Forest Research 17:697-704.

7. Akenson, J. J., and M. G. Henjum. 1994. Black bear den site selection in the Starkey study area. Natural Resource News 4(2):1-2. Blue Mountains Natural Resource Institute, La Grande, OR.

8. Allen, R. B., P.W. Clinton, and M. R. Davis. 1997. Cation storage and availability along a Nothofagus forest development sequence in New Zealand. Canadian Journal of Forest Research 27:323-330.

9. Amaranthus, M. P., and D.A. Perry. 1987. Effect of soil transfer on extomycorrhiza formation and the survival and growth of conifer seedlings on old, nonreforested clear-cuts. Canadian Journal of Forest Research 17:944-950.

10. ———, R. J. Molina, and J. M.Trappe. 1989 a. Long-term productivity and the living soil. Pages 36-52 in: D. Perry, D., R. Meurisse, B. Thomas, R. Miller, and J. Boyle, editors. Maintaining the long-term productivity of Pacific Northwest ecosystems. Timber Press. Portland, OR.

11. ———, D. S. Parrish, and D.A. Perry. 1989 b. Decaying logs as moisture reservoirs after drought and wildfire. Pages 191-194 in: E. B. Alexander, editor. Proceedings of Watershed '89: A Conference On The Stewardship Of Soil, Air, And Water Resources. U.S. Forest Service, Juneau, AK.

12. ———, J. F. Weigand, and R. Abbott. 1998. Managing high-elevation forests to produce American matsutake (Tricholoma magnivelare), high-quality timber, and non-timber forest products. Western Journal of Applied Forestry 13(4): 1-9.

13. ———, J. M.Trappe, L. Bednar, and D. Arthur. 1994. Hypogeous fungal production in mature Douglas-fir forest fragments and surrounding plantations and its relation to coarse woody debris and animal mycophagy. Canadian Journal of Forest Research 24:2157-2165.

14. Anderson, N. H. 1982. A survey of aquatic insects associated with wood debris in New Zealand streams. Mauri Ora 10:21-23.

15. Aplet, G. H., N. Johnson, J. T. Olson, and V.A. Sample, editors. 1993. Defining sustainable forestry. Island Press, Washington, D.C.

16. Apps, M. J. and D.T. Price. 1996. Forest ecosystems, forest management and the global carbon cycle. Springer, New York, NY.

17. Arnolds, E. 1991. Decline of ectomycorrhizal fungi in Europe. Agriculture, Ecosystems & Environment 35:209-244.

18. Arthur, M.A. and T. J. Fahey. 1990. Mass and nutrient content of decaying boles in an Engelmann spruce—subalpine fir forest, Rocky Mountain National Park, Colorado. Canadian Journal of Forest Research 20:730-737.

19. Asner, G. P., T. R. Seastedt, and A. R. Townsend. 1999. The decoupling of terrestrial carbon and nitrogen cycles. BioScience 47(4):226-234.

20. Aubry, K. B., Amaranthus, M. P., Halpern, C. B., White, J. D., Woodard, B. L., Peterson, C. E., Lagoudakis, C. A., and A. J. Horton. 1999. Evaluating the effects of varying levels and patterns of green-tree retention: experimental design of the DEMO study. Northwest Science 73: 12-26.

21. ———, J. G. Hallett, S. D. West, M. A. O'Connell, and D. A. Manuwal. 1997. Wildlife use of managed forests: a landscape perspective, Volume I. Timber Fish and Wildife Program, Report No. TFW-WL4-98-001.

22. ———, L. L. Jones, and P. A. Hall. 1988. Use of woody debris by plethodontid salamanders in Douglas-fir forest in Washington. Pages 32-37 in R. C. Szaro, K. E. Severson, and D. R. Patton, technical coordinators. Management of amphibians, reptiles, and small mammals in North America. Proceedings of a symposium, July 19-21 1988 in Flagstaff, AZ. U.S. Forest Service, Rocky Mountain Forest and Range Experiment Station, Fort Collins, CO. General Technical Report RM-GTR-166.

23. Ausmus, B. S., G. J. Dodson, and D. G. Todd. 1975. Microbial - invertebrate interactions: The mechanism of wood decomposition. Ecological Society of America Bulletin 56(2):42.

24. Avery, C. C., F.R. Larson, and G. H. Schubert. 1976. Fifty year records of virgin stand development in southwestern ponderosa pine. U.S. Forest Service, Rocky Mountain Forest And Range Experiment Station, Fort Collins, Colorado. General Technical Report RM-GTR-22.

25. Baker, F.A., Daniels, S. E., and C. A. Parks. 1996. Inoculating trees with wood decay fungi with rifle and shotgun. Western Journal of Applied Forestry 11(1): 13-15.

26. Bailey, J. D., and J. C. Tappeiner. 1998. Effects of thinning on structural development in 40- to 100-year-old Douglas-fir stands in western Oregon. Forest Ecology and Management 108: 99-113.

27. Bate, L. J., E.O. Garton, and M. J. Wisdom. 1999. Estimating snag and large tree densities and distributions on a landscape for wildlife management. U.S. Forest Service, Pacific Northwest Research Station, Portland, Oregon. General Technical Report PNW-GTR-425.

28. Bartels, R., J. D. Dell, R. L. Knight, and G. Schaefer. 1985. Dead and down woody material. Pages 172-186 in E.R. Brown, editor. Management of wildlife and fish habitats in forests of western Oregon and Washington. U.S. Forest Service, Pacific Northwest Region.

29. Bednarz, J. C., D. M. Juliano, M. J. Huss, A. A. Beall, and D. E. Varland. 1998. Interactions between fungi and cavity nesters: applications to managing wildlife and forests in western Washington: 1997 annual report. Unpublished report. Arkansas State University.

30. Beese, W. J., and A. B. Bryant. 1999. Effect of alternative silvicultural systems on vegetation and bird communities in coastal montane forests of British Columbia, Canada. Forest Ecology and Management 115: 231-242.

31. Benda, L. E. and T. Dunne. 1987. Sediment routing by debris flows. Pages 213-223 in R. L. Beschta, T. Blinn, G. E. Grant, G. G. Ice, and F. J. Swanson, editors. Erosion and sedimentation in the Pacific Rim. International Association of Hydrological Sciences, Publication 165. Oxfordshire, England.

31a. Berg, D. R. 1995. Forest harvest-setting design evaluation incorporating logger's preference. Ph.D. Dissertation. University of Washington, Seattle.

32. Beschta, R. L. and W. S. Platts. 1986. Morphological features of small streams: significance and function. Water Resources Bulletin 22:369-379.

33. Bilby, R. E. and P.A. Bisson. 1998. Function and distribution of large woody debris. Pages 324-346 in R. J. Naiman, and R. E. Bilby, editors. River ecology and management. Springer, New York, NY.

34. Binkley, D. and T. C. Brown. 1993. Management impacts on water quality of forests and rangelands. U.S. Forest Service, Rocky Mountain Research Station, General Technical Report RM-GTR-239. Fort Collins, CO.

35. ———, K. Cromack, Jr., and R. L. Fredricksen. 1982. Nitrogen accretion and availability in some snowbrush ecosystems. Forest Science 28(4):720-724.

36. Birch, K. 1999. First approximation report for sustainable forest management in Oregon. Oregon Department of Forestry. Salem, OR.

37. Bisson, P.A., J. L. Nielsen, R. A. Palmason, and L. E. Grove. 1982. A system of naming habitat types in small streams, with examples of habitat utilization by salmonids during low streamflow. Pages 62-73 in N. B. Armantrout, editor. Acquisition and utilization of aquatic habitat inventory information. Western Division, American Fisheries Society, Portland, Oregon. The Hague Publishing, Billings, MT.

38. ———, and R. E. Bilby. 1998. Organic matter and trophic dynamics. Pages 373-398 in R. J. Naiman, and R. E. Bilby, editors. River ecology and management. Springer, New York, NY.

39. ———, T. P. Quinn, G. H. Reeves, and S.V. Gregory. 1992. Best management practices, cumulative effects, and long-term trends in fish abundance in Pacific Northwest river systems. Pages 189-232 in R.J. Naiman, editor. Watershed management balancing sustainability and environmental change. Springer Verlag, New York, NY.

40. ———, R. E. Bilby, M. D. Bryant, C. A. Dolloff, G. B. Grette. R. A. House, M. L. Murphy, K.V. Koski, and J. R. Sedell. 1987. Large woody debris in forested streams in the Pacific Northwest: past, present, and future. Pages 143-190 in E. O. Salon and T. W. Cundy, editors. Streamside management: forestry and fishery interactions. Contribution 57, Institute of Forest Resources, University of Washington, Seattle, WA.

41. Bolsinger, C. L. and K. L. Waddell. 1993. Area of old-growth forests in California, Oregon, and Washington. U.S. Forest Service, Pacific Northwest Research Station, Resource Bulletin PNW-RB-197. Portland, OR.

42. ———, N. McKay, D. R. Gedney, and C. Alerich. 1997. Washington's public and private forests. U.S. Forest Service, Pacific Northwest Research Station, Resource Bulletin PNW-RB-218. Portland, OR.

43. Borchers, J. G. and D.A. Perry. 1990. Effects of prescribed fire on soil organisms. Pages 143-175 in J. D. Walstad, S. R. Radosevich, and D.V. Sandberg, editors. Natural and prescribed fire in Pacific Northwest forests. Oregon State University Press, Corvallis, OR.

44. ———, D.A. Perry, P. Sollins. et al. 1992. The influence of soil texture and aggregation on carbon and nitrogen dynamics in southwest Oregon forests and clear-cuts. Canadian Journal of Forest Research 22:298-305.

45. Bormann, B. T., P. G. Cunningham, M. H. Brookes, V. W. Manning, and M. W. Collopy. 1994. Adaptive ecosystem management in the Pacific Northwest. U.S. Forest Service, Pacific Northwest Research Station, General Technical Report PNW-GTR-341.

46. ———, H. Spaltenstein, M. H. McClellan, F.C. Ugolini, K. Cromack, Jr., S. M. Nay. 1995. Rapid soil development after windthrow disturbance in pristine forests. Journal of Ecology 83:1-20.

47. Breznak, J.A. 1982. Intestinal microbiota of termites and other xylophagous insects. Annual Review of Microbiology 36:323-343.

48. Brown, R. E. 1985. Management of wildlife and fish habitats in forests of western Oregon and Washington. U.S. Forest Service, Pacific Northwest Region.

49. Brown, T. K. 1996. Snags and wildlife tree preservation and enhancement. Pages 71-75 in P. Bradford, T. Manning, and B. l'Anson. Editors. Wildlife tree/stand-level biodiversity workshop proceedings. October 17-18, 1995, Victoria, BC, Canada.

50. Bull, E. L. 1996. Owl nests in brooms caused by dwarf mistletoes and Elytroderma disease. U.S. Forest Service, Pacific Northwest Research Station, Portland, OR. Unpublished report.

51. ———. 1995. Progress report on American marten home range and habitat use. U.S. Forest Service, Pacific Northwest Research Station, Portland, OR. Unpublished report.

52. ———, and C. T. Collins. 1993. Vaux's swift (*Chaetura vauci*). In A. Poole, and E. Gill, editors, The Birds Of North America. The Academy of Natural Sciences, Report No 77., Philadelphia, PA.

53. ———. and R. S. Holthausen. 1993. Habitat use and management of pileated woodpeckers in northeastern Oregon. Journal of Wildlife Management 57:335-345.

54. ———. and A. D. Partridge. 1986. Methods of killing trees for use by cavity nesters. Wildlife Society Bulletin 14:142-146.

55. ———, R. S. Holthausen, and M.G. Henjum. 1992. Roost trees used by pileated woodpeckers in northeastern Oregon. Journal of Wildlife Management. 56:786-793.

56. ———, C. G. Parks, and T. R. Torgerson. 1997. Trees and logs important to wildlife in the interior Columbia River Basin. U.S. Forest Service, Pacific Northwest Research Station, General Technical Report PNW-GTR-391.

57. ———, A. D. Partridge, and W. G. Williams. 1981. Creating snags with explosives. U.S. Forest Service, Pacific Northwest Research Station, Portland, OR. Research Note PNW-RN-393.

58. Burge, W. D. and F. E. Broadbent. 1961. Fixation of ammonia by organic soils. Soil Science Society of America Proceedings 25:199-204.

59. Burns, R. M., compiler. 1983. Silvicultural systems for the major forest types of the United States. U.S. Forest Service, Agriculture Handbook No. 445. Washington, D.C.

59a Bury, R. B., and P. S. Corn. 1988. Douglas fir forests in Oregon and Washington Cascades; relation of the herpetofauna to stand age and moisture. Pages 11-122 in R. C. Szaro, K. E. Severson, and D. R. Patton, technical coordinators. Management of amphibians, reptiles, and small mammals in North America. Proceedings of a symposium, July 19-21 1988 in Flagstaff, AZ. U. S. Forest Service, Rocky Mountain Forest and Range Experiment Station, Fort Collins, CO. General Technical Report RM-GTR-166.

60. Camp, A. E. (submitted). Demographics of dead trees on the eastern slope of the Washington Cascades from the Cascades crest to the Columbia Plateau. Northwest Science.

61. Carey, A. B. and R. O. Curtis. 1996. Conservation of biodiversity: a useful paradigm for forest ecosystem management. Wildlife Society Bulletin 24(4):610-620.

62. ———. and J. D. Gill. 1983. Direct habitat improvements: some recent advances. In J. W. Davis, G. A. Goodwin, and R. A. Ockerfells, technical coordinators. Snag habitat management: proceedings of a symposium. U. S. Forest Service, Washington, D.C. Technical Report RM-99.

63. ———. 1995. Sciurids in Pacific Northwest managed and old-growth forests. Ecological Applications 5(3):648-661.

64. ———. and M. L. Johnson. 1995. Small mammals in managed, naturally young, and old-growth forests. Ecological Applications 5(2): 336-352.

65. ———, and H. R. Sanderson. 1981. Routing to accelerate tree-cavity formation. Wildife Society Bulletin 9:14-21.

66. ———, D. R. Thysell, and A. Brodie. 1999. The forest ecosystem study: background, rationale, implementation, baseline conditions, and silvicultural assessment. U.S. Forest Service, Pacific Northwest Research Station, Portland, OR. General Technical Report PNW-GTR-457.

67. ———, C. Elliott, B. R. Lippke, J. Sessions, C. J. Chambers, C. D. Oliver, J. F. Franklin, and M. J. Raphael. 1996. A pragmatic ecological approach to small-landscape management. Washington Forest Landscape Management Project—report No.2. Washington Forest Landscape Management Project Report. No. 2, Washington Department of Natural Resources, Olympia, WA.

68. ———, D. R. Thysell, L. J. Villa, T. M. Wilson, S. M. Wilson, J. M. Trappe, W. Colgan III, E. R. Ingham, and M. Holmes. 1996. Foundations of biodiversity in managed Douglas-fir forests. Pages 68-82 in D. L. Pearson and C. V. Klimas, editors. The role of restoration in ecosystem management. Society For Ecological Restoration, Madison, WI.

69. ———, T. M. Wilson, C. C. Maguire, B. Biswell. 1997. Dens of northern flying squirrels in the Pacific Northwest. Journal of Wildlife Management 61(3): 684-699.

70. ———. 2000. Effects of new forest management strategies on squirrel populations. Ecological Applications: 10(1):248-257.

71. Caza, C. L. 1993. Woody debris in the forests of British Columbia: a review of the literature and current research. Ministry Of Forests, British Columbia, Canada.

72. Cederholm, C. J. and N. P. Peterson. 1985. The retention of coho salmon (Oncorhynchus kisutch) carcasses by organic debris in small streams. Canadian Journal of Fisheries and Aquatic Sciences 42: 1222-1225.

72a. CFR 1995. Logging Standard (29 CFR 1910.266) Code of Federal Regulations. Washington, DC.: US Government Printing Office, Office of the Federal Register.

73. Chambers, C. L., W. C. McComb, and J. C. Tappeiner, II. 1999. Breeding bird responses to three silvicultural treatments in the Oregon Coast Range. Ecological Applications 9(1): 171-185.

74. Chapman, D. W. and K. P. McLeod. 1987. Development of criteria for fine sediment in the Northern Rockies Ecoregion. U.S. Environmental Protection Agency, Water Division, Report 910/9-87-162, Seattle, WA.

75. Chappell, C., R. Crawford, J. Kagan, and P. J. Doran. 1997. A vegetation, land use, and habitat classification system for the terrestrial and aquatic systems of Oregon and Washington. Wildlife habitats and species associations in Oregon and Washington—progress report #3, Washington Department of Fish And Wildlife, Olympia.

76. Chen, H. 1989. Studies on tree mortality and log decomposition of main species in conifer-deciduous mixed forest on Changbai Mountain Reserve. Thesis. Academia Sinica, Shenyang, People's Republic of China.

77. Childs, T. W., K. R. Shea, J. L. Stewart. 1971. Elytroderma disease of ponderosa pine. U.S. Forest Service, Forest Insects And Disease Program, Leaflet No.42.

78. Christensen, N. L. et al. 1996. The report of the Ecological Society of America commitee on the scientific basis for ecosystem management. Ecological Applications 6:665-691.

79. Christiansen, E., R. H. Waring, and A. A. Berryman. 1987. Resistance of conifers to bark beetle attack: searching for general relationships. Forest Ecology and Management 22:89-106.

80. Christy, E. J. and R. N. Mack. 1984. Variation in demography of juvenile Tsuga heterophylla across the substratum mosaic. Journal of Ecology 72:75-91.

81. Cline, S. P. and C. A. Phillips. 1983. Coarse woody debris and debris-dependent wildlife in logged and natural riparian zone forests: a western Oregon example. Pages 33-39 in J. W. Davis, G. A. Goodwin, and R. A. Ockenfels, technical coordinators. Snag Habitat Management: Proceedings of the Symposium. U.S. Forest Service, Rocky Mountain Research Station, General Technical Report, RM-GTR-99.

82. ———, A. B. Berg, and H. M. Wight. 1980. Snag characteristics and dynamics in Douglas-fir forests, Western Oregon. Journal of Wildlife Management 44(4): 773-786.

83. Clinton, P. W., R. B. Allen, R. H. Newman, and M. R. Davis. 1997. Nitrogen availability and forest development in mountain beech. Soil Biology and Biochemistry: in press.

84. Conner, R. N., J. G. Dickson, and J. H. Williamson. 1983. Potential woodpecker nest trees through artificial innoculation with heart rots. In J. W. Davis, G. A. Goodwin, and R. A. Ockerfells, technical coordinators. Snag habitat management: proceedings of a symposium. U.S. Forest Service, Washington, D.C. Technical Report RM-99.

85. Cornaby, B. W., and J. B. Waide. 1973. Nitrogen fixation in decaying chestnut logs. Plant and Soil 39(2):445-448.

86. Coutts, M. P. and J. Grace. 1995. Wind and trees. Cambridge University Press, New York, NY.

87. Cromack, K., Jr., R. L. Todd, and C. D. Monk. 1975. Patterns of Basidiomycete nutrient accumulation in conifer and deciduous forest litter. Soil Biology And Biochemistry 7(4-5): 265-268.

88. ———, C. C. Delwiche, and D. H. McNabb. 1979. Prospects and problems of nitrogen management using symbiotic nitrogen fixers. Pages 210-223 in J. C. Gordon, C. T. Wheeler, and D. A. Perry, editors. Symbiotic nitrogen fixation in the management of temperate forests. Forest Research Lab, Corvallis, OR.

89. Crossley, D. A. Jr. 1976. The roles of terrestrial saprophagous arthropods in forest soils: current status of concepts. Pages 49-56 in W. J. Mattson, editor. The role of arthropods in forest ecosystems. Springer Verlag, New York, NY.

90. Cummins, K. W., J. R. Sedell, F. J. Swanson, G. W. Minshall, S. G. Fisher, C. E. Cushing, R. C. Petersen, and R. L. Vannote. 1982. Organic matter budgets for stream ecosystems: problems in their evaluation. Pages 299-353 in G. W. Minshall and J. R. Barnes, editors. Stream ecology: application and testing of general ecological theory. Plenum, New York, NY.

91. Curtis, R. O. 1995. Extended rotations and culmination age of coast Douglas-fir: Old studies speak to current issues. U.S. Forest Service, Pacific Northwest Research Station, Portland, Oregon. Research Paper PNW-RP-485.

92. ———. 1997. The role of extended rotations. Pages 165-170 in K. A. Kohm, and J. F. Franklin, editors. Creating a forestry for the 21st century: the science of ecosystem management. Island Press, Washington, D.C.

93. ———, and D. D. Marshall. 1993. Douglas-fir rotations: time for reappraisal? Western Journal of Applied Forestry 8(3): 81-85.

94. ——— D. S. DeBell, C. A. Harrington, D. P. Lavender, J. B. St. Clair, J. C. Tappeiner, and J. D. Walstad. et al. 1998. Silviculture for multiple objectives in the Douglas-fir region. U.S. Forest Service, Pacific Northwest Research Station, Portland, Oregon. General Technical Report PNW-GTR-435.

95. Daily, G. C. 1993. Heartwood decay and vertical distribution of red-naped sapsucker nest cavities. Wilson Bulletin 105(4):674-679.

96. DeBell, D. S., R. O. Curtis, C. A. Harrington, and J. C. Tappeiner. 1997. Shaping stand development through silvicultural practices. Pages 141-150 in K. A. Kohm, and J. F. Franklin, editors. Creating a forestry for the 21st century: the science of ecosystem management. Island Press, Washington, D.C.

97. Deyrup, M. A. 1975. The insect community of dead and dying Douglas-fir. 1: The Hymenoptera. Ecosystem Analysis Studies Bulletin 6, Coniferous Forest Biome. University of Washington, Seattle, WA.

98. ———. 1976. The insect community of dead and dying Douglas-fir : Diptera, Coleoptera, and Neuroptera. Dissertation, University of Washington, Seattle, WA.

99. ———. 1981. Deadwood decomposers. Natural History 90:84-91.

100. Diaz, S., J. P. Grime, J. Harris, and E. McPherson. 1993. Evidence of a feedback mechanism limiting plant response to elevated CO_2. Nature 364:616-617.

101. Dickson, B. A., and R. L. Crocker. 1953. A chronosequence of vegetation and soils near Mt. Shasta, California. II. The development of the forest floors and the carbon and nitrogen profiles of the soils. Journal of Soil Science 4:142-154.

102. Dixon, R. D. 1995. Ecology of white-headed woodpeckers in the central Oregon Cascades. Thesis, University of Idaho, Moscow, ID.

103. Doyle, A. T. 1985. Small mammal micro- and macrohabitat selection in streamside ecosystems. Dissertation, Oregon State University, Corvallis, OR.

104. ———. 1990. Use of riparian and upland habitat by small mammals. Journal of Mammalogy 71:14-23.

105. Edmonds, R. L. 1987. Decomposition rates and nutrient dynamics in small-diameter woody litter in four forest ecosystems in Washington, U.S.A. Canadian Journal of Forest Research 17:499-509.

106. ———. 1991. Organic matter decomposition in western United States. Pages 118-128 in A.E. Harvey and L. F. Neuenschwander, editors. Management and productivity of western-montane forest soils. U.S. Forest Service, Intermountain Research Station, Ogden, UT.

107. ———, and A. Eglitis. 1989. The role of Douglas-fir beetle and wood borers in the decomposition and nutrient release from Douglas-fir logs. Canadian Journal of Forest Research 19:327-337.

108. Entry, J. A., N. E. Martin, and K. Cromack, Jr. 1986. Light and nutrient limitation in Pinus monticola: seedling susceptibility to Armillaria infection. Forest Ecology and Management 17:189-198.

109. Everest, F. H., R. L. Beschta, J. C. Srivener, K. V. Koski, J. R. Sedell, and C. J. Cederholm. 1987. Fine sediment and salmon production: a paradox. Pages 98-142 in E. O. Salo, and T. W. Cundy, editors. Streamside management: forestry and fishery interactions, Institute of Forest Resources, Contribution 57. University of Washington, Seattle, WA.

110. Everett, R., C. Oliver, J. Saveland, P. Hessburg, N. Diaz, and L. Irwin. 1993. Adaptive ecosystem management. Pages 351-364 in M. Jensen and P. Bourgeron, editors. Ecosystem management: principles and application. U. S. Forest Service, Northern Region. Missoula, MT.

111. ———, L. Lehmkuhl, R. Schellhaas, P. Ohlson, D. Keenum, H. Riesterer, and D. Spurbeck. In press. Snag dynamics in a chronosequence of 26 wildfires on the east slope of the Cascade Range in Washington. International Journal of Wildland Fire.

112. ———, compiler. 1994. Volume IV: restoration of stressed sites, and processes. U.S. Forest Service, Pacific Northwest Research Station, Portland, Oregon. General Technical Report PNW-GTR-30.

113. Fahey, T. J., J. W. Hughes, M. Pu, and M. A. Arthur. 1988. Root decomposition and nutrient flux following whole-tree harvest of northern hardwood forest. Forest Science 34(3): 744-768.

114. Federal Ecosystem Management Analysis Team (FEMAT). 1993. Chapter III, Option Development and Description, and Chapter IV, Terrestrial Forest Ecosystem Assessment. Forest Ecosystem Management: An Ecological, Economic, and Social Assessment. U.S. Government Printing Office.

115. Fogel, R. and K. Cromack, Jr. 1977. Effect of habitat and substrate quality on Douglas-fir litter decomposition in western Oregon. Canadian Journal of Botany 55(12):1632-1640.

116. Fogel, R., M. Ogawa, and J. M. Trappe. 1973. Terrestrial decomposition: a synopsis. U.S. International Biological Program, Coniferous Forest Biome Internal Report 135. University of Washington, Seattle, WA.

117. Foster, F. R. and G. E. Lang. 1982. Decomposition of red spruce and balsam fir boles in the White Mountains of New Hampshire. Canadian Journal of Forest Research 12:617-626.

118. Francke-Grosmann, H. 1967. Ectosymbiosis in wood inhabiting insects. Pages 142-205 in S. M. Henry, editor. Symbiosis. Volume 2. Academic Press, New York, NY.

119. Frankland, J. C., J. N. Hedger, and M. J. Swift, editors. 1982. Decomposer basidiomycetes, their biology and ecology. Cambridge University Press, New York, NY.

120. Franklin, J. F. 1988. Pacific Northwest forests. Pages 104-130 in M. G. Barbour, and W. D. Billings, editors. North American terrestrial vegetation. Cambridge University Press, New York, NY.

121. ———. 1989. Toward a new forestry. American Forests 95 (11/12): 37-44.

122. ———. 1992. The scientific basis for new perspectives in forests and streams. Pages 25-72 in R. J. Naiman, editor. Watershed management balancing sustainability and environmental change. Springer Verlag, New York, NY.

123. ———, and C. T. Dyrness. 1988. Natural vegetation of Oregon and Washington. Oregon State University Press, Corvallis, OR.

124. ———, and C. Maser. 1988. Looking ahead: some options for public lands. Pages 113-122 in C. Maser, R. F. Tarrant, J. M. Trappe, and J. F. Franklin, editors. U.S. Forest Service, Pacific Northwest Research Station, General Technical Report PNW-GTR-229, Portland, OR.

125. ———, and T. A. Spies. 1991. Composition, function, and structure of old-growth Douglas-fir forests. Pages 71-82 in L. F. Ruggiero, K. B. Aubry, A. B. Carey, and M. H. Huff, technical coordinators. Wildlife and vegetation of unmanaged Douglas-fir Forests. U.S. Forest Service, Pacific Northwest Research Station, Portland, OR, General Technical Report, PNW-GTR-285.

126. ———, and T. A. Spies. 1991b. Ecological definitions of old-growth Douglas-fir forests. Pages 61-70 in L. F. Ruggiero. K. B. Aubry, A. B. Carey, and M. H. Huff, technical coordinators. Wildlife and vegetation of unmanaged Douglas-fir forests. U.S. Forest Service, Pacific Northwest Research Station, General Technical Report, PNW-GTR-285, Portland, OR.

127. ———, and R. H. Waring. 1980. Distinctive features of the Nortwestern coniferous forest: Development, structure, and function Pages 59-85 in R. H. Waring, editor. Forests: fresh perspectives from ecosystem analysis. Proceedings of the 40th Biology Colloquium (1979), Oregon State University. Press, Corvallis, OR.

128. ———, P. M. Frenzen, and F. J. Swanson. 1988. Re-creation of ecosystems at Mount St. Helens: contrasts in artificial and natural approaches. Pages 1-37 in J. Cairns, Jr., editor. Rehabilitating damaged ecosystems. Vol. 2. CRC Press, Boca Raton, FL.

129. ———, J. A. Mac Mahon, F. J. Swanson, and J. R. Sedell. 1985. Ecosystem responses to the eruption of Mt. St. Helens. National Geographic Research. 2: 198-216.

130. ———, D. A. Perry, T. A. Schowalter, M. E. Harmon, A. McKee, and T. A. Spies. 1989. Importance of ecological diversity to long-term site productivity. Pages 82-97 in D. A. Perry, R. Meurisse, B. Thomas, R. Miller, J. Boyle, J. Means, C. R. Perry, and R. F. Powers, editors. Maintaining the productivity of Pacific Northwest forest ecosystems. Timber Press, Portland, OR.

131. ———, K. Cromack, Jr., W. Denison, A. McKee, C. Maser, J. R. Sedell, F. J. Swanson, and G. Juday. 1981. Ecological characteristics of old-growth forests. U.S. Forest Service, Pacific Northwest Research Station, Portland, OR. General Technical Report, PNW-GTR-118.

132. ———, D. R. Berg, D. A. Thornburgh, and J. C. Tappeiner. 1997. Alternative silvicultural approaches to timber harvesting: variable retention harvest systems. 1997. Pages 111-139 in K.A. Kohm, and J. F. Franklin, editors. Creating a forestry for the 21st century: the science of ecosystem management. Island Press, Washington, D.C.

133. Furniss, R. L. and V. M. Carolin. 1977. Western Forest Insects. U.S. Forest Service Publication No. 1339.

134. Gartner, B. L. 1995. Plant stems: physiology and functional morphology. Academic Press, New York, NY.

135. Gashwiler, J. S. 1959. Small mammal study in west-central Oregon. Journal of Mammalogy 40(1): 128-139.

136. ———. 1970. Plant and mammal changes on a clear-cut in west central Oregon. Ecology 51(6):1018-1026.

137. Gast, W., D. Scatt, C. Schmitt, D. Clemens, S. Howes, C. G. Johnson, Jr., R. Mason, F. Mofr, and R. A. Clapp. 1991. Blue Mountains Forest Health Report: New Perspectives In Forest Health. U.S. Forest Service, Pacific Northwest Region, Malheur, Umatilla, and Wallowa-Whitman National Forest.

138. Gholz, H. L. 1982. Environmental limits on aboveground net primary production, leaf area, and biomass in vegetation zones of the Pacific Northwest. Ecology 63(2):469-481.

139. Gonor, J. J., J. R. Sedell, and P. A. Benner. 1988. What we know about large trees in estuaries, in the sea, and on coastal beaches. Pages 83-112 in: Maser, C., R. F. Tarrant, J. M. Trappe, and J. F. Franklin, technical editors. 1988. From the forest to the sea: a story of fallen trees. U.S. Forest Service, Pacific Northwest Research Station, Portland, OR. General Technical Report PNW-GTR-229.

140. Graham, R.L. 1982. Biomass dynamics of dead Douglas-fir and western hemlock boles in mid-elevation forests of the Cascade Range. Ph.D. Dissertation. Oregon State University, Corvallis, OR.

141. ———, and K. C. Cromack, Jr. 1982. Mass, nutrient, content, and decay rate of dead boles in rain forests of Olympic National Park. Canadian Journal of Forest Research 12:511-521.

142. Grant, G. E., F. J. Swanson, and M. G. Wolman. 1990. Pattern and origin of stepped-bed morphology in high-gradient streams, western Cascades, Oregon. Geological Society of America Bulletin 102:340-352.

143. Gregory, S.V. 1997. Riparian Management In The 21st Century. Pages 69-86 in K. A. Kohm, and J. F. Franklin, editors, Creating a forestry for the 21st century. The science of ecosystem management. Island Press, Washington, D.C.

144. ———, and P. A. Bisson. 1997. Degradation and loss of anadromous salmonid habitat in the Pacific Northwest. Pages 277-214 in D. J. Stouder, P. A. Bisson, and R. J. Naiman, editors. Pacific salmon & their ecosystems. Chapman and Hall, New York, NY.

145. ———, F. J. Swanson, W.A. McKee, and K. W. Cummins. 1991. An ecosystem perspective of riparian zones. BioScience 41:540-551.

146. Grier, C. C. 1978. A Tsuga heterophylla-Picea sitchensis ecosystem of coastal Oregon: decomposition and nutrient balances of fallen logs. Canadian Journal of Forest Research 8:198-206.

147. ———, and R. S. Logan. 1978. Old-growth Pseudotsuga menziesii communities of a western Oregon watershed: biomass distribution and production budgets. Ecological Monographs 47:373-400.

148. ———, K. M. Lee, and N. M. Nadkarni et al. 1989. Productivity of forests of the United States and its relation to soil and site factors and management practices: a review. U.S. Forest Service, Pacific Northwest Research Station, Portland, OR. General Technical Report PNW-GTR-222.

149. Hagan, J. M. and S. L. Grove. 1999. Coarse woody debris. Journal of Forestry 97(1):6-11.

150. Hagar, J. C., W. C. McComb, and W. H. Emmingham. 1996. Bird communities in commercially thinned and unthinned Douglas-fir stands of western Washington. Wildlife Society Bulletin 24(2):353-366.

151. Halpern, C. B., and T. A. Spies. 1995. Plant species diversity in natural and managed forests of the Pacific Northwest. Ecological Applications 5(4):913-934.

152. ———, S. A. Evans, C. R. Nelson, D. McKenzie, D. Liquori, D. E. Hibbs, and M. G. Halaj. 1999. Response of forest vegetation to varying levels and patterns of green-tree retention: an overview of a long-term experiment. Northwest Science 73: 27-43.

153. Hamer, T. E., and S. K. Nelson. 1995. Characteristics of marbled murrelet nest trees and nesting stands. Pages 69-87 in C. J. Ralph, G. L. Hunt Jr., M. G. Raphael, and J. F. Piatt, editors. Ecology and conservation of the marbled murrelet. U.S. Forest Service, Pacific Southwest Forest and Range Experiment Station, Albany, CA. General Technical Report PSW-GTR-152.

154. Hansen, A. J., Spies, T.A., Swanson, F. J., and J. L. Ohmann. 1991. Conserving biodiversity in managed forests. BioScience 41:382-392.

155. ———, W. C. McComb, R. Vega, M. G. Raphael, and M. Hunter. 1995. Bird habitat relationships in natural and managed forests in the west Cacades of Oregon. Ecological Applications 5(3):555-569.

156. ———, S. L. Garman, J. F. Weigand, D. L. Urban, W. C. McComb, and M. G. Raphael. 1995. Alternative silvicultural regimes in the Pacific Northwest: simulations of ecological and economic effects. Ecologcial Applications 5(3):535-554.

157. ———, S. L. Garman, B. Marks, and D. L. Urban. 1993. An approach for managing vertebrate diversity across multiple-use landscapes. Ecological Applications 3: 481-496.

158. Harmon, M.E. 1992. Long-term experiments on log decomposition at the H. J. Andrews Experimental Forest. U.S. Forest Service, Pacific Northwest Research Station, General Technical Report, PNW-GTR-280.

159. ———, and H. Chen. 1991. Coarse woody debris dynamics in two old-growth ecosystems: comparing a deciduous forest in China and a conifer forest in Oregon. BioScience 41:604-610.

160. ———, and J. F. Franklin. 1989. Tree seedlings on logs in Picea-Tsuga forests of Washington and Oregon. Ecology 70(1):48-59.

161. ——— W. K. Ferell, and J. F. Franklin. 1990. Effects on carbon storage of conversion of old-growth forests to young forests. Science 247:699-702.

162. ———, J. Sexton, B. A. Caldwell, and S. E. Carpenter. 1994. Fungal sporocarp mediated losses of Ca, Fe, K, Mg, N, P, and Zn from conifer logs in the early stages of decomposition. Canadian Journal of Forestry Research 24: 1883-1893.

163. ———. 1989. Effects of bark fragmentation on plant succession on conifer logs in Picea-Tsuga forests of Olympic National Park, Washington. American Midland Naturalist 121:112-124.

164. ———, J. F. Franklin, F. J. Swanson, P. Sollins, S. V. Gregory, J. D. Lattin, N. H. Anderson, S. P. Cline, N. G. Aumen, J. R. Sedell, G. W. Lienkaemper, K. Cromack, Jr., and K. W. Cummins. 1986. Ecology of coarse woody debris in temperate ecosystems. Advances in Ecological Research 15:133-302.

165. Harrod, R. J., W. L. Gaines, W. E. Hartl, and A. Camp. 1998. Estimating historical snag density in dry forests east of the Cascade Range. U.S. Forest Service, Pacific Northwest Research Station, Portland, Oregon, General Technical Report PNW-GTR-428.

166. Hartmann, P., M. Scheitler, and R. Fischer. 1989. Chapter 2-C in E.D. Schulze, O. L. Lange, and R. Oren, editors. Forest decline and pollution. Springer-Verlag, New York, NY.

167. Harvey, A. E. 1995. Soil and the forest floor: what it is, how it works, and how to treat it. Nat. Res. News (5)1. Blue Mountains National Research Institute, La Grande, OR. U.S. Deptartment of Agriculture, Forest Service, Pacific Northwest Research Station.

168. ———, M. J. Larsen, and M. F. Jurgensen. 1976. Distribution of ectomycorrhizae in a mature Douglas-fir/larch forest soil in western Montana. Forest Science 22:393-398.

169. ———, M. F. Jurgensen, R. T. Graham. 1987. The role of woody residues in soils of a symposium. Washington State University, Pullman, WA.

170. ———, J. M. Geist, G. I. McDonald, M. F. Jurgensen, P. H. Cochran, D. Zabowski, and R. T. Meurisse. 1994. Biotic and abiotic processes in eastside ecosystems: the effects of management on soil properties, processes, and productivity. U.S. Forest Service, Pacific Northwest Research Station, Portland, OR. General Technical Report PNW-GTR-323.

171. Hawksworth, F. G. and D. Wiens. 1996. Dwarf mistletoes: biology, pathology and systematics. U.S. Forest Service, Agricultural Handbook No. 709, Washington, D.C.

172. Haynes, R. W., R. J. Alig, and E. Moore. 1994. Alternative simulations of forestry scenarios involving carbon sequestration options: investigation of impacts on regional and national timber markets. U.S. Forest Service, Pacific Northwest Research Station, General Technical Report PNW-GTR-335, Portland, OR.

173. Heede, B. H. 1972. Influences of a forest on the hydraulic geometry of two mountain streams. Water Resources Bulletin 8:523-529.

174. Hedin, L. O., J. J. Armesto, and A. H. Johnson. 1995. Patterns of nutrient loss from unpolluted temperate old-growth forests: evaluation of biogeochemical theory. Ecology 76:493-509.

175. Hepting, G. H. 1971. Diseases of forest and shade trees of the United States. USDA Forest Service Agricultural Handbook 386.

176. Hergert, H. L. and E. F. Kurth. 1952. The chemical nature of cork from Douglas-fir bark. Tappi 35: 59-66.

177. Hessburg, P. F., R. G. Mitchell, and G. M. Filip. 1994. Historical and current roles of insects and pathogens in eastern Oregon and forested landscapes. U.S. Forest Service, Pacific Northwest rearch Station, Portland OR. General Technical Report PNW-GTR-327.

178. ———, B. G. Smith, and R. Brion Salter. 1999. Detecting change in forest spatial patterns from reference conditions. Ecological Applications 9(4):1232-1252.

179. Hixson, S. E., R. F. Walker, and C. M. Skau. 1990. Soil denitrification rates in four subalpine plant communities of the Sierra Nevada. Journal of Environmental Quality 19:717-620.

180. Holsten, E. H., R. A. Werner, and T. H. Laurent. 1980. Insects and diseases of Alaskan forests. USDA Forest Service, Alaska Region report No. 75.

181. Holthausen, R. S., M. G. Raphael, K. S. McKelvey, E. D. Forsman, E. E. Starkey, and D. Erran Seaman. 1996. The contribution of federal and non-federal habitat to persistence of the Northern Spotted Owl on the Olympic Peninsula, Washington: report of the reanalysis team. U.S. Forest Service, Pacific Northwest Research Station, Portland, OR. General Technical Report PNW-GTR-352.

182. Hope, S. M. 1987. Classification of decayed *Abies Amabalis* logs. Canadian Journal of Forest Research 17:559-564.

183. Huggard, D. J. 1999. Static life-table analysis of fall rates of subalpine fir snags. Ecological Applications 9(3):1009-1016.

184. Hulme, M. A. and J. K. Shields. 1970. Biological control of decay fungi in wood by competition for non-structural carbohydrates. Nature 227:300-301.

184a. Hunter, M. 1995. Residual trees as biological legacies. Management Communique no. 2. Corvallis, Oregon. Cascade Center for Ecosystem Management.

185. Ingham, E. R. 1995. Organisms in the soil: the functions of bacteria, fungi, protozoa, nematodes, and arthropods. National Research News (5)1. Blue Mountains Nat. Res. Institute, La Grande, OR. U.S. Forest Service, Pacific Northwest Research Station.

186. ———, and A. R. Moldenke. 1995. Microflora and microfauna on stems and trunks: diversity, food webs, and effects on plants. Pages 241-256 in B. L. Gartner editor. Plant stems: physiology and functional morphology. Academic Press, New York, NY.

187. Johnson, C. G., Jr., R. R. Clausnitzer, P. J. Mehringer, and C. D. Oliver. 1994. Biotic and abiotic processes of eastside ecosystems: the effects of management on plant and community ecology, and on stand and landscape vegetation dynamics. U.S. Forest Service, Pacific Northwest Research Station, Portland, OR. General Technical Report PNW-GTR-322.

188. Johnson, D. H., and T. A. O'Neil. 2001. Wildlife-habitat relationships in Oregon and Washington. Oregon State University Press, Corvallis, OR.

189. Johnson, D. W. 1992. Nitrogen retention in forest soils. Journal of Environmental Quality 21:1-12.

190. Jones, J. A. and G. E. Grant. 1996. Peak flow responses to clear-cutting and roads in small and large basins, western Cascades, Oregon. Water Resources Research 32: 959-974.

191. Karr, J. R. 1991. Biological integrity: a long-neglected aspect of water resource management. Ecological Applications 1:66-84.

192. ———, K. D. Fausch, P. L. Angermeier, P. R. Yant, and I. J. Schlosser. 1986. Assessing biological integrity in running waters: a method and its rationale. Illinois Natural History Survey, Special Publication 5, Champaign, IL.

193. Kauffman, J. B. 1990. Ecological relationships of vegetation and fire in Pacific Northwest Forests. Pages 39-54 in J. D. Walstad, S. R. Radosevich, and D. V. Sandberg, editors. Natural and prescribed fire in Pacific Northwest forests. Oregon State University Press, Corvallis, OR.

194. Kauffman, M. R. 1996. To live fast or not: growth, vigor, and longevity of old-growth ponderosa pine and lodgepole pine trees. Tree Physiology 16:139-144.

195. Kauffman, J. B. and R. E. Martin. 1985. A preliminary investigation on the feasibility of preharvest prescribed burning for shrub control. Pages 89-114 in Proceedings of The 6th Forest Vegetation Management Conference, Redding, CA.

196. Keenan, R. J., and J. P. Kimmins. 1993. The ecological effects of clear-cutting. Environmental Review 1:121-144.

197. Kellomaki, S., and T. Karjalainen. 1995. Sequestration of carbon in the Finnnish boreal forest ecosystem managed for timber production. Pages 59-68 in M. J. Apps and D. T. Price, editors. Forest ecosystems, forest management and the global carbon cycle. Springer, New York, NY.

198. Kimmins, J. P. 1997. Forest ecology: A foundation for sustainable management, 2nd edition. Prentice Hall, NJ.

199. ———. 1997. Balancing act. Environmental issues in forestry, second edition. UBC Press, Vancouver, B.C., Canada.

200. Koehler, G. M., and K. B. Aubrey. 1994. Pages 74-98 in L. F. Ruggiero, K. B. Aubrey, S. W. Buskirk, L. J. Lyon, and W. J. Zielinski, editors. The scientific basis for conserving forest carnivores: American marten, fisher, lynx, and wolverine. U.S. Forest Service, Rocky Mountain Forest and Range Experiment Station, Fort Collins, CO. General Technical Report RM-GTR-254.

201. Kohm, K. A., J. F. Franklin. 1997. Creating a forestry for the 21st century: The science of ecosystem management. Island Press, Washington D.C.

202. Korol, J. J., M. A. Hemstrom, W. J. Hann, and R. A. Gravenmier. In press. Snags and down wood in the Interior Columbia Basin Ecosystem Management Project. In P. J. Shea and W. Laudenslayer, editors. Proceedings of a conference on the ecology and management of dead wood in western forests, Reno, Nevada. USDA Forest Service, Pacific Southwest Research Station General Technical Report.

203. Kurth, E. F. 1948. The chemical analysis of western woods. Part I. Paper Trade. Journal 126:56-58.

204. ———, and F. L. Chan. 1953. Extraction of tannin and dihydroquercetin from Douglas-fir bark. Journal of American Leather Chemical Association 48:20-32.

205. Lambert, R. L., G. E. Lang, and W. A. Reiners. 1980. Loss of mass and chemical change in decaying boles of a subalpine fir forest. Ecology 61:1460-1473.

206. Larsen, M. J., M. F. Jurgensen, and A. E. Harvey. 1978. N2 fixation associated with wood decayed by some common fungi in western Montana. Canadian Journal of Forest Research 8:341-345.

207. Lattin, J. D. 1990. Arthropod diversity in Northwest old-growth forests. Wings 15 (2):7-10.

208. Law, B. E. and R. H. Waring. 1994. Combining remote sensing and climatic data to estimate net primary production. Ecological Applications 4:717-728.

209. Lee, K. N. 1999. Appraising adaptive management. Conservation Ecology 3(2):1-19.

210. Lehmkuhl, J. F., S. D. West, C. C. Chambers, W. C. McComb, D. A. Manuwal, K. B. Aubry, J. L. Erickson, R. A. Gitzen, and M. Leu. 1999. An experiment for assessing vertebrate response to varying levels and patterns of green tree retention. Northwest Science 73: 45-55.

211. Lehmkuhl, J. F., P. F. Hessburg, R. L. Everett, M. H. Huff, and R. D. Ottmar. 1994. Historical and current forest landscapes of eastern Oregon and Washington. Part 1: vegetation pattern and insect and disease hazards. U.S. Forest Service, Pacific Northwest rearch Station, Portland, OR. General Technical Report PNW-GTR-328.

212. Lewis, J. C. 1998. Creating snags and wildlife trees in commercial forest landscapes. Western Journal of Applied Forestry 13(3):97-101.

213. Little, S. N. and J. L. Ohmann. 1988. Estimating nitrogen lost from forest floor during prescribed fires un Douglas-fir/ western hemlock clearcuts. Forest Science 34:152-164.

214. Lofroth, E. C. 1993. Scale dependent analyses of habitat selection by marten in the sub-boreal spruce biogeoclimatic zone, British Columbia. M. S. Thesis. Simon Fraser University, Burnaby, B. C., Canada

215. Lutes, D. C. 1999. A comparison of methods for the quantification of coarse woody debris and identification of its spatial scale; a study from the Tenderfoot Creek Experimental Forest, Montana. M. S. thesis. University of Montana, Missoula, MT.

216. Machmer, M. M., and C. Steeger. 1995. The ecological roles of wildlife tree users in forest ecosystems. Land Management Handbook No. 35. Ministry of Forests, Victoria, B.C., Canada.

217. Maclean, C. D. 1979. Relative density: the key to stocking assessment in regional analysis—a Forest Survey viewpoint. U.S. Forest Service, Pacific Northwest rearch Station, Portland, OR. General Technical Report PNW-78.

218. MacMillan, P., J. Means, G. M. Hawk, K. Cromack, Jr., and R. Fogel. 1977. Log decomposition in an old-growth Douglas-fir forest. Northwest Science. Assoc. Program and Abstract. at 50th annual mtg. Pullman, WA, Washington State University Press, WA.

219. Magill, A. H., J. D. Aber, J. J. Hendricks, R. D. Bowden, J. M. Melillo, and P. A. Steudler. 1997. Biochemical response of forest ecosystems to simulated chronic nitrogen deposition. Ecological Applications 7(2):402-415.

220. Manion, P. D. 1981. Tree Disease Concepts, second edition. Prentice-Hall. Englewood Cliffs, NJ.

221. Marcot, B. G. 1992. Snag Recruitment Simulator, Rel. 3.1 [computer program]. USDA Forest Service, Pacific Northwest Region, Portland, OR.

222. ———. 1997. Biodiversity of old forests of the west: A lesson from our elders. Pages 87 - 105 in K.A. Kohm, and J. F. Franklin, editors, Creating a forestry for the 21st Century. The science of ecosystem management. Island Press, Washington, D.C.

223. ———. In press. An ecological functional basis for managing decaying wood for wildlife. In: P. J. Shea and W. Laudenslayer, editors. Proceedings of a Conference on The Ecology and Management of Dead Wood in Western Forests, Reno, Nevada. USDA Forest Service, Pacific Southwest Research Station General Technical Report.

224. ———, and M. Vander Heyden. 2001. Key ecological functions of wildlife species. Pages 168-186 in: D. Johnson and T. O'Neil, managing directors. Wildlife-habitat relationships in Oregon and Washington. Oregon State University Press, Corvallis OR.

225. ———, M.A. Castellano, J.A. Christy, L. K. Croft, J. F. Lehmkuhl, R. H. Naney, R. E. Rosentreter, R. E. Sandquist, and E. Zieroth. 1997. Terrestrial ecology assessment. Pages 1498-1713 in Quigley, T. M, and S. J. Arbelbide, technical editors. An assessment of ecosystem components in the interior Columbian basin and portions of the Klamath and Great Basins. U.S. Forest Service, Pacific Northwest rearch Station, Portland, OR. General Technical Report PNW-GTR-405.

226. ———, K. Mellen, S.A. Livingston, and C. Ogden. In press. The DecAID advisory model: wildlife component. In: P. J. Shea and W. Laudenslayer editors. Proceedings of a conference on The Ecology and Management of Dead Wood in Western Forests, Reno, Nevada. USDA Forest Service, Pacific Southwest Research Station General Technical Report.

227. Marra, J. L. and R. L. Edmonds. 1994. Coarse woody debris and forest floor respiration in an old-growth coniferous forest on the Olympic Peninsula, Washington. Canadian Journal of Forest Research 24:1811-1817.

228. Maser, C. 1988. The redesigned forest. R&E Miles, San Pedro, CA.

229. ———. 1994. Sustainable forestry, philosophy, science, and economics. St. Lucie Press. Delray Beach, FL.

230. ———, and J. M. Trappe, technical editors. 1984. The seen and unseen world of the fallen tree. U.S. Forest Service, Pacific Northwest Research Station, Portland, OR. General Technical Report, PNW-GTR-164.

231. ———, J. M. Trappe, and R. A. Nussbaum. 1978. Fungal-small mammal interrelationships with emphasis on Oregon coniferous forests. Ecology 59(6):799-809.

232. ———, ———, and D. Ure. 1978. Implications of small mammal mycophagy to the management of western coniferous forests. Pages 78-88 in Transactions of the 43rd North American Wildlife and Natural Resources Conference.

233. ———, R. F. Tarrant, J. M. Trappe, and J. F. Franklin. 1988. From The Forest To The Sea: A Story Of Fallen Trees. USDA Forest Service, Pacific Northwest Research Station, Portland, OR. General Technical Report PNW-GTR-229.

234. ———, R. G. Anderson, K. Cromack, Jr., J. T. Williams, and R. E. Martin. 1979. Pages 78-95 in J. W. Thomas, editor. Dead and down woody material. Wildlife Habitats in Managed Forests; the Blue Mountains of Oregon and Washington. USDA Forest Service, Washington, D.C.

235. Mason, D. T. and J. Koon. 1985. Habitat values of woody debris accumulations of the lower Stehekin River, with notes on disturbances of alluvial gravels. Final report to the National Park Service, Contract CX-9000-3-8066. Fairhaven College, Western Washington University, Bellingham, WA.

236. Mason, R. R., B. E. Wickman, R. C. Beckwith, and H. G. Paul. 1992. Thinning and nitrogen fertilization in a grand fir stand infested with western spruce budworm. Part I. Insect Response. Forest Science 38: 235-251.

237. Matson, P. A. and R. H. Waring. 1984. Effects of nutrient and light limitation on mountain hemlock: susceptibility to laminated root rot. Ecology 65:1517-1524.

238. McClelland, B. R., S. S. Frissell, S. S., W. C. Fischer, and C. H. Halvorsen. 1979. Habitat management for hole-nesting birds in forests of western larch and Douglas-fir. Journal of Forestry 77: 480-483.

239. McComb, W. C., J. Tappeiner, L. Kellogg, C. Chambers, R. Johnson. 1994. Stand management alternatives for multiple resources: integrated management experiments. Pages 71-86 in M. H. Huff, L. K. Norris, J. B. Nyberg, and N. L. Wilkin, coordinators. Expanding horizons of ecosystem management. U.S. Forest Service, Pacific Northwest Research Station, Portland, Oregon. General Technical Report PNW-GTR-336.

240. ———, and J. L. Ohmann. 1996. Snag Dynamics Projection Model (SDPM), [computer program], USDA Forest Service, Pacific Northwest Research Station, Corvallis, OR.

241. McDade, M. H., F. J. Swanson, W. A. McKee, and J. F. Franklin. 1990. Source distances for coarse woody debris entering small streams in western Oregon and Washington. Canadian Journal of Forest Research 20:326-329.

242. McFee, W. W. and E. L. Stone. 1966. The persistence of decaying wood in the humus layers of northern forests. Soil Science Society American Proceedings. 30:513-516.

243. McIntosh, B. A., J. R. Sedell, J. Smith, R. C. Wissmar, S. E. Clarke, G. H. Reeves, and L. A. Brown. 1994. Management history of eastside ecosystems: changes in fish habitat over 50 years, 1935 to 1992. U.S. Forest Service, Pacific Northwest rearch Station, Portland, OR. General Technical Report PNW-GTR-321.

244. McKee, A., G. Laroi, and J. F. Franklin. 1982. Structure, composition, and reproduction behavior of terrace forests, South Fork Hoh River, Olympic National Park. Pages 22-29 in E. E. Starkey, J. F. Franklin, and J. W. Matthews, editors. Ecological research in national parks of the Pacific Northwest. Forest Research Laboratory, Oregon State University, Corvallis, OR.

245. McNabb, D. H. and K. Cromack, Jr. 1990. Effects of prescribed fire on nutrients and soil productivity. Pages 125-142 *in* J. D. Walstad, S. R. Radosevich, and D. V. Sandberg, editors. Natural and prescribed fire in Pacific Northwest forests. Oregon State University Press, Corvallis, OR.

246. Mellen, T. K., and A.. Ager. 1998. Coarse Wood Dynamics Model (CWDM), [computer model], USDA Forest Service, Pacific Northwest Region, Portland, OR.

247. ———, and A. Ager. In press. A coarse wood dynamics model for the western Cascades. *In* P. J. Shea and W. Laudenslayer, editors). The Ecology and Management of Dead Wood *in* Western Forests. Reno, NV.

248. ———, B. G. Marcot, J. L. Ohmann, K. L. Waddell, E. A. Willhite, B. B. Hostetler, S. A. Livingson, and C. Ogden. In press. DecAID: A decaying wood advisory model for Oregon and Washington. *In*: P. J. Shea and W. Laudenslayer, editors. Proceedings of a conference on The Ecology and Management of Dead Wood in Western Forests, Reno, Nevada. USDA Forest Service, Pacific Southwest Research Station General Technical Report.

249. Moldenke, A. R. 1990. One hundred twenty thousand little legs. Wings 15(2):11-14.

250. ———, and J. D. Lattin. 1990. Density and diversity of soil arthropods as "biological probes" of complex soil phenomena. Northwest Environmental Journal 6:409-410.

251. Molina, R. and J. M. Trappe. 1982. Patterns of ectomycorrhizal host specificity and potential among the Pacific Northwest conifers and fungi. Forest Science 28:423-458.

252. Moore, K. M. S. and S. V. Gregory. Response of young-of-the-year cutthroat trout to manipulation of habitat structure in a small stream. Transactions of the American Fisheries Society 117:162-170.

253. Morrison, M. L., and M. G. Raphael. 1993. Modeling the dynamics of snags. Ecological Applications 3(2):322-330.

254. Moseley, M. P. 1981. The influence of organic debris on channel morphology and bedload transport in a New Zealand forest stream. Earth Surface Processes and Landforms 6:571-579.

255. Mote, P. 1999. Impacts of climate variability and change. National Oceanic And Atmospheric Administration, Office of Global Programs, University of Washington, JISAO/SMA Climate Impacts Group Contribution #715, Seattle, WA.

256. Mowrey, R. A., and J. C. Zasada. 1984. Den tree use and movements of northern flying squirrels in interior Alaska and implications for forest managers. Pages 351-356 *in* W. R. Meehan, T. A. Hanley, and A. Thomas, editors. Fish and wildlife relationships in old-growth forests: proceedings of a symposium: 1982 April 12-14 in Juneau, Alaska. American Institute of Fishery Research Biologists.

257. Murphy, M. L., K. V. Koski, J. Heifetz, S. W. Johnson, D. Kirchofer, and J. F. Thedinga. 1985. Role of large organic debris as winter habitat for juvenile salmonids in Alaska streams. Proceedings, W. Association of Fish and Wildlife Agencies 1984:251-262.

258. Nabuurs, G. J. and G. M. J. Mohren. 1993. Carbon fixation through forestation activities. A study of the carbon sequestration potential of selected forest types, commissioned by the FACE Foundation. IBN Natural resource report 93/4. FACE/Institute for Forestry and Nature Research (IBN-DLO), Arnhem, Wageningen, The Netherlands.

259. Naiman, R. J. and J. R. Sedell. 1979. Benthic organic matter as a function of stream order in Oregon. Archiv für Hydrobiologie 87:404-422.

260. ———, K. L. Fetherston, S. J. McKay, and J. Chen. 1998. Riparian forests. Pages 289-323 in R. J. Naiman and R. E. Bilby, editors. River ecology and management. Springer, New York, NY.

261. ———, and R. E. Bilby. 1998. River ecology and management. Lessons from the Pacific coastal region. Springer, New York, NY.

262. ———, T. J. Beechie, L. E. Benda, D. R. Berg, P. A. Bisson, L. H. MacDonald, M. D. O'Connor, P. L. Olson, and E. A. Steel. 1992. Fundamental elements of ecologically healthy watersheds in the Pacific Northwest coastal ecoregion. Pages 127-18 *in* R. J. Naiman, editor. Watershed Management: Balancing Sustainability and Environmental Change. Springer-Verlag, New York, NY.

263. Neitro, W. A., V. W. Binkley, S. P. Cline, R. W. Mannan, B. G. Marcot, D. Taylor, and F. F. Wagner. 1985. Snags (wildlife trees). Pages 129-169 *in* E. R. Brown, technical editor. Management Of Wildlife And Fish Habitats In Forests Of Western Oregon and Washington. USDA Forest Service, Pacific NW Region, Portland, OR. Publication No. R6-F&WL-192-1985.

264. Nihlgard, B. 1985. The ammonium hypothesis—an additional explanation to the forest dieback in Europe. Ambio 14:2-8.

264a. NIOSH. 1995. National Institute for Occupational Safety and Health. Preventing injuries and deaths of loggers. US Dept. of Health and Human Services, Public Health Service, Centers For Disease Control and Prevention, National Institute for Occupational Safety and Health, Fatality Assessment and Control Evaluation (FACE) Report No. 95-101.

265. Noble, W. O., E. C. Meslow, M. D. Pope. 1990. Denning habits of black bears in the central Coast Range of Oregon. Oregon State University, Department of Fisheries and Wildlife, Corvallis, OR.

266. Nommik, H. and K. Vahtras. 1982. Retention and fixation of ammonium and ammonia in soils. Pages 123-171 *in* F. J. Stevenson, editor. Nitrogen In Agricultural Soils. American Society of Agronomy. Madison, WI.

267. North, M. and J. F. Franklin. 1990. Post-disturbance legacies that enhance biological diversity in a Pacific Northwest old-growth forest. The Northwest Environmental Journal 6:427-429.

268. ———, J. Chen, G. Smith, L. Krakowiak, and J. F. Franklin. 1996. Initial response of understory plant diversity and overstory tree diameter growth to a green tree retention harvest. Northwest Science 70(1): 24-35.

269. ———. 1993. Stand structure and truffle abundance associated with northern spotted owl habitat. Ph.D. thesis, University of Washington, Seattle, WA.

270. Nowacki, G. J. and M. G. Kramer. 1998. The effects of wind disturbance on temperate rainforest structure and dynamics of southeast Alaska. USDA Forest Service, Pacific Northwest Research Station, Portland, OR. General Technical Report, PNW-GTR-42.

271. NRC (National Research Council). 1990. Forestry Research: A Mandate For Change. National Academy Press, Washington, D.C.

272. ———. 1998. Changing conditions of the forest. Pages 57-71 *in* Forested landscapes in perspective. Prospects and opportunities for sustainable management of America's nonfederal forests. Committee on prospects and opportunities for sustainable management of America's nonfederal forests. National Academy Press, Washington, D. C.

273. Odum, E. P. 1971. Fundamentals of ecology. W. B. Saunders Co., Philadelphia, PA.

274. Ohmann, J. L., and T. A. Spies. 1998. Regional gradient analysis and spatial pattern of woody plant communities of Oregon forests. Ecological Monographs 68:151-182.

275. ———, W. C. McComb, and A. A. Zumrawi. 1994. Snag abundance for cavity-nesting birds on nonfederal lands in Oregon and Washington. Wildlife Society Bulletin 22:607-620.

276. ———, and K. L. Waddell. In press. Regional patterns of dead wood in forested habitats of Oregon and Washington. *In*: P. J. Shea and W. Laudenslayer, editors. Proceedings of a conference on The Ecology and Management of Dead Wood in Western Forests, Reno, Nevada. USDA Forest Service, Pacific Southwest Research Station General Technical Report.

277. Oliver, C. D., and B. C. Larson. 1996. Forest stand dynamics. John Wiley and Sons, Inc.

278. ———, D. R. Berg, D. R. Larsen, and K. L. O'Hara. 1992. Integrating management tools, ecological knowledge, and silviculture. Pages 361-382 *in* R. J. Naiman, editor. Watershed Management. Balancing sustainability and environmental change. Springer Verlag, NY.

279. ———, L. L. Irwin, W. H. Knapp. 1994. Eastside forest management: historical overview, extent of their applications, and their effects on sustainability of ecosystems. U.S. Forest Service, Pacific Northwest Research Station, Portland, OR. General Technical Report PNW-GTR-324.

280. Olson, R. K., D. Binkley, and M. Bohm. 1992. The Responses of Western Forests to Air Pollution. Ecological Studies Vol. 97. Springer-Verlag, New York, NY.

281. O'Neill, R. V., W.F. Harris, B. S. Ausmus, and D. E. Reichle. 1975. A theoretical basis for ecosystem analysis with particular reference to element cycling. Pages 28-40 in F. G. Howell, J. B. Gentry, and M. H. Smith editors. Mineral Cycling in Southeastern Ecosystems. ERDA Symposium Series (Conf-740513).

282. Parks, C. G. 1996. Bear Trees - An Eastern Oregon Landscape Legacy. USDA Forest Service, Pacific Northwest Research Station, Portland, OR.

283. ———, and D. C. Shaw. 1996. Death and decay: a vital part of living canopies. Northwest Science. 70:46-53.

284. ———, E. Bull, G. M. Filip, and R. L. Gilbertson. 1996. Wood decay fungi associated with woodpecker nest cavities in living western larch. Plant Disease. 80:959.

285. ———, ———, and T. R. Torgersen. 1997. Field guide to the identification of snags and logs in the interior Columbia River basin. U.S. Forest Service, Pacific Northwest rearch Station, Portland, OR. General Technical Report PNW-GTR-390.

286. ———, D. A. Conklin, L. Bednar, and H. Maffei. 1999. Woodpecker use and fall rates of snags created by killing ponderosa pine infected with dwarf mistletoe. U.S. Forest Service, Pacific Northwest rearch Station, Portland, OR. General Technical Report PNW-GTR-515.

287. Parmeter, J. R. 1978. Forest stand dynamics and ecological factors in relation to dwarf mistletoe spread, impact, and control. Pages 16-31 in: Proceedings of the Symposium on Dwarf Mistletoe Control Through Forest Management. Pacific Southwest Research Station. PSW-GTR-31. Berkeley, CA.

288. Paulus, W. and A. Bresinsky. 1989. Soil fungi and other microorganisms. Chapter 2-A in E.-D. Schulze et al. editors. Forest decline and pollution. Springer-Verlag, New York, NY.

288a. Pecore. M. 1992. Menominee sustained yield management: a successful land ethic in practice. Journal of Forestry 90(7):12-16.

289. Perlin, J. 1989. A forest journey: the role of wood in the development of civilization. Harvard University Press, Cambridge, MA.

290. Perry, D. A. 1994. Forest Ecosystems. John Hopkins University Press, Baltimore and London.

291. ———, and M. P. Amaranthus. 1997. Disturbance, recovery, and stability. Pages 31-67 in K. A. Kohm, and J. F. Franklin, editors. Creating a forestry for the 21st century. Island Press, Washington, D.C.

292. ———, and J. Maghembe. 1989. Ecosystem concepts and current trends in forest management: time for reappraisal. Forest Ecology and Management 26:123-140.

293. ———, M. P. Amaranthus, J. G. Borchers, S. L. Borchers, and R. E. Brainerd. 1989. Bootstrapping in ecosystems. Bioscience 39:230-236.

294. ———, J. G. Borchers, D. P. Turner, S. V. Gregory, C. R. Perry, R. K. Dixon, S. C. Hart, B. Kauffman, R. P. Neilson, and P. Sollins. 1991. Biological feedbacks to climate change: terrestrial ecosystems as sinks and sources of carbon and nitrogen. The Northwest Environmental Journal 7:203-232.

295. Peterjohn, W. T., and D. L. Correll. 1984. Nutrient dynamics in an agricultural watershed: observations on the role of a riparian forest. Ecology 65:1466-1475.

296. Powell, D. S., J. L. Faulkner, D. R. Darr, Z. Zhu, and D. W. MacCleery,. 1993. Forest resources of the United States, 1992. General Technical Report RM-234. U.S. Department of Agriculture, Forest Service.

297. Powell, R. A., and W. J. Zielinski. 1994. Fisher. Pages 38-74 in L. F. Ruggiero and K. B. Aubrey, technical editors. The scientific basis for conserving forest carnivores. U.S. Forest Service, Rocky Mountain Forest and Range Experiment Station. Fort Collins, CO. General Technical Report, RM-GTR-254.

298. Powers, R. F. 1990. Do timber management practices degrade long-term site productivity?: what we know and what we need to know. Pages 87-106 in Proceedings of the Eleventh Annual Forest Vegetation Management Conference, November 8-9, 1989.

299. ———. 1990. A soils research approach to evaluating management impacts on long-term productvity. Pages 127-145 in W. J. Dyck and C. A. Mees, editors. Impact of intensive harvesting on forest site productivity. Proceedings IEA/BE A3 Workshop. South Island, New Zealand, March 1989. Forest Research Institute, Rotorua, New Zealand, FRO Bulletin No. 159.

300. Prescott, C. E., J. P. Corbin, and D. Parkinson. 1989. Input, accumulation, and residence times of carbon, nitrogen, and phosphorus in four Rocky Mountain coniferous forests. Canadian Journal of Forest Research 19:489-498.

301. Putz, F. E., P. D. Coley, K. Lu, A. Montalvo, and A. Aiello. 1983. Uprooting and snapping of trees: structural determinants and ecological consequences. Canadian Journal of Forest Research 13:1011-1020.

302. Quigley, T. M., R. W. Haynes, and R. T. Graham, technical editors. 1996. Integrated scientific assessment for ecosystem management in the interior Columbia basin and portions of the Klamath and Great Basins. U.S. Forest Service. Pacific Northwest Research Station, Portland, OR. General Technical Report PNW-GTR-382.

303. Quinn, T. P. and R. F. Tallman. 1987. Seasonal environmental predictability in riverine fishes. Environmental Biology of Fishes 18: 155-159.

304. Raedecke, K. J. 1988. Streamside management: Riparian wildlife and forestry interactions. Institute of Forest Resources, Contribution 59, Univ. of Washington, Seattle, WA.

305. Raphael, M. G. and L. C. Jones. 1997. Characteristics of resting and denning sites of American marten in central Oregon and western Washington. Pages 146-165 in G. Proulx, H. N. Bryant, and P. M. Woodard, editors. Martes: taxonomy, ecology, techniques and management. Provincial Museum of Alberta, Edmonton, Alberta, Canada.

306. ———, and M. L. Morrison. 1987. Decay and dynamics of snags in the Sierra Nevada, California. Forest Science 33(3):774-783.

307. ———, and M. White. 1984. Use of snags by cavity-nesting birds in the Sierra Nevada. Wildlife Monograph 86: 1-66.

308. Rastetter, E. B., R. B. McKane, G. R. Shaver, and J. M. Melillo. 1992. Changes in C storage by terrestrial ecosystems: How C-N interactions restrict responses to CO2 and temperature. Water, Air, And Soil Pollution. 64:327-344.

309. ———, G. I. Agren, and G. R. Shaver. 1997. Responses of N-limited ecosystems to increased CO2: A balanced-nutrition, coupled-element-cycles model. Ecological Applications 7(2): 444-460.

310. Rayner, A. D. M. and L. Boddy. 1988. Fungal decomposition of wood, its biology and ecology. John Wiley & Sons, New York, NY.

311. Rose, C. R. and P. S. Muir. 1997. Green-tree retention: consequences form timber production in forests of the western Cascades, Oregon Ecological Applications 7(1): 209-217.

311a. ———. 1994. Relationships of green-tree retention following timber harvest to forest growth and species composition in the western Cascade Mountains. M.S. Thesis. Oregon State University, Corvallis.

312. Regional Forest Nutrition Research Project and Stand Management Cooperative (RFNRP/SMC) 1989. Planning for 2001: Research in Forest Nutrition, Silviculture, and Wood Quality in the 21st Century. Proceedings of a Joint Review And Planning Meeting at Pack Forest. November, 1989.

313. Ruggiero, L. F., and K. B. Aubry. 1994. The scientific basis for conserving forest carnivores. U.S. Forest Service, Rocky Mountain Forest and Range Experiment Station, Fort Collins, CO. General Technical Report RM-GTR-254.

314. ———, ———, A. B. Cary, and M. H. Huff. 1991. Wildlife and vegetation of unmanaged Douglas-fir forests. USDA Forest Service, Pacific Northwest Research Station, Portland, OR. General Technical Report PNW-GTR-285.

315. Runyon, J., R. H. Waring, S. N. Goward, and J. M. Welles. 1994. Environmental limits on net primary production and light-use efficiency across the Oregon transect. Ecological Applications 4(2): 226-237.

316. Ruth, R. H., and R. A. Yoder. 1953. Reducing wind damage in the forests of the Oregon Coast Range. Tree. U.S. Forest Service, Pacific Northwest Research Station, Portland, OR. Research Paper, PNW-RP-7.

317. Ryan, M. G., D. Binkley, and J. H. Fownes. 1997. Age-related decline in forest productivity: pattern and process. Advances in Ecological Research 27:213-262.

318. ———, and B. J. Yoder. 1997. Hydraulic limits to tree height and tree growth. BioScience 47(4): 235-242.

319. Sachs, D. L., and P. Sollins. 1986. Potential effects of management practices on nitrogen nutrition and long-term productivity of western hemlock stands. Forest Ecology and Management 17:25-36.

320. Society of American Foresters (SAF). 1993. Task Force report on sustaining long-term forest health and productivity. Society of American Foresters, Bethesda, MD.

321. Salo, E. O., and T. W. Cundy, editors. 1987. Streamside management: Forestry and fishery interactions. Institute of Forest Resources, Contribution 57. University of Washington, Seattle, WA.

322 Scharpf, R. F. 1993. Diseases of Pacific coast conifers. U.S. Forest Service, Pacific Southwest Research Station, Washington, D.C. Agriculture Handbook No 521.

323. Scheffer, T. C. and E. B. Cowling. 1966. Natural resistance of wood to microbial deterioration. Annual Reviews of Phytopathology. 4:147-170.

324. Schowalter, T. D. 1989. Canopy arthropod community structure and herbivory in old-growth and regenerating forests in western Oregon. Canadian Journal of Forest Research 19:318-322.

325. ———. 1990. Invertebrate diversity in old-growth versus regenerating forest canopies. The Northwest Environmental Journal 6:403-404.

326. ———. 1995. Canopy arthropod communities in relation to forest age and alternative harvest practices in western Oregon. Forest Ecology and Management 78:115-125.

327. ———, E. Hansen, R. Molina, and Y. Zhang. 1997. Integrating the ecological roles of phytophagous insects, plant pathogens, and mycorrhizae in managed forests. Pages 171-189 in: K. A. Kohm and J. F. Franklin, editors. Creating a forestry for the 21st century: the science of ecosystem management. Island Press, Washington, D.C.

328. Schreiber, B. 1998. Long-term monitoring of wildlife leave trees in clearcut harvest units and the Siuslaw National Forest, Northwest Oregon, 1987-98. Report to the Siuslaw National Forest.

329. Schulze, E. D., O. L. Lange, and R. Oren. 1989. Forest decline and pollution. Springer-Verlag, New York, NY.

330. Sedell, J. R., and F. J. Swanson. 1984. Ecological characteristics of streams in old-growth forests of the Pacific Northwest. Pages 9-16 in W. R. Meehan, T. R. Merrell, and T. A. Hanley, editors. Fish And Wildlife Relationships In Old-growth Forests. Transaction of the Annual Management of American Institute Of Fisheries Biologists, Juneau, AK.

331. ———, G. H. Reeves, and P. A. Bisson. 1997. Habitat policy for salmon in the Pacific Northwest. Pages 375-388 in D. J. Stouder, P. A. Bisson, and R. J. Naiman, editors. Pacific salmon & their ecosystems. Chapman and Hall, New York, NY.

332. Sharp, R. F. and J. W. Milbank. 1973. Nitrogen fixation in deteriorating wood. Experientia 29:895-896.

333. Shaw, D. C. and J. Bible. 1996. An overview of forest canopy ecosystem functions with reference to urban and riparian systems. Northwest Science 70:1-6.

334. Shearer, C. A. and S. B. von Bodman. 1983. Patterns of occurrence of ascomycetes associated with decomposing twigs in a Midwestern stream. Mycologia 75:518-530.

335. Shaetzl, R. J., D. L. Johnson, S. F. Burns, and T. W. Small. 1989. Tree uprooting: a review of terminology, process, and environmental implications cations. Canadian Journal of Forest Research 19:1-11.

336. Silvester, W. B., P. Sollins, T. Verhoeven, and S. P. Cline. 1982. Nitrogen fixation and acetylene reduction in decaying conifer boles: effects of incubation time, aeration, and moisture content. Canadian Journal of Forest Research 12:646-652.

337. Sinton, D. S., J. A. Jones, J. L. Ohmann, and F. J. Swanson. 2000. Windthrow, disturbance, forest composition, and structure in the Bull Run basin, Oregon. Ecology, 81(9):2539-2556.

338. Sit, V., and B. Taylor. Editors. 1998. Statistical methods for adaptive management studies. BC Ministry of Forests, Land Management Handbook No. 2. Victoria, BC, Canada.

339. Sollins, P. 1982. Input and decay of coarse woody debris in coniferous stands in western Oregon and Washington. Canadian Journal of Forest Research 12: 18-28.

340. ———, S. P. Cline, T. Verhoeven, D. Sachs, and G. Spycher. 1987. Patterns of log decay in old-growth Douglas-fir forests. Canadian Journal of Forest Research 17(12):1585-1595.

341. ———, C. C. Grier, F. M. McCorrison, K. Cromack, Jr., R. Fogel, and R. L. Fredricksen. 1980. The internal element cycles of an old-growth Douglas-fir forest in western Oregon. Ecological Monographs 50(3):261-285.

342. Sperry, J. S. 1995. Limitations on water stem transport and their consequences. Pages 105-125 in B. L. Gartner editor. Plant stems: Physiology and functional morphology. Academic Press, New York, NY.

343. Spies, T. A. 1997. Forest stand structure, composition, and function. Pages 11-30 in K. A. Kohm, and J. F. Franklin, editors. Creating a forestry for the 21st century. Island Press, Washington, D.C.

344. ———, and S. P. Cline. 1988. Coarse woody debris in forests and plantations of coastal Oregon. Pages 5-24 in C. Maser, R. F. Tarrant, J. M. Trappe, and J. F. Franklin, editors. From the forest to the sea: A story of fallen trees. U.S. Forest Service, Pacific Northwest Research Station, Portrland, OR. General Technical Report PNW-GTR-229.

345. ———, and J. F. Franklin. 1991. The structure of natural young, mature, and old-growth Douglas-fir forests in Oregon and Washington. Pages 111-121 in L. F. Ruggiero, K. B. Aubry, A. B. Carey, and M. H. Huff, technical coordinators. Wildlife and vegetation of unmanaged Douglas-fir forests. U.S. Forest Service, Pacific Northwest Research Station, Portland, OR. General Technical Report PNW-GTR-285.

346. ———, ———, and T. B. Thomas. 1988. Coarse woody debris in Douglas-fir forests of western Oregon and Washington. Ecology 69(1): 689-702.

347. ———, W. J. Ripple, and G. A. Bradshaw. 1994. Dynamics and pattern of a managed coniferous forest landscape in Oregon. Ecological Applications 4(3): 555-568.

348. Starr, L., editor. 1995. Soil: the foundation of the ecosystem. National Research News (5)1. Blue Mountains National Research Institute, La Grande, OR. U.S. Department of Agriculture, Forestry Service, Pacific Northwest Research Station.

349. Stathers, R. J., T. P. Rollerson, and S. J. Mitchell. 1994. Windthrow handbook for British Columbia forests. Ministry of Forests. Research Progress Working Paper 9401, Victoria, BC, Canada.

350. Statzner, B., J. A. Gore, and V. H. Resh. 1988. Hydraulic stream ecology: observed patterns and potential applications. Journal of the North American Benthological Society 7: 307-360.

351. Steeger, C. and C. L. Hitchcock. 1998. Influence of forest structure and disease on nest-site selection by red-breasted nuthatches. Journal of Wildlife Management 62(4):1349-1358.

352. Stevens, V. 1997. The ecological role of coarse woody debris. An overview of the ecological importance of CWD in BC Forests. Ministry of Forests Research Program. BC, Canada.

353. Stone, E. C., and R. B. Vasey. 1968. Preservation of coast redwood on alluvial flats. Science 159:157-161.

354. Stouder, D. J., P. A. Bisson, and R. J. Naiman, editors. 1997. Pacific salmon and their ecosystems. Chapman and Hall, New York, NY.

355. Strickland, T. C., P. Sollins, N. Rudd, and D. S. Schimel. 1992. Rapid stabilization and mobilization of ^{15}N in forest and range soils. Soil Biology And Biochemistry 9:849-855.

356. Swanson, F. J. 1980. Geomorphology and ecosystems. Pages 159-170 in R. H. Waring, editor. Forests: Fresh perspectives from ecosystem analysis. Proceedings of the 40th Annual Biology Colloquium. Oregon State University Press, Corvallis, OR.

357. ———, and G. W. Lienkaemper. 1978. Physical consequences of large organic debris in Pacific Northwest streams. U.S. Forest Service, Pacific Northwest Research Station, Portland, OR. General Technical Report, PNW-GTR-69.

358. Swanston, D. N. and F. Swanson. 1976. Timber harvesting, mass erosion, and steepland, forest geomorphology in the Pacific Northwest. Pages 199-221 in D. R. Coates, editor. Geomorphology And Engineering. Hutchinson Ross, Stroudsburg, Pennsylvania.

359. Swanson, F. J., R. L. Graham, and G. E. Grant. 1985. Some effects of slope movements on river channels. Pages 273-278 in International Symposium on Erosion, Debris Flow and Disaster Prevention, Tsukuba, Japan.

360. ———, G. W. Lienkaemper, and J. R. Sedell. 1976. History, physical effects, and management and implications of large organic debris in western Oregon streams. U.S. Forest Service, Pacific Northwest Research Station, Portland, OR. General Technical Report, PNW-GTR-56.

361. ———, S. V. Gregory, J. R. Sedell, and A. G. Campbell. 1982. Land-water interactions: the riparian zone. Pages 267-291 in R. L. Edmonds, editor. Analysis of coniferous forest ecosystems in the western United States. US/ISB Synthesis Series. Hutchinson Ross Publ. Co., Stroudsberg, PA.

362. Sweeney, B. W. and R. L. Vannote. 1978. Size variation and the distribution of hemimetabolous aquatic insects: two thermal equilibrium hypotheses. Science 200:44-446.

363. Swezy, D. M., and K. K. Agee. 1991. Prescribed-fire effects on fine-root and tree mortality in old-growth ponderosa pine. Canadian Journal of Forest Research 21:626-634.

364. Tappeiner, J. C., D. Huffman, D. Marshall, T. A. Spies, and J. D. Bailey. 1997. Density, ages, and growth rates in old-growth and young-growth forests in coastal Oregon. Canadian Journal of Forest Research 27: 638-648.

365. Tesch, S. D. 1995. The Pacific northwest region. Pages 499-558 in J. W. Barrett, editor. Regional silviculture of the United States. Third edition., John Wiley and Sons, New York, NY.

366. Thomas, J. W. editor. 1979. Wildlife habitats in managed forests. The Blue Mountains of Oregon and Washington. U.S. Forest Service. Agriculture Handbook No. 553.

367. Thomas, T. L., et al. 1993. Viability assessments and management considerations for species associated with late successional and old-growth forests of the Pacific Northwest. The report of the scientific analysis team. U. S. Forest Service.

368. ———, and J. K. Agee. 1986. Prescribed fire effects on mixed conifer forest structure at Crater Lake, Oregon. Canadian Journal of Forest Research 16:1082-1087.

369. Thomas, J. W., R. G. Anderson, C. Maser, E. L. Bull. 1979. Snags: Pages 60-77 in J. W. Thomas, editor. Wildlife habitats in managed forests: the Blue Mountains of Oregon and Washington. USDA Forest Service, Agric. Handbook. 553. Washington, D.C.

370. ———., R. J. Miller, H. Black, J. E. Rodiek, and C. Maser. 1976. Guidelines for maintaining and enhancing wildlife habitat in forest management in the Blue Mountains of Oregon and Washington. Pages 452-476 in Transactions of 41st North American Wildlife and National Research Conference. Wildlife Management Institute, Washington, D.C.

371. Tinnin, R. O., and D. M. Knutson. 1982. Witches' broom formation in conifers infected by Arceuthobium spp: an example of parasitic impact upon community dynamics. American Midland Botanist. 107:351-359.

372. ———, and ———. 1985. How to identify brooms in Douglas-fir caused by dwarf mistletoe. U.S. Forest Service, Pacific Northwest rearch Station, Portland, OR. Research Note, PNW-RN-426.

373. Torgersen, T. R. and E. L. Bull. 1995. Down logs as habitat for forest dwelling ants—the primary prey of pileated woodpeckers in northeastern Oregon. Northwest Science 69:294-303.

374. Townsend, A. R., M. T. Sykes, M. J. Apps, I. Fung, S. Kellomaki, P. J. Martinkainen, E. B. Rastetter, B. J. Stocks, W. J. A. Volney, S. C. Zoltai. 1996. WG1 Summary: Natural and anthropogenically-induced variations in terrestrial carbon balance. Chapter 9 in M. J. Apps, and D. T. Price, editors. Forest ecosystems: Forest management and the global Carbon cycle. NATO ASI Series 1: Global Environmental Change, Vol. 40. Springer, New York, NY.

375. Trappe, J. M. and C. Maser. 1976. Germination of spores of Glomus macrocarpus (Endogonaceae) after passage through a rodent digestive tract. Mycologia 68:433-436.

376. Triska, F. J., and K. Cromack, Jr. 1980. The role of wood debris in forests and streams. Pages 171-190 in R. H. Waring, editor. Forests : fresh perspectives from ecosystem analysis. Proceedings of the 40th annual biology colloquium. Oregon State University Press, Corvallis, OR.

377. Ugolini, F. C. and D. H. Mann. 1979. Biopedological orgin of peatlands in southeastern Alaska. Nature 281:366-368.

378. Ure, D. C., and C. Maser. 1982. Mycophagy of red-backed voles in Oregon and Washington. Canadian Journal of Zoology 60:3307-3315.

379. U.S. Forest Service. 1996. Status of the Interior Columbia Basin. Summary of scientific findings. U.S. Forest Service. Pacific Northwest Research Station, Portland, Oregon. General Technical Report PNW-GTR-385.

380. Valiela, I., G. Collins, J. Kremer, K. Lajtha, M. Geist, B. Seely, J. Brawley, and C. H. Sham. 1997. Nitrogen loading from coastal watersheds to receiving estuaries: new method and application. Ecological Applications 7(2): 358-380.

381. Van Cleve, K. C., T. Dyrness, L. A. Viereck, J. Fox, F. S. Chapin, III, and W. C. Oechel. 1983. Taiga ecosystems in interior Alaska. Bioscience 33: 39-44.

382. Vannote, R. L., G. W. Minshall, K. W. Cummins, J. R. Sedell, and C. E. Cushing. 1980. The river continuum concept. Canadian Journal of Fisheries and Aquatic Science 37: 130-137.

383. Vitousek, P. M. and P. A. Matson. 1984. Mechanism of nitrogen retention in forest ecosystems: a field experiment. Science 225: 51-52.

384. Vitousek, P. M. and W. A. Reiners. 1975. Ecosystem retention: a hypothesis. Bioscience 5:376-381.

385. Vitousek, P. M. and P. A. Matson. 1985. Disturbance, nitrogen availability, and nitrogen losses in an intensively managed loblolly pine plantation. Ecology 66:1360-1376.

386. Vogt, K. A., J. C. Gordon, J. P. Wargo, D. J. Vogt, H. Asbjornsen, P. A. Palmiotto, H. J. Clark, J. L. O'Hara, T. Patel-Weynard, B. Larson, D. Tortoriello, J. Perez, A. Marsh, M. Corbett, K. Kaneda, F. Meyerson, and D. Smith. 1997. Ecosystems: Balancing science and management. Springer-Verlag, Inc., New York, NY.

387. Vonhof, M. J., and R. M. R. Barclay. 1997. Use of tree stumps as roosts by the western long-eared bat. Journal of Wildlife Management 61(3): 674-684.

388. Vogt, K. A., C. C. Grier, and D. J. Vogt. 1986. Production, turnover, and nutrient dynamics of above- and belowground detritus of world forests. Advances in Ecological research 15: 303-375.

389. Walker, B. H. 1992. Biodiversity and ecological redundancy. Conservation Biology 6:18-20.

390. Walstad, J. D., S. R. Radosevich, and D. V. Sandberg. 1990. Natural and prescribed fire in Pacific Northwest forests. Oregon State University Press, Corvallis, OR.

391. Walters, C. J. 1986. Adaptive management of renewable resources. McGraw-Hill, New York, NY.

392. Ward, G. W., K. W. Cummins, R. W. Speaker, A. K. Ward, S. V. Gregory, and T. L. Dudley. 1982. Habitat and food resources for invertebrate communities in South Fork Hoh River, Olympic National Park. Pages 9-14 in E. E. Starkey, J. F. Franklin, and J. W. Matthews, editors. Ecological research in national parks of the Pacific Northwest. Forest Research Laboratory, Oregon State University, Corvallis, OR.

393. Waring, R. H. 1987. Characteristics of trees predisposed to die. Bioscience 37:569-574.

394. ———. 1983. Estimating forest growth efficincy in relation to canopy leaf area. Advances in Ecological Research 13:327-354.

395. ———, and J. F. Franklin. 1979. Evergreen coniferous forests of the Pacific Northwest. Science 204:1380-1386.

396. ———, J. J. Landsberg, and M. Williams. 1998. Net primary production of forests: a constant fraction of gross primary production? Tree Physiology 18(2).

397. ———, and G. B. Pitman. 1985. Modifying lodgepole pine stands to change susceptibility to mountain pine beetle attack. Ecology 66:889-897.

398. ———, and W. E. Winner. 1996. Constraints on terrestrial primary production in temperate forests along the Pacific Coast of North America and South America. Pages 89-102 in R. G. Lawford, P. Alaback, and E. R. Fuentes, editors. High latitude rain forests and associated ecosystems of the west coast of the Americas: climate, hydrology, ecology, and conservation. Springer Verlag, New York, NY.

399. ———, and S. W. Running. 1998. Forest ecosystems: Analysis at multiple scales, 2nd edition. Academic Press.

400. ———, K. Cromack, Jr., P. A. Matson, R. D. Boone, and S. G. Stafford. 1987. Responses to pathogen-induced disturbance: decomposition, nutrient availability, and tree vigor. Forestry 60:219-227.

401. ———, T. Savage, K. Cromack, Jr., and C. Rose. 1992. Thinning and nitrogen fertilization in a grand fir stand infested with western spruce budworm. Part IV. An ecosystem management perspective. Forest Science 38:275-286.

401a. Washington Department of Labor and Industries. 1992. Guidelines For Selecting Reserve Trees. Report# P 417-092-000, Olympia.

401b. ———. 1999. Work-related traumatic Head and Brain Injuries in Washington State 1990-1997. Report # 57-1-1999. State and Health Assessment and Research for Prevention Program. Wasington State Dept. of Labor& Industries.

401c. Weigand, J.F. and A. L. Burditt. 1992. Economic implications for management of structural retention on harvest units at the Blue River Ranger District, Willamette National Forest, Oregon. USDA PNW Research Station, Research Note PNW-RN-510.

402. Weikel, J. M., and J. P. Hayes. 1999. The foraging ecology of cavity-nesting birds in young forests of the notthern coast range of Oregon. The Condor 101: 58-66.

403. Wickman, B. E., R. R. Mason, and H. G. Paul. 1992. Thinning and nitrogen fertilization in a grand fir stand infested with western spruce budworm. Part II: Tree growth response. Forest Science 38:252-264.

404. Wright, P. J. 1998. The effects of fire regime on coarse woody debris in the west central Cascades, Oregon. Master Thesis. Oregon State University, Corvallis, OR.

405. Yavitt, J. B. and T. J. Fahey. 1985. Chemical composition of interstitial water in decaying lodgepole pine bole wood. Canadian Journal of Forest Research 15:1149-1153

406. Zenner, E. K., S. A. Acker, and W. H. Emmingham. 1998. Growth reduction in harvest-age, coniferous forests with residual trees in the western central Cascade Range of Oregon. Forest Ecology and Management 102:75-88.

407. Ziller, W. G. 1974. The tree rusts of western Canada. Canadian Forestry Service Publ. 1329. Department of the Environment. Victoria, BC, Canada.

408. Zinke, P. J. A. Stangenberger, and W. Colwell. 1979. The fertility of the forest. California Agriculture. 33(5): 10-11.

25

Single Species, Multiple Species, or Ecosystem Management: A Perspective on Approaches to Wildlife Conservation

Hal Salwasser

The conservation of wildlife and of biological diversity at large has taken various approaches in the U.S.[4] Sometimes, the focus is on the provision of life requisites for a single species of plant or animal, such as spotted owls, elk, or grizzly bears. Sometimes it is on the provision of habitats for a suite of species, i.e., a guild or biological community, such as cavity-dependent or wetland-associated animals. And sometimes the focus is on ecosystems, i.e., integrated systems of land, water, and biota in contiguous areas, e.g., watersheds, landscapes, or regions. Given this variety of approaches, it is logical to pose a simple question: when do you take a single species, multiple species, or ecosystem approach?

The Filter Concept

One answer to this question has been offered through the so-called coarse filter—fine filter concept. In this approach (also known as macro—micro filter) the needs of most species are assumed to be met through the coarse filter provision of mosaics of habitats, i.e., multi-species or ecosystem approaches. The needs of certain species, usually endangered or game species, are then addressed through a fine filter of special habitat components and species-specific actions. This "filter concept" is often attributed to Hunter,[3] who provided a good articulation of the concept. But in practical use, the filter concept is richer than the coarse-fine dichotomy. And it has been around for a very long time in wildlife conservation.

The debate over single species, multiple species or ecosystem approaches is essentially a false debate. In reality, except for zoos, botanical gardens, and gene pool banks, there is no such thing as pure single species approach to wildlife conservation. This could even be said of multi-species approaches. Thus, the above question of when you use one or the other is not really a useful question to pursue; we need to use all of them. Even when we try to address the needs of individual species or groups of species in actual habitats or landscapes, it is ultimately ecosystems that are conserved, restored or managed. This has the result that what start out as single species or multi-species approaches, in the end evolve toward ecosystem approaches. Conversely, when we set out to conserve, restore, or manage ecosystems, one of the first questions that arises is this: how much of what kinds of habitat conditions are needed, and how should those habitat

conditions be distributed across landscapes? And where do we turn for answers? We usually turn to our knowledge of what certain plant or animal species need for population viability or levels of abundance that allow for sustainable uses of surplus individuals. The upshot, then, is that it seems to make little difference what approach we start with, species or ecosystems; we eventually end up addressing both.

A more germane question for wildlife or biodiversity conservation is this: how do we best take advantage of all the tools available for perpetuating desired conditions of select species, biological diversity, and ecosystems? To gain perspective on possible answers to this question, we need an historical understanding of what has been used.

First, consider the possibility that the coarse filter—fine filter idea is really just a subset of a much richer, longer-standing concept, that of multiple overlapping and nested filters (to continue the metaphor). Consider also that the so-called filters do not even belong to the same dimension, i.e., that they all deal with habitats. We can see this through a brief exposition of how the system of biodiversity conservation has unfolded in the United States during the past 250 years.

Think of each of the major factors affecting conservation as a kind of filter or overlay. The first overlay was ordinances and laws designed to protect certain species from excessive harvest, some as early as the 1600s in the first colonies and others as late as the early 1900s in the west. The next overlay was reservation of large tracts of land in public parks, forests, and refuges, beginning locally in the 1700s in the east and extensively in the late 1800s in the west. The next overlay was restrictions on environmental toxics, poisons, and traps used to reduce pests and predators, mostly not occurring until the mid- to late-1900s. In most of these cases, only crude models or analytical tools were used to accomplish the intended conservation purposes, and sometimes no tools at all. To the extent that these three filters or overlays were applied, rather significant positive effects were accomplished for wildlife conservation.[5, 6, 8, 13] We might think of such actions as mega or meta filters (again to continue the filter metaphor). In fact, by the present time, most populations of native species have been protected from excessive harvest or unregulated poisoning for at least several decades. And about one-third of the land area of the U.S.

is protected to some degree in public stewardship, managed in ways that are conducive to the maintenance of significant aspects of biological diversity. But these are only the direct actions that impinge upon wildlife populations and availability of habitats.

Early in the twentieth century, improvements in agricultural production and the advent of machines powered by internal combustion engines provided another overlay of conservation, allowing the nation to meet its food and fiber needs without continued conversion of forests and grasslands to agriculture.[2, 5] This had a substantial positive, indirect effect on wildlife conservation, as habitat conversion is often the most significant and permanent negative impact to populations. Then, several decades ago, another set of mega filters was put into place that has had profound but often little recognized beneficial effects on biological diversity: namely, the nation's clean air and clean water statutes. Think of these indirect factors as another set of overlays on the coarse and fine filters of habitat. Think also what the environment and habitats for many species might be like without agricultural efficiency, clean air, and clean water.

Late in the conservation game, so to speak, we saw the emergence of new policy overlays: endangered species legislation (1960s in principle and ESA of 1973 in particular (P.L. 93-205, 87 Stat. 884, as amended); environmental policy require-ments (NEPA in 1969; P.L. 91-190, 83 Stat. 852); and forest and rangeland conservation laws (in particular, FLPMA in 1976; P.L. 94-579, 90 Stat. 2743, as amended, and NFMA in 1976; P.L. 94-588, 90 Stat. 2949, as amended). These laws essentially refined conservation tools and approaches, generally requiring more quantitative assessments of populations and habitats, leading to the emergence of books and journal articles on wildlife-habitat-relationships tools.[12] They also increased accountability for achieving desired results for certain species. Eventually, highly quantitative, analytical tools came into being as a result of these laws, to assess such things as population viability, habitat capability, biological diversity and ecosystem integrity.[7, 11, 12, 14] The models and analytical procedures, such as those presented in this book and its predecessors, are among the tools we can now use to determine both the coarse and fine filters. They have a dual edge, however. On the one hand, they can give us systematic ways to address species and ecosystem conservation matter. On the other, they can create the illusion that we can, in fact, be comprehensive, accurate, or precise in how we set out to achieve desired end results. The reality is that we can never be very comprehensive, accurate, and precise, because nature is more complex than we can account for in models, and it is never going to be fully understandable or predictable.

Where Do Models Fit?

So it might be useful to step back a bit from our focus on analytical models such as WHR matrixes (Wildlife Habitat Relationships tables), HSI (Habitat Suitability Indices), PATREC (Pattern Recognition), PVA (Population Viability Analysis) and AI (Artificial Intelligence) and think about how each of these potential tools might serve us in the future. First, though, what will the future in which these tools must operate be like? We know from ecology and paleoecology that the future is going to be dynamic,[1] and thus we will have limited ability to predict with any precision. So we will need tools that help us deal with change and surprises. The future is also going to have many more human beings seeking greater access to the land, water, air, and habitat resources upon which other species also depend for livelihood.[9, 15] So we need tools that will help us mitigate or reduce potential human impacts on those resources, essentially to optimize potentially complementary land and resources uses. It also appears that information will be more readily available to people in the future through emerging communications technologies. So we will need tools that tap the information sources, such as geographic information systems, remotely sensed databases, and internet data, and tools that will help people interpret or translate that information into meaningful understanding of what is going on in nature and what is most likely to deliver desired outcomes, i.e., predictive tools that work with reasonable simulations of the future.[10] Going back to the human dimension, a realistic outcome of the growing population is that less space and resources will be left for wild things to occupy or use. Thus we will need tools that help us choose wisely where the greatest returns for investment in those habitats might exist. These are essentially decision analysis or risk analysis tools.

It might also be useful to look at the environment that the first hundred years of conservation has bequeathed to us, and to realize that 70%, maybe 80% (maybe even more) of what there was to accomplish (or what was possible given the starting point of the late 1800s) has already been done through the land systems and environmental and conservation laws and policies now set in place by federal, state, county and tribal governments. What is left to us with our models, concepts, and analyses, then, is fine-tuning around the edges of these land conservation systems and various environmental laws. This is not to denigrate new tools and their potential utility. Nor is it to say that significant accomplishments are a thing of the past. It is, rather, to put our new tools into their proper perspective, to recognize them for what they bring to the already mature conservation business. If nothing else, these tools will be needed to help us retain or restore the health and productive capacity of lands and environments, so they can accommodate the many more people who will occupy the future without the significant losses in biological diversity that would occur were we to not use these tools.

We must also acknowledge that just as some seemingly unrelated forces or events of the past had as much to do with conservation of biological diversity as the direct actions taken on its behalf, or perhaps even more, the same might also be true in the future. I remind the reader of one historic example in support of this point: the agricultural revolution of the early twentieth century in the U.S. that stemmed the tide of increasing conversion of wildlands to farms. No conservation law, land reserve system, or WHR model affected this outcome, yet the positive result for wild plants and animals has been enormous.

Given this context, and given that we will use single-species, multi-species, and ecosystem approaches to conservation, let us consider the relative strengths and weaknesses of these approaches. (The following borrows heavily from Johnson).[4]

Single-species Approaches to Conservation

Strengths
- Ability to selectively focus on species most valued for some particular purpose, i.e., endangered, game, commercially valuable
- Efficient proxy for protecting habitats or ecosystems
- More easily understood by public, can be based on charismatic species

Weaknesses
- Species taxonomy might not be clear
- We don't know much about most species, and nothing at all about many
- Monitoring can be difficult and costly; easier to measure habitats, for example
- Prime habitat for one species is not necessarily prime for others
- No species is a perfect surrogate for another or for an entire ecosystem

Ecosystem Approaches to Conservation

Strengths
- Ability to conserve many species and genetic diversity through adequate amount and distribution of different ecosystem types (need good ecosystem classifications)
- Ability to conserve diversity and ecological processes essential to long-term productivity and resilience, hence the foundations for species persistence and adaptation
- Most cost-effective way to conserve multiple species
- Effective for species conservation when little or nothing is known about many species

Weaknesses
- There is no ecosystem classification scheme with the national or international consistency of species classifications
- Can miss rare, poorly distributed, or potentially endangered species

As noted earlier, species and ecosystem conservation can no longer be done without full consideration of how

humans fit into the scheme. So we return to the fundamental conservation challenge. We have only one Earth to work with, and it is finite. It has many different kinds of places, i.e., ecosystems, each with differing capabilities to sustain desired conditions of environments and human well-being. All ecosystems are constantly changing, though many are still well within geo-climatic ranges of variability. But Earth and all its ecosystems must contend with more people, well beyond anything these ecosystems have ever experienced. Regardless of where we enter the stream of species conservation, we will need to address individual species, groups of species, and ecosystems in an integrated manner. Furthermore, because people are so numerous, and because they and their artifacts are so widely distributed in nature, we will have to integrate species and ecosystem concerns with increasing human needs and impacts. And we will have to operate in the knowledge that events, forces, and trends beyond our control or even our understanding may exert more influence on our desired outcomes for biological diversity than anything we purposefully set out to do.

Thus I close this perspective with a review of the strengths and weaknesses of integrative, ecosystem approaches to conservation as they are now often proposed as key parts of sustainable development and adaptive management schemes:

Integrative, Ecosystem Conservation, i.e., Sustainable Development

Strengths
- Plans can have higher likelihood of durable implementation because human needs and institutional factors have been incorporated with species and ecosystem conservation; yields broader social ownership in the approach
- Can link biological reasons for conservation with social and economic forces to increase political support for conservation
- Can put biological, economic, social, and political factors into a more explicit policy and decision making forum

Weaknesses
- Wildlife and biodiversity can be devalued compared to other social and economic goals
- Experimental, still not as well developed as historical species and ecosystem conservation strategies
- Social, economic and biological data may not all be available or in consistent quality at appropriate scales, and some factors may change at different rates, e.g., when an economy changes faster than ecosystem conditions

The challenge in the future for those who wish to conserve wildlife or biological diversity in our increasingly human-dominated ecosystems is threefold. First, it is to use the strengths of each approach, i.e., single species, multiple species, and ecosystem, to compensate for the weaknesses of the other approaches. Second, it is to adopt adaptive management as the overarching framework to allow us to learn from our actions and respond to both systematic change and surprise events. And third, it is never to fool others or ourselves by believing that we can figure natural systems out precisely or comprehensively. We will simply do the best we can with the data and tools at hand, and learn as we go.

Literature Cited

1. Botkin, D.B. 1990. Discordant harmonies: a new ecology for the 21st century. Oxford University Press. New York, NY.

2. Fedkiw, J. 1989. The evolving use and management of the nation's forests grasslands, croplands and related resources. USDA Forest Service General Technical Report RM-175, Fort Collins, CO.

3. Hunter, M.L., Jr. 1990. Wildlife, forests, and forestry. Prentice-Hall, Inc., Englewood Cliffs, NJ.

4. Johnson, N.C. 1995. Biodiversity in the balance: approaches to setting geographic conservation priorities. Biodiversity Support Program, World Wildlife Fund, Washington, D.C.

5. MacCleery, D.W. 1996. American forests: a history of resiliency and recovery. Forest History Society, Durham, NC.

6. Matthiessen, P. 1959. Wildlife in America. The Viking Press, New York, NY.

7. McCullough, D.R., and R.H. Barrett, editors 1992. Wildlife 2001: Populations. Elsevier Science Publishers, Ltd., London, England.

8. Rieger, J.F. 1986. American sportsmen and the origins of conservation. Oklahoma University Press, Norman, OK. Revised edition: Oregon State University Press, Corvallis, OR, 2000.

9. RNRF. 1999. Congress on human population growth: impacts on the sustainability of renewable natural resources. Renewable Natural Resources Journal, Special Report. Volume 16, Number 4, Washington, D.C.

10. Salwasser, H. 1993. Perspectives on modeling sustainable forest ecosystems. Pages 176-181 in D.C. LeMaster and R.A. Sedjo, editors. Modeling sustainable forest ecosystems. Forest Policy Center, American Forests, Washington, D.C.

11. Soule, M.E. (editor) 1987. Viable populations for conservation. Cambridge University Press, Cambridge, England.

12. Thomas, J.W. (editor) 1979. Wildlife habitats in managed forests: the Blue Mountains of Oregon and Washington. USDA Forest Service, Agricultural Handbook No. 553. Washington, D.C.

13. Trefethen, J.B. 1975. An American crusade for wildlife. Winchester Press and the Boone and Crockett Club, New York, NY.

14. Verner, J., M.L. Morrison, and C.J. Ralphs (editors) 1986. Wildlife 2000. University of Wisconsin Press, Madison, WI.

15. Wilson, E.O. (editor) 1988. Biodiversity. National Academy Press, Washington, D.C.

26

Pacific Salmon and Wildlife—Ecological Contexts, Relationships, and Implications for Management

C. Jeff Cederholm, David H. Johnson, Robert E. Bilby, Lawrence G. Dominguez,
Ann M. Garrett, William H. Graeber, Eva L. Greda, Matt D. Kunze,
Bruce G. Marcot, John F. Palmisano, Rob W. Plotnikoff, William G. Pearcy,
Charles A. Simenstad, & Patrick C. Trotter

Introduction

The landscapes of Washington and Oregon at first glance appear to have some disconnect between the terrestrial and ocean environments. Abundant rivers and streams flow from the interior to the coastal zones, actively connecting the freshwater, estuarine, and ocean systems (Photograph 1). Within these environments there are countless abiotic and biotic processes which form a highly integrated ecosystem. Key inhabitants include wild anadromous Pacific salmon (*Oncorhynchus* spp.) (anadromous fishes are those that spend much of their lives feeding in the ocean and migrate to freshwater to breed), 605 common vertebrate wildlife species, and numerous species of macroinvertebrates and other fishes. Complex relationships have evolved within and between anadromous salmon and other inhabitants that may be important for maintaining this ecosystem. Former highly exploitive fisheries and poor land uses, an over-reliance on salmon hatcheries, and a change in ocean environment have contributed to many salmon stock declines in these states.[367, 397, 32, 153] It has been suggested that future salmon conservation will need to take an ecosystem approach if wild stocks are to survive.[485] The purpose of this paper is to identify known relationships between wild salmon and wildlife, to discuss the ecological context of these relationships, and to suggest new ways of managing the salmon resource with an ecosystem perspective.

We define wild salmon as indigenous species that are the progeny of streambed spawners. This definition is used to distinguish wild salmon from hatchery (artificially) propagated salmon. The genus *Oncorhynchus* includes both salmon and trout; however, for our purpose, we collectively refer to them simply as salmon. Wildlife are divided into two main categories, indigenous macroinvertebrates (aquatic and terrestrial) and vertebrates (amphibians, reptiles, fishes, birds, and mammals) found in Washington and Oregon.

It is important to recognize that the ecosystem of Washington and Oregon salmon can be hemispheric in scale. It reaches from local inland watersheds, where spawning occurs, all the way to ocean feeding grounds north of the Aleutian Islands of Alaska, west to the Asian side of the Pacific Ocean, and back again. This ecosystem spans an area of freshwater and ocean habitat in excess of 4 million km². The essence of the salmon is that they link together what humans generally consider distant, diverse, and separate ecosystems, and relatively long time spans. Scientific knowledge of salmon in Washington and Oregon was preceded by a rich legacy of aboriginal culture, which wove them into everyday life.[98, 379, 431] (Photograph 2) Salmon were an important food staple and a basis of many legends of the native people of these states, particularly those that lived along rivers and marine areas. Salmon were consumed by natives in large quantities, for example, Craig and Hacker[109, cited in 379] calculate that pre-contact catches of salmon in the Columbia basin alone ranged between 4.5 and 5.6 million fish annually. Most of the salmon caught at that time were consumed within their

*Photograph 1. Hoh River on the western Olympic Peninsula.
(Photo by Jeff Cederholm)*

Photograph 2. Indians fishing for salmon at Celilo Falls on the lower Columbia River. (Photo courtesy of Archive of the Spokesman Review, Spokane, Washington)

respective river drainage and some were traded with distant tribes. Columbia River tribal records indicate that salmon were transported long distances inland, including trade routes over the Continental Divide.

> *The Wishram and Wasco (tribes along the lower Columbia River near Celilo Falls) seem to have been the focal point in the most extensive trade network in the plateau—one that reached to the mouth of the Columbia and out onto the plains east of the Rockies. They traded dried fish (salmon) for bison hides and other commodities that originated on the plains.*"[177, cited in 379]

Some of the earliest Euro-Americans to view Pacific salmon traveled to the Northwest with the Lewis and Clark expedition in 1805. Near the confluence of the Columbia and Snake rivers, they observed salmon in unimaginable abundance, as William Clark reported: "The number of dead Salmon on the Shores & floating in the river is incrediable to say...."[252 cited in 119]

European settlement and commercial development of Washington and Oregon brought significant habitat problems for the salmon; resulting in many physical, chemical, hydrological, and biological modifications to the environment (Photograph 3). Varied effects on salmon habitat are often interrelated in complex ways, and the effects of various activities and ecosystem modifications can be cumulative.[513] Some of the more harmful habitat losses caused by humans have been: river channel clearing and channelization, log driving and splash damming, extensive land clearing, major water diversions, livestock grazing, mining runoff pollution, logging road-associated erosion and removal of the old growth forest, filling and diking of wetlands and estuaries, hydroelectric dam development, urban runoff, water and sediment contamination with toxicants, and recently recognized

Photograph 3. Urbanization and habitat loss in the Puyallup River estuary and floodplain, Tacoma, Washington. (Photo courtesy of Washington Department of Natural resources Photo and Mapping Section)

Photograph 4. Salmon catch at the Seattle wharf. (Photo courtesy of Washington State Historical Society, Tacoma, Washington. Negative Number 1994.123.121)

human-induced oligotrophication of waterways.[16, 59, 101, 174, 208, 245, 296, 365, 453, 485, 543]

Fishery exploitation of Columbia River salmon by Euro-Americans became a major factor after the middle to late 1800s (Photograph 4). To ensure primary access to the salmon, commercial fisheries were strategically located downstream of popular Indian fishing grounds. The principle means used to catch the salmon were gillnets, traps, seines, and fish wheels.[121, 274, 458] It was reported that "...on a single spring day in 1913 the Seufert brothers' wheel no. 5 turned a record catch of 70,000 pounds."[103] After the 1870s and up to the early 1900s, the Columbia River salmon fishery grew from 1 to 40 canneries.[368 cited in 365, 478] (Photograph 5) Fish wheels were prohibited on the Columbia River after 1935.[365] Commercial landings of Columbia River salmon and steelhead peaked between 1880 and 1930, and then went into a long-term decline through to present times.[365] Depletion of the prime spring and summer chinook probably started earlier than this time frame, however, as the fishery shifted to the less desirable coho and fall chinook.[275] One estimate of annual pre-Euro-American salmon and steelhead run size for the Columbia River ranges between 8.2 and 16.3 million fish.[379]

Early attempts to increase salmon catches using salmon hatcheries began as early as the 1870s, when concerns about over fishing led the Oregon and Washington Fish Propagating Company to construct a salmon-breeding station on the Clackamas River.[365] By the 1960s, with the advent of the *Oregon Moist Pellet* medicated food, hatchery salmon production increased dramatically. Total annual Columbia-Snake River system hatchery production (Washington, Idaho, and Oregon) reached 216 million smolts in 1989.[390] By the middle 1990s there were well over 100 state, federal, tribal and private salmon hatcheries in Washington and Oregon. The history of artificial propagation reveals a recurring cycle of technological optimism followed by pessimism.

While many attempts have been made at remedying the threats of habitat loss, over-fishing, and hatchery impacts, they have not been enough to prevent the widespread decline of wild salmon stocks in these states. Recent publications have chronicled the low abundance of wild salmon stocks along the Pacific Coast in the lower forty-eight states.[53, 181, 367, 387, 542] In 1991, the American Fisheries Society[367] published a list of 214 naturally spawning stocks of salmon, steelhead, and cutthroat from California, Oregon, Idaho, and Washington, including: 101 stocks at high risk of extinction, 58 stocks at moderate risk of extinction, 54 stocks of special concern, and one stock classified as threatened under the Federal Endangered Species Act (ESA) of 1973. In spite of past salmon habitat degradation and over-fishing, however, some stocks remain healthy.[221] Since 1990, The National Marine Fisheries Service (NMFS) has received a number of petitions to list Pacific salmon stocks as threatened or endangered under ESA, and the first salmon stock of this area to be listed as endangered was the Snake River sockeye, in November 1991.[366a]

General Salmon Life History

There are seven species of Pacific salmon and trout of the genus *Oncorhynchus* in Washington and Oregon, and they include: chum (*O. keta*), pink (*O. gorbuscha*), sockeye (*O. nerka*), chinook (*O. tshawytscha*), and coho salmon (*O. kisutch*); and rainbow (called steelhead when anadromous) (*O. mykiss*) and coastal cutthroat trout (*O. clarki clarki*). Some of these species, including the sockeye salmon (kokanee) and rainbow and cutthroat trouts, have both anadromous and nonanadromous forms.

A typical anadromous salmon life history has five main stages: (1) spawning and egg incubation, (2) freshwater rearing, (3) seaward migration, (4) ocean rearing, and (5) return migration and deposition of marine-derived nutrients into the freshwater ecosystem (Figure 1). Each species has slightly different temporal phases of the anadromous life history. The chum, pink, sockeye, chinook, and coho salmon all die after spawning just once, a life history strategy known as *semelparity*.[322] This life

Photograph 5. Canned salmon at the Apex fish company (1913). (Photo courtesy of Washington State Historical Society. Negative number 27683)

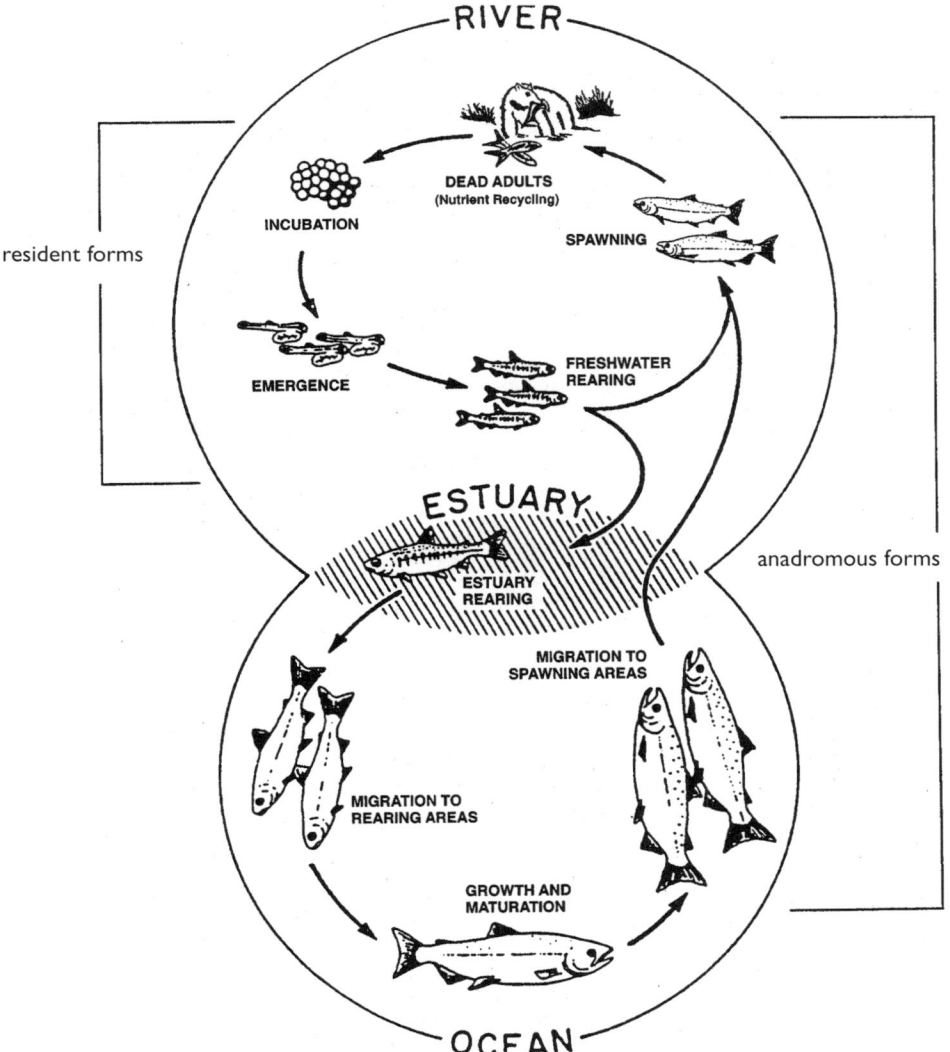

Figure 1. Generalized anadromous and nonanadromous (resident) Pacific salmon life histories, showing freshwater, estuary, and ocean components (the original diagram was from Nicolas and Hankin[372] and later modified by Spence et al.[485]

strategy has evolved because of the need to have a greater portion of the energy obtained from ocean feeding devoted to gamete production and juvenile survival. Consequently, survival after spawning no longer offered an advantage to these species.[322] The *iteroparous*, repeat spawning strategy, typical of the rainbow and cutthroat, probably occurred in the headwater reaches of larger rivers, where nonanadromous populations could be maintained year around. These fish generally were smaller in size, less fecund, and had sparser distribution and lower abundance than the anadromous forms. However, by retaining iteroparity, calamitous losses of young due to floods or drought, could be compensated for in subsequent breeding seasons.[322]

Chum, pink, sockeye, chinook, and coho salmon all spawn sometime from August through February, and cutthroat spawn between December and July. Pacific salmon are able to clean gravels by purging them of fine sand and silt particles during redd (spawning nest) excavation, but subsequent sediment transport processes and fine bedload flux tend to return this environment to the pre-spawning conditions.[414] After approximately 2-4 months of incubation, salmon fry swim up through the gravel and emerge into the stream. Emerging fry can vary widely in size at emergence, ranging from 20+ mm nonanadromous cutthroat to 35-40 mm chinook. Upon emergence, fry actively feed on a variety of aquatic insects, and for those that freshwater rear for extended periods of time (e.g., coho), the proportion of terrestrial food items in the diet may increase to over 30%.[338] Larger-sized juvenile salmon such as older aged rainbow and cutthroat prey on a mixed diet of aquatic and terrestrial macroinvertebrates, and may supplement their diet with occasional salmon eggs or fry.[498, 521] Sockeye fry are known to feed on cladocerens, copepods, and gammarid amphipods in lakes.[337]

After a summer of rearing in fresh water, juvenile coho average approximately 50 to 90 mm in length, and may weigh 2 to 5 g each.[87,128] Summer low-flow is a crucial time in the life of juvenile salmon that extended rear in freshwater. During this period the volume of aquatic habitat shrinks to a minimum, which can intensify inter-

and intra-specific competition.[9] Declining streamflow conditions also may cause some fish (e.g., chinook, coho) to emigrate to estuaries,[196, 522] where they continue to rear. Where species overlap in fresh water, a number of temporal and behavioral differences facilitate coexistence.[96, 282]

Upon the first rains and high waters of fall, coastal species (juvenile coho, steelhead, and cutthroat) make a directed migration to seasonally alternate rearing habitats. Juvenile coho and cutthroat exhibit major immigrations into side-channel swamps[80] and riverine ponds,[94, 398, 400, 162] located along river flood plains. Juvenile coho, steelhead, and cutthroat are also known to immigrate into small "runoff" tributaries (valley-wall tributaries) of rivers.[94, 271] Presumably these immigrations are to avoid high flows and turbidity of main rivers, as well as to take advantage of good feeding conditions.[398, 399] In contrast, interior (Idaho) juvenile chinook and trout are known to move out of tributaries and into main rivers to over-winter, probably to avoid winter ice conditions in the tributaries.[57]

Upon completing their freshwater stage, juvenile salmon of all anadromous forms undergo a physiological change called smoltification that includes osmoregulatory adjustments which prepare them to enter saltwater. For example, chum and pink salmon are nearly smolts upon emergence from the gravel, going directly to estuaries and the ocean;[259, 440] while chinook[197] and coho[442] may either go directly to sea the first summer of their life, or remain in freshwater for a whole year before smolting. Sockeye may rear in freshwater for one or two years before smolting,[77] and steelhead[283] and cutthroat[160, 162] may not smolt for two or three years or more.

Once in the estuary or ocean, most salmon prefer to feed on such prey as euphausiids, squid, herring, sandlance, rockfish, and anchovy.[83b, 197] While in the ocean most salmon species migrate long distances to feeding grounds along the North Pacific coast.[135, 136, 178, 194, 196] In contrast, anadromous cutthroat may only range several kilometers from their natal stream without overwintering in the ocean.[395]

During their anadromous life history salmon make important ecological contributions (as prey) to various

Table 1. Key sources of life history information for the seven salmon species of Washington and Oregon.

Species	Reference
chum	Bjornn and Reiser,[56] Everest et al.,[140] Koski,[259] Salo,[440] Wydoski and Whitney[573]
pink	Bjornn and Reiser,[56] Everest et al.,[140] Heard,[198] Wydoski and Whitney[573]
sockeye	Bjornn and Reiser,[56] Burgner,[77] Everest et al.,[140] Foerster,[146] Wydoski and Whitney[573]
chinook	Bjornn,[56] Bjornn,[57] Everest et al.,[140] Healey,[197] Wydoski and Whitney[573]
coho	Bjornn and Reiser,[56] Bustard and Narver,[80] Everest et al.,[140] Peterson and Reid,[401] Salo and Bayliff,[439] Sandercock,[442] Tagart,[510] Wydoski and Whitney,[573]
steelhead	Allee,[9] Bjornn and Reiser,[56] Cederholm and Scarlett,[94] Everest et al.,[140] Stolz and Schnell,[498] Winter,[560] Wydoski and Whitney[573]
cutthroat	Cederholm and Scarlett,[94] Everest et al.,[140] Fuss,[160] Garrett,[162] Glova,[166] Hall et al.,[181] Johnston,[238] Pearcy et al.,[395] Trotter,[521] Wydoski and Whitney[573]

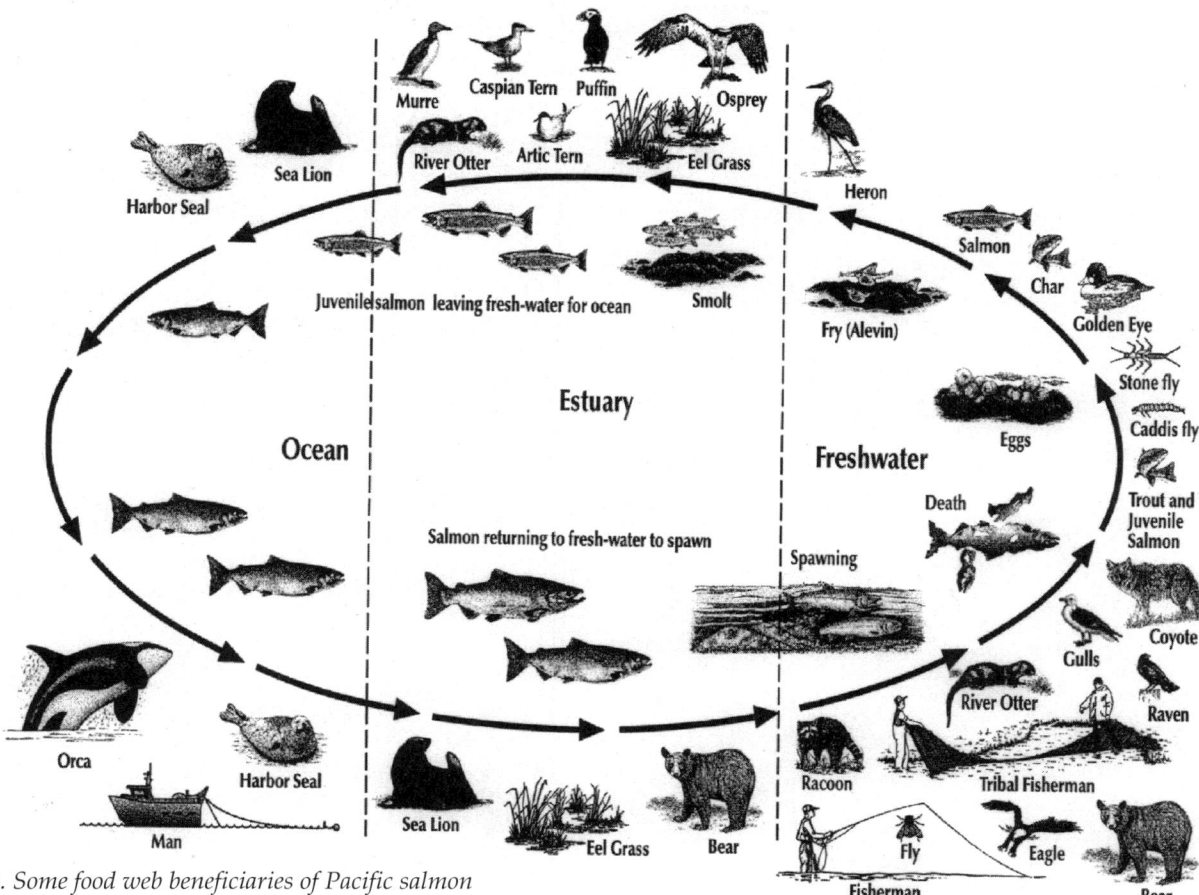

Figure 2. Some food web beneficiaries of Pacific salmon nutrient in freshwater, estuary, and ocean environments.

predators in the Pacific Northwest ecosystem, regardless of whether a particular individual salmon completes all life history stages or not (Figure 2). It is not uncommon for overall salmon survival rates to average 0.1% from egg to spawning adult. See Table 1 for more life history information on the seven salmon species of Washington and Oregon.

Freshwater and Terrestrial Habitat Relationships of Salmon

Freshwater Habitat

Freshwater habitat of salmon includes all the physical, chemical and biological elements within the aquatic environment. Geology, climate, topography, disturbance history, nutrients from returning salmon, and characteristics of the riparian vegetation typically govern the characteristics and the distribution of habitat types in a watershed.[33, 38, 332, 417] Components of freshwater habitat include:

Physical Characteristics: channel width and depth, substrate composition, pool and riffle frequency, pool types, and channel roughness.

Water Quality and Quantity: temperature, dissolved oxygen, dissolved nutrients, dissolved and particulate organic matter, hydrography.

Cover Factors: interstitial spaces (space between gravels), undercut banks, woody debris, water surface disturbance.

Biological Factors: food availability, salmon carcass nutrient inputs, competition, predation, disease, parasites, and functioning riparian conditions.

The abundance of fish in a stream is greatly affected by the stream's capacity to produce food. Many of the factors influencing stream productivity change predictably with changes in stream size, a pattern termed the *river continuum*.[529] Productivity is influenced by nutrient availability, input of organic matter from external sources, and the capacity for the channel to store and process organic matter and light. Differences in these factors can be very large and lead to a high degree of variability in production of fish, including salmon and trout populations.[223] Some of the highest freshwater production values have been reported for trout in New Zealand spring-fed streams, 54.7 g/m²/y;[10] however, production values from the Pacific Northwest are generally low when compared with other regions of the world, often below 1.0 g/m²/y and very rarely over 5.0 g/m²/y.[55]

Riparian Habitat

In this section, we offer only a summary of riparian aspects, as a chapter elsewhere in this book covers this topic more fully.

Many of the functional and structural attributes of stream habitat are created and maintained through interaction with riparian vegetation. Riparian areas constitute the interface between aquatic and terrestrial ecosystems,[171, 506] performing a number of vital functions that affect the quality of salmon habitats as well as providing habitat for a large variety of terrestrial plants and animals. Riparian areas influence streams and consequently salmon habitat in a variety of ways,[485] including:

Shade which dampens seasonal and diel fluctuations in stream temperature and controls primary and secondary production.

Streambank stabilization provides erosion resistant roots that bind soil particles together, thus facilitating bank building during high flow events by slowing the stream velocities.

Sediment control regulates sediment flow from upland areas by acting as a filter, or storing sediments in the primary floodplain.

Litter input contributes a significant amount of organic matter to streams, which acts as an important food resource for aquatic communities.

Large woody debris (LWD) provides important structure to the stream channel for energy dissipation, fish habitat, and salmon carcass retention.

Nutrients input: riparian zones mediate the flow of nutrients to the stream and are, therefore, important regulators of stream production. Some riparian species such as red alder (*Alnus rubra*) also fix atmospheric nitrogen and therefore augment N availability to the ecosystem.

Microclimate: streamside soils and vegetation can have a significant effect on moderating the climate within riparian zones.

Streamside vegetation moderates water temperature. This relationship is influenced by elevation, air temperature, stream width, water depth, and aspect.[41] Removal of riparian vegetation has been associated with increased maximum water temperatures, and diurnal fluctuations in water temperature during summer; and decreased winter water temperatures.[41] Small, low-elevation streams are the most susceptible to summer water temperature increases caused by canopy removal.[504] The biological consequences of elevated water temperature on aquatic communities are complex. There is little information indicating direct mortality of fishes as a result of temperature changes related to riparian canopy removal;[41] however, reductions in growth rate,[51, 572] changes in life history,[215] changes in competitive interactions between species,[418] reductions in fecundity of adults,[40] and

an increased susceptibility to disease,[363, 386] have all been documented. Some species of amphibians and aquatic macroinvertebrates also are thermally intolerant and elevated water temperatures may have detrimental impacts on their populations.[78, 137, 327]

Riparian vegetation increases streambank stability and resistance to erosion. Roots from woody and herbaceous vegetation bind soil particles together, helping to maintain bank integrity during erosive high-streamflow events.[485, 506] Riparian vegetation also facilitates bank-building during high flow events by slowing stream velocities, which in turn helps to filter sediments and debris from suspension. This combing action helps to stabilize and rebuild streambanks, allowing the existing channel to narrow and deepen, and increases the effectiveness of riparian vegetation in providing bank stability and shade.[132] During over-bank flows, water is slowed and fine silts are deposited in the flood plain, increasing future productivity of the riparian zone.[485]

Forested riparian areas generate much of the organic matter that provides the energy source for the trophic systems of small streams. In one study of forested headwater channels in Oregon, Sedell et al.[452] determined that over 90% of the in-channel organic matter was provided from the surrounding terrestrial environment. A 10-m wide stream in western Washington received over 75% of its annual organic matter supply from terrestrial sources.[46] Even though this source of organic matter decreases relative to autochthonous (biomass produced from within the stream) organic matter in larger channels, it remains vital to stream productivity.

Large woody debris (LWD) has been shown to be a critical structural component in Pacific Northwest streams, forming pools, waterfalls, and overhead cover; and it also regulates the transport of sediment, gravel and organic matter, for fish and other aquatic biota.[49, 50] (Photograph 6) In forested watersheds LWD provides the most common obstruction, often forming pools in various types of fluvial channels.[332] Without this material, pool abundance and size is decreased,[45] reducing habitat complexity and potentially reducing the diversity of the fish community. In addition to numerous habitat and morphological functions,[50] wood (organic debris) helps retain salmon carcasses in streams for many carnivorous wildlife species[91] and general biological activity.[90]

Riparian areas play a key role in determining the concentration of nutrients in stream water.[42, 520] The presence of even a narrow riparian buffer can profoundly influence stream water chemistry. Uptake and storage of various elements carried by groundwater can be considerable, even where input rates have been substantially altered as a result of upslope land uses.[284] Riparian vegetation composition can influence nitrogen input to streams. Early successional vegetation in riparian areas in the Pacific Northwest is often dominated by red alder, a nitrogen (N) fixing species. As a result, the N content of litter beneath these riparian stands is 1.5 to 3-fold higher than sites where conifer species are the dominant component of the over story.[128a] The result is

Photograph 6. Low gradient stream showing large woody debris formed habitat in Monroe Creek, Washington. (Photo by Jeff Cederholm)

higher N levels in the riparian soils[58] and increased delivery of N to the stream channel. Higher N levels in stream water may elevate primary production and decomposition (heterotrophy) in the channel and increase food availability for the invertebrate community. Increased invertebrate production may elevate food availability for stream-dwelling fishes, amphibians, and other insect feeders such as bats and flycatchers.

The riparian area may act as either a source or sink of organic matter and sediment during flood flows. The manner in which the stream and riparian area interact at these times depends upon the morphology and vegetation of the riparian zone and the intensity of the discharge event.[34] The structure and abundance of riparian vegetation plays a key role in moderating the movement of materials between the riparian area and the stream.[361] Vegetation in the riparian zone has been shown to be the single most important structural element for the retention of fluvially transported organic matter during high flow events.[484] Similarly, riparian vegetation promotes the storage of sediment,[216] that may provide germination sites for some species of riparian plants.[370] The variations in retentive capacity of different riparian areas for organic matter leads to large differences in the organic content of riparian soils, ranging from nearly all inorganic material in some locations to very high concentrations of organic matter in stream-adjacent swamps and wetlands.[71] This variation in substrate further contributes to riparian vegetation heterogeneity.[2]

The area in which water exchange between the channel and the underlying riparian soils occurs is termed the hyporheic zone.[491] The extent of the hyporheic zone varies as a function of site topography and soil characteristics. In riparian areas of low relief and porous soils, the hyporheic zone may extend as far as 3 km from the edge of the channel.[491] Riparian vegetation can influence the amount of water stored in the hyporheic zones and soil chemistry.

The attractiveness of riparian areas to wildlife likely reflects three main attributes: the presence of water, local microclimate condition, and more diverse plant assemblages found in riparian areas compared to uplands. Wildlife also congregate seasonally in riparian areas where salmon spawn, to take advantage of an abundant food supply of live fish[205] and carcass flesh.[91] The high value of riparian habitats to wildlife has been recognized by naturalists,[380] and considered a bridge between upland habitats and the aquatic environment. The combination of shape, moisture, deposition soils, and disturbance regime unique to riparian areas contributes to their exceptional productivity in terms of plant growth, plant diversity, and structural complexity of the vegetation.[237, 270, 329] Wildlife dependency and diversity peak at this terrestrial/aquatic boundary. Brown[69] reports that 359 of 414 (87%) species of wildlife in western Washington and western Oregon use riparian areas and wetlands during some season or part of their life cycle. In their detailed examination of wildlife and habitats for all of Washington and Oregon, Johnson and O'Neil[236] reported that 393 of 456 (86%) of the common terrestrial and freshwater wildlife species have seasonal use of riparian areas, wetlands, and streams. Of these 393 species, 110 were found to be closely *associated* (e.g., obligates) with eastside and westside riparian habitat types.

The close association may very well have evolved from the direct or indirect exploitation of the rich vegetative habitat provided by riparian areas.[255] Quantitative studies conducted during the past several decades have supported observations and have identified biological and physical attributes of riparian habitats which enhance their value to wildlife. Brinson et al.[64] and Oakley et al.[381, cited in 380] summarize these important biological and physical features of riparian areas:

• presence of surface water.
• increased humidity, high rates of transpiration, and greater air movement.
• complexity of biological and physical habitats.
• maximum edge effects with adjacent upland forests, which is beneficial for some species.
• food supply.
• thermal cover.

According to O'Connell et al.[380] stream type has a direct influence on the riparian habitat and its associated wildlife communities. In the smaller headwater streams the impacts of the upstream riparian vegetation on the streams is greater than downstream where flow volume increases, flooding is more widespread, and the impact of riparian vegetation on the stream is less. Brinson et al.[64] suggest that middle order perennial streams and associated riparian areas have the greatest wildlife use. Periodic flooding can enhance the availability of food for wildlife by creating new feeding areas.[64] Flooding can also make riparian habitat unsuitable for other species. Species abundance of riparian mammal communities has been related to the timing of recent hydrologic events; impoverished mammal populations have been attributed to recent flooding whereas more abundant populations have been observed in areas not subject to recent flooding.[64]

Habitat-forming Processes

Disturbance plays a major role in maintaining community diversity and productivity in many ecosystems,[105, 420] and is a key factor in creating and maintaining diverse stream habitat in the Pacific Northwest.[54] These disturbances range in severity from minor events, such as seasonal changes in flow, to less frequent high intensity events such as wildfire, debris torrents, and major floods. Riparian and channel conditions evolve as impacted areas recover from disturbance.[211] The result is a diverse set of riparian community types and stream habitat conditions that vary over both time and space.[2]

Disturbance contributes to both diversity of aquatic fauna and productivity of these communities when considered at a watershed level.[173] Aquatic communities associated with early-successional riparian areas typically exhibit low diversity, but high productivity for certain species. Removal of the channel shading canopy brings about dramatic increases in light and algal productivity; however, input of terrestrial litter decreases.[46, 172] Invertebrates that feed on algal material (e.g., grazers) typically dominate communities at recently disturbed sites.[137] These invertebrates form a major component of the diet of some salmon and trout[338] and can contribute to increased fish productivity following disturbance.[46, 52, 352] The increased productivity is typically observed during summer, and often does not extend into winter months when the availability of shelter from high flows for juvenile salmon becomes important.[94, 162, 401]

After forest canopy closure, primary productivity in streams decreases. The type of litter delivered to these systems and the physical characteristics of the channel differ from those at sites bordered by mature vegetation. Hardwood trees, especially red alder, often dominate the canopy at these sites. Litter from red alder trees decomposes much more rapidly than conifer litter, in part due to the higher N content.[451] The high N content of the litter improves its nutritional value for shredding macroinvertebrates but the high rate of decomposition causes it to be scarce at some times of the year.

Alder stands begin to die-out and provide LWD to channels after about 60 years.[175] Shade tolerant conifers, like western red cedar (*Thuja plicata*) and western hemlock (*Tsuga heterophylla*), colonize the site and begin to provide needles and other litter to the channel. Some stands of alder can persist and will repeat themselves several times (Slaney, pers. comm.). Woody debris amounts and average piece size increase for 100 or more years following conifer occupation of a site.[44, 506] The morphology of the channel and the routing of sediment and organic matter evolve slowly as the riparian community changes, ultimately creating channels which are highly complex structurally and support a macroinvertebrate community dominated by shredders.[11]

Forest practices and other land uses have accelerated the rate of occurrence of some types of disturbance. The acceleration in disturbance has led to the establishment of early successional communities in the majority of riparian areas on commercial forest land in the Pacific Northwest,[60, 86, 154] and has been shown to alter natural hydrology.[207] Practices such as splash damming of rivers to float logs to market,[550] and removing all trees to the channel's edge,[43] modify the riparian successional process. Timber harvest or roads constructed on unstable slopes or road drainage systems that were improperly maintained, dramatically increase the incidence of landslides.[93, 421, 456] Many hillslope failures enter stream channels and may move considerable distances downstream, removing streamside vegetation and soil. On the positive side, however, localized landslides also can input massive amounts of spawnable sized gravels and LWD into stream channels, where they may benefit salmon populations.[456] These disturbance events have affected a large proportion of the riparian areas bordering streams in the region over the last century, and have played a key role in determining channel form and habitat conditions.[507]

A study of fire history in Mt. Rainier National Park[201] reported that alluvial terraces and valley bottoms were often forested with old stands, and that every major river valley contained a stream-side old-growth corridor. These observations support an inference that the moist environment of riparian areas inhibits fire and reduces the fire return interval for riparian forests. The persistence of live trees in riparian forests may also provide a local seed source that facilitates a more rapid development of a multi-layered, conifer-dominated forest.[405]

Beaver

Beavers have long co-existed with salmon in the Pacific Northwest, and have had a important ecological relationship with salmon populations. The beaver created and maintained a series of beneficial aquatic conditions in many headwater streams, wetland, and riparian systems, which serves as juvenile salmon rearing habitat. Beavers have multiple effects on water bodies and riparian ecosystems that include altering hydrology, channel morphology, biochemical pathways, and stream productivity.[385] Beaver ponds were of special importance in more arid regions, but also had important roles in coastal systems.[365]

Beavers were once extremely abundant in the Pacific Northwest, but as far back as 1778 trapping expeditions into western North America began depleting their numbers. Between 1834 to 1837, pelts from 405,472 beavers from the area that would become southwest Washington and Oregon were shipped to Europe. Past excessive trapping, and subsequent unregulated land- and water-use activities, significantly reduced abundance of beaver and beaver ponds. Additionally, excessive livestock grazing in riparian areas has degraded habitat conditions for beaver.[385] Severe declines of beaver in Washington and Oregon have fundamentally altered important natural aquatic ecosystem processes such as nutrient cycling, flood plain development, and stream hydrology.

Beaver dams can obstruct channels and redirect channel flow and the flooding of streambanks and side channels. By ponding water, beaver dams create enhanced rearing and over-wintering habitat that protect juvenile salmon during high flow conditions.[365] Studies in Oregon coastal streams have suggested that where the amount of spawning is adequate, the winter survival of juvenile coho, which can be swept downstream in high winter flows, is limited by the presence of adequate slow-water habitat.[374] Beaver dams are often found associated with riverine ponds called "wall-base channels"[401] along main river flood plains, and these habitats are used heavily by juvenile coho salmon[94, 400] and cutthroat trout[94, 162] during the winter. Though their dams can occasionally block upstream migration by adult and juvenile salmon, studies of trout movement indicate that fish can pass over beaver dams during all seasons.[385] Beaver dams may temporarily keep salmon adults in the lower parts of spawning streams, where flows are greater and pools are deeper. Then when dam breaching flows occur, free passage to upstream areas is allowed.

Beaver foraging can cause a loss of woody riparian vegetation and an increase of fine sediments, but it also increases the input of large woody debris to streams, and beaver droppings may enrich pond productivity. Bank dens and channels can increase erosion potential, but because ponds fill with sediment to become wetlands over time, this helps to retard upstream erosion and retain sediments that otherwise could adversely alter downstream areas. In a Wyoming study of an area that had 10.5 beaver dams per km, each dam was found to retain 5,350 m³ of sediment. In another Wyoming study, sediment loads were reduced by 90% after flowing 8 km through an area with well developed riparian habitat and beaver dams.

Beaver ponds provide a sink for nutrients from tributary streams and create conditions that promote anaerobic decomposition and de-nitrification. These processes can cause nutrient enrichment and increased primary and secondary production downstream from the pond, while increasing nutrient retention time and enhanced invertebrate production in the pond.[365] These factors help increase salmon growth and survival, and also helps improve water quality. Beaver ponds can cause increased storage of water in the banks and flood plains,

and this increases the water table, enhances summer flows, adds cold water during summer, and causes more even stream flows throughout the year. During winter, beaver ponds in cold environments prevent anchor ice from forming and prevent super-cooling of the water. By storing spring and summer storm run-off, beaver ponds help to reduce downstream flooding and the damage from rapid increases in stream flows.[385]

Beavers also help shape riparian habitat. Beaver ponds increase the surface area of water several hundred times and thereby enhance the overall riparian habitat development.[385] They also enhance vegetation growth by increasing the amount of groundwater for use by riparian plants and wetland areas. The presence of beaver can have both positive and negative influence on salmon habitat, but on the whole, their presence is considered of great benefit to both water quality and salmon, particularly juvenile coho salmon and cutthroat trout, and to many other species of wildlife and invertebrates.

Estuary Habitat

By definition,[413] an estuary is a region where salt water of the ocean is measurably diluted by freshwater runoff from the land within a constricted body of water. In the Pacific Northwest, river flow plays a strong role in the structure and dynamics of estuarine circulation.[515] Estuaries of lowland watersheds along the Washington and Oregon coasts tend to exhibit high peak flows associated with winter storms, but often extremely low flows associated with the dry summers. This can frequently cause dramatic differences in the available estuarine habitat between winter and spring-summer periods, limiting summer rearing. In some southern Oregon and northern California estuaries river flow can decrease to the point that bars form across the estuaries' entrances, restricting juvenile salmon ocean emigration to extreme high (spring) tides.

Perhaps the most fundamental concept in understanding the estuarine ecology of juvenile salmon is that the salmon do not respond to singular habitats *per se*, but rather interact with a mosaic of habitats in response to changing migratory mandates, tidal cycles and freshwater runoff events (Figure 3). River flow and tide, physiological change, prey and predator distributions, and likely metapopulation genetic structure as well, all affect the rate of fish movement through the estuary. But the opportunity for juvenile salmon to exploit preferred habitats is just as likely dependent on the arrangement of key landscape features such as tidal-freshwater and brackish rearing zones, low-velocity refugia, migratory corridors and foraging patches. Although this is a relatively new topic of research, with few definitive experiments and tests, there is some emerging evidence that the edge of marsh vegetation in dendritic tidal channel and slough systems may relate directly to juvenile salmon production.[473]

Large watersheds with significant snow accumulations and extended melting periods can create prolonged spring freshets.[308] Spring and winter freshets, and winter "rain-on-snow" events associated with rapid snowmelt, produce

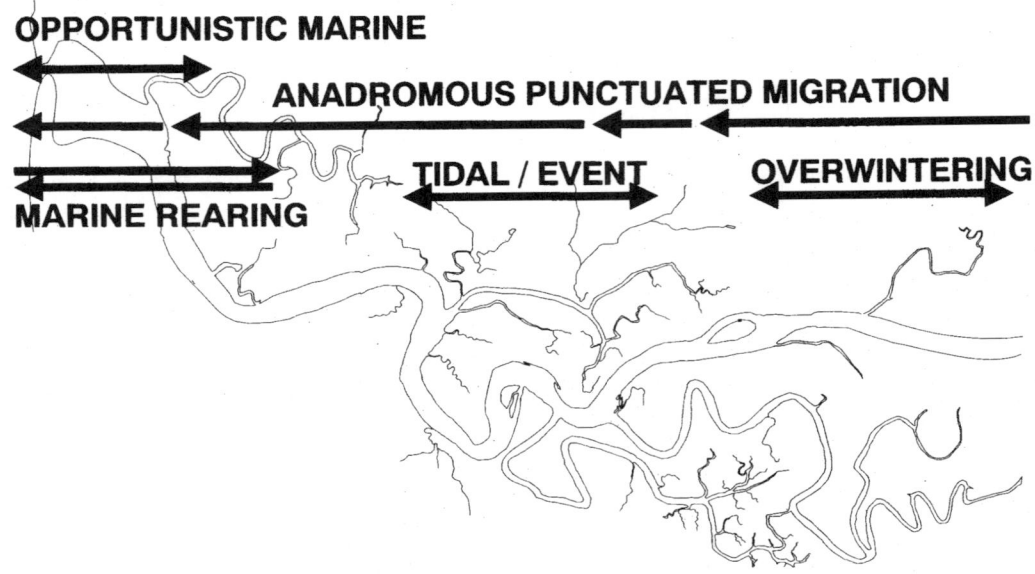

Figure 3. Movements and migrations of juvenile Pacific salmon across tidal-freshwater delta-estuarine landscapes. (Contributed by C. Simenstad, unpublished diagram)

flooding in tidal flood plains and estuaries that influences short-term and long-term productivity of juvenile salmon and their ecosystems.[360, 566] Although flood plain and estuarine wetland flooding increases flows in the main distributary channels, likely diminishing the ability of juvenile salmon to occupy them, considerable side-channel and other flood plain wetlands (i.e., ponds, relict side-channels) are inundated and become available for refuge and rearing. This flooding recruits organic detritus and dissolved nutrients from these peripheral wetlands and imports them to the estuary. While trapped in estuarine wetlands or circulation features such as estuarine turbidity maxima,[472] these materials contribute to primary and secondary production by supporting food web pathways to juvenile salmon.

The structure of the watershed and estuary, with the seasonal variability in river flow, shapes estuarine circulation, and strongly influences juvenile salmon residence time, habitat use and production. Except where the river has been extensively diked and channeled, the flood plain in the freshwater-tidal region is characterized by extreme habitat complexity, abrupt changes in water velocity and low-velocity off-channel habitats. As the "estuarine gateway," the tidal-freshwater mixing zone can be exceedingly important to juvenile salmon[469] because it: (1) provides habitat for overwintering chinook, coho and steelhead forced downstream during high river flows; (2) contains complex low-velocity refugia such as off-channel sloughs and LWD; (3) allows migrating juveniles to adapt physiologically as they encounter brackish waters of the upper estuary; (4) drifting insects are trapped and concentrated due to flow reversals, providing opportune feeding conditions;[523] and (5) is the first region of estuarine settling of suspended sediments and detritus, which can fuel soft-sediment habitat formation and detritus-based food webs exploited by salmon.

Estuaries are composed of both discrete and highly integrated habitat complexes and their associated plant and animal communities. Categorizing habitats to a large degree is a function of scale, as juvenile salmon can respond to habitat features (e.g., LWD or tidal channels) that are elemental to the broader habitats. Estuaries generally posses eight habitat components: (1) subtidal distributaries; (2) mud- and sand-flats; (3) gravel-cobble beaches; (4) low elevation emergent marshes; (5) high elevation emergent marshes; (6) forested and shrub swamps; (7) eelgrass; and (8) kelp. Salmon communities have been shown to utilize many of these habitat components. Juvenile coho (fry, fingerling) are often found rearing during winter and early spring in the tidal flood plains of many large rivers such as the Chehalis River.[323, 470, 471, 472] These fish are either staging for migration through the estuary or are moving back into freshwater for extended rearing. Work in British Columbia,[438, 522, 523] and Alaska[512] show that certain sub-populations have a minimal juvenile freshwater rearing phase of their life history ("ocean-type"), and spend extended periods of time either feeding in estuaries or in the ocean. Such subyearling migrant coho, may constitute significant portions (up to 50%) of the returning adult spawners.[523]

Natural disturbance regimes are responsible for creating and maintaining habitat complexes important to juvenile salmon. Erosive flooding, channel reconfiguration, and changes imposed by LWD all promote increased habitat complexity and heterogeneity. In the absence of disturbance, early successional habitats such as mudflats and low elevation estuarine marshes (e.g., *Carex lyngbyei* sedge) would not persist or would be relatively rare. Yet, these habitats can be some of the most productive and beneficial of salmon habitats and play unique roles for some salmon species. A number of important studies have described the association with specific migratory, rearing and residency times of salmon species in estuaries.[104, 192, 196, 197, 227, 247, 256, 273, 460, 466, 467, 565]

Ocean Habitat

Upwelling along the coast of the Pacific Northwest often results in high primary and secondary productivity, resulting in large standing stocks of fishes, seabirds, and marine mammals. The coastal upwelling domain extends from British Columbia to Baja California, and is located inshore of the equatorial flowing California Current. Coastal upwelling is driven by prevailing northwesterly winds during the spring and summer. Those winds result in offshore displacement of near-shore surface waters and vertical advection of deep, cool and often nutrient-rich waters into the euphotic zone along the coast and into estuaries. Rich blooms of phytoplankton are observed along the coast following episodic upwelling events.[286, 501, 514] Upwelling varies seasonally and over longer annual and semiannual cycles, with intensity generally increasing southward to northern California. Upwelling is most intense in regions of capes such as Cape Blanco in southern Oregon. There was a strong correlation between the intensity of coastal upwelling and the smolt-to-adult survival of hatchery coho salmon from the Oregon Production Index region south of the Columbia River from 1960 to 1981.[373] During the late 1970s, however, there was a major change in ocean climate in the North Pacific Ocean, called a regime shift or the Pacific Decadal Oscillation, which was correlated with a sharp decline in the survival of Oregon coho salmon between smolt release years 1975 and 1976.[32, 153, 397] After this regime shift the production of Oregon coho salmon has usually been low. The relationship between coastal upwelling and coho survival is no longer significant. The reason for this changed relationship is unclear, but is related to weak coastal upwelling, warm sea temperatures, high sea levels, and frequent El Niño events.[394] Upwelling has probably not been effective in injecting nutrient-laden water into the euphotic zone because of the deep lens of overlying warm, nutrient-depleted water along the coast.[195, 432] The persistence of warm, unproductive ocean conditions is a major reason for the decline of many stocks of anadromous fishes along the west coast, and for the very large variability in survival and reproduction of marine birds.

Although the mechanisms that have resulted in poor ocean survival of salmon are speculative, one hypothesis is that weak upwelling results in low growth and poor survival of zooplankton and forage organisms, and impacts juvenile salmon during their critical first summer in the ocean. This lack of forage and the narrow band of cool waters along the coast during weak upwelling years concentrates juvenile salmon near the coast where they are more vulnerable to predation by seabirds, marine mammals and fishes.[145] During warm years predators from southern waters, e.g., Pacific and jack mackerel, invade coastal waters and may either compete with or prey upon juvenile salmon.[394, 397]

The principal prey of juvenile salmon off the coast of Oregon and Washington during the spring and summer are fishes and crustaceans.[65b, 395] Salmon in the open ocean forage opportunistically on a diverse assemblage of pelagic organisms. The diets of maturing salmon in the North Pacific Ocean vary among species and sizes of fish, with season and year, and with location and proximity to the coast. Fishes, squids, amphipods, copepods, and pteropods are primary prey.[178, 268, 396]

Ecological Relationships of Salmon
Macroinvertebrates

Freshwater Macroinvertebrates and Salmon

Freshwater ecosystems are inhabited by a large variety of macroinvertebrates that play an integral part in the salmon's life history. They include insects, crustaceans, and other forms of macroinvertebrates (larger than 595 microns in their later instars or mature forms). Many species in their aquatic phase have been described from the hyporheic zone, or zone below the surface of the stream bottom.[492] Given the variety of physical habitat across the region's landscape, there is an opportunity for freshwater invertebrate species to form diverse and specialized communities.

Freshwater macroinvertebrates play a significant role in energy pathways of aquatic ecosystems. The consumption of algae, detritus, and bacteria is the basis for transfer of this energy. A few invertebrate species are known to actively derive their food base from higher life forms (e.g., small fish). The food source used by an invertebrate defines what function it performs in this food web.

Structure and Function in Macroinvertebrate Communities

The type and location of food in the aquatic environment consumed by invertebrates determines their functional designation. Headwater streams or heavily canopied streams are dominated by leaf litter input, allochthonous (biomass produced from outside the stream) material, which has been linked to significant shredder activity.[111, 536] Shredders comprise a group of aquatic insects that utilize coarse particulate organic matter, such as leaf litter, with a significant dependence on the associated microbial biomass.[529] Portions of a drainage where the riparian canopy opens can result in substantial autochthonous input (periphyton growth), and are consumed by scrapers like the mayfly family Heptageniidae.[315] Lower in a drainage the channel can accumulate large deposits of detritus. Invertebrates distributed here are mainly collector-gathers and may constitute the bulk of juvenile salmon diets.[196] The distribution of dominant food sources throughout a drainage are influenced by a continuum of physical changes as one travels from the steep headwater streams to the relatively low gradient flood plains.[529]

Invertebrate community structure in a stream or pond reflects physical characteristics of the living space. Numbers of species in a stream ecosystem are usually greater in physically diverse habitats. Structural attributes like species richness change along a disturbance gradient. Two investigations found that species richness was consistently higher in streams with intermediate disturbance of substrate.[116, 519] The effect of disturbance and physical change over a continuum results in species replacement and sometimes adjustments of the functional characteristics in the community.[325]

Physical and Chemical Influences on Macroinvertebrate Distribution

Factors that control distribution and abundance of macroinvertebrates are substrate, current velocity, temperature, predators, and food resources.[223] Substrate heterogeneity often promotes greater species richness.[326, 328] Interstitial spaces in stream gravels can serve as refuge from predators and physical disturbance, and entrap detritus. Water temperature in the interstitial microenvironment can be relatively constant and cooler than the overlying surface water.[557]

Early life stages of the salmon can be affected by substrate quality. Factors that favor survival of salmon eggs and fry (low levels of fine sands and silts) are coincident with requirements of aquatic invertebrates that have a narrow tolerance range to environmental fluctuations. Protection from natural physical disturbance is important for early life stages of salmon and mobile aquatic invertebrates. Stable stream bottoms during periods of flood or freshet reduce predation on dislodged animals. In some instances, salmon redd construction is a natural disturbance that reduces invertebrate density in localized areas of a stream.[324] This disturbance also opened niche space for other functional groups of aquatic insects, like blackflies, who feed on suspended particles and recolonized quickly along with stonefly nymphs and midge larvae.[324] Other invertebrates that enter the drift behaviorally or unintentionally from substrate disturbance are potential prey items for feeding salmon. Mayfly and stonefly density and richness can be reduced by physical alterations to the stream corridor. These changes may have significant implications to the salmon food base.

Invertebrate drift is either voluntary, a behavioral activity, or coincides with catastrophic stream conditions, especially during floods. Taxonomic groups prominent in behavioral drift are amphipods, Ephemeroptera (mayflies), Plecoptera (stoneflies), Tricoptera (caddisflies), and Simuliidae (blackfly larvae). Later stages of the nymph and larval forms are most active in the diel (24-hour cycle) drift.[547] Behavioral drift occurs with a diel periodicity, typically at two peaks in a 24-hour time frame. Most invertebrates that enter the drift are night-active, with photoperiods as the major cue. Fewer invertebrates are day-active and begin drifting by cues through change in water temperature.

Drifting invertebrates are a food source for certain species of fish that forage in the stream water column. Rader[415] determined that the mayfly genus, Baetis, whose drift propensity was high, was a significant food source to juvenile and adult salmon. Other studies indicated that food preference of juvenile fish was related to its abundance and location within the stream channel. Juvenile coho salmon diet varied seasonally depending on the type and abundance of invertebrates, salmon fry, or salmon eggs in the benthos or drift.[260]

Significance of Macroinvertebrate Life Cycles

There are two life strategies characteristic of freshwater macroinvertebrate species.[558] The simpler *hemimetabolous* strategy inherent in stonefly and mayfly species contain an egg, multiple nymph, and adult stages. A few of the stonefly species are long-lived (more than a year) in the aquatic nymphal form. Large-bodied stoneflies found in streams indicate adequate flow in channels that are key to survival of early salmon life stages and to some of the invertebrate fauna they will eventually consume.

The second life strategy contains representatives of the *holometabolous* invertebrates. Midges, blackflies, and caddisflies have egg, larva, pupa, and adult life stages. These types are mostly short-lived having one or many generations per year in a population. Aquatic environments that are seasonally stressed by high temperatures, low dissolved oxygen, or drought are primarily colonized by holometabolous invertebrates. These stressors increase the mortality of early life stages in salmon, but encourage dominance of holometabolous species in the aquatic invertebrate community.

Aquatic Macroinvertebrates as a Food Source for Salmon

Aquatic ecosystems are frequently inhabited by both hemimetabolous and holometabolous macroinvertebrates. The hemimetabolous species richness is greater in mid- to upper-drainage streams and play a larger role in the diet of juvenile chinook[197] and coho salmon.[442] Although holometabolous invertebrates are dominant in lower-drainages, species in the family Chironomidae are present in all habitats and are a significant food source to salmon in early freshwater stages.[100, 399, 440]

All species of salmon fry consume some life stages of dipterans, primarily Chironomidae, during the freshwater life phase.[178] Stonefly and mayfly nymphs are consumed by pink, chum, and chinook salmon fry. Coho fry are suspension and surface feeders whose diet is predominately terrestrial insects. Ecologically important freshwater invertebrates in coho natal habitat are emerging and flying insects such as mayflies, stoneflies, and midges (Chironomidae). The rapid migration of chinook fry to the river estuary introduces terrestrial homopterans (leaf hoppers and aphids) into their diet. Additional details of prey items during the freshwater cycle have been described.[95, 155, 178, 294, 338, 399, 449, 461, 466] The influence of riparian vegetation along streams and estuaries appears to be an important factor in determining abundance and type of terrestrial insects on which salmon forage.

Salmon as a Food Source for Aquatic Macroinvertebrates

Freshwater macroinvertebrates such as caddisflies, stoneflies, and midges are involved in processing the microbially conditioned salmon carcass. Bilby et al.[47] observed a significant contribution of nitrogen from spawning salmon to the collector-gatherer invertebrate community. (Photograph 7) Increases in aquatic invertebrate density from the introduction of salmon carcasses[564] stimulated feeding by early life stages of select

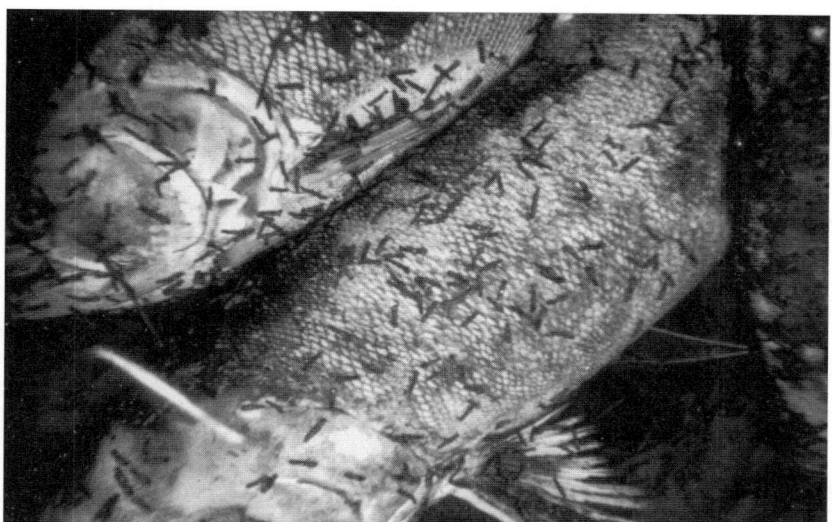

Photograph 7. Aquatic insects feeding on a salmon carcass. (Photo by Jason Walter and Brian Fransen)

salmon species.[48] Other stages of the salmon life history contribute to the invertebrate food base. Nicola[376] observed the stonefly nymph, *Alloperla*, scavenging dead pink and chum salmon embryos and alevins. Also, Elliott and Bartoo[131] found the midge, *Polypedilum* (Diptera), associated with dead pink salmon embryos and alevins.

Freshwater invertebrate shredder abundance increases in the presence of salmon carcasses.[564] Non-salmon-bearing streams support a limited abundance of shredders mediated through input of leaf matter. This organic food base must first be conditioned by the microbial community to increase palatability to shredders. Cool water temperatures characteristic of coastal streams slow the microbial decay of the leaf litter food source resulting in limitations in distribution and abundance of the shredder community. The appearance of salmon and the additional influx of biomass to streams appears to be a controlling factor for shredder species. However, the role of shredders in the presence of salmon carcasses continues to be investigated. Bilby et al.[47] found no significant concentrations of carbon contributed from decaying carcasses in the shredder community. Undigestable animal tissue consumed by shredders was excreted as fine particulate organic matter (FPOM). Nutritive food value for shredders may have been derived primarily from the microbial community on decaying carcasses. Aquatic insects of the collector-gatherers group typically benefit secondarily from the activity of shredders.[316, 47]

The relationship between invertebrates and salmon can be complex. Functions of invertebrates have not yet been fully defined, but we know they are essential to salmon survival. Invertebrates complete a loop beginning as recipients of food from adult salmon carcasses that, in turn, fuel the growth and survival of early stages in the salmon's life cycle.

Vertebrate Wildlife

Vertebrate Wildlife and Salmon

Anadromous salmon provide a rich, seasonal food resource that directly affects the ecology of both aquatic and terrestrial consumers, and indirectly affects the entire food-web that knits the water and land together. Wildlife species have likely had a very long, and probably co-evolutionary, relationship with salmon in the Pacific Northwest. In their *Natural History of Washington Territory and Oregon*, Suckley and Cooper[502] wrote of the California condor:

> The Californian vulture visits the Columbia river in fall, when its shores are lined with great numbers of dead salmon, on which this and the other vultures, besides crows, ravens, and many quadrupeds, feast for a couple of months.

The "Five Mile Rapids" prehistoric archaeological site along the banks of the Columbia River, five miles east of the Dalles, Oregon, yielded bones from at least 63 individual California condors, plus remains of turkey vultures, cormorants, bald eagles, and gulls.[110] Carbon-14 dating placed materials at this site from 10,000 to 7,500 years before present.[321, 474] Miller[321] suggested that these birds were attracted to the site by the presence of abundant living and dead salmon and human refuse resulting from fishing.

The life cycle stages of salmon (i.e., eggs, fry, smolts, adults, and carcasses) all provide direct or indirect foraging opportunities for terrestrial, freshwater, and marine wildlife (Photograph 8). While sometimes abundant and somewhat dependable from year to year, the availability of salmon to wildlife is largely seasonal in nature. The high seasonal variability in a particular food resource is reflected in the opportunistic foraging of many wildlife consumers—however, "opportunistic" is not a synonym for biological unimportance. Thus, one could hypothesize that while many wildlife species could develop important food-web relationships with salmon, few wildlife species would likely be able to form an ecological "dependance" on salmon. Only those species which are highly mobile, or are able to capture, consume, and store (in body tissues) substantial quantities of salmon biomass in a short period of time would be likely to develop a strong ecological dependance on salmon. It is more probable that the majority of wildlife which directly consume salmon will

Photograph 8. Juvenile glaucous-winged gull eating a chum salmon (O. keta) carcass at Kennedy Creek, Washington. (Photo by Jeff Cederholm)

have flexible foraging strategies, utilizing salmon when available, and alternate food sources during other times of the year.

Indirect relationships develop when a food resource is providing foraging opportunities to a secondary consumer. An example in our case is reflected by peregrine falcons which eat gulls that feed on salmon carcasses. As salmon are a concentrated resource, this will serve to concentrate otherwise dispersed wildlife species (e.g., bears). In this scenario, there may well be competition, parasitism, or other aggressive interactions between or among wildlife species. Some of these interactions, e.g., bald eagles disturbing common mergansers, serve to benefit salmon by reducing predation. The magnitude of the salmon-wildlife interaction warrants special examination and calls attention to the pervasive occurrence of these important ecological functions and linkages across the region. The loss or severe depletion of

anadromous fish stocks could have major effects on the population biology (i.e., age class, longevity, dispersal ability) of many species of wildlife, and thus, on the overall health and functioning of natural communities over the majority of the region.

Research on predator-prey interactions in which anadromous fish are the prey has strongly emphasized the effects of predation on the fish populations.[225, 331, 430, 559, 570, 571, 576] Many existing studies describe predatory species as competitors of human harvesters and attempt to control the rate of predation to maximize human consumption. We hope to reverse this perspective by focusing on the important interplay between salmon and wildlife populations. In the following sections, we discuss the relationships between salmon and their vertebrate consumers, and the salmon's role in enhancing ecological functions involving wildlife in terrestrial, freshwater, and marine systems.

Table 2. Relationship between Pacific salmon and 605 species of wildlife in Washington and Oregon.

| Salmon life stage | Relationship | | | | | |
	Strong, Consistent	Recurrent	Indirect	Rare	Unknown	None
Incubation—eggs and alevin(23)	2	10	1	10		
Freshwater Rearing— fry, fingerling, parr (49)	4	31	4	10		
Saltwater—smolts, immature adults, adults (63)	6	36	5	19		
Spawning (16)	5	10	0	1		
Carcasses (82)	5	28	22	38		
	(9)	(58)	(25)	(64)	(60)	(408)

There were 137 species with a positive relationship with salmon (i.e., combined total for species with *Strong, Consistent, Recurrent, Indirect*, and *Rare* relationships). The total number of individual wildlife species for columns and rows are shown in parenthesis; the number of species shown in the rows and columns may not equate to the numbers shown as totals as 19 species had more than one type of relationship with salmon, and 73 species are associated with salmon at more than one life stage.

Wildlife Species with a Relationship to Salmon

Johnson et al.[233] examined the relationships between the Pacific salmon and 605 species of terrestrial and marine mammals, birds, reptiles, and amphibians currently or historically common to Washington and Oregon. They found a positive relationship between salmon and 137 species of wildlife, the relationship was "unknown" for 60 species, and a determination of "no relationship" was made for 408 species (Table 2). Where a relationship existed, they identified both the type(s) of relationship and the stage(s) of the salmon life cycle to which it applied. Of the 137 species with a relationship to salmon, 9 species were categorized as having a *Strong, Consistent* relationship (Appendix I), 58 as *Recurrent* (Appendix II), 25 as *Indirect* (Appendix III), and 64 as *Rare* (Appendix IV). (This tally totals more than 137 because 19 species had more than one type of relationship with salmon.)

Of the 137 wildlife species, 88 were characterized as having a routine relationship (combination of species with *Strong, Consistent, Recurrent;* and *Indirect*) with salmon. Of these 88 species, there were 25 mammals (8 of these were marine mammals), 60 birds, 2 amphibians, and 1 reptile.

The relationship categories are briefly described as follows:

1) Strong, Consistent Relationship. Salmon play (or historically played) an important role in this species distribution, viability, abundance, and/or population status. The ecology of this wildlife species is supported by salmon, especially at particular life stages or during specific seasons. Timing of reproductive activities, and daily or seasonal movements often reflect salmon life stages. Relationship with salmon is direct (e.g., feeds on salmon, or salmon eggs) and routine. The relationship may be regional or localized to one or more watersheds. Examples: a significant portion of the diet of killer whales is adult salmon (*Saltwater* stage); common mergansers may congregate to feed on salmon fry (*Freshwater Rearing* stage) when they are available.

2) Recurrent Relationship. The relationship between salmon and this species is characterized as routine, albeit occasional, and often tends to be in localized areas (thus affecting only a small portion of this species population). While the species may benefit from this relationship, it is generally not considered to affect the distribution, abundance, viability, or population status of this species. The percent of salmon in the diet of these wildlife species may vary from 5% to over 50%, depending on the location and time of year. Example: turkey vultures routinely feed on salmon carcasses, but feed on many other items as well.

3) Indirect Relationship. Salmon play an important, routine, but *indirect* link to this species. The relationship could be viewed as one of a secondary consumer of salmon; for example, salmon support other wildlife that are prey of this species. This includes aspects such as salmon carcasses that support insect populations that are a food item for this species. Example: American dippers feed on aquatic insects that are affected by salmon-derived nutrients. The hypothesis of an *indirect* relationship

between an aerial insectivore and salmon was supported by the presence of two or more of the following characteristics of the insectivore: (1) riparian obligate or associate, (2) feeds below or near the canopy layer of riparian trees, (3) known or perceived to feed on midges, blackflies, caddisflies, stoneflies, or other aquatic insects that benefit from salmon-derived nutrients, and/or (4) feeds near the water surface. While this category includes general aspects of salmon nutrient cycling in stream/river systems, we are not including or examining the role of carcass-derived nutrient cycling on lentic system riparian and wetlands vegetation, and subsequent links to wildlife.

4) Rare Relationship. Salmon play a very minor role in the diet of these species, often amounting to less than 1 percent of the diet. Typically, salmon are consumed only on rare occasions, during a shortage of the usual food and may be especially evident during El Niño events. As salmon are often present in large quantities, they may be consumed on rare occasions by species that normally do not consume them. Examples: red-tailed hawks are known to consume salmon carcasses in times of distress; trumpeter swans are primarily vegetarians, but on rare occasions will consume eggs, parr, as well as salmon carcass tissue.

5) Unknown Relationship. A relationship between this species and salmon may exist, but there is not enough information to determine the scope or scale of the relationship at this time. Example: while it is logical to speculate that riparian feeding bats may feed on salmon-derived insects, aspects of seasonality of both bats and salmon carcasses are relevant, as is the nocturnal flight behavior of the insects. Do bats and salmon carcasses coincide seasonally, and if so, are salmon-derived insects actually available to feeding bats? At this time, the evidence for this relationship is inconclusive and remains to be examined.

6) No Relationship. There is no recognized or apparent relationship between salmon and this species.

As part of the same study, Johnson et al.[233] reported 60 species as having an "unknown" relationship with salmon (Appendix V), suggesting that the diets of these species in *Oregon and Washington,* were not understood well enough to characterize their relationship with salmon. Additional observations on the diets of these species will help determine whether the relationships of these species with salmon is routine, a rare and unusual event, or whether a relationship exists at all. Johnson et al.[233] identified 408 species as having "no relationship" to salmon.

Wildlife Response to Salmon Congregations

The numerical response of predators to salmon congregations is often substantial, sometimes spectacularly so. The ability of wildlife species to concentrate at salmon sites is more than opportunistic foraging, it provides for interaction of important ecological processes. Anadromous fishes (including their eggs) are a major source of high-energy food that allows for

Table 3. Wildlife species that have been observed or are perceived to aggregate at salmon congregations in Oregon and Washington.

American crow
American white pelican
Arctic tern
Bald eagle
Bank swallow
Barn swallow
Barrow's Goldeneye
Black bear[a]
Black-billed magpie
Brandt's cormorant
Brown pelican
California gull
California sea lion
Caspian tern
Clark's grebe
Cliff swallow
Common goldeneye
Common merganser
Common murre
Common raven
Common tern
Double-crested cormorant
Elegant tern
Forester's tern
Glaucous gull
Glaucous-winged gull
Grizzly bear[a]
Harbor seal
Herring gull
Killer whale
Northwestern crow
Northern rough-winged swallow
Northern (Steller) sea lion
Red-breasted merganser
Rhinoceros auklet
Ring-billed gull
Thayer's gull
Tree swallow
Tufted puffin
Turkey vulture
Violet-green swallow
Western grebe
Western gull

[a] now questionable; there may not be enough salmon in Oregon and Washington for bears to gather at a site.

successful reproduction and enhanced survival of adults and juveniles of many wildlife species, and support for long-distance migrant birds. Wildlife movements to salmon congregations can be seasonal (e.g., bald eagles along the Skagit River in Washington), or depending on the situation (e.g., hatchery fish released during an El Niño high food-stress seabird breeding season) can occur within a matter of hours. Perhaps as noteworthy, but much harder to detect, is that some wildlife species that have been reported to group at salmon sites in other areas (e.g., black bears in southeast Alaska) do not appear to be doing so with any regularity in Washington and Oregon. This may well be reflecting the depressed nature of some salmon stocks rather than the inherent behavior of the wildlife species. Of the 88 species with a link to salmon,[233] 43 species (37 birds, 6 mammals) concentrate or form loose aggregations at salmon sites (Table 3). Some reasons why other species do not congregate at salmon sites are: strong territoriality (e.g., great blue heron), foraging strategies which require above-water structures or perches (e.g., belted kingfisher), and limited movement capabilities (e.g., shrews).

That wildlife concentrate at salmon areas is well established for some species, and for others we may be witnessing signs of the impacts of salmon declines. We offer the following examples relevant for Oregon and Washington.

Bald Eagle. Suckley and Cooper[502] state:

This noble looking bird is exceedingly abundant in Oregon and Washington Territories, and in certain localities, especially during the salmon season, may be found in great numbers.

The North Fork of the Nooksack River (coastal Washington) currently hosts one of the largest and most visible concentrations of wintering bald eagles in the lower forty-eight states. Peak concentrations (100 or more eagles) occur along the Nooksack[263] and Skagit[218] rivers with December- and January-spawning chum salmon.

Caspian Tern. The first breeding record of Caspian terns along the Oregon/Washington coast was a colony of 50 pairs in Grays Harbor, Washington in 1957.[6] This, and other nesting colonies (mid-1950s through early 1990s) along the Washington coast in Grays Harbor, Willapa Bay, and near the mouth of the Columbia River, have been abandoned or destroyed by human actions.[429, 430, 479] A colony of Caspian terns originally settled on Rice Island, a dredge material disposal island in the lower Columbia River, in 1987. In 1997, an estimated 14,000 terns used this island for nesting and/or roosting during the 80-100 day (April-July) breeding season.[429, 430] This represents the largest known colony of Caspian terns in North America, and possibly the world. In 1997, the terns appeared to be largely dependent on juvenile salmon (roughly 75% of the diet), consuming an estimated 14.5 million smolts, the majority being hatchery fish.[429, 430] Tern nesting success was very low (roughly 5%) in 1997; predation on adult terns by bald eagles, and gull predation on tern eggs and chicks (caused by eagle and researcher disturbance) were the primary causes.[429, 430]

Common Murre. The common murre is a seabird that nests in large colonies along the Oregon coast; colonies in Washington have undergone significant declines in the last decade. It is only an occasional consumer of salmon; the vast majority of its diet is other small marine fishes. A severe El Niño event occurred in 1983 during the seabird nesting season along the eastern Pacific coast, with the majority of common murres (and other seabirds species) either not attempting to nest or abandoning their nests once initiated.[27, 210] Adult survival was also greatly reduced.[169] Oregon Aqua-Foods, Inc. had released a total of 2 million or more salmon smolts into the Yaquina Estuary at roughly 2.5 day intervals between June and August since 1977.[28] Murres were more numerous at the mouth of Yaquina Estuary for the first two days post-release in July of 1983 than in July of 1982 (a non-El Niño year), as they were drawn in to feed on the released salmon. First day post-release averages of murres were 3,710 in 1983 and 3,053 in 1982. In August of 1983 however, murre numbers were significantly less than in 1982 (average 1983=106; 1982=1,860), as murres had begun moving north earlier to feed on other food resources.[27] In summary, although not as a primary food resource, murres will make use of salmon resources during food-stress conditions.

Black Bear. Contrary to popular image, Washington and Oregon black bears only rarely congregate at salmon sites. Poelker and Hartwell[406] reported on three diet studies of black bears in western Washington for the time periods of 1952-54 and 1968, and found that fish represented 5.0% of the diet. In their treatise on land mammals, Verts and Carraway[535] describe black bears in Oregon as being largely herbivorous, and do not mention salmon as part of their diets. Cederholm et al.[91] found black bears on Washington's Olympic Peninsula to heavily consume salmon carcasses. In northern California, Kellyhouse[244] found evidence of salmon in 10% of black bear fecal samples analyzed from spawning areas. The California Department of Fish and Game[84] also reported black bears taking advantage of anadromous fish runs. The strong link between black bears and salmon was demonstrated in the Anas Creek drainage of southeastern Alaska (D. Chi, pers. comm.). This effort studied the behavior and activity patterns of black bears (n=40 individuals) which had established movement patterns according to salmon migrations. Bears arrived in the lower reaches of the creek in June to begin feeding on spawning salmon, and stayed through August and early September to feed on salmon carcasses. Thirteen of the bears were radio-marked and their movements indicated that they were moving in from at least eight miles (12.9 km) away. Other bears were assumed to be coming in from further away. The general lack of salmon in the published accounts of black bear diets across Washington and Oregon [exception: see91] is somewhat counter to the observed salmon use in adjacent regions. Radio-marked bears in Washington (G. Kohler,

pers. comm.) and Oregon[528] have been found to move to and congregate at higher elevations in the fall to feed on huckleberries (i.e., forming "traditional use areas"). Thus one could reasonably conclude that if salmon were to be found in substantial and predictable numbers, bears in Oregon and Washington, like those studied by Chi in Alaska, would also establish traditional use areas around salmon. Black bears in western Washington typically den by 1 November and emerge around 1 April, thus salmon runs occurring during the winter will not be available to bears. Recent bear studies in western Washington have included the Humptulips, Wishkah, Wynoochee, and Quinault Rivers, and while these hold low levels of hatchery-based salmon, bears do not congregate along them (G. Kohler, pers. comm.). D.H. Johnson (unpubl. data) summarized 1990-1998 hatchery return data of adult salmon for the Humptulips river system in western Washington. These fish return to the hatchery facility as early as late-September (most begin around mid-October), and as typical with most, were done spawning by mid-December. While there are additional fish in this system, an average number of 217 (range 95-320) chinook, 6,496 (range 177-10,195) coho, and 165 (range 51-339) steelhead returned annually to the hatchery. The substantial majority of these fish species return to spawn after 1 November (the average date of bear denning) and are not available to bears; an average of 73 chinook, 1,264 coho, and 0 (zero) returning steelhead were available to black bears. Here, "available" means simply present in the river system, and not located at spawning redds. In summary, bears have a strong relationship to salmon where they have access to them, but it appears that in substantial measure, current salmon populations do not represent a predictable food supply to bears in Washington and Oregon.

Review of Wildlife Relationships by Salmon Life Stages

For the 137 species with a relationship to salmon, Johnson et al.[233] identified the salmon life stage(s) involved for each species. In this study, the five general life history stages of salmon were identified as: (a) *Incubation* (egg and alevin); (b) *Freshwater Rearing* (fry, fingerling, and parr); (c) *Saltwater* (smolt, subadult, adult); and (d) *Spawner*; and (e) *Carcass*. The number of wildlife species associated with each (in parenthesis) were: *Incubation* (23); *Freshwater Rearing* (49); *Saltwater* (63); *Spawning* (16), and *Carcass* (82); this tally of wildlife species totals more than 137 because 73 species are associated with salmon at more than one life stage (Appendixes I, II, III). See Appendix VI for a list of published and unpublished observations of wildlife predators and scavengers on salmon.

Incubation Stage (egg and alevin). Twenty-three wildlife species are linked to salmon at this stage. Twenty-two wildlife species are direct consumers of "drift eggs" (eggs not buried in redds) or alevin (2 amphibians, 1 reptile, 19 birds, and 1 mammal); and 1 bird (bald eagle) is an indirect consumer of eggs/alevin, feeding on the waterfowl that consume eggs and alevin.

Photograph 9. Garter snake eating a salmon smolt, unknown Olympic Peninsula stream, Washington. (Photo by Jim Rozell, deceased)

Freshwater Rearing (fry, fingerling, and parr). Forty-nine wildlife species are linked to rearing salmon, including 2 amphibians, 5 reptiles, 34 birds, and 4 mammals. Forty-five of these species are direct consumers of salmon and 4 species (bald eagle, gyrfalcon, peregrine falcon, and snowy owl) are indirect consumers, feeding on terns, waterfowl, gulls, and other animals that eat rearing salmon (Photograph 9).

Saltwater (smolt, subadult, adult). Sixty-three wildlife species are consumers of salmon at this stage (51 birds and 12 marine mammals). Fifty-eight of these species are direct consumers of salmon and 5 species are indirect consumers. This list is somewhat expansive due to the geography being included, that is, the estuarine and all marine water habitats.

Spawner. Sixteen species of wildlife are consumers of spawning salmon (6 birds and 10 mammals). This list is relatively small, as few wildlife species are physically capable of capturing and handling live, adult fish. The gray wolf and grizzly bear are on this list, but both have undergone significant range contractions and declines in their abundance (i.e., both are extirpated from Oregon and significantly reduced in Washington).

Carcass. Carcasses are linked to the largest group of wildlife consumers of any salmon life cycle stage, with 82 species (1 reptile, 50 birds, and 31 mammals) being consumers of carcasses and/or carcass-derived insects. Body sizes of these animals range from shrews to grizzly bears. Seventy-two species of wildlife (1 reptile, 38 birds, and 31 mammals) are direct consumers of carcasses; 22 species (14 birds and 8 mammals) are consumers of carcass-derived insects; and 10 species (2 birds and 8 mammals) are consumers of both carcasses and carcass-derived insects.

Continued documentation of wildlife species—salmon interactions, especially of the 60 species having an "unknown" relationship, will provide vital information for ongoing developments in ecologically-based salmon spawner escapement research and prescriptions for riparian management practices. As Key Ecological Functions (KEFs) are identified through such research,[292] tools for informed decisions will be made available to fish and land managers operating under an ecosystem context.

Key Ecological Functions (KEFs) Provided to the Ecosystems through Salmon—Wildlife Interactions

In striving to manage for healthy and sustainable ecosystems, we simultaneously are striving to provide for the full range of ecological functions that these systems provide. Key ecological functions (KEFs) refers to the main ecological roles of a species (or group of species) that influence diversity, productivity, or sustainability of ecosystems.[334] A given KEF can be provided by a single species or shared by many species, and a given species can have several KEFs. Main categories of KEFs include trophic relations; herbivory; nutrient cycling; interspecies relations; disease; pathogen and parasite relations; soil relations; wood relations; water relations; and vegetation structure and composition relations. Building upon work by Marcot et al.,[291] Marcot and Vander Heyden[292] characterized the key ecological functions for each of the 605 common wildlife (i.e., terrestrial and marine birds, mammals, reptiles, and amphibians) species in Oregon and Washington. Several questions can be thus posed:

• In what way does providing for salmon also provide for a wider array of ecological functions of wildlife species associated with salmon?
• What are those functions?

- How do different kinds of salmon-wildlife relations, and different salmon life stages, provide for an array of ecological functions?

This somewhat innovative analysis describes the functional links among fish and wildlife species across aquatic and terrestrial communities. To conduct this analysis, we queried the database matrixes on salmon-wildlife relations, key ecological functions of wildlife species, and habitats used by wildlife.[236] The general conclusion is that salmon provide a causal mechanism for movement behaviors and a nutrient source for a variety of wildlife species which in turn perform a surprisingly broad array of ecological functions[292] across a wide span of habitats. For this analysis, one can think of the array of ecological functions performed by these wildlife species as a "functional web." It focuses on salmon in their various life stages, and extends well beyond the aquatic realm to influence the diversity, productivity, and ultimately sustainability of habitats and ecosystems throughout Washington and Oregon.

Wildlife with Strong Consistent Links to Salmon. The 9 species of wildlife with strong consistent links to salmon (bald eagle, American black bear, Caspian tern, common merganser, grizzly bear, harlequin duck, killer whale, osprey, and river otter) comprise a functional group of "salmon-eaters" with close affinities to salmon. There are 32 primary wildlife-habitats across Washington and Oregon[236]; Figure 4 summarizes the occurrence of these 9 wildlife species by habitat. Not surprisingly, most of these

9 species inhabit freshwater and marine habitats, but some of them also occur across the range of inland forest, woodland, shrubland, and grassland habitats. It is of interest that from 1-7 of these 9 species can be found in each of the 32 habitats. In this way, salmon provide for a set of wildlife species that occur well beyond just salmon-inhabited aquatic systems.

In addition, the full set of key ecological functions performed by these 9 species also extends beyond the aquatic system. Each of these species provides a set of ecological functions to the various array of habitats that they occur within. Figure 5 depicts the collective range of ecological functions that these 9 wildlife species provide to the number of habitats that they occupy. The functions range from various trophic, organismal, and wood and soil relations. Some functions are more widespread (occur in more habitats) than are other functions. Examples of some widespread functions are: potential control of vertebrate populations (through predation), carrion feeding, piscivory (fish-feeding), invertebrate feeding (including insectivory), omnivory, transportation or dispersal of seeds and animals, creation of terrestrial runways used by other species, and secondary use of burrows created by other species.

All Wildlife with Links to Salmon. What of the full set of species showing either strong consistent, recurrent, and/or indirect links to salmon? (Some species have more than one type of relation because they use more than one salmon life stage). Table 4 lists key ecological functions of

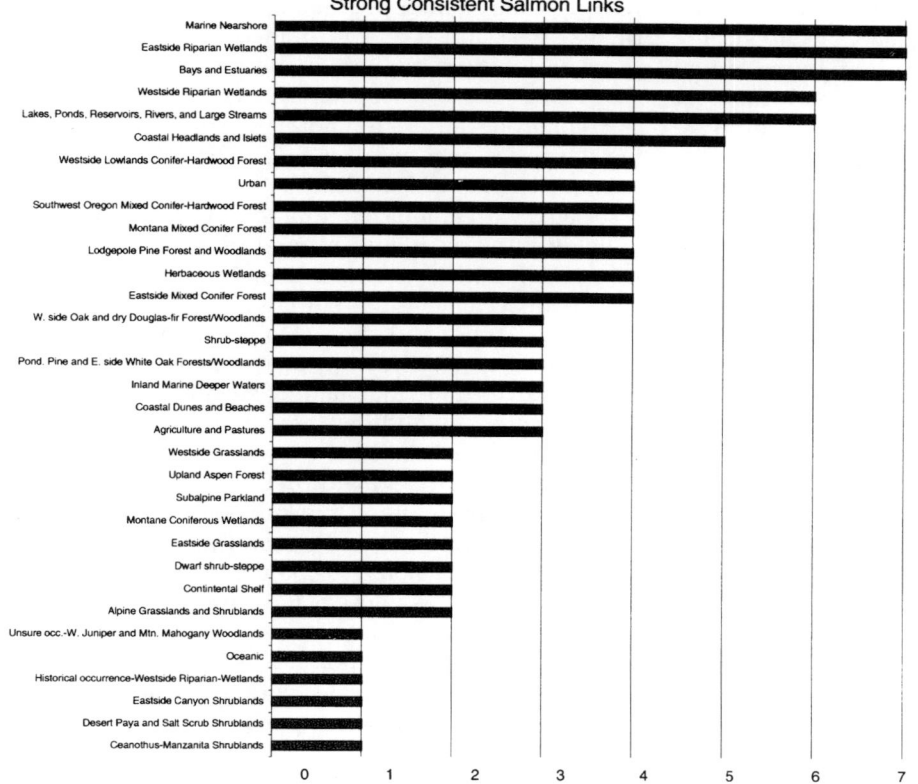

Species Richness by Habitat
Strong Consistent Salmon Links

Figure 4. Occurrence by number of vertebrate wildlife species in the 32 wildlife habitats in Washington and Oregon, as used by the 9 wildlife species with a strong consistent relationship to salmon.

Figure 5 . The array of key ecological functions performed by the 9 vertebrate wildlife species with a strong consistent relationship to salmon, across the 32 wildlife habitats in Washington and Oregon.

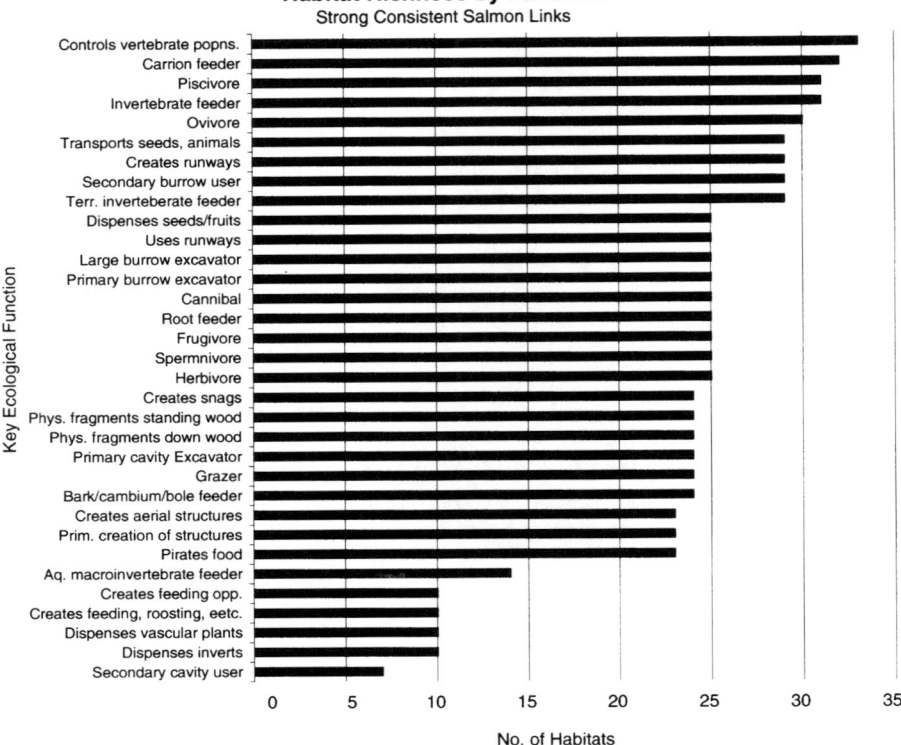

Habitat Richness by Function
Strong Consistent Salmon Links

wildlife more or less unique to each type of salmon-wildlife link. Each of the three types of relations provides for some unique set of ecological functions. For example, wildlife species indirectly linked to salmon can provide the following ecological functions: fungivory (fungus-eating); tertiary consumption or secondary predation; prey source; regulate insect populations through predation; serve as interspecific host for avian nest parasites; and create primary small ground burrows. These functions are either not performed, performed by far fewer wildlife species, or by the wildlife with occasional or strong consistent links to salmon.

What this means is that different degrees of salmon-wildlife relations provide for some unique kinds of wildlife ecological functions. Only the full set of all wildlife-salmon link relations can provide for all collective functions. Thus, to manage the full set of all ecological functions, one should not focus solely on those few wildlife species with strong consistent links to salmon, but on all types of links.

How Salmon Life Cycle Stages Provide for Ecological Functions. In a similar way, most of the five life cycle stages of salmon provide for a unique set of wildlife species and their ecological functions (Table 5). For example, wildlife associated with the incubation stage of salmon include secondary cavity users and primary excavators of small ground burrows; these two ecological functions are not provided, or only poorly provided, by wildlife species associated with any of the other salmon life cycle stages. Thus, to manage for the full set of ecological functions, one should focus on providing all life cycle stages of salmon.

Managing the Functional Web. So what is the manager to do with this information? For one, be aware that salmon can be viewed as the center of a broad "functional web" of wildlife and their ecological roles. Such roles extend well past the salmon populations and aquatic habitats themselves, and likely influence the structure and processes of the communities and ecosystems in which they reside, thus a "keystone" species.[559]

Second, one can use the information presented here and in the species data matrixes to list the collective set of habitat elements and conditions used by wildlife species associated with salmon. For example, one can link the list of wildlife associated with salmon life stages likely to be found in low order headwater streams, and determine the set of habitat elements used by this set of wildlife species, by habitat type, and then establish habitat-specific management guidelines to provide for such habitat elements over time. Maintaining such habitat elements and conditions would help maintain the full salmon-wildlife functional web.

Third, one can begin to predict—or, at least pose tentative management hypotheses about—which ecological functions may be in jeopardy if the wildlife that performs such functions are not maintained. That is, one can now determine which wildlife species may be influenced by altering salmon populations and habitats that imperil specific salmon life stages, and the set of ecological functions associated with such wildlife species. In some cases, other wildlife species not associated with salmon may also perform some ecological function, but never in exactly the same manner and in the same set of habitats and habitat elements.

Table 4. Importance of types of salmon-wildlife relations to key ecological functions. Listed are functions unique to each relationship category.

1. Strong Consistent Relationships are important for:
    trophic relations:
        primary consumption:
            spermivory
            grazing
            frugivory
            root feeding
    organismal relations:
        controlling vertebrate populations
        dispersing seeds, fruits, inverts, vasc plants
        creating feeding opportunities for other species
        primary cavity excavation in trees and snags
        primary creation of large ground burrows
        primary creation and secondary use of ground
        runways
    wood relations:
        fragmenting standing and down wood
        killing standing trees (creating snags)

2. Occasional Relationships are important for:
    trophic relations:
        pirating of food
    organismal relations:
        secondary use of aerial and aquatic structures
        created by other spp.
    disease relations:
        carrier of domestic animal disease
    soil relations:
        improving soil structure and aeration by digging and
        burrowing

3. Indirect Relationships are important for:
    trophic relations:
        primary consumption:
            fungivory
            tertiary consumption
    prey relations:
        providing prey for predators
    organismal relations:
        controlling insect populations
        serving as interspecific host for avian nest parasite
        primary creation of small ground burrows

Table 5. Importance of salmon life stages to key ecological functions. Listed are functions unique to each life stage category

1. Incubation Stage is important for:
    organismal relations:
        secondary cavity use
        primary excavation of small ground burrows

2. Freshwater Rearing Stage is important for:
    (no specific function is mostly supported by this stage)

3. Saltwater Stage is important for:
    organismal relations:
        creating aerial structures used by other spp.
        creating aquatic structures used by other spp.

4. Spawning Stage is important for:
    trophic relations:
        primary consumption:
            spermivory
            grazing
            frugivory
            root feeding
            bark/cambium/bole feeding
    organismal relations:
        controlling vertebrate populations
        creating feeding opportunities for other species
        primary cavity excavation
        primary excavation of large ground burrows
        primary creation of ground runways
        secondary use of ground runways
    wood relations:
        fragmenting standing and down wood
        killing standing trees (creating snags)

5. Carcass Stage is important for:
    trophic relations:
        primary consumption:
            fungivory
    organismal relations:
        controlling insect populations
        serving as interspecific host for avian nest parasite

Salmon Fisheries

Herein we provide only a very brief overview of fisheries as related to wildlife. A more thorough review of the magnitude and characteristics of Northwest Pacific Coast salmon fisheries and habitat issues are provided in a number of recommended readings.[103, 275, 365, 387, 390, 499]

Washington and Oregon salmon fisheries include commercial, recreational, and treaty harvests that occur in the states' rivers, inland lakes, inland marine waters, coastal embayments, and at sea. Fishing is an important source of mortality, both for immature fish in the ocean and mature fish on their return to freshwater to spawn. Understanding past fishery effects is important because man's increasing efficiency as a predator augments the mortality rates that salmon encounter in natural situations. Because of the presence of humans, salmon are significantly less available to a wide variety of natural consumers in both the freshwater, terrestrial, and marine environments, and thus these ecosystems are suffering. According to the NRC,[365] salmon habitat mortality factors caused by human activities and natural factors together usually exceed fishing mortality. Thus, although factors other than fishing have a major effect on the production of adult fish, fishing is still the easiest salmon mortality factor to control.[365]

The abundance of virtually all salmon species initially harvested in the Columbia River, Puget Sound, and coastal Washington and Oregon, have never recovered to their former levels, except for Puget Sound chum salmon (Figures 6 and 7). Increased hatchery production did enhance some fisheries, notably the rise of the coho salmon runs in the 1960s, and chum salmon runs in Puget Sound. In general, however, artificial propagation has failed to rebuild the runs to former levels,[394] and in some instances likely contributed to the further decline of wild stocks.[209]

Admittedly, managing salmon fisheries is a challenge, and past management approaches were generally commodity/extraction-based; however, this approach has significantly contributed to the decline of wild stocks.[544] Fishing activities that contribute to the problem include: indirect mortality due to catch and release of undersized fish (bycatch), out-of-state domestic and foreign interception, conflicts among user groups, and the mixed-stock fisheries.

A major dilemma that fishery resource agencies find themselves in is how to harvest hatchery salmon selectively, while still meeting spawning escapement goals of wild stocks of salmon. Hatchery-produced salmon co-mingle with wild salmon in ocean waters, and as a result, a mixed-stock fishery is created. If harvests are allowed in such mixed-stock fisheries, then wild and hatchery fish will be caught at rates that only hatchery fish can sustain. Wild fish cannot withstand the high hatchery exploitation rate because they are exposed to a full range of natural and human-caused selection pressures and mortalities. Hatchery fish are sheltered from mortality factors that normally occur during incubation and freshwater rearing. At smolt migration, many more progeny are still alive per hatchery female than per wild female. This condition enables hatchery stocks to withstand higher rates of

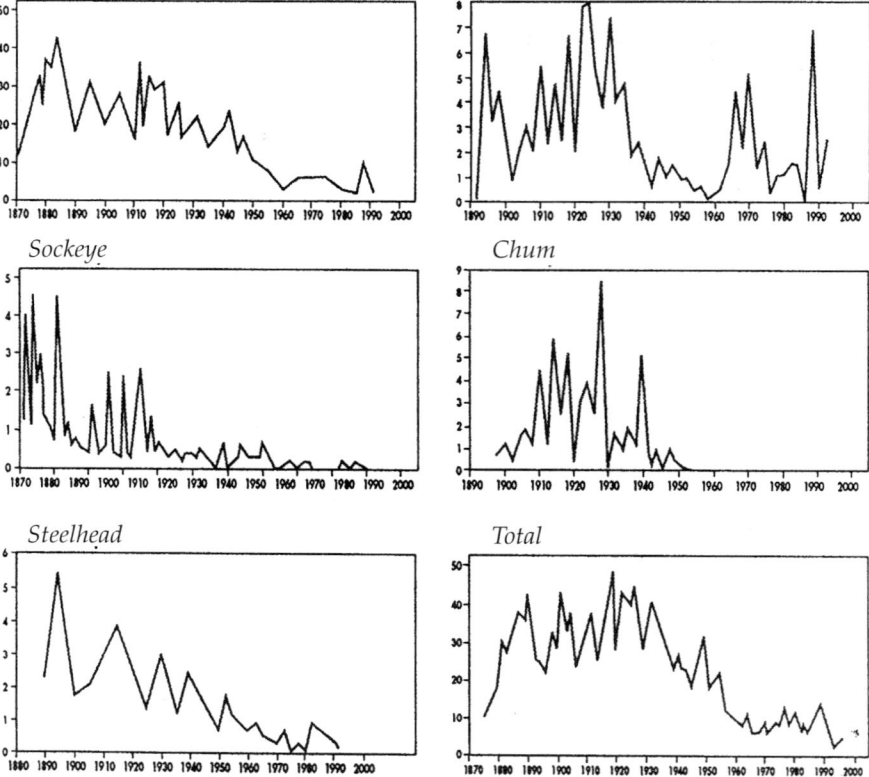

Figure 6. Total commercial catch, in millions of pounds, of Lower Columbia River salmon from 1870s to 1990s.[390, 379, 387b, 540]

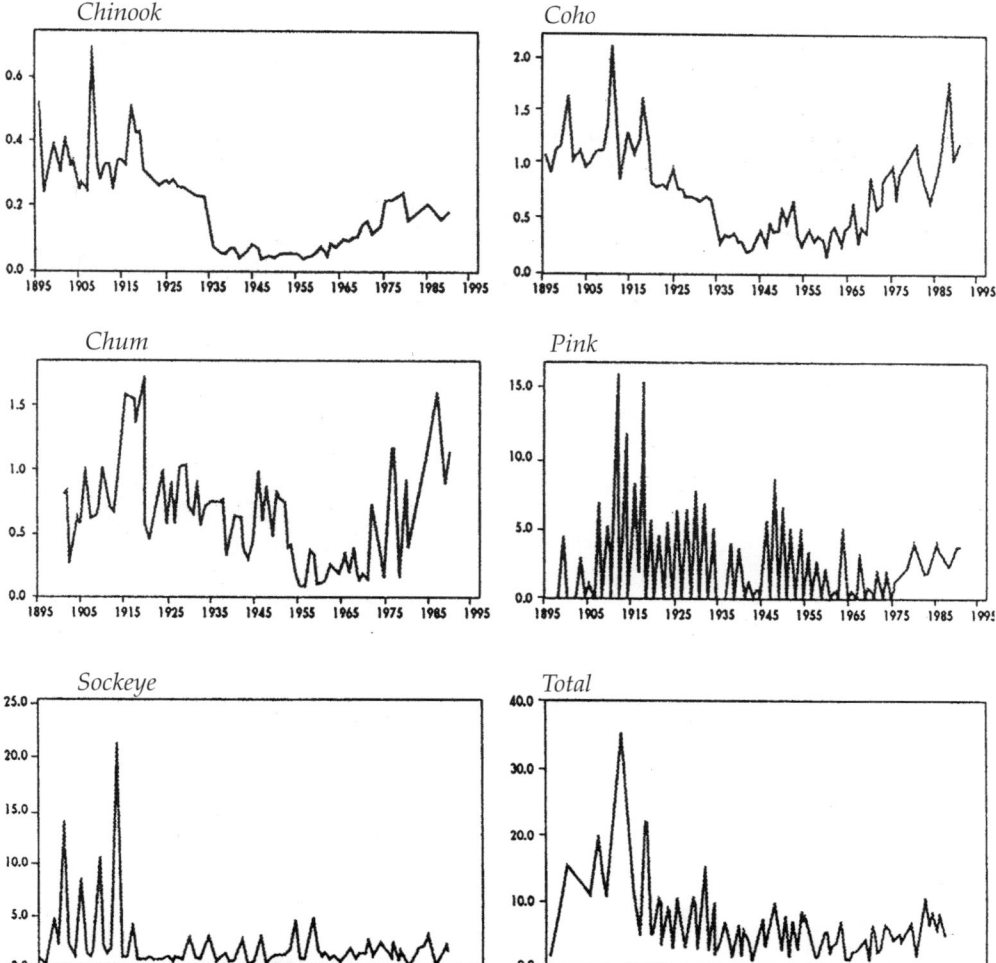

Figure 7. Estimated and reported total catch, in numbers, of Puget Sound salmon from 1890s to 1990s.[390, 57a, 538b]

harvest than wild stocks; however, even hatchery stocks eventually succumb to the high exploitation rates through changes to smaller adult size,[517] or different time of return.[190] Therefore, high fishing exploitation rates are associated with fishery resource management agency policies, and agency policies also determine hatchery policies and practices. The current demand by certain interests to protect wild stocks can be in direct conflict with the agency mandate to enhance or supplement current stocks with hatchery-produced fish to ensure sustainable harvest for the major user groups. This conflict will have an important bearing on future management of salmon fisheries and hatchery practices. Live capture selective fishing, including live release of wild unmarked fish and retention of marked hatchery fish, is potentially an option.[279]

Spawning Escapement Goals

Spawning escapement goals (the number of spawners required to perpetuate the population[254]) are set by fishery managers solely to determine the portion of the estimated returning adult population that can be harvested. No spawning escapement goals currently provide an identified portion of the escaping fish for wildlife or ecological functions (Photograph 10).

According to the Pacific Fisheries Management Council (PFMC)[388] the state of Washington has established annual escapement goals for coho, chum, and chinook salmon and steelhead, and include wild and hatchery fish. Some escapement goals exist for pink and sockeye salmon that are mostly for wild fish. No determination has been made of the spawning escapement needs of sea-run cutthroat trout. In Oregon, escapement goals have been established primarily for chinook and coho salmon, and include wild and hatchery fish. No determination of the spawning escapement needs has been made for wild steelhead, chum salmon, pink salmon, or sea-run cutthroat. Escapement goals may have been established for some of these stocks in more recent years. Spawning escapement goals have been established for 98 wild salmon stocks and 15 wild stocks of steelhead in Washington.[106, 390, 538a] Analysis of annual spawning escapement data by Konkel and McIntyre,[257] collected for naturally spawning salmon populations in the Pacific Northwest between 1969 and 1984, suggests that escapements are down for coho and chum; but up for chinook, sockeye, and pink salmon. The critical factor that salmon harvest managers need to face is how to reduce the annual salmon harvest, and achieve stock-by-stock ecosystem-based spawning escapement

Photograph 10. Chum salmon carcasses in Kennedy Creek, Washington. (Photo by Jeff Cederholm)

goals. With the exception of carcass supplementation programs, there has not been a concerted effort to manage salmon populations for the benefits they provide to the recovery of listed wildlife species (e.g., grizzly bears) or to the broader ecological systems. Salmon spawning escapement goals should not only replace a stock of salmon with sufficient numbers of recruits, but also meet the needs of the broader aquatic and terrestrial ecosystems that depend on salmon for nutrients and carbon influx.[47, 48, 92, 319] Salmon harvest managers could take a lesson from the worldwide conventions for herring harvest managers, where ecosystem function has been explicitly recognized through "...a precautionary, conservative approach to fisheries management."[541a]

There is increased interest in the Pacific Northwest in developing harvest management strategies that address nutrient delivery to freshwater ecosystems.[266, 544] However, there is relatively little information on which to base escapement targets that will meet this objective. Ideally, estimates of the number of spawners needed to fulfill this function would be determined by experimentally altering escapement levels for each stock and evaluating the impact on system productivity. However, conducting such an experiment on hundreds or thousands of stocks over a sufficient length of time is a daunting prospect and would not provide usable results for many years. Several other approaches to determining appropriate escapement levels are currently being investigated. One option being considered attempts to determine the amount of food required to support a population of rearing fish that fully utilizes the habitat available in a stream. Escapement levels would be established which would ensure sufficient nutrients and organic matter are returned to the stream to produce this level of food. Another alternative is to develop a relationship between spawner density and the proportion of marine-derived nutrients in the tissues of juvenile fish. This type of relationship may enable a "saturation level" for marine nutrients to be established and escapement goals set accordingly. However, these

approaches do not account for impacts associated with land use that drive down stock productivity (reduce survivals) and reduced habitat capacity, and decreased genetic diversity, nor do they incorporate any consideration of temporal variability in environmental conditions. Nonetheless, these approaches do represent a shift from MSY to more ecologically based stock management objectives.

Understanding Salmon Relationships

Salmon as a Key Linkage in Biodiversity And Productivity

Ecological processes have been so altered by human activities, especially in the more densely populated regions, that natural resource and environmental management will need to expand from current site- and case-specific methods, to landscape- and ecosystem-scale approaches. The struggle to develop the tools required for these scales of management has only just begun.[269, 305, 365, 460] The documents: *From the Forest to the Sea*,[295] *Forest Ecosystem Management: An Ecological, Economic, and Social Assessment*,[151] *Pacific Salmon and Their Ecosystems*,[499] *An Ecosystem Approach to Salmonid Conservation*,[485] *Fish Habitat Rehabilitation Procedures*,[477] and *River Ecology and Management—Lessons From Pacific Coastal Ecoregion*[362] are recommended readings that describe the understanding of natural systems and processes and take a holistic approach to rehabilitation and restoration of watersheds and ecosystems.

Anadromous salmon play an important role in maintaining an ecosystem's productivity. The seasonal migrations of millions of salmon between Pacific rim streams and the subarctic Pacific Ocean appear to increase overall terrestrial productivity. Key processes discussed here are the transport of materials, energy and nutrients between marine, aquatic, and terrestrial ecosystems, with emphasis on salmon as a transport vector. From a broad ecosystem management perspective, the status of salmon metapopulations is a powerful indicator of human adaptation to boreal biomes.[348, 419, 464] Sibatani[464] has suggested that salmon are the "canary in the mineshaft"; the mineshaft in this instance being entire subarctic ocean basin ecosystems. Cederholm et al.[92] review and discuss the mechanisms of salmon nutrient transport and the significance to terrestrial and freshwater ecosystems. The following discussion suggests a need to understand and apply information on the exchange of materials, energy, and nutrients between the aquatic and terrestrial ecosystems of the northern Pacific Basin. Such an application would occur at multiple geographic and temporal scales, examining both healthy and depressed salmon populations under varying conditions. Effectively applied, managers would be able to define and achieve long-term ecosystem management success not just for salmon, but for numerous other fish and wildlife resources and the overall health of the environment.

Biomass of Salmon Runs as an Energy Source

Organic matter that supports the trophic system of fresh water ecosystems is provided from both autochthonous and allochthonous sources. Common types of autochthonous sources are: algae, mosses, vascular plants, and phytoplankton. All of these factors are found in freshwater, and generate organic matter through the process of photosynthesis. Common types of allochthonous input include leaves, needles, wood and insects from the terrestrial environment and dissolved organic matter carried in groundwater that enters the water body. Salmon provide an important source of allochthonous organic matter for Pacific Northwest fresh water ecosystems.[47, 239, 248] Salmon spawning runs transport organic matter and nutrients from the northern Pacific Ocean to their natal spawning grounds. The organic matter

and nutrients carried in the biomass of the salmon runs is input to the trophic system through multiple pathways including direct consumption, excretion, decomposition, and primary production. Direct consumption may occur in the form of predation, parasitism, or scavenging on the live spawner, carcass, egg or fry cycle life stages. Carcass decomposition and the particulate and dissolved organic matter released by spawning fish (e.g., eggs and milt, excrement) delivers nutrients to primary producers. Potential nutrient or energy pathways and factors influencing biomass cycling of spawning salmon is graphically depicted in Figure 8.

Freshwater ecosystem productivity depends upon nutrient inputs and retention. Larkin and Slaney[266] and Munn et. al.[339] discuss nutrient cycling and the nutrient spiraling concept, whereby nutrients unidirectionally spiraling downstream can influence aquatic system functions for considerable distances. Larkin and Slaney[266] and Munn et. al.[339] also discuss the importance of instream habitat complexity (wood debris complexes) for increasing productivity by increasing salmon carcass retention; citing Cederholm and Peterson[90] and Cederholm et. al.[91] The discussion that follows emphasizes the necessity of retention mechanisms and how the physical and biological complexity of the aquatic, riparian, and wetland zones enhances this function.

Sportsmen and naturalists have long recognized the importance of salmon runs to the natural economy of streams, as this quotation by Haig-Brown[180] reveals:

The death of a salmon is a strange and wonderful thing, a great gesture of abundance. Yet the dying salmon are not wasted. A whole natural economy is built on their bodies. Bald eagles wait in the trees, bears hunt in the shallows and along the banks, mink and marten and coons come nightly to the feast. All through the winter mallards and mergansers feed in the eddies, and in

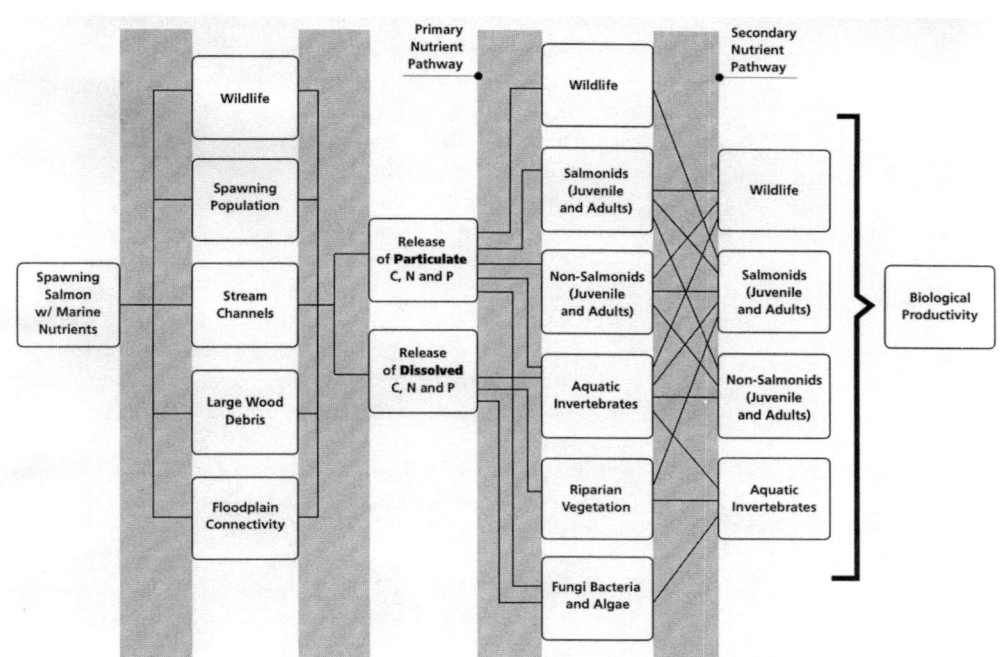

Figure 8. Factors of stream complexity and marine-derived nutrient pathways that influence biological productivity.[92]

*freshet time, the herring gulls come in to plunge down
on the swifter water and pick up the rotting drift.
Caddis larvae and other carnivorous insects crawl over
the carcasses that are caught in the bottoms of the pools
or against the rocks in the eddies. The stream builds its
fertility on this death and readies itself to support a new
generation of salmon.*

The scientific community has also recognized the
contribution of nutrients and organic matter from
spawning salmon for some time. Juday et al.[241] estimated
that sockeye salmon transported in excess of 2 million kg
of organic matter and 5000 kg of phosphorus to the Karluk
River system in Alaska in an average year. This recognition
resulted in sockeye lake fertilization programs in British
Columbia and Alaska, to replenish lost nutrients caused
by fish harvest.[496, 497] Over the last 10 years stable isotope
analysis has enabled direct measurement of marine-
derived nutrients in stream[47, 239, 248] and lake[249] ecosystems.
These studies have firmly established the need to consider
the importance of salmon biomass as a flow of energy and
nutrients into the freshwater and estuarine food webs of
the Pacific Northwest.

Nutrient levels in freshwater and estuarine systems can
be substantially enriched by the organic inputs of salmon
spawning runs. The majority of material transported to
freshwater by some species of anadromous salmon is of
marine origin. Mathisen et al.[298] demonstrated that over
95% of the body mass of some salmon species is produced
during ocean residence; the remainder represents mass
accumulated in freshwater prior to migration to the sea.
The species of salmon that spawn at high densities (e.g.,
chum, pink, sockeye) significantly alter nutrient loadings
and budgets in the freshwater systems where they spawn.
In Kamchatka, Krokhin[262] reported that 35-40% of the
yearly total phosphorus input to a lake was transported
by spawning sockeye salmon, as was much of the nitrogen
input to the system. Similar results have been reported
for the Iliamna Lake system in Alaska,[122, 249, 297, 298] and for
the Paratunka River basin, Kamchatka.[261, 262] Nutrients
from spawning pink and chum salmon have been shown
to enrich not only the freshwater and estuarine habitats
where they spawn, but also the estuarine habitats
downstream.[62, 503] Munn et. al.[339] consider changes in
nutrient loading and cycling and ecosystem productivity
that could result from restoration of historic salmonid
populations to the Elwha River system in Washington
state. The study indicates a potential 65-fold increase in
nitrogen and phosphorus loadings from salmon returns.
They concluded that restoration of the Elwha River system
salmon runs would have a profound effect on the primary
and secondary production in the system. Annual nutrient
inputs of 29.8 mt of nitrogen and 3.5 mt of phosphorus
are projected for a 980 mt biomass of restored salmon runs
based on Munn et al.[339] Average sizes of salmon by species
are from Ricker,[426] as quoted in Larkin and Slaney.[266]
Proportions of body composition for nitrogen and
phosphorus are from Larkin and Slaney.[266] These nutrient
input levels would be approximately 40% and 13%,
respectively, of the 74.5 mt of nitrogen and 27.3 mt of

phosphorus total loads to the Elwha River watershed
under existing conditions.[226] These preliminary estimates
as to the levels of nutrient loadings to the water column
and sediments, while crude, should be sufficient cause to
trigger some rethinking of what constitutes healthy
baseline water and sediment quality for Puget Sound and
Pacific Coast streams and estuaries.

Even low spawning densities can provide significant
contributions of nutrients to the system. Juvenile coho,
steelhead and cutthroat in a small stream in western
Washington obtained from 25% to 40% of their N and C
from dead coho salmon that spawned in the stream.[47]
Aquatic insects also contained high levels of marine-
derived N and C, and the foliage of plants growing along
the streams also contained nitrogen of marine origin.[47] The
direct feeding of salmon fry on salmon carcasses has been
known for a long time in the Amur River of Asia,[377]
however, the growth benefits for juvenile salmon has only
recently been documented in North America.[47]

Macroinvertebrate communities in streams receiving
salmon runs can change in response to spawning activity
and nutrient enrichment. In a Snoqualmie River tributary
and in Kennedy Creek, Washington, Minakawa[324] found
the presence of salmon carcasses and eggs produced a two-
fold or greater increase in total insect densities and biomass
compared to control reaches. Piorkowski[404] found insect
taxa richness and diversity to increase in response to
nutrient enrichment from salmon carcasses in southeast
Alaska, and suggested that insect colonization of carcasses
facilitated decomposition and subsequent nutrient release.
Bilby et al.[47] found all functional feeding groups except
insect shredders to be enriched with marine origin isotopes
of nitrogen and carbon in western Washington streams
after coho salmon spawning. Some aquatic invertebrates
such as stoneflies (Alloperla)[375] and Dipteran flies
(Chironomidae)[131] will scavenge for dead salmon eggs and
alevins within the gravel. Limnephilid caddisfly larvae
are attracted to recently expired salmon and have been
observed feeding directly on fish flesh.[324, 404]

Terrestrial insects including fly maggots (Diptera) have
also been observed feeding heavily on salmon carcasses
in streams in the Queen Charlotte Islands of British
Columbia,[422] but generally little work has been done to
systematically document these activities. Maggot larvae
have commonly been observed consuming beached
salmon carcasses during the warmer months of the
spawning season along the spawning reaches of several
Washington streams. Dead chum salmon along Kennedy
Creek in South Puget Sound often have their heads filled
with maggots (Cederholm, pers. obs.) (Photograph 11).
Chinook carcasses along Puget Sound Basin rivers can be
reduced to skeletal remains by maggots within a two week
period; fall freshets frequently have been observed to wash
the carcasses and masses of larvae back into the stream
where they are then available as food for juvenile salmon
and other organisms (W.Graeber, pers. obs.). Hornets have
also been observed to feed on carcass remains during
warm fall weather periods in the same areas; they are
especially attracted to exposed fresh flesh or blood
(W.Graeber, pers. obs.).

Photograph 11. Fly maggots eating a chum salmon carcass at Kennedy Creek, Washington. (Photo by Jeff Cederholm)

Quantitative measurements of salmon carcass consumption in the terrestrial environment has focused on their utilization by high profile species like bald eagles along the Skagit River, Washington;[490] and grizzly bears along the Columbia River.[205] Cederholm et al.[91] recorded 43 taxa of mammals and birds present on small Olympic Peninsula streams at a time when coho salmon carcasses were present, and found that 51% of those taxa had fed on carcasses. Skagen et al.[475] in their study of human disturbance on an avian scavenging guild, observed significant bird scavenging of chum and coho salmon carcasses along the North Fork of the Nooksack River, Washington. The primary bird scavengers were eagles, crows, and glaucous-winged gulls. Information from the expert panel process and the published literature on observations of wildlife predation and scavenging on salmon is presented in Appendix VI.

Cederholm et al.[91] also reported that black bears, raccoons, and river otters increase food availability for terrestrial species incapable of removing carcasses from the stream. The larger animals rarely completely consumed the carcasses they removed from the stream, and were often followed by an array of other smaller birds and animals who fed on the "leavings." A similar interaction occurred at McDonald Creek, Glacier National Park, Montana, where kokanee salmon captured by grizzly bears were incompletely consumed, leaving remains for birds and small mammals.[486]

As the above studies indicate, spawning salmon provide a source of carbon, nitrogen and phosphorus essential to maintaining the production of salmon juveniles and other trophic levels of the stream. Accumulating evidence suggests that spawning salmon populations are an important link to the adjacent riparian and terrestrial communities, and indeed, fortifies the role of salmon as a keystone species, wherein the integrity and persistence of the entire community is contingent upon the population's actions and abundance.[559]

Pacific Salmon Provide Key Ecological Functions

The key ecological functions that spawning salmon play within certain freshwater ecosystems may be illustrated with a well-documented case study from McDonald Creek in Glacier National Park, Montana. This stream is a principal spawning tributary for the Flathead Lake—Flathead River ecosystem. The triggering event in the series of changes that cascaded through this ecosystem was caused by the introduction of an exotic species, the opossum shrimp (*Mysis relicta*), from 1968 and 1975.[486] The shrimp were added to the lake as a food source for kokanee salmon, but behavioral patterns made them unavailable for consumption. Opossum shrimp are voracious predators of zooplankton, the principal food of the kokanee. The shrimp decimated the zooplankton in the lake, and by the late 1980s the lake and McDonald Creek spawning kokanee population had collapsed. These fish served as an important food source for various birds and mammals that had fed upon them in the spawning tributaries. Among the most prominent predators and scavengers utilizing this resource were bald eagles that gathered by the hundreds during the kokanee spawning period. In 1981, spawning kokanee in excess of 100,000 attracted 639 eagles, the densest eagle concentration south of Canada. Beginning in 1987 eagle numbers declined along with the kokanee, reaching a low of just 25 birds in 1989. It is feared that loss of the kokanee spawning run could lead to higher eagle mortality during migration or during winter, unless the birds can find alternate food resources, a prospect that is not likely in that ecosystem.[486] A number of other bird and mammal species that used the McDonald Creek kokanee also have been displaced.[486] Gulls, mergansers and mallards commonly fed on kokanee carcasses, while Barrows and common goldeneyes and dippers fed on loose eggs. Mammals that fed on spawning kokanee or carcasses along McDonald Creek, including grizzly bears, coyotes, mink, and river otters, are now less common along the creek.

Estuaries, where rivers and streams meet tidal influence and enter the ocean, act as traps for many of the nutrients washed from watersheds. Some species of Pacific salmon typically spawn near saltwater, even beginning within the upper reaches of estuaries. Spawning very often occurs at high densities. The effect of salmon carcasses on the nutrient dynamics and trophic productivity of estuarine systems is just beginning to be examined, Kline et al.[248] reported that approximately 30,000 pink salmon spawned within 1.2 km of the estuary of Sashin Creek in southeast Alaska. In southwestern Washington, the 5 km of Kennedy Creek accessible to anadromous fishes has supported as many as 80,000 spawning chum salmon (WDFW unpub. data). Using the size and body composition information previously cited (under the Elwha River discussion), we estimate that this peak escapement to Kennedy Creek delivered approximately 398 mt of salmon flesh containing 12 mt of nitrogen and 1.4 mt of phosphorus to 0.075 km^2 of stream channel area (5 km with an average channel width of 15 m). The nutrient loading per unit of channel area would be 160 mt/km^2 nitrogen and 18.7 mt/km^2 phosphorus. Many of these salmon become carcasses within the stream system or their dissolved nutrients may be carried to the estuary during freshets. Therefore, a nutrient link may function between adult salmon carcasses and juvenile salmon rearing in the estuary. For example, Fujiwara and Highsmith[159] found elevated stable isotope ratios of nitrogen in *Ulva* sp., an estuarine macroalga, following the decomposition of salmon carcasses in Seldovia Bay, Alaska. *Ulva* spp. are a major food source for harpacticoid copepods, which in turn are a preferred prey of juvenile chum salmon fry in the estuary. Thus the contribution of nutrients and organic matter from salmon carcasses may be a substantial source in some systems, and may be a key factor in promoting estuarine productivity. The importance of estuaries as nursery zones for anadromous salmon along the Pacific Northwest coast is well documented;[194, 196, 289, 355, 397, 454] the role carcasses play in maintaining productivity of these systems may be critical in supporting the health of salmon populations.[159]

The role salmon populations play as a key vector in the recycling of energy and nutrients inland from the North Pacific Ocean to aquatic and the terrestrial ecosystems is now gaining recognition as a critical component of ecosystem function.[348, 350, 463, 464] River and lake fertilization with inorganic nutrients has been undertaken with ecosystem restoration in mind in some British Columbia systems;[15, 16, 476, 496] however, artificially supplementing inorganic nutrients may not fully mitigate for the loss of the multiple pathway flow of energy and materials provided by naturally spawning salmon.

Salmon as Vectors in Broader Nutrient Cycling

The flux of nutrients is essential for the continuity and stability of any living system,[412] and nutrients provide a link between aquatic and terrestrial ecosystems.[61, 317] Biological vectors through which materials and energy are transported include migration of animals (i.e., mammals, birds, fish) that carry nutrients across ecosystem boundaries.[61, 349, 379, 463, 464] Therefore the role and importance of salmon in the freshwater and terrestrial ecosystems can be recognized within the context of broader nutrient cycling, and spawning migrations of salmon represent an obvious example of this process. Other means of moving nutrients upstream, such as the emergence of the adult stages of insects and meteorological vectors, are considered to be relatively insignificant compared to anadromous fish.[278] The nutrient-subsidy contributed by salmon in the North Pacific could potentially serve as a model for a more general, and global, aspect of nutrient circulation. Such studies may be supplemented by an assessment of the extent to which migratory birds, which travel long distances between boreal/subboreal zones and temperate and/or tropical regions, contribute to similar effects.

Using the discipline known as *Resource Physics*, Tsuchida[524] has explained how the hydrologic cycle and the convection of air are able to keep the earth in a low-entropy state.[350] The hydrologic cycle is in turn driven by solar thermal energy and the earth's gravity. Subsequently, inorganic salts (especially nitrates and phosphates), which are essential to formation and activities of animals, plants, and microorganisms, are eventually washed downstream.[525] Ultimately, these salts are dissolved in river water and transported to the ocean, where they attain the highest and most uniform concentration below the depth of 1,000 m, largely free from biological consumption in the absence of photosynthesis. However, due in part to ocean currents (local upwelling), these nutrients eventually find their way back to the surface waters. This occurs in northern temperate or subpolar oceans by the effective vertical mixing of seawater due to the approximation of water temperature between deep and shallow water, primarily during colder seasons. Finally, uptake of these salts by marine plants near the surface, where photosynthesis is possible during the warmer seasons, allows a means through which other animals are able to derive and transport the nutrients inland. For example, Tsuchida[525] speculates that in coastal regions, some birds will carry nutrients back to the land after deriving them from the consumption of marine organisms. Bird excrement is a fertilizer rich in inorganic matter, especially phosphates, as evidenced by the material deposits (Peruvian bird guano) on tropical sea islands. Sibatani[463, 464] points out that another, arguably more significant way that nutrients are transported back onto the land is by anadromous fish swimming up, spawning, and dying in the many rivers of the Asian and North American continents. Murota et al.[351] discussed how migratory fish move ocean nutrients inland and benefit the Siberian forest and its various wildlife inhabitants (Figure 9).

In the Edo era, some people in Japan started to notice that forests along seashores or rivers attracted fish towards them. It was considered that a forest could give benefits to fish in the forms of shadow as shelter, nutrients, and so on. This consideration remained in

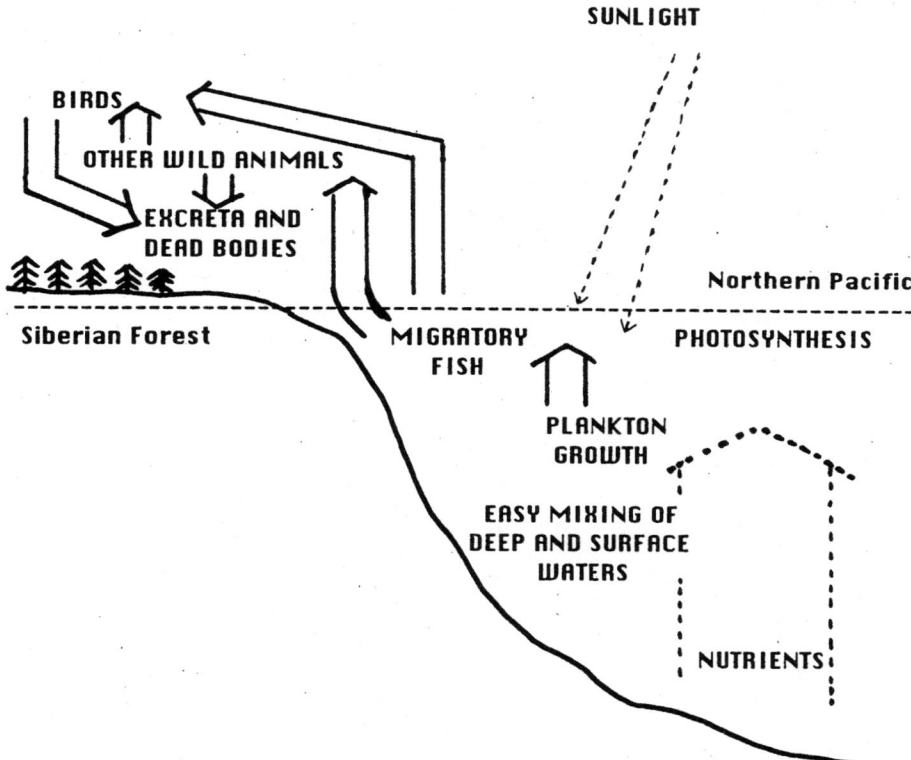

Figure 9. Anadromous fish and the Siberian forest.[351]

the minds of people living near waterfronts or forests after the Meiji Restoration (1868). When the first forest act was introduced at the beginning of the twentieth century, it contained the article ordering the conservation of uo-tsuki-rin, which literally meant "fish-attaching forest.'"This article is still valid in the present-day forest act of Japan.[348]

The theories of Sibatani,[463, 464] Murota,[348, 349] and Tsuchida[524, 525] cause one to reflect on the once bountiful salmon runs of the Columbia River. Before Europeans settled the Pacific Northwest, salmon and steelhead had access to over 20,000 km of main river and tributaries in the Columbia River basin.[379] The annual Columbia River salmon and steelhead run size was estimated to range between 8.2 and 16.3 million fish.[379] Using an average weight of 6.75 kg per salmon, spawning populations could have potentially contributed between 55,350 and 110,025 mt of nutrient annually. This amounts to potential average annual contributions of salmon nutrient in the anadromous area of the Columbia River on the order of 2.76 to 5.50 mt/km. But what of the native people of the Columbia River? Many traveled long distances to partake in the catch and consumption of salmon, and in doing so participated in the further cycling of nutrients over this vast watershed, and beyond. Some tribes of the upper Columbia were known to cross the Continental Divide to trade dried salmon for buffalo hides,[379] thus providing an additional mechanism for transfer of marine-derived nutrients to the inland land mass. Salmon, wildlife, and humans, therefore, may be the most prominent carriers of ocean nutrients to inland ecosystems.

With an Ecosystem Perspective in Mind— Where Do We Go from Here?

The need for an ecosystem approach to salmon management has inevitably grown in Washington and Oregon.[305, 365, 499] Terrestrial ecologists have recognized the influence human uses (and thereby disturbances) have had on terrestrial and aquatic ecosystems of large scales.[151, 295, 545] Some investigators have explored the significance of the salmon as a key ecological process vector on the broadest scales of energy and nutrient transport.[348, 464, 559] The use of salmon as an indicator of ecosystem health, or "the canary" is complicated by the fact that this canary is a food resource in high demand by humans.[464]

The magnitude of the role of salmon populations as keystone vectors in energy and nutrient cycling inland from the Northern Pacific to freshwater and terrestrial ecosystems is now gaining recognition as a critical component to an overall understanding of ecosystem functions.[92, 348, 349, 464] Application of this knowledge to understand the cumulative impacts of human land use practices and fisheries exploitation on ecosystem functions is only just beginning. New tools are necessary to make management actions toward regaining lost productivity and biodiversity objectives.[305, 365, 499]

Greater understanding of the hydrologic cycle has helped us to improve land and water uses to better adapt to our environment. Better understanding of geologic processes including the role of hydrogeology in shaping the landscape and controlling the rates at which sediments, and small and large organic debris cycle through watersheds is also leading to changes in views on land and water uses.[295, 434] Understanding nutrient cycling processes, pathways and the effects of nutrient loading has helped in managing water and sediment quality problems.[245, 435, 538] In much the same manner, an understanding of nutrient spiraling and cycling in streams[369] and the keystone role of anadromous salmon[266, 339] will be valuable, if not essential, to understand how we have affected coastal ecosystems and the processes that support them. Such an understanding will lead us to better identification of those management action options we may take to achieve desired future conditions.

Maser et al.[295] have added substantially to the literature on downstream and seaward movement of materials, energy and nutrient transport processes to and through aquatic systems. Their report compiled and presented a wealth of information on the inputs, fates, and effects of forest debris, particularly LWD, in freshwater, estuarine, and marine ecosystems. In essence, physical and chemical processes were well described, now the biological is considered. Adding estimates of the upstream flow of energy and nutrients via salmon to existing watershed processes literature, will provide a more complete picture of the large scale and long-term energy and nutrient cycles for entire watersheds. An understanding of the overall cycles and levels of productivity at this scale will provide the context for interpretation of local level trends in production and materials transport, utilization and storage. It has been well established that aquatic systems have metabolisms that function based upon physical processes, rates of loadings of materials, energy, and nutrients, rates of primary and secondary production, and resulting changes in standing stocks of fishes.

As a keystone species to the productivity and biodiversity of the ecosystems of the North Pacific basin, anadromous salmon closely link the management issues of the forests, the flood plains and lowlands, the estuaries and nearshore areas, and the ocean domains as a continuum. Materials, energy and nutrient budget analyses on an appropriate time scale will be necessary to estimate the potential effects that past land use practices have had on production and discern which ones persist. Budgets for the North Pacific Basin can theoretically be calculated in a similar manner to that already used on smaller systems. Then we can begin to use resulting information to provide a context for management decisions in various disciplines and forums that will affect materials, energy, and nutrient flows and stocks at various scales. To address the disconnects apparent in the current state of terrestrial and aquatic systems will require some estimate of many factors, including:

1. What is the status of the nutrient capital and rates of transport within the domains and the basin as a whole (nutrient budget)?
2. What is the range of the nutrient and materials capital and rates of transport (how does the current budget relate to the known ranges of standing stocks and rates of metabolism and transport)?

3. How have humans altered the nutrient budget?
4. What adaptive management actions might be warranted and feasible to push the terrestrial and aquatic systems toward the identified goals?
5. Are there some measures to employ in the interim until stocks of salmon can be restored?
6. What are the desired future conditions?

Early European settlement of the eastern North Pacific Rim territories provides many accounts of heavy extractions of forest resources.[70, 110a, 276, 295, 348, 365, 464, 554] Logging of the forests had the obvious effect of short-circuiting the prior cycles that supplied woody debris to freshwater, estuarine, and marine ecosystems. In recent decades, mechanized logging equipment combined with highly efficient ("clean") logging practices and slash burning to prevent wildfires and accelerate re-growth of planted conifers has further resulted in very little debris left on the site or entering streams. The store of vast quantities of nutrients in the form of the decayed woody debris and trapped detritus that serves as substrate for long-term nitrogen fixation and retention no longer exists. Large woody debris may continue to enter stream corridors, but not necessarily in the amount, size, and quality that it did in the past, thereby decreasing potential to provide stream structure and organic matter to food-webs.[299, 416]

The long-term loss of the function of LWD as a primary component of the floodplain waterways, resulting from land and channel clearing, may well be more significant than the loss of the wood material itself. The resistance of the abundant woody material slowed the flows of water, sediment, and smaller debris; resulting in very complex valley floor stream-ways composed of multiple highly sinuous channels that were generally well connected to off-channel wetland systems by sloughs and high water channels. Put this liquidation of natural resource capital into the context of long-term climate cycles[189] and Maser et al.'s[295] long-term geologic and successional cycles, and one can begin to formulate management goals for ecological processes.

A combination of development activities have diverted water from, shortened, straightened, cleared, dammed, diked, drained, filled and polluted the habitats of salmon. Early in settlement, logging and splash damming, land clearing for agriculture, and channel clearing for navigation appear to have had the most pronounced effects. The development and consumptive utilization of natural resources was highly dependent upon water-borne transportation and patterns of impacts are reflective of navigation-centered commerce. Continuing development for agriculture, industry, and urban growth has resulted in further losses through conversions to other uses. Impacts have become more pervasive throughout the landscape as transportation infrastructure and vehicle capabilities have increased. Releases of persistent toxins has contaminated coastal sediments. The available freshwater and estuarine salmon habitats also continue to be degraded by ongoing land uses.[174, 245, 295, 365, 454, 526, 554]

Desired Future Condition

The inability of the various interest groups to resolve conflicting and agreed-to goals or conditions has been identified as the fatal gap in salmon management.[54, 365] Describing the desired future condition of the terrestrial and aquatic ecosystems and identifying and defining human actions can influence movement of ecosystems toward those conditions. The institutional changes suggested by Lichatowich[277] will be necessary to implement the long-term management approach required. Some of those institutions are known and have been in use during other eras where human land and resource uses appeared to have been indefinitely sustainable.[464, 554] The essence of the Pacific salmon is that it links together what humans generally consider distant, diverse, and separate ecosystems and long time spans.

In spite of the high potential commodity value of the harvest and our knowledge that the Pacific salmon is a key driver for the biodiversity and productivity of the northern Pacific basin, we have not developed the strategies for effective long-term management of the resource's health. The widespread use of salmon hatcheries has also significantly reduced the amount of salmon carcass nutrients available for the aquatic food webs. Modern human culture has not fully adapted to the environment of the northern Pacific Basin. The environment of our region is showing the signs of stress all around us which indicates the failure of our past and current approaches. If humans are to thrive at present population levels within portions of the basin, we will need to look more at multiple scales of ecosystem management and at integrating that management to sustain productivity over very long time frames.

The challenge, then, for the people of Washington and Oregon and the whole North Pacific Rim is to recognize that the character and the health of the northern Pacific Basin sub-ecosystems depend upon the material and energy flow processes that link them. The flows of energy and of organic and inorganic materials among the various aquatic and upland ecosystems determines the productivity of each component and of the whole. In short, the whole is greater than the sum of the parts, symbiosis on a grand scale. Understanding the biological processes and the impacts of our management practices upon them will be necessary for our long-term (measured in generations) success in the region. To do that we will need to look beyond the plants, animals, and the habitats of a given smaller-scale ecosystem to the processes that link them together, perhaps into the large-scale ecosystem of the anadromous salmon. The flow of sediments, woody debris, detritus, and nutrients through a watershed determines the character and productivity of the entire watershed, estuary, the near shore zone and even the domains of the northern Pacific Ocean. The flow of energy and nutrients back upstream via the Pacific salmon and the ability of the watershed to retain them, in large measure determines the productivity of the entire watershed.

Acknowledgments

We thank the Washington Departments of Fish and Wildlife (WDFW) and Natural Resources (WDNR) for supporting this project. We appreciate greatly the thoughtful reviews given by Peter Bisson of the United States Forest Service; Jim Lichatowich, Private Consultant; Hal Michael, Jr., of the WDFW; Tom Mumford of WDNR; Pat Slaney of the British Columbia Watershed Restoration Program; and Mary F. Willson of the Great Lakes Program of The Nature Conservancy. We also thank Luis Prado of WDNR and Darrell Pruett of the WDFW for graphic art work on some of the figures in this report, and Billie Wyckoff of WDNR for her help.

Literature Cited

1. Adams, C. 1998. Determining the effects of marine derived nutrients on growth of juvenile coho salmon (*Oncorhynchus kisutch*) using stable isotopes. The 1998 Annual General Meeting of the North Pacific International Chapter of American Fisheries Society Meeting, held March 18-20, 1998 at Alderbrook, WA.

2. Agee, J. K. 1988. Successional dynamics of forest riparian zones. Pages 31-43 in: K. J. Raedeke, editor. Streamside Management: Riparian Wildlife and Forestry Interactions. Contrib. No. 59, Institute of Forest Resources, University of Washington, Seattle, WA.

3. Ainley, D. G. C. S. Strong T. M. Penniman, and R. J. Boekelheide. 1990. The feeding ecology of Farallon seabirds. Pages 51-127 in: D. A. Ainley and R. J. Boekelheide, editors. Seabirds of the Farallon Islands, Ecology, Dynamics, and Structure of an Upwelling-System Community. Stanford University Press, Stanford, CA.

4. ———, and D. W. Anderson. 1981. Feeding ecology of marine cormorants in southwestern North America. Condor. 83:120-131.

4a. ———, and G. A. Sanger. 1979. Trophic relations of seabirds in the northeastern Pacific Ocean and Bering Sea. Pages 95-122 in J.C. Bartonek and D.N. Nettleship, editors. Conservation of marine birds of northern North America. U.S. Department of Interior, Fish and Wildlife Service, Wildlife Research Report No. 11.

5. Akande, M. 1972. The foods of feral mink (*Mustela vison*) in Scotland. Journal of Zoology 167:475-479.

6. Alcorn, G.D. 1958. Nesting of the Caspian Tern in Gray's Harbor, Washington. Murrelet 39:19-20.

7. Alexander, G. R. 1979. Predators of fish in cold water streams. Pages 153-170 in: Predator-Prey Systems in Fisheries Management. Sport Fishing Institute, Washington, D.C.

8. Alexander, G. 1977. Food of vertebrate predators on trout waters in north-central lower Michigan. Michigan Academician 10:181-195.

9. Allee, B.A. 1982. The role of interspecific competition in the distribution of salmonids in streams. Pages 111-122 in: E. L. Brannon and E. O. Salo, editors. Proceedings of the Salmon and Trout Migratory Behavior Symposium. 1981, First International Symposium. School of Fisheries, University of Washington, Seattle, WA.

10. Allen, K. R. 1951. The Horokiwi Stream: a study of a trout population. New Zealand Dept. Fish. Bull. 10:1-238.

11. Anderson, N. H. 1992. Influence of disturbance on insect communities of Pacific Northwest streams. Hydrobiologia 248:79-92.

12. Anderson, M. E. 1977. Aspects of the ecology of two sympatric species of *Thamnophis* and heavy metal accumulation within the species. M.S. Thesis, University of Montana, Missoula, MT.

13. Antonelis, G. A. Jr. and M. A. Perez. 1984. Estimated annual food consumption by northern fur seals in the California current. CalCOFI Rep. 25:135-145.

14. Antonelli, A. L., R. A. Nussbaum, and S. D. Smith. 1972. Comparative food habits of four species of stream-dwelling vertebrates (*Dicamptodon ensatus*, *D. copei*, *Cottus tenuis*, *Salmo gairdneri*). Northwest Science 46:277-289.

15. Ashley, K., L. C. Thompson, D. C. Lasenby, L. McEachern, K. E. Smokorowski, and D. Sebastian. 1997. Restoration of an interior lake ecosystem: the Kootenay Lake fertilization experiment. Vol. 32: 295-323.

16. ———, and P.A. Slaney. 1997. Accelerating recovery of stream, river and pond productivity by low-level nutrient replacement. Pages 13-1 to 13-24 in: P.A. Slaney and D. Zaldokas, editors. Fish Habitat Rehabilitation Procedures. Watershed Restoration Program, Ministry of Environment, Lands and Parks, Vancouver, BC, Canada.

17. Ayles, G. B., J. G. I. Lark, J. Barica, and H. Kling. 1976. Seasonal mortality of Rainbow trout (Salmo gairdneri) planted in small eutrophic lakes of central Canada. Journal of the Fisheries Research Board of Canada 33:647-655.

18. Baird, P. H. 1990. Influence of abiotic factors and prey distribution on diet and reproductive success of three seabird species in Alaska (USA). Ornis Scandinavica . 21:224-235.

19. Baird, R. 1991a. W. The Risso's dolphin in British Columbia. Victoria Naturalist (Victoria, BC, Canada) 47:6-7.

20. ———. 1991b. Optimal foraging and intraspecific competition in the tufted puffin. Condor 93:503-515.

21. Baker, R. C., F. Wilke, and C. H. Baltzo. 1970 The Northern Fur Seal. United States Dept. of the Interior. U. S. Fish and Wildlife Service. Circular No. 336. Washington D.C.

22. Balcomb, K. C., J. R. Boran, and S. L. Heimlich. 1982. Killer whales in Greater Puget Sound. Pages 681-685 in: International Whaling Commission, Report of the Commission 32.

23. ———, J. R. Boran, J. R. Osborne, and N. J. Haenel. 1980. Observations of killer whales (Orcinus orca) in Greater Puget Sound, State of Washington. INTIS PB80-224728, U. S. Dept. of Comm., Springfield, VA.

24. Baltz, D. M., and G.V. Morejohn. 1977. Food habits and niche overlap of seabirds wintering on Monterey Bay, California. Auk 94:526-543.

25. Banci, V. 1987. Ecology and behavior of wolverine in Yukon. M.S. Thesis. Simon Fraser University. 178 pp.

26. Banfield, A. W. F. 1974. The mammals of Canada. University of Toronto Press, Toronto, Canada. 438 pp.

26a. Barr, J.F. 1973. Feeding biology of the common loon (Gavia immer) in oligotrophic lakes of the Canadian shield. Ph.D. dissertation, University of Guelph, Ontario, 192pp.

27. Bayer, R. D. 1986. Seabirds near an Oregon estuarine salmon hatchery in 1982 and during the 1983 El Niño. Fish Bulletin 84:279-286.

28. ———. 1986a. Nearshore flights of seabirds past Yaquina Estuary, Oregon, during the 1982 and 1983 summers. Western Birds 16:169-173.

29. ———. 1986b. Breeding success of seabirds along the mid-Oregon coast concurrent with the 1983 El Niño. Murrelet 67:23-26.

29a. ———. 1989. The cormorant/fisherman conflict in Tillamook County, Oregon. Studies in Oregon Ornithology, No. 6. Gahmken Press, Newport, OR. 101 pp.

30. Beach, U. S. 1937. The destruction of trout by fish ducks. Transactions of the American Fisheries Society 66:338-342.

31. Beach, R., A. C. Geiger, S. J. Jeffries, S. D. Treacy and B. L. Troutman. 1985. Marine mammals and their interactions with fisheries of the Columbia River and adjacent waters, 1980-1982. Third Annual Report. Washington Department of Wildlife, Wildlife Management Division. Olympia, WA.

32. Beamish, R. J., and R. Boullion. 1993. Pacific salmon production trends in relation to climate. Canadian Journal of Fisheries and Aquatic Sciences 50:1002-1016.

33. Beechie, T. J., and T. H. Sibley. 1990. Evaluation of the TFW stream classification system: stratification of physical habitat area and distribution. Timber/Fish/Wildlife Final Report TFW-16B-89-006, Washington Department of Natural Resources, Olympia, WA.

34. Bell, D.T., and S. K. Sipp. 1975. The litter stratum in the streamside forest ecosystem. Oikos 26: 391-397.

35. Ben-David, M., R. W. Flynn, and D.M. Schell. 1997. Annual and seasonal changes in diets of martens: evidence from stable isotope analysis. Oecologia 111:280-291.

36. Ben-David, M.T. 1997. Timing of reproduction in wild mink: the influence of spawning Pacific salmon. Canadian Journal of Zoology 75:376-382.

37. ———, A. Hanley, D. R. Klein, and D. M. Schell. 1997. Seasonal changes in diets of coastal and riverine mink: the role of spawning Pacific salmon. Canadian Journal of Zoology 75:803-811.

38. Benda, L., T. J. Beechie, R. C. Wissmar, and A. Johnson. 1992. Morphology and evolution of salmonid habitats in a recently deglaciated river basin, Washington State, USA. Canadian Journal of Fisheries and Aquatic Sciences 49:1246-1256.

39. Bent, A. C. 1921. Life histories of North American Gulls and Terns. United States National Museum. Bulletin No.1131:65-71.

39a. ———. 1926. Life histories of North American marsh birds. U.S. National Museum Bulletin No. 135.

40. Berman, C., and T. P. Quinn. 1991. Behavioural thermoregulation and homing by spring chinook salmon, Oncorhynchus tshawytscha Walbaum, in the Yakima River. Journal of Fish. Biology 39:301-312-

41. Beschta, R. L., R. E. Bilby, G. W. Brown, L. B. Holtby, and T. D. Hofstra. 1987. Stream temperature and aquatic habitat: fisheries and forestry interactions. Pages 191-232 in: E. O. Salo and T. W. Cundy, editors. Streamside Management: Forestry and Fishery Interactions. Institute of Forest Resources Contribution Number 57. University of Washington, Seattle, WA.

42. Bilby, R. E. 1988. Interactions between terrestrial and aquatic systems. Pages 13-29 in: K. J. Raedeke, editor. Streamside Management: Riparian Wildlife and Forestry Interactions. Contrib. No. 59, Inst. For. Res., University of Washington, Seattle, WA.

43. ———. 1984. Post-logging removal of woody debris affects stream channel stability. Journal of Forestry 82: 609-613.

44. ———, and J. W. Ward. 1989. Changes in characteristics and function of woody debris with increasing size of streams in western Washington. Transactions of American Fisheries Society 118:368-378.

45. ———, and ———. 1991. Characteristics and function of large woody debris in streams draining old-growth, clear-cut, and second-growth forests in southwestern Washington. Canadian Journal of Fisheries and Aquatic Sciences 48:2499-2508.

46. ———, and P.A. Bisson. 1992. Allochthonous versus autochthonous organic matter contributions to the trophic support of fish populations in clear-cut and old-growth forested streams. Canadian Journal of Fisheries and Aquatic Sciences 49:540-551.

47. ———, B.R. Fransen and P.A. Bisson. 1996. Incorporation of nitrogen and carbon from spawning coho salmon into the trophic system of small streams: evidence from stable isotopes. Canadian Journal of Fisheries and Aquatic Sciences 53:164-173.

48. ———, ———, ———, and J. W. Walter. 1998. Response of juvenile coho salmon (Oncorhynchus kisutch) and steelhead (Oncorhynchus mykiss) to the addition of salmon carcasses to two streams in Southwestern Washington, U.S.A. Canadian Journal of Fisheries and Aquatic Sciences 55:1909-1918.

49. ———, and P.A. Bisson. 1998. Function and distribution of large woody debris, pages 324-346 In: R. J. Naiman and R. E. Bilby, editors. River Ecology and Management: Lessons From The Pacific Coastal Ecoregion. Springer-Verlag. New York City, NY.

50. ———, R.E. Bilby, M.D. Bryant, C.A. Dolloff, G.B. Grette, R.A. House, M.L. Murphy, K.V. Koski, and J.R. Sedell. 1987. Large woody debris in forested streams in the Pacific Northwest: past, present, and future, pages 143-190 in: E.O. Salo and T.W. Cundy, editors. Streamside Management: Forestry and Fishery Interactions. Contribution Number 57, Institute of Forest Resources, University of Washington, Seattle, WA.

51. ——— and G. E. Davis. 1976. Production of juvenile chinook salmon, Oncorhynchus tshawytscha, in a heated model stream. NOAA Fisheries Bulletin 74:763-774.

52. ———, and J. R. Sedell. 1984. Salmonid populations in streams in clearcut vs. old-growth forests of western Washington. Pages 121-130 in: W. R. Meehan, T. R. Merrell, Jr., T.A. Hanley, editors. Proceedings of a Symposium: Fish and Wildlife Relationships in Old-growth Forests. April, 1982, Juneau, AK.

53. ———, T. Quinn, G. H. Reeves, and S.V. Gregory. 1992. Best management practices, cumulative effects, and long-term trends in fish abundance in Pacific Northwest river systems. Pages 223-265 in: R. J. Naiman, editor. Watershed management, Balancing Sustainability and Environmental Change. Springer-Verlag. New York, NY.

54. ——— G. H. Reeves, R. E. Bilby, and R. J. Naiman. 1997. Watershed management and Pacific salmon: Desired future conditions. Pages 447-474 in: Pacific Salmon and Their Ecosystems: Status and Future Options. Chapman and Hall, New York, NY.

55. ———— and R. E. Bilby. 1998. Organic matter and trophic dynamics. In: R. J. Naiman and R. E. Bilby, editors. River Ecology and Management: Lessons From the Pacific Coastal Ecoregion. Springer-Verlag. New York, NY.

56. Bjornn, T. C., and D. W. Reiser. 1991. Habitat requirements of salmonids in streams. Pages 83-138 in: W. R. Meehan, editor. Influences of Forest and Rangeland Management on Salmonid Fishes and Their Habitats. American Fisheries Society Special Publication No.19.

57. ————. 1971. Trout and salmon movements in two Idaho streams as related to temperature, food, stream flow, cover, and population density. Transactions of American Fisheries Society 100:423-438.

57a. Bledsoe, L. J. et al. The Puget Sound runs of salmon: An examination of the changes in run size since 1896. Pages 50-61 in: C. D. Levings et al., editors. Proceedings of the National Workshop on Effects of Habitat Alteration on salmonid stocks. Canadian Special Publication Fisheries and Aquatic Science. No. 105.

58. Bollen, W. B., and K. C. Lu. 1968. Nitrogen transformations in soils beneath red alder and conifers. Pages 141-148 in: J. M. Trappe, J. F. Franklin, R. F. Tarrant, and G. M. Hansen, editors. Biology of Alder. USDA Forest Service. Portland, OR.

59. Booth, D. B., and C. R. Jackson. 1997. Urbanization of aquatic systems: Degradation thresholds, stormwater detection, and the limits of mitigation. Journal of American Water Resources Association 33:1077-1090.

60. Booth. D. E. 1991. Estimating prelogging old-growth in the Pacific Northwest. J. For. 89:25-29.

61. Bormann, F. H., and G. E. Likens. 1967. Nutrient cycling. Science 155:424-429.

62. Brickell, D. C., and J. J. Goering. 1970. Chemical effects of salmon decomposition on aquatic ecosystems. Pages 125-138 in: R. S. Murphy, editor. Proceedings of the Symposium on Water Pollution Control in Cold Climates. U. S. Government Printing Office, Washington, DC.

63. Briggs, D. T. and C. W. Davis. 1972. A study of predation by sea lions on salmon in Montery Bay. California Fish Game 58:37-43.

64. Brinson, M. M., B. L. Swift, R. C. Plantico, and J. S. Barclay. 1981. Riparian ecosystems: Their ecology and status. USDI Fish and Wildlife Service, Biological Services Program. Kearneysville, WV.

65. Brittell, J. D. Sweeney and S. T. Knick. 1979. Washington bobcats: diet, population dynamics and movement. National Wildlife Federation Scientific and Technical Service No. 6: 107-110.

65a. Brodeur, R. D. 1989. Neutonic feeding by juvenile salmonids in coastal waters of the Northwest Pacific. Canadian Journal of Zoology 67:1995-2007.

65b. ————, and W. G. Pearcy. 1990. Trophic relations of juvenile Pacific salmon off the Oregon and Washington coasts. Fisheries Bulletin 88:617-630.

66. Brooks, A. 1928. Does the marbled murrelet nest inland? Murrelet 9:68.

67. Brown, R. F., S. Riemer, and S. J. Jeffries. 1995. Food of pinnipeds collected during the Columbia River area salmon gillnet observation program, 1991-1994. Oregon Dept. of Fish and Wildlife, Wildlife Diversity Program, Technical Report #95-6-01 (Available From Oregon Dept. of Fish and Wildlife, 2501 SW First, Portland, OR 97201).

67a. Brown, H. A., R. B. Bury, D. M. Darda, L. V. Diller, C. R. Peterson, and R. M. Storm. 1995a. Reptiles of Washington and Oregon. Seattle Audubon Society, Seattle, WA. 176pp.

68. Brown, R. F., and B. R. Mate. 1983. Abundance, movements, and feeding habits of harbor seals, *Phoca vitulina*, at Netarts and Tillamook Bays, Oregon. U. S. Fish and Wildlife Service Fish. Bull. 81(2):291-301.

69. Brown, E. R. 1985. Riparian zones and freshwater wetlands. Management of wildlife and fish habitats in forests of western Oregon and Washington, Part I—Chapter Narratives. USDA, Forest Service.

70. Brown, B. 1982. Mountain in the clouds: A search for wild salmon. Simon and Schuster, New York, NY.

71. Brown, S., M. M. Brinson, and A. E. Lugo. 1979. Structure and function of riparian wetlands. Pages 17-31 In: R.R. Johnson and J. F. McCormick, editors. Strategies for protection and management of floodplain wetlands and other riparian ecosystems. USDA For. Ser. Gen. Tech. Rep. WO-12.

72. Browne, P., R. L. DeLong, H. R. Huber, and J. L. Laake. 1997. Pinniped predation on endangered salmonids in Washington and Oregon: harbor seal food habits on the Columbia River. Pages 109-118 in: P. S. Hill and D. P. DeMaster, editors. Marine Mammal Protection Act and Endangered Species Act Implementation Program 1996. AFSC Processed Report 97-10.

73. Burcham, J. S. 1904. Notes on the habits of the water ousel (*Cinclus mexicanus*). Condor. 6:50.

74. Burger, A. E, R. P. Wilson, D. Garnier, and M. P. T. Wilson. 1993. Diving depths, diet, and underwater foraging of Rhinoceros Aucklets in British Columbia. Canadian Journal of Zoology 71:2528-2540.

75. Burgess, S. A. and J. R. Bider. 1980. Effects of stream habitat improvements on invertebrates, trout populations, and mink activity. Journal of .Wildlife Management 44:871-880.

76. Burger, A. E. 1993 (1994). Mortality of seabirds assessed from beached-bird surveys in southern British Columbia. Canadian Field-Naturalist 107:164-176.

77. Burgner, R. L. 1991. The life history of sockeye salmon (*Oncorhynchus nerka*). Pages 1-118 in: C. Groot and L. Margolis, editors. Pacific Salmon Life Histories. University of British Columbia Press, Vancouver, BC, Canada.

78. Bury, R. B., and P. S. Corn. 1988. Responses of aquatic and streamside amphibians to timber harvest: A review. Pages 165-188 in: K. J. Raedeke, editor. Streamside Management: Riparian Wildlife and Forestry Interactions. Contribution No. 59, Institute of Forest Resources, University of Washington, Seattle, WA.

79. Busby, P. J., T. C. Wainwright, G. J. Bryant, L. J. Lierheimer, R. S. Waples, F. W. Waknitz, and I. V. Lagomarsino. 1996. Status review of west coast steelhead from Washington, Idaho, Oregon, and California U.S. Dept. of Commerce. NOAA Tech. Memo. NMFS-NWFSC-27:261 pp.

80. Bustard, D. R., and D. W. Narver. 1975. Aspects of the winter ecology of juvenile coho salmon (*Oncorhynchus kisutch*) and steelhead trout (*Salmo gairdneri*). Canadian Journal of Fisheries and Aquatic Sciences. 32: 667-80.

81. Butler, R. L. 1991. Trout as predator. Pages 65-72 in: J. Stolz and J. Schnell, editors. Trout, the Wildlife Series. Stackpole Books. Harrisburg, PA.

82. Butler, R. W. 1973. Trumpeter swans eating salmon. Vancouver Nat. Hist. Soc. Discovery (New Ser.). 2:120.

83a. Brodeur, R. D. 1989. Neutonic feeding by juvenile salmonids in coastal waters of the Northeast Pacific. Canadian Journal of Zoology 67:1995-2007.

83b. ————, and W. G. Pearcy. 1990. Trophic relations of juvenile Pacific salmon off the Oregon and Washington coasts. Fisheries Bulletin 88:617-630.

84. California Department of Fish and Game. 1993. Final environmental document regarding bear hunting. Resource Agency, California Department of Fish and Game, Sacramento, CA.

85. Campbell, R. W. and N. K. Dawe. 1990. The birds of British Columbia, Volume 1, Nonpasserines: Loons through waterfowl. Royal British Columbia Museum, Victoria, BC, Canada.

85a. ————, ————, I. McTaggart-Cowan, J.M. Cooper, G.W. Kaiser, and M.C.E. McNall. 1990. The birds of British Columbia, volume III: Passerines, flycatchers through vireos. Royal British Columbia Museum, Victoria, BC, Canada.

86. Carlson, A. 1991. Characterization of riparian management zones and upland management areas with respect to wildlife habitat.. Washington Timber/Fish/Wildlife Rep. T/F/W-WLI-91-001. Washington Department of Natural Resources, Olympia, WA.

87. Carman, R. E., C. J. Cederholm, and E. O. Salo. 1984. A baseline inventory of juvenile salmonid populations and habitats in streams in Capitol Forest, Washington 1981-1982. Progress Report for Sampling During 1981 and 1982, University of Washington, Fisheries Research Institute. Report No. FRI-UW-8416, Seattle, WA.

88. Carter, H. R. and S. G. Sealy. 1984. Marbled murrelet mortality due to gill-net fishing in Barkley Sound, British Columbia. Pages 212-220 in: D. N. Nettleship, G. A. Sanger, and P. F. Springer, editors. Marine Birds: Their Feeding Ecology and Commercial Fisheries Relationships. Canadian Wildlife Service.

89. Case, D. J., and D. R McCullough. 1987. White-tailed deer forage on alewife. Journal of Mammalogy. 68:195-197.

90. Cederholm, C. J., and N. P. Peterson. 1985. The retention of coho salmon (Oncorhynchus kisutch) carcasses by organic debris in small streams. Canadian Journal of Fisheries and Aquatic Sciences 42:1222-1225.

91. ———, D. B. Houston, D. L. Cole, and W. J. Scarlett. 1989. Fate of coho salmon (Oncorhynchus kisutch) carcasses in spawning streams. Canadian Journal of Fisheries and Aquatic Sciences 46:1347-1355.

92. ———, Kunze, M. D., Murota, T., and A. Sibatani. 1999. Pacific salmon carcasses: Essential contributions of nutrients and energy for aquatic and terrestrial ecosystems. Fisheries 24 (10):6-15.

93. ———, and E. O. Salo. 1979. The effects of logging road landslide siltation on the salmon and trout spawning gravels of Stequaleho Creek and the Clearwater River Basin, Jefferson County, Washington. University of Washington, Fisheries Research Institute. Report No. FRI-UW-7915. Seattle, WA.

94. ———, and W. J. Scarlett. 1982. Seasonal immigrations of juvenile salmonids into four small tributaries of the Clearwater River, Washington, 1977-198. Pages 98-110 in: E. L. Brannon and E. O. Salo, editors. Proceedings of the Salmon and Trout Migratory Behavior Symposium, June 3-5, 1981. University of Washington, School of Fisheries, Seattle, WA.

95. Chapman, D. W., and T. C. Bjornn. 1969. Distribution of salmonids in streams, with special reference to food and feeding. Pages 153-176 in: T. G. Northcote, editor. Symposium On Salmon and Trout In Streams, H. R. MacMillan Lectures In Fisheries. February 22-24, 1968, the University of British Columbia. Vancouver, BC, Canada.

96. ———. 1962. Aggressive behavior in juvenile coho salmon as a cause of emigration. Canadian Journal of Fisheries and Aquatic Sciences 19:1047-1080.

97. Chi, D. K. 1999. The effects of salmon availability, social dynamics, and people on Black Bear (Ursus americanus) fishing behavior on an Alaskan salmon stream. Ph.D. Dissertation. Utah State University, Logan, UT.

98. Clark, E. E. 1984. Indian legends of the Pacific Northwest. University of California Press. Berkeley, CA. 225 pp.

99. Clemens, W. A. and G. V. Wilby. 1933. Food of the fur seal off the coast of British Columbia. Journal of Mammalogy 14:43-46.

100. Coffman, W. P., and L. C. Ferrington, Jr. 1996. Chironomidae. Pages 635-754 In: R. W. Merritt and K. W. Cummins, editors. Kendall/Hunt Publishing. Dubuque, IA.

101. Commencement Bay Natural Resource Trustees. 1997. Commencement Bay Natural Resource Damage Assessment Restoration Plan and Final Programmatic Environmental Impact Statement. Prepared by the U. S. Fish and Wildlife Service and the National Oceanic and Atmospheric Administration for the Commencement Bay Natural Resource Trustees and Cooperating Agencies. Olympia and Seattle, WA.

102. Conaway, C. H. 1952. Life history of the water shrew (Sorex palustris navigator). The American Midland Naturalist 48:219-248.

103. Cone, J., and S. Ridlington. 1996. The Northwest salmon crisis—a documentary history. Oregon State University Press, Corvallis, OR.

104. Congleton, J. L., S. K. Davis, and S. R. Foley. 1982. Distribution, abundance and outmigration timing of chum and chinook salmon fry in the Skagit salt marsh. Pages 153-163 In: E. L. Brannon and E. O. Salo, editors. Proceedings of the Salmon and Trout Migratory Behavior Symposium, June 3-5, 1981. University of Washington, School of Fisheries, Seattle, WA.

105. Connell, J. H. 1978. Diversity in tropical rain forests and coral reefs. Science 199:1302-1310.

106. Cooper, R., and T. H. Johnson. 1992. Trends in steelhead abundance in Washington and along the Pacific coast of North America. Report #92-90. Fisheries Management Division. Washington Department of Wildlife, Olympia, WA.

107. Cottam, C. 1939. Food habits of North American diving ducks. U.S. Departmen of Agriculture Technical Bulletin No. 643.

108. ———, and Uhler F. M. 1937. Birds in relation to fishes. Wildlife Research and Management Leaflet BS-83.

109. Craig, J. A., and R. L. Hacker. 1940. The history and development of the fisheries of the Columbia River. Bulletin of the Bureau of Fisheries, Vol. XLIX.

110. Cressman, L. S. 1960. Cultural sequences at the Dalles, Oregon. A contribution to Pacific Northwest prehistory. Transactions of the American Philosophical Society, Vol. 50, Part 10.

110a. Crutchfield, J. A., and G. Pontecorvo. 1969. The Pacific salmon fisheries: A study of irrational conservation. The John Hopkins Press, Baltimore, MD.

111. Cummins, K. W., M. A. Wilzbach, D. M. Gates, J. B. Perry, and W. B. Taliaferro. 1973. Shredders and riparian vegetation. Bioscience 39:24-30.

112. Dahm, C. C. 1981. Pathways and mechanisms for removal of dissolved organic carbon from leaf leachate in streams. Canadian Journal of Fisheries and Aquatic Science 38:68-76.

113. Day, J. W., Jr., C. A. S. Hall, W. M. Kemp, and A. Yáñez-Arancibia. 1989. Estuarine Ecology. John Wiley & Sons, New York, NY.

114. Dalquest, W. W. 1948. Mammals of Washington. University of Kansas Publication Volume 2, Museum of Natural History, University of Kansas, Lawrence, KS.

115. Dawson, W. L. 1909. The Birds of Washington. A Complete, Scientific and Popular Account of the 372 Species of Birds Found in the State. Seattle, WA.

116. Death, R. G., and M. J. Winterbourn. 1995. Diversity patterns in stream benthic invertebrate communities: the influence of habitat stability. Ecology 76:1446-1460.

117. DeGange, A. R. 1996. The marbled murrelet: a conservation assessment. General Technical Report PNW-GTR-388. U.S. Department of Agriculture, Forest Service, Pacific Northwest Research Station, Portland, OR.

118. DeShazo, J. J. 1980. Sea-run cutthroat trout management in Washington - An overview. Fisheries Management Division, Washington State Game Dept., Olympia, WA.

119. DeVoto, B. 1981. The journals of Lewis and Clark. Houghton Mifflin Company. Boston, MA.

120. Dolloff, C. A. 1993. Predation by river otters (Lutra canadensis) on juvenile coho salmon (Oncorhynchus kisutch) and Dolly Varden (Salvelinus malma) in southeast Alaska. Canadian Journal of Fisheries and Aquatic Science 50:312-315.

121. Donaldson, I. J., and F. K. Cramer. 1971. Fishwheels of the Columbia. Binfords and Mort, Publishers. Portland, OR.

122. Donaldson, J. R. 1967. The phosphorus budget of Iliamna Lake, Alaska as related to the cyclic abundance of sockeye salmon. Ph.D. Dissertation, University of Washington, School of Fisheries, Seattle, WA.

123. Drummond, H. 1983. Aquatic foraging in garter snakes: a comparison of specialists and generalists. Behaviour 86:1-30.

124. Dunstone, N. 1993. The mink. T & A. D. Poyser Ltd, London.

125. Dzinbal, A., and R. L. Jarvis. 1982. Coastal feeding ecology of harlequin ducks in Prince William Sound, Alaska, during summer. Pages 6-10 in: D. N. Nettleship, G. A. Sanger and P. F. Springer, editors. Marine Birds: Feeding Ecology and Commercial Fisheries Relationships. Proceedings of the Pacific Seabird Group Symposium, Seattle, WA.

126. Eagle, T. C., and J. S. Whitman. 1987. Mink. Pages 614-624 in: M. Novak, J. A. Baker, M. E. Obbard and B. Malloch, editors. Wild Furbearer Management and Conservation in North America. Ministry of Natural Resources, Ontario, Canada.

127. Eastman, D. E. 1996. Response of freshwater fish communities to spawning sockeye salmon (Oncorhynchus nerka). M. S. Thesis, University of Washington, School of Fisheries, Seattle, WA.

128. Edie, B. G. 1975. A census of the juvenile salmonids of the Clearwater River Basin, Jefferson County, Washington, in relation to logging. M. S. Thesis. University of Washington, School of Fisheries, Seattle, WA.

128a. Edmonds, R. L. 1980. Litter decomposition and nutrient release in Douglas-fir, red alder, western hemlock, and Pacific silver fir ecosystems in western Washington. Canadian Journal of Forestry Res. 10:327-337.

129. Ehinger, C. E. 1930. Some studies of the American Dipper or Water Ouzel. Condor 47:487-498.

130. Eipper, A. W. 1956. Differences in the vulnerability of the prey of nesting kingfishers. Journal of Wildlife Management 20:177-183.

131. Elliott, S. T., and R. Bartoo. 1981. Relation of larval *Polypedilum* (Diptera: Chironomidae) to pink salmon eggs and alevins in an Alaska stream. Prog. Fish-Culturist 43: 220-221.

132. Elmore, W. 1992. Riparian responses to grazing practices. Pages 442-457 in: R. B. Naiman, editor. Watershed Management: Balancing Sustainability and Environmental Change. Springer-Verlag. New York, NY.

133. Elphick, C.S., and T.L. Tibbitts. 1998. Greater Yellowlegs (*Tringa melaoleuca*). The Birds of North America, No. 355. A. Poole and F. Gill, editors. The Academy of Natural Sciences, Philadelphia, PA, and The American Ornithologists Union, Washington D.C.

134. Elson, P. F. 1962. Predator-prey relationship between fish-eating birds and Atlantic salmon. Fisheries Research Board Can. Bulletin No 133.

135. Emmett, R. L. 1997. Estuarine survival of salmonids: The importance of interspecific and intraspecific predation and competition. Pages 147-158 in: R.L. Emmett and M.H. Schiewe, editors. Estuarine and Ocean Survival of Northeastern Pacific Salmon. NOAA Technical Memorandum NMFS-NWFSC-29.

136. ———— and M. H. Schiewe. 1997. Estuarine and ocean survival of Northeastern Pacific salmon: Proceedings of the workshop. U.S. Dept. Commerce., NOAA Technical Memo. NMFS-NWFSC-29.

137. Erman, D. C., J. D. Newbold, and K. B. Roby. 1977. Evaluation of streamside bufferstrips for protecting aquatic organisms. California Water Resources Center, Contribution Number 165, University of California, Davis, CA.

138. Erickson, J. 1988. Competition for fish. Headlight Herald. June 4th(A-4).

139. Eriksson, M. O. G. D. Blomqvist M. Hake and O. C. Johansson. 1990. Parental feeding in the Red-throated Diver *Gavia stellata*. Ibis. 132:1-13.

140. Everest, F. H., N. B. Armantrout, S. M. Keller, W. D. Parante, J. R. Sedell, T. E. Nickelson, J. M. Johnston, and G. N. Haugen. 1985. Salmonids. Pages 199-226 in: R. Brown, editor. Management of Wildlife and Fish Habitats in Forests of Western Oregon and Washington. Volume 1. USDA, Forest Service. Pacific Northwest Region.

141. Everitt, R. D., P. J. Gearin, J. S. Skidmore, and R. L. DeLong. 1981. Prey items of harbor seals and California sea lions in Puget Sound, Washington. Murrelet 62:83-86.

142. Farley, L. C. 1980. The Behavioral-Ecology of Swans Wintering in Southeast Alaska. MS Thesis. Idaho State University.

143. Felleman, F. L. J. R. Heimlich-Boran and R. W. Osborne. 1991. The feeding ecology of killer whales (*Orcinus orca*) in the Pacific Northwest. Pages 113-147 in: K. Pryor, and K. S. Norris, editors. Dolphin Societies, Discoveries and Puzzles. University of California Press, Berkeley, CA.

144. Fiscus, C. H. and G. A. Baines. 1966. Food and feeding behavior of Steller and California sea lions. Journal of Mammalogy 47:195-200.

145. Fisher, J. P., and W. G. Pearcy. 1988. Growth of juvenile coho salmon (*Oncorhynchus kisutch*) in the ocean off Oregon and Washington, USA, in years of differing coastal upwelling. Canadian Journal of Fisheries and Aquatic Sciences 45:1036-1044.

146. Foerster, R. E. 1968. The sockeye salmon (*Oncorhynchus nerka*). Fisheries Research Board Can. Bulletin No. 162. Ottawa, Canada.

147. Fitch, H. S. 1984. *Thamnophis couchii*. Catalogue of American Amphibians and Reptiles. 351.1-351.3.

148. ————. 1941. The feeding habits of California garter snakes. California Fish and Game 27:2-32.

149. Fontaine, P. M. M. O. Hammill C. Barrette and M. C. Kingsley. 1994. Summer diet of the harbor porpoise (*Phocoena phocoena*) in the estuary and the Northern Gulf of St. Lawrence. Canadian Journal of Fisheries and Aquatic Science 5:172-178.

150. Forbes, L. C. and K. Simpson. 1982. Behavioral studies of great blue herons at Pender Harbor and Sechelt, British Columbia, in 1980. Unpublished Report., Canadian Wildlife Service, Delta, B.C, Canada.

151. Forest Ecosystem Management Assessment Team (FEMAT). 1993. Forest Ecosystem Management: An Ecological, Economic, and Social Assessment. USDA, Forest Service.

152. Frame, G. W. 1974. Black bear predation on salmon at Olsen Creek, Alaska. Zeitung der Tierpsychol. 35:23-38.

153. Francis, R. C., and S. R. Hare. 1994. Decadal-scale regime shifts in the large marine ecosystems of the Northeast Pacific: A case for historical science. Fish. Oceanogr. 3:279-291.

154. Franklin, J. F. 1992. Scientific basis for new perspectives in forests and streams. Pages 25-72 in: R. J. Naiman, editor. Watershed Management: Balancing Sustainability and Environmental Change. Springer-Verlag, New York, NY.

155. Friesen, W. 1990. Winter dietary studies of juvenile coho salmon (*Oncorhynchus kisutch*) utilizing two enhanced wall-base channels along the Clearwater River in Jefferson County, Washington. Master of Environmental Studies, The Evergreen State College, Olympia, WA.

156. Fraser, J. M. 1972. Recovery of planted brook trout, splake, and rainbow trout from selected Ontario lakes. Journal of Fisheries Research Board Can. 29:129-142.

157. ————. 1974. An attempt to train hatchery reared brook trout to avoid predation by the common loon. Transactions of the American Fisheries Society 103:815-818.

158. French, J. M. and J. R. Koplin. 1977. Distribution, abundance, and breeding status of ospreys in northwestern California. Trans. North American Osprey Research Conference U.S. National Park Service Transactions Proceedings Series No. 2: 223-240.

159. Fujiwara, M., and R. C. Highsmith. 1997. Harpacticoid copepods: Potential trophic link between inbound adult salmon and outbound juvenile salmon. Marine Ecology Progress Series 158:205-216.

160. Fuss, H. J. 1982. Age, growth, and instream movement of Olympic Peninsula coastal cutthroat trout (*Salmo clarki clarki*). M. S. Thesis. University of Washington, School of Fisheries, Seattle, WA.

161. Gabrielson, I. N., and S. G. Jewett. 1940. Birds of Oregon. Oregon State College, Corvallis, OR.

162. Garrett, A. M. 1998. Interstream movements of coastal cutthroat trout (*Oncorhynchus clarki clarki*) in the Clearwater River, Jefferson County, Washington. Master of Environmental Studies. The Evergreen State College, Olympia, WA.

163. Gaston, A. J and S. B. C. Dechesne. 1996. Rhinoceros Auklet, *Cerorhinca monocerata. In:* A. Poole and F. Gill, editors. The Birds of North America No. 212. The Academy of Natural Sciences, Philadelphia, PA, and The American Ornithologists' Union, Washington, D.C.

164. Gearin, P. J., S. R. Melin, R. L. DeLong, H. Kajimura, and M. A. Johnson. 1994. Harbor porpoise interactions with a chinook salmon set-net fishery in Washington state. Pages 427-438 in: W. F. Perrin, G. P. Donovan, and J. Barlow, editors. Gillnets and Cetaceans: Incorporating the Proceedings of the Symposium and Workshop on the Mortality of Cetaceans in Passive Fishing Nets and Traps. La Jolla, CA. October 1990. Whaling Commission Special Issue No.15.

165. ————, R. Pfeifer, S. J. Jeffries, R. L. DeLong, and M. A. Johnson. 1988. Results of the 1986-1987 California sea lion—steelhead trout predation control program at the Hiram M. Chittenden Locks. U. S. Department of Commerce. NWAFC Processed Report 88-30.

166. Glova, G. J. 1986. Interactions for food and space between experimental populations of juvenile coho salmon (*Oncorhynchus kisutch*) and coastal cutthroat trout (*Salmo clarki*) in a laboratory stream. Hydrobiologia 132:155-168.

167. Gould, V. E. 1934. A monograph of the belted kingfisher *Megaceryle alcyon* (Linnaeus).

168. Graf, W. 1949. Observations on the salamander Dicamptodon . Copeia 1:79-80.

169. Graybill, M.R., and J. Hodder. 1985. Effects of the 1982-1983 El Niño on reproduction of six species of seabirds in Oregon. Pages 205-210 in: W. S. Wooster and D. L. Fluharty, editors. El Niño North. Washington Sea Grant Program, University of Washington, Seattle, WA.

170. ————. 1981. Haul out patterns and diet of harbor seals, *Phoca vitulina*, in Coos County, Oregon. M.S. Thesis, University of Oregon, Eugene, OR.

171. Gregory, S.V., F. J. Swanson, and W. A. McKee. 1991. An ecosystem perspective of riparian zones. BioScience 40:540-551.

172. ————. 1980. Effects of light, nutrients, and grazing on periphyton communities in streams. Ph.D. Dissertation. Oregon State University, Corvallis, OR.

173. ————, G. A. Lamberti, D. C. Erman, K. V. Koski, M. L. Murphy, and J. R. Sedell. 1987. Influences of forest practices on aquatic production. Pages 233-255 *In:* E. O. Salo and T. Cundy , editors. Streamside Management—Forestry and Fishery Interactions. College of Forest Resources, University of Washington. Contribution No. 57. Seattle, WA.

174. ——, and P.A. Bisson. 1997. Degradation and loss of anadromous salmonid habitat in the Pacific Northwest. Pages 277-314 in: D. J. Stouder, P.A. Bisson, and R. J. Naiman, editors. Pacific Salmon and their ecosystems. Chapman and Hall Publishers, New York, NY.

175. Grette, G. 1985. The role of large organic debris in juvenile salmonid rearing habitat in small streams. M.S. Thesis, University of Washington, School of Fisheries. Seattle, WA.

176. Grinnell, J. J. Dixon and J. M. Linsdale. 1937. Fur-bearing mammals of California. Their natural history, systematic status, and relations to man. University of California Press, Berkeley, CA.

177. Griswold, G. 1953. Aboriginal patterns of trade between the Columbia Basin and the northern plains. M.A. Thesis, Montana State University. Missoula, MT.

178. Groot, C., and L. Margolis. 1991. Pacific Salmon Life Histories, University of British Columbia Press, Vancouver, BC, Canada.

179. Gross, M. R., R. M. Coleman, and R. M. McDowell. 1988. Aquatic productivity and the evolution of diadromous fish migration. Science 239:1291-1293.

180. Haig-Brown, R. 1946. A river never sleeps. Crown Publishers. New York, NY.

181. Hall, J. D., P.A. Bisson, and R. E. Gresswell. 1997. Sea-run cutthroat trout-biology, management , and future considerations. Proceedings of a Symposium, Reedsport, Oregon. October 12-14, 1995. Published by the Oregon Chapter of the American Fisheries Society. Corvallis, OR.

182. ——. 1986. Notes on the distribution and feeding behavior of killer whales in Prince William Sound, Alaska. Pages 69-83 in: B. C. Kirkevold, and J. S. Lockard, editors. Behavioral biology of killer whales. Alan R. Liss, Inc., New York. NY.

183. Hamilton, A. N. and W. R. Archibald. 1985. Grizzly bear habitat in the Kimsquit River Valley, coastal British Columbia: evaluation. Pages 50-57 in: G. P. Contreras, and K. E. Evans, editors. Proceedings—Grizzly Bear Habitat Symposium, Missoula, Montana, April 30 - May 2, 1985. Intermountain Research Station, Ogden, UT.

184. ——, and F. L. Bunnell. 1987. Foraging strategies of coastal grizzly bears in the Kimsquit River Valley, British Columbia. International Conference on Bear Research and Management 7:187-197.

185. Hampton, P. D. 1981. The wintering and nesting behavior of the trumpeter swan. M.S. Thesis. University of Montana.

186. Hansen, W. R. 1980. Western aquatic garter snakes in central California: an ecological and evolutionary perspective. M.S. Thesis, California State University, Fresno, CA.

187. Hansen, A. J., E. L. Boeker, J. I. Hodges, and D. R. Cline. 1984. Bald eagles of the Chilkat Valley, Alaska: ecology, behavior, and management. National Audubon Society. New York, NY.

188. Hanson, L. C. 1993. The foraging ecology of harbor seals, *Phoca vitulina*, and California sea lions, *Zalophus californianus*, at the mouth of the Russian River, California. M.A. Thesis, Sonoma State University, Rohnert Park, CA.

189. Hare, S. R., N. J. Mantua and R. C. Francis. 1999. Inverse production regions: Alaska and West Coast Pacific salmon. Fisheries 24: 6-14.

190. Hartman, W., and R. Raleigh. 1964. Tributary homing of sockeye salmon at Brooks and Karluk lakes, Alaska. Canadian Journal of Fisheries and Aquatic Science 21:485-504.

191. Harvey, J.T. and M. J. Weise. 1997. Impacts of California sea lions and Pacific harbor seals on salmonids in Monterey Bay, California. MLML Technical Publication No.97-03. Moss Landing Marine Laboratories.

192. Hasegawa, S., T. Hirano, T. Ogasawara, M. Iwata, T.Akiyama, and S.Arai. 1987. Osmoregulatory ability of chum salmon, *Oncorhynchus keta*, reared in fresh water for prolonged periods. Fish Physiol. Biochem. 4:101-110.

193. Hatler, D. F. 1976. The coastal mink on Vancouver Island, British Columbia. Ph. D. Dissertation, University British Columbia, Vancouver, BC, Canada.

194. Hayman, R.A., E. M. Beamer, and R. E. McClure. 1996. Fiscal Year 1995 Skagit River chinook restoration research. Final project performance report, National Marine Fisheries Service, Contract# 3311 for FY 1995. Skagit System Cooperative. La Conner, WA.

195. Hayward, T. L. 1993. Preliminary observations of the 1991-1992 El Niño in the California Current. California Cooperative Oceanic Fisheries Investigations Reports 34: 21-29.

196. Healey, M. C. 1982. Juvenile Pacific salmon in estuaries: The life support system Pages 315-341 in: V. S. Kennedy, editor. Estuarine Comparisons. Academic Press. New York, NY.

197. ——. 1991. Life history of chinook salmon (*Oncorhynchus tshawytscha*). Pages 311-393 in: C. Groot and L. Margolis, editors. Pacific Salmon Life Histories, University of British Columbia Press. Vancouver, BC, Canada.

198. Heard, W. R. 1991. Life history of pink salmon (*Oncorhynchus gorbuscha*). Pages 121 to 230 in: C. Groot and L. Margolis, editors. Pacific Salmon Life Histories, University of British Columbia Press. Vancouver, B.C.

199. Heimlich-Boran, J. R. 1986. Fishery correlations with the occurrence of killer whales in Greater Puget Sound. Pages 113-131 in: B. C. Kirkevold, and J. S. Lockard, editors. Behavioral biology of killer whales. Alan R. Liss, Inc., New York, NY.

200. ——. 1988. Behavioral ecology of killer whales (*Orcinus orca*) in the Pacific Northwest. Canadian Journal of Zoology 66:565-578.

201. Hemstrom, M.A., and J. F. Franklin. 1982. Fire and other disturbances of the forests in Mount Rainier National Park. Quaternary Research 18:32-51.

202. Henny, C. J. and M. R. Bethers. 1971. Population ecology of the great blue heron with special reference to western Oregon. Canadian Field-Naturalist 85:205-209.

203. Herder, M. J. 1983. Pinniped fishery interactions in the Klamath River system, July 1979 to October 1980. Southwest Fish. Centennial Administrative Report LJ-83-12C. (Available From Southwest Fisheries Science Center, National Marine Fisheries Services, NOAA, P.O. Box 271, La Jolla, CA 92038).

204. Hewson, R. 1995. Use of salmonid carcasses by vertebrate scavengers. Journal of Zoology 235:53-65.

205. Hildebrand, G.V., S. D. Farley, C.T. Robbins, T.A. Hanley, K. Titus, and C. Serveheen. 1996. Use of stable isotopes to determine diets of living and extinct bears. Canadian Journal of Zoology 74:2080-2088.

207. Hicks, B.J., R. L. Beschta, and R. D. Harr. 1991b. Long-term changes in streamflow following logging in western Oregon and associated fisheries implications. Water Resources Bulletin 27:217-226.

208. ——, J. D. Hall, P.A. Bisson, and J. R. Sedell. 1991a. Response of salmonids to habitat changes. Pages 483-518 in: W.R. Meehan, editor. Influences of Forest and Rangeland Management On Salmonid Fishes and Their Habitats. American Fisheries Society Special Publication 19. Bethesda, MD.

209. Hilborn, R. 1992. Hatcheries and the future of salmon in the Northwest. Fisheries 17:5-8.

210. Hodder, J., and M.R. Graybill. 1985. Reproduction and survival of seabirds in Oregon during the 1982-1983 El Niño. Condor 87:535-541.

211. Hogan, D.L., and J.W. Schwab. 1991. Stream channel response to landslides in the Queen Charlotte Islands, B.C.: changes affecting pink and chum salmon habitat. Pages 222-236 in: B.White and I. Gutherie, editors. Proceedings Of the 15th Northeast Pacific Pink and Chum Workshop. Canada Department of Fisheries and Oceans. Vancouver, BC, Canada.

212. Hoffman, T. and T. Hall. 1988. Tillamook trip report regarding smolt/cormorant problem. U. S. Dept. of Agriculture, Animal Damage Control Program Interoffice Memorandum About Their Cormorants Collection on 27 April 1988.

213. Holland, D. C. 1985a. *Clemmys marmorata* (western pond turtle). Feeding. Herpetol. Review 16:112-113.

214. ——. 1985b. An ecological and quantitative study of the Western Pond Turtle (*Clemmys marmorata*) in San Luis Obispo County, California. M.S. Thesis, California State University, Fresno, CA.

215. Holtby, L. B. 1988. Effects of logging on stream temperatures in Carnation Creek, British Columbia, and associated impacts on the coho salmon (*Oncorhynchus kisutch*). Canadian Journal of Fisheries and Aquatic Sciences 45:502-515.

216. Horner, R. R., and B.W. Mar. 1982. Guide for water quality impact assessment of highway operations and maintenance. Report to Washington Dept. of Transp. FHWA WA-RD-39.14. Dept. of Civil Engineering, University of Washington, Seattle, WA.

216a. Hughes. J. 1983. On osprey habitat and productivity: a tale of two habitats. Pages 269-273 in D.M. Bird, editor. Biology and management of bald eagles and ospreys. Proceedings of 1st International Symposium on Bald Eagles and Ospreys, Montreal, 28-29 October 1981. Harpell Press, Ste Anne de Bellevue, Quebec.

217. Hunt, W. G., B. S. Johnson, and R. E. Jackman. 1992. Carrying capacity for bald eagles wintering along a northwestern river. Journal of Raptor Research 26:49-60.

218. ———, J. M. Jenkings, R. E. Jackman, C. G. Thelander, and A. T. Gerstell. 1992. Foraging ecology of bald eagles on a regulated river. Journal of Raptor Research 26:243-256.

219. Hunt, W. A. 1993. Jasper National Park harlequin duck research project: 1992 pilot projects - interim results. Jasper Warden Service Biological Report Series, No. 1, Heritage Resource Conservation, Parks Canada, Jasper, Alberta, Canada.

220. Hunt, W. G., B. S. Johnson, and R. E. Jackman. 1992. Carrying capacity for bald eagles wintering along a northwestern river. Journal of Raptor Research 26(2):49-60.

221. Huntington, C. W., W. Nelhsen, and J. Bowers. 1996. A survey of healthy native stocks of anadromous salmonids in the Pacific Northwest and California. Fisheries 21(3).

222. Huntsman, S. G. 1941. Cyclical abundance of birds versus salmon. Journal of the Fisheries Research Board of Canada 5:227-235.

223. Hynes, H. B. N. 1970. The ecology of running waters. University of Liverpool Press. Liverpool, England.

224. Imler, R. H. and H. R. Sarber. 1947. Harbor seals and sea lions in Alaska. U. S. Fish. Wildlife Services Special Sci. Report No.28.

225. Science Team. 1998. Pinniped and seabird predation: implications for recovery of threatened stocks of salmonids in Oregon under the Oregon Plan for Salmon and Watersheds. Technical Report 1998-2 to the Oregon Plan for Salmon and Watersheds. Governor's Natural Resources Office, Salem, OR.

226. Inkpen, E. L., and S. S. Embrey. 1998. Nutrient transport in the major rivers and streams of the Puget Sound Basin, Washington. U. S. Department of the Interior-U. S. 1 Survey, National Water-Quality Assessment Program.

227. Iwata, M., T. Hirano, and S. Hasegawa. 1982. Behavior and plasma sodium regulation of chum salmon fry during transition into seawater. Aquaculture 28:133-142.

228. Jameson, R. J. and K. W. Kenyon. 1977. Prey of sea lions in the Rogue River, Oregon. Journal of Mammalogy 58: 672.

229. Jeffries, S. 1985. Marine mammals of the Columbia River Estuary. Washingotn. State Department of Game; 1984. From U.S. Government Reports 85(2):85. Available From NTIS As PB85-107050/GAR.

230. Jewett, S. G., W. P. Taylor, W. T. Shaw, and J. W. Aldrich. 1953. Birds of Washington State. University of Washington Press, Seattle, WA.

231. Johnson, W. E. and A. D. Hasler. 1954. Rainbow trout production in dystrophic lakes. Journal of Wildlife Management 18:113-134.

232. Johnson, J. H. and A. A. Wolman. 1984. The Humpback Whale, *Megaptera novaeangliae*. U. S. National Marine Fisheries Service Marine Fisheries Review 6:30-37.

233. Johnson, D. H., M. M. Hoover, E. L. Greda, and C. J. Cederholm. (In prep.). Relationships between Pacific salmon and 605 species of birds, mammals, reptiles, and amphibians in Oregon and Washington.

234. Johnson, O. W., M. H. Ruckelshaus, W. S. Grant, F. W. Waknitz, A. M. Garrett, G. J. Bryant, K. Neely, and J. J. Hard. 1999. Status review of coastal cutthroat trout in Washington, Oregon, and California. U.S. Dept. Commerce, NOAA Tech. Memo. NMFS-NWFSC-37.

235. Johnson, R. R., and S. W. Carothers. 1982. Riparian habitats and recreation: Interelationships and impacts in the Southwest and Rocky Mountain region. Eisenhower Consortium for West. Environmental For. Res. Bulletin 12: 1-31.

236. Johnson, D. H., and T. A. O'Neil. 2001. Wildlife—habitat relationships in Oregon and Washington. Oregon State University Press, Corvallis, OR.

237. Johnston, J. M. 1982. Life histories of anadromous cutthroat with emphasis on migratory behavior. Pages 123-127 in: E. L. Brannon and E. O. Salo, editors. Proceedings of the Salmon and Trout Behavior Symposium. School of Fisheries, University of Washington, Seattle, WA.

239. Johnston, N. T., J. S. MacDonald, K. J. Hall, and P. J. Tschaplinski. 1997. A preliminary study of the role of sockeye salmon (*Oncorhynchus nerka*) carcasses as carbon and nitrogen sources for benthic insects and fishes

in the 'Early Stuart' stock spawning streams, 1050 km from the ocean. Fisheries Project Report No. RD55, Fisheries Branch, Ministry of Environment, Lands, and Parks, Province of British Columbia, Canada.

240. Jones, R. E. 1981. Food habits of smaller marine mammals from northern California. Proceedings of the California Academy of Sciences 42:409-433.

241. Juday, C., W. H. Rich, G. I. Kemmerer, and A. Mean. 1932. Limnological studies of Karluk Lake, Alaska 1926-1930. Bulletin of the U. S. Bureau of Fisheries 47:407-436.

242. Jurek, R. M. 1974. Special wildlife investigations.: American River Green Heron study, 1974. California Department of Fish and Game.

243. Kajimura, H. 1983. Food of the Pacific White-sided Dolphin, *Lagenorhynchus obliquidens*, Dall's Porpoise, *Phocoenoides dalli*, and Northern Fur Seal, *Callorhinus ursinus*, off California and Washington with appendices of size and food of Dall's porpoise from Alaskan waters. U.S. National Marine Fisheries Service Technical Memorandum; F/NWC-2. June 1980. From Monthly Catalog of U.S. Government Publications 83-6731.

244. Kellyhouse, D. G. 1975. Habitat utilization by black bears in northern California. International Conference on Bear Research and Management 4:221-227.

245. Kennish, M. J. 1997. Practical handbook of estuarine and marine pollution. (CRC Press, Marine Science Series). CRC Press. Boca Raton, FL.

246. Kingery, H. E. 1996. American Dipper. The Birds of North America, No. 229 in: A. Poole and F. Gill, editors. The Academy of Natural Sciences, Philadelphia, PA, and The American Ornithologist's Union, Washington, D.C.

247. Kjelson, M. A., P. F. Raquel, and F. W. Fisher. 1982. Life history of fall-run juvenile chinook salmon, *Oncorhynchus tshawytscha*, in the Sacramento-San Joaquin estuary, California. Pages 393-411 in: V. S. Kennedy, editor. Estuarine Comparisons. Academic Press, New York, NY.

248. Kline, T. C., Jr., J. J. Goering, O. A. Mathisen, P. H. Poe, and P. L. Parker. 1990. Recycling of elements transported upstream by runs of Pacific salmon: I. $\partial 15N$ and $\partial 13C$ evidence in Sashin Creek, southeastern Alaska. Canadian Journal of Fisheries and Aquatic Science 47: 136-144.

249. ———, ———, ———, ———, ———, and R. S. Scalan. 1993. Recycling of elements transported upstream by runs of Pacific salmon: II. $\partial 15N$ and $\partial 13C$ evidence in the Kvichak River watershed, Bristol Bay, southwestern Alaska. Canadian Journal of Fisheries and Aquatic Sciences 50: 2350-2365.

250. Knick, S. T. S. J. Sweeney J. R. Alldredge and J. D. Brittell. 1984. Autumn and winter food habits of Bobcats in Washington State. Great Basin Naturalist 44:70-74.

251. Knight, R. L., P. J. Randolph, G. T. Allen, L. S. Young, and R. J. Wigen. 1990. Diets of nesting bald eagles, *Haliaeetus leucocephalus*, in western Washington. Canadian Field-Naturalist 104:545-551.

252. ———, and D. P. Anderson. 1990. Effects of supplemental feeding on an avian scavenging guild. Wildlife Society Bulletin 18:388-394.

253. Knight, S. K and R. L. Knight. 1983. Aspects of food finding by wintering bald eagles. Biology and management of bald eagles and ospreys. Pages 28-29 in: D. M. Bird, editor. Proceedings of First International Symposium on bald eagles and ospreys Montreal. October, 1981. Harpell Press, Ste Anne De Bellevue, Quebec, Canada.

254. Knudsen, E. E. 1999. Managing Pacific salmon escapements: the gaps between theory and reality. Pages 237-272 in: E. E. Knudsen, C. S. Steward, D. D. MacDonald, J. E. Williams, and D. W. Reiser, editors. Sustainable Fisheries Management: Pacific Salmon. C. R. C. Lewis Publishers, Boca Raton, FL.

255. Knutson, K. L., and V. L. Naef. 1997. Management recommendations for Washington's Priority habitats. Washington Department of Fish and Wildlife. Olympia, WA.

256. Kojima, H., M. Iwata, and T. Kurokawa. 1993. Development and temporal decrease in seawater adaptability during early growth in chum salmon, *Oncorhynchus keta*. Aquaculture 118:141-150.

257. Konkel, G. W., and J. D. McIntyre. 1987. Trends in spawning populations of Pacific anadromous salmonids. Technical Report 9, USDI—Fish and Wildlife Service.

258. Koplin, J. R., D. S. MacCarter, D. P. Garber, and D. L. MacCarter. 1977. Food resources and fledgling productivity of California and Montana ospreys. Pages 205-214 in: Trans. North American Osprey Research Conference U.S. National Park Service Transactions of Proceedings Series No. 2.

259. Koski, K V. 1975. The survival and fitness of two stocks of chum salmon (Oncorhynchus keta) from egg deposition to emergence in a controlled-stream environment at Big Beef Creek. Ph.D. Dissertation, University of Washington, School of Fisheries. Seattle, WA.

260. ———, and D. Kirchofer. 1984. A stream ecosystem in an old-growth forest in Southeastern Alaska. Part IV: Food of juvenile coho salmon in relation to abundance of drift and benthos. Pages 81-88 In: W. R. Meehan, T. R. Merrell, Jr., and J. W. Matthews, editors. Proceedings of a Symposium on Fish and Wildlife Relationships in Old-Growth Forests, Held April, 12-15, 1982. Juneau, AK.

261. Krokhin, E. M. 1967. Effect of the size of sockeye migration on the phosphate regime of spawning lakes. Izvestia, Pacific Scientific Institute of Fisheries and Oceanography 64: 353-364 (in Russian).

262. ———. 1975. Transport of nutrients by salmon migrating from the sea into lakes. Pages 153-156 in: A. D. Hasler, editor. Coupling of Land and Water Systems. Springer-Verlag. New York, NY.

263. La Tourrette, J. 1992. Washington Wildlife Viewing Guide. Falcon Publishing, Inc. Helena, MT.

263a. Lagler, K.F. 1943. Food habits and economic relations of the turtles of Michigan with special reference to fish management. The American Midland Naturalist 29(2):257-312.

264. ———, and J. C. Salyer II. 1945. Influence of availability on the feeding habits of the common garter snake. Copeia. 1945:100-107.

265. Lampman, B. H. 1947. A note on the predaceous habits of water shrews . 1947. Journal of Mammalogy 28:181.

266. Larkin, G. A., and P. A. Slaney. 1997. Implications of trends in marine-derived nutrient influx to south coastal British Columbia salmonid production. Fisheries 22 (11):16-24.

267. Leatherwood, S. and R. R. Reeves. 1983. The Sierra Club handbook of whales and dolphins. Tien Wah Press, Singapore, Malaysia.

268. LeBrasseur, R. J. 1966. Stomach contents of salmon and steelhead trout in the northeastern Pacific Ocean. Journal of Fisheries Research Board of Canada 23:85-100.

269. Lee, K. N. 1993. Compass and gyroscope. Integrating science and politics for the environment. Island Press. Washington, D.C.

270. Lee, L. C., T. A. Muir, and R. R. Johnson. 1987. Riparian ecosystems as essential habitat for raptors in the American West. Pages 15-26 in: Proceedings of the Western Raptor Management Symposium and Workshop. National Wildlife Fed. Washington, D.C.

271. Leider, S. A., M. W. Chilcote, and J. J. Loch. 1986. Movement and survival of presmolt steelhead in a tributary and the main stem of a Washington river. North American Journal of Fisheries Management 6:526-531.

272. Leonard, W. P. H. A. Brown L. L. C. Jones K. R. McAllister and R. M. Storm. 1993. Amphibians of Washington and Oregon. Seattle Audubon Society, Seattle, WA.

273. Levy, D. A., and T. G. Northcote. 1982. Juvenile salmon residency in a marsh area of the Fraser River estuary. Canadian Journal of Fisheries and Aquatic Sciences 39: 270-276.

274. Lichatowich, J. A., and L. E. Mobrand. 1996. Chinook salmon (Oncorhynchus tshawytscha) in the Columbia River: The components of decline. In: Applied Ecosystem Analysis—Background—Mobrand Biometrics, Inc., Prepared for US Department of Energy Bonniville Power Administration, Environmental Fish and Wildlife. Portland, OR.

275. ———. 1999. Salmon without rivers—A history of the Pacific salmon crisis. Island Press, Covelo, CA.

276. ———. 1998. Habitat alteration and changes in habitat of coho (Oncorhynchus kisutch) and chinook (O. tshawytscha) in Oregon's coastal streams. Pages 92-99 in: C. C. Levings, L. B. Holtby, and M.A. Henderson, editors. Proceedings of the National Workshop on Effects of Habitat Alteration on Salmonid Stocks. Canadian Special Publication Fisheries and Aquatic Science 105.

277. ———. 1997. Evaluating salmon management institutions: The importance of performance measures, temporal scales, and production cycles. Pages 69 to 90 in: D. J. Stouder, P. A. Bisson, and R. J. Naiman, editors. Pacific Salmon and Their Ecosystems. Chapman and Hall Publishers, New York, NY.

278. Likens, G. E., and F. H. Bormann. 1974. Linkages between terrestrial and aquatic ecosystems. BioScience 24: 447-456.

279. Link, R. M., and K. K. English. 1999. Long-term, sustainable monitoring of Pacific salmon populations using fish wheels to integrate harvesting, management, and research. Pages 667-674 in: E. E. Knudsen, C. S. Steward, D. D. MacDonald, J. E. Williams, and D. W. Reiser, editors. Sustainable Fisheries Management: Pacific Salmon. C. R. C. Lewis Publishers, Boca Raton, FL.

280. Lind, G. S. 1976. Production, nest site selection, and food habits of ospreys on Deschutes National Forest, Oregon. M.S. Thesis, Oregon State University, Corvallis, OR.

281. Lingle, G. R. 1977. Food habits and sexing-aging criteria of the white pelican at Chase Lake National Wildlife Refuge, North Dakota. M.S. Thesis, Michigan Technological University.

282. Lister, D. B., and H. S. Genoe. 1970. Stream habitat utilization by cohabiting under-yearlings of chinook (Oncorhynchus tshawytscha) and coho (Oncorhynchus kisutch) salmon in the Big Qualicum River, British Columbia. Journal of Fisheries Research Board of Canada 27:1215-1224.

283. Loch, J. J., S. A. Leider, M. W. Chilcote, R. Cooper, and T. H. Johnson. 1988. Differences in yield, emigration-timing, size, and age structure of juvenile steelhead from two small western Washington streams. California Fish and Game 74:106-118.

284. Lowrance, R., R. Todd, J. Fail Jr., O. Hendriksen Jr., R. Leonard, and L. Asmussen. 1984. Riparian forests as nutrient filters in agricultural watersheds. BioScience 34:374-377.

285. Loegering, J. P. 1997. Abundance, habitat association, and foraging ecology of American dippers and other riparian-associated wildlife in the Oregon Coast Range. Ph.D. Thesis, Oregon State University, Corvallis, OR.

286. Landry, M. R., J. R. Postel, W. K. Peterson, and J. Newman. 1989. Broad-scale distributional patterns of hydrographic variables on the Washington/Oregon shelf. Pages 1-40 In: M. R. Landry and B. M. Hickey, editors. Coastal Oceanography of Washington and Oregon. Elsevier Oceanography Series 47, Amsterdam, Netherlands.

287. MacCarter, D. L. 1972. Food habits of ospreys at Flathead Lake, Montana. M. S. Thesis. California State University. Humboldt, Arcata, CA.

288. Macdonald, J. S. C. D. Levings C. D. McAllister U. H. M. Fagerlund and J. R. McBride. 1988. A Field Experiment to Test the Importance of Estuaries for Chinook Salmon (Oncorhynchus tshawytscha) Survival: Short-Term Results. Canadian Journal Fisheries and Aquatic Sciences 45:1366-1377.

289. ———, I. K. Birtwell, and G. M. Kruzynski. 1987. Food and habitat utilization by juvenile salmonids in the Campbell River estuary. Canadian Journal of Fisheries and Aquatic Sciences 44:1233-1246.

290. Mace, P. M. 1983. Bird Predation on Juvenile Salmonids in the Big Qualicum Estuary, Vancouver Island. Can. Technical Report, Fisheries Aquatic Science 1176:1-77.

291. Marcot, B. G., M. A. Castellano, J. A. Christy, L. K. Croft, J. F. Lehmkuhl, R. . Naney, R. E. Rosentreter, R.E. Sandquist, and E. Zieroth. 1997. Terrestrial ecology assessment. Pages 1497-1713 in: T. M. Quigley, S. J. Arbelbide, and S. F. McCool, editors. An assessment of ecosystem components in the interior Columbia Basin and portions of the Klamath and Great Basins, USDA Forest Service General Technical Report PNW-GTR-405. Pacific Northwest Research Station, Portland, OR.

292. ———, and M. Vander Heyden. 2001. Key ecological functions of wildlife species. Pages 168-186 in: D. H. Johnson and T. A. O'Neil, managing directors. Wildlife—habitat relationships in Oregon and Washington. Oregon State University Press, Corvallis, OR.

293. Marquiss, M. and K. Duncan. 1993. Variation in the abundance of red-breasted Mergansers Mergus serrator on a Scottish river related to season, year, river hydrography, salmon density and spring culling. Ibis. 135:33-41.

294. Martin, D. J. 1985. Production of cutthroat trout (Salmo clarki) in relation to riparian vegetation in Bear Creek, Washington. Ph. D. Dissertation. University of Washington, School of Fisheries, Seattle, WA.

295. Maser, C., R. F. Tarrant, J. M. Trappe, and J. F. Franklin. 1988. From the forest to the sea: A story of fallen trees. USDA- Forest Service, and USDI- Bureau of Land Management. General Technical Report PNW-GTR-229.

296. ———, and J. R. Sedell. 1994. From the forest to the sea—the ecology of wood in streams, rivers, estuaries, and oceans. St. Lucie Press. Delray Beach, FL. 297. Mathisen, O.A. 1972. Biogenic enrichment of sockeye salmon lakes and stock productivity. Verh. Int. Ver. Limnol. 18:1089-1095.

298. Mathisen, O.A., P. L. Parker, J. J. Goering, T. C. Kline, P. H. Poe, and R. S. Scalan. 1988. Recycling of marine elements transported into freshwater systems by anadromous salmon. Verh. Int. Ver. Limnol. 23:2249-2258.

299. McHenry, M. L., E. Shott, R. H. Conrad, and G. B. Grette. 1998. Changes in the quantity and characteristics of large woody debris in streams of the Olympic Peninsula, Washington, U. S.A. (1982-1993). Canadian Journal of Fisheries and Aquatic Science 55:1395-1407.

300. Matkowski, S. M. D. 1989. Differential susceptibility of tree species of stocked trout to bird predation. North American Journal of Fisheries Management 9:184-187.

301. ———. 1984. Angler harvest and other causes of mortality of stocked salmonids in Duck Mountain Provincial Park, Maniotba. M.S. Thesis. University of Manitoba, Winnipeg, Canada.

302. Matthews, D. R. 1983. Feeding ecology of the common murre, Uria aalge, off the Oregon coast. M. S. Thesis, University of Oregon, Eugene, OR.

303. Mattson, D. J. B. M. Blanchard and R. R. Knight. 1991. Food habits of Yellowstone Grizzly Bears, 1977-1987. Canadian Journal of Zoology 69:1619-1629.

304. McClelland, B. R. 1973. Autumn concentrations of bald eagles in Glacier National Park. Condor 75:121-123.

305. McMurray, G. R., and R. J. Bailey. 1998. Change in Pacific Northwest Coastal Ecosystems: Science for Solutions, Pacific Northwest Coastal Ecosystems Regional Study (PNCERS). U. S. Department of Commerce. NOAA Coastal Ocean Program Decision Analysis Series No. 11.

306. McNeil, W. J., J. R. Gowan, and R. Severson. 1991. Offshore release of salmon smolts. American Fisheries Society Symposium, 10:548-553.

307. Meffe, G. K. 1992. Techno-Arrogance and halfway technologies: Salmon hatcheries on the Pacific coast of North America. Conservation Biology 6(3):350-354.

308. Melone, A. M. 1985. Flood producing mechanisms in coastal British Columbia. Canadian Water Research Journal 10:46-64.

309. Melquist, W. E., and A. E. Dronkert. River Otter. 1987. Wild Furbearer Management and Conservation in North America. Pages 626-641 in: M. J.A. Novak, M. E. Obbard, and B. Malloch, editors. Ministry of Natural Resources, Ontario, Canada.

310. ———, J. S. Whitman, and M. G. Hornocker. 1981. Resource partitioning and coexistence of sympatric mink and river otter populations. Proceedings of the Worldwide Furbearer Conference 1:187-220.

311. ———, and M. G. Hornocker. 1983. Ecology of river otters in west central Idaho. Wildlife Monographs No. 83:1-60.

312. ———, and D. R. Johnson. 1984. Additional comments on the migration of northern Idaho and eastern Washington ospreys. Journal of Field Ornithol. 55:483-485.

313. Mendall, H. L. 1944. Food of hawks and owls in Maine. Journal of Wildlife Management. 8:198-208.

314. ———. 1939. Food habits of the herring gull in relation to freshwater game fishes in Maine. Wilson Bulletin 41(223-226).

315. Merritt, R. W., and K. W. Cummins. 1996. An introduction to the aquatic insects of North America. Kendall/Hunt Publishing Company. Dubuque, IA.

316. ———, ———, and T. M. Burton. 1984. The role of aquatic insects in the processing and cycling of nutrients. Pages 134-163 In: V. H. Resh and D. M. Rosenberg, editors. The Ecology of Aquatic Insects. Praeger Publishers. New York, NY.

317. Meyer, J. I., and eight co-authors. 1988. Elemental dynamics in streams. Journal of North American Benthological Society 7:410-432.

318. Michael, J. H., Jr. 1995. Enhancement effects of spawning pink salmon on stream rearing juvenile coho salmon: Managing one resource to benefit another. Northwest Science 69:228-233.

319. ———. 1998. Pacific salmon spawner escapement goals for the Skagit River watershed as determined by nutrient cycling considerations. Northwest Science 72: 239-248.

320. Miegs, R. C. and C.A. Rieck. 1958. Mergansers and trout in Washington. Pages 306-318 in: Proceedings of 47th Annual Conference West Assoc. State Game Fish Commission.

321. Miller, L. 1957. Bird remains from an Oregon Indian midden. Condor 59:59-63.

322. Miller, R. J., and E. L. Brannon. 1982. The origin and development of life history patterns in Pacific salmon. Pages 296 to 309 in: E. L. Brannon, and E. O. Salo, editors. Proceedings of the Salmon and Trout Migratory Behavior Symposium. June 3-5, 1981, First International Symposium, School of Fisheries, University of Washington, Seattle, WA.

323. Miller, J.A., and C.A. Simenstad. 1997. A comparative assessment of a natural and created estuarine slough as rearing habitat for juvenile chinook and coho salmon. Estuaries 20:792-806.

324. Minakawa, N. 1997. The dynamics of aquatic insect communities associated with salmon spawning. Ph. D. Dissertation, University of Washington, School of Fisheries. Seattle, WA.

325. Minshall, G. W., R. C. Petersen, K. W. Cummins, T. L. Bott, J. R. Sedell, C. E. Cushing, and R. L. Vannote. 1983. Interbiome comparison of stream ecosystem dynamics. Ecological Monographs 53:1-25.

326. ———, and J. N. Minshall. 1977. Microdistribution of benthic invertebrates in Rocky Mountain (U.S.A.) Stream. Hydrobiologia 55:231-249.

327. ———. 1968. Community dynamics of the benthic fauna in a woodland springbrook. Hydrobiologia 32:305-337.

328. ———. 1984. Aquatic insect-substratum relationships. Pages 358-400 in: V. H. Resh and D. M. Rosenberg, editors. The ecology of aquatic insects. Praeger Publishers. New York, NY.

329. Mitsch, W. J., and J. G. Gosselink. 1986. Wetlands. Van Nostrand Reinhold, New York, NY.

330. Mizue, K. K. Yoshida and A. Takemura. 1966. On the ecology of the Dall's porpoise in the Bering Sea and the North Pacific Ocean. Fac. Fish., Nagasaki University, Bulletin 21:1-21.

331. Modde, T., A. F. Wasowicz, and D. K. Hepworth. 1996. Cormorant and grebe predation on rainbow trout stocked in a southern Utah reservoir. North American Journal of Fisheries Management 16:388-394.

332. Montgomery, D. R., and J. M. Buffington. 1993. Channel classification, prediction of channel response, and assessment of channel condition. Draft report to the Sediment, Hydrology, and Mass Wasting Committee of the Washington State Timber/Fish/Wildlife Agreement. Department of Geological Sciences and Quarternary Research Center, University of Washington. Seattle, WA.

333. Morejohn, G. V. J. T. Harvey and L. T. Krasnow. 1978. The importance of *Loligo opalescens* in the food web of marine vertebrates in Monterey Bay, California. California Department of Fish and Game Bulletin No. 169: 67-98.

334. Morrison, M. L., B. G. Marcot, and R. W. Mannan. 1998. Wildlife-Habitat Relationships. Second edition. University of Wisconsin Press, Madison, WI.

335. Mossman, A. S. 1959. Selective predation of Glaucous-winged gulls upon adult red salmon. Ecology 39:482-486.

335a. ———. 1958. Selective predation of glaucous-winged gulls upon adult red salmon. Ecology 39(3):482-486.

336. Moyle, P. 1966. Feeding behavior of the Glaucous-winged gull on an Alaskan salmon stream. Wilson Bulletin. 78:175-190.

337. Muir, W. D., and R. L. Emmett. 1988. The food habits of migrating salmonid smolts passing Bonneville Dam in the Columbia River, 1984. Re. Riv. Res. Man. 2:1-10.

338. Mundie, J. H. 1969. Ecological implications of the diet of juvenile coho in streams. Pages 135-152 in: T. G. Northcote, editor. Symposium On Salmon and Trout In Streams, H. R. MacMillan Lectures in Fisheries. Held February 22-24, 1968, at the University of British Columbia, Vancouver, BC, Canada.

339. Munn, M. D., R. W. Black, A. L. Haggland, M. A. Hummling, and R. L. Huffman. 1999. An assessment of stream habitat and nutrients in the Elwha River Basin: Implications for restoration. U.S. Geological Survey,

Water-Resources Investigations Report 98-4223. Prepared in cooperation with the Lower Elwha Tribe and National Park Service.

340. Munro, J. 1941. Studies of waterfowl in British Columbia. Greater scaup duck, lesser scaup. Canadian Journal Research Section D Zoological Sci. 19:113-138.

341. Munro, J.A. 1932. Food of the American Merganser (*Mergus merganser americanus*) in British Columbia. Canadian Field-Naturalist 46:166-168.

341a. ———. 1939. Food of ducks and coots at Swan Lake British Columbia. Canadian Journal of Research 17:178-186.

342. ———, and W.A. Clemens. 1939. The food and feeding habits of the Red-breasted Merganser in British Columbia. Journal of Wildlife Management 3:46-53.

343. ———. 1945. Observations of the loon in the Cariboo Parklands, British Columbia. Auk 62:38-49.

344. ———. 1923. A preliminary report on the relation of various ducks and gulls to the propagation of sockeye salmon at Henderson lake, Vancouver Island, BC. Canadian Field Naturalist 37:107-116.

345. ———. 1938. Studies of waterfowl in British Columbia: Barrow's goldeneye, American goldeneye. Transactions of the Royal Canadian Institute 22:259-318.

346. ———. 1938a. The northern bald eagle in British Columbia. Wilson Bulletin 50:38-35.

347. Murie, O. J. 1959. Fauna of the Alutian Islands and Alaska Peninsula. North American Fauna.

348. Murota, T. 1998a. Material cycle and sustainable economy. Pages 120-138 In: D. Bell, L. Fawcett, R. Keil, and P. Penz, editors. Political Ecology. Routledge, London & New York, NY.

349. ———. 1998b. Nutrient shadow cast by anadromous fishes: Perspectives in comparison with marine fishery and guano occurance. Paper presented at the Western Division American Fisheries Society Symposium, held in Anchorage, Alaska (Sept.-Oct., 1998).

350. ———. 1987. The environmental economics of the water planet earth. In: Environmental Economics—The Analysis of a Major Interface. Pages 185-199 in: G. Pilet and T. Murota, editors. R. Leimgruber, Geneva, Switzerland.

351. ———, and Faculty of Environmental Studies. 1994. Material cycle and sustainable economy: A thermodynamical approach to political ecology. A paper read at The Global Political Ecology Conference (Harold Innis Centenary Celebration). Held March 3-6, 1994. York University, Toronto, Canada.

352. Murphy, M. L., and J. D. Hall. 1981. Varied effects of clear-cut logging on predators and their habitat in small streams of the Cascade Mountains, Oregon. Canadian Journal of Fisheries and Aquatic Sciences 38:137-145.

353. Myers, G. L. and J. J. Peterka. 1976. Survival and growth of Rainbow trout (*Salmo gairdneri*) in four prairie lakes, North Dakota. Journal of the Fisheries Research Board of Canada 33:1192-1195.

354. Myers, K. W. 1980. An investigation of the utilization of four study areas in Yaquina Bay, Oregon, by hatchery and wild juvenile salmonids. M. S. Thesis. Oregon State University, Corvallis, OR.

355. ———, and H. F. Horton. 1982. Temporal use of an Oregon estuary by hatchery and wild juvenile salmon. Pages 377-392 in: V. S. Kennedy, editor. Estuarine comparisons. Academic Press, New York, NY.

356. Myers, J. M., R. G. Kope, G. J. Bryant, D. Teel, L. J. Lierheimer, T. C. Wainwright, W. S. Grant, F. W. Waknitz, K. Neeley, S. T. Lindley, and R. S. Waples. 1998. Status review of chinook salmon from Washington, Idaho, Oregon, and California. U.S. Dept. of Commerce, NOAA Tech. Memo. NMFS-NWFSC-35.

357. Myers, J. P. 1980. Territoriality and flocking by Buff-breasted Sandpipers: variations in non-breeding dispersion. Condor 82:241-250.

358. Nagorsen, D. W. Morrison K. F. and Forsberg J. E. 1989. Winter diet of Vancouver Island marten (*Martes americana*). Canadian Journal of Zoology 67:1394-1400.

359. ———, Campbell W. R. and Giannico G. R. 1991. Winter Food Habits of Marten, *Martes americana*, on the Queen Charlotte Islands. Canadian Field-Naturalist 105:55-59.

360. Naiman, R. J., and E. C. Anderson. 1997. Streams and rivers: their physical and biological variability. Pages 131-148 in: P. Schoomaker, B. von Hagen, and E. Wolf, editors. The Rain Forests of Home: Profile of a North American Bioregion. Ecotrust/Interain Pacific and Island Press.

361. ———., T. J. Beechie, L. E. Benda, P.A. Bisson, L. H. MacDonald, M. D. O'Connor, C. Oliver, P. Olson, and E.A. Steel. 1992. Fundamental elements of ecologically healthy watersheds in the Pacific Northwest Coastal Ecoregion. Pages 127-188 in: R. J. Naiman, editor. Watershed Management—Balancing Sustainability and Environmental Change. Springer-Verlog. New York, NY.

362. ———, and R. E. Bilby. 1998. River ecology and management: Lessons from the Pacific Coastal Ecoregion. Springer-Verlag, New York, NY.

363. Nakatani, R. E. 1969. Effects of heated discharge on anadromous fish. In: P.A. Krenkel and F. L. Parker, editors. Biological aspects of thermal pollution. Vanderbilt University Press. Nashville, TN.

364. National Marine Fisheries Service (NMFS). 1997. Investigation of scientific information on the impacts of California Sea Lions and Pacific Harbor Seals on salmonids and on the coastal ecosystems of Washington, Oregon, and California. U.S. Dep. Commer., NOAA Tech. Memo. NMFS-NWFSC-28.

365. National Research Council (NRC). 1996. Upstream - Salmon and society in the Pacific Northwest. National Academy Press. Washington, D.C.

366. National Marine Fisheries Service (NMFS). 1998. Essential Fish Habitat Advisory Report to the Pacific Fishery Management Council for Pacific salmon species of California, Oregon, and Washington. Draft report prepared by the Pacific State Marine Fisheries Commission. Gladstone, OR.

366a. ———. 1991. Final rule for endangered status of Snake River sockeye salmon. November 20, 1991. Federal Register Volume 56: 58619.

367. Nehlsen, W., J. E. Williams, and J.A. Lichatowich. 1991. Pacific salmon at the crossroads: stocks at risk from California, Oregon, Idaho, and Washington. Fisheries 16 (2):4-21.

368. Netboy, A. 1980. The Columbia River salmon and steelhead trout: Their fight for survival. University of Washington Press, Seattle, WA.

369. Newbold, J. D., J. W. Elwood, R. V. O'Neil, and W. V. Winkle. 1981. Measuring nutrient spiralling in streams. Canadian Journal of Fisheries and Aquatic Sciences 38:860-863.

370. Newton, M., B.A. El Hassen, and J. Zavitkovski. 1968. Role of red alder in western Oregon forest succession. Pages 73-84 in: J. M. Trappe, J. F. Franklin, R. F. Tarrant, and G. M. Hansen, editors. Biology of Alder. USDA Forest Service. Portland, OR.

371. Nichol, L. M. and D. M. Shackleton. 1996. Seasonal movements and foraging behavior of northern resident killer whales (*Orcinus orca*) in relation to the inshore distribution of salmon (*Oncorhynchus* spp.) in British Columbia. Canadian Journal of Zoology 74:983-991.

372. Nicholas, J. W., and D. G. Hankin. 1988. Chinook salmon populations in Oregon coastal river basins: Description of life histories and assessment of recent trends in run strengths. Oregon Department of Fish and Wildlife Fisheries Div. Info. Report, No. 88-1.

373. Nickelson, T. E. 1986. Influences of upwelling, ocean temperatures, and smolt abundance on marine survival of coho salmon (*Oncorhynchus kisutch*) in the Oregon Production Area. Canadian Journal of Fisheries and Aquatic Sciences 43:527-535.

374. Nickelson, T., J. Rodgers, S. Johnson, and M. Solazzi. 1992. Seasonal changes in habitat use by juvenile coho salmon (*Oncorhynchus kisutch*) in Oregon Coastal Streams. Canadian Journal of Fisheries and Aquatic Sciences 49:783-789.

375. Nicola, S. J. 1966. The relationship of Alloperla nymphs to incubating pink (*Oncorhynchus gorbuscha*) and chum (*Oncorhynchus keta*) salmon eggs, alevins, and pre-emergent fry in a southeastern Alaska stream. M. S. Thesis. University of Washington, School of Fisheries, Seattle, WA.

376. ——— 1968. Scavenging by *Alloperla* (Plecoptera: Chloroperlidae) nymphs on dead pink (*Oncorhynchus gorbuscha*) and chum (*Oncorhynchus keta*) salmon embryos. Canadian Journal of Zoology 46:787-796.

377. Nikol'skii, G. V. 1952. The type of dynamics of stocks and the character of spawning of the chum (*Oncorhynchus keta, Walb.*) and the pink salmon (*Oncorhynchus gorbuscha, Walb.*) in the Amur River. Doklady Akademii Nauk. SSSR, 86(4): 873-875.

378. Norris, K. S., and J. H. Prescott. 1961. Observations on Pacific cetaceans of California and Mexican waters. University of California. Publ. Zoo. 63:291-403.

378a North, M. R. 1994. Yellow-billed loon, *Gavia adamsii*. The Birds of North America No.121. *In*: A. Poole and F. Gill, editors. The Academy of Natural Sciences, Philadelphia, PA, and The American Ornithologist's Union, Washington, D.C.

379. Northwest Power Planning Council (NPPC). 1986. Compilation of information on salmon and steelhead losses in the Columbia River basin. Appendix D of the 1987 Columbia River Basin Fish and Wildlife Program. Portland, OR.

380. O'Connell, M.A., J. G. Hallett, and S. D. West. 1993. Wildlife use of riparian habitats: A Literature review. Washington Department of Natural Resources, Timber/Fish/Wildlife, Report TFW-WL1-93-001.

381. Oakley, A. L., J. A. Collins, L. B. Everson, D.A. Heller, J. C. Howerton, and R. E. Vincent. 1985. Riparian zones and freshwater wetlands. Pages 58-76 in: R. E. Brown, editor. Management of Wildlife and Fish Habitats in Forests of Western Oregon and Washington. U. S. For. Serv. Portland, OR.

382. Obermayer, K. E. Hodgson A. and M. F. Willson. 1999. American Dipper, *Cinclus mexicanus*, foraging on Pacific salmon, *Oncorhynchus* spp., eggs. Canadian Field-Naturalist 113:288-290.

383. Ofelt, C. H. 1975. Food habits of nesting bald eagles in southeast Alaska. Condor 77:337-338.

384. Olesiuk, P. F. 1993. Annual prey consumption by harbor seal (*Phoca vitulina*) in the Strait of Georgia, British Columbia. Fish. Bull. 91:491-515.

384a. ———, M.A. Bigg, G.M. Ellis, S. J. Crockford, and R.J. Wigen. 1990. An assessment of the feeding habits of harbour seals (*Phoca vitulina*) in the Strait of Georgia, British Columbia, based on scat analysis. Canadian Technical Report of Fisheries and Aquatic Sciences. No. 1730. Department of Fisheries and Oceans. Biological Sciences Branch. Pacific Biological Station. Nanaimo, British Columbia.

385. Olson, R. and W. Hubert. 1994. Beaver: Water resources and Riparian Habitat Manager. University of Wyoming. Laramie, WY.

386. Ordal, E., and R. E. Pacha. 1963. The effects of temperatures on disease in fish. in Water temperature: Influences, effects and control. Proceedings of the twelfth Pacific Northwest symposium on water pollution research. U.S. Public Health Service, Pacific Northwest Water Laboratory. Corvallis, OR.

387. Oregon Coastal Salmon Restoration Initiative (OCSRI). 1997. OCSRI Conservation Plan. Draft Revision February 24, 1997. Salem, OR.

388. Pacific Fishery Management Council (PFMC). 1998. Review of 1997 ocean salmon fisheries. A Report of the Pacific Fishery Management Council pursuant to National Oceanic and Atmospheric Administration, Award No. NA87FC0008. Portland, OR.

388a. ———. 1991. Review of 1990 ocean salmon fisheries. Portland, OR

388b. ———. 1992. Review of 1990 ocean salmon fisheries. 2130 SW Fifth Avenue, Suite 224, Portland, OR.

389. Palmer. 1962. Handbook of North American Birds 1. Yale University Press. New Haven, CT.

390. Palmisano, J. F., R.H. Ellis and V.W. Kaczynski. 1993. The impact of environmental and management factors on Washington's wild anadromous salmon and trout. Washington Forest Protection Association and State of Washington Department of Natural Resources. Olympia, WA.

391. Parker, M. S. 1994. Feeding ecology of stream-dwelling Pacific giant salamander larvae (Dicamptodon tenebrosus). Copeia 3:705-718.

392. ———. 1991. Relationship between cover availability and larval Pacific giant salamander density. Journal of Herpetology 25:355-357.

393. ——— 1993. Predation by Pacific Giant Salamander larvae on juvenile steelhead trout. Northwestern Naturalist 74:77-81.

394. Pearcy, W. G. 1997. What have we learned in the last decade? What are research priorities? Pages 271-277 in: R. L. Emmett and M. H. Schiewe, editors. Estuarine and Ocean Survival of Northeastern Pacific Salmon, NOAA Technical Memorandum NMFS-NWFSC-29.

395. ———, R. D. Brodeur, and J. P. Fisher. 1990. Distribution and ecology of juvenile cutthroat trout (*Oncorhynchus clarki clarki*) and steelhead (*Oncorhynchus mykiss*) in the ocean off Oregon and Washington. Fisheries Bulletin 88:697-711.

396. ———, ———, J. M. Shenker, W. W. Smoker, and Y. Endo. 1988. Food habits of Pacific salmon and steelhead trout, midwater trawl catches and oceanographic conditions in the Gulf of Alaska, 1980-1985. Bull. Ocean Res. Instit., Univ. of Tokyo 26(2):29-78.

397. ———. 1992. Ocean ecology of north Pacific salmonids. Washington Sea Grant Program, distributed by University of Washington Press. Seattle, WA.

398. Peterson, N. P. 1980. The role of spring ponds in the winter ecology and natural production of coho salmon (*Oncorhynchus kisutch*) on the Olympic Peninsula, Washington. M. S. Thesis, University of Washington, College of Fisheries. Seattle, WA.

399. ———. 1982a. Population characteristics of juvenile coho salmon (*Oncorhynchus kisutch*) overwintering in riverine ponds. Canadian Journal of Fisheries and Aquatic Science 39:1303-1307.

400. ———. 1982b. Immigration of juvenile coho salmon (*Oncorhynchus kisutch*) into riverine ponds. Canadian Journal of Fisheries and Aquatic Science 39:1308-1310.

401. ———, and L. M. Reid. 1984. Wall-base channels: their evolution, distribution, and use by juvenile coho salmon in the Clearwater River, Washington. Pages 215-226 in: J. M. Walton and D. B. Houston, editors. Proceedings of the Olympic Wild Fish Conference, March 23-25, 1983, Port Angeles, WA.

402. Phinney, L.A., and P. Bucknell. 1975. A catalog of Washington streams and salmon utilization. Washington Department of Fisheries, Olympia, WA.

403. Pike, G. C. 1950. Stomach contents of whales caught of the coast of British Columbia. Progress Report of the Pacific Coast Stations. Fisheries Research Board Can. 83:27-28.

404. Piorkowski, R. J. 1995. Ecological effects of spawning salmon several south-central Alaskan streams. Ph. D., University of Alaska, Fairbanks.

405. Poage, N. J., and T.A. Spies. 1996. A tale of two unmanaged riparian forests. COPE Report 9(1):6-9.

406. Poelker, R.J., and H.D. Hartwell. 1973. Black bear of Washington. Washington State Game Department. Biological Bulletin No. 14. Olympia, WA.

407. Pitcher, K.W. 1981. Prey of the Steller sea lion, *Eumetopias jubatus*, in the Gulf of Alaska. U. S. Fish and Wildlife Service Fishery Bulletin 79:467-472.

410. ———. 1977. Population Productivity and Food Habits of Harbor Seals in the Prince William Sound, Copper River Delta Area, Alaska. Report to U.S. Marine Mammal Commission for Contract MM5AC01.

411. ———, and D. G. Calkins. 1981. Reproductive biology of Steller sea lions in the Gulf of Alaska. Journal of Mammalogy 62:599-605.

412. Pomery, L. R. 1970. The strategy of mineral cycling. Annual Review Ecol. Syst. 1:171-190.

413. Pritchard, D.W. 1967. What is an estuary: Physical viewpoint. Pages 3-5 in: G. H. Lauff, editor. Estuaries. American Association of the Advancement of Science Publication No. 83, Washington, D.C.

414. Quinn, T. P., and N. P. Peterson. 1994. The effects of forestry practices on fish populations: Incubation Environment of Chum Salmon (*Oncorhynchus keta*) in Kennedy Creek - Part A. Persistence of Egg Pocket Architecture in Chum Salmon Redds. Washington Department of Natural Resources, Olympia, WA. Timber/Fish/Wildlife Report. TFW-FA-94-001.

415. Rader, R. B. 1997. A functional classification of the drift: traits that influence invertebrate availability to salmonids. Canadian Journal Fisheries Aquatic Sciences 54:1211-1234.

416. Ralph, S. C., G. C. Poole, L. L. Conquest, and R. J. Naiman. 1994. Stream channel morphology and woody debris in logged and unlogged basins of western Washington. Canadian Journal of Fisheries and Aquatic Sciences 51:37-51.

417. Reeves, G. H., P.A. Bisson, J. M. Dambacher. 1998. Fish Communities. Pages 200-234 in: R. J. Naiman and R. E. Bilby, editors. River Ecology and Management: Lessons From the Pacific Coastal Ecoregion. Springer-Verlag. New York, NY.

418. ———. F. H. Everest, and J.D. Hall. 1987. Interactions between the redside shiner (*Richardsonius balteatus*) and the steelhead trout (*Salmo gairdneri*) in western Oregon: the influence of water temperature. Canadian Journal Fisheries and Aquatic Science 44:1602-1613.

419. Regier, H. A. 1997. Old traditions that led to abuses of salmon and their ecosystems. Pages 17-28 In: D. J. Stouder, P. A. Bisson, and R. J. Naiman, editors. Pacific Salmon and Their Ecosystems: Status and Future Options. Chapman & Hall. New York, NY.

420. Reice, S. R. 1994. Nonequilibrium determinants of biological community structure. Amer. Sci. 82:424-435.

421. Reid, L. M. 1981. Sediment production from gravel-surfaced forest roads, Clearwater Basin, Washington. M. S. Thesis, University of Washington Department of Geological Sciences. Seattle, WA.

422. Reimchen, T. E. 1994. Further studies of predator and scavenger use of chum salmon in stream and estuarine habitats at Bag Harbour, Gwaii Haanas. Island Ecological Research, Queen Charlotte City, B.C., Prepared for Canadian Parks Service.

423. Riemer, S. D. and R. F. Brown. 1997. Prey of pinnipeds at selected sites in Oregon identified by scat (fecal) analysis, 1983-1996. Oregon Department of Fish and Wildlife, Wildlife Diversity Program, Technical Report No. 97-6-02.

424. Reinhart, D. P. and D. J. Mattson. 1989. Bear use of cutthroat trout spawning streams in Yellowstone National Park. International Conference of Bear Research and Management 8:343-350.

425. Richey, J. E., M. A. Perkins, and C. R. Goldman. 1975. Effects of kokanee salmon (Oncorhynchus nerka) decomposition on the ecology of a subalpine stream. Journal of Fisheries Research Board of Canada 32:817-820.

426. Ricker, W. 1980. Causes of the decrease in age and size of chinook salmon (Oncorhynchus tshawytscha). Canadian Technical Report Aquatic Science 944.

427. Robertson, I. 1973. Predation by fish-eating birds on stocks of Pacific herring, Clupea pallasi, in the Gulf Islands of British Columbia. Unpublished Report, Pacific Biol. Stn., Nanaimo, BC, Canada.

428. ———— 1974. The food of nesting Double-crested and Pelagic Cormorants at Mandarte Island, British Columbia, with notes on feeding ecology. Condor 76:346-348.

429. Roby, D. D., D. P. Craig, K. Collis, and S. L. Adamany. 1998. Avian predation on juvenile salmonids in the lower Columbia River—1997 Annual Report. Oregon Coop. Wildl. Res. Unit, Oregon State University, Corvallis, OR.

430. ————, ————, ————, and ————. 1998. (Unpublished) Avian predators on juvenile salmonids in the lower Columbia River. Annual Report to Bonneville Power Administration and U. S. Army Corps of Engineers.

431. Roche, J., and M. McHutchison. 1998. First fish/first people: Salmon tales of the North Pacific rim. University of Washington Press. Seattle, WA.

432. Roemmich, D., and J. McGowan. 1995. Climatic warming and the decline of zooplankton in the California Current. Science 267:1324-1326.

433. Roffe, T. J. and B. R. Mate. 1984. Abundances and feeding habits of pinnipeds in the Rogue River. Oregon. Journal of Wildlife Management 48:1262-1274.

434. Rosgen, D. 1996. Applied river morphology. Wildland Hydrology. Pagosa Springs, CO.

435. Ruttner, F. 1971. Fundamentals of limnology. English translation of Third Edition Copyright University of Toronto Press. Toronto, Canada.

436. Rowlett, R. A. 1980. Observarions of marine birds and mammals in the northern Chesapeake Bight. U. S. Fish Wildl. Serv., Biol. Serv. Prog., FWS/OBS-80/04.

437. Ruggerone, G. T. 1986. Consumption of migrating juvenile salmonids by gulls foraging below a Columbia River dam. Transactions of the American Fisheries Society 115:736-742.

438. Ryall, R., and C. D. Levings. 1987. Juvenile salmon utilization of rejuvenated tidal channels in the Squamish estuary, British Columbia. Can. Manuscript Rep. of Fisheries Aquatic Science 1904.

439. Salo, E. O., and W. H. Bayliff. 1958. Artificial and natural production of silver salmon, Oncorhynchus kisutch, at Minter Creek, Washington. State of Washington Department of Fisheries. Res. Bulletin No. 4.

440. ————. 1991. Life history of chum salmon. Pages 231-307 in: C. Groot and L. Margolis, editors. Pacific Salmon Life Histories, University of British Columbia Press. Vancouver, BC, Canada.

441. Salyer, J. C. and K. F. Lagler. 1940. The food and habits of the American merganser during winter in Michigan, considered in relation to fish management. Journal of Wildlife Management 4:186-219.

442. Sandercock, F. K. 1991. Life history of coho salmon (Oncorhynchus kisutch). Pages 397 to 445 in: C. Groot and L. Margolis, editors. Pacific Salmon Life Histories, University of British Columbia Press. Vancouver, BC, Canada.

442a. Sanger, G. A. 1983. Diets and Food Web Relationships of Seabirds in the Gulf of Alaska and Adjacent Marine Regions. Pages 631-771 in: Final Report Outer Continental Shelf Environmental Assessment Program, US Fish and Wildlife Service Denver Wildlife Research Center Migratory Bird Project.

443. Schlorff, R. W. 1978. Predatory ecology of the Great Egret at Humboldt Bay, California. Pages 347-353 in: A. J. Sprunt, J. C. Ogden, and S. Winkler, editors. Wading Birds. National Audubon Society Research Report No. 7.

444. Scheffer, V. B. and J. W. Slipp. 1944. The harbor seal in Washington State. American Naturalist 32:374-416.

444a. ————, and ————. 1948. The whales and dolphins of Washington State with a key to the cetaceans of the west coast of North America. American Midland Naturalist 39:257-337.

445. ————. 1950. The food of the Alaska Fur Seal. United States Dept. of the Interior, Fish and Wildlife Service. Wildlife Leaflet No. 329. Washington D.C.

446. Scheffer, T. H. and C. C. Sperry. 1931. Food habits of the Pacific Harbor Seal, Phoca richardii. Journal of Mammalogy 12:214-226.

447. Schwartz, J. E. II and G. E. Mitchell. 1945. The Roosevelt elk on the Olympic Peninsula, Washington. Journal of Wildlife Management 9:295-319.

448. Scott, J. M. 1973. Resource allocation in four synoptic species of marine diving birds. Ph.D. Dissertation, Oregon State University, Corvallis, OR.

449. Scott, W. B., and E. J. Crossman. 1973. Freshwater fishes of Canada. Fisheries Research Board Can., Bulletin 184.

450. Schuldt, and Hershey. 1995. Effect of salmon carcass decomposition on Lake Superior tributary streams. Journal of North American Benthological Society 14 (2) 259-268.

451. Sedell, J. R., F. J. Triska, and N. S. Triska. 1975. The processing of conifer and hardwood leaves in two coniferous forest streams. 1. Weight loss and associated invertebrates. Verh. Int. Ver. Limnol. 19:1617-1627.

452. ————, R. J. Naiman, K. W. Cummins, G. W. Minshall, and R. L. Vannote. 1979. Transport of particulate organic material in streams as a function of physical processes. International. Verh. Int. Ver. Limnol. 20:1366-1375.

453. ————, and K. J. Luchessa. 1981. Using the historical record as an aid to salmonid habitat enhancement. Pages 210 to 223 in: N. B. Armantrout, editor. Aquisition and Utilization of Aquatic Habitat Information. Proceedings of a Symposium held 28-30 October, 1981. Portland, OR.

454. Seliskar, D. M., and J. L. Gallagher. 1983. The ecology of tidal marshes of the Pacific Northwest coast: a community profile. USDI—Fish and Wildlife Service, Division of Biological Services. FWS/OBS-82/32. Washington, D.C.

455. Senn, H. G. 1958. Merganser depredation. Unpublished Report to Fisheries Management Division, State of Washington Department of Game, Olympia, WA.

456. Serdar, C. F. 1999. Description, analysis, and impacts of the Grouse Creek Landslide, Jefferson County, Washington, 1997-98. Master of Environmental Studies, The Evergreen State College. Olympia, WA.

457. Servheen, C. W. 1975. Ecology of the wintering Bald Eagle on the Skagit River, Washington. M.S. Thesis. University of Washington, Seattle, WA.

458. Seufert, F. 1980. Wheels of fortune. Oregon Historical Society.

459. Shea, D. S. 1973. White-tailed deer eating salmon. Murrelet 54:23.

459a. Shetter, D.S., and G.R. Alexander. 1970. Result of predator reduction on brook trout and brown trout in 4.2 milse (6.76 km) of the North Branch of the Au Sable River. Transactions of the American Fisheries Society 2:312-319.

460. Shreffler, D. K., and R. M. Thom. 1993. Restoration of Urban Estuaries: New approaches for site location and design. Prepared for the

Washington State Department of Natural Resources by Battelle/ Marine Sciences Laboratory. Sequim, WA.

461. Shreffler, D. A., C. A. Simenstad, and R. M. Thom. 1992. Temporary residence of juvenile salmon in a restored estuarine wetland. Canadian Journal of Fisheries and Aquatic Science 47:2079-2084.

462. Shuldt, J. A., and A. E. Hershey. 1995. Effect of salmon carcass decomposition on Lake Superior tributary streams. Journal of Wildlife Management 14:1-9.

463. Sibatani, A. 1992. Naze sake ha kawa wo sozyousuruka (Why do salmon go up rivers). In: Tyuuou Kouron/Chuo Koron (Tokyo) 1992(4): 286-295 (In Japanese).

464. ———. 1996. (English translation by: R. Davis). Why do salmon ascend rivers? Selected Papers on Entropy Studies 3:3-11.

465. Simenstad, C. A., W. J. Kinney, S. S. Parker, E. O. Salo, J. R. Cordell, and H. Buechner. 1980. Prey community structures and trophic ecology of outmigrating juvenile chum and pink salmon in Hood Canal, Washington: A synthesis of three years' studies, 1977-1979. Final Rep. to Washington Department of Fisheries. Fisheries Research Institute, University of Washington. FRI-UW-8026. Seattle, WA.

466. Simenstad, C. A., K. L. Fresh, and E. O. Salo. 1982. The role of Puget Sound and Washington coastal estuaries in the life history of Pacific salmon: An unappreciated function. Pages 343-364 In: V. S. Kennedy, editor. Estuarine Comparisons. Academic Press. New York, NY.

467. ———, and E. O. Salo. 1982. Foraging success as a determinant of estuarine and nearshore carrying capacity of juvenile chum salmon (Oncorhynchus keta) in Hood Canal, Washington. Pages 21-37 in: B. R. Melteff and R. A. Neve, editors. Proceedings of the North Pacific Aquaculture Symposium, 18-27 August 1980. Anchorage, Alaska and Newport, Oregon. Alaska Sea Grant Rep. 82-2. University of Alaska, Fairbanks, AK.

468. ———, B. S. Miller, C. F. Nyblade, K. Thornburgh, and L. J. Bledsoe. 1979. Food web relationships of northen Puget Sound and the Strait of Juan de Fuca. MESA Puget Sound Project, Eviron. Res. Labs. DOC/EPA Interagency Energy /Environment R&D Program Rep., EPA-600/7-79-259.

469. ———, M. Dethier, C. Levings, and D. Hay. 1997b. The Land-Margin Interface of Coastal Temperate Rain Forest Ecosytems: Shaping the Nature of Coastal Interactions. Pages 149-187 in: P. Schoonmaker, B. von Hagen, and E. Wolf, editors. The Rain Forests of Home: Profile of a North American Bioregion. Ecotrust/Interain Pacific and Island Press.

470. ———, J. R. Cordell, W. G. Hood, B. E. Feist, and R. M. Thom. 1997a. Ecological status of a created estuarine slough in the Chehalis River estuary: Assessment of created and natural estuarine sloughs, January-December 1995., Fisheries Research Institute, School of Fisheries, FRI-UW-9621, University of Washington, Seattle, WA.

471. ———, J. R. Cordell, J. A. Miller, W. G. Hood, and R. M. Thom. 1993. Ecological status of a created estuarine slough in the Chehalis River estuary: Assessment of created and natural estuarine sloughs, January-December 1992. FRI-UW-9305. Fish. Res. Inst., University of Washington, Seattle, WA.

472. ———, , W. G. Hood, J. A. Miller and R. M. Thom. 1992. Ecological status of a created estuarine slough in the Chehalis River estuary: Report of monitoring in created and natural estuarine sloughs, January-December 1991. Fisheries Research Institute, University of Washington, FRI-UW-9206. Seattle, WA.

473. ———, W. G. Hood, R. M. Thom, D. A. Levy and D. L. Bottom. In press. Landscape structure and scale constraints on restoring estuarine wetlands for Pacific Coast juvenile fishes. In: M. P. Weinstein and D. A. Kreeger, editors. Concepts and Controversies in Tidal Marsh Ecology, Kluwer Academic Publ., Dordrecht, Netherlands.

474. Simons, D. D. 1983. Interactions between California Condors and humans in prehistoric far western North America. Pages 470-494 in: S. R. Wilbur and J. A. Jackson, editors. Vulture Biology and Management. University of California Press, Berkeley, CA.

475. Skagen, S. K. R. L. Knight and G. H. Orians. 1991. Human disturbance of an avian scavenging guild. Ecological Applications 1:215-225.

476. Slaney, P. A., B. O. Rublee, C. J. Perrin, and H. Goldberg. 1994. Debris structure placements and whole-river fertilization for salmonids in a large regulated stream in British Columbia. Bulletin of Marine Science 55:1160-1180.

477. ———, and D. Zaldokas. 1997. Fish habitat rehabilitation procedures. Watershed Restoration Technical Circular No. 9. Watershed Restoration Program, Ministry of Environment, Lands and Parks, Vancouver, BC, Canada.

478. Smith, C. L. 1979. Salmon fishers of the Columbia. Oregon State University Press, Corvallis, OR.

479. Smith, M. R., P. W. Mattocks, Jr., M. K. Cassidy. 1997. Breeding birds of Washington State. In: K. M. Cassidy, C. E. Grue, M. R. Smith, and K. M. Dvornich, editors. Vol. 4. Washington State GAP Analysis - Final Report. Seattle Audubon Society, Publications in Zoology, No. 1. Seattle, WA.

480. Smith, J. L. and D. R. Mudd. 1978. Food of the Caspian Tern in Grays Harbor, Washington. Murrelet 59:105-106.

481. Smith, M. W. 1968. Fertilization and predator control to increase growth rate and yield of trout in a natural lake. Journal of Fisheries Research Board of Canada 25:2011-2036.

482. Spalding, D. J. 1964. Comparative feeding habits of fur seal, sea lion, and harbor seal on the British Columbia coast. Fisheries Research Board of Canada Bull. No. 146.

483. Spanier, E. 1980. The use of distress calls to repel night herons (Nycticorax nycticorax) from fish ponds. Journal of Appl. Ecol. 17:287-294.

484. Speaker, R. W., K. J. Luchessa, J. F. Franklin, and S. V. Gregory. 1988. The use of plastic strips to measure leaf retention by riparian vegetation in a coastal Oregon stream. American Midland Naturalist 120:22-31.

485. Spence, B. C., G. A. Lomnicky, R. M. Hughes and R. P. Novitzki. 1996. An ecosystem approach to salmonid conservation. Report prepared by ManTech. TR-4501-96-6057. Corvallis, OR.

486. Spencer, C. N., B. R. McClelland, and J. A. Stanford. 1989. Shrimp stocking, salmon collapse and eagle displacement: cascading interactions in the food web of a large aquatic ecosystem. BioScience 41:14-21.

487. Stalmaster, Mark V. 1980. Salmon carrion as a winter food source for red-tailed hawks. Murrelet 61(1):43-44.

488. Stalmaster, M. V. J. R. Newman and A. J. Hansen. 1979. Population dynamics of wintering bald eagles on the Nooksack River, Washington. Northwest Science 53:126-131.

489. Stalmaster, M. V. 1976. Winter ecology and effects of human activity on bald eagles in the Nooksack River Valley, Washington. M.S. Thesis. Western Washington State College, Bellingham, WA.

490. ———, and J. A. Gessaman. 1984. Ecological energetics and foraging behavior of overwintering bald eagles. Ecological Monographs 54:407-428.

491. Stanford, J. A., and J. V. Ward. 1988. The hyporheic habitat of river ecosystems. Nature 335:64-66.

492. Stanford, J. A., and A. R. Gaufin. 1974. Hyporheic communities of two Montana rivers. Science 185:700-702.

493. Steeger, C., H. Esselink, and R. C. Ydenberg. 1992. Comparative feeding ecology and reproductive performance of Ospreys in different habitats of southeastern British Columbia. Canadian Journal of Zoology 70:470-475.

494. Stenson, G. B., G. A. Badgero and H. D. Fisher. 1984. Food habits of the River Otter Lutra canadensis in the marine environment of British Columbia. Canadian Journal of Zoology 62:88-91.

495. Stewart, B. S. and S. Leatherwood. 1985. Minke whale. Balaenoptera acutorostrata Lacepede, 1804. Pages 91-136 in: S. H. Ridgway, and R. Harrison, editors. Handbook of Marine Mammals. Volume 3. The Sirenians and Baleen Whales. Academic Press, London & Orlando, FL.

496. Stockner, J. G., and E. A. MacIsaac. 1996. British Columbia lake enrichment program: two decades of habitat enhancement for sockeye salmon. Regulated Rivers: Research & Management 12: 547-561.

497. ———, and K. R. S. Shortreed. 1978. Enhancement of autotrophic production by nutrient addition in a coastal rainforest stream on Vancouver Island. Journal of Fisheries Res. Bd. Can. 35:28-34.

498. Stolz, J., and J. Schnell. 1991. Trout—The Wildlife Series. Stackpole Books. Cameron and Kelker Streets, Harrisburg, PA.

499. Stouder, D. J., P. A. Bisson, and R. J. Naiman, editors. 1997. Pacific salmon and their ecosystems: Status and future options. I. T. P. Chapman and Hall International Thomson Publishing. New York, NY.

500. Stroud, R. K. C. H. Fiscus and H. Kajimura. 1980. Food of the Pacific White-sided Dolphin, Lagenorhynchus obliquidens, Dall's Porpoise, Phocoenoides dalli and Northern Fur Seal, Callorhinus ursinus, off

California and Washington. U. S. Fish and Wildlife Service Fisheries Bulletin 78(4):951-959.

501. Strub, P.T., C. James, A. C. Thomas, and M. R. Abbott. 1990. Seasonal and nonseasonal variability of satelite-derived surface pigment concentration in the California Current. J. Geophy. Res. 95(C7):11503-11530.

502. Suckley, G. and J. G. Cooper. 1860. The natural history of Washington Territory and Oregon. Bailliere Brothers, New York, NY.

503. Sugai, S. F., and D. C. Burrell. 1984. Transport of dissolved organic carbon, nutrients, and trace metals from the Wilson and Blossom rivers to Smeaton Bay, southeast Alaska. Canadian Journal of Fisheries and Aquatic Sciences 41:180-190.

504. Sullivan, K., J., Tooley, K. Doughty, J. E. Caldwell, and P. Knudsen. 1990. Evaluation of prediction models and characterization of stream temperature regimes in Washington. Washington Department of Natural Resources. Timber/Fish/Wildlife Rep. No. TFW-WQ3-90-006. Olympia, WA.

505. Sumner, F. H. 1962. Migration and growth of coastal cutthroat trout in Tillamook County, Oregon. Transactions of American Fisheries Society 91:77-83.

506. Swanson, F. J., S.V. Gregory, J. R. Sedell, and A.G. Campbell. 1982. Land-water interactions: The riparian zone. Pages 267-291 In: R. L. Edmonds, editor. Analysis of coniferous forest ecosystems in the western United States. U.S. Int. Biolog. Prog. Synthesis Series 14, Hutchinson Ross. Stroudsburg, PA.

507. ———, L. E. Benda, S. H. Duncan, G. E. Grant, W. F. Megahan, L. M. Reid, and R. R. Ziemer. 1987. Mass failures and other processes of sediment production in Pacific Northwest forest landscapes. Pages 9-38 in: E. O. Salo and T. W. Cundy, editors. Streamside Management: Forestry and Fishery Interactions. College of Forest Resources, University of Washington. Institute of Forest Resources, Contribution No. 57. Seattle, WA.

508. Sweeney, S. J. 1978. Diet, reproduction and population structure of the bobcat (Lynx rufus fasciatus) in western Washington. M. S. Thesis. University of Washington, WA.

508a. Sweonson, J.E. 1978. Factors affecting status and reproduction of ospreys in Yellowstone National Park. Journal of Wildlife Management 43(3):595-601.

509. Sydeman, W. J. K. A. Hobson P. Pyle and E. B. McLaren. 1997. Trophic relationships among seabirds in central California: combined stable isotope and conventional dietary approach. Condor 99:327-336.

510. Tagart, J. V. 1976. The survival from egg deposition to emergence of coho salmon in the Clearwater River, Jefferson County, Washington. M.S. Thesis, University of Washington School of Fisheries, Seattle, WA.

510a. Tanner, W.W. 1949. Food of the wandering garter snake, Thamnophis elegans vagrans (Baird & Girard) in Utah. Herpetologica 5:85-86.

511. Taverner, P.A. 1934. Birds of Canada. Canada Dept. Mines, Natl. Mus. Bull. No. 72 (Biol. Ser. 19). Ottawa, Canada.

512. Thedinga, J. F., and K.V. Koski. 1984. A stream ecosystem in an old-growth forest in southeast Alaska. Part VI: The production of coho salmon, Oncorhynchus kisutch, smolts and adults from Porcupine Creek. Pages 99-108. in: W. R. Meehan, T. R. Merrel, Jr., and T.A. Hanley, editors. Proceedings from a Symposium on Fish and Wildlife Relationhships in Old-growth Forests. Am. Inst. Fish. Res. Biol., Juneau, AK.

513. Thom, R. M., and A. B. Borde. 1997. Human intervention in Pacific Northwest coastal ecosystems. Pages 5-37 in: G. R. McMurray and R. J. Bailey, editors. Change in Pacific Northwest Coastal Ecosystems. Science for Solutions, Pacific Northwest Coastal Ecosystems Regional Study (PNCERS). U.S. Department of Commerce. NOAA Coastal Ocean Program Decision Analysis Series No. 11.

514. Thomas, A. C., and P.T. Strub. 1989. Large scale patterns of phytoplankton pigment distribution during the spring transition along the west coast of North America. J. Geophy. Res. 94(C12):18-117.

515. Thomson, R. E. 1981. Oceanography of the British Columbia Coast. Can. Spec. Publ. Fisheries and Aquatic Science 56, Dept. Fish. Oceans. Sydney, BC, Canada.

516. Thut, R. N. 1970. Feeding habits of the Dipper in southwestern Washington. Condor 72:234-235.

517. Todd, I., and P. Larkin. 1971. Gillnet selectivity on sockeye (Oncorhynchus nerka) and pink salmon (O. gorbuscha) of the Skeena River system, British Columbia. Journal of Fisheries Research Board Can. 28:821-842.

518. Toweill, D. E. 1974. Winter food habits of River Otters in western Oregon. Journal of Wildlife Management 38:107-111.

519. Townsend, C. R., and M. R. Scarsbrook. 1997. The intermediate disturbance hypothesis, refugia, and biodiversity in steams. Limnol. Oceanogr. 42(5):938-949.

520. Triska, F. J., J. H. Duff, and R. J. Avanzino. 1990. Influence of exchange flow between channel and hyporheic zone on nitrate production in a small mountain stream. Canadian Journal of Fisheries and Aquatic Sciences 47:2099-2111.

521. Trotter, P. C. 1987. Cutthroat—Native Trout of the West. Colorado Associated University Press. Boulder, CO.

522. Tschaplinski, P. J. 1982. Aspects of the population biology of estuarine-reared and stream-reared juvenile coho salmon in Carnation Creek: A summary of current research. Pages 289-307 in: G. F. Hartman, editor. Proceedings of the Carnation Creek Workshop: A Ten-year Review. Malaspina College, Nanaimo, BC, Canada.

523. ———. 1987. The use of estuaries as rearing habitats by juvenile coho salmon. Pages 123-141 in: T. W. Chamberlin, editor. Proceedings of the Workshop: Applying 15 Years of Carnation Creek Results. Pacific Biol. Station. Nanaimo, BC, Canada.

524. Tsuchida, A. 1994. Resource physics and the limitations of nuclear fusion power generation. Selected Papers On Entropy Studies, Vol. 1.

525. ———. 1996. The importance of the "Cycle of Matter:" A discussion of nutrient cycling in basins, land, and human society from the viewpoint of entropy, pages 150-170. In Toyohashi, Proceedings of Clean Sea '96. International Workshop and Symposium On Environmental Restoration for Enclosed Seas.

526. United States Army Corps of Engineers (USACOE), U. S. Environmental Protection Agency, U. S. Fish and Wildlife Service, and National Marine Fisheries Service. 1993. Commencement Bay Cumulative Impact Study. U. S. Army Corps of Engineers, Seattle District. Seattle, WA.

527. Van Daele, L. J. and H. A. Van Daele. 1982. Factors affecting the productivity of ospreys nesting in west-central Idaho. Condor 84:292-299.

528. Vander Heyden, M. 1997. Female black bear habitat selection and home range ecology in the central Cascades of Oregon. M.S. Thesis, Oregon State University, Corvallis, OR.

529. Vannote, R. L., G. W. Minshall, K. W. Cummins, J. R. Sedell, and C. E. Cushing. 1980. The river continuum concept. Canadian Journal of Fisheries and Aquatic Sciences 37:130-137.

530. Vermeer, K. and K. Devito. 1986. Size, caloric content, and association of prey fishes in meals of nestling Rhinoceros Auklets. Murrelet 67:1-9.

531. ———, and S. J. Westrheim. 1984. Fish changes in diets of nestling rhinoceros auklets and their implications. Pages 96-105 in: D. N. Nettleship,, G. A. Sanger, and P. F. Springer, editors. Marine Birds: Their Feeding Ecology and Commercial Fisheries Relationships. Minister of Supply and Services, Canada.

532. ———. 1979. Nesting requirements, food and breeding distribution of rhinoceros auklets, Cerorhinca monocerata, and tufted puffins, Lunda cirrhata. Ardea 67:101-110.

533. ———. 1982. Comparison of the diet of the glaucous-winged gull on the east and west coasts of Vancouver Island. Murrelet 63:80-85.

533a. ———. 1992. The diet of birds as a tool for monitoring the biological environment. Pages 41-50 in K. Vermeer, R. W. Butler, and K.H. Morgan, editors. The ecology, status, and conservation of marine and shoreline birds on the west coast of Vancouver Island. Canadian Wildlife Service Occasional Papers, No. 75.

534. ———, and K. H. Morgan. 1992. Marine bird populations and habitat use in a fjord on the west coast of Vancouver Island. Canadian Wildlife Service Occasional Paper No. 75: 86-96.

535. Verts, B. J. and L. N. Carraway. 1998. Land Mammals of Oregon. University of California Press, Berkeley, CA.

536. Wallace, J. B., S. L. Eggert, J. L. Meyer, and J. R. Webster. 1997. Multiple trophic levels of a forest continuum concept. Canadian Journal of Fisheries and Aquatic. Sciences 37:130-104.

537. Watson, J. W. M. G. Garrett and R. G. Anthony. 1991. Foraging ecology of bald eagles in the Columbia River estuary. Journal of Wildlife Management 55:492-499.

538. Warren, C. E. 1971. Biology and water pollution control. The Department of Fisheries and Wildlife, Oregon State University, Corvallis, OR.

538a. Washington Department of Fisheries. 1992. Draft. Goals and objectives for stocks and fisheries. Olympia, WA.

538b. ———. 1992. Strategic plan for management of Washington's salmon resources. Olympia, WA.

539. Washington Department of Fish and Wildlife and Western Washington Treaty Indian Tribes. 1994. 1992 Washington State Salmon And Steelhead Stock Inventory. Appendix One, Puget Sound Stocks.

540. Washington Department of Fisheries (WDF) and Oregon Department of Fish and Wildlife (ODFW). 1992. Status report. Columbia River fish runs & fisheries, 1938-91. Olympia, WA.

541. ———. 1998. 1998 Washington salmonid stock inventory, coastal cutthroat trout appendix. DRAFT. Washington Deptartment of Fish and Wildlife, Olympia, WA.

541a. Washington Department of Fish and Wildlife (WDFW). 1998. Forage fish management plan - A plan for managing the forage fisheries in Washington. Adopted by the Washington Fish and Wildlife Commission on January 24, 1998.

542. Washington Department of Fisheries, Washington Department of Wildlife, and Western Washington Treaty Indian Tribes. 1992. 1992 Washington State Salmon And Steelhead Stock Inventory. Washington Department of Fish and Wildlife, Olympia, WA

543. Washington Department of Ecology. 1994. Inventory of dams in the state of Washington. Washington Department of Ecology, Water Resources Program, Dam Safety Division. Revised Edition, January 1994, Publication #94-16.

544. Washington Department of Fish and Wildlife's (WDFW). 1997. Wild Salmonid Policy. Washington Department of Fish and Wildlife. Olympia, WA.

545. Washington Department of Natural Resources (WDNR). 1997. Final Habitat Conservation Plan. Washington Department of Natural Resources, Jennifer Belcher, Commissioner of Public Lands. Olympia, WA.

546. Washington Department of Fish and Wildlife and Oregon Department of Fish and Wildlife. 1994. Status Report. Columbia fish runs and fisheries 1938-93. Olympia, WA.

547. Waters, T. F. 1969. Invertebrate drift - Ecology and significance to stream fishes. Pages 121-134 In: T. G. Northcote, editor. H. R. MacMillan Lectures In Fisheries. Held Feb. 22-24, 1968. Univ. British Columbia. Vancouver, BC, Canada.

548. Wehle, D. H. S. 1983. The food, feeding, and development of young tufted and horned puffins in Alaska. Condor 85:427-442.

549. Weitkamp, L. A., T. C. Wainwright, G. J. Bryant, G. B. Milner, D. J. Teel, R. G. Kope, and R. S. Waples. 1995. Status review of coho salmon from Washington, Oregon, and California. U.S. Dept. of Commerce. NMFS-NWFSC-24.

550. Wendler, H. O., and G. Deschamps. 1955. Logging dams on coastal Washington streams. Fish. Res. Pap. 1: 27-38, Washington Department of Fisheries, Olympia, WA.

551. White, H. C. 1936. The food of kingfishers and mergansers on the Margaree River, Nova Scotia. J. Biol. Bd. Can. 2:299-309.

552. ———. 1939. Bird control to increase the Margaree River salmon. Bull. Fisheries Research Board. Can. 58.

553. ———. 1957. Food and natural history of mergansers on salmon waters in the Maritime provinces of Canada. Fisheries Research Board Can., Bulletin No. 116.

554. White, R. 1992. Land use, environment, and social change: The shaping of Island County, Washington. University of Washington Press. Seattle, WA.

555. Whitman, J. S. 1981. Ecology of the mink (Mustela vison) in west-central Idaho. M.S. Thesis, University of Idaho.

555a. Wilke, F. A., and K. W. Kenyon. 1957. The food of fur seals in the eastern Bering Sea. Journal of Wildlife Management 21:237-238.

556. Williams, W. R., M. Laramie, and J. J. Ames. 1975. Catalog of Washington Streams and Salmon Utilization—Volume 1 Puget Sound. Washington Department of Fisheries. Olympia, WA.

557. Williams, D. D., and H. B. N. Hynes. 1974. The occurrence of benthos deep in the substratum of a stream. Freshwater Biology 4:233-256.

558. ———, and B. W. Feltmate. 1992. Aquatic insects. CAB International. United Kingdom.

559. Willson, M. F., and K. C. Halupka. 1995. Anadromous fish as keystone species in vertebrate communities. Conservation Biology 9:489-497.

560. Winter, B. D. 1992. Determinate migratory behavior of steelhead (Oncorhynchus mykiss) parr. Ph.D. Dissertation. University of Washington, School of Fisheries. Seattle, WA.

561. Willson, M. F., S. M. Gende, B. H. Marston. 1998. Fishes and the Forest: Expanding perspectives on fish-wildlife interactions. BioScience 48:455-462.

562. ———, Halupka K. C. 1995. Anadromous fish as keystone species in vertebrate communities. Conservation Biology 9:489-497.

563. Wilson, U. W and D. A. Manuwal. 1986. Breeding biology of the rhinoceros auklet in Washington. Condor 88:143-155.

564. Wipfli, M. S., J. Hudson, and J. Caouette. 1998. Influence of salmon carcasses on stream productivity: response of biofilm and benthic macroinvertebrates in southeastern Alaska, U.S.A. Canadian Journal of Fisheries and Aquatic Sciences 55:1503-1511.

565. Wissmar, R. C., and C. A. Simenstad. 1988. Energetic constraints of juvenile chum salmon (Oncorhynchus keta) migrating in estuaries. Canadian Journal of Fisheries and Aquatic Science. 45:1555-1560.

566. ———, and ———. 1998. Variability of estuarine and riverine ecosystem productivity for supporting Pacific salmon. Pages 253-301 in: G. R. McMurray and R. J. Bailey, editors. Change in Pacific Northwest Coastal Ecosystems. NOAA Coastal Ocean Prog., Decision Analysis Series No. 11, NOAA Coastal Ocean Office. Silver Spring, MD.

567. Wood, C. C. 1986. Dispersion of Common Merganser (Mergus merganser) breeding pairs in relation to availability of juvenile Pacific salmon in Vancouver Island streams. Canadian Journal of Zoology 64:756-765.

568. ———. 1985. Aggregative response of Common Mergansers (Mergus merganser): predicting flock size and abundance on Vancouver Island salmon streams. Canadian Journal of Fisheries and Aquatic Science 42:1259-1271.

569. ———, and C. M. Hand. 1985. Food searching behavior of the Common Merganser (Mergus merganser) I: functional responses to prey and predator density. Canadian Journal of Zoology 63:1260-1270.

570. ———. 1987a. Predation by the common merganser (Mergus merganser) on Eastern Vancouver Island. I: Predation during the seaward migration. Canadian Journal of Fisheries and Aquatic Sciences 44:941-949.

571. ———. 1987b. Predation by the common merganser (Mergus merganser) on Eastern Vancouver Island. II: Predation of stream-resident juvenile salmon by merganser broods. Canadian Journal of Fisheries and Aquatic Sciences 44:950-959.

572. Wurtsbaugh, W. A., and G. E. Davis. 1977. Effects of temperature and ration level on the growth and food conversion efficiency of Salmo gairdneri Richardson. J. Fish. Biol. 11:87-98.

573. Wydoski, R. S., and R. R. Whitney. 1979. Inland fishes of Washington. University of Washington Press, Seattle, WA.

574. Yoakum, J. 1964. Observations on bobcat—water relationships. Journal of Mammalogy 45:477-479.

575. Young, S. P. 1944. The Wolves of North America. Their history, life habits, economic status, and Control. The American Wildlife Institute, Washington, D.C.

576. Zarnowitz, J. E., and K. J. Raedeke. 1984. Winter predation on coho fingerlings by birds and mammals in relation to pond characteristics. College of Forest Resources, University of Washington. Service Contract No. 1480. Seattle, WA.

Appendix I. The nine wildlife species identified as having (or historically had) a *strong, consistent* relationship with salmon in Oregon and Washington.

	Incubation	FreshwaterRearing	Saltwater	Spawning	Carcass
Common Merganser	x	x	x		
Harlequin Duck	x		x		
Osprey		x	x	x	
Bald Eagle			x	x	x
Caspian Tern		x	x		
Black Bear				x	x
Grizzly Bear				x	x
Northern River Otter		x		x	x
Killer Whale			x		

An "x" identifies the life stage(s) of salmon applicable to the species.

Appendix II. The fifty-eight wildlife species identified as having (or historically had) a *recurrent* relationship with salmon in Oregon and Washington.

	Incubation	Freshwater Rearing	Saltwater	Spawning	Carcass
Cope's Giant Salamander	x	x			
Pacific Giant Salamander	x	x			
Pacific Coast Aquatic Garter Snake	x	x			
Red-throated Loonx		x	x		
Pacific Loon			x		
Common Loon		x	x		
Pied-billed Grebe		x			
Western Grebe		x	x		
Clark's Grebe			x		
American White Pelican		x			
Brandt's Cormorant		x	x		
Double-crested Cormorant		x	x		
Pelagic Cormorant		x	x		
Great Blue Heron		x	x		
Black-crowned Night-heron		x	x		
Turkey Vulture					x
California Condor					h
Common Goldeneye	x			x	x
Barrow's Goldeneye	x			x	x
Common Merganser					x
Red-breasted Merganser	x	x	x		
Golden Eagle				x	x
Bonaparte's Gull	x		x		x
Heermann's Gull			x		
Ring-billed Gull		x	x		x
California Gull			x		x
Herring Gull		x	x		x
Thayer's Gull			x		
Western Gull			x		x
Glaucous-winged Gull	x		x	x	x
Glaucous Gull			x		x
Common Tern		x	x		
Arctic Tern		x	x		
Forster's Tern		x	x		
Elegant Tern			x		
Common Murre			x		

	Incubation	Freshwater Rearing	Saltwater	Spawning	Carcass
Marbled Murrelet		x	x		
Rhinoceros Auklet			x		
Tufted Puffin			x		
Belted Kingfisher		x	x	x	
American Dipper	x	x			x
Steller's Jay					x
Black-billed Magpie		x			x
American Crow		x			x
Northwestern Crow		x	x		x
Common Raven		x		x	x
Virginia Opossum					x
Water Shrew	x	x			x
Coyote					x
Gray Wolf				x	x
Raccoon		x			x
Mink		x		x	x
Bobcat				x	x
Northern Fur Seal			x		
Northern (Steller) Sea Lion			x	x	x
California Sea Lion			x	x	
Harbor Seal			x	x	x
Pacific White-sided Dolphin			x		

An "x" identifies the life stage(s) of salmon applicable to the species currently, "h" represents a historical relationship with salmon.

Appendix III. The twenty-five wildlife species identified as having an *indirect* relationship with salmon in Oregon and Washington.

	Incubation	Freshwater Rearing	Saltwater	Spawning	Carcass
Harlequin Duck					x[a]
Bald Eagle	x[b]	x[b]	x[b]		x[b]
Gyrfalcon		x[b]	x[b]		x[b]
Peregrine Falcon		x[b]	x[b]		x[b]
Killdeer					x[a]
Spotted Sandpiper					x[a]
Snowy Owl		x[b]			
Willow Flycatcher					x[a]
Tree Swallow					x[a]
Violet-green Swallow					x[a]
Northern Rough-winged Swallow					x[a]
Bank Swallow					x[a]
Cliff Swallow					x[a]
Barn Swallow					x[a]
American Dipper					x[a]
Masked Shrew					x[a]
Vagrant Shrew					x[a]
Montane Shrew					x[a]
Fog Shrew					x[a]
Pacific Shrew					x[a]
Water Shrew					x[a]
Pacific Water Shrew					x[a]
Trowbridge's Shrew					x[a]
Harbor Porpoise			x[b]		
Dall's Porpoise			x[b]		

An "x" identifies the life stage(s) of salmon applicable to the species. Species have indirect relationship with salmon either through consuming insects associated with carcasses ([a]), or through feeding on gulls, terns, or waterfowl that eat live salmon or salmon carcasses ([b]).

Appendix IV. The sixty-four wildlife species identified as having (or historically had) a *rare* relationship with salmon in Oregon and Washington.

Incubtion	Freshwater Rearing	Saltwater	Spawning	Carcass
Snapping Turtle		x		
Western Pond Turtle		x		
Western Terrestrial Garter Snake		x		
Common Garter Snake		x		
Pacific Loon				x
Common Loon				x
Yellow-billed Loon			x	
Horned Grebe	x		x	
Red-necked Grebe			x	x
Western Grebe				x
Sooty Shearwater			x	
Brown Pelican			x	
Great Egret		x	x	
Snowy Egret		x		
Green Heron		x	x	
Trumpeter Swan	x	x		x
Mallard	x			x
Green-winged Teal	x			
Canvasback				x
Greater Scaup	x			x
Surf Scoter			x	x
White-winged Scoter			x	x
Common Goldeneye			x	x
Barrow's Goldeneye			x	
Hooded Merganser	x	x		x
Red-tailed Hawk				x
Greater Yellowlegs	x			
Franklin's Gull		x		
Mew Gull	x			
Black-legged Kittiwake			x	x
Pigeon Guillemot			x	
Ancient Murrelet			x	
Gray Jay				x
Winter Wren				x
American Robin	x			
Varied Thrush	x			x
Spotted Towhee				x
Song Sparrow				x
Masked Shrew				x
Vagrant Shrew				x
Montane Shrew				x
Fog Shrew				x
Pacific Shrew				x
Pacific Water Shrew				x
Torwbridge's Shrew				x
Douglas' Squirrel				x
Northern Flying Squirrel				x
Deer Mouse				x
Red Fox				x
Gray Fox				x
Ringtail				x
American Marten				x
Fisher				x
Long-tailed Weasel				x
Wolverine				x

Incubation	Freshwater Rearing	Saltwater	Spawning		Carcass
Striped Skunk					x
Mountain Lion				x	x
White-tailed Deer					x
Minke Whale			x		
Sperm Whale			x		
Humpback Whale			x		
Northern Right-whale Dolphin			x		
Dall's Porpoise			x		
Harbor Porpoise			x		

An "x" identifies the life stage(s) of salmon applicable to the species.

Appendix V. The sixty wildlife species identified as having an *unknown* relationship with salmon in Oregon and Washington.

Northwestern Salamander
Long-toed Salamander
Olympic Torrent Salamander
Columbia Torrent Salamander
Southern Torrent Salamander
Cascade Torrent Salamander
Rough-skinned Newt
Dunn's Salamander
Van Dyke's Salamander
Tailed Frog
Western Toad
Woodhouse's Toad
Red-legged Frog
Oregon Spotted Frog
Foothill Yellow-legged Frog
Northern Leopard Frog
Bullfrog
Painted Turtle
Red-eared Slider Turtle
Western Sandpiper
Dunlin
Short-billed Dowitcher
Least Flycatcher
Hammond's Flycatcher
Dusky Flycatcher
Pacific-slope Flycatcher
Cordilleran Flycatcher
Black Phoebe
Warbling Vireo
Purple Martin
Gray Catbird
European Starling
Brewer's Blackbird
Northern Waterthrush
Baird's Shrew
Merriam's Shrew
Pygmy Shrew
Shrew-mole
California Myotis
Western Small-footed Myotis
Yuma Myotis
Little Brown Myotis
Long-legged Myotis

Fringed Myotis
Keen's Myotis
Long-eared Myotis
Silver-haired Bat
Western Pipistrelle
Big Brown Bat
Hoary Bat
Spotted Bat
Townsend's Big-eared Bat
Pallid Bat
Brazilian Free-tailed Bat
Townsend's Chipmunk
Columbian Mouse
Townsend's Vole
Water Vole
Ermine
Western Spotted Skunk

Appendix VI. List of published and unpublished observations of wildlife predation and scavenging on salmon.

Species	Relationship to salmon	References	Location of study/report
Incubation—Eggs and Alevin			
Cope'sGiant Salamander	Recurrent	Johnson et al. in prep.	
Pacific Giant Salamander	Recurrent	Graf 1949	California
Pacific Coast Aquatic Garter Snake	Recurrent	Brown et al. 1995	Oregon
Common Merganser	Strong	Munro and Clemens 1939, Munro 1932, 1937	British Columbia
Hooded Merganser	Rare	Reimchen 1994	British Columbia
Red-breasted Merganser	Recurrent	Munro and Clemens 1939	Canada
Barrow's Goldeneye	Recurrent	Willson and Halupka 1995	Alaska
		Munro 1938, 1939	British Columbia
Common Goldeneye	Recurrent	Willson and Halupka 1995	Alaska
		Munro 1938, 1939	British Columbia
American Dipper	Recurrent	Ehinger 1930	Washington
		Munro 1923, Obermayer et al. 1999, Piorkowski 1995, Willson and Halupka 1995	Alaska
		Reimchen 1994, Burcham 1904	British Columbia
Glaucous-winged Gull	Recurrent	Baird 1990	Washington
		Reimchen 1994	British Columbia
		Mossman 1958, Moyle 1966	Alaska
Bonaparte's Gull	Recurrent	Moyle 1966	Alaska
Harlequin Duck	Strong	Dzinbal and Jarvis 1982, Obermayer et al. 1999	Alaska
Horned Grebe	Rare	Palmer 1962	Washington
		Munro 1941	British Columbia
Trumpeter Swan	Rare	Farley 1980	Alaska
Mallard	Rare	Willson and Halupka 1995	Alaska
Green-winged Teal	Rare	Ned Pittman pers. comm.	Washington
Greater Scaup	Rare	Munro 1941	British Columbia
Greater Yellowlegs	Rare	Elphick and Tibbitts 1998	Alaska
Mew Gull	Rare	Moyle 1966	Alaska
American Robin	Rare	Willson and Halupka 1995	Alaska
Varied Thrush	Rare	Ned Pittman pers. comm.	Washington
Water Shrew	Recurrent	Banfield 1974	Canada
Freshwater Rearing—Fry and Parr			
Cope's Giant Salamander	Recurrent	Antonelli et al. 1972	Washington
Pacific Giant Salamander	Recurrent	Antonelli et al. 1972	Washington
		Parker 1993, Parker 1994	California
Snapping Turtle	Rare	Johnson and Hasler 1954, Lagler 1943	Michigan
Western Pond Turtle	Rare	Johnson et al. in prep.	
Pacific Coast Aquatic Garter Snake	Recurrent	Brown et al. 1995a, Fitch 1941, 1984, Drummond 1983, Hansen 1980	Oregon, California
Western Terrestrial Garter Snake	Rare	Anderson 1977, Tanner 1949	Montana, Utah
Common Garter Snake	Rare	Lagler and Salyer 1945	Michigan
Red-throated Loon	Recurrent	Eriksson et al. 1990	Sweden
		Palmer 1962	Labrador
Common Loon	Recurrent	Palmer 1962, Johnson and Hasler 1954, Alexander 1977	Michigan
		Fraser 1972, 1974, Matkowski 1989, Matkowski 1984, Barr 1973, Smith 1968	Canada, British Columbia
		Willson and Halupka 1995	Alaska
		Munro 1945	British Columbia
Pied-billed Grebe	Recurrent	Zarnowitz and Raedeke 1984	Washington
Western Grebe	Recurrent	Modde et al. 1996	Utah
American White Pelican	Recurrent	Myers and Peterka 1976, Lingle 1977	North Dakota
		Palmer 1962	Wyoming

Species	Relationship to salmon	References	Location of study/report
Brandt's Cormorant	Recurrent	Johnson et al. in prep.	
Double-crested Cormorant	Recurrent	Modde et al. 1996	Utah
		Mayers and Peterka 1976	North Dakota
Pelagic Cormorant	Recurrent	Johnson et al. in prep.	
Great Blue Heron	Recurrent	Alexander 1977, 1979, Bent 1926, Johnson and Hasler 1954	Michigan
		Fraser 1972, Matkowski 1989, Smith 1968	Canada
		Dolloff 1993, Willson and Halupka 1995	Alaska
		Zarnowitz and Raedeke 1984	Washington
		Henney and Bethers 1971	Oregon
Great Egret	Rare	Johnson et al. in press	
Snowy Egret	Rare	Johnson et al. in press	
Green Heron	Rare	Jurek 1974	California
Black-crowned Night Heron	Recurrent	Spanier 1980	Israel
		Myers and Peterka 1976	North Dakota
Trumpeter Swan	Rare	Farley 1980	Alaska
		Hampton 1981	Montana
Common Goldeneye	Recurrent	Beach 1937	Michigan
		White 1939	Nova Scotia
Barrow's Goldeney	Recurrent	Munro 1938, Munro and Clemens 1939	British Columbia
Hooded Merganser	Rare	Zarnowitz and Raedeke 1984	Washington
Common Merganser	Strong	Alexander 1977, 1979, Beach 1937, Johnson and Halser 1954, Salyer and Lagler 1940, Shetter and Alexander 1970, Miegs and Rieck 1958, Senn 1958	Michigan
			Washington
		Fraser 1972, Huntsman 1941, Munro and Clemens 1939, Munro 1923, 1939, Smith 1968	Canada
		Willson and Halupka 1995	Alaska
		White 1936, 1957	Nova Scotia
Red-breasted merganser	Recurrent	Munro and Clemens 1939	Canada
		Marquiss and Duncan 1993	Scotland
Osprey	Strong	Swenson 1978	Wyoming
		Steeger et al. 1992	British Columbia
		MacCarter 1972	Montana
		Johnson and Hasler 1954	Michigan
		Van Daele and Van Daele 1982	Idaho
		French and Koplin 1977	California
		Hughes 1983	Alaska
		Lind 1976	Oregon
Franklin's Gull	Rare	Myers and Peterka 1976	North Dakota
Ring-billed Gull	Recurrent	Johnson et al. in prep.	
Caspian Tern	Strong	Johnson et al. in prep.	
Herring Gull	Recurrent	Mendall 1939	Maine
Forster's Tern	Recurrent	Ayles et al. 1976	Canada
Common Tern	Recurrent	Ayles et al. 1976	Canada
Arctic Tern	Recurrent	Mossman 1959, Willson and Halupka 1995	Alaska
Marbled Murrelet	Recurrent	Brooks 1928	British Columbia
		Carter and Sealy 1984	Alaska
Belted Kingfisher	Recurrent	Alexander 1977, 1979	Michigan
		Willson and Halupka 1995	Alaska
		Gould 1934, Eipper 1956	New York
		White 1936	Nova Scotia
		Elson 1962, Huntsman 1941	Canada
Black-billed Magpie	Recurrent	Willson and Halupka 1995	Alaska
American Crow	Recurrent	Willson and Halupka 1995	Alaska
Northwestern Crow	Recurrent	Willson and Halupka 1995	Alaska
Common Raven	Recurrent	Johnson et al. in prep.	
American Dipper	Recurrent	Dolloff 1993	Alaska
		Loegering 1997	Oregon
		Thut 1970	Washington
		Kingery 1996, Cottam and Uhler 1937, Munro 1923	British Columbia

Species	Relationship to salmon	References	Location of study/report
Water Shrew	Recurrent	Lampman 1947	Oregon
		Banfield 1974	Canada
		Conaway 1952	Montana
Raccoon	Recurrent	Alexander 1977	Michigan
Mink	Recurrent	Whitman 1981	Idaho
		Dunstone 1993	New York
		Banfield 1974, Burgess and Bider 1980, Fraser 1972	Canada
		Ben-David et al. 1997	Alaska
		Grinnell et al. 1937	California
		Alexander 1977, 1979	Michigan
		Akande 1972	Scotland
Northern River Otter	Strong	Zarnowitz and Raedeke 1984	Washington
		Banfield 1974, Stenson et al. 1984	British Columbia
		Dolloff 1993	Alaska
		Alexander 1979	Michigan

Saltwater—Smolt, Immature Adults, and Adults

Species	Relationship to salmon	References	Location of study/report
Pacific Loon	Recurrent	Mace 1983	British Columbia
Red-throated Loon	Recurrent	Johnson et al. in prep.	
Common loon	Recurrent	Palmer 1962	Alaska
		Mace 1983	British Columbia
Yellow-billed Loon	Rare	North 1994	Russia
Horned Grebe	Rare	Mace 1983, Vermeer et al. 1992	British Columbia
Red-necked Grebe	Rare	Mace 1983	British Columbia
Western Grebe	Recurrent	Vermeer 1992, Mace 1983	British Columbia
Clark's Grebe	Recurrent	Johnson et al. in prep.	
Sooty Shearwater	Rare	Emmett 1997, Bayer 1989	Oregon
Brown Pelican	Rare	Bayer 1986a, Emmett 1997, McNeil et al. 1991	Oregon
Brandt's Cormorant	Recurrent	Ainley and Sanger 1979	California
		Bayer 1986, Scott 1973	Oregon
Double-crested Cormorant	Recurrent	Bayer 1986, Bayer 1989, Erickson 1988, Hoffman and Hall 1988, Roby et al. 1998	Oregon
		Ainley and Anderson 1981, Mace 1993, Robertson 1974	British Columbia
Pelagic Cormorant	Recurrent	Bayer 1986, Scott 1973	Oregon
		Jewett et al. 1953	Washington
		Mace 1983	British Columbia
Great Blue Heron	Recurrent	Forbes and Simpson 1982, Mace 1983, Myers 1980	British Columbia
Green Heron	Rare	Johnson et al. in prep.	
Black-crowned Night Heron	Recurrent	Johnson et al. in prep.	
Great Egret	Recurrent	Johnson et al. in prep., Schlorff 1978	California
Harlequin Duck	Strong	Cottam 1939, Mace 1983	British Columbia
Surf Scoter	Rare	Mace 1983	British Columbia
White-winged Scoter	Rare	Mace 1983	British Columbia
Common Goldeneye	Rare	Mace 1983	British Columbia
Barrow's Goldeneye	Rare	Mace 1983	British Columbia
Common Merganser	Strong	Elson 1962	Canada
		Mace 1983, Macdonald et al. 1988, Wood and Hand 1985, Wood 1985, 1986, 1987a	British Columbia
Red-breasted Merganser	Recurrent	Mace 1983	British Columbia
Osprey	Strong	Bayer 1986a, Emmett 1997, Roby et al. 1998	Oregon
Bald Eagle	Recurrent	Knight et al. 1990, Watson et al. 1991	Washington
Bonaparte's Gull	Recurrent	Macdonald et al. 1988, Mace 1983	British Columbia
Heermann's Gull	Recurrent	Bayer 1986, 1989	Oregon
Ring-billed Gull	Recurrent	Bayer 1989, Roby et al. 1998	Oregon
		Ruggerone 1986	Washington
California Gull	Recurrent	Roby et al. 1998	Oregon
		Mace 1983	British Columbia
Herring Gull	Recurrent	Bent 1921	Washington
		Mace 1983, Macdonald et al. 1988	British Columbia

Species	Relationship to salmon	References	Location of study/report
Thayer's Gull	Recurrent	Mace 1983	British Columbia
Western Gull	Recurrent	Bayer 1986, 1986b, Roby et al. 1998	Oregon
Glaucous-winged Gull	Recurrent	Baird 1990	Washington
		Roby et al. 1998, Bayer 1986	Oregon
		Mace 1983, Vermeer 1982	British Columbia
Glaucous Gull	Recurrent	Sanger 1983	Alaska
Black-legged Kittiwake	Rare	Rowlett 1980, Sanger 1983	Alaska
		Simenstad et al. 1979	Oregon
Caspian Tern	Strong	Smith and Mudd 1978	Washington
		Roby et al. 1998	Oregon
Elegant Tern	Recurrent	Johnson et al. in press	
Common Tern	Recurrent	Johnson et al. in press, Simenstad et al. 1979	Oregon
Forster's Tern	Recurrent	Johnson et al. in press	
Arctic Tern	Recurrent	Simenstad et al. 1979	Oregon
		Willson and Halupka 1995	Alaska
Common Murre	Recurrent	Sydeman et al. 1997	California
		Bayer 1986, Bayer 1986b, Matthews 1983	Oregon
Common Murre	Recurrent		Oregon
		Ainley et al. 1990	Alaska
Pigeon Guillemot	Rare	Bayer 1986	Oregon
Marbled Murrelet	Recurrent	DeGange 1996	Alaska
		Mace 1983	British Columbia
Ancient Murrelet	Rare	Mace 1983	British Columbia
Rhinoceros Auklet	Recurrent	Wilson and Manuwal 1986	Washington
		Burger et al. 1993, Vermeer and DeVito 1986, Vermeer and Westrheim 1984, Vermeer 1979	British Columbia
		Gaston and Dechesne 1996, Sydeman et al. 1997	California
		Sanger 1983	Alaska
Tufted Puffin	Recurrent	Baird 1990, 1991b, Wehle 1983	Alaska
Belted Kingfisher	Recurrent	Bayer 1989	Oregon
		Mace 1983	British Columbia
Northwestern Crow	Recurrent	Mace 1983	British Columbia
Northern Fur Seal	Recurrent	Antonelis an Perez 1984, Baker et al. 1970, Kajimura 1983	Washington, Oregon
		Clemens and Wilby 1933, Spalding 1964	British Columbia
		Scheffer 1950, Wilke and Kenyon 1957	Bering Sea
		Banfield 1974	Canada
		Fiscus and Baines 1966, Imler and Sarber 1947, Pitcher 1981	Alaska
		Dalquest 1948	Washington
		Riemer and Brown 1997, Roffe and Mate 1984	Oregon
		Banfield 1974, Spalding 1964	British Columbia
California Sea Lion	Recurrent	Gearin et al. 1988, Jeffries 1985	Washington
		Jameson and Kenyon 1977, Riemer and Brown 1997, Roffe and Mate 1984	Oregon
		Baltz and Morejohn 1977, Harvey and Weise 1997, Jones 1981, NMFS 1997	California
Harbor Seal	Recurrent	Dalquest 1948, Everitt et al. 1981, Scheffer and Sperry 1931, Scheffer and Slipp 1944	Washington
		Beach et al. 1985, Brown and Mate 1983, Brown et al. 1995, Browne et al. 1997, Graybill 1981, Jeffries 1985, Riemer and Brown 1997, Roffe and Mate 1984	Oregon
		Olesiuk 1993, Olesiuk et al. 1990, Spalding 1964	British Columbia
		Imler and Sarber 1947, Pitcher and Calkins 1981, Pitcher 1977, 1981	Alaska
		Briggs and Davis 1972, Jones 1981, Herder 1983, Hanson 1993	California
Minke Whale	Rare	Banfield 1974	Canada
		Stewart and Leatherwood 1985	Atlantic Ocean

Species	Relationship to salmon	References	Location of study/report
Humpback Whale	Rare	Johnson and Wolman 1984	Northern Hemisphere
Pacific White-sided Dolphin	Recurrent	Kajimura 1983, Stroud et al. 1980	Washington
Northern Right-Whale Dolphin	Rare	Johnson et al. in prep.	
Killer Whale	Strong	Hall 1986	Alaska
		Scheffer and Slipp 1948	Washington
Killer Whale	Strong	Nichol and Shackleton 1996, Banfield 1974	British Columbia
		Balcomb et al. 1982, Heimlich-Boran 1986, 1988, Felleman et al. 1991	Pacific Northwest
Harbor Porpoise	Rare	Gearin et al. 1994	Washington
		Fontaine et al. 1994	Canada
Dall's Porpoise	Rare	Norris and Prescott 1961	Oregon
		Mizue et al. 1996	Washington
Sperm Whale	Rare	Leatherwood and Reeves 1983, Pike 1950	British Columbia
Osprey	Strong	Johnson et al. in prep.	
Bald Eagle	Strong	Hunt et al. 1992, Servheen 1975, Spencer et al. 1989, Stalmaster 1976	Washington
		Simons 1983	Oregon

Spawning

Species	Relationship to salmon	References	Location of study/report
Bald Eagle	Strong	Munro 1938a	British Columbia
		Ofelt 1975	Alaska
Golden Eagle	Recurrent	McClelland 1973	Montana
Glaucous-winged Gull	Recurrent	Mossman 1958	Alaska
Belted Kingfisher	Recurrent	Reimchen 1994	British Columbia
Common Raven	Recurrent	Reimchen 1994	British Columbia
Gray Wolf	Recurrent	Young 1944	Alaska
Black Bear	Strong	Kellyhouse 1975	California
		Reimchen 1994	British Columbia
		Reinhart and Mattson 1989	Idaho
		Chi 1999, Moyle 1966, Piorkowski 1995, Wilson et al. 1998	Alaska
		Banfield 1974	Canada
Grizzly Bear	Strong	Banfield 1974	Canada
Mink	Recurrent	Eagle and Whitman 1987	Canada
		Melquist et al. 1981	Idaho
		Hatler 1976	British Columbia
Northern River Otter	Strong	Melquist et al. 1981	Idaho
		Toweill 1974	Oregon
Mountain Lion	Rare	Ned Pittman pers. comm.	Washington
Bobcat	Recurrent	Yoakum 1964	Washington
Harbor Seal	Recurrent	Reimchen 1994	British Columbia
		Everitt et al. 1981	Washington
		Riemer and Brown 1997	Oregon
California Sea Lion	Recurrent	Jeffries 1985	Oregon
		Riemer and Brown 1997, Roffe and Mate 1984	Oregon
Northern (Steller) Sea Lion	Recurrent	Roffe and Mate 1984, Riemer and Brown 1997	Oregon
		Reimchen 1994	British Columbia

Carcasses

Species	Relationship to salmon	References	Location of study/report
Western Pond Turtle	Rare	Holland 1985a	California
Common Loon	Rare	Reimchen 1994	British Columbia
Western Grebe	Rare	Reimcher 1994	British Columbia
Pacific Loon	Rare	Reimchen 1994	British Columbia
Red-necked Grebe	Rare	Reimchen 1994	British Columbia
Turkey Vulture	Recurrent	Jewett et al. 1953	Washington
California Condor	Recurrent	Gabrielson and Jewett 1940, Simons 1983	Oregon
California Condor	Recurrent	Suckley and Cooper 1860	Washington
Trumpeter Swan	Rare	Butler 1973	British Columbia

Species	Relationship to salmon	References	Location of study/report
Mallard	Rare	Jewett et al. 1953	Washington
		Reimchen 1994	British Columbia
Canvasback	Rare	Jewett et al. 1953	Washington
Greater Scaup	Rare	Munro 1941, Reimchen 1994	British Columbia
Surf Scoter	Rare	Reimchen 1994	British Columbia
White-winged Scoter	Rare	Reimchen 1994	British Columbia
Common Goldeneye	Recurrent	Dawson 1909, Jewett et al. 1953, Servheen 1975	Washington
		Taverner 1934	Canada
Barrow's Goldeneye	Recurrent	Jewett et al. 1953	Washington
Hooded Merganser	Rare	Reimchen 1994	British Columbia
Common Merganser	Recurrent	Stalmaster and Gessaman 1984	Washington
		Munro and Clemens 1939, Munro 1932, 1937	British Columbia
Bald Eagle	Strong	Cederholm et al. 1989, Jewett et al. 1953, Knight and Knight 1983, Skagen et al. 1991, Stalmaster and Gessaman 1984	Washington
		Simons 1983	Oregon
		Munro 1938a	British Columbia
		Shea 1973	Montana
Red-tailed Hawk	Rare	Willson and Halupka 1995	Alaska
		Cederholm et al. 1989, Ned Pittman pers. comm., Stalmaster and Gessaman 1984, Stalmaster 1980	Washington
Golden Eagle	Recurrent	Stalmaster and Gessaman 1984	Washington
Ring-billed Gull	Recurrent	Johnson et al. in prep.	
California Gull	Recurrent	Johnson et al. in prep., Reimchen 1994	British Columbia
Bonaparte's Gull	Recurrent	Moyle 1966, Willson and Halupka 1995	Alaska
Western Gull	Recurrent	Johnson et al. in prep.	
Herring Gull	Recurrent	Reimchen 1994	British Columbia
Glaucous-winged Gull	Recurrent	Reimchen 1994	British Columbia
		Simons 1983	Oregon
		Moyle 1966, Bent 1921	Alaska
Glaucous-winged Gull	Recurrent	Servheen 1975, Skagen et al. 1991, Stalmaster and Gessaman 1984	Washington
Glaucous Gull	Recurrent	Johnson et al. in prep.	
Black-legged Kittiwake	Rare	Reimchen 1994	British Columbia
Gray Jay	Rare	Cederholm et al. 1989	Washington
Steller's Jay	Recurrent	Cederholm et al. 1989	Washington
		Willson and Halupka 1995	Alaska
Black-billed Magpie	Recurrent	Willson and Halupka 1995	Alaska
		McClelland 1973	Montana
American Crow	Recurrent	Cederholm et al. 1989, Skagen et al. 1991, Stalmaster and Gessaman 1984	Washington
Northwestern Crow	Recurrent	Campbell et al. 1990, Reimchen 1994	British Columbia
Common Raven	Recurrent	Murie 1959, Willson and Halupka 1995	Alaska
		Knight and Anderson 1990, Stalmaster and Gessaman 1984, Cederholm et al. 1989, Suckley and Cooper 1860	Washington
Winter Wren	Rare	Cederholm et al. 1989	Washington
		Reimchen 1994	British Columbia
American Dipper	Recurrent	Willson and Halupka 1995	Alaska
		Cederholm et al. 1989	Washington
Varied Thrush	Rare	Reimchen 1994	British Columbia
Spotted Towhee	Rare	Ned Pittman pers. comm.	Washington
Song Sparrow	Rare	Ned Pittman pers. comm.	Washington
Virginia Opossum	Recurrent	Johnson et al. in prep.	
Deer Mouse	Rare	Cederholm et al. 1989	Washington
Douglas' Squirrel	Rare	Cederholm et al. 1989	Washington
Northern Flying Squirrel	Rare	Cederholm et al. 1989	Washington
Water Shrew	Recurrent	Cederholm et al. 1989	Washington
		Conaway 1952	Montana

Species	Relationship to salmon	References	Location of study/report
Vagrant Shrew	Rare	Cederholm et al. 1989	Washington
Masked Shrew	Rare	Cederholm et al. 1989	Washington
Trowbridge's Shrew	Rare	Johnson et al. in prep.	
Pacific Water Shrew	Rare	Johnson et al. in prep.	
Pacific Shrew	Rare	Johnson et al. in prep.	
Montane Shrew	Rare	Johnson et al. in prep.	
Fog Shrew	Rare	Johnson et al. in prep.	
Coyote	Recurrent	Young 1944	Oregon
		Cederholm et al. 1989, Stalmaster and Gessaman 1984	Washington
Gray Wolf	Recurrent	Young 1944	Oregon, British Columbia, Alaska
Red Fox	Rare	Young 1944	Oregon
		Willson and Halupka 1995	Alaska
		Hewson 1995	Scotland
Gray Fox	Rare	Young 1944	Oregon
Black Bear	Strong	Young 1944	British Columbia
		Piorkowski 1995, Chi 1999	Alaska
		Cederholm et al. 1989, Hilderbrand et al. 1996	Washington
Grizzly Bear	Strong	Mattson et al. 1991	Idaho
		Hamilton and Archibald 1985, Hamilton and Bunnel 1987, Young 1944	British Columbia
		Banfield 1974	Canada
		Verts and Carraway 1998	Oregon
Ringtail	Rare	Johnson et al. in prep.	
Raccoon	Recurrent	Cederholm et al. 1989	Washington
American Marten	Rare	Reimchen 1994, Nagorsen et al. 1989, 1991, Hatler 1976	British Columbia
Striped Skunk	Rare	Cederholm et al. 1989	Washington
Long-tailed Weasel	Rare	Cederholm et al. 1989	Washington
Mink	Recurrent	Ben-David et al. 1997, Ben-David 1997, Willson and Halupka 1995	Alaska
		Cederholm et al. 1989	Washington
		Hatler 1976, Young 1944	British Columbia
		Eagle and Whitman 1987	Canada
		Melquist et al. 1981	Idaho
Fisher	Rare	Young 1944	British Columbia
Wolverine	Rare	Willson and Halupka 1995, Banci 1987	Alaska
Northern River Otter	Strong	Cederholm et al. 1989	Washington
		Young 1944	British Columbia
		Grinnell et al. 1937	California
Bobcat	Recurrent	Brittell et al. 1979, Cederholm et al. 1989, Knick et al. 1984, Schwartz and Mitchell 1945, Sweeney 1978	Washington
White-tailed Deer	Rare	Shea 1973, Cederholm et al. 1989	Washington
Black-tailed Deer	Rare	Cederholm et al. 1989	Washington
Northern (Steller) Sea Lion	Recurrent	Reimchen 1994	British Columbia
		Willson and Halupka 1995	Alaska
Harbor Seal	Recurrent	Reimchen 1994	British Columbia

27

An Introduction to Wildlife-Habitat Relationships CD-ROM

Marla Trevithick, Thomas A. O'Neil, & Charley Barrett

Introduction

The purpose of this chapter is to provide a brief introduction to the accompanying CD-ROM to help navigate through the Wildlife-Habitat Relationships (WHR) digital information, and to list the kinds of information found on it. The CD-ROM is divided into four main sections, *Introduction, Definitions, Data Queries*, and *Maps*, which can be found on the Main Navigation Bar. Specifically, the *Introduction* gives an overview that has 7 submenus that depict Acknowledgements, Authors who developed the digital information, CD-ROM Navigation, Trouble-shooting, Site Map MetaData, and copyright information. The *Definitions* section gives explanations of the fields found within each matrix. The detailed descriptions can actually be found on the CD-ROM. *Data Queries* allows the user to view the digital information in a pre-determined or canned query format. These canned queries were developed to answer the most commonly asked questions regarding the subject matter and to allow the user to view and print out the results. Finally, the *Maps* section illustrates several maps of wildlife-habitat types for Oregon and Washington that can be reviewed on screen and also printed.

The data presented on the CD-ROM are a compilation of 7 matrixes that focus on wildlife species relationships with: 1) Wildlife-Habitat Types, 2) Structural Conditions, 3) Habitat Elements, 4) Key Ecological Functions, 5) Life History, 6) Salmon-Wildlife Relationships, and 7) Management Activity Links. Each of these matrixes can be found as submenus under the Main Navigation Bar for *Data Queries*. In-depth discussions and descriptions about the CD-ROM data (i.e. metadata) actually reside in specific chapters within this book, and these chapters will be highlighted in the Guide to the Wildlife-Habitat Relationships on the CD-ROM.

Information presented on the CD-ROM has undergone an extensive Quality Control effort that was the equivalent of two people working on reviewing the integrity of the data for two years. Thus a significant attempt has been made to ensure that the information presented is accurate.

CD-ROM Navigation

System Requirements

The *Matrixes for Wildlife-Habitat Relationships in Oregon and Washington's* CD-ROM are optimized to operate on IBM compatible Personal Computers (PCs). See the ReadMe.txt file for Unix and Macintosh. To view this CD-ROM, you will need Microsoft® Internet Explorer 5.0 or later. Recommended minimum computer requirements are: a Pentium 100 processor (or its equivalent), 32 megabytes RAM (64+ is highly recommended), 4 megabyte video adapter, 4X CD-ROM drive or faster, and if copying to a hard drive you will need 100 megabytes of free space.

Getting Started

To view the CD-ROM, insert the disk into your CD-ROM drive. On PCs with a Microsoft® Windows operating system and Auto-Run enabled, the CD-ROM will automatically start. On other systems, start Internet Explorer and then click "File/Open" from the main menu. In the Open dialog box, either type "*CD-ROM Drive Letter [i.e. D:]*/START.HTM" into the "Open:" box or use the "Browse" button to locate the file "START.HTM" in the root directory of the CD-ROM. Next, click the "OK" button in the Open dialog box and the CD-ROM will start. Please note, if you experience problems, check the Troubleshooting section.

Appendix

Scientific and common names for 743 wildlife species found in Oregon and Washington

Class	Order	Family	Scientific Name (Other Scientific Name)	Common Name (Other Common Name)	Subspecies Name Oregon & Washington only
Amphibia	Caudata	Ambystomatidae	Ambystoma tigrinum	Tiger Salamander	melanosticum
Amphibia	Caudata	Ambystomatidae	Ambystoma gracile	Northwestern Salamander	
Amphibia	Caudata	Ambystomatidae	Ambystoma macrodactylum	Long-toed Salamander	"macrodactylum, columbianum, sigillatum"
Amphibia	Caudata	Dicamptodontidae	Dicamptodon copei	Cope's Giant Salamander	
Amphibia	Caudata	Dicamptodontidae	Dicamptodon tenebrosus (Dicamptodon ensatus)	Pacific Giant Salamander	
Amphibia	Caudata	Rhyacotritonidae	Rhyacotriton olympicus	Olympic Torrent Salamander (Olympic Seep Salamander)	
Amphibia	Caudata	Rhyacotritonidae	Rhyacotriton kezeri (Rhyacotriton olympicus)	Columbia Torrent Salamander (Columbia Seep Salamander)	
Amphibia	Caudata	Rhyacotritonidae	Rhyacotriton variegatus (Rhyacotriton olympicus)	Southern Torrent Salamander (Southern Seep Salamander)	
Amphibia	Caudata	Rhyacotritonidae	Rhyacotriton cascadae (Rhyacotriton olympicus)	Cascade Torrent Salamander	
Amphibia	Caudata	Salamandridae	Taricha granulosa	Rough-skinned Newt	granulosa
Amphibia	Caudata	Plethodontidae	Plethodon dunni (Plethodon gordoni)	Dunn's Salamander	gordoni
Amphibia	Caudata	Plethodontidae	Plethodon larselli	Larch Mountain Salamander	
Amphibia	Caudata	Plethodontidae	Plethodon vandykei	Van Dyke's Salamander	"idahoensis, vandykei"
Amphibia	Caudata	Plethodontidae	Plethodon vehiculum	Western Red-backed Salamander	
Amphibia	Caudata	Plethodontidae	Plethodon elongatus	Del Norte Salamander	
Amphibia	Caudata	Plethodontidae	Plethodon stormi (Plethodon elongatus)	Siskiyou Mountains Salamander	
Amphibia	Caudata	Plethodontidae	Ensatina eschscholtzii	Ensatina	"oregonensis, picta"
Amphibia	Caudata	Plethodontidae	Aneides ferreus	Clouded Salamander	
Amphibia	Caudata	Plethodontidae	Aneides flavipunctatus	Black Salamander	
Amphibia	Caudata	Plethodontidae	Batrachoseps wrighti	Oregon Slender Salamander	
Amphibia	Caudata	Plethodontidae	Batrachoseps attenuatus	California Slender Salamander	
Amphibia	Anura	Leiopelmatidae	Ascaphus truei	Tailed Frog (American Bell Toad)	
Amphibia	Anura	Pelobatidae	Scaphiopus intermontanus (Spea intermontana)	Great Basin Spadefoot	
Amphibia	Anura	Bufonidae	Bufo boreas	Western Toad ("Boreal Toad, Northwestern Toad")	halophilus
Amphibia	Anura	Bufonidae	Bufo woodhousii	Woodhouse's Toad	woodhousii
Amphibia	Anura	Hylidae	Pseudacris regilla (Hyla regilla)	Pacific Chorus (Tree) Frog	

Class	Order	Family	Scientific Name (Other Scientific Name)	Common Name (Other Common Name)	Subspecies Name Oregon & Washington only
Amphibia	Anura	Ranidae	*Rana aurora*	Red-legged Frog (Northern Red-Legged Frog)	*aurora*
Amphibia	Anura	Ranidae	*Rana cascadae*	Cascades Frog	
Amphibia	Anura	Ranidae	*Rana pretiosa*	Oregon Spotted Frog (Spotted Frog)	
Amphibia	Anura	Ranidae	*Rana luteiventris* (*Rana pretiosa*)	Columbia Spotted Frog (Spotted Frog)	
Amphibia	Anura	Ranidae	*Rana boylii*	Foothill Yellow-legged Frog	None
Amphibia	Anura	Ranidae	*Rana pipiens*	Northern Leopard Frog	None
Amphibia	Anura	Ranidae	*Rana catesbeiana*	Bullfrog	
Amphibia	Anura	Ranidae	*Rana clamitans*	Green Frog	
Reptilia	Testudines	Chelydridae	*Chelydra serpentina*	Snapping Turtle	
Reptilia	Testudines	Emydidae	*Chrysemys picta*	Painted Turtle (Western Painted Turtle)	*belli*
Reptilia	Testudines	Emydidae	*Clemmys marmorata*	Western Pond Turtle (Northwestern Pond Turtle)	*marmorata*
Reptilia	Testudines	Emydidae	*Trachemys scripta* (*Pseudemys scripta*)	Red-eared Slider Turtle (Slider)	*"elegans, scripta"*
Reptilia	Testudines	Cheloniidae	*Caretta caretta*	Loggerhead Sea Turtle	
Reptilia	Testudines	Cheloniidae	*Chelonia mydas*	Green Sea Turtle	
Reptilia	Testudines	Cheloniidae	*Lepidochelys olivacea*	Pacific Ridley Sea Turtle	
Reptilia	Testudines	Dermochelyidae	*Dermochelys coriacea*	Leatherback Turtle	
Reptilia	Squamata	Anguidae	*Elgaria coerulea* (*Gerrhonotus coerulea*)	Northern Alligator Lizard	*"principis, shastensis"*
Reptilia	Squamata	Anguidae	*Elgaria multicarinata* (*Gerrhonotus multicarinata*)	Southern Alligator Lizard	*scincicauda*
Reptilia	Squamata	Iguanidae	*Crotaphytus bicinctores*	Mojave Black-collared Lizard	
Reptilia	Squamata	Iguanidae	*Gambelia wislizenii*	Long-nosed Leopard Lizard	
Reptilia	Squamata	Iguanidae	*Phrynosoma douglassii*	Short-horned Lizard	*douglassii*
Reptilia	Squamata	Iguanidae	*Phrynosoma platyrhinos*	Desert Horned Lizard (Northern Desert Horned Lizard)	*platyrhinos*
Reptilia	Squamata	Iguanidae	*Sceloporus graciosus*	Sagebrush Lizard	*graciosus*
Reptilia	Squamata	Iguanidae	*Sceloporus occidentalis*	Western Fence Lizard	*"occidentalis, longipes"*
Reptilia	Squamata	Iguanidae	*Uta stansburiana*	Side-blotched Lizard	*stansburiana*
Reptilia	Squamata	Scincidae	*Eumeces skiltonianus*	Western Skink	*"utahensis, skiltonianus"*
Reptilia	Squamata	Teiidae	*Cnemidophorus tigris*	Western Whiptail	*tigris*
Reptilia	Squamata	Teiidae	*Cnemidophorus velox*	Plateau Striped Whiptail	
Reptilia	Squamata	Boidae	*Charina bottae*	Rubber Boa	
Reptilia	Squamata	Colubridae	*Coluber constrictor* (*Coluber mormon*)	Racer	*mormon*
Reptilia	Squamata	Colubridae	*Contia tenuis*	Sharptail Snake	

Class	Order	Family	Scientific Name (Other Scientific Name)	Common Name (Other Common Name)	Subspecies Name Oregon & Washington only
Reptilia	Squamata	Colubridae	*Diadophis punctatus* (*Diadophis amabolis*)	Ringneck Snake (Northwest Ringneck Snake)	*occidentalis*
Reptilia	Squamata	Colubridae	*Hypsiglena torquata*	Night Snake	
Reptilia	Squamata	Colubridae	*Lampropeltis getula*	Common Kingsnake	*californiae*
Reptilia	Squamata	Colubridae	*Lampropeltis zonata*	California Mountain Kingsnake	
Reptilia	Squamata	Colubridae	*Masticophis taeniatus*	Striped Whipsnake	*taeniatus*
Reptilia	Squamata	Colubridae	*Pituophis catenifer* (*Pituophis melanoleucus*)	Gopher Snake	"*catenifer, deserticola*"
Reptilia	Squamata	Colubridae	*Sonora semiannulata*	Western Ground Snake	None
Reptilia	Squamata	Colubridae	*Thamnophis atratus* (*Thamnophis couchii*)	Pacific Coast Aquatic Garter Snake	*hydrophilus*
Reptilia	Squamata	Colubridae	*Thamnophis elegans*	Western Terrestrial Garter Snake	"*vagrans, elegans, terrestris*"
Reptilia	Squamata	Colubridae	*Thamnophis ordinoides*	Northwestern Garter Snake	None
Reptilia	Squamata	Colubridae	*Thamnophis sirtalis*	Common Garter Snake	"*concinnus, fitchi, pickeringii*"
Reptilia	Squamata	Viperidae	*Crotalus viridis*	Western Rattlesnake	"*oreganus, lutosus*"
Aves	Charadriiformes	Alcidae	*Synthliboramphus antiquus*	Ancient Murrelet	None
Aves	Charadriiformes	Alcidae	*Ptychoramphus aleuticus*	Cassin's Auklet	None
Aves	Charadriiformes	Alcidae	*Aethia psittacula* (*Cyclorrhynchus psittacula*)	Parakeet Auklet	
Aves	Charadriiformes	Alcidae	*Cerorhinca monocerata*	Rhinoceros Auklet	None
Aves	Charadriiformes	Alcidae	*Fratercula corniculata*	Horned Puffin	None
Aves	Charadriiformes	Alcidae	*Fratercula cirrhata*	Tufted Puffin	None
Aves	Columbiformes	Columbidae	*Columba livia*	Rock Dove	
Aves	Columbiformes	Columbidae	*Columba fasciata*	Band-tailed Pigeon	*monilis*
Aves	Columbiformes	Columbidae	*Zenaida asiatica*	White-winged Dove	
Aves	Columbiformes	Columbidae	*Zenaida macroura*	Mourning Dove	*marginella*
Aves	Cuculiformes	Cuculidae	*Coccyzus erythropthalmus*	Black-billed Cuckoo	
Aves	Cuculiformes	Cuculidae	*Coccyzus americanus*	Yellow-billed Cuckoo	*occidentalis*
Aves	Strigiformes	Tytonidae	*Tyto alba*	Barn Owl (Common Barn Owl)	*pratincola*
Aves	Strigiformes	Strigidae	*Otus flammeolus*	Flammulated Owl	*idahoensis*
Aves	Strigiformes	Strigidae	*Otus kennicottii*	Western Screech-owl	"*kennicottii* (*brewsteri*), *bendirei* (*macfarlanei*)"
Aves	Charadriiformes	Scolopacidae	*Heteroscelus incanus*	Wandering Tattler	
Aves	Charadriiformes	Scolopacidae	*Heteroscelus brevipes*	Gray-tailed Tattler	
Aves	Charadriiformes	Scolopacidae	*Actitis macularia*	Spotted Sandpiper	None
Aves	Charadriiformes	Scolopacidae	*Bartramia longicauda*	Upland Sandpiper	
Aves	Charadriiformes	Scolopacidae	*Numenius phaeopus*	Whimbrel	*hudsonicus*
Aves	Charadriiformes	Scolopacidae	*Numenius tahitiensis*	Bristle-thighed Curlew	
Aves	Charadriiformes	Scolopacidae	*Numenius americanus*	Long-billed Curlew	*parvus*
Aves	Charadriiformes	Scolopacidae	*Limosa haemastica*	Hudsonian Godwit	

Class	Order	Family	Scientific Name (Other Scientific Name)	Common Name (Other Common Name)	Subspecies Name Oregon & Washington only
Aves	Charadriiformes	Scolopacidae	Limosa lapponica	Bar-tailed Godwit	baueri
Aves	Charadriiformes	Scolopacidae	Limosa fedoa	Marbled Godwit	"fedoa, beringiae"
Aves	Charadriiformes	Scolopacidae	Arenaria interpres	Ruddy Turnstone	"interpres, morinella"
Aves	Charadriiformes	Scolopacidae	Arenaria melanocephala	Black Turnstone	None
Aves	Charadriiformes	Scolopacidae	Aphriza virgata	Surfbird	
Aves	Charadriiformes	Scolopacidae	Calidris tenuirostris	Great Knot	None
Aves	Charadriiformes	Scolopacidae	Calidris canutus	Red Knot	"rogersi, canutus"
Aves	Charadriiformes	Scolopacidae	Calidris alba	Sanderling	
Aves	Charadriiformes	Scolopacidae	Calidris pusilla	Semipalmated Sandpiper	None
Aves	Charadriiformes	Scolopacidae	Calidris mauri	Western Sandpiper	None
Aves	Charadriiformes	Scolopacidae	Calidris ruficollis	Red-necked Stint	
Aves	Charadriiformes	Scolopacidae	Calidris minuta	Little Stint	
Aves	Charadriiformes	Scolopacidae	Calidris subminuta	Long-toed Stint	
Aves	Charadriiformes	Scolopacidae	Calidris minutilla	Least Sandpiper	None
Aves	Charadriiformes	Scolopacidae	Calidris fuscicollis	White-rumped Sandpiper	
Aves	Charadriiformes	Scolopacidae	Calidris bairdii	Baird's Sandpiper	
Aves	Charadriiformes	Scolopacidae	Calidris melanotos	Pectoral Sandpiper	None
Aves	Charadriiformes	Scolopacidae	Calidris acuminata	Sharp-tailed Sandpiper	
Aves	Charadriiformes	Scolopacidae	Calidris ptilocnemis	Rock Sandpiper	"tschuktschorum, couesi"
Aves	Charadriiformes	Scolopacidae	Calidris alpina	Dunlin	pacifica
Aves	Charadriiformes	Scolopacidae	Calidris ferruginea	Curlew Sandpiper	
Aves	Charadriiformes	Scolopacidae	Calidris himantopus	Stilt Sandpiper	None
Aves	Charadriiformes	Scolopacidae	Tryngites subruficollis	Buff-breasted Sandpiper	None
Aves	Charadriiformes	Scolopacidae	Philomachus pugnax	Ruff	
Aves	Charadriiformes	Scolopacidae	Limnodromus griseus	Short-billed Dowitcher	"caurinus, hendersoni"
Aves	Charadriiformes	Scolopacidae	Limnodromus scolopaceus	Long-billed Dowitcher	
Aves	Charadriiformes	Scolopacidae	Gallinago gallinago	Common Snipe (Wilson's Snipe)	delicata
Aves	Charadriiformes	Scolopacidae	Phalaropus tricolor	Wilson's Phalarope	None
Aves	Charadriiformes	Scolopacidae	Phalaropus lobatus	Red-necked Phalarope (Northern Phalarope)	None
Aves	Charadriiformes	Scolopacidae	Phalaropus fulicaria	Red Phalarope (Grey Phalarope)	None
Aves	Charadriiformes	Laridae	Stercorarius maccormicki	South Polar Skua (Mccormick's Skua)	None
Aves	Charadriiformes	Laridae	Stercorarius pomarinus	Pomarine Jaeger (Pomarine Skua)	None
Aves	Charadriiformes	Laridae	Stercorarius parasiticus	Parasitic Jaeger (Arctic Skua)	None
Aves	Charadriiformes	Laridae	Stercorarius longicaudus	Long-tailed Jaeger (Long-Tailed Skua)	None

Class	Order	Family	Scientific Name (Other Scientific Name)	Common Name (Other Common Name)	Subspecies Name Oregon & Washington only
Aves	Charadriiformes	Laridae	Larus atricilla	Laughing Gull	None
Aves	Charadriiformes	Laridae	Larus pipixcan	Franklin's Gull	
Aves	Charadriiformes	Laridae	Larus minutus	Little Gull	
Aves	Charadriiformes	Laridae	Larus ridibundus	Black-headed Gull (Common Black-Headed Gull)	
Aves	Charadriiformes	Laridae	Larus philadelphia	Bonaparte's Gull	None
Aves	Charadriiformes	Laridae	Larus heermanni	Heermann's Gull	None
Aves	Charadriiformes	Laridae	Larus canus	Mew Gull (Common Gull)	brachyrhynchus
Aves	Charadriiformes	Laridae	Larus delawarensis	Ring-billed Gull	None
Aves	Charadriiformes	Laridae	Larus californicus	California Gull	None
Aves	Charadriiformes	Laridae	Larus argentatus	Herring Gull	smithsonianus
Aves	Charadriiformes	Laridae	Larus thayeri	Thayer's Gull	
Aves	Charadriiformes	Laridae	Larus glaucoides	Iceland Gull	
Aves	Charadriiformes	Laridae	Larus schistisagus	Slaty-backed Gull	
Aves	Charadriiformes	Laridae	Larus occidentalis	Western Gull	occidentalis
Aves	Charadriiformes	Laridae	Larus glaucescens	Glaucous-winged Gull	None
Aves	Charadriiformes	Laridae	Larus hyperboreus	Glaucous Gull	"barrovianus, hyperboreus"
Aves	Charadriiformes	Laridae	Xema Sabini	Sabine's Gull	None
Aves	Charadriiformes	Laridae	Rissa tridactyla	Black-legged Kittiwake	pollicaris
Aves	Charadriiformes	Laridae	Rissa brevirostris	Red-legged Kittiwake	
Aves	Charadriiformes	Laridae	Rhodostethia rosea	Ross's Gull (Ross' Gull)	
Aves	Charadriiformes	Laridae	Pagophila eburnea	Ivory Gull	None
Aves	Charadriiformes	Laridae	Sterna caspia	Caspian Tern	None
Aves	Charadriiformes	Laridae	Sterna elegans	Elegant Tern	
Aves	Charadriiformes	Laridae	Sterna hirundo	Common Tern	hirundo
Aves	Charadriiformes	Laridae	Sterna paradisaea	Arctic Tern	None
Aves	Charadriiformes	Laridae	Sterna forsteri	Forster's Tern	None
Aves	Charadriiformes	Laridae	Sterna antillarum	Least Tern	
Aves	Charadriiformes	Laridae	Chlidonias niger	Black Tern	surinamensis
Aves	Charadriiformes	Alcidae	Uria aalge	Common Murre	"inornata, californica"
Aves	Charadriiformes	Alcidae	Uria lomvia	Thick-billed Murre	
Aves	Charadriiformes	Alcidae	Cepphus columba	Pigeon Guillemot	columba
Aves	Charadriiformes	Alcidae	Brachyramphus perdix	Long-billed Murrelet	
Aves	Charadriiformes	Alcidae	Brachyramphus marmoratus	Marbled Murrelet	"marmoratus, perdix"
Aves	Charadriiformes	Alcidae	Brachyramphus brevirostris	Kittlitz's Murrelet	
Aves	Charadriiformes	Alcidae	Synthliboramphus hypoleucus	Xantus's Murrelet (Xantus' Murrelet)	
Aves	Strigiformes	Strigidae	Bubo virginianus	Great Horned Owl	"saturatus, lagophonus, subarcticus"

Class	Order	Family	Scientific Name (Other Scientific Name)	Common Name (Other Common Name)	Subspecies Name Oregon & Washington only
Aves	Strigiformes	Strigidae	Nyctea scandiaca	Snowy Owl	None
Aves	Strigiformes	Strigidae	Surnia ulula	Northern Hawk Owl (Northern Hawk-Owl)	
Aves	Strigiformes	Strigidae	Glaucidium gnoma	Northern Pygmy-owl	"grinnelli, californicum"
Aves	Strigiformes	Strigidae	Athene cunicularia (Speotyto cunicularia)	Burrowing Owl	hypugaea
Aves	Strigiformes	Strigidae	Strix occidentalis	Spotted Owl	caurina
Aves	Strigiformes	Strigidae	Strix varia	Barred Owl	varia
Aves	Strigiformes	Strigidae	Strix nebulosa	Great Gray Owl	nebulosa
Aves	Strigiformes	Strigidae	Asio otus	Long-eared Owl	tuftsi
Aves	Strigiformes	Strigidae	Asio flammeus	Short-eared Owl	flammeus
Aves	Strigiformes	Strigidae	Aegolius funereus	Boreal Owl	richardsoni
Aves	Strigiformes	Strigidae	Aegolius acadicus	Northern Saw-whet Owl	acadicus
Aves	Caprimulgiformes	Caprimulgidae	Chordeiles minor	Common Nighthawk	"hesperis, minor"
Aves	Caprimulgiformes	Caprimulgidae	Phalaenoptilus nuttallii	Common Poorwill	californicus, nuttallii
Aves	Apodiformes	Apodidae	Cypseloides niger	Black Swift	borealis
Aves	Apodiformes	Apodidae	Chaetura vauxi	Vaux's Swift	vauxi
Aves	Apodiformes	Apodidae	Aeronautes saxatalis	White-throated Swift	saxatalis
Aves	Apodiformes	Trochilidae	Archilochus alexandri	Black-chinned Hummingbird	None
Aves	Apodiformes	Trochilidae	Calypte anna	Anna's Hummingbird	None
Aves	Apodiformes	Trochilidae	Calypte costae	Costa's Hummingbird	
Aves	Apodiformes	Trochilidae	Stellula calliope	Calliope Hummingbird	None
Aves	Apodiformes	Trochilidae	Selasphorus platycercus	Broad-tailed Hummingbird	platycercus
Aves	Apodiformes	Trochilidae	Selasphorus rufus	Rufous Hummingbird	None
Aves	Apodiformes	Trochilidae	Selasphorus sasin	Allen's Hummingbird	sasin
Aves	Coraciiformes	Alcedinidae	Ceryle alcyon	Belted Kingfisher	caurina
Aves	Piciformes	Picidae	Melanerpes lewis	Lewis's Woodpecker (Lewis' Woodpecker)	None
Aves	Piciformes	Picidae	Melanerpes formicivorus	Acorn Woodpecker	bairdi
Aves	Piciformes	Picidae	Sphyrapicus thyroideus	Williamson's Sapsucker	None
Aves	Piciformes	Picidae	Sphyrapicus varius	Yellow-bellied Sapsucker	
Aves	Piciformes	Picidae	Sphyrapicus nuchalis	Red-naped Sapsucker	None
Aves	Piciformes	Picidae	Sphyrapicus ruber	Red-breasted Sapsucker	"ruber, daggetti"
Aves	Piciformes	Picidae	Picoides nuttallii	Nuttall's Woodpecker	
Aves	Piciformes	Picidae	Picoides pubescens	Downy Woodpecker	"fumidus, gairdnerii, turati, leucurus"
Aves	Piciformes	Picidae	Picoides villosus	Hairy Woodpecker	"monticola, orius, harrisi, septentrionalis"
Aves	Piciformes	Picidae	Picoides albolarvatus	White-headed Woodpecker	albolarvatus
Aves	Piciformes	Picidae	Picoides tridactylus	Three-toed Woodpecker	fasciatus
Aves	Piciformes	Picidae	Picoides arcticus	Black-backed Woodpecker	None
Aves	Piciformes	Picidae	Colaptes auratus	Northern Flicker (Red-Shafted Flicker)	"cafer, collaris"

Class	Order	Family	Scientific Name (Other Scientific Name)	Common Name (Other Common Name)	Subspecies Name Oregon & Washington only
Aves	Piciformes	Picidae	Dryocopus pileatus	Pileated Woodpecker	abieticola (picinus)
Aves	Passeriformes	Tyrannidae	Contopus cooperi (Contopus borealis)	Olive-sided Flycatcher	cooperi
Aves	Passeriformes	Tyrannidae	Contopus sordidulus	Western Wood-pewee	"veliei, saturatus"
Aves	Passeriformes	Tyrannidae	Contopus virens	Eastern Wood-pewee	
Aves	Passeriformes	Tyrannidae	Empidonax traillii	Willow Flycatcher (Traill's Flycatcher)	"brewsteri, adastus"
Aves	Passeriformes	Tyrannidae	Empidonax minimus	Least Flycatcher	None
Aves	Passeriformes	Tyrannidae	Empidonax hammondii	Hammond's Flycatcher	None
Aves	Passeriformes	Tyrannidae	Empidonax wrightii	Gray Flycatcher	None
Aves	Passeriformes	Tyrannidae	Empidonax oberholseri (Empidonax wrighti)	Dusky Flycatcher	None
Aves	Passeriformes	Tyrannidae	Empidonax difficilis	Pacific-slope Flycatcher (Western Flycatcher)	difficilis
Aves	Passeriformes	Tyrannidae	Empidonax occidentalis (Empidonax difficilis)	Cordilleran Flycatcher (Western Flycatcher)	helmayi
Aves	Passeriformes	Tyrannidae	Sayornis nigricans	Black Phoebe	semiatra
Aves	Passeriformes	Tyrannidae	Sayornis phoebe	Eastern Phoebe	
Aves	Passeriformes	Tyrannidae	Sayornis saya	Say's Phoebe	"saya, yukonensis"
Aves	Passeriformes	Tyrannidae	Pyrocephalus rubinus	Vermilion Flycatcher	
Aves	Passeriformes	Tyrannidae	Myiarchus cinerascens	Ash-throated Flycatcher	cinerascens
Aves	Passeriformes	Tyrannidae	Tyrannus melancholicus	Tropical Kingbird	
Aves	Passeriformes	Tyrannidae	Tyrannus verticalis	Western Kingbird	None
Aves	Passeriformes	Tyrannidae	Tyrannus tyrannus	Eastern Kingbird	None
Aves	Passeriformes	Tyrannidae	Tyrannus forficatus	Scissor-tailed Flycatcher	
Aves	Passeriformes	Tyrannidae	Tyrannus savana	Fork-tailed Flycatcher	
Aves	Passeriformes	Laniidae	Lanius ludovicianus	Loggerhead Shrike	"mexicanus (gmbeli), exubitorides"
Aves	Passeriformes	Laniidae	Lanius excubitor	Northern Shrike	borealis
Aves	Passeriformes	Vireonidae	Vireo griseus	White-eyed Vireo	
Aves	Passeriformes	Vireonidae	Vireo bellii	Bell's Vireo	
Aves	Passeriformes	Vireonidae	Vireo flavifrons	Yellow-throated Vireo	
Aves	Passeriformes	Vireonidae	Vireo plumbeus	Plumbeous Vireo (Solitary Vireo)	plumbeus
Aves	Passeriformes	Vireonidae	Vireo cassinii (Vireo solitarius)	Cassin's Vireo (Solitary Vireo)	None
Aves	Passeriformes	Vireonidae	Vireo huttoni	Hutton's Vireo	obscurus
Aves	Passeriformes	Vireonidae	Vireo gilvus	Warbling Vireo	swainsoni (leucopolius)
Aves	Passeriformes	Vireonidae	Vireo philadelphicus	Philadelphia Vireo	
Aves	Passeriformes	Vireonidae	Vireo olivaceus	Red-eyed Vireo	carnividis
Aves	Passeriformes	Corvidae	Perisoreus canadensis	Gray Jay (Canada Jay)	"bicolor, obscurus (rathbuni, connexus, griseus)"

Class	Order	Family	Scientific Name (Other Scientific Name)	Common Name (Other Common Name)	Subspecies Name Oregon & Washington only
Aves	Passeriformes	Corvidae	Cyanocitta stelleri	Steller's Jay	"annectens, fronatlis, paralia"
Aves	Passeriformes	Corvidae	Cyanocitta cristata	Blue Jay	
Aves	Passeriformes	Corvidae	Aphelocoma californica (Aphelocoma coerulescens)	Western Scrub-Jay (Scrub Jay)	"californica (immanis, caurina), superciliosa, woodhouseii = nevadae,"
Aves	Passeriformes	Corvidae	Gymnorhinus cyanocephalus	Pinyon Jay	cassini
Aves	Passeriformes	Corvidae	Nucifraga columbiana	Clark's Nutcracker	None
Aves	Passeriformes	Corvidae	Pica hudsonia (Pica pica)	Black-billed Magpie	
Aves	Passeriformes	Corvidae	Corvus brachyrhynchos	American Crow	hesperis
Aves	Passeriformes	Corvidae	Corvus caurinus	Northwestern Crow	None
Aves	Passeriformes	Corvidae	Corvus corax	Common Raven	"sinuatus, clarionensis"
Aves	Passeriformes	Alaudidae	Alauda arvensis	Sky Lark (Eurasian Skylark)	arvensis
Aves	Passeriformes	Alaudidae	Eremophila alpestris	Horned Lark	"strigata, arcticola, alpina, merrilli, lamprochroma"
Aves	Passeriformes	Hirundinidae	Progne subis	Purple Martin	"arboricola, subis"
Aves	Passeriformes	Hirundinidae	Tachycineta bicolor	Tree Swallow	None
Aves	Passeriformes	Hirundinidae	Tachycineta thalassina	Violet-green Swallow	thalassina (lepida)
Aves	Passeriformes	Hirundinidae	Stelgidopteryx serripennis	Northern Rough-winged Swallow (Rough-Winged Swallow)	serripennis (aphractus)
Aves	Passeriformes	Hirundinidae	Riparia riparia	Bank Swallow (Sand Martin)	riparia
Aves	Passeriformes	Hirundinidae	Petrochelidon pyrrhonota (Hirundo pyrrhonota)	Cliff Swallow	"pyrrhonota, hypopolia, aprophata"
Aves	Passeriformes	Hirundinidae	Hirundo rustica	Barn Swallow	erythrogaster
Aves	Passeriformes	Paridae	Poecile atricapilla (Poecile atricapillus, Parus atricapillus)	Black-capped Chickadee	"fortuitus, occidentalis, nevadensis"
Aves	Passeriformes	Paridae	Poecile gambeli (Parus gambeli)	Mountain Chickadee	"baileyae (grinnelli, abbreviatus), inyoensis"
Aves	Passeriformes	Paridae	Poecile rufescens (Parus rufescens)	Chestnut-backed Chickadee	rufescens
Aves	Passeriformes	Paridae	Poecile hudsonica (Poecile hudsonicus, Parus hudsonicus)	Boreal Chickadee	columbianus (cascadensis)
Aves	Passeriformes	Paridae	Baeolophus inornatus (Parus inornatus)	Oak Titmouse (Plain Titmouse)	inornatus (sequestratus)
Aves	Passeriformes	Paridae	Baeolophus ridgwayi ("Parus inornatus, Baeolophus griseus")	Juniper Titmouse (Plain Titmouse)	zaleptus
Aves	Passeriformes	Aegithalidae	Psaltriparus minimus	Bushtit	"saturatus, minimus, californicus, plumbeus"
Aves	Passeriformes	Sittidae	Sitta canadensis	Red-breasted Nuthatch	None

Class	Order	Family	Scientific Name (Other Scientific Name)	Common Name (Other Common Name)	Subspecies Name Oregon & Washington only
Aves	Passeriformes	Sittidae	Sitta carolinensis	White-breasted Nuthatch	"aculeata, tenuissima"
Aves	Passeriformes	Sittidae	Sitta pygmaea	Pygmy Nuthatch	melanotis
Aves	Passeriformes	Certhiidae	Certhia americana	Brown Creeper	"montana, occidentalis, zelotes"
Aves	Passeriformes	Troglodytidae	Salpinctes obsoletus	Rock Wren	obsoletus
Aves	Passeriformes	Troglodytidae	Catherpes mexicanus	Canyon Wren	griseus
Aves	Passeriformes	Troglodytidae	Thryomanes bewickii	Bewick's Wren	"calophonus, drymoecus, atrestus"
Aves	Passeriformes	Troglodytidae	Troglodytes aedon	House Wren	parkmanii
Aves	Passeriformes	Troglodytidae	Troglodytes troglodytes	Winter Wren	"pacificus, salebrosus"
Aves	Passeriformes	Troglodytidae	Cistothorus palustris	Marsh Wren (Long-Billed Marsh Wren)	"pulverius, browningi, paludicola, plesius"
Aves	Passeriformes	Cinclidae	Cinclus mexicanus	American Dipper	unicolor
Aves	Passeriformes	Regulidae	Regulus satrapa	Golden-crowned Kinglet	"olivaceus, apache"
Aves	Passeriformes	Regulidae	Regulus calendula	Ruby-crowned Kinglet	"calendula, grinnelli"
Aves	Passeriformes	Sylviidae	Polioptila caerulea	Blue-gray Gnatcatcher	obscura
Aves	Passeriformes	Turdidae	Oenanthe oenanthe	Northern Wheatear	
Aves	Passeriformes	Turdidae	Sialia mexicana	Western Bluebird	occidentalis
Aves	Passeriformes	Turdidae	Sialia currucoides	Mountain Bluebird	None
Aves	Passeriformes	Turdidae	Myadestes townsendi	Townsend's Solitaire	townsendi
Aves	Passeriformes	Turdidae	Catharus fuscescens	Veery	salicicola
Aves	Passeriformes	Turdidae	Catharus minimus	Gray-cheeked Thrush	
Aves	Passeriformes	Turdidae	Catharus ustulatus	Swainson's Thrush	"ustulatus, swainsoni"
Aves	Passeriformes	Turdidae	Catharus guttatus	Hermit Thrush	"auduboni, slevini, jewetti, vaccinius, oromelus = dwighti, nanus = osgoodi, guttatus, verecundas"
Aves	Passeriformes	Turdidae	Hylocichla mustelina	Wood Thrush	
Aves	Passeriformes	Turdidae	Turdus migratorius	American Robin	"caurinus, propinquus, migratorius"
Aves	Passeriformes	Turdidae	Ixoreus naevius	Varied Thrush	"naevius, godfreii, meruloides, carlottae"
Aves	Passeriformes	Timaliidae	Chamaea fasciata	Wrentit	"phaea, margra"
Aves	Passeriformes	Mimidae	Dumetella carolinensis	Gray Catbird	ruficrissa
Aves	Passeriformes	Mimidae	Mimus polyglottos	Northern Mockingbird	None
Aves	Passeriformes	Mimidae	Oreoscoptes montanus	Sage Thrasher	None
Aves	Passeriformes	Mimidae	Toxostoma rufum	Brown Thrasher	
Aves	Passeriformes	Mimidae	Toxostoma redivivum	California Thrasher	
Aves	Passeriformes	Sturnidae	Sturnus vulgaris	European Starling	vulgaris
Aves	Passeriformes	Prunellidae	Prunella montanella	Siberian Accentor	
Aves	Passeriformes	Motacillidae	Motacilla flava	Yellow Wagtail	
Aves	Passeriformes	Motacillidae	Motacilla alba	White Wagtail	
Aves	Passeriformes	Motacillidae	Motacilla lugens	Black-backed Wagtail	
Aves	Passeriformes	Motacillidae	Anthus cervinus	Red-throated Pipit	
Aves	Passeriformes	Motacillidae	Anthus rubescens (Anthus spinoletta)	American Pipit (Water Pipit)	"pacificus, alticola, geophilus"

Class	Order	Family	Scientific Name (Other Scientific Name)	Common Name (Other Common Name)	Subspecies Name Oregon & Washington only
Aves	Passeriformes	Bombycillidae	Bombycilla garrulus	Bohemian Waxwing	pallidiceps
Aves	Passeriformes	Bombycillidae	Bombycilla cedrorum	Cedar Waxwing	larifuga
Aves	Passeriformes	Ptilogonatidae	Phainopepla nitens	Phainopepla	
Aves	Passeriformes	Parulidae	Vermivora pinus	Blue-winged Warbler	
Aves	Passeriformes	Parulidae	Vermivora chrysoptera	Golden-winged Warbler	
Aves	Passeriformes	Parulidae	Vermivora peregrina	Tennessee Warbler	
Aves	Passeriformes	Parulidae	Vermivora celata	Orange-crowned Warbler	"lutescens, orestera"
Aves	Passeriformes	Parulidae	Vermivora ruficapilla	Nashville Warbler	ridgwayi
Aves	Passeriformes	Parulidae	Vermivora virginiae	Virginia's Warbler	
Aves	Passeriformes	Parulidae	Vermivora luciae	Lucy's Warbler	
Aves	Passeriformes	Parulidae	Parula americana	Northern Parula	
Aves	Passeriformes	Parulidae	Dendroica petechia	Yellow Warbler	"morcomi, brewsteri"
Aves	Passeriformes	Parulidae	Dendroica pensylvanica	Chestnut-sided Warbler	
Aves	Passeriformes	Parulidae	Dendroica magnolia	Magnolia Warbler	
Aves	Passeriformes	Parulidae	Dendroica tigrina	Cape May Warbler	
Aves	Passeriformes	Parulidae	Dendroica caerulescens	Black-throated Blue Warbler	
Aves	Passeriformes	Parulidae	Dendroica coronata	Yellow-rumped Warbler (Audubon's Warbler)	auduboni
Aves	Passeriformes	Parulidae	Dendroica nigrescens	Black-throated Gray Warbler	"nigrescens, halseii"
Aves	Passeriformes	Parulidae	Dendroica virens	Black-throated Green Warbler	
Aves	Passeriformes	Parulidae	Dendroica townsendi	Townsend's Warbler	None
Aves	Passeriformes	Parulidae	Dendroica occidentalis	Hermit Warbler	None
Aves	Passeriformes	Parulidae	Dendroica fusca	Blackburnian Warbler	
Aves	Passeriformes	Parulidae	Dendroica dominica	Yellow-throated Warbler	
Aves	Passeriformes	Parulidae	Dendroica pinus	Pine Warbler	
Aves	Passeriformes	Parulidae	Dendroica discolor	Prairie Warbler	
Aves	Passeriformes	Parulidae	Dendroica palmarum	Palm Warbler	palmarum
Aves	Passeriformes	Parulidae	Dendroica castanea	Bay-breasted Warbler	
Aves	Passeriformes	Parulidae	Dendroica striata	Blackpoll Warbler	
Aves	Passeriformes	Parulidae	Mniotilta varia	Black-and-white Warbler	
Aves	Passeriformes	Parulidae	Setophaga ruticilla	American Redstart	None
Aves	Passeriformes	Parulidae	Protonotaria citrea	Prothonotary Warbler	
Aves	Passeriformes	Parulidae	Helmitheros vermivorus	Worm-eating Warbler	
Aves	Passeriformes	Parulidae	Seiurus aurocapillus	Ovenbird	
Aves	Passeriformes	Parulidae	Seiurus noveboracensis	Northern Waterthrush	notabilis
Aves	Passeriformes	Parulidae	Oporornis formosus	Kentucky Warbler	
Aves	Passeriformes	Parulidae	Oporornis philadelphia	Mourning Warbler	
Aves	Passeriformes	Parulidae	Oporornis tolmiei	Macgillivray's Warbler	tolmiei (intermedia)
Aves	Passeriformes	Parulidae	Geothlypis trichas	Common Yellowthroat	"arizela, campicola, occidentalis (idahonicola, oregonicola)"

Class	Order	Family	Scientific Name (Other Scientific Name)	Common Name (Other Common Name)	Subspecies Name Oregon & Washington only
Aves	Passeriformes	Parulidae	Wilsonia citrina	Hooded Warbler	
Aves	Passeriformes	Parulidae	Wilsonia pusilla	Wilson's Warbler	"pileolata, chryseola"
Aves	Passeriformes	Parulidae	Wilsonia canadensis	Canada Warbler	
Aves	Passeriformes	Parulidae	Icteria virens	Yellow-breasted Chat	auricollis (longicauda)
Aves	Passeriformes	Thraupidae	Piranga rubra	Summer Tanager	
Aves	Passeriformes	Thraupidae	Piranga olivacea	Scarlet Tanager	None
Aves	Passeriformes	Thraupidae	Piranga ludoviciana	Western Tanager	None
Aves	Passeriformes	Emberizidae	Pipilo chlorurus	Green-tailed Towhee	"oreganus, falcinellus, curtatus"
Aves	Passeriformes	Emberizidae	Pipilo maculatus (Pipilo erythrophthalmus)	Spotted Towhee (Rufous-Sided Towhee)	
Aves	Passeriformes	Emberizidae	Pipilo crissalis (Pipilo fuscus)	California Towhee (Brown Towhee)	bullatus
Aves	Passeriformes	Emberizidae	Spizella arborea	American Tree Sparrow	ochracea
Aves	Passeriformes	Emberizidae	Spizella passerina	Chipping Sparrow	"stridula, arizonae"
Aves	Passeriformes	Emberizidae	Spizella pallida	Clay-colored Sparrow	None
Aves	Passeriformes	Emberizidae	Spizella breweri	Brewer's Sparrow	"breweri, taverneri"
Aves	Passeriformes	Emberizidae	Spizella atrogularis	Black-chinned Sparrow	cana
Aves	Passeriformes	Emberizidae	Pooecetes gramineus	Vesper Sparrow	"affinis, confinis"
Aves	Passeriformes	Emberizidae	Chondestes grammacus	Lark Sparrow	strigatus (actitus)
Aves	Passeriformes	Emberizidae	Amphispiza bilineata	Black-throated Sparrow	deserticola
Aves	Passeriformes	Emberizidae	Amphispiza belli	Sage Sparrow	campicola = nevadensis
Aves	Passeriformes	Emberizidae	Calamospiza melanocorys	Lark Bunting	
Aves	Passeriformes	Emberizidae	Passerculus sandwichensis	Savannah Sparrow	"brooksi, nevadensis, crassus, sandwichensis, anthinus"
Aves	Passeriformes	Emberizidae	Ammodramus savannarum	Grasshopper Sparrow	perpallidus
Aves	Passeriformes	Emberizidae	Ammodramus leconteii	Le Conte's Sparrow	
Aves	Passeriformes	Emberizidae	Ammodramus nelsoni (Ammodramus caudacutus)	Nelson's Sharp-tailed Sparrow (Sharp-Tailed Sparrow)	
Aves	Passeriformes	Emberizidae	Passerella iliaca	Fox Sparrow	"fuliginosa, schistacea, olivacea, unalaschcensis, ridgwayi (insularis), sinuosa, annectens, townsendi, chilcatensis, megarhyncha"
Aves	Passeriformes	Emberizidae	Melospiza melodia	Song Sparrow	"merrilli, fisherella, morphna, kenaiensis, caurina, rufina, inexpectata, cleonensis, montana"
Aves	Passeriformes	Emberizidae	Melospiza lincolnii	Lincoln's Sparrow	"lincolnii, alticola"
Aves	Passeriformes	Emberizidae	Melospiza georgiana	Swamp Sparrow	ericrypta
Aves	Passeriformes	Emberizidae	Zonotrichia albicollis	White-throated Sparrow	None
Aves	Passeriformes	Emberizidae	Zonotrichia querula	Harris's Sparrow (Harris' Sparrow)	None

Class	Order	Family	Scientific Name (Other Scientific Name)	Common Name (Other Common Name)	Subspecies Name Oregon & Washington only
Aves	Passeriformes	Emberizidae	Zonotrichia leucophrys	White-crowned Sparrow	"pugetensis, gambelii, oriantha"
Aves	Passeriformes	Emberizidae	Zonotrichia atricapilla	Golden-crowned Sparrow	None
Aves	Passeriformes	Emberizidae	Junco hyemalis	Dark-eyed Junco	"oreganus, simillimus, thurberi, shufeldti (montanus, eumesus)"
Aves	Passeriformes	Emberizidae	Calcarius mccownii	McCown's Longspur	
Aves	Passeriformes	Emberizidae	Calcarius lapponicus	Lapland Longspur	alascensis
Aves	Passeriformes	Emberizidae	Calcarius ornatus	Chestnut-collared Longspur	
Aves	Passeriformes	Emberizidae	Emberiza rustica	Rustic Bunting	
Aves	Passeriformes	Emberizidae	Plectrophenax nivalis	Snow Bunting	Nivalis
Aves	Passeriformes	Emberizidae	Plectrophenax hyperboreus	Mckay's Bunting	
Aves	Passeriformes	Cardinalidae	Pheucticus ludovicianus	Rose-breasted Grosbeak	
Aves	Passeriformes	Cardinalidae	Pheucticus melanocephalus	Black-headed Grosbeak	"maculatus, melanocephalus"
Aves	Passeriformes	Cardinalidae	Guiraca caerulea	Blue Grosbeak	None
Aves	Passeriformes	Cardinalidae	Passerina amoena	Lazuli Bunting	
Aves	Passeriformes	Cardinalidae	Passerina cyanea	Indigo Bunting	
Aves	Passeriformes	Cardinalidae	Passerina ciris	Painted Bunting	
Aves	Passeriformes	Cardinalidae	Spiza americana	Dickcissel	
Aves	Passeriformes	Icteridae	Dolichonyx oryzivorus	Bobolink	None
Aves	Passeriformes	Icteridae	Agelaius phoeniceus	Red-winged Blackbird	"caurinus, nevadensis"
Aves	Passeriformes	Icteridae	Agelaius tricolor	Tricolored Blackbird	None
Aves	Passeriformes	Icteridae	Sturnella neglecta	Western Meadowlark	"confluenta, neglecta"
Aves	Passeriformes	Icteridae	Xanthocephalus xanthocephalus	Yellow-headed Blackbird	None
Aves	Passeriformes	Icteridae	Euphagus carolinus	Rusty Blackbird	
Aves	Passeriformes	Icteridae	Euphagus cyanocephalus	Brewer's Blackbird	"minusculus, brewsteri, cyanocephalus (aliastus)"
Aves	Passeriformes	Icteridae	Quiscalus quiscula	Common Grackle	
Aves	Passeriformes	Icteridae	Quiscalus mexicanus	Great-tailed Grackle	
Aves	Passeriformes	Icteridae	Molothrus ater	Brown-headed Cowbird	artemisiae
Aves	Passeriformes	Icteridae	Icterus spurius	Orchard Oriole	
Aves	Passeriformes	Icteridae	Icterus cucullatus	Hooded Oriole	
Aves	Passeriformes	Icteridae	Icterus pustulatus	Streak-backed Oriole	
Aves	Passeriformes	Icteridae	Icterus galbula	Baltimore Oriole (Northern Oriole)	
Aves	Passeriformes	Icteridae	Icterus bullockii (Icterus galbula)	Bullock's Oriole (Northern Oriole)	bullockii
Aves	Passeriformes	Icteridae	Icterus parisorum	Scott's Oriole	
Aves	Passeriformes	Fringillidae	Fringilla montifringilla	Brambling	
Aves	Passeriformes	Fringillidae	Leucosticte tephrocotis (Leucosticte arctoa)	Gray-crowned Rosy-Finch (Rosy Finch)	"wallowa, tephrocotis, littoralis"
Aves	Passeriformes	Fringillidae	Leucosticte atrata (Leucosticte arctoa)	Black Rosy-finch (Rosy Finch)	None

Class	Order	Family	Scientific Name (Other Scientific Name)	Common Name (Other Common Name)	Subspecies Name Oregon & Washington only
Aves	Passeriformes	Fringillidae	Pinicola enucleator	Pine Grosbeak	"flammula, carlottae, montanus, leucurus"
Aves	Passeriformes	Fringillidae	Carpodacus purpureus	Purple Finch	californicus (rubidus)
Aves	Passeriformes	Fringillidae	Carpodacus cassinii	Cassin's Finch	vinifer
Aves	Passeriformes	Fringillidae	Carpodacus mexicanus	House Finch	"frontalis (grinnelli), solitudinus"
Aves	Passeriformes	Fringillidae	Loxia curvirostra	Red Crossbill	"minor group, pusilla group"
Aves	Passeriformes	Fringillidae	Loxia leucoptera	White-winged Crossbill	leucoptera
Aves	Gaviiformes	Gaviidae	Gavia stellata	Red-throated Loon (Red-Throated Diver)	None
Aves	Gaviiformes	Gaviidae	Gavia pacifica	Pacific Loon (Black-Throated Diver)	
Aves	Gaviiformes	Gaviidae	Gavia immer	Common Loon (Great Northern Diver)	None
Aves	Gaviiformes	Gaviidae	Gavia adamsii	Yellow-billed Loon (White-Billed Diver)	None
Aves	Podicipediformes	Podicipedidae	Podilymbus podiceps	Pied-billed Grebe	podiceps
Aves	Podicipediformes	Podicipedidae	Podiceps auritus	Horned Grebe (Slavonian Grebe)	cornutus
Aves	Podicipediformes	Podicipedidae	Podiceps grisegena	Red-necked Grebe ("Holboll's Grebe, Holboell's Grebe, Gray-Cheeked Grebe")	holbollii
Aves	Podicipediformes	Podicipedidae	Podiceps nigricollis	Eared Grebe (Black-Necked Grebe)	californicus
Aves	Podicipediformes	Podicipedidae	Aechmophorus occidentalis	Western Grebe	occidentalis
Aves	Podicipediformes	Podicipedidae	Aechmophorus clarkii	Clark's Grebe	clarkii
Aves	Procellariiformes	Diomedeidae	Thalassarche cauta	Shy Albatross	
Aves	Procellariiformes	Diomedeidae	Phoebastria immutabilis (Diomedea immutabilis)	Laysan Albatross	None
Aves	Procellariiformes	Diomedeidae	Phoebastria nigripes (Diomedea nigripes)	Black-footed Albatross	None
Aves	Procellariiformes	Diomedeidae	Phoebastria albatrus (Diomedea albatrus)	Short-tailed Albatross (Steller's Albatross)	
Aves	Procellariiformes	Procellariidae	Fulmarus glacialis	Northern Fulmar (Arctic Fulmar)	rodgersii
Aves	Procellariiformes	Procellariidae	Pterodroma ultima	Murphy's Petrel	
Aves	Procellariiformes	Procellariidae	Pterodroma inexpectata	Mottled Petrel	
Aves	Procellariiformes	Procellariidae	Pterodroma cookii	Cook's Petrel	
Aves	Procellariiformes	Procellariidae	Puffinus creatopus	Pink-footed Shearwater	
Aves	Procellariiformes	Procellariidae	Puffinus carneipes	Flesh-footed Shearwater	
Aves	Procellariiformes	Procellariidae	Puffinus bulleri	Buller's Shearwater (New Zealand/Grey-Backed Shearwater)	None
Aves	Procellariiformes	Procellariidae	Puffinus griseus	Sooty Shearwater	

Class	Order	Family	Scientific Name (Other Scientific Name)	Common Name (Other Common Name)	Subspecies Name Oregon & Washington only
Aves	Procellariiformes	Procellariidae	Puffinus tenuirostris	Short-tailed Shearwater (Slender-Billed Shearwater)	None
Aves	Procellariiformes	Procellariidae	Puffinus puffinus	Manx Shearwater	
Aves	Procellariiformes	Procellariidae	Puffinus opisthomelas	Black-vented Shearwater	
Aves	Procellariiformes	Hydrobatidae	Oceanites oceanicus	Wilson's Storm-petrel	
Aves	Procellariiformes	Hydrobatidae	Oceanodroma furcata	Fork-tailed Storm-petrel	plumbea
Aves	Procellariiformes	Hydrobatidae	Oceanodroma leucorhoa	Leach's Storm-petrel	leucorhoa
Aves	Procellariiformes	Hydrobatidae	Oceanodroma melania	Black Storm-petrel	
Aves	Pelecaniformes	Phaethontidae	Phaethon aethereus	Red-billed Tropicbird	
Aves	Pelecaniformes	Sulidae	Sula nebouxii	Blue-footed Booby	
Aves	Pelecaniformes	Pelecanidae	Pelecanus erythrorhynchos	American White Pelican	None
Aves	Pelecaniformes	Pelecanidae	Pelecanus occidentalis	Brown Pelican	californicus
Aves	Pelecaniformes	Phalacrocoracidae	Phalacrocorax penicillatus	Brandt's Cormorant	None
Aves	Pelecaniformes	Phalacrocoracidae	Phalacrocorax auritus	Double-crested Cormorant	albociliatus
Aves	Pelecaniformes	Phalacrocoracidae	Phalacrocorax pelagicus	Pelagic Cormorant	resplendens
Aves	Pelecaniformes	Fregatidae	Fregata magnificens	Magnificent Frigatebird	
Aves	Ciconiiformes	Ardeidae	Botaurus lentiginosus	American Bittern	None
Aves	Ciconiiformes	Ardeidae	Ixobrychus exilis	Least Bittern	hesperis
Aves	Ciconiiformes	Ardeidae	Ardea herodias	Great Blue Heron	"herodias, treganzai, fannini"
Aves	Ciconiiformes	Ardeidae	Ardea alba (Casmerodius albus)	Great Egret	egretta
Aves	Ciconiiformes	Ardeidae	Egretta thula	Snowy Egret	brewsteri
Aves	Ciconiiformes	Ardeidae	Egretta caerulea	Little Blue Heron	
Aves	Passeriformes	Fringillidae	Carduelis flammea	Common Redpoll	flammea
Aves	Passeriformes	Fringillidae	Carduelis hornemanni	Hoary Redpoll	
Aves	Passeriformes	Fringillidae	Carduelis pinus	Pine Siskin	"vagans, pinus"
Aves	Passeriformes	Fringillidae	Carduelis psaltria	Lesser Goldfinch	psaltria (hesperophilus)
Aves	Passeriformes	Fringillidae	Carduelis lawrencei	Lawrence's Goldfinch	
Aves	Passeriformes	Fringillidae	Carduelis tristis	American Goldfinch	"pallida, jewetti"
Aves	Passeriformes	Fringillidae	Coccothraustes vespertinus	Evening Grosbeak	brooksi (californica)
Aves	Passeriformes	Passeridae	Passer domesticus	House Sparrow	domesticus
Aves	Ciconiiformes	Ardeidae	Egretta tricolor	Tricolored Heron	
Aves	Ciconiiformes	Ardeidae	Bubulcus ibis	Cattle Egret	ibis
Aves	Ciconiiformes	Ardeidae	Butorides virescens (Butorides striatus)	Green Heron (Green-Backed Heron)	anthonyi
Aves	Ciconiiformes	Ardeidae	Nycticorax nycticorax	Black-crowned Night-heron	hoactli
Aves	Ciconiiformes	Ardeidae	Nyctanassa violacea	Yellow-crowned Night-heron	None
Aves	Ciconiiformes	Threskiornithidae	Plegadis chihi	White-faced Ibis	
Aves	Ciconiiformes	Cathartidae	Cathartes aura	Turkey Vulture	meridionalis

Class	Order	Family	Scientific Name (Other Scientific Name)	Common Name (Other Common Name)	Subspecies Name Oregon & Washington only
Aves	Ciconiiformes	Cathartidae	*Gymnogyps californianus*	California Condor	
Aves	Anseriformes	Anatidae	*Dendrocygna bicolor*	Fulvous Whistling-Duck	
Aves	Anseriformes	Anatidae	*Anser albifrons*	Greater White-fronted Goose	"frontalis, gambeli"
Aves	Anseriformes	Anatidae	*Chen canagica*	Emperor Goose	
Aves	Anseriformes	Anatidae	*Chen Caerulescens*	Snow Goose (Lesser Snow Goose)	caerulescens
Aves	Anseriformes	Anatidae	*Chen rossii*	Ross's Goose (Ross' Goose)	None
Aves	Anseriformes	Anatidae	*Branta canadensis*	Canada Goose (Great Basin Canada Goose or Giant Canada Goose)	
Aves	Anseriformes	Anatidae	*Branta canadensis moffitti*	Western Canada Goose	
Aves	Anseriformes	Anatidae	*Branta canadensis maxima*	Giant Canada Goose	
Aves	Anseriformes	Anatidae	*Branta canadensis taverneri*	Taverner's Canada Goose	
Aves	Anseriformes	Anatidae	*Branta canadensis leucopareia*	Aleutian Canada Goose	
Aves	Anseriformes	Anatidae	*Branta canadensis minima*	Cackling Canada Goose	
Aves	Anseriformes	Anatidae	*Branta canadensis occidentalis*	Dusky Canada Goose	
Aves	Anseriformes	Anatidae	*Branta canadensis fulva*	Vancouver Canada Goose	
Aves	Anseriformes	Anatidae	*Branta canadensis parvipes*	Lesser Canada Goose	
Aves	Anseriformes	Anatidae	*Branta bernicla*	Brant (Black Brant)	nigricans
Aves	Anseriformes	Anatidae	*Cygnus olor*	Mute Swan	None
Aves	Anseriformes	Anatidae	*Cygnus buccinator*	Trumpeter Swan	None
Aves	Anseriformes	Anatidae	*Cygnus columbianus*	Tundra Swan	columbianus
Aves	Anseriformes	Anatidae	*Cygnus cygnus*	Whooper Swan	
Aves	Anseriformes	Anatidae	*Aix sponsa*	Wood Duck	None
Aves	Anseriformes	Anatidae	*Anas strepera*	Gadwall	None
Aves	Anseriformes	Anatidae	*Anas falcata*	Falcated Duck (Falcated Teal)	
Aves	Anseriformes	Anatidae	*Anas penelope*	Eurasian Wigeon	
Aves	Anseriformes	Anatidae	*Anas americana*	American Wigeon (Baldpate)	None
Aves	Anseriformes	Anatidae	*Anas rubripes*	American Black Duck	
Aves	Anseriformes	Anatidae	*Anas platyrhynchos*	Mallard	platyrhynchos
Aves	Anseriformes	Anatidae	*Anas discors*	Blue-winged Teal	discors
Aves	Anseriformes	Anatidae	*Anas cyanoptera*	Cinnamon Teal	eptentrionalium
Aves	Anseriformes	Anatidae	*Anas clypeata*	Northern Shoveler	None
Aves	Anseriformes	Anatidae	*Anas acuta*	Northern Pintail	acuta
Aves	Anseriformes	Anatidae	*Anas querquedula*	Garganey	
Aves	Anseriformes	Anatidae	*Anas formosa*	Baikal Teal	
Aves	Anseriformes	Anatidae	*Anas crecca*	Green-winged Teal	carolinensis

Class	Order	Family	Scientific Name (Other Scientific Name)	Common Name (Other Common Name)	Subspecies Name Oregon & Washington only
Aves	Anseriformes	Anatidae	Aythya valisineria	Canvasback	
Aves	Anseriformes	Anatidae	Aythya americana	Redhead	
Aves	Anseriformes	Anatidae	Aythya collaris	Ring-necked Duck	None
Aves	Anseriformes	Anatidae	Aythya fuligula	Tufted Duck	
Aves	Anseriformes	Anatidae	Aythya marila	Greater Scaup	mariloides
Aves	Anseriformes	Anatidae	Aythya affinis	Lesser Scaup	None
Aves	Anseriformes	Anatidae	Polysticta stelleri	Steller's Eider	
Aves	Anseriformes	Anatidae	Somateria spectabilis	King Eider	
Aves	Anseriformes	Anatidae	Histrionicus histrionicus	Harlequin Duck	pacificus
Aves	Anseriformes	Anatidae	Melanitta perspicillata	Surf Scoter	None
Aves	Anseriformes	Anatidae	Melanitta fusca	White-winged Scoter (Velvet Scoter)	deglandi
Aves	Anseriformes	Anatidae	Melanitta nigra	Black Scoter	americana
Aves	Anseriformes	Anatidae	Clangula hyemalis	Long-Tailed Duck (Oldsquaw)	epixanthum
Aves	Anseriformes	Anatidae	Bucephala albeola	Bufflehead	None
Aves	Anseriformes	Anatidae	Bucephala clangula	Common Goldeneye	americana
Aves	Anseriformes	Anatidae	Bucephala islandica	Barrow's Goldeneye	
Aves	Anseriformes	Anatidae	Mergellus albellus	Smew	
Aves	Anseriformes	Anatidae	Lophodytes cucullatus	Hooded Merganser	None
Aves	Anseriformes	Anatidae	Mergus merganser	Common Merganser (Goosander)	americanus
Aves	Anseriformes	Anatidae	Mergus serrator	Red-breasted Merganser	serrator
Aves	Anseriformes	Anatidae	Oxyura jamaicensis	Ruddy Duck	rubida
Aves	Falconiformes	Accipitridae	Pandion haliaetus	Osprey	carolinensis
Aves	Falconiformes	Accipitridae	Elanus leucurus (Elanus caeruleus)	White-tailed Kite (Black-Shouldered Kite)	majusculus
Aves	Falconiformes	Accipitridae	Haliaeetus leucocephalus	Bald Eagle	alascanus
Aves	Falconiformes	Accipitridae	Circus cyaneus	Northern Harrier	hudsonius
Aves	Falconiformes	Accipitridae	Accipiter striatus	Sharp-shinned Hawk	"velox, perobscurus"
Aves	Falconiformes	Accipitridae	Accipiter cooperii	Cooper's Hawk	None
Aves	Falconiformes	Accipitridae	Accipiter gentilis	Northern Goshawk	"atricapillus, laingi"
Aves	Falconiformes	Accipitridae	Buteo lineatus	Red-shouldered Hawk	elegans
Aves	Falconiformes	Accipitridae	Buteo platypterus	Broad-winged Hawk	
Aves	Falconiformes	Accipitridae	Buteo swainsoni	Swainson's Hawk	None
Aves	Falconiformes	Accipitridae	Buteo jamaicensis	Red-tailed Hawk	calurus
Aves	Falconiformes	Accipitridae	Buteo regalis	Ferruginous Hawk	None
Aves	Falconiformes	Accipitridae	Buteo lagopus	Rough-legged Hawk	sanctijohannis
Aves	Falconiformes	Accipitridae	Aquila chrysaetos	Golden Eagle	canadensis
Aves	Falconiformes	Falconidae	Falco sparverius	American Kestrel	sparverius
Aves	Falconiformes	Falconidae	Falco columbarius	Merlin	"columbarius, suckleyi, richardsoni"

Class	Order	Family	Scientific Name (Other Scientific Name)	Common Name (Other Common Name)	Subspecies Name Oregon & Washington only
Aves	Falconiformes	Falconidae	Falco rusticolus	Gyrfalcon	None
Aves	Falconiformes	Falconidae	Falco peregrinus	Peregrine Falcon	"pealei, tundrius, anatum"
Aves	Falconiformes	Falconidae	Falco mexicanus	Prairie Falcon	None
Aves	Galliformes	Phasianidae	Alectoris chukar	Chukar	chukar
Aves	Galliformes	Phasianidae	Perdix perdix	Gray Partridge (Hungarian Partridge)	perdix
Aves	Galliformes	Phasianidae	Phasianus colchicus	Ring-necked Pheasant (Chinese Pheasant)	"sabini, castanea, affinis, phaia"
Aves	Galliformes	Phasianidae	Bonasa umbellus	Ruffed Grouse	phaios
Aves	Galliformes	Phasianidae	Centrocercus urophasianus	Greater Sage-grouse	franklinii
Aves	Galliformes	Phasianidae	Falcipennis canadensis (Dendragapus canadensis)	Spruce Grouse	
Aves	Galliformes	Phasianidae	Lagopus leucurus	White-tailed Ptarmigan	"rainierensis, leucurus"
Aves	Galliformes	Phasianidae	Dendragapus obscurus	Blue Grouse	"sierrae, fuliginosus, pallidus, richardsonii"
Aves	Galliformes	Phasianidae	Tympanuchus phasianellus	Sharp-tailed Grouse	columbianus
Aves	Galliformes	Phasianidae	Meleagris gallopavo	Wild Turkey	"merriami, intermedia, silvestris"
Aves	Galliformes	Odontophoridae	Oreortyx pictus	Mountain Quail	"palmeri, pictus"
Aves	Galliformes	Odontophoridae	Callipepla squamata	Scaled Quail	pallida
Aves	Galliformes	Odontophoridae	Callipepla californica	California Quail (Valley Quail)	"californicus, brunnescens, orecta"
Aves	Galliformes	Odontophoridae	Colinus virginianus	Northern Bobwhite	"virginianus, texanus, taylori"
Aves	Gruiformes	Rallidae	Coturnicops noveboracensis	Yellow Rail	noveboracensis
Aves	Gruiformes	Rallidae	Rallus limicola	Virginia Rail	limicola
Aves	Gruiformes	Rallidae	Porzana carolina	Sora	None
Aves	Gruiformes	Rallidae	Gallinula chloropus	Common Moorhen	
Aves	Gruiformes	Rallidae	Fulica americana	American Coot	americana
Aves	Gruiformes	Gruidae	Grus canadensis	Sandhill Crane	"canadensis, tabida"
Aves	Charadriiformes	Charadriidae	Pluvialis squatarola	Black-bellied Plover	None
Aves	Charadriiformes	Charadriidae	Pluvialis dominica (Pluvialis dominicus)	American Golden-Plover	None
Aves	Charadriiformes	Charadriidae	Pluvialis fulva (Pluvialis dominica)	Pacific Golden-Plover (Lesser Golden-Plover)	None
Aves	Charadriiformes	Charadriidae	Charadrius mongolus	Mongolian Plover	
Aves	Charadriiformes	Charadriidae	Charadrius alexandrinus	Snowy Plover	nivosus
Aves	Charadriiformes	Charadriidae	Charadrius semipalmatus	Semipalmated Plover	None
Aves	Charadriiformes	Charadriidae	Charadrius melodus	Piping Plover	
Aves	Charadriiformes	Charadriidae	Charadrius vociferus	Killdeer	vociferus
Aves	Charadriiformes	Charadriidae	Charadrius montanus	Mountain Plover	
Aves	Charadriiformes	Charadriidae	Charadrius morinellus	Eurasian Dotterel	

Class	Order	Family	Scientific Name (Other Scientific Name)	Common Name (Other Common Name)	Subspecies Name (Oregon & Washington only)
Aves	Charadriiformes	Haematopodidae	Haematopus bachmani	Black Oystercatcher	None
Aves	Charadriiformes	Recurvirostridae	Himantopus mexicanus	Black-necked Stilt	None
Aves	Charadriiformes	Recurvirostridae	Recurvirostra americana	American Avocet	None
Aves	Charadriiformes	Scolopacidae	Tringa melanoleuca	Greater Yellowlegs	None
Aves	Charadriiformes	Scolopacidae	Tringa flavipes	Lesser Yellowlegs	
Aves	Charadriiformes	Scolopacidae	Tringa brythropus	Spotted Redshank	
Aves	Charadriiformes	Scolopacidae	Tringa solitaria	Solitary Sandpiper	"solitaria, cinnamomea"
Aves	Charadriiformes	Scolopacidae	Catoptrophorus semipalmatus	Willet	inornatus
Mammalia	Didelphimorphia	Didelphidae	Didelphis virginiana	Virginia Opossum	
Mammalia	Insectivora	Soricidae	Sorex cinereus	Masked Shrew (Commom Shrew)	
Mammalia	Insectivora	Soricidae	Sorex preblei	Preble's Shrew	
Mammalia	Insectivora	Soricidae	Sorex vagrans (Sorex trigonirostris)	Vagrant Shrew (Wandering Shrew)	
Mammalia	Insectivora	Soricidae	Sorex monticolus (Sorex monticola)	Montane Shrew (Dusky Shrew)	"setosus, obscurus"
Mammalia	Insectivora	Soricidae	Sorex bairdi (Sorex bairdii)	Baird's Shrew	"bairdi, permiliensis"
Mammalia	Insectivora	Soricidae	Sorex sonomae	Fog Shrew	"sonomae, tenelliodus"
Mammalia	Insectivora	Soricidae	Sorex pacificus	Pacific Shrew	"pacificus, cascadensis"
Mammalia	Insectivora	Soricidae	Sorex palustris	Water Shrew	navigator
Mammalia	Insectivora	Soricidae	Sorex bendirii	Pacific Water Shrew (Marsh Shrew)	"palmeri, bendirii, albiventer"
Mammalia	Insectivora	Soricidae	Sorex trowbridgii	Trowbridge's Shrew	"trowbridgii, mariposae, destructioni"
Mammalia	Insectivora	Soricidae	Sorex merriami	Merriam's Shrew	merriami
Mammalia	Insectivora	Soricidae	Sorex hoyi (Microsorex hoyi)	Pygmy Shrew	washingtoni
Mammalia	Insectivora	Talpidae	Neurotrichus gibbsii	Shrew-mole	"gibbsii, minor"
Mammalia	Insectivora	Talpidae	Scapanus townsendii	Townsend's Mole	"townsendii, olympicus"
Mammalia	Insectivora	Talpidae	Scapanus orarius	Coast Mole	"orarius, schefferi, yakimensis"
Mammalia	Insectivora	Talpidae	Scapanus latimanus	Broad-footed Mole	dilatus
Mammalia	Chiroptera	Vespertilionidae	Myotis californicus	California Myotis	"californicus, caurinus"
Mammalia	Chiroptera	Vespertilionidae	Myotis ciliolabrum ("Myotis leibii, Myotis subulatus")	Western Small-footed Myotis	melanorhinus
Mammalia	Chiroptera	Vespertilionidae	Myotis yumanensis	Yuma Myotis	"saturatus, sociabilis"
Mammalia	Chiroptera	Vespertilionidae	Myotis lucifugus	Little Brown Myotis	"carissima, alascensis"
Mammalia	Chiroptera	Vespertilionidae	Myotis volans	Long-legged Myotis	"longicrus, interior"
Mammalia	Chiroptera	Vespertilionidae	Myotis thysanodes	Fringed Myotis	thysanodes
Mammalia	Chiroptera	Vespertilionidae	Myotis keenii	Keen's Myotis	keenii
Mammalia	Chiroptera	Vespertilionidae	Myotis evotis	Long-eared Myotis	"evotis, pacificus"
Mammalia	Chiroptera	Vespertilionidae	Lasionycteris noctivagans	Silver-haired Bat	

Class	Order	Family	Scientific Name (Other Scientific Name)	Common Name (Other Common Name)	Subspecies Name Oregon & Washington only
Mammalia	Chiroptera	Vespertilionidae	Pipistrellus hesperus	Western Pipistrelle	hesperus
Mammalia	Chiroptera	Vespertilionidae	Eptesicus fuscus	Big Brown Bat	bernardinus
Mammalia	Chiroptera	Vespertilionidae	Lasiurus cinereus	Hoary Bat	cinereus
Mammalia	Chiroptera	Vespertilionidae	Euderma maculatum	Spotted Bat	
Mammalia	Chiroptera	Vespertilionidae	Corynorhinus townsendii (Plecotus townsendii)	Townsend's Big-eared Bat	"Townsendii, Pallescens"
Mammalia	Chiroptera	Vespertilionidae	Antrozous pallidus	Pallid Bat	"pacificus, pallidus"
Mammalia	Chiroptera	Molossidae	Tadarida brasiliensis	Brazilian Free-tailed Bat	mexicana
Mammalia	Lagomorpha	Ochotonidae	Ochotona princeps	American Pika	"brunnescens, fumosa, taylori, jewetti"
Mammalia	Lagomorpha	Leporidae	Brachylagus idahoensis	Pygmy Rabbit	
Mammalia	Lagomorpha	Leporidae	Sylvilagus bachmani	Brush Rabbit	"ubericolor, tehamae"
Mammalia	Lagomorpha	Leporidae	Sylvilagus floridanus	Eastern Cottontail	
Mammalia	Lagomorpha	Leporidae	Sylvilagus nuttallii	Nuttall's (Mountain) Cottontail	nuttalli
Mammalia	Lagomorpha	Leporidae	Oryctolagus cuniculus	European Rabbit	
Mammalia	Lagomorpha	Leporidae	Lepus americanus	Snowshoe Hare	"washingtoni, klamathensis, oregonus"
Mammalia	Lagomorpha	Leporidae	Lepus townsendii	White-tailed Jackrabbit	townsendii
Mammalia	Lagomorpha	Leporidae	Lepus californicus	Black-tailed Jackrabbit	"californicus, wallawalla"
Mammalia	Rodentia	Aplodontidae	Aplodontia rufa	Mountain Beaver	
Mammalia	Rodentia	Sciuridae	Tamias minimus (Eutamias minimus)	Least Chipmunk	
Mammalia	Rodentia	Sciuridae	Tamias amoenus (Eutamias amoenus)	Yellow-pine Chipmunk	"abiventris, amoenus, ludibundus, ochraceus, affinis, canicaudus, luteiventris, felix, caurinus"
Mammalia	Rodentia	Sciuridae	Tamias townsendii (Eutamias townsendii)	Townsend's Chipmunk	
Mammalia	Rodentia	Sciuridae	Tamias senex (Eutamias senex)	Allen's Chipmunk	
Mammalia	Rodentia	Sciuridae	Tamias siskiyou (Eutamias siskiyou)	Siskiyou Chipmunk	
Mammalia	Rodentia	Sciuridae	Tamias ruficaudus (Eutamias ruficaudus)	Red-tailed Chipmunk	
Mammalia	Rodentia	Sciuridae	Marmota flaviventris	Yellow-bellied Marmot	"avara, flaviventris"
Mammalia	Rodentia	Sciuridae	Marmota caligata	Hoary Marmot	
Mammalia	Rodentia	Sciuridae	Marmota olympus	Olympic Marmot	
Mammalia	Rodentia	Sciuridae	Ammospermophilus leucurus	White-tailed Antelope Squirrel	leucurus
Mammalia	Rodentia	Sciuridae	Spermophilus townsendii (Citellus townsendii)	Townsend's Ground Squirrel	"townsendii, nancyae, canus, vigillis, mollis"
Mammalia	Rodentia	Sciuridae	Spermophilus canus ("Citellus canus, Spermophilus vigilis")	Merriam's Ground Squirrel	"canus, vigilis"

Class	Order	Family	Scientific Name (Other Scientific Name)	Common Name (Other Common Name)	Subspecies Name Oregon & Washington only
Mammalia	Rodentia	Sciuridae	Spermophilus mollis (Citellus mollis)	Piute Ground Squirrel	mollis
Mammalia	Rodentia	Sciuridae	Spermophilus washingtoni (Citellus washingtoni)	Washington Ground Squirrel	
Mammalia	Rodentia	Sciuridae	Spermophilus elegans (Citellus elegans)	Wyoming Ground Squirrel (Richardson's Ground Squirrel)	"oregonus, creber"
Mammalia	Rodentia	Sciuridae	Spermophilus beldingi (Citellus beldingi)	Belding's Ground Squirrel	
Mammalia	Rodentia	Sciuridae	Spermophilus columbianus (Citellus columbianus)	Columbian Ground Squirrel	"columbianus, ruficaudus"
Mammalia	Rodentia	Sciuridae	Spermophilus beecheyi ("Citellus beecheyi, Otospermophilus beecheyi")	California Ground Squirrel	douglasii
Mammalia	Rodentia	Sciuridae	Spermophilus lateralis ("Citellus lateralis, Callospermophilus lateralis")	Golden-mantled Ground Squirrel	"chrysodeirus, trinitatis, connectens, trepidus, saturatus"
Mammalia	Rodentia	Sciuridae	Spermophilus saturatus ("Citellus saturatus, Callospermophilus saturatus")	Cascade Golden-mantled Ground Squirrel	
Mammalia	Rodentia	Sciuridae	Sciurus carolinensis	Eastern Gray Squirrel	pennsylvanicus
Mammalia	Rodentia	Sciuridae	Sciurus niger	Eastern Fox Squirrel	
Mammalia	Rodentia	Sciuridae	Sciurus griseus	Western Gray Squirrel	griseus
Mammalia	Rodentia	Sciuridae	Tamiasciurus hudsonicus	Red Squirrel	"richardsoni, streatori"
Mammalia	Rodentia	Sciuridae	Tamiasciurus douglasii	Douglas' Squirrel	"douglasii, mollipilosus, albolimbatus"
Mammalia	Rodentia	Sciuridae	Glaucomys sabrinus	Northern Flying Squirrel	"oregonensis, fuliginosus, klamathensis, bangsi, columbiensis, latipes"
Mammalia	Rodentia	Geomyidae	Thomomys talpoides	Northern Pocket Gopher	"columbianus, wallowa, quadratus, douglasii, immunis, limosus, shawi, yakimensis, aequalidnes, devexus, fuscus"
Mammalia	Rodentia	Geomyidae	Thomomys mazama	Western Pocket Gopher (Mazama Pocket Gopher)	"helleri, niger, hesperus, oregonus, mazama, nascius, couchi, glacialis, buiei, pugatensis, tacomensis, tumuli, yelmensis, melanops"
Mammalia	Rodentia	Geomyidae	Thomomys bulbivorus	Camas Pocket Gopher	
Mammalia	Rodentia	Geomyidae	Thomomys bottae (Thomomys umbrinus)	Botta's (Pistol River) Pocket Gopher	"laticeps, detumidus, leucodon"
Mammalia	Rodentia	Geomyidae	Thomomys townsendii	Townsend's Pocket Gopher	"townsendii, nevadensis"
Mammalia	Rodentia	Heteromyidae	Perognathus parvus	Great Basin Pocket Mouse	"parvus, mollipilosus, yakimensis, columbianus, lordi"

Class	Order	Family	Scientific Name (Other Scientific Name)	Common Name (Other Common Name)	Subspecies Name Oregon & Washington only
Mammalia	Rodentia	Heteromyidae	Perognathus longimembris	Little Pocket Mouse	nevadensis
Mammalia	Rodentia	Heteromyidae	Microdipodops megacephalus	Dark Kangaroo Mouse	oregonus
Mammalia	Rodentia	Heteromyidae	Dipodomys ordii	Ord's Kangaroo Rat	columbianus
Mammalia	Rodentia	Heteromyidae	Dipodomys microps	Chisel-toothed Kangaroo Rat	preblei
Mammalia	Rodentia	Heteromyidae	Dipodomys californicus (Dipodomys heermanni)	California Kangaroo Rat	californicus
Mammalia	Rodentia	Castoridae	Castor canadensis	American Beaver	"longicaudus, megalolis"
Mammalia	Rodentia	Muridae	Reithrodontomys megalotis	Western Harvest Mouse	"rubidus, gambelii, sonoriensis, artemisiae, alpinus, austerus"
Mammalia	Rodentia	Muridae	Peromyscus maniculatus	Deer Mouse	
Mammalia	Rodentia	Muridae	Peromyscus keeni (Peromyscus oreas)	Columbian Mouse (Forest Deer Mouse)	crinitus
Mammalia	Rodentia	Muridae	Peromyscus crinitus	Canyon Mouse	"gilberti, preblei"
Mammalia	Rodentia	Muridae	Peromyscus truei	Pinon Mouse	"durranti, brevicaudus"
Mammalia	Rodentia	Muridae	Onychomys leucogaster	Northern Grasshopper Mouse	nevadensis
Mammalia	Rodentia	Muridae	Neotoma lepida	Desert Woodrat	"fuscipes, monochroura"
Mammalia	Rodentia	Muridae	Neotoma fuscipes	Dusky-footed Woodrat	"fusca, aticola, occidentalis, pulla"
Mammalia	Rodentia	Muridae	Neotoma cinerea	Bushy-tailed Woodrat	"idahoensis, saturatus, cascadensis, occidentalis, caurinus, nivarius"
Mammalia	Rodentia	Muridae	Clethrionomys gapperi	Southern Red-backed Vole	"californicus, manzama, obscurus"
Mammalia	Rodentia	Muridae	Clethrionomys californicus (Clethrionomys occidentalis)	Western Red-backed Vole	
Mammalia	Rodentia	Muridae	Phenacomys intermedius (Phenacomys olympicus)	Heather Vole	"intermedius, oramontis"
Mammalia	Rodentia	Muridae	Phenacomys albipes (Arborimus albipes)	White-footed Vole	
Mammalia	Rodentia	Muridae	Phenacomys longicaudus ("Arborimus longicaudus, Phenacmys silvicola")	Red Tree Vole	"silvicola, longicaudus"
Mammalia	Rodentia	Muridae	Microtus pennsylvanicus	Meadow Vole	kincaidi
Mammalia	Rodentia	Muridae	Microtus montanus	Montane Vole	"montanus, nanus, micropus, canescens"
Mammalia	Rodentia	Muridae	Microtus canicaudus	Gray-tailed Vole	
Mammalia	Rodentia	Muridae	Microtus californicus	California Vole	eximius
Mammalia	Rodentia	Muridae	Microtus townsendii	Townsend's Vole	"townsendii, pugetti"
Mammalia	Rodentia	Muridae	Microtus longicaudus	Long-tailed Vole	"abditus, angusticeps, longicaudus, halli, macrurus"
Mammalia	Rodentia	Muridae	Microtus oregoni	Creeping Vole	"adocetus, bairdii, oregoni"
Mammalia	Rodentia	Muridae	Microtus richardsoni	Water Vole	"arvicoloides, macropus"
Mammalia	Rodentia	Muridae	Lemmiscus curtatus (Lagurus curtatus)	Sagebrush Vole	"pauperrimus, intermedius"

Class	Order	Family	Scientific Name (Other Scientific Name)	Common Name (Other Common Name)	Subspecies Name Oregon & Washington only
Mammalia	Rodentia	Muridae	Ondatra zibethicus	Muskrat	"occipitalis, osoyoosensis"
Mammalia	Rodentia	Muridae	Synaptomys borealis	Northern Bog Lemming	
Mammalia	Rodentia	Muridae	Rattus rattus	Black Rat (Roof Rat)	
Mammalia	Rodentia	Muridae	Rattus norvegicus	Norway Rat	
Mammalia	Rodentia	Muridae	Mus musculus	House Mouse	
Mammalia	Rodentia	Zapodidae	Zapus princeps	Western Jumping Mouse	"pacificus, oregonus"
Mammalia	Rodentia	Zapodidae	Zapus trinotatus	Pacific Jumping Mouse	"trinotatus, montanus"
Mammalia	Rodentia	Erethizontidae	Erethizon dorsatum	Common Porcupine	
Mammalia	Rodentia	Myocastoridae	Myocastor coypus	Nutria	bonariensis
Mammalia	Carnivora	Canidae	Canis latrans	Coyote	"lestes, umpquensis"
Mammalia	Carnivora	Canidae	Canis lupus	Gray Wolf	"fuscus, irremotus, columbianus"
Mammalia	Carnivora	Canidae	Vulpes vulpes ("Vulpes fulvus, Vulpes fulva")	Red Fox	"cascadensis, macroura"
Mammalia	Carnivora	Canidae	Vulpes velox (Vulpex macrotis nevadensis)	Kit Fox	macrotis
Mammalia	Carnivora	Canidae	Urocyon cinereoargenteus	Gray Fox	townsendi
Mammalia	Carnivora	Ursidae	Ursus americanus (Euarctos americanus)	Black Bear	"altifrontalis, cinnamomum"
Mammalia	Carnivora	Ursidae	Ursus arctos (Ursus chelan)	Grizzly Bear (Brown Bear)	"horribilis, arctos"
Mammalia	Carnivora	Procyonidae	Bassariscus astutus	Ringtail	raptor
Mammalia	Carnivora	Procyonidae	Procyon lotor	Raccoon	"pacificus, excelsus"
Mammalia	Carnivora	Mustelidae	Martes americana	American Marten	"caurina, vulpina"
Mammalia	Carnivora	Mustelidae	Martes pennanti	Fisher	pacific
Mammalia	Carnivora	Mustelidae	Mustela erminea	Ermine (Short-Tailed Weasel)	"streatori, muricus"
Mammalia	Carnivora	Mustelidae	Mustela frenata	Long-tailed Weasel	
Mammalia	Carnivora	Mustelidae	Mustela vison	Mink	energumenos
Mammalia	Carnivora	Mustelidae	Gulo gulo (Gulo luscus)	Wolverine	luscus
Mammalia	Carnivora	Mustelidae	Taxidea taxus	American Badger	jeffersonii
Mammalia	Carnivora	Mustelidae	Spilogale gracilis	Western Spotted Skunk	"saxatalis, latifrons"
Mammalia	Carnivora	Mustelidae	Mephitis mephitis	Striped Skunk	"major, notata, occidentalis, spissigrada"
Mammalia	Carnivora	Mustelidae	Lutra canadensis (Lontra canadensis)	Northern River Otter	pacifica
Mammalia	Carnivora	Felidae	Puma concolor (Felis concolor)	Mountain Lion	"oregonensis, californica, missoulensis"
Mammalia	Carnivora	Felidae	Lynx canadensis ("Felis canadensis, Felis lynx")	Lynx	canadensis

Class	Order	Family	Scientific Name (Other Scientific Name)	Common Name (Other Common Name)	Subspecies Name Oregon & Washington only
Mammalia	Carnivora	Felidae	Lynx rufus (Felis rufus)	Bobcat	
Mammalia	Perissodactyla	Equidae	Equus asinus	Wild Burro	
Mammalia	Perissodactyla	Equidae	Equus caballus	Feral Horse (Mustang)	
Mammalia	Artiodactyla	Suidae	Sus scrofa	Feral Pig ("Wild Pig, Wild Boar, Feral Swine, Wild Hog, European Wild Boar")	
Mammalia	Artiodactyla	Cervidae	Cervus elaphus roosevelti (Cervus canadensis)	Roosevelt Elk	roosevelti
Mammalia	Artiodactyla	Cervidae	Cervus elaphus nelsoni (Cervus canadensis)	Rocky Mountain Elk	nelsoni
Mammalia	Artiodactyla	Cervidae	Odocoileus hemionus columbianus	Black-tailed Deer	columbianus
Mammalia	Artiodactyla	Cervidae	Odocoileus hemionus hemionus	Mule Deer	
Mammalia	Artiodactyla	Cervidae	Odocoileus virginianus leucurus	Columbian White-tailed Deer	
Mammalia	Artiodactyla	Cervidae	Odocoileus virginianus ochrourus	White-tailed Deer (Eastside)	ochrourus
Mammalia	Artiodactyla	Cervidae	Alces alces	Moose	shirasi alces alces
Mammalia	Artiodactyla	Cervidae	Rangifer tarandus	Mountain Caribou (Woodland Caribou)	caribou
Mammalia	Artiodactyla	Antilocapridae	Antilocapra americana	Pronghorn Antelope	americana
Mammalia	Artiodactyla	Bovidae	Bos bison (Bison bison)	Bison (Buffalo)	athabascae
Mammalia	Artiodactyla	Bovidae	Oreamnos americanus	Mountain Goat	"americanus, missoulae"
Mammalia	Artiodactyla	Bovidae	Ovis canadensis canadensis	Rocky Mountain Bighorn Sheep	canadensis
Mammalia	Artiodactyla	Bovidae	Ovis canadensis californiana	California Bighorn Sheep	californiana
Mammalia	Carnivora	Otariidae	Callorhinus ursinus	Northern Fur Seal	None
Mammalia	Carnivora	Otariidae	Eumetopias jubatus	Northern (Steller) Sea Lion	None
Mammalia	Carnivora	Otariidae	Zalophus californianus	California Sea Lion	californianus
Mammalia	Carnivora	Phocidae	Phoca vitulina (Halicyon richardsi)	Harbor Seal ("Hair Seal, Spotted Seal, Common Seal")	richardsi..richardsi
Mammalia	Carnivora	Phocidae	Mirounga angustirostris	Northern Elephant Seal	None
Mammalia	Carnivora	Mustelidae	Enhydra lutris	Sea Otter	
Mammalia	Cetacea	Eschrichtiidae	Eschrichtius robustus (Eschrichtius glaucus)	Gray Whale (California Gray Whale)	
Mammalia	Cetacea	Balaenopteridae	Balaenoptera acutorostrata	Minke Whale (Little Piked Whale)	
Mammalia	Cetacea	Balaenopteridae	Balaenoptera borealis	Sei Whale	
Mammalia	Cetacea	Balaenopteridae	Balaenoptera musculus	Blue Whale (Sulphur Bottom)	musculus
Mammalia	Cetacea	Balaenopteridae	Balaenoptera physalus	Fin Whale (Finback Whale)	

Class	Order	Family	Scientific Name (Other Scientific Name)	Common Name (Other Common Name)	Subspecies Name (Oregon & Washington only)
Mammalia	Cetacea	Balaenopteridae	Megaptera novaeangliae	Humpback Whale	
Mammalia	Cetacea	Balaenidae	Balaena glacialis (Eubalaena glacialis)	Northern Right Whale ("Black Right Whale, Pacific Right Whale")	Glacialis
Mammalia	Cetacea	Delphinidae	Stenella coeruleoalba	Striped Dolphin	
Mammalia	Cetacea	Delphinidae	Delphinus delphis	Common Saddle-backed Dolphin ("Common Dolphin, White-Bellied Porpoise")	
Mammalia	Cetacea	Delphinidae	Lagenorhynchus obliquidens	Pacific White-sided Dolphin	
Mammalia	Cetacea	Delphinidae	Grampus griseus (Delphinus griseus)	Risso's Dolphin (Gray Grampus)	
Mammalia	Cetacea	Delphinidae	Pseudorca crassidens	False Killer Whale (False Pilot Whale)	
Mammalia	Cetacea	Delphinidae	Globicephala macrorhynchus ("Globicephala macrorhyncha, Globicephala sieboldii")	Short-finned Pilot Whale ("Pilot Whale, Pothead Whale")	
Mammalia	Cetacea	Delphinidae	Orcinus orca	Killer Whale (Orca)	
Mammalia	Cetacea	Delphinidae	Lissodelphis borealis	Northern Right-whale Dolphin (Pacific Right Whale Porpoise)	
Mammalia	Cetacea	Phocoenidae	Phocoena phocoena	Harbor Porpoise (Common Porpoise)	
Mammalia	Cetacea	Phocoenidae	Phocoenoides dalli	Dall's Porpoise (Dall Porpoise)	
Mammalia	Cetacea	Ziphiidae	Berardius bairdii	North Pacific Bottle-nosed Whale ("Baird's Beaked Whale, Giant Bottlenose Whale")	
Mammalia	Cetacea	Ziphiidae	Ziphius cavirostris	Goose-beaked Whale (Cuvier's Beaked Whale)	
Mammalia	Cetacea	Ziphiidae	Mesoplodon stejnegeri	Bering Sea Beaked Whale ("Stejneger's Beaked Whale, North Pacific Beaked Whale")	
Mammalia	Cetacea	Ziphiidae	Mesoplodon carlhubbsi	Arch-beaked Whale (Hubb's Beaked Whale)	
Mammalia	Cetacea	Kogiidae	Kogia breviceps	Pygmy Sperm Whale	
Mammalia	Cetacea	Physeteridae	Physeter macrocephalus (Physeter catodon)	Sperm Whale	
Mammalia	Cetacea	Monodontidae	Delphinapterus leucas	Beluga Whale (White Whale)	

Glossary

Abiotic: Non-living components of an ecosystem; basic elements, and compounds of the environment.

Abyssal: Pertaining to zones of great depth in the oceans or lakes into which light does not penetrate; occasionally restricted to depths below 2,000 meters, but more usually used for depths between 4,000 and 6,000 meters.

Accretion: Deposition of material by sedimentation which increases land area.

Active Layer: A seasonally thawed surface layer of soil in arctic or alpine regions that lies above permanently frozen ground and is between a few centimeters and about three meters thick.

Adaptive Radiation: The evolutionary diversification of a taxon into a number of different forms, usually as a result of encounters with new resources or habitats.

Adaptation: A genetically determined characteristic that enhances an organism's chances for survival and reproduction.

Adaptive Management: An adaptive approach to management where we use the best scientific knowledge and technologies, clearly recognize knowledge gaps, build shared expectations among those who have a stake in ecosystem outcomes, monitor actions, and adjust management actions accordingly.

Adventive Plant: A species of plant that is not native and has been introduced into the area but has not become permanently established.

Afforestation: The establishment of forest by natural succession or by the planting of trees on land where they did not grow formerly.

Albedo: A measure of surface reflectivity, usually expressed as a percentage, such as the proportion of solar radiation that is reflected back into space from the Earth, clouds, and atmosphere without heating the receiving surface.

Alcids: Any of the Alcidae family (Order Charadriiformes) of marine birds having a stout bill, short wings and tail, webbed feet, a large head and heavy body, and thick, compact plumage. Confined to the northern parts of the Northern Hemisphere, alcids include auks, guillemots, murres, and puffins.

Alevin: A young fish, particularly a young salmon that is still attached to the yolk sac.

Algae: The common name for the relatively simple type of unicellular or multicellular plant which is never differentiated into root, stem, and leaves, contains chlorophyll *a* as its photosynthetic pigment, has no true vascular system, and has no sterile layer of cells surrounding its reproductive organs.

Alluvial: Relating to river and stream deposits.

Alluvial Soil: Soil formed in material deposited by the action of running water, such as a floodplain or delta.

Alpine Tundra: A treeless region above the treeline of high mountains, characterized by cold winters and short, cool summers and having permafrost below a surface layer that may melt in summer.

Altricial: Naked or helpless when hatched; immobile, downless, eyes closed.

Alvar: A plant community dominated by mosses and herbs, occurring on shallow, alkaline limestone soils.

Amphidromous: Referring to the migratory behavior of fishes moving from fresh water to the sea and vice versa, not for breeding purposes but occurring regularly at some stage of the life cycle (such as feeding or overwintering).

Amphipod: Any of a large order of small, usually aquatic crustaceans with a laterally compressed body, for example, beach fleas.

Anadromous: Referring to the life cycle of fishes, such as salmon, in which adults travel upriver from the sea to breed, usually returning to the area where they were born.

Anaerobic: Referring to an environment in which oxygen is absent, or to a process, which occurs only in the absence of oxygen, or to an organism, which lives, is active, or occurs in the absence of oxygen, such as some yeasts or bacteria.

Annelids: Any of a phylum (Annelida) of usually elongated, segmented coelomate invertebrates, such as earthworms, various marine worms, and leeches.

Anoxic: Greatly deficient in oxygen; oxygenless.

Anthropogenic: Of, relating to, or resulting from the influence of humans on nature.

Anticyclonic: Referring to an area or system of high atmospheric pressure having a characteristic pattern of air circulation which usually induces settled weather conditions. Light winds flow clockwise in the northern hemisphere and counterclockwise in the southern hemisphere.

Aquatic Ecosystem: Any body of water, such as a stream, lake, or estuary, and all organisms and non-living components within it, and functioning as a natural system.

Aquatic Integrity: A mosaic of well connected, high-quality water and habitats that support a diverse assemblage of native and desired non-native species, the full expression of potential life histories and taxonomic lineages, and the taxonomic and genetic diversity necessary for long-term persistence and adaptation in a variable environment.

Arboreal: Living in the canopies of trees.

Archaebacteria: A taxonomic kingdom of bacteria, including sulphur-dependent bacteria, methane-producing bacteria, and halophilic bacteria.

Areas of Environmental Concern: Areas within the public lands where special management attention is required (when such areas are developed or used or where no development is required) to protect and prevent irreparable damage to important natural systems or processes, or to protect life and safety from natural hazards.

Arthropod: Invertebrate animals with a segmented body and jointed appendages, for example, spiders, bees, and crabs.

Aspect: The direction a slope faces with respect to the cardinal compass points.

Association: A stable grouping of two or more plant species that characterize or dominate a type of biotic community.

Autecology: A subdivision of ecology that deals with the relationship of individuals of a species to their environment.

Avalanche Chute: An area where periodic snow or rockslides prevent the establishment of forest conditions; typically shrub and herb dominated.

Avian: Relating to or derived from birds.

Avifauna: The birds of a specific region or period.

Barrens: A level area with poor, usually sandy or serpentine soils that is sparsely forested or unable to support normal vegetative cover and that generally has a low level of productivity. Barrens are frequently dominated by specialized groups of endemic plants.

Bathymetry: The measurement of the depth of the ocean floor from the water surface; the oceanic equivalent of topography.

Bathypelagic: Of, relating to, or living in the depths of the ocean, especially in the area between about 600 and 3,000 meters deep. The number of species and populations is relatively low in the bathypelagic zone, where no light source exists other than bioluminescence, temperature is uniformly low, and pressures are great.

Benthic: Occurring at the bottom of a body of water, for example, a seabed, riverbed, or lake bottom.

Benthos: In freshwater and marine ecosystems, the collection of organisms both attached to or resting on the bottom sediments and burrowed into the sediments.

Bight: A large indentation in a coastline or continental shelf margin forming an open bay.

Bioaccumulation (Also called *Biomagnification*): The process by which chemical contaminants become more concentrated in the tissues of organisms as they pass higher up the food chain. Heavy metals and pesticides such as DDT are stored in the fatty tissues of animals and are passed along to predators of those animals. The resulting concentrations eventually reach harmful levels in predators at the top of the food chain.

Biodiversity: The variety of organisms considered at all levels, from genetic variants belonging to the same species through arrays of genera, families, and still higher taxonomic levels, includes the variety of ecosystems, which comprise both the communities of organisms within particular habitats and the physical conditions under which they live.

Bioenvironments: Combinations of environmental factors to which the biota responds directly (e.g., temperature), or consume as resources (e.g., nutrients).

Biogeochemical: The flow and interactions of biological and chemical processes in relation to organisms and their environments.

Biogeographic: The spatial distribution patterns of organisms in relation to changes through time (paleoecological, historical, current, and future).

Biogeographical Region: Any geographical region characterized by distinctive flora or fauna (such as a biome or a province).

Biogeography: The science that deals with the geographical distribution of animals and plants.

Biologic Diversity: The full variety of living organisms and their assemblages; the genetic variation within and between populations of species, and the many processes that link organisms and their physical environments into ecological systems.

Biomass: The total mass of all living organisms or of a particular set of organisms in an ecosystem or at a trophic level in a food chain; usually expressed as a dry weight or as the carbon, nitrogen, or caloric content per unit area.

Biome: A major regional ecological community characterized by distinctive life forms and principal plant or animal species, such as a tropical rain forest, tundra, grassland, or a desert.

Bioregion: A territory defined by a combination of biological, social, and geographic criteria, rather than geopolitical considerations; generally, a system of related, interconnected ecosystems.

Biota: The plants and animals of a specific region or period, or the total aggregation of organisms in the biosphere.

Biotic Community: Any assemblage of populations living in a prescribed area or physical habitat; an aggregate of organisms, which form a distinct ecological unit.

Biotic Climax: A climax caused by a permanent influence or culmination of influences caused by one or more kinds of organisms, including humans.

Biotype: A group of individuals within a population occurring in nature, all with essentially the same genetic constitution. A species usually consists of many biotypes.

Bivalve: A mollusk whose body is enclosed by two hinged valves or shells.

Blowdown: An extensive toppling of trees by wind within a relatively small area, which significantly alters the small-scale climate within the ecosystem.

Boreal Forest: The circumpolar, subarctic forest of high northern latitudes that is dominated by conifers. It is found south of the tundra in the Northern Hemisphere and often contains peaty or swampy areas.

Brackish: Water that is saline but not as salty as seawater.

Braided Channel: A stream consisting of a network of interlacing small channels separated by bars, which may be vegetated and stable or barren and unstable.

Breeding Bird Survey: The North American Breeding Bird Survey (BBS) begun in 1966 to collect standardized data on bird populations along more than 3,400 survey routes across the continental United States and southern Canada for more than 250 species.

Broad Scale: Encompassing a wide area.

Broader Geographic Scales: (See landscape). Watershed, river basin, or other physiographic region suitable for analyzing management proposals relative to other proposals or activities; something larger than a stand, soil mapping unit, local landform, lake or stream.

Brood Parasitism: (Also called *nest parasitism* or *breeding parasitism*.) The laying of eggs by one bird species in the nest of another bird species and the subsequent brooding of the egg and raising of the young by the parasitized host, usually to the detriment to the host's young.

Bunchgrass: Any of several grasses, especially of the western United States that grow in tufts rather than forming turf, for example, the genus *Andropogon*.

Bycatch: Nontarget organisms that are caught in fishing or other harvest operations and are usually discarded.

Calcareous: Consisting of or containing calcium carbonate; a soil rich in calcium salts, derived from limestone or chalk. Also, an organism which has an affinity for such an alkaline or basic soil.

Candidate Species: A species being considered for listing as a federally endangered or threatened species.

Canopy: A layer of foliage in a forest stand; most often refers to the uppermost layer of foliage.

Canopy Closure: The degree to which the canopy blocks sunlight or obscures the sky. It can only be accurately determined from measurements taken under the canopy, as openings in the branches and crowns must be accounted for.

Capability: The potential of an area of land to produce resources, supply goods and services, and allow resource uses under an assumed set of management practices and at a given level of management intensity.

Carrying Capacity: The maximum population of a given organism that a particular environment or habitat can sustain; implies continuing yield without environmental damage; often denoted as *K*.

Catadromous: An organism, which lives in fresh water and goes to the sea to spawn, such as some eels.

Catchment: The area drained by a river or body of water.

Cetacean: Any of an order of aquatic, mostly marine mammals that include the whales, dolphins, porpoises, and related forms.

Chaetognaths: A group of small, active, transparent marine worms of uncertain systemic position with horizontal lateral and caudal fins and a row of moveable, curved spines around the mouth, for example, arrowworms.

Changed Biophysical Template: Biophysical systems that have the biotic or physical potentials of the historical range of variability (HRV), but have a different composition, structure, or disturbance regime than present during HRV.

Channelization: The straightening of rivers or streams by means of an artificial channel.

Chaparral: A vegetation type dominated by shrubs and small trees, especially evergreen trees with thick, small leaves.

Chironomids: Any of a family (Chironomidae) of midges that lack piercing mouthparts.

Chlorofluorocarbons (Also called greenhouse gases or CFCs.): A group of gaseous compounds that contain carbon, chlorine, fluorine, and sometimes hydrogen, and are aerosol propellants and in the manufacture of plastic foams.

Cirque: A steep hollow, often containing a small body of water, found at the upper end of a mountain valley.

CITES: Convention on International Trade in Endangered Species of Wild Fauna and Flora an agreement between 103 nations to restrict international commerce involving endangered and threatened species of animals and plants, such as tropical birds, rhinoceros horns, orchids, and ivory.

Cladocerans: Any of an order (Cladocera) of minute, freshwater brachiopod crustaceans, including the water fleas.

Clear-cut: An area where the entire stand of trees has been removed in one cutting.

Climate: Generalized statement of the prevailing weather conditions at a given place, based on statistics of a long period of record. Includes seasonality of temperature and moisture.

Climax: The final stage of succession in an ecosystem. Also, a community that reached a steady state under a particular set of environmental conditions.

Close-crowned: Descriptive of crowded forests where closely spaced trees have tops that touch or overlap.

Cluster Analysis: A method grouping those variables within a set of variables that are highly correlated and excluding from clusters those that are negatively correlated or uncorrelated.

Coarse Woody Debris (CWD): Portion of a tree that has fallen or been cut and left in the woods. Usually refers to pieces at least 20 inches in diameter.

Cohort: Individuals all resulting from the same birth-pulse, and thus all of the same age.

Commensal: Referring to the relationship between two kinds of organisms in which one obtains food or other benefits from the other without damaging or benefiting it.

Community: Any grouping of populations of different organisms that live together in a particular environment.

Connectivity: Condition, in which the spatial arrangement of land cover types allows organisms and ecological processes (such as disturbance) to move across the landscape. Connectivity is the opposite of fragmentation.

Conservation Biology: The body of knowledge that deals with the careful protection, utilization, and planned management of living organisms and their vital processes to prevent their depletion, exploitation, destruction, or waste.

Conservation Strategy: A management plan for a species, group of species, or ecosystem that prescribes standards and guidelines that if implemented provide a high likelihood that the species, groups of species, or ecosystem, with its full complement of species and processes, will continue to exist well-distributed throughout a planning area, i.e., a viable population.

Conspecific: Relating to the same species.

Continental Shelf: The shallow, gradually sloping seabed around a continental margin not usually deeper than 200 meters and formed by submergence of part of a continent.

Continentality: Tendency of large land areas in mid-latitude and high latitudes to impose a large annual temperature range on the air temperature cycle.

Copepods: Any of a large subclass (Copepoda) of usually minute freshwater and marine crustaceans that form an important element of the plankton in the marine environment and in some fresh waters.

Corridor: A more or less continuous connection between landmasses or habitats; a migration route that allows more of less uninhibited migration of most of the animals of one faunal region to another. In terms of conservation biology, a connection between habitat fragments in a fragmented landscape.

Corridors: The landscape elements that connect similar patches through a dissimilar matrix or aggregation of patches.

Corvids: A family of birds that includes the crows, ravens, jays, and magpies.

Cover: Vegetation used by wildlife for protection from predators, to mitigate weather conditions, or to reproduce. May also refer to the protection of soil and the shading provided to herbs and forbs by vegetation.

Crevasse: A breach in a levee along the bank of a river through which floodwater may flow and produce sheet-like deposits of gravel or sandy sediment; or a large, open fissure forming in a glacier as it moves and is deformed.

Critical Habitat: Under the Endangered Species Act, critical habitat is defined as the specific areas within the geographic area occupied by a federally listed species on which are found physical and biological features essential to the conservation of the species, and that may require special management considerations or protection, and specific areas outside the geographic area occupied by a listed species, when it is determined that such areas are essential for the conservation of the species.

Crown Fires: Fires that spread from tree crown to tree crown, usually indicative of particularly hot fires in dry conditions.

Crustacean: Any of a large class (Crustacea) of mostly aquatic mandibulate arthropods that have a chitinous of calcareous and chitinous exoskeleton, a pair of often modified appendages on each segment, and two pairs of antennae; includes lobsters, shrimps, crabs, wood lice, water fleas, and barnacles.

Cultivar: A variety of a plant produced and maintained by horticultural techniques and not normally found in wild populations.

Cyanobacteria: A large and varied group of bacteria which possess chlorophyll *a* and which carry out photosynthesis in the presence of light and air, producing oxygen. They were formerly regarded as algae and were called "blue-green" algae. The group is very old and is believed to have been the first oxygen-producing organisms on Earth.

Cyclonic: Referring to a region of low atmospheric sea level pressure; or, the wind system around such a low-pressure center that has a clockwise rotation in the Northern Hemisphere and a counterclockwise rotation in the Southern Hemisphere.

Debouch: To emerge or issue; often used in reference to rivers or streams.

Debris torrent: A flood of debris (branches, shrubs, rocks, mud, and so forth) and water rushing down a stream channel, caused by excessive rainfall or snowmelt. Debris torrents have a significant scouring effect on the stream ecosystem.

Deciduous: Plants having structures that are shed at regular intervals or at a given stage in development, such as trees that shed their leaves seasonally.

Decreaser Species: plant species of the potential vegetation that will decrease in relative amount when management-induced disturbance stresses (such as excessive livestock grazing pressure or alteration of fire frequency and severity) operate during drought and when management-induced disturbance stresses operate during and immediately after drought. The term is used particularly for plant species on rangelands.

Defoliators: Insects that feed on foliage and act to remove some or all of the foliage from a tree shrub, or herb.

Degradation: The breaking down of a substance into smaller or simpler parts, usually by erosion.

Delta: An alluvial deposit at the mouth of a river or tidal inlet. Deltas occur when a sediment-laden current enters an open body of water, at which point there is a reduction in the velocity of the current, resulting in rapid deposition of the sediment, as at the mouth of a river where the river discharges into the sea or a lake.

Demersal: Living at or near the sea floor but having the capacity for active swimming.

Demography: The quantitative analysis of population structure and trends; population dynamics.

Dendrochronology: The science of dating events and variations in the environment by the comparative study of annual growth rings of trees.

Desertification: The process by which an area or region becomes more arid through loss of soil and vegetative cover. The process is often accelerated by excessive continuous overstocking and drought.

Detritus: Debris or waste material, usually organic, such as dead or partially decayed plants and animals, often important as a source of nutrients; or, small particles of minerals from weathered rock, such as sand or silt.

Dewatering/Dewater: The removal of water from a stream/river network, typically for irrigation, industrial, or human use; commonly changes a network that developed by concentrating flows from stream/river branches to main-stems branching to canals, which reduces the flow in the main-stems.

Diel: A 24-hour period, usually encompassing one day and one night.

Dinoflagellates: Any of an order (Dinoflagellata) of chiefly marine, planktonic, usually solitary phytoflagellates (which have many characteristics in common with algae) that includes luminescent forms, forms important in marine food chains, and forms causing red tides.

Disjunct: Distinctly separate; a discontinuous range in which one or more populations are separated from other potentially interbreeding populations by a sufficient distance to preclude gene flow between them.

Dispersal: The movement, usually one-way and on any time scale, of plants or animals form their point of origin to another location where they subsequently produce offspring.

Distributary: A river branch flowing away from the main stream.

Disturbance: An effect of a planned human management activity, or unplanned native or exotic agent or event that changes the state of a landscape element, landscape pattern, or regional composition.

Disturbance Severity Classes:

 Lethal: Disturbance that causes morality to most of the upper layers of vegetation and changes the structure.

 Non-lethal: Disturbance does not cause mortality to a substantial portion of the upper layer, removes susceptible individuals from all layers, and maintains the structure.

 Mixed: Disturbance causes a fine-scale mosaic of lethal and non-lethal effects that result in clump/gap mosaics of changed and maintained or unaffected structure.

Disturbance Regime: The pattern of intervals between disturbance and severity of disturbance. For landscapes, this can be for a given disturbance, such as fire, or for a complex of disturbances.

Diurnal: Occurring or active only in daylight.

Diversity: The distribution and abundance of different plant and animal communities and species within the area covered by a land and resource management plan.

Doliolids: Any of a small family of oceanic tunicates.

Downwelling: The downward movement of surface waters caused by the convergence of different water masses or where surface waters flow toward the coast.

Drawdown: A lowering of the water level in a reservoir or other body of water.

Echinoderms: Any of a phylum (Echinodermata) of radially symmetrical coelomate marine animals including the starfishes, sea urchins, and related forms.

Ecological Approach: Natural resource planning and management activities that assure consideration of the relationship between all organisms (including humans) and their environment.

Ecological Disequilibria/Disequilibrium: A system that has unequal relationships of inputs and outputs that result in erratic (and unpredictable) successional patterns and associated responses to disturbance.

Ecological Element: The individual constituent of the whole. For example: vegetation patch, stream reach, road, city site, or large snag.

Ecological Function: The activity or role performed by an organism or element in relation to other organisms, element in relation to other organisms, elements, or the environment.

Ecological Integrity: The maintenance of native and desired non-native species and associated processes.

Ecological Predictability: Causal agents of disturbance (such as fire, erosion, floods, grazing, harvest, predation, insects, and pathogens) produce "expected" effects that are consistent with the limitations of the biophysical system and inherent disturbance regimes (based on native historical range of variability, or altered regime, as appropriate).

Ecological Process: A series of actions, changes, or functions that produce a resulting condition for biota, elements, or the environment. For example succession, decay, photosynthesis, food chain, fire, drought, or flood.

Ecological Resiliency: Ability of system biota and their environments to renew the cycle of functions and processes following disturbance. At a landscape scale the focus is on connections of the hydrologic and land system, carbon-nutrient system, food web, and evolutionary systems and their aggregate tendency towards equilibrium or disequilibrium in response to disturbance.

Ecological Simplification: Loss of inherent patterns of elements (such as native species, large trees, bunchgrasses, soil, or channel stability), mosaics (such as vegetation patch patterns, migration routes, or connected stream pools and runs), and succession/disturbance regimes (such as frequent non-lethal fire to infrequent lethal fire or high sinuosity streams to low sinuosity). Simplification results in loss of resiliency and associated predictability (or reliability) of system response.

Ecological Site: A specific location on the land, that is representative of an ecological type.

Ecological Succession: The chronological sequence of vegetation and associated animals in an area; or, continuous colonization, extinction, and replacement of species' populations at a particular site, due either to environmental changes or to the intrinsic properties of the plants and animals.

Ecological Type: A category of land having a unique combination of potential natural community, soil, landscape features, climate, and differing from other ecological types in its ability to produce vegetation and respond to management.

Ecology: The relationship of species, including humans, and their environment.

Ecoregion: A continuous geographic area in which the environmental complex, produced by climate, topography, and soil, is sufficiently uniform to develop characteristics of potential major vegetation communities.

Ecosystem: A community of organisms and their physical environment that interact as an ecological unit.

Ecosystem Degradation: Reductions in ecosystem sustainability because of natural or human effects.

Ecosystem Function (Processes): The major processes of ecosystems that regulate or influence the structure, composition, and pattern. These include nutrient cycles, energy flows, trophic levels (food chains), diversity patterns in time/space development and evolution, cybernetics (control), hydrologic cycles and weathering processes.

Ecosystem-based Management: The careful and skillful integration of ecological, economic, social, and managerial principles to conserve, enhance, and restore ecosystems (including their functions, processes, constituent species, and productive capacities) to maintain their long-term viability and integrity while seeking desired conditions for uses, products, values, and services.

Ecosystem Pattern: The structure that results from the distribution of organisms in, and their interaction with their environment. Includes zonation, stratification, activity, or periodicity, food-webs, reproductive, social, and stochastic.

Ecosystem Structure: The physical arrangement of the various components. Also, trophic structure; measured in standing crop or energy fixed per unit area per unit time. May be pyramids of numbers, biomass, or energy flows.

Ecosystem Viability: The ability to maintain diversity, productivity, resilience to stress health, renewability, and/or yields of desired values, resource used, products, or services from an ecosystem while maintaining the integrity of the ecosystem over time.

Ecosystems Approach: The ecosystem approach embodies three fundamental concepts: designating the physical boundary of the system and its parts; understanding the interactions of the parts as a functioning whole; and understanding the relation between the system and its context (external factors that influence the system and also internal information that must be synthesized to be understood at the scale of the defined system).

Ecotone: The boundary or transitional zone between adjacent communities containing the characteristic species of each, such as the edge of a woodland next to a field or lawn.

Ecotype: A locally adapted population of a species which has a distinctive limit of tolerance to environmental factors; a genetically uniform population of a species resulting from natural selection by the special conditions of a particular habitat factors.

Ecotypic Variation: The variation within a species that is adapted to the variety of different habitats across the range of the species.

Ectotherm: A cold-blooded animal, one having a body temperature determined primarily by the temperature of its surrounding environment. Terrestrial reptiles are ectotherms.

Edaphic: Pertaining to soil or to the physical, chemical, and biological properties of the soil or substratum, which influence associated biota, such as pH and organic matter content.

Edge Effect: The tendency for a transitional zone between communities (an *ecotone*) to contain a greater variety of species and more dense populations of species than either community surrounding it.

El Niño (Also called *El Niño—Southern Oscillation Event*, or *ENSO*): A warm water current which periodically flows southward along the coast of Ecuador, associated with the southern oscillation in the atmosphere, and which affects climate throughout the Pacific region.

Elasmobranches: Any of a subclass (Elasmobranchii) of cartilaginous fishes that have five to seven later gill openings on each side, comprising sharks, rays, skates, and extinct related fishes.

Electrophoresis: A technique for separating mixtures of organic molecules based on their different rates of travel in electric fields.

Emergent: An aquatic plant having most of its vegetative parts above water. Also, a tree which reaches or exceeds the level of the surrounding canopy.

Empirical: Originating in or based upon observation or experience; capable of being verified or disproved by observation or experiment.

Encroachment: Conditions where the succession/disturbance regimes have been changed to allow transition to dominance by species or structures that are not adapted to the biophysical succession/disturbance regime.

Endemic: Belonging or native to a particular people or geographic region; a genetically unique life form.

Endotherm: A warm-blooded animal, one that maintains a body temperature largely independent of the temperature of the environment. Mammals are endothermic.

Environment: The complex of climatic, soil and biotic factors that act upon an organism or ecological community and ultimately determine its form and survival.

Eolian: Pertaining to the action or effect of the wind; wind-borne..

Ephemeral streams: Streams that contain running water only sporadically, such as during and following storm events.

Epipelagic: The oceanic zone extending from the surface to about 200 meters, where enough light penetrates to allow photosynthesis.

Epiphyte: A plant that uses another plant (usually a tree) for support or anchorage but not for water or nutrients.

Epizootic: An outbreak of disease (an epidemic) in nonhuman animals, or pertaining to such an outbreak.

Equilibria/Equilibrium: A system that has cyclic successional patterns or multiple stable states, and associated response in disturbances.

Ericaceous: Of, relating to, or being a heath or of the heath family of plants, which are mostly shrubby, dicotyledonous, and often evergreen plants that thrive on open, barren soil that is usually acidic and poorly drained.

Escapement: The number of fish that are permitted to survive and spawn (as by adjustment of fishing season or by provision of fishways).

Estivation: Dormancy associated with warm/dry period of the year.

Estuary: A semi-enclosed coastal body of water which has a free connection with the open sea and where fresh water derived from land drainage (usually mouths of rivers) is mixed with seawater; often subject to tidal action and cyclic fluctuations in salinity.

Euryhaline: Able to live in waters with a wide range of salinity.

Eutrophication: The process by which a body of water acquires a high concentration of nutrients, especially phosphates and nitrates, which typically promote excessive growths of algae, decomposition of which depletes oxygen, causing the death of other organisms.

Evapotranspiration: Loss of water from the soil both by evaporation and by transpiration from plants.

Excessive livestock grazing pressure: Grazing pressure that results in a decline in physiological vigor of plants, typically observed as a decline in reproductive output (for example, tiller production of grasses) and below ground (for example, root growth) growth.

Exotic Species: Species which occur in a given place, area, or region as the result of direct or indirect, deliberate or accidental introduction of the species by humans, and for which introduction has permitted the species to cross a natural barrier to dispersal.

Extinction: The dying out of a species, or the condition of having no remaining living members; also, the process of bringing about such a condition.

Extirpation: The loss or removal of a species from one or more specific areas but not from all areas.

Fauna: The animal life of a region or geological period.

Fecundity: The potential reproductive capacity of an organism or population.

Fen: A marshy, low-lying wetland covered by shallow, usually stagnant, and often alkaline water that originates from groundwater sources.

Feral: Relating to plants or animals which have escaped from domestication, and to their descendants.

Fetch: The distance along open water or land over which the wind blows; the distance traversed by waves without obstruction.

Fire regime: The characteristic frequency, extent, intensity, severity, and seasonality of fires in an ecosystem.

Flora: Plant or bacterial life forms of a region or geological period.

Fluvial: Pertaining to rivers or streams and their action.

Food Web: The interlocking pattern of food chains in an ecosystem. A food chain is a transfer of food energy from plants through a series of animals.

Forb: An herbaceous plant which is not a grass.

Fragmentation: Breaking up of contiguous areas into progressively smaller patches of increasing degrees of isolation.

Gallery Forest: A narrow strip of forest along the margins of a river in an otherwise unwooded landscape.

Gap Analysis: The process of identifying and classifying components of biological diversity to determine which components already occur in protected areas and which are not present or are under-represented in protected areas.

Gastropod: Any of a large class (Gastropoda) of mollusks, usually with a univalve shell or no shell and a distinct head bearing sensory organs, such as snails and slugs.

Geographic Information System (GIS): A spatial type of information management system, which provides for the entry, storage, manipulation, retrieval, and display of spatially oriented data.

Geomorphology: The study of landforms on a plant's surface and of the processes that have fashioned them.

Glade: An open space in the forest.

Graminoids: Grasses and grass like plants, such as sedges.

Gravid: Carrying eggs or young; pregnant.

Grazing System: A specialization of grazing management that defines the periods of grazing and non-grazing.

Greenhouse Effect: Heating of the Earth's atmosphere that is loosely analogous to the glass of a greenhouse letting light in but not letting heat out.

Gregarious: Tending to form into groups, which possess a social organization, such as schools of fish, herds of mammals, flocks of birds.

Groundfish: A bottom-dwelling fish, especially one of commercial importance such as cod, haddock, pollock, or flounder.

Guild: A group of species having similar ecological resource requirements and foraging strategies and therefore having similar roles in the community.

Gymnosperm: A plant, such as a cycad or a conifer, whose seeds are not enclosed in an ovary (fruit).

Gyre: A circular or spiral system of movement, especially a giant circular oceanic surface current.

Habitat: The place, including physical and biotic conditions, where a plant or an animal usually occurs.

Habitat Connections: A network of habitat patches linked by areas of like habitat. The linkages connect habitat areas within the watershed to each other and to areas outside the watershed. These connections include riparian areas, mid-slopes, and ridges. In the case of old-growth forest habitat connections, each connection is planned to be sufficiently wide (at least 1,000 feet) to retain interior old-growth associated species.

Habitat Fragmentation: The breaking up of a habitat into unconnected patches interspersed with other habitat, which may not be inhabitable by species occupying the habitat that was broken up. The breaking up is usually by human action, as, for example, the clearing of forest or grassland for agriculture, residential development, or overland electrical lines.

Habitat Type: Place where an animal or plant normally lives, often characterized by a dominant plant form or physical characteristic.

Halophytic: Referring to a plant that can tolerate or thrive in alkaline soil rich in sodium or calcium salts; tolerant of saline conditions.

Hard Mast: Fruit of hardwood trees such as beech and oaks.

Heavy Grazing: A comparative term, which indicates that the stocking rate of a pasture is relatively greater than that of other pastures. Often erroneously used to mean overuse.

Heavy Metals: A metallic element of high specific gravity, such as antimony, bismuth, cadmium, copper, gold, lead, mercury, nickel, silver, tin, and zinc. These metals, which are toxic even in low concentrations, persist in the environment and can accumulate to levels that stunt plant growth and interfere with animal life.

Hectare (ha): A metric unit of measure for area, equal to 2.47 acres.

Hematocrit: The percentage of blood volume occupied by red blood cells.

Hermaphroditic: An individual that possesses both male and female sex organs.

Herptiles: Reptiles and amphibians collectively.

Heterogeneity: Variation in the environment over space and time.

Heterogeneous: Consisting of diverse or dissimilar parts; having non-uniform structure or composition.

Heterotrophic: An organism that is unable to manufacture its own food from simple chemical compounds and therefore consumes other organisms, living or dead, as its main source of carbon.

Heterozygous: Having two different alleles at a particular gene locus on a chromosome pair. Provides a measure of genetic variation either in a population or in an individual.

Hibernation: Dormancy associated with cold period of the year.

Historic: The approximate 1,000-year time period prior to Euro-American settlement (substantial effects in Oregon and Washington assumed to have begun by the mid 1800s).

Holocene: The present, post-Pleistocene geologic epoch of the Quaternary period, including the last 10,000 years; the most recent postglacial period.

Home Range: The geographic area within which an animal restricts its normal, daily activities.

Homeothermy: Of being warm blooded, able to maintain body temperature above that of the surroundings despite large variation in environmental temperature.

Human Dimension: An integral component of Ecosystem Management that recognizes people are part of ecosystems, that people's pursuits of past, present, and future desires, needs and values have and will continue to influence ecosystems and must be included in ecosystem management.

Hybridization: Any crossing of individuals of different genetic composition, often belonging to separate species, resulting in hybrid offspring.

Hydric: Characterized by, relating to, or requiring an abundance of moisture.

Hydrocarbon: A naturally occurring organic compound that contains carbon and hydrogen; may be gaseous, solid, or liquid, for example, natural gas, bitumens, and petroleum.

Hydrographic: Relating to the characteristic features of bodies of water, such as depth and flow.

Hydrological cycle: The movement of water from the sea through the air to the land and back to the sea.

Hydrology: The study of the movement of water from the sea through the air to the land and back to the sea; the properties, distribution, and circulation of water on or below the Earth's surface and in the atmosphere.

Hydromorphic: Descriptive of an intra-zonal soil formed under waterlogged or poorly drained conditions.

Hydroperiod: The duration and frequency of flooding.

Hyporheic Zone: The area under the stream channel and floodplain that contributes to the stream.

Hypoxic: Deficient in oxygen.

Impoundment. A natural or artificial body of water held back by a dam.

Increaser: Plant species that will increase in relative amount when management induced disturbance stresses (such as excessive livestock grazing pressure or alteration of fire frequency and severity) operate and when management-induced disturbance stresses operate during and immediately after drought. The term is used particularly for plant species on rangelands.

Indicator Species: An organism whose presence or state of health is used to identify a specific type of biotic community or as a measure of ecological conditions or changes occurring in the environment.

Indigenous: A species that occurs naturally in an area; native.

Integrated Pest Management: A pest management philosophy based on an understanding of forest growth and development, forest pest dynamics, and the interaction of the two.

Integrated Resources Management: The simultaneous consideration of ecological, physical, economic, and social aspects of lands, waters, and resources in developing and carrying multiple-use, sustained-yield management.

Intermediate Host: The host occupied by juvenile stages of a parasite prior to the definitive host and in which asexual reproduction often occurs.

Intermittent Stream: Any non-permanent flowing drainage feature having a definable channel and evidence of scour or deposition. This includes what are sometimes referred to as ephemeral streams if they meet these two criteria.

Intertidal: Relating to the littoral zone above the low-tide mark.

Invertebrate: An animal without a backbone, such as snails, worms, and insects.

Invertivore: An animal or plant that eats invertebrate animals.

Isobath: A line on a map or chart that connects all points having the same depth below the surface of a body of water; also, having constant depth.

Isopod: Any of a large order of sessile-eyed crustaceans with the body composed of seven free thoracic segments, each bearing s similar pair of legs.

Isotherm: A line on a map or chart of the Earth's surface connecting points having the same temperature at a given time or the same mean temperature for a given period.

Iteroparous: Repeat spawning strategy typical of the rainbow and cutthroat trout.

Karst: A limestone landscape that is characterized by sinks, underground streams, and caverns.

Keystone Species: Organisms that play dominant roles in an ecosystem and affect many other organisms. The removal of a keystone predator from an ecosystem causes a reduction of the species diversity among its former prey.

Krill: Planktonic crustaceans and larvae that constitute the primary food of baleen whales.

Krummholz: A discontinuous belt of stunted forest or scrub typical of windswept alpine regions close to treeline; a wind-deformed tree at high elevations.

Lacustrine: Pertaining to or living in lakes or ponds.

Lagoon: A shallow water body that is near or connected to a larger body of water.

Landscape: A spatially heterogeneous area with repeating patterns of elements and associated disturbance regimes, with similar climate and geomorphology.

Landscape Connectivity: The spatial contiguity with in the landscape: a measure of how easy or difficult it is for organisms to move through the landscape without crossing habitat barriers.

Landscape Contrast: The degree to which adjacent landscape elements differ from each other, with respect to species composition and physical attributes.

Landscape Ecology: The relationships of structure, function, and change in a heterogeneous land area composed of interacting ecosystems. Structure, function, and change refer to the patterns and processes of terrestrial, aquatic, hydrologic, social, and economic systems across space and through time.

Landscape Edge: The interface between landscape elements of different composition and structure, for example between an open clear-cut and a closed-canopy forest.

Landscape Grain: The average size of landscape elements: the "texture" of the landscape.

Landscape Heterogeneity: The variation in aggregations of landscape elements across a landscape.

Landscape Matrix: The most concentrated portion of the landscape, that is, the vegetation type that is most contiguous.

Landscape Patches: Areas of vegetation which are relatively homogeneous internally (with respect to composition, successional stage, etc.) but differ from what surrounds them (the matrix, or other patches).

Landscape Patchiness: The density of all types of patches within a landscape; considers the diversity among patches.

Landscape Porosity: The density of a particular type of patch within a matrix.

Landscape Stability: The likelihood a landscape structural element will change significantly (in composition, physical features, etc.) over time, and the rate of that change.

Landscape Unit: A continuous geographic area with fairly consistent landform, and vegetation communities.

Landscape Use by Wildlife:

  Contrast: Species requires the contrast between two major structural stages; uses two major structural stages in close proximity. Typically forages in open stages and breed/rests in large tree structural stages.

  Generalist: Species use all or many patch types. Patches not restricted to one structural stage.

  Mosaic: Species uses aggregates of patches of habitat.

  Patch: Species likely uses only one homogeneous patch during life cycle (for residents) or breeding/wintering period (for migrants).

Large Woody Debris (LWD): see Coarse Woody Debris (CWD).

Larva (Larvae): The wingless and often wormlike hatchlings of insects; also, the early form of an animal (such as a frog or sea urchin) which at birth or hatching is fundamentally unlike its parent and must metamorphose before assuming adult characteristics.

Leaching: The removal of readily soluble components, such as chlorides, sulfates, organic matter, and carbonates, from soil by percolating water. The remaining upper layer of leached soil becomes increasingly acidic and deficient in plant nutrients.

Legacy Tree (also called *Live Remnant Tree*): A live tree remaining from the previous stand. Does not necessarily have to be a mature or old growth tree.

Lek: A mating system among birds during which males display communally at a traditional site (one used year after year).

Lentic: Related to still waters such as ponds lakes, or swamps.

Levee: A raised embankment along the edge of a river channel, often constructed as protection against flooding. Natural levees result from periodic overbank flooding, when coarser sediment is immediately deposited because of a reduction in river velocity.

Lichen: A composite organism consisting of a fungus and algae or cyanobacteria living in symbiotic association.

Life History: The significant features of the life cycle through which an organism passes, with particular reference to strategies influencing survival and reproduction.

Limnic: Pertaining to lakes or to other bodies of standing fresh water; often used with reference only to the open water of a lake away from the bottom; limnetic.

Linkages: Route that permits movement of individual plant (by dispersal) and animals from a Landscape Unit and/or habitat type to another similar Landscape Unit and/or habitat type.

List of Endangered or Threatened Species: A listing of animals and plants administratively determined to meet legal criteria for protection under provisions of the U.S. Endangered Species Act.

Littoral Zone: The biogeographic zone in a body of fresh water where light penetration is sufficient for the growth of plants; the intertidal zone of the seashore.

Loess: Unconsolidated sediment deposited by wind. Loess is usually composed of unstratified fine sand or silt.

Lotic: Relating to or living in moving water, such as a river or stream.

Macroclimate: Climate that lies just beyond the modifying irregularities of landform and vegetation.

Macrofauna: Animals large enough to be seen with the naked eye.

Management Disturbances: Intentional, planned human disturbance that changes the structure and composition of a landscape element, landscape pattern, or regional composition, such as timber harvest, thinning, range improvements, livestock grazing, prescribed fire planned ignition, fire suppression, etc.

Management Region: The collective delineation of land that are modeled with similar assumptions relative to management objectives.

Marine Protected Areas (MPAs): Any area of intertidal or subtidal terrain, together with its overlying water and associated flora, fauna, historical, and cultural features, which has been reserved by law or other effective means to protect part or all of the enclosed environment (IUCN 1988).

Marsh: An ecosystem of more or less continuously waterlogged soil dominated by emersed herbaceous plants but without a surface accumulation of peat. A marsh differs from a swamp in that it is dominated by rushes, reeds, cattails, and sedges, with few if any woody plants, and differs from a bog in having soil rather than peat at its base.

Matrix: The most extensive and most connected landscape element type present, which plays the dominant role in landscape functioning. Also, a landscape element surrounding a patch.

Mature Forest: Generally used in an economic sense to indicate that a forest has attained harvest age.

Maximum Sustainable Yield: The maximum yield or crop which may be harvested year after year without damage to the system, or the theoretical point at which the size of a population is such as to produce a maximum rate of increase.

Megafauna: The largest size category of animals in a community.

Meiofauna: That part of the microfauna, which inhabits algae, rock fissures, and superficial layers of the muddy sea bottom. They are smaller than 1 millimeter but larger than 0.1 millimeter.

Melanism: A condition in which dark pigment produces dark color or blackness in scales, skin, or plumage.

Meristems: The undifferentiated, growing parts of plants, consisting of groups of cells capable of actively dividing.

Mesic: Neither wet (hydric) nor dry (xeric); intermediate in moisture, without extremes.

Mesopelagic: The ocean zone from 200 to 1,000 meters deep, where little light penetrates and the temperature gradient is even and gradual with little seasonal variation. This zone contains an oxygen minimum layer and usually has the maximum concentrations of the nutrients nitrate and phosphate. It overlies the *bathypelagic* zone and is overlain by the *epipelagic* zone.

Metabolite: A product of metabolism or a substance that is essential to the metabolism of an organism or to a metabolic process.

Metapopulation: A group of populations, usually of the same species, which exist at the same time but in different places.

Microclimate: The climate that prevails in a small area, usually in the layer near the ground.

Midden: A heap of refuse. Also, a pile of seeds or of various items that were gathered by a rodent, for example, by a squirrel or packrat.

Mitochondria (Mitochondrion): Organelles occurring in the cytoplasm of all aerobic cells of plants and animals and containing enzymes responsible for converting foods to usable energy. Mitochondria also contain double-stranded DNA encoding some of the genes functioning in the processing energy and protein synthesis.

Miocene: A geologic epoch within the Tertiary period (about 26 to 5 million years before the present).

Mode: Value occurring most frequently in a series of observations.

Model: An idealized representation or reality developed to describe, analyze, or understand the behavior of some aspect of it; a mathematical representation of the relationships under study. The term model is applicable to a broad class of representation, ranging from a relatively simple qualitative description of a system or organization to a highly abstract set of mathematical equations.

Mollusk: An organism in the phylum Mollusca (for example, snails, clams, or squids), characterized by soft, un-segmented body parts enclosed in a shell.

Monitoring: A process of collecting information to evaluate whether objectives of a management plan are being realized.

Montane: Of, relating to, growing in, or being the biogeographical zone of relatively moist, cool upland slopes below the timberline, often dominated by large coniferous trees.

Moraine: An accumulation of boulders, stones, or other debris carried and deposited by a glacier.

Morphology: The form and structure of organisms.

Mosaic: Heterogeneous ecological conditions on a landscape usually produced by the variable, patchy effects of disturbances; a patchwork of vegetation communities within a landscape as determined by environmental conditions.

Mustelid: One of a large, widely distributed family of small, lithe, carnivorous mammals, including weasels, otters, skunks, wolverines, and minks.

Mutagen: Any agent that produces a mutation or enhances the rate of mutation in an organism, for example, x-rays, gamma rays, and certain chemicals.

Mutualism: An interaction between members of two species, which benefits both; in strict terms, obligatory mutualism, in which neither species can survive under natural conditions without the other.

Mycorrhizae: The mutually beneficial association between a fungus and the roots of a plant; a mycorrhizal root takes up nutrients more efficiently than an uninfected root.

National Environmental Policy Act (NEPA): An act which encourages productive and enjoyable harmony between man and his environment; promotes efforts to prevent or eliminate damage to the environment and biosphere and stimulate the health and welfare of man; enriches the understanding of the ecological system and natural resources important to the nation; and establishes a Council on Environmental Quality.

National Forest Management Act (NFMA): A law passed in 1976 as amendments to the Forest and Rangeland Renewable Resources Planning Act that requires the preparation of regulations to guide that development.

Native: Plants or animals that are indigenous to a given place; the pre-Euro-American settlement system.

Natural Conditions: Plant and animal communities where people have not directly impacted either the plant community or the soil by such activities as logging, grazing, or cultivation.

Natural Variability: Range of the spatial, structural, compositional, and temporal characteristics of ecosystem elements during a period specified to represent "natural" conditions.

Nekton (Also spelled *necton*): Free-swimming organisms in aquatic ecosystems; unlike plankton, they are able to navigate at will (such as fishes, amphibians, and large swimming insects).

Nematode: Any of a phylum (Nematoda or Nemata) of elongated cylindrical worms parasitic in animals or plants, or free-living in soil or water.

Nemerteans: Any of a phylum (Nemertea) of often vividly colored marine worms, most of which burrow in the mud or sand along seacoast; often called ribbon worms.

Neotenic: Referring to an organism, which has attained sexual maturity while retaining juvenile characteristics.

Neotropical Migrant: A bird that nests in temperate regions and migrates to the Neotropical faunal region, which includes the West Indies, Mexico, Central America, and that part of South America within the tropics.

Neritic: Relating to or inhabiting the shallow water, or nearshore marine zone extending from the low-tide level to a depth of 200 meters. The neritic zone is populated by benthic organisms because of the penetration of sunlight to these shallow depths.

Nitrogen Fixation: The process of converting inorganic, atmospheric nitrogen into an organic form of nitrogen, ammonia. This process can be carried out by lightning, by photochemical fixation in the atmosphere, or by the action of microorganisms. Also, the chemical processes used in the manufacture of fertilizers.

Nival: Of, relating to, or growing under or in snow.

Nocturnal: Referring to organisms that are active or functional at night.

Nonindigenous (Also called *exotic, nonnative, introduced,* and *alien*): A plant or animal that is not native to the area in which it occurs; it was either purposely or accidentally introduced.

Nonpoint: Not from a single, well-defined site. Nonpoint sources are pollution-producing entities not tied to a specific origin, such as an individual smokestack; include runoff, which washes pollutants from roads into storm sewers and bodies of water or agricultural chemicals from lawns, fields, and golf courses.

Nutrient Cycling: Circulation or exchange of elements such as nitrogen and carbon between non-living and living portions of the environment. Includes all mineral and nutrient cycles involving mammals and vegetation.

Obligate: Essential, necessary; unable to exist in any other state, mode, or relationship; restricted to one particularly characteristic mode of life.

Obligate Species: A plant of animal that occurs only in a narrowly defined habitat such as tree cavity, rock cave, or wet meadow.

Old-growth: Referring to an ecosystem or community, particularly a forest, which has not experienced intense or widespread disturbance for a long time relative to the life spans of the dominant species and which has entered a late successional stage; usually associated with high diversity of species, specialization, and structural complexity.

Oligochaetes: Any of a class or order (Oligochaeta) of hermaphroditic terrestrial or aquatic annelid worms that lack a specialized head.

Oligotrophic: Waters or soils that are poor in nutrients and have low primary productivity.

Omnivore: Consuming a variety of plants and animals; neither plant nor animal food usually comprises less than one-third of diet.

Ontogenetic: Relative to the course of growth and development of an individual organism.

Osmerid: A member of the family of fishes (Osmeridae) to which the true smelts belong; smelts and smelt-like fishes.

Overgrazing: Continued heavy grazing that exceeds the recovery capacity of the plant community and creates a deteriorated range.

Ovigerous: Carrying eggs, or modified for carrying eggs.

Oxbow: A pond or wetland created when a river bend is cut off from the main channel of the river.

Palustrine: Pertaining to wet or marshy habitats.

Parasite: An organism that is intimately associated with and metabolically dependent on another living organism (the host) for completion of its life cycle, and which is typically detrimental to the host.

Passerine: Of or relating to the largest order (Passeriformes) of birds, which includes more than half of all living birds and consists primarily of perching songbirds, whose young are hatched in an immature and helpless condition

Patch: Ecosystem elements (e.g., areas of vegetation) that are relatively homogeneous internally and that differ from what surrounds them.

Patch Dynamics: The idea that communities are a mosaic of different areas (patches) within which non-biological disturbances (such as climate) and biological interactions proceed.

Pathogen: A specific causative agent of a disease, such as a bacterium or a virus.

Pelagic: Referring to or occurring in the open sea.

Percent Cover: In descriptions of plant communities, the proportion of ground, expressed as a percentage that is occupied by the perpendicular projection down onto it of the aerial parts of individuals of the species under consideration.

Perennial: A plant that normally lives for more than two seasons and which, often after an initial period, produces flowers annually.

Perennial Stream: A stream that typically has running water on a year-round basis.

pH: A Measure of acidity and alkalinity of a solution, taken by measuring the relative concentration of hydrogen ions in the solution.

Phenology: The study of the relationship between climate and the timing of periodic natural phenomena such as migration of birds, bud bursting, or flowering of plants.

Phenotype: The observable manifestation of a specific genetic makeup; those observable properties of structure and function of an organism as modified by genetic structure in conjunction with the environment.

Philopatry (also referred to as site fidelity or site tenacity): The tendency to return each season to the same nest or breeding colony.

Photic Zone: The surface zone of the sea or a lake having sufficient light penetration for photosynthesis.

Photoperiod: The length of time an organism is daily exposed to light, especially with regard to how that exposure affects growth and development.

Phylogenetic: Pertaining to the evolutionary history of a group or lineage, or the evolutionary relationships within and between taxonomic levels; the relationships of groups of organisms as reflected by their evolutionary history.

Physiographic Province: A region of the landscape with distinctive geographical features.

Physiography: Landform; physical geography.

Phytoplankton: One of two groups into which *plankton* are divided, the other being *zooplankton*. Phytoplankton comprises all the freely floating photosynthetic forms in the oceans.

Pinniped: Any of a suborder of aquatic carnivorous mammals with all four limbs modified into flippers; includes seals, sea lions, and walruses.

Pioneer: The first species or community to colonize or re-colonize a barren or disturbed area, thereby commencing a new biological succession.

Piscivores: Fish-eaters; those organisms that subsist exclusively or primarily on fish.

Plankton: Small aquatic organisms (animals and plants) that generally having no locomotive organs, drift with the currents. The animals in this category include protozoans, small crustaceans, and the larval stages of larger organisms.

Plant Association: Stands of vegetation with similar combinations of species united into abstract types; a basic unit in plant community classification.

Playa: A nearly level area at the bottom of an undrained desert basin, sometimes temporarily covered with water during wet periods. Playas are barren and usually saline.

Pleistocene: The earlier epoch of the Quaternary period or the corresponding system of rocks; 1.6 million-10,000 years ago; the "Ice Age."

Pluvial: Characterized by abundant rain.

Polychaetes: Any of a class (Polychaeta) of chiefly marine annelid worms (such as clam worms), usually with paired segmental appendages, separate sexes, and a free-swimming trochophore larva.

Polyandry: One female mates with two or more males.

Polychlorinated Biphenyls (PCBs): A group of toxic, carcinogenic organic compounds containing more than one chlorine atom; very stable compounds, fat-soluble; they therefore accumulate in ever-higher concentrations as they move up the food chain.

Polymerase Chain Reaction (PCR): A molecular biology technique allowing small and specific amounts of DNA sequences to be amplified by many thousand fold in an automated fashion.

Polygamy: Both polygyny and polyandry occur.

Polygyny: One male mates with two or more females.

Population: A group of organisms, all of the same species, which occupies a particular area. Also, the total number of individuals of a species within an ecosystem, or of any group of similar individuals.

Population Distribution by Wildlife:

Contiguous: Habitat is broadly distributed over the species' range with opportunity for continuous or nearly continuous occupation by the species; little or no limitation on population interaction.

Gaps: Habitat is broadly distributed over the species' range, but gaps exist within this distribution. Disjunct patches of habitat are typically large enough and close enough to other patches to permit dispersal among patches and to allow species to interact as a meta-population.

Isolated: Habitat locally distributed as isolated patches, causing strong limitations for population interaction among patches, and limited opportunity of dispersal among patches. Some local populations may be extirpated and rates of re-colonization will likely be slow.

Patchy: Habitat exists primarily as patches, some of which are small or isolated to the degree that species interactions are limited. Local subpopulations in most of the species' range interact as a meta-population, but some populations are so disjunct that subpopulations in those patches are essentially isolated from other populations.

Scarce: Habitat is very scarce throughout species' range with little or no possibilities for interactions between local populations, strong potential for extirpations, and little likelihood of re-colonization.

Population Dynamics: The aggregate of changes that occur during the life of a population. Included are all phases of recruitment and growth, senility, mortality, seasonal fluctuation in biomass, and persistence of each year class and its relative dominance, as well as the effects that any or all of these factors exert on the population.

Population Viability: Probability that a population will persist for a specified period across its range despite normal fluctuations in population and environmental conditions.

Potential Natural Community: The biotic community that would be established if all successional sequences of its ecosystem were completed without additional human-caused disturbances under present environmental conditions.

Precocial Young: Mobile, downy, follow parents, and find their own food.

Prenatal: Existing or occurring before birth.

Prescribed Fire: A fire burning under specified conditions that will accomplish certain planned objectives. The fire may result from planned or unplanned ignitions.

Primary Producer: An organism capable of using the energy derived from light or a chemical substance in order to manufacture energy-rich organic compounds, mainly green plants.

Primary Productivity: The rate at which biomass is produced by organisms which synthesize complex organic substances from simple inorganic substrates, such as in photosynthesis and chemosynthesis.

Primary Production: The biomass produced through photosynthesis and chemosynthesis in a community or group of communities.

Primiparity: Having a single offspring/pregnancy.

Progradation: The outward building of a sedimentary deposit, such as the seaward advance of a delta or shoreline, or the outbuilding of an alluvial fan.

Promiscuity: Males and females mate more or less indiscriminately.

Province: An area of land, less extensive than a region, having a characteristic plant and animal population.

Purse Seine: A large seine net designed to be set by two boats around a school of fish and so arranged that after the ends have been brought together, the bottom can be closed.

Radiation: In ecology, the spread of a group of organisms into new habitats.

Range (of a species): The area or region over which an organism occurs.

Range of Variability: The spectrum of conditions possible in ecosystem composition, structure, and function considering both temporal and spatial factors.

Rangeland: Land on which the native vegetation is predominantly grasses, grass-like plants, forbs, or shrubs. Includes lands re-vegetated naturally or artificially when routine management of that vegetation is accomplished mainly through manipulation of grazing.

Recovery Plan: A plan that lists the actions that must be taken and the objectives that must be reached before an organism is no longer endangered or threatened and may be removed from the list of endangered and threatened species.

Recruitment: The influx of new members into a population by reproduction or immigration.

Refugium: An isolated area where extensive changes, typically due to changing climate (such as glaciation) but also due to large-scale disturbances such as those caused by humans, have not occurred, and where plants and animals typical of a region may survive. Such a refuge is a center of *relict* forms, from which dispersion and speciation may take place after environmental readjustment.

Regime: A regular pattern of occurrence or action.

Region: The broadest scale of landscape ecology composed of a course-grained pattern of connected landscapes, with contrasting boundaries, that have a similar macroclimate and sphere of human activity and interest.

Rehabilitation: Returning of land to farm use or to productivity in conformity with a prior land use plan, including a stable ecological state that does not contribute substantially to environmental deterioration and is consistent with surrounding aesthetic values.

Relict: Persistent remnants of a formerly widespread species in certain isolated areas.

Remote Sensing: Methods for gathering data on a large or landscape scale which do not involve on-the-ground measurement, especially satellite photographs and aerial photographs; often used in conjunction with Geographic Information Systems.

Resilience: The ability of an ecosystem to maintain diversity, integrity, and ecological processes following disturbance.

Resource Partitioning: Division of some resource or resources among two or more con-occurring species; for example, eating slightly different foods.

Restoration:

Ecological: The reestablishment of predisturbance functions and related chemical, biological, and hydrological characteristics.

Passive (or Natural): The discontinuation of those activities that are causing degradation or preventing the ecosystem's recovery.

Retrogression: The return along the successional path to an earlier successional community or an altered simpler state.

Riffles: Shallow rapids where the water flows swiftly over completely or partially submerged obstructions to produce surface agitation, but where standing waves are absent.

Riparian: Relating to, living, or located on the bank of a natural watercourse (such as a river) or sometimes of a lake or tidewater.

Riparian Ecosystem: Ecosystems transitional between terrestrial and aquatic ecosystems. Also, streams, lakes, wet areas and adjacent vegetation communities and their associated soils, which have free water at or near the surface.

Riparian Zone: An area of vegetation adjacent to an aquatic ecosystem. It has a high water table, certain soil characteristics, and some vegetation that requires free (unbound chemically) water or conditions that are more moist than normal. This zone is transitional between aquatic and upland zones.

Riprap: A general term for large, blocky stones that are artificially placed to stabilize and prevent erosion along a riverbank or shoreline.

Risk Analysis: A qualitative assessment of the probability of persistence of wildlife species and ecological systems under various alternatives and management options; generally also accounts for scientific uncertainties.

Rookery: Breeding or nesting place for some gregarious mammals and birds.

Runoff: Precipitation on land that runs off to a body of water.

Runs/Glides: Areas of swiftly flowing water, without surface agitation or waves, which approximates uniform flow and in which the slope of the water surface is roughly parallel to the overall gradient of the stream reach.

Salinity: A measure of the total concentration of dissolved salts in water.

Salmonid: Any of a family of elongate bony fishes (such as salmon or trout) that have the last three vertebrae upturned.

Sanitation: The removal of dead or damaged trees, or trees susceptible to insect and disease attack, such as intermediate and suppressed trees, essentially to prevent the spread of pest or pathogens and to promote forest health.

Savanna (also spelled *savannah*): A grassland-woodland mosaic vegetation type found in tropical and subtropical regions with long dry periods and receiving more rain fall than desert areas but not enough to support complete forest cover.

Seamount: An underwater mountain (usually a submarine volcanic mountain peak) rising from the ocean floor whose summit is below the water's surface.

Secondary Production: The biomass production resulting from the assimilation of organic matter produced by a primary consumer: production by organisms (mainly animals), which consume primary producers (mainly plants).

Secondary Productivity: The rate of biomass production resulting from the assimilation of organic matter produced by a primary consumer; production by organisms (mainly animals), which consume primary producers (mainly plants).

Sediment: Materials that sink to the bottom of a body of water or materials that are deposited by wind, water, or glaciers.

Semelparous: A reproductive strategy that allows only one mating prior to death, typical of the Pacific salmon.

Senescence: The aging process in mature individuals; or, the period near the end of an organism's life cycle; in deciduous plants, the process that occurs before the shedding of leaves.

Sensitive Species: A species not formally listed as endangered or threatened, but considered to be at risk, as evidenced by: a significant current or predicted downward trend in population numbers or density; or a significant current or predicted downward trend in habitat capability that would reduce a species' existing distribution.

Seral: Relating to a phase in the sequential development of ecological communities formed in ecological succession in a particular habitat and leading to a particular climax association; intermediate communities in an ecological succession.

Sere: The series of stages that follow one another in an ecologic succession; a series of biotic communities that follow one another in time on any given area of the Earth's surface.

Serotinous Cones: Pinecones that remain on the tree for many years and are tightly closed until stimulated by the heat of a forest fire to open and release seeds.

Serpentine: A mineral rock consisting essentially of a hydrous magnesium silicate (chrysolite and antigorite and usually having a dull green color and often a mottled appearance; or, the usually infertile, excessively well-drained soil derived from serpentine.

Sessile: Permanently attached to a substrate or established; not free to move about. Also, attached without a stalk.

Short-stopping: The process of creating habitat improvements to hold geese or ducks throughout the winter in an area that was historically used only as a migratory stopover point en route to wintering grounds farther south.

Silviculture: The art and science of managing forest stands to provide or maintain structures, species composition, and growth rates that contribute to forest management goals.

Simulation: To project or estimate the future or historic conditions or outcomes of various attributes.

Sink: A sinkhole; or, an area with a demand for metabolic substances. For example, growing meristems are sinks for energy compounds from photosynthesis, mitochondria are oxygen sinks, and tropical rainforests, or deep oceans may act as carbon sinks, absorbing carbon dioxide from the atmosphere.

Site: The classification of land area based on its climate, physiographic (physical geography), edaphic (soil), and biotic factors that determine its suitability and productivity for particular species and silvicultural alternatives.

Site-potential Tree: A tree that has attained the maximum height possible given site conditions where it occurs.

Slough: A swamp, marsh, or muddy backwater.

Smolt: The stage in the life of salmon and similar fishes in which the subadult individuals acquire a silvery color and migrate down the river to begin their adult lives in the open sea.

Snag: A standing dead tree or stump that provides habitat for a broad range of wildlife, from beetle larvae (and the birds that feed upon them) to dens for raccoons.

Spawn: The eggs of certain aquatic organisms; also, the act of producing such eggs or egg masses.

Species: A group of organisms formally recognized as distinct from other groups; the taxon rank in the hierarchy of biological classification below genus; the basic unit of biological classification, defined by the reproductive isolation of the group from all other groups of organisms.

Species Diversity: See *Biological Diversity*.

Species Richness: The absolute number of species in an assemblage or community.

Staging Area: A traditional area, usually a lake, where birds that migrate in flocks rest and feed either immediately before or during migration. Many flocks may be gathered in such an area.

Stand Composition: The representation of tree species in a forest stand, expressed by some measure of dominance (e.g., percent of volume, number, basal area, cover).

Stand Structure: The physical and temporal distribution of plants in a stand.

Standard Error: In statistics, the standard deviation of the sampling distribution of a statistic; an estimate of the range by which the means of a number of sets of data deviate about the mean of those means.

Standing Stock: Biomass; the total mass of organisms comprising all or part of a population or other specified group or within a given area; measured as volume, mass, or energy.

Steppe: Specifically, the temperate, semiarid areas of treeless grassland in the mid-latitudes of Europe and Asia; more generally, any such grassland.

Stewardship: A land ethic for current and future generations that 1) encourages wise use and conservation of resources; 2) sustains and enhances productivity of resources; and 3) protects resources.

Stochastic: Random.

Stressors: Physical or biotic factors that stress individual organisms/communities.

Structural Stage: A stage of development of a vegetation community that is classified on the dominant processes of growth, development, competition, and mortality.

Structure: The various horizontal and vertical physical elements of the vegetation.

Subaerial: Occurring immediately above the surface of the ground.

Subalpine: The zone just below treeline on temperate mountains, usually dominated by a coniferous forest ecologically similar to boreal forest. The elevation of this zone increases with a decrease in latitude.

Subbasin: The fourth delineation within the hydrologic unit code system. Provides a delineation generally of a river, or group of rivers, that flow into a basin.

Sublittoral Zone: The deeper zone of a lake below the limit of rooted vegetation; the marine zone extending from the lower margin of the intertidal (littoral) to the outer edge of the continental shelf at a depth of about 200 meters.

Submersed: Pertaining to a plant or plant structure growing entirely underwater.

Subnivian: Beneath the snow cover; specifically, the interface between snow and the surface of the ground where small mammals are active in winter.

Subsidence: The process of sinking or settling of a land surface or a crustal elevation because of natural or artificial causes.

Subspecies: A race of a species that is granted a taxonomic name; rules for designating subspecies are subjective, but subspecies are generally geographically distinct and form populations (not merely morphs) which differ to some degree from other geographic populations of the species.

Substrate: The surface of medium that serves as a base for something.

Subterranean: Under the surface of the Earth.

Subtidal: Applied to that portion of a tidal-flat environment which lies below the level of mean low water for spring tides. Normally it is covered by water at all states of the tide. Often used as a general descriptive term for a subaqueous but shallow marine depositional environment.

Subtropical: The latitudinal zone between 23.50 and 34.00 in either hemisphere, bordering the tropical zone. Also can refer to vegetation, organisms, or weather typical of subtropical habitats.

Subwatershed: The sixth delineation within the hydrologic unit code system. Provides a delineation of a group of streams that flow into a watershed.

Succession: The development of biotic communities following disturbances that produce an earlier successional community.

Successional Path: The sequence of successional communities; may follow one path or multiple paths depending on competition, morality of individuals, and non-lethal disturbances. There may be one endpoint or multiple endpoints depending on the degree of change in climate and soil properties.

Successional Stage: One in a series of usually transitory communities or developmental stages that occur on a particular site or area over a period of time.

Succulent: A plant that has a specialized fleshy tissue in roots, stems, or leaves for the conservation of water. Most succulents are xerophytes, plants preferring dry climates, such as cactus or aloe, but some are halophytes, adapted for living in salty soils where water retention is a problem.

Suitability: The appropriateness of applying certain resource management practices to a particular area of land, as determined by an analysis of the economic and environmental consequences and the alternative uses foregone.

Sustainability: The ability to sustain diversity, productivity, resilience to stress, health, renewability, and/or yields of desired values, resources uses, products, or services from an ecosystem while maintaining the integrity of the ecosystem over time.

Sustainable Development: The use of land and water to sustain production indefinitely without environmental deterioration, ideally without loss of native biodiversity.

Sustainable Ecological System: Emphasizing and maintaining the underlying ecological processes that ensure long-term production of goods, services, and values without impairing productivity of the land.

Sympatric: Referring to populations, species, or taxa occurring together in the same geographical area; they may occupy the same habitat or different habitats within the same geographical area.

Synergistic: Pertaining to the cooperative action of two or more agencies such that the total is greater than the sum of the component actions; combined action or operation.

Syntopic: Relating to or displaying conditions, as they exist simultaneously over a broad area, as of the atmosphere or weather.

Talus: Broken rock forming a more or less continuous layer that may or may not be covered by duff and litter.

Taxon (Taxa): Any organism or group of organisms of the same taxonomic rank; for example, members of an order, family, genus, or species.

Tectonic Movement: The formation of faults and folds on the crust of a planet.

Temporal Niche: The functional position of an organism in its environment as determined by the periods of time during which it occurs and is active there.

Terrestrial Community Type (Also referred to as terrestrial vegetation type): A group of cover types in the same seral stages that have similar characteristics for interpretation habitat values.

Terrestrial Ecosytem: A land-based ecosystem (see ecosystem). An interacting system of soil, geology, and topography with plant and animal communities.

Territory: The area that an animal defends, usually during breeding season, against intruders of its own species.

Tertiary: The first period of the Cenozoic Era which began about 65 million years and lasted to 1.6 million years before the present, marked by formation of high mountains, the dominance of mammals on land, and angiosperms superseding gymnosperms as dominant plants.

Thaliaceans: Any of a small order of tunicates consisting of various aberrant, free-swimming pelagic forms, including those of the genera *Salpa* and *Doliolum*.

Threatened Species: Those plant or animal species likely to become endangered species throughout all or a significant portion of their range within the foreseeable future as identified by the Secretary of Interior as threatened, in accordance with the 1973 Endangered Species Act.

Threshold: The boundary between ecological states that, once crossed, is not easily reversible and results in the loss of capacity of produce commodities and satisfy values.

Topography: The natural and constructed relief of an area.

Transect: A line or narrow belt used in ecological surveys to provide a means of measuring and representing graphically the distributions of organisms across a given area.

Transpiration: The loss of water vapor from a plant to the outside atmosphere, mainly through the stomata of leaves and the lenticels of stems.

Treeline: The upper limits of tree growth in mountains or at high latitudes.

Trophic: Pertaining to nutrition or to a position in a food web, food chain, or food pyramid.

Tropical: Referring to the zone between the Tropic of Cancer (23E27'N) and the Tropic of Capricorn (23E27'S); characterized by a climate with high temperatures, humidity, and rainfall. Also can refer to vegetation, organisms, or weather typical of tropical conditions.

Tundra: A level or rolling treeless plain in the arctic or subarctic regions; the soil is black and mucky, the subsoil is permanently frozen, and the vegetation is dominated by mosses, lichens, herbs, and dwarf shrubs. A similar environment occurs in mountainous areas above the timberline.

Tunicate: Any of a subphylum (Urochordata or Tunicata) of marine chordate animals that have a thick secreted covering layer, a greatly reduced nervous system, a heart able to reverse the direction of blood flow, and a notochord in the larval stage.

Turbid: Having sediment or foreign particles stirred up or suspended; muddy.

Tussock: A compact tuft of grass or sedges, or an area of raised solid ground, which is held together by roots of low vegetation, found in a wetland or tundra.

Understory: The vegetation layer between the overstory or canopy and the groundcover of a forest community, usually formed by shade-tolerant species or young individuals of emergent species. May also refer to the groundcover if no tree or shrub layer is present.

Ungulate: Any four-footed, hoofed, grazing mammal (such as a ruminant, swine, deer, hippopotamus, horse, antelope, elk, elephant, or hyrax) that is adapted for running but is not necessarily related to other ungulates.

Upwelling: The upward movement of cold, nutrient-rich water from ocean depths, produced by wind or diverging currents.

Vascular Plants: A plant with a specialized conducting system (for the transport of water and nutrients) that includes xylem and phloem; includes familiar high plants such as trees, shrubs, and grasses.

Vertebrate: An animal with a backbone; includes mammals, birds, reptiles, amphibians, and fishes.

Viability: The likelihood of continued existence in an area for some specified period of time.

Viable Population: A population, which has adequate numbers and dispersion of reproductive individuals to ensure the continued existence of the species population on the planning area.

Volant: Flying or capable of flying.

Watershed: An area or a region that is bordered by a divide and from which water drains to a particular watercourse or body of water.

Watershed Analysis: A systematic procedure for characterizing watershed and ecological processes to meet specific management and social objectives. Watershed analysis is a stratum of ecosystem management planning applied to watersheds of approximately 20 to 200 square miles.

Wetland: A general term applied to land areas, which are seasonally or permanently waterlogged, including lakes, rivers, estuaries, and freshwater marshes; an area of low-lying land submerged or inundated periodically by fresh or saline water.

Wilderness: An area designated by congressional action under the 1964 Protection Act. Wilderness is defined as undeveloped federal land retaining its primeval character and influence without permanent improvements or human habitation. Wilderness areas are protected and managed to preserve their natural conditions, which generally appear to have been affected primarily by the forces of nature, with the imprint of human activity substantially unnoticeable; have outstanding opportunities for solitude or for a primitive and confined type of recreation; include at least 5,000 acres or a are of sufficient size to make practical their preservation, enjoyment, and use in an unimpaired condition; and may contain features of scientific, education, scenic, or historical value as well as ecological and geologic interest.

Woodland: A vegetation community that includes widely spaced large trees. The tree crowns are typically more spreading in form than those of forest trees and do not form a closed canopy. Grass, heath, or scrub may develop between the trees.

Xeric: Dry; tolerating or adapted to dry conditions.

Year-class: Fish of a given species spawned or hatched in a given year; for example, a three year-old fish caught in 1998 would be a member of the 1995 year-class.

Zoeae: The free-swimming, planktonic larval forms of many decapod crustaceans (especially crabs) that have a relatively large cephalothorax, conspicuous eyes, and fringed antennae and mouthparts.

Zooplankton: See *plankton*.

Managing Directors

To describe our role in creating this book and CD-ROM we use the term "Managing Directors" because of our overall involvement in the many facets of how the book and CD-ROM were designed and developed. Specifically, our roles included (1) creating and identifying the structure and roles of the five scientific teams that guided this project, (2) developing a concept (and for several cases an outline) for each of the chapters in the book, (3) identifying the structure and core fields of information for the matrixes on the CD-ROM, (4) organizing and conducting the 15 expert scientific panels to gather key wildlife habitat information, (5) raising funds to support Senior Staff and contractors for four years, and for the publication of this book, (6) fostering the development of several new wildlife and ecological concepts, (7) developing and publishing the CD-ROM, (8) coordinating and guiding the involvement of more than 600 people to help build a common understanding for management along with coordinating with adjacent states and provinces.

David H. Johnson David has been involved in the natural resource field for 24 years. He has held forestry, wildlife biologist, and habitat scientist positions with the Minnesota Department of Natural Resources, BLM, USFWS, ODFW, and WDFW. He has Associate degrees in Natural Resource Management and Civil Engineering, a B.A. in Biology with a Minor in Archaeology, and an M.S. in Wildlife Science. He has a passion for owls and is deeply involved with international owl conservation. He loves traveling and reading historical exploration accounts, and for therapy from it all, builds and paddles sea kayaks and cedar-strip canoes.

Thomas O'Neil Tom has spent 25 years working on ecological issues in the Pacific Northwest and has several degrees from the University of Montana and the University of Toledo. Currently, Tom is the Director of the Northwest Habitat Institute. Previously, he has worked as a Wildlife Ecologist for ODFW, Research Division; University of Chicago's Argonne National Laboratory; Montana Power Company; as a Wildlife Biologist for the USFS, and as an Administrator for the Jeep Corporation. Tom has also co-authored the book *Atlas of Oregon Wildlife*, and has written over 40 other publications or documents. Tom is a second generation Irish-American with three of four grandparents coming directly from Ireland. His father worked in the early stages of the U.S. Space Program. Tom is married and has 2 children. He has been a Certified Wildlife Biologist with The Wildlife Society for 18 years. In his spare time, Tom enjoys the outdoors, traveling, and working on his 1949 Willys Jeep.

Bringing all this together could not have been possible without a dedicated staff who devoted many hours wading through countless reports, articles, databases, and books, assisting with the scientific panels, as well as coordinating and working with numerous wildlife professionals through-out the region. These invaluable people, who helped us from development to completion of this book and digital matrixes and maps, are listed below:

Senior Staff

Charley Barrett. Charley has been specializing in geographic information systems (GIS) for nine years. He earned an M.S. degree in Geography at Oregon State University in 1998. He studied GIS, cartography, and rural planning while also teaching computer mapping and GIS labs. As an undergraduate at Towson University, Maryland, Charley majored in geography and economics with an emphasis in environmental planning. Currently a member of the Northwest Habitat Institute, he assists in mapping vegetation, habitat, and species at local to regional scales, as well as developing the Internet site. Charley enjoys motorcycling, bicycling, camping, and picking the five-string banjo in his spare time.

Kelly Anne Bettinger. Kelly received her B.S. in Wildlife Management from Virginia Tech in 1989 and her M.S. in Wildlife Biology from Oregon State University in 1996. She has worked on a variety of projects on both U.S. coasts studying screech owls, spotted owls, red-cockaded woodpeckers, sea snakes, sea turtles, and rare plants. Her current interests center on breeding bird community studies and she has spent the past eight field seasons surveying birds in various habitats throughout Oregon. Kelly has been involved in most aspects of the project including compiling literature, organizing and moderating expert panel sessions, filling out life history matrixes, reviewing the matrixes for quality control, providing literature and database information to chapter authors, and editing chapters. She also enjoys participating in Christmas bird counts, running a breeding bird survey in eastern Oregon, and collecting old natural history books.

Pat Doran. Pat earned M.S. degrees in Ecology and Environmental Science from Indiana University. His research focused on the relationship between land cover and the reproductive success of songbirds. While with our project, Pat was involved in database and chapter development, and scientific panels, as well as project planning. Pat left the project to pursue a Ph.D. in Ecology at Dartmouth College. In his free time, Pat enjoys bird watching, hiking and backpacking, traveling, and just about any athletic activity.

Eva Greda. Eva graduated from Western Washington Univer-sity with a B.S. degree in Cell Biology. While at Western she was involved in a research project attempting to find a sequence homology between the *Caudal* gene of the fruit fly *Drosophila melanogaster*, and cDNA of the wasp *Nasonia vitripenisis*. Eva has worked for more than a year with the WDFW and previously with the Washington State

Attorney General's office regarding tobacco litigation. Her contributions to this project included recording and entering the results from the scientific panel process, compiling a portion of the life history matrix, and co-authoring several chapters. Her interests are gardening, painting, playing the piano, and walking her dog, Dana.

Teri Guydish. Teri received her Clerk/Receptionist Certificate at South Puget Sound Community College (SPSCC) in Olympia, Washington after twenty-seven years of being a housewife and mother of four children. Through the cooperative work program at SPSCC, she found a position with the Washington Department of Fish and Wildlife. This opened the door for several opportunities in the Habitat Program. She contributed to this project by copying articles to support the development of the matrixes, doing data entry, and helping out with many other tasks. In her free time she enjoys spending time with her family, gardening, fishing, camping, and baking great cookies.

Chris Kiilsgaard. Chris received a B.S. in Botany from Montana State University and an M.S. in Biogeography from Oregon State University. He has eighteen years of experience in mapping vegetation and wildlife habitats in the Pacific Northwest. Chris is the Director for Mapping Research for the Northwest Habitat Institute. Presently he is compiling a map of wildlife habitats for the seven-state Columbia River Watershed and mapping upland vegetation in the Oregon Coast Range. His other professional interest is habitat restoration. In his spare time he enjoys tennis, vegetable gardening, and scheming to own a vineyard.

Derek Stinson. Derek received his B.S. in Biology at Framingham State College, and his M.S. in Zoology from Washington State University. His diverse interests and work experiences have included shrews, bats, galliforms, raptors, insects, and wetland birds. He spent four years working on endangered birds, fruit bats, and migrant shore and waterbirds in the Mariana Islands, and wrote the recovery plan for the Micronesian Megapode. More recently he has worked on rare species and conservation planning of forest wildlife in Washington. Derek's expertise was used to assist in compiling information for the life history matrix. He enjoys gardening, birding, painting, and other forms of celebrating and caring for creation.

Marla Trevithick. Marla received a B.A. in Computer Science from Western Oregon University (WOU) and worked with the ODFW and WDFW for three years. Currently, she is working with U.S. Geological Survey's Forest and Rangeland Ecosystem Science Center as a computer specialist in technical user support. She was the data systems and digital technologies co-lead for the project. Marla's interests include both indoor and outdoor soccer, downhill skiing, crafts, and traveling. She has also been an assistant coach for the WOU women's soccer team for two years. Having grown up in Alaska gave her an appreciation for the outdoors, and she enjoys exploring Oregon's landscapes.

Madeleine Vander Heyden Madeleine is a wildlife biologist currently living in North Bend, Oregon. She graduated with an M.S. in Wildlife Ecology from Oregon State University in 1997 and received her B.S. from the University of Wisconsin in 1987. She has worked on a variety of wildlife species, including research on spotted owls, small mammals, bald eagles, and black bears. Madeleine's experience contributed to the digitial matrixes for habitat elements, species key ecological functions, and management activity links to the habitat elements. Her special areas of interest include wildlife-habitat relationships, carnivore ecology, and public lands management.

While we were initiating this effort, Dr. Laurie Wunder helped tremendously with identifying and defining information to be considered in the *Wildlife-Habitat* types, *Structural Conditions*, *Habitat Elements*, and *Life History* matrixes, as well as assisting with the scientific panel process. Additional project staff whose efforts were also invaluable are Kelly Cruce, Jackie Enriquez, Connie Iten, Patricia Johnson, Kathryn LePome, Kathleen Morley, John Morgan, Russell Rodgers, Meg Shaughnessy, Daniel Stein, and Susan Tank.

A number of individuals who, although they did not work for the project directly (as employees), contributed importantly to key aspects of the effort: Joseph R. Evenson consistently offered his assistance and insights into marine mammal and bird aspects. A most talented individual, Terrance P. Johnson, provided critical GIS support by developing many of the species range maps, marine habitat type boundary determinations, marine species richness analysis, county occurrence determinations, and essential habitat-type analysis for the project. We are deeply indebted to the scientific insights and rigor, and all-around professional input of Dr. Bruce G. Marcot, who helped guide us at many important junctures in the project. We also thank Nikki Derringer, Monica Hoover, Karol McFarlane, Cindy J. Neff, Sharon Newton, Jeff Reams, Joann Smith, Lori Turner, Sue Vance, Sara Vickerman, and Karla Yeager for their important contributions to the success of this project.

Acknowledgments

This book, *Wildlife—Habitat Relationships in Oregon and Washington,* has been developed through the combined efforts of more than 600 people. The final tally shows that we received input from 246 questionnaire respondents, 225 people who attended the landscape modeling workshop in Olympia, 73 species specialists who participated in the 15 scientific panels, 40 people who accessed and contributed to our Internet site, 88 chapter authors, and another 30 people who guided our process by participating on one of five advisory teams.

Special thanks go to Bill Martin, the individual who designed the Northwest Coast Indian artwork for this book.

Bill is a fisherman, diver, and renowned carver and artist, and specializes in Northwest Coast Indian artwork. Bill is a Makah tribal member, and he and his wife, April, and their four children live in Neah Bay, Washington. His work can be found in many locations, including the Smithsonian Institute. A recent example of his talent is reflected in the eight-pole carving at the new marina in Neah Bay. During discussions of the artwork for this project, Bill was overheard saying that he can draw the Northwest Coast designs as easily as he breathes.

During the course of this project, project staff acquired over 9,000 hard copies of journal articles and related publications. The Washington State Library staff heroes on this monumental task were Diane Mitchell, Shaun Fuller, Don Gulliford, Ron Howard, Ingrid Morley, David Pope, Marina Rodriguez, Heidi Schroder, and Becky Stewart. Heartfelt thanks go to Diane for her undying support and peppy attitude, and to Heidi, who contributed more than two summers of her time working on the project's literature requests.

We are indebted to the primary contacts we had from our thirty-four Project Partners and Contributing Sponsors. Extra efforts came from the following folks: Robert Anthony, Kate Benkert, Tim Bodurtha, Jim Bottorff, Kay Brown, Charlie Bruce, Joe Buchanan, Eric Campbell, Alan Christensen, Pete Comaner, Katie Distler, Bob Falkenstein, Kelly Cassidy, Ron Escaño, Randy Fisher, Guy Green, Craig Hansen, Leslie Lehmann, Dave Marshall, Lisa Norris, Rose Owens, Peter Paquet, Russ Peterson, Claire Puchy, Jim Rochelle, Paul F. and Teresa Roline, J. Michael Scott, Tom Toman, Julie Thompson, Laura Todd, Bill Tweit, Cheryl Quade, Paul Wagner, Jeff Waldon, Tom Williams, and Lenny Young. Maria A. Hug consistently played a crucial and supportive role in the fiscal aspects of the project.

The process to create and develop the wildlife habitat information was guided by five teams. Team members gave freely of their time and the insights from their many years of experience proved essential to the integrity of the final products. *Science Team* members were Richard Holthausen, USFS; Donavin Leckenby, ODFW (retired); Charles Meslow, Wildlife Management Institute (retired); Martin Raphael, USFS; and James Rochelle, Weyerhauser Company (retired). *Species Habitat Team* members were Steve Kohlmann, Prairie Wings, Inc.; Eric Larsen, WDFW; Kim Mellen, USFS; Tom Mumford, WDNR; Doug Runde, Weyerhaeuser Company; and Matt Vander Haegen, WDFW. *Digital Products Team* members were Donavin Leckenby, ODFW (retired); Tom Owens, WDFW; Jon Sadowski, BLM; Barbara Wales, USFS; and Tim Young, WDNR. *Management Application Team* members were Heather Ballash, Washington Community, Trade, and Economic Development; Bonita Cleveland and Trevin Taylor, Quileute Indian Tribe; Cheryl Friesen, USFS; Cheryl Gruenthal, Boise Cascade; Joe Lint, BLM; Terry Luther, Confederated Tribes of the Warm Springs; Russ Peterson, USFWS; Steve Smith, ODFW; and Paul Wagner, Washington Department of Transportation. *Marine Team* members were Helen Berry, WDNR; Ken Warheit, WDFW; Jeff Skriletz, WDFW; Dave Nysewander, WDFW; and Joseph Evenson, WDFW.

People who contributed information via the Internet were Burr J. Betts, Ed Bowlby, R. Mark Brigham, Brian L. Biswell, Anna M. Bruce, Eric K. Cole, Michael G. Cope, L. Morris Eiffert, Joseph Engler, Joseph R. Evenson, Terry Farrell, Howard Ferguson, John H. Guetterman, Lisa A. Hallock, Aaron L. Holmes, Ronald J. Jameson, Martha Jordan, Thomas W. Keegan, Jeffrey C. Lewis, Joseph B. Lint, John P. Loegering, Michael Marsh, Hal Michael, Russell Morgan, Tim R. Mullican, Dave Nysewander, Thomas O'Neil, Joel E. Pagel, Mike Patterson, Dr. Mary Poss, Gary J. Roloff, Derek W. Stinson, Rebecca Thompson, Cynthia Trombino, Matthew Vander Haegen, Madeleine Vander Heyden, David Vesely, James W. Watson, and Simon Wray.

People who participated in the 15 scientific panels were: *Amphibians and Reptiles:* Doug Calvin, Marc Hayes, Richard Hoyer, Bill Leonard, Kelly McAllister, Richard Nauman, Chuck Peterson, and Alan St. John; *Bats:* Ralph Anderson, Burr Betts, Steve Cross, James Hallett, Barry Keller, Pat Ormsbee, Steve West, and Laurie Wunder; *Small Mammals:* Ralph Anderson, Brian Biswell, Charles Drabeck, Jim Hallett, Barry Keller, Tom Manning, Margaret O'Connell, Steve West, Laurie Wunder, and Eric Yensen; *Meso Mammals:* Robert Anthony, Keith Aubrey, Terry Ferrell, Tom O'Neil, Dale Toweill, and Don Whittaker; *Ungulates:* Ed Arnett, Lou Bender, Richard Pedersen, Dale Toweill, Madeleine Vander Heyden, Gary Witmer, and Don Whittaker; *Marine Mammals:* John Calambokidis, Greg Green, Jan Hodder; *Loons, Grebes and Waterfowl:* Kevin Blakely, Robert Jarvis, Dave Nysewander, and Tom O'Neil; *Pelicans, Herons, and Cranes:* Ken Popper, Joe Engler, Gary Ivey, Deborah Jaques, and Don Norman; *Shorebirds:* Joe Buchanan, Pat Doran, Jeff Gilligan, and Nils Warnock; *Seabirds:* Ken Warheit, Kim Nelson, Chris Thompson, and Craig Strong; *Raptors:* Eric Forsman, Rick Gerhardt, Denver Holt, Frank Isaacs, David Johnson, John Marzluff, Joel Pagel, and Brian Woodbridge; *Gallinaceous Birds, Pigeons, and Doves:* Ralph Anderson, John Crawford, Kathy Martin, Alan Sands, and Mike Schroeder; *Passerine and Near Passerine Birds (Group 1):* Paul Adamus, Bob Altman, Kelly Bettinger, Dan Gumtow-Farrior, Rick Lundquist, Dennis Vroman, Kim Mellen, Catherine Raley, and Russell Rogers; and *Passerine Birds (Group 2):* Bob Altman, Kelly Bettinger, Chris Chappell, Steve Dowlan, Joan Hagar, Matt Hunter, Rex Sallabanks, and Matt Vander Haegen. *Management Activities and Links:* Susan Tank, Marla Trevithick, Thomas O'Neil, David H. Johnson, Pat Chapman, Curt Leigh, Randy Carman, Cheryl Friesen, Donavin Leckenby, and Doug Runde, Derek Stinson, and Kelly Bettinger, Cheryl Broyles, Paul Wagner, and E. Charles Meslow.

We are deeply indebted to the chapter authors (their names are published with their respective chapters in this book), who brought forward their scientific best for this publication. We are also grateful to Sally Olson-Edge, who was a technical editor for many chapters, and to the staff at Oregon State University Press, especially Jo Alexander and Jeff Grass.

Lastly, the Managing Directors take full responsibility for any errors found within this publication and CD-ROM.

Credits for Habitat Photographs
(Chapter 2)

Agriculture (color)
1. Near Samish Bay, Washington. Photo: Joseph B. Buchanan
2. Sinlahekin Valley, Washington. Photo: Reid Schuller
3. Ellensburg area, Washington. Photo: Andrew Kratz
4. Wheat Field, Southeast Washington. Photo: Rollin Geppert
5. Christmas Tree farm, Benton County, Oregon. Photo: Kelly A. Bettinger
6. Scotch Creek Wildlife Area, Okanogan County, Washington. Photo: Jerry Benson

Alpine Grasslands and Shrublands (color)
1. Panhandle Gap, Mt. Rainier National Park, Washington. Photo: Chris Chappell

2. Steens Mt., Oregon. Photo: Chris Chappell
3. Berkeley Park, Mt. Rainier National Park, Washington. Photo: Chris Chappell
4. NE Olympics, Washington. Photo: Chris Chappell

Alpine Grasslands and Shrublands (black and white)
Example of Alpine grasslands and shrublands. Photo: Jeff Reams

Bays and Estuaries (color)
1. Estuary, Washington. Photo: WDFW
2. Dogfish Pt., Skagit County, Washington. Photo: Chris Chappell
3. Niawiakum River, Willapa Bay, Washington. Photo: Linda Kunze
4. Nisqually Delta, Washington. Photo: Linda Kunze
5. Niawiakum River, Willapa Bay, Washington. Photo: Mark Sheehan
6. Newport, Oregon. Photo: David H. Johnson

Bays and Estuaries (black and white)
Bays and Estuaries. Skokomish River, Hood Canal, Washington. Photo: Steve Jeffries.

Ceanothus Manzanita Shrublands (color)
1. Southwestern Oregon. Photo: James F. Harper
2. South of Shady Cove, Oregon. Photo: James F. Harper
3. Southwestern Oregon. Photo: Jeff Reams.
4. South of Shady Cove, Oregon. Photo: James F. Harper

Ceanothus Manzanita Shrublands (black and white)
Example of Ceanothus Manzanita Shrublands, southwestern Oregon. Photo: James F. Harper.

Coastal Headlands and Islets (color)
1 Near Cape Perpetua, Oregon. Photo: Jeff Ream
2 Heceta Head, Oregon. Photo: Jeff Ream
3. Near Yaquina Head, Oregon. Photo: David H. Johnson
4. Near Cape Perpetua, Oregon. Photo: Jeff Reams
5. Coastline near Quinalt River, Washington. Photo: Chris Chappell

Coastal Headlands and Islets (black and white)
Example of Coastal Headlands and Islets. Photo: Jeff Reams.

Coastal Dunes (color)
1. Florence, Oregon. Photo: Madeleine Vander Heyden
2. Dugualla Bay, Washington. Photo: DNR Aquatic Lands Program
3. Ocean Shores, Washington. Photo: David H. Johnson
4. Rocky Pt., Whidbey Island, Washington. Photo: Chris Chappell
5. Rocky Pt., Whidbey Island, Washington. Photo: Chris Chappell

Coastal Dunes (black and white)
Example of Coastal Dunes. Photo: Jeff Reams

Desert Playa (color)
1. Alvord Desert, Oregon. Photo: Jeff Rems
2. Alvord Desert, Oregon. Photo: Chris Chappell
3. Alvord Desert, Oregon. Photo: Chris Chappell
4. Harney Basin, Oregon. Photo: Joseph B. Buchanan
5. Alvord Desert, Oregon. Photo: Chris Chappell

Desert Playa (black and white)
Example of desert playa and salt scrub shrublands. Photo: Jeff Reams.

Dwarf Shrub-Steppe (color)
1. Castle Rock, Grand Coulee, Washington. Photo: Andy Kratz
2. Umtanum, Kittitas County, Washington. Photo: Chris Chappell
3. Saddle Mts., Washington. Photo: Chris Chappell
4. Sheldon N.R.W., Nevada. Photo: Michael Gregg

Dwarf Shrub-Steppe (black and white)
Example of Dwarf Shrub-Steppe. Moses Coulee, Douglas County, Washington. Photo: Rex Crawford

Eastside Canyon Shrublands (color)
1. Barker Mt. Okanogan County, Washington. Photo: Chris Chappell
2. H. J. Experimental Forest, Oregon. Photo: Chris Chappell
3. H. J. Experimental Forest, Oregon. Photo: Chris Chappell
4. Smith Canyon, Methow Valley, Washington. Photo: Chris Chappell
5. Klickitat County, Washington. Photo: Mark Sheehan

Eastside Canyon Shrublands (black and white)
Example of Eastside Canyon Shrublands. Steens Mt., Oregon. Photo: Madeleine Vander Heyden

Eastside Grasslands (color)
1. Gibraltar Mt., Ferry County, Washington. Photo: Rex Crawford
2. Palouse River, Washington. Photo: Rex Crawford
3. Dalles Mts., Washington. Photo: Rex Crawford
4. Hanford area, Washington. Photo: Chris Chappell

Eastside Grasslands (black and white)
Example of Eastside Grasslands. Franklin County, Washington. Photo: Andy Kratz

Eastside Mixed Conifer Forest (color)
1. Wenatche Mts., Kittitas County, Washington. Photo: Chris Chappell
2. Alice Mae Mt., Stevens County, Washington. Photo: Chris Chappell
3. Alice Mae Mt., Stevens County, Washington. Photo: Chris Chappell
4. Rainbow Creek, Research Natural Area, Blue Mts., Washington. Photo: Chris Chappell

Eastside Mixed Conifer Forest (black and white)
Example of Eastside Mixed Conifer Forest. Photo: Jeff Reams.

Eastside Riparian Wetlands (color)
1. Northrup Canyon, Washington. Photo: Reid Schuller
2. Little Pend Oreille River, Stevens County, Washington. Photo: Reid Schuller
3. Douglas Creek, Douglas County, Washington. Photo: Chris Chappell
4. Myers Creek, Okanogan County, Washington. Photo: John Gamon
5. Eastern Klickitat County, Washington. Photo: DNR Natural Heritage Program
6. Crimm's Creek, Washington. Photo: Cheryl Quade

Eastside Riparian Wetlands (black and white)
1. Example of Eastside Riparian Wetlands. Crab Creek, Lincoln County, Washington. Photo: Rex Crawford.

Herbaceous Wetlands (color)
1. Leuton Flat, Okanogan County, Washington. Photo: Elisabeth Rodrick.
2. Methow Marsh, Washington. Photo: Rex Crawford.
3. Pacific County, Washington. Photo: Robin Woodin
4. Lincoln County, Washington. Photo: Rex Crawford
5. Linn County, Oregon. Photo: Jeff Reams

Herbaceous Wetlands (black and white)
Example of Herbaceous Wetlands. Photo: Jeff Reams

Inland Marine Deeper Waters (color)
1. Squaxin Island, South Puget Sound, Washington. Photo: David H. Johnson
2. San Juan Islands, Washington. Photo: Coastal and Ocean Resources, Inc.

Inland Marine Deeper Waters (black and white)
Example of Inland Marine Deeper Waters. March Point, Skagit County, Washington. Photo: DNR, Nearshore Habitat Program.

Lodgepole Pine Forest and Woodlands (color)
1. Loomis State Forest, Washington. Photo: Cheryl Quade
2. Loomis State Forest, Washington. Photo: Cheryl Quade
3. Loomis State Forest, Washington. Photo: Cheryl Quade
4. Loomis State Forest, Washington. Photo: Cheryl Quade
5. Loomis State Forest, Washington. Photo: Cheryl Quade

Lodgepole Pine Forest and Woodlands (black and white)
Example of Lodgepole Pine Forest and Woodlands. Loomis, Washington. Photo: Cheryl Quade

Marine Nearshore (color)
1. Ben Ure Island, Island County, Washington. Photo: DNR, Nearshore Habitat Program
2. Cypress Island, Skagit County, Washington. Photo: DNR, Nearshore Habitat Program
3. Guemes Island, Washington. Photo: DNR, Nearshore Habitat Program
4. Rock North of Biz Pt., Skagit County, Washington. Photo: DNR, Nearshore Habitat Program
5. Near Cascade Head, Lincoln County, Oregon. Photo: Jeff Reams

Marine Nearshore (black and white)
Example of Marine Nearshore. North Samish Island, Skagit County, Washington. Photo: DNR, Nearshore Habitat Program

Montane Coniferous Wetlands (color)
1. Arlecho Creek, Washington. Photo: Chris Chappell
2. Roger Lake, Washington. Photo: Rex Crawford

3. Mt. Rainier, Washington. Photo: Chris Chappell
4. Boulder Creek, Washington. Photo: Chris Chappell

Montane Coniferous Wetlands (black and white)
Example of Montane Coniferous Wetlands. Photo: Jeff Reams.

Montane Mixed Conifer Forest (color)
1. Boulder Creek, Washington. Photo: Chris Chappell
2. Pend Oreille County, Washington. Photo: Reid Schuller
3. Bald Mt., Mt. Pilchuck, Washington. Photo: Chris Chappell
4. Arlecho Creek, Washington. Photo: Chris Chappell

Montane Mixed Conifer Forest (black and white)
1. Example of Mixed Conifer Forest. Photo: Jeff Reams

Open Water
1. Lake Ozette, Washington. Photo: Linda Kunze
2. Oxbow in lower Soleduck River, Washington. Photo: Chris Chappell
3. Grant County, Washington. Photo: James Hannah
4. Snake River, Washington. Photo: Rollin Geppert
5. Willamette River, Linn County, Oregon. Photo: Jeff Reams

Open Water (black and white)
Example of Open Water. Columbia River, Washington. Photo: Rollin Geppert.

Ponderosa Pine Forest and Woodlands (color)
1. Barker Mt., Okanogan County, Washington. Photo: DNR, Natural Heritage Program
2. Near Sisters, Oregon. Photo: Cathy Rose
3. Turnbull Wildlife Refuge, near Spokane, Washington. Photo: Cheryl Quade
4. Indian Ford, north of Sisters, Oregon. Photo: Chris Chappell
5. Briske Canyon, Washington. Photo: Rollin Geppert
6. Badger Gulch Natural Area Preserve, Klickitat County, Washington. Photo: Mark Sheehan

Ponderosa Pine Forest and Woodlands (black and white))
Example of Ponderosa Pine. Photo: Jeff Reams

Shrub-Steppe (color)
1. Yakima, Washington. Photo: Chris Chappell
2. Horse Heaven Hills near Prosser, Washington. Photo: Matt Vander Haegen
3. Yakima, Washington. Photo: Chris Chappell
4. Steens Mt, Oregon. Photo: Chris Chappell
5. Varnita, Washington. Photo: Rex Crawford

Shrub-Steppe (black and white))
Example of Shrub-Steppe. Goose Hill, Benton County, Washington. Photo: Reid Schuller

Southwest Oregon Mixed Conifer-Hardwood Forest (color)
1. Southwestern Oregon. Photo: James F. Harper
2. Jackson County, Oregon. Photo: Jeff Reams
3. Ruch, Oregon. Photo: Jeff Reams

Southwest Oregon Mixed Conifer-Hardwood Forest (black and white)
Example of Southwest Oregon Mixed Conifer Forest. Southwestern Oregon. Photo: James. F. Harper

Subalpine Parkland (color)
1. Grand Park, Mt. Rainier National Park, Washington. Photo: Chris Chappell
2. Strawberry Mt., Oregon. Photo: Madeleine Vander Heyden
3. Bald Mt., Mt. Pilchuck, Washington. Photo: Chris Chappell
4. Eagle Cap, Oregon. Photo: Madeleine Vander Heyden
5. Goat Rocks, Lewis County, Washington. Photo: Chris Chappell

Subalpine Parkland (black and white))
Example of Subalpine Parkland. Eagle Cap, Oregon. Photo: Madeleine Vander Heyden.

Upland Aspen Forest (color)
1. Sum Mt., Winthrop, Washington. Photo: Rollin Geppert
2. Hart Mt., Oregon. Photo: Joseph B. Buchanan
3. Steens Mt., Oregon. Photo: Chris Chappell
4. Hart Mt., Oregon. Photo: Madeleine Vander Heyden

Upland Aspen Forest (black and white))
Example of Upland Aspen Forest. Sun Mt., Winthrop, Washington. Photo: Rollin Geppert

Urban (color)
1. Example of high density urban, Seattle, Washington. Photo: Rollin Geppert
2. Example of medium density urban, Washington. Photo: DNR Nearshore Habitat Program
3. Example of low density urban, Benton County, Oregon. Photo: Kelly a. Bettinger
4. Example of low density urban, Wenatchee, Washington. Photo: Rollin Geppert

Urban (black and white)
Example of Urban. Maple Valey, Washington. Photo: Rollin Geppert

Westside Grasslands (color)
1. Scatter Creek, Washington. Photo: Chris Chappell.
2. Scatter Creek, Washington. Photo: Chris Chappell.
3. Bald Hill Natural Area Preserve, Washington. Photo: Mark Sheehan
4. Burrows Island, Washington. Photo: Chris Chappell.
5. Washington. Photo: Dave Hayes.

Westside Grasslands (black and white)
Example of Westside Grasslands. Smith Prairie, Washington. Photo: Chris Chappell.

Western Juniper and Mountain Mahogany Woodlands (color)
1. Sheldon National Wildlife Refuge, NV. Photo: Michael A. Gregg
2. Sheldon National Wildlife Refuge, NV. Photo: Michael A. Gregg
3. Jefferson County, Oregon. Photo: Jeff Reams
4. Klamath County, Oregon. Photo: Jeff Reams.

Western Juniper and Mountain Mahogany Woodlands (black and white)
Example of Western Juniper and Mountain Mahogany Woodlands. Photo: Jeff Reams.

Westside Lowlands Conifer-Hardwood Forest (color)
1. Goodman Creek, Oregon. Photo: David H. Johnson.
2. H. J. Andrews Experimental Forest, Oregon. Photo: Chris Chappell
3. Hoypus South, Washington. Photo: Chris Chappel
4. Capitol Forest, Olympia, Washington. Photo: Rollin Geppert
5. Upper Dungeness, Washington. Photo: Chris Chappell
6. Lake Quinalt, Washington. Photo: David H. Johnson

Westside Lowlands Conifer-Hardwood Forest (black and white)
Example of Westside Lowlands Conifer-Hardwood Forest. Skokomish, Washington. Photo: Rollin Geppert.

Westside Oak and Dry Douglas-Fir Forest and Woodlands (color)
1. Cormorant Bay, San Juan Islands, Washington. Photo: Chris Chappell
2. James Island, San Juan Islands, Washington. Photo: Chris Chappell
3. Cady Mt., Orcas Islands, Washington. Photo: Chris Chappell
4. Argonne Forest, Fort Lewis, Washington. Photo: Chris Chappell

Westside Oak and Dry Douglas-Fir Forest and Woodlands (black and white)
Example of Westside Oak and Dry Douglas-Fir Forest and Woodlands. Photo: Jeff Reams

Westside Riparian Wetlands (color)
1. Middle Stequaleho, WA. Photo: Chris Chappell
2. Cranberry Creek, WA. Photo: Linda Kunze
3. Maxfield Creek, WA. Photo: Chris Chappell
4. Cranberry Creek, Grays Harbor, WA. Photo: Linda Kunze
5. Quinalt River, WA. Photo: David H. Johnson

Westside Riparian Wetlands (black and white)
Example of Westside Riparian Wetlands. Tacoma Creek, Pierce County, Washington. Photo: Chris Chappell.

Index

Names of wildlife species are listed under their common names in inverted format (e.g., eagles, bald), except when directed by cross-reference. A complete listing of scientific and common names is provided in the Appendix (pages 687-711).